Encyclopedia of Cancer

Manfred Schwab

Editor

Encyclopedia of Cancer

Volume 2

Fourth Edition

C–E

With 1230 Figures and 260 Tables

 Springer

Editor
Manfred Schwab
German Cancer Research Center (DKFZ)
Tumorgenetik, Heidelberg, Germany

ISBN 978-3-662-46874-6   ISBN 978-3-662-46875-3 (eBook)
ISBN 978-3-662-47424-2 (print and electronic bundle)
DOI 10.1007/978-3-662-46875-3

Library of Congress Control Number: 2017933328

Printed on acid-free paper

This Springer imprint is published by Springer Nature
The registered company is Springer-Verlag GmbH Germany
The registered company address is: Heidelberger Platz 3, 14197 Berlin, Germany

Preface to the Fourth Edition

Welcome to the fourth edition of the *Encyclopedia of Cancer*. The third edition had appeared in 2011, and the tremendous response by the scientific community has encouraged us to prepare a subsequent edition that is now available. The past 5 years have seen an enormous progress in cancer research, with particular emphasis on the bench-to-bed paradigm and the application of personalized cancer medicine.

For this new edition, the multidisciplinary approach bridging basic science and clinical application was further developed. Numerous new entries by authorities from the international scientific community were added to meet the substantial progress in molecular cancer etiology, diagnostics, and therapy. Entries from the third edition were updated, and new entries were added addressing central areas of basic and clinical cancer research, such as personalized cancer medicine, immunotherapy, pediatric and adult oncology, and epigenetics.

The *Encyclopedia of Cancer*, fourth edition, will be available both in print and online version. The online version is designed as an interactive and dynamic database where authors at any time will be able to modify and update presentations in order to keep the content up to date. Additionally, new entries can be entered at any time, and contributors are encouraged to suggest new topics that they feel are insufficiently covered.

The technical preparation of the *Encyclopedia of Cancer* would not have been possible without the competent and dedicated input by Daniela Graf and Melanie Thanner. Their excellent and pleasant cooperation is highly appreciated. Thanks also to the publisher who has taken every effort to develop this prestigious *Encyclopedia of Cancer* into a useful instrument from which both basic scientists and clinicians may benefit.

Heidelberg, March 1, 2016
Manfred Schwab

Preface to the Third Edition

Recent developments in the rapidly developing field of cancer research are seeing a dynamic progress in basic and clinical cancer science, with translational research increasingly becoming a new paradigm. In particular, the identification of a large number of prognostic and predictive clinically validated biomarkers now allows exciting and promising new approaches in both personalized cancer medicine and targeted therapies to be pursued.

The third edition of the *Encyclopedia of Cancer* is now available 10 years after the first edition had come out in 2001. Numerous new entries addressing topics of basic cancer research have been added. As a major new feature, up-to-date and authoritative essays present a comprehensive picture of topics ranging from pathology, to clinical oncology and targeted therapies for personalized cancer medicine for major cancers types, such as breast cancer, colorectal cancer, prostate cancer, ovarian cancer, renal cancer, lung cancer, and hematological malignancies, leukemias, and lymphomas. This information source should be of great value to both the clinical and basic science community.

The *Encyclopedia of Cancer*, Third Edition, is available both in print and online versions. Contributors to the *Encyclopedia of Cancer* are encouraged to keep their presentations up-to-date by online editing. Clinical and basic scientists are encouraged to suggest new essays to the editor-in-chief.

The technical preparation of the *Encyclopedia of Cancer* would not have been possible without the competent input of Jutta Jaeger-Hamers, Melanie Thanner, and Saskia Ellis; their excellent and pleasant cooperation is highly appreciated.

Heidelberg, Germany
Manfred Schwab

Preface to the Second Edition

Given the overwhelming success of the first edition of the Cancer Encyclopedia, which appeared in 2001, and the amazing development in the different fields of cancer research, it has been decided to publish a second fully revised and expanded edition, following the principal concept of the first edition that has proven so successful.

Recent developments are seeing a dynamic merging of basic and clinical science, with translational research increasingly becoming a new paradigm in cancer research. The merging of different basic and clinical science disciplines toward the common goal of fighting against cancer has long ago called for the establishment of a comprehensive reference source both as a tool to close the language gap between clinical and basic science investigators and as a platform of information for advanced students and informed laymen alike. It is intended to be a resource for all interested in information beyond their own specific expertise.

While the first edition had featured contributions from approximately 300 scientists/clinicians in one volume, the second edition includes more than 1,000 contributors in four volumes with an A–Z format of more than 7,000 entries. It provides definitions of common acronyms and short definitions of both related terms and processes in the form of keyword entries. A major information source are detailed essays that provide comprehensive information on syndromes, genes and molecules, and processes and methods. Each essay is well structured, with extensive cross-referencing between entries. Essays represent original contributions by the corresponding authors, all distinguished scientists in their own field, editorial input has been carefully restricted to formal aspects.

A panel of field editors, each an eminent international expert for the corresponding field, has served to ensure the presentation of timely and authoritative Encyclopedia entries. These new traits are likely to meet the expectance that a wide community has toward a cancer reference work.

An important element in the preparation of the Encyclopedia has been the competent support by the Springer crew, Dr. Michaela Bilic, Saskia Ellis, and lately, Jana Simniok. I am extremely grateful for their excellent and pleasant cooperation.

The Cancer Encyclopedia, Second Edition, will be available both in print and online versions. Clinicians, research scientists, and advanced students will find this an amazing resource and a highly informative reference for cancer.

Heidelberg, Germany
Manfred Schwab

Preface to the First Edition

Cancer, although a dreadful disease, is at the same time a fascinating biological phenotype. Around 1980, cancer was first attributed to malfunctioning genes and, subsequently, cancer research has become a major area of scientific research supporting the foundations of modern biology to a great extent. To unravel the human genome sequence was one of those extraordinary tasks, which has largely been fueled by cancer research, and many of the fascinating insights into the genetic circuits that regulate developmental processes have also emerged from research on cancer.

Diverse biological disciplines such as cytogenetics, virology, cell biology, classical and molecular genetics, epidemiology, biochemistry, together with the clinical sciences, have closed ranks in their search of how cancer develops and to find remedies to stop the abnormal growth that is characteristic of cancerous cells. In the attempt to establish how, why, and when cancer occurs, a plethora of genetic pathways and regulatory circuits have been discovered that are necessary to maintain general cellular functions such as proliferation, differentiation, and migration. Alterations of this fine-tuned network of cascades and interactions, due to endogenous failure or to exogenous challenges by environmental factors, may disable any member of such regulatory pathways. This could, for example, induce the death of the affected cell, may mark it for cancerous development or may immediately provide it with a growth advantage within a particular tissue.

Recent developments have seen the merger of basic and clinical science. Of the former, particularly genetics has provided instrumental and analytical tools with which to assess the role of environmental factors in cancer, to refine and enable diagnosis prior to the development of symptoms, and to evaluate the prognosis of patients. Hopefully, even better strategies for causal therapy will become available in the future. Merging the basic and clinical science disciplines toward the common goal of fighting cancer calls for a comprehensive reference source to serve both as a tool to close the language gap between clinical and basic science investigators and as an information platform for the student and the informed layperson alike. Obviously this was an extremely ambitious goal, and the immense progress in the field cannot always be portrayed in line with the latest developments. The aim of the Encyclopedia is to provide the reader with an entrance point to a particular topic. It should be of value to both basic and clinical scientists working in the field of cancer research. Additionally, both students and lecturers in the life sciences should benefit highly from this database. I therefore hope that this Encyclopedia will become an essential complement to existing science resources.

The attempts to identify the mechanisms underlying cancer development and progression have produced a wealth of facts, and no single individual is capable of addressing the immense breadth of the field with undisputed authority. Hence, the "Encyclopedic Reference of Cancer" is the work of many authors, all of whom are experts in their fields and reputable members of the international scientific community. Each author contributed a large number of keyword definitions and in-depth essays and in so doing it was possible to cover the broad field of cancer-related topics within a single publication. Obviously this approach entails a form of presentation, in which the author has the freedom to set priorities and to promote an individual point of view. This is most obvious when it comes to nomenclature, particularly that of genes and proteins. Although the editorial intention was to apply the nomenclature of the Human Genome Organisation (HUGO), the more vigorous execution of this attempt has been left to future endeavors.

In the early phase of planning the Encyclopedia, exploratory contacts to potential authors produced an overwhelmingly positive response. The subsequent contact with almost 300 contributory authors was a marvelous experience, and I am extremely grateful for their excellent and constructive cooperation. An important element in the preparation of the Encyclopedia has been the competent secretarial assistance of Hiltrud Wilbertz of the Springer-Verlag and of Ingrid Cederlund and Cornelia Kirchner of the DKFZ. With great attention to detail they helped to keep track of the technical aspects in the preparation of the manuscript. It was a pleasure to work with the Springer crew, including Dr. Rolf Lange as the Editorial Director (Medicine) and Dr. Thomas Mager, Senior Editor for Encyclopedias and Dictionaries. In particular I wish to thank Dr. Walter Reuss, who untiringly has mastered all aspects and problems associated with the management of the numerous manuscripts that were received from authors of the international scientific community. It has been satisfying and at times comforting to see how he made illustration files come alive. Thanks also to Dr. Claudia Lange who, being herself a knowledgeable cell biologist, has worked as the scientific editor. Her commitment and interest have substantially improved this Encyclopedia. As a final word, I would like to stress that although substantial efforts have been made to compose factually correct and well-understandable presentations, there may be places where a definition is incomplete or a phrase in an essay is flawed. All contributors to this Encyclopedia will be extremely happy to receive possible corrections, or revisions, in order for them to be included in any future editions of the "Encyclopedic Reference of Cancer."

Heidelberg, Germany
Manfred Schwab

Editor-in-Chief

Manfred Schwab, Dr. rer. nat.

University-Professor of Genetics
Neuroblastoma Genomics B087
German Cancer Research Center (DKFZ)

Contributors

Apart from few editorial input, the respective authors are responsible for the content of their own texts.

Vesa Aaltonen Department of Ophthalmology, University of Turku, Turku, Finland

Trond Aasen Department of Pathology, Vall d'Hebron University Hospital, Barcelona, Spain

Cory Abate-Shen Herbert Irving Comprehensive Cancer Center, Columbia University Medical Center, New York, NY, USA

Phillip H. Abbosh Department of Pathology and Laboratory Medicine, Indiana University School of Medicine, Indianapolis, IN, USA

Kotb Abdelmohsen RNA Regulation Section, National Institute on Aging, National Institutes of Health, Biomedical Research Center, Baltimore, MD, USA

Fritz Aberger Department of Molecular Biology, University of Salzburg, Salzburg, Austria

Hinrich Abken Tumor Genetics, Clinic I Internal Medicine, University Hospital Cologne, and Center for Molecular Medicine Cologne, University of Cologne, Cologne, Germany

Amal M. Abu-Ghosh Department of Oncology and Pediatrics, Lombardi Comprehensive Cancer Center, Georgetown University, Washington, DC, USA

Rosita Accardi Infections and Cancer Biology Group, International Agency for Research on Cancer, Lyon, France

Filippo Acconcia Molecular and Cellular Oncology, The University of Texas MD Anderson Cancer Center, Houston, TX, USA

Christina L. Addison Cancer Therapeutics Program, Ottawa Hospital Research Institute, Ottawa, ON, Canada

Vaqar M. Adhami School of Medicine and Public Health, University of Wisconsin, Madison, WI, USA

Farrukh Afaq Department of Dermatology, University of Alabama at Birmingham, Birmingham, AL, USA

Chapla Agarwal SOP-Administration, University of Colorado Denver – Anschutz Medical Campus, Aurora, CO, USA

Garima Agarwal College of Pharmacy, The Ohio State University, Columbus, OH, USA

Rajesh Agarwal Department of Pharmaceutical Sciences, Skaggs School of Pharmacy and Pharmaceutical Sciences, University of Colorado, Aurora, CO, USA

Patrizia Agostinis Department of Cellular and Molecular Medicine, Cell Death Research and Therapy Lab, KU Leuven Campus Gasthuisberg, Leuven, Belgium

Terje C. Ahlquist Roche Norway, Oslo, Norway

Kazi Mokim Ahmed Department of Radiation Oncology, Houston Methodist Research Institute, Houston, TX, USA

Khalil Ahmed Minneapolis VA Health Care System and University of Minnesota, Minneapolis, MN, USA

Shahid Ahmed Department of Oncology, University of Saskatchewan, Saskatoon, SK, Canada

Joohong Ahnn Department of Life Science, Hanyang University, Seoul, South Korea

Cem Akin University of Michigan, Ann Arbor, MI, USA

Gada Al-Ani Department of Cancer Biology, University of Kansas Cancer Center, The University of Kansas Medical Center, Kansas City, KS, USA

Ami Albihn Department of Microbiology, Tumor and Cell Biology (MTC), Karolinska Institutet, Stockholm, Sweden

Adriana Albini IRCCS Multimedica, Milano, Italy

Jérôme Alexandre Faculté de Médecine Paris – Descartes, UPRES 18-33, Groupe Hospitalier Cochin – Saint Vincent de Paul, Paris, France

Amal Yahya Alhefdhi Department of Surgery – MBC 40, King Faisal Specialist Hospital and Research Center, Riyadh, Kingdom of Saudi Arabia

Shadan Ali Karmanos Cancer Institute, Wayne State University, Detroit, MI, USA

Malcolm R. Alison Centre for Diabetes and Metabolic Medicine, Queen Barts and the London School of Medicine and Dentistry, Institute of Cell and Molecular Science, London, UK

Catherine Alix-Panabieres University Medical Center, Lapeyronie Hospital, Montpellier, France

Alison L. Allan Cancer Research Laboratories, London Regional Cancer Program and Departments of Oncology and Anatomy and Cell Biology, Schulich School of Medicine and Dentistry, University of Western Ontario, London, ON, Canada

Paola Allavena Department of Immunology, Fondazione Humanitas per la Ricerca, Rozzano, Milan, Italy

Damian A. Almiron Departments of Pediatrics and of Genetics, Norris Cotton Cancer Center, Geisel School of Medicine at Dartmouth, Hanover, NH, USA

Angel Alonso Deutsches Krebsforschungszentrum, Heidelberg, Germany

Gianfranco Alpini Departments of Medicine and Medical Physiology, Texas A&M Health Science Center, College of Medicine, Central Texas Veterans Health Care System, Baylor Scott & White Health, Temple, TX, USA

Marie-Clotilde Alves-Guerra Molecular and Cellular Oncogenesis Program, The Wistar Institute, Philadelphia, PA, USA

Pierre Åman LLCR, Department of Pathology, Institute of Biomedicine, Sahlgrenska Academy, Goteborg University, Gothenburg, Sweden

Kurosh Ameri Department of Medicine, Division of Cardiology, Translational Cardiac Stem Cell Program, Eli and Edythe Broad Center of Regeneration Medicine and Stem Cell Research, Cardiovascular Research Institute, University of California San Francisco (UCSF), San Francisco, CA, USA

Mounira Amor-Guéret Institut Curie – UMR 3348 CNRS, Orsay Cedex, France

Grace Amponsah Department of Pathology, Comprehensive Cancer Centre, The Ohio State Medical Centre, Columbus, OH, USA

John W. Anderson Dream Master Laboratory, Chandler, AZ, USA

Kenneth C. Anderson Department of Medical Oncology, Jerome Lipper Multiple Myeloma Center, Dana-Farber Cancer Institute, Boston, MA, USA

Nicolas André Centre for Research in Oncobiology and Oncopharmacology, INSERM U911, Marseille, France

Metronomics Global Health Initiative, Marseille, France

Department of Pediatric Hematology and Oncology, La Timone Children's Hospital, Marseille, France

Peter Angel Division of Signal Transduction and Growth Control, Deutsches Krebsforschungszentrum, Heidelberg, Germany

Andrea Anichini Department of Experimental Oncology, Fondazione IRCCS Istituto Nazionale per lo Studio e la Cura dei Tumori, Milan, Italy

Talha Anwar Medical Scientist Training Program and Department of Pathology, University of Michigan Medical School, Ann Arbor, MI, USA

Peter D. Aplan Genetics Branch, Center for Cancer Research, National Cancer Institute, National Institutes of Health, Bethesda, MD, USA

Natalia Aptsiauri UGC Laboratorio Clínico Hospital Universitario Virgen de las Nieves Facultad de Medicina, Universidad de Granada, Granada, Spain

Rami I. Aqeilan He Lautenberg Center for General and Tumor Immunology, Department of Immunology and Cancer Research-Institute for Medical Research Israel-Canada, Hebrew University-Hadassah Medical School, Jerusalem, Israel

Tsutomu Araki Departments of Obstetrics and Gynecology, Nippon Medical School, Kawasaki and Tokyo, Japan

Sanchia Aranda School of Nursing, The University of Melbourne, Carlton, VIC, Australia

Diego Arango CIBBIM - Nanomedicina Oncologia Molecular, Vall d'Hebron Hospital Research Institute, Barcelona, Spain

David J. Araten NYU School of Medicine, Laura and Isaac Perlmutter Cancer Center and the New York VA Medical Center, New York, NY, USA

Laura Arbona Department of Biology, University of the Balearic Islands, Palma de Mallorca, Spain

Valentina Arcangeli Department of Oncology, Instituto Scientifico Romagnolo per lo s, Infermi Hospital, Rimini, Italy

Gemma Armengol Faculty Biosciences, U. Biological Anthropology, Universitat Autonoma de Barcelona, Barcelona, Spain

Elias S. J. Arnér Department of Medical Biochemistry and Biophysics, Karolinska Institutet, Stockholm, Sweden

Marie Arsenian-Henriksson Department of Microbiology, Tumor and Cell Biology (MTC), Karolinska Institutet, Stockholm, Sweden

Stefano Aterini Department of Experimental Pathology and Oncology, University of Firenze, Florence, Italy

Scott Auerbach Biomolecular Screening, National Toxicology Program, National Institute of Environmental Health Sciences (NIEHS), Research Triangle Park, NC, USA

Katarzyna Augoff Department of Gastrointestinal and General Surgery, Wroclaw Medical University, Wroclaw, Poland

Marc Aumercier CNRS, INRA, UMR 8576-UGSF-Unité de Glycobiologie Structurale et Fonctionnelle, Université de Lille, Villeneuve d'Ascq, France

Riccardo Autorino Clinica Urologica, Seconda Università degli Studi, Naples, Italy

Matias A. Avila Division of Hepatology, CIMA, University of Navarra, Pamplona, Spain

Hava Karsenty Avraham Division of Experimental Medicine, Beth Israel Deaconess Medical Center, Harvard Institutes of Medicine, Boston, MA, USA

Shalom Avraham Division of Experimental Medicine, Beth Israel Deaconess Medical Center, Harvard Institutes of Medicine, Boston, MA, USA

Sanjay Awasthi United States Longview Cancer Center, Longview, TX, USA

Yogesh C. Awasthi City of Hope, Duarte, CA, USA

Debasis Bagchi Department of Pharmacy Sciences, Creighton University Medical Center, Omaha, NE, USA

Xue-Tao Bai State Key Laboratory of Medical Genomics, Shanghai Institute of Hematology, Rui-Jin Hospital, Shanghai Jiao Tong University School of Medicine, Shanghai, People's Republic of China

Michael J. Baine Department of Radiation Oncology, Fred and Pamela Buffett Cancer Center, University of Nebraska Medical Center, Omaha, NE, USA

Jürgen Bajorath Department of Life Science Informatics, B-IT, University of Bonn, Bonn, Germany

Stuart G. Baker Biometry Research Group, National Cancer Institute, Bethesda, MD, USA

Elizabeth K. Balcer-Kubiczek Department of Radiation Oncology, Marlene and Stewart Greenebaum Cancer Center, University of Maryland School of Medicine, Baltimore, MD, USA

Enke Baldini Department of Experimental Medicine, University of Rome "Sapienza", Rome, Italy

Graham S. Baldwin Department of Surgery, Austin Health, The University of Melbourne, Heidelberg, VIC, Australia

Sherri Bale GeneDx, Rockville, MD, USA

Laurent Balenci INSERM Unité Mixte 873, Grenoble, France

Sushanta K. Banerjee Cancer Research Unit, Research Division, VA Medical Center, Kansas City, MO, USA

Michal Baniyash The Lautenberg Center for Immunology and Cancer Research, Israel-Canada Medical, Research Institute Faculty of Medicine, The Hebrew University, Jerusalem, Israel

Shyam S. Bansal Division of Cardiovascular Disease, University of Alabama at Birmingham, Birmingham, AL, USA

Nektarios Barabutis Frank Reidy Research Center for Bioelectrics, Old Dominion University, Norfolk, VA, USA

Aditya Bardia Massachusetts General Hospital Cancer Center and Harvard Medical School, Boston, MA, USA

Rafijul Bari Departments of Medicine and Molecular Sciences, Vascular Biology Center, Cancer Institute, University of Tennessee Health Science Center, Memphis, TN, USA

Nicola L. P. Barnes Department of Academic Surgery, South Manchester University Hospital, Manchester, UK

Robert Barouki Inserm UMR-S 1124, Université Paris Descartes, Paris, France

Juan Miguel Barros-Dios Department of Preventive Medicine and Public Health, School of Medicine, University of Santiago de Compostela, Santiago de Compostela, Spain

Harry Bartelink Department of Radiotherapy, The Netherlands Cancer Institute–Antoni van Leeuwenhoek Hospital, Amsterdam, The Netherlands

Stefan Barth Institute of Infectious Disease and Molecular Medicine and Department of Integrative Biomedical Sciences, University of Cape Town, Cape Town, South Africa

Helmut Bartsch Division of Toxicology and Cancer Risk Factors, German Cancer Research Center (DKFZ), Heidelberg, Germany

Thomas Barz Max-Panck-Institut für Psychiatrie, Munich, Germany

Holger Bastians Abt. Molekulare Onkologie, Universitätsmedizin Göttingen, Göttingen, Germany

Anna Batistatou Ioannina University Medical School, Ioannina, Greece

Surinder K. Batra Eppley Institute for Research in Cancer and Allied Diseases and Department of Biochemistry and Molecular Biology, University of Nebraska Medical Center, Omaha, NE, USA

Frederic Batteux Faculté de Médecine Paris – Descartes, UPRES 18-33, Groupe Hospitalier Cochin – Saint Vincent de Paul, Paris, France

Jacques Baudier INSERM Unité Mixte 873, Grenoble, France

Paul Bauer Pfizer Research Technology Center, Cambridge, MA, USA

Tobias Bäuerle Institute of Radiology, University Medical Center Erlangen, Erlangen, Germany

Asne R. Bauskin Department of Medicine, Centre for Immunology, St. Vincent's Hospital, University of New South Wales, Sydney, NSW, Australia

Boon-Huat Bay Department of Anatomy, National University of Singapore, Singapore, Singapore

Jean-Claude Béani Clinique Universitaire de Dermato-Vénéréologie, Photobiologie et Allergologie, Pôle Pluridisciplinaire de Médecine, CHU de Grenoble, Grenoble, France

Nicole Beauchemin Goodman Cancer Research Centre, McGill University, Montreal, QC, Canada

John F. Bechberger Department of Cellular and Physiological Sciences, The University of British Columbia, Vancouver, BC, Canada

Gerhild Becker Department of Palliative Care, University Hospital Freiburg, Freiburg, Germany

Katrin Anne Becker Department of Molecular Biology, University of Duisburg-Essen, Essen, Germany

Marie E. Beckner Department of Pathology, University of Pittsburgh School of Medicine, Pittsburgh, PA, USA

Roberto Bei Department of Clinical Sciences and Translational Medicine, Faculty of Medicine, University of Rome "Tor Vergata", Rome, Italy

Claus Belka Department of Radiation Oncology, University of Tübingen, Tübingen, Germany

Anita C. Bellail Department of Pathology and Laboratory Medicine, Henry Ford Health System, Detroit, MI, USA

Larissa Belov School of Molecular and Microbial Biosciences, University of Sydney, Sydney, NSW, Australia

P. Annécie Benatrehina College of Pharmacy, The Ohio State University, Columbus, OH, USA

Maurizio Bendandi Department of Clinical Medicine, School of Medicine, Ross University, Roseau, Commonwealth of Dominica

Yaacov Ben-David Division of Molecular and Cellular Biology, Sunnybrook Health Sciences Centre, Toronto, ON, Canada

Martin Benesch Division of Pediatric Hematology and Oncology, Department of Pediatrics and Adolescence Medicine, Medical University of Graz, Graz, Austria

Suzanne M. Benjes Cancer Genetics Research, University of Otago, Christchurch, New Zealand

Carmen Berasain Division of Hepatology, CIMA, University of Navarra, Pamplona, Spain

Alan Berezov Department of Pathology, Laboratory Medicine and Abramson Cancer Center, University of Pennsylvania, Philadelphia, PA, USA

Rob J. W. Berg University Medical Center Utrecht, Utrecht, The Netherlands

Corinna Bergelt Institute of Medical Psychology, University Medical Center Hamburg-Eppendorf, Hamburg, Germany

Rene Bernards The Netherlands Cancer Institute, Amsterdam, The Netherlands

Zwi Berneman Vaccine and Infections Disease Institute (VAXINFECTIO) Laboratory of Experimental Hematology, Faculty of Medicine and Health Sciences, University of Antwerp, Edegem, Belgium

Jérôme Bertherat Endocrinology, Metabolism and Cancer Department, INSERM U567, Institut Cochin, Paris, France

Saverio Bettuzzi Department of Biomedicine, Biotechnology and Translational Research, University of Parma, Parma, Italy

Arun Bhardwaj Department of Oncologic Sciences, Mitchell Cancer Institute, University of South Alabama, Mobile, AL, USA

Kumar M. R. Bhat Department of Anatomy, Kasturba Medical College, Manipal University, Manipal, Karnataka, India

Malaya Bhattacharya-Chatterjee University of Cincinnati and The Barrett Cancer Center, Cincinnati, OH, USA

Caterina Bianco Division of Extramural Activities, National Institutes of Health, Rockville, MD, USA

Tina Bianco-Miotto Robinson Research Institute and School of Agriculture, Food and Wine, The University of Adelaide, Adelaide, SA, Australia

Jean-Michel Bidart Department of Clinical Biology, Institut Gustave-Roussy, Villejuif, France

Jaclyn A. Biegel Department of Pathology and Laboratory Medicine, Children's Hospital of Los Angeles, Los Angeles, CA, USA

Margherita Bignami Istituto Superiore di Sanita', Rome, Italy

Irene V. Bijnsdorp Department of Medical Oncology, VU University Medical Center, Amsterdam, The Netherlands

Chen Bing Institute of Ageing and Chronic Disease, University of Liverpool, Liverpool, UK

Angelique Blanckenberg Department of Chemistry and Polymer Science, Stellenbosch University, Matieland, South Africa

Giovanni Blandino Translational Oncogenomic Laboratory, Regina Elena Cancer Institute, Rome, Italy

David E. Blask Laboratory of Chrono-Neuroendocrine Oncology, Department of Structural and Cellular Biology, Tulane University School of Medicine, New Orleans, LA, USA

Jonathan Blay Department of Pharmacology, Dalhousie University, Halifax, NS, Canada

Peter Blume-Jensen Xtuit Pharmaceuticals, Boston, MA, USA

Sarah Bocchini Department of Experimental Medicine, University of Rome "Sapienza", Rome, Italy

Ann M. Bode The Hormel Institute, University of Minnesota, Austin, MN, USA

Paolo Boffetta Icahn School of Medicine at Mount Sinai, New York, NY, USA

Stefan K. Bohlander Faculty of Medical and Health Sciences, The University of Auckland, Auckland, New Zealand

Valentina Bollati EPIGET - Epidemiology, Epigenetics and Toxicology Lab - Department of Clinical Sciences and Community Health, University of Milan, Milan, Italy

Subbarao Bondada Department of Microbiology, Immunology and Molecular Genetics, University of Kentucky, Lexington, KY, USA

Maria Grazia Borrello Department of Experimental Oncology and Molecular Medicine, Fondazione IRCCS Istituto Nazionale dei Tumori, Milan, Italy

Giuseppe Borzacchiello Department of Veterinary Medicine and Animal Productions, University of Naples "Federico II", Naples, Italy

Valerie Bosch Forschungsschwerpunkt Infektion und Krebs, F020, German Cancer Research Center (DKFZ), Heidelberg, Germany

Chris Boshoff Cancer Research Campaign Viral Oncology Group, Wolfson Institute for Biomedical Research, University College London, London, UK

Irina Bosman Institute of Pharmacy, University of Bonn, Bonn, Germany

Galina I. Botchkina Department of Pathology, Stony Brook University, Stony Brook, NY, USA

Institute of Chemical Biology and Drug Discovery, Stony Brook University, Stony Brook, NY, USA

Franck Bourdeaut Département de pédiatrie, INSERM 830, Biologie et génétique des tumeurs, Institut Curie, Paris, France

Jean-Pierre Bourquin Pediatric Oncology, University Children's Hospital Zurich, Zurich, Switzerland

Hassan Bousbaa Instituto Investigação Formação Avançada Ciências Tecnologias Saúde, CESPU – Cooperativa de Ensino Superior Politecnico e Universitario, Gandra PRD, Portugal

Norman Boyd Campbell Family Institute for Breast Cancer Research, Ontario Cancer Institute, Toronto, ON, Canada

Sven Brandau Department of Otorhinolaryngology, University Duisburg-Essen, Essen, Germany

Burkhard H. Brandt Institute of Clinical Chemistry, University Medical Centre Schleswig-Holstein, Kiel, Germany

Hiltrud Brauch Breast Cancer Susceptibility and Pharmacogenomics, Dr. Margarete Fischer-Bosch-Institute of Clinical Pharmacology, University of Tübingen, Stuttgart, Germany

Massimo Breccia Department of Cellular Biotechnologies and Hematology, Sapienza University, Rome, Italy

Samuel N. Breit Cytokine Biology and Inflammation Research Program, St Vincent's Centre for Applied Medical Research (AMR), St Vincent's Hospital, Sydney, NSW, Australia

Edwin Bremer Department of Pathology and Laboratory Medicine, Section Medical Biology, Laboratory for Tumor Immunology, University Medical Center Groningen, Groningen, The Netherlands

Catherine Brenner INSERM UMR-S 769, Labex LERMIT, Châtenay-Malabry, University of Paris Sud, Paris, France

David J. Brenner Department of Radiation Oncology, Columbia University, New York, NY, USA

Amanda E. Brinker Department of Cancer Biology, University of Kansas Cancer Center, The University of Kansas Medical Center, Kansas City, KS, USA

Nikko Brix Clinic for Radiotherapy and Radiation Oncology, LMU Munich, Munich, Germany

Katja Brocke-Heidrich Praxis für Naturheilkunde und ganzheitliche Therapie, Leipzig, Germany

Angela Brodie University of Maryland School of Medicine, Baltimore, MD, USA

Jonathan Brody Department of Surgery, Thomas Jefferson University, Philadelphia, PA, USA

Christopher L. Brooks Institute for Cancer Genetics, and Department of Pathology, College of Physicians and Surgeons, Columbia University, New York, NY, USA

Mai N. Brooks Surgical Oncology, School of Medicine, University of California, Los Angeles, CA, USA

David A. Brown St. Vincent's Centre for Applied Medical Research, St Vincent's Hospital, University of New South Wales, Sydney, NSW, Australia

Karen Brown Department of Cancer Studies, University of Leicester, Leicester, UK

Kevin Brown University of Florida, College of Medicine, Gainesville, FL, USA

Tilman Brummer Institut für Molekulare Medizin und Zellforschung, Zentrum für Biochemie und Molekulare Zellforschung (ZBMZ), Albert-Ludwigs-Universität Freiburg, Freiburg, Germany

Antonio Brunetti Department of Health Sciences, University of Catanzaro "Magna Græcia", Catanzaro, Italy

Andreas K. Buck Department of Nuclear Medicine, University of Würzburg, Würzburg, Germany

Laszlo Buday Department of Medical Chemistry, Semmelweis University Medical School, Budapest, Hungary

Marie Annick Buendia Hopital Paul Brousse, Inserm U785, Centre Hépatobiliaire, Villejuif, France

Ralf Buettner City of Hope National Medical Center and Beckman Research Institute, Duarte, CA, USA

Nigel J. Bundred Department of Academic Surgery, South Manchester University Hospital, Manchester, UK

Alexander Bürkle Department of Biology, University of Konstanz, Konstanz, Germany

Barbara Burwinkel Division Molecular Biology of Breast Cancer, University of Heidelberg, Department of Gynecology and Obstetrics, Heidelberg, Germany

Xavier Busquets Department of Biology, University of the Balearic Islands, Palma de Mallorca, Spain

Jagdish Butany Laboratory Medicine and Pathobiology, University Health Network/Toronto, Toronto, ON, Canada

Neville J. Butcher School of Biomedical Sciences, University of Queensland, St Lucia, QLD, Australia

Timon P. H. Buys Department of Cancer Genetics and Developmental Biology, British Columbia Cancer Research Centre, Vancouver, BC, Canada

Miguel A. Cabrita Cancer Therapeutics Program, Ottawa Hospital Research Institute, Ottawa, ON, Canada

Jean Cadet Département de Médecine Nucléaire et Radiobiologie, Faculté de Médecine et des Sciences de la Santé, Université de Sherbrooke, Sherbrooke, QC, Canada

Yi Cai Department of Pathology, Baylor College of Medicine, Houston, TX, USA

Yiqiang Cai Section of Nephrology, Yale University School of Medicine, New Haven, CT, USA

Bruno Calabretta Kimmel Cancer Institute, Thomas Jefferson University, Philadelphia, PA, USA

Daniele Calistri Molecular Laboratory, Istituto Scientifico Romagnolo per lo Studio e la Cura dei Tumori (I.R.S.T.), Meldola, Italy

Javier Camacho Department of Pharmacology, Centro de Investigación y de Estudios Avanzados del I.P.N., Mexico City, D.F., Mexico

William G. Cance Departments of Surgical Oncology, Roswell Park Cancer Institute, Buffalo, NY, USA

Amparo Cano Departamento de Bioquímica, Facultad de Medicina, UAM, Instituto de Investigaciones Biomédicas "Alberto Sols" CSIC-UAM IdiPAZ, Madrid, Spain

Anthony J. Capobianco Molecular and Cellular Oncogenesis Program, The Wistar Institute, Philadelphia, PA, USA

Emilia Caputo Institute of Genetics and Biophysics – ABT, Napoli, Italy

Salvatore J. Caradonna Department of Molecular Biology, Rowan University School of Osteopathic Medicine, Stratford, NJ, USA

Michele Carbone University of Hawaii Cancer Center, Honolulu, HI, USA

Vinicio Carloni University of Florence, Florence, Italy

Neil O. Carragher Drug Discovery Group, Edinburgh Cancer Research Centre, University of Edinburgh, Edinburgh, UK

Michela Casanova Pediatric Oncology Unit, Fondazione IRCCS Istituto Nazionale Tumori, Milano, Italy

Wolfgang H. Caselmann Medizinische Klinik und Poliklinik I, Rheinische Friedrich-Wilhelms-Universität, Bonn, Germany

Giuliana Cassinelli Molecular Pharmacology Unit, Department of Experimental Oncology and Molecular Medicine, Fondazione IRCCS Istituto Nazionale dei Tumori, Milan, Italy

Webster K. Cavenee Ludwig Institute for Cancer Research, UCSD, La Jolla, CA, USA

Esteban Celis Georgia Cancer Center, Augusta University, Augusta, GA, USA

Chiswili Chabu Howard Hughes Medical Institute, Yale University School of Medicine, New Haven, CT, USA

Wook-Jin Chae Department of Immunobiology, Yale University School of Medicine, New Haven, CT, USA

Ho Man Chan Division of Biochemistry and Molecular Biology, Davidson Building, University of Glasgow, Glasgow, UK

Shing Leng Chan Cancer Science Institute of Singapore, National University of Singapore, Singapore, Singapore

Dawn S. Chandler Department of Pediatrics, Columbus Children's Research Institute, Center for Childhood Cancer, The Ohio State University School of Medicine, Columbus, OH, USA

Guru Chandramouly Beth Israel Deaconess Medical Center, Harvard Medical School, Boston, MA, USA

Mau-Sun Chang Institute of Biochemical Sciences, National Taiwan University, Taipei, Taiwan

Mei-Chi Chang Biomedical Science Team, Chang Gung Institute of Technology, Taoyuan, Taiwan

Lung-Ji Chang Department of Molecular Genetics and Microbiology, College of Medicine, University of Florida, Gainesville, FL, USA

Jane C. J. Chao School of Nutrition and Health Sciences, Taipei Medical University, Taipei, Taiwan

Christine Chaponnier Department of Pathology and Immunology, University of Geneva, Geneva, Switzerland

Konstantinos Charalabopoulos Ioannina University Medical School, Ioannina, Greece

Malay Chatterjee Department of Pharmaceutical Technology, Jadavpur University, Calcutta, West Bengal, India

Sunil K. Chatterjee University of Cincinnati and The Barrett Cancer Center, Cincinnati, OH, USA

Gautam Chaudhuri Department of Molecular and Medical Pharmacology and Department of Obstetrics and Gynecology, David Geffen School of Medicine at UCLA, Los Angeles, CA, USA

M. Asif Chaudry University Department of Surgery, Royal Free and University College London Medical School, London, UK

Dharminder Chauhan Department of Medical Oncology, The Jerome Lipper Multiple Myeloma Center, Dana Farber Cancer Institute, Harvard Medical School, Boston, MA, USA

Jeremy P. Cheadle Institute of Medical Genetics, Cardiff University, Heath Park, Cardiff, UK

Ai-Ping Chen Department of Gynecology, Affiliated Hospital of Qingdao University, Qingdao, China

Chienling Chen Department of Molecular Genetics and Microbiology, College of Medicine, University of Florida, Gainesville, FL, USA

Herbert Chen Department of Surgery, University of Alabama - Birmingham (UAB) School of Medicine, UAB Hospital and Health System, University of Alabama Comprehensive Cancer Center, Birmingham, AL, UK

Jie Chen Department of Pharmacology and Pharmacy, The University of Hong Kong, Hong Kong, China

Sai-Juan Chen State Key Laboratory of Medical Genomics, Shanghai Institute of Hematology, Rui-Jin Hospital, Shanghai Jiao Tong University School of Medicine, Shanghai, People's Republic of China

Taosheng Chen Chemical Biology and Therapeutics, St. Jude Children's Research Hospital, Memphis, TN, USA

Wenxing Chen Department of Clinical Pharmacy, College of Pharmacy, Nanjing University of Chinese Medicine, Nanjing, China

Yingchi Chen Department of Molecular Genetics and Microbiology, College of Medicine, University of Florida, Gainesville, FL, USA

Zhu Chen State Key Laboratory of Medical Genomics, Shanghai Institute of Hematology, Rui-Jin Hospital, Shanghai Jiao Tong University School of Medicine, Shanghai, People's Republic of China

George Z. Cheng Harvard Medical School, Boston, MA, USA

Jin Q. Cheng Molecular Oncology Program and Research Institute, H. Lee Moffitt Cancer Center, University of South Florida College of Medicine, Tampa, FL, USA

Liang Cheng Department of Pathology and Laboratory Medicine, Indiana University School of Medicine, Indianapolis, IN, USA

Chun Hei Antonio Cheung Department of Pharmacology and Institute of Basic Medical Sciences, College of Medicine, National Cheng Kung University, Tainan, Taiwan, Republic of China

Ya-Hui Chi Institute of Biotechnology and Pharmaceutical Research, National Health Research Institutes, Zhunan, Taiwan

Martyn A. Chidgey School of Cancer Sciences, University of Birmingham, Birmingham, UK

Sudhakar Chintharlapalli Department of Veterinary Physiology and Pharmacology, Texas A&M University, College Station, TX, USA

Alexandre Chlenski Department of Pediatrics, Section of Hematology/Oncology, University of Chicago, Chicago, IL, USA

Daniel C. Cho Beth Israel Deaconess Medical Center, Boston, MA, USA

William Chi-Shing Cho Department of Clinical Oncology, Queen Elizabeth Hospital, Kowloon, Hong Kong

Michael Chopp Neurology Research, Henry Ford Health System, Detroit, MI, USA

Pei-Lun Chou Division of Allergy-Immunology-Rheumatology, Department of Internal Medicine, Lin Shin Hospital, Taichung, Taiwan

Claus Christensen Department of Cancer Genetics, Danish Cancer Society, Copenhagen, Denmark

Rikke Christensen Clinical Genetics, Aarhus University Hospital, Aarhus, Denmark

Gerhard Christofori Department of Biomedicine, University of Basel, Basel, Switzerland

Richard I. Christopherson School of Life and Environmental Sciences, University of Sydney, Sydney, NSW, Australia

Fong-Fong Chu Department of Cancer Biology, Beckman Research Institute of City of Hope, Duarte, CA, USA

Wen-Ming Chu Cancer Biology Program, University of Hawaii Cancer Center, Honolulu, HI, USA

Bong-Hyun Chung BioNanotechnology Research Center, Korea Research Institute of Bioscience and Biotechnology, Yuseong, Daejeon, Republic of Korea

Fung-Lung Chung Department of Oncology, Lombardi Comprehensive Cancer Center, Georgetown University Medical Center, Washington, DC, USA

Jacky K. H. Chung Department of Medical Genetics and Microbiology, University of Toronto, Toronto, ON, Canada

Sue Clark Imperial College London, London, UK

Pier Paolo Claudio The University of Mississippi, Medical Center Cancer Institute, Jackson, MS, USA

Elizabeth B. Claus Department of Epidemiology and Public Health, Yale University School of Medicine, New Haven, CT, USA

Pascal Clayette SPI-BIO, Service de Neurovirologie, CEA, CRSSA, EPHE, Fontenay aux Roses Cedex, France

Dahn L. Clemens Research Service, Veterans Administration Medical Center, Omaha, NE, USA

Steven C. Clifford Northern Institute for Cancer Research, Newcastle University, Newcastle upon Tyne, UK

Kevin A. Cockell Nutrition Research Division, Health Canada, Ottawa, ON, Canada

Susan L. Cohn Department of Pediatrics, Section of Hematology/Oncology, University of Chicago, Chicago, IL, USA

Graham A. Colditz Washington University in St. Louis, St. Louis, MO, USA

Paola Collini Anatomic Pathology Department, Fondazione IRCCS Istituto Nazionale Tumori, Milano, Italy

Andrew R. Collins Department of Nutrition, University of Oslo, Oslo, Norway

Nicoletta Colombo Istituto Europeo di Oncologia, Milan, Italy

Joan W. Conaway Stowers Institute for Medical Research, Kansas, MO, USA

Ronald C. Conaway Stowers Institute for Medical Research, Kansas, MO, USA

Bong-Hyun Chung: deceased.

Lellys Mariella Contreras Department of Cancer Biology, University of Kansas Cancer Center, The University of Kansas Medical Center, Kansas City, KS, USA

Amanda E. Conway Molecular Cancer Biology, Duke University Medical Center, Durham, NC, USA

Nathalie Cools Vaccine and Infections Disease Institute (VAXINFECTIO) Laboratory of Experimental Hematology, Faculty of Medicine and Health Sciences, University of Antwerp, Edegem, Belgium

Helen C. Cooney UCD School of Biomolecular and Biomedical Science, UCD Conway Institute, University College Dublin, Dublin, Ireland

Scott Coonrod Baker Institute for Animal Health, Department of Biomedical Sciences, School of Veterinary Medicine, Cornell University, Ithaca, NY, USA

Kumarasen Cooper Pathology and Laboratory Medicine, Perelman School of Medicine at the University of Pennsylvania, Philadelphia, PA, USA

Laurence J. N. Cooper Division of Pediatrics, Department of Immunology, MD Anderson Cancer Center, Houston, TX, USA

Michael K. Cooper Department of Neurology, Vanderbilt Medical Center, Nashville, TN, USA

Peter J. Coopman IRCM, INSERM U1194, Montpellier Cancer Research Institute, Montpellier, France

Lanfranco Corazzi Department of Experimental Medicine, University of Perugia, Perugia, Italy

Maria Paola Costi Department of Pharmaceutical Sciences, University of Modena and Reggio Emilia, Modena, Italy

Richard J. Cote Department of Pathology, Miller School of Medicine, University of Miami, Miami, FL, USA

Massimo Cristofanilli Division of Hematology and Oncology, Robert H Lurie Comprehensive Cancer Center, Chicago, IL, USA

Marcus V. Cronauer Department of Urology, University Hospital Schleswig-Holstein – Campus Lübeck, Lübeck, Germany

Sidney Croul Department of Pathology, UHN, University of Toronto, Toronto, ON, Canada

Ronald G. Crystal Division of Pulmonary and Critical Care Medicine, Weill Cornell Medical College, New York, NY, USA

Bruce D. Cuevas Department of Molecular Pharmacology and Therapeutics, Stritch School of Medicine, Loyola University Chicago, Maywood, IL, USA

Jiuwei Cui Jilin University, Changchun, Jilin, China

Edna Cukierman Basic Science/Tumor Cell Biology, Fox Chase Cancer Center, Philadelphia, PA, USA

Zoran Culig Department of Urology, Innsbruck Medical University, Innsbruck, Austria

David Cunningham Department of Medicine, The Royal Marsden NHS Foundation Trust, London, UK

David T. Curiel Division of Cancer Biology, Washington University, St. Louis, MO, USA

Franck Cuttitta NCI Angiogenesis Core Facility, National Cancer Institute, National Institutes of Health, Advanced Technology Center, Gaithersburg, MD, USA

Andrea Cziffer-Paul Department of Pathology, The Mount Sinai School of Medicine, New York, NY, USA

Massimino D'Armiento Department of Experimental Medicine, University of Rome "Sapienza", Rome, Italy

Yun Dai Department of Gastroenterology, Peking University First Hospital, Beijing, China

Yataro Daigo Institute of Medical Science, The University of Tokyo, Tokyo, Japan

Lokesh Dalasanur Nagaprashantha City of Hope National Medical Center, Duarte, CA, USA

Ashraf Dallol Centre of Innovation in Personalised Medicine, King Abdulaziz University, Jeddah, Kingdom of Saudi Arabia

Tamas Dalmay School of Biological Sciences, University of East Anglia, Norwich, UK

Ivan Damjanov Department of Pathology, University of Kansas School of Medicine, Kansas City, KS, USA

Vincent Dammai Dammai-Morgan Scientific Consultants LLC, Mount Pleasant, SC, USA

Chendil Damodaran University of Louisville, Louisville, KY, USA

Janet E. Dancey Canadian Cancer Trials Group, Queen's University, Kingston, ON, Canada

Nadia Dandachi Department of Internal Medicine, Division of Oncology, Medical University Graz, Graz, Austria

Chi V. Dang Division of Hematology, Department of Medicine, Johns Hopkins University School of Medicine, Baltimore, MD, USA

Alla Danilkovitch-Miagkova National Cancer Institute-FCRDC, Frederick, MD, USA

Kakoli Das Cancer and Stem Cell Biology Program, Duke-NUS Graduate Medical School, Singapore, Singapore

Kaustubh Datta Department of Urology Research, Department of Biochemistry and Molecular Biology, Mayo Clinic College of Medicine, Rochester, MN, USA

Pran K. Datta Departments of Surgery and Cancer Biology, Vanderbilt-Ingram Cancer Center, Vanderbilt University School of Medicine, Nashville, TN, USA
Department of Medicine, University of Alabama at Birmingham, Birmingham, AL, USA

Leonor David IPATIMUP (Institute of Molecular Pathology and Immunology of the University of Porto) and Medical Faculty of the University of Porto, Porto, Portugal

David Mark Davies Department of Oncology, South West Wales Cancer Centre, Swansea, UK

Juhayna Kassem Davis Carolinas HealthCare System, Charlotte, NC, USA

Alexey Davydov Fox Chase Cancer Center, Philadelphia, PA, USA

Shaheenah Dawood Department of Medical Oncology, Dubai Hospital, Dubai, United Arab Emirates

Robert Day Department of Surgery/Division of Urology, Institut de Pharmacologie, Faculté de Médecine et des sciences de la santé, Université de Sherbrooke, Sherbrooke, QC, Canada

Terry Day Head and Neck Tumor Program, Hollings Cancer Center, Medical University of South Carolina, Charleston, SC, USA

Suzane Ramos da Silva Department of Molecular Microbiology and Immunology, Keck School of Medicine, University of Southern California, Los Angeles, CA, USA

Enrique de Alava Institute of Biomedicine of Sevilla (IBiS), Virgen del Rocio University Hospital /CSIC/University de Sevilla, Seville, Spain

Diederik de Bruijn Department of Human Genetics, Radboud University Nijmegen Medical Centre, Nijmegen, The Netherlands

Floris Aart de Jong Amgen BV, Breda, The Netherlands

Vincenzo de Laurenzi Department of Experimental Medicine and Biochemical Sciences, University of Tor Vergata, Rome, Italy

Ben O. de Lumen Department of Nutritional Sciences and Toxicology, University of California at Berkeley, Berkeley, CA, USA

Elvira de Mejia Department of Food Science and Human Nutrition, University of Illinois, Urbana-Champaign, IL, USA

Ana Ramirez de Molina Nutritional Genomics and Cancer Unit, IMDEA Food Institute, Madrid, Spain

Christiane de Wolf-Peeters Department of Pathology, University Hospitals of K.U. Leuven, Leuven, Belgium

Jochen Decker Hematology Oncology Medical School Clinic III, University of Mainz, Mainz, Germany

P. Markus Deckert Zentrum für Innere Medizin II – Abteilung für Onkologie und Palliativmedizin, Klinikum Brandenburg, Brandenburg an der Havel, Germany

Francesca Degrassi Institute of Molecular Biology and Pathology IBMN c/o "Sapienza" University, Italian National Research Council CNR, Rome, Italy

Amir R. Dehdashti Division of Neurosurgery, University of Toronto, Toronto, ON, Canada

Maryse Delehedde R&D Lunginnov, Campus de l'Institut Pasteur de Lille, Lille, France

Olivier Dellis Signalisation Calcique et Interactions Cellulaires dans le Foie, INSERM UMR-S 1174, Université Paris-Sud 11, Orsay, France

Renée M. Demarest Molecular and Cellular Oncogenesis Program, The Wistar Institute, Philadelphia, PA, USA

Berna Demircan University of Florida, College of Medicine, Gainesville, FL, USA

Miriam Deniz Department of Obstetrics and Gynaecology, University of Ulm, Ulm, Germany

Samuel Denmeade The Sidney Kimmel Comprehensive Cancer Center, Johns Hopkins, Baltimore, MD, USA

David A. Denning Department of Surgery, Marshall University, Huntington, WV, USA

Sylviane Dennler Department of Molecular Cell Biology, Leiden University Medical Center, Leiden, The Netherlands

Channing J. Der University of North Carolina at Chapel Hill, Chapel Hill, NC, USA

Barbara Deschler Comprehensive Cancer Center Mainfranken, Clinical Trials Office, University of Würzburg, Würzburg, Germany

Chantal Desdouets Institut Cochin, Université Paris Descartes, CNRS, Paris, France

Peter Devilee Human Genetics, Leiden University Medical Center, Leiden, The Netherlands

Mark W. Dewhirst Department of Radiation Oncology, Duke University, Durham, NC, USA

Girish Dhall Division of Hematology-Oncology, Department of Pediatrics, Children's Hospital Los Angeles and the Keck School of Medicine, University of Southern California, Los Angeles, CA, USA

Danny N. Dhanasekaran Stephenson Cancer Center, University of Oklahoma Health Sciences Center, Oklahoma City, OK, USA

Pier Paolo Di Fiore IFOM, the FIRC Institute of Molecular Oncology, Milan, Italy

Giuseppe Di Lorenzo Cattedra di Oncologia Medica, Dipartimento di Endocrinologia e Oncologia molecolare e clinica, Università degli Studi "Federico II", Naples, Italy

Dario Di Luca Department of Medical Sciences, University of Ferrara, Ferrara, Italy

Marc Diederich College of Pharmacy, Seoul National University, Seoul, South Korea

Joseph DiFranza Department of Family Medicine and Community Health, University of Massachusetts Medical Center, Worcester, MA, USA

Martin Digweed Institute of Medical and Human Genetics, Charité – Universitätsmedizin Berlin, Berlin, Germany

Peter ten Dijke Department of Molecular Cell Biology, Leiden University Medical Center, Leiden, The Netherlands

Gerard Dijkstra University Medical Center Groningen, University of Groningen, Groningen, The Netherlands

Nathalie Dijsselbloem Lab of Eukaryotic Gene Expression, LEGEST-University Gent, Ghent, Belgium

Helen Dimaras The Hospital for Sick Children, Department of Ophthalmology and Vision Science, The University of Toronto, Toronto, ON, Canada

Jian Ding State Key Laboratory of Drug Research, Shanghai Institute of Materia Medica, Shanghai Institutes for Biological Sciences, Chinese Academy of Sciences, Shanghai, People's Republic of China

Zhaoxia Ding Department of Gynecology, Affiliated Hospital of Qingdao University, Qingdao, China

Jürgen Dittmer Klinik für Gynäkologie, Universität Halle-Wittenberg, Halle (Saale), Germany

Henrik J. Ditzel Department of Cancer and Inflammation Reserch, Institute fo Molecular Medicine, University of Southern Denmark, Odense C, Denmark

Dan Dixon Cancer Biology, University of Kansas Medical Center, Kansas City, KS, USA

Cholpon S. Djuzenova Klinik für Strahlentherapie der Universität Würzburg, Würzburg, Germany

Christian Doehn Urologikum Lübeck, Lübeck, Germany

Yasufumi Doi Department of Medicine and Clinical Science, Graduate School of Medical Sciences, Kyushu University, Fukuoka, Japan

Milos Dokmanovic Division of Monoclonal Antibodies, Office of Biotechnology Products, Office of Pharmaceutical Science, Center for Drug Evaluation and Research, U.S. Food and Drug Administration, Bethesda, MD, USA

Qihan Dong The University of Western Sydney, Sydney, NSW, Australia

Department of Endocrinology, Central Clinical School, Royal Prince Alfred Hospital, The University of Sydney, Sydney, NSW, Australia

Zigang Dong The Hormel Institute, University of Minnesota, Austin, MN, USA

Ben Doron Oregon Health and Science University, Portland, OR, USA

Qing Ping Dou The Prevention Program, Barbara Ann Karmanos Cancer Institute and Department of Pathology, School of Medicine, Wayne State University, Detroit, MI, USA

Thierry Douki Laboratoire "Lésions des Acides Nucléiques", Institute Nanosciences et Cryogénie, Grenoble, France

Harry A. Drabkin Division of Hematology-Oncology, Medical University of South Carolina and the Hollings Cancer Center, Charleston, SC, USA

Tommaso A. Dragani Fondazione IRCCS Istituto Nazionale Tumori, Milan, Italy

Kenneth Drake Department of Chemistry and Biochemistry, University of Texas at Arlington, Arlington, TX, USA

Martin Dreyling Department of Internal Medicine III, University of Munich, Großhadern, Munich, Germany

Nathalie Druesne-Pecollo UMR U1153 INSERM, U1125 INRA, CNAM, Université Paris 13, Centre de Recherche Epidémiologie et Statistique Sorbonne Paris Cité, Bobigny, France

Brian J. Druker Oregon Health and Science University Cancer Institute, Portland, OR, USA

Denis Drygin Pimera, Inc., San Diego, CA, USA

Raymond N. DuBois ASU Biodesign Institute, Tempe, AZ, USA

Dan G. Duda Steele Laboratories for Tumor Biology, Department of Radiation Oncology, Massachusetts General Hospital and Harvard Medical School, Boston, MA, USA

Jaquelin P. Dudley Department of Molecular Biosciences and Institute for Cellular and Molecular Biology, The University of Texas at Austin, Austin, TX, USA

Roy J. Duhé Department of Pharmacology and Toxicology, University of Mississippi Medical Center, Jackson, MS, USA

Department of Radiation Oncology, University of Mississippi Medical Center, Jackson, MS, USA

Ignacio Duran Department of Medical Oncology and Hematology, Robert and Maggie Bras and Family New Drug Development Program, Princess Margaret Hospital, Toronto, ON, Canada

Stephen T. Durant R&D, Oncology, Innovative Medicines, AstraZeneca, Little Chesterford, UK

Meenakshi Dwivedi Department of Life Science, Hanyang University, Seoul, South Korea

Madalene A. Earp Department of Health Sciences Research, Mayo Clinic College of Medicine, Rochester, MN, USA

Behfar Ehdaie Department of Surgery, Urology Service, Memorial Sloan-Kettering Cancer Center, New York, NY, USA

Justis P. Ehlers Cole Eye Institute, Cleveland Clinic, Cleveland, OH, USA

Gerhard Eisenbrand Department of Chemistry, Division of Food Chemistry and Toxicology, University of Kaiserslautern, Kaiserslautern, Germany

Mohamad Elbaz Department of Pathology, Comprehensive Cancer Centre, The Ohio State Medical Centre, Columbus, OH, USA

Patricia V. Elizalde Laboratory of Molecular Mechanisms of Carcinogenesis, Institute of Biology and Experimental Medicine (IBYME), CONICET, Buenos Aires, Argentina

Bassel El-Rayes Department of Hematology and Medical Oncology, Emory University School of Medicine, Atlanta, GA, USA
Winship Cancer Institute of Emory University, Atlanta, GA, USA

Mitsuru Emi Departments of Obstetrics and Gynecology, Nippon Medical School, Kawasaki and Tokyo, Japan

Steffen Emmert Clinic for Dermatology and Venereology, University Medical Center Rostock, Rostock, Germany

Caroline End Division of Molecular Genome Analysis, DKFZ, Heidelberg, Germany

Daniela Endt Department of Human Genetics, Biozentrum University of Würzburg, Würzburg, Germany

Rainer Engers Institute of Pathology, University Hospital Düsseldorf, Düsseldorf, Germany

Marica Eoli Unit of Clinical Neuro-Oncology, Istituto Neurologico Besta, Milan, Italy

Anat Erdreich-Epstein Division of Hematology-Oncology, Department of Pediatrics, Children's Hospital Los Angeles and the Keck School of Medicine, University of Southern California, Los Angeles, CA, USA

Süleyman Ergün Institut für Anatomie und Zellbiologie, Julius-Maximilians-Universität Würzburg, Würzburg, Germany

Pablo V. Escribá Department of Biology, University of the Balearic Islands, Palma de Mallorca, Spain

Nuria Están-Capell Service of Clinical Analysis, Dr. Peset University Hospital, Valencia, Spain

Konstantinos Evangelou Molecular Carcinogenesis Group, Laboratory of Histology-Embryology, Medical School, National and Kapodistrian University of Athens, Athens, Greece

Mark F. Evans Department of Pathology and Laboratory Medicine, University of Vermont, Burlington, VT, USA

B. Mark Evers Department of Surgery, The University of Texas Medical Branch, Galveston, TX, USA

Vera Evtimov Monash University, Melbourne, VIC, Australia

Jörg Fahrer Department of Toxicology, University Medical Center Mainz, Mainz, Germany

Cristina Maria Failla Experimental Immunology Laboratory, IDI-IRCCS, Rome, Italy

Marco Falasca Faculty of Health Sciences, School of Biomedical Sciences, Curtin University, Perth, WA, Australia

Fang Fan Department of Pathology, University of Kansas School of Medicine, Kansas City, KS, USA

Saijun Fan Long Island Jewish Medical Center, Albert Einstein College of Medicine, Bronx, NY, USA

Bingliang Fang Department of Thoracic and Cardiovascular Surgery, The University of Texas MD Anderson Cancer Center, Houston, TX, USA

Jinxu Fang Department of Chemical Engineering and Materials Science, Viterbi School of Engineering, University of Southern California, Los Angeles, CA, USA

Lei Fang Dermatology Branch, National Cancer Institute, National Institutes of Health, Bethesda, MD, USA

Valeria R. Fantin Merck Research Laboratories, Boston, MA, USA

Z. Shadi Farhangrazi Biotrends International, Denver, CO, USA

Omid C. Farokhzad Laboratory of Nanomedicine and Biomaterials, Department of Anesthesiology, Brigham and Women's Hospital, Boston, MA, USA

William L. Farrar National Cancer Institute – Frederick, Frederick, MD, USA

Alessandro Fatatis Department of Pharmacology and Physiology, Drexel University College of Medicine, Philadelphia, PA, USA

Andrew P. Feinberg Department of Medicine and Center for Epigenetics, Institute for Basic Biomedical Sciences, Johns Hopkins University School of Medicine, Baltimore, MD, USA

Mark A. Feitelson Department of Biology, Temple University, Philadelphia, PA, USA

Francesco Feo Department of Biomedical Sciences, Division of Experimental Pathology and Oncology, University of Sassari, Sassari, Italy

Félix Fernández Madrid Department of Internal Medicine, Division of Rheumatology, Wayne State University, Detroit, MI, USA

Paula Fernández-García Department of Biology, University of the Balearic Islands, Palma de Mallorca, Spain

Marie Fernet INSERM U612, Institut Curie-Recherche, Orsay, France

Audrey Ferrand INSERM U.858, Institut de Médecine Moléculaire de Rangueil, IFR150, Université Paul Sabatier, Toulouse, France

Andrea Ferrari Pediatric Oncology Unit, Fondazione IRCCS Istituto Nazionale Tumori, Milano, Italy

Stefania Ferrari Department of Pharmaceutical Sciences, University of Modena and Reggio Emilia, Modena, Italy

Robert A. Figlin Division of Hematology Oncology, Samuel Oschin Comprehensive Cancer Institute, Cedars-Sinai Medical Center, Los Angeles, CA, USA

Lorena L. Figueiredo-Pontes Medical School of Ribeirão Preto, University of São Paulo, Ribeirão Preto, Brazil

Cristina Fillat Institut d'Investigacions Biomèdiques August Pi i Sunyer (IDIBAPS) and Centro de Investigación Biomédica en Red de Enfermedades Raras (CIBERER), Barcelona, Spain

Daniel Finley Department of Cell Biology, Harvard Medical School, Boston, MA, USA

Gaetano Finocchiaro Unit of Experimental Neuro-Oncology, Istituto Nazionale Neurologico Besta, Milan, Italy

Paul B. Fisher Departments of Urology, Pathology and Neurosurgery, Columbia University Medical Center, College of Physicians and Surgeons, New York, NY, USA

James Flanagan Institute of Reproductive and Developmental Biology, Imperial College London, London, UK

Michael Fleischhacker Universitätsklinikum Halle (Saale), Klinik für Innere Medizin I, Schwerpunkt Pneumologie, Halle (Saale), Germany

Eliezer Flescher Department of Human Microbiology, Sackler Faculty of Medicine, Tel Aviv University, Tel Aviv, Israel

Jonathan A. Fletcher Albany Medical College, Albany, NY, USA

Barbara D. Florentine Department of Pathology, Henry Mayo Newhall Memorial Hospital, Valencia, CA, USA

CA and Keck School of Medicine, University of Southern California, Los Angeles, CA, USA

Tamara Floyd Cancer Vaccine Section, National Cancer Institute, National Institutes of Health, Bethesda, MD, USA

Riccardo Fodde Department of Pathology, Josephine Nefkens Institute, Erasmus MC, Rotterdam, The Netherlands

Judah Folkman Children's Hospital and Harvard Medical School, Boston, MA, USA

Hamidreza Fonouni Department of General, Visceral and Transplantation Surgery, University of Heidelberg, Heidelberg, Germany

Kenneth A. Foon The Pittsburgh Cancer Institute, Pittsburgh, PA, USA

Alessandra Forni Department of Occupational and Environmental Health "Clinica del Lavoro L. Devoto", University of Milan, Milan, Italy

David A. Foster Department of Biological Sciences, Hunter College of the City University of New York, New York, NY, USA

Paul Foster Department of Endocrinology and Metabolic Medicine, Imperial College Faculty of Medicine, St. Mary's Hospital, London, UK

Paul Fréneaux Département de Pathologie, Institut Curie, Paris, France

Rodrigo Franco Redox Biology Center, School of Veterinary Medicine and Biomedical Sciences, University of Nebraska-Lincoln, Lincoln, NE, USA

David A. Frank Dana-Farber Cancer Institute and Harvard Medical School, Boston, MA, USA

Stuart J. Frank Division of Endocrinology, Diabetes, and Metabolism, Department of Medicine, UAB, Endocrinology Section, Birmingham VAMC Medical Service, Birmingham VA Medical Center, University of Alabama, Birmingham, AL, USA

Stanley R. Frankel Merck Research Laboratories, Boston, MA, USA

Michael J. Franklin Division of Hematology, Oncology and Transplantation, University of Minnesota, Minneapolis, MN, USA

Aleksandra Franovic Department of Cellular and Molecular Medicine, Faculty of Medicine, University of Ottawa, Ottawa, ON, Canada

Ralph S. Freedman UT MD Anderson Cancer Center, Houston, TX, USA

Michael R. Freeman Urological Diseases Research Center, Children's Hospital Boston, Harvard Medical School, Boston, MA, USA

Emil Frei Dana-Farber Cancer Institute, Boston, MA, USA

Jean-Noël Freund INSERM U1113 and Fédération de Médecine Translationnelle de Strasbourg (FMTS), Université de Strasbourg, Faculté de Médecine, Strasbourg, France

Errol C. Friedberg University of Texas Southwestern Medical Center, Dallas, TX, USA

Steven M. Frisch Mary Babb Randolph Cancer Center and Department of Biochemistry, West Virginia University, Morgantown, WV, USA

Andrew M. Fry University of Leicester, Leicester, UK

Mark Frydenberg Department of Surgery, Monash University, Melbourne, VIC, Australia

Hendrik Fuchs Institute for Laboratory Medicine, Clinical Chemistry and Pathobiochemistry, Charité – Universitätsmedizin Berlin, Berlin, Germany

Atsuko Fujihara Department of Urology, Kyoto Prefectural University of Medicine, Kyoto, Japan

Hirota Fujiki Department of Clinical Laboratory Medicine, Faculty of Medicine, Saga University, Saga, Japan

Jiro Fujimoto Department of Obstetrics and Gynecology, Gifu University School of Medicine, Gifu City, Japan

Jun Fujita Department of Clinical Molecular Biology, Graduate School of Medicine, Kyoto University, Kyoto, Japan

Hiroshi Fukamachi Department of Molecular Oncology, Graduate School of Medical and Dental Sciences, Tokyo Medical and Dental University (TMDU), Tokyo, Japan

Kenji Fukasawa Molecular Oncology Program, H. Lee Moffitt Cancer Center and Research Institute, Tampa, FL, USA

Tomoya Fukui Department of Respiratory Medicine, Kitasato University School of Medicine, Sagamihara, Kanagawa, Japan

Simone Fulda Institute for Experimental Cancer Research in Pediatrics, Goethe-University Frankfurt, Frankfurt, Germany

Claudia Fumarola Department of Experimental Medicine, Unit of Experimental Oncology, University of Parma, Parma, Italy

Kyle Furge Van Andel Research Institute, Grand Rapids, MI, USA

Mutsuo Furihata Department of Pathology, Kochi Medical School, Kochi, Japan

Rhoikos Furtwängler Universitätsklinikum des Saarlandes, Klinik für Pädiatrische Onkologie und Hämatologie, Homburg/Saar, Germany

Bernard W. Futscher Department of Pharmacology and Toxicology, Arizona Cancer Center and College of Pharmacy, University of Arizona, Tucson, AZ, USA

Ulrich Göbel Clinic of Pediatric Oncology, Hematology and Immunology, Heinrich-Heine-University Düsseldorf, Düsseldorf, Germany

Tobias Görge Department of Dermatology, University of Münster, Münster, Germany

Ursula Günthert Institute of Pathology, University Hospital, Basel, Switzerland

Shirish Gadgeel Karmanos Cancer Institute, Wayne State University, Detroit, MI, USA

Jochen Gaedche Department of General, Visceral and Pediatric Surgery, University Medical Center, Göttingen, Germany

Federico Gago Departamento de Ciencias Biomédicas, Facultad de Medicina, Universidad de Alcalá, Alcalá de Henares, Madrid, Spain

William M. Gallagher UCD School of Biomolecular and Biomedical Science, UCD Conway Institute, University College Dublin, Dublin, Ireland

Bernard Gallez Biomedical Magnetic Resonance, Université Catholique de Louvain, Brussels, Belgium

Brenda L. Gallie The Hospital for Sick Children, Department of Ophthalmology and Vision Science, The University of Toronto, Toronto, ON, Canada

Antoine Galmiche EA4666, Université de Picardie Jules Verne (UPJV), Amiens, France

Service de Biochimie, Centre de Biologie Humaine (CBH), University Hospital of Amiens (CHU Sud), Amiens, France

Ramesh K. Ganju Department of Pathology, Comprehensive Cancer Centre, The Ohio State Medical Centre, Columbus, OH, USA

Ping Gao Division of Hematology, Department of Medicine, Johns Hopkins University School of Medicine, Baltimore, MD, USA

Dolores C. García-Olmo Unidad de Investigación, Complejo Hospitalario Universitario de Albacete, Albacete, Spain

Roy Garcia City of Hope National Medical Center and Beckman Research Institute, Duarte, CA, USA

Robert A. Gardiner School of Medicine, University of Queensland, Brisbane, QLD, Australia

Centre for Clinical Research, University of Queensland, Herston, QLD, Australia

Royal Brisbane and Women's Hospital, Brisbane, QLD, Australia

Edith Cowan University Western Australia, Joondalup, WA, Australia

Lawrence B. Gardner The NYU Cancer Institute, New York University School of Medicine, New York, NY, USA

Patricio Gariglio Genetic and Molecular Biology, CINVESTAV-IPN, Mexico City, México

Cathie Garnis MIT Center for Cancer Research, Cambridge, MA, USA

Andrei L. Gartel Department of Medicine, University of Illinois at Chicago, Chicago, IL, USA

Ronald B. Gartenhaus The University of Maryland Marlene and Stewart Greenebaum Cancer Center, Baltimore, MD, USA

Thomas A. Gasiewicz University of Rochester Medical Center, Rocheser, NY, USA

Patrizia Gasparini Tumor Genomic Unit, Department of Experimental Oncology, Istituto Nazionale Tumori, Milan, Italy

Zoran Gatalica Department of Pathology, Creighton University School of Medicine, Omaha, NE, USA

Grégory Gatouillat Laboratory of Biochemistry, IFR53, Faculty of Pharmacy, Reims, France

Adi F. Gazdar Hamon Center for Therapeutic Oncology Research and Departments of Pathology, Internal Medicine and Pharmacology, University of Texas Southwestern Medical Center, Dallas, TX, USA

Christian Geisler Department of Hematology, The Finsen Centre, Rigshospitalet, Copenhagen, Denmark

Klaramari Gellci Department of Biomedical Engineering, Wayne State University, Detroit, MI, USA

Eleni A. Georgakopoulou Department of Histology and Embryology, Faculty of Medicine, National and Kapodistrian University of Athens, Athens, Greece

Spyros D. Georgatos Department of Basic Sciences, The University of Crete, School of Medicine, Heraklion, Crete, Greece

Julia M. George Queen Mary University of London, London, UK

Kimberly S. George Parsons Department of Chemistry, Marietta College, Marietta, OH, USA

Armin Gerger Department of Internal Medicine, Division of Oncology, Medical University Graz, Graz, Austria

Ulrich Germing Klinik für Hämatologie, Onkologie und Klinische Immunologie, Heinrich-Heine-Universität, Düsseldorf, Germany

Jeffrey E. Gershenwald Department of Surgical Oncology, The University of Texas MD Anderson Cancer Center, Houston, TX, USA

Andreas J. Gescher Department of Cancer Studies, Cancer Biomarkers and Prevention Group, University of Leicester, Leicester, Leicester, UK

Christian Gespach Laboratory of Molecular and Clinical Oncology of Solid tumors, Faculté de Médecine, Université Pierre et Marie Curie-Paris 6, Paris, France

INSERM U. 673, Paris, France

B. Michael Ghadimi Department of General, Visceral and Pediatric Surgery, University Medical Center, Göttingen, Germany

Michelle Ghert Department of Surgery, Hamilton Health Sciences, Juravinski Cancer Centre, McMaster University, Hamilton, ON, Canada

Riccardo Ghidoni Laboratory of Biochemistry and Molecular Biology, San Paolo Medical School, University of Milan, Milan, Italy

Saurabh Ghosh Roy Department of Cell and Developmental Biology, University of California, Irvine, Irvine, CA, USA

Ronald A. Ghossein Department of Pathology, Memorial Sloan-Kettering Cancer Center, New York, NY, USA

Lorenzo Gianni Department of Oncology, Instituto Scientifico Romagnolo per lo s, Infermi Hospital, Rimini, Italy

Michael K. Gibson Case Western Reserve University, Cleveland, OH, USA

Michael Z. Gilcrease Department of Pathology, Breast Section, MD Anderson Cancer Center, Houston, TX, USA

M. Boyd Gillespie Head and Neck Tumor Program, Hollings Cancer Center, Medical University of South Carolina, Charleston, SC, USA

François Noël Gilly Department of Digestive Oncologic Surgery, Hospices Civils de Lyon–Université Lyon 1, Lyon, France

Thomas Gilmore Biology Department, Boston University, Boston, MA, USA

Oliver Gimm Department of Surgery, University Hospital, Linköping, Sweden

Alessio Giubellino Laboratory of Pathology, Center for Cancer Research, National Cancer Institute, Bethesda, MD, USA

Morten F. Gjerstorff Department of Oncology, Odense University Hospital, Odense C, Denmark

Shannon S. Glaser Department of Internal Medicine, Texas A&M Health Science Center, Central Texas Veterans Health Care System, Temple, TX, USA

Hansruedi Glatt Federal Institute for Risk Assessment (BfR), Berlin, Germany

Olivier Glehen Department of Digestive Oncologic Surgery, Hospices Civils de Lyon–Université Lyon 1, Lyon, France

Aleksandra Glogowska Department of Human Anatomy and Cell Science, College of Medicine, Faculty of Health Sciences, University of Manitoba, Winnipeg, MB, Canada

Thomas W. Glover Department of Human Genetics, University of Michigan, Ann Arbor, MI, USA

John C. Goddard Jacksonville Hearing and Balance Institute, Jacksonville, FL, USA

Andrew K. Godwin The University of Kansas Medical Center, Kansas City, KS, USA

Elspeth Gold Department of Anatomy, Otago School of Medical Sciences, Dunedin, New Zealand

Gary S. Goldberg Molecular Biology, University of Medicine and Dentistry of New Jersey, Stratford, NJ, USA

Itzhak D. Goldberg Long Island Jewish Medical Center, Albert Einstein College of Medicine, Bronx, NY, USA

Susanne M. Gollin Department of Human Genetics, University of Pittsburgh Graduate School of Public Health and the University of Pittsburgh Cancer Institute, Pittsburgh, PA, USA

Roy M. Golsteyn Department of Biological Sciences, University of Lethbridge, Lethbridge, AB, Canada

Rohini Gomathinayagam Stephenson Cancer Center, University of Oklahoma Health Sciences Center, Oklahoma City, OK, USA

Ellen L. Goode Department of Health Sciences Research, Mayo Clinic College of Medicine, Rochester, MN, USA

Gregory J. Gores Miles and Shirley Fiterman Center for Digestive Diseases, Division of Gastroenterology and Hepatology, Mayo Clinic College of Medicine, Rochester, MN, USA

Vassilis Gorgoulis Department of Histology and Embryology, Faculty of Medicine, National and Kapodistrian University of Athens, Athens, Greece

Noriko Gotoh Division of Cancer Cell Biology, Cancer Research Institute, Kanazawa University, Kanazawa city, Ishikawa, Japan

Lynn F. Gottfried LeClairRyan, Rochester, NY, USA

Stéphanie Gout Le Centre de recherche du CHU de Québec-Université Laval: axe Oncologie, Le Centre de recherche sur le cancer de l'Université Laval, Québec, QC, Canada

Ammi Grahn Department of Clinical Chemistry and Transfusion Medicin, Institute of Biomedicine, Sahlgrenska Academy at Göteborg University, Göteborg, Sweden

Galit Granot Felsenstein Medical Research Center, Beilinson Hospital, Sackler School of Medicine, Tel Aviv University, Petah Tikva, Israel

Denis M. Grant Department of Pharmacology and Toxicology, Faculty of Medicine, University of Toronto, Toronto, ON, Canada

Heidi J. Gray Gynecologic Oncology, University of Washington, Seattle, WA, USA

Peter Greaves Department of Cancer Studies, University of Leicester, Leicester, UK

John A. Green Department of Cancer Medicine, University of Liverpool, Liverpool, UK

Mark I. Greene Department of Pathology, Laboratory Medicine and Abramson Cancer Center, University of Pennsylvania, Philadelphia, PA, USA

Michael Greene Auburn University, Auburn, AL, USA

Arjan W. Griffioen Angiogenesis Laboratory, Department of Pathology, Maastricht University, Maastricht, The Netherlands

Dirk Grimm BIOQUANT, Cluster of Excellence Cell Networks, University of Heidelberg, Heidelberg, Germany

Matthew J. Grimshaw Breast Cancer Biology Group, King's College London School of Medicine, Guy's Hospital, London, UK

Stephen R. Grobmyer Department of Surgery, Division of Surgical Oncology, University of Florida, Gainesville, FL, USA

Bernd Grosche Department of Radiation Protection and Health, Bundesamt für Strahlenschutz (Federal Office for Radiation Protection), Oberschleissheim, Germany

Isabelle Gross INSERM U1113, Université de Strasbourg, Strasbourg, France

Michael Grusch Institute of Cancer Research, Department of Medicine I, Medical University of Vienna, Vienna, Austria

Wei Gu Institute for Cancer Genetics, and Department of Pathology, College of Physicians and Surgeons, Columbia University, New York, NY, USA

Francisca Guardiola-Serrano University of the Balearic Islands, Palma de Mallorca, Spain

Juliana Guarize Department of Thoracic Surgery, European Institute of Oncology, Milan, Italy

Valentina Guarneri Istituto Oncologico Veneto IRCCS, Division of Medical Oncology 2, Department of Surgery, Oncology and Gastroenterology, University of Padova, Padova, Italy

Tiziana Guarnieri Department of Biology, Geology and Environmental Sciences, Alma Mater Studiorum University of Bologna, Bologna, Italy

Liliana Guedez Immunopathology Section, National Eye Institute, Bethesda, MD, USA

Frederick Peter Guengerich Department of Biochemistry and Center in Molecular Toxicology, Biochemistry and Center in Molecular Toxicology, Vanderbilt University School of Medicine, Nashville, TN, USA

Abhijit Guha Division of Neurosurgery, University of Toronto, Toronto, ON, Canada

Katherine A. Guindon Department of Pharmacology and Toxicology, Queen's University, Kingston, ON, Canada

Erich Gulbins Department of Molecular Biology, University of Duisburg-Essen, Essen, Germany

Charles A. Gullo Microbiology NUS (Research), Duke/NUS GMS, Singapore, Singapore

Aparna Gupta Life Science Research Associate, Department of Gastroenterology and Hepatology, Stanford University School of Medicine, Stanford, CA, USA

Sonal Gupta Department of Pathology, The Sol Goldman Pancreatic Cancer Research Center, Johns Hopkins University School of Medicine, Baltimore, MD, USA

Murali Gururajan Department of Hematology and Oncology, Cedars-Sinai Medical Center, Los Angeles, CA, USA

Bristol-Myers Squibb & Co, Princeton, NJ, USA

James F. Gusella Center for Human Genetic Research, Massachusetts General Hospital, Boston, MA, USA

Graeme R. Guy Signal Transduction Laboratory, Institute of Molecular and Cell Biology, Singapore, Singapore

Manuel Guzmán Department of Biochemistry and Molecular Biology I, School of Biology, Complutense University, Madrid, Spain

Geum-Youn Gwak Department of Medicine, Samsung Medical Center, Sungkyunkwan University School of Medicine, Gangnam-gu, Seoul, South Korea

Guy Haegeman Lab of Eukaryotic Gene Expression, LEGEST-University Gent, Ghent, Belgium

Stephan A. Hahn University of Bochum, Bochum, Germany

Jörg Haier Comprehensive Cancer Center Münster, University Hospital Münster, Münster, Germany

Numsen Hail Department of Pharmaceutical Sciences, The University of Colorado at Denver and Health Sciences Center, Denver, CO, USA

Pierre Hainaut International Prevention Research Institute, Lyon, France

Brett M. Hall Department of Pediatrics, Columbus Children's Research Institute, The Ohio State University, Columbus, OH, USA

Janet Hall Centre de Recherche en Cancérologie de Lyon (CRCL), UMR Inserm 1052 - CNRS 5286, Lyon, France

Joyce L. Hamlin Department of Biochemistry and Molecular Genetics, University of Virginia School of Medicine, Charlottesville, VA, USA

Rasha S. Hamouda GeneDx, Rockville, MD, USA

Kelsey R. Hampton Department of Cancer Biology, Kansas University Cancer Center, Kansas City, KS, USA

The University of Kansas Medical Center, Kansas City, KS, USA

Lina Han Department of Leukemia, Section of Molecular Hematology and Therapy, The University of Texas MD Anderson Cancer Center, Houston, TX, USA

Ross Hannan Department of Cancer Biology and Therapeutics, John Curtin School of Medical Research, ANU College of Medicine, Biology and the Environment, Canberra, ACT, Australia

Chunhai Hao Department of Pathology and Laboratory Medicine, Henry Ford Health System, Detroit, MI, USA

J. William Harbour Bascom Palmer Eye Institute, University of Miami, Miami, FL, USA

Mark Harland Section of Epidemiology and Biostatistics, Cancer Research UK Clinical Centre, Leeds Institute of Molecular Medicine, St. James's University Hospital, Leeds, UK

Adrian L. Harris Weatherall Institute of Molecular Medicine, John Radcliffe Hospital, University of Oxford, Cancer Research UK, Headington, Oxford, UK

Randall E. Harris Director Center of Molecular Epidemiology, The Ohio State University, Columbus, OH, USA

Marion Hartley Ruesch Center for the Cure of Gastrointestinal Cancers, Lombardi Comprehensive Cancer Center, Georgetown University, Washington, DC, USA

Uzma Hasan CIRI, Oncoviruses and Innate Immunity, INSERM U1111, Ecole Normale Supérieure, Université de Lyon, CNRS-UMR5308, Hospices Civils de Lyon, Lyon, France

Mia Hashibe University of Utah, Salt Lake City, UT, USA

Masaharu Hata Division of Radiation Oncology, Department of Oncology, Yokohama City University Graduate School of Medicine, Yokohama, Kanagawa, Japan

Yosef S. Haviv Division of Nephrology, Hadassah-Hebrew University Medical Center, Department of Medicine, Jerusalem, Israel

John D. Hayes Medical Research Institute, Jacqui Wood Cancer Centre, University of Dundee, Dundee, UK

Nicole M. Haynes Cancer Therapeutics Program, Trescowthick Laboratories, Peter MacCallum Cancer Centre, East Melbourne, VIC, Australia

Hong He Department of Surgery, Austin Health, The University of Melbourne, Heidelberg, VIC, Australia

Lili He Molecular Oncology Program and Research Institute, H. Lee Moffitt Cancer Center, University of South Florida College of Medicine, Tampa, FL, USA

Li-Zhen He Memorial Sloan-Kettering Cancer Center, Weill Cornell Graduate School of Medical Sciences, New York, NY, USA

Ruth He Lombardi Comprehensive Cancer Center, Georgetown University, Washington, DC, USA

Yu-Ying He Medicine/Dermatology, University of Chicago, Chicago, IL, USA

Stephen S. Hecht The Cancer Center, University of Minnesota, Minneapolis, MN, USA

Ingrid A. Hedenfalk Department of Oncology, Clinical Sciences, Lund University, Lund, Sweden

Petra Heffeter Department of Medicine I, Institute of Cancer Research, Medical University of Vienna, Vienna, Austria

Ahmed E. Hegab Department of Geriatric and Respiratory Medicine, Tohoku University Hospital, Sendai, Japan

Axel Heidenreich Division of Oncological Urology, Department of Urology, University of Köln, Köln, Germany

Olaf Heidenreich Northern Institute for Cancer Research, Newcastle University, Newcastle upon Tyne, UK

Werner Held Ludwig Center for Cancer Research, Department of Oncology, University of Lausanne, Epalinges, Switzerland

Carl-Henrik Heldin Ludwig Institute for Cancer Research, Uppsala University, Uppsala, Sweden

Wijnand Helfrich Groningen University Institute for Drug Exploration (GUIDE), University Medical Center Groningen, Department of Pathology and Laboratory Medicine, Section Medical Biology, Laboratory for Tumor Immunology, University Medical Center Groningen, Groningen, The Netherlands

Debby Hellebrekers Department of Pathology, GROW-School for Oncology and Developmental Biology, Maastricht University Hospital, Maastricht, The Netherlands

Ingegerd Hellstrom Department of Pathology, University of Washington, Seattle, WA, USA

Karl Erik Hellstrom Department of Pathology, University of Washington, Seattle, WA, USA

Paul W. S. Heng Department of Pharmacy, National University of Singapore, Singapore, Singapore

Kai-Oliver Henrich DKFZ, German Cancer Research Center, Heidelberg, Germany

Rui Henrique Department of Pathology, Portuguese Oncology Institute-Porto, Porto, Portugal

Ellen C. Henry University of Rochester Medical Center, Rocheser, NY, USA

Elizabeth P. Henske Center for LAM Research and Clinical Care, Brigham and Women's Hospital, Harvard Medical School, Boston, MA, USA

Donald E. Henson Uniformed Services University of the Health Sciences, Bethesda, MD, USA

Serge Hercberg UMR U1153 INSERM, U1125 INRA, CNAM, Université Paris 13, Centre de Recherche Epidémiologie et Statistique Sorbonne Paris Cité, Bobigny, France

Meenhard Herlyn The Wistar Institute, Philadelphia, PA, USA

Heike M. Hermanns Med. Klinik II, Hepatologie, Universitätsklinikum Würzburg, Würzburg, Germany

Blanca Hernandez-Ledesma Instituto de Investigación en Ciencias de la Alimentación (CIAL, CSIC-UAM, CEI UAM+CSIC), Madrid, Spain

Wolfgang Herr Universitätsklinikum Regensburg, Regensburg, Germany

Erika Herrero Garcia Department of Pharmacology, University of Illinois College of Medicine, Chicago, IL, USA

Helen E. Heslop Center for Cell and Gene Therapy, Baylor College of Medicine, Texas Children's Hospital, and The Methodist Hospital, Houston, TX, USA

Jochen Hess Division of Signal Transduction and Growth Control, Deutsches Krebsforschungszentrum, Heidelberg, Germany

Dominique Heymann Physiopathologie de la Résorption Osseuse et Thérapie des Tumeurs Osseuses Primitives, University of Nantes, Nantes, France

Martha Hickey Obstetrics and Gynaecology, The University of Melbourne, Parkville, VIC, Australia

James Hicks Cold Spring Harbor Laboratory, Cold Spring Harbor, New York, USA

Kevin O. Hicks Auckland Cancer Society Research Centre, The University of Auckland, Auckland, New Zealand

Colin K. Hill Department of Radiation Oncology, Keck School of Medicine, University of Southern California, Los Angeles, CA, USA

Shawn Hingtgen Division of Molecular Pharmaceutics, UNC Eshelman School of Pharmacy, Biomedical Research Imaging Center, University of North Carolina, Chapel Hill, NC, USA

Isabelle Hinkel INSERM U1113, Université de Strasbourg, Strasbourg, France

Boaz Hirshberg Cardiovascular and Metabolic Diseases, Pfizer Inc, Groton, CT, USA

Ari Hirvonen Finnish Institute of Occupational Health, Helsinki, Finland

Ricardo Hitt Hospital Universitario Severo Ochoa, Madrid, Spain

Eiso Hiyama Natural Science Center for Basic Research and Development, Department of Pediatric Surgery, Hiroshima University Hospital, Hiroshima University, Hiroshima, Japan

Falk Hlubek Department of Pathology, Ludwig-Maximilians-University of Munich, Munich, Germany

Steven N. Hochwald Departments of Surgical Oncology, Roswell Park Cancer Institute, Buffalo, NY, USA

Mir Alireza Hoda Division of Thoracic Surgery, Medical University of Vienna, Vienna, Austria

Michael Hodsdon Department of Laboratory Medicine, Yale University School of Medicine, New Haven, CT, USA

Kasper Hoebe Division of Immunobiology, Cincinnati Children's Hospital Medical Center, Cincinnati, OH, USA

Markus Hoffmann Hals-, Nasen- und Ohrenheilkunde, Kopf- und Halschirurgie, Universitätsklinikum Schleswig-Holstein, Campus Kiel, Kiel, Germany

Michèle J. Hoffmann Department of Urology, Heinrich Heine University, Düsseldorf, Germany

Lorne J. Hofseth Department of Pharmaceutical and Biomedical Sciences, South Carolina College of Pharmacy, University of South Carolina, Columbia, SC, USA

Susanne Holck Department of Pathology, Copenhagen University Hospital, Hvidovre, Denmark

Stefan Holdenrieder Institute of Clinical Chemistry and Clinical Pharmacology, Universitatsklinikum Bonn, Bonn, Germany

James F. Holland Tisch Cancer Institute, Icahn School of Medicine at Mount Sinai, New York, NY, USA

Petra Den Hollander Department of Translational Molecular Pathology, The University of Texas MD Anderson Cancer Center, Houston, TX, USA

Caroline L. Holloway BC Cancer Agency, Vancouver Island Centre, Victoria, BC, Canada

Arne Holmgren Department of Medical Biochemistry and Biophysics, Karolinska Institutet, Stockholm, Sweden

Astrid Holzinger Tumor Genetics, Clinic I Internal Medicine, University Hospital Cologne, and Center for Molecular Medicine Cologne, University of Cologne, Cologne, Germany

Jun Hyuk Hong Division of Urologic Oncology, The Cancer Institute of NJ, Robert Wood Johnson Medical School, New Brunswick, NJ, USA

Adília Hormigo Department of Neurology, Medicine (Division Hematology Oncology) and Neurosurgery, Icahn School of Medicine at Mount Sinai and The Tisch Cancer Institute, New York, NY, USA

Joshua Hornig Head and Neck Tumor Program, Hollings Cancer Center, Medical University of South Carolina, Charleston, SC, USA

Michael R. Horsman Department of Experimental Clinical Oncology, Aarhus University Hospital, Aarhus, Denmark

Andrea Kristina Horst Inst. Experimental Immunology and Hepatology, University Medical Center Hamburg-Eppendorf, Hamburg, Germany

David W. Hoskin Departments of Pathology, and Microbiology and Immunology, Dalhousie University, Halifax, NS, Canada

Andreas F. Hottinger Departments of Clinical Neuroscience and Oncology, CHUV, Lausanne University Hospital, Lausanne, VD, Switzerland

Peter J. Houghton Greehey Children's Cancer Research Institute, UT Health Science Center, San Antonio, TX, USA

Anthony Howell CRUK Department of Medical Oncology, University of Manchester, Christie Hospital NHS Trust, Manchester, UK

Lynne M. Howells Department of Cancer Studies, University of Leicester, Leicester, UK

Chia-Chien Hsieh Department of Human Development and Family Studies (Nutritional Science and Education), National Taiwan Normal University, Taipei, Taiwan

Shie-Liang Hsieh Department of Microbiology and Immunology, National Yang-Ming University, Immunology Research Center, Taipei Veterans General Hospital; Genomics Research Center, Academia Sinica, Taipei, Taiwan

Wei Hu Departments of Gynecologic Oncology and Reproductive Medicine, The University of Texas MD Anderson Cancer Center, Houston, TX, USA

Cheng-Long Huang Department of Second Surgery, Kagawa University, Kagawa, Japan

Gonghua Huang Department of Immunology, St. Jude Children's Research Hospital, Memphis, TN, USA

Shile Huang Department of Biochemistry and Molecular Biology and Feist-Weiller Cancer Center, Louisiana State University Health Sciences Center, Shreveport, LA, USA

Kay Huebner Department of Molecular Virology, Immunology and Medical Genetics, Ohio State University Comprehensive Cancer Center, Columbus, OH, USA

Pere Huguet Department of Pathology, Vall d'Hebron University Hospital, Barcelona, Spain

Maureen B. Huhmann Department of Nutrition Sciences, School of Health Related Professions, Rutgers The State University, Newark, NJ, USA

Wen-Chun Hung National Institute of Cancer Research, National Health Research Institutes, Tainan Taiwan, Republic of China

Tony Hunter Salk Institute, Molecular and Cell Biology Laboratory, La Jolla, CA, USA

Teh-Ia Huo Institute of Pharmacology, School of Medicine, National Yang-Ming University, Taipei, Taiwan

Jacques Huot Le Centre de recherche du CHU de Québec-Université Laval: axe Oncologie, Le Centre de recherche sur le cancer de l'Université Laval, Québec, QC, Canada

Karen L. Huyck Department of Pathology, Brigham and Women's Hospital, Boston, MA, USA

Sam T. Hwang Dermatology Branch, National Cancer Institute, National Institutes of Health, Bethesda, MD, USA

Brandy D. Hyndman Department of Pathology and Molecular Medicine, Queen's University Cancer Research Institute, Queen's University, Kingston, ON, Canada

Maitane Ibarguren Department of Biology, University of the Balearic Islands, Palma de Mallorca, Spain

Takafumi Ichida Department of Hepatology and Gastroenterology, Juntendo University School of Medicine, Shizuoka Hospital, Shizuoka, Japan

Yoshito Ihara Department of Biochemistry, School of Medicine, Wakayama Medical University, Wakayama, Japan

Hitoshi Ikeda Department of Pediatric Surgery, Dokkyo Medical University Koshigaya Hospital, Koshigaya, Saitama, Japan

Landon Inge Norton Thoracic Institute, St. Joseph's Hospital and Medical Center, Phoenix, AZ, USA

Kazuhiko Ino Department of Obstetrics and Gynecology, Nagoya University Graduate School of Medicine, Nagoya, Japan

Juan Iovanna INSERM, Stress Cellulaire, Parc Scientifique et Technologique de Luminy, Marseille Cedex, France

Irmgard Irminger-Finger Molecular Gynecology and Obstetrics Laboratory, Department of Gynecology and Obstetrics, Geneva University Hospitals, Geneva, Switzerland

Meredith S. Irwin Cell Biology Program and Division of Hematology-Oncology Hospital for Sick Children, University of Toronto, Toronto, ON, Canada

Toshihisa Ishikawa Biochemistry, Molecular Biology, and Pharmacogenomics, NGO Personalized Medicine and Healthcare, Yokohama, Japan

Toshiyuki Ishiwata Department of Integrated Diagnostic Pathology, Graduate School of Medicine, Nippon Medical School, Tokyo, Japan

Mark A. Israel Departments of Pediatrics and of Genetics, Norris Cotton Cancer Center, Geisel School of Medicine at Dartmouth, Hanover, NH, USA

Antoine Italiano Early Phase Trials and Sarcoma Units, Institut Bergonie, Bordeaux, France

Norimasa Ito Departments of Surgery and Bioengineering, University of Pittsburgh, Pittsburgh, PA, USA

Michael Ittmann Department of Pathology, Baylor College of Medicine, Houston, TX, USA

Richard Ivell School of Biosciences and School of Veterinary Medicine and Science, University of Nottingham, Nottingham, UK

Antoni Ivorra Department of Information and Communication Technologies, Universitat Pompeu Fabra (UPF), Barcelona, Spain

Nobutaka Iwakuma Department of Surgery, Division of Surgical Oncology, University of Florida, Gainesville, FL, USA

Shai Izraeli Pediatric Hemato-Oncology, Sheba Medical Center and Tel Aviv University, Ramat Gan, Israel

Paola Izzo Department of Molecular Medicine and Medical Biotechnology, School of Medicine and Surgery, University of Naples Federico II, Naples, Italy

Mark Jackman Wellcome/CRC Institute, Cambridge, UK

Alan Jackson Centre for Imaging Sciences, University of Manchester, Manchester, UK

Deborah Jackson-Bernitsas Department of Systems Biology, The University of Texas MD Anderson Cancer Center, Houston, TX, USA

Stephan C. Jahn Department of Pharmacology and Therapeutics and the UF and Shands Cancer Center, University of Florida, Gainesville, FL, USA

David Jamieson School of Clinical and Laboratory Sciences, Newcastle University, Newcastle upon Tyne, UK

Siegfried Janz Department of Pathology, Carver College of Medicine, University of Iowa, Iowa City, IA, USA

Daniel G. Jay Tufts University School of Medicine, Boston, MA, USA

Gordon C. Jayson Cancer Research UK Department of Medical Oncology, Christie Hospital, Manchester, UK

Kuan-Teh Jeang National Institute of Allergy and Infectious Disease, NIH, Bethesda, MD, USA

Diane F. Jelinek Department of Immunology, Mayo Clinic, College of Medicine, Rochester, MN, USA

Jiiang-Huei Jeng Laboratory of Pharmacology and Toxicology, School of Dentistry, National Taiwan University Hospital and National Taiwan University Medical College, Taipei, Taiwan

Elwood V. Jensen National Institute of Health, Bethesda, MD, USA

Erika Jensen-Jarolim Institute of Pathophysiology and Allergy Research, Center of Pathophysiology, Infectiology and Immunology, Medical University Vienna, Vienna, Austria

The Interuniversity Messerli Research Institute, University of Veterinary Medicine Vienna, Medical University Vienna and University Vienna, Vienna, Austria

Carmen Jeronimo Research Center, Portuguese Oncology Institute-Porto, Porto, Portugal

Lin Ji Department of Thoracic and Cardiovascular Surgery, The University of Texas MD Anderson Cancer Center, Houston, TX, USA

Shuai Jiang Department of Biology and Biological Engineering, California Institute of Technology, Pasadena, CA, USA

Yufei Jiang Cancer Vaccine Section, National Cancer Institute, National Institutes of Health, Bethesda, MD, USA

Charlotte Jin Departments of Clinical Genetics, University Hospital, Lund, Sweden

Chengcheng Jin The David H. Koch Institute of Integrative Cancer Research, Massachusetts Institute of Technology, Cambridge, MA, USA

Andrew K. Joe Department of Medicine, Herbert Irving Comprehensive Cancer Center, New York, NY, USA

Manfred Johannsen Facharztpraxis Urologie Johannsen and Laux, Berlin, Germany

Kaarthik John Division of Microbiology, Tulane University, Covington, LA, USA

Alan L. Johnson Pennsylvania State University, State College, PA, USA

Sara M. Johnson Department of Surgery, The University of Texas Medical Branch, Galveston, TX, USA

Won-A Joo The Wistar Institute, Philadelphia, PA, USA

V. Craig Jordan Breast Medical Oncology, MD Anderson Cancer Center, Houston, TX, USA

Serene Josiah Cambridge, MA, USA

Richard Jove Vaccine and Gene Therapy Institute of Florida, Port Saint Lucie, FL, USA

Jaroslaw Jozwiak Department of Histology and Embryology, Medical University of Warsaw, Warsaw, Poland

Jesper Jurlander Department of Hematology, Rigshospitalet, Copenhagen, Denmark

Donat Kögel Experimental Neurosurgery, Center for Neurology and Neurosurgery, Goethe-University Hospital, Frankfurt am Main, Germany

Ralf Küppers Institute of Cell Biology (Cancer Research), University of Duisburg-Essen, Medical School, Essen, Germany

Chaim Kahana Department of Molecular Genetics, Weizmann Institute of Science, Rehovot, Israel

Bernd Kaina Department of Toxicology, University Medical Center Mainz, Mainz, Germany

Kiran Kakarala Departments of Otolaryngology-Head and Neck Surgery, University of Kansas Medical Center, Kansas City, KS, USA

Tadao Kakizoe National Cancer Center, Tokyo, Japan

Ganna V. Kalayda Institute of Pharmacy, University of Bonn, Bonn, Germany

Tuula Kallunki Unit of Cell Death and Metabolism, Danish Cancer Society Research Center, Copenhagen, Denmark

Takehiko Kamijo Research Institute for Clinical Oncology, Saitama Cancer Center, Ina, Saitama, Japan

Yasufumi Kaneda Department of Gene Therapy Science, Graduate School of Medicine, Osaka University, Suita, Osaka, Japan

Kazuhiro Kaneko Department of Gastroenterology, Endoscopy Division, National Cancer Center Hospital East, Chiba, Japan

Inkyung Kang Department of Surgery, University of California, San Francisco, San Francisco, CA, USA

Jayakanth Kankanala Center for Drug Design, Academic Health Center, University of Minnesota, Minneapolis, MN, USA

Yung-Hsi Kao Department of Life Sciences, College of Science, National Central University, Jhongli City, Taiwan

David E. Kaplan Division of Gastroenterology, University of Pennsylvania, Philadelphia, PA, USA

Niki Karachaliou Instituto Oncológico Dr. Rosell, Quiron-Dexeus University Hospital, Barcelona, Spain

Sophia N. Karagiannis St. John's Institute of Dermatology, Division of Genetics and Molecular Medicine, Faculty of Life Sciences and Medicine, King's College London, London, UK
NIHR Biomedical Research Centre at Guy's and St. Thomas' Hospitals, Guy's Hospital, King's College London, London, UK

Michalis V. Karamouzis Department of Biological Chemistry, Medical School, University of Athens, Goudi, Athens, Greece

Adam R. Karpf Department of Pharmacology and Therapeutics, Roswell Park Cancer Institute, Buffalo, NY, USA

Nilesh D. Kashikar Departments of Surgery and Cancer Biology, Vanderbilt-Ingram Cancer Center, Vanderbilt University School of Medicine, Nashville, TN, USA

Matilda Katan CRC Centre for Cell and Molecular Biology, Institute of Cancer Research, London, UK

William K. Kaufmann Department of Pathology and Laboratory Medicine, University of North Carolina at Chapel Hill, Chapel Hill, NC, USA

Manjinder Kaur Department of Pharmaceutical Sciences, School of Pharmacy, University of Colorado Health Sciences Center, Denver, CO, USA

Sukhwinder Kaur Department of Biochemistry and Molecular Biology, University of Nebraska Medical Center, Omaha, NE, USA

Ingo Kausch Department of Urology, Ammerlandklinik Westerstede, Westerstede, Germany

Koji Kawakami Department of Pharmacoepidemiology, Graduate School of Medicine and Public Health, Kyoto University, Kyoto, Japan

Frederic J. Kaye National Cancer Institute, NIH and National Naval Medical Center, Bethesda, MD, USA

Stanley B. Kaye Drug Development Unit, Institute of Cancer Research, The Royal Marsden Hospital, Sutton, UK

Evan T. Keller Departments of Urology and Pathology, University of Michigan, Ann Arbor, MI, USA

Daniel Keppler Department of Biological Science, College of Pharmacy, Touro University-CA, Vallyo, CA, USA

Santhosh Kesari Department of Translational Neuro-Oncology and Neurotherapeutics, John Wayne Cancer Institute, Providence St. John's Health Center, Santa Monica, CA, USA

Jorma Keski-Oja Departments of Pathology and of Virology, Haartman Institute, University of Helsinki, Helsinki, Finland

Khandan Keyomarsi Department of Experimental Radiation Oncology, Unit 1052, University of Texas MD Anderson Cancer Center, Houston, TX, USA

Abdul Arif Khan Department of Pharmaceutics, College of Pharmacy, King Saud University, Riyadh, Saudi Arabia

Shahanavaj Khan Department of Pharmaceutics, College of Pharmacy, King Saud University, Riyadh, Saudi Arabia

Chand Khanna Comparative Oncology Program, Center for Cancer Research, National Cancer Institute, Bethesda, MD, USA

Samir N. Khleif GRU Cancer Center, Augusta, GA, USA

Roya Khosravi-Far Department of Pathology, Harvard Medical School, Beth Israel Deaconess Medical Center, Boston, MA, USA

Tobias Kiesslich Department of Internal Medicine I, Paracelsus Medical University, Institute of Physiology and Pathophysiology, Paracelsus Medical University, Salzburg, Austria

Fumitaka Kikkawa Department of Obstetrics and Gynecology, Nagoya University Graduate School of Medicine, Nagoya, Japan

Nerbil Kilic Kantonspital St. Gallen, St. Gallen, Switzerland

Isaac Yi Kim Division of Urologic Oncology, The Cancer Institute of NJ, Robert Wood Johnson Medical School, New Brunswick, NJ, USA

Jung-whan Kim Department of Biological Sciences, The University of Texas at Dallas, Richardson, TX, USA

Miran Kim Division of Gastroenterology, Liver Research Center, Rhode Island Hospital and Warren Alpert Medical School of Brown University, Providence, RI, USA

Moonil Kim BioNanotechnology Research Center, Korea Research Institute of Bioscience and Biotechnology, Yuseong, Daejeon, Republic of Korea

Seong Jin Kim Laboratory of Cell Regulation and Carcinogenesis, National Cancer Institute, Bethesda, MD, USA

Su Young Kim Pediatric Oncology Branch, Center for Cancer Research, National Cancer Institute, National Institutes of Health, Bethesda, MD, USA

Adi Kimchi Department of Molecular Genetics, Weizmann Institute of Science, Rehovot, Israel

A. Douglas Kinghorn College of Pharmacy, The Ohio State University, Columbus, OH, USA

David Kirn Jennerex Biotherapeutics Inc., San Francisco, CA, USA

Youlia M. Kirova Department of Radiation Oncology, Institut Curie, Paris, France

Shinichi Kitada Burnham Institute for Medical Research, La Jolla, CA, USA

Karel Kithier Department of Pathology, Wayne State University School of Medicine, Detroit, MI, USA

Chikako Kiyohara Department of Preventive Medicine, Graduate School of Medical Sciences, Kyushu University, Fukuoka, Japan

Celina G. Kleer Department of Pathology and Comprehensive Cancer Center, University of Michigan Medical School, Ann Arbor, MI, USA

George Klein Microbiology, Tumor and Cell Biology, Karolinska Institute, Stockholm, Sweden

Michael J. Klein Department of Pathology and Laboratory Medicine, Hospital for Special Surgery, New York, NY, USA

Elena Klenova Department of Biological Sciences, University of Essex, Colchester, Essex, UK

Thomas Klonisch Department of Human Anatomy and Cell Science, College of Medicine, Faculty of Health Sciences, University of Manitoba, Winnipeg, MB, Canada

Elizabeth Knobler Department of Dermatology, Columbia College of Physicians and Surgeons, New York, NY, USA

Robert Knobler Department of Dermatology, Medical University of Vienna, Vienna, Austria

Beatrice Knudsen Cedars-Sinai, Los Angeles, CA, USA

Stefan Kochanek Division of Gene Therapy, University of Ulm, Ulm, Germany

Manish Kohli Medical Oncology, Mayo Clinic, Rochester, MN, USA

Katri Koli Translational Cancer Biology Program, University of Helsinki, Helsinki, Finland

Christian Kollmannsberger Division of Medical Oncology, British Columbia Cancer Agency, Vancouver Cancer Centre, University of British Columbia, Vancouver, BC, Canada

Yutaka Kondo Department of Epigenomics, Nagoya City University Graduate School of Medical Sciences, Nagoya, Japan

Lin Kong Department of Radiation Oncology, Fudan Universtiy Shanghai Cancer Center, Shanghai, China

Marina Konopleva Department of Leukemia and Department of Stem Cell Transplantation and Cellular Therapy, The University of Texas MD Anderson Cancer Center, Houston, TX, USA

Roland E. Kontermann Institute of Cell Biology and Immunology, University of Stuttgart, Stuttgart, Germany

Janko Kos Faculty to Pharmacy, University of Ljubljana, Ljubljana, Slovenia

Marta Kostrouchova Institute of Cellular Biology and Pathology, 1st Faculty of Medicine, Charles University, Prague, Czech Republic

Athanassios Kotsinas Molecular Carcinogenesis Group, Laboratory of Histology-Embryology, Medical School, National and Kapodistrian University of Athens, Athens, Greece

Evangelia A. Koutsogiannouli Department of Urology, Heinrich Heine University, Düsseldorf, Germany

Heinrich Kovar Children's Cancer Research Institute, Vienna, Austria

Craig Kovitz Department of Medical Oncology, University of Texas MD Anderson Cancer Center, Houston, TX, USA

Christian Kowol Institute of Inorganic Chemistry, University of Vienna, Vienna, Austria

Barnett S. Kramer Office of Disease Prevention, National Institutes of Health, Bethesda, MD, USA

Oliver H. Krämer Department of Toxicology, University Medical Center Mainz, Mainz, Germany

Barbara Krammer Department of Molecular Biology, University of Salzburg, Salzburg, Austria

Henk J. van Kranen National Institute of Public Health and Environment, Bilthoven, The Netherlands

Robert Kratzke Division of Hematology, Oncology and Transplantation, University of Minnesota, Minneapolis, MN, USA

Thomas Krausz Department of Pathology, University of Chicago, Chicago, IL, USA

Jürgen Krauter Medizinische Klinik III – Hämatologie und Onkologie, Klinikum Braunschweig, Braunschweig, Germany

Bernhard Kremens Department of Pediatric Hematology, Oncology and Respiratory Medicine, University Hospitals of Essen, Essen, Germany

Betsy T. Kren Minneapolis VA Health Care System and University of Minnesota, Minneapolis, MN, USA

Yasusei Kudo Department of Oral Molecular Pathology, Institute of Biomedical Sciences, Tokushima University Graduate School, Tokushima, Japan

Deepak Kumar Department of Biological and Environmental Sciences, University of the District of Columbia, Washington, DC, USA

Parvesh Kumar Department of Radiation Oncology, Keck School of Medicine, University of Southern California, Los Angeles, CA, USA

Rakesh Kumar Department of Biochemistry and Molecular Medicine, George Washington University, Washington, DC, USA

Hiroki Kuniyasu Department of Molecular Pathology, Nara Medical University School of Medicine, Kashihara, Nara, Japan

Siavash K. Kurdistani Department of Biological Chemistry, David Geffen School of Medicine at UCLA, Los Angeles, CA, USA

Elena Kurenova Departments of Surgical Oncology, Roswell Park Cancer Institute, Buffalo, NY, USA

Keisuke Kurose Departments of Obstetrics and Gynecology, Nippon Medical School, Kawasaki and Tokyo, Japan

Peter Kurre Department of Pediatrics, Oregon Health and Science University, Portland, OR, USA

Robert M. Kypta Cell Biology and Stem Cells Unit, CIC bioGUNE, Derio, Spain

Imperial College London, London, UK

Juan Carlos Lacal Instituto de Investigaciones Biomedicas, CSIC, Madrid, Spain

James C. Lacefield Departments of Electrical and Computer Engineering and Medical Biophysics, University of Western Ontario, London, ON, Canada

Stephan Ladisch Center for Cancer and Immunology Research, Children's Research Institute, Children's National Medical Center and The George Washington University School of Medicine, Washington, DC, USA

Hermann Lage Institute of Pathology, Charité Campus Mitte, Berlin, Germany

Charles P. K. Lai Department of Cellular and Physiological Sciences, The University of British Columbia, Vancouver, BC, Canada

Henry Lai Departments of Bioengineering, University of Washington, Seattle, WA, USA

Dale W. Laird Department of Anatomy and Cell Biology, University of Western Ontario, London, ON, Canada

Hilaire C. Lam Center for LAM Research and Clinical Care, Brigham and Women's Hospital, Harvard Medical School, Boston, MA, USA

Janice B. B. Lam Department of Pharmacology and Pharmacy, The University of Hong Kong, Hong Kong, China

Wan L. Lam Department of Cancer Genetics and Developmental Biology, British Columbia Cancer Research Centre, Vancouver, BC, Canada

Hui Y. Lan The Chinese University of Hong Kong, Hong Kong, China

Joseph R. Landolph, Jr. Department of Molecular Microbiology and Immunology, and Department of Pathology; Laboratory of Chemical Carcinogenesis and Molecular Oncology, USC/Norris Comprehensive Cancer Center, Keck School of Medicine; Department of Molecular Pharmacology and Pharmaceutical Sciences, School of Pharmacy, Health Sciences Campus, University of Southern California, Los Angeles, CA, USA

Ari L. Landon The University of Maryland Marlene and Stewart Greenebaum Cancer Center, Baltimore, MD, USA

Robert Langer Department of Chemical Engineering and Center for Cancer Research, Massachusetts Institute of Technology, Cambridge, MA, USA

Sigrid A. Langhans Nemours Center for Childhood Cancer Research, Alfred I duPont Hospital for Children, Wilmington, DE, USA

Cinzia Lanzi Molecular Pharmacology Unit, Department of Experimental Oncology and Molecular Medicine, Fondazione IRCCS Istituto Nazionale dei Tumori, Milan, Italy

Rosamaria Lappano Department of Pharmacy and Health and Nutritional Sciences, University of Calabria, Rende, Italy

Paola Larghi Department of Immunology, Fondazione Humanitas per la Ricerca, Rozzano, Milan, Italy

James M. Larner Department of Therapeutic Radiology and Oncology, University of Virginia School of Medicine, Charlottesville, VA, USA

Göran Larson Department of Clinical Chemistry and Transfusion Medicin, Institute of Biomedicine, Sahlgrenska Academy at Göteborg University, Göteborg, Sweden

Lars-Inge Larsson Department of Pathology, Copenhagen University Hospital, Hvidovre, Denmark

Susanna C. Larsson Division of Nutritional Epidemiology, Institute of Environmental Medicine, Karolinska Institutet, Stockholm, Sweden

Philippe Lassalle INSERM U774, Institut Pasteur de Lille, Lille, France

Antony M. Latham Endothelial Cell Biology Unit, Leeds Institute of Genetics Health and Therapeutics (LIGHT), University of Leeds, Leeds, UK

Farida Latif Institute of Cancer and Genomic Sciences, University of Birmingham, Edgbaston, Birmingham, UK

Paule Latino-Martel UMR U1153 INSERM, U1125 INRA, CNAM, Université Paris 13, Centre de Recherche Epidémiologie et Statistique Sorbonne Paris Cité, Bobigny, France

Kirsten Lauber Clinic for Radiotherapy and Radiation Oncology, LMU Munich, Munich, Germany

Béatrice Lauby-Secretan Section of the IARC Monographs, IARC/WHO, Lyon, France

Virpi Launonen Department of Medical Genetics, Biomedicum Helsinki, University of Helsinki, Helsinki, Finland

Martin F. Lavin University of Queensland Centre for Clinical Research at Royal Brisbane and Women's Hospital, The University of Queensland, Brisbane, QLD, Australia

Brian Law Department of Pharmacology and Therapeutics and the UF and Shands Cancer Center, University of Florida, Gainesville, FL, USA

Gwendal Lazennec INSERM, Montpellier, France

Pedro A. Lazo CSIC-Universidad de Salamanca, Instituto de Biología Molecular y Celular del Cáncer, Salamanca, Spain

Gail S. Lebovic Director of Women's Services, The Cooper Clinic, Dallas, TX, USA

David P. LeBrun Department of Pathology and Molecular Medicine, Queen's University Cancer Research Institute, Queen's University, Kingston, ON, Canada

Protein Function Discovery Group, Queen's University, Kingston, ON, Canada

Division of Cancer Biology and Genetics, Cancer Research Institute, Queen's University, Kingston, ON, USA

Sean Bong Lee Department of Pathology and Laboratory Medicine, Tulane University School of Medicine, New Orleans, LA, USA

Seong-Ho Lee Department of Nutrition and Food Science, University of Maryland, College Park, MD, USA

Stephen Lee Department of Cellular and Molecular Medicine, Faculty of Medicine, University of Ottawa, Ottawa, ON, Canada

William P. J. Leenders Department of Pathology, Radboud University Medical Center Nijmegen, Nijmegen, The Netherlands

Andreas Leibbrandt Institute of Molecular Biotechnology of the Austrian Academy of Sciences, Vienna, Austria

Manuel C. Lemos CICS-UBI, Health Sciences Research Centre, University of Beira Interior, Covilhã, Portugal

Eric Lentsch Head and Neck Tumor Program, Hollings Cancer Center, Medical University of South Carolina, Charleston, SC, USA

Derek LeRoith Division of Endocrinology, Diabetes and Bone Diseases, Mount Sinai School of Medicine, New York, NY, USA

Yun-Chung Leung Lo Ka Chung Centre for Natural Anti-cancer Drug Development and Department of Applied Biology and Chemical Technology, The Hong Kong Polytechnic University, Hong Kong, China

Francis Lévi Warwick Medical School, University of Warwick, Coventry, UK

Jay A. Levy University of California, School of Medicine, San Francisco, CA, USA

Benyi Li Department of Urology, The University of Kansas Medical Center, Kansas City, KS, USA

Guideng Li Institute for Immunology, School of Medicine, University of California, Irvine, CA, USA

Kaiyi Li Department of Surgery, Baylor College of Medicine, Houston, TX, USA

Yan Li Department of Immunology, Cleveland Clinic, Cleveland, OH, USA

Daiqing Liao Department of Anatomy and Cell Biology, UF Health Cancer Center, University of Florida College of Medicine, Gainesville, FL, USA

Yung-Feng Liao Institute of Cellular and Organismic Biology, Academia Sinica, Taipei, Taiwan

Emmanuelle Liaudet-Coopman IRCM, INSERM, UMI, CRLC Val d'Aurelle, Montpellier, France

Rossella Libè Endocrinology, Metabolism and Cancer Department, INSERM U567, Institut Cochin, Paris, France

Danielle Liddle Gray Institute for Radiation Oncology and Biology, Department of Oncology, University of Oxford, Oxford, UK

Jane Liesveld James P. Wilmot Cancer Center, University of Rochester, Rochester, NY, USA

Stephanie Lim Medical Oncology, Ingham Research Institute, Liverpool, NSW, Australia

Ke Lin Department of Haematology, Royal Liverpool University Hospital, Liverpool, UK

Sheng-Cai Lin Department of Biomedical Sciences, School of Life Sciences, Xiamen University, Xiamen, Fujian, China

Shiaw-Yih Lin Department of Systems Biology, The University of Texas MD Anderson Cancer Center, Houston, TX, USA

Wan-Wan Lin Department of Pharmacology, College of Medicine, National Taiwan University, Taipei, Taiwan

Yong Lin Molecular Biology and Lung Cancer Program, Lovelace Respiratory Research Institute, Albuquerque, NM, USA

Janet C. Lindsey Northern Institute for Cancer Research, Newcastle University, Newcastle upon Tyne, UK

Christopher A. Lipinski Melior Discovery, Waterford, CT, USA

Joseph Lipsick Stanford University, Stanford, CA, USA

Fei-Fei Liu Princess Margaret Cancer Centre, University Health Network, Toronto, ON, Canada

Department of Radiation Oncology, Princess Margaret Hospital, Toronto, ON, Canada

Department of Radiation Oncology, University of Toronto, Toronto, ON, Canada

Department of Medical Biophysics, University of Toronto, Toronto, ON, Canada

Tao Liu Department of Medicine, Harvard Medical School, Brigham and Women's Hospital, Boston, MA, USA

Wen Liu Division of Life Science, Hong Kong University of Science and Technology, Kowloon, Hong Kong

Xiangguo Liu School of Life Science, Shandong University, Jinan, Shandong, China

Yiyan Liu Department of Radiology, New Jersey Medical School, Rutgers University, New Brunswick, NJ, USA

Hui-Wen Lo Department of Cancer Biology, Wake Forest University School of Medicine, Winston-Salem, NC, USA

Ting Ling Lo Signal Transduction Laboratory, Institute of Molecular and Cell Biology, Singapore, Singapore

Victor Lobanenkov Section of Molecular Pathology, Laboratory of Immunopathology, NIAID, National Institutes of Health, Bethesda, MD, USA

Holger N. Lode Klinik und Poliklinik für Kinder und Jugendmedizin, Universitätsmedizin Greifswald, Greifswald, Germany

Lawrence A. Loeb University of Washington, Seattle, WA, USA

Robert Loewe Department of Dermatology, Division of General Dermatology, Medical University of Vienna, Vienna, Austria

Steffen Loft Department of Environmental Health, University of Copenhagen, Copenhagen, Denmark

Dietmar Lohmann Institut für Humangenetik, Universitätsklinikum Essen, Essen, Germany

Matthias Löhr Department of Clinical Science, Intervention and Technology (CLINTEC), Karolinska Institutet, Stockholm, Sweden

Vinata B. Lokeshwar Department of Biochemistry and Molecular Biology, Medical College of Georgia; Augusta University, Augusta, GA, USA

Alexandre Loktionov DiagNodus Ltd, Babraham Research Campus, Cambridge, UK

Elias Lolis Department of Laboratory Medicine, Yale University School of Medicine, New Haven, CT, USA

Pier-Luigi Lollini Laboratory of Immunology and Biology of Metastasis, Department of Experimental, Diagnostic and Specialty Medicine, University of Bologna, Bologna, Italy

Weiwen Long Department of Biochemistry and Molecular Biology, Wright State University, Dayton, OH, USA

David J. López University of the Balearic Islands, Palma de Mallorca, Spain

Miguel Lopez-Lazaro Department of Pharmacology, Faculty of Pharmacy, University of Seville, Seville, Spain

Ana Lopez-Martin Hospital Universitario Severo Ochoa, Madrid, Spain

Charles L. Loprinzi Department of Oncology, Mayo Clinic, Rochester, MN, USA

Jochen Lorch Dana Farlur Cancer Institute, Boston, MA, USA

Edith M. Lord Department of Microbiology and Immunology, University of Rochester School of Medicine and Dentistry, Rochester, NY, USA

Reuben Lotan Department of Thoracic Head and Neck Medical Oncology, The University of Texas MD Anderson Cancer Center, Houston, TX, USA

Ragnhild A. Lothe Department of Cancer Prevention, Rikshospitalet-Radiumhospitalet Medical Centre, Oslo, Norway

Michael T. Lotze Department of Surgery and Department of Immunology, University of Pittsburgh, Pittsburgh, PA, USA

Christophe Louandre EA4666, Université de Picardie Jules Verne (UPJV), Amiens, France

Service de Biochimie, Centre de Biologie Humaine (CBH), University Hospital of Amiens (CHU Sud), Amiens, France

Chrystal U. Louis Center for Cell and Gene Therapy, Baylor College of Medicine, Texas Children's Hospital, and The Methodist Hospital, Houston, TX, USA

Dmitri Loukinov Section of Molecular Pathology, Laboratory of Immunopathology, NIAID, National Institutes of Health, Bethesda, MD, USA

David B. Lovejoy Department of Pathology, University of Sydney, Sydney, NSW, Australia

José Lozano Department of Molecular Biology and Biochemistry, University of Málaga, Málaga, Spain

Guanning N. Lu Departments of Otolaryngology-Head and Neck Surgery, University of Kansas Medical Center, Kansas City, KS, USA

Jiade J. Lu Department of Radiation Oncology, Fudan Universtiy Shanghai Cancer Center, Shanghai, China

Jing Lu Departments of Molecular and Cellular Oncology, The University of Texas MD Anderson Cancer Center, Houston, TX, USA

Tzong-Shi Lu Division of Experimental Medicine, Beth Israel Deaconess Medical Center, Harvard Institutes of Medicine, Boston, MA, USA

Yuanan Lu Department of Public Health Science, University of Hawaii, Honolulu, HI, USA

Irina A. Lubensky National Cancer Institute, Division of Cancer Treatment and Diagnosis, National Institutes of Health, Rockville, MD, USA

Jared M. Lucas Divisions of Human Biology, Fred Hutchinson Cancer Research Center, Seattle, WA, USA

Andreas Luch German Federal Institute for Risk Assessment (BfR), Berlin, Germany

Maria Li Lung Department of Clinical Oncology, Li Ka Shing Faculty of Medicine, The University of Hong Kong, Hong Kong, China

Jian-Hua Luo Department of Pathology, University of Pittsburgh, Pittsburgh, PA, USA

Gary H. Lyman Public Health Sciences and Clinical Research Divisions, Hutchinson Institute for Cancer Outcomes Research, Fred Hutchinson Cancer Research Center, Seattle, WA, USA

Henry Lynch Department of Preventive Medicine and Public Health, Creighton University, Omaha, NE, USA

Elsebeth Lynge Institute of Public Health, University of Copenhagen, Copenhagen, Denmark

Scott K. Lyons Molecular Imaging Group, CRUK Cambridge Research Institute, Li Ka Shing Centre, Cambridge, UK

Wenjian Ma National Institute of Environmental Health Sciences (NIEHS), Research Triangle Park, NC, USA

Michael MacManus Department of Radiation Oncology, Peter MacCallum Cancer Centre, East Melbourne, VIC, Australia

Britta Mädge DKFZ, Heidelberg, Germany

Claudie Madoulet Laboratory of Biochemistry, IFR53, Faculty of Pharmacy, Reims, France

Rolando F. Del Maestro Montreal, QC, Canada

Marcello Maggiolini Department of Pharmacy and Health and Nutritional Sciences, University of Calabria, Rende, Italy

Brinda Mahadevan Abbott Nutrition, Regulatory Affairs, Abbott Laboratories, Columbus, OH, USA

Joseph F. Maher Cancer Institute, University of Mississippi Medical Center, Jackson, MS, USA

Csaba Mahotka Institute of Pathology, Heinrich Heine Universität, Düsseldorf, Germany

Sourindra N. Maiti Division of Pediatrics, Department of Immunology, MD Anderson Cancer Center, Houston, TX, USA

Isabella W. Y. Mak Department of Surgery, Hamilton Health Sciences, Juravinski Cancer Centre, McMaster University, Hamilton, ON, Canada

N. K. Mak Department of Biology, Hong Kong Baptist University, Kowloon Tong, Hong Kong, China

Jennifer Makalowski Tumor Genetics, Clinic I Internal Medicine, University Hospital Cologne, and Center for Molecular Medicine Cologne, University of Cologne, Cologne, Germany

Cédric Malicet INSERM, Stress Cellulaire, Parc Scientifique et Technologique de Luminy, Marseille Cedex, France

Alessandra Mancino Department of Immunology, Fondazione Humanitas per la Ricerca, Rozzano, Milan, Italy

Evelyne Manet CIRI-International Center for Infectiology Research, INSERM U1111, Université Lyon 1, ENS de Lyon, Lyon, France

Sridhar Mani Department of Medicine, Oncology and Molecular Genetics, Albert Einstein College of Medicine, New York, NY, USA

Marcel Mannens Academic Medical Centre, University of Amsterdam, Amsterdam, The Netherlands

Alberto Mantovani Department of Immunology, Fondazione Humanitas per la Ricerca, Rozzano, Milan, Italy

Ashley A. Manzoor Department of Radiation Oncology, Duke University, Durham, NC, USA

Selwyn Mapolie Department of Chemistry and Polymer Science, Stellenbosch University, Matieland, South Africa

Lucia Marcocci Department of Biochemical Sciences "A. Rossi Fanelli", Sapienza University of Rome, Rome, Italy

Maurie Markman Department of Medical Oncology, Eastern Regional Medical Center, Philadelphia, PA, USA

Dieter Marmé Tumor Biology Center, Institute of Molecular Oncology, Freiburg, Germany

Marie-Claire Maroun Department of Internal Medicine, Division of Rheumatology, Wayne State University, Detroit, MI, USA

Deborah J. Marsh Kolling Institute of Medical Research and Royal North Shore Hospital, University of Sydney, Sydney, NSW, Australia

John L. Marshall Lombardi Comprehensive Cancer Center, Georgetown University, Washington, DC, USA

Angela Märten National Centre for Tumour Diseases; Department of Surgery, University Hospital Heidelberg, Heidelberg, Germany

Francis L. Martin Centre for Biophotonics, Lancaster University, Lancaster, Lancashire, UK

Olga A. Martin Division of Radiation Oncology and Cancer Imaging, Molecular Radiation Biology Laboratory, Peter MacCallum Cancer Centre, Melbourne, VIC, Australia

The Sir Peter MacCallum Department of Oncology, The University of Melbourne, Melbourne, VIC, Australia

Victor D. Martinez British Columbia Cancer Research Centre, Vancouver, BC, Canada

Gaetano Marverti Department of Biomedical Sciences, Metabolic and Neural Sciences, University of Modena and Reggio Emilia, Modena, Italy

Edmund Maser Institute of Toxicology and Pharmacology for Natural Scientists, University Medical School, Kiel, Germany

Thomas E. Massey Department of Biomedical and Molecular Sciences, Queen's University, Kingston, ON, Canada

Noriyuki Masuda Department of Respiratory Medicine, Kitasato University School of Medicine, Sagamihara, Kanagawa, Japan

Atsuko Masumi Department of Safety Research on Blood and Biological Products, National Institute of Infectious Diseases, Tokyo, Japan

Yasunobu Matsuda Department of Medical Technology, Niigata University Graduate of Health Sciences, Niigata, Japan

Sachiko Matsuhashi Department of Internal Medicine, Saga Medical School, Saga University, Saga, Japan

Takaya Matsuzuka Department of Anatomy and Physiology, Kansas State University, Manhattan, KS, USA

Malgorzata Matusiewicz Department of Medical Biochemistry, Wroclaw Medical University, Wroclaw, Poland

Warren L. May Department of Health Administration, School of Health Related Professions, University of Mississippi Medical Center, Jackson, MS, USA

Arnulf Mayer Department of Radiooncology and Radiotherapy, University Medical Center Mainz, Mainz, Germany

Matthew A. McBrian Department of Biological Chemistry, David Geffen School of Medicine at UCLA, Los Angeles, CA, USA

Joseph H. McCarty MD Anderson Cancer Center, Houston, TX, USA

Molliane Mcgahren-Murray Department of Systems Biology, Unit 1058, University of Texas MD Anderson Cancer Center, Houston, TX, USA

Katherine A. McGlynn Division of Cancer Epidemiology and Genetics, National Cancer Institute, National Institutes of Health, Bethesda, MD, USA

W. Glenn McGregor University of Louisville, Louisville, KY, USA

Iain H. McKillop Department of General Surgery, Carolinas Medical Center, Charlotte, NC, USA

Margaret McLaughlin-Drubin Brigham and Women's Hospital, Boston, MA, USA

Roger E. McLendon Department of Pathology, Duke University Medical Center, Durham, NC, USA

Donald C. McMillan University Department of Surgery, Royal Infirmary, Glasgow, UK

David W. Meek Division of Cancer Research, Jacqui Wood Cancer Centre/CRC, University of Dundee, Dundee, UK

Annette Meeson Institute of Genetic Medicine and North East Stem Cell Institute, Newcastle University, International Centre for Life, Newcastle upon Tyne, UK

Kamiya Mehla The Eppley Institute for Research in Cancer and Allied Diseases, University of Nebraska Medical Center, Omaha, NE, USA

Arianeb Mehrabi Department of General, Visceral and Transplantation Surgery, University of Heidelberg, Heidelberg, Germany

Mohammad Mehrmohammadi Department of Biomedical Engineering, Wayne State University, Detroit, MI, USA

Anil Mehta Division of Cardiovascular Medicine, University of Dundee, Dundee, UK

Kapil Mehta The University of Texas MD Anderson Cancer Center, Houston, TX, USA

Rekha Mehta Regulatory Toxicology Research Division, Bureau of Chemical Safety, Food Directorate, HPFB, Health Canada, Ottawa, ON, Canada

Yaron Meirow The Lautenberg Center for Immunology and Cancer Research, Israel-Canada Medical, Research Institute Faculty of Medicine, The Hebrew University, Jerusalem, Israel

Bar-Eli Menashe Department of Cancer Biology, The University of Texas MD Anderson Cancer Center, Houston, TX, USA

Wenbo Meng Special Minimally Invasive Surgery, Hepatopancreatobiliary Surgery Institute of Gansu Province, Clinical Medical College Cancer Center, First Hospital of Lanzhou University, Lanzhou University, Lanzhou, Gansu, China

Deepak Menon Department of Biological Sciences, Hunter College of the City University of New York, New York, NY, USA

Heather Mernitz Alverno College, Milwaukee, WI, USA

Karl-Heinz Merz Department of Chemistry, Division of Food Chemistry and Toxicology, University of Kaiserslautern, Kaiserslautern, Germany

Enrique Mesri Viral Oncology Program, Sylvester Comprehensive Cancer Center and Development Center for AIDS Research, Department of Microbiology and Immunology, University of Miami Miller School of Medicine, Miami, FL, USA

Roman Mezencev Georgia Institute of Technology, School of Biology, Atlanta, GA, USA

Jun Mi Department of Therapeutic Radiology and Oncology, University of Virginia School of Medicine, Charlottesville, VA, USA

Dennis F. Michiel Biopharmaceutical Development Program, Leidos Biomedical Research, Inc., Frederick National Laboratory for Cancer Research, Frederick, MD, USA

Josef Michl Departments of Pathology, Molecular and Cell Biology, State University of New York, Downstate Medical Center, New York, NY, USA

Stephan Mielke Abteilung Hämatologie und Onkologie, Medizinische Klinik und Poliklinik II, Zentrum Innere Medizin (ZIM), Universitätsklinikum Würzburg, Würzburg, Germany

Oleg Militsakh Head and Neck Surgery, Nebraska Medical Center, Nebraska Methodist Hospital, Omaha, NE, USA

Mark Steven Miller Department of Cancer Biology, Comprehensive Cancer Center, Wake Forest School of Medicine, Winston-Salem, NC, USA

Takeo Minaguchi Department of Obstetrics and Gynecology, University of Tsukuba, Tokyo, Japan

Nagahiro Minato Department of Immunology and Cell Biology, Graduate School of Medicine, Kyoto University, Kyoto, Japan

Rodney F. Minchin School of Biomedical Sciences, University of Queensland, St Lucia, QLD, Australia

Lucas Minig Gynecologic Department, Valencian Institute of Oncology (IVO), Valencia, Spain

John D. Minna Hamon Center for Therapeutic Oncology Research and Departments of Pathology, Internal Medicine and Pharmacology, University of Texas Southwestern Medical Center, Dallas, TX, USA

Claudia Mitchell Institut Cochin, Université Paris Descartes, CNRS, Paris, France

Kazuo Miyashita Faculty of Fisheries Sciences, Department of Bioresources Chemistry, Hokkaido University, Hakodate, Hokkaido, Japan

Eiji Miyoshi Department of Molecular Biochemistry and Clinical Investigation, Osaka University Graduate School of Medicine, Suita, Japan

Jun Miyoshi Department of Molecular Biology, Osaka Medical Center for Cancer and Cardiovascular Diseases, Osaka, Japan

Toshihiko Mizuta Department of Internal Medicine, Imari Arita Kyoritsu Hospital, Saga, Japan

Omeed Moaven Department of Surgery, Massachusetts General Hospital, Harvard Medical School, Boston, MA, USA

K. Thomas Moesta Klinik für Chirurgie und Chirurgische Onkologie, Charité Universitätsmedizin Berlin, Berlin, Germany

Seyed Moein Moghimi Nanomedicine Research Group, Centre for Pharmaceutical Nanotechnology and Nanotoxicology, Faculty of Health and Medical Sciences, University of Copenhagen, Copenhagen, Denmark

Sunish Mohanan Baker Institute for Animal Health, Department of Biomedical Sciences, School of Veterinary Medicine, Cornell University, Ithaca, NY, USA

Sonia Mohinta Department of Medical Microbiology, Immunology and Cell Biology, Southern Illinois University, School of Medicine, Springfield, IL, USA

Jan Mollenhauer Molecular Oncology Group, University of Southern Denmark, Odense, Denmark

Michael B. Møller Department of Pathology, Odense University Hospital, Odense, Denmark

Bruno Mondovì Department of Biochemical Sciences "A. Rossi Fanelli", Sapienza University of Rome, Rome, Italy

Alessandra Montecucco Istituto di Genetica Molecolare CNR, Pavia, Italy

Ruggero Montesano International Agency for Research on Cancer, Lyon, France

Wolter J. Mooi Department of Pathology, VU Medical Center, Amsterdam, The Netherlands

Amy C. Moore Georgia Cancer Coalition, Atlanta, GA, USA

Malcolm A. S. Moore Department of Cell Biology, Memorial-SloanKettering Cancer Center, New York, NY, USA

Cesar A. Moran Department of Pathology, MD Anderson Cancer Center, Houston, TX, USA

Jan S. Moreb Department of Medicine, Division of Hematology/Oncology, College of Medicine, University of Florida, Gainesville, USA

Sergio Moreno Instituto de Biología Molecular y Celular del Cáncer, CSIC/Universidad de Salamanca, Salamanca, Spain

Fabiola Moretti Institute of Cell Biology and Neurobiology, National Council Research of Italy, Rome, Italy

Eiichiro Mori Department of Radiation Oncology, School of Medicine, Nara Medical University, Kashihara, Nara, Japan

Akira Morimoto Department of Pediatrics, Kyoto Prefectural University of Medicine, Kyoto, Japan

Pat J. Morin Laboratory of Molecular Biology and Immunology, National Institute on Aging, Baltimore, MD, USA

Department of Pathology, Oncology and Gynecology and Obstetrics, Johns Hopkins Medical Institutions, Baltimore, MD, USA

American Association for Cancer Research, Philadelphia, PA, USA

Christine M. Morris Cancer Genetics Research, University of Otago, Christchurch, New Zealand

Cynthia C. Morton Department of Pathology, Brigham and Women's Hospital, Boston, MA, USA

Gabriela Möslein Helios Klinik, Allgemein- und Viszeralchirurgie, Bochum, Germany

Justin L. Mott Department of Biochemistry and Molecular Biology, University of Nebraska Medical Center, Omaha, NE, USA

Spyro Mousses Cancer Genetics Branch, National Human Genome Research Institute, NIH, Bethesda, MD, USA

Pavlos Msaouel Jacobi Medical Center, Albert Einstein College of Medicine, Bronx, NY, USA

Sebastian Mueller Centre of Alcohol Research (CAR), University of Heidelberg, Heidelberg, Germany

Susette C. Mueller Lombardi Comprehensive Cancer Center, Georgetown University Medical Center, Washington, DC, USA

Subhajit Mukherjee Albert Einstein College of Medicine, New York, NY, USA

Hans K. Müller-Hermelink Institute of Pathology, University of Würzburg, Würzburg, Germany

Gabriele Multhoff Klinikum rechts der Isar, Department Radiation Oncology, TU München and CCG – "Innate Immunity in Tumor Biology", Helmholtz Zentrum München, Munich, Germany

Julia Münzker Division of Endocrinology and Diabetology, Department of Internal Medicine, Medical University of Graz, Graz, Austria

Ramachandran Murali Department of Biomedical Sciences, Cedars-Sinai Medical Center, Los Angeles, CA, USA

Kenji Muro Department of Neurological Surgery, Robert H. Lurie Comprehensive Cancer Center, Northwestern University Feinberg School of Medicine, Chicago, IL, USA

Mandi Murph Department of Pharmaceutical and Biomedical Sciences, Georgia Cancer Coalition Distinguished Cancer Scholar, University of Georgia and College of Pharmacy, Athens, GA, USA

Edward L. Murphy University of California, School of Medicine, San Francisco, CA, USA

Paul G. Murray CRUK Institute for Cancer Studies, Molecular Pharmacology, Medical School, University of Birmingham, Birmingham, UK

Ruth J. Muschel Radiation Oncology and Biology, University of Oxford, Oxford, UK

Markus Müschen Leukemia and Lymphoma Program, Norris Comprehensive Cancer Center, University of Southern California, Los Angeles, CA, USA

Antonio Musio Institute for Genetic and Biomedical Research, National Research Council, Pisa, Italy

Istituto Toscano Tumori, Firenze, Italy

Akira Naganuma Laboratory of Molecular and Biochemical Toxicology, Graduate School of Pharmaceutical Sciences, Tohoku University, Sendai, Japan

Shigekazu Nagata Osaka University Medical School, Osaka, Japan

Christina M. Nagle Cancer and Population Studies, Queensland Institute of Medical Research, Royal Brisbane Hospital, Brisbane, QLD, Australia

Rita Nahta Department of Pharmacology, Emory University, Atlanta, GA, USA

Akira Nakagawara Saga Medical Center KOSEIKAN, Tosu, Japan

Tetsuya Nakatsura Division of Cancer Immunotherapy, Explonatory Oncology Research and Clinical Trial Center, National Cancer Center, Kashiwa City, Chiba Prefecture, Japan

Hariktishna Nakshatri IU Simon Cancer Center, Indiana University School of Medicine, Indianapolis, IN, USA

Patrizia Nanni Laboratory of Immunology and Biology of Metastasis, Department of Experimental, Diagnostic and Specialty Medicine, University of Bologna, Bologna, Italy

Zvi Naor Department of Biochemistry and Molecular Biology, The George S. Wise Faculty of Life Sciences, Tel Aviv University, Tel Aviv, Israel

Mohd W. Nasser Department of Pathology, Comprehensive Cancer Centre, The Ohio State Medical Centre, Columbus, OH, USA

Christian C. Naus Department of Cellular and Physiological Sciences, The University of British Columbia, Vancouver, BC, Canada

Tim S. Nawrot Division of Lung Toxicology, Department of Occupational and Environmental Medicine (T.S.N.) and the Studies Coordinating Centre (J.A.S.), Division of Hypertension and Cardiovascular Rehabilitation, Department of Cardiovascular Diseases, University of Leuven, Leuven, Belgium

David F. Nellis Biopharmaceutical Development Program, SAIC-Frederick, Inc., National Cancer Institute-Frederick, Frederick, MD, USA

Kenneth P. Nephew School of Medicine, Indiana University, Bloomington, IN, USA

David M. Neskey Department of Otolaryngology and Head and Neck Surgery, Medical University of South Carolina, Charleston, SC, USA

Klaus W. Neuhaus School of Dental Medicine, Department of Preventive, Restorative and Pediatric Dentistry, University of Bern, Bern, Switzerland

Kornelia Neveling Department of Human Genetics, Radboud University Nijmegen Medical Centre, Nijmegen, The Netherlands

Brad Neville Head and Neck Tumor Program, Hollings Cancer Center, Medical University of South Carolina, Charleston, SC, USA

Calvin S. H. Ng Division of Cardiothoracic Surgery, Chinese University of Hong Kong, Hong Kong, China

Irene O. L. Ng Department of Pathology, The University of Hong Kong, Hong Kong, China

Duc Nguyen Howard Hughes Medical Institute, Yale University School of Medicine, New Haven, CT, USA

Carole Nicco Faculté de Médecine Paris – Descartes, UPRES 18-33, Groupe Hospitalier Cochin – Saint Vincent de Paul, Paris, France

Santo V. Nicosia H. Lee Moffitt Cancer Center, Tampa, FL, USA

Anne T. Nies Dr. Margarete Fischer-Bosch-Institut für Klinische Pharmakologie, Stuttgart, Germany

M. Angela Nieto Instituto de Neurociencias de Alicante CSIC-UMH, Sant Joan d'Alacant, Spain

Omgo E. Nieweg Melanoma Institute Australia, North Sydney, NSW, Australia

Jonas Nilsson Department of Clinical Chemistry and Transfusion Medicin, Institute of Biomedicine, Sahlgrenska Academy at Göteborg University, Göteborg, Sweden

Ewa Ninio INSERM UMRS, Université Pierre et Marie Curie-Paris, Paris, France

Douglas Noonan University of Insubria, Varese, Italy

Larry Norton Breast Cancer Medicine Service, Department of Medicine, Memorial Sloan-Kettering Cancer Center, New York, NY, USA

Francisco J. Novo Department of Biochemistry and Genetics, University of Navarra, Pamplona, Spain

Ruslan Novosiadly Department of Cancer Immunobiology, Eli Lilly and Company, New York, NY, USA

Noa Noy Department of Cellular and Molecular Medicine, Lerner Research Institute, Cleveland Clinic and Case Western Reserve University, Cleveland, OH, USA

Hala H. Nsouli Department of Epidemiology and Biostatistics, The George Washington University School of Public Health and Health Services, Washington, DC, USA

Lauren M. Nunez Department of Biological Science, College of Pharmacy, Touro University-CA, Vallyo, CA, USA

John P. O'Bryan Department of Pharmacology, University of Illinois College of Medicine, Chicago, IL, USA

Jesse Brown VA Medical Center, Chicago, IL, USA

James P. B. O'Connor Institute of Cancer Sciences, University of Manchester, Manchester, UK

Sarah T. O'Dwyer Colorectal and Peritoneal Oncology Centre, The Christie NHS Foundation Trust, University of Manchester, Manchester, UK

John O'Leary Departments of Obstetrics and Gynaecology/Histopathology, Trinity College Dublin, Trinity Centre for Health Sciences, Dublin, Ireland

Ruth M. O'Rega Winship Cancer Institute, Emory University, Atlanta, GA, USA

Sharon O'Toole Departments of Obstetrics and Gynaecology/Histopathology, Trinity College Dublin, Trinity Centre for Health Sciences, Dublin, Ireland

André Oberthür Department of Pediatric Oncology and Hematology, Children's Hospital, University of Cologne, Cologne, Germany

Takahiro Ochiya Division of Molecular and Cellular Medicine, National Cancer Center Research Institute, Tokyo, Japan

Stefan Offermanns Department of Pharmacology, Max-Planck-Institute for Heart and Lung Research, Bad Nauheim, Germany

Anat Ohali Cancer Vaccine Section, National Cancer Institute, National Institutes of Health, Bethesda, MD, USA

Takeo Ohnishi Department of Radiation Oncology, School of Medicine, Nara Medical University, Kashihara, Nara, Japan

Hitoshi Ohno Department of Internal Medicine, Faculty of Medicine, Kyoto University, Kyoto, Japan

Kevin R. Oldenburg MatriCal, Inc., Spokane, WA, USA

Magali Olivier Group of Molecular Mechanisms and Biomarkers, International Agency for Research on Cancer, World Health Organization, Lyon, France

Egbert Oosterwijk Laboratory of Experimental Urology, University Medical Centre Nijmegen, Nijmegen, The Netherlands

Gertraud Orend Department of Clinical and Biological Sciences, Institute of Biochemistry and Genetics, Center for Biomedicine, DKBW, University of Basel, Basel, Switzerland

Makoto Osanai Department of Pathology, Kochi University School of Medicine, Kochi, Japan

Eduardo Osinaga Departamento de Inmunobiología, Facultad de Medicina, Universidad de la República, Montevideo, Uruguay

German Ott Department of Clinical Pathology, Robert-Bosch-Krankenhaus, Stuttgart, Germany

Christian Ottensmeier CRC Wessex Oncology Unit, Southampton General Hospital and Tenovous Laboratory, Southampton University Hospital Trust, Southampton, UK

Sai-Hong Ignatius Ou Chao Family Comprehensive Cancer Center, University of California, Irvine, CA, USA

Iwata Ozaki Health Administration Center, Saga Medical School, Saga University, Saga, Japan

Shuji Ozaki Department of Hematology, Tokushima Prefectural Central Hospital, Tokushima, Japan

Mónica Pérez-Ríos Department of Preventive Medicine and Public Health, School of Medicine, University of Santiago de Compostela, Santiago de Compostela, Spain

Helen Pace Department of Molecular Virology, Immunology and Medical Genetics, Ohio State University Comprehensive Cancer Center, Columbus, OH, USA

Simon Pacey Cancer Research UK Center for Cancer Therapeutics, The Institute of Cancer Research, Sutton, Surrey, UK

Mabel Padilla Molecular Biology and Lung Cancer Program, Lovelace Respiratory Research Institute, Albuquerque, NM, USA

Sumanta Kumar Pal Department of Medical Oncology and Experimental Therapeutics, City of Hope Comprehensive Cancer Center, Duarte, CA, USA

Viswanathan Palanisamy Department of Oral Health Sciences, Medical University of South Carolina, Charleston, SC, USA

Pier Paolo Pandolfi Division of Genetics, Beth Israel Deaconess Medical Center, Boston, MA, USA

Klaus Pantel Universitäts-Krankenhaus Eppendorf, Hamburg, Germany

Melissa C. Paoloni National Cancer Institute, Center for Cancer Research, Comparative Oncology Program, Bethesda, MD, USA

Evangelia Papadimitriou Laboratory of Molecular Pharmacology, Department of Pharmacy, School of Health Sciences, University of Patras, Patras, Greece

Philippe Paparel Department of Urology, Lyon Sud University Hospital, Pierre Benite, France

Athanasios G. Papavassiliou Department of Biological Chemistry, Medical School, University of Athens, Goudi, Athens, Greece

Sabitha Papineni Department of Veterinary Physiology and Pharmacology, Texas A&M University, College Station, TX, USA

Benoit Paquette Department of Nuclear Medicine and Radiobiology, Faculty of Medicine and Health Sciences, Université de Sherbrooke, Sherbrooke, QC, Canada

Ben Ho Park The Sidney Kimmel Comprehensive Cancer Center, Johns Hopkins University, Baltimore, MD, USA

Geoff J. M. Parker Centre for Imaging Sciences, University of Manchester, Manchester, UK

Sarah J. Parsons University of Virginia, Charlotteville, VA, USA

Eddy Pasquier Centre for Research in Oncobiology and Oncopharmacology, INSERM U911, Marseille, France

Metronomics Global Health Initiative, Marseille, France

Children's Cancer Institute, Randwick, NSW, Australia

Oneel Patel Department of Surgery, Austin Health, The University of Melbourne, Heidelberg, VIC, Australia

Rusha Patel Otolaryngology, Medical University of South Carolina, Charleston, SC, USA

Shyam Patel Standford University, Palo Alto, CA, USA

Patrizia Paterlini-Bréchot Faculté de Médecine Necker Enfants Malades, INSERM Unit 1151, Team 13, Paris, France

Yvonne Paterson Department of Microbiology, Perelman School of Medicine, University of Pennsylvania, Philadelphia, PA, USA

Konan Peck Institute of Biomedical Sciences, Academia Sinica Taipei, Taiwan, Republic of China

Florence Pedeutour Laboratory of Solid Tumors Genetics, Faculty of Medicine, Nice University Hospital, Nice, France

Dan Peer Laboratory of Precision NanoMedicine, Department of Cell Research and Immunology, George S. Wise Faculty of Life Sciences, Tel Aviv University, Tel Aviv, Israel

Department of Materials Science and Engineering, The Iby and Aladar Fleischman Faculty of Engineering, Tel Aviv University, Tel Aviv, Israel

Center for Nanoscience and Nanotechnology, Tel Aviv University, Tel Aviv, Israel

Tobias Peikert Division of Pulmonary and Critical Care Medicine, Department of Internal Medicine, Mayo Clinic College of Medicine, Rochester, MN, USA

Miguel A. Peinado Institute of Predictive and Personalized Medicine of Cancer (IMPPC), Badalona, Barcelona, Spain

Angel Pellicer Department of Pathology, New York University School of Medicine, New York, NY, USA

Juha Peltonen Department of Anatomy, Institute of Biomedicine, University of Turku, Turku, Finland

Sirkku Peltonen Department of Dermatology, University of Turku, Turku, Finland

Josef M. Penninger Institute of Molecular Biotechnology of the Austrian Academy of Sciences, Vienna, Austria

Richard T. Penson Division of Hematology Oncology, Massachusetts General Hospital, Boston, MA, USA

Maikel P. Peppelenbosch Erasmus Medical Center, University Medical Center Rotterdam, Rotterdam, The Netherlands

Carlos Perez-Stable Geriatric Research, Education, and Clinical Center Research Service, Bruce W. Carter Veterans Affairs Medical Center, Miami, FL, USA

Francisco G. Pernas National Institute on Deafness and Other Communication, Disorders and National Cancer Institute, NIH, Bethesda, MD, USA

Silverio Perrotta Department of Pediatrics, Second University of Naples, Naples, Italy

Godefridus J. Peters Department of Medical Oncology, VU University Medical Center, Amsterdam, The Netherlands

Marleen M. R. Petit Department of Human Genetics, University of Leuven, Leuven, Belgium

Peter Petzelbauer Department of Dermatology, Division of General Dermatology, Medical University of Vienna, Vienna, Austria

Claudia Pföhler Department of Dermatology, Saarland University Medical School, Homburg/Saar, Germany

Michael Pfreundschuh Klinik für Innere Medizin I, Universität des Saarlandes, Homburg, Germany

Philip A. Philip Karmanos Cancer Institute, Wayne State University, Detroit, MI, USA

Marco A. Pierotti Molecular Genetics of Cancer, Fondazione Istituto FIRC di Oncologia Molecolare, Milan, Italy

Paola Pietrangeli Department of Biochemical Sciences "A. Rossi Fanelli", Sapienza University of Rome, Rome, Italy

Torsten Pietsch Institut für Neuropathologie, Kinderchirurgie, Universitätskliniken Bonn, Bonn, Germany

Sreeraj G. Pillai Department of Surgery, Washington University School of Medicine, St. Louis, MO, USA

Lorenzo Pinna Department of Biological Chemistry, University of Padua, Padua, Italy

Michael Pishvaian Lombardi Comprehensive Cancer Center, Georgetown University, Washington, DC, USA

Ellen S. Pizer Laboratory of Cellular and Molecular Biology, National Institute on Aging, NIH, Baltimore, MD, USA

Kristjan Plaetzer Laboratory of Photodynamic Inactivation of Microorganisms, Division of Physics and Biophysics, University of Salzburg, Salzburg, Austria

Christoph Plass German Cancer Research Center (DKFZ), Heidelberg, Germany

Jeffrey L. Platt Departments of Microbiology and Immunology and Department of Surgery, University of Michigan, Ann Arbor, MI, USA

Mark R. Player Johnson & Johnson Pharmaceutical Research and Development, Spring House, PA, USA

Isabelle Plo INSERM, U1170, Hématopoièse et cellules souches, Gustave Roussy–PR1, Villejuif, France

Stephen R. Plymate Department of Medicine, Division of Gerontology and Geriatric Medicine, University of Washington, Seattle, WA, USA

Klaus Podar Medical Oncology, National Center for Tumor Diseases (NCT), University of Heidelberg, Heidelberg, Germany

Beatriz G. T. Pogo Tisch Cancer Institute, Icahn School of Medicine at Mount Sinai, New York, NY, USA

Jeffrey W. Pollard MRC Centre for Reproductive Health, Queen's Medical Research Institute, The University of Edinburgh, Edinburgh, UK

Department of Developmental and Molecular Biology, Albert Einstein College of Medicine, New York, NY, USA

Simona Polo University of Milan, Medical School, Milan, Italy

Satyanarayana R. Pondugula Department of Anatomy, Physiology, and Pharmacology, Auburn University, Auburn, AL, USA

Auburn University Research Initiative in Cancer, Auburn University, Auburn, AL, USA

Sreenivasan Ponnambalam Endothelial Cell Biology Unit, School of Molecular and Cellular Biology, University of Leeds, Leeds, UK

Mirco Ponzoni Experimental Therapies Unit, Laboratory of Oncology, Istituto Giannina Gaslini, Genoa, Italy

Beatrice L. Pool-Zobel Nutritional Toxicology, Friedrich-Schiller-University of Jena, Jena, Germany

Annemarie Poustka Division of Molecular Genome Analysis, DKFZ, Heidelberg, Germany

Marissa V. Powers Cancer Research UK Cancer Therapeutics Unit, The Institute of Cancer Research, Sutton, London, UK

Garth Powis NCI-Designated Cancer Center, Sanford Burnham Prebys Medical Discovery Institute, La Jolla, CA, USA

Graziella Pratesi Fondazione IRCCS Istituto Nazionale dei Tumori, Milan, Italy

George C. Prendergast Department of Pathology, Anatomy and Cell Biology, Jefferson Medical School, Lankenau Institute for Medical Research, Wynnewood, PA, USA

Victor G. Prieto Department of Pathology, The University of Texas MD Anderson Cancer Center, Houston, TX, USA

Sharon Prince Department of Human Biology, Health Science Faculty, Division of Cell Biology, University of Cape Town, Rondebosch, South Africa

Kevin M. Prise Centre for Cancer Research and Cell Biology, Queen's University Belfast, Belfast, UK

Kathy Pritchard-Jones Institute of Cancer Research/Royal Marsden Hospital, Sutton, Surrey, UK

Tassula Proikas-Cezanne Autophagy Laboratory, Department of Molecular Biology, Interfaculty Institute for Cell Biology, Faculty of Science, Eberhard Karls University Tübingen, Tübingen, Germany

Ching-Hon Pui St. Jude Children's Research Hospital, Memphis, TN, USA

Karen Pulford Nuffield Division of Clinical Laboratory Sciences, University of Oxford, John Radcliffe Hospital, Oxford, UK

Teresa Gómez Del Pulgar Instituto de Investigaciones Biomedicas, CSIC, Madrid, Spain

Vinee Purohit The Eppley Institute for Research in Cancer and Allied Diseases, and Department of Pathology and Microbiology, University of Nebraska Medical Center, Omaha, NE, USA

Keith R. Pye Cell ProTx, Aberdeen, UK

Chao-Nan Qian Department of Nasopharyngeal Carcinoma, Sun Yat-sen University Cancer Center, Guangzhou, People's Republic of China

Jiahua Qian Qiagen, Frederick, MD, USA

Liang Qiao Storr Liver Centre, Westmead Millennium Institute for Medical Research, The University of Sydney at Westmead Hospital, Westmead, NSW, Australia

Hartmut M. Rabes Institute of Pathology, University of Munich, Munich, Germany

Bar-Shavit Rachel Department of Oncology, Hadassah-University Hospital, Jerusalem, Israel

Ronny Racine Department of Urology, University of Miami – Miller School of Medicine, Miami, FL, USA

Dirk Rades Department of Radiation Oncology, University Hospital Schleswig-Holstein, Campus Luebeck, Germany

Jerald P. Radich Clinical Research Division, Fred Hutchinson Cancer Research Center, Seattle, WA, USA

Norman S. Radin Department of Psychiatry, University of Michigan, Ann Arbor, MI, USA

Fulvio Della Ragione Department of Biochemistry and Biophysics, Second University of Naples, Naples, Italy

Ryan L. Ragland Department of Human Genetics, University of Michigan, Ann Arbor, MI, USA

Gilbert J. Rahme Departments of Pediatrics and of Genetics, Norris Cotton Cancer Center, Geisel School of Medicine at Dartmouth, Hanover, NH, USA

Nino Rainusso Department of Pediatrics, Section of Hematology-Oncology, Baylor College of Medicine, Texas Children's Cancer and Hematology Centers, Houston, TX, USA

Ayyappan K. Rajasekaran Nemours Center for Childhood Cancer Research, Alfred I duPont Hospital for Children, Wilmington, DE, USA

Jayadev Raju Regulatory Toxicology Research Division, Bureau of Chemical Safety, Food Directorate, HPFB, Health Canada, Ottawa, ON, Canada

Sundaram Ramakrishnan Department of Pharmacology, University of Minnesota, Minneapolis, MN, USA

Kota V. Ramana Department of Biochemistry and Molecular Biology, University of Texas Medical Branch, Galveston, TX, USA

Pranela Rameshwar Medicine-Hematology/Oncology, Rutgers, New Jersey Medical School, Newark, NJ, USA

Santiago Ramón y Cajal Department of Pathology, Vall d'Hebron University Hospital, Barcelona, Spain

Giorgia Randi Department of Epidemiology, Institute for Farmacological Research Mario Negri, Milan, Italy

Ramachandran Rashmi Department of Radiation Oncology, Washington University School of Medicine, St. Louis, MO, USA

Mariusz Z. Ratajczak Stem Cell Institute at James Graham Brown Cancer Center, University of Louisville, Louisville, KY, USA

Anke Rattenholl Applied Biotechnology Division, Department of Engineering and Mathematics, University of Applied Sciences Bielefeld, Bielefeld, Germany

Cocav A. Rauwerdink Lahey Center for Hematology/Oncology at Parkland Medical Center, Salem, NH, USA

Alberto Ravaioli Department of Oncology, Instituto Scientifico Romagnolo per lo s, Infermi Hospital, Rimini, Italy

Mira R. Ray The Prostate Centre at Vancouver General Hospital, University of British Columbia, Vancouver, BC, Canada

Roger Reddel Children's Medical Research Institute, The University of Sydney, Westmead, NSW, Australia

May J. Reed Department of Medicine, Division of Gerontology and Geriatric Medicine, University of Washington, Seattle, WA, USA

Eduardo M. Rego Medical School of Ribeirão Preto, University of São Paulo, Ribeirão Preto, Brazil

Reuven Reich Institute for Drug Research, School of Pharmacy, Faculty of Medicine, The Hebrew University of Jerusalem, Jerusalem, Israel

Jean-Marie Reimund Université de Strasbourg, Faculté de Médecine, INSERM U1113 and Fédération de Médecine Translationnelle de Strasbourg (FMTS), and, Hôpitaux Universitaires de Strasbourg, Hôpital de Hautepierre, Service d'Hépato-Gastroentérologie et d'Assistance Nutritive, Strasbourg, France

Celso A. Reis Institute of Molecular Pathology and Immunology, University of Porto, Porto, Portugal

Ling Ren Pediatric Oncology Branch, National Cancer Institute, Center for Cancer Research, Bethesda, MD, USA

Andrew G. Renehan Colorectal and Peritoneal Oncology Centre, The Christie NHS Foundation Trust, University of Manchester, Manchester, UK

Marcus Renner Division of Molecular Genome Analysis, DKFZ, Heidelberg, Germany

Paul S. Rennie The Prostate Centre at Vancouver General Hospital, University of British Columbia, Vancouver, BC, Canada

Domenico Ribatti Department of Basic Medical Sciences, Neurosciences and Sensory Organs, University of Bari Medical School, Bari, Italy

Raul C. Ribeiro Department of Oncology, St. Jude Children's Research Hospital, Memphis, TN, USA

Des R. Richardson Department of Pathology, University of Sydney, Sydney, NSW, Australia

Victoria M. Richon Merck Research Laboratories, Boston, MA, USA

Justin L. Ricker Merck Research Laboratories, Boston, MA, USA

Thomas Ried Genetics Branch, Center for Cancer Research, National Cancer Institute, NIH, Bethesda, MD, USA

Jörg Ringel Department of Medicine A, University of Greifswald, Greifswald, Germany

Carrie Rinker-Schaffer Department of Surgery, Section of Urology, The University of Chicago, Chicago, IL, USA

Francisco Rivero Centre for Cardiovascular and Metabolic Research, The Hull York Medical School, University of Hull, Hull, UK

Tadeusz Robak Department of Hematology, Medical University of Lodz, Lodz, Poland

Rita Roberti Department of Experimental Medicine, University of Perugia, Perugia, Italy

Fredika M. Robertson The University of Texas MD Anderson Cancer Center, Houston, TX, USA

Angelo Rodrigues Department of Pathology, Portuguese Oncology Institute-Porto, Porto, Portugal

Delvys Rodriguez-Abreu Hospital Universitario Insular, Las Palmas de Gran Canaria, Spain

Jose Luis Rodríguez-Fernández Departamento de Microbiología Molecular y Biología de las Infecciones, Centro de Investigaciones Biológicas, Madrid, Spain

Carlos Rodriguez-Galindo Dana-Farber Cancer Institute, Boston, MA, USA

Florian Roka Department of Surgical Oncology, The University of Texas MD Anderson Cancer Center, Houston, TX, USA

Cleofé Romagosa Department of Pathology, Vall d'Hebron University Hospital, Barcelona, Spain

Ze'ev Ronai Signal Transduction Program, Burnham Institute for Medical Research, La Jolla, CA, USA

Luca Roncucci Department of Diagnostic and Clinical Medicine, and Public Health, University of Modena and Reggio Emilia, Modena, Italy

Igor B. Roninson Department of Drug Discovery and Biomedical Sciences, South Carolina College of Pharmacy, Columbia, SC, USA

Jatin Roper Tufts Medical Center, Boston, MA, USA

Rafael Rosell Instituto Oncológico Dr. Rosell, Quiron-Dexeus University Hospital, Barcelona, Spain

Pangaea Biotech, Barcelona, Spain

Cancer Biology and Precision Medicine Program, Catalan Institute of Oncology, Hospital Germans Trias i Pujol, Badalona, Spain

Molecular Oncology Research (MORe) Foundation, Barcelona, Spain

Eliot M. Rosen Department of Oncology, Georgetown University School of Medicine, Washington, DC, USA

Department of Biochemistry, Molecular and Cellular Biology, Georgetown University School of Medicine, Washington, DC, USA

Department of Radiation Medicine, Georgetown University School of Medicine, Washington, DC, USA

Carol L. Rosenberg Boston Medical Center and Boston University School of Medicine, Boston, MA, USA

Steven A. Rosenzweig Department of Cell and Molecular Pharmacology and Experimental Therapeutics, Medical University of South Carolina, Charleston, SC, USA

Angelo Rosolen Department of Pediatrics, Hemato-oncology Unit, University of Padua, Padova, Italy

Jeffrey S. Ross Albany Medical College, Albany, NY, USA

Theodora S. Ross Department of Internal Medicine, University of Texas, Southwestern Medical Center, Dallas, TX, USA

Catalina A. Rosselló University of the Balearic Islands, Palma de Mallorca, Spain

Anita De Rossi Viral Oncology Unit and AIDS Reference Center, Section of Oncology and Immunology, Department of Surgery, Oncology and Gastroenterology, University of Padova, Padova, Italy

Alberto Ruano-Ravina Department of Preventive Medicine and Public Health, School of Medicine, University of Santiago de Compostela, Santiago de Compostela, Spain

Tami Rubinek Tel Aviv Medical Center and Tel Aviv University, Tel Aviv, Israel

Luca Rubino Department of Oncology, Humanitas Research Hospital, Humanitas Cancer Center, Rozzano, Milan, Italy

Marco Ruggiero Dream Master Laboratory, Chandler, AZ, USA

Francisco Ruiz-Cabello Osuna UGC Laboratorio Clínico Hospital Universitario Virgen de las Nieves Facultad de Medicina, Universidad de Granada, Granada, Spain

María Victoria Ruiz-Pérez Department of Microbiology, Tumor and Cell Biology (MTC), Karolinska Institutet, Stockholm, Sweden

Zoran Rumboldt Department of Radiology and Radiological Science, Medical University of South Carolina, Charleston, SC, USA

Erkki Ruoslahti Cancer Research Center, Sanford Burnham Prebys Medical Discovery Institute, La Jolla, CA, USA

Center for Nanomedicine and Department of Molecular Cellular and Developmental Biology, University of California, Santa Barbara, Santa Barbara, CA, USA

Dario Rusciano Friedrich Miescher Institute, Basel, Switzerland

Giandomenico Russo Istituto Dermopatico dell'Immacolata, Istituto di Ricovero e Cura a Carattere Scientifico, Roma, Italy

Irma H. Russo Breast Cancer Research Laboratory, Fox Chase Cancer Center, Philadelphia, PA, USA

Jose Russo Breast Cancer Research Laboratory, Fox Chase Cancer Center, Philadelphia, PA, USA

James T. Rutka The Arthur and Sonia Labatt Brain Tumour Research Centre, The Hospital for Sick Children, The University of Toronto, Toronto, ON, Canada

James Ryan Head and Neck Tumor Program, Hollings Cancer Center, Medical University of South Carolina, Charleston, SC, USA

Venkata S. Sabbisetti Renal Division, Department of Medicine, Brigham and Women's Hospital, The Harvard Clinical and Translational Science Center, Boston, MA, USA

Anne Thoustrup Saber National Institute of Occupational Health, Copenhagen, Denmark

Gauri Sabnis University of Maryland School of Medicine, Baltimore, MD, USA

Mohamad Seyed Sadr Montreal, QC, Canada

Guillermo T. Sáez Department of Biochemistry and Molecular Biology, Faculty of Medicine and Odontology-INCLIVA, University of Valencia, Valencia, Spain
Service of Clinical Analysis, Dr. Peset University Hospital, Valencia, Spain

Stephen Safe Department of Veterinary Physiology and Pharmacology, Texas A&M University, College Station, TX, USA

Xavier Sagaert Department of Pathology, University Hospitals of K.U. Leuven, Leuven, Belgium

Asim Saha University of Cincinnati and The Barrett Cancer Center, Cincinnati, OH, USA

Emine Sahin Institute for Physiology, Center for Physiology and Pharmacology, Medical University of Vienna, Vienna, Austria

Kunal Saigal National Institute on Deafness and Other Communication, Disorders and National Cancer Institute, NIH, Bethesda, MD, USA

Toshiyuki Sakai Department of Molecular-Targeting Cancer Prevention, Graduate School of Medical Science, Kyoto Prefectural University of Medicine, Kyoto, Japan

Bodour Salhia Cancer and Cell Biology Division, The Translational Genomics Research Institute, Phoenix, AZ, USA

Helmut Rainer Salih Department of Internal Medicine II, University Hospital of Tübingen, Eberhard-Karls-University, Tübingen, Germany

Beth A. Salmon Department of Pharmacology and Therapeutics, University of Florida, Gainesville, FL, USA

Howard W. Salmon Department of Radiation Oncology, North Florida Radiation Oncology, Gainesville, FL, USA

Raed Samar Cancer Vaccine Section, National Cancer Institute, National Institutes of Health, Bethesda, MD, USA

Julian R. Sampson Institute of Medical Genetics, Cardiff University, Heath Park, Cardiff, UK

Nianli Sang Department of Biology, Drexel University College of Arts and Sciences, Philadelphia, PA, USA

Manoranjan Santra Neurology Research, Henry Ford Health System, Detroit, MI, USA

Ehsan Sarafraz-Yazdi Division of Gynecologic Oncology, Department of OB/GYN, State University of New York, Downstate Medical Center, New York, NY, USA

Frank Saran Department of Radiotherapy and Paediatric Oncology, Royal Marsden Hospital NHS Foundation Trust, Sutton, Surrey, UK

Devanand Sarkar Department of Human and Molecular Genetics, Virginia Commonwealth University, VCU Medical Center, School of Medicine, Richmond, VA, USA

Fazlul H. Sarkar Karmanos Cancer Institute, Wayne State University, Detroit, MI, USA

Debashis Sarker Cancer Research UK Center for Cancer Therapeutics, The Institute of Cancer Research, Sutton, Surrey, UK

Ken Sasaki Department of Cancer Biology, University of Kansas Cancer Center, The University of Kansas Medical Center, Kansas City, KS, USA

Hiroyuki Sasaki Division of Epigenomics and Development, Medical Institute of Bioregulation, Kyushu University, Fukuoka, Japan

Tomikazu Sasaki Department of Chemistry, University of Washington, Seattle, WA, USA

A. Kate Sasser Department of Pediatrics, Columbus Children's Research Institute, The Ohio State University, Columbus, OH, USA

Aaron R. Sasson Department of Surgery, University of Nebraska Medical Center, Omaha, NE, USA

Robert L. Satcher Orthopaedic Oncology, University of Texas MD Anderson Cancer Center, Houston, TX, USA

Leonard A. Sauer Bassett Research Institute, Cooperstown, NY, USA

Christobel Saunders School of Surgery and Pathology, QEII Medical Centre, University of Western Australia, Crawley, WA, Australia

Constance L. L. Saw Department of Pharmaceutics, Rutgers, The State University of New Jersey, Ernest Mario School of Pharmacy, Piscataway, NJ, USA

Anurag Saxena Department of Pathology and Laboratory Medicine, Royal University Hospital, Saskatoon Health Region/University of Saskatchewan, Saskatoon, SK, Canada

Reinhold Schäfer Comprehensive Cancer Center, Charité Universitätsmedizin Berlin, Berlin, Germany

Amanda Schalk University of Illinois at Chicago, Chicago, IL, USA

Manfred Schartl Physiologische Chemie I, Biozentrum, Universität Würzburg, Würzburg, Germany

Huub Schellekens Department of Innovation Studies, Department of Pharmaceutical Sciences, Utrecht University, TD Utrecht, The Netherlands

Detlev Schindler Department of Human Genetics, Biozentrum University of Würzburg, Würzburg, Germany

Peter M. Schlag Comprehensive Cancer Center, Charité Campus Mitte, Berlin, Germany

Peter Schlosshauer Department of Pathology, The Mount Sinai School of Medicine, New York, NY, USA

Martin Schlumberger Department of Nuclear Medicine and Endocrine Oncology, Referral Center for Refractory Thyroid Tumors, Institut National du Cancer, Institut Gustave Roussy, Villejuif, France

Peter Schmezer Division Epigenomics and Cancer Risk Factors, German Cancer Research Center (DKFZ), Heidelberg, Germany

Annette Schmitt-Graeff Department of Pathology, University hospital Freiburg, Freiburg, Germany

Marc Schmitz Institut für Immunologie, Technische Universität Dresden, Dresden, Germany

Dominik T. Schneider Clinic of Pediatrics, Klinikum Dortmund, Dortmund, Germany

Katrina J. Schneider Research Service, Veterans Administration Medical Center, Omaha, NE, USA

Stefan W. Schneider Hauttumorzentrum Mannheim (HTZM), Universi-tätsmedizn Mannheim, Mannheim, Germany

Maria Schnelzer Department of Radiation Protection and Health, Bundesamt für Strahlenschutz (Federal Office for Radiation Protection), Oberschleissheim, Germany

Nathalie Scholler Center for Cancer, SRI Biosciences, Menlo Park, CA, USA

Axel H. Schönthal University of Southern California, Keck School of Medicine, Los Angeles, CA, USA

Bart H. W. Schreuder Department of Orthopaedics, Radboud University Medical Centre, Nijmegen, The Netherlands

Morgan S. Schrock Department of Molecular Virology, Immunology and Medical Genetics, Ohio State University Comprehensive Cancer Center, Columbus, OH, USA

Laura W. Schrum Department of Biology, The University of North Carolina at Charlotte, Charlotte, NC, USA

Wolfgang A. Schulz Department of Urology, Heinrich Heine University, Düsseldorf, Germany

Manfred Schwab German Cancer Research Center (DKFZ), Heidelberg, Germany

Markus Schwaiger Department of Nuclear Medicine, Technical University of Munich, Munich, Germany

Edward L. Schwartz Department of Medicine (Oncology), Albert Einstein College of Medicine, Bronx, NY, USA

Julie K. Schwarz Department of Radiation Oncology, Washington University School of Medicine, St. Louis, MO, USA

Rony Seger Department of Biological Regulation, The Weizmann Institute of Science, Rehovot, Israel

Gail M. Seigel Center for Hearing and Deafness, University at Buffalo, Buffalo, NY, USA

Hiroyuki Seimiya Division of Molecular Biotherapy, Cancer Chemotherapy Center, Japanese Foundation for Cancer Research, Koto-ku, Tokyo, Japan

Paule Seite UMR CNRS 6187 Pôle Biologie Santé, University of Poitiers, Poitiers cedex, France

Helmut K. Seitz Centre of Alcohol Research (CAR), University of Heidelberg, Heidelberg, Germany

Department of Medicine, Salem Medical Center, Heidelberg, Germany

Periasamy Selvaraj Department of Pathology, Emory University School of Medicine, Atlanta, GA, USA

Wolfhard Semmler Department of Medical Physics in Radiology, German Cancer Research Center, Heidelberg, Germany

Subrata Sen Department of Molecular Pathology (Unit 951), The University of Texas MD Anderson Cancer Center, Houston, TX, USA

Suvajit Sen Department of Obstetrics and Gynecology, Jonsson Comprehensive Cancer Center, David Geffen School of Medicine, University of California at Los Angeles, Los Angeles, CA, USA

Vitalyi Senyuk Department of Medicine (M/C 737), College of Medicine Research Building, University of Illinois at Chicago, Chicago, IL, USA

Nedime Serakinci Medical Genetics, Near East University, Nicosia, Northern Cyprus

Christine Sers Institute of Pathology, University Medicine Charité, Berlin, Germany

Marta Sesé Department of Pathology, Vall d'Hebron University Hospital, Barcelona, Spain

Vijayasaradhi Setaluri Department of Anatomy, Kasturba Medical College, Manipal University, Manipal, Karnataka, India

John F. Seymour Haematology Department, Peter MacCallum Cancer Centre, East Melbourne, VIC, Australia

University of Melbourne, Parkville, VIC, Australia

Girish V. Shah Department of Pharmacology, University of Louisiana College of Pharmacy, Monroe, LA, USA

Rabia K. Shahid Department of Medicine, University of Saskatchewan, Saskatoon, SK, Canada

Sharmila Shankar Department of Pathology and Laboratory Medicine, The University of Kansas Medical Center, Kansas City, KS, USA

Anand Sharma Head and Neck Tumor Program, Hollings Cancer Center, Medical University of South Carolina, Charleston, SC, USA

Narinder Kumar Sharma Department of Pharmacology, Toxicology and Therapeutics, and Medicine, The University of Kansas Medical Center, Kansas City, KS, USA

Jerry W. Shay University of Texas Southwestern Medical Center, Dallas, TX, USA

Shijie Sheng Department of Pathology and Oncology, Wayne State University School of Medicine, Karmanos Cancer Institute, Detroit, MI, USA

James L. Sherley Asymmetrex, LLC, Boston, MA, USA

Donna Shewach Department of Pharmacology, University of Michigan Medical School, Ann Arbor, MI, USA

Ie-Ming Shih Department of Pathology, Johns Hopkins University School of Medicine, Baltimore, MD, USA

Kentaro Shikata Department of Environmental Medicine, Graduate School of Medical Sciences, Kyushu University, Fukuoka, Japan

Yosef Shiloh Sackler School of Medicine, Tel Aviv University, Tel Aviv, Israel

Hyunsuk Shim Department of Hematology/Oncology, Winship Cancer Institute, Emory University, Atlanta, GA, USA

Yutaka Shimada Department of Surgery, Graduate School of Medicine, Kyoto University, Kyoto, Japan

Masahito Shimojo School of Medicine, Osaka Medical College, Takatsuki, Osaka, Japan

Yong-Beom Shin BioNanotechnology Research Center, Korea Research Institute of Bioscience and Biotechnology, Yuseong, Daejeon, Republic of Korea

Toshi Shioda Massachusetts General Hospital Center for Cancer Research, Charlestown, MA, USA

Janet Shipley The Institute of Cancer Research, Sutton, Surrey, UK

Girja S. Shukla Department of Surgery, Vermont Comprehensive Cancer Center, College of Medicine, University of Vermont, Burlington, VT, USA

Arthur Shulkes Department of Surgery, Austin Health, The University of Melbourne, Heidelberg, VIC, Australia

Antonio Sica Department of Immunology, Fondazione Humanitas per la Ricerca, Rozzano, Milan, Italy

Gene P. Siegal Department of Pathology, University of Alabama at Birmingham, Birmingham, AL, USA

Dietmar W. Siemann Department of Radiation Oncology, University of Florida, Gainesville, FL, USA

Christine L. E. Siezen National Institute of Public Health and Environment, Bilthoven, The Netherlands

Alexandra Silveira Ocular Molecular Genetics Institute, Harvard Medical School, Massachusetts Eye and Ear Infirmary, Boston, MA, USA

Martin J. Simard Le Centre de recherche du CHU de Québec-Université Laval: axe Oncologie, Le Centre de recherche sur le cancer de l'Université Laval, Québec, QC, Canada

Diane M. Simeone Department of Physiology, University of Michigan Medical Center, Ann Arbor, MI, USA

Hans-Uwe Simon Department of Pharmacology, University of Bern, Bern, Switzerland

Bryan Simoneau Le Centre de recherche du CHU de Québec-Université Laval: axe Oncologie, Le Centre de recherche sur le cancer de l'Université Laval, Québec, QC, Canada

Ajay Singh Department of Oncologic Sciences, Mitchell Cancer Institute, University of South Alabama, Mobile, AL, USA

Amrik J. Singh Department of Pathology, Harvard Medical School, Beth Israel Deaconess Medical Center, Boston, MA, USA

Harprit Singh De Montfort University, Leicester, UK

Kamaleshwar Singh The Institute of Environmental and Human Health (TIEHH), Texas Tech University, Lubbock, TX, USA

Narendra P. Singh Departments of Bioengineering, University of Washington, Seattle, WA, USA

Pankaj K. Singh The Eppley Institute for Research in Cancer and Allied Diseases, and Department of Pathology and Microbiology, and Department of Biochemistry and Molecular Biology, and Department of Genetic Cell Biology and Anatomy, University of Nebraska Medical Center, Omaha, NE, USA

Shalini Singh Department of Surgery, McMaster University, Hamilton, ON, Canada

Shree Ram Singh Basic Research Laboratory, National Cancer Institute at Frederick, Frederick, MD, USA

Vineeta Singh School of Surgery and Pathology, QEII Medical Centre, Sir Charles Gairdner Hospital, Nedlands, WA, Australia

Lillian L. Siu Department of Medical Oncology and Hematology, Robert and Maggie Bras and Family New Drug Development Program, Princess Margaret Hospital, Toronto, ON, Canada

Anita Sjölander Cell and Experimental Pathology, Department of Laboratory Medicine, Lund University, Malmö University Hospital, Malmö, Sweden

Judith Skoner Head and Neck Tumor Program, Hollings Cancer Center, Medical University of South Carolina, Charleston, SC, USA

Keith Skubitz Division of Hematology, Oncology and Transplantation, University of Minnesota Medical School, Minneapolis, MN, USA

Christopher Slape Genetics Branch, Center for Cancer Research, National Cancer Institute, National Institutes of Health, Bethesda, MD, USA

Keiran S. M. Smalley The Wistar Institute, Philadelphia, PA, USA

Lubomir B. Smilenov Department of Radiation Oncology, Columbia University, New York, NY, USA

Bruce F. Smith Scott-Ritchey Research Center, College of Veterinary Medicine, Auburn University, Auburn, AL, USA

Russell Spencer Smith Department of Pharmacology, University of Illinois College of Medicine, Chicago, IL, USA

Josef Smolle Department of Dermatology, Medical University Graz, Graz, Austria

Jimmy B. Y. So Department of Surgery, National University of Singapore, National University Hospital, Singapore, Singapore

Robert W. Sobol University of South Alabama Mitchell Cancer Institute, Mobile, AL, USA

Alexander S. Sobolev Department of Molecular Genetics of Intracellular Transport, Institute of Gene Biology, Russian Academy of Sciences, Moscow, Russia

Eric Solary Inserm Unité Mixte de Recherche (UMR) 1009, Institut Gustave Roussy, University Paris-Sud 11, Villejuif, France

Graziella Solinas Department of Immunology, Fondazione Humanitas per la Ricerca, Rozzano, Milan, Italy

Toshiya Soma Department of Surgery, Graduate School of Medicine, Kyoto University, Kyoto, Japan

Guru Sonpavde Texas Oncology and Veterans Affairs Medical Center and the Baylor College of Medicine, Houston, TX, USA

Anil K. Sood Departments of Gynecologic Oncology and Reproductive Medicine and Cancer Biology and The Center for RNA Interference and Non-Coding RNAs, The University of Texas MD Anderson Cancer Center, Houston, TX, USA

Henrik Toft Sørensen Department of Clinical Epidemiology, Aarhus University Hospital, Aarhus C, Denmark

Pavel Soucek Toxicogenomics Unit, Center for Toxicology and Health Safety, National Institute of Public Health, Prague, Czech Republic

Lorenzo Spaggiari University of Milan School of Medicine, Milan, Italy

Ulrich Specks Division of Pulmonary and Critical Care Medicine, Department of Internal Medicine, Mayo Clinic College of Medicine, Rochester, MN, USA

David W. Speicher The Wistar Institute, Philadelphia, PA, USA

Valerie Speirs Leeds Institute of Molecular Medicine, University of Leeds, Leeds, UK

Dietmar Spengler Max-Panck-Institut für Psychiatrie, Munich, Germany

Phillippe E. Spiess Department of Genitourinary Oncology, Moffitt Cancer Center, Tampa, FL, USA

Melanie Spotheim-Maurizot Centre de Biophysique Moleculaire, CNRS, Orleans, France

Cynthia C. Sprenger Department of Medicine, Division of Gerontology and Geriatric Medicine, University of Washington, Seattle, WA, USA

Lakshmaiah Sreerama Department of Chemistry and Biochemistry, St. Cloud State University, St. Cloud, MN, USA

Department of Chemistry and Earth Sciences, Qatar University, Doha, Qatar

Rakesh Srivastava Department of Pathology and Laboratory Medicine, The University of Kansas Medical Center, Kansas City, KS, USA

Satish K. Srivastava Department of Biochemistry and Molecular Biology, University of Texas Medical Branch, Galveston, TX, USA

M. Sharon Stack Northwestern University Medical School, Chicago, IL, USA

Jan A. Staessen Division of Lung Toxicology, Department of Occupational and Environmental Medicine (T.S.N.) and the Studies Coordinating Centre (J.A.S.), Division of Hypertension and Cardiovascular Rehabilitation, Department of Cardiovascular Diseases, University of Leuven, Leuven, Belgium

Eric Stanbridge Department of Microbiology and Molecular Genetics, University of California, Irvine, CA, USA

Barry Staymates Department of Pathology, Henry Mayo Newhall Memorial Hospital, Valencia, CA, USA

Stacey Stein Center for Advanced Biotechnology and Medicine, UMDMJ – Robert Wood Johnson Medical School, Piscataway, NJ, USA

Martin Steinhoff UCD Charles Institute of Dermatology, University College Dublin, Belfield, Ireland

Department of Dermatology School of Medicine and Medical Sciences, University College Dublin, Dublin, Ireland

Alexander Steinle Institute for Molecular Medicine, Centre for Molecular Medicine, Goethe University, Frankfurt am Main, Germany

Carsten Stephan Department of Urology, Charité, Universitätsmedizin, Campus Charité Mitte, Berlin, Germany

Peter L. Stern Cancer Research UK Manchester Institute, University of Manchester, Manchester, UK

William G. Stetler-Stevenson Extracellular Matrix Pathology Section, Cell and Cancer Biology Branch, National Cancer Institute, Bethesda, MD, USA

Richard G. Stevens University of Connecticut Health Center, Farmington, CT, USA

Freda Stevenson CRC Wessex Oncology Unit, Southampton General Hospital and Tenovous Laboratory, Southampton University Hospital Trust, Southampton, UK

William P. Steward Department of Cancer Studies, University of Leicester, Leicester, UK

Constantine A. Stratakis Program on Developmental Endocrinology of Genetics, NICHD, NIH, Bethesda, MD, USA

Alex Y. Strongin Burnham Institute for Medical Research, La Jolla, CA, USA

Deepa S. Subramaniam Georgetown University Hospital, Washington, DC, USA

Garnet Suck Health Sciences Authority, Centre for Transfusion Medicine, Singapore, Singapore

Paul H. Sugarbaker Washington Cancer Institute, Washington Hospital Center, Washington, DC, USA

Baocun Sun Department of Pathology, Tianjin Cancer Hospital and Tianjin Cancer Institute, Tianjin, People's Republic of China

Duxin Sun Department of Pharmaceutical Sciences, University of Michigan, Ann Arbor, MI, USA

Shi-Yong Sun School of Medicine and Winship Cancer Institute, Emory University, Atlanta, GA, USA

Zhifu Sun Department of Health Sciences Research, Mayo Clinic College of Medicine, Rochester, MN, USA

Saul Suster Department of Pathology, Medical College of Wisconsin, Milwaukee, WI, USA

Russell Szmulewitz The University of Chicago Medicine, Chicago, IL, USA

Thomas Tüting Laboratory for Experimental Dermatology, Department of Dermatology, University of Bonn, Bonn, Germany

Dirk Taeger Institute for Prevention and Occupational Medicine of the German Social Accident Insurance (IPA), Ruhr-University Bochum, Bochum, Germany

Masatoshi Tagawa Division of Pathology and Cell Therapy, Chiba Cancer Center Research Institute, Chiba, Japan

Stanley Tahara Keck School of Medicine, Department of Molecular Microbiology and Immunology, University of Southern California, Los Angeles, CA, USA

Yoshikazu Takada UC Davis School of Medicine, Sacramento, CA, USA

Akihisa Takahashi Heavy Ion Medical Center, Gunma University, Maebashi, Gunma, Japan

Tsutomu Takahashi Department of Environmental Health, School of Pharmacy, Tokyo University of Pharmacy and Life Sciences, Tokyo, Japan

Yoshimi Takai Faculty of Medicine, Osaka University Graduate School of Medicine, Suita, Japan

Tamotsu Takeuchi Department of Pathology, Kochi Medical School, Kochi, Japan

Constantine S. Tam Haematology Department, Peter MacCallum Cancer Centre, East Melbourne, VIC, Australia

University of Melbourne, Parkville, VIC, Australia

Luca Tamagnone Department of Oncology, University of Turin, Candiolo, Italy

Candiolo Cancer Center-IRCCS, University of Turin, Candiolo, Italy

Harald Tammen PXBioVisioN GmbH, Hannover, Germany

Masaaki Tamura Department of Anatomy and Physiology, Kansas State University, Manhattan, KS, USA

David S. P. Tan Department of Medical Oncology, National University Cancer Institute, Singapore (NCIS), National University Hospital, and Cancer Science Institute, National University of Singapore, Singapore, Singapore

Takuji Tanaka Department of Oncologic Pathology, Kanazawa Medical University, Kanazawa, Japan

Dean G. Tang Department of Carcinogenesis, Science Park-Research Division, The University of Texas MD Anderson Cancer Center, Smithville, TX, USA

Ya-Chu Tang Department of Life Sciences, College of Science, National Central University, Jhongli City, Taiwan

Nizar M. Tannir Department of Genitourinary Medical Oncology, University of Texas MD Anderson Cancer Center, Houston, TX, USA

Weikang Tao Department of Cancer Research, Merck Research Laboratories, West Point, PA, USA

Chi Tarn Department of Medical Oncology, Fox Chase Cancer Center, Philadelphia, PA, USA

Clive R. Taylor Department of Pathology, University of Southern California Keck School of Medicine, Los Angeles, CA, USA

Jennifer Taylor Committee on Cancer Biology, The University of Chicago, Chicago, IL, USA

Andrew R. Tee Institute of Medical Genetics, Cardiff University, Heath Park, Cardiff, UK

Ayalew Tefferi Division of Hematology, Mayo Clinic College of Medicine, Rochester, MN, USA

Bin T. Teh Cancer and Stem Cell Biology (CSCB), Duke-NUS, Graduate Medical School, Singapore, Singapore

Marie-Hélène Teiten Laboratoire de Biologie Moléculaire et Cellulaire du Cancer (LBMCC), Hôpital Kirchberg, Luxembourg, Luxembourg

Joseph R. Testa Fox Chase Cancer Center, Philadelphia, PA, USA

John Thacker Medical Research Council, Radiation and Genome Stability Unit, Harwell, Oxfordshire, UK

Rajesh V. Thakker Academic Endocrine Unit, Radcliffe Department of Medicine, Oxford Centre for Diabetes, Endocrinology and Metabolism (OCDEM), Churchill Hospital, University of Oxford, Oxford, UK

Nicholas B. La Thangue Department of Oncology, University of Oxford, Oxford, UK

Dan Theodorescu Department of Surgery, Urology, School of Medicine, University of Colorado Cancer Center, Aurora, CO, USA

Panayiotis A. Theodoropoulos Department of Basic Sciences, The University of Crete, School of Medicine, Heraklion, Crete, Greece

Frank Thévenod Private Universität Witten/Herdecke gGmbH, Witten, Germany

Karl-Heinz Thierauch Berlin, Germany

Megan N. Thobe University of Cincinnati College of Medicine, Cincinnati, OH, USA

Natalie Thomas Clinical Network Services Pty Ltd, St Albans, UK

Peter Thomas Departments of Surgery and Biomedical Sciences, Creighton University, Omaha, NE, USA

Sufi M. Thomas Departments of Otolaryngology-Head and Neck Surgery, University of Kansas Medical Center, Kansas City, KS, USA

Cancer Biology, University of Kansas Medical Center, Kansas City, KS, USA

Anatomy and Cell Biology, University of Kansas Medical Center, Kansas City, KS, USA

Sven Thoms University of Göttingen, Göttingen, Germany

Magnus Thörn Department of Surgery (MT), Karolinska Institutet, Stockholm, Sweden

Anna Tiefenthaller Clinic for Radiotherapy and Radiation Oncology, LMU Munich, Munich, Germany

Derya Tilki Martini-Klinik, Prostatakrebszentrum, Universitätsklinikum Hamburg-Eppendorf, Hamburg, Germany

Donald J. Tindall Department of Urology Research, Department of Biochemistry and Molecular Biology, Mayo Clinic College of Medicine, Rochester, MN, USA

Umberto Tirelli Department of Medical Oncology, National Cancer Institute, Aviano, PN, Italy

Martin Tobi Section of Gastroenterology, Detroit VAMC, Detroit, MI, USA

Philip J. Tofilon Radiation Oncology Branch, National Cancer Institute, Bethesda, MD, USA

Masakazu Toi Department of Surgery (Breast Surgery), Graduate School of Medicine, Kyoto University, Kyoto, Japan

Amanda Ewart Toland Division of Human Cancer Genetics, The Ohio State University, Columbus, OH, USA

Massimo Tommasino Infections and Cancer Biology Group, International Agency for Research on Cancer, Lyon, France

Antonio Toninello Department of Biological Chemistry, University of Padua, Padua, Italy

Jeffrey A. Toretsky Department of Oncology and Pediatrics, Lombardi Comprehensive Cancer Center, Georgetown University, Washington, DC, USA

Jorge R. Toro National Institutes of Health, Bethesda, MD, USA

Manuel Torres University of the Balearic Islands, Palma de Mallorca, Spain

Tibor Tot Department of Pathology and Clinical Cytology, Central Hospital Falun, Uppsala University, Falun, Sweden

Mathilde Touvier UMR U1153 INSERM, U1125 INRA, CNAM, Université Paris 13, Centre de Recherche Epidémiologie et Statistique Sorbonne Paris Cité, Bobigny, France

Philip C. Trackman Department of Molecular and Cell Biology, Boston University Henry M. Goldman School of Dental Medicine, Boston, MA, USA

Tiffany A. Traina Breast Cancer Medicine Service, Department of Medicine, Memorial Sloan-Kettering Cancer Center, New York, NY, USA

Luba Trakhtenbrot Molecular Cytogenetics Laboratory, Institute of Hematology, The Chaim Sheba Medical Center, Tel Hashomer, Israel

Janeen H. Trembley Minneapolis VA Health Care System and University of Minnesota, Minneapolis, MN, USA

Pierre-Luc Tremblay Le Centre de recherche du CHU de Québec-Université Laval: axe Oncologie, Le Centre de recherche sur le cancer de l'Université Laval, Québec, QC, Canada

Matthew Trendowski Department of Biology, Syracuse University, Syracuse, NY, USA

Edward L. Trimble Department of Health and Human Services, National Cancer Institute, National Institutes of Health, Bethesda, MD, USA

Jörg Trojan Universitätsklinikum Frankfurt, Medizinische Klinik 1, Frankfurt am Main, Germany

Alisha M. Truman Northeastern University, Boston, MA, USA

Gregory J. Tsay Department of Medicine, Institute of Immunology, Chung Shan Medical University, Taichung, Taiwan

Apostolia-Maria Tsimberidou Department of Investigational Cancer Therapeutics, Division of Cancer Medicine, The University of Texas MD Anderson Cancer Center, Houston, TX, USA

Kunihiro Tsuchida Division for Therapies Against Intractable Diseases, Institute for Comprehensive Medical Science (ICMS), Fujita Health University, Toyoake, Japan

Nobuo Tsuchida Department of Molecular Cellular Oncology and Microbiology, Tokyo Medical and Dental University, Bunkyo-ku, Tokyo, Japan

Florin Tuluc Department of Pediatrics, Perelman School of Medicine, University of Pennsylvania, Philadelphia, PA, USA

The Children's Hospital of Philadelphia, Philadelphia, PA, USA

Mehmet Kemal Tur Institute of Pathology, University Hospital, Justus-Liebig-University Giessen, Giessen, Germany

Greg Turenchalkb 454 Life Sciences, Branford, CT, USA

Andrew S. Turnell Cancer Research UK Institute for Cancer Studies, The Medical School, The University of Birmingham, Edgbaston, Birmingham, UK

Jeffrey Turner Prostate Oncology Specialists, Los Angeles, CA, USA

Michelle C. Turner McLaughlin Centre for Population Health Risk Assessment, University of Ottawa, Ottawa, ON, Canada

ISGlobal, Centre for Research in Environmental Epidemiology (CREAL), Barcelona, Spain

Universitat Pompeu Fabra (UPF), Barcelona, Spain

CIBER Epidemiología y Salud Pública (CIBERESP), Madrid, Spain

Guri Tzivion Cancer Institute, Department of Biochemistry, University of Mississippi Medical Center, Jackson, MS, USA

Salvatore Ulisse Department of Experimental Medicine, University of Rome "Sapienza", Rome, Italy

Nick Underhill-Day School of Biosciences, Swift Ecology Ltd, Warwickshire, UK

Rosemarie A. Ungarelli Boston Medical Center and Boston University School of Medicine, Boston, MA, USA

Gretchen M. Unger GeneSegues Inc., Chaska, MN, USA

Motoko Unoki Division of Epigenomics and Development, Medical Institute of Bioregulation, Kyushu University, Fukuoka, Japan

Markus Vähä-Koskela Molecular Cancer Biology Research Program, University of Helsinki, Helsinki, Finland

Antti Vaheri Medicum, Faculty of Medicine, University of Helsinki, Helsinki, Finland

Kedar S. Vaidya Global Pharmaceutical Research and Development, Abbott Laboratories, North Chicago, IL, USA

Ilan Vaknin The Lautenberg Center for Immunology and Cancer Research, Israel-Canada Medical, Research Institute Faculty of Medicine, The Hebrew University, Jerusalem, Israel

Anne M. VanBuskirk Takeda Oncology, Cambridge, MA, USA

Wim Vanden Berghe Epigenetic Signaling Lab PPES, Department Biomedical Sciences, University Antwerp, Antwerp, Belgium

Marry M. van den Heuvel-Eibrink Princess Maxima Center for Pediatric Oncology/Hematology, Utrecht, The Netherlands

Michael W. Van Dyke Department of Chemistry and Biochemistry, Kennesaw State University, Kennesaw, GA, USA

Casper H. J. van Eijck Department of Surgery, Erasmus MC, Rotterdam, The Netherlands

Manon van Engeland Department of Pathology, GROW-School for Oncology and Developmental Biology, Maastricht University Hospital, Maastricht, The Netherlands

Wilhelmin M. U. van Grevenstein Department of Surgery, Erasmus MC, Rotterdam, The Netherlands

Ad Geurts van Kessel Department of Human Genetics, Radboud University Nijmegen Medical Centre, Nijmegen, The Netherlands

Ron H. N. van Schaik Department of Clinical Chemistry, Erasmus University Medical Center, Rotterdam, The Netherlands

Viggo Van Tendeloo Vaccine and Infections Disease Institute (VAXINFECTIO) Laboratory of Experimental Hematology, Faculty of Medicine and Health Sciences, University of Antwerp, Edegem, Belgium

Alex van Vliet Department of Cellular and Molecular Medicine, Cell Death Research and Therapy Lab, KU Leuven Campus Gasthuisberg, Leuven, Belgium

Carter Van Waes National Institute on Deafness and Other Communication, Disorders and National Cancer Institute, NIH, Bethesda, MD, USA

Sakari Vanharanta Department of Medical Genetics, Biomedicum Helsinki, University of Helsinki, Helsinki, Finland

Roberta Vanni Department of Biomedical Science and Technology, University of Cagliari, Monserrato (CA), Italy

Judith A. Varner Moores UCSD Cancer Center, University of California San Diego, La Jolla, CA, USA

Aikaterini T. Vasilaki University Department of Surgery, Royal Infirmary, Glasgow, UK

Peter Vaupel Department of Radiooncology and Radiotherapy, University Medical Center Mainz, Mainz, Germany

Guillermo Velasco Department of Biochemistry and Molecular Biology I, School of Biology, Complutense University, Madrid, Spain

Marcel Verheij Department of Radiotherapy, The Netherlands Cancer Institute–Antoni van Leeuwenhoek Hospital, Amsterdam, The Netherlands

Mukesh Verma Division of Cancer Control and Population Sciences, National Cancer Institute (NCI), National Institutes of Health (NIH), Rockville, MD, USA

Rakesh Verma Prescient Healthcare Group, London, UK

Srdan Verstovsek Leukemia Department, University of Texas MD Anderson Cancer Center, Houston, TX, USA

René P. H. Veth Department of Orthopaedics, Radboud University Medical Centre, Nijmegen, The Netherlands

G. J. Villares Department of Cancer Biology, The University of Texas MD Anderson Cancer Center, Houston, TX, USA

Akila N. Viswanathan Brigham and Women's/Dana-Farber Cancer Center, Boston, MA, USA

Kris Vleminckx Department of Biomedical Molecular Biology and Center for Medical Genetics, Ghent University, Ghent, Belgium

Israel Vlodavsky Anatomy and Cell Biology, Technion Israel Institute of Technology, Cancer and Vascular Biology Research Center, Haifa, Israel

Martina Vockerodt Department of Pediatrics I, Children's Hospital, Georg-August University of Gottingen, Gottingen, Germany

Charles L. Vogel Sylvester Cancer Center, School of Medicine, University of Miami, Plantation, FL, USA

Tilman Vogel Department of Surgery, Krankenhaus Maria Hilf, Mönchengladbach, Germany

Ulla Vogel National Institute of Occupational Health, Copenhagen, Denmark

Daniel D. von Hoff Arizona Cancer Center, Tucson, AZ, USA

Silvia von Mensdorff-Pouilly Department of Obstetrics and Gynaecology, Vrije Universiteit Medisch Centrum (VUmc), Amsterdam, The Netherlands

Ingo Kausch von Schmeling Klinik für Urologie und Kinderurologie, Ammerland Klinik GmbH, Westerstede, Germany

Dietrich von Schweinitz Klinikum der Universität München, Kinderchirurgische Klinik im Dr. von Haunerschen Kinderspital, München, Germany

Alireza Vosough Department of Radiotherapy, Royal Marsden Hospital NHS Foundation Trust, Sutton, Surrey, UK

George F. Vande Woude Van Andel Research Institute, Grand Rapids, MI, USA

Tom Waddell GI/Lymphoma Research Unit, Royal Marsden Hospital, Surrey, UK

Christoph Wagener University Medical Center Hamburg-Eppendorf, Hamburg, Germany

Sabine Wagner Department of Pediatrics, Klinik St. Hedwig, Krankenhaus der Barmherzigen Brüder, Regensburg, Germany

Kristin A. Waite Genomic Medicine Institute, Lerner Research Institute, and Taussing Cancer Institute, Cleveland Clinic Foundation, Cleveland, OH, USA

Toshifumi Wakai Division of Digestive and General Surgery, Niigata University Graduate School of Medical and Dental Sciences, Niigata, Japan

Heather M. Wallace University of Aberdeen, Aberdeen, UK

Håkan Wallin National Institute of Occupational Health, Copenhagen, Denmark

Susan E. Waltz Cancer and Cell Biology, University of Cincinnati College of Medicine, Cincinnati Veteran's Administration Hospital, Cincinnati, OH, USA

Jack R. Wands Division of Gastroenterology, Liver Research Center, Rhode Island Hospital and Warren Alpert Medical School of Brown University, Providence, RI, USA

Bo Wang The Ohio State University, Columbus, OH, USA

Gang Wang Feil Brain and Mind Research Institute, Weill Cornell Medicine, Cornell University, New York, NY, USA

Helen Y. Wang Center for Inflammation and Epigenetics, Houston Methodist Research Institute, Houston, TX, USA

Hwa-Chain Robert Wang Molecular Oncology, Department of Biomedical and Diagnostic Sciences, The University of Tennessee, College of Veterinary Medicine, Knoxville, TN, USA

Jianghua Wang Department of Pathology, Baylor College of Medicine, Houston, TX, USA

Mingjun Wang Center for Inflammation and Epigenetics, Houston Methodist Research Institute, Houston, TX, USA

Rong-Fu Wang Center for Inflammation and Epigenetics, Houston Methodist Research Institute, Houston, TX, USA

Xianghong Wang Department of Anatomy, The University of Hong Kong, Hong Kong, China

Xiang-Dong Wang Jean Mayer USDA Human Nutrition Research Center on Aging at Tufts University, Boston, MA, USA

Yu Wang Department of Pharmacology and Pharmacy, The University of Hong Kong, Hong Kong, China

Zhu A. Wang Department of Genetics and Development, Columbia University Medical Center, Herbert Irving Comprehensive Cancer Center, New York, NY, USA

Patrick Warnat Department of Theoretical Bioinformatics, German Cancer Research Center, Heidelberg, Germany

Kounosuke Watabe Department of Medical Microbiology, Immunology and Cell Biology, Southern Illinois University, School of Medicine, Springfield, IL, USA

School of Medicine, Department of Cancer Biology, Wake Forest University, Winston-Salem, NC, USA

Dawn Waterhouse Experimental Therapeutics, BC Cancer Agency, Vancouver, BC, Canada

Catherine Waters The Ohio State University College of Medicine, Columbus, OH, USA

Valerie M. Weaver Department of Surgery, University of California, San Francisco, San Francisco, CA, USA

Lau Weber Department of Urology, Singapore General Hospital, Singapore, Singapore

Daniel S. Wechsler Pediatric Hematology-Oncology, Duke University Medical Center, Durham, NC, USA

Scott A. Weed Department of Neurobiology and Anatomy, Mary Babb Randolph Cancer Center, West Virginia University, Morgantown, WV, USA

Oliver Weigert Department of Internal Medicine III, University of Munich, Großhadern, Munich, Germany

Eugene D. Weinberg Biology and Medical Sciences, Indiana University, Bloomington, IN, USA

I. Bernard Weinstein Columbia University, New York, NY, USA

Ellen Weisberg Department of Medical Oncology, Dana Farber Cancer Institute, Boston, MA, USA

Lawrence M. Weiss Division of Pathology, City of Hope National Medical Center, Duarte, CA, USA

Danny R. Welch Department of Cancer Biology, University of Kansas Cancer Center, The University of Kansas Medical Center, Kansas City, KS, USA

Thilo Welsch Department of Visceral, Thoracic and Vascular Surgery, TU Dresden, Dresden, Germany

Sarah J. Welsh Harris Manchester College, University of Oxford, Oxford, UK

Tania M. Welzel Universitätsklinikum Frankfurt, Medizinische Klinik 1, Frankfurt am Main, Germany

Tamra E. Werbowetski-Ogilvie Regenerative Medicine Program, Biochemistry and Medical Genetics and Physiology and Pathophysiology, College of Medicine, Faculty of Health Sciences, University of Manitoba, Winnipeg, MB, Canada

Frank Westermann DKFZ, German Cancer Research Center, Heidelberg, Germany

Linda C. Whelan UCD School of Biomolecular and Biomedical Science, UCD Conway Institute, University College Dublin, Dublin, Ireland

Bruce A. White Department of Cell Biology, UConn School of Medicine, UConn Health, Farmington, CT, USA

Robert P. Whitehead Nevada Cancer Institute, Las Vegas, NV, USA

Theresa L. Whiteside University of Pittsburgh Cancer Institute and University of Pittsburgh School of Medicine, Pittsburgh, PA, USA

Christophe Wiart University of Nottingham, Nottingham, UK

Andreas Wicki Department of Medical Oncology, University Hospital, Basel, Switzerland

Carol Wicking Institute for Molecular Bioscience, The University of Queensland, St Lucia, QLD, Australia

Lisa Wiesmüller Department of Obstetrics and Gynaecology, University of Ulm, Ulm, Germany

Edwin van Wijngaarden Department of Public Health Sciences, University of Rochester School of Medicine and Dentistry, Rochester, NY, USA

Kandace Williams Department of Biochemistry and Cancer Biology, Health Science Campus, UT College of Medicine, Toledo, OH, USA

Elizabeth D. Williams Australian Prostate Cancer Research Centre – Queensland (APCRC-Q), Brisbane, QLD, Australia

Translational Research Institute, Institute of Health and Biomedical Innovation, Faculty of Health, School of Biomedical Sciences, Queensland University of Technology, Brisbane, QLD, Australia

Elizabeth M. Wilson Department of Pediatrics and Biochemistry and Biophysics, University of North Carolina at Chapel Hill, Chapel Hill, NC, USA

George Wilson Storr Liver Centre, Westmead Millennium Institute for Medical Research, The University of Sydney at Westmead Hospital, Westmead, NSW, Australia

Ola Winqvist Department of Medicine (OW), Karolinska Institutet, Stockholm, Sweden

Jordan Winter Department of Surgery, Thomas Jefferson University, Philadelphia, PA, USA

John Pierce Wise Department of Pharmacology and Toxicology, University of Louisville, Louisville, KY, USA

Christian Wittekind Department für Diagnostik, Institut für Pathologie, Universitätsklinikum Leipzig, Leipzig, Germany

Isaac P. Witz Department of Cell Research and Immunology, Tel Aviv University, Tel Aviv, Israel

Ido Wolf Division of Oncology, The Tel Aviv Sourasky Medical Center, Tel Aviv University, Tel Aviv, Israel

The Sackler Faculty of Medicine, Tel Aviv University, Tel Aviv, Israel

Roland C. Wolf Biomedical Research Centre, University of Dundee, Dundee, UK

Alice Wong University of Hong Kong, Hong Kong, China

Chun-Ming Wong Department of Pathology, The University of Hong Kong, Hong Kong, China

Yung H. Wong Division of Life Science, Biotechnology Research Institute, The Hong Kong University of Science and Technology, Kowloon, Hong Kong

Dori C. Woods Northeastern University, Boston, MA, USA

Paul Workman Cancer Research UK Center for Cancer Therapeutics, The Institute of Cancer Research, Sutton, Surrey, UK

Maria J. Worsham Department of Otolaryngology, Henry Ford Health System, Detroit, USA

Thomas Worzfeld Institute of Pharmacology, University of Marburg, Marburg, Germany

Department of Pharmacology, Max-Planck-Institute for Heart and Lung Research, Bad Nauheim, Germany

Jie Wu Department of Molecular Oncology, SRB-3, H. Lee Moffitt Cancer Center and Research Institute, Tampa, FL, USA

Mei-Yi Wu Department of Biochemistry and Molecular Medicine, The George Washington University, Washington, DC, USA

Ray-Chang Wu Department of Biochemistry and Molecular Medicine, The George Washington University, Washington, DC, USA

Shiyong Wu Edison Biotechnology Institute and Department of Chemistry and Biochemistry, Ohio University, Athens, OH, USA

Wen Jin Wu Division of Monoclonal Antibodies, Office of Biotechnology Products, Office of Pharmaceutical Science, Center for Drug Evaluation and Research, U.S. Food and Drug Administration, Bethesda, MD, USA

Xiaosheng Wu Department of Immunology, Mayo Clinic, College of Medicine, Rochester, MN, USA

Xifeng Wu Department of Epidemiology, The University of Texas MD Anderson Cancer Center, Houston, TX, USA

Yi-Long Wu Guangdong Lung Cancer Institute, Guangdong General Hospital and Guangdong Academy of Medical Sciences, Guangzhou, China

Christopher Xiao Department of Otolaryngology-Head and Neck Surgery, Medical University of South Carolina, Charleston, SC, USA

Guang-Hui Xiao Fox Chase Cancer Center, Philadelphia, PA, USA

Huajiang Xiong Department of Zoophysiology, Zoological Institute, Christian-Albrechts-University of Kiel, Kiel, Germany

Jianming Xu Department of Molecular and Cellular Biology, Baylor College of Medicine, Houston, TX, USA

Tian Xu Howard Hughes Medical Institute, Yale University School of Medicine, New Haven, CT, USA

Zhengping Xu Zhejiang University School of Medicine, Hangzhou, China

Jing Xue Stanford University School of Medicine, Stanford, CA, USA

Judy W. P. Yam Department of Pathology, The University of Hong Kong, Hong Kong, China

Sho-ichi Yamagishi Department of Pathophysiology and Therapeutics of Diabetic Vascular Complications, Kurume University School of Medicine, Kurume, Japan

Michiko Yamamoto Department of Respiratory Medicine, Kitasato University School of Medicine, Sagamihara, Kanagawa, Japan

Wei Yan Department of Cancer Biology, Beckman Research Institute of City of Hope, Duarte, CA, USA

Haining Yang University of Hawaii Cancer Center, Honolulu, HI, USA

Hong Yang Cancer Vaccine Section, National Cancer Institute, National Institutes of Health, Bethesda, MD, USA

Jia-Lin Yang Adult Cancer Program, Lowy Cancer Research Centre, Prince of Wales Clinical School, Faculty of Medicine, University of New South Wales, Sydney, NSW, Australia

Ping Yang Department of Health Sciences Research, Mayo Clinic College of Medicine, Rochester, MN, USA

Rongxi Yang Molecular Epidemiology Unit, German Cancer Research Center, Heidelberg, Germany

Libo Yao Department of Biochemistry and Molecular Biology, The Fourth Military Medical University, Xi'an, Shananxi, China

Masakazu Yashiro Department of Surgical Oncology, Osaka City University Graduate School of Medicine, Osaka, Japan

Nelson Yee Penn State Hershey Cancer Institute, Hershey, PA, USA

Yerem Yeghiazarians Department of Medicine, Division of Cardiology, Translational Cardiac Stem Cell Program, Eli and Edythe Broad Center of Regeneration Medicine and Stem Cell Research, Cardiovascular Research Institute, University of California San Francisco (UCSF), San Francisco, CA, USA

W. Andrew Yeudall Department of Oral Biology, College of Dental Medicine, Georgia Regents University, Augusta, GA, USA

Maksym V. Yezhelyev Winship Cancer Institute, Emory University, Atlanta, GA, USA

Ömer H. Yilmaz The David H. Koch Institute for Integrative Cancer Research, Massachusetts Institute of Technology, Cambridge, MA, USA

Açelya Yilmazer Aktuna Biomedical Engineering Department, Engineering Faculty, Ankara University, Golbasi, Ankara, Turkey

Anthony P. C. Yim Division of Cardiothoracic Surgery, Chinese University of Hong Kong, Hong Kong, China

John H. Yim Department of Surgery, City of Hope, Duarte, CA, USA

Chengqian Yin Department of Biology, Drexel University College of Arts and Sciences, Philadelphia, PA, USA

Helen L. Yin Department of Physiology, University of Texas Southwestern Medical Center, Dallas, TX, USA

Min-Jean Yin Oncology Research, Pfizer Worldwide R&D, San Diego, CA, USA

Xiao-Ming Yin Department of Pathology and Laboratory Medicine, Indiana University, Indianapolis, IN, USA

Zhimin Yin College of Life Science, Nanjing Normal University, Nanjing, People's Republic of China

George Wai-Cheong Yip Department of Anatomy, National University of Singapore, Singapore, Singapore

Kenneth W. Yip Princess Margaret Cancer Centre, University Health Network, Toronto, ON, Canada

Harry H. Yoon Mayo Clinic Comprehensive Cancer, Rochester, MN, USA

Jung-Hwan Yoon Department of Internal Medicine, Seoul National University College of Medicine, Chongno-gu, Seoul, South Korea

Kazuhiro Yoshida Department of Surgical Oncology, Gifu University School of Medicine, Gifu, Japan

Tatsushi Yoshida Department of Molecular-Targeting Cancer Prevention, Graduate School of Medical Science, Kyoto Prefectural University of Medicine, Kyoto, Japan

Kouichi Yoshimasu Department of Hygiene, School of Medicine, Wakayama Medical University, Wakayama, Japan

Anas Younes Lymphoma Service, Department of Medicine, Memorial Sloan Kettering Cancer Center, New York, NY, USA

Graeme P. Young Flinders Cancer Control Alliance, Flinders University, Adelaide, SA, Australia

Ken H. Young Department of Hematopathology, The University of Texas MD Anderson Cancer Center, Houston, TX, USA

Dihua Yu Departments of Molecular and Cellular Oncology, The University of Texas MD Anderson Cancer Center, Houston, TX, USA

Jian Yu Department of Pathology, University of Pittsburgh Cancer Institute, University of Pittsburgh School of Medicine, Pittsburgh, PA, USA

Yan Ping Yu Department of Pathology, University of Pittsburgh, Pittsburgh, PA, USA

Yu Yu Department of Pathology, University of Sydney, Sydney, NSW, Australia

Xiao Yuan Research and Development Center, Wuhan Botanical Garden, Chinese Academy of Science, Wuhan, Hubei, People's Republic of China

Anthony Po-Wing Yuen Division of Otorhinolaryngology, Department of Surgery, The University of Hong Kong, Hong Kong, SAR, China

Zhong Yun Department of Therapeutic Radiology, Yale School of Medicine, New Haven, CT, USA

Stefan K. Zöllner Department of Pediatric Hematology and Oncology, University Childrens Hospital Münster, Münster, Germany

Leo R. Zacharski VA Hospital, White River Junction, VT, USA

Gerard P. Zambetti Department of Biochemistry, Dana-Farber Cancer Institute, Boston, MA, USA

Behrouz Zand UT MD Anderson Cancer Center, Houston, TX, USA

Laura P. Zanello Department of Biochemistry, University of California-Riverside, Riverside, CA, USA

Uwe Zangemeister-Wittke Department of Pharmacology, University of Bern, Bern, Switzerland

Andrew C. W. Zannettino Myeloma Research Laboratory, School of Medicine, Faculty of Health Sciences, University of Adelaide, Adelaide, SA, Australia

Kamran Zargar-Shoshtari Department of Urology, Moffitt Cancer Center and Research Institute, Tampa, FL, USA

Laura Zavala-Flores Redox Biology Center, School of Veterinary Medicine and Biomedical Sciences, University of Nebraska-Lincoln, Lincoln, NE, USA

Berton Zbar Laboratory of Immunobiology, NIH – Frederick, Frederick, MD, USA

Herbert J. Zeh III UPMC/University of Pittsburgh Schools of the Health Sciences, Pittsburgh, PA, USA

Jason A. Zell Cancer Prevention Program, Division of Hematology/Oncology and Epidemiology, Department of Medicine, School of Medicine, Chao Family Comprehensive Cancer Center, University of California, Irvine, CA, USA

Danfang Zhang Department of Pathology, Tianjin Cancer Hospital and Tianjin Cancer Institute, Tianjin, People's Republic of China

Fengrui Zhang Michigan State University, East Lansing, MI, USA

Hao Zhang The University of Texas MD Anderson Cancer Center, Houston, TX, USA

Hong Zhang Biogen Idec, San Diego, CA, USA

Hui Zhang Department of Chemistry and Biochemistry, University of Nevada, Las Vegas, NV, USA

Jinping Zhang Departments of Pathology and Immunology, Center for Cell and Gene Therapy, Baylor College of Medicine, Houston, TX, USA

Ji-Hu Zhang Lead Discovery Center, Novartis Institute for Biomedical Research, Cambridge, MA, USA

Lin Zhang Biogen Idec, San Diego, CA, USA

Lin Zhang Department of Pharmacology and Chemical Biology, University of Pittsburgh Cancer Institute, University of Pittsburgh School of Medicine, Pittsburgh, PA, USA

Ruiwen Zhang University of Alabama at Birmingham, Birmingham, AL, USA

Shiwu Zhang Department of Pathology, Tianjin Cancer Hospital and Tianjin Cancer Institute, Tianjin, People's Republic of China

Yong Zhang Department of Neuroscience, Johns Hopkins University School of Medicine, Baltimore, MD, USA

Xin A. Zhang Departments of Medicine and Molecular Sciences, Vascular Biology Center, Cancer Institute, University of Tennessee Health Science Center, Memphis, TN, USA

Xuefeng Zhang Duke Pathology, Duke University School of Medicine, Durham, NC, USA

Yu-Wen Zhang Department of Oncology, Georgetown University Medical Center, Washington, DC, USA

Yuesheng Zhang Roswell Park Cancer Institute, Buffalo, NY, USA

Liang Zhong Le Centre de recherche du CHU de Québec-Université Laval: axe Oncologie, Le Centre de recherche sur le cancer de l'Université Laval, Québec, QC, Canada

Guang-Biao Zhou State Key Laboratory of Membrane Biology, Institute of Zoology, Chinese Academy of Sciences, Beijing, China

Jerry Zhou School of Molecular and Microbial Biosciences, University of Sydney, Sydney, NSW, Australia

Zeng B. Zhu Departments of Medicine, Pathology, Surgery, Obstetrics and Gynecology and the Gene Therapy Center, Division of Human Gene Therapy, University of Alabama at Birmingham, Birmingham, AL, USA

M. Zigler Department of Cancer Biology, The University of Texas MD Anderson Cancer Center, Houston, TX, USA

Margot Zoeller DKFZ, Heidelberg, Germany

Massimo Zollo Department of Molecular Medicine and Medical Biotechnology, University Federico II of Naples, Naples, Italy

Roberto T. Zori University of Florida, Gainesville, FL, USA

Enrique Zudaire NCI Angiogenesis Core Facility, National Cancer Institute, National Institutes of Health, Advanced Technology Center, Gaithersburg, MD, USA

Carsten Zwick Klinik für Innere Medizin I, Universität des Saarlandes, Homburg, Germany

C

C.elegans Cell Death 4 Homolog

▶ APAF-1 Signaling

C.I. 75300

▶ Curcumin

C/EBP-Epsilon-Regulated Myeloid-Specific Secreted Cysteine-Rich Protein (XCP1)

▶ Resistin

---

c-erb-B2

▶ HER-2/neu

C21H2006

▶ Curcumin

C33

▶ Metastasis Suppressor KAI1/CD82

© Springer-Verlag Berlin Heidelberg 2017
M. Schwab (ed.), *Encyclopedia of Cancer*,
DOI 10.1007/978-3-662-46875-3

Ca^{2+} Homeostasis

Olivier Dellis
Signalisation Calcique et Interactions Cellulaires dans le Foie, INSERM UMR-S 1174, Université Paris-Sud 11, Orsay, France

Definition

Ca^{2+} homeostasis is the dynamic equilibrium of Ca^{2+} ion concentration in the body or in the cell ("cellular Ca^{2+} homeostasis").

Characteristics

Even if 99% of the Ca^{2+} ions of the body are trapped in the bones and could be considered as stable, Ca^{2+} homeostasis refers to the control of the free Ca^{2+} ion concentrations at the extracellular ("$[Ca^{2+}]_{ex}$") and the intracellular ("$[Ca^{2+}]_i$") levels.

Control of $[Ca^{2+}]_{ex}$ is important because Ca^{2+} ions stabilize numerous ionic channels like the voltage-gated ones: thus a decrease of $[Ca^{2+}]_{ex}$ could induce spontaneous contraction of muscles, implying that $[Ca^{2+}]_{ex}$ must be tightly controlled in the blood. This is mainly done by parathyroid hormone that controls the uptake of Ca^{2+} ions by the gusts, the release by the bones, and the reabsorption by the kidneys.

Cellular Ca^{2+} Homeostasis

At the cellular level, $[Ca^{2+}]_i$ is an equilibrium between the Ca^{2+} ion concentration of the different compartments, mainly the cytosol ($[Ca^{2+}]_{cyt}$), endoplasmic reticulum ($[Ca^{2+}]_{ER}$), and mitochondria ($[Ca^{2+}]_m$). Even if $[Ca^{2+}]_{ER}$ is important for the synthesis and the right folding of proteins and $[Ca^{2+}]_m$ for the activity of the mitochondria, the control of $[Ca^{2+}]_{cyt}$ must be tightly done by the cell as it directly regulates almost all the cellular processes, from the egg fertilization to activation, neurotransmission, proliferation, and apoptosis. As depicted by Berridge et al. (2003), $[Ca^{2+}]_{cyt}$ is the result of an equilibrium between "on" reactions allowing entry of Ca^{2+} ions in the cytosol and "off" reactions extruding Ca^{2+} ions from the cytosol.

Indeed, the $[Ca^{2+}]_{ex}$ is commonly of 1–2 mM, and $[Ca^{2+}]_{cyt}$ must remain at 100 nM to avoid any uncontrolled activation of cellular process: due to this huge gradient of concentration, reinforced by the fact that cells are negatively charged, the Ca^{2+} electrochemical gradient is in favor of a massive Ca^{2+} ion entry. Thus, even in resting conditions, cells face a Ca^{2+} ion entry. To avoid the fast rise of $[Ca^{2+}]_{cyt}$, Ca^{2+} ions are rapidly exited from the cytosol by two ways:

(i)   Plasma membrane Ca^{2+} ATPases ("PMCA") pump the Ca^{2+} ions out of the cells, with the help of Na$^+$/Ca^{2+} exchangers ("NCX"),

(ii) And/or Ca^{2+} ions are mainly pumped in the lumen of the ER by the sarcoplasmic-endoplasmic reticulum Ca^{2+} ATPases ("SERCA"). However, according to the cell types, Ca^{2+} ions can be also uptake in the apparatus of Golgi by the secretory pathway Ca^{2+} ATPases ("SPCA").

Furthermore, the cytosol contains different type of proteins able to buffer the Ca^{2+} concentration.

Thus, each type of cells expresses different isoforms of these Ca^{2+} transporters and buffers allowing cell type control of resting $[Ca^{2+}]_{cyt}$.

Cellular Ca^{2+} Homeostasis During Cell Stimulation

During the cell stimulation, the Ca^{2+} homeostasis will reach a new equilibrium due to the entry of Ca^{2+} ions from the extracellular medium and/or the Ca^{2+} ions release by internal stores.

Ca^{2+} influx is allowed by a large variety of plasma membrane channels with different types of opening: membrane depolarization for voltage-operated channels (L, N, P/Q, R, and T types), fixation of a ligand for receptor-operated channels (NMDA receptors or ATP receptor P2X), fixation of a second messenger for cyclic nucleotide-gated (CNG) channels, interaction with an ER protein for Ca^{2+} release-activated channels (CRAC, allowing the Ca^{2+} influx known as store-operated calcium entry (SOCE), formally controlled by the ER), etc.

Ca^{2+} release by internal stores, mainly by the ER (or its derivative the sarcoplasm in muscle cells), is controlled by Ca^{2+} itself or by second messengers like inositol 1,4,5-triphosphate ("IP$_3$") allowing the opening of intracellular Ca^{2+} channels.

The expression spectra of this different kind of plasma membrane and internal channels in the different type of cells create different types of Ca^{2+} homeostasis change during cell stimulation, inducing different response from the cells.

For example, in cells like T lymphocytes, the stimulation of the T-cell receptor induces the synthesis of IP$_3$, allowing the release of Ca^{2+} ions by the ER through IP$_3$ receptors. This release induces the opening of the plasma membrane ORAI1, allowing the massive entry of extracellular Ca^{2+} ions. This increase activates different proteins like calmodulin and calcineurin, which one induces the translocation of the nuclear factor NFAT to the nucleus. Nuclear NFAT can then activates the transcription of various genes, like the one of interleukin-2. In the absence of this $[Ca^{2+}]_{cyt}$ rise due to a default in the signal transduction from the plasma membrane to the ER, or the non-expression of ORAI1 channels, T lymphocytes cannot be activated, and the immune system is shut down. Inhibition of calcineurin by cyclosporin impairs the signal transduction between the Ca^{2+} rise and the activation of NFAT.

Thus in every kind of cells, the defect of only one Ca^{2+} transporters could have huge impacts on the cell activity and properties.

Cellular Ca^{2+} Homeostasis in Cancer

Even if the role of Ca^{2+} ions is well established in a large number of cellular processes like proliferation and apoptosis, two processes implied in the appearance of cancer, its role in cancerogenesis and metastasis formation is surprisingly poorly documented.

However, since few years, some works have clearly implied Ca^{2+} ions in the proliferation of cancerous cells and metastasis. Thus, since the discovery of the ORAI1 channels in 2006 and its important role for the proliferation of non-excitable cells, it seems that ORAI1 channel activity may control the proliferation of cancerous cells. For example, inhibition of ORAI1 channel impairs the proliferation and migration of melanoma cells (Umemura et al. 2014) or the formation of bone metastasis of breast cancer cells (Chantôme et al. 2013). Some other compounds acting on ORAI1 seem to induce apoptosis.

To conclude, Ca^{2+} homeostasis appears modified in numerous cancer cells. In the near future, new molecules able to control Ca^{2+} channels to avoid the cancer cell proliferation and formation of metastases will probably appear.

Cross-References

▶ Apoptosis Induction for Cancer Therapy
▶ Ion Channels
▶ Membrane Transporters
▶ Signal Transduction

References

Berridge MJ, Bootman MD, Roderick HL (2003) Calcium signalling: dynamics, homeostasis and remodelling. Nat Rev Mol Cell Biol 4:517–529

Chantôme A, Potier-Cartereau M, Clarysse L, Fromont G, Marionneau-Lambot S, Guéguinou M, Pagès JC, Collin C, Oullier T, Girault A, Arbion F, Haelters JP, Jaffrès PA, Pinault M, Besson P, Joulin V, Bougnoux P, Vandier C (2013) Pivotal role of the lipid raft SK3-Orai1 complex in human cancer cell migration and bone metastases. Cancer Res 73(15):4852–4861

Umemura M, Baljinnyam E, Feske S, De Lorenzo MS, Xie LH, Feng X, Oda K, Makino A, Fujita T, Yokoyama U, Iwatsubo M, Chen S, Goydos JS, Ishikawa Y, Iwatsubo K (2014) Store-operated Ca2+ entry (SOCE) regulates melanoma proliferation and cell migration. PLoS One 9(2):e89292

See Also

(2012) Ca 2+ aTPase. In: Schwab M (ed) Encyclopedia of cancer, 3rd edn. Springer, Berlin, p 577. doi:10.1007/978-3-642-16483-5_763

(2012) Ca 2+ -release channels. In: Schwab M (ed) Encyclopedia of cancer, 3rd edn. Springer, Berlin/Heidelberg, p 577. doi:10.1007/978-3-642-16483-5_764

(2012) Calcineurin. In: Schwab M (ed) Encyclopedia of cancer, 3rd edn. Springer, Berlin/Heidelberg, p 584. doi:10.1007/978-3-642-16483-5_775

(2012) Cyclosporin A. In: Schwab M (ed) Encyclopedia of cancer, 3rd edn. Springer, Berlin/Heidelberg, p 1036. doi:10.1007/978-3-642-16483-5_1440

(2012) Inositol 1,4,5-trisphosphate. In: Schwab M (ed) Encyclopedia of cancer, 3rd edn. Springer, Berlin/Heidelberg, p 1870. doi:10.1007/978-3-642-16483-5_3071

(2012) T lymphocyte. In: Schwab M (ed) Encyclopedia of cancer, 3rd edn. Springer, Berlin/Heidelberg, p 3600. doi:10.1007/978-3-642-16483-5_5654

Ca^{2+}-Activated Phospholipid-Dependent Protein Kinase

▶ Protein Kinase C Family

CaBP3

▶ Calreticulin

Cachectin

▶ Tumor Necrosis Factor

Cachexia

Chen Bing
Institute of Ageing and Chronic Disease, University of Liverpool, Liverpool, UK

Definition

Cachexia came from the Greek "kakos" and "hexis" meaning "bad conditions." Cachexia is a complex metabolic syndrome characterized by progressive weight loss with extensive loss of skeletal muscle and adipose tissue, which is secondary to the growing malignancy.

Characteristics

Most cancer patients develop cachexia at some point during the course of their disease, and nearly one-half of all cancer patients have weight loss at diagnosis. Cachexia prevents effective treatments for cancer and predicts a poor prognosis because the severity of wasting inversely correlates with survival. The consequences of cachexia are detrimental and cachexia is considered to be the direct cause of about 20% of cancer deaths. The pathogenesis of cancer cachexia remains to be fully understood, but it is evidently multifactorial.

Weight Loss

Clinically, cachexia should be suspected if involuntary weight loss of more than five percent of premorbid weight occurs within a 6-month period. Weight loss is not simply caused by competition for nutrients between tumor and host as the tumor burden may be only 1–2% of total body weight. The frequency of weight loss varies with the type of malignancy, being more common and severe in patients with cancers of the gastrointestinal tract (▶ gastrointestinal stromal tumor) and lung (▶ lung cancer). Gastric and pancreatic cancer patients may lose large amounts of weight, up to 25% of initial body weight. Over 15% of weight loss in patients is likely to cause significant impairment of respiratory muscle function, which probably contributes to premature death. Weight loss can arise from several metabolic changes that take place during malignancy, for example, reduced food intake, increased energy expenditure, and tissue breakdown.

Poor Appetite

Loss of the desire to eat or lack of hunger is common in cancer patients. It can be related to the mechanical effect of the tumor such as obstructions (especially of the upper gastrointestinal tract), side-effects of chemotherapy or

radiotherapy (▶ chemoradiotherapy), and emotional distress. Some tumors may secrete products which act on the brain to inhibit appetite. Regulation of food intake involves the integration of the peripheral and neural signals in the hypothalamus and other brain regions. In the hypothalamus, the orexigenic signals such as neuropeptide Y (NPY), the most potent appetite stimulant, increase food intake, and the anorexigenic signals including the pro-opiomelanocortin/cocaine and amphetamine regulated transcript (POMC/CART) inhibit appetite. Dysregulation of NPY in the hypothalamic pathway can lead to decreased energy intake but higher metabolic demand for nutrients. It has been demonstrated that NPY-immunoreactive neurons in the hypothalamus are decreased in experimental model of cancer anorexia. In contrast, reduced food consumption can be restored to normal levels by blocking the POMC/CART pathway in tumor-bearing animals. High level of leptin, a hormone primarily secreted by adipocytes, inhibits the release of hypothalamic NPY. In cancer cachexia the leptin feedback loop appears to be deranged, altering the signaling pathway of NPY. Cytokines such as interleukin 1β (IL-1β), interleukin-6 (IL-6), and tumor necrosis factor-α (TNFα) are implicated to be involved in cancer anorexia, possibly by stimulating corticotrophin-releasing factor, a neurotransmitter which suppresses food intake at least in rodents, and/or by inhibiting neurons that produce NPY in the hypothalamus.

Increased Metabolism and Energy Expenditure

Maintaining normal body weight requires energy intake to equal energy expenditure. In some patients with cancer cachexia, energy balance becomes negative as reduced food intake is not accompanied by a parallel decrease in energy expenditure. For example, patients with lung and pancreatic cancers generally have higher resting energy expenditure (REE) compared with normal control subjects; however, REE is usually normal in patients with colorectal cancer. The mechanisms of increased energy expenditure are not clear although studies suggest that it might be through the upregulation of uncoupling proteins, a family of mitochondrial membrane proteins, which are proposed to be involved in the control of energy metabolism. Uncoupling protein-1 (UCP-1), which decreases the coupling of respiration to ADP phosphorylation thereby generating heat instead of ATP, is only expressed in brown adipose tissue (BAT). UCP-1 mRNA levels in BAT are increased in mice bearing the MAC16 colon adenocarcinoma. Although BAT is uncommon in adults, the prevalence of BAT has been found to be higher in cancer cachectic patients than the age-matched control subjects. mRNA levels of UCP-2 (expressed ubiquitously) and UCP-3 (expressed in skeletal muscle and BAT) in skeletal muscle are upregulated in rodent models of cancer cachexia. In humans, skeletal muscle UCP-3 mRNA levels are over fivefold higher in cachectic cancer patients compared with patients without weight loss and health controls. Elevated expression of UCP-2 and UCP-3 has been suggested to contribute to lipid utilization rather than whole-body energy expenditure. Cytokines such as TNFα and/or other tumor products may be responsible for the changes in UCP expression at least in rodents. Additional energy consumption could arise from the metabolism of tumor-derived lactate via "futile cycles" between the tumor and the host. The main energy source for many solid tumors is glucose, which is converted to lactate and transferred to the liver to convert back into glucose. This "futile cycle" requires large amount of ATP, resulting in an extra loss of energy in cancer patients.

Loss of Adipose Tissue

Fat constitutes 90% of normal adult fuel reserves, and depletion of adipose tissue together with hyperlipidemia becomes a hallmark of cancer cachexia. Computed tomography (CT) scanning has revealed that cachectic cancer patients with gastrointestinal carcinoma had significantly smaller visceral adipose tissue area than control subjects. Increased lipolysis is implicated in cancer-associated adipose atrophy. The activity of hormone-sensitive lipase, a rate-limiting enzyme of the lipolytic pathway, is increased in cancer cachectic patients, which causes elevated plasma levels of free fatty acids and triglycerides. Meanwhile, there is a fall in lipoprotein lipase (LPL) activity in white

adipose tissue, thus inhibiting cleavage of triglycerides from plasma lipoproteins into glycerol and free fatty acids for storage, causing a net flux of lipid into the circulation. Finally, glucose transport and de novo lipogenesis in the tissue are reduced in tumor-bearing state, leading to a decrease in lipid deposition. There is also evidence that loss of adipose tissue in cancer cachexia could be the result of impairment in the formation and development of adipose tissue. The expressions of several key adipogenic transcription factors including CCAAT/enhancer-binding protein alpha, CCAAT/enhancer-binding protein beta, peroxisome proliferator-activated receptor gamma, and sterol regulatory element-binding protein-1c are markedly reduced in adipose tissue of cancer cachectic mice.

Various factors produced by tumors or the host's immune cells responding to the tumor can disturb lipid metabolism. TNFα has been shown to affect adipose tissue formation by inhibiting the differentiation of new adipocytes, causing dedifferentiation of mature fat cells and suppressing the expression of genes encoding key lipogenic enzymes. TNFα has also been associated with increased lipolysis probably through suppression of LPL activity in adipocytes. In addition, both TNFα and IL-1β are able to inhibit glucose transport in adipocytes and consequently decrease the availability of substrates for lipogenesis. Certain prostate, gut, and pancreatic tumors secrete a lipid-mobilizing factor (LMF), also produced by a mouse adenocarcinoma model. LMF has been shown to be identical to the plasma protein zinc-α 2-glycoprotein (ZAG). It is found to be secreted by human adipocytes and upregulated in adipose tissue of mice with cancer cachexia. ZAG causes rapid lipolysis in vitro and in vivo, possibly through activation of intracellular cyclic AMP. ZAG also stimulates expression of UCPs in brown fat of mice, which may contribute to increased energy expenditure as well as lipid catabolism during cachexia. Moreover, ZAG expression and secretion by adipose tissue is enhanced weight loss patients with gastrointestinal cancer. Given its lipid-mobilising effect, ZAG could contribute to adipose tissue loss associated with cancer cachexia in humans.

Loss of Muscle Protein

Weakness, commonly seen in cancer cachectic patients, is directly related to wasting of muscle that accounts for almost half the body's total protein and bears the brunt of enhanced protein destruction. Reduced protein synthesis together with enhanced proteolysis has been observed in experimental animal models and in muscle biopsies from cancer patients with cachexia, and whole-body protein turnover can be markedly increased in cachectic cancer patients.

Some mediators and pathways of excessive protein breakdown have been incriminated in cancer cachexia. TNFα appears to be involved, as treatment with recombinant TNFα enhances proteolysis in rat skeletal muscle and activates the ubiquitin–proteasome system. Ubiquitin, an 8.6 kD peptide, is crucially involved in targeting of proteins undergoing cytosolic ATP-dependent proteolysis. There is an increase in ubiquitin gene expression in rat skeletal muscle after incubation with TNFα in vitro. Tumors also produce cachectic factors such as proteolysis-inducing factor (PIF), a 24 kD glycoprotein initially isolated from a cachexia-inducing tumor (MAC16) and the urine of cachectic cancer patients. PIF induces muscle protein breakdown by stimulation of the ubiquitin–proteasome proteolytic pathway.

There is increasing evidence that both cytokines and PIF cause protein degradation by activation of ▶ nuclear factor kappa B (NFκB), a transcription factor that regulates the expression of a number of proinflammatory cytokines. TNFα and PIF can upregulate components of the ubiquitin–proteasome pathway in an NFκB-dependent manner. Activation of NFκB by TNFα in murine muscle cells suppresses mRNA of the transcription factor MyoD, inhibiting skeletal muscle cell differentiation as well as preventing the repair of damaged skeletal muscle fibers.

Treatment

Current treatment designed to ameliorate cancer cachexia has limited benefit. Nutritional supplementation (oral or parenteral) alone has little effect and, critically, does not restore muscle mass and improve quality of life or prognosis in cancer patients. Appetite stimulants such as megestrol acetate and medroxyprogesterone acetate are commonly used at present in the treatment of anorexia and cachexia. These agents are believed to stimulate orexigenic peptide NPY in the hypothalamus and inhibit the synthesis and release of proinflammatory cytokines. Their effects on appetite and well being are short-termed and they do not influence lean body mass and survival. ▶ Cannabinoids have also been studied as potential appetite stimulants. However, dronabinol has failed to prevent progressive weight loss in patients with advanced cancer.

Therapeutic interventions include anticytokines such as thalidomide with multiple immunomodulatory properties. It suppresses the production of TNFα, IL-1β, IL-12, and cyclooxygenase-2, which is probably through inhibiting NFκB activity. Thalidomide has been shown to attenuate total weight loss and loss of lean body mass in cachectic patients with advanced pancreatic cancer.

Eicosapentaenoic acid (EPA), a polyunsaturated fatty acid from fish oil, has attracted attention as a potential anticachectic agent. EPA has been shown to attenuate the increased expression of the components of the ubiquitin–proteasome proteolytic pathway in skeletal muscle of mice with cancer cachexia, and EPA can block PIF-induced protein degradation in vitro. In randomized clinical trials, cachectic patients with unresectable pancreatic cancer receiving EPA have shown a stabilization in the rate of weight loss, fat and muscle mass, as well as the REE. Data from animal studies suggest that EPA combined with the leucine metabolite beta-hydroxy-beta-methylbutyrate seems to be more effective in the reverse of muscle protein wasting.

Cross-References

- ▶ Cannabinoids
- ▶ Chemoradiotherapy
- ▶ Gastrointestinal Stromal Tumor
- ▶ Lung Cancer
- ▶ Nuclear Factor-κB

References

Argiles JM, Busquets S, Lopez-Soriano FJ (2006) Cyto-kines as mediators and targets for cancer cachexia. Cancer Treat Res 130:199–217

Bing C, Brown M, King P et al (2000) Increased gene expression of brown fat uncoupling protein (UCP)1 and skeletal muscle UCP2 and UCP3 in MAC16-induced cancer cachexia. Cancer Res 60:2405–2410

Bing C, Russell S, Becket E et al (2006) Adipose atrophy in cancer cachexia: morphologic and molecular analysis of adipose tissue in tumour-bearing mice. Br J Cancer 95:1028–1037

Fearon KC, Moses AG (2002) Cancer cachexia. Int J Cardiol 85:73–81

Mracek T, Stephens NA, Gao D et al (2011) Enhanced ZAG production by subcutaneous adipose tissue is linked to weight loss in gastrointestinal cancer patients. Br J Cancer 104:441–7. doi: 10.1038/sj.bjc.6606083

Tisdale MJ (2002) Cachexia in cancer patients. Nat Rev Cancer 2:862–871

See Also

(2012) Brown adipose tissue. In: Schwab M (ed) Encyclopedia of cancer, 3rd edn. Springer, Berlin/Heidelberg, p 572. doi:10.1007/978-3-642-16483-5_742

(2012) Eicosapentaenoic acid. In: Schwab M (ed) Encyclopedia of cancer, 3rd edn. Springer, Berlin/Heidelberg, p 1212. doi:10.1007/978-3-642-16483-5_1836

(2012) Hyperlipidemia. In: Schwab M (ed) Encyclopedia of cancer, 3rd edn. Springer, Berlin/Heidelberg, p 1784. doi:10.1007/978-3-642-16483-5_2909

(2012) Lipogenesis. In: Schwab M (ed) Encyclopedia of cancer, 3rd edn. Springer, Berlin/Heidelberg, p 2055. doi:10.1007/978-3-642-16483-5_3378

(2012) MyoD. In: Schwab M (ed) Encyclopedia of cancer, 3rd edn. Springer, Berlin/Heidelberg, p 2440. doi:10.1007/978-3-642-16483-5_3942

(2012) Neuropeptide Y. In: Schwab M (ed) Encyclopedia of cancer, 3rd edn. Springer, Berlin/Heidelberg, p 2504. doi:10.1007/978-3-642-16483-5_4043

(2012) Resting energy expenditure. In: Schwab M (ed) Encyclopedia of cancer, 3rd edn. Springer, Berlin/Heidelberg, p 3264. doi:10.1007/978-3-642-16483-5_5060

(2012) Uncoupling protein-1. In: Schwab M (ed) Encyclopedia of cancer, 3rd edn. Springer, Berlin/Heidelberg, p 3846. doi:10.1007/978-3-642-16483-5_6106

Cachexia-Inducing Agent

▶ Leukemia Inhibitory Factor

Cadherin-1

▶ E-Cadherin

Cafe-Au-Lait Macule

▶ Cafe-Au-Lait Spots

Cafe-Au-Lait Spots

Synonyms

Cafe-au-lait macule

Definition

Coffee-with-milk-colored spots on the skin that are seen characteristically in the neurofibromatosis type 1 (NF1) syndrome.

Cajal Bodies

Vincenzo de Laurenzi
Department of Experimental Medicine and Biochemical Sciences, University of Tor Vergata, Rome, Italy

Synonyms

Coiled bodies

Definition

Small nuclear organelles (0.1–2.0 μM in diameter), present in all eukaryotic cells, involved in a number of different nuclear functions.

Characteristics

The nucleus of eukaryotic cells contains a number of different highly specialized organelles. Unlike cytoplasmic organelles these nuclear structures are not delimited by a membrane but are by all means compartments that contain a number of specific proteins. Most of the organelles can be clearly identified through immunostaining using antibodies directed against specific marker proteins; however, it should be kept in mind that these organelles are highly dynamic structures that often exchange components and therefore many proteins can be found in more than one organelle.

Among these organelles are Cajal bodies (CBs), described over a century ago by Ramon y Cajal. CBs were originally described in neuronal cells but have since been described in a variety of cell types, both in animals and in plants, suggesting that they are involved in some fundamental cellular process. Due to their characteristic ultrastructural appearance as a tangle of coiled fibrillar strands, they have also been called coiled bodies. They usually vary in size from 0.2 to 2 μM, but can be occasionally larger. The number of CBs is usually between 0 and 4 in normal diploid cells; however, many more can be found in some cancer cells. The number of CBs per cell is regulated during the cell cycle. Indeed, CBs disappear in prophase nuclei, to reappear in G1 at the same time of the nucleolus. Their number is then doubled, usually reaching the number of four, in the S phase. It has been suggested that in these cells the number of CBs depends on the ploidy of the cells or more specifically on the number of chromosomes 1 and 6. CBs can be found associated with specific gene loci such as snoRNA, snRNA, and histone gene clusters. In addition, CBs can also be found in association with other nuclear bodies such as cleavage bodies and PML bodies, suggesting that there is an exchange of components between the different nuclear organelles.

CBs have a heterogeneous composition, containing different small nuclear ribonucleoproteins (snRNPs), small nucleolar ribonucleoproteins (snoRNPs), cell cycle regulating proteins, and transcription factors, as well as other proteins, whose function still needs to be determined.

The generally recognized marker of CBs is p80 coilin. The function of this protein is still unknown; its deletion in mice results in reduced coilin −/− animal litters, suggesting a developmental defect; however surviving animals appear normal. Deletion of coilin results in residual bodies that still contain some components such as fibrillarin, Nopp140, and FLASH, but not others like splicing snRNPs.

While their function is still in part elusive, recent work suggests that they are involved in several nuclear functions. CBs are supposed to be the site of assembly of the three eukaryotic RNA polymerases (pol I, pol II, and pol III) with their respective transcription and processing factors that are then transported as multiprotein complexes to the sites of transcription. They are also involved in the modification of small nuclear RNAs (snRNAs) and small nuclear ribonucleoproteins (snRNPs), which are important for spliceosome formation. Indeed CBs contain newly assembled snRNPs and snoRNPs that later accumulate in speckles and nucleoli, and it has been suggested that CBs are sites of modification for snRNPs and particularly sites where $2'$-O-methylation and pseudouridine formation occur. This process requires a novel class of small CB-specific RNAs (scaRNAs) that pair with the snRNAs and function as guides for $2'$-O-methylation. The reaction is probably mediated by the fibrillarin, a CB, and nucleolar-associated protein with methyl transferase activity. CBs have also been implicated in replication-dependent histone gene transcription, and a subset of CBs is physically associated with histone gene clusters on chromosomes 1 and 6. Phosphorylation of a CB component p220/NPAT by cyclin E/Cdk2 is required for activation of histone transcription, exit from G1, and progression through S phase of the cell cycle. Moreover it has been shown that another CB component FLASH is essential for this function. Downregulation of FLASH results in structural alteration of CBs, reduction of replication-dependent histone gene transcription, and block of cells in the S phase (▶ S-phase damage-sensing checkpoints) of the cell cycle. In addition, CBs are involved in U7 snRNA-dependent cleavage of the 3′ end of histone pre-mRNA before the mature mRNA can be exported to the cytoplasm.

Finally, a role for CBs in regulating ▶ telomerase function has been proposed. Based on the presence of the RNA component of telomerase (hRT) in CBs of cancer cells, it has been suggested that CBs play a role in the maturation of hRT or in the assembly of the telomerase complex. However, CBs might represent only a site of accumulation of hRT; alternatively, this could be an altered localization only present in cancer cells; therefore further studies are required to clarify this potential CB function.

Alteration of CB structure, as well as other nuclear structure alterations, has been observed in various diseases; however, in most cases it is not clear if these defects are a consequence of altered nuclear functions or play a role in the disease pathogenesis.

CBs have been found associated with the aggregates formed in CAG triplet expansion diseases and ataxin-1; mutated in spinocerebellar ataxia type 1 (SCA1), it has been shown to interact with coilin. The role of these findings in the disease pathogenesis is yet to be established.

Spinal muscular atrophy (SMA) is an autosomal recessive disease characterized by motor neuron degeneration associated with muscular atrophy and paralysis; it is usually caused by mutations of the surviving motor neuron 1 (SMN1) gene. SMN is a 294 amino acid protein, ubiquitously expressed; it bears no homology to other known proteins and its function is still unknown. It is localized both in the cytoplasm and in the nucleus where it is found in two different nuclear organelles: Cajal bodies (CBs) and Gems (for Gemini of CBs). Pathogenesis of SMA is not clearly understood, but reduction of SMN levels results in an alteration of CB structure.

Alteration of CBs in cancer has not been thoroughly studied yet, and a role for these organelles in cancer has not been clearly established. However cancer cell lines often show an increased number of CBs, and some alteration of CBs can be found in specific cancers. In MLL-ELL leukemia (▶ acute myeloid leukemia) the presence of the MLL-ELL fusion protein results in alteration of CB structure and altered localization of coilin. The TLS/CHOP fusion protein generated by the t(12;16) translocation (▶ chromosomal translocations), found in liposarcomas, shows high transforming capacity and is in part localized in CBs.

In conclusion, while studies have started to shed light on the function of CBs and on the interrelationship between these organelles and other nuclear structures, more work is required to clearly understand the molecular mechanisms involved in their formation and clarify their different roles in nuclear function. This in turn will provide information on their potential role in the pathogenesis of a range of human diseases.

References

Cioce M, Lamond AI (2005) Cajal bodies: a long history of discovery. Annu Rev Cell Dev Biol 21:105–131

Gall JG (2000) Cajal bodies: the first 100 years. Annu Rev Cell Dev Biol 16:273–300

Ogg SC, Lamond AI (2002) Cajal bodies and coilin – moving towards function. J Cell Biol 159(1):17–21

CAK1 Antigen

▶ Mesothelin

Calcitonin

Girish V. Shah
Department of Pharmacology, University of Louisiana College of Pharmacy, Monroe, LA, USA

Definition

Calcitonin (CT) is a 32-amino acid peptide synthesized in mammals by the C cells of the thyroid gland. Several extrathyroidal sites including the ▶ prostate gland, gastrointestinal tract, thymus, ▶ bladder, ▶ lung, pituitary gland, and central nervous system (CNS) also produce this peptide molecule (Ball 2007).

Characteristics

Almost all cells of human body synthesize and secrete procalcitonin (proCT), a precursor of the CT peptide, in response to infection/▶ inflammation. Only cells of the thyroid and neuroendocrine organs can process proCT to produce mature CT molecule (Davies 2015). CT sequence among various species shows remarkable divergence (Steinwald et al. 1999). However, all sequences contain 32 amino acids, a carboxy-terminal proline amide, and a disulfide bridge between cysteine residues at positions 1 and 7. In addition to CT, other biologically and chemically diverse molecules such as CT gene-related peptide (CGRP), ▶ adrenomedullin (AFP-modern), and amylin (AMY) are considered as CT family of peptides because of their ability to interact with CT receptor (CTR) and induce biological response. Each of these peptides displays selective tissue distribution and distinct physiological effects (Findlay and Saxton 2003). For example, CGRP is predominantly present in central and peripheral nervous system and is important for neurotransmission and neuromodulation. ADM is relatively abundant in vascular space, and plays an important role in the regulation of cardiovascular and respiratory functions, and CT is essential for calcium balance. However, CT does not regulate calcium in extrathyroidal tissues but is implicated to play an important role in cell growth, cell differentiation, and other regulatory functions.

Biosynthesis

Four CT genes, CALC-I, CALC-II, CALC-III, and CALC-IV with significant nucleotide homologies have been identified. However, CT is encoded by only CALC-I gene. CALC-I and CALC-II encode two different forms of CGRP, CGRP-I and CGRP-2. CALC-III is thought to be a pseudo gene, and CALC-IV produces AMY. Human (h) CT (CALC-I) gene is located in the p14qter region of chromosome 11. CT gene encodes two distinct peptides CT and CGRP, which arise by tissue-specific alternative splicing of the same primary mRNA transcript. The primary mRNA transcript is spliced almost exclusively to CT mRNA in thyroid and to CGRP in the nervous system.

CT is synthesized as part of a larger precursor protein of 136 amino acids. The DNA sequence of the hCT gene predicts that the hormone is flanked in the precursor by N- and C-terminal peptides. Both N-terminal and C-terminal flanking peptides are detected in the plasma and thyroidal tissues of both normal and medullary thyroid carcinoma (MTC) patients. However, no biological function for either of these two peptides has been conclusively determined. Cyclic adenosine monophosphate (cAMP), pentagastrin, and progesterone are potent stimulators of CT gene expression. In contrast, testosterone and estrogens have inhibitory effect.

There is evidence for polymorphisms in CALC-I gene that leads to increased risk for ovarian ▶ cancer in carrier women (T-C 624 bp upstream of translation initiation codon); and 16 bp microdeletion polymorphism has been reported in a family with multiple cases of unipolar and bipolar depressive disorders.

Biological Actions

Actions on Bone

CT is the only hormone that inhibits bone resorption by direct action on osteoclasts in the bone. It is characterized by the rapid loss of osteoclast ruffled borders, reduced cytoplasmic spreading, decreased release of lysosomal enzymes, and inhibition of collagen breakdown. This role is physiologically more relevant at times of stress on skeletal calcium conservation such as pregnancy, lactation, and growth, when bone remodeling by osteoclasts and the consequent release of calcium stores in the bone need to be tightly regulated to prevent unnecessary bone loss. In normal adult humans, even large dose of CT has little effect on serum calcium. However, in pathologies created by increased bone turnover such as thyrotoxicosis, metastatic bone disease, or Paget's disease, CT treatment effectively inhibits bone resorption and lowers serum calcium.

Renal Actions

CT increases urinary excretion rate of sodium, potassium, phosphorus, and magnesium. CT also enhances 1-hydroxylation of 25-hydroxy vitamin D in the proximal straight tubule by stimulating the expression of 25-hydroxy vitamin D 1-hydroxylase.

Central Actions

Central administration of CT produces analgesia, affects sleep cycles producing insomnia, major reduction in slow wave sleep and long period of alteration of rapid eye movement (REM) sleep and wakening. The centrally mediated actions of CT correlate well with the location of CT binding sites. CT also demonstrates multiple hypothalamic actions such as modulation of hormone release, decreased appetite, gastric acid secretion, and intestinal motility. Administration of CT in clinical situations of bone pain is very effective in ameliorating the pain symptoms.

Other Actions

CT and its receptors have been identified in a large number of other cell types and tissue sites suggesting multiple roles for CT–CTR axis. CTR-binding sites have been identified in the kidneys, brain, pituitary, testis, prostate, spermatozoa, lung, and lymphocytes. There is evidence to suggest the involvement of CT in cell growth and differentiation, tissue development and tissue remodeling. CT appears to be important for blastocyst implantation and development of the early blastocyst.

CT in Cancer

Overexpression of CT has been reported in cancer-derived cells from thyroid, ► lung, ► Brms1, ► prostate, ► pancreas, pituitary, bone (osteoclastoma, osteogenic sarcoma), and embryonal carcinoma, suggesting the deregulation of CT expression is an important event in several malignancies. The results from our laboratory

have shown that CT and CTR are present in undifferentiated basal cells but absent in differentiated secretory cells of normal human prostate gland. However, CT and CTR become detectable in malignant secretory epithelium suggesting malignancy-associated deregulation of CT/CTR expression. CT and CTR transcripts in malignant human prostate become detectable as early as high-grade PIN and progressively increase with increase in tumor grade. In human pancreas, CTR is present in benign as well as malignant regions but CT is exclusively detected in malignant sections of multiple pancreatic carcinomas, including ductal adenocarcinomas.

Mechanism of CT Action

Receptors

CT acts by binding to receptors on the plasma membrane of responding cells. CTR cDNA has been cloned in multiple mammalian species. Analysis of the protein translated from CTR cDNA sequence reveals the size of approximately 500 amino acids, and the receptor belongs to the class B family of G protein-coupled receptors (GPCRs), which also includes numerous potentially important drug targets. The human CTR gene is located on chromosome 7 at 7q21.3. The CTR gene exceeds 70 Kb in length, comprises of at least 14 exons, separated by introns ranging in size from 70 nucleotides to >20,000 nucleotides.

Multiple polymorphic sites in CTR gene have been identified, and several of them lead to lower bone mineral density in postmenopausal women.

Receptor Isoforms

Human CTR (hCTR) is known to exist in mutiple isoforms that arise from alternative splicing of the same primary transcript. The two most common hCTR variants arise by alternative splicing of intracellular domain 1 (ondel 2000). The most common variant (type 1 hCTR) leads to the addition of a 16 amino acid insert in the first intracellular loop. Alternative splicing of this small exon leads to the expression of type 2 hCTR, which differs from type 1 hCTR (abundant in the brain and the kidneys) by the absence of a 16-amino acid insert in the first intracellular loop. Type

2 CTR is predominantly expressed in malignant prostate and pancreatic cells. It has been shown that the lack of 16-amino acid insert in first intracellular loop enables type 2 hCTR to coactivate both adenylyl cyclase and phospholipase C. In addition, another receptor referred as calcitonin receptor-like receptor (CRLR) has been reported (Barwell et al. 2012).

Modulation of CTR Specificity

CTR displays high affinity for CT but low affinities for other CT family peptides. However, the ligand specificity of CTR is significantly altered when it binds to a RAMP protein. Several RAMPs have been identified but three (RAMP1, RAMP2, RAMP3) are investigated. hCTR displays low affinity for AMY, but association with RAMPs enables hCTR to bind AMY with high affinity. Similarly, ligand specificity of CRLR depends on the complexed RAMP. For example, CRLR-RAMP1 serves as CGRP receptor whereas CRLR-RAMP2 acts as ADM receptor. CRLR-RAMP3 displays high affinity for ADM as well as CT. This phenomenon opens up a possibility that ligand specificity of CTRs can be regulated by modulation of RAMPs expression.

CTR Signaling

G Protein-Mediated Signaling

The intracellular mechanisms by which CTR produces biological effects are still being elucidated. However, the signaling pathways appear to vary with cell type as well as animal species. As with most other GPCRs, the CTRs show coupling with multiple G proteins, which also depends on the isoform of CTR. For example, type 1 CTR preferentially couples to Gαs, leading to the activation of adenylyl cyclase and elevation in the intracellular levels of cAMP. The inhibitory action of CT on osteoclasts is accompanied by increase in cAMP levels. Forskolin, a direct activator of adenylyl cyclase, as well as dibutyryl cAMP, which elevates intracellular cAMP levels independent of adenylyl cyclase, mimic CT actions on bone resorption. Similarly, CTR is known to activate adenylyl cyclase in kidney as well as in cancers of lung, breast, and bone.

Unlike type 1 CTR, type 2 CTR simultaneously couples with Gαs and Gαq, leading to the coactivation of adenylyl cyclase and phospholipase C. This results in the elevation of intracellular levels of cAMP, as well as inositol triphosphates, and thence increased cytosolic calcium levels. This, together with coliberated diacyl glycerols, activates protein kinase C. In brain tissue, CT couples to G proteins other than Gαs as indicated by limited activation of adenylyl cyclase in neural tissues. In hepatocytes, CT increases cytosolic calcium levels without activating adenylyl cyclase. In LLC-PK1 kidney cells, CT increases either intracellular cAMP levels or cytosolic calcium levels in a cell cycle-dependent manner; and in pituitary lactotrophs, CT inhibits TRH-induced increases in cytosolic levels and activation of protein kinase C. In ► prostate cancer cells, CT coactivates protein kinases A and C and pathways activated by these enzymes play an important role in CT-stimulated growth, invasiveness, and tumorigenicity of prostate cancer cells.

G Protein-Independent Signaling

Evidence suggests that GPCRs also activate G - protein-independent signaling by interacting with proteins referred as GPCR-interacting proteins (GIPs). GPCRs activate this signaling by binding to GIPs through one or more of structural interacting domains such as Src homology 2 (SH2) and SH3, plackstrin homology, PDZ and Eva/WASP homology (EVH) domains.

Examination of CTR sequence reveals that the last four amino acids at the extreme C-terminus of the C-tail form E-S-S-A tetramer (amino acids 447–50), which conforms to the canonical type I PDZ ligand. A single serine-to-alanine substitution in the PDZ ligand of prostate CTR almost abolished CT-elicited increase in invasiveness and tumorigenicity of PC-3 prostate cancer cell line, raising a strong possibility that metastasizing ability of CTR is dependent upon its ability to interact with intracellular proteins containing PDZ domain(s). CTR seems to induce ► metastasis by disassembly of tight junctions on prostate cancer cells, which leads to the loss of cell polarity, activation of proteases such as urokinase type plasminogen activator, matrix metalloproteinases 2 and 9. These results raise a possibility that the prevention of interaction between CTR and its intracellular partner through the PDZ ligand can be an effective strategy to prevent CT-mediated metastasis. With current advances in medicinal chemistry and peptide mimetics, it should be possible to design a small peptide of 4–6 amino acids or a small molecule to prevent this interaction.

CT also activates phosphoinositol-3-kinase (PI3K)–Akt–Survivin pathway and induces chemoresistance and ▶ apoptosis in multiple prostate cancer cell lines through as yet uncharacterized mechanism. CT activated protein kinase A plays a key role in multiple actions of CT on prostate cancer cell lines, suggesting that both, G protein-dependent and G protein-independent, actions of CTR may act in concert to increase oncogenicity of prostate cancer cell lines.

Significance of CT–CTR Axis in Cancer: Clinical Aspects

CT Is "Oncogene" for Prostate Cancer but "Tumor Suppressor" for Breast Cancer

Although growing body of evidence suggests elevated expression of CT and CTR in multiple cancers, extensive studies on CT actions have been conducted only in ▶ prostate cancer and ▶ breast cancer (▶ tumor suppressor genes) cell lines. Interestingly, CT displays sexual dimorphism in these two cancers, raising a possibility of the modulatory role of sex hormones on CT actions in these two organs. For example, CT is a potent ▶ oncogene for prostate cancer as indicated by the progressive increase in CT and CTR expression in primary prostate cancers with tumor progression, and potent stimulatory actions of CT on tumorigenicity of prostate cancer cell lines. In contrast, CT and CTR is constantly expressed in normal mammary ductal epithelium, the loss of CTR expression is associated with the progression of breast cancer to ▶ metastatic phenotype, and CT inhibits growth of some breast cancer cell lines. Although opposing actions of CT on prostate and breast cancer cell lines remain to be thoroughly investigated, initial studies in the author's laboratory suggest that CT protects junctional complexes in breast cancer cell lines, and estradiol attenuates the actions of CT in estradiol receptor-positive breast cancer cell lines (Han et al. 2006; Seck et al. 2005). These results emphasize the importance of CTR actions on junctional complexes in cancer and its significance in cancer cell growth and metastasis.

CT Is an Angiogenic Factor

▶ Angiogenesis, the process of new vessel formation or neovascularization, has aroused increasing interest over the last 25 years. Expansion of the tumor cell mass is dependent on both the degree of tumor vascularization and the rate of angiogenesis. Our results have demonstrated the presence of CTR in HMEC-1 cell line, and that CT stimulates in vivo angiogenesis in nude mice, and directly stimulates all major phases of in vitro angiogenesis including endothelial cell migration, invasion, proliferation, and tube morphogenesis. The stimulatory actions of CT on in vitro angiogenesis are comparable to the actions of ▶ vascular endothelial growth factor (VEGF). Importantly, silencing of CTR in HMEC-1 cells completely abolishes CT-induced tube morphogenesis. Furthermore, prostate and thyroid cancer cell lines expressing high levels of CT form large, highly vascular tumors. In contrast, the silencing of CT expression in these cell lines markedly reduces tumor growth and vascularity. These results may also explain the findings that malignancies displaying high levels of CT expression (such as MTCs and multiple endocrine neoplasias) also produce highly vascular tumors. Considering that therapeutic use of CT for pain relief is fairly widespread in cancers as well as other diseases, it will be important to consider oncogenic and angiogenic effects while determining CT therapy in these patients.

In summary, CT and CTR expression has been well investigated in breast and prostate carcinomas. CT is a potent stimulator of tumor growth, angiogenesis, and metastasis in prostate cancer cell lines. In contrast, CTR expression is lost with breast cancer progression, and CTR attenuates growth of breast cancer cell lines. Significant expression of CT and CTR has also been reported in MTCs, multiple endocrine neoplasias, and carcinomas of lung, ▶ pancreas, gastrointestinal tract, thymus, and ▶ bladder. However, the significance of CT–CTR axis in these carcinomas remains to be investigated.

Cross-References

▶ Bladder Cancer
▶ Lung Cancer
▶ Medullary Thyroid Cancer Targeted Therapy
▶ Metastasis
▶ Multiple Endocrine Neoplasia Type 1
▶ Pancreatic Cancer
▶ SH2/SH3 Domains
▶ Src
▶ Thyroid Carcinogenesis

References

Ball DW (2007) Medullary thyroid cancer: therapeutic targets and molecular markers. Curr Opin Oncol 19(1):18–23
Barwell J, Gingell JJ, Watkins HA, Archbold JK, Poyner DR, Hay DL (2012) Calcitonin and calcitonin receptor-like receptors: common themes with family B GPCRs? Brit J Pharmacol 166(1):51-65. doi:10.1111/j.1476-5381.2011.01525.x.
Davies J (2015) Procalcitonin. J Clin Pathol 68: 675–79
Findlay DM, Saxton PM (2003) Calcitonin. In: Henry HL, Normon AW (eds) Encyclopedia of hormones and related cell regulators. Academic, New York, pp 220–230
Han B, Nakamura M, Zhou G et al (2006) Calcitonin inhibits invasion of breast cancer cells: involvement of urokinase-type plasminogen actor and uPA receptor. Int J Oncol 28(4):807–814
ondel M (2000) Calcitonin and calcitonin receptors: bone and beyond. Int J Exp Pathol 81(6):405–422
Seck T, Pellegrini M, Florea AM et al (2005) The delta e13 isoform of the calcitonin receptor forms six transmembrane domain receptor with dominant negative effects on receptor surface expression and signaling. Mol Endocrinol 19(8):2132–2133
Steinwald PM, Whang KT, Becker KL, Snider RH, Nylen ES, White JC (1999) Elevated calcitonin precursor levels are related to mortality in an animal model of sepsis. Critical Care 3(1):11–16
Thomas S, Chigurupati S, Anbalagan M, Shah G (2006) Calcitonin increases tumorigenicity of prostate cancer cells: evidence for the role of protein kinase A and urokinase-type plasminogen receptor. Mol Endocrinol 20(8):1894–1911

Calcium-Binding Proteins

Meenakshi Dwivedi and Joohong Ahnn
Department of Life Science, Hanyang University, Seoul, South Korea

Definition

Calcium-binding proteins are proteins that participate in calcium signaling pathways by binding to Ca^{2+}. The most ubiquitous Ca^{2+}-binding protein, found in all eukaryotic organisms including yeasts, is calmodulin. With their role in signal transduction, Ca^{2+}-binding proteins contribute to all aspects of the cell's functioning, from homeostasis to ▶ cancer.

Characteristics

Normal cell cycle division is a highly coordinated progression of molecular events that is subject to control mechanisms from both outside and inside the cell. Commitment to cell cycle initiation is made from outside and occurs as a response to extracellular signals such as growth factors. Inside the cell, control mechanisms exist to determine the timing of intracellular events such as nuclear and cytoplasmic cleavage. Under normal conditions, growth-regulating mechanisms endeavor to maintain homeostasis. Homeostasis within a cell is regulated by the balance between proliferation, growth arrest, and ▶ apoptosis. Intracellular Ca^{2+} is an important modulator of a variety of biochemical processes associated with cell cycle progression. With few exceptions, the controls exerted by intracellular Ca^{2+} are transduced through site-specific interactions with specialized Ca^{2+}-binding proteins. There exist at least three main families of Ca^{2+}-binding proteins. The first of these is represented by proteins that possess one or more EF-hand helix–loop–helix structural motifs predicting Ca^{2+}-binding domains as typically found within calmodulin. The second class of Ca^{2+}-binding proteins is known generically as annexins. A possible third family of Ca^{2+}-binding proteins is the "calreticulin-like" group of proteins that include ▶ calreticulin, Grp78, endoplasmin, and protein disulfide isomerase. Ca^{2+} has also been implicated in cell growth under pathological states. An altered cellular response to extracellular calcium ion concentration is one of the earliest changes induced in mouse epidermal cells by chemical carcinogens. However, whereas some human breast cancer cell lines and leukemia cell lines exhibit Ca^{2+}-induced cell proliferation, other carcinoma cell lines exhibit retarded growth in the presence of Ca^{2+} or no sensitivity to Ca^{2+} at all. In a human breast cancer cell line, which is sensitive to Ca $^{2+}$, the administration of calcium channel antagonists lowered intracellular Ca^{2+} and inhibited cell proliferation. A sustained physiological elevation of intracellular calcium ion concentration (Ca^{2+}i) may be responsible for a loss of proliferative potential in neoplasmic keratinocytes. It appears from preliminary evidence that Ca^{2+} is not only important to cell cycling and growth in normal cells, but the abnormal regulation of Ca^{2+} may also contribute to changes in these processes in disease conditions like cancer.

EF-Hand Motif Calcium-Binding Proteins

The EF-hand protein structural motif was first discovered in the crystal structure of parvalbumin. It consists of two alpha helices positioned roughly perpendicular to one another and linked by a short loop region (usually about 12 amino acids) that often binds calcium ions. A consensus amino acid sequence for this motif has aided the identification of new members of this family that now has over 200 members. A few of these proteins are present in all cells, whereas the vast majorities are expressed in a tissue-specific fashion. Some members, like S100 family, calcineurin, calmodulin, etc., have proved to be useful therapeutic markers for a variety of cancers.

S100 family: The S100 ("Soluble in 100% saturated solution with ammonium sulfate") family, the largest family within the EF-hand protein, comprises at least 26 members, 19 of which (S100A1–16, profillagrin, trychohyalin, and repetin) are located in the epidermal differentiation complex situated at 1q21, while S100B, S100G, S100P, S100Z, and

S100A7L2–S100A7L4 are present at other genomic locations (21q22, xp22, 4p16, 5q14, and 1q22, respectively). Their gene structure is highly conserved, in general comprising three exons and two introns, of which the first exon is noncoding. The S100 family is a remarkable group of proteins that acquired highly specialized functions during their evolution, even though they are small proteins (9–13 kDa acidic proteins) with a single functional domain. S100 proteins, which exhibit dramatic changes in the expression, are involved in tumor progression and include S100A1, S100A4, S100A6, S100A7, and S100B (11–14), whereas S100A2 has been postulated to be a tumor suppressor (Table 1). Such changes might be caused by rearrangements and deletions in chromosomal region, which are frequently observed in tumor cells. Although in most cases, the function of S100 proteins in cancer cells is still unknown, the specific expression patterns of these proteins can be used as a valuable prognostic tool.

The other members of EF-hand motif Ca^{2+}-binding family involved in cancer are calcineurin, recoverin, calretinin, oncomodulin, etc. (Table 2).

Annexins and Other Non-EF-Hand Motif Proteins

Annexin (AnxA1) is a Ca^{2+}-binding and acidic phospholipid binding protein with anti-inflammatory properties. AnxA1 has been found in leukocytes, tissue macrophages, T-lymphocytes, and epithelial cells of the respiratory and urinary systems. Cellular functions of AnxA1 include regulation of membrane trafficking, cellular adhesion, cell signaling, and membrane fusion in exocytosis and endocytosis. The AnxA1 protein is involved in maintaining normal breast biology. The *AnxA1* gene expression may provide data about the future therapeutic plan of breast carcinomas. The decreased expression of *AnxA1* gene in normal histological sections of breast may warn the clinician that a malignant version of the cancer is about to form from the benign gland. This observation carries an important prognostic clinical value on microscopic reading of the surgical specimen, especially if these normal glands are adjacent to surgical margins. Similar result was reported as a prognostic

Calcium-Binding Proteins, Table 1 S100 proteins involved in cancer with characteristic features

S100 protein	Previous name	Cancer type	Characteristic features
S100A2	CAN19, S100L	Esophageal SCC	Function in carcinogenesis is dependent upon the context of tissue and associated tumor type as well as stage of malignancy
		Thyroid	
		Oral SCC	
		Laryngeal SCC	
		Melanoma	
		Skin tumors (other)	
		NSCLC	
		Lung SCC	Marker of patient prognosis
		Gastric	
		Lymphoma	
		Prostate	
		Ovarian	
		Breast	
S100A3	S100E	Astrocytomas	Level of S100A3 protein expression identified pilocytic astrocytomas
S100A4	CAPL, Calvasculin, MTS1, Metastasin, p9Ka, FSP1	Thyroid carcinoma	Stimulate angiogenesis
		NSCLC	Overexpression is associated with many different cancer types
		Colorectal	
		Gastric	
		Prostate	Poor patient prognosis in breast, colorectal, NSCLC, and bladder cancers
		Breast	
		Gallbladder	
		Bladder	
		Pancreas	
		Esophageal SCC	
		Melanoma	
		Oral SCC	
S100A5	S100D	Meningioma	Expression can be associated with prognostic value in recurrence of meningiomas
S100A6	CABP, CACY, Calcyclin, MLN 4, PRA, Prolactin receptor-associated protein	Thyroid	Upregulation of S100A6 appears to be an early event in progression towards pancreatic cancer
		Pancreas	
		Breast	
		Lung	
		Melanoma	Potentially act as a predictor of clinical outcome
		Colorectal	
S100A7	PSOR1, Psoriasin	Breast	Expression is restricted to keratinocytes and breast epithelial cells
		Esophageal SCC	
		Bladder SCC	
		Melanoma	Overexpression has a role in early breast tumor progression
		Skin SCC	
		Gastric	

(*continued*)

Calcium-Binding Proteins, Table 1 (continued)

S100 protein	Previous name	Cancer type	Characteristic features
S100A8	Calgranulin A, MRP-8, MIF, NIF, P8, CFAG, CGLA, CP-10, Calprotectin	Prostate	Upregulated in PIN and in prostatic adenocarcinomas
		Breast	
		Esophageal SCC	
		HNSCC	
		Gastric	
		Pancreas	Early involvement of the proteins in prostate cancer
		Bladder TCC	
		Endometrial	
		Ovarian	
		Colorectal	
S100A9	Calgranulin B, MRP-14, P14, Calprotectin	Prostate	Upregulated in PIN and in prostatic adenocarcinomas
		Breast	
		HNSCC	
		Esophageal SCC	
		Liver (hepatocellular)	
		Gastric	
		Lung	
		Ovarian	
S100A10	CAL1L, CLP11, GP11, p10, ANX2LG	NSCLC	Annexin 2 protein ligand
		Gastric	Overexpressed in human renal cell carcinoma
		Renal cell carcinoma	
		Lymphoma	
S100A11	Calgizzarin, MLN70, S100C	Breast	Downregulated at transcriptional level in malignant bladder cells
		Bladder	
		Prostate	
		Thyroid	Expression may be involved in tumor suppression and better prognosis
		Lymphoma	
		Gastric	
		Colon	Cytoplasmic staining pattern in papillary carcinomas
S100A12	Calgranulin C, MRP6, p6, CAAF1, ENRAGE	Esophageal SCC	Downregulated
S100A14	BCMP84, S100A15, S114	Esophageal SCC	Used for CTC monitoring in peripheral blood
		Circulating tumor cells	
S100A16	S100F, DT1P1A7	Circulating tumor cells	Used for CTC monitoring in peripheral blood
S100B	S100, S100 protein (beta chain), NEF	Melanoma	Putative cancer biomarker
S100P	S100E, MIG9	NSCLC	Isolated from placenta
		Pancreas	
		Prostate	Upregulation is early event in pancreatic cancer valuable marker for the prediction of clinically relevant early pancreatic lesions
		Breast	
		Colon	

SCC squamous cell carcinoma, *HNSCC* head and neck SCC, *NSCLC* nonsmall cell lung carcinoma, *TCC* transitional cell carcinoma, *CTCC* circulating tumor cells

Calcium-Binding Proteins, Table 2 Other EF-hand motif family members

EF-hand motif protein	Cancer type	Characteristic features
Recoverin	Cancer-associated retinopathy	Autoimmune response
Calretinin	Colon adenocarcinoma and mesothelioma of epithelial type	Belongs to the calbindin subgroup, autoantigen in a paraneoplastic disease
Sorcin	Ovarian carcinoma	Overexpression leads to paclitaxel resistance
Calcineurin B	Squamous cell carcinoma of cervix, pancreatic cancer	Calcineurin B subunit appears to be a significant biological response modifier due to its anticancer effects
Oncomodulin/parvalbumin	Carcinoma cell lines characterized by translocative activity	Related to motile behavior of carcinoma cells, can be a possible candidate for tumor marker
Calmodulin	Osteoclast apoptosis	Calcium ion receptor in neoplastic cells

factor with downregulation of AnxA1 and other Anxs in the development of the lethal prostatic carcinoma phenotype.

The other major protein that belongs to this category is clusterin (CLU). CLU is a disulfide-linked heterodimeric protein associated with the clearance of cellular debris and apoptosis. In prostate, breast, and colorectal cancers, the CLU was found to have anti- or proapoptotic activity regulated by calcium homeostasis. Reports so far suggest "two faces" of CLU activity: the calcium-dependent cytoplasmic localization of CLU positively correlates with cell survival, whereas nuclear translocation of this protein promotes cell death in calcium-deprived cells. The cytoplasmic retention and high level of the 50-kDa CLU protect tumor cells from apoptotic stimuli induced by chemotherapeutic drugs or natural ligands, such as FasL, whereas its nuclear

localization (nCLU) enhances cell apoptosis. The 50-kDa CLU isoform is mainly over-expressed in cancer cells and retained in the cytoplasm, promoting cancer progression and aggressiveness. Cytoplasmic CLU could easily translocate into the nucleus in the presence of various inducers, such as IR, chemotherapy, hormones, or cytokines, or depletion of cellular calcium. These findings support CLU as a valid therapeutic target in strategies employing novel multimodality therapy for advanced prostate cancer.

Calreticulin-Like Proteins

Calreticulin is a 46-kDa Ca^{2+}-binding chaperone protein found across a diverse range of species. The human gene for calreticulin is located on chromosome 19 at locus p13.3–p13.2 and the homologous gene in the mouse maps to chromosome 8. Calreticulin at the cell surface may play a role in cell adhesion, cell–cell communication, and apoptosis. Calreticulin has also been implicated in the pathology of some cancers. The protein also plays an important role in autoimmunity and cancer. For example, it appears that calreticulin might be an excellent molecular marker for prostate cancer. The expression of calreticulin is downregulated in metastatic melanoma and ▶ squamous cell carcinoma, whereas significantly upregulated in colon cancer. Further, the N-domain of the protein has been reported to have inhibitory effects on tumors and to inhibit ▶ angiogenesis on endothelial cells. This observation is of great interest because the development of angiogenesis inhibitors is currently a highly promising approach in anticancer therapy.

Cross-References

▶ Angiogenesis
▶ Apoptosis
▶ Calreticulin
▶ Cancer
▶ Squamous Cell Carcinoma

References

Donato R (2003) Intracellular and extracellular roles of S100 proteins. Microsc Res Tech 60:540–551

Kretsinger RH, Tolbert D, Nakayama S et al (1991) The EF-hand, homologs and analogs. In: Heizmann CW (ed) Novel calcium-binding proteins: fundamentals and clinical implications. Springer, New York, pp 17–37

Pfyffer GE, Haemmerlit G, Heizmann CW (1984) - Calcium-binding proteins in human carcinoma cell lines. Proc Natl Acad Sci U S A 81:6632–6636

Pfyffer GE, Humbel B, Strauli P et al (1987) Calcium-binding proteins in carcinoma, neuroblastoma and glioma cell lines. Virchows Arch 412:135–144

See Also

(2012) EF-Hand Proteins. In: Schwab M (ed) Encyclopedia of Cancer, 3rd edn. Springer Berlin Heidelberg, p 1211. doi: 10.1007/978-3-642-16483-5_1819

(2012) Proteins. In: Schwab M (ed) Encyclopedia of Cancer, 3rd edn. Springer Berlin Heidelberg, p 3099. doi: 10.1007/978-3-642-16483-5_4812

Calcium-Binding Reticuloplasmin of Molecular Weight 55 kDa

▶ Calreticulin

CALI

▶ Chromophore-Assisted Laser Inactivation

CALM

▶ PICALM

Calpain

Neil O. Carragher
Drug Discovery Group, Edinburgh Cancer Research Centre, University of Edinburgh, Edinburgh, UK

Definition

The calpains represent a unique class of intracellular protein degrading enzymes. This class of proteases was named "calpains" to reflect their dependency upon calcium ions for proteolytic activity, and homology to the papain family of cysteine proteases. In mammalian species, the calpain protein family is comprised of 15 members, of which nine are ubiquitously expressed in all tissues and the remainder are expressed in a tissue-specific manner. The ubiquitously expressed calpain 1 and calpain 2 are the most well characterized isoforms. Calpain 1 and calpain 2 function as heterodimeric enzymes composed of a large catalytic subunit (calpain 1 and calpain 2) bound to a small regulatory subunit (calpain 4). Calpain activity in vivo is tightly regulated by the ubiquitously expressed endogenous inhibitor calpastatin. Calpains result in the proteolysis of a broad spectrum of cellular proteins. No unique consensus amino acid sequence has been identified as a calpain-binding or -cleavage site, rather, it appears that calpains target substrates for cleavage by recognition of unidentified tertiary structure motifs. Another distinguishing feature of calpain proteases is their ability to confer limited cleavage of protein substrates into stable fragments, rather than complete proteolytic digestion. Thus, the calpain-calpastatin proteolytic system represents a major pathway of posttranslational modification of proteins that influences various aspects of cellular physiology. The application of pharmacological and molecular intervention strategies against calpain activity demonstrates a broad role for this class of proteases in the control of proliferation, ▶ migration, and ▶ apoptosis in most cell types.

Characteristics

Cell proliferation, migration, and apoptosis are key processes that have to be tightly regulated in order to maintain optimal tissue homeostasis, required for development and viability of multicellular organisms. Deregulation of any of these cellular processes will ultimately result in pathological outcomes, such as cancer. A number of studies have identified a correlative link between modulation of calpain gene expression and/or activity with cancer development and progression in vivo. For example, in human renal cell carcinomas, significantly higher levels of calpain 1 expression are found in tumors that metastasized to peripheral lymph nodes relative to tumors that had not metastasized. In addition, elevated calpain activity was detected in breast cancer tissues relative to normal breast tissues and was determined to be greater in estrogen receptor (ER)-positive tumors than ER negative tumors. Calpain-mediated proteolysis of the tumor suppressor protein neurofibromatosis type 2 (NF2 or ▶ Merlin) is associated with the development of schwannomas and menigiomas. Experimental studies performed in vitro demonstrate that total cellular calpain activity is elevated upon transformation induced by the *v-src*, *v-jun*, *v-myc*, *k-ras*, and *v-fos* oncogenes. Furthermore, calpain activity is necessary for full cellular transformation induced by such oncogenes. A number of intervention studies utilizing small molecule inhibitors or oligonucleotides that impair calpain activity have demonstrated a role for calpain during tumor cell progression in vitro and in vivo. Cell proliferation, migration, and apoptosis are controlled by a plethora of regulatory proteins that participate in complex biochemical signaling cascades, of which calpain is a pivotal regulator. Thus, targeting calpain activity may represent an effective strategy for cancer prevention and/or treatment.

Calpain and Cell Proliferation

Studies using pharmacological inhibitors of calpain activity, overexpression of calpastatin, and cells expressing depleted levels of calpain activity have all implicated calpain in the promotion of cell proliferation. Sequential progression through G1, S, G2, and M phases of the cell cycle is required for mitosis and cell proliferation. Several studies indicate that calpain can cleave a number of cell cycle control proteins such as ▶ cyclin D, cyclin E, and p27^{kip1} all of which regulate progression through G1 and S phase. Calpain has also been demonstrated to cleave upstream regulators of cell cycle control proteins such as p53 and p107. Consequently, elevated calpain levels and activity in tumors may contribute to cancer cell proliferation through cleavage of cell cycle control proteins and deregulation of normal cell cycle control.

More detailed mechanistic studies also demonstrate that calpain 2 is an important downstream component of many growth factor receptor and non-receptor tyrosine kinase signaling pathways that include signaling kinases such as, ▶ epidermal growth factor receptor (EGFr), ▶ platelet derived growth factor receptor (PDGFr), Src, and ▶ focal adhesion kinase (FAK). Such signaling molecules play an important role in transmitting extracellular signals to intracellular mediators that control cell proliferation and are often constitutively activated in cancer cells. Activation of receptor and non-receptor kinases subsequently leads to activation of the Ras/MAPK pathway, which results in ERK-mediated phosphorylation of calpain 2 on a serine residue (Ser 50). Phosphorylation of calpain 2 on Ser 50 initiates a conformational switch culminating in enhanced proteolytic activity. This evidence, together with pharmacological and molecular intervention studies targeting calpain activity, suggests that activation of calpain, in part, mediates growth factor receptor and non-receptor induced cell proliferation and migration of cancer cells.

Calpain and Cell Migration

The interaction between cell surface adhesion receptors known as integrins and their extracellular matrix substrates controls the migration of all cells. Integrin-linked focal adhesions are large complexes of structural and signaling proteins that provide both a structural and biochemical link between the extracellular environment and intracellular proteins. Dynamic spatial and temporal regulation of focal adhesion assembly and disassembly is required for optimal cell motility. Several studies indicate that calpains localize to integrin-associated adhesions. Furthermore, many of the protein components of focal adhesions are known substrates of calpain. Calpain-mediated cleavage of the focal adhesion components, FAK, paxillin, talin, and possibly others promotes the disassembly of these complexes, contributing to reduced cell adhesion and increased migration. In fact, calpain-mediated cleavage of talin has been reported to represent the rate-limiting step in adhesion turnover. In addition to mediating focal adhesion turnover, emerging evidence suggests a role for calpain in regulating components of the actin cytoskeleton involved in cell spreading and membrane protrusion, mechanisms that are also essential for persistent and directed cell migration. It is likely that calpain cleavage of actin-binding and actin-regulatory proteins such as, ezrin, rhoA, and cortactin influences the dynamic formation and retraction of membrane structures known as filipodia and lamellipodia thereby influencing cancer cell migration and ▶ invasion. Pharmacological and molecular inhibition of calpain activity has been shown to impair cancer cell migration across experimental two-dimensional substrates and invasion into three-dimensional extracellular matrix substrates in vitro. Furthermore, an intervention study demonstrates that antisense-mediated suppression of calpain 2 gene expression reduced the invasion of prostate cancer cells both in vitro and in a mouse model in vivo. Thus, evidence strongly indicates that calpain activity contributes to the invasion and ▶ metastasis of cancer cells.

Calpain and Apoptosis

Apoptosis is defined as the process of programmed cell death. Apoptosis often follows activation of the caspase family of cysteine proteases, which degrade numerous proteins that are essential for cell viability. Regulated apoptosis is critical for the development of multicellular organisms and also restricts the growth and spread of malignant cancer cells. Conflicting roles for calpain activity in the promotion and suppression of cell apoptosis have been proposed. Calpain activity has previously been shown to play a pro-apoptotic role through the activation of caspase 3 and caspase 12 and cleavage of Bax and ▶ Bid proteins to their pro-apoptotic forms. Enhanced calpain activity has also been implicated as the major proteolytic pathway resulting in breakdown of essential proteins during caspase-independent mechanisms of apoptosis. Conversely, calpain-mediated cleavage of caspase 7 and caspase 9 has been found to suppress their activity and subsequent apoptosis. In addition, calpain-mediated cleavage of IκBα can lead to activation of the NFκB transcription factor resulting in subsequent expression of anti-apoptotic survival proteins. Many chemotherapeutic agents such as cisplatin induce their tumoricidal effect via inducing apoptosis of cancer cells. Tumor cell resistance to cisplatin-induced apoptosis is a common feature frequently encountered during chemotherapy of cancer patients. Inhibition of calpain activity has been shown to sensitize resistant tumor cells to cisplatin-induced death, whereas other studies suggest that calpain potentiates cisplatin-induced cell death. Thus, the role of calpain during cell apoptosis is context dependent and determined by cell type, the apoptotic stimuli, and status of intrinsic regulators of cell apoptosis.

In contrast to the aforementioned studies suggesting a pro-tumorigenic role for calpain activity, an anti-tumorigenic role is also supported by studies indicating that calpain degrades a number of oncogene-generated protein products such as PDGFr, EGFr, c-Jun, c-Fos, c-Src, and c-Mos. Also, calpain-mediated cleavage of protein kinase C (PKC), a downstream effecter for tumor promoting phorbal esters, inhibits malignant transformation. Furthermore, specifically calpain 9 (nCL-4) activity contributes to the suppression of cell transformation in vitro and gastric tumors in vivo. Although the calpain 9 substrates that mediate this antitumor effect remain to be determined.

A substantial body of evidence has accumulated demonstrating that activity of the calpain family of proteases plays a broad and important role in the physiology of both normal and cancer cells. Further investigation into the complex and multifaceted role of calpain in cancer may lead to the discovery of novel therapeutic approaches targeting calpain activity that may impact on the development, progression, and prevention of cancer.

Cross-References

▶ Apoptosis
▶ Cyclin D
▶ Focal Adhesion Kinase
▶ Epidermal Growth Factor Receptor

- ▶ Invasion
- ▶ Merlin
- ▶ Metastasis
- ▶ Migration
- ▶ Platelet-Derived Growth Factor

References

Carragher NO, Frame MC (2002) Calpain: a role in cell transformation and migration. Int J Biochem Cell Biol 34:1539–1543

Franco SJ, Huttenlocher A (2005) Regulating cell migration: calpains make the cut. J Cell Sci 118(17):3829–3838

Goll DE, Thompson VF, Li H et al (2003) The calpain system. Physiol Rev 83:731–801

See Also

(2012) Calpastatin. In: Schwab M (ed) Encyclopedia of Cancer, 3rd edn. Springer, Berlin/Heidelberg, p 598. doi:10.1007/978-3-642-16483-5_783

(2012) EGFR. In: Schwab M (ed) Encyclopedia of Cancer, 3rd edn. Springer, Berlin/Heidelberg, p 1211. doi:10.1007/978-3-642-16483-5_1828

(2012) ERK. In: Schwab M (ed) Encyclopedia of Cancer, 3rd edn. Springer, Berlin/Heidelberg, pp 1307-1308. doi:10.1007/978-3-642-16483-5_1987

(2012) Extracellular Matrix. In: Schwab M (ed) Encyclopedia of Cancer, 3rd edn. Springer, Berlin/Heidelberg, p 1362. doi:10.1007/978-3-642-16483-5_2067

(2012) Filipodia. In: Schwab M (ed) Encyclopedia of Cancer, 3rd edn. Springer, Berlin/Heidelberg, p 1407. doi:10.1007/978-3-642-16483-5_2189

(2012) Focal Adhesion. In: Schwab M (ed) Encyclopedia of Cancer, 3rd edn. Springer, Berlin/Heidelberg, pp 1436-1437. doi:10.1007/978-3-642-16483-5_2227

(2012) Integrin. In: Schwab M (ed) Encyclopedia of Cancer, 3rd edn. Springer, Berlin/Heidelberg, p 1884. doi:10.1007/978-3-642-16483-5_3084

(2012) Lamellipodia. In: Schwab M (ed) Encyclopedia of Cancer, 3rd edn. Springer, Berlin/Heidelberg, p 1971. doi:10.1007/978-3-642-16483-5_3267

(2012) MAPK. In: Schwab M (ed) Encyclopedia of Cancer, 3rd edn. Springer, Berlin/Heidelberg, p 2167. doi:10.1007/978-3-642-16483-5_3532

(2012) Papain. In: Schwab M (ed) Encyclopedia of Cancer, 3rd edn. Springer, Berlin/Heidelberg, p 2781. doi:10.1007/978-3-642-16483-5_4366

(2012) Platelet-Derived Growth Factor Receptor. In: Schwab M (ed) Encyclopedia of Cancer, 3rd edn. Springer, Berlin/Heidelberg, p 2910. doi:10.1007/978-3-642-16483-5_4612

(2012) Proteases. In: Schwab M (ed) Encyclopedia of Cancer, 3rd edn. Springer, Berlin/Heidelberg, p 3080. doi:10.1007/978-3-642-16483-5_4786

Calreticulin

Yoshito Ihara
Department of Biochemistry, School of Medicine, Wakayama Medical University, Wakayama, Japan

Synonyms

CaBP3; Calcium-binding reticuloplasmin of molecular weight 55 kDa; Calsequestrin-like protein; CRP55; CRT; ERp60; HACBP; High affinity Ca^{2+}-binding protein; Reticulin

Definition

Calreticulin (CRT) is a Ca^{2+}-binding multifunctional ▶ molecular chaperone in the endoplasmic reticulum (ER). CRT is a 46-kDa soluble protein with a cleavable N-terminal amino acid signal sequence and the C-terminal sequence Lys-Asp-Glu-Leu (KDEL), a retrieval signal in the ER. Calnexin (CNX), a membrane-binding paralog of CRT, shares the ▶ chaperone function in the ER. CRT is expressed in a variety of tissues and organs, but its levels are particularly high in the pancreas, liver, and testis. It is also a highly conserved protein with over 90% amino acid identity in mammals including humans, rabbits, rats, and mice. The CRT gene has been mapped to human chromosome 19 at p13.2, and its expression is up-regulated by ▶ ER stress such as unfolded protein responses and deprivation of Ca^{2+} in the ER.

Characteristics

Structure of CRT

Based on structural and functional studies, CRT can be divided into three distinct domains: N-terminal [N], proline-rich [P], and C-terminal [C]. The proline-rich P-domain shows a characteristic structure with an extended and curved arm

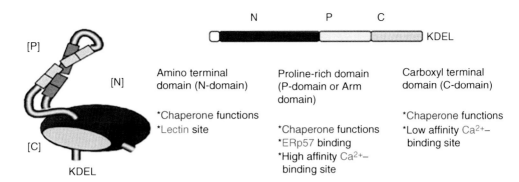

Calreticulin, Fig. 1 Schematic structure of calreticulin

connected to a globular N-domain. The N-terminal region encompassing the N- and P-domains of CRT interacts with misfolded proteins and glycoproteins, binds ATP, Zn^{2+}, and Ca^{2+} with high affinity and low capacity, and is likely to be involved in the chaperone function of the protein. The C-domain binds Ca^{2+} with high capacity and plays a role in the storage of Ca^{2+} in the ER in vivo, though no structural information is available at present (Fig. 1).

Functions of CRT in the Cell

CRT is involved in a number of biological processes including the regulation of glycoprotein folding, Ca^{2+} homeostasis and intracellular signaling, cell adhesion, gene expression, and nuclear transport (Fig. 2).

CRT, a Lectin-Like Molecular Chaperone in the ER
A molecular chaperone function of CRT has been reported for several protein substrates. In the biosynthesis of glycoproteins bearing N-linked glycans in the ER, the oligosaccharide Glc3Man9GlcNAc2 (Glc, glucose; Man, mannose; GlcNAc, N-acetylglucosamine) is attached to the Asn residue contained in the consensus sequence Asn-X-Ser/Thr, of newly synthesized polypeptides. CRT or CNX binds the Glc1Man5–9GlcNAc in glycoproteins after the processing of sugar chains. The N-domain of CRT and CNX is speculated to be the oligosaccharide-binding site (lectin site). If the glycoprotein is completely folded in the ER, the terminal glucose is removed by glucosidase-II

and the glycoprotein is released from the CNX/CRT chaperone cycle. However, if the glycoprotein is not properly folded, the terminal glucose is once again attached by the action of UDP-Glc: glycoprotein glucosyltransferase, which discriminates between folded and unfolded substrates. Together, CRT and CNX form a specific chaperone cycle for the biosynthesis of glycoproteins in the ER. Because of the preference of CNX/CRT for oligosaccharides as substrates, CNX and CRT are called "lectin-like chaperones." CRT and CNX function with the help of other chaperones such as ERp57 and BiP/GRp78. The binding site for ERp57 has been identified in the P-domain of CRT or CNX. As a chaperone, CRT plays an important role in the formation of major histocompatibility complex (MHC) class I to aid in antigen presentation.

CRT, a Regulator of Ca^{2+} Homeostasis in the ER
The ER is the main reservoir of intracellular Ca^{2+} and plays an important role in ► Ca^{2+} homeostasis. CRT has two Ca^{2+}-binding sites and this characteristic contributes to the function of the ER as a Ca^{2+} reservoir. Ca^{2+} is released from the ER by receptors for inositol-1,4,5-trisphosphate (IP3) and ryanodine, and taken up into the ER by sarcoplasmic and endoplasmic reticulum Ca^{2+}-ATPase (SERCA). With respect to the regulation of the Ca^{2+} level, the involvement of CRT and SERCA2b or the IP3-receptor has been reported. Furthermore, the store-operated release of Ca^{2+} from the ER was shown to be suppressed by overexpression of CRT protein. These findings indicate that CRT is not

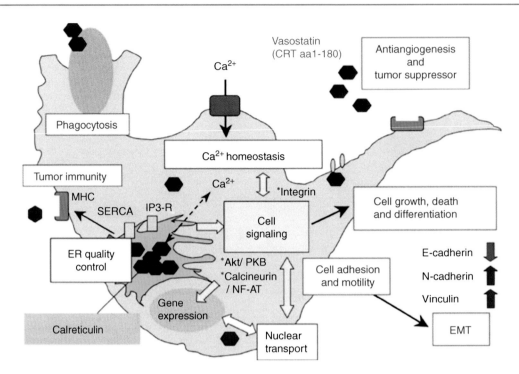

Calreticulin, Fig. 2 Functions of calreticulin in the cell. CRT is involved in a variety of cellular processes, including the quality control of glycoprotein synthesis in the ER, Ca^{2+} homeostasis, intracellular signaling, gene expression, and nuclear transport. In cancer cells, the altered expression of CRT may lead to alterations in cellular characteristics, such as growth, adhesion, motility, immune responses, and susceptibility to apoptosis. Furthermore, extracellular CRT fragments (i.e., vasostatin) elicit antiangiogenic or tumor-suppressing activities

only a reservoir of Ca^{2+} but also a regulator of Ca^{2+}-homeostasis in the ER.

Other Miscellaneous Functions of CRT In and Out of the ER

CRT is involved in cell ► adhesion by affecting integrin-related cell signaling. In CRT-deficient embryonic stem cells, integrin-mediated Ca^{2+} influx was impaired leading to a decrease in cell adhesion to fibronectin and laminin. It is still not clear whether CRT affects integrin directly or indirectly to regulate cell adhesion signaling.

Cell surface expression of CRT has also been reported in various cell types and may be related with cell adhesion and ► migration. The cell surface CRT may modulate cell adhesion by binding with extracellular matrix proteins, such as ► fibrinogen, ► laminin, and ► thrombospondin. Furthermore, extracellular CRT is implicated in the pathological processes of autoimmune diseases. ► Autoantibodies against CRT were found in ~40% of patients with systemic lupus erythematosus, patients with secondary Sjogrens syndrome, rheumatoid arthritis, celiac disease, complete congenital heart block, and halothane hepatitis. CRT is known to bind to complement, C1q, and compete with antibodies for binding to C1q and inhibition of C1q-dependent hemolysis. In autoimmune diseases, impairment of the classical pathway of compliment causes a failure to clear immune complex, resulting in progression of the disease. Therefore, extracellular CRT may contribute to the progression of autoimmune diseases by preventing the clearance of immune complex. Furthermore, it has been reported that cell surface CRT is involved in the mechanism for clearance of viable or apoptotic cells through the transactivation of LDL-receptor-related protein (LRP) on phagocytes. However, it is still controversial whether CRT is exported from necrotic cells or apoptotic cells under pathologic conditions.

Cytosolic CRT functions as an export factor for multiple nuclear hormone receptors, such as steroid hormone, nonsteroid hormone, and orphan receptors. This function is consistent with previous findings that CRT suppresses the transactivation of nuclear hormone receptors including ▶ androgen receptor and vitamin D. However, the mechanisms by which CRT molecules are transported into, and retained in, the cytosol/nucleus are not fully defined.

CRT and Development

CRT is essential for cardiac and neural development in mice. CRT-deficient embryonic cells showed an impaired nuclear import of nuclear factor of activated T cell (NF-AT3), a transcription factor, indicating that CRT functions in cardiac development as a component of the Ca^{2+}/calcineurin/NF-AT/GATA-4 transcription pathway. Actually, cardiac-specific expression of calcineurin reversed the embryonic lethality of CRT-deficient mouse.

CRT transgenic mice suffer a complete heart block and sudden death, and CRT-dependent cardiac block involves an impairment of both the L-type Ca^{2+} channel and gap junction ▶ connexins (Cx40 and Cx43). Phosphorylated Cx43 was also decreased in CRT transgenic heart, suggesting that the functions of protein kinases are altered via the regulation of Ca^{2+} homeostasis. Collectively, CRT plays a vital role in cardiac differentiation and function, though how has not been fully clarified.

CRT and Cancer

Expression of CRT in Cancer

In terms of the relationship between CRT and cancer, proteomic analysis has revealed a new functional role of CRT in the early diagnosis of cancers. CRT is proposed to be a new tumor marker of bladder cancer. In addition, it was reported that the expression of CRT is up-regulated in a variety of malignant cells or tissues including progressive fibrosarcoma cells, colorectal cancer cells, and pituitary adenomas. Furthermore, autoantibodies to CRT isoforms have utility for the early diagnosis of pancreatic

cancer. These reactions are not indicative of malignant properties of CRT, but rather are markers of immunogenicity and anticancer responses. On the other hand, another report demonstrated that CRT is overexpressed in the nuclear matrix in ▶ hepatocellular carcinoma, compared with normal liver tissue, suggesting a relationship between overexpressed CRT and malignant transformation. In contrast, it was also reported that CRT expression correlates with the differentiation of ▶ neuroblastomas to predict favorable patient survival.

Pathophysiological Relevance of CRT in Malignant Disease

Susceptibility to ▶ apoptosis is important in terms of cancer treatments including the use of antibiotics and irradiation. In embryonic fibroblasts from CRT knockout mice, susceptibility to apoptosis was significantly suppressed, indicating that CRT functions in the regulation of apoptosis. Furthermore, it was found that overexpression of CRT modulates the ▶ radiation sensitivity of human glioma U251MG cells by suppressing Akt/protein kinase B signaling for cell survival via alterations of cellular Ca^{2+} homeostasis. These findings suggest that the expression level of CRT is well correlated with the susceptibility to apoptosis. In contrast, overexpression of CRT provides resistance to oxidant-induced cells death in renal epithelial LLC-PK1 cells. The function of CRT in the regulation of apoptosis may differ in specific cell types and is still controversial.

As for cell adhesion, it was reported that CRT expression modulates cell adhesion by coordinating up-regulation of N-cadherin and vinculin. It has been reported that overexpression of CRT induces ▶ epithelial-mesenchymal transition (EMT)-like morphological changes and enhances cellular invasiveness in renal epithelial MDCK cells. The enhanced invasiveness mediated through ▶ E-cadherin gene repression was regulated by the gene repressor, Slug, via altered Ca^{2+} homeostasis caused by overexpression of CRT in MDCK cells. This study suggests that expression of CRT may play some causative role in the gain of invasiveness during the process of malignant transformation. In addition, it has been reported that cellular migration and binding to collagen type V are apparently suppressed in embryonic fibroblasts from CRT knockout mice, indicating that the cellular level of CRT is important for the regulation of cell motility.

Furthermore, CRT protein binds to GCN repeats in mRNA of the myeloid transcription factor CCAAT/enhancer-binding protein α (CEBPA), and thereby impedes translation of the CEBPA mRNA, suggesting that CRT plays a functional role in the differentiation block in ► acute myeloid leukemia through suppression of CEBPA by the leukemic ► fusion gene *AML1-MDS1-EVI1*. Together, these findings suggest that CRT is involved in the regulation of cancer characteristics, although the overall mechanisms are still not clear.

CRT as a Tool for Cancer Therapy

CRT can form complexes with peptides in vitro to elicit peptide-specific CD8+ ► T cell responses. In addition, peptide-bound CRT purified from tumor extracts elicits an antitumor effect specific to the source tumor. Antigen-specific cancer immunotherapy is an attractive approach to the eradication of systemic tumors at multiple sites in the body. It has been reported that vaccination with DNA encoding chimera for CRT and a tumor antigen, human papilloma virus type-16 (HPV-16) E7 [CRT/E7], resulted in a significant reduction in the number of lung tumor nodules in immunocompromised mice. All together, the use of CRT represents a feasible approach for enhancing tumor-specific T cell-mediated immune responses.

Therapeutic agents that target the tumor vasculature may prevent or delay tumor growth and even promote tumor regression or dormancy. As another approach to cancer therapy, CRT or a fragment thereof (amino acids 1–180) (i.e., vasostatin) inhibits ► angiogenesis and suppresses tumor growth. The combination of vasostatin and IL-12 as well as vasostatin and interferon-inducible protein-10 had a suppressing effect on the cell growth of Burkitt lymphoma and colon carcinoma in mouse metastasis models. Although this suggests some potential for use in cancer therapy, the molecular mechanism of CRT actions at the cell surface is not fully understood.

References

Eggleton P, Michalak M (2003) Introduction to calreticulin. In: Eggleton P, Michalak M (eds) Calreticulin, 2nd edn. Landes Biosciences/Eurekah. com/Kluwer/Plenum Publishers, Georgetown/New York, pp 1–8

Gelebart P, Opas M, Michalak M (2005) Calreticulin, a Ca-binding chaperone of the endoplasmic reticulum. Int J Biochem Cell Biol 37:260–266

Johnson S, Michalak M, Opas M et al (2001) The ins and outs of calreticulin: from the ER lumen to the extracellular space. Trends Cell Biol 11:122–129

Michalak M, Corbett EF, Mesaeli N et al (1999) Calreticulin: one protein, one gene, many functions. Biochem J 344:281–292

Williams DB (2006) Beyond lectins: the calnexin/calreticulin chaperone system of the endoplasmic reticulum. J Cell Sci 119:615–623

Calsequestrin-Like Protein

► Calreticulin

CAM

► Chorioallantoic Membrane

Campto®

► Irinotecan

Camptosar®

► Irinotecan

CAMs

► Cell Adhesion Molecules

CAMTA1

Kai-Oliver Henrich and Frank Westermann
DKFZ, German Cancer Research Center,
Heidelberg, Germany

Definition

CAMTA1 is a candidate ▶ tumor suppressor gene encoding a member of a protein family designated as calmodulin-binding transcription activators (CAMTAs). It resides within a distal portion of chromosomal arm 1p that is frequently deleted in a wide range of human malignancies.

Characteristics

CAMTA1 maps to 1p36.31-p36.23 and its 23 exons are spread over 982.5 kb. The 6,582 bp cDNA encodes a protein of 1,673 amino acids. The primary structure of protein contains a nuclear localization signal, two DNA-binding domains (CG-1 and TIG), a transcription activation domain, calmodulin binding motifs (IQ motifs), and ankyrin domains. Although the expression of *CAMTA1* is seen in various organs, the highest levels are found in neuronal tissues. Information on the physiologic roles of CAMTAs is scarce and most data derive from plant and drosophila studies.

CAMTAs are transcription factors that typically bind to CGCG boxes via their CG-1 domain. An alternative mechanism of transcriptional activation has been described for CAMTA2, the second human CAMTA homolog. It acts as a coactivator of another transcription factor, Nkx2-5, to stimulate gene expression. This function is inhibited by binding of class II histone deacetylases to the ankyrin-repeat region of CAMTA2. Upstream signaling components can activate CAMTA2 by promoting the export of class II histone deacetylases to the cytoplasm, relieving their repressive influence on CAMTA2.

The sole fly homolog of CAMTA1 induces the expression of an F-box gene, the product of which inhibits a Ca^{2+}-stimulating G-protein-coupled receptor (GPCR). The controlled deactivation of Ca^{2+}-stimulating GPCRs is needed to tune Ca^{2+}-mediated signaling and prevents abnormal cell proliferation. As CAMTA activity is increased by the Ca^{2+}-sensor calmodulin, the Ca^{2+}/calmodulin/CAMTA/F-box protein pathway may mediate a negative feedback loop controlling the activity of Ca^{2+}-stimulating GPCRs. This regulatory loop is of special interest taking into account the fundamental links between GPCR-mediated pathways and cancer biology.

Clinical Relevance

Deletions within 1p occur in various types of human malignancies, ranging from virtually all types of solid cancers to leukemias and myeloproliferative disorders. Functional evidence for a role of 1p in tumor suppression derives from experiments in which the introduction of 1p chromosomal material into ▶ neuroblastoma cells resulted in reduced tumorigenicity. In neuroblastoma and other cancers, deletion of 1p36 is a predictor of poor patient outcome. Therefore, it is widely assumed that distal 1p harbors a gene (or genes) with tumor suppressive properties. To define the DNA, deleted from 1p, more precisely in pursuit of identifying the gene(s) of interest, substantial mapping efforts have been undertaken with the most detailed picture being worked out for neuroblastoma. In this tumor entity, the combination of loss of heterozygosity (LOH) fine mapping studies allowed to considerably narrow down a smallest region of consistent deletion spanning only 261 kb at 1p36.3 and pinpointing the *CAMTA1* locus. Sequence analysis revealed no evidence for somatic mutations in the remaining *CAMTA1* copy of neuroblastomas with 1p deletion. However, a rare sequence variant leading to amino acid substitution within the ankyrin domain was seen in a subgroup of neuroblastomas. More importantly, low *CAMTA1* expression is significantly associated with markers of unfavorable tumor biology and is itself a marker of poor neuroblastoma patient outcome. Moreover, *CAMTA1* expression is a neuroblastoma predictor variable that is independent of the established molecular markers including 1p deletion. Thus, the measurement of this variable should allow an additional biological stratification of neuroblastomas and help to assign patients to the appropriate therapy.

Additional evidence for a role of CAMTA1 in tumor development comes from glioma and colon cancer in which 1p is frequently deleted. In glioma, a 1p minimal deleted region spans 150 kb and resides entirely within *CAMTA1*. In colorectal cancer, a genome-wide analysis of genomic alterations revealed that loss of a 2 Mb recurrently deleted genomic region encompassing *CAMTA1* has the strongest impact on survival when compared with other genomic changes. Furthermore, as in neuroblastoma, the low expression of *CAMTA1* is an independent marker of poor patient outcome. The high prevalence of *CAMTA1* deletion in neuroblastoma, glioma, and colorectal cancer together with the independent predictive power of low *CAMTA1* expression for neuroblastoma and colorectal cancer outcome is consistent with the idea that low CAMTA1 levels mediate a selective advantage for developing tumor cells.

Cross-References

► Neuroblastoma
► Tumor Suppressor Genes
► Ubiquitin Ligase SCF-Skp2

References

Barbashina V, Salazar P, Holland EC et al (2005) Allelic losses at 1p36 and 19q13 in gliomas: correlation with histologic classification, definition of a 150-kb minimal deleted region on 1p36, and evaluation of CAMTA1 as a candidate tumor suppressor gene. Clin Cancer Res 11(3):1119–1128

Bouche N, Scharlat A, Snedden W et al (2002) A novel family of calmodulin-binding transcription activators in multicellular organisms. J Biol Chem 277(24):21851–21861

Henrich KO, Fischer M, Mertens D et al (2006) Reduced expression of CAMTA1 correlates with adverse outcome in neuroblastoma patients. Clin Cancer Res 12(1):131–138

Henrich KO, Claas A, Praml C et al (2007) Allelic variants of CAMTA1 and FLJ10737 within a commonly deleted region at 1p36 in neuroblastoma. Eur J Cancer 43(3):607–616

Kim MY, Yim SH, Kwon MS et al (2006) Recurrent genomic alterations with impact on survival in colorectal cancer identified by genome-wide array comparative genomic hybridization. Gastroenterology 131(6):1913–1924

See Also

(2012) Glioma. In: Schwab M (ed) Encyclopedia of Cancer, 3rd edn. Springer Berlin Heidelberg, p 1557. doi:10.1007/978-3-642-16483-5_2423

(2012) G-protein Couple Receptor. In: Schwab M (ed) Encyclopedia of Cancer, 3rd edn. Springer Berlin Heidelberg, p 1587. doi:10.1007/978-3-642-16483-5_2294

(2012) Loss of Heterozygosity. In: Schwab M (ed) Encyclopedia of Cancer, 3rd edn. Springer Berlin Heidelberg, pp 2075-2076. doi:10.1007/978-3-642-16483-5_3415

Cancer

Manfred Schwab
German Cancer Research Center (DKFZ), Heidelberg, Germany

Definition

Cancer is a deregulated multiplication of cells with the consequence of an abnormal increase of the cell number in particular organs. Initial stages of the developing cancer are usually confined to the organ of origin whereas advanced cancers grow beyond the tissue of origin. Advanced cancers invade the surrounding tissues that are initially connected to the primary cancer. At a later stage, they are distributed via the hematopoetic and lymphatic systems throughout the body where they can colonize in distant tissues and form ► metastasis. The development of cancers is thought to result from the damage of the cellular genome, either due to random endogenous mechanisms or due to environmental influences.

The origin of cancers can be traced back to alterations of cellular genes. Genetic damage can be of different sorts:

• Recessive mutations in ► tumor suppressor genes

- Dominant mutations of ▶ oncogenes
- Loss-of-function mutations in genes, involved in maintaining genomic stability and ▶ repair of DNA (resulting in ▶ genomic instability)

History

Human cancer is probably as old as the human race. It is obvious that cancer did not suddenly start appearing after modernization or industrial revolution. The world's oldest documented case of cancer comes from ancient Egypt, in 1500 BC. The details were recorded on a papyrus, documenting eight cases of tumors occurring on the breast. It was treated by cauterization, a method to destroy tissue with a hot instrument called "the fire drill." It was also recorded that there was no treatment for the disease, only palliative treatment. The word cancer came from the father of medicine, Hippocrates, a Greek physician (460–370 BC). Hippocrates used the Greek words, carcinos and carcinoma to describe tumors, thus calling cancer "karkinos." The Greek terms actually were words to describe a crab, which Hippocrates thought a tumor resembled. Hippocrates believed that the body was composed of four fluids: blood, phlegm, yellow bile, and black bile. He believed that an excess of black bile in any given site in the body caused cancer. This was the general thought of the cause of cancer for the next 1,400 years. Autopsies done by Harvey in 1628 paved the way to learning more about human anatomy and physiology. By about the same time period, Gaspare Aselli discovered the lymphatic system, and this led to the end of the old theory of black bile as the cause of cancer. The new theory suggested that abnormalities in the lymph and lymphatic system as the primary cause of cancer. The lymph theory replaced Hippocrates' black bile theory on the cause of cancer. The discovery of the lymph system gave new insight to what may cause cancer, it was believed that abnormalities in the lymphatic system were the cause. Other theories surfaced, such as cancer being caused by trauma, or by parasites, and it was thought that cancer may spread "like a liquid" (Bentekoe, 1687, Heinrich Vierling, "personal communication"). The belief that cancer was composed of fermenting and degenerating lymph fluid was predominant.

The discovery of the microscope by Leeuwenhoek in the late seventeenth century added momentum to the quest for the cause of cancer. By late nineteenth century, with the development of better microscopes to study cancer tissues, scientists gained more knowledge about the cancer process. It wasn't until the late nineteenth century that Rudolph Virchow, the founder of cellular pathology, recognized that cells, even cancerous cells, derived from other cells. The early twentieth century saw great progress in our understanding of microscopic structure and functioning of the living cells. Researchers pursued different theories to the origin of cancer, subjecting their hypotheses to systematic research and experimentation. John Hill first recognized an environmental cause from the dangers of tobacco use in 1761 and published a book "Cautions Against the Immoderate Use of Snuff." Percivall Pott of London in 1775 described an occupational cancer of the scrotum in chimney sweeps caused by soot collecting under their scrotum. This led to identification of a number of occupational carcinogenic exposures and public health measures to reduce cancer risk. This was the beginning of understanding that there may be an environmental cause to certain cancers.

A virus causing cancer in chickens was identified in 1911 (Rous sarcoma virus). Existence of many chemical and physical carcinogens was conclusively identified during the later part of the twentieth century. The later part of the twentieth century showed tremendous improvement in our understanding of the cellular mechanisms related to cell growth and division. The identification of ▶ transduction of oncogenes with the discovery of the ▶ SRC gene, the transforming gene of Rous sarcoma virus, led to formulating the oncogene concept of tumorigenesis and can be viewed as the birth of modern molecular understanding of cancer development. Subsequently, tumor suppressor genes were identified. Many genes that suppress or activate the cell growth and division are known to date, their number is ever growing. It is conceivable that in the end the confusing situation may arise to recognize that all genes of the human genome, in one way or another, take part in signaling normal or cancerous cellular growth.

Characteristics

A large proportion of genetic changes appears to arise by mechanisms endogenous to the cell, such as by errors occurring during the replication of the $\sim 3 \times 10^9$ base pairs present in the human genome. Environmental factors have a major role as well, predominantly as:

- Chemical carcinogens (e.g. aflatoxin B1 in liver cancer) (▶ hepatocellular carcinoma molecular biology), tobacco smoke in ▶ lung cancer; (▶ tobacco carcinogenesis)
- Radiation (▶ radiation carcinogenesis)
- Viruses (such as ▶ hepatitis B virus in liver cancer, or human papillomavirus in ▶ cervical cancers)

Types of Genetic Damage
Damage to oncogenes and tumor suppressor genes can be of different sorts:

- Point mutations resulting in the activation of a latent oncogenic potential of a cellular gene (e.g. RAS) or in the functional inactivation of a tumor suppressor gene by generating an intragenic stop codon that leads to premature translation termination with the consequence of an incomplete truncated protein (e.g. p53) or the failure for maintaining genomic stability (mismatch repair genes in HNPCC).
- ▶ Amplification leading to an increase of the gene copy number beyond the two alleles normally present in the cell (copy number can reach 500 and more; example: MYCN in human ▶ neuroblastoma).
- Translocation, which is defined as an illegitimate recombination between nonhomologous chromosomes, the result being either a fusion protein (where recombination occurs between two different genes such as BCR-ABL in chronic myclogenous leukemia) or in the

disruption of normal gene regulation (where the regulatory region of a cellular gene is perturbed by the introduction of the distant genetic material such as MYC in Burkitt lymphoma (▶ Epstein-Barr virus)).
- Viral insertion by the integration of viral DNA into the regulatory region of a cellular gene. This integration can occur after a virus has infected a cell. Viral insertion is well documented in animal tumors (HBV integration in the vicinity of MYCN in liver cancer in experimental animals; liver cancer, molecular biology).

Cellular Aspects
Cancer in solid tissues (solid cancer) usually develops over long periods (often 20–30 years latency period) of time. An exception is solid cancers (such as neuroblastoma) in children, which often are diagnosed shortly after birth. Malignant cancers are characterized by their ability to develop metastasis (i.e. secondary cancers at distance from the primary tumor), often they also show multidrug resistance, which means that they hardly react to conventional chemotherapy. It is thought that the development of a normal cell to a metastatic cell is a continuous process driven by genetic damage and genomic instability, with the progressive selection of cells that have acquired a selective advantage within the particular tissue environment (▶ multistep development). Studies of colorectal cancers have identified 6–7 genetic events required for the conversion of a normal cell to a cell with metastatic ability. This is in contrast to leukemias, which usually require one genetic event, most often a translocation, for disease development.

Sporadic Versus Familial Cancer
The vast majority of cancers are "sporadic," which simply means that they develop in an individual. Descendants of this individual do not have an increased risk because the cellular changes that have resulted in cancer development are confined to this individual. In contrast, ~10% to 15% of cancer cases have a clearly recognizable hereditary background, they show familial clustering (prominent examples include retinoblastoma

(▶ retinoblastoma, cancer genetics), breast cancer, FAP (▶ APC gene in familial adenomatous polyposis) and HNPCC as familial forms of colorectal cancer (colon cancer), ▶ melanoma). This is not to say that "sporadic" cancers never are related to heredity. In fact, it is well possible that an undetermined fraction, if not all, "sporadic" cancers may be related to an individual inherited susceptibility that does not appear as a strong single gene determinant, but rather as a genetic constitution consisting of complex balance of polymorphic genes.

Familial cancers have been identified to result from germline mutation of genes. These germ line mutations do not always directly dictate cancer development, although they are considered "strong" hereditary determinants. They represent susceptibility genes that confer a high risk for cancer development to the gene carrier. The relative risk of the individual carrying the mutant gene can vary considerably. For instance, the risk of carriers of one of the breast cancer susceptibility genes *BRCA1* or *BRCA2* for breast cancer development can vary between approximately 60% and 90%. In reality, this means that the risk for cancer development is difficult to predict, and individuals may not develop cancer at all in spite of the presence of a mutated gene in their germ line. The molecular basis for the differences in risk is unknown. Formally the activity of modifying factors, either environmental or genetic, has been suggested. Such modifying factors appear to be less important for some other familial cancers, such as retinoblastoma, where the risk is constant between 90% and 95% for gene carriers.

Polygenic Determinants of Risk

The relative risk of the individual for cancer development can also be determined by the so-called weak genetic factors. Normal cells contain a number of genes involved in ▶ detoxification reactions. Different allelic variants of these genes exist in the human population that encodes proteins with slightly different enzymatic activities. Although the exact contribution of individual allelic variants to cancer development is difficult to assess, it is reasonable to assume that individuals that have inherited "weak" enzymatic activities in different detoxification systems are likely to have a higher risk. It is likely, therefore, that the risk for such cancers is "polygenic."

Cross-References

▶ Cohesins
▶ Toxicological Carcinogenesis

Cancer (or Tumor) Stroma

▶ Tumor Microenvironment

Cancer and Cadmium

Tim S. Nawrot and Jan A. Staessen
Division of Lung Toxicology, Department of Occupational and Environmental Medicine (T.S.N.) and the Studies Coordinating Centre (J.A.S.), Division of Hypertension and Cardiovascular Rehabilitation, Department of Cardiovascular Diseases, University of Leuven, Leuven, Belgium

Definition

Cadmium is a metal that has the symbol Cd and atomic number 48 in the periodic table. Cadmium has high toxic effects, an elimination half-life of 10–30 years, and accumulates in the human body, particularly the kidney. Roughly 15,000 t of cadmium is produced worldwide each year for nickel–cadmium batteries, pigments, chemical stabilizers, metal coatings, and alloys.

Characteristics

Urinary excretion of cadmium over 24 h is a biomarker of lifetime exposure. Exposure to cadmium occurs through intake of contaminated food or water, or by inhalation of tobacco smoke or polluted air. Occupational exposures can be found in industries such as electroplating, welding, smelting, pigment production, and battery manufacturing. Other exposures to cadmium can occur through inhalation of cigarette smoker. Gastrointestinal absorption of cadmium is estimated to be around 5–8%. Inhalation absorption is generally higher, ranging from 15% to 30%. Absorption after inhalation of cadmium fume, such as cigarette smoke, can be as high as 50%. Once absorbed, cadmium is highly bound to the metal-binding protein, metallothionein. Cadmium is stored mainly in the kidneys and also the liver and testes, with a half-life in the body of 10–30 years. In general, nonsmokers have urinary cadmium concentrations of 0.02–0.7 µg/g creatinine, which increase with age in parallel with the accumulation of cadmium in the kidney. Cadmium is a global environmental contaminant. Populations worldwide have a low-level intake through their food, causing an age-related cumulative increase in the body burden of this toxic metal. Environmental exposure levels to cadmium, that are substantially above the background, occur in areas with current or historical industrial contamination, for instance, in regions of Belgium, Sweden, the UK, Japan, and China.

As an environmental carcinogen, cadmium could have substantial health implications. Three lines of evidence explain why the International Agency for the Research on Cancer classified cadmium as a human carcinogen. First, as reviewed by Verougstraete and colleagues, several, albeit not all studies in workers, showed a positive association between the risk of lung cancer and occupational exposure to cadmium; discrepancies between these studies should not be ascribed to the better design of the studies. Verougstraete and colleagues suggested that such inconsistencies might be attributed to the high relative risk of cancer in the presence of coexposure to ► arsenic, nickel, or toxic fumes and that the increasingly stringent regulations with regard to levels of exposure permissible at work might be a confounding factor (► lead

exposure, ► nickel carcinogenesis). Second, data from rats showed that the pulmonary system is a target site for carcinogenesis after cadmium inhalation. However, exposure to toxic metals in animal studies has usually been much higher than those reported in environmentally exposed humans to toxic metals. Third, several studies done in vitro have shown plausible pathways, such as increased oxidative stress, modified activity of transcription factors, and inhibition of DNA repair. Most errors that arise during DNA replication can be corrected by DNA polymerase proof reading or by postreplication mismatch repair. In fact, inactivation of the DNA repair machinery is an important primary effect, because repair systems are required to deal with the constant DNA damage associated with normal cell functions. The latter mechanism might indeed be relevant for environmental exposure because Jin et al. found that chronic exposure of yeast to environmentally relevant concentrations of cadmium can result in extreme hypermutability. In this study the DNA-mismatch repair system is already inhibited by 28% at cadmium concentrations as low as 5 µM. For example, the prostate of healthy unexposed humans accumulates cadmium to concentrations of 12–28 µM, and human lungs of nonsmokers accumulate cadmium to concentrations of 0.9–6 µM. Further, in vitro studies provide evidence that cadmium may act like an estrogen, forming high-affinity complexes with estrogen receptors, suggesting a positive role in breast cancer carcinogenesis.

Along with this experimental evidence, two epidemiological studies in 2006 gave important positive input into the discussion on the role of exposure to environmental cadmium in the development of cancer in human beings. First, the results of a population-based case–control study noticed a significant twofold increased risk of breast cancer in women in the highest quartile of cadmium exposure compared with those in the lowest quartile. Second, we conducted a population-based prospective cohort study with a median follow-up of 17.2 years in an area close to three zinc smelters. Cadmium concentration in soil ranged from 0.8 to 17.0 mg/kg. At baseline, geometric mean urinary cadmium excretion was 12.3 nmol/day for people in the high-exposure area, compared with 7.7 nmol/day for those in the reference (i.e., low exposure) area. The risk of lung cancer was 3.58 higher than in a reference population from an area with low exposure. 24-h urinary excretion is a biomarker of lifetime exposure to cadmium. The risk for lung cancer was increased by 70% for a doubling of 24-h urinary cadmium excretion. Confounding by coexposure by arsenic could not explain the observed association. Epidemiological studies did not convincingly imply cadmium as a cause of prostate cancer. Of 11 cohort studies, only three (33%) found a positive association.

In conclusion, experimental and epidemiological studies strongly suggest environmental exposure to cadmium as a causal factor in the development of cancer of the lung and breast.

References

Jarup L, Berglund M, Elinder CG et al (1998) Health effects of cadmium exposure – a review of the literature and a risk estimate. Scand J Work Environ Health 24(Suppl 1):1–51

Jin YH, Clark AB, Slebos RJ et al (2003) Cadmium is a mutagen that acts by inhibiting mismatch repair. Nat Genet 34(3):326–329

McElroy JA, Shafer MM, Trentham-Dietz A et al (2006) Cadmium exposure and breast cancer risk. J Natl Cancer Inst 98(12):869–873

Nawrot T, Plusquin M, Hogervorst J et al (2006) Environmental exposure to cadmium and risk of cancer: a prospective population-based study. Lancet Oncol 7(2):119–126

Verougstraete V, Lison D, Hotz P (2003) Cadmium, lung and prostate cancer: a systematic review of recent epidemiological data. J Toxicol Environ Health B Crit Rev 6(3):227–255

Cancer Antigen 3

▶ NY-ESO-1

Cancer Causes and Control

Graham A. Colditz
Washington University in St. Louis, St. Louis, MO, USA

Synonyms

Cancer etiology; Cancer prevention; Prevention

Definition

The process of identifying causes of cancer and developing strategies to change cancer risk through health-care providers, regulations that reduce risk, or individual and community level changes.

Characteristics

Over six million people around the world die from cancer each year. There is overwhelming evidence that lifestyle factors impact cancer risk and that positive population-wide changes can significantly reduce the cancer burden. Current epidemiologic evidence links behavioral factors to a variety of diseases, including the most common cancers diagnosed in the developed world – ▶ lung cancer, colorectal cancer, ▶ prostate cancer, and ▶ breast cancer. These four cancers account for over 50% of all cancers diagnosed on western countries. As summarized in Fig. 1, tobacco causes some 30% of cancer, lack of physical activity 5%, obesity 15%, diet 10%, alcohol 5%, viral infections 5%, and UV light by excess sun exposure 3%. Because of the tremendous impact of modifiable factors on cancer risk, especially for the most common cancers, it has been estimated that at least 50% of cancer is preventable. Currently in the US not all risk factors are equally distributed across race and social class. Trends in risk factors should also be

Cancer causes

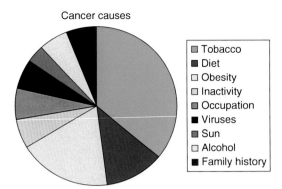

- ■ Tobacco
- ■ Diet
- □ Obesity
- □ Inactivity
- ■ Occupation
- ■ Viruses
- ■ Sun
- □ Alcohol
- ■ Family history

Cancer Causes and Control, Fig. 1 Causes of cancer

considered when assessing potential for prevention. To bring about dramatic reductions in cancer incidence, widespread lifestyle changes will be necessary.

Rose advocates the need for population approaches for prevention of chronic disease. He emphasizes that when the relation between a lifestyle factor and biological predictor of risk is continuous, the majority of cases attributable to the exposure will likely arise in those who are not classified as being at high risk. He illustrates this with examples of blood pressure and rates of coronary heart disease. Specifically, even small changes in blood pressure at the population level can translate into large reduction in the rates of coronary disease and stroke. To reduce the risk of disease in the population, substantial benefits can be achieved by a small reduction for all members of the society rather than just focusing on the high-risk groups. Because population wide trends in cardiovascular risk factors show continuing improvement, the rate of coronary heart disease incidence and mortality continues to decrease.

When we consider population approaches to cancer prevention, we must address the etiologic process, which covers a different time course and sequence from coronary heart disease. Although cardiovascular disease is the end point of the chronic process of atherosclerosis, treatment focuses on the reversal and subsequent prevention of the acute thrombotic process of myocardial infarction. Cancer, on the other hand, is the result of a long process of accumulating DNA damage (▶ multistep development), leading ultimately to

clinically detectable lesions such as in situ and invasive cancer. For example, studies of the progression in ▶ colon cancer from first mutation to invading malignancy suggest that DNA changes accumulate over a period of as long as 40 years. The goal of cancer prevention is to arrest this progression; different interventions interrupt carcinogenesis at different points in the process. Further, most cancers do not have a late "acute" event, analogous to thrombosis, which can be prevented with medical interventions.

The benefits of cancer prevention and control programs take time to be observed. The fact that different interventions will impact at different points along the pathway to cancer, that can stretch over nearly half a century, has implications for when we can expect to see pay-off in terms of lower cancer rates. Research has demonstrated that those who initiate smoking during early adolescence greatly increase risk of lung cancer even when one takes into account both the dose and duration of smoking. If we could delay the age at which most adolescents first start to smoke, we would probably substantially reduce lung cancer rates, but this benefit will not be observable for 20–40 years after the intervention. Adult cessation, on the other hand, reduces risk more rapidly, but fails to address the continuing recruitment of the next generation of smokers. Declines in the incidence of lung cancer among younger men and women in the United States reflect reductions in the rate of smoking among younger adults. Other lifestyle interventions may act as preventive early in the DNA pathway to cancer. For example, ▶ aspirin and folate appear to act early in the pathway inhibiting colon cancer.

Population-wide prevention strategies for cancer do work. For example, reductions in lung cancer rates in the United States mirror changes in cigarette smoking patters, with marked decreases seen first in young men, then older men, and finally in women. Introduction of the Papaniculou test for cervical cancer in the 1950s was followed by a dramatic decline in cervical cancer in those countries that made widespread screening available. The decline in Australian ▶ melanoma mortality for those born after 1950 is an additional example of effective intervention at the population level. Behavior change is possible and offers great potential for cancer prevention. The recommendations for cancer risk reduction include reducing tobacco use, increasing physical activity, maintaining a healthy weight, improving diet, limiting alcohol, avoiding excess sun exposure, utilizing safer sex practices, and obtaining routine cancer screening tests.

Age is the dominant factor that drives cancer risk; for all major malignancies, risk rises markedly with age. The importance of age is exemplified by the fact that the aging U.S. population together with projected population growth will result in a doubling of the total number of cancer cases diagnosed each year by the year 2050, assuming that incidence rates remain constant. With this estimated growth in cancer from 1.3 million to 2.6 million cases per year, it is expected that that both the number and proportion of older persons with cancer will also rise dramatically.

Tobacco

Tobacco is the major cause of premature death around the world accounting for some 5 M deaths each year. In the United States, adult smokers lose an average of 13 years of life because of smoking, and approximately half of all smokers die of tobacco-related disease. Smoking is well known to cause over 90% of ▶ lung cancers in addition to a range of other malignancies (▶ tobacco carcinogenesis; ▶ tobacco-related cancers). It causes about 30% of all the cancer in the developed world, including lung cancer, mouth cancer, larynx cancer, esophagus cancer, ▶ pancreas cancer, cervix cancer, kidney cancer, and ▶ bladder cancer. Smoking also confers increased risk of colorectal cancer, ▶ gastric cancer, ▶ liver cancer, and prostate cancer, as well as to leukemia. In addition, smoking leads to many other health problems, including heart disease, stroke, lung infections, emphysema, and pregnancy complications.

Tobacco may act on multiple stages of carcinogenesis; it delivers a variety of carcinogens, causes irritation and ▶ inflammation, and interferes with the body natural protective barriers. The health risks of tobacco use are not limited to cigarette smoking. Cigar and pipe use increase the

risk of disease, as does exposure to second-hand smoke and smokeless tobacco use (▶ lung cancer and smoking behavior).

Avoiding initiation of tobacco use clearly offers the greatest potential for disease prevention. However, for those who use tobacco products, there are substantial health benefits that come with quitting. There are numerous effective cessation methods, and in the past 25 years, 50% of all living Americans who have ever smoked, have successfully quit.

Quitting smoking has immediate and significant health benefits for men and women of all ages. For example, former smokers live longer than individuals who continue smoking. Those who quit before age 50 have approximately half the risk of dying in the next 15 years. This decline in mortality risk is measurable shortly after cessation and continues for at least 10–15 years.

Strategies to assist smoking cessation and decreasing youth initiation from both a population and clinical perspective are essential steps to reducing the burden of cancer.

Trends

Current smoking among US adults has remained steady over the past decade.

Once quite pronounced, gender disparity in smoking rates is now relatively small and has been stable since 1990. In 2002, 25.7% of men were current smokers compared to 20.8% of women. Given the profound impact of smoking on cancer, disparities in smoking rates and in access to effective cessation methods will continue to translate directly into differences in the burden of smoking-related cancers.

Physical Activity

Lack of physical activity causes over 2 M deaths each year around the world. People in the US and in other developed nations are extremely inactive – over 60% of the US adult population does not participate in regular physical activity, which includes 25% of adults who are almost entirely sedentary. Fortunately, the negative effects of a sedentary lifestyle are reversible: increasing one's level of physical activity, even after years of inactivity, can reduce mortality risk.

Lack of physical activity increases the risk of colon and breast cancer and likely endometrial cancer, as well as diabetes, osteoporosis, stroke, and coronary heart disease. Overall, sedentary lifestyles have been linked to 5% of deaths from cancer. Among both men and women, high levels of physical activity may decrease the risk of colon cancer by as much as half. Using a variety of measures of activity, studies have consistently shown higher physical activity lowers risk of colon cancer. Physical activity also appears to lower the risk of large adenomatous polyps, precursor lesions for colon cancer, suggesting that it may influence the early stages of the adenoma-carcinoma sequence. In addition, the relationship between physical activity and breast and colon cancer are seen across levels of obesity, indicating that physical activity and obesity have separate or independent effects on cancer incidence. Growing evidence suggests that physical activity may also be protective against lung and prostate cancer.

Several mechanisms have been proposed to explain these associations. Physical activity reduces circulating levels of insulin, a growth factor for colonic epithelial cells. Additionally, it is postulated that cancer risk is reduced through alterations in prostaglandin levels, improvement in immune function, and modification of bile acid metabolism.

Potential mechanisms for the reduction of breast cancer risk include physical activity's lowering of the cumulative lifetime exposure to circulating estrogens and improving immune pathways.

The benefits of physical activity include the prevention of cancer and a large number of other chronic diseases. Increasing levels of physical activity, even after years of inactivity, reduces mortality risk. As little as 30 min of moderate physical activity (such as brisk walking) per day significantly reduces disease risk.

Trends

One major determinant of activity level that has changed over time is the amount of activity required for work and daily living. With advances in technology and the development of labor-saving devices, there is now a greatly reduced need for physical activity for transportation, household tasks, and occupational requirements. Overall, the prevalence of physical inactivity in the United States is remarkably high; in 1996, about 28% of Americans reported absolutely no participation in leisure-time physical activity. In addition, physical activity in schools has declined, and almost half of young Americans between the ages of 12–21 are not vigorously active on a routine basis.

Given the trends in our society, it is unlikely that this decreasing energy expenditure will reverse rapidly. Accordingly, the burden of cancer due to lack of physical activity will increase in the years ahead unless new strategies to promote activity are rapidly implemented.

Weight Control and Obesity Prevention

Overweight and obesity (► Obesity and Cancer Risk) is increasing at epidemic rates in the United States, around the world, and is estimated to account for 2.6 M deaths each year. Currently almost 65% of American adults are overweight (body mass index (BMI) ≥ 25 kg/m^2), and over 30% are considered obese (BMI ≥ 25 kg/m^2).

Overweight and obesity cause a variety of cancers; colorectal cancer, postmenopausal ► breast cancer, ► endometrial cancer, ► renal cancer, and ► esophageal cancer. The proportion of cancer caused by obesity ranges from 9% for postmenopausal breast cancer to 39% for endometrial cancer. One large US study suggested that obesity influences an even broader range of cancers, increasing the risk of death from cancers of the colon and rectum, prostate, breast, esophagus, liver, gallbladder, pancreas, kidney, stomach, uterus, and cervix in addition to non-Hodgkin lymphoma and multiple myeloma. Overall, obesity causes 14% of cancer deaths among men and 20% of cancer deaths among women. Excess body fat may act by altering levels of hormones and tumor growth factors. It is clear that excess weight has severe health consequences.

In addition to raising the risk of cancer, overweight and obesity also increase the risk of a multitude of other diseases and chronic conditions, such as stroke, cardiovascular disease, type 2 diabetes, osteoarthritis, and pregnancy complications.

The International Agency for Research on Cancer has proposed a comprehensive set of recommendations to address the issue of weight control at multiple levels, including steps by health care providers, regulatory approaches to create adequate access to safe places for exercise (including school, worksite, and community), and family and community level actions.

Trends

In the USA, the prevalence of overweight and obesity has increased so dramatically and so rapidly, it is frequently referred to as an obesity epidemic. The trend is also being seen among children and adolescents.

This epidemic has affected people of all ages, races, ethnicities, socioeconomic levels, and geographic regions. Given limited long-term success in weight reduction programs, the cancer burden due to obesity will likely continue to follow the rising prevalence of this risk factor in the coming years.

Dietary Improvements

Fruit and vegetable intake has been most consistently evaluated as a cancer prevention strategy. The global burden of inadequate intake is estimated to account for over 2 M deaths each year. While evidence for cardiovascular benefits and reduced risk of diabetes are clear, evidence for cancer risk reduction has become less convincing with the results of numerous prospective cohort studies showing weaker associations with cancer risk. Low intake of fruits and vegetables are probably related to increased risk of pancreas, bladder, lung, colon, mouth, pharynx, larynx, esophagus, and stomach cancer.

Although the effect of fruit and vegetable consumption on the risk of prostate cancer has been examined in nearly twenty studies, data remain inconsistent. The majority of studies suggest that overall fruit and vegetable intake has a little effect if any on the risk of prostate cancer. However, individual fruits and vegetables may offer the potential for greater risk reduction, with tomatoes being the most promising, with a 40–50% reduction in risk among men who consumed large amounts of tomatoes and tomato products. The

carotenoid lycopene is hypothesized to be responsible for the protective effect.

A number of mechanisms have been suggested to explain the protective effect of fruits and vegetables, but it is not known if specific agents, such as ▶ carotenoids, folic acid, and vitamin C, or a special combination of factors create anticarcinogenic effects. It is also possible that diet in childhood and adolescence is more important than later in life in driving risk of cancer.

A number of studies have found that as folate intake increases, the risk of colorectal cancer (as well as polyps) decreases. The Nurses' Health Study found that a high intake of folate from fruits and vegetables was sufficient to lower risk but that supplementation with a multivitamin that contained folate offered even greater reductions. The underlying biologic role of folate and its interaction with the *MTHFR* gene add support to the causal relation between low folate and colon cancer.

In addition to the reduction in risk of colon cancer, growing evidence points to folate also reducing the adverse effect of alcohol on breast cancer. Based on this evidence and the benefits for prevention of neural tube defect and cardiovascular disease, use of a daily vitamin supplement containing folate is recommended.

Dietary Fat

Variations in international cancer rates have often been attributed to differences in total fat intake, yet evaluation has shown no clear link between dietary fat and breast, colon or prostate cancer. Although dietary fat overall does not appear to impact cancer risk, there is some evidence to suggest that certain types of fat, such as animal fat, may increase risk.

Fiber has been shown to reduce the risk of heart disease and diabetes, but it does not appear to offer protection against cancer. Long believed to help prevent colon cancer, the data do not support this hypothesis.

Red Meat

High intake of red meat, including beef, pork, veal, and lamb is associated with an elevated risk of colorectal cancer. The mechanism of this increased risk is not well understood, but it may be related to the high concentrations of animal fat or to carcinogens such as heterocyclic amines produced when the meat is cooked at high temperatures.

Calcium

Higher calcium intake has been linked to a reduced risk of colorectal adenomas and colorectal cancer. However, increased dietary calcium is also associated with an increased risk of prostate cancer. Research indicates that there may be a moderate intake of calcium that provides protection against colorectal cancer risk without causing a large increase in prostate cancer risk.

Excess Caloric Intake

One consistent dietary finding is that excess calories from any source result in weight gain and increased cancer risk. As the obesity epidemic continues to spread, the importance of balancing caloric intake with caloric expenditure becomes even more evident for the prevention of cancer and other chronic diseases.

Whole Grains

Although grain products in general have not been shown to affect cancer risk, whole-grain foods may provide some protection against stomach cancer. Grains such as wheat, rice, and corn form the basis for most diets worldwide. Some grain products, such as whole-wheat bread and brown rice, are consumed in the "whole-grain" form, while others, like white bread and white rice, are more refined. During the process of refining grain, most of the fiber, vitamins, and minerals are removed, thus whole-grain foods tend to be more nutrient-rich than refined foods and may offer more in terms of disease prevention. The benefits of whole-grain foods in reducing cardiovascular disease and ischemic stroke are well established.

Vitamin A and Carotenoids

Isolated vitamin A and carotenoids are not likely to play a large role in cancer prevention. Some observational data support a probable inverse relation with lung cancer risk, but randomized trials of beta-carotene intake found either no effect or an increased risk of lung cancer. It has also been suggested that beta-carotene impacts breast cancer risk, however, it seems that at best, there is only a small decrease in breast cancer risk associated with a high intake of carotenoids.

Selenium

Ecological studies have suggested that increased ▶ selenium intake is associated with decreased risk of colon and breast cancer. A randomized control trial of selenium for skin cancer prevention showed no effect of selenium on skin cancer incidence, however it did show a reduction in incidence of lung, colon, and prostate cancer. Despite these promising results, the impact of selenium remains unclear. Fortification of the soil in Finland in the mid-1980s led to higher blood selenium levels, but no decline in incidence or mortality has been noted for prostate or colon cancer.

Vitamin D

Growing evidence relates lower levels of ▶ vitamin D to increased risk of cancer and to poor survival after diagnosis.

Trends

The proportion of adults consuming the recommended five servings of fruits and vegetables a day varies between 8 and 32%. While these estimates are clearly low for the entire population, certain groups, as defined by gender, race/ethnicity, education, and income, are of particular concern.

Given constraints due to both financial resources and physical access to markets that provide fresh fruit and vegetables, it remains likely that SES gradients in diet will continue. Interventions are needed to overcome these existing barriers and make healthy foods readily available to all.

Limitation of Alcohol Use

Globally, alcohol intake in excess is responsible for 1.8 M deaths each year. Clear benefits of moderate alcohol intake have been shown in terms of reducing cardiac and diabetes risk, but alcohol remains a risk factor for cancer mortality.

Alcohol is a known carcinogen that may raise cancer risk in several ways. For example, it may act as an irritant, directly causing increased cell turnover, or it may allow for improved transport and penetration of other carcinogens into cells. Alcohol use is a primary cause of esophageal and oral cancer, and it is associated with an increased risk of breast, liver, and colorectal cancer. Multiple other risks are also associated with alcohol use, including the risk of hypertension, addiction, suicide, accident, and pregnancy complication.

To balance the cardiovascular benefits with the risks of cancer and other negative consequences, it is recommended that those who drink alcohol should do so only in moderation. Intake should be limited to less than one drink per day for women and less than two drinks per day for men.

Safer Sex and Decreased Viral Transmission

Unsafe sex is responsible for 2.9 M deaths each year, primarily due to the transmission of HIV. However, unprotected sexual contact also results in the spread of multiple other sexually transmitted infections including oncogenic viruses. Some of these viruses may also be spread through exposure to blood and blood products.

Human papillomavirus causes cervical cancer, vulvar, penile, and anal cancer; hepatitis B virus and ► hepatitis C virus cause hepatocellular cancer; human lymphotropic virus-type 1 is associated with adult T-cell leukemia (► human T-lymphotropic virus); human immunodeficiency virus-type 1 causes ► Kaposi sarcoma and non-Hodgkin lymphoma; and human herpes virus causes Kaposi sarcoma and body cavity lymphoma.

Prevention strategies to contain the spread of these viruses should include behavioral and educational interventions to modify sexual behavior, and structural and regulatory changes to promote safer sex and make condoms readily available. Biomedical interventions to administer vaccines are also needed. For example, it is estimated that vaccination programs could reduce the global burden of liver cancer by 60%. Additional strategies to prevent viral spread include needle exchange programs for intravenous drug users; regulation of tattooing and acupuncture; and screening of blood donors and the development of artificial blood products.

Trends and Disparities

Current U.S. data on the prevalence of these different viruses is not adequate to predict trends in cancer incidence. In addition, the development of new technologies such as vaccines against HPV suggest a new era in prevention of cervical cancer. However, for success, such vaccines must be available and accessible to the entire population. Assuring access remains a policy priority to maximize the potential benefit of this cancer prevention strategy.

Sun Protection

The American Cancer Society estimates over 50,000 melanoma diagnoses each year in the USA. The incidence of melanoma is rising more rapidly than that of any cancer in this country. Exposure to the sun (► UV radiation) is the major modifiable cause of melanoma and other skin cancers. For most people, the majority of lifetime sun exposure occurs during childhood and adolescence, and migrant studies clearly show that age at migration to high-risk countries has a strong impact on risk of this malignancy. For this reason, early intervention has the greatest potential for prevention.

The risk of melanoma and other less aggressive forms of skin cancer exists for all racial and ethnic groups, but skin cancers occur predominantly in the non-Hispanic white population. Constitutional characteristics including hair color, mole count, and family history contribute to risk of melanoma. However, studies show that established risk factors alone do not identify a sufficient proportion of cases to focus prevention efforts on only a subset of the population. Because identifying high-risk individuals will miss the majority of cases, population-based efforts provide greater protection. There is tremendous potential to substantially reduce the burden of this common malignancy through effective prevention efforts.

Screening

Screening for cancer can provide protection in several ways. In the case of colorectal and cervical cancers, screening can detect premalignant changes that can be treated to prevent cancer from developing. This primary prevention has the potential to substantially reduce the burden of cancer. With colorectal screening, the mortality from colon cancer is reduced by a half or more.

If cancer is already present, screening can act as a secondary prevention, such as mammography for breast cancer, facilitating early diagnosis and treatment, thereby decreasing morbidity and mortality. This type of prevention is an added benefit of colorectal and cervical screening, and is the main goal of breast cancer and prostate cancer screening.

Trends and Disparities

Trends in cervical cancer screening have been impacted by the breast cancer cervical cancer screening act which provided resources to states, via the Centers for Disease Control and Prevention, to bring screening services to low income women. Despite these efforts, national data suggest that low income and Hispanic women are less likely to be current with screening recommendations. Lack of access to care, defined as not having a usual source of health care, was associated with significantly lower compliance with cervical screening. Evidence from the 1998 Health Interview Survey indicates that all US born women have comparable and high compliance with screening for cervical cancer. Foreign born women, however, appear to be under-screened, accounting for the disparity among Hispanic women and suggesting a priority area for prevention as the USA continues to have a large immigrant population at risk of cancer. Surveys of colorectal screening suggest that the rates of screening have been rising and that Caucasians are more likely to be up to date with screening than other racial or ethnic groups.

Conclusion

Lifestyle changes offer tremendous potential for prevention of cancer and multiple other chronic conditions. This potential is often underestimated. To achieve the maximal benefit through behavioral change, interventions are necessary at multiple levels. Societal changes are needed to support

and encourage the behavior modification of individuals. Approaches are needed to target individuals, communities, and systems, and create an environment less inductive to high-risk lifestyles. Social systems and regulatory efforts must complement individual behavior changes if these changes are to be sustained and the benefits of reduced disease burden realized.

Overall, the major lifestyle factors considered here account for the majority of cancer and could be modified to prevent at least half of all cancers. However, the burden of cancer is not limited to just the major lifestyle factors considered here. For example, occupational and environmental exposures also account for a relatively small number of cancer cases compared to the lifestyle factors considered above. Yet the burden of exposure to these harmful agents may be disproportionately high among low-income populations, accentuating their cancer risk. In large part, these exposures can be prevented through adequate enforcement of regulatory changes, and this should remain a high priority.

Small individual changes can result in large population benefits, but efforts to create prevention programs for only certain members of our society limits the potential for prevention. We must largely reframe our approach to the issue. Identifying risk factors and setting goals for reduction is only the beginning. Research and policy must now focus on bringing about population-wide lifestyle change, addressing the issues of disparities, and leaving no group or community behind.

Cross-References

- ▶ Alcohol Consumption
- ▶ Aspirin
- ▶ Bladder Cancer
- ▶ Breast Cancer
- ▶ Carotenoids
- ▶ Cervical Cancers
- ▶ Colorectal Cancer
- ▶ Endometrial Cancer
- ▶ Esophageal Cancer

▶ Gastric Cancer

▶ Hepatitis C Virus

▶ Hematological Malignancies, Leukemias, and Lymphomas

▶ Hepatocellular Carcinoma

▶ Human T-Lymphotropic Virus

▶ Inflammation

▶ Kaposi Sarcoma

▶ Lung Cancer

▶ Lung Cancer and Smoking Behavior

▶ Multistep Development

▶ Obesity and Cancer Risk

▶ Pancreatic Cancer

▶ Prostate Cancer

▶ Prostate Cancer Clinical Oncology

▶ Renal Cancer Clinical Oncology

▶ Renal Cancer Genetic Syndromes

▶ Selenium

▶ Tobacco Carcinogenesis

▶ UV Radiation

▶ Vitamin D

References

Colditz GA, DeJong D, Emmons K et al (1997) Harvard report on cancer prevention. volume 2. Prevention of human cancer. Cancer Causes Control 8:1–50

Curry S, Byers T, Hewitt M (2003) Fulfilling the potential of cancer prevention and early detection. National Academy Press, Washington, DC

International Agency for Research on Cancer (2002) Weight control and physical activity, vol 6. International Agency for Research on Cancer, Lyon

Rose G (1981) Strategy of prevention: lessons from cardiovascular disease. Br Med J (Clin Res) 282:1847–1851

U.S. Department of Health and Human Services (1996) Physical activity and health: a report of the Surgeon General. US Department of Health and Human Services, Centers for Disease Control and Prevention, National Center for Chronic Disease Prevention and Health Promotion, Atlanta

See Also

(2012) Folate. In: Schwab M (ed) Encyclopedia of cancer, 3rd edn. Springer, Berlin/Heidelberg, p 1440. doi:10.1007/978-3-642-16483-5_2231

(2012) Hepatitis B Virus. In: Schwab M (ed) Encyclopedia of cancer, 3rd edn. Springer, Berlin/Heidelberg, p 1663. doi:10.1007/978-3-642-16483-5_2659

(2012) HIV. In: Schwab M (ed) Encyclopedia of cancer, 3rd edn. Springer, Berlin/Heidelberg, p 1706. doi:10.1007/978-3-642-16483-5_2764

(2012) Lycopene. In: Schwab M (ed) Encyclopedia of cancer, 3rd edn. Springer, Berlin/Heidelberg, p 2116. doi:10.1007/978-3-642-16483-5_3441

(2012) MTFR. In: Schwab M (ed) Encyclopedia of cancer, 3rd edn. Springer, Berlin/Heidelberg, pp 2383–2384. doi:10.1007/978-3-642-16483-5_6767

(2012) Renal cancer. In: Schwab M (ed) Encyclopedia of cancer, 3rd edn. Springer, Berlin/Heidelberg, pp 3225–3226. doi:10.1007/978-3-642-16483-5_6575

(2012) Tobacco. In: Schwab M (ed) Encyclopedia of cancer, 3rd edn. Springer, Berlin/Heidelberg, pp 3716–3717. doi:10.1007/978-3-642-16483-5_5844

(2012) Ultraviolet light. In: Schwab M (ed) Encyclopedia of cancer, 3rd edn. Springer, Berlin/Heidelberg, p 3841. doi:10.1007/978-3-642-16483-5_6101

(2012) Viruses. In: Schwab M (ed) Encyclopedia of cancer, 3rd edn. Springer, Berlin/Heidelberg, p 3924. doi:10.1007/978-3-642-16483-5_6203

(2012) Vitamin A. In: Schwab M (ed) Encyclopedia of cancer, 3rd edn. Springer, Berlin/Heidelberg, p 3925. doi:10.1007/978-3-642-16483-5_6205

Möslein G (2009) Colon cancer. In: Schwab M (ed) Encyclopedia of cancer, 2nd edn. Springer, Berlin/Heidelberg, pp 722–727. doi:10.1007/978-3-540-47648-1_126

Cancer Cell Cytotoxicity

▶ Artemisinin

Cancer Cell-Platelet Microemboli

▶ Tumor Cell-Induced Platelet Aggregation

Cancer Epidemiology

Paolo Boffetta
Icahn School of Medicine at Mount Sinai,
New York, NY, USA

Synonyms

Population-based cancer research

Definition

Knowledge about causes and preventive strategies for malignant neoplasms has greatly advanced during the last decades. This is largely attributable to the development of cancer epidemiology. In parallel to the identification of the causes of cancer, primary and secondary preventive strategies have been developed. A careful consideration of the achievements of cancer research, however, suggests that the advancements in knowledge about causes and mechanisms have not been followed by an equally important reduction in the burden of cancer. Part of this paradox is explained by the long latency occurring between exposure to carcinogens and development of the clinical disease. In addition, the most important risk factors of cancer are linked to lifestyle, and their modification entails cultural, societal, and economic consequences. The failure to identify valid biomarkers of cancer risk is another reason of the limited success in cancer control.

Cancer epidemiology investigates the distribution and determinants of cancer in human populations. Although the main tool in cancer epidemiology is the observational study, the intervention study, of experimental nature, is conducted to evaluate the efficacy of prevention strategies, such as screening programs and chemoprevention trials (clinical trials are usually considered outside the scope of epidemiology). Intervention studies follow the randomized trial design. Observational epidemiology can be broadly divided in descriptive epidemiology and analytical studies. Analytical studies can be based on data collected at the individual or population level. The former consist of ► cohort study, ► case–control study, and cross-sectional study (and a few variations on these themes), the latter of the so-called ecological study. Family-based studies are used in genetic epidemiology to identify hereditary factors. An additional useful distinction among etiological studies concerns the nature of the information on exposure: while some studies use information routinely collected for other purposes, such as censuses and medical records, in other circumstances exposure data are collected ad hoc following a variety of approaches including questionnaires, pedigrees, environmental measurements, and measurement of biological markers. A method-oriented (rather than subject- or design-oriented) approach has led to the identification of specific subdisciplines such as molecular epidemiology.

Characteristics

A distinctive feature of cancer epidemiology is the availability in many countries of a population-based cancer registry, which allow the calculation of valid and reliable estimate of the occurrence of cancer (incidence, mortality, prevalence, survival). Typically, registries collect routinely demographic data of patients, which are used to generate statistics according to period of diagnosis, age, sex, and other characteristics. These studies of descriptive epidemiology have been critical in developing etiological hypotheses. One particular type of descriptive studies concerns migrants from low- to high-risk areas or vice versa: the repeated demonstration of rapid changes in the risk of many cancers among migrants (from that prevalent in the area of origin toward that of the host area) provided very strong evidence of a predominant role of modifiable factors in the etiology of human cancer. While cancer registries were initially established in high-resource countries, a growing number of populations in middle- and low-resource countries are now covered by good-quality registries, thus providing a solid infrastructure for most ambitious research projects.

Other routinely collected data are used in epidemiology. Mortality statistics are available in many countries of the world and provide a good approximation of the incidence of the most fatal cancers. In a growing number of populations, automatic linkage is possible between incidence or mortality data and other population-based registries (e.g., hospital discharges, use of medications). Record-linkage study may represent an efficient alternative to investigations based on ad hoc collection of data.

The number of new cases of cancer which occurred worldwide in 2012 has been estimated at about 14,100,000. Of them, 7,400,000 occurred in men and 6,700,000 in women. About 6,100,000 cases occurred in more developed countries (North America, Japan, Europe including Russia, Australia, and New Zealand) and 8,000,000 in less developed countries. Among men, lung, prostate, colorectal, stomach, and liver cancers are the most common malignant neoplasms, while breast, colorectal, lung, cervical, and stomach cancers are the most common neoplasms among women. The number of deaths from cancer in 2012 was estimated at about 8,200,000 and that of 5-year prevalent cases at about 32,500,000.

Epidemiology has been instrumental to identify the causes of human cancer. In several cases, the epidemiological results preceded the elucidation of the underlying mechanisms. In other areas, however, epidemiological techniques are not sufficiently sensitive and specific to lead to conclusive evidence on the presence or absence of an increased risk. As for other branches of the discipline, the observational nature of epidemiology represents an opportunity for bias, including that generated by confounding, to generate spurious results. Techniques have been developed to prevent, control, and assess the presence and extent of bias in epidemiological studies.

Cancer epidemiology has led to the identification of tobacco smoking and use of smokeless tobacco products, chronic infections, overweight, alcohol drinking, and reproductive factors as major causes of human cancer. Other important causes include medical conditions, some drugs, perinatal factors, physical activity, occupational exposures, and ultraviolet and ionizing radiation. A role of diet in cancer risk has been suggested, but for very few dietary factors, there is a conclusive evidence of an effect on cancer risk. With a few exception of little relevance in most populations, the role of pollutants on cancer is not established. Tobacco smoking is the main single cause of human cancer worldwide. It is a cause of cancers of the oral cavity, pharynx, esophagus, stomach, liver, pancreas, nasal cavity, larynx, lung, cervix, kidney, and bladder and of myeloid leukemia. The proportion of cancers in a population attributable to tobacco smoking depends on the distribution of the habit a few decades earlier. Therefore, in populations in which the tobacco epidemic has not fully matured (e.g., men in many low-income countries and women in most European countries), the full effect of tobacco smoking on cancer burden is not yet observed.

The notion that genetic susceptibility plays an important role in human cancer is old, and genetic epidemiology studies that have characterized familial conditions entailing a very high risk of cancer have been identified, such as the Li–Fraumeni syndrome and the familial polyposis of the colon, and have identified high-risk cancer genes responsible for these syndromes. However, such high-risk conditions explain only a small fraction of the role of inherited susceptibility to cancer. The remaining fraction of genetic predisposition is likely explained by the combination of common variants in genes involved in one or more steps in the carcinogenic process, such as preservation of genomic integrity, repair of DNA damage. The identification of such low-penetrance susceptibility genes and of their interactions with exogenous factors (so-called gene–environment interactions) represents a challenge to genetic epidemiology.

Advances in molecular biology and genetics offer new tools for epidemiological investigations and have led to the development of new methodological approaches, broadly defined as molecular epidemiology. The application of biomarkers to epidemiology has led to advances in the identification of human carcinogens (e.g., the role of aflatoxin in liver cancer) and in the elucidation of mechanisms of carcinogenesis (e.g., *TP53* mutations in tobacco-related carcinogenesis).

Exposure to most known carcinogens – at least in theory – should be avoided or reduced. This is true in particular for tobacco smoking and chronic infections, the two major known causes of cancer. Tobacco control measures have been implemented in most countries, and effective vaccination is available today against two of the main carcinogenic viruses, hepatitis B and human papilloma. Control of workplace exposure to known and suspected carcinogens in high-resource countries is another example of successful primary prevention of cancer. In many instances, however, primary prevention of cancer would require major changes in lifestyle, which are difficult to achieve.

Detection of preclinical neoplastic lesions before they have developed the full malignant phenotype and notably the ability to metastasize is a highly appealing approach to control cancer. The effectiveness of screening has been demonstrated via epidemiological studies for cervical cancer (cytological smear), breast cancer (mammography), and colorectal cancer (colonoscopy). The development of effective strategies for the early detection of other neoplasms is an active area of research.

Cancer epidemiology exemplifies the strengths and the weaknesses of the discipline at large. Cancer epidemiology has the privilege of using complete and good-quality disease registries available in many populations and covering a broad spectrum of rates and exposures. In many occasions, cancer epidemiology has been the key tool to demonstrate the causal role of important cancer risk factors. The best example is the association between tobacco smoking and lung cancer, which led in the early 1960s to the establishment of criteria for causality in observational research. These findings have brought important regulatory and public health initiatives as well as lifestyle changes in many countries of the world. These epidemiological "discoveries" share two important characteristics: they involve potent carcinogens, and methods are available to reduce misclassification of exposure to the risk factor of interest and to major possible confounders. It has been therefore possible to demonstrate consistently an association in different human populations. Note that it is not necessary for the prevalence of exposure to be high (although this obviously has an impact on the population attributable risk): examples are the many occupational exposures and medical treatments for which conclusive evidence of carcinogenicity has been established on the basis of epidemiological studies conducted in small populations of individuals with well-characterized exposure.

When these conditions are not met, however, the evidence accumulated from epidemiological studies is typically inconsistent and difficult to interpret. The history of cancer epidemiology presents many examples of premature conclusions, which have not been confirmed by subsequent investigations and have damaged the reputation of the discipline. Exposure misclassification, uncontrolled confounding, emphasis of positive findings generated by chance, and inadequate statistical power are the most common limitations encountered in epidemiological studies. Several solutions have been proposed to overcome these problems. First, epidemiological studies should be very large in size. This is achieved either by conducting multicentric studies including thousands of cases of cancer or by performing pooled and meta-analyses of independent investigations. Second, as mentioned above, the use of biological markers of exposure and early effect might contribute to reduce exposure misclassification, increase the prevalence of the relevant outcomes, and shed light on the underlying mechanisms. Finally, guidelines that have been developed to improve and standardize the conduct are report of observational epidemiological studies.

Although relatively young, epidemiology has become a key component in cancer research. Most cancer centers have an epidemiological research group, and cancer is a major subject of research in most academic departments of epidemiology. Epidemiologists are more and more often invited to meetings of clinicians and basic researchers not only to provide an introduction to the distribution and the risk factors of a given cancer but to participate in interdisciplinary discussions on clinical, preventive, or mechanistic aspects of the disease. The strongest cancer epidemiology groups in the world are those combining different lines of expertise, from biostatistics to molecular biology and genetics to medical oncology. Despite its limitations, cancer epidemiology remains one of the most powerful tools at the disposal of the research community to combat cancer at all levels.

Cross-References

▶ Case Control Association Study
▶ Epidemiology of Cancer

References

Doll R, Peto R (2003) Epidemiology of cancer. In: Warrell DA, Cox TM, Firth TD (eds) Oxford textbook of medicine, 4th edn. Oxford University Press, London, pp 193–218

Ferlay J, Soerjomataram I, Ervik M, Dikshit R, Eser S, Mathers C, Rebelo M, Parkin DM, Forman D, Bray, F (2013) GLOBOCAN 2012 v1.0, Cancer Incidence and Mortality Worldwide. International Agency for Research on Cancer, Lyon, France

Last JM (1983) A dictionary of epidemiology. Oxford University Press, New York

STROBE Statement. STrengthening the Reporting of OBservational studies in Epidemiology. http://www.strobe-statement.org/

Cancer Epigenetics

Berna Demircan and Kevin Brown
University of Florida, College of Medicine,
Gainesville, FL, USA

Definition

Epigenetics is defined as chromatin modifications that can alter gene expression, are heritable during cell division, but do not involve a change in DNA coding sequence.

Characteristics

In the context of normal biological processes, epigenetic mechanisms establish regions within the genome containing transcriptionally active (termed euchromatin) and silent (termed heterochromatin) DNA. Further, epigenetic mechanisms are responsible for stably inherited patterns of gene expression such as X chromosome inactivation and genomic imprinting (i.e., selective expression of maternal or paternal alleles). Chromatin modifications that alter gene expression are

---

This entry was first published in the 2nd edition of the Encyclopedia of Cancer in 2009.

both changes to the methylation state of DNA and posttranslational modifications to histone complexes.

It is well recognized that genetic mutations occur in cancer cells and that these events can exert profound and disease-associated changes in gene expression and/or function. However, it is becoming widely accepted that cancer cells also exhibit aberrant epigenetic alterations and that these changes can play a prominent role in disease initiation and progression. Epigenetic changes are potentially as important as genetic mutations in causing cancer since chromatin alterations can exert an influence regional gene expression, thereby changing the transcriptional profile of multiple genes. In this chapter, we summarize the principal epigenetic alterations that occur in cancer cells: regional DNA hypermethylation and ▶ histone modifications and global DNA hypomethylation.

DNA Hypermethylation

Chromatin structure is influenced by cytosine ▶ methylation, the only known naturally occurring base modification in DNA. Cytosine methylation occurs at $5'$-CG-$3'$ dinucleotides (referred to as CpGs) and is catalyzed by a class of enzymes termed DNA methyltransferases (DNMTs). Several DNMTs have been characterized in mammalian cells including DNMT1, DNMT3a, and DNMT3b. These enzymes catalyze the transfer of a methyl group from S-adenosylmethionine (SAM) to the 5-carbon position of cytosine, forming 5-methycytosine. DNMT3a and DNMT3b appear to be principally involved in methylating previously unmodified cytosines (termed de novo methylation). In contrast, DNMT1 preferentially methylates hemimethylated DNA and is thus viewed as the DNA methyltransferase principally responsible for continuation of DNA methylation patterns in daughter cells (termed *maintenance* methylation).

From a statistical standpoint, the human genome is depleted in CpG dinucleotides; however, ∼60% of genes in our genome are associated with regions ranging from 200 to 4,000 bases in length containing high density of CpG dinucleotides relative to the bulk genome. These regions

Cancer Epigenetics, Table 1 Genes subject to epigenetic silencing in cancer

Gene	Function	Tumor types
APC	Regulation of β-catenin, cell adhesion	Colorectal, gastrointestinal
BRCA1	DNA repair	Breast, ovarian
CDH1 (E-cadherin)	Homotypic epithelial cell–cell adhesion	Bladder, breast, colon, liver
MGMT	DNA repair	Brain, colorectal, lung, head, and neck
MLH1	DNA repair	Colorectal, endometrial, ovarian
CDKN2A (p16)	Cell cycle control	Lung, brain, breast, colon, bladder, melanoma prostate
PTEN	Regulation of cell growth and apoptosis	Prostate, brain, endometrial, melanoma
VHL	Inhibits angiogenesis, regulates transcription	Renal cell carcinoma
ATM	DNA damage response	Breast, colorectal, head, and neck

are referred to as ► CpG islands and are usually located within upstream promoter regions or gene transcriptional start sites. In normal somatic cells, gene-associated CpG islands are usually unmethylated and associated with genes in a transcriptionally active euchromatic state. In cancer cells, hypermethylation of such CpG islands is strongly correlated with the transcriptional silencing of genes. Thus, through this epigenetic mechanism, tumor cells can dramatically downregulate expression of numerous genes, including ► tumor suppressor genes (TSGs). At present, numerous TSGs have been characterized as targets for epigenetic silencing through hypermethylation of associated CpG islands. Table 1 is a partial listing of characterized TSG whose promoter regions have been shown to be hypermethylated in various tumor types. Given the wide spectrum of tumor types that display epigenetic silencing of TSGs, mounting evidence clearly supports the assertion that epigenetic silencing is a prominent mechanism driving the process of tumorigenesis.

Cancer cells often overexpress DNMTs. Compared to normal tissues, the expression of DNMT1 is almost always increased in tumors. However, since DNMT1 expression is normally regulated during the cell cycle with increased abundance paralleling entry into S-phase, much of this increased expression may simply reflect increased cell proliferation within the tumor. Although demonstrable in model experiments, it remains unresolved if increased expression of DNMT1 is responsible for aberrant methylation in cancer cells. In contrast, increased expression of DNMT3a and DNMT3b observed in some tumors is likely significant, since these enzymes are normally expressed at low levels in somatic cells. However, it is still unclear to what extent overexpression of these enzymes is responsible for cancer-associated DNA hypermethylation especially when one considers that cancer cells exhibit overall genome hypomethylation (discussed below). Thus, it remains unclear how CpG islands associated with specific TSG are targeted for hypermethylation during the process of tumorigenesis.

One mechanism by which DNA methylation can negatively impact gene expression is by simply blocking the binding of essential transcription factors to gene promoter sequences. While several examples of this are documented, it is also apparent that CpG methylation is also capable of directing transcriptional repression through promoting additional layers of chromatin alteration. Specifically, several proteins have been characterized that bind to methylated CpG dinucleotides and are capable of promoting further chromatin condensation and consequential transcriptional repression through recruitment of chromatin-modifying activities.

Histone Modification

Owing to technical considerations, DNA methylation is the most widely analyzed type of epigenetic alteration in human tumors. However, another extremely important epigenetic modification capable of altering gene expression during carcinogenesis involves various types of histone modifications.

The fundamental packing unit of chromatin within the nucleus is termed the nucleosome.

A single nucleosome unit contains 146 base pairs of DNA wrapped around eight histone subunits (histone octamer). The histone octamer contains two copies each of histones H2A, H2B, H3, and H4. Structural studies have determined that each histone possesses an amino-terminal tail rich in the amino acid lysine. These lysine residues can undergo a variety of posttranslational modifications including acetylation, methylation, phosphorylation, and ubiquitination. Such modifications are recognized by various proteins and protein complexes, and combinations of histone modifications constitute a proposed histone code important in establishing a given gene's transcriptional profile.

Perhaps the best-studied histone modification is acetylation of the ε-amino group of lysine residues within the amino-terminal tail of H3 and H4 although acetylation of both H2A and H2B occurs as well. This is a reversible modification that is carefully controlled by two large enzyme families: histone acetyltransferases (HATs) and histone deacetylases (HDACs). The net positive charges carried by these lysine residues are proposed to contribute to the high affinity of histones for negatively charged DNA. Acetylation of lysine residues by HATs neutralizes this positive charge, thus decreasing histone/DNA interaction. This raises molecular access to DNA and promotes gene transcription. Conversely, HDACs promote transcriptional repression by supporting chromatin condensation into a heterochromatic conformation.

Several proteins that bind specifically to methylated CpG through a conserved methyl-binding domain (MBD) motifs have been discovered. The first such protein to be characterized, termed MeCP2, is capable of recruiting the corepressor molecule mSin3 to the sites of methylated DNA. In turn, mSin3 binds to HDAC1 and HDAC2, thus promoting localized histone deacetylation. The importance of HDAC activity in transcriptional repression is underscored by the observation that repression of several gene promoters can be partially relieved by HDAC inhibitors. A functionally similar complex termed MeCP1 binds to methylated DNA via an associated protein termed MBD2. The MeCP1 complex contains multiple subunits besides MBD2, including components of the NuRD complex, a characterized repressor complex containing both chromatin remodeling and HDAC activities.

In addition to acetylation, lysine residues within histone tails can be methylated and exist in either mono-, di-, or trimethylated states. Similar to the effects of histone acetylation/deacetylation, methylated lysine 4 of H3 (H3K4) is associated with transcriptionally active chromatin, while transcriptionally silent chromatin generally contains methylated lysine 9 of H3 (H3K9). H3K9 is methylated by a number of histone methyltransferases including ESET, Eu-HMTase, G9a, and the closely related methyltransferases SUV39-H1 and SUV39-H2. Histone methylation is likely a dynamic process since a histone demethylase, termed LSD1, was characterized. Methylated H3K9 binds to the chromodomain protein heterochromatin protein 1 (HP1) which promotes heterochromatin formation and gene silencing. Moreover, since H3K9 methylation cannot occur when this position is acetylated, it is clear that H3K9 acetylation and methylation represent opposing forces in determining chromatin conformation.

In a broad view, it is reasonable to propose that CpG methylation and histone acetylation/deacetylation act synergistically in the progressive silencing of genes. One model that accounts for tumor suppressor gene silencing by epigenetic mechanisms invokes abnormal hypermethylation of the promoter CpG island followed by recruitment of MBD proteins, including complexes such as MeCP2 and MeCP1 that recruit HDACs to the area of hypermethylation and promote further transcriptional repression through histone modification. An alternate model proposes that transcriptionally repressive histone modifications are the first event in gene silencing and subsequently promote CpG methylation, resulting in further transcriptional repression. Experimental evidence supports both of these models and may be reflective of the variety of gene promoters and model systems used for study. Less equivocal is the fact that DNA methylation appears as the dominant silencing mechanism since inhibition of DNA methylation generally restores gene expression, while HDAC inhibitors generally exert more modest effects on gene silencing.

DNA Hypomethylation

While CpG islands associated with gene promoters are generally unmethylated in normal adult somatic cells, the majority of CpG dinucleotides elsewhere in the genome are generally methylated. Moreover, despite the fact that many CpG islands are subject to hypermethylation in cancer cells, it is equally well documented that tumor cells display an overall loss of methylated cytosines compared to normal tissue. This tumor-associated global DNA hypomethylation predominantly occurs in repetitive DNA sequences within the human genome although the molecular mechanisms responsible for this loss of DNA methylation are poorly understood.

Recent work on a human genetic disorder has underscored an important role for DNA methylation in maintenance of genome stability. The immunodeficiency, centromeric region instability, facial anomalies (ICF) syndrome is a rare autosomal recessive disease characterized by germline mutation of the DNMT3B gene. Loss of DNMT3B activity in ICF leads to hypomethylation of repetitive satellite DNA sequences within heterochromatin adjacent to the centromeric region of chromosomes. The loss of methylation is most prominent within the pericentric regions of human chromosomes 1 and 16 and leads to multiple chromosomal abnormalities including chromatin decondensation, chromosomal translocations and deletion, and multiradial chromosomal structures. These observations, as well as those made on cells with engineered disruption of DNMTs, clearly support the view that DNA methylation is critical in maintaining normal chromosome structure. Since cancer cells often show chromosomal rearrangements, it is likely that cancer-associated DNA hypomethylation allows for heightened rates of chromosomal instability.

Retrotransposon sequences of the LINE (long interspersed nuclear element) and SINE (short interspersed nuclear element) classes as well as human endogenous retroviruses (HERVs) are major targets of tumor-associated DNA hypomethylation. Mobility of these DNA elements is kept in check in normal tissues owing, in part, to dense methylation of CpG dinucleotides within their genomic structure. It follows that increased mobility of these dormant mobile elements occurs as a result of cancer-associated DNA hypomethylation, and there have been reports of retrotransposition-like insertions involving LINE-1 sequences in tumors although retrotransposition of endogenous elements seemingly occurs more often in rodents than humans.

In addition to the hypomethylation of CpG dinucleotides present within repetitive DNA elements, cancer-associated hypomethylation also occurs in regions of the genome encoding single-copy genes. Dysregulation of allele-specific methylation will result in the loss of imprinting (LOI) and allow for both maternal and paternal gene expression. Perhaps the best-studied example of this is the insulin-like growth factor 2 (IGF2) gene where LOI occurs in primary tumors and in patients with the inherited, cancer-prone ▶ Beckwith–Wiedemann. While not as well-studied as gene silencing due to DNA hypermethylation, it is likely that additional examples of cancer-promoting increased gene expression stemming from DNA hypomethylation will be uncovered in the future.

Epigenetic Alterations as Targets for Diagnosis and Therapeutic Intervention

Sequencing data obtained from the human genome project are currently undergoing analysis to construct a human epigenetic map based on CpG content. This knowledge coupled with cross-species comparisons of the epigenome will be invaluable in deciphering the epigenetic elements involved in gene regulation. Epigenetic alterations typically occur early during the oncogenic process, and detection of such early abnormalities may aid in early diagnosis and/or preventing cancer progression through dietary alterations or pharmacological intervention. With increasing awareness of the importance of epigenetics in tumorigenesis, and the advent of sensitive laboratory approaches to analyze epigenetic alterations, it is likely that epigenetic profiles will ultimately be used in the clinical setting to provide information useful in predicting an individual's predisposition to cancer, assisting in tumor staging, and guiding optimal therapeutic approaches.

A promising feature of alterations in DNA methylation patterns and chromatin structure in cancer cells is their potential for reversibility, because these modifications occur without changing the primary nucleotide sequence. At present, two major pharmacological targets associated with these epigenetic changes are DNMTs and HDACs. The DNMT inhibitor 5-aza-deoxy-cytidine (5-azadC) and related compounds cause transcriptional reactivation of endogenous genes with hypermethylated promoters. This drug, also termed decitabine, is currently used to treat certain types of hematological malignancies, especially advanced ► myelodysplastic syndromes (MDS). HDAC inhibitors, such as trichostatin A and sodium butyrate, have been shown to increase the level of histone acetylation in cultured cells and to cause growth arrest, differentiation, and apoptosis. Based on these observations, the potent HDAC inhibitor suberoylanilide hydroxamic acid (SAHA) is in clinical trials.

Cross-References

- ► Beckwith-Wiedemann Syndrome
- ► Cancer Epigenetics
- ► CpG Islands
- ► Epigenetic
- ► Epigenetic Gene Silencing
- ► Histone Modification
- ► Methylation
- ► Myelodysplastic Syndromes
- ► Tumor Suppressor Genes

See Also

(2012) Acetyltransferase. In: Schwab M (ed) Encyclopedia of cancer, 3rd edn. Springer, Berlin/Heidelberg, p 17. doi:10.1007/978-3-642-16483-5_27

(2012) DNA-methyl-transferases. In: Schwab M (ed) Encyclopedia of cancer, 3rd edn. Springer, Berlin/Heidelberg, p 1147. doi:10.1007/978-3-642-16483-5_1681

Cancer Etiology

► Cancer Causes and Control

Cancer Germline Antigens

Adam R. Karpf
Department of Pharmacology and Therapeutics, Roswell Park Cancer Institute, Buffalo, NY, USA

Synonyms

Cancer-testis antigens; CG antigens

Definition

CG antigens (cancer-testis (CT) antigens) are a class of immunogenic tumor antigens encoded by genes expressed in gametogenic cells of the testis and/or ovary and in human cancer.

Characteristics

Identification

The main criterion for the classification of a gene as a CG antigen pertains to its expression pattern in gametogenic, somatic, and tumor tissues. A gene is generally considered to be a CG antigen if it is expressed in the gametogenic cells of the testis or ovary (including fetal ovary) and in some proportion of human cancers, but is expressed in two or fewer normal somatic tissues. CG antigen genes are also commonly expressed in trophoblast tissue. CG antigens were originally identified from searches for autoantigens expressed in human cancer. The original method for antigen screening used autologous typing, in which T-cells (T-lymphocyte) from a ► melanoma patient were screened for reactivity with tumor cells from the same patient; this method led to the identification of MAGE-A (named for ► melanoma antigens)/CT1 genes. In later studies, another immunological assay was developed to identify tumor antigens and this method, ► SEREX (Serological analysis of recombinant cDNA expression libraries), was used to successfully identify a variety of important CG antigen genes, including NY-ESO-1/CT6 and SSX/CT5. Recognition of the unique expression pattern of CG antigen genes has led to the use of gene expression analyses (including EST or SAGE database searching) to identify CG antigen genes. This method has led to the identification of additional CG genes, including XAGE-1/CT12 and SCP-1/CT8. Although formally classified as CG antigen genes, genes identified by the latter nonimmunological method may not be antigenic in cancer patients.

Nomenclature

A nomenclature system for CG antigens has been devised, which is based on their chronology of discovery, and also accounts for the numerous family members that exist for certain CG antigens. In this system, CG antigen genes are referred to by their original given names and also are assigned a separate CT identifier or CT#. Currently, over 40 CG antigen gene families are recognized, comprising more than 89 distinct mRNA transcripts. CG antigen genes have been assigned into two groups on the basis of chromosomal localization. CG-X antigens: These genes reside on the X-chromosome where, interestingly, close to 10% of the total number of genes encode CG antigens. CG-X genes are typically members of large multigene families, e.g., MAGE-A/CT1, MAGE-B/CT3, and MAGE-C/CT7. In normal tissues, CG-X genes are often expressed in premeiotic spermatocytes in the testis. All of the current important targets of CG antigen ► cancer vaccines are members of this group, including MAGE-A1/CT1.1, MAGE-A3/CT1.3, and NY-ESO-1/CT6.1. Nonseminomatous gene cell tumor: A number of CG antigen genes are located on autosomal chromosomes. Unlike CG-X genes, these genes are highly dispersed in the genome and do not exist in multigene families. In normal tissues, non-X CG genes are often expressed during meiosis, where some members play roles in DNA recombination, including SCP-1/CT8 and

SPO11/CT35. The members of this gene group do not include any currently validated cancer antigens, although certain members are expressed at high levels in cancer.

Regulation of Expression

Certain cancer types appear to expresses CG antigen genes frequently, while others rarely express them. Tumor types that frequently express CG antigens include melanoma, lung, ovary, and ► bladder cancer; tumors that rarely express CG antigens include colon cancer, renal cancer, and leukemia/lymphoma. CG antigen genes show coordinate expression in human cancer. That is, the great majority of tumors either do not express CG antigen genes or express two or more CG antigen genes simultaneously, while relatively few tumors express only one CG antigen gene. Another characteristic of CG antigen gene expression in cancer (revealed by immunohistochemical staining) is that tumors that express CG antigens show heterogeneous expression within the tumor: often only focal staining is observed. The coordinate but heterogeneous expression of CG antigens in cancer has led to the intriguing hypothesis that CG antigen expression is indicative of the activation of a normally dormant gametogenic program in tumor cells (possibly corresponding to tumor stem cells).

The observation of coordinate expression of CG antigen genes suggests that CG antigen gene activation may be controlled by a common molecular mechanism. Supporting this idea, a number of studies have suggested a key role for DNA methylation in regulating CG antigen gene expression. Promoter DNA hypermethylation has been observed to correlate with CG antigen gene repression in normal tissues and nonexpressing tumors, while treatment of tumor cell lines in vitro with DNA methyltransferase inhibitors such as ► 5-aza-2′-deoxycytidine (DAC) leads to CG antigen gene activation, coincident with promoter DNA hypomethylation. Conversely, tumor cell lines and tissues that endogenously express CG antigen genes often display promoter DNA hypomethylation. Many CG antigen genes have CpG-rich promoter regions that serve as targets for regulation by DNA methylation.

Other studies have shown that histone deacetylase (HDAC) inhibitors can either augment DAC-mediated CG antigen gene activation or activate CG antigen genes on their own. As DNA ▶ methylation and chromatin structure (in the form of histone modification status) are intimately linked, it is not surprising that both of these ▶ epigenetic mechanisms serve as important regulators of CG antigen gene expression. Consistent with the model that epigenetic mechanisms regulate CG antigen gene expression is the observation that DNA hypomethylation occurs during gametogenesis, which is the normal setting for CG antigen gene expression.

Function

CG antigens are a rare group of genes in that clinical studies designed to target these antigens for ▶ immunotherapy of cancer are more advanced than is our basic knowledge of the function of the gene products. However, some information about CG antigen gene function has come to light. As mentioned earlier, many non-X CG antigens have roles in germ cell maturation, including mediating the structure of synaptonemal complexes (SCP1/CT8), facilitating DNA recombination during meiosis (SPO11/CT35), and contributing to spermatid function (ADAM2/CT15, OY-TES-1/CT23). In tumors, the function of CG antigen genes is less clear, but studies of the MAGE-type antigens, which share a region referred to as the MAGE homology domain (MHD), indicate that these proteins might serve as transcriptional repressors via interactions with other transcriptional regulatory proteins that themselves recruit corepressors such as HDACs. CG antigen genes have also been reported to play a role in the evolution of ▶ chemotherapy resistance in cancer cell lines, suggesting that the CG antigen gene products could serve as viable targets for anticancer therapy. A report appears to link these two observations by showing that MAGE-A2/CT1.2 disrupts p53 function by recruiting HDAC3 to p53, leading to chemotherapy resistance in cancer cells.

Clinical Studies

The identification of CG antigens as tumor-specific antigens has led to a great deal of interest in treating cancer by targeting CG antigens via vaccine-based immunotherapy. In particular, MAGE-A1/CT1.1, MAGE-A3/CT1.3, and NY-ESO-1/CT6.1 have been developed as targets for this approach. In early studies, the antigenic peptides from CG antigens that elicited T-cell dependent responses were mapped, and these peptides were utilized for vaccination. Because responses to peptide-based vaccine formulations are limited by patient HLA type, vaccination approaches targeting CG antigens have utilized full-length recombinant proteins. These recombinant proteins can be introduced using viral vectors, including vaccinia and fowlpox viruses. Alternatively, recombinant CG antigen proteins can be assembled with adjuvants such as ISOMATRIX, which further enhances immune responses. A common finding in CG antigen vaccine clinical studies is that the treatment is safe and elicits both antibody and T-cell mediated immune responses in vivo. In particular, NY-ESO-1/CT6.1 vaccine trials have shown encouraging results, with durable and multifaceted immune responses, as well as suggestive data indicating clinical benefit, in terms of disease stabilization and prolonged time to recurrence. Many of the patients targeted in these clinical trials have had malignant melanoma, and a proportion of these patients displayed evidence of immune recognition to the target antigen prior to vaccine therapy. In virtually all cases, patients have been selected for inclusion in CG antigen vaccine trials based on the expression of the antigenic target in tumor biopsies. To expand the patient population that would benefit from this ▶ immunotherapy approach, a number of investigators have proposed using DNA methyltransferase and/or HDAC inhibitors (which are FDA approved and known to augment CG antigen gene expression) in combination with CG antigen directed vaccines. The potential benefit of this multimodality approach awaits clinical testing.

Cross-References

▶ 5-aza-2′ Deoxycytidine
▶ Bladder Cancer
▶ Cancer Vaccines

▶ Chemotherapy
▶ Epigenetic
▶ Immunotherapy
▶ Melanoma Antigens
▶ Methylation
▶ SEREX
▶ T-Cell Response
▶ Testicular Cancer

References

Davis ID, Chen W, Jackson H et al (2004) Recombinant NY-ESO-1 protein with ISOMATRIX adjuvant induces broad integrated antibody and CD4+ and CD8+ T cell responses in humans. Proc Natl Acad Sci U S A 101:10697–10702
Scanlan MJ, Gure AO, Jungbluth AA et al (2002) Cancer/testis antigens: an expanding family of targets for cancer immunotherapy. Immunol Rev 188:22–32
Scanlan MJ, Simpson AJG, Old LJ (2004) The cancer/testis genes: review, standardization, and commentary. Cancer Immun 4:1–15
Simpson AJG, Caballero OL, Jungbluth A et al (2005) Cancer/testis antigens, gametogenesis and cancer. Nat Rev Cancer 5:615–625

See Also

(2012) Adjuvant. In: Schwab M (ed) Encyclopedia of Cancer, 3rd edn. Springer Berlin Heidelberg, p 75. doi: 10.1007/978-3-642-16483-5_107
(2012) Antibody. In: Schwab M (ed) Encyclopedia of Cancer, 3rd edn. Springer Berlin Heidelberg, p 208. doi: 10.1007/978-3-642-16483-5_312
(2012) Antigen. In: Schwab M (ed) Encyclopedia of Cancer, 3rd edn. Springer Berlin Heidelberg, p 209. doi: 10.1007/978-3-642-16483-5_319
(2012) Autologous Typing. In: Schwab M (ed) Encyclopedia of Cancer, 3rd edn. Springer Berlin Heidelberg, p 317. doi: 10.1007/978-3-642-16483-5_482
(2012) Chromatin. In: Schwab M (ed) Encyclopedia of Cancer, 3rd edn. Springer Berlin Heidelberg, p 825. doi: 10.1007/978-3-642-16483-5_1125
(2012) DNA Methylation. In: Schwab M (ed) Encyclopedia of Cancer, 3rd edn. Springer Berlin Heidelberg, p 1140. doi: 10.1007/978-3-642-16483-5_1682
(2012) Gametogenic. In: Schwab M (ed) Encyclopedia of Cancer, 3rd edn. Springer Berlin Heidelberg, p 1493. doi: 10.1007/978-3-642-16483-5_2311
(2012) HLA. In: Schwab M (ed) Encyclopedia of Cancer, 3rd edn. Springer Berlin Heidelberg, p 1706. doi: 10.1007/978-3-642-16483-5_2765
(2012) Hypomethylation. In: Schwab M (ed) Encyclopedia of Cancer, 3rd edn. Springer Berlin Heidelberg, p 1791. doi: 10.1007/978-3-642-16483-5_2922

(2012) P53. In: Schwab M (ed) Encyclopedia of Cancer, 3rd edn. Springer Berlin Heidelberg, p 2747. doi: 10.1007/978-3-642-16483-5_4331
(2012) Recombinant. In: Schwab M (ed) Encyclopedia of Cancer, 3rd edn. Springer Berlin Heidelberg, p 3205. doi: 10.1007/978-3-642-16483-5_4991
(2012) T Cell. In: Schwab M (ed) Encyclopedia of Cancer, 3rd edn. Springer Berlin Heidelberg, p 3599. doi: 10.1007/978-3-642-16483-5_5645
(2012) Trophoblast. In: Schwab M (ed) Encyclopedia of Cancer, 3rd edn. Springer Berlin Heidelberg, p 3785. doi: 10.1007/978-3-642-16483-5_5989

Cancer of B-Lymphocytes

▶ B-Cell Tumors

Cancer of the Large Intestine

▶ Colorectal Cancer Clinical Oncology

Cancer of the Lung

▶ Lung Cancer Clinical Oncology

Cancer Prevention

▶ Cancer Causes and Control

Cancer Prevention with Green Tea

▶ Green Tea Cancer Prevention

Cancer Process of the Large Intestine

▶ Colon Cancer Genomic Pathways

Cancer Stem Cell Therapies

► Targeting Cancer Stem Cells

Cancer Stem Cells

► Stem-Like Cancer Cells

Cancer Stem Cells Targeted Drug Development

Galina I. Botchkina
Department of Pathology, Stony Brook
University, Stony Brook, NY, USA
Institute of Chemical Biology and Drug
Discovery, Stony Brook University, Stony Brook,
NY, USA

Definition

Cancer stem cells (CSCs) targeted drug development refers to the treatment strategy focused on the eradication or promotion of differentiation of CSCs and not only on tumor shrinkage.

Characteristics

Cancer Stem Cells

The consensus definition of a cancer stem cell (CSC) is "a cell within a tumor that possesses the capacity to self-renew and to cause the heterogeneous lineages of cancer cells that comprise the tumor," as determined at the AACR workshop in 2006. Cancer stem cells are thus defined by functional test: their ability to generate physiologically relevant tumors after serial transplantation into immunodeficient animals. This central CSC feature is reflected in the alternative terms, such as "tumor-initiating cell" and "tumorigenic cell," sometimes used in the literature to describe putative CSCs. Like normal tissue and embryonic stem cells, CSCs have the ability to undergo both symmetrical self-renewing cell division, producing identical daughter stem cells that retain self-renewal capacity, and asymmetrical self-renewing cell division, resulting in one stem cell and one non-stem committed progenitor cell. A third type of possible stem cell division is symmetrical division producing two non-stem committed progenitor cells, leading to stem cell depletion. It is of great therapeutic value to promote this last form of stem cell division.

Novel Concept of Carcinogenesis Suggests Alternative Paradigm of Cancer Treatment

For the majority of cancer types, tumor regression induced by standard anticancer therapies can lead to only an insignificant increase in the overall survival of patients. The low effectiveness of standard treatment modalities has been attributed to the existence of relatively rare stemlike cells that are highly resistant to therapies and possess the ability to induce and maintain tumor growth and spread. After the first discoveries of CSCs in hematological malignancies and solid tumors, it became increasingly evident that the majority, if not all, of tumors exhibit a hierarchical organization similar to normal tissues: immature pluripotent cells occupy the top of the hierarchy and give rise to the heterogeneous majority of tumor cells at different stages of their maturation (differentiation). To identify novel molecular targets for the development of effective anticancer drugs, the tumor-initiating and drug-resistant subset of cancer cells should be purified and characterized. CSC research is rapidly evolving and holds significant promise for the development of the next generation of anticancer therapeutics. Although presently there are no highly selective tools for the isolation of a pure population of CSCs, several cell surface markers or their combinations allow for initial enrichment and purification of the tumor-initiating cells. Thus, it was demonstrated that a single epithelial cell with a particular CSC-relevant phenotype can regenerate a whole tumor in an animal model. It is important to test the stemness state of all candidate cell subpopulations for their ability to serially induce tumors in immunodeficient animals after transplantation of a low number of cells of a particular phenotype.

Numerous studies on different cancer types have demonstrated that the tumorigenic cells expressing common stemness markers, in particular CD133 and CD44, are not only exceptionally resistant to conventional anticancer drugs, such as 5-FU, oxaliplatin, irinotecan, docetaxel and others, but may also significantly increase in number after treatment. Such an enrichment of the immature stemlike cells is usually manifested as a more drug-resistant and more aggressive recurrence or metastatic disease. In many cancers, the ratio of CSCs correlates with tumor aggressiveness, histologic grade, poor prognosis, and distant metastasis. All of the above indicate that purified CSCs represent a critical source for the identification and validation of novel molecular targets and the development of effective anticancer therapies.

Potential Molecular Targets

Targeting CSC Signaling Pathways

Theoretically, any molecule that is in high demand in CSCs and any step in stemness-relevant signaling pathways can potentially provide an opportunity for therapeutic intervention. However, the network of mechanisms that regulate stem cells and CSC renewal and carcinogenesis is extremely complex and not completely known. In many types of human cancers, CSCs express the majority of genes and transcription factors characteristic of embryonic stem cells. Thus, they possess the upregulated levels of several key markers of pluripotent stem cells, including Sox2, Oct-4, and c-Myc, compared to differentiated cancer cells. In a variety of systems, such as embryonic stem cells, induced pluripotency cells, normal adult tissue stem cells, and CSCs, a number of critical pathways are conserved, including Wnt, Notch, Hedgehog (Hh), TGFβ/BMP, JAK/STAT, FGF/MAPK/PI3K, and others. At least three of them, including Wnt, Notch, and Hh, are frequently deregulated in cancer (numerous reviews analyzed these pathways). Many signaling pathways have redundant functions and consist of multiple receptors, ligands, kinases, signal transducers, and transcription factors, which require complex multidisciplinary study. Nevertheless, several potential molecular targets have been

identified. In general, inhibition/modulation of signal transduction pathways can include (a) disruption of the receptor/ligand interaction; (b) abrogation of the cytosolic downstream cascade; and/or (c) abrogation of the nuclear signaling components using monoclonal antibodies, small molecules, and natural phytochemicals. Several CSC-targeted approaches to enhance responsiveness to systemic therapy are presently under development. They include: (i) CSC ablation using antitumor agents such as monoclonal antibodies, small molecules, engineered oncolytic viruses, or activated immune cells; (ii) blockade of CSC function; (iii) reversal of CSC resistance; (iv) CSC-directed differentiation therapy; and (v) targeting the CSC environment. Numerous preclinical therapeutics, as well as clinical trials involving Notch, Wnt and Hh pathway inhibitors, several kinase inhibitors, immunotherapy, and molecular chaperons, are currently under evaluation by many pharmaceutical companies and research laboratories. However, several major weaknesses limit their effectiveness, including poor penetration of solid tumors, inability to cross the blood–brain barrier, and systemic toxicities. Accumulated data indicate that successful normalization of the deregulated activities of multiple signaling pathways requires multiplex, combinatorial therapeutics involving protective natural phytochemicals with diverse, pleiotropic anticancer activities rather than single cytotoxic agents.

Targeting CSC Resistance to Therapies

Stem cells in general are evolutionary predisposed to be resistant to any unfavorable conditions, including lack of oxygen, deficit of nutrients, and presence of cytotoxic agents. Induction of chemo and radioresistance is mediated, in part differentially, through many target genes regulated and orchestrated by multiple transcription factors. CSC resistance to conventional chemo and radiation therapies is associated with the deregulation or sustained activation of multiple developmental pathways, including increased Wnt/b-catenin and Notch signaling, upregulation of antiapoptotic Bcl-2 family members, downregulation of proapoptotic machinery, and others. Resistance of CSCs to chemotherapy has been also attributed to the high levels of expression of multidrug ABC transporters family genes, resulting in a more efficient efflux of chemotherapeutic drugs, and the expression of multidrug resistance genes, including MDR1 and BRCP1. In addition, the expression of high cytoplasmic levels of aldehyde dehydrogenase (ALDH) enzyme activity, which inactivates the bioactive metabolic byproducts, as well as the relatively slow proliferation rate of CSCs (quiescence), allows them to escape toxicity by drugs in general and toxicity by cell cycle-dependent drugs targeting rapidly dividing cells in particular. CSC resistance to radiation is attributable to amplified DNA damage repair, although other common mechanisms of resistance can contribute significantly to the insensitivity of CSC to radiotherapy.

Targeting Angiogenesis

The contribution of CSCs to angiogenesis is a relatively new but actively investigated field. It was found that CSCs in brain tumors closely interact with endothelial cells and can promote the growth of new vessels. Moreover, CSCs can directly differentiate into endothelial cells, thereby generating the necessary vasculature for secondary tumors. On the other hand, when CSCs are transplanted together with endothelial cells, it enhances their stemness state and accelerates tumor initiation and growth. Although targeting tumor vasculature with conventional anti-VEGF therapies seems reasonable, it has shown only limited survival benefit. This data suggests that a combination of VEGF production suppression with specific targeting of CSCs might be more effective.

Targeting CSCs with Phytochemicals (Nutraceuticals)

Phytochemicals (also known as phytonutrients, plant compounds believed to have health-protecting qualities) and nutraceuticals (natural foods or supplements with therapeutic effects) are becoming increasingly common as a form of alternative or complementary anticancer therapy. It is well known that cancer rates in Asia are

significantly lower compared to those in Western countries, suggesting the existence of some epidemiological factors such as diet. Growing clinical and experimental data suggest that several natural phytochemicals, including curcumin, green tea, and soy isoflavones, which are consumed at high rates in Asia, exert multiple anticancer effects. As a rule, naturally derived compounds exert a fairly large spectrum of molecular mechanisms underlying their physiological activities. In particular, these agents induce antiproliferative, anti-invasive, anti-angiogenic, and proapoptotic effects on human cancer cells in vivo and in vitro without cytotoxic effects on healthy cells. For example, the pleiotropic effects of curcumin can be explained in part by its ability to scavenge, as well as to generate reactive oxygen species (ROS), induce apoptosis via rapid generation of ROS, and downregulate the expression of multiple anti-apoptotic proteins. Importantly, there is accumulating data that it can inhibit multiple stem cell-relevant signaling pathways, decrease the proliferative potential of CSCs, and/or induce their differentiation. Several clinical trials showed improved efficacy of conventional anticancer drugs, including 5-FU, dasatinib, and oxaliplatin, when used in combination with curcuminoids. Since curcumin has low biological availability and bioactivity, many laboratories are presently focused on the development of chemically modified or synthetic phytochemicals with improved characteristics. In this context, pan-inhibitory activities of the combination of new-generation toxoids (such as SBT-1214) with synthetic analogs of curcumin (such as CMC2.24) against multiple stemness-relevant genes and transcription factors are highly promising. Other phytochemicals, such as resveratrol, cyclopamine, piperine, and the combination of piperine with curcuminoids, also exert pleiotropic anticancer and CSC-targeted activities. Cyclopamine is a naturally occurring teratogenic steroidal alkaloid isolated from the corn lily (*Veratrum californicum*). Cyclopamine inhibits Hedgehog signaling at the level of Smoothened. The effect of cyclopamine was uncovered when lambs fed to ewes grazing on *Veratrum californicum* were born with cyclopia (single eye). A similar phenotype results from mutations in Sonic hedgehog (Shh). Resveratrol, a natural polyphenol found in grapes, berries, red wine, and peanuts, may inhibit the proliferation and tumorigenicity of the CD133$^+$ cells in vitro and in vivo and may enhance sensitivity to radiotherapy. In particular, the anticancer properties of resveratrol are realized by inhibiting the Notch pathway and induction of apoptosis via Fas-, p53-, and p21-mediated pathways. The steroidal alkaloid cyclopamine affects CSCs via downregulation of the Hh signaling. However, since each signaling pathway may contain multiple ligands with redundant functions, the combinatorial targeting of several pathways is more likely to successfully eliminate CSCs. Thus, neither inhibition of the Shh pathway with cyclopamine nor inhibition of mTOR signaling with rapamycin alone, but only a combination of inhibitors of these pathways, could deplete the CSCs pool. Importantly, in vivo studies have demonstrated that combined therapy with cyclopamine, rapamycin, and gemcitabine was well tolerated and resulted in tumor-free, long-time survival.

CSC-Targeted Preclinical Evaluation of the Anticancer Agents

Preclinical evaluation of the candidate anticancer agents is traditionally based on the use of two-dimensional cultures of the total, unselected cancer cells. However, this in vitro model ignores rare yet functionally significant, highly drug-resistant tumor-initiating cells and thereby has low relevance to the complexity and pathophysiology of in vivo tumors. Monolayer cultures have unnatural cell-to-cell and cell-to-matrix contacts, which can significantly affect their phenotype, signal transduction pathways, and drug response. Since they are directly exposed to medium content and to oxygen, which is a key signal for many important biological processes including stem cell self-renewal, apoptosis, differentiation, and migration, studies on two-dimensional cancer cell cultures may lead to inaccurate conclusions. This represents one of the reasons for the high rate of attrition of candidate anticancer agents. Thus, only 5% of agents that have anticancer activity in preclinical development are licensed. In addition,

even highly purified CSCs can undergo relatively fast differentiation after being placed in standard adherent culturing conditions. An alternative three-dimensional model of floating cancer spheroids was established by Sutherland and colleagues long before the isolation and characterization of CSCs. This model is largely used in cancer research, since it is more closely related to original tumors than to the cancer cell monolayers with respect to cell morphology, metabolic and proliferative gradients, oxygen and drug penetration, cell–cell junctions, kinase activation, and other parameters. Floating cancer spheroids are organized hierarchically; can be passaged for many generations, which suggests that they contain a population of cells with extensive self-renewal capacity; coexpress multiple CSC markers, including CD133, CD166, CD44, CD24, CD29, Msi-1, Lgr5, and nuclear localization of β-catenin; and possess an increased resistance to chemo and radiotherapy.

Therefore, it is conceivable that CSCs represent the most crucial target in the development of a new generation of anticancer drugs and that the search for effective therapeutic interventions should be focused on the evaluation of the status of cancer-specific tumor-initiating cells and not only on the bulk cancer cells or tumor shrinkage. However, the adequate cultivation and efficient propagation of CSCs in vitro and in vivo is critical for studying CSC biomolecular characteristics, as well as for high-throughput drug screening and rational drug development based on novel CSC markers and signaling pathways. Accumulated data indicate that early passage patient-derived mice tumor xenografts and early passage floating spheroids induced by purified patient-derived cancer-specific CSCs may be relatively appropriate models for CSC-targeted research and drug development.

Conclusion

Selective molecular targeting of cancer stem cells is a rapidly developing field in oncology and holds high promise for the efficient eradication of cancer. Progress in this direction depends on a better understanding of CSC biology and carcinogenesis, prompt and efficient translation of scientific findings into clinical research, and adequately organized clinical trials. The extreme heterogeneity of human cancers requires the establishment of patient-derived low-passage CSC lines and CSC-relevant in vivo and in vitro models, critical for the identification of novel clinically relevant molecular targets. On the other hand, new criteria for the efficacy of anticancer therapies should be based on comparative, multidisciplinary analyses of the posttreatment status of CSCs and CSC-relevant signaling pathways, as well as long-term follow-up for possible tumor recurrence, and not exclusively on the achievement of temporary tumor shrinkage.

References

Botchkina G (2013) Colon cancer stem cells–from basic to clinical application. Cancer Lett 338(1):127–140. doi: 10.1016/j.canlet.2012.04.006

Botchkina GI, Zuniga ES, Das M, Wang Y, Wang H, Zhu S, Savitt AG, Rowehl RA, Leyfman Y, Ju J, Shroyer K, Ojima I (2010) New-generation taxoid SB-T-1214 inhibits stem cell-related gene expression in 3D cancer spheriods induced by purified colon tumor-initiating cells. Mol Cancer 9:192–204

Clarke MF, Dick JE, Dirks PB, Eaves CJ, Jamieson CHM, Jones DL, Visvader J, Weissman IL, Wahl GM (2006) Cancer stem cells: perspectives on current status and future directions. AACR workshop on cancer stem cells. Cancer Res 66:9339

Harris PJ, Speranza G, Ullmann CD (2012) Targeting embryonic signaling pathways in cancer therapy. Expert Opin Ther Targets 16:49–66

Takebe N, Harris PJ, Warren RQ, Ivy SP (2011) Targeting cancer stem cells by inhibiting Wnt, Notch, and Hedgehog pathways. Nat Rev Clin Oncol 8:97–106

Zhang Y, Gu Y, Lee H-M, Hambardjieva E, Vranková K, Golub LM, Johnson F (2012) Design, synthesis and biological activity of new polyenolic inhibitors of matrix metalloproteinases: a focus on chemically-modified curcumins. Curr Med Chem 19:4348

See Also

(2012) Smoothened. In: Schwab M (ed) Encyclopedia of cancer, 3rd edn. Springer, Berlin/Heidelberg, p 3456. doi:10.1007/978-3-642-16483-5_5383

(2012) Sonic hedgehog. In: Schwab M (ed) Encyclopedia of cancer, 3rd edn. Springer, Berlin/Heidelberg, p 3471. doi:10.1007/978-3-642-16483-5_5420

(2012) Teratogenic. In: Schwab M (ed) Encyclopedia of cancer, 3rd edn. Springer, Berlin/Heidelberg, p 3651. doi:10.1007/978-3-642-16483-5_5731

Cancer Stem-Like Cells

Gaetano Finocchiaro
Unit of Experimental Neuro-Oncology, Istituto Nazionale Neurologico Besta, Milan, Italy

Synonyms

Tumor-initiating cells

Definition

Cancer stem(-like) cells are those cells that possess the capacity for self-renewal and for causing the heterogeneous lineages of cancer cells that comprise the tumor.

Characteristics

The definition follows a consensus at a workshop on cancer stem(-like) cells (CSC) organized by the American Association for Cancer Research (AACR). There is considerable debate and some controversy on the CSC concept, so that a consensus definition is required. The importance of the debate is proportional to its relevance to the change in our perception of cancer, intrinsic to the CSC paradigm, implying that not all cancer cells are equal but that only a small fraction of them is endowed with the properties of perpetuating the disease. This hierarchical model has not only important biological consequences but also relevant therapeutic implications, as we discuss in this essay.

The CSC paradigm fits in a model of cancer as a caricature of an organ that is already present in the literature as suggested by data published 30–40 years ago. In particular, Hamburger and Salmon established growth conditions for cancer cells in soft agar medium and found that tumor stem cell colonies, arising from different types of cancer with 0.001–0.1% efficiency, had differing growth characteristics and colony morphology. Studies by Dick and coworkers in the 1990s showed that in several forms of acute myeloid leukemia (AML) cells that could engraft in immunodeficient mice are restricted to a minority subpopulation defined as [CD34$^+$/CD38^{neg}]: these cells, therefore, shared a cell surface phenotype with normal human primitive hematopoietic progenitors, suggesting that they may have originated from normal stem cells rather than from committed progenitors. Also of interest was the observation that leukemic cells engrafted in NOD-SCID mice (nonobese diabetic-severe combined immunodeficiency: an immunodeficient mouse strain characterized by lack of B, T, and NK lymphocytes) showed similar phenotypic heterogeneity to the original donor: thus, [CD34$^+$, CD38^{neg}] retain the differentiating capacity necessary to give rise to CD38$^+$ and Lin$^+$ cells (lineage positive).

The presence of CSC has also been demonstrated in chronic myeloid leukemia (CML). This disease has a chronic phase and a terminal stage; the blast crisis and molecular events underlying this evolution are not completely understood. In the chronic phase, the chromosomal translocation t(9:22)BCR-ABL, a diagnostic marker of CML, can be detected in most circulating mature lineages. In the blast crisis, however, highly undifferentiated BCR-ABL$^+$ cells accumulate in the blood. In particular, an expansion of granulocyte-macrophage progenitors (GMP) is present in blast cells, showing aberrant acquisition of self-renewal properties and nuclear expression (i.e., activation) of beta-catenin, a key, positive regulator of stem cell self-renewal. These observations imply that during progression of CML the GMP subfraction of leukemic progenitors acquire stem cell characteristics. Thus, the functional hierarchy of CSC can be modified during the natural history of this tumor as a result of its progression.

The requirement for a periodical renovation is not only present in blood but also in the skin and epithelia of the respiratory, gastrointestinal, reproductive, and genitourinary systems. Other tissues like brain, previously considered as exclusively post-mitotic, contain stem cells that can be mobilized and activated under conditions of stress, such as hypoxia. Thus the CSC model could also be applied to solid tumors, and a series of papers

Cancer Stem-Like Cells, Table 1 Molecular markers of cancer stem-like cells

Tumor	Markers	References
Acute myeloid leukemia	CD34$^+$/CD38neg	Bonnet and Dick (1997)
Breast cancer	CD44$^+$/CD24-/neg	Al-Hajj et al. (2003)
Glioblastoma	CD133$^+$	Singh et al. (2003)
Myeloma	CD138neg	Matsui et al. (2004)
Prostate cancer	CD44$^+$/alpha2beta1 integrin high/CD133$^+$	Collins et al. (2005)
Melanoma	CD20+	Fang et al. (2005)
Lung cancer	Sca-1$^+$/CD45neg/Pecam neg/CD34pos	Kim et al. (2005)
Colon cancer	CD133$^+$	O'Brien et al. (2007), Ricci-Vitiani et al. (2007)
Pancreatic cancer	CD44$^+$/CD24$^+$/ESA (epithelial specific antigen)$^+$	Li et al. (2007)

report data supporting the identification of a stem cell population in different cancers (see Table 1). Initial data were gained in breast cancer where a small population of cells with a CD44$^+$/CD24$^{neg-low}$ phenotype appears exclusively capable of tumor initiation.

The most malignant of brain tumors, glioblastoma multiforme (GBM), was also found to contain a fraction of neoplastic cells identified and selected on the basis of CD133 expression. Not only could CD133$^+$ cells self-renew and differentiate into different neural lineages but also, in vivo, only the CD133$^+$ cells were able to reinitiate malignant gliomas with phenotype similar to the original tumor.

The CSC paradigm may also help explain intratumor heterogeneity, a frequent finding in most cancers: heterogeneity could be consequent to functional diversity of cells at different states of differentiation. On the other hand, the patterns of tumor heterogeneity and gene expression profiles can be highly similar in the original tumor and in distant metastasis.

Cancer Stem-Like Cells, Table 2 Therapeutic potential of cancer stem-like cells

Pathway/mechanism in CSC	Potential treatment	References
Angiogenesis-increased production of VEGF	Bevacizumab	Bao et al. (2006) Cancer Res
Increased resistance to radiation	Chk1 and Chk2 inhibitor	Bao et al. (2006) Nature
Specific patterns of expression	Dendritic cell targeting	Pellegatta et al. (2006)
Cell cycle deregulation	Bone morphogenetic protein 4	Piccirillo et al. (2006)
Resistance to chemotherapy	BCRP1/ MGMT inhibition (?)	Liu et al. (2006)

It is conceivable that the existence of cancer stem cells may provide novel therapeutic targets of increased effectiveness in contrasting or even eliminating cancer. Brain tumors have provided a highly fertile ground to start verifying this hypothesis, as outlined in Table 2. Data are piling up indicating that CD133$^+$ GBM CSC are highly proangiogenic, because of the high levels of VEGF expression, and have greater resistance to chemotherapy and radiotherapy. As a consequence, specific therapeutic strategies can be attempted and combined to overcome CSC. Upon radiotherapy CD133$^+$ GBM CSC activates checkpoint kinases 1 and 2 and repair mechanisms more effectively than CD133^{neg} cells. Resistance to chemotherapy can be linked to an intriguing aspect of the CSC phenotype, the side population (SP) phenotype. SP cells have the ability to extrude the DNA binding dye Hoechst 33342 via the drug transporter BCRP1/ABCG2. Interestingly, the BCRP1/ABCG2 pump can also effectively extrude chemotherapeutic drugs such as mitoxantrone.

Also related, although of less immediate relevance in the clinical setting, are the observations reported by Pellegatta et al. using glioma neurospheres as a target for dendritic cell (DC, the most potent of antigen presenting cells) immunotherapy. Normal neural stem cells may grow as neurospheres (NS) in the absence of serum and in the presence of two critical growth factors, EGF and bFGF. NS are enriched in neural stem cells but also contain partially committed progenitors as well as a differentiated progeny. Oncospheres with similar characteristics were obtained from GBM but also from other solid tumors like breast or colon carcinomas. Pellegatta et al. set up a murine model showing that DC loaded with GBM NS are much more effective in protecting mice against the GBM challenge than DC loaded with GBM cells where CSC are poorly represented. Thus, CSC targeting by immunotherapy is feasible and highly effective, opening new scenarios for clinical immunotherapy and supporting the idea that CSC are at the heart of malignant growth. Also of interest is the observation by Piccirillo et al. that treating GBM CSC with the differentiating factor BMP4 can block growth in vitro and avoid tumor formation in the majority of mice in vivo.

Given the increasing number of observations supporting the CSC paradigm in different tumors, it is expected that more therapeutically relevant observations will be proposed in the near future.

Together with therapeutic and clinical implications, the CSC concept seems to have important consequences for our understanding of tumor biology. Modern genetics and molecular biology have given a definition of cancer as a genetic disease in which a growing burden of mutations leads to a progressively more aggressive and ultimately lethal phenotype (Fig. 1). A Darwinian selection for these mutations, privileging those that can confer resistance to different challenges, like hypoxic stress or immune attack, appears to be the most plausible rationale for making sense of this evolutionary catastrophe. The hierarchical CSC model seems to introduce an element of rigidity in this highly flexible scenario, implying that only cells endowed with stem cell properties can afford tumor perpetuation (Fig. 1). Are these two models different or are they compatible? A convincing answer to this tough question will undoubtedly require a lot of robust science in the time to come but comments can be given on the basis of data that are already available. One important issue that the CSC model addresses is that of the cell of origin for cancer(s): stem cells,

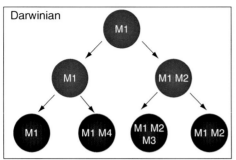
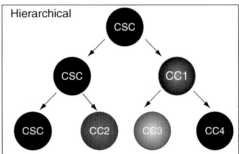
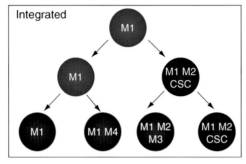

Cancer Stem-Like Cells, Fig. 1 Biological models for tumor evolution

because they are long-lived and self-renewing, are excellent candidates to play the "cell of origin" role. A stem cell hosting a critical mutation could be quiescent for years and then be engaged in a repair response requiring mobilization and proliferation. For example, hypoxic stress may activate the CXCR4 pathway that not only attracts stem cells but may also favor their proliferation, thus being the spark initiating the cancer fire. However, an initiating mutation could also arise in a more committed progenitor (see the integrated model in Fig. 1): acquisition of a stem-like phenotype could in this context be the consequence of environmental challenges; in vivo, for instance, hypoxia could play an important role in dedifferentiation; in vitro, the modification of growth factors could have similar consequences. Epigenetic changes could play important roles in mediating rapid and genome-wide changes that can substitute for genetic mutations and lead to dedifferentiation.

In the *Darwinian model*, different mutations (M1 through M4) accumulate during evolution and confer heterogeneity.

In the *Hierachical model*, tumor arises in a stem cell, thus becoming a cancer stem cell (CSC): heterogeneity is conferred by asymmetrical divisions creating different types of cancer cells (CC1 through 4).

In the *Integrated model*, a first mutation (M1) can arise in a progenitor or even a committed cell. During progression, though, external stimuli may give rise to a cancer stem cell that through asymmetric division will create other CSC as well as more differentiated tumor cells.

Cross-References

▶ Stem-Like Cancer Cells

References

Al-Hajj M, Wicha MS, Benito-Hernandez A, Morrison SJ, Clarke MF (2003) Prospective identification of tumorigenic breast cancer cells. Proc Natl Acad Sci U S A 100:3983–3988

Bao S, Wu Q, Sathornsumetee S, Hao Y, Li Z, Hjelmeland AB, Shi Q, McLendon RE, Bigner DD, Rich JN (2006) Stem cell-like glioma cells promote tumor angiogenesis through vascular endothelial growth factor. Cancer Res 66:7843–7848

Bonnet D, Dick JE (1997) Human acute myeloid leukemia is organized as a hierarchy that originates from a primitive hematopoietic cell. Nat Med 3:730–737

Clarke MF, Dick JE, Dirks PB et al (2006) Cancer stem cells – perspectives on current status and future directions: AACR workshop on cancer stem cells. Cancer Res 66:9339–9344

Collins AT, Berry PA, Hyde C, Stower MJ, Maitland NJ (2005) Prospective identification of tumorigenic prostate cancer stem cells. Cancer Res 65:10946–10951

Dalerba P, Cho RW, Clarke MF (2007) Cancer stem cells: models and concepts. Annu Rev Med 58:267–284

Fang D, Nguyen TK, Leishear K, Finko R, Kulp AN, Hotz S, Van Belle PA, Xu X, Elder DE, Herlyn M (2005) A tumorigenic subpopulation with stem cell properties in melanomas. Cancer Res 65:9328–9337

Feinberg AP, Ohlsson R, Henikoff S (2006) The epigenetic progenitor origin of human cancer. Nat Rev Genet 7:21–33

Jamieson CH, Ailles LE, Dylla SJ et al (2004) Granulocyte-macrophage progenitors as candidate leukemic stem cells in blast-crisis CML. N Engl J Med 351:657–667

Kim CFB, Jackson EL, Woolfenden AE, Lawrence S, Babar I, Vogel S, Crowley D, Bronson RT, Jacks T (2005) Identification of bronchioalveolar stem cells in normal lung and lung cancer. Cell 121:823–835

Li C, Heidt DG, Dalerba P, Burant CF, Zhang L, Adsay V, Wicha M, Clarke MF, Simeone DM (2007) Identification of pancreatic cancer stem cells. Cancer Res 67:1030–1037

Liu G, Yuan X, Zeng Z, Tunici P, Ng H, Abdulkadir IR, Lu L, Irvin D, Black KL, Yu JS (2006) Analysis of gene expression and chemoresistance of CD133+ cancer stem cells in glioblastoma. Mol Cancer 5:67

Matsui W, Huff CA, Wang Q, Malehorn MT, Barber J, Tanhehco Y, Smith BD, Civin CI, Jones RJ (2004) Characterization of clonogenic multiple myeloma cells. Blood 103:2332–2336

O'Brien CA, Pollett A, Gallinger S, Dick JE (2007) A human colon cancer cell capable of initiating tumour growth in immunodeficient mice. Nature 445:106–110

Pellegatta S, Poliani PL, Corno D, Menghi F, Ghielmetti F, Suarez-Merino B, Caldera V, Nava S, Ravanini M, Facchetti F et al (2006) Neurospheres enriched in cancer stem-like cells are highly effective in eliciting a dendritic cell-mediated immune response against malignant gliomas. Cancer Res 66:10247–10252

Piccirillo SGM, Reynolds BA, Zanetti N, Lamorte G, Binda E, Broggi G, Brem H, Olivi A, Dimeco F, Vescovi AL (2006) Bone morphogenetic proteins inhibit the tumorigenic potential of human brain tumour-initiating cells. Nature 444:761–765

Ricci-Vitiani L, Lombardi DG, Pilozzi E, Biffoni M, Todaro M, Peschle C, De Maria R (2007) Identification and expansion of human colon-cancer-initiating cells. Nature 445:111–115

Sanai N, Alvarez-Buylla A, Berger MS (2005) Neural stem cells and the origin of gliomas. N Engl J Med 353:811–822

Singh SK, Clarke ID, Terasaki M, Bonn VE, Hawkins C, Squire J, Dirks PB (2003) Identification of a cancer stem cell in human brain tumors. Cancer Res 63:5821–5828

Cancer Vaccines

Malaya Bhattacharya-Chatterjee[1], Sunil K. Chatterjee[1], Asim Saha[1] and Kenneth A. Foon[2]
[1]University of Cincinnati and The Barrett Cancer Center, Cincinnati, OH, USA
[2]The Pittsburgh Cancer Institute, Pittsburgh, PA, USA

Definition

A vaccine should activate a unique lymphocyte (B and/or T cell) response, which has an immediate antitumor effect as well as memory response against future tumor challenge (Fig. 1). The primary role of a cancer vaccine is the treatment of cancer or prevention of recurrence in a patient with surgically resected cancer, rather than "prevention" of cancer in a person who has never had cancer. Therefore, cancer vaccines are not thought of in the traditional sense of vaccines that are used for infectious diseases. If the current cancer vaccines prove to be useful in the above respects, then they may have a future role in preventing cancer in persons who have never had cancer but are at high risk for a particular type of cancer.

Characteristics

The first and most obvious types of vaccines are prepared from autologous or allogeneic tumor cells. Autologous refers to a setting in which the donor and recipient are the same person, for instance in blood transfusion or transplantation. Alternatively, membrane preparations from tumor cells may be used. In some instances, tumor cell vaccines have been combined with cytokines such as granulocyte macrophage-colony stimulating factor (GM-CSF) and interleukin-2 (IL-2). With advances in molecular biological approaches, gene modified-tumor cells expressing antigens designed to increase the immune response, or gene modified to secrete cytokines have been an additional tool used in vaccination. In addition,

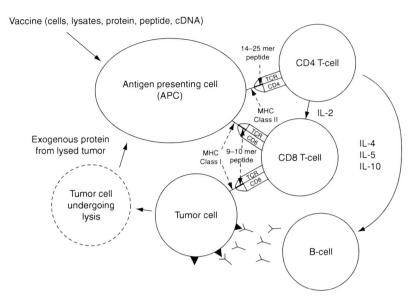

Cancer Vaccines, Fig. 1 T-cell activation. T-cells recognize antigens as fragments of proteins (peptides) presented with major histocompatibility complex (*MHC*) molecules on the surface of cells. The antigen presenting cell processes exogenous protein from the vaccine or from the lysed tumor cell in to a peptide and presents the 14/25 mer peptide to CD4 helper-T-cells on a class II molecule. There is also data that suggests exogenous proteins can be processed into 9/10 mer peptides that may be presented on MHC class I molecules to CD8 cytotoxic T cells. Activated Th1 CD4 helper T-cells secrete Th1 cytokines such as IL-2 that upregulate CD8 cytotoxic T cells. Activated Th2 CD4 helper T-cells secrete Th2 cytokines such as IL-4, IL-5, and IL-10 that activate B cells

increase in our knowledge of tumor associated antigens (TAA) have led to the use of purified TAAs, DNA-encoding protein antigens, and/or protein derived peptides. All of these approaches are currently being tested in the clinic.

Mechanistically, the ultimate aim of a vaccine is to activate a component of the immune system such as B lymphocytes, which produce antibodies or T lymphocytes, which directly kill tumor cells. Antibodies must recognize antigens in the native protein state on the cell's surface. Once bound, these molecules can mediate antibody-dependent cellular cytotoxicity (ADCC) or complement-mediated cytotoxicity, both mechanisms which are capable of destroying tumor cells. ADCC is a passive immune response in which the Fc fragment of a (therapeutic) monoclonal antibody binds or ligates activating immunoglobulin Fc receptors, e.g., Fc RI (CD64), Fc RIIa (CD32a), Fc RIIc (CD32c), or Fc RIII (CD16), present on monocytes, macrophages, granulocytes, and natural killer (NK) cells, driving cytotoxic effector functions to target membrane-associated antigens.

T lymphocytes, on the other hand, recognize proteins as fragments or peptides that vary in size, presented in the context of major histocompatibility (MHC) antigens on the surface of the cells recognized (Fig. 1). The proteins from which the peptides are derived may be cell surface or cytoplasmic proteins. MHC antigens are highly polymorphic, and different alleles have distinct peptide binding capabilities. The sequencing of peptides derived from MHC molecules have led to the discovery of allele-specific motifs that correspond to anchor residues that fit into specific pockets on MHC class I or II molecules.

T Lymphocytes

There are two types of T lymphocytes, helper T lymphocytes and cytotoxic T lymphocytes (CTLs), that recognize antigens through a specific T cell receptor (TCR) in close conjunction to the CD3 molecules, which is responsible for signaling. CD4 helper T cells recognize antigens in association with class II MHC gene products, and CD8 positive CTLs recognize antigens in association with class I MHC gene products. CD4 helper T cells are activated by binding via their TCR to class II molecules that contain 14–25 amino acid peptides in their antigen-binding cleft. Specialized antigen presenting cells (APCs), such as dendritic cells (DCs), macrophages, and B lymphocytes, capture extracellular protein antigens, internalize and process them, and display class II-associated peptides to CD4 helper T cells. The CD8 positive CTLs are activated by binding via their TCR to class I molecules that contain 9–10 amino acid peptides in their antigen-binding cleft. All nucleated cells can present class I-associated peptides, derived from cytosolic proteins such as viral and tumor antigens, to CD8 positive T cells.

There are two types of CD4 helper T cells capable of generating either antibody or cell-mediated immune responses, based on the type of signaling they receive. Th1 CD4 helper T cells stimulate cell-mediated immunity by activating CTLs through the release of cytokines such as IL-2. Th2 CD4 helper T cells mediate an antibody response through the release of cytokines such as IL-4 and IL-10.

Tumor Cells

The most straightforward means of immunization is the use of whole tumor cell preparations (either autologous or allogeneic tumor cells). The advantage of this approach is that the potential TAAs are presented to the immune system for processing and presentation to the appropriate T cell precursors. The difficulty with this approach lies in the availability of fresh autologous tumor material and the scarcity of well-characterized long-term tumor cell lines. Regardless, whole tumor cell vaccines have been an area of intense interest. A variety of trials using autologous tumors for colon cancer and malignant melanoma have been reported. In one trial, freshly thawed autologous colon cancer cells were inactivated with radiation, mixed with BCG (bacille Calmette-Guerin) and injected into patients who had their primary colon cancer resected but were at risk for recurrence. This study did reveal disease-free survival and overall survival trends in favor of the vaccine arm. In a melanoma study, autologous tumor cells were mixed with dinitrophenyl (DNP) and mixed with BCG. Promising results were reported for patients with metastatic disease and for patients with locally resected melanoma.

The weakness of autologous cell vaccines can be overcome with the allogeneic approach: First, an allogeneic vaccine is generic and developed from cell lines selected to provide multiple TAAs and a broad range of HLA expression. Second, allogeneic cells are more immunogeneic than autologous cells. Third, there is no requirement to obtain tumor tissue by surgical resection for a prolonged course of immunotherapy.

A polyvalent melanoma cell vaccine called CancerVax developed for allogeneic viable melanoma cell lines has demonstrated promising results for patients with resected metastatic disease and for resected local disease. Randomized phase III studies are ongoing in the United States comparing CancerVax plus BCG versus BCG for patients with stage III melanoma.

Another variation of cell vaccines is using "shed" antigen vaccines. These are vaccines that are prepared from the material shed by viable tumor cells into culture medium. The potential advantage is that it contains a broad range of antigens expressed on the surface of melanoma cells and the shed antigens are partially purified. Trials of such vaccines in melanoma patients have demonstrated specific humoral and cellular immune responses in patients and promising early clinical results.

Another approach to tumor cell vaccines is the introduction of foreign genes encoding cytokines such as IL-2 and GM-CSF into tumor cells. Alternatively, molecules designed to increase the immunogenicity of the tumor cell such as CD80 and CD86. Gene transfer can be accomplished by transfection of plasmid constructs (electroporation) or transduction using a viral vehicle such as a retrovirus or an adenovirus. Another option tested for gene transfer is physical gene delivery in which a plasmid or "naked" DNA is delivered directly into tumor cells. There are a number of mechanisms to carry this out including liposomes as gene carriers, use of a "gene gun," electroporation and calcium phosphate-mediated gene transfer. In one phase I trial, 21 patients with metastatic melanoma were vaccinated with irradiated autologous melanoma cells engineered to secrete human GM-CSF. Metastatic lesions resected after vaccination were densely infiltrated with T lymphocytes and plasma cells and showed extensive tumor destruction.

Peptides and Carbohydrates

An advantage to peptide vaccines is that they can be synthetically generated in a reproducible fashion. The major disadvantage is that they are restricted to a single HLA molecule and are not of themselves very immunogenic. To increase their immunogenicity, peptides may be injected with adjuvants, cytokines or liposomes or presented on DCs. Whole proteins have the advantage over peptides in that they can be processed for a wider range of MHC class I and II antigens.

Mucins such as MUC I are heavily glycosylated high molecular weight proteins abundantly expressed on human cancers of epithelial origin. The MUC I gene is over-expressed and aberrantly glycosylated in a variety of cancers including colorectal cancer. MUC 1 is being widely used as a focus for vaccine development.

Using expression-cloning techniques, several groups have cloned the genes encoding melanoma antigens recognized by T cells and have identified the immunogenic epitopes presented on HLA molecules. Ten different melanoma antigens have been identified. Direct immunization using the immunodominant peptides from the tumor antigens or recombinant viruses such as adenovirus, fowlpox, and vaccinia virus encoding the relevant genes have been pursued to immunize patients with advanced melanoma. Initial results have demonstrated increased antitumor T cell reactivity in patients receiving peptide immunization. Immunization in melanoma patients with melanoma antigens has been reported. One study showed that immunization of melanoma patients with MAGE-1 peptide pulsed on DCs induced melanoma-reactive and peptide specific CTL responses at the vaccination sites and at distant tumor deposits. Administration of the gp-100 molecule in conjunction with high-dose bolus

IL-2 to 31 patients with metastatic melanoma revealed an objective response of 42%. This is compared with the typical response of high-dose systemic IL-2 without peptide of only 15%. Based on these data, a randomized trial was initiated to compare the peptide vaccine plus IL-2 versus IL-2 alone in metastatic melanoma patients.

Immunization against tumor-associated carbohydrate antigens has also been attempted. Carbohydrate antigens typically bypass T cell help for B cell activation. Investigators demonstrated that some carbohydrates may activate an alternative T cell pathway. Vaccine studies have been reported using the GM-2 ganglioside vaccine. Patients were pretreated with low dose cyclophosphamide. After a minimum follow up of 72 months, there was a 23% increase in disease-free interval and a 17% increase in overall survival in patients who produced antibody against GM-2. This suggested a benefit to the GM-2 ganglioside vaccine which has led to a current phase III trial.

Recombinant Vaccines Expressing Tumor Antigens

The ▶ carcinoembryonic antigen (CEA) is highly expressed on ▶ colorectal cancer and on a variety of other epithelial tumors and is thought to be involved in cell-cell interactions. A recombinant vaccinia virus expressing human CEA (rV-CEA) stimulates specific T cell responses in patients. This was the first vaccine to demonstrate human CTL responses to specific CEA epitopes and class I HLA-2 restricted T-cell mediated lysis, and demonstrated the ability of human tumor cells to endogenously process CEA to present a specific CEA peptide in the context of a MHC for T-cell mediated lysis.

Anti-idiotype Vaccines

The idiotype network offers an elegant approach to transforming epitope structures into idiotypic determinants expressed on the surface of antibodies. According to the network concept, immunization with a given TAA will generate production of antibodies against these TAA, which are termed Ab1; the Ab1 is then used to generate a series of anti-idiotype antibodies against the Ab1, termed Ab2. Some of these Ab2 molecules can effectively mimic the three-dimensional structure of the TAA identified by the Ab1. These Ab2 can induce specific immune responses similar to those induced by the original TAA and therefore can be used as surrogate TAAs. Immunization with Ab2 can lead to the generation of anti-anti-idiotypic antibodies (Ab3) that recognize the corresponding original tumor-associated antigen identified by Ab1. The anti-idiotype antibody represents an exogenous protein that should be endocytosed by APCs and degraded to 14–25 mer peptides to be presented by class II antigens to activate CD4-helper T cells. Activated Th2 CD4-helper T cells secrete cytokines such as IL-4 that stimulate B cells that have been directly activated by Ab2 to produce antibody that binds to the original antigen identified by Ab1. In addition, activation of Th1 CD4-helper T cells secrete cytokines that activate T cells, macrophages, and natural killer cells that directly lyse tumor cells and, in addition, contribute to ADCC. Th1 cytokines such IL-2 also contribute to the activation of a CD8-CTL response. This represents a putative pathway of endocytosed anti-idiotype antibody. The anti-idiotype antibody may be degraded to 9/10 mer peptides to present in the context of class I antigens to activate CD8-cytotoxic T cells, which are also stimulated by IL-2 from Th1 CD4-helper T cells.

Anti-idiotype antibodies that mimic distinct TAAs expressed by cancer cells of different histology have been used to implement active specific immunotherapy in patients with malignant diseases including colorectal carcinoma, malignant melanoma, breast cancer, B cell lymphoma and leukemia, ovarian cancer, or lung cancer. A murine monoclonal anti-idiotype antibody, 3H1 or CeaVac, which mimics CEA was developed by the authors and was used in a phase I clinical trial. Among 23 patients with advanced colorectal cancer, 17 patients generated anti-anti-idiotypic Ab3 responses, and 13 of these responses were proven to be true anti-CEA responses. The median survival of 23 evaluable patients was 11.3 months, with 44% 1-year survival. Toxicity was limited to local swelling and minimal pain. In another clinical trial, 32 patients with resected colorectal cancer were randomized to treatment with CeaVac. All 32 patients entered into this trial generated high-titer IgG anti-CEA antibodies, and ~75% generated CEA specific T cell responses. These data demonstrated that 5-fluorouracil based chemotherapy regimens did not have any adverse effect on the immune response developed by CeaVac. TriGem, an anti-idiotype monoclonal antibody that mimics the disialoganglioside GD2, was used as a vaccine in clinical trial consisting of 47 patients with stage IV melanoma. Forty of 47 patients developed high-titer IgG anti-GD2 antibodies. Seventeen patients were stable on the study from 8 to 34 months. Disease progression occurred in 27 patients on the study from 1 to 9 months. For the 26 patients with soft tissue disease, the median overall survival has not been reached. For 18 patients with visceral metastasis, the median overall survival was 15 months. These results exceed historical controls with stage IV melanoma. Another anti-idiotype monoclonal antibody, TriAb, which mimics the human milk fat globule (HMFG) membrane antigen, is highly overexpressed on breast cancer cells and a variety of other cancer cells, including ovarian cancer, non small-cell lung cancer, and colon cancer. Immunizations with this anti-idiotype antibody elicited both anti-HMFG antibodies and idiotype specific T cell responses in patients with breast cancer in the adjuvant setting as well as in patients with advanced disease following autologous bone marrow transplantation. Although these initial clinical data are promising, active specific immunotherapy with anti-idiotype antibodies need to be tested in combination with other conventional and experimental therapies to overcome the multiple mechanisms by which tumor cells escape immune recognition and destruction. The anti-idiotype vaccine therapy for patients with minimal residual disease might be curative in the adjuvant setting and may improve the quality of patients' life.

Dendritic Cell-based Vaccines

DCs are the professional APCs of the immune system and are present in peripheral tissues, where they capture antigens. These antigens are subsequently processed into small peptides as the DCs mature and move toward the draining secondary lymphoid organs. There the DCs present the peptides to naïve T cells, thereby inducing a cellular immune response that involves both CD4 T helper 1 (Th1) cells and cytotoxic CD8 T cells. DCs are also important at inducing humoral immune response through their capacity to activate naïve and memory B cells. DCs can also activate natural killer (NK) cells and natural killer T (NKT) cells. Therefore, DCs can conduct all of the elements of the immune orchestra, and they are therefore a fundamental target and tool for vaccination.

The development of ex vivo techniques for generating large numbers of DCs in vitro from mouse bone marrow cells supplemented with either GM-CSF alone or GM-CSF plus IL-4 allowed the approach of DC-based tumor vaccination to be fully exploited. Numerous studies in mouse tumor models have shown that DCs pulsed with tumor antigens can induce protective and therapeutic antitumor immunity. In 1996, Hsu et al. reported the first DC-based clinical trial of follicular B cell lymphoma patients who were treated with peripheral blood-derived DCs pulsed with a tumor-specific idiotype (Id) protein. Of these ten patients, eight developed a proliferative cellular response to Id and one patient developed an Id-specific CTL response. However, tumor regression was not reported in these DC-vaccinated patients. In several other trials, a correlation between immunological and clinical outcome has been demonstrated. However, the efficacy of therapeutic DC-based vaccination has been modest and these trials have had similar clinical outcome: mainly, immunized patients often demonstrate significant activation of adaptive immunity to the targeted tumor antigen(s) as shown by various methods such as tetramer analysis, IFN-γ ELISPOT, and ^{51}Cr-release assay; but only a limited number of immunized patients demonstrate significant tumor regression.

The complexity of the DC system requires rational manipulation of DCs to achieve protective or therapeutic immunity. Further research is needed to analyze the immune responses induced in patients by distinct ex vivo generated DC subsets that are activated through different pathways. These ex vivo strategies should help to identify the parameters for in vivo targeting of DCs. Overall, we remain optimistic that improved cancer vaccines will ultimately yield favorable clinical results, particularly after these approaches have been modified in a manner that integrates progress related to the physiology of DCs and our improved understanding of how tumors and the host immune system interact with each other.

Conclusion

There exist several promising immunologic approaches to vaccine therapy of cancer. The challenge of immunotherapy research is to determine which combination of approaches leads to a favorable clinical response and outcome. Several studies have shown enhanced survival of patients receiving vaccines; however, a randomized phase III clinical trial has yet to show a statistically significant improvement in the survival of such patients.

Cross-References

▶ Carcinoembryonic Antigen
▶ Cancer Germline Antigens
▶ Colorectal Cancer
▶ Cytokine Receptor as the Target for Immunotherapy and Immunotoxin Therapy
▶ T-Cell Response

References

Bhattacharya-Chatterjee M, Chatterjee SK, Foon KA (2002) Anti-idiotype antibody vaccine therapy for cancer. Expert Opin Biol Ther 2:869–881

Dalgleish AG, Whelan MA (2006) Cancer vaccines as a therapeutic modality: the long trek. Cancer Immunol Immunother 55:1025–1032

Emens LA (2006) Roadmap to a better therapeutic tumor vaccine. Int Rev Immunol 25:415–443

Nestle FO, Farkas A, Conrad C (2005) Dendritic-cell-based therapeutic vaccination against cancer. Curr Opin Immunol 17:163–169

Saha A, Chatterjee SK, Mohanty K et al (2003) Dendritic cell based vaccines for immunotherapy of cancer. Cancer Ther 1:299–314

See Also

(2012) BCG. In: Schwab M (ed) Encyclopedia of Cancer, 3rd edn. Springer Berlin Heidelberg, p 356. doi: 10.1007/978-3-642-16483-5_560

(2012) FcR. In: Schwab M (ed) Encyclopedia of Cancer, 3rd edn. Springer Berlin Heidelberg, p 1386. doi: 10.1007/978-3-642-16483-5_2135

Cancer Without Disease

► Dormancy

Cancer/Testis Antigen1b

► NY-ESO-1

Cancer-Mediated Bone Loss

► Bone Loss Cancer Mediated

Cancers of Hormone-Responsive Organs or Tissues

► Endocrine Related Cancers

Cancer-Testis Antigen 1.11

► MAGE-A11

Cancer-Testis Antigens

► Cancer Germline Antigens

Candidate of Metastasis 1

► P8 Protein

Canine Transmissible Tumor (CTVT)

► Sticker Sarcoma

Cannabinoids

Guillermo Velasco and Manuel Guzmán
Department of Biochemistry and Molecular Biology I, School of Biology, Complutense University, Madrid, Spain

Synonyms

Endocannabinoids; Marijuana; Phyto-cannabinoids; Synthetic cannabinoids

Definition

Cannabinoids are a family of lipid molecules that comprises a series of metabolites produced by the hemp plant *Cannabis sativa* (the phyto-cannabinoids), several fatty-acid derivatives endogenously produced by most animals (the endogenous ligands for cannabinoid receptors), and different synthetic compounds structurally or functionally related with the natural cannabinoids. Activation of cannabinoid receptors by some of these molecules reduce the symptoms associated to cancer chemotherapy and inhibit the growth of tumor cells in culture and in animal models of tumor xenografts.

This entry was first published in the 2nd edition of the Encyclopedia of Cancer in 2009.

Cannabinoids,
Fig. 1 Cannabinoids, cannabinoid receptors, and their mechanisms of action. (**a**) Δ^9-tetrahydrocannabinol (*THC*), the main active component of marijuana, and the endocannabinoids anandamide and 2-arachidonoylglycerol are ligands of cannabinoid receptors. (**b**) Both CB_1 and CB_2 receptors belong to the family of G-protein-coupled receptors. Binding of cannabinoids to cannabinoid receptors leads, among other actions and depending on the cell context, to: inhibition of adenylyl cyclase, modulation of the activity of several ion channels, modulation of phosphatidylinositol-3 kinase (PI3K) and of mitogen activated protein kinase cascades, or stimulation of ceramide generation

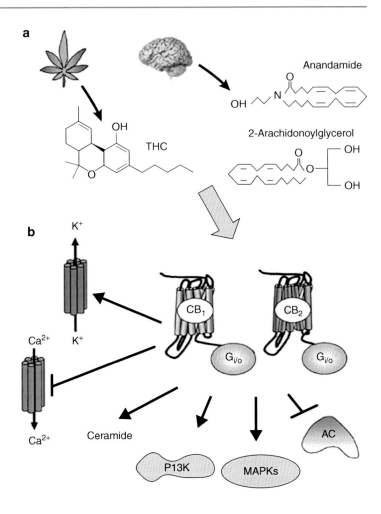

Characteristics

The hemp plant *Cannabis sativa* produces approximately 70 unique compounds known as cannabinoids, of which Δ^9-tetrahydrocannabinol (THC) is the most important owing to its high potency and abundance in cannabis. THC exerts a wide variety of biological effects by mimicking endogenous substances – the endocannabinoids anandamide and 2-arachidonoylglycerol – that bind to and activate specific cannabinoid receptors (Fig. 1a, b). So far, two cannabinoid-specific G-protein-coupled receptors have been cloned and characterized from mammalian tissues: The CB_1 receptor is particularly abundant in discrete areas of the brain, but is also expressed in peripheral nerve terminals and various extraneural sites. In contrast, the CB_2 receptor

was initially described to be present in the immune system, although it has been shown that expression of this receptor also occurs in cells from other origins including many types of tumor cells.

Signaling Pathways Modulated by Cannabinoid Receptors

Most of the physiological, therapeutic, and psychotropic actions of cannabinoids rely on the activation of CB_1 and CB_2 receptors (Fig. 1a, b). Extensive molecular and pharmacological studies have demonstrated that cannabinoids inhibit adenylyl cyclase through CB_1 and CB_2 receptors. The CB_1 receptor also modulates ion channels, inducing, for example, inhibition of N- and P/Q-type voltage-sensitive Ca^{2+} channels and activation of G protein-activated inwardly rectifying K^+ channels. Besides these well-established signaling events that mediate – among others – the neuromodulatory actions of the endocannabinoids, cannabinoid receptors also modulate several pathways that are more directly involved in the control of cell proliferation and survival, including extracellular signal-regulated kinase, c-Jun N-terminal kinase and p38 mitogen-activated protein kinase, phosphatidylinositol 3-kinase/Akt, and focal adhesion kinase. In addition, cannabinoids stimulate the generation of the bioactive lipid second messenger ceramide via two different pathways: sphingomyelin hydrolysis and ceramide synthesis de novo.

Palliative Effects of Cannabinoids in Cancer

Cannabinoids have been known for several decades to exert palliative effects in cancer patients, and nowadays capsules of THC (Marinol-TM) and its synthetic analog nabilone (Cesamet-TM) are approved to treat nausea and emesis associated with cancer chemotherapy. In addition, several clinical trials are testing other potential palliative properties of cannabinoids in oncology such as appetite stimulation and analgesia.

Mechanism Involved in the Antiemetic Effect of Cannabinoids

One of the most important physiological functions of the cannabinoid system is to modulate synaptic transmission. Thus, activation of cannabinoid receptors at presynaptic locations leads to reduced neurotransmitter release. As the CB_1 receptor is present in cholinergic nerve terminals of the myenteric and submucosal plexus of the stomach, duodenum and colon, it is likely that cannabinoid-induced inhibition of digestive tract motility is due to blockade of acetylcholine release in these areas. There is also evidence that cannabinoids act on CB_1 receptors localized in the dorsal vagal complex of the brainstem – the region of the brain that controls the vomiting reflex. In addition, endocannabinoids and their inactivating enzymes are present in the gastrointestinal tract and may play a physiological role in the control of emesis.

Mechanism Involved in Appetite Stimulation by Cannabinoids

The endogenous cannabinoid system may serve as a physiological regulator of feeding behavior. For example, endocannabinoids and CB_1 receptors are present in the hypothalamus, the area of the brain that controls food intake; hypothalamic endocannabinoid levels are reduced by leptin, one of the most prominent anorexic hormones; and blockade of tonic endocannabinoid signaling with the CB_1 antagonist rimonabant – inhibits appetite and induces weight loss. CB_1 receptors present in nerve terminals and adipocytes also participate in the regulation of feeding behavior.

Mechanism Involved in the Analgesic Effect of Cannabinoids

Cannabinoids inhibit pain in animal models of acute and chronic hyperalgesia, allodynia, and spontaneous pain. Cannabinoids produce antinociception by activating CB_1 receptors in the brain (thalamus, periaqueductal gray matter, rostral ventromedial medulla), the spinal cord (dorsal horn), and nerve terminals (dorsal root ganglia, peripheral terminals of primary afferent neurons). Endocannabinoids serve naturally to suppress pain by inhibiting nociceptive neurotransmission. In addition, peripheral CB_2 receptors might mediate local analgesia, possibly by inhibiting the release of various mediators of pain and inflammation, which could be important in the management of cancer pain.

Antitumoral Effects of Cannabinoids

Cannabinoids have been proposed as potential antitumoral agents on the basis of experiments performed both in cultured cells and in animal models of cancer. A number of plant-derived, synthetic, and endogenous cannabinoids are now known to exert antiproliferative actions on a wide spectrum of tumor cells in culture. More importantly, cannabinoid administration to nude mice curbs the growth of various types of tumor xenografts, including lung carcinoma, glioma, thyroid epithelioma, lymphoma, skin carcinoma, pancreatic carcinoma, and melanoma. The requirement of cannabinoid receptors for this antitumoral

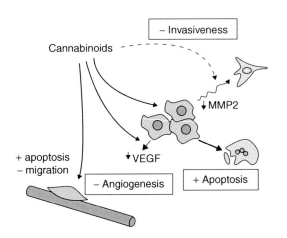

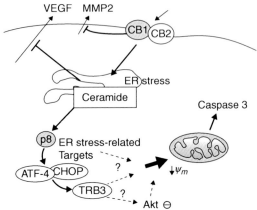

Cannabinoids, Fig. 2 Mechanism of cannabinoid antitumoral action. (**a**) Cannabinoid administration decreases the growth of tumors by several mechanisms, including at least: (i) reduction of tumor angiogenesis, (ii) induction of tumor cell apoptosis, and perhaps (iii) inhibition of tumor cell migration and invasiveness. (**b**) Cannabinoid treatment induces apoptosis of several types of tumor cells via ceramide accumulation and activation of an ER stress-related pathway. The stress-regulated protein p8 plays a key role in this effect by controlling the expression of ATF-4, CHOP, and TRB3. This cascade of events triggers the activation of the mitochondrial intrinsic apoptotic pathway through mechanisms that have not been unraveled as yet. Cannabinoids also decrease the expression of various tumor-progression molecules such as VEGF and MMP2

activity has been revealed by various biochemical and pharmacological approaches, in particular by determining cannabinoid receptor expression in the tumors and by using selective cannabinoid receptor agonists and antagonists.

Although the downstream events by which cannabinoids exert their antitumoral action have not been completely unraveled, there is substantial evidence for the implication of at least two mechanisms: induction of apoptosis of tumor cells and inhibition of tumor angiogenesis (Fig. 2a).

Induction of Apoptosis
Different studies have shown that the proapoptotic effect of cannabinoids on tumor cells relies on the stimulation of cannabinoid receptors and a subsequent activation of the proapoptotic mitochondrial intrinsic pathway. In glioma and pancreatic tumor cells, treatment with cannabinoids leads to accumulation of the proapoptotic sphingolipid ceramide which in turn leads to upregulation of the stress-regulated protein p8, which belongs to the family of HMG-I/Y transcription factors. The acute increase of p8 levels after cannabinoid treatment

triggers a cascade of events that involves the upregulation of several genes involved in the endoplasmic reticulum (ER) stress response including the activating transcription factor 4 (ATF-4) and the C/EBP-homologous protein (CHOP). These two transcription factors cooperate in the induction of the tribbles homologue 3 (TRB3), a pseudokinase that is involved in the induction of apoptosis (Fig. 2b).

The processes downstream of ER stress activation involved in the execution of cannabinoid-induced apoptosis of tumor cells are not completely understood yet but include inhibition of the antiapoptotic kinase Akt and activation of the mitochondrial intrinsic pathway.

Of interest, the proapoptotic effect of cannabinoids is selective of tumor cells. For instance, treatment of primary cultured astrocytes with these compounds does not trigger ceramide accumulation, induction of the aforementioned ER stress-related genes, or apoptosis. Furthermore, cannabinoids promote the survival of astrocytes, oligodendrocytes, and neurons in different models of injury, supporting the notion that cannabinoids activate opposite responses in transformed and nontransformed cells.

Inhibition of Tumor Angiogenesis

To grow beyond minimal size, tumors must generate a new vascular supply (angiogenesis) for purposes of cell nutrition, gas exchange and waste disposal, and therefore blocking the angiogenic process constitutes one of the most promising antitumoral approaches currently available. Immunohistochemical analyses in mouse models of glioma, skin carcinoma, and melanoma have shown that cannabinoid administration turns the vascular hyperplasia characteristic of actively growing tumors to a pattern of blood vessels characterized by small, differentiated, and impermeable capillaries. This is associated with a reduced expression of vascular endothelial growth factor (VEGF) and other proangiogenic cytokines such as angiopoietin-2 and placental growth factor, as well as of type 1 and type 2 VEGF receptors, in cannabinoid-treated tumors. Pharmacological inhibition of ceramide synthesis de novo abrogates the antitumoral and antiangiogenic effect of cannabinoids in vivo and decreases VEGF production by glioma cells in vitro and by gliomas in vivo, indicating that ceramide plays a general role in cannabinoid antitumoral action.

Other reported effects of cannabinoids might be related with the inhibition of tumor angiogenesis and invasiveness by these compounds (Fig. 2a, b). Thus, activation of cannabinoid receptors on vascular endothelial cells in culture inhibits cell migration and survival. In addition, cannabinoid administration to glioma-bearing mice decreases the activity and expression of matrix metalloproteinase-2, a proteolytic enzyme that allows tissue breakdown and remodeling during angiogenesis and metastasis. In line with this notion, cannabinoid intraperitoneal injection reduces the number of metastatic nodes produced from paw injection in lung, breast, and melanoma cancer cells in mice.

Therapeutic Potential of Cannabinoids as Antitumoral Agents

On the basis of these preclinical findings, a pilot clinical study of THC in patients with recurrent glioblastoma multiforme has been run. Cannabinoid delivery was safe and could be achieved without significant psychoactive effects. Also, although the limited number of patients involved in the trial did not permit the extraction of statistical conclusions, median survival of the cohort was similar to other studies performed in recurrent glioblastoma multiforme with temozolomide and carmustine, the drugs of reference for the treatment of these tumors. In addition, THC administration correlated with decreased tumor cell proliferation and increased tumor cell apoptosis.

The significant antiproliferative action of cannabinoids, together with their low toxicity compared with other chemotherapeutic agents and their ability to reduce symptoms associated to standard chemotherapies, might make these compounds promising new antitumoral agents.

Cross-References

▶ Angiogenesis
▶ Apoptosis
▶ Ceramide
▶ Endoplasmic Reticulum Stress
▶ Matrix Metalloproteinases
▶ P8 Protein
▶ Vascular Endothelial Growth Factor

References

Carracedo A, Lorente M, Egia A et al (2006) The stress-regulated protein p8 mediates cannabinoid-induced apoptosis of tumor cells. Cancer Cell 9:301–312

Guzman M (2003) Cannabinoids: potential anticancer agents. Nat Rev Cancer 3:745–755

Guzman M, Duarte MJ, Blazquez C et al (2006) A pilot clinical study of Delta9-tetrahydrocannabinol in patients with recurrent glioblastoma multiforme. Br J Cancer 95:197–203

Hall W, Christie M, Currow D (2005) Cannabinoids and cancer: causation, remediation, and palliation. Lancet Oncol 6:35–42

Mackie K (2006) Cannabinoid receptors as therapeutic targets. Annu Rev Pharmacol Toxicol 46:101–122

See Also

(2012) Allodynia. In: Schwab M (ed) Encyclopedia of cancer, 3rd edn. Springer, Berlin/Heidelberg, p 138. doi:10.1007/978-3-642-16483-5_193

(2012) G-protein couple receptor. In: Schwab M (ed) Encyclopedia of cancer, 3rd edn. Springer, Berlin/Heidelberg, p 1587. doi:10.1007/978-3-642-16483-5_2294

(2012) Hyperalgesia. In: Schwab M (ed) Encyclopedia of cancer, 3rd edn. Springer, Berlin/Heidelberg, p 1780. doi:10.1007/978-3-642-16483-5_2902

(2012) Mitochondrial intrinsic pathway. In: Schwab M (ed) Encyclopedia of cancer, 3rd edn. Springer, Berlin/Heidelberg, p 2333. doi:10.1007/978-3-642-16483-5_3766

(2012) Pseudokinase. In: Schwab M (ed) Encyclopedia of cancer, 3rd edn. Springer, Berlin/Heidelberg, p 3112. doi:10.1007/978-3-642-16483-5_4839

(2012) Tribbles homologue 3. In: Schwab M (ed) Encyclopedia of cancer, 3rd edn. Springer, Berlin/Heidelberg, p 3783. doi:10.1007/978-3-642-16483-5_5971

(2012) Tumor xenografts. In: Schwab M (ed) Encyclopedia of cancer, 3rd edn. Springer, Berlin/Heidelberg, p 3807. doi:10.1007/978-3-642-16483-5_6061

CAP20

▶ p21

CAR

▶ Chimeric Antigen Receptor on T Cells
▶ Constitutive Androstane Receptor

8-Carbamoyl-3-methylimidazo (5,1-d)-1,2,3,5-tetrazin-4(3H)-one

▶ Temozolomide

Carbohydrate Part of Glycoconjugates

▶ Glycosylation

Carbon Metabolism

Nianli Sang and Chengqian Yin
Department of Biology, Drexel University
College of Arts and Sciences, Philadelphia, PA, USA

Definition

Cells utilize reduced carbon sources including carbohydrates, lipids, and amino acids, to satisfy the basic needs for adenosine triphosphate (ATP), reducing power and building blocks, which are critical for cell survival, growth, and proliferation.

Characteristics

Three fundamental needs must be satisfied to support the robust proliferation of cancer cells: sufficient amount of ATP production to provide energy, rapid biosynthesis of biomolecules to support cell structure and function, and delicate maintenance of cellular redox status. Carbon metabolism plays an essential role in all three aspects (Yin et al. 2012; Cairns et al. 2011), and oncogenic signaling pathways activate the utilization of carbon sources to facilitate cell survival, growth, proliferation, and cancer progression.

There are three major types of carbon sources for cell metabolism: carbohydrates, lipids, and amino acids. Glucose is the major product of carbohydrates after digestion and absorption and is a universal carbon source utilized in cancer cells. Cancer cells have an extraordinary dependence on glucose as they consume a large amount of glucose for glycolysis and lactate fermentation even in the presence of ample oxygen, which is well known as Warburg effect. Upon entering cells via a glucose transporter, glucose is first phosphorylated to glucose 6-phosphate (G6P) by hexokinases. G6P is the common starting point of multiple metabolic pathways. In glycolytic pathway, G6P is finally converted to pyruvate. Glycolysis provides ATP independent of mitochondria and molecular oxygen. Some intermediates of the glycolytic pathway are important precursors for biosynthesis of other intermediary metabolites such as nonessential amino acids. Pyruvate can either be reduced to lactate by cytosolic nicotinamide adenine dinucleotide (NADH), reduced or enter the mitochondrion where it will be decarboxylated to acetyl-CoA or carboxylated to oxaloacetate and then enter the citric acid cycle. The citric acid cycle not only provides various precursors for the biosynthesis of many components such as heme and some amino acids but also transfers electrons to form reducing molecules NADH and flavin adenine dinucleotide (FADH2), reduced for ATP production via oxidative phosphorylation. In the pentose phosphate pathway (PPP), G6P is utilized to produce riboses and NADPH. Riboses are precursors for the biosynthesis of nucleotides and some cofactors, and NADPH is the most important reducing power for biosynthesis and maintenance of intracellular redox status.

Triacylglycerol and free fatty acids collectively represent another type of carbon source from diet. Free fatty acids can be oxidized into acetyl-CoA through β oxidation. In addition, absorbed free fatty acids can be directly used as precursors of phospholipids for biomembrane construction. Generally, free fatty acids have little role in the production of NADPH.

The carbon skeletons of amino acids also function as carbon source. Generally, carbon skeletons from proteinogenic amino acids can be converted to either ketone or glucose (by gluconeogenesis, except for lysine and leucine). Either way, they may end up in citric acid cycle and be used to produce ATP. Particularly, glutamine plays a vital role in cancer cell metabolism. Like glucose, glutamine is another nutrient which cancer cells have an extraordinary demand for. Through glutaminolysis, glutamine is degraded into glutamate and then α-ketoglutarate (α-KG). After entering the citric acid cycle, α-KG will be used either for energy production and anaplerosis or be converted to malate or isocitrate for NADPH generation. All anaplerotic amino acids may contribute to intracellular NADPH production via either the malate or isocitrate pathway.

Finally, in carbon metabolism, a key molecule that links catabolism to anabolism is acetyl-CoA. Catabolism of carbohydrates, lipids, or amino acids generates acetyl-CoA, which can be used for the biosynthesis of fatty acids, ketones, mevalonate, isoprenes, and other indispensible biomolecules derived from them, such as cholesterol, heme, quinone and dolichol.

Homeostatic Regulation of Carbon Metabolism

Cancer cells require continuous and abundant energy supply for their survival, growth, and proliferation. Therefore, cancer cells need a delicate energy sensing and regulating system to maintain the energy homeostasis. The AMP-activated protein kinase (AMPK) plays a crucial role in the process. The decreasing of the ATP level leads to the increasing of AMP concentration, which activates AMPK through allosteric activation and protection against dephosphorylation. AMPK contributes to maintaining the energy level through two aspects: enhancement of ATP production and inhibition of ATP consumption. Activated AMPK activates a lot of catabolic enzymes participating in glycolysis and fatty acid oxidation to increase ATP generation. AMPK is also reported to promote the translocation of glucose transporter 4 (GLUT4) in short term and upregulate the expression of GLUT4 in long term to increase the glucose uptake. On the other hand, the activation of AMPK inhibits the synthesis of many molecules such as fatty acids, cholesterol, glycogen, and proteins. In addition, activated AMPK causes the G1-phase cell-cycle arrest where a large amount of energy is required or even promotes apoptosis through activating the tumor suppressor p53 (Hardie 2011).

Besides the reduced biosynthesis of macromolecules, the reducing power is also indispensible for the maintenance of redox homeostasis in the cells. Various causes including normal metabolic processes, irradiation, and pharmaceutical agents result in the generation of reactive oxygen species (ROS). Although moderate levels of ROS are required and beneficial for certain cellular processes such as signal transduction, pathogen killing, and gene expression regulation, high levels of

ROS damages biomolecules and cell structures. ROS can attack and damage macromolecules including DNA, proteins, and lipids, which is implicated in apoptosis, genetic instability, cancer, and many other diseases. Therefore, cells should maintain adequate levels of reducing power which is usually in the form of NADPH to balance the oxidative and reductive levels. The ratio of [NADPH]/[NADP+] is dynamically regulated by oxidizing reduced carbon sources. NADPH can be regenerated from NADP+ in several metabolic pathways: the pentose phosphate pathway which oxidizes G6P, the oxidation of glutamate catalyzed by glutamate dehydrogenase, the reaction converting isocitrate to α-KG facilitated by isocitrate dehydrogenase 1 and 2 (IDH 1 and 2), and the malate oxidation catalyzed by malic enzyme 1 (Yin et al. 2012).

Another important function of carbon source is for the biosynthesis of building blocks. Particularly, biosynthesis of biomembranes has been found to be important for cancer progression. Important enzymes involved in the fatty acid synthesis are acetyl-CoA carboxylase and fatty acid synthase. The key rate-limiting enzyme of the cholesterol synthesis pathway is HMG-CoA reductase. The synthesis of fatty acids and cholesterol is regulated by sterol regulatory element (SRE) and SRE-binding proteins (SREBPs). Newly synthesized SREBP is inserted in the membranes of endoplasmic reticulum (ER), bound to the SREBP cleavage-activating protein (SCAP). When the intracellular sterol level is low, SREBP migrates to the Golgi apparatus, where SREBP is cleaved by site-1 and site-2 protease (S1P and S2P) activated by SCAP. The cleaved and activated SREBP then moves to the nucleus and upregulates more than 30 genes involved in the synthesis of cholesterol, fatty acids, and phospholipids, as well as the NADPH required for the reduced synthesis of these molecules (Horton et al. 2002). In addition to the biosynthesis of biomembrane, carbons are also needed for the generation of the skeletons of nonessential amino acids; some of them are expected to be actively synthesized as intermediary metabolites for the production of macromolecules such as proteins, DNA, and RNA.

At the organismal level, glucose homeostasis is regulated by insulin, glucagon, and other hormones. At the cellular level, two transcription complexes have been reported to regulate gene expression in response to high glucose concentrations: MondoA/Mlx and MondoB/Mlx. When the intracellular G6P level increases, the two transcription complexes upregulate metabolic enzymes to utilize or store the carbon source. The cellular response to low G6P levels has not been well studied. Since low glucose levels usually result in ATP depletion, AMPK pathway has been considered to play a role in cell response to low-glucose conditions (Yin et al. 2012).

References

Cairns RA, Harris IS, Mak TW (2011) Regulation of cancer cell metabolism. Nat Rev Cancer 11(2):85–95
Hardie DG (2011) AMP-activated protein kinase-an energy sensor that regulates all aspects of cell function. Genes Dev 25(18):1895–1908
Horton JD, Goldstein JL, Brown MS (2002) SREBPs: activators of the complete program of cholesterol and fatty acid synthesis in the liver. J Clin Invest 109(9):1125–1131
Yin C, Qie S, Sang N (2012) Carbon source metabolism and its regulation in cancer cells. Crit Rev Eukaryot Gene Expr 22(1):17–35

Carbonyl Metabolism

Lakshmaiah Sreerama
Department of Chemistry and Biochemistry, St. Cloud State University, St. Cloud, MN, USA
Department of Chemistry and Earth Sciences, Qatar University, Doha, Qatar

Definition

Carbonyl metabolism is a general term used to collectively describe the reactions in which either the carbonyl group is formed or carbonyl carbon is reduced and/or further oxidized (Fig. 1). These

ADH: Alcohol Dehydrogenase
ALDH: Aldehyde Dehydrogenase
AKR: Aldoketo Reductase

NAD⁺: Nicotinamide Adinine Dinucleotide
NADH: Reduced Nicotinamide Adinine Dinucleotide
NADP⁺: Nicotinamide Adinine Dinucleotide Phosphate
NADPH: Reduced Nicotinamide Adinine Dinucleotide Phosphate

Carbonyl Metabolism, Fig. 1 Metabolism of compounds containing carbonyl groups

reactions are catalyzed by three distinct families of enzymes:

- Alcohol dehydrogenases (ADHs)
- Aldehyde dehydrogenases (ALDHs)
- Aldo-keto reductases (AKRs)

Each of these enzyme families catalyzes NAD(P)⁺- or NAD(P)H-dependent oxidation and/or reduction of the carbonyl carbon (Fig. 1) present in a wide variety of endogenous and exogenous compounds. Some of the endogenous compounds that are substrates for these enzymes are generated as a result of oxidative stress, metabolism of mono- and polyamines, prostaglandins, vitamins, sugars, and steroids. Exogenous compounds that are known to serve as substrates for these enzymes include anticancer drugs, alcohols, and carcinogens. Since these enzymes are involved in redox reactions, they are often referred to as phase I drug metabolizing enzymes. Certain cytochrome P450s and aldehyde oxidase are also capable of carbonyl metabolism; however, their role appears to be minimal in this process, accordingly are not considered here.

Characteristics

Alcohol Dehydrogenases (ADHs)

ADHs are ubiquitous. They are present in many organisms as well as in most tissue to varying levels. ADHs catalyze NAD⁺-dependent oxidation of alcohols to aldehydes as well as NADH-dependent reduction of ketones and aldehydes to alcohols. In humans as well as animals, ADHs serve to bioactivate certain alcohols to their aldehydes, e.g., retinol àretinal, that are further metabolized to carboxylic acids which are important in cell growth and differentiation. They break down toxic alcohols and participate in the generation of useful aldehydes, ketones, or alcohol groups in various biosynthetic pathways. In certain organisms, including yeast, some plants, and many bacteria, ADHs catalyze the reduction of acetaldehyde to ethanol (part of fermentation process) to maintain a balance of NAD⁺/NADH ratio.

The ADHs are a superfamily of isozymes. The human ADHs are coded for by at least seven different genes, and the isozymes are classified into five classes (I–V). The class I ADH (liver forms) in humans consists of α, β, and γ subunits

Carbonyl Metabolism, Fig. 2 ADH- and ALDH-catalyzed oxidation of ethanol

that are encoded by the genes ADH1A, ADH1B, and ADH1C. The class II, III, IV, and V ADHs are encoded by ADH4, ADH5, ADH7, and ADH6, respectively. Each of the human ADH isozymes is a dimer, and each subunit has an active site with zinc ion (Zn^{2+}) associated with it. The Zn^{2+}ions are located at the catalytic site to aid in binding the hydroxyl group of an alcohol and are critical for its catalytic activity.

Class I ADHs are primarily responsible for the oxidation of ethanol to acetaldehyde (Fig. 2). Although the purpose for the presence of these ADHs is most likely to break down alcohols naturally present in foods or those produced during metabolism, the reaction shown in Fig. 2 allows us to consume ethanol-containing beverages and other products.

Another function of class I ADHs is to metabolize endogenous retinol that ultimately results in the formation of retinoic acid (Vitamin A) via carbonyl metabolism. It is also believed that ADHs primarily eliminate toxic levels of retinol. Class I ADHs are also responsible for bioactivation and toxicity associated with certain alcohols. For example, class I ADHs oxidize methanol, ethylene glycol, and many ethylene glycol ethers to their corresponding aldehydes. These aldehydes are known to cause various types of cancers in animal models.

The ADH levels are gender, age, and race specific. For example, men generally have higher levels ADH activity as compared to women. Young women are unable to process alcohol at the same rate as young men because their ADH levels are lower. The ADH levels are also different among various populations. Polymorphism in these enzymes has clinical significance in alcoholism. For example, the expression of slower alcohol metabolizing isozymes ADH2 and AHD3 poses increased risk for alcoholism. ADH polymorphism is also associated with drug dependence; however, this line of thought needs further investigations.

Aldehyde Dehydrogenases (ALDHs)

Like ADHs, ALDHs are also ubiquitous, present in most tissues as well as organisms. ALDHs function in conjunction with ADHs and catalyze NAD(P)-dependent oxidation (detoxification and/or bioactivation) of endogenous as well as exogenous aldehydes, Fig. 1. ALDHs exhibit relatively broad substrate specificity, and the substrates include straight- and branched-chain aliphatic and aromatic aldehydes. For example, the conversion of ethanol-derived acetaldehyde to acetic acid is considered detoxification (Fig. 2). The conversion of retinal to retinoic acid (physiologically active) and conversion of ethylene glycol ether-derived aldehydes to their corresponding acids (toxic and carcinogenic in certain cases) are considered bioactivation.

Similar to ADHs, the ALDHs also belong to a superfamily of isozymes. According to the latest literature reports, the human genome contains 19 ALDH functional genes and three pseudogenes. The ALDH isozymes are classified into at least 12 classes, and each of these classes has multiple members. Human ALDHs are homotetrameric enzymes with the exception of class 3 isozymes which are homodimers. The most well-investigated ALDH isozymes include class 1 (ALDH1A members), class 2 (ALDH2), and class 3 (ALDH3A members). ALDH1A and ALDH2 isozymes are constitutive forms, whereas ALDH3A isozymes are inducible in response to oxidative stress.

ALDH1A members are mainly responsible for retinal metabolism and thus play a significant role in vertebrate embryogenesis. ALDH1A members are expressed in stem cells and thus considered as markers in these cells. ALDH2 is mainly responsible for the detoxification of ethanol-derived acetaldehyde (Fig. 2). More than 40% of individuals from East Asian descent exhibit a functional polymorphism in ALDH2 gene (ALDH2*2; Glu487 has been replaced by a lysine) that leads to a partially inactive form of ALDH2. This results in acetaldehyde accumulation and an alcohol-induced flushing reaction, an increased sensitivity to alcohol and thus resulting in lower rates of alcoholism in this population. Polymorphism in ALDH2 in association with polymorphism in class I ADH isozymes is considered a risk factor for many cancers. ALDH2 has also been implicated in the bioactivation of nitroglycerin, a compound used to treat angina and heart failure. ALDH3A members are expressed in tumors, stomach, and cornea. They appear to be responsible for the maintenance of corneal transparency, protection of the lens crystallins by scavenging hydroxyl radicals, direct absorption of UV-light, and metabolism of cytotoxic aldehydes generated from UV-induced lipid peroxidation.

ALDH1A1 and ALDH3A1 isozymes catalyze detoxification of certain anticancer drugs, e.g., oxazaphosphorines such as cyclophosphamide and ifosfamide. They also detoxify/bioactivate many biologically and environmentally important aldehydes such as acrolein, chloroacetaldehyde, and 2-butoxyacetaldehyde. The latter aldehydes are implicated in carcinogenesis. Polymorphisms in other ALDHs play a significant role in hyperprolinemia; neurological disorders including mental retardation, ataxia, and seizures; and stress management in vital organs such as the kidney.

Aldo-Keto Reductases (AKRs)

AKRs are also a superfamily of isozymes. Like ADHs and ALDHs, AKRs are present in prokaryotes as well as eukaryotes and are ubiquitously expressed in various tissues. They catalyze NAD(P)H-dependent reduction of aldehydes or ketones to primary or secondary alcohols, respectively (Fig. 1). AKRs, like other carbonyl-metabolizing enzymes, exhibit broad substrate specificity. The compounds that serve as substrates for these enzymes include drugs, carcinogens, and reactive aldehydes. Many of the alcohols resulting from AKR-catalyzed reactions are further conjugated to sulfate or glucuronide for excretion (elimination reactions; detoxification) and some are bioactivated, e.g., tobacco carcinogens.

AKRs are implicated in the metabolism of certain cancer chemotherapeutics leading to their detoxification, and thus, AKRs are associated with anticancer drug resistance. AKRs convert tobacco carcinogens such as polycyclic aromatic *trans*-dihydrodiols to reactive and redox-active *o*-quinones (bioactivation). They detoxify nicotine-derived nitrosoamino ketones. AKRs are also known to detoxify exogenous toxins such as aflatoxin and endogenous toxins such as lipid peroxides.

More than 50 genes are known to code for AKR isozymes. AKR isozymes are mostly monomeric (~37 kDa) and cytosolic enzymes. They are categorized into 14 classes (families). Most of the human AKRs are placed into AKR1 class and are further subdivided into four subclasses, viz., (i) AKR1A (aldehyde reductases), (ii) AKR1B (aldose reductases), (iii) AKR1C (hydroxysteroid/dihydrodiol dehydrogenases), and (iv) AKR1D (steroid 5β-reductases).

AKR1A and AKR1B members utilize sugars, glycation products (methylglyoxal), and lipid aldehydes (4-hydoxy-2-nonenal) as substrates. One of the members of AKR1B subclass, viz., AKR1B10, prefers retinals as substrates. AKR1B10 is primarily expressed in small intestine and colon. Its levels are elevated in some liver cancers suggesting it may be involved in liver pathogenesis. AKR1C1 (human 20-α-hydroxysteroid dehydrogenase/dihydrodiol dehydrogenase 1) has been shown to be overexpressed (>50-fold) in non-small cell lung cancer (NSCLC). Elevated levels of AKR1C1 in NSCLC have been correlated with poor prognosis outcome in NSCLC, and it has been implicated in anticancer drug resistance.

References

Moreb JS (2008) Aldehyde dehydrogenase as a marker for stem cells. Curr Stem Cell Res Ther 3:237–246

Parkinson A (2001) Biotransformation of xenobiotics. McGraw-Hill, New York, pp 133–224

Penning TM (2015) The aldo-keto reductases (AKRs): Overview. Chemico-biological interactions, 234, 236–246

Sladek NE (2003) Human aldehyde dehydrogenases: potential pathological, pharmacological, and toxicological impact. J Biochem Mol Toxicol 17:7–23

Yokoyama A, Mizukami T, Yokoyama T (2015) Genetic polymorphisms of alcohol dehydrogense-1B and aldehyde dehydrogenase-2, Alcohol flushing, mean corpuscular Volume, and aerodigestive tract neoplasia in japanese drinkers. In: Biological basis of alcohol-induced cancer Springer international publishing, pp 265–279

Carbonyl Reductases

▶ Reductases

Carcinoembryonic Antigen

Peter Thomas
Departments of Surgery and Biomedical Sciences, Creighton University, Omaha, NE, USA

Synonyms

CD66e; CEA; CEACAM5

Definition

CEA is a glycoprotein of approximately 150–180 kDa. Its measurement in serum is used clinically as a biomarker for a number of cancers (pancreas, breast, stomach, ovary, lung, and medullary carcinoma of the thyroid), but its primary use is in monitoring cancers of the colon and rectum.

Characteristics

Protein Structure

CEA was discovered in 1965 in colon cancer and fetal tissue extracts and was described as an oncofetal antigen. Many of the advances in tumor marker research lead directly back to the discovery of CEA. The protein component of CEA is 79 kDa in size, and the balance of 70–100 kDa is made from up to 28 complex N-linked multi-antennary carbohydrate structures containing N-acetyl-glucosamine, mannose, galactose, fucose, and sialic acid. Low-resolution X-ray studies have shown an elongated monomeric structure that could be described as a bottlebrush. The molecule is composed of a series of six disulfide-linked immunoglobulin-like domains (IgC2-like) of either 93 (type A) or 85 (type B) amino acids and a seventh N-domain of 108 amino acids which is an IgV (variable antigen recognition domain) structure without the stabilizing disulfide bridge. CEA can attach to the cell membrane, and this is achieved by posttranslational modification of a small (26 amino acids) hydrophobic C-terminal domain to a glycosylphosphatidylinositol linkage (see Fig. 1a). Cleavage of this linkage by phospholipases releases CEA into the lumen of the intestine or other extracellular compartments.

The CEA Gene Family

The complete gene for CEA has been cloned, and it includes a promoter region that confers cell type-specific expression. The ▶ CEA gene family comprises 29 genes or pseudogenes located between the q13.1 and q13.3 regions of chromosome 19. The family can be divided into three groups: The CEA group of 12 genes, the pregnancy-specific glycoprotein (PSG) group of 11 genes, and a third group composed of 6 pseudogenes. Only 16 of the 29 genes are expressed. Sequence data has shown that the CEA family is a subset of the immunoglobulin supergene family. Comparative sequence studies of the CEA gene family from various species suggest that the CEA family has a common ancestry and arose relatively recently in evolution.

Function of CEA in Normal and Cancerous Tissue

In general, members of the CEA family subgroup have a ubiquitous distribution in adult tissues. However, CEA itself has a more restricted

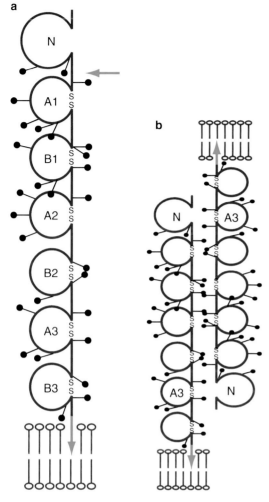

Carcinoembryonic Antigen, Fig. 1 CEA structure. (**a**) Insertion of CEA into the plasma membrane (*down arrow*), the Ig domain structure, and the position of the N-linked sugar chains. An *arrow marks* the position of the PELPK receptor recognition sequence. (**b**) Homotypic binding between two CEA molecules with attachment between the N and A3 domains (Structures are modified from the CEA homepage http://cea.klinikum.uni-meunchen.de)

expression being found only in the colon, pyloric mucus cells, epithelial cells of the prostate, sweat glands, and squamous cells in the tongue, cervix, and esophagus. In the colon CEA is located at the apical surface of colonic enterocytes and is associated with the glycocalyx or fuzzy coat. In the normal colon CEA is maximally expressed on columnar cells at the level of the free luminal surface. CEA is also found in goblet cells in association with mucins. The function of CEA in the normal individual is not well understood and has been the subject of much speculation. It has been estimated that the normal person can produce 70 mg or more of CEA a day and excrete it in the feces. CEA has been shown to bind to various fimbriated gut pathogens, and therefore, it has been suggested that it has a function in protecting the gut epithelia. In cancer cells CEA may perform a number of functions. Unlike the normal colonocyte where CEA expression is highly polarized in cancers, this polarity is lost and its expression occurs through the whole of the cell surface. It has been shown that CEA can act as a Ca^{2+}-independent homotypic adhesion molecule binding with itself through an interaction between the N and A3 domains (Fig. 1b) and causing aggregation of tumor cells. This allows the malignant epithelium to adopt a multilayered structure and may disrupt the normal pattern of differentiation. CEA can also bind heterotypically to other members of the gene family including the nonspecific cross-reacting antigen (NCA, CD66c, CEACAM6) and the biliary glycoprotein (BGP, CD66a, CEACAM1), of which seven different forms have been identified. It is unlikely that CEA functions as a ► cell adhesion molecule in the normal colon because of its apical expression. CEA is cleared from the circulation by the hepatic ► macrophages (► Kupffer cells). A cell surface receptor identical to the heterogeneous nuclear RNA-binding protein hnRNP M4 recognizes a pentapeptide (Pro–Glu–Leu–Pro–Lys (PELPK)) located at the hinge region between the N and the first immunoglobulin loop domain (A1) of CEA. Patients with a mutation in the region coding for this peptide have extremely high circulating CEA levels presumably due to the inability of Kupffer cells to clear the protein from the blood. CEA has also been implicated in the development of hepatic ► metastasis from colorectal cancers by the induction of a localized inflammatory response that affects retention and implantation in the liver. Cytokines produced also protect the tumor cells against the toxic effects of hypoxia. CEA-producing cells therefore have a selective advantage for growth in the liver. Studies have also shown that CEA can protect cancer cells from a form of programmed cell death called ► anoikis, and this also seems to involve the PELPK motif and inhibition of Trail-R2 (DR5) signaling. CEA is also protective against other forms of ► apoptosis including drug- and UV light-induced programmed cell death. The related protein CEACAM-1, however, is a proapoptotic protein. Research has shown that CEA may be involved in promoting angiogenesis (growth of new blood vessels) in colorectal cancer, by interacting with endothelial cells directly through its receptor. This raises the possibility of alternate therapeutic approaches and further emphasizes CEA as a multi-functional glycoprotein.

Clinical Aspects

The main clinical use for CEA is as a tumor marker especially for cancers in the colon and rectum and approximately 90% of these cancers produce CEA. CEA has also been used as a marker for breast and small cell cancer of the lung. Approximately 50% of breast and 70% of small cell cancers express CEA. Accurate immunoassays are commercially available for its measurement in body fluids. Immunohistochemistry on biopsy or resection specimens is also often carried out, for example, the intensity of CEA staining has been associated with a worse prognosis for breast cancer. Normal serum levels are <2.5 ng/ml unless the subject is a heavy smoker when the normal cutoff becomes <5 ng/ml. CEA levels can be elevated in a number of nonmalignant conditions such as pancreatitis, inflammatory bowel disease including ► Crohn disease, hereditary polyposis including Gardener syndrome, polycystic disease of the liver, and a variety of other liver diseases including cirrhosis and hepatitis and benign biliary duct obstruction. Rarely do these diseases result in a CEA elevation over 10 ng/ml. CEA levels above 20 ng/ml almost always indicate the presence of a malignant tumor. Patients with colorectal cancer who present with a CEA level of over 5 ng/ml have a poorer prognosis and are at higher risk for developing metastasis to the liver. However, because the CEA assay lacks sensitivity for early stage colorectal cancer, it cannot be used as a population screen. CEA is most useful for the early detection of liver metastasis in colorectal cancer patients. It is not as effective in detecting locoregional or pulmonary metastases. Elevated CEA levels that fall to normal following tumor resection are an indication of a successful surgery; however, a rising CEA level postoperatively indicates a progression or recurrence of the tumor. There is no clear agreement on how often CEA measurements should be taken following curative surgery. The guidelines put out by the American Society of Clinical Oncologists (ASCO) recommend measurements every 2–3 months for a minimum of 2 years. CEA measurements can give a lead time of up to a year before the onset of clinical symptoms of recurrence. Serial serum CEA measurements have been shown to be useful in the follow-up of patients with breast cancer and small cell cancer of the lung. The reverse transcriptase-polymerase chain reaction (RT-PCR) has been used to detect CEA-producing circulating cancer cells. A real-time PCR method has been developed for the quantitative detection of CEA mRNA transcripts in blood, peritoneal washings, and lymph nodes. These methods can be used for a more exact staging and prognosis in cancer patients. CEA has been used as a target antigen for both radioimmunodetection and ► radioimmunotherapy of cancers. Imaging after administration of radiolabeled anti-CEA antibodies provides information on the location and extent of disease. Radiolabeled anti-CEA antibodies have shown therapeutic effects in reducing tumor size in metastatic disease. Radiolabeled anti-CEA antibodies are also used to guide second-look surgery and can detect occult disease. CEA has been used as the antigen of choice for cancer vaccines against colorectal cancers. Clinical trials have been conducted using recombinant CEA-vaccine virus (vaccine constructed from a recombinant vaccine virus containing the human ► carcinoembryonic antigen gene) and recombinant ALVAC-CEA vaccines. The latter is a cancer vaccine constructed from canarypox virus (ALVAC) and combined with the human carcinoembryonic antigen (CEA) gene. The vaccines were well tolerated and elicited CEA-specific T-cell responses. This promises to be a useful addition to standard therapies.

Cross-References

▶ Anoikis

▶ Apoptosis

▶ CEA Gene Family

▶ Cell Adhesion Molecules

▶ Crohn Disease

▶ Kupffer Cells

▶ Macrophages

▶ Metastasis

▶ Oncofetal Antigen

▶ Radioimmunotherapy

References

Beauchemin N, Draber P, Dveksler G, Gold P, Gray-Owen S, Grunert F, Hammerstrom S, Holmes KV, Karlsson A, Kuroki M, Lin S-H, Lucka L, Najjar S.M, Neumaier M, Obrink B, Shivley JE, Skubitz K, Stanners CP, Thomas P, Thompson JA, Virji M, von Kleist S, Wagener C, Watt S, Zimmermann W (1999) Redefined nomenclature for members of the carcinoembryonic antigen gene family. Experimental Cell Res. 252:243–249

Gold P, Freedman SO (1965) Specific carcinoembryonic antigens of the human digestive system. J Exp Med 122:467–481

Goldstein MJ, Mitchell EP (2005) Carcinoembryonic antigen in the staging and follow up of patients with colorectal cancer. Cancer Invest 23:338–351

Hammerstrom S (1999) The carcinoembryonic antigen (CEA) family: structures, suggested functions and expression in normal and malignant tissues. Semin Cancer Biol 9:67–81

Jessup JM, Thomas P (1998) CEA and metastasis: a facilitator of site specific metastasis. In: Stanners C (ed) Cell adhesion and communication mediated by the CEA family: basic and clinical perspectives. Harwood Academic, Amsterdam, pp 195–222

Koppe MJ, Bleichrodt RP, Oyen WJG et al (2005) Radioimmunotherapy and colorectal cancer. Br J Surg 92:264–276

See Also

(2012) ALVAC-CEA. In: Schwab M (ed) Encyclopedia of Cancer, 3rd edn. Springer Berlin Heidelberg, p 150. doi: 10.1007/978-3-642-16483-5_214

(2012) Biliary Glycoprotein. In: Schwab M (ed) Encyclopedia of Cancer, 3rd edn. Springer Berlin Heidelberg, p 401. doi: 10.1007/978-3-642-16483-5_621

(2012) Biomarkers. In: Schwab M (ed) Encyclopedia of Cancer, 3rd edn. Springer Berlin Heidelberg, pp 408–409. doi: 10.1007/978-3-642-16483-5_6601

(2012) CEA-Vaccine Virus. In: Schwab M (ed) Encyclopedia of Cancer, 3rd edn. Springer Berlin Heidelberg, p 720. doi: 10.1007/978-3-642-16483-5_971

(2012) Cirrhosis. In: Schwab M (ed) Encyclopedia of Cancer, 3rd edn. Springer Berlin Heidelberg, p 869. doi: 10.1007/978-3-642-16483-5_1184

(2012) DR5. In: Schwab M (ed) Encyclopedia of Cancer, 3rd edn. Springer Berlin Heidelberg, p 1160. doi: 10.1007/978-3-642-16483-5_1726

(2012) Gardner Syndrome. In: Schwab M (ed) Encyclopedia of Cancer, 3rd edn. Springer Berlin Heidelberg, p 1503. doi: 10.1007/978-3-642-16483-5_2327

(2012) Glycocalyx. In: Schwab M (ed) Encyclopedia of Cancer, 3rd edn. Springer Berlin Heidelberg, p 1569. doi: 10.1007/978-3-642-16483-5_2445

(2012) Glycosyl Phosphatidyl Inositol. In: Schwab M (ed) Encyclopedia of Cancer, 3rd edn. Springer Berlin Heidelberg, p 1571. doi: 10.1007/978-3-642-16483-5_2456

(2012) HnRNP M. In: Schwab M (ed) Encyclopedia of Cancer, 3rd edn. Springer Berlin Heidelberg, p 1711. doi: 10.1007/978-3-642-16483-5_2777

(2012) Nonspecific Cross Reacting Antigen. In: Schwab M (ed) Encyclopedia of Cancer, 3rd edn. Springer Berlin Heidelberg, p 2551. doi: 10.1007/978-3-642-16483-5_4125

(2012) Phospholipase. In: Schwab M (ed) Encyclopedia of Cancer, 3rd edn. Springer Berlin Heidelberg, p 2867. doi: 10.1007/978-3-642-16483-5_4536

(2012) Radio-Immunodetection. In: Schwab M (ed) Encyclopedia of Cancer, 3rd edn. Springer Berlin Heidelberg, p 3148. doi: 10.1007/978-3-642-16483-5_4910

Carcinofetal Proteins

▶ Alpha-Fetoprotein

Carcinogen Metabolism

Frederick Peter Guengerich
Department of Biochemistry and Center in Molecular Toxicology, Biochemistry and Center in Molecular Toxicology, Vanderbilt University School of Medicine, Nashville, TN, USA

Definition

The transformation of chemicals is important in carcinogenesis both in terms of bioactivation and

Carcinogen Metabolism, Fig. 1 General paradigm for carcinogen metabolism, including both bioactivation and detoxication reactions

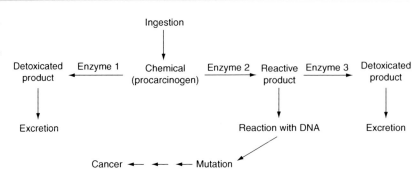

Carcinogen Metabolism, Fig. 2 Major events in the metabolism of the hepatocarcinogen aflatoxin B_1

detoxication. Most chemical carcinogens need to be activated within the body. Such reactive forms can then cause biological damage (Fig. 1). As an example for competing processes, aflatoxin B_1 was chosen (Fig. 2) (▶ Adducts to DNA). Exactly what proportion in human cancers is the result of chemical exposure is not clear. However, in most countries, at least one third of cancer cases are due to tobacco carcinogens (▶ Tobacco carcinogenesis, ▶ tobacco-related cancers). A significant number of cancer cases may be related to diet, although it is unknown exactly which chemicals in food cause or influence cancer. As a result of precautions adapted in the course of the last century, the number of cases due to industrial exposure seems to be very low.

Characteristics

History
In 1761, the London physician J. Hill associated the use of snuff with nasal cancers (Tobacco carcinogenesis, tobacco-related cancers). More than one hundred years later in 1895, Rehn and others reported a link of large-scale arylamine exposure of workers in the aniline dye industry in Germany and Switzerland to bladder cancer (▶ Aromatic amine). Aniline is an aromatic amine. It is a colorless, oily liquid, originally obtained from indigo, a blue dyestuff derived from several plants, by distillation. Today it is largely manufactured from coal tar or nitrobenzene as a base from which many brilliant dyes are made. In

Japan, Yamagiwa and Ichikawa were in 1915 the first to demonstrate the formation of tumors in rabbits exposed to coal tar, a mixture of polycyclic hydrocarbons (▶ Polycyclic aromatic hydrocarbons). The concept that metabolic processes are a necessity for the bioactivation of chemical carcinogens was primarily developed by J. A. and E. C. Miller at the University of Wisconsin in the early 1940s (▶ DNA damage). Over the next few decades, they and others provided further insight, defining metabolically derived carcinogenic products that react with DNA ("ultimate carcinogens") (Adducts to DNA). However, although the relationship between carcinogens and mutagenesis had been considered, it was not clearly defined. It was only after B. N. Ames developed a (still widely used) bacterial mutation system in which rat liver extracts are able to transform carcinogens into mutagens that the correlation between carcinogenesis and mutagenesis became obvious (▶ Genetic toxicology). Advances in enzymology and recombinant DNA technology made it possible to discern the role of individual human enzymes in various steps in carcinogen metabolism. Using inbred mouse strains and knockout mice, it was possible to demonstrate the critical role of mouse orthologues in carcinogen activation.

Metabolism

Metabolism of carcinogens occurs in many tissues throughout the body (▶ ADMET screen). Many in vitro studies utilize liver tissue samples because many enzymes of interest are concentrated there. However, for tumors that originate elsewhere, extrahepatic sites are of greater interest. The question of which kind of tissue is most important is related to the site of entry of a carcinogen, as well as how much of the activated form(s) of the carcinogen is able to circulate within the body before reacting with the target tissue. Examples of important carcinogens and their metabolism are given below (▶ Xenobiotics).

- Polycyclic aromatic hydrocarbons are systems of fused benzene rings that are found in carcinogenic soots, tars, and tobacco smoke (Polycyclic aromatic hydrocarbons). A widely studied member of this class of compounds is benzo[a]pyrene. It is widely believed that the main metabolic pathway involves the oxidation of benzo[a]pyrene by cytochrome P450 (P450) to an epoxide. The hydrolysis of this epoxide to a dihydrodiol is followed by another oxidation by P450 that generates highly reactive diol epoxides (▶ Cytochrome P450). The latter can either react with DNA or are detoxicated by glutathione transferase.

- Aflatoxin B_1 is a mycotoxin and a prominent contributor to human liver cancer (▶ Hepatocellular carcinoma). A critical feature of its metabolism is the formation of an epoxide by P450 enzymes (Fig. 2). The epoxide (with a half-life in water of $t_{1/2} = 1$ s) is able to react with DNA or can be conjugated with glutathione. P450 enzymes can also detoxicate aflatoxin B_1 by catalyzing several other oxidation steps (e.g., the oxidation to 3α- and 9α-hydroxylated products).

- Olefins (alkenes) can be oxidized to epoxides (▶ Alkylating agents). A member of this group is vinyl chloride, a carcinogenic substance that was shown to cause a rare liver hemangiosarcoma in people working in the rubber industry.

- Another problematic group of substances are N-nitrosamines. They can result from some industrial settings but are also produced endogenously from amines and nitrites in the acidic environment of the stomach. Sources are the so-called tobacco-specific nitrosamines as well as sodium nitrite that are used to preserve processed meats (Tobacco carcinogenesis, tobacco-related cancers). As in the examples stated above, P450 activates N-nitrosamines by oxidation. The formation of an alcohol on the adjacent carbon atom yields an unstable product that decomposes and alkylates DNA.

- Another group of chemicals of concern present in food and tobacco is heterocyclic amines, substances derived from creatinine and amino acids following pyrolysis (Aromatic amines). Amine activation involves its oxidation by a P450 enzyme to a hydroxylamine (–NHOH). An unstable compound (–NHOAc) is the result of the enzymatic transfer of an acetyl group (▶ Arylamine N-acetyltransferases (NAT) and cancer, biomarkers). It ultimately breaks down to a nitrenium ion ($-NH^+$) that can react with DNA. Detoxication involves other P450 enzymes, glutathione transferases, and UDP glucuronosyltransferases.

Mechanisms

Conjugation reactions (including those catalyzed by the enzyme *N*-acetyltransferase) (Arylamine *N*-acetyltransferases (NAT) and cancer) are usually involved in detoxication reactions; they can, however, also be part of bioactivation schemes (► Sulfotransferases). An example is the pesticide ethylene dibromide ($BrCH_2Cl_2Br$) and related compounds where the enzymatic conjugation of ethylene dibromide with endogenous tripeptide glutathione yields a molecule (in this case glutathione-CH_2CH_2Br) that can react with DNA (► Glutathione S-Transferase).

Cancers

Numerous studies support the important role of carcinogen metabolism in human cancers.

First, substances, such as aflatoxin B_1, whose metabolic products can cause cancers (Fig. 2) have been identified in foods (Xenobiotics). Second, it has been shown in animal models that either the absence or the induction of certain enzymes that are involved in carcinogen metabolism can have a dramatic effect on chemical-caused cancers. Third, humans are known to show great phenotypic variation in many enzymes involved in carcinogen metabolism. Dramatic effects on the metabolism of drugs have been demonstrated with these enzymes. Large international and other interindividual differences in cancer incidence, as well as the documented effects of diet on cancer, justify the considerable interest to study carcinogen metabolism, particularly in humans. Research in carcinogen metabolism and its applications can be divided into several areas. Investigating cancer cause and cancer etiology depends upon the understanding of basic chemistry, enzymology, and physiology of metabolic processes as well as how the chemicals react with DNA once they are activated. Molecular epidemiology utilizes information about carcinogen metabolism in order to establish their relevance in human cancer (► Biomarkers). A related topic is

risk assessment, which uses the knowledge of carcinogen metabolism derived from animal bioassay studies and sometimes epidemiology to determine critical exposure levels of environmental carcinogens in humans (► Cancer epidemiology).

Metabolism mechanisms play an important role in cancer safety assessment studies of prospective new drugs, including those used to treat cancer. Another important area is chemoprevention where beneficial effects of certain chemicals are investigated, e.g., their ability to change the metabolism of carcinogens.

Cross-References

► Adducts to DNA
► ADMET Screen
► Alkylating Agents
► Aromatic Amine
► Arylamine *N*-Acetyltransferases
► Biomarkers in Detection of Cancer Risk Factors and in Chemoprevention
► Cancer Epidemiology
► Clinical Cancer Biomarkers
► Cytochrome P450
► Detoxification
► DNA Damage
► Genetic Toxicology
► Glutathione S-Transferase
► Hepatocellular Carcinoma
► Polycyclic Aromatic Hydrocarbons
► Sulfotransferases
► Tobacco Carcinogenesis
► Tobacco-Related Cancers
► Xenobiotics

References

Guengerich FP (2000) Metabolism of chemical carcinogens. Carcinogenesis 21:345–351

Guengerich FP, Shimada T (1991) Oxidation of toxic and carcinogenic chemicals by human cytochrome P-450 enzymes. Chem Res Toxicol 4:391–407

Miller JA (1998) The metabolism of xenobiotics to reactive electrophiles in chemical carcinogenesis and mutagenesis. Drug Metab Rev 30:645–674

Searle CE (ed) (1984) Chemical carcinogens, vols 1 and 2. American Chemical Society, Washington, DC

See Also

(2012) Biomarkers. In: Schwab M (ed) Encyclopedia of cancer, 3rd edn. Springer, Berlin/Heidelberg, pp 408–409. doi:10.1007/978-3-642-16483-5_6601

Carcinogenesis

I. Bernard Weinstein
Columbia University, New York, NY, USA

Definition

Carcinogenesis is the process by which cancer develops in various tissues in the body.

Characteristics

In most cases, carcinogenesis occurs via a stepwise process that can encompass a major fraction of the lifespan (▶ multistep development). These progressive stages often include hyperplasia, dysplasia, metaplasia, benign tumors, and then, eventually, malignant tumors. Malignant tumors can also undergo further progression to become more invasive and metastatic, autonomous of hormones and growth factors, and resistant to chemotherapy or radiotherapy.

Causes

Known causes of carcinogenesis include various chemicals or mixture of chemicals present in several sources. This includes cigarette smoke; diet; workplace or general environment; ultraviolet and ionizing radiation; specific viruses, bacteria, and parasites; and endogenous factors (▶ DNA Oxidation Damage, DNA depurination, deamination). According to the International Agency for Research on Cancer (IARC), 69 agents, mixtures, and exposure circumstances are known to be carcinogenic to humans (group 1), 57 are probably carcinogenic (group 2A), and 215 are possibly carcinogenic to humans. Some of these agents,

---

Bernard Weinstein: deceased.

or their metabolites, form covalent ▶ adducts to DNA and are mutagenic. Others act at the epigenetic level by altering pathways of signal transduction and gene expression. These include tumor promoters, growth factors, and specific hormones. Dietary factors also play an important role. Fruits and vegetables often have a protective effect. Excessive fat and/or calories may enhance carcinogenesis in certain organs. Hereditary factors can also play an important role in cancer causation. Indeed, human cancers are often caused by complex interactions between these multiple factors. An example is the interaction between the naturally occurring carcinogen ▶ aflatoxin and the chronic infection with hepatitis B virus in the causation of liver cancer in regions of China and Africa.

Molecular Genetics

Studies indicate that the stepwise process of carcinogenesis reflects the progressive acquisition of activating mutations in dominant-acting ▶ oncogenes and inactivating recessive mutations in ▶ tumor suppressor genes. It is also apparent that epigenetic abnormalities in the expression of these genes also play an important role in carcinogenesis. Thus far over 100 oncogenes and at least 12 tumor suppressor genes have been identified. Tumor progression is enhanced by genomic instability due to defects in DNA repair and other factors. The heterogeneous nature of human cancers appears to reflect heterogeneity in the genes that are mutated and/or abnormally expressed. Individual variations in susceptibility to carcinogenesis are influenced by hereditary variations in enzymes that either activate or inactivate potential carcinogens, variations in the efficiency of DNA repair, and other factors yet to be determined. Age, gender, and nutritional factors also influence individual susceptibility.

Clinical Relevance

Prevention

Cancer is a major cause of death throughout the world. Therefore, the prevention of carcinogenesis is a major goal of medicine and public health. The carcinogenic process can be prevented by avoidance of exposure to various carcinogenic factors (i.e., cigarette smoking, excessive sunlight, etc.), dietary changes, early detection of precursor lesions, and chemoprevention.

Cross-References

▶ Adducts to DNA
▶ Aflatoxins
▶ DNA Oxidation Damage
▶ Multistep Development
▶ Nucleoporin
▶ Oncogene
▶ Toxicological Carcinogenesis
▶ Tumor Suppressor Genes

References

Kitchin KT (ed) (1999) Carcinogenicity, testing predicting and interpreting chemical effects. Marcel Dekker, New York/Basel
Weinstein IB (2000) Disorders in cell circuitry during multistage carcinogenesis: the role of homeostasis. Carcinogenesis 22:857–864
Weinstein IB, Santella RM, Perera FP (1995a) Molecular biology and molecular epidemiology of cancer. In: Greenwald P, Kramer BS, Weed DL (eds) Cancer prevention and control. Marcell Dekker, New York, pp 83–110
Weinstein IB, Carothers AM, Santella RM et al (1995b) Molecular mechanisms of mutagenesis and multistage carcinogenesis. In: Mendelsohn J, Howley PM, Israel MA, Liotta LA (eds) The molecular basis of cancer. Saunders, WB, Philadelphia, pp 59–85

See Also

(2012) Mutagenic. In: Schwab M (ed) Encyclopedia of cancer, 3rd edn. Springer, Berlin/Heidelberg, p 2412. doi:10.1007/978-3-642-16483-5_3909

Carcinogenesis in Colon

▶ Colorectal Cancer Nutritional Carcinogenesis

Carcinogenic Compounds in Food

▶ Food-Borne Carcinogens

Carcinoid

Synonyms

Carcinoid tumors

Definition

Carcinoids originate in hormone-producing cells of the gastrointestinal (GI) tract (i.e., esophagus, stomach, small intestine, colon), the respiratory tract (i.e., lungs, trachea, bronchi), the hepatobiliary system (i.e., pancreas, gallbladder, liver), and the reproductive glands (i.e., testes, ovaries).

The most common site of origin is the GI tract, and carcinoid tumors often develop in the appendix, the rectum, and the lower sections of the small intestine (i.e., the jejunum and the ileum). The large tubes that lead from the windpipe to the lungs (bronchi) are other common sites of origin.

Carcinoids are classified as ▶ neuroendocrine tumors. They develop in peptide- and amineproducing cells, which release hormones in response to signals from the nervous system. Excessive amounts of these hormones cause a condition called carcinoid syndrome in approximately 10% of patients with carcinoid tumors.

Carcinoids are slow growing, and tumors with the same site of origin often have different characteristics and growth patterns. They can be subdivided according to the following:

- Cellular growth pattern (e.g., trabecular, glandular, undifferentiated, mixed)
- Hormones produced (e.g., bradykinin, serotonin, histamine, ▶ prostaglandins)
- Site of origin – foregut (respiratory tract, pancreas, stomach, first section of the small intestine [duodenum]), midgut (jejunum, ileum, appendix, diverticulum, ascending colon), or hindgut (transverse colon, descending colon, rectum)

Incidence and Prevalence of Carcinoid Tumors
According to the American Cancer Society, approximately 5,000 carcinoid tumors are diagnosed each year in the United States. According to the National

Cancer Institute (NCI), approximately 74% of these tumors originate in the GI tract and 25% occur in the respiratory tract. Carcinoids are rare in children and are more common in patients older than the age of 50. They are twice as common in men. Carcinoid tumors of the appendix usually are benign and often occur between the ages of 20 and 40.

Cross-References

▶ Carcinoid Tumors
▶ Neuroendocrine Neoplasms
▶ Prostaglandins

See Also

(2012) Bradykinin. In: Schwab M (ed) Encyclopedia of cancer, 3rd edn. Springer, Berlin/Heidelberg, p 468. doi:10.1007/978-3-642-16483-5_701
(2012) Carcinoid syndrome. In: Schwab M (ed) Encyclopedia of cancer, 3rd edn. Springer, Berlin/Heidelberg, p 654. doi:10.1007/978-3-642-16483-5_846
(2012) Serotonin. In: Schwab M (ed) Encyclopedia of cancer, 3rd edn. Springer, Berlin/Heidelberg, p 3389. doi:10.1007/978-3-642-16483-5_5262
http://www.healthcommunities.com/carcinoid-malignancy/carcinoid-malignancy-overview.shtml

Carcinoid (Well-Differentiated Neuroendocrine Tumor (NET) of the Respiratory and Gastrointestinal Tract)

▶ Neuroendocrine Neoplasms

Carcinoid Tumors

Phillip H. Abbosh and Liang Cheng
Department of Pathology and Laboratory Medicine, Indiana University School of Medicine, Indianapolis, IN, USA

Synonyms

Argentaffin carcinoma; Carcinoid; NET; Neuroectodermal tumor; Neuroendocrine carcinoma

Definition

Carcinoid tumors represent a family of diseases derived from neuroendocrine cells. These tumors were first described by Langhans in 1867 but were not described in detail until Lubarsch described them in 1888. The name *karzinoide* was not used until 1907 by Oberndorfer and was chosen to reflect his idea that these were benign growths. However, these tumors have a wide range of clinical presentations and outcomes from benign to malignant. Clinicians must recognize the nature of carcinoid disease, because these often have a significantly different clinical course than typical carcinomas occurring within the body. Additionally, with few exceptions, neuroendocrine tumors (NETs) comprise a tiny fraction of tumors within any specific organ. These neoplasms cause <1% of all malignancies in the United States, currently occurring at a rate of 2.5–4.5/100,000 people.

NETs can arise in almost any tissue within the body, but most are derived from the embryonic foregut, midgut, and hindgut with over two-thirds of these occurring in the gastroenteropancreatic axis, with approximately 25% occurring in the foregut. NETs of the midgut (true carcinoids) are the only ones which secrete serotonin, and only these tissues give rise to tumors causing carcinoid syndrome. Carcinoid syndrome is the most recognized complication of carcinoid tumors, originally described in 1890 by Ransom. This manifestation was not recognized as an endocrine or paraneoplastic syndrome until 1914 by Gosset. Carcinoid syndrome is caused by carcinoid in the midgut which secretes serotonin. In its localized state, venous drainage of tumor secretions that are metabolized by the liver and serotonin is deactivated. However, upon metastasis to the liver, secretions are not as readily processed and serotonin not deactivated and so is released into the systemic circulation, causing the syndrome. This is often characterized by flush, diarrhea, and cramping.

Characteristics

Within the GI system, the small intestine is the most frequent site of NETs, followed by the rectum. Overall, 32–45% of NETs will spread, but the propensity for these tumors to metastasize also varies individually by the organ in which they develop. As with most neoplasms, the set of symptoms that each tumor will cause the patient is dependent on where the tumor begins, and thus, NETs will probably present in different stages depending on the site. For instance, >70% of NETs starting in the cecum and pancreas spread regionally or distantly, compared to NETs of the rectum and stomach, which spread less than 18% and 33% of the time, respectively. Overall, 5-year survival rates for NETs are 67%, with survival rates after spread to localized or distant sites exceeding 75% and below 40%, respectively. Accordingly, 5-year survival rates vary by site, with the highest and lowest rates being in the rectum (88%) and liver (18%), regardless of stage.

Risk for developing NETs can also be part of inherited familial cancer syndromes. Although rare, this most frequently occurs in type I multiple endocrine neoplasia and type I neurofibromatosis, with isolated case reports in other inherited cancer syndromes. However, the site of NETs in these syndromes is most often the duodenum or pancreas, and not the midgut, which is the site of most sporadic carcinoids. The molecular genetics of *MEN1* (the gene which is mutated MEN I) in carcinoids is complex but includes point mutation as well as loss of one allele. In lung NETs, for example, 4/11 tumors showed both point mutation and deletion of one allele. Additionally, *MEN1* is located on chromosome 11q13, which is frequently lost from NETs and will be discussed later. *MEN1* which encodes the protein menin is frequently altered in sporadic carcinoid, in addition to type I MEN. Menin behaves as a nuclear protein when consisting of the wild-type sequence but does not localize to the nucleus in many mutated forms. As with almost all genes which are implicated in familial cancer syndromes, menin predictably behaves as a tumor suppressor gene. It has been shown to dampen transcriptional transactivational activity of such oncogenes as NF-$\kappa\beta$ and JunD. However, menin is also known to behave as an oncogene and in

fact is required for mixed-lineage leukemogenesis in relevant models of the disease. Menin interacts with the mixed-lineage leukemia (MLL) protein, which is also known to regulate transcription. Indeed, deletion of the menin-interaction domain from MLL results in failed leukemogenesis in a validated model for this disease. MLL behaves as a histone methyltransferase for histone H3-K4, and H3-K4 methylation is known to be strongly associated with transcriptional activation. Therefore, it appears menin retains transcriptional activating as well as repressing activity and behaves in oncogenic as well as tumor-suppressive signaling pathways. Only a handful of other genes are known to be definitively mutated in NET and carcinoid, including succinate-ubiquinone oxidoreductase subunit D (*SDHD*) and β-catenin (► APC/β-Catenin Pathway). Notably, mutation of "classical" oncogenes and tumor suppressors like *RAS* and ► *TP53* is not present in NETs.

Much more work has been done to identify chromosomal losses in NETs or carcinoid tumors. Identification of regions of DNA which are consistently lost in carcinoid tumors implies that carcinoid- or NET-specific tumor suppressors may reside in these deleted regions. These regions most frequently include 11q, which contains *MEN1* as previously discussed, but this region may contain other tumor suppressors as well. Another chromosomal region frequently lost in these malignancies is 18q. This occurs in 33–88% of NETs and varies by NET site. For instance, lung NETs, like carcinoids, often have loss of 11q but rarely have loss of 18q. Several known tumor suppressors reside on 18q which include two transcription factors in the TGF-β signaling pathway (► Smad proteins in TGF-β signaling) (*SMAD2* and *SMAD4*) and *DCC*, all of which are known to be altered in pancreatic and colorectal cancer. At least one study has delimited the region of lost genomic DNA in midgut carcinoids to the sequence between 18q22 and the 18q telomere. Interestingly, *SMAD2*, *SMAD4*, and *DCC* do not reside in the region of recurrent loss in carcinoid, strongly suggesting that other genes in 18q22-qter are carcinoid tumor suppressors. Less evidence is available regarding the loss of 9p21 (which contains *p16/ARF*), 3p (which contains ▶ peroxisome proliferator-activated receptor, *RASSF1A*, ▶ von Hippel-Lindau tumor suppressor gene, *FHIT*, and ▶ retinoic acid), and 16q21.

The ability to measure chromosomal abnormalities and point mutation of specific genes in NETs combined with the fact that NETs are often multifocal has led to the question of whether each tumor focus originated as an independent event or if multifocality is secondary to local invasion and metastasis. This question can be answered in a variety of ways, but currently, the most definitive method is based on molecular analysis of the genomic content of each individual focus. By analyzing regions which are frequently lost and/or genes which are commonly mutated in each tumor individually, inferences regarding the clonality of multiple tumor foci can be drawn. If multiple tumors form independently, then one would expect that genetic alterations at specific loci would occur randomly. However, if the same genetic alterations are seen in a majority of tumors from the same patient, then it is likely that the tumors do not form independently. This was shown by Katona et al. using microdissected tissues from patients with multiple pancreatic NETs or carcinoids. Seventy-two tumors from 24 patients were analyzed using LOH analysis. The results of this study showed that indeed at least some tumors show identical patterns of allelic loss, indicating that carcinogenesis in these patients probably occurred once and that each tumor focus represents a locally invasive metastasis. However, the largest fraction of patients had tumors with nonidentical LOH pattern in each tumor, indicating that these tumors actually formed independently. Other groups have shown a similar pattern of evidence with different genetic markers implicating multiple independent foci in carcinoid carcinogenesis.

The implications of identifying the mechanism of multiple tumor formation are directly related to therapy, especially with regard to surgical intervention. This is directly related to the pathogenesis of NET formation. One would surmise that multiple tumors arise independently in an organ for one of a few reasons: Cells were exposed to an endogenous mitogen or trophic factor; cells were exposed to a genotoxin, causing genetic damage to accumulate in many cells which independently acquire further mutations; or intrinsic germline polymorphisms provide a propensity for tumors to form in multiple independent cells. In all three cases, the idea of a field defect, first described by Slaughter in 1953, plays a significant role. A field defect may be thought of as normal-looking tissue which, at a gene-by-gene level, is not normal. In other words, it is a pre-benign lesion. If the entire organ at the site of NET formation harbored premalignant genomic changes, then more aggressive surgical intervention would predictably result in better outcomes for patients, because the procedure would leave behind the tissue with a propensity to continue to undergo carcinogenesis. However, if multiple NETs arise in an organ due to local spread, then wider surgical margins would probably not result in higher survival rates, assuming the procedure was able to remove all of the malignant tissue.

Much has been learned about carcinoid since its original description over 100 years ago. Much is left to be learned about this disease as well, including exactly which signaling pathways are altered in the disease and identification of genes which are in chromosomal regions of frequent alteration. As yet, there are very few reports of experimental carcinoid models which to study, both at the cell line and animal level. Detailed epidemiological studies become difficult to perform due to the rare nature of this disease, which also complicates prospective studies. Overcoming or averting these obstacles will significantly speed the rate of progress in carcinoid research.

Cross-References

▶ Carcinoid
▶ Multiple Endocrine Neoplasia Type 1
▶ Neuroendocrine Carcinoma
▶ Neurofibromatosis 1
▶ Telomerase

References

Dreijerink KM, Hoppener JW, Timmers HM et al (2006) Mechanisms of disease: multiple endocrine neoplasia type 1-relation to chromatin modifications and transcription regulation. Nat Clin Pract Endocrinol Metab 2(10):562–570

Katona TM, Jones TD, Wang M et al (2006) Molecular evidence for independent origin of multifocal neuroendocrine tumors of the enteropancreatic axis. Cancer Res 66(9):4936–4942

Kytola S, Hoog A, Nord B et al (2001) Comparative genomic hybridization identifies loss of 18q22-qter as an early and specific event in tumorigenesis of midgut carcinoids. Am J Pathol 158(5):1803–1808

Modlin IM, Lye KD, Kidd M (2003) A 5-decade analysis of 13,715 carcinoid tumors. Cancer 97(4):934–959

Walch AK, Zitzelsberger HF, Aubele MM et al (1998) Typical and atypical carcinoid tumors of the lung are characterized by 11q deletions as detected by comparative genomic hybridization. Am J Pathol 153(4):1089–1098

Carcinoma In Situ

Synonyms

In Situ Carcinoma

Definition

CIS is a lesion that exhibits the cytologic changes of invasive carcinoma but that is limited to the epithelium with no invasion of the basement membrane. For instance, in colorectal cancer, CIS represents an early form of carcinoma that is restricted to the colon mucosa and only locally expands within the mucosa without expanding over the mucosa limits. There is a complete absence of invasion of the surrounding tissues. However, becoming bigger, CIS will then grow over the colon mucosa limits, reach the surrounding vessels, and thus become invasive. Invasive carcinomas, that are able to metastasize, are often the first clinical presentation of colon cancer (► Colorectal Cancer Clinical Oncology). Transitional cell CIS carries a high risk of progression to invasion. CIS also appears in breast cancer and cervical cancer.

Cross-References

► Breast Cancer
► Cervical Cancers
► Colorectal Cancer
► Colorectal Cancer Clinical Oncology
► Dormancy

Carcinoma of the Adrenal Cortex

► Adrenocortical Cancer

Carcinoma Pathogenesis

► Epithelial Tumorigenesis

Carcinoma with Amine Precursor Uptake Decarboxylation Cell Differentiation

► Extrapulmonary Small Cell Cancer

Carcinomatosis

François Noël Gilly and Olivier Glehen
Department of Digestive Oncologic Surgery, Hospices Civils de Lyon–Université Lyon 1, Lyon, France

Synonyms

Peritoneal carcinomatosis; Peritoneal malignancy; Peritoneal tumor

Definition

Until the 1980s, "carcinomatosis" was a condition typically characterized by widespread dissemination of malignant metastases throughout the body. It was then used to describe conditions with more limited spread as in "leptomeningeal carcinomatosis," "lymphangitic carcinomatosis" (which is diffuse malignant infiltration of the lungs with obstruction of the lymphatic channels that occurs most commonly in patients with carcinoma of the breast, lung, stomach, pancreas, prostate, cervix, or thyroid, as well as in patients with metastatic adenocarcinoma from an unknown primary site), and "peritoneal carcinomatosis." Since 2000, due to the worldwide diffusion of cytoreductive surgery combined with intra-abdominal chemotherapy, "carcinomatosis" is a word now describing tumoral spreading within the peritoneal cavity.

Peritoneum is the mesothelial tissue (serosa) that covers most of the organs in the abdominal cavity as well as the interior part of the abdominal wall (parietal peritoneum covers the abdominal walls, and visceral peritoneum covers intra-abdominal organs like the stomach, colon, gallbladder, spleen, liver, etc.). Peritoneum, which is now regarded as an organ by itself, could be the site for "peritoneal carcinomatosis" arising from primary peritoneal tumors or from metastatic nonperitoneal tumors.

Characteristics

Etiology of Peritoneal Carcinomatosis

Primary peritoneal carcinomatosis could arise:

(i) From pseudomyxoma: a rare borderline or malignant mucinous tumor, generally originating from the appendix, incidence of which is estimated 1/million/year and sometimes called the "jellylike fluid disease" – only the histopathologic analysis is able to determine grade 1 (peritoneal adenomucinosis, 84% 5-year survival), grade 3 (peritoneal mucinous adenocarcinoma, 7% 5-year survival), and grade 2 (intermediate, 37% 5-year survival).

(ii) From mesothelioma: peritoneal location of mesothelioma is not so frequent as pleural location, and about 250 new cases are diagnosed in the USA each year; contrary to pleural mesothelioma, its asbestosis exposure relation remains controversial mainly in women, and other carcinogenic agents were reported such as virus, abdominal irradiation, chronic peritonitis, mica, or thorium dioxide exposure.

(iii) From primary serous carcinoma: a very rare disease mostly diagnosed in women and sometimes related to a chromosome deletion.

(iv) From desmoplastic tumors.

(v) From psammocarcinoma: the estimated incidence for these three last diseases remains unknown and is probably $<1/10$ million/ year.

Metastatic peritoneal carcinomatosis is common; it mainly arises from colorectal cancer (detected in about 10% of patients at the time of primary cancer resection), ovarian cancer, and gastric or pancreatic cancers. The mechanisms causing carcinomatosis are multifactorial and include peritoneal dissemination of free cancer cells as a result of serosal involvement of the primary tumor, implantation of free cancer cells caused by the presence of adhesion molecules, and presence of cancer cells in lymph fluid or venous blood retained within the peritoneal cavity (the role of laparoscopic approach in malignant cell diffusion, as well as the role of surgeon during the tumor handling, is still controversial). Metastatic peritoneal carcinomatosis could also arise from extra peritoneal cancer (lymph and/or blood dissemination) such as breast cancer, uterus cancer, thyroid cancer, etc.

Natural history of metastatic peritoneal carcinomatosis is well known from three international prospective series; without curative treatment, the overall median survival is 3–7 months, according to the stage of the peritoneal carcinomatosis.

Diagnosis

Primary and metastatic peritoneal carcinomatosis have no specific symptoms; due to the absence of specific symptoms, clinical diagnosis could be a rather difficult one. Peritoneal carcinomatosis is often diagnosed during surgical exploration of a known primary tumor; if not, symptoms can include abdominal or pelvic pain, changes in bowel functions (up to intestinal obstruction), increase of abdominal volume (caused by ascitis or by tumoral volume itself), abdominal swelling and bloating, infertility (mainly in pseudomyxoma), loss of weight, anorexia, asthenia, etc. Clinical examination of the patient could reveal ascitis or malignant nodules detectable through the abdominal wall.

Abdominal ultrasonography could reveal ascitis or primary ovarian cancer but remains unuseful for detection of peritoneal lesion <1 cm in diameter. While CT scan and MRI help the diagnosis when peritoneal carcinomatosis is made of >5 mm in diameter lesions, PET scan is still under evaluation. Laparoscopic exploration is useful to perform large biopsies and to stage the peritoneal carcinomatosis using the Gilly staging system (Fig. 1) and Sugarbaker peritoneal cancer index (Fig. 2); combination of these two scoring systems was demonstrated as an independent survival predictive factor.

There is no biologic specificity; tumoral markers (CEA, CA 19-9, CA 125, etc.) could be increased in relation to or not with the primary tumor, and molecular markers are still being evaluated.

Microscopic examination of biopsies (or surgically removed tumor) is the key for the diagnosis and could need immune analysis using calretinin, B72.3, Ber EP4, estrogen, and progesterone receptors (mainly to differentiate mesothelioma and primary serous carcinoma).

Treatment

While peritoneal carcinomatosis was thought to be a terminal disease for a long time, most oncologists regarded it as a condition only to be palliated. Systemic chemotherapy (mainly 5-fluorouracil based, oxaliplatinum based, and irinotecan based) combined or not with antiangiogenic drugs achieves a 15-month median survival (most of the available trials are related to the liver and/or pulmonary metastases with only few information regarding peritoneal metastases). Since the

Stages carcinomatosis	
Stage 1	< 5mm, one part
Stage 2	< 5mm, diffuse
Stage 3	5mm to 2 cm
Stage 4	Large malignant cakes

Carcinomatosis, Fig. 1 Gilly staging system for peritoneal carcinomatosis

Carcinomatosis, Fig. 2 Sugarbaker peritoneal cancer index (PCI) for peritoneal carcinomatosis

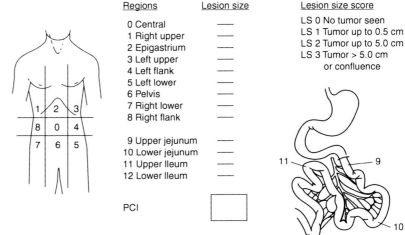

Peritoneal cancer index

Regions
0 Central
1 Right upper
2 Epigastrium
3 Left upper
4 Left flank
5 Left lower
6 Pelvis
7 Right lower
8 Right flank

9 Upper jejunum
10 Lower jejunum
11 Upper Ileum
12 Lower Ileum

PCI

Lesion size
—
—
—
—
—
—
—
—
—

—
—
—
—

Lesion size score
LS 0 No tumor seen
LS 1 Tumor up to 0.5 cm
LS 2 Tumor up to 5.0 cm
LS 3 Tumor > 5.0 cm
or confluence

Carcinomatosis,
Fig. 3 Operative view of right diaphragmatic cupula peritonectomy (the parietal peritoneum is stripped to remove all the macroscopic lesions)

C

1980s, a renewed interest in peritoneal surface malignant diseases developed through new multimodal therapeutic approaches, mainly with cytoreductive surgery combined with intraperitoneal chemotherapy (using mitomycin C, cisplatinum, oxaliplatinum, doxorubicin) with or without hyperthermia (42–43 °C). Despite these aggressive and multidisciplinary approaches are reserved to experienced teams, many phase II and phase III studies revealed a strong advantage for selected patients with colorectal carcinomatosis, peritoneal pseudomyxoma, and mesothelioma.

Cytoreductive surgery (also called peritonectomy procedure) aims to remove as much tumor as possible within the abdominal cavity. The objective is to clear the entire peritoneal cavity of all macroscopic detectable disease. Procedures for cytoreductive surgery (Fig. 3) have been described extensively by Sugarbaker: parietal peritonectomy (which is a stripping of the parietal peritoneum) combined with organ resections (where visceral peritoneum is involved) followed by immediate intraperitoneal chemotherapy with or without hyperthermia (using closed or opened technique) aims to clear all potential microscopic residual disease. Whatever the used technique is, intraperitoneal chemohyperthermia is defined as a heated fluid circulation with cytotoxic drugs for 30–90 min within the abdominal cavity under a nonstop control of core temperature as well as cardiac flow rate. At least, combination of optimal cytoreductive surgery and intraperitoneal chemohyperthermia is a long

surgical, oncologic, and anesthetic procedure (5–10-h long) which requires an experienced multidisciplinary team.

These therapeutic strategies need a strict patient selection (younger than 70 years who have not had cardiorespiratory or renal failure) and need to be done in experienced centers involved in the management of peritoneal surface malignancies. Combination of cytoreductive surgery and intraperitoneal chemotherapy leads to a 1–5% mortality rate and 30% morbidity rate (related to the extent of the carcinomatosis, the duration of surgery, and the number of digestive anastomoses performed). Postoperatively, the patients also received systemic chemotherapy, according to the primary tumor location.

Results are currently encouraging ones. For colorectal carcinomatosis, the Dutch randomized trial showed that 2-year survival was 43% using cytoreductive surgery and intraperitoneal chemohyperthermia versus 16% in the control group; the international registration (including more than 500 patients) showed that 5-year survival was 33% for patients treated by optimal cytoreductive surgery combined with intraperitoneal chemotherapy. Concerning pseudomyxoma (the natural history of this disease is not extensively documented, but the prognosis is better than that for colorectal carcinomatosis), the 5-year survival is 80% for patients with complete cytoreductive surgery (whatever the pathologic grade is). Concerning peritoneal mesothelioma (the median survival in the past was approximately 12 months), optimal cytoreductive surgery and intraperitoneal chemohyperthermia achieve a 5-year median survival. Peritoneal carcinomatosis arising from gastric or ovarian cancer and the use of intraperitoneal chemohyperthermia in a prophylactic way for high locoregional recurrence risk tumors are still under evaluation.

Cross-References

▶ Colorectal Cancer
▶ Cancer
▶ Gastric Cancer
▶ Gastric Cancer Therapy
▶ Hyperthermia
▶ Locoregional Therapy
▶ Mesothelioma
▶ Ovarian Cancer
▶ Ovarian Cancer Chemoresistance
▶ Pseudomyxoma Peritonei

References

Bakrin N et al (2012) Cytoreductive Surgery and Hyperthermic Intraperitoneal Chemotherapy (HIPEC) for Persistent and Recurrent Advanced Ovarian Carcinoma: A Multicenter, Prospective Study of 246 Patients. Ann Surg Oncol

Bakrin N et al (2013) Peritoneal carcinomatosis treated with cytoreductive surgery and Hyperthermic Intraperitoneal Chemotherapy (HIPEC) for advanced ovarian carcinoma: A French multicentre retrospective cohort study of 566 patients. Eur J Surg Oncol

Chua TC, et al (2011) Multi-institutional experience of diffuse intra-abdominal multicystic peritoneal mesothelioma. Br J Surg 98(1): 60–64

Chua TC et al (2012) Early- and Long-Term Outcome Data of Patients With Pseudomyxoma Peritonei From Appendiceal Origin Treated by a Strategy of Cytoreductive Surgery and Hyperthermic Intraperitoneal Chemotherapy. J Clin Oncol

Elias D et al (2010) Pseudomyxoma peritonei: a French multicentric study of 301 patients treated with cytoreductive surgery and intraperitoneal chemotherapy. Eur J Surg Oncol 36(5): 456–462

Glehen O, Mohamed F, Gilly FN (2004) Peritoneal carcinomatosis from digestive tract cancer: new management by cytoreductive surgery and intraperitoneal chemohyperthermia. Lancet Oncol 5:219–228

Glehen O et al (2010) Toward curative treatment of peritoneal carcinomatosis from nonovarian origin by cytoreductive surgery combined with perioperative intraperitoneal chemotherapy: a multi-institutional study of 1,290 patients. Cancer 116(24): 5608–5618

Glehen O et al (2014) GASTRICHIP: D2 resection and hyperthermic intraperitoneal chemotherapy in locally advanced gastric carcinoma: a randomized and multi-center phase III study. BMC Cancer 14: 183

Sadeghi B, Arvieux C, Glehen O (2000) Peritoneal carcinomatosis from non gynaecologic malignancies: results of the EVOCAPE 1 multicentric prospective study. Cancer 88:358–363

Sugarbaker PH (1995) Peritonectomy procedures. Ann Surg 221:29–42

Sugarbaker PH, Chang D (1999) Results of treatment of 385 patients with peritoneal surface spread of appendiceal malignancy. Ann Surg Oncol 6:727–731

Verwall VJ, Ruth S, van de Bree E (2003) Randomized trial of cytoréduction and hyperthermic intraperitoneal chemotherapy versus systemic chemotherapy and palliative surgery n patients with peritoneal carcinomatosis of colorectal cancer. J Clin Oncol 21:3737–3743

Carcinosarcoma

Definition

A malignant tumor that is a mixture of carcinoma (cancer of epithelial tissue, which is skin and tissue that lines or covers the internal organs) and sarcoma (cancer of connective tissue such as bone, cartilage, and fat).

Cardiac Tumors

Jagdish Butany
Laboratory Medicine and Pathobiology,
University Health Network/Toronto, Toronto,
ON, Canada

Definition

Cardiac tumors, like all other tumors, may be classified as primary and secondary and were previously considered as incidental curiosities seen at autopsy. In comparison to the incidence and range of neoplastic proliferations seen in other organs, tumors of the heart are uncommon and have a fairly limited morphologic spectrum. With the advent of innovative diagnostic techniques such as echocardiography, computed tomography (CT) scan, and magnetic resonance imaging (MRI) and better delineation, this has changed, and most tumors are now diagnosed antemortem. With new therapeutic techniques, both surgical and pharmacologic, patients with cancers live longer and show more evidence of cardiac involvement.

Characteristics

The heart was considered the "royal organ" and hence immune to damage, including tumors. We know now that the heart is as prone to disease as most other organs and that includes tumors. The prevalence of immunodeficiency states (especially human immune deficiency virus infection) has also led to an increase in some cancers, involving the heart.

Incidence
Secondary tumors or metastases to the heart and pericardium are 100–1,000 times more common than the primary cardiac tumors with an incidence of about 1.23%. In contrast, the frequency of primary neoplasms ranges from 0.001% to 0.030%, and three quarters of these are benign. The types of benign tumors vary with age. The myxoma and papillary fibroelastoma are more common in adults, while in children, the common ones are cardiac rhabdomyoma and fibroma. Primary cardiac cancers comprise the remaining quarter, and most of these are sarcomas.

Clinical Features
Cardiac tumors, in general, can have a varied clinical presentation, and the pattern depends on the location of the tumor. Most primary tumors, benign and malignant, usually produce intracavitary masses. Such lesions produce one or more patterns of the classic triad of constitutional symptoms, obstruction to the inflow and outflow of the blood within the cardiac chambers and/or complications related to tumor embolization. Embolization refers to the therapeutic introduction of various substances into the circulation to occlude vessels, by purposely introducing emboli. A treatment that clogs small blood vessels and blocks the flow of blood, such as to a tumor, aims either to arrest or prevent hemorrhaging; to devitalize a structure, tumor, or organ by occluding its blood supply; or to reduce blood flow to an arteriovenous malformation. Tumors with significant myocardial involvement may lead to arrhythmias or features related to coronary artery disease due to narrowing or obliteration of small intramyocardial coronary arteries. If the tumor is large or multifocal, it can by itself produce frank, symptomatic cardiac failure. Pericardial tumors are often associated with varying degrees of serous or hemorrhagic effusion with or without tamponade. This is also the presentation with metastatic tumors, which more commonly involve the pericardium. It is noteworthy that 12% of primary tumors and almost 90% of metastases to the heart are clinically silent and detected only at routine assessment or are a surprise at necropsy.

Primary Tumors

Benign
Cardiac myxoma is the commonest tumor encountered in adults and accounts for nearly half the cases. It is more common in females than in males. While no age is immune, it is more common in the 30–60-year-old group. A familial predisposition has been noted in some of these patients. In the familial form, the tumors appear at a younger age, involve the right side of the heart, are often multicentric, and have a high rate of recurrence. There is a germ line mutation in PRKARIA gene, and other associations such as pigmented adrenal micronodular hyperplasia and cutaneous melanocytic or neurogenic tumors are inherited as autosomal dominant disease. These are collectively known as the ▶ Carney complex. A majority of myxomas are solitary, arising as smooth-surfaced, firm, gray-white sessile polypoid masses on either side of the interatrial septum (Fig. 1a, b). Approximately 75% of myxomas are located in the left atrium, where they produce features related to the classical triad. Patients

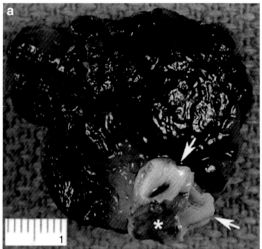
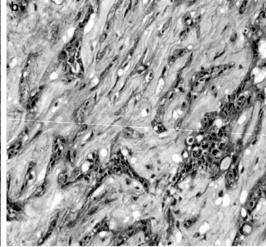

Cardiac Tumors, Fig. 1 (**a**) Polypoidal glistening hemorrhagic myxoma in the left atrium, resected with the underlying interatrial septal endocardium (*arrows*) and myocardium (***). (**b**) Myxoma cells in cords within a mucopolysaccharide (*greenish blue*)-rich stroma (**b**) (Stain Movat pentachrome; original magnification, ×10.0)

with this tumor often present with a typical complaint of "hearing a plop" as they bend forward, associated with severe shortness of breath, and that they hear a second "plop" and improvement of symptoms, as they bend backward. This is due to the tumor "plopping" into the mitral valve orifice and obstructing it. Systemic embolization is often related to the softer, gelatinous, blunt papillary fronds.

Superimposed thrombus or even infection can occur and these can embolize. The cut surface characteristically appears "wet," yellowish white, gelatinous, and translucent with foci of fibrosis, hemorrhage, calcification, and rarely ossification. Histologically, nestling amidst the mucopolysaccharide (jellylike) stroma are the myxoma cells or lipidic cells arranged in the form of cords or form vessel-like structures. Foci of hematopoiesis and intestinal glandular metaplasia may be seen occasionally.

The next common tumor is the papillary fibroelastoma, representing about 8% of cardiac tumors, with no gender predilection. The tumor arises from the endocardium, especially over the cardiac valves, particularly the aortic. The true incidence is not known as they are often small, escaping gross detection; besides, the surgically excised native valves are not always subjected to rigorous histological examination. The morphology of this tumor is best appreciated by holding the mass under water, when a short central stalk with multiple papillary fronds (up to 1 cm or more in length) is seen (Fig. 2a, b). The whole tumor resembles a sea anemone. These papillae are delicate and often break off and embolize. Additionally, thrombi can also occur over the surfaces of the papillae with subsequent embolization. In fact, one of the presentations of this tumor is the sudden development of blindness of one eye or the sudden development of chest pain. Sudden death has also been reported. The fronds have a soft myxoid core surrounded by collagen and elastic tissue fibers, lined by endothelial cells.

Rhabdomyoma, considered a hamartoma, is the most common tumor found in children, with an overall incidence of about 5%. Sporadic rhabdomyomas are mostly seen as solitary, small or large (0.1 cm or more), firm, opaque white, and well-circumscribed endocardial nodules. They are usually found in the left ventricle and the interventricular septum and produce obstructive symptoms. In contrast to the sporadic variants, about 50% of patients with tuberous sclerosis have multiple rhabdomyomas, leading to intrauterine hydrops fetalis and stillbirths. It is important to rule out the presence of tuberous sclerosis in not

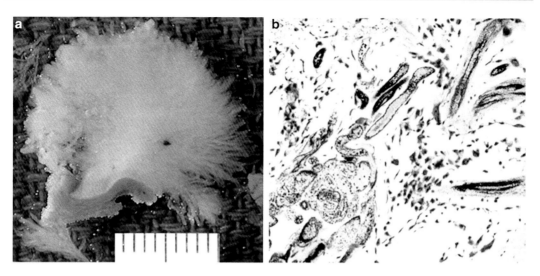

Cardiac Tumors, Fig. 2 (**a**) "Sea anemone"-like appearance of papillary fibroelastoma, photographed under water. (**b**) Delicate papillae with a core of collagen and elastic, lined by plump endothelial cells (**b**) (Stain hematoxylin and eosin; original magnification, ×10.0)

only multiple tumors but also in patients with solitary lesions. The rhabdomyomas are composed of ballooned out cardiac myocytes with clear vacuolated pink-staining cytoplasm, radiating from the centrally located nucleus to the periphery, and responsible for the "spider-cell" appearance.

Cardiac fibroma, the second common tumor in the pediatric population, is seen as a solitary, circumscribed, firm to hard, gray-white fasciculated tumor occurring in the ventricular chambers with a predilection to affect the interventricular septum. The mean age of presentation is around 13 years and these are the commonly resected tumors at that age. There is a proliferation of innocuous appearing fibroblasts, which characteristically entrap islands of myofibers (pseudo-invasion). With increasing age, the tumors are rendered paucicellular, with multifocal areas of calcification.

Both rhabdomyomas and fibromas are considered as congenital lesions that undergo gradual spontaneous regression or cease to be progressive. Other benign lesions include the hemangioma (vascular) and lipoma (fat-cell tumor). Brief mention must be made of the cystic tumor of the atrioventricular node, the smallest known tumor, associated with sudden death. It is composed of small fluid-filled endoderm-derived spaces in a connective tissue stroma. The spaces may be lined a variety of epithelial and epitheliod cells.

Malignant

Sarcomas are the most common malignant primary tumors of the heart, forming ~10% of the surgically resected cardiac neoplasms. Angiosarcoma is the commonest with an incidence of 35–37%, slight male predominance and occurs in the third to fifth decade of life. Most often, it is seen as an irregular, soft, friable, hemorrhagic mass in the right atrium, projecting into the cavity as well as infiltrating the wall and adjacent structures. Clinical presentation is often with recurrent, hemorrhagic pericardial effusion and/or symptoms related to metastases (seen in more than half of the patients), frequently pulmonary or rarely even distant. At times, the angiosarcomas may be purely pericardial, forming a sheetlike mass. Depending on the degree of differentiation, angiosarcomas show irregular, anastomosing vascular channels lined by plump, atypical endothelial cells with papillary projections or sheets of epithelioid or spindled cells. If areas of spindle-shaped cells predominate, the presence of red blood cells in the stroma and the presence of intracellular vacuoles offer

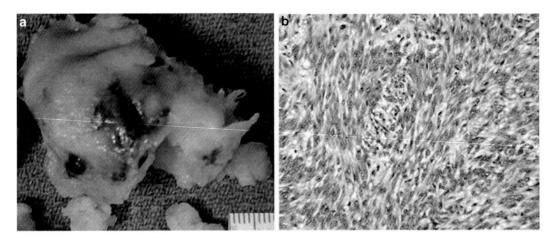

Cardiac Tumors, Fig. 3 (**a**) Left atrial sarcoma resected in pieces. (**b**) Malignant spindle-shaped cells arranged in intersecting bundles, suggestive of leiomyosarcoma (**b**) (Stain hematoxylin and eosin; original magnification, ×20.0)

important clues to the diagnosis. Areas of necroses and brisk mitotic activity are evident. Some of the tumors are also associated with chromosomal abnormalities as seen by cytogenetic analysis.

The next common sarcomas are a group of sarcomatous proliferation, which share many morphological and clinical features. These are designated as myofibroblastic sarcomas and include malignant fibrous histiocytoma, fibrosarcoma, fibromyxosarcoma, and myxosarcoma (Fig. 3a, b). Many of them in addition can show focal osteosarcomatous or chondrosarcomatous differentiation. The mean age of presentation is around 40 years of age with no gender predilection. They form bulky, lobulated, polypoidal graywhite masses projecting into the left atrium with the result that many of these patients present with the left-sided inflow tract obstruction. Most of them resemble their soft tissue counterparts, but often these tumors in their cardiac location show a prominent myxoid change. Hence, at times, many such tumors are misdiagnosed as myxomas.

Undifferentiated sarcomas constitute about 10–24% of these malignant cardiac tumors. They are designated undifferentiated, as they are not associated with specific, classifiable morphological, immunohistochemical, or ultrastructural features. These are present as large, lobulated, polypoidal masses, chiefly in the left-sided chambers in middle-aged adults. They can affect either gender, with a wide age range of occurrence.

Patients are often symptomatic at an early stage. The tumors are composed of proliferating spindle cells or epithelioid-looking cells and pleomorphic cells in varying pattern.

Rhabdomyosarcomas, which constitute about 5% of the tumors, are common in younger patients, especially children. They too form very bulky and infiltrative masses in either of the ventricular chambers. Majority are embryonal rhabdomyosarcomas. The other sarcomas include leiomyosarcoma and synovial sarcoma, though virtually any type of soft tissue sarcoma may occur.

Primary lymphomas of the heart are extremely rare. However, their incidence has increased, in patients who are HIV positive as well as those with other causes of immunosuppression. An important criterion in these patients is the demonstrable absence of lymphomatous proliferation or any other sarcoma at any other site, before they can be categorized as a primary cardiac tumor. These patients therefore require thorough imaging investigations. In this type of lymphoma, there is a slight male predominance, and multiple, soft to firm, creamy white nodules are seen, especially in the right atrium. They mostly exhibit the full range of the neoplastic B cell proliferation, though large cell lymphomas are common.

Sarcomas can also affect the great vessels. Leiomyosarcomas involve chiefly the veins especially the inferior vena and less commonly the

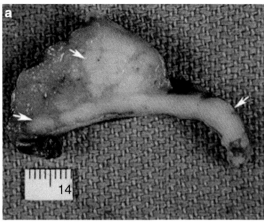 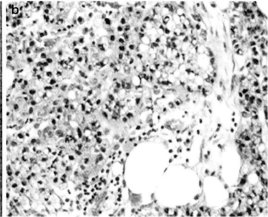

C

Cardiac Tumors, Fig. 4 (**a**) Cut surface (viewed en face) of an excised piece of pericardium showing metastatic tumor, with diffuse *gray-white* thickening and a large nodule, with some surrounding *yellow* adipose (fatty) tissue. (**b**) Histology shows a metastatic adenocarcinoma composed of clusters of mucin-secreting cells (Stain hematoxylin and eosin; original magnification, ×20.0)

superior vena cava and azygos vein. Women with a mean age of 49 years are affected, with clinical features of pain and venous obstructions. The growth may be intraluminal or extraluminal. Extraluminal tumors, despite their extension into the surrounding tissues, appear circumscribed and lobulated. Metastases usually involve the lungs, though other organs like the liver or kidney can also be affected, especially with inferior vena caval tumors. Sarcomas of the great arteries, pulmonary trunk, and aorta arise from the multipotential mesenchymal cells of the intima and are designated as intimal sarcomas. Despite the intimal origin, the tumor can have an intraluminal or mural growth patterns. The patients are usually elderly and the symptoms depend on the growth pattern. Aortic involvement is more common than pulmonary. The luminal tumors often produce sheetlike or plaque-like growths, which in due course of time can form an intraluminal polyp. There can be superimposed thrombi, which may form the major chunk over the tumor, leading to misleading diagnosis. The patients present with effects of obstruction or embolization. The mural growth pattern is uncommon and there is medial and adventitial infiltration with resultant local invasion. Most of them are of the undifferentiated type, which have 50% shorter survival as compared with the differentiated type. The latter includes myxofibrosarcoma, angiosarcoma, malignant fibrous histiocytoma, leiomyosarcoma, or myxoid chondrosarcoma.

Secondary or Metastatic Tumors

The commonest tumor seen in the heart is the metastatic or secondary tumor and is seen in at least 3% of patients with cancer, that is, at least 3% of patients with a malignancy (usually advanced ones) have cardiac metastases (Fig. 4a, b). Primary cancers can spread to the heart by direct extension from adjoining structures, hematogenous or lymphatic spread, and sometimes as extensions through the inferior vena cava and even the pulmonary veins. The noncardiac solid organ primary cancers may be placed in three groups, depending on their propensity to produce metastases: uncommon malignant tumors with a high incidence of cardiac metastases (malignant melanoma and malignant germ cell tumors), common cancers with an intermediate frequency of cardiac involvement (carcinoma of lung in males and breast cancer in females), and common cancers with rare metastases (cervical carcinoma).

Of all malignancies, leukemias have the highest incidence of cardiac involvement; however, this infiltrative process usually does not produce symptoms. There is a diffuse or patchy interstitial infiltrate of neoplastic cells. On the other hand, the solid tumor metastases produce multiple or single nodules over the epicardial surface. Occasionally, nodules may also be found in the myocardium or on the endocardial surface. The mode of presentation therefore depends on the location of the tumor. The tumors do not pose a diagnostic problem, as the histologic appearance is usually similar to that of the primary site.

Therapy and Prognosis

Benign tumors, which are symptomatic or located at sites that might lead to catastrophic complications, are resected. Primary cancers with limited growth and no evidence of metastases are treated by surgical resection with adjuvant chemotherapy and radiotherapy. The latter may be the only treatment option available in very large tumors which are not amenable to even palliative debulking. Autotransplantation or orthotopic transplantation may be an option in some cases. Primary lymphomas, on the other hand, are usually best treated by combined chemotherapy and radiotherapy. The prognosis in benign tumors after surgical excision is usually excellent, except for a rare case of recurrences, likely due to incomplete resection. The prognosis for cardiac sarcoma is extremely dismal with a mean survival of 3 months to a year.

Cross-References

▶ Aging
▶ AIDS-Associated Malignancies
▶ Carcinogenesis
▶ Carcinoid Tumors
▶ Carney Complex
▶ Staging of Tumors
▶ Targeting Cancer Stem Cells

References

Butany J, Leong SW, Carmichael K, Komeda M (2005a) A 30-year analysis of cardiac neoplasms at autopsy. Can J Cardiol 21(8):675–680

Butany J, Nair V, Naseemuddin A, Nair GM, Catton C, Yau T (2005b) Cardiac tumours: diagnosis and management. Lancet Oncol 6(4):219–228

Gazit AZ, Gandhi SK (2007) Pediatric primary cardiac tumors: diagnosis and treatment. Curr Treat Options Cardiovasc Med 9(5):399–406

Neragi-Miandoab S, Kim J, Vlahakes GJ (2007) Malignant tumours of the heart: a review of tumour type, diagnosis and therapy. Clin Oncol (R Coll Radiol) 19(10):748–756

Sarjeant JM, Butany J, Cusimano RJ (2003) Cancer of the heart: epidemiology and management of primary tumors and metastases. Am J Cardiovasc Drugs 3(6):407–421

Carney Complex

Jérôme Bertherat
Endocrinology, Metabolism and Cancer Department, INSERM U567, Institut Cochin, Paris, France

Synonyms

Atrial myxoma; Blue nevi; CNC; Ephelide; LAMB; Lentigineses; Mucocutaneous myxoma; Myxoid neurofibroma; NAME; Nevi

Definition

The Carney complex (CNC) is a dominant autosomal hereditary multiple neoplasia syndrome characterized mainly by cardiac myxomas (a benign tumor of the heart and the most common type of heart tumor in adults; cardiac myxomas can appear in an isolated case or in families), spotty skin pigmentation, and endocrine tumors. It was first described in 1985 by J. Aidan Carney, a pathologist at the Mayo Clinic.

Characteristics

The manifestations of CNC can be numerous and vary between patients. Even in the same kindred, phenotypic variability can be observed. The estimated frequencies of these manifestations are listed in Table 1. Endocrine, dermatologic, and cardiac anomalies are the main manifestations of the disease. The lentiginosis is observed in most patients and is so characteristic that can make the

Carney Complex, Table 1 Main manifestations of Carney complex

Main features of Carney complex	Frequency (%)
Primary pigmented nodular adrenocortical disease (PPNAD)	25–60
Cardiac myxoma	30–60
Skin myxoma	20–63
Lentiginosis	60–70
Multiple blue nevus	
Breast ductal adenoma	25
Testicular tumors (LCCSCT, large cell calcifying Sertoli cell tumor) (in male)	33–56
Ovarian cyst (in female)	20–67
Acromegaly	10
Thyroid tumor	10–25
Melanotic schwannoma	8–18
Osteochondromyxoma	<10

diagnosis. It appears as small brown to black macules typically located around the upper and lower lips and on the eyelids, ears, and the genital area. Multiple blue nevi and junctional or compound nevi may also be observed in CNC, as well as cutaneous myxomas. The skin myxomas present as nonpigmented subcutaneous nodules. Myxomas can also be located in the ear canal.

Table 1 lists the most frequent features of CNC and their estimated frequency. The incidence of each manifestation depends on its presentation and might not reflect true prevalence. For instance, according to autopsy studies, primary pigmented nodular adrenocortical disease (PPNAD) is a constant feature in CNC patients; however, reports of ▶ Cushing syndrome in the literature indicate that only 25–45% of CNC patients have PPNAD.

Cardiac myxoma is an important manifestation of CNC. It may be the cause of the high rate (16%) of sudden death historically reported in CNC families, thus, underlying the importance of its early diagnosis. In the past, underdiagnosis of cardiac myxomas may have accounted for the majority of deaths due to CNC. In contrast with sporadic myxoma, they can develop in any cardiac chamber and may be multiple. Cardiac myxoma can be the cause of stroke due to embolism and cardiac deficiency. It is therefore important to screen regularly (by ultrasound) patients with CNC for the presence of cardiac myxoma. In difficult cases, transesophageal ultrasound and cardiac magnetic resonance imaging (MRI) can be very helpful.

Endocrine tumors are also a major manifestation of the disease. Most characteristic is the adrenocorticotropic hormone (ACTH)-independent ▶ Cushing syndrome due to PPNAD observed in 30–60% of patients with CNC. The disease was named after the macroscopic appearance of the adrenals that is characterized by the small pigmented micronodules observed in the cortex. The disease is usually bilateral with primary involvement of both adrenals. Cushing syndrome due to PPNAD is most often observed in children and young adults, with a peak during the second decade of life. Diagnosis of Cushing syndrome due to PPNAD is often difficult because hypercortisolism can develop progressively over years. In contrast, a large and rapid burst of cortisol excess can be observed in some patients who might spontaneously regress. In some cases of PPNAD, clearly cyclic forms of hypercortisolism have been documented. PPNAD can also be diagnosed by systematic screening in patients with CNC, investigated for other clinical manifestations of the complex, or after familial screening. Despite the unusual time course of Cushing syndrome observed in some patients with PPNAD, clinical signs are quite similar to those observed in patients having other causes of hypercortisolism. Urinary cortisol is increased in most patients at the time of diagnosis of PPNAD, but its level can be highly variable. The circadian rhythm of cortisol secretion is usually completely abolished. As with ACTH-independent Cushing syndrome due to other causes, patients with PPNAD have low plasma levels of ACTH and show no stimulation of cortisol or ACTH secretion after corticotropin-releasing hormone (CRH) injection. In addition, dexamethasone fails to suppress cortisol secretion, even after high-dose administration. Pathological investigation reveals that adrenal glands from patients with PPNAD are usually normal in size and weight (between 4 and 17 g). In keeping with this finding, adrenals appear normal on computed tomography (CT) scan in one out of three patients. In the other patients, micronodules can be visible and, more rarely, macronodules (>1 cm diameter) in one or both glands. Iodocholesterol scintigraphy, when performed, usually shows a bilateral uptake despite ACTH suppression by endogenous hypercortisolism.

Acromegaly due to a pituitary GM-CSF-secreting tumor is not very frequent, but most patients with CNC present with a mild increase in GH and sometimes in ► prolactin (PRL) secretion.

Alterations in the rhythm of GH secretion are frequently observed. Thyroid tumors are most often benign, nontoxic adenomas, mostly of follicular type. Some patients present with papillary carcinoma that can be multiple and sometimes quite aggressive. Testicular tumors (large cell calcifying Sertoli cell tumors (LCCST)) are easily detected by ultrasound investigation as bilateral microcalcifications. They can be diagnosed by ultrasound. Ovarian cysts and cystadenoma have been observed in CNC patients.

Various other tumors, some of them quite specific for CNC, can be observed. Melanotic schwannoma is a rare tumor and occurs mainly in CNC. It is a pigmented tumor that can be misdiagnosed as a melanoma. This tumor can be observed in any peripheral nerve and can be, in rare cases, malignant. Breast ductal adenomas, breast myxomas, and osteochondromyxoma are among the tumors also observed in CNC.

CNC is an autosomal dominant hereditary disease, and at least two loci have been postulated: 2p16 and 17q22-24. The *CNC1* gene, located on 17q22-24, has been identified as the regulatory subunit (R1A) of the protein kinase A (*PRKAR1A*). PRKAR1A is a key component of the cAMP-signaling pathway that has been implicated in endocrine tumorigenesis. Heterozygous inactivating mutations of *PRKAR1A* have been detected in about 65% of CNC families. In CNC patients with Cushing syndrome, the frequency of *PRKAR1A* mutations is about 80%, suggesting that families with PPNAD are more likely to carry a 17q22-24 defect. Interestingly, patients with isolated PPNAD and no familial history of CNC may also carry a germ line de novo mutation in *PRKAR1A*. In the tumors of CNC patients, loss of heterozygosity (LOH) at 17q22-24 may be observed, suggesting that *PRKAR1A* is a tumor suppressor gene. Somatic mutation of *PRKAR1A* in a patient with PPNAD already carrying a germ line mutation may lead to inactivation of the wild-type allele. However, inactivation of the remaining wild-type allele by genetic alteration does not appear to be a constant step in PPNAD and CNC tumor development. In a mice transgenic model with heterozygous inactivation of *PRKAR1A*, tumors may develop without allelic loss. This suggests that the classic model of tumor suppressor gene with a germ line inactivating first allelic alteration, followed by a second genetic hit leading to inactivation of the remaining wild-type allele, might to some extent be applicable to PRKAR1A. It is also possible that in PPNAD, a general polyclonal expansion might be stimulated by haploinsufficiency due to the first germ line defect; a second genetic hit would then lead to the inactivation of the wild-type allele and further stimulate tumorigenesis and the development of adrenocortical nodules. PRKAR1A inactivation in transgenic models is associated with an increased PKA activity. Stimulation of the MAP kinase pathway as well as mTOR phosphorylation has been observed in experimental models of PRKAR1A inactivation and might be a mechanism for oncogenesis in CNC.

Considering the genetics of isolated PPNAD, the clinical manifestations in a subgroup of very young PPNAD patients may differ from those in older patients with CNC. In these patients, the classical pathological finding of pigmented nodules may be absent although micronodules are present. In this subgroup of very young PPNAD patients, Cushing syndrome may occur between birth and the age of 5 years. The main reason for differentiating this group of PPNAD or PPNAD-like patients is the lower rate of germ line inactivating mutation. This observation leads to the identification by genome-wide screen of a gene responsible for isolated PPNAD: the phosphodiesterase PDE11A4. The affected patient presents with a germ line heterozygous mutation of *PDE11A4* gene located at 2q31-35. An allelic loss at 2q31-35 is observed in adrenal tissue from these patients, suggesting also that PDE114 might be a tumor suppressor gene. Inactivating mutations lead to increased cAMP and cGMP levels in keeping with the observation that PDE11A4 is a dual phosphodiesterase.

Cross-References

► Cushing Syndrome
► Osteochondroma
► Prolactin

References

Carney JA, Gordon H, Carpenter PC et al (1985) The complex of myxomas, spotty pigmentation, and endocrine overactivity. Medicine (Baltimore) 64(4):270–283

Groussin L, Cazabat L, Rene-Corail F et al (2005) Adrenal pathophysiology: lessons from the Carney complex. Horm Res 64(3):132–139

Horvath A, Boikos S, Giatzakis C et al (2006) A genome-wide scan identifies mutations in the gene encoding phosphodiesterase 11A4 (PDE11A) in individuals with adrenocortical hyperplasia. Nat Genet 38(7):794–800

Kirschner LS, Carney JA, Pack SD et al (2000) Mutations of the gene encoding the protein kinase A type I-alpha regulatory subunit in patients with the Carney complex. Nat Genet 26(1):89–92

Veugelers M, Wilkes D, Burton K et al (2004) Comparative PRKAR1A genotype-phenotype analyses in humans with Carney complex and prkar1a haploinsufficient mice. Proc Natl Acad Sci U S A 101(39):14222–14227

See Also

(2012) Acromegaly. In: Schwab M (ed) Encyclopedia of Cancer, 3rd edn. Springer Berlin Heidelberg, p 18. doi: 10.1007/978-3-642-16483-5_39

(2012) Benign Tumor. In: Schwab M (ed) Encyclopedia of Cancer, 3rd edn. Springer Berlin Heidelberg, p 381. doi: 10.1007/978-3-642-16483-5_579

(2012) Cardiac Myxoma. In: Schwab M (ed) Encyclopedia of Cancer, 3rd edn. Springer Berlin Heidelberg, p 661. doi: 10.1007/978-3-642-16483-5_855

(2012) Corticotrophin-Releasing Hormone. In: Schwab M (ed) Encyclopedia of Cancer, 3rd edn. Springer Berlin Heidelberg, p 983. doi: 10.1007/978-3-642-16483-5_1343

(2012) Granulocyte-Colony Stimulating Factor. In: Schwab M (ed) Encyclopedia of Cancer, 3rd edn. Springer Berlin Heidelberg, p 1597. doi: 10.1007/978-3-642-16483-5_2505

(2012) Large Cell Calcifying Sertoli Cell Tumor. In: Schwab M (ed) Encyclopedia of Cancer, 3rd edn. Springer Berlin Heidelberg, p 1980. doi: 10.1007/978-3-642-16483-5_3278

(2012) Lentiginosis. In: Schwab M (ed) Encyclopedia of Cancer, 3rd edn. Springer Berlin Heidelberg, p 2000. doi: 10.1007/978-3-642-16483-5_3309

(2012) Melanotic Schwannoma. In: Schwab M (ed) Encyclopedia of Cancer, 3rd edn. Springer Berlin Heidelberg, p 2217. doi: 10.1007/978-3-642-16483-5_3616

(2012) PKA. In: Schwab M (ed) Encyclopedia of Cancer, 3rd edn. Springer Berlin Heidelberg, p 2895. doi: 10.1007/978-3-642-16483-5_4581

(2012) Primary Pigmented Nodular Adrenocortical Disease. In: Schwab M (ed) Encyclopedia of Cancer, 3rd edn. Springer Berlin Heidelberg, p 2988. doi: 10.1007/978-3-642-16483-5_4741

Carotenoids

Heather Mernitz[1] and Xiang-Dong Wang[2]
[1]Alverno College, Milwaukee, WI, USA
[2]Jean Mayer USDA Human Nutrition Research Center on Aging at Tufts University, Boston, MA, USA

Definition

Carotenoids are lipophilic plant pigments with polyisoprenoid structures that occur naturally in plants and other photosynthetic organisms. There are over 600 known carotenoids with chemical structures characterized by a large (35–40 carbon atoms) conjugated polyene chain, sometimes terminated by ring structures. Carotenoids are divided into two major groups: xanthophylls, oxygenated carotenoids including lutein, zeaxanthin, and β-cryptoxanthin, and carotenes, hydrocarbon carotenoids that are either cyclized, such as α-carotene and β-carotene, or linear like lycopene. The most abundant carotenoids in human plasma include lutein, lycopene, β-carotene, zeaxanthin, β-cryptoxanthin, and α-carotene. The two main mechanisms by which carotenoids may influence cancer risk are by exerting antioxidant effects and through interaction with ligand-dependent nuclear hormone receptors and their signaling pathways. The capacity of carotenoids to act as lipid-soluble antioxidants serves a functional and protective

role in plants during photosynthesis and may protect animals against free radical damage to lipid
membranes and DNA. In addition, carotenoids
may enhance cell-cell gap junctional communication and induce phase II detoxifying enzymes (see
"► Carcinogen Metabolism"). Provitamin
A carotenoids (e.g., β-carotene, α-carotene, and
β-cryptoxanthin) can be cleaved to generate vitamin A and other metabolites that interact with
signaling pathways controlling gene expression.
Non-provitamin A carotenoids (e.g., lutein, zeaxanthin, and lycopene) not only have significant
antioxidant activity, but their metabolites may
also exert effects on gene expression (Fig. 1).

Characteristics

► DNA damage by free radicals (including ► reactive oxygen species and reactive nitrogen species)
is thought to be a major contributor to ► carcinogenesis, and carotenoids contain an extended system of conjugated double bonds that make them
efficient scavengers of free radicals. While antioxidant supplements and diets high in antioxidant
nutrients, including carotenoids (lycopene,
β-carotene, lutein, β-cryptoxanthin), have been
shown to reduce DNA strand breaks and other
biomarkers of ► oxidative DNA damage, it is still
unclear whether these changes are sufficient to

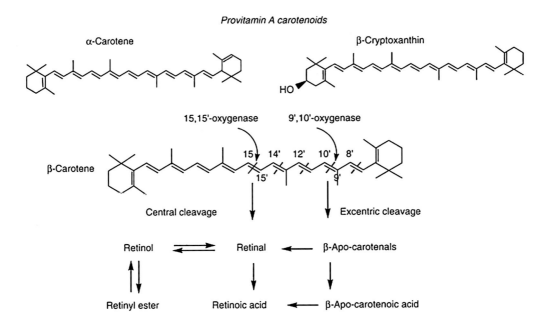

Carotenoids, Fig. 1 Metabolic pathway of β-carotene and chemical structures of provitamin A carotenoids (β-carotene, α-carotene, and β-cryptoxanthin) and non-provitamin A carotenoids (lutein, zeaxanthin, and lycopene)

lower cancer risk in humans with chronic exposure to low levels of carcinogenic compounds over a lifetime. Epidemiological studies suggest that a higher dietary intake of carotenoids and high levels of certain carotenoids in the plasma may offer protection against the development of certain cancers (e.g., lung, prostate, stomach, colon, breast), as well as other health conditions linked to oxidative damage (e.g., heart disease, macular degeneration, cataracts). However, two intervention trials, the Beta-Carotene and Retinol Efficacy Trial (CARET) and the Alpha-Tocopherol, Beta-Carotene Cancer Prevention Study (ATBC), have shown that supplementation with high-dose β-carotene, alone or in combination with vitamin A, does not reduce the risk of lung cancer and may even increase that risk in smokers and ► asbestos workers. These findings have led to an increased effort to better understand the role of carotenoids and ► retinoids (vitamin A and its derivatives) in the process of carcinogenesis, with special attention to dose and the oxidative environment at the tissue level. Based on the accumulated evidence, it appears that low-dose carotenoids (similar to the amounts consumed in a diet high in fruits and vegetables) may act as antioxidants and protect against cancer, whereas at high doses, carotenoids may lose their effectiveness as antioxidants, function as prooxidants, and/or interfere with retinoid signaling pathways, increasing cancer risk.

The series of conjugated double bonds in the central chain of carotenoids make them susceptible to oxidative cleavage and isomerization from *trans* to *cis* forms. Cleavage can result in the formation of potentially bioactive metabolites, such as retinoids and other biological compounds. For provitamin A carotenoids, such as β-carotene, α-carotene, and β-cryptoxanthin, central cleavage by β-carotene 15,15′-oxygenase, a nonheme iron oxygenase enzyme which can cleave carotenoids at their central 15,15′ double bond, is a major pathway leading to vitamin A formation. An alternative pathway for carotenoid metabolism into vitamin A in mammals is excentric cleavage or asymmetric cleavage. Characterization and study of β-carotene 9′,10′-oxygenase have demonstrated that this enzyme a mitochondrial protein can catalyze the excentric cleavage of both

provitamin A carotenoids and non-provitamin A carotenoids. Since disruption in retinoid metabolism and signaling may play a key role in the process of carcinogenesis, understanding the molecular details behind the actions of these carotenoid oxidative metabolites may yield insights into both physiological and pathophysiological processes in human health and disease, particularly the potential for beneficial effects of small quantities of carotenoids and harmful effects of large quantities of carotenoid metabolites.

Disruption of carotenoid and retinoid metabolism and signaling due to diet and lifestyle factors has been associated with increased risk of cancers at multiple sites (see "► Nutrition Status" and "► Hepatic Ethanol Metabolism"). Cigarette smoking is associated with substantially decreased plasma levels of carotenoids, despite only slightly lower intakes of carotenoids in smokers compared to nonsmokers. Several hypotheses have been proposed to explain this increased metabolism of carotenoids in the tissues of smokers, including increased induction of metabolic enzymes, excentric cleavage of β-carotene into harmful oxidative products, and oxidative degradation of cellular antioxidants (e.g., ascorbic acid, α-tocopherol) that normally serve to stabilize the reduced form of β-carotene. While initial cleavage of provitamin A carotenoids can lead to the generation of ► retinoic acid, a bioactive form of vitamin A, additional oxidation can lead to degradation into polar metabolites. With respect to lung cancer, laboratory studies have demonstrated that the oxidative cleavage products of β-carotene, when formed in large quantities in the cell after supplementation with high-dose β-carotene in the highly oxidative environment of the smoke-exposed lung, enhance catabolism of retinoic acid by their induction of cytochrome P450 enzymes (CYP enzymes) and facilitate the binding of carcinogen ► adducts to DNA. Lower retinoic acid levels then combine with smoke-induced changes to alter ► signal transduction pathways and promote lung carcinogenesis. On the other hand, low-dose β-carotene supplementation, particularly when combined with other antioxidants, inhibits tobacco smoke-induced changes in retinoic acid levels and signaling in the lung tissue, preventing the formation of smoke-induced ▶ preneoplastic lesions (see "▶ Tobacco Carcinogenesis").

Therefore, the effects of provitamin A carotenoids can be mediated by conversion to retinoic acid and transcriptional activation of a series of genes with distinct antiproliferative or proapoptotic activity or by induction of ▶ apoptosis, eliminating cells with unrepairable alterations in the genome or killing neoplastic cells. Certain carotenoids may also be able to interact directly or indirectly with transcription factors, such as retinoic acid receptors (RARs), peroxisome proliferator-activated receptors (PPARs), nuclear factor E_2-related factor 2 (Nrf2), or orphan receptors, or may indirectly influence transcriptional activity of redox-sensitive transcription systems, such as activator protein-1 (▶ AP-1), ▶ nuclear factor-κB (NF-κB), and the antioxidant response element (ARE). Greater understanding of the biological functions of carotenoids mediated via their oxidative metabolites through their effects on these important cellular signaling pathways and molecular targets, as well as their significance to cancer prevention, is needed. In considering the efficacy and complex biological functions of carotenoids in human cancer prevention, it seems that increasing consumption of vegetable and fruits rich in carotenoids as a part of a balanced diet would be an effective chemopreventive strategy against cancer development.

References

Palozza P, Serini S, Ameruso M et al (2009) Modulation of intracellular signaling pathways by carotenoids. In: Britton G, Liaaen-Jensen S, Pfander H (eds) Carotenoids: Vol 5: Nutrition and Health. Birkhauser Verlag, Basel, pp 211–234

Rock CL (2009) Carotenoids and cancer. In: Britton G, Liaaen-Jensen S, Pfander H (eds) Carotenoids: Vol 5: Nutrition and Health. Birkhauser Verlag, Basel, pp 269–286

Wang XD (2012) Carotenoids. In: Ross AC, Caballero B, Cousins RJ, Tucker KL, Ziegler TR (eds) Modern nutrition in health and disease. Lippincott Williams & Wilkins, Philadelphia, pp 427–439

Yeum KJ, Aldini G, Russell RM, et al (2009) Antioxidant/Pro-oxidant actions of carotenoids. In: Britton G, Liaaen-Jensen S, Pfander H (eds) Carotenoids: Vol 5: Nutrition and Health. Birkhauser Verlag, Basel, pp 235–268

Case Control Association Study

Ahmed E. Hegab
Department of Geriatric and Respiratory Medicine, Tohoku University Hospital, Sendai, Japan

Synonyms

Case–control association analysis; Genetic association study; Population candidate gene association study

Definition

Case–control association study aims to detect association between one or more genetic markers (usually a polymorphism but also may be a microsatellite) and a trait, which might be a disease (e.g., lung cancer), a quantitative characteristic (e.g., serum level of a ▶ cytokine), or a discrete attribute.

Characteristics

Several genetic methods are used for detecting genes responsible for the development of complex human diseases; these are nonparametric linkage analysis, case–control association analysis, and DNA microarray.

Case–control association analysis involves selecting genes that are likely to be associated with the pathogenesis of disease based on our understanding of its pathophysiology. Then genetic polymorphisms in these candidate genes are investigated in a large number of unrelated patients and healthy ethnically matched controls. Significant differences in genotype or allele frequencies between the two groups suggest either that (i) the polymorphism predisposes one to the disease, (ii) the polymorphism is in ▶ linkage disequilibrium with a disease susceptibility gene, or (iii) there is a confounding factor such as poor ethnic matching between the cases and controls.

When several markers are being examined for association with the same trait, it is advisable to check them for linkage disequilibrium and disease-associated genetic haplotypes.

Association studies have greater power than linkage analysis. They can detect genes with a relative risk of 1.5 at nearly 80% probability if several hundred samples are collected. However, since association studies examine much smaller regions than linkage analyses, many more markers would need to be typed to conduct a genome-wide association study. This is not possible with current technology. At present, association studies are limited to the investigation of candidate genes and regions identified in linkage analysis. As association studies are not comprehensive, the possibility that the most important genes have been overlooked cannot be excluded. Choosing candidate genes from areas spotted by linkage analysis might be the most fruitful practice.

When performing case–control association studies, factors such as study design, methods for recruitment of case and controls, selection of candidate genes, functional significance of polymorphisms chosen for study, and statistical analysis require close attention to ensure that only genuine associations are detected.

Potential Problems in Association Studies

Marked inconsistency can be observed between association studies in different or even the same ethnic groups. The association between tumor necrosis factor-alpha promoter polymorphism and gastric cancer is an example for inconsistency, while the association of head and neck cancers with M1 polymorphism of glutathione S-transferase gene is an example for the reproducible associations. This situation has led some commentators to question the value of genetic association studies, suggesting that association studies should be restricted to polymorphisms that have been shown to have a direct effect on gene function. Possible explanations for these inconsistent results include:

1. Different populations might have different genetic components for the same disease phenotype. Furthermore, the degree of association between the specific gene and disease is different between populations. It is also possible that the different environments to which every population is exposed will interact differently with the genetic components responsible for the development of the disease.

2. Most complex human diseases are heterogeneous disorders, e.g., leukemia. It may happen that patients diagnosed with leukemia in different ethnic groups have distinct types of polymorphisms causing a specific type of leukemia.

3. When the population under study consists of a mixture of two or more subpopulations that have different allele frequencies, associations between genotype and outcome could be confounded by population stratification.

4. Ascertainment errors include undiagnosed affected individuals in the control group or including patients with heterogeneous etiologies for the complex disease in the patient group.

5. Failure to check for Hardy–Weinberg equilibrium. The presence of disequilibrium in the control group can result from genotyping errors, inbreeding, small sample size, or mutation.

6. A possible reason for failure to replicate positive findings is that subsequent studies are underpowered.

7. Failure to exclude chance is the most likely explanation for difficulty in replication of reports of genetic associations with complex diseases. Applying a significance level of $p = 0.05$ leads to one false positive in 20 results. In order to avoid type I error, p values calculated from association studies must be corrected for the number of loci analyzed (x) and the number of alleles at each loci (y). In the Bonferroni correction, the required significance level should be divided by $x(y - 1)$. However, this method is too conservative because closely located loci are not usually independent. The appropriate correction for multiple comparisons in association studies remains unclear.

8. Publication bias should be considered. Negative results in association studies may not be submitted for publication.

Cross-References

▶ Cytokine
▶ Linkage Disequilibrium

References

Hegab AE, Sakamoto T, Sekizawa K (2005) Assessing the validity of genetic association studies. Thorax 60:882–883
Lohmueller KE, Pearce CL, Pike M et al (2003) Meta-analysis of genetic association studies supports a contribution of common variants to susceptibility to common disease. Nat Genet 33:177–182
Newton-Cheh C, Hirschhorn JN (2005) Genetic association studies of complex traits: design and analysis issues. Mutat Res 573:54–69
Risch N, Merikangas K (1996) The future of genetic studies of complex human diseases. Science 273:1516–1517

See Also

(2012) Allele. In: Schwab M (ed) Encyclopedia of Cancer, 3rd edn. Springer Berlin Heidelberg, p 137. doi:10.1007/978-3-642-16483-5_6570
(2012) DNA Microarray. In: Schwab M (ed) Encyclopedia of Cancer, 3rd edn. Springer Berlin Heidelberg, p 1140. doi:10.1007/978-3-642-16483-5_1683
(2012) Genetic Haplotype. In: Schwab M (ed) Encyclopedia of Cancer, 3rd edn. Springer Berlin Heidelberg, pp 1526–1527. doi:10.1007/978-3-642-16483-5_2378
(2012) Genetic Polymorphism. In: Schwab M (ed) Encyclopedia of Cancer, 3rd edn. Springer Berlin Heidelberg, p 1528. doi:10.1007/978-3-642-16483-5_2382
(2012) Genotype. In: Schwab M (ed) Encyclopedia of Cancer, 3rd edn. Springer Berlin Heidelberg, p 1540. doi:10.1007/978-3-642-16483-5_2396
(2012) Hardy–Weinberg Law. In: Schwab M (ed) Encyclopedia of Cancer, 3rd edn. Springer Berlin Heidelberg, pp 1631–1632. doi:10.1007/978-3-642-16483-5_2566
(2012) Linkage. In: Schwab M (ed) Encyclopedia of Cancer, 3rd edn. Springer Berlin Heidelberg, p 2043. doi:10.1007/978-3-642-16483-5_3367
(2012) Microsatellite. In: Schwab M (ed) Encyclopedia of Cancer, 3rd edn. Springer Berlin Heidelberg, p 2305. doi:10.1007/978-3-642-16483-5_3730
(2012) Polymorphism. In: Schwab M (ed) Encyclopedia of Cancer, 3rd edn. Springer Berlin Heidelberg, pp 2954–2955. doi:10.1007/978-3-642-16483-5_4673

Case–Control Association Analysis

▶ Case Control Association Study

Casein Kinase 2

▶ CK2

Casein Kinase II

▶ CK2

CASH

▶ FLICE-Inhibitory Protein

CASP-8

▶ Caspase-8

Caspase

Definition

Are protein degrading enzymes (proteases) that act as mediators of programmed cell death (▶ apoptosis). Proteins within the large family of these cell-death proteases are all similar to each other. Caspases are highly conserved during evolution and can be found in humans as well as in insects and worms and are even found in lower multicellular organisms. More than a dozen caspases have been identified in humans. Usually caspases selectively cleave a restricted set of target proteins in the primary sequence at one position, or at a few positions at most. Cleavage always occurs behind an aspartate amino acid. The caspase-mediated cleavage of specific substrates supplies an explanation for several characteristic features of apoptosis. Cleavage of the nuclear lamins, for instance, is required for nuclear shrinking. Cleavage of cytoskeletal proteins causes the overall loss of cell shape. In healthy cells, caspases normally lie dormant. In response to diverse stimuli, they become activated when cell death is required. Dormant caspases exist as precursor polypeptides or "proenzymes" that are largely activated by proteolytic processing. This involves cleaving of proenzymes at specific points to generate the large and small subunits that associate to the active caspase enzyme. The proenzymes have low protease activity themselves and can therefore process each other when brought into vicinity. This process starts when an external stimulus, a "death ligand," binds to a receptor (such as CD95/FAS/APO-1) on the cell surface. Ligand binding results in the aggregation of procaspase-8. The high density of caspase-8 proenzymes has the result that they mutually activate each other. Caspase-8 is an initiator caspase that can activate downstream procaspases, in particular procaspase-3, either by direct cleaving or indirectly by cleaving BID and inducing cytochrome C release from mitochondria.

An alternative mechanism of caspase activation in response to death stimuli involves procaspase-9. In this case, the adaptor molecule APAF-1 sequesters several procaspase-9 molecules that, within this complex (often referred to as apoptosome), are activated by a change in conformation, not by proteolysis. In response to that change, they can activate downstream caspases.

In short, initiator caspases become primarily activated by regulated protein-protein interaction, whereas downstream effector caspases are activated proteolytically. Besides caspase pathways, other death-inducing pathways must exist since developmental apoptosis is functional in mice that are defective in regard to the ▶ caspase-8 and caspase-9 pathways.

Cross-References

▶ APAF-1 Signaling
▶ Apoptosis
▶ Autophagy
▶ Caspase-8
▶ PUMA

Caspase Homologue

▶ FLICE-Inhibitory Protein

Caspase-8

Simone Fulda
Institute for Experimental Cancer Research in Pediatrics, Goethe-University Frankfurt, Frankfurt, Germany

Synonyms

CASP-8; FADD-like ICE; FLICE; Mach; Mch5

Definition

Caspase-8 belongs to the family of cysteine proteases called caspases that act as mediators of programmed cell death (▶ apoptosis). It is a protein of 480 amino acids and 55 kDa that is widely expressed in various tissues. Caspase-8 displays 20% identity to the *ced-3*-encoded protein of *Caenorhabditis elegans*. The gene maps to 2q33.

Characteristics

Structure and Physiological Functions of Caspase-8

Caspase-8 contains two death-effector domains (DED) in the N-terminal prodomain that serve as protein–protein interaction sites and a catalytic protease domain at the C-terminus consisting of a large and small subunit. The active caspase-8 molecule is composed of a heterotetramer of two of each of the large and small subunits (Fig. 1). The preferred substrate specificity for caspase-8 is I/L/V/E X D (where X is any amino acid).

Caspase-8 exists in different splice variants of which caspase-8a and caspase-8b are expressed in most cell lines and catalytically active. Caspase-

Caspase-8, Fig. 1 Caspase-8 structure. Caspase-8 is a 480 amino acid protein that consists of two death-effector domains (*DED*) and a catalytic protease domain with a large subunit (*p20*) and small subunit (*p10*)

8 L is generated by alternative splicing of intron 8 of the human *caspase-8* gene generating a 136 bp insertion between exon 8 and exon 9 of full-length caspase-8 mRNA. This produces a premature stop codon and a truncated protein that contains only the two N-terminal DED domains but lacks the C-terminal proteolytic domain.

Caspase-8 is an initiator caspase that is expressed as proenzyme (zymogen) in an inactive state and becomes activated during apoptosis through oligomerization in a multimeric complex. Cross-linking of ► death receptors such as CD95 or the agonistic TRAIL receptors TRAIL-R1 and TRAIL-R2 by their corresponding ligands CD95 ligand or TRAIL or by agonistic antibodies initiates receptor trimerization, clustering of the receptors death domains, and recruitment of adaptor molecules such as Fas associated with a death domain (FADD) through homophilic protein–protein interactions mediated by the death domains (Fig. 2). FADD in turn recruits caspase-8 to activated death receptors through interaction via the DED domains to form the death-inducing signaling complex (DISC). Oligomerization of caspase-8 upon DISC formation drives its activation through autoproteolysis. Once activated, caspase-8 cleaves downstream effector caspases such as caspase-3. For the CD95 signaling pathway, two distinct prototypic cell types have been identified. In type I cells, caspase-8 is activated upon CD95 ligation at the DISC in quantities sufficient to directly activate downstream effector caspases such as caspase-3. In type II cells, however, the amount of active caspase-8 generated at the DISC is insufficient to fully activate caspase-3. In these cells, a mito-chondrial amplification loop is required for com-plete activation of the caspase cascade involving caspase-8-mediated cleavage of BH3-interacting death domain agonist (Bid), which translocates to

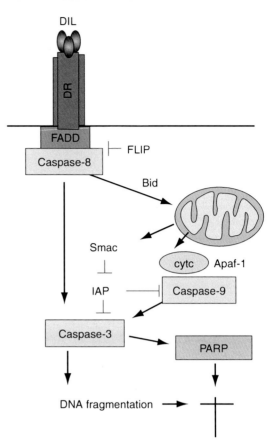

Caspase-8, Fig. 2 Apoptosis signaling pathways. Apo-ptosis pathways can be initiated by cross-linking of death receptors (*DR*), e.g., CD95 or TRAIL receptors, by death-inducing ligands (*DILs*) such as CD95 ligand or TRAIL followed by recruitment of the adaptor molecule FADD and caspase-8, which drives caspase-8 activation through autoproteolysis (receptor/extrinsinc pathway). In type I cells, caspase-8 is activated at the receptor level in quan-tities sufficient to directly activate effector caspase-3. In type II cells, caspase-8 initiates a mitochondrial amplifica-tion loop to activate effector caspases by cleaving Bid, which translocates to mitochondria to trigger the release of cytochrome *c* (*cytc*) and Smac. The mitochondrial (intrinsic) pathway is initiated by the release of cytochrome *c* or Smac from mitochondria into the cytosol. Cytochrome *c* triggers caspase-3 activation via formation of the cyto-chrome *c*/Apaf-1/caspase-9-containing apoptosome com-plex, while Smac neutralizes the inhibitor of apoptosis protein (*IAP*)-mediated inhibition of caspase-3 and caspase-9. Cellular ► FLICE-inhibitory protein (*c-FLIP*) inhibits apoptosis by blocking caspase-8 activation

mitochondria to trigger the release of apoptogenic proteins such as cytochrome *c* from mitochondria into the cytosol. Also, a similar cell-type-dependent organization (type I and type II) of the TRAIL signaling pathway has been described. Besides its activation at the DISC, caspase-8 can also be activated downstream of mitochondria upon initiation of the intrinsic apoptosis pathway, e.g., through cleavage by caspase-6.

In addition to its established role in apoptosis signaling, evidence indicates that caspase-8 can also exert several nonapoptotic functions. For example, caspase-8 is required to maintain homeostasis of peripheral T cells by controlling T-cell proliferation via regulation of IL-2 production. In addition, caspase-8 is involved in the regulation of differentiation and proliferation of B cells, NK cells, and hematopoietic progenitors. Also, caspase-8 has been reported to be important for NF-κB activation through the T-cell receptor, for CD95 clustering and internalization upon CD95 stimulation, as well as for the survival of endothelial cells. Furthermore, it has been shown that caspase-8 can promote cell motility by regulating activation of ▶ calpains, Rac, and lamellipodial assembly and regulates cell spreading by cleaving the cytolinker plectin, a component of hemidesmosomes and focal adhesion complexes. Loss-of-function mutation in caspase-8 is lethal to the mouse embryo around day 12.5, an indication of its critical role during normal development. Caspase-8 knockout mice die in utero as a result of defective development of heart muscle and display abdominal hemorrhage and fewer than normal hematopoietic progenitor cells. Together, these findings indicate that constitutive caspase-8 activity is relevant to normal physiology.

Caspase-8 and Cancer

It is well established that the evasion of apoptosis is one of the hallmarks of ▶ cancer. It is therefore not surprising that some cancers have used the inactivation of caspase-8 to avoid apoptotic signals suggesting that caspase-8 may act as tumor suppressor. In principle, caspase-8 expression and/or function can be altered through genetic or ▶ epigenetic mechanisms, or alternatively, caspase-8 function can be compromised in cancers. For example, caspase-8 expression can be impaired by mutations. Such mutant variants of caspase-8 may act in a dominant-negative

manner, e.g., by blocking the recruitment of wild-type caspase-8 to activated death receptors, thereby inhibiting apoptosis. Despite the key role of caspase-8 for cell death execution, caspase-8-mutations in human tumors have, however, only been identified at low frequency in some tumors, e.g., in colorectal, head and neck, or vulvar carcinoma. In addition, homo- or heterozygous genomic deletions were found in some neuroblastoma. In contrast to these rare genetic alterations, caspase-8 expression is frequently impaired by epigenetic mechanisms in cancer cells. To this end, caspase-8 expression was found to be inactivated by hypermethylation of a regulatory sequence of the *caspase-8* gene, which maps to the boundary between exon 3 and intron 3. Silencing of caspase-8 was detected in a variety of cancers, e.g., in ▶ neuroblastoma, medulloblastoma, malignant glioma, ▶ rhabdomyosarcoma, ▶ Ewing sarcoma, ▶ retinoblastoma, and small lung cell carcinoma both in cell lines and in primary tumor samples. Although this regulatory region of caspase-8 does not meet the criteria of a classical ▶ CpG island and shows no promotor activity, the ▶ methylation status of this domain correlated with caspase-8 expression in several human tumors. In addition, treatment with the demethylating agent 5-aza-2′deoxycytidine (5-AZA) resulted in demethylation of this regulatory sequence, which in turn led to increased caspase-8 promotor activity and re-expression of caspase-8. This suggests that demethylation of a *trans*-acting factor may be involved in controlling activity of the caspase-8 promotor. Another level of transcriptional regulation of caspase-8 in cancers is alternative splicing, for example, in leukemia or neuroblastoma cells. Alternative splicing of intron 8 of the *caspase-8* gene generates caspase-8 L that misses the catalytic site but retains the two N-terminal DED repeats. Thus, caspase-8 L is recruited to activated death receptors where it acts as a dominant-negative inhibitor of apoptosis in cancer cells by interfering with the recruitment of wild-type caspase-8.

In addition to genetic and epigenetic mechanisms, caspase-8 signaling can also be functionally impaired in cancer cells, e.g., by overexpression of antiapoptotic proteins that interfere with caspase-8 activation at the death receptor level. Examples are cellular ► FLICE-inhibitory protein (c-FLIP) or phosphoprotein enriched in diabetes/phosphoprotein enriched in astrocytes-15 kDa (PEST region) that exert their antiapoptotic function by blocking the recruitment of caspase-8 to activated death receptors. The adenoviral E1B19K early protein has similar properties.

The biological relevance of caspase-8-inactivation in cancers follows from its key role in the apoptotic machinery. Tumor cells with loss of caspase-8 were found to be resistant to death receptor-triggered apoptosis. Similarly, embryonic fibroblasts derived from caspase-8 knockout mice were completely resistant to apoptosis induced by death receptors including CD95, TRAIL receptors, or TNF receptor-1, whereas they retained sensitivity to other apoptotic stimuli such as UV irradiation, ► ceramide, or several anticancer drugs. These findings indicate that caspase-8 plays a necessary and nonredundant role in transducing the death signal from activated death receptor to intracellular effector caspases. In addition, chemotherapeutic drugs can initiate caspase-8 activation in a receptor-dependent and also in a receptor-independent manner. Of note, loss of caspase-8 expression has been reported to significantly correlate with unfavorable survival outcome in medulloblastoma patients, while no correlation with survival or established parameters of poor prognosis was found in neuroblastoma. Restoration of caspase-8 expression by gene transfer or by demethylation treatment in cancer cells where caspase-8 is epigenetically silenced also sensitized resistant tumor cells for death-receptor- or drug-induced apoptosis. In addition, treatment with the cytokine IFNγ caused transcriptional activation of caspase-8 in cancer cells lacking caspase-8 and enhanced expression of caspase-8 through interferon-sensitive response elements within the caspase-8 promotor and STAT-1.

Loss of caspase-8 fosters cancer ► metastasis and ► invasion by rendering cancer cells resistant to integrin-mediated cell death. Integrin receptors that are unable to find appropriate ligands can form a large molecular complex containing caspase-8, thereby initiating an apoptosis cascade. In this context, caspase-8 may function as metastasis suppressor gene that, together with integrins, regulates cell death of cancer cells that migrate from the primary tumor. Thus, loss of caspase-8 may promote metastasis of cancers by providing a survival advantage in foreign microenvironments.

Cross-References

- ► Apoptosis
- ► Calpain
- ► Cancer
- ► Ceramide
- ► CpG Islands
- ► Death Receptors
- ► Epigenetic
- ► Ewing Sarcoma
- ► FLICE-Inhibitory Protein
- ► Invasion
- ► Metastasis
- ► Methylation
- ► Migration
- ► Neuroblastoma
- ► Retinoblastoma
- ► Rhabdomyosarcoma
- ► Signal Transducers and Activators of Transcription in Oncogenesis
- ► TNF-Related Apoptosis-Inducing Ligand

References

Barnhart BC, Lee JC, Alappat EC et al (2003) The death effector domain protein family. Oncogene 22:8634–8644

Fulda S, Debatin KM (2004) Exploiting death receptor signaling pathways for tumor therapy. Biochim Biophys Acta 1705:27–41

Lahti JM, Teitz T, Stupack DG (2006) Does integrin-mediated cell death confer tissue tropism in metastasis? Cancer Res 66:5981–5984

Park SM, Schickel R, Peter ME (2005) Nonapoptotic functions of FADD-binding death receptors and their signaling molecules. Curr Opin Cell Biol 17:610–616

See Also

(2012) Alternative RNA Splicing. In: Schwab M (ed) Encyclopedia of Cancer, 3rd edn. Springer Berlin Heidelberg, p 148. doi:10.1007/978-3-642-16483-5_212

(2012) BH3-Interacting Death Domain Agonist. In: Schwab M (ed) Encyclopedia of Cancer, 3rd edn. Springer Berlin Heidelberg, p 389. doi:10.1007/978-3-642-16483-5_601

(2012) C-FLICE-like Inhibitory Protein. In: Schwab M (ed) Encyclopedia of Cancer, 3rd edn. Springer Berlin Heidelberg, p 753. doi:10.1007/978-3-642-16483-5_1039

(2012) Death Domain. In: Schwab M (ed) Encyclopedia of Cancer, 3rd edn. Springer Berlin Heidelberg, p 1065. doi:10.1007/978-3-642-16483-5_1534

(2012) Death-Effector-Domain. In: Schwab M (ed) Encyclopedia of Cancer, 3rd edn. Springer Berlin Heidelberg, p 1066. doi:10.1007/978-3-642-16483-5_1535

(2012) Death-Inducing Signaling Complex. In: Schwab M (ed) Encyclopedia of Cancer, 3rd edn. Springer Berlin Heidelberg, p 1066. doi:10.1007/978-3-642-16483-5_1536

(2012) Fas Associated with a Death Domain. In: Schwab M (ed) Encyclopedia of Cancer, 3rd edn. Springer Berlin Heidelberg, p 1379. doi:10.1007/978-3-642-16483-5_2123

(2012) Initiator Caspases. In: Schwab M (ed) Encyclopedia of Cancer, 3rd edn. Springer Berlin Heidelberg, p 1865. doi:10.1007/978-3-642-16483-5_3061

(2012) Microenvironment. In: Schwab M (ed) Encyclopedia of Cancer, 3rd edn. Springer Berlin Heidelberg, p 2296. doi:10.1007/978-3-642-16483-5_3720

(2012) PEST Sequence. In: Schwab M (ed) Encyclopedia of Cancer, 3rd edn. Springer Berlin Heidelberg, p 2828. doi:10.1007/978-3-642-16483-5_4478

(2012) Phosphoprotein Enriched in Diabetes/Phosphoprotein Enriched in Astrocytes-15kDa. In: Schwab M (ed) Encyclopedia of Cancer, 3rd edn. Springer Berlin Heidelberg, p 2870. doi:10.1007/978-3-642-16483-5_4543

(2012) Receptor for TNF-Related Apoptosis-Inducing Ligand. In: Schwab M (ed) Encyclopedia of Cancer, 3rd edn. Springer Berlin Heidelberg, p 3198. doi:10.1007/978-3-642-16483-5_4981

(2012) STAT. In: Schwab M (ed) Encyclopedia of Cancer, 3rd edn. Springer Berlin Heidelberg, p 3502. doi:10.1007/978-3-642-16483-5_5481

(2012) Tumor Suppressor. In: Schwab M (ed) Encyclopedia of Cancer, 3rd edn. Springer Berlin Heidelberg, p 3803. doi:10.1007/978-3-642-16483-5_6056

Caspase-Eight-Related Protein

▶ FLICE-Inhibitory Protein

Caspase-Independent Apoptosis

Chun Hei Antonio Cheung
Department of Pharmacology and Institute of Basic Medical Sciences, College of Medicine, National Cheng Kung University, Tainan, Taiwan, Republic of China

Synonyms

AIF-mediated cell death; Caspase-independent cell death; Nonclassical apoptosis

Definition

Caspase-independent apoptosis is defined as the process of apoptosis-like cell death, in which caspase activation does not contribute to the completion of this process.

Characteristics

Cellular and Molecular Characteristics of Apoptotic Cells

Cell death can be induced through necrosis and apoptosis, in which apoptosis is also called programmed cell death. Generally, apoptosis processing requires sequential activations of a specific family of proteases called caspases. At the molecular level, initial activation of initiator caspases such as caspase-9/caspase-8 and the subsequent activation of effector caspases such as caspase-3/caspase-7 play important roles in inducing apoptosis in cells. Activation of caspases induced proteolytic cleavages of a set of proteins that are important in maintaining cellular integrity and cell-to-cell attachment. Activation of caspases also activates various DNase enzymes (CAD, caspase-activated DNase), leading to the induction of DNA fragmentations (with fragment size of approximately 200 base pairs). Therefore, the appearance of cytoplasmic shrinkage and presence of DNA strand breaks/fragmentations are two morphological changes typically observed in apoptotic cells.

Molecules Involved in the Caspase-Independent Apoptosis Process

Noticeably, cells can also process apoptosis-like cell death without activating caspases (caspase-independent apoptosis). During caspase-independent apoptosis, apoptosis-inducing factor (AIF, mature form of approximately 57 kDa), which is a mitochondrial intermembrane flavoprotein, translocates from the mitochondria upon mitochondrial membrane depolarization into the nucleus. It binds to DNA strands through electrostatic interaction and induces chromosome condensation and large-scale DNA fragmentation (approximately 50 k base pairs), resulting in the promotion of cell death. Cells can also process apoptosis through the translocation and activation of endonuclease G (endoG) in the absence of caspase activation or during caspase dysfunction. EndoG is a sequence-unspecific DNase that also exhibits RNase activity. As similar to AIF, endoG is released from the intermembrane space of mitochondria upon mitochondrial membrane depolarization to the cytosol during caspase-independent apoptosis. It subsequently translocates into the nucleus and induces DNA fragmentations. Besides AIF and endoG, activation and nuclear translocation of another proapoptotic molecule called WOX1 (WW domain-containing oxidoreductase, WWOX) also play an important role in processing apoptosis without caspase activation. In addition, the activated WOX1 can bind to the tumor suppressor p53 and promotes the proapoptotic functions of p53.

Inter-regulations Between Caspase-Dependent and Caspase-Independent Apoptosis in Cancer Cells

The process of caspases, AIF, endoG, and WOX1-mediated apoptosis is not mutually exclusive in cells. Although cells can carry out programmed cell death through nuclear translocation of both AIF and endoG and the subsequent induction of DNA fragmentation without caspase activation, caspase activation is capable of triggering translocation and activation of these molecules under certain circumstances during the Bax-/Bak-mediated caspase-involved apoptosis. In addition, inter-switch between caspase-dependent apoptosis and caspase-independent apoptosis does exist in cells. Upon apoptotic stimulations, cells can switch to process caspase-independent apoptosis if caspase-3/caspase-7 is inhibited by a few caspase-specific inhibitors. Furthermore, overexpression of AIF can induce caspase-7 activation occasionally and inhibit protein synthesis. Cancer cells can inhibit both caspase-dependent apoptosis and caspase-independent apoptosis simultaneously in order to maintain their survival and induce drug resistance to various chemotherapeutic treatments. In fact, a few antiapoptotic molecules are overexpressed in cancer cells and capable of inhibiting both the caspase-dependent and caspase-independent apoptosis pathways. For example, survivin, a member of the inhibitor-of-apoptosis proteins (IAPs) family which is overexpressed in a variety of cancer cells but not in differentiated tissues, inhibits the activation of caspase-3 through direct and indirect mechanisms and also interferes with the translocation of AIF in cancer cells.

Cross-References

▶ Apoptosis
▶ Caspase-8
▶ Survivin

References

Candé C, Cecconi F, Dessen P, Kroemer G (2002) Apoptosis-inducing factor (AIF): key to the conserved caspase-independent pathways of cell death? J Cell Sci 115:4727–4734
Lorenzoa HK, Susinb SA (2004) Mitochondrial effectors in caspase-independent cell death. FEBS Lett 557:14–20

Caspase-Independent Cell Death

▶ Caspase-Independent Apoptosis

Caspase-Like Apoptosis-Regulatory Protein

▶ FLICE-Inhibitory Protein

CASPER

▶ FLICE-Inhibitory Protein

CASTing

▶ Combinatorial Selection Methods

Castrate-Resistant Prostate Cancer

Saurabh Ghosh Roy
Department of Cell and Developmental Biology,
University of California, Irvine, Irvine, CA, USA

Synonyms

Hormone-refractory prostate cancer; Metastatic prostate cancer

Keywords

Androgen deprivation therapy; Androgen receptor signaling

Definition

Prostate cancer as the name suggests is the cancer of the prostate, an essential gland of the male reproductive system. In general, prostate cancers are slow-growing cancers, and frequently they are not detected by PET scanning. The disease could be segregated into two stages. The initial stage which responds to castration and majority of the patients go disease-free for life or for a few years. The later stage is marked by aggressiveness, and overall survival is significantly reduced by this later stage of the disease. Castration decreases androgen levels in the body and thereby helps in the remission or overall survival during the initial stage of the disease, whereas castration has no effect on the later stage of the disease and many a times induces the rapid onset of the late stage of this disease giving the name: castration-resistant prostate cancer (CRPC). Prostate cancer that progresses despite castrate levels of serum testosterone is defined as "castrate resistant."

Characteristics

Prostate cancer is the most frequently diagnosed cancer in men aside from skin cancer. For reasons that remain unclear, incidence rates are 70% higher in African-Americans than in whites. With an estimated 29,720 deaths in 2013, prostate cancer is the second-leading cause of cancer death in men. For most patients, prostate cancer is a localized indolent disease that may be cured with surgery or radiation therapy, but the disease recurs in approximately 20–30% of patients. However, it is far more difficult to treat those patients with aggressive or metastatic form of the disease. Androgen deprivation therapy (ADT), the most common treatment after recurrence, is effective, but the disease eventually progresses in most patients who receive such treatment. For men with metastatic castration-resistant prostate cancer, the median survival in studies has ranged from 12.2 to 21.7 months.

The male body produces male hormones called androgens. Research has shown that androgens help fuel the prostate tumor. Androgen is produced from two different sources in the human body: primarily from the male testes and some from the cortical adrenal glands situated on top of the kidneys. However, in men, androgens could also be made from yet another source: the tumor tissue itself, thereby making them self-sufficient and does not depend on other sources of androgens.

Previously, the initial stage of the disease was referred to as hormone-dependent prostate cancer and the later stage as hormone-independent prostate cancer (as castration was thought to obliterate androgen signaling). Later, research showed that although androgen deprivation therapy (ADT) improved the tumor burden in the hormone-dependent stage of the disease, it had no effect during the ► hormone-independent stage. While the later stage of the disease was no longer responsive to castration or ADT therapy by either chemical or surgical means, these late-stage "hormone refractory" cancers still show reliance upon hormones for ► androgen receptor (AR) activation, hence the re-nomenclature as castration-resistant prostate cancer (CRPC). There is abundant evidence that the road to CRPC is paved by the reactivation of the AR and the re-expression of androgen responsive genes. Research suggests that AR is reactivated by a gain of function in a ligand-sensitized manner. Mutations also occur in the AR ligand binding domain which broadens the specificity for steroid hormone ligands. Briefly, several cellular and molecular alterations are related to this post-castration activation of AR, including incomplete blockade of AR ligand signaling, AR amplifications, AR mutations, AR splice variant expression, and aberrant AR co-regulator activities.

Genetic Causes of the Disease
Profiling studies of prostate cancers have shown that several receptor tyrosine kinases including HER kinase family (EGFR, HER2), PDGFR, c-met, and c-myc are expressed in a certain number of these cancers. One of the most frequent genetic alterations is the deletion of tumor suppressor PTEN (phosphatase and tensin homologue).

Diagnosis of the Disease
Diagnosis could be done by ultrasonography, MRI, as well as prostate biopsy. Perhaps the most widely used method for screening the disease is by monitoring the rising levels of PSA (prostate-specific antigen).

Survival
The majority (93%) of prostate cancers are discovered in the local or regional stages, for which the 5-year relative survival rate approaches 100%. Patients with metastatic castration-resistant prostate cancer (mCRPC) have a poor prognosis, and those patients with metastases are expected to survive ≤19 months.

Current Treatment Options
Until 2009, there were only few drugs approved for the treatment of CRPC, only one, docetaxel, that showed improvement in overall survival. Currently, the FDA has approved a number of novel drugs which targets the disease at different stages thereby improving overall progression-free survival.

As the androgen receptor signaling is still active in CRPC patients, several new agents which are both FDA approved and/or in development targets the AR activation by the following mechanisms.

1. Direct androgen receptor antagonists: enzalutamide (FDA approved) and ARN-509 (phase III clinical trials)
2. Androgen biosynthesis inhibitors: abiraterone (FDA approved) and TAK-700 (phase III clinical trials)
3. Androgen receptor coactivators: OGX-111 (phase III clinical trials) and OGX-427 (phase II clinical trials)
4. Immunologic therapy (vaccine therapy): sipuleucel-T (also known as Provenge; FDA approved), Prostvac-VF (phase III clinical trials), and ipilimumab (phase III clinical trials)
5. Tyrosine kinase inhibitors: cabozantinib (phase III clinical trials)
6. Radiopharmaceutical therapy: radium 223 (FDA approved)

Brief Description of Each Therapy
(A) Direct androgen receptor antagonists:
   (i) Enzalutamide is an oral androgen receptor signaling inhibitor that inhibits nuclear translocation of the androgen receptor hormone complex, DNA binding, and coactivator recruiting and induces cell ► apoptosis.
   (ii) ARN-509 is an oral competitive androgen receptor antagonist that impairs

androgen receptor binding to DNA and androgen receptor target gene modulation and induces apoptosis.

(B) Androgen biosynthesis inhibitors:
- (i) Abiraterone is a small molecule inhibitor of 17-alpha-monooxygenase (17-alpha-hydroxylase and C17,20-lyase, named as CYP17 complex), a member of the ► cytochrome P450 family that blocks androgen synthesis by the adrenal glands and testes and within the prostate tumor in a ligand-dependent fashion.
- (ii) TAK-700 is a selective, nonsteroidal potent CYP17 inhibitor that inhibits the 17,20-lyase activity of CYP17A1.

(C) Androgen receptor coactivators:
- (i) OGX-111 also known as clusterin is a chaperone protein involved in cell proliferation and survival. It is a stress-induced androgen receptor-regulated cytoprotective chaperone that is upregulated in cell death.
- (ii) OGX-427 also known as heat shock protein 27 (Hsp27) is a chaperone protein that regulates cell signaling and survival pathways involved in cancer progression and is uniformly expressed in metastatic CRPC.

(D) Immunologic therapies:
- (i) Sipuleucel-T is a personalized antigen-presenting cell-based immunotherapy product.
- (ii) Prostvac-VF is a prostate cancer vaccine consisting of a recombinant vaccinia virus expressing the entire PSA transgene as a primary vaccination, followed by multiple recombinant fowlpox booster vaccinations and a viral vector encoding three major costimulatory molecules.
- (iii) Ipilimumab is a monoclonal antibody that blocks the activity of the T-cell inhibitory receptor cytotoxic T-lymphocyte-associated antigen 4 (CTLA4)

(E) Tyrosine kinase inhibitors:
- (i) Cabozantinib is an orally bioavailable novel tyrosine kinase inhibitor with specific activity against c-MET and VEGF receptor 2 (VEGFR2).

(F) Radiopharmaceutical therapy:
- (i) Radium 223 is a novel alpha particle-emitting radiopharmaceutical targeting bone metastasis in CRPC.

Cross-References

- ► Cancer Vaccines
- ► Metastasis
- ► Prostate Cancer Clinical Oncology
- ► Prostate Cancer Diagnosis
- ► Prostate Cancer Hormonal Therapy
- ► Prostate Cancer Targeted Therapy
- ► Prostate-Specific Antigen

References

Acar O, Esen T, Lack NA (2013) New therapeutics to treat castrate resistant prostate cancer. Sci World J 379641

Dayyani F, Gallick GE, Logothetis CJ, Corn PG (2011) Novel therapies for metastatic castrate resistant prostate cancer. J Natl Cancer Inst 103(22): 1665–1675

Scher HI, Sawyers CL (2005) Biology of progressive, castration-resistant prostate cancer: directed therapies targeting the androgen receptor signaling axis. J Clin Oncol 23(32):8253–8261

Thoreson GR, Gayed BA, Chung PH, Raj GV (2014) Emerging therapies in castration resistant prostate cancer. Can J Urol 21(2 Suppl):98–105

Castration-Resistant Prostate Cancer

Definition

Androgen deprivation therapy (ADT) has been a mainstay in ► prostate cancer therapy. After an excellent clinical response to ADT, however, prostate cancer returns as a therapy resistant and deadly form, resulting in a short survival time of 18–24 months. This form of prostate cancer, now known as castration-resistant prostate cancer (CRPC), often metastasizes to the bone and results in the clinical symptoms of pain. A great challenge is the discovery of new chemotherapy drugs that can increase overall survival of patients with recurrent CRPC.

Cross-References

▶ Androgen Ablation Therapy
▶ Prostate Cancer
▶ Prostate Cancer Chemotherapy

Catechin

▶ Epigallocatechin

Cathepsin-D

Emmanuelle Liaudet-Coopman
IRCM, INSERM, UMI, CRLC Val d'Aurelle, Montpellier, France

Definition

Cathepsin-D (E.C. 3.4.23.5) is a ubiquitous lysosomal aspartic endo-proteinase cleaving preferentially -Phe-Phe-, -Leu-Tyr-, -Tyr-Leu-, and -Phe-Tyr- bonds in peptide chains containing at least five amino acids at an acidic pH.

Characteristics

Cathepsin-D is ubiquitously distributed in lysosomes. It was considered for a long time that the main function of cathepsin-D was to degrade proteins in lysosomes at an acidic pH. Apart from its function in general protein turnover, cathepsin-D can also activate precursors of biologically active proteins in pre-lysosomal compartments of specialized cells. Knock-out of cathepsin-D gene induces death shortly after birth with severe apoptotic and necrotic phenotypes. Its pH optimum depends on the enzyme source and on the substrate used for the determination of the activity and ranges between 2.8 and 5. No endogenous cathepsin-D tissue inhibitor is known in mammals. Pepstatin, a natural inhibitor of aspartic proteases isolated from various species of actinomycetes, inhibits its catalytic activity. Cathepsin-D, like other aspartic proteases, such as renin, chymosin, pepsinogen, has a bilobed organization. Crystal structures of native and pepstatin-inhibited forms of mature human cathepsin-D revealed a high degree of tertiary structural similarity with other members of the aspartic proteinase family (e.g., pepsinogen and human immunodeficiency virus protease). The human cathepsin-D gene containing nine exons is located in chromosome 11p15 and expresses a single transcript of 2.2 kb. Cathepsin-D is synthesized as a 52 kDa catalytically inactive precursor (Fig. 1). During its transport to lysosomes, cathepsin-D can be found in the endosomes where it is present as partially active 48 kDa single-chain intermediate (Fig. 1). This intermediate is subsequently transported to the lysosomes where it is converted into the fully active mature protease that is composed of a 34 kDa heavy and a 14 kDa light chain (Fig. 1). The human cathepsin-D catalytic site includes two critical aspartic residues (amino acids 33 and 231) located on the 34 and 14 kDa chains (Fig. 1a). Mannose-6-phosphate (M6P) receptors are involved in lysosomal routing of cathepsin-D and in the cellular uptake of the secreted pro-cathepsin-D. In ▶ breast cancer cell lines, over-expressed cathepsin-D is hyper-secreted in the extracellular environment and can be endocytosed (▶ Endocytosis) by both ▶ cancer cells and fibroblasts via M6P receptors and other as yet unidentified receptor(s) (Fig. 1b). Endocytosed pro-cathepsin-D also undergoes successive maturations leading to the 48 kDa and 34 + 14 kDa forms. In addition, secreted pro-cathepsin-D, like pepsinogen, is capable of acid-dependent auto-activation in vitro, resulting in a catalytically active pseudo-cathepsin-D, an enzyme species that retains 18 residues (27–44) of the pro-segment.

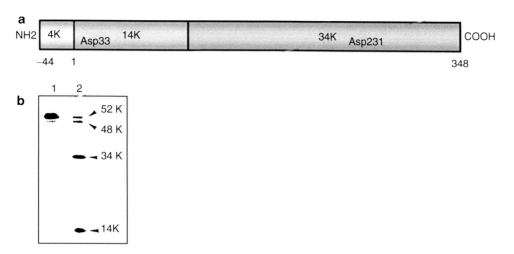

Cathepsin-D, Fig. 1 Cathepsin-D structure and expression in breast cancer cells (**a**) Schematic representation of the human 52 kDa pro-cathepsin-D sequence. Location of 4 kDa cathepsin-D pro-fragment, 14 kDa light and 34 kDa heavy mature chains are indicated. Intermediate 48 kDa form (not shown) corresponds to noncleaved 14 + 34 kDa chains. Number 1 corresponds to the first amino acid of the mature cathepsin-D. Position of the 2 aspartic acids of the catalytic site is shown. Molecular mass is shown in K (kDa). (**b**) Expression of Human cathepsin-D in MCF-7 breast cancer cell line. MCF-7 cells were metabolically labeled with [^{35}S]Methionine and human cathepsin-D immunoprecipitated from cell extract (*lane 2*) and medium (*lane 1*) was analyzed by SDS-PAGE

Apoptosis

Cathepsin-D is a key mediator of ► apoptosis induced by many apoptotic agents, such as IFN-gamma, FAS/APO, TNF-alpha, ► oxidative stress, ► adriamycin, etoposide, cisplatin and 5-fluorouracil, as well as staurosporine. The role of cathepsin-D in apoptosis has been linked to the lysosomal release of mature 34 kDa cathepsin-D into the cytosol, leading in turn to the mitochondrial release of cytochrome c into the cytosol and the activation of pro-caspases-9 and -3.

Regulation

Studies on ► estrogen receptor positive breast cancer cell lines revealed that this housekeeping enzyme is highly upregulated by estrogens (► Estradiol) and growth factors (i.e., IGF1, EGF). In estrogen receptor positive breast cancer cell lines, both estrogens and growth factors stimulate cathepsin-D protein and mRNA accumulation levels. The regulation of cathepsin-D mRNA accumulation by estrogens is mainly due to increased initiation of transcription. Estrogen-responsive elements have been defined in the proximal promoter region of the gene, and in conjunction with other regulatory sequences (e.g., SP1, AP1), they may be responsible for the stimulation of cathepsin-D gene expression. Studies in estrogen receptor negative breast cancer cell lines that are the more aggressive, invasive, and metastatic indicated a constitutive over-expression of cathepsin-D. The mechanism of this over-expression is still unknown but does not seem to involve gene amplification or major chromosomal rearrangements (► Chromosomal Translocations).

Cancer

Cathepsin-D over-expressed by cancer cells stimulates tumorigenicity and ► metastasis in nude mice. The direct role of cathepsin-D in cancer metastasis was first demonstrated in rat tumor cells in which transfection-induced cathepsin-D over-expression increased their metastatic potential in vivo. In this rat tumor model, the cathepsin-D mechanism responsible for metastasis stimulation seemed to be a positive effect on cell proliferation, favoring the growth of micro-metastases. Using an RNA antisense strategy, cathepsin-D was then shown to be a rate limiting factor for the outgrowth, tumorigenicity, and lung colonization of MDA-MB-231 breast cancer cells. Several reports have indicated that cathepsin-D stimulates cancer cell proliferation. Purified pro-cathepsin-D from MCF-7 breast cancer cells stimulated MCF-7 cell growth. Moreover, 3Y1-Ad12 rat cancer cells transfected with human cathepsin-D cDNA grew more rapidly both at low or high cell densities in vitro and showed an increased experimental metastatic potential in vivo. In addition, pro-cathepsin-D was also mitogenic for breast and prostate cancer cells.

Clinical Aspects

Different approaches, such as cytosolic immunoassay, ▶ immunohistochemistry, in situ hybridization, and Northern and Western blot analyses, have indicated that in most breast cancer tumors, cathepsin-D is over-expressed from 2- to 50-fold compared to its concentration in other cell types such as fibroblasts or normal mammary glands. Several independent clinical studies have shown that the cathepsin-D level in primary breast cancer cytosols is an independent prognostic parameter correlated with the incidence of clinical metastasis and shorter survival times. The major cathepsin-D producing cells appear to be epithelial cancer cells (Epithelial Tumors) and stromal ▶ macrophages. Cathepsin-D production by fibroblasts appears variable according to various publications. Certain studies have indicated that cathepsin-D production is low relative to cancer cells as shown by immunohistochemistry and in situ hybridization with antisense RNA. Other studies have indicated a prognostic role for cathepsin-D over-expression by reactive stromal cells. Pro-cathepsin-D is also increased in the plasma of patients with metastatic breast cancer, indicating that part of the pro-cathepsin-D secreted by tumors can be released into the circulation.

Cross-References

- ▶ Adriamycin
- ▶ Amplification
- ▶ Apoptosis
- ▶ Breast Cancer
- ▶ Cancer
- ▶ Chromosomal Translocations
- ▶ Endocytosis
- ▶ Epithelial Tumorigenesis
- ▶ Estradiol
- ▶ Estrogen Receptor
- ▶ Immunohistochemistry
- ▶ Macrophages
- ▶ Metastasis
- ▶ Oxidative Stress

References

Chwieralski CE, Welte T, Buhling F (2006) Cathepsin-regulated apoptosis. Apoptosis 11:143–149

Liaudet-Coopman E, Beaujouin M, Derocq D et al (2006) Cathepsin D: newly discovered functions of a long-standing aspartic protease in cancer and apoptosis. Cancer Lett 237:167–179

Rochefort H (1992) Cathepsin D in breast cancer: a tissue marker associated with metastasis. Eur J Cancer 28A:1780–1783

Westley BR, May FE (1999) Prognostic value of cathepsin D in breast cancer. Br J Cancer 79:189–190

See Also

(2012) Epithelial cell. In: Schwab M (ed) Encyclopedia of cancer, 3rd edn. Springer Berlin Heidelberg, pp 1291–1292. doi:10.1007/978-3-642-16483-5_1958

(2012) Estrogens. In: Schwab M (ed) Encyclopedia of cancer, 3rd edn. Springer Berlin Heidelberg, p 1333. doi:10.1007/978-3-642-16483-5_2019

(2012) Knock-out. In: Schwab M (ed) Encyclopedia of cancer, 3rd edn. Springer Berlin Heidelberg, p 1957. doi:10.1007/978-3-642-16483-5_3237

(2012) Lysosome. In: Schwab M (ed) Encyclopedia of cancer, 3rd edn. Springer Berlin Heidelberg, p 2128. doi:10.1007/978-3-642-16483-5_3472

(2012) Promoter. In: Schwab M (ed) Encyclopedia of cancer, 3rd edn. Springer Berlin Heidelberg, p 3004. doi:10.1007/978-3-642-16483-5_4768

(2012) Proteinase. In: Schwab M (ed) Encyclopedia of cancer, 3rd edn. Springer Berlin Heidelberg, p 3092. doi:10.1007/978-3-642-16483-5_4805

Cathepsins

Definition

Are mainly lysosomal cysteine proteases (human cathepsins B, C, F, H, K, L, O, S, V, X, and W), other cathepsins belong to the serine (cathepsin G) and the aspartic (cathepsins D, E) proteases.

Cathepsins were long believed to be involved in intracellular protein degradation; it has become evident that they are involved in a number of specific cellular processes and that their irregular function is associated with pathological conditions, including cancer. Cathepsins were originally defined as a group of digestive proteases present in lysosomes and involved in lysosomal protein breakdown. From a genetic, biochemical, and catalytic point of view, cathepsins constitute an extremely heterogeneous group of proteases. This diversity assures in most tissues complete degradation of ingested proteins. With the identification of select cathepsins in other vesicular compartments of the secretory and endosomal system, however, the definition of cathepsins has evolved to also take into account their capacity to act by limited proteolysis on certain proteins.

Cross-References

▶ Cystatins
▶ Stefins

Caudal Type Homeobox 2

▶ CDX2

Caveolins

Klaus Podar[1] and Kenneth C. Anderson[2]
[1]Medical Oncology, National Center for Tumor Diseases (NCT), University of Heidelberg, Heidelberg, Germany
[2]Department of Medical Oncology, Jerome Lipper Multiple Myeloma Center, Dana-Farber Cancer Institute, Boston, MA, USA

Definition

Caveolins are integral membrane proteins responsible for the formation of caveolae, small vesicular invaginations of the plasma cell membrane. They play a key role in membrane trafficking, ▶ signal transduction, mechano-sensing, and cell metabolism.

Characteristics

Caveolae ("little caves") are flask-shaped, "smooth," vesicular invaginations of the plasma membrane (50–100 nm in diameter) distinct from the larger electron-dense clathrin-coated pits. As a subset of detergent-resistant liquid-ordered lipid rafts, which are clustered protein microdomains within a "sea of homogeneously distributed lipids," they are uniquely enriched in cholesterol, sphingolipids, and phosphatidylethanolamine and additionally contain essential structural marker proteins termed caveolins, cavins, and pacsin-2. Specifically, caveolins are highly conserved hairpin loop-shaped (both the C-terminus and the N-terminus face the cytoplasmic side of the membrane), oligomeric, integral membrane proteins of 22–24 kDa with a typical short stretch of eight amino acids (FEDVIAEP), the "caveolin signature sequence." Three distinct caveolin genes have been identified: caveolin-1 or VIP-21 (Cav-1), caveolin-2 (Cav-2), and caveolin-3 (Cav-3). Cav-1 exists in two isoforms Cav-1α (containing residues 1–178) and Cav-1β (containing residues 32–178); Cav-2 exists in three isoforms, the full-length Cav-2α, and two truncated variants, Cav-2β and Cav-2γ. Cav-1 and Cav-2, which is proposed to function as an accessory protein to Cav-1, are co-expressed in most differentiated cells, including adipocytes, endothelial cells, pneumocytes, Schwann cells, and fibroblasts, whereas Cav-3 is found specifically in skeletal muscle, the diaphragm, and the heart. Apart from the plasma cell membrane, caveolins are also present in other cellular localizations including endocytic vesicles called caveosomes, mitochondria, the endoplasmic reticulum (ER), the Golgi/trans-Golgi network (TGN), and secretory vesicles. In addition, Cav-1 is secreted by some cells into the extracellular space.

Functionally, caveolae, caveolins, and cavins have been implicated in vesicular transport (transcytosis, pinocytosis, and clathrin-independent ► endocytosis), mechano-sensing, cholesterol homeostasis, and cell metabolism. Moreover, caveolins in general and Cav-1 in particular interact through the caveolin scaffolding domain (CSD) with a vast variety of proteins, thereby sequestering and organizing protein complexes and regulating multiple intracellular signaling pathways. Such molecules include ► Src family tyrosine kinases, ► G protein α subunits, G protein-coupled receptors, ► receptor tyrosine kinases (i.e., receptors for ► epidermal growth factor (EGFR), ► insulin-like growth factor (IGFR), placenta-derived growth factor (PDGFR), ► interleukin-6 (IL-6), ► vascular endothelial growth factor (VEGFR)), Ca^{2+} pumps, endothelial ► nitric oxide synthetase (eNOS), integrins, protein kinase C α, as well as components of the tumor growth factor β (TGFβ/SMAD), Wnt/β-catenin/Lef-1, and ► MAP Kinase (e.g., H-Ras, ► Raf kinase, p38) pathway. In addition to the CSD, SH2 domain-containing molecules (i.e., Grb7) interact with Cav-1 via the growth factor-/cytokine-triggered phosphorylation of Tyr 14. Dysregulation of caveolins is associated with the pathogenesis of several human diseases including type II diabetes, Alzheimer disease, atherosclerosis, muscular dystrophy, and ► cancer.

Clinical Aspects

The ability of Cav-1 to interact with and regulate the activity of proteins involved in cell transformation, growth, metabolism, invasion, and cytoskeletal rearrangement renders Cav-1 a key role in tumorigenesis. The effect of Cav-1 expression depends on whether it is expressed in tumor cells or stroma cells. Loss of Cav-1 in fibroblasts induces a cancer-associated fibroblast (CAF) phenotype, which has been consistently linked to higher tumor grade and poor patient outcome in a variety of malignancies including prostate cancer, esophageal squamous cell carcinoma, gastric cancer, pancreatic cancer, and melanoma. Based on these data, expression of Cav-1 together with expression of cavin-1 and CD36 in the tumor stroma has been suggested as prognostic biomarkers, i.e., in breast cancer. In addition, new therapeutic strategies aim to exploit the loss of stromal Cav-1 by targeting the tumor microenvironment.

In contrast to stromal Cav-1, the functional roles of Cav-1 and cavins in tumor cells depend on cancer cell types and conditions. While initial studies have demonstrated that Cav-1 negatively regulates signaling molecules in some tumor cells (i.e., head and neck cancer and extrahepatic biliary carcinoma cells) thereby mediating cell growth inhibition, several reports clearly show a positive correlation between high Cav-1 expression, ► tumor grade, ► progression, ► metastasis, and chemoresistance in other tumor cells. This dual role of Cav-1 may be caused by microenvironment-stimulated Cav-1 tyrosine and/or serine phosphorylations and the presence of a Cav-1 P132L dominant-negative point mutation, which counteract the growth inhibitory function of Cav-1. Moreover, the secreted form of Cav-1 (e.g., in prostate cancer) acts as a growth factor and an inhibitor of apoptosis, as well as a stimulator of angiogenesis. Increased Cav-1 expression has been linked to the progression of tumors including human ► prostate cancer, primary and metastatic human ► breast cancer, progression of thyroid cancer, high-grade ► bladder cancer, metastasis of the ► lung, ► pancreatic cancer, lymph node metastasis in esophageal ► squamous cell carcinoma, and ► multiple myeloma. Based on these proposed roles of Cav-1 in tumor progression, ongoing studies are now exploring caveolins as novel therapeutic targets in cancer therapies. High levels of Cav-1 expression in vascular endothelial cells additionally provide the rationale for using Cav-1-targeted therapy to inhibit tumor ► angiogenesis. Approaches to target caveolins in general and Cav-1 in particular include the use of Cav-1 antisense and Cav-1 ► siRNA, as well as the use of synthetic CSD, which competitively inhibits protein interactions with Cav-1. Further therapeutic strategies include attempts to inhibit or disrupt caveola formation using either statins (3-hydroxy-3-methylglutaryl-coenzyme A (HMG-CoA) reductase inhibitors), which block the production of the cholesterol intermediate mevalonate, or the cholesterol-binding agent methyl-β-cyclodextrin

(MβCD). Alternatively, caveolae might be used as a drug and gene delivery transport system to specifically target anticancer therapies to tumor cells, thereby reducing required dosages and overall toxicity.

Cross-References

- ▶ Angiogenesis
- ▶ Bladder Cancer
- ▶ Breast Cancer
- ▶ Cancer
- ▶ Endocytosis
- ▶ Epidermal Growth Factor Receptor
- ▶ G Proteins
- ▶ Grading of Tumors
- ▶ Insulin-Like Growth Factors
- ▶ Interleukin-6
- ▶ Lung Cancer
- ▶ MAP Kinase
- ▶ Metastasis
- ▶ Multiple Myeloma
- ▶ Nitric Oxide
- ▶ Pancreatic Cancer
- ▶ Platelet-Derived Growth Factor
- ▶ Progression
- ▶ Prostate Cancer
- ▶ Raf Kinase
- ▶ Receptor Tyrosine Kinases
- ▶ Signal Transduction
- ▶ SiRNA
- ▶ Squamous Cell Carcinoma
- ▶ Src
- ▶ Vascular Endothelial Growth Factor

References

Carver LA, Schnitzer JE (2003) Caveolae: mining little caves for new cancer targets. Nat Rev Cancer 3:571–581

Liu P, Rudick M, Anderson RG (2002) Multiple functions of caveolin-1. J Biol Chem 277:41295–41298

Martinez-Outschoorn UE, Sotgia F, Lisanti MP (2015) Caveolae and signalling in cancer. Nat Rev Cancer 15(4):225–237

Parton RG, del Pozo MA (2013) Caveolae as plasma membrane sensors, protectors and organizers. Nat Rev Mol Cell Biol 14(2):98–112

van Golen KL (2006) Is caveolin-1 a viable therapeutic target to reduce cancer metastasis? Expert Opin Ther Targets 10:709–721

See Also

(2012) Integrin. In: Schwab M (ed) Encyclopedia of cancer, 3rd edn. Springer, Berlin/Heidelberg, p 1884. doi:10.1007/978-3-642-16483-5_3084

(2012) Wnt. In: Schwab M (ed) Encyclopedia of cancer, 3rd edn. Springer, Berlin/Heidelberg, p 3953. doi:10.1007/978-3-642-16483-5_6255

C-BAS/HAS

▶ HRAS

CBFA2

▶ Runx1

CBP/p300 Coactivators

Andrew S. Turnell
Cancer Research UK Institute for Cancer Studies, The Medical School, The University of Birmingham, Edgbaston, Birmingham, UK

Definition

CBP is an acronym for cAMP-regulated-enhancer (CRE)-binding protein (CREB)-binding protein. p300 is a protein that is highly homologous to CBP and has been named according to its approximate molecular weight. Coactivators are a group of cellular proteins that enhance transcription factor-dependent transcriptional activation.

Characteristics

CBP was initially identified as an auxiliary cofactor required for the CREB-mediated activation of cAMP-stimulated gene transcription. CBP binds specifically, at CREs, to an activated CREB species which has been suitably modified through phosphorylation by the cAMP-responsive protein kinase, PKA. p300 was subsequently characterized, independently, upon the basis of its interaction with the protein product of the adenoviral transforming *E1A* gene and, like CBP, can function as a coactivator in CREB-mediated transcriptional activation. CBP, akin to p300, also binds to E1A. CBP and p300 are highly related at the amino acid sequence level, sharing approximately 60% identity, and both proteins have predicted molecular weights of 265 kDa (Goodman and Smolik 2000). Although CBP and p300 bind to a similar set of cellular proteins, share identical enzymatic activities (Fig. 1), and overlap functionally in regulating cell cycle and differentiation pathways, it is important to note that they also possess distinct biological functions. For example, discrete roles for CBP and p300 during retinoic acid-induced differentiation, cell cycle exit, and ▶ apoptosis of embryonal carcinoma

F9 cells have been identified. p300, but not CBP, was found to be required for both retinoic acid-induced differentiation and transcriptional upregulation of the cell cycle inhibitor p21$^{CIP1/WAF1}$. In contrast, CBP, but not p300, was required for transcriptional induction of p27^{KIP1}. Interestingly, both CBP and p300 were required for retinoic acid-induced apoptosis.

CBP and p300 function primarily as transcriptional coactivators for many sequence-specific transcription factors. In this capacity both CBP and p300 function as lysine (K)-directed acetyltransferases (ATs; Fig. 2a). They modify chromatin structure and function through acetylation of the core histones H2A, H2B, H3, and H4 at numerous sites within their N-terminal tail regions. Specific p300-directed acetylation sites within nucleosome-associated histones have been identified. p300 acetylates H2A upon K5; H2B upon K5, K12, K15, and K20; H3 upon K14 and K18; and H4 upon K5, K8, and K12. Histone acetylation by CBP and p300 facilitates further epigenetic histone modifications and the

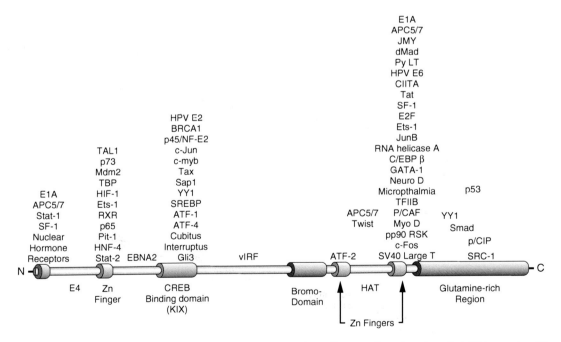

CBP/p300 Coactivators, Fig. 1 Schematic depiction of CBP/p300 primary sequence displaying conserved domains. The diagram shows the binding sites for a number of proteins including the APC/C subunits APC5 and APC7, p53, as well as the adenoviral E1A protein: *E4* ubiquitin E4 ligase activity, *HAT* histone-directed AT activity

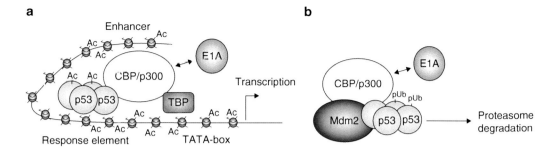

CBP/p300 Coactivators, Fig. 2 Role of CBP and p300 in acetylation and ubiquitylation. (**a**) CBP and p300 bind to enhancer and promoter regions and promote the acetylation of the core histones in order to promote the recruitment of transcription factors and auxiliary factors to sites of transcription. Acetylation of the transcription factor p53 promotes its binding to p53-response elements, Ac: acetylation (**b**) CBP and p300 accelerate Mdm2-mediated polyubiquitylation (pUb) of p53 promoting its degradation by the proteasome. The adenoviral E1A protein binds to CBP/p300 to regulate both acetylation and ubiquitylation activities

recruitment of other proteins involved in transcriptional activation to promoter/enhancer regions, potentially through reducing the affinity of histone tails for DNA. Interestingly, p300 AT activity itself is enhanced by autoacetylation of critical lysine residues in an activation loop motif found within its AT domain. Specifically, autoacetylation of critical residues K1499, K1549, K1554, K1558, and K1560 enhances AT activity.

CBP and p300 also enhance transcription through their ability to interact with and acetylate nonhistone proteins and regulate their cellular activities. Indeed, CBP and p300 acetylate a variety of transcription factors directly, including p53, E2F-1, NF-κB, and c-Myc. For example, p300 has been shown to enhance p53 transcriptional activity by promoting p53 sequence-specific binding to DNA through the acetylation of multiple residues in p53's C-terminal region. Lysine residues K370, K372, K373, K381, and K382 have all been found to be substrates for p300-directed acetylation in vitro. Consistent with these observations, K373 is acetylated in vivo in circumstances when p53 transcriptional activity is stimulated by UV and ionizing radiation. Interestingly, Mdm2, the E3 ubiquitin ligase that targets p53 for degradation, inhibits p300-mediated acetylation of p53. CBP and p300 can also function as transactivators independently of AT activity. Thus CBP and p300 mutants that lack the AT domain can still stimulate transcription. CBP and p300 function in this regard through specific binding to transcription factors such as nuclear receptors, or p53. p300 also possesses an N-terminal E4 ubiquitin ligase domain. It has been shown that this domain catalytically enhances the Mdm2-directed polyubiquitylation of p53, promoting degradation (Fig. 2b). E1A inhibits p300 function in this regard.

A role for CBP and p300 in cell cycle and cellular transformation was first established during early studies with E1A. E1A mutants incapable of binding to CBP and p300 were found to be defective in their ability to promote S phase and initiate DNA synthesis in baby rat kidney (BRK) cells; E1A was also shown to induce S phase by a redundant pathway through its interaction with the protein product of the *Retinoblastoma* gene, pRb. Interestingly, E1A's capacity to induce mitosis in BRKs requires its interaction with both pRb and CBP/p300. Moreover, the ability of E1A to transform primary rodent cells in tissue culture was found to be wholly dependent upon its interaction with CBP and p300, suggesting that both CBP and p300 might function as tumor suppressors. In vitro models suggest that E1A inhibits CBP/p300-directed AT activity and represses CBP/p300-dependent transcription programs. Alternatively, E1A could utilize CBP/p300 acetyltransferases during tumorigenesis to promote an altered program of gene expression. A role for the E3 ubiquitin ligase, the APC/C, in CBP/p300 function has been determined. E1A and APC/C subunits APC5 and APC7 share evolutionarily conserved CBP/p300-binding domains within their primary sequence. Studies have suggested that E1A deregulates CBP/p300 during tumorigenesis by disrupting CBP/p300-APC/C cell cycle function. Interestingly, E1A residue K239 is acetylated by CBP/p300 in vivo, and E1A associates with CBP/p300 AT activity from adenovirus-infected and adenovirus-transformed cells. Acetylation of E1A has been proposed to affect its interaction with the corepressor CtBP and alter its nuclear localization by disrupting E1A association with importin-α. Whether acetylation of E1A is required for transformation with either Ras or E1B is not known. The requirement for the CBP/p300 E4 ligase in E1A-mediated transformation is similarly not known.

There is increasing evidence to suggest that CBP and p300 might be functionally deregulated in ► cancer. In support of this notion, studies have indicated that both CBP and p300 genes are functionally deregulated in ► acute myeloid leukemia (AML). Specifically, chromosomal translocations occur during AML tumorigenesis where a significant portion of the gene encoding the monocytic leukemia zinc finger AT (MOZ) fuses with a large part of the CBP or p300 gene to form MOZ-CBP or MOZ-p300 chimeras. It is proposed that these chimeric proteins possess aberrant AT activity which is important in promoting tumorigenesis. Chromosomal rearrangements are more common for CBP than p300 in this regard. Mixed lineage leukemia (MLL), MLL-CBP, and MLL-p300 translocations have also been described. Studies have also indicated that somatic mutations in one p300 allele, accompanied by loss of heterozygosity (LOH) of the second wild-type allele, also occur in isolated cases of human colorectal and breast tumors. Similarly, biallelic somatic inactivation of CBP has been observed in ovarian tumors, esophageal squamous cell carcinomas, and some lung cancers, suggesting that both CBP and p300 might function as classical tumor suppressors in epithelial cancers.

In support of these findings, germ-line monoallelic inactivation of CBP is the genetic basis for Rubinstein-Taybi syndrome (RTS), a disease characterized by pleiotropic developmental abnormalities and an increased incidence of malignancies, usually childhood tumors of neural crest origin. Whether these tumors are characterized by LOH is, however, not known. Interestingly, mice displaying monoallelic inactivation of CBP also display characteristics of RTS, while mice-engineered heterozygous for CBP displays hematological developmental abnormalities, and with increased age develop a number of hematological malignancies, which in some instances are characterized by LOH. Germ-line monoallelic mutations in p300 also result in RTS. It is not known at present, however, whether these RTS patients also have an increased risk of developing tumors. However, mice heterozygous for p300 do not develop malignancies at a higher frequency. The ability of CBP and/or p300 to function as ► tumor suppressor genes may reside in their capacity to directly interact with tumor suppressor gene products and ► oncogene products, or through regulating, indirectly, multiple signaling pathways that coordinate cell cycle progression and/or differentiation programs.

Cross-References

- ► Acute Myeloid Leukemia
- ► Apoptosis
- ► Cancer
- ► Oncogene
- ► Tumor Suppressor Genes

References

Goodman RH, Smolik S (2000) CBP/p300 in cell growth, transformation and development. Genes Dev 14:1553–1577

Iyer NG, Ozdag H, Caldas C (2004) p300/CBP and cancer. Oncogene 23:4225–4231

Hennenkam RCM (2006) Rubinstein–Taybi syndrome. Eur J Hum Genet 14:981–985

Miller RW, Rubinstein JH (1995) Tumors in Rubinstein-Taybi syndrome. Am J Med Genet 56:112–115

Turnell AS, Mymryk JS (2006) Roles for the coactivators CBP and p300 and the APC/C E3 ubiquitin ligase in E1A-dependent cell transformation. Br J Cancer 95:555–560

See Also

(2012) Acetyltransferase. In: Schwab M (ed) Encyclopedia of cancer, 3rd edn. Springer Berlin Heidelberg, p 17. doi: 10.1007/978-3-642-16483-5_27

(2012) Cell cycle. In: Schwab M (ed) Encyclopedia of cancer, 3rd edn. Springer Berlin Heidelberg, p 737. doi: 10.1007/978-3-642-16483-5_994

(2012) Chromatin. In: Schwab M (ed) Encyclopedia of cancer, 3rd edn. Springer Berlin Heidelberg, p 825. doi: 10.1007/978-3-642-16483-5_1125

(2012) Differentiation. In: Schwab M (ed) Encyclopedia of cancer, 3rd edn. Springer Berlin Heidelberg, p 1113. doi: 10.1007/978-3-642-16483-5_1616

(2012) E3 ubiquitin ligase. In: Schwab M (ed) Encyclopedia of cancer, 3rd edn. Springer Berlin Heidelberg, p 1184. doi: 10.1007/978-3-642-16483-5_1771

(2012) E4 ubiquitin ligase. In: Schwab M (ed) Encyclopedia of cancer, 3rd edn. Springer Berlin Heidelberg, p 1184. doi: 10.1007/978-3-642-16483-5_1772

(2012) Loss of heterozygosity. In: Schwab M (ed) Encyclopedia of cancer, 3rd edn. Springer Berlin Heidelberg, pp 2075–2076. doi: 10.1007/978-3-642-16483-5_3415

(2012) P53. In: Schwab M (ed) Encyclopedia of cancer, 3rd edn. Springer Berlin Heidelberg, p 2747. doi: 10.1007/978-3-642-16483-5_4331

(2012) Transcription. In: Schwab M (ed) Encyclopedia of cancer, 3rd edn. Springer Berlin Heidelberg, p 3752. doi: 10.1007/978-3-642-16483-5_5899

(2012) Transformation. In: Schwab M (ed) Encyclopedia of Cancer, 3rd edn. Springer Berlin Heidelberg, pp 3757–3758. doi: 10.1007/978-3-642-16483-5_5913

(2012) Tumor suppressor. In: Schwab M (ed) Encyclopedia of cancer, 3rd edn. Springer Berlin Heidelberg, p 3803. doi: 10.1007/978-3-642-16483-5_6056

C-CAM

▶ CEA Gene Family

CCCTC-Binding Factor

Elena Klenova[1], Dmitri Loukinov[2] and Victor Lobanenkov[2]
[1]Department of Biological Sciences, University of Essex, Colchester, Essex, UK
[2]Section of Molecular Pathology, Laboratory of Immunopathology, NIAID, National Institutes of Health, Bethesda, MD, USA

Synonyms

CTCF

Definition

CTCF (acronym for a "CCCTC-binding factor") is a highly conserved and ubiquitous protein with multiple functions, which include regulation of transcription, chromatin insulation, and genomic imprinting.

Characteristics

The CTCF protein was originally identified for its ability to bind to a promoter element of the chicken c-myc gene. The sequence recognized by CTCF contained the CCCTC repeats and therefore the protein was defined as CTCF (the CCCTC-binding factor). However, it was later discovered that other CTCF-target sequences (or CTSs) were remarkably dissimilar, and the term "multivalent transcription factor" was coined for CTCF. Another unusual feature of the CTSs is their length: the analysis of binding patterns of CTCF to multiple sites demonstrated that CTCF requires about 50–60-bp-long sequence to form a complex with DNA.

The ability of CTCF to bind such diverse targets has been attributed to its DNA-binding domain, which is composed of 11 zinc fingers (ZFs), 10 of them of the C_2H_2 class and 1 ZF of C_2HC class (Fig. 1a, b). According to this model, the combinatorial utilization of different ZFs results in binding to diverse DNA targets. In addition, CTCF-DNA complex formation can be regulated by DNA ▶ methylation, if symmetrically methylated CpG dinucleotides present on both DNA-strands within any given CTS coincide with the DNA bases required for the CTS recognition by a particular subset of CTCF fingers. Not all CTCF-target sequences contain CpG bp that can be modified by methylation, nevertheless the capability of CTCF to distinguish differentially methylated DNA targets is one of the major features of CTCF with a broad spectrum of functional implications.

The CTSs have been identified in many genomic elements. It is estimated there may be well over 30,000 of CTSs in the human genome, with ~14,000 localized in potential insulators. Many

CCCTC-Binding Factor, Fig. 1 (**a**) Schematic drawing of the CTCF protein. The three domains of CTCF are depicted as follows: *N* N-terminal domain (*Patterned box*), *ZF* ZF domain (*box with half ovals* designating 11 Zinc Fingers; the *black half ovals* refer to the C_2H_2 class and the *gray half oval* refers to the C_2HC class), *C* C-terminal domain (*open box*). The amino acid numbers for the start and the end of each domain are indicated *above* the diagram. (**b**) The cartoon illustration of the wild-type human CTCF protein represents the N-terminal and C-terminal domains of CTCF and the DNA-binding domain of CTCF composed of 10 ZF of C_2H_2 class and 1 ZF of C_2HC class. (**c**) The locations of the tumor-specific mutations in the CTCF protein are shown. The mutations CTCFHR, KE, and RW are located in ZF3, and the mutation CTCFRQ is located in ZF7. The position of the 14 bp insertion is indicated

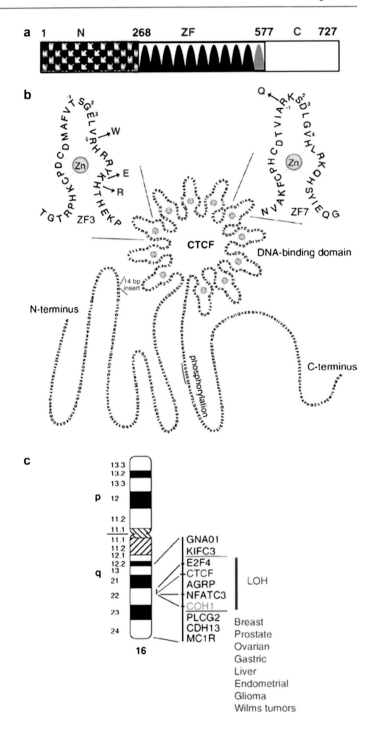

of these sites are methylation sensitive and map to promoter, intergenic and intragenic regions, and both exons and introns. Examples of CTCF-target promoters include 5′-noncoding regions of the c-*Myc* oncogene, chicken lysozyme, *IRAK2*, *BRCA1*, the amyloid precursor protein (*APP*), the exon regions of *hTERT*, and the intron regions of the serotonin transporter gene, *SLC6A4*. Other CTCF-driven regulatory elements include vertebrate enhancer-blocking elements (insulators), classic examples of which are chicken β-globin insulators that flank β-globin gene cluster. Such intergenic insulators seem to have a consensus binding motif for CTCF. CTCF sites are universally present in all mammalian differentially methylated domains/regions (DMD/DMR) or imprinting control regions (ICR), as exemplified by CTSs in ICRs of such imprinted gene clusters as *IGF2/H19*, *Rasgfr*, *KvDMR*, and other loci, deregulation of which through aberrant (biallelic) CTS-methylation or CTS-demethylation contributes to cancer.

CTCF has now been cloned from various organisms which include insects, fish, amphibians, birds, rodents, and primates. The comparison between the proteins revealed a high degree of homology between the CTCF from different organisms, especially in the ZF DNA-binding domain. Thus, this domain is 100% identical at the protein level among mouse, man, and chicken, whereas the full-length protein is 93% identical in those three species; the *Drosophila* CTCF protein has a 46% identity within the zinc-finger regions and 27% overall identity.

Typically for a transcriptional factor, CTCF is localized to the nucleus. It is ubiquitously expressed in various tissues and cells in different organisms. Such conservation in the protein composition and also wide representation in cells/tissues signifies the important and general cellular functions mediated by CTCF.

The size of the CTCF protein varies depending on the organism. For example, the human CTCF protein is composed of 727 amino acids, chicken CTCF of 728, and *Drosophila* CTCF of 818 amino acids. The structure of the human CTCF is shown in Fig. 1 (panels a and b). The ZF DNA-binding domain is positioned in the center of CTCF and accounts for about one third of the protein's size.

The N-terminal domain of human CTCF is composed of 268 amino acids and is rich in proline residues. The C-terminal domain is the smallest part of the molecule (150 amino acids) and is highly negatively charged. These CTCF domains play an important role in the modulation of CTCF functions in the regulation of transcription. In some cases, this regulation relies on posttranslational modifications. For example, the C-terminal domain contains the sites of phosphorylation by the protein kinase CK2 (former casein kinase II), whereas the N-terminal domain contains the sites for poly(ADP-ribosyl)ation by the PARP-1 (poly(ADP-ribose) polymerase-1). The sites for SUMOylation have been mapped to the N- and C-terminal domains of CTCF.

The posttranslational modifications and interactions with protein partners have been demonstrated to modulate important functions of CTCF. For example, specific phosphorylation of CTCF by CK2 and SUMOylation affect the CTCF functions in transcriptional regulation. Poly(ADP-ribosyl)ation was found to be important for insulator function of CTCF, CTCF-dependant nucleolar transcription, and barrier function. Posttranslational modifications of CTCF have also been implicated in human myeloid cell differentiation.

Regulation of CTCF-dependent molecular processes also involves CTCF associations with other proteins. Thus, CTCF interactions with sin3 and YB-1 are shown to modulate CTCF function as a transcriptional repressor. Cooperation of CTCF with nucleophosmin, Kaiso, and helicase protein CHD8 has been linked to the control of insulator function of CTCF and epigenetic regulation. Cohesins and CTCF have been shown to co-localize genome wide; this association has been implicated in the insulator function of CTCF. Interaction of CTCF with another transcription factor, YY-1, is required to control the X-chromosome inactivation, and cooperation of CTCF with RNA Polymerase II may be important for regulation of transcription.

A testis-specific paralogue of CTCF has been reported. This protein was termed ▶ *BORIS* (the acronym for Brother of the Regulator of Imprinted Sites). BORIS possesses the 11 ZF domain homologous to that of CTCF; the flanking N-and C terminal domain, on the other hand, are dissimilar. These structural features indicate that BORIS could recognize the same set of DNA targets as CTCF, while different flanking domains could be important for regulation of BORIS-specific functions.

CTCF Functions

A growing body of evidence suggests that CTCF is involved in the organization and regulation of a whole range of distinct genomic functions in three-dimensional nuclear space. They include gene activation, repression, and silencing; CTCF is also involved in the control of insulator function and imprinting. All vertebrate enhancer-blocking elements tested so far contain CTCF-binding sites. The importance of the insulator function of CTCF was further demonstrated in the regulation of CTG/CAG repeats in the DM1 locus and in the X-chromosome inactivation. It is now generally accepted that the molecular basis for the insulator function of CTCF lies in the ability of CTCF to influence chromatin architecture by mediating long-range chromatin looping and modification of histones. Such alterations then settle the balance between active and repressive chromatin and influence gene expression.

CTCF binding to many of its targets can be regulated by DNA methylation; the ability of CTCF to read such epigenetic marks contributes significantly to the versatility of CTCF functions. Several findings support the concept of CTCF being a ▶ tumor suppressor gene (TSG). Firstly, CTCF suppresses cell growth and proliferation, and, further, in some cell systems (for example, myeloid cells) induces cell differentiation. Secondly, the CTCF gene maps within the smallest region of overlap for loss of heterozygosity (LOH) that has been observed at chromosome 16q22.1 in breast, prostate, and Wilm's tumor (Fig. 1c). Finally, functionally significant, tumor-specific CTCF mutations in the ZF domain of CTCF were identified in various sporadic cancers including breast, prostate, and Wilm's tumor in the remaining allele (Fig. 1b). All four reported tumor-specific point mutations in the CTCF Zn finger domain result in a missense codon at a position predicted to be critical for ZF formation or DNA base recognition. Another reported tumor-specific mutation constituted of a 14 bp insertion in the N-terminal domain of CTCF (Fig. 1b). In familial non-*BRCA1/BRCA2* breast cancers, two sequence variants, G240A in the 5′ untranslated region and C1455T (S388S) in exon 4, were also identified.

The CTCF's function as a negative regulator of cell growth has been well documented on various cellular models. Thus, over-expression of CTCF leads to inhibition of cell growth and proliferation. Normal embryonic rat cells, made haploinsufficient for CTCF by the retroviral insertion into the intron upstream of the first coding exon, manifest all major features of cancerous transformation in vitro. The mechanism of this function of CTCF, at least in part, lies in the ability of CTCF to control genes responsible for regulation of cell growth and proliferation, negatively ▶ oncogenes and positively TSG. Examples of such CTCF-target genes include oncogenes ▶ *MYC, PIM-1, PLK, E2F1, TERT, IGF2* and TSGs *p19ARF(p16/INK4a)*, BRCA1, ▶ p53, ▶ p21, and *p27*. Based on these findings, CTCF emerges as a key versatile element linking genetics, epigenetics, development, and disease.

The ability of CTCF to interact with the repeated sequences and read epigenetic marks (DNA methylation) may provide a causal link not only to some forms of neoplasia but also to degenerative and neurological conditions. Epigenetic disturbances in these diseases are frequently associated with the instability of repeats, which is considered to be the hallmark of this pathology.

Clinical Aspects

A link between CTCF and the disease development has been generally recognized. Various genetic and epigenetic mechanisms that result in CTCF malfunction can lead to pathogenesis.

The tumor-specific mutations in CTCF can dramatically change the normal biological functions of the wild-type CTCF protein. The sets of the genomic targets of the mutant CTCF variants may alter due to the loss of binding to the usual CTCF targets and/or binding of the mutants to the new targets, especially if the wild-type allele is lost. Each ZF mutation abrogates CTCF binding to a subset of target sites within the promoters and/or insulators of certain genes involved in regulating cell proliferation but do not alter binding to the regulatory sequences of other genes. These observations suggest that CTCF may represent a novel tumor suppressor gene that displays tumor-specific "change of function" rather than complete "loss of function."

The 14 bp insertion in the N-terminal domain, on the other hand, most likely leads to the loss of function of CTCF as it creates a premature stop codon, thus generating a truncated CTCF protein. The significance of the sequence variants in the familial breast cancers, however, is not yet clear.

The genetic alterations in CTCF are rare events; therefore, considerable efforts are being currently made to identify epigenetic mechanisms responsible for inactivation of CTCF. The rationale behind these studies is that the binding of CTCF to its DNA targets is methylation sensitive, with the current view that the bound CTCF can protect the CpG islands of DNA against methylation. Indeed, it has been reported that derepression of the maternal *IGF2* allele is linked to abnormal methylation of the CTCF target sites within the *ICR H19* in a wide range of cancer types (breast, prostate, colorectal, Wilm's tumor). This has been explained by the inability of CTCF to bind to the methylated *ICR H19* and therefore its failure to establish the chromatin insulator function on the maternal allele thus leading to activation of *IGF2*.

There is a growing body of evidence to suggest that even mutations of a single CTCF site leads to dramatic biological consequences. For instance, mutations of the CTCF site in the Xist promoter that alter CTCF binding result in the skewed X-chromosome inactivation in affected families. Furthermore, deletions of CTCF sites in human *ICR H19* lead to predisposition to Wilm's tumors in families with Beckwith-Wiedemann Syndrome (BWS). Finally, a mutation of the single CTCF site in the homologous *ICR H19* predisposes the mice carrying such a mutation to colorectal cancer.

Epigenetic inactivation of a number of cancer genes due to aberrant methylation of the CpG islands within their promoters has also been established. Interestingly, many of these genes are regulated by CTCF. As in the case with the *ICR H19*, CTCF may be necessary to protect the promoters of the TSGs from unwanted DNA methylation. According to another, yet to be proven, model, CTCF may demarcate the boundary between methylated and unmethylated genomic domains, as may be the case for the BRCA1 promoter.

The utility of CTCF as a cancer ► biomarker is yet to be established, although there are indications that CTCF may be an interesting target for therapy in breast tumors where levels of CTCF were found elevated compared with breast cell lines with finite life span and normal breast tissues. Such upregulation of CTCF in breast cancer cells has been linked to resistance of these cells to apoptosis. The results of the experiments in breast cancer cell lines point to a possible link between CTCF expression and sensitivity to apoptosis; that is, higher levels of CTCF may be necessary to protect the more sensitive cancer cells from apoptotic stimuli. These findings may be relevant to the potential use of CTCF as a therapeutic target in breast cancers: reducing the levels of CTCF would then result in apoptotic cell death of cancer cells hopefully without affecting normal breast tissue; the effect of CTCF downregulation may be more dramatic in high grade breast tumors. On the other hand, elevated levels of CTCF in breast tumors may correlate with several clinical and/or pathological parameters, which make CTCF a potential prognostic marker. More research is needed to clarify the full potential of CTCF as a clinical target and a cancer biomarker.

Cross-References

► Biomarkers in Detection of Cancer Risk Factors and in Chemoprevention
► *BRCA1/BRCA2* Germline Mutations and Breast Cancer Risk
► Clinical Cancer Biomarkers
► MYC Oncogene

References

Klenova EM, Morse HC, III HC, Ohlsson R et al (2002) The novel BORIS + CTCF gene family is uniquely involved in the epigenetics of normal biology and cancer. Semin Cancer Biol 12:399–414

Ohlsson R, Renkawitz R, Lobanenkov V (2001) CTCF is a uniquely versatile transcription regulator linked to epigenetics and disease. Trends Genet 17:520–527

Ohlsson R, Lobanenkov V, Klenova E (2010) Does CTCF mediate between nuclear organization and gene expression? Bioessays 32:37–50

Phillips JE, Corces VG (2009) CTCF: master weaver of the genome. Cell 137:1194–1211

Recillas-Targa F, De La Rosa-Velazquez IA, Soto-Reyes E et al (2006) Epigenetic boundaries of tumour suppressor gene promoters: the CTCF connection and its role in carcinogenesis. J Cell Mol Med 10:554–568

CCI779

▶ Rapamycin

CCI-779

▶ Temsirolimus

CCRG-81045

▶ Temozolomide

CD Antigens

Richard I. Christopherson
School of Life and Environmental Sciences,
University of Sydney, Sydney, NSW, Australia

Synonyms

Cellular antigens; Cluster of differentiation antigens; Immunophenotypic determinants; Surface molecules

Definition

The human clusters of differentiation (CD) antigens are surface molecules originally detected on white blood cells (leukocytes) from peripheral blood. The first Human Leukocyte Differentiation Antigen (HLDA) workshop was held in Paris in 1982 where 15 surface molecules were assigned based upon the "clustering" of submitted antibodies whose reactivities were screened against a panel of cell lines. Different antibodies that showed similar or identical patterns of reactivity against the panel of cell types were considered to be reacting with the same surface molecule. This clustering of antibody reactivity enabled designation of a specific CD number for a particular surface molecule. The identification of CD antigens was facilitated by the prior development by Kohler and Milstein of a procedure for generation of monoclonal antibodies against a particular antigen. Meetings of the HLDA group were held approximately every 4 years, culminating in HLDA10 that was held at Wollongong (NSW, Australia) in December 2014. At that workshop, further CD antigens were added to the list to give a total of 371 CD antigens. The CD antigen organization has now been renamed Human Cell Differentiation Molecules (HCDM) in recognition that CD antigens are not found uniquely on leukocytes. Indeed CD antigens are found on all types of human cells in different repertoires controlled by the genetic program of the tissue.

Characteristics

The CD antigens are a diverse group of surface glycoproteins with a multitude of functions, providing the interface between a cell and the external environment that includes other cells. The CD antigens may be cell-cell or cell-matrix adhesion molecules, cytokine receptors, ion pores, or nutrient transporters. The CD antigens perform a variety of roles in immune system function. CD1, for example, presents lipids to T-cells and is essential for immunity against the mycobacterial infections that cause tuberculosis and leprosy. CD4 is a co-receptor in antigen-induced T-cell activation and is a receptor for HIV, CD35 is a complement receptor, CD40 is a member of the TNF receptor family with the ligand CD154, and CD54 is an intercellular adhesion molecule.

The method of discovery of CD antigens has classically involved testing monoclonal antibodies submitted to a workshop against a panel

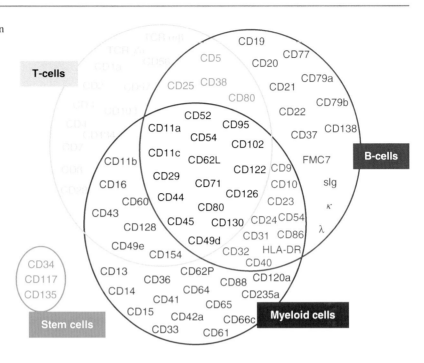

CD Antigens, Fig. 1 Venn diagram showing the differential expression of CD antigens on different categories of leukocytes

of 75 cell types using fluorescently tagged antibodies and ▶ flow cytometry. Hierarchical cluster analysis is then performed and a dendrogram plotted. Monoclonal antibodies that cluster show similar patterns of interaction with the panel of cells. With the development of sophisticated procedures for membrane proteomics, this clustering procedure is becoming outdated, and CD antigens may in the future be designated using different criteria. There are certainly several 1,000 cell surface proteins that could, in principle, be detected and characterized using methods of higher sensitivity. The discovery of further CD antigens will continue to involve raising monoclonal antibodies against antigens on intact cells in the traditional manner but will certainly utilize modern proteomic techniques such as two-dimensional gel electrophoresis and multidimensional chromatography with detection and identification of proteins using mass spectroscopy and extensive protein databases.

CD Antigens Provide Immunophenotypes of Leukocytes

The repertoires of surface CD antigens found on different types of leukocytes reflect the genetic programs that operate in particular cell types.

Thus, cells may be classified according to their cell surface profile (immunophenotype). This concept is illustrated in Fig. 1 as a Venn diagram for T-cells, B-cells, and myeloid cells. T-cells (yellow) express certain antigens uniquely such as CD2, CD3, and CD4; B-cells (blue) express CD19, CD20, CD21, and CD22; and myeloid cells (red) express CD13, CD14, CD15, and CD33. Certain CD antigens are shared between two lineages of leukocytes, for example, CD5 and CD38 (green) are shared between T-cells and B-cells. The so-called pan leukocyte markers are shared between all three categories of leukocytes and include well-known antigens such as CD44 and CD45. All leukocytes originate from stem cells via proliferation and differentiation of cells down lineages to form the many types of mature leukocytes. The stem cell antigen CD34 (black) is a marker of undifferentiated cells.

Classification of Leukemias Using CD Antigens

The principles described above for normal cells can also be applied to cancers such as leukemias. Most leukemias arise as mutations in precursors of leukocytes in the lineages of differentiation found in the bone marrow. A mutation will stop further differentiation of a precursor cell, and there is proliferation rather than differentiation. The resultant identical (monoclonal) cells accumulate in the circulation and the patient is eventually diagnosed with leukemia. Most leukemias are monoclonal, and the leukemic cells usually have a similar or identical surface expression profile (immunophenotype) to that of the precursor cell from which the leukemia arose. Thus, identification of a large number of CD antigens using flow cytometry or antibody microarrays may be sufficient to diagnose leukemia.

CD Antigens as Targets for Therapeutic Antibodies

These cell surface proteins are potential targets for therapeutic antibodies. Such antibodies may block the function of a receptor, selectively activate leukocyte subpopulations, carry a toxin or radioisotope, or act as a site for antibody-dependent cellular cytotoxicity (ADCC) or complement-dependent cytotoxicity (CDC) where the target cell is eliminated by cytotoxic cells such as neutrophils, monocytes, and natural killer cells. There are a number of therapeutic antibodies in clinical use for treatment of a variety of leukemias and lymphomas. For example, rituximab is specific for CD20 and is used to treat chronic lymphocytic leukemia (CLL) and non-Hodgkin lymphoma (NHL). Both are B-cell cancers that express CD20 (Fig. 1) and are killed by this antibody. Mylotarg is specific for CD33, contains a toxin, and is used to treat certain types of acute myeloid leukemia (AML). Campath-1H (alemtuzumab) binds to CD52 and is used to treat NHL. There are many more therapeutic antibodies in development, one of the most rapidly growing area of pharmaceuticals, where monoclonal antibodies are first made against the desired CD antigen and the characteristics of the antibody are then "engineered" to make it suitable for use in patients.

Methods for Identification of CD Antigens

Flow cytometry has been the "gold standard" for identification of a limited number of CD antigens on the surface of leukocytes. In this method, the leukocytes in suspension are mixed with a fluorescently labeled antibody that is specific for the extracellular portion (epitope) of a surface molecule thought to be expressed on the cells. The fluorescently labeled sample is aspirated into the flow cytometer, and the cells pass singly through a narrow aperture where a laser beam individually excites fluorescent antibodies bound to single cells. The emitted fluorescence is detected and data accumulates for a large number (e.g., 10,000 cells). Flow cytometry can detect three different fluorescent antibodies simultaneously;

CD Antigens,
Fig. 2 Capture of live leukocytes on the CD antibody microarray. The *red bars* across the cell membrane represent a CD antigen (e.g., CD20) that forms an initial interaction with antibodies against CD20 that are immobilized on a solid support as a dot in the microarray. Cell capture occurs progressively as CD20 moves in the membrane of the cell and becomes progressively captured by the antibodies on one side of the cell

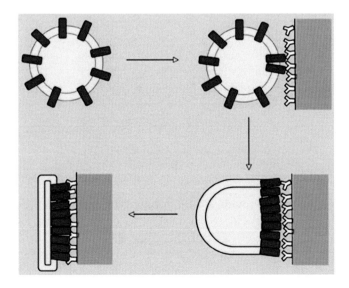

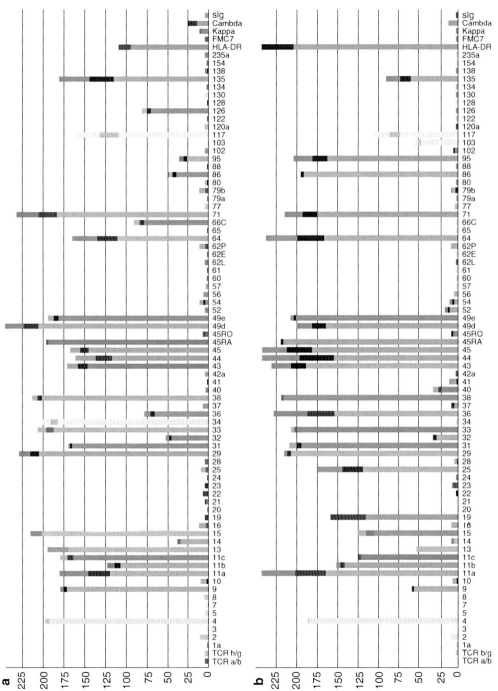

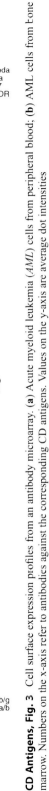

CD Antigens, Fig. 3 Cell surface expression profiles from an antibody microarray: (**a**) Acute myeloid leukemia (*AML*) cells from peripheral blood; (**b**) AML cells from bone marrow. Numbers on the x-axis refer to antibodies against the corresponding CD antigens. Values on the y-axis are average dot intensities

more sophisticated systems can detect eight and up to 17 CD antigens. To diagnose leukemias, 10–15 CD antigens are usually identified using several cycles of flow cytometry, and the information is combined with other criteria such as cell morphology, cell staining, an image of the chromosomes, and sometimes analysis of the DNA in the cells.

A CD antibody microarray has been developed that detects the presence of 147 different CD antigens on leukocytes in a single assay. This microarray called DotScan (Medsaic Pty Ltd, Eveleigh, NSW, Australia), consists of CD antibodies immobilized on a microscope slide. Live cells (three million) are placed on the microarray that is ~0.5 cm square and contains more than 300 antibody dots. Cells are captured by an immobilized antibody if the cell has the corresponding CD antigen on its surface (Fig. 2). After one hour, unbound cells are gently washed off and the resultant dot pattern is the immunophenotype (surface expression profile, disease signature) for the leukemia. The dot pattern for leukemia is stored as a digital image and may be analyzed with a variety of software to provide an expression profile (Fig. 3) that in many cases enables diagnosis of the type of leukemia.

Cross-References

▶ Flow Cytometry

References

Belov L, Mulligan SP, Barber N et al (2006) Analysis of human leukaemias and lymphomas using extensive immunophenotypes from an antibody microarray. Br J Haematol 135:184–197

Chattopadhyay PK, Price DA, Harper TF et al (2006) Quantum dot semiconductor nanocrystals for immunophenotyping by polychromatic flow cytometry. Nat Med 12:972–977

Köhler G, Milstein C (2005) Continuous cultures of fused cells secreting antibody of predefined specificity. J Immunol 174:2453–2455. Reprinted from Nature 256(5517):495–497 (1975)

Zola H, Swart B, Banham A et al (2006) CD molecules – human cell differentiation molecules. J Immunol Methods 319:1–5

See Also

(2012) CD Antibody Microarray. In: Schwab M (ed) Encyclopedia of Cancer, 3rd edn. Springer Berlin Heidelberg, p 689. doi:10.1007/978-3-642-16483-5_946

(2012) Clustering. In: Schwab M (ed) Encyclopedia of Cancer, 3rd edn. Springer Berlin Heidelberg, p 885. doi:10.1007/978-3-642-16483-5_1226

(2012) Immunophenotype. In: Schwab M (ed) Encyclopedia of Cancer, 3rd edn. Springer Berlin Heidelberg, p 1826. doi:10.1007/978-3-642-16483-5_3000

(2012) Leukocytes. In: Schwab M (ed) Encyclopedia of Cancer, 3rd edn. Springer Berlin Heidelberg, p 2028. doi:10.1007/978-3-642-16483-5_3330

(2012) Monoclonal Antibody. In: Schwab M (ed) Encyclopedia of Cancer, 3rd edn. Springer Berlin Heidelberg, p 2367. doi:10.1007/978-3-642-16483-5_6842

(2012) Proteomic Techniques. In: Schwab M (ed) Encyclopedia of Cancer, 3rd edn. Springer Berlin Heidelberg, p 3100. doi:10.1007/978-3-642-16483-5_4820

(2012) Surface Glycoproteins. In: Schwab M (ed) Encyclopedia of Cancer, 3rd edn. Springer Berlin Heidelberg, p 3571. doi:10.1007/978-3-642-16483-5_5593

CD156b Antigen

▶ ADAM17

CD184

▶ Chemokine Receptor CXCR4

CD246

▶ ALK Protein

CD26

▶ CD26/DPPIV in Cancer Progression and Spread

CD26/DPPIV in Cancer Progression and Spread

Jonathan Blay
Department of Pharmacology, Dalhousie
University, Halifax, NS, Canada

Synonyms

ADAbp; ADA-CP; CD26; Dipeptidyl-peptidase IV; DPPIV

Definition

CD26/DPPIV is a multifunctional protein in the outer membrane of normal and cancer cells that can (i) remove an amino-terminal dipeptide from many regulatory peptides, terminating their activity, (ii) bind the enzyme adenosine deaminase (ADA) from the extracellular fluid, and (iii) associate directly with proteins of the ▶ extracellular matrix. Levels of CD26/DPPIV are variable but typically decline as cancer develops, and this has been linked to disease progression and the shift to metastasis.

Characteristics

CD26/DPPIV is a molecule that has been known in different forms since the 1960s but whose key role in cancer has only been appreciated since the early 1990s when it was shown that the absence or presence of CD26/DPPIV in melanocytes determined whether or not those cells showed behavior that was characteristic of a cancer. Our understanding of CD26/DPPIV has an interesting history, as it reflects the collective findings of four different areas of research – in fact directly reflecting the multifunctional nature of the protein itself. The different aspects of the function of this molecule are illustrated in Fig. 1.

Some of the earliest data on this molecule were obtained in studies of the major binding protein for the enzyme adenosine deaminase (ADA) in gastrointestinal epithelia. When ADA was isolated from the tissue, it was found to exist in both high-molecular-weight and low-molecular-weight forms. The high-molecular-weight form was found to be a complex of ADA itself with a larger, 110-kDa protein, subsequently referred to as ADA-complexing protein (ADA-CP) or ADA-binding protein (ADAbp). This anchoring protein for ADA was later shown to be identical to CD26/DPPIV, the extracellular part of which has a region that acts to bind ADA from outside of the cell.

Some of the major substrates for this activity are listed in Table 1. Early studies on CD26/DPPIV also addressed its enzyme activity. The dipeptidyl-peptidase IV (DPPIV) activity is an intrinsic part of the molecule itself and was initially studied mostly at a biochemical level. This very selective form of enzyme activity removes just two amino acids from the N-(amino-)terminus of a peptide, which is why it is called a dipeptidase. The characteristic activity of DPPIV requires that the penultimate N-terminal amino acid has a particular identity, usually proline and less commonly alanine. This is a part of the peptide that often has effects on its stability within the body – the existence of a proline in that position typically confers greater stability. So the removal of this dipeptide by DPPIV is a means of regulating the persistence and bioactivity of important regulatory peptides.

The relative susceptibilities to cleavage of the substrates are given on an arbitrary scale based upon their specificity constants (k_{cat}/K_m). A high number indicates that the peptide is a good substrate for the dipeptidyl-peptidase IV activity of CD26/DPPIV.

The third area of research that led to our present knowledge of CD26/DPPIV involved the way in which lymphocytes become activated. Lymphocytes normally reside in the body within particular tissue structures – specialized structures called lymph nodes or at specific sites within the gut mucosa, for example – in numbers that are necessary to be able to respond to almost all of the threats that may be encountered. In the event of such a challenge, however, the cells that are most able to deal with the threat are mobilized, divide

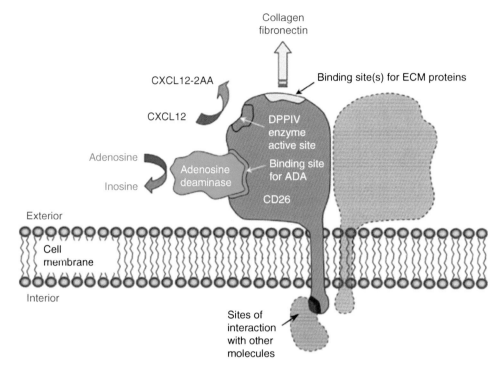

CD26/DPPIV in Cancer Progression and Spread, Fig. 1 The different domains and functions of CD26/DPPIV. The CD26 protein is anchored in the plasma membrane of the cell, with the bulk of its molecular structure on the outer face. The enzyme domain that underlies its dipeptidyl-peptidase IV activity, removing pairs of amino acids (*AA*) from substrates such as the chemokine CXCL12, comprises one of three functional sites in contact with the external environment. A separate domain acts as the major cellular binding site for another enzyme, adenosine deaminase (*ADA*), which is present in the extracellular fluid. There are also at least two potential sites for the binding of the extracellular matrix proteins collagen and fibronectin. CD26/DPPIV usually exists as a dimer; the second molecule is shown in outline. The intracellular portion of CD26/DPPIV is small and no functional domains have been identified. CD26/DPPIV must signal intracellularly by coupling with other cellular components

so as to make a larger population of specialized defenders, and become armed to respond in the appropriate way. As these cells become activated, various important proteins are produced at the cell surface. These "activation proteins" are given "CD" numbers as unique identifiers ("CD" refers to "cluster of differentiation" markers or antigens). The differentiation antigen designated CD26 has proven to be identical to the molecules ADAbp and DPPIV.

The last of the roles for CD26/DPPIV follows from its ability to bind to extracellular matrix molecules, primarily collagen and ▶ fibronectin. These are embedded within the molecular scaffold that surrounds all cells and which provides particular cues for cellular behavior in three dimensions. For the CD26/DPPIV that is present on cancer cells, this opens up the possibility that it may act as an additional anchor to tether cells to the extracellular matrix, along with dedicated cell adhesion molecules such as the integrins. The reverse situation may also be important during the process of metastasis. It has been shown that the CD26/DPPIV that is present at the surface of endothelial cells lining blood vessels can interact with a form of fibronectin that is deposited on the surface of cancer cells. This may cause arrest of circulating cancer cells that have become detached from the main tumor and help to seed the cancer at secondary sites like the lung.

The same molecule therefore has four different functions and has four different names that have been used over the years with greater or lesser frequencies. The designation CD26 is probably

CD26/DPPIV in Cancer Progression and Spread, Table 1 Some of the major substrates for the dipeptidyl-peptidase IV activity of CD26/DPPIV

Molecule	Full name and main function(s) in normal tissues	DPPIV sensitivity (k_{cat}/K_m)
CXCL12	SDF-1α (stromal cell-derived factor-1α): Involved in development of the nervous system, bone marrow, and intestine and in the homing of stem cells	100
CCL22	Macrophage-derived chemokine: Is an attractant for various types of white cells and functions in immune and inflammatory responses	80
GRP	Gastrin-releasing peptide: Released by nerves in the stomach to cause the production of gastrin from G cells in the mucosa	40
NPY	Neuropeptide Y: Peptide neurotransmitter found in the brain that has a role in regulating normal physiological processes	20
GLP-1	Glucagon-like peptide-1: Gut hormone secreted by L cells in the intestine has a role in control of insulin levels	4
CCL11	The chemokine eotaxin-1: Causes the recruitment of eosinophils into tissues and plays a role in allergic responses	1.6
CCL5	The chemokine RANTES ("regulated on activation, normal T expressed and secreted"): Selective attractant for memory T lymphocytes and monocytes	0.8
VIP	Vasoactive intestinal peptide: Peptide hormone produced by various tissues, with effects on blood vessels and secretory processes	0.2

the most neutral, because although CD proteins have been studied primarily in white cells, they also exist in other tissues, and the nomenclature has no link to function. The abbreviation "DPPIV"

refers to its enzyme activity and – given the other activities this talented component incorporates – is not a valid name for the overall molecule. However, as so as much research on this protein has focused upon its enzymatic role, and this facet of its action is of significance in certain diseases such as cancer and diabetes, the term "CD26/DPPIV" serves as a compromise.

CD26/DPPIV is found at the surface of the cells that form the functional barrier (epithelium) in most of the major sites that give rise to cancer in adults (e.g., intestine, lung, breast, and prostate). The levels detected in cancer (the "expression") vary from those of the corresponding normal tissue, but the pattern is not consistent across all cancers and within a single cancer type there may be variable findings. So, for example, while the prevailing change in adult solid cancers (e.g., lung and prostate cancer) is for CD26/DPPIV to decline, in certain less common cancers such as those of the thyroid and kidney, CD26/DPPIV levels actually increase. This suggests that the absence or presence of CD26/DPPIV does not universally favor or disfavor cancer progression but that its role depends very much on the tissue type, meaning that changes in CD26/DPPIV as a tissue becomes cancerous will depend very much on its normal role. Additionally, in some cancers (such as colorectal cancer), the expression of CD26/DPPIV is very variable, not just between different tumors but in different regions of the same cancer. This points to a likelihood that CD26/DPPIV levels can be regulated by factors that are generated within the developing cancer tissue.

The ability of CD26/DPPIV to bind the enzyme ADA seems to be part of a fundamental mechanism whereby cells can resist the actions of the purine nucleoside adenosine in certain disease situations. This helps them to resist a threat to their survival by high concentrations of adenosine or the risk of responding excessively to adenosine when it persists in the environment for an extended period. High concentrations of adenosine can occur persistently in the disorganized environment of a solid cancer (▶ Adenosine and tumor microenvironment). By retaining ADA close to the cell surface, the cell has a greater chance of scavenging adenosine near to the cell and preventing excessive action through adenosine receptors that are embedded in the cell membrane.

This dynamic situation involving extracellular adenosine production (from ATP breakdown and through cellular export) and breakdown (ADA bound to CD26/DPPIV) next to the cell surface provides substantial opportunity for the cell to modulate other signals that might be acting on it from other sources. Adenosine modulates many of the signals that are produced to act on leukocytes in inflammation and cancer, leaving CD26/DPPIV – as the docking site for ADA – in a unique position to act as one of the central determinants of the overall cellular response. In leukocytes, this seems to allow cells to resist somewhat the immunomodulatory effects of adenosine that may be produced during inflammation. Indeed, levels of CD26/DPPIV, either on the surface of leukocytes or in a soluble form (sCD26) that is shed from cells and can be recovered from blood plasma, have been used to indicate levels of inflammation.

In cancer, the status quo is altered by two things. Firstly, as indicated above, adenosine levels in solid cancers are persistently high. Secondly, cellular levels of CD26/DPPIV are altered from normal and (with the exception of a few specific cancers) are typically low. These factors will combine to leave cells within a cancer (tumor cells, supporting fibroblastic cells, and infiltrating leukocytes) more susceptible to the effects of adenosine.

The two factors may be linked, as it has been shown that persistently high adenosine levels can cause the amounts of CD26/DPPIV at the surface of cancer cells to decline precipitously. Adenosine, which is produced regionally within cancers, is likely a major factor responsible for the spatial variations in CD26/DPPIV expression within certain cancers.

Changes in CD26/DPPIV levels in cancer will also have an impact as a result of alterations in the DPPIV enzyme activity available. The substrates of this enzyme are typically hormones and other peptide regulators that are important in controlling the functions of epithelial and nervous cells, as well as cells involved in the body's defenses (Table 1). Among the most sensitive of the various mediators that are substrates for this enzyme is a chemokine molecule called CXCL12. (Chemokines are small peptide mediators that play an important role in controlling cellular arrangement in developing tissues and directing cell movement in the immune and inflammatory systems of our body's defenses.) CXCL12 is important in cancer because it seems to be one of the major factors that provides the "right environment" for cancer cells that have left the original tumor to settle into new locations in the process of metastasis. It provides a signal that activates a receptor on cancer cells called CXCR4 to facilitate their seeding and growth in such metastatic sites as the lungs, liver, and bone marrow (▶ Chemokine Receptor CXCR4).

Changes in CD26/DPPIV levels in cancer likely help cancers to grow by affecting the activities of these mediators that are substrates for the DPPIV enzyme activity. The result of excising the N-terminal two amino acids in most cases is to inactivate the mediator or cause it to be more rapidly degraded. In the common cancers in which CD26/DPPIV tends to have declined, there will therefore be a shift to higher levels of the active mediator(s). As mediators such as CXCL12 are strongly linked to cancer progression, this will be one of the many different ways in which cancers can act to encourage their own expansion.

Cross-References

▶ Extracellular Matrix Remodeling
▶ Integrin Signaling
▶ Melanocytic Tumors

CD314

▶ NKG2D Receptor

CD318 (Cluster of Differentiation 318)

▶ CDCP1

CD44

Ursula Günthert
Institute of Pathology, University Hospital, Basel, Switzerland

Synonyms

Cluster of differentiation 44; ECMRIII; gp90^Hermes; H-CAM; Homing receptor; Hyaluronan receptor; pgp-1; Phagocytic glycoprotein-1

Definition

CD44 is a type I transmembrane glycoprotein, which exists in a large number of isoforms. The gene contains 20 exons within a region of ~60 kb on chromosome 11p13 in humans and on chromosome 2 at 56 cM in mice. CD44 is in close proximity to the recombination-activating genes Rag-1 and Rag-2.

Characteristics

CD44 is the major receptor for hyaluronic acid and other ▶ extracellular matrix molecules (▶ fibronectin, laminin 5, collagen type IV, serglycin). The standard molecule is heavily glycosylated by N- and O-linked residues and chondroitin sulfate side chains, while some of the variant isoforms carry in addition heparan sulfate moieties, which can present various growth factors and ▶ chemokines (for local concentration and activation). The number of extracellular molecules that can associate with CD44 is ever growing, among them matrix metalloproteinase-7 (MMP-7) and matrix metalloproteinase-9 (MMP-9) inducing activation of latent transforming growth factor β (TGF-β) and hence promote ▶ invasion and ▶ angiogenesis. Further associating molecules are ErbB2 (HER-2/neu), EpCAM, E-selectin, CD8+ cytotoxic T cells, and VLA-4 (Integrin α4β1). While

c-met/▶ scatter factor receptor, c-kit/stem cell factor receptor, ▶ osteopontin (OPN), and CD95 have specifically been shown to associate with CD44 variant isoforms, association with the other molecules has not been specified to a CD44 isoform. The association between VLA-4 (integrin α4β1) and CD44 directs cells into inflammatory regions, while the c-met/CD44v6 interaction is required for c-met/scatter factor receptor signaling leading to ▶ RAS activation, and when CD44v6 associates with CD95, trimerization of the death receptor is prevented and hence apoptosis signaling is blocked (see Fig. 1).

Upon cellular activation, CD44 localizes to plasma membrane microdomains and associates (see Fig. 1) with nonreceptor tyrosine kinases lck and fyn, smad-1, membrane-bound OPN, and Rho. Via ezrin (▶ ERM protein), ankyrin, or annexin II, the cytoplasmic region of CD44 is linked to the cytoskeleton. CD44 is involved in the ▶ Wnt signaling pathway. ▶ P-glycoprotein, the product of the multidrug resistance (MDR) gene, has also been demonstrated to interact physically and functionally with CD44, thus promoting cell ▶ migration and invasion and possibly enforcing resistance to ▶ chemotherapy. The p-glycoprotein–CD44 interaction is the first hint of a functional association between MDR and ▶ metastasis formation, involving CD44. Further it is of importance that the presenilin-dependent γ-secretase cleaves off the intracellular domain (ICD) of CD44, which then translocates to the nucleus and acts as a transcription factor for genes containing TPA (12-O-tetradecanoyl phorbol 13-acetate) response elements in their promoter. The ICD of CD44 promotes the fusion of ▶ macrophages, is localized in the nucleus of macrophages, and promotes the activation of nuclear factor kappa (NF-κ) B.

Cellular and Molecular Regulation

The standard form of CD44 (CD44s) is expressed in almost all tissues and leukocytes and is encoded by exons s1–s10, yielding a product of 90 kDa. The variant isoforms (CD44v) are generated by alternative splicing of the nuclear RNA between exons s5 and s6 and are encoded by exons v2–v10

CD44, Fig. 1 Multiprotein complexes can be formed between CD44 and various membrane-linked (*top*) and intracellular molecules (*bottom*)

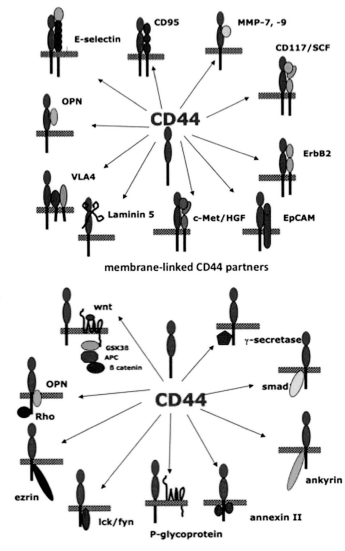

membrane-linked CD44 partners

intracellular CD44 partners

(exon v1 is silent in humans, but not in mice and rats). Combinations of different variant exons with the standard backbone result in numerous variant isoforms, with masses of 100–250 kDa. All the variant regions are located extracellularly and are highly hydrophilic. In contrast to the ubiquitous expression of CD44s, CD44v isoforms are expressed in a highly restricted manner in nonmalignant tissues: in early embryogenesis, stem cells of epithelia and hemopoiesis, activated leukocytes, and memory cells. However, in malignant tissues, CD44v isoforms are often upregulated, e.g., in carcinoma, various ▶ hematological malignancies, and in autoimmune lesions.

A positive feedback loop was identified which couples *RAS* activation with alternative splicing of the CD44 variant isoforms. The presence of CD44v6 then sustains Ras signaling, which is in turn important for cell cycle progression.

CD44 is implicated in various aspects of tumor progression: invasion, migration, and ▶ apoptosis blockade.

Clinical Relevance

Originally identified by its metastasizing potential in rats, CD44v isoform expression was identified in various human tumors and correlated with clinical relevance. *Upregulation* of CD44v correlates with poor prognosis in gastric and colorectal carcinoma, non-small cell lung tumors, ▶ hepatocellular carcinoma, ▶ pancreatic cancer, B-cell chronic lymphocytic leukemia, ▶ multiple myeloma, non-Hodgkin lymphoma, and acute myeloblastic leukemia. *Downregulation* of CD44v correlates with poor prognosis in esophageal squamous cell carcinoma, bronchial carcinoid tumors, ovarian neoplasms, uterine cervical tumors, transitional cell bladder tumors, and prostate cancers, while downregulation of CD44s correlates with amplification of MYCN and is indicative for an unfavorable outcome in ▶ neuroblastoma patients. In breast carcinoma, controversial data between CD44v expression and survival were established and need further evaluation.

Elevated serum levels of CD44v have prognostic value for gastric and colon carcinoma and non-Hodgkin lymphoma, which are indicative for a poor prognosis.

An emerging new field (although hypothesized some 150 years ago) is the area of cancer-initiating cells, also termed ▶ cancer stemlike cells. They exist as a small population in every tumor and determine the capability of the tumor to grow and propagate. In tumors of the ▶ Brms1, the pancreas, the prostate, the head and neck, the brain (glioblastoma), and in the blood system (leukemia), the cancer-initiating cells are CD44$^+$. A major goal currently is to identify specific markers (▶ stem cell markers) that enable to distinguish between normal, benign tissue stem cells and those that are cancer-initiating.

CD44 is also strongly upregulated in inflammatory lesions of patients with autoimmune diseases (▶ inflammatory bowel disease-associated cancer (Crohn disease), multiple sclerosis, rheumatoid arthritis).

Cross-References

- ▶ γ-Secretase
- ▶ Apoptosis
- ▶ Autoimmunity and Cancer
- ▶ Caspase
- ▶ Cell Adhesion Molccules
- ▶ Colorectal Cancer
- ▶ Crohn Colitis
- ▶ Death Receptors
- ▶ Embryonic Stem Cells
- ▶ EpCAM
- ▶ ERM Proteins
- ▶ E-Selectin-Mediated Adhesion and Extravasation in Cancer
- ▶ Extracellular Matrix Remodeling
- ▶ Gastric Cancer
- ▶ HER-2/neu
- ▶ Hyaluronidase
- ▶ *HRAS*
- ▶ Inflammation
- ▶ Kit/Stem Cell Factor Receptor in Oncogenesis
- ▶ Lipid Raft
- ▶ Matrix Metalloproteinases
- ▶ MET
- ▶ Mouse Models
- ▶ Osteopontin
- ▶ P-Glycoprotein
- ▶ *RAS* Activation
- ▶ Receptor Cross-Talk
- ▶ Receptor Tyrosine Kinases
- ▶ Scatter Factor
- ▶ Stem Cell Markers
- ▶ Wnt Signaling

References

Cheng C, Yaffe MB, Sharp PA (2006) A positive feedback loop couples Ras activation and CD44 alternative splicing. Genes Dev 20:1715–1720

Jin L, Hope KJ, Zhai O et al (2006) Targeting of CD44 eradicates human acute myeloid leukemic stem cells. Nat Med 12:1167–1174

Martin TA, Harrlison G, Mansel RE et al (2003) The role of the CD44/ezrin complex in cancer metastasis. Crit Rev Oncol Hematol 46:165–186

Ponta H, Sherman L, Herrlich PA (2003) CD44: from adhesion molecules to signalling regulators. Nat Rev Mol Biol 4:33–45

Ponti D, Zaffaroni N, Capelli C et al (2006) Breast cancer stem cells: an overview. Eur J Cancer 42:1219–1224

Zeilstra J, Joosten SP, van Andel H, Tolg C, Berns A, Snoek M, van de Wetering M, Spaargaren M, Clevers H, Pals ST (2014) Stem cell CD44v isoforms promote intestinal cancer formation in Apc(min) mice downstream of Wnt signaling. Oncogene 3(5):665-70

CD55

▶ Decay-Accelerating Factor

CD62 Antigen-Like Family Member E (CD62E)

▶ E-Selectin-Mediated Adhesion and Extravasation in Cancer

CD66a

▶ CEA Gene Family
▶ CEACAM1 Adhesion Molecule

CD66b

▶ CEA Gene Family

CD66c

▶ CEA Gene Family

CD66e

▶ Carcinoembryonic Antigen
▶ CEA Gene Family

CD82

▶ Metastasis Suppressor KAI1/CD82

2-CdA

▶ Cladribine

CdA

▶ Cladribine

CDA2

▶ Activation-Induced Cytidine Deaminase

CDCP1

Brian Law and Stephan C. Jahn
Department of Pharmacology and Therapeutics
and the UF and Shands Cancer Center, University
of Florida, Gainesville, FL, USA

Synonyms

CD318 (cluster of differentiation 318); CDCP1
(CUB domain-containing protein 1); gp140
(glycoprotein 140); SIMA135 (subtractive immunization M(+)HEp3 associated 135 kDa protein);
Trask (transmembrane and associated with Src
kinases)

Definition

CDCP1 is an 836-amino-acid protein that is present in cells as an apparent 140 kDa full-length
protein and an 80 kDa fragment. It is
overexpressed in some cancers and has been
implicated in ▶ invasion, ▶ metastasis, and
tumor ▶ progression.

Characteristics

Discovery

The CDCP1 gene was first discovered in 2001 when high levels of mRNA were found in colon cancer cells, and the protein was later identified in three separate instances. SIMA135 was described as an N-glycosylated and tyrosine phosphorylated membrane protein upregulated in metastatic human epidermoid carcinoma cells in 2003. It was later identified as glycoprotein 140, a protein that was highly phosphorylated when cells were cultured in suspension and could be cleaved to an 80 kDa fragment. Its final name, Trask, came in 2005 when it was discovered to be a substrate for the Src family kinases.

Protein Structure

The type 1 transmembrane glycoprotein contains a 29-amino-acid signal sequence on the amino terminus, a 636-amino-acid extracellular domain, a 21-amino-acid membrane spanning sequence, and a 150-residue intracellular domain. The predicted molecular weight is approximately 90 kDa; however, CDCP1 migrates nearer to 140 kDa on SDS-polyacrylamide gels due to high levels of glycosylation. The extracellular portion holds three CUB (complement protein subcomponents C1r/C1s, urchin embryonic growth factor, and bone morphogenetic protein 1) domains and contains 14 consensus N-glycosylation sites. It is structurally similar to membrane receptors, but no ligand has been identified. The intracellular domain is also posttranslationally modified, with 5 phosphorylatable tyrosine residues. Two proline-rich stretches make up SH3 ligand binding domains. CDCP1 protein structure is summarized in Fig. 1.

Cleavage

The 140 kD full-length protein is cleaved between R368 and K369 in some cancers, creating an 80 kDa fragment with a truncated extracellular

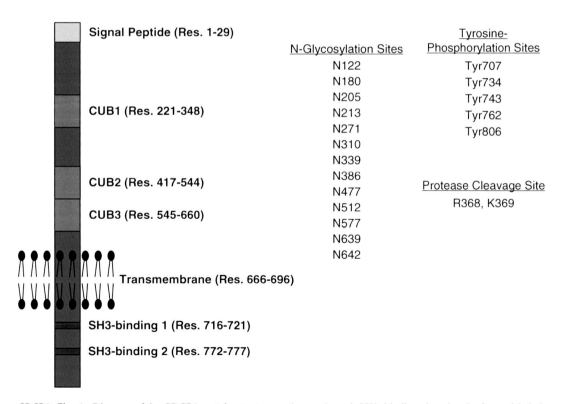

Signal Peptide (Res. 1-29)		
CUB1 (Res. 221-348)	N-Glycosylation Sites	Tyrosine-Phosphorylation Sites
CUB2 (Res. 417-544)	N122	Tyr707
CUB3 (Res. 545-660)	N180	Tyr734
	N205	Tyr743
	N213	Tyr762
	N271	Tyr806
	N310	
	N339	
	N386	
	N477	Protease Cleavage Site
	N512	R368, K369
	N577	
	N639	
	N642	

Transmembrane (Res. 666-696)

SH3-binding 1 (Res. 716-721)

SH3-binding 2 (Res. 772-777)

CDCP1, Fig. 1 Diagram of the CDCP1 protein structure showing the extracellular, transmembrane, and intracellular portions. The signal peptide (*yellow*), CUB domains (*orange*), and SH3 binding domains (*red*) are labeled. N-glycosylation, phosphorylation, and protease cleave sites are also listed

domain lacking the original N-terminus. Trypsin and Matriptase are capable of carrying out this cleavage in vitro at K277 and R368, respectively. This cleavage is primarily carried out in vivo by the serine protease Plasmin during early-stage colonization of ▶ metastatic cells. CDCP1 cleavage is initiated by cell detachment and leads to phosphorylation by Src Family Kinases and pro-invasive and pro-survival signaling.

CDCP1 Expression

CDCP1 is normally expressed in a small number of stem and progenitor cells but is also highly expressed in various cancers. Its expression levels are controlled by promoter methylation in both cases and by ▶ hypoxia-inducible factor 1 and 2 in renal cell carcinoma cells in vitro. High levels of CDCP1 expression in tumors correlate with a poor prognosis.

CDCP1 Signaling

Phosphorylation
CDCP1 is a heavily tyrosine phosphorylated protein and is a key target of the Src family of kinases (SFK). Tyr734 is phosphorylated by SFK, allowing SFK to bind and further phosphorylate Tyr762. In cancer cells, this phosphorylation is induced by detachment and is important in initiating signaling cascades responsible for invasion and metastasis.

Downstream Signaling
The phosphorylation of Tyr762 by SFK allows the binding and activation of Protein Kinase-C δ (PKCδ), and PKCδ signaling is responsible for the pro-tumorigenic effects of CDCP1, including cell invasion, resistance to ▶ anoikis, ▶ matrix metalloproteinase 9 (MMP-9) secretion, and invadopodia formation. The mechanisms of inducing MMP-9 secretion and invadopodia formation are not known; however, the activation of the CDCP1-PKCδ complex results in a reduction in phosphorylation of ▶ focal adhesion kinase, decreasing cell adhesion and increasing ▶ motility. This signaling process is outlined in Fig. 2.

Known Binding Partners
SFK, PKCδ, Yes, Integrins, P-cadherin, N-cadherin

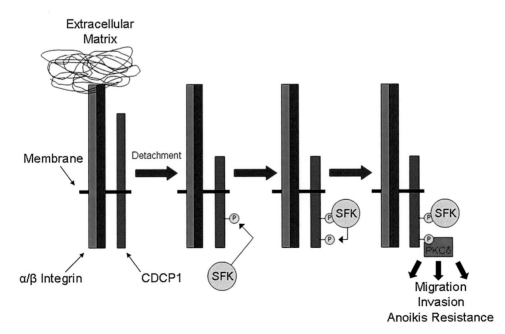

CDCP1, Fig. 2 Diagram of CDCP1-mediated signaling. Upon detachment, CDCP1 is phosphorylated by SFK leading to PKCδ recruitment and subsequent downstream signaling

Cross-References

► Adhesion
► Akt Signal Transduction Pathway
► Anoikis
► Cell Adhesion Molecules
► Focal Adhesion Kinase
► Glycosylation
► Hypoxia-Inducible Factor-1
► Invadosome
► Invasion
► Matrix Metalloproteinases
► Metastasis
► Motility
► Progression
► Proteinase-Activated Receptor-4
► SH2/SH3 Domains
► Src

References

Bhatt AS, Erdjument-Bromage H, Tempst P, Craik CS, Moasser MM (2005) Adhesion signaling by a novel mitotic substrate of src kinases. Oncogene 24:5333–5343

Brown TA, Yang TM, Zaitsevskaia T, Xia Y, Dunn CA, Sigle RO, Knudsen B, Carter WG (2004) Adhesion or plasmin regulates tyrosine phosphorylation of a novel membrane glycoprotein p80/gp140/CUB domain-containing protein 1 in epithelia. J Biol Chem 279:14772–14783

Hooper JD, Zijlstra A, Aimes RT, Liang H, Claassen GF, Tarin D, Testa JE, Quigley JP (2003) Subtractive immunization using highly metastatic human tumor cells identifies SIMA135/CDCP1, a 135 kDa cell surface phosphorylated glycoprotein antigen. Oncogene 22:1783–1794

Scherl-Mostageer M, Sommergruber W, Abseher R, Hauptmann R, Ambros P, Schweifer N (2001) Identification of a novel gene, CDCP1, overexpressed in human colorectal cancer. Oncogene 20:4402–4408

Uekita T, Ryuchi S (2011) Roles of CUB domain-containing protein 1 signaling in cancer invasion and metastasis. Cancer Sci 102:1943–1948

See Also

(2008) SDS-polyacrylamide gels. In: Rédei GP (ed) Encyclopedia of genetics, genomics, proteomics and informatics. Springer, Netherlands, p 1768. doi:10.1007/978-1-4020-6754-9_15185

(2008) YES1 oncogene. In: Rédei GP (ed) Encyclopedia of cancer, Encyclopedia of genetics, genomics, proteomics and informatics. Springer, Netherlands, p 2125. doi:10.1007/978-1-4020-6754-9_18335

(2012) Cadherins. In: Schwab M (ed) Encyclopedia of cancer, 3rd edn. Springer, Berlin/Heidelberg, pp 581 582. doi:10.1007/978-3-642-16483-5_770

(2012) CUB domain. In: Schwab M (ed) Encyclopedia of cancer, 3rd edn. Springer, Berlin/Heidelberg, p 1012. doi:10.1007/978-3-642-16483-5_1408

(2012) Glycoprotein. In: Schwab M (ed) Encyclopedia of cancer, 3rd edn. Springer, Berlin/Heidelberg, p 1570. doi:10.1007/978-3-642-16483-5_2451

(2012) Integrin. In: Schwab M (ed) Encyclopedia of cancer, 3rd edn. Springer, Berlin/Heidelberg, p 1884. doi:10.1007/978-3-642-16483-5_3084

(2012) Invadopodia. In: Schwab M (ed) Encyclopedia of cancer, 3rd edn. Springer, Berlin/Heidelberg, p 1904. doi:10.1007/978-3-642-16483-5_3132

(2012) Matriptase. In: Schwab M (ed) Encyclopedia of cancer, 3rd edn. Springer, Berlin/Heidelberg, p 2182. doi:10.1007/978-3-642-16483-5_3552

(2012) Plasmin. In: Schwab M (ed) Encyclopedia of cancer, 3rd edn. Springer, Berlin/Heidelberg, p 2904. doi:10.1007/978-3-642-16483-5_4604

(2012) Phosphorylation. In: Schwab M (ed) Encyclopedia of cancer, 3rd edn. Springer, Berlin/Heidelberg, p 2870. doi:10.1007/978-3-642-16483-5_4544

(2012) Progenitor cells. In: Schwab M (ed) Encyclopedia of cancer, 3rd edn. Springer, Berlin/Heidelberg, p 2990. doi:10.1007/978-3-642-16483-5_4752

(2012) Prognosis. In: Schwab M (ed) Encyclopedia of cancer, 3rd edn. Springer, Berlin/Heidelberg, p 2994. doi:10.1007/978-3-642-16483-5_4758

(2012) Promoter hypermethylation. In: Schwab M (ed) Encyclopedia of cancer, 3rd edn. Springer, Berlin/Heidelberg, p 3004. doi:10.1007/978-3-642-16483-5_4769

(2012) Renal-cell carcinoma. In: Schwab M (ed) Encyclopedia of cancer, 3rd edn. Springer, Berlin/Heidelberg, p 3252. doi:10.1007/978-3-642-16483-5_5023

(2012) SH3 domain. In: Schwab M (ed) Encyclopedia of cancer, 3rd edn. Springer, Berlin/Heidelberg, pp 3399–3400. doi:10.1007/978-3-642-16483-5_5281

(2012) Signal sequence. In: Schwab M (ed) Encyclopedia of cancer, 3rd edn. Springer, Berlin/Heidelberg, p 3403. doi:10.1007/978-3-642-16483-5_5297

(2012) Src family tyrosine kinase. In: Schwab M (ed) Encyclopedia of cancer, 3rd edn. Springer, Berlin/Heidelberg, p 3498. doi:10.1007/978-3-642-16483-5_5467

(2012) Trypsin. In: Schwab M (ed) Encyclopedia of cancer, 3rd edn. Springer, Berlin/Heidelberg, p 3786. doi:10.1007/978-3-642-16483-5_5996

CDCP1 (CUB Domain-Containing Protein 1)

► CDCP1

CDDP

► Cisplatin

CDK

► Cyclin-Dependent Kinases

Cdk1 Kinase

► Cyclin-Dependent Kinases

CDK2/Cyclin A-Associated Protein p45

► Ubiquitin Ligase SCF-Skp2

CDK4I

► CDKN2A

CDKN1A

► p21

CDKN2

► CDKN2A

CDKN2A

Mark Harland
Section of Epidemiology and Biostatistics,
Cancer Research UK Clinical Centre, Leeds
Institute of Molecular Medicine, St. James's
University Hospital, Leeds, UK

Synonyms

CDK4I; CDKN2; CMM2; Cyclin-dependent kinase inhibitor 2A; INK4A; MTS1; p16; $p16^{INK4}$; $p16^{INK4A}$; $p16^{INK4a}$

Definition

Cyclin-dependent kinase inhibitor 2A gene (CDKN2A), the first identified ► melanoma predisposition gene, encodes the tumor suppressor proteins p16 and ARF.

Characteristics

Identification of CDKN2A
The 9p21-22 chromosomal region was originally implicated in the development of melanomas through a combination of cytogenetic and loss of heterozygosity (LOH) studies. Subsequent linkage analysis in melanoma families indicated that this region harbored a melanoma predisposition locus. Homozygous deletions in cell lines derived from several different tumor types narrowed down the region significantly. This led to the isolation, by two independent groups, of the cell cycle regulatory gene encoding the cyclin-dependent kinase (CDK) inhibitor, p16, which had been previously identified in a yeast two-hybrid screen to identify proteins that bound to CDK4 (Fig. 1).

Gene Structure of CDKN2A
In the original description of human p16, the initiating methionine was incorrectly identified. It was later found that the protein included eight

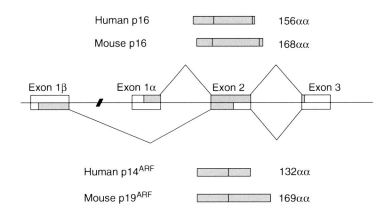

CDKN2A, Fig. 1 Alternative transcripts and products encoded by the CDKN2A locus. The exons of CDKN2A are shown as boxes and identified as exons 1β, 1α, 2, and 3. Alternative splicing occurs as indicated to give rise to two transcripts, exons that splice to encode p16 are shown above, and those that encode p14ARF are shown below. The sizes and composition of the respective mouse and human proteins are indicated

additional amino acids at its amino terminus, although these residues are not present in murine p16. Three exons, spread over approximately 7.2 kb of genomic DNA, encode the 156 amino acid protein with predicted molecular weight of 16,533 Da, designated p16. The primary structural feature of p16 is the four tandem ankyrin-like repeats that comprise approximately 85% of the protein. This domain is believed to facilitate protein-protein interactions (Fig. 2).

The sizes of the translated regions encoded by exon 1α, exon 2, and exon 3 are 150, 307, and 11 bp, respectively. The CDKN2A-locus also has the capacity to encode two distinct transcripts from two different promoters. This is achieved by alternative splicing and the use of different reading frames. Each transcript has a specific 5′ exon, exon 1α (E1α) or exon 1β (E1β), which is spliced onto common second (E2) and third (E3) exons. The E1α-containing transcript encodes p16, and the E1β-containing transcript encodes a protein translated into an alternate reading frame initiated in E1β, designated p19ARF in mice and p14ARF in humans. In contrast to p16, where the murine and human genes share 85% amino acid homology, the alternative reading frame (ARF) proteins share only 59% amino acid homology. The different sizes of the encoded proteins are brought about by the earlier truncation of the ARF transcript in exon 2 in humans.

Two different translation start sites have been reported for the ARF protein, which has lead to some confusion in the numbering of the ARF protein amino acids in publications.

Tumor Suppressor

CDKN2A is a tumor suppressor gene for multiple tumor types. The frequency of mutations at this locus in various cancers is rivaled only by mutations in TP53. As with other classical tumor suppressor genes, both alleles need to be abrogated for tumorigenesis to occur. A wide variety of mechanisms of inactivation of CDKN2A have been documented, including intragenic mutation, homozygous deletion, and transcriptional silencing through methylation of the promoter. Notably in melanomas, many of the intragenic mutations are C > T or tandem CC > TT transitions, implicating ultraviolet radiation (UVR) as the causal somatic mutagen. Although CDKN2A is inactivated in the majority of melanoma cell lines examined, deletions and interstitial mutations of CDKN2A are much less common in uncultured melanoma tumors. Present studies indicate that only 5–10% of uncultured melanomas demonstrate mutations in CDKN2A, a surprisingly low figure given the obvious importance of CDKN2A in familial melanoma and the frequency of LOH seen at chromosome 9p21 in melanomas.

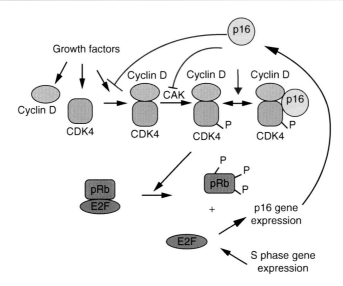

CDKN2A, Fig. 2 Schematic representation of the protein interactions in the cyclin D/CDK4/p16/pRb pathway. Through a complex system of signal transduction, growth factors lead to the assembly of cyclin D and CDK4. This complex is then activated through phosphorylation by the CDK-activating kinase (CAK), and cyclin D/CDK4 in turn phosphorylates pRb, leading to the release of transcription factors of the E2F family. These are then capable of transactivating the genes necessary for entry into S phase, and p16 has been shown to inhibit this process in several ways, by binding to the complex and inhibiting the kinase activity of CDK4, inhibiting CAK-dependent phosphorylation of CDK4, or inhibiting the assembly of the cyclin D/CDK4 complex, with the latter being the principal mechanism of inhibition in vivo. The scheme provided is necessarily simplistic; however, it appears that p16 may also inhibit the phosphorylation of pRb by indirectly inactivating other CDKs, e.g., CDK2, as a consequence of the redistribution of other CDK inhibitors, e.g., p27 and p21. There is also a feedback loop whereby the release of the E2F transcription factor results in the activation of p16 expression, although the absence of E2F binding sites in the CDKN2A promoter precludes direct transactivation by E2F. Aberration of this pathway through either deletion or mutation of pRb, the binding of viral oncogenes to pRb, overexpression or activation of CDK4 or cyclin D, or deletion or mutation of CDKN2A all can result in constitutive transactivation of S phase genes by E2F transcription factors

P16 Is a CDK Inhibitor

P16 is the archetype member of the ▶ INK4 (inhibitor of CDK4) family of CDK inhibitors, which is comprised of p16INK4A, p15INK4B, p18INK4C, and p19INK4D, encoded by CDKN2A, CDKN2B, CDKN2C, and CDKN2D, respectively. Each of the proteins inhibits CDK4- or CDK6-mediated phosphorylation of the ▶ retinoblastoma susceptibility gene product, pRb, thereby providing a powerful negative signal, or "brake," to progression through the cell cycle.

The ▶ cyclin D1/CDK4/p16/pRb signaling pathway is the major growth control pathway for entry into the cell cycle. For cells to progress through G1 into S phase they must pass the late G1 restriction point, which controls entry into S phase. For progression past this restriction point, cyclin D/CDK4 must phosphorylate the ▶ retinoblastoma protein pRb. During G0/G1 the Rb protein exists in a DNA-bound protein complex, where it is bound to the transactivation domain of E2F transcription factors, preventing transactivation of E2F target genes. The phosphorylation of pRb results in the disassociation of this protein complex and the release of E2F such that it can transactivate genes required for entry into S phase. Overexpression of p16 inhibits progression of cells through the G1 phase of the cell cycle by binding to CDK4/cyclin D complexes (or CDK6/cyclin D) and blocking the kinase activity of the holoenzyme. Given that p16 normally functions to inhibit CDK4, it is easy to understand how inactivation of this gene could result in uncontrolled cellular growth leading to

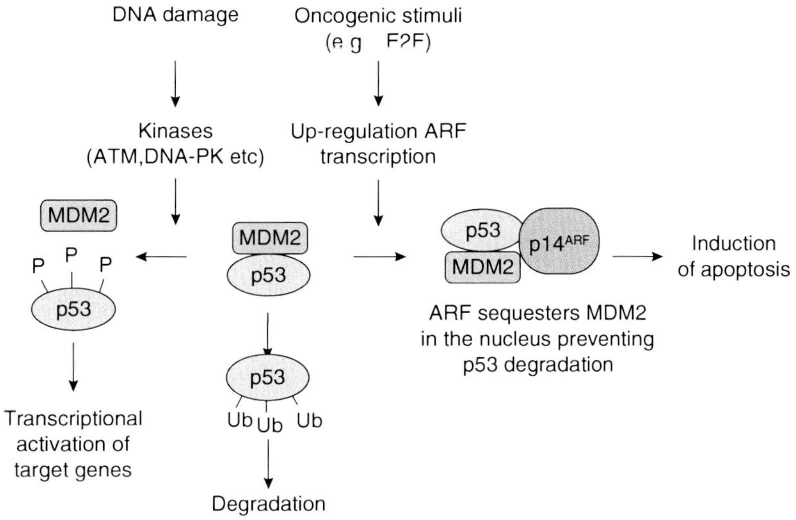

CDKN2A,
Fig. 3 Schematic
representation of the role of
ARF in p53 activation by
DNA damage and
oncogenic stimuli. ARF
functions to sequester
MDM2 in the nucleus
preventing
nucleocytoplasmic
shuttling of the MDM2/p53
complex; however, the
details have not yet been
fully elucidated and results
suggest the mechanism may
differ between humans
and mice

cancer. In many tumor types, an inverse correlation between mutations of p16 and pRb has been observed. Since p16 lies upstream of pRb, inactivation of both proteins would be redundant.

Role of the Alternative Reading Frame (ARF) Product

The ARF protein also regulates the G1/S phase transition via a distinct pathway involving the ▶ TP53 ▶ tumor suppressor gene product p53 and MDM2, which function upstream of p21 (a cyclin-dependent kinase inhibitor closely related to p16) and the CDK2/cyclin E complex (Fig. 3). p53 is a transcription factor that plays a major role in monitoring the integrity of the genome and can be activated to inhibit cell cycle progression or initiate apoptosis through two distinct pathways: (i) in response to a variety of cellular stresses including ▶ DNA damage and ▶ hypoxia and (ii) via overexpression of viral or cellular oncoproteins such as E1A and c-myc. In this way, cells prevent the repair of mutations in successive generations by inducing apoptosis in incipient cancer cells. ARF plays a crucial role in p53-induced apoptosis. Murine p19ARF is capable of inducing a p53-dependent G1 cell cycle arrest that is not mediated through the direct inhibition of known CDKs. Ectopic expression of ARF leads to stabilization of p53 in multiple cell types, but unlike other known upstream effectors of p53, this activation is not through phosphorylation. Instead, ARF binds to MDM2 and blocks both MDM2-mediated p53 degradation and the transactivational silencing of p53. MDM2 continuously shuttles between the nucleus and the cytoplasm. This shuttling is essential for its ability to promote p53 degradation, indicating that MDM2 must export p53 from the nucleus to the cytoplasm to target p53 to the cytoplasmic proteosome. ARF activates p53 by binding to MDM2 in the nucleus and blocking the transport of the MDM2/p53 complex out of this organelle. Results obtained with murine and human ARF are somewhat different. In murine cells results indicate that p19ARF sequesters MDM2 away from p53 into the nucleolus. In human cells p14ARF moves out from the nucleolus to form discrete nuclear bodies in conjunction with MDM2 and p53, thereby blocking their nuclear export and leading to p53 stabilization. The discovery that ARF transcription is induced by the overexpression of a variety of cellular and viral oncoproteins including c-myc, E1A, and E2F has provided the link by which hyperproliferative signals result in p53-dependent apoptosis.

To determine whether mutations in CDKN2A contribute to tumorigenesis via p19ARF in addition to p16, cDNAs carrying a variety of exon 2 mutations have been transfected into cell lines and cell cycle arrest monitored. These mutations have included several that are silent in p16 but caused missense mutations in p19ARF, as well as several deletion mutants that removed either exon 1β or various portions of exon 2. Results indicate that the majority of p19ARF activity is encoded by the exon 1β sequences, as all missense mutations in exon 2 of p19ARF remained fully active in blocking cell cycle progression, and removal of exon 2 sequences only marginally reduced the ability to induce arrest. In contrast, deletion of exon 1β resulted in a transcript that was incapable of inhibiting cell cycle progression. Missense mutations in exon 2 of the human p14ARF transcript similarly did not reduce the growth suppressive function of p14ARF.

Senescence

p16 is not normally expressed at detectable levels in most cycling cells; however, CDKN2A mRNA and p16 protein accumulate in late-passage non-immortalized cells, implicating a role for p16 in cellular ► senescence. This is supported by studies revealing that loss of p16 expression is a critical event in ► immortalization (the flip side to senescence) of a range of cell types. This conclusion was initially alluded to by finding that the frequency of deletions and intragenic mutations of CDKN2A in uncultured tumors was considerably lower than in immortalized cell lines. Growth and survival experiments using cells with impaired CDKN2A function suggest that a p16/pRb-dependent form of senescence may be particularly important in melanocytes. Individuals with defective p16INK4a have been found to have increased numbers of naevi, and it has been speculated that naevi are senescent clones of melanocytes.

Mouse Models

The generation of a CDKN2A "knockout" mouse, carrying a germline homozygous deletion encompassing exons 2 and 3 of the gene, revealed that p16 and p19ARF (since both proteins are eliminated by deletion of exon 2) were not essential for viability or organomorphogenesis. However, the mice did demonstrate abnormal extramedullary hematopoiesis, suggesting that p16 or p19ARF may regulate the proliferation of some hematopoietic lineages. In addition, the mice developed spontaneous tumors at an early age, specifically fibrosarcomas and B cell lymphomas, and were highly sensitive to carcinogens. In contrast to wild-type mouse embryonic fibroblasts (MEFs), cultured MEFs from Cdkn2a nullizygous mice (Cdkn2a$^{-/-}$) failed to undergo senescence crisis and could be transformed by oncogenic ras alleles. Although Cdkn2a$^{-/-}$ mice did not develop melanomas, transformation of Cdkn2a$^{-/-}$ MEFs by activated ras prompted experiments to cross the Cdkn2a$^{-/-}$ mice with a previously generated transgenic mouse in which an activated ras allele was targeted exclusively to melanocytes under the control of the tyrosinase promoter. These mice spontaneously developed melanomas at high frequency and with short latency.

To determine whether p16 or p19ARF was the principal mediator of the above effects, knockout mice strains with targeted deletions of p16 and p19ARF were generated. In general, p19ARF null animals were observed to develop a tumor spectrum more closely related to p53 null rather than p16 null mice. Tumors observed in p19ARF null mice included lymphomas and an increased incidence of soft tissue sarcomas, carcinomas, and osteosarcomas. Mice lacking p16 were found to develop soft tissue sarcomas, osteosarcomas, and melanomas. Mouse strains with specific inactivation of either p16 or p19ARF were tumor prone, but neither was as severely affected as animals lacking both p16 and p19ARF, suggesting cooperation between p16 and p19ARF loss in tumorigenesis.

Clinical Aspects

CDKN2A Mutations and Melanoma

Germline CDKN2A mutations have been observed in approximately 20–40% of melanoma families worldwide. However, melanoma appears to segregate with chromosome 9p markers in a far greater proportion of families than have been shown to carry mutations of CDKN2A. This suggests that melanoma predisposition in some of these families is caused by: (i) another gene in the vicinity of CDKN2A, (ii) mutations outside of the p16 coding region, and (iii) another gene somewhere else in the genome, with linkage to this region occurring simply by chance. The most parsimonious explanation is that a combination of all these possibilities is likely.

Overall, approximately 40% of pedigrees with three or more cases of melanoma have been found to harbor mutations in the CDKN2A gene. This figure varies with location and is lowest in regions of high ► UV radiation (UVR), e.g., Australia (20%), and higher in regions with low incident UVR, e.g., Europe (57%).

There is a significant increase in the yield of CDKN2A mutations with increasing number of affected cases in families with melanoma. In addition, an early age of diagnosis and the presence of family members with multiple primary melanomas or with ► pancreatic cancer have also been shown to be significantly associated with an increased likelihood of finding a CDKN2A mutation.

The population-based frequency of CDKN2A mutations in melanoma cases is of the order of 1–2%, even in those individuals that had developed multiple primary tumors, much lower than observed in families selected for multiple cases of melanoma.

Disease-associated mutations are distributed along the entire length of the p16 coding region. At least one mutation has been described in the promoter of the gene, and several putative mutations have been identified in the intronic sequences. The most frequent CDKN2A mutations identified to date are c.255_243del19 (also known as p16 Leiden), p.M53I, p.G101W, c.331_332insGTC (p.R112_L113insR) (all in exon 2), c.-34G > T (promoter), and c.IVS2-105A > G (intron). There are considerable differences in the frequencies and distribution of CDKN2A mutations across the world. Many mutations have been shown to arise from a common founder and are more frequent in particular geographic locations. For example, Sweden and the Netherlands have single predominant founder mutations (p.R112_L113insR and p16 Leiden, respectively) involving over 90% of families tested. The G101W mutation, common in Italy, France, and Spain, has been calculated to arise from a single genetic event approximately 93 generations ago. Many additional mutations have been repeatedly reported, and where analysis has been performed these have invariably been shown to be due to common founders. The only exception to this appears to be a 24 bp insertion in exon 1a, that has arisen multiple times, presumably because of DNA slippage over a 24 bp repeat region.

Mutation of ARF Germline mutations affecting ARF but not p16INK4a have been reported in a small number (∼3%) of melanoma families. Whereas the distribution of p16 mutation types (approximately 70% missense or nonsense, 23% insertion or deletion, 5% splicing, and 2% regulatory) is consistent with that observed in the Human Genome Mutation Database, the reported ARF-specific mutations are almost all either splicing mutations (affecting the 3' splice site of exon 1b) or large deletions.

Penetrance The pattern of susceptibility in melanoma pedigrees is consistent with the inheritance of autosomal dominant genes with incomplete penetrance. The overall penetrance of CDKN2A mutations in melanoma families has been estimated to be 0.30 by the age of 50 years and 0.67 by the age of 80 years. There is significant variation in the penetrance of CDKN2A mutations with geographical location. By the age of 50 years, penetrance was estimated to be 0.13 in Europe, 0.5 in the United States, and 0.32 in Australia and by the age of 80 years 0.58 in Europe, 0.76 in the United States, and 0.91 in Australia (Fig. 4).

This indicates that the CDKN2A mutation penetrance varies with melanoma population incidence rates, thus the same factors that effect population incidence of melanoma may also mediate CDKN2A penetrance.

Multiple Primary Melanoma

General characteristics of inherited susceptibility to many types of cancer are early age of onset and the development of multiple primary tumors. Hence the presence of multiple primary melanomas (MPM) in an individual may be a sign of them being a CDKN2A mutation carrier. This is the case for a small proportion (13/133, 10%) of MPM cases without a family history of the disease. In contrast, analysis of MPM cases with a

CDKN2A, Fig. 4 Age-specific penetrance estimates for CDKN2A mutations. Penetrance is shown for melanoma pedigrees from Australia, Europe, America, and all geographic locations combined

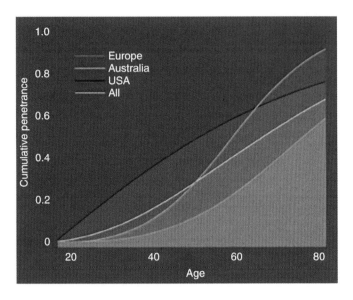

family history of disease yields CDKN2A mutations in 55/139 (40%) of samples tested. The proportion of CDKN2A mutations in sporadic MPM cases increases with increasing number of melanomas (10/119 (8.5%) of cases with two primary melanomas, compared to 11/83 (33%) cases with three or more primary tumors).

CDKN2A Mutations and Nonmelanoma Cancers
Since CDKN2A is a tumor suppressor found to be inactivated in a wide range of different tumors, one might expect individuals carrying germline mutations of CDKN2A to be prone to cancers other than melanoma. ▶ Brms1, prostate, colon, and ▶ lung cancers have been suggested to be associated with CDKN2A mutations; however, these common cancers may occur in CDKN2A-positive pedigrees by chance. Convincing evidence for susceptibility to another tumor type has been shown only for pancreatic cancer, which has been shown to be significantly associated with CDKN2A mutations in all regions except Australia, the reason for this is not yet understood.

There appears to be no evidence of an association between neural system tumors (NSTs) and CDKN2A mutations involving p16. However, there is marginal evidence for the association of NSTs with ARF-specific mutations.

Modifiers of Penetrance of CDKN2A Mutations
The MC1R gene (16q24) which encodes for the melanocyte-stimulating hormone has been shown to be a risk factor in families with segregating CDKN2A mutations. MC1R variants have been shown to act as modifier alleles, increasing the penetrance of CDKN2A mutations and reducing the age of onset of melanoma.

CDKN2a Polymorphisms as Low-Risk Factors
The A148T variant, located in exon 2 of the CDKN2A gene, has no observed effect on p16 function and does not segregate with disease in melanoma pedigrees. The contribution of this polymorphism to melanoma risk remains unclear; an association with increase in risk has been seen in some populations, but not in others.

The 500 C > G and the 540 C > T polymorphisms in the 3′ untranslated region of the CDKN2A gene have been shown to be associated with melanoma risk. The frequencies of the rare alleles at these loci have been shown to be higher in melanoma cases than in controls. It is possible that these variants might alter the stability of the CDKN2A transcript or the level of transcription, or that they may be in linkage disequilibrium with an unidentified variant which is directly responsible for melanoma predisposition. The contribution of these polymorphisms to melanoma risk is likely to be small in comparison to that of CDKN2A inactivating mutations.

CDKN2A and the Atypical Mole Syndrome

Since the description of the "B-K mole syndrome," much debate has ensued regarding the association between melanoma and the atypical mole syndrome (AMS). Several authors have concluded that atypical moles segregate independently of CDKN2A mutations, although individuals with high numbers of naevi in melanoma-prone families are three times more likely to be CDKN2A mutation carriers than those with a low number of naevi. Support for the notion that CDKN2A is naevogenic comes from a study of a large series of 12-year-old twins in which total naevus count was found to be tightly linked to CDKN2A. This finding has been corroborated by two independent genome wide association studies that have mapped loci responsible for naevi in twin cohorts. Both studies showed peaks of high linkage scores at 9p21 directly over the CDKN2A gene.

References

Bishop JN, Harland M, Randerson-Moor J et al (2007) Management of familial melanoma. Lancet Oncol 8(1):46–54
Goldstein AM, Chan M, Harland M et al (2007) Features associated with germline CDKN2A mutations: a GenoMEL study of melanoma-prone families from three continents. J Med Genet 44(2):99–106
Hayward NK (2003) Genetics of melanoma predisposition. Oncogene 22(20):3053–3056
Sharpless NE (2005) INK4a/ARF: a multifunctional tumor suppressor locus. Mutat Res 576(1–2):22–38
Sharpless E, Chin L (2003) The INK4a/ARF locus and melanoma. Oncogene 22(20):3092–3098

CDKN4

▶ p27

cDNA Chips

▶ Microarray (cDNA) Technology

CDX2

Isabelle Gross and Isabelle Hinkel
INSERM U1113, Université de Strasbourg, Strasbourg, France

Synonyms

Caudal type homeobox 2; CDX3; CDX-3

Definition

CDX2 is a member of the caudal-related homeobox transcription factor gene family. As a determinant of cell fate, CDX2 is critical for various aspects of embryonic development, including intestinal morphogenesis. In the adult, CDX2 expression is restricted to the gut and is required to maintain intestinal homeostasis. Altered CDX2 expression is associated with several types of cancer, namely, colon cancer and acute myeloid leukemia.

Characteristics

Structure

CDX1, CDX2, and CDX4 are the three members of the mammalian homeobox transcription factor gene family related to the Drosophila gene *caudal* and belong to the ▶ ParaHox gene cluster, a paralogue of the Hox gene cluster.

The human CDX2 gene is located on chromosome 13 at band q12.3 and consists of three exons encoding a 313 amino acid protein. The central region of the CDX proteins is the most conserved and corresponds to the homeodomain, a 60 amino acid sequence arranged in three alpha-helices, which binds to DNA. The N-terminal region of CDX2 acts as a transcriptional activator domain and together with the C-terminal region modulates its activity. Alternative splicing of the CDX2 gene can also generate miniCDX2 in which the N-terminal transactivation domain is replaced by a specific 13 amino acid extension.

Expression, Activity, and Mechanisms of Regulation

Nuclear CDX2 expression is detected at E3.5 in the murine ► trophectoderm and around E8.5 in several developing tissues of the embryo itself (posterior gut, tail bud, neural tube, etc.). By E12.5 onwards, CDX2 expression is restricted to the intestinal ► epithelium where it is maintained throughout life. Species- and stage-specific gradients of expression along the anteroposterior and dorsoventral axes have been described: for instance, CDX2 expression generally increases with differentiation in the small intestine but not in the colon.

The regulation of CDX2 transcription is highly dynamic, involving stage-specific promoter elements and possibly various transcription factors such as HNF4alpha, GATA6, TCF4/beta-catenin, NF-kappaB, SMAD, or CDX2 itself. The transcription of CDX2 can be modified by multiple extracellular factors (collagen I, Laminin 1, Wnt5A, sodium butyrate, etc.) and is highly sensitive to the cellular microenvironment.

CDX2 levels are also regulated by posttranslational modifications affecting the half-life of the protein. Indeed, phosphorylation of CDX2 by kinases implicated in cell cycle progression, such as ERK1/2 and CDK2, leads to its polyubiquitination and degradation by the ► proteasome. Conversely, in intestinal cells that start to differentiate, the ► cyclin-dependent kinase inhibitor p27Kip1 stabilizes CDX2 by preventing its phosphorylation by CDK2.

Posttranslational modifications are not only involved in the regulation of CDX2 protein levels but can also modulate the transcriptional activity of CDX2. For instance, the MAPK p38alpha phosphorylates CDX2 on a not yet identified residue in differentiated cells and this leads to enhanced transcription of CDX2 target genes. On the opposite, high levels of S60-phosphorylated CDX2 are detected in the proliferative crypt cells, and this phosphorylation actually inhibits CDX2 transcriptional activity: this might explain why CDX2 target genes are mainly activated in the upper third of the crypt, although no CDX2 expression gradient is observed in colonic ► crypts.

Finally, another way of regulating CDX2 activity was revealed with the detection in the proliferative ► crypt cells of a dominant negative isoform of CDX2 (miniCDX2), that lacks the transcription activator domain and whose fixation on the CDX2 binding sites inhibits transcription by full-length CDX2.

Structure Physiological Functions

The existence of a large panel of mice models provides us with considerable information about the biological functions of CDX2.

Ubiquitous and homozygous gene invalidation of CDX2 is lethal before gastrulation as CDX2 is required for ► trophectoderm maturation and consequently blastocyst implantation. In contrast, heterozygous $CDX2^{-/+}$ mice are viable and fertile and present no major dysfunctions despite morphological defects. Indeed, these $CDX2^{-/+}$ mice display anterior homeotic shifts of their axial skeleton, tail abnormalities, or stunted growth, illustrating the role of CDX2 in anteroposterior patterning and posterior axis elongation. In addition, these mice totally lose CDX2 expression in some regions of the proximal colon, which allows intercalary growth of more anterior gastrointestinal tissue types (esophageal, gastric), highlighting the role of CDX2 in intestinal identity. Accordingly, ectopic expression of CDX2 in the stomach of transgenic mice induces the conversion of gastric epithelial cells into enterocyte-like cells.

To circumvent the problem of embryonic lethality induced by complete CDX2 depletion, conditional inactivation of CDX2 was performed to study the consequences of CDX2 loss at different stages of development and in the adult. Because CDX1 and CDX2 can be functionally redundant, double knockout mice for CDX1 and CDX2 were sometimes analyzed using $CDX1^{-/-}$ mice, which are viable and only show alterations of the skeleton. For instance, ubiquitous inactivation of CDX2 post-implantation at E5.5 in $CDX1^{-/-}$ mice is lethal at E10: the mice present abnormal axis elongation, neural tube closure defects, and ► somite patterning alterations, demonstrating that the CDX genes are crucial for these events in early embryonic development.

CDX2 expression was also specifically suppressed in the developing intestine: strikingly, none of these mice survived longer than 2 days after birth because of severe abnormalities in the morphology and function of the gut. For instance, mice in which CDX2 is invalidated at E9.5 in the early endoderm fail to form a colon. In addition, the small intestine lacks most of the ▶ villi critical for nutrient absorption and displays more cycling cells, and many of the mutant cells resemble more to keratinocytes that constitute the esophageal ▶ epithelium than to differentiated intestinal cells. If ablation of CDX2 in the developing intestine is performed later at E13.5 or E15.5, colon formation occurs, but the ▶ epithelium of mutant mice is highly disorganized and ▶ villi are smaller. Inactivation of CDX2 at E13.5 leads to an upregulation of gastric markers (H^+/K^+-ATPase, ghrelin) and a downregulation of intestinal markers (I-FABP). Ablation of CDX2 at E15.5 generates enterocytes that display profound defects in their typical microvilli and disrupted apicobasal polarity, but no features of gastric/esophageal transdifferentiation.

Finally, specific ablation of CDX2 in the adult intestinal ▶ epithelium is also lethal, indicating that CDX2 expression is required throughout life to maintain a functional intestine. Indeed, mutant mice lose weight, have chronic diarrhea, and die of starvation (malabsorption) at the latest 3 weeks after CDX2 inactivation. The ▶ villi of these mice are smaller and the microvilli on absorptive cells are shorter, less dense, and disorganized compared to those of their wild-type littermates. Although conversion into stomach-like tissue is not observed, analysis of the gene expression profiles of $CDX2^{-/-}$ mice shows upregulation of stomach-specific markers.

Mode of Action at the Cellular Level

In line with the spectacular consequences of CDX2 depletion on intestinal cell differentiation in mice, numerous reports show that overexpression of CDX2 can induce various degrees of intestinal differentiation in vitro. For instance, undifferentiated colorectal cell lines can acquire a polarized, columnar shape with apical microvilli, produce various digestive enzymes, and form tight, adherens, and desmosomal junctions upon CDX2 expression. The effect of CDX2 on apicobasal polarity was demonstrated using a 3D culture system and was associated with defective apical transport. This effect is consistent with the formation of large cytoplasmic vacuoles and downregulation of genes involved in endolysosomal function in intestinal cells of conditional CDX2 knockout mice.

CDX2 expression can also reduce anchorage-dependent or anchorage-independent growth of normal, ▶ adenoma, and carcinoma epithelial cells. This may be achieved through reduced cell proliferation as CDX2 can block the G0/G1-S progression in intestinal cell lines. However, a proapoptotic effect of CDX2 can also be observed in various intestinal contexts and thus may also contribute to reduced cell numbers. Of note, the activity of CDX2 on cell growth appears to be dependent on the context and cell type: for instance, somatic knockout of CDX2 reduces anchorage-independent growth of LoVo intestinal cells, and shRNA silencing of CDX2 expression inhibits the proliferation of various human leukemia cell lines.

CDX2 can inhibit intestinal cell ▶ migration and ▶ invasion in Boyden chambers coated or not with Matrigel. These are hollow plastic chambers sealed at one end with a porous membrane and suspended in a well containing chemoattractants. Cells are placed inside the chamber and allowed to migrate through the pores to the other side of the membrane. CDX2 expression appears to influence chromosome segregation, as well as DNA damage repair in intestinal cells.

Mode of Action at the Molecular Level

As a bona fide transcription factor, the main function of CDX2 is to activate specific gene expression in the embryo and later in the intestinal ▶ epithelium. The consensus binding site of CDX2 is (C/TATAAAG/T), an AT-rich sequence typical of homeobox proteins, but CDX2 can also bind to sequences that are slightly different.

During early development, CDX2 regulates anteroposterior patterning by stimulating the expression of various HOX genes such as HOXA5. Later, in the developing of mature intestinal ▸ epithelium, CDX2 regulates a large number of genes involved in intestinal identity and in various intestinal functions. Indeed, CDX2 regulates the transcription of genes implicated in cell-fate decision, such as the Notch ligand DLL1, the transcription factors Math1 or KFL4, and even itself. Since CDX2 is critical for enterocyte maturation, the first direct target genes identified encoded digestive enzymes like sucrase-isomaltase, lactase, or phospholipase A/lysophospholipase. Many transporters, necessary for the absorption and secretion of nutrients by enterocytes, are also CDX2 target genes, for instance, the iron transporter hephaestin, the multidrug resistance 1 (MDR1/P-glycoprotein/ABCB1), or the solute carrier family 5, member 8 (SLC5A8). Other direct CDX2 target genes are involved in the modeling of the intestinal mucus-covered brush border: they encode, for example, the actin-binding protein villin 1 (important for microvilli architecture) and mucus constituents such as MUC2 and MUC4. Some CDX2 target genes encode adhesion molecules, potentially involved in intestinal barrier function and cell polarization: several members of the cadherin superfamily such as LI-cadherin, Mucdhl, and Desmocollin 2, but also claudin-2 and claudin-1. Finally, the transcription of the cyclin-dependent kinase inhibitor p21Cip1 can be stimulated by CDX2 and thus may contribute to the antiproliferative effect of CDX2.

CDX2 does not necessarily bind to DNA and use its properties of transcriptional activator to modulate gene expression. For instance, CDX2 affects the ▸ Wnt signaling pathway by direct interaction with beta-catenin, thereby inhibiting the formation of the beta-catenin/TCF4 complex and consequently Wnt target gene activation. Another example is the binding of CDX2 to the p65 subunit of NF-κB, which prevents its binding and activation of the COX-2 promoter. More unexpectedly, CDX2 can modulate the activity of proteins that are not involved in gene transcription: as an example, CDX2 interacts by its homeodomain with the protein complex Ku70/Ku80 and inhibits its activity of DNA repair by the nonhomologous end joining process. Furthermore, CDX2 can affect proteins without direct

interaction: indeed, CDX2 can stabilize the cyclin-dependent kinase inhibitor p27Kip1 by inhibiting its polyubiquitination and thereby reduce cell proliferation.

In some cases, the exact mechanism of CDX2 activity is not yet understood but might be important for intestinal homeostasis. One example is the potential repression of the mTOR pathway by CDX2, which might oppose cell cycle progression and chromosomal segregation defects. Another example is the enhanced trafficking of ▸ E-cadherin to the membrane of colon cancer cells, which strengthens Ca^{2+}-dependent adhesion and might be linked to the fact that CDX2 can reduce the phosphorylation of beta-catenin and p120 catenin.

Clinical Relevance for Colon Cancer

In colon ▸ adenocarcinomas, nuclear CDX2 expression is generally reduced, becoming sometimes diffuse and cytoplasmic, but there is a lot of heterogeneity in the level of reduction between different tumors or even between different areas within a tumor, which might explain why conflicting results have been obtained in separate studies. Reduced expression of CDX2 can be associated with high ▸ microsatellite instability (MSI) status, advanced tumor stage, higher tumor grade, lymph node metastasis, and reduced survival. In addition, CDX2 expression is more systematically decreased in cells located at the tumor front or disseminated in the adjacent stroma compared to the cells of the tumor center. Strikingly, most of the time, ▸ metastases (lymph nodes, liver) exhibit a similar level of CDX2 expression than the primary tumor, suggesting a dynamic expression pattern of CDX2 during tumor progression, with a specific but transient reduction in invasive cells.

Deletions or mutations at the CDX2 locus occur very rarely in colon tumors. Actually, most (chromosomal instability) CIN tumors present a gain of CDX2 copy number, but this gene amplification does not correlate with CDX2 expression. On the other hand, somatic cell hybrid experiments indicate that silencing of CDX2 expression was transferable upon cell fusion, suggesting a dominant repression mechanism. Since no epigenetic modifications of the CDX2 promoter have been detected in colon cancer cell lines, the existence of a transcriptional repression pathway is likely. Of note, such a regulatory mechanism would be consistent with a transient change of CDX2 expression in invasive cells. Several oncogenic signaling pathways (PI3K, Raf-MEK-ERK1/2) that are aberrantly activated in a large fraction of colon tumors can repress CDX2 expression in colon cancer cell lines. Transcriptional repressors inducing EMT (Slug, Snail, and Zeb1) can repress CDX2 transcription in vitro and may be involved in the systematic decrease of CDX2 expression in invasive cells. Several microenvironmental factors linked to tumor progression (▶ hypoxia, extracellular matrix, protein changes) can modify CDX2 transcription in colon cancer cell lines, and nude mice grafting experiments highlight the plasticity of CDX2 expression. However, all of the above data obtained with cell lines still await confrontation with cohorts of human colon tumors.

Given that CDX2 expression is downregulated in colon tumors and impacts on cell proliferation and ▶ migration, it is hypothesized that CDX2 acts as a tumor suppressor in the colon. Heterozygous CDX2$^{-/+}$ mice do not develop spontaneous tumors (the initially described "intestinal polyps" turned out to be nonneoplastic; see above), suggesting that the loss of CDX2 alone is not sufficient to initiate tumor formation, but only one allele is invalidated in these mice to allow survival. In contrast, upon tumor initiation, the tumor suppressor activity of CDX2 becomes obvious. Indeed, CDX2$^{+/-}$ mice treated with a colon carcinogen (azoxymethane) develop numerous ▶ adenocarcinomas in the distal colon much faster than their wild-type littermates. Similarly, when CDX2$^{+/-}$ mice are crossed with mice that spontaneously develop adenomatous polyps in the small intestine (APC$^{+/\Delta716}$ mice), they form six times more adenomatous polyps, and these are now located in the distal colon.

Finally, forced expression of CDX2 in colon cancer cells injected in nude mice correlates not only with reduced tumor size, but also with decreased metastasis incidence, suggesting that CDX2 opposes metastatic dissemination. Thus, even if CDX2 cannot be considered as a classic

tumor suppressor gene (no genomic alteration, no spontaneous tumor), it impacts on various cellular processes (proliferation, ▶ adhesion, polarity, ▶ migration; see above) involved in tumor growth and dissemination, and experimental evidences in mice indicate that reduced expression of CDX2 has important consequences for colon tumor (speed, number, location) and ▶ metastasis formation.

Clinical Relevance for Other Types of Cancer
Ectopic CDX2 expression is described in various types of ▶ adenocarcinomas, especially in those arising in the stomach, esophagus, and ovary.

More surprisingly, leukemia patients, and above all 90% of patients with ▶ acute myeloid leukemia (AML), exhibit ectopic CDX2 expression. The mechanism involved in this aberrant expression of CDX2 is not yet elucidated. Nevertheless, CDX2 expression represents a marker of bad prognosis and reduced survival for leukemia patients. In contrast to most intestinal cell lines, CDX2 stimulates the proliferation and the ability to form colonies of hematopoietic cells in vitro. In addition, ectopic CDX2 expression in transplanted hematopoietic cells was sufficient to induce AML in mice by perturbing the expression of HOX genes. The pro-oncogenic role of CDX2 in leukemia may be linked to the involvement of CDX genes in embryonic hematopoiesis described in zebrafish or murine pluripotent stem cells but awaits further investigation.

References

Aoki K et al (2011) Suppression of colonic polyposis by homeoprotein CDX2 through its nontranscriptional function that stabilizes p27Kip1. Cancer Res 71(2):593–602

Beck F, Stringer EJ (2010) The role of Cdx genes in the gut and in axial development. Biochem Soc Trans 38(2):353–357

Gao N, White P, Kaestner KH (2009) Establishment of intestinal identity and epithelial-mesenchymal signaling by Cdx2. Dev Cell 16(4):588–599

Lengerke C, Daley GQ (2012) Caudal genes in blood development and leukemia. Ann N Y Acad Sci 1266:47–54

Subtil C et al (2007) Frequent rearrangements and amplification of the CDX2 homeobox gene in human sporadic colorectal cancers with chromosomal instability. Cancer Lett 247(2):197–203

CDX3

▶ CDX2

CDX-3

▶ CDX2

CEA

▶ Carcinoembryonic Antigen

CEA Gene Family

Nicole Beauchemin
Goodman Cancer Research Centre, McGill
University, Montreal, QC, Canada

Synonyms

C-CAM; CD66a; CD66b; CD66c; CD66e;
CEACAM1 = BGP; CEACAM5 = CEA;
CEACAM6 = NCA; CEACAM7 = CGM2;
CEACAM8 = CGM6

Definition

The carcinoembryonic antigen (CEA) gene family
comprises 33 genes, 22 of which are expressed.
All family members share similar structural features encompassing immunoglobulin (Ig) variable
and/or constant domains and therefore constitute
members of the large immunoglobulin superfamily. These proteins are either secreted or membrane bound. Several CEACAMs function as
homophilic or heterophilic intercellular ▶ cell
adhesion molecules. CEA, CEACAM1,
CEACAM6, and CEACAM7 also play a significant role as regulators of tumor cell proliferation

and differentiation and their overexpression (CEA
and CEACAM6) or their downregulation
(CEACAM1 and CEACAM7) contributes to progression of many epithelial cancers and immune
dysfunctions.

Characteristics

The *CEA* gene family encodes a set of 22 genes
and 11 pseudogenes clustered in a 1.8 Mb region
on human chromosome 19q13.2 between the
CY2A and *D19S15* marker genes. The *CEA*
genes encompass an N-terminal Ig variable
domain followed by one to six Ig constant-like
domains. A striking characteristic of these proteins is their extensive ▶ glycosylation on asparagine residues with multiantennary carbohydrate
chains. CEA and CEACAM1 are further modified
by the addition of Lewis and sialyl-Lewis^x high-
mannose residues. The proteins differ, however,
in their C-terminal regions producing either
secreted entities such as the pregnancy-specific
glycoproteins (PSG1–11) or others, tethered to
the cell surface by either a glycosyl phosphatidylinositol linkage (CEA, CEACAM6–8) or a *bona
fide* transmembrane domain (CEACAM1,
CEACAM3, CEACAM4, CEACAM18–21)
(Fig. 1). The *CEACAM1* gene is unique in this
family in that it produces 12 different splicing
variants. More information on the structural features of the *CEA* gene family members is available
at http://www.carcinoembryonic-antigen.de/.
CEA is a monomeric protein adopting a β-barrel
cylindrical shape resembling a "bottle brush,"
whereas CEACAM1 is present as both a monomeric and dimeric protein.

Expression and Functions of CEA Family Members in Normal and Tumor Tissues
Although not ubiquitous, CEA family members
exhibit a wide tissue distribution. CEA and
CEACAM6 are found mainly in columnar epithelial and goblet cells of the colon in the early fetal
period and are maintained in adult life. In the
colonic brush border, CEA, CEACAM1, 6 and
7 demonstrate maximal expression at the free
luminal surface, although CEACAM1 and 7 are

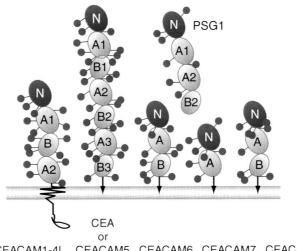

CEA Gene Family, Fig. 1 Schematic representation of some members of the CEA family. Most CEA family members, except the pregnancy-specific glycoproteins (*PSG*) that are secreted proteins, are associated with the cell membrane (depicted in *grey*). The immunoglobulin variable-like domains (the N domain) are shown in *blue* and the immunoglobulin constant-like domains are represented in *orange*. The N-linked glycosylation sites are indicated by *sticks* and *balls*, colored in *dark orange*. The glycosylphosphatidylinositol membrane anchors are represented by *arrows*. The *CEACAM1* gene expresses many splice variants. However, only the CEACAM-4L isoform containing four Ig domains and the longer cytoplasmic tail is shown here

also found at the lateral membrane. In addition to its expression in epithelia, CEACAM1 is located on granulocytes, lymphocytes, and endothelial cells, whereas CEACAM6 is also expressed on granulocytes and monocytes. CEACAM3 and 8 are found exclusively on granulocytes.

CEA, CEACAM1, and CEACAM6 are recognized as cell adhesion molecules contacting each other by antiparallel self-binding (homophilic). Some associations are exclusive, such as CEACAM8-CEACAM6. The first Ig domain is crucial in these interactions. Various CEA family members also act as heterophilic partners for E-selectin and galectin-3. Another striking feature of CEA family members is their ability to act as pathogen receptors binding to outer membrane proteins of *Neisseria* gonococci and *Haemophilus influenzae* as well as fimbriae of *Salmonella typhimurium* and *Escherichia coli*. In addition, CEACAM1 is the receptor for the mouse hepatitis viruses. The bacterial and viral adhesin functions of the CEA family members confer strong immunosuppressive activity in T and B lymphocytes, whereas they enhance integrin-dependent cell adhesion in epithelial cells with concomitant

increase of the TGF-β1 receptor CD105. Other functions for CEA and CEACAM6 include the inhibition of cellular differentiation as demonstrated in a number of cellular systems and inhibition of the apoptotic process of ▶ anoikis by activation of β1 integrins.

PSG1–11 are mainly expressed in syncytiotrophoblast during the first trimester of pregnancy where they act as immunomodulators and inhibit cell-matrix interactions.

CEA is abundantly expressed in tumors of epithelial origin such as colorectal, lung, mucinous ovarian, and endometrial adenocarcinomas. For these reasons, CEA has a long history as a marker of colonic, intestinal, ovarian, and breast tumor progression and its high expression is associated with poor prognostic and recurrence of disease postsurgically. High preoperative CEA levels are indicative of a poor prognosis whereas low levels are associated with increased survival of the patients. The tumorigenic potential of CEA and CEACAM6 was clarified by transgenic overexpression of a bacterial artificial chromosome fragment of 187 kb encoding the full *CEA*, *CEACAM6*, and *CEACAM7* genes. When the

CEABAC transgenic mice were treated with the azoxymethane carcinogen to induce colon cancers, expression of CEA and CEACAM6 was increased by 2–20 fold, a situation reminiscent to that observed in the human cancer. Information on CEACAM7 expression in tumors is more limited. It is downregulated in colorectal cancers, but increased in gastric tumors. CEACAM6, however, exhibits a broader distribution than in the cancers described above, as it is additionally found in gastric and breast carcinomas and ► acute lymphoblastic leukemias. In fact, overexpression of CEACAM6 in ► pancreatic cancer confers increased resistance to anoikis and increased metastasis. It also modulates chemoresistance to the ► gemcitabine agent, thereby suggesting that CEACAM6 determines cellular susceptibility to apoptosis.

Expression and Functions of CEACAM1

CEACAM1 expression is more complex. It is downregulated in colon, prostate, hepatocellular, bladder, endometrial, renal cell, and 30% of breast carcinomas, but overexpressed in gastric and squamous lung cell carcinomas, bladder cancer and ► melanomas. In thyroid carcinomas, CEACAM1 was shown to restrict tumor cell growth. However, it increases the thyroid cancer metastatic potential. Manipulation of CEACAM1 expression levels in colonic, prostatic, and bladder tumor cell lines, negative for CEACAM1, has indeed confirmed that expression of the longer variant, CEACAM1-4L, produces reduction of tumorigenic potential in vitro and inhibition of tumor growth in xenograft mouse models. The importance of cell surface CEACAM1 expression for maintenance of normal epithelial cellular behavior has been confirmed in vivo; a *Ceacam1*-null mouse exhibits a significantly increased colon tumor load compared to the wild-type littermates upon carcinogenic induction of colorectal cancer.

CEACAM1's role as a modulator of tumor progression depends on the involvement of its cytoplasmic domain in signaling via its tyrosine and serine phosphorylation. Two Tyr residues are positioned within immunoreceptor tyrosine-based inhibition motifs (ITIM). The membrane-proximal Tyr488 is a phosphorylation substrate of Src-like kinases as well as of the insulin and epidermal growth factor receptors. Upon Tyr phosphorylation, CEACAM1-L associates with the tyrosine phosphatases SHP-1 and SHP-2. The SHP-1-CEACAM1-L protein complex regulates its function in various tissues such as inhibition of epithelial cell growth, CD4$^+$ T cell activation, and insulin clearance from hepatocytes. CEACAM1-L tyrosine phosphorylation also stimulates its association with the cytoskeletal proteins G-actin, tropomyosin, and paxillin, thereby influencing cell adhesion, and with the β3 integrin, hypothesized to influence cell motility. The CEACAM1-L cytoplasmic domain also carries 17 serine residues most of which lie in consensus sequences recognized by serine kinases. However, little is known about their functional implications apart from the CEACAM1-S Thr/Ser452 and Ser456, shown to modulate direct binding to G- and F-actin, tropomyosin, and calmodulin, and CEACAM1-L's Ser503 whose mutation to an Ala residue enhances colonic or prostatic tumor development in xenograph models. Additionally, Ser503 renders permissive Tyr488 phosphorylation by the insulin receptor. Transgenic mice overexpressing a Ser503Ala CEACAM1-L mutant in the liver developed hyperinsulinemia, secondary insulin resistance, and defective insulin clearance. As a consequence of the decreased insulin receptor endocytosis and altered insulin signaling, the transgenic mice became obese demonstrating increased visceral adiposity, elevated serum free fatty acids and plasma and hepatic triglyceride levels.

CEACAM1-L also contributes to important functions in the immune system. It functions as an inhibitory coreceptor in T lymphocytes. Its conditional deletion in these cells amplified TCR-CD3 signaling, whereas overexpression in T cells was responsible for decreased proliferation, allogeneic reactivity, and cytokine production in vitro, with delayed type hypersensitivity and inflammatory bowel disease in vivo. Regulation of this function involves the ITIM motifs and the SHP-1 tyrosine phosphatase. A similar function and mechanism have been described in B lymphocytes and natural killer cells. Indeed,

CEACAM1-mediated intercellular adhesion between melanomas with increased CEACAM1 expression and NK cells allows inhibition of NK-cell-elicited killing, thereby conferring upon CEACAM1 a role in tumor immunosurveillance. Similarly, heterophilic engagement of CEACAM1 with CEA, overexpressed in many tumors, also inhibits lymphocyte-mediated and NK-cell-mediated killing having therefore detrimental effects on immune surveillance. In addition, increased expression of CEACAM1 on endothelial cells present in tumors in response to VEGF activation and/or hypoxia provokes a proangiogenic switch with increased endothelial tube formation and invasion. Therefore CEACAM1's contribution to cancer progression most likely depends on its positive or negative expression and signaling in epithelial tumor cells, on its systemic effects on metabolism and adiposity, on its role in immunosurveillance, and most probably on endothelial proliferation and invasion.

Transcriptional Regulation

The upstream promoters of the *CEA* and *CEACAM1* genes have been dissected to identify important binding sites responsible for their transcriptional regulation. These two genes do not encompass classical TATA and CAAT boxes and are considered members of the housekeeping gene family. Their distal promoter regions (> -500 bp) contain highly repetitive elements, whereas their proximal promoter regions are rich in GC boxes and SP1 binding sites. Five footprinted regions have been identified in the *CEA* promoter, the first three binding respectively, to the upstream stimulatory factor (USF) and SP1 and SP1-like factors. Similarly, the human *CEACAM6* promoter is regulated by the USF1 and USF2 as well as SP1 and SP3 transcription factors. A silencer element has also been located in its first intron. In contrast, the human *CEACAM1* promoter does not bind the SP1 factors, but associates with an AP-2-like factor and the USF and HFN-4 transcription factors. The gene is additionally controlled by the hormonal changes (estrogens and androgens) and can be induced by cAMP, retinoids, glucocorticoids, and insulin. Moreover, many genes of this large family are triggered by inflammation via interferons, tumor necrosis factors, and interleukins. It has been reported that expression of the *CEACAM1* gene is influenced by TPA and calcium ionophore in endometrial cancers, the expression of BCR/ABL in leukemias, the expression of the β3 integrin in melanomas, and VEGF and hypoxia in angiogenic situations. In prostate cancer, there is an inverse correlation between the downregulation of CEACAM1 and the increased expression of the transcriptional repressor Sp2 that acts to recruit histone deacetylase to the *CEACAM1* promoter.

The Next Frontier

The diversity of functions of the members of the CEA gene family and their dynamic expression patterns in normal and tumor tissues has slowed the development of effective targeted therapies. Effective strategies have been devised using vaccination with CEA peptide-loaded mature dendritic cells that induced potent CEA-specific T cell responses in advanced colorectal cancer patients. Effective protection from tumor development have also been seen with delivery of adenoviral vectors encoding CEA fused to immunoenhancing agents such as tetanus toxin or the Fc portion of IgG1. Likewise, targeting of CEACAM6 in pancreatic cancer may result in decreased tumor load. The therapeutic and selective targeting of CEACAM1 in melanomas, gastric and lung carcinomas as well as its location in tumor endothelia may prove to be a favorable avenue of future interventions.

References

Beauchemin N, Arabzadeh A (2013) Carcinoembryonic antigen-related cell adhesion molecules (CEACAMs) in cancer progression and metastasis. Cancer and Mets Rev 32:643–671

Beauchemin N, Draber P, Dveksler G, Gold P, Gray-Owen S, Grunert F, Hammarstrom S, Holmes KV, Karlsson A, Kuroki M, et al (1999) Redefined nomenclature for members of the carcinoembryonic antigen family. Exp Cell Res 252:243–249

Gray-Owen SD, Blumberg RS (2006) CEACAM1: contact-dependent control of immunity. Nat Rev Immunol 6:433–446

Hammarström S (1999) The carcinoembryonic antigen (CEA) family: structures, suggested functions and expression in normal and malignant tissues. Semin Cancer Biol 9:67–81

Horst A, Wagener C (2004) CEA-related CAMs. Handb Exp Pharmacol 165:283–341

Kuespert K, Pils S, Hauck CR (2006) CEACAMs: their role in physiology and pathophysiology. Curr Opin Cell Biol 18:1–7

Leung N, Turbide C, Marcus V et al (2006) Carcinoembryonic antigen-related cell adhesion molecule 1 (CEACAM1) contributes to progression of colon tumors. Oncogene 25:5527–5536

CEACAM1

▶ CEACAM1 Adhesion Molecule

CEACAM1 Adhesion Molecule

Andrea Kristina Horst[1] and Christoph Wagener[2]
[1]Inst. Experimental Immunology and Hepatology, University Medical Center Hamburg-Eppendorf, Hamburg, Germany
[2]University Medical Center Hamburg-Eppendorf, Hamburg, Germany

Synonyms

BGP; Biliary glycoprotein; CD66a; CEACAM1; CEA-related cell adhesion molecule 1; Cluster of differentiation antigen 66 a; NCA-160; Nonspecific cross-reacting antigen with a Mw of 160kD

Definition

CEACAM1 (CEA-related cell adhesion molecule 1) belongs to the CEA (▶ carcinoembryonic antigen, ▶ CEA gene family) family of cell surface glycoproteins, a subfamily of the immunoglobulin gene superfamily. The CEA family comprises two major groups, the CEA-related molecules and the PSG (pregnancy-specific glycoprotein)-related molecules. Additionally, a number of pseudogenes have been identified. To date, 29 genes are known, which are clustered on human chromosome 19 (19q13.1-19q13.2). The CEA-related members of the CEA family display a complex expression pattern on human healthy and malignant tissues. They are linked to the cell membrane via GPI anchors, or they are transmembrane proteins with a cytoplasmatic tail. The PSG-related molecules are soluble glycoproteins; their expression is restricted to the placenta, more specifically, to the syncytiotrophoblast, which is the outermost fetal component of the placenta. CEACAM1 has been structurally and functionally conserved in humans and rodents.

Characteristics

Properties of CEACAM1

Human CEACAM1 has been originally identified in human bile due to its crossreactivity with CEA-antisera. It was therefore named biliary glycoprotein I or nonspecific cross-reacting antigen at first. Amongst the cluster of differentiation antigens on human leukocytes, CEACAM1 used to be referred as CD66a. However, with the latest revision of the nomenclature for the CEA family, CD66a, BGP, or NCA-160 became CEACAM1. Its structural similarities to CEA and the immunoglobulin superfamily proteins became apparent, once the cDNA sequence for CEACAM1 became available.

CEACAM1 displays the broadest expression pattern amongst CEA family members; it has first been described as a cell–cell adhesion molecule on rat hepatocytes. CEACAM1 is expressed on epithelia, endothelia, and leukocytes.

CEACAM1 is a heavily glycosylated molecule that exists in 11 known isoforms emerging from differential splicing and proteolytic processing. The two major isoforms of CEACAM1 consist of four extracellular Ig-like domains, a transmembrane domain, and either a long or a short cytoplasmic tail, referred to as the long (CEACAM1-4L) and the short isoform (CEACAM1-4S), respectively. In addition to these transmembrane isoforms, soluble

CEACAM1 isoforms are found in body fluids, for example, in saliva, serum, seminal fluid, and bile. Glycans on the extracellular domains of CEACAM1 are linked to the protein backbone via N-glycosidic linkages. It is presently unknown whether all of the 19 motifs that may render N-linked ▶ glycosylation actually harbor sugar moieties. On human granulocytes, CEACAM1 is a major carrier of Lewis^x glycans that are implicated in cellular adhesion to cognate lectins on blood vessels, within the extracellular matrix, or antigen presenting cells. CEACAM1 also elicits cell–cell adhesion via self-association in a homomeric fashion or via formation of heteromers with other CEA-family members and different adhesion molecules that are either located on the same cell or on neighboring cells. The resulting adhesive properties are modulated by differential expression ratios between the long and short CEACAM1 isoform, respectively. Through its long and short cytoplasmic tail, CEACAM1 mediates molecular interactions with cytoskeletal components or adapter proteins, which are integral parts of various key signal transduction pathways (signal transduction, cell biology). These interactions are in part dependent on differential phosphorylation of the CEACAM1-4L cytoplasmic domain on tyrosine and serine residues. The overall phosphorylation status of the CEACAM1-4L cytoplasmic domain relays signals, which contribute to cellular motility and differentiation, and thus determine cell fate by promoting proliferation or cell death. Phosphorylation of CEACAM1-4L cytoplasmic tyrosines that are part of an imperfect ITIM (immune receptor tyrosine-based inhibition motif) and serine residues regulate the interaction with kinases, phosphatases, cellular receptors for insulin (▶ Insulin receptor), the epidermal growth factor (epidermal growth factor receptor ligand, epidermal growth factor receptor inhibitor), and other cellular adhesion molecules, for example, integrin $\alpha_v\beta_3$ (integrin signaling and cancer). These qualities make CEACAM1 an important tool for cellular communication and they illustrate why so many different biological functions have been attributed to CEACAM1 in different biological contexts (Fig. 1).

CEACAM1 in Cancer

The first report on CEACAM1, in the context of human pathological conditions, was on elevated serum levels of a biliary glycoprotein in patients with liver or biliary tract disease. Later, aberrant CEACAM1 expression in a broad variety of human malignancies has been reported. In the progression of malignant diseases, two general patterns in the changes of CEACAM1 expression levels have emerged. In the first group of tumors, CEACAM1 expression is downregulated in the course of progressing disease. In the second group of tumors, CEACAM1 expression appears to be upregulated; often, this upregulation of CEACAM1 expression is observed in the context with increased invasiveness (▶ invasion) of the primary tumor or is found on microvessels in progressing (▶ progression) tumor areas (Fig. 2).

Loss of CEACAM1 Expression in Tumorigenesis and Tumor Progression

Human cancers that show the downregulation of CEACAM1 expression in the course of tumor progression are carcinomas of the liver (▶ hepatocellular carcinoma), colon (colon cancer, colorectal premalignant lesions), kidney (renal cell carcinoma, renal carcinoma), urinary bladder (bladder cancer, bladder tumors), prostate (prostate cancer, clinical oncology), mammary gland (▶ breast cancer), and the endometrium (▶ endometrial cancer). In general, downregulation and subsequent loss of CEACAM1 expression is more frequent in high-grade tumors that are poorly differentiated and often associated with a larger tumor size.

On epithelia, especially those that form a lumen, CEACAM1 exhibits a pronounced apical expression, like in the entire gastrointestinal tract, breast, liver, prostate, bladder, and kidney. CEACAM1 expression has been implicated in morphogenesis of lumen formation. In the process of building an asymmetrical epithelium, lateral CEACAM1 expression on neighboring cells is lost and often becomes entirely apical once a lumen or a duct has been formed. The loss of CEACAM1 expression in the context of tumorigenesis has been studied most extensively in the context of breast, colonic, and prostate carcinomas.

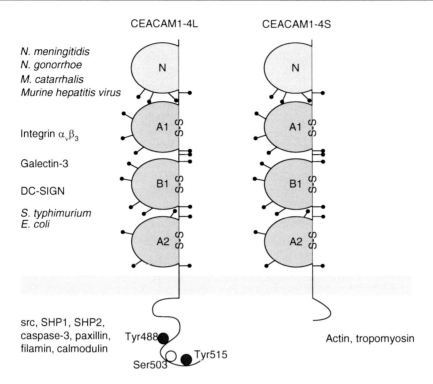

CEACAM1 Adhesion Molecule, Fig. 1 Schematic representation of CEACAM1-4L and CEACAM1–4S and their participation in extracellular and intracellular communication. The two major CEACAM1 isoforms consist of four extracellular immunoglobulin-like domains, a transmembrane domain and either a long or a short cytoplasmic tail. The N-terminal domain (N) resembles a variable-like Ig domain but lacks the cystin bond usually found in Ig members. The A1, B1, and A2 domain resemble constant I-type-like Ig domains. Motifs for N-linked glycosylation are represented by lollipops. With its extracellular domains, CEACAM1 mediates recognition of various pathogens, such as *Escherichia coli*, *Salmonella typhimurium*, *Moraxella catarrhalis*, *Neisseria gonorrhoeae*, and *Neisseria meningitidis*. The murine homologue of CEACAM1 is the receptor for the murine hepatitis virus: Additionally, CEACAM1 binds to galectin-3, DC-SIGN (dendritic cell ICAM3-grabbing nonintegrin), and integrin $\alpha_v\beta_3$. Tyrosine and serine residues involved in relaying CEACAM1-4L-mediated signal transduction are indicated by red and grey circles, respectively. Through its long cytoplasmic tail, CEACAM1-4L interacts with intracellular kinases of the SRC-family (▶ *SRC*), the tyrosine phosphatases SHP-1 and SHP-2, caspase-3 as well as with paxillin, filamin, and calmodulin. Differential phosphorylation of the CEACAM1-4L cytoplasmic domain is required for its interaction with the insulin receptor, regulating insulin receptor internalization and recycling, and for modulating immune responses elicited by lymphocytes, for example. The short cytoplasmic domain of CEACAM1–4S binds to actin and tropomyosin

A hallmark of carcinomatous lesions is the loss of polarity of their epithelial structures. In colonic epithelium, for example, loss of polarity is accompanied by the loss of apical CEACAM1 expression that occurs in early adenomas and carcinomas. In these tumors, the presence and absence of CEACAM1 correlate with normal and reduced apoptosis (apoptosis, apoptosis signals), respectively. Furthermore, the naturally occurring process of ▶ anoikis, once cells lose contact to their substratum, is compromised.

This observation and the fact that the *CEACAM1* gene is silenced in the course of aberrant cell growth prompted the hypothesis that CEACAM1 acts as a tumor suppressor. In intestinal cells, the presence of the long CEACAM1 isoform is required to suppress tumor growth, and the lack of CEACAM1-4L expression is accompanied by a decrease in proteins that inhibit cell cycle progression.

In human mammary epithelial cells, CEACAM1 expression is causally related to

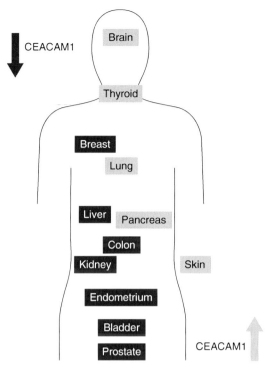

CEACAM1 Adhesion Molecule, Fig. 2 Dysregulation of CEACAM1 expression in human cancers. Changes of epithelial CEACAM1 expression in the course tumor progressison: In mammary carcinomas and carcinomas of the liver, colon, endometrium, kidney, bladder, and prostate, CEACAM1 expression is downregulated on tumor epithelium (epithelial cancers). Downregulation of CEACAM1 levels often correlates with dedifferentiation of the tumor and loss of tissue architecture. In carcinomas of the thyroid, ▶ non-small cell lung cancer (▶ lung cancer), pancreatic tumors (pancreas cancer, clinical oncology), and malignant melanomas, CEACAM1 is induced or upregulated in the course of tumor growth. Here, CEACAM1 expression is found on the invasive front of the tumors and is related to development of metastatic disease (▶ metastasis) and poor prognosis. In pancreatic cancer, CEACAM1 has been identified as a novel biomarker (biomarker, clinical cancer biomarker) that indicates the presence of malignant disease

lumen formation and differentiation. In mammary glands, CEACAM1-4S is the predominating isoform, and only the short cytoplasmic tail induces apoptosis of the central cells and subsequently leads to lumen formation in mammary morphogenesis. During tumor progression, CEACAM1-4S expression is lost and acinar polarity no longer can be observed.

However, since particular mutations or allelic loss of the *CEACAM1* gene in human cancers has not been described so far, it is likely that the dysregulation of CEACAM1 expression rather than irreversible loss of the *CEACAM1* gene are linked to tumorigenesis and tumor progression in vivo. Hence, gene silencing may attribute to the loss of the tumor suppressive qualities of CEACAM1. Though there are no changes in promoter ▶ methylation of the *CEACAM1* gene linked to tumor progression, CEACAM1 promoter activity appears to be regulated by binding of the transcription factor Sp2. In high-grade prostate carcinomas, Sp2 is highly abundant, whereas CEACAM1 expression is lost. Sp2 localizes to the CEACAM1 promoter and imposes repression of gene transcription by recruiting histone deacetylase.

Upregulation of CEACAM1 Expression in Malignant Diseases

Opposed to its tumor suppressive functions, certain tumors gain CEACAM1 expression in the course of cancer development. In the case of malignant melanomas and thyroid carcinomas, expression of CEACAM1 correlates with an increase of tumor invasiveness and development of metastatic disease. In primary cutaneous malignant melanomas, for example, CEACAM1 expression is found at the invasive front of the tumors, and its coexpression with integrin $\alpha_v\beta_3$ indicates that CEACAM1 may directly promote on cellular invasion. In a follow-up study, CEACAM1 was identified as an independent prognostic marker, predicting the development of metastatic disease and poor survival. In this context, it is noteworthy that CEACAM1 on melanoma cells forms homophilic cell–cell contacts with CEACAM1 molecules on tumor-infiltrating lymphocytes and leads to the inhibition of their cytolytic function. Similarly, in human non-small cell lung cancer, CEACAM1 expression correlates with advanced disease, whereas it is not expressed on the normal bronchiolar epithelium; this CEACAM1 neoexpression was identified as an independent prognostic marker, indicating lower incidence of relapse-free survival.

In pancreatic carcinomas, CEACAM1 has been identified as a novel serum biomarker, with an increased CEACAM1 expression on neoplastic cells of pancreatic adenocarcinomas and elevation of serum levels at the same time. Additionally, significant differences in CEACAM1 serum levels were found in patients with either pancreatic cancer or chronic pancreatitis. Opposed to the classical pancreatic tumor marker CA19-9, CEACAM1 was confirmed as an independent marker to distinguish between the presence of malignant disease and pancreatitis.

CEACAM1 and Tumor Angiogenesis

CEACAM1 expression on human blood vessels is restricted to newly formed vessels, and usually, no CEACAM1 is found on mature, large vessels. The first indication that CEACAM1 is related to ► angiogenesis was the description of CEACAM1 neoexpression on newly formed vessels in the human placenta. Furthermore, CEACAM1 is expressed on vessels in wound healing tissues and on tumor vessels of human bladder carcinomas, the prostate, hemangiomas, and ► neuroblastomas. CEACAM1 expression in endothelia is induced by VEGF (► vascular endothelial growth factor)-dependent pathways and appears to favor vessel maturation.

In human prostate carcinomas, CEACAM1 shows divergent expression on tumoral blood vessels and the tumor epithelium. The presence of epithelial CEACAM1 is observed in the context of poor tumoral blood vessel growth and loss of epithelial CEACAM1 expression parallels enhanced tumor angiogenesis. Especially in high-grade prostate carcinomas, tumor proximal vessels are expressing CEACAM1. Contrary to prostate carcinomas, microvessels in human neuroblastomas are CEACAM1-positive only during tumor maturation, but absent in undifferentiated, high-grade tumors. In ► Kaposi sarcomas, CEACAM1 upregulation is observed, indicating that CEACAM1 might be related to lymphatic reprogramming of the vasculature in these tumors.

Studying CEACAM1 in Cancer: Animal Models

In animal models investigating CEACAM1 function in tumorigenesis in vivo, the observations from human diseases could be confirmed. The focus of the mouse and rat models (► Mouse model) studied to date was set largely on the tumor-suppressive effects or enhancement of metastatic disease of CEACAM1-4L on the progression of colonic cancer, prostate cancer, hepatocellular carcinomas, and malignant melanomas. In CEACAM1-knockout mice, chemically induced colonic tumor growth was significantly increased in terms of tumor numbers and size opposed to CEACAM1-expressing wild type littermates. In syngeneic and xenotypic transplantation of tumor cells of the colon, prostate, and hepatocellular carcinomas, the tumor-suppressive effects of CEACAM1-4L expression could also be validated. After xenotransplantation of human CEACAM1-expressing melanoma cell lines into immune-deficient mice, enhanced metastasis was observed when compared to transplantation of CEACAM1-negative cell lines.

References

Beauchemin N, Draber P, Dveksler G et al (1999) Redefined nomenclature for members of the carcinoembryonic antigen family. Exp Cell Res 252:243–249

Gray-Owen SD, Blumberg RS (2006) CEACAM1: contact-dependent control of immunity. Nat Rev Immunol 6:433–446

Kuespert K, Pils S, Hauck CR (2006) CEACAMs: their role in physiology and pathophysiology. Curr Opin Cell Biol 18:565–571

Prall F, Nollau P, Neumaier M et al (1996) CD66a (BGP), an adhesion molecule of the carcinoembryonic antigen family, is expressed in epithelium, endothelium, and myeloid cells in a wide range of normal human tissues. J Histochem Cytochem 44:35–41

Singer BB, Lucka LK (2005) CEACAM1. UCSD-nature molecule pages. Nat Publ Group. doi:10.1038/mp. a003597.01

CEACAM1 = BGP

► CEA Gene Family

CEACAM5

▶ Carcinoembryonic Antigen

CEACAM5 = CEA

▶ CEA Gene Family

CEACAM6 = NCA

▶ CEA Gene Family

CEACAM7 = CGM2

▶ CEA Gene Family

CEACAM8 = CGM6

▶ CEA Gene Family

CEA-Related Cell Adhesion Molecule 1

▶ CEACAM1 Adhesion Molecule

CED

▶ Convection-Enhanced Delivery

Celastrol

Qing Ping Dou[1] and Xiao Yuan[2]
[1]The Prevention Program, Barbara Ann Karmanos Cancer Institute and Department of Pathology, School of Medicine, Wayne State University, Detroit, MI, USA
[2]Research and Development Center, Wuhan Botanical Garden, Chinese Academy of Science, Wuhan, Hubei, People's Republic of China

Synonyms

Quinone methide friedelane tripterene (2R,4aS,6a S,12bR,14aS,14bR)-10-hydroxy-2,4a,6a,9,12b,14a-hexamethyl-11-oxo-1,2,3,4,4a,5,6,6a,11,12b,13,14,14a,14b-tetradecahydropicene-2-carboxylic acid; Tripterine

Definition

Celastrol is a natural quinone methide tripterene, widely found in the plant genera *Celastrus*, *Maytenus*, and *Tripterygium*, all of which are present in China. For example, celastrol is one of the active components extracted from *Tripterygium wilfordii Hook F*, an ivy-like vine also known as "Thunder of God Vine," which belongs to the family of *Celastraceae* and has been used as a natural medicine in China for hundreds of years (Fig. 1).

Characteristics

Biological Properties

Celastrol has strong antifungal, anti-inflammatory, and antioxidant effects. It has been shown that celastrol isolated from the roots of *Celastrus hypoleucus* (Oliv) Warb f argutior Loes exhibited inhibitory effects against diverse phytopathogenic fungi. Celastrol was also found to inhibit the mycelial growth of *Rhizoctonia solani* Kuhn and *Glomerella cingulata* (Stonem) Spauld and Schrenk in vitro. Furthermore,

a

b

Celastrol, Fig. 1 The chemical structure and nucleophilic susceptibility of celastrol. (a) The chemical structure of celastrol is shown. (b) Nucleophilic susceptibility of celastrol analyzed using CAChe software. Higher susceptibility was shown at the C_2 and C_6 positions of celastrol

celastrol has good preventive effect and curative effect against wheat powdery mildew in vivo.

Celastrol in low nanomolar concentrations suppresses the production of the pro-inflammatory cytokines tumor necrosis factor-alpha (TNF-α) and interleukin-1 beta (IL-1β) by human monocytes and macrophages. Celastrol also decreases the induction of class II major histocompatibility complex (MHC) expression by microglia. In macrophage lineage cells and endothelial cells, celastrol decreases induction of nitric oxide (NO) production. Celastrol also suppresses adjuvant arthritis in the rat, demonstrating in vivo anti-inflammatory activity. Low doses of celastrol administered to rats could significantly improve the performance of these animals in memory, learning, and psychomotor activity.

In an isolated rat liver assay of lipid peroxidation, the antioxidant potency of celastrol (IC_{50} 7 μM) is 15 times stronger than that of α-tocopherol or vitamin E. Under in vitro conditions, celastrol was found to inhibit ▶ cancer cell proliferation and induce programmed cell death (or ▶ apoptosis) in a broad range of tumor cell lines, including 60 National Cancer Institute (NCI) human cancer cell lines. As a ▶ topoisomerase II inhibitor, celastrol was fivefold more potent than the well-known topoisomerase inhibitor etoposide to induce apoptosis in HL-60 leukemia cells. Celastrol was also found to be a

tumor ▶ angiogenesis inhibitor. In a sharp comparison, celastrol can block neuronal cell death in cultured cells and in animal models. These unique features of celastrol suggest potential use for treatment of cancer and neurodegenerative diseases accompanied by inflammation, such as Alzheimer disease.

Potential Molecular Targets

Celastrol is a naturally occurring potent inhibitor of the ▶ proteasome and nuclear factor kappa B (NFκB). Proteasome, or 26S proteasome, is a multicatalytic protease complex consisting of a 20S catalytic particle capped by two 19S regulatory particles. The ubiquitin-proteasome pathway is responsible for the degradation of most endogenous proteins involved in gene transcription, cell cycle progression, differentiation, senescence, and apoptosis. Inhibition of the proteasomal chymotrypsin-like but not trypsin-like activity is associated with induction of apoptosis in tumor cells.

Both computational and experimental data support the hypothesis that celastrol is a natural proteasome inhibitor. Atomic orbital energy analysis demonstrates high susceptibility of C_2 on A-ring and C_6 on B-ring of celastrol toward a nucleophilic attack. Computational modeling shows that celastrol binds to the proteasomal chymotrypsin site (β5 subunit) in an orientation and conformation that is suitable for a nucleophilic attack by the hydroxyl (OH) group of N-terminal

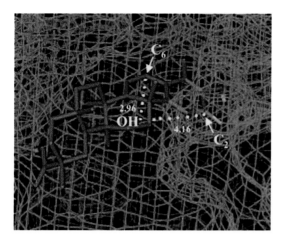

Celastrol, Fig. 2 Docking solution of celastrol. Celastrol was docked to S_1 pocket of β5 subunit of 20S proteasome. Celastrol was shown in pink while β5 subunit was shown in purple. The selected conformation with 92% possibility showed the distances to the OH group of N-Thr from C_6 and C_2 were 2.96 Å and 4.16 Å, respectively

threonine of β5 subunit. The distances to the OH of N-terminal threonine of β5 from the electrophilic C_6 and C_2 of celastrol are measured as 2.96 Å and 4.16 Å, respectively. Both carbons, more probably C_6, of celastrol potentially interact with N-terminal threonine of β5 subunit and inhibit the proteasomal chymotrypsin-like activity (Fig. 2).

Celastrol potently and preferentially inhibits the chymotrypsin-like activity of a purified 20S proteasome with an IC_{50} value 2.5 μM. Celastrol at 1–5 μM inhibits the proteasomal activity in intact human prostate cancer cells. The inhibition of the cellular proteasome activity by celastrol results in accumulation of ubiquitinated proteins and three natural proteasome substrates, IκB-α, Bax, and p27, leading to induction of apoptosis in ► androgen receptor (AR)-negative PC-3 cells. In AR-positive LNCaP cells, celastrol-mediated proteasome inhibition was accompanied by suppression of AR protein, probably by inhibiting ATP-binding activity of heat shock protein 90 (Hsp90) that is responsible for AR folding. Treatment of PC-3 tumor-bearing nude mice with celastrol (1–3 mg/kg/day, *i.p.*, for 1–31 days) resulted in significant inhibition (65–93%) of the tumor growth. Multiple assays using the animal tumor tissue samples from both early and end time points demonstrated in vivo inhibition of the proteasomal activity and induction of apoptosis after celastrol treatment.

Antitumor activity of celastrol was also observed in a breast cancer mouse model. Celastrol inhibited ∼60% tumor growth in breast cancer xenograft through NFκB inhibition. NFκB inhibition by celastrol includes inhibition of its DNA-binding activity and inhibition of IκBα degradation induced by TNF-α or phorbol myristyl acetate. Further investigation showed that the cysteine-179 in the IκBα kinase was a potential target of celastrol-suppressed IκBα degradation. Since the proteasome is required for the activation of NFκB by degrading IκBα, the proteasome inhibition may also contribute to the NFκB inhibition by celastrol.

TNF could send both anti-apoptotic and pro-apoptotic signals. The effects of celastrol on cellular responses activated by the potent pro-inflammatory cytokine TNF have also been investigated. Celastrol was able to potentiate the apoptosis induced by TNF and chemotherapeutic agents and inhibited invasion, both regulated by NFκB activation. TNF induced the expression of gene products involved in anti-apoptosis (IAP1, IAP2, ► Bcl2, Bcl-X_L, c-FLIP, and survivin), proliferation (cyclin D1 and COX-2), invasion (MMP-9), and angiogenesis (VEGF), and celastrol treatment suppressed the expression of these genes. Celastrol also suppressed both inducible and constitutive NFκB activation. Furthermore, celastrol was found to inhibit the TNF-induced activation of IκBα kinase, IκBα phosphorylation, IκBα degradation, p65 nuclear translocation and phosphorylation, and NFκB-mediated reporter gene expression. Therefore, celastrol potentiates TNF-induced apoptosis and inhibits invasion through suppression of the NFκB pathway.

Clinical Relevance

Due to its antioxidant or anti-inflammatory effects, celastrol has been effectively used in the treatment of autoimmune diseases (rheumatoid arthritis, systemic lupus erythematosus), asthma, chronic inflammation, and neurodegenerative diseases. As a bioactive component in Chinese traditional medicinal products from the extract of the roots of *Tripterygium wilfordii Hook F*, celastrol has been used since the 1960s in China for autoimmune diseases but has showed some side effects such as nausea, vomiting, etc. Celastrol has not been used solely as a medication product. Celastrol has antitumor activities via inhibition of the proteasome and NFκB activation, indicating that celastrol has a great potential to be used for cancer prevention and treatment. This finding can be applied to various human cancers and diseases in which the proteasome is involved and on which celastrol has an effect.

Cross-References

► Topoisomerases

References

Hieronymus H, Lamb J, Ross KN et al (2006) Gene expression signature-based chemical genomic prediction identifies a novel class of HSP90 pathway modulators. Cancer Cell 10:321–330

Sassa H, Takaishi Y, Terada H (1990) The triterpene celastrol as a very potent inhibitor of lipid peroxidation in mitochondria. Biochem Biophys Res Commun 172:890–897

Sethi G, Ahn KS, Pandey MK et al (2006) Celastrol, a novel triterpene, potentiates TNF-induced apoptosis and suppresses invasion of tumor cells by inhibiting NF-?B-regulated gene products and TAK1-mediated NF-?B activation. Blood 109:2727–2735

Setty AR, Sigal LH (2005) Herbal medications commonly used in the practice of rheumatology: mechanisms of action, efficacy, and side effects. Semin Arthritis Rheum 34:773–784

Yang HJ, Chen D, Cui QZC et al (2006) Celastrol, a triterpene extracted from the Chinese "Thunder of God Vine", is a potent proteasome inhibitor and suppresses human prostate cancer growth in nude mice. Cancer Res 66:4758–4765

Celebra

► Celecoxib

Celebrex

► Celecoxib

Celecoxib

Numsen Hail[1] and Reuben Lotan[2]
[1]Department of Pharmaceutical Sciences, The University of Colorado at Denver and Health Sciences Center, Denver, CO, USA
[2]Department of Thoracic Head and Neck Medical Oncology, The University of Texas MD Anderson Cancer Center, Houston, TX, USA

Synonyms

Celebra; Celebrex; 4-[5-(4-Methylphenyl)-3-(trifluoromethyl)-1H-pyrazol-1-yl] benzene sulfonamide

Characteristics

Celecoxib, a diaryl-substituted pyrazole drug, was developed by G. D. Searle & Company and is currently marketed by Pfizer Incorporated under the brand names Celebrex and Celebra. Celecoxib is a member of the class of agents known as ► non-steroidal anti-inflammatory drugs (NSAIDs). NSAIDs are the most commonly used therapeutic agents for the treatment of acute pain, fever, menstrual symptoms, osteoarthritis, and rheumatoid arthritis. Because of their ability to reduce tissue ► inflammation, which is often associated with tumorigenesis at various sites in the body (e.g., gastrointestinal tract and lung), celecoxib and certain other NSAIDs are also considered to have a potential in cancer chemoprevention as exemplified by their ability to prevent the formation and decrease the size of polyps in familial adenomatous polyposis (FAP) patients. Orally administered celecoxib exhibits good systemic bioavailability and tissue distribution with an estimated plasma half-life of approximately 11 h. Celecoxib binds to plasma albumin and is metabolized primarily by hepatic enzymes prior to excretion. In humans, long-term exposures to celecoxib taken for arthritis pain relief at 100 mg twice daily caused no biologically significant adverse reactions. However, higher doses of

Celecoxib, Fig. 1 The chemical structure of celecoxib

400 mg twice daily recommended for patients with FAP resulted in threefold increased risk of cardiovascular events (Fig. 1).

▶ *Cyclooxygenase Dependent Mechanisms for Cancer Chemoprevention by Celecoxib.* Cyclooxygenases are enzymes that are indispensable for the synthesis of ▶ prostaglandins. Prostaglandins are ▶ hormones generated from arachidonic acid, and they are found in virtually all tissues and organs. Prostaglandins typically act as short-lived local cell signaling intermediates that regulate processes associated with inflammation. In the early 1990s, cyclooxygenases were demonstrated to exist as two isoforms, cyclooxygenase-1 (COX-1) and cyclooxygenase-2 (COX-2). COX-1 is characterized as a constitutively expressed housekeeping enzyme that mediates physiological responses like platelet aggregation, gastric cytoprotection, and the regulation of renal blood flow. In contrast, COX-2 is recognized as the inducible cyclooxygenase isoform that is primarily responsible for the synthesis of the prostaglandins that are involved in pathological processes (e.g., chronic inflammation) in cells that mediate inflammation (e.g., macrophages and monocytes). COX-2 is inducible by oncogenes (e.g., RAS and ▶ SRC), interleukin-1, ▶ hypoxia, benzo[*a*]pyrene, ultraviolet light, epidermal growth factor, ▶ transforming growth factor β, and tumor necrosis factor α. Many of these inducers activate

nuclear factor kappa B (NF-κB), which controls COX-2 expression and has been associated with tumorigenesis in various cell types.

The COX-2 isoenzyme is frequently unregulated in cancer cells, as well as cells that constitute premalignant lesions, which are important targets for cancer chemoprevention. The expression of the inducible COX-2 is enhanced in 50% of colon adenomas and in the majority of human colorectal cancers, as opposed to COX-1, which typically remains unchanged. Thus, the increase in COX-2 expression, which is an early event in colon carcinogenesis, is believed to be necessary for tumor promotion. Aberrant COX-2 expression has also been implicated in tumorigenesis in the lung, prostate, esophagus, ▶ Brms1, liver, pancreas, and skin. The activity of COX-2 to produce arachidonic acid metabolites appears to enhance the proliferation of transformed cells and/or increases their survival through the suppression of ▶ apoptosis. Furthermore, COX-2 expression by tumor cells can stimulate ▶ angiogenesis at the tumor site and alter tumor cell adhesion to promote ▶ metastasis.

Celecoxib is a highly selective inhibitor of COX-2. Traditional NSAIDs (e.g., aspirin) inhibit both COX-1 and COX-2 isozymes. In contrast, celecoxib is approximately 20 times more selective for COX-2 inhibition compared to its inhibition of COX-1. This specificity allows celecoxib, and other selective COX-2 inhibitors, to reduce inflammation while minimizing adverse drug reactions (e.g., stomach ulcers and reduced platelet aggregation) that are common with non-selective NSAIDs. This selectivity for COX-2 is also intimately associated with the putative cancer chemopreventive activity of celecoxib, which has been demonstrated in colorectal cancer prevention. Epidemiological studies have shown that persons who regularly take aspirin have about a 50% lower risk of developing colorectal cancer. Celecoxib was the most effective NSAID in reducing the incidence and multiplicity of colon tumors in a rat colon carcinogenesis model. Moreover, in a clinical setting celecoxib has been used effectively to suppress the development and/or reduce the number of colorectal polyps in patients with FAP. This

inflammatory disease often predisposes individuals to the development of ▸ colorectal cancers. The anti-inflammatory mediated anticancer effects of celecoxib may be tissue-specific considering that celecoxib reduced lung inflammation in mice, but failed to inhibit the formation of chemically induced lung tumors in these animals.

Cyclooxygenase Independent Mechanisms for Cancer Chemoprevention by Celecoxib. The results of several in vitro and animal studies suggest the celecoxib may suppress tumorigenesis through several COX-2-independent mechanisms, which may account, at least in part, for celecoxib's anti-cancer effects in humans. For example, celecoxib inhibited the proliferation of various cancer cell types in vitro irrespective of their expression of COX-2, including transformed haematopoietic cells and immortalized and transformed human bronchial epithelial cells that were deficient in COX-2 expression. Celecoxib also inhibited the growth of human COX-2-deficient colon cancer cells that were transplanted as xenografts in nude mice. Thus, the chemopreventive effect of COX-2-specific inhibitors like celecoxib may be due to their effect on COX-2 as well as targets other than COX-2.

One putative COX-2 independent target for celecoxib is the phosphatidylinositol 3-kinase (PI3K) pathway, which is often deregulated in tumor cells. Celecoxib appears to directly inhibit the phosphoinositide-dependent kinase-1 (PDK1), and its downstream substrate protein kinase B/AKT, in the PI3K pathway. Protein kinase B/AKT inhibits apoptosis through the phosphorylation, and thus inactivation, of the proapoptotic ▸ BCL-2 family protein BAD. During apoptotic stimuli, BAD antagonizes BCL-2 and BCL-X_L activity, which can promote mitochondrial membrane permeabilization and cell death. The inhibition of the PI3K pathway by celecoxib is believed to be specific in its ability to promote apoptosis in transformed cells. For example, rofecoxib, another specific COX-2 inhibitor, had only marginal protein kinase B/AKT inhibitory activity in tumor cells during apoptosis induction.

Another presumed COX-2 independent target of celecoxib in tumor cells is sphingolipid metabolism. Celecoxib treatment increases the level of the sphingolipid ceramide in murine mammary tumor cells irrespective of COX-2 expression. This increase in ▸ ceramide was considered essential to apoptosis induction in these cells. Ceramide has been shown to mediate apoptosis in response to inflammatory cytokines like Fas and tumor necrosis factor α, and/or conditions associated with ▸ oxidative stress. During conditions of cell stress, the deregulation of ceramide generating and/or utilizing processes are believed to cause a net increase in cellular ceramide that is sufficient to trigger apoptosis induction via a mitochondrial membrane permeabilization mechanism.

Celecoxib treatment has also been shown to suppress the activity of the Ca ATPase located in the endoplasmic reticulum of human prostate cancer cells. The inhibition of the $Ca^{2\pm}$ ATPase by celecoxib disrupted Ca^{2+} homeostasis in the prostate cancer cells. This activity was highly specific for celecoxib and was not associated with the exposure to other COX-2 inhibitors, including rofecoxib. Microsome and plasma membrane preparations from the human prostate cancer cells showed that only the $Ca^{2\pm}$ ATPases located in the endoplasmic reticulum were the direct targets of celecoxib. The disruption of Ca^{2+} homeostasis played a central role in apoptosis induction in the prostate cancer cells because it was required for the activation of Ca^{2+}-dependent hydrolyses that carried out cellular degradation. Moreover, mitochondrial membrane permeabilization, which releases cytochrome *c* to activate cell death, is sensitive to elevations in intracellular free Ca^{2+}. Consequently, the celecoxib-induced inhibition $Ca^{2\pm}$ ATPases located in the endoplasmic reticulum may provide a link to mitochondrial membrane permeabilization for apoptosis induction much in the same way that celecoxib inhibition of the PI3K pathway can regulate BAD phosphorylation to trigger mitochondrial-mediated cell death.

It is apparent that the central hypothesis of a dominant role for COX-2 inhibition in cancer prevention by celecoxib may need re-examination. Furthermore, the COX-2 dependent and independent action of celecoxib in cancer prevention may be tissue specific. Since the aberrant expression of COX-2 is implicated in the pathogenesis of various types of human cancers, perhaps this inducible enzyme may be a useful surrogate biomarker of the anticancer activity of celecoxib when evaluating the chemoprevention of cancer at various sites in the body. Although the precise molecular mechanism for its chemopreventive effects are still fairly unknown, celecoxib may be still useful as a chemopreventive agent for a variety of malignancies, especially since it triggers less toxicity and adverse side effects during long-tern use when compared to traditional NSAIDs. Celecoxib may be useful when combined with other cancer chemopreventive/therapeutic agents to control the process of tumorigenesis.

References

Chun KS, Surh JY (2006) Signal transduction pathways regulating cyclooxygenase-2 expression: potential molecular targets for chemoprevention. Biochem Pharmacol 68:1089–1100

Grosch S, Maier TJ, Schiffmann S et al (2006) Cyclooxygenase-2 (COX-2)-independent anticarcinogenic effects of selective COX-2 inhibitors. J Natl Cancer Inst 98:736–747

Kismet K, Akay MT, Abbasoglu O et al (2004) Celecoxib: a potent cyclooxygenase-2 inhibitor in cancer prevention. Cancer Detect Prev 28:127–142

Psaty BM, Potter JD (2006) Risks and benefits of celecoxib to prevent recurrent adenomas. N Engl J Med 355:950–952

Schroeder CP, Kadara H, Lotan D et al (2006) Involvement of mitochondrial and akt signaling pathways in augmented apoptosis induced by a combination of low doses of celecoxib and N-(4-hydroxyphenyl) retinamide in premalignant human bronchial epithelial cells. Cancer Res 66:9762–9770

Cell Adhesion Molecules

Kris Vleminckx
Department of Biomedical Molecular Biology and Center for Medical Genetics, Ghent University, Ghent, Belgium

Synonyms

Adhesion molecules; CAMs

Definition

Cell ▶ adhesion molecules are transmembrane or membrane-linked glycoproteins that mediate the connections between cells or the attachment of cells to substrate (such as stroma or basement membrane). Dynamic cell-cell and cell-substrate adhesion is a major morphogenetic factor in developing multicellular organisms. In adult animals, adhesive mechanisms underlie the maintenance of tissue architecture, allow the generation of force and movement, and guarantee the functionality of the organs (e.g., to create barriers in secreting organs, intestines, and blood vessels) as well as the generation and maintenance of neuronal connections. Cell adhesion is also an integrated component of the immune system and wound healing. At the cellular level, cell adhesion molecules do not function just as molecular glue. Several signaling functions have been attributed to adhesion molecules, and cell adhesion is involved in processes such as contact inhibition, growth, and ▶ apoptosis. Deficiencies in the function of cell adhesion molecules underlie a wide variety of human diseases including cancer. By their adhesive activities and their dialogue with the ▶ cytoskeleton, adhesion molecules directly influence the invasive and metastatic behavior of tumor cells and by their signaling function they can be involved in the initiation of tumorigenesis.

Characteristics

At the molecular level, cell adhesion is mediated by molecules that are exposed on the external surface of the cell and are somehow physically linked to the cell membrane. In essence, there are three possible mechanisms by which such membrane-attached adhesion molecules link cells to each other (Fig. 1a). Firstly, molecules on one cell bind directly to similar molecules on the other cell (homophilic adhesion). Secondly, adhesion molecules on one cell bind to other adhesion receptors on the other cell (heterophilic adhesion). Finally, two different adhesion molecules on two cells may both bind to a shared secreted multivalent ligand in the extracellular

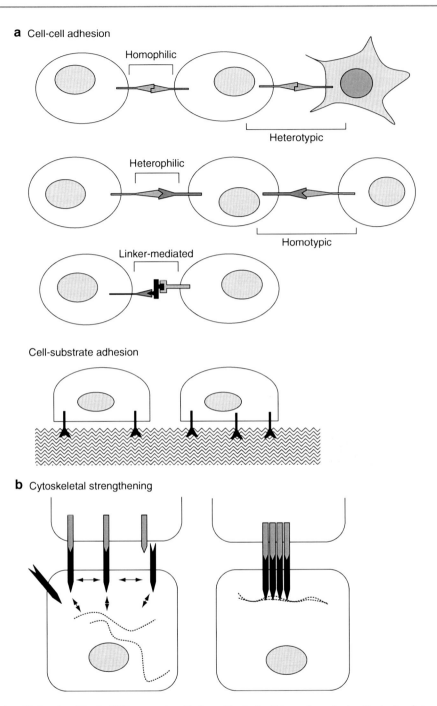

Cell Adhesion Molecules, Fig. 1 Different modes of cell-cell and cell-substrate adhesion and the mechanism of cytoskeletal strengthening. (**a**) Three possible mechanisms by which cell adhesion molecules mediate intercellular adhesion. A cell surface molecule can bind to an identical molecule (homophilic adhesion) on the opposing cell or can interact with another adhesion receptor (heterophilic adhesion). Alternatively, cell adhesion receptors on two neighboring cells can bind to the same multivalent, secreted ligand (linkermediated adhesion). Intercellular adhesion can take place between identical cell types (homotypic adhesion) or between cells of different origin (heterotypic adhesion), independently of the involved adhesion molecules. Cell-substrate adhesion molecules attach cells to specific compounds of the extracellular matrix. Cell-cell and cell-substrate adhesion can occur simultaneously. (**b**) Intercellular and cell-substrate adhesion can be strengthened by indirect intracellular linkage of the cytoplasmic tail of the adhesion molecules to the cytoskeleton and by lateral clustering in the membrane space. Also, cell-cell adhesion between two identical cells is called homotypic (cell) adhesion, while heterotypic (cell) adhesion takes place between two different cell types. In the case of cell-substrate adhesion, the adhesion molecules bind to the extracellular matrix (ECM).

Cell Adhesion Molecules and the Cytoskeleton

Adhesion molecules can be associated with the cell membrane either by a glycosylphosphatidylinositol (GPI) anchor or by a membrane-spanning region. In the latter case, the cytoplasmic part of the molecule often associates indirectly with components of the cytoskeleton (e.g., actin, intermediate filaments, or submembranous cortex). This implies that adhesion molecules, which by themselves establish extracellular contacts, can be structurally integrated with the intracellular cytoskeleton, and they are often clustered in specific restricted areas in the membrane, the so-called junctional complex (Fig. 1b). This combined behavior of linkage to the cytoskeleton and clustering, considerably strengthens the adhesive force of the adhesion molecules. In some cases, exposed adhesion molecules can be in a conformational configuration that does not support binding to its adhesion receptor. A signal within the cell can induce a conformational change that activates the adhesion molecule. Dynamic adhesion can also be mediated via regulated endocytosis of the adhesion molecules. These mechanisms of regulation allow for a dynamic process of cell adhesion that, amongst others, is required for morphogenesis during development and for efficient immunological defense.

Classification of Cell Adhesion Molecules

Based on their molecular structure and mode of interaction, five classes of adhesion molecules are generally distinguished; the cadherins, integrins, immunoglobulin (Ig) superfamily, selectins, and proteoglycans (Fig. 2).

Cadherins

Cadherins and protocadherins form a large and diverse group of adhesion receptors. They are Ca^{2+}-dependent adhesion molecules, involved in a variety of adhesive interactions both in the embryo and the adult. Cadherins play a fundamental role in metazoan embryos, from the earliest gross morphogenetic events (e.g., separation of germ layers during gastrulation) to the most delicate tunings later in development (e.g., molecular wiring of the neural network). The extracellular part of vertebrate classical cadherins consists of a number of cadherin repeats whose conformation is highly dependent on the presence or absence of calcium ions. Homophilic interactions can only be realized in the presence of calcium, usually by the most distal cadherin repeat. Classical cadherins are generally exposed as homodimers and their cytoplasmic domain can be structurally or functionally associated with the actin cytoskeleton. Cadherins are the major adhesion molecules in tissues that are subject to high mechanical stress such as epithelia (▶ E-cadherin) and endothelia (VE-cadherin). However, finer and more elegant intercellular interactions, such as synaptic contacts, also involve cadherins.

Integrins

Integrins are another group of major players in the field of cell adhesion. They are involved in various processes such as morphogenesis and tissue integrity, homeostasis, immune response, and inflammation. Integrins are a special class of adhesion molecules not only because they mediate both cell-cell and cell-substrate interactions (with components in the ECM such as laminin, fibronectin and collagen) but also because they function as heterodimers consisting of an α- and β-subunit. To date, at least 16 α-subunits and 8 β-subunits have been indentified. Of the theoretical 128 heterodimeric pairings, at least 21 are known to exist. While most integrin heterodimers bind to ECM components, some of them, more particularly those expressed on leukocytes, are heterophilic adhesion molecules binding to members of the Ig superfamily. The α-subunit mostly contains a ligand-binding domain and requires the binding of divalent cations (Mg^{2+}, Ca^{2+}, and Mn^{2+}, depending on the integrin) for its function. Interestingly, integrins may be present on the cell-surface in a nonfunctional and functional configuration. The cytoplasmic domain appears to be responsible for the conformational change that activates the integrin.

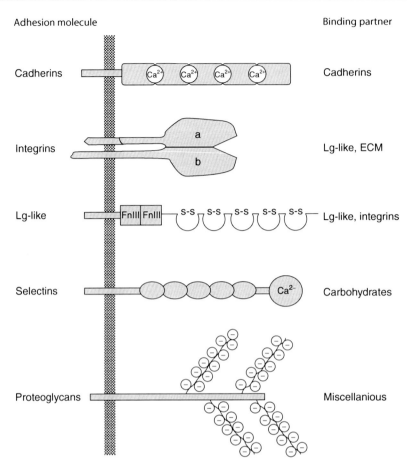

Cell Adhesion Molecules, Fig. 2 The five major classes of cell adhesion molecules and their binding partners. Cadherins are Ca^{2+}-dependent adhesion molecules that consist of a varying number of cadherin repeats (five in case of the classical cadherins). The conformation and activity of cadherins is highly dependent on the presence of Ca^{2+}-ions. In general, cadherin binding is homophilic. Integrins are functional as heterodimers and consist of an a- and b-subunit. They interact with members of the immunoglobulin superfamily or with compounds of the extracellular matrix (e.g., fibronectin, laminin). Members of the immunoglobulin superfamily (Ig-like proteins) are characterized by a various number of immunoglobulin-like domains (*open circles*). Membrane-proximal, fibronectin type III repeats are often observed (gray boxes). They can either bind to other members of the Ig-family (homophilic) or to integrins. Selectins contain an N-terminal Ca^{2+}-dependent lectin domain (*circle*) that binds carbohydrates, a single EGF-like repeat (*gray box*) and a number of repeats that are related to those present in complement-binding proteins (*ovals*). Proteoglycans are huge molecules that consist of a relatively small protein core to which long side chains of negatively charged glycosaminoglycans are covalently attached. They bind various molecules, including components of the extracellular matrix

The Ig Superfamily

Among the classes of adhesion molecules discussed here, the Ig superfamily is probably the most diverse. The main representatives are the neural cell adhesion molecules (NCAMs) and V(ascular)CAMs. As the name suggests, the members of this family all contain an extracellular domain consisting of different immunoglobulin-like domains. NCAMs sustain homophilic and heterophilic interactions that play a central role in regulation and organization of neural networks, specifically in neuron-target interactions and fasciculation. The basic extracellular structure consists of a number of Ig domains, which are responsible for homophilic interaction, followed by a discrete number of fibronectin type III repeats. This structure is linked to the membrane either by a GPI anchor or a transmembrane

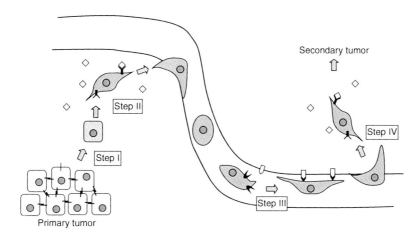

Cell Adhesion Molecules, Fig. 3 Cell adhesion processes involved in the metastatic cascade. A subset of cells (*gray*) growing in a primary tumor will reduce cell-cell contacts (*Step I*) and migrate in the surrounding stroma by increasing specific cell-substrate adhesion (*Step II*). These invasive tumor cells can extravasate into the circulation and, at distant sites, attach to the endothelial blood vessel wall through specific cell-cell interactions (*Step III*). Once these cells have extravasated through the vessel wall they use cell-substrate adhesion molecules to invade the surrounding stroma (*Step IV*). See text for details

domain. The VCAM subgroup, including I (ntercellular)CAMs and the mucosal vascular addressin adhesion molecule (MAdCAM), is involved in leukocyte trafficking (or homing) and extravasation. They consist of membrane-linked Ig domains that make heterophilic contacts with integrins. Other members of this family that are associated with cancer are carcinoembryonic antigen (CEA), "deleted in colon cancer" (DCC) and platelet endothelial (PE)CAM-1.

Selectins

These types of adhesion molecules depend on carbohydrate structures for their adhesive interactions. Selectins have a C-type lectin domain that specifically binds to discrete carbohydrate structures present on cell-surface proteins. Intercellular interactions mediated by selectins are of particular interest in the immune system, where they play a fundamental role in trafficking and homing of leukocytes.

Proteoglycans

Proteoglycans are large extracellular proteins consisting of a relatively small protein core to which long chains of glycosaminoglycans are attached. Although poorly documented, proteoglycans may bind to each other or may be the attachment site for other adhesion molecules.

Role of Adhesion Molecules in Cancer

The Metastatic Cascade

Cell adhesion molecules play an important role during the progression of tumors, more particularly in the metastatic cascade (Fig. 3). When a benign tumor becomes malignant, cells at the periphery of the tumor will lose cell-cell contact (step I) and invade the surrounding stroma (step II) (see also ► invasion). Cells then extravasate and enter the vasculature or lymphatic system, where they are further transported. A fraction of the circulating tumor cells survives and is arrested at a distant site, attaches to the endothelium (step III), and extravasates through the blood vessel wall and into the surrounding tissue (step IV). Here the tumor cells grow, attract blood vessels, and develop to a secondary tumor (► metastasis).

Adhesive Events in Metastasis

All the classes of cell adhesion molecules play a role in the metastatic cascade. During the first step, tumor cells need to disrupt intercellular junctions in order to detach from the primary tumor. This step often involves the suppression of cadherin function. The second step of ▶ migration through the stroma and into the blood or lymphatic vessels requires dynamic cell-substrate adhesion, mostly mediated by integrins. In the third step, where cells arrest in the circulation by aggregation with each other or attachment to platelets, leukocytes, and endothelial cells, critical roles have been attributed to cell adhesion molecules of the Ig superfamily, selectins, integrins, and specific membrane-associated carbohydrates. The fourth step is similar to step II and mostly involves integrins. Details on the adhesive events associated with metastasis are outlined below.

- In benign epithelial tumors, cells maintain firm intercellular adhesive contacts, mostly by formation of a junctional complex (including tight junctions, ▶ adherens junctions, and desmosomes). Establishment and maintenance of such a strong junctional complex requires expression and function of cadherins (more particularly E-cadherin). Loss of E-cadherin expression or function appears to be a hallmark of progression of a benign epithelial tumor (adenoma) to a malignant one (carcinoma). Epithelial tumor cells often acquire invasive properties by mutational inactivation of E-cadherin or one of its cytoplasmic binding partners (catenins). It is important to keep in mind that cadherin-mediated adhesion is a dynamic process and that E-cadherin can be temporarily inactivated at the functional level, for example by phosphorylation or other posttranslational modifications. E-cadherin and other molecules of the junctional complex are very often suppressed or functionally modulated in the epithelial-mesenchymal transitions (EMT), a hallmark of malignant tumor progression. EMT can be a tumor-intrinsic feature or can be induced by their microenvironment. Paracrine factors such as scatter factor or juxtacrine signaling via Ephrin/Eph receptor or via ▶ semaphorins/plexins can affect adhesion via direct activity on the cell adhesion molecules or via regulation of the cytoskeleton.
- Dynamic cell-substrate adhesion is a critical factor in the migration of invasive tumor cells into

the surrounding stroma. Integrins are instrumental in this process. Several studies have correlated the migratory behavior of tumor cells either with an increased or decreased expression of particular integrins. This apparent paradox may be explained by the fact that firm but temporary cell-substrate contacts are required for cells to migrate on a substrate. In order to crawl directionally through the stroma, a cell needs to "grab" the ECM, release after pulling itself forward and then has to establish the next contact. Both inhibiting adhesion and preventing release of the substrate contacts "locks" the cell in its position and prevents migration. It should be remembered that integrins may exist in two functional states and that signals passed through the cytoplasm determine whether membrane-exposed integrins are functional or not.

- In the third step of the metastatic cascade, cell-cell interactions are again the most determining. Homotypic interactions between circulating tumor cells promote formation of aggregates that are preferentially retained in the capillary network. PECAM-1 is a cell adhesion molecule potentially involved in this process. It should be pointed out that (re)expression of the invasion-suppressor molecule E-cadherin would actually promote metastasis formation. Besides these homotypic interactions, heterotypic interactions are also of major importance in the metastatic process. Tumor cells can attach to the blood-vessel wall either directly or indirectly through platelets and leukocytes. The adhesion molecules involved in this process are similar to those involved in the "multistep adhesion cascade" observed during homing and extravasation of leukocytes or trafficking of lymphocytes. Cell adhesion events include interactions of tumor-associated lectins with selectins expressed on platelets, leukocytes, and endothelium (P-, L-, and E-selectins, respectively). These adhesion molecules are also involved in the initial transient low-affinity interactions (rolling) of circulating leukocytes (and probably tumor cells) with the endothelium. Other and more stringent heterotypic heterophilic interactions in this metastatic stage include the binding of integrins on tumor cells to ICAMs expressed on the surface of the endothelial cells.

- The fourth step in the metastatic cascade is extravasation and invasion at a distant site. This process is very similar to step 2 and the same adhesion molecules are likely to be involved. Specific interactions of the tumor cells with molecules present on the endothelial cells (e.g., N-cadherin) will facilitate the extravasation process.

Other Cancer-Related Functions of Cell Adhesion Molecules

It has become clear that some cell adhesion molecules are involved in signaling processes that are relevant to cancer. Germline mutations in E-cadherin predispose patients to the development of diffuse gastric carcinomas, and in lobular breast carcinoma, E-cadherin seems to act as a tumor suppressor. Interestingly, β-catenin, a protein cytoplasmically linked to cadherins, has a central role in ► Wnt signaling and has oncogenic properties that are counteracted by the adenomatous polyposis coli (APC) gene product. Signaling by integrins can also be an important factor that prevents cells from undergoing apoptosis (apoptosis upon loss of cell adhesion is called ► anoikis), which might be critical when tumor cells are traveling in the circulation. Interdisciplinary research has revealed new unexpected functions for known cell adhesion molecules. The suspected tumor suppressor DCC, a member of the Ig superfamily of adhesion molecules, turned out to be the receptor for netrin-1, an axonal chemoattractant crucial in neuronal development. Other molecules known to have adhesive or repulsive activities in the axonal growth cone or in migrating neural crest cells, turn out to have similar activities in tumor cells (see also the chapters on ► EPH receptors, Ephrin signaling in cancer, ► semaphorins, and ► plexins).

Cross-References

- ► Adherens Junctions
- ► Adhesion
- ► Anoikis
- ► Apoptosis
- ► Carcinoembryonic Antigen
- ► Cytoskeleton
- ► E-Cadherin
- ► Eph Receptors
- ► Invasion
- ► Metastasis
- ► Migration
- ► Plexins
- ► Semaphorin
- ► Wnt Signaling

References

Cavallaro U, Christofori G (2004) Cell adhesion and signalling by cadherins and Ig-CAMs in cancer. Nat Rev Cancer 4:118–132

Chothia C, Jones EY (1997) The molecular structure of cell adhesion molecules. Annu Rev Biochem 66:823–862

Hynes RO (2000) Cell adhesion: old and new questions. Trends Cell Biol 9:M33–M37

Mizejewski GJ (1999) Role of integrins in cancer: survey of expression patterns. Proc Soc Exp Biol Med 222:124–138

Sanderson RD (2001) Heparan sulfate proteoglycans in invasion and metastasis. Semin Cell Dev Biol 12:89–98

See Also

(2012) Cadherins. In: Schwab M (ed) Encyclopedia of Cancer, 3rd edn. Springer Berlin Heidelberg, pp 581–582. doi:10.1007/978-3-642-16483-5_770

(2012) Contact Inhibition. In: Schwab M (ed) Encyclopedia of Cancer, 3rd edn. Springer Berlin Heidelberg, pp 973–974. doi:10.1007/978-3-642-16483-5_1323

(2012) E-Selectin. In: Schwab M (ed) Encyclopedia of Cancer, 3rd edn. Springer Berlin Heidelberg, p 1317. doi:10.1007/978-3-642-16483-5_1780

(2012) Extracellular Matrix. In: Schwab M (ed) Encyclopedia of Cancer, 3rd edn. Springer Berlin Heidelberg, p 1362. doi:10.1007/978-3-642-16483-5_2067

(2012) Homophilic and Heterophilic Adhesion. In: Schwab M (ed) Encyclopedia of Cancer, 3rd edn. Springer Berlin Heidelberg, p 1729. doi:10.1007/978-3-642-16483-5_2804

(2012) Integrin. In: Schwab M (ed) Encyclopedia of Cancer, 3rd edn. Springer Berlin Heidelberg, p 1884. doi:10.1007/978-3-642-16483-5_3084

(2012) Junctional Complex. In: Schwab M (ed) Encyclopedia of Cancer, 3rd edn. Springer Berlin Heidelberg, p 1929. doi:10.1007/978-3-642-16483-5_3188

(2012) Lectin. In: Schwab M (ed) Encyclopedia of Cancer, 3rd edn. Springer Berlin Heidelberg, p 1999. doi:10.1007/978-3-642-16483-5_3303

(2012) Proteoglycans. In: Schwab M (ed) Encyclopedia of Cancer, 3rd edn. Springer Berlin Heidelberg, p 3100. doi:10.1007/978-3-642-16483-5_4816

Cell Biology

Filippo Acconcia[1] and Rakesh Kumar[2]
[1]Molecular and Cellular Oncology, The University of Texas MD Anderson Cancer Center, Houston, TX, USA
[2]Department of Biochemistry and Molecular Medicine, George Washington University, Washington, DC, USA

Definition

Cell biology deals with all aspects of the normal and of the tumor cell, their normal and abnormal multiplication, their differentiation, their stem origins, and their regulated cell death.

Characteristics

The Cell

The intracellular environment is separated from the external environment by a lipid bilayer called plasma membrane. The plasma membrane controls the movement of substances in and out of the cell and it is important for the cell to sense the surrounding environment. Within the cell the nucleus occupies most of the space. The cell nucleus contains genes, which drive all cellular activities and processes. Genes are organized in chromosomes (i.e., genome) and are made of DNA. The genetic information is used to produce proteins, which are the critical effectors required for all cellular processes. The nucleus is separated from the rest of the cellular content by the nuclear membrane, which remains in contact with the cytoplasm as well as the nucleoplasm. In the cytoplasm, proteins are organized into specific functional structures and also connected with the structural network referred to as cytoskeleton network, which physically sustains the cell. Moreover several intracellular organelles are located in the cytoplasm (e.g., mitochondria, Golgi apparatus) and allow the cells to self sustain. To continuously adjust the intracellular processes and to promptly respond to the demands of the

extracellular environment, cells need to exchange matter, energy, and information with the external milieu.

Cell Division and Reproduction

One of the unique features of cell is its ability to divide and produce two daughter cells that are an exact copy of their parental cell, by a process called "mitosis." However, some differentiated cells undergo the process of meiosis. For simplicity, meiotic division can be considered as the sum of two successive mitotic divisions, which result in four daughter cells with half the number of chromosomes and rearranged genes. These specialized cells (i.e., gametes) serve as reproductive cells. The fusion of the female and male gametes (eggs and spermatozoa, respectively) results in a new cell called zygote. The zygote, by definition, is a stem cell. Following mitotic division, it becomes an embryo and, at the end of the embryonic development, results in a new organism.

Cell Proliferation

The physiological functions of an organ require maintenance of homeostasis, a process of regulated balance between cell proliferation and cell death (also known as ► apoptosis), in the differentiated tissue. Indeed, a variety of extracellular stimuli activate specific ► signal transduction pathways that affect the expression and activity of molecules involved in the control of cell proliferation or cell death. Thus, the balance between cell cycle progression and apoptosis defines the cell fate, and this process depends on genetic factors as well as the kinetics of signal transduction pathways in exponentially growing cells.

Cell Cycle

In mammalian cells, one cell cycle takes about 24 h in most cell types and can be schematically divided into two stages: mitosis and interphase. Mitosis (M phase) consists of a series of molecular processes that result in cell division. On the other hand, the interphase can be subdivided into three major gaps (G1, S, and G2 phase). The G1 phase of the cell cycle separates the M and S phases. In G1 phase, cells express a specific pattern of gene products required for the DNA synthesis; the G2 phase of the cell cycle resides in between the S and M phases and is important for the completion of processes that are necessary for mitosis. The G0 phase of the cell cycle is entered by the cells from the G1. In the G0 phase, cells are out of the cell cycle and into a quiescent state where they do not proliferate.

Regulation of Cell Cycle Progression

Cell cycle progression is achieved through a series of coordinated molecular events that allow the cells to transit across the restriction points, also known as cell cycle checkpoints. There are three main restriction points in the cell cycle (G2/M, M/G1, and G1/S, respectively). Broadly, these checkpoints are defined as points after which the cell is committed to progress to the next phase in a nonreversible manner. Therefore, the transition between the phases of the cell cycle is strictly regulated by a specific set of proteins. ▶ Cyclin-dependent kinases (CDK) act in various phases of the cell cycle by binding to its activating proteins called cyclins. For example, both ▶ cyclin D/CDK4 and cyclin E/CDK2 complexes regulate transition of the cells through G1/S phase whereas cyclin A/CDK1, cyclin A/CDK2, and cyclin B/CDK1 complexes are active during the rest of the cell cycle. On the other hand, another class of regulatory proteins, the cyclin-dependent kinase inhibitors (CKI) (e.g., $p21^{Cip/Kip}$; $p19^{Ink4d}$) antagonizes the activation of CDK activity, thus impeding the progression of the cell cycle.

Programmed Cell Death

Programmed cell death (PCD) is a physiological process of eliminating a living cell. The PCD involves activation of specific intracellular programs that commit cells to a "suicidal route." The process of PCD plays an important role in a variety of biological events, including morphogenesis, maintenance of tissue homeostasis, and elimination of harmful cells. To date, different forms of PCD have been described among which apoptosis, necrosis, and ▶ autophagy are the most common.

Apoptosis

One of the critical events in apoptosis is the activation of cystein proteases, called caspases, upon a given signal. The initiator caspases (▶ Caspase 8 and 9) are the first enzymes involved in the activation of the apoptotic cascade. Caspase 8 and 9 activate the downstream effector caspases (caspase 3, 6, and 7) by proteolytic cleavage which in turn results in the hydrolysis and inactivation of the enzymes involved in the processes of DNA repair such as by poly-ADP-ribose polymerase (PARP). Upon stimulation of apoptotic cascade, cells display a specific set of characters, which constitute the hallmark of apoptosis (DNA fragmentation, cell shrinkage, cytoplasmic budding, and fragmentation). The activation of caspases is achieved through two principle pathways – an extrinsic pathway that transduces signals from the plasma membrane directly to the caspases, and an intrinsic pathway that involves activation of caspases through a series of biochemical events leading to permeabilization of the mitochondrial membrane and release of cytochrome c (▶ Cytochrome P450) in the cytoplasm. Apoptotic cells are eventually eliminated by the immune system without the activation of inflammatory reactions (▶ Inflammation).

Necrosis

Necrosis results from a severe physical, mechanical, or metabolic cellular damage. The necrotic phenotype is very different from those of an apoptotic cells. Overall, the cell switches off its metabolic pathways and the DNA condenses at the margins of the nucleus and the cellular constituents start to degrade. In general, necrosis consists in a general swelling of the cell before it disintegrates. Furthermore, upon leakage of the intracellular content, necrotic cells stimulate an inflammatory response that usually damages the surrounding tissue.

Autophagy

Autophagy, i.e., autophagic cell death, occurs by sequestration of intracellular organelles in a double membrane structure termed autophagosome. Subsequently, the autophagosomes are delivered to the lysosomes and degraded. Autophagy is responsible for the turnover of dysfunctional organelles and cytoplasmic proteins and thus, contributes to cytosolic homeostasis. Autophagy can occur either in the absence of detectable signs of apoptosis or concomitantly with apoptosis. Indeed, autophagy is activated by signaling pathways that also control apoptosis.

Signal Transduction

Extracellular signals are transduced by the activation of a series of phosphorylation-dependent intracellular pathways initiated by cell surface receptors. Eventually, such signals feed into the nucleus, stimulate transcription factors, and regulate gene transcription.

Signaling Targets

Signaling pathways regulate gene transcription by triggering the promoter activity of the target gene. For example, regulation of cyclin D is critical for cell cycle progression. The extracellular signal-mediated activation of specific signal transduction pathways stimulates the activity of transcription factors such as AP-1, SP-1, and NF-κB, which coordinate the activation of the cyclin D1 promoter and thus lead to cyclin D1 expression. On the other hand, signaling molecules can also change the activity of a preexisting protein. For example, activation of p21-activated kinase (PAK) induces the phosphorylation of phosphoglucomutase (PGM) that stimulates its enzyme activity and the phosphorylation of ▶ estrogen receptor alpha (ERα) thus inducing its transcriptional activity. One of the most studied signaling pathways is the extracellular-regulated kinase (ERK) (▶ MAP kinase) cascade. It consists of three steps of sequential phosphorylations that impact on diverse cellular effectors. The ERK cascade is activated by mitogenic stimuli (e.g., growth factors (▶ Fibroblast growth factors)) and plays a critical role both in cell proliferation and cell survival. Indeed, activation of ERK induces the activation of AP-1 transcription factor, which, in turn, regulates cyclin D1 expression in addition to many of other proliferative molecules. Further, ERK activity leads to an increased expression of the antiapoptotic protein ▶ BCL-2 and inactivation of the proapoptotic protein Bad. Conversely, the JNK/SAPK (▶ JNK Subfamily) and the p38/MAPK (MAP kinase) pathways mediate stress and apoptotic stimuli (e.g., UV, ischemic-reperfusion

damage). Activation of JNK/SAPK and p38/MAPK often results in an increased expression of proapoptotic proteins (e.g., Bax), and in the activation of the caspase cascade and cytochrome c release from the mitochondria.

Systems Biology

Systems biology represents a new analytical tool that has begun to emerge for balanced comprehensive analyses of cellular pathways at the level of genes and proteins. Signal transduction pathways often cross-talk and influence each other, and the functionality of the effector molecule is influenced by the overall outcome of a set of signaling pathways. Thus, cells form a web of intracellular interactions that are critical for a timely and dynamic response. The intracellular signaling network is considered a complex system rapidly adapting to extracellular challenges. Therefore, an additional level of complication is the evaluation of the network as a whole, rather than the individual pathway.

Cell Motility and Migration

▶ Motility and ▶ migration are important components for the functionality of a variety of cell types and are involved in physiologic processes such as embryonic development, immune response, as well as in pathologic processes such as ▶ invasion and ▶ metastasis. Cell motility and migration are coordinated physiological processes that allow the cells to move or to invade the surrounding tissues, respectively. They occur as a result of a complex interplay between the focal ▶ adhesion sites (cell-to-substrate contacts) and the extracellular matrix (ECM) (substrate). Phenotypically, migratory cells develop motile structures such as pseudopodia, lamellipodia, and filopodia. An ordered sequence of events (protrusion of motile structures, formation and disruption of focal contacts) generate the traction forces that drive the cell movement. Moreover, when migration is required, cells secrete specific proteolytic enzymes (matrix metalloproteinases, MMPs) that digest the ECM, thus opening a passage across the substrate. Cytoskeleton is critical for the correct occurrence of cell motility and migration.

Cytoskeleton

Cytoskeleton is a network of cytoplasmic proteins, which define the cell "bones." Many different protein filaments are important for cytoskeleton functions. In particular, microtubules, built from different types of tubulin, originate from specific intracellular structures called microtubules organizing centers (MTOC). Dynamic changes in the polymerization and depolymerization of tubulin maintain microtubule integrity and resulting functions. Furthermore, actin microfilaments form a network of cytoskeleton-associated proteins and connect the focal adhesion with the intracellular cytoskeleton. The dynamic remodeling of microtubules and microfilaments has an impact on cell motility, migration and cell–cell adhesion, ▶ endocytosis, intracellular trafficking, organelle function, cell survival, gene expression, and cell division.

Signaling Regulation

At the focal adhesion sites, cells accumulate receptors (e.g., growth factor receptors), adaptors (e.g., vinculin), and signaling molecules, as well as structural and motor proteins (e.g., actin, myosin). Migration-specific stimuli (e.g., integrins engagement of ECM, growth factor stimulation, and mechanical stimuli) activate specific biochemical pathways. ▶ Focal Adhesion Kinase (FAK), integrin-linked kinase (ILK), PAK, and ▶ Src play key roles in modulating cell migration and invasion. The FAK/Src complex regulates the assembly and disassembly of focal contacts, F-actin cytoskeleton remodeling, and the formation of lamellipodia and filopodia through the activation of specific downstream cytoskeleton-associated signaling pathways. Further, ILK is also implicated in cell motility and migration by linking integrins with cytoskeleton dynamics through the ▶ PI3K signaling pathway. Also, PAK1 dynamically regulates cytoskeletal changes by coordinating upstream signaling with multiple effectors. By acting on actin reorganization, PAK1 drives directional cell motility and migration.

Tumor Biology

Cancer is a progressive disease that arises from the clonal expansion of a single transformed cell into a mass of uncontrolled proliferating cells. Tumorigenesis is a multistep process and involves progressive conversion of a normal cell into a malignant cell, which subsequently invades the surrounding tissues. The process of tumorigenesis consists of major steps (initiation, promotion, and progression), each involving specific molecular mechanisms, often interlaced with each other, that drive tumor development.

Initiation and Promotion

In general, initiation of tumorigenesis is referred to as the first oncogenic stimulus. However, such as initial event is not sufficient for tumor induction. In most cases, a second oncogenic stimulus must occur in a restricted time frame, thus promoting an irreversible effect. Chemical (e.g., aromatic compounds (▶ Polycyclic aromatic hydrocarbons)), physical (e.g., ▶ UV radiation), as well as biological (e.g., viruses as Human Papillomavirus) stress have impact on the cells and can induce DNA mutations (e.g., point mutations). In addition, gene deletion or duplication also alters gene function and contributes to the process of tumorigenesis. These genomic changes result in the production of proteins with altered functions or in the overexpression or downregulation of specific proteins, which affects the associated cellular functions.

Protooncogenes or oncogenes are genes that encode for proteins involved in the induction of cell proliferation (e.g., *cyclin D1*, *CDK*, *EGFR*, *Src*, *Ras*, etc.) and whose overexpression or hyperactivation leads to an uncontrolled cell proliferation. On the other hand, tumor suppressor genes are genes encoding for proteins that negatively regulate cell proliferation (e.g., *p53*, *PARP*, *CKI*, etc.). Inactivating mutations or downregulation of tumor suppressor genes are also critical for enhanced cell proliferation. In addition to DNA damage, oncogenes and tumor suppressor genes, abnormal changes in the epigenetic cellular information (e.g., DNA ▶ methylation) can also participate in clonal evolution of human cancers.

Progression

The modified balance between the growth-inhibitory programs and proliferative networks allows the cell to escape the physiological growth restrains. These selective growth advantages produce a population of more aggressive or transformed cells that resist clearance by the immune system (i.e., immune defense escape), and in turn, contributes to the accumulation of additional mutations and eventually, in tumor growth. In this context, an in situ tumor develops, that is the uncontrolled mass of transformed cells stays within the limit of the tissue in which the first cell resided. During this phase, tumor volume increases in parallel with an increased dedifferentiation of the cells that also secrete angiogenic factors (▶ Angiogenesis) to promote blood vessels formation in the tumor.

Metastasis

Metastasis is the process by which highly vascularized tumor cells acquire the ability to invade the blood-stream and seed in distant organs. Deregulation of cytoskeleton-associated proteins and secretion of protein factors play a critical role in the functionality of the metastatic cells.

Stem Cell Biology

In 1998, the group of Prof. James Thomson reported the isolation of a human embryonic stem cell line from the blastocyst stage of a human embryo. This cell line showed stability in a specifically developed culture medium and, upon transplantation in the nude mice, had the ability to form tumor-like structures made up of all the major human tissue types. This pioneer study opened the field of stem cell biology. Since then, enormous research efforts have been focused on the understanding of stem cell biology as well as their potential medical and therapeutic implications. Nonetheless, although the last 10 years witnessed an enormous progress, the field of stem cell research is in its infancy. The first controversy is the definition of stem cell itself. For simplicity, a stem cell is a clonal self-renewing entity that is multipotent and can generate several different cell types. This definition introduces three major characteristic of the stem cells: self-renewal, clonality, and potency.

Self-Renewal and Clonality

Self-renewal is the process by which a stem cell undergoes an asymmetric mitotic division that produces, rather than two identical daughter cells, one cell that is completely identical to the parental stem cell and another cell that is already committed to a more restricted developmental path and more specialized abilities. Thus, stem cells have both the ability to self-maintain their clonal cell population and to produce a population of clones with more differentiated characteristics. In this way, stem cells form a hierarchy of potency.

Potency

Stem cells have the ability to give rise to a population of daughter stem cells with a reduced differentiation. The totipotent cells are the first embryonic cells that can become any kind of cell type (e.g., zygote). These cells become pluripotent cells, which can differentiate into most but not all cell types (e.g., embryonic stem cells). Next, cells that are committed to produce only a certain lineage of cell types (e.g., ▶ adult stem cells) are the multipotent cells. Some multipotent cells can only generate one specific kind of terminally differentiated cell type and thus, such cells, are called unipotent cells.

Environmental Regulation

The molecular mechanism by which regulatory processes occur in stem cells are not clear but are believed to be tightly regulated to avoid imbalance in stem cell population or mutation that can lead to tumorigenesis. One possibility is that the asymmetric division produces two daughter cells and, because of intrinsic factors, such cells follow different fates in spite of residing in the same microenvironment. Alternatively, the two daughter cells become functionally different because they are exposed to different extrinsic factors. Most likely, both intrinsic and extrinsic factors are integrated in the milieu of the surrounding microenvironment, also known as the stem cell niche. Signals from the niche determine the type of gene regulation that allows the asymmetric division to take place. In this model, one daughter cell stays in the niche and the other one moves out. Indeed, the importance of the microenvironment in stem cell biology is highlighted by the ability of a particular stem cell to transdifferentiate or to dedifferentiate when put in a different niche. Although the concept of plasticity is debated in the literature, it is part of the "stemness" of a cell, which is the hallmark for a cell to be defined as a stem cell.

Social Implications

The ability to scientifically manipulate the human embryo or human adult stem cells has opened new perspectives for treatment of several human diseases. However, it has also initiated intense philosophical and political debates on the ethical issues associated with the use of such potential tools in medical practice.

Cross-References

▶ Adhesion
▶ Adult Stem Cells
▶ Angiogenesis
▶ Apoptosis
▶ Autophagy
▶ Bcl2
▶ Caspase-8
▶ Cyclin D
▶ Cyclin-Dependent Kinases
▶ Cytochrome P450
▶ Endocytosis
▶ Estrogen Receptor
▶ Fibroblast Growth Factors
▶ Focal Adhesion Kinase
▶ Inflammation
▶ Invasion
▶ JNK Subfamily
▶ MAP Kinase
▶ Metastasis
▶ Methylation
▶ Migration
▶ Motility
▶ PI3K Signaling
▶ Polycyclic Aromatic Hydrocarbons
▶ Signal Transduction
▶ Src
▶ UV Radiation

References

Feinberg AP, Tycko B (2004) The history of cancer epigenetics. Nat Rev Cancer 4:143–153

Gearhart J, Hogan B, Melton D et al (2006) Essential of stem cell biology. Academic, London

Lowe SW, Cepero E, Evan G (2004) Intrinsic tumour suppression. Nature 432:307–315

Pestell RG, Albanese C, Reutens AT et al (1999) The cyclins and cyclin-dependent kinase inhibitors in hormonal regulation of proliferation and differentiation. Endocr Rev 20:501–534

Potten C, Wilson J (2004) Apoptosis – the life and death of cells. Cambridge University Press, New York

See Also

(2012) Cell Cycle. In: Schwab M (ed) Encyclopedia of Cancer, 3rd edn. Springer Berlin Heidelberg, p 737. doi: 10.1007/978-3-642-16483-5_994

(2012) Extracellular Matrix. In: Schwab M (ed) Encyclopedia of Cancer, 3rd edn. Springer Berlin Heidelberg, p 1362. doi: 10.1007/978-3-642-16483-5_2067

(2012) Microenvironment. In: Schwab M (ed) Encyclopedia of Cancer, 3rd edn. Springer Berlin Heidelberg, p 2296. doi: 10.1007/978-3-642-16483-5_3720

Cell Cycle Checkpoint

Wenjian Ma
National Institute of Environmental Health Sciences (NIEHS), Research Triangle Park, NC, USA

Definition

Cell cycle checkpoints are the control mechanisms that stop cell progression during particular stage of the cell cycle to check and ensure the accurate completion of earlier cellular processes and faithful transmission of genetic information before cell division.

Characteristics

Cell growth and division proceeds through an ordered set of events called cell cycle, which is divided into four distinct phases namely G1 (the first gap phase), S (DNA synthesis), G2 (the second gap phase), and M (mitosis). G1 and G2 are two gap phases that accumulate nutrients, perform biosynthesis, and monitor cell state to get ready for DNA synthesis and mitosis, respectively. DNA replication occurs in S phase and the duplicated chromosomes are separated into two identical sets during mitosis (M phase). Followed by cytokinesis, the mother cell is divided into two daughter cells that are genetically identical to each other.

The cell cycle is highly regulated and each phase is monitored by surveillance mechanisms to maintain cellular integrity and faithful transmission of genetic information from mother cell to daughter cell. If a crucial process has not been completed or if a cell has sustained damage, progression into the next cell phase would be prevented. These mechanisms that capable of delaying the cell cycle at specific time points are now referred to as checkpoints, which were first identified in the late 1980s.

Various stresses can activate the checkpoint and cause cell cycle arrest, such as nutrient deprivation, mitogenic stimuli, and cytotoxins. However, the most important function of checkpoints is to monitor DNA damages and coordinate repair. Cells are under constant attack by DNA-damaging agents arising from endogenous or exogenous sources such as UV and the reactive oxygen species that inevitably generates during metabolism. These attacks can interfere with DNA replication, transcription, and other cellular functions and finally lead to genome instability. As repairing damaged DNA takes time, it is essential to activate specific checkpoint machinery to temporarily stall the cell cycle progression. In case the damages cannot be dealt with, the checkpoint can also activate other mechanisms such as apoptosis to target the cell for destruction.

Multiple checkpoints have been identified from lower eukaryotes to human. Despite variations in molecular details, the controlling mechanisms of different organism share some conserved features in that they are tightly regulated through the interaction of specific protein kinases and adaptor proteins. The transition from one phase of the cell cycle to the next is driven by a group of kinases called cyclin-dependent kinases (CDKs),

which become active when bound by their cyclin partners. CDKs phospharylate specific downstream substrates to alter their biochemical function and elicit specific cellular responses. The level of cyclins and CDKs fluctuate during the cell cycle that is controlled by complex negative-feedback loops. Through the oscillation of cyclin-CDKs, cellular processes within the cell cycle such as DNA replication, chromosome segregation, and cell division are precisely modulated.

Simple eukaryotes such as yeast has only one CDK (Cdc28 in *Saccharomyces cerevisae* and Cdc2 in pombe), whereas higher eukaryotes have multiple CDKs, and through different combination of CDKs and cyclins, to control different aspects of the cell cycle. For example, S-phase is controlled by cyclin A in combination with CDK2, whereas progression into mitosis is regulated by cyclin B-CDK1 in mammalian cells. So far 16 eukaryotic cyclins and up to nine CDKs have been discovered.

CDK activity is also negatively controlled by certain families of inhibitory proteins, and the cell-cycle progression is determined by the relative abundance of positive and negative regulators. The core cell cycle control protein/enzyme machineries sense stress/damage and trigger the cell cycle arrest are not conserved between different eukaryotes. Below describes the major checkpoints in mammalian cells as shown in Fig. 1.

G1 checkpoint

The G1 checkpoint is located at the end of the G1 phase that ensures everything is ready for DNA synthesis. It is the major restriction point to decide whether the cell continue for a further round of cell division. Under unfavorable environmental conditions, it signals the cell to temporally withdraw from the cell cycle and enter into a resting phase called G0. Once passing this checkpoint, the cell would tend to complete the whole cycle. During G1 phase, the cells may also irreversibly withdraw from the cell cycle into terminally differentiated or senescent states.

One of the control pathways acting in G1 checkpoint is through the regulation of the tumor suppressor retinoblastoma protein (Rb) and the transcription factor called E2F. The

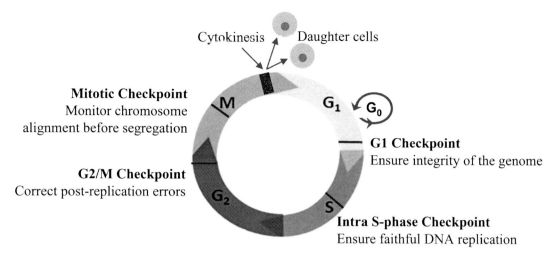

Cytokinesis   Daughter cells

Mitotic Checkpoint
Monitor chromosome
alignment before segregation

M G_1 G_0

G1 Checkpoint
Ensure integrity of the genome

G2/M Checkpoint
Correct post-replication errors

G_2 S

Intra S-phase Checkpoint
Ensure faithful DNA replication

Cell Cycle Checkpoint, Fig. 1 The cell cycle checkpoints in mammalian cells

hypophosphorylated form of Rb is active and represses cell cycle progression by inhibiting E2F, which is necessary for S phase entry. Phosphorylation of Rb blocks its inhibition on E2F and brings about the G1 phase progression or G1-S transition. In early G1 phase, increased expression of cyclin D in conjunction with CDK4 or CDK6 (depending on the cell types) leads to Rb phosphorylation. In late G1 phase, Rb is phospharylated by cyclin E/CDK2 complex. Phospharylation of Rb and subsequent release of E2F facilitates the transcription of late G1 genes to get ready for DNA synthesis and S-phase entry. Besides this positive regulation, G1 checkpoint is also negatively regulated by a family of proteins called cyclin-dependent kinase inhibitors (CKIs), which have a function in inhibiting the cyclin/CDK complexes. In mammalian cells, there are two major families of CKIs – INK4 family (selectively for CDK4 and CDK6) and the CIP/KIP family (has a broader range of inhibition).

In addition to the above pathway, another control of the G1 checkpoint is through the tumor suppressor p53 and its negative regulator MDM2. p53 Activation can cause G1 growth arrest via the CIP family member p21Cip1. This pathway, which also works in G2 checkpoint, plays an important regulatory role in DNA repair, senescence, and apoptosis.

Intra S-phase checkpoint

Strict control of S-phase is important to ensure the genome stability and precise transmission of genetic information. The intra S-phase checkpoints monitor DNA damage, coordinate DNA repair pathways, and cause transient and reversible inhibition of the DNA replication during the whole S phase. They are activated when the replication fork stalls which can help preventing the conversion of primary DNA damages into lethal lesions such as DNA double strand breaks.

There are two major checkpoint pathways in human that are initiated by the sensor proteins ATR or ATM, which delays the cell cycle either through the downstream signal cascade of Chk1(-Chk2)/cdc25a/CDK2 or ATM/MRN/SMC1. In the first pathway, it is often triggered by the formation of single-stranded DNA (ssDNA) in replication fork as a result of uncoupling between DNA unwinding and DNA synthesis. ssDNA signals the recruitment of ATR to the stalled forks then activates downstream mediator and transduces the signal to Chk1/2. Phospharylated Chk1/2 then activating other downstream proteins/factors, such as cdc25 and CDK2/cyclin A, to control several cellular processes including cell cycle delay, prevention of late replication origins from firing, and the activation of DNA repair pathways. In the second pathway, ATM is recruited to sites of DNA damage by a component of the double strand break repair complex MB-N. ATM then phosphorylates another component of the MRN complex called NBS1 as well as the cohesin complex SMC1 and lead to s-phase delay with mechanisms that are poorly understood.

G2/M checkpoint

G2/M checkpoint is located at the end of G2 phase, which controls the entry into mitosis. It checks a number of factors such as the completion of DNA replication and the genomic integrity before cell division starts. Genomic DNA often contains damaged parts prior to mitosis, which makes G2 checkpoint an important control in preventing transmission of damages to daughter cells. It is especially critical in repairing some lethal damages such as DNA double strand break, which can be repaired precisely by homologues recombination using the intact DNA sequences in sister chromatids as template.

The G2-M transition is regulated by the cdc2/Cyclin B complex. Under favorable conditions, it is activated by the mitosis promoting factor (MPF) for further cell progression. It is maintained in an inactive state by the tyrosine kinases Wee1 and Myt1 as the negative control. Once DNA damages are recognized by the sensory protein kinases DNA-PK or ATM, two parallel signal cascades can be activated and induce growth arrest by inactivating cdc2/Cyclin B. The first pathway is through Chk1/2 kinases and its downstream target cdc25, which prevents cdc2 activation and rapidly inhibits G2-M progression. A second pathway, which is slower, is through the tumor suppressor p53. P53 regulates multiple downstream players such as $p21^{cip1}$, 14-3-3, and Gadd45, which inactivate cdc2-cyclin B by different mechanisms, to arrest the cell cycle progression.

Mitotic checkpoint

The mitotic checkpoint, also known as the spindle assembly checkpoint, occurs at the metaphase/anaphase transition to ensure that all the chromosomes are aligned at the mitotic plate and a bipolar spindle is formed.

The central element in this checkpoint is the anaphase promoting complex (APC) which is highly conserved across different eukaryotes. In its activated form, APC can target many cyclins for degradation, which in turn triggers the signal cascade leading to the cuts of the cohesin complex that holds sister chromatids together. APC is negatively controlled by MAD1/2, BUB1/3, BUBR2, and the centromere protein E (CENP-E). Under favorable cellular conditions when chromosomes are correctly aligned, the checkpoint signal that inhibits APC is silenced and renders the latter to target cyclin B for destruction and inactivates CDK1, thereby promoting exit from mitosis and initiating anaphase. Followed by cytokinesis, the cell splits into two cells and the daughter cells enter into G1 to start a new cell cycle.

Checkpoints and cancer

Checkpoint failure often causes accumulation of genome mutations and rearrangements, which is a major factor in the development of many diseases including cancer. Cell cycle arrest with its vital role in maintaining genome stability is the most important barrier to prevent uncontrolled proliferation, the hallmarks of cancer. Many tumor suppressors are in fact components of the cell cycle checkpoints, such as p53, p16, ATM, and BRCA1/2. Mutation or loss of these tumor suppressors is common in cancer cell, which provides growth advantages over adjacent normal cells that are regulated by growth signals.

Cell cycle checkpoints are also one of the most important targets in cancer drug development, which can enhance the efficacy of DNA damage related therapies. Most tumor cells have defects in their G1 checkpoint pathway, and therefore rely more on the efficient S and G2 phase checkpoints for repairing DNA damages and cell survival. Modulating the S and G2/M checkpoints has emerged as an attractive therapeutic strategy for anticancer therapy. Various inhibitors selectively targeting the key players in S or G2/M but not G1 checkpoints, such as Chk1, have been developed and showed promising effects in enhancing the conventional chemotherapy and radiotherapy. Checkpoint inhibition has become an area of intense interest in cancer biology and is continue growing.

References

Malumbres M, Barbacid M (2009) Cell cycle, CDKs and cancer: a changing paradigm. Nat Rev Cancer 9(3):153–166

Reinhardt HC, Yaffe MB (2009) Kinases that control the cell cycle in response to DNA damage: Chk1, Chk2, and MK2. Curr Opin Cell Biol 21(2):245–255

Sclafani RA, Holzen TM (2007) Cell cycle regulation of DNA replication. Annu Rev Genet 41:237–280

Cell Locomotion

▶ Migration

Cell Motility

▶ Migration

Cell Movement

▶ Motility

β-Cell Tumor of the Islets

▶ Insulinoma

Cell-Free Circulating Nucleic Acids (cfDNA)

▶ Circulating Nucleic Acids

Cell-Free Nucleic Acids in Plasma and Serum (CNAPS), Circulating Tumor DNA (ctDNA)

▶ Circulating Nucleic Acids

Cellular Antigens

▶ CD Antigens

Cellular Immunotherapy

▶ Adoptive Immunotherapy

Cellular Self-cannibalism

▶ Autophagy

Cellular Self-digestion

▶ Autophagy

Central Neurocytoma

▶ Neurocytoma

Central Neurofibromatosis

▶ Neurofibromatosis 2

Centroblastic

▶ Diffuse Large B-Cell Lymphoma

Centrocytic (Mantle Cell) Lymphoma

▶ Mantle Cell Lymphoma

Centrosome

Kenji Fukasawa
Molecular Oncology Program, H. Lee Moffitt
Cancer Center and Research Institute, Tampa, FL,
USA

Synonyms

Major microtubule organizing center; MTOC;
SPB; Spindle pole body

Definition

The centrosome is a nonmembranous organelle
(1–2 μm in diameter) normally localized at the
periphery of nucleus, and its primary function is to
nucleate and anchor microtubules.

Characteristics

Structure and Function

The centrosome in mammalian cells consists of a
pair of centrioles and the surrounding protein
aggregates consisting of a number of different
proteins (known as pericentriolar material;
PCM). The centrioles in the pair structurally differ
from each other; one with a set of appendages at
the distal ends (mother centriole) and another
without appendages (daughter centriole). These
appendages are believed to be important for
nucleating and anchoring microtubules. The
daughter centriole acquires the appendages in
late G2-phase of the cell cycle. As the primary
function of the centrosome is to nucleate and
anchor microtubules, centrosomes organize the
cytoplasmic microtubule network during interphase, which is involved in vesicle transport,
proper distribution of small organelles, and establishment of cell shape and polarity. In mitosis,
centrosomes become the core structures of spindle
poles and direct the formation of mitotic spindles
(Fig. 1).

Centrosome Duplication

Upon cytokinesis, each daughter cell receives
only one centrosome. Thus, the centrosome, like
DNA, must duplicate once prior to the next mitosis. In other words, cells have either one
unduplicated or two duplicated centrosomes at
any given time point during the cell cycle. Since
DNA and centrosome are the only two organelles
that undergo semiconservative duplication once
in a single cell cycle, cells are equipped with a
mechanism that coordinates these two events,
likely to ensure these two organelles to duplicate
once, and only once. In late G1/early S-phase, the
centrosome initiates duplication by physical separation of the paired centrioles, which is followed
by the formation of a procentriole in the proximity
of each preexisting centriole. During S and G2,
the procentrioles elongate and two centrosomes
continue to mature by recruiting PCM. By late
G2, two mature centrosomes are generated.

The coupling of the initiation of DNA and
centrosome duplication is in part achieved by
late G1-specific activation of cyclin-dependent
kinase 2 (CDK2)/cyclin E. CDK2/cyclin
E triggers initiation of both DNA synthesis and
centrosome duplication. The activation of CDK2/
cyclin E is controlled by the late G1-specific
expression of cyclin E as well as the basal level
expression of p53 and its transactivation target
p21$^{Waf1/Cip1}$ (p21), a potent CDK inhibitor. Several potential targets of CDK2/cyclin E for centrosome duplication have been identified,
including nucleophosmin, Mps1 kinase, and
CP110. For instance, nucleophosmin localizes
between the paired centrioles, likely functioning
in the pairing of the centrioles. CDK2/cyclin
E-mediated phosphorylation promotes dissociation of nucleophosmin from the centriole pairs,
leading to physical separation of the paired centrioles (Fig. 2).

Abnormal Amplification of Centrosomes and Chromosome Instability in Cancer

The presence of two centrosomes at mitosis
ensures the formation of bipolar mitotic spindles.
Since chromosomes are pulled toward each spindle pole, the bipolarity of mitotic spindles is
essential   for   the   accurate   chromosome

a

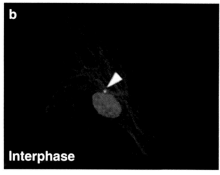

Appendages

Centriole pair

Daughter centriole

Mother centriole

Fibers

Pericentriolar Material (PCM)

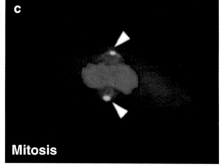

Interphase

Mitosis

Centrosome, Fig. 1 Structure and function of centrosomes. (**a**) The centrosome consists of a pair of centrioles and surrounding protein aggregates (*PCM*). (**b, c**) Mouse embryonic fibroblasts were immunostained for γ-tubulin (one of major centrosomal proteins, *green* – appearing in *yellow*) and α- and β-tubulin (primary constituents of microtubules, *red*). Cells were also counterstained for DNA with DAPI. Panel **b**: interphase cell, panel **c**: mitotic cell

segregation into two daughter cells during cytokinesis. Abrogation of the regulation underlying the numeral homeostasis of centrosomes (i.e., regulation of centrosome duplication) results in abnormal amplification of centrosomes (presence of >2 centrosomes), which in turn increases the frequency of mitotic defects (i.e., formation of >2 spindle poles) and chromosome segregation errors/chromosome instability (see Fukasawa 2005 for the full description of the mechanisms for generation of centrosome amplification). Chromosome instability has been recognized as a hallmark of cancer and contributes to multistep carcinogenesis by facilitating the accumulation of genetic lesions required for the acquisition of various malignant phenotypes. To date, a number of studies have shown that centrosome amplification is a frequent event in almost all types of solid tumors, including breast, bladder, brain, bone, liver, lung, colon, prostate, pancreas, ovary,

testicle, cervix, gallbladder, bile duct, adrenal cortex, and head and neck squamous cell, to name a few. Centrosome amplification has also been observed in certain cases of leukemia and lymphoma. Many studies have also shown the strong association between the occurrence of centrosome amplification and a high degree of aneuploidy. Thus, centrosome amplification can be reasonably considered as a major contributing factor for chromosome instability in cancer (Fig. 3).

Loss of Tumor Suppressor Proteins and Centrosome Amplification

In view of carcinogenesis, it is important to mention that loss or inactivating mutation of certain tumor suppressor proteins, most notably p53 and BRCA1, results in centrosome amplification. For both p53 and BRCA1, they were initially implicated in the control of centrosome duplication and

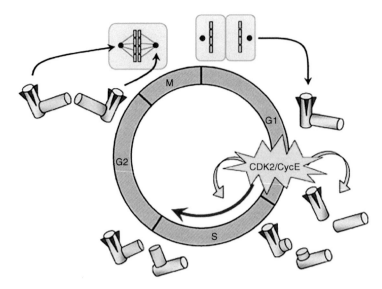

Centrosome, Fig. 2 The centrosome/centriole duplication cycle. Late G1-specific activation of CDK2/cyclin E triggers initiation of both DNA and centrosome duplication. Centrosome duplication begins with the physical separation of the paired centrioles, which is followed by formation of procentrioles. During S- and G2-phases, procentrioles elongate and two centrosomes progressively recruit PCM. In late G2, the daughter centriole of the parental pair acquires appendages (shown as *red* wedges), and two identical centrosomes are generated. During mitosis, two duplicated centrosomes form spindle poles and direct the formation of bipolar mitotic spindles. Upon cytokinesis, each daughter cell receives one centrosome

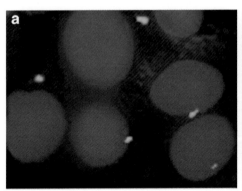

Normal bladder epithelium

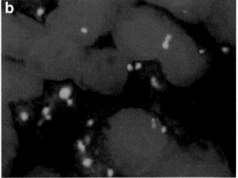

G3 bladder tumor

Centrosome, Fig. 3 Representative immunostaining images of centrosome amplification in human cancer. The touch preparations of G3 tumor grade bladder cancer specimens and the adjacent normal bladder epithelium samples were subjected to immunostaining for γ-tubulin (centrosome, *green*) and counterstained for DNA with DAPI (*blue*). No centrosome amplification can be seen in normal bladder epithelium (**a**), while a high frequency of centrosome amplification in the G3 tumors (**b**)

numeral homeostasis of centrosomes by the observations that centrosome amplification and consequential mitotic aberrations were frequent in the embryonic fibroblasts (as well as various tissues) of p53-null mice as well as mice harboring BRCA1 mutation, which implies that the destabilization of chromosomes due to centrosome amplification contributes to the cancer susceptibility phenotype associated with loss or mutational inactivation of p53 as well as BRCA1.

Centrosome Amplification and Cancer Chemotherapy

In cells inhibited for DNA synthesis (i.e., by exposure to DNA synthesis inhibitors such as aphidicolin (Aph) or hydroxyurea (HU)), centrosomes undergo multiple rounds of duplication in the absence of DNA synthesis, resulting in abnormal amplification of centrosomes. However, this phenomenon preferentially occurs when p53 is either mutated or lost. In the presence of wild-type p53, centrosome duplication is also blocked by exposure to DNA synthesis inhibitors; p53 is upregulated upon prolonged exposure to Aph or HU, leading to the transactivation of p21, which in turn blocks the initiation of centrosome duplication via continuous inhibition of CDK2/cyclin E. In contrast, in cells lacking p53, p21 fails to be upregulated in response to the cellular stress imposed by DNA synthesis inhibitors, allowing "accidental" activation of CDK2/cyclin E, which triggers the initiation of centrosome duplication. Considering the high frequency of p53 mutation in human cancer, it is important to address the effect of commonly used anticancer drugs targeting S-phase (DNA replication) on centrosomes. When p53-null cells were exposed to subtoxic concentrations of the S-phase targeting chemotherapeutic agents (i.e., 5-′-fluorouracil, arabinoside-C), centrosome amplification was efficiently induced. Moreover, after removal of drugs, these cells resumed cell cycling and suffered dramatic destabilization of chromosomes. This finding may be significant in the context of cancer chemotherapy using the S-phase targeting drugs. During chemotherapy, not all cells in tumors receive a maximal dose of drugs – such cells may not be killed, but only arrested for cell cycling. If these cells harbor p53 mutations, centrosome amplification occurs during the drug-induced cell cycle arrest. Upon cessation of chemotherapy, these cells resume cell cycling in the presence of amplified centrosomes and suffer significant mitotic aberrations and chromosome instability, which increases the risk of acquiring further malignant phenotypes. This may in part explain why the recurrent tumors after chemotherapy are often found to be more malignant than the original tumors. Many S-phase targeting anticancer drugs have been found to be effective, and there is no doubt that DNA duplication should be one of the major targets for future development of more effective anticancer drugs. However, the possibility that the S-phase targeting drugs may exacerbate a chromosome instability phenotype by inducing centrosome amplification should be taken into consideration.

Another important issue to be addressed is the concept of centrosome duplication as a target of cancer chemotherapy. Like DNA replication, centrosome duplication occurs only in proliferating cells. Inhibition of centrosome duplication will not only suppress centrosome amplification and chromosome instability but also block cell division and possibly induce cell death – cells with one centrosome fail to form bipolar mitotic spindles, and are often undergo cell death. Moreover, in contrast to genotoxic drugs which impose an increased rate of secondary mutations through interfering with DNA metabolisms, such side effects will likely be minimal in the protocol designed to block centrosome duplication.

Cross-References

▶ Genomic Imbalance
▶ Microtubule-Associated Proteins

References

Bennett RA, Izumi H, Fukasawa K (2004) Induction of centrosome amplification and chromosome instability in p53-null cells by transient exposure to sub-toxic levels of S-phase targeting anti-cancer drugs. Oncogene 23:6823–6829

Deng CX (2002) Roles of BRCA1 in centrosome duplication. Oncogene 21:6222–6227

Fukasawa K (2005) Centrosome amplification, chromosome instability and cancer development. Cancer Lett 230:6–19

Hinchcliffe EH, Sluder G (2002) Two for two: Cdk2 and its role in centrosome doubling. Oncogene 21:6154–6160

Tarapore P, Fukasawa K (2002) Loss of p53 and centrosome hyperamplification. Oncogene 21:6234–6240

Ceramide

Katrin Anne Becker and Erich Gulbins
Department of Molecular Biology, University of
Duisburg-Essen, Essen, Germany

Definition

Ceramide belongs to the group of sphingolipids and is constituted by the amide ester of the sphingoid base D-erythro-sphingosine and a fatty acid of C_{16} through C_{32} chain length. At present, the differential biological function of different ceramide species is unknown and, thus, the term ceramide is used collectively to represent all long-chain ceramide molecules.

Characteristics

Formation of Ceramide
Ceramide molecules are very hydrophobic and exclusively present in membranes. Sphingomyelin, the choline-ester of ceramide, is hydrolyzed by acid, neutral, and alkaline sphingomyelinases to release ceramide. Ceramide is also de novo synthesized via a pathway involving the serine-palmitoyl-CoA transferase and a variety of partly cell type-specific ceramide synthases. Under some circumstances, ceramide can be also formed from sphingosine by a reverse activity of the acid ceramidase.

Ceramide-Induced Changes of Biological Membranes
The formation of ceramide within biological membranes results in a dramatic change of the biophysical properties of the lipid bilayer. Ceramide molecules have the tendency to self-associate and to form small ceramide-enriched membrane microdomains. These membrane microdomains spontaneously fuse to large ceramide-enriched membrane macrodomains that constitute a very hydrophobic and stable membrane domain. Furthermore, ceramide molecules seem to compete with and displace cholesterol from membrane domains. Ceramide-enriched membrane platforms serve to reorganize and cluster/aggregate receptor molecules in the membrane resulting in a very high density of receptors within a small area of the cell membrane. At least for some receptors, the transmembranous domain of the receptor determines its preferential partitioning in ceramide-enriched membrane platforms.

Ceramide-enriched membrane macrodomains are also involved in the recruitment or exclusion, respectively, of intracellular signaling molecules that mediate transmission of signals into the cell via a particular receptor. In general, clustering of receptors in ceramide-enriched membrane domains serves to amplify a weak primary signal. For instance, it was shown that ceramide-enriched membrane platforms amplify CD95 signaling ~100-fold.

Ceramide in Receptor-Mediated Signaling
Death receptors, in particular CD95 or DR5, activate the acid sphingomyelinase and trigger the translocation of the enzyme onto the extracellular leaflet of the cell membrane. Translocation of the acid sphingomyelinase onto the extracellular leaflet of the cell membrane may occur by fusion of intracellular vesicles that are mobilized upon receptor stimulation with the cell membrane. Surface exposure and stimulation of the acid sphingomyelinase results in very rapid release of ceramide in the cell membrane. Ceramide forms membrane platforms and mediates clustering of the death receptors, which is required for the induction of cell death via these receptors (Fig. 1).

However, ceramide is not only involved in the mediation of apoptotic stimuli, but also many other stimuli trigger the release of ceramide including CD40, CD20, FcγRII, CD5, LFA-1, CD28, TNFα, Interleukin-1 receptor, PAF-receptor, infection with *P. aeruginosa*, *S. aureus*, *N. gonorrhoeae*, Sindbis-Virus, Rhinovirus, γ-irradiation, measles virus, UV-light, doxorubicin, cisplatin, gemcitabine, disruption of integrin-signaling, and some conditions of developmental death. In general, ceramide is often involved in acute and strong stress responses.

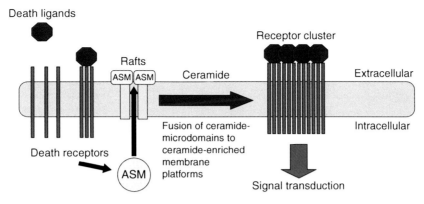

Ceramide, Fig. 1 Receptors cluster in ceramide-enriched membrane domain to transmit signals into cells. The interaction of a ligand with its receptor results in translocation of the acid sphingomyelinase onto the extracellular leaflet and a concomitant release of ceramide. Ceramide spontaneously forms ceramide-enriched microdomains that fuse to large ceramide-enriched macrodomains. These domains trap activated receptor molecules finally resulting in clustering of many receptor molecules within a small area of the cell membrane. The high density of receptor molecules and associated intracellular molecules amplifies the primarily weak signal, permits the generation of a strong signal and, thus, efficient transmission of the signal into the cell (Modified from A. Carpinteiro et al. Cancer Letters)

Signaling Molecules Regulated by Ceramide

Ceramide interacts with and activates phospholipase A_2, kinase suppressor of Ras (KSR; identical to ceramide-activated protein kinase), ceramide-activated protein serine-threonine phosphatases, protein kinase C isoforms, and c-Raf-1. Furthermore, ceramide inhibits the potassium channel Kv1.3 and calcium release activated calcium (CRAC) channels. Lysosomal ceramide specifically binds to and activates cathepsin D resulting in the translocation of cathepsin D into the cytoplasm and induction of cell death via the proapoptotic proteins Bid, Bax, and Bak.

Ceramide in Mitochondria and Cell Death

Besides a function of ceramide in the plasma membrane and lysosomes for the induction of cell death, ceramide is also generated in mitochondria via the de novo synthesis pathway, a reverse activity of the ceramidase and/or activity of the acid sphingomyelinase. Although at present the function of ceramide in the mediation of mitochondrial proapoptotic events is poorly defined, it was suggested that C_{16}-ceramide molecules form large channels in mitochondrial membranes that may permit the exit of cytochrome c from mitochondria to execute death. Further, ceramide has been shown to be involved in the apoptosis response of mitochondria.

Genetic Evidence for a Function of Ceramide in Apoptosis

The role of the acid sphingomyelinase and ceramide for CD95 and DR5-triggered apoptosis was evidenced by studies on acid sphingomyelinase-deficient cells or mice, respectively, that revealed a resistance of these cells to CD95- and DR5-triggered apoptosis, but also γ-irradiation- and UV-light- or P. aeruginosa-triggered cell death.

Ceramide in γ-Irradiation- and UV-A Light-Induced Apoptosis

The acid sphingomyelinase and ceramide are critically involved in the response of cells to γ-irradiation. Animals or cells lacking the acid sphingomyelinase are resistant to γ-irradiation-induced cell death. In particular, endothelial cells in acid sphingomyelinase-deficient mice are resistant to γ-irradiation.

Ceramide also plays a critical role for UV-light induced apoptosis. UV-A and UV-C light activate the acid sphingomyelinase, trigger the release of ceramide, and the formation of large ceramide-enriched membrane domains in the cell membrane to initiate death.

Ceramide and Chemotherapy

In addition to a central role of ceramide in γ-irradiation-induced cell death, ceramide is also critically involved in the induction of cell death by at least some chemotherapeutic drugs. Thus, doxorubicin-, cisplatin-, taxol-, and gemcitabine-induced cell death of malignant and nonmalignant cells requires expression of the acid sphingomyelinase, release of ceramide and/or the formation of ceramide-enriched membrane platforms to trigger death.

Short Chain Ceramide

Short chain ceramide molecules composed of a fatty acid chain with C_2 through C_{12} length are water soluble and, thus, very much differ from endogenous long ceramide molecules (C_{16}-C_{32}). However, they are very efficient reagents to kill tumor cells *in vitro*. Cationic pyridinium-ceramides seem to accumulate in mitochondria of tumor cells and may, thus, serve as a new class of antitumor reagents, although at present no convincing concepts are available to selectively target tumor cells *in vivo* and to avoid effects of short chain ceramide on normal cells.

Cross-References

▶ Adriamycin
▶ Cisplatin
▶ Death Receptors
▶ Docetaxel
▶ Gemcitabine

References

Fulda S, Debatin KM (2006) Extrinsic versus intrinsic apoptosis pathways in anticancer chemotherapy. Oncogene 25:4798–4811
Gulbins E, Kolesnick RN (2003) Raft ceramide in molecular medicine. Oncogene 22:7070–7077
Jaffrezou JP, Laurent G (2004) Ceramide: a new target in anticancer research? Bull Cancer 91:E133–E161
Kolesnick RN, Goni FM, Alonso A (2000) Compartmentalization of ceramide signaling: physical foundations and biological effects. J Cell Physiol 184:285–300
Ogretmen B, Hannun YA (2004) Biologically active sphingolipids in cancer pathogenesis and treatment. Nat Rev Cancer 4:604–616

Cervical Cancers

Jiro Fujimoto
Department of Obstetrics and Gynecology, Gifu University School of Medicine, Gifu City, Japan

Definition

The regions of the uterus are the corpus and the cervix. Cancer originating from the cervix is defined as cancer of the cervix. When cancers are simultaneously detected in the cervix and corpus, squamous cell carcinoma (SCC) is designated as a cancer of the cervix, and adenocarcinoma is designated as a cancer of the corpus. When cancer occupies both the cervix and vagina without the junctional area (the fornix), the cancer extending to the exocervix is recognized as a cancer of the cervix. Thus, cervical cancer is defined apart from cancer of the uterine corpus (cancer of the uterine endometrium) and cancer of the vagina.

Characteristics

The main gynecological cancers originate from the cervix, endometrium, and ovary. Among them, cervical cancer is the most common malignancy in women.

Main risk factors are the following:

- Young age at first intercourse, especially shortly after the menarche
- High number of sexual partners
- High number of sexual partners of the partner
- High number of children
- Excessive douching

Smoking appears to increase the incidence of SCC, but not of adenocarcinoma or adenosquamous carcinoma. Immunosuppression by smoke-derived nicotine and its metabolite cotinine in the cervical mucus may enhance the effects of sexually transmitted disease (STD) including human papillomavirus (HPV) infection. Most epidemiological risk factors for cervical cancer are associated with STDs. HPV induces an STD, human venereal condyloma, which is associated with cervical, vaginal, and vulvar dysplasia, and invasive carcinomas. HPV particles and DNA, especially HPV-16, HPV-18, and HPV-33, are detected in cervical and vulvar dysplasia and in invasive carcinomas. Additionally, it has been demonstrated that HPV transforms human cell lines. HPV infection of the cervix is a main etiology of cervical cancer.

Symptoms

Main symptoms of cervical cancer are the following:

- Vaginal bleeding, which may be recognized as postmenopausal bleeding, irregular menses, or postcoital bleeding
- Abnormal vaginal (watery, purulent, or mucoid) discharge

In advanced cases, corresponding local symptoms occur. A Pap smear even in unsymptomatic cases is useful for the early detection of cervical dysplasia and cancers. Among women over the age of 18 who have had sexual intercourse, high-risk women should be screened at least yearly.

Pathology

Histopathological types in cervical cancers are mainly SCC and adenocarcinoma, which account for about 90% of all cervical cancers (adenosquamous carcinoma, glassy cell carcinoma, adenoid cystic carcinoma, adenoid basal carcinoma, carcinoid, small cell carcinoma, and undifferentiated carcinoma also occur). SCCs are keratinizing or nonkeratinizing in most cases and may be verrucous, condylomatous, papillary, or lymphoepithelioma-like carcinomas in a few cases. Adenocarcinomas are classified into mucinous, endometrioid, clear cell, serous, and mesonephric adenocarcinomas; mucinous adenocarcinomas are subclassified with endocervical type into adenoma malignum and villoglandular papillary adenocarcinoma and intestinal type adenocarcinoma.

Staging

Clinical staging represents the degree of advancement of the tumor and is defined by the FIGO classification established in 1994 and by the TNM classification of malignant tumors set by the UICC in 1997 as follows (classified by FIGO [TNM]):

- Stage 0 (Tis): carcinoma in situ (preinvasive carcinoma).
- Stage I (T1): cervical carcinoma confined to the uterus.
- Stage II (T2): tumor invades beyond the uterus but not to the pelvic wall or to the lower third of the vagina.
- Stage III (T3): tumor extends to the pelvic wall and/or involves the lower third of the vagina and/or causes hydronephrosis or nonfunctioning kidney.
- Stage IVA (T4): tumor invades the mucosa of the bladder or rectum and/or extends beyond the true pelvis.
- Stage IVA (Ml): distant metastasis.

Stage IA (T1a) has been further classified by microinvasive depth and width into stage IA1 (T1a1) (depth of stromal invasion ≤ 3 mm, horizontal spread ≤ 7 mm) and stage IA2 (T1a2) (depth of stromal invasion >3 mm, ≤ 5 mm; horizontal spread ≤ 7 mm). Stage IB (T1b) has been further classified by tumor size into stage IB1 (T1b1) (greatest dimension ≤ 4 cm) and stage IB2 (T1b2) (greatest dimension >4 cm). In cases staged IA2 (T1a2) or less advanced, colposcopically directed biopsy in the transformation zone of the cervix, endocervical curettage, or cervical conization is required.

Prognosis

Unfavorable prognostic factors include younger age, advanced clinical stage, certain histopathological types, vessel permeation, large tumor volume, parametrium involvement, and lymph node metastasis. Nodal metastasis is an especially critical prognostic factor after curative resection. Vascular endothelial growth factor (VEGF)-C and osteopontin contribute to the aggressive lymphangitic metastasis in uterine cervical cancers. Platelet-derived endothelial cell growth factor (PD-ECGF) contributes to the advancement of metastatic lesions as an angiogenic factor. PD-ECGF, VEGF-C, and osteopontin levels in metastatic lesions are prognostic indicators. Furthermore, serum PD-ECGF level reflects the status of advancement of cervical cancers and is recognized as a novel tumor marker for both SCC and adenocarcinoma of the cervix, while the tumor marker SCC is well known only as an indicator for SCC of the cervix. VEGF-C and osteopontin contribute to the aggressive lymphangitic metastasis in uterine cervical cancers.

Therapy

The treatment for cervical cancer consists mainly of surgery and radiation. Chemotherapy is performed in combination with surgery and/or radiation for advanced cases, and immunotherapy is an adjuvant treatment for surgery, radiation, and chemotherapy. The standard treatment for carcinoma in situ is cervical conization or total hysterectomy. The standard treatment for microinvasive carcinoma stage IA (T1a) is modified radical hysterectomy regardless of regional lymphadenectomy. The standard surgical treatment for invasive carcinoma is radical hysterectomy with regional lymphadenectomy. Although oophorectomy can be avoided in some cases during the reproductive period, ovarian metastasis must be considered especially in adenocarcinoma of the cervix. When oophorectomy is avoided, the ovary is better shifted out of radiation area. For patients who undergo oophorectomy, hormone replacement therapy can be useful. In more advanced cases, extended radical hysterectomy or pelvic exenteration is appropriate. After surgery external irradiation is followed in some cases. The standard radiotherapy without surgery for invasive carcinoma is intracavitary and/or external irradiation. Neoadjuvant therapy (chemotherapy) has been tried in order to make surgery more successful, and concurrent radiochemotherapy has been tested for the purpose of enhancing the effect of radiation.

References

Fujimoto J, Sakaguchi H, Hirose R et al (1999) Clinical implication of expression of platelet-derived endothelial cell growth factor (PD-ECGF) in metastatic lesions of uterine cervical cancers. Cancer Res 59:3041–3044

Fujimoto J, Sakaguchi H, Aoki I et al (2000) The value of platelet-derived endothelial cell growth factor as a novel predictor of advancement of uterine cervical cancers. Cancer Res 60:3662–3665

Fujimoto J, Toyoki H, Sato E et al (2004) Clinical implication of expression of vascular endothelial growth factor-C in metastatic lymph nodes of uterine cervical cancers. Br J Cancer 91:466–469

Cetuximab

Definition

Trade name Erbitux® is a chimeric IgG1κ monoclonal antibody specifically binding to epidermal growth factor receptor (EGFR).

Epidermal growth factor receptor (EGFR) (EGFR, ErbB-1, HER1 in humans) is the cell-surface receptor for members of the epidermal growth factor family (EGF family) of extracellular protein ligands. Ligands which induce activation of EGFR are epidermal growth factor and transforming growth factor-α, for example. Upon activation by its growth factor ligands, EGFR undergoes a transition from an inactive monomeric form to an active homodimer. EGFR dimerization stimulates its intrinsic intracellular protein-tyrosine kinase activity resulting in activation of several signal transduction cascades which lead to DNA synthesis and cell proliferation. EGFR mutations can lead to EGFR overexpression or overactivity and consequently result in uncontrolled cell division. Mutations of EGFR have been identified in several types of cancer, such as lung cancer and colorectal cancer.

Cetuximab is approved for the treatment of irinotecan-refractory metastatic colorectal cancer (CRC) in combination with irinotecan and for the treatment of locoregional advanced head and neck cancer as monotherapy or in combination with radiation. Dermatological toxicity is a limiting factor to the use of cetuximab. Hypersensitivity to cetuximab (rash, urticaria, fever, dyspnea, and hypotension) is frequent in certain regions and has been related to the presence of IgE antibodies specific for an oligosaccharide, galactose-α-1,3-galactose, which is present on the Fab portion of the cetuximab heavy chain. Cetuximab is under investigation in combination with chemotherapy (carboplatin or irinotecan) in pretreated triple-negative breast cancer (TNBC) with advanced disease. TNBC is estrogen and progesterone receptor negative, as well as HER2 negative, and therefore is not amenable to treatment with hormonal therapy or with trastuzumab. Furthermore, several phase I/II studies with cetuximab in combination with cytotoxic agents or with other targeted therapies, such as trastuzumab, are currently ongoing.

Cross-References

▶ Epidermal Growth Factor Inhibitors

See Also

(2012) Monoclonal antibody therapy. In: Schwab M (ed) Encyclopedia of cancer, 3rd edn. Springer, Berlin/Heidelberg, pp 2367–2368. doi: 10.1007/978-3-642-16483-5_3823

CG Antigens

▶ Cancer Germline Antigens

CGP57148

▶ Imatinib
▶ STI-571

C-HA-RAS1

▶ *HRAS*

Charged Particle Therapy

▶ Proton Beam Therapy

Checkpoint with Forkhead and RING Finger Domain Protein

▶ CHFR

Chelation Therapy

▶ Chelators as Anti-Cancer Drugs

Chelators as Anti-Cancer Drugs

David B. Lovejoy, Yu Yu and Des R. Richardson
Department of Pathology, University of Sydney, Sydney, NSW, Australia

Synonyms

Chelation therapy

Definition

Iron is an element fundamental for life. Many vital cellular processes such as energy metabolism and DNA synthesis consist of reactions that require catalysis by iron-containing proteins. These proteins include cytochromes, and ribonucleotide reductase (RR). The latter is more significant in the context of cellular proliferation due to its role in catalyzing the rate-limiting step of DNA synthesis. Ultimately, the importance of iron is highlighted by the fact that iron-deprivation leads to G_1/S cell cycle arrest and ► apoptosis. Cancer cells, in particular, have a higher iron requirement because of their rapid rate of proliferation. In order to satisfy their iron requirement, some cancer cells have altered iron metabolism. In addition, iron chelators also demonstrate the ability to inhibit growth of aggressive tumors such as ► neuroblastoma. For these reasons, iron-deprivation through iron chelation is seen as an exploitable therapeutic strategy for the treatment of cancer.

Characteristics

Iron Metabolism in Cancer Cells

In order to attain more iron, cancer cells have higher numbers of the transferrin receptor-1 molecule (TfR1) on their cell surface. The TfR1 binds the serum iron transport protein, transferrin (Tf). Hence, cancer cells are able to bind more Tf and, thus, take up iron at a greater rate than their normal counterparts. This is reflected by the ability of tumors to be radiolocalized using a radioisotope of gallium, ^{67}Ga, which binds to the iron-binding site on Tf for delivery via TfR1. ^{67}Ga can bind to iron-binding sites of Tf due to the similar atomic properties between gallium(III) and iron(III). Additionally, gene therapy by the administration of antisense TfR1 targeted to the sequences of *TfR1* mRNA also showed selective anti-cancer activity, further demonstrating the importance of TfR1 in mediating cancer cell growth.

Apart from TfR1 up-regulation, the expression of the iron-storage protein ferritin is also often altered in neoplastic cells, especially neuroblastoma (NB) and breast carcinoma. In childhood NB, serum ferritin levels are elevated at stages III and IV of the disease. In a longitudinal study, it was found that the elevated level was associated with a markedly poorer prognosis of the disease. In addition, serum ferritin levels also exceeded the normal limit in ► hepatocellular carcinoma and were found to be directly related to axillary lymph node status, presence of metastatic disease (► metastasis), and clinical stages of breast cancer.

Desferrioxamine, an Iron Chelator with Some Anti-Cancer Activity

Desferrioxamine (DFO) is a natural ligand secreted by the bacterium *Streptomyces pilosus* to selectively sequester iron for biological use (Fig. 1). DFO is used clinically for the treatment of iron overload disorders such as the transfusion-related iron overload in β-thalassemia.

DFO is active against aggressive tumors including NB and leukemia in cell culture and clinical trials. The cytotoxicity of DFO in vitro was prevented by co-incubation of the cells with iron or iron saturated DFO, indicating that its antiproliferative activity was due to the depletion of cellular iron. Furthermore, DFO induces a block in cell cycle progression. Therefore, it was proposed that the mechanism of action of DFO involved the depletion of cellular iron, leading to the inhibition of ribonucleotide reductase for DNA synthesis and cell cycle arrest. In human NB cells, 5 days of exposure to DFO resulted in approximately 90% cell death. In contrast, the effect of DFO was minimal on non-NB cells, suggesting that it had selective anti-NB activity. A clinical trial showed that seven of nine NB patients had up to 50% reduction in bone marrow infiltration after a course of DFO administered for 5 days. Other clinical trials using DFO as a single agent and in combination with other chemotherapeutic drugs confirmed the anti cancer potential of this chelator. However, in some animal studies and clinical trials, DFO was found to exhibit limited or no activity.

DFO also suffers a number of limitations as a result of its highly hydrophilic nature. It has poor gastrointestinal absorption and a short plasma half-life of about 12 min due to rapid metabolism. As a result, DFO is not orally active and needs to be administered via subcutaneous infusion for prolonged periods ranging from 8 to 12 h for five to seven times per week. The prolonged infusion results in pain and swelling, which consequently leads to poor patient compliance. DFO is also expensive to produce.

Despite these limitations and mixed results in clinical trials, DFO nonetheless provides "proof of principle" that iron chelation therapy may be specific and useful for cancer treatment.

Chelators as Anti-Cancer Drugs, Fig. 1 Chemical structures of the iron chelators desferrioxamine (DFO), N, N',N''-tris(2-pyridylmethyl)-*cis,cis*-1,3,5-triaminocyclohexane (tachpyridine or tachpyr), 3-aminopyridine-2-carboxaldehyde-thiosemicarbazone (3-AP or Triapine®), 2-hydroxy-1-napthaldehyde isonicotinoyl hydrazone (311), and di-2-pyridylketone-4,4,-dimethyl-3-thiosemicarbazone (Dp44mT) showing coordination to iron (Fe) through pyridyl nitrogen, aldimine nitrogen, and thionyl sulfur donor atoms

Other Chelators with Anti-Cancer Potential

The limitations of DFO as an anti-cancer agent have encouraged the search for other active iron-chelating drugs against cancer.

Other experimental iron chelators include Triapine® (3-AP; Fig. 1), an iron-binding thiosemicarbazone-based drug currently in clinical trials for cancer therapy. Triapine® is a chelator that binds iron via sulfur and two nitrogen donor atoms and is suggested to be one of the most potent inhibitors of RR yet identified. In clinical trials, high doses of Triapine® (160 mg/m^2/day) resulted in dose-limiting toxicities, including reduction in white blood cells, jaundice, nausea, and vomiting. Lower doses of Triapine® administered as a 96-h iv infusion at 120 mg/m^2/day every 2 weeks was found to be well tolerated. In clinical trials with patients with advanced cancer, Triapine® was combined with the cytotoxic cancer drug gemcitabine, which also targets DNA synthesis. Of the 22 patients examined after treatment with gemcitabine and Triapine®, three were observed to have an objective response, and one patient had evidence of tumor reduction. In this trial, Triapine® was suggested to cause oxidation of hemoglobin to met-hemoglobin. This may have led to or contributed to the hypoxia, acute hypotension, and electrocardiogram changes in patients receiving this chelator. An asymptomatic myocardial infarction was also observed in one individual administered Triapine® and this may also be related to its oxidative effects. Triapine® continues to be examined in clinical trials, particularly in combination with standard chemotherapy drugs. However, these deleterious effects must be considered when designing future studies with compounds of this class.

Tachpyridine (or tachpyr; Fig. 1) is a novel chelator based upon the framework of the triamine *cis,cis*-1,3,5-triaminocyclohexane. Tachpyridine is cytotoxic to cultured bladder cancer cells with an activity approximately fifteen times greater than that of DFO. Although tachpyridine has the potential to chelate a number of metals, including calcium(II), magnesium(II), manganese(II), copper(II), and zinc(II), toxicity studies on tachypyridine complexes suggest that iron and zinc depletion mediates its cytotoxic effects.

Similar to Triapine®, Tachpyridine induces apoptotic cell death independent of functional p53 (see section "Iron Chelation and Cell Cycle Control Molecules," below) (▶ p53 family). In addition, tachpyridine-iron complexes produce toxic free radicals (▶ reactive oxygen species), which was also thought to contribute to its anti-tumor activity.

Tachpyridine arrests cells at the G_2 phase, whereas the majority of iron chelators arrest cells at the G_1/S phase due to the inhibition of ribonucleotide reductase. The G_2 phase stage of the cell cycle is particularly sensitive to the effects of radiation. Ionizing radiation increases the sensitivity of tumor cells to the action of tachpyridine. Currently, tachpyridine is in preclinical development with the National Cancer Institute, USA.

PIH Chelators

The most comprehensively assessed alternate chelators for cancer treatment are the pyridoxal isonicotinoyl hydrazone (PIH) analogues. This class of chelators binds iron through the carbonyl oxygen, imine nitrogen, and phenolic oxygen (Fig. 1).

Originally conceived for the treatment of iron overload disorders, several chelators of the PIH class were found to inhibit the growth of cancer cells. In fact, the chelator 311 (Fig. 1) was found to be highly active against a range of cancer cells. These compounds also showed marked ability to remove Fe from cells and prevent cellular Fe uptake from transferrin. The marked anti-cancer activity of chelator 311 was attributed to its relatively high lipophilicity, which facilitates entry into the cell. Indeed, a general trend observed with the PIH analogues was that anti-cancer activity increased as the chelator became more lipophilic. Mechanistically, PIH analogues have multiple modes of anti-cancer activity, aside from chelation of iron and inhibition of ribonucleotide reductase. Some members of the PIH class of chelators (e.g., see section "The DpT Chelators: Dp44mT," below) increase the generation of toxic free radicals (reactive oxygen species) in cancer cells and affect the expression of cell-cycle control molecules (see section "Iron Chelation and Cell Cycle Control Molecules," below). Additional studies with 311 have also shown that it can markedly induce the expression of the metastasis suppressor protein, Drg-1 in tumor cells. The Drg-1 protein is known to play a critical role in suppressing tumor growth and metastasis. Hence, induction of Drg-1 by potent iron chelators such as 311 may significantly contribute to the anti-cancer activity of these analogues.

The DpT Chelators: Dp44mT

The DpT class of chelators are structurally related to PIH analogues, but feature a sulfur donor atom instead of the hydrazone oxygen donor atom (Fig. 1). The chelator Dp44mT has been shown to be the most effective of the DpT series of ligands in terms of anti-cancer activity. It acts with selectivity against tumor cells and has much less effect on the growth of normal cells. Dp44mT also showed high iron chelation efficacy and prevented cellular uptake of iron from iron-labeled Tf. Another mechanism of its action involves the generation of toxic free radicals (reactive oxygen species) when Dp44mT interacts with cellular iron pools.

Initially, in vivo studies of Dp44mT in mice bearing chemotherapy-resistant M109 lung carcinoma showed a reduction in the size of the tumor by 53% after 5-days of treatment. A later investigation also found marked inhibition of the growth of human lung, neuroepithelioma, and melanoma xenografts growing in mice. In fact, a 7-week administration of Dp44mT in mice bearing human melanoma xenografts resulted in the decrease of tumor growth to 8% of that in untreated control mice. At the dose given, no hematological abnormalities were detected, although at a higher dose, myocardial fibrosis was identified. This side effect at a high dose may be due to the marked redox activity of the Dp44mT-iron complex. However, at a lower dose Dp44mT was well tolerated with no hematological abnormalities and less cardiotoxicity. Other studies with Dp44mT showed that it also markedly increased the expression of the metastasis suppressor protein, Drg-1 in tumor cells. Induction of Drg-1 could potentially be a very significant component of the anti-cancer mechanism of Dp44mT. Further development of DpT series chelators is currently underway.

Iron Chelation and Cell Cycle Control Molecules

Iron-deprivation generally leads to G_1/S phase cell cycle arrest as a result of inhibition of ribonucleotide reductase. This has prompted many studies assessing the effect of iron chelation by DFO and chelator 311 on the expression of many cell cycle control molecules, namely, cyclins, ► cyclin dependent kinases (cdks), cdk inhibitors, and P53 (p53 gene family). Consistently, these studies found that iron chelation markedly decreased the expression of ► cyclin D (D1, D2, and D3) and to a lesser extent cyclin A and B. The expression of cdk2 and cdk1, but not cdk4, were also decreased upon iron chelation. These effects were dependent on iron-deprivation, as iron-chelator complexes were unable to induce such effects.

Cyclins D, E, and A and cdks 2, 4, and 6 are involved in progression through the G_1 phase, although cyclin E, A, and cdk2 are also involved in S phase progression. The formation of the cyclin A-cdk2 complex is essential for G_1/S progression. Cyclin B and cdk1, on the other hand, are important for mitosis. During the G_1 phase, cyclin D and E bind to cdk4 and cdk2, respectively, to phosphorylate (phosphorylation) the retinoblastoma protein (pRb) (► retinoblastoma protein, biological and clinical functions). This results in the release of molecules such as the E2F transcription factor from pRb that promotes the expression of genes for S phase. The decrease in the expression of these cyclins upon iron chelation causes hypophosphorylation of pRb, which in turn leads to the G_1/S phase arrest.

In addition to cyclins and cdks, iron chelation also affects the expression of cell cycle modulatory molecules. In particular, iron chelators caused a marked increase in the expression of the cyclin-dependent kinase inhibitor $p21^{WAF1/CIP1}$ (► p21(WAF1/CIP1/SDI1)) at the mRNA level. $P21^{WAF1/CIP1}$ mediates G_1/S phase arrest by directly binding the cyclin-cdk complexes. It was speculated that the increased level of $p21^{WAF1/CIP1}$ upon iron chelation was consistent with its potential role in the G_1/S phase arrest. However, an increase of $p21^{WAF1/CIP1}$ expression only occurred at the mRNA level, with either no change or a decrease in $p21^{WAF1/CIP1}$ protein expression being observed. This was unexpected and it was subsequently demonstrated that $p21^{WAF1/CIP1}$ protein level could be controlled by

proteasomal (► proteasome) degradation after iron chelation.

In contrast, investigations examining p53 showed that its protein expression and DNA-binding activity were increased after chelation. P53 is a tumor suppressor and acts as a transcription factor that is involved in the transcription of a variety of genes involved in cell cycle arrest, differentiation, apoptosis, and DNA repair. An increase in p53 after iron chelation may be the result of a decrease in deoxyribonucleotide levels due to the inhibition of RR activity or changes in intracellular redox status. Despite the fact that $p21^{WAF1/CIP1}$ is a downstream effector of p53, elevated expression of $p21^{WAF1/CIP1}$ upon iron chelation occurs through a p53-independent pathway. The ability of chelators to potentially inhibit tumor cell growth by a p53-independent pathway is significant, since p53 is the most frequently mutated gene in cancer. This also explains why cells with wild type or mutant p53 are similarly sensitive to the growth inhibitory effects of iron chelators. However, the function of increased p53 expression after chelation remains a subject for further investigation.

Conclusions

The demonstration that some iron chelators may be clinically useful for cancer treatment followed on from initial observations that rapid cancer cell proliferation requires iron. Currently, the iron chelator, Triapine®, is being examined in a variety of clinical trials, with focus on a potential role in combination chemotherapy. The search for more effective anti-cancer Fe chelators than DFO has also led to the development of other potent Fe chelators, including Dp44mT and tachpyridine, and significant progress has been made toward understanding their molecular targets. However, further in vivo experiments and pre-clinical studies will be necessary to build upon the promise of these agents.

Cross-References

► Apoptosis
► Cyclin D

▶ Cyclin-Dependent Kinases
▶ Hepatocellular Carcinoma
▶ Metastasis
▶ Neuroblastoma
▶ P53 Family
▶ p21
▶ Proteasome
▶ Reactive Oxygen Species
▶ Retinoblastoma Protein, Biological and Clinical Functions

References

Buss JL, Greene BT, Turner J et al (2004) Iron chelators in cancer chemotherapy. Curr Top Med Chem 4:1623–1635

Kalinowski D, Richardson DR (2005) Evolution of iron chelators for the treatment of iron overload disease and cancer. Pharmacol Rev 57(4):1–37

Le NTV, Richardson DR (2004) Iron chelators with high anti-proliferative activity up-regulate the expression of a growth inhibitory and metastasis suppressor gene: a novel link between iron metabolism and proliferation. Blood 104:2967–2975

Whitnall M, Howard J, Ponka P et al (2006) A class of iron chelators with a wide spectrum of potent anti-tumor activity that overcome resistance to chemotherapeutics. Proc Natl Acad Sci U S A 103:14901–14906

Yu Y, Wong J, Lovejoy DB et al (2006) Chelators at the cancer coalface: desferrioxamine to triapine and beyond. Clin Cancer Res 12:6876–6883

See Also

(2012) Cell cycle. In: Schwab M (ed) Encyclopedia of cancer, 3rd edn. Springer Berlin Heidelberg, p 737. doi: 10.1007/978-3-642-16483-5_994

(2012) Cytochrome c . In: Schwab M (ed) Encyclopedia of cancer, 3rd edn. Springer Berlin Heidelberg, p 1043. doi: 10.1007/978-3-642-16483-5_1458

(2012) Drg-1. In: Schwab M (ed) Encyclopedia of cancer, 3rd edn. Springer Berlin Heidelberg, p 1160. doi: 10.1007/978-3-642-16483-5_1730

(2012) E2F transcription factor . In: Schwab M (ed) Encyclopedia of cancer, 3rd edn. Springer Berlin Heidelberg, p 1183. doi: 10.1007/978-3-642-16483-5_1770

(2012) Lipophilicity. In: Schwab M (ed) Encyclopedia of cancer, 3rd edn. Springer Berlin Heidelberg, p 2058. doi: 10.1007/978-3-642-16483-5_3384

(2012) Ribonucleotide reductase. In: Schwab M (ed) Encyclopedia of cancer, 3rd edn. Springer Berlin Heidelberg, p 3308. doi: 10.1007/978-3-642-16483-5_5102

(2012) Chelator. In: Schwab M (ed) Encyclopedia of cancer, 3rd edn. Springer Berlin Heidelberg, p 755. doi: 10.1007/978-3-642-16483-5_1052

(2012) Phosphorylation. In: Schwab M (ed) Encyclopedia of cancer, 3rd edn. Springer Berlin Heidelberg, p 2870. doi: 10.1007/978-3-642-16483-5_4544

Chemical Biology Screen

▶ Small Molecule Screens

Chemical Carcinogenesis

Joseph R. Landolph, Jr.
Department of Molecular Microbiology and Immunology, and Department of Pathology; Laboratory of Chemical Carcinogenesis and Molecular Oncology, USC/Norris Comprehensive Cancer Center, Keck School of Medicine; Department of Molecular Pharmacology and Pharmaceutical Sciences, School of Pharmacy, Health Sciences Campus, University of Southern California, Los Angeles, CA, USA

Definition

Chemical carcinogenesis (▶ carcinogenesis) is the process of the genesis of a tumor (carcinoma) and the series of sequential steps that occur when lower animals or humans are treated with **chemical carcinogens** that lead to tumor development. After all these steps are accomplished, the physiological mechanisms regulating the control of growth in the normal cells are degraded, and the normal cells are degraded and converted into tumor cells. The tumor cells then grow in an unregulated fashion and evade the host immune system, leading to development of visible tumors.

Characteristics

Normal Cell Types in Animals and the Tumors They Give Rise To

During embryogenesis in mammals (warm-blooded animals), there are three primary germ layers of the early embryo which develop into all the basic cell types, tissues, and organs in the body. These are the ectoderm, the endoderm, and the mesoderm. The ectoderm and endoderm are epithelial layers. Most of the epithelial organs in the body are derived from the endodermal and the ectodermal germ layers. The epidermis of the skin, the corneal epithelium, and mammary glands develop from the ectoderm. The endoderm layer develops into the liver, pancreas, stomach, and intestines. The mesoderm develops into the kidney and linings of male and female reproductive tracts. Three types of cells are important in chemical carcinogenesis. These cell types are (i) epithelial cells, which form the coverings and internal parts of organs; (ii) fibroblasts, which are connective tissue cells derived from primitive mesenchymal cells; and (iii) cells of the hemato-lymphopoietic series, which are derived from the blood-forming elements. These cell types all have special and specific characteristics.

In humans, 92% of the tumors that arise are derived from epithelial cells (epithelial cell tumors). These tumors are called carcinomas. The remaining 8% of the tumors are derived from a combination of tumors derived from fibroblasts, called sarcomas, and tumors derived from white blood cells, called leukemias and lymphomas.

Carcinogens

There are a group of molecules and radiations referred to as "carcinogens." A carcinogen is any molecule, or group of molecules, such as viruses (▶ Virology), or radiation (▶ Radiation carcinogenesis; ▶ radiation oncology), that can cause tumors in lower animals and humans, when they are exposed to this agent. This happens when carcinogens cause normal cells to transform or convert into transformed cells and tumor cells during experiments in vitro, called chemical transformation experiments.

Chemicals referred to as chemical carcinogens (chemical carcinogenesis) can cause tumors in lower animals and in humans exposed to them. Examples of chemical carcinogens are vinyl chloride, aflatoxin B1 (a metabolite and biocide of the fungus, *Aspergillus flavus*) (▶ Aflatoxins), benzo(a)pyrene (a polycyclic aromatic hydrocarbon formed when organic matter is pyrolyzed in the absence of oxygen) (▶ Polycyclic aromatic hydrocarbons), and beta-naphthylamine (an aromatic amine used to manufacture dyestuffs that causes bladder cancer in animals and humans) (▶ Aromatic amine). Nitrosamines are another class of chemical carcinogens. An example is dimethylnitrosamine (DMN). Many nitrosamines are synthetic compounds. Some are believed to form in the stomach of humans when amines (derived from fish in the diet) contact nitrous acid (formed from the nitrate from fertilizer that is used to grow foodstuffs) in the acidic conditions (acid pH) of the stomach. Chemicals in all these classes of carcinogens can cause tumors in humans and in lower mammals.

There are also a number of radiations that cause tumors in humans and lower animals. These include ionizing radiations, such as alpha particles (charged helium nuclei), beta particles (naked electrons), and gamma particles. There are also tumor viruses, consisting of RNA (RNA tumor viruses) and DNA (DNA tumor viruses). When animals are treated with these viruses, tumors are formed. Examples of RNA tumor viruses are the Rous sarcoma virus, the Abelson leukemia virus, and the Kirsten Ras virus. Examples of DNA tumor viruses are the polyoma virus, the SV40 (simian virus 40) (▶ SV40) virus, the ▶ Epstein-Barr virus, and the human papilloma viruses 16 and 18.

Mechanisms of Chemical Carcinogenesis

There are two broad mechanisms of chemical carcinogenesis. In the first type, which we refer to here as "complete ▶ carcinogenesis," a mammal is treated with a large dose of a chemical carcinogen, such as 7,12-dimethylbenz(a) anthracene, and the animals treated eventually develop tumors. Carcinogenesis with complete carcinogens is usually dose dependent, such that the higher doses of carcinogens that the animals are treated with, the high the yield of tumors per animal and in the percentage of animals with tumors.

The second mechanism of chemical carcinogenesis, discovered by Dr. Isaac Berenblum of the Weizmann Institute in Israel, is referred to as "two-step carcinogenesis," or "initiation and promotion." In initiation and promotion experiments, Berenblum treated mice on the skin of their shaved backs with chemical carcinogens at low doses and also with tumor promoters. Berenblum was testing the hypothesis that carcinogenesis was due to irritation and inflammation. Hence, he used croton oil, a product of the plant, *Euphorbia lathyris*, which the plant uses as a biocide against insects. Croton oil is a very irritating substance, which is important in the plant's use of it as a biocide against insects. When mice were treated with low doses of 7,12-dimethylbenz(a) anthracene (DMBA, a carcinogenic PAH), one time, they exhibited no tumors. A second group of animals was treated with the tumor promoter, croton oil, once per week, and the animals also exhibited no tumors. When the mice were treated with a low dose of DMBA, and then once weekly with croton oil, they developed many tumors. If the latter treatment was reversed, i.e., the animals were treated first with croton oil once per week and then later treated with a low dose of DMBA, the animals showed no tumors. If the animals were treated with a low dose of DMBA, then no treatment was performed for a significant amount of time and then the animals were treated with croton oil once per week, the animals also developed a high yield of tumors. In this system, treatment of the animals with the low dose of DMBA is referred to as the "initiation step," and later treatment with croton oil is called the "promotion" step. Initiation is believed to be a genotoxic event, likely a mutation, and is an irreversible step. Initiated cells can be promoted to tumors cells if they are treated with croton oil long enough. The promotion step is believed to be due to the binding of tetradecanoyl-phorbol acetate (TPA, the most active constituent of the mixture of phorbol esters in croton oil), to protein kinase C, triggering signal transduction and cell division in cells bearing mutations in proto-oncogenes. If promotion is interrupted, then tumorigenesis is reversible, i.e., the cellular death rate will equal the cellular growth rate, and the tumor will regress. If promotion is continued long enough, the tumor becomes fixed and will not regress. Eric Hecker of the German Cancer Research Center (Deutsch Krebs Forschung Zentrum) in Heidelberg, Germany, fractionated croton oil used by Berenblum, by high pressure liquid chromatography, and found that TPA was the most active tumor promoter in it.

From experiments with high doses of chemical carcinogens, and experiments with initiation and promotion, we now have evidence that chemical carcinogens such as DMBA cause mutations in proto-oncogenes, such as *RAS* genes, converting them into activated ► oncogenes. In complete carcinogenesis experiments, further mutations in other proto-oncogenes can also occur, leading to activation of additional oncogenes. In addition, activated metabolites of the carcinogens (formed in the animals/mammals by cytochrome P450 or other enzymes of metabolic activation) also cause mutational inactivation of ► tumor suppressor genes or breakage of chromosomes bearing them, leading to loss of these tumor suppressor genes. Together, activation of oncogenes and inactivation of tumor suppressor genes leads to the genesis of tumors in mammals.

Insights into Mechanisms of Chemical Carcinogenesis from Studies of Chemically Induced Neoplastic Transformation

Studies of the abilities of chemical carcinogens to convert normal cells into tumor cells in cell culture dishes have given us substantial insight into the molecular mechanisms of chemical carcinogenesis. In cell culture, normal fibroblasts and normal epithelial cells grow if they are fed properly, until they eventually fill the culture dish, and touch each other. Growth then ceases. This process is called contact inhibition of cell division. Cells can then be removed from the cell culture dish with a protease called trypsin, diluted, and replated into new cell culture dishes. This process can be repeated many times, until the population of total cells has undergone sixty population doublings. At this point, the cells senesce (► Senescence and immortalization) or die. This is due to progressive shortening of telomeres (► Telomerase), structures at the end of chromosomes, with each successive DNA replication and cell division. Telomere shortening acts as a cellular and molecular "clock," to mark the lifetime of the cell. This process aids in the control of the normal physiology of the organism, by removing old cells which accumulated many mutations, which could eventually lead to cancer.

▶ Chemically induced cell transformation is the process by which normal cells are treated with chemical carcinogens in vitro in a cell culture dish or flask, and then their growth control mechanisms degrade, converting or transforming them into transformed cells. There are two mechanisms by which cells can be converted by chemical carcinogens into transformed cells. Firstly, cells can be treated with genotoxic (DNA damaging) (▶ Genetic toxicology) chemical carcinogens. Many of these genotoxic carcinogens are mutagens (▶ Mutation rate). These carcinogens either already are direct mutagens (rare), or more commonly they are pre-carcinogens, and can be converted into mutagenic proximate carcinogens by cytochrome P450 enzymes or other enzyme systems that activate the pre-carcinogens into mutagens. The carcinogens benzo(a)pyrene, aflatoxin B1, and nitrosamines are all examples of pre-carcinogens that are metabolically activated into mutagens by various types of cytochrome P450 enzymes.

Most pre-carcinogens are hydrophobic (fat loving) compounds that would bioaccumulate in the body and cause alterations in the properties of enzymes and membranes in cells. Mammals must therefore derive strategies to eliminate hydrophobic pre-carcinogens. The cytochrome P450 enzyme systems, and other enzyme systems, have evolved in order to metabolize these pre-carcinogens, to make them water-soluble, so they can be excreted in the urine and removed from the body. Since these compounds are inherently chemically inert, a necessary first chemical reaction step has evolved, in which cytochrome P450 enzymes attack pre-carcinogens like benzo(a)pyrene (BaP) with molecular oxygen and reducing equivalents (NADPH and NADH) to generate epoxides and diol epoxides from it. These metabolites are mutagens, and this step results in "metabolic activation." In a second step, which is closely coupled to the first step, these active metabolites are reacted with and conjugated to, molecules of water by the enzyme, epoxide hydrolase, converting them to trans-dihydrodiols and tetraols, which are highly water-soluble, so they are excreted in the urine. The small amount of epoxides and diol epoxides derived from BaP

then bind covalently to DNA bases, resulting in mutations in proto-oncogenes, activating them into oncogenes, and mutations in tumor suppressor genes, inactivating them.

In a second mechanism of ▶ carcinogenesis, chemicals called "non-genotoxic carcinogens" transform normal cells into tumor cells in a different way, by non-mutagenic mechanisms. One example is the chemical, 5-azacytidine, a chemical analog of a normal base. 5-azacytidine binds to DNA methyltransferases (▶ Methylation), inhibiting them. This results in a loss of methylation of the cytidine in DNA. If this occurs in quiescent proto-oncogenes, then these can become transcriptionally activated, leading to cell transformation. Other examples of non-genotoxic carcinogens include hormones, such as testosterone and estrogen. Higher steady-state levels of testosterone and estrogen are believed to lead to aberrantly high numbers of cell divisions in the prostate and breast tissue. The resultant spontaneous mutations that occur are believed to lead to prostate cancer and breast cancer, respectively.

The process by which a normal cell is converted into tumor cells, or chemically induced neoplastic transformation (neoplastic cell transformation), occurs in four steps. In the first step, when cells are treated with mutagenic chemical carcinogens, there occur mutations in proto-oncogenes, activating them to oncogenes, and mutations in tumor suppressor genes inactivating them. The cells then develop the ability to grow in multilayers and form foci. This is particularly true for fibroblastic cells, less so for epithelial cells. This first step in cell transformation is called morphological cell transformation or focus formation. Further genetic changes occur in the transformed cells. The second step that occurs is that the cells become immortal and do not die or senesce. Some activated oncogenes (v-myc) can cause cells to become immortal. This step would be called transformation to cellular immortality. In the third step, cells develop the ability to grow in soft agar, in three-dimensional suspension. This step is called anchorage-independent cell transformation or transformation to anchorage independence. A final step that develops after further genetic change is that cells develop the ability to form tumors when injected into athymic (nude) mice. This step is called neoplastic transformation, or the ability of cells to be transformed so that they form neoplasms or new growths, which we call tumors. Often, a number of activated oncogenes, two or more, may cooperate together to perturb normal cellular physiology to cause neoplastic transformation of normal rodent or human cells in culture. It is now believed by scientists that activation of proto-oncogenes into oncogenes, and inactivation of tumor suppressor genes, such that approximately eight total genes are genetically altered, leads to the aberrant expression of approximately 150 genes or more in the tumor cells. This then leads to neoplastic transformation of cells in culture and hence to chemical cacinogenesis in the animal. We believe that chemically induced neoplastic transformation is a good model for how cells in the animal become converted (transformed) into tumor cells when the animal is treated with chemical carcinogens.

Significance of Chemical Carcinogenesis

The significance of the process of chemical carcinogenesis is twofold. Firstly, the assay for chemical carcinogenesis in lower animals, usually mice and rats, can be used to test chemicals to determine whether they are carcinogens by virtue of their ability to induce tumors in mice and rats. Those chemicals that are able to cause a reproducible, dose-dependent induction of tumors in mice and/or rats are presumed to be human carcinogens. This presumption is due first to the relationship that rodents and humans are both warm-blooded animals and mammals. As such, their biochemistry and physiology are similar. In addition, many chemical carcinogens were first found to be carcinogenic in rodent carcinogenesis bioassays and later found to be carcinogens in humans. Almost all carcinogens that have been shown to be carcinogenic in humans are also carcinogenic in rodents (aflatoxin B1, vinyl chloride, asbestos, cigarette smoke, asbestos, polycyclic aromatic hydrocarbons).

Secondly, the process of chemical carcinogenesis as studied in rodents has led to unique insights into the mechanisms of carcinogenesis. Investigators frequently use whole animal carcinogenesis bioassays to study how proto-oncogenes are activated into oncogenes, how tumor suppressor genes are inactivated by chemical carcinogens, and how oncogene activation and tumor suppressor gene inactivation lead to induction of tumors in mammals. Studying the mechanisms of carcinogenesis in rodents has also led to the identification of agents that interfere with this process and may eventually be used to prevent the induction of cancer in humans.

Cross-References

▶ Adenocarcinoma
▶ Aflatoxins
▶ Alkylating Agents
▶ Amplification
▶ Anchorage-Independent
▶ Aneuploidy
▶ Aromatic Amine
▶ Benzene and Leukemia
▶ Benzpyrene
▶ Bladder Cancer
▶ Cancer
▶ Cancer Causes and Control
▶ Cancer Epidemiology
▶ Carcinogen Metabolism
▶ Carcinogenesis
▶ Chemically Induced Cell Transformation
▶ Chromium Carcinogenesis
▶ Chromosomal Instability
▶ Class II Tumor Suppressor Genes
▶ Detoxification
▶ DNA Damage
▶ DNA Oxidation Damage
▶ Embryonic Stem Cells
▶ Endocrine Oncology
▶ Endocrine-Related Cancers
▶ Epidemiology of Cancer
▶ Epigenetic
▶ Epigenomics
▶ Epstein-Barr Virus
▶ Fibrosarcoma
▶ Genetic Toxicology
▶ Genomic Instability
▶ Helicobacter Pylori in the Pathogenesis of Gastric Cancer

▶ Hematological Malignancies, Leukemias, and Lymphomas

▶ Hepatitis B Virus

▶ Hepatitis B Virus x Antigen-Associated Hepatocellular Carcinoma

▶ Hepatocellular Carcinoma: Etiology, Risk Factors, and Prevention

▶ Hexavalent Chromium

▶ Hormonal Carcinogenesis

▶ Hypomethylation of DNA

▶ Inflammation

▶ Lung Cancer

▶ Lung Cancer Epidemiology

▶ Mesenchymal Stem Cells

▶ Methylation

▶ Mutation Rate

▶ Oncogene

▶ Oxidative Stress

▶ Polycyclic Aromatic Hydrocarbons

▶ Radiation Carcinogenesis

▶ Radiation Oncology

▶ Reactive Oxygen Species

▶ Renal Cancer Pathogenesis

▶ Repair of DNA

▶ Senescence and Immortalization

▶ SV40

▶ Telomerase

▶ Toxicological Carcinogenesis

▶ Tumor Suppressor Genes

▶ Virology

References

Landolph JR Jr, Xue W, Warshawsky D (2006) Whole animal carcinogenicity bioassays, Chapter 2. In: Warshawsky D, Landolph JR Jr (eds) Molecular carcinogenesis and the molecular biology of human cancer. CRC/Taylor and Francis Group, Boca Raton, pp 25–44

Verma R, Ramnath J, Clemens F et al (2005) Molecular biology of nickel carcinogenesis: identification of differentially expressed genes in morphologically transformed C3H/10T1/2 Cl 8 mouse embryo fibroblast cell lines induced by specific insoluble nickel compounds. Mol Cell Biochem 255:203–216

Warshawsky D (2006) Carcinogens and mutagens, Chapter 1. In: Warshawsky D, Landolph JR Jr (eds) Molecular carcinogenesis and the molecular biology of human cancer. CRC/Taylor and Francis Group, Boca Raton, pp 1–24

Warshawsky D, Landolph JR Jr (2006) Overview of human cancer induction and human exposure to carcinogens, Chapter 13. In: Warshawsky D, Landolph JR Jr (eds) Molecular carcinogenesis and the molecular biology of human cancer. CRC/Taylor and Francis Group, Boca Raton, pp 289–302

Weinberg RW (2007) Multi-step tumorigenesis, Chapter 11. In: Ram A (ed) The biology of cancer. Garland Science/Taylor and Francis Group, LLC, New York, pp 399–462

See Also

(2012) Carcinogen. In: Schwab M (ed) Encyclopedia of cancer, 3rd edn. Springer, Berlin/Heidelberg, p 644. doi:10.1007/978-3-642-16483-5_839

(2012) Cytochrome P450 enzymes. In: Schwab M (ed) Encyclopedia of cancer, 3rd edn. Springer, Berlin/Heidelberg, p 1043. doi:10.1007/978-3-642-16483-5_1465

(2012) Epithelial cell. In: Schwab M (ed) Encyclopedia of cancer, 3rd edn. Springer, Berlin/Heidelberg, pp 1291-1292. doi:10.1007/978-3-642-16483-5_1958

(2012) Fibroblasts. In: Schwab M (ed) Encyclopedia of cancer, 3rd edn. Springer, Berlin/Heidelberg, p 1398. doi:10.1007/978-3-642-16483-5_2176

(2012) Genotoxic. In: Schwab M (ed) Encyclopedia of cancer, 3rd edn. Springer, Berlin/Heidelberg, p 1540. doi:10.1007/978-3-642-16483-5_2393

(2012) Mutagen. In: Schwab M (ed) Encyclopedia of cancer, 3rd edn. Springer, Berlin/Heidelberg, p 2409. doi:10.1007/978-3-642-16483-5_3907

(2012) Mutation. In: Schwab M (ed) Encyclopedia of cancer, 3rd edn. Springer, Berlin/Heidelberg, p 2412. doi:10.1007/978-3-642-16483-5_3911

(2012) Neoplastic cell transformation. In: Schwab M (ed) Encyclopedia of cancer, 3rd edn. Springer, Berlin/Heidelberg, p 2474. doi:10.1007/978-3-642-16483-5_4013

(2012) Proto-oncogenes. In: Schwab M (ed) Encyclopedia of cancer, 3rd edn. Springer, Berlin/Heidelberg, pp 3107-3108. doi:10.1007/978-3-642-16483-5_6656

(2012) Tumor. In: Schwab M (ed) Encyclopedia of cancer, 3rd edn. Springer, Berlin/Heidelberg, p 3792. doi:10.1007/978-3-642-16483-5_6014

(2012) Tumor promoter. In: Schwab M (ed) Encyclopedia of cancer, 3rd edn. Springer, Berlin/Heidelberg, p 3800. doi:10.1007/978-3-642-16483-5_6047

(2012) Two-step carcinogenesis. In: Schwab M (ed) Encyclopedia of cancer, 3rd edn. Springer, Berlin/Heidelberg, p 3821. doi:10.1007/978-3-642-16483-5_6071

Chemical Genetic Screen

▶ Small Molecule Screens

Chemical Mutagenesis

▶ Genetic Toxicology

Chemically Induced Cell Transformation

Joseph R. Landolph, Jr.
Department of Molecular Microbiology and
Immunology, and Department of Pathology;
Laboratory of Chemical Carcinogenesis and
Molecular Oncology, USC/Norris
Comprehensive Cancer Center, Keck School of
Medicine; Department of Molecular
Pharmacology and Pharmaceutical Sciences,
School of Pharmacy, Health Sciences Campus,
University of Southern California, Los Angeles,
CA, USA

Definition

Chemically induced cell transformation is the series of sequential steps that occur when mammalian cells are treated with ► Chemical Carcinogenesis and converted into tumor cells.

The intermediate cell phenotypes (cell properties) are acquired one at a time, including first cellular immortality, then morphological transformation (change in cell shape, leading to crisscrossing of cells in abnormal patterns), then anchorage independence (growth of cells as colonies or balls of cells in three-dimensional suspension of agar, without attachment to the plastic dishes cells are usually grown on), and finally neoplastic transformation (neoplastic cell transformation), or the ability of cells to form tumors when injected into nude (athymic) mice.

Characteristics

Normal Growth of Normal Cells

In the mammalian organism (warm-blooded animal), there are many types of cells. In general, these cell types are divided into (i) epithelial cells, which form the coverings of organs; (ii) fibroblasts, which are connective tissue cells; and (iii) cells of the hemato-lymphopoietic series, which are derived from the blood-forming elements. These cell types all have special and specific characteristics.

These three general cell types can be grown outside the body in an artificial situation, in cell culture medium in plastic cell culture dishes. This constitutes a model system in which the physiology of cells can be studied outside of the complicated conditions of the body. When grown in cell culture, epithelial cells and fibroblastic cells attach to the cell culture dish, by virtue of the surface charge of the cell relative to that of the plastic of the cell culture dish. These normal fibroblastic and epithelial cells must anchor to the bottom inside of the cell culture dish in order to be able to replicate their DNA and divide. This is called anchorage dependence of cell growth. These cells continue to grow if fed properly with cell culture medium, containing 5–10% fetal calf serum and cell culture medium. Cell culture medium consists of sugars, amino acids, salts, and buffers, along with an indicator to detect the acidity of the culture medium (pH indicator), all dissolved in water.

In cell culture, the normal fibroblasts and normal epithelial cells continue to grow if they are fed properly, until they eventually fill the culture dish, and touch each other. Growth then ceases. This process is called contact inhibition of cell division. These cells can then be removed from the cell culture dish with a protease called trypsin, diluted and replated into new cell culture dishes. This process can be repeated many times, until the population of total cells has undergone approximately 60 population doublings. This is called the "Hayflick limit," after Dr. Leonard Hayflick, who discovered it. At this point, the cells undergo cellular senescence (► Senescence and immortalization) or die. This is due to progressive shortening of telomeres (► Telomerase), structures at the end of chromosomes that are progressively shortened with each successive DNA replication and cell division. Hence, telomere shortening acts as a cellular and molecular "clock," to mark the lifetime of the cell. This process is believed to aid in the control of the normal physiology of the organism, and to rid it of old cells which have many mutations, which could eventually lead to cancer. If these normal cells are injected into mice lacking an immune system (athymic or "nude" mice), they will not grow and will not form tumors.

In contrast, cells of the hemato-lymphopoietic series grow in three-dimensional suspension (the blood) in vivo. Hence, when grown in vitro (outside the body), these cells must also be grown in three-dimensional suspension. A common practice is to grow the cells in varying concentrations of agar. When injected into athymic or "nude" mice, these normal cells, whether cells of the hematopoietic (red blood cell) or lymphoid (white blood cell) lineages, will not form tumors.

Carcinogens

There are a group of chemical molecules, radiations, and viruses referred to as "carcinogens." A carcinogen is any chemical or group of molecules, such as viruses (▶ Virology) or radiation (▶ Radiation carcinogenesis; ▶ radiation oncology) that can cause tumors in lower animals when they are treated with this agent. These agents can also cause normal cells to transform (convert) into transformed cells and tumor cells.

There are a group of chemicals referred to as chemical carcinogens (▶ Chemical carcinogenesis). These are specific chemicals that can cause tumors in animals treated with them. Examples of these are vinyl chloride, aflatoxin B1 (a metabolite and biocide of the fungus, *Aspergillus flavus*) (▶ Aflatoxins), benzo(a)pyrene (a polycyclic aromatic hydrocarbon formed when organic matter is burned in the absence of oxygen) (▶ Polycyclic aromatic hydrocarbons), and beta-naphthylamine (an aromatic amine used to manufacture dyestuffs that causes bladder cancer in animals and humans) (▶ Aromatic amine). Another class of chemical carcinogens is called nitrosamines. An example is dimethylnitrosamine (DMN). Many nitrosamines are synthetic compounds. Some are believed to form in the stomach of humans when amines (derived from fish in the diet) contact nitrous acid (formed from the nitrate from fertilizer that is used to grow foodstuffs) in the acidic conditions (acid pH) of the stomach. Chemicals in all these classes of carcinogens can cause tumors in humans and in lower mammals.

There are also a number of radiations (radiation carcinogenesis) that can cause tumors in humans and lower animals. These include ionizing radiations, such as alpha particles (charged helium nuclei), beta particles (naked electrons), and gamma particles.

In addition, there are also tumor viruses, consisting of RNA (RNA tumor viruses) and DNA (DNA tumor viruses). When animals are treated with these viruses, tumors are formed. Examples of RNA tumor viruses are the Rous sarcoma virus, the Abelson leukemia virus, and the Kirsten Ras virus. Examples of DNA tumor viruses are the polyoma virus, the SV40 (simian virus 40) (▶ SV40) virus, the ▶ Epstein-Barr virus, and the human papilloma viruses 16 and 18.

Chemically Induced Cell Transformation: Description and Mechanisms

Chemically induced cell transformation is the process by which normal cells are treated with chemical carcinogens in vitro in a cell culture dish or flask, and they then convert or transform into transformed cells. There are two mechanisms by which cells can be converted by chemical carcinogens into transformed cells. Firstly, cells can be treated with genotoxic (DNA damaging) (▶ Genetic toxicology) chemical carcinogens. Many of these genotoxic carcinogens are mutagens (▶ Mutation rate). These carcinogens either already are direct mutagens (rare), or more commonly they are pre-carcinogens, and can be converted into mutagenic proximate carcinogens by cytochrome P450 enzymes or other enzyme systems that activate the pre-carcinogens into mutagens. The pre-carinogens benzo(a)pyrene, aflatoxin B1, and nitrosamines are all examples of pre-carcinogens that are metabolically activated into mutagens by various types of cytochrome P450 enzymes.

The perspective for this process is that most pre-carcinogens are hydrophobic (fat loving) compounds that would bioaccumulate in the body and cause alterations in the properties of enzymes and membranes in cells. Hence, the organism must derive a strategy to eliminate these hydrophobic pre-carcinogens. Therefore, the cytochrome P450 enzyme systems, and other enzyme systems, have evolved in order to metabolize these pre-carcinogens, to make them water-soluble, so they can be excreted in the urine and removed from the body. Since these compounds are inherently chemically inert, a necessary first chemical reaction step has evolved, in which cyto-chrome P450 enzymes first attack pre-carcinogens like benzo(a)pyrene (BaP) with molecular oxygen and reducing equivalents (NADPH and NADH) to generate epoxides and diol epoxides from it. These metabolites are mutagens, and this step results in "metabolic activation." In a second step, which is closely coupled to the first step, these active metabolites are reacted with and conjugated to, molecules of water by the enzyme, epoxide hydrolase, converting them to trans-dihydrodiols and tetraols, which are highly water-soluble, so they are excreted in the urine. The small amount of epoxides and diol epoxides derived from BaP then go on to bind covalently to DNA bases, resulting in mutations in proto-▶ oncogenes, acti-vating them into ▶ oncogenes, and mutations in ▶ tumor suppressor genes, inactivating them.

In a second mechanism of ▶ carcinogenesis, chemicals called "non-genotoxic carcinogens" transform normal cells into tumor cells in a differ-ent way, by non-mutagenic mechanisms. One example is the chemical, 5-azacytidine, a chemi-cal analog of a normal base. 5-azacytidine binds to DNA methyltransferases (▶ Methylation), inhibiting them. This results in a loss of methyla-tion of the cytidine in DNA. If this occurs in quiescent proto-oncogenes, then these can become transcriptionally activated, leading to cell transformation. Other examples of non-genotoxic carcinogens include hormones, such as testosterone and estrogen. Higher steady-state levels of testosterone and estrogen are believed to lead to aberrantly high numbers of cell divisions in the prostate and breast tissue. The resultant spontaneous mutations that occur are believed to lead to prostate cancer and breast cancer, respectively.

The process of chemically induced neoplastic transformation, or the process of generating a tumor cell, falls into at least four steps. In the first step, when cells are treated with mutagenic chemical carcinogens, there occur mutations in proto-oncogenes, activating them to oncogenes, and mutations in tumor suppressor genes, inactivating them. The cells then develop the abil-ity to grow in multilayers and form foci. This is particularly true for fibroblastic cells, less so for epithelial cells. This first step in cell transforma-tion is called morphological cell transformation or focus formation. Further genetic changes occur in the transformed cells. The second step that occurs is that the cells become immortal and do not die or senesce. Some activated oncogenes (v-myc) can cause cells to become immortal. This step would be called transformation to cellular immortality. A third step that occurs is that the cells develop the ability to grow in soft agar, in three-dimensional suspension. This step is called anchorage-independent cell transformation or transformation to anchorage independence. A final step that develops after further genetic change is that the cells develop the ability to form tumors when injected into athymic (nude) mice. This step is called neoplastic transformation, or the ability of the cell to be transformed so that it forms neoplasms or new growths, which we call tumors. Often, a number of activated oncogenes, two or more, may cooperate together to perturb normal cellular physiology to cause neoplastic transformation of normal rodent or human cells in culture.

Significance of Chemically Induced Neoplastic Transformation

The significance of the process of chemically induced neoplastic transformation is two-fold. Firstly, the assay for chemically induced morpho-logical cell transformation can be used an assay to detect chemical carcinogens. Those chemicals that have the ability to induce foci of morpholog-ically transformed cells are highly likely to be able to induce tumors in animals. Hence, this assay can detect chemical carcinogens by virtue of their ability to induce foci of morphologically transformed cells.

Secondly, the study of chemically induced morphological, anchorage-independent, and neo-plastic transformation in vitro is frequently used as a model system to study the process of chem-ical carcinogenesis. Investigators frequently use these assays to study how proto-oncogenes are activated into oncogenes, and how tumor suppres-sor genes are inactivated by chemical carcino-gens, and how oncogene activation and tumor suppressor gene inactivation leads to induction of morphological transformation, cellular immortality, anchorage-independent transformation, and neoplastic transformation.

Cross-References

▶ 5-aza-2′ Deoxycytidine
▶ Aflatoxins
▶ Anchorage-Independent
▶ Aromatic Amine
▶ Benzpyrene
▶ Cancer
▶ Carcinogen Metabolism
▶ Carcinogenesis
▶ Cervical Cancers
▶ Chemical Carcinogenesis
▶ Class II Tumor Suppressor Genes
▶ DNA Damage
▶ Epigenetic
▶ Epithelium
▶ Epstein-Barr Virus
▶ Estrogenic Hormones
▶ Genetic Toxicology
▶ *KRAS*
▶ Methylation
▶ Mutation Rate
▶ Oncogene
▶ Polycyclic Aromatic Hydrocarbons
▶ Radiation Carcinogenesis
▶ Radiation Oncology
▶ Senescence and Immortalization
▶ SV40
▶ Telomerase
▶ Tumor Suppressor Genes
▶ Virology

References

Kumar V, Abbas AK, Fausto N (2005) Neoplasia, Chapter 7. In: Robbins and Cotran's pathologic basis of disease, 7th edn. Elsevier Saunders, Philadelphia, pp 269–342

Landolph JR Jr (2006) Chemically induced morphological and neoplastic transformation in C3H/10T1/2 mouse embryo cells, Chapter 9. In: Warshawsky D, Landolph JR Jr (eds) Molecular carcinogenesis and the molecular biology of human cancer. CRC/Taylor and Francis Group, Boca Raton, pp 199–220

Pitot HC, Dragan YP (2001) Chemical carcinogenesis, Chapter 8. In: Klaassen CD (ed) Casarett and Doull's toxicology, the basic science of poisons, 6th edn. McGraw-Hill, New York, pp 239–320

Verma R, Ramnath J, Clemens F et al (2005) Molecular biology of nickel carcinogenesis: identification of differentially expressed genes in morphologically transformed C3H/10T1/2 Cl 8 mouse embryo fibroblast cell lines induced by specific insoluble nickel compounds. Mol Cell Biochem 255:203–216

Weinberg RW (2007) Multi-step tumorigenesis, Chapter 11. In: The biology of cancer. Garland Science/Taylor and Francis Group, LLC, New York, pp 399–462

See Also

(2012) Anchorage-independent cell transformation. In: Schwab M (ed) Encyclopedia of cancer, 3rd edn. Springer, Berlin/Heidelberg, p 173. doi:10.1007/978-3-642-16483-5_263

(2012) Carcinogen. In: Schwab M (ed) Encyclopedia of cancer, 3rd edn. Springer, Berlin/Heidelberg, p 644. doi:10.1007/978-3-642-16483-5_839

(2012) Cellular senescence. In: Schwab M (ed) Encyclopedia of cancer, 3rd edn. Springer, Berlin/Heidelberg, p 743. doi:10.1007/978-3-642-16483-5_1019

(2012) Cytochrome P_{450} enzymes. In: Schwab M (ed) Encyclopedia of cancer, 3rd edn. Springer, Berlin/Heidelberg, p 1043. doi:10.1007/978-3-642-16483-5_1465

(2012) Contact inhibition of cell division. In: Schwab M (ed) Encyclopedia of cancer, 3rd edn. Springer, Berlin/Heidelberg, p 974. doi:10.1007/978-3-642-16483-5_1324

(2012) Epithelial cell. In: Schwab M (ed) Encyclopedia of cancer, 3rd edn. Springer, Berlin/Heidelberg, pp 1291–1292. doi:10.1007/978-3-642-16483-5_1958

(2012) Fibroblasts. In: Schwab M (ed) Encyclopedia of cancer, 3rd edn. Springer, Berlin/Heidelberg, p 1398. doi:10.1007/978-3-642-16483-5_2176

(2012) Genotoxic. In: Schwab M (ed) Encyclopedia of cancer, 3rd edn. Springer, Berlin/Heidelberg, p 1540. doi:10.1007/978-3-642-16483-5_2393

(2012) Morphological cell transformation. In: Schwab M (ed) Encyclopedia of cancer, 3rd edn. Springer, Berlin/Heidelberg, p 2373. doi:10.1007/978-3-642-16483-5_3836

(2012) Mutagen. In: Schwab M (ed) Encyclopedia of cancer, 3rd edn. Springer, Berlin/Heidelberg, p 2409. doi:10.1007/978-3-642-16483-5_3907

(2012) Mutation. In: Schwab M (ed) Encyclopedia of cancer, 3rd edn. Springer, Berlin/Heidelberg, p 2412. doi:10.1007/978-3-642-16483-5_3911

(2012) Neoplastic cell transformation. In: Schwab M (ed) Encyclopedia of cancer, 3rd edn. Springer, Berlin/Heidelberg, p 2474. doi:10.1007/978-3-642-16483-5_4013

(2012) Transformation. In: Schwab M (ed) Encyclopedia of cancer, 3rd edn. Springer, Berlin/Heidelberg, pp 3757–3758. doi:10.1007/978-3-642-16483-5_5913

Chemoattractant Cytokine

▶ Chemokines

Chemoattraction

Jose Luis Rodríguez-Fernández
Departamento de Microbiología Molecular y
Biología de las Infecciones, Centro de
Investigaciones Biológicas, Madrid, Spain

Synonyms

Directed migration; Directed motility

Definition

Chemoattraction is the process whereby a cell
detects a chemical gradient of a ligand called
chemoattractant and, as a consequence, gets oriented and subsequently moves in the direction
from a low to a high concentration of the
chemoattractant. Chemoattraction is controlled by
specific chemoattractant receptors that are able to
detect selectively these ligands. Chemoattraction is
called *chemotaxis* or *haptotaxis* when the chemical
gradient of the chemoattractant is presented to the
cell either in a soluble or bound to a substrate form,
respectively. As it is not clear which one of these
two types of motile processes takes place in vivo, it
is more appropriate to refer to these directional
motile processes with the more general term of
chemoattraction.

Characteristics

Chemoattractants use specific chemoattractant
receptors to guide different migratory cell types
toward specific sites in the organism. These receptors, upon binding to the chemoattractant, transform the information of this ligand in intracellular
signals that result in the movement of the migratory

cell toward the positions where chemoattractant is
present at high concentration. Therefore, the analysis, in a specific context, in one hand, of the type
of chemoattractant receptors expressed by a certain
migratory cell and, on the other hand, the position
in the organism of the chemoattractants recognized
by these receptors, allow to make predictions on
the potential tissues where this cell can be attracted.
Upon arrival to the position where the
chemoattractant is at a high concentration, adhesive receptors may contribute to slow down
(function largely performed by selectin adhesive
receptors for cells in blood vessels) and eventually
attach (cells use integrin receptors for this function
in most cell types) the cells to these sites.

Chemoattractants can be conveniently classified according to the type of receptor that they
bind. In this regard, the first and the largest group
include chemoattractants that bind members of the
G protein-coupled receptor (GPCR) superfamily.
In this first group is included the family of
▶ chemokines. A second group is formed by
chemoattractants that bind tyrosine kinase receptors (e.g., epidermal growth factor (EGF),
platelet-derived growth factor (PDGF)). A third
group includes ligands that bind receptors different of the two aforementioned families (e.g., laminin and fibronectin, which bind integrin
receptors). This article deals mainly with the
chemokines because they have been the
chemoattractant family most studied in relation
to ▶ cancer and ▶ metastasis.

Chemokines

Chemokines (chemotactic chemokines) are a family of peptides (60–100 amino acids (aa)) that
includes some 50 members (Fig. 1). Based on
the number and spacing of the conserved cysteine
(C) residue in the N-terminus of the protein,
chemokines are subdivided into four families (C,
CC, CXC, CX_3C), where X is any intervening
amino acid between the cysteines. Chemokine
receptors transmit intracellular signals that can
control either chemoattraction or other functions
(Fig. 1). The chemokine receptors (some 20 members) are included in the G protein-coupled receptor (GPCR) superfamily. They are classified based
on the class of chemokines that they bind, i.e.,

Chemokines		Chemokine receptor
Common name	New name	
IL-8	CXCL8	
GCP-2	CXCL6	CXCR1
NAP-2	CXCL7	
ENA-78	CXCL5	CXCR2
GROα	CXCL1	
GROβ	CXCL2	
GROγ	CXCL3	
IP-10	CXCL10	
Mig	CXCL9	CXCR3
I-TAC	CXCL11	CXCR7
SDF-1α/β	CXCL12	CXCR4
BCA-1	CXCL13	CXCR5
	CXCL16	CXCR6
BRAK	CXCL14	Unknow
MCP-1	CCL2	
MCP-4	CCL13	CCR2
MCP-3	CCL7	
MCP-2	CCL8	
MIP-1β	CCL4	CCR5
MIP-1αS	CCL3	
MIP-1αP	CCL3LI	
RANTES	CCL5	
MPIF-1	CCL23	CCR1
HCC-1	CCL14	
HCC-2	CCL15	
HCC-4	CCL16	
Eotaxin-2	CCL24	
Eotaxin-3	CCL26	CCR3
Eotaxin	CCL11	
TARC	CCL17	
MDC	CCL22	CCR4
MIP-3α	CCL20	CCR6
ELC	CCL19	
SLC	CCL21	CCR7
I-309	CCL1	CCR8
TECK	CCL25	CCR9
CTACK	CCL27	CCR10
PARC	CCL18	Unknown
Lymphotactin	XCL1	
SCM-1β	XCL2	XCR1
Fractalkine	CX3CL1	CX3CR1

Chemoattraction, Fig. 1 Classical and new names of chemokines are included. *Red* identifies "inducible" or "inflammatory" chemokines, *green* "homeostatic" agonists, and *yellow* ligands belonging to both realms. *BCA* B cell-activating chemokine, *BRAK* breast and kidney chemokine, *CTACK* cutaneous T-cell attracting-chemokine, *ELC* Epstein-Barr virus-induced receptor ligand chemokine, *ENA-78* epithelial cell-derived neutrophil-activating factor (78 amino acids), *GCP* granulocyte chemoattractant

receptors that bind to C, CC, CXC, and CX$_3$C chemokines are called, respectively, CR, CCR, CXCR, and CX$_3$CR receptors. Based largely on studies performed in the immune system, chemokines have been classified in three functional groups: homeostatic, inducible, and dual function (Fig. 1). The first group, which includes chemokines constitutively produced by "resting cells" in specific organs or in tissues inside these organs, controls homeostatic migratory processes that determinate the correct location of different cell types in the organism under normal conditions. The second group is inducible or inflammatory chemokines, which are secreted in different tissues in emergency situations and serve to attract to these places' specialized cell types that contribute to the resolution of the emergency situation. The third group is formed by dual function chemokines, which can be either homeostatic or inducible depending on the context (Fig. 1). Although chemoattraction is the function most commonly regulated by chemokines, however, studies performed mainly on leukocytes have demonstrated that these peptides, acting through specific chemokine receptors, may control additional cellular functions, including proliferation, ► adhesion, ► motility, survival, or protease secretion, among other functions. By controlling these activities, chemokines may contribute to modulate the functions of leukocytes and other cell types.

Chemokines and Cancer

Cancer is a disease where cells have disrupted the mechanisms that regulate their normal growth

Chemoattraction, Fig. 1 (continued) protein, *GRO* growth-related oncogene, *HCC* human CC chemokine, *IP* IFN-inducible protein, *I-TAC* IFN-inducible T-cell α chemoattractant, *MCP* monocyte chemoattractant protein, *MDC* macrophage-derived chemokine, *Mig* monokine induced by gamma interferon, *MIP* macrophage inflammatory protein, *MPIF* myeloid progenitor inhibitory factor, *NAP* neutrophil-activating protein, *PARC* pulmonary and activation-regulated chemokine, *RANTES* regulated upon activation normal T cell expressed and secreted, *SCM* single C motif, *SDF* stromal cell-derived factor, *SLC* secondary lymphoid tissue chemokine, *TARC* thymus and activation-related chemokine, *TECK* thymus-expressed chemokine

and, consequently, proliferate without control. This affliction becomes life threatening when cancer cells become metastatic, that is, they acquire the ability to leave their original sites of growth (primary tumor) and invade other tissues or organs where the uncontrolled growing cells can form new colonies (▶ metastasis) that can interfere with vital functions. The process leading to metastasis formation has been divided into several steps. In the first step, the cancer cells detach from the substrate and from the neighboring cells and escape from the primary tumors. The second step involves the penetration of the cancer cells into the blood or lymphatic vessels and their ▶ migration through these vessels. In the case of cells that migrate through the afferent lymphatics, they migrate first to the lymph nodes from where they can exit through the efferent lymphatics, eventually ending up in the blood vessels. In the third stage, cancer cells extravasate from blood vessels and home into new sites in the organism where new metastatic colonies can be formed. During these migratory processes, the cells undergo changes in their adhesive properties that are regulated by modulation of the activities and/or levels of integrin receptors. Moreover, cancer cells and/or associated stromal cells secrete proteases which, by degrading extracellular matrix (ECM) proteins of connective tissues, facilitate the moving of the cells and the ▶ invasion of other tissues. Finally, at the metastatic sites, the cancer cells attach and grow as secondary colonies. In addition, they may secrete chemokines and other soluble factors that induce new vascular vessel formation (▶ angiogenesis) and contribute to maintain the growth of the metastatic cells. Although millions of cells may be shed into the blood from primary tumors, however, only a reduced percentage of these cells are able to form metastases, suggesting that metastatic cells develop mechanisms that increase their survival in the face of a hostile environment.

Chemoattraction: A Key Process to Attract Cancer Cells to New Biological Niches

Since the work of Stephen Paget in the second half of the nineteenth century, it is known that metastatic cells do not move randomly, displaying in contrast a marked tropism toward specific organs (Table 1). A variety of experimental data indicates that chemokines may play an important role in determining this bias of the metastatic cells. Analysis of the phenotype of multiple metastatic cell types shows that these cells express specific sets of chemokine receptors (Table 1). Furthermore, a clear correlation has been observed between the expression of a specific chemokine receptor by a metastatic cell and the presence of its respective ligands in the metastatic sites, suggesting the involvement of these receptors in the homing processes (Table 1). Finally, a direct role for chemokines and their receptors in the control of the tropism of metastatic cells is corroborated in studies that show that interference with the binding to the chemokine receptors impairs the ability to metastasize to specific organs. For instance, antibody neutralization of ▶ CXCR4 in breast cancer cells reduced the ability of these cells to form metastases in the lung, both upon intravenous injection and after orthotopic implantation of the cells. Conversely, overexpression of CCR7 in B16 melanoma resulted in a dramatic enhancement in the ability of these cells to form metastases in the draining lymph nodes upon intravenous injection of the cells in mice. From these studies it has also emerged that CCR7 and CXCR4 are the chemokine receptors most commonly expressed by metastatic cells. This finding contributes to explain the ability of multiple metastatic cell types that express these receptors to colonize the lymph node and other organs where CXCL12 (ligand for CXCR4 and CXCR7) and CCL19 and CCL21 (both ligands of CCR7) are expressed (Table 1).

Premetastatic niche is the name given to the specific regions, whose formation is induced by soluble factors released by primary tumor cells, which eventually become colonized by distant metastatic cells from the primary tumors. It has been shown that chemokine expression may confer premetastatic niches the ability to attract metastatic cells from the distant primary tumor. In this regard, it has been shown that chemokines S100A8 and S100A9, expressed by myeloid and endothelial in premetastatic niches in the lung, are responsible of attracting incoming

Chemoattraction, Table 1 Chemokine receptors involved in cancer metastases

Chemokine/s receptor/s/ligand/s	Site/s of metastases	Cancer cell types	Function/s regulated by chemokine receptor
CXCR3/CXCL9, CXCL10, CXCL11	Lung, bone, lymph node	Acute lymphoblastic leukemia, chronic myelogenous leukemia, colon, melanoma	Chemoattraction
CXCR4/ CXCL12	Lung, bone, lymph node	Breast, ovarian, prostate, glioma, pancreas, melanoma, esophageal, lung (small cell lung cancer), head and neck, bladder, colorectal, renal, stomach, astrocytoma, cervical cancer, squamous cell cancer, osteosarcoma, multiple myeloma, intraocular lymphoma, follicular center lymphoma, rhabdomyosarcoma, neuroblastoma, B-lineage acute lymphocytic leukemia, B-chronic lymphocytic leukemia, non-Hodgkin lymphoma, acute myeloid leukemia, thyroid cancer, acute lymphoblastic leukemia, chronic myelogenous leukemia	Chemoattraction, angiogenesis, survival, growth
CXCR5/ CXCL13	Lymph node	Head and neck, chronic myelogenous leukemia	Chemoattraction
CXCR7/ CXCL11, CXCL12	Lymph node	Breast, cervical carcinoma, glioma, lymphoma, lung carcinoma	Adhesion, survival, growth
CCR4/CCL17, CCL22	Skin	Cutaneous T-cell lymphoma	Chemoattraction
CCR7/CCL19, CCL21	Lymph node	Breast, melanoma, lung (non-small cell lung cancer), head and neck, colorectal, stomach, chronic lymphocytic leukemia	Chemoattraction
CCR9/CCL25	Small intestine	Melanoma, prostate	Chemoattraction
CCR10/CCL27	Skin	Melanoma, cutaneous T-cell lymphoma	Chemoattraction, growth, survival

Lewis lung carcinoma metastatic cells to these niches because neutralization of the chemokines with antibodies reduced the metastases in these areas. In sum, chemokine/chemokine receptor pairs are important factors that control the colonization of cancer cells to specific sites in the organism.

Other Biological Effects of Chemokines on Cancer Cells Apart from Chemoattraction

Chemokines may affect cancer not only by regulating chemoattraction but also by regulating other functions that control cancer progression.

Chemokines Can Contribute to Regulate the Growth of Cancer Cells

Uncontrolled growth is a hallmark of cancer cells. Considering that chemokines may control cell growth in different cell types, the effect of chemokines on the proliferation of cancer cells is not unexpected. The growth of tumor cells may be affected by chemokines that can be either released in an ▸ autocrine signaling fashion by the cancer cells or secreted by the stromal tissues associated to the cancer cells. As an example of the first case, it is known that CXCL1, CXCL2, CXCL3, and CXCL8, secreted as autocrine growth factors by melanoma, pancreatic, and liver cancer cells, regulate the proliferation of all these cell types. As an example of the second case, it has been reported that CXCL12, which is secreted in the lungs and lymph nodes, leads to the increase in the growth of glioma, ovarian, small cell lung, basal cell carcinoma, and renal cancer, all cancer cell types that colonize the aforementioned organs. The effects of chemokines on growth can be complex because, for instance, interference with CCR5 seems to increase the proliferation of xenografts of human breast cancer, suggesting that CCR5 inhibits the growth of this cancer cells.

Chemokines Can Contribute to Regulate the Survival of Cancer Cells

A reduced susceptibility to ▶ apoptosis, leading to a concomitant extended survival, is also an important factor to explain the uncontrolled growth and the ability of cancer cells to form metastases. Chemokines have been involved in regulating survival in leukocytes and other cells; therefore these ligands may potentially contribute to regulate the carcinogenic phenotype by modulating this function. Stimulation of melanoma B16 cells expressing CCR10 with its ligand CCL27 enhances the resistance of these cells to the apoptosis induced by stimulation of the death receptor CD95. These in vitro results are consistent with in vivo experiments that show that the neutralization of CCL27 ligand with antibodies results in the blocking of tumor cell formation. Also, stimulation of glioma cells with CXCL12 protects these cells from the apoptosis induced by serum deprivation. It has been shown that CXCR7, a novel second receptor for CXCL12, is expressed in a variety of cancer cells. It has been indicated that CXCR7 may regulate survival, growth, and adhesion. Thus, it is possible that CXCR7 may also contribute to control all these functions in cancer cells.

Chemokines Can Contribute to Regulate the Adhesion to New Sites in Cancer Cells

Migratory cancer cells experience changes in adhesion, including processes of attachment and detachment, as they move through the organism. Enhanced adhesion is particularly crucial at the final stages of cancer progression where these cells require attaching to the new metastatic sites. Stimulation of cancer cells with chemokines may change the adhesion of these cells either by increasing the activity of integrins or by inducing changes in the expression levels on the membrane of these receptors. As an example of the first case, it has been observed that stimulation of B16 melanoma cells with CXCL12 leads to an increase in the affinity of the β1 integrin by the ligand VCAM-1 both in in vitro and in in vivo

experiments. As an example of the second case, stimulation of prostate tumor cells with CXCL12 induces enhanced expression of the integrins α3 and β5.

Chemokines Can Contribute to Control Protease Secretion in Cancer Cells

Metalloproteins are largely responsible for ECM remodeling and play key roles in solid tumor cell invasion. In this regard, it has been shown that chemokines enhance in protease secretion in some cancer cell types. For instance, stimulation of myeloma cells with CXCL12 induces metalloproteinase secretion.

Chemokines Can Contribute to Control Angiogenesis in Cancer Cells

At metastatic sites cancer cells induce formation of new vessels (angiogenesis), which allow the nourishment of the metastatic colonies. Angiogenesis is a finely orchestrated process where endothelial cells proliferate, secrete proteases, change their adhesive properties, migrate, and, finally, differentiate into new vessels. Chemokines can act as positive or negative regulators of the angiogenesis in the tumor microenvironment. In this regard, the members of the CXC chemokine family play an important role during this process. The CXC family has been divided into two groups. The first group includes members that present the triplet glutamic acid-leucine-arginine (ELR) before the first Cys (ELR$^+$ CXC chemokines), and the second group includes the members that lack this three amino acids (ELR$^-$ CXC chemokines). Although there are exceptions, by and large, ELR$^+$ CXC chemokines (including CXCL1, CXCL2, CXCL3, CXCL5, CXCL6, CXCL7, and CXCL8) play pro-angiogenic roles, promoting vessel formation through the stimulation of the CXCR2 receptor. For instance, in human ovarian carcinoma, CXCL8 induces both angiogenesis and tumorigenesis. Furthermore, treatment of mice that bear CXCL8-producing non-small cell ▶ lung cancer cells with anti-CXCL8 antibodies blunted the growth of these tumors in the mice. Exceptions to the rule ELR$^+$ CXC=angiogenic chemokines are the ELR$^+$ CXC members CXCL1 and

CXCL2, which are angiostatic, i.e., they inhibit angiogenesis.

ELR⁻ CXC chemokines, including CXCL9, CXCL10, and CXCL11, are generally angiostatic. For instance, CXCL9 and CXCL10 inhibit Burkitt lymphoma tumor formation probably by blocking blood vessel formation. An exception to the rule ELR⁻ CXC=angiostatic chemokine is CXCL12 that is angiogenic, as suggested by CXCL12 and CXCR4 KO mice that display cardiovascular development defects. It is believed that the angiogenic effects of CXCL12 are mediated by the vascular endothelial growth factor (VEGF) that is secreted by endothelial cells upon stimulation with CXCL12. The latter chemokine can be secreted in the tumor microenvironment by both the cancer cells and associated stromal cells. Finally, apart from CXC chemokines, other chemokines families may also regulate angiogenesis. In this regard, the CC chemokine CCL21 is angiostatic. In contrast, three CC family members (CCL1, CCL2, CCL11) and one CX3C family member (CX3CL1) can induce angiogenesis. All these chemokines, secreted inside the tumor, may potentially regulate the growth of the metastatic cells.

Therapeutical Aspects

The multiple points at which chemokines may regulate cancer progression make them attractive targets to develop anticancer drugs. Several strategies have been adopted to harness the power of chemokines against cancer, including the use of antibodies against the overexpressed chemokine receptors in the target cancer cells to induce apoptosis of these cells. One common strategy has been the development of inhibitors to block the binding of the chemokines to the receptors and consequently the function of these receptors. The fact that chemokine receptors are on the membrane and that much information is available on the sequences, both on the ligands and on the receptors, necessary for receptor-ligand binding have enabled the development of numerous peptide or small molecule inhibitors that interfere with chemokine function. Some of these inhibitors have been developed against CCR1, CCR5, CXCR7, and CXCR4. Most of these inhibitors

relay on their ability to inhibit survival or angiogenesis in the target cells. As CXCR4 is one of the most broadly expressed chemokine receptor in cancer cells, at least six peptides or small molecule inhibitors of the function of CXCR4 have been developed and used in preclinical cancer models. CXCR4 is particularly interesting due to its pro-angiogenic functions. A variety of data indicate that the growth and persistence of tumors and their metastases depend on an active angiogenesis at the tumor sites. In this regard, interference with this process is a powerful strategy to inhibit tumor growth. Interference with CXCR4 has been used in several cancer models, including many of the cancers indicated in Table 1. Although peptide inhibitors of chemokine receptors may not have by itself tumoricidal affects, however, along with other strategies may be a powerful therapy against tumors.

Summary and Final Conclusions

Upon becoming carcinogenic and metastatic, a variety of cancer cells upregulate the expression of chemokine receptors. In this regard, the microenvironment conditions inside the tumors are also known to induce chemokine receptor expression in some cases. For instance, the low oxygen concentration (▶ hypoxia) inside a tumor induces CXCR4 expression which concomitantly leads to a more aggressive metastatic phenotype in cancer cells. Chemokine receptors endow cancer cells with "postal codes" that determine their migration to tissues where the ligands of these receptors are expressed and therefore are important for the metastatic ability of these cells. In addition, these receptors may confer or modulate cancer cells functions that, by regulating different steps in cancer progression, may contribute to the carcinogenic and metastatic phenotype of these cells. The case of the Kaposi sarcoma herpesvirus (KSHV), which induces cancer lesions similar to that of the Kaposi sarcoma, is a dramatic example that shows the important role that chemokines and their receptors may play in cancer. Interestingly, this virus encodes a constitutively active receptor that displays a high degree of sequence similarity to chemokine receptors CXCR1 and CXCR2 and which can even be further activated by the

CXCR2 ligands CXCL1 and/or CXCL8. KSHV is also pro-angiogenic and induces survival effects in the cancer cells where it is expressed. Further supporting a causative role of CXCR2 in cancer, a constitutive form of CXCR2, can induce cell transformation in susceptible cell types.

Cross-References

▶ Adhesion
▶ Angiogenesis
▶ Apoptosis
▶ Autocrine Signaling
▶ Cancer
▶ Chemokine Receptor CXCR4
▶ Chemokines
▶ G Proteins
▶ Hypoxia
▶ Invasion
▶ Lung Cancer
▶ Metastasis
▶ Migration
▶ Motility

References

Balkwill F (2004) Cancer and the chemokine network. Nat Rev Cancer 4:540–550
Ben-Baruch A (2006) The multifaceted roles of chemokines in malignancy. Cancer Metastasis Rev 25:357–371
Kakinuma T, Hwang ST (2006) Chemokines, chemokine receptors, and cancer metastasis. J Leukoc Biol 79:639–651
Sánchez-Sánchez N, Riol-Blanco L, Rodríguez-Fernández JL (2006) The multiples personalities of the chemokine receptor CCR7 in dendritic cells. J Immunol 176:5153–5159
Zlotnik A (2006) Chemokines and cancer. Int J Cancer 119:2026–2029

See Also

(2012) Chemotaxis. In: Schwab M (ed) Encyclopedia of Cancer, 3rd edn. Springer Berlin Heidelberg, p 793. doi: 10.1007/978-3-642-16483-5_1081
(2012) Glioma. In: Schwab M (ed) Encyclopedia of Cancer, 3rd edn. Springer Berlin Heidelberg, p 1557. doi: 10.1007/978-3-642-16483-5_2423
(2012) G-protein Couple Receptor. In: Schwab M (ed) Encyclopedia of Cancer, 3rd edn. Springer Berlin Heidelberg, p 1587. doi: 10.1007/978-3-642-16483-5_2294
(2012) Haptotaxis. In: Schwab M (ed) Encyclopedia of Cancer, 3rd edn. Springer Berlin Heidelberg, p 1631. doi: 10.1007/978-3-642-16483-5_2565
(2012) Integrin. In: Schwab M (ed) Encyclopedia of Cancer, 3rd edn. Springer Berlin Heidelberg, p 1884. doi: 10.1007/978-3-642-16483-5_3084
(2012) Orthotopic. In: Schwab M (ed) Encyclopedia of Cancer, 3rd edn. Springer Berlin Heidelberg, p 2661. doi: 10.1007/978-3-642-16483-5_4264
(2012) Xenograft. In: Schwab M (ed) Encyclopedia of Cancer, 3rd edn. Springer Berlin Heidelberg, p 3967. doi: 10.1007/978-3-642-16483-5_6278

Chemokine Receptor CXCR4

Jonathan Blay
Department of Pharmacology, Dalhousie University, Halifax, NS, Canada

Synonyms

CD184; Fusin; Receptor for CXCL12; Receptor for stromal cell-derived factor-1 alpha; SDF-1α

Definition

CXCR4 is a cell surface protein that acts as a receptor for the molecule CXCL12 (stromal cell-derived factor-1 alpha, SDF-1α). CXCL12 is one of a class of signaling molecules called chemokines that regulate the movement and other activities of cells throughout the body. Although CXCL12 and CXCR4 play major roles in regulating stem cells and cells of the immune system, CXCR4 is also found on many cancer cells and plays a part in metastasis, spread of the cancer cells being influenced by tissue levels of CXCL12.

Characteristics

Chemokines are a class of peptide mediators that play important roles in controlling cellular homing and migration both in embryonic development and in the regulation of cell populations in the adult. There are at least 40 different chemokines that fall into four classes depending upon their peptide structure. The different classes are "C," "CC," "CXC," and "CX$_3$C" chemokines, for which characteristic sequence motifs involve residues of the amino acid cysteine (C) either in sequence or separated by one or three other amino acids (X or X$_3$). The chemokines themselves are peptides that can exist freely in solution in biological fluids and act by binding to corresponding ▶ receptors. In the language of molecular interactions, a chemokine is therefore known as a ligand. Chemokines are denoted by the letter L within their name. CXCL12 is thus a ligand and a chemokine of the CXC class of chemokine mediators.

The chemokine receptors are named according to the chemokine class of their binding partner (or ligand), with the letter "R" to designate their receptor status. CXCR4 is therefore a receptor. As for chemokines, the numbers serve to distinguish individual members of the overall family. The partnership between chemokine receptors and the chemokines is not monogamous, and some chemokine receptors may bind as many as ten different chemokines. However, most receptors have between one and three distinct partners. With very few exceptions, these partnerships are within a particular chemokine class (e.g., CXCL chemokines bind selectively to certain CXCR receptors). At this point, the only chemokine factor known to bind to CXCR4 is CXCL12, although CXCL12 itself is able to bind to an alternate receptor (CXCR7, previously known as RDC-1) as well as to CXCR4.

Chemokine receptors such as CXCR4 are seven-transmembrane, G protein-coupled receptors. The protein chain of CXCR4 therefore winds back and forth across the outer membrane of the cell so that it crosses the membrane a total of seven times. One end of the protein chain (the amino terminus) protrudes from the outside of the cell. This region of the protein, together with certain parts of the three extracellular loops, forms the binding domain for CXCL12. The part of the receptor that protrudes from the inner face of the membrane (composed of the carboxy-terminus and three intracellular loops) contains the characteristics that allow it to provoke a cascade of events within the cell (Fig. 1). These steps are initiated firstly by a linkage to one or more of a

Chemokine Receptor CXCR4, Fig. 1 The cellular signaling pathways of CXCR4. When the chemokine ligand CXCL12 binds to its receptor CXCR4, one or more of several pathways can be activated through initial links involving G proteins that associate with the receptor. These pathways, which are shown only in outline, involve a further network of interactions that eventually lead to a cellular response that may ensure cell growth, migration or survival

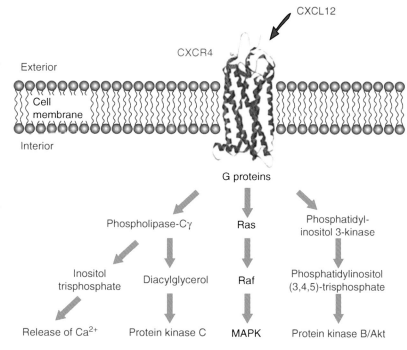

small family of proteins that interact directly with the receptor, called ▶ G proteins (in this case primarily $G_{\alpha i}$ and $G_{\alpha q}$). G protein involvement leads to the activation of three major signaling pathways: (i) the phospholipase C-diacylglycerol/IP$_3$ pathway, (ii) the Ras-Raf-MAP kinase pathway, and (iii) the PI3-kinase pathway.

CXCR4 is a crucially important member of the chemokine receptor family. If CXCR4 or CXCL12 is absent during embryonic development, the organism is unable to survive. The key dependence on CXCL12 and CXCR4 reflects the importance of this signal/receptor pair in marshaling the correct formation of cells as tissues are formed from their more rudimentary cellular precursors in the embryo. The CXCL12-CXCR4 axis, as it is often called, is a central part of the normal development of the central nervous system (the brain itself) and the exquisitely organized tissue that replenishes the different cells of the blood through adult life (the hematopoietic system). In addition, CXCR4 and CXCL12 seem to play a particular role in the development of the gut, and their participation is important for the proper development of the blood vessel system that is required for efficient intestinal function in the adult. In adult organisms, CXCR4 and CXCL12 partly reprise their developmental role during tissue damage by participating in repair processes.

Once the organism is fully formed, the most evident role for CXCR4 and CXCL12 in a normal individual is that of continued regulation of the hematopoietic system. This takes place mainly in the bone marrow, which acts as a reservoir for the ancestral cells (stem cells and other progenitor cells) that are needed for the continued production of various white cells (leukocytes) and other progeny that are required to ensure a proper defense against infection or injury or to deal with replacement and remodeling of damaged tissues. These stem cells – which need to be maintained safely by the body until required to respond – are located within the protected environment of the bone marrow and are supported and nourished by a specialized grouping of cells that together are referred to as the "microenvironmental niche."

These supporting cells or "stromal cells" secrete a number of factors that serve to nourish the stem cells and to keep them within a safe environment in their primitive and "resting" state.

Notable among these factors is CXCL12 (the "stromal cell-derived factor"), which can bind to CXCR4 on the stem cells. The binding of CXCL12 to its receptor has several effects on cell behavior, but the principal outcome is to attract cells toward the source of CXCL12. In the case of stem cells in the bone marrow, this results in retention within the microenvironmental niche or directs migrant stem cells back to this location. This ability of the CXCL12:CXCR4 axis to direct cell movement is what underlies its key role in orchestrating tissue development and repair. The phenomenon can be demonstrated in experiments using isolated cells, such that cells that have the CXCR4 receptor can be induced to migrate through pores in an artificial filter in response to an upward concentration gradient of CXCL12 in the fluid. This is a cellular response known as chemotaxis, and CXCL12 is referred to as a chemoattractant.

Unfortunately, this normal and very important process by which CXCL12 and CXCR4 assist directed cell movement has been subverted by cancer cells to assist the spread of a cancer or metastasis. Normal tissues that are not subject to inflammation or repair processes typically have very low levels of CXCR4. However, when cancers are formed the affected cells frequently experience a dramatic increase ("upregulation") of CXCR4. This has been shown for the common adult cancers (carcinomas of the breast, colon, lung, prostate, cervix, etc.), which arise in the membranous linings (epithelia) of certain organs; but CXCR4 levels are also elevated in cancers arising in the bone (e.g., osteosarcoma), muscle (e.g., rhabdomyosarcoma), nervous tissue (e.g., glioblastoma), or white cells (various leukemias).

This is such a consistent finding that in many cancers the level, or "expression," of CXCR4 can be used as cancer biomarker. The levels of CXCR4 that are present on the cells give an indication of how the cancer is likely to behave in the future and what therapeutic steps might need to be considered. Levels are assessed using a technique called immunohistochemistry. In this approach very thin slices or "sections" – no more than 0.005 mm thick – are taken from the suspect tissue onto glass slides. Special protein reagents called antibodies are used that recognize any molecules of CXCR4 in the tissue, and additional steps in the process generate color wherever the antibody has bound. The resulting picture under a microscope tells the pathologist not only about the architecture of the tissue and the characteristics of the cells but whether or not they have high levels of CXCR4. High levels (expression) of CXCR4 are associated with cancer aggressiveness, a likelihood that the cancer will spread or metastasize and means that the outlook for the patient is likely to be poorer.

The link between cancer aggressiveness/metastasis exists because the CXCL12:CXCR4 axis has a similar role of "directing traffic" in cancer as it does in normal circumstances. In this situation it is the cancer cells that possess the receptor – CXCR4 – and have levels at the cell surface that are much greater than are found on their normal counterparts. The exact reasons for these elevated levels of the chemokine receptor are not fully understood. Undoubtedly the genetic changes that are characteristic of cancer cells lead to alterations in transcription of the CXCR4 gene

that may provide certain subpopulations with greater amounts of the CXCR4 protein, and these cells have a selective advantage. However, there are also indications that factors within the environment of the tumor can make the situation worse by stimulating the cell to make even more CXCR4. The hypoxic nature of tumor tissue causes an increase in CXCR4 gene transcription through a pathway involving ▸ hypoxia-inducible factor 1 alpha (HIF-1α). Various small-molecular-weight and polypeptide mediators have also been shown to enhance the cellular expression of this chemokine receptor.

The cancer cells are therefore equipped to be attracted toward sources of CXCL12 and to be captured within environments that are high in concentrations of CXCL12. Thus, it is no coincidence that the tissues that are high in CXCL12 are also those in which cancers form secondary tumors or metastases. Such tissues include the lymph nodes – central filters in the system that drain fluid from all tissues – as well as the liver, lung, and bone marrow. CXCL12 is believed to be one of the major factors driving metastasis (Fig. 2). As a colorectal cancer develops in the large intestine, for example, and small groups of tumor cells are shed into the blood circulation and the lymphatic drainage, circulating cells will find

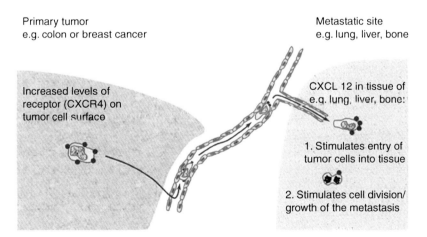

Chemokine Receptor CXCR4, Fig. 2 How CXCR4 and CXCL12 work together to facilitate metastasis. Tumor cells have increased levels of the receptor at their cell surface. When the tumor grows sufficiently for the cancer cells to find their way into the bloodstream, some cells lodge in tissues (e.g., lungs, liver, and bone marrow) that

have high concentrations of CXCL12, the molecule for which CXCR4 is the receptor. CXCL12 both encourages the entry of cells into the tissue and promotes growth of the cell population, facilitating metastatic spread. Tissues that have low levels of CXCL12 are much less likely to accept metastases

an attractive home as they encounter lymph nodes in the mesenteric fat around the intestinal wall, when they are delivered to the liver through the portal circulation or as they lodge in the capillary beds of the lung after traversing the systemic circulation. Conversely, they have a much reduced probability of taking up residence in sites such as the heart or skeletal (voluntary) muscle, which are low in CXCL12.

In addition to being attracted and retained in tissues that have high concentrations of CXCL12, the CXCR4-bearing cancer cells may respond in other ways. Although this may not be the case for all cancers, in some types (e.g., carcinomas of the colon and prostate), there is evidence that once the cells have settled in to their new location, the presence of CXCL12 acting through CXCR4 also enhances their ability to grow and colonize the tissue. In this way, CXCL12 can also be regarded as a growth factor, alongside other polypeptide growth stimulators that participate in tumor expansion.

One additional factor that makes CXCR4 of interest for many different clinicians and researchers is that it is one of the two major coreceptors by which the AIDS virus infects human cells. One of the proteins that is present within the outer surface of the HIV-1 virus, called gp120, binds to CXCR4, although at a slightly different site to CXCL12. When the virus binds to its major target (the CD4 protein) on susceptible cells, it requires a coreceptor in order to complete its cellular attack. This allows it to complete the molecular changes that allow it to infect the cell. Depending on the exact cell and viral type, the coreceptor may be CXCR4 or another chemokine receptor, CCR5. While the link with AIDS has limited direct relevance to most cancers, the two fields of research have synergized to extend our present understanding of CXCR4.

Cross-References

▶ G Proteins
▶ Hypoxia-Inducible Factor-1
▶ Receptors

See Also

(2012) Antibody. In: Schwab M (ed) Encyclopedia of Cancer, 3rd edn. Springer Berlin Heidelberg, p 208. doi: 10.1007/978-3-642-16483-5_312

(2012) Biomarkers. In: Schwab M (ed) Encyclopedia of Cancer, 3rd edn. Springer Berlin Heidelberg, pp 408–409. doi: 10.1007/978-3-642-16483-5_6601

(2012) Chemotaxis. In: Schwab M (ed) Encyclopedia of Cancer, 3rd edn. Springer Berlin Heidelberg, p 793. doi: 10.1007/978-3-642-16483-5_1081

(2012) G-protein Couple Receptor. In: Schwab M (ed) Encyclopedia of Cancer, 3rd edn. Springer Berlin Heidelberg, p 1587. doi: 10.1007/978-3-642-16483-5_2294

(2012) Hematopoietic System. In: Schwab M (ed) Encyclopedia of Cancer, 3rd edn. Springer Berlin Heidelberg, p 1645. doi: 10.1007/978-3-642-16483-5_2621

(2012) Ligands. In: Schwab M (ed) Encyclopedia of Cancer, 3rd edn. Springer Berlin Heidelberg, p 2040. doi: 10.1007/978-3-642-16483-5_3352

(2012) Microenvironmental Niche. In: Schwab M (ed) Encyclopedia of Cancer, 3rd edn. Springer Berlin Heidelberg, p 2296. doi: 10.1007/978-3-642-16483-5_3721

(2012) Stromal Cells. In: Schwab M (ed) Encyclopedia of Cancer, 3rd edn. Springer Berlin Heidelberg, p 3544. doi: 10.1007/978-3-642-16483-5_5535

(2012) Transcription. In: Schwab M (ed) Encyclopedia of Cancer, 3rd edn. Springer Berlin Heidelberg, p 3752. doi: 10.1007/978-3-642-16483-5_5899

Chemokines

Lei Fang and Sam T. Hwang
Dermatology Branch, National Cancer Institute, National Institutes of Health, Bethesda, MD, USA

Synonyms

Chemoattractant cytokine; Chemotactic cytokine

Definition

Chemokines are a large group of small proteins that play multiple biological roles, including stimulating directional migration (chemotaxis) of leukocytes and tumor cells via their membrane-bound receptors.

The name comes from "chemotactic cytokines," these small cytokines induce migration of diverse immune cells. The family of the chemokines is quite numerous, as are the chemokine receptors, and often there is "promiscuity," in that a single chemokine can activate multiple receptors and multiple chemokines can activate a single receptor. These molecules direct trafficking of leucocytes. Two chemokine receptors are also the principal coreceptors for HIV involved in viral entry: CCR5, expressed on monocytes and macrophages as well as other cells, and the more widely expressed CXCR4. The tropism of specific chemokine receptors is associated with HIV clinical effects, with CCR5 linked to infection and CXCR4 tropism linked to progression to AIDS.

Characteristics

Chemokines are divided into four subgroups (C, CC, CXC, and CX$_3$C) based on the spacing of the key cysteine residues near the N terminus of these proteins. The CC and CXC families represent the majority of known chemokines. Chemokines signal through seven-transmembrane-domain receptors, which are coupled to heterotrimeric G$_i$-proteins. Activation of phospholipase C (PLC) and phosphatidylinositol-3-kinase γ (PI3Kγ) by $\beta\gamma$ subunits of ► G-proteins is well established.

So far, approximately 50 chemokines and 18 chemokine receptors have been identified. Some chemokine receptors bind to multiple chemokines and vice versa, suggesting possible redundancies in chemokine functions. Chemokine receptors permit diverse cells to sense small changes in the gradient of soluble and extracellular matrix-bound chemokines, thus facilitating the directional migration of these cells toward higher relative concentrations of chemokines. While soluble chemoattractants can induce directional migration, chemokines (due to their net positive charges) will often be bound to and presented by negatively charged macromolecules such as endothelial cell-derived proteoglycans in vivo. Chemokine gradients bound to solid surfaces are capable of mediating haptotaxis of leukocytes and other cells. Chemokine receptor activation can also trigger conformational changes in membrane integrins, permitting strong cell–cell adhesion in the presence of appropriate integrin receptors. This signaling pathway is particularly relevant in triggering cellular integrins found on leukocytes and cancer cells to bind to their respective receptors (e.g., ICAM-1) on vascular endothelial cells, facilitating stable binding and spreading of cells to endothelium. The stable binding of metastatic tumor cells to vascular endothelial cells at distant sites of metastasis is likely to be a crucial early step in the process of ► metastasis.

Circumstantial evidence supports the idea that tumor cells use chemokines to promote their own survival and metastasis through multiple mechanisms. For example, certain chemokines secreted by tumor cells contribute to tumor growth and ► angiogenesis. Members of chemokines that contain an ELR motif (Glu–Leu–Arg) act as angiogenic factors, which are chemotatic for endothelial cells in vitro and can stimulate in vivo. In contrast, members without an ELR motif inhibit angiogenesis. Chemokine-mediated tumor cell activation through cellular kinases such as PI3K, ► Akt signal transduction pathway in oncogenesis), and other downstream mediators (Fig. 1) influences tumor cell resistance to apoptotic death. For example, activation of the chemokine receptor CCR10 prevents Fas-mediated tumor cell death induced by cytolytic antigen-specific T cells.

Selected chemokine receptors are upregulated in a large numbers of common human cancers, including breast, lung, prostate, colon, and melanoma. Chemokine receptors expressed on tumor cells coupled with chemokines preferentially expressed in a variety of organs are believed to play critical roles in cancer metastasis to vital organs as well as draining lymph nodes. CXCR4 is by far the most common chemokine receptor expressed on most cancers. In addition, CXCL12, the ligand for CXCR4, is highly expressed in lung, liver, bone marrow, and lymph nodes, which represent the common sites of metastasis of many cancers. Chemokine receptor expression

Chemokines,
Fig. 1 Chemokine receptor signaling. Upon stimulation by chemokine, βγ subunits of G-protein are dissociated from Gα$_i$ subunit. βγ subunits activate phospholipase C (*PLC*) and phosphatidylinositol 3 kinase γ (*PI3Kγ*), whereas Gα$_i$ subunit directly activates ► Src-like kinase

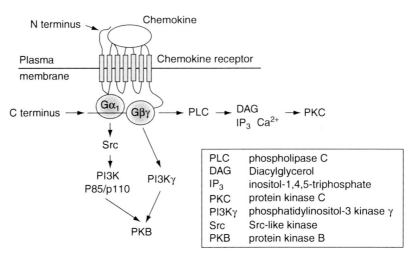

on cancer cells may influence the conversion of small, clinically insignificant foci of cancer cells at metastatic sites to rapidly growing, clinically serious secondary tumors. Cancers that upregulate CCR7 expression also facilitate their entry into lymphatic vessels, which strongly express the CCR7 ligand (CCL21), and subsequent retention within CCL21-rich secondary lymphoid organs. Upregulation of chemokine receptors such as CCR7 may be a major reason for efficient lymph node metastasis observed in many epithelial cancers.

Chemokines released by tumor cells have been shown to attract ► regulatory T cells, thus suppressing host responses to invasive tumors. Moreover, chemokine and their receptors are involved in ► dendritic cell maturation, B and T cell development, and T1 and T2 polarization of the T-cell response. These actions suggest the possibility that chemokines may play a role in altering the magnitude and polarity of host immune responses to cancer cells.

Although individual chemokine and chemokine receptor appear to affect many aspects of cancer cell survival, migration, angiogenesis, and the host response to cancer cells, it is still unclear which of these functions predominate in the multistep establishment of primary tumors and secondary metastases.

Cross-References

► Akt Signal Transduction Pathway
► Angiogenesis
► Chemokine Receptor CXCR4
► Dendritic Cells
► G Proteins
► Metastasis
► Regulatory T Cells
► Src

References

Kakinuma T, Hwang ST (2006) Chemokines, chemokine receptors, and cancer metastasis. J Leukoc Biol 79:639–651
Müller A, Homey B, Soto H et al (2001) Involvement of chemokine receptors in breast cancer metastasis. Nature 410:50–56
Murphy PM (2002) International Union of Pharmacology. XXX. Update on chemokine receptor nomenclature. Pharmacol Rev 54:227–229
Rossi D, Zlotnik A (2000) The biology of chemokines and their receptors. Annu Rev Immunol 18:217–242
Thelen M (2001) Dancing to the tune of chemokines. Nat Immunol 2:129–134

See Also

(2012) Chemotaxis. In: Schwab M (ed) Encyclopedia of cancer, 3rd edn. Springer Berlin Heidelberg, p 793. doi: 10.1007/978-3-642-16483-5_1081

(2012) FAS. In: Schwab M (ed) Encyclopedia of cancer, 3rd edn. Springer Berlin Heidelberg, p 1379. doi: 10.1007/978-3-642-16483-5_2121

(2012) Haptotaxis. In: Schwab M (ed) Encyclopedia of cancer, 3rd edn. Springer Berlin Heidelberg, p 1631. doi: 10.1007/978-3-642-16483-5_2565

(2012) Integrin. In: Schwab M (ed) Encyclopedia of cancer, 3rd edn. Springer Berlin Heidelberg, p 1884. doi: 10.1007/978-3-642-16483-5_3084

(2012) TH1 Cells. In: Schwab M (ed) Encyclopedia of cancer, 3rd edn. Springer Berlin Heidelberg, p 3600. doi: 10.1007/978-3-642-16483-5_5647

Chemokinesis

▶ Motility

Chemoprevention

Definition

Chemoprevention involves the use, in healthy people, of natural or laboratory-made substances to prevent cancer or reduce cancer risk both in high-risk individuals and in the general population. The aim is to reduce the cancer burden in humans. Most work is being done to reduce the risk for ▶ oral cancer, prostate cancer (see "▶ Prostate Cancer Clinical Oncology"), ▶ cervical cancer, ▶ lung cancer, ▶ colorectal cancer, and ▶ breast cancer. The first chemopreventive agent to reach the clinic – and possibly the best known – was ▶ tamoxifen, which has been shown to cut breast cancer incidence in high-risk women by 50%. It was followed by finasteride, found to reduce prostate cancer (see "▶ Prostate Cancer Clinical Oncology") incidence by 25% in men at high risk for the disease. However, the large-scale trials that confirmed these benefits brought to light a troublesome issue: the drugs caused serious side effects in some patients. This is an issue of particular concern when considering long-term administration of a drug to healthy people who may or may not develop cancer. Obviously, this is raising a number of ethical issues. An effective chemopreventive agent should not significantly alter quality of life and should be ideally inexpensive, safe, well tolerated, and effective in preventing more than one cancer.

Experience with ▶ celecoxib (Celebrex) and other COX-2 inhibitors illustrates the importance of an assessment of the risk/benefit ratio for patients. COX-2 inhibitors have shown impressive efficacy in the prevention of colon cancer and several other forms of cancer, but they also increase the risk of serious cardiovascular side effects.

Attention has focused on ▶ nutraceuticals and phytochemicals (see "▶ Phytochemicals in Cancer Prevention") as chemopreventive agents. ▶ Curcumin (found in the curry spice turmeric) has shown dramatic anticancer results in preclinical studies owing to its significant anti-▶ inflammation properties. Curcumin has been used for thousands of years in the diets of people in the Middle and Far East and therefore is believed to have a low probability of serious side effects. Under investigation for their potential in breast cancer chemoprevention are aromatase inhibitors (see "▶ Aromatase and Its Inhibitors"), a class of estrogen blockers, which are approved to treat metastatic breast cancer in postmenopausal women. While the idea of cancer chemoprevention is extremely attractive, much research remains to be done to make this a generally applicable option for reducing the human cancer burden. An important element will be to identify informative biomarkers to assess individual cancer risk and to possibly provide information of patient's tolerance toward individual chemopreventive agents.

Cross-References

▶ Aromatase and Its Inhibitors
▶ Breast Cancer
▶ Celecoxib
▶ Cervical Cancers
▶ Chemoprotectants
▶ Colorectal Cancer
▶ Curcumin
▶ Detoxification
▶ Inflammation

▶ Lung Cancer
▶ Nutraceuticals
▶ Oral Cancer
▶ Photochemoprevention
▶ Phytochemicals in Cancer Prevention
▶ Prostate Cancer Clinical Oncology
▶ Tamoxifen

See Also

(2012) Biomarkers. In: Schwab M (ed) Encyclopedia of cancer, 3rd edn. Springer, Berlin/Heidelberg, pp 408–409. doi:10.1007/978-3-642-16483-5_6601
(2012) Cyclooxygenase-2. In: Schwab M (ed) Encyclopedia of cancer, 3rd edn. Springer, Berlin/Heidelberg, p 1035. doi:10.1007/978-3-642-16483-5_1435
(2012) Estrogens. In: Schwab M (ed) Encyclopedia of cancer, 3rd edn. Springer, Berlin/Heidelberg, p 1333. doi:10.1007/978-3-642-16483-5_2019
(2012) Finasteride. In: Schwab M (ed) Encyclopedia of cancer, 3rd edn. Springer, Berlin/Heidelberg, p 1407. doi:10.1007/978-3-642-16483-5_2191

Chemoprotectants

Debasis Bagchi
Department of Pharmacy Sciences, Creighton University Medical Center, Omaha, NE, USA

Synonyms

Chemoprevention; Chemoprotection

Definition

Chemoprotectants are natural or synthetic chemical compounds which exhibit the ability to ameliorate, mimic, or inhibit the toxic or adverse effects of structurally different chemotherapeutic agents, radiation therapy, cytotoxic drugs, or naturally occurring toxins, without compromising the anticancer or antitumor potential of the chemotherapeutic drugs. Chemoprotectants should not affect the therapeutic efficacy of the chemotherapeutic agents, radiation, or drugs, disrupt the serum enzyme levels, or induce significant injury to the tissues/organs. These chemoprotectants include anticancer, antitumor, anti angiogenic, and antioxidant compounds and are used as an adjuvant in cancer ▶ chemotherapy.

Characteristics

According to the World Health Organization (WHO), cancer accounts for 7.6 million (or 13%) of all deaths in 2005, and the incidence of cancer is expected to rise with an estimated 9 and 11.4 million deaths from cancer in 2015 and 2030, respectively. Cancer chemotherapy and radiation therapy are the most promising choice available for the cancer patients. The global outlook of cancer therapy has made dramatic improvement since the discovery of various synthetic and natural chemoprotectants which slow down the progress of this deadly disease and enhance the life span of the cancer patients. Chemoprotectants may exert toxic effects. Thus, it is very important to determine the right dosage and exposure scenario for each chemoprotectant prior to the exposure to demonstrate adequate safety.

Synthetic Chemoprotectants

Amifostine. A white powder, water-soluble organic thiophosphate compound, chemically known as 2-[(3-aminopropyl)amino]-ethanethiol dihydrogen phosphate (ester) or 2-(3-aminopropylamino)ethylsulfanyl phosphonic acid or aminopropylaminoethyl thiophosphate (Fig. 1a), and used as a cytoprotective adjuvant in cancer chemotherapy to reduce the incidence of ▶ neutropenia-related fever and infection caused by DNA-binding chemotherapeutic agents including cyclophosphamide and cisplatin. Amifostine (empirical formula $C_5H_{15}N_2O_3PS$; molecular weight 214.22; trade name Ethyol, synonyms: ethiofos, ethanethiol, gammaphos, WR2721, NSC-296961) is used to decrease the cumulative nephrotoxicity caused by cisplatin in patients with ovarian or lung cancer, as well as to reduce the incidence of moderate to severe xerostomia (dry mouth) in patients undergoing radiotherapy

a Amifostine
(R)-2-acetamido-3-mercaptopropanoic acid

b Dexrazoxane
4-[1-(3, 5-dioxopiperazin-1-yl) propan-2-yl]
piperazine-2, 6-dione

c Glutathione
2-amino-5-{[2[(carboxymethyl)amino]-
1-(mercaptomethyl)-2-oxoethyl]amino}-
5-oxopentanoic acid

d Mesna
Sodium-2-sulfanylethane-
sulfaonate

e N-Acetylcysteine
(R)-2-acetamido-3-
mercaptopropanoic acid

Chemoprotectants, Fig. 1 Structures and IUPAC nomenclature of (**a**) amifostine, (**b**) dexazoxane, (**c**) glutathione, (**d**) mesna, and (**e**) *N*-acetylcysteine

for head and neck cancer. Amifostine is dephosphorylated by alkaline phosphatase in tissues to a pharmacologically active free thiol metabolite, which readily scavenge noxious reactive oxygen species (ROS) generated by exposure to either cisplatin or radiation, as well as detoxify reactive metabolites of platinum and other alkylating agents. Pharmacokinetic studies show that amifostine is rapidly cleared from the plasma with a distribution half-life of <1 min and an elimination half-life of approximately 8 min.

Ethyol is supplied in 500 mg vials and administered intravenously (i.v.). Amifostine-induced adverse side effects include nausea, vomiting, flushing, chills, dizziness, shortness of breath, fainting, seizures, cardiovascular problems, skin rash, hives, and swelling of the throat.

Dexrazoxane. A whitish crystalline powder, sparingly soluble in water, and chemically known as (S)-4,4′-(1-methyl-1,2-ethanediyl)bis-2,6-piperazinedione or 4-[1-(3,5-dioxopiperazin-1-yl) propan-2-yl]piperazine-2,6-dione (Fig. 1b). Dexrazoxane (empirical formula $C_{11}H_{16}N_4O_4$;

molecular weight 268.28; trade names: Zinecard®, ICRF-187, ADR-529, or NSC 169780, synonym: 2,6-piperazinedione) is a cyclic derivative of EDTA that readily penetrates cell membranes and a potent intracellular chelating agent. Dexrazoxane is used to protect the heart against the cardiotoxic side effects of anthracycline chemotherapy and to reduce the incidence and severity of cardiomyopathy associated with doxorubicin administration in women with breast cancer. Dexrazoxane is hydrolyzed by the enzyme dihydropyrimidine amidohydrolase in the liver and kidney to active metabolites, which have been shown to chelate both free and bound intracellular iron, thereby preventing the formation of cardiotoxic ROS and anthracycline-mediated cardiomyopathy. However, dexrazoxane may potentiate hematological toxicity induced by chemotherapy or radiation. Dexrazoxane is rapidly distributed into the body's tissues and fluids, while the highest concentration is found in the hepatic and renal tissues. Urinary excretion plays an important role in the elimination of dexrazoxane (half-life, $t_{1/2}$ 2–4 h). Forty-two percent of the 500 mg/m^2 dose of dexrazoxane is excreted in the urine.

Dexrazoxane is available in 250 and 500 mg for i.v. administration. Adverse effects include alopecia, nausea, vomiting, fever, fatigue, anorexia, urticaria, leucopenia, hematologic thrombocytopenia, and neurotoxicity.

Glutathione. A tripeptide, made of the amino acids γ-glutamic acid, cysteine, and glycine, is the predominant nonprotein thiol and functions as a redox buffer and exhibits diverse antioxidant activities and protects cells from oxidative stress. Glutathione (synonyms: reduced glutathione, monomeric glutathione, GSH; empirical formula $C_{10}H_{17}N_3O_6S$; and molecular weight 307.33) is chemically known as *N*-(*N*-L-γ-glutamyl-L-cysteinyl)glycine (Fig. 1c), while its dimer is known as oxidized glutathione, glutathione disulfide, diglutathione, and GSSG, chemically known as L-γ-glutamyl-L-cysteinyl-glycine disulfide (empirical formula $C_{20}H_{32}N_6O_{12}S_2$). The primary function of GSH is to act as a nonenzymatic reducing agent to help keep cysteine thiol side chains in a reduced state on the surface of proteins. GSH levels in intracellular fluids decline dramatically with advancing age, and thus, the ability to detoxify ROS diminishes. GSH is available as a single-ingredient dietary supplement or in combination products. Daily dosage ranges from 50 to 600 mg daily. No adverse effects were reported.

Mesna. A synthetic sulfhydryl compound, chemically known as sodium-2-mercaptoethane sulfonate or sodium-2-sulfanylethane-sulfonate, and forms a clear and colorless aqueous solution. Mesna ($HS-CH_2-CH_2SO_3-Na^+$; empirical formula $C_2H_5O_3S_2Na$; molecular weight 164.18; trade names: Uromitexan, Mesnex) (Fig. 1d) is a thiol uroprotective chemoprotectant used as an adjuvant in cancer chemotherapy to protect the bladder and kidneys from the urotoxic side effects of the chemotherapy drugs ifosfamide (Mitoxana, Ifex, and Holoxan), trofosfamide (Ixoten), and cyclophosphamide (Endoxan). It was developed as a prophylactic agent to reduce or detoxify the risk of hemorrhagic cystitis and hematuria (excretion of blood in urine) induced by

ifosfamide or cyclophosphamide and to decrease the incidence of ifosfamide-associated urothelial toxicity. Hematuria can also happen with higher doses of cyclophosphamide chemotherapy, but is less common. Higher doses of mesna are recommended if blood is detected in the urine. Ifosfamide or cyclophosphamide is converted to urotoxic metabolites such as acrolein and oxazaphosphorine metabolites, while mesna neutralizes these metabolites by binding through its sulfhydryl moieties and increases urinary cysteine excretion. Analogous to the physiological cysteine–cystine system, mesna is rapidly oxidized to its biologically inert disulfide metabolite, mesna disulfide or dimesna. Both mesna and dimesna are very hydrophilic and, therefore, remain in the intravascular compartment, where they are rapidly eliminated by the kidneys. In the kidney, the mesna disulfide is reduced to the free thiol compound, mesna, which reacts chemically with the urotoxic ifosfamide metabolites including acrolein and 4-hydroxy-ifosfamide, resulting in their detoxification. After oral administration, mesna has a bioavailability of 50–75%, and urinary mesna concentrations are approximately one half of those observed after i.v. infusion. The mean terminal half-life of mesna is 0.4 h, and the half-life of dimesna is 1.2 h.

Mesnex Injection contains 100 mg/ml mesna and is recommended for both oral and/or i.v. Adverse effects include nausea, vomiting, taste changes, headache, diarrhea, weakness, pain, skin rash, itching, irritation, and mood swings.

N-*Acetylcysteine.* It is a precursor of intracellular glutathione and cysteine and has an impressive array of mechanisms and protective effects toward DNA damage, carcinogenesis, and other mutation-related diseases. *N*-Acetylcysteine (empirical formula $C_5H_9NO_3S$; molecular weight 163.19; synonyms: LNAC, NAC, *N*-acetyl-L-cysteine; trade names: ACC, Mucomyst, Acetadote, Fluimucil, Parvolex) (Fig. 1e) is chemically known as (R)-2-acetamido-3-mercaptopropanoic acid and used mainly as a mucolytic (mucus dissolving) in a variety of respiratory conditions or in the management of paracetamol overdose. However, novel applications of

NAC, alone and in combination with other anti-cancer compounds, have been shown to be successful in treatment of tumor cell growth. First, it scavenges noxious ROS, and later, NAC is deacetylated in many tissues and cells to form L-cysteine, supporting glutathione biosynthesis that serves directly as an antioxidant or as a substrate in the glutathione redox cycle. Efficacy of different doses of NAC on potent carcinogens such as benzo(a)pyrene, 2-aminofluorene, and aflatoxin B_1 have been reported. NAC, a precursor of intracellular glutathione, is also capable of stimulating phase II enzymes in the glutathione cycle (GSH peroxidase, GSSG, reductase, GSH S-transferase). Repair of DNA damage has also been found to be stimulated by thiols like NAC and glutathione. In a rat hepatocarcinogenesis model, NAC administered by gavage inhibited the formation of carcinogen–DNA adducts. NAC (250–1500 mg/day) is well tolerated while mild gastrointestinal upset reported at very high doses.

ORG 2766. A neuroprotective chemo-protectant which slows down the neurotoxic effect or neuropathy of the cancer chemotherapy drug cisplatin, while leaving the antitumor activity of cisplatin unaffected. ORG 2766, a hexapeptide analog of ACTH-(4–9) [synonym: adrenocorticotropic hormone-(4–9)], prevents Taxol-induced neuropathy in rats and cisplatin-induced ototoxicity (ear poisoning). ORG 2766 is given subcutaneously in a dose of 0.25 mg/m^2 (low dose) or 1 mg/m^2 (high dose). No adverse effects were reported.

Natural Chemoprotectants

Berry Anthocyanin. Natural anthocyanins (synonyms: anthocyanins, anthocyanidins), including petunidin, malvidin, pelargonidin, peonidin, delphinidin, and cyanidin (Fig. 2), provide pigmentation (color) to fruits (especially berries), vegetables, and red wine and demonstrate novel chemotherapeutic, anticancer, anti-inflammatory, and antimutagenic properties. Blueberry, bilberry, cranberry, strawberry, lingonberry, tart cherry, black raspberry, and red raspberry as such, and their extracts, have exhibited potential cancer chemopreventive properties.

R_1 – H	R_2 – H	: Pelargonidin
R_1 = OH	R_2 = H	: Cyanidin
R_1 = OH	R_2 = OH	: Delphinidin
R_1 = OCH$_3$	R_2 = H	: Peonidin
R_1 = OCH$_3$	R_2 = OH	: Petunidin
R_1 = OCH$_3$	R_2 = OCH$_3$	: Malvidin

Chemoprotectants, Fig. 2 Structures of berry anthocyanins

Extensive studies were conducted on six edible berry extracts including wild blueberry, wild bilberry, cranberry, elderberry, raspberry, seed, and strawberry, and accordingly a novel synergistic combination of these six berry extracts known as "OptiBerry" was developed. The six berry extracts and OptiBerry demonstrated excellent antiangiogenic properties. OptiBerry was also shown to eradicate ▶ *Helicobacter pylori*, a causative factor for diverse gastrointestinal diseases including gastric cancer. Anthocyanins can be identified in human blood plasma and serum after consumption of berries.

Grape Seed Proanthocyanidins (GSP). OPC is the acronym for "oligomeric proanthocyanidins" [synonyms: procyanidins, grape seed extract, grape seed proanthocyanidins (GSP)], a class of polyphenolic bioflavonoids especially found in grape seeds and the bark of maritime pine trees. Catechin, epicatechin, and OPC dimers, trimers, and tetramers are shown in Fig. 3. GSP exhibited excellent free radical scavenging ability and provided significantly better protection as compared to vitamin C, vitamin E, and β-carotene in both in vitro and in vivo models. GSP exhibited significant protection against acetaminophen-induced hepato- and nephrotoxicity, amiodarone-induced

Chemoprotectants, Fig. 3 Structures of grape seed proanthocyanidins

pulmonary toxicity, dimethylnitrosamine (DMN)-induced splenotoxicity, cadmium chloride-induced nephrotoxicity, doxorubicin-induced cardiotoxicity, and O-ethyl S,S-dipropyl phosphorodithioate (MOCAP)-induced neurotoxicity in mice. GSP was shown to induce selective cytotoxicity toward cultured human MCF-7 breast cancer, A-427 lung cancer, and CRL-1739 gastric adenocarcinoma cells, while enhancing the growth and viability of normal human gastric mucosal cells and murine macrophage J774A.1 cells. The protective ability of GSP was assessed against chemotherapeutic drug-induced cytotoxicity toward normal human liver cells. Chang liver cells were treated with idarubicin (Ida) (30 nM) or 4-hydroxyperoxycyclophosphamide (4-HC) with or without GSP. GSP dramatically reduced the growth inhibitory effects of Ida and 4-HC on

liver cells. Thus, GSP can serve as a potential candidate to ameliorate the toxic effects associated with chemotherapeutic agents used in the treatment of cancer. Another study demonstrated that long-term exposure to GSP may serve as a potent barrier to all three stages of DMN-induced liver carcinogenesis and tumorigenesis by selectively altering oxidative stress, genomic integrity, and cell death patterns in vivo. No adverse effect is known.

Green Tea Catechins and Polyphenols. The history of tea as a beverage is traced by the Chinese to about 2700 BC. The green tea polyphenols (synonym: green tea extract) are composed of seven different kinds of catechin derivatives including (+)-catechin, (−)-epicatechin, (+)-gallocatechin, (−)-epigallocatechin, (−)-epicatechin-3-gallate, (−)-gallocatechin-3-gallate,

a Lycopene

(6E, 8E, 10E, 12E, 14E, 16E, 20E, 22E, 24E, 26E)-
2, 6, 10, 14, 19, 23, 27, 31-Octamethyldotriaconta-
2, 6, 8, 10, 12, 14, 16, 1B, 20, 22, 24, 30-tridecaene

b *trans-* Resveratrol   *cis-* Resveratrol

Trans-3, 4', 5-trihydroxystilebene; 3, 4', 5-stilbenetriol;
trans-resveratorl; (E)-5-(p-hydroxystyryl)resorcinol;
5-[(E)-2-(4-hydroxyphenyl)-ethenyl]benzene-1,3-doil

c Diadzein   Genistein

4', 7-Trihydroxyisoflavone   4', 5, 7-Trihydroxyisoflavone

Chemoprotectants, Fig. 4 Structure and IUPAC nomenclature of (**a**) lycopene, (**b**) *trans-* and *cis*-resveratrol, and (**c**) daidzein and genistein

and (−)-epigallocatechin-3-gallate (Fig. 3). Green tea catechins exhibit powerful antioxidant, antitumor, and anticancer properties. These help DNA molecules against oxidative damage, as well as eradicate *H. pylori*. High consumption of green tea and a low incidence of prostate and breast cancers have been reported in epidemiological studies. Case–control studies have demonstrated that high consumption of green tea, especially more than ten cups a day, is associated with cancer chemoprevention, while consumption of five cups lower the risk of esophageal, stomach, and gastric cancer. Green tea polyphenols increase the activity of both glutathione peroxidase and catalase in the intestines, liver, and lungs of mice and suppress spontaneous mutagenesis mediated by peroxide in the microenvironment of DNA following a substantial reduction of activated carcinogens. EGCG inhibits the growth and causes regression of human prostate and breast tumors. The advantages of cancer chemoprevention with green tea components are safety, economical, and early to mass-produce.

Lycopene. A bright red carotenoid pigment, chemically a terpene assembled from eight isoprene units, is a natural pigment biosynthesized by and accumulated in various fruits mostly in deep-red color of ripe tomatoes, vegetables, plants, and algae. Lycopene (empirical formula $C_{40}H_{56}$; molecular weight 536.87; synonyms: all-*trans* lycopene) (Fig. 4a) is one of the most potent carotenoid antioxidant in the human body, as well as a potent chemoprotectant. Lycopene exhibits anticancer properties by regulating cancer cell growth by interfering with cell cycle progression thereby inhibiting proliferation. An inverse association exists between intake of tomatoes and plasma levels of lycopene, and a lower risk for cancer was strongest for cancers of the lung, stomach, and prostate gland and was suggestive for cancers of the cervix, breast, oral cavity, pancreas, colorectum, and esophagus. Lycopene exhibited a cancer risk reduction of 30–40%. The red color of lycopene is due to many conjugated C = C double bonds, which absorb most of the visible spectrum. The antioxidant properties are responsible for the anticancer properties of lycopene. Lycopene in combination with vitamin D or vitamin E has been reported to inhibit cancer cell growth. It was shown that lycopene, with a half-maximal inhibitory concentration of 1–2 μM, more effectively impaired growth of select cancer cell types as compared with α-carotene or β-carotene. No adverse effects were reported.

Resveratrol. A phytoalexin, a natural antibiotic, conferring disease resistance in the plant kingdom and is produced under the conditions of UV radiation, fungal infection, and pathogenic attack. ► Resveratrol (3,4′,5-trihydroxystilbene, 5-[(E)-2-(4-hydroxyphenyl) ethenyl]benzene-1,3-diol; empirical formula $C_{14}H_{12}O_3$; molecular weight 228.25) (Fig. 4b) is mostly found in grapes, berries, nuts, red wine, and Japanese knotweed and produced with the help of the enzyme stilbene synthase. The *trans* configuration of resveratrol is the only naturally occurring isomer. Resveratrol was demonstrated to function as a potent antimutagen including the induction of phase II drug-metabolizing enzymes (antiinitiation activity), inhibition of cyclooxygenase and hydroperoxidase functions (antipromotion activity), and induction of human promyelocytic leukemic cell differentiation (antiprogression activity). Resveratrol was also found to possess chemopreventive activity by inhibiting ribonucleotide reductase and cellular events associated with cell proliferation, tumor initiation, promotion, and progression. Rapid absorption of resveratrol occurs at the intestinal level in both animals and humans and reaches the highest concentrations in the blood plasma approximately 1 h after administration. No adverse effects were reported.

Soy Isoflavonoids. Isoflavonoids, natural plant estrogens (► phytoestrogen), exhibit novel antioxidant and anticancer properties. Soybeans contain beneficial isoflavones such as daidzein (empirical formula $C_{15}H_{10}O_4$; molecular weight 254.25; chemical name 4′,7-dihydroxyisoflavone; synonyms: 7-hydroxy-3-(4-hydroxyphenyl)-4H-1-benzopyran-4-one, 4′,7-dihydroxyisoflavone) and ► genistein (empirical formula $C_{15}H_{10}O_5$; molecular weight 270.24; chemical name 4′,5,7-trihydroxyisoflavone; synonyms: 5, 7-dihydroxy-3-(4-hydroxyphenyl)-4H-1-benzopyran-4-one, 4′,5,7-trihydroxyisoflavone) (Fig. 4c). Being a weak form of estrogen, isoflavones can positively interact at estrogen receptor sites and maintain the requisite amount of estrogen in the blood level to reduce the risk factor for breast cancer and menopausal symptoms. Genistein reduces the risk factor for breast and prostate cancer and slows down the prostate cancer growth and renders prostate cancer cells to die. No adverse effects were reported.

Vitamins C and *E and β-carotene*. These antioxidants are associated with decreased risk of cancer.

Vitamin C. It is a water-soluble, highly bioavailable antioxidant and chemically known as 2-oxo-L-threo-hexono-1,4-lactone-2,3-enediol *or* (R)-3,4-dihydroxy-5-((S)-1,2-dihydroxyethyl) furan-2(5H)-one (Fig. 5a). Vitamin C (synonyms: L-ascorbate, L-ascorbic acid, L-xylo-ascorbic acid; empirical formula $C_6H_8O_6$; molecular weight 176.13) induces antioxidant efficacy in the biological systems. The low one-electron reduction potentials of ascorbate and the ascorbyl radical enable them to react with and reduce ROS and reactive nitrogen species. The ascorbyl radical may scavenge another radical or rapidly dismutates to form ascorbate and dehydroascorbic acid. Vitamin C acts as a coantioxidant by regenerating α-tocopherol from the α-tocopheroxyl radical. The recommended dietary allowance (RDA) for vitamin C is from 75 to 90 mg/day.

Vitamin E. A fat-soluble antioxidant present in cell membranes and lipoproteins and chemically known as (2R)-2,5,7,8-tetramethyl-2-[(4R,8R)-4,8,12-trimethyltridecyl]-3,4-dihydro-2H-chromen-6-ol (Fig. 5b). The term vitamin E (synonyms: α-tocopherol, tocopherol, (±)-α-tocopherol, 3,4-dihydro-2,5,7,8-tetramethyl-2-(4,8,12-trimethyltridecyl)-2H-1-benzopyran-6-ol; empirical formula $C_{29}H_{50}O_2$; molecular weight 430.69) describes a family of eight plant-derived antioxidants, α-, β-, γ-, and δ-tocopherol and α-, β-, γ-, and δ-tocotrienol. Only the *RRR*-stereoisomer occurs naturally. Vitamin E inhibits lipid peroxidation by reacting with lipid peroxyl radicals much faster than these radicals can react with polyunsaturated fatty acids to propagate the chain reaction of lipid peroxidation. The α-tocopheroxyl radical is relatively stable and can be reduced back by a coantioxidant such as vitamin C. The RDA for vitamin E is 15 mg/day.

β-Carotene. Another fat-soluble vitamin present in cell membranes and chemically known as 3,7,12,16-tetramethyl-1,18-bis(2,6,6-trimethyl-1-

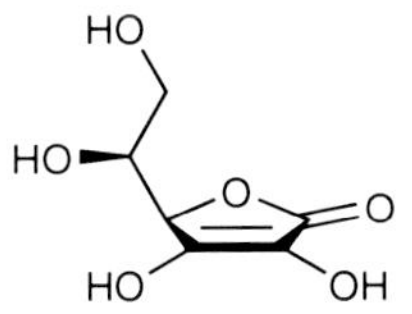

a Vitamin C

2-oxo-L-threo-hexono-1, 4-lactone-2, 3-enediol; (R)-3, 4-dihydroxy-5-((S)-1, 2-dihydroxyethyl)furan-2(5H)-one

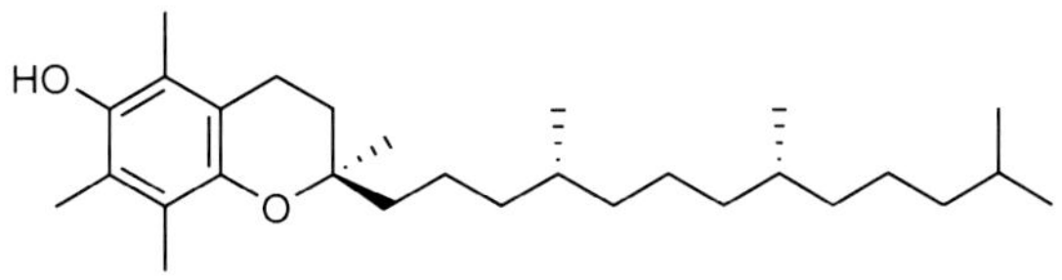

b Vitamin E

(R)-2, 5, 7, 8-tetramethyl-2((4R,8R)-4, 8, 12-trimethyltridecyl)chroman-6-ol

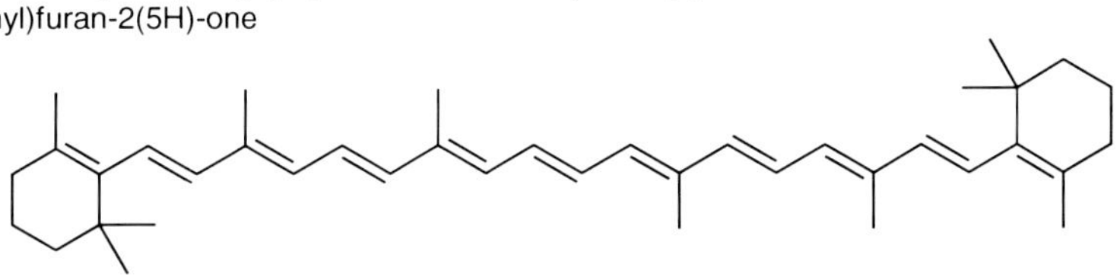

c β-Carotene

3, 7, 12, 16-tetramethyl-1, 18-bis(2, 6, 6-trimethyl-1-cyclohexenyl)-octadec; a-1, 3, 5, 7, 9, 11, 13, 15, 17-nonaene

Chemoprotectants, Fig. 5 Structure and IUPAC nomenclature of (**a**) vitamin C, (**b**) vitamin E, and (**c**) β-carotene

cyclohexenyl)-octadeca-1,3,5,7,9,11,13,15,17-nonaene (synonyms: β,β-carotene, provitamin A, β-cryptoxanthin, all-*trans* β-carotene; empirical formula $C_{40}H_{56}$; molecular weight 536.85) (Fig. 5c). Carotenoids, a class of carotenes, are a group of more than 600 naturally occurring pigments, of which only about 50 can be bioconverted to vitamin A. Carotenoids circulate in the blood with lipids in lipoproteins, while liver and adipose tissues are the major tissues where intact carotenoids accumulate. β-Carotene is the primary provitamin A carotenoid in the human diet. Vitamin A plays essential roles in visual function, immune system function, and cell growth and differentiation. The Institute of Medicine recommends 3–6 mg of β-carotene/day.

References

Bagchi D, Preuss HG (eds) (2004) Phytopharmaceuticals in cancer chemoprevention. CRC Press, Boca Raton

Block KI, Gyllenhaal C (2005) Commentary: the pharmacological antioxidant amifostine – implications of recent research for integrative cancer care. Integr Cancer Ther 4:329–351

Cvetkovic RS, Scott LJ (2005) Dexrazoxane: a review of its use for cardioprotection during anthracycline chemotherapy. Drugs 65:1005–1024

Jang M, Cai L, Udeani GO et al (1997) Cancer chemopreventive activity of resveratrol, a natural product derived from grapes. Science 275:218–220

Jones DP (2006) Redefining oxidative stress. Antioxid Redox Signal 8:1865–1879

Chemoprotection

▶ Chemoprotectants

Chemoradiation

▶ Chemoradiotherapy

Chemoradiotherapy

Marcel Verheij and Harry Bartelink
Department of Radiotherapy, The Netherlands Cancer Institute–Antoni van Leeuwenhoek Hospital, Amsterdam, The Netherlands

Synonyms

Bioradiotherapy; Chemoradiation; Combined modality treatment; Radiochemotherapy

Definition

Chemoradiotherapy refers to the combination of a cytostatic drug and external beam irradiation and can be applied sequentially or concurrently. There are several arguments to combine both modalities. While radiotherapy is aimed at controlling the primary tumor, chemotherapy is used to eradicate distant (micro-) metastases (spatial cooperation). Both modalities may be active against different tumor cell populations (independent cell-killing effect). In addition, chemotherapy may synchronize cells in a vulnerable phase for radiotherapy, decrease repopulation after radiotherapy, and enhance reoxygenation. It was also thought that shrinking a tumor with chemotherapy first should be advantageous for radiotherapy. However, this concept has failed in most clinical trials, probably due to fast repopulation of tumor cells after cytoreduction with chemotherapy before the start of radiotherapy. In contrast to sequential regimens, concurrent chemoradiotherapy exploits the ability of chemotherapeutic agents to sensitize radioresistant tumors to the lethal effect of ionizing irradiation. Optimal efficacy can be expected when the interaction between both modalities is synergistic.

Bioradiotherapy refers to the combination of radiotherapy with biological agents that specifically target deregulated pathways in tumor cells.

Characteristics

Several clinical trials carried out during the last decades clearly show that concurrent delivery of both chemotherapy and radiotherapy modalities significantly improves local control in a variety of advanced solid tumors. In most of these trials, ▶ cisplatin alone or in combination with other drugs has been used. This has led to improved survival rates in head and neck cancer, ▶ lung cancer, and ▶ cervical cancer. An additional important advantage of this combined treatment is the possibility to obtain a higher organ-preservation rate, such as in patients with advanced head and neck or anal cancer. Finally,

in the preoperative and ▶ adjuvant therapy setting, concurrent chemoradiation has contributed to a better outcome in terms of tumor downsizing/downstaging (▶ esophageal cancer and rectal cancer) and survival (▶ gastric cancer). Major further improvement can be expected from the combination with biological agents that are directed toward specific molecular targets in tumor cells (often referred to as bioradiotherapy). Examples include ▶ epidermal growth factor receptor (EGFR) inhibitors, antiangiogenic drugs, apoptosis modulators, and DNA repair-interfering agents.

Chemoradiotherapy for Non-small Cell Lung Cancer

For many years, radiotherapy has been the standard of care for inoperable stage III ▶ non-small cell lung cancer (NSCLC). These patients, however, show a poor outcome with long-term survival rates of 5–10%. Therefore, many groups have explored the possibility to improve these results by adding chemotherapy to the radiation treatment. The EORTC (European Organisation for Research and Treatment of Cancer) was the first in 1992 to report the results of a randomized phase III study of concomitant ▶ cisplatin (weekly or daily) and radiotherapy versus radiotherapy alone in patients with inoperable NSCLC. This combination of cisplatin with radiotherapy resulted in improved survival and control of local disease. The largest and significant benefit was seen in the treatment arm with radiotherapy and daily cisplatin. Two meta-analyses confirmed the benefit of concurrent cisplatin-based chemoradiotherapy compared with radiation alone and consolidated this regimen as standard treatment for stage III NSCLC. Whether chemoradiotherapy should be given sequentially or concurrently has also been the topic of several studies. Several randomized trials and ▶ meta-analyses have demonstrated that concurrent is superior over sequential chemoradiotherapy in terms of local control and survival, but is also associated with more, yet manageable acute toxicity, mainly esophageal. So far, no significant increase in late toxicity has been reported.

Chemoradiotherapy for Small Cell Lung Cancer

In the treatment of limited stage SCLC, the central role of chemotherapy has been widely recognized. To define an additional role of thoracic irradiation, several large phase III studies have been performed. These, together with subsequent meta-analyses, established the positive impact of thoracic irradiation in combination with chemotherapy in terms of local tumor control and survival. Regarding the timing of both treatment modalities, it has been shown that early thoracic irradiation during chemotherapy is superior to its late scheduling.

Chemoradiotherapy for Cervical Cancer

The introduction of chemoradiotherapy in the treatment of ▶ cervical cancer shows many similarities with that in NSCLC: until the 1980s, radiotherapy was the standard therapy for patients with locally advanced tumors. Despite modifications of the total radiation dose and overall treatment time, more than 70% of these patients developed a local regional recurrence. Therefore, improvements of these results were sought into the addition of chemotherapy to radiotherapy. In 1999, three articles were published reporting on studies comparing chemoradiotherapy with conventional radiotherapy for locally advanced cervical cancer. In all three studies, the combination of radiotherapy with cisplatin was significantly better than the control arms. An interesting observation came from Rose and colleagues, demonstrating that the single use of cisplatin was as effective as a combination of three drugs, the latter scheme being much more toxic. In the meantime, three additional trials on the concomitant use of cisplatin in cervical cancer have been published, demonstrating now in five out of six trials a significant improvement of local control and survival when concomitant cisplatin and irradiation was used. This is in contrast with eight out of nine phase III studies of neoadjuvant chemotherapy prior to radiotherapy, showing no benefit. Based on these results, concurrent chemoradiotherapy is nowadays the standard of care for cervical cancer. Several open questions persist, however. For example, it is unclear whether patients with very large tumors benefit as much from chemoradiotherapy as those with smaller tumors. Also, it remains to be established what the optimal chemotherapy or combination of cytostatic drug is for combined use with radiation and what role immunotherapy will play in combined modality strategies.

Chemo-Bioradiotherapy for Head and Neck Cancer

Several chemoradiation trials have been conducted in patients with previously untreated head and neck cancer using cisplatin alone, cisplatin and 6,4-photoproducts (▶ 5-FU), and other combinations. In eight single institutional studies, the average complete response to concomitant therapy was 67.5%. A ▶ meta-analysis performed by the MACH-NC group concerning the updated results of 63 randomized trials including 10,717 patients demonstrated a clear benefit of 8% ($p = 0.0001$) improved disease-free survival for the concomitant chemotherapy treatment. In the same analysis adjuvant and neoadjuvant chemotherapy showed no improvement. Subsequent trials confirmed that the concomitant use of ▶ cisplatin or carboplatin and irradiation leads to improved local cure and survival when compared with radiotherapy alone, including in the postoperative setting. The 3-arm GORTEC 99-02 randomized trial further showed that acceleration of radiotherapy cannot compensate for the absence of chemotherapy.

Despite these encouraging results, cisplatin-based chemoradiation protocols for advanced head and neck cancer are still associated with too many locoregional recurrences. Besides dose-escalation strategies, molecular targeted drugs represent a new and promising approach to further improve treatment results. One of these is the humanized monoclonal antibody directed against the ▶ EGFR which is frequently overexpressed in head and neck cancer and associated with chemo-/radioresistance and poor outcome. In 2006, a large multicenter randomized phase III study was published comparing radiotherapy alone with radiotherapy plus ▶ cetuximab in patients with locally advanced head and neck cancer. The results were very encouraging and demonstrated that the addition of cetuximab to radiotherapy significantly improved locoregional control and survival. Whether cetuximab (or other EGFR-blocking strategies) can further improve the results when added to standard chemoradiotherapy has been investigated in the RTOG 0522 randomized trial. Adding cetuximab to cisplatin-based chemoradiotherapy did not improve outcome, but was associated with significantly more acute toxicity. Whether HPV-positive head and neck cancer represents a separate disease entity requiring a differential approach is subject of ongoing trials.

Chemoradiotherapy for Esophageal Cancer

Surgical resection is currently the preferred treatment for ▶ esophageal cancer. Neoadjuvant chemotherapy may improve the results of surgery and may prevent patients from recurrent disease. However, a Cochrane meta-analysis based on seven phase III randomized trials with neoadjuvant chemotherapy failed to demonstrate such a beneficial effect. In a number of studies, sequential chemoradiotherapy or concurrent chemoradiotherapy was compared with radiotherapy alone. The RTOG 85-01 phase III study comparing radiotherapy alone with 5-FU/cisplatin-based chemoradiotherapy showed a statistically significant survival difference in favor of the chemoradiotherapy arm. Treatment-related toxicity was increased in the chemoradiotherapy arm, 44% severe and 20% life-threatening side effects versus 25% and 3% in the radiotherapy alone arm. Late toxicity was not increased as has been reported in other studies with concomitant chemoradiotherapy. Al-Sarraf reported on an additional group of patients treated with the same chemoradiotherapy regime. The 5-year survival was 26% versus 0% in the chemoradiotherapy arm and radiotherapy alone arm, respectively. These studies show that concurrent chemoradiation is recommended compared with radiotherapy alone. In most concurrent chemoradiotherapy studies, the classic 5-FU/cisplatin regimen has been used. Studies with taxanes as concurrently administered cytotoxic drugs showed promising results. In the Dutch randomized phase III CROSS trial, preoperative

chemoradiotherapy with a weekly schedule of ▶ paclitaxel and carboplatin was shown to result in significant tumor downstaging, high microscopically complete resection rate, and improved survival as compared to surgery only, with acceptable toxicity.

Chemoradiotherapy for Gastric Cancer

Surgical resection remains the cornerstone of curative treatment of ▶ gastric cancer. However, the long-term prognosis remains poor for patients with locally advanced disease. Therefore, different (neo-)adjuvant strategies have been evaluated in the past decades to improve these results. Adjuvant chemotherapy only resulted in a small survival benefit of 3–5% in Western populations as shown in multiple meta-analyses. Preoperative radiotherapy also showed a small, but significant improvement in survival. MacDonald et al. performed a randomized phase III study comparing surgery alone with surgery and postoperative adjuvant therapy, combining radiotherapy with 5-FU-leucovorin. In this study of 556 patients, a statistically and clinically significant reduced risk of relapse and improved survival were observed. Median overall survival in the surgery alone group was 27 months, compared with 36 months in the chemoradiation group. The 3-year survival rate was 41% versus 50% ($p = 0.005$), respectively. An update of the 10-year follow-up results in 2012 showed a strong persistent benefit from adjuvant chemoradiotherapy. In 2005, final results of the MAGIC study on perioperative chemotherapy have been presented. In this large multicenter study, patients were randomized between surgery only and three cycles of preoperative ECF (epirubicin, ▶ cisplatin, ▶ 5-FU) followed by surgery and then another three cycles of ECF chemotherapy. This regimen resulted in a 10% higher resectability rate and a significant survival benefit of 13% (23% vs. 36% at 5 years). Which of both strategies – postoperative chemoradiotherapy and perioperative chemotherapy – is superior remains to be determined. Since preoperative-combined chemoradiotherapy has shown a beneficial impact on surgical outcome in esophageal and rectal cancer, this is considered an attractive approach to explore in operable gastric cancer as well. Indeed, several phase I–II studies showed significant tumor downsizing, high R0, and pathological complete response rates by neoadjuvant chemoradiotherapy in locally advanced gastric cancer.

Chemoradiotherapy for Rectal Cancer

Surgical resection is the only curative treatment for ► colorectal cancer. However, following resection, local recurrence rate varies between 5% and 40%. Total mesorectal excision (TME), the standard surgical technique for primary resectable rectal cancer, has significantly improved the outcome of this disease, in particular, through the realization of free circumferential margins. The Dutch TME trial demonstrated that short-term preoperative radiotherapy is effective in preventing local recurrences, but not in patients with a positive resection margin. Although positive margins can be partly due to poor surgical techniques, they occur more often in locally advanced tumors. For these stages, a more aggressive (neo-)adjuvant approach is required. Postoperative chemoradiation has been mainly evaluated in the United States. The Gastrointestinal Tumor Study Group conducted a four-arm study: surgery only, postoperative chemotherapy, postoperative radiotherapy, and postoperative chemoradiotherapy (GITSG 71-75). Pairwise comparisons showed superior survival and local recurrence rates in the chemoradiation arm versus the surgery-only arm. The North Central Cancer Treatment Group compared radiotherapy with postoperative chemoradiation and demonstrated lower local and distant recurrence rates in the combined treatment arm. Survival was significantly increased (NCCTG 794751). The evidence that the addition of chemotherapy to preoperative radiotherapy improves local control rates has been provided by two separate trials. The EORTC 22921 trial has a two by two factorial design and randomized between preoperative radiotherapy and preoperative 5-FU-based chemoradiotherapy. A second randomization took place for postoperative chemotherapy versus no adjuvant treatment. The results demonstrated an increased local control rate for the chemoradiation arm: 92% versus 87%. A similar result was found in the French FFCD 9203 study, which randomized between preoperative radiotherapy and preoperative 5-FU-based chemoradiotherapy, with local recurrence rates of 16.5% and 8%, respectively. All studies that compare preoperative radiotherapy with preoperative chemoradiotherapy demonstrate an increase in toxicity in the combined modality arm.

It became clear that apart from cytotoxic agents, biological agents may play a role in the achievement of tumor response. In an experimental study, ► VEGF blockade enhanced radiotherapeutic activity, probably due to reduction of tumor vascular permeability and tumor interstitial pressure, thereby increasing the delivery of large therapeutic compounds to the tumor. In an early report on a small number of patients treated with the combination of an anti-VEGF monoclonal antibody, bevacizumab, 5-FU, and radiotherapy, significant downstaging occurred in all six patients.

Chemoradiotherapy for Anal Cancer

Over the past decades, the treatment of anal cancer has shifted from a surgical approach toward organ-sparing radiotherapy with or without concurrent chemotherapy. It was shown in two randomized studies that concomitant radiotherapy and 5-FU and ► mitomycin C (MMC) is superior to radiotherapy alone and significantly reduced the number of local recurrences. These anal cancer trials also clearly demonstrated the advantage of organ preservation by combined modality treatment as it results in an improved colostomy-free survival. The enhanced acute toxicity observed during these combined regimens did not translate in a significant increase in late side effects. MMC has contributed significantly to these results. In a RTOG study, patients were randomized to radiotherapy and 5-FU or radiotherapy, 5-FU, and MMC. The colostomy-free survival rate at 4 years was significantly better in patients who received both 5-FU and MMC compared with those who received 5-FU only (71% and 59%, respectively). In addition, others found that by deleting MMC from a comparable combined treatment protocol, the local tumor control rate at 2 years dropped from 87% to 58%. In order to minimize treatment-related toxicity, cisplatin has been evaluated as a replacement for MMC, with good results in nonrandomized studies. Randomized trials are now underway to confirm at least equal efficacy of cisplatin and MMC. The RTOG 98-11 phase III study compared 5-FU plus MMC and radiation to 5-FU plus cisplatin and radiation in 632 anal carcinoma patients. The results showed that the combination of radiotherapy with 5-FU plus MMC had a statistically significant, clinically meaningful impact on disease-free and overall survival, as compared to induction plus concurrent 5-FU and cisplatin. It was concluded that MMC, 5-FU (frequently given as oral capecitabine), and radiotherapy remain the standard of care for patients with anal canal carcinoma. Further improvements in treatment results are expected from the application of novel biological agents.

Chemoradiotherapy for Glioblastoma

Glioblastoma has a dismal prognosis with most patients dying within 2 years after diagnosis. Standard therapy consisted of surgical resection, followed by radiotherapy. Although a meta-analysis of 12 randomized trials suggested a small survival advantage by the addition of chemotherapy, Stupp et al. in 2004 were the first to demonstrate a clinically meaningful and statistically significant survival benefit from the addition of temozolomide (TMZ) to radiotherapy with minimal additional toxicity. The subsequent 5-year analysis of this EORTC-NCIC randomized phase III trial confirmed these results with the 2-year overall survival improving from 10.9% after surgery only to 27.2% with adjuvant TMZ and radiotherapy. The MGMT promoter methylation status of the tumor was identified as a prognostic biomarker, selecting those patients most likely to benefit from the addition of TMZ.

Concluding Remarks

The combination of radiotherapy and chemotherapy has resulted in a major step forward in the treatment of patients with advanced solid tumors. The recognition that concurrent chemoradiotherapy is superior to sequential regimens may be viewed as one of the major achievements in clinical oncology of the past decades. In general, the interaction between radiation and cytostatic agents is time-, dose-, and sequence-dependent as shown for cisplatin, the most widely used ► radiosensitizer. In the near future, the combination of radiotherapy with biological agents and a number of new cytostatic drugs will become available for testing in concomitant chemotherapy or biotherapy and radiotherapy approaches. These agents should be selected based upon their mechanisms of action. Given the results of many randomized clinical studies, it is quite likely that chemo- or bioradiotherapy will be the standard of care for an increasing number of advanced squamous cell cancers, but until the best regimen of each disease has been determined, there is now more than ever an urgent need to encourage treatment of patients within the framework of carefully controlled clinical trials.

Cross-References

► Adjuvant Therapy
► Cervical Cancers
► Cetuximab
► Cisplatin
► Colorectal Cancer
► Epidermal Growth Factor Inhibitors
► Epidermal Growth Factor Receptor
► Esophageal Cancer
► Fluorouracil
► Gastric Cancer
► Lung Cancer
► Meta-Analysis
► Mitomycin C
► Neoadjuvant Therapy
► Non-Small-Cell Lung Cancer
► Paclitaxel
► Radiosensitizer
► Vascular Endothelial Growth Factor

References

Bartelink H, Schellens JHM, Verheij M (2002) The combined use of radiotherapy and chemotherapy in the treatment of solid tumors. Eur J Cancer 38:216–222

Hennequin C, Favaudon V (2002) Biological basis for chemo-radiotherapy interactions. Eur J Cancer 38:223–230

John MJ (2004) Radiotherapy and chemotherapy. In: Leibel SA, Phillips TL (eds) Textbook of radiation oncology, 2nd edn. Saunders, Philadelphia, pp 77–100

Stewart FA, Bartelink H (2002) The combination of radiotherapy and chemotherapy. In: Steel GG (ed) Basic clinical radiobiology, 3rd edn. Hodder Arnold, London, pp 217–230

Tannock IF (1996) Treatment of cancer with radiation and drugs. J Clin Oncol 14:3156–3174

Verheij M, Vens C, van Triest B (2010) Novel therapeutics in combination with radiotherapy to improve cancer treatment: rationale, mechanisms of action and clinical perspective. Drug Resist Updat 13:29–43

See Also

(2012) Adjuvant. In: Schwab M (ed) Encyclopedia of cancer, 3rd edn. Springer, Berlin/Heidelberg, p 75. doi:10.1007/978-3-642-16483-5_107

(2012) Anti-angiogenic drugs. In: Schwab M (ed) Encyclopedia of cancer, 3rd edn. Springer, Berlin/Heidelberg, pp 207–208. doi:10.1007/978-3-642-16483-5_302

(2012) Carboplatin. In: Schwab M (ed) Encyclopedia of cancer, 3rd edn. Springer, Berlin/Heidelberg, p 641. doi:10.1007/978-3-642-16483-5_833

(2012) Concurrent. In: Schwab M (ed) Encyclopedia of cancer, 3rd edn. Springer, Berlin/Heidelberg, p 965. doi:10.1007/978-3-642-16483-5_6821

(2012) EGFR. In: Schwab M (ed) Encyclopedia of cancer, 3rd edn. Springer, Berlin/Heidelberg, p 1211. doi:10.1007/978-3-642-16483-5_1828

(2012) Epirubicin. In: Schwab M (ed) Encyclopedia of cancer, 3rd edn. Springer, Berlin/Heidelberg, p 1291. doi:10.1007/978-3-642-16483-5_1955

(2012) Leucovorin. In: Schwab M (ed) Encyclopedia of cancer, 3rd edn. Springer, Berlin/Heidelberg, p 2005. doi:10.1007/978-3-642-16483-5_3321

(2012) Neoadjuvant. In: Schwab M (ed) Encyclopedia of cancer, 3rd edn. Springer, Berlin/Heidelberg, p 2472. doi:10.1007/978-3-642-16483-5_4003

(2012) 6,4-Photoproduct. In: Schwab M (ed) Encyclopedia of cancer, 3rd edn. Springer, Berlin/Heidelberg, p 2881. doi:10.1007/978-3-642-16483-5_4557

(2012) VEGF. In: Schwab M (ed) Encyclopedia of cancer, 3rd edn. Springer, Berlin/Heidelberg, p 3906. doi:10.1007/978-3-642-16483-5_6174

Chemoresistance

▶ Drug Resistance

Chemosensibilization

Grégory Gatouillat and Claudie Madoulet
Laboratory of Biochemistry, IFR53, Faculty of Pharmacy, Reims, France

Synonyms

Chemosensitization; Resistance modulation; Resistance reversion

Definition

The sensitization to chemotherapeutic agents of resistant tumor cells is known as chemosensibilization.

Characteristics

Chemosensibilization is used when tumor cells no longer respond to chemotherapeutic drugs. This resistance can be inherent to tumor cells or can be acquired during ▶ chemotherapy treatment, leading to the inefficiency of a wide range of antineoplastic agents. This phenomenon is named multidrug resistance (MDR). Once MDR appears, chemotherapy is not efficient anymore even when using high doses of drugs, which stimulates the resistance mechanism and brings toxic side effects. To overcome this problem, several strategies aiming at restoring drug sensitivity are used. The main approach to achieve chemosensibilization is the use of substances capable of bypassing the resistance mechanism, named chemosensitizers or MDR modulators. A typical chemosensitizer should not really have inherent antitumor properties; however, most chemosensitizers exhibit antitumor activity and act synergistically with antineoplastic drugs to kill tumor cells. After the discovery of the mechanisms leading to chemoresistance, much effort has been devoted to discover such candidate molecules. Although numerous chemosensitizers have been developed, few of them have reached clinical trials. MDR is a multifactorial phenomenon, and the best characterized mechanisms are resistant to apoptosis and enhanced drug efflux due to the overexpression of ATP-binding cassette transporters (▶ ABC transporters) such as ▶ P-glycoprotein (P-gp), MRP, or BCRP in tumor cells.

Chemosensibilization of Tumors by Targeting P-gp

Inhibitors of P-gp Activity

The development of chemosensitizers to overcome chemoresistance mediated by ABC transporters mainly focuses on the inhibition of P-gp that is overexpressed in a number of malignancies (Table 1). Many chemosensitizers are administered simultaneously with anticancer drugs. They overcome drug resistance by functioning as competitive or noncompetitive inhibitors for P-gp and by binding either to drug modulation sites or to other modulator binding sites. Other reversing agents act by interfering with ATP hydrolysis required to P-pg activity.

The first chemosensitizers used to inhibit P-gp-mediated MDR were drugs that possess unrelated pharmacological functions. Among them, verapamil was the first substance that showed chemosensitizing activity. It was originally used as a calcium channel blocker in the treatment of heart disease. Verapamil is a substrate of P-gp and inhibits the transport of chemotherapeutic drugs in a competitive manner without interfering with its catalytic cycle. Cyclosporin A, a commonly used immunosuppressant for organ transplantation, can also sensitize MDR tumor cells by interfering with both substrate recognition and ATP hydrolysis. Unfortunately, the use of these first-generation chemosensitizers in clinical studies has been limited. These sensitizers reverse MDR at high concentrations, which brings toxic side effects due to their innate pharmacological function.

The search for nontoxic second-generation chemosensitizers resulted in newer analogs of the first-generation modulators that were more potent and considerably less toxic. Structural analogs of verapamil show increased reversal activity

Chemosensibilization, Table 1 Compounds used to chemosensitize P-gp-mediated MDR tumor cells

Antiarrhythmics	Quinidine, amiodarone, propafenone
Antibiotics	Cephalosporins
Antihistaminics	Terfenadine, azelastine
Antihypertensive	Reserpine
Antimalarials	Quinine, quinacrine, mefloquine
Calcium channel blockers	Verapamil, dexverapamil, nicardipine, azidopine
Calmodulin antagonists	Trifluorperazine, chlorpromazine
Immunosuppressants	Cyclosporin A, SDZ PSC 833, staurosporine, rapamycin
Neuroleptics	Phenothiazine, fluoxetine, haloperidol
P-gp-specific chemosensitizers	MS 209, GF 120918, XR 9576, VX-710, LY 335979
Steroid hormones and synthetic derivatives	Progesterone, tamoxifen
Alkaloids	Cyclopamine, tetrandrine, fangchinoline
Flavonoids and dietary compounds	Quercetin, genistein, curcumin, green tea polyphenols, ginsenoside Rg, indole-3-carbinol, diallyl sulfide
Anti-P-gp antibodies	Monoclonal antibodies, immunization-induced antibodies
siRNA, antisense oligonucleotides	
Anti-MDR1 ribozymes	

when used at lower doses and are less cardiotoxic than verapamil. PSC 833, a potent and non-immunosuppressive analog of cyclosporine, efficiently reverses MDR and has been used in combination with anticancer drugs in clinical studies. It restores sensitivity to chemotherapy by direct interaction with P-gp. Although these modulators are more efficient than the first-generation chemosensitizers, they influence the pharmacokinetics of anticancer drugs, elevating plasma concentration beyond acceptable toxicity.

Third-generation chemosensitizers were designed using structure-activity relationships specifically for high transporter affinity and low pharmacokinetic interaction. The latest synthetic compounds, including VX-710, LY 335979, GF-120918, and XR 9576, are currently used in clinical trials in association with anticancer drugs to sensitize tumor cells.

Several other compounds with different pharmacological functions possess chemosensitizing effects in several models of resistant tumor cells expressing P-gp (Table 1).

Other Strategies to Inhibit P-gp

Alternative strategies that include the use of chemicals as chemosensitizers can be applied to restore sensitivity of tumors:

1. *Anti-P-gp Antibodies.* Antibodies specific to P-gp have been developed and are capable of potent reversal of MDR by disrupting P-gp drug efflux activity. These antibodies can be generated by several ways. Monoclonal antibodies generated from hybridomas have been developed to target P-gp. The monoclonal antibody UIC2 recognizes a conformational epitope that involves several peptide fragments of the human P-pg. The binding of UIC2 to P-gp induces the blockade of conformational changes required to the activity of the efflux pump, leading to an increase in intracellular accumulation of drugs. Recombinant antibody fragments targeted to extracellular loops of P-gp are also used in vitro to sensitize MDR cells to chemotherapy, as well as antibodies induced by immunization with P-gp-derived peptides that allow the in vivo sensitization of tumor cells to anticancer drugs.

2. *Altered Levels of MDR1 mRNA.* Downregulation of the *MDR1* gene coding for P-gp is another way to overcome chemoresistance. It is based on the use of molecules such as antisense oligonucleotides (ASOs), small interfering RNAs (siRNAs), and ribozymes whose activity leads to altered level of a specific mRNA (▶ Antisense DNA therapy, ▶ RNA interference). They can specifically modulate the transfer of the genetic information from DNA to proteins. ASOs are short single-strand DNAs that hybridize to a unique mRNA sequence. Hybridization of

ASOs to mRNA leads to the formation of mRNA/DNA hybrid duplexes that become the target of RNase H, an enzyme that catalyzes the cleavage of RNA in RNA/DNA duplexes. siRNAs are small RNAs duplexes that assemble into an RNA-induced silencing complex (RISC). These complexes target a specific mRNA that is cleaved and degraded. The targeting of a unique mRNA can also be achieved by using catalytic RNAs called ribozymes. These small RNAs hybridize to a complementary sequence of mRNA and catalyze site-specific cleavage of the substrate. Specific ASOs, siRNAs, and ribozymes targeted to the MDR1 mRNA are used to prevent P-gp expression in tumor cells and can improve sensitivity toward chemotherapeutic drugs in resistant tumor cells.

Chemosensibilization of Tumors by Triggering Apoptosis

Tumor cells can become resistant to chemotherapy due to a reduced susceptibility to die by ▶ apoptosis, by the overexpression of antiapoptotic proteins and activation of prosurvival signaling pathways, or by the downregulation of proapoptotic proteins. Cancer treatment by chemotherapy kills cells principally by inducing apoptosis. Therefore, modulation of the key elements of apoptotic signaling directly influences therapy-induced tumor cell death and represents another way to sensitize tumor cells to apoptosis-inducing drugs. Given that antiapoptotic Bcl-2 family members are overexpressed in many types of cancers, specific ASOs directed against these proteins can be useful to improve drug sensitivity. The administration of Bcl-2 and Bcl-x_L antisenses in combination with anticancer drugs results in a decreased level of these antiapoptotic molecules and the subsequent improved efficiency of drugs. In many cases, complete cure of mice-bearing Bcl-2 or Bcl-x_L-overexpressing tumors occurs. Downregulation of other proteins inhibiting apoptosis including survivin and XIAP and silencing of prosurvival signaling mediated by the PI-3kinase/Akt pathway and NFκB can also sensitize tumor cells to programmed cell death.

A growing interest has been placed upon the use of dietary polyphenols, which induce apoptosis in cancer cells while they protect normal cells. These compounds not only have the capacity to trigger cell death when used as single agents but also enhance apoptosis triggered by numerous anticancer drugs in several tumor cell lines by interfering with multiple pathways leading to chemoresistance (▶ Polyphenols).

Cross-References

- ▶ ABC-Transporters
- ▶ Antisense DNA Therapy
- ▶ Apoptosis
- ▶ Chemotherapy
- ▶ P-Glycoprotein
- ▶ Polyphenols
- ▶ RNA Interference

References

Garg AK, Buchholz TA, Aggarwal BB (2005) Chemosensitization and radiosensitization of tumors by plant polyphenols. Antioxid Redox Signal 7:1630–1647
Shabbits JA, Hu Y, Mayer LD (2003) Tumor chemosensibilization strategies based on apoptosis manipulation. Mol Cancer Ther 2:805–813
Szakács G, Paterson JK, Ludwig JA et al (2006) Targeting multidrug resistance in cancer. Nat Rev Drug Discov 5:219–234

See Also

(2012) Monoclonal Antibody. In: Schwab M (ed) Encyclopedia of Cancer, 3rd edn. Springer Berlin Heidelberg, p 2367. doi: 10.1007/978-3-642-16483-5_6842

Chemosensitization

▶ Chemosensibilization

Chemotactic Cytokine

▶ Chemokines

Chemotaxis

▶ Motility

Chemotherapy

Emil Frei
Dana-Farber Cancer Institute, Boston, MA, USA

Definition

Chemotherapy is defined as the use of chemical agents for treatment. Chemotherapy as used for cancer generally refers to small molecules that damage proliferating cells. It represents systemic treatment in contrast to radiotherapy and surgery that represent local treatment. Classes of systemic agents may also include ▶ hormones, ▶ cytokines, and antitumor vaccines.

http://www.chemocare.com/whatis/important_ chemotherapy_terms.asp

Characteristics

The Challenge

Cancer is the most feared, morbid, and mortal of diseases. In the USA, five million people contract cancer per year, of whom one third, or almost half a million citizens, will die of their disease. Most cancers start in a specific location (e.g., breast, lung) and spread to regional lymph nodes; in ▶ breast cancer, spread is to the armpit and subsequent dissemination by the bloodstream to distant organs. For cancers that are diagnosed before such dissemination, local treatment with surgery and/or radiotherapy may be curative. Most patients who die of cancer die because of disseminated metastatic tumor. These are either clinically present at the time of diagnosis or occur months to

---

Emil Frei: deceased.

years after diagnosis because of microscopic clinically undetectable cancer that only becomes clinically evident following local treatment. Cancer chemotherapy along with hormone therapy (▶ Endocrine Therapy) and ▶ immunotherapy is designed to treat and ideally eradicate metastatic cancer.

The Agents

Most of the currently effective chemotherapeutic agents were discovered by serendipity and/or empiricism (by trial and error). For example, the first effective agent, nitrogen mustard (Nitrogen Mustards), was a derivative of chemical warfare studies conducted in World War I. Among the side effects of mustard gas was the suppression of normal bone marrow. Because of this, it was given to mice bearing a tumor derived from the bone marrow, i.e., leukemia, and found to be effective. Subsequent clinical trials affirmed this effectiveness. Analogs were synthesized and mustard-like compounds, termed ▶ alkylating agents, are effective in many forms of cancer.

The antimetabolites (Antimetabolite) are compounds that are similar to normal metabolites, such that they enter the same metabolic system but because of slight differences, inhibit or antagonize that system. For example, white cells consume high quantities of the vitamin folic acid, and this is particularly true of cancerous white cells, that is leukemic cells. Slight chemical modifications of folic acid have lead to the antifol class of compounds, and these have been found to be active in patients with leukemias as well as in many solid tumor patients.

Two important classes of compounds are

- The anthracyclines (▶ Anthracycline)
- The platinum analogs (▶ Platinum Complexes) were discovered by serendipity and developed largely through screening methods. All of the above compounds target DNA and therefore cell proliferation. Another class of compounds target the ▶ cytoskeleton of the tumor cell. These are derived from fungi and plants, and include the vinca alkaloids and ▶ taxol.

These examples of currently used agents, while varyingly effective against human cancers, have a significant limitation in the area of specificity. They attack not only the tumor but also certain rapidly growing normal tissues, such as the bone marrow and bowel, and hence produce dose-limiting toxicity relating to depression of the marrow (infection and bleeding), nausea and vomiting, and ulceration of the gastrointestinal tract.

The use of high dose combination chemotherapy with stem cell rescue for patients with breast cancer was the subject of considerable enthusiasm during the 1980s and early 1990s. However, in the late 1990s and particularly since the American Society of clinical Oncology (ASCO) reports in 1999 have been considered to be largely ineffective. This paper considers some of the reasons for this change and particularly on the basis of preclinical and clinical models, considers current and future directions. The intensification regimen may produce resistance that could compromise the important intensification component. Microenvironmental and clinical trials of adjuvant chemotherapy strongly indicate that one cycle of intensification is not enough and that two and perhaps three will be required. The components of the intensification regimen are reviewed with respect to dose response and with respect to mechanisms of resistance, cross resistance, and potential additive or synergistic effects.

To reiterate, the major limitation of classical cancer chemotherapy is the lack of specificity for the tumor as compared to normal tissue. This limitation is being addressed by basic science particularly relating to molecular biology; a summary of which follows:

Cancer is a genetic disease of somatic cells, following a series of mutations or genetic events incident to lifestyles such as smoking.

In ▶ tobacco carcinogenesis and genetic susceptibility, a sufficient number of events occur such that a cell becomes transformed into a cancer cell. The vast majority of cancers therefore derive from a single cell. However, the process that produces cancer also results in a marked increase in genetic instability such that daughter cells are variable. This variation permits selection of those daughter variants that have a survival advantage, such as resistance to certain drugs, a higher proliferative thrust, or a greater capacity to invade and metastasize. This clonal evolution to heterogeneity is adverse. However, it does lead to events that are unique to the cancer cell, that is, they are not present in normal cells in the same person. For example, ▶ chronic myelogenous leukemia is due to white cells being driven to cancer behavior by a product of the fusion of two genes. By advanced pharmacologic techniques including designer drug synthesis and high throughput screening, an agent termed STI571 was developed that inhibits the action of the fusion gene protein product. STI571 has been found to be capable of producing complete regression of leukemia in the majority of patients and, in contrast to essentially all chemotherapy, is non-toxic. There are a number of molecular targets in other tumors that have been identified, and academia and the pharmaceutical industry have given major priority to the development of agents capable of selectively attacking and inhibiting such molecular targets. It is this process more than any other that leads to optimism on the part of cancer investigators concerning the future of cancer treatment.

Clinical Strategies

Although there are numerous different types of cancer, they often share many biological and molecular processes, an important feature to keep in mind.

In the late 1950s, a series of integrated clinical trials were conducted. As a result, the cure rate for childhood leukemia increased from 0% to 70% and many scientific principles of cancer therapy were established. These were (in chronological order):

- The application of quantitative clinical trials, involving comparisons and randomizations.
- The use of agents in an appropriate combination can increase the complete remission rate from zero to more than 90%.
- The generation of complete remission as the most powerful discriminant for survival.

- The identification of active agents in an experimental model with the duration of complete remission (DCR) being central parameter of response.
- The use of the DCR model to develop and evaluate optimal doses, schedules, and combinations.
- It was observed that meningeal leukemia occurred with increasing frequency in patients with prolonged complete remission. Pharmacological and clinical trial studies found this be due to the failure of standard anti-leukemia agents to cross the blood–brain barrier. The introduction of intrathecal chemotherapy and radiotherapy to the brain markedly reduced this type of complication.
- The importance of supportive and symptomatic care, for example, platelet transfusions, antibiotics, and antiemetics, markedly reduced the morbidity and mortality of chemotherapy.

The result of these advances was an increase in the cure rate for childhood leukemia from 0% in 1955, to 35% by 1970, up to 80% within the last 15 years. This experience with childhood leukemia had a profound effect on the field of cancer chemotherapy in general. Most importantly, it established the position that it can be done, that is, systemic cancer could be cured by systemic therapy.

Most patients who die of cancer die because of disseminated metastatic tumor. Today, in major centers and cooperative groups the cure rate of childhood leukemia is close to 80%. It was hoped that the solid tumors would follow on closely behind the leukemias, but they have in the main proven more difficult to treat. This is unfortunate as adult solid tumors such as breast, bowel, and lung cancer constitute 80% of all cancers. A major strategy has been to use agents in combination. This is because solid tumor cells are heterogeneous and thus they have multiple targets. The second rationale for combination chemotherapy is that it works, with essentially all highly effective and certainly curative cancer chemotherapy involving combinations. The best of combinations produce partial responses in 30–50% of patients with, for example, metastatic cancer of bowel and lung. Complete tumor regression rarely occurs.

It has long been known that in experimental tumors, for example in mice, tumor burden is critical to chemotherapeutic effect. Thus, chemotherapy that has a minor effect on palpable tumor is often curative of the same tumor in microscopic form. This led to the strategy known as adjuvant chemotherapy. Here patients known to be of high risk of having micrometastatic cancer at the time of initial treatment are given chemotherapy immediately following surgery and/or radiotherapy. This increases the cure rate some 20% for breast and large bowel cancers. In a strategy (termed neoadjuvant chemotherapy), chemotherapy is given prior to surgery. This moves chemotherapy still further forward in the disease. It provides shrinkage of the primary tumor and thus facilitating the use and effectiveness of local treatment. It may decrease the need for radical surgery in certain tumors such as head, neck, and bladder cancer. Finally, combination chemotherapy can be given concurrently with radiotherapy as initial treatment. In addition to the above advantages, local control may be superior with many chemotherapeutic agents since some of them, particularly the platinum analogs and fluorouracil, are highly radiosensitizing.

In addition to combination chemotherapy, dose is a significant factor in cancer chemotherapy. The dose of certain chemotherapeutic agents, particularly the alkylating agents, can be substantially increased if one protects the bone marrow. Protection is provided by harvesting marrow stem cells before the high dose chemotherapy and returning the marrow to the patient following chemotherapy. Such peripheral blood stem cell rescue has been effective in the leukemias and lymphomas and is under study often in combination with some of the above-mentioned strategies in selected solid tumors.

Supportive Care

Bone marrow or peripheral blood stem cell transplantation is one form of supportive care. The first major advance in supportive care involved platelet transfusions for the treatment and prevention of thrombocytopenic hemorrhage, starting in the late 1950s and early 1960s. The nausea and vomiting associated with cancer, and some forms of cancer treatment, has been markedly reduced by the development of antiemetics. Pain control has markedly improved and radical surgery has been reduced by neoadjuvant approaches.

Long-Term Effects of Cancer Treatment

Perhaps the most worrisome long-term effect has been the development of secondary cancers (▶ Second Primary Tumors), particularly leukemia and leukemia-like illnesses. There is a latent period of 5–10 years for most of these secondary cancers, and for solid tumors it is even longer. Clinical treatment, environmental and genetic factors, and in vitro and in vivo laboratory models are being developed to study these events. The alkylating agents and X-ray are the chief offenders. ▶ Hodgkin disease was found to be curable by strategies similar to that of acute lymphocytic leukemia, but there was a cumulative long-term risk of secondary cancer. With this knowledge and the development of newer active agents for Hodgkin disease, the combination regimen that included alkylating agents has been modified without loss of effectiveness, but with major diminution in secondary cancers.

Conclusions

With a marked increase in support for cancer research including both basic and clinical and the extraordinary increase in molecular sophistication of such research, it is expected that major progress in the curative treatment of most, if not all, cancers will be achieved in the next decade. It is the clinical scientist who must translate this progress in basic research to the clinic. This ultimate challenge would require the most sophisticated of treatment methodology and must always be conducted in a setting where the primary beneficiary of such research is the patient.

References

Fearon ER, Vogelstein B (2000) Tumor suppressor gene defects in human cancer. Cancer Med 5:67–87

Holland JF, Frei E III, Kufe DW et al (2000) Principles of medical oncology. Cancer Med 5:503–510

Pollock RE, Morton DL (2000) Principles of medical oncology. Cancer Med 5:448–458

Ries LAG, Kosary CL, Hankey BF et al (eds) (1999) SEER cancer statistics review 1973–1996. National Cancer Institute, Bethesda

CHETK

▶ Choline Kinase

CHFR

Maria J. Worsham
Department of Otolaryngology, Henry Ford
Health System, Detroit, USA

Synonyms

Checkpoint with forkhead and RING finger domain protein; E3 ubiquitin protein ligase; FLJ10796; FLJ33629; RING finger protein 196; RNF116; RNF196

Definition

The *CHFR* (checkpoint with forkhead and RING finger domains) gene, located at 12q24.33, coordinates mitotic prophase by delaying chromosome condensation in response to a mitotic stress.

Characteristics

Chromosomal segregation at mitosis is preceded by a series of steps, including condensation of chromosome and separation of the centrosome, chromosomal alignment, and sister-chromatid separation (Shibata et al. 2002). To ensure fidelity of the replicated genetic material, mitosis is carefully choreographed and monitored by several checkpoint systems (Nigg 2001; Wassmann and Benezra 2001). Missteps in any one of these processes can result in aneuploidy or genetic instability, setting off any number of deleterious events such as unregulated cell growth, leading to neoplastic transformation and tumor progression. A mitotic checkpoint that delays chromosome condensation in response to mitotic stress induced by paclitaxel or nocodazole involves the *CHFR* gene which is associated with delaying prophase in human cells (Scolnick and Halazonetis 2000). Mitotic checkpoint function is impaired in a significant proportion of human cancer cell lines, often by genetic alterations, supporting a possible link between the impaired mitotic checkpoint and oncogenesis.

Activity in Cancer

CHFR expression has been found to be reduced in cancers due to the loss of gene copy number, possible interaction between *CHFR* and the DNA mismatch repair system, and promoter hypermethylation. The downregulation of *CHFR* is associated with microsatellite stable and instable tumors as well as chromosomal instability to varying degrees in multiple cancers. A statistically significant correlation has been found to exist between hypermethylation of *MLH1* and incidences of *CHFR* hypermethylation in primary colorectal cancer and gastric cancer. In HNSCC, comprehensive high-throughput methods have underscored the contribution of genetic and epigenetic events, often working together, in the development and progression of HNSCC. Epigenetic mechanisms involve DNA and histone modifications, resulting in the heritable silencing of genes without a change in their coding sequence. Gene transcriptional inactivation via hypermethylation at the CpG islands within the promoter regions is an important mechanism.

The association of promoter hypermethylation of *CHFR* and tumors was first reported in primary lung cancer where *CHFR* downregulation appeared to be associated with poor prognosis. The examination of the prevalence and pattern of *CHFR* inactivation in human tumors found CpG methylation-dependent silencing of *CHFR* expression in 40% of primary colorectal cancers,

53% of colorectal adenomas, and 30% of primary head and neck cancers.

Various cancers have demonstrated the inactivation of *CHFR* at different stages. In colorectal cancers, *CHFR* inactivation occurs at an early stage (Brandes et al. 2005), whereas in hepatocellular, non-small cell lung cancer (NSCLC), and oral squamous cell carcinoma, it occurs at a late stage. Attempts at determining a correlation between stage and *CHFR* inactivation have not demonstrated any clear pattern. In a pilot cohort of 28 HNSCCs, aberrant methylation of *CHFR* was detected in only stage IV tumors. This preliminary finding of promoter hypermethylation of *CHFR*, by MS-MLPA and MSP assays, only in late-stage tumors appears to suggest *CHFR* as a late event and a putative diagnostic biomarker for late-stage disease.

Biomarker Potential

CHFR methylation status may be useful as a diagnostic or prognostic marker. In smoking-related NSCLC, *CHFR* hypermethylation is associated with poorer outcomes, and in colorectal cancer it is better associated with the sporadic type of cancer. In gastric cancer, *CHFR* hypermethylation is a diagnostic marker of lymph node micrometastasis which could be useful in determining recurrence.

Conclusion

Treatment with the methyltransferase inhibitor 5-aza-2'-deoxycytidine induced re-expression of *CHFR*. In addition, because cancer cells that lack *CHFR* expression have been shown to be more susceptible to the microtubule inhibitor paclitaxel, silencing of *CHFR* by methylation can serve as a marker for predicting sensitivity to particular chemotherapeutic agents.

References

Brandes JC, van Engeland M, Wouters KA et al (2005) CHFR promoter hypermethylation in colon cancer correlates with the microsatellite instability phenotype. Carcinogenesis 26:1152–1156

Nigg EA (2001) Mitotic kinases as regulators of cell division and its checkpoints. Nat Rev Mol Cell Biol 2:21–32

Scolnick DM, Halazonetis TD (2000) Chfr defines a mitotic stress checkpoint that delays entry into metaphase. Nature 406:430–435

Shibata Y, Haruki N, Kuwabara Y et al (2002) Chfr expression is downregulated by CpG island hypermethylation in esophageal cancer. Carcinogenesis 23:1695–1699

Wassmann K, Benezra R (2001) Mitotic checkpoints: from yeast to cancer. Curr Opin Genet Dev 11:83–90

See Also

Baba S, Hara A, Kato K et al (2009) Aberrant promoter hypermethylation of the CHFR gene in oral squamous cell carcinomas. Oncol Rep 22:1173–1179

Baylin SB, Herman JG, Graff JR et al (1998) Alterations in DNA methylation: a fundamental aspect of neoplasia. Adv Cancer Res 72:141–196

Cahill DP, Lengauer C, Yu J et al (1998) Mutations of mitotic checkpoint genes in human cancers. Nature 392:300–303

Derks S, Postma C, Carvalho B et al (2008) Integrated analysis of chromosomal, microsatellite and epigenetic instability in colorectal cancer identifies specific associations between promoter methylation of pivotal tumour suppressor and DNA repair genes and specific chromosomal alterations. Carcinogenesis 29:434–439

Hernando E, Orlow I, Liberal V et al (2001) Molecular analyses of the mitotic checkpoint components hsMAD2, hBUB1 and hBUB3 in human cancer. Int J Cancer 95:223–227

Hiraki M, Kitajima Y, Sato S et al (2010) Aberrant gene methylation in the lymph nodes provides a possible marker for diagnosing micrometastasis in gastric cancer. Ann Surg Oncol 17:1177–1186

Homma N, Tamura G, Honda T et al (2005) Hypermethylation of Chfr and hMLH1 in gastric noninvasive and early invasive neoplasias. Virchows Arch 446:120–126

Joensuu EI, Abdel-Rahman WM, Ollikainen M et al (2008) Epigenetic signatures of familial cancer are characteristic of tumor type and family category. Cancer Res 68:4597–4605

Koga T, Takeshita M, Yano T et al (2011) CHFR hypermethylation and EGFR mutation are mutually exclusive and exhibit contrastive clinical backgrounds and outcomes in non-small cell lung cancer. Int J Cancer 128:1009–1017

Maruya S, Issa JP, Weber RS et al (2004) Differential methylation status of tumor-associated genes in head and neck squamous carcinoma: incidence and potential implications. Clin Cancer Res 10:3825–3830

Nigg EA (2001) Mitotic kinases as regulators of cell division and its checkpoints. Nat Rev Mol Cell Biol 2:21–32

Nomoto S, Haruki N, Takahashi T et al (1999) Search for in vivo somatic mutations in the mitotic checkpoint gene, hMAD1, in human lung cancers. Oncogene 18:7180–7183

Sakai M, Hibi K, Kanazumi N et al Aberrant methylation of the CHFR gene in advanced hepatocellular carcinoma. Hepatogastroenterology 52:1854–1857

Sanbhnani S, Yeong FM (2012) CHFR: a key checkpoint component implicated in a wide range of cancers. Cell Mol Life Sci 69:1669–1687

Sanchez-Cespedes M, Esteller M, Wu L et al (2000a) Gene promoter hypermethylation in tumors and serum of head and neck cancer patients. Cancer Res 60:892–895

Sanchez-Cespedes M, Okami K, Cairns P et al (2000b) Molecular analysis of the candidate tumor suppressor gene ING1 in human head and neck tumors with 13q deletions. Genes Chromosomes Cancer 27:319–322

Scolnick DM, Halazonetis TD (2000) Chfr defines a mitotic stress checkpoint that delays entry into metaphase. Nature 406:430–435

Shah SI, Yip L, Greenberg B et al (2000) Two distinct regions of loss on chromosome arm 4q in primary head and neck squamous cell carcinoma. Arch Otolaryngol Head Neck Surg 126:1073–1076

Shaw RJ, Liloglou T, Rogers SN et al (2006) Promoter methylation of P16, RARbeta, E-cadherin, cyclin A1 and cytoglobin in oral cancer: quantitative evaluation using pyrosequencing. Br J Cancer 94:561–568

Shibata Y, Haruki N, Kuwabara Y et al (2002) Chfr expression is downregulated by CpG island hypermethylation in esophageal cancer. Carcinogenesis 23:1695–1699

Sidransky D (2000) Circulating DNA. What we know and what we need to learn. Ann N Y Acad Sci 906:1–4

Smeets SJ, Braakhuis BJ, Abbas S et al (2006) Genome-wide DNA copy number alterations in head and neck squamous cell carcinomas with or without oncogene-expressing human papillomavirus. Oncogene 25:2558–2564

Soutto M, Peng D, Razvi M et al (2010) Epigenetic and genetic silencing of CHFR in esophageal adenocarcinomas. Cancer 116:4033–4042

Takahashi T, Haruki N, Nomoto S et al (1999) Identification of frequent impairment of the mitotic checkpoint and molecular analysis of the mitotic checkpoint genes, hsMAD2 and p55CDC, in human lung cancers. Oncogene 18:4295–4300

Takeshita M, Koga T, Takayama K et al (2008) CHFR expression is preferentially impaired in smoking-related squamous cell carcinoma of the lung, and the diminished expression significantly harms outcomes. Int J Cancer 123:1623–1630

Toyota M, Sasaki Y, Satoh A et al (2003) Epigenetic inactivation of CHFR in human tumors. Proc Natl Acad Sci U S A 100:7818–7823

Wassmann K, Benezra R (2001) Mitotic checkpoints: from yeast to cancer. Curr Opin Genet Dev 11:83–90

Worsham MJ, Pals G, Schouten JP et al (2003) Delineating genetic pathways of disease progression in head and neck squamous cell carcinoma. Arch Otolaryngol Head Neck Surg 129:702–708

Worsham MJ, Chen KM, Meduri V et al (2006a) Epigenetic events of disease progression in head and neck squamous cell carcinoma. Arch Otolaryngol Head Neck Surg 132:668–677

Worsham MJ, Chen KM, Tiwari N et al (2006b) Fine-mapping loss of gene architecture at the CDKN2B (p15INK4b), CDKN2A (p14ARF, p16INK4a), and MTAP genes in head and neck squamous cell carcinoma. Arch Otolaryngol Head Neck Surg 132:409–415

Childhood Adrenocortical Carcinoma

Raul C. Ribeiro[1], Carlos Rodriguez-Galindo[2] and Gerard P. Zambetti[3]

[1]Department of Oncology, St. Jude Children's Research Hospital, Memphis, TN, USA

[2]Dana-Farber Cancer Institute, Boston, MA, USA

[3]Department of Biochemistry, Dana-Farber Cancer Institute, Boston, MA, USA

Definition

Adrenocortical carcinoma (ACC) is a cancer of the cortex of the adrenal gland (▶ endocrine-related cancers). There are two types. In one type, the tumor continues to secrete the hormones normally produced by the cortex, including glucocorticoids, mineralocorticoids, and adrenal sex hormones. However, these steroids may be produced in excessive amounts, with negative effects on the body. In the other type, the tumor does not produce these hormones and may go undiscovered until it metastasizes.

Characteristics

Incidence

Cancer of the adrenal cortex is exceedingly rare: only about 300 cases are diagnosed in the United States each year. Adrenocortical tumors (ACT) represent only about 0.2% of all malignancies in children. The frequency of ACT is 0.4 per million during the first 4 years of life, 0.1 per million during the subsequent 10 years, and 0.2 per million during the late teens. These tumors also occur in adults, usually during the fourth to fifth decades of life. The incidence of ACT differs across

Childhood Adrenocortical Carcinoma, Table 1 Constitutional syndromes associated with adrenocortical tumors

Condition	Tumor types	Observations
Germline ▶ *TP53* mutations, including Li–Fraumeni syndrome (▶ IARC TP53 Database) (▶ p53 family) (p53 protein, biological and clinical aspects)	Adenomas, sarcomas, carcinomas	Penetrance of ACT is about 10% or less
Beckwith–Wiedemann syndrome (Beckwith–Wiedemann syndrome-associated childhood tumors)	Adenomas, carcinomas	ACT is the second most common tumor (~15% of children with this syndrome)
Hemihypertrophy	Adenomas, carcinomas	20% of these tumors are ACT
Congenital adrenal hyperplasia	Adenoma, carcinoma	Very rare occurrence of ACT
Carney complex	Primary pigmented nodular adrenocortical disease	ACT occurs in ~25% of patients; common in children
Multiple endocrine neoplasia I	Nodules, adenomas, carcinomas	ACT is very rare in children

geographic regions, ranging from 0.1 per million in Hong Kong and Bombay to 0.4 in Los Angeles and 3.4 in southern Brazil.

Causes

Predisposing genetic factors have been found in the majority of children and adolescents with ACT (Table 1).

The most common genetic abnormalities in young children with ACT are germline mutations in various exons of the *TP53* tumor suppressor (▶ Li–Fraumeni syndrome, or LFS). However, low-penetrant mutant alleles can also reduce rather than abrogate TP53 tumor suppressor activity and contribute to ACT without being associated with LFS. Strong evidence suggests that one such inherited *TP53* mutation (Arg337His) explains the extraordinarily high incidence of pediatric adrenocortical carcinoma in southern Brazil.

Adrenocortical tumors may also occur in the context of Beckwith–Wiedemann syndrome (BWS) (Beckwith–Wiedemann syndrome-associated childhood tumors), which is characterized by a loss of heterozygosity at chromosome 11p15, resulting in the overproduction of ▶ insulin-like growth factor II (IGF-II) (insulin-like growth factors) and diminished levels of the cyclin kinase inhibitor *p57* ^{Kip2}. ▶ Carney complex, hemihypertrophy, congenital adrenal hyperplasia, and multiple endocrine neoplasia type I (inherited mutations in the *MENIN* ▶ tumor suppressor gene) also give rise to pediatric ACT.

Normal Physiology and Tumor Biology

The human fetal adrenal cortex rapidly develops into two morphologically distinct zones (fetal and definitive) and is essentially adult size by midgestation. The outer, definitive zone exhibits high proliferative activity and is thought to be the germinal cell compartment from which the lipid-dense fetal zone cells migrate. Cortisol is also synthesized primarily by the definitive zone.

The inner, fetal zone produces large amounts of dehydroepiandrosterone sulfate (DHEA-S), which maintains placental function and integrity. Soon after birth, the fetal zone atrophies by undergoing massive cell death through an apoptotic mechanism. Remodeling of the cortex in the neonate results in three functional regions: (i) zona glomerulosa, (ii) zona fasciculata, and (iii) zona reticularis. The outer zona glomerulosa primarily produces aldosterone, and the zonae fasciculata and reticularis synthesize corticosteroids and androgens, respectively.

Adrenocortical tumorigenesis in children, in contrast to adults, often results in the hyperproduction of steroids, which is readily apparent by marked physical changes (see below and Fig. 1).

The extensive growth of the adrenal cortex during gestation and its postnatal resolution are likely important in the sensitivity of this gland to small losses in TP53 tumor suppressor activity in those carriers of germline *TP53* mutations. An adrenocortical cell that was destined to cease

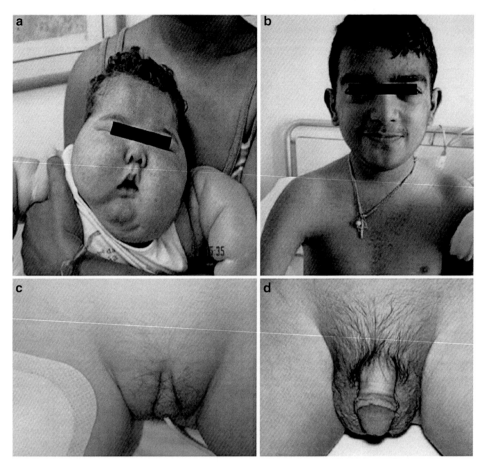

Childhood Adrenocortical Carcinoma, Fig. 1 Clinical signs of adrenocortical tumors. (**a**) Typical facies of a patient with hypercortisolism. (**b**) Patient with hirsutism. (**c**) Precocious pseudopuberty (clitorimegaly) in a girl with adrenocortical carcinoma. (**d**) Precocious pseudopuberty in a boy with adrenocortical carcinoma

proliferation during the expansion phase or was targeted to die after birth could continue to survive and divide. The timing of tumor development by 3 or 4 years of age is consistent with this hypothesis. The fetal zone or the zona reticularis have been implicated as the source of cells that contribute to ACT formation, although definitive proof is lacking.

Pediatric adrenocortical tumorigenesis relies on the acquisition of multiple genetic hits. In addition to frequent germline *TP53* mutations, *steroidogenic factor-1* (*SF1*), which encodes a transcription factor required for normal adrenal gland development, is amplified on chromosome 9q34 and overexpressed in ~90% of patients with ACT. Inactivation of the TGFβ-related *Inhibinα*

(INHA) on chromosome 2q33 by mutation of one allele and deletion of the remaining wild-type allele is also common in ACT. The gross overexpression of IGF-II in pediatric ACT, regardless of the genetic predisposing factors (LFS or BWS), has been convincingly established. Adult ACT shares this biochemical alteration, and clinical trials of drugs to inactivate IGF-II signaling in both childhood and adult ACT are being designed.

Clinical Manifestations

Features of virilization, including pubic hair, facial acne, clitorimegaly, voice change, facial hair, hirsutism, muscle hypertrophy, growth acceleration, and increased penis size, are the most

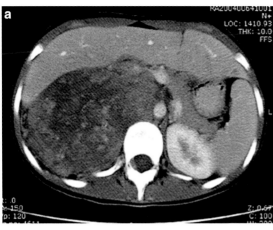
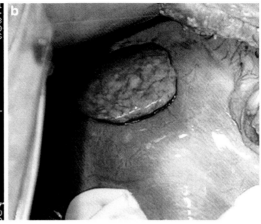

Childhood Adrenocortical Carcinoma, Fig. 2 (**a**) Axial computed tomographic image of a large adrenocortical tumor showing a central area of stellate appearance caused by hemorrhage, necrosis, and fibrosis. Small calcifications are also seen. (**b**) Regional relapse is apparent adjacent to the liver surface. This patient's initial surgery was complicated by tumor rupture and spillage

common clinical manifestations of ACC. Virilization can be observed either alone (virilizing tumors, 40% of patients) or with clinical manifestations resulting from the overproduction of other adrenal cortical hormones, including glucocorticoids, androgens, aldosterone, or estrogens (mixed type, 45%; Fig. 1). About 10% of patients show no clinical evidence of an endocrine syndrome at presentation (nonfunctional tumors). Finally, overproduction of glucocorticoids alone (▶ Cushing syndrome) is evident in about 3% of patients. Primary hyperaldosteronism (Conn syndrome) and pure feminization can also occur. In some circumstances, clinical manifestations of ACT can be present at birth. Conn syndrome, first reported by Jerome W. Conn, is a result of an increased production of aldosterone, a hormone produced by the zona glomerulosa of the adrenal cortex. This hormone causes the retention of water and sodium and excretion of potassium. The clinical manifestations include high blood pressure, headaches, and muscle cramps. A small proportion of children with ▶ adrenocortical cancer overproduce aldosterone.

Testing

Routine laboratory evaluation for suspected ACT includes measuring urinary 17-ketosteroids (17-KS), 17-hydroxycorticosteroid (17-OH), and free cortisol, as well as plasma cortisol, DHEA-S, testosterone, androstenedione, 17-hydroxyprogesterone, aldosterone, renin activity, deoxycorticosterone, and other 17-deoxysteroid precursors. Most patients with ACT who are tested have elevated levels of 17-KS. Plasma DHEA-S levels are abnormal in ~90% of cases. Elevated glucocorticoid and androgen levels are strong indications of an adrenal tumor.

Several different imaging modalities are used to diagnose ACT. Computed tomography (CT; Fig. 2), sonography, magnetic resonance imaging (MRI), and ▶ positron emission tomography (PET) are the most commonly used.

Ultrasound is useful for evaluating tumor extension into the inferior vena cava and right atrium. On CT imaging, ACT is usually well demarcated, with an enhancing peripheral capsule. Large tumors usually have a central area of stellate appearance that is caused by hemorrhage, necrosis, and fibrosis. Calcifications are common. Because ACT is metabolically active, fluorodeoxyglucose-positron emission tomography (FDG-PET) imaging is frequently used in patients with ACT. PET imaging can also detect tumor recurrence in areas that routine follow-up CT imaging may miss.

The definitive diagnosis of ACT is made on the basis of the gross and histological appearance of tissue obtained surgically. Tumors are classified as adenoma or carcinoma, although even an experienced pathologist can find it difficult to differentiate between benign and malignant tumors.

Treatment

Surgery is the mainstay of treatment for ACT. A curative, complete resection may be attempted in patients with local or regional disease (70–75% of cases). En bloc resection, including the adjacent structures invaded by the tumor, is required for good local control. Nephrectomy and resection of liver segments and portions of the pancreas may be included. Because of tumor friability, rupture of the capsule with resultant tumor spillage is common (Fig. 2). When ACT is suspected, laparotomy and a curative procedure are recommended rather than fine-needle aspiration, to avoid the risk of tumor rupture.

Infiltration of the vena cava by tumor thrombus occurs in 20% of patients and may make radical surgery difficult; a combined thoracic and abdominal approach may be required in those cases. The pattern of recurrence is locoregional (15–25%), combined local and distant (25–30%), or distant alone (50%). Chemotherapy with mitotane is indicated for unresectable and recurrent disease, although it has a small impact on overall outcome. At low doses, mitotane suppresses the secretion of adrenal steroids, providing symptomatic improvement and partial regression of endocrine dysfunction in most patients with functional tumors. Higher doses (>3 g/day) are required for an adrenolytic effect.

Although responses to mitotane alone may occur in 20–30% of cases, most responses are transient, and the prospect for long-term survival is uncertain. The antitumor effect of mitotane is influenced by its pharmacokinetics and by the duration of its therapeutic exposure. Serum concentration plateaus after 8–12 weeks of treatment and antitumor responses occur only when a serum concentration of at least 14 µg/mL is maintained for a prolonged period. The severe gastrointestinal (nausea, vomiting, diarrhea, and abdominal pain) and neurologic (somnolence, lethargy, ataxia, depression, and vertigo) toxic effects of mitotane reduce patient adherence. Because mitotane is adrenolytic, all patients receiving this agent should be considered to have severe adrenal insufficiency and treated accordingly. ► Cisplatin-based regimens, usually including etoposide and doxorubicin, are used in combination with mitotane, although less than 40% of patients respond.

The use of radiotherapy in pediatric ACT has not been consistently investigated, although ACT is generally considered to be radioresistant. Furthermore, because many children with ACT carry germline *TP53* mutations that predispose them to cancer, radiation may increase the incidence of secondary tumors. For most patients with metastatic or recurrent disease that is unresponsive to mitotane and chemotherapy, repeated surgical resection is the only alternative. However, given the infiltrative nature of the disease, complete resection is difficult. Image-guided tumor ablation with radiofrequency currently offers a valid alternative for these patients.

Prognosis

Complete tumor resection is the single most important prognostic indicator. Patients who have distant or local with gross or microscopic residual disease after surgery have a dismal prognosis. Long-term survival (5 years or more after the diagnosis) is about 75% for children after complete tumor resection. Among those who undergo complete tumor resection, tumor size has prognostic value. The estimated event-free survival is 40% for those with tumors weighing more than 200 g and 80% for those with smaller tumors. Children whose tumors produce excess glucocorticoid appear to have a worse prognosis than children who have pure virilizing manifestations. Classification schemes or disease staging systems (Table 2) are still evolving. Prognosis will likely be further refined by adding other predictive factors, including those from gene expression studies.

Concluding Remarks

Adrenocortical tumors remain difficult to treat, and little progress has been made in developing effective chemotherapeutic regimens. The rarity of ACT hinders the opportunity to conduct

Childhood Adrenocortical Carcinoma, Table 2 Staging criteria for childhood adrenocortical tumor

Stage	Description
I	Tumor totally excised, tumor size <100 g or <200 cm^3, absence of metastasis, and normal hormone levels after surgery
II	Tumor totally excised, tumor size ≥ 100 g or ≥ 200 cm^3, absence of metastasis, and normal hormone levels after surgery
III	Unresectable tumor, gross or microscopic residual tumor, tumor spillage during surgery, persistence of abnormal hormone levels after surgery, or retroperitoneal lymph node involvement
IV	Distant tumor metastasis

adequately powered clinical trials, including biological studies. Therefore, efforts must be coordinated and resources must be consolidated to advance our understanding and treatment of ACT. In this regard, a long-standing international ACT registry and tissue bank has been established. Short-term goals are to establish tissue culture, xenograft transplants, and genetically engineered mouse models to explore novel therapies. Clinical investigators, physicians, and basic scientists are encouraged to participate in these studies.

References

Michalkiewicz E, Sandrini R, Figuciredo B et al (2004) Clinical and outcome characteristics of children with adrenocortical tumors: a report from the International Pediatric Adrenocortical Tumor Registry. J Clin Oncol 22:838–845

Ribeiro RC, Sandrini F, Figueiredo B et al (2001) An inherited p53 mutation that contributes in a tissue-specific manner to pediatric adrenal cortical carcinoma. Proc Natl Acad Sci U S A 98:9330–9335

Pinto EM, Chen X, Easton J, Finkelstein D, Liu Z, Pounds S, Rodriguez-Galindo C, Lund TC, Mardis ER, Wilson RK, Boggs K, Yergeau D, Cheng J, Mulder HL, Manne J, Jenkins J, Mastellaro MJ, Figueiredo BC, Dyer MA, Pappo A, Zhang J, Downing JR, Ribeiro RC, Zambetti GP (2015) Genomic landscape of paediatric adrenocortical tumours. Nat Commun 6:6302. doi:10.1038/ncomms7302

Pinto EM, Morton C, Rodriguez-Galindo C, McGregor L, Davidoff AM, Mercer K, Debelenko LV, Billups C, Ribeiro RC, Zambetti GP (2013) Establishment and characterization of the first pediatric adrenocortical carcinoma xenograft model identifies topotecan as a potential chemotherapeutic agent. Clin Cancer Res 19 (7):1740–7

Ribeiro RC, Pinto EM, Zambetti GP, Rodriguez-Galindo C (2012) The International Pediatric Adrenocortical Tumor Registry initiative: contributions to clinical, biological, and treatment advances in pediatric adrenocortical tumors. Mol Cell Endocrinol 351(1):37–43

Childhood Cancer

Stefan K. Zöllner[1], Amal M. Abu-Ghosh[2] and Jeffrey A. Toretsky[2]

[1]Department of Pediatric Hematology and Oncology, University Childrens Hospital Münster, Münster, Germany

[2]Department of Oncology and Pediatrics, Lombardi Comprehensive Cancer Center, Georgetown University, Washington, DC, USA

Definition

Childhood cancer, also known as pediatric cancer, describes a cancerous tumor burden, which can occur anywhere in the body originating from cells with the propensity to invade surrounding tissue and to spread from its primary site of occurrence, the latter referred to as ▶ metastasis, and specifically affects children and adolescents.

Characteristics

Incidence

In general, cancer in children and teenagers is uncommon, representing between 0.5% and 4.6% of all cancer. The incidence of childhood cancers such as leukemia (▶ Hematological Malignancies, Leukemias, and Lymphomas) and tumors of the brain and central nervous system (CNS) (see ▶ Brain Tumors, ▶ Pediatric Brain Tumors, ▶ Neuro-Oncology: Primary CNS Tumors) varies between countries with higher overall rates in industrialized countries, while, for example, populations in sub-Saharan Africa

Childhood Cancer,
Fig. 1 SEER delay-
adjusted incidence and US
mortality 1975–2012, all
childhood cancers, under
20 years of age, both sexes,
all races

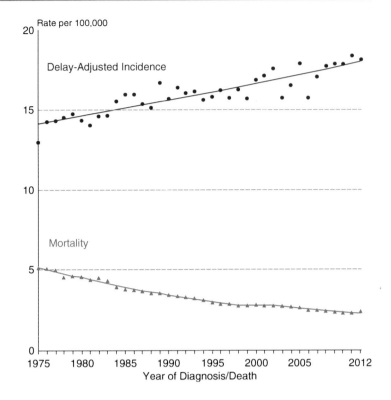

have higher incidence rates of lymphomas (► Hematological Malignancies, Leukemias, and Lymphomas) than other regions. These variations may reflect differences in diagnostic techniques or registration or in the distribution of possible risk factors.

The overall incidence of childhood cancer varies between 50 and 200 per million children across the world. In the United States of America (USA), an estimated 10,380 children (younger than 15) and about 5,000 adolescents aged 15–19 will be diagnosed with cancer per year. An estimated 69,212 adolescents and young adults (AYAs) aged 15–39 were diagnosed with cancer in 2011. Reports from the SEER (Surveillance Epidemiology and End Results) program on childhood cancer since the 1970s have shown a gradual increase in incidence, although it appears to have leveled off in the past decade (Fig. 1).

The most common childhood cancer diagnoses are leukemia followed by brain and CNS tumors and lymphoma (Fig. 2). The incidence and type of cancer varies with age with certain cancers occurring mostly in childhood (<15 years) (► Neuroblastoma (NB), ► nephroblastoma (a.k. a. ► Wilms tumor), ► retinoblastoma (RB), ► hepatoblastoma) and others at later ages (>15 years) (lymphoma, soft tissue, and bone tumors (see ► Bone Tumors, ► Non-Rhabdomyosarcoma Soft Tissue Sarcomas, ► Rhabdomyosarcoma, ► Osteosarcoma, ► Ewing Sarcoma, ► Synovial Sarcoma), germ cell tumors (see ► Ovarian Germ Cell Tumors, ► Germinoma, ► Ovarian Stromal and Germ Cell Tumors, ► Testicular Germ Cell Tumors), thyroid cancer (see ► Follicular Thyroid Tumors, ► Thyroid Carcinogenesis, ► Papillary Thyroid Carcinoma)). Other tumors affect both children and adolescents (Leukemia, brain, and CNS tumors). Tumors arising from postnatally persistent embryonal remnants or rests are referred as embryonal tumors and include NB, Wilms tumor, and brain tumors such as medulloblastoma, etc. Scientific evidence suggests that with developmental transition in AYAs, the spectrum of cancer reflects a similar transition harboring unique genetic and

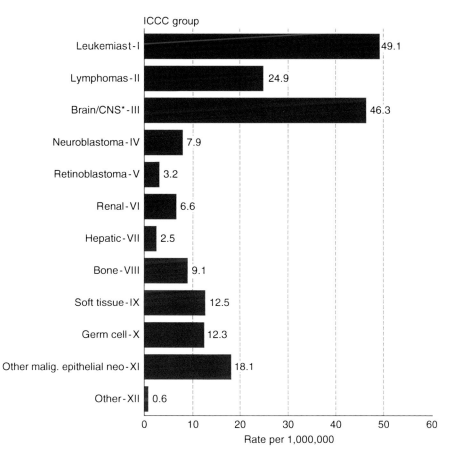

Childhood Cancer, Fig. 2 SEER incidence rates 2008–2012 by ICCC group (leukemias include myelodysplastic syndromes and brain/CNS include benign brain tumors), under 20 years of age, both sexes, all races

biological features which led to the concept of AYA-focused oncology to meet both the distinct cancer profiles (Hodgkin lymphoma (► Hodgkin Disease) (HD), ► melanoma, testicular cancer (see ► Testicular Germ Cell Tumors, ► Testicular Cancer), thyroid cancer, and sarcomas) and age-appropriate needs of these patients (Fig. 3).

Most children and adolescents diagnosed with cancer can be treated successfully. While survival differs between cancer types, the 5-year survival after childhood cancer in general has dramatically improved over the last 30 years, reaching 83%, which is mainly due to increased survival of childhood ► acute lymphoblastic leukemia (ALL) with introduction of multiagent chemotherapy and its risk-adapted application including CNS prophylaxis. Additionally, improvements in supportive

care have contributed to a better prognosis in all types of childhood cancer.

Despite improved survival rates, approximately 20% of children with cancer will die each year of their disease. From 1970 to 2011, the number of deaths from childhood cancer has decreased steadily by 67% in the USA. However, cancer remains the second leading cause of death in children 0–14 years of age after accidents. It is estimated that 1,250 deaths per year from cancer will occur in children in this age group, and 600 deaths from cancer will occur in teens aged 15–19 in the USA. In 2011, cancer was the leading cause of disease-related death in the AYA population.

Causes and Risk Factors
The causes (► Cancer Causes and Control) of childhood cancer have been systematically

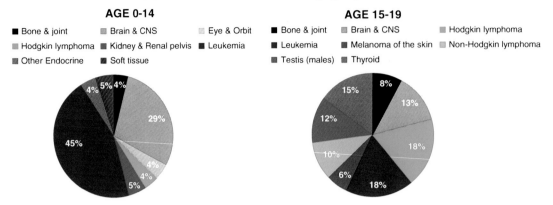

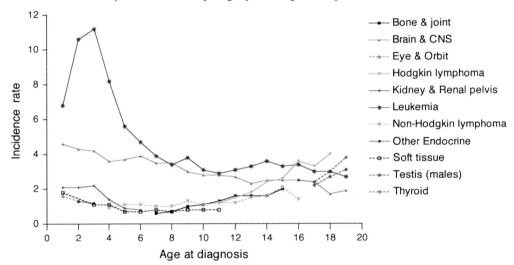

Childhood Cancer, Fig. 3 SEER age-specific incidence rates 2008–2012 (leukemia includes myelodysplastic syndromes and brain and CNS include benign brain tumors), under 20 years of age, both sexes, all races

studied for decades but apart from high-dose radiation (see ▶ Ionizing Radiation Therapy, ▶ Radiation Carcinogenesis) and prior ▶ chemotherapy there are few strong external risk factors. Most of the time, there is no known cause for childhood cancers. However, studying epidemiologic patterns (see ▶ Cancer Epidemiology, ▶ Epidemiology of Cancer) can shed some light over possible causes occurring alone or as part of multiple risk factors that work together and lead to cancer. Inherent risk factors including birth weight, parental age, and congenital anomalies are consistently associated with most types of pediatric

cancer. The contribution of common genetic variation to etiology has come into focus through genome-wide association studies; the genetic sequence analyses might contribute insights into inherited components of etiology. Finally, exposure to a variety of chemical carcinogens (see ▶ Chemical Carcinogenesis, ▶ Benzene and Leukemia) shortly before conception, during pregnancy, and/or after birth appear to increase the risk of some childhood cancer types, especially leukemia. Some other important and established risk factors are discussed below.

Socioeconomic and Ethnic Factors

From approximately 200,000 children and adolescents diagnosed with cancer every year worldwide, 80% live in low- and middle-income countries (LMICs), which account for 90% of cancer-related deaths. Outcome of children with cancer in LMICs is dictated by late presentation, underdiagnosis, high abandonment rates, and high prevalence of malnutrition; in therapy suboptimal supportive and palliative care as well as limited access to curative therapies further hinder morbidity and mortality. The rising proportion of cases in these countries, especially in AYAs, is caused by population growth and aging, combined with reduced mortality from infectious diseases. Meanwhile, there is a dramatic inequity in the distribution of resources for cancer care and control worldwide; the consequence limits options for patients in resource-limited countries. The lack of quality population-based cancer registries in LMICs limits knowledge of the epidemiology of pediatric cancer but available information show variations in incidence of leukemia and some embryonal tumors, possibly related to environmental factors and geographical and ethnic patterns.

Differences in the incidence of embryonal tumors between countries and ethnic groups have been consistently reported, particularly for NB and RB. While an inverse correlation between the incidence of RB and socioeconomic index has been described, the opposite is true for NB, with higher incidence in regions with high socioeconomic status. Conversely, in the USA, black and Native American patients with NB have a higher prevalence of high-risk disease. Further examples for geographical variations in childhood cancers include Wilms tumor occurring at the lowest rates in East Asia, and Ewing Sarcoma (ES) which is extremely rare in black and East Asian patients compared to Caucasians.

Infections

Several viruses are linked with childhood cancer development. The relative cellular tropism of viral infection presumably predetermines a specific cancer risk. Some viruses may cause more than one cancer, while some cancers may be caused by

Childhood Cancer, Table 1 Virus-associated cancer entities

Virus type	Cancer type/location
HPV	Cervical, penis, anus, vagina, vulva, mouth, throat
EBV	Lymphomas (Burkitt, Hodgkin), nasopharyngeal, stomach
HBV	Liver
HCV	Liver, non-Hodgkin lymphoma
HIV	Kaposi sarcoma, cervical, lymphoma (Hodgkin, non-Hodgkin), anal, lung, mouth, throat, skin, liver
HHV-8 (a.k.a. KSHV)	Kaposi sarcoma
HTLV-1	T-cell leukemia/lymphoma
MCV	Merkel cell carcinoma (skin)
Simian virus 40 (SV40)[a]	Mesothelioma, brain, bone, lymphomas

[a]Even though ▶ SV40 causes cancer in some lab animals, the evidence so far suggests that it does not cause cancer in humans

more than one virus. However, only a proportion of persons infected by oncogenic viruses will develop cancer. ▶ Epstein-Barr virus (EBV), human papilloma virus (HPV), human herpes virus 8 (HHV-8, a.k.a. Kaposi sarcoma herpes virus, KSHV), human T-cell lymphotropic virus type 1 (▶ Human T-Lymphotropic Virus) (HTLV-1), and Merkel cell polyomavirus (MCV) are the main viruses associated with the development of cancer and are considered direct carcinogens. ▶ Hepatitis B virus (HBV) and ▶ hepatitis C virus (HCV) represent indirect carcinogens through chronic inflammation. Besides being a main cause of immunodeficiency, which in turn contributes to oncogenesis by increasing the susceptibility to other infections, human immunodeficiency virus type 1 (▶ Pediatric HIV/AIDS) (HIV-1) also exerts a direct oncogenic effect (Table 1).

In general, persistent infection and high viral load are important risk predictors of virus-caused cancers. Some viruses display distinct geographical distribution due to endemic infection and, in combination with other unidentified factors such as genetic predisposition, may account for ethnic differences. Socioeconomic conditions further correlate with the risk for virus-induced cancer as in LMICs prevention programs including screening tests, and vaccine availability if applicable (HBV, HPV) are underdeveloped.

Epidemiological studies of the incidence of HD worldwide have shown a strong association of poor socioeconomic status and EBV infection which is detected in more than 50% of cases of HD worldwide. ► Burkitt lymphoma (BL) has also been strongly linked with EBV infection and malaria as a cofactor as found in all BL cases in tropical Africa and in Papua New Guinea. EBV infection further occurs in 50–70% of BL in North Africa and South America and 20% in Europe and North America. ► Nasopharyngeal carcinoma has the highest incidence in North Africa and is also associated with EBV infection. ► Kaposi sarcoma (KS) is the most common soft tissue sarcoma among children in sub-Saharan countries, likely originated in endemic HIV and heightened infection with HHV-8. ► Hepatocellular carcinoma (HCC) is a rare hepatic tumor in Europe and North America, but is still the most common childhood liver tumor in Saharan Africa, East and Southeast Asia, and Melanesia, and is associated with chronic hepatitis B infection.

Radiation Exposure

Following the Chernobyl explosion, areas in Belarus, Russia, and the Ukraine contaminated with radiation fallout reported a significant increase of more aggressive forms of thyroid carcinoma. The incidence fell to pre-Chernobyl levels in children conceived after the contamination. Few studies reported increased numbers of leukemia cases in children living in the vicinity of nuclear power plants but the emitted ionizing radiation is generally considered too small to cause cancer. Other studies could not confirm suspected associations between nonionizing radiation exposure (e.g., cell phones, AM and FM radio, televisions, and microwaves) and childhood brain tumors. In contrast, high background radiation from terrestrial gamma and cosmic rays may contribute to the risk of cancer in children, including leukemia and CNS tumors. Numerous epidemiologic cohort studies of childhood exposure to radiation for treatment of nonmalignant diseases have demonstrated radiation-related

risks of cancer of the thyroid, breast, brain and skin, as well as leukemia. It is noteworthy that the cancer risk following diagnostic radiation exposure in children is not increased.

UV Light Exposure

Overexposure to ultraviolet radiation (► UV Radiation) (UVR) during childhood is a major risk factor for skin cancer in adulthood since 40–50% of total UVR by age 60 occurs before age 20. Still, the incidence of childhood melanoma has increased and 98% of melanoma cases are diagnosed in adolescents and young adults. ► Melanoma in children appears to have similar epidemiologic characteristics (family history) to the adult form of the disease, being associated with a cluster of phenotypic attributes (large congenital or numerous nevi, heavy facial freckling, and an inability to tan on exposure to light), indicating cutaneous sensitivity to the effects of sun exposure. In consequence, the highest rate in childhood melanoma is seen in Oceania.

Genetic and Chromosomal Syndromes

Genetic factors play a crucial role in the development of cancer. There are several genetic conditions that are associated with increased risk of cancers in children, including hereditary cancer predisposition syndromes, bone marrow failure syndromes, primary immunodeficiency disorders, and numerical chromosomal abnormalities, which will be discussed in a separate chapter.

Clinical Presentation and Diagnosis

Cancer diagnosis in children is challenging, as (I) childhood cancers may behave very differently from adult cancers, even when they start in the same part of the body, (II) infants and children are often either motorically or mentally incapable of depicting their physical problems or describing their mental status, (III) early signs and symptoms of childhood cancers can be nonspecific, and (IV) symptoms often mimic those of common childhood diseases. Different cancers have different symptoms related to the site of disease and type of cancer. General signs and symptoms can be summarized with the mnemonic CHILD

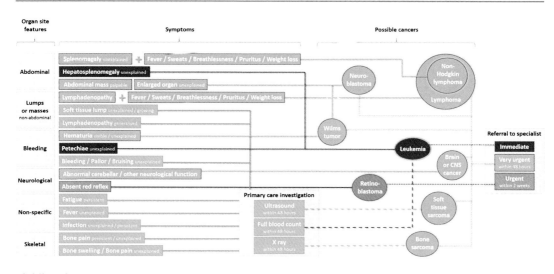

Childhood Cancer, Fig. 4 Diagnostic workflow from symptom to diagnosis (Adapted from Hamilton et al. 2015)

CANCER (provided by The Pediatric Oncology Resource Center):

> Continued, unexplained weight loss
> Headaches, often with early morning vomiting
> Increased swelling or persistent pain in the bones, joints, back, or legs
> Lump or mass, especially in the abdomen, neck, chest, pelvis, or armpits
> Development of excessive bruising, bleeding, or rash
> Constant, frequent, or persistent infections
> A whitish color behind the pupil
> Nausea that persists or vomiting without nausea
> Constant tiredness or noticeable paleness
> Eye or vision changes that occur suddenly and persist
> Recurring or persistent fevers of unknown origin

Primary care physicians are often the first to suspect a cancer diagnosis. Although many of the described symptoms frequently go along with common, benign childhood diseases, it is the duration, unusual associations, course, and quality of such symptoms that alerts the physician to the possibility of an underlying oncologic disorder. A full patient's history with the help of parents and a thorough clinical examination are indispensable. Childhood cancers can often be detected with blood counts or radiographic imaging. However, persistent, symptoms require referral for more intensive radiologic and pathological investigation (Fig. 4).

A diagnosis of cancer may be suspected based on history and physical examination but requires confirmation by tumor biopsy and histopathologic examination. In addition to diagnosing childhood cancer, the following tests may be used to stage the cancer disease, i.e., determine the status of metastasis: blood tests including complete blood count, liver and kidney function tests, and tumor-specific markers if applicable (also used as course parameter). Surgical excision or biopsy of the suspected lesions is performed to confirm the diagnosis by microscopic examination of the tumor tissue, and define a tumor subtype through histopathological techniques like tissue staining. Bone marrow examination and biopsy are required to assess the histopathology of hematopoiesis and screen for bone marrow metastases. Lumbar puncture (spinal tab) will assess metastatic involvement of the cerebrospinal fluid. Radiological evaluation such as ultrasound of both the lesion and other organ sites, computed tomography (CT) scan and magnetic resonance imaging (MRI) of different body compartments, positron emission tomography (PET) scan, etc., are necessary: (I) to characterize the exact tumor localization and (II) to identify all involved organ sites by cancer metastases.

Treatment

Childhood cancer is rare and the exact treatment (see ► Combinatorial Cancer Therapy,

▶ Adjuvant Therapy, ▶ Neoadjuvant Therapy) protocols for each malignancy are complex and constantly evolving which makes it unreasonable for primary care physicians to initiate care without consulting a pediatric oncologist. Moreover, diagnostic tools and pillars of cancer treatment such as chemotherapy require certain technical, logistical, and medical prerequisites only available at specialized centers. Children with cancer are therefore best cared for at a pediatric comprehensive cancer center that includes a multidisciplinary team of pediatric oncologists, specialized nurses, surgeons of different disciplines, radiation oncologists, pathologists, and supportive care services such as child life specialists, nutritionists, physical and occupational therapists, social workers, and counselors, for the patients and their families.

Children younger than 15 years old with cancer are often treated as part of a clinical trial, referred to as a research study that compares standard treatments (the best proven treatments available) with newer approaches to treatments that may be more effective. Clinical trials may test such approaches as a new drug, a new combination of standard treatments, or alternative regimens of current therapies to reduce long-term toxicity. The 5-year overall survival rate in countries where such infrastructures are established has significantly improved from <20% before 1950 to >80% since 1995.

The treatment of childhood cancer depends on several factors, including the type and stage of cancer, possible side effects, the family's preferences, and the child's overall health. Treatment may include single or combinatorial ▶ chemotherapy, radiation therapy (see ▶ Ionizing Radiation Therapy, ▶ Radiation Oncology), and surgery. The choice of chemotherapeutic agents depends on the disease and previous experiences from clinical trials. A chemotherapy regimen (schedule) (see ▶ Induction Chemotherapy, ▶ Maintenance Chemotherapy) usually consists of a specific number of cycles given over a set period of time. Side effects include hematological toxicities (anemia, ▶ neutropenia, and thrombocytopenia); nausea and vomiting; alopecia; infections; and cardiac, renal, and liver toxicities.

At several stages during anticancer therapy, the response to treatment (▶ Minimal Residual Disease) is assessed and the treatment may be adjusted based upon tumor response and patient side effects. Attention to psychosocial and financial issues may reduce existing delays in initiating therapy and also the fraction of patients that abandon therapy, especially in AYA patients.

Outcome

Advances in diagnostic precision, treatment strategies, and supportive care have resulted in significantly improved outcome for children with cancer over the past 60 years, with greater than 80% of patients today becoming 5-year survivors (Fig. 1). Several decades ago, a diagnosis of ALL was almost always fatal, but a child diagnosed with it today has about a 90% chance of long-term survival (>5 years). The use of CNS prophylaxis and multidrug regimens is largely responsible for this success. Delayed intensification and the use of maintenance intrathecal methotrexate have contributed to the dramatic improvement in survival rate.

Despite this progress, cancer remains the leading cause of death from disease in children in the US, and significant short-term and long-term treatment toxicities continue to impact the majority of children with cancer. Therefore, one of the major ongoing treatment objectives is to preserve the quality of life of cured patients through reduction and monitoring of treatment-related toxicities by risk-adapted treatment and standardized follow-up protocols, respectively. The characteristics of treatment-related toxicities in children will be discussed in a separate chapter.

Cross-References

- ▶ Acute Lymphoblastic Leukemia
- ▶ Adjuvant Therapy
- ▶ Benzene and Leukemia
- ▶ Bone Tumors
- ▶ Brain Tumors
- ▶ Burkitt Lymphoma

- ► Cancer Causes and Control
- ► Cancer Epidemiology
- ► Chemical Carcinogenesis
- ► Chemotherapy
- ► Combinatorial Cancer Therapy
- ► Epstein-Barr Virus
- ► Ewing Sarcoma
- ► Follicular Thyroid Tumors
- ► Germinoma
- ► Hematological Malignancies, Leukemias, and Lymphomas
- ► Hepatitis B Virus
- ► Hepatitis C Virus
- ► Hepatoblastoma
- ► Hepatocellular Carcinoma
- ► Hodgkin Disease
- ► Human T-Lymphotropic Virus
- ► Induction Chemotherapy
- ► Ionizing Radiation Therapy
- ► Kaposi Sarcoma
- ► Maintenance Chemotherapy
- ► Metastasis
- ► Minimal Residual Disease
- ► Nasopharyngeal Carcinoma
- ► Neoadjuvant Therapy
- ► Nephroblastoma
- ► Neuroblastoma
- ► Neuro-Oncology: Primary CNS Tumors
- ► Neutropenia
- ► Non-Rhabdomyosarcoma Soft Tissue Sarcomas
- ► Osteosarcoma
- ► Ovarian Germ Cell Tumors
- ► Ovarian Stromal and Germ Cell Tumors
- ► Papillary Thyroid Carcinoma
- ► Pediatric Brain Tumors
- ► Pediatric HIV/AIDS
- ► Radiation Carcinogenesis
- ► Radiation Oncology
- ► Retinoblastoma
- ► Rhabdomyosarcoma
- ► SV40
- ► Synovial Sarcoma
- ► Testicular Cancer
- ► Testicular Germ Cell Tumors
- ► Thyroid Carcinogenesis
- ► UV Radiation
- ► Wilms' Tumor

References

Chen CJ et al (2014) Epidemiology of virus infection and human cancer. Recent Results Cancer Res 193:11–32

Ferlay J et al (2015) Cancer incidence and mortality worldwide: sources, methods and major patterns in GLOBOCAN 2012. Int J Cancer 136(5):E359–E386

Hamilton W et al (2015) Suspected cancer (part 1 – children and young adults): visual overview of updated NICE guidance. BMJ 350:h3036

Kaatsch P (2010) Epidemiology of childhood cancer. Cancer Treat Rev 36(4):277–285

Marshall GM et al (2014) The prenatal origins of cancer. Nat Rev Cancer 14(4):277–289

Ries LAG, Harkins D, Krapcho M, Mariotto A, Miller BA, Feuer EJ, Clegg L, Eisner MP, Horner MJ, Howlader N, Hayat M, Hankey BF, Edwards BK (eds) (2006) SEER cancer statistics review, 1975–2003. National Cancer Institute, Bethesda. http://seer.cancer.gov/csr/1975_2003/. Based on Nov 2005 SEER data submission, posted to the SEER web site

Rodriguez-Galindo C et al (2013) Global challenges in pediatric oncology. Curr Opin Pediatr 25(1):3–15

Young G et al (2000) Recognition of common childhood malignancies. Am Fam Physician 61:2144–2154

Childhood Cancer and Pediatric Cancer Predisposition Syndromes

Stefan K. Zöllner[1] and Jeffrey A. Toretsky[2]
[1]Department of Pediatric Hematology and Oncology, University Childrens Hospital Münster, Münster, Germany
[2]Department of Oncology and Pediatrics, Lombardi Comprehensive Cancer Center, Georgetown University, Washington, DC, USA

Definition

Genetic factors play a crucial role in the development of cancer. There are several genetic conditions which are associated with increased risk of cancers, termed hereditary cancer syndromes. In adults, the percentage of cancer attributed to underlying, inherited genetic mutations is estimated at 5–10%. However, the rate of cancer predisposition in children might be significantly higher, and potentially as high as a third of all new pediatric cancer diagnoses may be due to an inherited genetic cause. Identification of at-risk individuals leads to early tumor detection and better prognosis for some cancers. Familial neoplastic syndromes, bone marrow failure syndromes, inherited immunodeficiency (see ▶ Immunosuppression and Cancer), and numerical chromosome abnormalities are presented.

Characteristics

Hereditary Cancer Predisposition Syndromes

Although hereditary cancer predisposition syndromes are rare, the list of well-defined inherited cancer predisposition syndromes is steadily growing. Malignancies arising in this context involve a large variety of organ systems and affect individuals of all ages. Many children with newly diagnosed cancer and suspicion of an underlying cancer predisposition syndrome have a family history of syndrome-specific symptoms and/or cancer, but some are diagnosed "de novo," without any clinical or genetic evidence in family members. Cancer predisposition is caused by alterations in one of the three groups of genes: ▶ tumor suppressor genes (TS), ▶ oncogenes (OG), and DNA stability genes (SG) (see ▶ DNA Damage Response Genes). Germline mutations coupled with acquired somatic mutations in the same genes confer cellular homozygosity, loss of tumor suppressor activity, and malignant transformation. Herein, we discuss syndromes of inherited cancer predisposition based on the primarily affected organ site of cancer which is associated with the syndrome. The attached table gives a comprehensive overview on associated tumor types and frequency, related genes including type (TS, OG, SG) and mutation frequency, mode of inheritance, time of tumor onset, symptoms, and survival of the different cancer predisposition syndromes in childhood.

Central Nervous System Cancer Predisposition Syndromes

▶ Retinoblastoma (RB) is a tumor that occurs in heritable (25–30%) and nonheritable (70–75%) forms. Heritable RB is the classic example of the "two-hit-hypothesis" of tumor suppressor genes where two mutational events of the *RB1* gene (see ▶ Retinoblastoma Protein, Biological and Clinical Functions) locus are necessary for development of RB. Patients with heritable RB often develop bilateral or multifocal tumors and are younger (median age 11 months) at diagnosis. In heritable tumors, one mutation is inherited through the germline and the second occurs in somatic cells, whereas in the sporadic tumors, both mutational events occur in somatic cells in utero. Patients with mutant *RB1* are at an increased risk for osteosarcomas in their first three decades of life as well as other types of cancers with advancing age including melanoma and carcinomas of the lung (see ▶ Lung Cancer, ▶ Non-Small Cell Lung Cancer) and bladder (see ▶ Bladder Cancer).

The ▶ rhabdoid tumor predisposition syndrome is an autosomal dominant cancer syndrome predisposing to renal or extrarenal malignant rhabdoid tumors (MRT) and to a variety of central nervous system (CNS) tumors (see ▶ Neuro-Oncology: Primary CNS Tumors), including choroid plexus carcinoma, medulloblastoma, and central primitive neuroectodermal tumors. In the CNS, rhabdoid tumors may be pure rhabdoid tumors or a variant that has been designated as atypical teratoid tumor (AT/RT). Up to one-third of patients with rhabdoid tumors harbor *SMARCB1* (see ▶ HSNF5/INI1/SMARCB1 Tumor Suppressor Gene) germline-inactivating mutations at 22q11.2 and rarely mutations in a 2nd locus of the SWI/SNF complex, the *SMARCA4* gene.

Paragangliomas and ▶ pheochromocytomas are rare tumors of parasympathetic or sympathetic ganglia, respectively. Characterized familial paraganglioma syndromes (types 1–4) include those related to perturbation in the succinate dehydrogenase (*SDH*) genes. Other familial paraganglioma conditions include von Hippel-Lindau (*VHL* gene) (see ▶ Von Hippel-Lindau Disease, ▶ Von Hippel-Lindau Tumor Suppressor Gene), multiple endocrine neoplasia type 2 (*RET*; see below), neurofibromatosis (*NF1*; see below), and germline perturbation of tumor suppressor *TMEM127*.

Hereditary Gastrointestinal Malignancies

Familial adenomatous polyposis (FAP) is one of the most common hereditary syndromes caused by germline mutations in adenomatous polyposis coli (*APC*) gene (see ▶ APC Gene in Familial Adenomatous Polyposis) in chromosome 5q21. FAP is associated with an increased risk of ▶ colorectal cancer and extracolonic neoplasms including gastric (see ▶ Gastric Cancer) and thyroid cancer (see ▶ Follicular Thyroid Tumors, ▶ Thyroid Carcinogenesis, ▶ Papillary Thyroid Carcinoma) and ▶ hepatoblastoma.

▶ Peutz-Jeghers syndrome (PJS) is a condition characterized by the association of gastrointestinal polyposis, mucocutaneous pigmentation, and cancer predisposition. The majority of patients that meet the clinical diagnostic criteria have a causative mutation in the *STK11* gene, which is located at 19p13.3. The cancer risks in this condition are substantial, particularly for a wide variety of epithelial malignancies (see ▶ Epithelial Tumorigenesis) (colorectal, gastric, pancreatic [see ▶ Pancreatic Cancer], breast [see ▶ Breast Cancer], lung, and ▶ ovarian cancers).

Juvenile polyposis syndrome (JPS) is a similar hereditary condition, which is characterized by the presence of hamartomatous polyps in the digestive tract. Hamartomas are noncancerous (benign) masses but exhibit a neoplastic potential for gastrointestinal cancers (gastric, small intestinal, colorectal, pancreatic cancer). Both sporadic and familial cases are found with mutations in *SMAD4* and *BMPR1A*. JPS has significant malignant potential, but unlike PJS, extraintestinal cancers are not prominent.

Genitourinary Cancer Predisposition Syndromes

A germline pathogenic variant is thought to be the cause of about 10–15% of ▶ Wilms tumor (see ▶ Nephroblastoma). *WT1* germline variants give rise to WAGR (Wilms tumor-aniridia-genital anomalies retardation), Denys-Drash syndrome (DDS), Frasier syndrome (FS), and isolated Wilms tumor, i.e., Wilms tumor with no evidence of an underlying syndrome. Other Wilms tumor predisposition genes have been mapped to 17q (locus name *FWT1*) and 19q (locus name *FWT2*).

Epigenetic (gain or loss of methylation of imprinting center region 1, *ICR1*) and genomic (*CDKN1C* mutations or 11p15 paternal uniparental isodisomy) alterations give rise to ▶ Beckwith-Wiedemann syndrome (BWS), an overgrowth syndrome. BWS exhibits an increased risk for embryonic tumors (see ▶ Beckwith-Wiedemann Syndrome Associated Childhood Tumors). BWS is associated with alterations in two distinct imprinting domains at 11p15: a telomeric domain containing *H19* and *IGF2* and a centromeric domain including *KCNQ1*, *KCNQ1OT1*, and *CDKN1C*. The types of tumors observed in children with telomeric defects (mainly Wilms tumors) are different from those observed in cases with aberrations limited to the centromeric domain (▶ rhabdomyosarcoma and gonadoblastoma).

Simpson-Golabi-Behmel syndrome (SGBS) is a complex congenital overgrowth syndrome with an increased risk of embryonal cancers. Most cases of SGBS appear to arise as a result of either deletions or point mutations within the glypican-3 (*GPC3*) gene at Xq26. Similar to BWS, patients with SGBS have an increased risk of developing Wilms tumor and ▶ neuroblastoma (NB). However, unlike BWS, patients with SGBS also appear to have an increased risk of ▶ hepatocellular carcinoma and medulloblastoma (see ▶ Testicular Germ Cell Tumors, ▶ Testicular Cancer).

Endocrine Cancer Predisposition Syndromes

Multiple endocrine neoplasia type 1 (MEN1) is a rare cancer syndrome presented mostly by endocrine-characterizing tumors (see ▶ Endocrine-Related Cancers) of the parathyroids, endocrine pancreas, and anterior pituitary. Other endocrine and non-endocrine lesions, such as adrenal cortical tumors (see ▶ Adrenocortical Cancer, ▶ Childhood Adrenocortical Carcinoma); carcinoids (see ▶ Carcinoid Tumors) of the bronchi, gastrointestinal tract, and thymus, lipomas, angiofibromas, collagenomas, and meningiomas, have been described. The responsible gene, *MEN1*, maps on chromosome 11q13.

Multiple endocrine neoplasia type 2 (MEN2) is characterized by the presence of medullary thyroid carcinoma (MTC), unilateral or bilateral

► pheochromocytoma (PHEO), and other hyperplasia and/or neoplasia of different endocrine tissues within a single patient. Predisposition to MEN2 is caused by germline-activating mutations of the *c-RET* proto-oncogene (see ► RET) on chromosome 10q11.2.

Sarcoma Predisposition Syndromes

► Li-Fraumeni Syndrome (LFS) is a rare syndrome, characterized by early onset of bone and soft tissue sarcomas (see ► Bone Tumors, ► Non-Rhabdomyosarcoma Soft Tissue Sarcomas, ► Rhabdomyosarcoma), brain tumors (see ► Brain Tumors, ► Pediatric Brain Tumors), breast cancer, leukemia (see ► Hematological Malignancies, Leukemias, and Lymphomas), adrenocortical carcinoma (see ► Adrenocortical Cancer), and other tumors, as well as multiple primary tumors in a single individual. The majority of families (77%) with classic LFS have an inherited or de novo ► *TP53* germline mutation.

Hereditary multiple exostoses (HME) is a skeletal disorder with mutations in *EXT-1* or *EXT 2* gene, characterized by the development of several benign tumors in the form of ► osteochondromas, which eventually become malignant.

Werner syndrome (WS) is a genetic instability and progeroid (premature aging) syndrome caused by loss-of-function mutations in the *WRN* gene. It is associated with an elevated risk of ► osteosarcoma but also thyroid neoplasms, malignant ► melanoma, meningioma, and leukemia. WS is estimated to affect 1:200,000 individuals in the USA, but occurs more often in Japan, affecting 1:20,000 to 1:40,000 individuals.

Genodermatoses (Inherited Genetic Skin Disorders) with Cancer Predisposition

Genodermatoses consign to inherited skin disorders that often present with multisystem involvement leading to increased morbidity and mortality.

► Naevoid basal cell carcinoma syndrome (NBCCS), also known as Gorlin syndrome, is a hereditary condition which is caused by mutations in the *PTCH1* gene. NBCCS accounts for less than 1% of all NBCCS diagnoses. The main clinical manifestation includes multiple ► basal cell carcinomas, but about 5–10% of NBCCS patients develop medulloblastomas.

Neurofibromatosis (NF) consists of three genetic disorders that primarily cause tumors to grow around the nerves: neurofibromatosis type 1 (*NF1*) (see ► Neurofibromatosis 1), neurofibromatosis type 2 (*NF2*) (see ► Neurofibromatosis 2), and schwannomatosis, which is a genetic condition, but unlike NF1 and NF2, does not have a clear pattern of inheritance. Benign tumors of NF1 (neurofibroma and optic pathway glioma), NF2 (schwannoma, ependymoma, and meningioma), and schwannomatosis (schwannoma) can cause significant morbidity. Schwannomatosis typically does not develop into malignancies, in contrast to NF1 and NF2 where nearly 10% of tumors will develop into malignant cancers. Malignant tumors commonly associated with the most diagnosed type, NF1, include malignant peripheral nerve sheath tumor (MPNS), CNS tumors (optic pathway glioma, ► astrocytoma, and brain stem glioma), soft tissue sarcoma, rhabdomyosarcoma, ► gastrointestinal stromal tumors, and leukemia.

► Tuberous sclerosis complex (TSC) is a neurocutaneous, multisystem disorder characterized by benign hamartomas in multiple organ systems, predominantly the skin (fibromas [see ► Aggressive Fibromatosis in Children]), brain (tubers, nodules), kidney (angiomyolipoma), and heart (cardiac rhabdomyoma [see ► Cardiac Tumors]). Although the overall cancer risk associated with TSC is low, patients with TSC have an increased risk of subependymal giant cell astrocytoma and renal cell carcinoma (see ► Renal Cancer Clinical Oncology, ► Renal Cancer Genetic Syndromes). The genetic cause is mutations in the *TSC1* gene, found on chromosome 9q34, and *TSC2* gene, found on chromosome 16p13.

► Xeroderma pigmentosum (XP) is a rare disorder, based on a genetic defect, in the DNA repair system. XP is characterized by photosensitivity, pigmentary changes, premature skin aging, and malignant tumor development. In consequence, patients with XP have a nearly 100% risk of developing multiple ► skin cancers, and the first diagnosis commonly occurs in childhood. Genetically, XP is differentiated into seven complementation groups (XP-A to XP-G) and the xeroderma pigmentosum variants (XP-V). XP is estimated to affect about 1:1,000,000 people in the USA and Europe, but is more common in Japan, North Africa, and the Middle East.

Rothmund-Thomson syndrome (RTS) is a rare inherited disorder, characterized by a poikilodermatous rash starting in infancy, skeletal abnormalities, and predisposition to specific cancers, particularly osteosarcoma as well as nonmelanoma skin cancers. The gene defect in two-thirds of cases is due to mutations in *RECQL4*.

Leukemia/Lymphoma Predisposition Syndromes

► Bloom syndrome (BS) is an inherited genomic instability disorder caused by disruption of the *BLM* helicase which confers an extreme cancer predisposition. The cancer predisposition is characterized by (I) broad spectrum, including leukemia, lymphomas, and adenocarcinomas, (II) early age of onset relative to the same cancer in the general population, (III) frequency, as more than half of the BS patients develop cancer, and (IV) multiplicity, that is, synchronous or metachronous cancers (see ► Second Primary Tumors). BS patients with five independent primary cancers have been described. Only a few hundred affected individuals have been described in the literature, about one-third of whom are of Central and Eastern European (Ashkenazi) Jewish background.

► Fanconi anemia (FA) is a chromosomal instability syndrome characterized by the presence of pancytopenia, congenital malformations, and cancer predisposition. Mutations in at least 15 genes can cause FA, but 80 to 90% of cases of DFA are due to mutations in one of three genes, *FANCA*, *FANCC*, and *FANCG* (SG, recessive). The hallmark neoplastic events in FA cases are myeloid leukemia (see ► Acute Myeloid Leukemia, ► Chronic Myeloid Leukemia), liver tumors, head and neck carcinomas (see ► Head and Neck Cancer), and gynecologic malignancies. FA is more common among people of Ashkenazi Jewish descent, the Roma population of Spain, and black South Africans.

► Nijmegen breakage syndrome (NBS) is a syndrome of chromosomal instability mainly characterized by microcephaly at birth, combined immunodeficiency, and predisposition to malignancies, predominantly of lymphoid origin including non-Hodgkin lymphoma (NHL), Hodgkin lymphoma (HD; see ► Hodgkin Disease), and leukemia but also brain tumors such as medulloblastoma and glioma and soft tissue sarcomas like rhabdomyosarcoma. Of all the chromosomal instability syndromes, the incidence of cancer in NBS patients is one of the highest. Due to a founder mutation in the underlying *NBN* gene, the disease is encountered most frequently among Slavic populations.

Ataxia-telangiectasia (A-T) is characterized by progressive cerebellar ataxia beginning between ages one and four years, telangiectasias of the conjunctivae, immunodeficiency, and an increased risk for malignancy, particularly leukemia and lymphoma, usually of the B-cell type (see ► B-Cell Lymphoma, ► Diffuse Large B-Cell Lymphoma, ► Marginal Zone B-Cell Lymphoma). The gene associated with A-T is *ATM* (see ► ATM Protein), meaning ataxia-telangiectasia mutated.

Fanconi anemia, ataxia-telangiectasia, xeroderma pigmentosum, Bloom syndrome, Werner syndrome, Rothmund-Thomson syndrome, and Nijmegen breakage syndrome form a class of cancer predisposition syndromes which consist of autosomal recessive disorders of DNA repair (see ► Repair of DNA, ► DNA Damage Response, ► DNA Damage Response Genes). Failure to diagnose hereditary cancer predisposition syndromes with impaired DNA repair pathways may result in the use of chemotherapeutic agents or radiation therapy with conventional dosages, which are contraindicated in these patients.

Bone Marrow Failure Syndromes

Fanconi anemia can also be classified with dyskeratosis congenita (DC), Diamond-Blackfan anemia (DBA), Shwachman-Diamond syndrome (SDS), severe congenital neutropenia (SCN), and amegakaryocytic thrombocytopenia (CAMT) to a second group comprising the inherited bone marrow failure syndromes which predispose to lymphoreticular malignancies. These disorders have diverse genetic mechanisms, including

Childhood Cancer and Pediatric Cancer Predisposition Syndromes, Table 1

Cancer predisposition syndrome		Synonyms	Malignancy risk and associated tumor types including index tumors (bold) and frequencies	Frequency
Central nervous system cancer predisposition syndromes	Hereditary retinoblastoma syndrome		(Bilateral) **Retinoblastoma**, pineoblastoma, malignant midline primitive neuroectodermal tumor, osteosarcoma, melanoma, lung, bladder cancer	1:15,000–20,000
	Rhabdoid tumor syndrome		**Rhabdoid tumors** (renal, CNS: pure or atypical teratoid rhabdoid tumor (AT/RT)), choroid plexus carcinoma, medulloblastoma, central primitive neuroectodermal tumors	<1:1Mio
	Hereditary pheochromocytoma-paraganglioma syndrome	Familial cerebelloretinal angiomatosis	**Pheochromocytoma** (=PCC), **paraganglioma** (=PGL; abdomen, head, neck, trunk) → malignant PCC, PGL (extra-adrenal)	1:500,000 (PCC)-1Mio (PGL)
Hereditary gastrointestinal malignancies	Familial adenomatous polyposis syndrome	Familial polyposis coli	Polyps → 100% **colorectal**, 0.5–12% gastrointestinal, 2% thyroid (papillary), 2% pancreatic, adrenal, bile duct cancer, 10–20% desmoid tumors, 1.5% hepatoblastoma, < 1% medulloblastoma	1:7,000–22,000
	Peutz-Jeghers syndrome	Hamartomatous intestinal polyposis / Polyps-and-spots syndrome	93%; hamartomatous polyps (jejunal) → 40% **colorectal**, 12–30% gastrointestinal, 50% **breast**, 36% pancreatic, 10% cervix, 21% ovarian, < 10% testicular, 15% lung cancer (non-small cell)	1:25,000–300,000
	Juvenile (gastrointestinal) polyposis syndrome		Juvenile polyps (colonal) → 9–50% **gastrointestinal cancer**	1:15,000–100,000

Related genes including type (tumor suppressor *TS*, oncogene *OG*, DNA stability gene *SG*) and frequency	Inheritance	Tumor onset (years)	Cancer and syndrome symptomology	Cancer therapy to avoid	Survival/life expectancy
RB1 (TS; chr13) (overall and particular for each gene)	ADI	<1–5	Leukocoria, strabismus, pain, redness, irritation, poor vision → blindness	UV, ionizing radiation (IR)	10 years: > 95% (post-enucleation)
30% *SMARCB1* (TS; chr22), *SMARCA4* (TS; chr19)	ADI	<2	Cancer site specific		<1 year/5 years: 25% (AT/RT)
SDHD (type I; TS; chr11), *SDHAF2* (type II; TS; chr11), *SDHC* (type III; TS; chr1), *SDHB* (type IV; TS; chr1)	ADI	<45	Cancer site specific; catecholamine production by sympathetic paragangliomas → neurologic symptoms related to hypertension		5 years: < 50% (malignant PCC, PCG)
APC (TS; chr5), *MUTYH* (TS; chr1)	ADI (*APC*), ARI (*MUTYH*)	1–7	Cancer site specific; gastrointestinal bleeding, (jaw) osteomata, teeth (extra, missing, unerupted), skin (epidermoid cysts and fibromas) abnormalities, congenital hypertrophy of the retinal pigment epithelium		63–70 years
80-94% *STK11* (*LKB1*; TS; chr19)	ADI	40	Cancer site specific; obstruction, intussusception, mucocutaneous hyperpigmentation		45–58 years
20% *BMPR1A* (TS; chr10), 20% *SMAD4* (TS; chr18)	ADI	24–47	Diarrhea, rectal prolapse, protein-losing enteropathy; cardiovascular (valvular heart disease), urogenital abnormalities, macrocephaly, cleft palate		56 years

(continued)

Childhood Cancer and Pediatric Cancer Predisposition Syndromes, Table 1 (continued)

Cancer predisposition syndrome		Synonyms	Malignancy risk and associated tumor types including index tumors (bold) and frequencies	Frequency
Genitourinary cancer predisposition syndromes	WAGR(O) syndrome	Deletion/monosomy 11p13	33–57%; **Wilms tumor = nephroblastoma**, gonadoblastoma	1:500,000-1Mio
	Denys-Drash syndrome		**Nephroblastoma** (multiple tumors in one or both kidneys), gonadoblastoma	~150 cases
	Frasier syndrome			<1:1Mio
	Beckwith-Wiedemann syndrome	Exomphalos-macroglossia-gigantism	3–43%; nephro**blastoma**, hepato**blastoma**, pancreato**blastoma**, neuro**blastoma**, gonado**blastoma**, adrenocortical carcinoma, rhabdomyosarcoma, gastric teratoma	1:14,000
	Simpson-Golabi-Behmel syndrome	X-linked dysplasia gigantism	8–10%; nephro**blastoma**, neuro**blastoma**, gonado**blastoma**, hepato**blastoma**, hepatocarcinoma, medulloblastoma	~250 cases

Related genes including type (tumor suppressor *TS*, oncogene *OG*, DNA stability gene *SG*) and frequency	Inheritance	Tumor onset (years)	Cancer and syndrome symptomology	Cancer therapy to avoid	Survival/life expectancy
WT1 (TS; chr11), *PAX6* (TS; chr11; responsible for eye features), *BDNF* (OG/TS; chr11; responsible for obesity)	ADI	1–3	**W**ilms tumor, **a**niridia (± cataract, glaucoma, nystagmus), **g**enitourinary (cryptorchidism, streak ovaries, bicornate uterus) abnormalities, mental **r**etardation, **o**besity; pancreatitis, scoliosis, autism, asthma, proteinuria, glomerulosclerosis (focal segmental)		4 years: 95% 27 years: 48%
WT1 (TS; chr11)	ADI		Glomerulosclerosis (Denys-Drash, diffuse mesangial; Frasier, focal segmental), pseudohermaphroditism		5 years: 90% (Wilms tumor)
Mutation or deletion of imprinted genes (*CDKN1C, H19, KCNQ1OT1*), hypermethylation and variation in the *H19/IGF2*-imprinting control region (ICR1) on chr 11p15.5; *NSD1* (TS; chr5)	ADI	<8	Cancer site specific; macrosomia, macroglossia, midline abdominal wall defects (omphalocele, umbilical hernia, diastasis recti), hemihyperplasia, visceromegaly, cleft palate, kidney abnormalities, posterior helical ear pits; hypoglycemia, nevus flammeus, prematurity		Normal/life-threatening complications/cancer specific
GPC3 (type I; TS; chrX), *OFD1* (*CXORF5*; type II; TS; chrX)	XRI	<10	Cancer site specific; macrosomia, craniofacial (coarse facies, macrocephaly, cleft palate, macroglossia), cardiovascular, skeletal (syn- and polydactyly), abdominal (visceromegaly, umbilical/diaghragmatic hernia) abnormalities, hypotonia		

(*continued*)

Childhood Cancer and Pediatric Cancer Predisposition Syndromes, Table 1 (continued)

Cancer predisposition syndrome		Synonyms	Malignancy risk and associated tumor types including index tumors (bold) and frequencies	Frequency
	Von Hippel-Lindau syndrome		40%; 90% **hemangioblastoma** (50% retinal, 60–80% CNS), cystadenoma (epididymis, uterus broad ligament, pancreas, endolymphatic sac), **PCC**, 70% **renal cell carcinoma**	1:30,000–40,000
Endocrine cancer predisposition syndromes	Multiple endocrine neoplasia type I	Wermer syndrome (type IV)	**Adenoma** (95% **parathyroid**, 15–90% **pituitary**, 20–40% adrenocortical), 30–80% **pancreatic islet cell tumor**, 10% carcinoid, 25% thyroid cancer, 88% angiofibroma (facial), 20–30% (angiomyo-)lipoma, leiomyoma, ependymoma (spinal cord), 60% meningioma, > 70% collagenoma (facial)	1:10,000–30,000
	Multiple endocrine neoplasia type II(A/B)		80% **medullary thyroid carcinoma** (=MTC), 80–100% **PCC** (bilateral), 80–100% **parathyroid adenoma**	1:35,000
Sarcoma predisposition syndromes	Li-Fraumeni syndrome	SBLA cancer syndrome	90%; *s*arcoma (osteosarcoma), *b*reast cancer, *b*rain tumor (glioblastoma), *l*eukemia, *a*drenocortical carcinoma, melanoma, nephroblastoma, gonadal germ cell tumor, gastric, pancreatic, colorectal, prostate cancer, choriocarcinoma	1:5,000–25,000
	Hereditary multiple exostoses/ osteochondromas	Diaphyseal aclasis	0.5–5%; **exostoses/ osteochondromas** (benign, metaphyseal) → chondrosarcoma, osteosarcoma	1:50,000
		Bessel-Hagen-disease		

Related genes including type (tumor suppressor *TS*, oncogene *OG*, DNA stability gene *SG*) and frequency	Inheritance	Tumor onset (years)	Cancer and syndrome symptomology	Cancer therapy to avoid	Survival/life expectancy
VHL (TS; chr3)	ADI	1–20	Cancer site specific		48–59 years
65-90%; *MEN1* (TS; chr11), *CDKN1B* (type IV; TS; chr12)	ADI	5–20	Cancer site specific; hyperparathyroidism, tuberous sclerosis-like skin abnormalities		20 years: 64% (normal)
RET (OG; chr10)	ADI	1–3	Cancer site specific; 70–80% type IIA, hyperparathyroidism, cutaneous lichen amyloidosis, M. Hirschsprung; 5% type IIB, no hyperparathyroidism, ganglioneuromas (lip, tongue, colon), marfanoid habitus, scoliosis		10 years: 68–75% (MEN IIA with MTC)
CHEK2 (TS; chr22), 77% *TP53* (TS; chr17)	ADI	30	Cancer site specific	UV, IR	<40 years
70–95%; 56–78% *EXT1* (TS; chr8), 21–44% *EXT2* (TS; chr11)	ADI	30	Cancer site specific; growth perturbation		10 years: 29–83%

(continued)

Childhood Cancer and Pediatric Cancer Predisposition Syndromes, Table 1   (continued)

Cancer predisposition syndrome		Synonyms	Malignancy risk and associated tumor types including index tumors (bold) and frequencies	Frequency
	Werner syndrome	Adult progeria	**Sarcoma** (osteosarcoma), **melanoma**, myelodysplastic syndrome (=MDS), leukemia, meningioma, thyroid, liver cancer, malignant fibrous histiocytoma	1: 20,000 (Japan)– 200,000 (USA)
Genodermatoses with cancer predisposition	Nevoid basal cell carcinoma syndrome	Gorlin-Goltz syndrome	90% **basal cell carcinoma** (face, chest, back), 5–10% **medulloblastoma, fibromas** (cardiac, ovarian), fibrosarcoma, rhabdomyosarcoma, leiomyosarcoma, 75% keratocystic odontogenic tumor	1:18,000– 250,000
	(Peripheral) Neurofibromatosis type 1	Von Recklinghausen disease	100% **neurofibromas** → (neuro-) fibrosarcoma, 3–15% malignant peripheral nerve sheath tumors, malignant schwannoma, 15% **glioma** (optic pathway, iridial Lisch nodules), PCC, leukemia, brain tumors, squamous cell carcinoma, rhabdomyosarcoma (genitourinary), gastrointestinal stromal tumors	1:2,500–4,000
	(Central) Neurofibromatosis type 2	MISME syndrome; familial acoustic neuroma/neurinoma/ vestibular schwannoma	100% *m*ultiple *i*nherited *s*chwannomas (vestibular = acoustic neuroma), 60% *m*eningioma, 20% *e*pendymoma; astrocytoma, malignant schwannoma	1:33,000– 60,000

Related genes including type (tumor suppressor *TS*, oncogene *OG*, DNA stability gene *SG*) and frequency	Inheritance	Tumor onset (years)	Cancer and syndrome symptomology	Cancer therapy to avoid	Survival/life expectancy
90%; *WRN* (*RECQL2*; SG; chr8)	ARI	25–64	Cancer site specific; progeroid syndrome with skin (facial wrinkling, subcutaneous calcification, ulcers), hair (alopecia, premature graying, thinning) abnormalities, cataract (bilateral), short stature; high-pitched voice, premature atherosclerosis, diabetes mellitus type 2, impaired fertility, osteoporosis	DNA-damaging agents	30–50 years
85% *PTCH1* (TS; chr9)	ADI	1–5 years	Cancer site specific; intracranial calcifications, cysts (epidermoid, conjunctival, jaw, bone, abdominal, genital), palmar-plantar pits, macrocephaly, cleft palate, skeletal (bifid, fused ribs, vertebrae), ocular (cataract, hypertelorism) abnormalities	UV, IR	70–81 years
NF1 (TS; chr17)	ADI	1–50	Cancer site specific; neurofibromas (cutaneous, subcutaneous, plexiform), ≥6 café-au-lait spots, freckling (axillary, inguinal), skeletal (scoliosis, short stature, sphenoid dysplasia) abnormalities, hypertension, learning disability	UV, IR	54–74 years
NF2 (TS; chr22)	ADI	1–20	Cancer site specific; fewer café-au-lait spots and neurofibromas than NF1, cataract; neurofibromatous neuropathy, multifocal meningioangiomatosis	UV, IR	62–73 years

(*continued*)

Childhood Cancer and Pediatric Cancer Predisposition Syndromes, Table 1 (continued)

Cancer predisposition syndrome		Synonyms	Malignancy risk and associated tumor types including index tumors (bold) and frequencies	Frequency
	Tuberous sclerosis complex	Bourneville syndrome	**Hamartoma, angiofibroma** (facial = adenoma sebaceum, brain, lung, renal angiomyolipoma, cardiac rhabdomyoma, coloboma), 14% **giant cell astrocytoma** (subependymal, retinal), 4% **renal cell carcinoma**, chordoma, PCC	1:6,000–10,000
	Xeroderma pigmentosum		100%; **basal cell carcinoma, squamous cell carcinoma** (head and neck), melanoma, brain tumors, lung, eye, tongue cancer, leukemia	1:1Mio
	Rothmund-Thomson syndrome	Poikiloderma of Rothmund-Thomson	30% **osteosarcoma**, 5% **squamous cell carcinoma, basal cell carcinoma**, MDS, leukemia	~300 cases
Leukemia/ lymphoma predisposition syndrome	Bloom syndrome	**Bloom**-Torre-Machacek **syndrome**	**Leukemia, lymphoma, adenocarcinoma** (breast, gastrointestinal, urogenital), nephroblastoma, osteosarcoma	<300 cases

Related genes including type (tumor suppressor *TS*, oncogene *OG*, DNA stability gene *SG*) and frequency	Inheritance	Tumor onset (years)	Cancer and syndrome symptomology	Cancer therapy to avoid	Survival/life expectancy
70%; *TSC1* (TS; chr9), *TSC2* (TS; chr16; more severe phenotype)	ADI	<1	Cancer site specific; ≥ 2 (angio-) fibromas, hypomelanotic macules (= ash leaf spots), tubers (= thick, firm, pale gyri), lymphangiomyomatosis (pulmonary); enamel pits, 60% epilepsy, 50% mental retardation		50 years
XPC (SG; chr3), *ERCC2* (SG; chr19), *POLH* (SG; chr6)	ARI	10	Cancer site specific; skin (dry, pigmented, increased sensitivity to sunlight) abnormalities, progressive psychomotoric impairment, hearing loss	UV	20 years: ≤ 40%
RECQL4 (SG; chr8)	ARI	14–34	Cancer site specific; progeroid syndrome with skin (poikiloderma, atrophy, telangiectases, hyper- and hypopigmentation), skeletal (short stature, saddle nose, osteoporosis, radial ray defect), dental, gastrointestinal abnormalities, alopecia, dystrophic nails, cataract	Hydroxyurea, camptothecin, doxorubicin, cisplatin, UV, IR	5 years: 60–70% (osteosarcoma)/ normal (no cancer)
BLM (SG; chr15)	ARI	4–46	Cancer site specific; high-pitched voice, "birdlike" facial (narrow face, prominent nose, ears, mandibular hypoplasia), skin (sun-sensitive telangiectatic erythema, poikiloderma, reduced subcutaneous fat) abnormalities, growth retardation with short stature, bilateral optic nerve hypoplasia, hypogonadism, immunodeficiency (IgM and IgA deficiency)	UV	50 years/28 years (cancer)

(*continued*)

Childhood Cancer and Pediatric Cancer Predisposition Syndromes, Table 1 (continued)

Cancer predisposition syndrome	Synonyms	Malignancy risk and associated tumor types including index tumors (bold) and frequencies	Frequency
Fanconi anemia	Fanconi pancytopenia	20–60%; 32% BMF → MDS, 10–30% **leukemia** (acute myeloid leukemia = AML), lymphoma, 14% head and neck, esophageal, colorectal, anogenital, breast, skin cancer, brain tumors	1:160,000
Nijmegen breakage syndrome	Berlin breakage syndrome; ataxia-telangiectasia, variant 1; immunodeficiency-microcephaly-chromosomal-instability-lymphoreticuloma	42%; **lymphoreticular malignancies** (non-Hodgkin lymphoma, leukemia), glioma, medulloblastoma, rhabdomyo sarcoma	1:100,000

Related genes including type (tumor suppressor *TS*, oncogene *OG*, DNA stability gene *SG*) and frequency	Inheritance	Tumor onset (years)	Cancer and syndrome symptomology	Cancer therapy to avoid	Survival/life expectancy
80–90%; 65% *FANCA* (SG; chr16), 15% *FANCC* (SG; chr9), 10% *FANCG* (SG; chr9); *FANCB* (chrX), *BRCA2* (*FANCD1*; TS/SG, chr13), *FANCD2* (SG; chr3), *FANCE* (SG; chr6), *FANCF* (SG; chr11), *FANCI* (SG; chr15), *BRIP1* (*FANCJ*; TS/SG; chr17), *FANCL* (SG; chr2), *FANCM* (chr14), *PALB2* (*FANCN*; TS/SG; chr16), *RAD51C* (*FANCO*; TS/SG; chr17), *SLX4* (*FANCP*; TS/SG; chr16), *ERCC4* (*FANCQ*; SG; chr16)	ARI, rarely XRI	16–34	Cancer site specific; BMF (macrocytosis, anemia, thrombocytopenia, neutropenia, hypocellular marrow), skeletal (growth retardation with short stature, radial ray defect, thenar hypoplasia), urogenital (hypoplastic genital organs, hypogonadism, ectopic, horseshoe, hypoplastic, double kidney, hydronephrosis), skin (hypo- and hyperpigmentation, café-au-lait spots), cardiopulmonary (valvular stenosis, cardiomyopathy), ocular (microphthalmia, strabismus, cataract), gastrointestinal organ atresia, microcephaly, conductive deafness	Mitomycin C (MMC), diepoxybutane (DEB), IR	20–33 years/5 years: 94% (after HSCT)
NBN (SG; chr8)	ARI	10	Cancer site specific; skeletal (growth retardation with short stature, thenar hypoplasia), "birdlike" facial (sloping forehead, long, beaked, upturned nose, mandibular hypoplasia, large ears), skin (café-au lait spots, vitiligo), urogenital (ectopic, horseshoe, hypoplastic, double kidney, premature ovarian insufficiency) abnormalities, CNS (microcephaly), immunodeficiency (both cellular and humoral)	IR, bleomycin, MMC, DEB	11 years/ 20 years: 85% (no cancer), 35% (cancer)

(continued)

Childhood Cancer and Pediatric Cancer Predisposition Syndromes, Table 1 (continued)

Cancer predisposition syndrome		Synonyms	Malignancy risk and associated tumor types including index tumors (bold) and frequencies	Frequency
	Ataxia-telangiectasia	Louis-Bar syndrome	25–40%; 85% **lymphoreticular malignancies** ((T cell) **leukemia**, (B cell) **lymphoma**), breast, thyroid, liver cancer, gastric mucinous adenocarcinoma, medulloblastoma, glioma	1:40,000–100,000
	Dyskeratosis congenita	Zinsser-Engman-Cole syndrome	10–50%; 22–80% bone marrow failure (=BMF) → MDS, **leukemia** (AML), Hodgkin lymphoma, 40% **squamous cell carcinoma** (head and neck), **anogenital**, pancreatic cancer	~1:1Mio
	Diamond-Blackfan anemia	Aase(-Smith II) syndrome; congenital pure red cell aplasia	6%; BMF → MDS, **leukemia** (AML), **osteosarcoma**	1:150,000

Related genes including type (tumor suppressor TS, oncogene OG, DNA stability gene SG) and frequency	Inheritance	Tumor onset (years)	Cancer and syndrome symptomology	Cancer therapy to avoid	Survival/life expectancy
90% *ATM* (SG; chr11)	ARI	5–20	Cancer site specific; CNS (progressive cerebellar ataxia, oculomotor apraxia, neuropathy) abnormalities, oculocutaneous telangiectases, gonadal dysfunction, growth retardation with short stature, immunodeficiency (both cellular and humoral)	IR, bleomycin, MMC, DEB	24 years (no cancer), 15 years (cancer)
60–70%; 40% *DKC1* (TS; chrX), 5% *TERC* (TS; chr3), *TERT* (OG; chr5), *TINF2* (TS; chr14), *NHP2* (TS; chr5), *NOP10* (TS; chr15), *RTEL1* (TS; chr20), *WRAP53* (OG; chr17), *USB1* (TS; chr16), *CTC1* (OG; chr17)	XRI (*DKC1*), ADI (*TERC, TERT, TINF2, RTEL1*), ARI (*TERT, CTC1, RTEL1, WRAP53, NHP2, NOP10*)	29–37	Cancer site specific; mucocutaneous (lacy reticular pigmentation, skin atrophy (chest, head and neck), palmar, plantar hyperkeratosis, pigmentation, nail dystrophy, oral leukoplakia) abnormalities; pulmonary fibrosis, immunodeficiency (both cellular and humoral), growth retardation with short stature, CNS (microcephaly, cerebellar hypoplasia, developmental delay), gastrointestinal (gastrointestinal tract stenosis, enteropathy), retinal abnormalities		46 years (no cancer), 39 years (cancer)
40–60%; 25% *RPS19* (TS; chr19), 9% *RPL5* (TS; chr1), 6.5% *RPL11* (TS; chr1), 7% *RPS26* (TS; chr12), 1–3%: *RPS7* (TS; chr2), *RPS17* (TS; chr15), *RPS24* (TS; chr10), *RPS10* (TS; chr6), *RPL35a* (TS; chr3), *GATA1* (TS; chrX)	ADI	56	Cancer site specific; BMF (macrocytosis, anemia with reticulocytopenia, erythroid hypoplasia in marrow) skeletal (growth retardation with short stature, upper limb malformations), craniofacial (Pierre Robin syndrome, cleft palate, cataract, glaucoma, strabismus), cardiac, urogenital (hypospadia) abnormalities		40 years: 75%

(*continued*)

Childhood Cancer and Pediatric Cancer Predisposition Syndromes, Table 1 (continued)

Cancer predisposition syndrome		Synonyms	Malignancy risk and associated tumor types including index tumors (bold) and frequencies	Frequency
	Shwachman-(Bodian-) Diamond syndrome	Pancreatic insufficiency and bone marrow dysfunction	40% BMF $\rightarrow$ 8–19% MDS, **leukemia** (AML); breast cancer, dermatofibrosarcoma	1:77,000– 350,000
	Congenital amegakaryocytic thrombocytopenia (CAMT)		BMF $\rightarrow$ MDS, **leukemia** (AML)	<100 cases
Primary immuno deficiency disorders with cancer predis position	Common variable immuno deficiency (CVID)	Primary hypogammaglobu-linemia	7–13%; 16% **gastrointestinal**, breast, bladder, cervix cancer, 8% **lymphoid malignancies** ((non-)Hodgkin lymphoma)	1:10,000– 100,000

Related genes including type (tumor suppressor *TS*, oncogene *OG*, DNA stability gene *SG*) and frequency	Inheritance	Tumor onset (years)	Cancer and syndrome symptomology	Cancer therapy to avoid	Survival/life expectancy
90% *SBDS* (TS; chr7)	ARI	4–19	Cancer site specific; BMF (macrocytosis, neutropenia, myeloid hypoplasia in marrow), exocrine pancreatic insufficiency with steatorrhea, skeletal (growth retardation with short stature, chondrodysplasia or congenital thoracic dystrophy), skin, dental, cardiac abnormalities, hepatomegaly	Cyclophos phamide, busulfan	20 years: 85%
MPL (type I and II; OG; chr1), *RUNX1* (type III; OG; chr21)	ARI	>2–3	Cancer site specific; BMF (hypomegakaryocytic thrombocytopenia, ± pancytopenia), cardiac (septal defects), neurological (cerebral, cerebellar hypoplasia) abnormalities, strabismus, psychomotor retardation		Mortality: 30% of bleeding, 20% after HSCT
ICOS (CVID 1; T-cell defect; OG; chr2), *TNFRSF13B* (*TACI*; CVID 2; TS/OG; chr17), *CD19* (CVID 3; B-cell defect; OG; chr16), *TNFRSF13C* (*BAFFR*; CVID 4; OG; chr22), CD20 (*MS4A1*; CVID 5; OG; chr11), *CD81* (CVID 6; TS/OG; chr11), *CD21* (*CR2*; CVID 7; TS/OG; chr1), *LRBA* (CVID 8; OG; chr4), *NFKB2* (CVID 10; OG; chr10), *IL21* (CVID 11; TS; chr4), *NFKB1* (CVID 12; OG; chr4)	80% ADI, 20% ARI	23–27	Cancer site specific; immunodeficiency (hypogamma globulinemia, recurrent 98% bronchitis, sinusitis, otitis, pneumonia with subsequent bronchiectasia), 25% autoimmune phenomena (immune thrombocytopenic purpura (ITP) and autoimmune hemolytic anemia (AIHA), rheumatoid arthritis), generalized lymphadenopathy ± splenomegaly, granulomas		43 years

(continued)

Childhood Cancer and Pediatric Cancer Predisposition Syndromes, Table 1 (continued)

Cancer predisposition syndrome		Synonyms	Malignancy risk and associated tumor types including index tumors (bold) and frequencies	Frequency
	Wiskott-Aldrich syndrome	Eczema-thrombocytopenia-immunodeficiency	13–22% **lymphoreticular malignancies** (MDS, B-cell lymphoma, leukemia)	<1:100,000
	Severe congenital neutropenia (SCN)	Kostman's disease (ARI disease)	9–15% **lymphoreticular malignancies** (MDS, leukemia (AML))	1:250,000-1Mio
	Severe combined immunodeficiency (SCID)		**Lymphoid malignancies** (30% non-Hodgkin lymphoma)	1:50,000–100,000

Related genes including type (tumor suppressor *TS*, oncogene *OG*, DNA stability gene *SG*) and frequency	Inheritance	Tumor onset (years)	Cancer and syndrome symptomology	Cancer therapy to avoid	Survival/life expectancy
WAS (TS; chrX)	XRI	10	Cancer site specific; immunodeficiency (both cellular and humoral, recurrent (middle ear) infections), microthrombocytopenia with hematochezia, mucosal bleeding and/or petechiae, eczema (chronic, acute), 40% autoimmune phenomena (AIHA, neutropenia, vasculitis, inflammatory bowel disease, nephropathy, arthritis)		8–15 years, > 80% after HSCT
60%; 50–60% *ELANE* (*ELA2*; SCIN1; TS; chr19), *GFI1* (SCIN2; chr1), 4–30% *HAX1* (SCIN3, OG; chr1), *G6PC3* (SCIN4; TS; chr17), *VPS45* (SCIN5, chr1), *JAGN1* (SCIN6, chr3), *WAS* (SCNX; TS; chrX)	ADI (*ELA2, GFI1*), ARI (*HAX1, G6PC3, VPS45, JAGN1*), XRI (*WAS*)	10	Cancer site specific; immunodeficiency (maturation arrest of myeloid precursors at promyelocyte stage, granulocytopenia with recurrent bacterial and mycotic infections, stomatitis, ear, nose, throat, pulmonary); osteoporosis		20 years: 82%
69%; *IL2RG* (TS; chrX), *JAK3* (TS/OG; chr19) and others	XRI (*IL2RG*), ARI (15 genes)	6	Cancer site specific; immunodeficiency (both cellular and humoral with lack of functional, peripheral T lymphocytes, recurrent (opportunistic) infections, absent lymph nodes), CNS (sensorineural deafness, microcephaly, neurodevelopmental deficit), skin (rash, alopecia), hepatic abnormalities		10 years: > 66–90% (after HSCT)

(continued)

Childhood Cancer and Pediatric Cancer Predisposition Syndromes, Table 1 (continued)

Cancer predisposition syndrome		Synonyms	Malignancy risk and associated tumor types including index tumors (bold) and frequencies	Frequency
	Selective immunoglobulin A deficiency		**Gastrointestinal cancer, lymphoid malignancies**	1:700–2,000
Numerical chromosomal abnormalities with cancer predisposition	Down syndrome	Trisomy 21	**Leukemia** (60% acute lymphoblastic leukemia, 40% AML; lymphoma, (extra-)gonadal germ cell tumor, retinoblastoma)	1:700–1,000
	Klinefelter syndrome	47,XXY syndrome	**Breast, lung, testicular cancer, non-Hodgkin lymphoma, germ cell tumor** (mediastinal)	1:500–1,000 males

Related genes including type (tumor suppressor *TS*, oncogene *OG*, DNA stability gene *SG*) and frequency	Inheritance	Tumor onset (years)	Cancer and syndrome symptomology	Cancer therapy to avoid	Survival/life expectancy
TNFRSF13B (TS/OG; chr17), *IGHA1* (chr14), *IGHA2* (chr14)	ARI	<40	Cancer site specific; 8590% patients asymptomatic; immuno-deficiency with recurrent infections, food intolerance, celiac disease, 40% allergic disorders (rhinitis, conjunctivitis, asthma, atopic dermatitis), 55% autoimmune phenomena (ITP, AIHA, rheumatoid arthritis, thyroiditis, diabetes mellitus type 1)		Normal/progression to CVID/cancer specific
Numerical chromosomal abnormalities	NA	<5	Cancer site specific; variable intellectual disability, muscular hypotonia, joint laxity, characteristic facial dysmorphism, cardiac, gastrointestinal (duodenal atresia, celiac disease), endocrine (hypothyroidism, diabetes mellitus type 1) abnormalities, Alzheimer disease, short stature, cataract, conductive hearing loss	DNA-damaging agents	>55 years
		15–30	Cancer site specific; motor, cognitive, behavioral dysfunction, vascular disease, primary hypogonadismus with severe endocrine (delayed, incomplete puberty, eunuchoid habitus, gynecomastia, infertility), reproductive (hypospadia, small testes, phallus or cryptorchidism) abnormalities		Normal/cancer specific

(continued)

Childhood Cancer and Pediatric Cancer Predisposition Syndromes, Table 1 (continued)

Cancer predisposition syndrome		Synonyms	Malignancy risk and associated tumor types including index tumors (bold) and frequencies	Frequency
	Turner syndrome	45,X syndrome	2–28% **gonadoblastoma, meningioma, childhood brain tumors**, Wilms tumor, neuroblastoma, retinoblastoma, bladder, uterus, colon cancer, melanoma, leukemia	1:2,500 females
	Edwards syndrome	Trisomy 18	1% **Wilms tumor, hepatoblastoma**	1:3,600–10,000
	Pleuropulmonary blastoma syndrome	DICER1 syndrome	**Pleuropulmonary blastoma**, 9% cystic nephroma, sarcoma, medulloblastoma, Sertoli-Leydig cell tumor, Hodgkin lymphoma, leukemia, thyroid cancer	~350 cases

chr chromosome, *TS* tumor suppressor gene, *OG* oncogene, *SG* DNA stability gene/DNA damage response gene, *ADI* autosomal dominant inheritence, *ARI* autosomal recessive inheritence, *XRI* X-linked recessive inheritance, *HSCT* hematopoietic stem cell transplantation, *UV* ultraviolet radiation

Related genes including type (tumor suppressor *TS*, oncogene *OG*, DNA stability gene *SG*) and frequency	Inheritance	Tumor onset (years)	Cancer and syndrome symptomology	Cancer therapy to avoid	Survival/life expectancy
		<15	Cancer site specific; short stature, ovarian failure (infertility), skin (lymphedema, multiple nevi), cardiovascular (coarctation aortae, aortic dissection, dilatation, valve anomalies), renal, hepatic, metabolic (osteoporosis, diabetes mellitus type 2) abnormalities, otitis media ± conductive deafness, autoimmune phenomena (inflammatory bowel disease, thyroiditis)		Reduced by 13 years/cancer specific
		1–5	Cancer site specific; skeletal (growth retardation, clenched fist with overriding fingers, nail hypoplasia), facial (microretrognathia, microphthalmia), cardiac (septal defects, patent ductus arteriosus, polyvalvular disease), urogenital, CNS (choroid plexus cysts, dolichocephaly, microcephaly, hypotonia, psychomotor and cognitive disability) abnormalities		>1 year: 10%
97% *DICER1* (TS; chr14)	ADI	2	Type I (cystic), type II (cystic/solid), type III (solid); symptoms of pneumonia, pneumothorax		5 years: 90% (type I), 71% (type II), 53% (type III)

autosomal recessive (DC, SDS, SCN), autosomal dominant (DC, DBA, SCN), and X-linked (DC) inheritance patterns. Within each bone marrow failure syndrome, the composition and severity of the physical phenotype vary widely, but there is overlap in features such as poor growth, radial ray anomalies, and involvement of skin, eyes, renal, cardiac, skeletal, and other organs. There is also a wide spectrum to the hematologic picture. ► Acute myeloid leukemia (AML) has been observed in FA, DBA, DC, SDS, SCN, and CAMT. Solid tumors are also appearing in patients whose underlying disease involves hematopoiesis and physical development. These tumors occur at much younger ages than in the general population and have patterns that are characteristic to the syndrome, such as head and neck and gynecologic cancers in FA and DC and osteogenic sarcomas in DBA. The other syndromes have not yet been reported to have a propensity for solid tumors.

Primary Immunodeficiency Disorders

Only a few of the more than 150 subtypes of primary immunodeficiency disorders (PIDDs) are associated with elevated risks for different types of cancer. The overall risk for cancer developing in children with PIDD is estimated to range from 4% to 25%. There seems to be a complex relationship between PIDD, the viral infections to which patients with PIDD are susceptible, and the development of cancer. As support for this assumption, the most common cancer subtypes in immunodeficient patients are NHL and HD, representing immune system-related malignancies. Further, cancer development in immunocompromised patients frequently correlates with either de novo, reactivated, or chronic infection, in particular with oncogenic viruses, such as EBV (see ► Epstein-Barr Virus) and HHV-8. In most PIDD cases with cancer, B-cell function is at least partially defective, whereas T-cell function might be unaffected. More than half of PIDD-related cancer cases have been reported in patients with ataxia-telangiectasia (A-T) and common variable immunodeficiency (CVID). One-third is associated with Wiskott-Aldrich syndrome (WAS), severe combined immunodeficiency (SCID), and

selective IgA deficiency, with NHL being the predominant malignancy in A-T, CVID, WAS, and SCID. Bloom syndrome and Nijmegen breakage syndrome are further recognized as immunodeficiency syndromes by their particular clinical or immunological features.

Numerical Chromosomal Abnormalities

Besides familial neoplastic syndromes, inherited immunodeficiency, and bone marrow failure syndromes, several numerical chromosome abnormalities are associated with childhood cancer. Down syndrome (trisomy 21) accounts for the largest number of cases including leukemia (60% ► acute lymphoblastic leukemia (ALL) and 40% AML) with a 50-fold risk in the first 5 years of life and tenfold risk in the next 10 years. Less commonly, Down syndrome patients are diagnosed at a higher frequency with germ cell tumors (see ► Ovarian Germ Cell Tumors, ► Germinoma, ► Ovarian Stromal and Germ Cell Tumors, ► Testicular Germ Cell Tumors), lymphomas, and RB. Patients with trisomy 18 have an increased risk of Wilms tumor. Female patients with Turner syndrome (45, X; other rare forms) are at increased risk for NB and Wilms tumor. Patients with Klinefelter syndrome (47, XXY; other rare forms) display an increased risk of breast and lung cancer, and germ cell tumors.

Miscellaneous

Pleuropulmonary blastoma (PPB) is a rare embryonal cancer affecting the lungs of infants and young children, and it is suspected that approximately 60–70% of PPBs are due to germline *DICER1* mutation which functions as a haploinsufficient tumor suppressor (Table 1).

Cross-References

► Acute Lymphoblastic Leukemia
► Acute Myeloid Leukemia
► Adrenocortical Cancer
► Aggressive Fibromatosis in Children
► APC Gene in Familial Adenomatous Polyposis

▶ Astrocytoma
▶ ATM Protein
▶ Basal Cell Carcinoma
▶ B-Cell Lymphoma
▶ Beckwith-Wiedemann Syndrome
▶ Beckwith-Wiedemann Syndrome Associated Childhood Tumors
▶ Bladder Cancer
▶ Bloom Syndrome
▶ Bone Tumors
▶ Brain Tumors
▶ Breast Cancer
▶ Carcinoid Tumors
▶ Cardiac Tumors
▶ Childhood Adrenocortical Carcinoma
▶ Chronic Myeloid Leukemia
▶ Colorectal Cancer
▶ Diffuse Large B-Cell Lymphoma
▶ DNA Damage Response
▶ DNA Damage Response Genes
▶ Endocrine-Related Cancers
▶ Epithelial Tumorigenesis
▶ Epstein-Barr Virus
▶ Fanconi Anemia
▶ Follicular Thyroid Tumors
▶ Gastric Cancer
▶ Gastrointestinal Stromal Tumor
▶ Germinoma
▶ Head and Neck Cancer
▶ Hematological Malignancies, Leukemias, and Lymphomas
▶ Hepatoblastoma
▶ Hepatocellular Carcinoma
▶ Hodgkin Disease
▶ HSNF5/INI1/SMARCB1 Tumor Suppressor Gene
▶ Immunosuppression and Cancer
▶ Li-Fraumeni Syndrome
▶ Lung Cancer
▶ Marginal Zone B-Cell Lymphoma
▶ Multiple Endocrine Neoplasia Type 1
▶ Naevoid Basal Cell Carcinoma Syndrome
▶ Nephroblastoma
▶ Neuroblastoma
▶ Neurofibromatosis 1
▶ Neurofibromatosis 2
▶ Neuro-Oncology: Primary CNS Tumors
▶ Nijmegen Breakage Syndrome

▶ Non-Rhabdomyosarcoma Soft Tissue Sarcomas
▶ Non-Small-Cell Lung Cancer
▶ Oncogene
▶ Osteochondroma
▶ Osteosarcoma
▶ Ovarian Cancer
▶ Ovarian Germ Cell Tumors
▶ Ovarian Stromal and Germ Cell Tumors
▶ Pancreatic Cancer
▶ Papillary Thyroid Carcinoma
▶ Pediatric Brain Tumors
▶ Peutz–Jeghers Syndrome
▶ Pheochromocytoma
▶ Renal Cancer Clinical Oncology
▶ Renal Cancer Genetic Syndromes
▶ Repair of DNA
▶ RET
▶ Retinoblastoma
▶ Retinoblastoma Protein, Biological and Clinical Functions
▶ Rhabdoid Tumor
▶ Rhabdomyosarcoma
▶ Second Primary Tumors
▶ Skin Cancer
▶ Testicular Cancer
▶ Testicular Germ Cell Tumors
▶ Thyroid Carcinogenesis
▶ TP53
▶ Tuberous Sclerosis Complex
▶ Tumor Suppressor Genes
▶ Von Hippel-Lindau Disease
▶ Von Hippel-Lindau Tumor Suppressor Gene
▶ Wilms' Tumor
▶ Xeroderma Pigmentosum

References

Alter BP et al (2010) Malignancies and survival patterns in the National Cancer Institute inherited bone marrow failure syndromes cohort study. Br J Haematol 150(2):179–188

Garber JE, Offit K (2005) Hereditary cancer predisposition syndromes. J Clin Oncol 23(2):276–292

Schiffman JD et al (2013) Update on pediatric cancer predisposition syndromes. Pediatr Blood Cancer 60(8):1247–1252

Shapiro RS (2011) Malignancies in the setting of primary immunodeficiency: Implications for hematologists/oncologists. Am J Hematol 86(1):48–55

Stiller CA (2004) Epidemiology and genetics of childhood cancer. Oncogene 23(38):6429–6444

Strahm B, Malkin D (2006) Hereditary cancer predisposition in children: genetic basis and clinical implications. Int J Cancer 119(9):2001–2006

Childhood Cancer and Treatment-Related Toxicities

Stefan K. Zöllner[1] and Jeffrey A. Toretsky[2]
[1]Department of Pediatric Hematology and Oncology, University Childrens Hospital Münster, Münster, Germany
[2]Department of Oncology and Pediatrics, Lombardi Comprehensive Cancer Center, Georgetown University, Washington, DC, USA

Definition

Treatment-related toxicities of childhood cancer encompass both the physical and non-physical burden of children and young adolescents diagnosed with cancer which result from invasive and non-invasive medical treatment modalities and the psycho-socio-economic effects of being a cancer patient, occurring during and/or after the disease-related treatment.

Risk awareness

The overall survival rates of many pediatric cancers continue to improve with each decade due to new advances in therapy. As this trend continues, the focus and importance of minimizing acute and long-term toxicity associated with treatment is paramount; significant treatment-related toxicities continue to impact the majority of children with cancer. Awareness of short- and long-term health risks is important, and careful follow-up of long-term survivors is essential. Children with cancer and their families are affected not only by the disease and treatments but also by significant effects on the child's physical and emotional development. The risk of specific health-related outcomes in childhood cancer survivors is extremely varied and may be influenced by characteristics of the individual, the childhood cancer diagnosis, and the therapeutic regimen. Long-term survivors are at high risk for developing complications, such as subsequent malignant neoplasms, cardiovascular disease, and endocrinopathies, years after completion of cancer treatment. Apart from direct organ toxicities, other treatment-related side effects significantly influence the psychosocial well-being of patients, including the risk of infertility, physical disability and stigmatization, and cognitive deficits.

Risk prevention

For many of the complications, there are well-established associations between therapeutic exposures and adverse health-related outcomes, which set the stage for primary prevention (avoidance of certain treatments, when possible) and secondary prevention (screening for and treatment of asymptomatic disease) strategies in these survivors. Some of these side effects were related to the cumulative dose of chemotherapy agents and resulted in dose modification or replacement strategies in newer regimens to minimize immediate and late toxicity without affecting efficacy. Similar modifications have been made with radiation therapy by instituting techniques to improve delivery of the radiation to the tumor and minimizing damage to neighboring healthy tissues (involved field radiotherapy) as well as reducing the total dose of radiation delivered. To ameliorate and enable the follow-up care for patients and families and physicians, specific recommendations for screening of treatment-related complications have been elaborated.

Death

Although the survival rate for most pediatric cancer patients is high, approximately 20% of children with cancer will die each year of the disease. Any patient's death is not only challenging for relatives but poses a unique set of emotional and practical difficulties for caregivers and health-care professionals alike. Specific studies revealed that caring for terminal patients (see ▶ Palliative

Therapy) can result in feelings of distress and burnout in health-care professionals when patients die. The findings point to the complexity of working with children where parents are included in the decision-making processes around a child's treatment. In contrast, parents of children diagnosed with cancer are at risk for the development of posttraumatic stress symptoms, especially at the end of treatment. Some bereaved parents develop post-traumatic stress disorder up to 5 years after the end of treatment or child's death.

Secondary Malignant Neoplasms

The occurrence of secondary and subsequent malignant neoplasms (SMNs) has been recognized for many years as late sequelae of childhood cancer therapy. SMNs should be distinguished from metastases or recurrences of the primary tumor. They are histologically different from the first primary and they occur after the first primary. While a majority of SMNs develop within 10 years, the incidence of SMNs in childhood cancer survivors increases with sustained age, with the cumulative incidence exceeding 20% at 30 years after diagnosis of the primary cancer. These survivors have an up to sixfold risk of SMNs when compared with age- and sex-matched general population. In detail, the excess risk of SMNs is highest for Hodgkin lymphoma (HD; see ▶ Hodgkin Disease) and ▶ Ewing sarcoma (ES) survivors (8.7-fold and 8.5-fold, respectively). Importantly, these patients are even at a higher risk to develop additional tumors, as within 20 years from diagnosis of the first SMNs, nearly one half (47%) will develop a subsequent neoplasm.

The main therapeutic risk factors for development of SMNs include chemotherapeutic agents (see ▶ Chemotherapy), exposure to ionizing radiation (see ▶ Radiation Oncology, ▶ Ionizing Radiation Therapy), and undergoing allogeneic hematopoietic stem cell transplantation (see ▶ Allogeneic Cell Therapy). The distinct differences in the onset and role of specific therapeutic exposures have resulted in the classification of SMNs into two distinct categories: (I) radiation-related solid SMNs, commonly including breast (see ▶ Breast Cancer), thyroid (see ▶ Follicular Thyroid Tumors, ▶ Thyroid Carcinogenesis, ▶ Papillary Thyroid Carcinoma), skin (see ▶ Skin Cancer), and brain cancer (see ▶ Brain Tumors, ▶ Pediatric Brain Tumors) and characteristically occurring after a latency of 10 years after radiation exposure, and (II) chemotherapy-related ▶ myelodysplastic syndrome/▶ acute myeloid leukemia (MDS/AML), which are notable for their shorter latency, i.e., less than 5 years from primary cancer diagnosis, and association with ▶ alkylating agent and/or ▶ topoisomerase II inhibitor chemotherapy. Success in treating children with cancer should not be overshadowed by the incidence of SMNs, but patients and health-care providers must be aware of risk factors for SMNs so that surveillance is focused and early prevention strategies are implemented.

Cardiovascular Disease

Childhood cancer survivors treated with cardiotoxic substances (▶ anthracycline chemotherapy and/or chest radiation) are at risk for developing cardiovascular complications including cardiomyopathy/heart failure, coronary artery disease, valvular disease, conduction abnormalities, and pericardial disease. The risk for cardiovascular complications is highest in survivors of HD, kidney tumors (see ▶ Renal Cancer Therapy, ▶ Renal Cancer Treatment), and ES. Apart from the fact that childhood cancer survivors have a sevenfold increased risk of cardiac death when compared with the general population, there is a long latency between cancer treatment and onset of clinically overt cardiovascular disease, e.g., asymptomatic cardiomyopathy eventually progresses to symptomatic heart failure. Therefore, survivors treated with anthracyclines and/or radiation therapy involving the heart region should be aware of the risk of cardiomyopathy, and their left ventricular systolic function should be monitored by echocardiography.

Endocrine Complications

Endocrine complications of cancer therapy are reported in over 40% of childhood cancer survivors. The main endocrine complications include disorders of the hypothalamic-pituitary axis, disorders of pubertal development, thyroid dysfunction, gonadal dysfunction, decreased bone mineral density (see ▶ osteoporosis), obesity, and alterations in glucose metabolism. Determining an individual child's risk for developing endocrine abnormalities requires a thorough analysis of the child's chemotherapeutic regimen, radiation dose, fractionation of radiotherapy, age during therapy, pubertal status during therapy, and length of time since completion of therapy.

Thyroid Disorders

Primary hypothyroidism frequently occurs in survivors treated with radiation to the thyroid gland, including nasopharyngeal, cervical, mantle/supraclavicular, and craniospinal fields, as well as those exposed to total body irradiation. The risk of primary hypothyroidism is observed beginning at 10 Gy of thyroidal irradiation, and it increases as the dose of radiation increases above this dose. At-risk childhood cancer survivors should have their thyroid function serially monitored, generally at least annually.

Gonadal Dysfunction

Loss of opportunity for fertility is a prime concern in both male and female cancer survivors. Endocrine effects of gonadal damage are also central to long-term health and well-being, and therefore preservation of gonadal function is an important priority at treatment. All approaches to fertility preservation have specific challenges in children and teenagers, including ethical, practical, and scientific issues. For both sexes, fertility preservation involves an invasive procedure. Decision-making for fertility often preservation needs assessment of the individual's risk of fertility loss and is made at a time of emotional distress but, in the end, must be offered to any child with cancer even if maturation of immature germ cells is uncertain.

Exposure of the sperm-producing cells to radiation and/or alkylating chemotherapeutics including mechlorethamine, ▶ cyclophosphamide, ifosfamide, procarbazine, busulfan, and melphalan may result in impaired spermatogenesis. The risk that is dose dependent, but individual responses vary greatly. Sperm production may be impaired at radiation doses as low as 0.15 Gy,

with permanent sterility common following testicular doses of more than 6 Gy, especially in men treated with total body irradiation. Chemotherapy alone rarely results in Leydig cell failure and subsequent impairment of testosterone production, but most prepubertal males who receive radiation doses of 24 Gy or greater to the testis will develop pubertal delay or arrest, if it occurs before or during puberty, and reduced libido, erectile dysfunction, decreased bone mineral density, and decreased muscle mass, if it occurs after completion of normal puberty.

The interdependence of sex steroid-producing cells and oocytes within the ovarian follicle leads to ovarian failure resulting in impairment of both sex hormone production and fertility. Ovarian dysfunction may result from treatment with gonadotoxic chemotherapy (typically treatment with high-dose alkylators), radiation affecting the ovaries (doses exceeding 10 Gy), or surgical removal of the ovaries. If ovarian failure occurs before pubertal onset, delayed puberty and primary amenorrhea will result. If ovarian function is lost during or after puberty, pubertal arrest, secondary amenorrhea, and symptoms of menopause will occur.

Female survivors treated with high-dose alkylating agents and/or pelvic irradiation have been shown to experience fewer pregnancies when compared with siblings; women who do achieve pregnancy after treatment with chemotherapy alone do not appear to have adverse pregnancy outcomes. Female cancer survivors treated with pelvic irradiation, who become pregnant, however, are at a higher risk of stillbirth or neonatal death and having offspring who are premature, have low birth weight, or are small for gestational age. In contrast, their offspring do not appear to be at increased risk of congenital anomalies or genetic defects.

Neurocognitive Impairment

A robust literature has developed documenting neurocognitive late effects in up to 40% of survivors of childhood cancer. Patterns of late effects include deficits in attention and concentration, working memory, processing speed, executive function, contributing to declines in intellectual and academic abilities, and ultimately affect employment, independent living, and health-care use in adult survivors of childhood cancer. In particular, patients who have undergone treatment for brain tumors or received prophylactic cranial radiation therapy for leukemia treatment are at high risk for neurocognitive impairment. For leukemia, the doses of cranial radiation therapy are usually much lower than they are for brain tumors, thereby resulting in less severe neurocognitive sequelae. In order to ameliorate the neurocognitive outcome of survivors, prophylactic interventions should be implemented during or immediately after therapy to minimize the effects associated with neurotoxic therapies. As not all children with similar diagnoses and treatment show identical neurocognitive outcomes, studies are needed to identify, early in their treatment, those patients at most risk for the greatest declines in cognitive damage.

Socioeconomic Burden and Health Status

The survival rate improvement of childhood cancer patients also affects their educational and social outcome. Returning to school after cancer diagnosis can be both academically, i.e., school performance, and socially, i.e., (re)socialization into a peer group, challenging for children with cancer and is influenced by (I) type of tumor (primarily brain tumors but also hematological malignancies (see ▶ Hematological Malignancies, Leukemias, and Lymphomas) and ▶ bone tumors), (II) age at diagnosis (very young children and adolescent), (III) treatment modalities (neurotoxic treatments, hematopoietic stem cell transplantation), and (IV) social or educational status of the parents.

Cancer treatment may cause financial stress for pediatric oncology families but also affects the quality of life of children with cancer in lower-income families which may have fewer resources to cope with their child's disease. In consequence, significant socioeconomic disparities exist in the quality of life of these children.

In general, the prevalence of poor health status is higher among survivors than siblings, increases rapidly with age particularly among female patients, and is related to an increasing burden of chronic health conditions.

Long-term follow-up is an important part of care for survivors of pediatric cancers, guided by the cancer diagnosis and the type of treatment the child received. Despite standardization in disease assessments, curative interventions, and follow-up care, palliative assessments and psychosocial interventions require improvement. The ultimate goal of cancer therapy today is not simply medical cure but cure that results in the survivors' healthy, long-term neurocognitive outcome and optimum quality of life.

Cross-References

- ▶ Acute Myeloid Leukemia
- ▶ Alkylating Agents
- ▶ Allogeneic Cell Therapy
- ▶ Anthracyclines
- ▶ Bone Tumors
- ▶ Brain Tumors
- ▶ Breast Cancer
- ▶ Chemotherapy
- ▶ Cyclophosphamide
- ▶ Ewing Sarcoma
- ▶ Follicular Thyroid Tumors
- ▶ Hematological Malignancies, Leukemias, and Lymphomas
- ▶ Hodgkin Disease
- ▶ Ionizing Radiation Therapy
- ▶ Myelodysplastic Syndromes
- ▶ Osteoporosis
- ▶ Palliative Therapy
- ▶ Papillary Thyroid Carcinoma
- ▶ Pediatric Brain Tumors
- ▶ Radiation Oncology
- ▶ Renal Cancer Therapy
- ▶ Renal Cancer Treatment
- ▶ Skin Cancer
- ▶ Thyroid Carcinogenesis
- ▶ Topoisomerases

References

Armenian SH, Kremer L.C, Sklar C (2015) Approaches to reduce the long-term burden of treatment-related complications in survivors of childhood cancer. Am Soc Clin Oncol Educ Book 196–204

Bhatia S, Sklar C (2002) Second cancers in survivors of childhood cancer. Nat Rev Cancer 2(2):124–132

Landier W, Armenian S, Bhatia S (2015) Late effects of childhood cancer and its treatment. Pediatr Clin North Am 62(1):275–300

Lipshultz SE et al (2013) Long-term cardiovascular toxicity in children, adolescents, and young adults who receive cancer therapy: pathophysiology, course, monitoring, management, prevention, and research directions: a scientific statement from the American Heart Association. Circulation 128(17):1927–1995

Oeffinger KC et al (2006) Chronic health conditions in adult survivors of childhood cancer. N Engl J Med 355(15):1572–1582

Prasad PK et al (2015) Psychosocial and neurocognitive outcomes in adult survivors of adolescent and early young adult cancer: a report from the childhood cancer survivor study. J Clin Oncol 33(23):2545–2552

Children's Brain Tumors

▶ Pediatric Brain Tumors

Chimeric Antigen Receptor (CAR)

Astrid Holzinger, Jennifer Makalowski and Hinrich Abken
Tumor Genetics, Clinic I Internal Medicine, University Hospital Cologne, and Center for Molecular Medicine Cologne, University of Cologne, Cologne, Germany

Synonyms

Chimeric immune receptor; Immunoreceptor; T-body

Definition

Chimeric antigen receptors are recombinant transmembrane receptor molecules which are composed of an extracellular binding domain, mostly derived from an antibody, and an intracellular signaling domain, mostly derived from the T cell receptor (TCR) complex, to initiate immune cell activation upon engagement of the receptor ligand.

Characteristics

Chimeric antigen receptors (CARs) are designed to redirect immune cells, preferentially T cells, with predefined specificity toward target cells for the use in the ▶ adoptive immunotherapy. The strategy thereby takes advantage of the power of the immune cell response to target predefined cells in patients for the treatment of various diseases, mostly malignant diseases.

Technically, patient's T cells are ex vivo engineered with the CAR by ▶ viral vector-mediated gene transfer, amplified to clinically relevant numbers and readministered to the patient. Currently, γ-retroviral and lentiviral vectors are mostly used; RNA and DNA-mediated transfer techniques are applied as well.

Apart from the entire T cell population, T cell subsets like CD8$^+$ T cells, CD4$^+$ T cells or ▶ regulatory T (Treg) cells, cytokine activated killer (CIK) cells, or NK-T cells are also used for CAR-based cell therapy. Tregs are explored in experimental models to treat autoimmunity. Other immune cells like natural killer (NK) cells or monocytes were also redirected by CARs to drive receptor-signaling initiated effector functions.

The efficacy of CAR-mediated T cell activation depends on various parameters including the CAR expression level on the T cell surface, binding affinity, the targeted antigen, the accessibility of the targeted epitope within the antigen, the density of the cognate antigen on the target cell, the CAR signaling, and others.

The Basic CAR Design

Based on the similarity of the primary structure and the spatial conformation of the variable regions of immunoglobulins (Ig) and TCR α and β chain molecules, antibody-derived binding regions V_H and V_L for antigen were grafted onto the constant domain of the TCR α and β chains, respectively, in earlier studies. The need to simultaneously express two modified TCR chains and

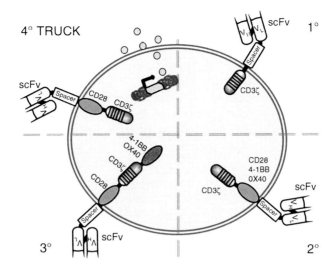

Chimeric Antigen Receptor (CAR), Fig. 1 The generations of chimeric antigen receptors (CARs). The chimeric antigen receptor consists in the extracellular part of a heavy (VH) and light chain (VL) variable region single chain antibody (scFv) for binding linked by a spacer to a transmembrane and cytoplasmic signaling domain. The first generation CAR (1°) has the CD3ζ chain derived from the T cell receptor (TCR) to mediate T cell activation, the second generation CAR (2°) has additionally a co-stimulatory signaling domain, mostly derived from CD28, 4-1BB, or OX40, the third generation CAR (3°) has two costimulatory domains in tandem. Multiple variations of these prototype CARs were reported. The fourth generation CAR (4°), also called TRUCKs (T cells redirected for universal cytokine killing), are CAR T cells which are additionally engineered with a CAR inducible expression cassette for a transgenic product, e.g., a cytokine, which is released upon CAR signaling

the risk of unintended hetero-dimerization with the endogenous TCR chains lead to the development of a single chain format which is now known as chimeric antigen receptor, CAR.

A typical CAR consists of one polypeptide chain which is comprised of a single chain variable fragment (scFv) antibody, optionally an extracellular spacer domain, a transmembrane domain, and one or more intracellular signaling domains (Fig. 1; Eshhar et al. 1993; Bridgeman et al. 2010). Multiple variations of the basic design were described. The prototypic CAR has entered clinical exploration with substantial success.

The Binding Domain

The CAR binds to cognate target by an antibody-derived domain, the scFv, which is engineered by joining the V_H and V_L immunoglobulin regions by a flexible peptide linker, e.g., $(Gly_4Ser)_3$, resulting in a continuous polypeptide chain (V_H-linker-V_L or V_L-linker-V_H). Due to the antibody-derived binding domain, CARs can target a broad variety of antigens of any chemical composition or conformation as far as a scFv antibody is available. Other binding domains, for instance, derived from physiological receptors, were also applied in this context.

In contrast to the TCR, CARs mediate T cell activation independent of MHC recognition. This particular property provides CAR-modified cells some advantages, including the targeting of nonclassical T cell antigens like carbohydrates and of antigens that are not properly processed or presented due to the downregulated antigen processing and presentation machinery in cancer cells. On the other hand, only surface proteins can be targeted with the exception to use antibodies that recognize the processed peptide in the context of MHC and thereby allow antigen recognition in a MHC-dependent fashion.

Instead of engineering a panel of CARs with different specificities, a CAR may alternatively be used that binds the immunoglobulin Fc region, e.g., via CD16, and binds an added antibody which in turn provides the specificity for the target. Thereby, the CAR strategy combines the T cell with the antibody therapy (Kudo et al. 2014).

CARs of multiple specificities can be engineered by linking scFvs to each other. Bispecific CARs with two linked scFvs of different specificities may be advantageous in targeting cancer cell variants with downregulated antigens.

The Spacer Domain

The requirement of a spacer domain in the extracellular part between the scFv and the transmembrane domain of the CAR depends in parts on the intracellular signaling moiety and on the accessibility of the targeted epitope for CAR binding. The spacer is typically derived from IgG1 or IgG4 CH2-CH3 (Fc) domains with or without hinge or from CD4 or CD8. Lengthening or shortening the spacer can optimize binding and subsequent CAR-mediated T cell activation.

The Transmembrane Domain

Various transmembrane regions, including those of CD3ζ, CD4, CD8, or OX40, are used. CARs with CD3ζ transmembrane domain incorporate into the endogenous TCR/CD3 complex and may have the advantage of more stable expression and more robust signaling compared to others.

The Signaling Domains

The CAR intracellular signaling domain is most frequently derived from CD3ζ of the TCR/CD3 complex or the γ-chain of the IgE Fc receptor-I (FcεRI). The CD3ζ provides three immunoreceptor tyrosine activation motifs (ITAMs), the γ-chain one ITAM. Upon CAR engagement of antigen the ITAMs become phosphorylated and serve as specific adaptors for downstream signaling proteins resulting in a cascade of T cell activation events. In this context, signaling moieties of downstream kinases like lck or fyn can also be used as CAR activation domains.

CAR-mediated downstream signaling promotes T cell activation resulting in the amplification of T cells, secretion of proinflammatory ► cytokines, and cytolysis of antigen-positive target cells. By using different primary signaling and costimulatory domains, the ► T-cell response can be modulated with some impact on the T-cell response, persistence, and maturation.

The Four Generations of CARs

First-generation CARs contain a primary signaling domain, mostly derived from CD3ζ or (FcεRI) γ-chain (Fig. 1). T cells with those CARs, however, failed to produce therapeutic efficacy in vivo which is thought to be due to the lack of a costimulatory signal needed for full T cell activation and prolonged persistence. Second (2°)- and third (3°)-generation CARs additionally provide costimulation through one (2°) or two (3°) costimulatory signaling domains, respectively. Different costimulatory signals produce different activation profiles, T cell functions, and cytokines released upon activation. T cells with second-generation CARs are currently evaluated in early phase trials and have shown increased downstream signaling potency and improved clinical efficacy in the treatment of leukemia/lymphoma.

T cells with CARs of the fourth generation (4°), so-called TRUCKs, are CAR T cells with an additional payload (i.e., the constitutive or inducible release of a transgenic product, for instance, a cytokine) (Chmielewski et al. 2014). Technically, the CAR T cells are additionally engineered with a construct for the transgenic payload directed under control of a T cell activation responsive promoter which becomes activated upon successful signaling of the CAR. Such activated TRUCKs deposit the transgenic product, e.g., IL-12, in the targeted tissue. In the example of an IL-12 TRUCK, the released cytokine recruits components of the innate immune response, like NK cells and ► macrophages, which in turn attack antigen-negative tumor cells, thereby broadening the overall antitumor response.

CAR T Cell Therapy: Challenges and Safety

Engineering T cells with defined specificity against cancer cells is a significant advancement toward a more specific and individualized cell therapy. The CAR T cell therapy has some advantages over other ► immunotherapies including the use of autologous ► cytotoxic T cells that actively penetrate tissues, execute their killing toward cognate target cells in a repetitive fashion, and activate by release of proinflammatory factors the entire cellular defense system in trans. As "living drugs" the CAR T cells amplify, circulate, and persist over long periods of time, in some cases over years, and may provide an antigen-specific memory. Thereby, the CAR T cell therapy is anticipated to eliminate disseminated cancer cells, like (micro-) ▸ metastasis, as well as ▸ circulating tumor cells including leukemia.

▸ Clinical trials are currently ongoing in several centers using CAR T cells directed toward CD19 or other antigens to treat ▸ hematological malignancies, most trials with substantial success (June et al. 2014). Patients with refractory ▸ chronic lymphocytic leukemia (CLL) and ▸ acute lymphocytic leukemia (ALL) experienced complete and enduring remission. Therapy of solid tumors, however, is still more challenging which is thought to be due to the stroma barrier which hampers CAR T cell penetration into the tumor lesion and the immunosuppressive ▸ tumor microenvironment which represses the antitumor response.

A number of parameters on the CAR T cell side, the patient side, and the pretreatment schedule need to be considered when translating the CAR T cell strategy into a clinical efficacious and safe treatment of malignant diseases. These include on the CAR side an accurate target antigen selection, a sufficient binding affinity and specificity, an appropriate CAR design, use of primary and costimulatory signaling domains, a sufficient CAR expression level, and others. The cell product itself requires an adequate T cell subset with depletion of suppressor T cells, an appropriate preactivation and maturation level, and ex vivo expansion capacities. The patient treatment includes the preconditioning like lymphodepletion, reducing the bulk of tumor mass prior therapy, the route of administration, systemic cytokine administration, and clinical managing of comorbidities and toxicities. The prediction of CAR T cell-related toxicities is still in its infant stage; some toxicities are briefly mentioned.

The "on-target on-tumor" toxicity includes the tumor lysis syndrome which is mediated through the rapid destruction of a large tumor mass in response to therapy. The release of tumor cell components into circulation leads to electrolyte and metabolic disturbances and, at worst, multiorgan failure. Vascular leakage syndrome (VLS) and the cytokine release syndrome (CRS) due to extensive T cell activation are associated with fever, nausea, and supraphysiological serum levels of proinflammatory cytokines; in particular IFN-γ and ▸ Interleukin-6, are life-threatening and demand intensive care treatment and the application of an IL-6 receptor blocking antibody. While the production of cytokines is beneficial in tumor elimination through the direct killing of target cells and recruitment of innate immune cells, the fine-tuning between the antitumor effect and the cytokine-related toxicity is currently a major hurdle.

"On-target off-tumor" toxicity occurs when CAR T cells recognize healthy tissues with physiological expression of the cognate antigen. In the case of B cell malignancies, B cell aplasia occurred due to CD19 targeting on normal B cells which is, however, clinically manageable. Targeting of healthy lung epithelial cells by ▸ HER-2/neu specific third-generation CAR T cells during treatment of a metastatic cancer patient resulted in fatal toxicity soon after adoptive transfer.

CAR T cells may produce "off-target off-tumor" toxicity due to target unrelated immune cell activation. For instance, the extracellular Fc spacer in the CAR may bind to and activate Fc receptor (FcR) expressing cells of the innate immune system including NK cells or macrophages which may mediate a systemic inflammatory response. Modification of the IgG1 Fc domain or the use of the IgG4 domain is aiming at avoiding this situation. In addition, CAR T cells may cause anaphylaxis due to the induction of IgE antibodies against the binding domain or other CAR moieties.

To avoid such toxicities, truly tumor-restricted antigens or physiologically embryonic antigens which are re-expressed by cancer cells may therefore be ideal antigens for CAR T cell targeting. Such tumor-selective antigens are mostly not available. Currently targeted antigens are rather tumor associated; some healthy cells express the cognate antigen, albeit at lower levels than on cancer cells. Several approaches are aiming at increasing the selectivity for cancer cells including the coexpression of two CARs, each CAR recognizing a different antigen. One CAR provides the "signal 1" for primary and suboptimal T cell activation; the other CAR provides "signal 2" for costimulation and prolongation of the response. While cosignaling through both CARs initiates full T cell effector functions, each CAR alone is not capable to initiate a lasting and sufficient T cell activation. When both coexpressed CARs recognize their cognate targets on a cancer cell, the suboptimal activation by each individual CAR is rescued by the complementation of both signals. Such combinatorial antigen recognition increases the selectivity in targeting cancer cells which is based on the reduced likelihood of two tumor-associated antigens being simultaneously expressed on healthy tissues in sufficient amounts to exceed the threshold for T cell activation.

In the case of toxicity, the depletion of CAR T cells may be required which, however, will result in the abrogation of their therapeutic potential. Strategies to selectively eliminate CAR T cells include the coexpression of a suicide gene, e.g., inducible caspase 9 (iCasp9), herpes simplex thymidine kinase (▶ HSV-TK), or others. A less selective approach to deplete modified T cells is the administration of high dose corticosteroids which proved successful in a clinical trial.

Cross-References

▶ Acute Lymphoblastic Leukemia
▶ Adoptive Immunotherapy
▶ Chronic Lymphocytic Leukemia
▶ Circulating Tumor Cells
▶ Clinical Trial
▶ Cytokine
▶ Cytotoxic T Cells
▶ Hematological Malignances, Leukemias, and Lymphomas
▶ HER-2/neu
▶ HSV-TK-/Ganciclovir-Mediated Toxicity
▶ Immunotherapy
▶ Interleukin-6
▶ Macrophages
▶ Metastasis
▶ Regulatory T Cells
▶ T-Cell Response
▶ Tumor Microenvironment
▶ Viral Vector-Mediated Gene Transfer

References

Bridgeman JS, Hawkins RE, Hombach AA, Abken H, Gilham DE (2010) Building better chimeric antigen receptors for adoptive T cell therapy. Curr Gene Ther 10:77–90

Chmielewski M, Hombach AA, Abken H (2014) Of CARs and TRUCKs: chimeric antigen receptor (CAR) T cells engineered with an inducible cytokine to modulate the tumor stroma. Immunol Rev 257:83–90

Eshhar Z, Waks T, Gross G, Schindler DG (1993) Specific activation and targeting of cytotoxic lymphocytes through chimeric single chains consisting of antibody-binding domains and the gamma or zeta subunits of the immunoglobulin and T-cell receptors. Proc Natl Acad Sci U S A 90:720–724

June CH, Maus MV, Plesa G, Johnson LA, Zhao Y, Levine BL, Grupp SA, Porter DL (2014) Engineered T cells for cancer therapy. Cancer Immunol Immunother 63:969–975

Kudo K, Imai C, Lorenzini P, Kamiya T, Kono K, Davidoff AM, Chng WJ, Campana D (2014) T lymphocytes expressing a CD16 signaling receptor exert antibody-dependent cancer cell killing. Cancer Res 74:93–103

See Also

(2012) Antibody. In: Schwab M (ed) Encyclopedia of cancer, 3rd edn. Springer, Berlin/Heidelberg, p 208. doi:10.1007/978-3-642-16483-5_312

(2012) Autoimmunity. In: Schwab M (ed) Encyclopedia of cancer, 3rd edn. Springer, Berlin/Heidelberg, p 312. doi:10.1007/978-3-642-16483-5_478

(2012) CD4. In: Schwab M (ed) Encyclopedia of cancer, 3rd edn. Springer, Berlin/Heidelberg, p 698. doi:10.1007/978-3-642-16483-5_914

(2012) CD4+ T-Cells. In: Schwab M (ed) Encyclopedia of cancer, 3rd edn. Springer, Berlin/Heidelberg, pp 698–699. doi:10.1007/978-3-642-16483-5_915

(2012) CD8. In: Schwab M (ed) Encyclopedia of cancer, 3rd edn. Springer, Berlin/Heidelberg, p 703. doi:10.1007/978-3-642-16483-5_917

(2012) Corticosteroids. In: Schwab M (ed) Encyclopedia of cancer, 3rd edn. Springer, Berlin/Heidelberg, pp 982–983. doi:10.1007/978-3-642-16483-5_1341

(2012) Costimulation. In: Schwab M (ed) Encyclopedia of cancer, 3rd edn. Springer, Berlin/Heidelberg, pp 983–984. doi:10.1007/978-3-642-16483-5_1348

(2012) ScFv. In: Schwab M (ed) Encyclopedia of cancer, 3rd edn. Springer, Berlin/Heidelberg, p 3340. doi:10.1007/978-3-642-16483-5_5170

(2012) Suicide gene. In: Schwab M (ed) Encyclopedia of cancer, 3rd edn. Springer, Berlin/Heidelberg, p 3555. doi:10.1007/978-3-642-16483-5_5557

(2012) T cell. In: Schwab M (ed) Encyclopedia of cancer, 3rd edn. Springer, Berlin/Heidelberg, p 3599. doi:10.1007/978-3-642-16483-5_5645

(2012) T cell receptor complex. In: Schwab M (ed) Encyclopedia of cancer, 3rd edn. Springer, Berlin/Heidelberg, p 3599. doi:10.1007/978-3-642-16483-5_5646

(2012) Treg. In: Schwab M (ed) Encyclopedia of cancer, 3rd edn. Springer, Berlin/Heidelberg, p 3782. doi:10.1007/978-3-642-16483-5_5967

(2012) Tumor-associated antigen. In: Schwab M (ed) Encyclopedia of cancer, 3rd edn. Springer, Berlin/Heidelberg, pp 3807–3808. doi:10.1007/978-3-642-16483-5_6017

(2012) Tumor lysis syndrome. In: Schwab M (ed) Encyclopedia of cancer, 3rd edn. Springer, Berlin/Heidelberg, p 3796. doi:10.1007/978-3-642-16483-5_6035

(2012) Epithelial cell. In: Schwab M (ed) Encyclopedia of cancer, 3rd edn. Springer, Berlin/Heidelberg, pp 1291–1292. doi:10.1007/978-3-642-16483-5_1958

(2012) Epitope. In: Schwab M (ed) Encyclopedia of cancer, 3rd edn. Springer, Berlin/Heidelberg, p 1297. doi:10.1007/978-3-642-16483-5_1966

(2012) FcR. In: Schwab M (ed) Encyclopedia of cancer, 3rd edn. Springer, Berlin/Heidelberg, p 1386. doi:10.1007/978-3-642-16483-5_2135

(2012) Immunoglobulin. In: Schwab M (ed) Encyclopedia of cancer, 3rd edn. Springer, Berlin/Heidelberg, p 1819. doi:10.1007/978-3-642-16483-5_2990

(2012) Interferon gamma. In: Schwab M (ed) Encyclopedia of cancer, 3rd edn. Springer, Berlin/Heidelberg, p 1888. doi:10.1007/978-3-642-16483-5_3092

(2012) ITAM. In: Schwab M (ed) Encyclopedia of cancer, 3rd edn. Springer, Berlin/Heidelberg, p 1921. doi:10.1007/978-3-642-16483-5_3165

(2012) MHC. In: Schwab M (ed) Encyclopedia of cancer, 3rd edn. Springer, Berlin/Heidelberg, p 2281. doi:10.1007/978-3-642-16483-5_3700

(2012) Monocyte. In: Schwab M (ed) Encyclopedia of cancer, 3rd edn. Springer, Berlin/Heidelberg, p 2371. doi:10.1007/978-3-642-16483-5_3825

(2012) NK. In: Schwab M (ed) Encyclopedia of cancer, 3rd edn. Springer, Berlin/Heidelberg, p 2529. doi:10.1007/978-3-642-16483-5_4096

Chimeric Antigen Receptor on T Cells

Laurence J. N. Cooper and Sourindra N. Maiti
Division of Pediatrics, Department of Immunology, MD Anderson Cancer Center, Houston, TX, USA

Synonyms

CAR

Definition

A chimeric antigen receptor (CAR) consists of an extracellular antigen-binding exodomain, typically derived from a single-chain antibody fragment (scFv) of a monoclonal antibody (mAb), a spacer (such as an antibody Fc region), a transmembrane region, and one or more intracellular signaling endodomains, which can be genetically introduced into hematopoietic cells, such as T cells, to redirect specificity for a desired cell-surface antigen.

Characteristics

Background on Manipulating T-Cell Responses to Cancer

Adoptive transfer of tumor-specific T cells in mouse models can function as potent anticancer biological agents leading to the elimination of established malignancies. Continued advances in tumor immunology support the premise and promise for ▶ adoptive immunotherapy as a treatment for human malignancies. Yet, infusion of tumor-specific T cells has only been partially successful in clinical oncology trials. Indeed, most of these trials demonstrate the safety and feasibility of infusing T cells, but with the exception of treating melanoma and chronic myelogenous leukemia (CML), only occasionally show a sustained anti-tumor effect. In contrast, infusing viral-specific T cells has successfully treated and protected patients from opportunistic diseases associated with adenovirus, CMV, and EBV. Why is it that augmenting an immune response against neoplasms by infusing tumor-specific T cells has proven more challenging than engendering an effective anti-viral response? The answer is partly due to the relative inability of T cells to recognize, via an endogenous T-cell receptor (TCR), poorly immunogenic tumor-associated antigens (TAAs) compared with the highly immunogenic/stimulatory viral antigens presented in the context of human leukocyte antigen (HLA). While TAAs generally have a little or no expression in normal postnatal tissues outside of sanctuary sites, naturally arising T cells are typically not reactive to tumors expressing a TAA due to immunologic tolerance. However, investigators have been able to manipulate T cells into recognizing TAA in the context of HLA molecules. This has been exploited by injection/infusion of (i) vaccines presenting TAA to overcome tolerance and stimulate T-cell immunity to tumors, (ii) tumor-specific T cells which have been culled from the patient and massively expanded in the laboratory, and (iii) T cells that have redirected specificity for tumor by genetically introducing predefined tumor-specific immunoreceptor genes. Clinical trials are currently evaluating all three approaches.

While vaccines will likely have application for cancer prevention (► cancer vaccines), vaccination to eradicate most established tumors is a difficult challenge at this time due in part to iatrogenic damage of the underlying immune system from chemotherapy and accumulation of large tumor masses in sanctuaries which are functionally protected from immune recognition and response. Rather than vaccinate, or addition to vaccination, investigators have chosen to augment T-cell response through infusion of effector T cells (adoptive immunotherapy). Indeed, isolation, ex vivo-expansion of autologous tumor-infiltrating cytotoxic T-lymphocytes (CTLs), and subsequent transfer of these CTLs in lymphodepleted patients can mediate regression of metastatic melanoma. The widespread therapeutic success of adoptive immunotherapy using T cells that have not been genetically manipulated is, however, limited because of (i) difficulties in obtaining sufficient number of tumor-reactive CTLs from patients, (ii) TAAs are typically poor immunogens, (iii) existence of immune-regulatory mechanisms that prevent T-cell dependent reactivity against TAA (such emergence of tumor escape variants with loss of HLA), and (iv) the requirement that patients have preexisting tumor-reactive cells that can be expanded ex vivo.

To overcome these hurdles, investigators have combined the endogenous effector function of CTLs with redirected antigen specificity that results from genetic introduction of an immunoreceptor. This introduced immunoreceptor transgene can be engineered ex vivo to provide nonphysiologic signaling via a chimeric antigen receptor (CAR) that co-opts the ability of recombinant antibody to bind to tumor targets leading to T-cell activation. The development of CAR$^+$ T cells can overcome the relative inability of antibodies to localize and penetrate into tumor masses and provides a self-renewing source of chimeric antibody linked to T-cell effector function.

Tumor-Specific CAR Development

Generally, a CAR consists of an extracellular domain composed of a single chain variable fragment (scFv) derived from a monoclonal antibody (mAb) against a cell-surface tumor-antigen which is typically a lineage specific molecule, such as CD19 on B cells. The scFv is typically suspended from the cell surface by a spacer (e.g., mAb Fc region) and uses a transmembrane (TM) region (e.g., from CD4 or CD28) to affix the scFv–Fc to the cell surface. This TM region is in turn fused in frame to one or more signaling modules that are normally present in an endogenous TCR signaling complex, such as the CD3-ζ chain (Fig. 1). The CAR can confer scFv-mediated antigen-recognition to T cells that is independent of HLA on tumor cells and endogenous TCR. T cells genetically modified to express a CAR can be propagated in vitro and demonstrated to exert robust CAR-dependent effector function. Upon antigen-mediated cross-linking of the CAR, the intracellular signaling domain or domains initiate cellular activation which can result in proliferation, cytokine secretion, and specific cytolysis of the antigen-expressing target cells. Several requirements need to be met to enable T cells expressing CAR to exert a pronounced therapeutic effect. These include (i) expression of the CAR at sufficient density to activate effector function upon binding antigen, (ii) generation of clinically meaningful numbers of T cells suitable for infusion, (iii) traffic T cells to and within tumor stroma, and (iv) conditional activation upon antigen binding, manifesting appropriate T-cell effector mechanisms such as cytokine secretion, proliferation, and cytolysis leading to tumor

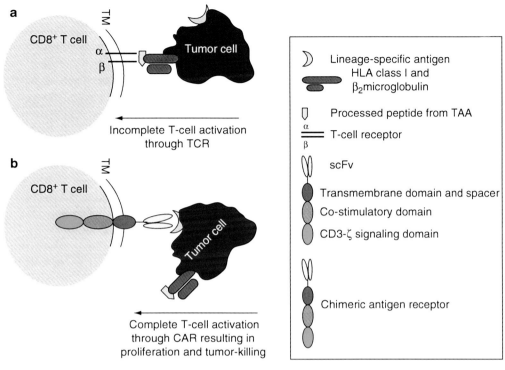

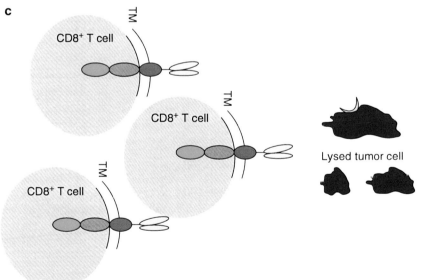

Chimeric Antigen Receptor on T Cells, Fig. 1 (**a**) Incomplete activation of T cells recognizing TAA through αβ TCR in context of HLA class I. (**b**) Fully competent activation signal by T cells recognizing lineage-specific antigen (e.g., CD19) independent of HLA by introduced CAR. (**c**) Complete CAR-mediated activation results in T-cell proliferation and tumor lysis destruction. We and others have generated a panel of CARs and demonstrated that T cells (and NK cells) equipped with CAR mediate a highly efficient antitumor immune response against antigen-defined tumor target cells in vitro as well as in vivo. These CAR-grafted T cells thus are appealing candidates for adoptive immunotherapy.

Clinical Application of CAR-Specific T Cells

Adoptive transfer of T cells for treatment of tumors is an attractive therapeutic option as it has the potential to cure disease refractory to conventional therapies. Successful allogeneic hematopoietic stem-cell transplantation (HSCT) with the engraftment of donor-derived tumor-specific T cells and adoptive transfer of T cells genetically rendered specific for melanoma antigens currently provide the two cornerstones for the rational application of adoptive transfer of T cells genetically modified to express CAR. Along with other investigators, we have chosen to develop CD19-specific CAR, since the CD19 molecule is widely expressed on most cancers arising from B cells. B-cell tumors are a class poorly immunogenic with few described TAA, thus isolating and expanding endogenous T cells with specificity for malignant B cells has proven difficult in the context of allogeneic HSCT and near-impossible in the autologous setting. Thus, to target B-lineage neoplasms, an initial CD19-specific CAR (designated CD19R) was generated from the variable regions of a mouse mAb specific for CD19, and T-cell activation was achieved through chimeric CD3-ζ endodomain. The CD19 molecule is a 95-kDa membrane glycoprotein found on human B lymphocytes at all stages of maturation, although it typically disappears upon differentiation to terminally differentiated plasma cells. It is expressed on B-lineage acute leukemias and lymphomas as well as chronic lymphocytic leukemia and is rarely lost during the process of neoplastic transformation. It is not expressed on hematopoietic progenitor cells or on normal tissues outside the B-lineage and is not thought to be shed into the circulation. An advantage of CD19-directed therapy for lymphomas over targeting CD20, another B-lineage antigen, is that unbound

rituximab (CD20-specific therapeutic mAb) will not interfere with binding of CD19-specific T cells. A clinical-grade DNA plasmid vector coding for CD19R and bifunctional hygromycin (Hy) phosphotransferase selection gene fused to thymidine kinase (TK) suicide/imaging fusion gene (HyTK) was developed and produced by the National Gene Vector Laboratory. Using this vector, a gene therapy trial was opened (BB-IND 11411, ClinicalTrials.gov Identifier: NCT00182650) to determine the feasibility and safety of infusing autologous T cells coexpressing CD19R and HyTK transgenes, along with exogenous low dose recombinant human IL-2 (as a surrogate T_h-response) in patients with refractory lymphoma.

Improved Therapeutic Potential of CAR$^+$ T Cells

Ongoing projects in our laboratory to improve therapeutic efficacy have focused on prolonging the survival of the infused CAR$^+$ T cells mainly via three approaches.

1. *To produce T cells those are capable of endogenous IL-2 production.* To generate an effective antitumor response for CD19-redirected effector T cells in tumor microenvironment, we introduced a chimeric CD28 T cell costimulatory molecule into the CD19R CAR. This is based on the rationale that T-cell binding of CD28 to B7 molecules on target cells generates critical regulatory signals necessary for full T-cell activation and preventing T-cell apoptosis after CAR engagement. However, massively ex vivo-expanded genetically modified T cells may lose endogenous CD28 cell-surface expression. Therefore, to provide genetically modified CD28^{neg} T cells with tandem activation and costimulation upon engagement with B7^{neg}CD19$^+$ B-lineage tumors, the CD19R CAR was modified to include CD28-signalling domain. This second generation CAR, designated CD19RCD28, has been expressed in primary T cells and shown to activate genetically modified T cells for killing, IFN-γ, and IL-2 cytokine production, and improve

in vivo survival of adoptively transferred CD19RCD28$^+$ T cells resulting in a greater antitumor effect, compared with first-generation CD19R$^+$ T cells.

2. *To target IL-2 to the tumor micro-environment.* In addition to engineering CAR for T-cell production of IL-2, a cytokine that ex vivo-expanded T cells typically depend on for continued in vivo persistence, we have directed exogenous IL-2 cytokine to the B-cell tumor microenvironment using a CD20-mAb fused to IL-2 (immunocytokine), and this combination immunotherapy enhanced the antitumor effect of infused T cells.

3. *To produce T cells with improved biologic potential.* Shortening the ex vivo manufacturing time may improve therapeutic efficacy, as extensively propagated T cells differentiate in vitro into cytolytic effectors lacking desired homing receptors and a tendency to undergo replicative senescence. Therefore, we have developed a new rapid propagation technology using a CD19$^+$ immortalized artificial antigen presenting cell (aAPC) that can be lethally irradiated and used in coculture to numerically expand cytolytic CD19-specific T cells. To develop noninvasive biomarkers and to evaluate adoptively transferred T-cell distribution and function, we have used radionuclides that are metabolized by TK which acts as a reporter gene for the detection of T cells by positron emission tomography (PET).

Future Challenges

There has been a paucity of published data on the safety and feasibility of infusing T cells expressing CAR, and at this time there are no reports describing sustained clinical response of adoptive transfer of CAR$^+$ T cells. This will soon change as multiple clinical trials are currently underway world-wide using adoptive cellular immunotherapy with T cells expressing CAR. To maximize therapeutic efficacy, future trials will likely infuse CAR$^+$ T cells combined vaccine to deliver a T-cell activation signal though endogenous $\alpha\beta$ TCR, or CAR$^+$ T cells combined with cytokine, such as immunocytokine, to deliver costimulatory signal through endogenous cytokine receptors. One of the major advantages of the strategy infusing CAR$^+$ T cells lies in the modular composition of the CAR molecule that combines an antigen-binding domain with signaling domains for effector-cell activation. This allows investigators to swap the CAR exodomain with a scFv or ligand that recognizes or binds to a desired cell-surface antigen. Future experiments will build on the catalog of antigen-targeting receptors to evaluate expression level, density, and stability of the CAR expression on the T-cell surface, as well as affinity of the binding domain for antigen in the context of tumor targets with varying antigen density. These studies will impact CAR-mediated immunotherapy since low-density antigen-positive tumor cells or low affinity CAR may lead to emergence of tumor escape. Just as genetic engineering can be used to alter the CAR exodomain, so the transmembrane, or endodomain may be altered to provide a fully competent antigen-dependent T-cell activation signal. The majority of the CARs generated harbor CD3-ζ signaling chain which is currently considered more efficient in activating T cells for cytolysis, compared to chimeric FcϵRI-γ. Other CARs have been generated activating T cells through syk, lck, but the full-effect of these receptors with respect to cellular activation and stability of receptor expression on the cell surface has yet to be determined. Similarly, the CAR endodomain has been altered to express a costimulatory signaling molecule, such as chimeric CD28, to provide a coordinated signal with CD3-ζ to provide a more complete T-cell activation signal than achieved by signaling through CD3-ζ alone. Because different types of costimulation may result in different patterns of cellular activation, it will likely be beneficial to explore alternative costimulatory pathways in CAR-mediated T-cell activation. Increasingly, investigators are developing the tools to genetically modify T cells using techniques that cause minimal manipulation of the product leaving intact the full range of T-cell homing and proliferation potentials. Finally, genetic modification is being used to accomplish more than redirect T-cell specificity. For example, experiments are underway to render the T cells resistant to the anti-inflammatory and deleterious effects of iatrogenic glucocorticoids and TGF-β secreted by tumor cell.

Conclusion

Although limitations remain, genetic modifications enable investigators to engineer T cells with augmented therapeutic potential. It is expected that infusion of genetically modified T cells will be a center-piece of personalized medicine rivaling the influence of therapeutic mAbs as a treatment for malignancies.

References

Cooper L, Topp MS, Serrano LM et al (2003) T-cell clones can be rendered specific for CD19: toward the selective augmentation of the graft-versus-B-lineage leukemia effect. Blood 101:1637–1644

Eshhar Z, Waks T, Bendavid A et al (2001) Functional expression of chimeric receptor genes in human T cells. J Immunol Methods 248(1–2):67–76

Morgan RA, Dudley ME, Wunderlich JR et al (2006) Cancer regression in patients after transfer of genetically engineered lymphocytes. Science 314:126–129

Park JR, Digiusto DL, Slovak M et al (2007) Adoptive transfer of chimeric antigen receptor re-directed cytolytic T lymphocyte clones in patients with neuroblastoma. Mol Ther 15(4):825–833

Rossig C, Brenner MK (2003) Chimeric T-cell receptors for the targeting of cancer cells. Acta Haematol 110(203):154–159

Chimeric Genes

▶ Fusion Genes

Chimeric Immune Receptor

▶ Chimeric Antigen Receptor (CAR)

Chimeric Oncogenes

▶ Fusion Genes

Chinese Medicine

▶ Chinese Versus Western Medicine

Chinese Versus Western Medicine

William Chi-Shing Cho
Department of Clinical Oncology, Queen
Elizabeth Hospital, Kowloon, Hong Kong

Synonyms

Chinese medicine; Traditional chinese medicine

Definition

Chinese medicine is a patient-oriented medical system that treats the patients instead of the diseases. It is believed that qi (the Chinese term for vital energy) supports the functional activities and blood supplies nutriments for the whole body. There exists a system of channels within the human body, through which the vital energy and blood circulate, and by which the internal organs are connected with superficial organs and tissues, and the body is made an organic whole. Using these holistic and harmonic approaches, Chinese medicine emphasizes to strengthen the body resistance. It attaches importance to the self-healing ability of human body to remove pathogenic factors and recover health. Many of its ▶ cancer therapies, such as Chinese medication (including medicinal decoction, patent medicine, and proprietary medicine), medicated diet, acupuncture and moxibustion, as well as qigong and massage, are employed for enhancing this power.

Western medicine is an evidence-based medical system, it is the science and practice of the diagnosis, treatment, and prevention of disease. The clinical problems faced by oncologists include overcoming the inherent or acquired resistance of the malignant cell to therapy, ameliorating the toxicities of aggressively applied therapies, as well as exploiting the synergistic potency of surgery, radiotherapy, and ► chemotherapy. In the postgenomic era, targeted therapy and novel therapeutic strategies are applied to complement the conventional treatment for an achievement of optimal anticancer results.

Characteristics

Developing History

The Huangdi's Internal Classic is believed to be the earliest medical monograph in China, which appeared during the Warring States period (475–221 BC), first defined the etiology of tumor. Since the ancient times, Chinese medicine has made a great contribution to the health of the Asian people. Based on empirical and clinical experience, Chinese medicine has been systematized and theorized in complex practice. Many safe and effective methods have been developed to diagnose and treat cancer over the past thousands of years.

Hippocrates (460–375 BC), the father of Western medicine, first attributed the origin of cancer to natural causes. Improved microscopes, stimulated cancer researches, and important discoveries in human and animal studies have resulted in a better understanding of neoplasia. The last few decades has witnessed spectacular progress in describing the fundamental molecular basis of cancer following the advent of molecular biology and genetics, which allows the device of advanced or targeted therapy for cancer.

Theories and Principles

The philosophical theories and fundamental principles of Chinese medicine include the theory of yin and yang, the phases concept, the physiological functions of viscera and bowels, the conception of vital energy and blood, as well as the theory of the channels and collaterals. The five phases (*synonym* five elements; wood, fire, earth, metal, and water) concept is a philosophical theory developed in ancient China to explain the composition and phenomena of the physical universe. It is used in Chinese medicine to expound the unity between human and nature, as well as the

physiological and pathological relationship and interconnection among the internal organs. The five phases match the five viscera, in which liver, heart, spleen, lung, and kidney correspond to wood, fire, earth, metal, and water, respectively. The five phase concept explains the interpromoting and interacting relations, as well as the encroachment and violation in illness condition between the five viscera. According to the basic theories, the physical structure and physiological phenomena of human body as well as the pathological changes are in adaptive conformity with the variations of the natural environment. Hence considerations of personalized cancer medicine are based on the patient's constitution, geographical localities, climatic, and seasonal conditions. The therapeutic principle of cancer is to treat the disease by looking into both its root cause and symptoms. Sometimes different treatments are applied to the same kind of cancer in the light of different physical reactions and clinical manifestations, whereas the same therapy can be used to treat different cancers if they are alike in clinical manifestations and pathogenesis.

According to the theories of Western medicine, cancer is not one illness but a variety of disorders with different pathophysiology that can arise from and spread to almost every organ and tissue in the body. Mechanism of the cancer development varies according to the site of the malignant disease and the precipitating cause. Each cancer has its unique pattern of presentation and approach to diagnosis and treatment. Therapy must be directed not only toward cure of the cancer and control of potential ► metastasis, but also to optimize the quality of life.

Etiology and Pathogenesis

Based on Chinese medicine, there are two main categories of etiological factors for cancer, which include exogenous and endogenous factors. The exogenous factors refer to the six excessive and untimely atmospheric influences (wind, cold, summerheat, dampness, dryness, and fire), as well as unhealthy diet. The endogenous factors refer to the excessive emotional changes (joy, anger, thought, anxiety, sorrow, fear, and fright) and the deficiency of functional organ. In general, the pathogenesis of cancer can be summarized as accumulation of phlegm dampness, internal noxious heat due to accumulation of pathogenic heat, blood stasis due to vital energy stagnancy, dysfunction of internal organs, vital energy, and blood deficiency, as well as yin and yang imbalance.

Extensive epidemiological and prospective studies have allowed Western medicine to identify two categories of etiological factors for cancer, which include environmental and genetic factors. The environmental factors refer to smoking, alcohol, unhealthy diet, ultraviolet light, ionizing radiation, carcinogens, certain viruses, and infections. The genetic factors refer to monogenic and polygenic disorders. In general, cancer is a clonal disease arising by the ▶ multistep development of genetic or epigenetic changes in oncogenes, ▶ tumor suppressor genes, and caretaker genes that favor expansion of the new clone over the old. These changes allow a normal cell to achieve the hallmark features of cancer, which include the capacity to proliferate irrespective of exogenous mitogen, the refractoriness to growth inhibitory signal, the resistance to ▶ apoptosis, the potential to reactivate ▶ telomerase resulting in unrestricted proliferation, the capacity to recruit a vasculature, as well as the ability to invade surrounding tissue and eventually metastasize.

Diagnostic Methods

Chinese medicine views the human body as a unity, and a malignant disease reflects both the interior and exterior of the body. The diagnosis of cancer is based on an overall analysis and differentiation of the patient's signs and symptoms, which include observation of the patient's mental state and inspection of the tongue, auscultation and olfaction, interrogation, as well as pulse taking and palpation.

Western medicine diagnoses cancer based on tissue diagnosis, which include tissue biopsy, diagnostic medicine, cytology, histopathology, and immunocytochemistry. The tumor-node-metastases classification is applied to establish the anatomical staging for most cancers. In addition, there are a number of specific tumor markers which are useful in diagnosis, such as α-fetoprotein for hepatocellular carcinoma, β-human chorionic gonadotrophin for choriocarcinoma, and prostate-specific antigen for prostate cancer.

Treatment Modalities

The principal methods of cancer treatment by Chinese medicine involve Chinese medication (derived from plant, animal, and mineral substances), medicated diet, acupuncture and moxibustion, as well as qigong and massage. In most cases, the patients are treated with medications to strengthen their resistance and dispel the invading pathogenic factors of cancer. Sometimes the malignancy is treated with poisonous medication to combat poison with poison, such as the application of arsenic trioxide. The cancer remedy should be made up in accordance with the physique of an individual, pathologic changes occur in the course of cancer, as well as the geographical and seasonal conditions. The prescription for cancer usually consists of various medicinal ingredients with the purpose to produce the desired therapeutic effect in unison and reduction of toxicity or side effects. The principal ingredient provides the principal curative action, the adjuvant ingredient helps to strengthen the principal action, the auxiliary ingredient relieves the secondary symptom or tempers the strong action of the principal ingredient, and the conductant directs the action to the affected channel or site. Over thousands of species of Chinese pharmaceuticals have been reported to treat cancer. As there are up to thousands of compounds in a medication, multitargets are exhibited to the malignant disease and some of the compounds may exert a synergistic anticancer effect.

Using the combined modality approach, the principal methods of cancer treatment by Western medicine involve the combined application of surgery, radiotherapy, combination chemotherapy, hormonal therapy, biological therapy, palliative care, and symptom control. Surgery alone is curative in many early stage neoplastic tumors. Radiotherapy is often used after surgery to reduce the chance of recurrence. It can also be used on its own with curative intent or as a palliative treatment. Chemotherapy is a systemic treatment that can reach any part of the body with an adequate blood supply, and therefore it is normally used to treat disseminated cancer. It can also be used as an ► adjuvant therapy to reduce the volume of advanced cancer, with intent to prolong life and relieve symptom. The development of new less toxic chemotherapeutic drugs and more effective antiemetics have reduced many adverse effects of chemotherapy. Hormonal therapy is performed to manipulate the hormone level, which can result in the regression of a number of cancers, particularly for the breast cancer, endometrial cancer, and prostate cancer. Biological therapy exploits insight into the nature of tumor antigen, the molecular and cellular requirement for immune activation, the role of cytokine in amplifying the immune response, as well as the evolution of recombinant DNA approach to introduce the genetic material into eukaryotic cell. Palliative care aims to achieve the best possible quality of life for patient and their family by controlling the physical symptom, as well as recognizing the psychological, social, and spiritual problems.

Western medicine is currently used as the primary therapy for cancer, and Chinese medicine is employed as a supplementary therapy in some Asian nations. There is considerable evidence for the promising benefits of Chinese medicine in alleviating the toxic effect of radiotherapy and chemotherapy, strengthening the anticancer activity, as well as enhancing the immune function. Chinese and Western medicines are obviously two distinct medical systems with different diagnostic and therapeutic methods for cancer. However, with a common goal to eradicate this systemic disease, it may be feasible for a certain degree of complementation and integration of these two medical systems in the realm of clinical practice.

References

Beauchamp EM, Ringer L, Bulut G, Sajwan KP, Hall MD, Lee YC, Peaceman D, Ozdemirli M, Rodriguez O, Macdonald TJ, Albanese C, Toretsky JA, Uren A (2011) Arsenic trioxide inhibits human cancer cell growth and tumor development in mice by blocking Hedgehog/GLI pathway. J Clin Invest 121:148–1602

Cho WC (2010) Supportive cancer care with Chinese medicine. Springer, New York, US3

Cho WC (2011) Evidence-based anticancer materia medica. Springer, New York, US4

Cho WC, Leung KN (2007) In vitro and in vivo anti-tumor effects of *Astragalus* membranaceus. Cancer Lett 252:43–54

Lee KH, Lo HL, Tang WC, Hsiao HH, Yang PM (2014) A gene expression signature-based approach reveals the mechanisms of action of the Chinese herbal medicine berberine. Sci Rep 4:6394

2-Chloro-2′-deoxyadenosine

► Cladribine

2-Chlorodeoxyadenosine

► Cladribine

ChoK

► Choline Kinase

Cholangiocarcinoma

Justin L. Mott[1] and Gregory J. Gores[2]
[1]Department of Biochemistry and Molecular Biology, University of Nebraska Medical Center, Omaha, NE, USA
[2]Miles and Shirley Fiterman Center for Digestive Diseases, Division of Gastroenterology and Hepatology, Mayo Clinic College of Medicine, Rochester, MN, USA

Synonyms

Bile duct carcinoma; Cholangiocellular carcinoma

Definition

Cholangiocarcinoma refers to malignancy within the biliary duct system (Bile Duct Neoplasms) and is distinct from ▶ gallbladder cancer. Cholangiocarcinomas generally have features of biliary tract epithelium, such as a glandular appearance, small regular nuclei, and scant cytoplasm. Tumor cells often express cytokeratins, mucin, and cancer-associated antigen 19-9, a carbohydrate antigen that is used as a tumor marker in serum. It is thought to be a sialylated Lewis blood group antigen. CA19-9 levels are elevated in many gastrointestinal malignancies including cholangiocarcinoma and pancreatic cancer, as well as some nonmalignant conditions such as cholangitis and peritoneal ▶ inflammation/infection. Patients who have a genetic deficiency in a fucosyltransferase specified by the Le gene are Lewis^{a-b-} and are unable to make this antigen; thus CA19-9 testing in Lewis^{a-b-} patients can be falsely negative.

- 95% of bile duct tumors are adenocarcinoma.
- The remainder is comprised of squamous, carcinoid, sarcomatous, mixed cholangiocellular/hepatocellular tumors, Kaposi sarcoma, and lymphoma.
- The characteristics described here refer to the adenocarcinoma cell type of either intrahepatic or extrahepatic bile duct origin.
- Extrahepatic cholangiocarcinoma is referred to as ▶ Klatskin tumor.

Characteristics

Etiology

Cholangiocarcinomas are the second most common primary liver cancer, representing 10% of all primary liver tumors. Cholangiocarcinomas occur with an approximate 1:100,000 incidence in the Western world but are more common in Southeast Asia and Japan. There is a slight male preponderance, with tumors occurring most frequently in the seventh decade of life. More than 90% occur sporadically in an otherwise normal liver. Risk factors for the development of cholangiocarcinoma include primary sclerosing cholangitis (PSC), congenital dilatation of the biliary tree (i.e., Caroli disease), hepatolithiasis and chronic infestation of the biliary tree by the parasitic liver flukes *Opisthorchis viverrini* and *Clonorchis sinensis*, a human liver fluke of the class *Trematoda*, Phylum Platyhelminthes. This parasite is found mainly in the common bile duct and gall bladder and feeds on bile. Infection with these parasites is endemic in some parts of Southeast Asia and in part may contribute to the higher incidence of cholangiocarcinoma in this region. These conditions all have in common chronic inflammation of the biliary tree, biliary stasis, or both. The inflammatory milieu likely contributes to two of the cardinal features of cancer, resistance to ▶ apoptosis, and continuous proliferation.

The cellular origin of malignant cells is not certain. Biliary epithelial cells are long lived under normal circumstances and thus survive long enough to accumulate carcinogenic mutations. Alternatively, multipotent progenitor cells that reside in the canals of Hering – small ductules lined in part by cholangiocytes and in part by hepatocytes, hepatocytes secrete bile into bile canaliculi which in turn drain into the canals of Hering – are not terminally differentiated and have the proliferative capacity that may facilitate malignant transformation. Finally, gland-like structures along the bile ducts can show hyperplasia and dysplasia adjacent to malignant cholangiocarcinoma. It is possible that each of these cellular compartments contributes depending on tumor focus and carcinogenic insult.

Biology

Grossly, cholangiocarcinomas usually develop at the hilum (60%) and less frequently develop as distal extrahepatic or as intrahepatic tumors (Fig. 1). Lesions can be categorized based on growth pattern as mass-forming, periductal-infiltrating, and intraductal. Microscopically, tumors have relatively few cancer cells in a desmoplastic stroma. This fibrosis leads to a low diagnostic yield of random sampling and the sensitivity of routine cytology from biliary brushings is usually reported in the range of 4–26%.

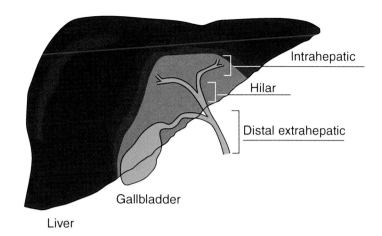

Intrahepatic

Hilar

Distal extrahepatic

Gallbladder

Liver

Cholangiocarcinoma, Fig. 1 Diagram of the liver and biliary tree. Cholangiocarcinoma can arise anywhere along the biliary tree, including intrahepatic bile ducts (peripheral cholangiocarcinoma), the hilum (Klatskin tumors), and more distally (distal extrahepatic cholangiocarcinoma). Hilar cholangiocarcinoma can be further subdivided using the Bismuth-Corlette classification depending on the involvement of only the hepatic duct (type I), the common hepatic duct and the confluence of the left and right hepatic ducts (type II), the common hepatic duct and either the left or right hepatic duct (type III), or the common, left, and right hepatic ducts or multifocal tumors (type IV)

The mechanistic link between inflammation and the development and progression of cholangiocarcinoma is of interest. Several inflammatory mediators have been shown to stimulate cholangiocyte proliferation and inhibit DNA repair or block apoptosis – all factors important in carcinogenesis. Indeed, interleukin-6 (IL-6) is an effective cholangiocyte mitogen and drives proliferation of this cell type via the MAP-kinase pathway. Additionally, IL-6-driven activation of signal-transducer and activator of transcription 3 (STAT3) increases the expression of Mcl-1, a potent antiapoptotic protein. Mcl-1 acts to protect cancerous cells and may contribute to the refractory nature of cholangiocarcinoma to chemotherapy. Proinflammatory cytokines induce cholangiocyte DNA damage and inhibit DNA repair by a nitric oxide-dependent mechanism. In addition, immunohistochemical studies of human cholangiocarcinoma specimens have shown the ubiquitous presence of inducible nitric oxide synthase (iNOS). Thus, IL-6 and iNOS with nitric oxide generation in chronically inflamed tissues likely contribute to the initiation and progression of cholangiocarcinoma.

The epidermal growth factor receptor (EGFR) has been implicated in numerous cancers, and activating mutations of EGFR have been described in cholangiocarcinoma. Further, bile acids stimulate signaling through the EGFR, including increasing Mcl-1 protein levels. EGFR activation leads to activation of the survival kinase Akt which is inhibited by phosphatase and tensin homolog deleted on chromosome 10 (PTEN). Experimental PTEN deficiency in mice combined with the loss of SMAD4 (a mediator of TGF-beta signaling) leads to cholangiocarcinoma by 4–5 months of age. Thus, dysregulation of EGFR signaling may play a prominent role in cholangiocarcinoma.

Diagnosis

Most patients with cholangiocarcinoma present with signs and symptoms of biliary obstruction, including jaundice, pruritis, chalk-colored stools, and dark-colored urine. This is the result of partial or complete blockage of the biliary tract leading to cholestasis and applies to Klatskin tumors and distal extrahepatic tumors. Intrahepatic tumors do not cause clinically significant biliary stasis.

Because the majority of tumors are not mass-forming, diagnosis by imaging can be difficult. Still, MRI with magnetic resonance cholangiopancreatography (MRCP) can be used to define the location and extent of a lesion. MR angiography is useful for the assessment of vascular involvement. CT and CT angiography can also be used to assess tumor location and vascular involvement, as well as lymph node enlargement. Ultrasound is often the initial study used in the evaluation of obstructive jaundice. While nonspecific, ultrasound can visualize biliary duct dilatation proximal to the obstruction and potentially can visualize an intrahepatic mass.

Endoscopic retrograde cholangiography (ERC) in the case of obstruction is necessary and can be diagnostically and therapeutically useful. ERC can demonstrate the site and extent of a stricture, intraluminal brushings taken within the stricture can provide cells for cytologic and advanced cytologic evaluation, and stenting relieves symptoms due to obstruction. Transluminal or percutaneous biopsy of hilar lesions or lymph nodes is not recommended due to the risk of tumor seeding.

Histologic diagnosis is the gold standard, however often diagnosis is based on the overall clinical picture. Cytologic evaluation has high specificity when positive for malignancy, but due to the desmoplastic nature of the tumor has low sensitivity. Advanced methods of cytologic analysis (digital image analysis and fluorescence in situ hybridization) have been used to improve sensitivity without compromising specificity. Serum CA19-9 level can aid in the diagnosis.

Patients with PSC represent a significant diagnostic challenge, as noncancerous stricture formation is the norm in this disease. Still, with a lifetime incidence of cholangiocarcinoma in PSC patients of 10–20%, a high index of suspicion must be maintained. Stable asymptomatic patients can be surveyed by noninvasive techniques such as MRCP and CA19-9 serum testing.

Treatment and Prognosis

Surgical resection for intrahepatic cholangiocarcinoma is curative for a minority of patients. For extrahepatic tumors (including Klatskin tumors), resectability depends on the extent of biliary and vascular involvement. Involvement of the left and right hepatic lobar structures (bilateral portal vein, main portal vein, or bilateral hepatic ducts) precludes resection, as does underlying liver disease or PSC. In resectable cases, partial hepatectomy improves outcomes. Selected patients with unresectable extrahepatic tumors benefit from liver transplantation; liver transplantation for intrahepatic cholangiocarcinoma is contraindicated due to disease recurrence.

Only a minority of patients present with disease amenable to surgical resection or transplantation. For the remaining patients, palliative treatment improves quality of life. The most important intervention involves relief of biliary obstruction generally by endoscopic approach or percutaneously. Drainage of one functional lobe is sufficient to relieve obstructive symptoms. Photodynamic therapy in conjunction with stent placement can improve drainage and increase survival by ~1 year. Palliative chemotherapy with gemcitabine leads to only limited responses.

Overall, the prognosis for cholangiocarcinoma remains poor. Five year survival ranges from 10% to 45%, while periampullary tumors have slightly higher 5-year survival (50–60%). Future improvements may come from rationally designed therapy based on the tumor biology, for instance targeting IL-6, EGFR, or Mcl-1.

Cross-References

▶ Adenocarcinoma
▶ Akt Signal Transduction Pathway
▶ Bile Acids
▶ Bile Duct Neoplasms
▶ Carcinoid Tumors
▶ Clinical Cancer Biomarkers
▶ Epidermal Growth Factor Receptor
▶ Gallbladder Cancer
▶ Gemcitabine
▶ Hepatocellular Carcinoma Molecular Biology
▶ Inflammation
▶ Interleukin-6
▶ Kaposi Sarcoma
▶ Klatskin Tumors
▶ Nitric Oxide
▶ Palliative Therapy
▶ Serum Biomarkers
▶ Smad Proteins in TGF-Beta Signaling
▶ Squamous Cell Carcinoma
▶ STAT3

References

de Groen PC, Gores GJ, LaRusso NF et al (1999) Biliary tract cancers. N Engl J Med 341:1368–1378

Patel T (2006) Cholangiocarcinoma. Nat Clin Pract Gastroenterol Hepatol 3:33–42

Roskams T (2006) Liver stem cells and their implication in hepatocellular and cholangiocarcinoma. Oncogene 25:3818–3822

Xu X, Kobayashi S, Qiao W et al (2006) Induction of intrahepatic cholangiocellular carcinoma by liver-specific disruption of Smad4 and Pten in mice. J Clin Invest 116(7):1843–1852

Cholangiocellular Carcinoma

▶ Cholangiocarcinoma

Choline Kinase

Ana Ramirez de Molina
Nutritional Genomics and Cancer Unit, IMDEA Food Institute, Madrid, Spain

Synonyms

CHETK; ChoK; Choline phosphokinase; Choline-ethanolamine kinase; CK

Definition

Choline kinase is the first enzyme in the Kennedy pathway of biosynthesis of phosphatidylcholine (PC), one of the major components of eukaryotic membranes (Fig. 1). This enzyme catalyzes the phosphorylation of choline (Cho) or ethanolamine (ETA) in the presence of ATP and magnesium as a cofactor to yield phosphocholine (PCho) or phosphoethanolamine (PEta), respectively (Fig. 2).

Characteristics

Choline kinase exists in mammalian cells in at least three isoforms, CK-alpha1, CK-alpha2, and CK-beta, encoded by two separate genes: choline kinase alpha (CHKA), located in chromosome 11q13.2, and choline kinase beta (CHKB), located in chromosome 22q13.33. Choline kinase is not active in its monomeric form; the active enzyme acts as a hetero- or homodimeric (or oligomeric) protein, which proportions have been proposed to be tissue specific.

CK-alpha1 and 2 are two splicing variants of the primary mRNA, in which the latter differs only in an additional stretch of 18 aa starting at position 155 from CK-alpha1. No differences have been found to date about the biological role of these two variants. By contrast, CK-beta is structurally and biologically different to CK-alpha. The first conspicuous difference lies in their sequence, differing in approximately 40%. In addition, whereas mice lacking the alpha isoform die during embryogenesis, mice lacking the beta isoform survive to adulthood though affected of muscular dystrophy and bone deformity. The involvement of each isoform in distinct biochemical pathway has been also suggested since CK-alpha presents a dual choline/ethanolamine function in vivo, whereas CK-beta seems to display only ethanolamine kinase activity in the same conditions.

Choline Kinase Alpha and Human Tumorigenesis

The most fundamental feature of cancer cells involves their ability to maintain chronic proliferation. In this process, cancer cells require the synthesis of new cell membranes as well as reprogramming their energy metabolism. CK is an essential enzyme in the generation of the major membrane phospholipids and, subsequently, displays a structural and energetic function within the cell, which confers this enzyme a relevant role in cell division. Accordingly, CK-alpha plays a role in the regulation of cell growth and is involved in malignant transformation.

Increased levels of CK-alpha have been associated with proliferation, tumorigenesis, and oncogenic signaling through Ras GTPases. CK-alpha has also been implicated in numerous cancers. It has been found overexpressed in a high percentage of cell lines derived from human

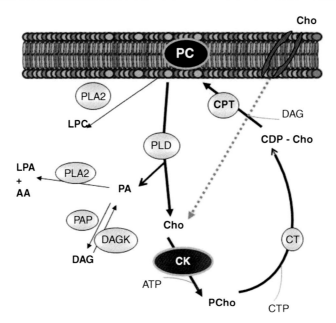

Choline Kinase, Fig. 1 Kennedy pathway (also called choline pathway) of biosynthesis of PC. Choline, an essential nutrient whose main function underlies in the synthesis of PC, crosses the plasmatic membrane into the cell by passive diffusion or choline-specific transports. Within the cell, choline is phosphorylated by choline kinase (CK, EC 2.7.1.32) to yield phosphocholine (PCho), which in turn is converted into CDP-choline by CDP-phosphorylcholine cytidyltransferase (CT, EC 2.7.7.15). Then, through the action of DAG-choline phosphoryltransferase (CPT, EC 2.7.8.2), DAG is incorporated in the reaction resulting in phosphatidylcholine (PC). PC is also catabolized through phospholipase D (PLD, EC 3.1.4.4) to choline and phosphatidic acid (PA). PA can be hydrolyzed to generate DAG by phosphatidic acid phosphohydrolase (PAP, EC 3.1.3.4), or deacetylated to lysophosphatidic acid (LPA) and arachidonic acid (AA) by phospholipase A2 (PLA2, EC 3.1.1.4). The generation of these metabolites has resulted in the consideration of PC as a precursor of lipid-related second messengers involved in cell proliferation and mitogenesis

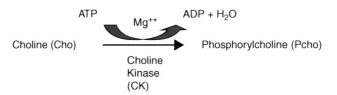

Choline Kinase, Fig. 2 Choline kinase. Choline (*Cho*), obtained by dietary source through the catabolism of PC, is phosphorylated to phosphocholine (PCho) catalyzed by choline kinase (*CK*) in the presence of ATP and magnesium ions. This reaction constitutes the first step in the cycle of biosynthesis of PC

tumors, as well as in different breast, lung, colorectal, prostate, bladder, and ovarian tumor tissues. An increase of its product, PCho, has been also frequently found in different tumor types using nuclear magnetic resonance (NMR) techniques. Since choline-containing compounds are detected by noninvasive magnetic resonance spectroscopy, increased levels of these compounds have been proposed as a noninvasive biomarker of tumorigenesis. In this sense, increased levels of CK-alpha activation have also been associated to increased levels of malignancy in several types of cancer, and CK-alpha expression has been proposed as a prognostic molecular biomarker for patients with lung cancer.

Due to the relevance of CK-alpha in human carcinogenesis, the inhibition of this enzyme has been proposed as an efficient targeted antitumor strategy. In this sense, small interfering RNA (siRNA) technology has been used to target ChoK showing antitumoral activity. In addition, numerous compounds have been synthesized mostly based on the structure of hemicholinium-3, a known competitive inhibitor of the enzyme. Promising results have been obtained with these compounds in terms of antitumoral activity, specificity, and associated toxicity. In this sense, one of these inhibitors, TCD-717, is currently in Phase I of clinical trials in humans as a "first-in-class" antitumoral agent (http://clinicaltrials.gov/ct2/show/NCT01215864).

References

Aoyama C, Liao H, Ishidate K (2004) Structure and function of choline kinase isoforms in mammalian cells. Prog Lipid Res 43(3):266–281

Glunde K, Bhujwalla ZM, Ronen SM (2011) Choline metabolism in malignant transformation. Nat Rev Cancer 11(12):835–848

Lacal JC (2001) Choline kinase: a novel target for antitumor drugs. IDrugs 4(4):419–426

Ramírez de Molina A, Sarmentero-Estrada J, Belda-Iniesta C et al (2007) Expression of choline kinase alpha to predict outcome in patients with early-stage non-small-cell lung cancer: a retrospective study. Lancet Oncol 8(10):889–897

Wu G, Vance DE (2010) Choline kinase and its function. Biochem Cell Biol 88(4):559–564

Choline Phosphokinase

▶ Choline Kinase

Choline-Ethanolamine Kinase

▶ Choline Kinase

Chorioallantoic Membrane

Domenico Ribatti
Department of Basic Medical Sciences, Neurosciences and Sensory Organs, University of Bari Medical School, Bari, Italy

Synonyms

CAM; Chorioallantois

Definition

The chorioallantoic membrane – synonym chorioallantois – (abbreviated CAM) is a vascular membrane in eggs of some amniotes, such as birds and reptiles. The chick embryo chorioallantoic membrane is an extraembryonic adnexa that mediates gas exchanges between the avian embryo and the outer environment. Due to its extensive vascularization and easy accessibility, the CAM has been used as an experimental in vivo model to study angiogenesis and antiangiogenesis. Moreover, because of the lack of a fully developed immunocompetent system, the CAM represents a host tissue for tumor engrafting suitable to study various aspects of tumor angiogenesis and metastatic potential.

Characteristics

Angiogenesis
The classical assays for studying angiogenesis in vivo include the hamster cheek pouch, rabbit ear chamber, the dorsal skin and air sac, the chick embryo chorioallantoic membrane (CAM), and the iris and the avascular corneal of rodent eye. The allantois is an extraembryonic membrane, derived from the mesoderm in which primitive blood vessels begin to take shape on day 3 of incubation. On day 4, the allantois fuses with the chorion and forms the chorioallantois. Until day 8 of incubation, primitive vessels continue to

proliferate and to differentiate into an arteriovenous system and thus originate a network of capillaries that migrate to occupy an area beneath the chorion and mediate gas exchanges with the outer environment. The CAM vessels grow rapidly up to day 11, after which the endothelial cell mitotic index decreases just as rapidly, and the vascular system attains its final setup on day 18 of incubation, just before hatching.

A variety of growth factors and normal cells have been reported to induce CAM angiogenesis. In studies of angiogenesis inhibitors, on the other hand, there are two approaches which differ in the target vessels, i.e., those which examine the response in the rapidly growing CAM and those that evaluate the inhibition of growth induced by an angiogenic cytokine. Several vehicles have been proposed to analyze the effect of the test substance: synthetic polymers, methylcellulose disks, agarose and millipore disks, and collagen gels. Gelatin sponges treated with a stimulator or an inhibitor of blood vessel formation have been implanted on the CAM surface on day 8 of incubation. The blood vessels growing vertically into the sponge and at the boundary between sponge and surrounding CAM mesenchyme are counted on day 12. The gelatin sponge is also suitable for the delivery of tumor cell suspensions, as well as of any other cell type, onto the CAM surface and for the evaluation of their angiogenic potential. Many techniques can be applied within the constraints of paraffin and plastic embedding, including histochemistry and immunohistochemistry. Electron microscopy can also be used in combination with light microscopy. Moreover, tissue specimens can be utilized for chemical studies, such as the determination of DNA, protein, and collagen content, as well as for RT-PCR analysis of gene expression by infiltrating cells, including endothelial cells. Moreover, the study of intracellular signaling pathways mediating the angiogenic response to growth factors and cytokines has been successfully performed.

Besides in ovo experimentation, a number of shell-less culture techniques have been described, that facilitate experimental access and continuous observation of the growing embryo.

Tumor-Induced Angiogenesis

The first evidence of the tumor-induced angiogenesis in vivo by using the CAM assay was dated 1913. The CAM has long been a favored system for the study of tumor angiogenesis and metastasis, because at this stage the chick's immunocompetent system is not fully developed and the conditions for rejection have not been established. All studies of mammalian neoplasms in the CAM have utilized bioptic tumor specimens and cell suspensions derived from tumors. Compared with mammalian models, where tumor growth often takes between 3 and 6 weeks, assays using chick embryos are faster. Between 2 and 5 days after tumor cell inoculation, the tumor xenografts become visible and are supplied with vessels of CAM origin. Tumors grafted onto the CAM remain non-vascularized for a couple of days, after which they can be penetrated by new blood vessels and begin a phase of rapid growth. Tumor cells can be identified in the CAM, as well as in the internal organs of the embryo.

Spontaneous metastasis model is based on grafting human tumor cells on the CAM surface. Within 5–7 days, tumor cells develop sizeable tumors, and tumor cells actively migrate from the chorionic epithelium through the mesenchyme, actively enter the vasculature, and colonize secondary tissues and organs. In tumor tissues grafts, preexisting blood vessels in the tumor graft disintegrated within 24 h after implantation, and revascularization occurred by penetration of proliferating host vessels into the tumor tissue. By contrast, preexisting vessels did not disintegrate in the embryo tissues grafts and anastomosed to the host vessels with almost no neovascularization. In adult tissue grafts, preexisting graft vessels disintegrated and did not stimulate capillary proliferation in the host.

Many innovative studies of L. Ossowski and A. Chambers have consolidated the CAM as a useful model to study metastasis. Several methods for semiquantitative analysis of metastasis in the chick embryo have been developed including morphometric assessment of individual metastasized cells, detection of microscopic tumor colonies, detection of human urokinase-type plasminogen activator (uPA) within secondary organs of the embryo, the use of green fluorescent protein, and in vivo videomicroscopy. Because the human genome is uniquely enriched in *Alu* sequences, PCR-mediated amplification of human-specific Alu sequences was used for semiquantitative detection of intravasated tumor cells in the CAM and within chicken tissues, followed by sensitive real-time Alu PCR assay. CAM has been successfully used in experimental metastasis studies, providing comprehensive information about other critical steps of metastatic process including (a) survival of cancer cells in the systemic circulation and transport to target organs, (b) arrest of cancer cells in the microcirculation, (c) migration of cancer cells through the vessel wall into the interstitial space (extravasation), and (d) proliferation of cancer cells in the target organs (colonization).

Cancer Progression

The majority of animal model systems to study cancer progression involve the use of immunocompromised mice and rats for hetero- or orthotopic transplantations of human tumor cells. Being naturally immunodeficient, the chick embryo accepts transplantation from various tissues and species without specific or nonspecific immune responses. Other advantages are the following: (a) CAM is particularly rich in blood vessels, allowing rapid vascularization, survival, and development of tumor cells or tissues placed on its surface; (b) CAM is connected to the embryo through a continuous circulatory system that is readily accessible from experimental manipulations and observations; (c) CAM allows to observe by in vivo microscopy the real-time changes in morphology of cancer cells arrested in its microcirculation; (d) in contrast to standard mouse models, most cancer cells arrested in the CAM microcirculation survive without significant cell damage and a large number of them eventually complete extravasation, (e) and the simplicity and low cost of the CAM assay strengthen the use of this experimental model.

There are some limitations in the use of the CAM assay. Nonspecific inflammatory reactions represent the main limitation of the chick CAM assay. However, nonspecific inflammatory reactions are much less frequent when implants are made relatively early in CAM development, when the host's immune system is relatively immature. Because the duration of the assay is limited to a 7–9-day window available before the chick hatches, most tumor cells cannot produce macroscopically visible colonies in secondary organs before the termination of the assay. The reduced number of specific antibodies available with specificity for chicken tissues limits the fine characterization of the different steps by using this experimental model. The complete characterization of the chick embryo genome (www.nhgri.gov/11510730) will be helpful to synthesize a broad panel of antibodies with high specificity for chicken tissues, especially for blood and lymphatic endothelial cells and stroma components. This aspect could be useful to better characterize the interactions between implanted human and/or mouse tumors and chicken tissues.

Metastatic cancer cells resemble stem cells in their ability to self-renew and to derive a diverse progeny. Embryonic microenvironments have been shown to inhibit the tumorigenicity of a variety of cancer cell lines. In this context, the embryonic microenvironment of the chick CAM could give an opportunity to study its influence of such on metastatic properties of cancer cells.

In the last years, retroviral, lentiviral, and adenoviral vectors have been used to infect the CAM (as well as the whole chick embryo), leading to the expression of the viral transgene. This allows the long-lasting presence of the gene product that is expressed directly by CAM cells, and this makes feasible the study of the effects of intracellular or membrane-bound proteins as well as of dominant-negative gene products. This approach will shed new lights on the study of the metastatic process by using the CAM assay.

Cross-References

▶ Angiogenesis
▶ Antiangiogenesis

References

Cimpean AM, Raica M, Ribatti D (2008) The chick embryo chorioallantoic membrane as a model to study tumor metastasis. Angiogenesis 11:311–319

Ribatti D (2008) Chick embryo chorioallantoic membrane as a useful tool to study angiogenesis. Int Rev Cell Mol Biol 270:181–224

Ribatti D (2010) The chick embryo chorioallantoic membrane in the study of angiogenesis and metastasis. Springer Science, Dordrecht, pp 1–124

Ribatti D (2014) The chick embryo chorioallantoic membrane as a model for tumor biology. Exp Cell Res 328:314–324

Chorioallantois

▶ Chorioallantoic Membrane

Choriocarcinoma

▶ Ovarian Tumors During Childhood and Adolescence

Choroidal Melanoma

▶ Uveal Melanoma

C-HRAS

▶ HRAS

Chromate

▶ Hexavalent Chromium

Chromatin Modification

▶ Chromatin Remodeling

Chromatin Remodeling

Yutaka Kondo
Department of Epigenomics, Nagoya City University Graduate School of Medical Sciences, Nagoya, Japan

Synonyms

Chromatin modification; Nucleosome remodeling

Definition

Chromatin remodeling is regulated by reorganization of nucleosome position by ATP-dependent nucleosome remodeling factors (ADNR) and covalent modifications of histone proteins. Because chromatin structure affects the binding of proteins including transcription factors to DNA, it is involved in many essential cellular processes.

Characteristics

Nucleosome consists of 147 bp DNA wrapped around the histone octamer comprising histone proteins, H2A, H2B, H3, and H4. Since the position of nucleosomes on DNA and the chromatin structure can affect the binding of proteins to DNA, chromatin remodeling is required for all the key processes such as gene expression (▶ epigenetic gene silencing), DNA replication, repair, chromosomal recombination, and mitosis. ADNR factors, SWI-/SNF-type factors, which are the part of multiprotein complexes, shuffle the nucleosomes and change chromatin structure and organization (nucleosome mobility), leading to either activation or repression of gene expression.

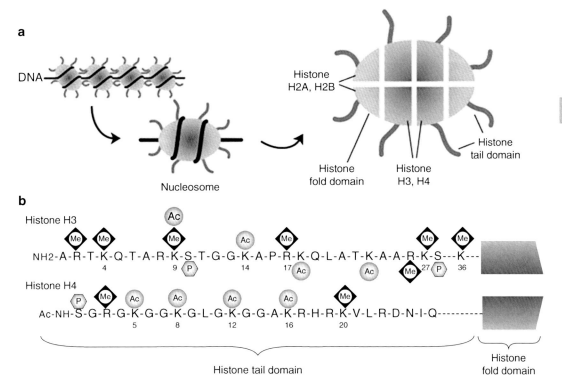

Chromatin Remodeling, Fig. 1 (**a**) DNA is compacted in the nucleus through a hierarchy of histone-dependent interactions. The fundamental repeating unit of chromatin is the nucleosome, which consists of 147 bp of DNA wrapped around an octamer of core histone proteins, H2A, H2B, H3, and H4. (**b**) Core histone proteins consist of less structured amino–terminal tails (histone tail domain) protruding from the nucleosome and globular carboxy-terminal domains making up the nucleosome scaffold (histone fold domain). Histone tail domain could be a target of a variety of posttranslational modifications, including acetylation (*Ac*), methylation (*Me*) and ▶ ubiquitination of lysine (*K*) residues, phosphorylation (*P*) of serine (*S*) and threonine (*T*) residues, and methylation of arginine (*R*) residues

The N-terminal domains of all core histones are subjected to chemical modifications, such as acetylation, methylation, and phosphorylation at certain residues (Fig. 1). Histone-modifying enzymes (histone deacetylases; HDACs) bring complexity of posttranslational modifications that can either activate or repress transcription, depending on the type of chemical modification and its location in the histone protein. The modification pattern of histone has been functionally linked to transcription and acts as a "histone code," which alters the structure of higher-order chromatin and helps recruit effector molecules. Various observations have suggested a connection between nucleosome remodeling and covalent histone modifications.

Recent study suggested that histones at transcriptionally active loci can be selectively replaced in a manner that is independent of DNA replication, that is, replacement of canonical histone H3 with variant histone H3.3, which is a highly conserved histone variant.

Mutation or dysfunction of these "epigenetic mechanisms" has been implicated in human cancers.

Mechanisms and Clinical Aspects

Chromatin-Remodeling Complex
In order to obtain the access of proteins to the DNA inside nucleosomes, chromatin-remodeling ATPases are required to unwrap the nucleosomal

b

SNF2 family proteins in mammals

Subfamily	Protein	Characteristic domain	Function
SWI2/SNF2	BRG1/SMARCA4	Bromo	Activation and repression
	BRM/SMARCA2	Bromo	Activation and repression
ISW1	SNF2H/SMARCA5	Sant-like	Chromosome structure
	SNF2L/SMARCA1	Sant-like	Chromosome structure
CHD1	CHD1	Chromo	Activation
	CHD3	Chromo	Repression
	CHD4	Chromo	Repression
INO80	INOB0	DBINO	DNA repair and gene expression
RAD54	ATRX		Regulation of transcription, DNA methylation
DDM1	HELLS/PASG/Lsh		Repression

Chromatin Remodeling, Fig. 2 (**a**) The nucleosome is a substrate of ATP-dependent nucleosome remodeling factor (*ADNR*). Nucleosome mobility is regulated by ADNR. (**b**) The SWI2/SNF2 family of ATP-dependent nucleosome remodeling proteins in mammals is classified into different subfamilies. SWI2/SNF2 family contains one or more domains in addition to helicase-like and ATPase domains. SANT (SWI3, ADA2, N-CoR, TFIIIB), DBINO (DNA-binding domain of INO80), Bromo (bromodomain), and Chromo (chromodomain) might interact with surfaces on the nucleosomes

DNA or to slide the nucleosome along the DNA to expose the buried sequences (Fig. 2). SWI-/SNF-type chromatin-remodeling factors have been shown to be required to this process. These factors are multiprotein complexes containing a central nucleic acid substrate stimulated ATPase belonging to the SWI2/SNF2 family. SWI2/SNF2 family is thought to be involved in gene expression, although their regulation and functions are not fully understood. The mammalian homologues, *Brahma* gene (*BRM*) and *Brahma*-related gene 1 (*BRG1*), are major components with ATPase enzymatic activities in the nucleosome remodeling SWI/SNF complex. BRM and BRG1 have a high degree of homology and either one of them might be contained in each SWI/SNF complex. Mutations or lack of expression of BRG1 has been identified in pancreatic, breast, lung, and prostate cancer cell lines. Germline and somatic mutations in *SNF5* (also called *INI1*), which is another mammalian SWI/SNF complex component but itself does not possess a chromatin-remodeling function, cause malignant rhabdoid tumor. Although BRM and BRG1 make a chromatin-remodeling complex with SNF5, mutation in SNF5 resulted in more sever phenotype than the tumor harboring either BRM or BRG1 mutations. There could be a degree of functional redundancy between BRM and BRG1, though the functions of SWI/SNF complexes containing BRG1 or BRM might not be interchangeable in some cases. BRG1 is involved in preventing cell cycle progression through its interaction with RB that has been shown to function as a brake on the cell cycle at least in part by establishing stable epigenetic silencing of the target genes. The SWI2/SNF2 family also contains generally one or more domains in addition to helicase-like and ATPase domains. A number of these domains have been shown to interact with surfaces on the nucleosome, which are often the targets of posttranslational modifications (see below). For example, bromodomain and chromodomain could bind acetylated lysine and methylated lysine, respectively.

Histone Modifications

Histone modifications are important in transcriptional regulation and are stably maintained during cell division. The less structured N-terminal domains of all core histones protrude from the nucleosomes and are subjected to chemical modifications, such as acetylation, ▶ methylation, and phosphorylation at certain residues. The modification patterns of histone have been functionally linked to transcription and act as a "histone code," which implies that transcription states can be predicted simply by deciphering this code. Generally, acetylation of lysine (K) residues on histone H3 and H4 leads to the formation of an open chromatin structure, with transcription factors accessible to promoters. Phosphorylation on serine 10 and acetylation on K14 on histone H3 work antagonistically to K9 methylation on H3 leading to the gene activation. Methylation at lysine is considered as a stable modification. There are the extra complications that histone lysine methylation can be either activating (e.g., H3K4 and

H3K36) or repressing (e.g., H3K9, H3K27, and H4K20), and the respective enzymes vary in their potential to induce mono-, di-, or trimethylation. Trimethylation at K9 on histone H3 or K20 on histone H4 has been shown to be a marker of heterochromatin from yeast to human. Dimethylation at K9 is associated with inactivation of gene expression. Trimethylation at histone H3K27 is a distinct histone modification involved in the regulation of homeotic (Hox) genes (▶ homeobox genes) expression and in early steps of X-chromosome inactivation in women. Di- or trimethylation on K4 on histone H3 localizes to sites of active transcription and this modification may be stimulatory for transcription. These different combinations of histone tail modifications influence transcription by affecting chromatin structure. Modifications on the lateral surface of core histone could also affect the histone–DNA interactions as well. Control of nucleosome mobility could be regulated by the valance of modifications of acetylation, methylation, and phosphorylation on the lateral surface amino acid residues.

DNA methylation is a crucial epigenetic mechanism for silencing tumor suppressor genes in human cancers, which also affect the chromatin structures. The link between DNA methylation and the histone modifications is mediated by a group of proteins with methyl–DNA binding activity, including MeCP2, MBD1, and Kaizo; these proteins localize to DNA-methylated promoters and recruit a protein complex that contains HDACs and histone methyltransferases. The DNA methyltransferases may also play a role in direct repression of transcription through cooperation with HDACs in late S-phase. While evidences for interactions between the DNA methylation and histone modifications are accumulating, the critical initiating events in silencing remain to be defined. In fungi, mutations of a histone H3K9 methyltransferase reduced DNA methylation indicate a simple linear model in which H3K9 methylation acts as an upstream epigenetic mark which signals to DNA methylation. However, in mammalian cells, DNA methylation inhibition could also rapidly changes histone methylation, and in plants histone and DNA methylation play distinct roles depending on the locus studied. The interactions between DNA methylation and histone H3K9 methylation currently best fit a model whereby these two changes form a reinforcing silencing loop, and this may explain why silencing is less stable in organisms that lack DNA methylation.

Histone Variant

At transcriptionally active loci, histone H3.3 variant substitutes for the canonical H3 histones. This replacement is independent of DNA replication. Histone modifications that may change histone–DNA or histone–histone contacts also affect catalyzing histone variant exchange. This replacement process is also catalyzed by ATP-dependent nucleosome-remodeling complexes. Histone replacement, which presumably is associated with activating transcription factors on the promoter region, offers an explanation for gene reactivation that were previously silenced via histone methylation. Whether histone replacement might be perturbed in cancer cells remains an open question.

Epigenetic Therapy

The most promising aspect of this field is the restoring gene function silenced by epigenetic changes including chromatin remodeling in cancer. "Epigenetic therapy" has the potential of "normalizing" cancer cells, which may lead to differentiation, senescence, or apoptosis. This could have a novel impact on the prevention and treatment of human cancers. HDAC inhibitors (► valproic acid) lead to the accumulation of acetylation in histones resulted in changes of chromatin status and of transcriptional activity to a normal state. P21 is a good example that is induced by HDAC inhibitors. However, exact mechanism through which the HDAC inhibitors mediate antitumor activity remains to be unclear, although these agents have been known to induce apoptosis and to inhibit angiogenesis and metastasis.

In a mouse model of colonic tumorigenesis, reducing DNA methylation genetically and/or pharmacologically has been shown to have tumor-preventive effects, a finding which was confirmed. The cytosine analogues, 5-azacytidine, and 5-aza-2′-deoxycytidine are powerful inhibitors of DNA methylation, which are incorporated into DNA during cell division and trap DNA methyltransferases and lead to cell differentiation and growth repression. Indeed these demethylating agents have been widely studied in hematological diseases and received FDA approval for the treatment of myelodysplastic syndrome. Further, the synergistic effects between DNA-demethylating agent and HDAC inhibitors suggest clinical trials of this approach to restore gene function silenced by aberrant chromatin changes in cancers.

Cross-References

- ► Cancer Epigenetics
- ► Epigenetic
- ► Epigenetic Gene Silencing
- ► Epigenetic Therapy
- ► Homeobox Genes
- ► Histone Modification
- ► hSNF5/INI1/SMARCB1 Tumor Suppressor Gene
- ► Methylation
- ► Ubiquitination
- ► Valproic Acid

References

Cosgrove MS, Boeke JD, Wolberger C (2004) Regulated nucleosome mobility and the histone code. Nat Struct Mol Biol 11:1037–1043

Gibbons RJ (2005) Histone modifying and chromatin remodelling enzymes in cancer and dysplastic syndromes. Hum Mol Genet 14:R85–R92

Mellor J (2005) The dynamics of chromatin remodeling at promoters. Mol Cell 19:147–157

Yoo CB, Jones PA (2006) Epigenetic therapy of cancer: past, present and future. Nat Rev Drug Discov 5:37–50

Chromium Carcinogenesis

Joseph R. Landolph, Jr.
Department of Molecular Microbiology and
Immunology, and Department of Pathology;
Laboratory of Chemical Carcinogenesis and
Molecular Oncology, USC/Norris Comprehensive
Cancer Center, Keck School of Medicine;
Department of Molecular Pharmacology and
Pharmaceutical Sciences, School of Pharmacy,
Health Sciences Campus, University of Southern
California, Los Angeles, CA, USA

Synonyms

Chromium-induced carcinogenesis; Chromium-induced cell transformation; Chromium tumorigenesis; Hexavalent chromium-induced carcinogenesis

Definition

Chromium carcinogenesis (▶ carcinogenesis) is the process of the genesis of tumors (carcinomas) by specific carcinogenic chromium compounds, containing ▶ hexavalent chromium Cr(VI). It includes the series of sequential steps that occur when lower animals or humans are exposed to specific hexavalent chromium-containing compounds that leads to tumor development. After all these steps are accomplished, the physiological mechanisms regulating control of growth in the normal cells are degraded. Hence, the normal cells are degraded and converted into tumor cells. The tumor cells then grow autonomously in an unregulated fashion and evade the host immune system, leading to development of visible tumors. Chromium carcinogenesis encompasses carcinogenesis by both insoluble and soluble hexavalent chromium-containing compounds.

Characteristics

Chromium: The Chemical Element and Its Ionic Species and Chemical Compounds

Pure chromium, the element, is a white, hard, lustrous, brittle metal with a high melting point ($1,903\ ^\circ$C). Pure chromium metal is resistant to mild corrosive agents. Therefore, pure chromium metal is used as a protective coating that can be electroplated onto other metals to protect them from corrosive. Chromium can be found in chemicals, in the +6, +5, +4, +3, +2, and 0 oxidation states. Cr(V) is usually a transient state, which can be best detected by ESR spectroscopy during the reduction of Cr(VI).

Chromium is most commonly found in nature in the ore called chromite or ferrous chromite ($FeCr_2O_4$). In chromite, chromium is in the +3 valence state, a very reduced state. The mineral, chromite, has Cr(III) at octahedral sties and Fe(II) at tetrahedral sites. Chromite can be reduced with carbon in a furnace. This reduction yields iron, chromium, and carbon monoxide in a carbon-containing alloy called ferrochromium.

High-purity chromium metal can be obtained by treating ferrous chromite ore with molten alkali and oxygen. This oxidizes Cr(III) to chromate, containing Cr(VI). The chromate can then be dissolved in water and precipitated as sodium dichromate. The sodium dichromate can then be reduced with carbon to Cr(III) oxide (Cr_2O_3). The chromium(III) oxide can then be reduced with aluminum metal to Cr(0) metal and aluminum oxide.

Important Commercial Uses of Chromium and Chromium Compounds

Chromium is also a very important and useful metal commercially. Chromium is used in large quantities in many important alloys, such as stainless steel (consisting of iron, nickel, and chromium) and ferrochromium (consisting of iron and chromium). Chromium is also used in the manufacture of paints and pigments, where various chromium compounds impart yellow and orange colors to the paints and pigments. Such compounds as lead chromate (PbCrO), strontium chromate, and barium chromate have been used in paints to paint aircraft due to the anti-corrosive properties of these compounds. Lead chromate is very toxic and hence has been replaced by strontium and barium chromates. Because chromium metal is resistant to corrosive agents, it is extensively used as a protective coating that is delivered onto other metals by an electroplating process.

Biological Aspects of the Essentiality of Chromium(III) in Mammalian Cells

In mammals, Cr(III) is considered an essential trace element. The required daily uptake of Cr (III) is from 50 to 200 µg/day. Cr(III) is usually ingested by humans from the diet. Cr(III) is considered an essential trace element because it is a component of the complex called glucose tolerance factor, which aids in glucose and lipid metabolism. Glucose tolerance factor, or "low-molecular-weight Cr-binding substance," is an oligopeptide. It is thought that a tetranuclear Cr (III) carboxylate complex may be present in glucose tolerance factor. Glucose tolerance factor is believed to aid insulin in mediating the uptake of glucose into mammalian cells. This area of nutrition requires further investigation.

Exposure of Humans to Chromium Compounds

Cr(VI)-containing compounds are very toxic to humans, both insoluble and soluble Cr(VI)-containing compounds. In terms of toxicology, humans are commonly exposed to chromium as Cr(VI)-containing compounds, primarily during the refining of chromium from chromite ore and also from the manufacture of chromium compounds for use in pigments and paints and during chromium electroplating. In the past, exposures of workers manufacturing chromate compounds to Cr(VI) compounds resulted in higher incidences of nasal and respiratory cancers. In workers who conducted chromium electroplating, there were also increased incidences of nasal and respiratory cancer. There were also increased incidences of various cancers in workers who were employed at tanneries, where they tanned cow hides, where chromate was also used to enhance the tanning process.

When inhaled, insoluble Cr(VI)-containing compounds can enter the airways and be phagocytosed by macrophages and by normal airway epithelial cells. This leads to deposition of a bolus of Cr(VI) inside the cells. When inhaled, soluble Cr(VI) compounds, in the form of chromate $(CrO_4)^{-2}$, bind to and enter mammalian cells on the sulfate-phosphate anion transport carrier. This anion transport carrier is somewhat nonspecific, and chromate can bind to it in place of phosphate and sulfate. Hence, inhalation of both soluble and insoluble Cr(VI)-containing compounds can cause cancer of the respiratory tract, insoluble Cr(VI) compounds via phagocytosis and soluble Cr(VI) compounds by entering cells on the anion transport carrier.

To date, we also know that ingestion of soluble Cr(VI)-containing compounds, or drinking them when they are added to or contaminate drinking water, can lead to these compounds being taken up into cells of the alimentary tract and entering cells on the sulfate-phosphate anion carrier. This can pose a risk for stomach cancer and cancer of the intestines, as well as a cancer risk to many other internal organs, in both lower animals (rodents) and humans.

Genotoxicity of Chromium Compounds

Chromium(VI) compounds are found in both the soluble and insoluble forms. Soluble and insoluble chromium compounds are both taken up by mammalian cells. Insoluble chromium(VI) compounds are taken up into mammalian cells by phagocytosis, and soluble Cr(VI) compounds are taken up into mammalian cells by the phosphate-sulfate transport carrier. Lead chromate, strontium chromate, and barium chromate are examples of Cr(VI)-containing compounds that are only sparingly soluble in water. Calcium chromate and potassium dichromate are examples of Cr(VI) compounds that are highly water soluble.

There are many insoluble chromium(VI) compounds, such as lead chromate, barium chromate, and strontium chromate. If these insoluble Cr(VI)-containing chromium compounds are found in particle sizes of <10 µm, they are phagocytosed by mammalian cells. Phagocytosis involves an invagination of the plasma membrane around the particles to form a phagocytic vesicle. The phagocytic vesicle then is internalized into the cell with its entrapped particle of insoluble Cr(VI)-containing compound. Hence, by the process of phagocytosis, a bolus of insoluble Cr(VI)-containing compound can be taken up into mammalian cells. The phagocytosed insoluble Cr(VI) compounds then enter the lysosomal network. The resultant chromium compounds then dissolve into soluble chromate ions and counterions. The chromate ions then migrate through the cell in an attempt to establish chemical equilibrium. Eventually, some chromate ions will travel into the nucleus. Eventually, the Cr(VI) in the chromate ions will be reduced to Cr(V), Cr(IV), and then to Cr(III). In addition, it is thought that Cr(VI) can act as a pseudo-Fenton reagent that can generate hydroxyl radical from superoxide radical (formed during normal metabolism) plus ▶ hydrogen peroxide (also formed during normal cellular metabolism).

Soluble Cr(VI)-containing compounds can enter mammalian cells on the sulfate-phosphate anion transport carrier, because they are competitive substrates and therefore inhibitors of sulfate and phosphate uptake into mammalian cells. Following phagocytosis of insoluble Cr(VI) compounds, or uptake of soluble Cr(VI) compounds on the phosphate-sulfate anion transport carrier, the resultant intracellular Cr(VI) ions cause cytotoxicity, DNA strand breaks, DNA-protein cross-links, and chromosomal aberrations, in the forms of gaps, breaks, fragments, dicentrics, and satellite associations, to mammalian cells.

Among the soluble Cr(VI) compounds, calcium chromate, sodium chromate, chromium trioxide, and potassium dichromate all induce mutation to 6-thioguanine resistance in diploid human fibroblasts. Lead chromate, an insoluble Cr(VI) compound, induced a weak yield of mutation to 6-thioguanine resistance in cultured diploid human fibroblasts. Induction of mutation to 6-thioguanine resistance by Cr(VI) compounds, both soluble and insoluble, occurred over the range of 0.05–1.00 μM. For Cr(III) compounds, weak mutation was induced by soluble chromium chloride and insoluble chromium chloride and Cr(III) oxide but only at the very high concentrations of 50–1,000 μM, which were 1,000-fold higher than those used to induce mutation by Cr(VI) compounds. Hence, Cr(VI) compounds are 1,000-fold more mutagenic, and cytotoxic, than Cr(III) compounds. Cr(III) compounds are not considered significantly toxic at low concentrations, particularly since Cr(III) is considered an essential nutrient and a component of "glucose tolerance factor."

Once inside mammalian cells, Cr(VI) ions are reduced to Cr(V), Cr(IV), and Cr(III) ions. Cr(III) ion binds to the DNA and thereby induces mutation to mammalian cells. There is also some evidence that intracellular Cr(VI) ions and more reduced Cr species can generate ▶ reactive oxygen species (ROS) intracellularly in mammalian cells. In mammalian cells, superoxide radicals arise from normal cellular oxidative metabolism. Following the dismutation of two superoxide radicals by superoxide dismutase, hydrogen peroxide is formed, particularly in mitochondria (mitochondrial DNA and cancer). There is some evidence that intracellular Cr(VI) ions can catalyze pseudo-Fenton reactions, in which superoxide radical, hydrogen peroxide, and Cr(VI) ions can cause the reaction of superoxide radical and hydrogen peroxide to generate hydroxyl radicals and hydroxyl ions. The resultant hydroxyl radicals are able to cause formation of 8-hydroxydeoxyguanosine and therefore mutations in DNA. However, this latter pathway has not been demonstrated conclusively with rigorous experimentation.

Chromium-Induced Cell Transformation

As noted above, insoluble Cr(VI)-containing compounds can be phagocytosed into mammalian cells, leading to generation of intracellular Cr(VI)-containing ions. Soluble chromium compounds generate soluble Cr(VI)-containing ions, which are taken up on the sulfate-phosphate anion transport carrier and enter mammalian cells. These intracellular Cr(VI) ions can be reduced by intracellular reductants, such as glutathione, which will then generate Cr(V)-, Cr(IV)-, and Cr(III)-containing ions. These Cr(III) ions can then induce DNA-DNA cross-links, DNA-protein cross-links, DNA strand breaks, mutations, and chromosomal aberrations, which are genotoxic events. Many Cr(VI)-containing compounds, including insoluble lead chromate and soluble Cr(VI) compounds, can induce morphological and neoplastic transformation of mammalian cells, and in particular in rodent cells, in culture. Lead chromate has been shown to induce morphological, anchorage-independent, and neoplastic transformation of C3H/10 T1/2 mouse embryo fibroblastic cells.

It is thought that the mechanism of Cr(VI)-induced morphological transformation is due to genotoxic events, including the mutations and chromosomal aberrations that Cr(VI) causes in mammalian cells. These chromosomal aberrations are thought to lead to loss of regions of chromosomes bearing ▶ tumor suppressor genes, or disrupt areas of chromosomes bearing tumor suppressor genes, which can lead to loss of tumor suppressor genes from cells, contributing to morphological, anchorage-independent, and neoplastic cell transformation. Similarly, intracellular Cr(VI) species can also generate reactive oxygen species, which can cause mutations in proto-oncogenes, leading to activation of oncogenes, which is a part of the mechanism of Cr(VI)-induced morphological and neoplastic cell transformation. In addition, generation of intracellular Cr(III) ions due to reduction of intracellular Cr(VI) by intracellular reductants, such as glutathione and other reductants, is thought to lead to mutations in proto-oncogenes, leading to activation of oncogenes, which also contributes to cell transformation. It is likely a combination of all these events that leads to Cr(VI)-induced morphological, anchorage-independent, and neoplastic cell transformation.

Chromium Carcinogenesis

Specific insoluble Cr(VI)-containing compounds, and also soluble Cr(VI) compounds, are carcinogens in lower animals such as rodents by the inhalation route. Similarly, both insoluble and soluble Cr(VI) compounds are also carcinogenic in humans when humans are exposed to them by the inhalation route. This is particularly true in the context of chrome electroplating and during the manufacture of chromates, where chromates are contacted by inhalation.

Increasing attention has been paid to the risk of cancer when Cr(VI) compounds are ingested or taken in with the drinking water. This is because Cr(VI) compounds have been used to paint aircraft in order to utilize the anti-corrosive properties of the Cr(VI) compounds. This has led to

contamination of drinking water sources in various states in the United States, such as California. In addition, Cr(VI) compounds have been used in the water of cooling towers. Discharge of this water into the drinking water sources has led to contamination of the drinking water sources with Cr(VI) compounds. When Cr(VI) enters humans or lower animals through the drinking water route, a significant amount of the Cr(VI) is reduced to Cr(III) extracellulary, reducing its toxicity, mutagenicity, and carcinogenicitiy. However, while this extracellular reduction is going on, there is a simultaneous, competitive uptake of Cr(VI) by the anion transport carrier into cells. This leads to a significant fraction of Cr(VI) getting into cells in a competitive mechanistic scheme [(i.e., A → B (cells) or C (enzymes of reduction)]. In rodents, administration of Cr(VI) by the oral route leads to stomach and intestinal tumors. There is weaker epidemiological data which also indicates that when humans drink water containing Cr(VI), the incidence of various internal tumors is increased. This includes a small increased incidence of stomach tumors, tumors of the bones, leukemias, kidney tumors, and liver tumors. Hence, while the cancer risk due to inhalation of Cr(VI) compounds is large, the cancer risk due to ingestion or drinking water containing Cr(VI) is smaller than the inhalation risk but is still of appreciable significance from the standpoint of induction of human cancer.

Mechanisms of Chromium Carcinogenesis

At present, the molecular mechanisms of Cr(VI) carcinogenesis appear to be due largely to genotoxic effects. These genotoxic effects include the ability of intracellular Cr(VI) ions to be reduced to Cr(V), Cr(IV), and Cr(III) ions and the ability of Cr(III) ions to bind to DNA and cause mutations. There is also some thinking that Cr(VI) ions can act as redox catalysts to cause pseudo-Fenton reactions. The reduction of Cr(VI) ions is believed to lead to generation of reactive oxygen species, such as superoxide, hydrogen peroxide, and hydroxyl radicals, which can cause 8-hydroxyguanosine in DNA, leading to mutations. Generation of 8-hydroxydeoxyguanosine and the binding of Cr(III) to

DNA would be expected to lead to mutations in many genes, including proto-oncogenes and tumor suppressor genes. This would be expected to result in the activation of proto-oncogenes into oncogenes and the mutational inactivation of tumor suppressor genes, leading to neoplastic cell transformation and eventually carcinogenesis.

Cross-References

▶ Aneuploidy
▶ Cancer
▶ Cancer Causes and Control
▶ Cancer Epidemiology
▶ Carcinogen Metabolism
▶ Carcinogenesis
▶ Class II Tumor Suppressor Genes
▶ DNA Damage
▶ Epidemiology of Cancer
▶ Gastric Cancer
▶ Genetic Toxicology
▶ Hexavalent Chromium
▶ Hydrogen Peroxide
▶ Lung Cancer Epidemiology
▶ Mesenchymal Stem Cells
▶ Nickel Carcinogenesis
▶ Oncogene
▶ Oxidative Stress
▶ Reactive Oxygen Species
▶ Renal Cancer Pathogenesis
▶ Repair of DNA
▶ Tumor Suppressor Genes

References

Biedermann KA, Landolph JR (1987) Induction of anchorage independence in human diploid foreskin fibroblasts by carcinogenic metal salts. Cancer Res 47:3815–3823

Biedermann KA, Landolph JR (1990) Role of valence state and solubility of chromium compounds in induction of cytotoxicity, mutagenesis, and anchorage independence in diploid human fibroblasts. Cancer Res 50:7835–7842

Cotton FA, Wilkinson G, Murillo CA et al (1999) The elements of the first transition series, Chapter 17. Section 17-c. Chromium: group 6. In: Advanced inorganic chemistry, 6th edn. Wiley-Interscience/Wiley, New York, pp 736–757

Landolph JR (1999) Role of free radicals in metal-induced carcinogenesis, Chapter 14. In: Sigel A, Sigel H (eds) Metal ions in biological systems. Interrelations between free radicals and metal ions in life processes, vol 36. Marcel Dekker, New York, pp 445–483

Patierno SR, Banh D, Landolph JR (1988) Transformation of C3H/10T1/2. Mouse embryo cells to focus formation and anchorage independence by insoluble lead chromate but not soluble calcium chromate: relationship to mutagenesis and internalization of lead chromate particles. Cancer Res 48:5380–52999

See Also

(2012) Chromosomal aberrations. In: Schwab M (ed) Encyclopedia of cancer, 3rd edn. Springer, Berlin/Heidelberg, p 838. doi:10.1007/978-3-642-16483-5_1138

(2012) Lead chromate. In: Schwab M (ed) Encyclopedia of cancer, 3rd edn. Springer, Berlin/Heidelberg, p 1992. doi:10.1007/978-3-642-16483-5_3295

(2012) Phagocytosis. In: Schwab M (ed) Encyclopedia of cancer, 3rd edn. Springer, Berlin/Heidelberg, p 2840. doi:10.1007/978-3-642-16483-5_4493

(2012) Proto-oncogenes. In: Schwab M (ed) Encyclopedia of cancer, 3rd edn. Springer, Berlin/Heidelberg, pp 3107–3108. doi:10.1007/978-3-642-16483-5_6656

(2012) Superoxide radical. In: Schwab M (ed) Encyclopedia of cancer, 3rd edn. Springer, Berlin/Heidelberg, p 3563. doi:10.1007/978-3-642-16483-5_5580

Chromium Tumorigenesis

▶ Chromium Carcinogenesis

Chromium(VI)

▶ Hexavalent Chromium

Chromium-Induced Carcinogenesis

▶ Chromium Carcinogenesis

Chromium-Induced Cell Transformation

▶ Chromium Carcinogenesis

Chromophore-Assisted Laser Inactivation

Daniel G. Jay
Tufts University School of Medicine, Boston, MA, USA

Synonyms

CALI

Definition

Technology to address protein function in situ. CALI uses laser light of 620 nm, targeted via specific malachite green-labeled non-function-blocking antibodies, which generate short-lived protein-damaging free radicals (Fig. 1). This wavelength is not absorbed by cells, such that nonspecific light damage does not occur. The short lifetime of the free radicals generated restricts the damage largely to the bound antigen (~15 Å) such that even neighboring proteins are not significantly affected. Micro-CALI focuses the laser light through microscope optics such that proteins within a 10 μ spot may be inactivated.

Characteristics

The advent of complete genomic information will herald a new revolution in molecular biology to develop a mechanistic understanding of how proteins function together in the living cell. This increased understanding will provide insight into cancer (as well as other diseases) and potentially will help define protein targets for drug discovery. Target validation of proteins of disease relevance (the site of most drugs) is the limiting first step in obtaining drugs of clinical value. In particular, identifying proteins that have essential roles in cancer-relevant cellular processes remain a major challenge. There is a current lack of technology that addresses protein function directly and

rapidly, as most functional inactivation approaches target genes or mRNAs. One useful tool to address this is chromophore-assisted laser inactivation (CALI). CALI uses targeted laser light to inactivate proteins of interest via a dye-labeled antibody that by itself does not block function. CALI provides a high degree of temporal and spatial resolution to acutely perturb protein function in situ.

Advantages

The major advantages of CALI compared to other functional inactivation approaches are its unprecedented temporal and spatial resolution. The range of area for CALI-based inactivation is localized to regions within the beam, between microns to millimeters depending on the focus of the laser. Inactivation occurs acutely upon initiating laser irradiation. Because the loss of protein function is acute and transient, CALI does not appear to be subject to genetic compensation that is occasionally observed in chronic deletion strategies such as gene knockouts in mice. CALI is particularly useful for systems lacking genetic methods. For example, CALI may be used for human tissue culture cells that are disease relevant. As such, it is not necessary to extrapolate target validation from model systems. Many proteins of interest to cancer research have essential roles in early development and are thus difficult to address by gene knockout. CALI may be used to study the roles of these essential gene products in cellular processes after development is completed. The coupling of micro-CALI with real time imaging has been used for studying dynamic cellular processes. The application of micro-CALI on part of a single cell generates a transient asymmetry of function across a cell, and this has been particularly useful in addressing proteins required for cell motility and migration. The capacity and relative ease of multiplex approaches for CALI (compared to gene knockouts) may lend itself well to whole proteome approaches and high-throughput screens.

Limitations

As with any technology, CALI has its limitations, and a clear understanding of these is required for

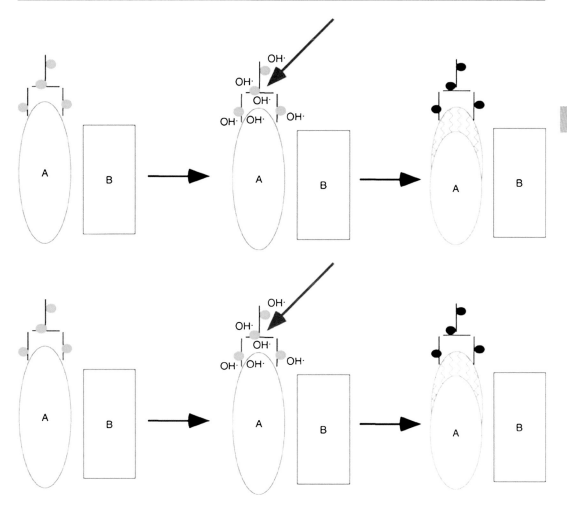

Chromophore-Assisted Laser Inactivation, Fig. 1 The principle of CALI. Specific proteins (**a**) in cells are bound by an antibody and labeled with the chromophore malachite green (*MG*). Irradiation with pulsed laser light of 620 nm (*red arrow*) generates short-lived hydroxyl radicals. These radicals selectively inactivate the bound protein by oxidative damage because their half-maximal radius is about 15 Å due to their short life time in cells. Neighboring proteins (**b**) are not significantly affected

its judicious application. Inactivation is dependent on quality, specificity and site of antibody binding and also on the susceptibility of the targeted protein to free radical damage. It should be noted that only the protein (not the gene) is inactivated and hence recovery is dependent on de novo synthesis. This usually allows a loss of function of hours to perhaps a day. The loss of function may not be complete so that activity is "knocked down" as opposed to "knocked out." As such, residual activity may obscure a potential phenotype. In general, a negative result is difficult to interpret for CALI as with most other inactivation approaches.

Application

CALI has been used for over 50 proteins and been successful in ~90% of these cases. CALI has precisely mimicked Drosophila genetic loss of function mutations in several direct comparisons. The proteins studied in many cell and animal systems span a diverse array that includes membrane receptors, cytoskeletal proteins, signal transduction molecules, and transcription factors.

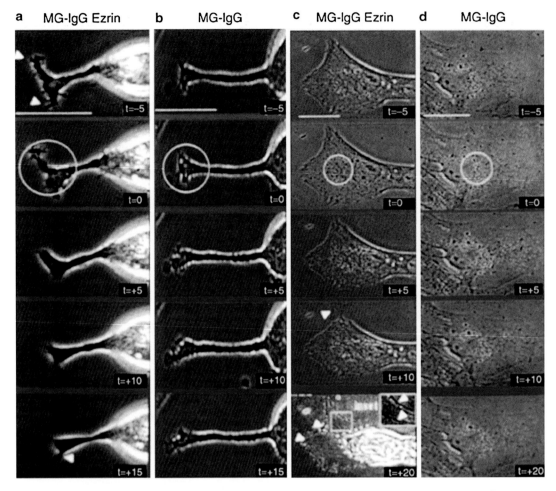

a MG-IgG Ezrin   **b** MG-IgG   **c** MG-IgG Ezrin   **d** MG-IgG

Chromophore-Assisted Laser Inactivation, Fig. 2 Micro-CALI of Ezrin affects fibroblast shape and motility. Fibroblasts transformed with v-fos, change their shape and motile behavior, and show an increase in the expression and phosphorylation of the actin-associated ERM protein, Ezrin. We applied micro-CALI of Ezrin within the circled areas to v-fos-transformed fibroblasts (A and B) and normal fibroblasts (C and D). Micro-CALI of Ezrin in v-fos transformed cells caused a loss of membrane ruffling (*arrowheads*) and pseudopodial retraction (**a**) while laser irradiation of cells injected with malachite green-labeled non-immune IgG had no effect on motility (**b**). Micro-CALI of Ezrin in normal fibroblasts caused a marked collapse of the leading edge (**c**) with filaments remaining attached to the substratum (*arrowheads* in *inset* of panel labeled $t = +20$). Irradiation of cells, injected with malachite green-labeled non-immune IgG had no effect on cell shape (**d**). Scale bars $= 10 \mu m$; time is in minutes

The understanding of how proteins function in the nerve growth cone has been a major area of study that has utilized CALI. The functional roles of proteins such as NCAM, L1, calcineurin, talin, vinculin, myosin V and Ib, radixin, and tau have been addressed. For these studies, methods for introducing antibodies into living cells have been optimized including electroporation, trituration, and microinjection.

Of particular relevance to cancer research is the application of CALI to proteins that have roles in cancer cell migration. One example of the application of CALI to a protein of cancer relevance is the prototypic ERM-family (► ERM proteins) member, ezrin. Ezrin is an actin-associated protein that shows increased expression and phosphorylation upon fos-mediated transformation of fibroblasts that is correlated with a change in cell shape

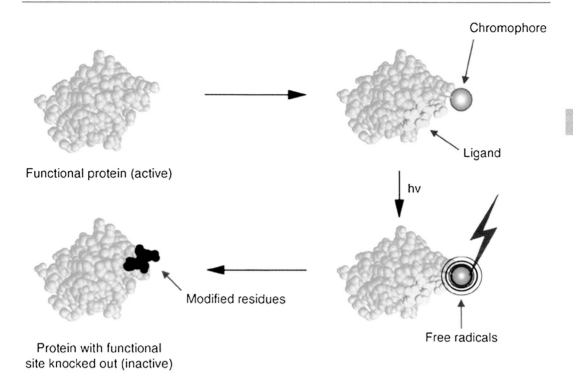

Chromophore-Assisted Laser Inactivation, Fig. 3 Principle of Xplore. Xplore uses CALI and high resolution mass spectroscopy to map regions of functional importance on proteins. After CALI inactivation, sites of oxidative damage are mapped by high resolution mass spectroscopy, providing a correlation of a protein function with specific domains of the targeted protein

(from flat to rounded) and motility (from lamellipodial to pseudopodial). CALI of ezrin in transformed fibroblasts causes a decrease in membrane ruffling and pseudopodial retraction. CALI of ezrin in normal fibroblasts causes a marked collapse of the leading edge lamellipodia (Fig. 2). These studies implicate ezrin in cell shape and motility and suggest that ezrin has a critical role in the shape and motility changes associated with oncogenic transformation. A second protein of interest is the tumor supressor ▶ hamartin. Hamartin binds to ERM proteins and its function is regulated by the small GTPase, Rho. CALI was used to show a role for hamartin in cell adhesion and suggests it might be involved in a rate-limiting step in tumor formation.

CALI is currently being combined with advances in dynamic imaging to visualize subcellular changes in response to the loss of function of specific proteins. We view that a major application of CALI for cancer research will be in target validation. As CALI lends itself well to combinatorial approaches and high throughput methods, it may be a powerful tool in addressing function in a proteome wide manner. A new use for CALI is in refining drug discovery screens to direct them against binders of a single domain on the target protein. CALI causes localized oxidative damage to modify residues of the protein near the antibody-binding site. By combining CALI with high resolution mass spectrometry to map those sites of damage, it may be possible to correlate loss of function with particular domains on a protein (Fig. 3).

A major development for CALI has been the application of endogenously encoded photosensitizers for the acute inactivation of fusion proteins, thus combining light induced loss of function with molecular genetics. While work in the 2000s established this using Green Fluorescent Protein, more efficient generators of oxygen radicals such as KillerRed, SuperNova and miniSOG expand CALI's potential and ease of use.

Developments

A major development for CALI has been the application of endogenously encoded photosensitizers for the acute inactivation of fusion proteins, thus combining light induced loss of function with molecular genetics. While work in the 2000s established this using Green Fluorescent Protein, more efficient generators of oxygen radicals such as KillerRed, SuperNova and miniSOG expand CALI's potential and ease of use.

Conclusions

CALI is a means for the inactivation of specific proteins in situ with a high degree of spatial and temporal resolution. CALI converts a binding reagent (such as an antibody) into an functional inhibitor. A large number of studies have demonstrated the potential of CALI in addressing cellular processes. It has been employed to address cellular mechanisms of cancer, and we believe that this technology is poised to contribute significantly to target validation and drug discovery for cancer-relevant processes.

Cross-References

► ERM Proteins
► Hamartin

References

Beermann AE, Jay DG (1994) Chromophore-assisted laser inactivation of cellular proteins. Methods Cell Biol 44:716–732

Bulina ME, Lukyanov KA, Britanova OV, Onichtchouk D, Lukyanov S, Chudakov DM (2006) Chromophore-assisted light inactivation (CALI) using the phototoxic fluorescent protein KillerRed. Nat Protoc 1:947–953

Ilag LL, Ng JH, Jay DG (2000) Chromophore-assisted laser inactivation (CALI) to validate drug targets and pharmacogenomic markers. Drug Dev Res 49:65–73

Lamb RF, Ozanne BW, Roy C et al (1997) Essential functions of ezrin in maintenance of cell shape and lamellipodial extension in normal and transformed fibroblasts. Curr Biol 7:682–688

Lamb RF, Roy C, Diefenbach TJ et al (2000) The TSC1 tumor suppressor hamartin regulates cell adhesion through ERM proteins and the GTPase Rho. Nat Cell Biol 2:281–287

Lin JY, Sann SB, Zhou K, Nabavi S, Proulx CD, Malinow R, Jin Y, Tsien RY (2013) Optogenetic inhibition of synaptic release with chromophore-assisted light inactivation (CALI). Neuron 79(2):241–53. doi: 10.1016/j.neuron.2013.05.022

Takemoto K, Matsuda T, Sakai N, Fu D, Noda M, Uchiyama S, Kotera I, Arai Y, Horiuchi M, Fukui K, Ayabe T, Inagaki F, Suzuki H, Nagai T (2013) SuperNova, a monomeric photosensitizing fluorescent protein for chromophore-assisted light inactivation. Sci Rep 3:2629. doi:10.1038/srep02629

Wang FS, Jay DG (1996) Chromophore-assisted laser inactivation (CALC): probing protein function in situ with a high degree of spatial and temporal resolution. Trends Cell Biol 6:444–447

See Also

(2012) High Throughput Screens. In: Schwab M (ed) Encyclopedia of Cancer, 3rd edn. Springer Berlin Heidelberg, p 1695. doi:10.1007/978-3-642-16483-5_2732

(2012) Laser. In: Schwab M (ed) Encyclopedia of Cancer, 3rd edn. Springer Berlin Heidelberg, p 1984. doi:10.1007/978-3-642-16483-5_3284

(2012) Proteome. In: Schwab M (ed) Encyclopedia of Cancer, 3rd edn. Springer Berlin Heidelberg, p 3100. doi:10.1007/978-3-642-16483-5_4819

Chromosomal Fluorescence In Situ Hybridization

Synonyms

FISH

Definition

Detection of specific chromosome structures by hybridization of fluorescence dye-conjugated probes to DNA. The FISH technique relies on the hybridization of DNA probes which identify specific chromosomal structures. Probes can be used which are specific for the centromere region of particular chromosomes, for genes, or for complete chromosomes. The DNA of both the applied probe and of the patient sample are denaturated, i.e., both DNA strands of the double helix are separated. During the following renaturation, the DNA probes attach to the complementary section of the patient's DNA (hybridization). The DNA probes are either directly conjugated to a fluorescent dye or are analyzed using fluorescence conjugated antibodies. The respective chromosome structures therefore are assessable as fluorescence signals.

A significant advantage of the method lies in its applicability not only to metaphases but also to interphase nuclei. A disadvantage is that information is obtained only on chromosomes and genes for which probes are used.

Interphase FISH

▶ Interphase Cytogenetics; Due to the multitude of different chromosome aberrations, which are observed particularly in acute leukemias, a screening based on FISH on interphase nuclei covers only a fraction of potentially present aberrations and therefore cannot substitute the classic chromosome analysis. However, if a specific question should be answered, e.g., the detection of the ▶ chromosomal translocation t(15;17)(q22;q12) when ▶ acute promyelocytic leukemia is suspected, the FISH technique represents a fast and reliable method, providing a result within 4 h.

In follow-up assessments during therapy, the FISH technique can be used for the detection of residual disease if at diagnosis aberrations have been found by chromosome analysis for which FISH probes are available. The sensitivity for this method is higher than for the chromosome analysis; however, it is lower than for PCR.

Metaphase FISH

In addition to the probes applicable to interphase nuclei, so-called chromosome painting probes can be applied to metaphases which specifically bind to the complete DNA of a chromosome. This technique is used mainly for the confirmation of the conventional chromosome analysis in difficult cases.

The 24-color-FISH method allows the display of all 22 different pairs of chromosomes as well as of the sex chromosomes in one single hybridization. It is applicable to metaphase chromosomes only and helps in identifying complex structural aberrations.

Cross-References

▶ Acute Promyelocytic Leukemia
▶ Chromosomal Translocations
▶ Interphase Cytogenetics
▶ Minimal Residual Disease

See Also

(2012) Centromere. In: Schwab M (ed) Encyclopedia of cancer, 3rd edn. Springer, Berlin/Heidelberg, p 744. doi:10.1007/978-3-642-16483-5_1028

(2012) Chromosome. In: Schwab M (ed) Encyclopedia of cancer, 3rd edn. Springer, Berlin/ Heidelberg, p 848. doi:10.1007/978-3-642-16483-5_1145

(2012) FISH. In: Schwab M (ed) Encyclopedia of cancer, 3rd edn. Springer, Berlin/Heidelberg, pp 1415–1416. doi:10.1007/978-3-642-16483-5_2197

(2012) Fluorescence in situ hybridisation. In: Schwab M (ed) Encyclopedia of cancer, 3rd edn. Springer, Berlin/Heidelberg, p 1436. doi:10.1007/978-3-642-16483-5_6740

(2012) PCR. In: Schwab M (ed) Encyclopedia of cancer, 3rd edn. Springer, Berlin/Heidelberg, p 2803. doi:10.1007/978-3-642-16483-5_4417

Chromosomal Instability

Susanne M. Gollin
Department of Human Genetics, University of Pittsburgh Graduate School of Public Health and the University of Pittsburgh Cancer Institute, Pittsburgh, PA, USA

Synonyms

CIN

Definition

Chromosomal instability is the gain and/or loss of whole chromosomes or chromosomal segments at a higher rate in a population of cells, such as cancer cells, compared to their normal counterparts (normal cells). In some cancers, each cell within the tumor has a different chromosomal constitution (karyotype) due to chromosomal instability, which may be defined in practical terms as numerical and/or structural chromosomal alterations that vary from cell to cell. Although the terms chromosomal instability and genomic instability have been used interchangeably, this is technically incorrect, as they refer to different forms of genetic instability.

Characteristics

Chromosomal instability is a characteristic of cancer cells, especially solid tumors (rather than most hematologic (blood cell) malignancies). Several cellular mechanisms lead to numerical and structural chromosomal instability in cancer cells, including defects in (i) chromosomal distribution to the daughter cells (chromosome segregation), (ii) cell cycle checkpoints that protect against proliferation of abnormal cells, (iii) telomere (specialized structures that cap the ends of chromosomes) stability, and (iv) the DNA damage response. Although in the past, these mechanisms were thought to be unrelated, it has become clear that they are intimately intertwined, connecting the complex network of cellular pathways. Human papillomavirus and other oncogenic viruses interfere with these processes, causing chromosomal instability and tumor formation in the cells that they infect. Chromosomal instability plays an important role in cancer by creating large-scale genetic changes in as little as one cell generation, leading to rapid cancer cell evolution. The rate of discoveries about the mechanisms leading to chromosomal instability in cancer cells is accelerating, improving our understanding of how cells become cancer cells and how cancer cells become more dangerous to the patient by progressing and/or metastasizing.

Both clonal numerical and structural chromosomal alterations and chromosomal instability are common features of human cancers. Aneuploidy is the condition in which the chromosome number in a cell, population of cells, or person is not an exact multiple of the usual haploid chromosome number ($N = 23$ for humans). Aneuploidy results from numerical chromosomal alterations. Cancers with chromosomal instability are characterized by aneusomy, a condition in which a population of cells contain different numbers of chromosomes. In tumor cells, gains and losses of chromosomal segments arise as a result of structural chromosomal alterations, including reciprocal and nonreciprocal chromosomal translocations, homogeneously staining regions (in which a cassette of contiguous genes, including at least one oncogene or growth-related gene, is tandemly repeated (amplified) at least five times on a diploid background), other forms of gene amplification (e.g., double minute chromosomes), insertions, and deletions. Structural alterations may result in a further imbalance in gene expression, resulting in chromosomal instability. In some tumors, each cell within the tumor has a different karyotype due to chromosomal instability.

Historical Background

Chromosomal instability is thought to be the means by which cells develop the features that enable them to become cancer cells. In spite of the presence of cell-to-cell chromosomal instability, the tumor karyotype is thought to be quite stable over time, probably because advanced tumors have evolved a genetic makeup (genotype) optimized for growth, making it less likely that additional genetic alterations will confer an additional growth advantage. Chromosomal alterations and karyotypic instability in human tumor cells have been investigated for nearly a century. David von Hansemann first identified abnormal dividing cells in tissue sections of tumors, including cell divisions that appeared to have asymmetric spindles or multiple spindle poles (multipolar spindles) that would lead to unequal distribution of the chromosomes to the daughter cells, and chromosomes stretched between the two spindle poles late in cell division (anaphase bridges). Theodor Boveri (Hanahan and Weinberg 2011), while studying chromosomal segregation in *Ascaris* worms and *Paracentrotus* sea urchins in the early 1900s suggested that malignant tumors arise from a single cell with an abnormal genetic constitution acquired as a result of defects in the mitotic spindle apparatus. Today we know that numerical chromosomal instability arises as a result of chromosome segregational defects, most frequently resulting from multipolar spindles. Structural chromosomal instability results from chromosome breakage and rearrangement due to defects in cell cycle checkpoints, the DNA damage response, and/or loss of telomere integrity. Structural chromosomal instability frequently results from breakage-fusion-bridge (BFB) cycles, first described in maize by geneticist Barbara McClintock in 1938. In this process, a chromatid break occurs, exposing an unprotected chromosomal end which, after replication, is thought to fuse with either another broken chromatid or its sister chromatid to produce a dicentric chromosome. During the anaphase stage of mitosis, the two centromeres are pulled to opposite poles, forming a bridge which breaks, resulting in more unprotected chromosomal ends, and thus the cycle continues. Our studies of cancer cells suggest that structural chromosomal instability, including gene amplification, can occur by BFB cycles. The basis for these BFB cycles is not entirely clear, although studies of chromosomal fragile site breakage, some of which occurs as a result of cigarette smoking and leads to induction of BFB cycles, telomere dynamics, and the DNA damage response, suggest that these critical cellular processes play major roles in the development of structural chromosomal instability. In this contribution, defects in chromosomal segregation, cell cycle checkpoints, telomere function, and the DNA damage response and their role in mechanisms leading to chromosomal instability are introduced and literature citations (References) are provided for the interested reader.

Chromosome Segregational Defects Lead to Chromosomal Instability

One of the fundamental processes required in the life of a cell, whether from a unicellular or multicellular organism, is chromosome segregation. Fidelity of chromosome segregation, whether in meiosis or mitosis, is necessary for genomic stability and the continuation of life as

we know it. Abnormal chromosome segregation results in aneuploidy, abnormal numbers of chromosomes being distributed to daughter cells, such that the daughter cells don't match each other or their mother cell. This is the essence of chromosomal instability. Studies have shown that several factors can result in segregation defects, including abnormal chromosome-spindle interactions, premature chromatid separation, centrosome amplification, multipolar spindles, and abnormal cytokinesis (cell division). Chromosomal segregational defects (multipolar spindles, lagging chromosomes at metaphase and anaphase, and anaphase bridges) in cancer cell lines are an intrinsic, heritable trait in the general tumor cell population. Tumor cells expressing chromosomal instability cannot be "cloned," as they continue to express numerical and structural chromosomal instability generation after generation. In some cancers, ongoing chromosomal instability is a feature of both primary tumors in the patient and cell lines cultured in the laboratory from biopsies removed from those tumors. Many studies of proteins involved in the process of chromosome segregation, spindle function, and cytokinesis are in progress in numerous laboratories. The role of these proteins in chromosomal instability and implications in the diagnosis, prognosis, and therapy of human tumors will be revealed in the next few years.

A Defective Response to DNA Damage Leads to Chromosomal Instability

For many years, cytogeneticists (scientists who examine chromosomes) have known that patients with "chromosome breakage" syndromes express chromosomal instability. Yet, until recently, features of these syndromes have not been utilized to define defects in the DNA damage response in cancer cells. Causes of DNA damage include attack by ultraviolet light, ionizing radiation (X-rays), or environmental chemicals, and cellular errors, such as "spelling errors" (base pair mismatch) during DNA replication, replication fork collapse, or defects caused by naturally occurring reactive oxygen species. One type of DNA damage is the double strand break, which leads to a cascade of cellular events (the DNA damage response) that usually results in repair of the damage or cell death. Failure in the DNA damage response and double strand break repair can lead to genetic alteration or chromosomal instability, which can result in transformation from a normal cell to a cancer cell.

The DNA damage response involves the sensing of DNA damage followed by transduction of the damage signal to a network of cellular pathways, from those involved in the cellular survival response, including cell cycle checkpoints, DNA repair, and stress responses to telomere maintenance, and the apoptotic pathway. To make a simple analogy in an effort to describe the complex DNA damage response to double strand breaks, we can say that our cellular instruction book for all of the activities that go on in our cells and in our bodies is made up of 23 chapters, the chromosomes, and for safety's sake, we have two copies of the book, one from our mother and one from our father, although they aren't exact copies (e.g., the set of eye color genes from your mother may code for blue eyes and the one from your father, brown). The genes are like sentences in a chapter, made up of three letter words composed of the four letters of DNA, A, T, C, and G. The 23 different chromosomes in the cells, composed of many genes, are equivalent to the 23 chapters in the book, made up of many sentences. The total genome is equivalent to the whole instruction book for the cells, and the instructions code for proteins, the molecules that do the work in our bodies. So in total, we have 46 chromosomes, two copies of each one. Sometimes, this very long set of DNA instructions becomes damaged (like the pages in the book can become torn or fall out) from smoking, chemicals, X-rays, oxidants that occur naturally in our bodies (why some of us take "antioxidant" vitamins), or other insults. Although our DNA is a code of letters like words in a book, it really looks like a ladder or even like railroad tracks. To more easily think about DNA repair, we need to visualize it as railroad tracks. Like the railroad company, which has special vehicles that check the integrity of the tracks, we have proteins that check our DNA (sensor proteins and checkpoint proteins).

If the sensor protein spots a double strand break or another defect that might derail the train or cause a defect in the cellular instructions (mutation), she then tells the communications officer (signal transducer) to call headquarters which in turn calls the repair team. This happens in our cells, in which case the repair team is a series of proteins that carry out sequential multistep assessment and repair of the damage (the DNA damage response). If they find that a cargo train has already been instructed by the defect to race out of control, analogous to a cell proliferating in an uncontrolled fashion, making more and more copies of itself, on the way to making a cancer, they kill that cell. But, what if the protein that has the job of pushing the kill switch is sick that day, the cell cannot be killed and a cancer ensues. In our cells, this DNA damage response pathway is carried out by about 50 proteins in a carefully choreographed process. With the advances in the human genome project, we are learning more about the proteins in this pathway and how defects (mutations) in them can cause predisposition to cancer.

Loss, mutation, or altered function of the genes that code for some of the DNA damage response proteins cause familial cancer syndromes and in some cases, chromosomal breakage syndromes, which may affect heterozygous gene carriers or affected (homozygous) individuals. Although not clear at this time, the role of these critical DNA damage response genes in chromosomal instability merits further investigation. The DNA damage response genes involved in known familial cancer syndromes include *ATM*, *TP53*, *BRCA1*, *BRCA2*, *FANC*, *CHEK2*, *BLM*, and *MRE11A*. The involvement of the DNA damage response genes, *BRCA1* and *BRCA2*, in familial breast and ovarian cancer is well known. Both genes also appear to be associated with an increased risk of prostate cancer, and *BRCA2* is involved in familial pancreatic cancer. Germline *TP53* mutation carriers have Li-Fraumeni syndrome which is associated with a high risk of breast and brain tumors, sarcomas (muscle tumors), leukemia (blood cell tumors), laryngeal (voice box) and lung cancer, and other tumors. Germline *CHEK2* mutation carriers may present with a Li-Fraumeni-like syndrome and may have an increased risk for a wide range of tumors including breast, prostate, and colorectal (intestinal) cancer. Patients with ataxia telangiectasia, the autosomal recessive genetic disorder characterized by a defective *ATM* gene, manifest progressive cerebellar ataxia (staggering gait), telangiectases ("blood shot" eyes and skin), immune dysfunction, chromosomal instability, increased sensitivity to ionizing radiation (X-rays), and predisposition to cancer, especially leukemia. Heterozygous *ATM* carriers (both human and mouse) of dominant-negative (interfering) missense mutations are at increased risk for solid tumors, including breast cancer. Fanconi anemia (FA) is a rare genetic cancer susceptibility syndrome characterized by skeletal abnormalities, skin pigmentation abnormalities, bone marrow failure, chromosomal instability in the form of rearrangements between nonhomologous chromosomes, and sensitivity to DNA crosslinking agents. FA patients are predisposed to developing cancer, primarily leukemia and epithelial tumors, especially squamous cell carcinoma of the mouth and throat (called head and neck cancer) or cervical cancer. The risk of solid tumors in FA patients is ~50-fold higher for all solid tumors compared to the general population, but about 700-fold higher for head and neck cancers. Bloom syndrome is an autosomal recessive disorder characterized by growth deficiency, sun-sensitive facial redness, hypo- and hyper-pigmented skin, sterility in males, reduced fertility in females, predisposition to a variety of malignancies, and chromosomal instability. Thus, patients with cancer predisposition and "chromosomal breakage" syndromes will continue to educate us about the cellular processes that lead to chromosomal instability and cancer.

Telomere Dysfunction May Lead to Chromosomal Instability

Telomere loss or dysfunction is a cause of chromosomal instability in the laboratory mouse. Telomere loss can result from DNA damage or occur spontaneously in cancer cells which often have a high rate of telomere loss due to telomere shortening with each cell division. Telomere alterations in certain genetically engineered mice mirror those in human epithelial tumors, lending support to the hypothesis that telomere defects drive chromosomal instability in cancer cells and age-related epithelial carcinogenesis. Thus, in mouse and man, telomere dysfunction leads to chromosomal instability, as shown by studies of telomere dysfunction in the mouse, chromosomal breakage patterns in human tumors, and the observation that cancer predisposition syndromes can lead to both telomere dysfunction and chromosomal instability. Consistent with this hypothesis, both telomere shortening and cancer incidence increase with age. Telomeres play an important role in chromosomal instability, but the exact details remain under active investigation.

Cell Cycle Disturbances Result in Chromosomal Instability

Oncogenic (cancer causing) viruses, such as human papillomavirus (a sexually transmitted disease which causes cervical cancer in women, penile cancer in men, and oral and anal cancer in both men and women), recapitulate the abnormalities, including defects in chromosome segregation, centrosome dynamics, telomere mechanics, the DNA damage response, cell cycle regulation, and cell cycle checkpoints, that appear to play important roles in the development and maintenance of chromosomal instability. The primary impact of chromosomal instability is cancer. In addition, chromosomal instability is a major cause of tumor evasion of or resistance to therapy. Therefore, a complete understanding of the biological basis of chromosomal instability is essential for developing therapies targeted against the defects in cancer cells.

Cross-References

▶ ATM Protein
▶ *BRCA1/BRCA2* Germline Mutations and Breast Cancer Risk
▶ Cell Cycle Checkpoint
▶ DNA Damage Response
▶ DNA Damage Response Genes
▶ P53 Family
▶ Repair of DNA
▶ TP53

References

Boveri T (1929) The origin of malignant tumors (trans: Boveri M). The Williams and Wilkins, Baltimore, p 119

Hanahan D, Weinberg RA (2011) Hallmarks of cancer: the next generation. Cell 144:646–674

Murnane JP (2010) Telomere loss as a mechanism for chromosome instability in cancer cells. Cancer Res 70:4255–4259

Thompson SL, Bakhoum SF, Compton DA (2010) Mechanisms of chromosomal instability. Curr Biol 20: R285–R295

Chromosomal Translocation t(8;21)

Olaf Heidenreich[1] and Jürgen Krauter[2]
[1]Northern Institute for Cancer Research, Newcastle University, Newcastle upon Tyne, UK
[2]Medizinische Klinik III – Hämatologie und Onkologie, Klinikum Braunschweig, Braunschweig, Germany

Synonyms

AML1/ETO; AML1/MTG8; RUNX1/CBFA2T1; RUNX1/RUNX1T1; t(8;21); t(8;21)(q22;q22)

Definition

The ► chromosomal translocation t(8;21) is associated with ► acute myeloid leukemia. The resultant fusion gene AML1/MTG8 (AML1/ETO, RUNX1/CBFA2T1, RUNX1/RUNX1T1) is a repressor of gene transcription. In this chapter, the fusion gene is named *AML1/MTG8* and the corresponding fusion protein AML1/MTG8.

Characteristics

Cytogenetics and Morphology

Almost 50% of all cases of acute leukemia are associated with recurrent chromosomal changes such as inversions or translocations of material from one chromosome to the other. t(8;21)(q22; q22) marks a chromosomal translocation, where the chromosomes 8 and 21 exchanged their long arms (the q arms) from band 22 till the telomere. This translocation is exclusively associated with acute myeloid leukemia (AML). Most commonly, standard cytogenetic analysis is used to detect the t(8;21). In addition, molecular techniques such as FISH (► fluorescence in situ hybridization) or reverse transcriptase-polymerase chain reaction (RT-PCR) are increasingly used for the identification of t(8;21) positive patients. Several studies comparing the sensitivity of PCR techniques and standard cytogenetics for the detection of t(8;21) have found AML1/MTG8 transcripts also in patients with no cytogenetic evidence of this aberration. These findings indicate that the sensitivity for the detection of a t(8;21) can be increased by molecular screening of all AML patients.

The leukemic blasts of t(8;21)-positive AML patients are often large and display characteristic morphological features such as abundant cytoplasm, numerous granules, and single needlelike Auer rods. In most cases, the leukemic cells express the stem cell marker antigen CD34 on their surface. In contrast to most solid tumors, the amount of additional chromosomal changes is rather limited in t(8;21)-positive leukemia. The t(8;21) is significantly associated with the loss of a sex chromosome. Other additional chromosomal changes include a trisomy of chromosome 8 and a deletion of chromosome 9q.

AML1/MTG8

The translocation t(8;21) affects two genes. The *AML1* (► *RUNX*1) gene located on chromosome 21 codes for a transcription factor which is essential for hematopoiesis. The *MTG8* gene on chromosome 8 encodes a corepressor able to interact with several histone deacetylases (HDACs). Because of its reciprocal nature, the translocation t(8;21) generates two fusion genes, the derivative 8, *MTG8/AML1*, and the derivative 21, *AML1/MTG8*. However, leukemic cells express only *AML1/MTG8*; MTG8/AML1 protein has not been identified yet. In the case of the AML1/MTG8 fusion protein, the DNA-binding domain of AML1 (the runt homology domain, RHD) is linked to the almost complete MTG8 (Fig. 1). As a consequence, the transcriptional modulator AML1 is converted into a constitutive repressor. However, since only one of the

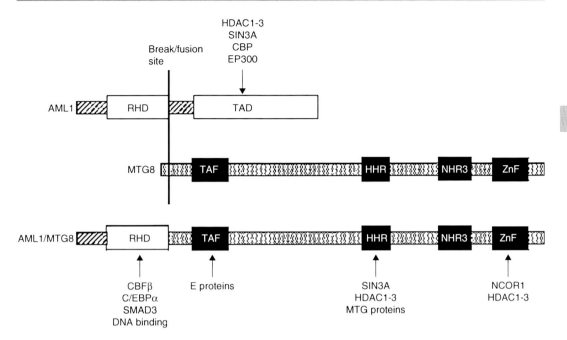

Chromosomal Translocation t(8;21), Fig. 1 Primary structure of AML1/MTG8, AML1b, and MTG8. The translocation t(8;21) fuses the N-terminal part of AML1 to the almost complete MTG8 protein. Functions and interacting proteins for the different domains are indicated. A line marks the fusion site. *RHD* runt homology domain, *TAD* transactivation domain, *TAF* TATA box binding protein-associated factor homology domain, *HHR* hydrophobic heptad repeat, *NHR3* nervy homology region 3, *ZnF* zinc-finger region

two copies of chromosome 8 and 21 are affected by the translocation, each t(8;21)-positive cell still contains one intact copy of these chromosomes and, thus, expresses nonfused wild-type AML1 in addition to AML1/MTG8.

AML1/MTG8 acts as a transcriptional repressor. Via the SIN3A and NCOR1 (N-CoR) bridging proteins, AML1/MTG8 recruits HDACs to genes, which contain AML1-binding sites in their promoters, thus leading to the deacetylation of histones and, consequently, silencing of the target gene (Fig. 2). Established target genes include cytokine and growth factor receptors such as the gene for M-CSF receptor (*CSF1R)* or cell cycle control genes such as p14^{ARF} (*CDKN2A*). Moreover, AML1/MTG8 interferes with hemopoietic differentiation by sequestering factors essential for these processes such as C/EBPα (CEBPA), SMADs, or vitamin D receptor (VDR).

AML1/MTG8 in Leukemogenesis

The translocation t(8;21) is most likely an initiating event in the development of leukemia. AML1/ MTG8 expression has been found in blood samples of newborn children at a much higher incidence than the probability to develop leukemia. Furthermore, some of the cured AML patients remain positive for AML1/MTG8. Moreover, AML1/MTG8 supports the expansion of hemopoietic stem cells both in cell culture and in animal models. In conclusion, AML1/MTG8 generates and maintains a pool of preleukemic cells but is not sufficient to induce leukemia. Additional genetic changes such as mutations in growth factor receptors (e.g., c-kit) or in p53 are required for full leukemic transformation. Nevertheless, leukemic persistence requires the continuous expression of AML1/MTG8 as shown by RNA interference experiments. Notably, a C-terminally truncated version lacking a binding domain for NCOR-HDAC complexes has a much higher transforming capacity in a leukemia mouse model than the full-length AML1/MTG8 protein. Interestingly, similar splice variants of *AML1/ MTG8* have been identified in patients suffering of t(8;21)-positive leukemia. Because of its

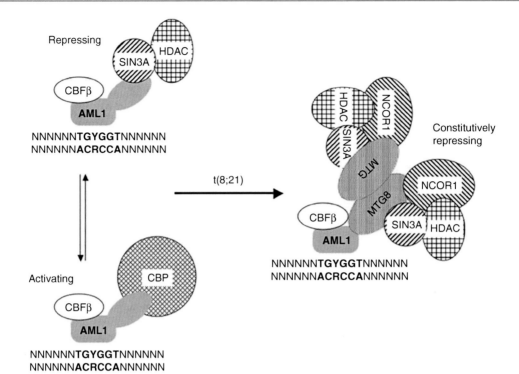

Chromosomal Translocation t(8;21), Fig. 2 AML1 and AML1/MTG8. Dependent on cellular signaling events, AML1 can switch from a repressive mode (complexed with SIN3A and HDACs) to an activating mode (complexed with the transcriptional activators CBP or EP300). Replacement of the transactivation domain by MTG8 results in oligomerization with other MTG proteins, recruitment of histone deacetylases, and, consequently, a constitutive repression of AML1 target genes. The DNA-binding site is indicated in bold. CBFβ, core-binding factor β (cofactor of AML1)

essential role in maintaining leukemia, and due to its exclusive expression in preleukemic and leukemic cells, *AML1/MTG8* might provide a promising target for leukemia-specific therapeutic approaches.

Clinical Relevance and Therapy

The translocation t(8;21) is found in about 10% of adult acute myeloid leukemia (AML) patients. Patients with t(8;21) are generally younger than 60 years. Most cases of t(8;21) positive AML show a FAB M2 or, less often, a M1 subtype that is with (M2) or with minimal (M1) signs of maturation. This translocation marks a subgroup of patients, which responds well to standard chemotherapy and, thus, has a rather good prognosis. Standard induction chemotherapy consisting of cytarabine (5-azacytidine) and an anthracycline achieves a very high complete remission (CR) rate of approximately 90% in patients with t(8;21). Moreover, an intensive consolidation with high-dose cytarabine or autologous stem cell transplantation yields an overall survival of approximately 50–70%. A low white blood cell count and high platelets at diagnosis are favorable prognostic factors, whereas the loss of the Y chromosome in male patients has an adverse prognostic effect. In t(8;21)-positive patients in CR, minimal residual disease can be detected by RT-PCR for AML/MTG8 fusion transcripts. As mentioned earlier, some of the patients remain positive for AML1/MTG8 even in long-term CR or after **allogeneic** stem cell transplantation most probably due to the persistence of nonleukemic t(8;21)-positive multipotent progenitors. However, it has been shown that serial quantification of *AML1/MTG8* transcript levels by quantitative RT-PCR might identify patients at high risk for relapse.

Cross-References

▶ Acute Myeloid Leukemia
▶ Allogeneic Cell Therapy
▶ Chromosomal Translocations
▶ Runx1

References

Downing JR (1999) The AML1-ETO chimaeric transcription factor in acute myeloid leukaemia: biology and clinical significance. Br J Haematol 106:296–308
Hug BA, Lazar MA (2004) ETO interacting proteins. Oncogene 23:4270–4274
Peterson LF, Zhang DE (2004) The 8;21 translocation in leukemogenesis. Oncogene 23:4255–4262

See Also
(2012) CCAAT/Enhancer-Binding Protein α. In: Schwab M (ed) Encyclopedia of Cancer, 3rd edn. Springer Berlin Heidelberg, pp 687–688. doi:10.1007/978-3-642-16483-5_901
(2012) FAB Classification. In: Schwab M (ed) Encyclopedia of Cancer, 3rd edn. Springer Berlin Heidelberg, p 1371. doi:10.1007/978-3-642-16483-5_2087
(2012) FISH. In: Schwab M (ed) Encyclopedia of Cancer, 3rd edn. Springer Berlin Heidelberg, pp 1415–1416. doi:10.1007/978-3-642-16483-5_2197
(2012) P53. In: Schwab M (ed) Encyclopedia of Cancer, 3rd edn. Springer Berlin Heidelberg, p 2747. doi:10.1007/978-3-642-16483-5_4331
(2012) SIN3A. In: Schwab M (ed) Encyclopedia of Cancer, 3rd edn. Springer Berlin Heidelberg, p 3411. doi:10.1007/978-3-642-16483-5_5309
(2012) SMAD. In: Schwab M (ed) Encyclopedia of Cancer, 3rd edn. Springer Berlin Heidelberg, p 3440. doi:10.1007/978-3-642-16483-5_5360

Chromosomal Translocations

Francisco J. Novo
Department of Biochemistry and Genetics, University of Navarra, Pamplona, Spain

Definition

A chromosomal translocation is a type of rearrangement between two chromosomes (usually nonhomologous) that involves breakage of each chromosome at a specific point called breakpoint, followed by fusion of the fragments generated by these breaks. A causative role has been demonstrated for some chromosomal translocations in various cancer types.

Characteristics

Instability of the genome, ▶ chromosomal instability in particular, is one of the hallmarks of cancer. Therefore, chromosomal rearrangements are very common in cancer cells. A frequent type of rearrangement is the translocation of genomic fragments between different chromosomal regions. The simplest case is a reciprocal translocation between two chromosomes, but translocations can also involve three or more chromosomes. If no genetic material is lost in the process, translocations are said to be "balanced." A well-known example of a reciprocal chromosomal translocation in cancer is the t(9;22) implicating the *ABL1* gene on chromosome 9 and the *BCR* gene on chromosome 22, which is found in most patients with ▶ Chronic Myeloid Leukemia.

Chromosome translocations are found both in solid tumors and in ▶ hematological malignancies, leukemias, and lymphomas. Solid tumors usually display complex karyotypes with many different translocations and other types of chromosomal rearrangements such as deletion, ▶ amplification, or inversion. In contrast, a frequent feature of most types of leukemias and lymphomas is the presence of a single or a few translocations, many of which are recurrent (i.e., found in different patients with the same type of cancer, or even in different tumor types). For this reason, chromosomal translocations have been best characterized in hematological cancers.

Biology
The mechanism underlying the presence of chromosomal translocations in cancer cells is the subject of active research. Various lines of evidence over the past years have identified several requirements for the generation of chromosome translocations. First of all, two breaks must be created

in different chromosomes at the same time. Additionally, the free ends must be close to each other within the cell nucleus. Finally, some DNA repair pathway must join the broken ends together, and the resulting molecule must provide some proliferative advantage to the cell.

With respect to the initial step, it is now generally accepted that chromosomal translocations are the result of DNA double-strand break, a type of ▶ DNA damage in which both strands of the double-helix are broken. Double-strand breaks are created throughout the genome by oxidative damage, radiation, replication over a single-strand break, genotoxic chemicals or physiological processes, such as the assembly of active immunoglobulin and T-cell receptor genes during lymphocyte development through ▶ V(D)J recombination. There are three main pathways that repair double-strand breaks in mammalian somatic cells:

- ▶ Homologous recombination repair, which relies on the presence of an intact homologous template in order to repair the DNA lesion
- Single-strand annealing, which requires some homology at both sides of the break, usually in the form of direct repeats
- Nonhomologous end-joining, which results in the religation of the ends without the requirement for a template.

The general consensus is that, in cancer cells, chromosomal translocations are the result of the repair of double-strand breaks via nonhomologous end-joining.

Work in the field of chromosome localization has shown that chromosomes occupy specific chromosomal territories inside the cell nucleus, and that a substantial amount of intermingling takes place between chromatin loops from neighboring territories. For example, loops from different territories can colocalize if genes present in those loops are transcribed at the same time and utilize the same transcription factory. Thus, two chromatin loops (from different chromosomes) that sustain a double-strand break simultaneously and are localized in close proximity are more likely to be involved in a translocation event.

This could explain the recurrence of certain chromosomal translocations in specific cancer types.

However, it is possible that many of the translocations generated in a cell never lead to the development of cancer. Chromosomal translocations are associated with cancer only when the resulting fusion products possess some oncogenic property that favors the clonal expansion of those cells. In this regard, there are two main mechanisms by which chromosomal translocations disrupt normal cellular processes. In one type of translocations, a gene is separated from its regulatory elements (promoter, enhancers) and juxtaposed to the regulatory elements of a different gene. As a result, the pattern of expression of the gene is altered and this leads to the acquisition of growth or survival advantage to those cells. Translocations involving immunoglobulin genes are the best example of this mechanism. For instance, the t(8;14) found in Burkitt lymphoma fuses the ▶ MYC oncogene to the regulatory elements of the gene coding for immunoglobulin heavy chains, resulting in deregulated and constitutive expression of MYC in lymphoid cells. In the second type of translocations, the oncogenic phenotype is the result of a ▶ fusion gene that is translated into a chimeric oncoprotein. This hybrid fusion protein brings together functional domains that were present in both original proteins, and this results in some gain of function which helps the cell to escape normal control mechanisms. For example, the t(15;17) found in patients with ▶ acute promyelocytic leukemia fuses part of the PML gene on chromosome 15 to part of retinoic acid receptor A (RARA) gene on chromosome 17. The chimeric fusion protein lacks RARA's responsiveness to retinoic acid, a consequence of which is that some bone marrow progenitor cells cannot undergo the normal process of differentiation.

Breakpoints are sometimes clustered in specific regions of the genes involved in a translocation. Such nonrandom distribution of breakpoints might be the result of functional selective pressures, so that even if double-strand breaks were generated randomly, not all the potential products would be functional: only those fusion transcripts that keep an intact reading frame and bring together specific functional domains will confer a proliferative advantage and will be found in cancer cells. Alternatively, specific structural DNA elements or sequence motifs might be responsible for the observed nonrandom distribution of translocation breakpoints in some tumor types. In this regard, recent work has shown that binding of nuclear receptors to DNA can bring specific genomic regions into close proximity within the cell nucleus and sensitize these regions to genotoxic stress. However, these two alternative explanations are not mutually exclusive, and it is likely that nonrandom clustering of translocation breakpoints in some cancers is the result of both processes.

Clinical Relevance

The fact that some chromosomal translocations are associated with specific malignancies is also important from the clinical point of view. A complete collection of published chromosomal translocations and the cancer types in which they were detected can be found in the Mitelman Database of Chromosome Aberrations in Cancer (http://cgap.nci.nih.gov/Chromosomes/Mitelman).

In some cases, especially in ▸ hematological malignancies, leukemias, and lymphomas, the diagnosis of the disease relies on the detection of a particular chromosomal translocation. The laboratory tools most frequently used in the diagnostic setting are conventional karyotyping (G-banding), fluorescence in situ hybridization (FISH), and PCR-based molecular techniques. Analysis of cancer patients has also shown that the clinical course of the disease sometimes depends on the presence of specific translocations. Therefore, the detection of chromosomal translocations is also important to estimate the probability of response to therapy or the risk that the cancer will recur after treatment. For this reason, specific translocations are part of the international classification system proposed by the World Health Organization for various types of malignancies. Importantly, the detection of specific chromosomal translocations is also used to assess the efficacy of treatment, since it provides a rational way to follow the evolution of the tumor clone and to confirm (or rule out) the presence of ▸ minimal residual disease.

Finally, the identification of chromosomal translocations has been instrumental in designing new effective therapies against some types of cancer. The best example of this is the new generation of drugs, like ▸ Imatinib (STI-571) against ▸ chronic myeloid leukemia and other malignancies characterized by the presence of chromosomal translocations involving tyrosine kinase genes. The finding that these tumors are the result of deregulated tyrosine kinase activity has led to the development of specific inhibitors and a dramatic increase in response to therapy and in survival rates in those patients.

Cross-References

▸ DNA Oxidation Damage
▸ Fusion Genes

References

Aplan PD (2006) Causes of oncogenic chromosomal translocation. Trends Genet 22:46–55

Lin C, Yang L, Tanasa B, Hutt K, Ju BG, Ohgi K, Zhang J, Rose DW, Fu XD, Glass CK, Rosenfeld MG (2009) Nuclear receptor-induced chromosomal proximity and DNA breaks underlie specific translocations in cancer. Cell 139:1069–1083

Meaburn KJ, Misteli T, Soutoglou E (2007) Spatial genome organization in the formation of chromosomal translocations. Semin Cancer Biol 17:80–90

van Gent DC, Hoeijmakers JH, Kanaar R (2001) Chromosomal stability and the DNA double-stranded break connection. Nat Rev Genet 2:196–206

Zhang Y, Rowley JD (2006) Chromatin structural elements and chromosomal translocations in leukemia. DNA Repair 5:1282–1297

Chronic Granulocytic Leukemia

▸ Chronic Myeloid Leukemia

Chronic Idiopathic Myelofibrosis

▸ Primary Myelofibrosis

Chronic Lymphocytic Leukemia

Jesper Jurlander
Department of Hematology, Rigshospitalet,
Copenhagen, Denmark

Synonyms

CLL

Definition

CLL is a chronic form of leukemia with accumulation of small mature B lymphocytes that express the surface membrane proteins CD5, CD19, and CD23.

Characteristics

Diagnosis

A diagnosis of CLL requires persistent absolute B lymphocytosis of 5×10^9/l or more, with a characteristic immunophenotype and cytomorphology. Usually the diagnosis can be made based on a blood sample. In the identical lymphoma disorder, small lymphocytic lymphoma, the level of circulating tumor cells in the blood and bone marrow may be very low or absent. The diagnosis of SLL therefore may require a lymph node biopsy.

Epidemiology

CLL is the most common leukemia in the Western world, with an incidence of approximately 5 new cases per 100,000 persons per year. The median age at diagnosis is 70 years, and the incidence increases with age. CLL may be seen in younger adults but never in children.

Etiology

The causes of CLL are largely unknown. Unlike other leukemias, there is no relation to exposure to chemotherapy or ionizing radiation.

Pathogenesis

The forces that drive the relentless expansion of the CLL clone are unknown. Models of the pathogenesis of CLL have focused on the B-cell receptor. Like other mature B-cells, CLL cells express immunoglobulin in the cell membrane, structurally ordered inside the B-cell receptor (BCR) complex. The immunoglobulin molecules in the CLL BCR complex are unique in several ways: (i) The repertoire of Ig-genes used by CLL cells is skewed, compared to normal B-cells. (ii) The genes encoding the heavy-chain variable segments only show signs of somatic hypermutation in about half of cases. (iii) The BCR complex is expressed at much lower densities, than on normal or other malignant B-cells. (iv) The three-dimensional structure of the immunoglobulin molecules encoded by B-cells is remarkably stereotypic. Taken together, these observations suggest that the BCRs of CLL cells may be activated and transduce signals, by a limited and restricted set of (auto)-antigens, that drive the expansion and survival of the CLL clone. Abnormal expression of certain molecules involved in signal transduction from the BCR, for example, overexpression of ZAP-70 or CD38 or low expression of $p72^{syk}$, may further modify the signaling capacity of the BCR in CLL cells. The result of this altered signaling is extended survival of B-CLL cells and perhaps even increased proliferation. Furthermore, the most common cytogenetic lesion in CLL cells results in deletion of a segment on chromosome 13q14 encoding the two ▶ microRNAs miR-15 and miR-16. These miRs can target and destroy Bcl-2 mRNA transcripts, leading to abandoned expression of Bcl-2 protein. When miR-15 and miR-16 are lost, due to the 13q14 deletion, the absence of the negative regulation of Bcl-2 may extend the longevity of CLL cells. Given the extended life cycle of a CLL cell, the risk of acquiring additional chromosomal aberrations is increased and may, for example, result in losses at 11q22 (the *ATM* gene) or 17p (the *p53* gene). Both of these aberrations will further destabilize the negative regulation of Bcl-2 and, furthermore, decrease the DNA damage response, in particular, the ability of p53 to induce cell cycle arrest upon DNA damage. In this

Chronic Lymphocytic Leukemia, Table 1 Risk prediction in CLL

Predictor	Low-risk CLL	High-risk CLL
Immunoglobulin heavy-chain gene mutations or usage	Mutated (less than 98% homology to the germ-line sequence)	Unmutated (more than 98% homology to the germ-line sequence). VH3-21 gene, regardless of mutational status
Cytogenetics by FISH	Del13q14 as sole abnormality	Del17p and/or Del11q22 and/or trisomy 12
Clinical stage	Lymphocytosis only	Bone marrow failure (nonimmune-mediated anemia/thrombocytopenia)
ZAP-70 protein expression	<20% positive cells by flow cytometry	>20% positive cells by flow cytometry
CD3	<30% positive cells by flow cytometry	>30% positive cells by flow cytometry
CLLU1 mRNA expression	<40-fold upregulation by quantitative RT-PCR	>40-fold upregulation by quantitative RT-PCR

way, the loss of control over apoptosis, induced by Bcl-2 overexpression, becomes linked to loss of control, the G1 restriction point of the cell cycle, resulting in increased proliferation and transformation to the truly malignant and aggressive form of leukemia, seen in the end stages of advanced chemotherapy refractory CLL.

Thus, the highly variable clinical course observed for many years in CLL is reflected by an equally variable spectrum of molecular aberrations detected in CLL cells. About half the cases of CLL have few molecular aberrations and are characterized by very slow expansion of the clone, resulting in an indolent form of leukemia that may not affect the mortality or morbidity of the patient. The other half of the patients show molecular features of aggressive disease and follow a clinical course that sooner or later develops into aggressive, refractory, and lethal leukemia.

Risk Prediction in CLL

Traditionally, the estimation of prognosis in CLL patients has relied on clinical staging systems, the two most widely used systems being those of Rai and Binet. Both systems use practical measurements of tumor size and bone marrow failure for prognostic estimation in CLL. Patients presenting with lymphocytosis alone generally have a favorable prognosis. Patients with lymphocytosis and enlarged lymph nodes, liver, or spleen have an intermediate prognosis. Patients presenting with signs of bone failure, i.e., anemia or thrombocytopenia not caused by autoimmunity, have a poor

prognosis. However, the clinical staging systems are static and can only describe the patient status at presentation. The increased usage of standard blood tests in the clinic results in identification of CLL patients at earlier stages of the disease, thereby eroding the informativeness of clinical staging systems. Today, more than 75% of patients are diagnosed by chance, usually because of examination for a non-CLL-related condition, at a stage where lymphocytosis is the only manifestation of the disease.

The identification of biological risk predictors now allows definition of low-risk and high-risk cases, based on molecular features at the time of diagnosis. The major risk predictors are seen on Table 1.

Patients with no high-risk features have an expected median survival of more than 15 years. Patients with some high-risk features have an expected median survival around 10 years. Patients with many high-risk features have an expected median survival of 5 years or less.

Symptoms and Signs of Active CLL

There is no chemotherapy treatment that can cure CLL at present. Non-myeloablative allogeneic hematopoietic cell transplantation may do so, but the considerable morbidity and mortality associated with this treatment makes it an option only for patients in advanced FC-refractory stages. Therefore, the treatment strategy is to await signs of active disease, before treatment is initiated. The signs of active CLL were defined by the NCI working group on CLL, also known as the

Cheson criteria, and updated in 2008 (Hallek criteria):

- Progressive bone marrow failure:
  - Development or worsening of nonimmune-mediated anemia
  - Development or worsening of nonimmune-mediated thrombocytopenia
- Massive or progressive lymph node enlargement
- Massive or progressive enlargement of the spleen
- Progressive lymphocytosis:
  - More than 50% increase in lymphocyte count in less than 2 months
  - Lymphocyte doubling time of less than 6 months
- Disease symptoms:
  - Weight loss (>10%) in less than 6 months
  - Fever of unknown origin for more than 2 weeks
  - Extreme fatigue
  - Night sweats
- Steroid-resistant autoimmune cytopenia

Treatment Strategies in CLL

Once the patient has developed active CLL, it is necessary to prepare a long-term treatment strategy, ensuring that the relevant options are available for the patient at the inevitable subsequent relapses. The goal of the treatment must be accessed, i.e., tumor reduction, tumor control, or tumor eradication, and developed in the context of patient age, comorbidity, and biological risk prediction. Elderly patients, or patients with significant comorbidity, may not benefit from aggressive treatment, not at least due to toxicity. Tumor control using single-agent alkylating regimens, such as chlorambucil, may be sufficient. Standard treatment, aiming at tumor control, for medically fit patients is the combination of (oral) ▶ fludarabine and ▶ cyclophosphamide (FC). For certain patients, the addition of the monoclonal antibody rituximab to FC (RFC) may further improve the result, and new data from controlled clinical trials suggest that patients with active CLL should be offered rituximab, in first line, second line, or both lines. The initiation of FC therapy in younger

patients is also the evaluation of an ultimative risk predictor. FC nonresponders, or early (<1 year) relapses, have a particularly unfavorable clinical course, with a median overall survival of less than 2 years. Therefore, patients younger than 70 years of age, started on treatment with a ▶ fludarabine-containing regimen, should have their options for allogeneic hematopoietic cell transplantation (allo-HCT allogeneic cell therapy) accessed at start of the treatment, primarily by tissue typing and identification of potential sibling allo-HCT donors. The expected median event-free survival is 1 year for alkylating agents, 2 years for fludarabine monotherapy, and 3 years or more for fludarabine combination regimens. None of these strategies will cure CLL, and relapses are inevitable. If the recurrence of active CLL occurs later than the expected time point, the initial treatment may successfully be repeated. If not, it may be considered to escalate to a more aggressive regimen, again considering the age and comorbidity of the patient.

Given the incurable nature of the disease, the described treatment strategy will eventually select for patients who are resistant to fludarabine-containing regimens. These patients constitute a growing and very significant challenge in CLL centers. First of all, their disease at this point has every sign of aggressive leukemia, with bone marrow failure and constitutional symptoms being the most important signs of disease. Secondly, fludarabine refractory patients more often than not present with a very severe immunodeficiency. It is opportunistic infections, more than the leukemia, that is threatening in advanced stages of CLL. Treatment options at this point, for example, monoclonal anti-CD52 antibodies (alemtuzumab), may reduce and control the tumor but will inevitably worsen the immunodeficiency. The only way out of this situation, with aggressive leukemia and disabled immune function, is allo-HCT.

Allo-HCT in CLL

The effectiveness of allogeneic HCT relies on two principles, with the aim to eradicate the tumor: (i) to deliver disease effective high-dose chemo- or radiotherapy that can eradicate the tumor and will eradicate normal bone marrow function, which however can be restored by reinfusion of normal bone marrow precursors and (ii) to develop alloreactivity that will target and destroy the leukemia, without targeting the patient. Standard transplantation evokes both principles. Non-myeloablative or reduced-intensity transplantation, using modern immunosuppression to allow the new immune system to develop, is focused on the second principle. This second form of allo-HCT appears to be particularly effective in CLL, however, at a certain cost. The introduction of a new immune system will create the risk that the new graft (immune system) will not only target the leukemia but several tissues in the engrafted recipient. The risk is development of graft-versus-host disease (GVHD). The GVH disease follows an acute and a chronic phase. The acute phase is responsible for a treatment-related mortality of approximately 10%. The deaths caused by chronic GVH occur at the same frequency, and living with chronic GVH disease causes severe reduction in life quality. Therefore, allo-HCT cannot be considered an option for the general CLL population. However, in younger patients, with FC-refractory disease or deletions at 17p, allo-HCT may be the only way to survive the disease.

In summary, the development of biological risk predictors, new effective chemoimmunotherapy combinations and the possibility for allo-HCT for at least some patients, has changed the management of CLL considerably since and will continue to do so over the next years.

Cross-References

▶ Allogeneic Cell Therapy
▶ Clinical Cancer Biomarkers
▶ Cyclophosphamide
▶ Fludarabine
▶ MicroRNA

See Also

(2012) Alemtuzumab. In: Schwab M (ed) Encyclopedia of cancer, 3rd edn. Springer, Berlin/Heidelberg, p 127. doi:10.1007/978-3-642-16483-5_177

(2012) CD52. In: Schwab M (ed) Encyclopedia of cancer, 3rd edn. Springer, Berlin/Heidelberg, p 702. doi:10.1007/978-3-642-16483-5_932

(2012) Cytogenetics. In: Schwab M (ed) Encyclopedia of cancer, 3rd edn. Springer, Berlin/Heidelberg, pp 1050–1051. doi:10.1007/978-3-642-16483-5_1470

(2012) Graft-versus-host disease. In: Schwab M (ed) Encyclopedia of cancer, 3rd edn. Springer, Berlin/Heidelberg, p 1597. doi:10.1007/978-3-642-16483-5_2502

(2012) Immunoglobulin. In: Schwab M (ed) Encyclopedia of cancer, 3rd edn. Springer, Berlin/Heidelberg, p 1819. doi:10.1007/978-3-642-16483-5_2990

(2012) Non-myeloablative. In: Schwab M (ed) Encyclopedia of cancer, 3rd edn. Springer, Berlin/Heidelberg, p 2538. doi:10.1007/978-3-642-16483-5_4116

(2012) Thrombocytopenia. In: Schwab M (ed) Encyclopedia of cancer, 3rd edn. Springer, Berlin/Heidelberg, p 3678. doi:10.1007/978-3-642-16483-5_5792

Chronic Myelogenous Leukemia

▶ Chronic Myeloid Leukemia

Chronic Myeloid Leukemia

Massimo Breccia
Department of Cellular Biotechnologies and Hematology, Sapienza University, Rome, Italy

List of Abbreviations

ALL	Acute lymphoblastic leukemia
AML	Acute myeloid leukemia
AP	Accelerated phase
BP	Terminal blastic phase
CCyR	Conventional cytogenetic analysis
CHR	Complete hematologic remission
CP	Chronic phase
MMR	Major molecular response
OS	Overall survival
Ph	Philadelphia chromosome
RQ-PCR	Real quantitative polymerase chain reaction
RT-PCR	Real-time polymerase chain reaction

Synonyms

Chronic granulocytic leukemia; Chronic myelogenous leukemia; CML; Ph-positive chronic leukemia

Definition

Chronic myeloid leukemia (CML) is a clonal disorder caused by a malignant transformation of a hematopoietic stem cell. Mature granulocytes and precursors proliferate and increase in bone marrow and peripheral blood.

Characteristics

The annual incidence is 1–2 cases every 100,000 inhabitants/year and increases with age, with a male prevalence. It account for 15–20% of all cases of leukemia in adult Western population. The disease is characterized by a reciprocal translocation t(9;22) (q34;q11) called Philadelphia chromosome (Ph). The Ph chromosome is present in more than 90% of adult CML patients, in 15–30% of adult acute lymphoblastic leukemia (ALL), and in 2% of acute myeloid leukemia (AML). Diagnosis is often based on morphological analysis of peripheral blood that showed increased mature myeloid cells, eosinophils, and basophils. Characterization of disease is performed with cytogenetic analysis and molecular analysis with real-time polymerase chain reaction (RT-PCR) or real quantitative polymerase chain reaction (RQ-PCR), which detects, respectively, Philadelphia chromosome and the presence and amount of *BCR/ABL1* mRNA.

CML is a progressive neoplasm that normally comprises three clinically recognized phases: approximately 90% of patients are diagnosed during the

- Typically indolent chronic phase (CP), which is followed by an
- Accelerated phase (AP) and finally a
- Terminal blastic phase (BP).

Although evolution through all stages is most common, 20% to 25% of patients progress directly from CP to BP. The time course for progression can also be extremely varied.

Chronic phase is often asymptomatic diagnosed incidentally with elevated white blood cell count on routine laboratory test. Symptoms at diagnosis include fever, fatigue, sweating, weight loss, joint pain, and enlargement of the spleen, liver, or both. The enlarged spleen may cause early satiety, decreased food intake, and abdominal fullness. Advanced phases of disease may present with bleeding, petechiae, ecchymoses, bone pain, fever, and infections. The mechanisms behind CML progression are not fully understood: it has been demonstrated that Src family kinases are involved in CML progression through the induction of cytokine independence and protection from apoptosis. Diagnosis of accelerated phase according to WHO criteria includes:

- 10–19% myeloblasts in blood or bone marrow
- Platelet count <100.000/mmc, unrelated to therapy, or >1.000.000/mmc unresponsive to therapy
- >20% basophils in the blood or bone marrow
- Additional cytogenetic abnormalities
- Increasing splenomegaly unresponsive to therapy

Diagnosis of blastic phase according to WHO criteria requires more than 20% of blast cells in bone marrow or peripheral blood, or extramedullary blast proliferation or large clusters of blasts in bone marrow biopsy.

Pathophysiology

The specific cytogenetic aberration associated with the disease consisting of a reciprocal translocation between the long arms of chromosomes 22 and 9. For the first time, it was described by Nowell and Hungerford, and subsequently called the Philadelphia (Ph1) chromosome. As a result, a region of chromosome 22 called "breakpoint cluster region" fused with the ABL gene on chromosome 9. This translocation results in the expression of the constitutively active protein BCR-ABL1 with tyrosine kinase activity.

Different molecular weight isoforms are generated based on different breakpoints and mRNA splicing. Most CML patients have a fusion protein of 210 kDa, while approximately 30% of Ph + ALL cases and a few CML cases express a 190 kDa BCR-ABL1 protein. The protein is able to phosphorylate several proteins involved in cell cycle control, apoptosis, and adhesion to the stromal layer. The discovery of the *BCR-ABL1*-mediated pathogenesis of CML provided the rationale for the design of specific inhibitory agents targeting BCR-ABL1 kinase activity.

Prognosis

Before the advent of tyrosine kinase inhibitors, median survival of CML patients was estimated to be 3–5 years. Now it has been estimated that overall survival (OS) is higher than 90% for patients who achieved a cytogenetic response with first-line tyrosine kinase inhibitor. Prognostic scores at diagnosis may be used to categorize the relative risk of overall survival. Sokal risk was based on an equation that considers age, spleen size, platelet count, and percentage of blood blast in peripheral blood. The score is able to identify three categories of risk (low, intermediate, and high) with different probabilities to achieve a complete cytogenetic response (absence of Ph-positive metaphases at conventional cytogenetic analysis, CCyR) and overall survival. In the interferon era, another prognostic score was developed, the so-called Hasford or Euro score that considered the same prognostic features as Sokal plus peripheral eosinophil and basophil count: this score also is able to identify three categories of risk. It has been reported that both scores were able to categorize patients even if treated with tyrosine kinase inhibitors. A new prognostic score was proposed, the so-called Eutos score, created on a large series of patients treated with front-line imatinib. It was based only on two features at baseline (spleen size and basophil count), and stratified patients in two categories, low and high risk.

Therapy of CML

Chemotherapy Agents and Interferon Alpha
Therapy of CML evolved over time on the basis of increased knowledge of pathogenesis. Myelosuppressive agents (hydroxyurea and busulfan) were initially used to obtain a haematological control of the disease, but did not induce cytogenetic or molecular remissions. In the 1980s, interferon alpha (IFN-alpha) entered the clinical practice, but allowed only limited cytogenetic responses and now is no longer considered as first-line therapy.

Imatinib
Imatinib mesylate, synonyms STI571, Gleevec, is currently approved as first-line treatment of CP-CML. The phase III international randomized study of IFN-alpha versus STI571 (IRIS) trial compared imatinib and IFN-alpha plus cytarabine in 1,106 CML patients in early CP. Eight-year follow-up showed that the cumulative CCyR rate is 83%, the estimated survival rate is 85%, and freedom from progression rate (FFP) is 92%, with an event-free survival of 81%. Disease progression occurs early, highlighting the importance of close monitoring during the early stages of treatment. Most of all events or disease progression reported during the 8-year follow-up occurred during the first 2–3 years. Achievement of CCyR in the first 12 months of therapy is associated with optimal long-term outcome: with the high rate of complete cytogenetic responders, the goal of therapy has become achieving molecular responses, as measured by the reduction or elimination of the *BCR-ABL1* transcript with RQ-PCR. Major molecular response (MMR) in the IRIS trial was defined initially as a >3 log reduction in transcript from baseline and then according to an International Scale (IS) as a *BCR-ABL1*/ABL ratio <0.1%. Early molecular responses were found to be predictive of a better outcome: progression of disease correlated with failure to achieve a 1 log reduction in the transcript level by 3 months and a 2 log reduction by 6 months. Obtaining an MMR within 18 months was associated with significantly better long-term event-free survival (98%) at 8 years. Adherence to therapy in the long-term is predictive of response. In a single-institution study of patients with newly diagnosed CML-CP who had achieved a CCyR with first-line imatinib treatment, patients with lower adherence to imatinib treatment had a significantly lower

probability of achieving a MMR and complete molecular response (undetectable residual disease or 4-log reduction, CMR) than patients with higher treatment adherence. The most frequent adverse events reported were oedema, nausea, muscle cramps, skin rash, and diarrhea.

A minority of CML patients in CP are refractory to imatinib or become insensitive to treatment with the agent after initial response to therapy, and consequently experienced relapse. In 2006, the European LeukemiaNet panel of experts proposed recommendations for monitoring CML patients treated with imatinib, and these recommendations were updated in 2009: three categories of patients were defined, optimal, suboptimal, and failure, at different scheduled time points (3, 6, 12, and 18 months). Despite the excellent results with imatinib, the 8-year follow-up results from IRIS study provided data on imatinib failure: primary resistance accounted for 17% of patients, whereas secondary resistance was approximately 15%. There are various mechanisms that could contribute to imatinib resistance, including increased expression of *BCR–ABL1* through gene amplification, decreased intracellular drug concentrations caused by drug efflux proteins, clonal evolution, and over-expression of the Src kinases (Lyn, Hck) involved in *BCR–ABL1*-independent activation of alternative pathways. However, 40–60% of resistance can be attributed to the emergence of clones expressing mutated forms of *BCR–ABL1* with amino acid substitutions in the ABL-kinase domain that impair imatinib binding, either through disruption of the critical contact point or by inducing a switch from the inactive to the active conformation. The drug was tested also in advanced phase and was able to induce 50–80% of complete hematologic remission (CHR), and about 30% of CCyR, unfortunately sustained in a minority of patients. Higher doses resulted in the improvement of responses and overall survival.

Dasatinib and Nilotinib as Second-Line Tyrosine Kinase Inhibitors

Second-generation tyrosine kinase inhibitors (nilotinib and dasatinib) were developed to improve results obtained with imatinib and to overcome different mechanisms of resistance.

Nilotinib is structurally related to imatinib, 30-fold more potent and selectively active against BCR-ABL1 protein. Like imatinib, nilotinib binds the inactive conformation of ABL but unlike imatinib, nilotinib is not a substrate for efflux and intake transporters. It is active against different mutations affecting imatinib response, except the T315I gatekeeper mutation. It was found less active against mutations with an IC50 > 150 nM such as E255V, Y253H, and F359. The drug was tested in a phase II trial that enrolled 321 CML patients with resistance or intolerance to imatinib treated with nilotinib at the standard dose of 400 mg twice daily. At a minimum follow-up of 4 years, 94% of patients rapidly reached a CHR and overall and 44% of patients reached a CCyR. MMR was obtained in 28% of patients with an overall survival of 87%. The most frequent nonhematological adverse events reported with the drug were skin rash, headache, and laboratory abnormalities (increased bilirubin, transaminases, glucose level, and pancreatic enzymes).

Dasatinib is an oral dual tyrosine kinase inhibitor active against ABL and Src-family kinases: the structure is based on a different chemical scaffold of imatinib, and it has a 325-fold greater potency. Dasatinib binds both the inactive and the active conformation of the ABL kinase domain. Also dasatinib was not able to inhibit T315I mutation and was found less active against mutations with IC50 > 3 nM, such as V299L, T315A, and F317. Several studies tested the efficacy and safety of dasatinib in resistant/intolerant patients to imatinib in different phases of disease. Based on the preliminary results of phase III trial, which enrolled 667 patients, the approved initial dose for CP patients in resistance, suboptimal response, or intolerance after imatinib therapy was changed from 70 mg twice daily to 100 mg once a day. At last follow-up of 7 years, 92% of patients achieved CHR, 50% achieved CCyR, and 44% of these reached MMR. Dasatinib induces more frequently hematologic adverse events (grade 3/4 neutropenia and thrombocytopenia) and nonhematological side effects, such as headache, diarrhea, and pleural/pericardic effusions.

Second-Generation Tyrosine Kinase Inhibitors for Newly Diagnosed Patients

After the results of dasatinib and nilotinib in imatinib resistant or intolerant patients, both were tested as single agent, or in phase III trials compared to imatinib, in newly diagnosed patients. Both are able to induce rapid CCyR and MMR, independently from risk category at baseline (evaluated by Sokal or Euro risk) and higher rates of complete molecular responses (considered as 4.5-log reduction or $MR^{4.5}$). Rapid achievement of responses reduces the rate of nonoptimal responders during the first years of treatment and reduces the rate of progression in BP. FDA has approved as possible treatment strategy in first line, both the drugs.

Third-Line Inhibitors

Bosutinib is 30 times more potent than imatinib, developed to be active against BCR-ABL1 signalling and Src family kinases with less activity against other receptors (c-KIT and PDGFR), active also against all imatinib-resistant mutations, with the exception of the T315I mutation. The results of a phase II trial in patients resistant or intolerant to imatinib were reported: CCyR rate was 50% and MMR rate 51%. Bosutinib was tested also in patients with newly diagnosed chronic-phase CML and compared to imatinib in a phase III study (BELA trial): the results of this trial showed that the drug allowed higher MMR rate compared to imatinib. The most frequent adverse events reported were diarrhoea, nausea, vomiting, and skin rash.

Ponatinib (AP24534) is an oral tyrosine kinase inhibitor for the treatment of CML and Ph-positive acute lymphoblastic leukemia (ALL). It is a multitargeted tyrosine-kinase inhibitor with a broad spectrum of action, active against all type of mutations, including T315I mutation. A phase II study, namely the PACE study, showed activity of ponatinib, in resistant or intolerant CML patients in different phases of disease and in Ph-positive ALL.

Allogeneic Bone Marrow Transplantion

Hematopoietic stem cell transplantation (HSCT) remains the only potential curative option for CML. In a period of 20 years, the estimated survival rate reported was 34% for HSCT from HLA-identical sibling. It has been reported that prior treatments with tyrosine kinase inhibitor do not affect negatively the outcome of HSCT. At the present time, the place of HSCT after imatinib treatment is debated. European LeukemiaNet recommendations indicated HSCT at diagnosis for patients presenting in AP or BP, even if a treatment with tyrosine kinase inhibitors is recommended, or after imatinib failure for patients who have experienced progression to AP/BP or carrying the T315I mutation after a second-line treatment. It is recommended that the search for a donor should be initiated at appropriate times, according to the stage of disease, response to treatment, and characteristics of the patient. Transplantation-related mortality ranges from 5% to 50% depending on factors including age, donor origin (related versus unrelated), degree of HLAmatching, host cytomegalovirus status, use of conditioning regimens, and institutional expertise. Potential risks of HSCT include the graft-versus-host disease (GVHD), life-threatening infections, risk of secondary malignancies, and possible relapses.

References

Baccarani M, Cortes J, Pane F, Niederwieser D, Saglio G, Apperley J, Cervantes F, Deininger M, Gratwohl A, Guilhot F, Hochhaus A, Horowitz M, Hughes T, Kantarjian H, Larson R, Radich J, Simonsson B, Silver RT, Goldman J, Hehlmann R (2009) Chronic myeloid leukemia: an update of concepts and management recommendations of European LeukemiaNet. J Clin Oncol 27:6041–6051

Breccia M, Alimena G (2011a) Activity and safety of dasatinib as second-line treatment or in newly diagnosed chronic phase chronic myeloid leukemia patients. BioDrugs 25:147–157

Breccia M, Alimena G (2011b) Nilotinib for the treatment of newly diagnosed Philadelphia chromosome-positive chronic myeloid leukemia: review of the latest clinical evidence. Clin Investig 1:707–719

Druker BJ, Talpaz M, Resta DJ et al (2001) Efficacy and safety of a specific inhibitor of the BCR-ABL tyrosine kinase in chronic myeloid leukemia. N Engl J Med 344:1031–1037

Hehlmann R, Hochhaus A, Baccarani M (2007) European LeukemiaNet. Chronic myeloid leukaemia. Lancet 370:342–350

Chronic Obstructive Pulmonary Disease and Lung Cancer

Juhayna Kassem Davis[1] and Ronald G. Crystal[2]
[1]Carolinas HealthCare System, Charlotte, NC, USA
[2]Division of Pulmonary and Critical Care Medicine, Weill Cornell Medical College, New York, NY, USA

Definition

Chronic obstructive pulmonary disease (COPD) and ▶ lung cancer have rising prevalence worldwide with estimates of significant increases in mortality over the next few decades. Both diseases are directly linked to cigarette smoking, environmental exposures and old age, and both cluster in families, suggesting genetic links (Genetic polymorphisms) to disease susceptibility. But irrespective of tobacco history (tobacco carcinogenesis; ▶ tobacco-related cancers) and environment, there is growing evidence supporting an increased incidence of lung cancer in individuals with COPD.

Primary lung cancer is usually divided into two broad classes, small-cell lung cancer and non-small-cell lung cancer (tobacco carcinogenesis; ▶ cancer causes and control; ▶ tobacco-related cancer), which includes adenocarcinoma, squamous cell carcinoma, and large-cell carcinoma.

Characteristics

COPD

COPD is a chronic disorder characterized by airflow limitation that may be accompanied by hyperreactivity but is not fully reversible and is usually progressive. The disease is primarily caused by cigarette smoking and, to a lesser extent, by other noxious particles and gases. Despite smoking cessation, once COPD is established, the disorder continues to progress, albeit at a slower rate. Although COPD is a lung disorder, it is also associated with significant systemic abnormalities.

COPD is a general category that includes chronic bronchitis, an inflammatory airway disorder characterized by a daily, productive cough for at least 3 months in two successive years, and emphysema, an abnormal, permanent airspace dilation with destruction of the alveolar walls without evidence of fibrosis. Most patients with COPD have components of both chronic bronchitis and emphysema, and some have superimposed asthma, with airway hyperactivity and reversible limitation to expiratory airflow.

Most cases of COPD start with abnormalities in the small airways, with epithelial changes, ▶ inflammation in the airway walls, and narrowing of the airway lumen, limiting airflow. As the disease progresses, the airflow limitation is manifested by abnormalities in spirometry, a physiologic test that measures inhaled and exhaled volumes of air independently and as a function of time. The most commonly used parameters relevant to COPD are the forced vital capacity (FVC, the volume of air that can be forcibly exhaled following maximal inspiration) and the forced expiratory volume in one second (FEV1, the volume of air that is exhaled in the first second of that same maneuver). A reduction in FEV1 directly correlates with the degree of airway disease from inflammation, fibrosis, or intraluminal exudates that characterize COPD. Following the use of inhaled bronchodilators, if the FEV1/FVC is <70% and the FEV1 is <80% of expected values, airflow limitation is present. The airflow limitation that characterizes COPD results from the intrinsic airway disease per se, as well as the loss of airway structural support from the destruction of the alveolar walls. The alveolar wall destruction also results in a reduction in the diffusing capacity for carbon monoxide and can be visualized by computerized tomography of the chest.

Lung Cancer

Lung cancer is the leading cause of cancer-related deaths worldwide. Like COPD, most (85–90%) of all lung cancers are a result of exposure to smoking. The risk of developing lung cancer, even after smoking cessation, persists for years. In the United States, the rate of lung cancer in former smokers is now equal to the rate in current smokers and is expected to increase even more in the next decades.

Link to Decline in FEV1

The incidence of both lung cancer and COPD increases with >20 pack-year smoking exposure. Smoking one pack of cigarettes per day can increase the normal decline in FEV1 from an expected 30 to 50–60 ml/year. Overall increases in mortality associated with COPD are directly associated with declining FEV1. The decrease in FEV1 associated with COPD is also directly linked to lung cancer, as well as to an increased risk of cardiovascular disease, including coronary heart disease and stroke.

COPD and Lung Cancer Associations

While COPD and lung cancer are both related to smoking, there is data from as early as the mid-1960s to support that the two diseases are even more closely related. Van Der Wal et al. were the first to demonstrate a high incidence of lung cancer in patients with "chronic nonspecific lung disease," now recognized as COPD. Goldstein et al. in 1968 reported a 32 times higher rate of lung cancer in patients with radiographic evidence of bullous emphysema in hospitalized patients with cancer as compared to other hospitalized patients who were used as controls. These data were supported by several subsequent studies in the 1970s and 1980s.

In 1976, Davis et al. suggested that a reduced FEV1 itself might be an independent risk factor for the development of lung cancer, demonstrating a four to fivefold increase in lung cancer in their patients with COPD as compared to lung cancer rates in previously reported series of smokers without COPD. A prospective study by Skillrud et al., comparing 113 people with COPD to 113 matched controls without COPD followed over a 10-year period, found that all-cause mortality increased with COPD, a decreased FEV1 was directly linked to a decreased time to death from any cause, and,

despite shorter survival data for patients with COPD, there was a clear, increased risk for the development of lung cancer in the COPD population. Overall, the risk of developing lung cancer in patients with airway obstruction is 4.4 times greater than in those without obstruction to expiratory airflow.

In 1997, a retrospective review of data collected from two large prospective study groups, the Intermittent Positive Pressure Breathing Trial sponsored by the National Heart, Lung, and Blood Institute and the Johns Hopkins Lung Project, was conducted by the National Cancer Institute's Cooperative Early Lung Cancer Detection Program. The goal was to evaluate the association between the degree of airway obstruction and the development of lung cancer. This study demonstrated that the frequency of lung cancer was proportional to the degree of airflow obstruction and that the risk of lung cancer was more closely linked to a decline in FEV1 than to older age or degree of tobacco use.

The 1994 study by Islam et al. reviewed prospectively data collected over a 28-year period in a community-based study in Tecumseh, Michigan, in 3,900 subjects. They determined that the initial FEV1 and the rate of decline in FEV1 were each independent predictors of lung cancer development. When loss of FEV1 reached 100 ml/year, the risk of lung cancer reached as high as 30 times the rate seen in matched controls.

A study by Mannino, published in 1993, reviewed patients in the First National Health and Nutrition Examination Survey database who had at least a 22-year follow-up. They demonstrated that moderate to severe obstructive disease was associated with a higher incidence of lung cancer and that there was no difference in rates of lung cancer when comparing current to former smokers.

Possible Mechanisms of the COPD: Lung Cancer Risk

The relationship between reduced FEV1 and the development of primary lung cancer is not clear, although many theories have been advanced. It is generally accepted that both disorders result from a combination of genetic and environmental factors (▶ Cancer Causes and Control; ▶ Cancer Epidemiology). However, while the link to smoking is clear, the commonalities of the smoke components that cause COPD and lung cancer are not clear nor are the genetic differences linking a susceptibility to both diseases.

One gene of common interest is ▶ vascular endothelial growth factor (VEGF). While COPD has been theorized to be associated with decreased availability of VEGF leading to capillary ▶ apoptosis in the lung, lung cancer, as with other cancers, is associated with an increased expression of VEGF supporting ingrowth of capillaries into the developing tumor. Linkage studies in both humans and mice have suggested that allelic loss in some regions of chromosomes 6q and 12q is associated with lung cancers as well as with COPD. Finally, excess amounts of ▶ matrix metalloproteinases (MMPs), enzymes that degrade ▶ extracellular matrix, are associated with a decline in lung function, as well as an increased risk of lung cancer.

Chronic inflammation (▶ Inflammation) has long been linked to cancer in many organs, as evidenced by the development of esophageal adenocarcinoma (▶ esophageal cancer) following chronic gastric reflux. This concept has led to the theory that COPD and lung cancer risks increase in the face of chronic inflammation. Consistent with that concept, cigarette smoking (▶ Tobacco Carcinogenesis; Tobacco-Related Cancers) has been shown to cause sustained changes in gene expression of respiratory epithelial cells. Some of these changes have been noted to persist, regardless of smoking cessation, for decades. Some of the inflammatory changes are likely associated with the development of both COPD and lung cancer. One example is the loss of the tumor suppressor gene ▶ TP53, the protein it encodes generally inhibits inflammation. Allelic loss of TP53 is known to lead to inflammatory responses including increases in ▶ nuclear factor-κB, a transcription factor linked to both COPD and cancer, via inflammatory pathways. Other ▶ oncogenes and other ▶ tumor suppressor genes have also been suggested as key risk factors to the development of COPD and lung cancer.

Summary

Many challenging questions remain unanswered regarding COPD, lung cancer, and the relationships between them. More information on their associations with environmental exposures (including but not limited to tobacco), genetic susceptibility, and inflammatory processes is still needed. It is evident that these connections are not simply based on tobacco use and that there is a genetic predisposition to the development of each. While smoking cessation will clearly reduce the incidence of COPD and lung cancer, it will not eliminate either disease for many decades to come. Further well-formulated studies are required to continue to evaluate the connections between environmental exposures, inflammation, genetic expression, and the development of COPD and lung cancer, which remain two pulmonary diseases with the highest morbidity and mortality in the world.

Cross-References

▶ Extracellular Matrix Remodeling

References

Buist AS et al (2007) Global strategy for the diagnosis, management, and prevention of chronic obstructive pulmonary disease. Global Initiative for Chronic Obstructive Lung Disease. http://www.goldcopd.org/Guidelineitem.asp?|1=2%26|2=1%26intld=989. Last accessed 26 Feb 2008

Harvey B-G, Heguy A, Leopold PL et al (2007) Modification of gene expression of the small airway epithelium in response to cigarette smoking. J Mol Med 86:39–53

Islam SS, Schottenfeld D (1994) Declining FEV1 and chronic productive cough in cigarette smokers: a 25-year prospective study of lung cancer incidence in Tecumseh, Michigan. Cancer Epidemiol Biomarkers Prev 3:289–296

Lundback B, Lindberg A, Lindstrom M, et al (2003) Not 15 but 50% of smokers develop COPD? Report from the Obstructive Lung Disease in Northern Sweden Studies. Respir Med 97:115–122

Mannino DM, Aguayo SM, Petty TL et al (2003) Low lung function and incident lung cancer in the United States. Arch Intern Med 163:1475–1480

Skillrud DM, Offord KP, Miller RD (1986) Higher risk of lung cancer in chronic obstructive pulmonary disease. A prospective, matched, controlled study. Ann Intern Med 105:503–507

Ciliary Body Melanoma

▶ Uveal Melanoma

CIN

▶ Chromosomal Instability

Cip1

▶ p21

Cip-Interacting Zinc Finger Protein 1 Ciz1

Petra Den Hollander[1] and Rakesh Kumar[2]
[1]Department of Translational Molecular
Pathology, The University of Texas MD
Anderson Cancer Center, Houston, TX, USA
[2]Department of Biochemistry and Molecular
Medicine, George Washington University,
Washington DC, USA

Definition

Ciz1 is a member of the Matrin 3 protein family and also called NP94 (Nuclear Protein 94). At the N-terminal region, Ciz1 contains poly-glutamine repeats and glutamine rich regions. Additionally, the C-terminus contains three zinc-finger motifs, an acidic region and a matrin 3-homologous domain 3 (MH3 domain), which shows 40.4% and 37.7% identity (55.8% and 64.2% similarity) to NP220 and matrin 3 (two other members of the Matrin 3 family), respectively (Fig. 1). Ciz1 is expressed in a wide variety of tissues, with highest expression in pancreas, testis, kidney, and brain (with the highest expression in cerebellum) and is mainly localized in the nucleus. Ciz1 is expressed in a number of cell lines from multiple organ systems. None of the genes of Yeast, *C. elegans* or *Drosophila* showed any overall structural similarities to Ciz1, suggesting that Ciz1 is a unique protein only found in vertebrates.

Characteristics

Ciz1 directly interacts with ▶ p21^{WAF1} and is predominantly located in the nucleus. However, upon co-overexpression of Ciz1 and p21^{WAF1}, an enhanced cytoplasmic localization of both proteins was detected. Ciz1 is a DNA-binding protein, and its DNA consensus sequence was determined by a modified selected and amplified binding (SAAB) sequence method (ARYSR(0–2) YYAC). Ciz1 is a binding partner of Dynein Light Chain 1 (DLC1), shown by using a modified ▶ proteomics technique. Ciz1 influences the cell cycle progression at the G_1-S transition by affecting the kinase activity of Cdk2. A reduced localization of p21^{WAF1} in nuclei of DLC1 overexpressing cells strengthened the hypothesis that Ciz1 (together with DLC1) is important for the sequestration of p21^{WAF1} in the cytoplasm, which will then release the repression of the Cdk2 kinase complex and induce the G_1-S transition. Also, cell-free experiments have demonstrated that Ciz1 has a role in mammalian DNA replication. The addition of Ciz1 protein increases the number of nuclei that initiate DNA replication, and when mutating the potential ▶ cyclin-dependent kinase (Cdk) phosphorylation sites, Ciz1 functions were compromised in vitro. Ciz1 co-localizes with PCNA in foci in the nucleus, and Ciz1-depleted cells are unable to replicate their DNA.

Besides significant role in the cell cycle progression, Ciz1 is responsible for potentiating the transactivation activity of the ▶ Estrogen Receptor alpha (ER). Ciz1 is a coregulator of ER by enhancing ER transactivation activity and recruitment to target gene chromatin. Ciz1 induces the hypersensitivity of ▶ breast cancer cells to ▶ estradiol and induces the expression of ER target gene ▶ Cyclin D1 at a femto-molar dose of estrogen, with likely downstream effects on G1

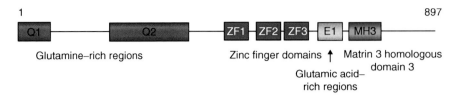

Cip-Interacting Zinc Finger Protein 1 Ciz1, Fig. 1 Schematic representation of protein domains in Ciz1

progression and DNA replication. In addition, overexpression of Ciz1 promotes the growth-rate, anchorage-independence, and tumorigenesis of breast cancer cells in a xenograft model. Interestingly, Ciz1 is also an estrogen-inducible gene, suggesting that overexpression of Ciz1 in breast cancer cells may support hormone hypersensitivity.

Ciz1 has a functional MH3 domain since its C-terminal region is essential for the anchoring of Ciz1 to the nuclear matrix, and this immobilization is cell cycle dependent and occurs probably during late G1 or early S phase. Although the C-terminus is sufficient for the immobilization of the Ciz1 protein to the nuclear matrix, the N-terminal region is required for the focal localization of Ciz1 in the nucleus. Using biotinylated-dUTP and GFP-tagged Ciz1, Ciz1 was shown to be colocalized with the newly synthesized DNA. Thus, Ciz1 has multiple functions in cell cycle progression and cell proliferation control and has a potential role in tumorigenesis.

Cross-References

▶ Breast Cancer
▶ Cyclin D
▶ Cyclin-Dependent Kinases
▶ Estradiol
▶ p21
▶ Proteomics

References

Coverley D, Marr J, Ainscough J (2005) Ciz1 promotes mammalian DNA replication. J Cell Sci 118:101–112

den Hollander P, Kumar R (2006) Dynein light chain 1 contributes to cell cycle progression by increasing cyclin-dependent kinase 2 activity in estrogen-stimulated cells. Cancer Res 66:5941–5949

den Hollander P, Rayala SK, Coverley D et al (2006) Ciz1, a novel DNA-binding coactivator of the estrogen receptor alpha, confers hypersensitivity to estrogen action. Cancer Res 66:11021–11029

Mitsui K, Matsumoto A, Ohtsuka S et al (1999) Cloning and characterization of a novel p21(CiP1/Waf1) interacting zinc finger protein, Ciz1. Biochem Biophys Res Commun 264:457–464

Warder DE, Keherly MJ (2003) Ciz1, Cip1 interacting zinc finger protein 1 binds the consensus DNA sequence ARYSR(0–2)YYAC. J Biomed Sci 10:406–417

See Also

(2012) Cell Cycle. In: Schwab M (ed) Encyclopedia of Cancer, 3rd edn. Springer Berlin Heidelberg, p 737. doi:10.1007/978-3-642-16483-5_994

Circadian Clock Induction

Francis Lévi
Warwick Medical School, University of Warwick, Coventry, UK

Definition

The circadian timing system efficiently orchestrates the physiology of living organisms to match environmental or imposed 24-h cycles. Most cells in the brain and peripheral tissues contain a molecular clock consisting of at least 12 specific clock genes in mammals. This molecular clock rhythmically controls the transcriptional activity of nearly 10% of the genome, ~10% of which are proliferation-related genes, over the 24 h. Among the 12 genes which constitute the molecular clock, *Per2*, *Bmal1*, and *Rev-erbα* play a central role. Thus, a null mutation in these genes results in profound alterations of the circadian phenotype. The intracellular clock mechanisms involve interacting positive and negative transcriptional feedback loops that drive recurrent rhythms in the RNA levels of these key components. High levels of Bmal1 mRNA and protein promote the formation of BMAL1:CLOCK heterodimers that bind to the ▶ E-box sequences in the promoter of clock genes *Per2* and *Rev-erbα* and activate their transcription. In turn, Rev-erbα negatively regulates Bmal1 transcription, while PER2:CRY1 complexes inhibit the transcription of their own genes by interfering with CLOCK:BMAL1. Phosphorylation of PER and CRY proteins through casein kinase (CK) Iδ/ε activity plays a key role in the regulation of clock proteins degradation and length of precise circadian period.

The circadian organization of drug metabolism pathways as well as ▶ cell cycle, ▶ DNA repair, and ▶ apoptosis is responsible for dosing time dependencies in drug ▶ pharmacokinetics and pharmacodynamics. As a result, about 24-h changes in anticancer drug tolerability and efficacy call for chronotherapeutics, i.e., the delivery of anticancer treatments according to circadian (and other) rhythms. Dedicated ▶ drug delivery systems can actually administer chemotherapeutic drugs at specific optimal times and/or according to an optimal circadian pattern to cancer patients.

Conversely, destruction of the circadian pacemaker in the brain, iterative alterations of the environmental cycles, or clock gene mutations usually produce severe modifications within the circadian timing system that result in the uncoupling of coordinated biological functions. Such circadian disruptions can influence cancer processes.

Characteristics

Relevance of Circadian Disruption for Cancer Processes

The relative risk of developing breast, colorectal, or prostate cancer is enhanced 50–300% in populations exposed to prolonged shift work, frequent transmeridian flights, or chronic light exposure at night. These latter conditions profoundly alter the circadian timing system and the downstream control it exerts on cellular proliferation, DNA repair, apoptosis, and metabolism. These circadian alterations can consist of a decrease in rhythm amplitude, a phase shift and/or a modification, or a suppression of the circadian period. Circadian physiology monitoring reveals relevant rhythm alterations in nearly one third of patients with metastatic cancer. In patients with advanced or metastatic cancer, disrupted circadian rhythms in rest-activity or cortisol secretion are associated with poor quality of life and increased risk of an earlier death. In experimental models, the growth rate of transplanted tumors is accelerated with the disruption of the circadian timing system through the ablation of the hypothalamic circadian pacemaker or chronic jet lag produced by 8-h advances of light onset every 2 days.

Circadian Clock Control of Cell Cycle

At least three molecular mechanisms link molecular circadian clock with the cycling of cell division. The molecular clock controls Wee1 transcription through an E-box-mediated mechanism. WEE1 negatively controls the activity of CDK1/cyclin B1, which facilitates the ▶ G2/M transition (Fig. 1). In addition, the BMAL1:CLOCK heterodimers repress c-Myc transcription through E-box-mediated reactions in the *c-Myc* gene P1 promoter, and PER 2 can suppress c-Myc expression indirectly. In general, knocking down the molecular clock modifies transcription patterns of genes involved in the cell cycle regulation. This can translate into genomic instability, thus favoring malignant ▶ progression or growth. Both *Per1* and *Per2* as well as possibly other clock genes also control DNA repair through interactions with ▶ *ATM* and *mdm2*.

Circadian Disruption in Cancer Tissues

The circadian expression pattern of core clock genes *Per2*, *Bmal1*, and *Rev-erbα* can be severely altered in experimental tumors in mice. In experimental Glasgow osteosarcoma, a transplantable mouse tumor with a doubling time of 2–3 days, the transcriptional rhythms in these three clock

Circadian Clock Induction, Fig. 1 Schematic representation of the interactions between the circadian timing system and the cell division cycle. *SCN* suprachiasmatic nuclei, the main circadian pacemaker in the hypothalamus

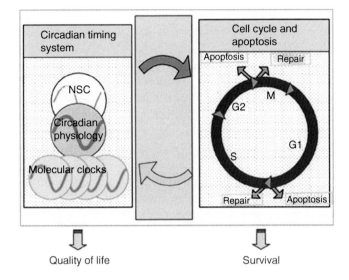

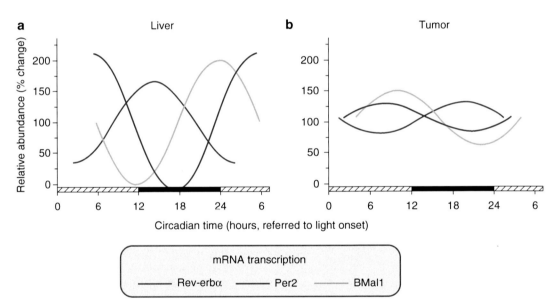

Circadian Clock Induction, Fig. 2 Cosinor-based model of molecular clock in mouse healthy liver (**a**) or tumor Glasgow osteosarcoma (**b**)

genes displayed altered amplitudes and phase at an early stage of growth that subsequently evolved toward arhythmicity (Fig. 2).

Physiologic Clock

The adjustment of a cosine curve with a 24-h period to mRNA expression of Rev-erbα (blue line), Per2 (red line), and Bmal1 (green line) allows to build a model of the circadian clock in healthy liver of B6D2F$_1$ mice. This model establishes physiological dynamic relations with maxima occurring near Circadian time 6 (CT6) for Rev-erbα, CT15 for Per2, and CT23 for Bmal1.

Tumor Clock

The same method is applied to clock genes data for control tumors. No clock gene transcription rhythm is validated.

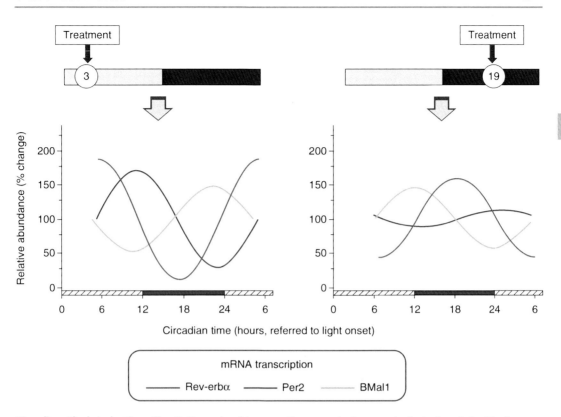

Circadian Clock Induction, Fig. 3 Example of tumor circadian clock induction with CDKI seliciclib as illustrated with cosinor-based model of molecular clock. Physiologic phase relations between expression patterns of the three core clock genes, similar to those in healthy liver, are induced with seliciclib treatment in the early rest span (at Zeitgeber Time 3, ZT3), but not if the drug is given during darkness (at ZT19)

The ablated rhythms in clock gene expression in tumor tissue could result from an altered synchronization of the molecular clocks in each malignant cell or from an impairment of the molecular clock within each individual cancer cell. The first possibility would happen, for instance, if the internal rhythm in each cell beats with a period that differs by minutes or hours from those of its neighbors. Alternatively, clock gene defects could result in nonfunctional molecular clocks through deletions, mutations, or ▶ methylation of promoter region. In several human cancers, decreased expressions of the Per1, Per2, or Per3 genes were found at a single time point, in comparison with reference tissues.

Circadian Clock Induction

Peripheral synchronizers, such as feeding schedules, and drugs, such as ▶ cyclin-dependent kinase inhibitor (CDKI) seliciclib, can induce rhythmic clock gene expression patterns in malignant tumors with otherwise disrupted or uncoordinated molecular clocks. Indeed, exposure of tumor-bearing mice to chronic jet lag severely disrupts the already altered temporal patterns of core clock gene expressions in tumor. Programming food availability to 6-h daily induces near normal molecular clock in the tumor and slows down malignant growth. The administration of seliciclib also induces near normal molecular clock in mice dosed in the early light span, but has no such effect if treatment is applied during darkness (Fig. 3). Feeding or pharmacologic induction of the tumor clock is associated with antitumor effect. Both clock induction and antitumor effects depend upon the circadian time of application.

Circadian clock induction improves control of ▶ G2/M gating through an enhancement of Wee1 transcription, a gene that is unidirectionally controlled by CLOCK-BMAL1. In the case of seliciclib, the induction of the molecular clock involves inhibition of CK1δ/ε, a key determinant of circadian period. In turn, inhibition of CK1δ/ε impairs PER2 degradation and nuclear translocation and results in increased *Bmal1* transcription. Since highly coordinated sequential transcription is a major mechanism of circadian rhythms, programmed food availability or daily seliciclib act as strong resetters of tumor cells that have lost synchrony in functional clocks, through transient inhibition of CK1δ/ε or other pathways within permissive time windows.

In conclusion, while malignant tumors tend to display disrupted molecular clocks and to lose circadian coordination, feeding schedules and drugs like CDKI seliciclib can slow down tumor growth through circadian clock induction. The mechanisms underlying these favorable effects involve the cross talk between the circadian clock and the cell cycle, two biological oscillators whose interactions represent a new relevant dynamic target for cancer therapeutics. Furthermore, the extent of circadian clock induction and that of antitumor activity depend upon the circadian time of drug administration, revealing the relevance of chronotherapeutics of CDKI for improving cancer control.

Cross-References

▶ ATM Protein
▶ Cell Cycle Checkpoint
▶ Drug Delivery Systems for Cancer Treatment
▶ Pharmacokinetics and Pharmacodynamics in Drug Development
▶ Repair of DNA

References

Filipski E, Innominato PF, Wu MW et al (2005) Effects of light and food schedules on liver and tumor molecular clocks. J Natl Cancer Inst 97:507–517
Iurisci I, Filipski E, Reinhardt J et al (2006) Improved tumor control through circadian clock induction by seliciclib, a cyclin-dependent kinase inhibitor. Cancer Res 66(22):10720–10728
Lévi F, Schibler U (2007) Circadian rhythms: mechanisms and therapeutic implications. Annu Rev Pharmacol Toxicol 47:593–628
Mormont MC, Lévi F (2003) Cancer Chronotherapy: principles, applications and perspectives. Cancer 97:155–169

Circulating Cancer Cells (CCC)

▶ Circulating Tumor Cells

Circulating Nucleic Acids

Michael Fleischhacker
Universitätsklinikum Halle (Saale), Klinik für Innere Medizin I, Schwerpunkt Pneumologie, Halle (Saale), Germany

Synonyms

Cell-free circulating nucleic acids (cfDNA); Cell-free nucleic acids in plasma and serum (CNAPS), circulating tumor DNA (ctDNA); CNAs; Extracellular nucleic acids

Definition

Circulating nucleic acids (CNAs), i.e., DNA and RNA, are isolated from cell-free plasma/serum and other circulating body fluids like lymphatic fluid. For nucleic acids obtained from other body fluids like liquor, ascites, milk, bronchial lavage fluids, urine, stool, or from cell-free supernatants of in vitro cultivated cells, the terms extracellular or cell-free nucleic acids are better suited. CNAs are found in animals, most of these nucleic acids seem to come from dead cells (▶ apoptosis, ▶ necrosis), but there are probably also mechanisms leading to an active release.

Characteristics

Basic Aspects

The presence of cfDNA in plasma samples of humans was reported by Mandel and Metais in 1948. Leon et al. were the first to describe a relationship between CNAs in cancer patients and their clinical parameters, and the group of Drs. Stroun and Anker clearly demonstrated that at least part of CNAs in cancer patients results from the tumor itself (Stroun and Anker 2005). The amount of CNAs in healthy subjects is very low (~1–20 ng DNA/ml plasma), while in patients with a benign or a malignant disease a higher quantity might be found (up to several orders of magnitude more).

CNAs are not naked but are released in complexes that make them resistant against attacks of nucleolytic enzymes. So far all cellular nucleotide sequences have also been detected in plasma/serum and other cell-free body fluids. Therefore, CNAs are probably a mirror of all cellular nucleic acids. The majority of cell-free circulating DNA molecules are of low molecular weight (~150–500 bp) but fragments of up to > 10 kb can also be detected.

The half-life of CNAs seems to be very short, in the range of minutes. The clearance mechanisms in animals, including humans, are probably mainly an excretion into urine and stool or an uptake by the liver. While DNA is a rather stable molecule, RNA is much more fragile and prone to degradation by the ubiquitous presence of RNA degrading enzymes. It is known that the serum RNAse concentration is elevated in cancer patients, but nevertheless it is possible to isolate and quantify EBV-associated RNA, mRNA and microRNAs (miRNA) coding for a variety of different genes from plasma/serum of tumor patients.

In plants, the traffic of DNA, and RNA, plays an important role in cell–cell communication (like spreading of a primary crown gall, a plant tumor, to "metastatic sites"), and in the animal kingdom CNAs might be involved in horizontal gene transfer (HGT).

When the genetic material from cells obtained with a biopsy and extracellular DNA are sequenced, frequently discordant results are found in the different specimens. This is due to the fact that extracellular nucleic acids are a mix of normal and tumor-derived nucleic acids. The molecular genetic characterization of human solid tumors using newly developed methods like deep sequencing demonstrated that the cells obtained by a biopsy are remarkably heterogenous in their genetic makeup (Gerlinger et al. 2012). In addition, the majority of solid tumors are genetically heterogeneous and contain subclones that harbor different genetic alterations. Since only part of the material is analyzed in a routine pathological examination of a biopsy, it is assumed that this leads to a loss of information about the genetic composition of the tumor. To solve this problem the method of „liquid biopsy" (or alternatively „liquid profiling") was established which comprises the analysis of genetic alterations found in CNAs and/or in ▶ circulating tumor cells. The rationale for this approach is the assumption that the characterization of extracellular nucleic acids gives a complete picture of all alterations present in a tumor and its metastases.

Clinical Aspects

When it had been demonstrated unambiguously that part of the extracellular DNA in cancer patients was released from tumor cells it was also shown that in these patients a higher concentration of plasma DNA could be measured. This lead to the concept that an increased plasma DNA concentration might be a useful and universal tumor biomarker. Later studies demonstrated that this increase in the amount of CNAs in diseased patients is very unspecific, not related to any malignancy in particular, and can be observed in trauma patients as well as in patients with chronic inflammatory conditions like ulcerative colitis.

It was demonstrated that the DNA integrity index of cfDNA (i.e. the amount of short vs. long DNA fragments) might be a useful biomarker for the identification of patients suffering from a tumor of the breast or the liver (Wang et al. 2003; Jiang et al. 2015), however these results could not be corroberated when lung and prostate cancer patients were examined. Several laboratories reported on the association of the DNA integrity index and clinical parameters like prognosis or survival of tumor patients, but these analyses yielded conflicting results. One of the reasons for these discordant results is the fact that in most studies the control group consisted of healthy subjects instead of patients with a benign disease affecting the same organ or patients with premalignant conditions. The tumor-associated genetic alterations found in CNAs include DNA ▶ amplifications, inversions, point mutations (as in ▶ oncogenes), deletion of a ▶ tumor suppressor gene, microsatellite alterations, hypermethylation of ▶ CpG islands in promoter regions, mutations in ▶ mitochondrial DNA, the presence of viral nucleic acids, and the increased amount of mRNA and microRNA for genes that are overexpressed in tumors (Fleischhacker and Schmidt 2007). A relationship between an increased total quantity of cell-free DNA and a higher concentration of DNA containing tumor-associated alterations in plasma/serum and the presence of a malignancy or other clinical parameters (tumor staging, tumor grading, size of the tumor, presence/absence of ▶ metastases, etc.) respectively, has previously been established.

The detection of mutations, frequently found in tumor cells (such as in TP53 gene family, KRAS, NRAS, APC, BRAF), has been described and in some cases a correlation between the presence of KRAS and TP53 mutations in CNAs and the survival of the affected patients could be established. The detection of a ▶ microsatellite instability (i.e., the loss of heterozygosity (LOH) or band shifts) in CNAs has also been reported. The presence of cell-free viral nucleic acids was reported before CNAs became popular as new tumor markers, and the strong correlation of a virus infection and the development of certain types of cancer make the detection of viral nucleic acids an interesting field. For example, there are strong correlations between the development of a nonnasopharyngeal head and neck carcinoma (NNHNC) and an infection with the ▶ Epstein-Barr virus (EBV); between ▶ cervical cancer and an infection with the human papilloma virus (HPV), especially types 16 and 18; between ▶ Hodgkin disease and other lymphoproliferative diseases and EBV infection. It has been shown that determining the quantity of circulating viral nucleic acids before and/or after therapy might be a good marker to evaluate the response to therapy or to predict an overall and disease-free survival. Also the presence of hypermethylated sequences in extracellular DNA has been demonstrated. These hypermethylated sequences are frequently located around the transcription start sites of human genes (promoter regions). Such a promoter hypermethylation is commonly found in tumor suppressor genes, leading to a shutdown of these genes (▶ epigenetic gene silencing). Developed

methods (mostly based on real-time PCR) show a high test sensitivity - one single hypermethylated allele can be detected against a background of up to 10,000 unmethylated alleles - and also allow a real-time quantification of methylated alleles. When extracellular DNA was examined for the presence of methylated sequences as a biomarker for a malignant disease, these alterations were found in APC, RASSF1A, p15, and p16INK4A genes. The quantification of methylated SHOX2 DNA in plasma from ▶ lung cancer patients was established as a sensitive and specific biomarker for therapy monitoring in these patients.

Targeted therapies are based on the observation that cancer cells are genetically different from normal cells leading to the development of substances (small molecules, monoclonal antibodies, etc.) that attack the tumor cells directly and in a more specific way than standard ▶ chemotherapy does. When ▶ non-small cell lung cancer patients are diagnosed they are screened for the presence of an activating mutation of the ▶ Epidermal Growth Factor Receptor gene (EGFR). In case of an EGFR mutation they will be treated with a variety of different ▶ tyrosine kinase inhibitors (TKIs). Patients with metastatic ▶ melanoma harboring a ▶ BRAF somatic alteration are treated with BRaf protein inhibitors targeting the ▶ MAP kinase signaling pathway in melanoma. In both diseases it has been demonstrated that the genetic analysis of extracellular DNA can detect these activating mutations at the time of diagnosis. Additionally, it has been shown that the use of sensitive methods like a digital PCR for follow-up analysis makes it possible to not only quantify the number of tumor-associated alterations (like mutations in the EGFR or BRAF genes) but to relate these results directly to the clinical parameters (such as therapy response, remission, or relapse detection). When patients receiving an anti-tumor therapy were examined with a digital PCR, the development of a drug resistance could be detected in their plasma DNA before a clinical relapse became visible.

In any case a thorough validation of the potential markers is a conditio sine qua non, since some of the tumor-associated genetic alterations are not only found in the tumor itself, but have also been detected in histologically normal tissue of cancer patients, in people having a certain risk for the development of a tumor, or even in cells obtained from presumably healthy subjects. This was demonstrated for the p53 tumor suppressor gene whose alterations are found in progenitor lesions of the lung, esophagus, head, and neck, and colon. With the advent of very sensitive and specific techniques like deep sequencing or digital PCR, which is able to count single DNA molecules absolutely (i.e. there is no standard necessary to compare with), it is essential to set threshold values in order to detect alterations that are reliable indicators for the presence of a tumor. Some of the tumor-associated changes seen in cellular DNA were also detected in the plasma DNA of non-tumor patients. These include microsatellite alterations in plasma DNA from patients with benign respiratory diseases and a tumor-associated gene promoter methylation in cell-free plasma DNA from women who have never smoked. In addition, patients with long-standing ulcerative colitis frequently test positive for mutations of the *KRAS* and *TP53* genes even if there are no signs of a tumor.

Finally, some important technical issues need to be resolved like determining the influence of pre-analytical factors on the result, standardization of the different methods, choice of an optimal marker panel, etc. before the analysis of CNAs can be a useful and clinically meaningful tool.

Cross-References

- ► Amplification
- ► Apoptosis
- ► *BRAF* Somatic Alterations
- ► Cervical Cancers
- ► Chemotherapy
- ► Circulating Tumor Cells
- ► CpG Islands
- ► Epidermal Growth Factor Receptor
- ► Epigenetic Gene Silencing
- ► Epstein-Barr Virus
- ► Hodgkin Disease
- ► Lung Cancer
- ► MAP Kinase
- ► Metastasis
- ► Microsatellite Instability
- ► Mitochondrial DNA
- ► Necrosis

- ► Non-Small-Cell Lung Cancer
- ► Oncogene
- ► Tyrosine Kinase Inhibitors

References

Fleischhacker M, Schmidt B (2007) Circulating nucleic acids (CNAs) and cancer – a survey. Biochim Biophys Acta 1775(1):181–232

Gerlinger M et al (2012) Intratumor heterogeneity and branched evolution revealed by multiregion sequencing. N Engl J Med 366(10):883–892

Jiang P et al (2015) Lengthening and shortening of plasma DNA in hepatocellular carcinoma patients. Proc Natl Acad Sci USA 112(11):E1317–1325

Mandel P, Metais P (1948) Les acides nucléiques du plasma sanguin chez l'homme. C R Seances Soc Biol Fil 142(3-4):241–243

Stroun M, Anker P (2005) Circulating DNA in higher organisms: cancer detection brings back to life an ignored phenomenon. Cell Mol Biol (Noisy-le-Grand) 51(8):767–774

Wang BG et al (2003) Increased plasma DNA integrity in cancer patients. Cancer Res 63(14):3966–3968

Circulating Tumor Cells

Patrizia Paterlini-Bréchot
Faculté de Médecine Necker Enfants Malades, INSERM Unit 1151, Team 13, Paris, France

Synonyms

Circulating cancer cells (CCC)

Keywords

Tumor invasion; Epithelial to mesenchymal transition; Metastases formation; Liquid biopsy; Tumor cell plasticity

Definition

Circulating tumor cells (CTC) are tumor cells spread in blood and/or lymphatic vessels from solid tumors, thus including all types of tumor

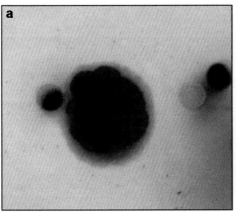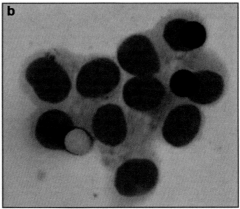

Circulating Tumor Cells, Fig. 1 CTC and CTM enriched by ISET and diagnosed by cytopathology. (**a**) CTC from a patient with mesothelioma (May–Grünwald–Giemsa staining, 100×). (**b**) CTM from a patient with breast cancer (May–Grünwald–Giemsa staining, 100×)

cells except those derived from leukemia and lymphoma. CTC may also circulate as aggregated tumor cells, which are defined as circulating tumor microemboli or "collective tumor cell migration (CTM)" (Fig. 1).

Characteristics

Tumor cells may circulate in blood spontaneously, i.e., because of their invasive capabilities or for other causes of cell spreading. Spontaneous circulation of tumor cells represents the early hallmark of the invasive behavior of a proportion of cancer cells and the first step of the process leading to the formation of ► metastases, which are known to account for 90% of cancer-related morbidity and mortality. Nonspontaneous circulation of tumor cells may derive from iatrogenic invasive procedures (biopsy, surgical intervention, etc.), tumor compression, and tumor inflammation.

The process by which tumor cells spreading from solid tumors give rise to metastases includes the following steps (Fig. 2): tumor cell growth involving genetic and epigenetic changes and tumor-induced microenvironment reprogramming (Meseure et al. 2014), ► angiogenesis, tumor cell detachment, ► epithelial to mesenchymal transition (EMT), motility, intravasation, survival in vessels and embolization, collective tumor cell migration (CTM), possible extravasation,

mesenchymal to epithelial transition (► MET), formation of ► micrometastases, and growth of macrometastases. Epithelial–mesenchymal cell plasticity is thought to play a central role in cancer progression, generation of cancer stem cells (CSC), and metastasis formation (Ye and Weinberg 2015).

Growing cells rapidly outstrip the supply of nutrients and oxygen and suffer from stress and hypoxia. Hypoxia-inducible factor (HIF), which mediates the transcriptional response to hypoxia, is a strong promoter of tumor growth and ► invasion and controls angiogenesis via two key angiogenic factors (VEGF-A and angiopoietin-2). Hypoxia determines cell necrosis and release of inflammatory mediators such as cytokines and ► chemokines which recruit, among other cells, leukocytes and ► macrophages. These, in turn, stimulate angiogenesis, extracellular matrix breakdown, and tumor cell motility (Noman et al. 2014). Local production of basic fibroblast growth factor (bFGF), epidermal growth factor (EGF), hepatocyte growth factor (HGF), and transforming growth factor beta (TGF-beta) mediates the control of tumor cell survival/► apoptosis balance and of E-cadherin downregulation leading to reduced cell adhesion and increased tumor cell invasiveness.

Hypoxia, acting through LOX induction and Snail activation, leads to E-cadherin repression, a crucial feature of the EMT. Furthermore, platelets may induce CTC to undergo EMT (Meseure

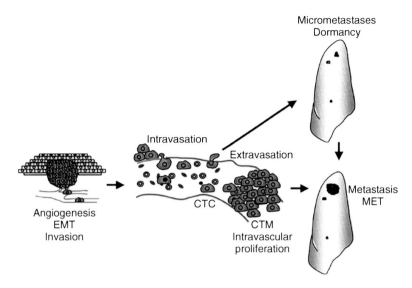

Circulating Tumor Cells, Fig. 2 Main steps leading to development of metastases (From Paterlini-Brechot 2007). Growing tumor cells outstrip oxygen supply and activate angiogenesis. Invading tumor cells undergo the phenotype switch "epithelial to mesenchymal transition (EMT)": they progressively lose epithelial antigens, acquire mesenchymal antigens, and motile propensities (like fibroblasts). After entering blood vessels (intravasation), circulating tumor cells (CTC) undergo apoptosis or circulate as isolated CTC. After extravasation to distant organs, CTC remain as dormant solitary cells or undergo limited proliferation (micrometastases). Unrestrained CTC proliferation gives rise to metastases, via phenotype reversion "mesenchymal to epithelial transition (MET)" and angiogenesis. Circulating tumor microemboli (CTM) represent "collective tumor cell migration" of tumor cells. They cannot extravasate, but arrest in capillaries and proliferate, rupturing the capillary walls and giving rise to metastases

et al. 2014; Ye and Weinberg 2015). During EMT, Twist may need to activate antiapoptotic programs in order to allow epithelial cells to convert to a mesenchymal fate while avoiding ▶ anoikis.

It is noteworthy that, in the journey of the tumor cells from the tumor tissue to the bloodstream and to the metastatic site, the cross talk of tumor cells with the microenvironment of the "primary" tissue, of blood, and of the metastatic tissue and the plasticity of tumor cells, i.e., their ability to shift from a differentiated to an undifferentiated phenotype and vice versa, are thought to play a central role in metastasis formation (Meseure et al. 2014; Ye and Weinberg 2015; Noman et al. 2014).

Epithelial cancer cells have very low survival rates in circulation. Animal studies, in which tumor cells are directly introduced into the systemic circulation, have established that approximately 1/40 CTC give rise to micrometastases and only approximately 0.01% proliferate into macrometastasis (Luzzi et al. 1998). The fate of

intravasated tumor cells includes a rapid phase of intravascular cancer cell disappearance related to sheer forces, detection by immune system, and "anoikis." Many cancer cell types with increased metastatic potential are resistant to anoikis and to elimination by the immune system compared with the parental cells, a tumor cell behavior related to the expression of apoptosis inhibitors and CTC chaperoning by endothelial cells and/or platelets.

Metastatic inefficiency is principally determined by CTC susceptibility to apoptosis, failure of solitary cells extravasated in distant organs to initiate growth, and failure of early micrometastases in distant organs to stimulate angiogenesis and continue growth into macrometastases.

Both solitary cells in organs, defined as disseminated tumor cells (DTC), and micrometastases may remain in "▶ dormancy" for years (Sosa et al. 2014). The immune system and angiogenesis have been shown to play a role in tumor cell dormancy, as well as extracellular and stromal microenvironments, autophagy, tumor cell epigenetics, and heterogeneity (Sosa et al. 2014). Finally, it has been suggested that any factor that tips the balance between proliferation and apoptosis may result in tumor progression or regression.

The mechanisms involved in the preferential choice of a target organ for metastatic tumor cell proliferation (▶ "seed and soil" theory) are still not completely understood, but include the close interaction between tumor cells (the "seeds") and the microenvironment of the "soil" (Langley and Fidler 2011). Organ-specific attractant molecules (chemokines) can stimulate migrating tumor cells to invade the walls of blood vessels and enter specific organs. Tumor–endothelial interaction, appropriate adhesion molecules expressed by endothelial cells in distant organs, and local growth factors can drive metastatic tumor cell proliferation. Once the target organ is reached, mesenchymal-like CTC may need to reverse to epithelial-like tumor cells via MET in order to regain the ability to proliferate.

Tumor cells can also invade as multicellular aggregates or clusters, a process known as "collective tumor cell migration." Multicellular aggregates of tumor cells, also called circulating tumor microemboli (CTM), are thought to have potential advantages for survival, proliferation, and establishment of micrometastatic lesions in distant organs. Actually, it has been shown that CTM may bring their own soil and give rise to metastasis without extravasation, by proliferating within the vasculature (Fig. 2). Thus, it is generally accepted that the presence of CTM in blood is a marker of highly metastatic potential (Paterlini-Bréchot 2014).

Convergent results have led to the present knowledge that invasion can be early and sometimes clinically dormant (Sosa et al. 2014). Tumor cell dissemination may precede evident primary tumor outgrowth by many years (Sosa et al. 2014; Kohn and Liotta 1995). The capacity to metastasize may be preordained by the spectrum of mutations acquired early in tumorigenesis, which means that some cancers start out "on the wrong foot." In fact, it has been demonstrated that cancer cells in the primary tumor may harbor a gene expression signature matching that observed in the metastatic colony and that this signature can help to predict whether the tumor will remain localized or not (Soundararajan et al. 2015).

CTC Detection and Characterization

The challenge of CTC/CTM detection is related to the requirement of high sensitivity combined with high specificity. Since invasion can start very early during tumor development, identification and counting of CTC when they are very rare (few CTC/CTM per 10 ml of blood, which means few CTC/CTM mixed with approximately 50–100 million leukocytes and 50 billion erythrocytes) could alert the oncologist about a developing tumor invasion process (Paterlini-Bréchot 2014).

Specificity is also an absolute requirement in this field. In fact, a wrong identification of circulating epithelial non-tumor cells as "tumor cells" is expected to generate wrong clinical and therapeutical choices with bad impact on cancer patients' survival.

Indirect methods to detect CTC do not provide a diagnostic identification of CTC (Paterlini-Bréchot 2014) as they target epithelial cells and/or use organ-specific markers which identify cells from organs but do not demonstrate their tumorous nature. These include immune-mediated methods and RT-PCR (reverse transcriptase polymerase chain reaction) methods. Since antigens or transcripts completely specific for CTC are not known (i.e., antigens or transcripts expressed by all tumor cells from a solid tumor type and not expressed by leukocytes nor by other circulating non-tumor cells), epithelial-specific or organ-specific antigens have been used to identify CTC (for instance, EpCAM, BerEP4, cytokeratins).

However, due to the lack of tumor specificity, epithelial-specific antibodies and transcripts have been proven to generate false-positive results through the biased detection of circulating non-tumor epithelial cells. Furthermore, epithelial-specific antibodies and transcripts can generate false-negative results since they cannot detect invasive circulating tumor cells which have lost their epithelial antigens due to the EMT process. Finally, CTM cannot be reliably detected by immune-mediated and RT-PCR approaches (as multiple cell labeling tends to dissociate tumor cell aggregates and RT-PCR methods destroy cell membranes). Thus, it appears that a reliable unbiased isolation and diagnostic identification of CTC and CTM cannot be based on the expression of epithelial-specific antigens or transcripts.

Accordingly, since the term circulating tumor cells (CTC) has been referred to circulating cells detected with methods using epithelial-specific antigens or transcripts, which have been demonstrated to generate false-positive and false-negative results, the terms of circulating cancer cells (CCC) and circulating cancer microemboli (CCM) have been introduced in 2014 to indicate circulating cells and microemboli isolated from blood without antibody-dependent bias and diagnostically (i.e., virtually without false-positive and false-negative results) identified by cytopathology (Paterlini-Bréchot 2014).

Direct methods, in particular density gradient isolation and ISET (isolation by size of tumor cells), which do not rely on the use of antibodies, isolate all types of CTC from blood without introducing bias of selection, thus without losing the most invasive tumor cells which have lost epithelial antigens. When they are followed by cytopathological analysis, they provide the diagnostic identification of CCC (Paterlini-Bréchot 2014).

CTC molecular characterization has revealed genetic heterogeneity of CTC and may detect potentially ▶ theranostic genetic abnormalities, useful to select targeted therapies and/or to detect escape mutants. Genotyping of CTC can be performed by several approaches including FISH (fluorescence in situ hybridization), CGH (comparative genomic hybridization), and NGS (next-generation sequencing). Analyses of oncogene abnormalities (e.g., HER2, ALK, BRAF) can be performed by FISH or by quantitative PCR. Immunolabeling of cancer cells isolated without using antibodies is an interesting approach to identify mutated oncogenic proteins (e.g., ALK, BRAF) and to characterize their invasive potential, for instance, through the expression of HER-2, metalloproteinases, EGF-R, uPAR, and alpha-fetoprotein.

Detection of apoptotic cells (for instance, by TUNEL (TdT-uridine nick end labeling) analysis) may be relevant before and after anticancer therapy, in order to assess the proapoptotic effect of therapeutic programs. However, the method used to prepare the cells for analysis may induce apoptotic cell death in cells made fragile by blood storage, multiple manipulations, and magnetic particles.

CTC culture, although potentially useful for CTC characterization and drug sensitivity studies, has been shown to be difficult, inconsistent, and with low efficiency up to now (Paterlini-Bréchot 2014).

CTC characterization assays are expected to expand our knowledge of the invasion process and generate new data aimed at improving cancer patients' diagnosis, follow-up, and treatment. However, it is noteworthy that addressing characterization studies only to a proportion of CTC isolated by antibody-dependent approaches is susceptible to generate biased results and false conclusions potentially leading to harmful clinical choices.

Clinical Impact of CTC Detection

Several studies have shown the potential of CTC/CTM detection and counting for cancer prognosis and follow-up. However, the clinical impact of CTC detection is not completely established because a substantial number of studies do not meet essential criteria for quality assurance, stressing the need for a gold standard assay based on a highly sensitive, unbiased isolation of CTC and their diagnostic cytopathologic detection (i.e., a gold standard method for CCC detection) (Paterlini-Bréchot 2014). This approach is crucial to assess the clinical impact of CCC by performing large clinical trials focused on patients with different types of solid cancers at different clinical stages. These trials are expected to generate reliable results and provide guidelines to the clinical use of CCC. In this setting, it is noteworthy that CCC diagnosed by a direct cytopathological assay (ISET) have been demonstrated to detect lung cancer before CT scan leading to its early diagnosis and surgical eradication (Ilie et al. 2014).

Glossary

Cell plasticity Capacity of cells to adopt the biological properties (gene expression profile, phenotype, etc.) of other undifferentiated (stem) or differentiated types of cells.

Escape mutants Mutated forms of a microorganism or tumor cell which escape the attack of immune system or selected therapy.

Liquid biopsy Detection of circulating tumor cells and/or cell-free molecules (DNA, RNA, miRNA) shed in blood by the primary tumor and/or metastases.

Theranostic A form of diagnostic testing employed for selecting targeted therapy.

Cross-References

▶ Angiogenesis
▶ Anoikis
▶ Apoptosis
▶ Chemokines
▶ Dormancy
▶ Epithelial-to-Mesenchymal Transition
▶ Invasion
▶ Macrophages
▶ MET
▶ Metastasis
▶ Micrometastasis
▶ "Seed and Soil" Theory of Metastasis
▶ Theranostics

References

Ilie M, Hofman V, LongMira E, Selva E, Vignaud JM, Padovani B, Mouroux J, Marquette CH, Hofman P (2014) "Sentinel" circulating tumor cells allow early diagnosis of lung cancer in patients with chronic obstructive pulmonary disease. PlosOne 9(10): e111597 1–7

Kohn EC, Liotta LA (1995) Molecular insights into cancer invasion: strategies for prevention and intervention. Cancer Res 55:1856–1862

Langley RR, Fidler IJ (2011) The seed and soil hypothesis revisited – the role of tumor-stroma interactions in metastasis to different organs. Int J Cancer 128(11):2527–2535

Luzzi KJ, MacDonald IC, Schmidt EE (1998) Multistep nature of metastatic inefficiency: dormancy of solitary cells after successful extravasation and limited survival of early micrometastases. Am J Pathol 153:865–873

Meseure D, Alsibai KD, Nicolas A (2014) Pivotal role of pervasive neoplastic and stromal cells reprogramming in circulating tumor cells dissemination and metastatic colonization. Cancer Microenviron 7:95–115

Noman MZ, Messai Y, Muret J, Hasmim M, Chouaib S (2014) Crosstalk between CTC, immune system and hypoxic tumor microenvironment. Cancer Microenviron 7:153–160

Paterlini-Brechot P and Benali N (2007) Circulating tumor cells (CTC) detection: Clinical impact and future directions. Cancer Letters, 253:180–204

Paterlini-Bréchot P (2014) Circulating tumor cells: who is the killer? Cancer Microenviron 7:161–176

Sosa MS, Bragado P, Aguirre-Ghiso JA (2014) Mechanisms of disseminated cancer cell dormancy: an awakening field. Nat Rev Cancer 14:611–622

Soundararajan R, Paranjape AN, Barsan V, Chang JT, Mani SA (2015) A novel embryonic plasticity gene signature that predicts metastatic competence and clinical outcome. Sci Rep 5:11766. doi:10.1038/srep11766

Ye X, Weinberg RA (2015) Epithelial–mesenchymal plasticity: a central regulator of cancer progression. Trends Cell Biol 25:675–686

See Also

(2012) Cancer stem cells. In: Schwab M (ed) Encyclopedia of cancer, 3rd edn. Springer, Berlin/Heidelberg, p 626. doi:10.1007/978-3-642-16483-5_815

(2012) Circulating tumor microemboli. In: Schwab M (ed) Encyclopedia of cancer, 3rd edn. Springer, Berlin/Heidelberg, pp 868–869. doi:10.1007/978-3-642-16483-5_1183

(2012) Extravasation. In: Schwab M (ed) Encyclopedia of cancer, 3rd edn. Springer, Berlin/Heidelberg, p 1370. doi:10.1007/978-3-642-16483-5_2080

(2012) Intravasation. In: Schwab M (ed) Encyclopedia of cancer, 3rd edn. Springer, Berlin/Heidelberg, p 1901. doi:10.1007/978-3-642-16483-5_3125

(2012) Macrometastasis. In: Schwab M (ed) Encyclopedia of cancer, 3rd edn. Springer, Berlin/Heidelberg, p 2130. doi:10.1007/978-3-642-16483-5_3483

(2012) Targeted therapy. In: Schwab M (ed) Encyclopedia of cancer, 3rd edn. Springer, Berlin/Heidelberg, p 3610. doi:10.1007/978-3-642-16483-5_5677

cis-Diamminedichloroplatinum

▶ Cisplatin

cis-Dichlorodiammineplatinum(II)

▶ Cisplatin

Cisplatin

Lin Ji
Department of Thoracic and Cardiovascular
Surgery, The University of Texas MD Anderson
Cancer Center, Houston, TX, USA

Synonyms

CDDP;   cis-Diamminedichloroplatinum;
cis-Dichlorodiammineplatinum(II); cis-Platinum
II; DDP

Definition

Cisplatin is classified as a platinum compound and
an alkylating cytotoxic agent. Much of our current
understanding of the unique properties of platinum drugs has come from studies of cisplatin,
especially its antitumor activity. The antitumor
activity of platinum (II) complexes requires several unique chemical properties including the
presence of chloride, bromide, oxalate, or
malonate as leaving group and the neutral complex with inert carrier ligands such as NH_3 groups.
Minor variations in the structure of these ligands
may have a profound effect on the antitumor
activity and toxicity of platinum compounds.
The cis conformation is required for a complex
to be a biologically effective agent and has significant cytotoxic properties, while the transisomer
does not.

Characteristics

Since the discovery of the antitumor potential of
cisplatin   by   Rosenberg   and   coworkers,
therapeutic efficacy of cisplatin as an anticancer
agent has been established in a variety of preclinical animal tumor models and in clinical human
cancers. Cisplatin has now been one of the most
widely used chemotherapeutic agents for the treatment of many human cancers. The success of
cisplatin in cancer treatment has been due to its
many unique properties: a wide spectrum of
antitumor activity against drug-sensitive as well
as drug-resistant human tumors; a potent inhibition against tumors with varied proliferation and
growth characters; effectiveness on both solid and
disseminated tumors; and broad cytotoxic activity
against   viral-induced,   chemical-induced,   and
transplantable tumors with no strain or species
specificity.

Mechanisms of Action
The biochemical and biological properties of cisplatin rely on the relative ease of substitution of
the chlorine ligands with nucleophilic species
such as nucleic acid bases of a DNA strand. It is
now widely accepted that cisplatin is similar to the
bifunctional alkylating agents and its primary target is DNA. After cisplatin enters the cells, the
chloride ligands are replaced by water molecules.
This reaction results in the formation of positively
charged platinum complexes that form covalent
bounds with nucleopholic sites on guanine bases
in a DNA strand using intrastrand and interstrand
cross-links and create cisplatin–DNA adducts.
The most prevalent and unique form of
cisplatin–DNA adducts is the 1,2-intrastrand
cross-link that cannot form with the inactive isomer of ciaplatin, trans-DDP, suggesting that such
an adduct might be responsible for the biological
activity of cisplatin. Other platinum–DNA
adducts form a distinct structural element that
interacts with DNA differently. The formation of
these DNA adducts disrupts DNA function and
prevents DNA, RNA, and protein synthesis. Regulatory mechanisms that detect the abnormal
DNA activate a chain of cellular response to correct or repair the faulty DNA and this ultimately
leads to programmed cell death (▶ apoptosis).
Cisplatin-mediated cell killing is believed to be
cell cycle phase nonspecific, although there is
now much evidence that it may be most effective

Cisplatin, Fig. 1 Cisplatin and DNA adduct formation

in G1 phase. Cisplatin also has immunosuppressive, radiosensitizing, and antimicrobial properties (Fig. 1).

Cisplatin and Cancer Treatment

Cisplatin is an effective chemotherapeutic drug against a wide spectrum of human cancers. It has been primarily used in the treatment of epidermoid carcinomas of the head and neck, lymphoma, nonHodgkin and Hodgkin disease sarcoma, mesothelioma, osteosarcoma, and adrenal carcinoma and of bladder, brain, and cervical, esophageal, gastric, lung, nasopharyngeal, ovarian, prostate, and testicular cancers. It has also proved to be of benefit in the treatment of other cancers of anal, kidney, liver, breast, penile, and thyroid and of choriocarcinoma, lymphomas, and melanoma. The effectiveness of ciaplatin is however mostly due to the inclusion of other antineoplastic agents into the chemotherapy regimens. For example, such combination therapy of cisplatin along with vinblastine and bleomycin produces complete remission in more than 70% of patients with testicular cancers and substantially improves the survival rate of patients with ovarian cancer. This high success rate is mostly due to synergistic effects, where multidrug combination prevents the drug-induced resistance in tumor cells and, in addition, to the reduced toxic effects of the combination therapy with respect to the total toxicity of each equivalent single agent. A marked therapeutic synergy has been shown in combination of cisplatin with a wide variety of other chemotherapeutic agents, such as 5-fluorouracil and cytarabine. In general, it is not uncommon for the therapeutic effect of two anticancer drugs on a particular cancer to be greater than the effect of each drug treatment alone or the sum of the individual effects. The presence of one drug enhances the effects of the second. This is called a synergistic effect or synergy, and the drugs are sometimes described as showing anticancer synergism.

Cisplatin is supplied for clinical use as a lyophilized powder in vials that contain 10 mg of the drug, a diuretic, usually mannitol, and salt, or as a 1 mg mL^{-1} aqueous solution. The powder is reconstituted with sterile water to a concentration of 1 mg mL^{-1} and followed by further dilution with saline for intravenous (i.v.) administration. The standard method of administration of cisplatin is as a single slow i.v. injection or infusion every 3–4 weeks. Cisplatin has been shown to be more effective when given locally to the site of the tumors. The most common method is intraperitoneal (i.p.) administration, and this type of therapy is most effective for ovarian cancers. The specific dose of cisplatin will vary from patient to patient and depends on a number of criteria.

Cisplatin-Induced Resistance

Even though cisplatin has proven to be a highly effective chemotherapeutic agent for treating various types of cancers, one of the significant limitations toward the successful treatment of malignant cancers with cisplatin and other platinum-based drugs is the emergency of drug resistance. Drug resistance has significant clinical implications and accounts for the failure of a single platinum agent-mediated chemotherapy in curing the majority of cancer patients. When cells become resistant to cisplatin, a large dose escalation has to be applied, which can lead to severe multiorgan toxicities such as failures of the kidneys and bone marrow, intractable vomiting, and deafness. Cellular resistance to these drugs consists of complex mechanisms involving multiple biological pathways. The acquisition or intrinsic presence of resistance significantly undermines the curative potential of these drugs against many human malignant cancers. Although the precise mechanisms by which cells develop resistance to cisplatin are still not well known, several cellular processes have been identified or suggested attributing to invulnerability to cisplatin-induced cytotoxicicty.

Inhibition of Drug Uptake and Decreased Intracellular Accumulation

In order for cisplatin to exercise its cytotoxic effect on tumor cell, it must be taken and accumulated inside of cancer cells to reach and bind to the DNA and cause cell death. The cancer cell, however, has to develop mechanisms either to keep cisplatin out of the cell or to remove cisplatin from the cell to survive. Alterations in cellular pharmacology, including inhibition of cisplatin uptake and reduced cisplatin accumulation by cancer cells, have been observed in numerous model systems and appear to be a major form of acquired resistance. An increase in the production of cellular thiols, such as metallothione and glutathione, has been shown to block the formation of cisplatin–DNA adducts and sequester cisplatin and remove it from the cell.

Increased DNA Repair

Cancer cells can also become resistant to cisplatin by an enhanced ability to remove cisplatin-DNA adducts and to repair cisplatin-induced ▶ DNA damages through an upregulated expression and activity of certain DNA repair proteins. For example, a nuclear protein called XPE-BF (xeroderma pigmentosum group E binding factor) has been shown to be upregulated early in the development of cisplatin resistance and be able to repair cisplatin damaging. Another example of a DNA repair protein that may be involved in the recognition of cisplatin damage is ERCC1, one of the essential components of the mammalian nucleotide excision repair (NER) pathway. A higher level of ERCC1gene expression is observed in cisplatin-resistant cells than in cells that are sensitive to cisplatin and in tumor tissues from patients who were clinically resistant to cisplatin therapy than those who responded favorably to the treatment. In addition, an increased level of ERCC1 expression was also found in patients who developed resistance after initial cisplatin treatment. It has been shown that an enhanced capacity to tolerate cisplatin-induced damage may also contribute to cisplatin resistance. Alterations in proteins that recognize cisplatin–DNA damage (mismatch repair and high-mobility group (HMG) family proteins) and in pathways that determine sensitivity to apoptosis may contribute to damage tolerance. Furthermore, ▶ tumor suppressor genes have also been linked to the ability of DNA repair to confer cisplatin sensitivity. Interruption of p53 ▶ tumor suppressor gene by dysfunctional mutations found in breast, ovarian, and lung cancer cells may increase tumor cells' sensitivity to cisplatin, possibly by a decrease in p53-mediated DNA repair.

NPRL2, a novel tumor suppressor gene identified in human chromosome 3p21.3 region, has been suggested to be involved in DNA mismatch repair, cell cycle checkpoint signaling, and regulation of the apoptotic pathway. The loss of NPRL2 protein expression was significantly correlated to cisplatin resistance in human nonsmall cell lung cancer cells. However, it remains to be determined whether any of these mechanisms contribute significantly to resistance in the clinical setting. Ongoing biochemical modulation and translational correlative trials should clarify which specific mechanisms are most relevant to clinical cisplatin resistance. Such investigations have the potential to improve the ability to predict likelihood of response and should identify potential targets for pharmacological or molecular intervention.

Development of Cisplatin Analogs

Besides a remarkable therapeutic efficacy in a series of solid tumors and outstanding activity of cisplatin, the platinum-based therapy is in part accompanied by a set of severe toxic side effects. Analogs or second-generation platinum drugs have being designed and developed to exhibit an exclusive tumor selectivity, enhance the efficacy, improve the toxicity profile, overcome resistance of the original drug, and to be able to be taken orally. Many second-generation analogs of cisplatin have been made. Some have been found to produce the same therapeutic effects as cisplatin but with lower required doses and reduced side effects. Three of these analogs are carboplatin, spiroplatin, and iproplatin. Carboplatin has proven to be the most useful of these three analogs and was approved by the FDA for the treatment of ovarian cancers and for first line lung cancer treatment. Carboplatin and cisplatin have been shown to form an identical type of adduct with DNA and have similar activities against ovarian and lung tumors but is less toxic to the peripheral nervous system and the kidneys. Carboplatin works in some cases when cisplatin has failed. The decreased toxicity of carboplatin and the activity of carboplatin against cisplatin-resistant tumors have led to greater use of carboplatin which has resulted in carboplatin becoming the greater

moneymaker of the two drugs. In addition to carboplatin and other second-generation cisplatin analogs, several third-generation drugs have been synthesized and tested, such as platinum (IV) dicarboxylates. These analogs can be taken orally, a significant improvement over cisplatin which can only be administered intravenously. These new platinum complexes and their promising therapeutic strategies in terms improved accumulation and activation at the tumor site are demonstrating a stepwise approach toward the "magic bullet" to human cancer therapy.

References

Barnes KR, Lippard SJ, Barnes KR et al (2004) Cisplatin and related anticancer drugs: recent advances and insights. Met Ions Biol Syst 42:143–177

Matsusaka S, Nagareda T, Yamasaki H et al (2005) Does cisplatin (CDDP) function as a modulator of 5-fluorouracil (5-FU) antitumor action? A study based on a clinical trial. Cancer Chemother Pharmacol 55:387–392

McEvoy GK (ed) (2004) AHFS 2004 drug information. American Society of Health-System Pharmacists, Bethesda, pp 929–945

Ueda K, Kawashima H, Ohtani S et al (2006) The 3p21.3 tumor suppressor NPRL2 plays an important role in cisplatin-induced resistance in human non-small-cell lung cancer cells. Cancer Res 66:9682–9690

Wang D, Lippard SJ (2005) Cellular processing of platinum anticancer drugs. Nat Rev Drug Discov 4:307–320

Cisplatin-Refractory Germ Cell Tumors

▶ Platinum-Refractory Testicular Germ Cell Tumors

Cisplatin-Resistant Germ Cell Tumors

▶ Platinum-Refractory Testicular Germ Cell Tumors

cis-Platinum II

▶ Cisplatin

c-Jun Activation Domain-Binding Protein-1

▶ JAB1

c-Jun N-Terminal Kinase

▶ JNK Subfamily

CK

▶ Choline Kinase

CK2

Denis Drygin[1] and Lorenzo Pinna[2]
[1]Pimera, Inc., San Diego, CA, USA
[2]Department of Biological Chemistry, University of Padua, Padua, Italy

Synonyms

Casein kinase 2; Casein kinase II; CKII

Definition

CK2 is a serine/threonine (Ser/Thr) protein kinase that can operate as a tetrameric holoenzyme consisting of two catalytically active subunits α and/or α' and a dimer of regulatory β-subunits or as a monomeric catalytic subunit (α or α'). Its

original name "casein kinase 2," which was derived from CK2 being identified as a kinase capable of phosphorylating casein in vitro, proved to be a misnomer since casein was found not to be a physiologic substrate of CK2.

Characteristics

CK2 was the first protein kinase to be identified in 1954 by Eugene Kennedy, who described the enzyme's ability to catalyze the phosphorylation of protein substrate by ATP. CK2 possesses several distinct features that distinguish it from the majority of other kinases. It is a highly acidophilic Ser/Thr kinase that atypically phosphorylates tyrosine (Tyr) residues in addition to Ser and Thr and can also utilize GTP, besides ATP, as a phosphate donor. CK2 is expressed in all cellular compartments, including the cellular membrane. It is extremely pleiotropic in nature with hundreds of substrates identified to date and is responsible for the generation of a substantial proportion of the human phosphoproteome (possibly up to 20%). CK2 is a very well conserved kinase with corresponding subunits having >98% sequence identity between human and mouse.

Regulation of Expression and Activity
The presence of fully functioning CK2α and CK2β subunits is absolutely required for embryonic development, with CK2α being a critical regulator of mid-gestational morphogenetic processes and loss of CK2β leading to early embryonic lethality. In contrast, CK2α'-deficient mice are viable, albeit suffering from defects in spermatogenesis. Even though they are encoded by distinct genes that reside on separate chromosomes, the two catalytic subunits possess a high level of homology to each other, with the exception of a C-terminal extension that is present only in CK2α. Furthermore, CK2α and CK2α' share the highly acidic recognition motif: X_{-1}-[S/T]X_{+1}-X_{+2}-[E/D/pS]-X_{+4}-X_{+5}, where S/T designates a phospho-acceptor residue, X_{-1}–X_{+5} are preferably acidic residues, and X_{+1} cannot be a proline. Based on this consensus, highly specific peptide substrates have been developed which

enable the detection and evaluation of CK2 activity in crude biological preparations and in cell lysates. In addition to redundant phosphorylation, CK2α and CK2α′ are also known to have independent substrates and functions. For example, siRNA-mediated knockdown of CK2α, but not CK2α′, was shown to downregulate the expression of IL-6. In contrast, CK2α′ exhibits a striking preference for caspase-3 phosphorylation in cells as compared to CK2α. The exact mechanism of the regulation of CK2 expression and activity is not well characterized. Both CK2 holoenzyme and monomeric catalytic subunits are constitutively active; in particular, they do not require any posttranslational modification for their function. In the past, several reports have described the increase in CK2 activity and/or expression in response to exogenous stimuli such as pro-inflammatory mediators or growth factors; presently, however, these results are being disputed. Interestingly, while there are multiple examples of changes in CK2 protein level under pathologic conditions, these do not necessarily correlate with the alteration in mRNA level, consistent with the occurrence of posttranscriptional and/or posttranslational regulation. Two molecular mechanisms contribute to constitutive activity: unique interactions between activation loop and N-terminal segment of CK2α/CK2α′ and association of catalytic subunits with regulatory subunits. In general, both the holoenzyme and individual catalytic subunits are capable of phosphorylating the same substrates, with the catalytic activity of the holoenzyme being somewhat higher ("class I" substrates). However, there are notable exceptions to this rule. Phosphorylation of certain protein substrates critically relies on the presence of β-subunits ("class III" substrates). In other cases, only individual subunits, but not the holoenzyme, were shown to be active ("class II" substrates). It should also be noted that association with β-subunits stabilizes CK2 activity.

Role in Carcinogenesis

Elevated levels and activity of CK2 have been associated with malignant transformation and poor prognosis in multiple types of cancers. Furthermore, forced expression of CK2 by genetic manipulation in mice leads to the development of lymphoma and breast cancer, particularly in the presence of c-Myc or Tal oncogene overexpression or loss of p53. However, in contrast to classical oncogenes, no gain-of-function mutations in any subunit of CK2 have been identified to date. While there are clearly several pathways that can contribute to cancer-specific alterations in CK2 expression, in general, they are not all well characterized. One example is that loss of miR-125b in breast cancer was shown to cause 40–56% increase in CK2α expression. Hypoxia, a condition that is commonly associated with tumors, was reported to increase CK2β protein levels and to enhance overall CK2 activity; however, the exact mechanism of this enhancement is not understood. While not an oncogene itself, CK2 plays an important role in maintaining an oncogenic phenotype by positively regulating multiple hallmarks of cancer. CK2 is involved in sustaining proliferative signaling, as it is known to regulate both G1/S and S/G2 transitions. CK2 was shown to phosphorylate multiple proteins involved in cell cycle regulation, including cyclin H, p53, p21, p27, Cdc34, and Cdk1. CK2 promotes the resistance of cancer cells to apoptosis through multiple mechanisms, including positive regulation of expression and stability of antiapoptotic proteins such as members of Bcl-2 family, as well as inhibitors of apoptosis (IAPs); by interfering with caspase activity, either directly or by rendering caspase targets refractory to cleavage; as well as by suppressing the signaling cascades triggered by death receptors. Another signaling pathway that plays an important role in suppression of apoptosis, PI3K-Akt-mTOR, is also regulated by CK2 at multiple nodules, including enhanced phosphorylation of Akt and inactivation of PTEN, as well as in a broader manner through activation of Hsp90/Cdc37 chaperone machinery. CK2 has been implicated in the promotion of tumor neo-angiogenesis by activating Hif-1α-dependent transcription through the HDAC-pVHL/p53 axis and by regulating the ability of the endothelium to produce and support neovasculature.

Epithelial–mesynchemal transition (EMT) is known to play an important role in activation of invasion and metastasis. Overexpression of CK2 has been demonstrated to induce a mesynchemal phenotype, in part through the direct phosphorylation and thus stabilization of Snail and in part through activation of Wnt signaling by phosphorylation of disheveled, Lef-1 and β-catenin, while inhibition of CK2 was shown to block TGFβ1-induced EMT. Inflammation has been recognized as an important contributor to carcinogenesis. CK2 controls one of the most important master regulators of the inflammatory response, NF-κB, by both activating its transcriptional activity through direct phosphorylation of the p65 subunit of NF-κB and by relieving the negative regulation through phosphorylation and inactivation of IkBα. Furthermore, CK2 has been shown to regulate the expression of several pro-inflammatory cytokines including IL-6 and TNFα. In addition to NF-κB, CK2 was shown to positively regulate other transcriptional factors/activators that are known to be involved in carcinogenesis, e.g., STAT3, Gli1, and Notch1.

Role in Drug Resistance

CK2 plays a significant a role in chemoresistance. Drug efflux by ATP-binding cassette (ABC) transporters represents one of the major mechanisms by which cancer cells negate the activity of chemotherapeutics. CK2 phosphorylates and thus positively regulates the activity and stability of two major members of the ABC transporter family: ABCB1/MDR1 (multidrug resistance protein 1) and ABCC1/MRP1 (multidrug resistance-associated protein 1). Suppression of CK2 by small molecule inhibitors or siRNA can counteract the activity of these transporters and restore the chemosensitivity of cancer cells. Another mechanism by which cancer cells circumvent the effect of treatment is the suppression of apoptosis. Upregulation of the expression of antiapoptotic proteins, as well as blockade of caspase-dependent proteolysis by CK2, prevents the activation of apoptosis in response to treatment and thus drives resistance. The efficacy of DNA-damaging drugs can be significantly diminished by DNA damage repair processes that cancer cells often employ to remove therapy-caused DNA lesions before they become irreversibly cytotoxic. CK2 phosphorylates multiple members of the DNA repair machinery and is known to play a prominent role in multiple types of DNA damage repair, including base excision repair, homologous recombination, and nonhomologous end joining. The additional mechanisms of CK2-dependent chemoprotection include EMT and inflammation. Targeting CK2 with either siRNA or small molecule inhibitors was shown to increase the antitumor potency of multiple types of cancer therapeutics, including DNA-damaging drugs, ionizing radiation, EGFR-targeting agents, proteasome inhibitor bortezomib, and BCR-Abl inhibitor imatinib.

Targeting CK2

Both catalytic subunits of CK2 belong to the CMGC kinase subfamily; have a bilobular topology with a β-strand-rich small lobe, an α-helix-rich large lobe, and a short-hinge region that contains the ATP-binding site; and display several main structural motifs that are common to all human protein kinases, e.g., phosphate-binding loop, catalytic loop, activation loop, and substrate binding site. They also possess several distinct features that separate them from most other kinases. The region that surrounds the substrate recognition site of both subunits contains an unusually high number of basic residues, which explains CK2's affinity for acidic substrates. In addition, three unique amino acids are present in the ATP/GTP-binding cleft of CK2: Val66, which is responsible for the resistance of CK2 to pan-kinase inhibitor staurosporine, Ile174, and Met163. The presence of these bulky amino acids creates a pocket that is smaller and more hydrophobic than the ones present in most other protein kinases. By replacing these hydrophobic residues with alanine, CK2 mutants have been generated which are less sensitive to a number of ATP site-directed inhibitors and are proving useful to demonstrate that functional alterations promoted by these compounds are actually mediated by CK2.

Multiple types of CK2 inhibitors have been described in the literature. In general they fall into three categories: (1) small molecule ATP-competing inhibitors of CK2 catalytic activity; (2) small molecule allosteric inhibitors; and (3) CK2-targeting "biologics."

Several classes of natural compounds and their derivatives have been shown to inhibit CK2 in an ATP-competitive manner, including anthraquinones emodin (Ki = 1.5 μM) and quinalizarin (Ki = 60 nM); flavonoids apigenin (IC$_{50}$ = 1.72 μM), quercetin (IC$_{50}$ = 510 nM), and luteolin (IC$_{50}$ = 860 nM); coumarin DBC (Ki = 60 nM); polyphenol ellagic acid (Ki = 20 nM), phenoxazine resorufin (Ki = 800 nM); and phytoestrogen coumestrol (IC$_{50}$ = 228 nM). The first synthetic CK2 ATP-mimetic inhibitor DRB (IC$_{50}$ = 23 μM), identified in 1986, belongs to the chemical class of benzimidazoles. Several improved analogs were developed including TBB (IC$_{50}$ = 160 nM), one of the most commonly used CK2 inhibitor in the literature. Numerous pharmaceutical and biotechnology companies also made a contribution to designing ATP-competitive inhibitors of CK2 including indoloquinazolinone IQA (IC$_{50}$ = 390 nM) from Novartis, pyrazole–triazine series from Polaris Pharmaceuticals that contains several compounds with Ki = 0.2–0.3 nM, pyrazole–pyrimidines from AstraZeneca (e.g., AZ285 IC$_{50}$ = 1.3 nM), 4'-hydroxyflavones from Otava, Ltd. (e.g., FNH79 IC$_{50}$ = 4 nM), and several series of inhibitors from Cylene Pharmaceuticals including pyrimidine–quinolines CX-5011 (IC$_{50}$ = 3 nM), CX-5279 (IC$_{50}$ = 9 nM), and pyrido-quinoline CX-4945 (IC$_{50}$ = 1 nM). To note that unlike the majority of ATP site-directed CK2 inhibitors, which have been shown to be quite promiscuous, CX-4945 and its analogs CX-5011 and CX-5279 are endowed with unprecedented selectivity. On the other hand, the advantage of developing dual and eventually multi-kinase inhibitors hitting CK2 and other kinase(s) that cooperate with CK2 to enhance the tumor phenotype has been highlighted and exemplified by the cytotoxic potency of CK2/PIM1 dual inhibitors on cancer cells. One of these, a deoxyribofuranosyl derivative of tetrabromo-benzimidazole (TDB) displays toward cancer cells a cytotoxic efficacy higher than that of CX-4945 and is evident also in a multidrug resistance background.

In addition to ATP-mimetics, several types of allosteric inhibitors of CK2 have been reported. These include hematein (IC$_{50}$ = 550 nM) that displays noncompetitive inhibition against ATP and mixed inhibition against a peptide substrate, W16, a podophyllotoxin indolo-analog (IC$_{50}$ = 20 μM) that interferes with CK2α(α')/CK2β interaction, and inorganic polyoxometalate $[P_2Mo_{18}O_{62}]^{6-}$ (IC$_{50}$ = 1.4 nM) for which the exact mechanism of action is still unclear. The utility of some of these compounds however is hampered by their lack of cell permeability.

Targeting protein expression with an antisense approach has been extensively tested in the preclinical setting as well as in the clinic, with at least two antisense-based drug securing regulatory approval. Nano-encapsulated anti-CK2 oligonucleotides (e.g., GS-10, an s50-encapsulated RNAi from GeneSegues Therapeutics) were shown to be effective in producing significant antitumor activity both in vitro and in vivo in several preclinical models of cancer. In addition to nucleic acid-based inhibitors, several cyclic peptides were shown to successfully inhibit the activity of CK2. One such cyclic peptide, CIGB-300, designed to target substrate phosphor-acceptor domains to abrogate phosphorylation by CK2, has been tested in patients with cervical malignancies demonstrating signs of clinical benefit, as was evidenced by significant decrease of the tumor lesion area and histological examination, while being well tolerated. Since this peptide fails to prevent the phosphorylation of peptide and protein substrates by CK2 in vitro, the biochemical features underlying its therapeutic potential are still unclear.

As of the fourth quarter of 2014, the only selective small molecule inhibitor of CK2 to be tested in clinical trials is CX-4945, which has completed two phase I clinical trials, one enrolling patients with advanced solid tumors, Castleman disease, or multiple myeloma and another focused solely on the relapsed or refractory multiple myeloma (http://clinicaltrials.gov, NCT00891280 and

NCT01199718). Two dosing regiments were investigated. CX-4945 was administered orally twice or four times daily for the first three consecutive weeks of the 4-week cycle. The drug was well tolerated with reversible dose-limiting toxicity being diarrhea and hypokalemia. The pharmacokinetic profile was generally linear and dose dependent. Inhibition of CK2 was evaluated by measuring phosphoproteins in peripheral blood mononuclear cells (Akt S129, Akt S473, and p21 T145), assessing plasma levels of interleukins 6 and 8 (IL-6 and IL-8), and by monitoring circulating tumor cells (CTCs). Twenty percent of the treated patients presented signs of stable disease for at least 16 weeks, with the most durable stabilization in patients with the highest percentage decreases in IL-6 and IL-8 levels. Reduction in monitored phosphoproteins and CTCs was also observed, but in general did not correlate with the clinical outcome. Thus, CX-4945 provided the first evidence that CK2 can be safely inhibited in humans. In June 2014, Senhwa Biosciences, which acquired CX-4945 from Cylene Pharmaceuticals, announced the initiation of phase II clinical trial testing CX-4945 in combination with gemcitabine and cisplatin for the frontline treatment of patients with bile duct cancers (cholangiocarcinoma) (http://clinicaltrials.gov, NCT02128282).

Cross-References

▶ Apoptosis
▶ Epithelial-to-Mesenchymal Transition
▶ Hypoxia
▶ Inflammation
▶ PI3K Signaling

References

Cozza G, Pinna LA, Moro S (2013) Kinase CK2 inhibition: an update. Curr Med Chem 20:671–693
Drygin D (2013) CK2 as a logical target in cancer therapy: potential for combining CK2 inhibitors with various classes of cancer therapeutic agents. In: Pinna L (ed) The Wiley-IUBMB series on biochemistry and molecular biology: protein kinase CK2. Wiley-Blackwell, Ames, pp 383–439
Guerra B, Issinger OG (2008) Protein kinase CK2 in human diseases. Curr Med Chem 15:1870–1886
Ruzzene M, Pinna LA (2010) Addiction to protein kinase CK2: a common denominator of diverse cancer cells? Biochim Biophys Acta 1804:499–504
Trembley JH, Chen Z, Unger G, Slaton J, Kren BT et al (2010) Emergence of protein kinase CK2 as a key target in cancer therapy. Biofactors 36:187–195

CKII

▶ CK2

c-Kit

▶ Kit/Stem Cell Factor Receptor in Oncogenesis

C-*KRAS*

▶ *KRAS*

Cladribine

Tadeusz Robak
Department of Hematology, Medical University of Lodz, Lodz, Poland

Synonyms

2-CdA; 2-Chloro-2′-deoxyadenosine; 2-Chlorodeoxyadenosine; Biodribin; CdA; Leustatin; NSC-10514-F

Definition

Cladribine is a purine nucleoside analog (PNA) synthesized by a simple substitution of a chlorine atom with a hydrogen atom at the position 2 of the purine ring of deoxyadenosine and resistant to deamination by adenosine deaminase (ADA) (Fig.1).

Characteristics

2-CdA is a prodrug and its intracellular phosphorylation is necessary for cytotoxic effect to occur. It is phosphorylated by deoxycytidine kinase (dCK) and accumulates as 2-chlorodeoxyadenosine triphopsphate (2-CdATP). High activity of this enzyme in lymphocytes along with their low 5 nucleotidase (5′-NT) activity explains its relatively high selectivity for lymphoid cells. The nucleoside that is formed does not readily exit from the cells through the cell membrane and therefore is accumulated inside the cell. This metabolite disrupts cell metabolism by incorporating into the DNA of the actively dividing cells and freezes cell cycles at S phase. In contrast to other antineoplastic drugs, 2-CdA is cytotoxic to both proliferating and quiescent cells. In quiescent cells 2-CdATP interferes with proper repair of DNA and leads to a total disruption of cellular metabolism via accumulation of breaks in DNA strand, which in turn lead to p53 expression and consequently to induction of apoptosis. Apoptosis induced by 2-CdA can be mediated either via DNA damage and p53 protein expression or directly via mitochondrial permeability transition pore. Inhibition of DNA repair and accumulation of DNA breaks lead to p53 expression, which plays a key role in control of apoptosis and cell cycle and influences the bcl-2 protein family with antineoplastic properties, as well as bcl-2 like proteins such as bax, bcl-xs, and bak, which have proapoptotic action (Fig. 2).

Administration and Pharmacokinetics

Clinical pharmacokinetics of 2-CdA have been evaluated in patients with lymphoproliferative diseases and acute leukemia. The drug is usually administrated i.v. in a dose of 0.12–0.14 mg/kg/day for 5–7 days in continuous infusion or 2-h infusion. Oral and subcutaneous method of administration can be also used. This routes result in substantial improvement of the quality of life in disorders that require repeated courses of treatment.

After administration of 2-CdA at a dose 0.14 mg/kg as a 2-h i.v. infusion, the mean maximum plasma concentration of the drug is 198 mol/L (range 70–381 mol/L). The steady-stage drug concentration during 24-h continuous infusion of 2-CdA at a dose of 0.14 mg/kg is 23 mol/L. The areas under the concentration time curves (AUC) are similar for both the 2-h (588 mol/L) and 24-h (552 mol/L) infusion. Following administration of 2-CdA at a dose of 0.12 mg/kg as a 2-h i.v. infusion or continuous 2-h infusion, the mean cellular concentration of 2-CdA nucleotides are 12.2 mol/L and 10.8 mol/L, respectively. Cellular concentration of the drug exceeded plasma concentration 128–373 times. There is a linear dose relationship for 2-CdA between 0.2 and 2.5 $mg/m^2/h$ and elimination followed by two compartment model. The two compartment model showed a half life $(T1/2\alpha)$ of 35 ± 12 min and $T1/2\beta$ of 6.7 ± 2.5 h. The mean apparent volume of distribution (Vdr) is 9.2 ± 5.4 L/kg.

Clinical Activity

2-CdA was approved by the FDA for the treatment of hairy cell leukemia (HCL), and in some European countries for the treatment of refractory/relapsed chronic lymphocytic leukemia (CLL). Moreover, several clinical trials continued the value of this agent in low-grade non-Hodgkin lymphoma (LG-NHL), Waldenström

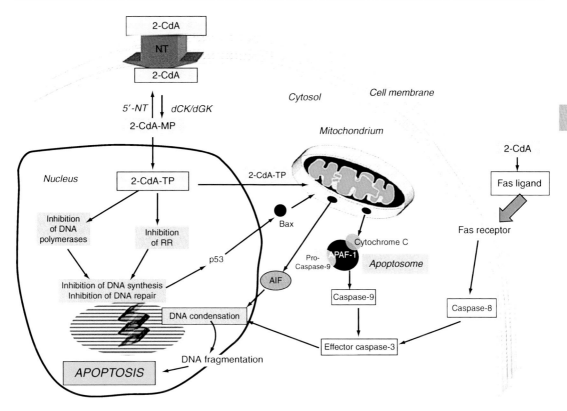

Cladribine, Fig. 2 Schematic presentation of cladribine (2-CdA) pathways. (2-CdA-MP-cladribine monophosphate; 2-CdA-DP-cladribine diphosphate; 2-CdA-TP cladribine triphosphate; dCK- deoxycytidine kinase; dGk-deoxyguanine kinase; 5′-NT - 5′ - nucleotidase)

macroglobulinemia (WM), cutaneous T-cell lymphoma (CTCL), Langerhans cell histiocytosis (LCH), and systemic mastocytosis. 2-CdA has also some activity in acute myeloid leukemia (AML) and idiopathic myelofibrosis (IM).

1. *Hairy cell leukemia.* 2-CdA induces durable and unmaintained complete response (CR) in about 80% of patients with HCL after a single course of therapy. However, patients in an apparent clinical and hematological remission following a single course of 2-CdA may have residual disease and 20–30% of them relapse at 10 years follow-up. However, 2-CdA may be equally effective in the reinduction therapy. Moreover, this agent may be also effective in patients with HCL who did not enter remission after splenectomy, interferon-α or even pentostatin.

2. *Chronic lymphocytic leukemia.* 2-CdA used alone or in combination with other cytotoxic drugs showed good efficacy and acceptable toxicity profile in CLL. The drug is more effective in the previously untreated patients than in the patients refractory or relapsed after conventional therapy with alkylating agents. The overall response (OR) rate ranges from 75% to 85% and CR from 10% to 47% when 2-CdA is used in first line therapy. In the pretreated patients, OR rate ranges first 30% to 70% and CR from 0% to 30%. The combination of 2-CdA with cyclophosphamide (CC), cyclophosphamide, and mitoxantrone (CMC) or rituximab (RC) can be more effective than 2-CdA alone. Randomized studies indicate that 2-CdA alone and CC used as the first-line therapy give similar OR and CR and are of comparable toxicity as fludarabine alone or

fludarabine combined with cyclophosphamide, respectively.

3. *Waldenström's macroglobulinemia.* 2-CdA is a reasonable choice for the first line treatment of WM patients. In this disease, 2-CdA has been shown to be active in 64–100% of the previously untreated patients and 14–78% of the refractory or relapsed patients. The median time of response to this agent in the previously untreated patients varied between 13 and 28 months. The response rate is higher and the duration of response is longer when 2-CdA is given to the patients with primary refractory disease or to the patients relapsing of therapy rather than to the patients with the disease in resistant relapse.

4. *Cutaneous T-cell lymphoma.* 2-CdA showed some activity in advanced CTCL patients, including Sezary syndrome, and *mycosis fungoides*. This drug produces 25% OR in CTCL. However, high incidence of septic complications and significant treatment related mortality was observed.

5. *Other lymphoid malignancies.* 2-CdA showed remarkable activity in both previously treated and untreated patients with low grade non Hodgkin lymphoma (LG-NHL). In relapsed/refractory LG-NHL, the drug-induced durable response with OR rates range from 36% to 56% and CR rates between 10% and 20%. 2-CdA is also effective in combination with alkylating agents and/or mitoxantrone in the treatment of refractory or relapsed advanced stage LG-NHL. High activity and low toxicity was reported in the patients with LG-NHL and mantle cell lymphoma treated with 2-CdA combined with rituximab (RC regimen) or rituximab and cyclophosphamide (RCC).

6. *Langerhans cell histiocytosis.* 2-CdA has a major clinical activity in both pediatric and adult patients with LCH. Clinical response was observed in 60–100% patients with a median disease free survival of 33–50 months. Combination with cyclophosphamide may be even more effective in the treatment of LCH.

7. *Systemic mastocytosis.* 2-CdA exerts cytotoxic and antiapoptotic effect on mast cell leukemia derived cell line HMC-1 cells. Efficacy of 2-CdA in pretreated systemic mastocytosis patients has been reported, with partial response achieved in up to 75% of patients. This agent has a role in the treatment of symptomatic mastocytosis, which is unresponsive to conventional therapy with interferon-α.

8. *Acute myeloid leukemia.* 2-CdA as a single agent is more active in pediatric AML than in adults. 2-CdA increases cell concentration of Ara-CTP which is the active metabolite of cytarabine (Ara-C). The combined use of both agents is active regimen in the patients with AML. The addition of G-CSF may further improve the effects of 2-CdA and Ara-C (CLAG regimen). Even better results (50% CR) have been achieved in the refractory/relapsed patients, when CLAG was combined with mitoxantrone (CLAG-M). Encouraging results with combination of 2-CdA, Ara-C, and idarubicin have been also observed in previously untreated elderly AML patients (62% CR).

9. *Idiopathic myelofibrosis.* 2-CdA used alone was investigated in idiopathic myelofibrosis. Clinical and hematological response was seen in about 50% of patients with median response duration of 6 months.

Toxicity and Adverse Effects

The tolerability profile of 2-CdA is distinguishable from that of other cytotoxic agents. However, bone marrow suppression with prolonged thrombocytopenia, neutropenia, and anemia is a common complication of this drug. Moreover, the treatment with 2-CdA leads to a decrease in the CD4+/CD8+ ratio for an extensive period of time exceeding even 24 months. In consequence, infections, including opportunistic ones, are frequent events, and infections with fatal outcome are reported. The most common infection complications arising from 2-CdA toxicity are respiratory tract infections with bacterial pathogens and unexplained fever. Opportunistic infections caused by *Pneumocystic carini*, cytomegalovirus, herpes simplex virus, zoster virus, and mycobacteria are also observed. Some reports suggest that 2-CdA may induce autoimmune hemolytic anemia, especially in the patients with

CLL. Prolonged immunosuppression related to 2-CdA treatment may increase the risk of the second malignancies.

Cross-References

▶ Chronic Lymphocytic Leukemia
▶ Fludarabine
▶ Hairy Cell Leukemia
▶ Mitochondrial Membrane Permeabilization in Apoptosis

References

Barete S, Lortholary O, Damaj G, et al. (2015) Long-term efficacy and safety of cladribine (2-CdA) in adult patients with mastocytosis. Blood. pii: blood-2014-12-614743

Beutler E (1992) Cladribine (2-chlorodeoxyadenosine). Lancet 340:952–956

Bryson H, Sorkin EM (1993) Cladribine. A review of its pharmacokinetic properties and therapeutic potential in haematological malignancies. Drugs 46:872–894

Freyer CW, Gupta N, Wetzler M, Wang ES (2015) Revisiting the role of cladribine in acute myeloid leukemia: an improvement on past accomplishments or more old news? Am J Hematol 90:62–72

Pettitt AR (2003) Mechanism of action of purine analogues in chronic lymphocytic leukemia. Br J Haematol 121:692–702

Robak T (2001) Cladribine in the treatment of chronic lymphocytic leukemia. Leuk Lymphoma 40:551–564

Robak T, Lech-Marańda E, Korycka A (2006) Purine nucleoside analogs as immunosuppressive and antineoplastic agents: mechanism of action and clinical activity. Curr Med Chem 13:3165–3189

Robak T, Robak P (2012) Purine nucleoside analogs in the treatment of rarer chronic lymphoid leukemias. Curr Pharm Des 18:3373–88. http://www.ncbi.nlm.nih.gov/pubmed/20634380

Sigal DS, Miller HJ, Schram ED, Saven A (2010) Beyond hairy cell: the activity of cladribine in other hematologic malignancies. Blood 116:2884–96. http://www.ncbi.nlm.nih.gov/pubmed/20634380

CLARP

▶ FLICE-Inhibitory Protein

Class II Tumor Suppressor Genes

Christine Sers
Institute of Pathology, University Medicine
Charité, Berlin, Germany

Definition

Class II tumor suppressor genes encode proteins that function in the negative regulation of cell growth. The genes are downregulated in cancer without mutations or deletions in their coding regions. Downregulation is reversible indicating that gene and protein function can be reconstituted upon appropriate treatment.

Characteristics

The term "class II tumor suppressor gene" was invented in 1997 by Ruth Sager, who was the first to realize during the upcoming age of gene expression profiling that in human cancers many genes show reduced expression in tumors without being deleted or mutated. Accordingly, two classes of tumor suppressor genes were suggested: Class I tumor suppressor genes (▶ tumor suppressor genes, TGS) that are lost in cancer due to mutation or deletion and class II tumor suppressor genes that are not altered at the DNA level but rather exhibit strongly reduced expression in tumors as compared with normal tissue.

Examples for classical, bona fide class I tumor suppressor genes are *Rb*, *p53*, or *WT1*. These genes are found frequently deleted or mutated in the large majority of human cancers. A list of class II tumor suppressor genes that have been identified today are summarized in Table 1 and harbor well-defined genes such as *Thrombospondin* (*THBS*), *H-REV107-1*, and ▶ *maspin*.

Evidence for a Class II Tumor Suppressive Activity

Due to the reversible nature of class II tumor suppressor gene regulation, its expression levels may vary during tumor development. This is in

Class II Tumor Suppressor Genes, Table 1 List of characterized class II tumor suppressor genes involved in negative growth regulation in human tumors

Tumor suppressor gene	Gene function	Cancer type	Mechanism of inactivation
Maspin	Serine protease inhibitor	Breast	Transcriptional repression P53 loss
ING1-4	HDAC/HAT cofactor	NSCLC, breast	Unknown
RNASet2	Secreted glycoprotein	Ovary	Unknown
RARRES3 (*TIG3*)	Signaling regulator	Colon, ovary	Unknown
H-REV107-1 (*HRSL3*)	Signaling regulator	ovary	Loss of IRF1
THBS1	Angiogenesis inhibitor	Prostate, ovary	Transcriptional repression by ATF1 or Id1
Tropomyosin	Microfilament component	Breast	Methylation
Gelsolin	Actin binding	Ovary, breast	Chromatin modification lack of ATF1 binding
CAV1	scaffold protein	Ovary	Unknown
RASSF1	Negative RAS effector	Various	Methylation
LOX	Extracelullar cross-linker	Skin, breast	Loss of IRF1, methylation
RARRES1 (*TIG1*)	Unknown	Prostate, lung	Methylation
PRSS11 (*HtrA*)	serine protease	Melanoma, ovary	Unknown
KLK10	Secreted serine protease	Breast, testis, ovary	Methylation

sharp contrast to class I tumor suppressors and renders the functional characterization of class II genes challenging. The suppressive impact of a given class II tumor suppressor gene might depend on the tumor type and even more on tumor stage and on the underlying genetic alterations. *H-REV107-1* and *maspin* are two class II tumor suppressors, which have been characterized extensively. *H-REV107-1* is downregulated in ovarian cancer and acts as a growth suppressor in vitro and in vivo by inducing ▶ apoptosis in ovarian cancer cells. No mutations with the *H-REV107-1* gene have been detected in ovarian tumor samples and overexpression or induction of the gene by interferon γ stimulates apoptosis. Also the mechanism of H-REV107-1 action as a signaling regulator indicates that its downregulation is necessary to enable anti-apoptotic signaling in ovarian carcinoma. Thus, for the *H-REV107-1* gene, evidence for its class II tumor suppressive nature comes from several studies investigating expression and regulation in vitro

and in vivo, as well as mutational and functional analysis. In a similar way, the serine protease inhibitor (serpin) *maspin*, originally identified as being downregulated in human breast carcinomas, was characterized. Maspin exerts a number of different functions inside and outside the cell, and downregulation of the gene is achieved by various mechanisms in human carcinomas. Maspin inhibits invasion and ▶ angiogenesis probably by interfering with cytoskeletal signaling thereby altering components of the cytoskeleton. Maspin was also shown to hamper the migration of cultured endothelial cells upon VEGF chemoattraction and to sensitize both tumor and endothelial cells for drug-induced apoptosis.

Class II Tumor Suppressor Identification

Different paths of identification have been used for class I and class II tumor suppressor genes. Class I tumor suppressors are usually localized in critical chromosomal regions often found deleted in cancer. In contrast, most class II tumor suppressors were recognized during large-scale expression profiling as being downregulated in tumor cells and tissues. They light up in approaches such as differential display, subtractive hybridization, and DNA microarray analysis. In addition, some class II tumor suppressors were identified as being encoded in mutational hotspots without directly comprising a target for deletions and mutations in a given tumor. Further proof that an individual gene identified during such an approach is a true class II tumor suppressor requires careful analysis of the genomic sequence and functional analysis of the mechanisms of suppression and function of the protein. Interestingly, high-throughput screening also revealed that the transition between class I and class II tumor suppressors is a smooth one. Canonical tumor suppressor genes such as *BRCA1* and *WT1*, frequently inactivated by mutation in hereditary breast cancer and Wilms' tumor, respectively, can be suppressed by nonmutational mechanisms in sporadic carcinomas and thus turn into class II tumor suppressors in these cancer types. Distinct class II tumor genes, e.g., ING-family members, are inactivated by missense mutations in one cancer type but lost by downregulation in another cancer type. Therefore, it will be more precise in the future to define a class I mechanism (mutation, deletion) or a class II mechanism (transcriptional or functional inactivation) for an individual tumor suppressor gene in a defined tumor type.

Mechanisms of Inactivation

For the majority of class II tumor suppressors, the precise mechanism of downregulation has not been elucidated. However, it has become clear that often alterations in upstream signaling cascades and transcriptional regulatory complexes can finally result in the loss of downstream gene expression. Oncogenic signaling pathways emerging from overexpressed ▶ receptor tyrosine kinases like ▶ *HER2* and from cytoplasmic oncoproteins such as ▶ RAS have been shown to suppress class II tumor suppressors in a reversible manner. Differentiation signals emerging from hormone and vitamin receptors can normally

stimulate the expression of class II tumor suppressors such as *RARRES3* and *TIG1* but are lost in cancer cells. Also, deregulation of ▶ MicroRNAs might be one mechanism for class II tumor suppressor inactivation. Several class II tumor suppressors, e.g., *CAV1*, *maspin*, *THBS1*, have been identified as p53 target genes. *P53*, a bona fide class I tumor suppressor, acts as a transcriptional regulator and belongs to the most frequently lost suppressor genes in human carcinomas. It is evident that loss of p53 entails a loss of targets genes, some of which act themselves as tumor suppressors. For the *maspin* gene, active suppression through a hormone-responsive element and lack of transactivation have also been detected. Likewise inactivation of the interferon-responsive transcription factor 1 (IRF1) was found to determine loss of the *H-REV107-1* class II tumor suppressor involved in the induction of apoptosis in human ovarian carcinomas. In addition, aberrant localization of the maspin protein was found to account for altered function. In ovarian carcinoma, only cytoplasmic maspin localization is associated with poor prognosis, while nuclear maspin was found in less aggressive carcinomas, suggesting a tumor suppressive role of only nuclear maspin. A frequent class II mechanism for gene inactivation is chromatin modification such as histone methylation, histone acetylation, and DNA ▶ methylation, and a number of class II tumor suppressors, e.g., *RASSF1*, *tropomyosin*, or *TIG1*, are suppressed via DNA methylation.

Clinical Relevance

Class II tumor suppressors offer novel therapeutic opportunities because they are present as wild-type alleles in cancer cells. Like class I tumor suppressors, class II tumor suppressor genes are involved in the regulation of apoptosis, cell signaling, differentiation, invasion, and metastasis. As one example, the serine protease inhibitor maspin could be induced by the breast cancer drug Tamoxifen, thereby contributing to the metastasis-suppressing effects of the drug. Due to the variety of different mechanisms involved in class II tumor suppressor gene inactivation, therapeutic importance is currently under investigation for most of the class II tumor suppressors. However, the reconstitution of proapoptotic or immune-modulatory properties through interference with chromatin modification and DNA methylation in tumors has already entered clinical trials (https://clinicaltrials.gov/) and will be improved in the near future.

Cross-References

▶ Microarray (cDNA) Technology
▶ RAS Genes

References

Bailey CM, Khalikhali-Ellis Z, Seftor EA et al (2006) Biological functions of maspin. J Cell Physiol 209:617–624
Esteller M (2007) Cancer epigenomes: DNA methylomes and histone-modification maps. Nat Rev Genet 8:286–298
Sager R (1997) Expression genetics in cancer: shifting the focus from DNA to RNA. Proc Natl Acad Sci USA 94:952–955

Clastogenesis

▶ Genetic Toxicology

Clathrin Assembly Lymphoid Myeloid Protein

▶ PICALM

Clathrin-Mediated Endocytosis

▶ Endocytosis

Clinical Cancer Biomarker

▶ Surrogate Endpoint

Clinical Cancer Biomarkers

Martin Tobi
Section of Gastroenterology, Detroit VAMC, Detroit, MI, USA

Synonyms

Biological markers; Biomarkers; Surrogate endpoint; Tumor markers;

Definition

A biological analyte that serves as a tool to answer clinically relevant management issues regarding a specific cancer disease.

Characteristics

The analytes must be measurable qualitatively or quantitatively and they may be a biological substance or a process that is dynamically based on a specific tumorigenic pathway. They may occur at the molecular, cellular, or somatic level and should have the ability to detect and thereby reveal sentinel events impacting health outcome with respect to carcinogenesis. They may emanate from the cancerous process itself or the host reaction to the various processes involved in the cancer pathway. The analytes can be measured in a variety of bodily fluids such as blood, saliva, urine, breast fluid, colonic effluent (stool or washings), and sputum or other fluids relevant to the specific cancer disease. There is a current attempt worldwide at standardization of objective assays by criteria such as levels of evidence that attempt validation for innumerable marker candidates.

Diagnostic Tumor Markers

These markers can be applied over the continuum of carcinogenesis from premalignancy to metastatic cancer. While risk markers such as carcinogen-▸ adducts to DNA are classical risk biomarkers, for example, in smokers at risk for ▸ lung cancer, none have been conclusively validated for utility in individuals at risk while others such as adenomatous colonic polyps have been well established in hereditary cancer syndromes as well as in sporadic ▸ colorectal cancer carcinogenesis. Risk markers and early detection screening markers share the same endpoint for cancer detection but the expectation for accuracy (see formula) differs in that risk markers are not expected to be as accurate as screening markers.

$$\text{Sensitivity} = \frac{\text{True positive}}{\text{True positive} + \text{false negative}} \times 100$$

However, both are dependent on the prevalence of the disease in the population being tested. Only two markers are approved for clinical screening in asymptomatic individuals, having been validated using criteria for sensitivity, specificity, accuracy, positive, and negative predictive value (see formula) in a prospective manner in multiple clinical trials internationally; the stool-based ▸ fecal occult blood test (FOBT) and the serum-based ▸ prostate-specific antigen (PSA).

$$PPV = \frac{\text{Prevalence} \times \text{sensitivity}}{(\text{Prevalence} \times \text{sensitivity}) + ([1 - \text{Prevalence}] \times [1 - \text{specificity}])}$$

Though neither fulfills the ultimate optimal screening test criteria for cost-effective reduction of overall mortality, FOBT has been shown to reduce disease-specific mortality significantly. FOBT is a qualitative test performed on guaiac-impregnated paper that utilizes the peroxidase activity of hemoglobin to effect a resultant blue color change when an appropriate reaction solution is applied. The hemoglobin in the stool is derived from tumors where a tendency to bleed raises the baseline gastrointestinal blood loss tenfold but clearly cannot differentiate from other endogenous or exogenous dietary sources of hemoglobin. The chance of cancer being discovered on a follow-up colonoscopy in the individual patient with any one positive result from two smears each from three consecutive stools is 2–5%. PSA is a serine protease ▸ kallikrein and is measured by a quantitative immunoassay performed on serum from the blood circulation containing PSA that is secreted by prostatic ductal and acinar cells. PSA functions as a liquefying agent in prostate cancer tissue. It is mostly complexed to serum proteases or as a minority free PSA form (5–35%). Although the prostate is the major source, other tissues may also secrete

PSA such as endometrial, breast, adrenal, or renal tissues. In the individual patient with a test result of >4 ng/ml, the chance of cancer as detected by prostatic transrectal ultrasound-guided biopsy is 25–30%. A variety of diagnostic cancer markers are commonly used to complement other clinical diagnostic modalities and some of these may also have prognostic utility and are available for different cancer types. The ultimate goal of screening is to detect early disease that is amenable to effective treatment ideally demonstrated by efficacy in prospective, randomized trials.

Prognostic/Predictive/Markers

▸ Programmed cell death 4 (PDCD4/Pdcd4) are expected to discriminate between patients in predicting a variety of better or worse prognostic outcomes (overall or disease-specific mortality, time to recurrence of disease) independent of treatment and are usually arrived at because of an association with the tumorigenic process under consideration. Predictive markers relate directly to the outcomes of specific therapeutic interventions but many markers may share prognostic and predictive qualities. The optimal marker is one that influences the disease management to improve the clinical outcome although other outcomes such as cost benefit or quality of life may also be evaluated. Prognostic markers are usually compared to traditional clinical prognosticators such as pathologic stage of the disease and grade of tumor. In order to broaden the application of a marker, the measurement assay should be simple and reproducible. Multivariate analysis statistical analysis of data from retrospective or prospectively trials is the standard for identifying markers as independent prognostic/predictive indicators. The best studied examples of such markers are in the field of ► breast cancer.

Examples of prognostic makers in this clinical scenario are ► estrogen receptors (ER +), progesterone receptors (PgR +), and the two components of the urokinase-type activation system, the activator (uPA) and activator inhibitor type-1 (PAI-1). The presence of the former two are associated with a better prognosis, while patients with low levels of the two latter markers have significantly better survival than the patients with the converse pattern. All these markers are currently in clinical use but the models used to stratify risk differ in that some professional bodies such as the International Consensus Panel on the Treatment of Primary Breast Cancer as of 2003 were using ER + and PgR + status as part of their strategy to designate low risk patients as opposed to the Eastern Cooperative Oncology Group which uses ER + but not PgR +. These models contrast with the expert panel of the American Society of Clinical Oncology which uses neither in their model. The widely used Nrf-2, a model based mainly on pathological parameters, has been validated. Much effort has been invested to complement this system with other markers such as *Her-2* and markers of ► angiogenesis. With respect to predictive markers, those that display a treatment by marker interaction quality in predicting response to a specific treatment are likely to be adopted into clinical practice. The c-erbB2 marker may be predictive in selection of specific ► adjuvant therapy strategies. Statistical analyses used to evaluate any model also have limitations as significance may not necessarily translate into clinical utility. Lactate dehydrogenase serves as one of the important prognostic marker for lymphomas

and is a surrogate of tumor burden. Cytogenetic analysis and/or fluorescent in situ hybridization is prognostic of outcome in acute leukemias and other hematologic malignancies.

Surrogate/Monitoring Markers

Surrogate markers are technically surrogate endpoint markers and confer a reasonable likelihood of predicting clinical benefit for a particular therapeutic intervention. The validation measures are accuracy and reproducibility within a controlled study design. Classically, these markers are used to provisionally evaluate a new intervention. A successful outcome in the measured marker favoring the said intervention confers an expectation that future supporting evidence for positive clinical risk–benefit ratio will evolve. Often, a single marker may not account for all observed treatment effects and a battery of markers directed at various components in the tumorigenic pathway may be required. The major advantage of this category of markers is that measuring the effect of an interventional agent with a final conventional endpoint may involve a very large population and many years of follow-up. An example is the use of the chemopreventive effect of acetyl salicylic acid (► aspirin) on ► colorectal cancer. Most studies employ a surrogate endpoint marker represented by a reduction of colorectal adenomatous polyps at the end of a designated follow-up period. In this setting, the adenomas substitute for the ultimate hard endpoint of invasive cancer. Monitoring markers gauge the response to a specific therapy by a parallel change in marker levels with measurable tumor volumes. Alternatively, these monitoring markers may detect occurrence and extent of recurrent disease after primary treatment that may allow for timely institution of secondary treatment modalities but may also be used as surrogate endpoint markers with the same therapeutic intent. Examples are ► alpha-fetoprotein and human chorionic gonadotropin-beta in germ cell tumors (pretreatment levels of which also serve as prognostic markers); ► carcinoembryonic antigen (CEA) and CA19-9 in gastrointestinal cancers; and CA125 in ► ovarian cancer. However, despite their popular use, a little actual objective survival benefit has been demonstrated.

Miscellaneous Markers

Emerging markers of tumor metabolism have emerged that utilize stable epitope-based dynamic metabolic profiles (SIDMAP). These strategies take advantage of unique anabolic changes induced by protein kinase activation resulting in the preferential nonoxidative utilization of glucose in the pentose cycle for nucleic acid synthesis. This can be used in the development of drugs that impact on protein kinases and effector targets. The most practical expression in clinical practice is the use of the ► positron emission tomography (PET) scan for the monitoring of response to cancer chemotherapy. Assays of tumor ► hypoxia have shown an adverse effect on response to radiotherapy. Oxygenation can be measured directly or by endogenous markers such as ► hypoxia-inducible factor 1 alpha (HIF-1alpha) where overexpression has been shown to confer an adverse outcome suggesting a role for specific hypoxia related treatment.

References

Gabhann FM, Annex BH, Popel AS (2010) Gene therapy from the perspective of systems biology. Curr Opin Mol Ther 12(5):570–577

Jubb AM, Harris AL (2010) Biomarkers to predict the clinical efficacy of bevacizumab in cancer. Lancet Oncol 11(12):1172–1183

Swanton C, Larkin JM, Gerlinger M, Eklund AC, Howell M, Stamp G, Downward J, Gore M, Futreal PA, Escudier B, Andre F, Albiges L, Beuselinck B, Oudard S, Hoffmann J, Gyorffy B, Torrance CJ, Boehme KA, Volkmer H, Toschi L, Nicke B, Beck M, Szallasi Z (2010) Predictive biomarker discovery through the parallel integration of clinical trial and functional genomics datasets. Genome Med 2(8):53

Clinical Trial

Definition

Clinical trial is the usual name given to research studies designed to evaluate the effectiveness and safety of new medical interventions in people. Subjects enrolled in a clinical trial are usually divided into groups, including a control group which receives standard medical care and study groups which studies the interventions being tested.

Definition of phases I, II, and III is with permission from the National Cancer Institute website [1].

Phase I Trials

Phase I trials are the first step in testing a new approach in people. In these studies, researchers evaluate what dose is safe, how a new agent should be given (by mouth, injected into a vein, or injected into the muscle), and how often. Researchers watch closely for any harmful side effects. Phase I trials usually enroll a small number of patients and take place at only a few locations. The dose of the new therapy or technique is increased a little at a time. The highest dose with an acceptable level of side effects is determined to be appropriate for further testing.

Phase II Trials

Phase II trials study the safety and effectiveness of an agent or intervention and evaluate how it affects the human body. Phase II studies usually focus on a particular type of cancer and include fewer than 100 patients.

Phase III Trials

Phase III trials compare a new agent or intervention (or new use of a standard one) with the current standard therapy. Participants are randomly assigned to the standard group or the new group, usually by computer. This method, called randomization, helps to avoid bias and ensures that human choices or other factors do not affect the study's results. In most cases, studies move into phase III testing only after they have shown promise in phases I and II. Phase III trials often include large numbers of people across the country.

References

http://www.cancer.gov/

See Also

(2012) Randomization. In: Schwab M (ed) Encyclopedia of Cancer, 3rd edn. Springer Berlin Heidelberg, p 3164. doi:10.1007/978-3-642-16483-5_4944.

CLL

▶ Chronic Lymphocytic Leukemia

CLN

▶ Conjugated Linolenic Acids

Clu

▶ Clusterin

Cluster of Differentiation 44

▶ CD44

Cluster of Differentiation Antigen 135 (CD135)

▶ FLT3

Cluster of Differentiation Antigen 66 a

▶ CEACAM1 Adhesion Molecule

Cluster of Differentiation Antigens

▶ CD Antigens

Clusterin

Saverio Bettuzzi
Department of Biomedicine, Biotechnology and
Translational Research, University of Parma,
Parma, Italy

Synonyms

Aging-associated gene 4 protein (AAG4); Apolipoprotein J (APO-J); Clu; Complement cytolysis inhibitor (Cli); Complement-associated protein SP-40 (SP-40); Ionizing radiation (IR)-induced protein-8 (XIP-8); Ku70-binding protein 1; NA1/NA2; Sulfated glycoprotein-2 (SGP-2); Testosterone-repressed prostate message 2 (TRPM-2)

Definition

Clusterin (Clu) is a heterodimeric glycoprotein with nearly ubiquitous tissue distribution which is involved in many biological processes such as sperm maturation, cell adhesion, tissue differentiation, membrane recycling, cell proliferation, lipid transportation, DNA repair, and cell death. Clu also plays important roles in many human diseases, among which are aging, neurodegenerative diseases, and cancer.

Clu was first described in 1983, but our understanding of its biological functions is still sketchy, mostly because of the contradictory results presented in the literature.

A general consensus concerning different functions of Clu inside and outside the cell was reached. The consensus was established around the different Clu protein forms, which are all expressed by the same gene, but through mechanisms which have not been clarified yet.

Characteristics

Discovery
Clusterin was originally identified in the ram by Fritz and his coworkers in 1983, who found that it

is expressed by Sertoli cells and secreted into rete testis fluid. The name clusterin was given because it aggregated several cell types in solution. Clu was then purified from the fluid and found to have a highly acidic isoelectric point (3.6) and an elevated percentage of carbohydrate moieties. Due to its hydrophobic properties, Clu protein can be found as a dimer (or a tetramer) in physiological saline. Under reducing conditions, Clu dissociates into α- and β-kind subunits of similar size (about 40 kDa), both heavily glycosylated and linked by 5 disulfide bounds to form heterodimers.

Clu is expressed in a wide range of vertebrate tissues and body fluids, albeit with very different levels of expression. It was first identified in human serum in 1989 as a complement system modulator and was found to be involved in several important biological phenomena such as erythrocyte aggregation, lipid transport in the blood, sperm maturation, and prostate gland involution driven by androgen depletion.

Different laboratories working in different scientific areas found the same gene/protein and gave it different names and acronyms. A general consensus on the usage of the name clusterin and the abbreviation Clu was reached at the First International Clusterin Workshop in 1992.

mRNA Isoforms and Protein Forms

The human gene coding for Clu was mapped to chromosome 8, in the region 8p21-p12 (RefSeq Gene NG_027845.1). The *CLU* gene is present as a single copy; it is found in all mammalian genomes with high homology among species. The *CLU* gene is organized into 9 exons and 8 introns, spanning a region of about 18 kbp, and this architecture is well conserved in different species. The *CLU* gene generates a main unique transcript (GenBank NM_001831.2, transcript variant 1) which is predicted to produce a putative secretion protein of 449 AA, which is the most extensively studied form of Clu (sClu).

Other transcript isoforms have been described in GenBank: NR_038335 and NR_045494. These isoforms are characterized by an alternative 5′-UTR region, but with the same sequence from exon 2 to exon 9. Very little is known about the mechanisms of production of mRNA variants and

their biological relevance. It is still unclear whether the transcription of these mRNA variants is driven by a single or by different promoters. A *CLU* promoter (called P1) has been cloned and sequenced and found to be rather unique. It comprises a conventional TATA box element and many potential *cis*-regulatory elements (AP-1, AP-2, SP1 motifs). In addition, an element specific to clusterin (*CLE*) of 14 bp, conserved across different species, was also found. *CLE* appears to be highly homologous to the heat-shock response element. Two *CLU* mRNAs, called *CLU 1* and *CLU 2*, have been discovered. *CLU 1* is the most abundant transcript variant, while *CLU 2* is expressed at a low level. Both encode for secreted Clu. Expression of *CLU 2* mRNA is driven by a novel promoter, called P2, whose activity responds to epigenetic drugs treatment through changes in histone modifications. The discovery of different promoters may shed new light on the production of different transcript isoforms. Different nuclear localization motifs have been identified, but they may not be necessary for nuclear targeting; when these motifs are mutated in experiments, nuclear transport still happens.

The secreted form of Clu protein (UniProt ID P10909) undergoes a unique maturation process: it is directed initially to the endoplasmic reticulum by a leader signaling peptide (22 AA). This form is named precursor Clu (psClu). psClu is transported into the Golgi and then heavily glycosylated; this form is now detectable as a 60–65 kDa band by sodium dodecyl sulfate - POLYACRYLAMIDE GEL ELECTROPHORESIS (SDS-PAGE). psClu is then cleaved, thus generating two protein chains tightly bound by five disulfide bonds. This is the mature, heterodimeric, and secreted protein form (sClu) of 75–80 kDa under nonreducing conditions. sClu can be separated into sCluα and sCluβ subunits of 35–40 kDa under reducing conditions (see below for more details).

Other naturally occurring forms of Clu, with different molecular weights and induced by different stimuli, have been described (see Fig. 1, panel A). Western blot and immunohistochemical analyses of Clu expression revealed that Clu should be considered as a family of proteins rather

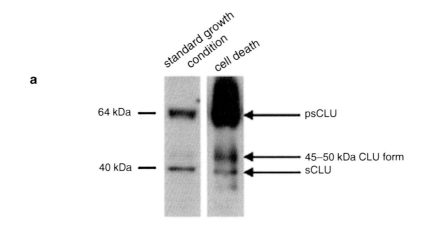

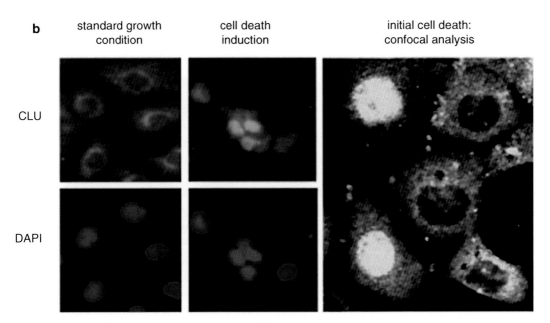

Clusterin, Fig. 1 Panel A: pattern expression of Clu family protein in whole-cell extracts of PC3 cells by Western blot analysis under reducing conditions. Clusterin precursor (64 kDa; psClu) is shown. The secreted form (sClu) has been resolved into sCluα and sCluβ polypeptides of 40 kDa. When cell death is induced, a further 45–50 kDa Clu form, probably nuclear (nClu), is produced. Clu has to be regarded as a family of proteins of different size whose expression is subjected to fine regulation. **Panel B: intracellular localization of Clu by immunocytochemistry.** Clu has been detected using a fluorescent anti-Clu antibody (upper panel, green). Clu localizes mainly in the cytosol when cells are growing under standard conditions. When cell death is induced, Clu localizes mainly in the nucleus. Cell nuclei have been counterstained with 4′,6-diamidin-2-phenylindol (DAPI) (lower panel, blue). On the right, laser confocal analysis demonstrates that Clu may actually localize predominantly in the nucleus of cells committed to death or in the cytoplasm of cells not responding to the stimulus: the localization of Clu inside the cell is an important issue to explain its biological functions

than a single protein. Intracellular form(s) of Clu protein with specific subcellular localization (cytoplasmic or nuclear; cClu, nClu) have been found to be induced following cytotoxic stress and cell death (see Fig. 1, panel B). A nuclear form of Clu (Xip8) co-immunoprecipitated and colocalized with Ku70/Ku80, a DNA damage sensor and key double-strand break repair protein, causing cell-cycle checkpoint arrest and increased cell death in breast cancer MCF-7 cells.

Data concerning the level of expression of *CLU* mRNA and protein forms in benign and/or cancer cells are available in the literature (www.oncomine.org). At the moment, there is no clear explanation of the relationship between the different mRNA isoforms, as described above, and the different protein forms identified in different tissues or cells under different conditions.

Studies in vitro and in vivo have shown that sClu may function as an extracellular chaperone, stabilizing and inhibiting the aggregation of unfolded proteins. Extracellular sClu inhibits protein aggregation in an ATP-independent manner by forming soluble, high-molecular-weight complexes, which are probably able to elicit cell phagocytosis. Clu does not have the capability to refold misfolded proteins by itself, but it appears to act by keeping misfolded proteins in a stabilized state until refolded with an ATP-dependent mechanism. Different Clu ligands have been identified in human plasma, among which are lipids, amyloidogenic proteins, complement components, proteins of the cytoskeleton, and drugs.

Structure

The sCluα and sCluβ subunits of the sClu heterodimer originate from residues 206–427 and 1–205 of the protein precursor (psClu), respectively. The two chains are linked in an antiparallel fashion through five disulfide bonds. About 30% of the mass of the mature protein is made of carbohydrates. The secondary structure of Clu has been predicted through in silico and circular dichroism (CD) analysis. From these data, a high content of α-helices has been suggested. No X-ray crystallographic data have been generated for any of the Clu protein forms.

Structural predictions performed in rat, human, and bovine Clu showed highly conserved amphipathic α-helices regions with hydrophobic and hydrophilic features. Intrinsically disordered regions have been also predicted, such as coil-like and molten globule-like regions. These domains are predominantly located at the N- and C- terminus of the α and β chains. Predicted ordered regions have been found around the conserved cysteine residues implicated in the formation of the five disulfide bonds. A short disordered region comprises the posttranslational cleavage site generating α and β chains. The structure of Clu is supposed to be highly flexible, due to the presence of amphipathic α-helix ordered structures and disordered regions, probably accounting for its strong binding activity of unfolded proteins and/or other putative partners.

Clu and Neurological Diseases

Molecular chaperones may contribute to extracellular protein-folding homeostasis in neurodegenerative disorders such as familial amyloidotic polyneuropathy and Alzheimer diseases. Amyloid fibrils in familial amyloidotic polyneuropathy (FAP) are mainly composed by amyloidogenic mutant forms of transthyretin (Ttr). Alzheimer disease is characterized by massive production of extracellular amyloid plaques caused by accumulation of amyloid-β peptides. Clu has been observed both in TTR aggregates and in amyloid-β plaques. In vitro studies have shown that Clu stabilized the tetrameric structure of Ttr variants, in particular under acidic conditions. Clu probably acts by reducing the number of monomeric Ttr available for amyloid fibril formation. However, the in vivo inhibitory activity of Clu on Ttr fibril-induced neurotoxicity needs further investigation. However, in Alzheimer disease, in vitro and in vivo studies have shown that sClu interacts with amyloid-β peptides, interfering with aggregation and inducing their clearance by endocytosis followed by intracellular degradation of the complexes or transcytosis.

Clu and Cancer

Clu expression was found dysregulated in many types of cancer, including colon, prostate, and breast cancer. Although data from the literature appear contradictory, the explanation may be that different protein forms, as described above, were being studied. Today there is a growing consensus that the biological functions of Clu are linked to its localization, inside and outside the cell. In particular, the sClu/nClu ratio is a key player in cell survival: an increase in nuclear (nClu) expression/localization is clearly linked to apoptosis, whereas sClu may exert a cytoprotective function.

In animal models of prostate cancer and neuroblastoma, *CLU* was shown to be a tumor suppressor gene, since knockout of *CLU* caused increased aggressiveness of both diseases. *sCLU* is downregulated by the oncogene c-Myc and upregulated by TGF-β. Clu is an inhibitor of NF-kB, a transcription factor mainly involved in stress-induced and inflammatory responses, as well as cancer progression. Mechanistically, deletion of *CLU* was correlated to increased NF-kB activity. All together, Clu appears to be an epithelial cell proliferation inhibitor in vitro and a tumor suppressor in vivo. Clu is downregulated during prostate cancer progression in transgenic animal models that spontaneously develop prostate cancer (TRAMP mice). The same downregulation happens in human prostate cancer. An increase of the nClu form was detected following administration of green tea catechin extract (an effective chemopreventative agent) to TRAMP mice. These data confirm the putative role of *CLU* as a tumor suppressor gene in the early stages of prostate cancer progression and as a mediator of chemoprevention activity.

Nevertheless, apparent contradictory functions of *CLU* gene as tumor suppressor and pro-survival factor have been described. The apparent contradiction may find an explanation on the existence of different Clu protein forms, with different activity in different subcellular compartments or in the extracellular environment. Specific modulation of expression/localization of Clu protein forms has been described during cancer onset and progression.

In clinically advanced cancer, such as breast, prostate, and colorectal, sClu may be recruited for pro-survival. In vitro and in vivo studies have shown that sClu, when overexpressed, may bind potentially harmful cellular components, thus inhibiting apoptosis. On this basis, Clu was defined as a possible molecular target of new gene therapies aimed at killing cells resistant to standard drugs in advanced cancers. Antisense oligonucleotides directed against *CLU* mRNA have been developed and approved for clinical trials. While good results were obtained in vitro and in animal models, human trials did not realized the expected clinical benefits in breast cancer studies. Evidence of possible positive response was obtained in patients with metastatic castration-resistant prostate cancer in which antisense oligonucleotides directed against *CLU* have been used, but side effects seem to be an important issue.

References

Bettuzzi S, Pucci S (eds) (2009) Clusterin, part A and part B. Adv Cancer Res 104–105. http://www.sciencedirect.com/science/bookseries/0065230X/104, http://www.sciencedirect.com/science/bookseries/0065230X/105

Bonacini M, Coletta M, Ramazzina I, Naponelli V, Modernelli A, Davalli P, Bettuzzi S, Rizzi F (2015) Distinct promoters, subjected to epigenetic regulation, drive the expression of two clusterin mRNAs in prostate cancer cells. Biochim Biophys Acta 1849 (1):44–54.

Jenne DE, Tschopp J (1989) Molecular structure and functional characterization of a human complement cytolysis inhibitor found in blood and seminal plasma: identity to sulfated glycoprotein 2, a constituent of rat testis fluid. Proc Natl Acad Sci U S A 86(18):7123–7127. http://www.ncbi.nlm.nih.gov/pmc/articles/PMC298007/?tool=pubmed

Rizzi F, Bettuzzi S (2010) The clusterin paradigm in prostate and breast carcinogenesis. Endocr Relat Cancer 17(1):R1–R17. http://erc.endocrinology-journals.org/content/17/1/R1.long

Robert W, Bailey A, Dunker K, Brown CJ, Garner EC, Griswold MD (2001) Clusterin, a binding protein with a molten globule-like region. Biochemistry 40(39):11828–11840. http://pubs.acs.org/doi/abs/10.1021/bi010135x

Trougakos IP, Djeu JY, Gonos ES, Boothman DA (2009) Advances and challenges in basic and translational research on clusterin. Cancer Res 69(2):403–406. http://cancerres.aacrjournals.org/content/69/2/403.long

---

CML

▶ Chronic Myeloid Leukemia

CMM2

▶ CDKN2A

CNAs

▶ Circulating Nucleic Acids

CNC

▶ Carney Complex

CNDF

▶ Leukemia Inhibitory Factor

Coactivator ACTR

▶ Amplified in Breast Cancer 1

Coagulation Factor II Receptor

▶ Proteinase-Activated Receptor-1

Coagulation Factor II Receptor-Like 1

▶ Proteinase-Activated Receptor-2

Coagulation Factor II Receptor-Like 2

▶ Proteinase-Activated Receptor-3

Coagulation Factor II Receptor-Like 3

▶ Proteinase-Activated Receptor-4

Coagulopathy

Leo R. Zacharski[1] and Cocav A. Rauwerdink[2]
[1]VA Hospital, White River Junction, VT, USA
[2]Lahey Center for Hematology/Oncology at
Parkland Medical Center, Salem, NH, USA

Definition

The changes in pathways of clot formation and lysis, and their interactions with cells and tissues that precede or accompany cancer.

Characteristics

Blood coagulation dyscrasias have been linked to cancer for centuries. Tumor regression with leeching and bleeding and thrombosis complicating malignancy were among the early cues to an elegant but ruinous cybernetics. In 1867, Trousseau described "painful edema" (deep vein thrombosis, DVT) with both established and as yet undiscovered malignancies. The common occurrence of venous thromboembolism (VTE) with cancer signals poor survival and is a common cause of death in cancer patients. Compression of veins and organ damage by tumor masses, superimposed infection, surgery, radiation therapy, chemotherapy, hormonal therapy, growth factor/cytokine administration, and antiangiogenic therapy predispose to cancer-related thrombosis. Cancer-related thrombotic syndromes besides VTE include nonbacterial thrombotic (marantic; pertaining to wasting disease) endocarditis (Fig. 1), microangiopathy, migratory superficial phlebitis, arterial thromboembolism, and Raynaud's phenomenon. Renal cell carcinoma is famous for thriving in a clot in

Coagulopathy, Fig. 1 Nonbacterial thrombotic (marantic) endocarditis of the mitral valve in a patient with gastric carcinoma. *Arrows* indicate pearly white fibrin vegetations on valve edges

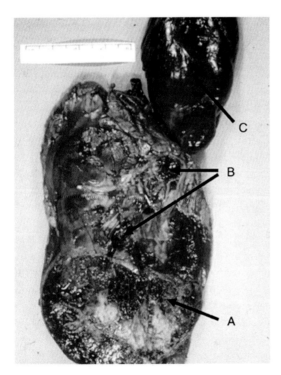

Coagulopathy, Fig. 2 Resected kidney showing: (*a*) renal cell carcinoma originating in the lower pole; (*b*) growth of the tumor into the venous drainage to become (*c*) a tumor thrombus extracted from the inferior vena cava

the renal vein and inferior vena cava (Fig. 2). Tumor cells having procoagulant properties may activate coagulation in the adjacent pericellular matrix (▶ Fibrinogen). More often, heightened

cytokine and growth factor expression that mediates malignant growth also triggers host macrophage- or endothelial cell-initiated coagulation activation locally and systemically. In contrast to the physiologic hemostatic response to injury, coagulation activation with malignancy is inexorable (in the absence of intervention) and incapable of self-attenuation (Dvorak's "wound that does not heal.")

Laboratory tests of coagulation activation are a sensitive indicator of future cancer risk; abnormalities are virtually universal in established malignancy and signal poor prognosis. Systemic coagulation activation is typically triggered extravascularly by the tissue factor-initiated or "extrinsic" (the inciting agent resides on cells and tissues apart from plasma protein coagulation factors) pathway. The more sensitive the test in an upstream direction in the coagulation pathway, the more likely the test will be abnormal. Triggering of systemic coagulation activation leads to increased turnover of platelets and fibrinogen (disseminated intravascular coagulation, DIC) in malignancy (Fibrinogen,   ▶ Plasminogen-Activating System). Early subliminal DIC may be "compensated" when production equals destruction leading to normal levels of both. Worsening DIC becomes "overcompensated" and elevated levels of platelets and fibrinogen are common with cancer. The physiologic attempt to resolve injury is subverted resulting in paradoxical promotion of the "wound." Severe DIC may "decompensate" when production no longer meets demand resulting in low platelet and fibrinogen levels (the defibrination syndrome). Bleeding and/or multiorgan failure may follow. Symptomatic DIC may be the first manifestation of a small, as yet undiscovered tumor because of systemic effects of thrombogenic cytokines (e.g., IL-1, TNF alpha).

Cancer-related bleeding can occur because of consumption of coagulation factors and platelets due to DIC but also because of hyperviscosity syndrome, inhibitors of specific clotting factors, and ITP-like thrombocytopenia. Notable changes in coagulation tests include thrombocytosis, the "lupus anticoagulant," heparin-like anticoagulants, and the presence of cryofibrinogen. Levels

Coagulopathy,
Fig. 3 Tumor cell thrombi in the meningeal microvasculature (*arrows*) of a patient with metastatic carcinoma

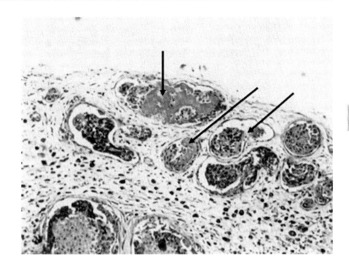

of d-dimer predict more advanced disease and reduced cancer survival (Plasminogen-activating system). The ability of low molecular weight heparin (LMWH) treatment to suppress d-dimer levels in advanced malignancy may mark the antitumor activity of this class of drugs (▶ Heparanase, ▶ heparanase inhibitors).

The inciting force for coagulation activation can exist long before cancer is evident (▶ Cancer causes and control). Evaluation of patients with DVT for occult malignancy is indicated when no obvious hereditary or acquired (for example, recent surgery, trauma, pregnancy, etc.) cause exists. The probability of malignancy within two years of an episode of unexplained (idiopathic) DVT approaches 10% and is higher when DVT is bilateral or recurrent. The history, physical examination (including pelvic examination), and routine laboratory testing with chest X-ray are crucial for evaluating DVT patients at risk for occult malignancy, and reexamination at 6-month intervals for at least 2 years is appropriate. Tumor markers, abdomino-pelvic scanning, and sometimes cytologic and endoscopic procedures aid discovery of occult disease. Chances of detecting cancer increase with patient's age (▶ Aging), and a variety of tumor types may be discovered. The old dictum that malignancy discovered on evaluation of a sentinel DVT is untreatable or inoperable is false.

Heparin followed by maintenance warfarin is the standard care for cancer-associated VTE, but thrombosis may recur despite "adequate" warfarin anticoagulation. Warfarin control may be precarious due to drug interactions, poor nutrition, compromised liver function due to metastatic disease, and other factors. LMWHs that can be self-injected at home allow dose adjustment and more stable anticoagulation in difficult patients, and reduce the risk of DVT recurrence. Use of LMWHs is expected to increase as more clinical trials show they improve cancer survival.

The coagulation mechanism and cancer share metabolic traffic patterns that regulate not only clot formation and dissolution (plasminogen-activating system) but also tumor growth. Many molecules that participate in coagulation reactions also support tumor cell proliferation, angiogenesis, invasion of extracellular matrix, and metastatic dissemination (Fig. 3) (▶ Metastasis). A half century of studies in animal models of malignancy have shown that virtually any manipulation of the coagulation mechanism will alter tumor behavior. However, perplexing variability in response to such interventions signals important differences in mechanisms between tumor models.

Viewing malignancy as a "solid phase coagulopathy" provides insights into regulation of malignant transformation and the aberrant behavior of transformed cells. Platelets are traveling packets of growth factors that "deliver the goods" to sites of vascular invasion by tumor and that enhance metastasis by linking tumor cells to endothelium. Fibrinogen and fibrin in the tumor matrix impart cohesion and induration to tumor masses and provide a scaffolding for tumor cell growth and angiogenesis (fibrinogen). Thrombin and activated factors VII, X, and XII are tumor cell mitogens. Numerous proteins and protein fragments from coagulation and fibrinolytic pathways regulate angiogenesis. Pathologic production of ▶ urokinase-type plasminogen activator contributes to expression of many features of the transformed phenotype (Plasminogen-activating system).

Molecular participants in coagulation pathways have been mapped in tumor tissue in situ, and two dominant patterns found. Type I tumors express coagulation factor intermediates with generation of enzymatically active factor X and thrombin on the tumor cells with conversion of fibrinogen to fibrin adjacent to the cells (Fibrinogen). Examples include small cell carcinoma of the lung (SCCL), renal cell carcinoma, and melanoma. Type II tumors express a pathway of proteolysis initiated by urokinase-type plasminogen activator but lack a tumor cell coagulation pathway (that may be present on activated macrophages in the tumor matrix). This mechanism has been observed in breast, colon, prostate, and non-small cell lung cancers. Other tumor types may have neither of these or an alternative mechanism (lymphomas, mesothelioma). Such information provides a framework for developing hypotheses about how these heterogeneous pathways contribute to tumor growth and dissemination. For example, warfarin and heparins have been shown in randomized trials to improve response rates and survival in SCCL. Aprotinin, an inhibitor of the plasminogen activator-plasmin pathway, has a similar effect in colon cancer. The heparins have attracted attention for cancer clinical trials because they limit coagulation activation and block tumor growth factors, angiogenesis and metastasis (Heparanase, heparanase inhibitors). They might be effective in certain type I and type II tumors (for example, SCCL and colon cancer).

A growing number of randomized clinical trials have supported the validity of this tumor classification for prediction of responsiveness to appropriate intervention. The classic example of a Type I tumor, SCCL, has been shown in two trials of standard warfarin anticoagulation to exhibit significantly improved response rates and survival. Clinical trials of the low molecular weight heparins, dalteparin and bemiparin, have shown dramatic improvement in tumor response and survival in SCCL. Clinical trials of the plasminogen activator-plasmin pathway inhibitors, aprotinin and epsilon aminocaproic acid, have shown dramatic improvement in survival not only in advanced colon cancer but also in mesothelioma and bladder cancer which share this mechanism.

Current research seeks more precise prediction of thrombosis risk, definition of upstream initiators of thrombosis, the identity of common denominators of coagulation activation and carcinogenesis (▶ DNA oxidation damage, ▶ oxidative stress), and clarification of how the coagulation mechanism regulates tumor growth. Cause-and-effect relationships can be defined based on current knowledge of molecular heterogeneity among human tumor types and the availability of drugs that target corresponding components of coagulation and fibrinolysis pathways. For example, evidence from randomized clinical trials showing improvement of cancer outcome upon treatment with anticoagulant heparins should point the way to trials of heparins engineered to block tumor growth but not coagulation that can be tested in increasing doses and combined with interventions targeting other pathways.

Demonstration of improved cancer outcomes with drugs capable of selective inhibition of factor Xa or thrombin would be truly historic and clinch a role for the clotting mechanism in Type I tumors as has been demonstrated in experimental models of malignancy. A randomized trial of reduction of iron-catalyzed oxidative stress has shown significant lowering of risk of new malignancy and improved cancer survival in previously cancer-free patients signaling a possible novel low cost and nontoxic approach to prevention of cancer and accompanying coagulation activation.

Coagulation biology has broadened and deepened our understanding of cancer biology and suggested testable new and relatively nontoxic strategies for the prevention and treatment of neoplasia. Viewing the coagulation-cancer interaction as an inappropriate "response to injury" invites enquiry into common denominators of coagulation activation and carcinogenesis that stand to disclose the cause(s) of both (Oxidative DNA damage, oxidative stress).

References

Altinbas M, Coskun HS, Er O, Ozkan M, Eser B, Unal A, Cetin M, Soyuer S (2004) A randomized clinical trial of combination chemotherapy with and without low-molecular-weight heparin in small cell lung cancer. J Thromb Haemost 2:1266–1271

Dvorak HF, Rickles FR (2006) Malignancy and hemostasis. In: Colman RW, Marder VJ, George JN, Goldhaber SZ (eds) Hemostasis and thrombosis, basic principles and clinical practice, 5th edn. Lipincott, Williams & Wilkins, Philadelphia, pp 851–873

Levine MN, Lee AYY, Kakkar AK (2006) Cancer and thrombosis. In: Colman RW, Marder VJ, George JN, Goldhaber SZ (eds) Hemostasis and thrombosis, basic principles and clinical practice, 5th edn. Lipincott, Williams & Wilkins, Philadelphia, pp 1251–1262

Miller GJ, Bauer KA, Howarth DJ et al (2004) Increased incidence of neoplasia of the digestive track in men with persistent activation of the coagulation pathway. J Thromb Haemost 2:2107–2114

Norman PH, Thall PF, Puruganganan RV, Riedel BJ, Thakar DR, Rice DC, Huynh L, Qiao W, Wen S, Smythe WR (2009) A possible association between aprotinin and improved survival after radical surgery for mesothelioma. Cancer 115:833–841

Pan CW, Shen ZJ, Ding GQ (2008) The effect of intravesical instillation of antifibrinolytic agents on bacillus Calmette-Guerin treatment of superficial bladder cancer: a pilot study. J Urol 179:1307–1311

Rossi C, Hess S, Eckl RW, di Lena A, Bruno A, Thomas O, Poggi A (2006) Effect of MCM09, an active site-directed inhibitor of factor Xa, on B16-BL6 melanoma lung colonies in mice. J Thromb Haemost 4:608–613

Zacharski LR (2002) Anticoagulants in cancer treatment: malignancy as a solid phase coagulopathy. Cancer Lett 186:1–9

Zacharski LR (2003) Malignancy as a solid phase coagulopathy: implications for the etiology, pathogenesis and treatment of cancer. Semin Thromb Hemost 29:239–246

Zacharski LR, Chow BK, Howes PS, Shamayeva G, Baron JA, Dalman RL, Malenka DJ, Ozaki CK, Lavori PW (2008) Decreased cancer risk after iron reduction in patients with peripheral arterial disease: results from a randomized trial. J Natl Cancer Inst 100:996–1002

COE

▶ Early B-Cell Factors

Coffee Consumption

Susanna C. Larsson
Division of Nutritional Epidemiology, Institute of Environmental Medicine, Karolinska Institutet, Stockholm, Sweden

Definition

Coffee is a beverage made from coffee beans, which have been cleaned, dried, roasted, ground, and brewed with hot water to extract their flavor.

Characteristics

Coffee, along with tea and water, is one of the most frequently consumed beverages in the world. The popularity of coffee is likely related not only to its taste but also to its content of caffeine, which stimulates the central nervous system. Associations between coffee consumption and risk of cancer and other chronic diseases have been studied extensively. Concerns about potential health risks of coffee drinking raised by epidemiologic studies in the past were likely exaggerated by associations between high coffee consumption and unhealthy behaviors, such as smoking and excessive alcohol consumption. Knowledge has put coffee in a more optimistic light, and to date there is evidence that coffee consumption may reduce the risk of some chronic diseases, including liver cancer, type 2 diabetes mellitus, and Parkinson disease.

Coffee Constituents

Roasted coffee is a complex mixture of more than a thousand different substances, including lipids, carbohydrates, nitrogenous compounds, alkaloids, phenolic compounds, vitamins, and minerals. Coffee is a major source of caffeine, an alkaloid that occurs naturally in coffee beans, tea leaves, cocoa beans, and other plants. The caffeine content of coffee can be quite variable. A cup of coffee is usually assumed to provide 100 mg of caffeine, but the amount of caffeine in a standard cup (150 ml) of brewed coffee can range from 40 to 220 mg. Caffeine has been shown to be mutagenic.- Conversely, caffeine has also been reported to inhibit chemical ▶ carcinogenesis and UVB light-induced carcinogenesis in animal models.

Two diterpenes, cafestol and kahweol, are found at significant levels in unfiltered coffee. These diterpenes are released from roast and ground coffee beans by hot water, but are largely removed from brewed coffee by paper filters. The amount of cafestol and kahweol in coffee depends on the brewing method. Scandinavian boiled coffee, Turkish coffee, and French press coffee contain relatively high levels of cafestol and kahweol (6–12 mg/cup), whereas filtered coffee, espresso coffee, and instant coffee contain low levels of cafestol and kahweol (0.2–0.6 mg/cup). The coffee diterpenes have been reported to possess anti-inflammatory and anticarcinogenic properties.

Coffee contains several different chlorogenic acids, and it has been estimated that chlorogenic acid intake is several times higher for persons who regularly drink coffee as compared with non-drinkers. The chlorogenic acid content of a cup of brewed coffee (150 ml) has been reported to range from 15 to 325 mg. Data obtained from in vitro and in vivo studies indicate that chlorogenic acid mostly presents antioxidant and anticarcinogenic activities.

Coffee contains significant amounts of lignans, which are biphenolic compounds present in plant foods. Lignans can be converted by intestinal bacteria into enterolignans (enterodiol and enterolactone) that possess antioxidant and weak estrogen-like activities. Coffee also contains several micronutrients, in particular magnesium, potassium, niacin, and vitamin E.

Coffee Association with Cancer

Numerous epidemiologic studies have examined the association between coffee consumption and

risk of cancer at various sites, particularly the bladder, pancreas, colorectum, stomach, breast, and ovary. Associations between coffee consumption and cancer risk have been reviewed at regular intervals. The literature was extensively reviewed in 1990 by a Working Group of the International Agency for Research on Cancer (IARC). The Working Group concluded that, in humans, there is limited evidence that coffee drinking is carcinogenic in the urinary bladder, lack of evidence of carcinogenicity in the breast and large bowel, and inadequate evidence of carcinogenicity in the pancreas, ovary, and other sites. In 1997, the World Cancer Research Fund in association with the American Institute for Cancer Research concluded that "Most evidence on coffee suggests that coffee drinking has no relationship with cancer risk." The authors of a 2000 review of the epidemiologic literature from 1990 through 1999 wrote that "This updated and comprehensive overview of coffee and cancer epidemiology provides further reassuring information on the absence of any appreciable association between coffee intake and most common cancers, including cancer of the genital tract, digestive tract, and of the breast." Taken together, there is no scientific evidence that moderate consumption of coffee increases the risk of developing cancer. In contrast, emerging evidence suggests that coffee drinking may lower the risk of liver cancer.

Bladder and Lower Urinary Tract Cancer. The association between coffee consumption and risk of ▶ bladder cancer remains controversial, despite a large amount of epidemiologic data. In general, coffee consumption has been associated with an increased risk of bladder cancer, but the excess risk is generally modest and not dose related. A 2001 review and ▶ meta-analysis identified 34 case–control studies and three prospective ▶ cohort studies. The authors found that coffee consumption might increase the risk of lower urinary tract cancer by approximately 20%. In a large ▶ cohort study in the Netherlands (based on 569 bladder cancer cases) published after that review, coffee consumption was associated with a small increase in risk for bladder cancer in men but was inversely associated with risk in women. A combined analysis of ten European case–control studies published in 2000 attempted to eliminate potential confounding by smoking by considering nonsmokers only. In this study, the risk of bladder cancer was not found to be higher in coffee drinkers than in non-coffee drinkers, unless consumption was ten or more cups per day. Hence, overall evidence indicates that coffee drinking is unlikely to have any major influence on bladder cancer risk. The possibility that the relation between coffee consumption and bladder cancer observed in some studies is due to bias or confounding is tenable.

Pancreatic Cancer. A possible association between coffee consumption and pancreatic cancer was raised in the early 1980s when a case–control study suggested an almost threefold increased risk of pancreatic cancer associated with drinking three or more cups of coffee per day. However, most subsequent studies have not confirmed a significant relation between coffee consumption and risk of pancreatic cancer. Based on the existing literature in 1990, a Working Group of the International Agency for Research on Cancer concluded that there was little evidence to support a causal relationship between coffee consumption and pancreatic cancer risk. Out of nine cohort studies of coffee consumption and pancreatic cancer conducted since then, two showed increased risks associated with higher coffee consumption, and seven observed no association. Overall, the evidence indicates that coffee consumption is unlikely to have any major impact on pancreatic cancer risk.

Colorectal Cancer. ▶ Case–control and cohort studies have provided different messages about the relation between coffee consumption and risk of colorectal cancer. While most case–control studies have reported an inverse association between coffee drinking and risk of colon cancer or colorectal cancer, no association has been found in large prospective cohort studies. However, results of two large cohort studies in the United States showed that men and women who drank two or more cups of decaffeinated coffee per day had approximately half the risk of rectal cancer than those who did not drink decaffeinated coffee. Overall, despite convincing findings in case–control studies, it remains unclear whether coffee consumption reduces the risk of colon or rectal cancer.

Gastric Cancer. In the literature to date, there are at least 16 case–control studies on coffee consumption and risk of ▶ gastric cancer; a significant association (inverse) was reported in only two of these studies. Of seven prospective cohort studies, a significant increase in risk of gastric cancer associated with higher coffee consumption was found in two studies, including a cohort of Hawaiian-Japanese and a cohort of Swedish women. The remaining five cohort studies observed no substantial association. Thus, it appears unlikely that coffee plays a major role in gastric carcinogenesis.

Breast Cancer. Out of 14 case–control studies on coffee consumption and risk of breast cancer, three reported an inverse association, and 11 observed no association. A Finish case–control study found that coffee consumption was inversely associated with breast cancer risk in postmenopausal (but not in premenopausal) women. A large case–control study in the United States also reported a decreased risk with higher coffee consumption, but the association was limited to premenopausal breast cancer. In a multicenter case–control study of women at high risk for breast cancer due to BRCA gene mutations, breast cancer risk was significantly lower among women who habitually drank six or more cups of coffee per day than among non-coffee drinkers. None of at least eight prospective cohort studies have found any significant association between coffee consumption and risk of breast cancer. Taken as a whole, there is no consistent evidence for a link between coffee drinking and breast cancer risk.

Ovarian Cancer. Of 13 case–control studies on coffee consumption and risk of ▶ ovarian cancer, a significant association was found in five studies, with four reporting an increased risk and one a decreased risk with higher coffee consumption. No significant relation between coffee consumption and ovarian cancer was observed in three prospective cohort studies, including a cohort of Seventh-day Adventists based on 51 fatal cases, a Norwegian cohort based on 93 incident cases, and a Swedish cohort based on 301 incident cases.

Overall, it appears unlikely that coffee consumption has a major impact on the risk of ovarian cancer.

Prostate Cancer. Of five ▸ case–control and six prospective cohort studies, none has reported any significant relation between coffee consumption and the risk of ▸ prostate cancer. Thus, it can be concluded that coffee consumption probably has no relationship with prostate cancer risk.

Kidney Cancer. Seven of eight case–control studies have found no association between coffee consumption and risk of kidney cancer (specifically, renal cell carcinoma). A case–control study in Los Angeles reported a significant increase in risk among women (but not men) who drank more than five cups of coffee per day. A Norwegian cohort study observed no significant association. Likewise, coffee consumption was not associated with risk of renal cell carcinoma in a study consisting of two large prospective cohorts of about 90,000 women and 48,000 men in the United States. Thus, it can be concluded that coffee consumption probably has no relationship with the risk of renal cell carcinoma.

Liver Cancer. To date, there are nine published epidemiologic studies on coffee consumption and risk of primary liver cancer or ▸ hepatocellular carcinoma, the major type of primary liver cancer. In a prospective cohort study of more than 90,000 Japanese men and women, those who consumed coffee on a daily or almost daily basis had half the risk of developing hepatocellular carcinoma than those who almost never drank coffee. The risk decreased with increasing amounts of coffee consumed; compared with non-coffee drinkers, those who consumed five or more cups per day had a 76% lower risk of hepatocellular carcinoma. Likewise, in a cohort study consisting of approximately 111,000 Japanese men and women, the risk of death due to hepatocellular carcinoma was 50% lower among drinkers of one or more cups of coffee per day than among non-coffee drinkers. Two smaller cohort studies in Japan and five case–control studies (two in Japan and three in Europe) have confirmed an inverse association between coffee consumption and risk of primary liver cancer or hepatocellular carcinoma.

Furthermore, coffee consumption has been associated with a decreased risk of chronic liver disease and hepatic cirrhosis. Compounds in coffee, including caffeine and chlorogenic acid, have been found to inhibit chemically induced hepatic carcinogenesis in animal models. Overall, the evidence suggests that coffee drinking may reduce the risk of liver cancer.

Cross-References

- ▸ Bladder Cancer
- ▸ Carcinogenesis
- ▸ Case Control Association Study
- ▸ Cohort Study
- ▸ Gastric Cancer
- ▸ Hepatocellular Carcinoma
- ▸ Meta-Analysis
- ▸ Ovarian Cancer
- ▸ Prostate Cancer

References

Higdon JV, Frei B (2006) Coffee and health: a review of recent human research. Crit Rev Food Sci Nutr 46:101–123

International Agency for Research on Cancer (1991) IARC monographs on the evaluation of carcinogenic risks to humans. Coffee, tea, mate, methylxanthines and methylglyoxal, vol 51. International Agency for Research on Cancer, Lyon

Larsson SC, Bergkvist L, Giovannucci E et al (2006) Coffee consumption and incidence of colorectal cancer in two prospective cohort studies of Swedish women and men. Am J Epidemiol 163:638–644

Tavani A, La Vecchia C (2000) Coffee and cancer: a review of epidemiological studies, 1990–1999. Eur J Cancer Prev 9:241–256

World Cancer Research Fund in Association with American Institute for Cancer Research (1997) Food, nutrition and the prevention of cancer: a global perspective. WCRF/AICR, Washington, DC

See Also

(2012) Antioxidant. In: Schwab M (ed) Encyclopedia of Cancer, 3rd edn. Springer Berlin Heidelberg, p 216. doi:10.1007/978-3-642-16483-5_328

(2012) Bias. In: Schwab M (ed) Encyclopedia of Cancer, 3rd edn. Springer Berlin Heidelberg, p 390. doi:10.1007/978-3-642-16483-5_607

(2012) Case-Control Study. In: Schwab M (ed) Encyclopedia of Cancer, 3rd edn. Springer Berlin Heidelberg, p 674. doi:10.1007/978-3-642-16483-5_870

(2012) Confounding. In: Schwab M (ed) Encyclopedia of Cancer, 3rd edn. Springer Berlin Heidelberg, p 968. doi:10.1007/978-3-642-16483-5_1304

(2012) Epidemiologic Studies. In: Schwab M (ed) Encyclopedia of Cancer, 3rd edn. Springer Berlin Heidelberg, p 1269. doi:10.1007/978-3-642-16483-5_1929

(2012) Mutagenic. In: Schwab M (ed) Encyclopedia of Cancer, 3rd edn. Springer Berlin Heidelberg, p 2412. doi:10.1007/978-3-642-16483-5_3909

Cohesin

▶ Cohesins

Cohesin Pathway Dysregulation

▶ Cohesins

Cohesins

Antonio Musio
Institute for Genetic and Biomedical Research, National Research Council, Pisa, Italy
Istituto Toscano Tumori, Firenze, Italy

Synonyms

Cancer; Cohesin; Cohesin pathway dysregulation; Gene mutations

Definition

Cohesin is a chromosome-associated multi-subunit protein complex that is evolutionarily conserved from yeast to humans. In mammalian cells, the mitotic cohesin core complex consists of four proteins, SMC1A, SMC3, RAD21, and either STAG1 or STAG2. SMC proteins are members of the "Structural Maintenance of Chromosomes" (SMC) family whose members are large ATPases with a peculiar domain organization. In fact, they contain two coiled coil domains separated by a flexible globular hinge domain (Fig. 1a). The hinge domains of SMC1A and SMC3 interact with each other to create a heterodimer, while their head domains connect with RAD21, forming a closed ring-like structure. Finally, the RAD21 subunit associates with STAG (Fig. 1b). Meiotic isoforms of cohesin core subunits also exist. In mammals, SMC1A and STAG1/STAG2 can be replaced by SMC1B and STAG3, respectively. In addition, in germinal cells, RAD21 is substituted by the meiosis-specific paralog called REC8.

Characteristics

The cohesin ring plays a pivotal role in correct chromosome segregation during mitosis and meiosis. In fact, cohesin's canonical role is holding sister chromatids together from the time of replication in S phase until their separation in anaphase. This activity requires the involvement of several other proteins contributing to its functions (Table 1). The deposit of cohesin onto chromatin requires the activity of the NIPBL/MAU complex. The establishment of physical bridges between sister chromatids during S phase requires ESCO2, whereas PDS5A and PDS5B interact with cohesin for its establishment and maintenance, ensuring proper sister chromatid pairing. Sister chromatid separation is finely regulated and it is also determined by the coordinated activity of the spindle assembly checkpoint; a lagging chromosome can activate this checkpoint, inhibiting the progression of the cell cycle. When chromosomes are correctly bi-oriented on the mitotic spindle, the anaphase-promoting complex/cyclosome (APC/C) triggers the metaphase-anaphase transition. Cohesion dissolution occurs in two steps. At first, cohesin is taken away from the chromosome arms following the phosphorylation of RAD21 and STAG by polo-like kinase 1 (PLK1), whereas centromeric cohesion is preserved by SGO1. This step also requires the regulation of the WAPL factor, which interacts directly with RAD21 and STAG. Thereafter, the

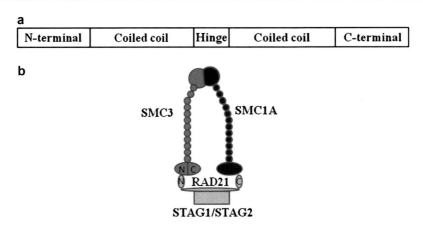

Cohesins, Fig. 1 (**a**) SMC proteins contain five distinct domains: N-terminal (or P-loop NTPase), a hinge motif flanked by two coiled coil regions, and a C-terminal (or P-loop NTPase). The protein length is not in scale. (**b**) The hinge domains of SMC1A and SMC3 bind tightly to each other, whereas the ATPase heads of both proteins are physically connected by the RAD21 subunit. The latter is associated with the fourth subunit of the cohesin core complex, called STAG. In mammals, two forms of STAG exist, STAG1 and STAG2. Cohesin complex contains either STAG1 or STAG2, but never both. The cohesin ring form supports the model where interaction between cohesin and DNA is topological. In fact, according to the so-called ring model, cohesin topologically encircles sister chromatids

Cohesins, Table 1 Components of cohesin pathway

Gene	Role
ESCO2	Cohesin regulatory, sister chromatid cohesion
NIPBL	Cohesin regulatory, cohesin loading
PDS5A	Cohesin regulatory, sister chromatid cohesion
PDS5B	Cohesin regulatory, sister chromatid cohesion
RAD21	Core cohesin member
REC8	Core cohesin member
Securin	Cohesin regulatory, sister chromatid dissociation
Separase	Cohesin regulatory, sister chromatid dissociation
SGOL1	Cohesin regulatory, sister chromatid cohesion
SMC1A	Core cohesin member
SMC1B	Core cohesin member
SMC3	Core cohesin member
STAG1	Core cohesin member
STAG2	Core cohesin member
STAG3	Core cohesin member
WAPL	Cohesin regulatory, cohesin dissociation

APC/C prompts the degradation of PTTG1 (also known as securin) during anaphase so that ESPL1(also called separase) can remove centromeric cohesin by cleaving RAD21.

Noncanonical Role of Cohesin

Beyond proper chromosome segregation, cohesin takes part in both gene transcription regulation and genome stability surveillance. Cohesin mediates the interaction between promoter and enhancer by chromatin looping at multiple gene loci important for differential gene expression during development. This interaction requires the insulator activity of the CCCTC binding factor (CTCF), a zinc finger DNA-binding protein that is evolutionarily conserved. In mammalian cells, the number of cohesin binding sites ranges from about 20,000 to about 100,000 depending on cell type and methodological approaches. About 60% of cohesin binding sites co-localize with CTCF. Most CTCF-free cohesin binding sites overlap with enhancer regions and genes showing a cell-specific expression. Altogether, these observations point out a possible role of cohesin as an insulator, inhibiting enhancer-promoter communication. However, experimental evidence suggests that the way cohesin influences gene expression is more complicated. In fact, data obtained in model organisms and Cornelia de Lange syndrome (CdLS; see below) cells show that cohesin is able to exert opposing effects on gene expression, both positive and negative.

Genome integrity is incessantly challenged by exogenous and endogenous stimuli, and the ability of cells to preserve genome integrity is important for avoiding the development of human diseases, including cancer. Considerable evidence supports the notion that cohesin participates in genome stability maintenance. The silencing of cohesin core members by RNA interference (iRNA) leads to genome instability, manifesting as chromosome gain or loss and chromosome aberrations. Furthermore, cohesin is specifically recruited at double-strand breaks (DSB) where de novo connection with DNA has to be established. Cohesion aids in the use of the sister chromatid as a template to repair DSB through homologous recombination-mediated repair. Finally, SMC1A is phosphorylated by ATM and ATR protein kinases in response to irradiation and replication stress. This phosphorylation represents the molecular signal to block the cell cycle, allowing the recruitment of the enzymatic apparatus responsible for DNA repair.

Cohesin and Human Diseases

Germinal mutations in cohesin and cohesin regulatory factor genes are responsible for human diseases collectively named cohesinopathies. CdLS, with an occurrence of 1:10,000 live births, is the most frequent among these human diseases resulting from cohesin dysfunction. Patients display pre- and postnatal growth delay, cognitive impairment, dysmorphism, hirsutism, upper extremity malformations, and other dysfunctions that affect cardiac, renal, and gastroesophageal systems. CdLS results from mutations in the NIPBL, SMC1A, SMC3, RAD21, and HDAC8 genes. Gene expression profiles show that CdLS cell lines and CdLS mouse models display many up- and downregulated genes.

Instead, mutations in ESCO2 cause Roberts syndrome (RBS) and the SC phocomelia syndrome. RBS is characterized by microcephaly, craniofacial defects, cognitive impairment, and tetraphocomelia, whereas SC phocomelia syndrome is a milder clinical variant of RBS. Warsaw breakage syndrome (WABS), another cohesinopathy characterized by microcephaly, growth retardation, and abnormal skin pigmentation, is caused by mutations in the DDX11 gene.

The DDX11 gene is a member of the XPD helicase family involved in sister chromatid cohesion. Mutations in the SGOL1 gene have been identified in a newly described disorder termed chronic atrial and intestinal dysrhythmia (CAID) syndrome characterized by the co-occurrence of sick sinus syndrome (SSS) and chronic intestinal pseudo-obstruction (CIPO). CdLS, RBS, and WABS cell lines show disturbances in DNA repair that lead to chromosome aneuploidy, chromosome aberrations, and sensitivity to genotoxic treatments. Notwithstanding this, cohesinopathy patients do not show predisposition to cancer.

Cohesin and Cancer

Due to its role in genome stability preservation, it is not surprising that somatic mutations in cohesin and cohesin regulatory factor genes have been identified in human cancers. Mutations in NIPBL, STAG3, SMC1A, and SMC3 have been identified in colorectal carcinoma characterized by chromosomal instability, whereas mutations in STAG2 have been detected in many cancer types including glioblastoma, melanoma, and bladder cancer. Instead, RAD21 is mutated at significant rates in acute myeloid leukemia (Table 2). The number of manuscripts reporting mutations in cohesin pathway genes is increasing continuously. In addition to mutations, cohesin gene expression is also dysregulated in cancer. The downregulation of STAG2, STAG3, PDS5B, RAD21, and SMC1A has occurred in ovarian cancer, acute myeloid leukemia, and in chronic myelomonocytic leukemia. On the contrary, ESCO2, ESPL1, PTTG1, RAD21, SMC3, and WAPL have been found upregulated in many cancer cell lines including breast, prostate, melanoma, colon, and cervical (Table 2). Cohesin dysfunction could trigger cancer development by increasing genome instability due to chromosome missegregation and defective DNA repair or replication. Chromosome imbalance might be the first event in cancer development, allowing different combinations of chromosomes to take place and hence providing a substrate for further selection based on a specific set of chromosome content. Again, gene expression changes of crucial oncogenes or tumor suppressor would be another obvious effect. Chromosome gain might lead to

Cohesins, Table 2 Cohesin pathway genes and their involvement in human cancer

Gene	Disease
ESCO2	Melanoma cancer
	Acute myeloid leukemia
NIPBL	Colorectal cancer
	Acute myeloid leukemia
PDS5A	Astrocytic cancer
	Renal cancer
PDS5B	Gastric cancer
	Colorectal cancer
	Esophageal squamous cell carcinoma
	Acute myeloid leukemia
RAD21	Breast cancer
	Prostate cancer
	Acute myeloid leukemia
	Oral squamous cell carcinoma
	Chronic myelomonocytic leukemia
REC8	Acute myeloid leukemia
Securin	Pituitary cancer
	Adrenocortical cancer
	Esophageal squamous cell carcinoma
Separase	Breast cancer
	Osteosarcoma
	Prostate cancer
SGOL1	Gastric cancer
	Colorectal cancer
SMC1A	Colorectal cancer
	Acute myeloid leukemia
SMC3	Colorectal cancer
	Acute mycloid leukemia
STAG1	Acute myeloid leukemia
STAG2	Bladder cancer
	Glioblastoma
	Acute myeloid leukemia
	Ewing's sarcoma
STAG3	Colorectal cancer
	Epithelial ovarian cancer
WAPL	Cervical cancer
	Acute myeloid leukemia

chromatid cohesion and proper chromosome segregation, cohesin takes part in many additional biological processes such as gene transcription regulation, cell cycle monitoring, DNA damage response, and safeguarding genome stability. The identification of mutations in cohesin pathway genes raises the question of how cohesin dysfunction contributes to cancer development. Missegregation and/or dysregulation of tumor-promoting gene expression due to cohesion defects could trigger the genetic environment favorable to tumorigenesis. In the long run, our increasing knowledge of the biological roles of cohesin might contribute to improving cancer diagnosis and patient treatment.

Cross-References

▶ Acute Myeloid Leukemia
▶ ATM Protein
▶ Bladder Cancer
▶ Bladder Cancer Pathology
▶ DNA Damage Response Genes
▶ RNA Interference

References

Dorsett D, Merkenschlager M (2013) Cohesin at active genes: a unifying theme for cohesin and gene expression from model organisms to humans. Curr Opin Cell Biol 25:327–333
Losada A (2014) Cohesin in cancer: chromosome segregation and beyond. Nat Rev Cancer 14:389–393
Mannini L, Musio A (2011) The dark side of cohesin: the carcinogenic point of view. Mutat Res 728:81–87
Mannini L, Menga S, Musio A (2010) The expanding universe of cohesin functions: a new genome stability caretaker involved in human disease and cancer. Hum Mutat 31:623–630
Nasmyth K, Haering CH (2009) Cohesin: its roles and mechanisms. Annu Rev Genet 43:525–558

enhanced expression of proto-oncogenes. On the other hand, chromosome loss might result in the removal of tumor suppressor genes, leading to cancer development if the first allele is inactivated.

Conclusions

It is now evident that cohesin is an emerging player in many games. In fact, in addition to sister

Cohort Study

Definition

In a cohort study, a defined population with known information on exposure is followed up. Information on diseases or causes of death is collected. Finally, the correlation of these outcomes with the exposure is analyzed. Subjects who presently have a certain condition and/or receive a particular treatment are followed over time and compared with another group who are not affected by the condition under investigation. For research purposes, a cohort is any group of individuals who are linked in some way or who have experienced the same significant life event within a given period. There are many kinds of cohorts, including birth (e.g., all those who born between 1970 and 1975), disease, education, employment, family formation, etc. Any study in which there are measures of some characteristic of one or more cohorts at two or more points in time is cohort analysis.

In some cases, cohort studies are preferred to randomized experimental design. For instance, since a randomized controlled study to test the effect of smoking on health would be unethical, a reasonable alternative would be a study that identifies two groups, a group of people who smoke and a group of people who do not, and follows them forward through time to see what health problems they develop.

In general, cohort analysis attempts to identify cohort effects: Are changes in the dependent variable (health problems in this example) due to aging, or are they present because the sample members belong to the same cohort (smoking vs. non smoking)? In other words, cohort studies are about the life histories of sections of populations and the individuals who comprise them. They can tell us what circumstances in early life are associated with the population's characteristics in later life – what encourages the development in particular directions and what appears to impede it. We can study such developmental changes across any stage of life in any life domain: education, employment, housing, family formation, citizenship, and health.

Cross-References

▶ Cancer Epidemiology
▶ Coffee Consumption
▶ Obesity and Cancer Risk
▶ Uranium Miners

Coiled Bodies

▶ Cajal Bodies

Cold Surgery

▶ Cryosurgery

Collapsin

▶ Semaphorin

Collier-Olf-EBF

▶ Early B-Cell Factors

Colloid Carcinoma

▶ Appendiceal Epithelial Neoplasms

Colon Cancer

▶ Colorectal Cancer

Colon Cancer Carcinogenesis in Human and in Experimental Animal Models

Takuji Tanaka
Department of Oncologic Pathology, Kanazawa Medical University, Kanazawa, Japan

Definition

Colon cancer (often subsumed under the term colorectal cancer (CRC), Fig. 1) develops as a result of the pathologic transformation of normal

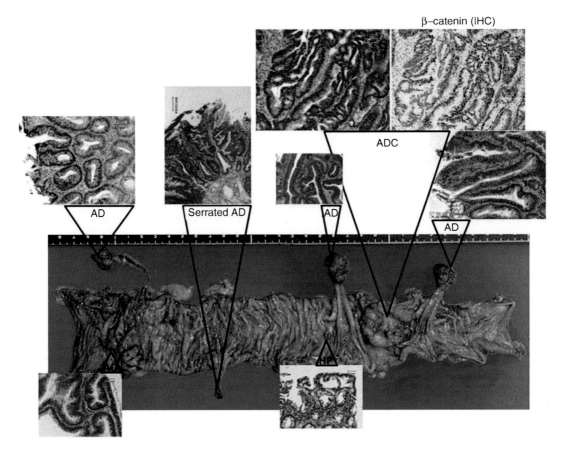

Colon Cancer Carcinogenesis in Human and in Experimental Animal Models, Fig. 1 A colonic carcinoma, histopathologically diagnosed as well-differentiated tubular adenocarcinoma (*ADC*) that developed in a 60-year-old Japanese male. Note: ADC shows beta-catenin-positive reactivity. There are four adenomas (*AD*), a serrated AD, and a hyperplastic polyp (*HP*) surrounding the carcinoma

colonic epithelium to an adenomatous polyp and ultimately an invasive cancer. The ▶ multistep development requires years and possibly decades and is accompanied by a number of genetic alterations.

Characteristics

Precursor and Premalignant Lesions of CRC
Aberrant crypt foci (ACF, Fig. 2), particularly dysplastic ACF, are postulated to be the earliest identifiable potential precursors of CRC in rodents and humans. ACF are postulated to be the early pathological and molecular changes that precede the development of an ▶ adenoma to CRC. Therefore, ACF may eventually evolve

into polyps and, subsequently, CRC in the case of the "adenoma-carcinoma" sequence. Beta-catenin-accumulated crypts (BCAC or microadenoma, Fig. 2) and mucin-depleted foci (MDF) are also early lesions that develop into CRC, but the detection of these lesions from human specimens is not easy by routine laboratory techniques. ACF and MDF are microscopically visible in the un-sectioned colon stained with methylene blue and high-iron diamine Alcian blue, respectively. Adenomas (Figs. 1 and 2) are considered to be premalignant lesions for CRC. Besides adenomatous polyps, other types of similar lesions deserve some comment either for their frequency (hyperplastic polyp, Fig. 1) or for their potential malignant evolution (flat adenoma) or for their peculiar histological features (serrated

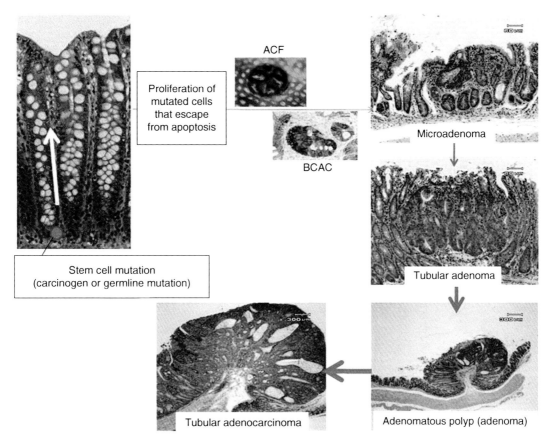

Colon Cancer Carcinogenesis in Human and in Experimental Animal Models, **Fig. 2** Interactions of colon crypt cells with blood-borne or luminal agents. (a) Mutation of APC in stem cell, as a result of blood-borne agents or germline mutation, produces abnormalities in cell proliferation, migration, and adhesion. (b) Abnormal cells accumulate, an ACF at the *top* of the crypt or a BCAC in the mucosa forms. (c) Other mutations are more likely with the contact of proliferating cells with fecal mutagens, an adenoma forms by sequential clonal expansion, and then an adenocarcinoma develops

adenoma, Fig. 1). Patients with ulcerative colitis (UC) and ▸ Crohn colitis of ▸ Crohn disease are at increased risk for colorectal malignancies. The risk increases with the duration of the disease and the extent of colorectal involvement. The morphological basis of tumor occurrence is the development of dysplastic changes in flat mucosa or in polypoid lesions. Colonic dysplasia in UC may grow as a flat lesion (dysplasia) or as a dysplasia-associated lesion or mass (DALM). The clinical distinction between DALM and sporadic adenoma that may occur in patients with ▸ inflammatory bowel disease (IBD) is extremely important. DALM arises as the result of a chronic inflammatory stimulus in a patient with UC or Crohn colitis.

Carcinogenic Pathways

There are at least four carcinogenic pathways associated with colorectal carcinogenesis, the "adenoma-carcinoma" sequence type, "de novo" type, hereditary nonpolyposis colorectal cancer (HNPCC) type, and colitic cancer ("ulcerative colitis dysplasia carcinoma" sequence) type.

Adenoma-Carcinoma Sequence Type

This pathway is an extension of the ▸ multistep development genetic model of colorectal carcinogenesis by Vogelstein et al. (1988). This model assumes that the APC gene mutation (▸ APC gene in familial adenomatous polyposis) occurs at the first stage of the carcinogenesis process. The APC gene has been identified as a causative gene of familial adenomatous polyposis (FAP) and is involved in the regulation of beta-catenin, ► cytoskeleton organization, ► apoptosis, ► cell cycle control, and cell ► adhesion. *APC* mutations occur in up to 80% of adenomas and adenocarcinomas, and in 4.3% of ACF. APC protein translated from the *APC* gene is a main factor in the ► Wnt signaling pathway, and APC regulates cell proliferation by binding and degrading beta-catenin protein that promotes cell proliferation. However, the mutant APC protein cannot bind and degrade beta-catenin protein and, as a result, beta-catenin protein translocates to the nucleus and binds to T-cell factor/lymphocyte enhancer (TCF/LEF) factor transcription factor that targets the ► MYC oncogene, and the ► *Cyclin D* 1 and *JUN* genes, and promotes cell proliferation. The *KRAS* gene (see ► RAS Genes) mutation occurs after the APC gene in the CRC. *RAS* gene mutations occur in 58% of adenomas larger than 1 cm and in 47% of CRCs. However, K-ras mutations are found in 9% of adenomas less than 1 cm in size. The mutation of the *KRAS* gene occurs during the early stage of carcinogenesis and increases the size of the tumor. More than 90% of primary CRCs with loss of heterozygosity (LOH) of chromosome 18q show a deletion in the deleted in colorectal carcinoma (DCC) gene included in the region of allelic loss. The most important point that determines the borderline between the adenoma and adenocarcinoma is a mutation of the ► p53 gene (► p53 family). The p53 gene mutation or LOH is present in about 75% of CRCs and is a conversion point from adenoma to adenocarcinoma.

HNPCC Type Pathway

HNPCC, also termed ► Lynch syndrome, is the most common of the hereditary CRC syndromes, accounting for approximately 1–6% of all colorectal malignancies. This disorder is characterized by early onset of CRC and other adenocarcinomas in the endometrium, ovary, stomach, and urinary tract. Individuals with HNPCC have an 80% lifetime risk of CRC. HNPCC is inherited in an autosomal dominant fashion. Mutations in five mismatch repair (MMR) genes (► mismatch

repair in genetic instability) are currently implicated in the pathogenesis of this disorder. hMSH2 on chromosome 2p and hMLH1 on chromosome 3p account for the majority of genetically defined cases. Fifteen to 60% of families with the clinical diagnosis of HNPCC are found to have mutations in these genes. Other less commonly implicated genes are hPMS1, hPMS2, and hMSH6. The DNA MMR proteins recognize and correct small sequence errors that occur during DNA replication. Mutations in both copies of a DNA mismatch repair gene lead to the accumulation of DNA sequence errors predominantly in segments of DNA containing multiple, short, repeated sequences known as a microsatellite. When these errors occur in critical growth-regulatory genes, carcinogenesis may ensue. Tumors in patients with HNPCC characteristically demonstrate ► microsatellite instability (MSI) – the widespread expansion or contraction of these short sequences of DNA. In HNPCC, approximately 90% of CRCs and 80% of adenomas demonstrate MSI, findings that can aid in the diagnosis of this syndrome.

"De Novo" Type Pathway

In the 1980s, several Japanese researchers began to report flat-type carcinomas (less than 10 mm in diameter), arising "de novo." This flat-type carcinoma shows fewer mutations of the APC and *KRAS* genes than polypoid-type carcinoma, though mutations of the p53 genes are seen the same level as in polypoid-type carcinoma. However, the ► epigenetic inactivation (► epigenetic gene silencing) of Ras-associated factor (RASSF) 1A (► RAS-association domain family 1) by hypermethylation of the promoter region is frequently detected in flat-type carcinoma. The inactivation of RASSF1A causes an aberration in the ras-signaling pathway without involving the *KRAS* gene mutation. The RASSF1A plays an important role with p53 in the "de novo" type of carcinogenesis pathway.

Colitic Cancer Pathway (Ulcerative Colitis Dysplasia Carcinoma Sequence)

This pathway to CRC is via ulcerative colitis (UC), one type of ► inflammatory bowel disease (IBD). Although this disease is a minor contributor to the overall population burden of CRC, individuals with UC have about a 20-fold excess risk. CRC occurs at a high rate in UC patients with extensive colitis of greater than 8–10 years duration. This pathway involves a dysplasia–carcinoma sequence, more like adenocarcinoma developing in ► Barrett esophagus. In UC patients, chronic inflammation can result in genetic alterations, which can progress to dysplasia and subsequently to cancer. In UC, APC mutations are uncommon (fewer than 10% of cases) and p53 loss can occur early, appearing even in diploid histologically normal colonic mucosa. There is no need to grow a polyp. Flat dysplasia or DALM thought to be a precancerous lesion (Colon Cancer Premalignant Lesions) occurs in the inflamed colonic mucosa of UC. ► Microsatellite instability (MSI) may occur in the absence of DNA ► mismatch repair (MMR) defects even in normal-appearing mucosa.

Experimental Colon Carcinogenesis

The ability to reliably induce colon tumors in animals has provided the opportunity to study various aspects of the carcinogenesis process. These models have provided information on the initiation, promotion, and progression of tumors, including detailed information on cellular transformation and the subsequent events leading to the formation of neoplastic lesions. The established models can be used for ► chemoprevention studies as well (► chemoprotectants). There are animal models that are chemically induced and genetically modified, including multiple intestinal neoplasia (Min) ► mouse models. Carcinogenesis studies using these models have also elucidated the role of genetic and environmental factors, and other influences on the various aspects of this complex disease. Animal colon cancer models have also been used to evaluate immunological, chemical, and surgical therapy regimens.

Colonic Carcinogens

There are chemical agents (► chemical carcinogenesis) that have been used to induce colon tumors in animals, mostly rats and mice. Many researchers deal mainly with experiments involving 1,2-dimethylhydrazine (DMH), and its metabolites azoxymethane (AOM), and methylazoxymethanol (MAM) acetate in the rodent colon tumor model. AOM is the most widely used colon carcinogen. Pursuing the argument that blood-borne carcinogens may well produce somatic mutations in APC and perhaps K-ras, important sources of known carcinogens with established links to risks to CRC include cooked meat (e.g., ► heterocyclic amines and nitrosamines). Sugimura and Sato originally proposed that specific heterocyclic amines, such as 2-amino-1-methyl-6-phenylimidazo[4,5-b]pyridine (PhIP), are important in the etiology of colon cancer. Several separate classes of these compounds have been identified and have been shown to be carcinogenic in animals, including having a direct effect on APC. DMH and its metabolite, AOM, are procarcinogens that require metabolic activation to form DNA-reactive products. Metabolism of these compounds involves multiple ► xenobiotics-metabolizing enzymes, which proceeds through several N-oxidation and hydroxylation steps, including the formation of MAM following hydroxylation of AOM. The reactive metabolite, MAM, readily yields a methyldiazonium ion, which can alkylate macromolecules in the liver and colon, including the addition of methyl groups at the O6 or N7 position of guanine. MAM is a substrate of the nicotinamide adenine dinucleotide-dependent dehydrogenase present in the colon and liver, suggesting that the active metabolite of MAM might be the corresponding aldehyde. A direct role for the alcohol-inducible ► cytochrome P450 isoform, CYP2E1, in activation of AOM and MAM has been established.

AOM/Dextran Sodium Sulfate (DSS) Model for Colitis-Related Colon Carcinogenesis

Perhaps the most commonly used mouse model that recapitulates many of the features associated with UC employs DSS. In a colitis related mouse CRC model (a two-stage mouse colon carcinogenesis model) that was reported in 2003, mice initiated with a low dose of a variety of colonic carcinogen (AOM, DMH, or PhIP) develop a number of colon tumors after a relatively short-term DSS exposure. The subsequent dysplasias and neoplasms are positive for β-catenin, cyclooxygenase-2 (cyclooxygenase-2 in colorectal cancer), and inducible ▶ nitric oxide synthase. In the model, neoplasms develop through dysplastic lesions. Many colon tumors develop from dysplastic crypts within 5 weeks in male and female ApcMin/+ mice when they receive 2% DSS in drinking water for 7 days. The tumor promotion effect of DSS is dose dependent, occurring at a concentration of 1% or higher, and the effect corresponds to the degree of inflammation and nitrosative stress within the colonic mucosa. Beta-catenin mutations within codons 33 and 34 are caused by AOM exposure, whereas codon 32 mutations result from DSS exposure. Altered gene expression in the inflamed colon of mice that receive a combination of AOM and DSS includes Wif1, Plat, Myc, Plscr2, Pparbp, and Tgfb3. Rats that receive AOM followed by DSS also develop a number of colonic tumors within a 20-week experimental period.

Cross-References

▶ Inflammatory Bowel Disease-Associated Cancer

References

Rosenberg DW, Giardina C, Tanaka T (2009) Mouse models for the study of colon carcinogenesis. Carcinogenesis 30(2):183–196. Epub 26 Nov 2008. Review. PubMed PMID: 19037092; PubMed Central PMCID: PMC2639048

Takahashi M, Wakabayashi K (2004) Gene mutations and altered gene expression in azoxymethane-induced colon carcinogenesis in rodents. Cancer Sci 95(6):475–80. Review. PubMed

Taketo MM, Edelmann W (2009) Mouse models of colon cancer. Gastroenterology 136(3):780–798. Review. PubMed

Tanaka T (2009) Colorectal carcinogenesis: review of human and experimental animal studies. J Carcinog 8:5. PubMed

Tanaka T, Kohno H, Suzuki R, Yamada Y, Sugie S, Mori H (2003) A novel inflammation-related mouse colon carcinogenesis model induced by azoxymethane and dextran sodium sulfate. Cancer Sci 94(11):965–73. PubMed

Yasui Y, Kim M, Oyama T, Tanaka T (2009) Colorectal carcinogenesis and suppression of tumor development by inhibition of enzymes and molecular targets. Curr Enzyme Inhib 5:1–26

Colon Cancer Classification

Definition

There are two classifications that are used separately, the Dukes and the TNM (tumor, lymphnodes, metastases) classification (Colon Cancer Clinical Oncology).

The **Dukes-Classification** is preferred in the US and UK and describes the following stages:

1. Dukes A Growth limited to wall, nodes negative
2. Dukes B Growth beyond muscularis propia, nodes negative
3. Dukes C1 Nodes positive and apical negative
4. Dukes C2 Apical node positive
5. Dukes D Growth beyond originating organ

The **TNM-staging** (suggested by the Union internationale contre le cancer, UICC) is preferred in European countries and distinguishes between the stages listed below. T stands for the expansion of the primary tumor; N for the lack or the presence of metastases of the lymph nodes; M for the lack or the presence of distant metastases. Numbers indicate the extent of malignant processes; p, postoperative (Table 1).

Colon Cancer Classification, Table 1 TNM-staging

pT1	Local invasion of submucosa
pT2	Local invasion of the muscularis propia
pT3	Local invasion beyond the muscularis propia
pT4	Tumor cells have reached peritoneal surface or invaded adjacent organs
pN0	No lymphnodes affected by metastases
pN1	One to three lymphnodes affected by metastases
pN2	Four or more lymphnodes affected by metastases
pM0	No distant metastasis
pM1	Distant metastasis

See Also

(2012) UICC. In: Schwab M (ed) Encyclopedia of Cancer, 3rd edn. Springer Berlin Heidelberg, p 3836. doi:10.1007/978-3-642-16483-5_6512.

Colon Cancer Genomic Pathways

Audrey Ferrand
INSERM U.858, Institut de Médecine
Moléculaire de Rangueil, IFR150, Université
Paul Sabatier, Toulouse, France

Synonyms

Cancer process of the large intestine

Definition

Colon carcinogenesis is thought to result from genomic changes – mutation, deletion, chromosome translocation, or ▶ epigenetic modifications – in genes controlling the normal balance between proliferation (▶ oncogene) and cell death (▶ Apoptosis) (▶ tumor suppressor gene), which appears in normal the epithelial cell of the colon and leading to a deregulation of the cell homeostasis. The mutated cells can first constitute preneoplastic lesions (▶ Colorectal Cancer Premalignant Lesions), then ▶ adenoma (benign tumors), and later on carcinoma able to lead to ▶ metastasis of the liver.

Characteristics

Histopathological Characteristics

In the colon adenoma-carcinoma sequence, the earliest identifiable lesion corresponds to small dysplastic lesions within the colonic epithelium named aberrant crypt foci (ACF). These ▶ preneoplastic lesions have been defined as crypts that: (i) are larger than the normal crypts in the field, (ii) have increased pericryptal space that separates them from the normal crypts, (iii) have a thicker layer of epithelial cells that often stains darker when stained with methylene blue and marked with permanent ink, and (iv) generally have oval rather than circular openings. A "top-down" model and a "bottom-up" model have been proposed to explain the origin and the development of these dysplastic ACF. The "top-down" model suggests that a mutation occurs in a cell localized at, or near, the top of the crypt. The cell then spreads laterally and downward to form new crypts. The "bottom-up" model proposes that the mutated cell is initially in the base of the crypt, within the stem cell zone, and expands upwards so the entire crypt might become dysplastic. Both models are still discussed.

ACF grow over time to generate macroscopically visible benign adenomatous polyps. The transition from benign (▶ adenoma) to malignant (carcinoma) growth is thought to be progressive. Adenomas first evolve into ▶ carcinoma in situ that will locally expand in the colon mucosa. These carcinomas will then grow over the colon mucosa limits and become invasive. Invasive carcinomas, which are able to metastasize, are often the first clinical presentation of colon tumors.

Genomic Alterations Involved

Carcinogenesis requires a ▶ genomic instability that leads to the accumulation of multiple genetic or epigenetic alterations in initially normal epithelial cell types (Worthley et al. 2007). These alterations confer the cells a growth advantage. This advantage can be the consequence of an increased proliferation rate, an impaired ▶ apoptosis, or both. Genomic instability is crucial for the carcinogenesis since it allows a high mutation rate. Two main types of genomic instability are found in colon cancer. The most common one −75–85% of colon cancers is the ▶ chromosomal instability (CIN) corresponding to an accumulation of numerical or structural chromosomal abnormalities (▶ aneuploidy). The second one is the ▶ microsatellite instability (MSI) resulting from impaired recognition and mismatch repair (▶ Mismatch Repair in Genetic Instability) in the daughter strand of DNA generated during the

DNA replication process. Although CIN and MSI seem to be mutually exclusive, in any case genomic instability is necessary and sufficient to induce colon carcinogenesis.

Chromosomal Instability Pathway

Comprehensive sequencing of 13,023 human genes in 11 colorectal tumors revealed that individual tumors accumulate an average of 90 mutant genes but that only a subset of these mutations actually contribute to the neoplastic process. The most well-known mutations in colon carcinogenesis are the following ones.

APC

The adenomatous polyposis coli (► APC Gene in Familial Adenomatous Polyposis) gene is a tumor suppressor mutated in most colon cancers. The APC protein has multiple domains able to interact with beta-catenin, axin, and glycogen synthase kinase-3 alpha to form a large protein complex. After formation of this complex, beta-catenin is phosphorylated, ubiquitinylated (► Ubiquitination), and broken-down by the ► proteasome system resulting in a negative regulation of the ► Wnt signaling pathway, a major player in colon development and carcinogenesis. As a consequence, beta-catenin cannot play its transcription factor role and activate its target genes. Most of APC mutations in colon cancer concern the beta-catenin binding domain. Thus, APC fails to inhibit Wnt signaling and the proliferative events it activates. Moreover, APC is a kinetochore-bound microtubule-associated protein that is important for the correct segregation of the chromosomes between the daughter cells. Loss of functional APC might also interfere with the normal regulation of mitosis and thus contribute to chromosomal instability (CIN). APC mutation status is different depending on the type of colon cancer. APC is frequently mutated in familial adenomatous polyposis (FAP) while notably absent from sporadic aberrant crypt foci (ACF). However, mutations in Wnt pathway (APC or beta-catenin) can be detected in nearly 80% of early adenomas suggesting that these alterations play a part in the ACF-adenoma transition and thus favor colon carcinogenesis. However, these mutations are not present in all the type of adenomas or cancers comforting the idea that APC is not the only major mutation involved in colon carcinogenesis.

KRAS

This GTP-binding protein activates signaling pathways regulating cell proliferation and ► apoptosis. Mutation of the *KRAS* gene (► *RAS* Genes) is frequent in cancers. Regarding colon carcinogenesis, while *KRAS* mutation is mainly absent in FAP dysplastic ACF, it is frequently found (60–80%) in sporadic ACF. However, only 35–50% of colon cancers display activating KRAS mutations suggesting that, despite the growth advantage conferred to the mutated cells, this alteration does not seem to be sufficient or even necessary to the advanced colon tumor process. Thus, it might be possible that most APC wild-type/KRAS mutated ACF will never develop into adenomas. Regarding the incidence of this mutation in the colon cancer treatment, a study reports that in metastastic colorectal cancer, somatic mutation of KRAS (30–50%), already known to be a factor of bad prognostic for the patient, is significantly associated with an absence of response to treatments based on anti-EGFR (► Epidermal Growth Factor Receptor) monoclonal antibody therapy suggesting that KRAS mutational status should be tested and considered before considering this type of therapy.

SMAD2, SMAD4 and DCC

SMAD2 and SMAD4 are downstream effectors of the TGF-Beta signaling pathway (► Smad Proteins in TGF-Beta Signaling) known to regulate cell proliferation and ► apoptosis. DCC (deleted in colorectal carcinoma) is a transmembrane receptor promoting apoptosis. The three genes coding for these proteins are located at the same locus on the chromosome 18 often subject to allelic loss (60% of colon cancer); however mutations seem to occur more often on SMAD4.

p53

The p53 protein, encoded by the ► TP53 tumor suppressor gene is a transcription factor that positively regulates the expression of genes retarding the cell cycle allowing more time for ► repair of DNA. In case of too much ► DNA damage, p53 activates pro-apoptotic genes. While APC and KRAS mutations are found early in the colon cancer process, mutations in the TP53 tumor-suppressor gene are detected at later stages. The frequency of either mutation or loss of heterozygosity (LOH) correlates with the progression of the pathology. Adenomas only display 4–26% of p53 abnormalities, while adenoma with invasive foci present up to 50%. This percentage goes up to 75% in established colon cancers.

Microsatellite Instability (MSI) Pathway

Microsatellite sequences are repetitive DNA genetic loci with 1–5 base pairs repeated 15–30 times, found in great number spread out over the whole DNA sequence. During DNA replication, these short sequences are subject to frameshift mutation and base-pair substitutions resulting in loss of function. This is referred to as the ► microsatellite instability (MSI) pathway (Soreide et al. 2006). MSI is present in virtually all the cases of hereditary nonpolyposis colorectal cancer (HNPCC; ► Lynch Syndrome) and approximately 15% of sporadic cases. These errors are normally controlled and repaired by the DNA mismatch repair (MMR) system composed of seven main proteins: MLMH1 and 3, MSH2, 3 and 6, PMS1 and 2. MSI in hereditary and sporadic colon cancer occurs through two different mechanisms. In more than 90% of HNPCC, MSI results from a germline mutation in the MutS homologue 2 and 6 (MSH2, MSH6) and MutL homologue 1 (MLH1) mismatch repair genes. In sporadic cancer, the main cause of MSI seems to be the loss of expression of a mismatch repair, most commonly MLH1, after ► epigenetic gene silencing such as biallelic or hemiallelic ► methylation of cytosine residues of the cytosine and guanine (CpG)-rich promoter sequences (► CpG Islands) of MLH1. However "mutator" and "methylator" phenotypes seem to be mutually exclusive. Aberrant methylation arises very early in the colon and may thus be part of the age-related field defect found in sporadic colon cancer but is also involved in later stage through the CpG island methylator phenotype (CIMP).

Animal Models for the Study of Colon Carcinogenesis

Carcinogen-Induced Models

Carcinogens can be divided into two groups: the nongenotoxic and the genotoxic carcinogens. Nongenotoxic carcinogens, such as ► hormones or organic compounds, do not directly alter DNA. Genotoxic carcinogens induce irreversible genetic alterations by directly interacting with DNA. Genotoxins include chemical or nonchemical agents such as ► UV radiation or ionizing radiation.

AOM

The most used carcinogen in rodent models of CRC is certainly the azoxymethane (AOM). The advantages of using this chemical agent include high potency and reproducibility, simple mode of application, and low price. AOM initiates cancer as a member of the ► alkylating agents of DNA. Although some tumor promotion activity has been reported for this chemical agent, it is mainly a tumor initiation agent. The AOM-induced carcinogenesis model has been successfully used in studies on factors modulating the tumor initiation and progression. AOM-based rodent models of carcinogenesis seem to be valuable tools to predict ► chemoprevention efficacy in humans. A ► meta-analysis of colon tumor chemoprevention has been done in rats, mice, and men (Corpet and Pierre 2005). This database can be used to perform detailed comparisons between various chemoprevention studies in different inducible models of CRC in rodents (http://www.inra.fr/reseau-nacre/sci-memb/corpet/indexan.html). Another advantage of AOM-induced colon carcinogenesis over other CRC models is that tumors frequently develop in the distal part of the colon corresponding to the predominant localization of spontaneous CRC in man.

DSS

Another chemical agent, dextrane sodium sulfate (DSS), has been used alone or in combination with AOM to study colitis-induced colon carcinogenesis. Repeated treatments with DSS, administrated by addition in the drinkable water, induce an ► inflammation of the colon mucosa generating colitis (Ulcerative Colitis). However, DSS alone is not as effective as AOM and the tumor development takes a long time. People have developed AOM-DSS treatment, the prior use of AOM accelerating the tumor process.

Genetic Models

► Mouse models displaying a mutation in the APC gene have been really helpful in understanding the colon carcinogenesis process (Femia and Caderni 2008).

APCMin Min (acronym for multiple intestinal neoplasia); these mice have a nonsense mutation at the codon 850 within the mouse APC gene leading to the development of tumors in the small intestine, associated with anemia and thus short life span. However, these mice develop only few tumors in the colon and are not the best model to study human colon carcinogenesis.

APC Transgenics These transgenic mice carry a targeted truncation at codon 716. Their phenotype is quite similar to the one of the APCMin mice although they develop more adenoma and do not present extraintestinal manifestations.

APC1638N Compared to the other APC models described above, animals carrying a mutation at codon 1638 N only display 5–6 adenoma/carcinoma, so have an attenuated tumor phenotype but more extraintestinal manifestations than APCMin mice.

APC(Delta14/+) A new model of APC germline mutation leading to inactivation by exon 14 deletion has been generated. The main phenotypic difference between these mice and the APCMin mice is the shift of the tumors in the distal colon and rectum, often associated with a rectal prolapse. All lesions, including early lesions, revealed APC LOH and loss of Apc gene expression. The APC(Delta14/+) model is an interesting tool to investigate the molecular mechanisms of colon carcinogenesis.

Perspective

Within the last decade, a new hypothesis in the development and recurrence of cancers has emerged, the stem cell hypothesis in cancer (CSC) (Reya et al. 2001). These cells form a subpopulation within the tumor able to generate tumors through self-renewal and differentiation into multiple cell types. CSCs, also called tumor-initiating cells or ► cancer stem-like cells, are tumorigenic in contrast to other nontumorigenic cancer cells. CSCs have a high survival rate and are responsible of the tumor relapse and ► metastasis. The origin of the CSC is unclear. Their characterization, the understanding of their capacity and the identification of their Achilles' heel could lead to major breakthroughs in cancer therapy.

Cross-References

► Adenoma

► Alkylating Agents

► Aneuploidy

► APC Gene in Familial Adenomatous Polyposis

► Apoptosis

► Cancer Stem-Like Cells

► Carcinoma in Situ

► Chemoprevention

► Chromosomal Instability

► Colorectal Cancer Premalignant Lesions

► CpG Islands

► DNA Damage

► Epidermal Growth Factor Receptor

► Epigenetic

► Epigenetic Gene Silencing

► Genomic Instability

► Hormones

► Inflammation

► Lynch Syndrome

► Meta-Analysis

► Metastasis

► Methylation

► Microsatellite Instability

► Mismatch Repair in Genetic Instability

► Mouse Models

► Oncogene

► Preneoplastic Lesions

► Proteasome

► *RAS* Genes

► Repair of DNA

▶ Smad Proteins in TGF-Beta Signaling
▶ TP53
▶ Tumor Suppressor Genes
▶ Ubiquitination
▶ UV Radiation
▶ Wnt Signaling

References

Corpet DE, Pierre F (2005) How good are rodent models of carcinogenesis in predicting efficacy in humans? A systematic review and meta-analysis of colon chemoprevention in rats, mice and men. Eur J Cancer 41(13):1911–1922

Femia AP, Caderni G (2008) Rodent models of colon carcinogenesis for the study of chemopreventive activity of natural products. Planta Med 74(13):1602–1607

Reya T et al (2001) Stem cells, cancer, and cancer stem cells. Nature 414(6859):105–111

Soreide K et al (2006) Microsatellite instability in colorectal cancer. Br J Surg 93(4):395–406

Worthley DL et al (2007) Colorectal carcinogenesis: road maps to cancer. World J Gastroenterol 13(28):3784–3791

See Also

(2012) Aberrant Crypt Foci. In: Schwab M (ed) Encyclopedia of Cancer, 3rd edn. Springer Berlin Heidelberg, pp 13–14. doi:10.1007/978-3-642-16483-5_6531

(2012) Allelic Loss. In: Schwab M (ed) Encyclopedia of Cancer, 3rd edn. Springer Berlin Heidelberg, p 137. doi:10.1007/978-3-642-16483-5_186

(2012) Axin. In: Schwab M (ed) Encyclopedia of Cancer, 3rd edn. Springer Berlin Heidelberg, p 324. doi:10.1007/978-3-642-16483-5_496

(2012) Azoxymethane. In: Schwab M (ed) Encyclopedia of Cancer, 3rd edn. Springer Berlin Heidelberg, p 329. doi:10.1007/978-3-642-16483-5_507

(2012) Beta-Catenin. In: Schwab M (ed) Encyclopedia of Cancer, 3rd edn. Springer Berlin Heidelberg, p 385. doi:10.1007/978-3-642-16483-5_889

(2012) Carcinogen. In: Schwab M (ed) Encyclopedia of Cancer, 3rd edn. Springer Berlin Heidelberg, p 644. doi:10.1007/978-3-642-16483-5_839

(2012) Cell Cycle. In: Schwab M (ed) Encyclopedia of Cancer, 3rd edn. Springer Berlin Heidelberg, p 737. doi:10.1007/978-3-642-16483-5_994

(2012) DCC. In: Schwab M (ed) Encyclopedia of Cancer, 3rd edn. Springer Berlin Heidelberg, pp 1063–1064. doi:10.1007/978-3-642-16483-5_1524

(2012) Deleted in Colorectal Carcinoma. In: Schwab M (ed) Encyclopedia of Cancer, 3rd edn. Springer Berlin Heidelberg, p 1073. doi:10.1007/978-3-642-16483-5_6544

(2012) Deletion. In: Schwab M (ed) Encyclopedia of Cancer, 3rd edn. Springer Berlin Heidelberg, p 1080. doi:10.1007/978-3-642-16483-5_1553

(2012) Epithelial Cell. In: Schwab M (ed) Encyclopedia of Cancer, 3rd edn. Springer Berlin Heidelberg, pp 1291–1292. doi:10.1007/978-3-642-16483-5_1958

(2012) Field Defect. In: Schwab M (ed) Encyclopedia of Cancer, 3rd edn. Springer Berlin Heidelberg, pp 1406–1407. doi:10.1007/978-3-642-16483-5_2186

(2012) Frameshift Mutation. In: Schwab M (ed) Encyclopedia of Cancer, 3rd edn. Springer Berlin Heidelberg, p 1454. doi:10.1007/978-3-642-16483-5_2266

(2012) Genotoxic. In: Schwab M (ed) Encyclopedia of Cancer, 3rd edn. Springer Berlin Heidelberg, p 1540. doi:10.1007/978-3-642-16483-5_2393

(2012) Germline Mutation. In: Schwab M (ed) Encyclopedia of Cancer, 3rd edn. Springer Berlin Heidelberg, p 1544. doi:10.1007/978-3-642-16483-5_2404

(2012) Glycogen Synthase Kinase-3. In: Schwab M (ed) Encyclopedia of Cancer, 3rd edn. Springer Berlin Heidelberg, p 1570. doi:10.1007/978-3-642-16483-5_2448

(2012) GTP-Binding. In: Schwab M (ed) Encyclopedia of Cancer, 3rd edn. Springer Berlin Heidelberg, p 1613. doi:10.1007/978-3-642-16483-5_2531

(2012) Initiation. In: Schwab M (ed) Encyclopedia of Cancer, 3rd edn. Springer Berlin Heidelberg, p 1865. doi:10.1007/978-3-642-16483-5_3057

(2012) Ionizing Radiation. In: Schwab M (ed) Encyclopedia of Cancer, 3rd edn. Springer Berlin Heidelberg, p 1907. doi:10.1007/978-3-642-16483-5_3139

(2012) Kinetochore. In: Schwab M (ed) Encyclopedia of Cancer, 3rd edn. Springer Berlin Heidelberg, p 1944. doi:10.1007/978-3-642-16483-5_3224

(2012) Loss of Heterozygosity. In: Schwab M (ed) Encyclopedia of Cancer, 3rd edn. Springer Berlin Heidelberg, pp 2075-2076. doi:10.1007/978-3-642-16483-5_3415

(2012) Methylator Phenotype. In: Schwab M (ed) Encyclopedia of Cancer, 3rd edn. Springer Berlin Heidelberg, p 2280. doi:10.1007/978-3-642-16483-5_3688

(2012) Microsatellite. In: Schwab M (ed) Encyclopedia of Cancer, 3rd edn. Springer Berlin Heidelberg, p 2305. doi:10.1007/978-3-642-16483-5_3730

(2012) Microtubule. In: Schwab M (ed) Encyclopedia of Cancer, 3rd edn. Springer Berlin Heidelberg, p 2308. doi:10.1007/978-3-642-16483-5_3734

(2012) Min. In: Schwab M (ed) Encyclopedia of Cancer, 3rd edn. Springer Berlin Heidelberg, p 2318. doi:10.1007/978-3-642-16483-5_3750

(2012) Monoclonal Antibody Therapy. In: Schwab M (ed) Encyclopedia of Cancer, 3rd edn. Springer Berlin Heidelberg, pp 2367–2368. doi:10.1007/978-3-642-16483-5_3823

(2012) Mutation. In: Schwab M (ed) Encyclopedia of Cancer, 3rd edn. Springer Berlin Heidelberg, p 2412. doi:10.1007/978-3-642-16483-5_3911

(2012) Nonsense Mutation. In: Schwab M (ed) Encyclopedia of Cancer, 3rd edn. Springer Berlin Heidelberg, p 2546. doi:10.1007/978-3-642-16483-5_4124

(2012) P53. In: Schwab M (ed) Encyclopedia of Cancer, 3rd edn. Springer Berlin Heidelberg, p 2747. doi:10.1007/978-3-642-16483-5_4331

(2012) Proliferation. In: Schwab M (ed) Encyclopedia of Cancer, 3rd edn. Springer Berlin Heidelberg, p 3004. doi:10.1007/978-3-642-16483-5_4766

(2012) Repetitive DNA. In: Schwab M (ed) Encyclopedia of Cancer, 3rd edn. Springer Berlin Heidelberg, p 3254. doi:10.1007/978-3-642-16483-5_5030

(2012) Replication. In: Schwab M (ed) Encyclopedia of Cancer, 3rd edn. Springer Berlin Heidelberg, p 3254. doi:10.1007/978-3-642-16483-5_5031

(2012) SMAD. In: Schwab M (ed) Encyclopedia of Cancer, 3rd edn. Springer Berlin Heidelberg, p 3440. doi:10.1007/978-3-642-16483-5_5360

(2012) Stem Cell Hypothesis in Cancer. In: Schwab M (ed) Encyclopedia of Cancer, 3rd edn. Springer Berlin Heidelberg, p 3508. doi:10.1007/978-3-642-16483-5_5489

(2012) Transcription Factor. In: Schwab M (ed) Encyclopedia of Cancer, 3rd edn. Springer Berlin Heidelberg, p 3752. doi:10.1007/978-3-642-16483-5_5901

(2012) Transgenic. In: Schwab M (ed) Encyclopedia of Cancer, 3rd edn. Springer Berlin Heidelberg, p 3763. doi:10.1007/978-3-642-16483-5_5919

(2012) Translocation. In: Schwab M (ed) Encyclopedia of Cancer, 3rd edn. Springer Berlin Heidelberg, p 3773. doi:10.1007/978-3-642-16483-5_5942

(2012) Transmembrane. In: Schwab M (ed) Encyclopedia of Cancer, 3rd edn. Springer Berlin Heidelberg, p 3773. doi:10.1007/978-3-642-16483-5_5947

(2012) Tumor Promotion. In: Schwab M (ed) Encyclopedia of Cancer, 3rd edn. Springer Berlin Heidelberg, p 3800. doi:10.1007/978-3-642-16483-5_6048

(2012) Tumor Suppressor. In: Schwab M (ed) Encyclopedia of Cancer, 3rd edn. Springer Berlin Heidelberg, p 3803. doi:10.1007/978-3-642-16483-5_6056

(2012) Ulcerative Colitis. In: Schwab M (ed) Encyclopedia of Cancer, 3rd edn. Springer Berlin Heidelberg, p 3836. doi:10.1007/978-3-642-16483-5_6095

Colon Cancer Molecular and Targeted Experimental Therapy

Yun Dai[1], George Wilson[2] and Liang Qiao[2]
[1]Department of Gastroenterology, Peking University First Hospital, Beijing, China
[2]Storr Liver Centre, Westmead Millennium Institute for Medical Research, The University of Sydney at Westmead Hospital, Westmead, NSW, Australia

Definition

Colorectal cancer (CRC) is the third most common cancer and the third leading cause of cancer death in the USA. CRC incidence rates have rapidly increased in several regions where the incidence was historically low, including parts of East Asia and Eastern Europe. In 2014, it was estimated that 136,830 new cases and 50,310 deaths have occurred in the USA and over 1.2 million new cases and some 600,000 deaths have occurred worldwide. The prognosis of CRC declines rapidly with the advanced staging. According to the American Joint Committee on Cancer (AJCC) staging system, the overall 5-year survival of patients with stage I CRC was greater than 90%, whereas that of the stage IV was less than 10%. Unfortunately, surgical resection is not suitable for the vast majority of CRC patients who have metastatic disease. As a result, more effective treatment approaches for CRC are needed.

Characteristics

Molecularly Targeted Therapies

Advances in the understanding of cellular and molecular mechanisms underlying the tumorigenesis and metastasis of CRC have enabled the development of novel approaches for treatment including the so-called *molecularly targeted approach* or *biological therapeutic approach*. These new approaches have demonstrated improved among CRC patients in a range of studies. In contrast to traditional cancer therapeutics that mainly kill the rapidly replicating cells, the new targeted therapies mainly act by influencing the processes that control cell proliferation, apoptosis, angiogenesis, and tumor spread.

Targeting Cell Proliferation Signaling as an Anticancer Approach

Deregulated proliferation is one of the major factors that contribute to tumor growth. Thus, any approaches that can suppress the growth of malignant cells would be a useful option for treating cancers. In most solid tumors, cell proliferation is controlled by multiple pathways, like epidermal growth factor (EGF)/epidermal growth factor receptor (EGFR) pathway, mitogen-activated protein kinase (MAPK) pathway, phosphatidylinositol 3-kinase (PI3K) signaling, and MET/hepatocyte growth factor (HGF) signaling. Thus, these pathways can be principal targets for cancer therapy.

Targeting EGF/EGFR Signaling

Approximately 85% of CRCs have increased expression of EGF, and about 36% of CRCs overexpress both EGF and EGFR. Constitutive activation of EGF/EGFR signaling is related to deregulated cancer cell proliferation, apoptosis, and tumor-induced angiogenesis and has thus been validated as a relevant therapeutic target in several human cancers, including CRC.

Monoclonal Antibodies Against EGFR The anti-EGFR monoclonal antibodies (MABs) cetuximab and panitumumab have been approved by FDA for treating EGFR-expressing metastatic CRC (mCRC). Data from clinical studies have demonstrated that cetuximab and panitumumab are effective anticancer drugs as single agent in heavily pretreated patients and patients with chemoresistant mCRC and that cetuximab is also effective in combination with standard chemotherapies as a first-line therapy. Pertuzumab, another MAB for EGFR, has been shown to mildly inhibit cell cycle progression of colon cancer cells and suppress the growth of xenograft tumors derived from these cells. A Phase I study has shown that the combination of pertuzumab and cetuximab could exert anticancer effects against the treatment-resistant CRCs. However, such a combination was associated with intolerable overlapping toxicities.

However, anti-EGFR MABs are only effective in a subset of patients. In CRC patients, activating mutations in KRAS, which are present in approximately 35–40% of patients with mCRC, can result in the activation of EGFR-independent signaling. Therefore, it is recommended that only patients with wild-type KRAS tumors should receive EGFR-targeted treatment. Moreover, preclinical evidence and preliminary clinical studies have suggested that other biomarkers in addition to KRAS status could be useful in predicting the response to anti-EGFR therapy. High levels of specific ligands for the EGFR (epiregulin and amphiregulin), as well as an increase in EGFR copy number, seem to predict higher efficacy of cetuximab in patients with wild-type KRAS tumors. Additionally, mutations in BRAF and PIK3CA and loss of PTEN (phosphatase and tensin homolog) protein expression have been suggested as novel biomarkers of resistance to anti-EGFR treatment in CRC patients. However, none of these markers have sufficient validation to be considered for routine clinical use.

Tyrosine Kinase Inhibitors Aberrant activation of receptor tyrosine kinases (RTKs) and non-RTKs contributes to both tumor initiation and progression, as well as promoting resistance to cancer therapy. Aberrant activation of RTKs, such as EGFR, hepatocyte growth factor (HGF) receptor MET, platelet-derived growth factor receptor (PDGFR), and vascular endothelial growth factor receptor (VEGFR), engages intracellular kinase cascades such as Ras/B-Raf/ERK and PI3K/AKT/mTOR. Thus, TKIs are potentially useful agents for cancer treatment.

TKIs targeting EGFR have been approved by the FDA for treatment of non-small cell lung cancer (NSCLC) or breast cancer but are under various stages of clinical trials for CRC. Preclinical studies using TKIs targeting EGFR alone have shown antitumor effects both in vitro and in vivo. In addition, a series of Phase I studies have shown that some TKIs targeting EGFR (vandetanib, lapatinib, erlotinib, EKB-569, and gefitinib), when combined with classic chemotherapy, appeared to enhance the antitumor effect but at the same time increase the cytotoxicity in patients with CRC.

Targeting PI3K/AKT/mTOR Signaling Dysregulation of the PI3K/AKT/mTOR pathway is seen in 40–60% of patients with CRC. Buparlisib (BKM120), a pan-Class I PI3K inhibitor, showed some antitumor activity with good safety profile in patients with mCRC in Phase I studies. Everolimus, an inhibitor of mTOR pathway, was tested in a Phase II clinical trial in patients with mCRC, and the results showed that this agent did not confer meaningful efficacy in patients with mCRC who were previously treated with bevacizumab-, fluoropyrimidine-, oxaliplatin-, and irinotecan-based regimens. Another Phase

I study of PI3K/mTOR inhibitor BGT226 in patients with advanced solid tumors including CRC showed limited preliminary antitumor activity, potentially due to low systemic exposure.

Targeting MEK/MAPK Signaling Many solid tumors including CRC display activation of the MAPK signaling pathway. MAPKs are phosphorylated and activated by MAPK kinases (MAPKKs, including MEK1/MEK2, MEK5, MKK3/MKK6, and MKK4/MKK7). Preclinical studies showed that MEK inhibitor exhibited potent antitumor activity in xenograft models. Oral MEK inhibitor CI-1040 was generally well tolerated but only showed limited antitumor activity in patients with advanced CRC as revealed in a Phase II study. A pilot study testing the effect of PD 0325901, a second-generation MEK inhibitor, was prematurely terminated because of an unexpected high incidence of musculoskeletal and neurological adverse events. A Phase II study of another MEK inhibitor, selumetinib (AZD6244, ARRY-142886), plus irinotecan as second-line therapy in patients with KRAS-mutated CRC showed promising results in improving progression-free survival (PFS). BAY 86-9766 (also termed RDEA119 or VRX-621119), a highly selective and potent inhibitor of MEK1/MEK2, also showed some benefits in patients with advanced CRC in Phase I studies. Further investigations of MEK inhibition in the treatment of mCRC are warranted.

Complex interactions between MAPK and PI3K/AKT/mTOR signaling pathways exist, and coactivation of these pathways is also seen in many cancers including CRC, thus providing a rationale for combining the therapeutic agents that simultaneously target both pathways. Preclinical studies have shown that combinatorial use of selumetinib and AZD8055 (a potent specific inhibitor of mTOR kinase) showed enhanced antitumor activity in xenograft CRC models derived from human CRC cells compared to either agent alone. However, the clinical relevance of these findings needs to be further evaluated.

MET/Hepatocyte Growth Factor (HGF) Signaling Pathway Aberrant MET/HGF signaling is a common feature in CRC. Downstream signaling effects are transmitted via MAPK, PI3K/AKT, signal transducer and activator of transcription proteins (STAT), and NF-κB, and the terminal effector components of these pathways lead to increased cell proliferation, survival advantage, and increased cell motility and invasive capacity. Activation of MET/HGF signaling appears to be a mechanism for resistance to anti-EGFR therapy in CRC patients and a negative prognostic indicator. Several MABs against MET and its ligand HGF as well as MET inhibitors have been tested in CRC patients. A Phase I/II trial with rilotumumab (anti-HGF antibody) in combination with panitumumab showed promising outcomes compared to panitumumab alone in chemoresistant CRC patients. The efficacy of anti-MET MAB onartuzumab (MetMAB) combined with bevacizumab plus mFOLFOX6 is currently being investigated in a Phase II study in mCRC patients. Combination of tivantinib, a selective MET TKI, with standard treatment was also investigated in Phase I/II trials, which showed a higher response rate and a slight improvement in PFS. The effect of combinatorial use of tivantinib with cetuximab is currently under investigation in a Phase II study in patients with advanced CRC who failed to respond to anti-EGFR therapy.

Targeting Angiogenesis as an Anticancer Approach

Angiogenesis plays a crucial role in tumor growth and metastasis and is considered an important target in cancer therapy. A number of growth factors and their cognate receptors regulate the process of angiogenesis, such as platelet-derived growth factor (PDGF), fibroblast growth factor (FGF), and transforming growth factor alpha (TGFα). However, the most well-studied pro-angiogenic pathway is vascular endothelial growth factor (VEGF) and its receptor VEGFR. Anti-VEGF therapy does not appear to have significant activity as a single agent in CRC and is generally used in combination with other anticancer regimens.

Monoclonal Antibodies Against VEGF Bevacizumab (Avastin) was the first antiangiogenic drug approved by the FDA for treating mCRC in combination with FOLFIRI or FOLFOX therapy, based on its ability to prolong survival. Ramucirumab (IMC-1121B) is a human MAB against VEGFR-2. Preclinical and Phase I clinical studies have demonstrated the antitumor efficacy of ramucirumab. The planned Phase III study of ramucirumab in mCRC will investigate its role as second-line treatment in combination with FOLFIRI. Other MABs against VEGF, such as HuMV833 and IMC-18F1, are currently under preclinical and Phase I clinical studies.

VEGF Trap Aflibercept is a fully humanized recombinant soluble fusion protein that consists of extracellular domains of VEGFR-1 and VEGFR-2. It binds VEGF-A and VEGF-B with high affinity, prevents their binding to native VEGF receptors, and therefore inhibits angiogenesis. Based on the results of the VELOUR study, aflibercept with FOLFIRI therapy prolonged overall survival (OS) and PFS in mCRC. Aflibercept has also been approved by the FDA for mCRC as a second-line therapy agent.

Tyrosine Kinase Inhibitor Regorafenib (BAY 73-4506) is a multikinase inhibitor against selected tyrosine kinases (VEGFR-2, VEGFR-3, TIE-2, PDGFR, FGFR, RET, and c-Kit) as well as an inhibitor of the RAF/MEK/MAPK pathway. A Phase III study has shown that regorafenib monotherapy increased OS and PFS in refractory mCRC patients, and as a result, regorafenib has been approved by FDA as the third-line treatment for mCRC. Vargatef (BIBF1120) blocks the activity of human VEGFR-1, VEGFR-3, FGFR-1, and FGFR-3. The Phase II study of Vargatef as the first-line treatment in combination with mFOLFOX6 is ongoing. Other TKIs against VEGFR signaling including brivanib (BMS582664), cediranib (ADZ2171), sunitinib (SU11248), vatalanib (PTK787/ZK222584), and semaxanib (SU5416) have shown promising preclinical activity in CRC. Unfortunately, Phase III trials where these agents were used as monotherapy or in combination with chemotherapy in CRC patients failed to show any improvement in clinical outcomes.

Targeting Apoptosis Signaling as an Anticancer Approach

Resistance to apoptosis is a hallmark of all malignancies, providing many potential targets for drug development. Apoptosis occurs through two major pathways: the intrinsic pathway and extrinsic pathway. The intrinsic apoptotic pathway is triggered by DNA damage, deregulated oncogenes, and growth factor deprivation and is largely regulated by the Bcl-2 family and mitochondria. The extrinsic apoptosis pathway is triggered by ligand-induced activation of cell death receptors (DRs) on the surface of the cell membrane. Three major death ligand-receptor systems have been identified: tumor necrosis factor (TNF)-tumor necrosis factor receptor (TNFR), FAS ligand-FAS, and TRAIL-TRAIL receptors. Several strategies have been devised to induce cell killing by manipulating apoptotic regulators directly.

Targeting Bcl-2 Family Under physiological conditions, the pro-apoptotic (Bax, Bak, Bok, Bad, and Bid) and anti-apoptotic members (Bcl-2, Bcl-XL, and Mcl-1) of the Bcl-2 family can cooperate to maintain a dynamic balance. Overexpression of Bcl-2 in CRC is a negative prognostic factor. Extensive preclinical studies have been conducted on various small-molecule Bcl-2 antagonists (HA14-1, antimycin A, obatoclax, ABT-737, and ABT-263) and a Bcl-2 antisense oligo (oblimersen) in experimental therapy for CRC. Only oblimersen has entered Phase I/II clinical trials for CRC. Preliminary data suggests that most Bcl-2 antagonists have limited efficacy in CRC as a single agent but can significantly sensitize cancer cells to other therapeutic agent-induced apoptoses.

IAP Antagonists Overexpression of inhibitors of apoptosis proteins (IAPs) is closely correlated to an elevated apoptotic threshold of cancer cells and has become an attractive target for cancer therapy. SMAC/Diablo is an endogenous IAP inhibitor and can relieve the anti-apoptotic activities of several IAPs. SMAC mimetics (SM-122, SM-164, and JP-1201) are being developed as a new class of anticancer therapies. Preclinical studies have demonstrated that the combination of SM-164 with TRAIL could induce significant apoptosis in CRC cells and induce rapid tumor regression in a xenograft model of CRC. The SMAC mimetic JP-1201 was also able to sensitize CRC cells to ionizing radiation both in vivo and in vitro. Small-molecule inhibitors for XIAP (such as embelin and 1396-12) have shown promise in preclinical studies in CRC. Because the single-agent activity of IAP antagonists is very limited, rational combinational regimes with these agents represent a viable strategy for their clinical development.

Targeting the Extrinsic Apoptosis Pathway Preclinical data suggest that recombinant human TRAIL (rhTRAIL) and agonistic MABs to TRAIL receptors such as mapatumumab (targeting DR4/TRAIL-R1), PRO95780, and apomab (targeting DR5/TRAIL-R2) have the ability to selectively induce apoptosis in a variety of human cancer cells. These agents have advanced into Phase I/II clinical trials for several cancers including CRC, showing some efficacy in combination regimens. The clinical outlook of Fas ligands (APO010 and Fasaret) or TNF-α (TNFerade) is limited due to severe side effects.

Targeting CRC Stem Cells
Cancer stem cells (CSCs) are defined as cells that are endowed with both self-renewal and multilineage differentiation potential. CSCs are considered to be responsible for tumor initiation, growth, and relapse. The failure of conventional radio- or chemotherapy in curing cancers is at least partially attributed to the inability of these approaches to eradicate CSCs, leading to treatment resistance and relapse after therapy. Thus, elimination of CSCs constitutes a promising approach for cancer therapy. Several signaling pathways are involved in maintaining CSCs including Wnt, Hedgehog, and Notch signaling pathways.

Targeting Wnt Pathway Components Approaches inhibiting Wnt/β-catenin signaling could of great therapeutic potential for cancers.

The Wnt canonical pathway is mainly regulated at the level of β-catenin, a protein kept under low cytosolic concentrations by the destruction complex. Wnt ligands bound to receptors trigger dissolution of the complex, leading to the translocation and accumulation of β-catenin in the cytosol and then in the nucleus where it converts TCF into a transcriptional activator. Most Wnt/β-catenin inhibitors are still in preclinical testing or in the developmental stage for CRC therapy. Likewise, targeting the TCF/β-catenin nuclear complex also holds promise for successful therapy. In this aspect, ICG-001, an inhibitor for Wnt/β-catenin/TCF-mediated transcription, was found to specifically bind to and inhibit the transcriptional coactivator element-binding protein (CBP) and prevent the subsequent activation of Wnt/β-catenin/TCF signaling. Treatment of CRC cells bearing APC or β-catenin mutations with ICG-001 causes dose-dependent cell death, whereas normal colonic epithelial cells are exempt. The effect is also seen in the APC^{min} mouse model and in tumor xenografts. ICG-001 is expected to shortly enter in clinical Phase I trials.

Targeting Notch Pathway Components An important ligand of the Notch pathway is delta-like 4 (DLL4). Inhibition of DLL4 with the human monoclonal antibody 21M18 in CRC xenografts led to reduced tumor growth and CSC frequency. Moreover, combination of irinotecan with anti-DLL4 has shown a synergistic effect for reducing the tumor growth and CSC frequency.

Colorectal Cancer Immunotherapy
Immunotherapies against tumor include active, passive, or immunomodulatory strategies. Active immunotherapies elicit the patient's own immune response to recognize tumor-associated antigens (TAAs) and attack cancer cells. In contrast, passive immunotherapy involves administration of exogenous lymphocytes or antibodies, to mediate an immune response. Immunotherapy alone or in combination with conventional therapy may be effective for treating advanced CRC and preventing relapse.

Cancer Vaccines

Cancer vaccines are active therapeutic approaches and include whole-tumor cell vaccines, peptide vaccines, dendritic cell-based vaccines, and viral vector-based vaccines. The identification of a suitable TAA is one of the most important steps in developing a cancer vaccine. The most widely studied TAAs in CRC for vaccine-based immunotherapy include CEA, MUC1, and guanylyl cyclase C (GUCY2C, GCC).

Autologous Tumor Cell Vaccines The autologous tumor cell vaccines comprise all tumor antigens, potentially eliciting adaptive antitumor immunity to multiple antigens. Phase III trials in CRC patients using OncoVAX, an irradiated autologous tumor cell vaccine with BCG adjuvant, have revealed that this agent could improve the recurrence-free survival and OS, although these effects were largely limited to the patients with stage II/III disease. Another cell-based vaccine for the treatment of CRC was developed using Newcastle disease virus-infected autologous tumor cells (ATV-NDV). In Phase II/III clinical trials, ATV-NDV was found to improve the metastasis-free survival and OS of CRC patients.

Peptide Vaccines Peptide vaccine employs the smallest possible unit of a vaccine: the 8–11-amino acid epitope of an antigen that is recognized by effector T cells. Increased expression of TAAs such as RNF43, TOMM34, and KOC1 was found in 80% of CRC specimens. Synthesized peptide vaccines derived from HLA-A2402-restricted epitopes of RNF43 and TOMM34, as well as HLA-A24-restricted peptides from RNF43, TOMM34, KOC1, and VEGFR-1 and VEGFR-2, have reached Phase I clinical trials. Future Phase II and III studies are needed to further define the efficacy of these approaches for CRC.

Dendritic Cell-Based Vaccines Dendritic cells (DCs) play a critical role in presentation of immunogenic peptides and activation of T cells. DCs can be collected from patients, pulsed with tumor epitopes, matured ex vivo, and transferred back into patients as a cancer vaccine to elicit antitumor immunity. Early-phase clinical trials have shown that CRC patients who were vaccinated with DCs pulsed with CEA peptides or CEA mRNA demonstrated positive CEA-specific T-cell responses and induction of stable disease. In a new Phase II trial, survival for all vaccinated mCRC patients who received autologous DCs modified with a poxvector encoding CEA and MUC1 was longer than the patients who did not receive active immunotherapy.

Viral Vector-Based Vaccines Viral vectors for cancer vaccines include recombinant lentiviruses, poxviruses, adenoviruses, and retroviruses. These vectors can effectively deliver the antigen and at the same time can provide sufficient adjuvant effects to elicit an immune response. Phase II trials using vaccinia and fowl pox encoding CEA and containing three co-stimulatory molecules B7-1, ICAM-1, and LFA-3 have demonstrated induction of anti-CEA-specific T-cell responses and production of prolonged disease stabilization in 40% of patients with metastatic cancers, including CRC. Another adenoviral vector encoding the CEA and Ad5 [E1-, E2b-]-CEA(6D) was shown to induce cell-mediated immunity in 61% of patients with advanced CRC. GUCY2C-targeted viral vector vaccine is in early stages of development. An adenoviral vector expressing GUCY2C (Ad5-GUCY2C) induces prophylactic and therapeutic immunity against mCRC in mice without adverse effects. A Phase I trial examining the effect of Ad5-GUCY2C in early-stage CRC patients is under design.

Adoptive Cell Therapy

Adoptive cell therapy (ACT) is one form of passive immunotherapy. In ACT, autologous T cells are removed from patients, activated and expanded ex vivo, and transferred back into patients for a therapeutic effect. Although ACT is successful in targeting melanoma and leukemia, this approach has failed to demonstrate safety and efficacy in CRC patients. Alternative approaches to minimize toxicities by identifying appropriate antigen or interventions that reduce the side effects will be necessary for this therapy to achieve success.

Immunomodulatory Strategy

In the last decade, it has become clear that the immune regulatory pathway composed of programmed death-1 (PD-1, CD279), a receptor expressed on activated T and B cells, plays an integral role in the down-modulation of antitumor immunity. In this perspective, S-936558 (MDX-1106/ONO-4538), a blocking monoclonal antibody specific for human PD-1, has shown promising results in treating a diversity of solid tumors including CRC.

Another Phase III trial has demonstrated that combination of chemotherapy and immune-adjuvant cytokines may represent a novel reliable option for mCRC. In this study, GOLF regimen (gemcitabine, oxaliplatin, levofolinate, and 5-fluorouracil) has been combined with the GM-CSF to activate peripheral DCs and recombinant IL-2 to promote the maturation of DCs and enhance the expansion of antitumor-specific CTL precursors. This regimen showed superiority over FOLFOX in terms of PFS and response rate with a trend to longer survival.

Acknowledgment This work was partially supported by the Robert W. Storr Bequest to the Sydney Medical Foundation, University of Sydney, a National Health and Medical Research Council of Australia (NHMRC) Project grant to LQ (ID: APP1047417), and two Cancer Council NSW grants to LQ (ID: APP1070076) and LH (ID: APP1069733).

Cross-References

- ► Angiogenesis
- ► Apoptosis
- ► Breast Cancer
- ► Cetuximab
- ► Chemotherapy
- ► DNA Damage
- ► Epidermal Growth Factor Receptor
- ► Erlotinib
- ► Fluorouracil
- ► Gefitinib
- ► Irinotecan
- ► Metastasis
- ► Non-Small-Cell Lung Cancer
- ► PI3K Signaling
- ► Receptor Tyrosine Kinases
- ► Tumor Necrosis Factor
- ► Tyrosine Kinase Inhibitors
- ► Vascular Endothelial Growth Factor

References

Camidge DR, Herbst RS, Gordon MS, Eckhardt SG, Kurzrock R, Durbin B, Ing J, Tohnya TM, Sager J, Ashkenazi A, Bray G, Mendelson D (2010) A phase I safety and pharmacokinetic study of the death receptor 5 agonistic antibody PRO95780 in patients with advanced malignancies. Clin Cancer Res 16(4):1256–1263

Correale P, Botta C, Rotundo MS, Guglielmo A, Conca R, Licchetta A, Pastina P, Bestoso E, Ciliberto D, Cusi MG, Fioravanti A, Guidelli GM, Bianco MT, Misso G, Martino E, Caraglia M, Tassone P, Mini E, Mantovani G, Ridolfi R, Pirtoli L, Tagliaferri P (2014) Gemcitabine, oxaliplatin, levofolinate, 5-fluorouracil, granulocyte-macrophage colony-stimulating factor, and interleukin-2 (GOLFIG) versus FOLFOX chemotherapy in metastatic colorectal cancer patients: the GOLFIG-2 multicentric open-label randomized phase III trial. J Immunother 37(1):26–35

de Sousa EM, Vermeulen L, Richel D, Medema JP (2011) Targeting Wnt signaling in colon cancer stem cells. Clin Cancer Res 17(4):647–653

Holt SV, Logie A, Davies BR, Alferez D, Runswick S, Fenton S, Chresta CM, Gu Y, Zhang J, Wu YL, Wilkinson RW, Guichard SM, Smith PD (2012) Enhanced apoptosis and tumor growth suppression elicited by combination of MEK (selumetinib) and mTOR kinase inhibitors (AZD8055). Cancer Res 72(7):1804–1813

Kemper K, Grandela C, Medema JP (2010) Molecular identification and targeting of colorectal cancer stem cells. Oncotarget 1(6):387–395

Lipson EJ, Sharfman WH, Drake CG, Wollner I, Taube JM, Anders RA, Xu H, Yao S, Pons A, Chen L, Pardoll DM, Brahmer JR, Topalian SL (2013) Durable cancer regression off-treatment and effective reinduction therapy with an anti-PD-1 antibody. Clin Cancer Res 19(2):462–468

Martinelli E, Troiani T, Morgillo F, Orditura M, De Vita F, Belli G, Ciardiello F (2013) Emerging VEGF-receptor inhibitors for colorectal cancer. Expert Opin Emerg Drugs 18(1):25–37

Melero I, Gaudernack G, Gerritsen W, Huber C, Parmiani G, Scholl S, Thatcher N, Wagstaff J, Zielinski C, Faulkner I, Mellstedt H (2014) Therapeutic vaccines for cancer: an overview of clinical trials. Nat Rev Clin Oncol 11(9):509–524

Shimizu T, Tolcher AW, Papadopoulos KP, Beeram M, Rasco DW, Smith LS, Gunn S, Smetzer L, Mays TA, Kaiser B, Wick MJ, Alvarez C, Cavazos A, Mangold GL, Patnaik A (2012) The clinical effect of the dual-targeting strategy involving PI3K/AKT/mTOR and

RAS/MEK/ERK pathways in patients with advanced cancer. Clin Cancer Res 18(8):2316–2325

Smyth EC, Sclafani F, Cunningham D (2014) Emerging molecular targets in oncology: clinical potential of MET/hepatocyte growth-factor inhibitors. Onco Targets Ther 12(7):1001–1014

Tang PA, Cohen SJ, Kollmannsberger C, Bjarnason G, Virik K, MacKenzie MJ, Lourenco L, Wang L, Chen A, Moore MJ (2012) Phase II clinical and pharmacokinetic study of aflibercept in patients with previously treated metastatic colorectal cancer. Clin Cancer Res 18(21):6023–6031

Colon Cancer Pathology of Hereditary Forms

Zoran Gatalica
Department of Pathology, Creighton University School of Medicine, Omaha, NE, USA

Definition

Approximately 5% of all colorectal carcinomas are due to a defined single genetic defect causing hereditary disease. Gross and microscopic pathologic examination of the resection or biopsy specimen can help identify an unsuspected case of hereditary colorectal carcinoma due to the characteristic morphologic findings seen in some syndromes. Additional immunohistochemical and molecular studies can then provide a definitive diagnosis. Furthermore, due to the germline nature of mutations in these syndromes, various extracolonic manifestations may be the first sign of the disease, and knowledge of such associations can greatly improve the quality of care for these patients.

Characteristics

Colorectal Cancer in Lynch Syndrome (Hereditary Nonpolyposis Colorectal Cancer)

The most common form of hereditary colorectal carcinoma is ▶ Lynch syndrome (*synonym* hereditary nonpolyposis colorectal cancer, HNPCC) caused by inactivating mutation(s) in genes for DNA mismatch repair (MMR) enzymes (▶ mismatch repair in genetic instability). It is characterized by proximally located tumors frequently showing mucinous and medullary type histologic features. The lack of functional MMR enzymes leads to genetic instability (which is reflected in high-frequency ▶ microsatellite instability, MSI-H).

Medullary colon carcinoma (Fig. 1), characterized by the sheets and nests of small to medium cells with variable cytoplasm, vesicular nuclei, and prominent nucleoli and with distinct stromal and intraepithelial lymphocytes (Fig. 2), is characteristic of microsatellite instability-high

Colon Cancer Pathology of Hereditary Forms, Fig. 1 Medullary carcinoma of the colon characterized by nests of medium-sized epithelial cells with variable cytoplasm, vesicular nuclei, and prominent nucleoli and with distinct stromal and intraepithelial lymphocytes

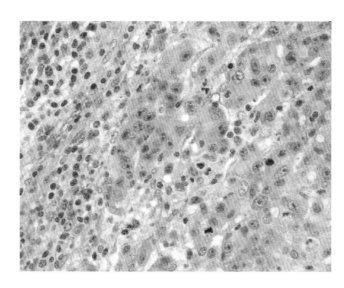

**Colon Cancer Pathology
of Hereditary Forms,
Fig. 2** Tumor-infiltrating
lymphocytes.
Immunohistochemical stain
highlighting increased
numbers of CD3-positive
T lymphocytes (*brown*) is
characteristically seen in
MSI-H colorectal cancer

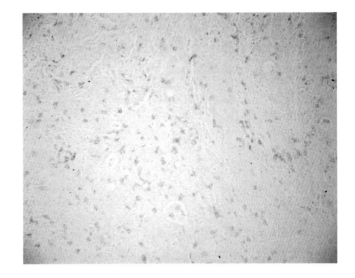

**Colon Cancer Pathology
of Hereditary Forms,
Fig. 3** Loss of expression
of MLH1 mismatch repair
protein in signet ring colon
carcinoma (*lower left*).
Nuclear expression (*brown*)
is retained in normal
lymphocytes (*upper right*)

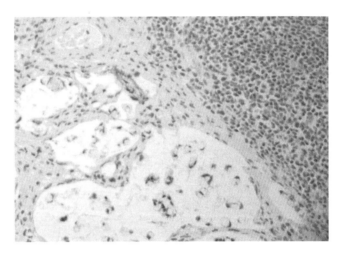

(MSI-H) phenotype, but it has very low (4%) overall prevalence.

Other histologic types of colorectal carcinomas of the Lynch syndrome are also characterized by the presence of increased number of intraepithelial lymphocytes. This feature may be the single most helpful morphologic characteristic of colorectal carcinoma caused by a deficiency in MMR proteins (e.g., Lynch syndrome). However, this feature alone does not discriminate between tumors caused by germline mutations in one of the MMR genes (Lynch syndrome) from sporadic colorectal carcinoma due to inactivation through promoter ▶ methylation of MLH-1 (major DNA mismatch repair gene/protein; Fig. 3).

▶ Immunohistochemistry tests for expression of MMR proteins in colorectal carcinoma (loss of expression in Lynch syndrome) may provide additional information about genetic events underlying MSI-H tumor phenotype. A loss of expression of mismatch repair proteins serves as a reasonably reliable test of mismatch repair deficiency if antibodies to hMLH1, hMSH2, hMSH6, and hPMS2 are employed. Patients with Lynch syndrome can sometimes present initially with extracolonic malignancies including small bowel carcinomas, sebaceous cutaneous tumors, and, in women, endometrial adenocarcinoma (▶ endometrial cancer). In a published study, 1.8% of all newly diagnosed endometrial cancer patients had Lynch syndrome.

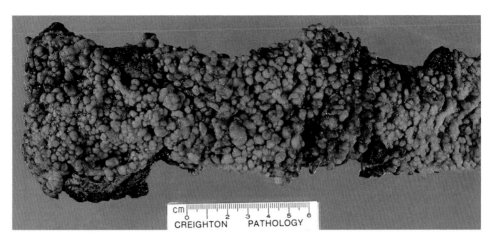

Colon Cancer Pathology of Hereditary Forms, Fig. 4 Colectomy specimen from a patient with classic familial adenomatous polyposis. The colon is carpeted by the hundreds of small polyps

Colorectal Cancer in Familial Polyposis Syndromes

Hereditary colorectal carcinoma may also arise in various familial polyposis syndromes which include familial adenomatous polyposis (FAP), "attenuated FAP," and other multiple adenomas syndromes as well as various types of hamartomatous polyposis syndrome. All of these rare conditions have characteristic clinical presentation and histopathologic features of polyps, and most of them have a defined genetic abnormality.

Familial adenomatous polyposis (FAP) is characterized by numerous (>100, usually several hundreds in fully developed cases) adenomatous colorectal polyps (Fig. 4).

FAP is an autosomal dominant hereditary cancer syndrome caused by a germline mutation in the APC gene. Several genotype-phenotype variations have been consistently observed between mutations in specific sites of the APC gene on chromosome 5q and disease manifestations. In fully developed cases, the colonic mucosa is carpeted by hundreds of mostly sessile polyps. Scattered pedunculated polyps are less numerous. Typically, polyps are evenly distributed along the whole large bowel. Microscopically, most adenomas display tubular architecture, and adenomas and adenocarcinomas in FAP are identical to sporadic counterparts. In the stomach, gastric fundic gland polyps develop in almost 90% of adult FAP patients.

Extra-gastrointestinal manifestations may be of importance for practicing pathologists in the diagnosis of unsuspected FAP. ► Desmoid tumor (fibromatosis) is rare in the general population, but it is commonly seen in FAP and can be the first manifestation of disease. In FAP, such desmoid tumors typically develop in retroperitoneal tissues or in the abdominal wall following surgical trauma (abdominal desmoids), while fibromatosis unrelated to FAP is more common in extra-abdominal localizations. ► Papillary thyroid carcinoma and its rare cribriform-morular variant can be associated with FAP, and this could lead to detection of unsuspected FAP. The risk of ► hepatoblastoma in children of patients with FAP is increased, and new germline mutations can be identified in 10% of cases.

► MUTYH-associated colorectal polyposis (MAP) is a recessively inherited disorder, and patients have homozygous or compound heterozygous germline mutations of the *MUTYH* gene (compound heterozygosity). Biallelic mutations in the germline in *MUTYH* were found in 18% of APC gene mutation-negative patients with attenuated phenotype (less than a 100 polyps in colon, presentation later in life, carcinomas of the left side of the colon).

Hamartomatous Polyposis Syndromes

These include ▶ Peutz-Jeghers syndrome, juvenile polyposis, Cronkhite-Canada syndrome, and Cowden disease/Bannayan-Riley-Ruvalcaba syndrome. Cronkhite-Canada syndrome is a nonfamilial syndrome characterized by epithelial disturbances in the gastrointestinal tract (hamartomatous polyposis syndrome) and skin (alopecia, onychodystrophy, and hyperpigmentation). The disease primarily affects adults (mean age at the onset is 59 years). All of these syndromes are characterized by hamartomatous polyps and associated risk of the development of gastrointestinal and extraintestinal carcinomas.

Peutz-Jeghers Syndrome

Peutz-Jeghers syndrome is characterized by mucocutaneous pigmentation and gastrointestinal hamartoma, which occurs anywhere from the stomach to the anus. Germline mutations in the serine/threonine kinase gene (STK11/LKB1) cause Peutz-Jeghers syndrome in about half of the affected families. A prototypic Peutz-Jeghers syndrome polyp is a hamartoma of the muscularis mucosae. Therefore, the core of the polyp consists of smooth muscle covered by lamina propria and mature glandular epithelium which gives rise to a characteristic arborizing smooth muscle core of the polyp. These polyps could be sessile, but are frequently pedunculated.

Juvenile Polyposis Coli Syndrome

Juvenile polyposis coli syndrome is caused by mutations in the SMAD4/MADH4 and in the BMPR1A genes. Patients develop numerous hamartomatous colorectal polyps, which are characterized by dilated crypts that do not include muscularis mucosae (a feature of Peutz-Jeghers polyps). The diagnosis of juvenile polyposis syndrome is made when multiple (3–10) juvenile polyps are found in the gastrointestinal tract. Adenomatous ▶ dysplasia can develop within a juvenile polyp, which carries an increased risk of malignancy.

Hamartomatous, Juvenile-Type Polyps

Hamartomatous, juvenile-type polyps can also be found in Cowden/Bannayan-Riley-Ruvalcaba syndrome (caused by germline mutations in PTEN) and Gorlin syndrome (germline patched gene (*PTCH*) mutations).

References

Atard TM, Young RJ (2006) Diagnosis and management of gastrointestinal polyps. Gastroenterol Nurs 29:16–22

Boland CR (2006) Decoding hereditary colorectal cancer. N Engl J Med 354:2815–2817

Burgart LJ (2005) Testing for defective mismatch repair in colorectal carcinoma. A practical guide. Arch Pathol Lab Med 129:1385–1389

Gatalica Z, Torlakovic E (2008) Pathology of the hereditary colorectal carcinoma. Fam Cancer 7:15–26

Colon Cancer Risk

▶ Colorectal Cancer Nutritional Carcinogenesis

Colorectal Cancer

Synonyms

Colon cancer; Rectal cancer

Definition

Cancer of the colon or rectum is also called colorectal cancer. In the United States, it is the fourth most common cancer in men and women. Caught early, it is often curable. It is more common in people over 50, and the risk increases with age.

Risk factors include:

1. Polyps – growths inside the colon and rectum that may become cancerous
2. A diet that is high in fat
3. A family history or personal history of colorectal cancer
4. Ulcerative colitis or ▶ Crohn disease

Symptoms can include blood in the stool, narrower stools, a change in bowel habits, and general stomach discomfort. However, even in the absence of symptoms, screening is important. Everyone who is 50 or older should be screened for colorectal cancer. Colonoscopy is one screening method for colorectal cancer. Treatments for colorectal cancer include surgery, chemotherapy, radiation, or a combination.

Cross-References

▸ APC Gene in Familial Adenomatous Polyposis
▸ Colon Cancer Carcinogenesis in Human and in Experimental Animal Models
▸ Colon Cancer Genomic Pathways
▸ Colon Cancer Molecular and Targeted Experimental Therapy
▸ Colon Cancer Pathology of Hereditary Forms
▸ Colorectal Cancer Chemoprevention
▸ Colorectal Cancer Clinical Oncology
▸ Colorectal Cancer Nutritional Carcinogenesis
▸ Colorectal Cancer Pathology
▸ Colorectal Cancer Premalignant Lesions
▸ Colorectal Cancer Therapeutic Antibodies
▸ Colorectal Cancer Vaccine Therapy
▸ COX-2 in Colorectal Cancer
▸ Crohn Disease
▸ KRAS in Colorectal Cancer Therapy
▸ Lynch Syndrome
▸ MUTYII-Associated Colorectal Polyposis
▸ Regorafenib

See Also

(2012) Colonoscopy. In: Schwab M (ed) Encyclopedia of cancer, 3rd edn. Springer, Berlin/Heidelberg, p 915. doi:10.1007/978-3-642-16483-5_1266
(2012) Polyp. In: Schwab M (ed) Encyclopedia of cancer, 3rd edn. Springer, Berlin/Heidelberg, p 2955. doi:10.1007/978-3-642-16483-5_6524
(2012) Ulcerative colitis. In: Schwab M (ed) Encyclopedia of cancer, 3rd edn. Springer, Berlin/Heidelberg, p 3836. doi:10.1007/978-3-642-16483-5_6095

Colorectal Cancer Chemoprevention

Lynne M. Howells and Karen Brown
Department of Cancer Studies, University of Leicester, Leicester, UK

Synonyms

Adenomatous polyps; COX-2 inhibitors; Therapeutic prevention

Definition

Chemoprevention involves the use of drugs (of synthetic or natural origin) in order to delay, reverse, or prevent disease progression. The chemoprevention concept is not a new medical intervention and has been used for decades with remarkable success as part of risk reduction approaches for cardiovascular disease. Diseases with a long latency period, such as colorectal cancer, provide an ideal scenario for the application of chemopreventive strategies with a considerable potential to improve patient outcomes. However, in order to determine whether such a strategy may be truly beneficial within this paradigm, two research areas need to experience significant advances. Firstly, our understanding of the natural progression of the disease and underlying molecular and genetic changes needs to be improved further, and secondly, robust and easily measurable biomarkers need to be discovered and developed for several purposes: pharmacodynamic markers are required to identify agents with activity in humans, and to ascertain the optimal dose and scheduling; surrogate biomarkers are needed for monitoring disease progression and agent efficacy; lastly, in the current era of personalized medicine, there is increasing interest in the discovery of biomarkers that will help tailor preventive interventions to individuals.

Characteristics

Within colorectal carcinogenesis, as for other solid cancers, the advent of tumour formation is preceded by early changes to the epithelia (dysplasia), which arise through genetic instability. Hyperproliferation within the crypts and subsequent expansion of dysplastic crypts lead to gross early histological changes, termed aberrant crypt foci (ACF). These ACF precede formation of overt adenomas, and agent-induced changes to ACF are often used as an early biomarker readout of putative chemopreventive efficacy. Despite the fact that adenomas represent a notable premalignant change, it is important to bear in mind that not all adenomas progress to cancer, with this process being governed by a multitude of factors including the profile of genetic alterations, degree of adenoma differentiation, and overall adenoma size. Even following successful removal of adenomas, certain features such as a site of origin and adenoma histology may allude to recurrence and future cancer risk. The determination of specific progression-associated risk factors for adenomas has the advantage of allowing the identification of target populations that may benefit most from preventive intervention strategies.

There are several facets to chemopreventive strategies for colorectal cancer, each targeting specific patient cohorts. Primary prevention aims to prevent disease in healthy individuals at high risk, such as those who may harbor germline mutations. People with Lynch syndrome (caused by mutations in mismatch repair genes) and familial adenomatous polyposis (FAP) (associated with adenomatous polyposis coli gene mutations) represent the most common groups with increased familial risk, accounting for approximately 5% of all colorectal cancers. Secondary prevention targets individuals in whom precursor lesions/adenomas already exist, and tertiary prevention primarily aims to prevent disease recurrence in individuals within an adjuvant setting.

As chemopreventive agents have to be administered over a long period of time, they must possess several characteristics which are key factors in deciding whether they warrant further clinical development: very low/no toxicity; possibility of oral dosage form, low cost; proven efficacy; known target/mechanism of action. There are a number of agents that have been or are currently being evaluated for chemopreventive efficacy against colorectal cancer. These compounds encompass a range of well-characterized repurposed drugs, in most cases with a long history of use in humans, and often agents of dietary origin. Each of the agents discussed below embodies at least three of the above five key requirements characterizing suitable chemopreventive agents.

Aspirin

Aspirin is one of the most commonly utilized analgesics and is recommended for prophylactic use against cardiovascular diseases. Preliminary evidence for a chemopreventive benefit in colorectal cancer originally arose from trials designed to assess the potential efficacy of aspirin in cardiovascular disease. More compelling support for a protective effect against colorectal cancer came from long-term follow-up of these trials. Five such randomized trials were subjected to 10- and 20-year follow-up, and effects of aspirin on incidence and mortality of colorectal cancer were assessed (Rothwell et al. 2010). It was revealed that 75 mg daily aspirin reduced both colon cancer incidence and mortality, with risk reduction observed in the proximal rather than distal colon. Furthermore, a reduced risk was also apparent for rectal cancer, where aspirin was taken for 5 years or more. Benefit for both sites increased with prolonged treatment duration. Retrospective analyses of these cardiovascular cohorts and the implications of aspirin-derived benefit for colorectal cancer prevention led to an increasing number of prospective population-targeted studies. The cancer prevention trials CAPP1 and CAPP2 have added further weight to the potential protective effects of aspirin, targeting cohorts with FAP and Lynch syndrome, respectively (Burn et al. 2011a, b). While there was only a trend toward benefit in FAP patients, there was a significant reduction in overall cancer incidence in those with Lynch syndrome.

Mechanistically, one of the primary targets through which aspirin is purported to exert its anticancer efficacy is via inhibition of the cyclo-oxygenase (COX) enzyme. COX enzymes exist in two major isoforms; COX-1 is constitutively expressed, and COX-2 is an inducible form. COX-2 can be induced by inflammatory cyto-kines and is frequently upregulated in both colo-rectal adenomas and cancers. The COX enzymes catalyze conversion of arachidonic acid into pros-taglandins and thromboxanes, which play a key role in proliferative and procarcinogenic pro-cesses. Aspirin is the only nonsteroidal anti-inflammatory that can irreversibly inhibit both COX isoforms, but irreversible inhibition of COX-1 may contribute to aspirin-related toxic-ities such as gastrointestinal bleeding. In order to overcome these limiting gastrointestinal toxic-ities, several trials are now underway investigat-ing the efficacy and toxicity of aspirin in combination with proton pump inhibitors such as esomeprazole, for stomach acid reduction. Other mechanisms engaged by aspirin in mediating its colorectal cancer preventive efficacy have yet to be fully delineated, but preclinical models suggest the potential importance of inhibition of a variety of signaling pathways including nuclear factor kappa B (NF-κB), wnt/β-catenin, and mitogen activated protein kinase (MAPK) (Stolfi et al. 2013).

Sulindac

Sulindac, like aspirin, is another nonsteroidal anti-inflammatory drug (NSAID) that has shown effi-cacy for colorectal cancer prevention in preclini-cal models, and also clinically. The active metabolite of the drug, sulindac sulfide is a potent yet reversible inhibitor of COX-1 and COX-2. Sulindac has shown efficacy in both FAP patients and sporadic colorectal cancer patients, reducing the size and number of polyps. Similarly to aspi-rin, gastrointestinal bleeding and peptic ulcer dis-ease have been reported as side effects. However, in addition to standard NSAID side effects, prolonged sulindac use has the potential to cause hepatic injury, and thus sulindac has fallen by the wayside offering limited use within a prevention setting.

Coxibs

Selective COX-2 inhibitors were developed as more targeted anti-inflammatory drugs in the treatment of osteoarthritis, to avoid the gastro-intestinal side effects attributed to the COX-1 inhibiting properties of NSAIDs. Celecoxib and rofecoxib are examples of COX-2 inhibitors that have shown benefit in patients with FAP or spo-radic adenomas, reducing the size and number of polyps. Celecoxib was licensed by the US Food and Drug Administration (FDA) for adjuvant intervention in FAP within a tertiary prevention setting. However, use in this context was later suspended, as increased risk of cardiovascular events emerged, particularly at doses that had previously proven most efficacious in polyp pre-vention. Rofecoxib was completely withdrawn from the market in 2004 due to severe cardiovas-cular toxicities. Causal factors in cardiotoxiciy may arise from inhibition of prostacyclins leading to enhanced platelet aggregation and atheroscle-rotic plaque formation.

5-Aminosalicilate (5-ASA)

Colorectal cancer risk is increased by up to sixfold in individuals with inflammatory bowel disease (IBD), with the highest risk observed in sufferers of ulcerative colitis. 5-Aminosalicilate-based drugs such as sulfasalazine and mesalamine are already used extensively and have proven efficacy in the management of IBD due to their anti-inflammatory properties. They are structurally related to NSAIDs, but as they are rapidly metab-olized and inactivated, the gastrointestinal toxic-ities typically observed with NSAIDs are avoided. 5-ASA inhibits COX-2 and activates peroxisome proliferator-activated receptor-γ (PPAR-γ), which has been shown to inhibit ACF formation and intestinal tumourigenesis in mouse models. How-ever, definitive proof of clinical chemopreventive efficacy remains ambiguous, largely due to the lack of randomized controlled trials. As 5-ASA drugs are the gold standard intervention in IBD they cannot ethically be withheld. Furthermore the role of IBD as a colorectal cancer risk factor is a minor one as it accounts for only ~1% of all colorectal cancers.

Metformin

Onset of type 2 diabetes shares several risk factors with the development of sporadic colorectal cancer, including an increased body mass index (BMI) and poor diet. Type-2 diabetic patients are commonly treated with the oral biguanide drug metformin, which acts to decrease the hyperinsulaemic state by reducing insulin-like growth factor signaling, downregulating mammalian target of rapamycin signaling (mTOR) and activating the $5'$ adenosine monophosphate-activated protein kinase (AMPK) pathway. Mechanistically, alterations to these pathways decrease proproliferative signaling and have proven anticancer effects in vitro and preclinically. Meta-analyses assessing overall cancer risk in type 2 diabetics show a significant reduction in those individuals receiving metformin (Gandini et al. 2014) with evidence for a specific benefit in colorectal cancer prevention. While data are promising, there are currently no recommendations for drug repurposing of metformin within an oncological or therapeutic prevention setting, but numerous trials are ongoing to expand the evidence base.

Vitamin D

A potential role for vitamin D in the chemoprevention of colorectal cancer is supported by strong epidemiological, in vitro, and preclinical evidence. More importantly, several meta-analyses suggest a significant inverse correlation between serum 25-hydroxyvitamin D levels and the propensity to develop colorectal cancer. To date there have only been a limited number of randomized controlled trials of vitamin D intervention to decrease the risk of colorectal cancer. Such trials have failed to recapitulate the significant inverse correlations between serum levels and risk seen in the meta-analyses, but a trend toward benefit has been observed. These interventions may have suffered from uncertainties concerning optimal conditions in terms of dose and duration of intervention.

Selenium

The diet constituent selenium is a non-nutritive essential trace element and antioxidant, with preclinical anticancer efficacy mediated via its ability to inhibit oxidative DNA damage and the inflammatory response. There have been a number of meta-analyses investigating the correlation between serum selenium levels and colorectal cancer risk, but outcomes were ambiguous. Evidence from the European prospective investigation of cancer and nutrition cohort study (EPIC) adds weight to the concept that suboptimal levels of selenium in European populations are associated with higher colorectal cancer risk, which is more evident in women than in men (Hughes et al. 2015). Indeed, the notion that there is a threshold concentration in plasma or target tissues below which individuals may benefit from pharmacological intervention, and above which little benefit is likely to be derived, is gaining acceptance across a variety of indications.

The Future

Undoubtedly, the most convincing evidence of chemopreventive effect in colorectal cancer (both familial and sporadic) to date arises from studies of NSAIDs, and of aspirin in particular. Despite exhaustive evaluation of the many studies that have involved aspirin use, not all outcomes show benefit from long-term use, and there are very few randomized controlled trials that have investigated cancer mortality as a primary endpoint. However, numerous observational studies have found aspirin to be of benefit in colorectal cancer cohorts, in terms of both recurrence and progression. Despite the common aspirin-induced side effect of gastrointestinal bleeds, aspirin has the added advantage of exerting cardioprotective benefit. Evidence to assess whether there are improvements in the side effect profile by concurrent administration of proton pump inhibitors is currently being garnered within the AspECT trial.

Development of future chemoprevention strategies must take a number of critical factors into consideration in order that chemoprevention can eventually be incorporated into standard clinical practice for colorectal cancer, in analogy to its successful application in the management of cardiovascular disease. For the most promising agents, the lowest effective doses and optimal treatment durations have still to be identified. The populations for whom chemopreventive benefits outweigh the risk of side effects must be clearly defined, and a consensus on biomarkers that are able to act as the most suitable intermediate readouts for long-term outcome measures must be established. Last but not least, there are a number of dietary constituents that warrant clinical evaluation based on a wealth of encouraging preclinical data. One such agent is exemplified by the curry constituent curcumin. Future clinical trials of putative chemopreventive agents should conceivably be undertaken in conjunction with an aspirin intervention arm, which represents the current gold standard with regard to proven efficacy.

Cross-References

▶ Adenocarcinoma
▶ Adenoma
▶ Anti-inflammatory Drugs
▶ Aspirin
▶ Cancer
▶ Cancer Causes and Control
▶ Carcinogenesis
▶ Celecoxib
▶ Chemoprevention
▶ Cohort Study
▶ Colorectal Cancer
▶ Curcumin
▶ DNA Damage
▶ Insulin-like Growth Factors
▶ Lynch Syndrome
▶ MAP Kinase
▶ Meta-Analysis
▶ Mismatch Repair in Genetic Instability
▶ Mitogen-Activated Protein Kinase Kinase Kinases
▶ Nonsteroidal Anti-inflammatory Drugs
▶ Nuclear Factor-κB
▶ Progression
▶ Selenium
▶ Surrogate Endpoint
▶ Vitamin D

References

Burn J et al (2011a) A randomized placebo-controlled prevention trial of aspirin and/or resistant starch in young people with familial adenomatous polyposis. Cancer Prev Res (Phila) 4(5):655–665

Burn J et al (2011b) Long-term effect of aspirin on cancer risk in carriers of hereditary colorectal cancer: an analysis from the CAPP2 randomised controlled trial. Lancet 378(9809):2081–2087

Gandini S et al (2014) Metformin and cancer risk and mortality: a systematic review and meta-analysis taking into account biases and confounders. Cancer Prev Res (Phila) 7(9):867–885

Hughes DJ et al (2015) Selenium status is associated with colorectal cancer risk in the European prospective investigation of cancer and nutrition cohort. Int J Cancer 136(5):1149–1161

Rothwell PM et al (2010) Long-term effect of aspirin on colorectal cancer incidence and mortality: 20-year follow-up of five randomised trials. Lancet 376(9754):1741–1750

Stolfi C et al (2013) Mechanisms of action of non-steroidal anti-inflammatory drugs (NSAIDs) and mesalazine in the chemoprevention of colorectal cancer. Int J Mol Sci 14(9):17972–17985

See Also

(2012) Aberrant crypt foci. In: Schwab M (ed) Encyclopedia of cancer, 3rd edn. Springer, Berlin/Heidelberg, pp 13–14. doi:10.1007/978-3-642-16483-5_6531

(2012) Acetylsalicylic acid. In: Schwab M (ed) Encyclopedia of cancer, 3rd edn. Springer, Berlin/Heidelberg, p 17. doi:10.1007/978-3-642-16483-5_26

(2012) Adenomatous polyposis coli. In: Schwab M (ed) Encyclopedia of cancer, 3rd edn. Springer, Berlin/Heidelberg, p 48. doi:10.1007/978-3-642-16483-5_86

(2012) Analgesic. In: Schwab M (ed) Encyclopedia of cancer, 3rd edn. Springer, Berlin/Heidelberg, p 168. doi:10.1007/978-3-642-16483-5_251

(2012) Antioxidant. In: Schwab M (ed) Encyclopedia of cancer, 3rd edn. Springer, Berlin/Heidelberg, p 216. doi:10.1007/978-3-642-16483-5_328

(2012) Atherosclerosis. In: Schwab M (ed) Encyclopedia of cancer, 3rd edn. Springer, Berlin/Heidelberg, p 299. doi:10.1007/978-3-642-16483-5_432

(2012) Beta-catenin. In: Schwab M (ed) Encyclopedia of cancer, 3rd edn. Springer, Berlin/Heidelberg, p 385. doi:10.1007/978-3-642-16483-5_889

(2012) Biomarkers. In: Schwab M (ed) Encyclopedia of cancer, 3rd edn. Springer, Berlin/Heidelberg, pp 408–409. doi:10.1007/978-3-642-16483-5_6601

(2012) Cardiotoxicity. In: Schwab M (ed) Encyclopedia of cancer, 3rd edn. Springer, Berlin/Heidelberg, p 666. doi:10.1007/978-3-642-16483-5_859

(2012) Concurrent. In: Schwab M (ed) Encyclopedia of cancer, 3rd edn. Springer, Berlin/Heidelberg, p 965. doi:10.1007/978-3-642-16483-5_6821

(2012) Coxibs. In: Schwab M (ed) Encyclopedia of cancer, 3rd edn. Springer, Berlin/Heidelberg, p 990. doi:10.1007/978-3-642-16483-5_1358

(2012) Risk factor. In: Schwab M (ed) Encyclopedia of cancer, 3rd edn. Springer, Berlin/Heidelberg, p 3310. doi:10.1007/978-3-642-16483-5_5111

(2012) Secondary cancer prevention. In: Schwab M (ed) Encyclopedia of cancer, 3rd edn. Springer, Berlin/Heidelberg, p 3347. doi:10.1007/978-3-642-16483-5_5198

(2012) Tertiary cancer prevention. In: Schwab M (ed) Encyclopedia of cancer, 3rd edn. Springer, Berlin/Heidelberg, p 3651. doi:10.1007/978-3-642-16483-5_5735

(2012) Toxicity. In: Schwab M (ed) Encyclopedia of cancer, 3rd edn. Springer, Berlin/Heidelberg, p 3731. doi:10.1007/978-3-642-16483-5_5868

(2012) Ulcerative colitis. In: Schwab M (ed) Encyclopedia of cancer, 3rd edn. Springer, Berlin/Heidelberg, p 3836. doi:10.1007/978-3-642-16483-5_6095

(2012) Wnt. In: Schwab M (ed) Encyclopedia of cancer, 3rd edn. Springer, Berlin/Heidelberg, p 3953. doi:10.1007/978-3-642-16483-5_6255

(2012) Cyclooxygenase-2. In: Schwab M (ed) Encyclopedia of cancer, 3rd edn. Springer, Berlin/Heidelberg, p 1035. doi:10.1007/978-3-642-16483-5_1435

(2012) Cyclooxygenases. In: Schwab M (ed) Encyclopedia of cancer, 3rd edn. Springer, Berlin/Heidelberg, pp 1035–1036. doi:10.1007/978-3-642-16483-5_1434

(2012) Epidemiologic studies. In: Schwab M (ed) Encyclopedia of cancer, 3rd edn. Springer, Berlin/Heidelberg, p 1269. doi:10.1007/978-3-642-16483-5_1929

(2012) Familial adenomatous polyposis. In: Schwab M (ed) Encyclopedia of cancer, 3rd edn. Springer, Berlin/Heidelberg, p 1373. doi:10.1007/978-3-642-16483-5_2106

(2012) Histology. In: Schwab M (ed) Encyclopedia of cancer, 3rd edn. Springer, Berlin/Heidelberg, p 1697. doi:10.1007/978-3-642-16483-5_2748

(2012) HNPCC. In: Schwab M (ed) Encyclopedia of cancer, 3rd edn. Springer, Berlin/Heidelberg, p 1711. doi:10.1007/978-3-642-16483-5_2776

(2012) In vitro. In: Schwab M (ed) Encyclopedia of cancer, 3rd edn. Springer, Berlin/Heidelberg, p 1839. doi:10.1007/978-3-642-16483-5_3023

(2012) MAPK. In: Schwab M (ed) Encyclopedia of cancer, 3rd edn. Springer, Berlin/Heidelberg, p 2167. doi:10.1007/978-3-642-16483-5_3532

(2012) Mesalazine. In: Schwab M (ed) Encyclopedia of cancer, 3rd edn. Springer, Berlin/Heidelberg, p 2238. doi:10.1007/978-3-642-16483-5_3638

(2012) Metabolite. In: Schwab M (ed) Encyclopedia of cancer, 3rd edn. Springer, Berlin/Heidelberg, p 2258. doi:10.1007/978-3-642-16483-5_3661

(2012) Mutation. In: Schwab M (ed) Encyclopedia of cancer, 3rd edn. Springer, Berlin/Heidelberg, p 2412. doi:10.1007/978-3-642-16483-5_3911

(2012) Omeprazole. In: Schwab M (ed) Encyclopedia of cancer, 3rd edn. Springer, Berlin/Heidelberg, p 2609. doi:10.1007/978-3-642-16483-5_4215

(2012) Personalised medicine. In: Schwab M (ed) Encyclopedia of cancer, 3rd edn. Springer, Berlin/Heidelberg, p 2828. doi:10.1007/978-3-642-16483-5_4476

(2012) Polyp. In: Schwab M (ed) Encyclopedia of cancer, 3rd edn. Springer, Berlin/Heidelberg, p 2955. doi:10.1007/978-3-642-16483-5_6524

(2012) Primary cancer prevention. In: Schwab M (ed) Encyclopedia of cancer, 3rd edn. Springer, Berlin/Heidelberg, p 2985. doi:10.1007/978-3-642-16483-5_4731

Colorectal Cancer Clinical Oncology

Gabriela Möslein
Helios Klinik, Allgemein- und Viszeralchirurgie, Bochum, Germany

Synonyms

Cancer of the large intestine; Malignant neoplastic changes of the colon

Definition

Colon cancer refers to malignant neoplasia of the large intestine. The demarcation line to the more distal rectal cancer is defined as being proximal to 16 cm of the anocutaneous line.

Characteristics

Due to the slow development of precursor lesions in the form of adenomatous polyps or dysplastic lesions, no other tumor offers as many possibilities and as much time for preventive measures. Diagnosis of colon cancer in early stages highly increases the probability of curative resection. A 5-year survival rate of patients diagnosed with stage one colon cancer (limited to the bowel wall) is 90%, which is decreased to 35–60% in patients with a positive nodal status (stage III) and drops to less than 10% in the metastatic disease (stage IV).

Screening Strategies for the Average-Risk Population

Early colon cancer detection programs have been suggested to asymptomatic population of a certain age. The World Health Organization (WHO), the American Cancer Society, and the Agency for Health Care Policy and Research (AHCPR) recommended an annual ▶ fecal occult blood test (FOBT) and a 5-yearly sigmoidoscopy for the asymptomatic population older than 50. In Germany this proposal has been extended to annual FOBT and rectal digital examination beginning at the age of 45. In 1994, however, this was followed by only 44.1% of women and 14.4% of men. To date it has not been demonstrated that the rectal digital examination by itself is an efficient means for the early detection of rectal cancer. It therefore seems unreasonable to replace the sigmoidoscopic examination by the rectal digital examination, although it is an essential part of every physical examination in patients older than 50.

FOBT (▶ Fecal Occult Blood Test)

Three large randomized studies carried out in the USA, Denmark, and Great Britain in a period of 8–13 years have demonstrated the benefits of FOBT in early colon cancer detection and in the reduction of mortality by 15–33%. Although FOBT is more specific without rehydration, best results were achieved when the test was carried out once a year and included rehydration. FOBT is an adequate screening modality for early cancer detection, reducing mortality rates as well as treatment costs. It is in itself, however, not the appropriate means in cancer prevention since it implies the removal of neoplastic changes even before the event of malignant transformation.

Sigmoidoscopy

Periodic sigmoidoscopy from age 50 onward reduces the mortality of rectosigmoidal cancers by 60%, and usually a control interval of 5 years is sufficient. The risk of developing colon cancer proximal to the splenic flexure, however, remains unaffected. Compared to FOBT alone, the combination of the two procedures increases the cancer-preventive effect by a factor of 2.2.

Colonoscopy

In the age group of 55- to 64-year-old asymptomatic persons, the combination of FOBT and sigmoidoscopy result will, in the case of positive test results, lead to the recommendation of performing colonoscopy. The question may therefore be raised, if a baseline colonoscopy is a suitable alternative for this age group. In approximately one third of all patients, polyps will be detected (and removed) such that this strategy would imply a true cancer prevention. On the other hand, 70% of persons with a negative colonoscopy would not require any further screening modalities for a period of 5 years.

Double-Contrast Barium Enema

This radiologic examination cannot replace colonoscopy since the sensitivity is significantly lower (83% vs. 95%). The probability of overlooking a small cancer is increased fourfold compared to the endoscopic procedure. Small polyps, however, are also frequently not recognized during the endoscopic examination. In a prospective study, every fourth adenoma under the size of 5 mm was overlooked. The detection of adenomas larger than 1 cm in diameter was reproducible in 94% of the cases. The reliability of colonoscopy depends much on the experience of the person performing the examination.

Preoperative   Diagnosis   of   Colon Cancer Required examinations:

- History (including family history).
- Physical examination (including recta digital examination).
- Colonoscopy with biopsy or double-contrast barium enema with subsequent biopsy of a pathological alteration. If the barium enema does not show a pathological lesion, it may entirely replace endoscopy.
- If a stenosis cannot be surpassed preoperatively, colonoscopic examination in the first 3 months after operation is warranted.

- Ultrasound sonography of the abdomen.
- Radiologic thorax examination.
- Tumor marker CEA (carcinoembryonic antigen).
- MRI (as an alternative or extended examination).
- CT scan of the thorax if in doubt about lung metastases.
- In case of sigmoid cancers, urine sedimentation and CT scan. If ultrasound examination suggests infiltration of the urinary tract or if red blood cells are demonstrated in the urine sedimentation, cystoscopy is recommended to investigate bladder infiltration. Gynecological examination, if infiltration of the uterus or the ovaries is suspected.

Preoperative (Neoadjuvant) Therapy To date no benefits have been shown in using ▶ neoadjuvant therapy in colon cancer.

Surgical Therapy (with the Aim to Cure) Surgery aims at the curative resection of the tumor-bearing segment of the colon together with the regional lymph nodes. In addition, the (partial) resection of adjacent organs, if these are infiltrated by tumor (multivisceral resection), may be necessary. Colon cancers usually have a circular growth pattern. In order to remove the intramural tumor cell spread, a minimal margin of 2 cm suffices. A regional lymph node involvement is more widespread. Lymph nodes show a tangential metastatic involvement (up to 10 cm away from the macroscopic tumor), their preferred distribution being towards the center.

Cancers of the Cecum and Ascending Colon

Generally, a right-sided hemicolectomy is the treatment of choice in such patients, including the radicular removal of the lymph nodes of the right colic artery and the ileocolic vessels. The large omentum of the colon is removed together with the colon segment. If the dissection of the gastrocolic ligament is considered, one might be confronted with contrasting opinions regarding the right gastroepiploic artery. Some authors recommend the preservation of the vessel, while others do not.

Cancers of the Right Flexure and the Proximal Transverse Colon

As a rule of thumb, extended right-sided hemicolectomy is warranted if the right colic artery is dissected at its origin, out of the superior mesenteric artery. The distal resection lies close to the splenic flexure, allowing circulation. If the blood supply of the distal transverse segment appears insufficient, the additional resection of this segment becomes necessary. The large omentum is completely removed, together with the gastroepiploic ligament and the right gastroepiploic vessels (resection of potentially involved lymph nodes above the pancreas).

Cancers of the Transverse Colon

Cancers in the mid-transverse colon are treated with an entire resection of the segment, including the flexures. The omentum as well as the gastroepiploic ligament and arcade is removed together with the colonic specimen. If the cancers are close to the flexures, hemicolectomy extended to the right and the left is the procedure of choice.

Tumors of the Left Colonic Flexure

Suggested is an extended left-sided hemicolectomy, together with the removal of the lymph nodes of the medial colic vessel and the inferior mesenteric vessels. Equally radical is the central ligature of the left colic artery at its origin, leaving the central part of the inferior mesenteric vessels intact. Under these circumstances the superior rectal vessels remain unaffected, such that the circulation in the remaining sigmoid colon is not impaired. Depending on the exact tumor localization and blood supply, the right colonic flexure may be preserved. Lymph nodes along the central portion of the superior mesenteric vessels should always be removed for diagnostic evaluation.

Tumors of the Descending Colon and Proximal Sigmoid Colon

Generally, left-sided hemicolectomy is recommended, with radicular ligation of the inferior mesenteric vessel. The distal margin of resection lies in the upper part of the rectum and usually the left flexure has to be removed. In order to obtain a tension-free anastomosis, sometimes the medial colic artery has to be sacrificed.

Tumors of the Middle and Distant Sigmoid Colon

In this case, radicular segmental sigmoid resection is the preferred option. The inferior mesenteric artery is ligated either centrally or distally, relative to the origin of the left colic artery. The inferior mesenteric vein should be ligated at the lower edge of the pancreas.

Further Constellations Influencing Surgical Strategy

Multivisceral resections: If adjacent structures are inherent to the tumor, these should, in addition to the lymph node resection, be resected "en bloc." In contrast, biopsies to confirm tumor infiltration of adjacent organs are to be avoided since cell dissemination might be initiated.

Distant metastases: The resection of synchronous or metachronous metastases of the liver, lung, etc. is indicated only, if this resection has curative intent and complies with the oncological principles. If the metastases are unresectable, "palliative measures" apply.

Multiple colonic primaries: The extent of resectional surgery depends on additional lymph node dissections recommended for each tumor. As a result colectomy with ileorectal anastomosis may be indicated.

Synchronous occurrence of colonic polyps: Adenomas that are not removable endoscopically should be resected during colon cancer surgery. In this case a margin of 2 cm applies. Extended resection to the segmental lymph node is not necessary.

Cancer diagnosis in endoscopically removed polyps: If, unexpectedly, the histological examination reveals malignancy, oncological resection of the colonic segment is indicated. This may be neglected only in the case of a polyp with tumor-free stem which is confined to the submucosa with "low risk" (pT1, G1–2, no lymph vessel involvement).

Segmental resection: In patients with metastatic disease, radical resection of the colonic segment with lymph node removal may not be indicated. Very poor physical condition or the high age of some patients may justify colonic surgery which does not follow oncological principles.

Emergency operation: Nevertheless, high-urgency surgery, unavoidable due to bowel obstruction or tumor or colon perforation, should comply with oncological principles.

Laparoscopic surgery: To date, no data are available that document the operative outcome in patients that underwent laparoscopic colon cancer surgery. Future consideration of ongoing studies with long follow-up periods is therefore necessary for the optimal treatment of colon cancer. Nevertheless, there are no objections to carry out laparoscopic colon cancer surgery in a palliative setting.

Ulcerative colitis and familial adenomatous polyposis (FAP (▶ APC gene in familial adenomatous polyposis)): Cancers of this type require proctocolectomy, if possible continence preserving. Especially in an early stage, cancer in the proximal two thirds of the rectum is not a contraindication for ileoanal pouch surgery.

Surgery of hereditary nonpolyposis colorectal cancer (HNPCC): In the event of this autosomal-dominant syndrome, many authors suggest extended cancer surgery in the form of prophylactic bowel removal (colectomy and ileorectal anastomosis or restorative proctocolectomy). Occurrence of metachronous colorectal cancer and the observation of so-called interval cancers are significant. However, the benefit of prophylactic colon removal (without the evidence of neoplasia) remains uncertain, especially if considering a reduced penetrance of ~80%.

Intra- and Postoperative Histopathological Diagnosis Due to technical complications the immediate pathological classification of a tumor/polyp in a frozen section is not an option. Pathological evaluation after oncological surgery is, however, of prognostic significance in the locoregional resection (R-classification), the depth of invasion (pT classification), and the grading and lymphnodal status (pN classification) and forms the basis in the decision process concerning ▶ adjuvant therapy. The total number of resected lymph nodes and metastatic lymph node

Colorectal Cancer Clinical Oncology, Table 1 Dukes classification

Dukes A	Growth limited to wall, nodes negative
Dukes B	Growth beyond muscularis propria, nodes negative
Dukes C1	Nodes positive and apical negative
Dukes C2	Apical node positive
Dukes D	Growth beyond originating organ

Colorectal Cancer Clinical Oncology, Table 2 TNM staging

pT1	Local invasion of submucosa
pT2	Local invasion of the muscularis propria
pT3	Local invasion beyond the muscularis propria
pT4	Tumor cells have reached peritoneal surface or invaded adjacent organs
pN0	No lymph nodes affected by metastases
pN1	One to three lymph nodes affected by metastases
pN2	Four or more lymph nodes affected by metastases
pM0	No distant metastasis
pM1	Distant metastasis

occurrence is therefore of essential relevance. Perforation of the tumor during surgery is of prognostic significance and must be documented.

Microsatellite tumor instability is of special relevance in the setting of HNPCC and of increasing interest since response rates to adjuvant therapy have shown that stable and unstable tumors differ in their biological response to chemotherapeutic agents. Although the natural course of unstable tumors is more benign than the natural course of stable tumors, the biological response to conventional chemotherapy in stable tumors is better.

Classification of Colorectal Cancers

There are two classifications that are used separately, the Dukes and the TNM (tumor, lymph nodes, metastases) classification.

The Dukes classification (Table 1) is preferred in the USA and UK and describes the following stages:

The TNM staging (suggested by the Union internationale contre le cancer (UICC)) (Table 2) is preferred in European countries and distinguishes between the stages listed below. T stands for the expansion of the primary tumor; N for the lack or the presence of metastases of the lymph nodes; M for the lack or the presence of distant metastases. Numbers indicate the extent of malignant processes; p, postoperative.

Adjuvant Therapy

- In order to recommend adjuvant therapy, complete removal of all regional and metastatic lesions (R0 resection) in addition to the tumor removal is necessary. Recommending adjuvant therapy is based on the pathohistological classification of the tumor, specifically of the pN status. In order to define the lymph node status, a minimum of 12 regional lymph nodes should be examined. Immunocytological studies of isolated tumor cells in either bone marrow aspiration biopsies or the peritoneal fluid should not, at least at this point, be referred to in the decision for or against adjuvant therapy since the impact of "minimal residual disease" remains to be established.

- Patients with early-stage colorectal cancer (stage I or II) and patients after R0 resection of distant metastases should receive adjuvant therapy in the setting of controlled studies only.

- The benefits of adjuvant therapy in UICC stage III cancers (all pT stages, pN1-2, M0) remain to be established. A quality-controlled surgical treatment with and without adjuvant therapy is currently under evaluation. Outside these studies, adjuvant therapy is recommended for stage III cancers.

- Adjuvant chemotherapy for stage III colon cancers: One-year administration of 5-FU (5-fluorouracil) and levamisole proved to be as effective as a 6-monthly administration of 5-FU and folinic acid. Although there is variation between different adjuvant protocols, general contraindications for adjuvant therapy are listed below:
  - General physical condition under the score of 2 (WHO)
  - Uncontrolled infection
  - Liver cirrhosis

- Severe coronary heart disease, cardiac insufficiency (NYHA III and IV)
- Preterminal and terminal renal insufficiency
- Limited bone marrow function
- Unavailability for regular control check-ups

To date there appears to be no benefit in the administration of monoclonal antibody treatment (17 1A) in addition to conventional chemotherapy.

Follow-Up Due to the low rate recurrence rate, no major prognostic advantage from follow-ups is expected for patients with early cancer (UICC I) and R0 resection. The advice to perform two colonoscopies, 2 and 5 years post colon cancer surgery, is aimed towards an early identification of second primaries. An intensified surveillance of individual cases is justified in circumstances that lead to suspect a higher recurrence rate, i.e., tumor perforation, G3 and G4 tumors, or histologically verified pericolic vessel infiltration. After palliative tumor resection (R2), a symptomatic follow-up is recommended.

Following R0 resections of tumor stages II and III, the main benefits of follow-up strategies can be expected, if the general physical condition of the patient does not object to recurrent surgical intervention. Two specific follow-ups are recommended within 5 years of primary surgery (year 2 and year 5) and include physical examination, CEA level, abdominal ultrasound, X-ray of the thorax, and colonoscopy. Intensified follow-up is recommended for patients with an increased hereditary risk.

Cellular and Molecular Features

Colorectal tumors provide an excellent system in which to search for and study genetic alterations involved in the development of neoplasia. It appears that most, if not all, malignant colorectal tumors arise from preexisting benign ▸ adenomas. These precursor lesions can be removed and studied at various stages of development. Colorectal tumors develop as a result of

▸ oncogene mutations in combination with the mutated ▸ tumor suppressor genes, the latter being predominant.

Human colorectal tumors, including very small adenomas, have a monoclonal composition. Adenomas therefore arise from a single or a small number of cells which initiate the process of neoplasia by clonal expansion. Genetic alterations within the majority of neoplastic cells studied so far suggest an impaired regulation of cell growth that enables those cells to become the predominant cell type, eventually constituting the neoplasm.

The development process in patients with sporadic cancer (as opposed to familial cancer) occurs over a period of decades. The series of genetic alterations involves ▸ oncogenes such as *RAS* as well as ▸ tumor suppressor genes (particularly those on chromosome 5q, 17p, and 18q). In general, the three stages are represented by increasing tumor size, dysplasia, and villous content. The mutation of the ▸ *RAS* gene (usually KRAS) appears to occur within a single cell of a preexisting small adenoma followed by clonal expansion which produces a larger and more dysplastic tumor. Deletions of chromosome 17p and 18q generally arise at a later stage of tumorigenesis than deletions of chromosome 5q or *RAS* gene mutations. In this ▸ multistep development, the total number of genetic alterations rather than their order of occurrence determines the biological properties of neoplasia.

Tumors continue to progress once cancers have formed, and the cumulative loss of tumor suppressor genes on different chromosomes correlates with the ability of the tumor to metastasize and to cause death.

Approximately 25% of randomly selected colorectal cancers have ▸ microsatellite instability, a phenomenon used as an independent prognostic factor in colorectal cancer. In addition, the loss of heterozygosity (LOH) at a chromosome 8p marker (termed allelic imbalance) has been related with a poor patient outcome. It may therefore be expected that the molecular characterization of colorectal tumors will increasingly affect the individual risk assessment and the suitable intervention strategies.

The identification and characterization of molecular mechanisms underlying tumor development and tumor growth offer new opportunities in cancer treatment. An intricate genetic scenario is responsible for a complex human neoplastic condition. The relevance of individual steps within might open doors for therapeuticals that specifically target essential, although malfunctioning, check-points.

Perspective

In developing countries, especially Asia, incidences of colon cancer are rapidly rising. In the USA and in Germany, ~130,000 and 50,000 patients, respectively, are diagnosed with colorectal cancer every year. Colorectal cancer comes second in the group of tumor-related deaths. The lifetime risk of developing colorectal cancer in Germany is 4–6%, with the majority of these cancers occurring in people aged 50 and above. Gaining new insight into the molecular pathogenesis of colorectal cancer will allow progress in many facets of disease control. They include the identification of genetically predisposed groups for targeted surveillance and/or chemoprevention, prognosis for patients with established cancer, predictions of treatment efficacy, and the development of novel treatment strategies.

References

Kronborg O, Fenger C, Olsen J et al (1996) Randomised study of screening for colorectal cancer with faecal-occult-blood test. Lancet 348:1467–1471

Levin B, Bond JH (1996) Colorectal cancer screening: recommendations of the U.S. preventive services task force. Gastroenterology 111:1381–1384

Rex DK, Cummings OW, Helper DJ et al (1996) 5-Year incidence of adenomas after negative colonoscopy in asymptomatic average-risk persons. Gastroenterology 111:1178–1181

Rex DK, Rahmani EY, Hasemann JH et al (1997) Relative sensitivity of colon of colorectal cancer in clinical practice. Gastroenterology 112:17–23

Selby JV, Friedmann GD, Quesenberry CP et al (1992) A case-control study of screening sigmoidoscopy and mortality from colorectal cancer. N Engl J Med 326:653–657

Colorectal Cancer Nutritional Carcinogenesis

Paule Latino-Martel, Nathalie Druesne-Pecollo, Serge Hercberg and Mathilde Touvier
UMR U1153 INSERM, U1125 INRA, CNAM, Université Paris 13, Centre de Recherche Epidémiologie et Statistique Sorbonne Paris Cité, Bobigny, France

Synonyms

Carcinogenesis in colon; Colon cancer risk; Diet; Food; Modulation of colon carcinogenesis; Nutritional factors; Nutritional prevention of colon cancer

Keywords

Nutritional status

Definition

The ability of nutritional factors (including food, alcohol, nutritional status, and physical activity) to increase or decrease the risk of ▶ colorectal cancer.

Characteristics

The etiology of colorectal cancer is multifactorial. Generally, it results from an interaction between genetic and environmental factors. As a part of the digestive tract, the colon is exposed to many native or metabolized dietary compounds. Gut microbiota, which metabolizes various dietary constituents and modifies their bioavailability and effects on the host, is also involved in colon ▶ carcinogenesis. The development of colon cancer is modulated by several nutritional factors, some acting as risk factors and others as protective factors.

Relation Between Nutrition and Colon Cancer: Evaluation by an International Expert Group

To evaluate the relationship between a protective factor or a risk factor and a cancer site, it is necessary to perform several types of studies: epidemiological studies like case-control studies, cohort studies and, when possible

1. Intervention studies (for potential protective factors only)
2. Mechanistic studies using cellular and animal models.

Each type of study has both advantages and limitations. In the field of nutrition and cancer (including colorectal cancer), an international working group of experts from the World Cancer Research Fund (WCRF) and the American Institute for Cancer Research (AICR) performed a comprehensive evaluation of the available scientific literature, in 1997 and 2007. Thereafter, the information has regularly been updated by cancer site, in the framework of the WCRF/AICR Continuous Update Project. In the last evaluation focused on colorectal cancer and published in 2011, a panel of international experts examined both the results of systematic reviews and meta-analyses of epidemiological studies on nutritional factors and colorectal cancer and the biological plausibility to grade the strength of evidence. Nutritional factors for which the evidence showed a convincing or a probable causal relationship with colorectal cancer are gathered in Table 1 and presented in the following sections.

Nutritional Factors Increasing the Risk of Colon Cancer

Red Meat and Processed Meat

The consumption of red meat and processed meat is associated with an increased risk of colorectal cancer. It has been estimated that the risk of colorectal cancer is increased by 17% per 100 g of red meat consumed per day and 18% per 50 g serving of processed meats consumed per day. Several mechanisms may explain the increased risk of colorectal cancer associated with the consumption of red meat and processed meat, nitrite salt provided by certain meats, production of N-nitroso carcinogens in the stomach and by the gut microbiota (▶ Carcinogen Metabolism), production of free radicals and pro-inflammatory ▶ cytokines associated with an excess of heme iron, and production of ▶ heterocyclic amines related to cooking at high temperatures. Among all these potential pro-carcinogenic bioactive compounds, heme iron is probably playing a key role, as demonstrated by mechanistic research. The increased risk of colorectal cancer by eating red meat and processed meat is considered convincing.

Alcoholic beverages

The consumption of alcoholic beverages is associated with an increased risk of colorectal cancer as well as other cancers. A significant dose–response relationship is observed. The percentage of increase in risk of colon cancer is estimated to 8% per 10 grams of ethanol per day, provided by one standard glass of alcoholic beverage consumed per day (▶ Alcoholic Beverages Cancer Epidemiology). Several mechanisms may explain the increased risk of colon cancer associated with the consumption of alcoholic beverages. Some mechanisms are common to several sites of cancer: the most important of these is the production of mutagenic metabolites from ethanol. Ethanol is metabolized into acetaldehyde (a highly reactive molecule with respect to DNA, known as a human carcinogen), mainly by alcohol

Colorectal Cancer Nutritional Carcinogenesis, Table 1 Strength of the evidence on the relationship between nutritional factors and colorectal cancer

	Convincing	Probable
Increase of the risk	Red meat, processed meat, alcoholic drinks (men), body fatness, abdominal fatness, adult attained height	Alcoholic drinks (women)
Decrease of the risk	Foods containing dietary fibre, Physical activity (colon)	Garlic, milk, calcium

dehydrogenase (ADH), expressed in several tissues including colon, and by the gut microbiota. Acetaldehyde is gradually eliminated by acetaldehyde dehydrogenase (ALDH2), which converts it to acetate. Other mechanisms seem to be more specific to colon cancer: for example, chronic use of alcohol induces folate deficiency, which itself contributes to the development and progression of colorectal cancer. The relationship between the consumption of alcoholic beverages and the risk of colon cancer is considered convincing in men and probable in women.

Body and Abdominal Fatness

The increase in body fatness is associated with an increased risk of colon cancer and of several other cancers. The percentage of increase in the risk of colon cancer is estimated to 3% per increase of body mass index (BMI) of 1 kg/m^2. A significantly increased risk is also observed with increasing abdominal fatness, the latter being estimated by waist circumference or the waist-to-hip ratio. Several mechanisms common to all cancer sites may explain the epidemiological association described between overweight, ▶ obesity, and the risk of colon cancer. For example, excess fat increases insulin resistance. The resultant chronic hyperinsulinemia induces the production of insulin growth factor-1 (IGF-1), which promotes cell proliferation. Furthermore, obesity induces a chronic inflammatory state, by increasing blood levels of pro-inflammatory factors such as ▶ tumor necrosis factor-alpha, ▶ interleukin-6, C-reactive protein, and leptin, which also induces cell proliferation. The increased risk of colorectal cancer associated with overweight and obesity is considered convincing.

Adult Attained Height

A higher adult attained height is associated with the risk of colon cancer (9% increase of cancer risk per 5 cm of height). Several factors that lead to greater adult attained height, such as early-life nutrition, altered hormone profiles, and the rate of sexual maturation, could plausibly influence colon cancer risk. The evidence that the factors leading to greater adult attained height, or its consequences, are a cause of colorectal cancer is convincing.

Nutritional Factors Decreasing the Risk of Colorectal Cancer

Physical Activity

Physical activity is associated with a decreased risk of various cancers including colorectal cancer. In the case of colorectal cancer, the percentage of decrease in risk is estimated to 8% for an increase of total physical activity of 5 MET.h/day, the MET (metabolic equivalent of task) being a physiological measure expressing the energy cost of physical activities. The main mechanisms that could explain the beneficial effect of physical activity on risk of cancer could be related to its effects on circulating levels of various ▶ hormones and growth factors: reduction of the plasma levels of insulin and ▶ insulin-like growth factor-1 (IGF-1) which are increased in overweight and ▶ obesity and promote cell proliferation. Physical activity may specifically reduce the risk of colorectal cancer through the acceleration of intestinal transit, thereby reducing the time of exposure of the colon mucosa to food-borne carcinogens. In addition, physical activity decreases the risk of weight gain, overweight, and obesity, thereby contributing also to the reduction of cancer risk, indirectly. The reduction of risk of colon cancer associated with physical activity of all types (occupational, household, transport and recreational) is considered convincing.

Foods Containing Dietary Fibre

Foods containing dietary fibre are associated with a decreased risk of colorectal cancer. It has been estimated that the risk of colon cancer is decreased by 11% per 10 grams of dietary fibre per day. Fibre exerts various effects on the gut: it decreases transit time, and then the exposure of the colon mucosa to foodborne carcinogens. Fibre is metabolized by the gut microbiota to short chain fatty acids, such as butyrate, which induce apoptosis, cell cycle arrest, and differentiation in experimental studies. It also reduces insulin-resistance and inflammation. The reduction of the risk of colorectal cancer associated with the intake of foods containing dietary fiber is considered convincing.

Other Factors

Garlic, milk, and calcium decrease colorectal cancer risk with a level of evidence judged as probable. For these nutritional factors, the level of evidence is lower for different reasons: possibility of residual confounding, low number of studies, and lack of ► cohort studies.

Conclusion

Increasing the intake of foods containing dietary fibre, increasing physical activity, avoiding overweight and obesity, and reducing the consumption of alcoholic beverages and of red meat and processed meat are the main actions in the field of nutrition that may contribute to the prevention of colorectal cancer and should be promoted by public health policies. The preventability of colon cancer by appropriate food, nutrition, physical activity, and body fatness has been estimated to 50% in the USA, 47% in the UK, 41% in Brazil, and 22% in China.

Cross-References

- ► Alcohol Consumption
- ► Alcoholic Beverages Cancer Epidemiology
- ► Cancer Causes and Control
- ► Carcinogenesis
- ► Carcinogen Metabolism
- ► Cohort Study
- ► Colorectal Cancer
- ► Cytokine
- ► Heterocyclic Amines
- ► Hormones
- ► Inflammation
- ► Insulin-Like Growth Factors
- ► Interleukin-6
- ► Obesity and Cancer Risk
- ► Tumor Necrosis Factor

References

Baan R, Straif K, Grosse Y et al (2007) Carcinogenicity of alcoholic beverages. Lancet Oncol 8:292–293

Calle EE, Kaaks R (2004) Overweight, obesity and cancer: epidemiological evidence and proposed mechanisms. Nat Rev Cancer 4:579–591

Fridenreich CM, Orenstein MR (2002) Physical activity and cancer prevention: etiologic evidence ad biological mechanisms. J Nutr 132:3456S–3464S

WCRF/AICR (2011) Continuous Update Project Report. Food, Nutrition, Physical Activity, and the Prevention of Colorectal Cancer. WCRF/AICR, Washington, DC. 40 pages

WCRF/AICR (2009) Policy and action for cancer prevention. Food, nutrition, and physical activity: a global perspective. AICR, Washington, DC. 190 pages

See Also

(2012) Acetaldehyde. In: Schwab M (ed) Encyclopedia of cancer, 3rd edn. Springer, Berlin/Heidelberg, p 16. doi:10.1007/978-3-642-16483-5_22

(2012) Alcohol dehydrogenase. In: Schwab M (ed) Encyclopedia of cancer, 3rd edn. Springer, Berlin/Heidelberg, p 120. doi:10.1007/978-3-642-16483-5_6732

(2012) Alcohol-mediated cancer. In: Schwab M (ed) Encyclopedia of cancer, 3rd edn. Springer, Berlin/Heidelberg, p 126. doi:10.1007/978-3-642-16483-5_170

(2012) Carcinogen. In: Schwab M (ed) Encyclopedia of cancer, 3rd edn. Springer, Berlin/Heidelberg, p 644. doi:10.1007/978-3-642-16483-5_839

(2012) Confounding. In: Schwab M (ed) Encyclopedia of cancer, 3rd edn. Springer, Berlin/Heidelberg, p 968. doi:10.1007/978-3-642-16483-5_1304

(2012) C-reactive protein. In: Schwab M (ed) Encyclopedia of cancer, 3rd edn. Springer, Berlin/Heidelberg, p 993. doi:10.1007/978-3-642-16483-5_1369

(2012) Folate. In: Schwab M (ed) Encyclopedia of cancer, 3rd edn. Springer, Berlin/Heidelberg, p 1440. doi:10.1007/978-3-642-16483-5_2231

(2012) Free radicals. In: Schwab M (ed) Encyclopedia of cancer, 3rd edn. Springer, Berlin/Heidelberg, p 1454. doi:10.1007/978-3-642-16483-5_2267

(2012) Insulin. In: Schwab M (ed) Encyclopedia of cancer, 3rd edn. Springer, Berlin/Heidelberg, pp 1873-1874. doi:10.1007/978-3-642-16483-5_3075

(2012) Leptin. In: Schwab M (ed) Encyclopedia of cancer, 3rd edn. Springer, Berlin/Heidelberg, p 2001. doi:10.1007/978-3-642-16483-5_6734

(2012) Proliferation. In: Schwab M (ed) Encyclopedia of cancer, 3rd edn. Springer, Berlin/Heidelberg, p 3004. doi:10.1007/978-3-642-16483-5_4766

(2012) Risk factor. In: Schwab M (ed) Encyclopedia of cancer, 3rd edn. Springer, Berlin/Heidelberg, p 3310. doi:10.1007/978-3-642-16483-5_5111

Colorectal Cancer Pathology

Christian Wittekind
Department für Diagnostik, Institut für
Pathologie, Universitätsklinikum Leipzig,
Leipzig, Germany

Definition

Colorectal carcinoma comprises all malignant epithelial tumors of the colon and rectum, the most frequent of which is colorectal adenocarcinoma. Unlike in the stomach or small intestine, only tumors that have invaded through the muscularis mucosae into the submucosa are considered malignant in the colon and rectum thus being called invasive carcinoma by the World Health Organization (WHO) classification (Bosman et al. 2010; UICC 2009). This is due to the fact that in the colon and rectum, a neoplasm has malignant potential only after invasion of the submucosa, where lymphatic vessels are located. Unfortunately, the term "carcinoma" is not used uniformly. This especially applies to the intermediate lesions between intraepithelial neoplasia (dysplasia) and invasive carcinoma that show invasion into the lamina propria mucosae or between the fibers of the muscularis mucosae without invading the submucosa. Hence, one has to be careful to avoid being misled by statistics and reports using data that do not distinguish between invasive carcinoma and high-grade intraepithelial neoplasia (high-grade dysplasia) (Wittekind 2007), as both Western and Eastern reports do not limit the use of the term carcinoma to invasion of at least the submucosa. Possible variations in classification between carcinoma in situ and invasive carcinoma are shown in Table 1 (Fig. 1).

Characteristics

Colorectal carcinoma is of great importance for health care systems considering the fact that it constitutes the fourth most common cancer in

Colorectal Cancer Pathology, Table 1 Variations in classification between carcinoma in situ and invasive carcinoma of the colorectum (Bosman et al. 2010; UICC 2009; Wittekind 2007)

pT category	Tumor entities	ICD-O M code
pTis	Severe dysplasia High-grade dysplasia High-grade intraepithelial neoplasia Intramucosal carcinoma	8140/2
pT1	Invasive carcinoma	8140/3

men and the third most common cancer in women worldwide. Colorectal cancer pathology covers a broad field beginning with etiology and pathogenesis including genetics and ranging to diagnosis and treatment as well as metastasis and causes of death of colorectal cancer patients (► Colorectal Cancer Clinical Oncology). The management of colorectal cancer is a team process, in which the pathologist plays an important role. The vast majority of colorectal cancers are carcinomas (malignant tumors of epithelial origin) and most of these are ► adenocarcinomas (malignant epithelial tumors evolving from glandular tissue).

Localization

A problem constitutes the definition of the colon and especially the rectum, which may vary greatly despite the fact that the pathological characteristics of carcinoma of the colon and rectum are in some ways quite similar. Clinically, the rectum extends 16 cm from the anal verge when measured below with a rigid rectosigmoidoscope. A tumor is classified as rectal if its inferior margin lies less than 16 cm from the anal verge or if any part of the tumor is at least partly within the supply of the superior rectal artery (Washington et al. 2009). The rectum itself is further subdivided in thirds with the upper third ranging from 12 to 16 cm, the middle third from 6 to less than 12 cm, and the lower third within less than 6 cm. Colon cancer risk varies by subsite within the colon (Fenoglio-Preiser et al. 2008), raising awareness of an increasing proportion of proximal, i.e., left-sided

Colorectal Cancer Pathology, Fig. 1 Early invasive carcinoma of the colorectum

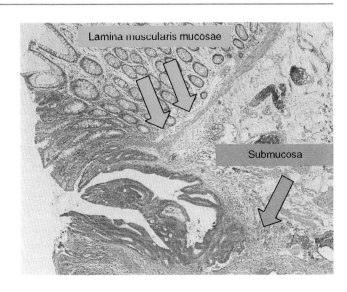

carcinomas as opposed to those located in the distal part of the large intestine, namely, the right-sided colon and rectum.

Macroscopy/Gross Morphology

Gross morphology varies and is influenced both by the phase of growth at the time of detection and the location in the large intestine. Small colorectal carcinomas grossly resemble adenomas and are mostly polypoid protruding into the lumen, pedunculated, semipedunculated, or sessile, but flat or even depressed types occur (Fenoglio-Preiser et al. 2008). In advanced carcinomas, the following growth patterns exist:

- Polypoid or exophytic with predominantly intraluminal growth.
- Ulcerative or endophytic with predominantly intramural growth, further subdivided into subgroups with sharply demarcated margins (the most common kind of growth pattern) or without definite borders.
- Diffusely infiltrating (linitis plastica): this growth pattern is characterized by a distinct desmoplastic reaction resulting in a rigid and thickened wall of the large intestine.

Carcinomas can involve only part of or the whole circumference of the large intestine, resulting in annular or circular proliferation and causing constriction of the lumen (Bosman et al. 2010) (Fig. 2).

Microscopy/Histomorphology

Histological Typing

In the following section, the assignment of some of these histological types is further explained (Table 2).

- Adenocarcinoma: this type is by far the most common and is defined by the presence of glandular epithelium.
- Medullary carcinoma: also a rare variant characterized by minimal or no glandular differentiation and by sheets of malignant cells showing vesicular nuclei with prominent nucleoli and abundant cytoplasm. Other characteristic features include prominent infiltration by intraepithelial lymphocytes, an invariable association with high-level microsatellite instability (MSI-H), and a favorable prognosis.
- Mucinous adenocarcinoma: this designation applies if >50% of the tumor is composed of extracellular mucin. Pools of mucin contain malignant epithelium in acinar, strip-like, or single-cell structures. Sometimes, the term mucoid or colloid adenocarcinoma is still used.
- Serrated adenocarcinoma: this rare variant resembles serrated polyps showing glandular serration.

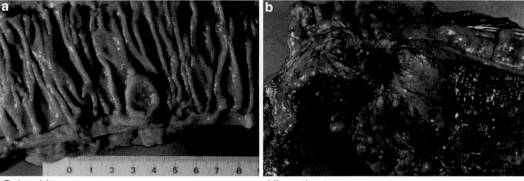

Polypoid type   *Ulcerative type*

Colorectal Cancer Pathology, Fig. 2 Gross morphology of colorectal carcinoma

Colorectal Cancer Pathology, Table 2 Histological typing: WHO classification (Bosman et al. 2010)

Histological type	ICD-O M code
Adenocarcinoma	8140/3
Cribriform comedo-type adenocarcinoma	8201/3
Medullary carcinoma	8510/3
Micropapillary carcinoma	8265/3
Mucinous adenocarcinoma	8480/3
Serrated adenocarcinoma	8213/3
Signet ring cell adenocarcinoma	8490/3
Adenosquamous carcinoma	8560/3
Spindle cell carcinoma	8032/3
Squamous cell carcinoma	8070/3
Undifferentiated carcinoma	8020/3

- Signet ring cell adenocarcinoma: this variant is defined by the presence of >50% signet ring cells, i.e., tumor cells with prominent intracytoplasmic mucin and displaced nuclei. As a primary tumor of the colorectum, this type is rare compared to the stomach and is associated with a bad prognosis.
- Adenosquamous carcinoma: unusual tumor of the colorectum with features of both adeno- and squamous cell carcinoma. This diagnosis requires more than just occasional small foci of squamous differentiation that constitute a not completely uncommon finding in adenocarcinoma.

- Squamous cell carcinoma: this constitutes a very uncommon type of tumor in the colorectum characterized by the presence of intercellular bridges and keratohyalin granules.
- Undifferentiated carcinoma: this type shows no glandular structure or other evidence of differentiation beyond that of an epithelial tumor (Bosman et al. 2010) (Fig. 3).

Histologic Grading

Histopathological grading serves as a prognostic indicator due to its correlation with the aggressiveness of a tumor and may influence decisions concerning treatment. It is based on architectural features, in case of adenocarcinoma of the colorectum on the extent of glandular appearance. Traditionally the four-step grading system also used by the International Union Against Cancer (UICC) classification is used (UICC 2009):

G1: well differentiated, uniform glandular structure consisting of well-formed glands

G2: moderately differentiated, intermediate stage between G1 and G3

G3: poorly differentiated, solid tumors composed of irregular, barely recognizable glands or single cells arranged in small or major clusters, partly with mucin production and remnants of gland formation

G4: undifferentiated

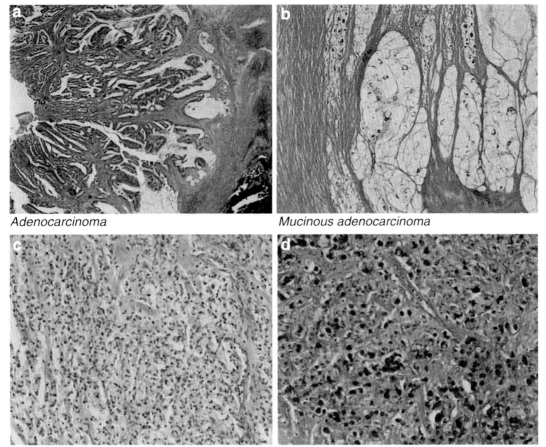

Adenocarcinoma

Mucinous adenocarcinoma

Signet-ring cell adenocarcinoma, HE

Signet-ring cell adenocarcinoma, PAS

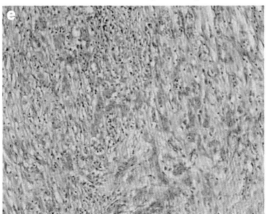

Undifferentiated carcinoma

Colorectal Cancer Pathology, Fig. 3 Histological types of colorectal carcinoma

Varying percentages of gland formation are ascribed to these four different grades by different authors. The WHO provides a two-step grading system that fulfills clinical requirements and is credited with higher reproducibility (Bosman et al. 2010):

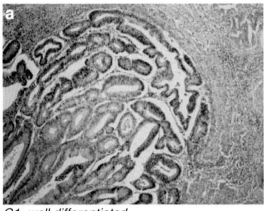
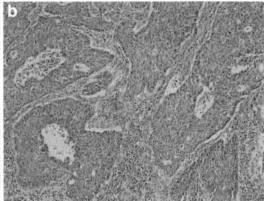

G1, well differentiated   *G3, poorly differentiated*

Colorectal Cancer Pathology, Fig. 4 Grading colorectal carcinoma

Low grade: encompassing G1 and G2
High grade: encompassing G3 and G4

In cases of carcinoma showing heterogeneous
differentiation, grading should be based on the
least differentiated component, i.e., assigning the
higher grade. Disorganized glands representing
poor differentiation are a common finding at the
advancing edge of tumors, and these tumors
should not be regarded as high-grade malignancies based on this finding only (Fig. 4).

Tumor Spread
In general, tumor spread is caused by tumor cells
escaping normal cellular growth control, thus
leading to tumor ► invasion and ► metastasis.
Knowledge of the mechanisms of tumor spread
is of prime importance, as they help in guiding
treatment strategies, especially concerning surgical resection.

Local Spread
Tumors of the colorectum can show continuous
intramural growth, both in a longitudinal and vertical direction. The latter is associated with invasion deeper into the bowel wall, eventually
resulting in penetration of the full thickness of
the bowel wall. Mural penetration allows direct
extension to adjacent organs or tissues or perforation of the serosal surface, thus leading to the

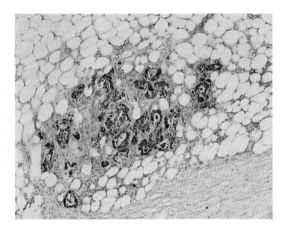

Colorectal Cancer Pathology, Fig. 5 Satellite

possibility of peritoneal metastasis, or both,
depending on the anatomic site. Especially in
rectal carcinomas, discontinuous spread in the
perimuscular tissues of the mesorectum in the
form of microscopic tumor nodules without residual lymphatic tissue, so-called satellites, has to be
taken into account. These can be found even some
centimeters distal to the lower margin of the tumor
(UICC 2009; Washington et al. 2009). Rarely,
tumors of the colon are reported to spread superficially along the mucosa or intraluminally
(Bosman et al. 2010; Fenoglio-Preiser
et al. 2008) (Fig. 5).

Lymphatic Spread

After penetration of the muscularis mucosae, colorectal carcinomas are able to enter the lymphatics, eventually resulting in lymph node metastasis. Invasion of ► lymphatic vessels by a colorectal tumor is classified as L1 in the L classification (in contrast to L0, no lymphatic invasion) (UICC 2009). Usually, the first lymph nodes to become involved are those closest to the tumor in the bowel wall, following normal lymphatic flow to farther regional lymph nodes and as the case may be non-regional lymph nodes. Via lymphatic drainage into the thoracic duct and caval vein, this may lead to lung metastasis (Fenoglio-Preiser et al. 2008).

Venous Spread

A feature present in a considerable proportion of colorectal carcinomas is venous permeation. According to the V classification, this state is classified as V1, i.e., microscopic venous invasion, or V2, i.e., macroscopic venous invasion (in contrast to V0, no venous invasion) (UICC 2009). Following drainage via the portal vein, tumor cells may cause liver metastasis. Only tumors of the lower rectum draining into the iliac and caval veins are also able to induce lung metastasis by the way of venous spread. Further distant metastases involve the brain, the bone, or the ovary (Fenoglio-Preiser et al. 2008).

Perineural Spread

An additional way of tumor spread is perineural or intraneural extension. Invasion of perineural spaces is classified according to the Pn classification as Pn1 (as opposed to Pn0, no perineural invasion) (UICC 2009) (Fig. 6).

Extent of Resection and Lymph Node Dissection

The different types of spread determine the extent of surgical resection, which is affected by the extent of lymph node dissection and the vascular supply in colon carcinoma. In rectal carcinoma, the distal and circumferential margin may be critical, too. Low anterior resection for tumors of the middle and lower rectal third therefore has to include total mesorectal excision down to the pelvic floor, while the margin of clearance within the rectal wall may be limited to 1 or 2 cm in cases of high-grade tumors (Wittekind 2007). During surgical procedures, local tumor cell spillage because of (iatrogenic) tumor perforation has to be avoided, as it may result in locoregional recurrence in the original tumor bed or peritoneal metastasis. This may even make a multivisceral resection necessary in case of adherence of tumor to adjacent organs, which may be caused either by inflammation or tumor invasion, if a curative operation is intended. Inspection or palpation may not settle the matter, and any biopsy from the site of adherence should be avoided as in case of tumor invasion, the result is local tumor spillage (Hermanek 2002). Lymphatic drainage determines the extent of lymph node dissection. This fact is taken into account by the definition of regional lymph nodes for the different sections of the colon and rectum provided by the current UICC TNM classification shown in Table 3.

Most parts of the colon show unidirectional lymphatic drainage through the lymphatic channels alongside major arteries. However, for both flexures and the right and left third of the transverse colon, lymphatic drainage is bidirectional, with the hepatic flexure and right third of the transverse colon draining into the nodes alongside both the right and middle colic artery and with the left third of the transverse colon and the splenic flexure draining into the nodes both alongside the middle and left colic artery (Wittekind 2007). For tumors of these locations, an extended right and left hemicolectomy respectively is therefore required. Rectal carcinomas metastasize to perirectal lymph nodes at the level of the primary tumor and above. Low rectal carcinomas, however, may involve both mesenteric and inguinal lymph nodes (Fenoglio-Preiser et al. 2008). Controversial reports exist on the probability of lateral spread to lymph nodes outside the mesorectum.

Staging

Every attempt to describe and define the anatomical extension of colorectal carcinoma is

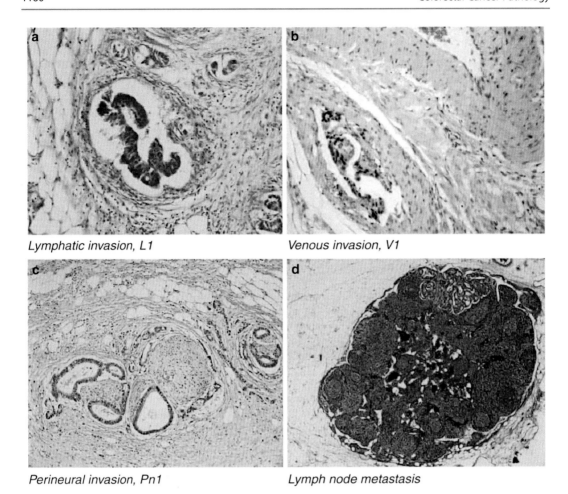

Lymphatic invasion, L1

Venous invasion, V1

Perineural invasion, Pn1

Lymph node metastasis

Colorectal Cancer Pathology, Fig. 6 Lymphatic spread, venous and perineural invasion

motivated by its known impact on prognosis. Cuthbert Esquire Dukes was the first to introduce a pathological staging classification, whose application is no longer advisable despite its remarkable historical value, and many others have followed since. Nowadays, the UICC and the American Joint Committee on Cancer (AJCC) TNM system is recommended for general use. It is based on the depth of tumor invasion and eventual invasion of adjacent structures concerning the primary tumor, the number of regional lymph nodes involved, and the presence or absence of distant metastases. The present seventh edition system is shown in Table 4.

Explanatory Notes Concerning the Application of the UICC TNM Classification

General Aspects

The TNM system constitutes a dual system in view of the fact that it comprises a clinical classification (TNM or cTNM) and a pathological classification (pTNM). The cTNM corresponds to a classification based on evidence acquired before treatment by way of physical examination, imaging, biopsies, and other means. The pTNM classification contains evidence acquired from surgical treatment and pathological examination (UICC 2009). The definitions of pTNM correspond to those of cTNM.

Colorectal Cancer Pathology, Table 3 Regional lymph nodes for each anatomical site or subsite (UICC 2009)

Cecum	Ileocolic, right colic
Ascending colon	Ileocolic, right colic, middle colic
Hepatic flexure	Right colic, middle colic
Transverse colon	Right colic, middle colic, left colic, inferior mesenteric
Splenic flexure	Middle colic, left colic, inferior mesenteric
Descending colon	Left colic, inferior mesenteric
Sigmoid colon	Sigmoid, left colic, superior rectal, inferior mesenteric, and rectosigmoid
Rectum	Superior, middle and inferior rectal, inferior mesenteric, internal iliac, mesorectal, lateral sacral, presacral, sacral promontory (Gerota)

Primary Tumor

Tis or pTis includes cancer cells confined within the glandular basement membrane (intraepithelial) or lamina propria (intramucosal) with no extension through the muscularis mucosae into the submucosa. For T4/pT4 further ramifications exist using suffixes (e.g., pT4a and pT4b). In T4/pT4, direct invasion includes invasion of other segments of the colorectum by way of the serosa, e.g., invasion of the sigmoid colon by a carcinoma of the cecum. Tumor that is adherent to other organs or structures, macroscopically, is classified as cT4. However, if no tumor is present in the adhesion, microscopically, the classification should be pT1–pT3 depending on the anatomical depth of wall invasion (UICC 2009; Wittekind et al. 2003).

Regional Lymph Nodes

Direct extension of primary tumor into regional lymph nodes is classified as lymph node metastasis. N1c in the N category refers to tumor deposits (satellites), which represent grossly or microscopically detectable nests or nodules in pericolorectal adipose tissue in the lymphatic drainage area of a primary tumor without histological evidence of residual lymph node. They may correspond to discontinuous spread, venous invasion, or

Colorectal Cancer Pathology, Table 4 The UICC TNM classification of tumors of the colon and rectum (UICC 2009; Wittekind et al. 2003)

Primary tumor (T)		
TX	Primary tumor cannot be assessed	
T0	No evidence of primary tumor	
Tis	Carcinoma in situ: intraepithelial or invasion of lamina propria	
T1	Tumor invades submucosa	
T2	Tumor invades muscularis propria	
T3	Tumor invades through the muscularis propria into the subserosa, or into non-peritonealized pericolic or perirectal tissue	
T4	Tumor directly invades other organs or structures and/or perforates visceral peritoneum:	
	T4a	Perforation of the visceral peritoneum
	T4b	Tumor directly invades other organs or structures
Regional lymph nodes (N)		
NX	Regional lymph nodes cannot be assessed	
N0	No regional lymph node metastasis	
N1	Metastasis in 1–3 regional lymph nodes:	
	N1a	1 node
	N1b	2–3 nodes
	N1c	Satellites in subserosa, without regional nodes
N2	Metastasis in 4 or more regional lymph nodes:	
	N2a	4–6 nodes
	N2b	7 or more nodes
Distant metastasis (M)		
M0	No distant metastasis	
M1	Distant metastasis:	
	M1a	One organ
	M1b	More than one organ or peritoneum

complete permeation of regional lymph nodes by the tumor. Presence of satellites in the subserosa without regional lymph nodes is classified as N1c. If such tumor nodules are considered lymph nodes totally replaced by tumor by the pathologist, usually because they have the form and smooth contour of a lymph node, they should be classified as regional lymph node metastases in the N category, with every nodule counted as a metastatic lymph node. Non-regional metastatic lymph nodes are classified as distant metastasis. The minimum number of lymph nodes to be removed and examined in case of carcinomas of the colorectum recommended by the UICC is 12 lymph nodes (UICC 2009).

Colorectal Cancer Pathology, Table 5 Stage grouping for colorectal carcinoma (UICC 2009)

Stage		T	N	M
0		Tis	N0	M0
I		T1, T2	N0	M0
II		T3, T4	N0	M0
	IIA	T3	N0	M0
	IIB	T4a	N0	M0
	IIC	T4b	N0	M0
III		Any T	N1, N2	M0
	IIIA	T1, T2	N1	M0
		T1	N2a	M0
	IIIB	T3, T4a	N1	M0
		T2, T3	N2a	M0
		T1, T2	N2b	M0
	IIIC	T4a	N2a	M0
		T3, T4b	N2b	M0
		T4b	N1, N2	M0
IV		Any T	Any N	M1
	IVA	Any T	Any N	M1a
	IVB	Any T	Any N	M1b

Colorectal Cancer Pathology, Table 6 R classification (UICC 2009)

RX	The presence of residual tumor cannot be assessed
R0	No residual tumor
R1	Microscopic residual tumor
R2	Macroscopic residual tumor

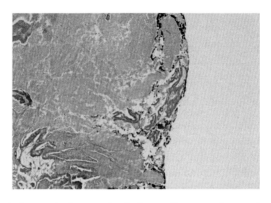

Colorectal Cancer Pathology, Fig. 7 Microscopic residual tumor directly at the resection margin marked with black color

Distant Metastasis

For the M category, MX or pMX does not exist. M0 or cM0 is used for cases that show no distant metastasis clinically. pM0, however, is to be employed only at autopsy. Therefore, if a clinically suspected metastatic node (cM1) is biopsied and is proven to be negative, it becomes M0. pM1 represents distant metastasis proven histologically (UICC 2009).

Stage Grouping

While the clinical TNM classification serves as a basis for treatment decisions, pTNM is the basis for estimating prognosis. For this purpose, the T, N, and M category of a carcinoma are summarized in stage groups 0 to IV, as shown in Table 5.

Residual Tumor Classification

Description of the anatomical extent of cancer before (mostly surgical) treatment is the purpose of the TNM and pTNM classification. In contrast, the auxiliary residual tumor (R) classification deals with tumor status after treatment. It reflects the efficacy of treatment, i.e., the completeness of tumor removal, influences the optional application of further treatment, and is a strong indicator of prognosis. The R classification indicates the presence or absence of residual tumor after treatment, specifying the amount of residual tumor as macroscopic or microscopic. In this respect, residual tumor at the primary site, in regional lymph nodes, and/or at distant sites is taken into account (UICC 2009; Wittekind et al. 2003). The R classification categories are listed in Table 6.

Thus, the R classification is based on clinical as well as pathological findings. After resection, the pathologist has to examine the resection margins in order to obtain a reliable R classification, supplemented by clinical information, e.g., on residual distant metastasis. In colorectal carcinoma, these margins are the proximal, distal, and circumferential (lateral/radial) margins. In colon and even more so in rectal cancer, the main area of concern with respect to prognosis is the circumferential margin (CRM), i.e., the resection margin of the mesocolon and mesorectum, respectively (Washington et al. 2009) (Fig. 7).

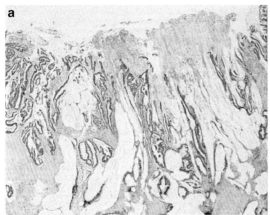 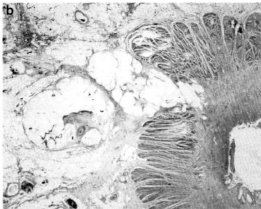

Minimal regression   *Subtotal regression*

Colorectal Cancer Pathology, Fig. 8 Tumor regression after ▶ neoadjuvant therapy

Tumor Regression Score

Preoperative, i.e., neoadjuvant, radiotherapy and/or ▶ chemotherapy is considered standard practice in advanced rectal cancer. In such cases, the pathologist is requested to assess the tumor response to therapy in terms of the extent of tumor regression. Unfortunately, a standardized, internationally accepted regression grading system does not exist yet. In Germany, a proposal by Dworak and its modification is widely used, among other regression grading systems (Neid et al. 2008). In colorectal carcinoma, complete regression is possible but not too common, with numbers ranging from 0% to 30% for rectal carcinomas in different reports. This has at least partially to be ascribed to the fact that no standardized procedure exists for these cases, which also pertains to pathological workup. Therefore, great care is necessary for workup of the tumor region supplemented by detailed statements on workup (e.g., number of blocks taken and number of levels cut) in order to allow an appraisal of the reliability of findings. With respect to TNM, the post-therapy stage is marked by the prefix "y," i.e., ycTNM and ypTNM, respectively (UICC 2009) (Fig. 8).

Prognosis

Variables affecting outcome, i.e., overall survival, disease-free survival, recurrence rate, and

Colorectal Cancer Pathology, Table 7 Prognostic factors in colorectal carcinoma (Bosman et al. 2010; Fenoglio-Preiser et al. 2008; Compton et al. 2000)

Staging
pT category: extent of local invasion of carcinoma
pN category: regional lymph node metastasis and number of lymph nodes with metastasis
pM category: distant metastasis
Features of the primary tumor
Histological grade
Histological type
Tumor border configuration (invasive margin)
Blood or lymphatic vessel invasion, perineural invasion
Tumor perforation
Consequences of surgical technique
Distance between resection margin and tumor (extent of resection)
Radial/circumferential margin (especially rectum)
Residual tumor classification (R classification)
Iatrogenic tumor perforation
Evidence of host response
Peritumoral fibrosis (desmoplasia)
Peritumoral inflammatory reaction
Reactive changes in regional lymph nodes
Molecular alterations
Chromosome 18q loss of heterozygosity (LOH)
MSI-H

response to treatment, are called prognostic factors. A multitude of prognostic factors has been proposed for colorectal carcinoma. One way of

Colorectal Cancer Pathology, Table 8 Minimum data to be included in surgical pathology reports on colorectal carcinoma specimens (Wittekind 2007; Washington et al. 2009; Fenoglio-Preiser et al. 2008; Compton et al. 2000; Wittekind and Koch 2005)

Incisional (endoscopic) biopsy	
Macroscopic examination	Number of pieces
	Largest dimension of each piece
Microscopic evaluation	Histological type
	Histological grade
	Extent of invasion, as appropriate
Polypectomy	
Macroscopic examination	Number of pieces
	Configuration (e.g., flat/sessile/pedunculated)
	Size (preferably three dimensions)
	If pedunculated, length of stalk
Microscopic evaluation	
Tumor	Histological type
	Histological grade
	Extent of invasion (e.g., intraepithelial/intramucosal/invasion of submucosa)
	Lymphatic/venous invasion
Margins	Minimal distance of tumor from margin (possibly differentiating between adenomatous and carcinomatous parts of a lesion)
Local excision (submucosal, full thickness)	
Macroscopic examination	
Specimen	Number of pieces
	Dimensions
Tumor	Dimensions
	Configuration
	Distance to margins
Microscopic evaluation	
Tumor	Histological type
	Histological grade
	Extent of invasion (pT)
	Lymphatic/venous (extramural?)/perineural invasion
Margins	Involvement or minimal distance of tumor from mucosal and deep margins
Additional findings	e.g., adenoma
Segmental resection specimen	
Macroscopic examination	
Specimen	Organ(s) included (parts of colorectum removed, adjacent organs if applicable)
	Number of pieces received (resection en bloc or not)
Tumor	Number of tumors
	Location/site of tumor(s)
	In rectal carcinomas
	Relation of tumor to peritoneal reflection
	Macroscopic intactness of mesorectum
	For abdominoperineal resection specimens: distance of tumor from dentate line/anal verge
	Dimensions of tumor(s), at least greatest dimension
	Tumor perforation (spontaneous/iatrogenic), if present
Margins	Minimum distance from proximal/distal margins
	Minimum distance from circumferential (radial/lateral) margin
	Specify method of measurement

(continued)

Colorectal Cancer Pathology, Table 8 (continued)

Microscopic evaluation			
Tumor	Histological type		
	Histological type		
	Microscopic tumor invasion (pT)		
	Lymphatic/venous (extramural?)/perineural invasion		
	Tumor deposits		
	Treatment effects, applicable to carcinomas after neoadjuvant therapy		
	Pattern of growth at invasive margin		
Margins	Proximal margin	}	Involvement or minimal distance
	Distal margin	}	
	Circumferential/lateral margin		
	Doughnuts, if applicable		
Metastatic spread	Number of regional lymph nodes examined		
	Number of regional lymph nodes involved (pN)		
	Histologically confirmed distant metastasis (pM)		
Additional findings	e.g., adenoma or other polyps, chronic inflammatory bowel disease		
Pathological staging	pTNM		
	Additional descriptors: L, V, Pn classification		
	Stage group		
	R classification		
	Response to neoadjuvant therapy, if applicable (tumor regression grading)		

organizing them is to assign them to certain categories reflecting the strength of the published evidence demonstrating their prognostic value (Compton et al. 2000). The following list of prognostic factors in colorectal carcinoma mainly focuses on those factors associated with pathological workup. Factors credited with strong evidence are in italics (Table 7).

Tumor stage is the most significant prognostic feature, with regional lymph node metastasis being second only to distant metastatic disease in importance (Compton et al. 2000). Both the number of lymph nodes removed by resection of the colon or rectum and the number of lymph nodes examined may vary considerably due to several influencing factors. One of these influencing factors constitutes the pathologist himself. Therefore, due diligence on the pathologist's behalf is essential concerning search for nodes in the resected specimen in order to retrieve at least the 12 lymph nodes recommended by the UICC (Wittekind et al. 2012).

Histological grade is also an important prognostic factor. In this respect, distinguishing between "low grade" and "high grade" is enough to retain prognostic significance (Compton et al. 2000). The impact of histological type on prognosis is not judged uniformly. Signet ring cell carcinoma, which is high grade by definition, is associated with a worse prognosis, and medullary carcinoma with a better prognosis (Compton et al. 2000). A histological feature correlated with worse prognosis is an infiltrative pattern of the invasive edge of the tumor as opposed to a pushing or expansile pattern (Bosman et al. 2010; Compton et al. 2000).

Invasion of both lymphatic and venous vessels is associated with worse prognosis, which increases with stage and grade. This aggravates if extramural blood vessels are affected by way of an increased risk for distant metastasis, especially to the liver (Fenoglio-Preiser et al. 2008; Compton et al. 2000). Tumor perforation resulting from extensive tumor invasion of the bowel wall is linked to poor prognosis (Fenoglio-Preiser et al. 2008). However, perforation may also be iatrogenic, thus linking up with another group of prognostic factors closely connected to surgical technique and performance. An important factor is the distance of the tumor to the resection margin, in a longitudinal direction and particularly in carcinomas of the rectum with a predominantly non-peritonealized surface also in a radial direction, i.e., the circumferential margin. The risk for local recurrence increases both with margin involvement and with decreasing distance between tumor and radial margin (Fenoglio-Preiser et al. 2008; Compton et al. 2000). Another feature of prognostic significance is the residual tumor (R) classification. Positive margins should be interpreted as the counterpart of residual tumor in the patient, unless proven otherwise, increasing the risk of local recurrence (Compton et al. 2000). Peritumoral inflammatory reaction and fibrosis are regarded as equivalents of host response to infiltrating tumor associated with better prognosis (Bosman et al. 2010; Fenoglio-Preiser et al. 2008). Although a multitude of molecular markers has been defined in colorectal carcinoma, most of them still lack strong evidence of prognostic significance. Among those being up for discussion are high level of ▶ microsatellite instability (MSI-H; correlated with a good prognosis) and loss of heterozygosity (LOH) of chromosome 18q (adverse prognostic marker) (Bosman et al. 2010; Compton et al. 2000).

Histopathological Report

The pathologist plays an important role in a multidisciplinary team, thus contributing to the quality of diagnosis, treatment, and estimation of prognosis. Different team members represented by surgeons, radiologists, or oncologists present the pathologist with different requirements. To meet all these needs, the pathological report has to contain a minimum number of data listed in Table 8 that are based on recommendations of national pathologists' associations.

Cross-References

▶ Colorectal Cancer

References

Bosman FT, Carneiro F, Hruban RH et al (eds) (2010) World Health Organization (WHO) classification of tumors. Pathology & genetics. Tumors of the digestive system. IARC Press, Lyon

Compton CC, Fielding LP, Burgart LJ et al (2000) Prognostic factors in colorectal cancer. College of American Pathologists Consensus Statement 1999. Arch Pathol Lab Med 124:979–994

Fenoglio-Preiser CM, Noffsinger AE, Stemmermann GN (2008) Gastrointestinal pathology: an atlas and text, 3rd edn. Lippincott Williams & Wilkins, Philadelphia

Hermanek P (2002) Pathology of colorectal cancer. In: Bleiberg H, Kemeny N, Rougier P et al (eds) Colorectal cancer. A clinical guide to therapy. M. Dunitz, London, p 55 et seq

Neid M, Tannapfel A, Wittekind C (2008) Gastrointestinal tumours. Histological regression grading after neoadjuvant therapy. Onkologe 14:409–418

UICC (2009) In: Sobin LH, Gospodarowicz MK, Wittekind C (eds) TNM classification of malignant tumours, 7th edn. Wiley-Blackwell, Oxford

Washington MK, Berlin J, Branton P et al (2009) Protocol for the examination of specimens from patients with primary carcinoma of the colon and rectum. Arch Pathol Lab Med 133:1539–1551

Wittekind C (2007) Pathology. In: Cassidy J, Johnston P, van Cutsem E (eds) Colorectal cancer. Informa Healthcare USA, New York, p 103 et seq

Wittekind Ch, Koch HK (2005) Empfehlungen zur pathologisch-anatomischen Diagnostik des kolorektalen Karzinoms des Berufsverbandes Deutscher Pathologen e.V. und der Deutschen Gesellschaft für Pathologie e.V., Version 1.0. http://www.bv-pathologie.de/mitgliederbereich/download.php?f=leit-linie_kolo_karzinom.pdf. Accessed Feb 2010

Wittekind C, Compton CC, Brierley J et al (2012) UICC TNM supplement. A commentary on uniform use, 4th edn. Wiley-Blackwell, Oxford

Colorectal Cancer Premalignant Lesions

Luca Roncucci
Department of Diagnostic and Clinical Medicine, and Public Health, University of Modena and Reggio Emilia, Modena, Italy

Synonyms

Adenomas; Adenomatous polyps; Microadenomas

Definition

▶ Colorectal cancer premalignant lesions are focal lesions which precede cancer development. There are different entities: ▶ adenoma is the only one for which the available scientific evidence of its premalignant nature is more convincing.

Colorectal adenomas are polypoid, flat, or depressed lumps of epithelial origin, which can be found throughout the large bowel, whose main histological feature is ▶ dysplasia. They may be single or multiple, rarely hundreds or even thousands, as in patients with familial adenomatous polyposis (FAP). Adenomas, if left untreated, grow and finally may become malignant. However, only a fraction of colorectal adenomas acquire malignant features, and sometimes may also regress. Polypoid adenomas are the most frequent and can be sessile (with a large base) or pedunculated (with a stalk attached by a narrow base on the colorectal mucosal surface).

Other kinds of colorectal lesions may be premalignant; they are hyperplastic polyps, hamartomatous polyps (Hamartoma), aberrant crypt foci, and ▶ dysplasia in inflammatory bowel disease (IBD). Anatomically, crypts are narrow and deep invaginations into a larger structure. The small intestine and the colon are typically formed by glandular crypts invaginate deep in the submucosa. In the small intestine, crypts are called crypts of Lieberkühn and are organized around finger-like protrusions (villi). Stem cells are located at the bottom of each crypt and generate actively dividing progenitors (the transit amplifying cells) that differentiates into specialized epithelial cell types (enterocytes, colonocytes, goblet cells, enteroendocrine cells) as they migrate upward.

Hyperplastic polyps are round, sessile, and pale lesions usually with a diameter of less than 1 cm. Histology shows elongated, dilated, and typically saw-like crypts, lined by a single layer of colonic epithelium, and usually no feature of ▶ dysplasia. Thus, hyperplastic polyps are not considered premalignant. However, a variant of hyperplastic polyps showing areas of both dysplasia and hyperplasia in the same lesion (referred to as mixed polyps) or others where, in a hyperplastic architecture, cytological signs of dysplasia are present (referred to as serrated adenomas), can be considered premalignant.

Hamartomatous polyps are rare colorectal polypoid lesions whose main histological feature is the presence in the same lesion of different normal colorectal tissues (mainly epithelial glands, connective tissue, and smooth muscle) showing marked alteration of the whole architecture.

Aberrant crypt foci are microscopic focal lesions which can be observed under a light or dissecting microscope at 30–40× magnification, or even during magnifying colonoscopy, on the colorectal mucosal surface after staining with a vital dye, methylene blue, in patients with familial adenomatous polyposis (FAP), cancer, or benign diseases of the large bowel. They were originally described in mice exposed to colonic carcinogens and then identified also in humans. They appear as clusters of altered colonic crypts, enlarged, and deeply stained than normal at topological view, often showing abnormal-shaped lumens and slightly bulging on the level of the normal mucosal surface.

In patients with inflammatory bowel disease (IBD) (ulcerative colitis and ▶ Crohn disease), dyplasia may be observed in biopsies taken during colonoscopy as flat lesions or polypoid masses. Dysplasia in IBD is a premalignant feature, which deserves further clinical management.

Characteristics

Adenomas

Adenomas are rare in Africa and in most Asian countries, while they are very frequent in many developed countries, where 30–40% of individuals over 50 years harbor an adenoma of any size in the colon. The widespread use of colonoscopy contributed to the increased prevalence of adenomas in these populations. The geographical distribution of adenomas lends support to the contention that dietary and lifestyle factors in western countries play some role in their etiology. Among them, a high intake of meat (especially red meat), a low intake of vegetables and fiber, a low physical activity, and insulin resistance are the most consistent. It is well-established that their prevalence increases with age. Adenomas are more frequent in the left colon (distal to the splenic flexure) and rectum, reflecting the anatomical distribution of carcinoma, though an increased prevalence in the right colon due to the use of pancolonoscopy has been reported. Adenomas are usually less than 1 cm in diameter, sometimes less than or equal to 0.5 cm (diminutive polyps). When larger, more often they show a villous architecture and proclivity to malignant transformation. Adenomas are single in most cases. In about half of the individuals, two or more lesions can be observed at endoscopy. The number of adenomas is strictly related to the risk of developing cancer; in FAP, cancer incidence approaches 100%. According to the main architecture of the lesion, adenomas may be divided into three histological types: tubular, tubulovillous, and villous, the latter having the highest probability of becoming malignant. Tubular adenomas show a prevalent glandular architecture in more than 80% of the whole lesion. Villous adenomas have more than 80% of villous

architecture, characterized by long fronds of papillary epithelium arising from the mucosal surface of the colon (Fig. 1). Tubulovillous adenomas show both components, each less than 80%. In all colorectal adenomas, dysplasia is graded into three different levels of severity, i.e., mild, moderate, and severe. The current grouping of grades into two categories: "low-grade" (mild and moderate) and "high-grade" (severe) is justified by the poor reproducibility in the separation between mild and moderate grades. This uncertainty stems from the definition of dysplasia, term with a high level of subjectivity. More advanced is the overall grade, higher is the risk of cancer development, which is evident when malignant cells pass through the muscularis mucosae and invade the submucosal layer of the bowel wall, identifying a malignant polyp. This is the earliest form of carcinoma with metastatic potential; the risk of lymph node ▶ metastasis has been estimated around 10% overall.

Flat adenomas are short lumps of mucosa with a reddish surface, sometimes with a central area of depression. At histology, these adenomas show the same three types of architecture described for

**Colorectal Cancer
Premalignant Lesions,
Fig. 1** Histological section
of a villous adenoma with
typical papillary epithelium

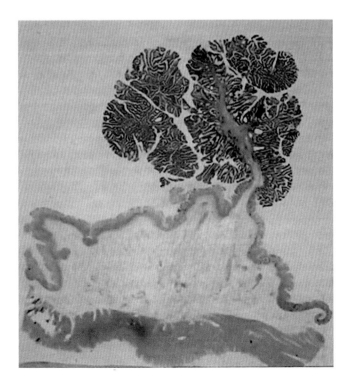

polypoid adenomas, but they progress to cancer with a higher frequency, providing further evidence for the adenoma–carcinoma sequence.

Sometimes adenomas bleed, but most remain asymptomatic, and are discovered by chance during colonoscopy. Flat and depressed adenomas are not easily seen during endoscopy, but it is important to identify and remove them, due to their higher malignant potential. The optimal treatment for adenomas (and for colorectal polyps in general) is its removal, usually during colonoscopy, or seldom by surgical operation, especially large adenomas of the rectum. The pathologic diagnosis is then mandatory for planning the future endoscopic follow-up of the patient. Estimating the risk of developing other lesions after removal of adenomas is not an easy task. The current evidence suggests that larger, multiple, and villous adenomas do recur more frequently and require a closer endoscopic follow-up.

Adenoma–Carcinoma Sequence

Overwhelming evidence suggests that colorectal carcinomas develop from preexisting adenomas, which is referred to as the so-called adenoma–carcinoma sequence hypothesis based upon the following observations:

1. Focal adenomatous areas have been frequently observed in colorectal carcinomas; on the other hand, focal infiltration by carcinomatous tissue (beyond the muscularis mucosae) can be observed in adenomas.

2. If adenomas are not removed, cancer may develop at the polyp site, and accurate surveillance with removal of polyps leads to a reduction of the expected number of invasive carcinomas.

3. Adenomas and carcinomas share a similar distribution along the large bowel, being more frequent in the descending and sigmoid colon and in the rectum.

4. Age-specific incidence rate of colorectal adenomas shows a peak that precedes that of cancer by about 5 years, in keeping with the estimated time lag required for malignant transformation.

5. Adenomas have an hyperproliferative epithelium, as carcinomas.

6. Similar patterns of genetic and epigenetic alterations have been reported for adenomas and carcinomas, though usually they are less frequent and severe in adenomas than in carcinomas.

The adenoma–carcinoma sequence has been revisited by a genetic point of view and definite steps of activation of oncogenes and inactivation by allelic loss of tumor suppressor genes, which underlies colorectal carcinogenesis (▶ colon cancer carcinogenesis in human and in experimental animal models), have been identified (Fig. 2). In particular, APC inactivation [*APC* gene in familial

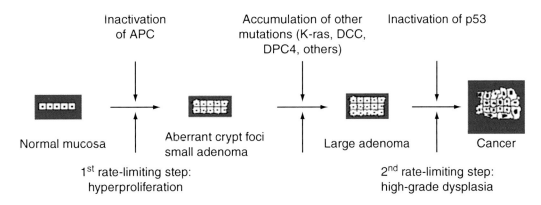

Colorectal Cancer Premalignant Lesions, Fig. 2 Genetic model of colorectal carcinogenesis, including the main molecular alterations underlying the development of colorectal lesions

adenomatous polyposis] seems one of the earliest events causing hyperproliferation of epithelial cells and microadenoma formation, then accumulation of mutations in other genes [*KRAS* (ras), DCC, and others] are responsible of adenoma growing. Finally, later inactivation of the tumor suppressor gene *p53* [p53 gene family] seems the key event for malignant transformation. Moreover, in some cases, ▶ microsatellite instability (MSI) (especially in tumors from patients with ▶ Lynch syndrome) and hypermethylation of ▶ CpG islands in the promoter region of several genes, causing ▶ epigenetic gene silencing, have been demonstrated in adenomas.

Hyperplastic Polyps

The premalignant potential of hyperplastic polyps is not supported by most evidence. However, some observations seem to point to the contrary. Polyps with mixed but distinct features of hyperplasia and dysplasia may become malignant. Their nature is not yet understood. It is possible that adenomatous tissue develops within a hyperplastic polyp. Also serrated adenomas have premalignant potential because of some cytological features of epithelial cells lining the crypts; evident signs of dysplasia have been identified in 30–40% of serrated adenomas. Moreover, it should be mentioned as rare hyperplastic polyposis, a syndrome which seems to have malignant potential, acquired through mutator pathways. Some biomolecular alterations similar to those found in adenomas have been reported in hyperplastic polyps, including *KRAS*, p53 mutations, and ▶ microsatellite instability.

Hamartomatous Polyps

Hamartomatous polyps are observed in two diseases: ▶ Peutz–Jeghers syndrome and juvenile polyposis. They are not considered precancerous lesions, although some authors have reported the development of adenocarcinoma in hamartomatous polyps, and others the coexistence of adenomatous and carcinomatous tissues in these polyps. Finally, in Peutz–Jeghers syndrome a frequent occurrence of gastrointestinal as well as extradigestive cancers has been observed.

Aberrant Crypt Foci

Aberrant crypt foci (ACF) are smaller than adenomas and usually not visible to the naked eye. Luminal openings of aberrant crypts show various shapes, each corresponding to definite histological alterations. Indeed, histologically ACF are rather heterogeneous. They may show various alterations, from hypertrophy (only dilated crypt with no cell alteration), to hyperplasia, to severe dysplasia. Only a minor fraction (10–30%) of the ACF examined is defined as dysplastic (referred to as microadenomas). Dysplasia may be focal in an aberrant crypt focus, and even in a single aberrant crypt, suggesting that a transition from hyperplasia to dysplasia in aberrant crypts is possible. Dysplasia seems more frequent in larger and in proximal colonic ACF. ACF with dysplasia can be considered true premalignant lesions, whereas for ACF with other histological features the matter is questionable, and the mechanisms and steps of the progression remain unclear. The main points in favor of their premalignant potential are as follows. The density of ACF is higher in patients with FAP. In these patients, the ACF examined show definite dysplasia at histology in 75–100% of cases. In patients with cancer and benign diseases of the large bowel, the density of ACF is lower, but higher in the distal colon and rectum. Older individuals harbor a higher number of ACF in their colon. Epidemiological evidence suggests that density of ACF in patients with colon cancer is related to environmental factors as it happens for adenomas and carcinomas. ACF have a hyperproliferative epithelium. In particular, an extension of the proliferative compartment to the upper portion of aberrant crypts is evident in dysplastic ACF. The premalignant potential of ACF is further supported by the finding of several genetic and epigenetic alterations (epigenetics), similar to those reported in adenomas and sometimes in hyperplastic polyps, including DNA ▶ microsatellite instability and DNA hypermethylation.

Inflammatory Bowel Diseases

Patients with ulcerative colitis or ▶ Crohn disease may have an increased risk of developing colorectal cancer, depending on the duration of the disease and the extent of the involved mucosa.

The risk is related to dysplasia which may be documented in the colorectal mucosa of these patients as flat areas or dysplasia-associated lesions or masses which resembles an adenoma. Dysplasia is the consequence of chronic ► inflammation typical of these diseases. The lesions are at high risk of becoming malignant, and their presence requires close follow-up and should guide surgical treatment.

Cross-References

► Colorectal Cancer Pathology

References

Fenoglio GM, Lane N (1974) The anatomical precursor of colorectal carcinoma. Cancer 34:819–823

Morson BC, Bussey HJR, Day DW (1983) Adenomas of the large bowel. Cancer Surv 3:451–477

Muto T, Bussey HJR, Morson BC (1975) The evolution of cancer of the colon and rectum. Cancer 36:2251–2270

Pretlow TP, Barrow BJ, Ashton WS et al (1991) Aberrant crypts: putative preneoplastic foci in human colonic mucosa. Cancer Res 51:1564–1567

Roncucci L, Medline A, Bruce WR (1991) Classification of aberrant crypt foci and microadenomas in human colon. Cancer Epidemiol Biomarkers Prev 1:57–60

Colorectal Cancer Therapeutic Antibodies

Larissa Belov[1], Jerry Zhou[1] and
Richard I. Christopherson[2]
[1]School of Molecular and Microbial Biosciences, University of Sydney, Sydney, NSW, Australia
[2]School of Life and Environmental Sciences, University of Sydney, Sydney, NSW, Australia

Definition

► Colorectal cancer (CRC) may occur in the colon, rectum, or appendix. It is the third most common form of cancer and the third leading cause of cancer-related death in the Western

world. CRC is the fourth most common cancer in men and the third in women, though significant international variations in the distribution of CRC have been observed. Worldwide, nearly 1.2 million new cases of CRC were diagnosed in 2007, resulting in about 6,30,000 deaths (8% of all cancer deaths). Many CRC are thought to arise from adenomatous polyps in the colon, which are usually benign, but may develop into cancer over time. Localized CRC is generally detected by colonoscopy. Therapy is usually through surgery, often followed by ► chemotherapy. Antibody-based therapies have also been used.

Antibody therapy is the use of "engineered" monoclonal antibodies (MAbs) to specifically target cells or soluble molecules important for disease progression. There are several ways MAbs can be used for therapy, e.g., to target cancer cells, which may stimulate the patient's immune system to attack these cells via ► antibody-dependent cellular cytotoxicity (ADCC) or ► complement-dependent cytotoxicity (CDC), or MAbs can be used to deliver lethal chemical or radioactive doses directly to the cancer cells. Alternatively, MAbs can be used to inhibit the disease-inducing biological activities of soluble factors by competitively blocking their interaction with receptors. It is possible to produce a MAb specific to almost any soluble protein or cell surface target, and a large amount of research and development is being carried out to create MAbs for the treatment of diseases such as leukemia and lymphoma, rheumatoid arthritis, multiple sclerosis, and various types of solid tumors including CRC.

Characteristics

Treatment of metastatic colorectal cancer

The standard treatment of metastatic CRC is ► chemotherapy using a combination of 5-fluorouracil/leucovorin with ► irinotecan (FOLFIRI) or oxaliplatin/► 5-fluorouracil/leucovorin (FOLFOX). The treatment options have been expanded by the development of several therapeutic antibodies. In 2004, the US Food and Drug Administration (FDA) approved the use of ► bevacizumab (Avastin ®; Genentech, Inc.,

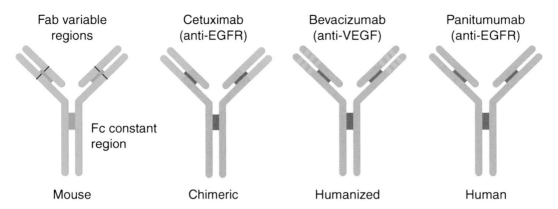

Colorectal Cancer Therapeutic Antibodies, Fig. 1 Engineered CRC therapeutic antibodies used against human CRC. Mouse protein sequence is *green* and human sequence is *blue*

South San Francisco, CA, USA) and ▶ cetuximab (Erbitux®; ImClone Systems, Inc., New York, USA). A third agent, panitumumab (Vectibix®; Amgen Inc., Thousand Oaks, CA, USA) was approved in 2007. Another agent, ramucirumab (CYRAMZA®, Eli Lilly and Company), was approved in April 2015. The main advantage of antibodies over older cytotoxic chemotherapy agents is that they more specifically target components of biologic pathways important for the growth and survival of cancer. ▶ Cetuximab and panitumumab inhibit growth-promoting signal transduction pathways in CRC cells by blocking the binding of epidermal growth factor (EGF) to its receptor (▶ epidermal growth factor receptor; EGFR). ▶ Bevacizumab inhibits the activity of ▶ vascular endothelial growth factor (VEGF), important for the development of new blood vessels (▶ angiogenesis) in tumors. Ramucirumab binds to the human vascular endothelial growth factor- receptor 2 (VEGF-R2), preventing the interaction of VEGF-R2 to its ligands. CRC therapeutic antibodies may also sensitize cancer cells to chemotherapy and are generally used in combination with FOLFOX or FOLFIRI as a first- or second-line therapy for advanced CRC.

MAbs generated by mouse hybridoma technology would cause serious side effects if injected into humans and therefore need to be altered by genetic engineering (Fig. 1). ▶ Cetuximab is a chimeric antibody generated by the insertion of the Fab variable, antigen-binding regions of the murine EGFR MAb into the human IgG1 Fc constant (framework) regions. ▶ Bevacizumab is a humanized antibody, generated by the insertion of mouse complementarity-determining regions (CDRs) into human IgG1 constant and variable domain frameworks. Panitumumab is a fully human IgG2 MAb generated using the XenoMouse, a transgenic mouse genetically engineered with a "humanized" humoral immune system by introducing almost the complete human immunoglobulin loci into the germ line of mice with inactivated mouse antibody machinery.

Cetuximab (Erbitux) and Panitumumab (Vectibix): Epidermal Growth Factor Receptor (EGFR) Inhibitors

EGFR is a membrane glycoprotein overexpressed on the surface of CRC cells in 70–80% of patients and known to be involved in carcinogenesis. Binding of EGF to EGFR induces homo- and heterodimerization of EGFR with other members of the EGF family of type I ▶ receptor tyrosine kinases, e.g., HER2 (c-Erb-B2, neu). This triggers a chain of downstream events including the activation of ▶ *KRAS*, BRAF, MAPK, ▶ PI3K, AKT, and STAT signal transduction pathways, leading to cell proliferation, ▶ angiogenesis, and survival (Fig. 2a). Treatment with ▶ cetuximab or panitumumab blocks EGF/EGFR binding, inhibiting the activation of these pathways (Fig. 2b). It slows tumor growth, ▶ invasion, ▶ metastasis, and ▶ angiogenesis and sensitizes

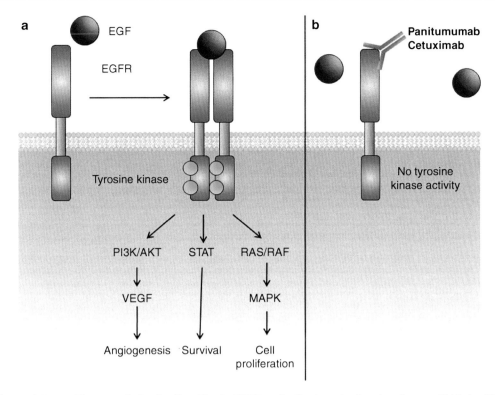

Colorectal Cancer Therapeutic Antibodies, Fig. 2 EGFR antibodies (cetuximab and panitumumab) bind to EGFR on the surface of CRC cells, preventing the binding of EGF

cancer cells to chemotherapy. In general, anti-EGFR antibodies are well tolerated, the most common side effect being a skin rash, effectively treated with a cream containing urea and vitamin K1.

▶ Cetuximab has shown promising results in patients with metastatic CRC, either as a single agent or in combination with ▶ chemotherapy, producing higher response rates and longer times to disease progression. There are no data showing that addition of cetuximab to chemotherapy significantly prolongs overall survival, but clinical trials are ongoing. Panitumumab is active against metastatic CRC when added to "best ▶ supportive care" in patients with metastatic CRC refractory to chemotherapy, with minimal side effects. Combined treatment with panitumumab and chemotherapy has also yielded promising results.

Anti-EGFR antibodies have anticancer activity in patients with CRC expressing EGFR together with "wild-type" KRAS protein. However, the therapeutic effects of anti-EGFR antibodies are abrogated in 30–40% of all CRC patients due to specific mutations in KRAS. These mutations enable constitutive EGFR signaling (Fig. 3) despite blocked EGF binding. In July 2009, the FDA limited the use of anti-EGFR MAbs to patients with CRC with wild-type KRAS.

Bevacizumab (Avastin): Vascular Endothelial Growth Factor (VEGF) Inhibitor

The therapeutic antibody ▶ bevacizumab inhibits ▶ angiogenesis by inhibiting the action of ▶ vascular endothelial growth factor (VEGF), a growth factor released by tumors, by blocking the binding of VEGF to VEGF receptor on stromal endothelial cells, thereby inhibiting the activation of signal transduction pathways leading to ▶ angiogenesis (Fig. 4). Bevacizumab was originally used for first-line treatment of patients with advanced CRC but may also be effective as second- or third-line therapy, with chemotherapy (FOLFOX or FOLFIRI). For advanced metastatic CRC, bevacizumab in combination with

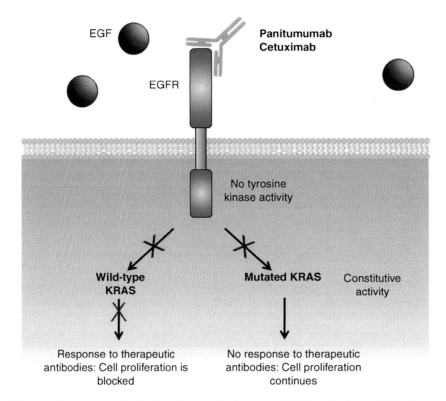

Colorectal Cancer Therapeutic Antibodies, Fig. 3 Resistance to EGFR antibody by CRC with mutant KRAS mutation is caused by constitutive activation of the KRAS/BRAF/MAPK signaling pathway

chemotherapy may extend overall survival by 33–52% compared to chemotherapy alone. However, only patients with elevated blood levels of the CRC biomarker CA19-9 benefit from bevacizumab, compared to chemotherapy alone. In patients with early stage CRC, treatment with bevacizumab plus chemotherapy after tumor resection does not improve disease-free survival.

Adverse side effects of bevacizumab are generally minor and reversible, but some patients may suffer from shortness of breath, dizziness, high blood pressure, weight loss, muscle aches and pains, and occasionally gastrointestinal perforation, hemorrhage, proteinuria, thromboembolism, or congestive heart failure. Combination of radiotherapy with bevacizumab has produced promising results for locally advanced and recurrent CRC. However, the combined use of bevacizumab with anti-EGFR antibodies and chemotherapy may be harmful, with a decrease in progression-free survival and a poorer quality of life.

Ramucirumab (CYRAMZA): Vascular Endothelial Growth Factor- Receptor 2 (VEGF-R2) inhibitor
Ramucirumab is a recombinant human monoclonal IgG1 antibody that inhibits angiogenesis by binding to VEGF-R2, preventing the interaction of VEGF-R2 with its ligands. It has been used in combination with FOLFIRI for the treatment of patients with metastatic CRC, whose disease had progressed during or after first-line treatment with bevacizumab, oxaliplatin and fluoropyrimidine. Patient survival was improved regardless of tumor KRAS mutation status, supporting the use of ramucirumab for second-line therapy to continue the inhibition of the VEGF-VEGF-R pathway. The most common grade ≥3 adverse events included neutropenia, hypertension, diarrhea and fatigue.

Other Antibodies with Anti-CRC Activity
HuA33 is a humanized MAb directed against the A33 protein, expressed at high levels in 95% of CRC metastases with only restricted expression in normal colonic mucosa. ▶ Radioimmunotherapy

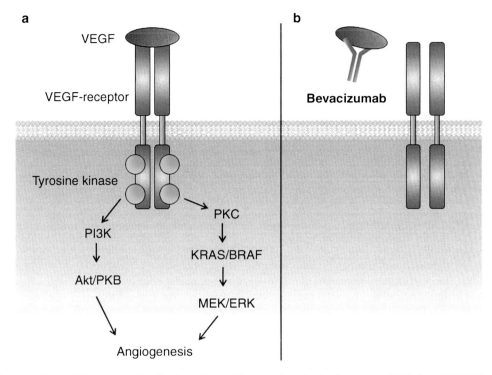

Colorectal Cancer Therapeutic Antibodies, Fig. 4 Therapeutic antibody (bevacizumab) binds to VEGF, blocking binding to the VEGF receptor and activation of signal transduction leading to angiogenesis

using 131 iodine-coupled huA33 (131I-huA33) shows promise in targeting CRC without harmful side effects. HuA33 has also been used in combination with chemotherapy in Phase I clinical trials for patients with advanced CRC, but no data are yet available on the efficacy of this combination. Several other novel CRC-targeting antibodies, e.g., chimeric SC104 and RAV12, have shown anticancer activity in preclinical studies and are undergoing further evaluation.

Conclusions

Anti-EGFR (cetuximab and panitumumab) and anti-VEGF (Bevacizumab) therapeutic antibodies exert their anticancer activity by blocking biological pathways important for the growth and survival of CRC. The benefits of these therapeutic antibodies are augmented by concurrent administration of chemotherapy or radiotherapy. However, improvements in patient outcomes in randomized trials have been more modest than anticipated. This may be due to tumor heterogeneity within patient cohorts, as demonstrated for tumor marker expression (e.g., CA19-9) and the mutational status of signaling pathways (e.g., KRAS). A better understanding of the molecular mechanisms that drive CRC and the genetic makeup of each patient may facilitate the selection of the most appropriate treatment. In the meantime, research and development of new antibodies for CRC continues.

Cross-References

▶ Akt Signal Transduction Pathway
▶ PI3K Signaling

References

Carter PJ (2006) Potent antibody therapeutics by design. Nat Rev Immunol 6:343–357

Ciprotti M, Chong G, Gan HK, Chan A, Murone C, MacGregor D, Lee FT, Johns TG, Heath JK, Ernst M, Burgess AW, Scott AM (2014) Quantitative intratumoural microdistribution and kinetics of (131) I-huA33 antibody in patients with colorectal carcinoma. EJNMMI research 4:22

Ferrara N, Hillan KJ, Novotny W (2005) Bevacizumab (Avastin), a humanized anti-VEGF monoclonal antibody for cancer therapy. Biochem Biophys Res Commun 333:328–335

Lievre A, Samalin E, Mitry E, Assenat E, Boyer-Gestin C, Lepere C, Bachet JB, Portales F, Vaillant JN, Ychou M, Rougier P (2009) Bevacizumab plus FOLFIRI or FOLFOX in chemotherapy-refractory patients with metastatic colorectal cancer: a retrospective study. BMC Cancer 9:347

Martinelli E, De Palma R, Orditura M, De Vita F, Ciardiello F (2009) Anti-epidermal growth factor receptor monoclonal antibodies in cancer therapy. Clin Exp Immunol 158:1–9

Siena S, Sartore-Bianchi A, Di Nicolantonio F, Balfour J, Bardelli A (2009) Biomarkers predicting clinical outcome of epidermal growth factor receptor-targeted therapy in metastatic colorectal cancer. J Natl Cancer Inst 101:1308–1324

Tabernero J, Yoshino T, Cohn AL, Obermannova R, Bodoky G, Garcia-Carbonero R, Ciuleanu TE, Portnoy DC, Van Cutsem E, Grothey A, Prausova J, Garcia-Alfonso P, Yamazaki K, Clingan PR, Lonardi S, Kim TW, Simms L, Chang SC, Nasroulah F (2015) Ramucirumab versus placebo in combination with second-line FOLFIRI in patients with metastatic colorectal carcinoma that progressed during or after first-line therapy with bevacizumab, oxaliplatin, and a fluoropyrimidine (RAISE): a randomised, double-blind, multicentre, phase 3 study. The Lancet. Oncology 16:499–508

Colorectal Cancer Vaccine Therapy

Peter M. Schlag
Comprehensive Cancer Center, Charité Campus
Mitte, Berlin, Germany

Definition

The term "cancer vaccination" (CV) (vaccine therapy) or "active specific immunization" (ASI) subsumes various strategies to induce an effective immune response against tumor cells. Like vaccinations against infectious pathogens, all strategies are based on the presentation of typical antigens (tumor-associated antigen (TAA)) to the immune system in a context favoring the induction of a cellular and, as some investigators propose, also a humoral immune response against the antigen-bearing cells (▸ cancer vaccines). The antigen has to be presented within a major histocompatibility complex (MHC) class I molecule (signal 1) in the presence of co-stimulatory molecules (signal 2) and in the context of a danger signal (mostly ▸ toll-like receptor (TLR)-mediated signals) and/or of soluble factors favoring the development and the polarization of the immune response (e.g., various types of ▸ cytokine). Several approaches have been developed and applied in ASI against colorectal cancer in order to fulfill these requirements.

Characteristics

Treatment of ▸ colorectal cancer relies actually on the three major columns: surgery, chemotherapy (chemotherapy of cancer), and radiotherapy (colon cancer clinical oncology). However, these treatment options are limited by the low efficiency against disseminated/residual tumor cells and by the lack of specificity as well as by toxicity, respectively. With the expansion of our understanding of tumor immunology and the development of increasingly sophisticated immunological methods, the development of potent ▸ immunotherapy strategies against colorectal cancer became conceivable as a further treatment option.

Since the second half of the last century, there is growing evidence from animal studies as well as from epidemiological surveys on human tumors that the immune system is implicated in the defense of the organism against malignant disease, finally leading to the creation of the concept of immunosurveillance by Burnet in 1970 (Burnet 1970). During the following decades it became increasingly clear that tumors dispose of various strategies to circumvent immunological mechanisms of tumor destruction and rejection. In advanced stages, tumors may even actively suppress an effective antitumoral immune response by producing immunosuppressive cytokines (▸ transforming growth factor beta, interleukin-10) or the expression of lymphotoxic molecules (e.g., Fas ligand). A strategy to counteract and to overcome tumor ▸ immune escape is

Colorectal Cancer Vaccine Therapy, Table 1 Clinical phase II/III vaccine trials in colorectal cancer patients in an adjuvant setting

Vaccine type	Trial type	Vaccine/adjuvant	Cancer type/ disease stage	Patient numbers	Clinical outcome	Author/ year/ references
Whole tumor cells	RCCT	Autologous TC/BCG	Colon; rectum/ stages II and III	80	Colon cancer: DFS↑ and OS↑ Rectal cancer: no effect	Hoover et al. (1993)
	RCCT	Autologous TC/BCG	Colon/ stages II and III	254	In stage II patients: DFS↑ In stage II patients: no effect	Vermorken et al. (1999)
	RCCT	Autologous TC/BCG	Colon/ stages II and III	412	No statistically significant effect	Harris et al. (2000)
	RCCT	Autologous TC/NDV	Colon; rectum/ stages I–IV	567	MSP↑	Liang et al. (2003)
	RCCT	Autologous TC/NDV	Colon; rectum/ metastatic	50	Colon cancer: DFS↑ and OS↑ Rectal cancer: no effect	Schulze et al. (2009)
Anti-idiotype vaccination	RCCT	Anti-id mimicking EpCAM/Alumn	Colon; rectum/ metastatic	39	No statistically significant effect, only in immunological responders: OS↑	Samonigg et al. (1999)
	RCCT	Anti-id mimicking EpCAM/Alumn	Colon; rectum/ stage IV	54	No statistically significant effect, only in immunological responders: OS	Loibner et al. (2004)
	PMIT	Anti-id mimicking CEA and human milk fat globule	Colon; rectum/ metastatic	52	No statistically significant effect	Posner et al. (2008)

RCCT randomized controlled trial, *PMIT* prospective multi-institutional trial, *TC* tumor cell, *BCG* bacillus Calmette-Guérin, *NDV* newcastle disease virus, *DFS* disease-free survival, *OS* overall survival, *MSP* median survival period

active immunization of patients against their tumor

Whole-Cell Vaccines (WCV)

WCV are composed of living but inactivated autologous tumor cells of the patient. The major advantage is the presence of multiple tumor antigens on the tumor cell without prior need to identify individual TAAs. On the other side, the absence of defined TAAs complicates a sophisticated monitoring of the induced immune response. However, as yet, the majority of randomized controlled clinical ASI trials are based on this strategy. Inactivated tumor cells were mixed with various types of adjuvant, e.g., ▶ bacillus Calmette-Guérin (BCG) or Newcastle disease virus (NDV) (Table 1). Using inactivated autologous tumor cells mixed with BCG, Vermorken et al. reported a significant prolongation of recurrence-free survival and a trend toward improved overall survival in a phase III trial including 254 colon cancer patients (Vermorken et al. 1999). A randomized phase III trial by Hoover et al. assessing the value of active specific vaccination with a tumor cell (BCG) vaccine in 80 stage II and III colorectal cancer patients showed a significant overall survival and disease-free survival benefit only in colon cancer patients, whereas no benefit was seen in patients with rectal cancer (Hoover et al. 1993). However, the Eastern Cooperative Oncology Group Study E5283, which was designed on the basis of the results of the Hoover study, failed to detect a significant clinical benefit of ASI after surgical resection of stage II or III colon cancer in 412 patients (Harris et al. 2000). Liang et al. reported an increased overall survival in the vaccine arm in a trial comprising 567 patients suffering from stage I–IV colorectal cancer vaccinated with inactivated autologous tumor cells infected with NDV after surgical resection of the primary tumor (Liang et al. 2003). Finally, in a phase III trial including 50 colorectal cancer patients with liver metastasis, Schulze et al. found an increased overall survival in colon but not in rectum cancer patients after resection of liver metastasis and vaccination with NDV-transformed inactivated tumor cells (Schulze et al. 2009). In total, these trials suggest that selected patient populations with colorectal cancer may benefit from ASI. Since the monitoring of the immune response in these studies was often limited to the assessment of the DTH reaction, there is actually little data for a prospective identification of patients with a relevant antitumoral response.

Heat Shock Proteins (HSP)

Heat shock proteins (heat shock protein 70, ► Hsp90) are intracellular proteins acting as a chaperone and thus harboring a broad range of cellular proteins. ► Dendritic cells (DCs) possess a special HSP receptor (CD91; ► CD antigens). Engagement of this receptor induces activation of the DC and effective processing and presentation of the associated tumor cell protein. As WCV, isolated HSP are a polyvalent vaccine and do not require previous identification of individual TAA. HSP-based vaccines have shown promising results in animal models and phase I human trials, but no phase III trials in colorectal cancer patients have been reported yet.

Peptide and Protein Vaccines

Peptide and protein vaccination strategies (► peptide vaccines for cancer) require the prior identification of an appropriate TAA. Peptides are presented within their specific MHC class I complex. Therefore, the success of a peptide vaccine in an individual patient depends not only on the expression of this peptide on the tumor but also on the presence of a permissive MHC haplotype. These constraints limit the potentially universal use of an identified peptide TAA as vaccine to those patients with the adequate HLA genotype. Several TAA peptides suitable for colorectal cancer patients have been identified, but their efficacy in human patients has not yet been assessed in phase III trials. In contrast to peptides, protein TAA must be cleaved and processed by the antigen-presenting cell before presentation in the MHC complex. Thus, they do not require a specific HLA type. Furthermore, a defined protein can give rise to multiple epitopes inducing the antitumoral immune response. Although this type of ASI has been largely applied in other tumor entities, only little data has been generated in colorectal cancer patients, and no clinical phase III trials are reported to date. The major advantage of both peptide and protein ASI is that the defined TAA used for vaccination allows for easy and relevant monitoring of the immune response.

Anti-idiotype Vaccines

According to the conventional network theory, in response to the administration of a TAA-directed antibody (Ab1), a mirror-image antibody (Ab2) directed against the paratope of the initial antibody will be generated that thus resembles strongly to the initial TAA. This anti-idiotype antibody can now be used for vaccination instead of the naturally occurring epitope (► idiotype vaccination). This method allows for relatively easy vaccine production also against nonprotein targets. Anti-idiotype antibodies mimicking various TAAs (e.g., ► EpCAM, CD55 (decay-accelerating factor), CEA (► CEA gene family)) have been tested in colorectal cancer in several clinical studies with variable clinical success (Table 1).

DNA Vaccines

DNA vaccines (► DNA vaccination) consist of a bacterial expression plasmid containing the coding sequence of a specific TAA. As a theoretical advantage, coding sequences for immunological adjuvants, e.g., cytokines and co-stimulatory molecules, or oligo-CpG can be added on the plasmid in order to enhance the elicited immune response. Phase III studies assessing the effect of this approach in colorectal cancer patients have not yet been reported.

Recombinant Virus Vaccines

The observation that virus proteins are presented within the MHC class I molecules on the surface of infected cells leads to the usage of viruses as TAA vectors. Viral vectors based on ► adenovirus or vaccinia virus can directly infect and activate antigen-presenting cells. The design of recombinant viral vectors can include not only sequences of TAAs but also those encoding co-stimulatory molecules and immunological adjuvants. Despite of these theoretically appealing assumptions, the clinical testing of viral vector-based vaccination strategies showed up to now only disappointing clinical results in colorectal cancer patients.

Dendritic Cell-Based Vaccines

► Dendritic cells (DCs) are the most potent antigen-presenting cells. By providing co-stimulatory molecules and by secreting soluble factors like interleukin-12, they are crucially implicated in the generation of the adaptive immune response (► adaptive immunity). TAA-pulsed DCs and DCs fused to tumor cells have been used in order to immunize colorectal cancer patients in small clinical phase I and II trials without convincing clinical effect. However, generation of larger quantities of DCs required for DC vaccination protocols is labor and cost intensive and has as yet precluded larger phase III clinical studies.

Outlook

The optimal clinical setting for ASI seems to be the adjuvant setting. A review assessing ASI in colorectal cancer patients with a measurable tumor burden revealed an objective clinical response rate of <1% and signs of immunological response in 50% of patients (Nagorsen and Thiel 2006). In sharp contrast, very promising clinical results have been obtained with WCV and to a certain degree with anti-idiotype vaccines in

colorectal cancer patients in the adjuvant setting. This indicates that patients with maximum benefit are probably those with totally resected disease but a high risk of tumor recurrence. The promising results of phase III studies with WCV and anti-idiotype vaccinations and the interesting results of preclinical studies involving other ASI protocols give new ground for optimism. With our increasing understanding of the mechanisms of peripheral tolerance induction, of the role of ► regulatory T cells, and of the innate immune systems in the induction of an antitumoral immune response and with the identification of several types of good immunological surrogate marker (► surrogate endpoint) for therapeutic monitoring and improved patient selection, ASI will acquire greater clinical effectiveness.

References

Burnet FM (1970) The concept of immunological surveillance. Prog Exp Tumor Res 13:1

Harris JE, Ryan L, Hoover HC Jr, Stuart RK, Oken MM, Benson AB 3rd, Mansour E, Haller DG, Manola J, Hanna MG Jr (2000) Adjuvant active specific immunotherapy for stage II and III colon cancer with an autologous tumor cell vaccine: Eastern Cooperative Oncology Group Study E5283. J Clin Oncol 18:148

Hoover HC Jr, Brandhorst JS, Peters LC, Surdyke MG, Takeshita Y, Madariaga J, Muenz LR, Hanna MG Jr (1993) Adjuvant active specific immunotherapy for human colorectal cancer: 6.5-year median follow-up of a phase III prospectively randomized trial. J Clin Oncol 11:390

Liang W, Wang H, Sun TM, Yao WQ, Chen LL, Jin Y, Li CL, Meng FJ (2003) Application of autologous tumor cell vaccine and NDV vaccine in treatment of tumors of digestive tract. World J Gastroenterol 9:495

Loibner H, Eckert H, Eller N, Groiss F, Himmler G, Rosenkaimer F, Salzberg M, Samonigg H, Schuster M, Settaf A (2004) A randomized placebo-controlled phase II study with the cancer vaccine IGN101 in patients with epithelial solid organ tumors (IGN101/2-01). J Clin Oncol (Meeting Abstracts) 22:2619

Nagorsen D, Thiel E (2006) Clinical and immunologic responses to active specific cancer vaccines in human colorectal cancer. Clin Cancer Res 12:3064

Posner MC, Niedzwiecki D, Venook AP, Hollis DR, Kindler HL, Martin EW, Schilsky RL, Goldberg RM (2008) A phase II prospective multi-institutional trial of adjuvant active specific immunotherapy following curative resection of colorectal cancer hepatic metastases: cancer and leukemia group B study 89903. Ann Surg Oncol 15:158

Samonigg H, Wilders-Truschnig M, Kuss I, Plot R, Stoger H, Schmid M, Bauernhofer T, Tiran A, Pieber T, Havelec L, Loibner H (1999) A double-blind randomized-phase II trial comparing immunization with antiidiotype goat antibody vaccine SCV 106 versus unspecific goat antibodies in patients with metastatic colorectal cancer. J Immunother 22:481

Schulze T, Kemmner W, Weitz J, Wernecke KD, Schirrmacher V, Schlag PM (2009) Efficiency of adjuvant active specific immunization with Newcastle disease virus modified tumor cells in colorectal cancer patients following resection of liver metastases: results of a prospective randomized trial. Cancer Immunol Immunother 58:61

Vermorken JB, Claessen AM, van Tinteren H, Gall HE, Ezinga R, Meijer S, Scheper RJ, Meijer CJ, Bloemena E, Ransom JH, Hanna MG Jr, Pinedo HM (1999) Active specific immunotherapy for stage II and stage III human colon cancer: a randomised trial. Lancet 353:345

Combinatorial Cancer Therapy

Lokesh Dalasanur Nagaprashantha
City of Hope National Medical Center, Duarte, CA, USA

Synonyms

Multimodality cancer therapy

Definition

The evidence-based combination of therapeutic interventions to optimize the mechanistic spectrum and specificity of cancer therapy to achieve effective tumor clearance, prevent recurrence and minimize off-target cytotoxicity.

Characteristics

Introduction

A tumor in any part of the body can be characterized by six classical hallmarks.

- Self-sufficiency in growth signals
- Insensitivity to antigrowth signals
- Limitless potential for cell division
- Evasion of ► apoptosis or resistance to cell death
- Sustained ► angiogenesis or the ability to form blood vessels
- Tissue ► invasion and ► metastasis, or distant spread

These hallmark features collectively differentiate the tumor from its normal surrounding tissues. Hence, the initial interventional approaches were aimed at targeting these hallmarks in cancers or inhibiting the expression and activity of individual molecules that contribute to acquisition and maintenance of one or more of these six hallmarks in respective cancer cells. Given its presence as an abnormal and harmful tissue in the body, surgical removal or "surgical therapy" of the tumor from its original and metastatic sites was one of the fundamental approaches employed to remove the tumor present in surgically accessible parts of the body. Further pharmacological inhibition of the expression and activity of specific proteins that are selectively expressed or overexpressed in tumors, relative to respective normal tissues, leads to the evolution of "chemotherapy," while radiological ablation of the tumor cells gave rise to "radiotherapy."

Each of the disciplines of conventional interventions in the management of cancers served their purpose varyingly depending on the nature of targeted tumor, local or distant spread, time of diagnosis, and other highly variable factors that determine patient compliance like age and presence of associated comorbid diseases. An essential feature of the tumors is their inherent ability to rapidly adapt to therapeutic interventions resulting in the recurrence of tumors and emergence of drug resistance. These factors necessitated more dynamic and interdisciplinary approaches that would maximize the antitumor effects while minimizing or desirably abolishing the ability of tumor cells to develop resistance to clinical interventions. The efforts at the development of such integrated approaches were greatly facilitated by rapid advances in the understanding of distinct profiles corresponding to the gross, histological and molecular behavior of various organ tumors. Thus, based on an integrated and multidisciplinary approach to serve the clinical goal of enhanced or complete tumor clearance, the age of "combinatorial cancer therapy" began.

Major Types of Combinatorial Cancer Therapy

Combinatorial Cancer Chemotherapy

The advances in the fields of molecular biology, pharmaceutical technology and identification of various compounds capable of targeting specific tumor molecules as discovered during the late part of twentieth century and the early phase of twenty-first century provided great thrust to the development of combinatorial chemotherapy. Consequently, many classes of anticancer agents like alkylating agents (e.g., oxaliplatin), antimetabolites (e.g., mercaptopurine), topoisomerase inhibitors (e.g., etoposide), growth factor receptor antibodies (e.g., Herceptin – epidermal growth factor antibody), and multi-specific tyrosine kinase inhibitors (e.g., sorafenib) were developed to target enabling proteins of specific importance in many tumors. In many cases, repeated doses of respective drugs were required which lead to the induction and potentiation of off-target effects on nonmalignant cells leading to commonly reported and dose-limiting effects of impaired immune function, loss of hair, bone loss, and severe toxicity to heart, brain, kidneys, and liver.

Clinical studies revealed that the combination of cyclophosphamide, methotrexate, and 5- fluorouracil [CMF regimen] provided better response which was further improved consequent to treatment with doxorubicin along with CMF regimen in breast cancer. Docetaxel and cisplatin together provided better response in head and neck squamous cell cancers than either of the single drugs alone. Weekly administration of relatively low doses of paclitaxel and carboplatin offered better response compared to three-weekly schedules of same drugs at high doses of in the head and neck cancers.

An important observation during the course of chemotherapy was that the tumors exposed to anticancer drugs developed resistance to subsequent treatment with same agents. The development of resistance to initial chemotherapy drugs was mediated by multiple factors. A major contributing factor for such multidrug resistance was the ability of tumors to rapidly export the anticancer agents out of the cells and activate parallel signaling proteins, so that the dependency of tumors on targeted protein/pathway is no more essential for tumor survival. In the quest to enhance the efficacy of chemotherapy, researchers started working on strategies to inhibit drug-transport proteins like P-glycoprotein and RLIP76 which are overexpressed in multiple cancers. The studies on the combination of drug-transport protein inhibitors and chemotherapy drugs are promising and awaiting large-scale testing in human populations.

With extensive investigations on gene mutations, the dependency of tumors on specific proteins and pathways have facilitated the development of advanced classes of chemotherapy drugs. Multi-specific tyrosine kinase inhibitors like sorafenib target multiple tyrosine kinase proteins that are commonly associated with driving the proliferative pathways of tumors. Melanomas overexpress a mutant form of $BRAF^{V600E}$, which has glutamine in the place of valine at codon 600. Vemurafenib is a drug that inhibits $BRAF^{V600E}$ thereby specifically targeting melanomas. Crizotinib, an approved drug, is known to selectively inhibit lung tumors with abnormal activation of ALK gene that is associated with tumor progression. Ipilimumab is a human monoclonal antibody which is known to activate immune system and has been approved by the US Food and Drug Administration (FDA) in the year 2011 for melanomas. Though various advanced classes chemotherapy drugs like sorafenib, vemurafenib, crizotinib, and ipilimumab have been developed and marketed, the dynamic adaptability of tumors is a factor which necessitates network-targeted therapies.

In order to evolve targeted and personalized therapies, research teams with multidisciplinary expertise are striving to characterize the networks of tumor dependency in respective cancers. This area of research represents an advanced and contemporary focus in combinatorial chemotherapy as it aims to integrate the various drugs into well-directed and tailored regimens that can be dynamically calibrated in terms of dosage, timing, and therapy response based on changing molecular profile of tumors in respective individuals over the course of treatment. In order to achieve this complex yet highly essential and beneficial task, various studies have been focusing on characterizing the global alternations in the signaling profiles of tumors in specific clinical types and stages using advanced tools including mass spectrometry-based proteomics and computer-guided network analyses. Such studies have yielded the characteristic patterns of up- and downregulation of critical signaling nodes and their interactions with cellular regulators of proliferation and drug resistance in specific tumors.

Based on the specific primary signaling proteins overexpressed and possible feedback nodes activated consequent to targeting a primary signaling protein in tumors, various networks have been identified as potential therapeutic targets. The mTOR signaling network which regulates the proliferative and apoptotic signals from growth factor receptors, energy sensor proteins like AMP kinase and PI3K/Akt pathway is an example of contemporary focus in network-targeted therapies. The preclinical studies with a novel EGFR ½ (that drives tumor proliferation) and VEGFR ½ (that drives tumor blood vessel formation) inhibitor called AEE788 have shown significant benefits in combination with the mTOR inhibitor RAD 001 (first approved by FDA in 2009 for cancer therapy) in advanced brain tumors called glioblastomas. The mTOR inhibition can result in the feedback activation of MEK in MAPK kinase pathway. Studies have shown that combination of mTOR inhibitors with MEK inhibitors together provides better antiproliferative effects in glioblastomas. Further translational advances in network-targeted therapies have immense promise as future and advanced strategies of combinatorial chemotherapy.

Combined Modality Cancer Therapy

In the initial period of clinical management of cancers, surgical therapy, chemotherapy, and radiotherapy were developed and practiced as possibly exclusive modes of targeting different types of cancers. The first initiatives toward interdisciplinary approaches were encouraged by experimental trials which showed that the combination of interventions at specific intervals during the course of treatment could be more beneficial compared to treatment with any single modality of cancer therapy. Thus, the strategy of preoperative chemotherapy/radiotherapy (neoadjuvant therapy to shrink the tumor), surgery, and postoperative chemotherapy/radiotherapy (adjuvant therapy to kill any residual tumor cells) often collectively called "combined modality cancer therapy" has been practiced in different stages of clinical cancer care.

Combined modality therapy was initially pioneered in the field of childhood tumors like Wilms' tumors, which showed significantly improved therapeutic response to radiotherapy and chemotherapy following surgical resection of primary tumors. Later, the combined modality therapy was applied to various cancers like lung cancer, colon cancer metastatic to the liver, and nasopharyngeal carcinoma. With the advances in the drug development, surgical and radiological interventions, various combinations of combined modality therapy are being investigated and employed to achieve effective tumor clearance and control. The encapsulation of chemotherapy drugs in magnetic nanoparticles which enable the targeting of chemotherapy drugs specifically toward tumors following application of external magnetic fields is one of the emerging areas of interventions. The development of intensity-modulated radiotherapy (IMRT) using photons has led to precise radiation dose calibration and specific targeting of tumor volumes, particularly in the head and neck cancers and prostate cancers, sparing surrounding normal tissues. The combination of gold-coated nanoparticles with radio-sensitizer drugs like doxorubicin is being investigated for image-guided radiotherapy (IGRT) which can combine the benefits of chemotherapy and targeted radiotherapy. In a study on 29 patients with locoregionally advanced nasopharyngeal carcinoma who underwent initial treatment with the EGFR antibody cetuximab

7–10 days before receiving concurrent IMRT, weekly cisplatin and cetuximab, 25 patients showed a complete response corresponding to a 96% response rate during therapy. Intensity-modulated proton therapy (IMPT) is another area of investigations which has shown promise in advanced non-small cell lung cancer. In one of the studies, concurrent administration of IMPT with chemotherapy in locally advanced non-small cell lung cancer had less off-target toxicities like reduction in neutrophils, lymphocytes, and hemoglobin levels compared to IMRT and chemotherapy.

Conclusion

Given the complexity of contributing factors that drive the tumor progression and the inherent nature of tumors to continuously adapt to changing microenvironment in which they live, further evolution of combinatorial cancer therapy using multi-targeting drugs, improved surgical techniques, and advanced modes of radiological ablation will greatly serve to enhance the response to clinical interventions. The integration of such rapid advances in the areas of chemotherapy, surgical therapy, and radiotherapy of various types of tumors is a current focus of translational cancer research where leading multidisciplinary experts are striving to develop effective combinatorial cancer therapies.

Cross-References

► Angiogenesis
► Apoptosis
► Invasion
► Metastasis

References

Arrigo S et al (1990) Better survival after combined modality care for adults with Wilms' tumor. A report from the National Wilms' Tumor Study. Cancer 66:827–830
Awasthi S et al (2000) Novel function of human RLIP76: ATP-dependent transport of glutathione conjugates and doxorubicin. Biochemistry 39:9327–9334
Bonadonna G et al (1995) Sequential or alternating doxorubicin and CMF regimens in breast cancer with more than three positive nodes. Ten-year results. JAMA 273:542–547
Devita VT et al (2008) DeVita, Hellman, and Rosenberg's cancer: principles & practice of oncology, 8th edn. Lippincott Williams & Wilkins, Philadelphia
Goudar RK et al (2005) Combination therapy of inhibitors of epidermal growth factor receptor/vascular endothelial growth factor receptor 2 (AEE788) and the mammalian target of rapamycin (RAD001) offers improved glioblastoma tumor growth inhibition. Mol Cancer Ther 4:101–112
Hanahan D, Weinberg RA (2000) The hallmarks of cancer. Cell 100:57–70
Paternot S, Roger PP (2009) Combined inhibition of MEK and mammalian target of rapamycin abolishes phosphorylation of cyclin-dependent kinase 4 in glioblastoma cell lines and prevents their proliferation. Cancer Res 69:4577–4581
http://www.mdanderson.org/newsroom/news-releases/2008/proton-therapy-and-concurrent-chemotherapy-may-reduce-bone-marrow-toxicity-in-advanced-lung-cancer-patients.html

See Also

(2012) Cell Death. In: Schwab M (ed) Encyclopedia of Cancer, 3rd edn. Springer Berlin Heidelberg, p 737. doi:10.1007/978-3-642-16483-5_6724

Combinatorial Selection Methods

Michael W. Van Dyke
Department of Chemistry and Biochemistry, Kennesaw State University, Kennesaw, GA, USA

Synonyms

CASTing; In vitro genetics; REPSA; SAAB; SELEX; TDA

Definition

Combinatorial selection methods refer to a series of reiterative approaches involving large pools of randomized oligonucleotides, a selection process, and PCR amplification for identifying preferred ligand-binding sites on nucleic acid receptors.

Combinatorial Selection Methods, Fig. 1 Steps within a cycle of CASTing. (1) Primary complex formation between ligand and selection template mixture, (2) secondary complex formation following antibody binding, (3) secondary complex purification by immunoprecipitation, and (4) PCR amplification of selected DNAs. Steps 1–4 are repeated until a population of selection templates with desired properties is isolated. *Open* and *filled* regions on selection templates refer to defined and randomized sequences, respectively

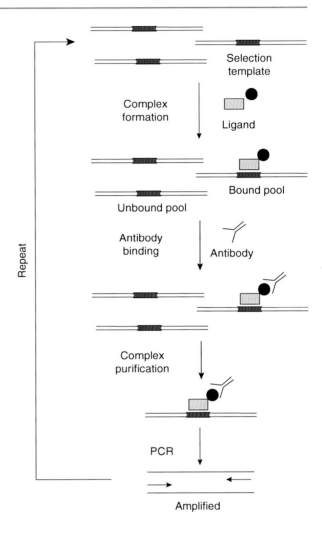

Characteristics

Combinatorial selection methods are reiterative in vitro methods used to find the preferred nucleic acid-binding sequences of many ligand types. Examples of combinatorial selection methods include cyclic amplification and selection of targets (CASTing), in vitro genetics, restriction endonuclease protection, selection, and amplification (REPSA), and systematic evolution of ligands by exponential enrichment (SELEX). Typically these combinatorial methods involve large populations of nucleic acids containing a region of randomized sequence, a selection process, a means of amplifying the selected subpopulation, and the ability to cyclically repeat selection and amplification steps to obtain

sequences that bind with high affinity to the selecting ligand. Figures 1, 2, and 3 show the basics of these combinatorial selection methods. Note: ▶ phage display, which uses randomized sequences within a bacteriophage genome to allow the expression of a variety of viral coat fusion proteins for use in the selection of peptides that interact with particular ligands, may also be formally considered a combinatorial selection method.

Ligands investigated by combinatorial selection methods include proteins, peptides, nucleic acids, and various small molecules (molecular mass <1000 Da). Receptors are usually DNA or RNA oligonucleotides. Selections are typically performed in vitro and require the physical separation of ligand-bound from unbound

Combinatorial Selection Methods, Fig. 2 Steps within a cycle of REPSA. (1) Primary complex formation between ligand and selection template mixture, (2) IISRE binding and selection template cleavage, and (3) PCR amplification of protected DNAs. Steps 1–3 are repeated until a population of cleavage-resistant selection templates emerges. *Open* and *filled* regions on selection templates refer to defined and randomized sequences, respectively

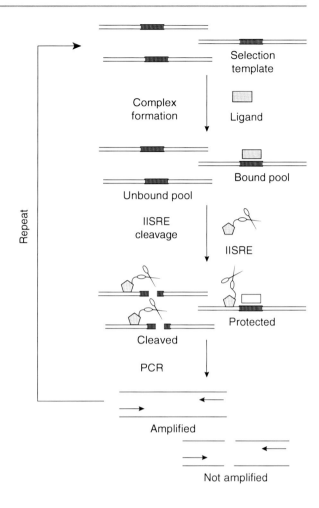

nucleic acids. Under optimal selection conditions, the selected subpopulation constitutes only a tiny fraction of the input nucleic acid. Thus, amplification of the selected nucleic acid, typically by a polymerase chain reaction (PCR) method, is necessary to acquire workable quantities of material. Amplifications can be performed under highly stringent conditions, to maintain nucleic acid sequence integrity. Alternatively, amplifications may be performed under less stringent conditions (e.g., through the use of a low-fidelity reverse transcriptase), thereby allowing the introduction of mutations that could provide even higher affinity in subsequent selection rounds. Finally, because of limitations in the selection process, often a single round of selection does not yield the highest possible affinity species. Thus, multiple cycles of selection and amplification are often

used to obtain the desired results. Progress toward isolating the highest affinity species can be determined after each cycle. Alternatively, many investigators proceed *sola fide*, with the hope that after a certain number of cycles, useful material will be obtained.

Combinatorial selection methods have been used to identify consensus DNA-binding sequences for specific transcription factors, isolate RNA aptamers for high-throughput proteomic applications, and identify preferred DNA-binding sequences for antineoplastic agents. They have become standard tools in biochemical and molecular biology research, as well as in the development of new drugs and diagnostics, especially in the cancer field. Descriptions of some of the more commonly used combinatorial selection methods and their

Combinatorial Selection Methods, Fig. 3 Steps within a cycle of SELEX. (1) Preparation of RNA transcript mixture from template DNA, (2) primary complex formation between ligand and selection transcript mixture, (3) secondary complex formation following antibody binding, (4) secondary complex purification by immunoprecipitation, (5) reverse transcription of selected RNAs, and (6) PCR amplification of selected cDNAs. Steps 1–6 are repeated until a population of selection templates with desired properties is isolated. *Open* and *filled* regions on selection templates refer to defined and randomized sequences, respectively. Raised, *rightward pointing arrows* on SELEX selection templates indicate transcription start sites

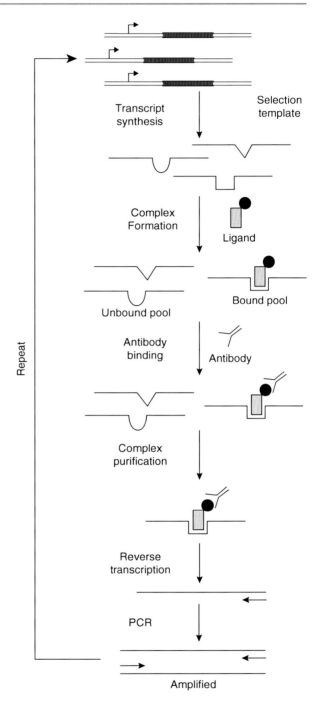

applications in cancer research and treatment are provided below.

CASTing

CASTing is an archetype for several related combinatorial selection methods, including selected and amplification binding (SAAB) and target detection assay (TDA). The steps involved in CASTing are shown in Fig. 1. Typically, these methods use randomized double-stranded DNA oligonucleotides as the receptor or selection template and proteins as ligands, although other ligands including small molecules and nucleic acids have also been used successfully. Selection templates include a central element (often 6–20 bp in length) containing a degenerate or randomized sequence flanked by two defined sequences of suitable length and base composition to allow for efficient amplification by PCR. Binding reactions are performed in vitro between a ligand and a population of selection templates sufficient for a good representation of all possible sequences. Ligand-bound oligonucleotides are then physically separated from unbound oligonucleotides by various methods, either through capitalizing on the different physical properties of ligand-DNA complexes compared with free oligonucleotides and standard biochemical techniques (e.g., reduced electrophoretic mobility and electrophoretic mobility shift assays (EMSA), increased hydrophobicity, and nitrocellulose filter binding) or through affinity methods (e.g., immunoprecipitation, affinity chromatography). Of course, application of these separation methods requires a priori knowledge of the ligand and its DNA complex and/or physical modification of the ligand (e.g., epitope-tagged proteins, biotinylated small molecules). Amplification of the selected oligonucleotides is achieved through direct PCR. Cycles of binding, separation, and amplification are often repeated four to eight times, depending on the length of the randomized region within the selection template and the efficiency of the selection process. The resulting selected oligonucleotides can then be individually subcloned and sequenced or the entire pool directly sequenced, and the resulting information can be used to derive a consensus.

Combinatorial selection methods such as CASTing have been used primarily to determine the specific DNA-binding sites of proteins such as eukaryotic transcription factors. Many of these proteins play important roles in cancer as ▶ oncogenes or as ▶ tumor suppressors (e.g., c-myc, ▶ p53).

REPSA

REPSA is a combinatorial selection method similar to CASTing in that it identifies preferred ligand-binding sites on double-stranded DNAs. The steps involved in REPSA are shown in Fig. 2. As in CASTing, a REPSA selection template also contains an element of randomized sequence flanked by two defined sequences. However, in REPSA selection templates, the defined flanking sequences also contain type IIS restriction endonuclease (IISRE) binding sites oriented so that the IISREs cleave sites within the randomized cassette. Type IISREs differ from conventional type II restriction endonucleases in that they do not cleave DNA directly at their binding site but rather at a fixed distance from their binding site. They also cleave duplex DNA without regard to sequence specificity; thus, they are powerful probes of ligand binding in the randomized cassette. After binding reactions are performed, the mixture of ligand, selection templates, and their complexes is subjected to cleavage by an IISRE. Unbound selection templates are preferentially cleaved, rendering them incapable of serving as templates in subsequent PCR amplifications. Intact templates are amplified, however, and serve as the input of reiterative rounds of selection, cleavage, and amplification. Subcloning, sequencing, and analysis are identical to those in CASTing.

Because REPSA does not require physical separation between ligand-bound and unbound selection templates, it is more versatile than many other combinatorial selection methods. Almost any ligand that can inhibit IISRE cleavage is suitable for use with REPSA. Ligands including proteins, nucleic acids, and both noncovalent- and covalent-binding small molecules have had their preferred duplex DNA-binding sequences successfully determined by REPSA. REPSA can also be used with mixtures of ligands, with unknown and uncharacterized ligands, and with native, unmodified ligands. Given the limitations of alternative methods, REPSA has been proven most effective in determining small molecule binding specificity, especially for those modular DNA-binding small molecules used to target specific genes (e.g., hairpin polyamides). In addition, REPSA has been proven highly effective in determining the binding specificity of several ▶ small molecule drugs, including ▶ alkylating agents and topoisomerase poisons (e.g., ▶ adriamycin and ▶ irinotecan). REPSA can also be used to identify the preferred sites of antineoplastic agent (e.g., ▶ cisplatin and other platinum drugs) and carcinogen-macromolecular adducts to DNA. It is envisioned that REPSA should be useful in the development of new anticancer drugs that target specific DNA sequences and genes and in better understanding chemical carcinogenesis.

SELEX

SELEX differs from CASTing and REPSA in that it primarily uses single-stranded RNA and DNA oligonucleotides as the receptor for various ligands. These single-stranded oligonucleotides adopt sequence-dependent three-dimensional structures in solution on the basis of base pairing (both through Watson-Crick and noncanonical hydrogen bonding schemes), base stacking, and other interactions. These structures then have the potential to interact with a variety of ligands, including proteins and small molecules. Thus, SELEX affords the possibility of identifying potential receptors within a larger conformational space than usually explored by other combinatorial approaches.

The steps involved in SELEX are shown in Fig. 3. The nucleic acids used in SELEX binding reactions are usually derivatives of double-stranded DNA templates (e.g., single-stranded RNA transcripts). Thus, whereas the flanking regions of the selecting transcripts need to be of sufficient length to allow PCR amplification, the complete templates are often longer to provide information for transcript generation (e.g., a bacteriophage promoter region). In addition, since the single-stranded nucleic acids need to be of sufficient length to adopt a stable three-dimensional structure, the randomized region of SELEX templates tends to be considerably longer (20–100 nucleotides) than that of other combinatorial selection methods. An initial step in SELEX involves the production of selection transcripts from the parent template. The binding and separation steps are comparable to those in CASTing. After selection, RNA transcripts need to be converted into complementary DNA strands by reverse transcription before amplification by PCR. As with other combinatorial selection methods, cycles of transcription, binding, separation, reverse transcription, and amplification can be repeated until desired results are obtained. Note that

the long length of some SELEX randomized regions makes it highly unlikely that all possible sequences are well represented in the initial selection. Thus, either the reverse transcription or PCR amplification steps are performed with enzymes having relatively low fidelity. This allows the introduction of mutations into the selection templates, which provides the opportunity to identify even higher-affinity oligonucleotides than were present in the initial selection. Subcloning, sequencing, and analysis are comparable to those in other combinatorial selection methods.

SELEX, being a combinatorial method that uses single-stranded nucleic acids, has the unique ability to present a large variety of different conformational shapes for selection, rather than just different duplex DNA sequences. This allows SELEX to identify linear nucleic acid sequences that adopt structures capable of interacting with a variety of ligands, including those not normally believed to interact with natural nucleic acids. Oligonucleotides containing these SELEX-selected sequences are known as aptamers, and aptamers have been identified that bind to a variety of proteins and small molecule ligands with specificities and affinities rivaling those of antibodies. Thus, a considerable number of uses for aptamers have been found in microarray-based proteomic analyses, especially in medical diagnostics. In addition, although unmodified nucleic acids have a relatively short half-life in vivo, chemically modified oligonucleotides can persist for several days. Thus, modified aptamers (e.g., ▶ aptamer bioconjugates) are being developed as targeted therapeutic agents for both acute and chronic diseases, including cancer.

Cross-References

▶ Combinatorial Selection Methods
▶ Tumor Suppressor Genes

References

Gold L, Polisky B, Uhlenbeck O et al (1995) Diversity of oligonucleotide functions. Annu Rev Biochem 64:763–797

Ouellette MM, Wright WE (1995) Use of reiterative selection for defining protein–nucleic acid interactions. Curr Opin Biotechnol 6:65–72

Szostak JW (1993) In vitro genetics. TIBS 17:89–93

Van Dyke MW, Van Dyke N, Sunavala-Dossabhoy G (2007) REPSA: general combinatorial approach for identifying preferred ligand–DNA binding sequences. Methods 42:118–127

Combined Modality Treatment

▶ Chemoradiotherapy

Comet Assay

Peter Schmezer
Division Epigenomics and Cancer Risk Factors, German Cancer Research Center (DKFZ), Heidelberg, Germany

Synonyms

SCGE; Single-cell gel electrophoresis assay; Single-cell microgel electrophoresis assay

Definition

The comet assay is a sensitive electrophoresis technique for studying ▶ DNA damage and ▶ repair of DNA in individual cells. It has become one of the standard methods for testing of genotoxic stress, and it is also frequently utilized in environmental and human ▶ biomonitoring, molecular ▶ cancer epidemiology, as well as fundamental cancer research.

Characteristics

The assay can be applied to all cell types including human, animal, or plant cells, whether in culture or isolated from organs or tissues. In general, single-cell suspensions are prepared, and cells are subsequently embedded in a thin agarose gel on microscope slides. Consecutive cell lysis with detergent and high salt removes cellular and nuclear membranes and proteins and liberates DNA in the form of a compact, nucleus-like structure which is also called nucleoid. The gel-embedded DNA is then subjected to electrophoresis. DNA migrates toward the anode in a way that is dependent on its size. Thereby, DNA migration corresponds to the number of DNA lesions, i.e., increased electrophoretic migration is correlated with an increased amount of DNA damage present in the cell. DNA is stained by a fluorescent DNA-binding dye and so visualized by fluorescence microscopy as to show DNA migration. Several imaging software programs are commercially available to analyze the microscopic pictures and to quantify DNA migration (Fig. 1).

Modifications

Different modifications of the assay have been developed. The most common version applies alkaline electrophoretic conditions in concert with an alkaline pretreatment of DNA. This leads to the conversion of so-called alkaline labile sites to strand breaks and increases the spectrum of DNA lesions that can be detected. Before electrophoresis, additional incubation of DNA with damage-specific DNA repair enzymes, such as DNA glycosylases or endonucleases, can enhance both sensitivity and specificity of the assay. These enzymes recognize DNA lesions with high specificity and convert them to strand breaks which increase DNA migration. Examples are endonuclease III to detect oxidized pyrimidines, formamidopyrimidine DNA glycosylase to detect 8-oxoguanine and other altered purines, and T4 endonuclease V to detect ▶ UV radiation-induced pyrimidine dimers. Neutral electrophoresis conditions facilitate the detection of DNA double-strand breaks.

Areas of Application

Genotoxicity Testing
International guidelines have been published for the application of the comet assay in genotoxicity

**Comet Assay,
Fig. 1** Comet assay
micrographs of (**a**)
undamaged cells and (**b**)
cells with DNA damage;
DNA stained with SYBR®
Green; observation at 250×
magnification using a
fluorescence microscope;
the term comet assay refers
to the comet-like structure
present in damaged cells
after lysis and gel
electrophoresis

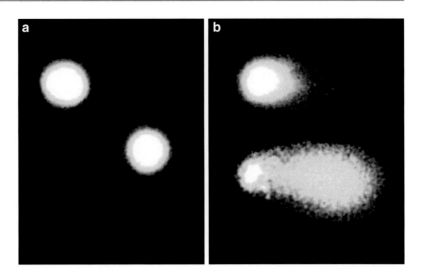

testing and biomonitoring. The assay is widely used in this research area due to several advantages such as (i) sensitivity to detect low levels of DNA damage, (ii) requirement for relatively small amounts of cells and test substances, and (iii) possibility to perform high-throughput analyses implying automated imaging. The technique is able to detect a broad spectrum of DNA damaging agents including both ionizing and UV radiation, ▸ alkylating agents, chemicals that form free radicals or ▸ adducts to DNA, and various metal compounds. Modified versions of the assay have been developed to measure the genotoxic effects of DNA–DNA or DNA–protein cross-linking agents such as ▸ cisplatin. Agents that directly induce DNA strand breaks are readily detectable. Other DNA lesions such as bulky DNA adducts, e.g., formed by ▸ polycyclic aromatic hydrocarbons, do not increase DNA migration by itself. Here, DNA breaks occur only as intermediates during their repair process when these adducts are eliminated from cellular DNA by ▸ nucleotide excision repair. These intermediate breaks are normally short-lived, especially in dividing cells. They are however effectively detectable by the comet assay when the breaks are accumulated by delaying DNA synthesis with specific inhibitors.

DNA Repair

The comet assay is increasingly used to measure DNA repair. Here, the removal of afore induced DNA lesions is monitored over time, and both clearance of alkaline labile sites and repair-mediated rejoining of strand breaks can be observed as a decrease in DNA migration. The repair activity of intact cells as well as of cell extracts has been successfully analyzed. The repair of specific DNA lesions can be followed (i) by utilizing agents that induce a well-characterized type of DNA damage and (ii) by pretreatment of the gel-embedded nucleoids with lesion-specific enzymes. An important and prominent example is the detection of 8-oxoguanine, a mutagenic base by-product which occurs as a result of exposure to reactive oxygen. Studying repair of oxidized bases by pretreatment of nucleoids with the enzyme Ogg1 (8-oxoguanine DNA N-glycosylase 1) or its bacterial counterpart Fpg (formamidopyrimidine DNA glycosylase) has revealed considerable variation among subjects. Furthermore, the application of the comet assay in human intervention trials showed that the level of oxidative DNA damage can be modulated, e.g., the supplementation of diet with antioxidant-rich fruit increased the antioxidant status of lymphocytes and enhanced DNA repair activity. The comet assay as described provides evidence of DNA damage and repair in the whole genome of the analyzed cell. These measurements can also be focused on specific genomic regions when the analysis is combined with fluorescence in situ hybridization (FISH).

Biomonitoring

The induction of DNA lesions is considered to be a crucial event in ▶ carcinogenesis and, in the case of absent or imperfect DNA repair, might lead to mutations, a key driving force in cancer development. The comet assay is applied in human biomonitoring studies, e.g., to detect genotoxic environmental or occupational exposures in human white blood cells, and it can be used as a tool to characterize hazards in risk assessment studies. The potential of the assay to detect DNA lesions in peripheral blood lymphocytes has also been exploited in cancer patients receiving antineoplastic chemo- or radiotherapy. It is however important to note that the level of DNA damage that is detected in these studies was shown to be – at least in some studies – influenced by age, gender, and a variety of additional environmental or lifestyle factors such as exposure of subjects to air pollution, sunlight, dietary components, smoking, or excessive physical exercise.

Cancer Susceptibility

In addition to its use as a ▶ biomarker of exposure to genotoxic carcinogens, it can also serve as a biomarker for cancer susceptibility. By analyzing human cell samples, e.g., peripheral blood lymphocytes, it allows estimation of interindividual differences in response to genotoxic carcinogens and facilitates the identification of susceptible subjects. When applied in molecular epidemiological studies, individual differences both in the extent of induced DNA damage (mutagen sensitivity) and in the ability to repair DNA lesions (DNA repair capacity) can be monitored. The assay was successfully utilized in such studies, e.g., to demonstrate that cells from lung cancer patients showed significantly increased mutagen sensitivity and reduced DNA repair capacity as compared to cells from control subjects. These data, and further results from studies using comparable assays, emphasize the importance of mutagen sensitivity and DNA repair capacity as host factors which are strongly associated with the risk of developing cancer and other diseases. Furthermore, family studies and studies in monozygotic and dizygotic twins provide strong and direct evidence that mutagen sensitivity is highly heritable. Overall, epidemiological studies revealed a positive and consistent association between these at risk phenotypes and cancer occurrence with an increased risk ranging from 2 to 10. Application of the comet assay, e.g., in prospective cohort studies and multi-laboratory trials, will further contribute to its validation and, if successful, will offer new possibilities to improve cancer screening programs, to prevent tumor initiation, and to intervene in tumor progression in a patient-tailored manner.

Cross-References

▶ Adducts to DNA
▶ Alkylating Agents
▶ Biomarkers in Detection of Cancer Risk Factors and in Chemoprevention
▶ Biomonitoring
▶ Cancer Epidemiology
▶ Carcinogenesis
▶ Cisplatin
▶ Clinical Cancer Biomarkers
▶ DNA Damage
▶ Nucleotide Excision Repair
▶ Polycyclic Aromatic Hydrocarbons
▶ Repair of DNA
▶ UV Radiation

References

Collins AR (2004) The comet assay for DNA damage and repair: principles, application, and limitations. Mol Biotechnol 26:249–261

Møller P (2006) The alkaline comet assay: towards validation in biomonitoring of DNA damaging exposures. Basic Clin Pharmacol Toxicol 98:336–345

Schmezer P, Rajaee-Behbahani N, Risch A et al (2001) Rapid screening assay for mutagen sensitivity and DNA repair capacity in human peripheral blood lymphocytes. Mutagenesis 16:25–30

Speit G, Hartmann A (2006) The comet assay: a sensitive genotoxicity test for the detection of DNA damage and repair. In: Henderson DS (ed) DNA repair protocols: mammalian systems. Methods in molecular biology. Humana Press, Totowa, pp 275–286

Tice RR, Agurell E, Anderson D et al (2000) Single cell gel/comet assay: guidelines for in vitro and in vivo genetic toxicology testing. Environ Mol Mutagen 35:206–221

See Also

(2012) Biomarkers. In: Schwab M (ed) Encyclopedia of cancer, 3rd edn. Springer, Berlin/Heidelberg, pp 408–409. doi:10.1007/978-3-642-16483-5_6601

(2012) Genotoxic. In: Schwab M (ed) Encyclopedia of cancer, 3rd edn. Springer, Berlin/Heidelberg, p 1540. doi:10.1007/978-3-642-16483-5_2393

Comparative Oncology

Melissa C. Paoloni[1] and Chand Khanna[2]
[1]National Cancer Institute, Center for Cancer Research, Comparative Oncology Program, Bethesda, MD, USA
[2]Comparative Oncology Program, Center for Cancer Research, National Cancer Institute, Bethesda, MD, USA

Definition

Comparative oncology is the discipline that includes spontaneous, naturally occurring cancers seen in companion (pet) animals into studies of cancer biology and therapy.

Characteristics

An underutilized group of animal models in the study of cancer biology and therapy includes companion animals, primarily dogs, which naturally develop cancer. In the United States, there are ~60–70 million pet dogs. Based on crude incidence rates, it is estimated that over one million new cases of cancer are diagnosed in pet dogs each year. Due to increasing emphasis on the human-animal bond, the pet owning public is motivated to seek out new and effective treatment options for their pet animals with cancer. This population provides a platform to study cancer biology and therapy in a natural system that can serve as an intermediary step between ▶ mouse models and human patients. Naturally occurring cancers in pet dogs and humans share many features, including histological appearance, tumor genetics, biological behavior, and response to conventional therapies. Tumor initiation and progression are influenced by the same factors for both human and canine cancer, including age, nutrition, sex, reproductive status, and environmental exposures. Several genetic alterations and molecular signaling pathways known to be important in human cancers have been defined and shown to be relevant in cancers of pet dogs.

Some malignant histologies of comparative interest include:

- Canine ▶ osteosarcoma
- Canine NH lymphoma
- Canine prostate carcinoma
- Canine mammary carcinoma
- Canine melanoma
- Canine lung carcinoma
- Canine head and neck carcinoma
- Canine soft tissue sarcoma
- Canine bladder carcinoma
- Canine renal cystadenocarcinoma

History

The value of naturally occurring cancers seen in companion animals as models of human cancer has been recognized for over 30 years. Early studies in the field of bone marrow transplantation utilized dogs with non-Hodgkin lymphoma (▶ malignant lymphoma: hallmarks and concepts) to define optimal preparatory regimens for bone marrow transplant. Since then, the activity and optimal use of a wide variety of anticancer agents have benefited from information derived from studies in these large animal naturally occurring cancer models. These have included:

- Limb-sparing techniques for osteosarcoma
- Cytotoxic chemotherapy
- Inhalation therapy for pulmonary malignancies
- Immunotherapy
  - Peptide vaccine (▶ peptide vaccines for cancer)
  - DNA vaccine
  - Cell-based immunotherapy
- Anti-angiogenic therapy
- Isolated perfusion techniques
- Small molecule inhibitors (▶ small molecule drugs)

Scientific Advancements

A recognized and long-standing weakness of comparative models was a limited opportunity to investigate the biological basis of an anticancer agent's activity or lack of activity. However, with the public release of a high-quality draft sequence covering 99% of the canine genome (2.5 billion base pairs), it is now possible to apply many of the same methodologies used to interrogate human cancer in the dog. The genome sequencing suggests that all ~19,000 genes in the dog have a similar gene in the human genome. The dog and human lineages are more similar than the rodent lineage both in terms of nucleotide divergence and rearrangements. Also the single-nucleotide polymorphism (SNP) frequency in dogs is similar to SNP frequency in the human population, even with the diversity of breed phenotypes. Thus, the genomes of dog and human are similar enough to suggest that genomic information learnt about one species can be easily transferred to and be applicable for the other.

Coupled with this more available genomic data, reductions in the cost of generating biological reagents have contributed to the development of novel investigative platforms for the study of canine tissues. Evidence of this includes the availability of commercially available canine oligonucleotide microarray, optimized conditions for proteomic studies, validated canine-specific antibodies, and characterization of human antibodies that cross-react with canine epitopes. Collectively, the opportunity now exists to conduct detailed and biologically intensive studies in dogs that have cancer that can evaluate cancer-associated target genes/proteins and pathways important in cancer biology and therapy.

Need for New Models of Drug Development

Cancer drug development (▶ drug design) is costly, linear, and inefficient. Costs associated with development incrementally rise as the path proceeds. The two most common causes of drug failure are toxicity or lack of efficacy. These failures are most costly once a drug enters phase I human trials but are even more costly if they occur after phase II trials and beyond (▶ clinical trial). It is therefore essential that "go-no-go" decisions focus on the issue of toxicity and efficacy as early in the development path as possible. An information gap has historically existed between preclinical studies and phase I human trials; however, with the development of novel noncytotoxic anticancer agents, this gap is now equally evident later in the drug development path. After successful completion of phase I human clinical trials, the design of phase II trials often has to take place without sufficient information including biological dose, schedule, and regimen for many of these novel agents. For both cytotoxic and, more importantly, novel noncytotoxic agents, additional model systems are needed. The "model" advantages of companion animal cancers provide an opportunity to integrate studies that include companion animals into the development paths of new cancer drugs. The outcome will include earlier assessment of agent activity and toxicity and the validation of biological end points and surrogate markers critical to the design of more informed phase I and phase II human clinical trials.

Comparative oncology models are well suited for integration into preclinical cancer drug development efforts for several reasons. By their nature, companion animal cancers are characterized by inter-patient and intra-tumoral heterogeneity, the development of recurrent or resistant disease, and metastasis to relevant distant sites. In these ways, companion animal cancers capture the "essence" of the problem of cancer in ways not seen in other animal model systems. The lack of gold standard treatments for canine cancer patients allow for the early and humane testing of novel therapies. The shortened life span of companion animal patients and their early metastatic failure allow rapid completion of clinical trials of novel agents. A further rationale for the use of these models in nonclinical efficacy studies is the immune competence of the host, relevant and species-concordant tumor-microenvironment interactions, spontaneous development of tumors, and more importantly spontaneous development of resistance patterns to standard therapies within an individual animal. Additional attributes of comparative oncology include the opportunity to gather serial biopsies from target and nontarget lesions and repeated body fluid collection (serum, whole blood, urine) from the same animal during exposure to an investigational agent. This serial sampling allows for the identification of tumoral and surrogate pharmacodynamics endpoints that can be uniquely correlated to respond in ways that are often deemed unacceptable in human trials (▶ preclinical testing).

Clinical Applications

Efforts to perform coordinated, multicenter preclinical cancer trials in companion (pet) dogs exist in a number of different formats allowing for the rapid evaluation of cancer drugs in biologically intensive trials. Hallmarks of comparative oncology trials include good clinical practice (GCP) trial conduct, computer-based data management and reporting, and multiple study end points. The ability to capture data contemporaneously in preclinical dog trials allows for the reporting of toxicities, if they are to occur, in a rapid and systemic fashion.

Mechanisms and procedures for integration of data from preclinical trials in tumor-bearing dogs into the regulatory pathway are currently being defined. However, the use of clinical trials using pet dogs will not only provide information important to the initiation of early human clinical trials (phase I) but also will inform the appropriate design of later development trials (phase II/phase III). It is expected, through the integration of comparative oncology modeling, that the cancer drug development path will become more informed, efficient, and less costly.

Cross-References

- ▶ Clinical Trial
- ▶ Drug Design
- ▶ Malignant Lymphoma: Hallmarks and Concepts
- ▶ Mouse Models
- ▶ Osteosarcoma
- ▶ Peptide Vaccines for Cancer
- ▶ Preclinical Testing
- ▶ Small Molecule Drugs

References

Hansen K, Khanna C (2004) Spontaneous and genetically engineered animal models; use in preclinical cancer drug development. Eur J Cancer 40:858–880

Lindblad-Toh K et al (2005) Genome sequence, comparative analysis and haplotype structure of the domestic dog. Nature 438(7069):803–819

Porrello A, Cardelli P, Spugnini EP (2006) Oncology of companion animals as a model for humans, an overview of tumor histotypes. J Exp Clin Cancer Res 25(1):97–105

Vail DM, MacEwen EG (2000) Spontaneously occurring tumors of companion animals as models for human cancer. Cancer Invest 18:781–792

Complement Cytolysis Inhibitor (Cli)

▶ Clusterin

Complement-Associated Protein SP-40 (SP-40)

▶ Clusterin

Complement-Dependent Cytotoxicity

Synonyms

Complement-mediated cytotoxicity

Definition

Complement-dependent cytotoxicity (CDC), refers to the lysis of a target cell in the presence of complement system proteins. The complement activation pathway is initiated by the binding and fixation of the first component of the complement system (CIq) to the fragment crystalline (Fc) region of a (therapeutic) antibody complexed with a cognate antigen. The end result is a membrane attack complex that generates a hole in the cell membrane, ultimately causing cell lysis and death.

Cross-References

▶ Diabody
▶ Immunotherapy
▶ Interferon-Alpha

See Also

(2012) Complement. In: Schwab M (ed) Encyclopedia of Cancer, 3rd edn. Springer Berlin Heidelberg, p 963. doi:10.1007/978-3-642-16483-5_1284.

Complement-Mediated Cytotoxicity

▶ Complement-Dependent Cytotoxicity

Compound Screen

▶ Small Molecule Screens

Conditionally Replicating Adenovirus

▶ Oncolytic Adenovirus

Conditionally Replicative Adenovirus

▶ Oncolytic Adenovirus

Confocal Laser-Scanning Microscopy In Vivo

Armin Gerger[1] and Josef Smolle[2]
[1]Department of Internal Medicine, Division of Oncology, Medical University Graz, Graz, Austria
[2]Department of Dermatology, Medical University Graz, Graz, Austria

Definition

Confocal laser-scanning microscopy (CLSM) in vivo represents a novel imaging tool that allows the noninvasive examination of skin cancer morphology in real time at a resolution for viewing microanatomic structures and individual cells.

Characteristics

In recent decades, enormous strides have been made in noninvasive imaging of cancer tissues with the development and refinement of computerized axial tomography, magnetic resonance imaging, and positron emission tomography, to name a few. Progress in noninvasive skin cancer imaging, however, has been slower than in other specialties due in part to the ease with which skin is visually examined and biopsied. Early detection of malignant skin tumors is essential and still one of the most challenging problems in clinical oncology. Although surgical excision in early stages of tumor development is almost always curative, delayed recognition of skin malignancies puts the patient at risk for destructive growth and death from disease once the tumor has progressed to competence for metastasis. The early diagnosis of malignant skin tumors by naked-eye examination, however, is still rather poor. Technological advancements have led to the development and investigation of imaging tools to provide information to the clinician that can improve the diagnostic performance for early diagnosis and assist in the management of cutaneous malignancies. Among novel noninvasive imaging techniques, CLSM stands out because of its high resolution. CLSM provides for the first time in vivo imaging of individual cancer cells and offers windows on living tissue. Investigations of CLSM for in vivo examination of human skin were first published in 1995; 2 years later, the first commercially available in vivo confocal reflectance microscope was introduced to the research community. Since then, valuable experience has been gained from research labs and hospitals around the world. Imaging is based on the detection of backscattered light with contrast due to naturally occurring refractive index variations of tissue microstructures. A confocal digital imager consists of a point source of light that illuminates a small spot within the biological specimen. The illuminated spot is then imaged onto a detector through a pinhole aperture. This aperture acts as a spatial filter, rejecting light that is reflected from the out-of-focus portions of the object, so the resultant image has the high contrast of a thin-section image. The light source, illuminated spot, and detector have the same foci, or are placed in conjugate focal planes, and are therefore confocal to each other. The diameter of the detector aperture is matched to the illuminated spot through the intermediate optics. Because a small spot is illuminated and then detected through a small aperture, only the plane in focus within the specimen is imaged. Light originating from out-of-focus planes is prevented from entering the detector. A confocal digital imager thus allows imaging of thin slices of tissue, or optical sectioning, with high axial resolution and contrast. The confocal digital imager illuminates and images only a small spot at a time. To view the whole specimen, the illumination spot is scanned over the desired field of view. The illumination spot is raster-scanned optomechanically to sweep the entire area. The specimen is illuminated point by point, and then the image is created in the corresponding manner. Real-time confocal imaging of human skin involves laser scanning rather than white-light tandem scanning. Laser scanning has the benefits of bright, higher contrast imaging, higher magnifications, and deeper penetration. Infrared lasers coupled to a fast scanner allow video rate imaging of skin to maximum depths

of 300 µm. The thickness of the in vivo optical slice obtained is 2–5 µm. To minimize blurring, a skin-to-microscope contact device stabilizes the skin to within ±2 cells. Live images are displayed on a video monitor. With in vivo imaging, the virtual sectioning occurs in the horizontal plane, which correlates to en face sections as opposed to the vertical sections of routine histology. Contrast in the image correlates to naturally occurring variations in refractive index of organelles and microstructures within the skin. Epidermal keratin, for example, varies in refractive index depending on the state of differentiation of the keratinocyte. As the keratinocytes mature within the epidermis and the molecular weight of the keratins within an epidermal keratinocyte increases, the keratinocytes become more refractile, thus causing an increase in refractive index. As a result, the confocal images become brighter and the keratinocytes within the epidermis are well defined. The pigment melanin within the epidermis also has a high refractive index, in fact higher than keratin. Visually, melanin has a characteristic brown-black appearance because of the absorption of visible light. When illuminated with infrared light, however, the absorption is greatly reduced. This reduced absorption, combined with the intrinsic high refractive index, causes enhanced backscattering of reflected light that is collected by the confocal microscope. Higher concentrations of melanin cause an increase of backscattering to occur. Consequently, what appears as brown-black to the naked eye will appear white or bright in a confocal image.

The main advantage of CLSM is the unique opportunity to image thin sections of living tissue at a resolution equal to that of conventional microscopes used to view histology slides. Cellular and architectural details can be examined without having to excise and process the tissue as in standard histology. When the objective lens is placed onto an adapter ring, which is fixed on the tumor, real-time images can be obtained in seconds at the bedside. As a limitation in the current state of technological confocal microscopy development, it has to be addressed that assessment of microanatomic structures can only be done to a depth of 300 µm, which corresponds to the

Confocal Laser-Scanning Microscopy In Vivo, Fig. 1 Confocal in vivo image of malignant melanoma

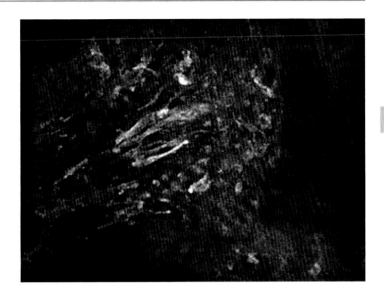

papillary dermis. Thus, processes in the reticular dermis and tumor invasion depth cannot be reliably evaluated at the present state of confocal imaging technology.

Clinical Relevance

Initial research has concentrated on the most clinically relevant cutaneous malignancies. Tumors that have been imaged and characterized include melanocytic skin tumors and nonmelanoma skin cancer. Confocal images have been qualitatively and quantitatively correlated to corresponding horizontal histology sections. The primary goal was to define and understand skin cancer morphology as seen with a confocal microscope in vivo. These preliminary studies were helping to develop an ability to understand and interpret confocal images of skin cancer. In the present state of investigations, the focus lies on the diagnostic accuracy of the method and consequently the integration into clinically routine procedures. Of all the cancers, ► melanoma of the skin represents one of the greatest challenges in early or preventative detection. Using CLSM, distinct morphologic features can be described for the differentiation of benign common nevi and malignant melanoma (Fig. 1). For example, in general, progression from monomorphic features in benign common nevi to increasing pleomorphism and architectural disarray in dysplastic nevi and melanomas was found.

Melanocyte cytology shows round to oval, bright, and monomorphic cells in benign nevi, whereas melanomas tend to present polymorphic and irregularly shaped cells. Nevus cell nests can be clearly seen in benign common nevi, but are less defined in dysplastic nevi. Disarray of architecture can be found in melanoma. Keratinocyte cell borders can be readily detected in benign common nevi, show focal absence in dysplastic nevi, and are poorly defined or absent in melanoma. Dendrite-like structures with a complex branching pattern are frequently seen in melanoma, but less frequently in benign nevi, where they are smaller and more delicate. In a published study, excellent sensitivity and specificity achieved for the diagnosis of melanoma using confocal microscopy in vivo, based on distinct morphologic features, has been described. Of note, the independent observer received only a 30 min presentation that instructed them in the confocal morphologic features of melanocytic skin tumors. Moreover, statistical analysis showed excellent to perfect inter- and intraobserver agreement for the confocal morphologic attributed studied. Nonmelanoma skin cancers are the most common malignancies among the Caucasian population. The most frequent of these are ► basal cell carcinomas and squamous cell carcinomas. Based on several studies, confocal images of both basal cell carcinomas and squamous cell carcinomas show relevant cellular and architectural features comparable to standard pathology. Moreover, a high diagnostic accuracy, prior to naked-eye and other noninvasive imaging techniques, could be achieved by the confocal microscope. Another potential use of CLSM in vivo is presurgical margin detection for skin cancer surgery. Surgical management of amelanotic melanomas as well as melanomas, basal cell carcinomas, and squamous cell carcinomas with ill-defined borders presents a significant clinical challenge currently addressed by serial excisions. In these settings, CLSM provides a much improved first approximation of the lateral borders between the tumor and normal skin. The cumulative experience with CLSM by different investigators clearly holds promise for this technology in the future. The results of several studies indicate that in vivo examination of skin tumors by CLSM can provide useful diagnostic information. CLSM represents an opportunity for clinicians to add useful and reliable information in their diagnostic decisions and therefore may spare some patients a biopsy or excision procedure and save time and costs.

Cross-References

▶ Melanocytic Tumors
▶ Skin Cancer

References

Gerger A, Koller S, Kern T et al (2005) Diagnostic applicability of in vivo confocal laser scanning microscopy in melanocytic skin tumors. J Invest Dermatol 124(3):493–498
Halpern AC, Rajadhyaksha M, Toledo-Crow R (2005) Bringing histology to the bedside. J Invest Dermatol 124(3):viii–x
Nori S, Rius-Diaz F, Cuevas J et al (2004) Sensitivity and specificity of reflectance-mode confocal microscopy for in vivo diagnosis of basal cell carcinoma: a multicenter study. J Am Acad Dermatol 51(6):923–930
Rajadhyaksha M, Grossman M, Esterowitz D et al (1995) In vivo confocal scanning laser microscopy of human skin: melanin provides strong contrast. J Invest Dermatol 104:946–952
Tannous Z, Torres A, Gonzalez S (2003) In vivo real-time confocal reflectance microscopy: a noninvasive guide for Mohs micrographic surgery facilitated by aluminum chloride, an excellent contrast enhancer. Dermatol Surg 29(8):839–846

Congenic Mice

▶ Mouse Models

Congenital Mesoblastic Nephroma

▶ Mesoblastic Nephroma

Congenital Telangiectatic Erythema

▶ Bloom Syndrome

Conjugated Linolenic Acids

Kazuo Miyashita
Faculty of Fisheries Sciences, Department of Bioresources Chemistry, Hokkaido University, Hakodate, Hokkaido, Japan

Synonyms

CLN

Definition

Conjugated linolenic acid (CLN) is a general term for the geometrical and positional isomers of octadecatrienoic (18:3) acid with three conjugated double bonds. Conjugated linolenic acids occur in several terrestrial plants (mainly seed oils). They include α-eleostearic acid (9cis(c),11$trans$(t),13t-18:3), catalpic acid (9t,11t,13c-18:3), punicic acid (9c,11t,13c-18:3), calendic acid (8t,10t,12c-18:3), and jacaric acid (8c,10t,12c-18:3) (Fig. 1). High contents of calendic acid, punicic acid, and α-eleostearic acid are found in seed oils of pot marigold, pomegranate, and tung/bitter gourd, respectively.

Conjugated Linolenic Acids, Fig. 1 Structure of conjugated linolenic acids (CLN)

α-Eleostearic acid [9c,11t,13t-18:3]

Catalpic acid [9t,11t,13c-18:3]

Punicic acid [9c,11t,13c-18:3]

Calendic acid [8t,10t,12c-18:3]

Characteristics

In Vitro Studies

Conjugated linolenic acid (CLN) shows cytotoxic effect on mouse tumor cell (SV-T2). However, there is a difference in the toxicity between CLN isomers. Fatty acid from pot marigold (8t,10t,12c-18:3; 33.4%) has no effect on the cell line up to 250 μM, but other kinds of fatty acids from seed oils are cytotoxic to SV-T2 cells below 20 μM. The same effect is observed in the case of human monocytic leukemia cell (U-937). Generally, 9,11,13-CLN and all trans-CLN are more cytotoxic than 8,10,12-CLN and CLN containing cis-configuration, respectively. The higher cytotoxicity of 9,11,13-CLN or all trans-CLN isomers is partly due to the different susceptibilities of these CLN isomers to lipid peroxidation. On the other hand, the inhibitory effect of CLN on the growth of colon cancer cells is related to the regulation of peroxisome proliferator-activated receptor gamma (PPAR)γ. PPARγ ligands such as troglitazone and 15-d-prostaglandin (PG) J₂ cause growth inhibition and induce ▶ apoptosis in cancer cells. CLN shows a higher ligand activity on PPARγ than troglitazone. ▶ BCL-2, GADD45, and ▶ p53 are known as an important molecular target in apoptosis-inducing pathways. In Caco-2 cell treated with 9c,11t,13t-CLN, Bcl-2 expression is downregulated, while GADD45 and p53 expressions are upregulated. Therefore, two possible mechanisms of the anticarcinogenic activity of CLN can be hypothesized, viz., induction of apoptosis via lipid peroxidation and regulation of target gene and protein.

In Vivo Studies

CLN from bitter gourd seed oil (BGO) significantly reduces the frequency of colonic aberrant crypt foci (ACF) in rat as a precursor of colon carcinogenesis. In this case, the proliferating cell nuclear antigen (PCNA)-labeling indices in ACF and normal-appearing crypts also decreases by dietary feeding of CLN. Furthermore, feeding of CLN enhances apoptotic cells in ACF without affecting the surrounding normal-appearing crypts. ▶ Chemopreventive ability of BGO on rat colon cancer can be found in a long-term in vivo assay. Dietary administration of BGO rich in CLN (9c,11t,13t-18:3) significantly inhibits the development of colonic adenocarcinoma induced by azoxymethane (AOM) in male F344 rats without causing any adverse effects. In addition, BGO intake significantly reduces the multiplicities of colorectal carcinoma (number of carcinomas/rats) in rats. Other CLN isomer (9c,11t,13c-18:3) from pomegranate seed oil (PGO) also shows the chemopreventive effect on rat colon cancer induced by AOM. Dietary feeding of PGO suppresses progression of adenoma to malignant neoplasm in post-initiation phase of colon cancer. Dietary feeding of BGO and PGO enhances PPARγ expression in nonlesional colonic mucosa. Synthetic ligands for PPARα and PPARγ effectively inhibit AOM-induced ACF in rats. Therefore, it may be possible that BGO and PGO suppress colon carcinogenesis by means of altering PPARγ expression in colonic mucosa.

References

Kohno H, Suzuki R, Noguchi R et al (2002) Dietary conjugated linolenic acid inhibits azoxymethane-induced colonic aberrant crypt foci in rats. Jpn J Cancer Res 93:133–142

Narayan B, Hosokawa M, Miyashita K (2006) Occurrence of conjugated fatty acids in aquatic and terrestrial plants and their physiological effects. In: Shahidi F (ed) Nutraceutical and specialty lipids and their co-products. CRC Taylor & Francis, New York, pp 201–218

Suzuki R, Noguchi R, Ota T et al (2001) Cytotoxic effect of conjugated trienoic fatty acids on mouse tumor and human monocytic leukemia cells. Lipids 36:477–482

Yasui Y, Hosokawa M, Sahara T et al (2005) Bitter gourd seed fatty acid rich in 9c,11t,13t-conjugated linolenic acid induces apoptosis and up-regulates the GADD45, p53 and PPARγ in human colon cancer Caco-2 cells. Prostaglandins Leukot Essent Fatty Acids 73:212–219

Yasui Y, Hosokawa M, Kohno H et al (2006) Growth inhibition and apoptosis induction by all-trans-conjugated linolenic acids on human colon cancer cells. Anticancer Res 26:1855–1860

Connexins

Dale W. Laird
Department of Anatomy and Cell Biology, University of Western Ontario, London, ON, Canada

Definition

Connexins are proteins which assemble into channels that allow for small molecules to pass directly from one cell to another.

Characteristics

The family of connexin (Cx) proteins is composed of 21 members in humans. All connexins (Cx32 and Cx26) share common features of assembling into connexons also called "hemichannels" consisting of six subunits of the same or different connexins (Fig. 1). Hemichannels from apposing cells dock and the resulting channels cluster into a junctional complex known as a ▶ Gap junction or often referred to as a gap junction plaque. Gap junctions allow for the direct intercellular exchange of secondary messengers and other small molecules, a process termed gap junctional intercellular communication (GJIC) (Fig. 1). Gap junctions have a ubiquitous distribution in human tissues, and these specialized intercellular channels are essential for normal cell function, proper cell differentiation, tissue development, metabolic transport, ion transfer, and cell growth control. Importantly, each connexin is assembled into a channel with unique properties that are thought to reflect distinct physiological roles for the different gap junction channel types. In numerous diseases, connexins are either not produced or mislocalized (e.g., many cancers) or contain mutations that inhibit the normal function of the resulting channels. Mutations of connexin genes are linked with human diseases including neurodegeneration, skeletal abnormalities, keratodermas, and hereditary sensorineural deafness. For example, mutations in the gene encoding Cx26 are the most common cause of congenital hearing loss.

Connexins as Tumor Suppressors

The evidence that connexins play a role in cell growth control and early tumorigenesis ranges from circumstantial to direct linkages. First, most soft tissue tumors typically have reduced gap junctions due to either decreased connexin expression or an inability to efficiently assemble connexins into gap junctions. Second, tumor promoters, mitogens, and ▶ oncogenes are known to reduce GJIC. Third, reexpression of connexins in tumor cells frequently reverts cancer cells to a less aggressive cell type and slows cell growth, highlighting the tumor suppressive behavior of connexins. Fourth, mice lacking one member of the connexin family (Cx32) are 10–25 times more at risk of developing chemically or radiation-induced liver or lung tumors. The evidence that mutations in the genes encoding connexins lead to increased susceptibility to cancer is sparse and thus connexins are more commonly thought of as conditional tumor suppressors reflecting the consequence of reduced expression or ability to make gap junctions. Collectively, convincing data suggest that connexins play a role in ▶ epithelial tumorigenesis particularly at disease onset, progression, and early events associated with ▶ metastasis. The role of connexins in cancer cells that enter the blood stream and proceed to

Connexins,
Fig. 1 Connexins (e.g., Cx32 or Cx26) are gap junction proteins that thread through the lipid bilayer of cell membranes four times. Connexins of the same type (*orange* or *green* rods) or different types (mixtures of *orange* and *green* rods) assemble into hexameric arrangements with a central pore known as connexons or "hemichannels." Connexons from opposing cells dock and tightly cluster into gap junction plaques allowing for bidirectional exchange of small molecules, a process known as gap junctional intercellular communication (GJIC)

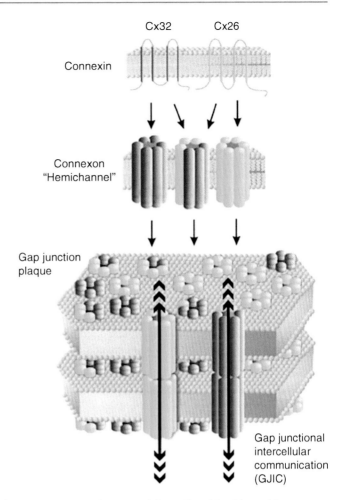

break through the endothelial wall in later stage metastasis is less clear. In fact, considerable evidence would support the position that connexin expression in later stage disease favors the ability of cancer cells to enter and propagate at new tissue sites. Consequently, a working paradigm is that connexins protect cells from becoming cancerous, act to suppress the growth of primary tumors, and play an inhibitory role in cells breaking away from the primary tumor. However, in advanced disease where tumor cells were successful in escaping the primary tumor, connexin reexpression may facilitate the cancer cells exiting the blood system to enter and populate a second tissue site.

Connexins as a Therapeutic Target

The likelihood of connexins being a good target for combination therapy for primary tumors remains promising. Considerable evidence supports the notion that up-regulation of connexins alone with only minimal increases in GJIC may be sufficient to suppress the growth and expansion of the primary tumor. However, it is likely that such a putative treatment would need to be combined with drugs designed to induce cell death. Unfortunately, there are no drugs in clinical cancer trials that specifically target the regulation of connexins. The reason for this is primarily due to the lack of a nontoxic drug that will specifically up-regulate connexins within the tumor. The need for a tumor specific drug is critical as many studies indicate that a system wide up-regulation of connexins and gap junctions in nondiseased organs would likely lead to pathological side effects. Consequently, any drug or gene therapy development would need to target the cancer cells only to avoid detrimental side effects. In addition to increasing connexin content in tumor cells, it is almost a certainty that any curative strategy would require combinational therapy where connexin up-regulation and increases in GJIC would be accompanied with a second therapeutic strategy. Evidence has suggested that gap junctions could act as conduits for delivery of pro-drugs deep into the tumor allowing for a more effective cell kill throughout the tumor. Again, such treatment strategies would necessitate good gene targeting or specific drug treatments that restrict their effects to the primary tumor. At a minimum, the increased presence of connexins and gap junctions would be expected to provide a decrease in tumor expansion while a patient is exposed to repeated treatment protocols designed to kill the tumor cells. The importance of connexins as a possible target in the treatment of metastatic disease is relatively unknown. Based on findings in animal models, additional precautions must be considered as connexins have been reported to enhance the movement of tumor cells from the blood to vital organ tissues. Additional research is also necessary to determine what role connexins play in facilitating or inhibiting the interaction of the tumor cell with the surrounding milieu of cells and the extracellular matrix that become the "soil" for metastatic tumor cell growth.

In summary, the role of connexins in tumorigenesis and metastatic disease may in fact be two fold. First, the bulk of the evidence would support members of the connexin family as acting as inhibitors of cancer onset, primary tumor growth, and early stage events associated with metastasis. As such, this highlights connexins as a viable target for cancer prevention and treatment of primary disease. Second, a paradigm is developing where connexin expression may favor later stage metastatic properties of at least some tumor cell types. Consequently, intentions where connexins are targeted and downregulated only in tumor cells circulating in the blood may serve as an advantage in treatment strategies of more advanced disease. Clearly more information involving better experimental models is necessary to resolve the full function of connexins in tumorigenesis and disease progression.

Cross-References

▶ Chemoprevention
▶ Epithelial Tumorigenesis
▶ Peroxisome Proliferator-Activated Receptor

References

Laird DW (2006) Life cycle of connexins in health and disease. Biochem J 394:527–543
Mesnil M, Crespin S, Avanzo JL et al (2005) Defective gap junctional intercellular communication in the carcinogenic process. Biochim Biophys Acta 1719: 125–145
Petersen MB, Willems PJ (2006) Non-syndromic, autosomal-recessive deafness. Clin Genet 69:371–392

Constitutive Androstane Receptor

Kaarthik John
Division of Microbiology, Tulane University, Covington, LA, USA

List of Abbreviations

CITCO	6-(4-chlorophenyl)imidazo[2,1-b] [1,3]thiazole-5-carbaldehyde O-[3,4-dichlorobenzyl)oxime
TCPOBOP	1,4-Bis [2-(3,5-dichloropyridyloxy)] benzene

Synonyms

CAR; NR1I3

Definition

The constitutive androstane receptor (CAR), a member of the nuclear receptor subfamily 1 (NR1I3- nuclear receptor subfamily 1, group I, member 3), initially identified as a xenosensor (sensor of xenobiotics), is a functionally pleiotropic, liver-enriched transcription factor with roles in various cellular processes and diseases.

Characteristics

Human CAR, first cloned in 1994 (and called MB67 at that point of time), was originally regarded as an "orphan" nuclear receptor, with no apparent endogenous ligand. It is expressed to the highest extent in the liver and small intestine. Additionally, unlike many nuclear receptors requiring the presence of a ligand to activate and cause its translocation from the cytoplasm to the nucleus, CAR, in heterodimerization with 9-cis retinoic acid receptor (RXR), transactivates and exhibits "basal" activity even in the absence of ligand. Hence, it initially came to be known as the "constitutively active receptor." Later, the identification of two endogenous testosterone metabolites, 5α-androstan-3α-ol (androstanol) and 5α-androst-16-en-3α-ol (androstenol), capable of acting as "inverse agonists" of the receptor, led to its current name, the **constitutive androstane receptor**. The identification of physiological ligands for this receptor has also now led to its classification as an "adopted orphan receptor," exhibiting a low level of affinity to these ligands. Additionally, given its role in Phase I, Phase II, and Phase III processes of xenobiotic metabolism, it is also regarded as a "master" xenobiotic receptor.

Structure of CAR

The human CAR gene, mapping on chromosome 1q23.3, comprises nine exons. Translation starts only from exon 2 with exon 1 being a noncoding exon. Exons 2, 3, and part of 4 encode the DNA binding and hinge regions of the protein. The remaining part of exon 4, along with exons 5–9, encode the ligand binding domain of the protein. A number of SNPs and splice variants have been reported for CAR. Among the SNPs, as of now, five are nonsynonymous with the remaining mostly being synonymous or occurring in noncoding regions of the gene. Despite the variable functional impact of these SNPs, their frequency in the general population, however, is rather low (<2%). With regard to splice variants, most appear to code for variant proteins or ones with a premature stop codon. Deletion appears to be a common mechanism. A couple of splice variants, designated CAR2 (SV23) and CAR3 (SV24), contain 12 and 15 bp nucleotide insertions between exons 6 and 7 and 7 and 8, respectively. These change the constitutive expression of wild-type CAR1 to ligand inducible expression in case of CAR2 and CAR3. In essence, the various splice variants possess different ligand specificities and differentially impact the final expression of CAR.

With regard to the protein, the original CAR clone encoded 348 amino acids, with the protein possessing a "modular" structure, typical of nuclear receptors. However, unlike other nuclear receptors, it lacks part of the N terminal A/B regulatory domain, housing the ligand-independent activation function 1 (AF-1) motif, and part of the C terminal E/F domain, following the ligand-dependent AF-2 transactivation motif (Fig. 1). Four different crystal structures of human

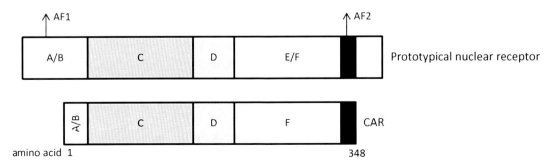

Constitutive Androstane Receptor, Fig. 1 Modular structure of CAR in contrast with that of a prototypical nuclear receptor. CAR lacks part of the A/B domain at the N terminal beyond the AF1 motif and part of the E/F domain beyond the AF2 motif. A/B = N terminal domain, C = DNA binding domain, D = hinge region, E/F = ligand binding/dimerization domain

and mouse CAR ligand binding domain, complexed with either two different human CAR ligands (in case of human CAR) or two different mouse CAR ligands (in case of mouse CAR), have also revealed a great deal of information about various structural aspects of the receptor. This in turn has also provided clues to certain properties of the receptor, such as its ligand independent constitutive activation.

Mechanism of Action

In primary hepatocytes and the liver, in the absence of a ligand, phosphorylated CAR (phosphorylated at Thr38 in human CAR and Ser202 in mouse CAR by protein kinase C) remains tethered in the cytoplasm, complexed with several chaperone proteins. These chaperones include heat shock protein 90 (Hsp90), cytoplasmic CAR repetition protein (CCRP), the protein phosphatase 1 regulatory subunit 16A PP1β (PPP1R16A), and possibly other unknown proteins. Exposure to a direct CAR activator (e.g., CITCO, a synthetic compound) or an indirect activator (such as phenobarbital (PB), an anticonvulsant) causes the recruitment of protein phosphatase 2A (PP2A) to the tethering complex and subsequent release of the receptor from the complex, mediated by PP2A directed dephosphorylation of the receptor. The released receptor subsequently undergoes nuclear translocation. Immortalized cancer cell lines lack such cytoplasmic tethering mechanism, with CAR constantly being localized in the nucleus. Additionally, while direct CAR activators and most inverse agonists physically bind the receptor causing its nuclear translocation, indirect activators such as PB and bilirubin do so by mechanisms other than direct receptor binding thereby leading to ligand independent mechanisms of nuclear translocation. The majority of the CAR activators cause nuclear translocation of the receptor through indirect PB-like mechanisms.

Nuclear translocation of most nuclear receptors, in general, relies on the presence of a nuclear localization signal (NLS), a short basic amino acid-rich sequence used to target the protein to the nucleus. Translocation of CAR, however, appears to occur even independent of a strong NLS or AF-2 motif in the ligand binding domain. However, the presence of a leucine-rich motif (LXXLXXL), termed as the xenobiotic response sequence/signal (XRS) in the C terminal ligand binding domain of the receptor, appears to be necessary. Once in the nucleus the receptor heterodimerizes with the retinoid X receptor (RXR), another nuclear receptor, and binds to specific DNA response elements in the promoter/enhancer regions of target genes. The PB response enhancer module (PBREM) and xenobiotic response enhancer module (XREM) constitute two such response elements identified in *CYP2B6*, a downstream target of CAR. These response elements typically consist of two direct repeats of a consensus hexameric sequence (AGGTCA), interspersed by direct repeats (DR) of four to five nucleotides (DR-4 and DR-5). Despite greatest affinity for this consensus sequence, binding to several other target elements is also possible. Examples include pregnane X receptor (PXR) responsive DR-3 and everted repeat 6 (ER-6) elements in *CYP3A* and peroxisome proliferator activated receptor (PPAR) responsive DR-1 elements. Once bound, several coregulators of transcriptional activity also get recruited and regulate transcriptional expression. These regulators could include several coactivators such as apoptotic speck protein-2 (ASC-2), glucocorticoid receptor interacting protein-1/transcriptional intermediary factor 2 (GRIP1/TIF2), PPAR gamma coactivator-1 (PGC-1), structural maintenance of chromosome-1 (SMC-1), steroid receptor coactivator 1 (SRC-1), **r**eceptor **a**ssociated **c**oactivator 3 (RAC3), PPAR binding protein (PPARBP), transcription factor Sp1, or corepressors such as nuclear receptor corepressor (NCoR), **sm**all heterodimer partner-**i**nteracting **le**ucine zipper protein (SMILE), **d**osage-sensitive sex reversal, **a**drenal hypoplasia critical region, on chromosome **X**, gene 1 (DAX1), **s**ilencing **m**ediator of **r**etinoic acid and **t**hyroid hormone receptor (SMRT), and several others, depending on the chemical environment. A schematic representation of CAR mediated gene regulation is provided in Fig. 2.

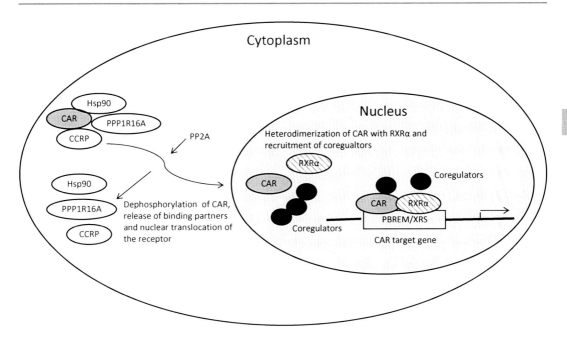

Constitutive Androstane Receptor, Fig. 2 Canonical model of CAR mediated gene regulation. Please refer to section "Mechanism of Action" for details

Regulators of CAR

CAR appears to bind various ligands and also be modulated by several chemicals. These include various drugs, pesticides, plasticizers, toxicants, certain natural and synthetic compounds, and several endogenous metabolites. These chemicals have been shown to function either as activators (by serving as direct or indirect agonists) or as repressors/inhibitors (by serving as inverse agonists) of CAR activity. In some cases their precise mechanism of modulation appears to be unknown. These chemicals also exhibit species specificity with some being specific human CAR modulators and certain others being rodent CAR modulators and some exhibiting activation/inhibition across different species. A representative list of some of the human CAR modulators along with their functional impact on the receptor is shown in Table 1.

CAR Targets

Various members of the Phase I, Phase II (bioactivation and detoxification), and Phase III (transporters) battery of xenobiotic metabolism genes in addition to certain members involved in endobiotic metabolism and cell cycle control constitute downstream targets of CAR. This in turn impacts on the absorption, distribution, metabolism, elimination, pharmacokinetics, pharmacodynamics, toxicokinetics, and toxicodynamics of various chemicals, drugs, and xenobiotics metabolized by these genes. Table 2 lists some of the human CAR target genes, associated with these various stages of biotransformation. A more extensive list of targets has been reported in rodents.

CAR and Carcinogenesis

CAR is a functionally pleiotropic receptor with roles in various physiological and pathological processes. Some of these processes include xenobiotic metabolism, drug metabolism, drug transport, glucose metabolism, hepatic bile metabolism, lipid and energy homeostasis, and endocrine regulation. Additionally, it has also been implicated in several diseases such as obesity, type 2 diabetes, dyslipidemia, certain neurological disorders, atherosclerosis, and cancer.

With respect to cancer, CAR exhibits a behavior that appears to be "species and context dependent."

Constitutive Androstane Receptor, Table 1 List of some human CAR activators and inhibitors

Compound	Nature of the compound	Effect on hCAR
Acetaminophen[a]	Drug	Activator
Artemisinin	Drug/herbal extract	Activator
Atorvastatin[a]	Drug	Activator
Butylated hydroxy anisole (BHA)	Food additive	Activator
2,2′,4,4′-Tetrabromodiphenyl ether (*BDE-47*)[a]	Flame retardant	Activator
Carbamates (benfuracarb)	Pesticide	Activator
Carbamazepine	Drug	Activator
Cerivastatin[c]	Drug	Activator
Cyclophosphamide	Drug	Activator
Diazepam	Drug	Activator
Di-(2ethylhexyl phthalate) (DEHP)	Plasticizer	Activator of hCAR2
Dehydroepiandrosterone (DHEA)	Hormone precursor	Activator
Dichlorodiphenyltrichloroethane (DDT)[a]	Pesticide	Activator
Ellagic acid, resveratrol[a]	Polyphenols	Activator
Fluvastatin[c]	Drug	Activator
FL81, UM104	Synthetic compound	Activator
Ginko biloba[a]	Herbal extract/ flavonoids	Activator
Myclobutanil	Pesticide	Activator
Nevirapine	Drug	Activator
Organochlorines (methoxychlor, PCB153)[b]	Pesticide	Activator
Pyrethroids (permethrin, cypermethrin)	Pesticide	Activator
5b-pregnane-3,20-dione	Hormone precursor	Activator
Sulfanilamides	Drug	Activator
Sulfonamides	Drug	Activator
Simvastatin[c]	Drug	Activator
Valproic acid[b]	Drug	Activator
Yin Zhi Huang	Herbal extract	Activator
CITCO	Synthetic compound	Direct Activator
tri_p_methylphenyl phosphate (TMPP)	Synthetic compound	Direct Activator
triphenyl phosphate	Plasticizer/Fire retardant	Direct Activator
6,7-dimethylesculetin	Herbal extract	Indirect Activator
Galangin, chrysin[a]	Herbal extract/flavonoids	Indirect Activator
Phenobarbital[a]	Drug	Indirect Activator
Phenytoin[a]	Drug	Indirect Activator
5α-Androstan-3α-ol	Steroid	Inverse agonist
17α-ethynyl-3,17 β-estradiol (EE2)	Steroid	Inverse agonist
Clotrimazole	Drug	Inverse agonist
Ketokonazole[a]	Drug	Inverse agonist
Metformin	Drug	Inverse agonist
PK11195, S07662, T0901317	Synthetic compound	Inverse agonist

[a]Activates/regulates mouse CAR as well
[b]Activates/regulates rat CAR as well
[c]Activates/regulates mouse and rat CAR as well

In rodents, CAR has been implicated in liver tumor development, following prolonged exposure to direct (e.g., TCPOBOP, a mouse CAR activator) or indirect CAR activators (e.g., PB, certain pesticides). Absence of CAR, as in CAR null mice, abrogates this response. Liver tumors formed following exposure to hepatic tumor initiators, such as N-Nitrosodiethylamine/diethylnitrosamine (DEN) or N-2-fluorenylacetamide (FAA), were maintained following exposure to the CAR activator, PB, or other CYP2B inducers. While some studies have implicated both hypertrophy and hyperplasia to play a role in the development of hepatocellular carcinoma, certain others have pointed to hypertrophy as causal mechanism. Additionally, PB also promotes the expansion of hepatocellular adenomas harboring activating

Constitutive Androstane Receptor, Table 2 List of certain human CAR target genes reported to be associated with different phases of biotransformation

Phase I					
CYP2A6	CYP2B6[a]	CYP2B10	CYP2C9[a]	CYP2C19[a]	CYP3A4[a]
CYP3A5[a]	CYP3A11				
Phase II					
UGT1A1[a]	GSTM1[b]				
Phase III					
ABCB1/MDR1A/p-glycoprotein			MRP2/ABCC2		MRP3/ABCC3

[a]Certain genes have been known to be regulated by PXR as well
[b]Predicted to be regulated by hCAR

β-catenin mutations with β-catenin null mice exhibiting no tumor formation despite proliferative responses in the liver. Some mechanistic studies have revealed CAR to enhance the expression of certain cell cycle genes, cyclin D1 and cdk2, an antiapoptotic factor, GADD45B, and a negative regulator of p53, Mdm2, and thereby contribute to some of the above tumorigenic effects. Others have pointed to induction of certain cytochrome P450s (downstream targets of CAR), enhanced replicative DNA synthesis, epigenetic changes in certain genes relevant to angiogenesis, migration, invasion, altered liver foci, and enhanced apoptosis to also play a role. Some studies observed knocking out certain other genes such as c-Jun, a transcription factor, connexin 32, a cellular gap junction protein, or N-acetylglucosaminyltransferase III, a regulatory enzyme involved in biosynthesis of certain membrane sugars, to result in subdued PB induced tumor formation in mice. However, upon further investigation, these did not turn out as targets of PB.

In contrast to rodents, the relevance of CAR to cancer in humans, as ascertained from humanized mouse models or from studies involving various human cancer cells, appears to be inconclusive. PB induces CAR in human hepatocytes and in humanized CAR transgenic mice (hCAR mice – mice where mouse CAR has been replaced with human CAR). In tune with the above observations, along with findings from rodents, some studies have reported similarities in hepatic transcriptomic profiles between wild-type mice (WT), harboring mouse CAR (mCAR), versus those double humanized for the receptors CAR and PXR (hCAR/hPXR), thereby suggesting the findings from rodents to be relevant to humans as well. Further, hCAR/hPXR mice were also reported to support tumor promotion, albeit to a lesser extent than WT mice, following administration of PB, post tumor initiation with DEN. In contrast, however, another study reported hypertrophy sans hyperplasia in hCAR/hPXR mice (i.e., no tumor formation) versus hypertrophy and hyperplasia mediated hepatic tumor formation in WT mice. In keeping with the latter observations, except for a few stray reports of PB mediated hepatocarcinogenesis in humans, epidemiological data from long-term PB usage among epilepsy patients fail to point to an enhanced incidence of hepatocarcinoma in humans. Hence overall, data from humanized animal models appears to be inconsistent. Interspecies differences in some of the receptor's functions, downstream target genes, and signaling cascades, along with possibilities of lower PB doses in humans, relative to rodents, constitute some of the plausible arguments put forth to account for the observed differences.

With respect to human cancer cells, based on certain reports, the choice of either activating or inhibiting the receptor as a therapeutic strategy appears to be context specific. For example, stimulation of CAR has been shown to potentially benefit in certain cases where treatment of the cancer relies on the metabolism of an administered prodrug (e.g., cyclophosphamide) to its pharmacologically active form (four hydroxy cyclophosphamide) by certain CAR target genes. In such instances, induction of CAR was found to turn on certain downstream CAR targets, such as CYP2B6, in turn aiding in the metabolism of the prodrug to its active form. The utility of this approach has been demonstrated in certain lab studies involving cyclophosphamide mediated treatment of leukemia cells and paclitaxel mediated control of lung tumors. In another study involving brain tumor stem cells (BTSCs), given the low expression of CAR in these cells, activation of the receptor was reported to control the expansion of BTSCs by inducing cell cycle arrest and apoptosis. In contrast, other studies have shown the inhibition of CAR to be a potentially useful cancer therapeutic approach. CAR has been known to regulate multidrug resistance (MDR). In such instances the activation of the receptor contributes to chemoresistance by enhancing the expression of MDR genes. Consequently, inhibition of the receptor sensitizes the cells to drugs by downregulating the expression of MDR genes. This strategy has been demonstrated in neuroblastoma and ovarian cancer cells.

Conclusions

CAR, originally identified only as orphan receptor, binds several compounds, both endogenous and exogenous, regulates several genes, and has roles in various physiological and pathological processes. Various modulators of the receptor, including activators and inhibitors, now exist. Given its functionally pleiotropic nature, more roles of the receptor, in erstwhile unreported processes and diseases, are likely to come to light with more research. Targeting CAR, as a therapeutic approach, could constitute a novel ancillary strategy to be used in future in conjunction with existing therapies. But this would need to be further validated. Details to do with some of the isoforms of the receptor that might be expressed in different diseases, including different cancers, need to be more completely understood. Additionally, a lot of information about the receptor is derived from rodents. Though its relevance to humans initially appeared equivocal and questionable, newer information might suggest otherwise and call for more studies to address its relevance in humans.

Glossary

Inverse agonist An agent capable of binding to the same receptor binding-site as the agonist

but causing a pharmacologically opposite effect as that of the agonist in a constitutively active receptor (i.e., a receptor exhibiting a certain level of intrinsic basal activity). In other words they cause a reduction of the basal expression. Depending on the affinity of the ligand to the receptor, inverse antagonism could be complete or partial.

Pregnane X receptor (PXR) A member of the nuclear receptor subfamily 1, group I (NR1I2), along with the constitutive androstane receptor (CAR) and vitamin D receptor (VDR). PXR, similar to CAR, has also been predominantly found to be expressed in the liver. It primarily serves as a xenobiotic and steroid hormone sensor. Human PXR was initially reported as a steroid and xenobiotic receptor and hence also called SXR. Along with CAR, PXR shares several similarities in its mechanism of action, along with overlapping roles in many of the physiological and disease processes, including inflammation and cancer.

Nuclear receptor structure Nuclear receptors are modular in structure with five-six domains, N to C terminus. These include an N-terminal A/B regulatory region with transcriptional activation function (AF-1) domain, a C region with a DNA-binding domain (DBD), a hinge D region and an E region with a ligand-binding domain (LBD) containing an activation function (AF-2) at its end.

Cross-References

▶ Aryl Hydrocarbon Receptor
▶ Peroxisome Proliferator-Activated Receptor
▶ Steroid X Receptor (SXR)
▶ Xenobiotics

References

Cherian MT, Chai SC, Chen T (2015) Small-molecule modulators of the constitutive androstane receptor. Expert Opin Drug Metab Toxicol 11(7):1099–1114
Molnar F, Küblbeck J, Jyrkkärinne J, Prantner V, Honkakoski P (2013) An update on the constitutive

androstane receptor (CAR). Drug Metabol Drug Interact 28(2):79–93

Omiecinski CJ, Vanden Heuvel JP, Perdew GH, Peters JM
(2011) Xenobiotic metabolism, disposition, and regulation by receptors: from biochemical phenomenon to
predictors of major toxicities. Toxicol Sci 120(S1):
S49–S75

See Also

(2012) Nuclear receptor. In: Schwab M (ed) Encyclopedia
of cancer, 3rd edn. Springer, Berlin/Heidelberg, p 2571.
doi:10.1007/978-3-642-16483-5_4157

Constitutive Photomorphogenic-9 (COP9) Signalosome 5 (CSN5)

► JAB1

Contact Normalization

Gary S. Goldberg
Molecular Biology, University of Medicine and
Dentistry of New Jersey, Stratford, NJ, USA

Synonyms

Heterologous growth control

Definition

The process by which nontransformed cells force
tumor cells to assume a normal morphology and
phenotype.

Characteristics

Transformed cells often survive medical treatment
and lay dormant for many years before they
emerge to cause relapse in a patient.
Nontransformed cells can force tumor cells to
assume a normal morphology and phenotype by
a process called "contact normalization" (Fig. 1).

Contact normalization is a powerful global
phenomenon. Cells transformed by a variety of
chemicals, viral agents, and oncogenes can be
normalized by contact with nontransformed
cells. This process is dramatically exemplified
by malignant tumor cells that form normal adult
organs when injected into mouse blastocysts.

Contact normalization is an important process
in vivo. Genetically transformed cells can assume
a normal morphology and reside in many organs
including the skin, breast, and intestine. Moreover, since these "occult tumor" cells are phenotypically normal, they tend to resist
chemotherapy. As stated above, contact normalization is a powerful process; transformed
keratinocytes that comprise up to 4% of epidermal

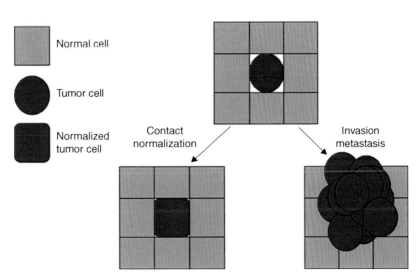

Contact Normalization, Fig. 1 Contact normalization. Tumor cells are normalized by contact with nontransformed cells. Cancer arises when this process is unsuccessful. Intercellular junctions are disrupted in invasive and metastatic tumor cells but should be stabilized during contact normalization

Normal cell

Tumor cell

Normalized tumor cell

Contact normalization

Invasion metastasis

volume can be controlled in the human skin for decades.

In at least some cell systems, contact normalization requires direct contact between transformed cells and nontransformed cells. Thus, intercellular junctional proteins such as ► connexins and cadherins may augment the ability of nontransformed cells to normalize the growth of adjacent tumor cells. Since these junctions are often disrupted in malignant and metastatic tumor cells, they should be stabilized in tumor cells undergoing contact normalization in the microenvironment.

Cross-References

► Adhesion
► Cell Adhesion Molecules
► E-Cadherin

References

Alexander DB, Ichikawa H, Bechberger JF et al (2004) Normal cells control the growth of neighboring transformed cells independent of gap junctional communication and SRC activity. Cancer Res 64:1347–1358

Glick AB, Yuspa SH (2005) Tissue homeostasis and the control of the neoplastic phenotype in epithelial cancers. Semin Cancer Biol 15:75–83

Naus CC, Goldberg GS, Sin WC (2005) Connexins in growth control and cancer. In: Winterhager E (ed) Gap junctions in development and disease. Springer, Berlin/Heidelberg/New York

Rubin H (2003) Microenvironmental regulation of the initiated cell. Adv Cancer Res 90:1–62

Rubin H (2006) What keeps cells in tissues behaving normally in the face of myriad mutations? Bioessays 28:515–524

Contralateral Breast Cancer

Definition

Bilateral primary breast cancer: increasing breast cancer incidence rates, improved prognosis, and growing life expectancy have resulted in an increasing number of women at risk of developing bilateral primary breast cancer. In the USA alone, there are approximately 2.2 million living diagnosed at sometime with breast cancer.

Contralateral breast cancers are divided in those that are synchronous, that is, when breast cancers are diagnosed in both breasts simultaneously, and metachronous, normally defined as diagnosed more than 6 months from the first cancer. The age-specific incidence of synchronous breast cancer mimics that of unilateral breast cancer. In contrast, the risk of being diagnosed with a metachronous bilateral cancer is higher among women diagnosed before the age of 45 compared to those older at time of diagnoses of the first breast cancer. The risk of metachronous bilateral breast cancer is elevated throughout the entire life of a woman and is approximately 0.5–1% annually.

The incidence of synchronous bilateral breast cancer has shown a steady increase in the last 30 years, although somewhat leveling off the last years. This is in sharp contrast to the metachronous contralateral breast cancers where the incidence has decreased by approximately 30% over the last 30 years. This is most likely a function of the increasing use of postoperative ► adjuvant therapy.

A woman with a synchronous bilateral breast cancer has a higher mortality rate compared to women with unilateral cancer; this is particularly evident before the age of 50 where a synchronous cancer entails about a two-time higher mortality rate. Women who develop a metachronous bilateral cancer within 5 years of the first cancer and before the age of 50 are at a four times higher risk of dying from breast cancer compared to an age-matched woman with unilateral cancer. Time since diagnoses of first cancer also influences the prognosis of the second cancer. Young women with an early metachronous cancer have a particularly bad prognosis, while women who were diagnosed more than 10 years after the first cancer has a prognosis similar of that of a unilateral breast cancer.

Cross-References

► Adjuvant Therapy

Convection-Enhanced Delivery

Kenji Muro
Department of Neurological Surgery, Robert
H. Lurie Comprehensive Cancer Center,
Northwestern University Feinberg School of
Medicine, Chicago, IL, USA

Synonyms

CED; High-flow microinfusion; Interstitial
microinfusion; Intracerebral clysis; Intracerebral
microinfusion

Definition

Convection-enhanced delivery (CED) is a novel
delivery method that allows direct drug infusion
into the brain in a locoregional manner. The delivery is accomplished through surgically implanted
catheters in the brain that are connected to external
drug infusion pumps that generate a positive infusion pressure. This positive pressure begins the
process of convection, which is the augmentation
and maintenance of the brain's normal physiologic bulk flow of interstitial fluid. The enhanced
bulk flow through the interstitial space acts as the
carrier of the desired agent.

Characteristics

High-grade primary ▶ brain tumors, such as glioblastoma multiforme, remain one of the most
challenging diagnoses to treat effectively. Despite
a range of therapeutic options and their combinations, including surgical resection, external beam
radiation therapy, and ▶ chemotherapy, the
median survival remains an astounding 12–14
months. Due to the compartmentalized distribution of functional areas within the brain and the
consequence to the patient's independence with
compromise of these functions, surgical resection
must be tempered. Radiation therapy has proven
benefit; however, there is a finite limit to the

brain's tolerance to radiation effects. Therefore,
focus has shifted toward maximizing the role of
chemotherapeutic agents in the treatment of primary brain tumors.

Several factors have contributed to the failure
to substantially improve survival among patients
with primary brain tumors. While some are attributable to the biology of the disease, others are due
to the limited activity of many agents and the
obstacles to effective delivery of therapeutic
agents within the brain. Advances in the understanding of the pathogenesis of primary brain
tumors have also resulted in the development of
novel therapeutic ▶ drug designs, many with high
specificity, due to the incorporation of structures
such as monoclonal antibodies to their structures.
Although highly specific in their targeting and
activity, these new agents also possess characteristics such as high molecular weight and polarity
and therefore may not be suitable for traditional
routes of drug delivery.

One physiological barrier that must be overcome is the ▶ blood–brain barrier (BBB). The
BBB is both a physical and a metabolic barrier
that allows entry of selected substance from the
circulation into the brain. Specifically, substances
with lipid solubility cross the cell membrane, as
well as those with specific transport systems. Substances with a molecular weight >500 Da cannot
cross through the BBB. This limitation may
exclude the use of otherwise highly bioactive
chemotherapeutic agents currently in development; in other words, antitumor agents must be
small in size and lipophilic in nature in order to
reach the brain following enteral or parenteral
administration.

Free concentration gradients of substances are
the driving force for diffusion as a passive transport mechanism. Diffusion of substances in the
interstitial space of the brain is largely dependent
on the molecular weight of the compound, with
higher-molecular-weight compounds resulting in
less diffusion than of smaller-molecular-weight
compounds. Diffusion is also a very slow process
and the desired agent is subject to many forces that
may limit its diffusive capacity, such as capillary
uptake and metabolism. Due to these physical
limitations, in order to achieve a therapeutically

meaningful concentration of a drug, very high concentrations, often supratherapeutic, must be delivered to ensure passage of meaningful concentrations of drug beyond the immediate delivery site. Realization that diffusion results in only millimeter distances of drug penetration through the interstitial space, when the biology of primary brain tumors dictates that regions, an order of magnitude greater needs to be covered with the desired agent, which led to interest in utilizing the brain's physiology, interstitial bulk flow, as a mechanism for drug delivery to the brain.

Although the brain itself lacks a lymphatic system, the interstitial space is a dynamic compartment where bulk flow of fluid occurs under normal physiologic conditions. Augmenting this bulk flow, by the initiation of a point source of positive iatrogenic pressure generated by a pump connected to a surgically implanted catheter in the brain, results in fluid convection. When desired agents are dissolved in the diluents and delivered through the catheter, the positive hydrostatic pressure results in distribution of the agents through the interstitial space in a radial, thus spherical, direction. CED has consistently resulted in centimeter-radius volumes of distribution and is able to distribute these agents, independent of their molecular weight, polarity, and concentration. In fact, CED results in a less than one-log decrease in concentration of the delivered agent at the "leading edge" compared to its concentration at the catheter site, unlike the distribution achieved by using diffusion. With the direct infusion of the desired agent into the interstitial space, the BBB is circumvented and theoretically acts as a barrier to keep agents from entering the circulation, which decreases systemic toxicity. To that end, small, lipophilic agents are not deemed favorable agents for CED application.

Clinical Application

CED is a novel technique that allows locoregional drug distribution for the treatment of a locoregional disease such as primary brain tumors; however, CED is not yet considered a clinical standard of care and is practiced only in research settings. However, preclinical, phase I, phase II, and now phase III clinical trials have

demonstrated CED to be well tolerated and safe for drug delivery into the brain. Adverse effects, seen in ~30% of patients undergoing CED, appear to be related to the increased cerebral edema (the accumulation of fluid in the brain) following drug infusion. Neurological deficits often respond to medical techniques aimed at reducing cerebral edema and thus are transient.

While the volume of distribution achieved by CED increases linearly with the volume of infusion, several technical factors have also been found to influence the efficacy of CED. As the delivery device situated within the brain, the catheter has received particular attention. One of the difficulties encountered with CED is the phenomenon of reflux, or leak back, of the agent along the catheter. This is encountered with high infusion rates and large catheter diameters. To combat reflux, research has focused on catheter design in an effort to create a reflux-proof catheter.

With regard to infusion rates, one of the disadvantages of CED is the need to infuse the desired agent over a protracted period of time, often lasting several days in duration. This necessity stems from the reflux problem mentioned above which, should it occur, limits the volume of distribution of the agent. Therefore, a rate of infusion is chosen which exceeds the rate at which the brain can remove fluid from the interstitial space but is less than the rate at which leak back may occur.

The anatomic complexity of the brain and of the region afflicted by a primary brain tumor also affects the efficacy of CED. Due to the presence of tumor tissue, white matter, and gray matter, the interstitial space is not uniform. Furthermore, the pattern of gyri and sulci and, therefore, the subarachnoid space, the proximity of the ependymal layer that lines the ventricle space, and the presence of tumor necrosis complicate catheter positioning, which must be accomplished after thorough presurgical planning. The subarachnoid space, ependymal layer, and tumor necrosis all represent low-resistance areas that would lead to the potential loss of convection. Radiographic correlate of effective convection is detected on T2-weighted magnetic resonance imaging sequence as an increase in the fluid signal within the targeted region.

Currently, there are several targeted therapeutic agents in advanced clinical development. The most advanced along its development is the agent IL13-PE38QQR, which is a chimeric protein based on the fusion of IL-13 as a ligand and the *Pseudomonas aeruginosa* exotoxin as the cytotoxic agent. The IL-13 receptor is known to be overexpressed in high-grade primary brain tumors, while the exotoxin is a potent inducer of cell death by arresting cellular protein synthesis.

IL13-PE38QQR has undergone rigorous testing thus far, and the results of three phase I/II trials were presented. In aggregate, 74 patients were enrolled in these trials, which determined the maximally tolerated dose of the agent. In addition, patient outcomes were compared when CED was conducted with the delivery catheters located in the peritumoral region compared to the intratumoral space. Improved survival was seen among patients undergoing peritumoral infusions; within this group of patients receiving peritumoral infusions, those who had more than two "optimally" placed catheters had a significantly improved median survival.

A phase III trial was initiated in 2004 which compared IL13-PE38QQR, delivered by CED, to another local drug delivery technique that relied on diffusion for drug dispersion. Three hundred patients were randomized to this trial, and the final results remain to be reported.

Future Directions

While CED represents a novel drug delivery technique, it also remains a field in evolution. Several lines of research are currently focusing on areas for continued improvement. Questions remain regarding the optimal placement and number of CED catheters. In phase I/II trials of IL13-PE38QQR, patients with "optimally" placed catheters had better outcomes. Whether this will be confirmed in the phase III trial will be of great interest. With regard to pharmacokinetic parameters, efforts are underway to increase the half-life of therapeutic agents, allowing the agents to remain available long after CED is halted. Investigators have demonstrated that encapsulation of their agents in liposomes is one such strategy. Another consideration is that, depending on the pharmacokinetic characteristics of the therapeutic agent, prolonged infusions may yield greater clinical efficacy. In that situation, alternative delivery methods of CED, such as implanted pumps housed entirely under the skin, may provide protection from infectious complications yet retain the drug delivery advantage.

Cross-References

▸ Blood-Brain Barrier
▸ Brain Tumors
▸ Chemotherapy
▸ Drug Design
▸ Locoregional Therapy

References

Krauze MT, Forsayeth J, Park JW et al (2006) Real-time imaging and quantification of brain delivery of liposomes. Pharm Res 23:2493–2504

Muro K, Das S, Raizer JJ (2006) Convection-enhanced and local delivery of targeted cytotoxins in the treatment of malignant gliomas. Technol Cancer Res Treat 5:201–213

Vogelbaum MA (2007) Convection enhanced delivery for treating brain tumors and selected neurological disorders: symposium review. J Neurooncol 87:97–109

See Also

(2012) Cerebral Edema. In: Schwab M (ed) Encyclopedia of Cancer, 3rd edn. Springer Berlin Heidelberg, p 750. doi:10.1007/978-3-642-16483-5_1034

(2012) Diffusion. In: Schwab M (ed) Encyclopedia of Cancer, 3rd edn. Springer Berlin Heidelberg, p 1116. doi:10.1007/978-3-642-16483-5_1620

(2012) Ependyma. In: Schwab M (ed) Encyclopedia of Cancer, 3rd edn. Springer Berlin Heidelberg, p 1267. doi:10.1007/978-3-642-16483-5_1926

(2012) Gyrus. In: Schwab M (ed) Encyclopedia of Cancer, 3rd edn. Springer Berlin Heidelberg, p 1620. doi:10.1007/978-3-642-16483-5_2542

(2012) Interstitial Space. In: Schwab M (ed) Encyclopedia of Cancer, 3rd edn. Springer Berlin Heidelberg, p 1899. doi:10.1007/978-3-642-16483-5_3110

(2012) Liposomes. In: Schwab M (ed) Encyclopedia of Cancer, 3rd edn. Springer Berlin Heidelberg, p 2063. doi:10.1007/978-3-642-16483-5_3388

(2012) Monoclonal Antibody. In: Schwab M (ed) Encyclopedia of Cancer, 3rd edn. Springer Berlin Heidelberg, p 2367. doi:10.1007/978-3-642-16483-5_6842

(2012) Pharmacokinetics. In: Schwab M (ed) Encyclopedia of Cancer, 3rd edn. Springer Berlin Heidelberg, p 2845. doi:10.1007/978-3-642-16483-5_4500

(2012) Subarachnoid Space. In: Schwab M (ed) Encyclopedia of Cancer, 3rd edn. Springer Berlin Heidelberg, p 3552. doi:10.1007/978-3-642-16483-5_5546

(2012) Sulcus. In: Schwab M (ed) Encyclopedia of Cancer, 3rd edn. Springer Berlin Heidelberg, p 3555. doi:10.1007/978-3-642-16483-5_5559

Coordination Compound

► Metal Drugs

COP9/CSN5

► JAB1

Core Binding Factor A2

► Runx1

Corin (TMPRSS10)

► Serine Proteases (Type II) Spanning the Plasma Membrane

Cortactin

Scott A. Weed
Department of Neurobiology and Anatomy, Mary Babb Randolph Cancer Center, West Virginia University, Morgantown, WV, USA

Synonyms

Amplaxin; EMS1; Src8

Definition

Cortactin is a protein that is a component of the cortical actin cytoskeleton, where it participates in regulating the assembly and organization of filamentous actin in protrusive structures generated during cellular movement. Cortactin gene ► amplification and overexpression is found in several cancer types, where it contributes to enhanced tumor cell ► motility, ► invasion, and ► metastasis.

Characteristics

Cortactin is an actin-binding protein and kinase substrate that is intimately associated with the microfilament network underlying the plasma membrane in most cells. It plays an important role in ► signal transduction pathways that mediate chemotactic cues from the extracellular environment that initiate and maintain cell ► migration. Activation of growth factor receptors or ► adhesion molecules results in the phosphorylation of cortactin at several tyrosine and serine residues. Cortactin phosphorylation is coincident with changes in plasma membrane architecture that occur during the initial phases of cellular movement, including the formation of lamellipodia and circular dorsal ruffles that are required for the extension of a cell's leading edge. Circular dorsal ruffles are transient regions of cell membrane that extend from the surface of cells as they initially respond to growth factor stimulation; they are thought to supply membrane and protein components required for sustained lamellipodia formation, as well as regulating growth factor receptor internalization. Cortactin is also enriched in invadopodia, ventral protrusive structures that contain membrane-bound proteases and enhance cellular invasion by facilitating the focal degradation of extracellular matrix. In addition to its role in cell motility, cortactin is also associated with various intracellular membrane compartments, including endosomal vesicles and the Golgi apparatus, and plays an important role in the early events of ► endocytosis and in vesicle trafficking.

Cortactin, Fig. 1 Domain structure of cortactin and associated binding proteins. This is a simplified representation showing domain organization, binding proteins, and regulatory signaling pathways. See text for details

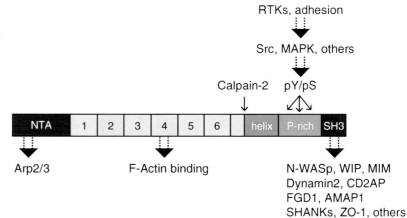

Structure and Binding Partners

Cortactin is expressed in all tissues except cells of myeloid lineage, where it is functionally replaced by the related protein HS1. Based on primary sequence analysis, cortactin is subdivided into several distinct domains (Fig. 1). The amino-terminal domain (NTA) contains a series of acidic residues and a binding motif that interacts with the Arp2/3 complex. The NTA domain is followed by a series of 37 amino acid tandem repeats, six complete and one incomplete in the predominant isoform. The repeat region interacts with F-actin, with binding activity centered around the fourth repeat. Alternative splicing in some cells is responsible for two additional isoforms that lack either the sixth complete or fifth and sixth complete repeat segments. These forms bind F-actin at reduced affinities. Following the repeats region is an alpha-helical domain that is the site of cleavage by the protease ► calpain 2. This is followed by a proline-rich region that harbors serine, threonine, and tyrosine residues that serve as the primary sites of phosphorylation. An SH3 domain is found at the extreme carboxyl terminus that binds to proline-rich sequences on a variety of proteins including the actin regulatory proteins N-WASp, WASp-interacting protein and the missing in metastasis protein, the endocytic proteins dynamin 2 and CD2AP, the small ► GTPase regulatory proteins FGD1 and AMAP1, scaffolding proteins of the SHANK family, and the ► tight junction protein ZO-1. These structural parameters and binding partners allow cortactin to function as a molecular scaffold by linking a wide variety of diverse regulatory molecules to sites of Arp2/3-mediated actin assembly.

Function

The function of cortactin has been best defined in regard to cell motility. Downregulation of cortactin protein expression reduces cellular movement, while overexpression of cortactin enhances this process. Biochemical studies have determined that cortactin activates Arp2/3 complex actin nucleation activity through the NTA domain, and its localization within lamellipodia indicates that cortactin contributes to the formation of the dendritic cortical actin network responsible for lamellipodia protrusion. Important in this aspect is the ability of cortactin to stabilize Arp2/3-produced actin networks, a feature unique among Arp2/3-activating proteins that serves to prolong the half-life of branched F-actin filaments at the cell periphery. Accordingly, cortactin depletion reduces the ability of extended lamellipodia to persist and inhibits efficient leading edge dynamics. Cortactin can effect Arp2/3-mediated actin polymerization by additional alternative mechanisms, most notably by activation of the Arp2/3 regulatory protein N-WASp through association with the cortactin SH3 domain. Cortactin fragments lacking the NTA but containing the SH3 domain are capable of stimulating motility, suggesting that the NTA and SH3 domains can function independently with regard to promoting actin-based cell movement. The interaction of cortactin with dynamin 2 is also noteworthy in that cortactin is recruited to subpopulations of clathrin-coated pits by dynamin 2 and is important for driving the scission of invaginating pits to produce intracellular endocytic vesicles. The cortactin-dynamin complex is also important in regulating cell morphology, invadopodia function, and the genesis of vesicles from the trans-Golgi network.

Regulation

Evidence to date indicates that phosphorylation on tyrosine and serine residues is the main factor involved in regulating cortactin function, although the precise mechanisms are unclear. Activation of ▶ receptor tyrosine kinases or adhesion molecules leads to phosphorylation of three tyrosine sites in the proline-rich domain that are required for efficient cell migration. These sites are direct targets of ▶ Src and related non-receptor tyrosine kinases and are hyperphosphorylated by oncogenic variants (i.e., v-Src). Tyrosine-phosphorylated cortactin is enriched within lamellipodia and invadopodia, indicating a potential role in regulating cortical actin dynamics and has been shown to influence F-actin architecture. Cortactin is also phosphorylated on two serine residues by ▶ MAP kinase in the proline-rich domain, and dual phosphorylation of cortactin by Src and MAP kinase has opposing effects on the ability of the cortactin SH3 domain to interact with and activate N-WASp. This has led to the proposal of a regulatory phosphorylation switch mechanism predicated by cortactin initially existing in an autoinhibited closed conformation, with the SH3 domain binding back and interacting with motifs in the proline-rich domain. Phosphorylation of cortactin by MAP kinase induces a conformation change that renders the SH3 domain accessible for binding and activating N-WASp, whereas phosphorylation of cortactin by Src causes disassociation of N-WASp from the SH3 domain and subsequent downregulation of N-WASp activity. This proposal remains theoretical in part since it is derived primarily from biochemical analysis and evidence for an intramolecular cortactin interaction is lacking. In addition

to phosphorylation indirectly regulating N-WASp activity, the serine/threonine kinase PAK1 phosphorylates cortactin within the first tandem repeat, resulting in reduced F-actin binding. Subsequent work has identified over 17 additional phosphorylation sites in every domain except the SH3, but the responsible signaling pathways and functional significance of these modifications are currently unknown. Besides phosphorylation, cortactin is also regulated by the calcium-dependent protease calpain 2, which cleaves cortactin between the repeats and alpha-helical domain and is important in limiting the extent of lamellipodia protrusion.

Role in Cancer

The cortactin gene (*CTTN*, formerly *EMS1*) maps to chromosome 11q13.3, a region that is frequently amplified in a number of cancers with inherently high invasive and metastatic potential, including ▶ Brms1, head and neck, ovarian, bladder, and ▶ hepatocellular carcinomas. 11q13 and *CTTN* amplification is associated with poor pathological outcome parameters including increased tumor recurrence, advanced disease stage, poor histological differentiation, increased lymph node metastasis, and reduced disease-specific survival. Mechanistically, cortactin overexpression as a result of *CTTN* amplification increases tumor cell motility and invasion as well as preventing the internalization and ubiquitylation-mediated degradation of EGF receptor, a receptor tyrosine kinase often overexpressed in carcinomas that is a potent activator of Src and MAP-K. Sustained EGF receptor activity as a result of *CTTN* amplification and cortactin overexpression promotes increased cortactin tyrosine phosphorylation, which has been shown to enhance distant metastasis of breast carcinoma cells in ▶ mouse models. EGF receptor inhibitors suppress tumor cell invasion and cortactin tyrosine phosphorylation, providing further support for the clinical relevance of cortactin phosphorylation in human cancer. Specific functions for cortactin in tumor cell invasion have been identified, most notable being its absolute role in the signaling and structural requirements governing the formation and function of invadopodia (Fig. 2). Cortactin is required to recruit and sequester the main

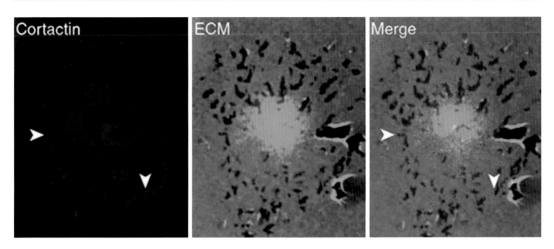

Cortactin, Fig. 2 Cortactin localization in invadopodia corresponds to sites of extracellular matrix degradation. Shown is a cell from a head and neck squamous cell carcinoma tumor containing invadopodia, visualized by immunofluorescent staining for cortactin in red as focal dots within the cell cytoplasm. The cell was grown on a *green* fluorescent extracellular matrix (*ECM*), and sites of matrix degradation are visualized as cleared dark regions against the *green* background. When merged, these areas correspond with cortactin-labeled invadopodia and are highlighted with *arrowheads*

invadopodial ▶ matrix metalloproteinase MT1-MMP into sites of newly initiated invadopodia. Cortactin tyrosine phosphorylation levels within invadopodia correlate to the degree of extracellular matrix degradation activity, but the functional significance of cortactin phosphorylation in invadopodia is currently undefined. Cortactin in invadopodia forms a complex with the focal adhesion protein ▶ paxillin and other signaling proteins. Often present in invasive carcinomas is amplification and overexpression of AMAP1, which physically links paxillin and cortactin together in promoting tumor invasion. Targeting of the trimeric paxillin-AMAP1-cortactin complex with competitive peptides mimicking the AMAP1 binding site for the cortactin SH3 domain suppresses carcinoma invasion and may show potential value in antimetastatic therapy. In addition to its role in promoting tumor cell invasion and metastasis, cortactin has been shown to be a prominent tumor antigen and is present at high levels in the sera of subsets of breast cancer patients. Recent work has determined that cortactin is an extracellular ligand for TEM7, a transmembrane receptor expressed primarily on the surface of tumor endothelial cells. While the function of TEM7 is currently unknown, related TEM proteins promote endothelial cell growth and survival, raising the possibility that cortactin released into the circulation from necrotic or damaged tumor cells, especially tumors with *CTTN* amplification, may serve an unexpected role by promoting or maintaining tumor ▶ angiogenesis.

References

Artym VV, Zhang Y, Seillier-Moiseivitsch F et al (2006) Dynamic interactions of cortactin and membrane type 1 matrix metalloproteinase at invadopodia: defining the stages of invadopodia formation and function. Cancer Res 66:3034–3043

Bryce NS, Clark ES, Leysath JL et al (2005) Cortactin promotes cell motility by enhancing lamellipodial persistence. Curr Biol 15:1276–1285

Cosen-Binker LI, Kapus A (2006) Cortactin: the gray eminence of the cytoskeleton. Physiology 21:352–361

Rossum AG, van Schuuring-Scholtes E, van Buuren-van Seggelen V et al (2005) Cortactin overexpression results in sustained epidermal growth factor receptor signaling by preventing ligand-induced receptor degradation in human carcinoma. Breast Cancer Res 7:235–237

Rothschild BL, Shim AH, Ammer AG et al (2006) Cortactin overexpression regulates actin-related protein 2/3 complex activity, motility and invasion in carcinomas with chromosome 11q13 amplification. Cancer Res 66:8017–8025

Cowden Disease

▶ Cowden Syndrome

Cowden Syndrome

Deborah J. Marsh[1] and Roberto T. Zori[2]
[1]Kolling Institute of Medical Research and Royal North Shore Hospital, University of Sydney, Sydney, NSW, Australia
[2]University of Florida, Gainesville, FL, USA

Synonyms

Cowden disease; Multiple hamartoma syndrome; PTEN hamartoma-tumor syndrome

Definition

Cowden syndrome (CS, OMIM#158350), along with Bannayan-Riley-Ruvalcaba syndrome (BRR, OMIM#153480), Peutz-Jeghers syndrome (PJS, OMIM#175200), and juvenile polyposis syndrome (JPS, OMIM#174900), is a member of a group of rare autosomally dominant inherited conditions classified as the hamartoma syndromes. Proteus syndrome (PS, OMIM#176920), although most frequently of sporadic presentation, has also been classified as part of this hamartoma-tumor syndrome spectrum. CS takes its name from the proposita of the first family described. It is characterized by an increased risk of developing breast, thyroid, and endometrial cancer along with the presence of hamartomas in multiple organ systems. The most frequently mutated susceptibility gene for both CS and BRR is the tumor-suppressor *PTEN* (alternatively named *MMAC1* or *TEP1*).

Characteristics

CS displays variable expressivity within families, which however usually presents in the third decade. CS hamartomas are present in tissues derived from all three germ cell layers, specifically the breast, thyroid, skin, central nervous system, and gastrointestinal tract. The incidence of CS has been estimated to be 1 in 200,000 individuals. Ninety-nine percent of CS patients display the hallmark CS hamartoma known as trichilemmomas in addition to mucocutaneous papules. Seventy percent of female CS patients develop breast fibroadenomas (benign tumor characterized by the proliferation of fibrous (stromal) and glandular tissue), 40–60% have thyroid adenomas, while gastrointestinal polyps occur in 35–40% of CS patients. Breast cancer develops in 25–50% of female CS patients and on occasion also in males, while thyroid cancer develops in 3–10% of all affected individuals. Disease of the central nervous system, most often benign but in some cases malignant, occurs in 40% of cases. Lhermitte-Duclos disease, a condition of dysplastic gangliocytoma of the cerebellum manifesting as seizures, tremors, and poor coordination, has been reported in conjunction with CS. Megencephaly or macrocephaly occurs in approximately 38% of CS patients. Other abnormalities including those of the genitourinary tract may also be present. Given the subtle and poorly recognized physical findings in individuals with CS, it is thought that this condition may often be underdiagnosed. The International CS Consortium has developed and revised diagnostic criteria to aid in the identification of this syndrome. Further, a clinical scoring system for selection of patients for PTEN mutation testing has been proposed, known as the PTEN Cleveland Clinic Score, or CC score.

CS shows partial clinical overlap with BRR as patients with either syndrome may develop intestinal hamartomatous polyps, macrocephaly (in nearly 100% of cases of BRR), and lipomas (occurring frequently in BRR but in a minority of patients with CS). Other features of BRR include very early age of onset, pigmented macules of the glans penis ("speckled penis") in males, hemangiomas, mild mental retardation, and developmental delay. Both the unique and overlapping clinical features of CS and BRR are described in the Table 1. A number of anecdotal cases of

Cowden Syndrome, Table 1 Clinical features seen in Cowden syndrome and Bannayan-Riley-Ruvalcaba syndrome

	Cowden syndrome	Bannayan-Riley-Ruvalcaba
CNS	Lhermitte-Duclos disease	Developmental delay[a], seizures, myopathy
Endocrine	Multinodular goiter, adenoma, thyroid anomalies, thyroiditis, hypothyroidism	Hashimoto's thyroiditis[a], diabetes mellitus
Growth disturbances		
Generalized	Macrocephaly	Macrosomia at birth, enlarged penis and testes, localized overgrowth, macrocephaly
Skin	Facial trichilemmoma[b], acral keratoses, hemangioma, mucosal lesions, papillomatous papules[b], hypertrichosis, vitiligo, pseudoacanthosis nigricans, skin cancers (basal cell, squamous cell, and melanoma)	Penile lentigines, acanthosis nigricans, verruca vulgaris type facial changes, tongue polyps, Café au lait spots, angiokeratoma, lipoma/lipomatosis[a]
Gastrointestinal	Polyps in entire gastrointestinal tract, gastrointestinal cancers	Polyps in distal gastrointestinal tract
Breast	Fibrocystic breast disease, adenocarcinoma	–
Other benign tumors	Uterine leiomyoma, ovarian cysts, fibroma, meningioma, glioma, neuroma	Meningioma, angiolipoma, hemangioma[a] (especially intracerebral and bony) lymphangioma
Other malignant tumors	Thyroid (non-medullary)[b], cervix, uterus, bladder, liver, renal, acute myelogenous leukemia, non-Hodgkin lymphoma, liposarcoma, trichilemmomal carcinoma, renal cell carcinoma	–

[a]Common in BRR, occasionally seen in CS
[b]Reported in CS, occasionally seen in BRR

malignancy affecting the thyroid and brain have been reported in BRR, however, while malignancy is well described in CS, it is not part of the classic BRR phenotype. Nevertheless, because of the apparent increased risk of developing cancer in BRR recommendations for cancer screening have been made. In addition, a number of families have been reported in which both CS and BRR are present. In these families, CS is generally present in the parental generation, while BRR appears in the younger generation, suggesting a form of anticipation (Figs. 1 and 2).

Molecular Features

PTEN, the first protein tyrosine phosphatase shown to function as a tumor suppressor, is the most frequently mutated susceptibility gene for both CS and BRR. A processed pseudogene is located on chromosome band 9p21, missing the initiating methionine present in PTEN but sharing greater than 98% homology with the *PTEN* coding region.

PTEN contains nine exons encoding a dual-specificity ▸ phosphatase mapped to 10q23.3, with homology to the cytoskeletal proteins tensin and auxillin. Residues 122–132 located in exon 5 encode the classic phosphatase core motif (I/V) HCXXGXXR(S/T)G. The COOH-terminus contains three potential tyrosine phosphorylation sites at residues 240, 315, and 336, as well as two potential serine phosphorylation sites at residues 335 and 338. It also contains a potential PDZ binding domain encoded by the last four amino acids (ITKV) that may have a role in its subcellular localization and/or substrate interactions.

PTEN has been shown to reduce tyrosine phosphorylation of focal adhesion kinase (FAK) in vitro suggesting a role in cell migration and invasion. However, the major endogenous substrate of PTEN would seem to be phosphatidylinositol 3,4,5-trisphosphate (Ptd-Ins(3,4,5)P$_3$), a phospholipid in the phosphatidylinositol 3-kinase (PI-3 kinase) pathway and an important second messenger in cell growth regulation. In this pathway, growth factors such as insulin, platelet-derived growth factor, and fibroblast growth factor stimulate the enzyme PI3-kinase to phosphorylate Ptd-Ins(4,5)P$_2$ to produce Ptd-Ins(3,4,5)P$_3$. PTEN acts as a 3-phosphatase to dephosphorylate Ptd-Ins(3,4,5)P$_3$ to Ptd-Ins(4,5)P$_2$. When PTEN is

**Cowden Syndrome,
Fig. 1** Female with
Cowden syndrome. (**a**)
Macrocephaly (head
circumference of 59½ cm,
greater than the 97th
percentile). (**b**) Multiple
small papules of the tongue
and mouth

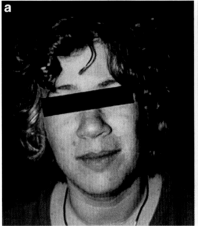
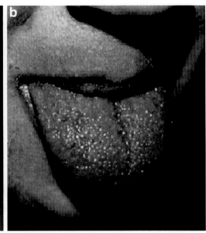

**Cowden Syndrome,
Fig. 2** Male with
Bannayan-Riley-Ruvalcaba
Syndrome (son of female in
Fig.1). (**a**) Macrocephaly
(head circumference of
59 cm, greater than the 97th
percentile). (**b**) Multiple
hyperpigmented macules of
the penis

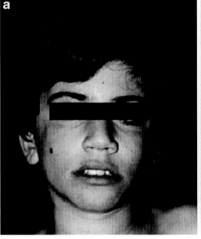
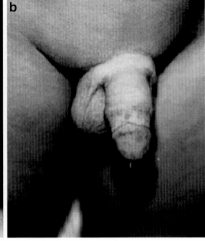

mutant, Ptd-Ins(3,4,5)P_3 accumulates and activates protein kinase B (PKB)/AKT to function as an oncogene, thus causing the tumorigenic state. AKT is a serine-threonine kinase and a known cell survival (antiapoptotic) factor. Thus, apoptosis is a likely mechanism for PTEN-induced growth suppression. However, PTEN is also able to cause cell cycle arrest in cells in the G_1 phase, possibly via modulation of levels of RB phosphorylation.

Elucidation of the crystal structure of PTEN has revealed a wider and deeper phosphatase active site than is usually described in other dual-specificity phosphatases that allows the accommodation of Ptd-Ins (3,4,5)P_3. Further, the makeup of the residues in this pocket causes it to have a positive charge consistent with the negative charge of Ptd-Ins (3,4,5)P_3 and with the preference displayed by PTEN for highly acidic polypeptide substrates. One particular germline mutant found only in CS, G129E, has been shown to have normal phosphatase activity against nonphospholipid substrates in vitro and in cell lines but has no phosphatase activity against Ptd-Ins(3,4,5)P_3. It is believed that mutation of this residue reduces the size of the active pocket so that it can no longer accommodate the phospholipid substrate Ptd-Ins(3,4,5)P_3.

However, catalysis of the smaller substrates of phospho-tyrosine, serine, and threonine is not disrupted. Furthermore, a C2 domain is present in the C-terminal domain and associates over an extensive interface with the phosphatase domain creating interdomain hydrogen bonds between conserved residues. This interphase region provides strong evidence that the C2 domain not only functions to recruit substrate but also optimally positions it available to the phosphatase catalytic domain. Germline mutations of conserved residues involved in the creation of this interphase, including serine at position 170, have been reported in BRR. Thus, there is strong evidence that the lipid phosphatase activity of PTEN is essential for its tumor-suppressor activity. Germline variants have been identified in PTEN mutation negative CS and CS-like patients, including in the genes *SDHB, SDHC, SDHD, AKT1, PIK3CA* and *SEC23B*.

Clinical Aspects

PTEN is mutated in the germline of up to 80% of patients with CS and up to 60% of patients with BRR. Mutations are scattered largely along the entire gene with the exception of exon 1, including point mutations, insertions, deletions, deletion-insertions, and splice site mutations. Germline mutations and deletions have also been reported in the promoter region of *PTEN* in patients with CS and BRR. In BRR alone, gross hemizygous deletions and also a balanced translocation likely affecting the *PTEN* gene have been reported. Further, loss of the wild-type allele has been identified in hamartomas from a subset of CS individuals with *PTEN* mutation, providing additional evidence that PTEN is functioning as a classic tumor suppressor according to Knudson's two-mutation model. However, there are many cases where loss of the wild-type allele is not observed in affected CS tissue. As is suggested by one of the $Pten^{+/-}$ mouse models, PTEN haploinsufficiency may be all that is required for the presence of the characteristic developmental

defects and tumor formation seen in CS and BRR. It was first thought that none of the 3 $Pten^{+/-}$ mouse models described developed the classic benign and malignant tumors of CS and BRR, although the presence of colonic microscopic hamartomatous polyps not dissimilar to what is seen in CS and BRR was reported. However, a study of $Pten^{+/-}$ mice older than 6 months reported the development of a range of tumors more similar to the spectrum of tumors observed in CS patients, specifically breast tumors in 50% of females, 100% of females with endometrial hyperplasia and a high incidence of endometrial cancer, prostate and adrenal neoplasia, and tumors of the gastrointestinal tract.

PTEN germline mutations in CS and BRR have been found to cluster in exons 5, 7, and 8, with the great majority occurring in exon 5. This may be a function of the fact that exon 5 is the largest exon of this gene, constituting 20% of the coding region, but this exon also contains the protein tyrosine phosphatase (PTPase) core motif. Of note in CS, most mutations that occur in the core motif are non-truncating, suggesting the importance of this functional domain.

Identical mutations, including Q110X, R130X, R233X, and R335X, have been reported in both CS and BRR, making the presence of other genetic and/or epigenetic factors such as modifier loci highly likely in the determination of phenotype. Furthermore, a number of families have been reported with CS diagnosed in the older generation and BRR present in the younger generation suggesting some form of anticipation that is currently not well understood. From this, it could be concluded that BRR and CS are different presentations of a single syndrome with broad clinical expression. In fact, it has been suggested that *PTEN* mutation-positive CS and BRR patients should be clinically grouped as a single entity and classified as the "PTEN hamartoma-tumor syndrome" (PHTS).

DNA-based predictive testing programs can now be incorporated as part of the clinical management of CS and BRR individuals. At the level of clinical management, cancer surveillance coupled with genetic counseling becomes important for CS and BRR patients as well as their first-degree relatives.

Genotype-phenotype correlations have been reported for both CS and BRR and a number of trends observed. Firstly, in a study of BRR and CS/BRR overlap families, the correlation of a germline *PTEN* mutation with the presence of lipomas and also with any cancer or breast fibroadenoma was determined. In CS families, a number of correlations were observed including an association between the presence of a *PTEN* mutation and breast involvement and the presence of a missense mutation and the involvement of all five organ systems (i.e., the breast, thyroid, gastrointestinal tract, central nervous system, and skin). It is possible that this latter trend may in fact be a positional effect given that the majority of missense mutations occur in the PTPase core motif. One study states that the presence of a *PTEN* mutation in either CS alone, BRR alone, or CS/BRR overlap families predisposes individuals to the presence of tumors, whether they be benign such as the lipomas seen predominantly in BRR or malignant such as the breast, thyroid, and uterine carcinomas seen in CS or CS/BRR overlap families. Confirmation of these findings requires analysis of a larger number of families before they can be directly transferred to the clinic.

Clinical cancer surveillance of patients with CS is recommended. All patients with a *PTEN* mutation should undergo careful annual physical examinations with special attention to the skin and thyroid from the teens. For females, breast self-examination from 18 years, annual clinical breast examination from age 25, and annual mammography/MRI at age 30–35 years are recommended. Thyroid ultrasound is recommended at 18 years with annual thyroid ultrasounds thereafter. Annual surveillance of the endometrium, with biopsies of the endometrium from the thirties and annual transvaginal ultrasound examination with biopsy of suspicious areas after menopause, should be performed. Urine should be checked annually for blood. All specific cancer screening should be started at least 5–10 years earlier than the earliest appearance of the specific cancer in the family.

In addition to being mutated in the germline of patients with CS and BRR, *PTEN* has been described as "…the most highly mutated tumor-suppressor gene in the post-p53 era…." It is mutated in a spectrum of human malignancies including glioblastoma (where *PTEN* mutation would seem to be a late event in tumor progression), endometrial hyperplasias (likely an early event) and carcinomas, prostate cancer, and malignant melanoma and less commonly in thyroid neoplasias and breast and colon cancer. Thus, it is likely that syndromic hamartomas and cancers in CS and BRR develop on a background created by loss of the tumor-suppressor function of PTEN. Furthermore, *PTEN* is a highly significant gene in the development of a wide range of sporadic human cancers.

References

Cristofano A, DiPesce B, Cordon-Cardo C et al (1998) *Pten* is essential for embryonic development and tumour suppression. Nat Genet 19:348–355

Marsh DJ, Kum JB, Lunetta KL et al (1999) *PTEN* mutation spectrum and genotype-phenotype correlations in Bannayan-Riley-Ruvalcaba syndrome suggest a single entity with Cowden syndrome. Hum Mol Genet 8:1461–1472

Pilarski R, Eng C (2004) Will the real Cowden syndrome please stand up (again)? Expanding mutational and clinical spectra of the PTEN hamartoma tumour syndrome. J Med Genet 41:323–326

Tan M-H, Mester J, Peterson C et al. (2011) A clinical scoring system for selection of patients for PTEN mutation testing is proposed on the basis of a prospective study of 3042 probands. Am J Hum Genet 88:42–56

Waite KA, Eng C (2002) Protean PTEN: form and function. Am J Hum Genet 70:829–844

Yehia L, Niazi F, Ni Y et al. (2015) Germline heterozygous variants in SEC23B are associated with Cowden syndrome and enriched in apparently sporadic thyroid cancer. Am J Hum Genet 97:661–676

Zhou XP, Waite KA, Pilarski R et al (2003) Germline PTEN promoter mutations and deletions in Cowden/Bannayan-Riley-Ruvalcaba syndrome result in aberrant PTEN protein and dysregulation of the phosphoinositol-3-kinase/Akt pathway. Am J Hum Genet 73:404–411

COX

▶ Arachidonic Acid Pathway

COX-2

▶ COX-2 in Colorectal Cancer

COX-2 in Colorectal Cancer

Raymond N. DuBois
ASU Biodesign Institute, Tempe, AZ, USA

Synonyms

COX-2; Cyclooxygenase (prostaglandin endoperoxide synthase 2)

Definition

COX-2 (PTGS2) is an enzyme that converts arachidonic acid to prostaglandin H2 (PGH2), an unstable endoperoxide intermediate. PGH2 is further converted to prostaglandins such as PGE2, PGD2, PGF2a, and PGI2 and thromboxane A2.

Characteristics

▶ Colorectal cancer remains a significant health concern for much of the industrialized world, even though mortality rates are beginning to decline in the USA. Diagnosis often occurs at a late stage in the progression of this disease, which reduces the likelihood of treatment being effective. Current treatment strategies often include a combination of surgical resection and adjuvant chemotherapy (▶ adjuvant therapy). Because of the unsatisfactory outcome of existing treatment methods, much emphasis has been placed on developing new treatment, prevention and screening strategies. Numerous population based studies indicate that use of ▶ nonsteroidal anti-inflammatory drugs (NSAIDs) reduce the risk for colorectal cancer and decrease the incidence of adenomatous polyps. NSAIDs induce polyp regression in familial adenomatous polyposis [FAP (▶ APC gene in Familial Adenomatous Polyposis)] patients and reduce tumor burden in animal models of colorectal cancer.

COX-2 and Colorectal Cancer Prevention

COX-2 mRNA and protein levels are increased in intestinal tumors that develop in rodents following carcinogen treatment and in adenomas taken from multiple intestinal neoplasia (Min) mice. When intestinal epithelial cells are forced to express COX-2 constitutively, they develop phenotypic changes that include increased ▶ adhesion to extracellular matrix (ECM) and resistance to butyrate-induced ▶ apoptosis. Both of these phenotypic changes are consistent with an increased tumorigenic potential. COX-2 expression has been detected in 80–90% of colorectal ▶ adenocarcinoma but in only 40–50% of premalignant adenomas. These data suggest that elevation of COX-2 expression is secondary to other initiating events such as dysregulation of the APC signaling pathway (▶ APC Gene in Familial Adenomatous Polyposis) and/or dysfunction of other genes affected during the adenoma to carcinoma sequence (▶ multistep development).

The observation of elevated COX-2 expression in three different models of colorectal carcinogenesis has led to consideration of the possibility that COX-2 expression may be related to colorectal tumorigenesis in a causal way. Studies have demonstrated a significant reduction in premalignant and malignant lesions in carcinogen-treated rats that were given a selective COX-2 inhibitor.

Tumor growth requires the maintenance and expansion of a vascular network. It has been demonstrated using in vitro assays that COX-2 can influence ▶ angiogenesis, and treatment with selective COX-2 inhibitors blocks angiogenesis. COX-2 appears to contribute to tumor vascularization, and there seems to be a link between COX-2 and regulation of VEGF expression.

Summary

Both preclinical and clinical data indicate that selective COX-2 inhibitors have antineoplastic activity. The precise role of COX-2 and the ▶ prostaglandins produced by this enzymatic pathway in carcinogenesis remains to be clearly delineated. Overexpression of COX-2 in epithelial cells leads to inhibition of ▶ apoptosis and increased adhesiveness to extracellular matrix. Inhibition of COX-2 activity leads to a marked reduction of tumor growth in a number of experimental models. Treatment with selective COX-2 inhibitors has been clearly shown to inhibit tumor-induced angiogenesis. The most effective role for selective COX-2 inhibitors for prevention and treatment of human cancers is currently under investigation.

Cross-References

- ▶ Adenocarcinoma
- ▶ Adhesion
- ▶ Adjuvant Therapy
- ▶ Angiogenesis
- ▶ APC Gene in Familial Adenomatous Polyposis
- ▶ Apoptosis
- ▶ Colorectal Cancer
- ▶ Multistep Development
- ▶ Nonsteroidal Anti-Inflammatory Drugs
- ▶ Prostaglandins
- ▶ Vascular Endothelial Growth Factor

References

DuBois RN, Abramson SB, Crofford L et al (1998) Cyclooxygenase in biology and disease. FASEB J 12:1063–1073

Smalley W, DuBois RN (1997) Colorectal cancer and non steroidal anti-inflammatory drugs. (August T, Anders MW, Myrad F, Coyle JT, eds). Adv Pharmacol 39:1–20

Williams CS, Mann M, DuBois RN (1999) The role of cyclooxygenases in inflammation, cancer and development. Oncogene 18:7908–7916

See Also

(2012) Epithelial Cell. In: Schwab M (ed) Encyclopedia of Cancer, 3rd edn. Springer Berlin Heidelberg, pp 1291–1292. doi:10.1007/978-3-642-16483-5_1958

(2012) Extracellular Matrix. In: Schwab M (ed) Encyclopedia of Cancer, 3rd edn. Springer Berlin Heidelberg, p 1362. doi:10.1007/978-3-642-16483-5_2067

(2012) Min. In: Schwab M (ed) Encyclopedia of Cancer, 3rd edn. Springer Berlin Heidelberg, p 2318. doi:10.1007/978-3-642-16483-5_3750

(2012) VEGF. In: Schwab M (ed) Encyclopedia of Cancer, 3rd edn. Springer Berlin Heidelberg, p 3906. doi:10.1007/978-3-642-16483-5_6174

COX-2 Inhibitors

▶ Colorectal Cancer Chemoprevention

CpG Islands

Christoph Plass
German Cancer Research Center (DKFZ), Heidelberg, Germany

Synonyms

HpaII tiny fragments islands; HTF islands

Definition

CpG islands are short stretches of DNA sequences with an unusually high GC content and a higher frequency of CpG dinucleotides as compared to the rest of the genome. Together CpG islands account for about 1–2% of the genome, and their location is mainly in the 5′ regulatory regions of all housekeeping genes as well as up to 40% tissue specifically expressed genes.

Characteristics

With the rapid accumulation of sequencing data, it became obvious that the distribution of the four bases, adenine (A), cytosine (C), guanine (G), and thymine (T), in the genomic sequence is not even. Normal DNA has an average GC content of 40%

and an AT content of 60%. Early work by Bird et al. (1985) identified stretches of genomic sequence characterized by an unusual high number of HpaII restriction sites (restriction site: C^CGG). These sequences were initially called "HpaII tiny fragments (HTF) islands." Careful inspection of those sequences indicated that the ratio of CpG dinucleotides is higher than in the rest of the genome. Normal DNA sequence contains only 25% of the CpG dinucleotides expected from the base composition. These stretches of DNA sequence, with a high GC content and a frequency of CpG dinucleotides that is close to the expected value, are now called CpG islands.

The following three criteria, established by Gardiner-Garden and Frommer (1987), are commonly used to define CpG islands: First, the sequence is longer than 200 bp but can be up to several kilo base pairs in size. Second, the GC content is above 50%, while the rest of the genome is at about 40%. Third, the CpG ratio (observed/expected) is above 0.6, while the rest of the genome is 0.2. Changes to this definition have been proposed using slightly modified criteria.

The human genome contains about 29,000 CpG islands, and the estimated number in the mouse genome is slightly less. The majority of these sequences are located in the 5′ region (promoter and or exon 1) of all housekeeping genes and a large number of tissue specifically regulated genes. However, CpG islands in the 3′ end of genes or in intronic sequences have been found. The preferential location of CpG islands in 5′ regions of genes can be used for the identification of novel genes. Rare cutting restriction enzymes with GC-rich recognition sequences such as NotI (GC^GGCCGC), AscI (GG^CGCGCC), BssHII (G^CGCGC), and EagI (C^GGCCG) can be used for the restriction mapping of large genomic clones. Clusters of those restriction enzyme cutting sites would indicate the presence of a CpG island. It is unknown why the mouse genome has fewer CpG islands than the human genome, when the estimated number of genes in both genomes is expected to be very similar. One possible explanation is that the rate of CpG dinucleotide loss due to deamination (see below) is higher in the mouse than it is in the

human genome. The location of CpG islands is in the early replicating, less condensed, and GC-rich R-bands of chromosomes.

Preservation of CpG Islands

The origin of CpG islands in the vertebrate genomes is closely associated with DNA ▸ methylation and a process called deamination. DNA methylation in the vertebrate genomes is found mainly in CpG dinucleotides. Those CpG dinucleotides that are located within CpG islands are usually unmethylated. However, CpG dinucleotides located outside of CpG islands are methylated at the 5′ position of the cytosine. Methylation of CpG dinucleotides makes these sites vulnerable to spontaneous deamination leading to a transition of the 5-methyl-cytosine to thymine, a process that is believed to be the cause for depletion of CpG dinucleotides from the genome.

Clinical Relevance

Although CpG islands are usually unmethylated, there are a few important exceptions. Methylation in CpG islands has been correlated with the transcriptional silencing of the adjacent genes. The detailed molecular process controlling this inactivation, however, is not known. Protein complexes, containing the methyl CpG-binding protein and corepressor enzymes modifying histone tails as major components, are able to bind to methylated promoters. These protein complexes induce histone deacetylation, which mediates the formation of transcription-repressing chromatin. In in vitro experiments, re-expression could be achieved by adding trichostatin A (TSA), a specific inhibitor of histone deacetylases. Two general methylation events in CpG islands can be distinguished: first, the developmentally regulated process of CpG island methylation found in the inactive X-chromosomes, in promoter regions of genes that are regulated in a tissue-specific manner, and in imprinted genes and, second, the aberrant CpG island methylation in cancer.

Normal, Developmentally Regulated, CpG Island Methylation

- Most CpG islands in the inactive X-chromosome of females are densely methylated. This process of X-chromosome inactivation is linked to the transcriptional silencing of genes on the inactive X-chromosome (phosphoglycerate kinase 1 (PGK1), glucose 6-phosphate dehydrogenase (G6PD), or androgen receptor (AR). Exceptions are found in a few number of CpG islands in genes that escape X inactivation (e.g., STS, ZFX, or UBE1)).

- Some CpG islands become methylated in other normal developmental processes including cell differentiation and aging. The result of this methylation is the selective inactivation of genes in specific tissues or at certain developmental stages (e.g., estrogen receptor).
- Genes that are expressed from either the paternal or the maternal allele are called imprinted genes. These genes are found to have CpG island methylation of one allele. While methylation usually occurs in the inactive allele, CpG island methylation was found in some instances in the active allele. This feature of allele-specific methylation in a CpG island was used as a tag for the identification of novel imprinted genes in the mouse using the ► restriction landmark genomic scanning (RLGS) technique for a genome wide scan for patterns of allele-specific methylation.

Aberrant CpG Island Methylation in Cancer

Hypermethylation of CpG islands in various cancers has been observed and is correlated with the transcriptional inactivation of tumor suppressor genes and other cancer-related genes. It was shown that methylation in a CpG island can serve as one of the two "hits" needed for the inactivation of a tumor suppressor gene. While CpG island methylation in some tumors is restricted to a small number of CpG islands, other tumors show a methylation phenotype with up to 10% methylated CpG islands. A subset of CpG islands is methylated in a tumor-type specific matter, while other CpG islands can be methylated in different tumor types. It was also shown that many of the genes associated with methylated CpG islands could be reactivated in cell lines by experimental demethylation using 5′-aza-2′deoxycytidine.

Cross-References

► Epigenetic Gene Silencing

References

Baylin SB, Herman JG, Graff JR et al (1998) Alterations in DNA methylation: a fundamental aspect of neoplasia. Adv Cancer Res 72:141–196

Bird A, Taggart M, Frommer M et al (1985) A fraction of the mouse genome that is derived from islands of nonmethylated, CpG-rich DNA. Cell 40:91–99

Costello JF, Frühwald MC, Smiraglia DJ et al (2000) Aberrant CpG island methylation has non-random and tumor type specific patterns. Nat Genet 25:132–138

Gardiner-Garden M, Frommer M (1987) CpG islands in vertebrate genomes. J Mol Biol 196:261–282

CPT-11

► Irinotecan

Cr(VI)

► Hexavalent Chromium

Cr⁶⁺

Cr^{6+}

► Hexavalent Chromium

cRaf

► Raf Kinase

C-Raf

► Raf Kinase

Cripto-1

Caterina Bianco
Division of Extramural Activities, National
Institutes of Health, Rockville, MD, USA

Synonyms

TDGF-1; Teratocarcinoma-derived growth factor-1

Definition

Human Cripto-1 is a cell membrane-associated
protein important for embryonic development,
stem cell renewal, and tumorigenesis.

Characteristics

Structure of Cripto-1, a Member of the EGF-CFC Protein Family

Human Cripto-1 (CR-1), originally identified
from a human embryonal carcinoma cDNA
library, is the founding member of the epidermal
growth factor (EGF)-CFC (Cripto in humans,
FRL1 in *Xenopus*, and Cryptic in mice) family
of proteins identified only in vertebrates. The
EGF-CFC protein family includes monkey
Cripto-1, mouse Cripto-1 (Cr-1), chicken
Cripto-1, zebra fish *one-eyed pinhead* (*oep*),
Xenopus FRL1, and mouse and human Cryptic.
EGF-CFC proteins contain multiple domains
consisting of an amino-terminal signal peptide, a
modified EGF-like domain, a cystein-rich CFC
motif, and a short hydrophobic carboxy-terminus
containing, in some cases, consensus sequences
for a glycosylphosphatidylinositol (GPI) anchorage site that serves to attach the protein to the cell
membrane (Fig. 1). EGF-CFC proteins are mostly
found to be cell membrane associated. However,
human CR-1 can also be detected in the conditioned medium of several cancer cell lines and in
the plasma of colon and breast cancer patients,
probably by cleavage of the glycosylphosphatidylinositol (GPI) linkage by GPI-specific enzymes.

An overall sequence identity of approximately
30% exists between the EGF-CFC members
across different species. Within the EGF-like
domain, there is a 60–70% sequence similarity,
whereas in the CFC motif, the similarity ranges
from 35 to 48%. The modified EGF-like domain
corresponds to a region of approximately
40 amino acids containing six cysteine residues.
Whereas the canonical EGF-like domain that is
present in the EGF family of growth factors
(▶ Epidermal Growth Factor Receptor Ligands),
such as EGF, transforming growth factor-α
(TGF-α), and heregulins, contains three loops
(A, B, and C) due to the presence of three intramolecular disulfide bonds, the variant EGF-like
domain in the EGF-CFC proteins lacks the
A loop, has a truncated B loop, and possesses a
complete C loop. The presence of this unusual
EGF-like domain explains the observation that
CR-1 does not directly bind to any of the known
*erb*B type I tyrosine kinase receptors including the
EGF receptor, *erb*B2, *erb*B3, and *erb*B4.
EGF-CFC proteins are glycoproteins that range
from 171 to 202 amino acids with an unmodified
core protein of 18–21 kDa in size. The native
mouse and human Cripto-1 proteins are 24, 28,
and 36 kDa in size, although proteins ranging in
size from 14 to 60 kDa have been identified in
mouse and human normal tissues. This variation
in size could be due to the removal of the hydrophobic signal peptide and to posttranslational
modifications of the core protein. In fact, all the
members of the EGF-CFC family, except for *oep*,
are glycoproteins that contain a single *N*-▶ glycosylation site and potential *O*-glycosylation sites
(Fig. 1). A single O-linked ▶ fucosylation site
(Fig. 1) has been identified within the EGF-like
domain of human CR-1, and a single point mutation in the fucosylation consensus sequence
results in the loss of Cripto-1-dependent Nodal
signaling (see below).

Cripto-1 During Embryonic Development and in Embryonic Stem (ES) Cells

Cripto-1 functions as a co-receptor for the TGF-β
family ligands, Nodal and Vg1/growth, and differentiation factor 1 and 3 (GDF1 and 3), during
early vertebrate embryogenesis. Genetic studies

Cripto-1,
Fig. 1 Schematic diagram
of the human Cripto-1
protein domains. Sites of
glycosylations are indicated
by *arrows*

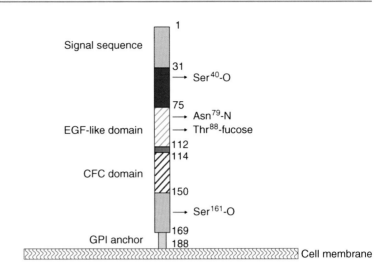

in zebra fish and mice have defined an essential role for Nodal that functions through Cripto-1 in the formation of the primitive streak, patterning of the anterior/posterior axis, specification of the mesoderm and endoderm during gastrulation, and establishment of left/right asymmetry of developing organs. Cripto-1-dependent Nodal signaling depends upon the Activin type II (Act RII) and type I (Alk4) serine/threonine kinase receptors that activate the Smad-2/Smad-3 intracellular signaling pathway (▶ Smad Proteins in TGFβ Signaling). Evidence from several studies suggests that Cripto-1 recruits Nodal to the Act RII/Alk4 receptor complex by interacting with Nodal through the EGF-like domain and with Alk4 through the CFC domain. Cr-1 null mice die at day 7.5 due to their inability to gastrulate and form appropriate germ layers. Disruption of Cr-1 in Cr-1$^{-/-}$ embryos results in the formation of embryos that possess a head without a trunk, demonstrating that there is a severe deficiency in mesoderm and endoderm without a loss of anterior neuroectoderm formation. Homozygous knock out of the Cr-1 gene in pluripotential ▶ embryonic stem cells (ES cells) impairs their ability to differentiate in vitro into cardiomyocytes without affecting the ability of ES cells to differentiate in other cell types. In fact, Cr-1$^{-/-}$ ES cells show extensive neuronal differentiation in vitro and in vivo, suggesting that Cripto-1 could represent a key molecule required for both induction of cardiomyocyte

differentiation and repression of neural differentiation. In this regard, Cr-1$^{-/-}$ ES cells, when transplanted in vivo at low doses, generate a pool of dopaminergic cells that are able to induce behavioral and anatomical recovery in animal models of Parkinson's disease. It has been established that Cripto-1 is a ▶ stem cell marker in mouse and human ES cells and in conjunction with Nanog, Nodal, Oct3/4, and GDF3 is involved in maintaining self-renewal and pluripotentiality of ES cells. Since malignant ES cells are probably the most appropriate targets for therapy in cancer, stem cell markers could be used as a signature to identify adult tissue cancer stem cells.

Cripto-1 in Mammary Gland Development

Ovarian hormones and several growth factors, such as TGF-β, TGF-α, EGF, and insulin-like growth factor have been shown to play a crucial role in the regulation of the development and maturation of the mammary gland. In the mouse mammary gland, Cr-1 is detected during different stages of postnatal mammary gland development. In fact, Cr-1 protein has been detected in 4–12-week-old virgin, midpregnant, and lactating mouse mammary gland. In the virgin pubescent mammary gland, Cr-1 expression is observed in the cap stem cells of the growing terminal end buds and in the ductal epithelial cells from pregnant and lactating mice. Expression of Cr-1 in ductal epithelial cells is enhanced by approximately 3- to 5-fold during pregnancy and lactation. Further support for CR-1 regulation of mammary epithelial cells derives from data showing that CR-1 can modulate milk protein expression in HC-11 mouse mammary epithelial cells. HC-11 mouse mammary epithelial cells express the milk protein β-casein after exposure to the lactogenic hormones dexamethasone, insulin, and prolactin (DIP). Prior treatment of HC-11 cells with exogenous CR-1 during logarithmic growth induces a competency response to DIP with respect to the induction of the milk protein β-casein. In contrast, simultaneous treatment of HC-11 cells with CR-1 in the presence of DIP inhibits β-casein expression. This inhibitory effect of CR-1 on milk protein expression may be biologically significant since soluble CR-1 protein can be found in human milk.

Cripto-1 in Transformation, Tumorigenesis, and Angiogenesis

A first clue to the biological activity of Cripto-1 derives from studies demonstrating the ability of human CR-1 to transform mouse NIH-3 T3 fibroblasts, mouse NOG-8, and mouse CID-9 mammary epithelial cells in vitro. However, NOG-8 and CID-9 transformed cells are unable to form tumors in nude mice, suggesting that additional genetic alterations are necessary to complete the tumorigenic phenotype in vivo. Further support for the transforming potential of Cripto-1 derives from studies showing increased expression of Cripto-1 in cells transformed by different oncogenes. In this regard, Ha-*ras* (► Ras) has been shown to upregulate Cr-1 expression in rat CREF embryo fibroblasts or rat FRLT-5 thyroid epithelial cells. Also, v-*ras*/Smad-7 transformed keratinocytes develop skin tumors that overexpress CR-1 and TGF-α, suggesting that Smad-7 induces tumor formation through upregulation of CR-1 and other EGF-related peptides. Overexpression of Cr-1 in EpH4 mouse mammary epithelial cells increases cell proliferation, anchorage-independent growth in soft agar, and the formation of branching structures when the cells are cultured in a three-dimensional type I collagen gel matrix. Furthermore, EpH4 Cr-1

cells show an increase in their migratory behavior in Boyden chamber studies and in wound-healing assay. Exogenous CR-1 protein is also able to stimulate chemotaxis of wild-type EpH4 cells and can induce scattering of NOG-8 mouse mammary epithelial cells grown at low density as colonies on plastic. The scattering effect is characterized by a change in morphology of the epithelial cells to a more fibroblastic-like phenotype and by a decrease in cell-cell adhesion due to reduction in ► E-cadherin expression. These findings suggest that Cripto-1 may play a role in inducing ► Epithelial to Mesenchymal Transition (EMT) of mammary epithelial cells. In fact, MCF7 breast cancer cells overexpressing CR-1 show increased invasion through matrix-coated membranes and mammary hyperplasias, and tumors from MMTV-CR-1 ► transgenic mice (see below) show a dramatic reduction in the levels of expression of the adhesion molecule E-cadherin, whereas the mesenchyme cell cytoskeleton component, vimentin, is significantly increased. Regulation of cell proliferation, cell motility, and survival by CR-1 is dependent upon activation of two major intracellular signaling pathways, the *Ras/Raf*/mitogen-activated protein kinase (MAPK) (► Map Kinase) and phopshatidylinositol 3′ kinase (PI3K)/Akt (► AKT Signal Transduction Pathway) signaling pathways. Activation of these two intracellular signaling pathways is independent of Nodal and Alk4, since CR-1 can activate MAPK and Akt in EpH4 mammary epithelial cells and MC3T3-E1 osteoblast cells that lack Nodal and Alk4 expression, respectively. Activation of these two signaling pathways is mediated by binding of CR-1 to the GPI-linked heparan sulfate proteoglycan Glypican-1, which can then activate the cytoplasmic tyrosine kinase c-Src triggering activation of MAPK and Akt. Finally, an intact c-Src kinase is required by CR-1 to induce in vitro transformation and enhance migration in mammary epithelial cells.

In addition to regulating cell proliferation and transformation, CR-1 plays an essential role in tumor ► angiogenesis. In fact, CR-1 has a strong angiogenic activity in vitro in cultured human umbilical vein endothelial cells (HUVECs), stimulating proliferation, migration, invasion, and differentiation of HUVECs into vascular-like structures when the cells are grown in Matrigel. Furthermore, recombinant CR-1 protein stimulates new blood vessel formation in silicone cylinders filled with Matrigel implanted under the skin of nude mice, and microvessel formation in response to CR-1 is significantly inhibited in vivo by an anti-CR-1 blocking mouse monoclonal antibody. Finally, tumor xenografts that develop from CR-1 overexpressing MCF-7 breast cancer cells in the cleared mammary fat pad of nude mice have a significantly higher microvessel density than tumor xenografts that form from control MCF7 cells.

Transgenic Mouse Models Overexpressing CR-1 in the Mammary Gland

Transgenic mouse models have shown that overexpression of a human CR-1 transgene in the mouse mammary gland under the control of the mouse mammary tumor virus (MMTV) or whey acidic protein (WAP) promoter results in mammary hyperplasias and adenocarcinomas. Virgin MMTV-CR-1 transgenic mice exhibit enhanced ductal branching, intraductal hyperplasias, and hyperplastic alveolar nodules. Approximately 30–40% of multiparous female mice develop papillary adenocarcinomas. The relatively long latency period suggests that additional genetic or regulatory alterations are required to facilitate mammary tumor formation in conjunction with CR-1. Unlike the MMTV promoter that starts to be active in the virgin mammary gland, the WAP promoter is maximally expressed at mid-pregnancy and lactation. Approximately 50% of old nulliparous WAP-CR-1 mice develop multifocal intraductal hyperplasias and more than half of multiparous WAP-CR-1 female mice develop multifocal mammary tumors of mixed histological subtypes. These tumors are a mixture of regions containing glandular, papillary, and undifferentiated carcinoma, as well as myoepithelioma and adenosquamous carcinoma. Mammary tumors of mixed histology are normally phenotypes that are associated with transgenic mice that have alterations in the canonical Wnt/β-catenin pathway (► Wnt Signaling).

In fact, increased expression of an activated β-catenin has been found in the mammary tumors of WAP-CR-1 transgenic mice, suggesting that a canonical Wnt pathway may be activated in these tumors.

Expression of CR-1 in Human Carcinomas and Premalignant Lesions

CR-1 is overexpressed, relative to noninvolved adjacent tissue, in ~50–90% of carcinomas that arise in the colon, breast, stomach, pancreas, lung, gall bladder, testis, bladder, ovary, endometrium, and cervix. Furthermore, enhanced expression of CR-1 has also been detected in premalignant lesions, such as colon adenomas, intestinal metaplasia of the gastric mucosa, and ductal carcinoma in situ of the breast. In this respect, the frequency and level of CR-1 expression in colon adenomas and intestinal metaplasia in the stomach are directly correlated with the size, histological subtype, and degree of dysplasia in these lesions, suggesting that CR-1 might be an early marker for malignant transformation in these tissues. CR-1 expression has also been detected in approximately 60% of normal colon mucosa specimens from individuals with a high incidence of colon carcinomas but only in 20% of colon mucosa from low-risk individuals. In addition, expression of CR-1 in the adjacent noninvolved colon epithelium surrounding colon tumors is significantly correlated with increased lymph node involvement and with a higher rate of recurrence of colorectal tumors. Although no significant correlations have been found between CR-1 expression and prognosis, a study suggests that CR-1 is an independent prognostic factor in breast cancer. In fact, in more than 100 invasive breast cancers, overexpression of CR-1 has been found more often in high grade and poor prognosis tumors compared to low grade and good prognosis breast cancers and is significantly associated with decreased patient survival. Another study has also demonstrated a significant increase in the plasma levels of CR-1 protein in patients affected by colon and breast carcinomas, suggesting that CR-1 might represent a novel serological marker for breast and colon cancer.

Cross-References

▶ Epidermal Growth Factor Receptor
▶ Epidermal Growth Factor-like Ligands
▶ Serum Biomarkers

References

Bianco C, Normanno N, Salomon DS et al (2004) Role of the cripto (EGF-CFC) family in embryogenesis and cancer. Growth Factors 22:133–139
Bianco C, Strizzi L, Normanno N et al (2005) Cripto-1: an oncofetal gene with many faces. Curr Top Dev Biol 67:85–133
Strizzi L, Bianco C, Normanno N et al (2005) Cripto-1: a multifunctional modulator during embryogenesis and oncogenesis. Oncogene 24:5731–5741

c-R*mil*

▶ BRaf-Signaling

Crohn Colitis

Synonyms

Granulomatous colitis; Crohn disease

Definition

Originally described as a small bowel process, is now known to involve the large bowel in approximately 40% of all cases, with or without a concomitant ileal component. Crohn colitis is ▶ inflammation that is confined to the colon. Abdominal pain and bloody diarrhea are the common symptoms. Anal fistulae and peri-rectal abscesses also can occur. This disease is histopathologically characterized by transmural and granulomatous inflammation. Obstruction and perforation, abscesses, fistulae, and intestinal bleeding are complications of Crohn colitis. Massive distention or dilatation of the colon

(megacolon), and rupture (perforation) of the intestine are potentially life-threatening complications. Extra-intestinal complications involve the skin, joints, spine, eyes, liver, and bile ducts. There is an increased risk of cancer of the small and large intestine in patients with long-standing Crohn disease (▶ Colon Cancer Carcinogenesis in Human and in Experimental Animal Models).

Cross-References

▶ Colon Cancer Carcinogenesis in Human and in Experimental Animal Models
▶ Crohn Disease
▶ Inflammation

Crohn Disease

Gerard Dijkstra[1] and Maikel P. Peppelenbosch[2]
[1]University Medical Center Groningen, University of Groningen, Groningen, The Netherlands
[2]Erasmus Medical Center, University Medical Center Rotterdam, Rotterdam, The Netherlands

Synonyms

Inflammatory Bowel Disease (also includes Ulcerative Colitis)-related Cancer; Nonprimary APC mutation Colorectal Cancer

Definition

Crohn disease (also: regional enteritis) and ulcerative colitis area chronic, inflammatory conditions of the gastrointestinal tract. Although the inflammation of the mucosa is usually episodic in nature, these Inflammatory bowel diseases (IBD) are associated with an increased risk of developing colorectal cancer (inflammatory bowel disease). Well-managed patients suffering from IBD have an approximately two times higher chance for contracting CRC as the population at large.

Characteristics

Inflammatory bowel disease (IBD) patients, which include the two related conditions of Crohn disease and ulcerative colitis, have an increased risk of developing colorectal cancer (CRC) (Itzkowitz and Yio 2004). For patients with Crohn disease [CD], the excess risk for contracting CRC has been estimated at 1.9, whereas the risk for small bowel cancer is 27.1 (Jess et al. 2005), for UC the risk appears considerably higher with a standardized incidence ratio of 2.4 (Munkholm 2003). The risk depends on disease duration, extent of inflammation, presence of primary sclerosing cholangitis, a positive family history of CRC, age of onset, and the degree of endoscopic and histologic activity (▶ Inflammation in Cancer). Furthermore, CRC accounts for about 15% of deaths related to IBD; however, IBD related colorectal carcinoma accounts for only 1–2% of all cases of CRC. Although the number of IBD-related CRC of the total cases of CRC is low, the mortality rate in patients with a diagnosis of CRCs in the setting of IBD is higher than for those afflicted with sporadic cases of CRC (▶ Cancer Epidemiology). Additionally, the risk for CRC is not related to disease activity, patients who are clinically quiescent do not have a lower risk for developing CRC compared to patients who suffer from a more active disease history. Moreover, during the last decades, the incidence of IBD has continued to rise worldwide, reaching incidence rates of 16.6/100,000 in North America and 9.8/100,000 in Europe. IBD, together with the hereditary syndromes of familial adenomatous polyposis (FAP), and hereditary nonpolyposis colorectal cancer (HNPCC), are the top three of high-risk conditions for CRC (colon cancer/gastrointestinal tumors/▶ Peutz-Jeghers-Syndrome).

Chemoprevention of IBD-Related CRC

Cancer chemoprevention is based on the arrest of one or several steps in the multistep carcinogenesis process, which attempts to block, reverse, or delay carcinogenesis before the development of invasive disease (▶ Chemoprotectants). Mesalazine (5-aminosalicylic acid (5-ASA)), for which there is long-term clinical experience in the treatment of patients with IBD, is well tolerated, has limited systemic side effects, and has no gastrointestinal toxicity (Schroeder 2002) (▶ Anti-Inflammatory Drugs). In a rodent model of colorectal cancer, mesalazine inhibits tumor growth and reduces the number of aberrant crypt foci (Brown et al. 2000) (▶ Colorectal Cancer Premalignant Lesions), whereas in patients with sporadic polyps or cancer of the large bowel mesalazine induces apoptosis and decreases proliferation in the colorectal mucosa. Epidemiological data from the chemopreventive action of mesalazine in IBD are inconsistent because of heterogeneity (Nguyen et al. 2012). Our own meta-analysis however supports a chemopreventive role for mesalazine in ulcerative colitis-associated colorectal cancer (Fig. 1). An important question is whether the chemopreventive effect of mesalazine is restricted to IBD-CRC or can be extrapolated to sporadic CRC as well.

Mechanistic Explanations

Confusingly many models as to the chemopreventive action for 5-ASA have been put forward, hampering rational decisions in this area. Reinacher-Schick et al. (2000) show that mesalazine induces apoptosis and decreases proliferation in colorectal mucosa in patients with sporadic polyps of the large bowel. In addition, Brown and coworkers (2000) show that mesalazine inhibits tumor growth and reduces the number of aberrant crypt foci in a rodent model of colorectal cancer. We ourselves show in a panel of human colon cancer cell lines that 5-ASA inhibits the Wnt/beta-catenin pathway through protein phosphatase 2A (Bos et al. 2006), but we also provided evidence that interferes with proliferation of colorectal cancer cells via inhibition of PLD-dependent generation of PA and loss of mTOR signaling (Baan et al. 2012). Schwab c.s. provide evidence for the involvement of PPARgamma in pro-apoptotic and antiproliferative actions of 5-ASA also in various CRC cell lines (Schwab

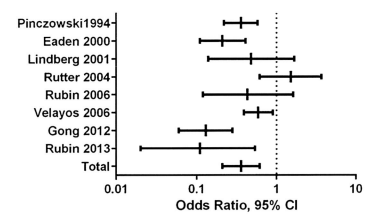

Crohn Disease, Fig. 1 Meta-analysis of the effect of mesalazine on IBD-associated CRC. Pubmed and Web of Science were searched for studies on the possible chemopreventive effect of mesalamine in UC-patients. Search terms: PubMed: ("Mesalamine"[Mesh] OR "Sulfasalazine"[Mesh]) AND "Colorectal Neoplasms"[Mesh] AND ("Colitis, Ulcerative"[Mesh] AND "Inflammatory Bowel Diseases"[Mesh]). Web of Science: ((mesalamine OR 5-asa OR mesalazine OR 5-aminosalicylic acid OR sulfasalazine OR sulphasalazine) AND (ulcerative colitis OR IBD) AND (colorectal cancer OR colorectal neoplasms OR colonic cancer OR colonic carcinoma)). Studies were excluded if they were reviews, in vitro studies, studies in any language but English, and studies using an animal model. Review manager 5 was used to conduct the meta-analyses. 274 abstracts were screened for eligibility. Full text review was performed on 25 studies. 8 studies were found suitable for inclusion in this meta-analysis, including a total of 5.467 patients. All studies are included, the data show that use of 5-ASA preparation conferred an *odds ratio of 0.36* (95% CI; 0.21–0.62) for development of CRC in IBD patients. Studies included are listed in Pinczowski et al. (1994), Eaden et al. (2000), Lindberg et al. (2001), Rutter et al. (2004), Rubin et al. (2006), Velayos et al. (2006), Gong et al. (2012), Rubin et al. (2013)

et al. 2008). Using both human and murine CRC lines, Monteleone c.s. show that mesalazine negatively regulates CDC25A protein expression, thus delaying CRC cell progression, independent of Chk1 and Chk2 (Stolfi et al. 2008), although in earlier work these authors also propose that 5-ASA disrupts EGFR signaling by enhancing SH PTP2 activity as a mechanism by which 5-ASA interferes with CRC growth (Monteleone et al. 2006). Alternatively, Chu et al. propose that downregulation of c-Myc is the key to its chemopreventive action in CRC (Chu et al. 2007). Gasche c.s. provide data that demonstrate that 5-ASA causes cells to reversibly accumulate in S phase and activate an ATR-dependent checkpoint (Luciani et al. 2007). In conclusion, it is fair to say that a plethora of mechanisms possibly mediating 5-ASA effects in chemoprevention of CRC has been proposed, and for this field to move forward a large-scale long-term study in patients will be essential.

Concluding Remarks

Chronic inflammation of the Colon, as observed in Crohn disease and ulcerative colitis, is closely linked to the development of CRC. The pathogenesis of this disease is different from the classical pathway, involving loss of APC function. Chemoprevention of CRC-development, using mesalazine, seems to be a promising chemopreventive strategy.

Cross-References

▶ Anti-inflammatory Drugs
▶ Cancer Epidemiology
▶ Chemoprotectants
▶ Colorectal Cancer Premalignant Lesions
▶ Crohn Colitis
▶ Inflammation
▶ Peutz-Jeghers Syndrome

References

Baan B, Dihal AA, Hoff E, Bos CL, Voorneveld PW, Koelink PJ, Wildenberg ME, Muncan V, Heijmans J, Verspaget HW, Richel DJ, Hardwick JC, Hommes DW, Peppelenbosch MP, van den Brink GR (2012) 5-Aminosalicylic acid inhibits cell cycle progression in a phospholipase D dependent manner in colorectal cancer. Gut 61(12):1708–1715

Bos CL, Diks SH, Hardwick JC, Walburg KV, Peppelenbosch MP, Richel DJ (2006) Protein phosphatase 2A is required for mesalazine-dependent inhibition of Wnt/beta-catenin pathway activity. Carcinogenesis 27:2371–2382

Brown WA, Farmer KC, Skinner SA, Malcontenti-Wilson C, Misajon A, O'Brien PE (2000) 5-aminosalicyclic acid and olsalazine inhibit tumor growth in a rodent model of colorectal cancer. Dig Dis Sci 45:1578–1584

Chu EC, Chai J, Ahluwalia A, Tarnawski AS (2007) Mesalazine downregulates c-Myc in human colon cancer cells. A key to its chemopreventive action? Aliment Pharmacol Ther 25:1443–1449

Eaden J, Abrams K, Ekbom A, Jackson E, Mayberry J (2000) Colorectal cancer prevention in ulcerative colitis: a case–control study. Aliment Pharmacol Ther 14(2):145–153

Gong W, Lv N, Wang B, Chen Y, Huang Y, Pan W, Jiang B (2012) Risk of ulcerative colitis-associated colorectal cancer in China: a multi-center retrospective study. Dig Dis Sci 57(2):503–507

Itzkowitz SH, Yio X (2004) Inflammation and cancer IV. Colorectal cancer in inflammatory bowel disease: the role of inflammation. Am J Physiol Gastrointest Liver Physiol 287:G7–G17

Jess T, Gamborg M, Matzen P, Munkholm P, Sørensen TI (2005) Increased risk of intestinal cancer in Crohn's disease: a meta-analysis of population-based cohort studies. Am J Gastroenterol 100:2724–2729

Lindberg BU, Broomé U, Persson B (2001) Proximal colorectal dysplasia or cancer in ulcerative colitis. The impact of primary sclerosing cholangitis and sulfasalazine: results from a 20-year surveillance study. Dis Colon Rectum 44(1):77–85

Luciani MG, Campregher C, Fortune JM, Kunkel TA, Gasche C (2007) 5-ASA affects cell cycle progression in colorectal cells by reversibly activating a replication checkpoint. Gastroenterology 132:221–235

Monteleone G, Franchi L, Fina D, Caruso R, Vavassori P, Monteleone I et al (2006) Silencing of SH-PTP2 defines a crucial role in the inactivation of epidermal growth factor receptor by 5-aminosalicylic acid in colon cancer cells. Cell Death Differ 13:202–221

Munkholm P (2003) Review article: the incidence and prevalence of colorectal cancer in inflammatory bowel disease. Aliment Pharmacol Ther 18(Suppl 2):1–5

Nguyen GC, Gulamhusein A, Bernstein CN (2012) 5--aminosalicylic acid is not protective against colorectal cancer in inflammatory bowel disease: a meta-analysis

of non-referral populations. Am J Gastroenterol 107(9):1298–1304

Pinczowski D1, Ekbom A, Baron J, Yuen J, Adami HO (1994) Risk factors for colorectal cancer in patients with ulcerative colitis: a case–control study. Gastroenterology 107(1):117–20

Reinacher-Schick A, Seidensticker F, Petrasch S, Reiser M, Philippou S, Theegarten D et al (2000) Mesalazine changes apoptosis and proliferation in normal mucosa of patients with sporadic polyps of the large bowel. Endoscopy 32:245–254

Rubin DT, Huo D, Kinnucan JA, Sedrak MS, McCullom NE, Bunnag AP, Raun-Royer EP, Cohen RD, Hanauer SB, Hart J, Turner JR (2013) Inflammation is an independent risk factor for colonic neoplasia in patients with ulcerative colitis: a case–control study. Clin Gastroenterol Hepatol 11(12):1601–8.e1–1601–8.e4

Rubin DT, LoSavio A, Yadron N, Huo D, Hanauer SB (2006) Aminosalicylate therapy in the prevention of dysplasia and colorectal cancer in ulcerative colitis. Clin Gastroenterol Hepatol 4(11):1346–1350

Rutter M, Saunders B, Wilkinson K, Rumbles S, Schofield G, Kamm M, Williams C, Price A, Talbot I, Forbes A (2004) Severity of inflammation is a risk factor for colorectal neoplasia in ulcerative colitis. Gastroenterology 126(2):451–459

Schroeder KW (2002) Role of mesalazine in acute and long-term treatment of ulcerative colitis and its complications. Scand J Gastroenterol Suppl 236:42–47

Schwab M, Reynders V, Loitsch S, Shastri YM, Steinhilber D, Schroder O et al (2008) PPARgamma is involved in mesalazine-mediated induction of apoptosis and inhibition of cell growth in colon cancer cells. Carcinogenesis 29:1407–1414

Stolfi C, Fina D, Caruso R, Caprioli F, Fantini MC, Rizzo A et al (2008) Mesalazine negatively regulates CDC25A protein expression and promotes accumulation of colon cancer cells in S phase. Carcinogenesis 29:1258–1266

Velayos FS, Loftus EV Jr, Jess T, Harmsen WS, Bida J, Zinsmeister AR, Tremaine WJ, Sandborn WJ (2006) Predictive and protective factors associated with colorectal cancer in ulcerative colitis: a case–control study. Gastroenterology 130(7):1941–1949

Crow-Fukase Syndrome

▶ POEMS Syndrome

CRP55

▶ Calreticulin

CRP-Ductin (Mouse)

▶ Deleted in Malignant Brain Tumors 1

CRT

▶ Calreticulin

Cryosurgery

René P. H. Veth and Bart H. W. Schreuder
Department of Orthopaedics, Radboud University
Medical Centre, Nijmegen, The Netherlands

Synonyms

Cold surgery; Freeze surgery

Definition

Operative cutting of tissue or the targeted destruction of pathological tissue by induced cold necrosis at temperatures down to $-196\,°C$.

Characteristics

Cryobiology
Cryobiology deals with the physical effects of low temperatures and the changing of temperatures in living tissues. The state or phase (vapor, liquid, or solid) of water depends on temperature, pressure, and volume. The liquid and solid phase of pure water are in equilibrium at atmospheric pressure and $0\,°C$. By increasing the pressure, this temperature ($0\,°C$) or freezing point can be lowered. This phenomenon is known as supercooling.

When its temperature is lowered, water will show vitrification or crystallization. Very rapid cooling of pure water will induce vitrification that entails the formation of amorphous, transparent, glasslike structures rather than crystals. Crystallization requires initiating nuclei, for instance, an insoluble crystalline impurity). Slow cooling rates of water ($<1\,°C/min$) will induce large crystals around a few nuclei. During fast cooling rates, many small crystals are formed which are thermodynamically unstable and tend to join each other by recrystallization to minimize their surface energies.

During freezing of solutions, ice crystals remove more and more pure water from the solution, elevating the dissolved solute concentration and lowering the vapor pressure of water to that of ice at the same temperature. In this situation, solid and liquid phase coexist, and this supercooled phase ends with a sudden rise of the temperature due to dissipation of latent heat generated by the recrystallization of the thermodynamically unstable small crystals. This phenomenon takes place at a eutectic temperature.

The nature of the tissue responding to low temperatures varies with the intensity of the induced cold. A minor cryogenic injury produces only an inflammatory response; a greater injury will produce tissue destruction. The effects of every physical state on living tissue can be divided in immediate and delayed effects. Immediate destructive properties of cryosurgery are the result of mechanical damage due to the formation of ice, whereas the delayed effects are due to progressive failure of the microcirculation (vascular stasis), tissue ischemia, and ultimately cell death. When tissue temperature is lowered without reaching subzero temperatures, cell metabolism is reduced. This is a reversible process and used to its benefit in cardiac surgery. However, if living tissue is continuously subjected to low, but nonfreezing temperatures, cell death will occur.

The freezing of tissue is more complicated since its solvent (water) is divided by cell membranes into extracellular and intracellular compartments. Cell membranes in general easily allow the passage of water, but far less readily allow passage of other solutes. When tissue is subjected to a constant slow lowering of temperature, it first enters a supercooled phase. Temperatures of 10–15 °C below zero will initiate ice formation in the extracellular compartment. The intracellular compartment remains unfrozen because it contains substances with high and low molecular weight, which lower freezing temperatures. Due to the freezing of water in the extracellular compartment, concentration of solutes will rise, creating an osmotic pressure-induced transport of water from the intra- to the extracellular compartment. This loss of water will lead to shrinkage of the cell, accompanied by higher concentrations of the solutes, which further prevent the formation of ice in the intracellular compartment.

The shrinkage and high concentration of solutes, especially of salts, may be responsible for cell injury. This phenomenon seems especially of importance during slow freezing rates. Very rapid cooling induces intracellular ice formation, because there is insufficient time for water leaving the cell to maintain osmotic equilibrium across the cell membrane. Intracellular ice formation is believed to be lethal to the cell. Based on histological investigations, it has been shown that intracellular ice causes mechanical damage to the membrane and disturbs the function of mitochondria and other cell organelles and membranes.

Furthermore, masses of frozen cells, closely packed, will be subjected to shearing forces of ice formation that will injure the tissue structure. Propagating ice will induce cell damage, regardless of the fact that ice is intra- or extracellular. Intracellular ice has to been shown to propagate from one cell to another via intercellular channels.

During thawing the "behavior" of the ice crystals is dependent on the rate of thawing. In contrast to rapid thawing, slow thawing is accompanied by recrystallization and the crystals can grow to damaging sizes. The damaging effect of these intracellular ice crystals, only formed during rapid freezing can therefore be exploited a second time, if slow thawing is allowed, thereby enhancing recrystallization. On the other hand, if tissues have been cooled slowly, causing shrinkage and intracellular dehydration, rapid thawing may be damaging because the cells are exposed to high electrolyte concentrations.

After thawing there is typically a brief period of vasodilation. Additionally the endothelium of blood vessels is particularly sensitive to freeze-thawing, leading to increased permeability of vascular walls, interstitial edema, slowing of circulation, and platelet aggregation. Capillary obstruction and vascular stasis ensue resulting in tissue ischemia and cell death.

Tumors Suitable for Cryosurgical Treatment

A number of benign and malignant ▶ bone tumors can be treated by cryosurgery. These include aneurysmal bone cyst, symptomatic enchondroma, borderline chondrosarcoma, low-grade chondrosarcoma, chondroblastoma, chordoma, and giant cell tumor of bone. In addition radio- and chemotherapy-resistant bone metastases may also be effectively treated. The same goes for benign aggressive soft tissue tumors.

Indications for Cryosurgery

Active or aggressive benign and low-grade malignant bone tumors are ideally treated by extralesional excision (marginal excision or wide excision). For tumors located in expendable bones, like ribs, this is the treatment of choice. However, since most benign and low-grade malignant bone tumors tend to occur in the metaphysis and/or epiphysis (the end of a bone that lies between the joint surface on one side and the epiphyseal plate (growth plate) on the other side) of long bones, marginal or wide excision would imply segmental loss of bone, compromising normal growth in children and loss of articular surface. Therefore, intralesional excision (curettage), combined with a powerful local adjuvant, is advocated in tumors in which this combination is equivalent to at least a marginal excision.

Cryosurgery is a powerful adjuvant therapy, and the main advantage is that the surgeon is in charge of the local extent beyond the surgical margin (7–12 mm) and is able to customize the treatment for a specific benign or low-grade malignant bone tumor.

Cryosurgical Technique

After sufficient exposure of the tumor, thorough curettage of the tumor is performed. To monitor the intralesional temperature and the local extent

Cryosurgery,
Fig. 1 Drawing of
cryosurgical technique for
bone tumors

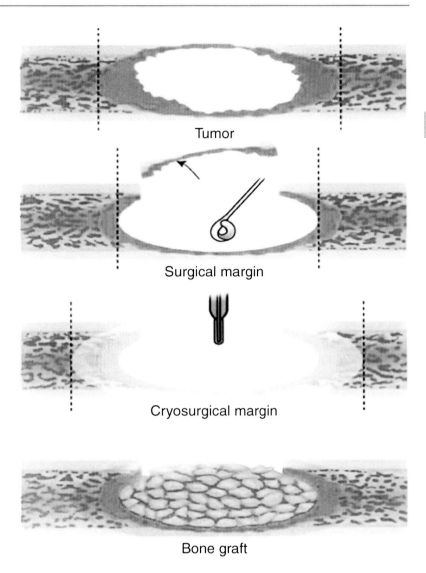

Tumor

Surgical margin

Cryosurgical margin

Bone graft

of the freeze, thermocouples are positioned in and around the lesion. Liquid nitrogen is sprayed in the cavity in every direction, until the whole cavity is wetted and becomes frosted. The duration of the freeze is based on the temperature readings and visual observation. Intralesional temperatures of at least minus 50 °C are pursued. After spontaneous thawing, two more cycles of freezing and thawing are done, to destroy tumor cells, which may have survived the previous cycle. Finally, the cavity is filled with allograft bone chips, and when feasible, the defect is reinforced with osteosynthesis to prevent pathological fracture of the bone.

Figure 1 shows this type of cryosurgical therapy for bone tumors.

Results of Treatment

All results will be presented according to the functional evaluation system of the Musculoskeletal Tumor Society. Veth et al. reported on 302 patients who had been treated by cryosurgery for a variety of bone and soft tissue tumors. At follow-up, 298 of these patients showed NED (no evidence of disease) or CDF (continuously free of disease), whereas 5 were AWD (alive with disease) and 2 DOD (dead of disease). The minimal follow-up period for this review was 2 years.

Of these cryosurgically treated patients, 43 had been diagnosed with giant cell tumor of bone, 15 with chondroblastoma, 73 with borderline chondrosarcoma, and 44 with chondrosarcoma grade 1. Chordoma was diagnosed in seven cases.

Most studies on giant cell tumor report on a rate of local recurrences varying from 0% to 47%. The risk for local recurrence for this tumor is greatly influenced by the method of first surgery. Primary surgery of a giant cell tumor in a non-bone tumor hospital is most likely to induce several local recurrences. The local recurrence rates in chondroblastoma and low-grade chondrosarcoma are, respectively, 7% and 3%. Figures for chordoma are small and rates of local recurrence vary to a great deal with the extent of the tumor. The functional results according to the MSTS system (MSTS functional evaluation system) for the first three tumors are good to excellent in 80% of cases.

Comparing the results, it appears that the risk for local recurrence is small after cryosurgery, compared to different types of treatment, and the functional results are at least similar. However, an import feature of cryosurgery is that tumor excision is never followed by prosthetic implants; thus, the patient keeps a biological reconstruction.

Complications

Possible complications of cryosurgery are:

Wound Infection

Cryosurgery appears to be accompanied by a deep infection rate of about 4%, but this differs between institutions. Sacral lesions are prone for developing an infection.

The following items are of importance in order to avoid infection: intraoperative broad-spectrum antibiotics, adequate drainage of wound fluids, avoidance of accidental freezing of the skin, and wound closure with sufficient soft tissue coverage.

Venous Gas Embolism

During cryosurgery liquid nitrogen is either sprayed or poured into the bony cavity, and since its boiling point is $-195\,^{\circ}\mathrm{C}$ nitrogen, gas bubbles are rapidly produced at room temperature. In general, whenever a gas is introduced into a body cavity, there is the hazard of intravascular introduction of gas bubbles especially when pressure is allowed to develop. Gas emboli in the vascular circulation can cause serious hemodynamic complications.

The risk is increased when the site of the tumor is located in a richly vascularized area such as the metaphysis of the long bones. Unfortunately, this is the location of preference for many bony tumors suitable for cryosurgical treatment.

Fracture

The tumor itself and the surgical exposure and resection jeopardize the structural integrity of the bone. Cryosurgery is said to further diminish bone strength by inducing necrosis of the local bone stock often leading to postoperative fractures.

In the late 1960s, the pioneers, who used cryosurgery for bone tumors, reported rather high fracture rates (up to 10%).

Fractures are most likely to occur 4–8 weeks after the cryosurgical treatment, but they can occur even after 8 months. Diaphyseal lesions are most prone for fracture. Therefore, prophylactic internal fixation is advised. Plate and screws are often used, which protects the bone especially from rotating forces. Intramedullary instrumentation is ill advised, because it has the risk of contaminating the entire intramedullary compartment with tumor cells. Titanium alloys are preferred because these implants induce little interference on MRI making tumor follow-up less difficult. Partial weight bearing is usually necessary until 3 months after the operation.

Experience and improvements in technique have reduced the fracture rate to an acceptable level of 1–2%.

Epiphyseal Damage

Benign bone tumors, especially simple- and aneurysmal bone cysts, tend to occur in patients of immature skeletal age. Furthermore, these tumors are commonly observed in the metaphysis, often adjacent or very close to the epiphysis. Damage of the epiphysis either by the tumor itself or the use of cryosurgery occurs and may result in arrest or disturbance of normal growth.

Whether an epiphysis is damaged by the bone tumor or by the treatment will not always become clear and in many cases may be the result of both.

Degenerative Osteoarthritis

Some bone tumors like giant cell tumor and chondroblastoma occur almost always extremely close to major joints. Damage of the articular surface either by the tumor itself (intra-articular fracture) or treatment (cryosurgery) may be the result, thus resulting in osteoarthritis.

Damage to Nerves

Nerve palsy is a complication of cryosurgery, which was recognized at the very early beginning of the introduction of cryosurgery for bone tumors.

If nerves are frozen, their function is only temporarily impaired. Most neuropraxias resulting from freezing will resolve in 6 weeks to 6 months. Very likely regenerating nerve fibers can grow down the nerve sheaths since they are left intact. Furthermore, the vital nerve cell nucleus is located away in the dorsal root ganglion. Tourniquets should not be used, in order to keep nerves and skin vascularized and thereby protect them from a freeze injury. Veth et al. saw in 302 cryosurgical procedures 10 nerve palsies; only one peroneal nerve failed to regain its function, however, not the cryosurgery but surgical traction was very likely the cause of this persistent palsy.

Prospects to the Future

Starting in the mid-1960s, cryosurgery has evolved from a medical tool with limited usefulness for treatment of all kind of tumors into a reliable technique, even for bone tumor patients. A study showed that two instead of three freeze-thaw cycles would be sufficient for tumor control. A study by Baust identified apoptosis as a cryosurgery-related mechanism of cell death, additive with ice-related cell damage and posttreatment coagulative necrosis. This may provide a possible route to molecular-based optimization of cryosurgical procedures and better results.

The use of cryosurgery in multi-recurrent schwannoma of peripheral nerves, chordoma, or other sacral tumors as well as its use in bone tumor areas where a peripheral nerve is often involved (proximal fibula) has shown that these nerves may dysfunction after cryosurgery for a period up to 6 months but in the end mostly recover completely. Additional study on the behavior of nerve tissue during cryosurgery is warranted in order to optimize the temperature for tumor cell kill and reduce the period of nerve dysfunction.

Cross-References

▶ Bone Tumors

References

Enneking WF, Dunham W, Gebhart MC (1993) A system for the functional evaluation of reconstructive procedures after surgical treatment of tumours of the musculoskeletal system. Clin Orthop Relat Res 286:241–246

Gage A, Baust J (1998) Mechanisms of tissue injury in cryosurgery. Cryobiology 37(3):171–186

Robinson D, Halperin N, Nevo Z (2001) Two freezing cycles ensure interface sterilization by cryosurgery during bone tumour resection. Cryobiology 43:4–10

Schreuder HWB (2001) Cryosurgery for bone tumors. In: Korpan NN (ed) Basics of cryosurgery. Springer, Wien, pp 231–253

Veth R, Schreuder B, van Beem H et al (2005) Cryosurgery in aggressive benign and low-grade malignant bone tumours. Lancet Oncol 6(1):25–34

See Also

(2012) Epiphysis. In: Schwab M (ed) Encyclopedia of cancer, 3rd edn. Springer Berlin Heidelberg, p 1291. doi:10.1007/978-3-642-16483-5_1952

(2012) Extralesional excision. In: Schwab M (ed) Encyclopedia of cancer, 3rd edn. Springer Berlin Heidelberg, p 1366. doi:10.1007/978-3-642-16483-5_2073

(2012) Intralesional excision. In: Schwab M (ed) Encyclopedia of cancer, 3rd edn. Springer Berlin Heidelberg, p 1900. doi:10.1007/978-3-642-16483-5_3122

(2012) Marginal excision. In: Schwab M (ed) Encyclopedia of cancer, 3rd edn. Springer Berlin Heidelberg, p 2168. doi:10.1007/978-3-642-16483-5_3538

(2012) MSTS functional Evaluation System. In: Schwab M (ed) Encyclopedia of cancer, 3rd edn. Springer Berlin Heidelberg, p 2383. doi:10.1007/978-3-642-16483-5_3865

(2012) Osteosynthesis. In: Schwab M (ed) Encyclopedia of cancer, 3rd edn. Springer Berlin Heidelberg, p 2670. doi:10.1007/978-3-642-16483-5_4290

(2012) Wide excision. In: Schwab M (ed) Encyclopedia of cancer, 3rd edn. Springer Berlin Heidelberg, p 3947. doi:10.1007/978-3-642-16483-5_6246

Cryptotanshinone

Shile Huang[1] and Wenxing Chen[2]
[1]Department of Biochemistry and Molecular
Biology and Feist-Weiller Cancer Center,
Louisiana State University Health Sciences
Center, Shreveport, LA, USA
[2]Department of Clinical Pharmacy, College of
Pharmacy, Nanjing University of Chinese
Medicine, Nanjing, China

Synonyms

(R)-1,2,6,7,8,9-hexahydro-1,6,6-trimethyl-
phenanthro(1,2-b)furan-10,11-dione; 1,2,6,7,8,9-
hexahydro-1,6,6-trimethyl-(R)-phenanthro
(1,2-b)furan-10,11-dione; 1,2,6,7,8,9-hexahydro-
1,6,6-trimethyl[1,2-*b*]furan-10,11-dione;
Tanshinone C

Definition

Cryptotanshinone is a cell-permeable diterpene
quinone and natural product isolated from the
root of *Salvia miltiorrhiza* Bunge (Fig. 1a), a
widely used herb in China for treatment of cardiovascular and cerebrovascular diseases.

Characteristics

Cryptotanshinone is one of the major tanshinones
(including tanshinone I, tanshinone IIA,
dihydrotanshinone, and cryptotanshinone)
derived from the traditional Chinese medicine
Salvia miltiorrhiza Bunge (simplified Chinese, ;
traditional Chinese, ; pinyin, *dānshēn*), also
known as red sage, Chinese sage, Dan Shen,
Dan-Shen, or Danshen (molecular formula,
$C_{19}H_{20}O_3$; molecular weight, 296.36; CAS registry number, 35825-57-1). The molecular structure
of cryptotanshinone is shown in Fig. 1b.

Cryptotanshinone is an orange-brown powder.
It is not water soluble, but soluble in dimethyl
sulfoxide (DMSO) or ethanol (~10 mM). It
should be stored under dry and cool (2–8 °C)
condition and protected from light.

Danshen has been used to treat a variety of
human diseases, including coronary artery disease, hyperlipidemia, acute ischemic stroke,
chronic renal failure, chronic hepatitis, and
Alzheimer disease. Danshen can also be used for
treatment of uterine myomas and menstrual problems in women. Studies have revealed that
cryptotanshinone not only has potential to prevent
ischemia, atherosclerosis, and Alzheimer disease
but also possesses diverse actions, such as
anti-inflammation, antidiabetes and antiobesity,
antiangiogenesis, antilymphangiogenesis, anti-
bacterial, and anticancer activities.

Cryptotanshinone,
Fig. 1 (**a**) Molecular
structure of
cryptotanshinone. (**b**)
Salvia miltiorrhiza Bunge
(http://bbs.zhong-yao.net/)

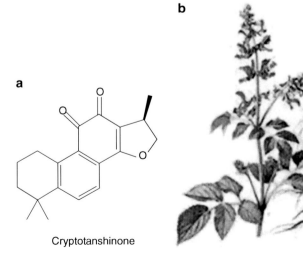

Cryptotanshinone

Antiangiogenesis

Cryptotanshinone inhibits proliferation in bovine aortic endothelial cells (BAECs) in culture. Also cryptotanshinone inhibits *basic fibroblast growth factor* (bFGF)-stimulated invasion and tube formation in BAECs. These data suggest that cryptotanshinone has antiangiogenic activity.

Antilymphangiogenesis

Cryptotanshinone inhibits tube formation in murine lymphatic endothelial cells (LEC), suggesting an antilymphangiogenic effect. This is partly attributed to downregulating protein expression of vascular endothelial growth factor (VEGF) receptor 3 (VEGFR-3), leading to decreased phosphorylation of the extracellular signal-related kinase 1/2 (ERK1/2). Additionally, cryptotanshinone also inhibits protein expression and activities of the small GTPases, such as Rac1 and Cdc42.

Anticancer Activities

Cryptotanshinone is also a potential anticancer agent. Cryptotanshinone inhibits cell proliferation and induces cell death in cancer cells. Cryptotanshinone inhibits cell proliferation by arresting cell cycle in G_1 or G_2/M phase depending on cell lines. It has been described that cryptotanshinone inhibits protein expression of cyclin D1 and phosphorylation of retinoblastoma protein (Rb), resulting in G_1 cell-cycle arrest, which is attributed to inhibition of the mammalian target of rapamycin (mTOR). Also, cryptotanshinone inhibits DU145 prostate cancer cell proliferation through inhibiting phosphorylation of the signal transducer and activator of transcription 3 (STAT3) (Tyr705). Cryptotanshinone inhibition of STAT3 is through the Janus-activated kinase 2 (JAK2)-independent mechanism, but by blocking the dimerization, nuclear translocation, and transcription activity of STAT3. In addition, cryptotanshinone induces reactive oxygen species (ROS), which activates c-jun N-terminal kinase (JNK)/p38 mitogen-activated protein kinase (MAPK) and inhibits extracellular signal-regulated kinases 1/2 (Erk1/2), leading to caspase-independent cell death in cancer cells. Moreover, cryptotanshinone downregulates androgen receptor (AR) signaling by inhibition of lysine-specific demethylase 1 (LSD1)-mediated demethylation of histone H3 lysine 9 (H3K9) and suppresses the transcriptional activity of AR in AR-positive prostate cancer cells. Cryptotanshinone also enhances the anticancer effects of TNF-α, Fas, cisplatin, etoposide, or 5-fluorouracil through inducing ER stress.

References

Chen W, Luo Y, Liu L et al (2010) Cryptotanshinone inhibits cancer cell proliferation by suppressing mammalian target of rapamycin-mediated cyclin D1 expression and Rb phosphorylation. Cancer Prev Res (Phila) 3:1015–1025

Luo Y, Chen W, Zhou H et al (2011) Cryptotanshinone inhibits lymphatic endothelial cell tube formation by suppressing VEGFR-3/ERK and small GTPase pathways. Cancer Prev Res (Phila) 4:2083–2091

Park IJ, Kim MJ, Park OJ et al (2010) Cryptotanshinone sensitizes DU145 prostate cancer cells to Fas (APO1/CD95)-mediated apoptosis through Bcl-2 and MAPK regulation. Cancer Lett 298:88–98

Shin DS, Kim HN, Shin KD et al (2009) Cryptotanshinone inhibits constitutive signal transducer and activator of transcription 3 function through blocking the dimerization in DU145 prostate cancer cells. Cancer Res 69:193–202

Crypts

Isabelle Gross
INSERM U1113, Université de Strasbourg, Strasbourg, France

Definition

Anatomically, crypts are narrow and deep invaginations into a larger structure. The small intestine and the colon are typically formed by glandular crypts that invaginate deep in the submucosa. In the small intestine, crypts are called crypts of Lieberkühn and are organized around fingerlike protrusions (villi). Stem cells are located at the bottom of each crypt and generate actively dividing progenitors (the transit-amplifying cells) that differentiate into specialized epithelial cell types (enterocytes, colonocytes, goblet cells, enteroendocrine cells, etc.) as they migrate upward.

Cross-References

▶ Villi

CSC

▶ Stem-Like Cancer Cells

c-Src

▶ Src

CT4

▶ GAGE Proteins

CT6.1

▶ NY-ESO-1

CTAG

▶ NY-ESO-1

CTAG1

▶ NY-ESO-1

CTAG1A

▶ NY-ESO-1

CTAG1B

▶ NY-ESO-1

CTCF

▶ CCCTC-Binding Factor

CTCFL (Stands for CTCF-Like)

▶ BORIS

CTCF-T (Stands for CTCF-Testis Specific)

▶ BORIS

CTLA-8 in Rodents

▶ Interleukin-17

CTLO

▶ *HRAS*

Cucurbitacin B

Christophe Wiart
University of Nottingham, Nottingham, UK

Definition

Cucurbitacin B is a bitter oxygenated triterpene produced principally by plant species classified in the cucumber family (Cucurbitaceae).

Characteristics

Chemical names
(2S,4R,23E)-2,16Beta,20-trihydroxy-9beta,10,14-trimethyl-1,11,22-trioxo-4,9-cyclo-9,10-secocholesta-5,23-dien-25-yl acetate

   1,2-Dihydro-alpha-elaterin

   2Beta,16alpha,20,25-tetrahydroxy-9-methyl-19-nor-9beta,10alpha-lanosta-5,23-diene-3,11,22-trione,25-acetate

   2Beta,16alpha,20,25-tetrahydroxy-9-methyl-3,11,22-trioxo-19-nor-9beta,10alpha-lanosta-5,23-dien-25-yl acetate

Activities on Cancer Cells
Cucurbitacin B has the ability to kill multiple types of cancer cell lines at very low concentration in vitro. Cucurbitacin B from *Helicteres isora* L. (family Sterculiaceae) abrogated the survival of nasopharyngeal carcinoma (KB) cells with an IC_{50} value at a concentration below 10^{-5} µg/mL. Cucurbitacin B isolated from *Begonia tuberhybrida* Voss var. *alba* (family Begoniaceae) and *Bryonia alba* L. (family Cucurbitaceae) inhibited the growth of human nasopharyngeal carcinoma (KB) cells. Cucurbitacin B isolated from *Marah oregana* (Torr. & A. Gray) Howell (family Cucurbitaceae) was cytotoxic against human nasopharyngeal carcinoma (KB) cells with an IC_{50}

value equal to 2.5×10^{-6} µg/mL. Cucurbitacin B from *Cucurbita andreana* Naudin (family Cucurbitaceae) compromised the proliferation of human colorectal carcinoma (HCT-116), human breast adenocarcinoma (MCF-7), and human large-cell lung cancer (NCI-H460) cells by 81.5%, 87%, and 96%, at a dose of 0.1 µM and reduced the number of human glioblastoma (SF268) cells by 92% at 0.05 µM. Besides, cucurbitacin B killed human prostate cancer (PC-3) cells with an IC_{50} value equal to 35 nM. Cucurbitacin B destroyed human leukemic lymphoblast (CCRF-CEM), human erythromyeloblastoid leukemia (K562), human malignant T-lymphoblastic (MOLT-4), human multiple myeloma(RPMI8226), and human large-cell immunoblastic lymphoma (SR) cells with IC_{50} values equal to 20 nM, 15 nM, 35 nM, 30 nM, and 20 nM. Of particular interest is the fact that cucurbitacin B is cytotoxic to PaCa-2, PL45, Panc-1, AsPC-1 SU86.86, Panc-03.27, and Panc-10.05 human pancreatic cancer cells with an IC_{50} value equal to 10^{-7} mol/L.

Mechanism of Action
Cucurbitacin B induces apoptosis in cancer cells by suppressing the activation of signal transducers and activators of transcription 3 (STAT3) and, to lesser degree, anti-Janus kinase 2 (JAK2) levels in cancer cells. Signal transducers and activators of transcription 3 (STAT3) is a leading target for cancer therapy which is activated in a variety of human solid tumors and hematological malignancies where it promotes growth, survival, and angiogenesis. Aberrant activation of STAT3 in pancreatic cancer is able to confer to these cells resistance to both radiation and chemotherapies.

The inhibition STAT3 by cucurbitacin B results in the upregulation of the expression of pro-apoptotic protein (p53), activation of caspase 3 and cyclin-dependent kinase inhibitors p21, and decreased level of anti-apoptotic Bcl2 protein and survivin. Activation of the Fas/CD95 receptor in Jurkat human T-cell lymphoblast-like cells increases the phosphorylation of eukaryotic translation initiation factor 2 subunit kinase (eIF2K) and therefore progression of cell death. In human erythromyeloblastoid leukemia (K562) cells, cucurbitacin B inhibited STAT3 activation at a dose of 50 μM with inhibition of c-Raf, mitogen-activated protein kinase (MEK), and extracellular signal-regulated kinase (ERK) indicating that cucurbitacin B inhibits the Raf/MEK/ERK pathway. The phosphorylation of STAT3 is abrogated by the inhibition of the extracellular signal-regulated kinase (ERK) activity. Besides, cucurbitacin B-induced apoptosis in human breast adenocarcinoma (SKBR-3), human breast adenocarcinoma (MCF-7), human ductal breast epithelial tumor (T47D), and human breast epithelial (HBL100) cells with IC_{50} values equal to 3.2 μg/mL, 63 μg/mL, 88.4 μg/mL, and 32.7 μg/mL with reduction of c-Myc and telomerase reverse transcriptase (hTERT) and therefore inhibition of telomerase activity via STAT3 inhibition since c-Myc is a downstream effector of transducer and activator of transcription 3 (STAT3) signaling. Additionally, human breast adenocarcinoma (SKBR-3) cells exposed to 10 μg/mL of cucurbitacin B became apoptotic via inhibition of galectin-3 resulting in decrease in phosphorylated glycogen synthase kinase 3 beta(GSK3beta), increase in beta-catenin degradation, and reduction of beta-catenin/TCF4 in the nucleus and impaired transcription of cyclin D1 and c-Myc.

Position of Cucurbitacin B

Cucurbitacin B is a first-line candidate for the treatment of pancreatic cancer. Pancreatic cancer is the fourth leading cause of cancer-related mortality in the United States. The survival rate of patients diagnosed with pancreatic cancer is low because the cells resist chemotherapy, radiotherapy, and immunotherapy on account of aberrant activation of STAT3. Cucurbitacin B has profound antipancreatic cancer activity in vitro and in vivo. It induces cell cycle arrest and apoptosis in pancreatic cancer cell lines by inhibiting the JAKT/STAT pathway. Besides, cucurbitacin potentiates the activity of the nucleoside analogue gemcitabine (2'-deoxy-2',2'-difluorocytidine) which is the standard drug for pancreatic cancer.

Adverse Effects

Toxicological examination of five-week-old female nu/nu athymic mice receiving 1 mg/Kg of cucurbitacin B showed no signs of liver fibrosis or renal damage. Total blood counts and glutamate oxaloacetate transaminase (GOT), glutamate pyruvate transaminase (GPT), g-glutamyl transpeptidase (gGT), total bilirubin, creatinine, glucose levels, and cholesterol as well as alkaline phosphatase, phosphate, chloride, calcium, potassium, total protein, albumin, uric acid, creatine kinase, and triglycerides were in the normal range for all the mice.

Conclusion

Cucurbitacin B is today a first-line candidate for the treatment of pancreatic cancer via STAT3 inhibition.

Cross-References

▶ Angiogenesis
▶ Apoptosis
▶ Bcl2
▶ Breast Cancer
▶ Chemotherapy
▶ Cyclin-Dependent Kinases
▶ Extracellular Signal-Regulated Kinases 1 and 2
▶ Glioblastoma Therapy
▶ Hematological Malignancies, Leukemias, and Lymphomas
▶ Immunotherapy
▶ Lung Cancer
▶ Multiple Myeloma
▶ Nasopharyngeal Carcinoma
▶ p21
▶ Pancreatic Cancer
▶ Prostate Cancer Clinical Oncology
▶ Signal Transduction
▶ STAT3
▶ Survivin
▶ Telomerase

References

Hoennissen NH, Iwanski GB, Doan NB, Okamoto R, Lin P, Abbassi S, Jee HS, Yin D, Toh M, Xie WD, Said JW, Koeffler HP (2009) Cucurbitacin B induces apoptosis by inhibition of the JAK/STAT pathway and potentiates antiproliferative effects of gemcitabine on pancreatic cancer cells. Cancer Res 69(14):5876–5884
Iwanski GB, Lee DH, En-Gal S, Doan NB, Castor B, Vogt M, Toh M, Bokemeyer C, Said JW, Thoennissen

NH, Koeffler HP (2010) Cucurbitacin B, a novel in vivo potentiator of gemcitabine with low toxicity in the treatment of pancreatic cancer. Br J Pharmacol 160(4):998–1007

Jian CC, Ming HC, Rui LN, Cordel GA, Qiuz SX (2005) Cucurbitacins and cucurbitane glycosides: structures and biological activities. Nat Prod Rep 22(3):386–399

Sun C, Zhang M, Shan X, Zhou X, Yang J, Wang Y, Li–Ling J, Deng Y (2010) Inhibitory effect of cucurbitacin e on pancreatic cancer cells growth via STAT3 signaling. J Cancer Res Clin Oncol 136(4):603–610

Zhang M, Sun C, Shan X, Yang X, Li–Ling J, Deng Y (2010) Inhibition of pancreatic cancer cell growth by cucurbitacin b through modulation of signal transducer and activator of transcription 3 signaling. Pancreas 39(6):923–992

Cullin Ubiquitin E3 Ligases

Hui Zhang
Department of Chemistry and Biochemistry, University of Nevada, Las Vegas, NV, USA

Synonyms

Cullin-containing ubiquitin E3 ligases; Cullin-RING ubiquitin E3 ligases (CRLs)

Definition

The cullin-RING ubiquitin E3 ligases (CRLs) are the largest family of multi-subunit ubiquitin E3 ligases in eukaryotes. CRLs share a structural archetype characterized by the presence of interchangeable substrate receptors assembled onto a core catalytic complex consisting of a cullin family member and a RING protein, RBX1 (ROC1 or Hrt1) or RBX2 (ROC2). The members of the human cullin family comprise cullin 1 (CUL1, CDC53), cullin 2 (CUL2), cullin 3 (CUL3), cullin 4A (CUL4A), cullin 4B (CUL4B), cullin 5 (CUL5), cullin 7 (CUL7), and cullin 9 (CUL9), each serves as a scaffold protein to assemble a subfamily of CRL E3 ligases. A cullin uses a conserved globular C-terminal domain (cullin homology domain) to recruit the catalytic RING protein and a series of N-terminal repeats of a five-helix bundle (cullin repeats) to modularly interact with different adaptor proteins and substrate receptors to assemble a unique CRL ubiquitin E3 ligase. The RING protein RBX1 or RBX2 contains a cysteine- and histidine-rich domain that coordinates two zinc metal ions to provide a binding site for an ubiquitin E2 conjugating enzyme that facilitates the transfer of the activated ubiquitin from E2 to a lysine residue in protein substrates to form a polyubiquitinated chain in a reaction that involves the ubiquitin E1 activating enzyme and ATP. The polyubiquitinated chain on the substrates are subsequently recognized and degraded by the 26S proteome. To recognize and interact with specific substrate proteins, each cullin-RING complex assembles a dedicated family of substrate receptor proteins, which are individually attached to the N-terminal domain of the cullin scaffold through a bridging molecule, the adaptor protein. The substrate receptors possess various protein-protein interaction domains to recognize and bind specific sequences of substrates, defined as "degrons," to mediate the selective ubiquitination and degradation of specific substrates under various physiological conditions. The eight cullin-RING core catalytic complexes, combined with many adaptor proteins and hundreds of substrate receptors, form the largest ubiquitin E3 ligase family that targets numerous protein substrates for degradation in response to cellular physiology and environmental cues.

Characteristics

CRL1: The prototype of CRL ubiquitin E3 ligases was established by the analysis of the founding member of CRLs, the SCF protein complex, which is now known as the CRL1 ubiquitin E3 ligase complex. An SCF (SKP1, CUL1/CDC53, F-box protein) or CRL1 ubiquitin E3 ligase complex consists of CUL1 as the scaffold protein, RBX1 as the catalytic RING subunit, SKP1 (S-phase kinase-associated protein 1) as the adaptor, and one of the "F-box" proteins as the substrate receptor (Fig. 1). In this complex,

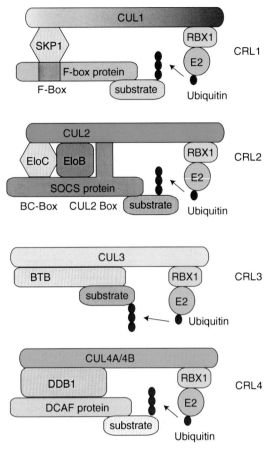

Cullin Ubiquitin E3 Ligases, Fig. 1 Illustration of the structures of CRL1, CRL2, CRL3, and CRL4 ubiquitin E3 ligase complexes

CUL1 exhibits an elongated crescent scaffold structure with an N-terminal stalk-like domain and a globular C-terminal domain. SKP1 interacts with the N-terminal domain of CUL1 through two conserved α-helices, while RBX1 associates with the C-terminal domain of CUL1. An F-box protein docks to SKP1 to form the SCF complex through a conserved F-box motif (first identified in cyclin F), which contains about 40 amino acid residues that are present in many otherwise unrelated proteins, the F-box proteins. In additional to the conserved F-box motif, each F-box protein possesses an independent and diversified protein-protein interaction domain to specifically interact with various degron motifs in the protein substrates. Sixty-nine human F-box proteins have been identified, and each of them is capable of

interacting with multiple protein substrates for ubiquitin-dependent proteolysis. The F-box proteins are usually characterized based on their substrate interaction domains. In human, 12 WD40 domain (FBXW or FBW), 21 leucine-rich repeat (LRR) domain (FBXL or FBL), and 36 F-box only but with variable potential substrate interaction domain-containing F-box proteins have been identified. However, the function of only a few F-box proteins has been well characterized. The canonical recognition and interaction between the F-box protein and its cognate substrate is triggered by the site-specific phosphorylation of serine/threonine residues. Typically, the phosphorylated substrates are directly recognized by the WD40- or leucine-rich repeat domains of F-box proteins or with the assistance of an accessory protein.

The best examples of various SCF complexes are F-box proteins SKP2, FBXW7, and beta-TRCP. SKP2 (S-phase kinase-associated protein 2) regulates the G1/S cell cycle transition by targeting the CDK inhibitor p27^{Kip1} for SCF-dependent degradation. In G1, p27 protein level is high to bind the cyclin/CDK kinases to inhibit the premature activation of these cell cycle kinases, which is required for S phase entry. At the end of the G1 phase, p27 is phosphorylated at threonine 187 by cyclin/CDKs, and this specific phosphorylation is recognized by CKS1, an accessory protein that binds to SKP2, which facilitates the binding of the LRR domain of SKP2 to the phosphorylated p27, promoting the degradation of p27, activation of cyclin/CDKs, and the cell cycle entry into S phase. In addition to p27^{Kip1}, SKP2 also targets other CDK inhibitors including p21^{Cip1} and p57^{Kip2} for SCF-dependent degradation. Because these CDK inhibitors act as tumor suppressors and SKP2 usually serves as a rate-limiting step for the degradation of these substrates, SKP2 usually acts as an oncogene that is overexpressed in many cancers including prostate cancer. Indeed, in most of cancers, p27 is inactivated by overexpression of SKP2, but not by genetic p27 mutations. Another well-known F-box protein is FBXW7 (human CDC4 or SEL10, FBW7). FBXW7 is a WD40 domain-containing F-box protein that binds to the phosphorylated Thr-Pro-Pro-X-Ser motif (both Thr and Ser are phosphorylated, X: any amino acid residue) in the substrates, which include cyclin E, Myc, Aurora A, Jun, and Notch. By degrading these important proteins that are often overexpressed in cancers, FBXW7 usually acts as a tumor suppressor. In T-cell acute lymphoblastic leukemia (T-ALL), the FBXW7 gene is mutated about 31%, but mutations are also found in solid tumor cancers of the breast, intestine, and bone in about 6% of all cancers. The loss of FBXW7 also induces resistance to certain chemotherapeutic drugs, such as anti-tubulin therapy. The F-box protein β-TRCP (beta-transducin repeat protein, βTRCP, FBXW1, and FBW1A) binds the consensus sequence of the Asp-Ser-Gly-X-X-Ser motif with both serines phosphorylated in substrates and targets many cell cycle and non-cell cycle proteins for degradation. The substrates of β-TRCP include CDC25A, β-catenin, EMI1, IκB, REST, Deptor, BIMEL, and circadian cycle proteins PER1 and PER2. β-TRCP1 or its close homologue β-TRCP2 is overexpressed in many cancers, including colorectal, pancreatic, hepatoblastoma, breast, melanoma, and gastrointestinal cancers. Although canonical F-box proteins interact with phosphorylated protein substrates, studies on F-box proteins with no recognizable WD40 or LRR domains indicate that these F-box proteins interact with unmodified degrons in substrates to target them for degradation. One example is cyclin F. Unlike other cyclins, the cyclin-homology domain of cyclin F does not bind to CDKs. Instead, this domain recognizes an Arg-X-Ile/Leu motif in substrates such as centrosomal protein CP110 and ribonucleotide reductase subunit M2 (RRM2) for their SCF-dependent degradation.

CRL2 and CRL5: CUL2 and CUL5 bind to the heterodimeric Elongin B and C (Elongin B/C, originally identified as the components of a transcription elongation factor complex that also contain Elongin A) protein complex as the adaptor protein. Elongin B/C in turn interacts with the substrate receptors that usually contain the SOCS (suppressor of cytokine signaling) box to assemble the CRL2 or CRL5 complexes. However, there are three specific features in CRL2 or

CRL5 that are different from SCF. While Elongin C bears structural similarities to SKP1 of the SCF/CRL1 complex, Elongin B is an ubiquitin-like protein that is absent from SCF/CRL1. In addition, although both CUL2 and CUL5 employ the Elongin B/C complex as the adaptor, CUL2 and CUL5 assemble into distinct CRL2 and CRL5 complexes, each with a subset of SOCS proteins. This is because different SOCS proteins can also specifically interact with CUL2 or CUL5 through an additional domain (CUL2 or CUL5 box) in the SOCS proteins (Fig. 1). Furthermore, CUL2 usually interacts with RBX1, whereas CUL5 typically docks RBX2 as the catalytic subunit.

One of the most characterized SOCS proteins is pVHL, the product of the von Hippel-Lindau tumor suppressor. The pVHL protein regulates the stability of the hypoxia-inducible transcription factor HIF-1-3α transcription factors in response to oxygen levels. Germline inactivation of the VHL gene causes the von Hippel-Lindau hereditary cancer syndrome, and somatic mutations of VHL are associated with the development of sporadic hemangioblastomas and clear renal cell carcinomas. The pVHL contains a "VHL box," which is composed of a BC box and a CUL2 box and conserved among the SOCS proteins that interact with CUL2-RBX1 complex. Under the normoxia (normal oxygen) condition, proline (prolyl) residues in the oxygen-dependent degradation (ODD) domain of HIF-1α, as well as that of other HIF-1α homologues such as HIF-2α and HIF-3α, is hydroxylated by hypoxia-inducible factor prolyl hydroxylase Egl-9 homologues including EGLN1-3 (PHD1-3). The hydroxylation-modified prolyl residues are subsequently recognized by pVHL and its associated CUL2-Elongin B/C complex. The binding of CRL2^{pVHL} leads to the proteolytic degradation of HIF1-3α proteins. Under the low-oxygen (hypoxia) conditions, the HIF-1α protein is stabilized and dimerizes with constitutively expressed HIF-1β (aryl hydrocarbon receptor nuclear receptor translocator, ARNT). The dimeric HIF complex then translocates to the nucleus to induce the expression of downstream hypoxia-regulated genes including vascular endothelial growth factor A (VEGFA) and platelet-derived growth factor-β (PDGFB). Inactivation of VHL prevents the oxygen-dependent degradation of HIF-1α and other related HIF proteins, resulting in the constitutive expression of HIF-dependent genes and causing the von Hippel-Lindau syndrome. In addition to HIF-1-3α, CRL2^{VHL} also polyubiquitinates and degrades Sprouty2 (Spry2) through prolyl hydroxylation, which is implicated in regulating tumor growth and progression. Other pVHL targets include epidermal growth factor receptor (EGFR) and the atypical PKCs, PKCλ, and PKCξII. Other CRL2 E3 ligase complexes include CRL2^{LRR-1}, which contains a VHL box and targets *C. elegans* Cip/Kip CDK inhibitor CKI-1 for degradation. Human LRR-1 homologue is also implicated in regulating p21^{Cip1} and cofilin, an actin-dependent protein that negatively controls cell mobility.

The endogenous CUL5 specifically interacts with endogenous RBX2, suggesting that RBX1 and RBX2 may be functionally distinct. The CRL5-Elongin B/C complexes interact with the CIS/SOCS proteins that consist of the family of suppressor of cytokine signaling (SOCS) and cytokine-inducible Src homology 2 (SH2)-domain-containing protein CIS (CISH). So far, eight CIS/SOCS family proteins have been identified, which are CIS and SOCS1-7; they contain a CUL5 box to interact specifically with the CUL5-Elongin B/C complex. These family members bind to Janus kinases (JAKs), certain cytokine receptors, or signaling molecules to suppress cell signaling. SOCS1 polyubiquitinates JAK2, Vav, IRS1, and IRS2. However, SOCS1 contains an incomplete CUL5 box, so no direct interaction between CUL5 and SOCS1 has been detected. Other CRL5 complexes include CRL5$^{Elongin\ A}$, CRL5^{SSB}, CRL5^{ASB}, CRL5$^{RAB-40C}$, CRL5^{Vif}, and CRL5^{WSB1} complexes that participate in various biological processes including transcriptional regulation, endocytosis, nitric oxide production, and HIV-1 infection.

CRL3: CUL3 uses the BTB (bric-a-brac, tramtrack, and broad complex, also called POZ for poxvirus and zinc finger) domain-containing proteins as the substrate-specific receptors. The BTB domain shares a conserved structural fold with SKP1 and Elongin C and interacts with the N-terminal domain of CUL3. A unique feature of the BTB domain is that it can dimerize, so the CRL3 ubiquitin E3 ligases often exist as a dimer with two CUL3 complexes. In addition, the BTB proteins also possess an additional protein-protein interaction domain as the substrate receptor domain for substrate binding. Thus, a single BTB polypeptide usually acts as both SKP1-like adaptor and substrate-recognition module (Fig. 1). The human genome encodes more than 150 proteins with recognizable BTB domains, often in combination with MATH, Kelch, or other domains. The BTB proteins that contain MATH (meprin and TRAF homology) and Kelch domains are typically associated with CUL3. The substrate interaction domain usually includes the C-terminal Kelch, PHR (PAM, Highwire, and RMP-1), or zinc finger domains, as well as an N-terminal MATH in the BTB protein SPOP. The most common substrate-interacting domain in CRL3 is the Kelch β-propeller domain, which occurs C-terminal to the BTB and BACK (BTB and C-terminal Kelch) domain. One of well-characterized CRL3-associated BTB-Kelch proteins is KEAP1, which regulates the protein stability of the transcription factor Nrf2 (nuclear factor, erythroid 2-like; Nfe2l2), a master regulator of the antioxidant response and cytoprotective genes. KEAP1 requires two Kelch domains to form a dimeric CRL3 architecture that simultaneously engages two distinct epitopes in Nrf2 for ubiquitination. Under normal physiological conditions, cytoplasmic Nrf2 interacts with Keap1, which requires the actin cytoskeleton in order to efficiently bind and target Nrf2 for proteolysis by CRL3. Under oxidative stress, the cysteine residues of Keap1 become oxidized, inducing a conformational change of the Keap1-Nrf2 complex to prevent the ubiquitination of Nrf2, which facilitates the translocation of Nrf2 to the nucleus to transcriptionally activate the antioxidant response genes. The Keap1-Nrf2 is central to carcinogenesis and is part of chemoprevention therapeutics. However, only limited BTB proteins were identified for CRL3 activity. For example, the CUL3-RBX1 complex interacts with Kelch-like 3 (KLHL3), which regulates the polyubiquitination and degradation of WNK4 kinase, a protein kinase that is mutated in familial hypertensive syndromes (Gordon syndrome or familial pseudohypoaldosteronism type II, PHAII). Both KLHL3 and CUL3 are mutated in PHAII. KLHL12 interacts with dopamine D4 receptor for ubiquitination and degradation. KLHL22 forms an E3 ligase with the CUL3-RBX1 complex to monoubiquitinate PLK1, a cell cycle kinase that regulates mitosis, to release PLK1 from kinetochores. CUL3-associated KLHL9 and KLHL13 also regulate the dynamic localization of Aurora kinase on mitotic chromosomes. CUL3 is also implicated in PLZF- and BCL6-dependent leukemia and lymphomas.

CRL4A and CRL4B: Both CUL4A and CUL4B interact with DDB1 (DNA damage-binding protein 1) with their N-terminal domain. DDB1 acts as an adaptor protein for CUL4A- and CUL4B-RBX1 complexes to interact with the substrate-recognition subunits, a subset of WD40 repeat proteins, the DDB1-CUL4-associated factors (DCAFs, also known as CDWs or DWDs), and other DDB1-associated proteins (Fig. 1). DDB1 (p127) was initially isolated with DDB2 (p48) as a heterodimer that recognizes the UV-induced DNA lesions in the nucleotide excision repair pathway for the global genome repair. Mutations of DDB2 cause xeroderma pigmentosum complementation group E, a disease with characteristics of increased sensitivity to UV light and high incidence of skin cancer. DDB1 also forms a complex with CSA (Cockayne syndrome group A protein) that contributes to the transcription coupled repair pathway. DDB1 contains three 7-bladed WD40 β-propeller domains (BPA, BPB, and BPC) and a C-terminal helical domain. The BPA and BPC domains interact with the DCAF substrate receptor proteins, whereas the BPB domain associates with the N-terminal region of CUL4A or CUL4B. The best characterized DCAF proteins of the CRL4 ubiquitin E3 ligases is CDT2 (DTL, L2DTL, DCAF2, CDW1), which was initially identified to target DNA replication licensing protein CDT1 for degradation. CDT2 is a WD40 repeat-containing protein that displays two WDXR motifs located at the end of two consecutive WD40 repeats, which appear to be commonly present in many DCAFs. Inactivation of CDT2 stabilizes CDT1 and causes the re-replication of DNA genome in a single cell cycle, producing a partial polyploid nucleus. The degradation of CDT1 and all other CDT2 substrates, such as CDK inhibitor $p21^{Cip1}$, SET8 (PR-set7 or SETD8), p12 of DNA polymerase δ, and thymine DNA glycosylase (TDG), requires the binding of PCNA (proliferating cell nuclear antigen, a DNA clamp that acts as a processivity factor for DNA polymerase δ), through a PCNA-interacting protein (PIP) box in the substrate proteins, which only occurs when PCNA is trimerized and recruited onto DNA during DNA replication or DNA repair. The orphan nuclear receptor, RORα, is mono-methylated by EZH2, a methyltransferase that normally catalyzes the trimethylation of histone H3 at lysine 27 (H3K27). The mono-methylated lysine 38 (K38) of RORa creates a degradation signal that is recognized by a chromo-domain of DCAF1 (VprBP) in the $CRL4^{DCAF1}$ complex. However, studies showed that DCAF1 is also involved in methylation-independent degradation pathways for CRL4 and other ubiquitin E3 ligases. In addition to initially identified 50 DCAFs or related proteins, the CUL4-DDB1 core complexes also interact with cereblon (CRBN) to form $CRL4^{CRBN}$ E3 ligase complex. CRBN is a highly conserved target for the immune-modulatory drugs (IMiDs) thalidomide and related compounds lenalidomide and pomalidomide. IMiDs were previously used to treat "morning sickness." However, they were found to cause teratogenicity during embryonic development. The IMiDs have been used for treatment of a variety of inflammatory, autoimmune, and neoplastic diseases. A seven-α-helical bundle domain (HBD) in CRBN interacts with the cavity between the BPA and BPC propeller domains of DDB1, which defines a new DDB1-interacting domain other than the canonical WD40 repeated domain. While IMiDs block the binding of CRBN to endogenous substrates such as the homeobox transcription factor MEIS2, the enantioselective interaction of IMiDs with CRBN redirects the substrate-specificity of CRBN to promote the ubiquitination and degradation of the IKAROS family B-cell-specific transcription factors IKZF1 and IKZF3 by CRL4^{CRBN}. IMiDs are critical in the treatment of multiple myeloma which is dependent on the presence of CRBN. These compounds are also effective in treatment of myelodysplasia, chronic lymphocytic leukemia, and some non-Hodgkin lymphomas. However, the function of many DCAFs remains uncharacterized.

CRL7 and CRL9: The CUL7 and CUL9 are atypical cullin proteins that contain additional protein domains. The CUL7-RBX1 complex binds to SKP1 and the F-box protein FBXW8 to form an SCF-like complex to ubiquitinate protein substrates, such as GORASP1, IRS1, and MAP4K1/HPK1, to regulate microtubule dynamics and genome integrity. CUL7 also contains additional domains that mediate its interaction with p53, glomulin (GLMN), or CUL9 (PARC). Mutations of CUL7 cause the 3-M syndrome and the Yakuts short stature syndrome, both of which are characterized by intrauterine and postnatal growth retardation. The CUL9-RBX1 protein complex mediates the ubiquitination and subsequent degradation of BIRC5 and is required to maintain microtubule dynamics and genome integrity. The binding of CUL7 to CUL9 is required to inhibit CUL9 activity.

Regulation

All cullin proteins except CUL7 are modified with the ubiquitin-like protein NEDD8 (Rub in yeast) at a conserved lysine residue in the C-terminal region. Like ubiquitination, the covalent NEDD8 conjugation (neddylation) to cullins is mediated by a NEDD8 activation enzyme (E1), which is composed of a heterodimer of NAE1 (APPBP1) and UBA3. Neddylation is also mediated by a NEDD8 E2-conjugating enzyme (UBCH12, also called UBE2M). The NEDD8 modification is removed by COP9 signalosome complex or by NEDP1, DEN1, and SENP8. Neddylation is required for CRL activity. One function of neddylation is to promote the dissociation of CAND1, a protein that only binds to unneddylated cullins when they are not complexed with adaptor and substrate receptor subunits. Neddylation is proposed to help reassemble various cullin-containing E3 ligase complexes with their adaptors and substrate receptors during ubiquitination cycles. In addition, neddylation changes the conformation of cullins to facilitate the transfer of ubiquitin from RING protein-bound E2 to substrates. MLN4924, a small molecular inhibitor of Nedd8 E1 activating enzyme, has been developed. MLN4924 disrupts cullin-RING E3 ligase-mediated protein turnover in many cancer cells, leading to apoptotic cell death by deregulation of S-phase DNA synthesis.

Conclusion

The CRL ubiquitin E3 ligases constitute the largest ubiquitin E3 ligase family in eukaryotes. By a modular assembly strategy to recruit multiple adaptor proteins and dedicated families of substrate receptors for substrate selectivity, the CRL ubiquitin E3 ligases employ limited cullin scaffold proteins and RING catalytic proteins to target numerous substrates in almost every biological process/pathway for selective degradation in response to various internal and external signals and cues. With their critical roles in regulating a wide variety of important biological processes, dysregulation of various CRL complexes is implicated in many human diseases. They also serve as potential targets for various therapeutic treatments.

Cross-References

► Androgen-independent Prostate Cancer
► Aneuploidy
► Beckwith-Wiedemann Syndrome Associated Childhood Tumors
► Breast Cancer
► Breast Cancer Multistep Development
► Breast Cancer Prognostic and Predictive Biomarkers
► Cancer
► Castration-Resistant Prostate Cancer
► Cell Cycle Checkpoint
► Chemotherapy
► Circadian Clock Induction
► Cyclin D

- ► Cyclin-dependent Kinases
- ► Cyclins
- ► Elongin BC Complex
- ► Genomic Instability
- ► Hormone-Refractory Prostate Cancer
- ► MYC Oncogene
- ► Oxygen Sensing
- ► p21
- ► p27
- ► Prostate Cancer Clinical Oncology
- ► Prostate Cancer Diagnosis
- ► Proteasome
- ► Proteasome Inhibitors
- ► Renal Cancer Clinical Oncology
- ► Renal Cancer Diagnosis
- ► Renal Cancer Genetic Syndromes
- ► Renal Cancer Pathogenesis
- ► S-phase Damage-Sensing Checkpoints
- ► Structural Biology
- ► Tumor Suppression
- ► Tumor Suppressor Genes
- ► Ubiquitin Ligase SCF-SKP2
- ► Ubiquitination
- ► UV Radiation
- ► Von Hippel-Lindau Disease
- ► Von Hippel-Lindau Tumor Suppressor Gene
- ► Xeroderma Pigmentosum

References

Jackson S, Xiong Y (2009) CRL4s: the CUL4-RING E3 ubiquitin ligases. Trends Biochem Sci 34:562–570

Jin J, Cardozo T, Lovering RC, Elledge SJ, Pagano M, Harper JW (2004) Systematic analysis and nomenclature of mammalian F-box proteins. Genes Dev 18:2573–2580

Petroski MD, Deshaies RJ (2005) Function and regulation of cullin-RING ubiquitin ligases. Nat Rev Mol Cell Biol 6:9–20

Stogios PJ, Downs GS, Jauhal JJ, Nandra SK, Prive GG (2005) Sequence and structural analysis of BTB domain proteins. Genome Biol 6:R82

Cullin-Containing Ubiquitin E3 Ligases

► Cullin Ubiquitin E3 Ligases

Cullin-RING Ubiquitin E3 Ligases (CRLs)

► Cullin Ubiquitin E3 Ligases

3D Culture

► Three-Dimensional Culture

Curcumin

Marie-Hélène Teiten[1] and Marc Diederich[2]
[1]Laboratoire de Biologie Moléculaire et Cellulaire du Cancer (LBMCC), Hôpital Kirchberg, Luxembourg, Luxembourg
[2]College of Pharmacy, Seoul National University, Seoul, South Korea

Synonyms

[HOC6H3(OCH3)CH:CHCO]2CH2; 1,7-bis [4-hydroxy-3-methoxyphenyl]-1,6-heptadiene-3,5-dione; Brilliant yellow S; C.I. 75300; C21H2006; Diferuloylmethane; E100; Natural yellow 3; Turmeric yellow

Definition

Curcumin (MW 368.3862) (Fig. 1a) is the active ingredient of the natural Indian spice curcuma purified from the root of the plant *Curcuma longa* L., also called turmeric, which is a member of the ginger family (Zingiberaceae) (Fig. 1b). In Ayurvedic medicine, turmeric is considered to possess health-promoting and anti-inflammatory properties. This natural product plays also a curative role in neurodegenerative, cardiovascular, pulmonary, metabolic, autoimmune, and neoplastic diseases. Curcumin, a multifunctional molecule, modulates several intracellular signaling

Curcumin, Fig. 1 (a) Molecular structure of curcumin (diferuloylmethane). (**b**) *Curcuma longa* L. (© 1995–2004 Missouri Botanical Garden, http://ridgwaydb.mobot.org/mobot/rarebooks)

Curcumin

pathways and exhibits strong antioxidant and antiproliferative properties (Teiten et al. 2010).

Characteristics

Structure-Activity Relationships

Structure-activity relationship of curcumin was intensely investigated (Reddy et al. 2013) and results indicated that its anti-inflammatory and antitumor potential was related to low levels of hydrogenation, high levels of unsaturation of the diketone moiety, and high levels of methoxylation. *Ortho-methoxy* substitutions and the level of hydrogenation of the heptadiene moiety manage its radical-scavenging potential.

Bioavailability

Due to its low water solubility, curcumin presents reduced bioavailability limiting its clinical use. To improve its therapeutic index by increasing its bioavailability and metabolism, novel delivery strategies have been developed such as combination with adjuvants and formulation in micelles, liposomes, nanoparticles, or phospholipid complexes as well as the design of synthetic analogs and hybrid molecules (Teiten et al. 2014).

Modulation of Cancer Cell Signaling Pathways

Curcumin inhibits tumor initiation, promotion, and progression through the regulation and modulation of several intracellular signaling pathways (Teiten et al. 2010):

- Curcumin was described as a potent modulator of inflammatory cell signaling pathways (Teiten et al. 2009). Curcumin inhibits nuclear factor-kappa B (NF-κB) signaling and thereby downregulates the transcription of NF-κB target genes (Reuter et al. 2009). Inhibition of IκB kinase (IKK) by curcumin blocks both IκBα phosphorylation and NF-κB p65 nuclear translocation and leads to NF-κB inhibition. Curcumin also inhibits interleukin (IL)-1α-, tumor necrosis factor (TNF)α-, 12-O-tetradecanoylphorbol-13-acetate (TPA)-, lipopolysaccharide (LPS)-, and thrombin-induced NF-κB activation. Curcumin also inhibits Janus kinase (JAK) and signal transducer and activator of transcription (STAT)-3 phosphorylation and activation as well as AP-1 (activating protein-1) transcription factor. Finally, pro-inflammatory enzymes like COX-2 (cyclooxygenase-2), LOX (lipoxygenase), and iNOS (inducible nitric oxide synthase), as well as the protein kinases cJun N-terminal kinase (JNK), protein kinase A (PKA), and NF-κB-inducing kinase (NIK), were inhibited by this natural product.

- Curcumin inhibits tumor cell proliferation through cell cycle arrest and blocks the invasion potential of cancer cells. In fact, curcumin induces the expression of cyclin-dependent kinase (CDK) inhibitors such as p16, p21, and p27, whereas it inhibits the expression of cyclins E and D1 and reduces the expression levels of oncogenes including c-jun, c-fos, and c-myc. The impact on cyclins is related to its impact on the Wingless (wnt signaling) pathway as curcumin was shown to decrease the nuclear level of β-catenin and T-cell factor (Tcf)-4 that leads to the inhibition of β-catenin/Tcf-mediated transactivation (Teiten et al. 2011). Moreover, this natural product targets phosphatidylinositol 3-kinase (PI3K)/ Akt signaling through modulation of the expression and phosphorylation of PI3K/Akt/ mammalian target of rapamycin (mTOR). It also inhibits signal transduction pathways leading to JNK, mitogen-activated protein kinase (MAPK), or extracellular-regulated kinase (ERK) activation. Proliferation of solid tumors is targeted by curcumin through the modulation of the metastatic, invasive, and proangiogenic potential of various cancer cell types by altering the expression of epidermal growth factor (EGF), vascular endothelial growth factor (VEGF), angiopoietin 1 and 2, tyrosine kinase Flk-1/KDR (VEGF receptor-2), and matrix metalloproteinase (MMP).

- Curcumin was also described as a regulator of epigenetic events and genomic modulations (Teiten et al. 2013). It modulates DNA methylation and covalent modifications of histones and alters miRNA expression patterns. Curcumin was described as a DNA-hypomethylating agent as it downregulates DNA methyltransferase (DNMT1) at both mRNA and protein levels and is able to reverse CpG methylation of the promoter region of several genes regulating tumorigenesis. Moreover, this natural product was described to act as a strong modulator of histone acetyltransferases (HATs) and histone deacetylases/lysine deacetylases (HDACs/ KDACs) as it promotes proteasome-dependent degradation of p300 and of the closely related CREB-binding protein (CBP). Curcumin also acts as a potent modulator of the miRNA signature profiles of cancer as it can modulate the expression of many miRNAs implicated in tumorigenesis including miR199a, miR186, and miR22 (Teiten et al. 2012). A link was established between the hypomethylating potential of curcumin and its impact on specific miRNAs. Finally, telomerase activity could be inhibited by curcumin leading to downregulation of the human telomerase reverse transcriptase (hTERT), the catalytic subunit of telomerase, which leads to the suppression of cell viability and induction of apoptosis.

Induction of Cancer Cell Death

- Many signaling pathways modulated by curcumin lead to the induction of cancer cell death by apoptosis, mitotic catastrophe, or autophagy. Curcumin is mainly described as an inducer of intrinsic and extrinsic apoptosis signaling in cancer cells without affecting normal cell viability (Reuter et al. 2008). This natural compound was reported to induce the downregulation of anti-apoptotic proteins (inhibitor of apoptosis (XIAP), B-cell lymphoma (Bcl)-2, Bcl-xL) and the upregulation of pro-apoptotic proteins from the Bcl-2 family (Bim, Bax, Bak, Puma, and Noxa) that lead to the opening of mitochondrial permeability transition pores, release of cytochrome c, and activation of the apoptosome and subsequent caspase cleavage that finally lead to cell death. However, the level of the suppressor gene p53 remains unchanged.

- On the other hand, the extrinsic apoptotic pathway is induced by curcumin through the activation of cell membrane receptors (Fas, TRAIL) that leads to the assembly of the death-inducing signaling complex (DISC) containing Fas, FAD, and caspase-8 and caspase-10. This extrinsic pathway converges to the intrinsic one by the induction of Bid cleavage and the subsequent cytochrome c release.

- Another mechanism of cell death induced by curcumin corresponds to mitotic catastrophe that results from aberrant mitosis through disruption of the mitotic spindle structure, appearance of micronucleation, accumulation of cells in G2/M, and increased levels of cyclin B1.

- Finally curcumin treatment could lead to cancer cell death through autophagy by inhibition of the Akt/mTOR kinase, activation of ERK1/2, and increased LC-3 conversion.

Clinical Trials

According to numerous published preclinical studies demonstrating and strongly supporting the fact that curcumin presents chemopreventive and chemotherapeutic potential, with minimal side effects, in a large variety of malignancies (e.g., pancreatic, breast, gastric, head and neck, hepatic, lung, ovarian, prostate, and colon cancers as well as lymphoma and leukemia), several clinical trials are underway or have been completed (http://clinicaltrials.gov/ct/gui/action/GetStudy). Most of the ongoing clinical trials explore the preventive or curative effect of curcumin used alone or in combination with conventional chemotherapeutic agent (e.g., 5-fluorouracil, oxaliplatin, gemcitabine, celecoxib, irinotecan, capecitabine, taxotere, etc.) or radiotherapy. Other clinical studies investigated optimized formulations of curcumin (e.g., liposomal or nanoparticle formulation, conjugation with plant exosomes or bioperine, etc.) in order to increase its solubility and subsequent bioavailability in patients. The Food and Drug Administration and World Health Organization also approved the safety of curcumin.

Cross-References

- ▶ Akt Signal Transduction Pathway
- ▶ Angiogenesis
- ▶ Antiangiogenesis
- ▶ Anti-Inflammatory Drugs
- ▶ Apoptosis
- ▶ Apoptosis Induction for Cancer Therapy
- ▶ Autophagy
- ▶ Bcl2
- ▶ Histone Modification
- ▶ Hypomethylation of DNA
- ▶ MAP Kinase
- ▶ MicroRNA
- ▶ Mitotic Catastrophe
- ▶ Natural Products
- ▶ Nuclear Factor-κB
- ▶ Phytochemicals in Cancer Prevention
- ▶ Signal Transducers and Activators of Transcription in Oncogenesis
- ▶ Wnt Signaling

References

Reddy AR, Dinesh P, Prabhakar AS, Umasankar K, Shireesha B, Raju MB (2013) A comprehensive review on sar of curcumin. Mini Rev Med Chem 13:1769–1777

Reuter S, Eifes S, Dicato M, Aggarwal BB, Diederich M (2008) Modulation of anti-apoptotic and survival pathways by curcumin as a strategy to induce apoptosis in cancer cells. Biochem Pharmacol 76:1340–1351

Reuter S, Charlet J, Juncker T, Teiten MH, Dicato M, Diederich M (2009) Effect of curcumin on nuclear factor kappab signaling pathways in human chronic myelogenous k562 leukemia cells. Ann N Y Acad Sci 1171:436–447

Teiten MH, Eifes S, Reuter S, Duvoix A, Dicato M, Diederich M (2009) Gene expression profiling related to anti-inflammatory properties of curcumin in k562 leukemia cells. Ann N Y Acad Sci 1171:391–398

Teiten MH, Eifes S, Dicato M, Diederich M (2010) Curcumin-the paradigm of a multi-target natural compound with applications in cancer prevention and treatment. Toxins (Basel) 2:128–162

Teiten MH, Gaascht F, Cronauer M, Henry E, Dicato M, Diederich M (2011) Anti-proliferative potential of curcumin in androgen-dependent prostate cancer cells occurs through modulation of the wingless signaling pathway. Int J Oncol 38:603–611

Teiten MH, Gaigneaux A, Chateauvieux S, Billing AM, Planchon S, Fack F, Renaut J, Mack F, Muller CP, Dicato M et al (2012) Identification of differentially expressed proteins in curcumin-treated prostate cancer cell lines. Omics 16:289–300

Teiten MH, Dicato M, Diederich M (2013) Curcumin as a regulator of epigenetic events. Mol Nutr Food Res 57:1619–1629

Teiten MH, Dicato M, Diederich M (2014) Hybrid curcumin compounds a new strategy for cancer treatment. Molecules 19:20839–20863

Cushing Syndrome

Constantine A. Stratakis
Program on Developmental Endocrinology of Genetics, NICHD, NIH, Bethesda, MD, USA

Definition

Cushing syndrome is a rare disease entity. The overall incidence of Cushing syndrome is $\sim$2–5 new cases per million people per year. Approximately 10% of the new cases each year occur in children. As in adult patients, in children with Cushing syndrome, too, there is a female to male predominance, which decreases with younger age; there might even be a male to female predominance in infants and young toddlers with Cushing syndrome. The most common cause of endogenous Cushing syndrome is the overproduction of adrenocorticotropin (ACTH) from the pituitary; this is called Cushing disease. It is usually caused by an ACTH-secreting

pituitary microadenoma and, rarely, a macroadenoma. ACTH secretion occurs in a semiautonomous manner, maintaining some of the feedback of the HPA axis. Cushing's disease accounts for $\sim$75% of all cases of Cushing syndrome in children over 7 years. In children under 7 years, Cushing disease is less frequent; adrenal causes of Cushing syndrome (adenoma, carcinoma, or bilateral hyperplasia) are the most common causes of the condition in infants and young toddlers. Ectopic ACTH production is almost unheard of in young children; it also accounts for less than 1% of the cases of Cushing syndrome in adolescents. Sources of ectopic ACTH include small cell carcinoma of the lungs, carcinoid tumors in the bronchus, pancreas or thymus, medullary carcinomas of the thyroid, pheochromocytomas, and other neuroendocrine tumors.

Rarely, ACTH overproduction by the pituitary may be the result of oversecretion of corticotropin-releasing-hormone (CRH) by the hypothalamus or by an ectopic CRH source. However, this cause of Cushing syndrome has only been described in a small number of cases and never in young children. Its significance lies in the fact that diagnostic tests that are usually used for the exclusion of ectopic sources of Cushing syndrome have frequently misleading results in the case of CRH-induced ACTH oversecretion.

Autonomous secretion of cortisol from the adrenal glands, or ACTH-independent Cushing syndrome, accounts for $\sim$10–15% of all the cases of Cushing syndrome. However, although adrenocortical tumors are rare in older children, in younger children they are more frequent. Adrenocortical neoplasms account for 0.6% of all childhood tumors.

Cushing syndrome is a manifestation of approximately one third of all adrenal tumors. A number of adrenal tumors presenting with Cushing syndrome can be malignant: the majority of patients present under age 5, contributing thus to the first peak of the known bimodal distribution of adrenal cancer across the life span. As in adults, in pediatric adrenal cancer, too, there is a female to male predominance. The tumors usually occur unilaterally; however, in 2–10% of patients they occur bilaterally.

Bilateral nodular adrenal disease has been appreciated as a rare cause of Cushing syndrome. Primary pigmented nodular adrenocortical disease (PPNAD) is a genetic disorder with the majority of cases associated with Carney complex, a syndrome of multiple endocrine gland abnormalities in addition to lentigines (benign lesions that occur on the sun-exposed areas of the body, the backs of hands and face are common areas; the lesions tend to increase in number with age, making them common among the middle age and older population, they can vary in size from 0.2 to 2 cm;. these flat lesions usually have discrete borders, are dark in color, and have an irregular shape) and myxomas. The adrenal glands in PPNAD are most commonly normal or even small in size with multiple pigmented nodules surrounded by an atrophic cortex. The nodules are autonomously functioning resulting in the surrounding atrophy of the cortex. Patients with PPNAD frequently have periodic Cushing syndrome.

Massive macronodular adrenal hyperplasia (MMAD) is another rare disease, which leads to Cushing syndrome. The adrenal glands are massively enlarged with multiple, huge nodules that are typical, yellow-to-brown cortisol-producing adenomas. Most cases of MMAD are sporadic, although few familial cases have been described; in those, the disease appears in children. In some patients with MMAD, cortisol levels appear to increase with food ingestion (food-dependent Cushing syndrome). These patients have an aberrant expression of the GIP receptor (GIPR) in the adrenal glands. In the majority of patients with MMAD, however, the disease does not appear to be GIPR-dependent; aberrant expression of other receptors might be responsible.

Bilateral macronodular adrenal hyperplasia can also be seen in McCune Albright syndrome (MAS). In this syndrome, there is a somatic mutation of the *GNAS* gene leading to constitutive activation of the Gsα protein and continuous, non-ACTH-dependent stimulation of the adrenal cortex. Cushing syndrome in MAS is rare and usually presents in the infantile period (before 6 months of age); interestingly, a few children have had spontaneous resolution of their Cushing syndrome.

Characteristics

In most patients, the onset of Cushing syndrome is rather insidious. The most common presenting symptom of the syndrome is weight gain, although it is not universally present. Almost pathognomonic for Cushing syndrome in childhood is weight gain associated with growth retardation; the combination of the two is among the most consistent and frequently encountered signs. Other common problems reported in patients include facial plethora, headaches, hypertension, hirsutism, amenorrhea, and delayed sexual development. Other patients may present with virilization. Skin manifestations, including acne, violaceous striae, bruising, and acanthosis nigricans, are also common. In comparison to adult patients with Cushing syndrome, symptoms that are less commonly seen include sleep disruption, weakness, and mental changes.

Diagnostic Guidelines

The appropriate therapeutic interventions in Cushing syndrome depend on accurate diagnosis and classification of the disease. The history and clinical evaluation, including growth charts in children, are important to make the initial diagnosis of Cushing syndrome. Upon suspicion of the syndrome, laboratory and imaging confirmations are necessary. An algorithm of the diagnostic process is presented in the attached diagram (Fig. 1). The first step in the diagnosis of Cushing syndrome is to document hypercortisolism. This step is usually done in the outpatient setting. Because of the circadian nature of cortisol and ACTH, isolated cortisol and ACTH measurements are not of great value in diagnosis. One excellent screening test for hypercortisolism is a 24-h urinary free cortisol (UFC) excretion corrected for body surface area. A normal 24-h UFC value is <70 $\mu g/m^2/day$ by radio immunoassay. A 24-h urine collection is often difficult for parents to do in children and may be done incorrectly, especially in the outpatient setting. Falsely high UFC may be obtained because of physical and emotional stress, chronic and severe obesity, pregnancy, chronic exercise, depression, alcoholism, anorexia, narcotic withdrawal, anxiety,

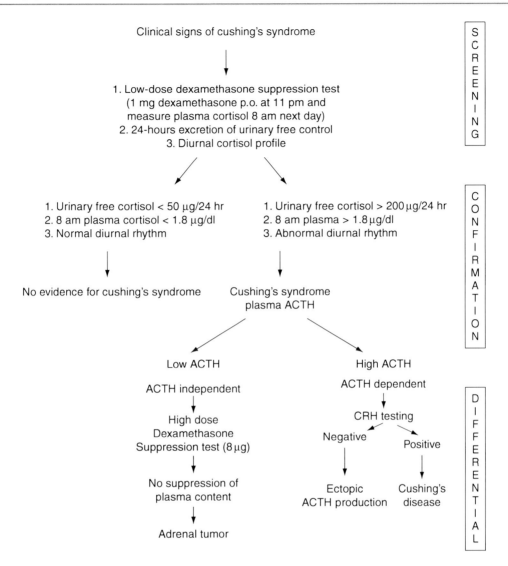

Cushing Syndrome, Fig. 1 Algorithm of the diagnostic process

malnutrition, and excessive water intake (more than 5 l/day). These conditions may lead to sufficiently high UFCs to cause what is known as pseudo-Cushing syndrome. On the other hand, falsely low UFC may be obtained mostly with inadequate collection. Another baseline test for the establishment of the diagnosis of Cushing syndrome is a low dose dexamethasone suppression test. This test involves giving a 1 mg of dexamethasone at 11 p.m. (adjusted for weight for children <70 kg by dividing the dose by 70 and multiplying by the weight of the child) and measuring a serum cortisol level the

following morning at 8 a.m. The problem with this test is that it has not been evaluated extensively in children; for adult patients, the cortisol cut-off level should be <1.8 μg/dl (50 nmol/l). If it is greater than 1.8 μg/dl, further evaluation is necessary. This test has a low percentage of false normal suppression; however, at our institution, very rarely we obtain the 1-mg test for screening for Cushing syndrome in children. It should also be noted that the 1-mg overnight test (like the 24-h UFCs) does not distinguish between hypercortisolism from Cushing syndrome and other hypercortisolemic states.

If the response to both the 1-mg dexamethasone overnight suppression test and the 24-h urinary free cortisol are both normal, a diagnosis of Cushing syndrome may be excluded with the following caveat: 5–10% of patients may have intermittent or periodic cortisol hypersecretion and may not manifest abnormal results to either test. If periodic or intermittent Cushing syndrome is suspected, continuous follow up of the patients is recommended. Diurnal plasma cortisol variation, including midnight cortisol values, is a fairly good test for the establishment of the diagnosis of Cushing syndrome. In our institution, it has become the test of choice for the confirmation of endogenous hypercortisolemia and is routinely done in patients with confirmed elevated urinary cortisol levels on the outside. There are several caveats for the interpretation of the test of which the most important ones are as follows: (i) The venous catheter has to be placed at least two hours before the test; and (ii) if the patient comes from another time zone, a 1 h per day adjustment should be taken into account prior to obtaining the test. In general, serum cortisol levels are drawn at 11:30 p.m. and 12:00 midnight and at 7:30 a.m. and 8:00 a.m., while the patient is lying in bed and asleep; midnight cortisol levels above 5 µg/dl are abnormal and confirm the diagnosis of Cushing syndrome, whereas an inverted diurnal rhythm is seen in PPNAD and some other adrenal tumors.

If one of the tests is suggesting Cushing syndrome or, if there is any question about the diagnosis, tests that distinguish between pseudo-Cushing states and Cushing syndrome may be obtained. One such test is the combined dexamethasone-CRH test. In this test, the patient is treated with low dose dexamethasone (0.5 mg adjusted for weight for children <70 kg by dividing the dose by 70 and multiplying by the weight of the child) every 6 h for eight doses prior to the administration of CRH (ovine CRH – oCRH) the following morning. ACTH and cortisol levels are measured at baseline and every 15 min for one hour after the administration of oCRH. The patient with a pseudo-Cushing state will exhibit low or undetectable basal plasma cortisol and ACTH and have a diminished or no response to oCRH stimulation. Patients with Cushing syndrome will have higher basal cortisol and ACTH levels and will also have a greater peak value with oCRH stimulation. The criterion used for the diagnosis of Cushing syndrome is a cortisol level of greater than 38 nmol/l (~1.4 µg/dl) 15 min after oCRH administration; all other patients (<1.4 µg/dl) may suffer from a pseudo-Cushing state.

Once the diagnosis of Cushing syndrome is confirmed there are several tests to distinguish ACTH-dependent disease from the ACTH-independent syndrome. A spot plasma ACTH may be measured; if this measurement is <5 pmol/l it is indicative of ACTH-independent Cushing syndrome, although the sensitivity and specificity of a single ACTH measurement are not high because of the great variability in plasma ACTH levels and the instability of the molecule after the sample's collection. Even if one assumes that the sample was collected and processed properly (collected on ice and spun down immediately in a refrigerated centrifuge for plasma separation; the sample should then be immediately processed or frozen at −20 °C), ACTH levels that are between 5 and 20 pmol/l are not informative in this era of high sensitivity assays; levels above 20 pmol/l are more suggestive of an ACTH-dependent condition, but again that is not a certainty until single ACTH levels are repeatedly over 30 pmol/l.

The standard six-day low- and high-dose dexamethasone suppression test (Liddle's test) is used to differentiate Cushing disease from ectopic ACTH secretion and adrenal causes of Cushing syndrome. In the classic form of this test, after 2 days of baseline urine collection, 0.5 mg of dexamethasone (adjusted per weight for children <70 kg by dividing the dose by 70 and multiplying by the weight of the child) every 6 h are given per os starting at 6.00 a.m. on day 3 ("low dose" phase of the test) for a total of eight doses (2 days); this is continued with a 2-mg dose of dexamethasone per os (adjusted per weight for children <70 kg by dividing the dose by 70 and multiplying by the weight of the child) on day 5 ("high dose" phase of the test) given every 6 h for another 8 doses (final 2 days) (13–15). Urinary free cortisol and 17-hydroxysteroid excretion are measured at baseline, during, and 1 day after the end of the dexamethasone administration. Approximately

90% of patients with Cushing disease will have suppression of cortisol and 17-hydroxysteroid values, whereas less than 10% of patients with ectopic ACTH secretion will have suppression. Urinary free cortisol values should suppress to 90% of baseline value and 17-hydroxysteroid excretion should suppress to less than 50% of baseline value. The criteria are similar if one uses serum cortisol values obtained at 8 a.m. of the morning after the last dose of dexamethasone, e.g., serum cortisol on day 7 should be 90% of baseline serum cortisol values (obtained at 8 a.m. the day before dexamethasone administration). An increase of urinary free cortisol values of 50% or more over baseline during Liddle test has been used in the differential diagnosis of PPNAD and other micronodular adrenocortical disease versus other causes of adrenal causes of Cushing syndrome.

The Liddle test has been modified to (i) giving 2 mg every 6 h (without the preceding low-dose phase); (ii) administering dexamethasone intravenously over 5 h at a rate of 1 mg/h; or (iii) giving a single high dose of dexamethasone (8 mg, in children adjusted for weight <70 kg) at 11 p.m. and measuring the plasma cortisol level the following morning. This overnight high-dose dexamethasone test has sensitivity and specificity values similar to those of the classic Liddle test. A 50% suppression of serum cortisol levels from baseline is what differentiates Cushing disease (more than 50% suppression) from other causes of Cushing syndrome (adrenal or ectopic ACTH production) (less than 50% suppression).

An oCRH stimulation test may also be obtained for the differentiation of Cushing disease from ectopic ACTH secretion. In this test, 85% of patients with Cushing disease respond to oCRH with increased plasma ACTH and cortisol production. Ninety-five percent of patients with ectopic ACTH production do not respond to the administration of oCRH. The criterion for diagnosis of Cushing disease is a mean increase of 20% above baseline for cortisol values at 30 and 45 min and an increase in the mean corticotropin concentrations of at least 35% over basal value at 15 and 30 min after oCRH administration. When the oCRH and high-dose dexamethasone (Liddle or overnight) tests are used together, diagnostic accuracy improves to 98%.

Another important tool in the localization and characterization of Cushing syndrome is diagnostic imaging. The most important initial imaging when Cushing disease is suspected is pituitary magnetic resonance imaging (MRI). The MRI should be done in thin sections with high resolution and always with contrast (gadolinium). The latter is important since only macroadenomas will be detectable without contrast; after contrast, an otherwise normal-looking pituitary MRI might show a hypoenhancing lesion, usually a microadenoma. More than 90% of ACTH-producing tumors are hypoenhancing, whereas only about 5% are hyperenhancing after contrast infusion. However, even with the use of contrast material, pituitary MRI may detect only up to ~30% of ACTH-producing pituitary tumors, although with the use of new modalities (e.g., SPGR-MRI), this percentage may be as high as 60%. Computed tomography (CT) (more preferable than MRI) of the adrenal glands is useful in the distinction between Cushing disease and adrenal causes of Cushing syndrome, mainly unilateral adrenal tumors. The distinction is harder in the presence of bilateral hyperplasia (MMAD or PPNAD) or bilateral adrenal carcinoma (conditions, however, that are rare). Most patients with Cushing disease have ACTH-driven bilateral hyperplasia, and both adrenal glands will appear enlarged and nodular on CT or MRI. Most adrenocortical carcinomas are unilateral and quite large by the time they are detected. Adrenocortical adenomas are usually small, less than 5 cm in diameter and, like most carcinomas, they involve one adrenal gland. MMAD presents with massive enlargement of both adrenal glands, whereas PPNAD is more difficult to diagnose radiologically because it is usually associated with normal or small-sized adrenal glands, despite the histologic presence of hyperplasia.

Ultrasound may not be used to image the adrenal glands for the diagnostic work up of Cushing syndrome, because its sensitivity and accuracy is much less than CT or MRI. A CT or MRI scan of the neck, chest, abdomen, and pelvis may be used for the detection of an ectopic source of ACTH production. Labeled-octreotide scanning and venous sampling may also help in the localization of an ectopic ACTH source. Since up to 50% of pituitary ACTH-secreting tumors and many of ectopic ACTH tumors can not be detected on routine imaging, and often laboratory diagnosis is not completely clear, catheterization studies must be used to confirm the source of ACTH secretion in ACTH-dependent Cushing syndrome (Magiakou, et al. 1994; Nieman 2002; Orth 1995). Bilateral inferior petrosal sinus sampling (IPSS) may also be used for the localization of a pituitary microadenoma (although not with great accuracy or sensitivity). IPSS is an excellent test for the differential diagnosis between ACTH-dependent forms of Cushing syndrome with a diagnostic accuracy that approximates 100%, as long as it is performed in an experienced clinical center. IPSS, however, may not lead to the correct diagnosis, if it is obtained when the patient is not sufficiently hypercortisolemic or, if venous drainage of the pituitary gland does not follow the expected, normal anatomy. In brief, sampling from each inferior petrosal sinuses is taken for the measurement of ACTH concentration simultaneously with peripheral venous sampling. ACTH is measured at baseline and at 3, 5, and 10 min after oCRH administration. Patients with ectopic ACTH secretion have no gradient between either one of the two sinuses and the peripheral sample. On the other hand, patients with an ACTH-secreting pituitary adenoma have at least a 2-to-1 at baseline, and 3-to-1 central-to-peripheral gradient after stimulation with oCRH.

References

Kirk JM et al (1999) Cushing's syndrome caused by nodular adrenal hyperplasia in children with McCune-Albright syndrome. J Pediatr 134:789–792

Magiakou MA (1994) Cushing's syndrome in children and adolescents: presentation, diagnosis and therapy. N Engl J Med 331:629–636

Nieman LK (2002) Diagnostic tests for Cushing's syndrome. Ann N Y Acad Sci 970:112–118

Orth DN (1995) Cushing's syndrome. N Engl J Med 332:791–803

Stratakis CA, Kirschner LS (1998) Clinical and genetic analysis of primary bilateral adrenal diseases (micro- and macronodular disease) leading to Cushing Syndrome. Horm Metab Res 30:456–463

Cutaneous Desmoplastic Melanoma

Florian Roka[1], Victor G. Prieto[2] and Jeffrey E. Gershenwald[1]

[1]Department of Surgical Oncology, The University of Texas MD Anderson Cancer Center, Houston, TX, USA

[2]Department of Pathology, The University of Texas MD Anderson Cancer Center, Houston, TX, USA

Synonyms

Desmoplastic melanoma

Definition

Desmoplastic (desmoplasia) melanoma (DM) is a rare variant of invasive melanoma of the skin composed of spindle cells surrounded by various degrees of sclerotic stroma. DM has a variable clinical appearance and may mimic several lesions.

Characteristics

Epidemiology

Desmoplastic melanoma (DM) is an uncommon variant of melanoma representing 2–4% of all melanoma cases. Like many melanomas, DM tends to occur in sun-exposed regions of the skin with highest incidence in the head and neck region (>40%), thus suggesting a predilection for chronically sun-damaged skin. Nevertheless, DM may occur anywhere, including acral and mucosal sites. DM predominantly affects elderly individuals; the mean age of onset is consistently 10 years later (~60 years) than it is for patients with conventional (i.e., non-DM) melanoma. DM has a male predominance ratio of ~2:1.

Microscopic Features

Microscopically, DM presents as a poorly circumscribed neoplasm of variable size with dermal and frequently subcutaneous infiltration. There is often an accompanying atypical or frankly malignant melanocytic proliferation in the epidermis. The tumor is present in a sarcoma-like pattern with elongated and usually amelanotic, hyperchromatic fibroblast-like spindle cells arranged singly or in thin fascicles separated from each other by fibrotic stroma. Epithelioid cells may be recognized in the superficial areas of the lesion, while in deeper areas the cells are predominantly spindle cell with fibroblast-like features. The spindle cell population varies greatly in appearance: it may be hypocellular and bland with minimal mitotic figures (differential diagnosis: scar, fibromatosis, dermatofibroma) or associated with nuclear atypia and high mitotic activity (differential diagnosis: high-grade sarcoma or sarcomatoid carcinoma). If these latter areas are more than focal, such lesions are not classified as DM but rather as spindle cell melanoma (see below). Focal lymphocytic aggregates are very frequently observed and may aid in diagnosis. By differentiating along Schwannian lines, DM may also mimic neurofibroma, neurotized melanocytic nevus, or nerve sheath myxoma. Variants of DM include the "neurotropic" melanoma showing a "neuroma-like" growth pattern with invasion of cutaneous nerves, usually in a spindle cell vertical component with fibrosis. The myxoid (or myxofibrous) variant displays abundant mucinous stroma. DM has been defined as either "pure" if the overwhelming majority of the invasive tumor is desmoplastic or as "combined" or "mixed" DM if the desmoplastic areas constitute less than 90% of the tumor.

The immunohistological profile of DM is different from that of epithelioid melanoma. While S100 is typically positive in 80–94% of cases, other markers often helpful in diagnosing melanoma (▶ melanoma antigens), such as HMB45 antigen, tyrosinase, and Melan-A/MART-1, are typically only focally positive or completely negative in DM. Nevertheless, absence of such immunoreactivity does not exclude the diagnosis when the clinical picture and/or histology are characteristic. Inconsistent results regarding the labeling pattern as well as the staining intensity have been noted for antigens such as CD68, NSE (neuron-specific enolase), CD34, and smooth muscle alpha-actin (SMA); as such these markers are usually not helpful in the diagnosis of DM.

Clinical Features

In most cases, DM presents as a slow-growing, painless, and usually amelanotic palpable plaque or nodule that may be associated with a lentigo maligna lesion. The clinical appearance, however, can be variable. Due to its sometimes innocuous appearance and the common lack of pigmentation, it is uncommonly diagnosed at an early stage and thus may be confused with a variety of benign (scars, dermatofibroma, melanocytic nevus) and malignant (carcinoma, sarcoma) lesions. DM commonly presents as a deeply invasive tumor at the time of diagnosis.

Studies reported a median tumor thickness between 2.5 and 6.5 mm, and the majority of lesions showed invasion of the reticular dermis or subcutis (Clark level IV/V).

Clark level refers to the degree of melanoma invasion, describing the maximal depth of melanoma cell penetration using the anatomical structures of the skin.

Clark level I: melanoma cells are confined to the epidermis (and appendages); equivalent to melanoma in situ

Clark level II: melanoma cells extending to the papillary dermis

Clark level III: extension of the tumor cells filling and expanding the papillary dermis

Clark level IV: invasion of the reticular dermis

Clark level V: invasion of the subcutaneous fat

Although the level of invasion has historically been shown to correlate with prognosis, its utility in staging primary cutaneous melanoma has been supplanted by Breslow depth, except for thin primary melanomas (i.e., up to 1 mm Breslow depth; AJCC/UICC T1) where both Clark level and Breslow depth are included.

This observation, in conjunction with a commonly described invasion along neural structures (neurotropism), was considered the main reason for the seemingly locally aggressive biologic behavior and high incidence of local recurrence of 11–55% reported in the literature. Nevertheless, data involving large patient datasets suggest that this may be due to frequent positive, unknown, or relatively narrow resection margins, particularly in the head and neck region. Excision of DM with a 2 cm margin (e.g., for melanomas over 1 mm in depth) appears to be associated with a reduced rate of local recurrence. This concept has been further substantiated by a study involving 65 patients with DM demonstrating no local recurrences after wide local excisions with 2 cm margins.

There is solid evidence to indicate that the incidence of lymph node metastasis in patients with DM is lower than in patients with conventional melanoma of similar thickness, ranging from 0% to 18.8%. With the widespread use of sentinel node biopsy for melanoma, regional lymph node status in patients with DM has often been assessed at the time of diagnosis. So far, all studies assessing SLN status in patients with pure DM showed a lower rate of lymph node involvement as compared to conventional melanoma. Specifically, synchronous microscopic stage III disease among patients with DM of the "pure" histologic subtype is extremely uncommon – i.e., below 3% – and suggests that the behavior of "pure" DM may be more like soft tissue sarcoma, which in general is also associated with a low rate of regional lymph node metastasis. In contrast, DM of the "mixed" phenotype showed a similar rate (~15%) of positive SLN nodes as is generally observed for conventional (i.e., non-DM) melanoma.

Although patients with DM usually present with thicker tumors, survival of patients with DM has repeatedly reported to be equal or better than for other forms of melanoma. The largest study at present reported similar survival rates among patients with desmoplastic and conventional melanoma, with a 5-year survival rate of 75%.

Management

Like any invasive melanoma, wide local excision – with margins appropriate for tumor thickness – represents the mainstay of treatment of the primary DM tumor location. As nerve involvement may be observed, adequate wide excision is in particular advisable in patients with DM whenever possible; such an approach likely enhances local control. The low incidence of regional lymph node metastases in patients with DM suggests that elective (prophylactic) lymph node dissection is not indicated in these patients. While sentinel node biopsy is regarded as the standard of care for patients with intermediate- and high-risk melanoma, only a subset of patients with pure DM may benefit from this procedure due to the expected low yield of positive regiona l lymph nodes.

In view of studies demonstrating significant local disease control by performing wide local excision with adequate margins, adjuvant postoperative radiation with the aim to reduce the rate of local recurrence may be dispensable in many patients with DM who have had an appropriate wide local excision.

Cross-References

▶ Melanoma Antigens

References

Busam KJ (2005) Cutaneous desmoplastic melanoma. Adv Anat Pathol 12(2):92–102

Lens MB, Newton-Bishop JA, Boon AP (2005) Desmoplastic malignant melanoma: a systematic review. Br J Dermatol 152(4):673–678

Pawlik TM, Ross MI, Prieto VG et al (2006) Assessment of the role of sentinel lymph node biopsy for primary cutaneous desmoplastic melanoma. Cancer 106(4):900–906

Prieto VG, Woodruff J (1997) Expression of HMB45 antigen in spindle cell melanoma. J Cutan Pathol 24:580–581

Quinn MJ, Crotty KA, Thompson JF et al (1998) Desmoplastic and desmoplastic neurotropic melanoma: experience with 280 patients. Cancer 83(6):1128–1135

See Also

(2012) Breslow depth. In: Schwab M (ed) Encyclopedia of cancer, 3rd edn. Springer Berlin Heidelberg, p 566. doi:10.1007/978-3-642-16483-5_729

Cutaneous Neoplasms

▶ Skin Cancer

Cyclin D

Rene Bernards
The Netherlands Cancer Institute, Amsterdam,
The Netherlands

Definition

D-Type cyclins belong to a family of related proteins that bind to and activate several protein kinases named ▶ cyclin-dependent kinases (CDKs), which are involved in the regulation of the cell division cycle.

Characteristics

D-Type cyclins are encoded by three closely related genes (cyclins D1, D2, and D3) that are expressed in a tissue-specific fashion. Biochemically, D-type cyclins act as regulatory subunits of a group of related protein kinases (CDKs), primarily the CDKs 4 and 6. Cyclin D/CDK4/6 complexes, together with cyclin E/CDK2, cause phosphorylation of the family of retinoblastoma proteins (pRb, p107, and p130) in the G1 phase of the cell cycle, resulting in abrogation of their growth inhibitory activity. Phosphorylation of the retinoblastoma proteins leads to release of E2F transcription factors from the retinoblastoma proteins and to progression to the S phase of the cell cycle (Fig. 1).

Regulation of D Cyclins

D-Type cyclins are major downstream targets of extracellular signaling pathways, which act to transduce mitogenic signals to the cell cycle machinery. Transcriptional induction of D-type cyclins occurs in response to a wide variety of mitogenic stimuli, including the Ras signaling cascade and the APC-β-catenin-Tcf/Lcf pathway. In addition, cyclin D1 protein turnover and subcellular localization is highly regulated during the cell cycle. Phosphorylation of cyclin D1 by GSK-3β in resting cells renders the protein a target for rapid destruction by the ▶ proteasome. In contrast, mitogenic stimulation of cells leads to inhibition of GSK-3β and stabilization of cyclin D1 protein. In response to DNA damage, cells initiate an immediate G1 arrest, which is caused by rapid proteolysis of cyclin D1. Together with activation of the p53 tumor suppressor protein, cyclin D1 destruction causes a fast withdrawal from the cell cycle to allow repair of the damaged DNA before DNA synthesis resumes.

Binding of D-type cyclins to their CDK partner is antagonized by the INK4 family of CDK inhibitors (CKI). INK4 proteins bind to CDK4 and 6 and thereby prevent association of D-type cyclins to these CDKs (Fig. 2). The most prominent member of this family is p16INK4A. Mutations in p16INK4A (also known as ▶ CDKN2A) are found in a variety of spontaneous tumors, and heterozygosity for p16INK4A in the germ line predisposes to melanoma. A second family of CKIs consists of three related proteins that bind to cyclin/CDK complexes. Members of this family include p21cip1 and p27kip1. This class of CKIs has quite divergent effects on the different cyclin/CDK complexes. Whereas cyclin E/CDK2 is inhibited by both p21cip1 and p27kip1, cyclin D/CDK4/6 complexes are active when complexed with this class of inhibitors (Fig. 2). In fact, formation of active cyclin D/CDK4/6 complexes requires the presence of p21cip1 or p27kip1 to act as "assembly factors" of cyclin D/CDK complexes. These opposing effects of p21cip1 and p27kip1 on cyclin E/CDK2 and cyclin D/CDK4 complexes endow cyclin D/CDK4 complexes with an important second, non-catalytic function during the G1 phase of the cell cycle. Synthesis of cyclin D1 by mitogenic stimulation leads to the absorption of p21cip1 or p27kip1 into active ternary complexes, thereby facilitating activation of cyclin E/CDK2 by removal of inhibitors.

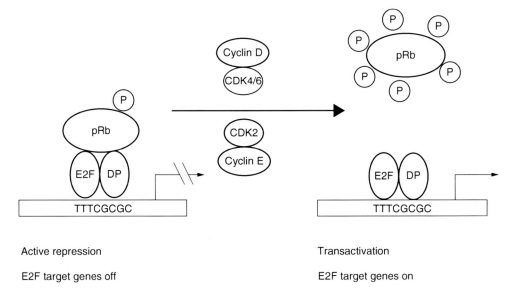

Active repression   Transactivation

E2F target genes off   E2F target genes on

Cyclin D, Fig. 1 Regulation of E2F activity through pRb phosphorylation. In the G1 phase of the cell cycle, the retinoblastoma protein pRb is hypophosphorylated, allowing it to bind E2F transcription factors. E2F/pRb complexes are able to bind DNA but are inactive in transcription activation. Phosphorylation of pRb by cyclin D/CDK4 and cyclin E/CDK2 complexes causes the release of E2F from pRb. Free E2F is then able to activate transcription of E2F target genes (genes with TTTCGCGC-like E2F sites in their promoters), allowing cells to enter the DNA synthesis phase (S phase) of the cell cycle

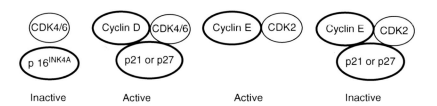

Inactive   Active   Active   Inactive

Cyclin D, Fig. 2 Effect of CDK inhibitors on cyclin/CDK complexes. CDKs 4 and 6 are activated by binding of D-type cyclins. Association of cyclin D to CDKs 4 and 6 is prevented by p16INK4A that binds with high affinity to these CDKs. Thereby, binding of cyclin D to these CDKs is prevented. The CDK inhibitors p21cip1 and p27kip1 bind both to cyclin E/CDK2 and to cyclin D/CDK4 complexes, although with different consequences. Even though these inhibitors antagonize cyclin E/CDK2 activity, they are required for proper assemblage and activity of cyclin D/CDK4/6 complexes

CDK-Independent Activities of D-Type Cyclins

Apart from their role in the activation of CDKs, D-type cyclins can have several profound effects on cellular physiology independent of their CDK partners. In ▶ breast cancer, cyclin D1 can bind directly to the estrogen receptor, thereby causing hormone-independent activation of the estrogen receptor. This activity of cyclin D1 may contribute to resistance to antihormonal therapy that is often seen in the clinic. In addition, D-type cyclins can modulate the activity of Myb transcription factors.

In this respect is the ▶ Myb-like transcription factor DMP1, which has antiproliferative activity. The expression of cyclin D inhibits this effect on cell proliferation of DMP1 through direct binding to DMP1, which prevents DNA binding by DMP1.

Clinical Relevance

Because of their critical role in linking cytoplasmic signals to nuclear responses, it is perhaps not surprising that D-type cyclins are frequently

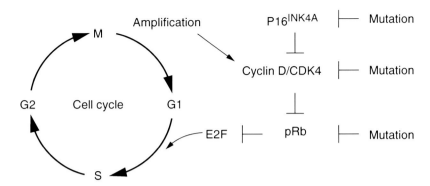

Cyclin D, Fig. 3 The p16-cyclin D-pRb pathway: a frequent target in human cancer. E2F transcription factors contribute to G1-S phase progression through the activation of specific target genes. E2F activity is negatively regulated by its binding to the retinoblastoma tumor suppressor gene product, pRb. The ability of pRb to bind E2F is regulated by cyclin D/CDK complexes. The activity of cyclin D/CDK complexes in turn is negatively regulated by p16INK4A that is encoded by the CDKN2A tumor suppressor gene

deregulated in several types of cancer. Cyclin D1 ► amplification or overexpression is found in a number of human malignancies, the most prominent being breast cancer, in which up to 50% of all cases have elevated levels of cyclin D1 protein. Chromosomal translocations involving cyclin D1 are found in parathyroid adenoma and in mantle cell lymphoma.

Not only is cyclin D1 itself often directly mutated in human cancer, its upstream regulators such as p16INK4A and its downstream target pRb are frequent targets in human carcinogenesis as well. It is generally believed that this p16INK4A-cyclin D1-pRb pathway is deregulated in virtually all human cancers (Fig. 3).

References

Bernards R (1997) E2F, a nodal point in cell cycle regulation. Biochim Biophys Acta 1333:M33–M40

Bernards R (1999) CDK-independent activities of D type cyclins. Biochim Biophys Acta 1424: M17–M22

Peeper DS, Bernards R (1997) Communication between the extracellular environment, cytoplasmic signalling cascades and the nuclear cell-cycle machinery. FEBS Lett 410:11–16

Sherr CJ, Roberts JM (1999) CDK inhibitors: positive and negative regulators of G1-phase progression. Genes Dev 13:1501–1512

Cyclin G-Associated Kinase

Mira R. Ray and Paul S. Rennie
The Prostate Centre at Vancouver General Hospital, University of British Columbia, Vancouver, BC, Canada

Synonyms

Auxilin-2; GAK

Definition

GAK/auxilin-2 is the ubiquitously expressed form of the neuronal-specific protein Auxilin-1. GAK is a member of the Ark/Prk serine/threonine protein kinases family and has an important role in ► endocytosis, un-coating clathrin-coated vesicles (CCVs) in nonneuronal cells, and clathrin-dependent trafficking from the *trans*-Golgi network. GAK has been implicated as a transcriptional coactivator of the ► androgen receptor (AR) and may have a role in prostate cancer

This entry was first published in the 2nd edition of the Encyclopedia of Cancer in 2009.

Cyclin G-Associated Kinase, Fig. 1 Human GAK (hGAK) is a 144 kDa protein that has three functional domains: an NH$_2$-terminal Ser/Thr kinase domain, a central Auxilin/Tensin homology domain, and a COOH-terminal J-domain

progression to androgen-independent disease. Expression patterns during androgen-withdrawal therapy suggest a prognostic role for GAK in advanced prostate cancer.

Characteristics

GAK is a large, 140 kDa protein that has three functional domains: (i) an NH$_2$-terminal Ser/Thr kinase domain, (ii) a central Auxilin/Tensin homology domain, and (iii) a COOH-terminal J-domain (Fig. 1). Although GAK was initially identified by its association with cyclin G, which is a downstream transcriptional target and a negative regulator of the p53 tumor suppressor protein, subsequent studies suggest that GAK has an important role in endocytosis and un-coating CCVs in nonneuronal cells.

Kinase assays demonstrated that GAK is one of two kinases present in clathrin-coated vesicles and that the Ser/Thr kinase activity of GAK was directed toward the μ2 component of CCVs. The protein homologue that is responsible for un-coating CCVs in neuronal cells is Auxilin. Consequently, the central Auxilin/Tensin homology domain of GAK is essential for its role in vesicle un-coating and has clathrin-binding motifs that allow for association with CCVs. GAK differs from Auxilin by its kinase domain, and GAK was classified as an Ark kinase family member due to its homology with actin-regulating kinase (ARK)-1.

The COOH-terminal J-domain of GAK is known to interact with the molecular chaperone Hsc70, which is the constitutively expressed form of Hsp70. Both GAK and Auxilin are DnaJ homologues that possess J-domains. The J-domain recruits ATP-bound Hsc70 to CCVs. The

mechanism of clathrin dissociation was first elucidated for Auxilin and then shown to be comparable for GAK. Recruitment of ATP-bound Hsc70 activates dormant ATPase activity of the chaperone and destabilizes clathrin-clathrin interactions in CCVs. Hsc70, now in its ADT state, remains tightly bound to clathrin while Auxilin is recycled for another round of un-coating. GAK activity in nonneuronal cells also involves Hsc-70 recruitment and ATPase-dependent destabilization of clathrin interactions. The kinase domain of GAK is indeed functional and, as described above, phosphorylates components of CCVs.

GAK and the Androgen Receptor

The impact of androgens, which are male steroid hormones, on prostate growth is a major consideration when treating advanced prostate cancer. Androgens carry out their function through the androgen receptor (AR), which is a ligand-dependent transcription factor. Activated AR regulates genes that are involved with growth and differentiation of the prostate gland. Dihydrotestosterone (DHT) has up to fivefold greater affinity for AR than testosterone (T) and, consequently, is up to 2.5-times more active as a hormone. For this reason, 5α-reductase inhibitors that prevent conversion of T to DHT were developed for use in prevention and treatment of prostate diseases. Another commonly used therapeutic strategy is removal of androgens by surgical and/or chemical methods to prevent activation of AR. Hormone therapy reduces testosterone levels significantly, down to ~10% of the normal level; low-levels of circulating androgens resulting from adrenal secretions are still present. Removal of hormone by androgen ablation therapy prevents the growth-promoting effects of androgens, leads to ► apoptosis of cancer cells, and ultimately results in tumor regression.

The decrease in tumor burden following androgen ablation occurs during the androgen-dependent (AD) stage of prostate cancer when the tumor still requires androgens for survival and growth. Unfortunately, this form of therapy offers limited aid, and the average range of overall survival is only 23–37 months. For reasons that are not fully understood, prostate cancer cells switch from an AD state to one that is androgen-independent (AI), in which cells are able to bypass requirement for the androgenic growth signal and grow in an uncontrolled fashion. As a result, tumor burden and prognostic disease markers such as ▶ Prostate-Specific Membrane Antigen (PSMA) increase dramatically.

Transcriptional coactivators of AR interact directly with the receptor to modify AR-mediated transactivation and gene expression. Originally, this class of proteins carried out their role at the level of the promoter, either by altering DNA accessibility to transcription machinery or by bridging AR to the basal transcription machinery. Several of these *classical*, or *Type I*, coactivators have indeed been associated with AR. However, a class of *nonclassical*, or *Type II*, coactivators was formed to differentiate their mechanism of action. Nonclassical coactivators enhance AR activity by altering other facets of the AR transactivation process. These include: (i) stability of inactive and active AR, (ii) nuclear translocation, and (iii) DNA binding. In addition, nonclassical coactivator activity may involve posttranslation modification of AR by phosphorylation, acetylation, or sumoylation. Unfortunately, the role of coactivators in directing AR activity is far from clear.

One possible mechanism for AI activation of AR in advanced prostate cancer is related to aberrant activity or expression of transcriptional coactivators. The majority of studies on AR coactivators have been carried out in vitro and the relevance of these accessory proteins has yet to be determined in vivo. However, transactivation assays with many coactivators have demonstrated that ectopic overexpression of these proteins results in enhanced AR transactivation and increased expression of AR-regulated genes.

GAK was identified as a putative AR-interacting protein by using the AR's NH_2-terminal transactivation domain as molecular bait in a modified yeast two-hybrid system. Subsequent studies, including immunoprecipitation and GST-pull-down assays, not only confirmed the interaction between AR and GAK in prostate carcinoma cell lines but also detailed the regions of interaction between the two protein molecules. GAK interacted with all three AR domains (NH_2-terminal transactivation, central DNA binding, and COOH-terminal ligand-binding domains), whereas only the Auxilin/Tensin homology domain of GAK is crucial for its interaction with AR.

Transactivation assays, which assess the impact of GAK on AR-mediated transcription, demonstrated that GAK could enhance AR activity from three- to fivefold in the presence of androgens. Importantly, however, overexpression of GAK increased AR activity by up to eightfold in low-androgen conditions, suggesting that increased GAK expression can result in increased AR sensitivity to low concentrations of androgens and may serve as a mechanism for prostate cancer progression to androgen independence.

The exact mechanism by which GAK serves as a transcriptional coactivator of AR is unknown. Also, its role as a Ser/Thr kinase is still being investigated. Upon activation by its ligand, AR is phosphorylated. Additionally, activated AR is sensitive to phosphorylation by signal transduction pathways. Whether these phosphorylation events are integral to the transcriptional activity of AR and whether phosphorylation of AR by GAK or any other kinase has a role in androgen-independent activation of AR remains to be determined.

GAK and Prostate Cancer Progression

The growing incidence of prostate cancer over the past two decades is due, in part, to increased detection through the introduction of PSA screening. PSA is a useful molecular marker for assessment of disease progression, particularly for emergence of AI disease in patients undergoing androgen ablation therapy, and is used in conjunction with digital rectal exam (DRE) and transrectal ultrasound for diagnosis. In addition, histological grading of prostate biopsies by Gleason score and TMN staging are important prognostic tools. Early detection is the most powerful weapon against prostate cancer since the disease is often curable in its early stages. However, patients with advanced disease have poor prognosis and an average life expectancy of 30 months with androgen ablation therapy.

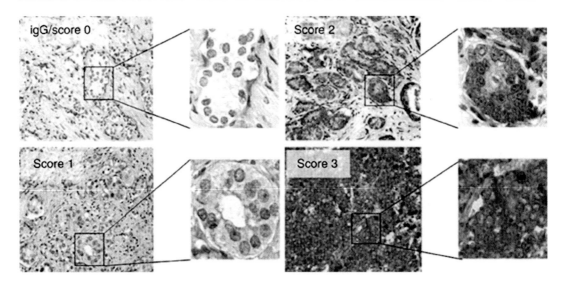

Cyclin G-Associated Kinase, Fig. 2 A NHT tissue microarray was stained with an antibody that recognizes GAK (Santa Cruz Biotechnologies, Inc.). Staining intensity was scored from 0 to 3 by a pathologist. Examples of IgG control and GAK staining with *score = 1*, *score = 2*, and *score = 3* are shown. Slides were visualized under × 40 magnification and further magnification of delineated areas is shown

Significant research has been carried out to identify diagnostic and prognostic molecular markers for disease progression as well as molecular targets for therapeutic strategies. Since GAK was identified as an AR-interacting protein with coactivator properties, it may have a role in inappropriate activation of AR in advanced disease. High throughput immunohistochemical analysis of neoadjuvant hormone therapy (NHT) tissue microarray was used to assess GAK's potential as a prognostic molecular marker of prostate cancer.

For the NHT arrays, a total of 112 samples were obtained and sampled in triplicate. Most tissues were radical prostatectomy specimens, while AI tissues were obtained either from transurethral resections from patients with hormone refractory disease or from warm autopsy samples of metastatic tissues. Specimens were chosen so as to represent various treatment durations of androgen-withdrawal therapy prior to radical prostatectomy ranging from no treatment ($n = 21$), ≤ 6 months ($n = 49$), and more than 6 months ($n = 28$). AI tumors were also identified ($n = 14$).

GAK expression was assessed by immunohistochemistry and staining intensity was scored visually by a pathologist on a scale from 0 to 3, ranging from no staining (score 0) to very intense staining (score 3) (Fig. 2). While GAK expression decreases slightly in response to androgen withdrawal, suggesting that GAK itself may be an AR-regulated gene, the proportion of samples with elevated GAK expression (score 2–3) increases during continued NHT treatment. In AI disease samples, elevated levels of GAK expression are considerably higher than during the AD phase. Since over 95% of AI tumor biopsies exhibit very high levels of GAK (visual score ≥ 2), the use of GAK as a prognostic marker for disease progression merits further investigation.

The mechanism by which GAK may encourage AI growth of prostate cancer is still unknown. GAK acts as a coactivator of AR and renders AR more responsive to lower androgen levels. Androgen ablation therapy, including NHT, reduces the level of circulating androgens. However, low levels of circulating androgens from adrenal secretions are still present. In prostate cancer cells, it is possible that increased levels of GAK are able to sensitize AR so that the receptor is responsive to these circulating low-level adrenal androgens, as observed in vitro. This would allow for AR activation and expression of AR-regulated growth and survival genes.

In summary, findings have identified GAK as a putative coactivator of AR and a possible molecular marker for prostate cancer progression. GAK expression may be a useful prognostic tool for assessing prostate cancer progression in patients undergoing androgen ablation therapy. An increase in GAK expression during the therapeutic course would suggest that the cancer has adapted to androgen-withdrawn conditions and are progressing to an AI stage. These early warnings of disease progression are crucial in allowing physicians the opportunity to reassess appropriate strategies for disease treatment.

Cross-References

- ► Androgen Receptor
- ► Apoptosis
- ► Endocytosis
- ► Prostate-Specific Membrane Antigen

References

Greener T, Zhao X, Nojima H et al (2000) Role of cyclin G-associated kinase in uncoating clathrin-coated vesicles from non-neuronal cells. J Cell Sci 275(2):1365–1370
Heinlein CA, Chang C (2002) Androgen receptor (AR) coregulators: an overview. Endocr Rev 23(2):175–200
Ray MR, Wafa LA, Cheng H et al (2006) Cyclin G-associated kinase: a novel androgen receptor-interacting transcriptional coactivator that is overexpressed in hormone refractory prostate cancer. Int J Cancer 118(5):1108–1119

See Also

(2012) P53. In: Schwab M (ed) Encyclopedia of cancer, 3rd edn. Springer, Berlin/Heidelberg, p 2747. doi:10.1007/978-3-642-16483-5_4331

Cyclin-Dependent Kinase Inhibitor 1A

► p21

Cyclin-Dependent Kinase Inhibitor 1B (CDKI1B)

► p27

Cyclin-Dependent Kinase Inhibitor 2A

► CDKN2A

Cyclin-Dependent Kinases

Roy M. Golsteyn
Department of Biological Sciences, University of Lethbridge, Lethbridge, AB, Canada

Synonyms

CDK; Cdk1 kinase; Maturation-promoting factor; MPF

Definition

A class of enzymes that add phosphates to other proteins.

Characteristics

Cyclin-dependent kinases (CDKs) catalyze one of the most important biological events in eukaryotic cells – cell proliferation. When cells proliferate, they follow the four phases of the ► cell cycle that enable them to duplicate their DNA and deliver exact copies into two daughter cells. Many enzymes orchestrate these steps, but the key enzymes that control the phases of the cell cycle are CDKs (Fig. 1).

CDKs are enzymes in the serine/threonine protein kinase family. They phosphorylate (add

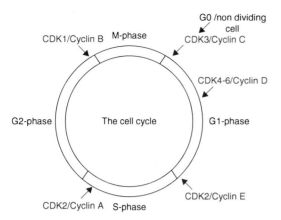

Cyclin-Dependent Kinases, Fig. 1 CDKs and the cell cycle. The role of the major CDKs (cyclin-dependent kinases) is shown relative to the four phases of the cell cycle. Each transition from one phase to another during the cell cycle, including entry into the cell cycle from G0 or nondividing cells, requires the formation and activation of a CDK

phosphate) to substrate proteins on hydroxy amino acids, serine, and threonine but not on tyrosine, which distinguishes CDKs from the tyrosine protein kinase family. Once a substrate is phosphorylated by a protein kinase, the substrate often has different biological properties relative to its nonphosphorylated counterpart. For example, a phosphorylated protein may increase or reduce its catalytic activity, change its cellular localization, bind to or dissociate from other proteins, and increase or decrease its biochemical half-life. The prototype CDK, Cdk1, phosphorylates many proteins during the G2/M-phase transition of the cell cycle. The change in the phosphorylation state of these proteins causes cells to organize their DNA into chromosomes and begin the process of mitosis and cytokinesis.

CDKs are composed of two protein subunits: a protein kinase catalytic domain and a regulatory cyclin subunit. The catalytic domains of protein kinases can be divided into 11 subdomains, each with a specific sequence that permits the classification of protein kinases into families. Among the 518 protein kinases encoded in the human genome, 13 are members of the CDK family. The signature sequences for cyclin subunits are not as well defined as those of the catalytic subunit, however, it is estimated that there are at

least 25 different human cyclin subunits. The one-to-one pairing of a catalytic domain to a cyclin posits that one catalytic subunit can bind to more than one cyclin subunit, although not at the same time. The exact composition of all possible CDKs is currently not known, but it is an important parameter because the pairing combination defines its substrates, its cellular location, and the timing of activation.

Cdk1 is the best-known member of the CDK family. It is composed of the 34 kD Cdk1 catalytic domain that binds to A- or B-type cyclin subunits of 50–55 kD. The Cdk1 catalytic domain is inactive as a monomer and its protein levels do not change during the cell cycle. As cells progress toward mitosis, the cyclin B protein levels increase and bind to the catalytic subunit to form the CDK complex. This complex is now competent to become active, but its activity is constrained by a coordinated network of three other enzymes: Wee1, Cdc25C, and CAK. Wee1 phosphorylates the Cdk1 catalytic subunit on threonine 14 and tyrosine 15, two amino acids that are within the ATP binding site and therefore directly block the interaction between the catalytic subunit and its ATP substrate. Cdc25C, a member of a protein ▶ phosphatase family, can dephosphorylate or remove the phosphates from the ATP binding site. In addition, access to the ATP binding and protein substrate binding sites are hindered by a peptide within the catalytic domain known as the T-loop. Upon phosphorylation of this peptide on threonine by CAK (another member of the CDK family), the domain is displaced thus permitting access of ATP and protein substrates to the catalytic center of Cdk1. The sequential formation of Cdk1 from its subunits followed by phosphorylation by Wee1 enables cells to accumulate a form of Cdk1 that can be rapidly activated. If a number of other cellular events are completed, such as DNA synthesis, then Cdc25C dephosphorylates Cdk1 and the T-loop is phosphorylated by CAK. This system of regulation enables a cell to achieve maximum Cdk1 catalytic activity in a very short time and engages a cell to enter mitosis.

The catalytic activity of Cdk1 is maintained until the cyclin subunit is selectively degraded by a protease. Once cyclin B1 is degraded, the

Cdk1 catalytic subunit returns to its inactive state, the cell exits from mitosis and forms two daughter cells. Cdk1 participates in its activation and inactivation by phosphorylating the enzymes that regulate it. This creates an autoactivation loop that results in an all-or-none activity of Cdk1, which is coherent with the all-or-nothing entry into mitosis during the cell cycle.

The Discovery of CDKs

Cell-division-cycle 2 (Cdc2) kinase was the original name of the prototype Cdk1, based upon the discovery of the role of Cdc2 in mitosis. A convention for naming this family of protein kinases was established by consensus at the Cold Spring Harbor Symposium on the Cell Cycle at Cold Spring Harbor, NY, USA in 1991. The role of Cdk1 in the cell cycle was originally identified by experiments with cells of lower eukaryotic species. The catalytic subunit was first identified by genetic studies in the fission yeast *Schizosaccharomyces pombe* as the Cdc2 gene product. Cyclin proteins were first identified in sea urchin eggs. After fertilization, sea urchin eggs enter mitosis in a synchronous manner. This natural synchrony and the experimental methods used to detect a new protein synthesis permitted the discovery of B-cyclins, whose levels of cycle are relative to mitosis. Evidence that cyclins and Cdc2 family members form a complex came from biochemical studies of *Xenopus* (toad) oocytes and starfish oocytes. Copurification of MPF with the Cdc2 subunit and cyclin B demonstrated that Cdc2/Cyclin B were the major components of an essential complex for mitosis. The confirmation that Cdc2 was part of a pathway conserved in all eukaryotic cells was made by genetic complementation studies in which yeast cells could proliferate after replacement of the yeast Cdc2 gene with the human CDC2 gene. Human and yeast genomes are separated by 1 billion years of genetic isolation, yet their functional activity has been conserved. Leland Hartwell, Tim Hunt, and Paul Nurse were awarded the Nobel Prize in physiology or medicine in 2001 for their role in the discovery and the characterization of Cdc2 and cyclins in the cell cycle.

Members of the CDK Family

Other CDKs, besides Cdk1, have important roles in progression through the cell cycle. CDK3 with its putative partner cyclin C is activated as cells start the cell cycle from a resting state (G_0). CDK4 and CDK6, which are associated with D-type cyclins, respond to extracellular growth signals and permit cells to continue the G1 phase of the cell cycle. CDK4 and CDK6 complexes phosphorylate the ▶ retinoblastoma protein, which controls the expression of genes required for the G1/S-phase transition and S-phase progression. The CDK2/cyclin E complex also participates in the G1/S-phase transition by phosphorylating replication factors. During S phase, CDK2/cyclin A phosphorylates different substrates allowing DNA replication and participates in the inactivation of G1 transcription factors. In mouse models, using gene knockout technology, it has been shown that CDK2 is not essential. By contrast CDK1 and CDK5 are essential genes. In the early phases of the cell cycle, the activity of CDKs is regulated by members of a small protein family, such as ▶ p21/WAF1 and ▶ p16/INK4. These proteins bind to the CDK complex and inhibit it. In the final phases of the cell cycle, after DNA synthesis is completed, the Cdk1 complex is activated and triggers the G2/M-phase transition. The cell ends M phase and enters G1 by phosphorylation of the anaphase-promoting complex (APC) by Cdk1. These successive waves of CDK/cyclin activation and inactivation drive the major phases of the cell cycle.

The role of CDK enzymes is not limited to the cell cycle. Some CDKs regulate DNA transcription, including the transcription of genes that are essential for cell cycle progression, therefore, these CDKs have a role that lies at the interface of cell cycle control and basal cell activity. CDK7, which binds to cyclin H, is a component of the transcription factor TFIIH. CDK7 has a dual role because it can phosphorylate RNA polymerase II, which is required for RNA elongation, and it can phosphorylate the Cdk1 and Cdk2 T-loop, which is required for CDK1 and CDK2 activation. The role of CDK8 overlaps with that of CDK7 in transcriptional regulation. CDK9/cyclin T is a component of a transcription elongation factor.

CDK activity can be detected in cells that undergo ▸ apoptosis (programmed cell death). In some cases, this activity is related to DNA damage, resulting in a type of mitosis that is known as ▸ mitotic catastrophe, which eventually leads to cell death. Activation of CDKs can occur in nondividing neuronal tissue after chemical shock and may be required for apoptosis. The role of CDKs in apoptosis has implications in chronic inflammation. Neutrophil, a specialized cell that participates in inflammatory response, may be made dependent upon CDKs to engage apoptosis under experimental conditions.

Relative to other CDKs, CDK5 has a specialized role in nondividing neural tissue. Its activity is important for neurite outgrowth and neuronal development and myogenesis and somite organization in embryos. The cyclin subunit, p35, is one of the smallest members of the cyclin family (35 kD) and it can be processed to exist in a smaller form of 25 kD. The three-dimensional structure of p35 reveals that it has similar protein folds to other cyclins, which provides a rationale for its capacity to activate CDK5.

The crystal structure of several CDKs has been resolved, which gives insight to the organization of the substrate binding site and the role of the regulatory cyclin subunit. In the case of CDK2, which is composed of a Cdk2 catalytic subunit and a cyclin A subunit, it was revealed the kinase complex consists of an amino-terminal lobe rich in β-sheets and a carboxyl-terminal lobe that is larger and mostly α-helical. Between these two lobes is a deep pocket that harbors the ATP binding site, which contains the conserved amino acids that participate in catalysis. The T-loop, which blocks substrate access, moves away from the catalytic cleft after cyclin binding and is then accessible for phosphorylation by CAK. This highlights the role of cyclin and CAK in CDK activation. Cyclin binding also reorientates amino acids within the ATP pocket to permit phosphate transfer from ATP to the protein substrate. The resolution of CDK5 and CDK6 revealed structures that are similar to that of CDK2, suggesting that CDK structures of the entire family are similar.

CDKs in Cancer and Other Human Diseases

CDKs are implicated in a broad range of human diseases. Their essential role in cell proliferation, especially in the case of Cdk1, has led to the proposal that these proteins play an important role in human cancers. Although it is known that cancer cells require Cdk1 to proliferate, it remains to be demonstrated that the function of Cdk1 in cancer cells is different from its role in normal, proliferating cells. The overexpression of cyclin E is frequently found in both precancerous and cancerous lesions in human tissue. It is believed that CDK activation by cyclin E may cause genomic instability. There is much evidence to link CDK5 activity to cytoskeletal abnormalities that can lead to neuronal cell death. This may be, in part, caused by the conversion of its cyclin partner p35 to p25, which leads to activation of CDK5 and alteration of its cellular localization. Uncontrolled activation of CDK5 causes the phosphorylation of neuronal proteins, such as tau, and may be linked to Alzheimer disease. Chronic inflammatory diseases such as gout, arthritis, and Crohn disease may be partly due to misregulation of CDKs. CDKs also participate in virus replication in infected cells. In the example of HIV, which is linked to the cause of AIDS, CDK9 is recruited by the ▸ HIV tat protein which enhances transcription of viral genes.

The involvement of CDKs in disease and the detailed knowledge of CDK atomic structure have led to the identification of potent chemical inhibitors that have the potential to become ▸ small molecule drugs. Many of these have been cocrystallized with CDKs, which gives insight to the molecular docking of inhibitors in the ATP pocket. Pharmacological inhibitors of CDKs are being evaluated for therapeutic use in major human diseases such as cancer, neurodegenerative disorders, cardiovascular disorders, viral infections, and parasitic infections.

Cross-References

▸ Cell Cycle Checkpoint
▸ Cyclin D

▶ Retinoblastoma Protein, Biological and Clinical Functions

▶ Retinoblastoma Protein, Cellular Biochemistry

▶ TAT Protein of HIV

References

Dorée M, Hunt T (2002) From Cdc2 to Cdk1: when did the cell cycle kinase join its cyclin partner? J Cell Sci 115:2461–2464

Knockaert M, Greengard P, Meijer L (2002) Pharmacological inhibitors of cyclin-dependent kinases. Trends Pharmacol Sci 23:417–425

Manning G, Whyte DB, Martinez R et al (2002) The protein kinase complement of the human genome. Science 298:1912–1934

Morgan DO (1995) Principles of CDK regulation. Nature 374:131–134

Cyclins

Definition

Cyclins are a family of proteins involved in the cell cycle progression, cooperating with its catalytic partner ▶ cyclin-dependent kinase (CDK), which activates the protein kinase function. The most important substrate of cyclin/CDK complex is ▶ retinoblastoma protein (Rb).

Cross-References

▶ Cyclin D
▶ Cyclin-Dependent Kinases
▶ Early B-Cell Factors
▶ INK4a
▶ Retinoblastoma Protein, Biological and Clinical Functions

Cyclooxygenase

Definition

Cyclooxygenases are proteins (enzymes) that are rate limiting in the production of thromboxanes and ▶ prostaglandins from arachidonic acid (arachidonic acid pathway). The constitutive form of these proteins (COX-1) is essential for maintaining homeostasis, while the inducible form (COX-2) is expressed in leukocytes in response to an ▶ inflammation stimulus, resulting in production of prostanoids. The activity of cyclooxygenases is inhibited by ▶ nonsteroidal anti-inflammatory drugs, such as ▶ aspirin, and corticosteroids interfere with gene expression of COX-2. COX-3 is the third discovered cyclooxygenase and is actively being studied for its similarities and differences between the COX-1 and COX-2 enzymes. It expresses a retained intron sequence from the COX-1 transcript that may decrease its enzymatic potential to generate prostaglandin E2. Currently, one hypothesis suggests that COX-3 exhibits the same role in prostaglandin synthesis.

COX-1 was first purified in 1976, and the gene was cloned in 1988. COX-1 is the key enzyme in the synthesis of ▶ prostaglandins (PGs) from arachidonic acid. In 1991, several laboratories identified a product from a second gene with COX activity and called it COX-2. However, COX-2 is inducible, and the inducing stimuli included pro-inflammatory ▶ cytokines and growth factors, implying a role for COX-2 in both ▶ inflammation and control of cell growth. COX-1 and COX-2 are almost identical in structure but have important differences in substrate and inhibitor selectivity and in their intracellular locations. Protective prostaglandins, which preserve the integrity of the stomach lining and maintain normal renal function in a compromised kidney, are synthesized by COX-1. In addition to the induction of COX-2 in inflammatory lesions, it is present constitutively in the brain and spinal cord, where it may be involved in nerve transmission, particularly that for pain and fever. ▶ Prostaglandins made by COX-2 are also important in ovulation and in the birth process. The discovery of COX-2 has made possible the design of drugs that reduce inflammation without removing the protective PGs in the stomach and kidney by COX-1. These highly selective COX-2 inhibitors may not only be anti-inflammatory but may also be active in ▶ colorectal cancer (▶ COX-2 in colorectal cancer) and Alzheimer disease.

Tumor development in several different tissues is frequently associated with overexpression of COX-2 in both premalignant and malignant stages, indicating that activation of COX-2 may be an early event in carcinogenesis. This overexpression often starts in tissues adjacent to the transformed epithelium giving rise to "activated" stroma. Expression of COX-2 is induced by numerous growth factors, ▶ cytokines and ▶ oncogenes, and regulated both transcriptionally and posttranscriptionally, especially through increased mRNA stability. Both genetic and pharmacologic studies support a causal role of COX in cancer development. Genetic inactivation of COX-2 strongly reduces tumor formation in several animal model systems including the classical two-stage mouse skin cancer model and in Apc mutant Min mouse models. These effects are not limited to COX-2 but also apply in part to COX-1. These data are corroborated by pharmacologic intervention studies using both nonselective COX inhibitors, like ▶ aspirin and other ▶ nonsteroidal anti-inflammatory drugs (NSAIDs) and COX-2 selective inhibitors. The subsequent reduction in tumor formation has been documented in numerous experimental animal model studies but also in patients with familial adenomatous polyposis (FAP) and supports the outcome of many epidemiological studies suggesting a chemoprotective effect of longtime regular use of these drugs.

Cross-References

- ▶ Arachidonic Acid Pathway
- ▶ Aspirin
- ▶ Colorectal Cancer
- ▶ COX-2 in Colorectal Cancer
- ▶ Cytokine
- ▶ Inflammation
- ▶ Nonsteroidal Anti-Inflammatory Drugs
- ▶ Oncogene
- ▶ Prostaglandins

See Also

(2012) Arachidonic acid. In: Schwab M (ed) Encyclopedia of cancer, 3rd edn. Springer, Berlin/Heidelberg, p 260. doi:10.1007/978-3-642-16483-5_379
(2012) Corticosteroids. In: Schwab M (ed) Encyclopedia of cancer, 3rd edn. Springer, Berlin/Heidelberg, pp 982–983. doi:10.1007/978-3-642-16483-5_1341
(2012) Familial adenomatous polyposis. In: Schwab M (ed) Encyclopedia of cancer, 3rd edn. Springer, Berlin/Heidelberg, p 1373. doi:10.1007/978-3-642-16483-5_2106
(2012) Min. In: Schwab M (ed) Encyclopedia of cancer, 3rd edn. Springer, Berlin/Heidelberg, p 2318. doi:10.1007/978-3-642-16483-5_3750
(2012) Prostanoid. In: Schwab M (ed) Encyclopedia of cancer, 3rd edn. Springer, Berlin/Heidelberg, p 3009. doi:10.1007/978-3-642-16483-5_4780

Cyclooxygenase (Prostaglandin Endoperoxide Synthase 2)

▶ COX-2 in Colorectal Cancer

Cyclophosphamide

Definition

Cyclophosphamide is a bifunctional nitrogen mustard that is a most commonly used drug in combination chemotherapy and is a DNA alkylating agent that is used as an immunosuppressive drug. It acts by killing rapidly dividing cells, including lymphocytes proliferating in response to antigen.

Cross-References

- ▶ Alkylating Agents

Cylindromatosis

Synonyms

Turban tumor syndrome

Definition

Cylindromatosis is a condition where mutations in the *CYLD* tumor suppressor gene predispose to benign tumors arising in hair follicles and in cells of sweat and scent glands, collectively called epitheliomas.

Cross-References

▶ Arachidonic Acid Pathway

CYP 450$_{arom}$

▶ Aromatase and Its Inhibitors

CYP450

▶ Cytochrome P450

Cystadenocarcinoma

▶ Appendiceal Epithelial Neoplasms

Cystadenoma

▶ Ovarian Tumors During Childhood and Adolescence

Cystatins

Lauren M. Nunez and Daniel Keppler
Department of Biological Science, College of Pharmacy, Touro University-CA, Vallyo, CA, USA

Synonyms

Thiol-protease inhibitors; Thiostatins; TPI

Definition

Cystatins were originally defined as endogenous inhibitors of thiol- or cysteine-proteases. Later, the discoveries of several proteins whose primary sequence revealed substantial homology to the typical cystatin domain led to the definition of a cystatin superfamily. With the identification of other types of intracellular cysteine-proteases, however, it became clear that cystatins inhibit mainly cysteine-proteases present in endosomes and lysosomes. Additionally, many of the newer members of the cystatin superfamily have less than 30% homology to classical cystatins and do not inhibit lysosomal cysteine proteases. Moreover, several proteins with little apparent amino acid sequence homology fold into typical three-dimensional structures attributed to cystatins. Thus, the term "cystatins" now refers to a heterogeneous group of proteins still lacking both a uniform identity and a cohesive definition.

Characteristics

Domain Structure

The primary structure of the archetypal cystatin domain consists of a polypeptide 100–120 amino acids in length. This polypeptide folds into a five-stranded β-sheet, which partly wraps around an inner α-helix, resulting in a three-dimensional structure commonly referred to as the "hot-dog fold." Members of the cystatin superfamily have been categorized into several distinct families according to the following criteria: (i) amino acid sequence homology; (ii) location of disulfide bonds; (iii) presence of a signal peptide in the primary translation product; and (iv) number of cystatin-like domains. Based on these criteria, roughly five different families can be distinguished in humans today: stefins, cystatins, latexins, fetuins, and kininogens (Fig. 1). Whether or not these diverse families and all of their respective members are phylogenetically linked remains to be clearly determined. Some members (i.e., fetuins and latexins) may indeed have acquired "cystatin"-like properties through other mechanisms such as convergent evolution.

Cystatins,
Fig. 1 Composition of the
cystatin superfamily.
Schematic ball and stick
representation of members
of the cystatin superfamily
with one, two, or three
cystatin-like domains. k,
kinin moiety; s, signal
peptide

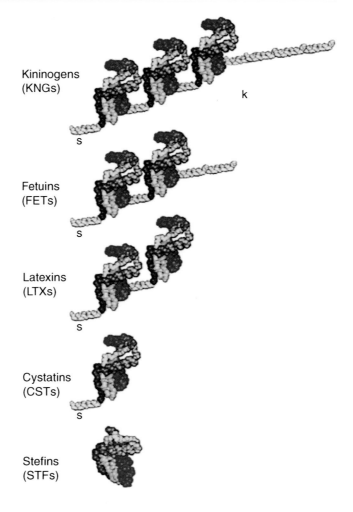

Kininogens
(KNGs)

Fetuins
(FETs)

Latexins
(LTXs)

Cystatins
(CSTs)

Stefins
(STFs)

The stefins (STFs) are 11-kDa cytoplasmic cystatins each containing a single cystatin-like domain and lacking typical features of secreted proteins such as a signal peptide, disulfide bonds, and glycosylation. The cystatins (CSTs) are secreted proteins 14–24-kDa in mass, which are distinguishable by a single cystatin domain, a typical signal peptide, and disulfide bonds. Some CSTs are *N*- or *O*-glycosylated or exhibit Ser/Thr-phosphorylation. The secreted latexins (LTXs) contain two typical cystatin domains, but share little sequence homology to other cystatins and seem to lack cysteine protease inhibitory activity. Plasma fetuins (FETs) also contain two cystatin-like domains, but share <30% sequence homology to other cystatins and also lack cysteine protease inhibitory activity. Finally, the kininogens (KNGs) are plasma proteins with three cystatin-

like domains, domain 1 being inactive, and domains 2 and 3 being active as cysteine protease inhibitors.

Gene Structure and Evolution

At the gene level, a typical cystatin domain is generally the pr\oduct of three coding exons. Consequently, the genes of most members of the cystatin superfamily are composed of 3+ (STFs and CSTs), 6+ (LTXs and FETs), and 9+ (KNGs) coding exons. Although the existence of a common evolutionary origin of the individual cystatin genes is still uncertain, a hypothetical model for the evolution of the superfamily has been proposed. According to this model, the various members arose from an ancestral STF-like cytoplasmic inhibitor. Upon evolution, the archetype gene acquired additional DNA elements such as, for example, the coding sequence for a signal peptide. This may have resulted in an archetypal gene for a secreted cystatin, which produced members with two and three cystatin domains upon successive gene duplication events. It is also possible that alternative evolutionary mechanisms have contributed to the level of diversity amongst the present members of the cystatin superfamily.

Prokaryotes do not seem to harbor any genes even remotely similar to cystatins. Cystatins, instead, seem to have emerged with the development of more complex forms of life in eukaryotes. Typical cystatin-like genes have indeed been identified in various plants, some yeast strains, various unicellular parasites, as well as in worms, insects, fishes, frogs, birds, snakes, and mammals.

Function

The primary function of cytoplasmic cystatins, such as STFA (stefin A/CSTA/ACPI/α-TPI) and STFB (stefin B/CSTB/NCPI/β-TPI), is generally assumed to be part of a safeguard mechanism protecting against the transient disruption of the integrity of lysosomes. Lysosomes are intracellular digestive factories filled with hazardous hydrolytic enzymes, including proteases generally referred to as ▶ cathepsins. Complexes between stefins and lysosomal cysteine-proteases form instantly and can be readily detected when the integrity of cells and lysosomes is compromised. Shear stress, fever, stress-induced oxidation or glycation of proteins, infection by pathogens, and other stress conditions all contribute to repeated episodes of lysosomal leakage, which must be concealed in space and time for cell survival. In proliferating tissues, these events bear less consequence than in postmitotic tissues such as cardiac and nervous tissue. Unsurprisingly, STFB deficiencies have been linked to hereditary forms of monoclonus epilepsies, in which cerebellar granule cells undergo cell death mediated by leakage of lysosomal cysteine-proteases.

Secreted cystatins such as CST3 (cystatin C/BCPI/γ-trace/post-γ-globulin) were long believed to function exclusively as inhibitors of lysosomal cysteine proteases, despite early studies proposing additional or alternate functions.

Before CST3 was coined a cystatin, it was believed to be a neuroendocrine peptide hormone. Over the years, five observations have greatly contributed to the need to reconsider the accepted model of CST function: (i) The number of CSTs, which seem to lack cysteine protease inhibitory activity, represents more than 50% of all members of the CST family in mammals. (ii) Several CST functions, such as the modulation of DNA synthesis, cell proliferation, and immune function, do not require cysteine protease inhibitory activity. (iii) Despite the fact that some CSTs form extremely tight molecular complexes with purified proteases in vitro, there is still little evidence for the occurrence of such complexes in vivo. (iv) Finally, under physiological and healthy conditions, most CSTs would be useless under the current model, as they are secreted proteins and their target enzymes are confined intracellularly.

Based on analyses of gene structures, a new model was proposed in which CSTs and ▶ chemokines evolved from a common ancestral gene and may have conserved similar integrated functions. Comparison of the crystal structures of CSTs and chemokines further underscores this relationship. Besides their similar gene structure (both are generally composed of three exons and two introns), CSTs share similarities in their basic three-dimensional fold: disordered *N*- and/or *C*-terminus, β-pleated sheet comprised of three to five anti-parallel β-strands, one β-strand with a bulge, a central α-helix, and two or three disulfide bonds to stabilize the core domain. CSTs and chemokines also share other intriguing features: Both types of proteins are secreted proteins, of low molecular mass (<24 kDa), which undergo *N*- or *C*-terminal processing by serine-proteases, have a tendency to dimerize, and exhibit multiple biological activities. To act like chemokines, however, CSTs would have to bind to specific ▶ G-protein coupled receptors (GPCRs). Such an interaction was proposed for the cystatin-like protein monellin from African berries, which, upon binding to a GPCR involved in taste perception, elicits an extremely sweet sensation in humans. Further direct evidence is required to establish an unambiguous relationship between CSTs and chemokines.

CSTs may play important roles in the modulation of the humoral and cellular immune responses. As inhibitors of lysosomal cysteine proteases, CSTs can block degradation of perforins and complement component C3, thus, preventing cytotoxic T cell- and complement-mediated cell lysis, respectively. In antigen presenting cells, lysosomal cysteine proteases play critical roles in major histocompatibility complex class II-mediated antigen processing and presentation. It has, however, not been conclusively determined whether CSTs are able to penetrate inside endosomes and/or lysosomes of antigen presenting cells to modify either process. As potential chemokines, CSTs have been shown to modulate the production of cytokines by various cell types such as macrophages and lymphocytes, as well as the release of nitric oxide, oxidative burst, and phagocytosis in neutrophils.

One of the functions of LTXs, as well as of other related members of the cystatin superfamily with two tandem cystatin domains, seems to be the inhibition of some serine- and/or metallo- but not cysteine proteases. LTX is the only known mammalian protein inhibitor of zinc-dependent carboxypeptidases and, thus, could potentially perform an important role in the regulation of peptide hormone activity during tissue growth and differentiation. The FETs regroups major plasma glycoproteins such as FETA (fetuin A/AHSG), FETB (fetuin B), and HRG (histidine- or histidine-and proline-rich glycoprotein). FETA is synthesized in the liver and constitutes a major plasma protein that accumulates in the matrix of bone and teeth. The protein also blocks calcium phosphate precipitation in the blood, thus preventing the spontaneous generation of systemic apatite crystals. This property is shared by HRG but not by FETB. A role for FETA in physiological and pathological mineralization is also evident from FETA knockout mice.

KNGs represent major plasma glycoproteins with three cystatin-like domains, and, like FETs, are also produced in the liver. One important function of KNGs is to serve as precursors of vasoactive kinins, such as bradykinin. Because bradykinin is extremely short-lived, this small peptide needs to be proteolytically released from the precursor molecule in the immediate vicinity of the GPCRs for bradykinin (BKRs). The third cystatin domain in KNGs is assumed to tether KNGs to the cell surface and thus bring the prokinin moiety close to the BKRs. With analogy to the interaction of chemokines with their respective receptors, one could postulate that KNGs perhaps also bind to BKRs in a two-step mode: an initial step in which there is a loose binding of the prokinin moiety, followed by proteolytic excision of the kinin moiety, tight binding, and activation of the receptor. Another important function of KNGs is to scavenge excess lysosomal cysteine-proteases that are locally released from damaged tissue during various injuries, infections, and inflammatory reactions. This buffering capacity of KNGs may limit further tissue damage and, hence, favor tissue remodeling and repair.

Role in Cancer

Lysosomal cysteine proteases have been implicated in multiple steps of tumor progression and recurrence, including early steps of immortalization and transformation, intermediate steps of tumor invasion and ▸ angiogenesis, and late steps of ▸ metastasis and drug resistance. During all these steps, tumor cells must actively escape immune surveillance. This may be accomplished, in part, through complete antigen degradation, which would leave no identifiable immunogenic peptides. Alternatively, exocytosis and cell surface binding of lysosomal cysteine proteases could lead to the degradation of the third component of complement C3 or the pore-forming protein perforin and, thus, interfere with immune cell-mediated lysis and the killing of target tumor cells. The importance of lysosomal cysteine proteases in the development of tumors from benign growths to aggressive lesions suggests that cystatins, i.e., STFs and CSTs, may in many ways safeguard against tumor progression. In spite of this, the scenario is not always as simplistic as some CSTs appear to have both tumor promoting and tumor suppressing activities, while others promote metastasis. Thus, STFs and CSTs play important but distinct roles in our current understanding of tumor formation and progression. There is a little or no information on LTXs and, the putative roles of FETs and KNGs in tumor neovascularization have been reviewed elsewhere.

In prostate and breast tissue, STFA is produced in basal and myoepithelial cells, respectively. STFA expression is thus lost with the loss of these cells during progression of most prostate and breast cancers. STFA expression is also lost during skin tumorigenesis, metastasis of oral/pharyngeal squamous cell carcinoma, and lung cancer progression. There is increasing evidence suggesting that STFs regulate initiation or propagation of the lysosomal cell death pathway. This cell death pathway is triggered by many different stimuli, including ▶ cytokines, p53 activation, and ▶ retinoids. In many tumor cells, the lysosomal cell death pathway involves cathepsin B as a major downstream executioner. Overexpression of STFA in various cancer cell lines indeed reduces their susceptibility to cell death-inducing agents. In addition, administration of STFA to mice bearing myeloid leukemias prolongs their mean survival. Exogenous treatments with STFA also reduce tumor cell ▶ motility. STFB is a far more ubiquitous and abundant protein than STFA, but like STFA, it also inhibits motility of tumor cells when exogenously administered.

CST3, unlike most other secreted cystatins, is considered a housekeeping-type gene. The mean concentration of this 15-kDa protein in normal human serum is about 77 nM (or 1.16 µg/ml). However, there are considerable differences in the levels of CST3 in other body fluids, suggesting that different tissues exhibit different accumulation rates of the protein or renewal rates of the extracellular fluid. The picture is further complicated by the fact that glomerular filtration is often impaired in patients with advanced cancer. CST3 has been extensively studied as a potential tumor marker, yet clinical studies tend to dismiss this cystatin as a useful diagnostic/prognostic marker. In spite of these clinical data, CST3 displays several noteworthy effects on tumor cells. As a potent inhibitor of lysosomal cysteine proteases, CST3 very efficiently inhibits in vitro tumor cell-mediated degradation and invasion of an extracellular matrix. In addition, the overexpression of CST3 in human glioblastoma

cells resulted in little apparent intracerebral tumor take compared to parental and mock controls. CST3, however, can also promote in vitro DNA synthesis and long-term proliferation of various cell types including stem cells. This may explain the fact that most established cell lines express and secrete this protein. Lung colony formation assays, in agreement with a mitogenic function of CST3, reveal that tail vein injection of highly metastatic melanoma cells results in a sevenfold reduction of lung colonies in CST3-null mice, compared to wild-type littermates.

CST6 (cystatin E/M) is expressed in a variety of normal human tissues, but its expression is lost in most established cancer cell lines and tumor tissues. In cancers of the breast, cervix, lung, and brain, the CST6 gene is epigenetically silenced by promoter hypermethylation rather than deleted or mutated. Overexpression of CST6 in breast and lung cancer cells results in a reduction of colony formation and cell proliferation, suggesting that this cystatin may hold some tumor-suppressing capabilities. This has been confirmed in vivo, after inoculation of tumor cell clones into mammary fat pads of severe combined immunodeficient mice. CST6 expression in breast cancer cells strongly reduces tumor growth during the first six to seven weeks after inoculation, but has only a minor effect on incidence of metastasis and number of lesions in the lungs. The overall metastatic burden in the lungs and the liver, however, is significantly smaller in the CST6 when compared to the control group. Further studies using the experimental lung colonization assay demonstrated that CST6 expression has no effect on the initial seeding and survival of tumor cells in the lungs, but does reduce the expansion of established lung colonies. Thus, CST6 is a *bona fide* ▶ tumor suppressor gene for breast cancer.

CST7 (cystatin F/leukocystatin/CMAP) is a highly tissue-specific cystatin, which explains its initial cloning and annotation as leukocystatin. The gene is expressed predominantly in cells of the hematopoietic lineage and not in fetal tissues. CST7 is highly expressed in peripheral blood cells, such as T-cells and monocytes, as well as in stem cell-derived dendritic cells. Most T- and

B-cell-derived cell lines and the promyelo-monocytic cell line U-937 secrete fair amounts of CST3, but little to no CST7. A significant amount of the CST7 produced by U-937 cells is in fact either retained intracellularly or is reabsorbed by the cells immediately following secretion. CST7 has also been cloned by RNA differential display as CMAP, a cystatin-like metastasis-associated protein. This gene is preferentially expressed in murine tumor cell lines that metastasize to the liver. Transfection of highly metastatic cells with CST7 antisense cDNA led to a reduction of experimental colonization of the liver and spleen by several transfection clones and improved survival of such tumor-bearing mice more than twofold. Additionally, CST7 expression was detected in several human tumor cell lines, particularly in those with high propensity to metastasize to the liver (i.e., colon, pancreas, and lung cancer cell lines). Other tumor cell lines, which demonstrated high levels of CST7 expression, included cells from melanomas, glioblastomas, and osteosarcomas. Based upon a multivariate analysis of 79 patients with colorectal cancer (including 17 cases with liver metastases), high expression levels of CST7 were identified as the strongest independent factor for liver metastasis. In addition, the 5-year survival rate was significantly lower in patients with high CST7 expression than it was for those who displayed low expression levels.

Conclusion

From the above considerations and hypotheses, it is clear that a much better understanding of the dual nature of cystatins could lead to the advancement of novel anticancer strategies. Further research could reveal that their roles as protease inhibitors and potential chemokines could be combined to signal anticancer immune responses, while also inhibiting tumor growth, invasion, and neovascularization. The literature on CST3 and CST7 suggests that chemokine-like function may need to be disrupted for efficient protease inhibitor-mediated tumor and metastasis suppression. In contrast, the example of CST6 shows that the two functions greatly contribute to tumor suppression.

Cross-References

▶ Angiogenesis
▶ Cathepsins
▶ Chemokines
▶ Cytokine
▶ Epigenetic Gene Silencing
▶ G Proteins
▶ Immunosurveillance of Tumors
▶ Metastasis
▶ Motility
▶ Retinoids
▶ Tumor Suppressor Genes

References

Keppler D (2006) Towards novel anti-cancer strategies based on cystatin function. Cancer Lett 235:159–176

Mohamed MM, Sloane BF (2006) Cysteine cathepsins: multifunctional enzymes in cancer. Nat Rev Cancer 6:764–775

Muller-Esterl W, Iwanaga S, Nakanishi S (1986) Kininogens revisited. Trends Biochem Sci 11:336–339

Turk V, Bode W (1991) The cystatins: protein inhibitors of cysteine proteinases. FEBS Lett 285:213–219

Zabel BA, Zuniga L, Ohyama T et al (2006) Chemoattractants, extracellular proteases, and the integrated host defense response. Exp Hematol 34:1021–1032

See Also

(2012) Complement. In: Schwab M (ed) Encyclopedia of Cancer, 3rd edn. Springer Berlin Heidelberg, p 963. doi:10.1007/978-3-642-16483-5_1284

(2012) Cytokinesis. In: Schwab M (ed) Encyclopedia of Cancer, 3rd edn. Springer Berlin Heidelberg, p 1055. doi:10.1007/978-3-642-16483-5_1477

(2012) Extracellular Matrix. In: Schwab M (ed) Encyclopedia of Cancer, 3rd edn. Springer Berlin Heidelberg, p 1362. doi:10.1007/978-3-642-16483-5_2067

(2012) Immortalization. In: Schwab M (ed) Encyclopedia of Cancer, 3rd edn. Springer Berlin Heidelberg, p 1811. doi:10.1007/978-3-642-16483-5_2969

(2012) Inhibitors. In: Schwab M (ed) Encyclopedia of Cancer, 3rd edn. Springer Berlin Heidelberg, p 1864. doi:10.1007/978-3-642-16483-5_3055

(2012) Lung Colony Formation Assay. In: Schwab M (ed) Encyclopedia of Cancer, 3rd edn. Springer Berlin Heidelberg, p 2115. doi:10.1007/978-3-642-16483-5_3436

(2012) Lysosome. In: Schwab M (ed) Encyclopedia of Cancer, 3rd edn. Springer Berlin Heidelberg, p 2128. doi:10.1007/978-3-642-16483-5_3472

(2012) N- and O-Glycosylation. In: Schwab M (ed) Encyclopedia of Cancer, 3rd edn. Springer Berlin Heidelberg, p 2447. doi:10.1007/978-3-642-16483-5_2459

(2012) N- or C-Terminal Processing. In: Schwab M (ed) Encyclopedia of Cancer, 3rd edn. Springer Berlin Heidelberg, p 2447. doi:10.1007/978-3-642-16483-5_5733

(2012) Neovascularization. In: Schwab M (ed) Encyclopedia of Cancer, 3rd edn. Springer Berlin Heidelberg, p 2474. doi:10.1007/978-3-642-16483-5_4016

(2012) P53. In: Schwab M (ed) Encyclopedia of Cancer, 3rd edn. Springer Berlin Heidelberg, p 2747. doi:10.1007/978-3-642-16483-5_4331

(2012) Perforin. In: Schwab M (ed) Encyclopedia of Cancer, 3rd edn. Springer Berlin Heidelberg, p 2814. doi:10.1007/978-3-642-16483-5_4443

(2012) Phagocytosis. In: Schwab M (ed) Encyclopedia of Cancer, 3rd edn. Springer Berlin Heidelberg, p 2840. doi:10.1007/978-3-642-16483-5_4493

(2012) Proteases. In: Schwab M (ed) Encyclopedia of Cancer, 3rd edn. Springer Berlin Heidelberg, p 3080. doi:10.1007/978-3-642-16483-5_4786

(2012) Ser/Thr-Phosphorylation. In: Schwab M (ed) Encyclopedia of Cancer, 3rd edn. Springer Berlin Heidelberg, p 3380. doi:10.1007/978-3-642-16483-5_5250

(2012) Severe Combined Immunodeficient Mice. In: Schwab M (ed) Encyclopedia of Cancer, 3rd edn. Springer Berlin Heidelberg, p 3395. doi:10.1007/978-3-642-16483-5_5271

(2012) Transfection. In: Schwab M (ed) Encyclopedia of Cancer, 3rd edn. Springer Berlin Heidelberg, p 3757. doi:10.1007/978-3-642-16483-5_5912

(2012) Transformation. In: Schwab M (ed) Encyclopedia of Cancer, 3rd edn. Springer Berlin Heidelberg, pp 3757–3758. doi:10.1007/978-3-642-16483-5_5913

Cystatins A, B

► Stefins

Cystic Fibrosis

Definition

Cystic fibrosis (CF) is a hereditary disease caused by mutations within the ► ABC transporter-encoding gene ABCC7 (CFTR).

Cystic fibrosis symptoms do not follow the same pattern in all patients but affect different people in different ways and to varying degrees. However, the basic problem is the same – an abnormality in the glands, which produces or secretes sweat and mucus. Sweat cools the body; mucus lubricates the respiratory, digestive, and reproductive systems and prevents tissues from drying out, protecting them from infection. People with cystic fibrosis lose excessive amounts of salt when they sweat. This can upset the balance of minerals in the blood, which may cause abnormal heart rhythms. Going into shock is also a risk. Mucus in cystic fibrosis patients is very thick and accumulates in the intestines and lungs. The result is malnutrition, poor growth, frequent respiratory infections, breathing difficulties, and eventually permanent lung damage. Lung disease is the usual cause of death in most patients. Loss of lung function is a major medical problem in most patients with CF. The average person with CF experiences a gradual worsening of lung function each year due to infection and inflammation. In people with CF, loss of lung function primarily is caused by blockage of air passages with infected mucus. The thick mucus plugs the air passages of the lungs and must be broken up and removed. The repeated lung infections also can cause permanent scarring of the lungs. Many adults with CF also develop symptoms of chronic sinus infections.

Cross-References

► ABC-Transporters

Cystic Nephroma

Definition

Cystic nephroma is a benign multicystic renal tumor. Contrary to the cystic partially differentiated ► nephroblastoma, it possesses no blastemal component in histology and can be found in adults too.

Cross-References

► Mesoblastic Nephroma
► Nephroblastoma

Cytochalasin B

Definition

The alkaloid cytochalasin B is a cell-permeable mycotoxin which inhibits cytoplasmic division by blocking the formation of contractile actin filaments. Cytochalasin B shortens actin filaments by blocking monomer addition at the fast-growing end of actin polymers. It blocks cytokinesis thus preventing the separation of daughter cells after mitosis that leads to the formation of binucleated cells.

Cross-References

▶ Microcell-Mediated Chromosome Transfer
▶ Micronucleus Assay

See Also

(2012) Cytokinesis. In: Schwab M (ed) Encyclopedia of cancer, 3rd edn. Springer, Berlin/Heidelberg, p 1055. doi:10.1007/978-3-642-16483-5_1477

Cytochrome P450

Ron H. N. van Schaik
Department of Clinical Chemistry, Erasmus University Medical Center, Rotterdam, The Netherlands

Synonyms

CYP450

Definition

Cytochrome P450 stands for a superfamily of closely related proteins that protect the individual against potentially harmful substances by modifying these substances by oxidation, hydroxylation, dealkylation, or dehalogenation, thereby increasing polarity and solubility and thus facilitating excretion from the body.

Characteristics

The body of an organism has a powerful system to facilitate the excretion of potentially harmful substances: the cytochrome P450 system. Cytochrome P450 stands for a superfamily of more than 50 closely related heme-containing enzymes, which can be divided into families (with at least 40% sequence homology on the amino acid level) and subfamilies (with at least 55% sequence homology), each with their own substrate specificity. The designation P450 is derived from the specific spectral absorbance at 450 nm of these proteins. The nomenclature is "CYP," followed by an Arabic number to indicate the family, then a capital letter indicating the subfamily if more members exist, and finally an Arabic number to indicate the specific enzyme (i.e., CYP2C9 is a member of the CYP2 family and belongs to the CYP2C subfamily) (Fig. 1). Family members include CYP1, CYP2, CYP3, CYP4, CYP11, CYP17, CYP19, CYP21, CYP26, and CYP51. These phase I monooxidases may carry out hydroxylation, dealkylation, oxidation, or dehalogenation reactions on a variety of lipophilic compounds of exogenous or endogenous origin, which are often followed by conjugation reactions (Fig. 2). The purpose is to increase aqueous solubility by increasing polarity of these molecules and thus facilitating their elimination from the body. Most of these enzymes are expressed in the liver, but also extrahepatic expression occurs (i.e., intestine, kidney). Within the cell, the enzymes are located in the endoplasmic reticulum. Besides the systemic action of the CYP450 system, also tumor cells may express CYP450 enzymes, especially those of the CYP1 family, thereby potentially affecting effective treatment with anticancer drugs. Antibodies against the highly expressed CYP1B1 in many tumor types have been developed, which potentially can be used as specific cancer detection. The high

Cytochrome P450,
Fig. 1 Nomenclature for
cytochrome P450 enzymes

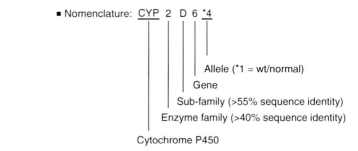

Cytochrome P450,
Fig. 2 Reactions catalyzed
by cytochrome P450
enzymes

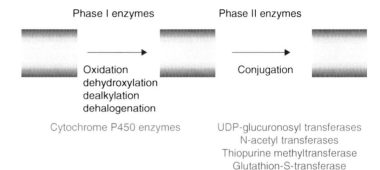

CYP1B1 expression in many tumors metabolically inactivates drugs like ▶ paclitaxel, ▶ docetaxel, doxorubicin, mitoxantrone, and ▶ tamoxifen and, as a consequence, mediates anticancer drug resistance. Besides detoxification reactions, part of the CYP450 family is involved in synthesis of endogenous compounds like steroids. The CYP11A1 enzyme catalyzes the conversion of cholesterol to pregnenolone, while CYP21A2 (21-decarboxylase) is involved in the formation of 11-deoxycortisol from 15α-hydroxyprgesterone. If this latter enzyme is deficient, it causes congenital adrenal hyperplasia. The enzyme CYP19 is involved in estradiol formation from androstenedione, making local inhibition of this enzyme in tumors an interesting target in breast cancer therapy.

Detoxification

Other important exogenous substrates of the CYP450 enzymes are environmental chemicals, plant toxins, and medical drugs. Metabolism by CYP450 enzymes mostly deactivates these compounds and facilitates urinary or fecal excretion. The best studied members of these detoxifying enzymes are the CYP2D6 enzyme, the CYP3A subfamily, and the CYP2C subfamily. The CYP3A subfamily (CYP3A4, CYP3A5, CYP3A7, CYP3A43), with CYP3A4 being the most important member, accounts for 50% of the total CYP450 protein in the liver in adults. Expression increases during the first year of life, replacing the fetal CYP3A7 expression. CYP3A5 expression is only found in 20% of Caucasians, and the enzyme has considerable substrate overlap with the CYP3A4 enzyme. The role of CYP3A43 is thought to be minimal. The enzyme CYP3A4 is involved in the metabolism of >50% of prescribed drugs, while CYP2D6, accounting only for 2% of total protein, is involved in the metabolism of 25% of these compounds (Fig. 3). Together with the CYP2C family (CYP2C8, CYP2C9, CYP2C19), which metabolizes 5–10% of drugs, these three subfamilies take credit for the majority of phase I reactions of prescribed drugs. Most anticancer agents are substrates for CYP3A4, like docetaxel, vinca alkaloids, paclitaxel, cyclophosphamide, ifosfamide, irinotecan, imatinib, gefitinib, etoposide, and teniposide. Important contributions of other CYP450 enzymes are, for instance, CYP2D6, which

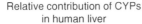

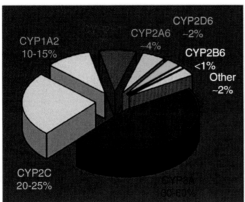
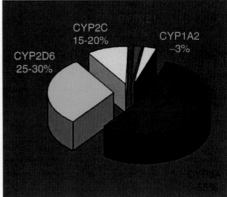

Cytochrome P450, Fig. 3 Percentage contribution of cytochrome P450 subfamilies to total protein in the human liver (*left panel*) or to drug metabolism (*right panel*)

activates tamoxifen, CYP2B6 for activation of cyclophosphamide, and the prominent role of CYP2C8 in the metabolism of paclitaxel. Furthermore, thalidomide (used in the treatment of chronic myeloid leukemia and prostate cancer) is metabolized by CYP2C19, while the prodrug of 5-fluorouracil, tegafur, needs activation by CYP2A6. All cytochrome P450 enzymes, with the exception of CYP2D6, are highly susceptible to induction or inhibition by numerous compounds, including drugs and environmental factors. Grapefruit juice, for instance, decreases CYP3A4 activity dramatically, the herb St. John's wort increases CYP3A4 activity, and smoking increases the transcription and activity of CYP1A2. These inducing and inhibiting events may affect the ability to metabolize drugs and thus interfere with the expected relationship between dose, blood concentration, and effect.

Using CYPs to Improve Therapy

The enzyme CYP1B1 was identified as the main CYP450 present in a wide range of human cancers of different histological types, among which are prostate, lung, and renal cell tumors. It is expressed in primary tumors as well as in metastases and is regarded as a biomarker of the neoplastic phenotype. It thus is an interesting target in the development of novel anticancer prodrugs that need specific activation by CYP1B1. In this way,

locally high concentrations of anticancer agents are generated without the high exposure of the total individual, although special precautions are necessary to prevent the premature elimination of those tumor cells expressing the CYP1B1 enzyme. Several patents on this approach have been filed in the last 5 years. Another, novel CYP450 enzyme was discovered: CYP2W1, which is expressed during fetal life but can also be found in transformed tissues. This high expression in tumors and the low expression found in normal cells also make CYP2W1 an interesting target for developing new anticancer therapies.

Genetic Polymorphisms

The enzymes of the CYP450 superfamily involved in detoxification reactions are highly polymorphic: genetic variants, usually present as single nucleotide polymorphisms (SNPs), may encode enzymes with reduced activity. Because these variant alleles can be determined prior to starting therapy, they can be used to predicting aberrant pharmacokinetics. This discipline of analyzing DNA to predict or explain drug metabolism is called pharmacogenetics. These variant alleles are depicted using an asterisk, in which *1 stands for the most common (wild-type) active allele and subsequent variants receive an increasing number (e.g., *CYP2C9*2*, *CYP2C9*3*) (Fig. 1). The power of these predictions depends on the pharmacokinetics of the drug (how much is it dependent on just one CYP450, or is there redundancy?), comedication, kidney and liver function, and environmental factors. For clinical use, these predictions are most valuable for drugs having a narrow therapeutic window, in which the consequences of subtherapeutic treatment, on the one hand, and toxic side effects, on the other hand, are high. Cancer therapy is therefore an important field to explore the potential contribution of pharmacogenetics, in which prediction of metabolism in increasing effectivity and avoiding extreme toxicity are of vital importance.

CYP2B6

Enzymatic activity of CYP2B6 shows a considerable interindividual variation, but in addition, also a difference between males and females exists, the latter having a 1.7-fold higher activity. The *CYP2B6* gene has shown to be quite polymorphic, with various variant alleles in the population, encoding CYP2B6 with altered activities. The correlation between specific variations in the DNA and assignment to specific variant alleles has been quite confusing because increasing knowledge showed that certain SNPs proved not as unique as anticipated for a specific variant allele. The most investigated SNPs at this moment are the 415A > G (Lys139Gln), 516G > T (Gln172His), 785A > G (Lys262Arg), and 1459C > T (Arg487Cys), which can be present in several allelic variants. The prominent role of CYP2B6 in the activation of cyclophosphamide was confirmed in some studies in which the 516G > T polymorphism correlated with a 1.5- to 2.0-fold increased clearance. Remarkably, this same SNP causes a reduced rather than increased enzymatic activity with respect to another CYP2B6 substrate, efavirenz, stressing the caution for extrapolating conclusions from one substrate to another. The frequency of the variant alleles containing this SNP is 23% for Caucasians and 16% for Japanese individuals, demonstrating that this may affect a substantial part of the patients treated with cyclophosphamide.

CYP2C8

The *CYP2C8*2* (805A > T, Ile269Phe) and *CYP2C8*3* (416G > A, Arg139Lys; 1196A > G, Lys399Arg) genetic variants, occurring with frequencies of 0% (*2) and 13% (*3) in Caucasians, and 18% (*2) and 2% (*3) in African Americans, were tested for their enzymatic activity on paclitaxel. This anticancer drug depends for 85% of its metabolism on CYP2C8 and for 15% on CYP3A4. Both CYP2C8 variant proteins showed a decreased activity on paclitaxel, implying a possible use in paclitaxel metabolism. However, further studies are needed to demonstrate the clinical use of genotyping for paclitaxel therapy.

CYP2C9

CYP2C9 is the major enzyme of the CYP2C family. The *CYP2C9* gene has numerous variant alleles, yet the *CYP2C9*2* (430C > T, R144C) and *CYP2C9*3* (1075A > C, I359L) are regarded as the most important variant alleles encoding decreased activity. The involvement of CYP2C9 in the metabolism of the anticoagulation drugs warfarin and acenocoumarol, affecting anticoagulation therapy, has resulted in major attention for this enzyme. Allele frequencies are quite high, with 11% (*2) and 7% (*3) in Caucasians, while the frequencies of these alleles in Asians (0% for *2 and 3% for *3) and Africans (4% for *2 and 2% for *3) are much lower. Although implicated in the metabolism of cyclophosphamide, no correlations between CYP2C9 variant alleles and cyclophosphamide pharmacokinetics are apparent, suggesting a very modest role for this enzyme.

CYP2C19

Population differences in CYP2C19 activity were first described around 1980 in an experiment using mephenytoin. One subject experienced such extreme sedation after taking mephenytoin for several days that he terminated his participation to the study. Based on the amounts of *R*- and *S*-mephenytoin in his urine, this individual was shown to be defective in *S*-mephenytoin 4-hydroxylation. Hereditary defective *S*-mephenytoin 4-hydroxylation is caused by the

Cytochrome P450,
Fig. 4 Distribution of
CYP2D6 activity in the
Caucasian population,
showing ultrarapid (UMs),
extensive (normal) (EMs),
and poor metabolizers
(PMs) (Adapted from
Guerich G (2003) Mol
Intern 3:194–204)

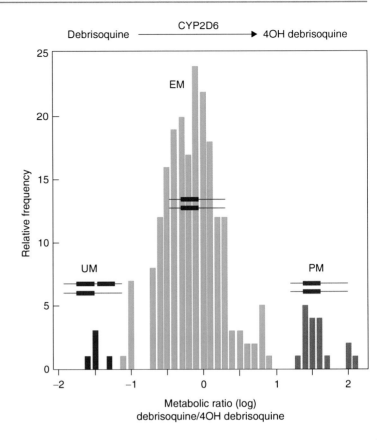

inheritance of two defective *CYP2C19* alleles. Approximately 2–3% of Caucasians is a CYP2C19 poor metabolizer, 3–5% of blacks, and 10–24% of Asians. The most common defective alleles are CYP2C19*2 (681G > A; splicing defect) and CYP2C19*3 (636G > A; W212X, incomplete protein). Known substrates for CYP2C19 are the proton pump inhibitors (omeprazole, pantoprazole), in which the treatment of individuals for *Helicobacter pylori* was found to be less effective in patients with two active alleles compared to patients having one or two deficient alleles. In cancer treatment, a correlation between CYP2C19 genotype and the activation of cyclophosphamide was demonstrated, thereby affecting survival, implicating potential use in cyclophosphamide therapy. Because of the involvement of CYP2C19 in the metabolism of thalidomide, pharmacogenetic analyses for this enzyme in the treatment of chronic myeloid leukemia or prostate cancer might be an interesting option.

CYP2D6

CYP2D6 activity displays a trimodal distribution in Caucasians, differentiating between poor metabolizers (PMs), extensive (normal) metabolizers (EMs), and ultrarapid metabolizers (UMs) (Fig. 4). Also, an intermediate metabolizer group can be distinguished, although phenotypically this group displays substantial overlap with the extensive metabolizers and is usually characterized on the genetic level. For CYP2D6, 5–10% of the Caucasian population is practically deficient, which was discovered using the probe drugs sparteine and debrisoquine. This deficiency is due to inheritance of two defective *CYP2D6* alleles, the most common polymorphism being a G to A conversion on position 1846 of the *CYP2D6* gene, leading to a RNA splicing defect, characteristic for the *CYP2D6*4* variant allele. The second most prominent CYP2D6 allele is the *5 allele, in which the whole *CYP2D6* gene is deleted. Over 50 variant alleles have been described until now for CYP2D6, which can be divided in those encoding no activity (null alleles), decreased activity (decreased function alleles), or normal activity (functional alleles). The frequency of variants in the population depends very much on ethnicity. The *CYP2D6*17*, for instance, is mainly found in Africans, while in Asians the decreased activity allele *CYP2D6*10* is found much more frequent than in Caucasians. Patients with a CYP2D6 deficiency who are treated with tamoxifen for breast cancer showed decreased effectiveness of therapy, due to decreased activation. In addition to genetic deficient alleles, *CYP2D6* can also be present as gene duplications in certain individuals, bringing the total number of CYP2D6 alleles to 3, or even higher. In Sweden, a family was identified that had 13 copies of the *CYP2D6* gene. This gene duplication is thought to be a compensating mechanism invented by evolution to circumvent the poor inducibility of CYP2D6. The frequency of this gene duplication displays an interesting north–south gradient, with duplications being present in 1–2% of individuals in Sweden, 3.6% in Germany, 7–10% in Spain, 10% in Italy, 20% in Saudi Arabia, and 29% in Ethiopia. The frequency of poor metabolizers shows the reverse, with 1–2% of poor metabolizers in Ethiopia to 5–10% in Northern Europe (Fig. 5). This gene duplication is accompanied by an increased activity and leads to an ultrarapid metabolizer phenotype. Individuals having these gene duplications may experience severe toxic effects with codeine, which is converted in the liver to morphine by CYP2D6; fatalities have been documented where the breast milk of a mother having codeine contained extreme high levels of morphine, leading to fatal morphine exposure in the child. This high morphine concentration in the milk was a result of the mother being a CYP2D6 ultrarapid metabolizer. In the treatment of cancer, CYP2D6 also plays a minor role in the metabolism of the new anticancer agent imatinib, but thus far the clinical implications of being a poor metabolizer for this kind of therapy are not known.

CYP3A4

This cytochrome is regarded as the most important enzyme and is involved in the metabolism of over 50% of commonly described drugs. It is highly susceptible to induction and inhibition, with a resulting large interindividual variability in activity. Although extensive research has been done identifying genetic polymorphisms in this enzyme, most SNPs and the 20 variant alleles described until now proved to have low frequencies (<1%) in the different populations. An exception is the promoter variant *CYP3A4*1B* ($-392A > G$), which has an allele frequency of 2–9% in Caucasians, 35–67% in blacks, but was rarely found in Asians. In vitro experiments demonstrated a slightly higher transcription, and thus CYP3A4 activity, but the clinical consequences of this are not clear yet. Interestingly, this *CYP3A4* polymorphism was shown to be linked genetically to a polymorphism in the *CYP3A5* gene. It was suggested that effects, attributes to the *CYP3A4* polymorphism, were, in fact, due to the *CYP3A5* polymorphism. However, several studies have shown correlations with the *CYP3A4*1B* allele without any correlation with CYP3A5 variant alleles. In study on cyclophosphamide, a decreased metabolism and a decreased median survival in breast cancer patients were demonstrated having the *CYP3A4*1B* allele. Another variant, the *CYP3A4*16B* ($554C > G$; Thr185Ser) allele, was indirectly shown by the use of tagging SNPs to be correlated with a 20% decrease in paclitaxel metabolism in Japanese patients, a population in which the allele frequency of this variant was found to be 1.7%.

CYP3A5

For a long time, the interindividual expression of CYP3A5 was poorly understood. However, its expression in only 20% of Caucasians appeared to be caused by a highly frequent genetic polymorphism, $6986A > G$, which is characteristic of the *CYP3A5*3* allele, and causes aberrant splicing. About 80% of Caucasians, but only 30% of the African American population, are deficient for this enzyme, most of them having the *CYP3A5*3/*3* genotype. Because of the substantial substrate overlap with CYP3A4, it is not always clear to what extent CYP3A5 contributes to the metabolism of certain drugs. This polymorphism does have an impact in the metabolism of

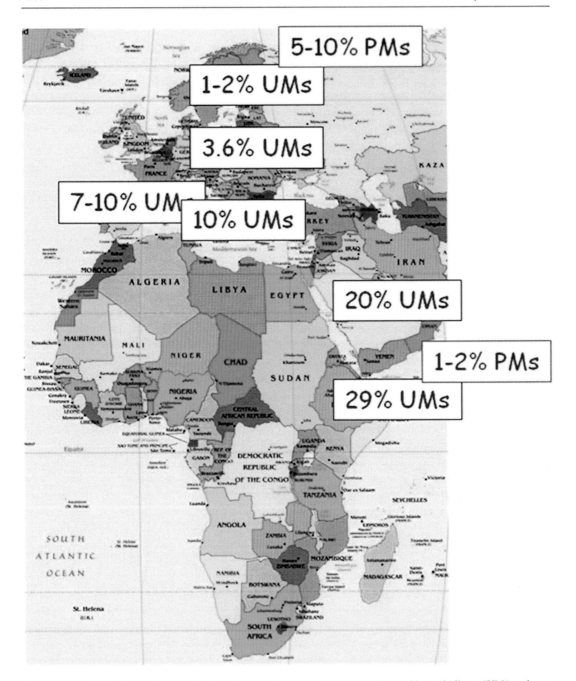

Cytochrome P450, Fig. 5 Geographical variation in frequency of CYP2D6 ultrarapid metabolizers (*UMs*) and poor metabolizers (*PMs*)

the immunosuppressive drug tacrolimus, which was slower in *CYP3A5*3/*3* individuals.

Cancer Incidence

A number of the detoxifying CYP450 enzymes have been associated with cancer incidence, mostly because of their involvement in the conversion of procarcinogens to carcinogenic metabolites (Table 1). The xenobiotic benzo(a)pyrene, for instance, is a common environmental pollutant produced from the burning of coal, from the combustion of tobacco products, from food barbeque

Cytochrome P450, Table 1 Correlations described for cytochrome P450 enzymes and the incidence of various cancers, either directly or in combination with environmental exposure

	1A1	1A2	2A6	1B1	2C9	2C19	2D6	2E1	3A4	3A5	17	19
Bladder						•						
Breast	•										•	•
Colorectal	•	•		•		•		•				
Dermatologic	•											
Gastric	•		•					•				
Gynecological				•							•	
Head and neck				•				•				
Hematologic												
Hepatocellular	•		•	•			•	•				
Lung	•		•			•		•		•		
Esophageal				•				•		•		
Prostate	•			•			•		•	•	•	•
Renal				•								

on charcoal briquettes, and from industrial processing. It is a weak carcinogen, but is converted to a potent carcinogen by CYP1A1, CYP1A2, and CYP1B1, making that variable activities in these enzymes correlate with cancer incidence. CYP1A1 alleles have been implicated in lung, gastric, colorectal, hepatocellular, breast, prostate, and dermatologic cancer, while CYP1B1 variant alleles correlated with colorectal, hepatocellular, prostate, head and neck, renal, and gynecological cancers. For CYP3A4, the *CYP3A4*1B* allele was discovered because it correlated with a more aggressive form of prostate cancer. In nonaggressive bladder cancer, a lower frequency of CYP2C19 poor metabolizers is found, while, in contrast, for squamous cell carcinoma, a higher frequency of CYP2C19 poor metabolizers was apparent. Also correlations with lung cancer and colorectal cancer have been reported, while CYP2C9 variant alleles showed correlation with colorectal cancer. *CYP2E1* polymorphisms were associated with an increased risk of lung, esophageal, head and neck, gastric, colorectal, and hepatocellular cancer, while CYP2D6 poor metabolizers appeared to be relatively protected for hepatocellular carcinoma but were at some increased risk for prostate cancer in patients who were using tobacco. Although not all mechanisms behind the correlation between variant CYP alleles and cancer incidence are known, it is obvious from the described observations that CYP450

enzymes not only play a role in the treatment of cancer but may also be involved in the development of cancer.

Cross-References

▶ Docetaxel
▶ Paclitaxel
▶ Tamoxifen

References

http://www.cypallele.ki.se

Ingelman-Sundberg M, Oscarson M, McLellan R (1999) Polymorphic human cytochrome P450 enzymes: an opportunity for individualized drug treatment. Trends Pharmacol Sci 20:342–349

Nelson DR, Koyman L, Kamataki T et al (1996) P450 superfamily: update on new sequences, gene mapping, accession numbers and nomenclature. Pharmacogenetics 6:1–42

Solus JF, Arietta BJ, Harris JR et al (2004) Genetic variation in eleven phase I drug metabolism genes in an ethnically diverse population. Pharmacogenomics 5:895–931

van Schaik RHN (2005) Cancer treatment and pharmacogenetics of cytochrome P450 enzymes. Invest New Drugs 23:513 522

See Also

(2012) Single nucleotide polymorphism. In: Schwab M (ed) Encyclopedia of cancer, 3rd edn. Springer, Berlin/Heidelberg, p 3412. doi:10.1007/978-3-642-16483-5_5316

(2012) SNP. In: Schwab M (ed) Encyclopedia of cancer, 3rd edn. Springer, Berlin/Heidelberg, p 3460. doi:10.1007/978-3-642-16483-5_5395

(2012) Variant allele. In: Schwab M (ed) Encyclopedia of cancer, 3rd edn. Springer, Berlin/Heidelberg, p 3885. doi:10.1007/978-3-642-16483-5_6152

(2012) Wild-type. In: Schwab M (ed) Encyclopedia of cancer, 3rd edn. Springer, Berlin/Heidelberg, p 3947. doi:10.1007/978-3-642-16483-5_6247

(2012) Pharmacogenetics. In: Schwab M (ed) Encyclopedia of cancer, 3rd edn. Springer, Berlin/Heidelberg, p 2840. doi:10.1007/978-3-642-16483-5_4496

Cytokine

Definition

Cytokine is any of a variety of secreted polypeptides that control the development, differentiation, and proliferation of hematopoietic cells. The effects of cytokines on lymphocytes are usually mediated through membrane-bound cytokine receptors and are especially critical during immune responses. A group of proteins mainly functioning as soluble signal transmitters in the immune system are primarily released by immune cells, but many other cell types can also release cytokines. Cytokines direct and mediate various functions of the adaptive and innate immunity and modulate functions of immune cells and other responsive cell types. Their local of systemic effects, elicited by binding to specific cell surface receptors, contribute mainly to innate and adaptive immune responses, so that their production is involved in a range of infectious, immunological, and inflammatory diseases. To this class belong interleukins, interferons, TNF-like molecules, etc. These molecular signals are similar to hormones and neurotransmitters and are used to allow one cell to communicate with another.

Cross-References

► Cytokine Receptor as the Target for Immunotherapy and Immunotoxin Therapy
► Signal Transducers and Activators of Transcription in Oncogenesis

Cytokine Receptor

► Cytokine Receptor as the Target for Immunotherapy and Immunotoxin Therapy

Cytokine Receptor as the Target for Immunotherapy and Immunotoxin Therapy

Koji Kawakami
Department of Pharmacoepidemiology, Graduate School of Medicine and Public Health, Kyoto University, Kyoto, Japan

Synonyms

Cancer vaccines; Cytokine receptor; Immunotherapy; Immunotoxins

Definition

Cancer ► immunotherapy is the treatment of cancer by improving the ability of a tumor-bearing individual to reject the tumor immunologically. In the case of immunotherapy targeting cytokine receptors, delivery of tumor antigen proteins induces immune response against cancer cells bearing such tumor antigen to eliminate them. Cytokine receptor-targeting immunotoxins are proteins containing a bacterial toxin or chemical compound along with an antibody or a ligand that binds specifically to its target receptor. Immunotoxins then internalize through the receptors and achieve cytotoxic effect derived from the toxin moiety.

Characteristics

When cancer cells are generated in a body, a variety of mechanisms cooperate in order to kill the cancer cells. Among these mechanisms, the immune system has an important role to eliminate tumor tissues. First, immunocytes like ▶ macrophages and natural killer (NK) cells are activated to attack the newly formed foreign body. If they were unable to completely remove the cancer cells, T-cells and B-cells are the next players to fight against cancer. Many scientists have tried to find how the natural immune system works in order to control and to beat cancer. As cancer cells are known to express specific antigens or cytokine receptors on their cell surface, these molecules may be utilized as a target for tumor immunotherapy and immunotoxin therapy.

Cytokine Receptor as the Target for Immunotherapy

Cancer immunotherapy attempts to stimulate the immune system to reject and destroy tumors. For example, administration of interferon can activate the systemic immunity. Therapeutic cancer vaccine is also included in this category, which is designed to activate the host immune system against tumor cells. Cancer vaccines take advantage of the fact that certain molecules on the surface of cancer cells are either unique or more abundant than those found on normal or noncancerous cells. These molecules act as antigens, stimulating the immune system to evoke a specific immune response. There are few licensed therapeutic vaccines to date. However, several cancer vaccines are in large-scale ▶ clinical trials.

The Her2/neu (c-erbB2), the target protein of ▶ Herceptin, which is the world's first therapeutic antibody to treat ▶ breast cancer, has long been studied as a target of cancer immunotherapy. The immunization methodology is variable, including administration of plasmid DNA, recombinant protein, or intracellular domain (ICD) peptide; virus vector delivery; dendritic cell pulse therapy; and combination therapy with adjuvant like granulocyte-macrophage colony stimulating factor (GM-CSF) or with anticancer reagents. Her2/neu immunotherapy has been tested in Phase II clinical trials to treat breast cancer, and in Phase I for various cancer types including ▶ ovarian cancer, ▶ prostate cancer, and ▶ nonsmall-cell lung carcinoma. In the Phase II trial contributed by Washington University, the ICD of HER-2

vaccine immunizes breast cancer patients with CD4 + helper T epitopes derived from the HER-2 protein. During ▶ preclinical testing, the rat neu peptide vaccine is effective because it circumvents tolerance to rat neu protein and generates rat neu-specific immunity. In Phase I study, patients underwent intradermal immunization once a month for a total of six immunizations with GM-CSF as an adjuvant. The endpoint of the study was to evaluate the toxicity and both the cellular and humoral HER-2/neu-specific immunity when the vaccinations were completed. The majority of patients (24 of 27 patients, 89%) developed HER-2/neu ICD-specific T-cell immunity. Out of 27 patients, 22 patients (82%) also developed HER-2/neu-specific IgG antibody immunity, and over half of the assessable patients retained HER-2/neu-specific T-cell immunity 9–12 months after the completion of immunizations.

▶ Vascular endothelial growth factor receptor-2 (VEGFR2) is highly expressed in neovascular endothelial cells in a tumor tissue. The epitope peptides of VEGFR2 were identified, and stimulation using these peptides induces CTLs with potent cytotoxicity in the ▶ HLA class I-restricted fashion against not only peptide-pulsed target cells but also endothelial cells endogenously expressing VEGFR2. In A2/Kb transgenic mice expressing α1 and α2 domains of human HLA-A*0201, vaccination using these epitope peptides in vivo was associated with significant suppression of the tumor growth and prolongation of the animal survival. A clinical trial has been initiated to verify its effectiveness on breast and ▶ gastric cancer at University of Tokyo, Japan.

Another approach attempts to avoid immunologic tolerance to cancer vaccines. Cancer immunotherapy utilizes the host immune system, and tolerance is one of the major causes weakening vaccine efficiency. If immunologic tolerance could be controlled, it is expected to enhance the effect of a cancer vaccine. There are a variety of other strategies which have shown antitumor activity in mouse model of cancer, e.g., dendritic cells pulsed with EphA2 (ephrin-A2, an angiogenic factor) epitope peptide, KLH-bound EGF receptor variant III peptide, VEGFR1 epitope peptide, and plasmid DNA of IL-13 receptor α2.

Cytokine Receptor as the Target for Immunotoxin Therapy

Immunotoxins are protein toxins connected to a cell binding ligand or antibody. Classically, immunotoxins were created by chemically conjugating an antibody to a whole protein toxin, or, for more selective activity, by using a protein toxin devoid of its natural binding domain. Immunologic proteins that are smaller than monoclonal antibodies (MAbs), like growth factors and ▶ cytokines, have also been chemically conjugated and genetically fused to protein toxins. Targeted cancer therapy such as immunotoxin therapy which targets tumor-specific cell surface receptors is one of the most effective strategies against cancer. The targeted agents require a threshold level of receptor expression on the cancer cells to achieve their antitumor activity.

At present, only one agent targeting cytokine receptor, which contains human interleukin (IL)-2 and truncated diphtheria toxin (ONTAK), is approved for use in cutaneous T-cell lymphoma (CTCL). ONTAK, or denileukin diftitox, is a fusion protein designed to direct the cytocidal action of diphtheria toxin to cells which express the IL-2 receptor (IL-2R). Among three forms of IL-2R, the high affinity form consisting of CD25/CD122/CD132 subunits is usually found only on activated hemocytes as T lymphocytes, B lymphocytes, and macrophages. A Phase III randomized, double-blind clinical trial was conducted in 71 patients with recurrent or persistent Stage Ib to IVa CTCL whose malignant cells express the CD25 component of the IL-2R. Administered with 9 or 18 µg/kg/day of ONTAK as an intravenous infusion daily for 5 days every 3 weeks (median: 6 courses), 7 patients (10%) achieved a complete response and 14 patients (20%) achieved a partial response. In 1999, the US FDA approved ONTAK indicated for the treatment of patients with persistent or recurrent CTCL whose malignant cells express CD25 component of the IL-2R.

In the case of the IL-13 receptor (IL-13R) system, these receptors are constitutively overexpressed on a variety of human solid cancer cells including renal cell carcinoma, glioma, AIDS-associated ▶ Kaposi sarcoma, head and neck cancer, ovarian cancer, and prostate cancer. To target IL-13R, a recombinant fusion IL-13 cytotoxin termed IL13-PE38QQR, or cintredekin besudotox, has been developed that is composed of IL-13 and a mutated form of *Pseudomonas* exotoxin. In early-phase studies, cintredekin besudotox was administered via intraparenchymal ▶ convection-enhanced delivery (CED) after resection of supratentorial recurrent malignant glioma. CED is a novel approach for the delivery of small and large molecules in solid tissues, utilizing a pressure gradient to distribute macromolecules to clinically significant volumes of tissue by bulk flow. The CED of cintredekin besudotox was fairly well tolerated, with a reasonable benefit/risk profile for treatment of patients with glioma. It has received orphan drug designation in Europe and the USA as well as fast-track drug development program status from the FDA (Fig. 1).

The ▶ interleukin-4 receptor (IL-4R), which is related to the IL-13R, is also expressed by a variety of solid tumors and ▶ hematologic malignancies. The IL-4 cytotoxin, IL4(38–37)-PE38KDEL, which is composed of circular permuted IL-4 and a mutated form of *Pseudomonas* exotoxin, is highly active in killing IL-4R-expressing tumor cells in vitro and in vivo. In a Phase I/II trial of 31 glioma patients, tumor necrosis was observed in 71% of patients, and one patient experienced long-term survival. The Phase II intratumoral study had been completed, but further development has stalled because of severe adverse events as infusion concentration of drug was possibly too high.

TransMID™, a modified diphtheria toxin conjugated to transferrin, is currently in Phase III clinical trials indicated for the treatment of glioblastoma multiforme. Transferrin receptors are particularly prevalent on rapidly dividing cells, and the high level of transferrin receptor expression on glioma cells makes it an ideal target for brain cancer. Phase I and Phase II clinical trials for TransMID™ have been successfully completed in patients suffering from inoperable, recurrent high grade gliomas who have failed to all other forms of treatment. In Phase II study, a reduction in tumor size of 50% or more was noted in 35% of evaluable patients, with a corresponding increase

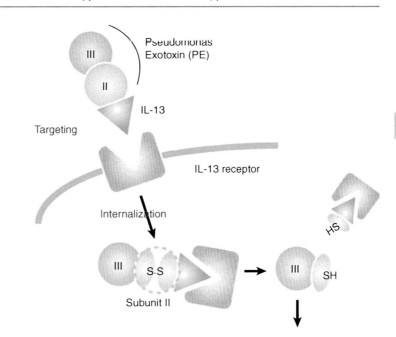

Cytokine Receptor as the Target for Immunotherapy and Immunotoxin Therapy, Fig. 1 Model for cytokine receptor-targeted cancer therapy. In this model, IL-13R expressing cancer cells are treated with IL-13 immunotoxin. When IL-13 moiety of this immunotoxin binds to IL-13 receptor, the receptor–immunotoxin complex immediately internalizes into cytosol. Then domain II of *Pseudomonas* exotoxin (PE) is degraded in endosome and domain III of PE shows cytotoxic effect through its irreversible inhibition of eF2

in life expectancy in those patients that did respond. In this study, median survival for patients receiving TransMID™ was ~37 weeks. TransMID™ received fast-track status from the FDA, and orphan drug designation in the USA, Europe, and Japan.

TP-38 is a recombinant chimeric targeted toxin composed of the EGFR binding ligand TGF-α and a genetically engineered form of the *Pseudomonas* exotoxin, PE38. A Phase I trial was conducted to define the maximum tolerated dose (MTD) and dose limiting toxicity of TP-38 delivered by CED in patients with recurrent malignant brain tumors. The Phase II studies, in which TP-38 was administered to patients with recurrent glioblastoma using CED, have shown initial encouraging results.

DT388GMCSF has been developed for the treatment of ► acute myeloid leukemia (AML). This molecule is composed of amino acids 1–388 of diphtheria toxin (DT), a histidine–methionine linker, and amino acids 1–124 of human GM-CSF. Phase I clinical trial has been completed in patients with AML, and 4 of 37 patients showed clinical remission of disease.

Another unique DT-based approach is VEGF121-DT385 and VEGF165-DT385 immunotoxins, which are a chemical conjugate containing VEGF165, VEGF121, and truncated DT. ► Vascular endothelial growth factor (VEGF) is the most critical inducer of blood vessel formation. In vivo animal studies have shown that these molecules are able to inhibit angiogenesis and tumor growth.

Cross-References

- ► Acute Myeloid Leukemia
- ► Breast Cancer
- ► Clinical Trial
- ► Convection-Enhanced Delivery
- ► Cytokine
- ► Gastric Cancer
- ► Hematological Malignancies, Leukemias, and Lymphomas
- ► Herceptin
- ► HLA Class I
- ► Immunotherapy
- ► Interleukin-4

- ▶ Kaposi Sarcoma
- ▶ Macrophages
- ▶ Non-Small-Cell Lung Cancer
- ▶ Ovarian Cancer
- ▶ Preclinical Testing
- ▶ Prostate Cancer
- ▶ Vascular Endothelial Growth Factor

References

Kawakami K, Nakajima O, Morishita R et al (2006) Targeted anticancer immunotoxins and cytotoxic agents with direct killing moieties. Scientific World J 6:781–790

Pastan I, Hassan R, Fitzgerald DJ et al (2006) Immunotoxin therapy of cancer. Nat Rev Cancer 6:559–565

Schuster M, Nechansky A, Kircheis R (2006) Cancer immunotherapy. Biotechnol J 1:138–147

See Also

(2012) CTL. In: Schwab M (ed) Encyclopedia of Cancer, 3rd edn. Springer Berlin Heidelberg, p 1012. doi:10.1007/978-3-642-16483-5_1406

(2012) Cytokine Receptor. In: Schwab M (ed) Encyclopedia of Cancer, 3rd edn. Springer Berlin Heidelberg, p 1052. doi:10.1007/978-3-642-16483-5_1475

(2012) Diphtheria Toxin. In: Schwab M (ed) Encyclopedia of Cancer, 3rd edn. Springer Berlin Heidelberg, p 1124. doi:10.1007/978-3-642-16483-5_1636

(2012) Pseudomonas Exotoxin. In: Schwab M (ed) Encyclopedia of Cancer, 3rd edn. Springer Berlin Heidelberg, p 3112. doi:10.1007/978-3-642-16483-5_4841

(2012) Renal Cancer. In: Schwab M (ed) Encyclopedia of Cancer, 3rd edn. Springer Berlin Heidelberg, pp 3225-3226. doi:10.1007/978-3-642-16483-5_6575

Cytokine Toxin Fusions or Conjugates

- ▶ Immunotoxins

Cytoplasmic Scaffolding Apoptotic Protease Activating Factor

- ▶ APAF-1 Signaling

Cytoskeleton

Francisco Rivero[1] and Huajiang Xiong[2]
[1]Centre for Cardiovascular and Metabolic Research, The Hull York Medical School, University of Hull, Hull, UK
[2]Department of Zoophysiology, Zoological Institute, Christian-Albrechts-University of Kiel, Kiel, Germany

Definition

The cytoskeleton is a complex network of interconnected filaments and tubules that extends mainly throughout the cytosol, reaching from the nuclear envelope to the inner surface of the plasma membrane. It gives shape to the cell, mediates anchoring to the substrate and to other cells, facilitates cell movements and movement of organelles, and is necessary for cell division. Although mainly cytosolic, some cytoskeleton components play roles within the nucleus.

Typical eukaryotic cells possess three cytoskeleton systems (Fig. 1) that can be distinguished on the basis of their diameter: microfilaments (MFs) (or actin filaments), intermediate filaments (IFs), and microtubules (MTs).

Characteristics

All three cytoskeleton systems are filamentous polymers built from small subunits. Contrary to what its name may suggest, the cytoskeleton is a highly dynamic structure: filaments grow and shrink, resulting in continual remodeling. This constant assembly and disassembly is subject to regulation in response to both intracellular and environmental cues. Moreover, none of the three cytosketon systems exists in isolation: they are connected to each other and to other cellular components (nucleus, organelles, plasma membrane). Numerous classes of associated proteins are responsible for the dynamic properties and the connectivity of the cytoskeleton. They also constitute the targets of signaling networks that control the spatial

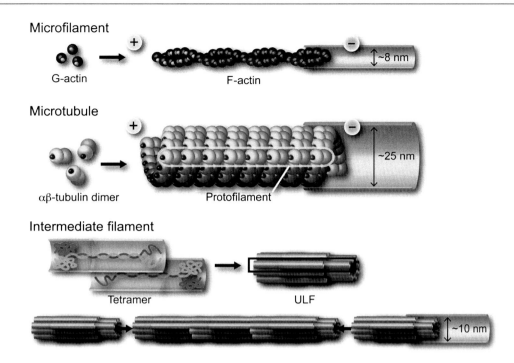

Cytoskeleton, Fig. 1 Schematic depiction of the composition of the three cytoskeleton systems. Microfilaments (actin filaments) are composed of filamentous (F-) actin, linear polymers of globular (G-) actin. Each actin monomer binds one ATP nucleotide (depicted as *yellow spheres*). ATP hydrolyses when the filament ages (depicted as *grey spheres*). Actin filaments are polar, with a fast growing (plus or barbed) end and a slow growing (minus or pointed) end. Microtubules are hollow tubules built from stable αβ-tubulin heterodimers. Both tubulins bind a GTP nucleotide, but only the one bound to β-tubulin (depicted as *red spheres*) can be hydrolyzed and exchanged. GTP hydrolyzes when the microtubule ages (depicted as *grey spheres*). Microtubules are polar, with a fast growing (plus) end and a slow growing (minus) end. Intermediate filament proteins form parallel dimers that then assemble to staggered antiparallel tetramers, the basic subunits. Eight tetramers associate laterally into unit-length filaments (ULF) that can extend from both ends (intermediate filaments lack polarity). The resulting filaments compress radially (not depicted) into mature 10 nm intermediate filaments

and temporal remodeling of the cytoskeleton. Figure 2 depicts schematically the three cytoskeleton systems and some common specialized assemblies of cytoskeleton-associated components.

Microfilaments (Actin Filaments)

Microfilaments are about 8 nm in diameter and constitute the smallest of the three cytoskeleton polymers (Fig. 1). They are linear polymers of actin, a 42 kDa globular protein. Actin is the most abundant protein (1–5% of the total weight of nonmuscle cells and up to 20% in muscle). Actins are encoded by six genes and are distributed in three classes, α-actin predominantly in muscle cells and β and γ-actin in nonmuscle cells. Several actin-related proteins (Arps) also exist that are similar in overall structure to actin and play distinct roles.

In its monomeric form, actin is referred to as G-actin (G stands for globular). The monomers polymerize to yield F-actin (F stands for filamentous). Most cells contain almost equal proportions of G- and F-actin. Actin monomers polymerize end-to-end, giving rise to a polar filament with a fast growing (plus or "barbed") and a slow growing (minus or "pointed") end. Each actin monomer binds one ATP nucleotide and it is in its ATP-bound form that actin polymerizes preferentially. Hydrolysis of ATP destabilizes the filament at the minus end and facilitates depolymerization.

MFs contribute to developing and maintaining cell shape and attachment to other cells and to the

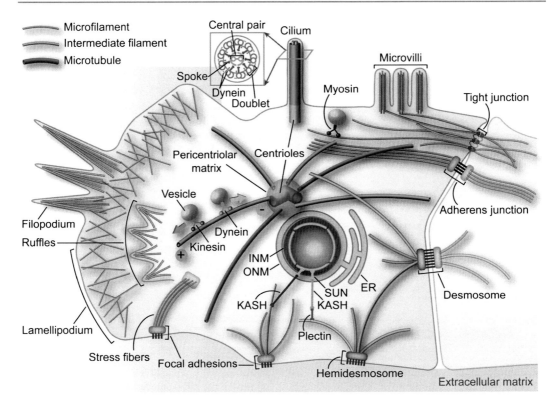

Cytoskeleton, Fig. 2 Schematic of the three cytoskeleton systems and some specialized assemblies. The hypothetical cell depicted here displays features of an epithelial cell on the right half and of a mesenchymal cell on the left side. Examples of microtubule and microfilament-associated motor proteins carrying a vesicle are depicted. Examples of components of LINC complexes (SUN and KASH proteins) that attach the nuclear lamina to the cytoskeleton are also provided. See text for details. ONM, outer nuclear membrane; INM, inner nuclear membrane; ER, endoplasmic reticulum

extracellular matrix and are essential for cell movements (cell migration, muscle contraction), movement of organelles, and completion of cell division. With the help of **actin-binding proteins** (ABPs), actin filaments are organized in networks, bundles, and complex structures. Dozens of ABPs have been described and have been classified according to functional or structural criteria, although many display more than one activity on actin. Some examples will be considered.

Several ABPs control remodeling of MFs by affecting their formation, stabilization or disassembly and constitute critical elements in signaling cascades that precisely modulate where and when actin remodeling takes place. Many ABPs are regulated by Ca^{2+} ions, binding to phospholipids (in particular phosphoinositides) or phosphorylation. A prominent role in the regulation of the actin cytoskeleton is played by small GTPases of the Rho family.

De novo formation of MFs is an unfavorable process if left to occur spontaneously. This situation is overcome in vivo by nucleation factors that stabilize the formation of the so-called nuclei or seeds (dimeric and trimeric complexes) that then elongate rapidly. There are three classes of nucleation factors, the Arp2/3 complex, formins, and spire proteins. The Arp2/3 complex produces branched filaments that form dendritic networks. This complex requires activation by proteins like ▶ cortactin and members of the WASP/Scar family that are themselves ABPs. Formins and spire, by contrast, produce filaments that are not branched. Monomeric ABPs like thymosin β4 and profilin sequester G-actin and maintain a pool of monomeric actin. Profilin also promotes actin polymerization by facilitating the exchange of ADP by ATP and binding to nucleation factors, ensuring a supply of fresh monomers to growing filaments. ▶ Gelsolin and cofilin/actin depolymerizing factor control the length of filaments by severing them and blocking the plus end or promoting the dissociation of actin monomers at the minus end. Capping proteins bind to and stabilize the end of the filament. CapG and the heterodimeric protein CapZ bind to the plus end and prevent elongation of the filament. Tropomodulin, on the contrary, caps the minus end and prevents the loss of actin subunits. Proteins like tropomyosin and nebulin bind along the actin filament and stabilize it.

Myosins are motor proteins that move along MFs using the energy provided by ATP hydrolysis. All myosins consist of a conserved motor domain that binds one or more light chains with regulatory function, and a variable tail. Many myosins have been described and have been distributed into classes on the basis of sequence similarities of the motor domain. The best know myosin is myosin II, which in muscle cells makes up the thick filaments of the sarcomere and powers muscle contraction. Other myosins are involved in association of the actin cytoskeleton with the plasma membrane (myosin I, myosin VII) and organelles (myosin V).

Actin filaments are organized in networks and bundles by actin crosslinking and bundling proteins. In networks, the MFs criss-cross and are loosely packed, whereas in bundles they are densely packed in parallel or antiparallel arrays. Filamin, spectrin, and α-actinin are examples of crosslinking proteins. Fascin, fimbrin (plastin), villin, and espin are common bundling proteins. Actin networks and bundles are found in a variety of protrusive structures. In motile cells, MFs are prevalent at cell edges, particularly in filopodia, pseudopodia, lamellipodia, and ruffles. Filopodia are thin fingerlike protrusions supported by parallel bundles of MFs. Lamellipodia and ruffles are sheetlike protrusions supported by MF networks. Pseudopodia are thick fingerlike protrusions characteristic of cells that exhibit amoeboid movement, like white blood cells. In cells that adhere tightly to the substrate actin, filaments organize in bundles called stress fibers that attach to focal adhesions, specialized complexes where the actin cytoskeleton is anchored through talin and other ABPs to integrins, which in turn bind to proteins of the extracellular matrix. MFs organize in antiparallel bundles at the cleavage furrow, a structure that, powered by myosin, constricts the cytoplasm of the dividing cell and contributes to separate the daughter cells. Networks of actin also assemble around endocytic structures at the plasma membrane like phagocytic cups and around endosomes and vesicles of the Golgi apparatus.

Cells of the intestinal and renal epithelia display fingerlike protrusions called microvilli that cover their apical pole and contribute to extend the cellular surface for digestion and transport of metabolites. Like filopodia, microvilli are built from parallel arrays of actin filaments, but they are more densely bundled and are attached to the plasma membrane by specialized myosins. Much thicker and longer bundles of MFs are characteristic of the stereocilia of hair cells of the inner ear, involved in sensing tiny movements caused by sound vibrations and changes of posture.

Other ABPs like spectrin, dystrophin, talin, proteins of the ERM (ezrin, radixin, moesin) family, and some unconventional myosins are responsible for the association of MFs to the plasma membrane. In association with short actin filaments, spectrin forms a network underneath the plasma membrane of the erythrocyte. In muscle cells, dystrophin is responsible for attaching the sarcomere to a glycoprotein complex which is anchored to the extracellular matrix. Actin also assembles at cell-cell contacts mediated by cadherin, the ▶ adherens junctions or by occludins, the tight junctions. In epithelial cells, these contacts constitute a circumferential band at the boundary between the apical and basolateral poles of the cell. Nesprins are ABPs of the outer nuclear membrane that connect the cytoplasmic cytoskeleton to the nuclear lamina through SUN-domain proteins and contribute to positioning of the nucleus and stability of the nuclear envelope.

The sarcomere, the contractile unit of the muscle cell, constitutes one of the most complex assemblies of cytoskeletal proteins. In the sarcomere, MFs and associated proteins build up the thin filaments and are arranged in a hexagonal pattern around thick filaments, constituted by bundles of myosin. Numerous structural proteins are responsible for maintaining the precise organization of the sarcomere, including some IF proteins (desmin, syncoilin, and synemin). Sliding of thin filaments past thick filaments is powered by ATP hydrolysis by myosin and results in shortening of the sarcomere and, on a large scale, muscle contraction.

Numerous MF-targeting agents have being used to perturb the assembly of actin into MFs (Table 1). Although several of them have demonstrated in vitro and in vivo anticancer efficacy, none of them has been approved for cancer therapy, usually because of their high toxicity, and only one, MKT-077 has been clinically evaluated. This compound crosslinks F-actin and produces aberrant MFs but part of its actions is due to the inhibition of mitochondrial Hsp70 and induction of apoptosis.

Microtubules

Microtubules are stiff hollow tubes of 25 nm diameter (Fig. 1). They are built from stable dimers of α and β tubulin, two similar globular proteins of approximately 55 kDa. Each tubulin monomer binds a GTP nucleotide but only the one bound to β-tubulin is exchangeable. Tubulin dimers polymerize end-to-end, giving rise to a polar protofilament with a fast growing (plus) and a slow growing (minus) end. On a cross section, a singlet MT is composed of 13 protofilaments around a hollow core, but other arrangements exist in centrioles (triplets) and the axonemes of cilia and flagella (doublets). GTP is needed for MT assembly. A GTP-tubulin cap prevents the peeling away of subunits from the plus end of a growing MT. Hydrolysis of GTP by β-tubulin results in an unstable tip and facilitates rapid depolymerization, an event called MT catastrophe. If the MT regains a GTP-tubulin cap it can resume growth, an event called MT rescue. Globally, this behavior is called dynamic instability.

There are two assemblies of MTs, cytoplasmic and axonemal. **Cytoplasmic MTs** build a very dynamic network that extends from the ▶ centrosome at the vicinity of the nucleus toward the cell periphery. They are mainly used as tracks for directional movement of organelles and also contribute to maintaining cell shape and polarity. During cell division, cytoplasmic MTs arrange into the mitotic spindle, a cage-like structure that captures the chromosomes and pulls the chromatids apart toward their respective daughter cell. The centrosome constitutes the MT organizing center (MTOC) of cytoplasmic MTs. It is composed of two centrioles surrounded by a cloud of amorphous pericentriolar material. The centrioles are two cylinders placed at right angles to each other. Each centriole is made of a radial array of 9 triplet MTs and contains additional tubulin isoforms (δ, ε). Interphase cells possess one centrosome that duplicates when the cell undergoes mitosis. Cytoplasmic MTs are not initiated in the centrioles, but in the pericentriolar material, form a protein complex called the γ-tubulin ring complex (γ-TuRC), that contains one more tubulin isoform, γ-tubulin, and 5 or 6 additional proteins.

Axonemal MTs are stable and found in specialized fingerlike appendages called cilia and flagella. Cilia are short (2–10 mm) and numerous and are characteristic of epithelial cells of the respiratory tract epithelium, where they produce an oarlike pattern of beating that contributes to carry mucus and foreign matter out of the airway. The flagellum is large (10–200 mm) and usually unique and is responsible for the undulatory beating that propels the sperm cells. Cilia and flagella have the same structure, with a core of MTs arranged in a radial array of nine double MTs around a core of two single microtubules in association with more than 100 accessory proteins. This structure, the axoneme, is an extension of a modified centriole, the basal body. Most mammalian cells possess a unique so-called primary cilium that emerges from one of the centrioles after cell division. The primary cilium is usually nonmotile and functions as a sensory antenna for many signal transduction pathways. The outer segments of rod and cone photoreceptors in the eye are formed from derivatives of primary cilia and possess a basal body and a vestigial axoneme.

Cytoskeleton, Table 1 Microfilament, microtubule, and intermediate filament targeting agents. The anticancer effects of most of the agents listed have been investigated in vitro and in vivo. Compounds "under investigation as cancer drug" indicates that they have been or are being tested in clinical studies

Agent family	Examples	Origin	Mechanism of action	Use
Microfilaments				
Cytochalasins	Cytochalasin B, cytochalasin D	Several species of fungi	Bind to the barbed end and inhibit elongation	Cell biology research
Chaetoglobosins	Chaetogobosin A, chaetoglobosin K	Several species of fungi	Similar to cytochalasins	Under investigation as anticancer drug
Jasplakinolide	Jasplakinolide	Marine sponge (*Jaspis johnstoni*)	Induces polymerization and stabilizes MFs	Cell biology research
Latrunculins	Latrunculin A	Marine sponge (*Latrunculia magnifica*)	Sequesters actin monomers	Cell biology research
MKT-077	MKT-077	Synthetic	Crosslinks MFsInhibits Hsp70	Under investigation as anticancer drug
Staurosporine	Staurosporine	Bacterium (*Streptomyces staurosporeus*)	Thinning and loss of MFsProtein kinase inhibitor	Biochemistry research
Scytophycins	Tolytoxin	Cyanobacteria (Scytonemataceae)	Inhibit actin polymerization and depolymerize MFs	Research
Phalloidin	Phalloidin	Death cap fungus (*Amanita phalloides*)	Stabilizes MFs by preventing depolymerization	Cell biology research
Microtubules				
Vinca alkaloids	Vincristine, vinblastine, vindesine, vinorelbin	Madagascar periwinkle (*Catharanthus roseus*)	Bind tubulin and inhibit assembly into MTs	Approved for cancer therapy
Dolastatins	Dolastatin 10, dolastatin 15, monomethyl auristatins	Sea hare (*Dolabella auricularia*)	Similar to vinca alkaloids	Approved for cancer therapy
Eribulin	Eribulin, halichondrin B	Marine sponge (*Halicondria okadai*)	Similar to vinca alkaloids	Approved for cancer therapy
Colchicine	Colchicine, colcemid	Meadow saffron (*Colchicum autumnale*)	Similar to vinca alkaloids	Approved as anti-inflammatoryCytogenetics
Cryptophycin	Cryptophycin 1	Cyanobacteria (*Nostoc*)	Similar to vinca alkaloids	Under investigation as anticancer drug
Nocodazole	Nocodazole	Synthetic	Similar to vinca alkaloids	Cell biology research
Taxanes	Paclitaxel, docetaxel	Pacific yew tree (*Taxus brevifolia*)	Bind and stabilize MTs	Approved for cancer therapyCell biology research
Epothilones	Ixabepilone, epothilone B (patupilone), sagopilone	Myxobacterium (*Sorangium cellulosum*)	Similar to taxanes	Approved for cancer therapy

(*continued*)

Cytoskeleton, Table 1 (continued)

Agent family	Examples	Origin	Mechanism of action	Use
Peloruside	Peloruside A	Marine sponge (*Mycale hentscheli*)	Similar to taxanes	Under investigation as anticancer drug
Laulimalide	Laulimalide	Marine sponge (*Cacospongia mycofijiensis*)	Similar to taxanes	Under investigation as anticancer drug
Discodermolide	Discodermolide	Deep-sea sponge (*Discodermia dissoluta*)	Similar to taxanes	Under investigation as anticancer drug
Intermediate filament				
Withanolides	Withaferin A	Winter cherry (*Withania somnifera*)	Bind vimentin at a Cys residue and inhibit assembly	Under investigation as anticancer drug

As with MFs, numerous MT-associated proteins (MAPs) control remodeling of MTs, formation of bundles, formation of networks with other cytoskeletal systems, and association to cellular structures. Like in the case of MFs, de novo formation of MTs is an unfavorable process that in vivo occurs in an organized manner at the MTOC or at basal bodies that function as nucleation centers. Most MAPs bind along the length of the MTs and stabilize them. They are particularly abundant in nervous tissue. Examples include the tau family (tau, MAP2, MAP4), MAP1A and 1B, STOP, tektin, and ▶ doublecortin. Some of these are restricted to specialized cells while others are expressed widely. Phosphorylated, proteolytically modified, and cross-linked tau is the main component of the neurofibrillary tangles that characterize Alzheimer disease. Other MAPs promote the destabilization of MTs. Stathmin/Op18 (oncoprotein 18), a protein expressed at unusually high levels in many malignant cells; sequesters tubulin heterodimers, preventing them from polymerizing. Catastrophin (MCKA, mitotic centromere associated kinesin) is an unusual kinesin that binds the plus ends of MTs and promotes the detachment of subunits. Katanins sever MTs into short fragments that then depolymerize. A group of proteins capture and stabilize the plus ends at the cell periphery, like CLIP-170 and EB1, or at kinetochores, like MAST/Orbit.

MT-associated motor proteins use the energy provided by the hydrolysis of ATP to walk along MTs. They belong to two major families, dynein and kinesin. In general, members of the large family of kinesins move toward the plus end and are involved in the transport of multiprotein complexes and organelles and in various phases of the cell division. Dyneins move toward the minus end and can be grouped into two classes, cytoplasmic (involved in transport of organelles and cell division) and axonemal (responsible for flagellar and cilliary beating).

Drugs that affect MT dynamics (Table 1) are used as effective antitumor drugs because blocking reorganization of the mitotic spindle preferentially affects rapidly dividing cells. This blocks the progression of mitosis and prolonged activation of the mitotic checkpoint triggers apoptosis. Vinca alkaloids like Vincristine (Oncovin™) and vinblastine bind to tubulin and cause it to aggregate, thereby preventing the assembly of MTs. They are used in combination with other cytostatic drugs. Vinblastine is used to treat Hodgkin lymphoma, nonsmall cell lung cancer, breast cancer, head and neck cancer, and testicular cancer. It is also used to treat Langerhans cell histiocytosis. Vincristine is mainly used to treat nonHodgkin's lymphoma, acute lymphoblastic leukemia, and nephroblastoma. Monomethyl auristatin E - (Vedotin) is more potent than vinca alkaloids but due to its toxicity it is linked to monoclonal antibodies that direct the drug to cancer cells. It is used to treat relapsed or refractory Hodgkin lymphoma and relapsed or refractory systemic anaplastic large cell lymphoma (Brentuximab

Vedotin). Taxanes stabilize MTs and protect them from disassembly, and as a result interfere with their normal dynamic breakdown during cell division. Paclitaxel (▶ Taxol™) is used to treat patients with lung, ovarian, breast and head and neck cancer, and advanced forms of Kaposi sarcoma. The clinical effectiveness of taxanes is often limited by primary or acquired resistance. This has led to the introduction of novel MT targeting agents like epothilones and eribulin. Epothilones have higher efficacy and lower toxicity than taxanes. Ixabepilone is used in the treatment of aggressive metastatic or locally advanced breast cancer no longer responding to currently available chemotherapies. Eribulin (Halaven™) is used to treat patients with metastatic breast cancer who have received at least two prior chemotherapy regimens for late-stage disease. Due to its relatively low therapeutic index, colchicine is not used in tumor therapy, but continues to be used for its anti-inflammatory properties for the treatment of acute flares of gout and familial Mediterranean fever.

Intermediate Filaments

Intermediate filaments owe their name to their diameter, 10 nm, intermediate between MFs and MTs (Fig. 1). In contrast to these, IFs vary largely in composition and size among different cell types. However, the subunits that make up IFs are structurally related rod-shaped proteins with a central α-helical coiled-coil domain that forms the core of the filaments and variable globular domains on either side.

IF proteins assemble into parallel coiled-coil dimers where the central domains are intertwined. Dimers assemble into antiparallel half-staggered tetramers (protofilaments). Eight tetramers associate laterally into unit-length filaments that anneal longitudinally to form 16 nm filaments which, unlike MFs and MTs, lack polarity. It follows a compaction phase during which filaments are compressed radially into mature 10 nm IFs. Assembly of IFs does not require nucleotide hydrolysis.

IFs are flexible, more stable than MFs and MTs and more abundant than these in neurons and epidermal cells. IFs form a meshwork that spans the whole cytoplasm, where they play a major structural role allowing cells to withstand stretching and shearing forces, although roles in cell signaling, epithelial polarity, growth, and apoptosis are emerging. IFs also make up the nuclear lamina that reinforces the nuclear envelope. IFs are not involved in cell movements. Although biochemically stable, IF networks undergo rearrangements, particularly during cell spreading, wound healing, cell division, and in response to environmental mechanical stresses.

The IF family consists of 73 gene products that have been grouped into six major types or sequence homology classes. Types I–IV are cytoplasmic, type V comprises lamins, present in the nucleus of all cells, and type VI comprises proteins found exclusively in the eye lens. Mutations in almost all genes encoding IF proteins have been found to cause or predispose to human disease, with more than 80 conditions described.

Type I and II IFs are acidic and basic keratins (or cytokeratins), respectively, that always associate into heterodimers of one acidic with one basic keratin. The keratin family comprises 54 functional genes, 37 epithelial keratins, and 17 hair keratins. Keratins are characteristic of epithelial cells of the skin and digestive, respiratory and urinary tracts and are an important component of structures that grow from skin, like hair and nails. Each type of epithelium expresses a characteristic combination of type I and type II keratins. For example, in the skin basal keratinocytes express K5/K14, whereas suprabasal keratinocytes express K1/K10.

Keratins are widely used tumor markers when diagnosis using conventional microscopic techniques is inconclusive, based on the fact that tumor cells retain the IF proteins characteristic of the tissue of origin. Staining with a pan-keratin specific antibody, for example, distinguishes epithelial from nonepithelial tumors. The precise pattern of keratin expression of metastases allows prediction of the origin of the primary tumor and can also be useful in predicting tumor prognosis. Circulating keratins and fragments thereof released form apoptotic and necrotic tumor and nontumor cells and have been used as markers of disease progression, response to treatment, and recurrence. The most commonly used markers are TPA (tissue polypeptide antigen, a mixture of K8, K18, and K19), TPS (tissue polypeptide-specific antigen, derived from K18), and CYFRA 21–1 (cytokeratin fragment 21–1, derived from K19).

Type III IFs comprise vimentin, desmin, glial fibrillary acidic protein (GFAP), peripherin, and syncoilin. Desmin is found in muscle and is responsible for stabilizing the sarcomere, where it has a fixed position and anchoring it to the plasma membrane. Syncoilin is also found at stress points in muscle cells. Vimentin is expressed in leukocytes, endothelial cells, and mesenchimal cells. GFAP is found in glial cells and peripherin in the peripheral nervous system.

Type IV IFs include neurofilaments (a triplet of low, mid, and high molecular weights), α-internexin, nestin, and synemin. Neurofilaments are found in neurons, where they determine axon diameter. Nestin and α-internexin are also expressed in neurons at early stages of development. Synemin is localized to stress points in muscle cells.

Type V IFs are made of lamins. They do not form ropes, but a planar meshwork underneath the inner nuclear membrane, the nuclear lamina. Three genes encode several alternatively spliced lamins that can be grouped into B-type (expressed in all cell types) and A-type (restricted to more differentiated cell types) lamins. Lamins interact with numerous proteins of the inner nuclear membrane, like lamin B receptor, emerin, and LAP (lamin-associated protein). The lamina is connected to the cytoskeleton through LINC (linker of nucleoskeleton and cytoskeleton) complexes composed of SUN proteins of the inner nuclear membrane interacting with nesprins of the outer nuclear membrane. LINC complexes are important for nuclear positioning and migration. Lamins become phosphorylated and disassemble at the onset of mitosis, concomitantly with the breakdown of the nuclear envelope, and reassemble after mitosis.

Type VI IFs group two proteins, Bfsp1 (filensin) and Bfsp2 (phakinin), responsible for the exceptional biochemical stability and optical clarity of the eye lens.

Unlike MFs and MTs, which are found in all cells, cytoplasmic IFs display complex patterns of cell and tissue distribution and the expression of some of them is markedly regulated during development and cell differentiation. Diverse **IF-associated proteins** (IFAPs) crosslink IFs to each other into bundles and networks, or link them to other cytoskeleton networks or specialized structures at the plasma membrane. As opposed to MFs and MTs, no IFAPs have been identified that possess IF-dependent nucleating, capping, or severing activity. IF remodelling is regulated primarily (albeit not exclusively) by phosphorylation. IFs also do not function as tracks for motor proteins, although IF particles can be themselves cargo of MT- and MF-dependent motor proteins.

The most widespread IFAPs are the plakins, a family of large multifunctional proteins. Prominent examples of this family are plectin, BPAG1 (bullous pemphigus antigen 1), epiplakin, periplakin, envoplakin, and desmoplakin. Plectin is a very large protein that crosslinks IFs to each other, to the plasma membrane, and to MFs and MTs. BPAG1 cross-links neuronal IFs to MFs. BPAG1 and plectin connect keratin filaments to hemidesmosomes. The other plakins mentioned above connect keratin filaments to desmosomes. ▶ Desmosomes and hemidesmosomes are specialized structures at the plasma membrane that allow the transmission of mechanical forces from one cell to another and to the extracellular matrix, respectively. Examples of proteins that crosslink IFs are filaggrin and KAPs (keratin associated proteins), which interact with keratin in the epidermis and the hair shaft.

The wide diversity of IF proteins complicates the search for potential therapeutic agents that specifically target IFs (Table 1). Currently, withaferin A is under investigation as an inhibitor of type III IFs, particularly vimentin. Withaferin A binds covalently to a conserved cysteine residue in vimentin and inhibits the assembly of the IF.

Cross-References

▶ Adherens Junctions
▶ Centrosome

- ▶ Cortactin
- ▶ Desmosomes
- ▶ Doublecortin
- ▶ Gelsolin
- ▶ Taxol

References

Campellone KG, Welch MD (2010) A nucleator arms race: cellular control of actin assembly. Nat Rev Mol Cell Biol 11:237–251

Godsel LM, Hobbs RP, Green KJ (2007) Intermediate filament assembly: dynamics to disease. Trends Cell Biol 18:28–37

Wade RH (2009) On and around microtubules: an overview. Mol Biotechnol 43:177–191

Cytosolic Transglutaminase

- ▶ Transglutaminase-2

Cytotactin

- ▶ Tenascin-C

Cytotoxic T Cells

Marc Schmitz
Institut für Immunologie, Technische Universität Dresden, Dresden, Germany

Definition

$CD8^+$ cytotoxic T cells (CTLs) are major cellular components of the adaptive immune system. They efficiently recognize and destroy virus-infected host cells or tumor cells, which expose antigenic peptides bound to major histocompatibility complex (MHC) class I molecules.

Characteristics

Dendritic Cells Play a Major Role in the Induction of $CD8^+$ T Cell Responses

$CD8^+$ CTLs can be efficiently activated by "professional" antigen-presenting cells (APCs) such as dendritic cells (DCs), which display the appropriate MHC class I-peptide complexes on their surface. DCs are the most effective APCs for stimulating naïve T cells that have not recognized and responded to antigens. Besides their extraordinary capability to initiate $CD8^+$ T cell responses, DCs also essentially contribute to the maintenance and regulation of previously activated $CD8^+$ CTLs. Macrophages and B lymphocytes also function as APCs, but mostly for previously stimulated T cells rather than for naïve T cells.

DCs originate from bone marrow progenitor cells, circulate through the blood, and migrate into various tissues such as the epidermis of the skin and the epithelia of the gastrointestinal and respiratory tracts. These so-called immature DCs are ideally located at common sides of pathogen entry to perform a sentinel function in the immune defense. They are characterized by the expression of different cell membrane receptors, which mediate an efficient endocytosis of pathogen-associated antigens. Furthermore, they also internalize antigens from the extracellular fluid receptor-independent by pinocytosis.

For $CD8^+$ T cell activation, DCs utilize the "classical" MHC class I pathway for processing and presentation of cytosolic proteins. In this context, ubiquitinated cytosolic proteins such as viral or tumor-associated proteins are degraded by proteasomes representing multiprotein enzyme complexes. The generated cytosolic peptides are translocated by a specialized transporter, termed transporter associated with antigen processing, into the endoplasmic reticulum (ER). Inside the ER, the peptides bind to the cleft of newly synthesized and assembled MHC class I molecules consisting of an α-chain and b2-microglobulin. Subsequently, stable MHC class I-peptide complexes move through the Golgi complex and are transported to the cell surface by vesicles. In addition to the "classical" MHC class I pathway, DCs have the extraordinary capacity to ingest virus-infected or tumor cells and

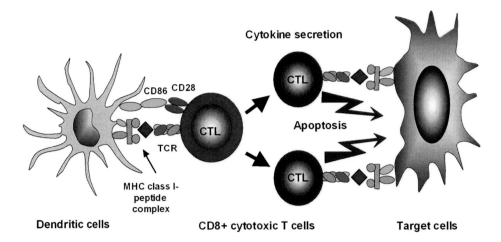

Cytotoxic T Cells, Fig. 1 DCs display an extraordinary capacity to induce and maintain CD8$^+$ T cell responses. Mature DCs are characterized by the high surface expression of MHC class I-peptide complexes and costimulatory molecules. Therefore, they are well equipped to efficiently activate and expand CD8$^+$ CTLs. Activated CD8$^+$ CTLs possess a profound capability to recognize and destroy pathogen-infected or tumor cells. Furthermore, they secrete important cytokines such as IFN-gamma, which can stimulate additional immune cell populations

present peptides derived from extracellular viral or tumor antigens in association with MHC class I molecules. This alternative pathway is called cross-presentation.

In contrast to their pronounced capacity to capture pathogen-associated antigens, immature DCs are not able to efficiently initiate CD8$^+$ T cell responses, which is based on the low expression of MHC, adhesion and costimulatory molecules on their surface. DC maturation is induced by pathogen-associated molecular patterns such as lipopolysaccharide that are recognized by Toll-like receptors and other pattern recognition receptors. Furthermore, pro-inflammatory cytokines produced by various immune cells promote the maturation of DCs. Mature DCs migrate from the tissues into draining lymph nodes via afferent lymphatic vessels, a process which is mediated by the interaction of the chemokine receptor CCR7 on DCs and the chemokines CCL 19 and CCL 21 that are produced in the T cell zone of lymph nodes. This leads to a colocalization of antigen-bearing DCs with naïve T cells. Naïve T lymphocytes circulate throughout the body in a nonstimulated state and perform their functional capabilities only after activation. Due to the high surface expression of

MHC class I-peptide complexes as well as adhesion and costimulatory molecules, mature DCs are able to efficiently induce CD8$^+$ T cell responses.

Activation of CD8$^+$ T cells

Cytokine secretion and clonal expansion of CD8$^+$ T cells as well as the differentiation of naïve T cells into effector and memory T lymphocytes require antigen recognition, costimulation, and cytokines that are provided by "professional" APCs, such as DCs, and by the T cells themselves. Antigen recognition is the important first signal for the activation of CD8$^+$ T lymphocytes, ensuring that the resultant immune response remains specific for the antigen (Fig. 1). In lymphoid organs, naïve CD8$^+$ T cells recognize via their T cell receptors (TCRs) antigen-derived peptides displayed by MHC class I molecules on the surface of DCs. In contrast to naïve CD8$^+$ T cells, differentiated effector T cells can respond to antigens presented by cells other than DCs. In addition to antigen recognition, naïve CD8$^+$ T cells require a second signal provided by costimulatory molecules on DCs (Fig. 1). In the absence of costimulation, T cells that encounter antigens either fail to respond and undergo apoptosis or enter a state of unresponsiveness called anergy.

The best characterized costimulatory pathway in T cell activation involves the T cell surface receptor CD28, which binds the costimulatory molecules B7-1 (CD80) and B7-2 (CD86) expressed on activated DCs. Previously activated effector CD8$^+$ T cells are less dependent on costimulation than naïve T cells. This property enables effector CD8$^+$ T cells to kill other cells that do not express costimulatory molecules. CD28 engagement together with TCR-mediated antigen recognition promotes survival and proliferation of T cells as well as the differentiation of naïve T cells into effector and memory T lymphocytes. Another costimulatory pathway for T cell activation is based on the interaction between inducible costimulator (ICOS) expressed on T cells and its ligand, ICOS-L, which is displayed on the surface of various cell types such as DCs. In the meantime, additional activating and inhibitory molecules have been identified that can regulate T cell responses. In addition to antigen recognition and costimulation, the activation of naïve CD8$^+$ T cells and their differentiation into CD8$^+$ effector and memory T cells may require the participation of CD4$^+$ T helper cells. Thus, CD4$^+$ T helper cells can improve the capacity of DCs to induce CD8$^+$ CTLs by the interaction between CD40 on DCs and CD40 ligand on activated CD4$^+$ T cells. Furthermore, CD4$^+$ T cells provide help for the expansion of CD8$^+$ CTLs by secreting cytokines such as interleukin (IL)-2.

Antigen-stimulated CD8$^+$ T cells are characterized by the expression of various surface molecules such as CD69 and CD25. The expression of CD25, a component of the IL-2 receptor, enables T cells to respond to the growth-promoting cytokine IL-2. In addition, activated CD8$^+$ T cells produce IL-2 that promotes the survival, proliferation, and differentiation of T cells. Furthermore, antigen recognition, costimulation, and growth factors such as IL-2 induce T cell proliferation. The result of this proliferation is clonal expansion, which generates a large number of CD8$^+$ T cells from a small pool of naïve antigen-specific CD8$^+$ T lymphocytes.

Antigen recognition and other activating stimuli also promote the differentiation of naïve T cells into effector and memory T lymphocytes. CD8$^+$ memory T cells have an ability to survive for long periods, persist in the circulating lymphocyte pool, mucosal tissues, skin, and lymphoid organs, respond rapidly to subsequent encounter with the same antigen, and generate new CD8$^+$ effector cells that contribute to antigen elimination. They mount larger and more rapid responses than naïve T cells to antigen challenge. Activated CD8$^+$ effector T cells, which are characterized by a strong capacity to lyze target cells and to produce cytokines such as interferon (IFN)-gamma after antigenic stimulation, migrate to peripheral tissues and essentially contribute to the elimination of pathogen-infected or tumor cells.

Cytotoxic Potential of Activated CD8$^+$ T Cells

CD8$^+$ effector CTLs efficiently destroy targets such as pathogen-infected cells, which express the same MHC class I-antigenic peptide complexes on their surface that triggered the proliferation and differentiation of naive CD8$^+$ T cells (Fig. 1). This specificity of CTL function ensures that normal cells are not killed by CTLs reacting against infected cells. In addition, the target cell killing is highly specific because an immunologic synapse is formed at the site of contact of the CTL and the antigen-expressing target, and the molecules that perform the killing are secreted into the synapse and cannot diffuse to other nearby cells.

The process of CD8$^+$ CTL-mediated killing of targets consists of antigen recognition, CTL activation, delivery of the lethal hit that kills the target cells, and release of the CTLs. CD8$^+$ CTLs recognize their target cells by the interaction of the TCR with the appropriate MHC class I-antigenic peptide complex and the accessory molecule leukocyte function antigen-1 with intercellular adhesion molecule-1. CTL recognition leads to activation and target cell killing. One important mechanism of CTL-mediated killing is the delivery of cytotoxic proteins stored within cytoplasmic granules to the target cell, thereby triggering apoptosis. The cytotoxic proteins in the granules include granzymes, which are serine proteases and the membrane-perturbing molecule perforin. Activation results in the release of granule contents from the CTL into the target cell through the area of contact. CTLs also use granule-independent mechanisms of killing that is mediated by interactions of membrane molecules on CTLs and target cells. For example, CTLs can express Fas ligand that binds to the death receptor Fas on target cells. This interaction also results in apoptosis of target cells. After delivering the lethal hit, the CTL is released from the target cell and kill other target cells. In addition to their high cytotoxic potential, activated CTLs can produce important proinflammatory cytokines such as IFN-gamma, which can stimulate various immune cell populations. Due to their functional properties, CD8$^+$ CTLs can essentially contribute to the elimination of pathogen-infected cells and tumor cells. In addition, CD8$^+$ CTLs emerged as promising candidates for cancer immunotherapy. In this context, clinical trials revealed that the adoptive transfer of CD8$^+$ T cells can mediate tumor regression in cancer patients.

Cross-References

▶ Adoptive T-Cell Transfer
▶ Chimeric Antigen Receptor on T Cells
▶ Dendritic Cells
▶ HLA Class I
▶ Proteasome

References

Andersen MH, Schrama D, Thor Straten P, Becker JC (2006) Cytotoxic T cells. J Invest Dermatol 126:32–41

Banchereau J, Briere F, Caux C, Davoust J, Lebecque S, Liu YJ, Pulendran B, Palucka K (2000) Immunobiology of dendritic cells. Annu Rev Immunol 18:767–811

Dudley ME, Wunderlich JR, Robbins PF, Yang JC, Hwu P, Schwartzentruber DJ, Topalian SL, Sherry R, Restifo NP, Hubicki AM, Robinson MR, Raffeld M, Duray P, Seipp CA, Rogers-Freezer L, Morton KE, Mavroukakis SA, White DE, Rosenberg SA (2002) Cancer regression and autoimmunity in patients after clonal repopulation with antitumor lymphocytes. Science 298:850–854

Mollderm JJ, Lee PP, Wang C, Felio K, Kantarjian HM, Champlin RE, Davis MM (2000) Evidence that specific T lymphocytes may participate in the elimination of chronic myelogenous leukemia. Nat Med 6:1018–1023

Rosenberg SA, Restifo NP, Yang JC, Morgan RA, Dudley ME (2008) Adoptive cell transfer: a clinical path to effective cancer immunotherapy. Nat Rev Cancer 8:299–308

Russell JH, Ley TJ (2002) Lymphocyte-mediated cytotoxicity. Annu Rev Immunol 20:323–370

Williams MA, Bevan MJ (2007) Effector and memory CTL differentiation. Annu Rev Immunol 25:171–192

Cytovillin

▶ ERM Proteins

D

3-(1,3-Dihydro-3-oxo-2H-indol-2-ylidene)-1,3 dihydro-2H-indol-2-one

▶ Indirubin and Indirubin Derivatives

DAC

▶ 5-Aza-2′ Deoxycytidine

Dacogen

▶ 5-Aza-2′ Deoxycytidine

Dactinomycin

Definition

Dactinomycin is a trade name for actinomycin D. It is a chemotherapy drug that is given as a treatment for some types of cancer, most commonly some cancers that occur in children such as ▶ Wilms' tumor and germ cell tumors, although it may sometimes be used for adults.

Cross-References

▶ Wilms' Tumor

See Also

(2012) Actinomycin D. In: Schwab M (ed) Encyclopedia of cancer, 3rd edn. Springer, Berlin/Heidelberg, p 19. doi:10.1007/978-3-642-16483-5_46

(2012) Germ cell tumors. In: Schwab M (ed) Encyclopedia of cancer, 3rd edn. Springer, Berlin/Heidelberg, p 1541. doi:10.1007/978-3-642-16483-5_6905

DAF

▶ Decay-Accelerating Factor

Damage Response

▶ Stress Response

DANCE, EVEC, UP50, FBLN-5

▶ Fibulins

© Springer-Verlag Berlin Heidelberg 2017
M. Schwab (ed.), *Encyclopedia of Cancer*,
DOI 10.1007/978-3-662-46875-3

Dap160 (Dynamin-Associated Protein of 160 kDa)

▶ Intersectin

DAP6

▶ Daxx

DARC

▶ Duffy Antigen Receptor for Chemokines

Daxx

Sheng-Cai Lin
Department of Biomedical Sciences, School of Life Sciences, Xiamen University, Xiamen, Fujian, China

Synonyms

BING2; DAP6; Death-associated protein 6; MGC126245; MGC126246

Definition

Daxx was originally identified as a protein factor that binds to a transmembrane receptor called Fas (FAS/APO-1/CD95), one member of the ▶ tumor necrosis factor receptor superfamily (Yang et al. 1997). The extracellular region of Fas is where its ligand, FasL, binds. The intracellular tail shares sequence similarity with another member of the TNF receptor family, TNF receptor I (TNFRI). The shared sequence is termed death domain for its critical role in signaling cell death upon ligand binding.

Characteristics

Since its identification in 1997, Daxx has been intensively investigated for its biological functions. However, up to date the three-dimensional crystal structure of Daxx has not been resolved. Human Daxx protein is a polypeptide of 740 amino acid residues in length. As depicted in the schematic diagram (Fig. 1), Daxx contains several putative domains and many binding sites for interacting with a wide spectrum of proteins including transcription factors, kinases, tumor suppressors, and chromatin remodeling factors. It is worth to note that for the influence from its acidic region, Daxx displays an aberrant molecular weight during electrophoretic migration on gels, being 120 kDa instead of the calculated 81.3 kDa (Yang et al. 1997).

Basic Information

Human *Daxx* gene is mapped to 6p21.3 in the major histocompatibility complex (MHC) region and transcribes a single 2.4 Kb RNA. It is ubiquitously expressed in human tissues and is highly conserved in mammals with 69% identity between the human and mouse proteins. Currently at least 35 molecules have been known to interact with Daxx, which were identified by various methods including yeast two-hybrid screening, coimmunoprecipitation, and immunofluorescent costaining in the cell (Salomoni and Khelifi 2006). Most notably, Daxx interacts with the tumor suppressor ▶ p53, Axin (Axis inhibitor), ubiquitin ligase Mdm23 (mouse double minute 2), the Ser/Thr kinase HIPK2 (homeodomain-interacting protein kinase 2), and the product of the promyelocytic leukemia (*PML*) gene.

Subcellular Distribution of Daxx Protein

Daxx is predominantly localized in the nucleus. Daxx and PML interact with each other and are colocalized in nuclear punctuate structures known as nuclear bodies (NBs) that have been involved in various processes such as transcriptional regulation, apoptosis, genome stability, and tumor suppression (Salomoni and Khelifi 2006). PML protein is an important tumor suppressor that is essential for the formation of the nuclear bodies,

Daxx, Fig. 1 Schematic diagram showing various domains of Daxx. Human Daxx is a protein with 740 amino acid residues in length that contains several putative domains: two N-terminal paired amphipathic helices (PAH1 and PAH2), a coiled-coil region, an acidic domain (D/E), and a Ser/Pro/Thr-rich domain (S/P/T). Axin and HIPK2 associate with Daxx through binding sites on the N-terminal portion. A region between amino acid 501 and 625 is required for ASK1 binding and the subsequent activation of JNK. The Ser/Pro/Thr-rich domain contains sites for interaction with a series of proteins, such as Fas, PML, Ubc9, and CENP-C

Daxx, Fig. 2 Schematic representation of two possible pathways through which Daxx mediates cell death. Upon stimulation by stress signals such as UV and nutrition deprivation, Daxx becomes associated with ASK1, leading to activation of JNK MAP kinase that is required for cell death induction. In parallel, UV irradiation causes dissociation of p53-destabilizing complex to release p53 and Daxx which then assemble into a p53-activating complex consisting of p53, Axin, Daxx, and HIPK2. Upon complex formation, HIPK2 is activated and activates p53 by phosphorylating it at Ser46. As a result, p53 target genes related to apoptosis are activated

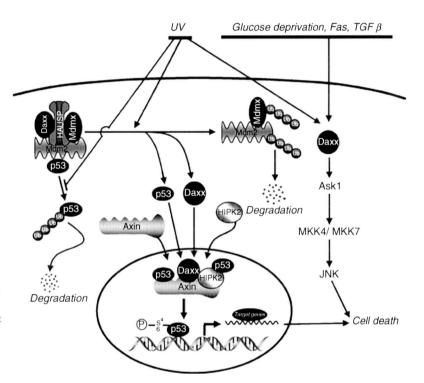

as deletion of the *PML* gene by genetic knockout can fully abrogate the formation of these structures. In $pml^{-/-}$ cells that lack PML-NB structures, the ability of Daxx to induce cell death is abolished, underscoring the importance of PML for cell death induction by Daxx-mediated signaling.

Daxx Mediates Stress-Induced Cell Death

Daxx has been implicated in the modulation of apoptosis induced by a wide range of stimuli, including ultraviolet radiation, hydrogen peroxide, arsenic trioxide, Fas ligand, transforming

growth factor beta (TGF-β), and interferon-γ. As summarized in Fig. 2, two main apoptotic pathways are thought to be mediated by Daxx.

• The first reported mechanism for Daxx to induce apoptosis is through association and subsequent activation of apoptotic signal-regulating kinase 1(ASK1), an upstream kinase of JNK MAP kinase that leads to activation of c-jun N-terminal kinase (JNK) (Salomoni and Khelifi 2006). Glucose deprivation as a stress can also stimulate the association of Daxx to ASK1 and activate

Daxx-dependent ASK1 activation. TGF-β is a multifunctional growth factor that has principal roles in growth control. In a yeast two-hybrid screening using the cytoplasmic domain of the type II TGF-β receptor as bait, Daxx was found to interact with this receptor and mediate TGF-β-induced apoptosis via facilitating JNK activation. Interference of Daxx function by truncated Daxx mutants or by knockdown of Daxx expression can inhibit TGF-β-, UV-, or oxidative stress-induced apoptosis.

- Another means adopted by Daxx to induce cell death is through Axin/HIPK2/p53 complex (Li et al. 2007). Currently, the best characterized role of Daxx is its ability to modulate the function of the tumor suppressor p53, which is mutated in over 50% of all tumor tissues. p53 exerts its tumor suppressional roles by acting as a transcription factor. Upon DNA damage, p53 is activated, which results in cell cycle arrest or apoptosis by inducing expression of a series of proteins such as p21, GADD45 (growth arrest and DNA damage), and ► PUMA (p53-upregulated modulator of apoptosis). HIPK2, an upstream serine/threonine kinase of p53, can specifically phosphorylate p53 at serine 46 upon UV irradiation and plays an important role in UV-induced p53-mediated apoptosis. Axin is a negative regulator of Axis formation in the development of mouse embryos, its deficiency can lead to axis duplication. Accumulating evidence demonstrates that this protein functions as a tumor suppressor. It has been shown that UV irradiation can enhance the association of Daxx and Axin. Further investigation demonstrates that Axin tethers Daxx to p53 and cooperates with Daxx to stimulate HIPK2-mediated Ser46 phosphorylation and transcriptional activity of p53. Axin is required for Daxx-induced p53 activation and apoptosis, as in $Axin^{-/-}$ cells the ability of Daxx to induce cell death is dramatically attenuated. Downregulation of HIPK2 by specific siRNA remarkably suppresses Daxx-induced apoptosis, suggesting that HIPK2 plays crucial role in Daxx-induced cell death. All above evidence indicates that formation of a new complex containing Daxx/Axin/p53/HIPK2 is important for Daxx-induced cell death. Knockdown of Axin and Daxx by ► siRNA can strongly attenuate UV-induced cell death, demonstrating that Daxx and Axin complex plays a critical role in UV-triggered cell death.

In Unstressed Cells, Daxx Seems to Play an Antiapoptotic Role

Daxx may also destabilize p53 by promoting the function of Mdm2, a ubiquitin ligase E3 that facilitates p53 ► ubiquitination and degradation (Tang et al. 2006). In unstressed cells, Daxx simultaneously interacts with Mdm2 (heterodimerized with MdmX) and the deubiquitinase HAUSP (herpes-associated ubiquitin-specific protease). In this complex, Daxx can stimulate the stabilizing effect of Hausp on Mdm2, which results in accumulation of Mdm2 in the cell. Moreover, Daxx enhances the intrinsic E3 activity of Mdm2 toward p53 leading to the inhibition of the antiproliferative effects of p53. Upon DNA damage, Daxx, HAUSP, and p53 dissociate from Mdm2, which allows Mdm2–MdmX complex to undergo autoubiquitination and subsequent degradation. The dissociated Daxx and p53 may then form an apoptotic complex with HIPK2 and Axin in the nucleus, in which Daxx and Axin cooperatively promote the HIPK2 kinase activity toward p53 at Serine 46. Therefore, in response to stress signals, Daxx enhances the death-inducing function of p53, suggesting that Daxx in fact may exert opposing roles in controlling p53 activity depending on cellular states. However, the detailed molecular mechanisms as to how Daxx dissociates from the Daxx/Mdm2/HAUSP/Mdmx complex and recycles to form an apoptotic complex upon DNA damage remain to be clarified in the future.

Cross-References

- ► MDM2
- ► p53 Family
- ► TP53
- ► Tumor Suppressor Genes
- ► Ubiquitination

References

Li Q, Wang X, Wu X et al (2007) Daxx cooperates with the Axin/HIPK2/p53 complex to induce cell death. Cancer Res 67:66–74

Salomoni P, Khelifi AF (2006) Daxx: death or survival protein. Trends Cell Biol 16:97–104

Tang J, Qu L-K, Zhang J et al (2006) Critical role for Daxx in regulating Mdm2. Nat Cell Biol 8:855–862

Yang X, Khosravi-Far R, Chang HY et al (1997) Daxx, a novel Fas-binding protein that activates JNK and apoptosis. Cell 89:1067–1076

DCIS

▶ Ductal Carcinoma In Situ

DcR3

▶ Decoy Receptor 3

DCX

▶ Doublecortin

DDP

▶ Cisplatin

DDS

▶ Drug Delivery Systems for Cancer Treatment

Death

▶ Necrosis

Death Receptors

Definition

Death receptors belong to the ▶ tumor necrosis factor (TNF) receptor gene superfamily, which is defined by similar, cysteine-rich extracellular domains. The death receptors also contain a homologous cytoplasmic sequence termed the "death domain." Death domains typically enable death receptors to engage the cell's apoptotic machinery, but in some instances they mediate functions that are distinct from or even counteract apoptosis. Some molecules that transmit signals from death receptors contain death domains themselves.

Cross-References

▶ FLICE-Inhibitory Protein
▶ Tumor Necrosis Factor

See Also

(2012) Death domain. In: Schwab M (ed) Encyclopedia of cancer, 3rd edn. Springer, Berlin/Heidelberg, p 1065. doi:10.1007/978-3-642-16483-5_1534

Death-Associated Protein 6

▶ Daxx

Decatenation G2 Checkpoint

William K. Kaufmann
Department of Pathology and Laboratory Medicine, University of North Carolina at Chapel Hill, Chapel Hill, NC, USA

Definition

A surveillance system that ensures that mitosis does not begin when topoisomerase IIα is locked in the closed clamp conformation by topoII catalytic inhibitors.

Characteristics

During DNA replication, daughter DNA duplexes become catenated or entangled through knots and links. These knots and links must be removed before sister chromatids can be properly segregated during mitosis. Type II DNA ▸ topoisomerases are enzymes that remove knots and links through a concerted set of reactions in which a protein-associated DNA double-strand break is generated, another segment of DNA is passed through the break, and the protein-associated double-strand break is reversed. The decatenation G2 checkpoint is an active signal transduction pathway that monitors topoisomerase IIα catalytic status and controls the location and/or activity of mitosis-promoting factor to delay the onset of mitosis.

Molecular Biology

Drugs that inhibit topoisomerase II fall into two groups known as poisons and catalytic inhibitors. Topoisomerase II poisons stabilize the enzyme when in a covalent complex with cleaved DNA strands. Proteolysis of the protein produces frank DNA double-strand breaks which are potentially lethal lesions. Topoisomerase poisons such as doxorubicin and etoposide are among the most widely used and successful of cancer chemotherapeutic drugs. The catalytic inhibitors of topoisomerase II prevent the enzyme from forming protein-associated DNA double-strand breaks; by preventing formation of protein-associated DNA double-strand breaks, catalytic inhibitors are able to block the toxicity of the poisons.

There are two type II topoisomerases in mammalian cells. Topoisomerase IIβ is a nonessential enzyme that is constitutively expressed in quiescent and proliferating cells. Topoisomerase IIα is an essential enzyme whose level of expression is tightly regulated in the cell division cycle. Maximal expression occurs in G2 and M when topoisomerase II is required to separate linked sister chromatids and condense mitotic chromosomes. Catalytic inhibitors of topoisomerase II include

ICRF-193 and ICRF-187. ICRF-193 causes mammalian cells to delay progression from G2 to mitosis and impedes separation of sister chromatids at anaphase in mitosis. This behavior suggested that cells monitor the state of chromatid catenation in G2 and delay entry to mitosis until sister chromatids are sufficiently decatenated. A test of this hypothesis indicated that chromatid catenations did not induce G2 arrest in human fibroblasts. Further biochemical analysis demonstrated that ICRF-193 inhibited topoisomerase IIα in a conformation with phospho-ser1524 accessible for interaction with MDC1. Binding of MDC1 was required for ICRF-193-induced G2 arrest. As MDC1 binds ATM and ATM is both activated by ICRF-193 and required for ICRF-193-induced G2 arrest, a current model holds that the decatenation G2 checkpoint involves activation of ATM at sites of catalytically inhibited topoisomerase IIα.

Both ICRF drugs cause cells to delay progression from G2 to mitosis. The G2 delay was established as an active checkpoint when various loss-of-function mutations or genetic alterations were shown to reverse the delay. The decatenation G2 checkpoint requires signaling by ATM, BRCA1, and WRN to block cellular entry to mitosis. Override of decatenation G2 checkpoint function using caffeine induces chromosome aberrations (breaks and exchanges) as mitotic cells attempt to condense and segregate knotted and linked chromatids.

The transition from G2 to mitosis (▸ G2/M Transition) is controlled by mitosis-promoting factor, a ▸ cyclin-dependent kinase made up of a catalytic subunit, CDK1, and a regulatory subunit, cyclin B1. Cyclin B1/CDK1 complexes can initiate all steps of mitosis including nuclear envelope breakdown, chromosome condensation, and formation of the mitotic spindle apparatus. Cyclin B1 is co-regulated with topoisomerase IIα, with peak levels of expression in G2 and mitosis. The activity of mitosis-promoting factor is regulated by phosphorylation/dephosphorylation. Inhibitory phosphates in the enzyme catalytic site are removed at the onset of mitosis by CDC25B/C. Cyclin B1/CDK1 complexes also are actively exported from the nucleus during G2, but at the ▸ G2/M transition this exportation is inhibited, intranuclear transport is enhanced, and mitosis-promoting factor accumulates in the nucleus. The decatenation G2 checkpoint blocks cellular progression from G2 to mitosis by preventing the accumulation of mitosis-promoting factor in the nucleus. The decatenation G2 checkpoint appears to function by inhibiting the activity of the polo-like kinase PLK1 that stimulates nuclear accumulation of mitosis-promoting factor.

Clinical Implications

The decatenation G2 checkpoint blocks progression from G2 to M when topoisomerase IIα is inhibited in a closed clamp conformation. Override of decatenation G2 checkpoint function using drugs such as caffeine is expected to cause instability of chromosome numbers and structure (► Chromosomal instability) as cells attempt mitosis with intertwined and insufficiently condensed chromatids. Linked chromatids may break under the force of chromatid condensation or segregation producing deletions, amplifications, and rearrangements. Cells with defective decatenation G2 checkpoint function due to defects in ATR, ATM, BRCA1, and WRN display chromosomal instability.

The breast cancer susceptibility gene BRCA1 is not only required for decatenation G2 checkpoint function, it also regulates the decatenatory activity of DNA topoisomerase IIα. BRCA1 contains an ubiquitin-ligase activity that is expressed in the presence of the cofactor BARD1. BRCA1 regulates ► ubiquitination of topoisomerase IIα, and ubiquitination stimulates decatenatory activity of topoisomerase IIα. Thus, breast cancers with inactivation of BRCA1 may display less chromatid decatenation by topoisomerase IIα. BRCA1 also serves as a mediator in DNA damage and decatenation G2 checkpoints, enhancing phosphorylation by checkpoint kinases of their downstream targets. Breast cancers with defects in BRCA1 are less able to decatenate daughter chromatids and less able to delay mitosis when topoisomerase IIα is inhibited.

► Lung cancer lines with defects in decatenation G2 checkpoint function display hypersensitivity to killing by ICRF-187. This suggests that topoisomerase IIα catalytic inhibitors may be clinically useful for selected cancers.

Cancer stem cells represent immortal clones that drive malignant progression (Stem Cells and Cancer). Mouse embryonic stem cells and human hematopoietic stem cells when grown in cell culture display a defect in the decatenation G2 checkpoint. Because decatenation G2 checkpoint function preserves chromosomal stability during cell division, a checkpoint defect may promote ► chromosomal instability in cancer stem cells.

Cross-References

- ► Chromosomal Instability
- ► Cyclin-Dependent Kinases
- ► G2/M Transition
- ► Lung Cancer
- ► Topoisomerases

References

Bower JJ, Karaca GF, Zhou Y, Simpson DA et al (2010) Topoisomerase IIα maintains genomic stability through decatenation G(2) checkpoint signaling. Oncogene 29:4784–4799

Damelin M, Sun YE, Sodja VB et al (2005) Decatenation checkpoint deficiency in stem and progenitor cells. Cancer Cell 8:479–484

Downes CS, Clarke DJ, Mullinger AM et al (1994) A topoisomerase II-dependent G2 cycle checkpoint in mammalian cells. Nature 372:467–470

Lou Z, Minter-Dykhouse K, Chen J (2005) BRCA1 participates in DNA decatenation. Nat Struct Mol Biol 12:589–593

Luo K, Yuan J, Chen J, Lou Z (2009) Topoisomerase IIα controls the decatenation checkpoint. Nat Cell Biol 11:204–210

Decay-Accelerating Factor

Burkhard H. Brandt
Institute of Clinical Chemistry, University Medical Centre Schleswig-Holstein, Kiel, Germany

Synonyms

Antigen of the Cromer blood group; CD55; DAF; Decay-accelerating factor for complement

Definition

Decay-accelerating factor (DAF) participates in the regulation of complement system activity by accelerating the decay of the C3/C5 convertase of the classic as well as of the alternative pathway. The highly polymorphic 50–100 kDa protein facilitates complement regulation by cysteine rich complement control protein repeats (CCP). Most of the DAF isoforms are linked to the cell membrane by a glycosyl phosphatidylinositol (GPI) anchor following the CCPs.

Characteristics

DAF has been detected in all mammalians. Physiologically it is expressed in all cells contacting the complement system enclosing cells within the peripheral blood and epithelial as well as endothelial cells. Soluble DAF is detectable in plasma, tears, saliva, and urine, as well as in synovial and cerebrospinal fluids. Besides its function as a regulator of the complement system, DAF inhibits natural killer cells, is a ligand of the CD97 receptor, and is a receptor for viruses and microorganisms. DAF is also involved in the development of spermatozoa and their survival within the female genital tract.

DAF is a glycoprotein appearing as different isoforms depending partially on different post-translational glycosylation patterns and on alternative splicing. The DAF protein possesses four units for the control of complement activation following each other and that are designated as CCP 1–4. Therefore, the molecular weight of the mature protein varies between 50 and 100 kDa in different cell types.

DAF has been detected in different malignancies, e.g., CLL, CML, ALL, AML, ► colorectal cancer, ► gastric cancer, ► thyroid cancer, medullary thyroid carcinoma, malignant glioma, ► breast cancer, renal cancer, ► non-small cell lung cancer, ► ovarian cancer, ► cervical cancer, and also partially in metastases of colorectal carcinomas. Furthermore, DAF is frequently overexpressed within the stroma of colorectal tumors suggesting that DAF comes from the tumor cells and is either cleaved from the cell membrane into the environment or is secreted by the tumor cells as a soluble form. Expression of more than one DAF isoform originated from different glycosylation patterns (colon cancer), or alternative splicing (breast cancer) has been detected.

Actions of DAF in Cancer

DAF expression in cancer cells is upregulated by interleukins (IL-4, IL-1α, IL-1β), cytokines (TNF-α, IFN-γ), growth factors (EGF, bFGF), ► prostaglandins (E2), and complement regulatory protein protectin (CD59) (Table 1).

DAF decreases complement deposition on tumor cell membranes as well as complement-mediated lysis in melanomas, renal tumors, thyroid, lung, squamous cell, and cervical carcinoma as well as in some hematological malignancies. DAF decreases cell adhesion of T-lymphocytes to leukemic cells as well as inhibiting effect on NK cells which could impair immune surveillance of cancer cells.

The GPI-anchored and membrane-linked form of DAF is part of a signal transduction cascade. In consequence, tyrosine phosphorylation in malignant tumors functioning within signal transduction goes far beyond immune-modulating effects. For example, the tyrosine kinase p56lck participates in a signal cascade conveying active motility of breast cancer cells.

DAF has been identified as ligand of the surface receptor CD97 (EGF-TM7), which belongs to the family of class B seven-span transmembrane (TM7) receptors. Predominantly expressed in hematopoietic cells, CD 97 is ectopic expressed in human thyroid, colorectal, gastric, pancreatic, esophageal, and oral squamous cell carcinomas. CD97 promotes invasive growth, ► migration, and ► angiogenesis. DAF, being the ligand of CD97, is synthesized and secreted by colorectal carcinoma cells leading to an autocrine stimulation of invasive growth, migration, and ► metastasis of these cells. A similar mechanism takes place in breast cancer cells as HER2-positive mammary carcinoma cells showing increased transendothelial invasiveness selectively overexpress and secrete a 45 kDa splice variant of DAF (Table 1).

Decay-Accelerating Factor, Table 1 Versatile actions of decay-accelerating factor in cancer

Action	Mechanisms
▶ Carcinogenesis	Upregulation of DAF in cancer and precancerous lesions (factors regulating DAF expression: IL-4, IL-1α, EGF, PGE2, TNFα, IFNγ, thrombin, βFGF, VEGF, CD59)
Complement inhibition	Decay acceleration of the C3/C5 convertase leading to decrease in complement deposition and in complement-mediated lysis of cancer cells
Inhibition of natural killer (NK) cells	Downregulation of NK cell-mediated cell lysis
Oncogenic tyrosine kinase pathway activation	Signal transduction through GPI-anchored form of DAF (tyrosine phosphorylation of src-kinases, TCR-zeta chain, and ZAP-70)
Angiogenesis	Increase of DAF synthesis by VEGF
Invasive growth, Metastasis	Increase in DAF expression, ligand binding of DAF to CD97 (autocrine stimulation by tumor-specific DAF isoforms)
Cell migration	Ligand binding of DAF to CD97, tyrosine phosphorylation of p56lck
Survival and apoptosis	Induction of apoptosis through monoclonal SC-1-antibody against tumor-specific DAF isoform (upregulation of c-myc, activation of caspase-6 and caspase-8, cleavage of cytokeratin 18)

Perspectives in Cancer Therapy

Removal of DAF from the cell membrane or neutralization of the protein increases the intensity of a potential local inflammatory reaction as well as complement-mediated cell lysis and could also possibly improve response to special therapeutic strategies (Table 2). In consequence, DAF was used as a target for molecular cancer therapy. Monoclonal antibody SC-1 binding an isoform of DAF expressed in gastric carcinoma induced ▶ apoptosis of these cells. In first clinical trials,

Decay-Accelerating Factor, Table 2 DAF targeted therapies

Targeted cancer	Therapeutic investigational design and benefit
Stomach adenocarcinoma of diffuse type	Monoclonal antibody SC-1 against a gastric cancer-specific 82 kD isoform of DAF; induced apoptosis in primary tumors as compared to pretreatment biopsy material in up to 90% of the cases, regression of tumor mass up to 50%
Metastasizing gastric cancer in nude mice	Monoclonal antibody SC-1 against a gastric cancer-specific 82 kD isoform of DAF; reduce the number of disseminated tumor cells in the bone marrow
Non-Hodgkin lymphoma	Rituximab (chimeric anti-CD20 monoclonal antibody); enhancement of complement-dependent killing activity of rituximab, by additional application of monoclonal anti-DAF antibody
Cervical cancer	Monoclonal antibody against DAF showed the widest range of specific reactivity
Renal cancer	Bispecific antibodies binding DAF as well as renal tumor-associated antigen G250; decrease of unwanted side effects
Osteosarcoma of children	DAF as a cancer vaccine additionally applied to myelosuppressive chemotherapy; induction of T-cell proliferation 71% and antigen-specific gammaIFN secretion in 59% of the cases; vaccination was well tolerated
Melanoma xenografts in immunodeficient mice	Coxsackievirus A21 infection; rapid viral oncolysis

patients with poorly differentiated stomach adenocarcinoma of diffuse type have been treated primarily with the SC-1 antibody followed by gastrectomy and lymphadenectomy. A significant induction of apoptotic activity in primary tumors as compared to pretreatment biopsy material in up to 90% of the cases and a significant regression of tumor mass in up to 50% was observed. Application of SC-1 antibody therapy in nude mice with metastasizing gastric cancer reduced the number of disseminated tumor cells in bone marrow. Other studies could show that the complement-dependent killing activity of ▶ rituximab, a chimeric anti-CD20 monoclonal antibody, in the treatment of non-Hodgkin lymphoma cells is enhanced by additional application of monoclonal anti-DAF antibody. The DAF is also expressed in many physiological human cells. Therefore, the use of bispecific antibodies binding DAF as well as another tumor-specific antigen like the renal tumor-associated antigen G250 might be a solution to decrease unbeneficial side effects. Additional application of DAF as a cancer vaccine for ▶ osteosarcoma under myelosuppressive chemotherapy induced T-cell proliferation gammaIFN secretion. DAF as a receptor for viruses enhanced the systemic effect on metastatic melanoma cells by an infection with coxsackie-A21 viruses (Table 2).

Cross-References

- ▶ Acute Lymphoblastic Leukemia
- ▶ Acute Myeloid Leukemia
- ▶ Angiogenesis
- ▶ Apoptosis
- ▶ Breast Cancer
- ▶ Cancer Vaccines
- ▶ Carcinogenesis
- ▶ Cervical Cancers
- ▶ Chronic Lymphocytic Leukemia
- ▶ Chronic Myeloid Leukemia
- ▶ Colorectal Cancer
- ▶ Complement-Dependent Cytotoxicity
- ▶ Gastric Cancer
- ▶ HER-2/neu
- ▶ Interleukin-4
- ▶ Metastasis
- ▶ Migration
- ▶ Natural Killer Cell Activation
- ▶ Non-Small-Cell Lung Cancer
- ▶ Osteosarcoma
- ▶ Ovarian Cancer
- ▶ Papillary Thyroid Carcinoma
- ▶ Prostaglandins
- ▶ Rituximab
- ▶ Thyroid Carcinogenesis
- ▶ Tumor Necrosis Factor

References

Brandt B, Mikesch JH, Simon R et al (2005) Selective expression of a splice-variant of decay-accelerating factor (DAF) in c-erbB-2-positive mammary carcinoma cells showing increased transendothelial invasiveness. Biochem Biophys Res Commun 329:319–324

Fishelson Z, Donin N, Zell S et al (2003) Obstacles to cancer immunotherapy: expression of membrane complement regulatory proteins (mCRPs) in tumors. Mol Immunol 40:109–123

Mikesch JH, Buerger H, Simon R et al (2006) Decay-accelerating factor (CD55): a versatile acting molecule in human malignancies. Biochim Biophys Acta 1766:42–52 (Review)

Niehans GA, Cherwitz DL, Staley NA et al (1996) Human carcinomas variably express the complement inhibitory proteins CD46 (membrane cofactor protein), CD55 (decay-accelerating factor), and CD59 (protectin). Am J Pathol 149:129–142

Spendlove I, Li L, Carmichael C et al (1999) Decay accelerating factor (CD55): a target for cancer vaccines? Cancer Res 59:2282–2286

See Also

(2012) BFGF. In: Schwab M (ed) Encyclopedia of cancer, 3rd edn. Springer, Berlin/Heidelberg, p 388. doi:10.1007/978-3-642-16483-5_596

(2012) EGF. In: Schwab M (ed) Encyclopedia of cancer, 3rd edn. Springer, Berlin/Heidelberg, p 1211. doi:10.1007/978-3-642-16483-5_1824

(2012) Glioma. In: Schwab M (ed) Encyclopedia of cancer, 3rd edn. Springer, Berlin/Heidelberg, p 1557. doi:10.1007/978-3-642-16483-5_2423

(2012) IFN. In: Schwab M (ed) Encyclopedia of cancer, 3rd edn. Springer, Berlin/Heidelberg, p 1806. doi:10.1007/978-3-642-16483-5_2949

(2012) Interleukin. In: Schwab M (ed) Encyclopedia of cancer, 3rd edn. Springer, Berlin/Heidelberg, p 1892. doi:10.1007/978-3-642-16483-5_3094

(2012) Medullary thyroid carcinoma. In: Schwab M (ed) Encyclopedia of cancer, 3rd edn. Springer, Berlin/Heidelberg, pp 2199-2200. doi:10.1007/978-3-642-16483-5_3600

(2012) Non-hodgkin lymphoma. In: Schwab M (ed) Encyclopedia of cancer, 3rd edn. Springer, Berlin/Heidelberg, p 2537. doi:10.1007/978-3-642-16483-5_4110

(2012) Renal cancer. In: Schwab M (ed) Encyclopedia of cancer, 3rd edn. Springer, Berlin/Heidelberg, pp 3225–3226. doi:10.1007/978-3-642-16483-5_6575

Decay-Accelerating Factor for Complement

▶ Decay-Accelerating Factor

Decitabine

▶ 5-Aza-2′ Deoxycytidine

Decoy Receptor

Definition

Decoy receptors recognize certain growth factors (such as ▶ vascular endothelial growth factor) or ▶ cytokines with high affinity and specificity but are structurally incapable of signaling or presenting the agonist to signaling receptor complexes. They act as a molecular trap for the agonist and for signaling receptor components. A decoy receptor, or sink receptor, is a receptor that binds a ligand, inhibiting it from binding to its normal receptor. For instance, the receptor VEGFR-1 can prevent ▶ vascular endothelial growth factor (VEGF) from binding to the VEGFR-2.

Cross-References

▶ Cytokine
▶ Decoy Receptor 3
▶ Vascular Endothelial Growth Factor

Decoy Receptor 3

Shie-Liang Hsieh[1] and Wan-Wan Lin[2]
[1]Department of Microbiology and Immunology, National Yang-Ming University, Immunology Research Center, Taipei Veterans General Hospital; Genomics Research Center, Academia Sinica, Taipei, Taiwan
[2]Department of Pharmacology, College of Medicine, National Taiwan University, Taipei, Taiwan

Synonyms

DcR3

Definition

DcR3 is a member of the ▶ TNF receptor superfamily (TNFRSF6B), and the DcR3 gene is mapped to chromosome 20q13.3. DcR3 is a glycosylated protein of 300 amino acids and 33 kD. The receptor lacks a transmembrane domain and exists as soluble protein.

Characteristics

Tissue Expression and Decoy Function of DcR3

DcR3 is generally undetectable in normal tissues but is highly expressed in several human malignant tissues, such as adenocarcinomas of the esophagus, stomach, colon, rectum, pancreas and lung, glioblastomas, and lymphomas. In addition, high serum levels of DcR3 have been detected in many cancer patients. Clinical data indicate the significance of detecting serum DcR3 as a novel parameter for the diagnosis, treatment, and prognosis of malignancies. Several lines of evidence suggest a significant role for DcR3 in immune suppression and tumor progression. DcR3 has been regarded to function as decoy receptor for

Decoy Receptor 3,
Fig. 1 This figure shows
the biological activities of
DcR3. Tumor-secreted
DcR3 can neutralize FasL,
LIGHT, and TL1A and also
bind to HSPG of monocytes
to trigger signal cascades,
resulting in differentiation
of M2 macrophages,
osteoclasts, and Th2-
dominant immune response

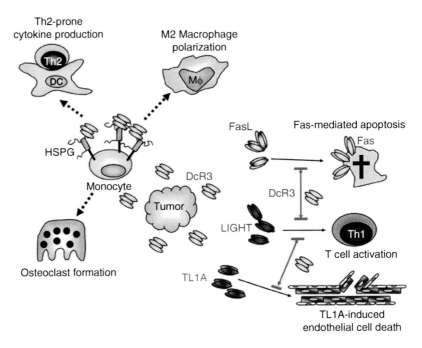

FasL, LIGHT (homologous to *l*ymphotoxins, shows *i*nducible expression and competes with herpes simplex virus *g*lycoprotein D for *h*erpesvirus entry mediator, a receptor expressed by *T* lymphocytes), and TL1A, which are three cytokine members of the TNF family. FasL and LIGHT are expressed by activated T cells and can induce tumor cell death through signaling pathways mediated by Fas and lymphotoxin β receptor (LTβR), respectively. LIGHT is also a T cell co-stimulator, and this action is mediated by receptor herpesvirus entry mediator (HVEM). Therefore, Fas and LIGHT are two cytokines contributing to the host immune surveillance. TL1A is an angiostatic cytokine releasing from endothelial cells. The cytokine neutralizing actions confer DcR3 function as a tumor molecule to decrease T-cell-mediated immunity and stimulate ▶ angiogenesis. Tumor cells engineered to release a higher amount of DcR3 may escape from FasL-induced apoptotic cell death.

Decoy Unrelated Novel Actions of DcR3
In addition to neutralizing bioactive cytokines, DcR3, which is also an effector molecule, directly modulates the activities of many cell types. DcR3 can regulate ▶ dendritic cells

(DCs) differentiation and downregulate several co-stimulatory molecules, leading to Th2 polarization (Fig. 1). Moreover, DcR3 induces actin reorganization and adhesion of monocytes as well as reducing phagocytic activity and proinflammatory cytokine production in ▶ macrophages. Osteoclast formation is promoted by the addition of DcR3 to monocyte/macrophage lineage precursor cells. Finally, DcR3 increases monocyte adhesion to endothelial cells via ▶ NF-κB activation, leading to the transcriptional up-regulation of adhesion molecules and IL-8 in endothelial cells. Thus DcR3 has pleiotropic effects to modulate inflammation and ▶ osteoporosis. As chronic ▶ inflammation has long been associated with tumorigenesis, DcR3-induced inflammation might provide a beneficial microenvironment for tumor growth. Increased ▶ osteoclast activity and decreased bone density are observed in DcR3 transgenic mice; therefore, DcR3 might also play an important role in bone erosion and destruction in cancer patients.

New Action Mechanisms of DcR3
It is a mystery why DcR3 has such diverse immunomodulatory functions independent of the neutralizing FasL, LIGHT, and TL1A (Fig. 1).

A glycosaminoglycan-binding domain of DcR3 has been identified as binding and cross-linking heparan sulfate proteoglycans (HSPG), such as syndecan 2 and CD44v3, to induce monocyte adhesion via activating PKC. Recombinant protein, comprising the HSPG-binding domain (HBD) of DcR3 and Fc portion of human IgG1 (HBD.Fc), has a similar effect as DcR3.Fc to modulate the activation and differentiation of DCs, macrophages, and induce osteoclast differentiation. Even though Fc stabilizes dimeric DcR3 and enhances cross-linking activity, transgenic mice overexpressing DcR3 also attenuates Th1 differentiation and enhances osteoclast formation. This indicates that the Fc portion is dispensable, and the biological effects of DcR3 in vivo are not restricted to its neutralizing effects on FasL, LIGHT, and TL1A.

Cross-References

▶ Angiogenesis
▶ Dendritic Cells
▶ Inflammation
▶ Macrophages
▶ Nuclear Factor-κB
▶ Osteoclast
▶ Osteoporosis
▶ Tumor Necrosis Factor

References

Chang YC, Chan YH, Jackson DG et al (2006) The glycosaminoglycan binding domain of decoy receptor 3 is essential for induction of monocyte adhesion. J Immunol 176:173–180
Chang YC, Hsu TL, Lin HH et al (2004) Modulation of macrophage differentiation and activation by decoy receptor 3. J Leukoc Biol 75:486–494
Hsu TL, Chang YC, Chen SJ et al (2002) Modulation of dendritic cell differentiation and maturation by decoy receptor 3. J Immunol 168:4846–4853
Yang CR, Hsieh SL, Teng CM et al (2004) Soluble decoy receptor 3 induces angiogenesis by neutralization of TL1A, a cytokine belonging to TNF superfamily and exhibiting angiostatic action. Cancer Res 64:1122–1129
Yang CR, Hsieh SL, Ho FM et al (2005) Decoy receptor 3 increases monocyte adhesion to endothelial cells via NF-κB-dependent up-regulation of intercellular adhesion molecule-1, VCAM-1, and IL-8 expression. J Immunol 174:1647–1656

See Also

(2012) HVEM. In: Schwab M (ed) Encyclopedia of cancer, 3rd edn. Springer Berlin Heidelberg, pp 1766-1767. doi:10.1007/978-3-642-16483-5_2872

Deferasirox

Galit Granot
Felsenstein Medical Research Center, Beilinson Hospital, Sackler School of Medicine, Tel Aviv University, Petah Tikva, Israel

Synonyms

Exjade®; ICL670A

Definition

Deferasirox is a rationally designed orally administered iron chelator.

Characteristics

Pharmacological Properties

Deferasirox is a rationally designed oral iron chelator.

Deferasirox is indicated for the treatment of chronic iron overload in patients who are receiving long-term blood transfusions for conditions such as β-thalassemia and other chronic anemias. It was approved by the United States Food and Drug Administration (FDA) in November 2005. It is the first oral medication approved in the USA for this purpose.

Its low molecular weight and high lipophilicity allows the drug to be taken orally unlike deferoxamine (DFO) which has to be administered by IV route. The half-life of deferasirox is between 8 and 16 h allowing once a day dosing. Deferasirox is capable of removing iron from cells

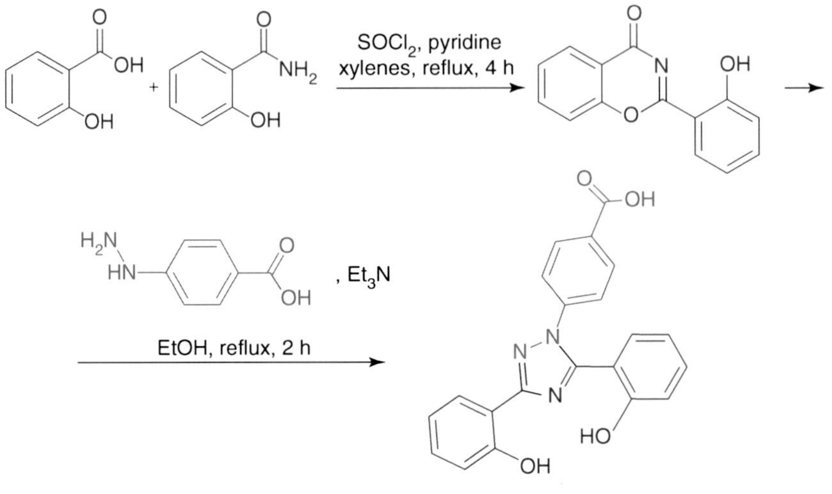

Deferasirox, Fig. 1 Synthesis of deferasirox (Tiwari. et al Journal of Analytical Sciences, Methods and InstrumentationVol.3 No.4(2013), Article ID:38319

as well as removing iron from the blood. Deferasirox is a tridentate iron chelator with high affinity for iron as Fe^{3+}; two molecules of deferasirox bind to one Fe^{3+}. This complex is excreted in bile and eliminated primarily via the feces.

Tolerability

Deferasirox is associated with a clinically manageable tolerability profile. The most common adverse events associated with deferasirox treatment include transient gastrointestinal adverse events (nausea, vomiting, abdominal pain, constipation, and diarrhea) and skin rash. Other adverse events include elevated serum creatinine and, less frequently, elevated ALT levels and visual and auditory disturbances. Regular monitoring and careful management are recommended. Postmarketing cases of renal failure and cytopenia have been reported.

Synthesis

Deferasirox is prepared from commercially available salicylic acid, salicylamide, and 4-hydrazinobenzoic acid in the following two-step synthetic sequence (Fig. 1):

The condensation of salicyloyl chloride (formed in situ from salicylic acid and thionyl chloride) with salicylamide under dehydrating reaction conditions results in formation of 2-(2-hydroxyphenyl)-1,3(4H)-benzoxazin-4-one. This intermediate is isolated and reacted with 4-hydrazinobenzoic acid in the presence of base to give 4-(3,5-bis(2-hydroxyphenyl)-1,2,4-triazol-1-yl)benzoic acid (deferasirox).

Iron

Iron is essential for life. Hemoglobin and myoglobin both contain iron; therefore, it plays a crucial role in oxygen transport and storage. Iron is equally as vital for processes such as ATP generation, effective DNA synthesis, and cell cycle progression. Iron homeostasis is strictly regulated at the level of intestinal absorption and any deviation from this tight regulation can have deleterious effects. This is best exemplified by the clinical syndrome of iron deficiency anemia, and on the other hand, since the body is incapable of effective iron excretion, excess systemic iron (as in the iron storage disease hereditary hemochromatosis or in iron overload conditions such as β-thalassemia and Friedreich's ataxia) leads to iron deposition in a number of organs including the heart, pancreas, and liver. This in turn causes irreversible tissue damage and fibrosis through the action of

► reactive oxygen species (ROS) generated by the iron. As such, excess iron has been implicated in a number of diseases including ischemic heart disease, diabetes mellitus, and neurodegenerative disorders.

Iron and Cancer

Iron-induced malignant tumors were first reported in 1959 by repeated intramuscular injection of iron dextran complex in rats. About 20 years later, sarcomas were shown to develop after intramuscular injection of iron. At that time, epidemiological reports had also associated increased iron exposure with elevated cancer risk. Although epidemiological studies on the association of iron with cancer remain inconclusive, the majority of these reports support the role of iron in human cancer. There has been data supporting the notion that iron overload is a risk factor for ► liver cancer, kidney cancer, lung cancer, ► stomach cancers, and ► colorectal cancer.

Iron Chelation in Cancer Therapy

Tumor cells in a highly proliferative state demonstrate elevated iron uptake and metabolism and have a high density of transferrin receptors. The dependency of tumor cells on high levels of iron and their sensitivity to iron depletion is believed to be the Achilles' heel of many types of cancers, and thus, it drove researchers to implement Fe chelators, initially developed to treat iron overload, as therapeutic antitumoral agents. It has been shown that iron chelators achieve their anticancer effect by targeting molecules that are critical for the regulation of both ► cell cycle and apoptotic processes. Indeed, iron chelation (mainly by DFO) has shown antiproliferative activity against leukemia and ► neuroblastoma cells in vitro, in vivo, and in ► clinical trials. Transferrin receptor antisense cDNA was shown to reduce transferrin receptor expression and subsequently to inhibit the growth of human ► breast carcinoma cells. Monoclonal antibodies against transferrin receptor severely restricted the growth of lymphoma tumors in mice. These data suggest that iron deprivation may be a useful anticancer strategy.

Deferasirox in Cancer Therapy

Given the apparent association between excess iron and cancer, there is significant interest in the investigation and development of iron-chelating drugs as antineoplastic agents.

Over the past few years, the potential for deferasirox to act as a cytotoxic agent has been investigated. At present results are predominantly limited to in vitro and in vivo laboratory studies. Human data currently comprises only small case series and anecdotal case reports.

Deferasirox has been shown to decrease cell viability, inhibit DNA replication, and induce DNA fragmentation in the human ► hepatoma cell line (HUH7) and human hepatocyte cultures. Interestingly, in contrast to other iron chelators, deferasirox also induces cell cycle blockade during S phase rather than the G1 phase. Importantly, higher concentrations of deferasirox are necessary to induce cytotoxicity in primary hepatocyte cultures compared to hepatoma cells suggesting that malignant cells may be more sensitive to deferasirox than benign ones.

Deferasirox was found to significantly reduce esophageal and lung tumor xenograft size with no marked alterations in normal tissue histology. Deferasirox was shown to increase the expression of the metastasis suppressor protein ► N-myc downstream-regulated gene 1. In addition, it was shown to upregulate the cyclin-dependent kinase inhibitor ► p21 while decreasing ► cyclin D1 levels. Moreover, this agent was shown to induce the expression of apoptosis markers, including cleaved caspase-3 and cleaved PARP.

Deferasirox and Mantle Cell Lymphoma

The biological and clinical heterogeneity of ► mantle cell lymphoma (MCL) is a substantial obstacle in treating and overcoming resistance in this disease. Several studies have represented new therapeutic approaches and strategies to target MCL by exploiting its unique biological features such as the ► mTOR and Bruton's tyrosine kinase (BTK) inhibitors, ► everolimus and ibrutinib, respectively. However, the emerging role of cancer metabolism has unveiled new potential targets to treat cancer in general and MCL in particular. As mentioned above, malignant cells exhibit an elevated proliferative potential and rapid growth rates and accordingly elevated iron uptake and metabolism. Interestingly, relapsed MCL tumors exhibit increased transferrin receptor mRNA levels when compared to primary tumors. In that concern, the antihuman transferrin receptor monoclonal antibody, A24, was shown to block MCL cell proliferation, to induce cell apoptosis, and to repress tumor growth in mice. These findings not only suggest an essential role of iron in MCL but also point out the vulnerability of MCL cells to iron deprivation making iron deprivation a useful approach for complementing current first-line therapies in this lymphoma.

Although currently DFO is considered the gold standard among iron chelation therapeutic approaches, deferasirox has shown promising results in MCL.

Deferasirox exhibited antitumoral activity against the MCL cell lines HBL-2, Granta-519, and Jeko-1. Deferasirox was shown to induce apoptosis mediated through caspase-3 activation and decreased cyclin D1 protein levels resulting from increased proteasomal degradation. In addition, exposure to deferasirox led to reduced phosphorylation of RB (Ser780), which resulted in increasing levels of the E2F/RB complex and G1/S arrest. The effect of deferasirox on these MCL cell lines was dependent on its iron-chelating ability. These data indicate that deferasirox, by downregulating cyclin D1 and inhibiting its related signals, may constitute a promising adjuvant therapeutic molecule in the strategy for MCL treatment.

Deferasirox and the PI3K/AKT/mTOR Pathway

The mTOR pathway is known to demonstrate aberrant activity in a number of cancers including ovarian, breast, colon, brain, lung and MCL. Overactivity of this pathway culminates in deregulation of the cell cycle and increased cellular proliferation. Conversely, inhibition of the mTOR pathway results in cytostatic effects. Deferasirox was shown to represses the mTOR signaling pathway in myeloid leukemia cells by enhancing the expression of REDD1. REDD1 is a stress response gene strongly induced by hypoxia. REDD1 can activate the TSC2 protein which is composed of TSC1 and TSC2. It has been shown that the major function of the TSC1/2 complex is to inhibit the checkpoint protein kinase mTOR (a major regulator of cell death and proliferation). mTOR enhances translational initiation in part by phosphorylating two major targets, the eIF4E-binding protein (4E-BPs) and the ribosomal protein S6 (S6K1 and S6K2) that cooperate to regulate translational initiation rates.

Deferasirox and the Wnt Signaling Pathway

Excessive signaling from the Wnt pathway is associated with numerous human cancers, most notably colorectal. In the absence of Wnt, cytoplasmic β-catenin protein is constantly degraded by the Axin complex, which is composed of the Axin protein, the tumor suppressor ▶ adenomatous polyposis coli gene product (APC), casein kinase 1 (CK1), and glycogen synthase kinase 3 β (GSK-3β). CK1 and GSK-3β sequentially phosphorylate β-catenin, resulting in the recognition of β-catenin by the E3 ubiquitin ligase, β-Trcp, leading to β-catenin ubiquitination and subsequent proteasomal degradation. This process prevents β-catenin from reaching the nucleus, enabling the Wnt target genes to be repressed by the DNA-bound T cell factor/lymphoid enhancer factor (TCF/LEF) family of proteins. The Wnt/β-catenin pathway is activated when a Wnt ligand binds to a seven-pass transmembrane Frizzled (Fz) receptor and its co-receptor, low-density lipoprotein receptor-related protein 5 (LRP5) or to LRP6. When this complex recruits the Dishevelled (Dvl) protein, LRP6 is phosphorylated and activated and the Axin complex is recruited to the receptors. These events lead to inhibition of Axin-mediated β-catenin phosphorylation and thereby to the stabilization of β-catenin, which accumulates and travels to the nucleus to form complexes with TCF/LEF and activates Wnt target gene expression. Deferasirox has been shown to attenuate Wnt signaling and cell growth in colorectal cancer cell lines with constitutive Wnt signaling through an iron-dependent mechanism.

Deferasirox and Nuclear Factor-κB (NF-κB)

NF-κB is a protein known to regulate several fundamental cellular processes such as apoptosis, proliferation, differentiation and migration. Incorrect regulation of NF-κB plays a pivotal role in the pathogenesis of a number of cancers. It has been shown to be constitutively active in many tumor cell lines and tissue samples. Deferasirox was found to dramatically reduce NF-κB activity in both leukemia cell lines and blood samples from patients with myelodysplastic disorders. In these experiments, the addition of ferric hydroxyquinoline during incubation did not reinstate NF-κB activity suggesting that deferasirox's effect may be independent of its iron-chelating abilities. NF-κB pathway inhibition was witnessed solely with deferasirox (it did not occur with DFO or deferiprone).

Interesting Case Report

A 73-year-old Japanese man diagnosed with acute monocytic leukemia (AML) was refractory to conventional chemotherapies. After declining further chemotherapy, he started receiving red blood cell transfusion to maintain a pretransfusional hemoglobin level. In parallel, iron chelation therapy with deferasirox was initiated. Less than 2 years later, bone marrow aspiration and biopsy revealed hematological and cytogenetic complete remission; the bone marrow was normocellular without leukemic monoblasts and myelofibrosis; cytogenetic abnormalities had disappeared in conventional karyotype analysis. Till the time when the paper was written (November 2010), normal blood cell counts have been maintained without transfusion. Remarkably, complete remission was achieved after iron chelation therapy with deferasirox in this AML patient, suggesting that deferasirox may have an antileukemic effect in the clinical setting.

Conclusion

Malignant cells exhibit an elevated proliferative potential and rapid growth rates and accordingly elevated iron uptake and metabolism. Therefore, most tumor cells are dependent on high levels of iron and are consequently sensitive to iron depletion. Deferasirox is an emerging drug that features a variety of significant biochemical, pharmaceutical and clinical advantages over other traditional iron chelators and is considered an effective antitumor agent against solid tumors, leukemias and lymphomas.

Cross-References

▶ Acute Myeloid Leukemia
▶ Cell Cycle Checkpoint
▶ Clinical Trial
▶ Colorectal Cancer
▶ Cullin Ubiquitin E3 Ligases
▶ Everolimus
▶ Gastric Cancer
▶ Hepatocellular Carcinoma
▶ Hepatocellular Carcinoma: Etiology, Risk Factors, and Prevention
▶ Mantle Cell Lymphoma
▶ Neuroblastoma
▶ N-myc Downstream-Regulated Gene
▶ P21
▶ Reactive Oxygen Species

References

Bedford MR, Ford SJ, Horniblow RD, Iqbal TH, Tselepis C (2013) Iron chelation in the treatment of cancer: a new role for deferasirox? J Clin Pharmacol 53:885–891

Chantrel-Groussard K, Gaboriau F, Pasdeloup N, Havouis R, Nick H, Pierre JL, Brissot P, Lescoat G (2006) The new orally active iron chelator ICL670A exhibits a higher antiproliferative effect in human hepatocyte cultures than O-trensox. Eur J Pharmacol 541:129–137

Fukushima T, Kawabata H, Nakamura T, Iwao H, Nakajima A, Miki M, Sakai T, Sawaki T, Fujita Y, Tanaka M, Masaki Y, Hirose Y, Umehara H (2011) Iron chelation therapy with deferasirox induced complete remission in a patient with chemotherapy-resistant acute monocytic leukemia. Anticancer Res 31:1741–1744

Huang X (2003) Iron overload and its association with cancer risk in humans: evidence for iron as a carcinogenic metal. Mutat Res 533:153–171

Vazana-Barad L, Granot G, Mor-Tzuntz R, Levi I, Dreyling M, Nathan I, Shpilberg O (2013) Mechanism of the antitumoral activity of deferasirox, an iron chelation agent, on mantle cell lymphoma. Leuk Lymphoma 54:851–859

See Also

(2001) Hepatoma. In: Schwab M (ed) Encyclopedic reference of cancer. Springer, Berlin/Heidelberg, p 401. doi:10.1007/3-540-30683-8_738

(2001) Reactive oxygen species. In: Schwab M (ed) Encyclopedic reference of cancer. Springer, Berlin/Heidelberg, pp 755–756. doi:10.1007/3-540-30683-8_1440

(2004) Retinoblastoma (Rb) gene. In: Offermanns S, Rosenthal W (eds) Encyclopedic reference of molecular pharmacology. Springer, Berlin/Heidelberg, p 815. doi:10.1007/3-540-29832-0_1404

(2005) Transferrin receptor. In: Vohr HW (ed) Encyclopedic reference of immunotoxicology. Springer, Berlin/Heidelberg, p 658. doi:10.1007/3-540-27806-0_1499

(2006) Adenomatous polyposis coli. In: Ganten D et al (eds) Encyclopedic reference of genomics and proteomics in molecular medicine. Springer, Berlin/Heidelberg, p 20. doi:10.1007/3-540-29623-9_6056

(2006) Apoptosis. In: Ganten D et al (eds) Encyclopedic reference of genomics and proteomics in molecular medicine. Springer, Berlin/Heidelberg, pp 84–85. doi:10.1007/3-540-29623-9_6191

(2006) Cyclin D. In: Ganten D et al (eds) Encyclopedic reference of genomics and proteomics in molecular medicine. Springer, Berlin/Heidelberg, p 363. doi:10.1007/3-540-29623-9_6677

(2006) Dishevelled. In: Ganten D et al (eds) Encyclopedic reference of genomics and proteomics in molecular medicine. Springer, Berlin/Heidelberg, p 409. doi:10.1007/3-540-29623-9_6774

(2008) Btk (Bruton's tyrosine kinase). In: Rédei GP (ed) Encyclopedia of genetics, genomics, proteomics and informatics. Springer, Netherlands, p 241. doi:10.1007/978-1-4020-6754-9_2087

(2008) Frizzled. In: Offermanns S, Rosenthal W (eds) Encyclopedia of molecular pharmacology, 2nd edn. Springer, Berlin/Heidelberg, p 511. doi:10.1007/978-3-540-38918-7_5730

(2008) GSK3 (glycogen synthase kinase 3β). In: Rédei GP (ed) Encyclopedia of genetics, genomics, proteomics and informatics. Springer, Netherlands, pp 827–828. doi:10.1007/978-1-4020-6754-9_7189

(2008) Iron Chelator. In: Offermanns S, Rosenthal W (eds) Encyclopedia of molecular pharmacology, 2nd edn. Springer, Berlin/Heidelberg, p 665. doi:10.1007/978-3-540-38918-7_6014

(2008) Myelofibrosis. In: Baert AL (ed) Encyclopedia of diagnostic imaging. Springer, Berlin/Heidelberg, p 1183. doi:10.1007/978-3-540-35280-8_1626

(2008) Myoglobin. In: Rédei GP (ed) Encyclopedia of genetics, genomics, proteomics and informatics. Springer, Netherlands, p 1314. doi:10.1007/978-1-4020-6754-9_11113

(2008) S6K1. In: Offermanns S, Rosenthal W (eds) Encyclopedia of molecular pharmacology, 2nd edn. Springer, Berlin/Heidelberg, p 1101. doi:10.1007/978-3-540-38918-7_6664

(2009) Iron. In: Manutchehr-Danai M (ed) Dictionary of gems and gemology. Springer, Berlin/Heidelberg, p 473. doi:10.1007/978-3-540-72816-0_11678

(2012) Anemia. In: Schwab M (ed) Encyclopedia of cancer, 3rd edn. Springer, Berlin/Heidelberg, p 178. doi:10.1007/978-3-642-16483-5_269

(2012) Axin. In: Schwab M (ed) Encyclopedia of cancer, 3rd edn. Springer, Berlin/Heidelberg, p 324. doi:10.1007/978-3-642-16483-5_496

(2012) βTrCP. In: Schwab M (ed) Encyclopedia of cancer, 3rd edn. Springer, Berlin/Heidelberg, p 3777. doi:10.1007/978-3-642-16483-5_5963

(2012) Beta-Catenin. In: Schwab M (ed) Encyclopedia of cancer, 3rd edn. Springer, Berlin/Heidelberg, p 385. doi:10.1007/978-3-642-16483-5_889

(2012) Caspase-3. In: Schwab M (ed) Encyclopedia of cancer, 3rd edn. Springer, Berlin/Heidelberg, p 675. doi:10.1007/978-3-642-16483-5_874

(2012) Cell Cycle. In: Schwab M (ed) Encyclopedia of cancer, 3rd edn. Springer, Berlin/Heidelberg, p 737. doi:10.1007/978-3-642-16483-5_994

(2012) E3 Ubiquitin ligase. In: Schwab M (ed) Encyclopedia of cancer, 3rd edn. Springer, Berlin/Heidelberg, p 1184. doi:10.1007/978-3-642-16483-5_1771

(2012) Hemoglobin. In: Schwab M (ed) Encyclopedia of cancer, 3rd edn. Springer, Berlin/Heidelberg, p 1646. doi:10.1007/978-3-642-16483-5_2632

(2012) Hepatocyte. In: Schwab M (ed) Encyclopedia of cancer, 3rd edn. Springer, Berlin/Heidelberg, p 1677. doi:10.1007/978-3-642-16483-5_2665

(2012) Hereditary hemochromatosis. In: Chen H (ed) Atlas of genetic diagnosis and counseling. Springer, pp 1025–1031. doi:10.1007/978-1-4614-1037-9_116

(2012) Leukemia. In: Schwab M (ed) Encyclopedia of cancer, 3rd edn. Springer, Berlin/Heidelberg, p 2005. doi:10.1007/978-3-642-16483-5_3322

(2012) Liver Cancer. In: Schwab M (ed) Encyclopedia of cancer, 3rd edn. Springer, Berlin/Heidelberg, p 2063. doi:10.1007/978-3-642-16483-5_3393

(2012) LRP5/6. In: Schwab M (ed) Encyclopedia of cancer, 3rd edn. Springer, Berlin/Heidelberg, p 2077. doi:10.1007/978-3-642-16483-5_3425

(2012) Lymphoma. In: Schwab M (ed) Encyclopedia of cancer, 3rd edn. Springer, Berlin/Heidelberg, p 2124. doi:10.1007/978-3-642-16483-5_3463

(2012) Medullary breast carcinoma. In: Schwab M (ed) Encyclopedia of cancer, 3rd edn. Springer, Berlin/Heidelberg, p 2199. doi:10.1007/978-3-642-16483-5_3599

(2012) MTOR. In: Schwab M (ed) Encyclopedia of cancer, 3rd edn. Springer, Berlin/Heidelberg, p 2384. doi:10.1007/978-3-642-16483-5_3867

(2012) Poly(ADP-Ribose) Polymerase. In: Schwab M (ed) Encyclopedia of cancer, 3rd edn. Springer, Berlin/Heidelberg, p 2935. doi:10.1007/978-3-642-16483-5_4655

(2012) Renal Cancer. In: Schwab M (ed) Encyclopedia of cancer, 3rd edn. Springer, Berlin/Heidelberg, pp 3225–3226. doi:10.1007/978-3-642-16483-5_6575

(2012) Sarcoma. In: Schwab M (ed) Encyclopedia of cancer, 3rd edn. Springer, Berlin/Heidelberg, p 3335. doi:10.1007/978-3-642-16483-5_5161

(2012) TCF/LEF. In: Schwab M (ed) Encyclopedia of cancer, 3rd edn. Springer, Berlin/Heidelberg, p 3625. doi:10.1007/978-3-642-16483-5_5705

(2012) Wnt. In: Schwab M (ed) Encyclopedia of cancer, 3rd edn. Springer, Berlin/Heidelberg, p 3953. doi:10.1007/978-3-642-16483-5_6255

(2013) Diabetes Mellitus. In: Gebhart GF, Schmidt RF (eds) Encyclopedia of pain, 2nd edn. Springer, Berlin/Heidelberg, p 954. doi:10.1007/978-3-642-28753-4_200571

(2013) Ischemic heart disease. In: Gebhart GF, Schmidt RF (eds) Encyclopedia of pain, 2nd edn. Springer, Berlin/Heidelberg, pp 1680–1681. doi:10.1007/978-3-642-28753-4_201097

Angelini C (2009) Neurodegenerative disorders. In: Lang F (ed) Encyclopedia of molecular mechanisms of disease. Springer, Berlin/Heidelberg, pp 1453–1455. doi:10.1007/978-3-540-29676-8_248

Galanello R, Origa R (2010) Beta-thalassemia. Orphanet J Rare Dis 5:11. doi:10.1186/1750-1172-5-11

Gilbert P (1996) Friedrich's ataxia. In: Gilbert P (ed) The A-Z reference book of syndromes and inherited disorders, 2nd edn. Springer, pp 123–126. doi:10.1007/978-1-4899-6918-7_32

Deleted in Liver Cancer 1

▶ DLC1

Deleted in Malignant Brain Tumors 1

Caroline End[1], Marcus Renner[1],
Jan Mollenhauer[2] and Annemarie Poustka[1]
[1]Division of Molecular Genome Analysis, DKFZ, Heidelberg, Germany
[2]Molecular Oncology Group, University of Southern Denmark, Odense, Denmark

Synonyms

Apactin (mouse); CRP-ductin (mouse); DMBT1; Ebnerin (rat); Glycoprotein-340 (gp-340 human); H3 (rhesus monkey); Hensin (rabbit); Muclin (mouse); Salivary agglutinin (SAG, human); Vomeroglandin (mouse)

Definition

DMBT1 located at chromosome 10q26.13 2 was initially identified as a gene that shows frequent homozygous deletions in malignant ▶ brain tumors. It codes for an extracellular glycoprotein of the evolutionary highly conserved group of scavenger receptor cysteine-rich (SRCR) proteins, and it is widely expressed in human tissues with strongest expression in epithelia and associated glands. Based on genomic alterations and loss of expression in several cancer types, *DMBT1* has been proposed to play a role in tumorigenesis. The currently available data suggests two physiological functions for DMBT1: a function in innate immunity/mucosal protection and a function in epithelial/stem cell differentiation and regenerative processes.

Characteristics

The human DMBT1-locus is built up by tandem-arrayed repeats with extensive homologies in the coding and noncoding region. The locus spans a region of about 103 kb with 55 exons with the last exon coding for a putative transmembrane domain (TMD). In contrast to the DMBT1 homologues in mouse and rat, the TMD exon has not yet been found to be contained in human transcripts. DMBT1 contains a signal peptide at the N-terminus and ▶ immunohistochemistry demonstrated that it is expressed and secreted mainly by epithelial cells and glands. The mode of secretion is thought to determine its functions: lumenally secreted DMBT1 is part of the mucus (the protective layer of epithelial surfaces) and is assumed to play a role in pathogen defense, while DMBT1 secreted to the extracellular matrix (ECM) may trigger epithelial and stem cell differentiation.

The largest (wild-type) variant of DMBT1 contains 13 highly homologous (87–100% identical) SRCR domains. The SRCR domains are separated by so-called SRCR-interspersed domains (SIDs) harboring potential sites for *O*-▶ glycosylation (Fig. 1). The SRCR domains have been shown to function in pathogen binding and in mediating interactions with endogenous proteins involved in pathogen defense.

Functions or ligands for the two C1r/C1s Uegf Bmp1 (CUB) domains and the 14th SRCR

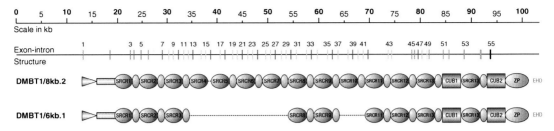

Deleted in Malignant Brain Tumors 1, Fig. 1 Genomic organization and domain structure of DMBT1. Topline represents scale in kilobases followed by the exon–intron structure of the DMBT1 gene consisting of 55 exons. The exons are colored according to the domain that they code for (see below). The presence of a further exon (*black box*) with coding potential for a transmembrane domain and a short cytoplasmic tail is predicted by homology searches with the cDNA sequences of the rodent homologues. The *bottom* two lines depict the domain organization of the wild-type protein (DMBT1/ 8 kb.2) isolated from human adult trachea and one of the 6 kb transcripts isolated from human fetal lung (DMBT1/ 6 kb.1). The two variants could represent alternative splice products, but simultaneously also correspond to the largest and the smallest allelic variants identified so far. *Pink triangle*, signal peptide; *blue box*, unknown motif; *red circles*, SRCR domains; *orange circles*, SIDs, threonine- and threonine–serine–proline-rich domain, respectively; *purple boxes*, CUB domains; *green circle*, ZP domain; *green* letters, ebnerin homologous domain

domain, which shares less homology to the 13 C-terminal ones, have not been identified so far. The second CUB domain is followed by a so-called zona pellucida (ZP) domain (Fig. 1). This type of domain mediates the formation of protein oligomers in other molecules, which is likely the case for DMBT1 as well. Genetic polymorphisms give rise to DMBT1 alleles with copy number variations resulting in a reduced number of SRCR domains and SIDs. Several DMBT1 alleles were identified coding for transcripts with sizes ranging from 6 to 8 kb (Fig. 1). A hemizygous deletion of SRCR4-SRCR7 was found in approximately 25% of the normal population.

Functional Characteristics

DMBT1 and Innate Immunity/Mucosal Protection

There is increasing evidence that DMBT1 is part of the innate immune system against bacteria and viruses. The protein is present at most mucosal surfaces and binds to a broad range of bacteria including ▶ *Helicobacter pylori*, a human pathogen, which is linked to the development of peptic ulcer and ▶ gastric cancer. Evidence that the SRCR domains could mediate the binding to bacteria has been provided by the identification of a small peptide sequence present in these domains, which exerts a broad bacteria-binding and

bacteria-agglutinating activity. However, several hints have also suggested an involvement of DMBT1-attached glycosyl residues in pathogen interactions. Furthermore, DMBT1 binds to viruses like HIV-1 and influenza A and is able to inhibit viral infection in vitro. Besides its interactions with microorganisms, DMBT1 is known to interact with mucosal and circulating defense factors among these surfactant proteins A and D, (s) IgA, ▶ trefoil factor 2, lactoferrin, and complement component C1q. This suggests a potential cooperative action of DMBT1 and its endogenous ligands in host defense. Further DMBT1 may activate the complement cascade and ▶ macrophage chemokinesis and regulate neutrophil respiratory burst response in vitro. Upregulation of DMBT1 has been reported in inflammatory processes, for example, following tissue injury or infection. Except for a role in innate mucosal protection, DMBT1 expression in lymphoid organs may point to a participation in the regulation of the acquired immune system, as has been reported for other proteins of the SRCR superfamily.

DMBT1 and Epithelial/Stem Cell Differentiation and Regenerative Processes

Functional studies in vitro and in vivo suggest a role for DMBT1 and its rodent homologues in epithelial/stem cell differentiation. The rabbit homologue of DMBT1 (hensin) was found to be responsible for the switch of cell polarity and the induction of terminal differentiation of intercalated epithelial cells in the kidney. These processes are triggered by the deposition and polymerization of hensin in the extracellular matrix. Localization to the ECM in human adult multilayered and fetal developing epithelia, and expression along the proliferation–differentiation axis in the intestine, may support such function. Rabbit DMBT1 has further been reported to trigger in vitro the differentiation of mouse embryonic stem cells toward columnar epithelial cells. DMBT1-deficient mice (termed hensin−/− mice) have been reported to display early embryonic lethality. Based on the upregulation of rat DMBT1 in liver stem cells during liver regeneration, a participation in stem cell-mediated regenerative processes has been proposed.

DMBT1 and Tumorigenesis

Genomic alterations in *DMBT1* and a loss or a reduction of its expression in several cancer types have led to the proposal that *DMBT1* might represent a ▶ tumor suppressor gene. Inactivating point mutations have not been found within *DMBT1*. Homozygous deletions (deletions concerning both copies of the gene) of/or within *DMBT1* initially have been observed in a substantial number of malignant ▶ brain tumors, ▶ lung cancer, and ▶ esophageal cancer. Due to the finding that *DMBT1* is a highly polymorphic gene with frequent intragenic deletions within the repetitive SRCR/SID exon-containing region, however, it has been proposed that the majority of the previously identified deletions may represent preexisting polymorphisms unmasked by a loss of heterozygosity of the wild-type allele. Hence, inactivation of *DMBT1* at the transcriptional/posttranscriptional level rather than genomic alterations may represent the predominant mechanism. Evidence has been gained that ▶ epigenetic silencing by DNA ▶ methylation may account for DMBT1 silencing in a smaller subset of tumors.

Contrary to other cancer types, prostate cancer and pancreatic cancer display highly elevated DMBT1 expression levels, so that DMBT1 has been discussed as potential biomarker for these cancer types. A 29 amino acid C-terminal peptide of DMBT1 has further been found at high levels in pancreatic cancer and has been suggested as tumor biomarker.

These data have led to the suggestion of a complex involvement of DMBT1 in tumorigenesis. Through protection from carcinogenic agents and/or pathogens and/or via regulation of the inflammatory response, DMBT1 may counteract tumor initiation and/or progression. Alternatively or additionally, loss of DMBT1 function may interfere with processes of differentiation and promote tumorigenesis.

References

Kang W, Reid KBM (2003) DMBT1 a regulator of mucosal homeostasis through the linking of mucosal defence and regeneration? FEBS Lett 540(1–3):21–25

Mollenhauer J, Wiemann S, Scheurlen W et al (1997) DMBT1, a new member of the SRCR superfamily, on chromosome 10q25.3-q26.1 is deleted in malignant brain tumours. Nat Genet 17:32–39

Mollenhauer J, Holmskov U, Wiemann S et al (1999) The genomic structure of the DMBT1 gene: evidence for a region with susceptibility to genomic instability. Oncogene 18:6233–6240

Mollenhauer J, Helmke B, Muller H et al (2002) An integrative model on the role of DMBT1 in epithelial cancer. Cancer Detect Prev 26(4):266–274

Vijayakumar S, Takito J, Gao X et al (2006) Differentiation of columnar epithelia: the hensin pathway. J Cell Sci 119(Pt 23):4797–4801

Deleted in Pancreatic Carcinoma Locus 4

Stephan A. Hahn
University of Bochum, Bochum, Germany

Synonyms

MADH4; Mother against decapentaplegic, drosophila, homolog of 4; SMA- and MAD-related protein 4; SMAD4; DPC4

Definition

Deleted in pancreatic carcinoma locus 4 (DPC4) belongs to the class of tumor suppressor genes (▶ tumor suppressor genes). It was identified within chromosomal band 18q21.1, which is frequently deleted in pancreatic carcinoma. DPC4 is a component of the transcription complex (transcriptional complex) that mediates cell surface signals to the nucleus which are initiated by transforming growth factor β [TGF-β]-related growth and differentiation factors.

Characteristics

The open reading frame of DPC4 spans 1,656 nucleotides and comprises 11 exons that code for

a 60 kD protein (552 amino acids). DPC4 belongs to the highly conserved family of Smad genes that has been identified by protein sequence homology studies. The founding member Mad (Mother against decapentaplegic) was identified in *Drosophila melanogaster*.

Mad and its homologs mediate signals from cell surface receptors to the cell nucleus (▶ Smad proteins in TGFβ signaling). Involved in this signaling cascade are serine/threonine kinase receptors of the TGFβ family that become activated upon binding of polypeptides of the TGFβ cytokine family (Fig. 1). At least 25 different cytokines from various species are currently known. They include TGF-β, activin, inhibin, bone morphogenic protein (BMP), and Müllerian-inhibiting substance and control, among others, important biological functions such as embryonic

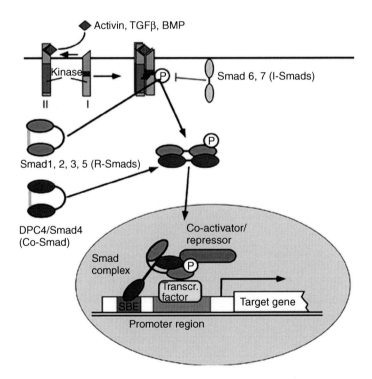

Deleted in Pancreatic Carcinoma Locus 4, Fig. 1 A model for DPC4 signaling. In response to ligand binding to the TGFβ receptor complex, receptor-regulated Smads (also called R-Smads) become C-terminally phosphorylated through the receptor kinase. The phosphorylated Smads change their folding pattern and form a heterodimeric or trimeric complex with DPC4. The newly formed Smad complex is then translocated into the

nucleus. In the nucleus, the Smad complex will make contact to transcription factors as well as bind directly to DNA through the Smad-binding element (SBE), thus stabilizing the higher order DNA-binding complex. In addition, transcriptional coactivators or corepressors may be recruited into the complex ultimately leading to either activation or repression of target gene expression

development, cell growth and cell differentiation, modulation of immune responses, and bone formation.

The number of Smad genes identified in humans has grown to a total number of eight (Table 1).

Common to all of them is a characteristic three-domain structure (Fig. 2): a highly conserved region, the Mad homology domain 1 (MH1), is located at the amino (N)-terminal end. Next to it is a poorly conserved, proline-rich linker, region which is followed by a second highly conserved

Deleted in Pancreatic Carcinoma Locus 4, Table 1 Summary of functional classes of human Smads

Classes of smad proteins		
Receptor-regulated	Common	Inhibitory
Smad 1 ⎫	DPC 4/Smad 4	Smad 6 Smad 7
Smad 5 ⎬ BMP		
Smad 8 ⎭		
Smad 2 ⎫ TGF β, Activin		
Smad 3 ⎭		

domain (MH2) located at the carboxy (C)-terminal end of the protein.

A link between the DPC4 protein and TGFβ signaling cascade was initially established by sequence homology studies of the proteins DPC4 and Mad. At the time of DPC4 discovery, its potential tumor suppressor function was hypothesized; it was believed that the loss of DPC4 function in tumors is the reason for the observed resistance towards TGFβ-mediated growth inhibition of many tumor types. Although this hypothesis has only been partly proved, a wealth of information regarding the signaling pathway has been collected. This gave rise to a model of DPC4 protein function within the TGFβ signaling pathway.

TGFβ-Smad Signaling Cascade

The Smad signaling cascade involves three different classes of Smad proteins: the receptor-regulated Smads (R-Smads), the common-mediator Smad (Co-Smad), and the inhibitory Smads (I-Smads).

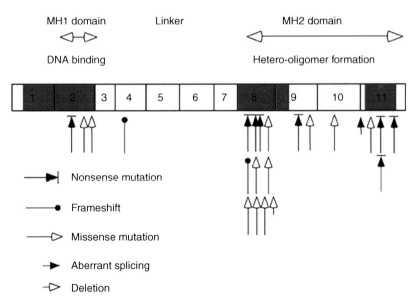

Deleted in Pancreatic Carcinoma Locus 4, Fig. 2 Functional domains and sites of identified DPC4 mutations. In addition to the mad homology domains MH1 and MH2, DPC4 carries a nuclear localization signal (NLS) domain and a nuclear export signal (NES) domain responsible for constant shuttling of DPC4 between nucleus and cytoplasm, thus helping the cell to constantly sense the TGFβ receptor activation state. A number of candidate phosphorylation target sites (P) for kinase pathways such as the MAPK (► MAP-Kinase) pathway have been described within the linker region which for example may modify nuclear accumulation rates of DPC4. The numbered squares forming the schematic DPC4 molecule also depict the 11 known exons within the DPC4 transcript

Upon ligand induced TGFβ receptor stimulation, R-Smads can transiently interact with the type I receptor. They become C-terminally phosphorylated by the receptor kinase and, once phosphorylated, are able to form a heterodimeric or trimeric complex with the "common-mediator" DPC4/Smad4 which then translocates into the nucleus. Here it can either up- or downregulate the transcription levels of target genes by interacting with other nuclear factors and by recruiting transcriptional coactivators or corepressors.

This signaling cascade can be negatively regulated by the I-Smads (Smad6 and Smad7). Whereas Smad7 acts as a more general inhibitor of TGFβ family signaling, Smad6 seems to preferentially block BMP signaling. I-Smads can compete with R-Smads for type I receptor binding and can therefore prevent the phosphorylation-dependent activation of R-Smads. Furthermore, Smad7 can interact with the E3 ubiquitin ligases Smurf 1 and 2. Once the Smad7/Smurf complex is bound to the TGFβ receptor, it induces TGFβ receptor degradation. Direct binding of I-Smads to R-Smads has also been shown, yielding R-Smads inactive. The expression of I-Smads appears to be regulated by TGFβ and BMP via an autoregulatory feedback loop. Furthermore, it has been shown that Interferon-γ (interferon-γ) (IFN-γ) via the Jak1/STAT1 pathway, and tumor necrosis factor alpha (TNFα) and interleukin 1 through NFκB (▶ Nuclear factor κB) /RelA can induce the expression of I-Smads to antagonize TGFβ signaling.

Transcriptional Regulation Through Smads

Since Smad proteins have no intrinsic enzymatic activity, they exert their effector function as transcriptional regulators either directly by binding to specific promoter consensus sequences termed Smad-binding elements (SBE) and/or indirectly by associating with transcription factors already bound to the promoter. Therefore, many but not all Smad responsive promoters have two adjacent DNA sequences. One provides the binding site for transcription factors that are cooperating with the Smad complex; the other allows direct binding of the Smad complex to the DNA. While R-Smads provide the interface for the binding to transcription factors, DPC4 makes the contact to SBE elements. DPC4 thereby stabilizes the formation of a higher order DNA-binding complex which is able to recruit transcriptional coactivators or corepressors.

Many of the factors that cooperate with the Smad complex are regulated independently by other signaling cascades. The function of an active Smad complex can therefore be described as a co-modulator of transcription. It can modulate gene expression positively as well as negatively by integrating various incoming signals, including those mediated by the TGFβ ligand family. Therefore, it is not surprising that currently more than 1,000 genes are described to be either directly or indirectly regulated by DPC4. In addition, many of these DPC4 target genes can only be found in a certain cell type and growth state, again illustrating how much the cellular differentiation and signaling state determines the net gene expression regulation pattern of a DPC4 containing transcription complex.

What Makes DPC4/Smad4 Unique Among the Other Smad Family Members?

- DPC4 is the only human Co-Smad that is currently known.
- It seems particular to DPC4 that it is, almost without exception, essential for the establishment of a functional active Smad complex, a fact that emphasizes its role as a "master switch" in the regulation of TGFβ-like signals.
- Most somatic and all germ line mutations in human Smad genes identified to date target DPC4. Only very few somatic mutations were found in the human Smad2 gene; none were found in the other members of the human Smad gene family.

Which Human Tumors Show Alterations of the DPC4 Gene?

Changes, resulting in the inactivation of the DPC4 gene, were found in approximately 50% of pancreatic carcinomas (▶ pancreas cancer). Research carried out in a variety of other cancer types suggested that DPC4 may contribute primarily to the formation of pancreatic neoplasia and to a lesser extent to ▶ colorectal cancer, ▶ cervical cancer, and biliary cancer (▶ bile duct neoplasia) as well as the induction of nonproducing ▶ neuroendocrine tumors. However, such changes appear to play only a minor role in the development of other tumor types such as head and neck cancer, ▶ lung cancer, ▶ ovarian cancer, ▶ breast cancer, and ▶ bladder cancer. The frequency of DPC4 mutations is markedly increased in metastatic colorectal carcinoma (35%) compared to nonmetastatic colorectal carcinomas (7%). Furthermore, during pancreatic carcinoma development, a high incidence of biallelic DPC4-inactivation is generally not present before the carcinoma-in-situ stage, suggesting the loss of DPC4 function is critical for the tumor cell to develop characteristics such as the ability to invade into the surrounding tissue and to form metastasis.

In addition, germ line mutations of the DPC4 gene have been identified in patients with familial juvenile polyposis, an autosomal dominant disorder that is characterized by a predisposition to hamartomatous polyps as well as an increased risk for gastrointestinal carcinomas.

How Are Naturally Occurring DPC4 Mutations Interfering with the Smad Signaling Cascade

Most DPC4 mutations identified to date are located within the C-terminal MH2 domain. Functional studies identified the MH2 domain as providing the binding properties to R-Smads, the latter being important for a functionally active Smad complex. It is therefore likely that compromising mutations of the MH2 domain structure restrict the formation of a functional Smad complex, thus preventing signal transduction to downstream components. In addition, a few mutations have been identified within the N-terminal MH1 domain which was shown to mediate the direct binding of DPC4 to DNA promoter sequences. Such mutations might interfere with Smad signaling by rendering the formation of the higher order Smad-DNA complex unstable. Furthermore, some DPC4 missense mutations targeting the MH1

domain result in an instable protein due to a mutation-induced poly-ubiquitination of DPC4 and its subsequent proteasomal degradation.

Does DPC4 Contribute to the Familial Risk for Pancreatic Cancer?

Although the DPC4 gene is most frequently altered in sporadic pancreatic carcinoma, to date no germ line mutations were found in families with an increased risk of this type of carcinoma. DPC4 is therefore unlikely to play an important role as a heritable genetic risk factor in pancreatic carcinoma.

How Does DPC4 Contribute to Tumor Formation?

Although Smad signaling (including DPC4/Smad4) is regarded as central to the TGFβ pathway, there are now numerous examples illustrating that DPC4 inactivation is not simply abolishing TGFβ responsiveness and thus providing the cell a growth advantage. This can partly be explained by the ability of TGFβ to modulate also Smad-independent pathways such as the Ras (▶ RAS)-ERK, PI3K (▶ PI3K Signaling)–AKT (▶ AKT Signal Transduction Pathway), and Rac/Rho pathways. Thus, loss of DPC4 function is not able to completely abrogate TGFβ signaling rather than shifting the balance between DPC4/Smad-dependent and DPC4/Smad-independent TGFβ signaling pathways towards the DPC4/Smad-independent pathways. The output of the latter is dependent on the successful activation of the latent form of TGFβ ligands and intactness of the TGFβ receptors. The cellular context will further modulate the signaling state of the DPC4/Smad-independent pathways through regulating the activity status of their pathway target genes by integrating signals from other signaling pathways. Thus, loss of DPC4 function has been shown in a cell specific manner to be involved in altering a number of different cell behaviors relevant to tumor formation such as, cell growth rate by modulating the cell cycle and/or the rate of ▶ apoptosis, altering the extracellular matrix components (▶ extracellular matrix remodeling), the cell adhesion (▶ adhesion) properties, and supporting epithelial to mesenchymal transition, thereby facilitating tumor ▶ invasion and ▶ metastasis. Furthermore, other experiments provided evidence that loss of DPC4 function might promote tumor ▶ angiogenesis by causing an increase in the concentration of angiogenic factors and/or a decrease in its corresponding inhibitors.

Additional insight of DPC4 function was provided by targeted mutagenesis in mice. Mice with two mutated alleles for DPC4 die at embryonic day 7.5, a result that underlines the importance of DPC4 in early embryonic development. DPC4 heterozygous mice develop gastric and duodenal polyps which resemble human juvenile polyps. Furthermore, knockout mice experiments have demonstrated a functional cooperation between the DPC4 and the APC (APC) (adenomatous polyposis coli) gene. In mice that were carrying defect copies of both genes, compared to mice carrying only the mutated APC gene, the induced colonic tumors displayed a much more aggressive phenotype. Lastly, data from primary human tumors and from mice experiments provided evidence that haploinsufficiency of the DPC4 locus may also contribute to progression of cancer. These data clearly support the importance of DPC4 in the suppression of tumorigenesis.

Cross-References

▶ Neuroendocrine Neoplasms

References

Alberici P, Jagmohan-Changur S, De Pater E et al (2005) Smad4 haploinsufficiency in mouse models for intestinal cancer. Oncogene 25:1841–1851

Hahn SA, Schutte M, Hoque AT et al (1996) DPC4, a candidate tumor suppressor gene at human chromosome 18q21.1. Science 271:350–353

Howe JR, Roth S, Ringold JC et al (1998) Mutations in the smad/dpc4 gene in juvenile polyposis. Science 280:1086–1088

Miyaki M, Iijima T, Konishi M et al (1999) Higher frequency of Smad4 gene mutation in human colorectal cancer with distant metastasis. Oncogene 18:3098–3103

Takaku K, Oshima M, Miyoshi H et al (1998) Intestinal tumorigenesis in compound mutant mice of both Dpc4 (Smad4) and Apc genes. Cell 92:645–656

Dendritic Cells

Nathalie Cools, Viggo Van Tendeloo and Zwi Berneman
Vaccine and Infections Disease Institute (VAXINFECTIO) Laboratory of Experimental Hematology, Faculty of Medicine and Health Sciences, University of Antwerp, Edegem, Belgium

Definition

Dendritic cells are a special subset of leukocytes that form a complex network of antigen-presenting cells (APC) throughout the body. They play a principal role in the initiation of immune responses to invading microorganisms (bacteria, fungi, and viruses), malignant cells, and allografts by activating naïve lymphocytes, by interaction with innate cells, and by the secretion of cytokines. At certain developmental stages they grow branched projections, the dendrites, hence the cell's name.

Characteristics

Origin and Function

Dendritic cells (DC) were characterized for the first time by Steinman in 1973 based on their distinct morphology with different cytoplasmic extensions, such as dendrites, pseudopodia, and lamellipodia, which give the cell its star-shaped feature. Due to their pronounced morphology, DC have a large surface, ensuring close contact with neighboring cells.

Variations among the tissue distribution of DC and differences in their phenotype and function indicate the existence of heterogenous populations of DC. DC originate from different hematopoietic lineages in the bone marrow (Table 1). A myeloid progenitor cell can differentiate in vivo to different DC populations: Langerhans cells that migrate to the skin epidermis and interstitial DC that migrate to the skin dermis and various other tissues (airways, liver, and intestine). Circulating, or migrating, DC are

Dendritic Cells, Table 1 Different subsets of dendritic cells

CD34+ hematopoietic stem cell				
	Myeloid progenitor cell			Lymphoid progenitor cell
	Monocyte-derived DC	Langerhans cells	Interstitial DC	Plasmacytoid DC
Phenotype				
CD11c	+	+	+	–
CD1a	±	+	–	–
CD123	–	–	–	+
Birbeck granules	–	+	–	–
Factor XIIIa	±	–	+	–
Function				
Endocytosis	+	+	+	+
IL-10*	+	+	+	+
IL-12*	+	–	+	+
IFN-α*	–	–	–	+

found in the blood and in the afferent lymphatics, respectively (the latter called veiled cells). Interdigitating DC are found in the paracortex of lymph nodes in close proximity with T cells. In addition, monocytes represent an abundant source of DC precursors during physiological stress. Another subset of DC, plasmacytoid DC (pDC) originate from a lymphoid progenitor cell in lymphoid organs. By contrast, follicular DC (FDC) are probably not of hematopoietic origin, despite similar morphology and function to the abovementioned subsets of DC. FDC are APC of the B cell follicles in lymph nodes and central players in humoral immunity.

After application of danger signals, DC express several different types of membrane molecules that determine their phenotypic and functional characteristics:

1. DC display a high surface density of antigen-presenting molecules, such as CD1a, major histocompatibility complex (MHC) class I and class II molecules. The level of expression of these molecules is 10- to 100-fold higher compared to other APC (e.g., B cells).
2. In addition, mature DC have high expression levels of costimulatory and adhesion molecules: CD40, ICAM-1/CD54, ICAM-3/CD50, LFA-3/CD58, B7-1/CD80, and B7-2/CD86. Binding of these molecules with their respective receptors on T cells results in T cell activation and subsequently stimulates the expression of cytokines, cytokine receptors, and genes for cell survival.

3. Several members of the integrin family are expressed by DC. Cadherins contribute to the generation of cell contacts and selectins are important for the motility of DC.
4. DC also express pathogen-recognition receptors, e.g., DEC-205, a macrophage-mannose receptor capable of binding bacterial carbohydrates and ► toll-like receptors (TLR), recognizing a variety of pathogen-associated molecular patterns (PAMP), such as carbohydrates, nucleic acids, peptidoglycans, and lipoteichoic acids.
5. Cytokine and chemokine receptors are also important for DC function, since growth, differentiation, and migration of DC as well as antigen processing and presentation are tightly regulated by cytokines and/or chemokines.

The widespread distribution of DC and their expression of a variety of membrane molecules underline their sentinel function: they patrol the body to capture invading pathogens and certain malignant cells in order to induce efficient antimicrobial or anti-tumor ► T cell responses. In their *in vivo* steady-state condition, immature DC are specialized in capturing antigens, i.e., they efficiently take up pathogens, apoptotic cells, and antigens from the environment by phagocytosis, macropinocytosis, or ► endocytosis. However, immature DC remain tissue resident, expressing only small amounts of (MHC) class II and of costimulatory molecules, which leads to T cell unresponsiveness. After encounter of a "danger" signal (e.g., TLR ligand) immature DC mature and migrate to the secondary lymphoid organs. Mature DC are considered to be immunogenic, mainly due to the marked upregulation of MHC class II and costimulatory molecules. This maturation step is believed to be a crucial event to regulate DC function and makes DC potent inducers of T cell immunity.

Dendritic Cell-Based Immunotherapy

Despite our immune system's function to protect us from malignant cells, tumor cells grow undisturbed and, unless treated, are fatal to the host. The reasons for the failure to eliminate tumor burden in a majority of patients can be the consequence of different tumor escape mechanisms. For example, tumor-derived inhibitory factors (e.g., PD-L1/2 IL-10 and/or TGF-β) or tumor cell-induced T regulatory cells (Treg) might be involved in downregulating or altering immune function. The goal of cancer ▸ immunotherapy is to resolve or circumvent these problems and generate tumor-specific immune responses. It is important to realize that immunotherapies will likely only be successful after reducing tumor mass via primary therapies: surgery and radio- and/or chemotherapy, i.e., in a ▸ minimal residual disease (MRD) setting.

Because of their pivotal immune-stimulatory capacity and their ability to activate naïve tumor-specific T cells, DC-based ▸ cancer vaccines could have important applications in the future treatment of cancer. For this, it was necessary to cultivate DC with high yields. Several cultivation protocols were developed for in vitro generation of DC. First, DC can be differentiated from CD34+ hematopoietic progenitor cells using granulocyte-monocyte colony stimulating factor (GM-CSF), tumor necrosis factor (TNF-α), stem cell factor (SCF), interleukin (IL)-3, and ▸ interleukin-6. Second, DC can be generated starting from monocytes using GM-CSF and ▸ Interleukin-4. Finally, DC can be directly

harvested from the peripheral blood of a patient, where they reside at low percentages (0.1%).

Next, cultivated DC can be loaded with the tumor antigen of importance in different ways:

1. DC can be grown in vitro in the presence of tumor-associated antigens (TAA). This technique is called peptide pulsing and results in direct binding of the immunodominant epitope on an empty MHC class I molecule on the DC membrane. This circumvents the need for antigen uptake and processing and ensures the stimulation of tumor-specific cell-mediated cytotoxicity. However, the number of known TAA is still restricted and highly dependent on the human leukocyte antigen (HLA) haplotype of the patient.

2. DC can also be fused with the patient's tumor cells in vitro or pulsed with tumor cell lysates. The former method combines sustained tumor antigen expression with the antigen-presenting and immunostimulatory capacities of DC. DC-tumor cell hybrids will also stimulate an active antitumoral immune response.

3. Tumor antigen can also be loaded on DC using plasmid DNA transfection or ▸ viral vector mediated gene transfer. The former method results in only low transfection efficiencies. On the other hand, viral transduction, for example by using adenoviral or lentiviral, vectors is very effective with regard to transfection efficiency. However, the immunogenic character of the viral vector itself is a serious disadvantage. In both cases, DC will transcribe and process the tumor antigen. This will result in a cytotoxic immune response, necessary for immunological defense against cancer cells.

4. It is also possible to transfect DC using in vitro transcribed mRNA coding for tumor antigens or total tumor RNA. It has been shown that electroporation of RNA is the most effective nonviral transfection method for DC (▸ Nonviral Vector for Cancer Therapy). mRNA is brought directly into the cytoplasm and the cell's metabolism will translate mRNA

into proteins, which can be presented onto MHC class I molecules after processing. This will guarantee a specific cell-mediated antitumoral immune response.

In a clinical context, in vitro cultured and activated DC loaded with appropriate tumor antigens could be administered to cancer patients in a therapeutic setting (active specific immunotherapy). The aimed generation of anti-tumor immunity, mediated by DC, could be of importance for both treatment (as adjuvant to conventional therapies) and to prevent relapse in an MRD setting. On the other hand, tumor antigen-loaded DC can also be used for the ex vivo generation of tumor-specific cytotoxic T lymphocytes (CTL) in an autologous system. These tumor-specific CTL can, in their turn, be administered to the patient to exert a direct cytotoxic effect on the patient's cancer cells (passive or adoptive immunotherapy).

The impact of a DC-based cancer vaccine is clear: an antigen-specific anti-tumor vaccine would influence both morbidity and mortality of various cancers. Currently, several phase I–II or III ► clinical trials using TAA-loaded DC are ongoing worldwide in order to stimulate the patient's immune system against tumor antigens. A number of these trials demonstrated some clinical and immunological responses (as evidenced by T cell proliferation, IFN-γ ELISPOT, and delayed type hypersensitivity [DTH] reaction) without any significant toxicity. However, despite the presence of expanded antigen-specific T cells in patients after vaccination, only a minor population of these patients showed a beneficial biologically relevant clinical response, i.e., tumor regression and increased disease-free survival. Clinical trials using DC have shown moderate success. To date, the combination of a targeted therapy exploiting the capacity of DC to stimulate the patient's own immune system against cancer with so-called immune checkpoint inhibitors is being examined in ongoing and future trials in order to eliminate tumor burden in patients.

Dendritic Cells in Cancers

DC can also infiltrate human tumors where they are involved in the induction of anti-tumor immune responses. It is likely that the establishment of tumor-specific immune responses depends on the migratory capacity of DC from the tumor microenvironment to the draining lymph nodes, where tumor antigen presentation to T cells takes place. Moreover, by their expression of costimulatory molecules and several cytokines, such as IFN-α and IL-12, DC also mediate T cell survival by preventing T cell ► apoptosis. In addition, mature DC have been reported to cause direct lysis, apoptosis, as well as cell cycle arrest of cancer cells through the secretion of soluble factors. As a consequence, the presence of a high number of DC in the tumoral or peritumoral area, as well as in the draining lymph nodes of various human tumors, has been shown to correlate with patients' survival and a better prognosis. Decreased numbers or dysfunction (e.g., decreased expression of costimulatory molecules) of DC is reported in poor-prognosis tumors. Furthermore, tumor cells can secrete certain factors (e.g., IL-10 and TGF-β) that counteract DC maturation and migration and thus actively contribute to DC dysfunction.

Occasionally, neoplasms of accessory immune cells (antigen-presenting cells, dendritic cells) can occur. These are primarily found in lymph nodes and extranodal lymphoid tissues (lymph node interdigitating cell sarcoma), but are also reported from other sites such as the skin (► Langerhans Cell Histiocytosis). The incidence of dendritic cell tumors is very rare: until now, only a few dozens of cases have been reported in literature.

Cross-References

- ► Apoptosis
- ► Cancer Vaccines
- ► Clinical Trial
- ► Endocytosis
- ► Immunotherapy
- ► Interleukin-4

► Interleukin-6
► Langerhans Cell Histiocytosis
► Macrophages
► Minimal Residual Disease
► Nonviral Vector for Cancer Therapy
► T-Cell Response
► Toll-Like Receptors
► Viral Vector-Mediated Gene Transfer

References

Banchereau J, Steinman RM (1998) Dendritic cells and the control of immunity. Nature 392:245–252

Gilboa E (2007) DC-based cancer vaccines. J Clin Invest 117:1195–1203

Lotze MT, Thomson AW (2001) Dendritic cells, 2nd edn. Academic, London

Ponsaerts P, Van Tendeloo VF, Berneman ZN (2003) Cancer immunotherapy using RNA-loaded dendritic cells. Clin Exp Immunol 134:378–384

Van Tendeloo VF, Van Broeckhoven C, Berneman ZW (2001) Gene-based cancer vaccines: an ex vivo approach. Leukemia 15:545–558

See Also

(2012) Adhesion molecules. In: Schwab M (ed) Encyclopedia of cancer, 3rd edn. Springer, Berlin/Heidelberg, p 66. doi:10.1007/978-3-642-16483-5_96

(2012) Antigen-presenting cells. In: Schwab M (ed) Encyclopedia of cancer, 3rd edn. Springer, Berlin/Heidelberg, pp 209–210. doi:10.1007/978-3-642-16483-5_321

(2012) Cadherins. In: Schwab M (ed) Encyclopedia of cancer, 3rd edn. Springer, Berlin/Heidelberg, p pp 581–582. doi:10.1007/978-3-642-16483-5_770

(2012) Delayed type hypersensitivity reaction. In: Schwab M (ed) Encyclopedia of cancer, 3rd edn. Springer, Berlin/Heidelberg, p 1073. doi:10.1007/978-3-642-16483-5_1550

(2012) ELISPOT. In: Schwab M (ed) Encyclopedia of cancer, 3rd edn. Springer, Berlin/Heidelberg, p 1217. doi:10.1007/978-3-642-16483-5_1850

(2012) Lamellipodia. In: Schwab M (ed) Encyclopedia of cancer, 3rd edn. Springer, Berlin/Heidelberg, p 1971. doi:10.1007/978-3-642-16483-5_3267

(2012) Langerhans cell. In: Schwab M (ed) Encyclopedia of cancer, 3rd edn. Springer, Berlin/Heidelberg, p 1975. doi:10.1007/978-3-642-16483-5_3272

(2012) Leukocytes. In: Schwab M (ed) Encyclopedia of cancer, 3rd edn. Springer, Berlin/Heidelberg, p 2028. doi:10.1007/978-3-642-16483-5_3330

(2012) Major histocompatibility complex. In: Schwab M (ed) Encyclopedia of cancer, 3rd edn. Springer, Berlin/Heidelberg, p 2137. doi:10.1007/978-3-642-16483-5_3500

(2012) Treg. In: Schwab M (ed) Encyclopedia of cancer, 3rd edn. Springer, Berlin/Heidelberg, p 3782. doi:10.1007/978-3-642-16483-5_5967

(2012) Tumor-associated antigen. In: Schwab M (ed) Encyclopedia of cancer, 3rd edn. Springer, Berlin/Heidelberg, pp 3807–3808. doi:10.1007/978-3-642-16483-5_6017

Dental Pulp Neoplasms

Klaus W. Neuhaus
School of Dental Medicine, Department of Preventive, Restorative and Pediatric Dentistry, University of Bern, Bern, Switzerland

Definition

Are tumors that are located in the dental pulp.

Characteristics

Dental pulp neoplasms (DPNs) are rare tumors of the dental pulp tissue which is not exposed to the oral cavity. Two types of DPNs can be distinguished: Type 1 originates from the dental pulp itself (primary DPN) and type 2 originates from tissue outside of the tooth (secondary DPN). Most DPNs are somewhat incidental findings in patients with a known tumor anamnesis. Therefore the number of histologic examples of DPNs is rather limited, and one also has to take into account articles from old literature in order to draw a complete clinical picture.

History

In the late nineteenth century, when systematic dental care and oral hygiene in general were considerably more deficient than today, dentists encountered numerous teeth with deep caries and sometimes massive exposed pulp tissue. This phenomenon was called "pulpitis chronica sarcomatosa," a chronically inflamed dental pulp supposedly caused by a sarcoma. Later it was found out that this pulpal alteration was in fact nothing to do with a sarcoma but rather was the result of colonization of the exposed dental pulp by free epithelium cells of the gums. This entity is a so-called dental pulp polyp. However, the first true description of a type 1 DPN was made in 1904 by V. A. Latham from Rogers Park, Illinois. He presented the case of a 56-year-old woman presenting with an upper right canine with a greenish-white tinge. The tooth was vital, symptomless, and caries-free, i.e., the dental pulp was not exposed to the oral environment. After tooth extraction (for prosthodontic reasons) and histological processing, this canine proved to have an epithelioma of the pulp. The extraction socket was curetted and subsequently cleaned with iodine and carbolic acid. According to his report, Latham thus seems to have cured the patient from a tumor by simply extracting the tooth. Until today, descriptions of type 1 DPNs are very rare.

First descriptions of type 2 DPNs also date back to the early twentieth century where reports of involvement of dental pulps in patients with ▶ breast cancer, lymphoma, or neuroma have been given. Three to thirty percent of tumors of the head and neck region (HNR) are associated with involvement of the dental pulp. Carcinomas are more likely to be associated with DPNs than sarcomas or any other type of tumors of the HNR. The maximum incidence of DPNs lies between the fifth and sixth decade of life.

Inflammatory Pulp Reactions

A DPN causes inflammatory reactions (▶ inflammation) in the dental pulp. Chronic inflammation of the pulp may either lead to calcification of parts of the dental pulp tissue or to resorption of the surrounding hard tissue, i.e., dentin. Calcifications – as regularly observed histological findings in dental pulps with a neoplasm – can be explained by the behavior of primary and secondary odontoblasts. These cells are determined to secrete dental hard substance. If a bacterial impact is directed toward the pulp (as is the case with dental caries), the primary odontoblasts immediately start to produce tertiary dentin in the targeted area. Thus increasing the distance between the bacteria and the pulp, an early opening of the pulp chamber in the course of the carious process is evaded. Meanwhile, the chronic

inflammation of the dental pulp in slowly progressing caries may lead to the calcification of parts of the pulpal tissue via secondary odontoblasts. These particular cells are differentiations of former pulpoblasts. Pulpoblast differentiation can be modified by bone morphogenetic protein (BMP) 2, 4, and 11, GDF, TGF-β, or high calcium concentrations, all of which are present in dentin. It can also be modified by certain medicaments, which originally were only used in periodontal regenerative therapy but have now also been introduced in endodontic therapy as well as dental traumatology as a means of direct pulp capping.

Radiotherapy

As an additional point of discussion the possibility of therapeutically induced DPNs by radiotherapy has to be mentioned. It is a given fact that one of the risks of radiotherapy of the HNR consists in radiation-induced tumors (▶ radiation-induced sarcomas after radiotherapy). Establishing a causal connection is often difficult due to a latency period of several years. However, teeth after radiotherapy sometimes show calcifications of the dental pulps, which can be detected in postirradiation radiographs. Since there has not been a study distinguishing between bacterial and abacterial calcifications as signs of chronic inflammations of the dental pulp in postirradiated cases, no predication can be given about a higher risk of DPN after radiotherapy.

Animal Investigations

As to DPN-cases, the pulp tissue reaction with respect to calcifications seems to be the same as in cases of dentin caries. At this point, observations made in experimental animal models become of interest: After several days calcification of the dental pulp tissue (with a simultaneous breakdown of the odontoblast layer) can be detected when inoculating the dental pulps of rodents with virulent sarcoma cells. Regular findings in these studies consist in massive development of intrapulpal dental hard substance like denticles, osteoids, or pulp stones. Also destruction of pulpal cells, particularly of the odontoblasts, by tumor tissue has been described in an animal study. In none of these investigations do the dental pulps survive longer

D

than 3 weeks. Nevertheless, it is a matter of speculation whether this effect is really due to the sarcoma cells or rather to the increased extravasal pressure of the inflamed pulpal tissue. In these animal models, the sarcomata are able to infiltrate the dental pulps and to proliferate to adjacent tissue like periodontium, mandibular bone, and masseteric muscle. In later stages, ▶ metastasis in the regional lymph nodes as well as in the sublingual, submandibulary, and parotid glands can be found.

The fact that rodent teeth are substantially different from human teeth must not be neglected. While rodent teeth are growing lifelong and have a largely open apex, human tooth formation literally comes to an endpoint in a constriction at the tip of the root(s).

Clinical Relevance

Since systematic autopsies of the jaws are no longer common, the entity of DPNs have somewhat moved out of the focus of scientific attention.

Type 1 DPNs are certainly of small clinical relevance. In the dental pulp, fibroblasts, subodontoblastic progenitor cells, pericytes, stem cells, and, occasionally, Malassez epithelium remainders of the Hertwig root sheath are cells with mitotic competence and thus are able to undergo neoplastic alteration. A relatively high grade of differentiation of the pulpal tissue limits further differentiation of purported neoplasms.

Due to the restricted anatomical macroenvironment of a tooth and possibly further due to microenvironmental interactions, a type 1 DPN is more or less self-limited. Concerning the formation of a DPN, the capability of the dental pulp to regularly form calcifications under certain circumstances as well as the fact that one encounters a terminal blood supply in the pulp plays a crucial role. Growth of a neoplasm will increase extravasal pressure within the dental pulp and thus stimulate secondary odontoblasts to secrete irritation dentin. A large amount of irritation dentin might influence the blood supply of the dental pulp and thus will probably lead to a hemorrhagic infarct. A growing tumor in the root canal

system will contribute to this effect. While becoming necrotic in such a way, the dental pulp does not necessarily have to show clinical symptoms (such as tooth ache). Teeth with necrotic pulps will normally receive endodontic treatment (i.e., root canal therapy) or they will be "cured" by tooth extraction. It can be acclaimed that the specialty about a type 1 DPN lies in its possibility to be removed successfully and in a relatively easy way. The anatomic prerequisite of the root canal system presents the unique fact that while growing a tumor is already limiting its further existence.

The risk of metastasis of a DPN is not given in normal-sized teeth. The volume of the dental pulp chamber and the root canal system do not provide sufficient space for a tumor to gain a critical cell mass in order to disseminate clonal cells. Only teeth with incomplete root formation (as in children or adolescents) or taurodonts, i.e., teeth with an abnormally large crown and roots, might provide enough space allowing a tumor to gain a critical cell mass. Large animal teeth, whose pulp chambers can surely provide enough space for a tumor (for instance in large mammalians), are not systematically screened for dental pulp diseases.

Type 2 DPNs seem to be mere incidental findings in patients with tumors mainly of the HNR. This leads to the assumption that DPNs are normally symptomless and of relatively small clinical relevance. Nevertheless, type 2 DPNs may also lead to tooth-related symptoms (pain) as has been described in single case reports.

Far more often (and clinically more important) is the opposite case when seemingly healthy teeth with no sign of caries, filling, or a positive trauma history mimic toothache. The projected toothache is thus drawing off the attention of a true HNR tumor, which often leads to unnecessary root canal treatment or tooth extraction. Therefore, apart from regular or ofacial neuropathic or nociceptive pain conditions, differential diagnosis therefore should always consider a neoplasm in the HNR.

In common classifications of dental pulp diseases, inflammation of the dental pulp due to neoplasms are neglected. However, animal tumor models (▶ mouse models) should be reinvestigated for changes within the dental pulp.

Cross-References

▶ Breast Cancer
▶ Inflammation
▶ Metastasis
▶ Mouse Models
▶ Radiation-Induced Sarcomas After Radiotherapy
▶ Transforming Growth Factor-Beta

References

Neuhaus KW (2007) Teeth: malignant neoplasms in the dental pulp? Lancet Oncol 8:75–78
Stewart EE, Stafne EC (1955) Involvement of the dental pulp by malignant tumors of the oral cavity. Oral Surg Oral Med Oral Pathol 8:842–855
Zajewloschin MN, Libin SI (1934) Histologische Untersuchungen der Zähne bei Neubildungen der Kiefer. Virchows Arch Pathol Anat Physiol Klin Med 293:365–380

See Also
(2012) BMP. In: Schwab M (ed) Encyclopedia of cancer, 3rd edn. Springer, Berlin/Heidelberg, p 441. doi:10.1007/978-3-642-16483-5_675
(2012) Caries. In: Schwab M (ed) Encyclopedia of cancer, 3rd edn. Springer, Berlin/Heidelberg, p 666. doi:10.1007/978-3-642-16483-5_861
(2012) Dentin. In: Schwab M (ed) Encyclopedia of cancer, 3rd edn. Springer, Berlin/Heidelberg, p 1087. doi:10.1007/978-3-642-16483-5_1560
(2012) GDF. In: Schwab M (ed) Encyclopedia of cancer, 3rd edn. Springer, Berlin/Heidelberg, p 1516. doi:10.1007/978-3-642-16483-5_2349
(2012) Microenvironment. In: Schwab M (ed) Encyclopedia of cancer, 3rd edn. Springer, Berlin/Heidelberg, p 2296. doi:10.1007/978-3-642-16483-5_3720
(2012) Odontoblasts. In: Schwab M (ed) Encyclopedia of cancer, 3rd edn. Springer, Berlin/Heidelberg, p 2598. doi:10.1007/978-3-642-16483-5_4193
(2012) Periodontium. In: Schwab M (ed) Encyclopedia of cancer, 3rd edn. Springer, Berlin/Heidelberg, p 2815. doi:10.1007/978-3-642-16483-5_4449
(2012) Taurodont. In: Schwab M (ed) Encyclopedia of cancer, 3rd edn. Springer, Berlin/Heidelberg, p 3614. doi:10.1007/978-3-642-16483-5_5685
(2012) TGF-β. In: Schwab M (ed) Encyclopedia of cancer, 3rd edn. Springer, Berlin/Heidelberg, p 3661. doi:10.1007/978-3-642-16483-5_5753

2′-Deoxy-5-azacytidine

▶ 5-Aza-2′ Deoxycytidine

Deoxyazacytidine

▶ 5-Aza-2′ Deoxycytidine

Dephosphorylating Enzyme

▶ Phosphatase

DES

▶ Diethylstilbestrol

Designer Foods

▶ Nutraceuticals

Desmoglein-2

Masakazu Yashiro
Department of Surgical Oncology, Osaka City University Graduate School of Medicine, Osaka, Japan

Synonyms

Dsg2

Definition

Dsg2 is one of the calcium-binding transmembrane glycoprotein components of the cell-cell ► adhesion molecules of the ► desmosomes. Dsg2 is one of the cadherin cell adhesion molecule superfamily in vertebrate epithelial cells.

Characteristics

Cell Junctions

Epithelial cell-cell junctions consist of four junctions: ► tight junctions, ► adherens junctions, ► desmosomes, and ► gap junctions (Fig. 1). Two adhering-type junctions, the adherens junctions and the desmosomes, are responsible for strong cell-cell adhesion. Each of these junctions consists of a transmembrane cadherin and a complex cytoplasmic plaque that serve to link cadherin to actin microfilaments or the intermediate filament cytoskeleton.

Desmosome

Intercellular junctions known as desmosomes are multimolecular membrane domains that provide intercellular adhesion and membrane anchors for the intermediate filament cytoskeleton. Desmosomes are essential adhesion structures in most epithelia that link the intermediate filament network of one cell to its neighbor, thereby forming a strong bond. Desmosomes contain the desmosomal cadherins, desmoglein (Dsg) and desmocollin (Dsc), that are linked to the intermediate filament cytoskeleton through interactions with plakoglobin and desmoplakin (Fig. 2).

Desmoglein and Cancer Epithelial cell-cell adhesion is important in tumor development. Dsgs are transmembrane glycoproteins of the desmosome, a cell-cell adhesive structure prominent in epithelial tissues, which have been reported to be associated with tumor development. cDNA and protein studies have revealed that there are subfamilies of Dsg (types 1, 2, and 3) and Dsc (types 1, 2, and 3) (Buxton et al. 1993). Dsg2 and Dsc2 are widely expressed and are found together in desmosomes of the basal layer of stratified epithelia, simple epithelia, and nonepithelial cells such as in the myocardium of the heart and lymph node follicles, whereas Dsg3/Dsc3 and Dsg1/Dsc1 are more restricted to complex epithelial tissues. Although considerable overlap is exhibited in the distribution of these isoforms in

Desmoglein-2,
Fig. 1 Cell junctions. Epithelial cell-cell junctions consist of four junctions: tight junctions, adherens junctions, desmosomes, and gap junctions

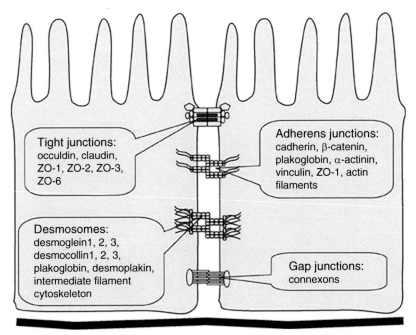

Tight junctions:
occuldin, claudin,
ZO-1, ZO-2, ZO-3,
ZO-6

Adherens junctions:
cadherin, β-catenin,
plakoglobin, α-actinin,
vinculin, ZO-1, actin
filaments

Desmosomes:
desmoglein1, 2, 3,
desmocollin1, 2, 3,
plakoglobin, desmoplakin,
intermediate filament
cytoskeleton

Gap junctions:
connexons

Desmoglein-2,
Fig. 2 Desmosomes. Desmosomes contain the desmosomal cadherins, desmoglein and desmocollin, that are linked to the intermediate filament cytoskeleton through interactions with plakoglobin and desmoplakin

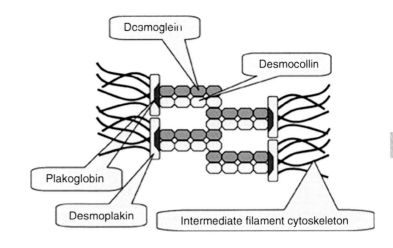

stratified tissues, their expression is clearly differentiation dependent. Dsg2, but not Dsg1 or Dsg3, is expressed in stomach epithelia. ▶ Gastric cancers have been classified into two histological types: intestinal type and diffuse type. Diffuse-type gastric cancers show decreased cell-cell adhesion, which is associated with metastatic potential. These histological features indicate that a decrease in adhesive junctions may be involved in the emergence of diffuse-type gastric cancers. A decrease in E-cadherin has been reported to be one cause of the decrease in adhesive junctions, but not all diffuse-type gastric cancers show such a decrease. Decreased expression of Dsg2 is associated with diffuse-type gastric cancers and poor prognosis in gastric carcinoma.

Adherens Junctions

The adherens junction is composed of a classic cadherin (e.g., E-, P-, or N-cadherin) linked to β-catenin or plakoglobin (Yashiro et al. 2006). Thus, plakoglobin is found in both adherens junctions and desmosomes, while β-catenin is restricted to the adherens junction. Alpha-catenin links the cadherin/catenin complex to the actin cytoskeleton through interactions with α-actinin, vinculin, ZO-1, and actin filaments. Lost or reduced plakoglobin expression has been observed in tumor tissues and metastatic lesions and has been linked to poor prognosis in a variety of tumors (Rieger-Christ et al. 2005; Holen et al. 2012; Pantel et al. 1998).

Cross-References

▶ APC/β-Catenin Pathway
▶ E-Cadherin

References

Buxton RS, Cowin P, Franke WW et al (1993) Nomenclature of the desmosomal cadherins. J Cell Biol 121:481–483

Holen I, Whitworth J, Nutter F, et al (2012) Loss of plakoglobin promotes decreased cell-cell contact, increased invasion, and breast cancer cell dissemination in vivo. Breast Cancer Res 14:R86

Jou TS, Stewart DB, Stappert J et al (1995) Genetic and biochemical dissection of protein linkages in the cadherin-catenin complex. Proc Natl Acad Sci U S A 92:5067–5071

Pantel K, Passlick B, Vogt J, et al (1998) Reduced expression of plakoglobin indicates an unfavorable prognosis in subsets of patients with non-small-cell lung cancer. J Clin Oncol 16:1407–1413

Rieger-Christ KM, Ng L, Hanley RS, et al (2005) Restoration of plakoglobin expression in bladder carcinoma cell lines suppresses cell migration and tumorigenic potential. Br J Cancer 92:2153–2159

Tselepis C, Chidgey M, North A et al (1998) Desmosomal adhesion inhibits invasive behavior. Proc Natl Acad Sci U S A 95:8064–8069

Wahl JK, Nieset JE, Sacco-Bubulya PA et al (2000) The amino- and carboxyl-terminal tails of (beta)-catenin reduce its affinity for desmoglein 2. J Cell Sci 113(Pt 10):1737–1745

Yashiro M, Nishioka N, Hirakawa K (2006) Decreased expression of the adhesion molecule desmoglein-2 is associated with diffuse-type gastric carcinoma. Eur J Cancer 42:2397–2403

Desmoid Tumor

Sue Clark
Imperial College London, London, UK

Synonyms

Aggressive fibromatosis; Gardner syndrome; Mesenteric fibromatosis

Definition

Desmoid (meaning tendon-like) tumors are a heterogeneous group of rare connective tissue neoplasms, which can occur at almost any anatomical location.

Desmoids have been classified as fibromatoses, along with pathologies such as palmar fasciitis, which are due to proliferation of well-differentiated fibroblasts and are locally infiltrative and tend to recur after excision but do not metastasize.

Characteristics

Desmoids are rare, accounting for less than 0.1% of all tumors, and have an annual incidence of two to four per million. While most occur sporadically, 2% are associated with ▶ familial adenomatous polyposis (FAP), an autosomal dominantly inherited cancer predisposition syndrome due to mutation of the tumor suppressor gene *APC* (▶ APC gene in familial adenomatous polyposis). Desmoids are over 1,000 times more common in individuals with FAP than in the population in general, occurring in about 10–20% of them, and are an important cause of death in this group.

It is useful to classify desmoid tumors as being either sporadic or FAP associated and by their location into intra-abdominal, abdominal wall, or extra-abdominal.

Pathology

Both sporadic and FAP-associated desmoids have been shown to be clonal proliferations of myofibroblasts. Those associated with FAP result from acquired mutations in the wild-type copy of *APC*. Somatic loss of the *β-catenin* gene has been described in sporadic desmoids, and *APC* mutation has also been identified in some cases ▶ APC/β-catenin pathway. Thus abnormal activation of the Wnt pathway seems to have an important role in desmoid tumorigenesis ▶ Wnt Signaling. A variety of complex chromosomal abnormalities, including trisomy 8 and gain of 1q21, has also been found in some tumors.

There is no true capsule, and the desmoid compresses and infiltrates surrounding tissues as it grows. Desmoids range in size from a few centimeters to large masses weighing several kilograms. A photograph of a mesenteric desmoid tumor taken at surgery can be found in ▶ APC gene in familial adenomatous polyposis. Growth rates are very variable. There have been reports of spontaneous resolution, and some desmoids grow relentlessly. The majority, however, either display cycles of growth and resolution or stabilize.

The cut surface is usually pale and whorled. There may be central hemorrhage, necrosis, or cystic degeneration. Histologically desmoids consist of mature, highly differentiated spindle shaped fibroblasts in an abundant collagen matrix. The histological appearances are not necessarily diagnostic and need to be interpreted in the light of the macroscopic findings.

Etiology

Trauma, sex hormones, and genetics have all been implicated in their etiology. Many sporadic abdominal wall desmoids seem to arise in women in pregnancy, perhaps as a result of low-grade trauma of stretching, coupled with high levels of female sex hormones. There have been numerous reports of desmoids arising at sites of surgical wounds, although many, particularly at extra-abdominal sites, seem to occur in the absence of any previous trauma.

The higher incidence of desmoids in females, association with pregnancy, presence of estrogen receptors, and results of some experimental studies on desmoid cell lines all suggest that estrogens may have a role in stimulating desmoid development.

Desmoids are very much more common in individuals with FAP. Within this group some familial clustering has been observed, in part explained by a genotype-phenotype correlation in which families with an *APC* mutation 3′ of codon 1444 have an attenuated colorectal phenotype but a high risk of desmoid development. There is also evidence of the influence of as yet unidentified modifier genes.

Clinical Features

Desmoids most commonly occur in young adults (mean age of onset around 30 years) but have been described in children and even babies. Sporadic desmoids are more frequent in women than men (reported gender ratio 2–5:1) but in FAP there is a less marked gender difference.

Sporadic desmoids are found predominantly in the abdominal wall (50%) and at extra-abdominal sites (40%), whereas about 80% of desmoids associated with FAP are within the abdomen, mostly in the small bowel mesentery. It is not uncommon for an individual to develop desmoids at multiple sites.

Intra-abdominal desmoids characteristically arise in the small bowel mesentery. Potential "desmoid precursor lesions," consisting of small plaques of peritoneal thickening, have been observed in patients with FAP. It is thought that these enlarge, causing a diffuse thickening and puckering of the mesentery which can be seen on CT scans. In some cases, a frank desmoid mass develops.

Most extra-abdominal desmoids cause symptoms because of their bulk and resulting mechanical effects. At some sites, for example in the neck, they can compress nerves and blood vessels. The overlying skin may ulcerate and abdominal wall desmoids occasionally adhere to and erode abdominal organs. Intra-abdominal desmoids can cause major morbidity and even death, usually due to ureteric obstruction, bowel obstruction, or perforation, either due to direct erosion or to compromise of the vascular supply.

CT and MRI are the most useful imaging modalities, showing both tumor size and relationship to neighboring structures. Signal intensity on T2 weighted MRI may reflect cellularity and is correlated with tumor growth.

Treatment

The treatment of desmoids is difficult and controversial. There are numerous case reports and small uncontrolled series in the literature, but these are difficult to interpret, particularly as the natural history of these tumors is so variable.

The drugs most widely used are ▶ nonsteroidal anti-inflammatory drugs (NSAIDs) (particularly sulindac) and antiestrogens (▶ tamoxifen or toremifene). Overall the response rates to a variety of drugs in these classes are claimed to be in the region of 50% but in reality is likely to be considerably less than this. There have been a handful of reports of acute desmoid necrosis, with abscess formation or bowel perforation in some, occurring in the weeks after initiation of drug treatment.

As NSAIDs have little in the way of adverse effects they are often used as first-line treatment. The mechanism of action in this setting is not clear, but there is some evidence that Cox-2 inhibition may inhibit desmoid growth, ▶ celecoxib, and ▶ cyclooxygenase-2 in colorectal cancer. There have been no trials of Cox 2 inhibitors used therapeutically. Antiestrogens can be used alone or in combination with NSAIDs.

Surgery is widely accepted as the first-line treatment for extra-abdominal and abdominal wall tumors. Recurrence rates are high (20–80%) but unaffected by use of prosthetic mesh in reconstruction. Serious morbidity and mortality rates are generally very low, although some sites, such as the neck, pose particular challenges. There are some reports suggesting that radiotherapy given postoperatively might reduce recurrence rates.

Excision of intra-abdominal desmoids is also associated with frequent recurrence but carries a substantial risk of perioperative mortality or major morbidity. The commonest reason for this is that the tumors lie close to or encase the superior mesenteric blood vessels, so that the blood supply of a large part of the intestine may be damaged or deliberately sacrificed during surgery. This may result in the need for lifelong parenteral nutrition and in a handful of cases small bowel transplantation has been performed in these circumstances. Careful case selection, using CT angiography and multiplanar reconstruction, together with accumulation of expertise in a specialized institution has been shown to produce better surgical results in the last 10 years. Generally, however, major resection of intra-abdominal desmoids should be avoided. Ureteric obstruction can be successfully overcome by stenting, and intestinal obstruction or fistulation may be managed in many cases, at least acutely, by defunctioning.

Cytotoxic chemotherapy has been used to treat life-threatening desmoids. Response rates of 50% have been obtained using doxorubicin and dacarbazine in combination and also with a less toxic regimen of methotrexate and vinblastine. In view of the potential toxicity of this type of treatment, it should probably be reserved for progressive, inoperable desmoid tumors in which other treatments have failed.

Cross-References

▶ Aggressive Fibromatosis in Children

References

Clark SK, Phillips RKS (1996) Desmoids in familial adenomatous polyposis. Br J Surg 83:1494–1504
Hosalkar HS, Fox EJ, Delaney T et al (2006) Desmoid tumours and current status of management. Orthop Clin North Am 37:53–63
Okuno S (2006) The enigma of desmoid tumours. Curr Treat Options Oncol 7:438–443
Reitamo JJ, Scheinin TM, Hayry P (1986) The desmoid syndrome. New aspects in the cause, pathogenesis and treatment of the desmoid tumour. Am J Surg 151:230–237
Sturt NJH, Clark SK (2006) Current ideas in desmoid tumours. Fam Cancer 5:275–285

Desmoplasia

A. Kate Sasser and Brett M. Hall
Department of Pediatrics, Columbus Children's Research Institute, The Ohio State University, Columbus, OH, USA

Synonyms

Scirrhous; Stroma; Stromal cell response

Definition

Desmoplasia is the formation of fibrous connective tissue by proliferation of fibroblasts. Desmoplasia is a key component of solid tumor stroma (Fig. 1).

Characteristics

Tumors have many parallels to wounds, including similar inflammatory and desmoplastic responses, and fibroblasts are the key cellular component in the development of desmoplasia. Fibroblasts are recruited into the wound or tumor, secrete and remodel extracellular matrix (ECM) (▶ extracellular matrix remodeling), and serve as scaffolding for other cell types in connective

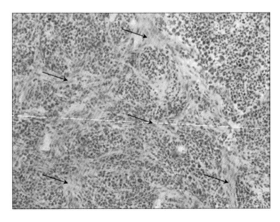

Desmoplasia, Fig. 1 Hematoxylin and eosin (*H&E*) stain. Tumor fibroblasts (i.e., desmoplasia) appear pink

tissue. As fibroblasts incorporate into a tumor environment, they undergo a phenotypic change and acquire an "activated fibroblast" appearance, which are also known as myofibroblasts or tumor-associated fibroblasts. Myofibroblasts have similar markers to fibroblasts, but myofibroblasts upregulate proteins such as α-smooth muscle actin (α-SMA), fibroblast activation protein (FAP-1), and ▶ fibronectin fibrils. During wound repair, the number of myofibroblasts returns to a normal level upon wound resolution. In contrast to wound repair, ▶ tumor microenvironments simulate a chronic wound in many ways. Thus, local fibroblasts and those that were recruited into the expanding stroma are continuously exposed to activation signals. Activated fibroblasts expand and contribute to an increased stromal response known as desmoplasia. Desmoplasia can be associated with increased tumor stage and poor prognosis in ▶ breast cancer patients, but it is unclear whether fibroblasts are *active* inducers or *passive* participants in cancer progression. It is clear, however, that activated fibroblasts play a large role in the expanding tumor stroma (▶ stromagenesis).

Fibroblastic stromal cells and desmoplasia have been linked to several activities that promote cancer growth and ▶ metastasis (▶ semaphorin) including ▶ angiogenesis, ▶ epithelial to mesenchymal transition (EMT), and progressive genetic instability. Additionally, fibroblastic stromal cells can dysregulate antitumor immune responses, as exemplified by experiments demonstrating that allogeneic murine tumor cells, when co-injected with fibroblastic stromal cells, can engraft across immunologic barriers. Together, these studies suggest that tissue-specific fibroblasts are influential players in progression of metastatic cancer. However, with the exception of promoting epithelial to mesenchymal transition, the direct biological impact on cancer cells themselves has been difficult to distinguish from indirect mechanisms such as enhanced support for angiogenesis or recruitment of inflammatory cells.

The origins of desmoplastic fibroblasts are not fully understood. Some studies have suggested that stromal cell fibroblasts are recruited to the expanding tumor mass from local tissue fibroblasts. However, other experimental evidences support that additional tumor-associated fibroblasts can be recruited from peripheral fibroblast pools, such as bone marrow-derived mesenchymal stem cells (MSC) or fibrocytes. It has been shown that once fibroblasts are recruited into the expanding stroma, they change their phenotype and may also undergo selective genetic alterations, which may drive additional tumor growth. Desmoplastic tumor fibroblasts have also been shown to carry unique genetic lesions when compared to those found in expanding tumor cells. These observations offer an additional insight into potential mechanisms for how genetic lesions can induce tumor cell expansion.

The mechanism for recruitment of desmoplastic fibroblasts into a developing tumor remains poorly defined. Yet several groups have shown that ▶ platelet-derived growth factor (PDGF) can contribute to the formation of desmoplasia. In a xenograft model using the human breast carcinoma cell line MCF-7 expressing the cellular oncogene, c-ras, investigators demonstrated that blocking tumor PDGF inhibited the formation of desmoplasia. Others have shown that blocking TGF-α, TGF-β, IGF-I, and IGF-II had no effect on the desmoplastic response. Since these models used murine xenografts, it remains unclear whether PDGF is as critical for the development of desmoplasia in human carcinomas (▶ epithelial tumorigenesis).

One important way that desmoplastic fibroblasts can contribute to tumor growth and metastasis is through the production of multiple growth factors (▶ fibroblast growth factors). Paracrine growth factors such as the stroma-derived factor 1 (SDF-1/CXCL12) (▶ angiogenesis), ▶ vascular endothelial growth factor (VEGF) (angiogenesis), ▶ fibroblast growth factor (FGF) family, hepatocyte growth factor (HGF), ▶ transforming growth factor beta (TGF-β) family, ▶ interleukin-6 (IL-6), and epidermal growth factor (EGF) have all been linked to increased tumor growth. Desmoplastic fibroblasts also contribute to tumor stroma through the production of fibrous connective tissues and extracellular matrix proteins (▶ fibronectin) (▶ focal adhesion kinase (FAK)). Collagen production is a hallmark feature of desmoplasia. As fibroblasts convert to

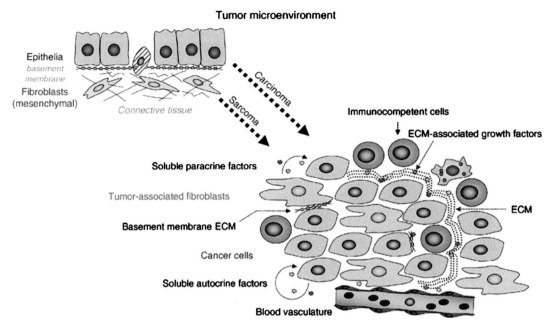

Desmoplasia, Fig. 2 The tumor microenvironment is composed of many cell types that support tumor cell growth and survival

myofibroblasts or tumor-associated fibroblasts, parallel increases in production of collagen are observed. A pathologist can readily visualize increased levels of tumor collagen using standard histology procedures (▶ pathology), and collagen types I and IV are the most prevalent forms of collagen found within most desmoplastic reactions. Collagen bundles interact with extracellular matrix and cell surface proteins such as integrins (▶ cell adhesion molecules) (focal adhesion kinase (FAK)) to influence the stiffness of a given tumor microenvironment.

Desmoplasia varies extensively between tumors and even within the same tumor. Some studies have suggested that desmoplasia is a defensive mechanism used to wall off the expanding tumor, but other data demonstrate that desmoplasia is associated with increased tumor growth, invasion, and metastasis. It is unclear, however, which underlying mechanisms determine the extent to which desmoplasia may promote tumor progression. As investigators continue to recognize the importance of the tumor microenvironment (Fig. 2), more detailed studies will allow clarification of the biological impact of desmoplasia in tumor development, survival, and metastasis.

Cross-References

▶ Angiogenesis
▶ Breast Cancer
▶ Cell Adhesion Molecules
▶ Cutaneous Desmoplastic Melanoma
▶ Epithelial-to-Mesenchymal Transition
▶ Epithelial Tumorigenesis
▶ Extracellular Matrix Remodeling
▶ Fibroblast Growth Factors
▶ Fibronectin
▶ Focal Adhesion Kinase
▶ Interleukin-6
▶ Metastasis
▶ Pathology
▶ Platelet-Derived Growth Factor
▶ Semaphorin

▶ Stem Cell Plasticity

▶ Stromagenesis

▶ Transforming Growth Factor-Beta

▶ Tumor Microenvironment

▶ Vascular Endothelial Growth Factor

References

Bhowmick NA, Neilson EG, Moses HL (2004) Stromal fibroblasts in cancer initiation and progression. Nature 432:332–337

Kunz-Schughart LA, Knuechel R (2002) Tumor-associated fibroblasts (part I): active stromal participants in tumor development and progression? Histol Histopathol 17(2):599–621

Mahadevan D, Von Hoff DD (2007) Tumor-stroma interactions in pancreatic ductal adenocarcinoma. Mol Cancer Ther 6(4):1186–1197

Walker RA (2001) The complexities of breast cancer desmoplasia. Breast Cancer Res 3:143–145

Zipori D (2006) The mesenchyme in cancer therapy as a target tumor component, effector cell modality and cytokine expression vehicle. Cancer Metastasis Rev 25:459–467

See Also

(2012) Allogeneic. In: Schwab M (ed) Encyclopedia of Cancer, 3rd edn. Springer Berlin Heidelberg, p 138. doi:10.1007/978-3-642-16483-5_194

(2012) Bone Marrow-Derived Mesenchymal Stem Cells. In: Schwab M (ed) Encyclopedia of Cancer, 3rd edn. Springer Berlin Heidelberg, p 446. doi:10.1007/978-3-642-16483-5_681

(2012) Collagen. In: Schwab M (ed) Encyclopedia of Cancer, 3rd edn. Springer Berlin Heidelberg, p 895. doi:10.1007/978-3-642-16483-5_1260

(2012) Extracellular Matrix. In: Schwab M (ed) Encyclopedia of Cancer, 3rd edn. Springer Berlin Heidelberg, p 1362. doi:10.1007/978-3-642-16483-5_2067

(2012) Fibroblasts. In: Schwab M (ed) Encyclopedia of Cancer, 3rd edn. Springer Berlin Heidelberg, p 1398. doi:10.1007/978-3-642-16483-5_2176

(2012) Genetic Instability. In: Schwab M (ed) Encyclopedia of Cancer, 3rd edn. Springer Berlin Heidelberg, pp 1527–1528. doi:10.1007/978-3-642-16483-5_2380

(2012) HGF In: Schwab M (ed) Encyclopedia of Cancer, 3rd edn. Springer Berlin Heidelberg, p 1693. doi:10.1007/978-3-642-16483-5_2710

(2012) Integrin. In: Schwab M (ed) Encyclopedia of Cancer, 3rd edn. Springer Berlin Heidelberg, p 1884. doi:10.1007/978-3-642-16483-5_3084

(2012) Microenvironment. In: Schwab M (ed) Encyclopedia of Cancer, 3rd edn. Springer Berlin Heidelberg, p 2296. doi:10.1007/978-3-642-16483-5_3720

(2012) Myofibroblasts. In: Schwab M (ed) Encyclopedia of Cancer, 3rd edn. Springer Berlin Heidelberg, pp 2440–2441. doi:10.1007/978-3-642-16483-5_3944

(2012) Paracrine. In: Schwab M (ed) Encyclopedia of Cancer, 3rd edn. Springer Berlin Heidelberg, p 2783. doi:10.1007/978-3-642-16483-5_4380

(2012) Xenograft. In: Schwab M (ed) Encyclopedia of Cancer, 3rd edn. Springer Berlin Heidelberg, p 3967. doi:10.1007/978-3-642-16483-5_6278

Desmoplastic Melanoma

▶ Cutaneous Desmoplastic Melanoma

Desmoplastic Small Round Cell Tumor

Sean Bong Lee
Department of Pathology and Laboratory Medicine, Tulane University School of Medicine, New Orleans, LA, USA

Synonyms

Malignancy of small round blue cell type; Small round cell tumor

Definition

DSRCT is a rare and highly aggressive tumor occurring mostly in the abdominal peritoneal cavity of adolescents and young adults. In rare cases, the tumors can also be found in other sites such as pleural cavity, pelvis, bone, and head and neck region. DSRCT belongs to a group of undifferentiated small round cell tumors, which include ▶ Ewing sarcoma/primitive peripheral neuroectodermal tumor (PNET)/Askin's tumor and ▶ rhabdomyosarcoma. DSRCT is invariably defined by a ▶ chromosomal translocation involving chromosomes 11 and 22, t(11;22)(p13;q12), leading to a fusion of two unrelated genes, *EWS* and *WT1*, into a single chimeric gene.

Characteristics

Clinical and Pathological Features

DSRCT was first described in 1989 and is a poorly understood cancer that primarily affects young adults in their second and third decades of life. DSRCT occurs predominantly in males than females, but the reason for this is unknown. Symptoms of DSRCT are usually associated with abdominal pain or pain in the primary site of tumor involvement, distention, and palpable mass. Local invasion or metastasis to the liver, lungs, and bone is commonly found at diagnosis. DSRCT displays distinct histological and immunological features. Most of DSRCT cases are presented as tumors in the serosal surface of abdominal cavity, displaying nests of tumor cells surrounded by dense stromal components (hence the term desmoplastic) containing spindle-shaped fibroblasts and hyperplastic blood vessels. Though rare, the primary tumors in sites other than abdominal region have been documented. The tumors are positive for various cell lineage markers, such as epithelial membrane antigen, keratin (epithelial), desmin (muscle), and neuron-specific enolase (neural). Thus, the tumor cell origin of DSRCT is not known.

DSRCT is a clinically aggressive tumor with a high risk of recurrence and an overall poor prognosis. A report on the comparison of different treatments of DSRCT patients suggests that compared to patients who received conventional treatments, a multimodal therapy, which include high-dose multiagent chemotherapy, aggressive debulking surgery, and radiotherapy, can prolong overall survival at 3 years (55%) and may provide a possibility of achieving a long-term survival, albeit at a low rate. The two key elements of the multimodal approach are the use of high-dose polychemotherapy, so-called P6 protocol, and greater than 90% removal of tumor by surgery. P6 protocol consists of seven courses of high-dose alkylating agents ▶ cyclophosphamide, doxorubicin, vincristine, ifosfamide, and ▶ etoposide.

This is followed by aggressive debulking surgery, which was shown to be the major determinant in patient survival. Postoperative radiotherapy also contributed to improved survival. Although the multimodal therapy can improve survival at 3 and 5 years, the prognosis of DSRCT still remains extremely low (median survival of 2.5 years).

Molecular Diagnosis

Although clinical, histological, and immunological features of DSRCT are distinct, a definitive diagnosis of DSRCT can be provided by genetic techniques. FISH technique, using fluorescently labeled genomic DNA probes derived from *EWS* and *WT1*, can be used to identify the specific t(11;22)(p13;q12) translocation of DSRCT. Alternatively, a definitive DSRCT diagnosis can be made with the use of reverse transcriptase-polymerase chain reaction (RT-PCR) technique to amplify and detect the DSRCT-specific EWS/WT1 hybrid mRNA transcripts using DNA primers specific for *EWS* and *WT1* genes. This is an extremely sensitive detection method that can provide accurate diagnosis with limiting tumor materials.

Molecular Genetics

Molecular genetic studies revealed that all cases of DSRCT harbor a balanced reciprocal chromosomal translocation, t(11;22)(p13;q12) (reciprocal translocation) (Fig. 1). The breakpoint in chromosome 22 has been mapped to the intron 7 of Ewing sarcoma gene, *EWS* (breakpoints in other sites of *EWS*, such as in introns 8 and 10, have also been observed in rare cases), while the other breakpoint in chromosome 11 has been invariably mapped to the intron 7 of ▶ Wilms' tumor gene *WT1*. This DSRCT-specific chromosomal translocation between *EWS* and *WT1* results in a fusion of the N-terminal domain (NTD) of *EWS* to the C-terminal DNA-binding domain of *WT1*.

EWS gene was first isolated from the Ewing sarcoma chromosomal breakpoint, where the translocation generates a fusion between *EWS* and an ETS-family transcription factor gene *FLI-1*. *EWS* encodes a putative RNA-binding protein with presumptive roles in transcription and

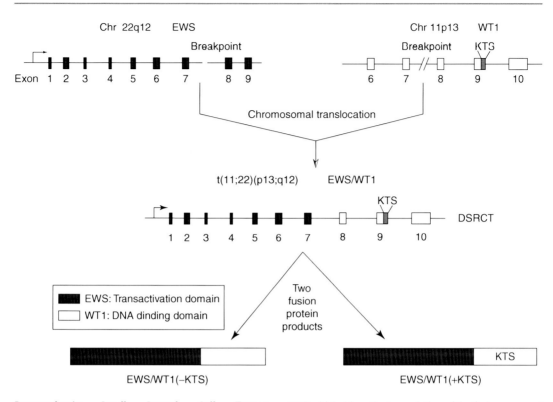

D

Desmoplastic Small Round Cell Tumor, Fig. 1 Schematic representation of DSRCT-specific chromosomal translocation. A reciprocal balanced chromosomal translocation that results in the fusion of *EWS* gene to *WT1* gene is shown. The arrow indicates the promoter of *EWS* which drives the transcription of the fusion gene and the boxes mark the exons. Alternative KTS splicing (*gray box*, KTS) within the exon 9 of *WT1* is shown. Two isoforms of the fusion product are shown separately. See text for details

splicing. The NTD of EWS mediates potent transcriptional activation when fused to a heterologous DNA-binding domain, while its C-terminal domain, which is lost in the translocation gene product, is involved in RNA recognition. *WT1* encodes a transcription factor which is mutated in a subset of Wilms' tumor, a childhood kidney cancer. *WT1* encodes four Cys_2-His_2 zinc fingers in the C terminus that mediate sequence-specific DNA binding and the NTD containing both transcriptional activation and repression domains. *WT1* is subjected to two alternative RNA splicing events, one of which involves the usage of two alternative splice donor sites at the end of exon 9, leading to inclusion or exclusion of three amino acids, lysine, threonine, and serine (termed KTS), between the zinc fingers 3 and 4 (Fig. 1). The KTS insertion leads to a markedly decreased DNA-binding affinity of WT1. In all *EWS/WT1*

translocations examined, only the last 3 exons of *WT1* (exons 8–10) encoding the last three zinc fingers are fused to the NTD of EWS (Fig. 1), while the first zinc finger of WT1 is invariably lost. The alternative KTS splicing of *WT1*, however, is preserved. As a result, *EWS/WT1* produces two isoforms, EWS/WT1(−KTS) and (+KTS), that differ in the DNA-binding affinity and specificity (Fig. 1). In vitro study has shown that only the EWS/WT1(−KTS) isoform, but not the EWS/WT1(+KTS), possesses the oncogenic activity in NIH3T3 transformation assay.

DSRCT is a rare disease and has been recognized as a distinct cancer type. Therefore, not much is known about the mechanisms of DSRCT, but molecular details are starting to emerge. The novel fusion protein EWS/WT1(-±KTS) acts as an aberrant transcription factor to presumably initiate the oncogenic process. To date, a number of direct transcriptional targets of EWS/WT1(−KTS) have been identified, which include *PDGF-A* (platelet-derived growth-factor A), *IGFR1* (insulin-like growth-factor receptor 1), *IL2RB* (interleukin 2 receptor beta), *BAIAP3* (BAI1-associated protein 3), a potential regulator of growth-factor release, and *TALLA*-1 (T-cell acute lymphoblastic leukemia-associated antigen 1), a gene encoding a tetraspanin-family protein. There is only one target gene identified for EWS/WT1(+KTS), which is *LRRC15* (leucine-rich repeat containing 15), a gene implicated in cell invasion. All of these target genes are not transcribed by the native WT1 and thus represent EWS/WT1-specific transcripts. Identification of the EWS/WT1 target genes may provide clues to the molecular and cellular pathways that are central to DSRCT. For example, the expression of IGFR1 and IL2RB can promote proliferation and survival of the tumor cells, while the expression of PDGF-A and BAIAP3 by the tumor cells can enhance recruitment and proliferation of surrounding fibroblasts and stromal tissues, which may further enhance the growth of the tumor cells and may explain the dense stroma (desmoplastic feature) associated with DSRCT. Some of these target genes may also have diagnostic and therapeutic values, but it will require further evaluation.

Cross-References

▶ Chromosomal Translocations
▶ Cyclophosphamide
▶ Etoposide
▶ Ewing Sarcoma
▶ Fusion Genes
▶ Rhabdomyosarcoma
▶ Wilms' Tumor

References

Gerald WL, Haber DA (2005) The EWS-WT1 gene fusion in desmoplastic small round cell tumor. Semin Cancer Biol 15:197–205

Gerald WL, Rosai J (1989) Desmoplastic small round cell tumor with divergent differentiation. Pediatr Pathol 9:177–183

Gerald WL, Ladanyi M, de Alava E et al (1998) Clinical, pathologic, and molecular spectrum of tumors associated with t(11;22)(p13;q12): desmoplastic small round cell tumor and its variants. J Clin Oncol 16:3028–3036

Ladanyi M, Gerald WL (1994) Fusion of the EWS and WT1 genes in the desmoplastic small round cell tumor. Cancer Res 54:2013–2840

Lal DR, Su WT, Wolden SL et al (2005) Results of multimodal treatment for desmoplastic small round cell tumors. J Pediatr Surg 40:251–255

See Also

(2012) Alternative RNA splicing. In: Schwab M (ed) Encyclopedia of cancer, 3rd edn. Springer, Berlin/Heidelberg, p 148. doi:10.1007/978-3-642-16483-5_212

(2012) Cytoreductive surgery. In: Schwab M (ed) Encyclopedia of cancer, 3rd edn. Springer, Berlin/Heidelberg, p 1057. doi:10.1007/978-3-642-16483-5_1489

(2012) Desmoplastic. In: Schwab M (ed) Encyclopedia of cancer, 3rd edn. Springer, Berlin/Heidelberg, p 1095. doi:10.1007/978-3-642-16483-5_1581

(2012) Doxorubicin. In: Schwab M (ed) Encyclopedia of cancer, 3rd edn. Springer, Berlin/Heidelberg, p 1159. doi:10.1007/978-3-642-16483-5_1722

(2012) EWS. In: Schwab M (ed) Encyclopedia of cancer, 3rd edn. Springer, Berlin/Heidelberg, p 1352. doi:10.1007/978-3-642-16483-5_2045

(2012) NIH-3T3 cells. In: Schwab M (ed) Encyclopedia of cancer, 3rd edn. Springer, Berlin/Heidelberg, p 2520. doi:10.1007/978-3-642-16483-5_4084

(2012) P6 protocol. In: Schwab M (ed) Encyclopedia of cancer, 3rd edn. Springer, Berlin/Heidelberg, p 2752. doi:10.1007/978-3-642-16483-5_4319

(2012) Reciprocal translocation. In: Schwab M (ed) Encyclopedia of cancer, 3rd edn. Springer, Berlin/Heidelberg, p 3204. doi:10.1007/978-3-642-16483-5_4989

(2012) RT-PCR. In: Schwab M (ed) Encyclopedia of cancer, 3rd edn. Springer, Berlin/Heidelberg, p 3322. doi:10.1007/978-3-642-16483-5_5129

(2012) Surgical debulking. In: Schwab M (ed) Encyclopedia of cancer, 3rd edn. Springer, Berlin/Heidelberg, p 3575. doi:10.1007/978-3-642-16483-5_5598

(2012) Transformation. In: Schwab M (ed) Encyclopedia of cancer, 3rd edn. Springer, Berlin/Heidelberg, pp 3757–3758. doi:10.1007/978-3-642-16483-5_5913

(2012) Translocation reciprocal. In: Schwab M (ed) Encyclopedia of cancer, 3rd edn. Springer, Berlin/Heidelberg, p 3773. doi:10.1007/978-3-642-16483-5_5945

(2012) Vincristine. In: Schwab M (ed) Encyclopedia of cancer, 3rd edn. Springer, Berlin/Heidelberg, p 3908. doi:10.1007/978-3-642-16483-5_6188

(2012) WT1. In: Schwab M (ed) Encyclopedia of cancer, 3rd edn. Springer, Berlin/Heidelberg, p 3958. doi:10.1007/978-3-642-16483-5_6265

Desmoplastic Tumor Microenvironment

▶ Stromagenesis

Desmosomes

Martyn A. Chidgey
School of Cancer Sciences, University of Birmingham, Birmingham, UK

Synonyms

Macula adherens; Maculae adherentes

Definition

Desmosomes are intercellular junctions that mediate cellular ▶ adhesion and maintain tissue integrity. They are found in epithelial cells, myocardial and Purkinje fiber cells of the heart, arachnoid cells of brain meninges, and follicular dendritic cells of lymph nodes.

Characteristics

Desmosomes are localized at sites of close cell-cell contact (Fig. 1a). They are less than 1 μm in diameter, have a highly organized structure at the ultrastructural level, and act as anchoring points for intermediate filaments of the cell cytoskeleton (Fig. 1b). By linking intermediate filaments of adjacent cells, desmosomes confer structural continuity and mechanical strength on tissues. Desmosomes are particularly prevalent in tissues, such as the epidermis and heart, that experience mechanical stress. The proteins that form desmosomes belong to three families, the desmosomal cadherins, the armadillo family, and the plakin family of cytolinkers.

Desmosomal Cadherins

The desmosomal cadherins are the membrane spanning cell adhesion molecules of desmosomes. In humans, there are seven, four desmogleins (Dsg1–4) and three desmocollins (Dsc1–3). Each desmoglein and desmocollin is encoded by a distinct gene that is located in the desmosomal cadherin gene cluster on chromosome 18q21. All three desmocollin genes encode a pair of proteins, a larger "a" protein and a smaller "b" protein, that are generated by alternative splicing of mRNA. All desmosomes contain at least one desmoglein and one desmocollin and both are required for adhesion. The desmosomal cadherins show tissue-specific patterns of expression with Dsg2 (desmoglein-2 adhesion molecule) and Dsc2 ubiquitously expressed in tissues that produce desmosomes and the others largely restricted to stratified epithelial tissues. The extracellular domains of desmosomal cadherins produced by adjacent cells interact in the intercellular space. Within the cell desmosomal cadherin, cytoplasmic domains associate with armadillo proteins (Fig. 1c).

Armadillo Family

Armadillo proteins that are found in desmosomes include plakoglobin (gamma-catenin) and plakophilins. Plakoglobin is indispensable for desmosome function and interacts with desmosomal cadherins, plakophilins, and desmoplakin. Plakoglobin is also found in adherens junctions where it is interchangeable with a closely related armadillo protein, beta-catenin. In addition to its structural role in adherens junctions, beta-catenin acts as a signaling molecule in the APC/beta-catenin pathway. There is a strong possibility that plakoglobin also has a signaling function in this pathway although its role has yet to be fully defined. There are three plakophilins (PKP1–3), each of which is encoded by a distinct gene. Two PKP1 and two PKP2 isoforms are known; in each case, a shorter "a" variant and a longer "b" variant are generated by alternative splicing. The plakophilins exhibit complex tissue-specific patterns of expression, and all three show dual localization in desmosomes and in the nucleus. The

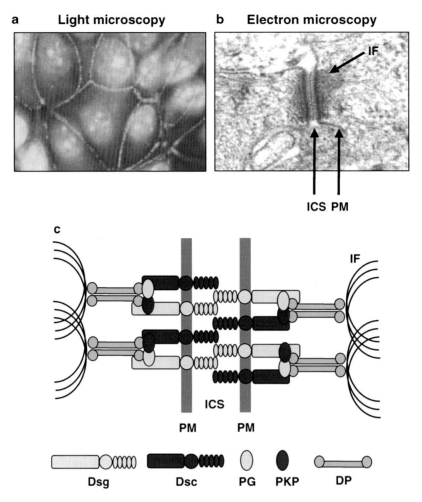

a   Light microscopy   **b   Electron microscopy**

c

Desmosomes, Fig. 1 Appearance of desmosomes by light and electron microscopy, and a schematic representation of desmosome structure. a) By light microscopy desmosomes have a punctate appearance and delineate the borders of adjacent cells. In this image desmosomes are highlighted by staining for the cytoplasmic desmosomal protein desmoplakin (green) and nuclei are stained blue. b) Desmosomes act as anchoring sites for intermediate filaments and at high magnification appear to rivet together the plasma membranes of adjacent cells. c) The minimum complement of proteins required for normal desmosomal adhesion include a desmoglein (Dsg), a desmocollin (Dsc), plakoglobin (PG), a plakophilin (PKP) and the intermediate filament (IF) binding protein desmoplakin (DP). For simplicity the Dsc 'b' protein and DPII are not shown. ICS, intercellular space; PM, plasma membrane. [Electron micrograph courtesy of M.Berika and D.Garrod, Manchester, UK]

plakophilins have an important structural role in desmosomes, and because of their nuclear localization and similarity to other armadillo proteins, it is possible that they act as signaling molecules.

Plakin Family

Plakin family proteins bind intermediate filaments and several, including desmoplakin, plectin, envoplakin and periplakin, localize to desmosomes. Of these, only the presence of desmoplakin is obligatory for normal desmosomal adhesion. It is a dumbbell-shaped molecule with two globular domains separated by a coiled-coil rod domain and is thought to exist as a homodimer. The desmoplakin gene encodes two proteins (DPI and DPII) that are generated by alternative splicing and differ only in the length of their central rod domain; the role of DPII, the smaller of these proteins, is unclear. The N-terminal end of desmoplakin binds to plakoglobin and plakophilins, whereas its C-terminal end binds to intermediate filaments. In epithelial tissues, desmoplakin anchors keratin intermediate filaments to the membrane, but in myocardial and Purkinje fiber cells, it interacts with desmin intermediate filaments, and in arachnoid and follicular dendritic cells, it associates with vimentin intermediate filaments.

Null Mutations in Mice

Genetic ablation studies in mice have shown the importance of desmosomes for embryonic development and normal tissue biology. Knock-out mice of either Dsg2, Dsc3, or desmoplakin display early embryonic lethality at around implantation or before. Mice without either plakoglobin or PKP2 survive longer but die during mid-gestation as a result of heart defects. Embryonic survival is not affected by the absence of either Dsg3, Dsg4, Dsc1, or PKP3, but loss of these molecules does result in defects in keratinocyte adhesion and skin and hair abnormalities.

Clinical Relevance

Loss of desmosomal adhesion can result in skin blistering diseases. Pemphigus is an autoimmune blistering disease that is caused by pathogenic autoantibodies against desmogleins. Staphylococcal scalded skin syndrome is caused by toxins with serine protease activity that are released by the bacterium *Staphylococcus aureus* and specifically cleave Dsg1. Mutations in DNA encoding the desmosomal proteins Dsg2, Dsc2, plakoglobin, PKP2, and desmoplakin can result in arrhythmogenic right ventricular cardiomyopathy, a heart muscle disorder associated with ventricular arrhythmias, heart failure, and sudden death. Mutations in desmosomal genes can also result in skin disorders such as palmoplantar keratoderma and hair loss.

Mutations in plakoglobin, concomitant with strong nuclear accumulation, have been linked to the pathogenesis of prostate cancer. Nuclear accumulation and improper activation of transcriptional targets as a result of a failure to degrade cytoplasmic β-catenin have been implicated in FAP (▶ APC gene in familial adenomatous polyposis), a familial syndrome that predisposes to ▶ colon cancer and sporadic colon cancer. It remains to be seen whether plakoglobin has, in common with β-catenin, pro-proliferative effects in cancer. In many cancers, loss of expression of plakoglobin has been observed, and it may be that plakoglobin is antiproliferative in some cell types. There is little doubt that plakoglobin plays a role in cancer, but whether this is related to its participation in desmosomes remains unclear.

To date, no mutations in desmosomal cadherin, plakophilin, or desmoplakin genes have been found in cancer. However, many reports have documented altered levels of expression of desmosomal proteins in carcinogenesis. Loss of expression of desmosomal cadherins, plakophilins, and desmoplakin has been reported in some types of cancer. Perhaps surprisingly, elevated expression of desmosomal cadherins and plakophilins has also been reported in certain cancers. It is difficult to come to any sort of firm conclusion about the role of desmosomes in cancer at present. However, it may be that the importance of desmosomes in cancer is twofold. Firstly, as mediators of cell-cell adhesion, reduced expression of desmosomal constituents could lead to loss of cell-cell adhesion, epithelial-mesenchymal transition, increased invasiveness, and metastasis. Secondly, desmosomes may act as signaling centers, and variations in expression levels of desmosomal proteins could trigger intracellular signaling cascades that contribute to cancer pathogenesis.

Cross-References

▶ Adherens junctions
▶ Adhesion
▶ APC/β-catenin pathway
▶ APC gene in Familial Adenomatous Polyposis
▶ Cell adhesion molecules
▶ Colorectal Cancer Clinical Oncology
▶ Cytoskeleton
▶ Desmoglein-2

References

Chidgey M, Dawson C (2007) Desmosomes: a role in cancer? Br J Cancer 96:1783–1787

Garrod D, Chidgey M (2008) Desmosome structure, composition and function. Biochim Biophys Acta 1778:572–587

Getsios S, Huen AC, Green KJ (2004) Working out the strength and flexibility of desmosomes. Nat Rev Mol Cell Biol 5:271–281

Green KJ, Simpson CL (2007) Desmosomes: new perspectives on a classic. J Invest Dermatol 127:2499–2515

Kottke MD, Delva E, Kowalczyk AP (2006) The desmosome: cell science lessons from human diseases. J Cell Sci 119:797–806

See Also

(2012) Epithelial Cell. In: Schwab M (ed) Encyclopedia of Cancer, 3rd edn. Springer, Berlin Heidelberg, pp 1291–1292. doi:10.1007/978-3-642-16483-5_1958

(2012) Plakin Family. In: Schwab M (ed) Encyclopedia of Cancer, 3rd edn. Springer, Berlin Heidelberg, p 2899. doi:10.1007/978-3-642-16483-5_4592

Detachment-Induced Cell Death

▶ Anoikis

Determination of Tumor Extent and Spread

▶ Staging of Tumors

Detoxication

▶ Detoxification

Detoxification

John D. Hayes
Medical Research Institute, Jacqui Wood Cancer Centre, University of Dundee, Dundee, UK

Synonyms

Carcinogen metabolism; Detoxication; Drug metabolism; Xenobiotic biotransformation; Xenobiotic metabolism

Definition

Metabolic and transport processes used to chemically inactivate noxious compounds and eliminate them from cells for subsequent excretion from the body.

Characteristics

Humans are continuously exposed to foreign chemicals (▶ xenobiotics) through administration of medicines, the consumption of food and drink, and air breathed. Protection against the detrimental effects of xenobiotics is achieved by the concerted actions of a battery of proteins that metabolize, transport, and ultimately pump out of cells modified forms of the compounds originally encountered. This process is called detoxification or detoxication (in instances where no toxicity occurs). Although detoxication occurs primarily in the liver, all cells possess some capacity to metabolize and eliminate unwanted chemicals. The xenobiotics subject to this process are numerous and include mycotoxins, phytoalexins, pesticides, herbicides, environmental pollutants, cytotoxic anticancer agents, and many pharmacologically active drugs. Detoxication processes also confer protection against harmful compounds of endogenous origin, many of which arise as a consequence of interaction with reactive oxygen species, such as the superoxide anion, produced normally in the body.

Detoxication is achieved in two distinct stages, the first involving metabolism of the xenobiotic and the second involving energy-dependent efflux of the xenobiotic from the cell. Historically, description of xenobiotic biotransformation has been divided into phase 1 and phase 2 metabolism, and consequently efflux of xenobiotics is referred to as phase 3 of detoxication.

- Phase 1 drug metabolism involves an initial chemical modification of the xenobiotic that results in the introduction, or exposure, of a functional chemical group (e.g., $-OH$, $-NH_2$, $-SH$, $-COOH$) into the compound. This usually entails enzyme-catalyzed oxidation reactions by ▶ cytochrome P450 (CYP) or flavin monooxygenase.

- Phase 2 drug metabolism often involves a second chemical alteration of the xenobiotic, usually at the same region of the molecule where the functional group was introduced. This is performed by enzymes catalyzing conjugation reactions (such as ▶ glutathione S-transferase (GST), *N*-acetyltransferase (NAT), ▶ sulfotransferase (SULT), and UDP-glucuronosyltransferase (UGT)). It should be noted that the use of the terms phase 1 and phase 2 to define the detoxication enzymes is somewhat arbitrary and does not necessarily reflect the pathway of biotransformation of all chemicals. Thus, a number of xenobiotics are subject to several modifications by the phase 1 CYP isoenzymes before serving as substrates for the phase 2 enzymes. Alternatively, some xenobiotics do not require modification by phase 1 enzymes before metabolism by phase 2 enzymes, and others are subject to modification by more than one phase 2 drug-metabolizing enzyme. As a result of differences in drug metabolism, the group of enzymes catalyzing reduction of hydrolysis reactions (e.g., ▶ aldehyde dehydrogenase (ADH), aldo-keto reductase (AKR), epoxide hydrolase (EPHX), and NAD(P)H-quinone oxidoreductase (NQO)) are variously referred to as phase 1 or phase 2 detoxication, depending on the individual xenobiotic being considered and the preferences of research workers. Clearly, these enzymes provide a highly flexible metabolic defense that has evolved to protect against a diverse spectrum of chemicals.
- Finally, phase 3 of detoxication involves ATP-dependent elimination of the parent compound or modified xenobiotic by proteins that are drug efflux pumps (e.g., multidrug resistance protein (MDR) and multidrug resistance-associated protein (multidrug resistance protein) (MRP)). As a consequence of the combined actions of phase 1 and phase 2 enzymes, a diverse spectrum of xenobiotics acquires a limited number of molecular "tags" (i.e., acetate, glutathione, glucuronide, or sulfate moieties) that are recognized by the MRP transmembrane pumps. Furthermore, the

xenobiotic metabolites produced by phase 1 and phase 2 are usually more soluble, and easily excreted, than the parent compound.

While the ability of CYP to oxidize xenobiotics is generally desirable, as it facilitates further metabolism and elimination of harmful chemicals, it can sometimes result in the generation of highly reactive products that may not be readily detoxified. In such instances, modification of intracellular macromolecules will occur resulting in necrosis, ▶ apoptosis, or malignant transformation. As an example of the interplay between toxification and detoxification reactions, a scheme depicting metabolism of ▶ aflatoxin B1 (AFB1), modification of macromolecules by AFB1 metabolites, and efflux of the AFB1-glutathione conjugate from a cell is shown in the illustration (Fig. 1).

Genetic Variation

Numerous proteins have evolved that detoxify drugs, and certain of the families listed above comprise over twenty genes. In total, the human probably possesses between 100 and 150 genes encoding detoxication proteins. Substantial variation can occur in the levels of these proteins in tissues from different individuals, and this can result in increased sensitivity of cells to chemical insult. In part, this interindividual variation is due to genetic polymorphisms. By definition, such differences must be present in at least 1% of the population in order to be considered a genetic polymorphism. In some instances, the variation involves deletion of detoxication genes with complete loss of specific functions, whereas in other instances, point mutations result in alteration of protein structure, causing only a modest attenuation of activity. In other cases, mutations alter the regulatory regions of genes, causing altered expression of normal protein. Detoxication genes that are polymorphic in the human include those for the enzymes CYP3A4, CYP2C9, CYP2C19, CYP2D6, CYP2E1, AKR1C4, GSTM1, GSTP1, GSTT1, NAT2, SULT1A1, SULT1E1, SULT2A1, UGT1A1, UGT1A4, UGT1A6 and UGT2B7, EPHX and NQO1, as well as the

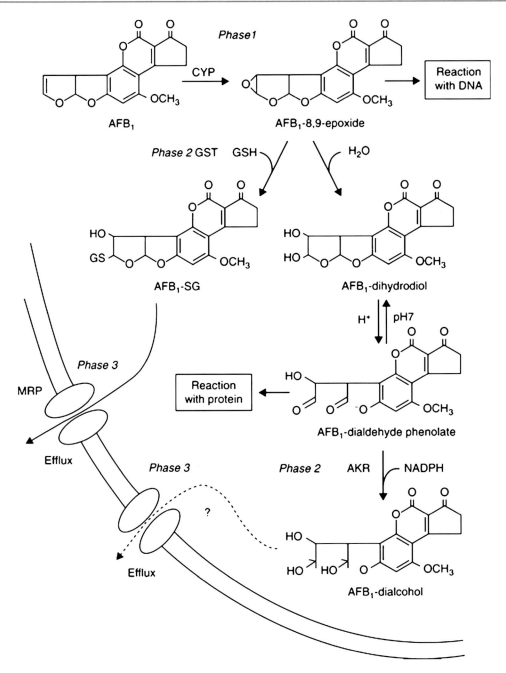

Detoxification, Fig. 1 Detoxification pathways for afla-toxin B1. The mycotoxin is converted to the ultimate carcinogen AFB1-8,9-epoxide, by the actions of the hepatic phase 1 CYP enzyme system. The epoxidated AFB1 is highly reactive, and if it is not detoxified, it will form DNA adducts that may cause hepatocarcinogenesis. The phase 2 GST enzymes can achieve detoxification of this unstable intermediate, and the resulting AFB1-glutathione conjugate is eliminated from the liver cell by MRP. In addition, AFB1-8,9-epoxide can rearrange to form a dialdehyde-containing metabolite which will cova-lently modify proteins by forming Schiff's bases. The dialdehyde can be reduced by phase 2 AKR to yield a dialcohol that may be a substrate for SULT or UGT before being transported out of the cell, presumably by MRP

MRP2 efflux pump. It is clear additional polymorphisms remain to be identified.

Cellular Regulation

In addition to genetic polymorphisms, induction of detoxication proteins by xenobiotics and environmental agents is a further mechanism that can cause interindividual differences in detoxification capacity. Induction of detoxication proteins represents an adaptive response to chemical and ► oxidative stress, which can be brought about by synthetic drugs or by naturally occurring compounds such as coumarins, indoles, and isothiocyanates that are found in edible plants. Increased expression provides short-term resistance to toxic xenobiotics. Enzyme induction also results in increased metabolism of therapeutic drugs. Many of the enzymes and pumps such as CYP, GST, ADH, AKR, NQO, and MRP are inducible, often by transcriptional activation of genes encoding the proteins. The promoters of these genes contain enhancers that enable a transcriptional response to a diverse spectrum of chemical agents. The enhancers that are involved in induction of detoxication proteins include ► AP-1 binding sites, the antioxidant responsive element, the xenobiotic responsive element, the phenobarbital-responsive enhancer module, progesterone X receptor, and peroxisome proliferator-activated receptor enhancer.

Clinical Relevance

It is apparent from studies into the mechanisms of selective toxicity between species that variation in the activity of detoxication proteins influences sensitivity to chemical insult. Increasing evidence suggests that genetic polymorphisms in detoxication enzymes can confer an inherited predisposition to a number of malignant diseases that are influenced by environmental factors (e.g., lung and colorectal cancer). They may also confer a predisposition to adverse drug reactions.

Induction of some phase 2 detoxication systems is believed to represent a major mechanism of cancer ► chemoprevention and is thought to explain in part the epidemiological data suggesting that consumption of diets rich in fruit and vegetables protects against certain malignant diseases.

Acquired ► drug resistance to chemotherapy is a major problem in the treatment of many cancers. There is overwhelming evidence that the overexpression of several detoxication proteins, particularly GST, MDR, and MRP, contributes to the drug-resistant phenotype.

References

Borst P, Evers R, Kool M et al (2000) A family of drug transporters: the multidrug resistance-associated proteins. J Natl Cancer Inst 92:1295–1302

Dinkova-Kostova AT, Massiah MA, Bozak RE et al (2001) Potency of Michael reaction acceptors as inducers of enzymes that protect against carcinogenesis depends on their reactivity with sulfhydryl groups. Proc Natl Acad Sci U S A 98:3404–3409

Guengerich FP, Shimada T (1991) Oxidation of toxic and carcinogenic chemicals by human cytochrome P-450 enzymes. Chem Res Toxicol 4:391–407

Hayes JD, McLellan LI (1999) Glutathione and glutathione-dependent enzymes represent a coordinately regulated defence against oxidative stress. Free Radic Res 31:273–300

Hayes JD, Pulford DJ (1995) The glutathione S-transferase supergene family: regulation of GST and contribution of the isoenzymes to cancer chemoprotection and drug resistance. Crit Rev Biochem Mol Biol 30:445–600

Klaassen CD (ed), Amdur MO, Doull J (eds emeriti) (1996) Casarett and Doull's toxicology: the basic science of poisons. McGraw-Hill, New York

Deubiquitinating Enzymes (DUBs)

► Herpesvirus-Associated Ubiquitin-Specific Protease De-ubiquitinase

Development Lymph Vessel

► Lymphangiogenesis

Development of New Lymphatic Vessels

▶ Lymphangiogenesis

Dezocitidine

▶ 5-Aza-2′ Deoxycytidine

D-factor

▶ Leukemia Inhibitory Factor

dFdC

▶ Gemcitabine

DIA

▶ Leukemia Inhibitory Factor

Diabody

Shuji Ozaki
Department of Hematology, Tokushima
Prefectural Central Hospital, Tokushima, Japan

Synonyms

Engineered antibody; Multimeric antibody fragments; Single-chain Fv dimer

Definition

Diabody is a noncovalent dimer of single-chain Fv (scFv) fragment that consists of the heavy-chain variable (V_H) and light-chain variable (V_L) regions connected by a small peptide linker. Another form of diabody is single-chain $(Fv)_2$ in which two scFv fragments are covalently linked to each other.

Characteristics

Advances in antibody technology are enabling the design of antibody-based reagents for specific purposes in cancer diagnosis and monoclonal antibody therapy. First, to minimize the immunogenicity and enhance the efficacy in human use, mouse monoclonal antibodies are engineered to chimeric antibodies or humanized antibodies by grafting to the human constant region or framework. Moreover, fully human antibodies are developed by the use of transgenic mice (see ▶ Transgenic Mouse) or phage display technology. Second, monoclonal antibodies are designed as immunoconjugates to deliver the cytotoxic agents such as chemotherapeutic drugs, toxins, enzymes, and radioisotopes. These therapeutic antibodies have emerged as potent agents and are used worldwide for cancer therapy. Engineered antibody fragments have been investigated as alternative reagents because of their unique properties resulting from the structure.

Structure

A variety of antibody fragments are developed including single V_H domain, Fab, scFv, and multimeric formats such as multivalent scFvs (diabody, triabody, and tetrabody), bispecific scFv, and minibody (scFv–CH3 dimer) (Fig. 1). In scFv fragments, the V_H domain binds to its attached V_L domain when the linker is flexible and long enough (a length of at least 12 amino acids). For example, the linker sequence of $(Gly_4Ser)_3$ provides sufficient flexibility for the V_H and V_L domain to form Fv comparable to the parent antibody. In contrast, when the linker is shortened to less than 12 residues (e.g., five

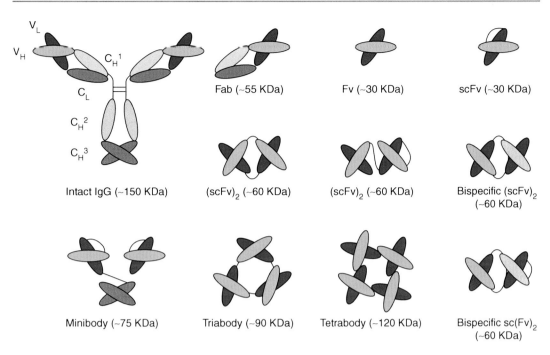

Diabody, Fig. 1 Schematic structure of intact IgG antibody and engineered antibody fragments. The variable regions of heavy (V_H) and light chains (V_L) contribute to the antigen binding. The V_H and V_L domains can be connected by a peptide linker to form single-chain Fv (scFv). The scFv fragments can form multimers such as $(scFv)_2$ diabody, triabody, and tetrabody depending on the linker length and the V-domain orientation. The bivalent sc $(Fv)_2$ can be generated by connecting two scFvs covalently. Bispecific diabodies can also be engineered by using two different Fv domains

amino acids of (Gly_4Ser)), the V_H and V_L domains are unable to bind each other, and instead the scFv fragment forms a noncovalent dimer by another scFv molecule [$(scFv)_2$ diabody]. Shortening of the linker length between the V_H and V_L domains (less than three residues) promotes the assembly of trimeric or tetrameric structures (triabody or tetrabody). However, this multimer formation also depends on the V-domain orientation either V_H–V_L or reverse V_L–V_H orientation in the scFv constructs. The bivalent Fv fragment can also be designed by linking two scFv domains covalently as a single-chain version [$sc(Fv)_2$ diabody]. The $sc(Fv)_2$ version is more stable than $(scFv)_2$, and this structure may form a noncovalent dimer [$sc(Fv)_2]_2$.

The capacity of multivalent binding of these fragments offers a significant opportunity to design multifunctional antibody reagents. The diabody structure is used to form ▶ bispecific antibodies by linking different V_H and V_L domains of two antibodies (e.g., $V_H A$–$V_L B$ and $V_H B$–$V_L A$). However, when two different polypeptides are produced within a single cell, purification steps are necessary to obtain the active heterodimeric antibody among the inactive homodimers. Therefore, bispecific $sc(Fv)_2$ version is developed by connecting two different scFv domains with the middle-length linker (e.g., $V_H A$–$V_L B$–$V_H B$–$V_L A$ or $V_H A$–$V_L A$–$V_H B$–$V_L B$).

Pharmacokinetics and Distribution

The pharmacokinetics of these antibody fragments is markedly different from intact IgG antibodies that exhibit prolonged circulation ($t_{1/2}$ of up to 3 weeks). The lower molecular weight constructs (below than 60 kDa) are subject to be excreted by renal clearance, resulting in a shorter serum half-life than intact IgG. In most cases, the $t_{1/2}$ values of scFv and diabody are extremely short such as 2 and 6 h, respectively. This rapid pharmacokinetics is the most favorable for imaging applications and ▶ radioimmunotherapy because of the lower background levels in normal tissues. Drug biodistribution studies of radio-labeled scFv and sc(Fv)$_2$ have shown high tumor-to-blood ratios in xenograft models compared with intact IgG antibodies. The fast blood clearance of antibody fragments contributes to avoid undesired toxicity, and these fragments have the remarkable advantage for ▶ targeted drug delivery of toxins or radioisotopes.

In addition, antibody fragments show better penetration into the tumor mass, but these smaller constructs have shorter retention to tumor cells at the same time. Thus, the valance of penetration and retention of antibodies is an important factor for therapeutic use, especially in solid tumors. The Fab and scFv fragments are monovalent and exhibit poor retention on target cells, but multivalent forms of these fragments such as diabody, triabody, and tetrabody exhibit dramatically increased affinity and high tumor retention compared with the parent scFv. The ideal tumor-targeting reagents are intermediate-sized multivalent antibodies such as bivalent diabodies that show a longer half-life as well. In another approach, the Fc portion is fused to antibody fragments to control the serum levels of the antibody. The scFv–Fc or scFv–CH3 fusion antibodies (minibodies) are expected to have a more prolonged half-life and increased tumor accumulation in vivo. The serum half-life of antibody fragments can also be extended by modification such as linkage to polyethylene glycol (PEG).

Agonistic Activity

In terms of mechanism of action, intact IgG antibodies kill tumor cells mainly by Fc-mediated effector functions such as antibody-dependent cell-mediated cytotoxicity (ADCC) and ▶ complement-dependent cytotoxicity. In contrast, antibody fragments have the compact structures without the Fc portion and have unique characteristics for using cancer treatment. The two binding sites of diabodies are located at a distance of about 70 Å less than half for those of intact IgG antibodies. Therefore, diabodies can place the antigens more closely to each other

than by the parent IgG antibodies, which efficiently induces the ligation of target molecules on the cell surface. When the targets are functional receptors, the diabody can mediate a direct effect or signal transduction in tumor cells, including the stimulation of ▶ apoptosis or cell death. For example, we and our collaborators have generated (scFv)$_2$ and sc(Fv)$_2$ diabodies that recognize CD47 or ▶ HLA class I molecules. These diabodies can cross-link the target antigens and show the enhanced cytotoxic activities against hematological malignancies such as leukemia, lymphoma, and myeloma cells when compared with the original IgG antibodies. Thus, enhancement of the cross-linking potential is one of the important bioactivities of antibody fragments.

Application of Diabodies

A variety of target antigens have been evaluated for therapeutic purposes including CD19, CD20, CD22, epithelial cell adhesion molecule (Ep-CAM), epidermal growth factor receptor (EGFR), HER2, MUC1, and ▶ carcinoembryonic antigen (CEA). Several types of diabodies and minibodies are engineered for targeting these candidate antigens on tumor cells. Immunotoxins are also constructed to deliver the cytotoxic agents, radio-isotopes, enzymes, cytokines, and liposomes by using antibody fragments. Previous studies have shown the effectiveness of these reagents in preclinical and clinical trials.

▶ Bispecific antibodies that comprise two different binding specificities have been studied extensively in cancer diagnosis and therapy. Most of the bispecific reagents are designed for the retargeting of effector cells such as cytotoxic T lymphocytes and NK cells. Recombinant bispecific diabodies such as anti-CD19 x anti-CD3, anti-Ep-CAM x anti-CD3, and anti-HER2 x anti-CD3 have been used in the immunotherapy of ▶ B-cell lymphoma, Breast (see ▶ Breast Cancer), ovarian (see ▶ Ovarian Cancer) and colorectal cancer (see ▶ Colorectal Cancer Clinical Oncology). Another strategy of bispecific antibodies is the recruitment of effector molecules including toxins, drugs, prodrugs, cytokines, and radionuclides in vivo. First, the tumor cells are targeted by the tumor-specific binding site of the diabody. After the unbound diabody is cleared from the serum, cytotoxic drugs or radiolabeled hapten is administered to be captured by another binding site of the bound diabody.

Future Directions

In principle, selection of target molecules and modification of antibody constructs are key issues of antibody-based strategies in clinical utility. Based on the properties of pharmacokinetics, biodistribution, and manufacturing production, engineered antibody fragments have been investigated as alternative reagents to target cancer cells. As more small-sized antibody than diabody, single-domain antigen-binding fragments from camelid heavy-chain antibodies (called V_HH or nanobody, ~15 kDa) are also used. Although the efficacy of these antibody fragments needs to be evaluated in clinical settings, the drastic potential of agonistic activity or multivalent activity of these reagents will provide new promises for development of the next generation of antibody drugs in cancer diagnosis and treatment.

Cross-References

▶ Apoptosis
▶ B-cell Lymphoma
▶ Bispecific Antibodies
▶ BRMS1
▶ Carcinoembryonic Antigen
▶ Colorectal Cancer Clinical Oncology
▶ Complement-Dependent Cytotoxicity
▶ HLA Class I
▶ Ovarian Cancer
▶ Radioimmunotherapy
▶ Targeted Drug Delivery
▶ Transgenic Mouse

References

Batra SK, Jain M, Wittel UA et al (2002) Pharmacokinetics and biodistribution of genetically engineered antibodies. Curr Opin Biotechnol 13:603–608
Beckman RA, Weiner LM, Davis HM (2007) Antibody constructs in cancer therapy. Cancer 109:170–179

Holliger P, Hudson PJ (2005) Engineered antibody fragments and the rise of single domains. Nat Biotechnol 23:1126–1136
Jain M, Kamal N, Batra SK (2007) Engineering antibodies for clinical applications. Trends Biotechnol 25:307–316

See Also

(2009) Antibody-dependent cell mediated cytotoxicity. In: Schwab M (ed) Encyclopedia of cancer, 3rd edn. Springer, Berlin/Heidelberg, p 188. doi: 0.1007/978-3-540-47648-1_313
(2012) Chimeric antibodies. In: Schwab M (ed) Encyclopedia of cancer, 3rd edn. Springer, Berlin/Heidelberg, p 806. doi:10.1007/978-3-642-16483-5_1091
(2012) Cytokine. In: Schwab M (ed) Encyclopedia of cancer, 3rd edn. Springer, Berlin/Heidelberg, p 1051. doi:10.1007/978-3-642-16483-5_1473
(2012) Drug biodistribution. In: Schwab M (ed) Encyclopedia of cancer, 3rd edn. Springer, Berlin/Heidelberg, p 1160. doi:10.1007/978-3-642-16483-5_1732
(2012) Humanized antibodies. In: Schwab M (ed) Encyclopedia of cancer, 3rd edn. Springer, Berlin/Heidelberg, p 1760. doi:10.1007/978-3-642-16483-5_2863
(2012) Monoclonal antibody therapy. In: Schwab M (ed) Encyclopedia of cancer, 3rd edn. Springer, Berlin/Heidelberg, pp 2367–2368. doi:10.1007/978-3-642-16483-5_3823
(2012) Pharmacokinetics. In: Schwab M (ed) Encyclopedia of cancer, 3rd edn. Springer, Berlin/Heidelberg, p 2845. doi:10.1007/978-3-642-16483-5_4500

Diagnostic Pathology

▶ Pathology

Dibasic Processing Enzyme

▶ Furin

Diet

▶ Colorectal Cancer Nutritional Carcinogenesis

Dietary Carcinogens

► Food-Borne Carcinogens

Dietary Essential Minerals

► Mineral Nutrients

Diethylstilbestrol

Rosemarie A. Ungarelli and Carol L. Rosenberg
Boston Medical Center and Boston University
School of Medicine, Boston, MA, USA

Synonyms

DES

Definition

Diethylstilbestrol is a synthetic nonsteroidal estrogen with biological properties similar to endogenous estrogens such as estradiol-17-beta and estrone (► Estradiol).

Characteristics

Pharmacology
Diethylstilbestrol is administered orally, is lipid-soluble, and readily absorbed from the proximal gastrointestinal tract. It is metabolized via the hepatic microsomal system to dienestrol and quinone and epoxide intermediates. It crosses the placenta and is thought to be metabolized by the fetus.

Initial Use and Early Epidemiologic Studies
Diethylstilbestrol (DES), first manufactured by Dodds and associates in London in 1938, was used to treat several gynecologic conditions. In particular, it was prescribed for the treatment of frequent or threatened miscarriages. As early as 1953, Dieckmann and colleagues demonstrated that DES did not improve pregnancy outcomes. (In fact, in a later reanalysis of these data in 1978, Brackbill and Berendes showed that women exposed to DES had higher risks of premature births, perinatal death, and miscarriages than women who were given placebo.) Other studies also found DES to be ineffective in preventing adverse pregnancy outcomes, but physicians continued to prescribe the drug to try to maintain high-risk pregnancies because its use seemed logical and it was well established. It was administered to approximately 5–10 million pregnant women in the United States between 1940 and 1971. It remained in use in Europe until the early 1980s.

In 1971, Herbst and colleagues established a strong connection between DES exposure in utero and subsequent development of clear-cell adenocarcinoma of the vagina and cervix in young women, aged 14–21 years. The incidence of this cancer in women whose mothers had been administered DES during pregnancy (DES daughters) is estimated to range from 1.4 cases per 1,000 exposed to one case per 10,000 exposed persons (► Cervical cancer). Previously, clear-cell adenocarcinoma of the vagina and cervix had been observed only rarely, primarily in postmenopausal women (over age 50) not exposed to DES. Consequently, the US Food and Drug Administration issued a drug bulletin recognizing DES as a ► transplacental carcinogen and banned its use during pregnancy. Since then, DES exposure has been observed to cause a range of teratogenic and neoplastic changes in humans and animals. It is used now mainly for treatment of a small subset of hormonally responsive refractory cancers. However, DES exposure can serve as a model for evaluating the potential effects of xenoestrogens (► Hormonal carcinogenesis; ► estrogenic hormones). Therefore, its investigation remains important and should not be limited to the study of the consequences of a specific, unintentionally deleterious administration.

Diethylstilbestrol, Table 1 Effects of diethylstilbestrol exposure

	Nonneoplastic effects	Neoplastic effects
DES mothers	Increased risk of • Adverse pregnancy outcomes	Increased risk of • Breast cancer
DES daughters	Increased risk of • Adverse pregnancy outcomes • Infertility • Reproductive tract structural abnormalities • Vaginal adenosis	Increased risk of • Clear-cell adenocarcinoma of the vagina and cervix • Breast cancer over age 40
DES sons	Increased risk of • Epididymal cysts • Cryptorchidism • Testicular hypoplasia • Semen and sperm abnormalities	No increased risk of hormone-related cancers

Animal Studies

DES has been studied extensively in animal models. In the Syrian hamster model, exposure to DES induces neoplasms of the liver and kidney, as well as aneuploidy (particularly chromosome gains) in the renal neoplasms (▶ Aneuploidy, chromosome instability). DES exposure in rats elicits tumors of the reproductive tract and pituitary and mammary glands. In addition, Green and colleagues have observed that DES metabolites produce DNA adducts (▶ Adducts to DNA) and, ultimately, cancer in the breast of female ACI rats. Tumors of the reproductive tract and mammary glands are seen in the murine model as well, along with alterations in the genetic pathways governing uterine differentiation. Data from Newbold and colleagues suggest that an increased susceptibility to tumor formation is transmitted along the maternal lineage to subsequent generations, both male and female. Thus, not only is the developing organism sensitive to the endocrine-disrupting chemical, but transgenerational effects are plausible as well.

Neoplastic Effects in Humans

The only increased risk of hormone-dependent cancers observed in women who took DES during pregnancy (DES mothers) is ▶ breast cancer (Table 1). Hatch and colleagues found that DES mothers had a 30% increased rate of breast cancer compared to the general population. In contrast, DES daughters have an increased risk of two types of hormone-dependent cancers, clear-cell adenocarcinoma of the vagina and cervix (mentioned above) and breast cancer (▶ Breast cancer). Clear-cell adenocarcinoma of the vagina and cervix generally presents in these women when they are in their teens and twenties; however, because

some women have been diagnosed in their thirties and forties, concerns have arisen about whether another increase in risk will occur as DES daughters approach the age at which this type of cancer is seen in the general population (the postmenopausal period).

There was a paucity of information on breast effects in DES daughters. However, the National Cancer Institute's Continuation of Follow-Up of DES-Exposed Cohorts has proven a rich source of information to study the long-term effects of in utero estrogen exposure in humans. These cohorts include over 4,000 women who had documented in utero exposure to DES and more than 2,000 unexposed women from the same record sources; all have been followed from 1994 or earlier. As women exposed in utero to DES have begun to reach the ages at which breast cancer is more common, it appears that these women have an increased risk. In the data from the National Cancer Institute collaborative follow-up study of DES health effects (Continuation of Follow-Up of DES-Exposed Cohorts), Palmer and colleagues found that at ages 40 and older, DES daughters had double the risk of breast cancer of unexposed women; no association was seen prior to age 40.

To date, males exposed in utero to DES (DES sons) do not exhibit an increase risk of developing hormone-related cancers. However, DES sons do display nonneoplastic abnormalities (see below) that place them at increased risk of developing testicular cancer regardless of DES exposure (▶ Testicular cancer).

Nonneoplastic Effects in Humans

In utero exposure to DES has been shown to elicit a variety of nonneoplastic reproductive tract abnormalities, including structural cervical, vaginal, or uterine abnormalities, in addition to changes in the vaginal epithelium such as adenosis (see Table 1). DES dosage and the stage of pregnancy during which the drug was administered appears to be directly linked to the severity of the adenosis, with the most severe manifestations seen in DES daughters whose mothers took the drug during their first trimesters. It is unclear whether these areas of adenosis progress to vaginal clear-cell adenocarcinoma. DES daughters have an increased risk of poor pregnancy outcomes, such as ectopic pregnancy or miscarriage, and also have a higher incidence of infertility than the general population.

DES sons are more likely than unexposed men to exhibit genital abnormalities such as epididymal cysts, cryptorchidism, and testicular hypoplasia. Although DES sons show an increased incidence of semen and sperm abnormalities, they have not demonstrated an increased risk of infertility, but this is still under investigation.

Mechanism of Toxicity

The mechanism by which estrogens in general, and DES in particular, exert their toxic and carcinogenic effects is not fully understood. Both proliferative and genotoxic mechanisms have been postulated. Classically, estrogen exerts its effects through interaction with the estrogen receptor α (ERα), which stimulates cell proliferation and inhibits apoptosis (▶ Estrogen receptor). ERα may also interact with other receptors (such as ERβ, insulin-like growth factor 1 receptor, and epidermal growth factor receptor) to influence proliferation or may act through non-genomic pathways, since it is found in nonnuclear subcellular fractions such as the plasma membrane and the mitochondria. An alternative or additional mechanism that may mediate estrogen's and perhaps DES' toxic effects is through the metabolites' genotoxic capacity. Estrogens, and specifically DES, can be oxidatively metabolized into potentially genotoxic intermediates. Estrogen and its metabolites have been reported to induce DNA damage (▶ DNA damage), manifesting as ▶ allele imbalance and DNA amplification in

human breast epithelial cells in vitro (▶ Amplification). DES metabolites have been reported to produce DNA adducts and cancer in mammary glands of female rats. DES is a strong mitotic inhibitor in cell lines, blocking equatorial plate formation, tubulin polymerization, and spindle assembly. This may, in turn, induce ▶ aneuploidy. Tumors associated with DES exposure exhibit genetic instability, such as whole and partial chromosome gains in vitro and in vivo, in the Syrian hamster model. ▶ Microsatellite instability has been reported in human vaginal clear-cell adenocarcinomas associated with in utero DES exposure, as well as in murine endometrial carcinomas after DES treatment.

In contrast to the substantial microsatellite instability seen in human vaginal clear-cell adenocarcinomas associated with in utero DES exposure, breast neoplasms in DES daughters do not exhibit an increased amount of microsatellite instability. Breast tumors of DES mothers have not been investigated. In fact, little microsatellite instability was observed in breast tumors in both exposed and unexposed women, which is consistent with previous results from unselected human breast cancers, confirming that microsatellite instability is unusual in human breast cancers and suggesting that prenatal DES exposure does not affect DNA mismatch repair mechanisms in the breast. Similarly, equivalent amounts of allele imbalance have been observed in breast tissue regardless of exposure, which differs from findings in animal models and in vitro systems. Therefore, the effect of in utero DES exposure may be tissue, timing, and/or species specific, as is the case with other hormonal agents such as ▶ tamoxifen, which has variable effects on human endometrium and mammary tissue. It remains under investigation as to whether the potential effects of in utero DES exposure on human breast carcinogenesis are mediated by enhanced proliferation, by alternative genotoxic effects, or by other pathways entirely.

Summary

The unfortunate consequences of DES administration to pregnant women have yielded clinical and scientific insights into estrogen's effects on developing and mature tissues. Its continued investigation should provide additional clinical and mechanistic information about these effects and have relevance to understanding the effects of exposure to xenoestrogens.

Cross-References

▶ Adducts to DNA
▶ Allele Imbalance
▶ Amplification
▶ Aneuploidy
▶ Breast Cancer
▶ Cervical Cancers
▶ Chromosomal Instability
▶ DNA Damage
▶ Estradiol
▶ Estradiol
▶ Estrogen Receptor
▶ Estrogenic Hormones
▶ Hormonal Carcinogenesis
▶ Microsatellite Instability
▶ Tamoxifen
▶ Testicular Cancer
▶ Transplacental Carcinogenesis

References

Giusti RM, Iwamoto K, Hatch EE (1995) Diethylstilbestrol revisited: a review of the long-term health effects. Ann Int Med 122:778–788

Larson PS, Ungarelli RA, De Las Morenas A et al (2006) In utero exposure to diethylstilbestrol (DES) does not increase genomic instability in normal or neoplastic breast epithelium. Cancer 107(9): 2122–2126

Newbold RR, Padilla-Banks E, Jefferson WN (2006) Adverse effects of the model environmental estrogen diethylstilbestrol are transmitted to subsequent generations. Endocrinology 147(Suppl 6): S11–S17

Schrager S, Potter BE (2004) Diethylstilbestrol exposure. Am Fam Physician 69:2395–2402

Yager JD, Davidson NE (2006) Estrogen carcinogenesis in breast cancer. New Eng J Med 354:270–282

See Also

(2012) Carcinogen. In: Schwab M (ed) Encyclopedia of cancer, 3rd edn. Springer Berlin Heidelberg, p 644. doi:10.1007/978-3-642-16483-5_839

(2012) DNA amplification. In: Schwab M (ed) Encyclopedia of cancer, 3rd edn. Springer Berlin Heidelberg, p 1129. doi:10.1007/978-3-642-16483-5_1665

(2012) DNA mismatch repair mechanism. In: Schwab M (ed) Encyclopedia of cancer, 3rd edn. Springer Berlin Heidelberg, p 1140. doi:10.1007/978-3-642-16483-5_1684

(2012) Estrogens. In: Schwab M (ed) Encyclopedia of cancer, 3rd edn. Springer Berlin Heidelberg, p 1333. doi:10.1007/978-3-642-16483-5_2019

(2012) Genetic instability. In: Schwab M (ed) Encyclopedia of cancer, 3rd edn. Springer Berlin Heidelberg, pp 1527-1528. doi:10.1007/978-3-642-16483-5_2380

Diferuloylmethane

▶ Curcumin

Differentiation-Inducing Factor

▶ Tumor Necrosis Factor

Diffuse

▶ Mantle Cell Lymphoma

Diffuse Astrocytoma

▶ Astrocytoma

Diffuse Intrinsic Pontine Glioma

▶ Astrocytoma

Diffuse Large B-Cell Lymphoma

Ken H. Young[1] and Michael B. Møller[2]
[1]Department of Hematopathology, The
University of Texas MD Anderson Cancer Center,
Houston, TX, USA
[2]Department of Pathology, Odense University
Hospital, Odense, Denmark

Synonyms

KIEL classification: Centroblastic,
B-immunoblastic, B-anaplastic; WHO: Diffuse
large cell; Working Formulation: Diffuse large
cell, Large cell immunoblastic, Diffuse mixed
small and large

Definition

Diffuse large B-cell lymphoma is an entity of
non-Hodgkin lymphoma composed of malignant
large lymphoid cells with blastic morphologic
features, expression of B-cell markers, and a diffuse growth pattern. The postulated cells of origin
are germinal or post germinal center B-cells. This
lymphoma entity is morphologically, clinically,
and molecularly heterogeneous.

Characteristics

Diffuse large B-cell lymphoma is the most common type of non-Hodgkin lymphoma comprising
30–40% of non-Hodgkin lymphomas in adult and
approximately 20% in childhood and adolescence. Diffuse large B-cell lymphoma can be
seen in all age groups, but the incidence increases
with age. The median age at diagnosis is approximately 65 years. There is a slight male preponderance (1.2:1). In most patients, the tumor
resides in lymph nodes, but 30–40% of patients
have primary extranodal disease. Approximately
70% of patients have at least one and 30% have
multiple extranodal involvements. Virtually any
extranodal site may be involved, but the most

frequently involved organs include the gastrointestinal tract, soft tissue, thyroid, skin, central
nervous system, testis, liver, bladder, bone,
gonads, breast, ovary, cervix, kidney, pancreas,
lung, and salivary glands. Transformed diffuse
large B-cell lymphomas arise from indolent lymphomas such as ▶ chronic lymphocytic leukemia/
small lymphocytic lymphoma, ▶ marginal zone
B-cell lymphoma, lymphoplasmacytic lymphoma
and ▶ follicular lymphoma. The etiology of most
diffuse large B-cell lymphoma cases is unclear.
However, patients with immunodeficiency such
as patients with human immunodeficiency virus,
Epstein-Barr virus, or human herpes virus 8 infection; patients receiving immunosuppressive therapy; and those having chronic inflammatory
disease are at increased risk of developing lymphoma or high-grade transformation.

Diagnosis

The typical clinical presentation of diffuse large
B-cell lymphoma patients with nodal disease is
enlarged lymph nodes. Patients with extranodal
presentation of the disease often have a mass with
rapid growth or symptoms related to dysfunction
of the involved organ(s). One third of the patients
have B symptoms. Approximately half of the
patients have localized lymphoma, i.e., Ann
Arbor stage I or II, and the remainder have disseminated or high stage disease.

The morphologic diagnosis is based on the
World Health Organization Classification. Diffuse
large B-cell lymphoma typically consists of a
diffuse proliferation of medium to large neoplastic
B-lymphoid cells with a nucleus at least twice the
size of a normal lymphocyte or similar to and
exceeding the size of the macrophage nuclei.
These large cells are a mixture of cells that resemble either the centroblasts or the immunoblasts
that normally reside in reactive germinal centers.
The diffuse large B-cell lymphoma entity is morphologically heterogeneous with several morphologic variants. The two most common variants are
the centroblastic variant which is dominated by
centroblasts and the immunoblastic variant with
>90% being immunoblasts. In the T-cell/
histiocyte rich variant, the majority of cells are
small T-cells and histiocytes and less than 10% of

the large cells are large neoplastic B-cells. An anaplastic variant of diffuse large B-cell lymphoma is recognized, which has a similar morphology and CD30 expression as the T-cell lymphoma anaplastic large cell lymphoma. However, anaplastic diffuse large B-cell lymphoma is clinically and genetically unrelated to anaplastic large cell lymphoma. Other rare variants include plasmablastic diffuse large B-cell lymphoma and diffuse large B-cell lymphoma with expression of full-length *ALK*, CD30 and *IRF4* rearrangement.

Diffuse large B-cell lymphoma cells usually express CD45 and B-lymphoid markers such as CD19, CD20, CD22, CD79a, and PAX5. The proliferation rate is high with most cases expressing the proliferation associated marker Ki-67 in >40% of the tumor cells. In some tumors, 95–100% of the malignant cells express Ki-67.

The prognosis is variable with a 5-year overall survival rate of 45–50% for all patients. The International Prognostic Index is widely used for prognostication of diffuse large B-cell lymphoma. It consists of five clinical factors (age, stage, performance score, serum lactate dehydrogenase, number of extranodal sites involved), each with independent prognostic value regarding overall survival. The index allocates 35–40% of the patients to the low risk group with a 5-year overall survival rate of >70%, while 15–20% of the patients have high risk lymphoma and a 5-year overall survival rate of <30%. By adding the anti-CD20 antibody ► rituximab to the treatment regimens, the survival rates cited above may increase with up to 10% or more.

Therapy

Multidrug chemotherapy and radiotherapy represent the mainstay of diffuse large B-cell lymphoma treatment. For several decades, CHOP (► Cyclophosphamide, Doxorubicin, Vincristine, and Prednisone) or CHOP-like regimens have been the gold standard chemotherapy regimens with the achievement of complete remission rates of 70–80% and cure rates of 40–50%. Trials have documented that the addition of rituximab to CHOP or CHOP-like regimens significantly increases event-free and overall survival for most diffuse large B-cell lymphoma patients.

The treatment strategy is influenced by the age and the International Prognostic Index risk group of the patient, as well as the location and stage of the disease. Patients with stage I/II lymphoma may be treated with three cycles of CHOP(-like) chemotherapy and radiotherapy of the involved field and possibly with the addition of rituximab. Consolidation radiotherapy is also used for patients with bulky or extranodal lesions with either limited-stage or advanced disease at presentation. Patients with higher stage disease are typically treated with six to eight cycles of CHOP (-like) chemotherapy with the addition of rituximab. The role of autologous stem cell transplantation is not well defined, but may be beneficial to younger patients with high risk disease who are unlikely to do well with standard therapy.

Approximately 30–40% of the diffuse large B-cell lymphoma patients either do not achieve complete remission or develop a relapse after remission, and most of these patients require second-line therapy. A variety of second-line chemotherapy regimens are used, and these regimens typically have a response rate of 40–50%. For patients with chemotherapy-sensitive relapse, autologous stem cell transplantation is considered the treatment of choice. Current immunotherapy with PD-1 pathway drugs and CAR-T regimens provide a novel avenue for those refractory and resistant patients.

Genetics

The cellular origin of diffuse large B-cell lymphoma is germinal or postgerminal center B-cells. Two major pathogenic mechanisms related to normal germinal centre function contribute to the lymphomagenesis of diffuse large B-cell lymphoma: aberrant somatic hypermutation and ► chromosomal translocation often involving the immunoglobulin genes. Virtually all diffuse large B-cell lymphoma cases have rearranged immunoglobulin genes, and most cases are somatically hypermutated in the immunoglobulin variable region genes.

Somatic hypermutation is a physiologic process in the germinal center that target immunoglobulin genes and thereby creates antibody diversity. However, aberrant somatic hypermutation is considered an important transformation mechanism in diffuse large B-cell lymphoma. In most diffuse large B-cell lymphoma cases, aberrant somatic hypermutation targeting nonimmunoglobulin genes results in hypermutation of multiple proto-oncogenes such as *BCL6* and ▶ *MYC* oncogene. This process is thought to contribute to the clinical and biological heterogeneity of diffuse large B-cell lymphoma.

Another important mechanism involved in the pathogenesis of diffuse large B-cell lymphoma is chromosomal translocations bringing a proto-oncogene under the influence of an active locus such as the immunoglobulin gene, thereby causing deregulated overexpression of the oncogene. These translocations may occur as (accidental) by-products of the immunoglobulin remodeling processes, i.e., ▶ V(D)J recombination, somatic hypermutation, and class switching, which take place in the germinal centers and require DNA strand breaks. The t(14;18) brings the antiapoptotic ▶ *BCL2* gene in juxtaposition with the joining segment of the immunoglobulin heavy chain gene and is a hallmark of follicular lymphoma. About 20 to 30% of diffuse large B-cell lymphomas also carry this translocation, and these cases are probably evolved from clinical or subclinical follicular lymphoma. The presence of the *BCL2* rearrangement is associated with disseminated, nodal disease, and bcl-2 overexpression. Bcl-2 overexpression can also be achieved by gene ▶ amplification (in 10% of the cases) and activation of B-cell receptor signaling. Approximately one third of diffuse large B-cell lymphoma cases have a rearrangement of the proto-oncogene *BCL6*. Deregulation of bcl-6 expression in germinal center cells may contribute to lymphomagenesis by functional inactivation of p53, thereby suppressing p53-mediated ▶ apoptosis. *MYC* is rearranged, often as a translocation to an immunoglobulin gene, in 10–15% of diffuse large B-cell lymphoma. Patients carrying concurrent *MYC* and *BCL2* translocations, known as *MYC+/BCL2+* "double-hit" lymphomas occurring in about 5% of all DLBCL patients, have an extremely poor survival.

Other frequent genetic lesions include ▶ *TP53* mutations in 20–25% and *CD95* mutations in 20% of the tumors. *TP53* mutations are associated with poor response to treatment and poor prognosis,

whereas *CD95* mutations correlate with extranodal disease and autoimmune phenomena. Several recurrent mutations are also reported, including those related to chromatin modification (MLL2, ARID1A and MEF2B), NF-κB (MYD88, CD79A/B, CARD11 and TNFAIP3), PI3 kinase (PIK3CD, PIK3R1, and MTOR), B-cell lineage (IRF8, POU2F2, and GNA13), and WNT signaling (WIF1). In transformed diffuse large B-cell lymphoma, genetic alterations are found in pathways deregulating cell-cycle progression, DNA damage responses (*CDKN2A/B, MYC*, and *TP53*) and aberrant somatic hypermutation.

Gene expression profiling studies with microarray technology (see "▶ Microarray (cDNA) Technology") have identified two major subgroups of diffuse large B-cell lymphoma. One type has an expression profile that resembles normal germinal centre B-cells (GCB-like), and the profile of the other group is similar to activated B-cells (ABC-like). In addition to having expression profiles with hallmarks related to B-cells of different stages of differentiation, these two subgroups of diffuse large B-cell lymphoma also have distinct genetic lesions. For example, *BCL2* is dysregulated by way of the t(14;18) in one third to half of the GCB-like cases, but this translocation is rarely found in ABC-like cases. Instead, as an alternative mechanism of *BCL2* activation, amplification of the *BCL2*-containing chromosome region 18q21 is present in 20–30% of ABC-like tumors, but it is rare in GCB-like cases. Furthermore, 20–26% of ABC-like cases have gain on chromosome 3q and/or trisomy 3, whereas these lesions are not identified in GCB-like lymphomas. Moreover, a major difference is that ABC-like diffuse large B-cell lymphoma has a constitutively active NF-κB pathway which is a potentially important therapeutic target. These molecular differences translate into clinical differences, as GCB-like diffuse large B-cell lymphoma has a better prognosis than ABC-like subtype.

Subtypes of Diffuse Large B-Cell Lymphoma

From the heterogeneous group of diffuse large B-cell lymphoma, several subtypes have sufficiently distinct clinical, morphological and genetic features to be recognized as entities.

Primary Mediastinal (Thymic) Large B-Cell Lymphoma

Primary mediastinal large B-cell lymphoma presumably arises from thymic B-cells. These cases comprise up to 10% of all diffuse large B-cell lymphomas. It is a disease with a female predominance. The patients are younger, in general in the third to fifth decade. Typically the lymphoma presents at diagnosis as a bulky and locally invasive anterior mediastinal mass which often causes airway compression and vena cava superior syndrome. In contrast to other diffuse large B-cell lymphoma subtypes, mediastinal large B-cell lymphomas seldom harbor rearrangements of *BCL2*, *BCL6*, or *MYC*, but a high frequency of *PD-L1* and *PD-L2* gene amplification and expression. Mediastinal large B-cell lymphoma has a gene expression profile distinct from ABC-like and GCB-like diffuse large B-cell lymphoma. However, in common with ABC-like diffuse large B-cell lymphomas, mediastinal large B-cell lymphoma has a constitutively active NF-κB pathway.

Intravascular Large B-Cell Lymphoma

Intravascular large B-cell lymphoma is a rare neoplasia characterized by proliferation of large CD20-positive lymphoid cells primarily in the lumina of small to medium-sized vessels. The disease is often widespread at diagnosis causing a highly variable clinical picture. The symptoms are mainly caused by occlusion of capillaries in the involved organs. The prognosis is very poor and the biological mechanisms are unclear.

Primary Effusion Lymphoma

Primary effusion lymphomas occur most often in immunocompromised patients and the outcome is extremely poor. Patients usually present with serous effusions without overt tumor masses. Characteristically the lymphomas cells are pleomorphic and lack B-cell-associated antigens such as CD19 and CD20. Instead they often express CD30 and the plasma cell-associated marker CD138. Reflecting the etiology, the lymphoma cells contain ► Kaposi sarcoma herpes virus/human herpes virus 8 and some patients are positive for ► Epstein-Barr virus.

Lymphomatoid Granulomatosis

Lymphomatoid granulomatosis is a rare Epstein-Barr virus-associated angiocentric lymphoproliferative neoplasm involving extranodal tissues, most commonly the lung, skin, and brain. The lesions are composed of large EBV-positive B-cells in a background of small reactive T-cells plasma cells and reactive histiocytes. Based on the proportion of EBV-positive large B-cells, lymphomatoid granulomatosis is graded from grade I (few cells) to grade III (numerous cells). Grade III lymphomatoid granulomatosis fulfill the morphologic criteria of a diffuse large B-cell lymphoma subtype and should be treated accordingly.

Cross-References

► Amplification
► Apoptosis
► Bcl2
► Chromosomal Translocations
► Chronic Lymphocytic Leukemia
► Cyclophosphamide
► Epstein-Barr Virus
► Follicular Lymphoma
► Kaposi Sarcoma
► Marginal Zone B-Cell Lymphoma
► Microarray (cDNA) Technology
► MYC Oncogene
► Nuclear Factor-κB
► Rituximab
► TP53
► V(D)J Recombination

References

Armitage JO, Mauch PM, Harris NL et al (2004) Diffuse large B-cell lymphoma. In: Mauch PM, Armitage JO, Coiffier B, Dalla-Favera R, Harris NL (eds) Non-Hodgkin's lymphomas. Lippincott Williams & Wilkins, Philadelphia, pp 427–453

Gatter KC, Warnke RJ (2001) Diffuse large B-cell lymphoma. In: Jaffe ES, Harris NL, Stein H, Vardiman JW (eds) World Health Organization classification of tumours. Pathology and genetics. Tumours of haematopoietic and lymphoid tissues. IARC Press, Lyon, pp 171–174

Hu S, Xu-Monette ZY, Balasubramanyam A, Manyam GC, Visco C, Tzankov A, Liu WM, Miranda RN, Zhang L,

Montes-Moreno S, Dybkær K, Chiu A, Orazi A, Zu Y, Bhagat G, Richards KL, Hsi ED, Choi WW, Han van Krieken J, Huang Q, Huh J, Ai W, Ponzoni M, Ferreri AJ, Zhao X, Winter JN, Zhang M, Li L, Møller MB, Piris MA, Li Y, Go RS, Wu L, Medeiros LJ, Young KH (2013) CD30 expression defines a novel subgroup of diffuse large B-cell lymphoma with favorable prognosis and distinct gene expression signature: a report from the International DLBCL Rituximab-CHOP Consortium Program Study. Blood 121(14):2715–2724, 4/2013

Küppers R (2005) Mechanisms of B-cell lymphoma pathogenesis. Nat Rev Cancer 5:251–262

Ok CY, Papathomas TG, Medeiros LJ, Young KH (2013) EBV-positive diffuse large B-cell lymphoma of the elderly. Blood 122(3):328–340

Staudt LM, Dave S (2005) The biology of human lymphoid malignancies revealed by gene expression profiling. Adv Immunol 87:163–208

Stein H, Warnke RA, Chan WC et al (2008) Diffuse large B-cell lymphoma, not otherwise specified. In: Swerdlow SH, Campo E, Harris NL et al (eds) WHO classification of tumours of haematopoetic and lymphoid tissues, 4th edn. International Agency for Research on Cancer (IARC), Lyon, pp 233–261

Visco C, Li Y, Xu-Monette ZY, Miranda RN, Green TM, Li Y, Tzankov A, Wen W, Liu WM, Kahl BS, d'Amore ES, Montes-Moreno S, Dybkær K, Chiu A, Tam W, Orazi A, Zu Y, Bhagat G, Winter JN, Wang HY, O'Neill S, Dunphy CH, Hsi ED, Zhao XF, Go RS, Choi WW, Zhou F, Czader M, Tong J, Zhao X, van Krieken JH, Huang Q, Ai W, Etzell J, Ponzoni M, Ferreri AJ, Piris MA, Møller MB, Bueso-Ramos CE, Medeiros LJ, Wu L, Young KH (2012) Comprehensive gene expression profiling and immunohistochemical studies support application of immunophenotypic algorithm for molecular subtype classification in diffuse large B-cell lymphoma: a report from the International DLBCL Rituximab-CHOP Consortium Program Study. Leukemia 26(9):2103–2113

Xia Y, Medeiros LJ, Young KH (2016) Signaling pathway and dysregulation of PD1 and its ligands in lymphoid malignancies. Biochim Biophys Acta 1865(1):58–71

Young KH, Medeiros LJ, Chan WC (2014) Diffuse large B-cell lymphoma. In: Orazi A, Weiss LM, Foucar K, Knowles DM (eds) Neoplastic hematopathology. Lippincott Willaims & Wilkins, Philadelphia, pp 502–565

See Also

(2012) B symptoms. In: Schwab M (ed) Encyclopedia of cancer, 3rd edn. Springer, Berlin/Heidelberg, p 331. doi:10.1007/978-3-642-16483-5_511

(2012) BCL-6. In: Schwab M (ed) Encyclopedia of cancer, 3rd edn. Springer, Berlin/Heidelberg, pp 363–364. doi:10.1007/978-3-642-16483-5_566

(2012) Bulky. In: Schwab M (ed) Encyclopedia of cancer, 3rd edn. Springer, Berlin/Heidelberg, p 574. doi:10.1007/978-3-642-16483-5_752

(2012) CD20. In: Schwab M (ed) Encyclopedia of cancer, 3rd edn. Springer, Berlin/Heidelberg, p 693. doi:10.1007/978-3-642-16483-5_922

(2012) CD30. In: Schwab M (ed) Encyclopedia of cancer, 3rd edn. Springer, Berlin/Heidelberg, p 697. doi:10.1007/978-3-642-16483-5_926

(2012) CD95. In: Schwab M (ed) Encyclopedia of cancer, 3rd edn. Springer, Berlin/Heidelberg, p 703. doi:10.1007/978-3-642-16483-5_939

(2012) Class switching. In: Schwab M (ed) Encyclopedia of cancer, 3rd edn. Springer, Berlin/Heidelberg, p 879. doi:10.1007/978-3-642-16483-5_1203

(2012) Doxorubicin. In: Schwab M (ed) Encyclopedia of cancer, 3rd edn. Springer, Berlin/Heidelberg, p 1159. doi:10.1007/978-3-642-16483-5_1722

(2012) Germinal center. In: Schwab M (ed) Encyclopedia of cancer, 3rd edn. Springer, Berlin/Heidelberg, p 1541. doi:10.1007/978-3-642-16483-5_2401

(2012) Immunoglobulin genes. In: Schwab M (ed) Encyclopedia of cancer, 3rd edn. Springer, Berlin/Heidelberg, p 1819. doi:10.1007/978-3-642-16483-5_2992

(2012) Ki-67. In: Schwab M (ed) Encyclopedia of cancer, 3rd edn. Springer, Berlin/Heidelberg, p 1943. doi:10.1007/978-3-642-16483-5_3213

(2012) Non-Hodgkin lymphoma. In: Schwab M (ed) Encyclopedia of cancer, 3rd edn. Springer, Berlin/Heidelberg, p 2537. doi:10.1007/978-3-642-16483-5_4110

(2012) Prednisone. In: Schwab M (ed) Encyclopedia of cancer, 3rd edn. Springer, Berlin/Heidelberg, p 2972. doi:10.1007/978-3-642-16483-5_4720

(2012) Somatic hypermutation. In: Schwab M (ed) Encyclopedia of cancer, 3rd edn. Springer, Berlin/Heidelberg, pp 3466–3467. doi:10.1007/978-3-642-16483-5_5410

(2012) Vincristine. In: Schwab M (ed) Encyclopedia of cancer, 3rd edn. Springer, Berlin/Heidelberg, p 3908. doi:10.1007/978-3-642-16483-5_6188

Diffuse Large Cell

▶ Diffuse Large B-Cell Lymphoma

Diffuse Mixed Small and Large

▶ Diffuse Large B-Cell Lymphoma

Diffuse or Nodular

▶ Mantle Cell Lymphoma

2′,2′-Difluoro-2′-deoxycytidine

▶ Gemcitabine

Dihydrogen Dioxide

▶ Hydrogen Peroxide

3,4-Dihydro-3-methyl-4-oxoimidazo (5,1-d)-as-tetrazine-8-carboxamide

▶ Temozolomide

3,4-Dihydro-3-methyl-4-oxoimidazo [5,1-d]-as-tetrazine-8-carboxamide

▶ Temozolomide

3,4-Dihydro-3-methyl-4-oxoimidazo (5,1-d)-1,2,3,5-tetrazine-8-carboxamide

▶ Temozolomide

Dihydrotestosterone Receptor

▶ Androgen Receptor

Dimethylfumarate

Peter Petzelbauer and Robert Loewe
Department of Dermatology, Division of General
Dermatology, Medical University of Vienna,
Vienna, Austria

Definition

Fumaric acid is one of two isomeric unsaturated dicarboxylic acids with the formula $HO_2CCH=CHCO_2H$ (the other being maleic acid with carboxylic acid groups in cis). Reactions with alcohol create fumaric acid esters (FAE) and water.

Chemistry and Pharmacodynamics

Dimethyl fumarate (DMF) is a lipophilic ester of fumaric acid (Fumaric acid is a dicarbonic acid and a component of the intracellular citric acid cycle). DMF is rapidly hydrolyzed to methyl hydrogen fumarate (MHF) in aqueous solutions at physiological pH.

In both DMF and MHF, nucleophiles can be added to the double bonds in the center of the molecule by a so-called Michaelis-type addition. Thus, they interact with thiols, as given in the tripeptide glutathione (GSH) or with cystein residues in larger proteins. This reactivity of DMF is higher than that of MHF (Fig. 1).

Characteristics

Biological Effects of FAEs In vitro

Due to the chemical structure of FAEs as low-molecular-weight electrophiles, much work has focused on their antioxidant effects. FAEs directly react with GSH. They initially deplete intracellular GSH levels, which return to normal or even supranormal levels 6–18 h thereafter. Possibly as a consequence of this GSH-FAE interaction, FAEs induce transcription and activity of so-called phase 2 enzymes, which are important

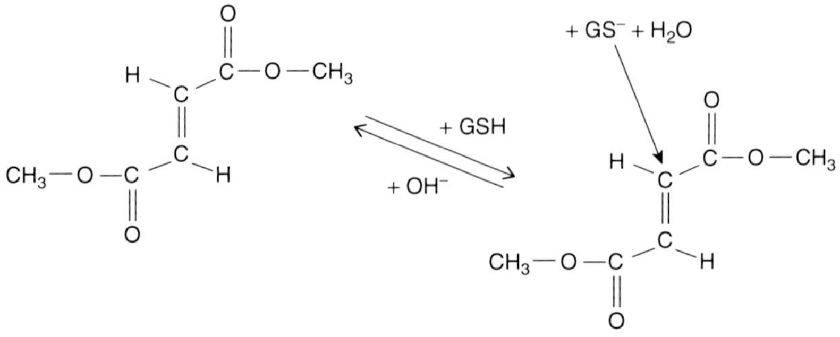

Dimethylfumarate,
Fig. 1 Reaction between
DMF and thiols (Modified
from Schmidt et al. 2007)

in the intermediate cellular metabolism for ► detoxification. Induction of Phase 2 enzymes, as, e.g., DT-diaphorase (NQO1 = NAD(P)H:quinone oxidoreductase I) and cytochrome b5 reductase, enhances cellular metabolism and decreases sensitivity toward environmental toxins. Phase 2 enzymes generally share similar antioxidant or electrophilic response elements (ARE/EpREs) within their promotor sequences. Interestingly, the induction of phase 2 enzymes by DMF is unevenly distributed within different tissues. In astrocytes, DMF leads to the induction of genes containing ARE/EpREs within their promotor sequences. Also in the gut and to a lesser degree in the liver, DMF increases activity of phase 2 enzymes, whereas in prostate tissue, no phase 2 enzymes are induced by DMF. It has been shown that DMF induces heme oxygenase 1 (HO-1) in human peripheral blood mononuclear cells. The induction of HO-1 expression by DMF was paralleled by the DMF-induced depletion of intracellular GSH, but a direct proof for a causal connection is pending.

A second approach toward identification of FAE effects was based on the clinical observation that FAEs reduce tumor necrosis factor (TNF)-induced cell activation. This effect of FAEs was identified in various different cells such as endothelial cells, fibroblasts, or ► melanoma cells. FAEs exert these effects by interacting with the ► nuclear factor-κB (NF-κB) pathway downstream the Iκ-kinases (IKK); they retain activated NF-κB proteins in the cytosol and prevent their nuclear translocation. As a result, NF-κB cannot

bind to DNA thereby preventing NF-κB-mediated gene transcription. In these experiments, DMF proved superior to MHF. Although the mode of this inhibition is not yet completely elucidated, it has been speculated that this is also mediated through interaction of FAEs with the redox status of the cell.

Drug Formulations of Currently Used Fumaric Acid Esters (FAEs)

1. Fumaderm®, a mixture of FAEs (120 mg DMF, 87 mg Ca^{2+} ethyl hydrogen fumarate, 5 mg Mg^{2+} ethyl hydrogen fumarate, 3 mg Zn^{2+} ethyl hydrogen fumarate) was registered in Germany in 1994. It is used for oral treatment of psoriasis, a chronic inflammatory skin condition. Off-label use revealed anecdotic efficacy in other inflammatory skin conditions such as granuloma annulare, cutaneous sarcoidosis, or necrobiosis lipoidica. Moreover, Fumapharm has initiated a phase IIa clinical trial with Fumaderm® for the treatment of multiple sclerosis. This study showed a significant reduction of gadolinium-enhancing lesions on T1-weighted magnetic resonance imaging (MRI) brain scans. This study could also show that exacerbations of the disease could be reduced by low-dose FAE therapy.

2. DMF as a monosubstance, named BG12®, was used for a European placebo-controlled multicenter phase III trial in psoriasis patients. In this study, treatment with BG12 for 16 weeks led to clinical response and improvement in 68% of the BG12 treated patients, whereas

placebo treated patients improved only for 10% Biogen has now initiated a phase III study in patients with multiple sclerosis using BG12.

In dendritic cells and keratinocytes, FAEs possess antiproliferative and proapoptotic properties. It is yet not clear whether this is based on the ability of FAEs to alter the redox status of the cell and/or by a direct interaction with the NF-kB pathway.

In malignancies in vitro, effects of DMF have been evaluated in ► lung cancer, ► breast cancer, ► neuroblastoma, and glioblastoma. DMF has been shown to increase cytotoxicity of chemotherapeutics when given in combination with for example ► mitomycin C or streptonigrin. This was shown for glioblastoma, lung cancer, and mammary cancer cell lines. DMF also sensitizes fibroblasts, neuroblastoma cells, as well as human ► bladder cancer cell lines to irradiation and therefore increases their ► radiation sensitivity.

Biological Effects of FAEs In vivo

Anti-inflammatory effects of oral treatment with Fumaderm® and BG12 in psoriasis are well documented in human patients. Although DMF is more active than MHF in vitro, it is still unclear whether DMF or MHF mediate biological effects in vivo. This is because (i) DMF hydrolyses into MHF and (ii) pharmacokinetic studies failed to detect DMF in human plasma, whereas MHF was measurable. However, these studies did not rule out that DMF was bound to cystein-containing plasma proteins or – due to its lipophilia – has rapidly entered cells and thus escaped detection by HPLC techniques.

For malignant disease, only preclinical data in animal models exist. DMF has been tested in a model of chemically induced colon cancer, where DMF reduced cancer ► progression and ► invasion. In humanized SCID mouse models for ► melanoma, oral treatment with DMF has reduced tumor growth and lymphogenic metastatic spread. In this model, beneficial effects of DMF were based on its antiproliferative and proapoptotic properties on melanoma cells.

Clinical Aspects and Future Perspectives

In many human malignancies the NF-κB pathway is constitutive active. This causes increased expression of many tumor progression genes (regulating cell cycle and apoptosis of tumor cells and the composition of the tumor matrix). Indeed, animal models have provided a direct link between activation of NF-κB and tumor progression. Moreover, NF-κB-dependent genes are clearly involved in the development of chemoresistance; inhibition of NF-κB has been shown to reduce chemoresistance.

DMF fulfills several criteria to justify clinical trials in human malignancies; first, it has been shown to be beneficial in animal models for malignant diseases; second, it inhibits NF-κB-dependent gene expression, which would reduce tumor growth and ► metastasis and also may reduce the development of chemoresistance; third, DMF has been clearly shown to be safe even in long-term treatment in human psoriasis patients.

References

https://clinicaltrials.gov/ct2/show/NCT02337426
https://clinicaltrials.gov/ct2/show/NCT02546440
Kastrati I, Siklos MI, Calderon-Gierszal EL et al (2016) Dimethyl Fumarate Inhibits the Nuclear Factor κB Pathway in Breast Cancer Cells by Covalent Modification of p65 Protein. J Biol Chem. 291(7):3639–47. doi:10.1074/jbc.M115.679704. Epub 2015 Dec 18
Lehmann J, Listopad J, Rentzsch C et al (2007) Dimethylfumarate induces immunosuppression via glutathione depletion and subsequent induction of heme oxygenase 1. J Invest Dermatol 127:635–645
Loewe R, Holnthoner W, Gröger M et al (2002) Dimethylfumarate inhibits TNF-induced nuclear entry of NF-κB/p65 in human endothelial cells. J Immunol 168:4781–4787
Loewe R, Valero T, Kremling S et al (2006) Dimethylfumarate impairs melanoma growth and metastasis. Cancer Res 66:11888–11896
Pereira MA, Barnes LH, Rassman VL et al (1994) Use of azoxymethane-induced foci of aberrant crypts in rat colon to identify potential cancer chemopreventive agents. Carcinogenesis 15:1049–1054
Schmidt TJ, Ak M, Mrowietz U (2007) Reactivity of dimethyl fumarate and methylhydrogen fumarate towards glutathione and N-acetyl-L-cysteine – preparation of S-substituted thiosuccinic acid esters. Bioorg Med Chem 15:333–342

Dioxin

Thomas A. Gasiewicz and Ellen C. Henry
University of Rochester Medical Center,
Rocheser, NY, USA

Synonyms

2,3,7,8-TCDD;   2,3,7,8-Tetrachlorodibenzo-*p*-
dioxin; TCDD

Definition

Dioxin is an unwanted by-product of a number of
chemical and industrial processes. It was first
observed as a contaminant formed in the synthesis
of trichlorophenols used for the production of
certain herbicides. Later it was recognized that
small amounts of dioxin (as well as other dioxins)
can be produced during different types of combustion processes including the burning of
chlorine-containing materials such as chemical
and hospital wastes and sewage sludge. Dioxin
may also be formed during manufacturing processes utilizing chlorine. These include the
bleaching of pulp and paper and chlorine-
dependent regeneration of metal catalysts. Dioxin
and numerous dioxin-like chemicals are ubiquitously present in trace amounts in the environment
(Fig. 1).

Characteristics

Based on studies in experimental animals, dioxin
is considered to be one of the most potent tumorigenic agents known. The results of epidemiologic studies are more equivocal in terms of

Dioxin, Fig. 1 Chemical structure of dioxin

carcinogenic potency in humans. Carcinogenic
and other toxic responses elicited by dioxin are
mediated by a unique mechanism involving binding to a transcription factor (aryl hydrocarbon
receptor, AhR) and subsequent modulation of
gene expression. The exact relationships between
particular gene alterations and ultimate carcinogenic responses to dioxin remain obscure. However, the growing recognition of the critical
involvement of this transcription factor in diverse
cells/tissues and in response to many possible
endogenous ligands suggests that dioxin likely
impacts many physiological processes.

Carcinogenic Activity

Dioxin is considered to be one of the most toxic
synthetic compounds known with acute lethal
doses in several animal species being in the
range of µg/kg body weight. In addition, numerous experimental studies have consistently shown
dioxin to be a potent carcinogen. Although there
are differences in sensitivity among species, some
rodents exhibit significantly increased and dose-
dependent tumor incidence with average daily
exposures of 100 ng/kg body weight and higher.
A variety of tumor types in several tissues including liver, lung, and thyroid have been reported.
The findings that in most experimental systems
dioxin fails to induce mutations, dioxin-derived
▶ adducts to DNA have not been detected, and
dioxin enhances the incidence of various tumor
types following initiation with known carcinogens, are consistent with dioxin acting primarily,
if not exclusively, as a tumor promoter. However,
although there is little or no evidence for dioxin to
be directly genotoxic, it may appear to be a complete carcinogen due to its exceptionally high
potency as a tumor promoter.

Mechanisms

The toxic and carcinogenic effects of dioxin are
mediated by its binding to and activation of a
transcription factor termed the ▶ aryl hydrocarbon receptor (AhR). This belongs to a family of
proteins containing a basic-helix-loop-helix
(bHLH) – PAS (Per-Arnt-Sim) domain structure.
The known bHLH-PAS proteins are involved in
regulating responses to signals in the tissue

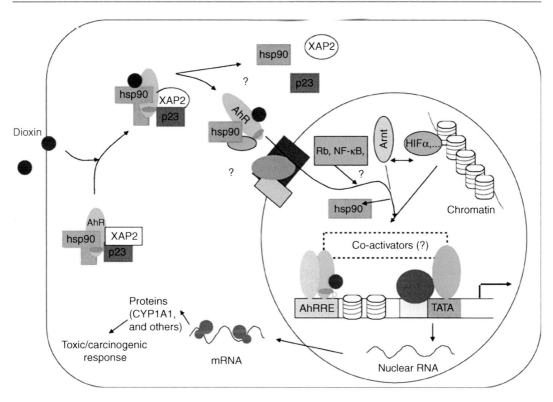

Dioxin, Fig. 2 AhR signaling pathway. Dioxin binds to the AhR to initiate a series of events leading to modulation of gene expression

environment such as oxygen tension and circadian rhythms and serve regulatory roles in development and cellular differentiation. Like many of these proteins, the AhR has been conserved through evolution. However, unlike most other proteins in this family, the AhR is ligand activated and its exact normal function in tissue processes is not clearly resolved. Evidence that the AhR mediates most, if not all, biological responses to dioxin is based in part on studies examining structure-activity relationships of other dioxin-like chemicals, and genetic studies in mice that have a mutant AhR protein. In particular, mice in which expression of a functional AhR has been "knocked out" (knockout mice) are insensitive to the toxic effects of dioxin. In the absence of ligand, the AhR appears to be localized primarily in the cellular cytosol complexed with several proteins including the chaperone protein ► hsp90, co-chaperone p23 protein, and an immunophilin-like protein XAP2 (Fig. 2). Upon ligand binding, the AhR is translocated to the cell

nucleus, dimerizes with another bHLH-PAS protein, Arnt (also known as ► hypoxia inducible factor (HIF) β), and binds to specific response elements (AhR response elements; AhREs) (also called xenobiotic response elements (XREs) or dioxin response elements (DREs)) within regulatory domains of responsive genes to modulate the expression of these genes. The AhR has been shown to interact with several proteins involved in other signaling pathways. These include the ► estrogen receptor and NF-κB. As such, some responses elicited by dioxin may not be dependent on the ability of the AhR-Arnt complex to bind AhREs. Nevertheless, the toxic and carcinogenic actions of dioxin are believed to result from its interference with unknown endogenous AhR ligands and/or the inappropriate and prolonged alteration of AhR-responsive genes and signaling pathways.

There are several postulated mechanisms whereby dioxin may act at stages involved in cancer initiation and progression to facilitate the development of malignancies. Of the identified AhR-responsive genes, many encode enzymes (e.g., ► cytochrome P450 (CYP) 1A1 and 1B1) known to metabolize and activate carcinogens to intermediates that cause DNA mutations. Although dioxin is resistant to metabolism, the induction of these pathways enhances the likelihood of bioactivation of other potential carcinogens that are in the environment. It has also been proposed that induction of the CYPs may generate ► reactive oxygen species that can directly damage DNA. Several studies have observed dioxin exposure to result in the expression of genes such as *p53* and *p21Waf1* indicative of DNA damage or cellular stress. Other studies have linked high CYP1A1 enzyme activity with increased cancer risk, and several carcinogens, including benzo[a] pyrene, induce tumors in normal mice but not AhR knockout mice. Increased expression of these genes by dioxin is also known to affect tissue levels of several hormones and growth factors, including estrogen and thyroxine, as well as their receptors. The progression of several tumors is known to be hormone dependent. Dioxin has been shown to alter several signaling pathways regulating cellular differentiation, proliferation, and cell death (► apoptosis). Furthermore the modulation of these cellular processes, intrinsic to metastatic progression, are affected in a variety of cell types by dioxin. In particular, many studies demonstrate a role, although as yet not clearly defined, of the AhR in regulating the cell cycle. This may occur through several complex pathways involving AhR-dependent activation of cellular kinases, e.g., ERK, or direct AhR interaction with cell cycle regulatory proteins including the tumor suppressor ► Retinoblastoma Protein (Rb). The latter interaction has been suggested to regulate key cell cycle components mediated by an alteration of E2F transcription factors. Data suggest that dioxin may affect the ability of tumor cells to migrate and metastasize.

Immune surveillance mechanisms are important in the control of malignant cell growth. In experimental animals, dioxin has consistently been shown to be a potent immunosuppressive agent affecting both humoral and cell-mediated immune pathways. Furthermore, dioxin exposure has been shown to increase tumor cell survival following engraftment. Prolonged inflammation is also associated with increased cancer development, and the AhR is known to interact with NF-kB proteins that are involved in regulating inflammatory responses. In addition, several proinflammatory genes (e.g., cyclooxygenase-2) are AhR-responsive and liver tumor promotion by dioxin in mice has been shown to depend on signaling by the inflammatory cytokines TNF/IL-1 as well as AhR. There is also evidence that the AhR may be involved in the regulation of ► angiogenesis important for tumor growth; AhR knockout animals exhibit defects in tissue vascularization. This has been postulated to be related to cross-talk between the AhR and HIF signaling pathways, since Arnt is also a dimeric partner for HIFα. Ultimately, the exact mechanisms underlying the carcinogenic activity of dioxin are poorly understood. As indicated above, there are likely many complex, and possibly tissue-specific, mechanisms whereby dioxin exhibits potent tumor promoter activity. Clearly, however, the multifaceted nature of the responses to dioxin reflect an importance of the AhR signaling pathway in fundamental processes, deregulation of which may contribute to altered cellular phenotype and uncontrolled growth.

Clinical Relevance

Several epidemiologic studies link dioxin exposure with cancer in human populations exposed occupationally or environmentally following accidental release. Among the increased cancers reported were lung cancer, lymphomas, and leukemias. The strongest evidence for dioxin carcinogenicity in humans was for "all cancers combined" in these high-exposure cohorts. However, in all of these studies the body burden exposure levels were very high, being often greater than several hundred μg/kg body weight. In many studies the increased cancer incidence was relatively small and the association, although significant, was weak. Based on limited human evidence and sufficient evidence in animals, the International Agency for Research on Cancer (IARC) and the National Toxicology Program classified dioxin as a human carcinogen. General population exposures to dioxin and dioxin-like chemicals are much lower, and occur mainly through food consumption. Average background tissue levels of dioxin are in the range of 2-3 ng/kg-fat, and have decreased two- to three-fold during the last 30 years (as dioxin emissions from manufacturing and incineration processes have decreased). These tissue concentrations are many-fold lower than reported in epidemiological studies in which a positive association with increased cancer incidence was observed. The actual risk for cancer development at low-level exposure remains uncertain. Several studies indicate that the human AhR has lower affinity for dioxin than the AhR in most animal species. Identified human AhR genetic polymorphisms have not yet been clearly associated with increased or decreased cancer risk. Nevertheless, the half-life of dioxin, a molecule resistant to metabolism, appears to be considerably longer in humans, ranging from 7 to 10 years. Furthermore, a variety of human cells and tissues respond to dioxin exposure in a manner similar to those of animals, at least in terms of some acute biochemical and cellular responses (e.g., *CYP1A1* induction and altered cellular differentiation). The full clinical significance of dioxin exposure may not be realized until we have a greater understanding of the physiological role of the AhR, its interactions with other signaling pathways, and the influence of the diverse putative endogenous ligands. The findings that dioxin exposure and/or altered activity of the AhR are associated with modulation of particular gene batteries resulting in altered immune responses, modulated angiogenesis, and deregulated cell cycle and cell proliferation, as well as the recognition that these responses may be tissue-specific and/or ligand-specific, suggest the possibility of new and unique therapeutic targets for treatment of various cancers.

Cross-References

▶ Adducts to DNA
▶ Apoptosis
▶ Angiogenesis
▶ Aryl Hydrocarbon Receptor
▶ Cytochrome P450
▶ Estrogen Receptor
▶ Hsp90
▶ Nuclear Factor-κB
▶ P21
▶ Reactive Oxygen Species
▶ Retinoblastoma Protein, Biological and Clinical Functions

References

Dietrich C, Kaina B (2010) The aryl hydrocarbon receptor (AhR) in the regulation of cell-cell contact and tumor growth. Carcinogenesis 31:1319–1328

Fernandez-Salguero PM (2010) A remarkable new target gene for the dioxin receptor. The Vav3 proto-oncogene links AhR to adhesion and migration. Cell Adh Migr 4:172–175

Huang G, Elferink CJ (2005) Multiple mechanisms are involved in Ah receptor-mediated cell cycle arrest. Mol Pharmacol 67:88–96

IARC (2012) 2,3,7,8-Tetrachlorodibenzo-para-dioxin, 2,3,4,7,8-pentachlorodibenzofuran, and 3,3',4,4',5-pentaclorobiphenyl. Monogr Eval Carcinog Risks Hum 100F: 339-378

Knerr S, Schrenk D (2006) Carcinogenicity of 2,3,7,8-tetrachlorodibenzo-*p*-dioxin in experimental models. Mol Nutr Food Res 50:897–907

Murray IA, Patterson AD, Perdew GH (2014) Aryl hydrocarbon receptor ligands in cancer: friend and foe. Nat Rev Cancer 14:801-814

Tuomisto J, Tuomisto JT (2012) Is the fear of dioxin cancer more harmful than dioxin? Toxicol Lett 210:338-344

See Also

(2012) BHLH-PAS proteins. In: Schwab M (ed) Encyclopedia of cancer, 3rd edn. Springer, Berlin/Heidelberg, p 389. doi:10.1007/978-3-642-16483-5_605

(2012) Carcinogen. In: Schwab M (ed) Encyclopedia of cancer, 3rd edn. Springer, Berlin/Heidelberg, p 644. doi:10.1007/978-3-642-16483-5_839

(2012) Chaperone. In: Schwab M (ed) Encyclopedia of cancer, 3rd edn. Springer, Berlin/Heidelberg, p 754. doi:10.1007/978-3-642-16483-5_1046

(2012) Cyclooxygenase-2. In: Schwab M (ed) Encyclopedia of cancer, 3rd edn. Springer, Berlin/Heidelberg, p 1035. doi:10.1007/978-3-642-16483-5_1435

(2012) Domain structure. In: Schwab M (ed) Encyclopedia of cancer, 3rd edn. Springer, Berlin/Heidelberg, p 1150. doi:10.1007/978-3-642-16483-5_1703

(2012) Gene battery. In: Schwab M (ed) Encyclopedia of cancer, 3rd edn. Springer, Berlin/Heidelberg, p 1522. doi:10.1007/978-3-642-16483-5_2364

(2012) Genetic polymorphism. In: Schwab M (ed) Encyclopedia of cancer, 3rd edn. Springer, Berlin/Heidelberg, p 1528. doi:10.1007/978-3-642-16483-5_2382

D

(2012) Half-life. In: Schwab M (ed) Encyclopedia of cancer, 3rd edn. Springer, Berlin/Heidelberg, p 1625. doi:10.1007/978-3-642-16483-5_2554

(2012) Hypoxia inducible factor. In: Schwab M (ed) Encyclopedia of cancer, 3rd edn. Springer, Berlin/Heidelberg, p 1796. doi:10.1007/978-3-642-16483-5_2927

(2012) Immunophilins. In: Schwab M (ed) Encyclopedia of cancer, 3rd edn. Springer, Berlin/Heidelberg, p 1827. doi:10.1007/978-3-642-16483-5_3003

(2012) Inflammatory response. In: Schwab M (ed) Encyclopedia of cancer, 3rd edn. Springer, Berlin/Heidelberg, pp 1858–1859. doi:10.1007/978-3-642-16483-5_3048

(2012) Knock-out mouse. In: Schwab M (ed) Encyclopedia of cancer, 3rd edn. Springer, Berlin/Heidelberg, p 1957. doi:10.1007/978-3-642-16483-5_3239

(2012) Leukemia. In: Schwab M (ed) Encyclopedia of cancer, 3rd edn. Springer, Berlin/Heidelberg, p 2005. doi:10.1007/978-3-642-16483-5_3322

(2012) Ligands. In: Schwab M (ed) Encyclopedia of cancer, 3rd edn. Springer, Berlin/Heidelberg, p 2040. doi:10.1007/978-3-642-16483-5_3352

(2012) Lymphoma. In: Schwab M (ed) Encyclopedia of cancer, 3rd edn. Springer, Berlin/Heidelberg, p 2124. doi:10.1007/978-3-642-16483-5_3463

(2012) P53. In: Schwab M (ed) Encyclopedia of cancer, 3rd edn. Springer, Berlin/Heidelberg, p 2747. doi:10.1007/978-3-642-16483-5_4331

(2012) Response elements. In: Schwab M (ed) Encyclopedia of cancer, 3rd edn. Springer, Berlin/Heidelberg, p 3264. doi:10.1007/978-3-642-16483-5_5058

Dioxin Receptor

► Aryl Hydrocarbon Receptor

Dipeptidyl-Peptidase IV

► CD26/DPPIV in Cancer Progression and Spread

Dipropylacetic Acid

► Valproic Acid

Directed Migration

► Chemoattraction

Directed Motility

► Chemoattraction

Disintegrin Metalloproteases

► ADAM Molecules

Disordered Domain

► Intrinsically Unstructured Proteins

Distal Intestinal Serine Protease (Mouse Only, TMPRSS8)

► Serine Proteases (Type II) Spanning the Plasma Membrane

Distant Bystander Effect

► Bystander Effect

Distant Bystander Effects

► Abscopal Effects

Distributed Stem Cells

► Tissue Stem Cells

DLC1

Judy W. P. Yam, Chun-Ming Wong and Irene O. L. Ng
Department of Pathology, The University of Hong Kong, Hong Kong, China

Synonyms

ARHGAP7; Deleted in liver cancer 1; Rho-GTPase-activating protein 7; STARD12; START domain-containing protein

Definition

DLC1 is a ▶ tumor suppressor gene frequently underexpressed in various types of cancers. Genomic deletion and promoter hypermethylation may account for the inactivation of *DLC1* in human cancers. Functionally, overexpression of *DLC1* can result in the ▶ tumor suppression in cancer cells and regulation of cytoskeleton organization.

Characteristics

Deleted in liver cancer 1 (*DLC1*), also known as *Rho-*▶ *GTPase-activating protein 7* (*ARHGAP7*) and *START domain-containing protein* (*STARD12*), is a putative tumor suppressor gene identified from primary ▶ hepatocellular carcinoma (HCC) in 1998. Human *DLC1* gene was mapped to chromosome region 8p21.3–22. The full-length *DLC1* mRNA is 6 kb long and consists of 14 exons which encode a protein of 1,091 amino acids. In silico analysis indicated that *DLC1* shares 86% sequence homology with rat *p122 RhoGAP* and is likely to be its human homologue. DLC1 protein contains three major functional domains, namely, SAM (sterile alpha motif), RhoGAP (GTPase-activating protein for Rho-like GTPases), and START (steroidogenic acute regulatory (StAR)-related lipid transfer) (Fig. 1). The RhoGAP domain is a characteristic of RhoGAP family proteins, functions to catalyze the intrinsic GTPase activity of Rho proteins.

Functions of DLC1

RhoGAP protein converts the active GTP-bound Rho to the inactive GDP-bound form. On the other hand, guanine nucleotide exchange factor (GEF) acts to activate Rho proteins. Rho proteins are important regulators in the remodeling of actin cytoskeleton, regulation of transcription, cell proliferation, metastasis, and tumorigenesis (Fig. 2). The best characterized members of Rho proteins are RhoA, Rac1, and Cdc42. RhoGAP serves tumor suppressor function by downregulating Rho GTPase activity. Therefore, negative regulation of Rho proteins by RhoGAP activity contributes to the functional role of DLC1.

Negative Regulation of Cytoskeleton

The in vitro GAP activities of DLC1 specific for RhoA and Cdc42 have previously been demonstrated. In cell model, expression of DLC1 resulted in inhibition of stress fiber formation and extensive cell rounding. These morphological changes were similar to the effects of Rho inhibitor, C3 exoenzyme, implicating that DLC1 negatively regulates cytoskeletal organization via inhibition of Rho proteins. The importance of RhoGAP of DLC1 was further supported by failure of DLC1 RhoGAP mutant to induce morphological changes. Tensin2, a focal adhesion protein, has been identified as the first and novel binding partner of DLC1. DLC1–tensin2 protein complex

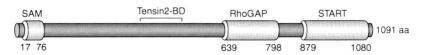

DLC1, Fig. 1 Schematic representation of DLC1 protein. Structural domains of DLC1, namely, sterile alpha motif (*SAM*), Rho-GTPase-activating protein (*RhoGAP*), steroidogenic acute regulatory (*StAR*)-related lipid transfer (*START*), and the tensin2 binding domain (*Tensin2-BD*) are shown

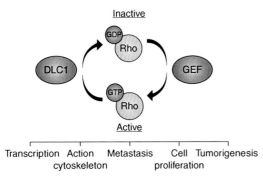

DLC1, Fig. 2 Functional role of DLC1 protein. Rho proteins exist as active GTP-bound form and inactive GDP-bound form. RhoGAP functions to convert the active Rho to the inactive state. On the other hand, guanine nucleotide exchange factor (*GEF*) acts to activate Rho proteins. Rho proteins are important regulators in the remodeling of actin cytoskeleton, regulation of transcription, cell proliferation, metastasis, and tumorigenesis. Negative regulation of Rho proteins by RhoGAP activity may contribute to the functional role of DLC1

localizes in caveolae where Rho GTPases are concentrated. It is postulated that DLC1–tensin2 complex brings DLC1 in close proximity to Rho GTPases which are enriched in caveolae and facilitates the inactivation of Rho proteins by DLC1.

Mouse model of DLC1 revealed that *DLC1* $^{+/-}$ heterozygous deletion mutant did not display any phenotypic abnormalities, while *DLC1* $^{-/-}$ homozygous deletion mutant (knock-out mice) was embryonic lethal. *DLC1* $^{-/-}$ embryos showed defects in several organs, including neural tube, brain, heart, and placenta. Fibroblasts isolated from *DLC1* $^{-/-}$ homozygous deletion embryos displayed aberrant cytoskeletal organization with diminished actin stress fiber formation and focal adhesions.

Tumor Suppressive Effect

Besides the role in the regulation of cytoskeletal reorganization, DLC1 also possesses growth inhibitory activity in cancer cell lines. The growth inhibitory effect of DLC1 was first demonstrated in HCC cell lines. Loss of growth suppressive effect with the DLC1 RhoGAP mutant demonstrated that the RhoGAP activity of DLC1 was associated with its growth suppressive effect in tumor cell lines. Expression of *DLC1* in HCC cells resulted in significant inhibition in cell proliferation, anchorage independent growth, and in vivo tumorigenicity in nude mice. Apart from HCC cells, the tumor suppressive effect of DLC1 has also been demonstrated in breast cancer and non-small cell lung carcinoma cells. Moreover, stable expression of *DLC1* resulted in inhibition of migratory and invasive abilities of HCC and breast cancer cells. Furthermore, expression of *DLC1* in HCC cells induced caspase-3-mediated apoptosis. The tumor suppressive role of *DLC1* has also been implicated in genome-wide expression profiling approach. Transcriptional profiling analysis of breast cancer cells that exhibit different metastatic efficiencies using microarrays identified *DLC1* as one of the differentially expressed genes. Functional analysis has further demonstrated that *DLC1* acts as a metastasis suppressor in breast cancer cells.

Clinical Relevance

DLC1 was first discovered from a subtractive hybridization screening study in primary HCC. Studies have indicated that *DLC1* possesses tumor-suppressing function and is implicated in human ▶ carcinogenesis. *DLC1* mRNA is frequently underexpressed in various human cancers, particularly in non-small cell lung carcinoma and nasopharyngeal carcinoma, and downregulation of *DLC1* mRNA expression has been observed in more than 90% of primary tumor samples. The inactivation of *DLC1* in human cancers could be due to both genetic and epigenetic alterations.

Genetic Alteration

DLC1 is mapped at chromosome 8p21.3–22, which is one of the most commonly deleted regions in human cancers. As indicated by its name, *DLC1* gene is frequently deleted in human HCC. Loss of heterozygous and single nucleotide polymorphism (SNP) analyses have revealed that hemizygous deletion of *DLC1* gene is present in approximately half of primary HCC and almost 90% of head and neck squamous cell carcinomas. Although not at a high frequency, homozygous deletion has also been reported in human HCC and medulloblastoma. Somatic mutation is rarely found in *DLC1* gene, and SNP on *DLC1* also shows no association with human cancers.

Epigenetic Silencing

In established cancer cell lines, frequent loss of *DLC1* gene expression, even in the absence of gene deletion, has been observed. In these cancer cell lines, *DLC1* expression can be significantly restored by administrating demethylating drugs or histone deacetylase inhibitors to the cells. This indicates that in addition to gene deletion, epigenetic alterations including DNA methylation and histone deacetylation also play important roles in *DLC1* inactivation. In all human cancer types tested, recurrent DNA ► methylation on the promoter region of *DLC1* gene has been detected in the primary cancer samples, and *DLC1* promoter methylation is closely associated with loss of gene expression. In ► cervical cancer, ► nasopharyngeal carcinoma, and ► multiple myeloma, DNA methylation accounts for more than 80% of *DLC1* gene inactivation in these diseases. Interestingly, *DLC1* promoter methylation has also been found in 71% of benign prostatic hyperplasia and was closely associated with the serum prostate-specific antigen level. These findings suggest that *DLC1* promoter methylation may occur in the early stage of carcinogenesis and has a potential value for early diagnosis of ► prostate cancer.

Cross-References

- ► Carcinogenesis
- ► Cervical Cancers
- ► GTPase
- ► Hepatocellular Carcinoma
- ► Methylation
- ► Multiple Myeloma
- ► Nasopharyngeal Carcinoma
- ► Prostate Cancer
- ► Tumor Suppression
- ► Tumor Suppressor Genes

References

Ng IOL, Liang ZD, Cao L et al (2000) DLC1 is deleted in primary hepatocellular carcinoma and exerts inhibitory effects on the proliferation of hepatoma cell lines with deleted DLC1. Cancer Res 60:6581–6584
Wong CM, Lee JMF, Ching YP et al (2003) Genetic and epigenetic alterations of DLC1 gene in hepatocellular carcinoma. Cancer Res 63:7646–7651
Wong CM, Yam JWP, Ching YP et al (2005) Rho GTPase activating protein DLC1 (deleted in liver cancer) suppresses cell proliferation and invasion in hepatocellular carcinoma. Cancer Res 65:8861–8868
Yam JWP, Ko FCF, Chan CY et al (2006) Interaction of DLC1-tensin2 complex with caveolin-1 and implications in tumor suppression. Cancer Res 66:8367–8372
Yuan BZ, Miller MJ, Keck CL et al (1998) Cloning, characterization, and chromosomal localization of a gene frequently deleted in human liver cancer (DLC1) homologous to rat RhoGAP. Cancer Res 58:2196–2199

See Also

(2012) Actin cytoskeleton. In: Schwab M (ed) Encyclopedia of cancer, 3rd edn. Springer, Berlin/Heidelberg, p 19. doi:10.1007/978-3-642-16483-5_43
(2012) Caveolae. In: Schwab M (ed) Encyclopedia of cancer, 3rd edn. Springer, Berlin/Heidelberg, p 682. doi:10.1007/978-3-642-16483-5_894
(2012) DNA methylation. In: Schwab M (ed) Encyclopedia of cancer, 3rd edn. Springer, Berlin/Heidelberg, p 1140. doi:10.1007/978-3-642-16483-5_1682
(2012) Homozygous deletion. In: Schwab M (ed) Encyclopedia of cancer, 3rd edn. Springer, Berlin/Heidelberg, p 1729. doi:10.1007/978-3-642-16483-5_2807
(2012) Knock-out mouse. In: Schwab M (ed) Encyclopedia of cancer, 3rd edn. Springer, Berlin/Heidelberg, p 1957. doi:10.1007/978-3-642-16483-5_3239
(2012) Loss of heterozygosity. In: Schwab M (ed) Encyclopedia of cancer, 3rd edn. Springer, Berlin/Heidelberg, pp 2075-2076. doi:10.1007/978-3-642-16483-5_3415
(2012) Nude mice. In: Schwab M (ed) Encyclopedia of cancer, 3rd edn. Springer, Berlin/Heidelberg, p 2584. doi:10.1007/978-3-642-16483-5_4172
(2012) Promoter. In: Schwab M (ed) Encyclopedia of cancer, 3rd edn. Springer, Berlin/Heidelberg, p 3004. doi:10.1007/978-3-642-16483-5_4768
(2012) Promoter hypermethylation. In: Schwab M (ed) Encyclopedia of cancer, 3rd edn. Springer, Berlin/Heidelberg, p 3004. doi:10.1007/978-3-642-16483-5_4769
(2012) Single nucleotide polymorphism. In: Schwab M (ed) Encyclopedia of cancer, 3rd edn. Springer, Berlin/Heidelberg, p 3412. doi:10.1007/978-3-642-16483-5_5316
(2012) Tensin2. In: Schwab M (ed) Encyclopedia of cancer, 3rd edn. Springer, Berlin/Heidelberg, p 3647. doi:10.1007/978-3-642-16483-5_5726

DMBT1

► Deleted in Malignant Brain Tumors 1

DNA Damage

Stephen T. Durant
R&D, Oncology, Innovative Medicines,
AstraZeneca, Little Chesterford, UK

Synonyms

DNA lesion

Definition

DNA damage refers to the myriad of chemical or structural perturbations that can affect the function of genes encoded along the DNA macromolecule of the cell. The continuous exposure of all cells to many types of DNA damaging agents can alter the backbone of the DNA double helix or the nucleotide base components that connect the two strands together and make up the four-lettered genetic code – guanine (G), cytosine (C), thymine (T), and adenine (A). Un-repaired or inappropriately repaired DNA damage can therefore alter genetic information controlling cell function, resulting in mutation, ▶ chromosomal instability, or ▶ aneuploidy – the hallmarks of all cancers.

Characteristics

DNA resides in both the nucleus and mitochondria of human cells, and both cellular regions contain significant levels of potentially damaging agents, derived either from the environment (exogenous) or from metabolic processes continually occurring inside the cell (endogenous). Although DNA is heavily protected by a complex packaging arrangement of supercoiled strands, wrapped around histone proteins within protective chromosomal structures, it is estimated that anything from 1,000 to 1,000,000 DNA lesions occur in any given cell per day. Because genomic DNA inside the nucleus contains the genetic code for expressing most of the proteins required for normal cell function, unrepaired or incorrectly repaired damage to nuclear DNA threatens the integrity of the genome and can result in disruption of cellular processes that control cellular growth and differentiation and therefore the development of cancer. DNA damage and subsequent mutagenesis in somatic cells also drive the processes of aging, degenerative diseases, and evolution.

Constant exposures to a barrage of environmental and cellular metabolic DNA damaging agents create enormous selective pressures on all prokaryotic and eukaryotic cells on Earth, from bacteria to human, to evolve a complex protective array of DNA damage sensing, signaling, and repair proteins. Subsequently, most of these proteins are conserved from single-cell organisms to humans. These enzymes and molecular scaffolds constantly interact to scan, recognize, alert, and repair many forms of DNA damage. They represent crucial cellular defense pathways that not only protect the integrity of the genome of a single cell but, coupled to abortive programmed cell death (or ▶ apoptosis) pathways, allow whole organisms to clear themselves of problematic and potentially malignant tissue. This is triggered when levels of DNA damage pass a certain threshold and this defines the repair capacity of a cell. This can vary from cell to cell and tissue to tissue, and some DNA repair pathways are more error prone and less efficient than others. Some repair pathways may become dysfunctional or deregulated and indirectly promote ▶ carcinogenesis, due largely to the activities of inappropriate cellular responses rather than the damage itself. Not surprisingly, individuals who inherit defects or mutations in specific DNA damage sensing and/or repair proteins are predisposed to and significantly more likely to develop early onset cancers, premature aging, and degenerative disorders. There are many types of DNA damage, and these are dealt with by several pathways of DNA repair that specialize in recognizing and repairing particular types of lesion.

Agents Causing Bulky, Structurally Distorting DNA Adducts

1. ▶ UV radiation. Predominantly UV-B and UV-C, with UV-B representing the wavelength

DNA Damage,
Fig. 1 Examples of DNA damage

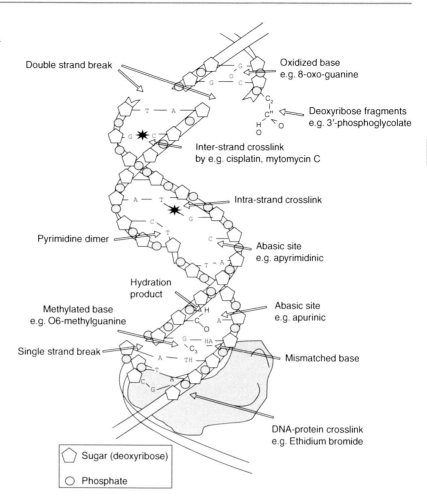

Double strand break

Oxidized base
e.g. 8-oxo-guanine

Deoxyribose fragments
e.g. 3'-phosphoglycolate

Inter-strand crosslink
by e.g. cisplatin, mytomycin C

Intra-strand crosslink

Pyrimidine dimer

Abasic site
e.g. apyrimidinic

Hydration
product

Methylated base
e.g. O6-methylguanine

Abasic site
e.g. apurinic

Single strand break

Mismatched base

DNA-protein crosslink
e.g. Ethidium bromide

⬠ Sugar (deoxyribose)

○ Phosphate

from solar ultraviolet light that can penetrate the Earth's ozone layer. This form of radiation reacts with DNA to form 6,4-photoproducts and cyclobutane pyrimidine dimers (CPDs) (Fig. 1). Chronic exposure of skin cells to solar radiation dramatically increases the cellular load of these adducts and increases the risk of ▶ skin cancers such as malignant melanoma.

An illustration showing examples of structural and chemical alterations to the deoxyribose backbone and the nucleotide bases in between the two strands of DNA. Chemical cross-links are indicated by bold asterisk, and a large protein–DNA interaction is shown as a blue globular protein distorting the nucleotide structure.

2. ▶ Polycyclic aromatic hydrocarbons (PAHs). Atmospheric pollutants found in smoke, soot, and tar. The adducts formed by PAH predominantly result in the conversion of G to T (called transversions) (Fig. 1). This results in the mutagenic transformation of affected genes and is responsible for the majority of smoke-related ▶ lung cancer. Tobacco cigarette smoke contains significant amounts of PAHs but also contains extremely high concentrations of acrolein that is one of many carcinogens also stimulating G to T transversions. Cigarette smoke is also known to oxidize bases (see "Nucleotide Base Damage" in next section).

3. Antibiotics such as Mitomycin C and ▶ Adriamycin. These can induce DNA–protein and DNA–DNA inter-/

intrastrand cross-links (Fig. 1) that covalently join together the nucleotide bases on one/both strands of DNA, respectively.

4. Genotoxic chemotherapeutic drugs such as ► cisplatin and camptothecin. These are also classed as DNA cross-linking agents, and their cytotoxic properties are exploited by oncologists to kill tumor cells such as testicular carcinomas. ► Cisplatin can also induce intrastrand cross-links (Fig. 1).

Bulky adducts are typically removed by the ► nucleotide excision repair (NER) system. This removes a stretch of DNA harboring the lesion (typically about 30 bases long) and replaces it with the correct sequence of nucleotide bases. Examples of NER enzymes are XPA, XPG, and TFIIH. Individuals with genetic defects that affect the function of these components are at a higher risk of developing cancers associated with ► xeroderma pigmentosum.

Nucleotide Base Damage

Many environmental as well as endogenous factors can induce smaller chemical alterations to the DNA sequence-encoding bases – G, C, T, and A. Although they are small alterations in comparison, they can have dramatic effects upon the integrity of the genetic sequence encoded along the particular stretch of DNA that harbors such damage. Examples include:

1. *Oxidative DNA damage.* The most common form of base damage that can stimulate the conversion of, e.g., guanine to 8-oxoguanine (Fig. 1). Oxidizing agents such as ► reactive oxygen species (ROS) are frequently produced during normal cellular metabolic processes, and these are responsible for a significant proportion of base oxidation. ROS are present in the nucleus and cytoplasm and especially inside the mitochondria where energy production from ATP hydrolysis produces high levels. Subsequently, mitochondrial DNA is particularly vulnerable to such damage. Chemicals such as ► hydrogen peroxide (H_2O_2) can also oxidize nucleotide bases and represent

powerful DNA damaging agents. Vinyl chloride and chromium (► chromium carcinogenesis) represent industrial chemicals that damage DNA by inducing, among others, etheno-DNA adducts and base oxidization, respectively. Occupational health hazards such as these are well documented to cause mutations and cancer among exposed workers. Antioxidants such as circulating uric acid, the enzyme superoxide dismutase, and those provided in many types of food, e.g., vitamin C - (ascorbic acid) and ► curcumin, work to absorb the free radicals that can oxidize DNA and are generally considered to be important agents that protect cells from cancer-causing oxidants.

2. *DNA alkylating agents and DNA methylation agents.* These can stimulate the conversion of, e.g., cytosine to O_6-methylcytosine by adding a methyl group ($-CH_3$) (Fig. 1). DNA alkylating agents like MNNG and 6-TG have been developed as chemotherapeutic agents to kill proliferating tumor cells by damaging their DNA. Unrepaired methylated bases are therefore very toxic to cells.

3. *Mismatched bases.* Unrepaired methylated lesions are also subject to secondary repair by the mismatch repair system that normally functions to replace the occasional mismatched base (Fig. 1) with its correct partner (A with T and C with G) (defects associated with ► microsatellite instability, ► colorectal cancer, and ► Lynch syndrome). In doing so, the repair system can become "fooled" by the structural change exerted by ► methylation and, if not sent round a futile cycle of toxic repair, can promote the fixation of potentially carcinogenic mutations.

The base excision repair (BER) system is responsible for recognizing and replacing many types of altered bases, and there are several components involved. Examples include the APE-1 endonuclease and Nth1 glycosylase enzymes that excise apurinic and apyrimidinic bases and Ogg1, MUTHY, and MTH1 that recognize and replace oxidized bases such as 8-oxoguanine.

Single- and Double-Strand Breaks

DNA damaging agents such as ionizing radiation (X-rays, α and β particles) and free radicals can physically break the covalent bonds between the phosphates, ribose, and bases and result in single- and double-strand DNA breaks (SSB and DSB) (Fig. 1). SSBs are recognized and processed efficiently by dedicated SSB repair proteins such as PARP-1 (poly(ADP-ribose) polymerase-1) and XRCC1 (which also functions in BER). DSBs are by far the most toxic form of DNA damage, and one unrepaired DSB inside a cell is capable of killing that cell. Almost immediately after a DSB forms, i.e., after exposure to ionizing radiation, a particular variant of the histones, H2AX, which coats the DNA along the entire genome, becomes phosphorylated upstream (5′) and downstream (3′) of a DSB, some million base pair distances away, thus amplifying the signal and initiating a global cellular response. Other histone modifications are now being linked to DNA damage responses in what seems to be a histone code of responses.

DSBs represent highly reactive substrates for DNA end repair proteins. These are components of the nonhomologous end-joining (NHEJ) and ▶ homologous recombination repair systems. Defects in the protein components of these DSB repair pathways predispose individuals to a range of neurodegenerative disorders and many forms of cancer, e.g., ATM (ataxia-telangiectasia), BLM (▶ Bloom syndrome), ligase IV, MRE11, NBS (lymphoma, ▶ Nijmegen breakage syndrome, ▶ multiple myeloma), BRCA1, BRCA2, and Fanc D (▶ Fanconi anemia, ▶ breast cancer). DSBs are therefore potentially extremely mutagenic lesions, capable of translocating and rearranging gene sequences that can give rise to cancer. A classic example of malignancy resulting from ▶ chromosomal translocation and rearrangement is ▶ Burkitt lymphoma.

The deoxyribose sugars at the ends of SSB and DSBs (3′ or 5′, depending on the direction of the gene transcript) often exist in fragmented forms, particularly if the breaks result from free-radical attack. Examples of these are phosphoglycans. These have to be removed by specific repair

enzymes before SSB and DSB repair can proceed. This extensive end-processing can result in loss of base sequence information, loss of heterozygosity (LOH), and cancer-causing mutations if not tightly regulated.

SSBs and DSBs are also actively induced during normal cellular processes such as ▶ V(D)J recombination and cell division, but the way in which these DSBs are discriminated from bona fide DNA damage is not fully understood. DSBs can also result from the collapse of regions undergoing DNA replication called replication forks. Collapse can result from preexisting DNA lesions that interrupt the DNA replication process.

Furthermore, existing DSBs in the genome represent easy target sites where foreign DNA, such as invading viral DNA, can insert, integrate, and rearrange genetic sequences. These random, sequence-independent events are called insertional mutagenic events and can be extremely carcinogenic. Indeed, this process was responsible for the unfortunate development of leukemia in a significant proportion of X-SCID patients who underwent ▶ gene therapy clinical trials to correct their immunodeficiency. The corrective gene was incorrectly inserted into a region close to the ▶ oncogene LMO-2 and resulted in uncontrolled proliferation. Metnase is a protein which promotes NHEJ and the random integration of foreign DNA into human cells.

A cell requires intricate signaling proteins that alert the presence of DNA damage and activate the appropriate responses. For example, if a cell does not repair DNA damage before it is replicated during cell division, daughter cells will inherit DNA with mutations around that region and the next generation or progeny of cells will be defective. Therefore, appropriate responses such as DNA replication checkpoints or cell cycle arrest pathways must be activated to prevent permanent alterations from being passed on. If this is not activated, then damaged cells can lose control of proliferation and lead to cancer. Key regulators of this response are the p53, ▶ MDM2, and Rb (retinoblastoma protein) that lie at the interface between cell cycle arrest and cell death responses.

Therapeutic Approaches

The DNA of cancer cells remains one of the most effective targets to damage and destroy cancerous tissue in patients. Traditional DNA damaging ▶ chemotherapy and radiotherapy approaches have served cancer patients very well over the past four to five decades. However, long-term survival rates under such management across the cancer spectrum have not improved that much over the same period of time, clearly defining the need for more advanced techniques.

As we learn more about how cells respond to such damage, it is not surprising that proteins that encode DNA damage response genes such as ATM, ATR, DNA-PK, PARP and p53 have been singled out as potentially powerful targets to treat human cancers. The PI3-kinase-like proteins ATR, ATM, DNA-PK and mTOR as well as the Poly-ADP-ribise polymerase (PARP) protein are all targets that potent and selective inhibitors have been designed against to block their catalytic activities. Many clinical trials and preclinical drug discovery programmes have been initiated to target certain cancers particulary sensitive to such inhibition. Indeed, olaparib (Lynparza) is the first of such drugs that blocks the function of PARP to be registered to treat ovarian cancer patients with BRCA mutations, another DNA damage response protein that, when not functioning, renders cells extremely sensitivie to PARP inhibition. Several other trials are now showing promising utility in our cancers where DANN repair defects and other vulnerabilities may exist. Restoring the functions of tumor suppressors such as p53 and suppressing oncogenes and DNA repair proteins that can promote carcinogenesis and cancer cell survival, respectively, are all validated approaches to kill tumor cells and are under current investigation. Nutlin-3 is a small molecule belonging to the cis-imidazoline nutlin analogs that blocks the interaction between p53 and its regulator Mdm2 and thus prolongs p53 expression/activity by preventing its degradation. Nutlin-3 has been suggested as a potential agent to kill p53-positive tumors. However, major hurdles remain to be overcome such as maintaining a therapeutic window by not overly activating wild-type p53 in normal cells as well as achieving acceptable pharmacokinetic profiles.

▶ Gene therapy holds great promise for treating all genetically tractable diseases including many cancers. Identifying individuals with DNA repair polymorphisms or mutations and replacing or enhancing the expression of wild-type protein may help restore appropriate responses to DNA damage and protect against cancer development. Improvements in precise genome editing technologies such as the CRISPR Cas9 platform will help in this endeavor. Furthermore, randomly positioned DNA DSB sites in the genome represent preferential regions where viruses insert their genomes and integrate. Inhibiting DNA repair pathways that promote this random integration may suppress viral infection and reduce the chances of insertional mutagenesis that has so far plagued the promising use of ▶ viral vector-mediated gene transfer for gene therapy in clinical trials.

Cross-References

- ▶ Lynch Syndrome
- ▶ MDM2
- ▶ Methylation
- ▶ Microsatellite Instability
- ▶ Mismatch Repair in Genetic Instability
- ▶ Multiple Myeloma
- ▶ Nijmegen Breakage Syndrome
- ▶ Nucleotide Excision Repair
- ▶ Oncogene
- ▶ Polycyclic Aromatic Hydrocarbons
- ▶ Radiotherapy
- ▶ Reactive Oxygen Species
- ▶ Repair of DNA
- ▶ Retinoblastoma Protein, Biological and Clinical Functions
- ▶ Skin Cancer
- ▶ UV Radiation
- ▶ V(D)J Recombination
- ▶ Viral Vector-mediated Gene Transfer
- ▶ Xeroderma Pigmentosum

References

Durant ST, Nickoloff JA (2005) Good timing in the cell cycle for precise DNA repair by BRCA1. Cell Cycle Rev 4(9):1216–1222

Friedberg EC (2001) How nucleotide excision repair protects against cancer. Nat Rev Cancer 1(1):22–33

Karran P (2000) DNA double strand break repair in mammalian cells. Curr Opin Genet Dev 10(2):144–150

Lindahl T (2016) The Intrinsic Fragility of DNA (Nobel Lecture). Angew Chem Int Ed Engl. 55(30):8528–8534

Mateo J et al (2015) DNA-repair defects and olaparib in metastatic prostate cancer. N Engl J Med 373(18):1697–1708

O'Connor MJ (2015) Targeting the DNA damage response in cancer. Mol Cell 60(4):547–560

Rouse J, Jackson SP (2002) Interfaces between the detection, signaling, and repair of DNA damage. Science 297:547–551

Tyteca S, Legube G, Trouche D (2006) To die or not to die: a HAT trick. Mol Cell 24(6):807–808

Tucker H, Charles Z, Robertson J, Adam J (2016) NICE guidance on olaparib for maintenance treatment of patients with relapsed, platinum-sensitive, BRCA mutation-positive ovarian cancer. Lancet Oncol 17(3):277–278. doi:10.1016/S1470-2045(16)00062-0

Vassilev LT, Vu BT, Graves B, Carvajal D, Podlaski F, Filipovic Z, Kong N, Kammlott U, Lukacs C, Klein C, Fotouhi N, Liu EA (2004) In vivo activation of the p53 pathway by small-molecule antagonists of MDM2. Science 303(5659):844–848. doi:10.1126/science.1092472

See Also

(2012) Base excision repair. In: Schwab M (ed) Encyclopedia of cancer, 3rd edn. Springer, Berlin/Heidelberg, p 349. doi:10.1007/978-3-642-16483-5_536

(2012) Camptothecin. In: Schwab M (ed) Encyclopedia of cancer, 3rd edn. Springer, Berlin/Heidelberg, p 603. doi:10.1007/978-3-642-16483-5_791

(2012) Carcinogen. In: Schwab M (ed) Encyclopedia of cancer, 3rd edn. Springer, Berlin/Heidelberg, p 644. doi:10.1007/978-3-642-16483-5_839

(2012) Cell cycle. In: Schwab M (ed) Encyclopedia of cancer, 3rd edn. Springer, Berlin/Heidelberg, p 737. doi:10.1007/978-3-642-16483-5_994

(2012) Checkpoint. In: Schwab M (ed) Encyclopedia of cancer, 3rd edn. Springer, Berlin/Heidelberg, pp 754–755. doi:10.1007/978-3-642-16483-5_1049

(2012) DNA methylation. In: Schwab M (ed) Encyclopedia of cancer, 3rd edn. Springer, Berlin/Heidelberg, p 1140. doi:10.1007/978-3-642-16483-5_1682

(2012) Nucleotide base. In: Schwab M (ed) Encyclopedia of cancer, 3rd edn. Springer, Berlin/Heidelberg, p 2581. doi:10.1007/978-3-642-16483-5_4170

(2012) 8-Oxoguanine. In: Schwab M (ed) Encyclopedia of cancer, 3rd edn. Springer, Berlin/Heidelberg, p 2734. doi:10.1007/978-3-642-16483-5_4313

(2012) P53. In: Schwab M (ed) Encyclopedia of cancer, 3rd edn. Springer, Berlin/Heidelberg, p 2747. doi:10.1007/978-3-642-16483-5_4331

(2012) Poly(ADP-Ribose) polymerase. In: Schwab M (ed) Encyclopedia of cancer, 3rd edn. Springer, Berlin/Heidelberg, p 2935. doi:10.1007/978-3-642-16483-5_4655

(2012) Replication. In: Schwab M (ed) Encyclopedia of cancer, 3rd edn. Springer, Berlin/Heidelberg, p 3254. doi:10.1007/978-3-642-16483-5_5031

(2012) Tobacco. In: Schwab M (ed) Encyclopedia of cancer, 3rd edn. Springer, Berlin/Heidelberg, pp 3716–3717. doi:10.1007/978-3-642-16483-5_5844

(2012) Vitamin C. In: Schwab M (ed) Encyclopedia of cancer, 3rd edn. Springer, Berlin/Heidelberg, p 3925. doi:10.1007/978-3-642-16483-5_6206

DNA Damage Response

Mau-Sun Chang
Institute of Biochemical Sciences, National
Taiwan University, Taipei, Taiwan

Definition

Damage to DNA invokes several cellular responses that enable cells either to eliminate the damage or activate programmed cell death. These responses include activation of checkpoints to arrest progression through the cell cycle; removal or repair of damaged DNA to prevent the transmission of damaged chromosomes to daughter cells; induction of genes required for DNA repair, cell cycle arrest, and ► apoptosis; and apoptosis to eliminate cells too seriously damaged or deregulated to repair.

Characteristics

Sources of Damage

Endogenous attack by ► reactive oxygen species can induce several types of damage to DNA bases, including oxidation and the generation of DNA strand interruptions; alkylation, depurination, and depyrimidination; and wrongly mismatched bases coupled to a new DNA strand. Exogenous damage is caused by a variety of external agents. Ultraviolet light from the sun causes cross-linking between adjacent thymine bases, creating pyrimidine dimers. Ionizing radiation from X-rays and gamma rays results in breaks in the DNA double strands. DNA intercalating compounds create a huge variety of DNA adducts.

Checkpoint Pathway

Checkpoint responses trigger cell cycle arrest, providing time to repair damaged chromosomes before they enter mitosis, thus maintaining genomic integrity. There are five major components of this system. First are checkpoint sensors which include the Rad9–Hus1–Rad1 (9-1-1) complex, the Rad17–RFC clamp loading complex, and the Mre11–Rad50–Nbs1 (MRN) complex. Second, checkpoint mediators include BRCA1, MDC1, 53BP1, and Claspin. Third are the apical signal transducing kinases including ATM and ATR kinases. Fourth, distal signal transducing kinases are represented by Chk1 and Chk2. Finally, there are checkpoint effectors encompassing cell cycle regulators such as the cdc25 phosphatases, various DNA repair proteins, transcription factors p53 or E2F1, and chromatin components such as histone H2AX (Fig. 1).

G1 Checkpoint

To prevent a cell with damaged DNA from entering the S phase, cells activate checkpoint transducing kinases ATM/ATR and Chk1/Chk2, which target Cdc25A and p53 at the G1 checkpoint. When DNA is damaged, p53 is phosphorylated at Ser15 and Ser20 in its trans-activation domain. This stabilizes p53 by inhibiting its interaction with ► Mdm2, thus stabilizing p53. The key transcriptional target of p53 is p21 which inhibits cyclin E–cdk2 activity and prevents the phosphorylation of Rb by cyclin D–cdk4. This sequence of events results in suppression of the Rb/E2F pathway, prolonging transition from G1 to S. The phosphorylation of cdc25A by Chk kinases creates binding sites on cdc25A with ► 14-3-3 proteins and excludes them from the nucleus. Lack of active cdc25A in the nucleus results in the accumulation of inactive cdk4. The inactive molecule is incapable of loading cdc45 onto chromatin, a step required for initiation of replication origin firing.

Intra-S-phase Checkpoint

When a double-strand break or DNA nick occurs during the S phase, the intra-S checkpoint, involving ATM, the MRN complex, and BRCA1, is activated. The ATM–Chk2–Cdc25A–Cdk2 pathway is responsible for the checkpoint response. In contrast, when DNA is damaged by UV radiation or chemicals that create bulky base lesions, the main damage sensor is the ATR–ATRIP complex. Activated ATR phosphorylates Chk1, which in turn phosphorylates and downregulates cdc25A, thus inhibiting replication origin firing.

G2/M Checkpoint

The key downstream target of the G2/M checkpoint is the mitosis-promoting activity of the cyclin B/Cdk1 kinase. After DNA damage, ATM/ATR and Chk1/Chk2 kinases downregulate cdc25A and cdc25C. Upregulated Wee1 and downregulated cdc25 collaboratively inhibit the activity of cyclin B/Cdk1 kinase at the G2/M boundary. Again, the phosphorylated cdc25A binds 14-3-3 proteins and is sequestered in the cytoplasm, promoting

DNA Damage Response,
Fig. 1 Signal transduction
of the DNA checkpoint
responses in human cells.
The DNA damage is
detected by sensors that,
with the aid of mediators,
transmit the signal to
transducers. The
transducers activate or
inactivate effectors that
directly participate in
inhibiting the G1/S, S, or
the G2/M transition

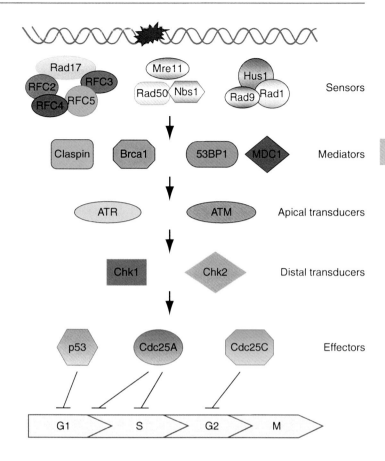

DNA Repair Responses

DNA damage alters the spatial structure of the helix, and such alterations can be detected by the cell. Once damage is located, specific DNA repair molecules are recruited to the site, inducing other molecules to bind and form a complex that enables actual repair to take place.

Excision Repair

Base excision repair corrects damage to a single nucleotide. In this process, DNA N-glycosylases recognize oxidized, alkylated, or deaminated bases and catalyze the hydrolytic removal of the mutated bases. Lesions are then repaired by DNA Pol $\beta/\delta/\varepsilon$, DNA ligase III, PCNA, and FEN1.

degradation by the ubiquitin–proteasome pathway. Y15-phosphorylated cdk1 accumulates, and mitosis is arrested.

Nucleotide excision repair amends longer strands of 2–30 damaged bases. This process, involving RPA, XPA, and XPC, recognizes bulky, helix-distorting changes such as thymidine dimers as well as single-strand breaks. TFIIH, XPG, and XPF–ERCC1 are recruited to the site to form an incision complex. The gap is filled by DNA Pol δ/ε with the aid of replication accessory proteins PCNA and RFC. Mismatch repair fixes errors of DNA replication and recombination that result in mispaired nucleotides.

Double-Strand Break Repair

Double-strand breaks are produced by ▶ reactive oxygen species, ionizing radiation, and certain antineoplastic chemicals. Two mechanisms exist to repair this damage. The nonhomologous end-joining pathway directly joins the two ends of the broken DNA strands without a template.

The Ku heterodimer binds to the two ends of a double-strand break and recruits DNA-PKcs and the ligaseIV–XRCC4 heterodimer to ligate the two duplex termini. The second mechanism, ► homologous recombination repair, requires an identical or a nearly identical sequence as a template. This pathway allows a damaged chromosome to be repaired using the newly created sister chromatid as a template. A key intermediate in this pathway is the Holliday junction in which the two recombining duplexes are joined covalently by single-strand crossovers.

Translesion Synthesis Repair

Translesion synthesis is an error-prone method of DNA repair. This pathway is mediated by DNA polymerases that insert extra bases at the site of damage and thus allow replication to bypass the damaged base.

Interstrand DNA Cross-Link Repair

Patients with ► Fanconi anemia have a predisposition to cancer. Their cells are hypersensitive to agents that cause interstrand cross-links. Repair of this type of damage is poorly understood but may involve excision repair and recombination.

Arrest, Repair, or Die?

p53-Dependent Mechanisms

A key determinant as to whether cells with DNA damage live or die is p53. ATM phosphorylates p53 at Ser15 and ► MDM2 at Ser395, while Chk1 and Chk2 phosphorylate p53 at Ser20. These modifications prevent the p53-MDM2 association that targets p53 for proteolysis, thereby allowing increments in cellular levels of p53. The consequences of p53 activation include transcriptional activation of DNA repair activities, cell cycle inhibitors (p21 and 14-3-3σ), and proapoptotic factors (► PUMA, FAS/APO-1/CD95, and Apaf-1).

p53-Independent Mechanisms

One possible mechanism for p53-independent apoptosis involves E2F1, under the control of Chk1- and Chk2-mediated activation. The transcriptional targets of E2F1 include proteins associated with DNA repair (BRCA1, Msh2, Msh6, RFC, and PCNA), cell cycle checkpoints (ARF and Chk1), and apoptosis (caspase-7, Apaf-1). Thus, like p53, E2F1 may determine whether cells live or die.

Cancer Susceptibility

Loss of ATM strongly predisposes humans to lymphoma and, to a lesser degree, to other malignancies. Patients with mutations in NBS1 or MRE11, for example, are predisposed to develop cancer. Seckel syndrome, in which there is a low level of ATR expression, involves chromosome instability after ► mitomycin C exposure. Inherited mutations in one allele of p53 and Chk2 are found in families with the extremely cancer-prone ► Li-Fraumeni syndrome. The inheritance of a single mutated allele of either BRCA1 or BRCA2 markedly increases the incidence of breast and ovarian cancers in women.

Nucleotide excision repair, homologous recombination, and translesion synthesis are all thought to be defective in ► Fanconi anemia cells. Nucleotide excision repair is defective in ► xeroderma pigmentosum and Cockayne syndrome, while mismatch repair malfunctions in children with ► Turcot syndrome and in tumor cells derived from hereditary nonpolyposis colorectal cancers (HNPCC). Translesion synthesis repair is defective in patients with a variant of xeroderma pigmentosum.

Cross-References

- ► Apoptosis
- ► BARD1
- ► *BRCA1/BRCA2* Germline Mutations and Breast Cancer Risk
- ► CDKN2A
- ► Chromosomal Instability
- ► Fanconi Anemia
- ► γH2AX
- ► Homologous Recombination Repair
- ► Li-Fraumeni Syndrome
- ► MDM2
- ► Mismatch Repair in Genetic Instability
- ► Mitomycin C
- ► 14-3-3 Proteins

▶ PUMA
▶ Reactive Oxygen Species
▶ Repair of DNA
▶ Turcot Syndrome
▶ UV Radiation
▶ Xeroderma Pigmentosum

References

Harrison JC, Haber JE (2006) Surviving the breakup: the DNA damage checkpoint. Annu Rev Genet 40:209–235

Kastan MB, Bartek J (2004) Cell-cycle checkpoints and cancer. Nature 432:316–323

Kennedy RD, D'Andrea AD (2006) DNA repair pathways in clinical practice: lesion from pediatric cancer susceptibility syndromes. J Clin Oncol 24:3799–3808

Sancar A, Lindsey-Boltz LA, Ünsal-Kaçmaz K et al (2004) Molecular mechanisms of mammalian DNA repair and the DNA damage checkpoints. Annu Rev Biochem 73:39–85

Su TT (2006) Cellular responses to DNA damage: one signal, multiple choices. Annu Rev Genet 40:187–208

See Also

(2012) APAF-1. In: Schwab M (ed) Encyclopedia of Cancer, 3rd edn. Springer Berlin Heidelberg, p 231. doi:10.1007/978-3-642-16483-5_344

(2012) ATR. In: Schwab M (ed) Encyclopedia of Cancer, 3rd edn. Springer Berlin Heidelberg, p 302. doi:10.1007/978-3-642-16483-5_443

(2012) Base Excision Repair. In: Schwab M (ed) Encyclopedia of Cancer, 3rd edn. Springer Berlin Heidelberg, p 349. doi:10.1007/978-3-642-16483-5_536

(2012) CDC25. In: Schwab M (ed) Encyclopedia of Cancer, 3rd edn. Springer Berlin Heidelberg, p 704. doi:10.1007/978-3-642-16483-5_953

(2012) Cell Cycle. In: Schwab M (ed) Encyclopedia of Cancer, 3rd edn. Springer Berlin Heidelberg, p 737. doi:10.1007/978-3-642-16483-5_994

(2012) Checkpoint. In: Schwab M (ed) Encyclopedia of Cancer, 3rd edn. Springer Berlin Heidelberg, pp 754-755. doi:10.1007/978-3-642-16483-5_1049

(2012) CHK1. In: Schwab M (ed) Encyclopedia of Cancer, 3rd edn. Springer Berlin Heidelberg, p 817. doi:10.1007/978-3-642-16483-5_1101

(2012) Cockayne Syndrome. In: Schwab M (ed) Encyclopedia of Cancer, 3rd edn. Springer Berlin Heidelberg, p 890. doi:10.1007/978-3-642-16483-5_1246

(2012) FAS. In: Schwab M (ed) Encyclopedia of Cancer, 3rd edn. Springer Berlin Heidelberg, p 1379. doi:10.1007/978 3-642-16483-5_2121

(2012) DNA Repair. In: Schwab M (ed) Encyclopedia of Cancer, 3rd edn. Springer Berlin Heidelberg, p 1141. doi:10.1007/978-3-642-16483-5_1687

(2012) G2/M Checkpoint. In: Schwab M (ed) Encyclopedia of Cancer, 3rd edn. Springer Berlin Heidelberg, p 1481. doi:10.1007/978-3-642-16483-5_2466

(2012) Hereditary Nonpolyposis Colorectal Cancer. In: Schwab M (ed) Encyclopedia of Cancer, 3rd edn. Springer Berlin Heidelberg, pp 1683–1684. doi:10.1007/978-3-642-16483-5_2683

(2012) Ionizing Radiation. In: Schwab M (ed) Encyclopedia of Cancer, 3rd edn. Springer Berlin Heidelberg, p 1907. doi:10.1007/978-3-642-16483-5_3139

(2012) Lymphoma. In: Schwab M (ed) Encyclopedia of Cancer, 3rd edn. Springer Berlin Heidelberg, p 2124. doi:10.1007/978-3-642-16483-5_3463

(2012) Mitosis. In: Schwab M (ed) Encyclopedia of Cancer, 3rd edn. Springer Berlin Heidelberg, p 2342. doi:10.1007/978-3-642-16483-5_3774

(2012) MRE11. In: Schwab M (ed) Encyclopedia of Cancer, 3rd edn. Springer Berlin Heidelberg, pp 2381–2382. doi:10.1007/978-3-642-16483-5_3853

(2012) NBS1. In: Schwab M (ed) Encyclopedia of Cancer, 3rd edn. Springer Berlin Heidelberg, p 2468. doi:10.1007/978-3-642-16483-5_3984

(2012) P53. In: Schwab M (ed) Encyclopedia of Cancer, 3rd edn. Springer Berlin Heidelberg, p 2747. doi:10.1007/978-3-642-16483-5_4331

(2012) PCNA. In: Schwab M (ed) Encyclopedia of Cancer, 3rd edn. Springer Berlin Heidelberg, p 2803. doi:10.1007/978-3-642-16483-5_4415

(2012) RAD50. In: Schwab M (ed) Encyclopedia of Cancer, 3rd edn. Springer Berlin Heidelberg, p 3133. doi:10.1007/978-3-642-16483-5_4894

(2012) Seckel Syndrome. In: Schwab M (ed) Encyclopedia of Cancer, 3rd edn. Springer Berlin Heidelberg, p 3342. doi:10.1007/978-3-642-16483-5_5190

(2012) Sister-Chromatids. In: Schwab M (ed) Encyclopedia of Cancer, 3rd edn. Springer Berlin Heidelberg, p 3418. doi:10.1007/978-3-642-16483-5_5329

(2012) Translesion Synthesis. In: Schwab M (ed) Encyclopedia of Cancer, 3rd edn. Springer Berlin Heidelberg, p 3772. doi:10.1007/978-3-642-16483-5_5939

(2012) Ultraviolet Light. In: Schwab M (ed) Encyclopedia of Cancer, 3rd edn. Springer Berlin Heidelberg, p 3841. doi:10.1007/978-3-642-16483-5_6101

DNA Damage Response Genes

Kandace Williams
Department of Biochemistry and Cancer Biology, Health Science Campus, UT College of Medicine, Toledo, OH, USA

Definition

DNA damage response genes encompass all genes that encode proteins required for either direct or indirect response to DNA damage. The

DNA Damage Response Genes, Table 1 Human hereditary cancer syndromes due to autosomal recessive inherited defects in DNA damage response genes

Syndrome	Gene	Cancer	Disabled process
Xeroderma pigmentosum	XPA à XPG, XPV	UV-induced skin cancer	NER, TLS
Hereditary nonpolyposis colon cancer	MSH2, MLH1, MSH6, PMS2	Colon cancer	MMR, MMR-induced DNA damage signaling
MYH-associated polyposis	MYH	Colon cancer	BER
Hereditary breast and ovarian cancer	BRCA1, BRCA2	Breast, ovarian cancer	HR
Fanconi anemia	FANCA à FANCP, RAD51C	Leukemia	DNA cross-link repair
Ataxia-telangiectasia	ATM	Leukemia, lymphoma, breast cancer	DNA DSB-induced damage signaling, G_2 checkpoint
Nijmegen break syndrome	NBS	Leukemia, lymphoma	DNA DSB repair
Seckel syndrome	ATR	Acute myeloid leukemia	DNA SSB-induced damage signaling, G_2 checkpoint
Li-Fraumeni	TP53	Multiple cancers	DNA damage signaling, G_1 checkpoint
Werner syndrome	WRN	Multiple cancers	Helicase and exonuclease functions
Bloom syndrome	BLM	Multiple cancers	Helicase functions, HR
Rothmund-Thomson	RECQL4	Osteosarcoma, skin cancer	Helicase functions ?

proteins encoded by these genes either act to help repair the damaged DNA within replicating cells or if damage is sufficient, to activate cell cycle checkpoints and if too severe, to activate cell death pathways. Each of the above enzymatic pathways also includes posttranslational modifiers such as protein kinases, phosphorylases, ubiquitin ligases, acetylators, sumoylators, methylators, scaffolding proteins required for proteasome complexes, and epigenetic gene silencing pathways. MicroRNA-mediated transcriptional regulation will likely be included in the future, as more information is gathered about this gene regulation system. The number of individual pathways and genetic sequences contributing to the major pathways above are unknown but likely encompass several 100 or more.

Characteristics

This essay will focus primarily on mutated genes associated with DNA damage response activities that have been identified as responsible for inherited autosomal recessive syndromes with increased incidence of human cancer (Table 1). These genes have been found mutated or epigenetically altered in sporadic tumors as well.

DNA Replication Polymerase Alterations in Cancer

DNA polymerases are essential for both replication and protection against genomic damage. Human cells contain at least 15 different DNA polymerases (Pols); Pols a, d, and e replicate chromosomal DNA in a 5′ to 3′ direction. High fidelity replication activity from these polymerases is essential for cellular viability. Pols d and e are responsible for the bulk of chromosomal replication and contain 3′ to 5′ exonuclease proofreading activity for removal of errors during replication. Pol a is responsible for initiating DNA synthesis on the leading strand and priming of Okazaki fragments during lagging strand DNA synthesis. These polymerases also perform fill-in synthesis during gap repair for specific DNA repair pathways, such as mismatch repair (MMR) and nucleotide excision repair (NER). Replicative DNA polymerases will not synthesize DNA over a damaged template, however. At least seven of the remaining mammalian DNA poly-merases now have defined activity as translesion synthesis (TLS) polymerases. The TLS Pols have decreased fidelity of replication with no proof-reading exonuclease activity. The function of these Pols is putatively to avoid replicating poly-merases stalling before unrepaired template lesions at the fragile replication fork, which would otherwise cause DNA strand breaks and genomic instability. Each of the TLS Pols pos-sesses a unique type of DNA damage bypass activity. The only DNA polymerase for which an inherited deficiency, in the form of a disabling mutation, predisposes humans to cancer is the TLS DNA Polymerase DNA Pol h and is one of several different mutated genes that result in xeroderma pigmentosum (XP). XP is an inherited disorder with eight complementation groups (see below), one of which is XPV – a lack of DNA Pol h activity. This disorder is associated with up to a 1,000-fold increased risk for sunlight-induced skin cancers. Pol h synthesizes correctly over a DNA template containing TT-cyclopyrimidine dimers (CPDs), the most common lesion resulting from ultraviolet (UV) light exposure. Surprisingly, whole genome sequencing of thousands of genomes from several different types of cancers has yet to reveal specific cancer-associated alter-ations in any other DNA polymerase gene, nor has gene silencing been observed by epigenetic mech-anisms. Several different mouse models have been developed that express different DNA Pol muta-tions that are predisposed to various tumors, however.

DNA Repair Pathway Alterations in Cancer

Chromosomal DNA is susceptible to errors by DNA replication machinery during every cell cycle in addition to constant attack by mutagens produced by endogenous metabolism as well as exogenous sources. DNA repair machinery must be able to correctly repair such damage quickly and correctly in replicating cells or undergo the risk of mutations that alter cellular phenotype. Cancer genome sequencing has confirmed that cancer cells bear up to 100,000 more mutational events than found within the genomes of normal cells. However, differences in mutation

prevalence between individual cancers and the vast number of these mutations reveal consider-able information about the development of neo-plasia. The somatic mutation signatures of individual tumors often carry imprints character-istic of mutagenic exposure or DNA repair defi-ciencies. For example, the mutational pattern in skin cancer reveals overwhelming exposure to UV light or tobacco carcinogens in lung cancer or an inherited deficiency of DNA mismatch repair in colorectal cancer.

Development of carcinogenesis during clonal expansion requires a subset of "driver mutations" to fall within a key set of "cancer genes." These mutations confer abilities for clonal growth, inva-sion, and metastasis, as well as impairment of programmed cell death or senescence pathways. The vast majority of "passenger" mutations, by definition, do not confer growth advantage. The increased mutational load within neoplastic cells also indicates that cancer is a process requiring many cell cycles and that the mutation rate in cancer cells is much higher than in normal cells; therefore, a malfunctioning DNA repair system is a required "driver" mutation at some point during carcinogenesis. Several inherited autosomal recessive cancer syndromes (mutated gene in germ cells) demonstrate the importance of DNA repair for normal growth and division of somatic cells. For example, XP is a complex genetic dis-order that places a person at greatly increased risk of sunlight-induced skin cancer (see above). Seven of the eight complementation groups dis-covered in this inherited syndrome are due to disabling mutations within the NER pathway (XPA-G). This DNA repair pathway is required to repair bulky DNA damage inflicted by exoge-nous agents such as CPDs produced by UV light from the sun. Indeed, the mutational signature in XP cells clearly derives from unrepaired CPD lesions.

Hereditary nonpolyposis colon cancer (HNPCC) is a familial cancer syndrome representing 2–3% of all colon cancer cases. This cancer syndrome is due to germ line muta-tions in DNA mismatch repair genes, most often hMSH2 or hMLH1. Genomic instability within cancer cells from this inherited syndrome is of a strikingly different pattern than the majority of tumors. Most human tumors display increased chromosomal instability (CIS) with many types of large chromosomal aberrations, such as large insertions, deletions, or translocations evident by cytogenetic analyses. HNPCC cells instead display microsatellite instability (MSI) evident only by sequencing of specific genomic locations. Microsatellite sequences are regions of homopolymeric stretches of nucleotides that replicating DNA polymerases have more difficulty synthesizing through, requiring frequent corrective input by the MMR system. Without this backup repair pathway, each replication cycle leaves more contraction and expansion errors within these sequences by the "slipping and sliding" errors made by the replicating polymerases. Many driver mutations in MMR defective tumors are within genes having microsatellite repeats nested in their sequences. Approximately 15% of sporadic gastric, colorectal, and endometrial tumors also exhibit MSI rather than CSI. The majority of these sporadic tumors have defective MMR because of epigenetic silencing via promoter hypermethylation of hMLH1.

An additional function of the MMR system, less well understood than DNA mismatch repair, is the ability of MMR proteins to recognize and bind to specific types of DNA damage, such as O^6methyldeoxyguanine (O^6meG), as a signal for the DNA damage response system to initiate cell cycle arrest either for subsequent repair or apoptosis if the DNA damage cannot be repaired. Monofunctional alkylating agents, such as N-methyl-N′-nitro-N-nitrosoguanidine (MNNG) or the clinical equivalent temozolomide (TMZ), produce several different alkylated DNA adducts in addition to O^6meG, the majority of which have low mutagenic potential and are repaired efficiently by the base excision repair pathway (BER). The O^6meG modification, however, is not repaired by BER but instead by a one-step enzymatic reaction that directly and covalently transfers the methyl group from the O^6meG position to methylguanine methyltransferase (MGMT), thus rendering this enzyme useless for further reactions and earning its alternate name as

the "suicide enzyme." MGMT can be rapidly depleted in the face of highly alkylated DNA. As well, MGMT is frequently expressed at low levels or epigenetically silenced in different tissues (normal and malignant). Low or absent MGMT expression in tumor cells can be important if alkylation chemotherapeutics, such as TMZ, are to be effectively used. For example, if the cell undergoes DNA replication before repair of O^6meG can occur, because of low or absent MGMT, there is an elevated likelihood of misinsertion of thymine opposite the damaged guanine leading to increased GàA transition mutations. A sufficient level of O^6meG within chromosomal DNA, however, will trigger an MMR-induced DNA damage response and subsequent apoptosis. To further complicate matters, both TLS and homologous recombination (see below) have been strongly implicated to also have roles in the DNA damage response to O^6me-G. Conversely, cells deficient in both MMR and MGMT expression do not trigger a DNA damage response or undergo apoptosis. MMR- and MGMT-deficient cells therefore demonstrate a significantly increased mutation rate, because of increased cell survival as well as lack of both mismatch and O^6meG repair within cells exposed to alkylating agents.

The base excision repair pathway (BER) removes small DNA lesions and mistakes that are frequently the result of endogenous damaging events, such as alkylation or oxidation damage, or abasic sites. BER has not been found to have inherited deficiencies in function that significantly increase susceptibility to cancer, with the exception of MYH, a glycosylase that removes misinserted adenines opposite 8-oxo-deoxyguanine as a repair step during oxidative damage repair. MYH-associated polyposis (MAP) is an inherited colon cancer syndrome resulting from disabling mutations within the MYH gene. The BER pathway has come under increasing scrutiny because of its major role in the repair of chemotherapeutic alkylation DNA damage. There are now numerous mouse models with altered BER protein expression that exhibit increased or decreased susceptibility to alkylation-induced cancer. Specific inhibitors of particular BER enzymes have been shown to potentiate the toxicity of clinical alkylators such as TMZ.

Perhaps the most highly recognized inherited defects in DNA repair within the lay public are those associated with BRCA1 and BRCA2 genes, conferring an increased susceptibility to breast and ovarian cancer. BRCA1 (but not BRCA2) is also frequently found epigenetically silenced in various sporadic cancers. These gene products are part of the homologous recombination (HR) pathway for repair of DNA double-strand breaks that is active in late S and G_2 phases of the cell cycle. Tumors lacking expression of BRCA1 or BRCA2, thus lacking a functional HR pathway, have created excitement as a target for "synthetic lethal" approaches to chemotherapy. Tumor cells that lack one DNA repair pathway are often highly susceptible to inhibitors of other DNA repair pathways used by the cell to compensate for the deficient pathway. For example, inhibitors to Poly ADP-ribose polymerase (PARP) effectively inhibit BER and have been discovered to be highly toxic to cells also lacking HR repair when combined with DNA damaging chemotherapy or radiation therapy. This type of synthetic lethal approach is proving effective for TMZ-induced alkylation damage as well. The principal idea is to increase nonrepairable DNA damage in the tumor cells to unsustainable levels in a very targeted approach, using the cell's own genetic deficiency as a chemotherapy-directed tool against itself. There is now great excitement in the field as more synthetic lethal approaches are sought within other DNA repair pathways.

Fanconi anemia (FA) has up to 15 genes now associated with this inherited disorder (FANCA-P, RAD51C). Disabling mutations in any of these genes results in an inability to repair DNA damaged by cross-linking agents and a highly increased susceptibility to acute myelogenous leukemia. The complex FA pathway coordinates NER, HR, and TLS pathways to resolve interstrand cross-links during replication. There is little knowledge in regard to epigenetic inactivation of this gene family in cancer, although some cases involving epigenetic inactivation of FancF have been reported.

DNA Damage Signaling Pathways in Cancer

Gene products such as ATM, ATR, and p53 are frontline defenses that initiate cell cycle checkpoint arrest at specific parts of the cell cycle for repair of DNA damage, or if damage is too severe, these gene products trigger apoptosis.

Ataxia-telangiectasia (AT) and Nijmegen breakage syndrome (NBS) are both inherited disorders with increased sensitivity to ionizing radiation and other treatments inducing DNA double-strand breaks. Both syndromes have increased risk of several different lymphatic tumors. ATM (ataxia-telangiectasia mutated) is the mutated gene that is inherited in AT. AT individuals also have an increased risk for breast cancer that has been ascribed to ATM's interaction and phosphorylation of BRCA1 following DNA damage. ATM is a protein kinase that normally initiates G_2 arrest after DNA double-strand breaks occur. In Nijmegen breakage syndrome, NBS is the inherited gene that is mutated. NBS codes for nibrin (NBS1), a protein that is normally part of a trimeric complex (NBS1/MRE11/RAD50) responsible for coordinating repair of DNA double-strand breaks and recruiting ATM to the strand breaks to initiate cell cycle arrest.

ATR (ATM and RAD3 related) initiates G_2 cell cycle arrest in response to persistent single-strand DNA breaks. ATR is required for cell viability, although certain inherited mutations have been found to be responsible for Seckel syndrome, which is an autosomal recessive disorder characterized by dwarfism, intrauterine growth retardation, bird-like facies, microcephaly, and mental retardation. ATR-Seckel syndrome has been found in patients with mutations in ATR. Cell lines from patients with Seckel syndrome, who are normal for ATR, show defective ATR signaling, suggesting that Seckel syndrome can be caused by mutations in other components of the ATR pathway.

Li-Fraumeni syndrome is the inherited deficiency of p53 due to germ line mutations in the gene TP53. Several different types of tumors are associated with this syndrome. The TP53 gene is also mutated in approximately 50% of all sporadic human tumors. Cells expressing mutated p53 do not exhibit G_1 cell cycle arrest, nor do they undergo p53-dependent apoptosis.

Werner syndrome presents as an accelerated aging process. Increased telomere attrition and genomic instability is observed in this syndrome and in sporadic tumors that express the dysfunctional protein from this mutated gene. The WRN gene responsible for this syndrome codes for a RECQ helicase family member (RECQL2). The WRN protein has both helicase and exonuclease functions that participate in many aspects of DNA metabolism, including maintenance of telomere structure and homology-dependent recombination. This gene has also been reported as epigenetically silenced in a large number of sporadic tumors.

Bloom syndrome is due to mutations of the BLM gene, which belongs to the DExH box-containing RecQ helicase subfamily. Cells harboring this mutated gene have a spontaneous mutation rate ten times higher than normal with highly increased rates of chromosome breakage and excessive HR activity. This helicase protein is thought to play a major role in DNA replication and HR.

Rothmund-Thomson syndrome is a very rare disorder that can severely affect many parts of the body, with increased incidence of osteosarcoma and skin cancer predominating. The RECQL4 helicase gene is mutated in this disorder. The gene product appears to play some role in replicating and repairing DNA, although little is understood currently about its function.

References

Fu D, Calvo JA, Samson LD (2012) Balancing repair and tolerance of DNA damage caused by alkylating agents. Nat Rev Cancer 12:104–120

Lahtz C, Pfeifer GP (2011) Epigenetic changes of DNA repair genes in cancer. J Mol Biol 3:51–58

Lange SS, Takata K-I, Wood RC (2011) DNA polymerases and cancer. Nat Rev Cancer 11:96–110

Polo SE, Jackson SP (2011) Dynamics of DNA damage response proteins at DNA breaks: a focus on protein modifications. Genes Dev 25:409–433

Stratton MR (2011) Exploring the genomes of cancer cells: progress and promise. Science 331:1553–1558

See Also

(2012) ATR. In: Schwab M (ed) Encyclopedia of cancer, 3rd edn. Springer, Berlin/Heidelberg, p 302. doi:10.1007/978-3-642-16483-5_443

DNA Damage Tolerance

Xianghong Wang
Department of Anatomy, The University of Hong Kong, Hong Kong, China

Synonyms

DNA lesion bypass; Postreplication repair; Replicative DNA lesion bypass

Definition

DNA damage tolerance is a biological mechanism in response to DNA damage which overcomes arrested DNA replication as a result of unrepaired DNA damage, leading to elimination of its potential lethal effects.

Characteristics

DNA is frequently damaged by endogenous and environmental factors. Base damage in DNA template strands blocks transcription to allow time to activate transcription-coupled repair pathways and eliminate DNA base damage. However, some lesions are persistent during replication therefore causing replication blockage and cell death. In order to overcome this problem, cells have evolved a damage tolerance system to allow complete replication in the presence of DNA damage. This process bypasses, rather removes, DNA damage, therefore, it is also named replicative bypass. It enables the cell to tolerate DNA damage and promote cell survival at the expense of high mutation rate. In fact, this process is responsible for most of the damage-induced point mutations and particularly important for oncogenesis. On the

DNA Damage Tolerance, Fig. 1 Schematic summary of consequences of DNA damage in mammalian cells. Note that DNA damage tolerance mechanism promotes cell survival, and defects in error-free translesion DNA synthesis result in predisposition to cancer

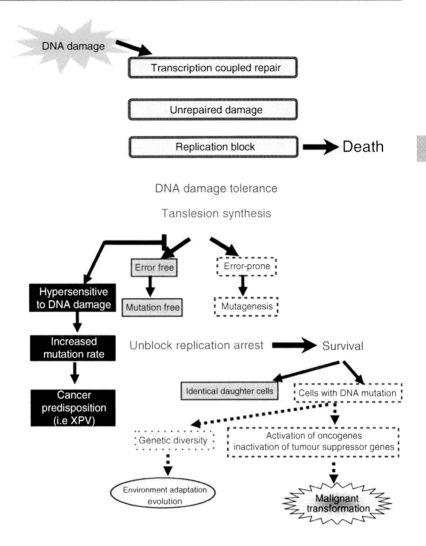

other hand, mutations are essential for evolution and adaptation which help a species to survive in a changing environment. The schematic summary of consequences of DNA damage is shown in Fig. 1.

Mechanism

Nonreplicative DNA Polymerases

A number of DNA polymerases have been identified specifically responsible for overcoming damage-induced replication arrest in human cells, which are also called specialized DNA polymerase (or bypass polymerase). So far ten of those polymerases have been identified in humans including Rev1, Polyη, κ, ι, λ, μ, β, θ, υ, and ζ.

Unlike replicative DNA polymerases, which can only use the opposite strand as template to initiate DNA synthesis, the specialized polymerases are able to promote stable incorporation of nucleotides opposite the lesion when replication is blocked by a damage. In addition to the conventional domains commonly found in all DNA polymerases such as "fingers," "thumbs," and "palm" domains, many of the specialized DNA polymerases have a "little finger" domain and catalytic sites which provide a flexible active site and allow the replicative bypass of various types of template structure. During an arrested replication, these polymerases take over temporarily from the replicative DNA polymerases and use either the damaged DNA strand or the newly synthesized

DNA strand as a template to proceed with replication. This event is referred as "polymerase switching." This step is essential for the specialized polymerases to transiently occupy the primer template for initiating synthesis to bypass the damaged site. Unlike the high-fidelity replicative DNA polymerases, these specialized polymerases have low fidelity of DNA synthesis and are responsible for mutagenesis in the genome. It is speculated that the reason that these polymerases are excluded from normal DNA replication is to maintain genomic stability in normal cells.

Translesion DNA Synthesis

The predominant and most well-studied DNA tolerance mechanism is known as translesion synthesis (TLS). This process allows tolerance of DNA damage by employing specialized DNA polymerases to synthesize DNA directly in order to bypass template DNA damage. The outcome of this process can be both error-free and error-prone. When normal DNA synthesis is blocked by a lesion on one of the template strands, the specialized DNA polymerase(s) use the newly synthesized daughter strand as template to proceed DNA synthesis. Therefore, copying of the damaged site of DNA template is avoided and the DNA replication continues. Because the newly synthesized daughter strand, instead of the damaged strand, is used as template, this process is also named "template switching." In this process, the correct nucleotide is incorporated opposite the damage site; therefore, it is error-free. In contrast, when specialized polymerase(s) use the damaged template to proceed with DNA synthesis, errors may occur. This process is called lesion bypass and usually error-prone because of lack of correct template. However, if the correct nucleotide is incorporated opposite to the damage site, it can be error-free. When an incorrect nucleotide is incorporated opposite the damage site and subsequently extended, a base mutation occurs. Therefore, the error-free lesion bypass is a mutation-avoiding mechanism, and error-prone lesion bypass is a mutation-generating mechanism. In humans, the poly η is demonstrated to be one of the error-free specialized polymerases, while Pol ι, κ, ζ, and Rev1 are found to be mutagenic

specialized polymerases. The error-prone translesion synthesis contributes a major mechanism of DNA damage-induced mutagenesis in humans.

Association with Cancer

Although the majority of polymerase errors are corrected by ► mismatch repair mechanism, the repair system may not function with 100% efficiency; therefore, errors are likely to escape correction and extended during subsequent replication. In addition, some unrepaired DNA damage, spontaneously formed or induced by environmental agents, will be processed by an error-prone lesion bypass mechanism leading to mutagenesis. Therefore, mutations are generated every time the cell replicates itself. Accumulation of mutations in DNA results in activation of proto-oncogenes and inactivation of ► tumor suppressor genes resulting in malignant transformation. The significance of translesion synthesis in the development of human cancer comes from the identification of germ line mutations of the *Poly η* gene in the form of a hereditary disease named ► xeroderma pigmentosum variant (XPV). Poly η is a highly specific bypass polymerase in repairing UV-induced thymidine dimers in an error-free manner. Loss of its function leads to increased UV-induced mutagenesis and hypersensitivity to sunlight. These XPV patients are hypermutable by ► UV radiation and develop cancer on sun-exposed skin at a very young age. Skin abnormalities are caused by defects in bypassing UV-induced DNA damage leading to cell death or malignancy.

References

Cleaver JE (2005) Cancer in xeroderma pigmentosum and related disorders of DNA repair. Nature 5:564–573

Friedberg EC (2004) The role of endogenous and exogenous DNA damage and mutagenesis. Curr Opin Gene Dev 14:5–10

Friedberg EC (2005) Suffering in silence: the tolerance of DNA damage. Nat Rev 6:943–953

Hoeijmakers JH (2001) Genome maintenance mechanisms for preventing cancer. Nature 411:366–411

Wang Z (2001) DNA damage-induced mutagenesis: a novel target for cancer prevention. Mol Interv 1:269–281

DNA Damage-Induced Apoptosis

Bernd Kaina
Department of Toxicology, University Medical
Center Mainz, Mainz, Germany

Synonyms

DNA damage-triggered death signaling pathways; DNA repair and damage processing

Definition

▶ DNA damage-induced cell death is executed by apoptosis, necrosis, parthanatos, mitotic catastrophe, and overactivated autophagy. From these different forms of cell inactivation, ▶ apoptosis is the main route of death following DNA damage. Cells undergo apoptosis upon genotoxic stress via the death receptor and/or the intrinsic mitochondrial pathway. DNA damage-induced apoptosis is thought to be a mechanism protecting against cancer because it eliminates genetically damaged cells. This is most obvious in sunburned skin in which p53 upregulation initiates the apoptotic process in response to light-induced DNA damage.

Characteristics

Summary

Not every type of DNA damage induces apoptosis. Many DNA lesions are tolerated by the cell, some are mutagenic without being toxic and some are more toxic than mutagenic. Apoptosis-inducing lesions are O^6-methylguanine, O^6-chloroethylguanine, base N-alkylations, cyclobutane pyrimidine dimers (CPDs) and (6–4) photoproducts, benzo[a]pyrene guanine adducts, cisplatin and mustard-induced crosslinks, and notably DNA double-strand breaks (DSBs). It is reasonable to suppose that any transcription and DNA replication blocking lesion bears apoptotic potential, and that transcription blockade and DNA replication fork stalling are apoptosis initiating events, triggering the DNA damage response pathway. Apoptosis signaling competes with DNA repair processes that either remove apoptosis-inducing lesions from DNA or tolerate them by replication bypass. Extended replication inhibition may lead to collapse of replication forks and the generation of DSBs, which are supposed to be the ultimate critical downstream apoptotic lesions for many, if not all, genotoxins. Primary DSBs induced by ionizing radiation and radiomimetic drugs and secondary DSBs derived from chemical DNA adducts during DNA replication are sensed by ATM (ataxia telangiectasia mutated) and ATR (ataxia telangiectasia related) protein, respectively. These PI3-like kinases phosphorylate a large number of proteins including p53 that regulates apoptosis by upregulating the death receptor FAS/APO-1/CD95, ▶ PUMA, NOXA, BAK und BAX, and translocating BAX to the mitochondrial membrane. Another mechanism implicated in apoptosis signaling rests on sustained upregulation of the transcription factor ▶ AP-1 by c-Fos and c-Jun activation, thus triggering the Fas ligand expression. The apoptotic pathway that finally becomes activated is cell type- and genotoxin- specific, depending on the p53 status, FAS/APO-1/CD95 responsiveness, and DNA repair capacity. DNA damage-induced apoptosis is executed by caspases, with caspase-8 activated by death receptors and caspase-9 by mitochondrial damage. The main executing ▶ caspases are caspase-3 and -7. These caspases cleave the inhibitor of caspase-activated DNase (CAD) that in turn cleaves DNA in the typical 180–200 bp nucleosomal fragments.

Physiological Role of Apoptosis Induced by DNA Damage

DNA is subject to spontaneous and induced damage by environmental genotoxins. These include ▶ alkylating agents, ▶ polycyclic aromatic hydrocarbons, biphenyls, ▶ heterocyclic amines, ultraviolet light, and ionizing radiation. DNA is also the main target for most anticancer drugs that react directly with DNA or interfere with DNA metabolism. Two cellular strategies have evolved for coping with DNA damage: (i) the damage is repaired or tolerated and (ii) cells harboring DNA damage are removed from the population by death. DNA damage has harmful consequences, which manifests as chromosomal changes, mutations, gene ▶ amplification and misregulation, cell death, and malignant transformation. Cells respond to genotoxins in a complex way by evoking numerous cellular responses that may ultimately lead to damage repair, damage fixation as mutations, or damage elimination by various routes of cell death (▶ necrosis, parthanatos (PARP-1 dependent cell death), apoptosis, reproductive cell death, interphase death, overactivated ▶ autophagy, and ▶ mitotic catastrophe). Programmed cell death (apoptosis) following DNA damage is based on a complex enzymatic machinery. Apoptosis occurs continuously in the body, notably in changing tissues, and it appears that DNA damage-triggered cell death utilizes the normal cellular suicide program as a strategy to eliminate genetically damaged cells.

DNA-damaging agents. There are three lines of evidence that apoptosis induced by genotoxic agents is related to DNA damage: (i) Inability of cells to repair DNA lesions results in hypersensitivity to the killing effect (an exception to the rule is DNA mismatch repair, see below). This has been shown for mutants defective in O^6-methylguanine-DNA methyltransferase (MGMT the repair protein is also known as alkyltransferase), base excision repair (BER), nucleotide excision repair (NER), DSB repair, and DNA cross-link repair. (ii) Modified nucleotides such as 6-thioguanine or gancyclovir incorporated in DNA induce apoptosis. (iii) DSBs induced by restriction enzymes in the cellular genome induce a strong apoptotic response. They hardly induce necrosis. The apoptotic pathways activated by a single DNA lesion, O^6methylguanine (O^6MeG), as well as its interplay with DNA repair pathways have been studied in great detail. O^6MeG does not block DNA replication. It mispairs with thymine, giving rise to GC to AT point mutations. In the presence of ▶ mismatch repair (MMR) the O^6MeG/thymine mismatch is recognized by MutSα (consisting of MSH2 and MSH6) and MutLα (heterodimer of

MLH1 and PMS2) that provoke a futile mismatch repair cycle leading to the formation of DSBs and apoptosis signaling. MMR deficient cells are highly resistant to simple alkylating agents because they do not undergo apoptosis. They tolerate O^6MeG adducts at the expense of mutations. Therefore, MMR driven apoptosis eliminates premutated cells from the population, reducing the mutation load (Mismatch Repair in Genome Stability).

Apoptotic pathways. The apoptotic pathways employed by cells following the exposure to O^6-methylating agents and other genotoxins differ depending on their p53 status. In p53 mutated cells the mitochondrial apoptotic pathway becomes activated, characterized by a decline in BCL-2 that increases the BAX/BCL-2 ratio and allows for the release of cytochrome c from the mitochondria. This release of cytochrome c leads to the activation of the apoptosome consisting of Apaf-1, ATP, cytochrome c, and procaspase-9. In turn, caspase-3 becomes activated that cleaves the inhibitor of caspase-activated DNase (ICAD) that degrades DNA, resulting in the characteristic apoptotic DNA fragmentation. In p53 wild-type cells, e.g., human lymphocytes, DNA damage activates the death receptor pathway (Fas/CD95/Apo-1). There are cell types, such as glioma cells, wild type for p53 in which both the Fas receptor and the mitochondrial damage pathway becomes activated in response to DNA damage. However, a higher level of DNA damage is required in order to induce apoptosis by activating the mitochondrial pathway compared to the death receptor pathway.

Another role for p53 has been ascribed to the regulation of PUMA. DNA damage localizes p53 to the nucleus and transcribes PUMA. PUMA in the cytoplasm liberates cytoplasmic p53 from Bcl-xL, thereby freeing p53 to activate BAX and facilitate mitochondrial apoptosis. Another player in DNA damage-induced apoptosis is caspase-2, the only nuclear localized caspase that seems to act upstream of the mitochondria.

DNA damage-triggered signaling: sustained JNK/p38 kinase activation. Some DNA-damaging agents provoke the activation of stress-activated protein kinase/c-Jun N-terminal kinase (SAPK/JNK) and p38 kinase, which results in an increase in c-Jun level and ▶ AP-1 activity. This sustained activation of AP-1 is accompanied by transcriptional activation of the Fas-L gene. NER repair defective mutants display a higher level of sustained JNK/p38 kinase activation, indicating DNA damage is responsible for the response. Together with DNA damage-induced p53 upregulation that triggers transcription of the Fas receptor, the upregulation of the Fas ligand via AP-1 is effective in driving apoptosis upon DNA damage.

The ATM/ATR/p53 connection. DSBs are the most lethal DNA lesions and, therefore, cells have to be equipped with sensors that recognize DSBs immediately upon formation. These sensors signal cell cycle blockage and ameliorate DNA repair or, if this fails, cell death. DSBs are recognized by the MRN complex consisting of MRE11, NBS1 and RAD50, and activate in turn DNA damage dependent kinases. The most important players are ataxia telangiectasia mutated (ATM) protein and the ATM- and Rad3-related (ATR) protein. ATM is mostly activated by DSBs, whereas ATR becomes activated by single-stranded DNA at blocked replication forks. Once activated, ATM and ATR phosphorylate various downstream substrates such as NBS1. Phosphorylated NBS1 acts as an adapter molecule and supports ATM dependent phosphorylation of CHK2 which activates Cdc25 responsible for S phase checkpoint control. Other substrates of ATM are p53, MDM2, CHK1, H2AX, and BRCA1. While ATM is activated by ionizing radiation-induced DSBs, ATR is activated in response to UV light and presumably all chemical agents that give rise to stalled DNA replication forks. ATM/ATR is implicated in three crucial functions: regulation and stimulation of DSB repair (homologous recombination), signaling cell cycle checkpoints, and signaling apoptosis via p53. The phosphorylation of p53 by ATM leads to its stabilization, nuclear translocation, and upregulation of p21 that triggers G1/S arrest. Low levels of DSBs activate only a minor fraction of p53 that is sufficient to drive the p21 gene, causing cell cycle arrest. With a higher level of DSBs, p53 becomes activated strongly and thus drives proapoptotic genes such as *bax* and the *fas receptor*. Therefore, the ATM/p53 activation level is very important in DNA damage-triggered apoptosis. Although AT cells (cells in which the ATM gene is mutated) are more sensitive to ionizing radiation, they exhibit less apoptosis after ionizing radiation than normal cells. ATM knockout mice also showed a lower apoptotic response after ionizing radiation than the corresponding wild-type mice, indicating the importance of the ATM/p53 pathway in triggering DNA damage-induced apoptosis. However, in a p53 mutated background, ATM knockout fibroblasts are more sensitive to apoptosis induced by alkylating agents than corresponding lines expressing ATM. Also, downregulation of ATM (and ATR) sensitizes cancer cells to anticancer drugs. This suggests that ATM (and ATR) is not indispensable for triggering apoptosis in response of DNA damage but fulfil also a significant protective function. It thus appears that DSBs, notably those induced by chemical genotoxins, are able to trigger apoptosis very efficiently also by another mechanism not involving ATM/p53 signaling. The high level of resistance of ATM wild-type cells is explained by the role of ATM in DSB repair.

p53 is considered a major player in the apoptotic response, which is in line with the finding that p53 knockout mouse are resistant to the toxic effect of ionizing radiation, which is largely due to impaired death of thymocytes. While p53 deficient thymocytes are more resistant to ionizing radiation, p53 deficient mouse fibroblasts are more sensitive. This indicates that in fibroblasts, p53 is not required for inducing apoptosis in response to DSBs. It rather exerts a protective effect, which is most likely due to its involvement in DNA repair and inhibition of DNA synthesis by blocking cells in G1/S, transition. The decision of whether p53 exerts a pro- or anti-apoptotic effect appears to be cell type-specific, and the conditions determining whether p53 stimulates or protects against apoptosis have yet to be explored in detail. However, there are data to indicate that the phosphorylation status of p53 is involved in the decision of whether p53 exerts pro- or antiapoptotic activity. Thus, p53 phosphorylated on Ser 15 was described to activate pro-survival genes, including the repair gene DDB2, while p53 phosphorylated at Ser 46 drives pro-death genes and activates the pro-apoptotic PTEN pathway. In this scenario the kinase HIPK2, which becomes activated in response to DNA damage, appears to play a key role.

Cross-References

▶ Alkylating Agents
▶ Amplification
▶ AP-1
▶ Apoptosis
▶ Autophagy
▶ BARD1
▶ Caspase
▶ Desmoglein-2
▶ DNA Damage
▶ γH2AX
▶ Heterocyclic Amines
▶ Mismatch Repair in Genetic Instability
▶ Mitotic Catastrophe
▶ Necrosis
▶ Polycyclic Aromatic Hydrocarbons
▶ PUMA

References

Kaina B (2003) DNA damage-triggered apoptosis: critical role of DNA repair, double-strand breaks, cell proliferation and signaling. Biochem Pharmacol 66(8):1547–1554

Lavin MF et al (2005) ATM signaling and genomic stability in response to DNA damage. Mutat Res 569:123–132

Ljungman M, Zhang F (1996) Blockage of RNA polymerase as a possible trigger for UV light-induced apoptosis. Oncogene 13:823–831

Mansouri A et al (2003) Sustained activation of JNK/p38 MAPK pathways in response to cisplatin leads to Fas ligand induction and cell death in ovarian carcinoma cells. J Biol Chem 278(21):19245–19256

Roos WP, Kaina B (2006) DNA damage-induced cell death by apoptosis. Trends Mol Med 12:440–450

Roos WP, Thomas AD, Kaina B (2016) DNA damage and the balance between survival and death in cancer biology. Nature Rev Cancer, 16 (1): 20–33.

See Also

(2012) Ataxia Telangiectasia. In: Schwab M (ed) Encyclopedia of Cancer, 3rd edn. Springer Berlin Heidelberg, p 298. doi:10.1007/978-3-642-16483-5_426

(2012) ATR. In: Schwab M (ed) Encyclopedia of Cancer, 3rd edn. Springer Berlin Heidelberg, p 302. doi:10.1007/978-3-642-16483-5_443

(2012) Caspase-3. In: Schwab M (ed) Encyclopedia of Cancer, 3rd edn. Springer Berlin Heidelberg, p 675. doi:10.1007/978-3-642-16483-5_874

(2012) CD95. In: Schwab M (ed) Encyclopedia of Cancer, 3rd edn. Springer Berlin Heidelberg, p 703. doi:10.1007/978-3-642-16483-5_939

(2012) CHK1. In: Schwab M (ed) Encyclopedia of Cancer, 3rd edn. Springer Berlin Heidelberg, p 817. doi:10.1007/978-3-642-16483-5_1101

(2012) Cytochrome c. In: Schwab M (ed) Encyclopedia of Cancer, 3rd edn. Springer Berlin Heidelberg, p 1043. doi:10.1007/978-3-642-16483-5_1458

(2012) DNA Double Strand Breaks. In: Schwab M (ed) Encyclopedia of Cancer, 3rd edn. Springer Berlin Heidelberg, p 1139. doi:10.1007/978-3-642-16483-5_1675

(2012) FAS. In: Schwab M (ed) Encyclopedia of Cancer, 3rd edn. Springer Berlin Heidelberg, p 1379. doi:10.1007/978-3-642-16483-5_2121

(2012) G1/S Transition. In: Schwab M (ed) Encyclopedia of Cancer, 3rd edn. Springer Berlin Heidelberg, p 1484. doi:10.1007/978-3-642-16483-5_2291

(2012) Ionizing Radiation. In: Schwab M (ed) Encyclopedia of Cancer, 3rd edn. Springer Berlin Heidelberg, p 1907. doi:10.1007/978-3-642-16483-5_3139

(2012) JUN. In: Schwab M (ed) Encyclopedia of Cancer, 3rd edn. Springer Berlin Heidelberg, p 1929. doi:10.1007/978-3-642-16483-5_3186

(2012) Knock-Out Mouse. In: Schwab M (ed) Encyclopedia of Cancer, 3rd edn. Springer Berlin Heidelberg, p 1957. doi:10.1007/978-3-642-16483-5_3239

(2012) MRN Complex. In: Schwab M (ed) Encyclopedia of Cancer, 3rd edn. Springer Berlin Heidelberg, p 2382. doi:10.1007/978-3-642-16483-5_6622

(2012) NBS1. In: Schwab M (ed) Encyclopedia of Cancer, 3rd edn. Springer Berlin Heidelberg, p 2468. doi:10.1007/978-3-642-16483-5_3984

(2012) P53. In: Schwab M (ed) Encyclopedia of Cancer, 3rd edn. Springer Berlin Heidelberg, p 2747. doi:10.1007/978-3-642-16483-5_4331

(2012) Replication. In: Schwab M (ed) Encyclopedia of Cancer, 3rd edn. Springer Berlin Heidelberg, p 3254. doi:10.1007/978-3-642-16483-5_5031

(2012) Transcription. In: Schwab M (ed) Encyclopedia of Cancer, 3rd edn. Springer Berlin Heidelberg, p 3752. doi:10.1007/978-3-642-16483-5_5899

(2012) Ultraviolet Light. In: Schwab M (ed) Encyclopedia of Cancer, 3rd edn. Springer Berlin Heidelberg, p 3841. doi:10.1007/978-3-642-16483-5_6101

DNA Damage-Triggered Death Signaling Pathways

▶ DNA Damage-Induced Apoptosis

DNA Demethylation

▶ Hypomethylation of DNA

DNA Lesion

▶ DNA Damage

DNA Lesion Bypass

▶ DNA Damage Tolerance

DNA Oxidation Damage

Andrew R. Collins
Department of Nutrition, University of Oslo,
Oslo, Norway

Definition

Represents free radical damage to DNA. Oxidation essentially involves the addition of oxygen or removal of hydrogen atoms from a molecule. Oxidation of DNA may simply result in a small change to one of the bases, or a deoxyribose in the backbone of the molecule may be altered to such an extent that the continuity of the backbone is broken. Single-strand breaks are more common than double-strand breaks.

Characteristics

DNA is thought of as a very stable molecule and yet it readily undergoes damage, by a variety of agents that can be either endogenous or exogenous in origin. Ionizing radiation (e.g., X-rays), ultraviolet radiation, and various chemicals, including some present in tobacco smoke (▶ tobacco carcinogenesis), cause the release of free radicals, and if DNA is not protected, oxidation can occur. Free radicals (▶ reactive oxygen species) also occur within the cells of the body, arising as a minor product during the cycle of oxidation of carbohydrates in the mitochondria. The hydroxyl radical, *OH, is particularly reactive with DNA.

In addition to single- and double-strand breaks, many different oxidation products of the four bases (▶ adducts to DNA) have been identified in DNA treated with radiation or other free radical-generating chemicals. Some of these modified bases are potentially capable of giving rise to a mutation. For instance, 8-oxoguanine, if present in the DNA when it is replicating, may lead to the incorporation of adenine rather than cytosine into the newly synthesized complementary strand, thus changing the DNA sequence.

Some oxidation is detectable in the DNA of normal human cells. In the past it was usually measured by gas chromatography (GC-MS) or high performance liquid chromatography (HPLC). Normally for GC-MS, the DNA is acid-hydrolyzed to bases (guanine etc.), while for HPLC the DNA is enzymically hydrolyzed to nucleosides (deoxyguanosine (dG) etc.). Both of these methods have given relatively high values for the extent of conversion of guanine to 8-oxoguanine, with up to 300 or more 8-oxo-guanines for every 10^6 unaltered guanines. However, guanine tends to oxidize during preparation of samples for analysis and so the early estimates of 8-oxoguanine are considered to be excessive. Values as low as 3 per 10^6 have been reported from HPLC analysis of anaerobically prepared samples.

There is another approach to measuring oxidized bases using bacterial repair endonucleases, which recognize the damage and make a corresponding break in the DNA. The enzyme FPG (formamidopyrimidine DNA glycosylase) recognizes 8-oxoguanine. DNA breaks can then be measured in various ways, including the ▶ comet assay (Fig. 1). This approach gives values for 8-oxoguanine that are even lower, with around 0.5 per 10^6 guanines.

The extent of background DNA oxidation in normal cells remains an important question.

Methodological problems must be solved before a consensus can be reached.

Mechanisms

Measurement is made of the steady-state level of DNA damage, which is a dynamic equilibrium between input of damage and its repair (Fig. 2). In the case of oxidative damage, input is controlled by antioxidant defenses, which are intrinsic compounds (e.g., albumin, glutathione, uric acid) and enzymes (catalase, superoxide dismutase) that can convert, scavenge, or inactivate free radicals (oxidative DNA damage). Dietary antioxidants, such as carotenoids and vitamins E and C, aid this process. The tripeptide glutathione is present at high concentration in the nucleus and "mops up" free radicals before they can cause damage. Superoxide dismutase and catalase are enzymes that convert superoxide and hydrogen peroxide (two reactive forms of oxygen) ultimately to non-harmful products. Other enzymes combine various organic free radicals with glutathione, thus inactivating them. Fruits, vegetables, and grains in the diet are a source of antioxidants, including vitamin C, vitamin E, ► carotenoids, and flavonoids. These ► natural products have the ability to quench or scavenge free radicals; whether they act as antioxidants in vivo depends on whether they are taken up from the gut in sufficient amounts and has been the subject of human intervention trials. In general, it is possible to detect a significant decrease in the steady-state level of base oxidation (and/or an increased resistance to in vitro oxidation of DNA) in white blood cells of volunteers taking individual antioxidant supplements or antioxidant-rich foods, ranging from fried onions to kiwifruit.

However effective the antioxidant defenses, some DNA oxidation does occur. The turnover of this damage is achieved by ► repair of DNA. Small base damage, which includes base oxidation, is repaired primarily by base excision repair. Here, the damaged base is removed, followed by the base-less sugar-phosphate residue and perhaps a few neighboring nucleotides. A small

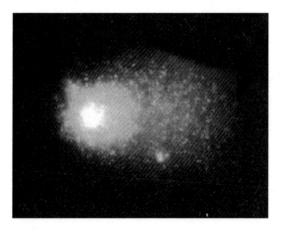

DNA Oxidation Damage, Fig. 1 The comet assay. Cells are embedded in agarose on a microscope slide, lysed, and electrophoresed at high pH. This view is of the DNA from one cell, stained with DAPI and visualized by fluorescence microscopy. The percentage of DNA fluorescence in the tail of the "comet" is proportional to the frequency of breaks – such as breaks introduced at 8-oxoguanine sites by the enzyme FPG

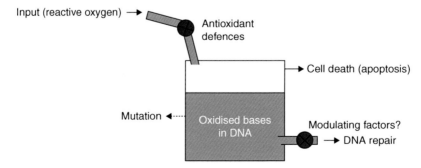

DNA Oxidation Damage, Fig. 2 DNA damage: a steady state. There is a constant inflow of damage, caused by free radical attack and attenuated by antioxidants. This is normally balanced by cellular DNA repair processes, which remove the damage and restore the DNA sequence. Little is known about what modulates repair activity

repair patch of new nucleotide(s) is inserted and ligated.

As well as being present in cellular DNA, 8-oxodG is detectable in urine as free nucleoside. The idea that urinary 8-oxodG represents the accumulated product of DNA repair in all the cells of the body is attractive but flawed; base excision repair releases the base and not the nucleoside. Even if this 8-oxodG originates in the oxidation of broken down DNA from dead cells passing through the kidneys, or from oxidised DNA precursor nucleotides in the cellular pool it still reflects ▶ oxidative stress, and it has given useful information about oxidative stress related to exercise, smoking, and nutrition. Most impressively, consumption of 300 g a day of Brussels sprouts led to a decrease of 28% in urinary 8-oxodG concentration.

Clinical Aspects

It is commonly stated that oxidative damage to DNA is a significant cause of cancer and that fruits and vegetables protect against cancer because the antioxidants they contain decrease the amount of base oxidation in the cellular DNA. However, the evidence for this is rather weak. In two large-scale human intervention trials, smokers and/or ▶ asbestos workers were given β-carotene (a carotenoid antioxidant) daily for several years. The ▶ lung cancer incidence was actually higher in these subjects compared with those taking placebo (or other supplement). Other intervention trials have shown no beneficial or harmful effect of supplementation with antioxidant micronutrients in terms of cancer risk – in spite of their ability to decrease oxidative damage.

In an experimental animal system, a high level of oxidative DNA damage is not necessarily marked by an elevated cancer risk. In a knockout mouse model, which is defective in the murine equivalent of FPG, there is a slight increase in the steady-state level of 8-oxoguanine, but no increase in cancer incidence. It seems that there is a backup repair pathway that deals (more slowly, but adequately) with oxidative damage.

Oxidative stress is a feature of many other diseases, including heart disease, diabetes, cataract, and rheumatoid and arthritic conditions. It may be a cause of the clinical condition or an effect. A common theory of aging argues that the accumulation of free radical-induced damage to biomolecules – lipids, proteins, and nucleic acids – is responsible for the general cellular dysfunction and deterioration of body processes in later life, but the evidence for accumulation of oxidative DNA damage is inconsistent.

The importance of fruits and vegetables in a healthy diet is not in doubt. But it is clear that antioxidants are not their only feature, and we should be looking at other effects that phytochemicals might have on metabolism to account for their capacity to prevent disease (Dolara et al. 2012).

Cross-References

▶ Adducts to DNA
▶ Asbestos
▶ Carotenoids
▶ Comet Assay
▶ Lung Cancer
▶ Natural Products
▶ Oxidative Stress
▶ Reactive Oxygen Species
▶ Repair of DNA
▶ Tobacco Carcinogenesis
▶ UV Radiation

References

Collins AR (1999) Oxidative DNA damage, antioxidants, and cancer. BioEssays 21:238–246

Collins AR (2005) Antioxidant intervention as a route to cancer prevention. Eur J Cancer 41:1923–1930

Dolara P, Bigagli E, Collins A (2012) Antioxidant vitamins and mineral supplementation, life span expansion and cancer incidence: a critical commentary. Eur J Nut 51:769–781

ESCODD, Gedik CM, Collins A (2005) Establishing the background level of base oxidation in human lymphocyte DNA: results of an interlaboratory validation study. FASEB J 19:82–84

Lindahl T (1993) Instability and decay of the primary structure of DNA. Nature 362:709–714

See Also

(2012) Antioxidant. In: Schwab M (ed) Encyclopedia of Cancer, 3rd edn. Springer Berlin Heidelberg, p 216. doi:10.1007/978-3-642-16483-5_328

(2012) Antioxidant Defenses. In: Schwab M (ed) Encyclopedia of Cancer, 3rd edn. Springer Berlin Heidelberg, p 216. doi:10.1007/978-3-642-16483-5_330

(2012) GC-MS. In: Schwab M (ed) Encyclopedia of Cancer, 3rd edn. Springer Berlin Heidelberg, p 1515. doi:10.1007/978-3-642-16483-5_2344

(2012) Knock-Out Mouse. In: Schwab M (ed) Encyclopedia of Cancer, 3rd edn. Springer Berlin Heidelberg, p 1957. doi:10.1007/978-3-642-16483-5_3239

(2012) Mutation. In: Schwab M (ed) Encyclopedia of Cancer, 3rd edn. Springer Berlin Heidelberg, p 2412. doi:10.1007/978-3-642-16483-5_3911

(2012) 8-Oxoguanine. In: Schwab M (ed) Encyclopedia of Cancer, 3rd edn. Springer Berlin Heidelberg, p 2734. doi:10.1007/978-3-642-16483-5_4313

DNA Repair and Damage Processing

▶ DNA Damage-Induced Apoptosis

DNA Undermethylation

▶ Hypomethylation of DNA

DNA Vaccination

Holger N. Lode
Klinik und Poliklinik für Kinder und Jugendmedizin, Universitätsmedizin Greifswald, Greifswald, Germany

Synonyms

Genetic immunization

Definition

Vaccination with deoxyribonucleic acid (DNA) against cancer is the most basic type of vaccination that, rather than consisting of the tumor-associated antigen itself, provides genes encoding for the antigen. Once produced in vivo following DNA delivery, the antigen is presented to the immune system inducing an antigen-specific immune response. This response is augmented by the immunological properties of the DNA itself, mediated by unmethylated CpG sequences. This essay reviews accomplishments and challenges in this area.

Characteristics

DNA vaccination represents a young field in cancer immunotherapy. It started with the observation that injection of plasmid DNA into a mammal resulted in the synthesis of the encoded protein. The unformulated or "naked" plasmid DNA containing a simple expression cassette, consisting of a promoter functioning in mammalian cells and of a gene encoding for a protein antigen, was injected into the muscle of mice. The subsequent induction of antigen-specific CD8+ ▶ cytotoxic T cells and antibodies was effective in protecting mice from challenges with the pathogenic agent expressing the antigen. This observation was surprising, given the low amount of antigen produced, the apparent lack of transfection of professional antigen-presenting cells (APC), and the absence of any replicative step. The robustness of the technology was demonstrated for a variety of disease models.

Mechanisms of Action

The method of DNA delivery critically affects the mechanisms involved in the induction of an immune response. Intramuscular injection of plasmid DNA leads to in vivo transfection of myocytes. Mechanistic studies revealed that antigen-specific immune responses following intramuscular injection of DNA are a result from cross priming. This mechanism describes production of the antigen by the myocyte and subsequent uptake and presentation by professional APCs. This was clearly demonstrated in bone marrow chimeric mice and in experiments with transfected myoblasts. In both systems, the induction of an antigen-specific immune response depended on the antigen presentation by APCs and not by the myocyte. The transfer of the antigen from myocyte to APC follows different routes, ranging from uptake of secreted protein, processed peptide alone or with heat-shock proteins or apoptotic bodies by APCs. The transfer of DNA into the myocyte and subsequent induction of an antigen-specific immune response can be largely improved by in vivo ► electroporation. The technique involves the application of an electric field around the DNA injection site. There are also needle-free systems available, injecting DNA in solution using high-pressure liquid jets. Clinical devices were developed by the pharmaceutical industry for application in humans, which are well tolerated.

Bombardment of the epidermis with plasmid loaded onto gold particles using the gene gun directly transfers DNA into APCs of the skin called ► Langerhans cells. Once the protein antigen is expressed, professional antigen presentation is mediated by this cell type. Efficient induction of an immune response occurs after migration of these APCs from the skin into regional lymph nodes.

Gene transfer of plasmid DNA into APCs is also accomplished by the use of live attenuated bacteria such as *Salmonella typhimurium* or *Listeria monocytogenes*. In both cases, these microorganisms are infectious, but not pathogenic, and therefore serve as in vivo carrier systems for plasmid DNA vaccines. After in vivo application, Peyer patches and the spleen become infected. Subsequently, the carrier microbes die due to distinct mutations in their genome and liberate multiple copies of the plasmid DNA vaccines in these secondary lymphoid organs. There, the DNA vaccines are expressed by APCs leading to the induction of an antigen-specific immune response.

A central role for the induction of an immune response by DNA vaccines is antigen expression by APCs and subsequent presentation to $CD8^+T$ cells, CD4 T cells, and B cells (Fig. 1).

Adjuvant Activity of DNA
Plasmid DNA derived from bacterial expression systems naturally contain unmethylated DNA sequences called CpG motifs. These sequences bind to Toll-like receptor 9 and are strong activators of ► innate immunity. This receptor is also expressed on APCs leading to improved antigen processing and presentation as well as the release of pro-inflammatory cytokines and chemokines that help to shift ► adaptive immunity responses from Th2 immune response to Th1 immune response. Th1 responses are required for most effective antitumor immunity. Therefore, CpG motifs in the DNA vaccine backbone can be considered endogenous adjuvants linking innate immunity with ► adaptive immunity, which provides for robust and long-lasting antigen-specific immune responses.

Tailoring Immune Responses by DNA Vaccine Design
In order to improve antigen-specific immune responses, the versatility of DNA vaccine design allows for the simultaneous expression of antigen, co-stimulatory molecules, and chemoattractants. These include cytokines, chemokines, and molecules of the B7 family and CD40 ligand. The design of the protein antigen itself can be altered to be secreted for the induction of B-cell responses or to be targeted into the endoplasmic reticulum or the proteasomal degradation pathway for the generation of T-cell epitopes. Protein antigens can be redesigned as mini genes only encoding for immunodominant peptide antigens. In summary, the versatility of DNA vaccines allows for specific tailoring of an optimized immune response following a rational vaccine design.

Formulation
The formulation of DNA vaccines to improve antigen-specific immune responses includes transfection-facilitating lipid complexes, nanoparticles, and classical adjuvants. Lipid complexes are varying combinations of DNA with cationic lipids. Microparticles are generated with DNA entrapped in biodegradable poly-lactide-co-glycolactide or complexed with nonionic block copolymers or polycations. Among the classical adjuvants, aluminum phosphate is noteworthy for its effectiveness and simplicity of preparation. Microparticles appear to improve the trafficking

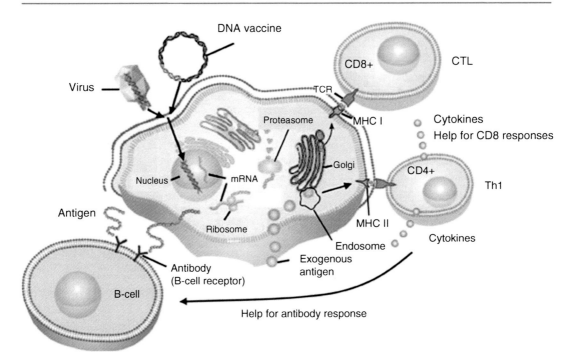

DNA Vaccination, Fig. 1 Mechanisms involved in the generation of antigen-specific humoral and cellular immune responses upon DNA vaccination. Antigen-specific activation of cytolytic T lymphocytes (CD8+T cells) occurs after proteasome-dependent antigen processing of intracytoplasmic proteins into peptides associated with newly synthesized MHC class I molecules. MHC class I/peptide complexes are presented on the surface of APCs in conjunction with costimulatory molecules to CD8+T cells. The activation of CD4T cells is primarily achieved by exogenous protein antigens taken up by the endolysosomal compartment. After degradation, peptides associate with MHC class II molecules which are then translocated to the cell surface. Specific CD4+ helper T cells recognize these MHC class II/peptide complexes and are activated to produce cytokines. These cytokines have multivarious activities helping B cells to mature into antibody-producing plasma cells and CD8+T cells to transform into cytolytic effector cells. For antibody responses, B cells recognize and respond to antigens that are either present extracellularly or exposed extracellularly by being transmembrane proteins

of DNA to APCs by facilitating the transfer of DNA into regional lymph nodes.

Mixed Modality Vaccines

A very promising strategy that is entering clinical trials is to combine DNA vaccines with other gene delivery systems. This is based on observations that if DNA encoding an antigen is given as a prime followed by another gene-based vector system as a boost such as recombinant viruses encoding the same antigen, most optimal immune responses and protection are achieved. The responses are significantly greater than using DNA or the virus for both the prime and the boost or if the order of the administration is reversed.

Results from Clinical Trials

First-generation DNA vaccines have been evaluated clinically as a therapeutic vaccine approach for cancer. These vaccines encoded for viral epitopes from transforming viruses, self-antigens expressed on tumors, and tumor-specific antigens. In most trials so far, the plasmid DNA was injected intramuscularly, intradermally, or intranodally. Antigen-specific humoral and cellular responses were observed in human cancer trials. However, this did not translate into clinical responses in the trial patient populations characterized by large tumor burden and progressive disease. The clinical trials so far have proved the principle that immune responses can be generated in humans. They also highlight the need to apply strategies to increase the potency of the technology as outlined above and to generate second-generation DNA vaccines for future application in cancer patients.

In summary, DNA vaccines hold great potential as immunotherapeutic tools to prevent and treat human cancer. Their advantages include cost-effectiveness, versatility, safety, stability, ease of construction and mass production, and most importantly ability to induce robust humoral and cellular immune responses. Lack of success in early clinical trials so far is similar to early clinical results with treatments based on monoclonal antibodies which are now established cancer therapeutics. To push for success, the next generation of DNA vaccines will have to incorporate multiple strategies to enhance plasmid DNA immunogenicity. Additional avenues may involve exploring the possibilities of combining adoptive cell therapies with DNA vaccines such as ex vivo gene transfer into autologous dendritic cells.

References

Donnelly JJ, Wahren B, Liu MA (2005) DNA vaccines: progress and challenges. J Immunol 175:633–639

Fest S, Huebener N, Weixler S et al (2006) Characterization of GD2 peptide mimotope DNA vaccines effective against spontaneous neuroblastoma metastases. Cancer Res 66:10567–10575

Liu MA, Ulmer JB (2005) Human clinical trials of plasmid DNA vaccines. Adv Genet 55:25–40

Lowe DB, Shearer MH, Jumper CA et al (2007) Towards progress on DNA vaccines for cancer. Cell Mol Life Sci 64:2391–2403

DNA-Bound Carcinogens

▶ Adducts to DNA

DnaJ (Hsp40) Homolog Subfamily C Member 15

▶ Methylation-Controlled J Protein

DNAJC15

▶ Methylation-Controlled J Protein

DNAJD1

▶ Methylation-Controlled J Protein

DNF15S2

▶ Macrophage-Stimulating Protein

Docetaxel

Ricardo Hitt and Ana Lopez-Martin
Hospital Universitario Severo Ochoa, Madrid, Spain

Synonyms

Taxotere

Definition

Docetaxel is a new class of anticancer agent that exerts the cytotoxic effects on microtubules. It is a semisynthetic drug with significant activity in a broad range of tumor types that are generally refractory to conventional therapies, including chemotherapy-resistant epithelial ▶ ovarian cancer, ▶ breast cancer, ▶ non-small cell lung cancer, head and neck cancer, ▶ bladder cancer, and ▶ gastric cancer.

Characteristics

Docetaxel is derived semisynthetically from 10-deacetyl-baccatin III, is more water soluble than ▶ paclitaxel, and is more potent antimicrotubule agent in vitro.

Mechanism of action: Docetaxel induces polymerization of tubulin, microtubule bundling in cells, and formation of numerous abnormal mitotic asters. The cytotoxic effects of docetaxel are also severalfold greater than paclitaxel in vitro and in tumor xenografts. Docetaxel inhibit proliferation of cells by inducing a sustained mitotis block at the metaphase–anaphase boundary at much lower concentrations than those required to increase microtubule polymer mass and microtubule bundle formation. These inhibitory effects at low drug concentrations are associated with the formation of an incomplete metaphase plate of chromosomes and an arrangement of spindle microtubules resembling the abnormal organization that occurs at low concentrations of the vinca alkaloids.

Docetaxel primarily blocks cell-cycle traverse in the mitotic phases and prevents the transition from Go to S phase. The inhibitory effects in the nonmitotic cell-cycle phases include the disruption of tubulin in the cell membrane and direct inhibitory effects on the disassembly of the interphase cytoskeleton. These effects may result in the disruption of many vital cell functions such as locomotion, intracellular transport, and transmission of proliferative transmembrane signals.

After disruption of microtubules and other processes by docetaxel, the precise means by which cell death occurs is not clear. Morphologic features and a DNA fragmentation pattern (nucleosomal DNA fragments) that are characteristic of programmed cell death, or apoptosis, in docetaxel-treated cells indicate that this taxane triggers apoptosis as do many other chemotherapeutic agents. Whether docetaxel-induced ▶ apoptosis requires a functional p53 pathway is unclear and probably depends on the cell line under study. The consensus seems to be that in most cell lines, disruption of p53 has little effect on drug sensitivity.

Mechanism of resistance: Selection of taxane-resistant cells in vitro is associated with changes in β-tubulin isotype expression. Six different isotypes of β-tubulin are expressed in nonmalignant tissues, with the class I isotype comprising 80–99% of cellular β-tubulin. The β-III isotype increases the dynamic instability of microtubules, impairs rates of microtubule assembly, and increases resistance to taxanes.

A second mechanism of acquired taxane resistance fits the general pattern of MDR. The particular species of Pgp found in taxane-resistant murine ▶ macrophages is similar, but not identical, to that found in vinblastine- and colchicine-resistant cells derived from the same parental line. These cells are cross-resistant with many other natural products, and resistance to docetaxel conferred by MDR-1 can be reversed by many classes of drugs, including ▶ tamoxifen, cyclosporine A, and antiarrhytmic agent.

Other changes in tumor cells selected for drug resistance have included upregulation of ▶ caveolin-1, a principal component of membrane-derived vesicles involved in transmembrane transport of small molecules and in intracellular signaling.

Pharmacokinetics

The single 1-h infusion every 3 weeks is the most common administration of docetaxel. The pharmacokinetics behavior on 1- or 2-h schedules is linear at doses of 115 mg/m^2 or less and optimally fits a three-compartment model. Docetaxel binds rapidly and avidly to plasma proteins (>90%), especially to albumin, α1-acid glycoprotein, and lipoproteins. In addition, peak plasma concentrations generally exceed levels required to induce relevant biologic effects in vitro. Limited information is available about the distribution of docetaxel in humans. Immediately after treatment, tissue uptake of radioactivity is highest in the liver, bile, and intestines, a finding that is consistent with substantial hepatobiliary extraction and excretion. High levels of radioactivity are also found in the stomach, which indicates the possibility of gastric excretion, as well as in the spleen, bone marrow, myocardium, and pancreas. Docetaxel has hepatic metabolism, and biliary excretion and urinary excretion account only 2%. Approximately 80% of the administered dose of total radioactivity is excreted in the feces within 7 days after treatment, with the majority of excretion occurring in the first 48 h. In the hepatic

▶ cytochrome P450-mixed function, oxidases are responsible for the bulk of drug metabolism, and CYP3A, CYP2B, and CYP1A isoforms may play major roles in biotransformation. The main metabolic pathway consists of oxidation of the tertiary butyl group on the side chain at the C-13 position of the taxane ring as well as cyclization of the side chain.

Toxicity

▶ Neutropenia is the principal toxicity of docetaxel. At dose of 100 mg/m^2, neutrophil count nadirs are <500/mcl in 50–80% of courses in previously untreated patients. The onset of neutropenia is early, with nadirs usually on day 9 and complete recovery by days 15–21. Neutropenia is not cumulative.

Hypersensitivity reaction (HSR) with premedication with dexamethasone is present in less of 10% of the cycles. Another possible toxicity is fluid retention that is cumulative doses, characterized by edema, weight gain, ▶ pleural effusion, and ▶ Aurora A. Neither hypoalbuminemia nor cardiac, renal, or hepatic dysfunction is found, and capillary filtration studies suggest that docetaxel causes a capillary permeability abnormality. Fluid retention is not usually significant at cumulative doses of <400 mg/m^2.

The incidence of this effect may be reduced by premedication with dexamethasone 8 mg orally twice daily for 5 days, beginning from day 1.

Between 50% and 75% of patients receiving docetaxel develop skin toxicity, typically characterized by an erythematous pruritic maculopapular rash that affects the forearms and hands. Other cutaneous effects include onychodystrophy, onycholysis, and soreness and brittleness of the fingernails. Docetaxel also induces palmar–plantar erythrodysesthesia and alopecia. With respect to neurotoxicity has been reported in 40% of previously untreated patients and may be higher in patients who were previously treated with ▶ platinum antitumor compounds. The peripheral neuropathy predominately affects large-fiber sensory function.

Asthenia has been observed in as many as 60–70% of patients and usually is moderate in severity, and when it is severe, it warrants dosage reduction or discontinuation of treatment. Nausea, vomiting, and severe diarrhea are rare.

Dosage

Docetaxel is indicated at a dose range of 60–100 mg/m^2 over 1 h. Although untreated or minimally pretreated patients generally tolerate docetaxel at a dose of 100 mg/m^2 without severe toxicity, emerging data indicate poorer tolerance in more heavily pretreated patients, in whom 75 mg/m^2 may be a more reasonable dose from a toxicologic perspective. Although weekly drug administration has no clear benefits in terms of antitumor activity, hematologic toxicity is much less than conventional schedules, and an unacceptable degree of asthenia and neurotoxicity is evident at doses that exceed 36 mg/m^2 per week.

Indications

Docetaxel has demonstrated antitumor activity in patients with metastatic breast cancer as first-line or salvage treatment, recurrent ovarian cancer, non-small cell lung cancer, adjuvant breast cancer, squamous cell head and neck cancer, and gastric cancer. In addition to these indications, docetaxel has demonstrated activity in previously treated patients with carcinomas of ▶ endometrial cancer, ▶ esophageal cancer, ▶ bladder cancer, ▶ prostate cancer, ▶ small-cell lung cancer, as well as lymphomas and other neoplasms.

Cross-References

- ▶ Apoptosis
- ▶ Aurora Kinases
- ▶ Bladder Cancer
- ▶ Breast Cancer
- ▶ Caveolins
- ▶ Cytochrome P450
- ▶ Endometrial Cancer
- ▶ Esophageal Cancer
- ▶ Gastric Cancer
- ▶ Macrophages
- ▶ Neutropenia
- ▶ Non-Small-Cell Lung Cancer
- ▶ Ovarian Cancer
- ▶ Paclitaxel

► P-Glycoprotein
► Platinum Complexes
► Pleural Effusion
► Prostate Cancer
► Small Cell Lung Cancer
► Tamoxifen
► Taxotere
► Endometrial Cancer

References

Cortes JE, Pazdur R (1995) Docetaxel. J Clin Oncol 13:2643–2655

Rowinsky EK, Donehower RC (1999) Antimicrotubule agents. In: Chabner BA, Longo DL (eds) Cancer chemotherapy. Lippincott-Raven, Philadelphia, pp 263–296

Sparreboom A, Tellingen O, Van Scherrenburg EJ et al (1996) Isolation, purification and biological activity of major docetaxel metabolites from human feces. Drug Metab Dispos 24:655

See Also

(2012) Colchicine. In: Schwab M (ed) Encyclopedia of cancer, 3rd edn. Springer, Berlin/Heidelberg, p 895. doi:10.1007/978-3-642-16483-5_1257

(2012) Cyclosporin A. In: Schwab M (ed) Encyclopedia of cancer, 3rd edn. Springer, Berlin/Heidelberg, p 1036. doi:10.1007/978-3-642-16483-5_1440

(2012) Lymphoma. In: Schwab M (ed) Encyclopedia of cancer, 3rd edn. Springer, Berlin/Heidelberg, p 2124. doi:10.1007/978-3-642-16483-5_3463

(2012) Microtubule. In: Schwab M (ed) Encyclopedia of cancer, 3rd edn. Springer, Berlin/Heidelberg, p 2308. doi:10.1007/978-3-642-16483-5_3734

(2012) P53. In: Schwab M (ed) Encyclopedia of cancer, 3rd edn. Springer, Berlin/Heidelberg, p 2747. doi:10.1007/978-3-642-16483-5_4331

(2012) Pharmacokinetics. In: Schwab M (ed) Encyclopedia of cancer, 3rd edn. Springer, Berlin/Heidelberg, p 2845. doi:10.1007/978-3-642-16483-5_4500

(2012) Tubulin. In: Schwab M (ed) Encyclopedia of cancer, 3rd edn. Springer, Berlin/Heidelberg, p 3792. doi:10.1007/978-3-642-16483-5_6011

(2012) Vinblastine. In: Schwab M (ed) Encyclopedia of cancer, 3rd edn. Springer, Berlin/Heidelberg, p 3907. doi:10.1007/978-3-642-16483-5_6186

(2012) Vinca alkaloids. In: Schwab M (ed) Encyclopedia of cancer, 3rd edn. Springer, Berlin/Heidelberg, p 3908. doi:10.1007/978-3-642-16483-5_6187

Docking Proteins

► Scaffold Proteins

Dormancy

Judah Folkman
Children's Hospital and Harvard Medical School, Boston, MA, USA

Synonyms

Cancer without disease; In situ cancer; In situ carcinoma; Occult cancer

Definition

Tumor dormancy describes human tumors with three characteristics:

(i) Visible only under a microscope and, therefore, cannot be detected by conventional diagnostic imaging methods and may have an average diameter the size of a pinhead but range from 0.1 to ~2 or 3 mm
(ii) Usually do not expand or spread to other organs
(iii) Usually asymptomatic and harmless but have the potential to resume growth and eventually to be fatal to their host

Characteristics

Virtually all adult humans have dormant cancers, as determined from autopsies of individuals who died of trauma (e.g., auto accidents), but who did not have a diagnosis of cancer during their lifetime. From these autopsies, pathologists report microscopic-sized cancers in different organs, often called carcinoma in situ. In women 40–50 years old, 39% have dormant in situ carcinomas in their breasts (► preneoplastic lesions), but only 1 out of 100 ever develops ► breast cancer during a normal lifetime (► ductal carcinoma in situ). Forty-six percent of men from 60 to 70 have carcinoma in situ of the prostate, but only 1 out of 100 in this age range is ever diagnosed with prostate cancer. In contrast, estimated 98% of

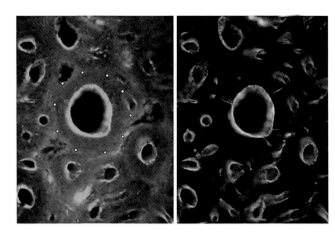

Dormancy, Fig. 1 *Left panel*: Blood vessels in a human breast cancer (MCa-IV). Tumors were transplanted into mice, and subsequently the tumors were perfused with fixative so that the microvessels would not be compressed. The *white dots* show the outer layer of multiple layers of tumor cells surrounding a new capillary blood vessel (lectin binding). *Right panel*: Endothelial cells are labeled by antibody to CD31 antigen. This section also shows new capillary sprouts in the breast cancer (Courtesy of Donald McDonald, University of California, San Francisco)

individuals age 50–70 harbor carcinoma in situ of the thyroid gland, but only 1 out of 1,000 develops thyroid cancer. Dormant carcinoma in situ can be found nestled among established capillary blood vessels, but such dormant tumors do not recruit new blood vessels. In other words, most dormant cancers in situ are non-angiogenic.

Blocked Angiogenesis

Therefore, one mechanism of tumor dormancy is blocked ▸ angiogenesis, i.e., the inability of emerging early tumors to recruit new blood vessels. Cancer arises from a single cell, for example, a liver cell. Normal liver cells rarely divide. A cancerous liver cell, however, can continue to divide without restraint until it has accumulated offspring of up to ~1 million tumor cells. Nevertheless, such a microscopic tumor becomes dormant when its further expansion is arrested by the limits of oxygen diffusion from the nearest open capillary blood vessel. This oxygen diffusion limit is ~180–200 μ (about 0.2 mm) for tumor cells and significantly less for normal cells. Virtually every normal cell lives either directly adjacent to a capillary blood vessel or at least not more than two cell widths from a capillary. However, tumor cells can surround a capillary vessel with multiple cells (Fig. 1). Once a microscopic tumor becomes oxygen depleted, tumor cell death (▸ apoptosis)

increases to match tumor cell proliferation. This "balance" between proliferating and dying tumor cells is one of the hallmarks of the dormant, microscopic, in situ cancer which virtually all humans harbor (▸ progression).

Maintenance of Tumor Dormancy by Endogenous Angiogenesis Inhibitors

The majority of new tumors and metastases remain dormant for prolonged periods of time (sometimes for years), in part because of endogenous angiogenesis inhibitors. At this writing, 29 angiogenesis inhibitors have been discovered in the body; none were known before 1980 (▸ anti-angiogenesis). Most of these angiogenesis inhibitors are proteins, such as ▸ thrombospondin-1, platelet factor 4, ▸ maspin, angiostatin, ▸ endostatin, tumstatin, canstatin, interleukin-12, SPARC, and others. Endostatin and tumstatin are under the control of p53. While some inhibitors circulate at low concentrations in the plasma, others are stored in platelets, white blood cells, bone marrow cells, fibroblasts of tissues throughout the body, or in the collagen basement membranes in the stroma underlying most tissue cells. Endothelial cells produce collagen type XVIII basement membrane. Tumor cells themselves also express certain angiogenesis inhibitor proteins directly, such as thrombospondin-1, or express enzymes that

Dormancy,
Fig. 2 Angiogenesis in rat sarcoma. In this micrograph, blood vessels grow toward a sarcoma (*dark area* at *right*) in rat muscle. This contrasts with the normal grid-like pattern of blood vessels that appears at the *upper left*. Tumor cells that have begun to surround the capillary vessels are not shown here (Courtesy of L. Heuser and R. Ackland, University of Louisville, USA)

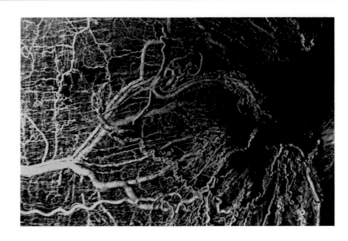

mobilize anti-angiogenic peptide fragments from larger proteins. Examples of the latter are angiostatin from plasminogen and endostatin from collagen XVIII. The gene for collagen XVIII is on chromosome 21. Individuals with Down syndrome have three copies of this chromosome (trisomy). As a result, elevated levels of endostatin are found in these individuals. They are protected against abnormal angiogenesis. Down syndrome individuals with diabetes are protected against neovascularization in the retina and also from neovascularization in atherosclerotic plaques. They are also protected against cancer. In fact, of the ~200 types of cancer, these individuals only develop ▶ testicular cancer and a rare but mild form of leukemia. For the other types of cancer, individuals with Down syndrome have <0.1 the expected incidence even though they live into middle age and beyond. In mice genetically engineered to overexpress collagen XVIII so that their endostatin level is increased by ~1.6-fold to mimic individuals with Down syndrome, implanted tumors are poorly neovascularized and grow 300–400% slower. Conversely, tumors grow threefold more rapidly in mice that lack endostatin. The genes for mental retardation are unrelated to endostatin. Furthermore, a second putative anti-angiogenic gene, called DSCR1 (Down syndrome critical region), has been identified on chromosome 21. Although this phenomenon is correlative and not yet proved to be causal, it provides a thought-provoking clinical clue that suggests the question – do individuals

with Down syndrome harbor equivalent numbers of microscopic dormant cancers as the rest of the population but a lower incidence of tumors that become angiogenic?

Escape from Tumor Dormancy by a Switch to the Angiogenic Phenotype

Miles of capillary blood vessels, thinner than a hair, supply every tissue in the body. A pound of fat contains ~1 mile of capillary blood vessels. Vascular endothelial cells which line the inside of these blood vessels normally proliferate only infrequently, to replace lost endothelial cells. The entire endothelial lining is replaced or "turned over" in ~3 years, in contrast to the turnover of intestinal epithelial cells that is measured in days. During physiological angiogenesis, such as reproduction, development, or wound repair, endothelial cells can proliferate rapidly, i.e., with a turnover measured in days or weeks. Physiologic angiogenesis is, however, self-limited. Angiogenesis during ovulation is turned off after a few days and in wounds after ~2 weeks. In contrast, once a tumor has become angiogenic, endothelial cells in the tumor bed proliferate continuously, the beginning of the switch to the angiogenic phenotype. As new capillary blood vessels grow toward the dormant tumor, tumor cells grow around them and the tumor mass now expands rapidly (Fig. 2). Angiogenic tumors become detectable by conventional imaging methods, cause symptoms, and metastasize to other organs. Angiogenic tumors are potentially fatal (▶ progression).

Experimental Analysis of the Angiogenic Switch in Human Dormant Cancers

The escape of human dormant tumors to the angiogenic phenotype has been studied in immunodeficient mice (SCID or nude mice), by cloning single tumor cells from tumor specimens discarded from the operating room or obtained as tumor cell lines from the American Type Culture Collection. When these clones are expanded in vitro, and reimplanted in immuno-deficient mice, ~3–5% form *non-angiogenic* tumors of ~1 mm diameter and remain dormant with a high proliferation rate of tumor cells balanced by a high apoptotic rate. The microscopic dormant tumors can be visualized in mice if tumors are stably infected either with luciferase or green fluorescent proteins (e.g., green fluorescent protein). For each tumor type, there is a predictable percentage of non-angiogenic tumors that will undergo a spontaneous switch to the angiogenic state (called phenotype), at a predictable time. For example, ~80% of the non-angiogenic tumors from a given type of human breast cancer become angiogenic at ~4 months. For liposarcoma (a cancer of fat tissue), the angiogenic switch occurs at 4 months in 95% of non-angiogenic tumors. A human brain tumor (glioblastoma) becomes angiogenic at 8 months in 60% of tumors. In human bone tumors (▶ osteosarcoma), the angiogenic switch does not occur until after 1 year and in only 5–15% of mice. After tumors become angiogenic, they escape tumor dormancy and form lethal tumors in 100% of mice regardless of cancer type. Thus, human tumors studied so far contain a subpopulation of non-angiogenic tumor cells that can form dormant tumors. For some tumor types, the majority of non-angiogenic tumors become angiogenic and escape from dormancy (▶ progression) (e.g., liposarcoma). For other tumor types, only a minority of non-angiogenic tumors become angiogenic and the rest remain non-angiogenic and dormant indefinitely.

Molecular Mechanisms of Angiogenic Switching in Dormant Cancers

If human tumors could be restricted to the non-angiogenic dormant phenotype, or if angiogenic tumors could be reversed to the non-angiogenic dormant phenotype, a novel anti-cancer therapy could be possible (▶ anti-angiogenesis). Therefore, molecular mechanisms are being studied. For example, transfection of a non-angiogenic human osteosarcoma with the *RAS* oncogene causes dormant tumors to become angiogenic and escape from dormancy within weeks to 1 month, in contrast to the spontaneous angiogenic switch which can take up to 1 year. Furthermore, after *RAS* transfection, 100% of dormant non-angiogenic tumors become angiogenic and grow rapidly, whereas spontaneous escape from dormancy occurred in only 5–15% of tumors. *RAS* transfection is followed by a 30% increased expression of vascular endothelial growth factor (i.e., ▶ adrenomedullin) (VEGF, a potent proangiogenic protein) and is accompanied by a 50% decrease in the expression of thrombospondin-1, a potent angiogenesis inhibitor. Many oncogenes induce increased expression of a proangiogenic proteins and suppression of an anti-angiogenic protein (▶ anti-angiogenesis). This common pattern could lead to a general molecular mechanism of escape from tumor dormancy by activation of the switch to the angiogenic phenotype.

Other Forms of Tumor Dormancy

Immune Surveillance as a Cause of Tumor Dormancy

Many experimental models reveal that the immune system can maintain a reduced load of cancer cells based on the manipulation of cytotoxic T lymphocytes (CD8+) which can kill tumor cells expressing specific antigens. Furthermore, escape from tumor dormancy in some models is based on tumor evasion of the immune system. In other experimental systems, growth of mouse lymphoma cells can be suppressed by a T-cell-mediated mechanism. It remains to be determined if therapeutic blockade of angiogenesis, by restricting expansion of a dormant tumor population, will synergize immune suppression.

Hormonal Depletion as a Cause of Tumor Dormancy

Certain hormonally dependent tumors, such as prostate cancer, become dormant when the hormone (e.g., androgen) is blocked or depleted. It has been shown that depletion of testosterone from a prostate cancer decreases VEGF expression and thus reduces tumor angiogenesis. However, in clinical practice, this therapy is temporary, and within 1 year or more, androgen-independent prostate cancer cells often emerge.

Dormancy of Single Metastatic Tumor Cells

It has been shown in experimental animals that single metastatic tumor cells can exit the circulation at a future metastatic site, for example, the liver or lungs, and survive for long periods near a capillary blood vessel without proliferating (the G_0 state).

Future Directions

Biomarkers to Detect Dormant Tumors

Many laboratories are developing a variety of molecular biomarkers in the blood to detect the presence of cancer. Some of these may be useful for detecting the presence of microscopic-sized tumors that cannot be located anatomically by conventional imaging methods. This would include non-angiogenic dormant tumors or those just beginning to switch to the angiogenic phenotype. If a biomarker of high sensitivity could be validated in the clinic, then it could eventually be used to guide nontoxic anti-angiogenic therapy, or anti-▶ telomerase therapy, or ▶ immunotherapy, to prevent recurrence of cancer years before symptoms or before anatomical location was possible.

Cross-References

- ▶ Adipose Tumors
- ▶ Adrenomedullin
- ▶ Antiangiogenesis
- ▶ Apoptosis
- ▶ Angiogenesis
- ▶ Breast Cancer
- ▶ Endostatin
- ▶ Ductal Carcinoma In Situ
- ▶ Immunotherapy
- ▶ Maspin
- ▶ Osteosarcoma
- ▶ *RAS* Genes
- ▶ Preneoplastic Lesions
- ▶ Progression
- ▶ Secreted Protein Acidic and Rich in Cysteine
- ▶ Telomerase
- ▶ Testicular Cancer
- ▶ Thrombospondin
- ▶ Vascular Endothelial Growth Factor

References

Aquirre-Ghiso JA (2007) Models, mechanisms and clinical evidence for cancer dormancy. Nat Rev Cancer 7:834–846

Finn OJ (2006) Human tumor antigens, immunosurveillance and cancer vaccines. Immunol Res 36:73–82

Folkman J (2007) Angiogenesis: an organizing principle for drug discovery? Nat Rev Drug Discov 6:273–286

Holmgren L, O'Reilly MS, Folkman J (1995) Dormancy of micrometastases: balanced proliferation and apoptosis in the presence of angiogenesis suppression. Nat Med 1:149–153

Naumov GN, Folkman J (2007) Strategies to prolong the nonangiogenic dormant state of human cancer. In: Davis DW, Herbst RS, Abbruzzese JL (eds) Antiangiogenic cancer therapy. CRC Press, Boca Raton/London/New York, pp 3–22

See Also

(2012) Angiostatin. In: Schwab M (ed) Encyclopedia of cancer, 3rd edn. Springer Berlin Heidelberg, p 187. doi:10.1007/978-3-642-16483-5_280

(2012) Basement membrane. In: Schwab M (ed) Encyclopedia of cancer, 3rd edn. Springer Berlin Heidelberg, p 349. doi:10.1007/978-3-642-16483-5_537

(2012) Biomarkers. In: Schwab M (ed) Encyclopedia of cancer, 3rd edn. Springer Berlin Heidelberg, pp 408–409. doi:10.1007/978-3-642-16483-5_6601

(2012) Cytotoxic T lymphocytes. In: Schwab M (ed) Encyclopedia of cancer, 3rd edn. Springer Berlin Heidelberg, p 1058. doi:10.1007/978-3-642-16483-5_1501

(2012) Down syndrome. In: Schwab M (ed) Encyclopedia of cancer, 3rd edn. Springer Berlin Heidelberg, p 1159. doi:10.1007/978-3-642-16483-5_6489

(2012) Glioblastoma. In: Schwab M (ed) Encyclopedia of cancer, 3rd edn. Springer Berlin Heidelberg, p 1554. doi:10.1007/978-3-642-16483-5_2421

(2012) Interleukin-12. In: Schwab M (ed) Encyclopedia of cancer, 3rd edn. Springer Berlin Heidelberg, p 1892. doi:10.1007/978-3-642-16483-5_3101

(2012) Leukemia. In: Schwab M (ed) Encyclopedia of cancer, 3rd edn. Springer Berlin Heidelberg, p 2005. doi:10.1007/978-3-642-16483-5_3322

(2012) Nude mice. In: Schwab M (ed) Encyclopedia of cancer, 3rd edn. Springer Berlin Heidelberg, p 2584. doi:10.1007/978-3-642-16483-5_4172

(2012) P53. In: Schwab M (ed) Encyclopedia of cancer, 3rd edn. Springer Berlin Heidelberg, p 2747. doi: 10.1007/978-3-642-16483-5_4331

(2012) Transfection. In: Schwab M (ed) Encyclopedia of cancer, 3rd edn. Springer Berlin Heidelberg, p 3757. doi:10.1007/978-3-642-16483-5_5912

(2012) VEGF. In: Schwab M (ed) Encyclopedia of cancer, 3rd edn. Springer Berlin Heidelberg, p 3906. doi:10.1007/978-3-642-16483-5_6174

Doublecortex

▶ Doublecortin

Doublecortin

Manoranjan Santra and Michael Chopp
Neurology Research, Henry Ford Health System, Detroit, MI, USA

Synonyms

DCX; Doublecortex

Definition

Doublecortex (*DCX*) is a brain-specific gene, deficient in glioma tumor cells and suppresses tumor in brain. DCX is a substrate of Jun N-terminal Kinase (JNK) and highly phosphorylated by JNK. The phosphorylated DCX interacts with a potent tumor suppressor, spinophilin neurabin II, and then associates with the protein serine/threonine phosphatase, protein phosphatase-1 (PP1). This association of protein complex of DCX, neurabin II, and PP1 leads to inhibition of anchorage-independent growth of glioma tumor cells and eventually suppresses tumor development.

Characteristics

DCX is a ▶ microtubule-associated protein (MAP), which directly interacts with microtubules (MTs) without any additional mediators. This interaction results in stabilization and bundling of MTs both in vivo and in vitro. DCX is a basic protein with an isoelectric point of 10, typical of other MT-binding proteins. DCX is an evolutionarily conserved MT binding protein that maintains almost identical amino acid as well as nucleic acid sequences from mouse to human. It is located on chromosome Xq22.3-q23. From Affymetrix gene chip analysis, *DCX* is one of the three dominant genes found in relation to glioma patient survival. Other two genes are osteonectin and ▶ semaphorin. All of them regulate cellular motility. De novo expression of *DCX*, a gene which is absent in wild-type glioma cells, leads to an arrest in the G2 phase of the cell cycle. During this process, the neoplastic cells cease to proliferate and lose their ability to form large colonies in semisolid medium and to induce tumor xenograft in immunocompromised hosts. This growth suppression can be reversibly abrogated by blocking *DCX* transcription via its sequence-specific inhibitor DCX ▶ siRNA. The absence of the *DCX* gene may increase propensity to develop tumors; thus, boosting endogenous *DCX* gene expression may be therapeutically beneficial and, alternatively, pharmacologic delivery of recombinant DCX may have beneficial therapeutic effects in the treatment of glioma.

Regulation

Regulation of many cellular functions depends on protein phosphorylation and dephosphorylation of multiple substrates by protein kinases and phosphatases. DCX, a substrate of JNK interacts with both JNK and JNK-interacting protein and is highly phosphorylated by JNK in glioma cells. Aside from ▶ epidermal growth factor receptor

(EGFR), JNK is the only other kinase that is almost exclusively expressed and basally active in primary human glial tumors. The phosphorylated DCX interacts with spinophilin/neurabin II, a tumor suppressor. Neurabin II inhibits anchorage-independent growth of human and mouse cancer cell lines, regardless of ▶ p53 and p14ARF (ARF) status. Ectopic expression of ARF is ineffective for suppression of colony formation in osteosarcoma cancer cells, Saos-2, whereas, neurabin II and ARF coexpression and interaction synergistically inhibits anchorage-independent growth. DCX and neurabin II interaction inhibits proliferation and anchorage-independent growth in glioma cells. In contrast, DCX-mediated growth suppression is lost in neurabin II null HEK 293 T cells and is reversed by knocking down neurabin II with siRNA (small interfering RNA). DCX and neurabin II interaction may contribute to strong antitumorigenicity to glioma.

Phosphorylation and dephosphorylation lead to association and dissociation of many proteins such as a phosphoprotein encoded by mouse a4 and the catalytic subunit of protein phosphatase 2A that regulates their activation. Neurabin II belongs to this class of regulators because it negatively regulates the PP1 catalytic subunit activity. Native neurabin II associates with PP1. DCX, neurabin II, and PP1 are found in the same protein complex when PP1 is pulled down with PP1-specific microcystin-agarose beads and also when DCX is immunoprecipitated from mouse brain extracts. JNK inhibitors and JNK site-directed mutagenesis (T331, S334) in DCX reduced the interaction between DCX and neurabin II in glioma cells and between DCX. JNK activation by MKK7–JNK1 increases DCX phosphorylation and its interaction with neurabin II in glioma cells. Phosphatase inhibitors reduce the interaction between DCX and neurabin II and also inhibit DCX–neurabin II–PP1 complex formation. Thus, phosphorylation and dephosphorylation both may be required for interaction between DCX and neurabin II and DCX–neurabin II–PP1 complex formation. Interaction between phosphorylated DCX and neurabin II induces association of DCX, neurabin II, and PP1 in vivo. PP1, one of the key

eukaryotic serine/threonine protein phosphatases, is involved in the mitotic dephosphorylation of the ▶ Retinoblastoma gene (pRb), as well as in the dephosphorylation of specific residues of p53, and regulates the control of cell cycle progression. PP1 dephosphorylates DCX specifically on amino acid residues S331 and T334 both in vitro and in vivo. Microinjection of PP1-neutralizing antibodies and PP1 inhibitors such as ▶ okadaic acid blocks mitosis and alters the progression of the cell cycle by accumulating at the nucleus to associate with chromatin during G2 and M phases. DCX overexpression blocks the G2-M phase of the cell cycle in glioma cells. DCX-mediated growth arrest in the G2-M phase of the cell cycle in glioma cells may be through inactivation of PP1 by neurabin II–DCX interaction. The expression of DCX and neurabin II is dynamic, and they are coexpressed in migrating neurons. Overexpressing the coiled-coil domain of neurabin II leads to interaction with DCX, recruits the endogenous neurabin II with PP1, and induces dephosphorylation of DCX on one of the JNK phosphorylated sites. In vitro, DCX is site-specifically dephosphorylated by PP1 without the presence of neurabin II. Overexpression of phosphorylated DCX, therefore, itself may competitively inhibit PP1 and block G2-M phase of cell cycle progression.

Phosphatase and tensin homologue deleted on chromosome 10 (PTEN) gene, also known as mutated in multiple advanced cancers (MMAC1), and transforming growth factor-β-regulated and epithelial cell-enriched phosphatase (TEP-1), is a tumor suppressor gene located at chromosome 10q23.3. Deletions of all or part of chromosome 10 are the most common genetic alterations in high-grade gliomas. Loss of heterozygosity on the long arm of chromosome 10 is found in 75–90% of high grade gliomas. PTEN maps to chromosome region 10q23 and is mutated by the most common genetic alteration of loss of heterozygosity in high-grade gliomas. Transfection of wild-type PTEN into PTEN-deficient glioma cells causes in vitro and in vivo growth suppression. PTEN induces DCX expression in mouse brain subventricular zone precursor cells and even in PTEN-deficient wild-type glioma cells. In addition, DCX siRNA treatment significantly reduces the growth suppression effect in PTEN-overexpressing glioma cells. These data indicate that involvement of DCX in PTEN-mediated tumor suppression is a novel mechanism.

DCX expression induces cell adhesion proteins such as E-, V-, and N-cadherin in glioma U87 cells at mRNA and protein level. One cell adhesion protein, E-cadherin is widely used as a negative marker of tumor invasion. E-cadherin is typically reduced in levels or absent in invasive tumors, as a result of mutation, transcriptional silencing, or protease-mediated ectodomain shedding. Such loss of E-cadherin function by proteolysis has been demonstrated for several types of protease, including matrix metalloproteinases MMPs, ADAM10, and plasmin. Glioma cells highly express MMPs, particularly MMP-2 and MMP-9. Their high expressions create an environment in favor of ▶ angiogenesis, cell growth, motility, survival, and eventually proinvasive functions that enable the general spreading of glioma cells into brain. MMPs are proteolytic enzymes that degrade cell–cell adhesion proteins and remodel the ▶ extracellular matrix (ECM). Induction of E-cadherin by DCX could therefore reverse the effect of invasion of glioma cells. Glioblastomas are the most malignant brain cancers, with a median survival of 10–12 months. Gliomas are the most common type of intracranial tumors and these rapidly invade the brain. Even drastic treatments for glioma such as surgery, radiotherapy, and chemotherapy fail. These treatments have only minimally altered the median survival time of patients with glioma, who eventually die within a year. The identification of the mechanisms of glioma cells invasion into the brain will possibly point toward the therapeutic approach to target the invading cells more specifically and reduce further spreading.

Clinical Relevance

Fundamental questions are how this microtubule-associated protein DCX modifies the growth rate and tumorigenic potential of transformed cells, and whether this growth suppression is also operational in normal cells. These MAPs including DCX are either absent or mutated in tumor cells.

The MTs play a critical role during mitosis. The polymerization of MTs into α/β tubulin heterodimers produces a complex structure known as the "mitotic spindle." The process of chromosome segregation is mediated by these mitotic spindles. The depolymerization of MTs is required to segregate sister chromatids on the mitotic spindle. The dynamic instability of MTs is a characteristic property of MTs, which allows them to switch abruptly between state of elongation (polymerization) and rapid shortening (depolymerization). The transition from a state of growth to a state of shrinkage is called "catastrophe" and the transition from a state of shrinkage to a state of growth is called "rescue." The polymerization and depolymerization of MTs depends on activities of two major classes of proteins, the MT-stabilizing and -destabilizing proteins and synchronizes spatially and temporally separation of sister chromatids. The MAPs stabilize the assembled MTs by suppressing catastrophe. The MT-destabilizing proteins are members of the kinesin superfamily and the MT-severing proteins. All these proteins can destabilize the assembled MTs by increasing the catastrophe rate of the polymers. These two classes of cellular polymerizers and depolymerizers of MTs play a critical role in the control of cell division. Mutation or absence of either of them in tumor cells disrupts the dynamic instability of microtubules, arrest cell growth in G2-M phase in cell cycle, inhibit cell division, and prevent or delay tumor progression. Similarly, induction of either of them in tumor cells interrupts the dynamic instability of microtubules, and causes accumulation of cells in the G2/M phases of the cell cycle, and inhibition of growth and tumor progression. Either an increase or decrease in the level of expression of polymerizers and depolymerizers causes mitotic arrest in the cell cycle. To complete mitosis in the cell cycle, the level of expression of polymerizers and depolymerizers are therefore required to be balanced. Overexpression of DCX as well as MAP2 leads to microtubule stabilization, cell cycle arrest in G2-M phase, and growth inhibition. Induction of MAPs including DCX in the primary tumor disrupts MT dynamics and spindle check point. Disruption of dynamic instability of MTs can lead to mitotic block and cell cycle arrest. Agents that target MTs such as DCX are, therefore, ideal for treatment of cancer.

References

Demuth T, Berens ME (2004) Molecular mechanisms of glioma cell migration and invasion. J Neurooncol 70(2):217–228 (Review)

Furnari FB, Lin H, Huang HS et al (1997) Growth suppression of glioma cells by PTEN requires a functional phosphatase catalytic domain. Proc Natl Acad Sci U S A 94:12479–12484

Gadde S, Heald R (2004) Mechanisms and molecules of the mitotic spindle. Curr Biol 14(18):R797–R805 (Review)

Santra M, Zhang X, Santra S et al (2006a) Ectopic doublecortin gene expression suppresses the malignant phenotype in glioblastoma cells. Cancer Res 66(24):11726–11735

Santra M, Lui XS, Zhang J et al (2006b) Ectopic expression of doublecortin protects adult mouse progenitor cells and human glioma cells from severe oxygen and glucose deprivation. Neuroscience 92:7016–7020

Santra M, Santra S, Roberts C et al (2009) Doublecortin induces mitotic microtubule catastrophe and inhibits glioma cell invasion. J Neurochem 108(1):231–45. http://www.ncbi.nlm.nih.gov/pubmed/19094064

Doxorubicin

▶ Adriamycin

DPC4

▶ Deleted in Pancreatic Carcinoma Locus 4

DPPIV

▶ CD26/DPPIV in Cancer Progression and Spread

DR4 Antibodies

▶ TRAIL Receptor Antibodies

DR5 Antibodies

▶ TRAIL Receptor Antibodies

Draining Lymph Node

▶ Sentinel Node

DRF

▶ Intrinsically Unstructured Proteins
▶ Leukemia Inhibitory Factor

Drug Carriers

▶ Liposomal Chemotherapy

Drug Delivery Systems for Cancer Treatment

Kunihiro Tsuchida
Division for Therapies Against Intractable Diseases, Institute for Comprehensive Medical Science (ICMS), Fujita Health University, Toyoake, Japan

Synonyms

DDS

Keywords

Tumor neovasculature; Drug delivery; Nanoparticles

Definition

Drug delivery systems (DDS) are defined as effective systems that deliver optimal amounts of drugs or chemicals to target tissues, enhancing drug efficacy and reducing adverse effects.

Characteristics

Drugs or chemicals for treatment of diseases are often tablets or solutions that are designed to enhance absorption *in vivo*. Efficient absorption, distribution, metabolism, and excretion during the process of drug delivery are major issues of drug therapies. However, diffusion of drugs before they reach to the target tissues occurs in many cases. Therefore, an amount of drugs properly targeting the desired tissues become low, and more importantly it is difficult to control the drug concentration in the disease foci. Furthermore, drugs must be administered in large amounts to exert their beneficial effects; therefore they affect normal cells, and adverse effects may occur.

Drug delivery systems are designed to enhance drug efficacy and reduce adverse effects by controlling drug concentration to a minimum needed at the target tissues. Regulation of the rate of drug release and targeting to specific tissues where the drugs are delivered are important objectives of DDS. If drugs can be administered by a sustained release in disease foci, regulation of the release and amounts of the drug to the targeted site becomes possible, and pharmacological therapy can be optimized.

Drug delivery systems are being actively investigated to develop novel therapies and to optimize therapeutic effects. Drug deliveries using macromolecules, liposomes, microcapsules, nanoparticles, and secreted extracellular vesicles have been developed and characterized as carriers for drugs, chemicals, and nucleic acids. Advances in developments of DDS would be enhanced with interdisciplinary cooperation among pharmaceutical and medical sciences, cell biology, biopolymer and material sciences, nanomedicine, tissue engineering, and clinical practices.

Characteristics of Vascular Walls of Cancer

Malignant cancer cells arise from normal cells and/or cancer stem cells. They show atypical cell morphology with high nucleus/cytoplasm ratio and invade to adjacent tissues and distant ► metastasis through blood, lymph node, and dissemination. To support cancerous growth, tumor neovasculization is very important. Basically, new vessels are derived from precursors such as pericytes and form endothelium. Newly developing vascular walls are often incomplete, and there are multiple fairly wide slits with widths of several hundred nanometers (nm). Although anticancer drugs adsorbed to micelles or nanoparticles do not penetrate endothelial cells and do not leak through normal vessel walls, they do penetrate vascular walls in cancer tissues (Fig. 1). Thus, drug carriers of less than several hundred nm sizes that are designed for DDS are likely to penetrate to cancer foci through leaky vasculature (Ying et al. 2015; Khawar et al. 2015). ► Lymphatic vessels that excrete drugs are also poorly developed in cancers, resulting in the retention of administered drugs for longer periods. This biological phenomenon is called the enhanced permeability and retention (EPR) effect.

Carriers for Drugs, Chemicals, and Nucleic Acids

Drugs, chemicals, and therapeutic nucleic acids are administered either by blood infusion, intradermal delivery, or through the digestive system. To achieve targeted delivery of drugs, multiple carriers such as nanoparticles, micelles, and secreted extracellular vesicles have been developed.

Particles between 1 and 100 nm have novel electric, optical, and structural properties. Nanoparticles can be used to conjugate various drugs and biomaterials and the resultant conjugated drugs customized as ► targeted drug delivery vehicles. This makes it possible to deliver sufficient amounts of therapeutic agents into cancer cells while protecting normal cells from exposure to these drugs.

Secreted extracellular vesicles, especially exosomes, are secreted from many cell types and

Drug Delivery Systems for Cancer Treatment, Fig. 1 Vasculature in cancer. Enhanced permeability and retention (EPR) effect. (**a**) Vasculature in normal tssues is tight, preventing drug carriers from penetrating and extravasating. (**b**) Vascular walls and endothelial cells in tumor tissues have multiple narrow slits and hyperpermeable to drugs with several hundred nm.

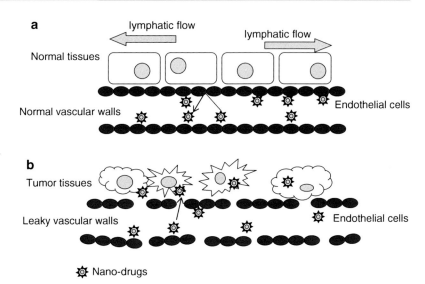

circulate (Srivastava et al. 2015). They are carriers for miRNA and mRNA and can be engineered to conjugate and release molecules. Therefore, engineered exosomes sized 100–200 nm in diameter would become good drug delivery vehicles for nucleic acids and ▶ gene therapy.

Engineered drug delivery vehicles can be further modified with targeting molecules such as peptides and antibodies. Several cancers are known to have unique peptides and/or abundant receptors on cell surface. For example, epidermal growth factor receptors (EGFs) are involved in tumor progression in many types of epithelial cells. Active targeting is achieved by conjugating nanoparticles with a targeting moiety such as antibodies to membrane receptors, antigens, and carbohydrates of cancer cells, thus allowing preferential accumulation of the drug at the cancer loci (Kulhari et al. 2015).

Conjugation of drugs and an antibody changes toxicity and pharmacodynamics *in vivo*. ▶ Positron emission tomography (PET) is particularly useful as an imaging system of cancer cells through the labeling of tumor-associating antibodies using radioisotopes or fluorodeoxyglucose (FDG).

Liposomes, micelles, and microcapsules also serve as suitable carriers of drugs and have been extensively studied.

Targeted Drug Delivery

Passive targeting takes advantage of the enhanced permeability of tumor vascular walls. Tumor-activated prodrug therapy is based on the tumor-specific microenvironment that activates prodrugs in tumor cells. As mentioned above active targeting using cancer-specific molecules can achieve active targeting.

Cross-References

▶ Gene Therapy
▶ Lymphatic Vessels
▶ Metastasis
▶ Positron Emission Tomography
▶ Targeted Drug Delivery

References

Khawar IA, Kim JH, Kuh HJ (2015) Improving drug delivery to solid tumors: priming the tumor microenvironment. J Control Release 201:78–89

Kulhari H, Pooja D, Rompicharla SV, Sistla R, Adams DJ (2015) Biomedical applications of trastuzumab: as a therapeutic agent and a targeting ligand. Med Res Rev 35(4):849–876

Srivastava A, Filant J, Moxley KM, Sood A, McMeekin S, Ramesh R (2015) Exosomes: a role for naturally occurring nanovesicles in cancer growth, diagnosis and treatment. Curr Gene Ther 15(2):182–192

Ying M, Chen G, Lu W (2015) Recent advances and strategies in tumor vasculature targeted nano-drug delivery systems. Curr Pharm Des 21(22):3066–3075

See Also

(2012) Liposomes. In: Schwab M (ed) Encyclopedia of cancer, 3rd edn. Springer, Berlin/Heidelberg, p 2063. doi:10.1007/978-3-642-16483-5_3388

(2012) Nanoparticles. In: Schwab M (ed) Encyclopedia of cancer, 3rd edn. Springer, Berlin/Heidelberg, p 2449. doi:10.1007/978-3-642-16483-5_3964

(2012) Neovascularization. In: Schwab M (ed) Encyclopedia of cancer, 3rd edn. Springer, Berlin/Heidelberg, p 2474. doi:10.1007/978-3-642-16483-5_4016

(2012) Prodrug. In: Schwab M (ed) Encyclopedia of cancer, 3rd edn. Springer, Berlin/Heidelberg, p 2989. doi:10.1007/978-3-642-16483-5_4751

(2012) Sustained release. In: Schwab M (ed) Encyclopedia of cancer, 3rd edn. Springer, Berlin/Heidelberg, p 3586. doi:10.1007/978-3-642-16483-5_5609

Drug Delivery Vehicles

▶ Liposomal Chemotherapy

Drug Design

Roy J. Duhé
Department of Pharmacology and Toxicology, University of Mississippi Medical Center, Jackson, MS, USA
Department of Radiation Oncology, University of Mississippi Medical Center, Jackson, MS, USA

Synonyms

Drug development; Targeted drug design

Definition

The process of discovering, inventing, improving, and/or testing a new therapeutic agent for the treatment of a specific disease.

Characteristics

Most of the drugs used in classical cancer chemotherapy were discovered by serendipity or by trial and error. A problem with such drugs arises because, in addition to having a desirable cytotoxic effect on cancer cells, many of these drugs can adversely affect rapidly dividing normal cells, such as gastric mucosal cells lining the stomach or hematopoietic stem cells in the bone marrow. This lack of specificity is a major contributing factor for the severity of side effects of classical cancer chemotherapy, which can include immunosuppression, nausea, vomiting and other gastrointestinal toxicities, and damage to the brain (neurotoxicity), heart (cardiotoxicity), liver (hepatotoxicity), kidney (renal toxicity), and lungs (pulmonary toxicity). The trend in anticancer drug development emphasizes the rational design of targeted chemotherapy (▶ Personalized Cancer Medicine), in which drugs are designed to inhibit a critical molecular target that is essential for the growth and survival of the tumor cell but which, ideally, does not exist in the normal cell.

The prototypical targeted chemotherapeutic agent is ▶ imatinib, a low molecular weight inhibitor of the protein-tyrosine kinase BCR/ABL. BCR/ABL is an oncoprotein produced from the fusion of the *bcr* and *abl* genes resulting from a t (9;22)(q34;q11) ▶ Chromosomal Translocation. This chromosomal translocation is observed in approximately 95% of adult patients with Chronic myelogenous leukemia (CML); this subset of patients can be identified by the presence of the Philadelphia Chromosome. While imatinib selectively inhibits BCR/ABL, it is not truly specific for this enzyme. In this instance, the lack of specificity benefits patients with ▶ gastrointestinal stromal tumor (GIST) because they can be successfully treated with imatinib by virtue of the fact that imatinib also inhibits the tyrosine kinase c-kit (▶ Kit/Stem Cell Factor Receptor in Oncogenesis). The inappropriate expression of *c-kit* contributes to the development of GIST and therefore provides another target for imatinib. In comparison to classical cancer chemotherapy agents such as ▶ cisplatin or ▶ adriamycin, imatinib has a relatively low toxicity profile which can include periorbital edema, dermatitis, or nausea. Following the introduction of imatinib, other protein ► tyrosine kinase inhibitors have been approved for the treatment of cancers. These include the use of ► epidermal growth factor receptor (EGF Receptor; EGFR), tyrosine kinase inhibitors – ► gefitinib and ► erlotinib, for the treatment of ► nonsmall cell lung cancer. ► Sorafenib, a dual inhibitor of both protein kinase Raf and VEGFR protein-tyrosine kinase used in the treatment of advanced renal cell carcinoma, is another example of successful targeted drug design.

Identification and Validation of Molecular Targets

In principle, drug design entails a progression of distinct stages. The first stage is to identify and validate one or more molecular targets for disease therapy. A molecular target for chemotherapy typically has a causal role in the etiology of the cancer, although some targets have supportive roles in the survival or progression of cancers. Just as ► Koch's Postulates are applied to validate a particular biological agent as a pathogen, so too must analogous postulates be applied in the validation of molecular pathogens. One popular approach to the identification and validation of molecular pathogens is to compare the complex genetic (Genomics), proteinaceous (► Proteomics) and metabolic (Metabolomics) differences between normal and diseased cells in order to find a pathogenic "needle" in a "haystack" of biological information by Bioinformatics. This is a discipline covering all aspects of biological information acquisition, processing, storage, visualization, distribution, as well as analysis and interpretation that combines the tools of mathematics, computer science, and biology to advance the scientific understanding of the biological significance of huge amount of data. It involves the creation and advancement of algorithms, computational and statistical techniques, as well as the theories to solve formal and practical problems inspired from the management and analysis of biological data. Genomics and proteomics analyses generate expression information

from tremendous amount of genes or proteins, which have to be organized, stored, and analyzed with the aid of bioinformatics.

However, these studies rarely yield a single unambiguous molecular target. Bioinformatics have typically provided a more complex molecular disease signature involving clusters of biomolecules, and these signatures have identified hitherto unrecognized distinctions amongst cancers which were not distinguishable through conventional diagnostic methods. The recognition of distinct subtypes of diffuse large ► B-cell lymphoma is an excellent example. Bioinformatics theory and its attendant technology and computational methodology are merging in the evolution of a new scientific discipline called systems biology. Systems biology examines not only how cellular components interact to result in cellular structure and function, but also the larger question of how cells interact to result in higher level organization and function. The great potential of these new computational and analytical tools is that they can not only yield new diagnostic and prognostic biomarkers (Biological Markers), but that they can also be used to validate a molecular pathogen and predict the physiological repercussions of targeting that molecule with a rationally designed drug. The major pitfall is, of course, is that the quality of the results will be strictly dependent on the quality of the data and the validity of the assumptions used to construct the predictive and analytical models. In reality, several of the signal transduction pathways which are the subject of contemporary targeted drug design are imperfectly understood, and this ignorance, far from being synonymous with bliss, has been a root cause for the failure of certain "targeted" drug candidates. In retrospect, it is clear that the success of imatinib against BCR/ABL-dependent CML (Chronic Myeloid Leukemia) occurred because BCR/ABL was thoroughly validated as a molecular pathogen in this disease.

Characterization of Molecular Targets

The second stage of drug design is the detailed characterization of the molecular target. The goal of this effort is to determine whether the target is a "druggable" target. Designers must not only determine whether a target *can* be affected by a drug (in other words, "is it possible?"), but also whether a target *should* be affected by a drug (in other words, "is it prudent?"). Failure to do so will result in the development of a drug that will be ineffective at treating cancer, likely to exhibit unacceptable toxicities, or at worst, both. The molecular pathogens that are best suited for targeted drug design are those that are biochemically distinct from their normal counterparts. Point mutations, chimeric fusions due to chromosomal translocations, posttranslational modifications, or other unique structural modifications are most likely to yield a good candidate target. The ideal characterization of a molecular target will include the determination of its three-dimensional structure by such techniques as nuclear magnetic resonance (NMR) spectroscopy and x-ray Crystallography, ▶ Structural Biology which will provide useful insights into the physical interaction between the drug and its target. Designers must also determine whether the candidate target has a functionally distinct role in the cancer cell in comparison to its role in the normal cell (Pathway Addiction), which further elaborates the information gleaned from the identification and validation phase. By thoroughly characterizing the role of the target in both normal and pathological cellular biology, designers can anticipate potential toxicities that would prove costly if discovered in later development stages. And finally, although it might seem quite obvious, it is important to characterize the biochemical behavior of the target. Most targets are either receptors or enzymes, and one must first understand their kinetic properties before one can compare the efficacy of candidate drugs on the target.

Compound Screening and Lead Optimization

The third and central stage of drug design is that of compound screening and ▶ lead optimization. Before entering this stage, the designer should consider which type of therapeutic agent would be most likely to succeed. The vast majority of drugs are low-molecular weight compounds; once approved for market use, these drugs are typically cheaper to produce. On the other hand, low-molecular weight compounds tend to interact with several other nontargeted biomolecules, which commonly causes undesirable adverse effects and toxicities. New therapeutic agents include recombinant therapeutic proteins and recombinant humanized monoclonal antibodies (see also Monoclonal Antibody Therapy). The complex structures of these agents provides the basis for their target specificity, such that if the biology of the target is clearly understood, then these agents are least likely to produce unanticipated adverse effects and toxicities and are most likely to succeed in late stage ▶ clinical trials. Their primary disadvantage is that, at present, they are exorbitantly expensive. In theory, ▶ gene therapy holds the promise of correcting the root cause of cancers and other diseases. In reality, the utility of gene therapy is severely hampered by the fact that current methods of gene delivery are inefficient, nonspecific, poorly understood at the mechanistic level, and have consistently resulted in unanticipated adverse effects in human subjects.

The goal of compound screening and lead optimization is to rationally determine which structural attributes of a molecule are most important for its usefulness with respect to the target and to modify the molecule until the best possible drug candidate has been generated. Many targeted anticancer drugs serve as inhibitors of crucial enzymes (▶ Tyrosine Kinase Inhibitors, ▶ Proteasome Inhibitors), so it is simplest to discuss compound screening and lead optimization in the context of inhibitor design. Compound screening, also referred to as the discovery phase, involves the rapid systematic evaluation of candidates contained in natural products libraries or combinatorial libraries. Given the large numbers of compounds present in such libraries, robotic processing is required for high-throughput screening (HTS). High-throughput enzyme assays, interaction assays, and cell-based assays must be designed with careful attention to accuracy, precision, sensitivity, specificity, and reliability. Methodologies such as homogeneous time-resolved fluorescence (HTRF), ▶ surface plasmon resonance, and luciferase reporter assays are commonly incorporated into high-throughput screening assays because of such considerations.

Ultimately, the quality of the data depends on the quality of the assay design, and miniaturized design can result in maximized errors. Statistical parameters, such as the ▶ Z-factor, have been developed to evaluate the quality and statistical reliability of high-throughput assays. In addition to considering the accuracy and sensitivity of cell-based HTS, one must ensure that the HTS strategy can distinguish between generalized toxicity versus target-selective effects before identifying active compounds (or "hits") in a drug library. Once a set of "hits" has been identified, the process of ▶ lead optimization begins. Lead compound optimization and preclinical evaluation form a reiterative loop (which is a formal acknowledgement of the need to "go back to the drawing board"). The chemical structures of the "hits" are ranked according to their HTS-ascribed functional profiles in an effort to define one or more appropriate ▶ pharmacophores, which provides a general molecular scaffold containing core functional domains that determine the drug's physicochemical and biochemical properties. When the structure of the target is known, computational algorithms such as DOCK can be applied to predict the conformation and orientation of a drug (or pharmacophore) bound to its target, although these predictions should be confirmed by biophysical methods such as Nuclear Magnetic Resonance (NMR), x-ray Crystallography, etc. Computational ▶ Quantitative Structure Activity Relationship (QSAR) analysis can then be applied to predict which structural modifications might improve the pharmacophore's properties. While a great deal of emphasis is placed on improving the potency of a drug, other properties can have a greater impact on the success of a drug. These properties include aqueous solubility, lipophilicity, pK_a, and metabolic stability, which will determine its pharmacokinetics/pharmacodynamics (PK/PD) profile. They also include the drug's specificity or selectivity for a given target, as well as its safety profile. There is a great interest in predicting the ADMET profile (Absorption, Distribution, Metabolism, Excretion, Toxicity, ▶ ADMET screen) of a drug before engaging in expensive and time-consuming clinical evaluations. It has been estimated that half of all drug development failures are attributable to unsatisfactory pharmacokinetic/pharmacodynamic properties and unacceptable animal toxicities, so the development of cost-effective HTS assays and computational methods to accurately predict ADMET profiles are of the utmost importance to the pharmaceutical industry. Ultimately, the efficacy of a drug is more important than its potency, and the overall drug development process will be most efficient if well-designed lead compounds emerge from this stage.

Preclinical Optimization

The fourth stage of targeted drug design, preclinical evaluation, should be considered separately from the screening and optimization stage, although the results of initial preclinical evaluations often require a reoptimization of the lead compound. This is especially true when unsatisfactory pharmacokinetic/pharmacodynamic properties or unacceptable toxicities (Pharmacokinetics/Pharmacodynamics), ▶ ADMET Screen are encountered during animal testing studies. Preclinical evaluation of a lead compound is the assessment of whether the compound can be safely administered to an animal to induce a therapeutic effect on the disease of interest. In rare instances, animals can be predisposed to develop a disease through genetic engineering or the disease can be chemically induced. If the disease is communicable, it can be induced into the animal by exposure to the appropriate pathogen, as is sometimes the case with cancers caused by viruses. The most commonly used animal in contemporary preclinical evaluations of anticancer drugs are immunosuppressed, SCID, or nude mice containing a transplanted tumor of human origin, also known as a xenografted mouse (Xenograft). This allows one to determine whether the drug can be administered to the animal in a fashion that will lead to regression or eradication of the tumor without causing unacceptable adverse or toxic effects. However, given the compromised immune status of xenografted mice, which forces them to live in an unrealistic sterile environment (not to mention the fact that mice have whiskers and a tail not normally found in humans), such preclinical studies do not provide a sufficient evidence to judge whether the lead compound should be safe and efficacious in humans. If a lead compound still appears promising after preclinical evaluation in xenografted mice, then it should be administered to other animals to improve the ▶ ADMET screen characterization profile. Other animals commonly used in preclinical evaluations of drugs include rat, rabbit, dog, and monkey. Theoretical and practical aspects are taken into consideration in designing such experiments; the cost per animal and the physiological similarity of a given animal to humans are two of the most important considerations. Internationally accepted ▶ Good Laboratory Practices (GLP) have improved the consistency of preclinical toxicology data as well asthe level of confidence in data generated by external laboratories. Such preclinical evidence will be carefully scrutinized before regulatory agencies will approve the administration of the first dose of a new drug in humans.

Clinical Evaluation

The fifth and final stage of drug design is the clinical evaluation of the drug. Once the drug has been demonstrated to be safe and efficacious in the treatment of humans, the manufacturer can petition the appropriate government authority for approval to market the drug. The design of clinical trials and the analysis of data from these trials must be impeccable, because clinical trial evidence is of paramount importance in the decision to approve or disapprove the marketing of a drug. Furthermore, sloppy design and analysis of clinical trials can be used in litigation against a manufacturer if a drug is shown to cause harm to patients after it has been approved for market use. Several issues must be considered before clinical trials can begin. First, it is essential to manufacture the drug under current ▶ Good Manufacturing Practices (GMP) to ensure that it is of consistent quality and purity before administering it to humans. Second, proposed protocols must be independently reviewed to ensure the protection of human subjects, and all investigational personnel must be properly trained in and committed to ethical conduct. Finally, if it has not already been addressed, biomarkers for patient

inclusion or exclusion should be incorporated into the clinical trial design. By requiring the detection of ▶ HER-2/neu in breast cancer biopsies, rather than merely the diagnosis of breast cancer, as an inclusion criterion for human test subjects, investigators were able to demonstrate that ▶ Herceptin was efficacious in the treatment of certain breast cancers. Retrospective analysis showed that such a demonstration would not have been possible with less stringent inclusion criteria. Similarly, the presence of the Philadelphia Chromosome provided a crucial biomarker for patient selection in the evaluation of imatinib for the treatment of CML (▶ Chronic Myeloid Leukemia). Clinical trials are conducted in three phases. Phase I clinical trials are the initial tests of a new drug in humans.

- Phase I trials are primarily designed to establish the dose-limiting toxicities of a new drug and to determine its maximum tolerable dose (MTD) as a gauge of the dose to be used in the next study phase. *Phase II trials are designed to determine whether a new drug is effective against the disease in question and to establish its effective dose. There is a growing recognition that the biologically effective dose (BED, i.e., the quantity of a drug that results in a therapeutic benefit; the accurate measurement of the BED is essential for the clinical evaluation of cytostatic molecularly targeted drugs) is a more relevant parameter than the MTD for the evaluation of molecularly targeted drugs and of the need for useful biomarkers to assess the BED.
- Phase III trials are designed to compare the efficacy of the new drug relative to the best established therapy for a given disease and to examine whether there are any unexpected adverse effects or toxicities in the general patient population. Phase III studies typically involve multiple study arms and are the volunteers divided into "control" and "experimental" groups.

In each of the above, the decision to progress to an advanced phase or to discontinue the evaluation process is predicated on the success or failure in an earlier study phase, and in each case the pool of participating human volunteers will increase in size to accommodate the increasingly rigorous statistical requirements of advanced phase trials.

Ideally, success in each of the preceding five stages of drug design will facilitate the approval of a new anticancer drug for use in the medical market. The work of the drug designer is not complete at this stage, however. Seemingly inevitable problems of drug resistance (Chemoresistance) will emerge, and if the cause of resistance is known, this problem can be overcome by returning to the stage of lead optimization. Such was the case with the development of dasatinib for the treatment of patients with CML (▶ Chronic Myeloid Leukemia) resistant to ▶ imatinib. Also, as new drugs are administered to larger and more genetically diverse patient populations, subtle population-based differences in drug efficacy and toxicities begin to emerge. It is hoped that these differences will be positively exploited to achieve the ideal goal of ▶ personalized medicine, which will require the application of Pharmacogenomics to the process of drug design.

On a final note, it should be remembered that while the process of drug design has been described as a logical strategy, serendipity still plays a major role both in the discovery of new drugs and in the discovery of new applications for old drugs. As Louis Pasteur once said, "Dans les champs de l'observation le hasard ne favorise que les esprits prepares" ("In the fields of observation, chance favors only the prepared mind").

Cross-References

- ▶ ADMET Screen
- ▶ Adriamycin
- ▶ B-cell Lymphoma
- ▶ Chromosomal Translocations
- ▶ Chronic Myeloid Leukemia
- ▶ Cisplatin
- ▶ Clinical Cancer Biomarkers
- ▶ Clinical Trial
- ▶ Epidermal Growth Factor Receptor
- ▶ Erlotinib
- ▶ Gefitinib

- ▶ Gene Therapy
- ▶ Good Laboratory Practices
- ▶ Good Manufacturing Practices
- ▶ HER-2/neu
- ▶ Herceptin
- ▶ Imatinib
- ▶ Kit/Stem Cell Factor Receptor in Oncogenesis
- ▶ Koch's Postulates
- ▶ Lead Optimization
- ▶ Luciferase Reporter Gene Assays
- ▶ Non-Small-Cell Lung Cancer
- ▶ Oncogene Addiction
- ▶ Personalized Cancer Medicine
- ▶ Pharmacophore
- ▶ Proteasome Inhibitors
- ▶ Proteomics
- ▶ Quantitative Structure Activity Relationship
- ▶ Small Molecule Screens
- ▶ Sorafenib
- ▶ Structural Biology
- ▶ Surface Plasmon Resonance
- ▶ Time-Resolved Fluorescence Resonance Energy Transfer Technology in Drug Discovery
- ▶ Tyrosine Kinase Inhibitors
- ▶ Z-Factor

References

Faivre S, Djelloul S, Raymond E (2006) New paradigms in anticancer therapy: targeting multiple signaling pathways with kinase inhibitors. Semin Oncol 33(4):407–420

Lord CJ, Ashworth A (2010) Biology-driven cancer drug development: back to the future. BMC Biol 8:38

Rowinsky EK (2003) Challenges of developing therapeutics that target signal transduction in patients with gynecologic and other malignancies. Clin Oncol 21(10 Suppl):175s–186s

See Also

(2012) Bioinformatics. In: Schwab M (ed) Encyclopedia of cancer, 3rd edn. Springer Berlin Heidelberg, pp 403–404. doi:10.1007/978-3-642-16483-5_631

(2012) Biologically effective dose. In: Schwab M (ed) Encyclopedia of cancer, 3rd edn. Springer Berlin Heidelberg, p 404. doi:10.1007/978-3-642-16483-5_638

(2012) Biomarkers. In: Schwab M (ed) Encyclopedia of cancer, 3rd edn. Springer Berlin Heidelberg, pp 408–409. doi:10.1007/978-3-642-16483-5_6601

(2012) Chemoresistance. In: Schwab M (ed) Encyclopedia of cancer, 3rd edn. Springer Berlin Heidelberg, p 790. doi:10.1007/978-3-642-16483-5_1076

(2012) Dasatinib. In: Schwab M (ed) Encyclopedia of cancer, 3rd edn. Springer Berlin Heidelberg, p 1060. doi:10.1007/978-3-642-16483-5_1518

(2012) Humanized monoclonal antibody. In: Schwab M (ed) Encyclopedia of cancer, 3rd edn. Springer Berlin Heidelberg, p 1760. doi:10.1007/978-3-642-16483-5_6844

(2012) Maximum tolerable dose. In: Schwab M (ed) Encyclopedia of cancer, 3rd edn. Springer Berlin Heidelberg, p 2188. doi:10.1007/978-3-642-16483-5_3566

(2012) Monoclonal antibody therapy. In: Schwab M (ed) Encyclopedia of cancer, 3rd edn. Springer Berlin Heidelberg, pp 2367–2368. doi:10.1007/978-3-642-16483-5_3823

(2012) Personalized medicine. In: Schwab M (ed) Encyclopedia of cancer, 3rd edn. Springer Berlin Heidelberg, p 2828. doi:10.1007/978-3-642-16483-5_4476

(2012) Pharmacodynamics. In: Schwab M (ed) Encyclopedia of cancer, 3rd edn. Springer Berlin Heidelberg, p 2840. doi:10.1007/978-3-642-16483-5_4495

(2012) Pharmacokinetics. In: Schwab M (ed) Encyclopedia of cancer, 3rd edn. Springer Berlin Heidelberg, p 2845. doi:10.1007/978-3-642-16483-5_4500

(2012) Philadelphia chromosome. In: Schwab M (ed) Encyclopedia of cancer, 3rd edn. Springer Berlin Heidelberg, p 2864. doi:10.1007/978-3-642-16483-5_4520

(2012) Systems biology. In: Schwab M (ed) Encyclopedia of cancer, 3rd edn. Springer Berlin Heidelberg, p 3598. doi:10.1007/978-3-642-16483-5_5643

(2012) Tyrosine kinase. In: Schwab M (ed) Encyclopedia of cancer, 3rd edn. Springer Berlin Heidelberg, p 3822. doi:10.1007/978-3-642-16483-5_6079

(2012) Xenograft. In: Schwab M (ed) Encyclopedia of cancer, 3rd edn. Springer Berlin Heidelberg, p 3967. doi:10.1007/978-3-642-16483-5_6278

Drug Development

▶ Drug Design

Drug Discovery

Min-Jean Yin
Oncology Research, Pfizer Worldwide R&D, San Diego, CA, USA

Definition

In the field of cancer biology and oncologic pharmacology, drug discovery is the process by which cancer drugs are discovered and developed. The process of drug discovery involves identification of cancer targets, chemical or biological design and synthesis of molecules against a given target, and screening and characterization of these molecules for maximal therapeutic efficacy and minimal toxicity. Once a molecule has shown activity in these tests preclinically, then the clinical development process will begin.

Characteristics

The process of cancer drug discovery involves multiple steps, beginning with selection of appropriate cancer targets, complex design and screening of potential molecules, and completion of drug characterization in disease relevant preclinical models to evaluate anticancer efficacy and toxicity tolerance prior to advancing to clinical development. Each step will be discussed in more detail in the following sections.

Cancer Targets

The biology of cancer development and progression is very complex and still under intensive investigation. It is common for multiple pathways or molecules to interact in the promotion of tumor progression, and this cooperation is dependent on the genetic backgrounds of specific tumors versus the host microenvironment. Therefore, making a decision to pursue the right target is the most vital step in the drug discovery process. In general, cancer targets can be categorized as follows:

Fusion Genes, Oncogenes, and Tumor Suppressors
Genetic alterations resulting in constitutively hyperactivation of gene products and oncogenic addiction directly drive tumorigenesis of normal human cells into malignant tumor cells. Therefore, targeting genetic alterations in cancers has proven to yield the most successful treatments for cancer patients. Successful examples are imatinib (Gleevec) targeting the BCR-ABL fusion gene in myelogenous leukemia (CML), crizotinib (Xalkori) targeting the EML4-ALK fusion gene in non-small cell lung cancer (NSCLC), erlotinib (Tarceva) and gefitinib (Iressa) targeting mutant

EGFR in NSCLC, vemurafenib (Zelboraf) targeting the B-Raf V600E mutation in melanoma, and sunitinib (Sutent) targeting the c-KIT mutation in gastrointestinal stromal tumors (GIST).

Tumor suppressor genes such as p53, PTEN, and others are frequent mutated and drive tumorigenesis of human cancers; however, targeted therapy options have not been successful partly due to the nature of the loss of function of these genes.

Pathway Activations

Tumorigenesis is complex, involving multistep processes of cancer hallmarks in addition to host-tumor microenvironment regulation (Hanahan and Weinberg 2011). Therefore, drugs targeting disrupted pathways in tumor cells have also provided treatment options for cancer patients.

Most chemotherapy or cytotoxic agents (microtubule inhibitors, alkylating agents, antimetabolites, anthracyclines, and topoisomerase inhibitors) inhibit cell proliferation by taking advantage of a more progressive cell cycle in tumor cells versus normal cells. Although targeted therapy is the current trend for cancer drugs, chemotherapy agents have been the standard of care in the clinic to prolong the life of cancer patients and even cure some childhood cancers (Mariotto et al. 2009).

Dysregulation of hormone binding to their corresponding nuclear receptors triggers signaling cascades that mediate cancer initiation, progression, and metastasis, especially in breast and prostate cancers. Hormone therapy in breast cancers has provided clinical success in hormone receptor-positive breast cancer patients. This includes agents such as aromatase inhibitors (e.g., anastrozole, exemestane, and letrozole), selective estrogen receptor modulators (e.g., tamoxifen), and estrogen receptor downregulators (e.g., fulvestrant). Similarly, hormone therapy in prostate cancers by reducing testosterone (e.g., Zoladex) or blocking testosterone binding to androgen receptor (e.g., Casodex) can prolong the life of prostate cancer patients.

Tumor angiogenesis mediated by growth factors and their receptors is the major hallmark of certain tumor types. This includes VEGF/VEGFR (vascular endothelial growth factor receptor) and other receptor tyrosine kinases (RTKs). Therefore, targeting angiogenesis activation is a proven successful strategy as demonstrated by sunitinib (Sutent) and sorafenib (Nexavar) in renal cell carcinoma (RCC) via targeting VEGFR and other RTKs, bevacizumab (Avastin) in colon and rectal cancer (CRC) via targeting VEGF, trastuzumab (Herceptin) in Her2-positive breast cancers by targeting Her2 overexpression, and cetuximab (Erbitux) in CRC via targeting EGFR overexpression.

Furthermore, there are a number of agents in preclinical or clinical stages with the intent to target molecules involved in tumor metabolism, epigenetics, or immune modulation and other novel pathways/molecules through cancer genome approaches. We will have to wait for the results of clinical trials to determine whether these are successful strategies.

Drug Types

There are different types of cancer drugs including chemical compounds from natural products (e.g., rapamycin and taxanes) or small molecular weight synthetic compounds (most targeted therapy agents), humanized monoclonal antibodies (e.g., Herceptin and Avastin), and antibody-drug conjugates (ADC, e.g., brentuximab vedotin). Other approaches such as cancer vaccines, anti-sense RNA, and peptide technology also have been developed but have not yet shown clinical success necessary for FDA approval. In general, an antibody approach can be used for cancer targets that are expressed on the cell surface or are secreted from the cells, chemical compounds are preferred for cytoplasmic and nuclear proteins, and ADCs are preferred for cell membrane proteins with fast internalization characteristics.

Design and Screen the Drug Candidates

The decision on the type of anticancer drug will affect how to implement the design, screening, and characterization of the drug candidates in the process of cancer drug discovery.

For chemical compounds, a compound library comprised of a huge number of multiple various structure scaffolds will be screened against the target of interest in a high-throughput manner to determine if certain structures of compounds can be identified as "leads," followed by multiple modifications of these initial compound hits in order to improve the antitumor efficacy, target selectivity, drug properties, and toxicity profiles in the appropriate preclinical models.

For biologic agents, the initial design step involves special or patented technologies to make humanized antibodies or ADCs with specificity to the target of interest, followed by modifications and a screening process to choose the agent with the best specificity, affinity, and activity against the target.

Characterize the Drug Candidates in Preclinical Models

Once the drug candidates are identified, it is critical to characterize antitumor activity in appropriate cell lines and animal models that sufficiently recapitulate human disease development and tumor progression. Together with the pharmacokinetic and drug safety characterization in preclinical species (usually rodents and dogs), the estimated human predicted doses and therapeutic window of the drug candidate can be determined and used as reference guide to start clinical trials in healthy volunteers or selected cancer patients. However, often times, the perfect human cancer cells or preclinical tumor models that reflect the human diseases are not available, but genetically engineered cell lines or genetically modified animal models can be used to determine antitumor activity with the caution that these preclinical data may not be perfectly translatable to clinical outcomes.

Conclusions

The process of drug discovery is complex and involves multiple steps, including identification of cancer targets, chemical or biological design, screening and synthesis of molecules, and characterization of molecules for maximal therapeutic efficacy and minimal toxicity. Each step is critical and holds scientific and investment risk in addition to the intrinsic heterogeneity of human cancer progression (Kamb et al. 2007). Therefore, despite advances in modern technology, better understanding of the biological mechanisms of cancer biology, and massive information on the human genome, cancer drug discovery is still a lengthy, expensive, and difficult process with a low success rate as measured by new cancer drug approvals.

References

Hanahan D, Weinberg RA (2011) Hallmarks of cancer: the next generation. Cell 144:646–673

Kamb A, Wee S, Lengauer C (2007) Why is cancer drug discovery so difficult? Nat Rev Drug Discov 6:115–120

Mariotto AB, Rowland JH, Yabroff KR, Scoppa S, Hachey M, Ries L, Feuer EJ (2009) Long-term survivors of childhood cancers in the United States. Cancer Epidemiol Biomarkers Prev 18:1033–1040

Drug Metabolism

▶ Detoxification

Drug Resistance

Bo Wang
The Ohio State University, Columbus, OH, USA

Synonyms

Chemoresistance; Resistance to chemotherapy

Definition

Drug resistance refers to the biochemical mechanisms by which cancer cells fail to respond to chemotherapy, such that there is growth of cancer despite therapy. Drug resistance is a major obstacle for the successful treatment of many cancers.

Characteristics

Based on the nature of drug resistance, there are mainly two classes of drug resistance, de novo and acquired resistance. De novo drug resistance means that resistance is present before drug exposure and selection, while acquired drug resistance occurs after prolonged drug treatment and cancer cells develop several derangements to overcome the toxic effect of chemotherapy. De novo resistance may contribute to the failure to eradicate residual minimal disease and facilitate the development of acquired resistance.

De Novo Drug Resistance

To date, several different forms of de novo resistance have been described. One of these forms derives from the mutations in some genes or their downstream signaling pathways that are targeted by chemotherapeutic agents. Several studies have shown that oncogenic mutations in the epidermal growth factor receptor (EGFR) signaling pathway, especially those activating extracellular signal-regulated kinase 1/2 (ERK1/2) signaling, such as mutations in KRAS, BRAF, and NRAS, result in de novo clinical resistance to cetuximab-based therapy, which was developed to target EGFR.

Another form of de novo resistance involves the interaction between cancer cells and tumor microenvironment, namely, environment-mediated drug resistance (EMDR). In EMDR, cancer cells are transiently protected from apoptosis induced by the toxicity of chemotherapy. This form of resistance is induced by either soluble factors, including cytokines, chemokines, and growth factors secreted by tumor stroma, or the adhesion of cancer cells to the stromal fibroblast or extracellular matrix (ECM). Soluble factors, such as IL-6 or stromal cell-derived factor 1 (SDF1), have been shown to increase the transcription of antiapoptotic proteins and confer the resistance in cancer cells to apoptosis. Unlike soluble factor-mediated drug resistance (SFM-DR), cell adhesion-mediated drug resistance (CAM-DR) is mainly mediated by non-transcriptional mechanisms that are not well understood yet. Some studies suggested that CAM-DR involves the proteasomal degradation of proapoptotic proteins in cancer cells triggered by interaction with integrins or ECM components, fibronectin, collagen, and laminin. CAM-DR may also be mediated by inducing redistribution or modulating the activity of apoptosis-related proteins.

Acquired Drug Resistance

Several mechanisms have been attributed to the development of acquired drug resistance, including altered drug metabolism; increased drug efflux; intracellular changes that overcome the toxic effects of drugs, such as overexpression or mutation of drug targets; reduced cell apoptosis; increased DNA damage repair; and deranged cytoskeleton organization.

Drug Metabolism

Most drugs are metabolized through a two-phase process composed of phase I and phase II reactions. Metabolic enzymes involved in these reactions may determine individual drug response and resistance. Phase I reactions are mainly carried out by cytochrome P450 (CYP450), a superfamily of mixed-function oxidative enzymes. The different expressions and genetic polymorphisms of CYP450 family members are major determinants in the organ-specific and individual variable response to a given chemotherapeutic agent. Phase II reactions, also known as conjugation reactions, involve the formation of conjugates between substances from phase I reaction and glutathione, glucuronic acid, or sulfate catalyzed by glutathione S-transferase (GST), UDP-glucuronosyltransferase, and sulfotransferase, respectively. Unlike phase I reactions, which may either detoxify/inactivate or toxify/activate substrates, phase II reactions usually inactivate the substances. Elevated expression of phase II detoxificating enzymes, especially GST, has been shown to be responsible for resistance to certain drugs, such as alkylating agents.

Drug Efflux

Increased efflux of chemotherapeutic agents leading to the reduction of intracellular drug concentrations is another well-established mechanism underlying acquired drug resistance, especially multidrug resistance (MDR). The ATP-binding cassette (ABC) transporter superfamily, the largest family of transmembrane proteins, is responsible for ATP-dependent transportation of a variety of xenobiotics, including drugs. There are 7 subfamilies of ABC transporters designated A to G based on DNA sequence and protein structure. In the 1970s, ABCB1 (or MDR1) gene, which encodes the P-glycoprotein (PGP) transporter, was first shown to be induced in cancer cells selected by in vitro culture with anticancer drugs and resulted in multidrug resistance. Thereafter, several other subfamily members of ABC transporters, including ABCC1 (also known as multidrug resistance-associated protein 1 (MRP1)) and ABCG2 (also known as mitoxantrone resistance (MXR) gene or breast cancer resistance protein (BCRP)), have been associated with multidrug resistance. The drug resistance mediated by these subfamily members shows different tissue specificities and drug specificities, probably due to different expression patterns and structural variations. Other ABC transporters may also confer resistance to certain drugs in different cancer cells.

Intracellular Changes in Drug Resistance

Apart from altered drug metabolism and increased drug efflux, some other mechanisms have been identified in cancer cells resistant to the toxic effects of chemotherapeutic agents. In the past several decades, the role of signaling pathways in the initiation and progression of tumors has been extensively studied, and new chemotherapeutic strategies targeting signal transduction from cell surface to nucleus have been developed. Monoclonal antibodies and small molecules targeting receptor tyrosine kinases (RTKs) including EGFR and IGF-1R have been widely used to treat a variety of cancers. But some patients develop resistance to these drugs over time. One major reason is that some receptors activate multiple downstream signaling pathways, and the overlapping or cross talking among these pathways prevents the effect of a given drug targeting one receptor. In addition, compensatory activation

of other receptors or mutations in the targeted receptors also contribute to the resistance to targeted therapeutics.

Cell apoptosis plays an important role in cancer treatment as most chemotherapeutic agents aim to induce cell death through different mechanisms. Consequently, drug efficacy depends not only on the damages they cause to cancer cells but also on the cellular response to these damages by triggering apoptotic machinery in cancer cells. The susceptibility of cancer cells to drug-induced apoptosis is determined by the balance between proapoptotic and antiapoptotic signals. Mutations or downregulation of proapoptotic genes, such as p53, PTEN, Apaf-1, and Bcl-2 family members promoting apoptosis, as well as activation or overexpression of pro-survival genes, such as IAPs and antiapoptotic Bcl-2 family members, is frequently associated with drug resistance in many cancers.

The cytotoxic effect of some anticancer drugs, such as alkylating agents and platinum drugs, relies on their ability to cause DNA damage. The response to DNA damage in cells may include DNA repair, damage tolerance, or apoptosis depending on the nature of DNA damage. These different responses have fundamental influence on drug resistance. Activation of DNA repair machinery may remove the potentially lethal DNA lesions caused by chemotherapy and confer resistance to these drugs. On the other hand, DNA damage tolerance provides an alternative mechanism for drug resistance. Cancer cells with mismatch repair deficiency, such as loss of MLH1 expression, may fail to recognize the DNA damage and allow cells to tolerate severe DNA damage without activating cell apoptosis, thus conferring drug resistance.

Microtubules are cytoskeletal structures that play important roles in various cellular processes such as cell division and migration and intracellular trafficking. Microtubules are composed of α–β-tubulin heterodimers that undergo dynamic polymerization and depolymerization. Considering the importance of tubulins in cellular functions, a variety of drugs targeting tubulin and microtubules have been developed. Tubulin-binding agents (TBAs) are a class of drugs that can stabilize or destabilize tubulin, cause the delay or block of mitosis progression, and eventually lead to cell death. Mutations in the predominant βI-tubulin isotype or abnormally high expression of other tubulin isotypes, in particular βIII-tubulin, has been established to be associated with drug resistance in various cancer types. Microtubule-associated proteins (MAPs), such as Tau, MAP2, and MAP4, can bind to and stabilize microtubules against depolymerization. Aberrant expression of MAPs is also responsible for some resistance to TBAs.

Drug Penetration

The above-described mechanisms focus on the intracellular changes that cause drug resistance. But we cannot ignore another important issue in chemotherapeutic treatment, especially for solid tumors, the drug penetration. Anticancer drugs must have access to all the cancer cells to exert the most robust effect. However, many drugs have limited distribution in solid tumors because of disorganized vascular network, the composition of ECM, hypoxic environment, excessive drug binding and retention in the microenvironment, cell–cell adhesion, and so on. All these obstacles limit drug effectiveness. Therefore, developing new drug delivery methods and improving drug design to increase penetration are essential to overcome drug resistance and increase therapeutic efficiency.

References

Fletcher JI, Haber M, Henderson MJ, Norris MD (2010) ABC transporters in cancer: more than just drug efflux pumps. Nat Rev Cancer 10(2):147–156

Fodale V, Pierobon M, Liotta L, Petricoin E (2011) Mechanism of cell adaptation: when and how do cancer cells develop chemoresistance? Cancer J 17(2):89–95

Gottesman MM, Fojo T, Bates SE (2002) Multidrug resistance in cancer: role of ATP-dependent transporters. Nat Rev Cancer 2(1):48–58

Kavallaris M (2010) Microtubules and resistance to tubulin-binding agents. Nat Rev Cancer 10(3):194–204

Meads MB, Gatenby RA, Dalton WS (2009) Environment-mediated drug resistance: a major contributor to minimal residual disease. Nat Rev Cancer 9(9):665–674

Minchinton AI, Tannock IF (2006) Drug penetration in solid tumours. Nat Rev Cancer 6(8):583–592

Drug Responses

▶ Lung Cancer Pharmacogenomics

Drug Targeting

▶ Targeted Drug Delivery

Dry Eye Syndrome

▶ Sjögren Syndrome

Dry Mouth Syndrome

▶ Sjögren Syndrome

Dsg2

▶ Desmoglein-2

DS-ML

▶ Acute Megakaryoblastic Leukemia

Ductal Carcinoma In Situ

Nicola L. P. Barnes and Nigel J. Bundred
Department of Academic Surgery, South Manchester University Hospital, Manchester, UK

Synonyms

DCIS; In situ breast cancer; Intraductal carcinoma; Noninvasive breast cancer; Preinvasive breast cancer

Definition

Ductal carcinoma in situ (DCIS) is a preinvasive ► breast cancer. The malignant cells remain confined behind an intact basement membrane. The tumor can spread locally along the breast ducts (up to 20 or so) that form the breast, but DCIS does not possess the ability to invade into surrounding structures or spread to distant sites. If left, over time, a proportion of cases will develop into an invasive breast cancer, which does have the ability to spread and metastasize.

Characteristics

Diagnosis

Over 90% of DCIS that is diagnosed cannot be felt on breast examination and is completely asymptomatic; it is often only detected at screening mammograms. Approximately 70% of these mammographically detected cases present as microcalcifications. Atypical mammographic features include circumscribed nodules, ill-defined masses, duct asymmetry, and architectural distortion. These screening-detected cases are frequently small (<4 cm) and localized and are able to be treated by breast-conserving surgery. The remaining 10% of cases are symptomatic, presenting with either a palpable breast lump, nipple discharge, or Paget disease of the nipple. If these symptoms are present, the underlying disease is usually extensive and will require a mastectomy.

Diagnosis can only be confirmed by core biopsy, which takes a cylindrical whole core of tissue through a biopsy gun. It is performed under local anesthetic and can be done at an outpatient clinic. This is preferred to fine-needle aspiration cytology, where a smear of cells are obtained through a finer needle, as this gives no information on whether the basement membrane is intact or if there is evidence of invasion (indicating an invasive breast cancer, not DCIS). X-ray or ultrasound image guidance, to ensure the accuracy of sampling of these mainly impalpable lesions, is crucial. If the area of DCIS is extensive (>4 cm in size), multiple areas of the lesion need to be biopsied preoperatively to ensure that there is no invasive component to the disease and that all the microcalcifications are truly DCIS.

If the definitive diagnosis cannot be made with core biopsy, then a vacuum assisted biopsy should be performed. If a diagnosis can still not be made then a surgical open biopsy will be necessary, usually under a general anesthetic. The area of concern is localized with fine wires or a radioactive marker injection placed preoperatively under X-ray or ultrasound guidance, to direct the surgeon to the area requiring excision. The excised specimen is X-rayed immediately, after careful orientation with radiopaque clips, to confirm that all the microcalcifications have been excised. These wire-guided localization procedures are often therapeutic rather than diagnostic, and no further procedure is necessary.

Risk Factors

Risk factors for developing DCIS include a family history of breast cancer, older age at first childbirth, and never being pregnant. There is no conclusive evidence to date that either the oral contraceptive pill or ► hormone replacement therapy (HRT) increases the risk of DCIS.

As DCIS itself has no ability to metastasize, the main reason to treat DCIS is that a significant proportion will progress to become an invasive cancer. It is a subsequent invasive cancer that has the ability to spread into local tissue, the lymphatic system, the blood, and distant organs, unlike pure DCIS. Screening mammography may detect some cases of DCIS that would not progress to an invasive cancer, but this number is small compared to the potential benefit of detecting more aggressive diseases that would in time become an invasive breast cancer and develop the potential to spread.

Classification

DCIS is classified into two major subtypes according to the presence or absence of comedo necrosis. Comedo necrosis is a type of necrosis occurring with glands in which there is central luminal ► inflammation with devitalized cells, usually occurring in the breast in ► ductal carcinoma in situ (DCIS). A case of DCIS is termed to be comedo necrotic when necrotic material is seen to fill at least one duct when the specimen is looked at under the microscope. The term "necrosis" refers to cells that are dying. When cells die, they often attract other chemicals in the body and form something called "comedo necrosis," which is basically the residual, leftover dead cells. Comedo necrosis can often attract calcium from the blood and can form calcium deposits, and that is why the little flecks of calcium appear as ▶ mammographic density.

Non-comedo tumors encompass all the other subtypes of DCIS:

- Solid – where tumor fills extended duct spaces
- Micropapillary – where tufts of cells project into the duct lumen perpendicular to the basement membrane
- Papillary – where the projecting tufts are larger than in micropapillary and contain a fibrovascular core
- Cribriform – where the tumor takes on a fenestrated/sievelike appearance
- Clinging (flat) DCIS – where there are variable columnar cell alterations along the duct margins

Rarer subtypes also exist including neuroendocrine, encysted papillary, apocrine, and signet cell. In addition, the UK- and EU-funded breast-screening programs use the system of low, intermediate, and high nuclear grade to classify DCIS. This definition is based on the characteristics of the lesion as seen with a high-power microscope lens (×40) and uses a comparison of tumor nuclear size with normal epithelial and red blood cell size. If a lesion contains areas of varying grade, it is awarded the highest grade present.

Poorly differentiated high-grade comedo DCIS has low ▶ estrogen receptor (ER) expression, high rates of cell proliferation, and high rates of ▶ apoptosis (programmed cell death) and overexpresses c-erbB-2 (HER-2/neu) and EGFR. Low-grade lesions have high ER expression, with lower rates of cell proliferation and apoptosis than high-grade lesions, and they rarely express c-cerbB-2.

Progesterone receptor (PR) expression correlates with ER expression in both low- and high-grade tumors. In comparison, the normal breast epithelium has a low expression of ER and PR and a very low rate of apoptosis and c-erbB-2 expression.

Treatment

Breast-conserving surgery is now the treatment of choice for small, localized areas of DCIS, <4 cm in diameter, where only the segment of breast containing the tumor is excised. When the specimen is examined under a microscope, if the DCIS extends close (<1 mm) to the edges of the specimen, the patient should undergo cavity re-excision, as clear margin status is a key factor in predicting a good prognosis.

The multidisciplinary consensus conference on the treatment of DCIS (1999) recommends mastectomy for patients with large areas of DCIS (>4 cm), for multicentric disease, and for patients where radiotherapy is contraindicated. Women should also be offered mastectomy if the excision margins are persistently involved following breast-conserving surgery and cavity re-excision. The recurrence rate following mastectomy for DCIS is <1%.

The incidence of macroscopic lymph node metastasis in DCIS is <1%, and formal axillary staging (removal of lymph nodes from the armpit) in women with DCIS should be avoided.

In the USA all patients who have undergone breast-conserving surgery for DCIS are recommended to receive a course of radiotherapy. In Europe, adjuvant radiotherapy is recommended for all high-grade DCIS. Intermediate- and low-grade DCIS is selected for adjuvant radiotherapy on an individual patient basis.

The use of drug therapy following treatment for DCIS is not yet standardized. The antiestrogen ▶ tamoxifen can be used in ER-positive women who have undergone breast-conserving surgery; its use is decided upon on an individual patient basis. There is little point in using tamoxifen in ER-negative cases. ▶ Chemotherapy is not offered to women with pure DCIS regardless of type.

Following primary treatment for DCIS and radiological and pathological confirmation that there has been complete excision of all suspicious microcalcifications with clear margins, patients should be given the opportunity to participate in clinical trials. Follow-up in outpatient clinics, after the initial postoperative reviews, should be by annual bilateral mammography to detect recurrence. Although most DCIS recurrence is impalpable, clinical examination is still important to detect invasive recurrences.

DCIS in Men

DCIS accounts for up to 15% of breast cancers in men. It mainly presents clinically with symptoms of a cystic-type mass behind the areola or bloody nipple discharge. The standard treatment for DCIS of the male breast is total mastectomy with excision of the nipple-areola complex. The percentage of men with DCIS that go onto develop an invasive cancer is not known.

Recurrence

Patients with a recurrence of DCIS, where the primary was treated with breast-conserving surgery alone, can be offered re-excision (ensuring clear margins) followed by postoperative radiotherapy. Patients who have already received radiotherapy following their primary excision should be advised to have completion mastectomy. A skin-sparing mastectomy with latissimus dorsi myocutaneous flap or a free Deep Inferior Epigastric Perforator (DIEP) flap, give excellent results. The management of invasive recurrence is again dependent on the initial therapy for DCIS. If the patient did not receive radiotherapy after initial DCIS excision, then wide local excision and radiotherapy may still be an option depending on the size and location of the invasive tumor. If wide local excision is not an option, then mastectomy and axillary staging is the treatment of choice, with adjuvant therapy dictated by standard protocol as for primary invasive cancers.

The overall recurrence for breast-conserving surgery alone is ~ 25% at 8 years follow-up, with up to 50% of recurrences (i.e., 12.5% of all cases) being invasive disease. The women who develop invasive disease are at risk of metastatic spread. The remaining 50% of women develop further in situ tumors, which by definition does not metastasize.

A fundamental risk factor for recurrence is inadequate excision following breast-conserving surgery. This is judged as either close (<1 mm) or involved margins and by failure to remove all suspicious microcalcifications.

High-grade tumors and tumors showing comedo necrosis are independent risk factors for recurrence. The degree of tumor differentiation is also predictive of local recurrence. A further risk factor for recurrence irrespective of tumor grade or type is a young age (<40 years) at diagnosis. None of the major trials have found any statistical significance between recurrence and tumor size.

References

Ernster VL, Barclay J (1997) Increases in ductal carcinoma in situ (DCIS) of the breast in relation to mammography: a dilemma. J Natl Cancer Inst Monogr 22:151–156

Schwartz GF, Solin LJ, Olivotto IA et al (2000) The consensus conference on the treatment of in situ ductal carcinoma of the breast, 22–25 April 1999. Cancer 88(4):946–954

Silverstein M (2002) Ductal carcinoma in situ of the breast, 2nd edn. Lippincott Williams and Wilkins, Philadelphia. ISBN 0-781732-23-9

The Steering Committee on Clinical Practice Guidelines for the Care and Treatment of Breast Cancer. The management of Ductal Carcinoma in Situ (DCIS) (1998) Can Med Assoc J 158(Suppl):S27–S34

Duffy Antigen Receptor for Chemokines

Sonia Mohinta[1] and Kounosuke Watabe[1,2]
[1]Department of Medical Microbiology, Immunology and Cell Biology, Southern Illinois University, School of Medicine, Springfield, IL, USA
[2]School of Medicine, Department of Cancer Biology, Wake Forest University, Winston-Salem, NC, USA

Synonyms

DARC; gp-Fy

Duffy Antigen Receptor for Chemokines, Fig. 1 Mechanism of DARC–KAI1 interaction

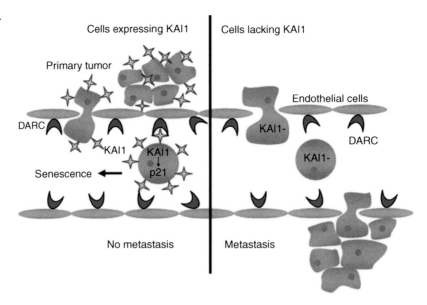

Definition

DARC or Duffy antigen receptor for ► chemokines is a seven-transmembrane protein, which has a molecular weight of ~45 k Da. DARC is expressed on vascular endothelium (plasma membrane and caveolae) as well as on red blood cells. It binds to chemokines of both C–C and C–X–C motif chemokines, but signal transduction through the G-protein-coupled receptor or Ca^{2+} flux is not observed. The genomic organization of the *DARC* gene consists of two alleles Fya and Fyb that are associated with a polymorphism at the 44-amino acid residue.

Characteristics

DARC is a receptor for malaria parasite *Plasmodium vivax*, and 70% of the West African population that lack DARC expression on their erythrocytes are resistant to malaria infection. A novel function of DARC has been identified where it has been found to interact with the metastasis suppressor *KAI1* CD82. When cancer cells expressing KAI1 intravasate and come in contact with the endothelial lining of small blood vessels during dissemination, KAI1 binds to the DARC expressed on the endothelial cells. The

physiologic outcome of this interaction is associated with the inhibition of cell proliferation and induction of cellular senescence. DARC constitutes two alleles, Fya and Fyb, and the Fyb allele of DARC is found to be associated with the KAI1–DARC interaction (Fig. 1). This KAI1–DARC interaction led to the cellular senescence by modulation of the expression of *TBX2* and *p21*(WAF1/CIP1/SDI1) genes. Furthermore, the metastasis suppression of KAI1 was significantly compromised in DARC-knockout mice, which show that DARC is essential for the function of KAI1 as a metastasis suppressor gene. The majority of West African population, which lacks DARC expression and is resistant to malaria, showed a significantly higher incidence of both prostate cancer and ► breast cancer as well as higher rate of metastatic disease than their white counterparts. DARC also serves as a promiscuous receptor for chemokines and is believed to function as "decoy" of excess chemokines. Therefore, DARC is also proposed to have an antitumorigenic role by clearing the angiogenic CXC chemokines. More evidence of DARC's antitumorigenic property has been observed, where DARC-knockout mice were shown to have a significant increase in incidence of prostate tumor in comparison to the wild-type mice. Overexpression of DARC in breast cancer cells significantly suppressed spontaneous lung metastasis. These lines of evidence indicate that DARC functions as an antitumorigenic as well as an antimetastatic protein, and development of drugs that can mimic its function may have a potential utility for the prevention and treatment of cancer.

References

Bandyopadhyay S et al (2006) Interaction of KAI1 on tumor cells with DARC on vascular endothelium leads to metastasis suppression. Nat Med 12(8):933–938

Iiizumi M, Bandyopadhyay S, Watabe K (2007) Interaction of Duffy antigen receptor for chemokines and KAI1: a critical step in metastasis suppression. Cancer Res 67(4):1411–1414

Iwamoto S et al (1995) Genomic organization of the glycoprotein D gene: Duffy blood group Fya/Fyb alloantigen system is associated with a polymorphism at the 44-amino acid residue. Blood 85(3):622–626

Mashimo T et al (2000) Activation of the tumor metastasis suppressor gene, KAI1, by etoposide is mediated by p53 and c-Jun genes. Biochem Biophys Res Commun 274(2):370–376

Miller LH et al (1975) Erythrocyte receptors for (*Plasmodium knowlesi*) malaria: Duffy blood group determinants. Science 189(4202):561–563

DUG (Rat Pdcd4)

▶ Programmed Cell Death 4

Dynamic Contrast-Enhanced Magnetic Resonance Imaging

James P. B. O'Connor[1], Geoff J. M. Parker[2] and Alan Jackson[2]
[1]Institute of Cancer Sciences, University of Manchester, Manchester, UK
[2]Centre for Imaging Sciences, University of Manchester, Manchester, UK

Definition

▶ Dynamic contrast-enhanced magnetic resonance imaging (DCE-MRI) is a noninvasive quantitative method of evaluating microvascular structure and function in tumors.

Characteristics

Solid tumors develop a circulatory blood supply of their own by ▶ angiogenesis to enable growth and survival. Since the vascular networks produce are structurally and functionally abnormal, they offer a potential target for novel anticancer therapy. This has led to worldwide interest in identifying and validating ▶ biomarkers of tumor vascular function that assist diagnosis, inform prognosis, or enable the monitoring of treatment effects for both conventional ▶ chemoradiotherapy and antiangiogenic agents.

T_1 and T_2* DCE-MRI Techniques

Imaging techniques allow repeated measurements of the tumor vasculature to be made in a noninvasive manner. Magnetic resonance imaging, unlike ▶ positron emission tomography and x-ray computed tomography, does not produce ionizing radiation and is thus considered a safe technique for evaluating functional changes in tumor blood vessels. Two main types of examination can be performed in DCE-MRI. In T-weighted techniques, repeated images are acquired while an intravenous bolus of gadolinium contrast agent traverses the tumor microvasculature. Gadolinium ions are paramagnetic and therefore interact with nearby hydrogen nuclei to shorten T_1 relaxation times in local tissue water, increasing signal intensity on T_1-weighted images. Signal enhancement within each imaging voxel is dependent on tissue perfusion, capillary surface area, capillary permeability, and the volume of the extracellular extravascular leakage space (EES), as well as the contrast agent dose, concentration of contrast agent in the artery supplying the tumor vascular bed and native tissue T_1 times. T_1-weighted DCE-MRI is therefore used to estimate the flow, volume, and permeability of blood vessels in tumors.

T*-weighted techniques (also known as dynamic susceptibility contrast MRI) offer an alternative method of investigating the microvasculature. Here, rapid loss of signal intensity observed on $T_2{}^*$-weighted images is used to calculate the change in concentration of contrast agent for each individual voxel. This method is particularly sensitive to changes in blood flow and volume and assumes that the gadolinium contrast agent remains within the tumor vessels throughout the examination, ignoring the effect of vessel permeability. For this reason, $T_2{}^*$-weighted methods have generally been limited to studies of the brain where it is assumed that the intact blood–brain barrier limits permeability. While this assumption is untrue in high-grade malignancy, numerous DCE-MRI studies of brain tumors employ $T_2{}^*$-weighted analysis techniques.

Data Acquisition

DCE-MRI protocols are complex. They require considerable technical expertise and interaction between basic scientists and clinicians. Most examinations are performed on conventional clinical scanners at a magnetic field strength of 1.5 T.

In general, three types of imaging data are acquired in T_1-weighted DCE-MRI. Initial anatomical images are followed by sequences that allow calculation of baseline tissue T_1 values. Finally, dynamic T_1-weighted images are acquired every few seconds over a period of approximately 5–10 min. Most investigators use gradient echo sequences for the dynamic series, as they allow good contrast medium sensitivity, high signal-to-noise ratio, adequate anatomical coverage, and rapid data acquisition. Policies for quality control are mandatory to achieve reliable, reproducible data, since DCE-MRI suffers from errors in T_1 calculation, significant motion artifact, and difficulties in accurate definition of the tumor region of interest.

Gradient echo pulse sequences are also used in dynamic susceptibility contrast MRI. Images are ideally acquired every 1–2 s to allow calculation of blood flow and volume. $T_2{}^*$-weighted techniques are limited by measurement errors and, unlike positron emission tomography, usually produce measurements that are relative to large cerebral vessels, rather than absolute values.

Data Analysis

DCE-MRI data analysis may be performed using one or more methods of varying complexity. Most examinations limit the area scanned to around 20 cm or less. Selection criteria for target lesions include size (tumors greater than 2 cm), adequate contrast with background tissue, and organ site free of excessive motion. Initially, a region of interest is defined to cover the entire tumor (e.g., in three-dimensional volume imaging) or part of the tumor (e.g., one or more two-dimensional slices).

Measures of tumor function are extracted based upon one or more of the following:

1. Simple features of the signal intensity–time curve (such as fraction of enhancing voxels, gradient, or time to 90% peak enhancement).
2. Conversion of signal intensity data into more robust parameters that describe the shape of the contrast agent concentration–time curve. These represent a combination of flow, blood volume, vessel permeability, and EES volume (such as the initial area under the gadolinium contrast agent concentration–time curve, IAUGC).
3. Fitting data from the contrast agent concentration–time curve to a pharmacokinetic model. This enables parameters that are independent of the data acquisition sequence to be derived.

Modeled parameters enable estimates of physiological characteristics such as flow and capillary endothelial permeability within a tumor; they are more "physiologically meaningful" than either signal intensity or IAUGC measurements. Many related parameters have been described in the literature making comparison between studies difficult. Current expert consensus recommends that the volume transfer coefficient of contrast agent between the blood plasma and the EES (K^{trans}), and the size of the EES should be preferred in ► clinical trials of

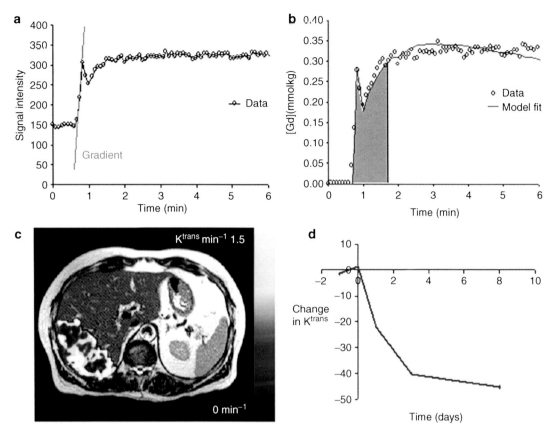

Dynamic Contrast-Enhanced Magnetic Resonance Imaging, Fig. 1 Panels (**a–c**) illustrate T_1-weighted DCE-MRI data analysis in a patient with a single colorectal liver ▶ metastasis prior to treatment with an antiangiogenic agent – analysis options include (**a**) measures from the signal intensity–time curve such as initial gradient, (**b**) calculated gadolinium contrast agent concentration–time curve with the integrated area under the curve at 60 s after injection (*IAUGC*), and (**c**) parametric map of median K^{trans} values for each voxel superimposed on an anatomical image. The effect of an antiangiogenic agent is demonstrated by (**d**) change in tumor median K^{trans}

angiogenesis inhibitors using T_1-weighted DCE-MRI, along with IAUGC. Cerebral blood volume (CBV) and cerebral blood flow (CBF) are typically derived in brain tumor studies that incorporate T_2^*-weighted techniques. Many of the aforementioned quantities have been investigated as potential biomarkers of tumor biology and drug effect but are yet to be validated as ▶ surrogate end points.

Finally, a choice of parameter analysis must be made. Median values are most commonly quoted, since both signal intensity and contrast agent concentration within the tumor voxels assume non-normal distributions. Single values may adequately quantify characteristics of tumor tissues, for example, median CBV can accurately predict histological tumor grading in glioblastoma multiforme and discriminate between high grade (higher CBV) and low grade (lower CBV) and identify areas of dedifferentiation. However, imaging drug effects in clinical trials of antiangiogenic agents requires changes in K^{trans} and IAUGC to be calculated from baseline values. Here, percentage change from baseline is frequently quoted. Other forms of data analysis, such as parametric (or calculated) maps, demonstrate the spatial heterogeneity of DCE-MRI parameters. Examples of DCE-MRI analyses are illustrated in Fig. 1.

Clinical Findings

DCE-MRI has been used to evaluate tumor microvasculature in numerous clinical studies. Comparison between study methodology and conclusions is difficult since most clinical studies have employed measures of signal intensity, rather than modeled parameters to quantify the microvasculature. Simple signal intensity measures, such as the initial gradient, shape of the enhancement curve, and maximum enhancement at 60 s, have been shown to reliably distinguish malignant ► breast cancer from benign breast lesions. High and/or increasing relative signal intensity of contrast enhancement in patients with ► cervical cancer before ► radiotherapy has predicted a low incidence of local recurrence. Some centers use DCE-MRI to distinguish between benign bladder wall thickening and ► bladder cancer, to characterize breast cancer and ► hepatocellular carcinoma, and to target areas of high-grade malignancy in glioblastoma multiforme. Despite these encouraging results, DCE-MRI has few clinical indications at present.

Evaluation of Angiogenesis Inhibitors

One major area of interest in imaging has been evaluating magnetic resonance-based ► biomarkers of changes in tumor microvasculature. Over 100 clinical trials of ► angiogenesis inhibitors have incorporated DCE-MRI and other forms of imaging in their study design. DCE-MRI is particularly attractive in this setting since antiangiogenic agents do not typically induce large changes in tumor size, as assessed by traditional imaging end points such as ► radiological response criteria.

Significant dose-dependent reductions in K^{trans}, IAUGC, and similar parameters have been demonstrated in a number of studies with anti-► vascular endothelial growth factor antibodies and inhibitors of tyrosine kinase ► signal transduction. For example, dose-dependent reduction in a parameter similar to K^{trans} was demonstrated in the study of patients with liver metastases from ► colon cancer, following administration of the VEGF receptor tyrosine kinase inhibitor vatalanib (PTK787/ZK222584; Novartis, Basel, Switzerland), which correlated with response rate and disease progression. Similar findings have been demonstrated in other studies, where DCE-MRI has been utilized alongside other functional imaging, histology, and ► serum biomarkers of tumor pathophysiology. Other studies, for example, of the VEGF receptor tyrosine kinase inhibitor sorafenib (Bayer; Leverkusen, Germany) in patients with metastatic renal cell cancer have shown high baseline values of K^{trans} related to good prognosis, measured by progression-free survival. Finally, evaluation of the anti-VEGF monoclonal antibody bevacizumab (Genentech, California) has shown how DCE-MRI can provide in-depth information regarding the temporal changes induced by antivascular drugs on the tumor microvasculature and how these functional and structural changes relate to downstream tumor shrinkage.

These studies highlight the potential use of DCE-MRI, namely, that change in tumor vasculature can be followed at regular intervals in a noninvasive manner. However, it also identifies the shortfalls of the technique – the precise mechanism of action cannot be determined by a technique that measures changes in composite processes at the millimeter scale. Furthermore, while changes in DCE-MRI parameters may be necessary to identify successful therapeutic compounds, they are not sufficient to identify success in a phase III randomized controlled trial – vatalanib has not produced an improvement in overall survival in patients with metastatic colon cancer when combined with cytotoxic chemotherapy.

Conclusion

The parameters obtained in DCE-MRI analysis undoubtedly provide valuable insight into how the ► tumor microenvironment responds to chemoradiotherapy and novel therapies, but the quantities produced are estimates of composite processes, rather than measurements of physiological mechanisms. Further work is required to validate their role in diagnosis, prognosis, and follow-up of solid tumors and as biomarkers of drug efficacy and mechanism in clinical trials.

References

Hahn OM, Yang C, Medved M et al (2008) Dynamic contrast-enhanced magnetic resonance imaging pharmacodynamic biomarker study of sorafenib in metastatic renal carcinoma. J Clin Oncol 26:4572–4578

Leach MO, Brindle KM, Evelhoch JL et al (2005) The assessment of antiangiogenic and antivascular therapies in early-stage clinical trials using magnetic resonance imaging: issues and recommendations. Br J Cancer 92:1599–1610

Morgan B, Thomas AL, Drevs J et al (2003) Dynamic contrast-enhanced magnetic resonance imaging as a biomarker for the pharmacological response of PTK787/ZK 222584, an inhibitor of the vascular endothelial growth factor receptor tyrosine kinases, in patients with advanced colorectal cancer and liver metastases: results from two phase I studies. J Clin Oncol 21:3955–3964

O'Connor JP, Carano RA, Clamp AR et al (2009) Quantifying antivascular effects of monoclonal antibodies to vascular endothelial growth factor: insights from imaging. Clin Cancer Res 15:6674–6682

O'Connor JP, Jackson A, Parker GJ et al (2012) Dynamic contrast-enhanced MRI in clinical trials of antivascular therapies. Nat Rev Clin Oncol 9:167–177

Tofts PS, Brix G, Buckley DL et al (1999) Estimating kinetic parameters from dynamic contrast-enhanced T-weighted MRI of a diffusable tracer: standardized quantities and symbols. J Magn Reson Imaging 10:223–232

Dysgerminoma

▶ Testicular Germ Cell Tumors

Dysgerminomas – Dysgerminoma

▶ Ovarian Tumors During Childhood and Adolescence

Dysmyelopoietic Syndrome

▶ Myelodysplastic Syndromes

Dysplasia

Definition

Dysplasia is an alteration found principally in epithelial tissues, characterized by a number of changes including loss of single cells uniformity and architectural orientation. Dysplastic cells are highly pleomorphic (variation in size and shape) and often exhibit hyperchromatic nuclei and increase in the nucleus/cytoplasm ratio. Mild-to-moderate dysplasia is often reversible following the removal of causative stimuli. However, severe dysplasia becomes irreversible and may be considered a premalignant lesion due to its ability to progress to carcinoma. Dysplastic epithelium may be precursor of carcinoma and may per se be malignant when associated with direct invasion beyond the muscularis mucosae in the submucosa of the bowel wall. Dysplasia is identified on the basis of a series of microscopic features including architectural alteration of the tissue and cytologic abnormalities, mainly nuclear pleomorphism and hyperchromatism, loss of nuclear polarity, and marked stratification of nuclei. On the skin, dysplasia may appear as a red patch on the mucosa that is not attributable to any obvious cause. Generally, these lesions have a well-defined border and a soft, velvet-like appearance. Their atrophic nature contributes to the red coloration, as underlying vasculature is more prominent. Around 90% show signs of severe dysplasia or ▶ carcinoma in situ and may progress to invasive squamous cell carcinoma.

Cross-References

▶ Carcinoma in Situ
▶ Colorectal Cancer Premalignant Lesions
▶ Preneoplastic Lesions
▶ Squamous Cell Carcinoma

E

E Motif

▶ E-Box

E T V6

Stefan K. Bohlander
Faculty of Medical and Health Sciences,
The University of Auckland, Auckland,
New Zealand

Synonyms

ETS variant gene 6

Definition

A gene encoding an *ets* domain transcription factor located on human chromosome 12 band p13. It is a frequent target of ▶ chromosomal translocations.

Characteristics

Discovery
The short arm of chromosome 12 is a hot spot for chromosomal rearrangements in diverse types of hematological malignancies. These rearrangements include balanced translocations with a great number of different chromosome partner bands as well as unbalanced translocations and deletions. The latter two rearrangements lead to the loss of genetic material from 12p. Molecular cytogenetic studies showed that more than half of the observed balanced translocations of 12p have breakpoints that involve the *ETV6* gene. There are currently more than 40 different 12p translocations described that involve the *ETV6* gene.

ETV6 was originally identified as the fusion partner of the ▶ platelet-derived growth factor receptor beta gene (*PDGFRB*) in a balanced t(5;12)(q31:p13) translocation from a case with chronic myelomonocytic leukemia (CMMoL). Initially, *ETV6* was called *TEL* (translocation *ets* leukemia gene) but was later renamed as *ETV6* (*ets* variant gene 6) to avoid confusion with the abbreviation for telomere.

Protein Domains
ETV6 is a member of the *ets* (E-26 transforming specific) family of transcription factors. All *ets* family proteins share a very conserved protein domain of about 88 amino acids in length, the so-called *ets* domain (Fig. 1). The *ets* domain is a sequence-specific DNA-binding domain but also mediates protein-protein interaction. It is evolutionarily highly conserved and found in invertebrates such as *Drosophila* and *C. elegans*. The *ets* domain of *ETV6* is more closely related to the *ets* domain of the *Drosophila* protein *yan* than to *ets* domain of the human *ETS1* or *SPI1* (PU.1) genes.

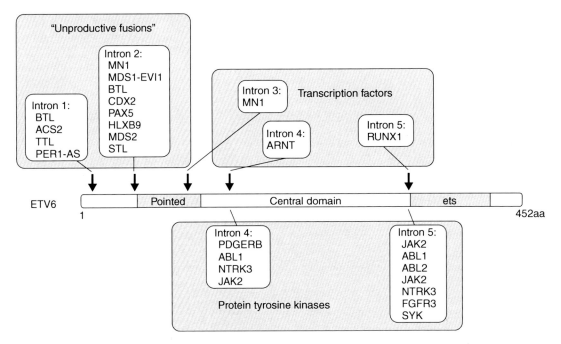

E T V6, Fig. 1 Diagrammatic representation of the ETV6 protein with the position of the breakpoints of the various fusion partner genes

The other evolutionarily conserved domain in ETV6 is the N-terminally located *pointed* or SAM (sterile alpha motif) domain. This domain is even more highly conserved in evolution than the *ets* domain and is found in many *ets* family members as well as in many other transcription factors and signal transduction proteins. The *pointed* domain serves as a homo- and heterodimerization module.

ETV6 Fusion Partners in Cancer

After the initial cloning of *ETV6*, several other fusion partners of *ETV6* were identified in quick succession. There are now well over 20 fusion partners of *ETV6* described, and the list is still growing.

The various *ETV6* fusions can be assigned to three groups: (i) protein tyrosine kinases (PTK), (ii) transcription factors and others, and (iii) "unproductive" fusions, i.e., fusions that do not result in an obvious fusion protein.

Protein Tyrosine Kinase Fusion Partners of ETV6

The first identified fusion partner of *ETV6* was the protein tyrosine kinase (PTK) platelet-derived growth factor receptor beta gene. The ETV6/ PDGFRB fusion protein is a constitutively active PTK and is the critical product of this translocation. In the ETV6/PDGFRB fusion protein, the N-terminal portion of ETV6, which includes the *pointed* domain, is fused to the C-terminal two thirds of the PDGFRB protein, which includes the tyrosine kinase domain of PDGFRB. This general structure, i.e., the *pointed* domain of ETV6 in the N-terminal half and the tyrosine kinase domain of the fusion partner in the C-terminal half of the fusion protein, is characteristic of all ETV6/PTK fusions. Many studies have shown that the *pointed* domain of ETV6 serves as a dimerization module for the fusion protein. Dimerization of the fusion protein leads to the constitutive activation of the PTK domain, which results in autophosphorylation of the fusion protein as well as phosphorylation of cellular proteins like rasGAP, Shc, SH-PTP2, SH-PTP1, CRK-L, CBL, paxillin, and STATs. Expression of ETV6/ PTK fusions in the interleukin 3-dependent hematopoietic cell line Ba/F3 leads to factor-independent growth. Several ETV6/PTK fusion proteins have also been assayed in murine bone

E T V6, Table 1 Translocation partners of ETV6

ETV6 fusion partner	Translocation	Disease
Tyrosine kinases		
PDGFRB	t(5;12)(q31;p13)	CMMoL
ABL1	t(9;12)(q34;p13)	AML, ALL
ABL2	t(1;12)(q25;p13)	AML-M3,-M4, T-ALL
JAK2	t(9;12)(p24;p13)	Pre-B-cell ALL, T-ALL
NTRK3	t(12;15)(p13;q25)	Congenital fibrosarcoma, mesoblastic nephroma, secretory breast carcinoma, AML
FGFR3	t(4;12)(p16;p13)	Peripheral T-cell lymphoma
▶ SYK	t(9;12)(q22;p12)	MDS
Transcription factors and cofactors		
RUNX1	t(12;21)(p13;q22)	ALL
MN1	t(12;22) (p13;q11)	AML and MDS
ARNT	t(1;12)(q21;p13)	AML-M2
"Unproductive fusions"		
MDS1/EVI1	t(3;12)(q26;p13)	MDS, CML
BTL	t(4;12) (q11-q12;p13)	AML-M0
CDX2	t(12;13)(p13;q12)	AML
PAX5	t(9;12)(q11;p13)	ALL
HLXB9	t(7;12)(q36;p13)	AML
MDS2	t(1;12)(p36.1;p13)	MDS
STL	t(6;12)(p13;q23)	B-ALL
ACS2	t(5;12)(q31;p13)	MDS-RAEB, AML, AEL
TTL	t(12;13)(p13;q14)	ALL
PER1 antisense	t(12;17)(p13;p13)	AML

marrow transplantation or transgenic mouse models where they lead to various hematological diseases like myelo- or lymphoproliferative syndromes. The different protein tyrosine kinase genes that have been found to be fused to ETV6 are listed in Table 1. It should be noted that most of these fusions are rather rare. For several partners just one or a handful of cases have been described in the literature.

ETV/PTK fusions are not only found in various types of leukemia, but the ETV6/NTRK3

fusion has also been found in solid tumors such as congenital fibrosarcoma, mesoblastic nephroma, and secretory ▶ breast cancer.

Transcription Factors and Other Fusion Partners of *ETV6*

There are only two fusions of *ETV6* with non-PTKs for which the transforming potential of the fusion protein could be established. This is the very common ETV6/▶ RUNX1 and the much rarer MN1/ETV6 fusion.

The ETV6/RUNX1 Fusion

The ETV6/RUNX1 (TEL/AML1) fusion was the second fusion of ETV6 that was identified. It was soon recognized that the ETV6/RUNX1 fusion is the most common fusion gene found in childhood ▶ acute lymphoblastic leukemia, which is present in up to 25% of all childhood B-ALL cases.

The ETV6/RUNX1 fusion protein, which is the critical fusion in this translocation, comprises the *pointed* domain of ETV6 as well as the DNA-binding and transactivation domain of RUNX1.

Many ETV6/RUNX1-positive ALLs have interstitial deletions (detectable by fluorescence in situ hybridization, FISH) of the short arm of the non-rearranged chromosome 12, which encompass the ETV6 locus. Even in cases in which no interstitial deletion of the non-rearranged ETV6 allele can be detected by FISH, there is no expression of the wild-type ETV6 allele suggesting a very important role for ETV6 loss of function in the pathogenesis of ALL.

Little is known about mechanisms by which the ETV6/RUNX1 fusion protein causes leukemia. Several attempts to establish ETV6/RUNX1 transgenic or bone marrow transplant leukemia models have failed or yielded leukemia only after a very long latency. There is some evidence that leukemogenesis by ETV6/RUNX1 is accelerated if cell cycle regulation is compromised by other mutations.

Some hints as to the function of ETV6/RUNX1 have been gleaned from reporter gene assays. In these experiments, the ETV6/RUNX1 fusion protein behaves as a strong transcriptional repressor on AML1 target genes. The *pointed* domain and the central domain of ETV6 have been shown to recruit transcriptional corepressors like SMRT, N-CoR, and mSin3A.

The MN1/ETV6 Fusion

The MN1/ETV6 fusion is found in some patients with ► acute myeloid leukemia (AML) or ► myelodysplastic syndrome with a t(12;22) (p13;q11) translocation. *MN1* was originally found as a gene disrupted by a translocation in meningioma. The critical MN1/ETV6 fusion protein, which is able to transform murine NIH3T3 fibroblasts in vitro (NIH3T3 transformation assay), contains transcriptional activation domains derived from MN1 and the *ets* DNA-binding domain of ETV6. This is the only known example in which the *ets* domain of ETV6 seems to be critical for transformation.

The ETV6/ARNT Fusion

Another rare ETV6/transcription factor fusion is the ETV6/ARNT fusion gene reported in one patient with AML and a t(1;12)(q21;p13), which links the N-terminal portion of ETV6 including the *pointed* domain to almost the complete ► aryl hydrocarbon receptor nuclear translocator (ARNT).

"Unproductive" *ETV6* Fusions

A large number of chromosomal translocations involving *ETV6* have been cloned, in which the breakpoints lie in intron 1 or 2 of *ETV6*. These translocations result in *ETV6* fusions that contain only the 54 amino-terminal amino acids of ETV6 and lack any of the important protein domains of ETV6 like the *pointed* or *ets* domain (Table 1). Almost all of these translocations have been identified in only one or at the most a handful of leukemia cases.

It is very likely that most, if not all, of these translocations do not result in the formation of a transforming ETV6/other fusion protein but rather that they lead to the transcriptional upregulation of genes adjacent to the translocation breakpoints. This has been elegantly shown for the t(12;13) (p13;q12) which results not only in the formation of a fusion between *ETV6* and the caudal-related

homeobox gene *CDX2* but also in the upregulation of a transcript that codes for the full-length CDX2 protein. In a bone marrow transplant model, the upregulation of the wild-type CDX2 protein and not the expression of the ETV6/CDX2 fusion protein is critical for the development of leukemia in a murine bone marrow transplantation model.

Putative Tumor Suppressor Gene and Physiological Function

There are several lines of evidence that suggest that *ETV6* might function as a ► tumor suppressor gene. In up to 70% of childhood ALL cases with ETV6/RUNX1 fusions, there is concomitant deletion of the non-rearranged *ETV6* allele. Deletions of the short arm of chromosome 12 are frequently found in a broad spectrum of hematological malignancies, and the common region of deletion was mapped to a small genomic region including *ETV6* and *CDKN1B*. In addition, several studies showed that even if no deletion of the *ETV6* locus can be detected, there was no expression of *ETV6* at the mRNA level or absence of the ETV6 protein.

In vivo and in vitro studies provide additional evidence that *ETV6* might function as a tumor suppressor gene. *ETV6* expression inhibits growth in soft agar of *RAS* transformed NIH3T3 fibroblasts cells and leads to the differentiation of erythroleukemia cells into erythrocytes. Additionally, the expression of *ETV6* in serum-starved NIH3T3 cells induces apoptosis. Reporter gene assays have demonstrated that ETV6 is a strong transcriptional repressor, which requires corepressors like N-Cor, mSin3, and SMRT.

Targeted deletion of the murine *Etv6* gene demonstrated that *Etv6* is essential for yolk sac angiogenesis and the establishment of definitive hematopoiesis in the bone marrow, while hematopoiesis in the yolk sac and fetal liver was not affected. $Etv6^{-/-}$ mice die between day 10.5 and 11.5 of embryonic development due to a defect in yolk sac ► angiogenesis and widespread apoptosis of mesenchymal and neural cells. Furthermore, *Etv6* could be shown to be an essential regulator for the maintenance of hematopoietic stem cells in adult murine bone marrow.

References

Dohlander SK (2005) ETV6: a versatile player in leuke-mogenesis. Semin Cancer Biol 15:162–174

Golub TR, Barker GF, Lovett M et al (1994) Fusion of PDGF receptor β to a novel ets-like gene, tel, in chronic myelomonocytic leukemia with t(5;12) chromosomal translocation. Cell 77:307–316

Hock H, Meade E, Medeiros S (2004) Tel/Etv6 is an essential and selective regulator of adult hematopoietic stem cell survival. Genes Dev 18:2336–2341

Rawat VPS, Cusan M, Deshpande A et al (2004) Ectopic expression of the homeobox gene Cdx2 is the transforming event in a mouse model of t(12;13)(p13; q12) acute myeloid leukemia. Proc Natl Acad Sci U S A 101:817–822

E100

▶ Curcumin

E11 Antigen

▶ Podoplanin

E2A-PBX1

Brandy D. Hyndman[1] and David P. LeBrun[1,2,3]
[1]Department of Pathology and Molecular Medicine, Queen's University Cancer Research Institute, Queen's University, Kingston, ON, Canada
[2]Protein Function Discovery Group, Queen's University, Kingston, ON, Canada
[3]Division of Cancer Biology and Genetics, Cancer Research Institute, Queen's University, Kingston, ON, USA

Definition

E2A-PBX1 is an oncogenic transcription factor expressed in neoplastic cells consequent to a somatic chromosomal translocation (t(1;19)) in some cases of ▶ acute lymphoblastic leukemia (ALL).

Characteristics

Leukemia cells commonly contain somatic chromosomal translocations that contribute to the induction and, presumably, perpetuation of the disease. They do this by altering proto-oncogenes that reside close to the involved breakpoints on the participating chromosomes. Translocation (1;19) (q23;p13.3) is the second most frequently observed recurrent translocation in ALL, detectable in approximately 5% of cases using conventional cytogenetics. In the vast majority of instances, t(1;19) fuses coding regions from two genes, *E2A* and *PBX1*, that reside respectively on chromosomes 19 and 1. Subsequent transcription and pre-mRNA splicing lead to expression of E2A-PBX1, an abnormal, chimeric transcription factor with potent oncogenic activity. Abundant evidence supports the notion that E2A-PBX1 contributes to the abnormal accumulation of primitive lymphoid progenitors that characterizes ALL by deregulating the transcription of key target genes.

Clinical Aspects

ALL is primarily a pediatric disease, so most patients with t(1;19)-positive and E2A-PBX1-expressing leukemia are children. Although the neoplastic cells in ALL most often have immunophenotypic (▶ flow cytometry) and genotypic (▶ molecular pathology) features characteristic of the very early stages of B-lymphoid development, prior to the initiation of immunoglobulin gene rearrangement or expression, the vast majority of cells associated with t(1;19) generally manifest characteristics typical of more mature "pre-B cells," including rearrangement of the immunoglobulin heavy-chain gene locus and expression of cytoplasmic, but not surface, immunoglobulin heavy-chain protein. Other clinical features associated with t(1;19)-positive ALL include especially high leukocyte counts at presentation, non-Caucasian race, and central nervous system involvement. Although the translocation was originally associated with

E

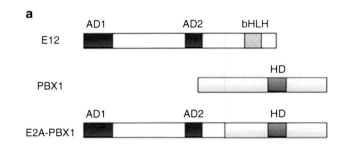

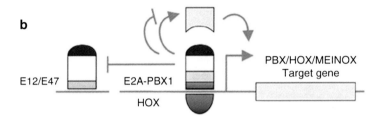

E2A-PBX1, Fig. 1 Structure and function of E2A-PBX1. (**a**) Translocation 1;19 results effectively in fusion of the amino-terminal two-thirds of the E2A proteins (this portion is identical in the E12 and E47 isoforms) with most of PBX1. The *vertical line* indicates the point of fusion. AD1 and AD2, transcriptional activation domains 1 and 2; bHLH, basic helix-loop-helix domain; HD, homeodomain.

(**b**) A model illustrating hypothetical mechanisms of neoplastic transformation by E2A-PBX1. The oncoprotein may deregulate the expression of critical genes normally controlled by PBX/HOX/MEINOX complexes, alter the function of transcriptional co-activators, or impair the function of wild-type E12/E47

unfavorable clinical outcomes, the more intensive treatment regimens used currently have largely or completely abrogated this association.

Wild-Type E2A and PBX1 Gene Products

The *E2A* gene resides at chromosome band 19p13.3 and encodes two proteins, called E12 and E47, which are generated by alternative splicing of exons (Fig. 1). E12 and E47 (or "E2A proteins") possess a C-terminal basic helix-loop-helix (bHLH) domain that mediates homo- or hetero-dimerization as well as binding to DNA at sites that contain the consensus sequence CANNTG (the ► E-box). E12 and E47 function as transcriptional activators, inducing the transcription of genes that lie in the general vicinity of the E-boxes to which they bind. The E2A proteins have important roles in regulating lineage-specific cellular differentiation; their contributions to various aspects of lymphocyte development have been especially well delineated.

The *PBX1* gene, at chromosome band 1q23, was identified through its involvement in the 1;19 translocation. Its protein products, PBX1a and PBX1b, contain a homeodomain, an evolutionarily ancient domain involved in DNA binding and protein-protein interactions (► homeobox genes). PBX1 binds to DNA and regulates the transcription of target genes in cooperation with other homeodomain-containing proteins of the HOX and MEINOX classes. These physical and functional interactions have important roles in embryonic development and tissue homeostasis. In particular, important roles in hematopoiesis are well documented.

Structure and Function of E2A-PBX1

The recombination of exons brought about by t(1;19) essentially fuses the amino-terminal two-thirds of the E2A proteins (a portion that is invariant between E12 and E47) to most of PBX1. The *E2A*-encoded portion includes two transcriptional activation domains capable of inducing target gene transcription by recruiting transcriptional co-activators (▶ chromatin remodeling), whereas the *PBX1*-derived portion includes the DNA-binding homeodomain.

E2A-PBX1 can function as a potently transforming oncoprotein in several cellular lineages. For example, enforced expression induces lethal lymphoproliferative diseases in transgenic mice and aggressive myeloproliferative diseases in a murine bone marrow transplantation model. The impaired cellular differentiation and accelerated proliferation that are associated with E2A-PBX1 expression result from the cumulative or cooperative effects of physical or functional interactions between E2A-PBX1 and other macromolecules. The available experimental evidence supports a general model, the basis of which was originally suggested by the nature of the functional domains that are brought together in the oncoprotein (Fig. 1). Largely by means of its *PBX1*-encoded portion, E2A-PBX1 retains the ability to bind cooperatively with HOX proteins and cognate PBX/HOX binding sites on the DNA. This probably results in the abnormal expression of target genes whose transcription is normally regulated by PBX/HOX/MEINOX complexes. The retention of potent transcription-inducing potential by the E2A portion of the oncoprotein and the documented involvement of HOX proteins (and, by implication, the target genes that they regulate) in normal and leukemia-associated hematopoiesis are consistent with this mechanism. In a perhaps complementary mechanism, the interaction with E2A-PBX1 may alter the function of transcriptional co-activator proteins, including the histone acetyltransferases p300 and CREB-binding protein (CBP) (chromatin remodeling in cancers), so as to promote neoplastic transformation in a manner at least superficially analogous to oncoproteins encoded by DNA tumor viruses such as Simian virus 40. Finally, E2A-PBX1 may exert dominant inhibitory effects on the wild-type E2A proteins, as these possess tumor suppressor activities.

The *E2A* locus is involved in another translocation, t(17;19)(q22;p13), seen relatively rarely in cases of ALL. Here, a similar portion of the E2A proteins is fused to a portion of the transcription factor hepatic leukemia factor (HLF), including the DNA binding domain. These observations indicate the promiscuity of the *E2A* locus with respect to translocation partners and suggest the involvement of common oncogenic mechanisms in cases of ALL associated with t(1;19) or t(17;19).

References

Abramovich C, Humphries RK (2005) Hox regulation of normal and leukemic hematopoietic stem cells. Curr Opin Hematol 12:210–216

LeBrun DP (2003) E2A basic helix-loop-helix transcription factors in human leukemia. Front Biosci 8:206–222

Murre C (2005) Helix-loop-helix proteins and lymphocyte development. Nat Immunol 6(11):1079–1086

Pui CH, Relling MV, Downing JR (2004) Acute lymphoblastic leukemia. N Engl J Med 350:1535–1548

E3 Ubiquitin Protein Ligase

▶ CHFR

Eag

▶ Ether à-go-go Potassium Channels

Eag1

▶ Ether à-go-go Potassium Channels

EAP1

▶ Securin

Early B-Cell Factors

Daiqing Liao
Department of Anatomy and Cell Biology, UF
Health Cancer Center, University of Florida
College of Medicine, Gainesville, FL, USA

Synonyms

COE; Collier-Olf-EBF; Early B-cell factors; EBF;
O/E; Olf; Olfactory neuronal transcription factor;
Olfactory/early B-cell factors

Definition

Early B-cell factors are a group of DNA-binding
transcription factors containing the helix-loop-
helix (HLH) domain. Within their highly conserved DNA-binding domain (DBD), a sequence
motif consisting of an atypical zinc finger (H-X_3-
C-X_2-C-X_5-C) is unique to this family of proteins.

Characteristics

Early B-cell factor 1 (EBF1) was initially isolated
in 1993 from the nuclear extracts of a murine
pre-B-cell line through oligonucleotide affinity
chromatography. The DNA sequence of the oligonucleotide was derived from the promoter of
the *mb-1* gene, encoding an immunoglobulin-
associated protein that is only expressed in the
early stages of B-lymphocyte differentiation.
EBF1 is essential for B-cell development, as
mice lacking *EBF1* gene do not produce functional B cells and immunoglobulins. The cDNA
encoding a related factor called Olf-1 was identified in the same year by screening rat cDNA
clones that could activate reporter gene in yeast
under the control of a synthetic promoter
containing DNA elements derived from olfactory
specific genes.

The EBF family is found only in the animal
kingdom from *Caenorhabditis elegans* to
humans. In the mouse and human genomes,
there are four paralogous genes of the *EBF* family
(EBF1–4). In humans, they localize at chromosomes *5q34*, *8p21.2*, *10q26.3*, and *20p13* for
EBF1–4, respectively. Like a typical
DNA-binding transcription factor, EBF proteins
contain well-defined modular structural and functional domains (Fig. 1). The DBD near the
N-terminus consists of approximately 200 residues whose sequence is exceedingly well conserved throughout evolution with >75%
sequence identity between distant species. This
family of proteins binds directly to DNA
sequences with a consensus of 5′-*CCCNNGGG*-3′
as homo- or heterodimers. The region following
the DBD resembles the conserved domains called
TIG-IPT (*i*mmunoglobulin-like fold, *p*lexins,
*t*ranscription factors, or *t*ranscription factor *i*mmu-
noglobulin). The function of this region in EBF
has not been determined, although the TIG-IPT
domain may be involved in homo- or heterodimerization in other transcription factors
containing this sequence. Located next to
TIG-IPT is the HLH domain that includes two
helices with similar sequence. This feature distinguishes the EBF family from other
HLH-containing transcription factors such as
Myc, Max, and MyoD that usually contain two
dissimilar amphipathic helices. The EBF HLH
domain is probably involved in dimerization, as
proteins with the deletion of this domain could not
form stable dimer in solution. The C-terminal
domain is highly rich in serine, threonine, and
proline residues. It is less conserved but is important for transcriptional activation; nonetheless,
mutant without this domain could still activate
transcription.

Widespread expression of *EBF* is detected in
diverse cell types such as adipocytes and neuronal
cells and during different developmental stages
such as limb buds and the developmental forebrain. The EBF orthologs in both *Drosophila
melanogaster* and *Caenorhabditis elegans* are
implicated in neurogenesis. During mouse
embryogenesis, *EBF* members are expressed in
early postmitotic neurons from midbrain to spinal
cord and at specific sites in the embryonic

Early B-Cell Factors,
Fig. 1 The structural
features of EBF family of
transcription factors.
Shown are the four paralogs
(EBF1–4) and their
corresponding accession
numbers (GenBank or
SwissProt database).
Specific domains are shown
in different color. Numbers
refer to the position of
amino acid residue in each
protein. The signature zinc
finger motif of the EBF
family of proteins in
complex with a zinc ion is
also depicted. The different
domains are *DBD*
DNA-binding domain, *TAD*
transactivation domain,
COE the signature sequence
of EBF family, *ZBM* zinc-
binding motif, *IPT/TIG* the
immunoglobulin-like fold
domain, and *HLH* helix-
loop-helix motif

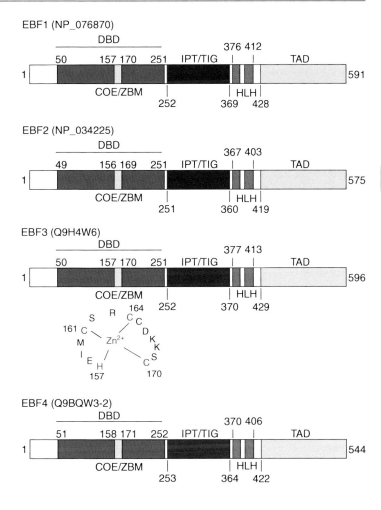

forebrain, indicating that EBF proteins may be involved in regulating neuronal maturation in the central nervous system (CNS). Interestingly, EBF1 is abundantly expressed in the striatonigral medium spiny neurons (MSNs). *EBF1* deficiency in mice results in markedly reduced number of striatonigral MSNs at postnatal day 14 (P14), although such neurons are properly specified in *EBF1* $^{-/-}$ mice by P0. Thus, EBF1 appears to be a lineage-specific transcription factor essential to the differentiation of striatonigral MSNs. Targeted deletion of mouse *EBF2* results in defects in peripheral nerve morphogenesis, migration of hormone-producing neurons, projection of olfactory neurons, and cerebellar development. In addition to B-cell development and neuronal differentiation, EBF transcription factors are

involved in other developmental processes. For example, EBF2 is a regulator of osteoblast-dependent differentiation of osteoclasts, and *EBF2*-deficient mice have reduced bone mass. Furthermore, EBF1 induces adipogenesis in NIH-3 T3 fibroblasts. The expression of EBF proteins in multiple tissues and their involvement in diverse developmental pathways suggest that they have fundamental cellular functions and their roles in lineage determination may be achieved through cooperation with other tissue-restricted factors.

The four paralogs of EBF members are highly similar to each other at the amino acid sequence level. It is therefore surprising that they have quite distinct functions, as *EBF3*-deficient mice exhibit neonatal lethality before postnatal day 2, and

EBF2-null mice are much smaller than their wild-type littermates, with body weight of the former being less than one half of that of the latter 30 days after birth. The N-terminal 50 amino acid residues and the entire C-terminal transactivation domain are the most divergent regions among these paralogs, although these regions are essentially identical among corresponding orthologs in different mammalian species. Therefore, functional specificity of individual EBF member may be determined by these divergent domains or through specific regulation of their expression. Indeed, the promoter sequences of the four paralogs appear quite distinct.

Potential Roles in Cancer

The *EBF3* locus at chromosome *10q26.3* is biallelically altered by genomic deletion and/or promoter hypermethylation in most cases of high-grade brain tumors. In a small number of examined clinical samples, the *EBF3* locus is inactivated in 50% of grade II, 83% of grade III, and 90% of grade IV brain tumors. *EBF3* is expressed in normal brain cells but is silenced in brain tumor cells. Thus, it is likely that EBF3 may be a tumor suppressor in the brain. EBF3 might also restrict abnormal proliferation in cancer of other tissue origins, as epigenetic silencing of *EBF3* occurs in cancer cell lines derived from the breast, colon, bone, and liver. Consistent with epigenetic silencing of the *EBF3* locus, *EBF3* expression can be reactivated by treating cells with 5-aza-2′-deoxycytidine, a demethylating agent, and trichostatin A, an inhibitor of histone deacetylases. Ectopic expression of *EBF3* causes cell-cycle arrest and apoptosis in cancer cells through regulating the expression of genes involved in cell-cycle control. Specifically, EBF3 directly activates genes encoding the *Cip/Kip* family of inhibitors of cyclin-dependent kinases such as $p21^{cip1/kip1}$ and $p27^{kip1}$. Conversely, EBF3 can repress the expression of genes responsible for cell proliferation and survival such as ▶ *cyclins A* and *B*, *CDK2*, *Daxx*, and *Mcl-1*. Therefore, EBF3 may act as a tumor suppressor by regulating expression of specific set of genes.

In mouse B-cell lymphomas, retroviral insertions occur frequently in two genetic loci encoding Evi3 (ecotropic viral integration site 3) and EBFAZ (EBF-associated zinc finger protein, also known as OAZ). Such viral integration results in heightened expression of EBFAZ or Evi3. These two proteins are highly similar to each other, with each containing 30 Krüppel-type zinc fingers. Via several zinc fingers near its C-terminus, EBFAZ or Evi3 binds to EBF, and it is suggested that overexpression of EBFAZ or Evi3 in B-cell leukemias and lymphomas causes aberrant expression of EBF1 target genes that might contribute to tumorigenesis in B cells. In support of this notion, *Evi3* is significantly expressed in most human acute myelogenous leukemias. *EVi3* expression is abundant in human hematopoietic progenitors and declines rapidly during cytokine-driven differentiation. Interestingly, Evi3 or EBFAZ could either repress or activate EBF-mediated transcription, depending on cell type and gene promoter. Therefore, the precise implications of the EBFAZ–EBF pathway in cancer development remain to be determined. Finally, focal deletions of the*EBF1* locus have been detected in significant cases of B-progenitor ALL (acute lymphoblastic leukemia).

References

Liberg D, Sigvardsson M, Akerblad P (2002) The EBF/Olf/Collier Family of transcription factors: regulators of differentiation in cells originating from all three embryonal germ layers. Mol Cell Biol 22:8389–8397

Zardo G, Tiirikainen MI, Hong C et al (2002) Integrated genomic and epigenomic analyses pinpoint biallelic gene inactivation in tumors. Nat Genet 32:453–458

Zhao LY, Niu Y, Santiago A et al (2006) An EBF3-mediated transcriptional program that induces cell cycle arrest and apoptosis. Cancer Res 66:9445–9452

Early Detection

▶ Prostate Cancer Diagnosis

Early Genes of Human Papillomaviruses

Massimo Tommasino[1], Rosita Accardi[1] and
Uzma Hasan[2]
[1]Infections and Cancer Biology Group,
International Agency for Research on Cancer,
Lyon, France
[2]CIRI, Oncoviruses and Innate Immunity,
INSERM U1111, Ecole Normale Supérieure,
Université de Lyon, CNRS-UMR5308, Hospices
Civils de Lyon, Lyon, France

Definition

Human Papillomavirus (HPV) 16 has six early
genes that are transcribed from the same DNA
strand. As in other viruses, the function of the
early proteins is to alter several cellular events to
guarantee the completion of the virus life cycle. In
addition, three early proteins, E5, E6, and E7, are
also involved in the induction of malignant transformation of the infected cells. The cutaneous
HPV types have a similar organization of the
early region, with exception that the majority of
these types lack the E5.

Characteristics

HPV constitutes of a heterogeneous group of
viruses from the *Papillomaviridae* family. The
HPV phylogenetic tree has been designed based
on the homologous nucleotide sequence of the L1
capsid protein. So far, 92 HPV types have been
fully sequenced, of which 60 belong to the alpha
genera and the other 32 belong to the beta and
gamma genera. Based on their tissue tropism,
HPVs can be divided in cutaneous and mucosal
HPV types. The mucosal HPV types are included
in the genus alpha together with certain benign
cutaneous HPV types, whereas the beta and the
gamma genera exclusively consist of cutaneous
HPV types. Biological and epidemiological studies have clearly demonstrated that certain mucosal

HPV types are associated with ▶ Cervical cancer.
On the contrary, the role of cutaneous types in
carcinogenesis is still under debate, although several lines of evidence support their association
with nonmelanoma skin cancer (NMSC). The
mucosal HPV type 16 (HPV16) is the most frequently found HPV genotype in cervical cancers
worldwide, and thus its early gene products are
the best studied and characterized (Fig. 1).

*HPV16 E6 and E7 are the major transforming
proteins*. Three different lines of evidence have
demonstrated the involvement of E6 and E7 in
cervical carcinogenesis:

1. The first indication came from the analysis of
   HPV-infected cells, which showed that viral
   DNA is randomly integrated in the genome of
   the majority of cervical carcinomas. Integration leads to the disruption of several viral
   genes with preservation of only the E6 and
   E7, which are actively transcribed.
2. The discovery that E6 and E7 proteins are able
   to induce cellular transformation in vitro confirmed their oncogenic role. Immortalized
   rodent fibroblasts can be fully transformed by
   expression of HPV16 E6 or E7 protein. These
   rodent cells acquire the ability to grow in an
   anchorage-independent manner and to be
   tumorigenic when injected into nude mice. In
   addition, HPV16 E6 and E7 together are able
   to immortalize primary human keratinocytes,
   the natural cellular host of the virus. In agreement with the in vitro assays, transgenic mice
   coexpressing both viral genes exhibit epidermal hyperplasia and various tumors. Similar to
   the mucosal HPV types, E6 and E7 from certain cutaneous HPV types of the genus beta
   (e.g., HPV8 and 38) display transforming
   properties in in vitro and in vivo models. Independent studies have demonstrated that
   HPV16 E6 and E7 proteins do not induce cellular transformation via a "hit and run" mechanism, but continuous expression of both
   proteins is required for the maintenance of the
   malignant phenotype.
3. Finally, biochemical studies have clarified the
   mechanism of action of E6 and E7. The viral

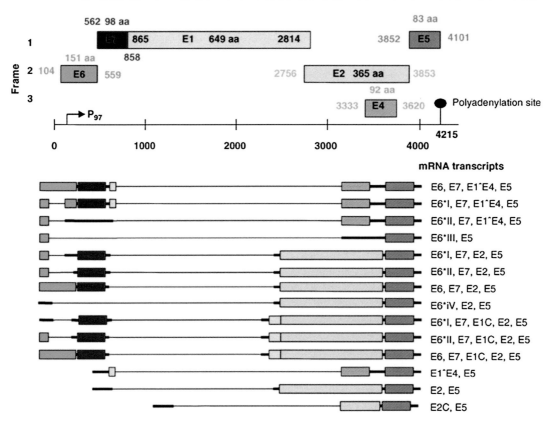

Early Genes of Human Papillomaviruses, Fig. 1 Organization of the early region of HPV16. The six early genes are represented by using different colors. The nucleotide positions of each early gene and the predicted size in amino acids are also shown. Several polycistronic transcripts have been identified, which comprise of 2–3 early genes in different combinations and are most likely transcribed from the P_{97} promoter. The function of each transcript is not known, but some of them are only transcribed at different stages of differentiation. Several lines of evidence indicate that the alternative splicing in E6 and E7 transcripts play an important role in the translational regulation of the viral proteins

oncoproteins are able to form stable complexes with cellular proteins and alter, or completely neutralize, their normal functions. These events lead to the loss of control of cell cycle checkpoints, apoptosis, and cellular differentiation.

E6 protein. HPV16 E6 is a small basic protein of 151 amino acids. The major structural characteristic of E6 is the presence of two atypical zinc fingers. At the base of these zinc fingers are two motifs containing two cysteines (Cys-X-X-Cys), which are conserved in all E6 HPV types. The best characterized HPV16 E6 activity is its ability to induce degradation of the tumor suppressor protein p53 via the ubiquitin pathway. This cellular

protein is a transcription factor that can trigger cell cycle arrest or apoptosis in response to stress or DNA damage. E6 binds to a 100 kDa cellular protein, E6AP (E6-protein), which functions as an ubiquitin protein ligase (E3). The E6/E6AP complex then binds p53, which becomes very rapidly ubiquitinated and as a consequence is targeted to proteasomes for degradation. Since the major role of p53 is to safeguard the integrity of the genome by inducing cell cycle arrest or apoptosis, cells expressing HPV16 E6 show chromosomal instability, which greatly increases the probability that HPV-infected cells will evolve toward malignancy. Additional findings have demonstrated that HPV16 E6 also associates with the transcriptional regulators, CBP and

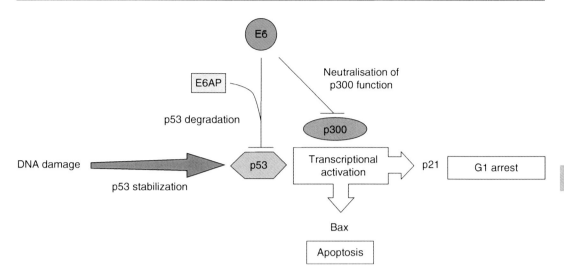

Early Genes of Human Papillomaviruses, Fig. 2 p53 pathways targeted by HPV16 E6. When cells are exposed to DNA-damaging agents, e.g., X-rays, the half-life of p53 can be greatly increased by posttranslational modifications (phosphorylation). In turn, p53 can either activate the transcription of the cyclin-dependent kinase (CDK) inhibitor, $p21^{WAF1/CIP1}$, leading to a G1 arrest and DNA repair before replication, or activate the transcription of the proapoptotic gene Bax with consequent induction of apoptosis. Cells expressing HPV16 E6 protein are resistant to the cell cycle arrest or apoptosis induced by DNA-damaging agents

p300, with resulting inhibition of p53-driven transcription. Thus, HPV16 E6 neutralizes p53 by two distinct mechanisms; the first is mediated by the p300/CBP association, while the second occurs via binding to E6AP to promote p53 degradation (Fig. 2).

Interestingly, the E6 protein from cutaneous HPV types is not able to induce degradation of p53. In fact, it has been shown that the beta HPV38 can inactivate p53 function by inducing accumulation of its antagonist, ΔNp73. HPV16 E6, as well as E6 from certain cutaneous HPV types, can also interfere with the apoptotic pathways via its association with Bak, a member of the Bcl-2 family. Analogously to its effect on p53, E6 induces Bak degradation via the ubiquitin-mediated pathway.

Several p53-independent cellular pathways, which are altered by the E6 molecule, have been identified. HPV16 E6 is able, through its association with E6AP, to promote the degradation of the transcriptional repressor NFX1-91 and consequently to activate the transcription of the *hTERT* (*human telomerase reverse transcriptase*) gene encoding the catalytic subunit of the telomerase complex. This effect directly results in telomerase activity upregulation, a key event in the immortalization of primary keratinocytes. In addition, HPV16 E6 is able to interfere with cell mobility, through interaction with the human homologue of the *Drosophila* discs large protein (DLG). Also in this case, E6 binding leads to degradation of the cellular protein. However, different E6 domains are required to induce degradation of p53 and DLG. Deletion of the carboxy terminus of HPV16 E6 abolishes its binding to DLG without influencing its ability to promote p53 destabilization.

Another cellular target of HPV16 E6 is paxillin, a protein involved in transducing signals from the plasma membrane to focal adhesions and the actin cytoskeleton. The fact that E6 from the oncogenic HPV16, but not E6 from the low-risk HPV types 6 and 11, is able to bind paxillin suggests that this interaction has a role in the carcinogenesis of HPV infection. Several factors have been described to minimize or prevent exposure of HPV to the immune system. It has also been shown that HPV16 E6 interacts with interferon regulatory factor-3 (IRF-3), a positive transcriptional regulator of the INF-β promoter, which is activated in response to virus infection.

E6 binding inhibits IRF-3 transactivation function. Thus, this E6-induced event may enable the virus to circumvent the antiviral response of the infected cell. The adhesion between keratinocytes and antigen presenting cells, i.e., Langerhans cells, in the epidermis is mediated by E-cadherin. It has been shown that E6 can reduce the levels of cell surface E-cadherin on keratinocytes, thereby limiting the presentation of viral antigens to the Langerhans cells and promoting viral survival. Furthermore, HPV16 E6 has been shown to suppress innate immune responses mediated by a family of toll-like receptors (TLRs) that are key sensors of evading pathogens. E6 is able to block the promoter of TLR9 which has been identified to recognize dsDNA sequences. Finally, E6 associates with ERC 55, a putative calcium-binding protein located in the endoplasmic reticulum. However, the biological significance of this interaction is still unclear.

E7 protein. HPV16 E7 is an acidic phosphoprotein of 98 amino acids, which is structurally and functionally related to a gene product of another DNA tumor virus, the adenovirus E1A protein. On the basis of the similarity in primary structure between the two viral proteins, they can be divided into three domains: conserved region 1–3 (CR1–3). Mutational analysis of HPV16 E7 has demonstrated that all three regions are important for the in vitro transforming activity of the molecule. CR3 contains two CXXC motifs involved in zinc binding and essential for the stability of the protein. Independent studies have demonstrated that this viral protein is located in the nucleolus, nucleus, and cytoplasm. Indeed, it has been shown that the E7 molecule associates with cytoplasmic and nuclear proteins. The best understood interaction of E7 with a cellular protein is that involving the "pocket" proteins, pRb, p107, and p130. The pocket proteins are central regulators of cell cycle division. They negatively regulate, via direct association, the activity of several transcription factors, including members of the E2F family (E2F1–5), which are associated with their partners, DPs. Under normal cell cycle regulation, phosphorylation of pRb, which is mediated by cyclin-dependent kinase (CDK) activity, leads to the disruption of pRb/E2F complexes, with consequent activation of E2Fs. HPV16 E7 binds the pocket proteins and, analogously to the phosphorylation, results in the release of active E2Fs which in turn activate the transcription of a group of genes encoding proteins essential for cell cycle progression, such as cyclin E and cyclin A. As described for the interaction between E6 and p53, HPV16 E7 protein is able to promote the destabilization of pRb through the ubiquitin–proteasome pathway (Fig. 3). Similarly, the beta HPV38 E7 binds to pRB and promotes its degradation. This property is not shared by all E7s from the different HPV genotypes. Indeed, E7 from the benign HPV1 can efficiently associate with pRb without inducing its degradation. It is likely that the E7-induced pRb degradation represents a more effective way to neutralize the function of the cellular protein. The other two members of the pocket protein family, p107 and p130, are involved in controlling additional cell cycle checkpoints; p130 exerts its transcriptional regulatory function during the G0/G1 transition, while p107 is active in the G1/S transition and in the G2 phase. Analogously to pRb, HPV16 E7 protein associates with p107 and p130, inactivating key cell cycle checkpoints.

Besides targeting the pocket proteins, E7 can alter cell cycle control by additional mechanisms. The HPV16 E7 protein is able to associate with the CDK inhibitors $p21^{WAF1/CIP1}$ and $p27^{KIP1}$ causing neutralization of their inhibitory effects on the cell cycle. Cells coexpressing HPV16 E7 and $p21^{WAF1/CIP1}$ or $p27^{KIP1}$ are still able to enter S phase, while in the absence of E7 cells are arrested in G1 phase. HPV16 E7 can also directly and/or indirectly interact with cyclin A/CDK2 complex. The biological function of this interaction remains to be elucidated, but it is possible that E7 may act by redirecting the kinase complexes to a different set of substrates.

Other cellular proteins involved in transcriptional regulation have been identified as HPV16 E7 targets. HPV16 E7 binds the TATA box-binding protein (TBP) and the TBP-associated factor TAF110, indicating that the viral protein is able to interfere with the basic transcriptional machinery of the host cell. Furthermore, E7

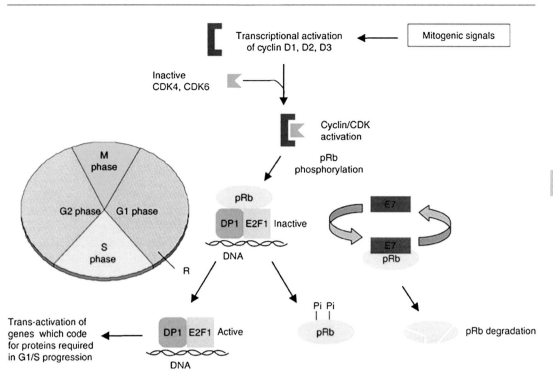

Early Genes of Human Papillomaviruses, Fig. 3 Deregulation of the restriction point (R) by HPV16 E7. E2F transcription factors form heterodimer complexes with members of the DP family and regulate the transcription of several genes during the cell cycle. In quiescent cells, pRb is present in a hypophosphorylated form and associates with E2F molecules, thereby inhibiting their transcriptional activity. When quiescent cells are exposed to mitogenic signals, genes encoding the G1-specific D-type cyclins (D1, D2, and D3) are activated. Subsequently, cyclins associate with a catalytic subunit, CDK4 or 6, and after transport into the nucleus, the kinase complexes phosphorylate pRb in mid-G1 phase causing release of active E2F/DP1 heterodimer complexes and progression through the restriction point (R). E7 binding to pRb mimics its phosphorylation. Thus, E7-expressing cells can enter S phase in the absence of a mitogenic signal

associates with AP1 complex, activating its transcriptional activity.

Similar to other stimulators of proliferation (e.g., c-myc or adenovirus E1A), the HPV16 E7 protein, besides the ability to deregulate the cell cycle, also promotes apoptosis. Expression of HPV16 E7 in normal human fibroblasts (NHF) or in human keratinocytes results in a cytocidal response, which displays the typical features of apoptosis and is much more evident in the absence of mitogenic signals. This E7-induced apoptosis requires pRb inactivation and is mediated by p53-dependent and -independent pathways. It is likely that the E7-induced apoptosis represents a cellular response elicited by the loss of cell cycle control. Interestingly, E6 protein is able to completely abrogate the E7 activity in promoting apoptosis. Thus, both viral proteins are required to

induce transformation of the host cells. Although E7 is constitutively expressed in HPV16-associated lesions and therefore appears as candidate antigen for a specific immune response, the immune system fails to produce an efficient defense against tumor outgrowth in affected patients. As in the case of E6, E7 has also evolved to escape immune surveillance. E7 has the ability to bind and prevent the transcription factor IRF-1 from activating the INF-α and INF-β promoters. E7 is also capable of downregulating innate responses via suppressing TLR9 expression. Furthermore, in HPV16 E6 and E7 transgenic mice, the transgene product E7 does not induce an immune response. However, upon vaccination with E7, anti-E7 antibodies were produced without causing signs of autoimmune disease. In contrast, E7-specific cytotoxic T lymphocytes (CTL) were not detected after immunization. Therefore, the E7 transgene expression induces specific immunological tolerance at the CTL level.

E5 protein plays an early role in the HPV-induced transformation. Studies on ▶ bovine papillomavirus (BPV) have provided evidence that E5 is a potent oncoprotein. It has also been shown that HPV16 E5 is able to induce cellular transformation, although with less efficiency than BPV E5. HPV16 E5 is a small hydrophobic protein, which is located in the endoplasmic reticulum (ER), nuclear membrane, and cytoplasmic vesicles. E5 is able to enhance growth factor-mediated signal transduction to the nucleus, resulting in stimulation of cellular proliferation. BVP1 and HPV16 E5 associate with the 16 kDa subunit c of the vacuolar H + -ATPase, which is responsible for acidification of membrane-bound organelles, such as Golgi, endosomes, and lysosomes. It has been shown that BPV1 E5 induces alkalinization of Golgi, and that this activity is linked to its in vitro transforming activity. Mutations in E5 that abolish the interaction with 16 kDa subunit c abrogate Golgi alkalinization and cellular transformation. HPV16 E5-mediated immune evasion also involves suppressing the expression of the major histocompatibility complex class I (MHC I) and antigen processing via the TAP pathway, reflecting the lack of antigen presentation to CTL.

Since the integration of viral DNA, which occurs in tumor cells, results in a loss of *E5* gene expression, it is clear that E5 is involved in early events during the multistep process of cervical carcinogenesis and that its function is no longer required after the establishment of the transformed phenotype.

E1 and E2 are involved in the regulation of viral DNA replication. The function of E1 is to control the replication of viral DNA. E1 contains the cyclin-binding RXL motif and is able to associate with cyclin E and A. Consistent with these findings, E1 is phosphorylated by cyclin E or A-associated kinase. Mutation of E1 phosphorylation sites results in a reduction of HPV DNA replication, supporting the idea that the E1/cyclin association plays an important role in viral DNA replication. Moreover, it has been shown that E1 has an ATPase and helicase activity. E1 forms a stable complex with E2 and binds to the replication origin of HPV in order to recruit cellular factors essential for DNA replication. Furthermore, E1 has been found associated with components of the cellular DNA replication machinery, e.g., DNA polymerase α.

E2 regulates the transcription of the early genes. In addition to controlling viral DNA replication together with E1, E2 is able to negatively or positively regulate the transcription of the early genes. Like all transcription regulatory factors, E2 has an amino-terminal transacting domain and a carboxy-terminal DNA-binding domain, which recognizes four *cis* elements (ACCN6GGT) in the long control region (LCR) of the HPV genome. These two domains are separated by a central region (hinge), which is important, together with the amino-terminal domain, for the nuclear localization of the molecule. Whether E2 binding results in repression or activation of the promoter of the early genes is dependent on the position of the E2-binding site in the LCR. E2 binding to the promoter-distal or -proximal elements leads to a positive or negative regulation of the promoter, respectively.

E4 is probably involved in virus maturation and/or replication. E4 is a late protein expressed from the early region of the genome. Most of the studies have been performed on E4 from the cutaneous HPV type 1. In these lesions E4 is present at very high levels and several E4-derived proteins have been detected. The primary product is a 17 kDa protein, which is expressed from E1^E4 transcript. E4 associates with and disrupts the cytoplasmic keratin network. The biological significance of this E4-induced event is not fully understood. It has been proposed that E4 plays a role in the productive phase of the infection establishing a favorable condition for viral maturation.

Clinical Relevance

HPV16 infection results in the induction of a benign proliferation, which, after a long latent period, can progress to invasive cancer. Persistent HPV infection is necessary for the development of the malignant lesion. This requirement is explained by the fact that viral proteins, in order to induce full malignant transformation of the host cells, have to cooperate with an activated cellular oncogene. Accumulation of mutations in cellular genes, which possibly lead to activation of an oncogene, requires continuous proliferation. This is achieved by the abilities of E6 and E7 to respectively neutralize apoptotic pathways and to induce unscheduled proliferation. Therefore, a possible approach to induce regression of an HPV-positive lesion is to target the biological functions of E6 and E7. This possibility is supported by findings, which clearly demonstrate that continuous expression of the two viral genes is necessary for the maintenance of the host cell transformed phenotype. Thus, we can predict that a blocking of the activity of E7 can lead to a rapid exit from the cell cycle. Neutralization of E6 function should result in an even more efficient way to induce regression of the HPV lesion. As described above, E7 has a dual activity, being able to induce proliferation and apoptosis. E6, acting upon p53-dependent and -independent pathways, completely abolishes the E7-induced apoptosis. Thus, we could imagine that in cells expressing *E6* and *E7* genes, the block of only E6 functions may push the balance between proliferation and apoptosis in favor of the latter causing regression of the HPV lesion.

Alternative targets are E1 and E2, which are involved in viral transcription and replication. Several approaches to neutralize the early viral proteins are under investigation. These include strategies to block the transcription or translation of viral genes or to identify small molecules able to specifically associate with and inactivate the viral proteins.

In addition, HPV E6 and E7 are able to downregulate the innate and adaptive immunity. Identification of strategies to reactivate the immune response in HPV-infected cells may favor the clearance of the infection preventing the development of cervical diseases.

References

Campo MS (2006) Papillomavirus research: from natural history to vaccine and beyond. Caister Academic Press, Norfolk

Munger K, Baldwin A, Edwards KM et al (2004) Mechanisms of human papillomavirus-induced oncogenesis. J Virol 78:11451–11460

O'Brien PM, Campo MS (2003) Papillomaviruses: a correlation between immune evasion and oncogenicity? Trends Microbiol 11:300–305

Tommasino M (ed) (1997) Human papillomaviruses in human cancer: the role of E6 and E7 oncoproteins. Molecular Biology Intelligence Unit, Landes Company, Austin

zur Hausen H (2002) Papillomaviruses and cancer: from basic studies to clinical application. Nat Rev Cancer 2:342

Early-Stage Ovarian Cancer

Behrouz Zand and Ralph S. Freedman
UT MD Anderson Cancer Center, Houston, TX, USA

Synonyms

Stage IA–C ovarian cancer

Definition

Early-stage ► ovarian cancer generally refers to malignancy of the ovary that is confined to one or both ovaries (International Federation of Gynecology and Obstetrics [FIGO] stage I A and B).

- Stage IA is growth limited to one ovary.
- Stage IB is growth limited to both ovaries.

Once the tumor involves the surface of an ovary, or there is capsule rupture, or presence of ascites or pelvic washings containing malignant cells (IC), the cancer is more advanced and prognosis for the patient is less favorable.

Characteristics

► Ovarian cancer is a relatively common gynecologic malignancy. Approximately 90% of ovarian cancers are of the epithelial cell type developing from the ovarian surface epithelium. Epithelial ovarian cancer peaks in incidence between 55 and 65 years of age and over 60% has serous histology. Less common epithelial cell cancers include primary mucinous, endometrioid, clear cell, and undifferentiated carcinomas, which comprise about 20–30% of the epithelial types. Non-epithelial types of ovarian cancer including germ cell tumors and sex cord stromal tumors comprise about 10–20% of all ovarian cancer cases. Non-epithelial ovarian cancers have overall a better prognosis than epithelial types and also occur at higher proportions in younger patients. Non-epithelial ovarian cancers mainly include germ cell types (dysgerminoma, immature teratoma, endodermal sinus tumor) and sex cord stromal cell types (▶ granulosa cell tumors, Sertoli-Leydig cell tumor). Overall they have a better prognosis than the epithelial types in part because they tend to be diagnosed at an early stage. They are also more commonly seen in women in their reproductive years. The non-epithelial malignancies particularly the germ cell tumors are more chemosensitive. A variety of other rare types of non-epithelial ovarian cancers including sarcomas, lipoid cell, and small cell or ▶ neuroendocrine tumors have poor prognosis but only contribute to 0.1% of all ovarian cancers.

Unfortunately, most epithelial ovarian cancers are diagnosed at a more advanced stage (stage II to IV). This makes ovarian cancer the most lethal gynecological cancer, except in developing countries where ▶ cervical cancer remains a major contributor to cancer deaths among women. Approximately 20% of patients with epithelial ovarian cancer can be diagnosed at an earlier stage of disease. Surgery is usually the first component in the treatment of most ovarian cancers of any histological type. Patients who are suspected of having an ovarian malignancy should undergo surgical exploration by a gynecological oncologist to confirm the diagnosis and to determine the stage and histological grade, followed by a rigorous effort to remove as much of the malignancy as possible. The type and duration of any required subsequent chemotherapy are determined in most cases by the accuracy and completeness of the surgical procedure and the histological type and grade. In both early-stage epithelial and non-epithelial cancers, there are certain situations where postoperative chemotherapy may not be indicated, hence the importance of obtaining adequate staging and pathological assessment in early-stage disease.

Preoperative Workup of an Adnexal Mass

The diagnosis of ovarian cancer remains a clinical and technological challenge despite the development of some new approved tests. Patients found to have a pelvic mass on physical examination should be evaluated with transvaginal pelvic sonography and a measurement performed of their ▶ serum biomarkers cancer antigen 125 (CA125) levels. CA125 is a mucin-like protein of high molecular mass estimated at 200–20,000 kDA. CA125 cell surface expression is upregulated when cells undergo metaplastic differentiation into a Müllerian epithelium. CA125 is the most extensively studied biomarker for possible use in ▶ ovarian cancer ▶ early detection. Its expression is elevated in some cases of endometriosis. Characteristics of masses that are more suggestive of malignancy include: size greater than 8 cm, a solid as opposed to a uniform cystic, consistency, immobility with an irregular shape, bilaterality, and the presence of ▶ ascites. A malignancy may be found with any one of these findings. A serum CA125 level of >35 U/mL combined with these features would suggest malignancy, although an elevated CA125 on its own is not definitive, as an elevated CA125 value can also occur commonly in patients with a variety of benign conditions such as ovarian hemorrhagic cysts, endometriosis, uterine fibroids, adenomyosis, ectopic pregnancy, pelvic inflammatory disease, appendicitis, and colitis. Specificity and positive predictive values for CA125 level measurements are consistently higher in postmenopausal women compared with premenopausal women in whom there is a higher frequency of functional and nonfunctional ovarian cystic or tubal inflammatory conditions that could confound the diagnosis. In a postmenopausal woman, any CA125 elevation accompanied by a palpable or transvaginal ultrasound documented pelvic mass should increase the suspicion for pelvic malignancy although there are medical conditions such as hypothyroidism, congestive heart failure, and cirrhosis that can sometimes contribute to false-positive results. In some instances it may be necessary to repeat the CA125 test for confirmation. The same assay and laboratory should be used for all repeat values. Another *caveat* related to CA125 is the fact that only 50% of women with stage I disease have an elevated CA125; therefore, a patient with a normal CA125 and adnexal mass does not exclude an early ovarian malignancy. This serves to tell us that there is no substitute for a good history, physical examination, and clinical judgment. Most early ovarian cancers are asymptomatic. Some non-epithelial tumors can present with pain due to tumor complications such as hemorrhagic necrosis or torsion.

Additionally, β-hCG, L-lactate dehydrogenase (LDH), and ▶ alpha-fetoprotein (AFP) levels may be elevated in the presence of certain malignant germ cell tumors, while inhibin A and B levels can be elevated in some granulosa cell tumors of the ovary. These latter markers should also be measured when surgical removal of an ovarian mass is being considered in any women during the reproductive period.

Surgical Staging

If the suspicion of malignancy is high, a computed tomography (CT) scan of the abdomen and pelvis with intravenous and rectal contrast can be helpful preoperatively for identifying the extent of disease in the abdomen and pelvis. It can also provide useful information for discussion with the patient regarding the nature and extent of surgical procedures that may be needed as well as any potential fertility-sparing procedures if appropriate. It is recommended that a gynecologic oncologist be involved in these discussions and performs the surgical procedures whenever possible. Some oncologists find it helpful to place ureteric catheters on the day of surgery to aid in the location of the ureters, especially where the mass is very large and appears to be immobile or there has been a prior history of endometriosis or pelvic inflammatory disease. Surgical pathological staging should be undertaken regardless of the patient's age or

desire for fertility. The diagnosis can usually be confirmed at frozen section on the affected ovary removed. It may happen that a younger patient desirous of preserving her fertility status has to undergo emergent surgery in a situation where either surgical expertise or frozen section and gynecologic pathology expertise are unavailable. In such a situation where there is also a suspicion of malignancy, the affected ovary should be removed completely for diagnosis and the patient referred to a center of expertise. There is not universal agreement as to the extent of the surgical staging that is required in early ovarian cancer. At a minimum the procedure should include peritoneal washings; careful inspection of the abdominal cavity to search for indications of spread outside of the ovaries, metastatic deposits which should be biopsied; and random peritoneal biopsies from the pelvis, paracolic gutters, the diaphragm, and the omentum. Either pelvic and para-aortic lymph node sampling or full ▶ lymphadenectomy can provide additional information that can result in upstaging if lymph node metastases are found. A woman who has completed her family should undergo a total hysterectomy and bilateral salpingo-oophorectomy in addition to the surgical staging.

Premenopausal patients who are interested in fertility conservation should be offered the option of unilateral salpingo-oophorectomy along with routine staging, with preservation of the contralateral ovary and the uterus – provided that these appear normal at the macroscopic level.

In early-stage disease, it is very important to establish that the disease has not spread beyond the ovaries or the pelvis since accurate staging will help determine the prognosis for the individual patient, the type and intensity and duration of postoperative chemotherapy, and in certain cases whether adjuvant chemotherapy is even indicated. Today with appropriate staging and appropriate adjuvant chemotherapy, 90% of patients with stage I ovarian cancer generally survive 5 years and more. ▶ Meta-analysis of the largest and best conducted randomized trials has confirmed the beneficial effects of adjuvant chemotherapy in "early-stage" ovarian cancer, but any benefit of chemotherapy after "optimal" surgical staging on progression-free or overall survival could not be shown. Optimal staging, even when defined, was performed in only a minority of these patients; thus, these studies were not designed or powered to address the question. A reasonable default position would be to offer adjuvant chemotherapy treatment to patients who were adequately staged and who have high risk factors (stage IA or B, with grade 3 or clear cell cancers, or stage IC) and to offer restaging to those patients whose staging were suboptimal or accept a higher bar for not using chemotherapy in such patients or to initiate treatment at first recurrence.

Although staging (▶ Staging of Tumors) for early-stage epithelial ovarian cancer has traditionally been performed by open laparotomy, there is an increasing body of evidence supporting the alternative use of a laparoscopic procedure. These studies have shown that laparoscopic staging and laparotomy staging result in similar surgical outcomes and staging adequacy in terms of nodal yield, omental size, and accuracy of identifying metastatic disease and safety. At present there are no prospective, collaborative trials comparing the use of laparoscopy staging with laparotomy staging for presumed early-stage epithelial ovarian cancer; therefore, it is difficult to draw any definite conclusions about laparoscopic staging in early ovarian cancer. In the hands of suitably trained and experienced laparoscopic gynecologic oncologists, there is good reason to believe that adequate and safe staging and surgical treatment by interventional laparoscopy may become possible in the future.

First-line adjuvant chemotherapy for epithelial cell ovarian cancer generally includes a combination of carboplatin and a taxane, either ▶ paclitaxel or ▶ docetaxel, usually for six cycles. In the USA both drugs are frequently utilized in early-stage disease. In Europe and the UK, single-agent carboplatin may be preferred because of concerns regarding toxicity particularly chronic toxicity such as peripheral neuropathy. Some have suggested that the number of cycles of chemotherapy should be reduced from six to three; there are no adequately powered studies to determine

which is better for early-stage disease. Since some patients will fail either due to understaging or to resistance to chemotherapy and ovarian cancer recurrences are rarely curable, a reasonable rationale is provided for treating patients with the standard six cycles that is used for higher-stage tumors. Monitoring patients closely for development of toxicities that could become irreversible from the chemotherapy is important.

Non-epithelial Ovarian Cancer: Surgical Staging and Treatment

Germ Cell Tumors

Dysgerminoma is the most common germ cell malignancy and accounts for 30–40% with 65% in stage I. Dysgerminomas have a higher rate of developing bilaterally (10–15%) than other germ cell tumors; therefore, careful inspection of the contralateral ovary for tumor is mandatory. Careful surgical staging should be done to determine the extent of occult metastatic disease. The minimum treatment is a unilateral salpingo-oophorectomy. If future fertility is desired, then the contralateral ovary, tube, and uterus can be left in situ even when metastasis is present, as these tumors are highly chemosensitive. Five percent of dysgerminomas occur in patients with abnormal XY karyotype (i.e., gonadal dysgenesis). In the latter patients, both gonads should be removed although the uterus may be left in situ for possible embryo transfer. Any suspicious lesion on the contralateral ovary should be biopsied and resected if found to be malignant (preserving some normal ovary) due to higher rate of bilaterality, but routine biopsy or wedge resection of a normal contralateral ovary is not indicated. A patient with stage IA dysgerminoma has a 5-year survival of 95% after a unilateral salpingo-oophorectomy alone.

Immature Teratoma

These also require full staging and can be treated with fertility-sparing surgery or total hysterectomy and bilateral salpingo-oophorectomy. Very few are bilateral and therefore contralateral biopsy is not indicated. Typically these tumors occur in combination with other germ cell tumors as mixed germ cell tumors, and less than 1% account for pure immature teratoma in all ovarian cancers. Overall the 5-year survival rate for patients with all stages of pure immature teratomas is 70–80% and 90–95% for patients with surgical stage I lesions.

Endodermal Sinus Tumor

EST; Also called "yolk sac" tumors, ESTs have a median age at diagnosis of 18 years. The primary treatment consists of unilateral salpingo-oophorectomy with frozen sections and removal of any gross disease. As with other germ cell tumors, fertility-sparing surgery is appropriate and all patients will need chemotherapy.

The malignant germ cell tumors as a group, dysgerminoma, immature teratoma, and endodermal sinus tumor, are very sensitive to a variety of chemotherapy combinations such as BEP (► bleomycin, ► etoposide, ► cisplatin), VBP (vinblastine, bleomycin, cisplatin), and VAC (vincristine, actinomycin D, ► cyclophosphamide).

Sex Cord Tumors

Granulosa cell tumors (GCT) are bilateral in only 2% of patients, so a unilateral salpingo-oophorectomy is indicated for stage IA tumors in children or in women of reproductive age. Surgical staging should be performed, and any suspicious ovarian lesion or enlargement should be biopsied. In postmenopausal women or those who do not desire future fertility, total abdominal hysterectomy-bilateral salpingo-oophorectomy (TAH-BSO) should be performed. If uterus is left in situ, preoperative pathologic assessment of the endometrium should be performed to exclude endometrial carcinoma or atypical hyperplasia as GCT typically secrete unopposed estrogen that can affect the endometrium. The prognosis of GCT is dependent on the stage. Stage I patients may be cured. However, late recurrences are sufficiently frequent to warrant close follow-up. Effective adjuvant

chemotherapy for early-stage disease has not been determined.

Sertoli-Leydig Cell Tumors

These occur in the 3rd and 4th decades of life and account for 0.2% of all ovarian cancers. They are typically low-grade malignancies, although some may be poorly differentiated. These tumors secrete androgens and can lead to virilization in 70–85% of patients. These tumors have a less than 1% chance of being bilateral, and so the treatment of choice is a unilateral salpingo-oophorectomy and evaluation of contralateral ovary in women in the reproductive period. Postmenopausal women may undergo TAH-BSO. The 5-year survival rate is 70–90% and recurrences are uncommon unless it is not poorly differentiated.

Fertility Preservation

In young patients with non-epithelial ovarian carcinoma or low malignant potential epithelial ovarian tumors (LMPT) confined to a single ovary, the surgical treatment should in most cases be conservative with preservation of the uterus and contralateral normal-appearing ovary even with invasive implants that are localized to the affected ovary. Routine biopsies of the macroscopically normal-appearing contralateral ovary identify a low rate of occult disease (<2%). Furthermore, routine ovarian biopsies can contribute to adhesions that might affect tubal function or to a painful ovarian entrapment syndrome. Ovarian biopsies should therefore be limited to suspicious lesions of the contralateral ovary when fertility preservation is being considered.

Fertility preservation of early-stage epithelial cancer has become more accepted since studies have shown excellent outcomes for adequately staged low-risk patients (stage IA and grade 1 or possibly 2). Other requirements for the conservational surgical approach include younger age patients with low parity; encapsulated tumor with no adhesions; no invasion of the capsule, lymphatics, or mesovarium; favorable histopathologic features; negative peritoneal washings; and close postoperative surveillance. The biological behavior of these early-stage epithelial tumors has only recently become better understood, and therefore, only relatively small numbers of successful pregnancy outcome reports are in the literature. It might therefore be useful to establish a national or international registry of patients with successful pregnancy outcomes.

Conclusion

Early-stage ovarian cancer following optimal surgical staging and pathological evaluation can contribute to excellent survival outcomes for most epithelial and non-epithelial ovarian cancers. Access to complete information enables more informed decision making for patients and their doctors; this is of particular importance when a conservative approach to surgery is indicated or considered feasible. Fertility-sparing surgery might be considered for young nulliparous patients or those desirous of keeping their reproductive potential, with stage IA grade 1 or 2 epithelial cancers and most non-epithelial tumors. Patients with more advanced disease would in general be counseled against conservative surgery unless they had non-epithelial tumors. It is equally important that all patients with early ovarian cancer, as with more advanced disease, have access to the best surgical and pathological expertise to ensure their best outcomes in terms of both survival and life quality.

References

ACOG (2007) ACOG practice bulletin number 83: management of adnexal masses. Obstet Gynecol 110:201–214

Berek JS, Hacker NF (2005) Practical gynecologic oncology, 4th edn. Lippincott Williams & Wilkins, Philadelphia

Eifel PJ, Gershenson DM, Kavanagh JJ, Silva EG (2006) In: Buzdar AU, Freedman RS (eds) MD Anderson Cancer Care Series: gynecologic cancer. Springer, New York

Park JY, Bae J, Lim MC et al (2008) Laparoscopic and laparotomic staging in stage I epithelial ovarian cancer:

a comparison of feasibility and safety. Int J Gynecol Cancer 18:1202–1209

Winter-Roach BA, Kitchener HC, Dickinson HO (2009) Adjuvant (post-surgery) chemotherapy for early stage epithelial ovarian cancer (Review). The Cochrane Collaboration. Wiley

EBF

▶ Early B-Cell Factors

Ebnerin (RAt)

▶ Deleted in Malignant Brain Tumors 1

E-Box

Britta Mädge
DKFZ, Heidelberg, Germany

Synonyms

E motif

Definition

E-box is the collective term for DNA motifs with the consensus sequence CANNTG. E-boxes appear in a broad variety of promoters and enhancers and serve as protein-binding sites. Proteins with affinity to this motif belong to the basic helix-loop-helix (bHLH) class of transcription factors, which can act as activators of transcription as well as repressors. More than 240 bHLH proteins are known to date in eukaryotes ranging from yeast to human, and the number continues to grow. Their binding specificity depends on both the nature of the "NN" nucleotides and sequences in the vicinity of the E-box. Genes containing E-boxes are activated during important developmental processes, and some are described to be involved in cancer development and/or progression.

Characteristics

The DNA sequence "CAGGTGGC" was originally identified in 1985 within the sequence of the immunoglobulin enhancers, the first enhancers found to be activated in a tissue-specific manner. Later the name E-box (with "E" for enhancer) was used as a common term for all motifs with the consensus sequence "CANNTG," and in 1989 the first two E-box binding proteins, E12 and E47, were identified. Both bind their recognition site as dimers, and dimerization is promoted by the helix-loop-helix (HLH) domain. HLH domains were subsequently found in other transcription factors such as MyoD and Myc proteins.

Genes Containing E-Boxes and the bHLH Proteins That Control Them

To date an overwhelming number of E-box containing genes, regulated by bHLH proteins, have been identified (Table 1), which include:

- Muscle (smooth and skeletal)-specific genes that control myogenesis and are regulated by the bHLHs of muscle regulatory factors (Mrf).
- Genes involved in heart development that are regulated by the gene products of the Hand gene family, Hand1 and Hand2.
- Genes involved in neuronal development and differentiation are regulated by different bHLH proteins such as the Achaete-Scute family, Atonal family, E12/E47 family, and Hen family as well as by Nex, Hairy/E(Spl), and Id (Id proteins).
- Insulin-inducible genes controlling pancreatic development are regulated by a member of the Atonal family, NeuroD. The latter is also a key player in neurogenesis.

- Genes involved in B- and T-cell development, regulated by bHLHs of the E12/E47 family.

Genes and Gene Products That Are Related to Growth Control and Cancer

- The proto-oncogenes BCL6 and FOS that encode transcription factors.
- CDC2 and cyclin B1 gene products, involved in cell-cycle regulation.
- Enzymes cathepsin B and cathepsin D, fatty acid synthase, and HMG-CoA reductase either appear to be involved in the process of tumor cells invading healthy tissue (cathepsins) or seem to support the growth of cancer cells (fatty acid synthase and HMG-CoA reductase).
- Overexpression of the HB-EGF gene, encoding a growth factor, promotes cancer cell growth.
- COX-2 and SP-A gene products may be involved in tumor promotion. Underlying mechanisms are still unknown.

Many E-box-containing genes that are thought to be involved in cancer development provide binding for members of the Myc protein family, which play an important role in many cancer types. Being transcription factors they can activate transcription of the genes *CAD*, *CDC25A*, *cyclin B1*, *ornithine decarboxylase*, *prothymosin alpha*, *RCC1*, *ID2*, and *TERT*.

Members of the USF (upstream transcription factor) protein family also contain the bHLH motif and bind to a variety of genes correlated with cancer. However, to date no oncogenic behavior has been described for these proteins, and since USFs are ubiquitously expressed, they may only play a less specific role in mechanisms of gene regulation (Table 2).

How Do bHLH Proteins Bind to E-Boxes?

bHLH proteins bind to the E-box only as dimers, either as homodimers or heterodimers with other members of the bHLH protein family. The HLH motif within the protein allows dimerization: Two amphipathic α-helices are separated by a short

E-Box, Table 1 Genes that contain E-boxes and that are involved in cell growth control or cancer development

Gene	Function/role in cancer development	E-box regulating protein
BCL6	Proto-oncogene coding for Krüppel-like zinc finger transcription factor, genetic rearrangements frequently found in lymphoma	?
Carbamoyl phosphate synthetase/aspartate carbamoyltransferase/ dihydroorotase (CAD)	Enzyme involved in de novo pyrimidine biosynthesis, essential for cell growth	Myc
Cathepsin B	Protease, overexpression causes ability of cancer cells to metastasize	?
Cathepsin D	Estrogen-induced protease, the same mechanism as cathepsin B but mainly in breast cancer	USF1, USF2
CDC2	Kinase, associates with cyclin A and cyclin B1, regulator of cell-cycle progression through G2 and M phase	USFs, myogenin
CDC25A	Oncogene coding for CDK-activating phosphatase, cell-cycle progression	Myc
c-Fos	Proto-oncogene coding for a transcription factor	E12, E47, myogenin
Cyclin B1	Activates cdc2 kinase, which regulates cell-cycle progression through G2 and M phase, increased expression often found in cancer cell lines	Myc
Cyclooxygenase-2 (COX-2)	Enzyme involved in prostaglandin synthesis in inflammatory processes; overexpression is correlated with tumor promotion	USFs
Fatty acid synthase (FAS)	Main enzyme in lipogenesis, overexpressed in a wide variety of cancer types, mainly breast and prostate cancer, identical with prognostic molecule OA-519	USF
Heparin-binding epidermal growth factor-like growth factor (HB-EGF)	Growth factor, member of the EGF family, postulated role in development of hepatocellular carcinoma, prostate cancer, breast cancer, esophageal cancer, and gastric cancer	MyoD
3-hydroxy-3-methylglutaryl-coenzyme A reductase (HMG-CoA reductase)	Rate-limiting enzyme in isoprenoid biosynthesis; inhibition blocks *RAS* activation and inhibits growth of Ras-transformed cells	SREBP
Ornithine decarboxylase (ODC)	Rate-limiting enzyme in polyamine synthesis, essential for cell growth	Myc, Mycn
Prothymosin alpha	Protein of unclear function, related to cell growth	Myc, Mycn
Pulmonary surfactant protein A (SP-A)	Lung-specific phospholipid-associated glycoprotein, mediates pathogen defense, frequently overexpressed in lung adenocarcinomas	USF1
Regulator of chromosome condensation 1 (RCC1)	Guanine exchange factor, necessary for cell proliferation	Myc
Telomerase reverse transcriptase (TERT)	Catalytic subunit of telomerase which maintains chromosome ends and is an immortalizing enzyme	Myc

stretch of amino acids, forming one or more β-turns, the "loop." Similar to the leucine-zipper motif, hydrophobic amino acid residues on one face of the helix interact with similar, also hydrophobic, residues of the helix from a second protein, thus stabilizing the dimer. Some proteins possess a leucine-zipper as a second dimerization motif; these include members of the Mad, Srebp, Tfe, and Myc family (Fig. 1).

Dimerization is necessary but not sufficient for DNA binding. A sequence-specific E-box recognition is mediated by a pattern of basic amino acids (Table 3). Since this basic region is localized N-terminal to the HLH region, it was this order

E-Box, Table 2 List of bHLH protein families in animals and their E-box binding specificity

E-box sequence	bHLH family	Family members	Function
CAG (CTG)	Achaete-Scute	Achaete, Scute, Mash-2, Hash-2, Fash-2	Neurogenesis
	Atonal	Atonal, Lin-32, Hath1, Math1, Math2, NeuroD, NeuroD2, NeuroD3, NeuroM	Neurogenesis, pancreatic development
	Delilah	Delilah	Differentiation of epidermal cells into muscle in *Drosophila*
	Hand	Hand1, Hand2	Cardiac morphogenesis, trophoblast cell development, neural crest development
	E12/E47	E12, E47, Pan1, Pan2	Ubiquitously expressed, myogenesis, neurogenesis, immunoglobulin gene expression
	Hen	Hen1, Hen2, Nhlh1, Nhlh2	Neurogenesis
	Lyl	Lyl-1, Nscl1, Nscl2, Scl/Tal-1, Tal-2	Hematopoietic proliferation and differentiation
	Mrf	MyoD, Myogenin, Myf5, Myf6, Mist1, Mrf4	Myogenesis
	Nex	Nex-1	Neurogenesis
	Twist	Twist, Paraxis, Scleraxis, Dermo-1, Tcf21	Specification of mesoderm lineages, myogenesis
CAC (GTG)	bHLH/PAS	Arnt, Arnt2, trh, Hif-1α, Sim, ahr, BMAL1, clock, tim, per	Reaction to aromatic hydrocarbons, regulation of circadian rhythm, hypoxia response
	Hairy/E (Spl)	Hairy, E(Spl), Deadpan, Hes1, Hes5, Her1, Her4, Hesr-1, Sharp-1, Sharp-2	Neurogenesis, segmentation
	Mad	Mad, Mad3, Mad4, Mnt, Mxi1	Regulation of cell proliferation
	Myc	Myc, Mycn, Mycl, Max	Cell proliferation, differentiation
	Srebp/Add	Srepb1, Srebp2, Add1, HLH106	Cholesterol homeostasis, sterol synthesis, adipocyte determination
	Tfe	Tfe3, Tfeb, Tfec, Mitf, Mi	Placenta vascularization, development of melanocytes, osteoclasts, mast cells
	USF	USF1, USF2, SPF1	Ubiquitous transcription factors
No binding	Id	Id1, Id2, Id3, Id4, TId1, TId2, XIdl, XIdx, XId2, emc	Negative regulators of myogenesis, neurogenesis

Extended, updated, and modified after Atchley and Fitch (1997)

which gave the name to a whole class of proteins, the "bHLH proteins." bHLH proteins that rather bind to the "CAGCTG" DNA motif have different sets of basic residues than those recognizing the DNA sequence "CACGTG." The change of only a single amino acid residue can alter binding specificity. The MyoD basic region usually recognizes the DNA sequence "CAGCTG." Replacement of the leucine at position 13 by an arginine within the basic MyoD region changed its binding preference to the "CACGTG" sequence.

Regulation of E-box-containing genes is also determined by the composition of the bHLH dimer bound to the DNA. Max, a member of the Myc family, can heterodimerize with Myc and activate target genes. If Myc is replaced by one of the Mad family members, the resulting heterodimer now represses the same target genes. Whereas Max is a ubiquitously expressed protein, the expression pattern of its potential partners is tissue, developmental, or cell cycle dependent. Consequently, transcriptional activation or repression of the target gene is determined by the ratio of Myc to Mad proteins, competing for Max dimerization.

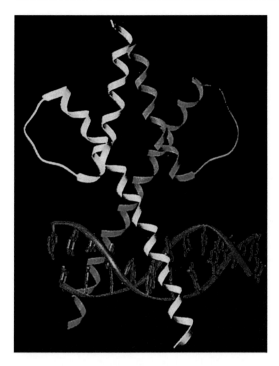

E-Box, Fig. 1 Shown is an E47 bHLH dimer bound to an E-box element (sequence: CACCTG). Each monomer consists of two α-helices that are separated by a loop; hydrophobic interactions between the α-helices stabilize dimerization. The basic region of each bHLH monomer makes contact with the major groove of the DNA molecule, each covering one half of the provided DNA binding site [From Ellenberger et al. (1994) with permission]

E-Box, Table 3 Amino acid sequence comparison of bHLH proteins that bind the DNA sequences CAC GTG and CAG CTG

CAC GTG	bHLH	BB—N—ER—R—
	Myc	KRRTHNVLERQRRNE
	Max	KRAHHNALERKRRDH
	USF	RRAQHNEVERRRRDK
	Tfe3	KKDNHNLIERRRRFN
CAG CTG	MyoD	RRKAATMRERRRLSK
	E12	RRVANNARERLRVRD
	E47	RRMANNARERVRVRD

B: basic residue

References

Atchley WR, Fitch WM (1997) A natural classification of the basic helix-loop-helix class of transcription factors. Proc Natl Acad Sci U S A 94:5172–5176

Ellenberger T, Fass D, Arnaud M et al (1994) Crystal structure of transcription factor E47: E-box recognition by a basic region helix-loop-helix dimer. Genes Dev 8:979–980

Massari ME, Murre C (2000) Helix-loop-helix proteins: regulators of transcription in eukaryotic organisms. Mol Cell Biol 20:429–440

Murre C, McCaw PS, Baltimore D (1989) A new DNA binding and dimerization motif in immunoglobulin enhancer binding, daughterless, MyoD, and myc proteins. Cell 56:777–783

See Also

(2012) BCL-6. In: Schwab M (ed) Encyclopedia of cancer, 3rd edn. Springer Berlin Heidelberg, pp 363-364. doi:10.1007/978-3-642-16483-5_566

(2012) BHLH. In: Schwab M (ed) Encyclopedia of cancer, 3rd edn. Springer Berlin Heidelberg, p 389. doi:10.1007/978-3-642-16483-5_604

(2012) FOS. In: Schwab M (ed) Encyclopedia of cancer, 3rd edn. Springer Berlin Heidelberg, p 1446. doi:10.1007/978-3-642-16483-5_2252

(2012) HAND gene family. In: Schwab M (ed) Encyclopedia of cancer, 3rd edn. Springer Berlin Heidelberg, p 1627. doi:10.1007/978-3-642-16483-5_2559

(2012) Id proteins. In: Schwab M (ed) Encyclopedia of cancer, 3rd edn. Springer Berlin Heidelberg, p 1803. doi:10.1007/978-3-642-16483-5_2940

EBV

▶ Epstein-Barr Virus

EC 2.7.11.1

▶ BRaf-Signaling
▶ Raf Kinase

E-Cadherin

Andreas Wicki[1] and Gerhard Christofori[2]
[1]Department of Medical Oncology, University Hospital, Basel, Switzerland
[2]Department of Biomedicine, University of Basel, Basel, Switzerland

Synonyms

Cadherin-1; Epithelial cadherin; Uvomorulin

Definition

E-cadherin, a 120 kDa molecule, is a prototypical member of the classical cadherin family of single-pass transmembrane proteins that mediate calcium-dependent cell-cell adhesion. Normally, epithelial cells are tightly interconnected through several junctional structures, including ▶ adherens junctions, ▶ tight junctions, and ▶ desmosomes. E-cadherin is the main ▶ adhesion molecule of the adherens junctions of epithelial cells, and via catenins it is linked to the underlying actin cytoskeleton. Two other members of the cadherin family, desmoglein and desmocollin, mediate adhesion in desmosomes. Via plakoglobin and desmoplakin, they are connected to intermediate cytoskeletal filaments. Thus, E-cadherin and other members of the cadherin family play a crucial role in establishing and maintaining the integrity of epithelial tissues.

The majority of human cancers (80–90%) are of epithelial origin. Loss of E-cadherin expression correlates with late-stage tumorigenesis characterized by tumor cell dedifferentiation, invasive tumor growth, and metastasis. Moreover, studies in a transgenic mouse model of carcinogenesis have demonstrated that the loss of E-cadherin is a rate-limiting step in the transition from benign tumors to malignant tumors and the subsequent formation of metastases. In addition, germline mutation of E-cadherin predisposes to diffuse ▶ gastric cancer.

Characteristics

The extracellular region of E-cadherin consists of five highly conserved cadherin (CAD) domains that combine with calcium ions to form rodlike structures (Fig. 1). The outermost CAD domain contains a conserved His-Ala-Val (HAV) motif, which is thought to mediate homotypic binding *in trans* to an E-cadherin molecule on the surface of an adjacent cell. Dimerization of E-cadherin molecules *in cis* appears to be mediated by their transmembrane domains. Depletion of calcium ions results in the disassembly of E-cadherin-mediated adhesive structures and the loss of cell-

cell adhesion. Critical for E-cadherin-mediated cell-cell adhesion is the interaction of its cytoplasmic domain with catenins. β-Catenin and γ-catenin/plakoglobin associate directly with the cytoplasmic domain of E-cadherin. They bind to α-catenin, which in turn connects the cadherin cytoplasmic complex to the actin cytoskeleton (Fig. 1). An additional catenin, p120 catenin, has been originally identified as a substrate of the non-receptor tyrosine kinase $pp60^{c-src}$ and binds to the juxtamembrane region in the cytoplasmic tail of classical cadherins. Depending on the cell type, p120 catenin has been found to positively or negatively modulate the strength of cell-cell adhesion.

E-Cadherin and the Formation of Adhesion Junctions

Upon contact of the cell membrane with adjacent cells, preexisting E-cadherin is recruited to the site of the interaction, and the junctional complex is formed. Once in place, E-cadherin-mediated cell-cell adhesion can be rapidly dismantled by growth factor-mediated signals. p120 catenin and to a lesser extent β-catenin and γ-catenin are phosphorylated upon treatment of cells with the growth factors hepatocyte growth factor (HGF), epidermal growth factor (EGF) (Growth factor), or ▶ platelet-derived growth factor – either directly by the respective receptor tyrosine kinases or indirectly by downstream effector kinases, such as $pp60^{c-src}$. Phosphorylation of the catenins causes their release from the cell adhesion complex resulting in the disruption of cell adhesion and eventually cell scattering. Similarly, treatment of cells with potent inhibitors of tyrosine phosphatases induces dissociation of the cytoplasmic cell adhesion complex. Deletion of the cytoplasmic catenin-binding domain or any disruption of the intracellular E-cadherin-catenin complex results in loss of cell-cell adhesion. Hence, mutations in genes other than E-cadherin itself might also affect E-cadherin function.

The mechanism through which β-catenin and α-catenin increase the adhesive strength of E-cadherin is not clear. Studies indicate that α-catenin may not simultaneously bind E-cadherin/β-catenin and actin, but rather

E-Cadherin,
Fig. 1 E-cadherin-
mediated cell-cell adhesion.
E-cadherin homodimers on
the plasma membranes of
adjacent cells interact in a
zipper-like fashion. The
most N-terminal CAD
domain on each E-cadherin
molecule contains a HAV
motif thought to interact
with an E-cadherin
molecule on an adjacent
cell. The cytoplasmic
cadherin complex, which
consists of α-catenin,
β-catenin, γ-catenin, and
p120 catenin, mediates the
interaction between
E-cadherin and the actin
cytoskeleton

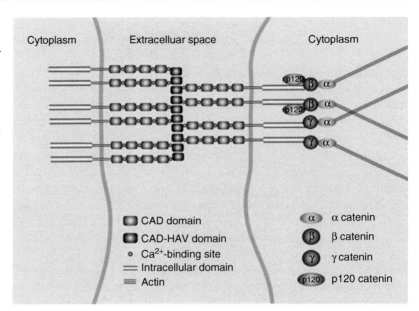

influence the structure of the actin cytoskeleton
via Arp2/3. The formation of adhesive junctions is
also regulated by Rho-GTPases (▸ GTPase).
Forced expression of constitutive-active forms of
Rac or Cdc42 inhibits HGF-induced scattering of
epithelial cells. Similarly, overexpression of
Tiam-1, a guanine nucleotide exchange factor
(GEF) for Rac, enhances E-cadherin-mediated
adhesion.

Signals Elicited by the Loss of Cell Adhesion
Besides being a major component of the cytoplasmic cadherin complex, β-catenin also takes part in
▸ Wnt signaling. In the absence of a Wnt signal,
free β-catenin is rapidly phosphorylated by glycogen synthetase kinase-3β (GSK3β) in the adenomatous polyposis coli (APC)-GSK3β complex
and subsequently degraded by the ubiquitin-
proteasome pathway. If the tumor suppressor
APC is nonfunctional or if GSK3β activity is
blocked by the activated Wnt signaling pathway,
β-catenin accumulates at high levels in the cytoplasm. Subsequently, it translocates to the
nucleus, where it binds to a member of the TCF/
LEF-1 family of transcription factors and modulates the expression of TCF/LEF-1 target genes.
Loss of function mutations of APC is a hallmark
of colon cancers, and stabilizing mutations of
β-catenin are frequently found in a number of

cancer types. However, in several human cancer
types and experimental tumor models, tumor progression induced by the loss of E-cadherin was
shown to be independent of β-catenin. Therefore,
β-catenin that is released from the E-cadherin
adhesion complex does not necessarily feed
directly in the Wnt signaling pathway, unless
β-catenin degradation is inhibited.

α-Catenin is not only connects adherens
junctions with the actin cytoskeleton but also
interacts with several actin-binding proteins,
including vinculin, ZO-1 (▸ zonula occludens
(ZO) protein-1), and α-actinin. It is also involved
in the communication between cell adhesion and
changes in cellular phenotype. As outlined above,
E-cadherin and the junctional complex can also
interact with Rho-GTPases. Experiments suggest
a direct influence of Rac and Cdc42 on the
cadherin-catenin complex via the protein
IQGAP1 (▸ IQGAP1 protein). IQGAP1 accumulates at contact sites of cells expressing
E-cadherin, where it directly interacts with the
N-terminus of β-catenin and the cytoplasmic tail
of E-cadherin. Active Rac and Cdc42 prevent the
interaction of IQGAP1 with β-catenin by sequestration of IQGAP1. Upon release of IQGAP1
from GDP-bound, inactive Rac and Cdc42, however, IQGAP1 displaces α-catenin from
β-catenin, resulting in the dissociation of

α-catenin from the cytoplasmic cadherin complex and concomitant loss of E-cadherin-mediated cell adhesion. Thus, Rac and Cdc42 seem to positively regulate cell adhesion by suppression of IQGAP1 activity. Finally, p120 catenin was shown to translocate to the nucleus, interact with Kaiso (a transcriptional repressor), and thereby modulate gene expression. Furthermore, evidence suggests that Frodo, a functional regulator of disheveled, may link p120 catenin to the Wnt pathway as well.

Apart from signaling through the catenins, the intracytoplasmic tail of E-cadherin can be cleaved by presenilin1, a γ-secretase, and participate in the regulation of cellular transcription and lysosomal degradation of amyloidogenic proteins.

E-Cadherin in Animal Models of Cancer

The effect of loss of E-cadherin on tumor progression was studied in two animal models of human cancer. In the Rip1Tag2 mouse model of insulinoma, the expression of a dominant-negative E-cadherin (resulting in a complete ablation of E-cadherin-mediated adhesion) promoted the formation of carcinomas and induced lymph node metastasis (Fig. 2). Forced expression of E-cadherin in the same model arrested tumor progression at the stage of an adenoma. Furthermore, the transgenic expression of ▶ podoplanin, a small mucin-like glycoprotein, led to the formation of carcinomas in the presence of E-cadherin and the other components of the zonula adherens.

In the p53 knockout (K14Cre;Trp53$^{F/F}$) mouse model of breast cancer, somatic inactivation of E-cadherin shifted the invasion pattern of carcinomas from an expansive (ductal-like) to an infiltrative (lobular-like) invasion pattern and induced the formation of distant metastasis. Interestingly, isolated knockout of E-cadherin in the mammary epithelium was not sufficient to induce breast cancer formation.

Taken together, loss of E-cadherin increases the formation of carcinomas, shifts the invasion pattern from collective to single-cell invasion, and thereby promotes metastasis. However, in certain animal models, the formation of carcinomas was possible in the presence of E-cadherin, and those carcinomas which expressed E-cadherin tended to

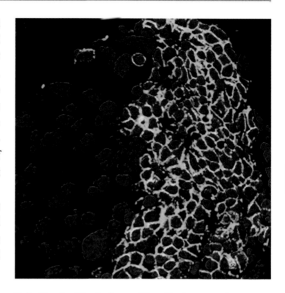

E-Cadherin, Fig. 2 Loss of E-cadherin expression during tumor progression. *Right hand*: E-cadherin (*green*) is expressed at the plasma membranes of tumor cells with benign, epithelial phenotype and normal nuclei (*red*). *Left hand*: In contrast, E-cadherin expression is focally downregulated within the portion of the tumor that exhibits invasive growth and atypical nuclei (*red*)

invade by adopting an expansive/collective invasion pattern.

E-Cadherin in Human Cancer

E-cadherin's adhesive function can be lost during the development of human epithelial cancers, including ▶ breast cancer, ▶ colorectal cancer, ▶ prostate cancer, ▶ gastric cancer, liver cancer, ▶ esophageal cancer, skin cancer, renal cancer, and ▶ lung cancer. In general, decreased E-cadherin function correlates with dedifferentiation of tumor cells, infiltrative tumor growth, metastasis, and poor prognosis. Several different mechanisms appear to cause the loss of E-cadherin function in human tumors. Hereditary mutations in the E-cadherin gene are evident in cases of familial gastric cancers. Patients with a germline mutation of E-cadherin have a 70–80% lifetime risk to develop diffuse gastric cancer, indicating that the penetrance of the E-cadherin mutation is similar to that of BRCA1 carriers in breast cancer.

In animal models of cancer, mutation of E-cadherin predisposes to neoplasia, but the mutation alone is not sufficient to induce the disease.

Apart from mutations of E-cadherin, expression of truncated α-catenin or truncated β-catenin (Beta-Catenin) abrogates E-cadherin function. Whereas mutations in the α-catenin gene or reduced α-catenin protein levels thus far have only been found in cultured tumor cell lines, mutations in the β-catenin gene are evident in many primary human tumors, including melanoma, ► colorectal cancer, ► gastric cancer, and ► prostate cancer.

► Epigenetic mechanisms such as chromatin rearrangement and loss of transcription factor binding also coincide with suppression of E-cadherin promoter activity in invasive carcinoma cells. In many tumor types, hypermethylation of the regulatory region of the E-cadherin gene and thus transcriptional silencing of the gene appears to be a major mechanism underlying E-cadherin loss of function. In several cellular systems, the epigenetic silencing of E-cadherin is mediated by the activation of transcriptional repressors. One of these repressors, Snail (► Snail transcription factors) has been shown to downregulate E-cadherin expression along with claudin-3, a protein involved in the formation of tight junctions. Finally, proteases that are upregulated during tumor progression are able to degrade the extracellular portion of the E-cadherin molecule resulting in the disruption of cell-cell adhesion and the cytoplasmic cadherin complex.

Other family members of classical cadherins are expressed in epithelia. It was observed that in several cancer types, expression of N-cadherin was upregulated during tumor progression concomitant with the loss of E-cadherin function. This phenomenon was termed "cadherin switch," and is related to ► epithelial-to-mesenchymal transition (EMT), a process implicated in the transition from benign neoplasia to malignant cancer. In these cases, N-cadherin appeared to induce tumor cell invasion and metastasis.

Clinical Relevance

E-cadherin is used as a diagnostic and prognostic marker for several human cancers. In particular, expression of E-cadherin is lost in most cases of lobular ► breast cancer, making the lack of E-cadherin a diagnostic criterium for the disease. Moreover, detection of mutations in the E-cadherin gene is diagnostic for hereditary diffuse ► gastric cancer. Therapeutic approaches that are based on the loss of E-cadherin function will have to await the identification of downstream effector genes that may be more amenable to therapeutic intervention.

Cross-References

- ► Adherens Junctions
- ► Adhesion
- ► BARD1
- ► Breast Cancer
- ► Colorectal Cancer
- ► Desmosomes
- ► Epigenetic
- ► Epithelial-to-Mesenchymal Transition
- ► Esophageal Cancer
- ► Gastric Cancer
- ► GTPase
- ► IQGAP1 Protein
- ► Lung Cancer
- ► Platelet-Derived Growth Factor
- ► Podoplanin
- ► Prostate Cancer
- ► Snail Transcription Factors
- ► Tight Junction
- ► Wnt Signaling
- ► Zonula Occludens Protein-1

References

Barrallo-Gimeno A, Nieto MA (2005) The Snail genes as inducers of cell movement and survival: implications in development and cancer. Development 132:3151–3161

Brembeck FH, Rosario M, Birchmeier W (2006) Balancing cell adhesion and Wnt signaling, the key role of beta-catenin. Curr Opin Genet Dev 16:51–59

Cavallaro U, Christofori G (2004) Cell adhesion and signalling by cadherins and Ig-CAMs in cancer. Nat Rev Cancer 4:118–132

Fitzgerald RC, Caldas C (2004) Clinical implications of E-cadherin associated hereditary diffuse gastric cancer. Gut 53:775–778

Perl AK, Wilgenbus P, Dahl U et al (1998) A causal role for E-cadherin in the transition from adenoma to carcinoma. Nature 392:190–193

See Also

(2012) Benign Tumor. In: Schwab M (ed) Encyclopedia of Cancer, 3rd edn. Springer Berlin Heidelberg, p 381. doi:10.1007/978-3-642-16483-5_579

(2012) Beta-Catenin. In: Schwab M (ed) Encyclopedia of Cancer, 3rd edn. Springer Berlin Heidelberg, p 385. doi:10.1007/978-3-642-16483-5_889

(2012) Epithelial Cell. In: Schwab M (ed) Encyclopedia of Cancer, 3rd edn. Springer Berlin Heidelberg, pp 1291–1292. doi:10.1007/978-3-642-16483-5_1958

(2012) Germline Mutation. In: Schwab M (ed) Encyclopedia of Cancer, 3rd edn. Springer Berlin Heidelberg, p 1544. doi:10.1007/978-3-642-16483-5_2404

(2012) Growth Factor. In: Schwab M (ed) Encyclopedia of Cancer, 3rd edn. Springer Berlin Heidelberg, pp 1607–1608. doi:10.1007/978-3-642-16483-5_2520

(2012) Hypermethylation. In: Schwab M (ed) Encyclopedia of Cancer, 3rd edn. Springer Berlin Heidelberg, p 1784. doi:10.1007/978-3-642-16483-5_2910

(2012) Malignant Tumor. In: Schwab M (ed) Encyclopedia of Cancer, 3rd edn. Springer Berlin Heidelberg, p 2150. doi:10.1007/978-3-642-16483-5_3519

(2012) Renal Cancer. In: Schwab M (ed) Encyclopedia of Cancer, 3rd edn. Springer Berlin Heidelberg, pp 3225–3226. doi:10.1007/978-3-642-16483-5_6575

(2012) Transcription Factor. In: Schwab M (ed) Encyclopedia of Cancer, 3rd edn. Springer Berlin Heidelberg, p 3752. doi:10.1007/978-3-642-16483-5_5901

ECMRIII

▶ CD44

ECSA

▶ Erythropoietin

Ecteinascidin 743

▶ Trabectedin

Ectonucleotide Pyrophosphatase/ Phosphodiesterase 2

▶ Autotaxin

Eczema

▶ Allergy

EDF

▶ Activin

Edible Salt

▶ Salt Intake

EDN

▶ Endothelins

Efferocytosis

Nikko Brix, Anna Tiefenthaller and Kirsten Lauber
Clinic for Radiotherapy and Radiation Oncology, LMU Munich, Munich, Germany

Keywords

Apoptosis; Necrosis; Phagocytes; Macrophages; Dendritic cells; Immune response; Inflammation; Immune tolerance; Autoimmunity; Cancer; "Find-me" signals; "Eat-me" signals; "Don't-eat-me" signals

Definition

The term *efferocytosis* (from *effere*, Latin for "to take away", "to carry to the grave", "to bury") has been defined as the process of apoptotic cell removal, which involves phagocyte recruitment, dying cell recognition, and engulfment.

Characteristics

The Efferocytic Process

Apoptotic Cell Death

In higher multicellular organisms, the removal of dying cells is a common event: It is estimated that one million cells undergo apoptosis per second in a human adult, and fundamental biological processes, such as embryogenesis, the resolution of inflammation, or homeostatic cell turnover involve apoptosis and subsequent clearance of dying cells. This is performed either by neighboring cells (when they are endowed with "amateur" phagocyte capacity) or by professional phagocytes, such as macrophages and immature dendritic cells (DCs), respectively. Macrophages and DCs serve as "undertakers" of dying cells with different tasks. Whereas macrophages can powerfully engulf and degrade huge amounts of dying cell material, DCs act as sentinels, which capture antigen and (cross-)present it to T cells, thus sculpting adaptive immune responses.

The entire process of apoptosis is finely controlled, swift, and innoxious for the surrounding tissue. These characteristics starkly distinguish it from the events being observed during necrosis, which is considered to be a nonphysiological form of cell death where plasma membrane rupture and the uncontrolled release of cytosolic danger signals occur.

Importantly, apoptosis is an immunogenically silent event. It is morphologically characterized by cell shrinking, chromatin condensation, nuclear fragmentation, and plasma membrane blebbing. On the molecular level, a set of proteases termed caspases commonly initiate and execute this form of programmed cell death. Besides, caspase-independent forms of apoptosis have been reported. The integrity of organelles and the plasma membrane is preserved until late stages of apoptosis. Thus, the liberation of pro-inflammatory intracellular molecules,

including heat shock proteins, high mobility group box 1 protein (HMGB1), S100 proteins, and uric acid, which may damage neighboring cells and induce inflammation, is prevented.

Phagocyte Recruitment by Apoptotic Cells

In order to "orchestrate" their own burial, apoptotic cells send out soluble chemotactic factors to trigger monocyte/macrophage recruitment, which is a premise for efficient efferocytosis. These "find-me" signals comprise biomolecules of very different classes, such as nucleotides (mainly ATP and UTP), phospholipids (e.g. lysophosphatidylcholine and sphingosine-1-phosphate), and proteins (for instance, the ectodomain of the IL-6 receptor and soluble fractalkine), which are recognized by the corresponding phagocyte receptors.

Recognition of Apoptotic Cells by Phagocytes

Upon attraction to the apoptotic site, phagocytes must precisely distinguish between healthy and dying cells. This substep of the efferocytic cascade is guided by the exposure of "eat-me" signals on the apoptotic cell surface. Translocation of phosphatidylserine (PS) to the outer leaflet of the plasma membrane is the best-known and probably most important example. This phospholipid is recognized by *bona fide* PS receptors on the phagocytes (e.g. brain angiogenesis inhibitor 1 (BAI-1), T cell immunoglobulin mucin domain (TIM) family members 1, 3, and 4, and the stabilins 1 and 2) or indirectly via soluble bridging proteins, which bind to both PS and their respective phagocytic cell surface receptors. Examples of bridging proteins secreted by phagocytes are milk fat globule EGF factor 8 (MFG-E8) and developmental endothelial locus 1 (Del-1). Moreover, bridging proteins derived from apoptotic cells, such as annexin A1, or from interstitial body fluids (β_2-glycoprotein, growth arrest specific gene 6 (GAS6), and protein S) have been described.

Furthermore, additional "eat-me" signals apart from PS exist. Examples are the surface exposure of ICAM-3 on apoptotic cells being recognized by CD14 and altered sugars, which are detected by lectins on the phagocyte. Besides, the inactivation and/or the lack of "don't-eat-me" signals (e.g. CD47 and CD46) contribute to efficient dying cell engulfment. These proteins are expressed on healthy cells and thus protect them from being accidentally ingested by phagocytes. Taken together, a plethora of different signals mediate the engulfment of apoptotic cells, and it is currently being unraveled which receptor-ligand axis dominates in which tissue or organ.

Apoptotic Cell Engulfment

The signaling mechanisms that are initiated upon ligation of phagocytic "eat-me" signal receptors are far from being fully understood. Yet, it is known that the ingestion of apoptotic cells requires massive modifications in the phagocyte's actin cytoskeleton. These are regulated by small GTPases of the RHO family, such as RHOA, RAC, and CDC42. RHOA activation has been described to inhibit apoptotic cell engulfment via binding and thereby activating its downstream effector Rho-associated coiled-coil containing protein kinase (ROCK). ROCK activation may in turn alter the phosphorylation status of myosin light chain and thus the phagocytic actin cytoskeleton structure and contractility. Inversely, RHOA inhibition was shown to promote dying cell engulfment. In contrast to RHOA, RAC activation positively affects apoptotic cell engulfment. For instance, the PS receptor BAI-1 is known to activate RAC by recruiting the adaptor protein engulfment and cell motility 1 (ELMO1) and its binding partner dedicator of cytokinesis 180 (DOCK180).

The Post-Phagocytic Immune Response

Apoptotic cell death itself is not only immunogenically silent, it also shapes the post-phagocytic immune response. Unlike the uptake of necrotic cell material by macrophages, apoptotic cell engulfment induces the secretion of anti-inflammatory cytokines including interleukin-10 (IL-10), transforming growth factor β (TGF-β), and prostaglandin E_2 (PGE$_2$). The removal of apoptotic cell material by DCs that have been educated in this milieu leads to tolerogenic (cross-)presentation of antigens in the draining lymph nodes. Among the "tolerate-me" signals,

which are involved in the immunomodulatory processes being initiated upon apoptotic cell clearance, exposure of PS is of crucial importance. However, since PS exposure is also observed during necrosis after plasma membrane rupture, it is evident that several other biomolecules might contribute to sculpting the immunological outcome of dying cell phagocytosis. Moreover, the cause of apoptotic cell death, the predominating dying cell type, and the quantity of apoptotic cells naturally modify the post-phagocytic immune response. The latter point is of particular importance: Although efferocytosis is performed rapidly under physiological conditions, excessive apoptotic cell death and/or inefficient clearance may overwhelm the phagocytic capacity: For instance, accumulating, uncleared apoptotic cells can be observed in the context of autoimmune diseases or after tumor radio(chemo)therapy as discussed below. As a result, these apoptotic cells may transit into secondary necrosis culminating in plasma membrane rupture and liberation of pro-inflammatory intracellular components. Considering their disparate origin, the soluble mediators released from primary and secondary necrotic cells encode for different immunological outcomes but commonly induce pro-inflammatory processes, which may attenuate the anti-inflammatory milieu at a formerly apoptotic site.

Diseases Caused by Malfunctioning Efferocytosis

The fact that apoptotic cells are rarely seen in homeostatic tissues testifies the efficiency of dying cell removal by professional and "amateur" phagocytes. In case of malfunctioning dying cell clearance, insufficient immunological quiescence causes recruitment of inflammatory cells into the tissue. Accumulation of this cell debris might initiate immunological reactions against self-antigens resulting in different types of chronic inflammatory diseases.

Rheumatic Diseases

The term "rheumatism" summarizes a wide range of autoimmune diseases affecting different organs. Their common pathogenesis is associated to autoimmune mechanisms – immune reactions against the body's own tissues.

From an immunological point of view, defective efferocytosis after tissue destruction leads to accumulation of cell corpses, which progress into secondary necrosis and release their intracellular contents (e.g., DNA, nucleosomes, and histones) upon cell membrane rupture. The immune system generates antibodies against these intracellular antigens, which in complex with their cognate antigens can stimulate an exacerbation of the inflammatory cascade. The immune reaction against own tissue structures and intracellular components results in general symptoms such as fever, pain, fatigue and progressive organ destruction when not adequately treated.

Systemic lupus erythematosus (SLE) is a systemic autoimmune disease that results from genetic and environmental stimuli, including ultraviolet light. As autoimmunity targets diverse intracellular components, several parts of the body are affected, especially inner organs (e.g., lungs, kidneys, liver, and the central nervous system) and the skin (butterfly rash). Although the etiopathogenesis of SLE is not completely understood, it is considered to involve multifactorial events and genetic predispositions. Alterations in regulators of apoptosis and efferocytosis, including C-reactive protein (CRP), pentraxins, and the complement system (especially C1q), lead to increased occurrence of secondary necrotic cell debris. The accumulation of secondary necrotic cell debris in the germinal centers of the lymph nodes favors complement-mediated immune complex formation. Blood phagocytes subsequently engulf these strongly pro-inflammatory structures, and the inflammatory reaction is further fueled by inappropriate cytokine release and production of inflammatory proteins, such as CRP. Clinically, these chronic inflammatory processes become manifest in fever, fatigue, and immune-mediated progressive organ destruction.

Even though defective apoptotic cell clearance has not yet been proven to contribute directly to the onset of rheumatoid arthritis, there exists some indirect evidence that this disease of progressive joint destruction could be caused by compromised efferocytosis as well. The extracellular debris located in inflamed joints contains – among other factors – histones and HMGB1, which both can inhibit efferocytosis. Eliminating these factors reportedly results in improved phagocytosis and could therefore be an interesting strategy for antirheumatic therapies.

Atherosclerosis

Atherosclerosis as one of the widest-spread cardiovascular diseases is considered to be a chronic inflammatory disorder, which is linked – at least in part – to lacking engulfment of apoptotic macrophages by their healthy counterparts. Tissue-resident macrophages undergo apoptotic cell death after excessive uptake of oxidized lipids. They remain in the vessel wall and transit into secondary necrosis due to defective clearing mechanisms. These secondary necrotic cells then stimulate chronic inflammation, further impairing phagocytosis and forming atherosclerotic plaques. In addition, malfunctioning or dead macrophages lack the ability of producing antiphlogistic cytokines (e.g., TGF-β, IL-10) and therefore do not contribute to atheroprotection.

Chronic Lung Diseases

Several chronic lung diseases are related to apoptosis and defective cell clearance, e.g., *chronic obstructive pulmonary disease (COPD)*, *asthma*, *pulmonary fibrosis*, *cystic fibrosis (CF)*, and *acute respiratory distress syndrome (ARDS)*. In all of these diseases, phagocytosis is reduced, mainly by neutrophil-produced reactive oxygen species (ROS) and subsequent activation of the efferocytosis inhibitor RHOA. Today, the common treatment of these diseases includes corticosteroid therapy. The mechanisms underlying corticosteroid action involve the inhibition of pro-inflammatory cytokine production and enhanced induction of apoptosis as well as efferocytosis.

Allograft Tolerance

Allotransplant patients are commonly treated with immunosuppressive agents in order to suppress allograft rejection. In mouse models, apoptotic cells appear to be a major contributor to immunogenic tolerance after graft transplantation. Hence, the question arises, if classic immunosuppressive treatment could be complemented by apoptotic cell transfer in order to exploit the apoptosis-dependent immunotolerance inducing effects.

Infection and Microbes

In case of tissue infection by microbes, pathogen-associated molecular patterns (PAMPs, e.g., nucleic acids, lipopolysaccharides, bacterial endotoxins), and danger signals derived from the host (e.g., ATP, HMGB1, heat shock proteins, uric acid, mitochondrial DNA) induce local and to some extent also systemic inflammation. This involves invasion of immune cells – neutrophils in the early phase, monocytes and macrophages later on. After pathogen engulfment, neutrophils locally undergo a specific form of apoptosis, the so-called phagocytosis-induced cell death (PICD), thus stimulating their internalization by macrophages and ensuring a second round of degradation of the internalized pathogens. Accordingly, efferocytosis appears to be a fundamental, well-established, and conserved mechanism of pathogen defense in mammals. However, in some cases, pathogens hijack this way of phagocytic clearance for infection. For *Chlamydia pneumonia*, it has been reported that efferocytosis of apoptotic, infected neutrophils, and subsequent lysosomal escape lead to macrophage infection – a phenomenon known as Trojan-horse strategy. Alternatively, exposure of PS is utilized as a means of apoptotic mimicry by certain viruses (vaccinia virus, HIV, and others) and parasites (*Leishmania* and *Trypanosoma species*). They expose PS on their membranes in order to be engulfed by macrophages and exploit the macrophages' tolerogenic, post-efferocytic state for pathogen spread and immune escape.

Apart from the disorders mentioned above, there are many more, which can be linked to defects or disturbances in efferocytosis. Consequently, the complexity of cell death, dying cell clearance, and subsequent immune reactions require further research in order to unravel the mechanisms of pathogenesis and to develop novel treatment approaches.

The Role of Efferocytosis in Cancer

Remarkably, apoptosis and efferocytosis are also highly relevant with regard to our understanding of cancer development and treatment. On the one hand, resistance to apoptosis is known to facilitate tumor cell growth and has therefore been defined as a hallmark of cancer. For instance, overexpression of anti-apoptotic proteins, such as B-cell lymphoma 2 (BCL-2) and B-cell lymphoma extralarge (BCL-XL), may prevent apoptotic cell death. Similarly, loss of p53 – a key DNA damage sensor endowed with the capacity to induce apoptosis – is a very common mechanism of tumor cell death evasion, and there are many other examples of apoptosis resistance in tumors.

However, on the other hand, findings suggest that the consequences of tumor cell apoptosis for a growing tumor may be more complex. Apoptotic cell death implies tumor cell removal and tolerogenic antigen (cross-)presentation by tumor-associated macrophages (TAMs) and intratumoral DCs, respectively. Astonishingly, this efferocytic process may be one of the fundamental elements in the progression of certain tumors as demonstrated in lymphomas: In order to enforce angiogenesis and to escape the immune system, a moderate level of tumor cell apoptosis might be beneficial for overall tumor growth by providing a pro-angiogenic, non-inflammatory milieu. In this context, TAMs may release tumor-protective factors, which activate angiogenesis, matrix remodeling, and wound healing.

Importantly, these findings do not contradict the well-established concept of apoptosis resistance as a hallmark of cancer. Evading apoptosis may be of crucial importance to initiate and establish the malignant phenotype. However, in later stages of tumor development, which are often accompanied by intratumoral hypoxia, an increase in angiogenesis and an immunotolerogenic milieu induced by TAM-dependent efferocytic processes may be beneficial for overall tumor growth.

Conversely, the growth of certain tumors might be impeded by altering the amount and especially the quality of tumor cell death: Novel cancer therapeutics may induce a delay in apoptotic cell clearance or block apoptotic signaling. For instance, masking the "eat-me" signal PS on apoptotic cells by annexin V has been shown to mediate antitumor immunity in vivo – presumably by retarding efferocytosis. Moreover, established cancer therapy approaches, such as radio- and chemotherapy, may be fine-tuned (and combined with additional therapeutics) in order to prevent a preponderance of apoptosis and subsequent tolerogenic immune responses. In summary, therapeutic induction of apoptosis should undoubtedly be considered as a key mechanism to induce tumor regression, but the ambivalent character of apoptosis with regard to efferocytosis and its immunological implications should also be taken into account.

Prospects

Extensive research on the process of dying cell clearance within the last decades has demonstrated that efferocytosis is much more than a simple method of "waste disposal" to eliminate dead cell corpses. Not only apoptotic cell death, phagocyte recruitment, and apoptotic cell engulfment but especially the post-phagocytic immunological consequences require further investigation as numerous patients may profit from a better understanding of the basic principles underlying the efferocytic process. New approaches aiming at modifying the quality and quantity of cell death and modulating the removal of dying cells may strongly improve the treatment of several severe diseases as these therapeutic strategies would not only control symptoms but target the origin of disease.

Cross-References

- ▶ Angiogenesis
- ▶ Apoptosis
- ▶ Autoimmunity and Cancer
- ▶ BCL2
- ▶ Cancer

- ▶ Dendritic Cells
- ▶ Hypoxia
- ▶ Inflammation
- ▶ Macrophages
- ▶ Prostaglandins
- ▶ Transforming Growth Factor-Beta
- ▶ Tumor-Associated Macrophages

References

deCathelineau AM, Henson PM (2003) The final step in programmed cell death: phagocytes carry apoptotic cells to the grave. Essays Biochem 39:105–117

Lauber K, Ernst A, Orth M, Herrmann M, Belka C (2012) Dying cell clearance and its impact on the outcome of tumor radiotherapy. Front Oncol 11(2):116

Muñoz LE, Lauber K, Schiller M, Manfredi AA, Herrmann M (2010) The role of defective clearance of apoptotic cells in systemic autoimmunity. Nat Rev Rheumatol 6(5):280–289

Poon IK, Lucas CD, Rossi AG, Ravichandran KS (2014) Apoptotic cell clearance: basic biology and therapeutic potential. Nat Rev Immunol 14(3):166–180

Willems JJ, Arnold BP, Gregory CD (2014) Sinister self-sacrifice: the contribution of apoptosis to malignancy. Front Immunol 4(5):299

See Also

(2012) Acute respiratory distress syndrome. In: Schwab M (ed) Encyclopedia of cancer, 3rd edn. Springer, Berlin/Heidelberg, p 36. doi:10.1007/978-3-642-16483-5_64

(2012) Allograft. In: Schwab M (ed) Encyclopedia of cancer, 3rd edn. Springer, Berlin/Heidelberg, p 142. doi:10.1007/978-3-642-16483-5_199

(2012) Allograft rejection. In: Schwab M (ed) Encyclopedia of cancer, 3rd edn. Springer, Berlin/Heidelberg, p 143. doi:10.1007/978-3-642-16483-5_200

(2012) Atherosclerosis. In: Schwab M (ed) Encyclopedia of cancer, 3rd edn. Springer, Berlin/Heidelberg, p 299. doi:10.1007/978-3-642-16483-5_432

(2012) Bcl-XL. In: Schwab M (ed) Encyclopedia of cancer, 3rd edn. Springer, Berlin/Heidelberg, p 368. doi:10.1007/978-3-642-16483-5_569

(2012) Caspase. In: Schwab M (ed) Encyclopedia of cancer, 3rd edn. Springer, Berlin/Heidelberg, pp 674–675. doi:10.1007/978-3-642-16483-5_873

(2012) Cdc42. In: Schwab M (ed) Encyclopedia of cancer, 3rd edn. Springer, Berlin/Heidelberg, p 704. doi:10.1007/978-3-642-16483-5_955

(2012) Chronic obstructive pulmonary disease. In: Schwab M (ed) Encyclopedia of cancer, 3rd edn. Springer, Berlin/Heidelberg, pp 853–855. doi:10.1007/978-3-642-16483-5_1158

(2012) Complement. In: Schwab M (ed) Encyclopedia of cancer, 3rd edn. Springer, Berlin/Heidelberg, p 963. doi:10.1007/978-3-642-16483-5_1284

(2012) C-Reactive protein. In: Schwab M (ed) Encyclopedia of cancer, 3rd edn. Springer, Berlin/Heidelberg, p 993. doi:10.1007/978-3-642-16483-5_1369

(2012) Cystic fibrosis. In: Schwab M (ed) Encyclopedia of cancer, 3rd edn. Springer, Berlin/Heidelberg, p 1042. doi:10.1007/978-3-642-16483-5_1454

(2012) Cytokine. In: Schwab M (ed) Encyclopedia of cancer, 3rd edn. Springer, Berlin/Heidelberg, p 1051. doi:10.1007/978-3-642-16483-5_1473

(2012) Endotoxins. In: Schwab M (ed) Encyclopedia of cancer, 3rd edn. Springer, Berlin/Heidelberg, p 1258. doi:10.1007/978-3-642-16483-5_1906

(2012) Germinal center. In: Schwab M (ed) Encyclopedia of cancer, 3rd edn. Springer, Berlin/Heidelberg, p 1541. doi:10.1007/978-3-642-16483-5_2401

(2012) High Mobility Group. In: Schwab M (ed) Encyclopedia of cancer, 3rd edn. Springer, Berlin/Heidelberg, p 1694. doi:10.1007/978-3-642-16483-5_2729

(2012) Immune response. In: Schwab M (ed) Encyclopedia of cancer, 3rd edn. Springer, Berlin/Heidelberg, p 1815. doi:10.1007/978-3-642-16483-5_2977

(2012) Immune system. In: Schwab M (ed) Encyclopedia of cancer, 3rd edn. Springer, Berlin/Heidelberg, p 1815. doi:10.1007/978-3-642-16483-5_2980

(2012) Infection. In: Schwab M (ed) Encyclopedia of cancer, 3rd edn. Springer, Berlin/Heidelberg, p 1848. doi:10.1007/978-3-642-16483-5_3041

(2012) Interleukin. In: Schwab M (ed) Encyclopedia of cancer, 3rd edn. Springer, Berlin/Heidelberg, p 1892. doi:10.1007/978-3-642-16483-5_3094

(2012) Lymph node. In: Schwab M (ed) Encyclopedia of cancer, 3rd edn. Springer, Berlin/Heidelberg, p 2116. doi:10.1007/978-3-642-16483-5_3443

(2012) Lymphoma. In: Schwab M (ed) Encyclopedia of cancer, 3rd edn. Springer, Berlin/Heidelberg, p 2124. doi:10.1007/978-3-642-16483-5_3463

(2012) Necrosis. In: Schwab M (ed) Encyclopedia of cancer, 3rd edn. Springer, Berlin/Heidelberg, p 2469. doi:10.1007/978-3-642-16483-5_3994

(2012) P53. In: Schwab M (ed) Encyclopedia of cancer, 3rd edn. Springer, Berlin/Heidelberg, p 2747. doi:10.1007/978-3-642-16483-5_4331

(2012) Pathogen-associated molecular pattern. In: Schwab M (ed) Encyclopedia of cancer, 3rd edn. Springer, Berlin/Heidelberg, p 2792. doi:10.1007/978-3-642-16483-5_4404

(2012) Phagocyte. In: Schwab M (ed) Encyclopedia of cancer, 3rd edn. Springer, Berlin/Heidelberg, p 2839. doi:10.1007/978-3-642-16483-5_4491

(2012) Phosphatidylserine. In: Schwab M (ed) Encyclopedia of cancer, 3rd edn. Springer, Berlin/Heidelberg, p 2866. doi:10.1007/978-3-642-16483-5_4531

(2012) Rac. In: Schwab M (ed) Encyclopedia of cancer, 3rd edn. Springer, Berlin/Heidelberg, p 3133. doi:10.1007/978-3-642-16483-5_4891

(2012) Rho. In: Schwab M (ed) Encyclopedia of cancer, 3rd edn. Springer, Berlin/Heidelberg, p 3302. doi:10.1007/978-3-642-16483-5_5099

(2012) Systemic lupus erythematosus. In: Schwab M (ed) Encyclopedia of cancer, 3rd edn. Springer, Berlin/Heidelberg, p 3598. doi:10.1007/978-3-642-16483-5_5639

Eg5

► KSP Mitotic Spindle Motor Protein

EGF

► Epidermal Growth Factor-Like Ligands

EGF Family

► Epidermal Growth Factor-Like Ligands

EGF-Like Ligands

► Epidermal Growth Factor-Like Ligands

EGFR

► Epidermal Growth Factor Receptor

EGP-2

► EpCAM

EGP34

► EpCAM

EHSH1 (EH Domain/SH3 Domain-Containing Protein)

▶ Intersectin

Eicosanoid Signaling

▶ Arachidonic Acid Pathway

Eicosanoids

Definition

Eicosa-, Greek for "twenty," are signaling molecules derived by oxygenation from omega-3 (ω-3) or omega-6 (ω-6) fatty acids. They exert complex control over many bodily systems, especially in inflammation, immunity, and as messengers in the central nervous system. The networks of controls that depend upon eicosanoids are among the most complex in the human body. There are four families of eicosanoids – the ▶ prostaglandins, the prostacyclins, the thromboxanes, and the leukotrienes. For each, there are two or three separate series, derived either from an ω-3 or ω-6 essential fatty acid.

Cross-References

▶ Arachidonic Acid Pathway
▶ Leukotrienes
▶ Lipid Mediators
▶ Prostaglandins

ELA2

▶ Neutrophil Elastase

Elastase

▶ Neutrophil Elastase

ELAV1

▶ HuR

Electrolytes

▶ Mineral Nutrients

Electromagnetic Fields

Marco Ruggiero[1] and Stefano Aterini[2]
[1]Dream Master Laboratory, Chandler, AZ, USA
[2]Department of Experimental Pathology and Oncology, University of Firenze, Florence, Italy

Definition

Magnetic fields are generated by the movement of any electrical charge. A continuous electric current passing through a conductor creates a static magnetic field, while an electric current changing in time creates a variable magnetic field, which radiates electromagnetic waves spreading around the surrounding space at light speed. These electromagnetic fields enter the living tissue but are known as non-ionizing radiation since they are weak and unable to break molecular bonds. Metals such as iron, zinc, manganese, and cobalt are sensitive to electromagnetic fields that may exert their effects on proteins and cellular components containing these metallic elements.

Characteristics

Few environmental issues are as contentious as the question of whether exposure to electromagnetic fields affects biological systems. Considering the exponentially increasing widespread use of electromagnetic radiation-generating applications such as radio, television, wireless devices, and cell phones, the continuing change in the frequencies used, the health hazard implications of any connection between electromagnetic fields, and the cancer risk have raised a growing interest in the potential biological effects of electromagnetic fields on the mammalian cell growth, viability, and response to genotoxic injury. This topic is still a subject of repeated argument, and caution in the interpretation of the effects of static and variable magnetic fields on cellular behavior and individuals' health needs to be claimed. A measurable magnetic field is created even by the residential electric current. It is noteworthy that we are pervaded by the Earth's static magnetic field that is hundreds of times greater than the low-frequency electromagnetic fields created by current within homes.

Epidemiological and Clinical Evidence

The first connection between human disease and electromagnetic fields was suggested by the observation of a higher incidence of cancer in children (childhood cancer) living near power distribution lines. Afterward, major power lines have been held responsible for the occurrence of different cancer varieties. Results of different studies of a possible link between exposure to electromagnetic fields and childhood cancer, namely, leukemia, have been rather inconsistent. One large study found no association between electromagnetic field exposure and an increased risk of childhood leukemia, in contrast to previous reports showing that the exposure to electromagnetic fields resulted in nearly a 20% increase in the risk of leukemia. This case–control investigation (case–control association study) did not find a significant link between the risk of childhood leukemia and the actual measurement of magnetic fields in children's current and former homes, including homes their mothers lived in during pregnancy of the affected subjects. Electromagnetic field exposure has also been associated with the risk of breast cancer, mainly in men. Epidemiological studies have shown that in industrialized countries, where the electromagnetic field-generating devices are in use on a large scale, breast cancer risk is higher. It has been suggested that electromagnetic field exposure might promote breast neoplasm through inhibition of melatonin release. Different occupational epidemiological studies have shown an increased incidence of breast neoplasm in women employed in occupations with high electromagnetic field exposure as well as in male electrical workers. However, other investigations, not producing any significant correlation, failed to confirm these suggestive data from occupational studies. In a large Swedish cohort study, an ~10% increase in the risk of cancer was documented in people in the medium- and high-exposure levels. Several types of cancer, including skin, digestive, respiratory, reproductive, and urinary organs, were linked with occupational magnetic field exposure, suggesting an involvement of the endocrine and immune systems. Discrepancies in epidemiological studies dealing with this matter have involved different estimates of electromagnetic field exposure, measurement, and characteristics; the statistical analysis performed with data obtained in such epidemiological reports is another Achilles heel, and considerable biases can create misleading conclusions. Higher exposure has been associated with an increase in the cancer risk even though care needs to be taken in drawing any conclusion, because no dose–response relation has been documented so far. New technologies have been introduced on a large scale, and the possible short lag period between exposure and disease manifestation needs to be considered when examining the available data. Children are increasingly heavy users of communication sources (mobile phones), and they are likely to accumulate many years of exposure during their lives. They should be thoroughly monitored in the study population to detect

E

possible effects involving long induction periods or effects from long-term exposure.

Experimental Evidence

In case of high-frequency magnetic fields, biological effects and health risks are related to the thermal effect associated with sources emitting fields high enough to cause a significant temperature rise in living tissue. Carcinogenesis is a multistep process of accumulating mutations and promoting events. It has been proposed that electromagnetic field exposure might enhance the effects of other carcinogens, provided that both exposures are chronic. The potential for genotoxicity of electromagnetic fields has been investigated, and several negative studies in several exposure categories have presented sound and independent, reproducible data. Using in vivo animal models of carcinogenesis, the assessment of the potential carcinogenic activity of electromagnetic fields has yielded negative results in different studies, while using the rat mammary carcinoma model results seem to be conflicting. According to available data, it is unlikely that long-term exposure to electromagnetic fields is carcinogenic per se in animal models. However, a promoting effect in the development of cancer under certain exposure conditions cannot be ruled out. Since exposure conditions vary widely in the different models thus far proposed, independent replication of experimental results is absolutely crucial. Exposure to electromagnetic field, alone or in combination with ionizing radiation, appears to induce an insult at the cellular level, to inhibit DNA synthesis and the growth of human tumor cell lines in vitro. However, controversies still exist about the possibility that electromagnetic fields may influence tumor promotion. Different in vitro studies have failed to demonstrate any detectable effect of electromagnetic fields on the rate of DNA synthesis and cultured cell growth. Moreover, exposure of cultured mammalian cells to electromagnetic fields has not resulted in the production of detectable DNA lesions and has not affected intracellular ATP levels, suggesting that electromagnetic fields are not genotoxic and cytotoxic. On the other hand, investigating the genotoxic potential of electromagnetic fields using in vitro experiments, statistically significant and suggestive positive results have been reported. Following electromagnetic field exposure, enzymatic activity induction, DNA mutation in human and nonhuman cells, and DNA strand breaks in rat brain cells have been demonstrated. The static magnetic field has been shown to induce a remodeling and differentiation of human neuronal cells in the absence of any alteration of DNA, thus ruling out a direct effect of the magnetic field on DNA stability. Investigating the effects of a static magnetic field on the ability to proliferate of human breast cancer cells in vitro, it has been observed that magnetic field exposure only temporarily slows down cellular growth, which then eventually fully recovers. The reduced cell growth caused by the magnetic field could be explained by a temporary effect on some cellular metabolic events leading to the reduced DNA synthesis. Alternatively, it could be ascribed to a transient cellular differentiation, since induction of differentiated phenotype often correlates with decreased cell proliferation. These results are consistent with the observation that magnetic field induces time-dependent developmental effects on the process of differentiation of the chick cerebellar cortex. Human skin fibroblasts exposed to electromagnetic fields, generated by mobile phones, show alterations in cell morphology and increased expression of mitogenic signal transduction genes (MAP kinase kinase 3 [MAP kinase], G2/mitotic-specific cyclin G1 [cyclin G-associated kinase]), cell growth inhibitors (transforming growth factor beta [transforming growth factor]), and genes controlling apoptosis (bax) [apoptosis signaling]; a significant increase in DNA synthesis and intracellular mitogenic second messenger formation [signal transduction] matches the high expression of MAP kinase family genes. A report (Blank and Goodman 2011) proposes the interesting interpretation of DNA as an antenna able to respond to electromagnetic fields in different frequency ranges. The authors found that electromagnetic field interactions with DNA were similar over a range of non-ionizing frequencies, from extremely low to radio frequencies, and concluded that DNA shows the functional characteristic of a fractal antenna, since it is endowed with two structural characteristics of fractal antennas, i.e., electronic conduction and self-symmetry. This concept of DNA as an antenna opens fascinating perspective that goes far beyond the field of oncology. In fact, if an antenna is capable of receiving signals, i.e., to interact with electromagnetic fields, it is conceivable that it is also capable of sending signals. The existence, nature, and meaning of such hypothetical signals transmitted by DNA are not yet even imagined, but we feel confident that this field of research will see important developments in the future. In fact, endogenous electromagnetic fields are continuously generated by electrically active cells within a body, and the intensity of such electromagnetic fields is enough to alter gene expression. For example, it was demonstrated that the myoelectric activity of the gut induces the expression of heat-shock protection mechanisms in the cells of gut epithelium as well as in gastrointestinal microorganisms thus de facto altering the delicate balance of the gut microbiome (Laubitz et al. 2006). Considering that the human microbiome is involved in the development and function of all other organs and systems and most notably the immune system (Palm et al. 2015), we hypothesize that the alteration of the microbiome may be one of the mechanisms through which electromagnetic fields, both endogenous and exogenous, exert their biological effects. Thus, the effects of electromagnetic fields on the human microbiome open a new perspective in assessing the risks for health and in preventing them. So far, most of the research was concentrated on assessing the effects of electromagnetic fields on those areas of the body that were exposed to their energies. In other words, all the effects of electromagnetic fields on human health were ascribed to their interaction with the human cells of our bodies. However, since we have learned that electromagnetic fields, even of minimal intensity such as the endogenous electromagnetic fields, modify the human microbiome, their effects might be much more complex and far ranging. In fact, microbes and the microbiome may amplify or mitigate carcinogenesis, responsiveness to cancer

therapeutics, and cancer-associated complication (Garrett 2015), and, therefore, electromagnetic fields modifying the microbiome may interfere with all such cancer-related responses. It is foreseeable that the development of functional foods containing probiotics for the prevention and treatment of cancer will have to take into account the effects of endogenous as well as exogenous electromagnetic fields on the human microbiome.

Clinical Relevance

Although some studies have given no consistent or convincing evidence for a causal relation between electromagnetic fields and cancer, other reports suggest that the exposure to electromagnetic fields brings about a weak increase in the risk estimates of neoplasm. In general, many of the earlier studies lack statistical power and show methodological deficiencies. However, larger studies and meta-analyses describing an epidemiological association between exposure to different types of electromagnetic fields and various types of cancer were conducted with the goal of overcoming these shortcomings and providing precautionary public health protection strategies. In fact, ever increasing use of wireless devices and cell phones prompted researchers to further investigate the association between commonly encountered electromagnetic fields and most common tumors. For example, a review on the association between childhood leukemia and power-frequency magnetic fields concluded that still there is no clear indication of harm at the field levels analyzed and that if the risk is real, its impact is likely to be small. This notwithstanding, according to the authors, "a precautionary approach suggests that low-cost intervention to reduce exposure is appropriate." Another review analyzed various adverse health outcomes associated with cell phone use, describing the epidemiology of tumors such as of glioma, acoustic neuroma, meningioma, testicular and salivary gland tumors, malignant melanoma of the eye, intratemporal facial nerve tumor, breast cancer, and non-Hodgkin lymphoma. In this review, the authors conclude that a precautionary principle should be used and invite health authorities to

E

revise current standard of exposure to microwave during mobile phone use. We are sure that this field will be the object of many more investigations, and therefore this entry will be continuously updated. At this point in time, however, on the basis of available evidence, we too are convinced that a precautionary principle should be used, in particular, considering that millions people use cell phones, with children among the heaviest users. Until further evidence is collected, we believe that the safety recommendations put forward by Dubey et al. (2010) should be taken into consideration.

Cross-References

▶ Adaptive Immunity
▶ Genomic Instability
▶ Gut Microbiota
▶ Mutation Rate

References

Blank M, Goodman R (2011) DNA is a fractal antenna in electromagnetic fields. Int J Radiat Biol 87:409–415
Campion EW (1997) Power lines, cancer and fear. N Engl J Med 337:44–462
Dubey RB, Hanmandlu M, Kumar Gupta S (2010) Risk of brain tumors from wireless phone use. J Comput Assist Tomogr 34:799–807
Garrett WS (2015) Cancer and the microbiota. Science 348:80–86
Hietanen M (2006) Health risks of exposure to non-ionizing radiation – myths or science-based evidence. Med Lav 97:184–1886
ICNIRP (International Commission for Non-Ionizing Radiation Protection) Standing Committee on Epidemiology, Ahlbom A, Green A, Kheifets L et al (2004) Epidemiology of health effects of radiofrequency exposure. Environ Health Perspect 112:1741–17545
Laubitz D, Jankowska A, Sikora A et al (2006) Gut myoelectrical activity induces heat shock response in *Escherichia coli* and Caco-2 cells. Exp Physiol 91:867–875
Maslanyj M, Lightfoot T, Schüz J et al (2010) A precautionary public health protection strategy for the possible risk of childhood leukaemia from exposure to power frequency magnetic fields. BMC Public Health 10:6737
McCann J, Kavet R, Rafferty CN (2000) Assessing the potential carcinogenic activity of magnetic fields using animal models. Environ Health Perspect 108(Suppl 1):79–1003
Pacini S, Ruggiero M, Sardi I et al (2002) Exposure to global system for mobile communication (GSM) cellular phone radiofrequency alters gene expression, proliferation and morphology of human skin fibroblasts. Oncol Res 13:19–244
Palm NW, de Zoete MR, Flavell R (2015) Immune-microbiota interactions in health and disease. Clin Immunol pii: S1521-6616(15)00199-0. doi:10.1016/j.clim.2015.05.014

Electropermeabilization

▶ Electroporation

Electroporation

Cristina Fillat[1] and Antoni Ivorra[2]
[1]Institut d'Investigacions Biomèdiques August Pi i Sunyer (IDIBAPS) and Centro de Investigación Biomédica en Red de Enfermedades Raras (CIBERER), Barcelona, Spain
[2]Department of Information and Communication Technologies, Universitat Pompeu Fabra (UPF), Barcelona, Spain

Synonyms

Electropermeabilization

Definition

Electroporation is the phenomenon in which cell membrane permeability to ions and macromolecules is increased by exposing the cell to short high electric field pulses. Such increase in permeability can be either reversible – resulting in viable cells after the period of increased permeability – or irreversible when cells die due to excessive permeabilization. Both outcomes have important applications in biotechnology and in medicine.

Reversible electroporation is now a routine technique in microbiology laboratories for in vitro gene transfection. Clinically, in vivo electroporation is employed for gene transfection ("electrogenetherapy") and for enhancing

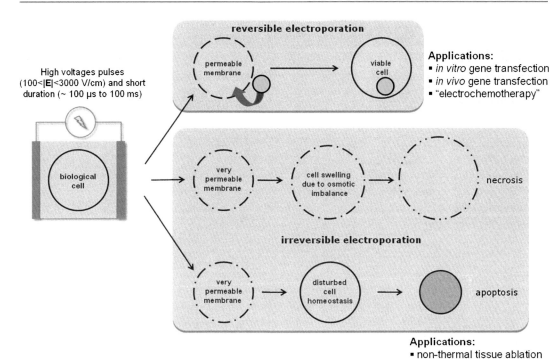

Electroporation, Fig. 1 Schematic illustration of electroporation outcomes

penetration of chemotherapeutic drugs ("electrochemotherapy").

Irreversible electroporation is used as a sterilization method for liquids, and it has been proposed and demonstrated as a tissue ablation method with significant advantages over thermal ablation techniques (Fig. 1).

Characteristics

Physical Aspects

The increase of cell membrane permeability is believed to be related to the formation of nanometric pores in the membrane, from which the suffix "-poration" stems. Those pores have not been observed directly, but molecular dynamics simulations show that, in a few nanoseconds, water channels resembling pores are indeed created in a bilipid membrane under a high electric field.

It is known that electroporation is a dynamic phenomenon that depends on the local transmembrane voltage. Commonly it is stated that, for a given pulse duration and shape, a specific transmembrane voltage threshold exists for the manifestation of the electroporation phenomenon (from 0.2 to 1 V). As a matter of fact, the degree of permeabilization shows very step gradualness with the transmembrane voltage.

Required transmembrane voltages for electroporation (larger than 0.2 V) do not occur under physiological conditions. These voltages are only reached when electric fields are artificially applied, in general through electrodes. The applied fields superimpose an artificially induced transmembrane voltage to the natural resting transmembrane voltage. In most cases, the applied electric field magnitude has to be significantly larger than 100 V/cm for achieving electroporation.

When electroporation occurs, cell membrane permeability is very high during electric field delivery, and although it very rapidly decreases after delivery cessation, membrane permeability can be significantly large for some minutes after. Presumably, this condition of prolonged permeability after electric field delivery is responsible for most physiological effects related to electroporation.

Electric fields for electroporation are delivered as short pulses (from 10 microseconds to 100 milliseconds) in order to avoid thermal effects and electrochemical reactions at the electrodes. It has been found that electroporation exhibits a memory effect: delivery of a second pulse produces larger increases in permeability than a single pulse, even if the single pulse duration combines the duration of both pulses. Because of that, electric fields are delivered as trains of pulses, typically 8 pulses for reversible electroporation and 80 pulses for irreversible electroporation. Pulse repetition frequencies range from 0.01 to 10 kHz.

Experimental Electroporation for Nucleic Acid Delivery

Gene transfer by electroporation is being used in in vitro studies to introduce DNA both into prokaryotic and eukaryotic cells. This procedure is in general terms very efficient and facilitates *E. coli* transformation and eukaryotic cell transfection. They constitute nowadays basic procedures in molecular biology laboratories.

In ovo electroporation has been applied mostly in developmental biology to study the role of specific genes or gene cascades in early developmental events. Plasmid DNAs are injected into the neural tube of stages 10–22 chick embryos, and a 25–40 V/25–50 ms square pulse is applied five times. Expression of the introduced DNA is observed 2 h after electroporation, peaks around 20 h after electroporation, and then weakens.

Although plasmid DNA is the most common material delivered in electrogene transfer studies, other nucleic acids with interest as anticancer agents are acquiring relevance. Electrotransfer of siRNA molecules has been shown to be feasible to induce gene silencing in melanoma cells, involving a clear specific mechanism related to the physicochemical properties of the molecules and different from plasmid DNA.

Clinical Uses of Electroporation

Electrochemotherapy

The term electrochemotherapy defines a therapeutic maneuver – currently in clinical use – consisting in enhancing the efficiency of chemotherapeutic drugs by means of electroporation. In practice, low-permeant drugs, such as bleomycin and cisplatin, are administered locally or systemically, and high-voltage pulses are delivered through electrodes at the site of interest – a solid tumor – so that significant uptake of the non-permeant drug will only occur in the cells exposed to a high enough electric current for electroporation. Electroporation-enhanced toxicity of bleomycin for subcutaneous tumors in mice has been reported to be above 10,000-fold.

This therapeutic approach is employed for treatment of cutaneous and subcutaneous tumors. Its use for deep-seated tumors, such as liver and brain tumors, is currently under research.

Electrogenetherapy

Electrogenetherapy was initially described as the in vivo electroporation of DNA to tissues with therapeutic purposes. Direct delivery into the tumor has been shown to generate an antitumor effect. Preclinical data has demonstrated that the optimization of the electric pulses, the voltage, and the type of electrodes are crucial elements to achieve high DNA transfection, and this could probably have slight variations depending on the tumor type. Moreover, in tumors with dense extracellular matrix, pretreatment of tumors with molecules that can modify the extracellular matrix demonstrated an increase in transfection efficiency.

Delivery to alternative sites has also been explored for the production of cancer vaccines, the reduction of tumor angiogenesis, or the induction of tumor cell apoptosis and demonstrated therapeutic effect against certain cancers. Striking differences on electroporation conditions are found between individual tissues, likely related to cell size and tissue organization.

Clinical trials of DNA vaccines and immunotherapy for cancer treatment using in vivo electroporation have been initiated in patients with melanoma and prostate cancer. Delivery of a plasmid DNA encoding interleukin-12 (IL-12) using electroporation has entered clinical trials in patients with metastatic melanoma and demonstrated to be safe with no systemic toxicity and only transient pain after electroporation as the major adverse event reported. Delivery of IL-12 into the metastatic melanoma lesions with electroporation resulted in significant necrosis of melanoma cells in the majority of treated tumors and significant lymphocytic infiltrates. In addition, clinical evidence of responses in untreated lesions suggested the induction of a systemic response following therapy.

A DNA vaccine delivered by intramuscular injection followed by electroporation has been evaluated as a method to stimulate humoral responses against prostate tumors. In a phase I/II clinical trial, in patients with recurrent prostate cancer, it was observed that the DNA vaccine tested with electroporation was a safe and well-tolerated procedure. Increased antibody responses were observed which persisted the 18 months of follow-up.

Many other clinical trials are currently under way, most of them exploring novel DNA vaccines in colorectal cancer, skin tumors, cervical cancer, or hematological malignancies.

Irreversible Electroporation

In the context of clinical applications, the term irreversible electroporation (IRE) defines a nonthermal tissue ablation modality consisting in applying high-voltage pulses through electrodes so that cell death is produced by excessive electroporation. It has been introduced into the surgical armamentarium, and currently, it is clinically being tested for ablation of solid tumors of, for example, the liver, prostate, kidney, and lungs. Other tumor locations addressed in animal experimentation are the bones, brain, and pancreas. In addition, other ablation targets are being explored such as coronary artery restenosis prevention or cardiac ablation for arrhythmia management. Typically, needle-shaped electrodes are inserted into or around the target tissue, and a train of high-voltage pulses (e.g., 80 pulses of 100 microseconds at a repetition frequency of 1 Hz and over 2,000 V) is delivered so that large cell membrane permeabilization is caused by electroporation.

The mechanism leading to cell death has not been elucidated so far, but it probably combines immediate lysis by osmotic imbalance and apoptosis triggering due to cell homeostasis disturbance and delayed necrosis and apoptosis due to ischemia caused by a vascular lock phenomenon which in turn seems to be caused by destruction of capillaries.

Cross-References

- ▶ Apoptosis
- ▶ Bleomycin
- ▶ Cancer
- ▶ Cancer Vaccines
- ▶ Chemoradiotherapy
- ▶ Chemotherapy
- ▶ Cisplatin
- ▶ Colon Cancer Molecular and Targeted Experimental Therapy
- ▶ Drug Delivery Systems for Cancer Treatment
- ▶ Electroporation
- ▶ Head and Neck Cancer
- ▶ Neoadjuvant Therapy
- ▶ Pancreatic Cancer
- ▶ Pancreatic Cancer Desmoplastic Reaction and Metastasis
- ▶ Photothermal Therapy
- ▶ Prostate Cancer Chemotherapy
- ▶ Prostate Cancer Experimental Therapeutics
- ▶ Radiation Oncology
- ▶ Renal Cancer Therapy
- ▶ Renal Cancer Treatment
- ▶ Suicide Gene Therapy

References

Krassowska W, Filev PD (2007) Modeling electroporation in a single cell. Biophys J 92(2):404–417

Mir LM, Morsli N, Garbay JR, Billard V, Robert C, Marty M (2003) Electrochemotherapy: a new treatment of solid tumors. J Exp Clin Cancer Res 22(4 Suppl):145–148

Murakami T, Sunada Y (2011) Plasmid DNA gene therapy by electroporation: principles and recent advances. Curr Gene Ther 11(6):447–456

Rubinsky B (2007) Irreversible electroporation in medicine. Technol Cancer Res Treat 6(4):255–260

Sardesai NY, Weiner DB (2011) Electroporation delivery of DNA vaccines: prospects for success. Curr Opin Immunol 23(3):421–429

E

Eliminate Gynecomazia

▶ Gynecomastia

ELISA

Definition

*E*nzyme-*l*inked *i*mmunosorbent *a*ssay is a biochemical analysis technique used to detect the presence of a protein in a fluid sample. The assay uses two antibodies, one that specifically detects the protein of interest and a second antibody that reacts to the first antibody-protein complex. The second antibody is usually linked to an enzyme that produces a chromogenic or fluorogenic substrate, thus providing the readout for the assay. ELISAs are often used to measure cancer biomarkers in the blood of cancer patients.

Cross-References

▶ Osteopontin

Elongin BC Complex

Ronald C. Conaway and Joan W. Conaway
Stowers Institute for Medical Research, Kansas, MO, USA

Definition

The mammalian elongin BC complex is a heterodimer, composed of the 118 amino acid ubiquitin-like elongin B protein and the 112 amino acid elongin C protein. The elongin BC complex interacts through a short, degenerate BC-box motif with multiple proteins, including the transcription factor elongin A, the von Hippel–Lindau (VHL) tumor suppressor protein, and members of the SOCS-box protein family.

Characteristics

The elongin B and C proteins were initially identified as positive regulatory subunits of the three-subunit elongin complex, which is one of several transcription factors capable of controlling the activity of the RNA polymerase II elongation complex. The three-subunit elongin complex was originally purified from rat liver nuclei by its ability to activate the overall rate of elongation by RNA polymerase II, by suppressing transient polymerase pausing at many sites within transcribed sequences. The elongin complex is composed of a transcriptionally active A subunit of approximately 770 amino acids and the elongin BC subcomplex. The latter positively regulates the activity of the elongin complex by binding to a BC-box motif in the elongin A elongation activation domain and potently inducing elongin A transcriptional activity. The BC-box is an ~10 amino acid sequence motif with consensus $Lxxx(C,A,S)xxx(A,I,L,V)$. Elongins B and C perform different functions in regulation of elongin A transcriptional activity. Elongin C functions as the inducing ligand and is capable of binding directly to the BC-box and maximally activating elongin A transcriptional activity in the absence of elongin B. Elongin B binds to elongin C and promotes stable binding of elongin C to elongin A.

Shortly after the discovery of the elongin complex, a large collection of BC-box motif-containing proteins were identified and found to function together with the elongin BC complex as integral components of members of two closely related classes of multiprotein E3 ubiquitin ligases (▶ ubiquitination). In addition to a BC-box protein and the elongin BC complex, these E3 ubiquitin ligases include a heterodimeric submodule composed either of Cullin family member Cul2 and RING-H2 finger protein Rbx1 (also known as ROC1 or Hrt1) or the related Cul5 and Rbx2 proteins. In the context of these E3 ubiquitin ligases, the BC-box protein functions as the substrate recognition subunit. The elongin BC complex functions as an adaptor that links the BC-box protein to the Cullin/Rbx submodule, which in turn functions to recruit E2 ubiquitin-conjugating enzymes of the Ubc5 family to the ubiquitin ligase complex and to activate ubiquitination of target proteins. The interaction of elongin BC with the BC-box is governed by the interaction of a highly conserved leucine found at the N-terminus of the BC-box with a hydrophobic pocket created by residues in the C-terminal half of elongin C. A short N-terminal elongin C region binds to the N-terminal ubiquitin-like domain of elongin B. Additional sequences near the BC-box motif in BC-box proteins specify selection of the particular Cul2/Rbx1 or Cul5/Rbx2 submodule present in a given elongin BC-containing E3 ubiquitin ligase.

The VHL Tumor Suppressor Complex

The founding member of the family of elongin BC-containing E3 ubiquitin ligases is the VHL tumor suppressor (▶ Von Hippel-Lindau Tumor Suppressor Gene, ▶ Tumor Suppression) complex, which is composed of the VHL BC-box protein, elongin BC, and a Cul2/Rbx1 submodule. The VHL gene on chromosome 3p25.5 is mutated in the majority of sporadic clear cell renal carcinomas, in familial erythrocytosis-2, and in the VHL disease, an autosomal dominant familial cancer syndrome that predisposes affected individuals to a variety of tumors including clear cell renal carcinomas, cerebellar hemangioblastomas and hemangiomas, retinal angiomas, and pheochromocytomas. A substantial fraction of VHL mutations found in sporadic clear cell renal carcinomas and in VHL kindreds results in mutation or deletion of the VHL BC-box and disruption of the VHL–elongin BC interaction and of the VHL ubiquitin ligase complex.

A major function of the VHL E3 ubiquitin ligase complex is to regulate transcription of genes such as vascular endothelial growth factor (VEGF), platelet-derived growth factor β (PDGFβ), and E-cadherin by controlling the cellular levels of ▶ hypoxia-inducible transcription factors 1–3 by ubiquitin-dependent proteolysis. Under normoxic cell growth conditions, HIF transcription factors are rapidly ubiquitinated and destroyed by the proteasome. Under hypoxic conditions, ubiquitination of HIF transcription factors is inhibited and their concentrations rise to sufficient levels to activate hypoxia-inducible genes. Targeting of HIF transcription factors by the VHL ubiquitin ligase requires that HIFs are first posttranslationally modified by hydroxylation of a critical proline in the HIF oxygen-dependent degradation domain (ODD) by a prolyl hydroxylase enzyme of the dioxygenase family. In light of evidence that the prolyl hydroxylase uses molecular oxygen as the oxygen donor, it has been proposed that the prolyl hydroxylase might serve as an oxygen sensor (or oxygen-dependent switch) for the regulation of expression of HIF-dependent hypoxia-inducible genes.

Although it is well established that the VHL protein plays a crucial role in the regulation of transcription of hypoxia-inducible genes through its control of cellular HIF levels, renal carcinoma cells expressing VHL mutants lacking a functional BC-box exhibit a collection of additional phenotypes, including:

- Defects in the regulation of mRNA stability resulting in misregulation and constitutive expression of hypoxia-inducible genes
- Cell-cycle defects resulting at least in part from misregulation of the degradation of the Cdk inhibitor p27
- Defects in ubiquitin-dependent degradation of improperly processed or folded proteins
- Extracellular matrix defects resulting from failure of cells to secrete and/or properly assemble fibronectin

At least some of these phenotypes result from defects in a VHL-dependent, HIF-independent pathway that remains to be defined. In addition, future studies are needed to determine whether these additional phenotypes are due to defects in the VHL E3 ubiquitin ligase complex or whether they result from other VHL defects not related to its function in the ubiquitin–proteasome pathway.

SOCS-Box Proteins and the Elongin BC Complex

In addition to the VHL E3 ubiquitin ligase complex, which includes a Cul2/Rbx1 submodule, a growing collection of elongin BC-containing E3 ubiquitin ligases, which include Cul5/Rbx2 submodules, have been identified and found to play roles in human disease. Among the BC-box proteins that assemble with elongin BC and Cul5/Rbx2 are the so-called SOCS-box family of proteins and various virally encoded proteins such as the Adenovirus E4orf6 and HIV-1 VIF proteins. In addition, elongin A assembles with elongin BC and Cul5/Rbx to form a potential E3 ubiquitin ligase that still possesses RNA polymerase II elongation factor activity.

SOCS-box proteins include the SH2 domain-containing suppressors of cytokine signaling (SOCS) proteins and more than 30 additional members of the ras, WD repeat, ankyrin repeat, and SPRY domain families. SOCS-box proteins are modular and are composed of an N-terminal SH2, ras, WD repeat, ankyrin repeat, or SPRY domain and a C-terminal SOCS-box. The SOCS-box is an ~50 amino acid motif composed of an N-terminal consensus BC-box and a short C-terminal L/P-rich region that helps specify interaction of SOCS-box proteins with Cul5/Rbx2. The founding members of the SOCS-box protein family were the SH2 domain-containing SOCS proteins, which assemble with elongin BC and Cul5/Rbx2 to form E3 ubiquitin ligases that function as negative regulators of cytokine-induced Jak/STAT signaling, at least in part by targeting Jak or receptor tyrosine kinases and inhibiting phosphorylation and activation of STATs.

The Adenovirus E4orf6 and HIV-1 VIF proteins function together with cellular elongin BC and Cul5/Rbx2 as substrate recognition subunits of E3 ubiquitin ligases that promote virus growth. The E4orf6 ubiquitin ligase assembles shortly after Adenovirus infection of susceptible cells and targets the p53 tumor suppressor for ubiquitination and degradation by the proteasome, thus inhibiting apoptosis of infected cells and promoting viral infection. The VIF ubiquitin ligase targets the cellular APOBEC3G cytidine deaminase for ubiquitination and degradation by the proteasome. The APCBEC3G enzyme acts as a cellular antiretroviral factor by inducing hypermutations in newly synthesized HIV-1 minus-strand DNA.

Clinical Relevance

Although mutation of the elongin A or SOCS genes have not yet been identified in human disease, mutation of the VHL gene on chromosome 3p25.5 is responsible for the majority of sporadic clear cell renal carcinomas, for familial erythrocytosis-2, and for VHL disease, an autosomal dominant familial cancer syndrome that predisposes affected individuals to a variety of tumors including clear cell renal carcinomas, cerebellar hemangioblastomas and hemangiomas, retinal angiomas, and pheochromocytomas. Given their contributions to the adenoviral and retroviral life cycles, the E4orf6 and Vif ubiquitin ligases offer potential therapeutic targets for antiviral agents.

Cross-References

► Hypoxia-Inducible Factor-1
► Tumor Suppression
► Ubiquitination
► Von Hippel-Lindau Tumor Suppressor Gene

References

Alexander WS, Hilton DJ (2004) The role of suppressors of cytokine signaling (SOCS) proteins in regulation of the immune response. Annu Rev Immunol 22:503–529

Kamura T, Sato S, Haque D et al (1998) The Elongin BC complex interacts with the conserved SOCS-box motif present in members of the SOCS, ras, WD-40 repeat, and ankyrin repeat families. Genes Dev 12:3872–3881

Kamura T, Koepp DM, Conrad MN et al (1999) Rbx1, a component of the VHL tumor suppressor complex and SCF ubiquitin ligase. Science 284:657–661

Ohh M (2006) Ubiquitin pathway in VHL Cancer Syndrome. Neoplasia 8:623–629

Stebbins CE, Kaelin WG, Pavletich NP (1999) Structure of the VHL-ElonginC-ElonginB complex: implications for VHL tumor suppressor function. Science 284:455–461

Embryonal Carcinoma

▶ Ovarian Tumors During Childhood and Adolescence

Embryonal Serum Alpha-Globulin

▶ Alpha-Fetoprotein Diagnostics

Embryonal Serum α-Globulin

▶ Alpha-Fetoprotein Diagnostics

Embryonic Stem Cells

Definition

Primitive cells from the embryo with pluripotential capacity. These cells are capable of proliferating indefinitely in a pluripotent state and have the potential to differentiate into all somatic cell types.

Cross-References

▶ Adult Stem Cells
▶ Stem Cell Markers

Embryo-Specific Alpha-Globulin

▶ Alpha-Fetoprotein Diagnostics

Embryo-Specific α-Globulin

▶ Alpha-Fetoprotein Diagnostics

Empirin

▶ Aspirin

EMS1

▶ Cortactin

Endocan

Philippe Lassalle[1] and Maryse Delehedde[2]
[1]INSERM U774, Institut Pasteur de Lille, Lille, France
[2]R&D Lunginnov, Campus de l'Institut Pasteur de Lille, Lille, France

Synonyms

Endocan; Endothelial cell-specific molecule-1; ESM-1

Definition

Endocan is a soluble proteoglycan of 50 kDa carrying a single dermatan sulfate chain. This dermatan sulfate proteoglycan has been shown to be overexpressed in cancers and in diseases when endothelium gets challenged. Endocan/ESM-1 is upregulated in the presence of pro-angiogenic molecules, and VEGF-induced endocan secretion is abolished in the presence of antiangiogenic drug. Clearly, endocan/ESM-1 expression is biomarker of endothelial dysfunction and/or vascular remodeling in inflammatory diseases and in cancer.

Characteristics

Endocan, with synonym endothelial cell-specific molecule-1 (ESM-1), was originally cloned from

a human endothelial cell cDNA library and
described as being secreted by human umbilical
vein endothelial cells (HUVEC). Endocan transcripts were detected from a variety of cultured
endothelial cells of different origins, i.e., in human
coronary artery endothelial cells (HCAEC),
human pulmonary artery endothelial cells
(HPAEC), human dermal microvascular endothelial cells (HDMVEC), and human capillary endothelial cells (HUCE) purified from adipose
tissues. Moreover, there have been rising evidences of endocan expression in cultured cell
lines that are not of endothelial origin as in
human adipocytes, in melanoma cells involved
in ▸ vasculogenic mimicry process, in
liposarcoma cells, and in ▸ glioblastoma cells.
Structurally, endocan/ESM-1 is a soluble proteoglycan of 50 kDa, constituted of a mature polypeptide of 165 amino acids and a single dermatan
sulfate chain covalently linked to the serine residue at position 137. This dermatant/chondroitin
sulfate chain has been sequenced. Experimental
evidence implicates endocan/ESM-1 as a key
player in the regulation of major processes such
as cell ▸ adhesion or proliferation in ▸ inflammation disorders and tumor progression. Endocan/
ESM-1 was shown to inhibit the interaction
between intercellular adhesion molecule-1
(ICAM-1) and the integrin (lymphocyte
function-associated antigen-1) LFA-1 on leukocytes. Endocan can then modulate LFA-1-
mediated leukocyte functions, such as the firm
adhesion of leukocytes to the endothelium and
the leukocyte transmigration. Inflammatory cytokines such as ▸ TNF-alpha and growth factors
involved in ▸ angiogenesis process such as
▸ vascular endothelial growth factor (VEGF)
and ▸ fibroblast growth factor (FGF-2) strongly
increased the expression, synthesis, or release of
endocan by human endothelial and/or tumor cells.
Endocan has then been shown to be clearly
overexpressed in various human tumors, with elevated serum levels being observed in late-stage
▸ lung cancer patients, in clear cell renal carcinoma patients, and in hepatocarcinoma patients
(as measured by enzyme-linked immunoassay)
and with its overexpression in vessels within the

tumors being evident by immunohistochemistry.
Interestingly, endocan/ESM-1 has been described
as a biomarker of recurrence in pituitary tumors.
Upregulation of endocan/ESM-1 has also been
recognized as a significant molecular signature
of a bad prognosis in several types of cancers. In
mouse xenograft models of human cancers,
endocan/ESM-1 was identified to be one of the
genes involved in the switch from dormant to
angiogenic tumors. Endocan/ESM-1 has been
shown to be a biomarker of tip cells, the highly
specialized cells involved in neoangiogenesis. In
vitro, the VEGF-induced endocan secretion is
completely abolished in the presence of
antiangiogenic drugs (i.e., antibodies, small molecules such as tyrosine kinase inhibitors) raising
the pertinence of endocan as a biomarker to follow
antiangiogenic drug therapy.

References

Bechard D, Gentina T, Delehedde M et al (2001) Endocan
is a novel chondroitin sulfate/dermatan sulfate proteoglycan that promotes hepatocyte growth factor/scatter
factor mitogenic activity. J Biol Chem
276:48341–48349
Lassalle P, Molet S, Janin A et al (1996) ESM-1 is a novel
human endothelial cell-specific molecule expressed in
lung and regulated by cytokines. J Biol Chem
271:20458–20464
Recchia FM, Xu L, Penn JS et al (2010) Identification of
genes and pathways involved in retinal neovascularization by microarray analysis of two animal models
of retinal angiogenesis. Invest Ophthalmol Vis Sci
2:1098–1105
Sarrazin S, Adam E, Lyon M et al (2006) Endocan
or endothelial cell specific molecule-1 (ESM-1):
a potential novel endothelial cell marker and a new
target for cancer therapy. Biochim Biophys Acta
1765:25–37
Sarrazin S, Maurage CA, Delmas D et al (2010a) Endocan
as a biomarker of endothelial dysfunction in cancers.
J Cancer Sci Ther 2:47–52
Sarrazin S, Lyon M, Deakin JA et al (2010b) Characterization and binding activity of the chondroitin/dermatan
sulfate chain from Endocan, a soluble endothelial proteoglycan. Glycobiology 20(11):1380–1388
Strasser GA, Kaminker JS, Tessier-Lavigne M (2010)
Microarray analysis of retinal endothelial tip cells identifies CXCR4 as a mediator of tip cell morphology and
branching. Blood 115:5102–5110

Endocannabinoids

▶ Cannabinoids

Endocrine Neoplasms

▶ Neuroendocrine Neoplasms

Endocrine Oncology

Pavlos Msaouel
Jacobi Medical Center, Albert Einstein College of
Medicine, Bronx, NY, USA

Definition

Endocrine oncology is a field of medicine that combines the two disciplines of oncology and endocrinology. Endocrine oncology covers (i) tumors derived from organs or tissues related to the endocrine system such as neoplasms of the gonads, the adrenal glands, the endocrine pancreas and gut, the thyroid, the parathyroid, the pituitary, and the hypothalamus and (ii) tumors that are responsive to hormonal manipulations such as prostate and breast cancer. (iii) Endocrine oncology may also involve the effects of tumors of endocrine or non-endocrine origin and/or their respective therapies on the endocrine system. For example, lung cancers are neoplasms of non-endocrine origin that may however produce paraneoplastic hormonal syndromes that affect endocrine homeostasis. Another example is the damage to the gonadal endocrine axis from radiation therapy and cytotoxic chemotherapy.

Characteristics

Whereas endocrine-responsive tumors such as prostate and breast cancer are very common, endocrine malignancies are more rare. The most common endocrine tumor is ovarian cancer which is the sixth most frequent cancer among women. In contrast to tumors arising from the female gonads, testicular cancer is uncommon with incidence rates ranging from 3.1 to 10/100,000/year among white men and even lower rates among Africans and Asians. Thyroid cancer is the most common cancer of the endocrine glands and accounts for 92% of all endocrine gland cancers in the United States. Parathyroid tumors are the most common benign endocrine neoplasms with US incidence rates of 21/100,000/year recorded on 1993–2001. On the other hand, parathyroid malignancies are extremely rare, accounting for less than 5% of all primary hyperparathyroidisms.

The population prevalence of pituitary adenomas is 14% on CT studies and 23% on autopsy studies with incidence estimates ranging from 1 to 2/100,000/year. Most pituitary adenomas are actually nonsecretory and thus do not overproduce pituitary hormones. The most common secretory pituitary tumors are prolactinomas, followed by growth hormone (GH)-secreting tumors, while adrenocorticotropic hormone (ACTH)-secreting tumors are the least common of the three.

Adrenal cortical tumors have a prevalence of 0.35–4.6% on CT studies, and approximately 12% of such incidentalomas are found to be malignant. Functioning adrenal cortical tumors may secrete aldosterone (Conn syndrome), corticosteroids (causing ACTH-independent Cushing syndrome), or androgens (resulting in virilization). Pheochromocytomas have an incidence of approximately 1.5–2/1,000,000/year and follow the "rule of 10's" whereby 10% are malignant, 10% are bilateral, 10% are extra-adrenal (paragangliomas), 10% are familial (associated with multiple endocrine neoplasia syndromes, neurofibromatosis type 1, von Hippel–Lindau disease, and familial pheochromocytoma), and 10% are not associated with hypertension. Neuroendocrine tumors, which include carcinoid tumors and other gastrointestinal endocrine neoplasms (e.g., gastrinomas, insulinomas, glucagonomas, VIPomas, and somatostatinomas), are very rare with the incidence rates of all carcinoid tumors ranging from 0.5 to 2/100,000/year. The incidence rate of insulinomas is approximately 1.2/1,000,000/year, while all other gastrointestinal endocrine tumors are even more rare with incidence rates of less than 1/1,000,000/year.

Hereditary Syndromes Associated with Endocrine Tumors

A number of inherited medical syndromes are also known to be associated with endocrine tumors. These include:

(I) The autosomal dominant multiple endocrine neoplasia (MEN) type 1 syndrome that is caused by inactivating germline mutations of the *MEN1* gene and is characterized by neoplasias of the anterior pituitary gland, the parathyroid glands, the adrenal glands, and the pancreatic islets, as well as the occurrence of neuroendocrine tumors in the thymus, lungs, and stomach.

(II) The autosomal dominant multiple endocrine neoplasia (MEN) type 2A syndrome manifested by medullary thyroid cancer (MTC) in over 90% of cases, pheochromocytomas in 30–50% of cases, and hyperparathyroidism in 10–30% of cases. The hyperparathyroidism seen in MEN 2A is frequently less severe compared to MEN 1 or to sporadic cases.

(III) Similar to MEN type 2A, patients with MEN type 2B have an inherited autosomal dominant predisposition to MTC and pheochromocytomas. However, parathyroid hyperplasia is not commonly seen in MEN 2B patients who also develop distinct phenotypic abnormalities such as large lips, mucosal neuromas of the lips and tongue, intestinal ganglioneuromas, and, in some cases, a Marfanoid habitus. MEN 2B patients suffer the most aggressive form of MTC compared to all other hereditary syndromes.

(IV) Familial medullary thyroid cancer (FMTC) is considered to be a rare variant of MEN 2A characterized by a strong autosomal dominant predisposition to MTC but none of the other clinical manifestations of either MEN 2A or 2B. FMTC patients have the most indolent form of MTC compared to other MEN 2 variants. All three MEN 2 variants (MEN 2A, MEN 2B, and FMTC) are caused by germline missense gain of function mutations in the *RET* proto-oncogene. This signifies a notable difference compared to almost all other hereditary cancer syndromes that are caused by inactivating mutations (as opposed to activating "gain of function" mutations).

(V) von Hippel–Lindau (VHL) disease is an autosomal dominant syndrome associated with various benign and malignant neoplasias such as retinal, cerebellar, and other central nervous system hemangioblastomas, clear cell renal cell carcinomas, and renal cysts. VHL disease is also associated with a variety of endocrine tumors such as pheochromocytomas. The syndrome is caused by inactivating germline mutations of the *VHL* tumor suppressor gene. Pheochromocytomas are found in approximately 10% of VHL disease cases, although the prevalence of pheochromocytoma within affected VHL families can range from zero to 92%. Types IIA–C of VHL disease are defined by the presence of pheochromocytomas, while type I VHL patients do not develop these tumors. Other endocrine manifestations of VHL disease include pancreatic cystic disease and islet cell neuroendocrine tumors. Most of the pancreatic manifestations of VHL disease are asymptomatic, and the majority of the pancreatic islet cell tumors are clinically nonfunctioning.

(VI) Neurofibromatosis type 1 (originally known as von Recklinghausen disease) is an autosomal dominant disorder with an incidence of 30–38 cases per 100 000 live births that is characterized by café au lait spots (patchy hyperpigmentations of the skin), multiple neurofibromas, iris Lisch nodules (hamartomas), and various other manifestations such as optic gliomas, neurofibrosarcomas, seizures, bone abnormalities, vascular dysplasias, and learning disabilities. Neurofibromatosis type 1 is caused by loss of function mutations in the *NF1* gene and is associated with pheochromocytomas, which are found in 0.5–2% of cases. Patients with the less frequent neurofibromatosis type 2 syndrome do not develop pheochromocytomas or other endocrine oncology manifestations. Because neurofibromatosis type 1 is significantly more frequent than MEN 2 or VHL disease, neurofibromatosis is actually the most common hereditary cancer syndrome with predisposition to pheochromocytoma. Pheochromocytomas appear later in life (at approximately 50 years of age) in neurofibromatosis type 1 patients compared to MEN 2 or VHL disease patients. Neuroendocrine tumors of the gastrointestinal tract are also found in 1–3% of neurofibromatosis type 1 patients.

(VII) Cowden syndrome is a rare autosomal dominant syndrome characterized by hamartomatous overgrowth of tissues originating from all three germ cell layers and a predisposition for benign and malignant neoplasms of the breast, thyroid, and endometrium. Approximately 80% of patients with Cowden syndrome have detectable inactivating mutations in the *PTEN* tumor suppressor gene. Thyroid lesions such as goiter, adenomas, and thyroglossal cysts develop in about 60% of patients, whereas non-medullary thyroid cancer is found in approximately 10% of patients. Around 75% of Cowden syndrome patients will develop breast lesions including fibrocystic dysplasia and fibroadenomas, while breast cancer is found in 25–50% of cases. Endometrial cancer is also found in 5–10% of Cowden syndrome patients.

(VIII) Carney complex is a very rare autosomal dominant familial lentiginosis syndrome that is most commonly caused by inactivating mutations in the *PRKAR1A* gene. It is most commonly manifested by spotty skin pigmentation (freckling, café au lait spots, blue nevi) and lentigines, endocrine gland tumors, and myxomas of the heart, skin, and breast. The most frequent endocrine disorders are ACTH-independent Cushing syndrome secondary to bilateral primary pigmented nodular adrenocortical disease, acromegaly due to GH-producing pituitary adenomas or pituitary somatotropic hyperplasia, testicular tumors (most commonly large cell calcifying Sertoli cell tumors), ovarian cysts that may sometimes progress to ovarian cancer, and thyroid gland tumors with a predisposition to non-medullary thyroid cancer.

(IX) Other rare hereditary syndromes predisposing to endocrine tumors include the hereditary pheochromocytoma–paraganglioma syndrome (an autosomal dominant syndrome resulting from inactivation of the *SDHB*, *SDHC*, and *SDHD* genes; *SDHD* penetrant mutations are most often paternally inherited due to maternal imprinting), the hyperparathyroidism–jaw tumor syndrome (an autosomal dominant disorder resulting from inactivating mutations of the *HRPT2* gene and characterized by the occurrence of ossifying jaw fibromas and hyperparathyroidism due to parathyroid hyperplasia, parathyroid adenoma, or parathyroid cancer), and the pituitary adenoma predisposition (PAP) autosomal dominant disorder that results from inactivating mutations in the *AIP* tumor suppressor gene and is manifested by isolated familial predisposition for developing pituitary adenomas, the majority of which are GH secreting somatotropinomas, although some may also be paired with prolactinomas.

Endocrine Tumor Markers

Endocrine tumor markers are usually more site specific compared to other tumor markers and are frequently used in the diagnosis and follow-up of endocrine neoplasms. Carcinoids produce 5-hydroxytryptamine (5-HT) which is metabolized to 5-HIAA in the liver and lungs and excreted by the kidneys. Thus, measurement of the 24-h urinary excretion of 5-hydroxyindoleacetic acid (5-HIAA) is the most useful initial diagnostic test for carcinoid syndrome. Serum 5-HT is less stable and more difficult to measure and demonstrates a nonlinear relationship with tumor mass which limits its utility in tumor monitoring. In those rare tumors, mostly of the foregut (gastroduodenal and bronchial tumors), that produce 5-hydroxytryptophan (5-HTP) instead of 5-HT, measurement of urinary 5-hydroxytryptamine (5-HT) may also be used diagnostically. This is because 5-HTP is converted to 5-HT in the kidneys and subsequently excreted in the urine. Serum chromogranin A (CG-A) can be used in the diagnosis and monitoring of various neuroendocrine tumors such as carcinoids, pheochromocytomas, neuroblastomas, functioning or nonfunctioning islet cell tumors, medullary thyroid carcinomas, and pituitary tumors. CG-A is elevated in nearly all cases of patients with carcinoid syndrome due to neuroendocrine hindgut tumors and in 80–90% of cases due to midgut or foregut tumors. The above markers are usually elevated only after hepatic metastasis of the carcinoids has occurred.

Plasma and urine metanephrines are preferred to catecholamines, homovanillic acid (HVA), or vanillylmandelic acid (VMA) for the diagnosis and monitoring of pheochromocytomas and of related tumors such as paraganglionomas. Urine HVA and VMA measurements are today preferred only for the diagnosis of neuroblastomas, as metanephrines have not been sufficiently validated in these tumors. Calcitonin levels are associated with disease burden and predict outcome in patients with MTC and are used for diagnosing and monitoring this disease. Early data also indicate the utility of procalcitonin in the diagnosis and follow-up of MTC. Measurement of serum thyroglobulin (TG) is commonly used to monitor treatment response and disease progression or recurrence in patients with follicular cell-derived thyroid cancers. On the other hand, the utility of TG as a primary diagnostic tool is compromised by the fact that other thyroid lesions such as diffuse or nodular goiter or the autoimmune follicular destruction seen in Hashimoto thyroiditis (a thyroid autoimmune disease in which the immune system attacks and destroys the cells of the thyroid gland; the thyroid helps set the rate of metabolism, which is the rate at which the body uses energy) or Graves disease may also increase serum TG levels. TG levels may also be low in small thyroid malignancies. Furthermore, TG use in the follow-up of thyroid cancer may be limited by the development of anti-TG autoantibodies found in 15–30% of thyroid cancer patients.

Other endocrine tumor markers include: (i) ACTH which is increased in Cushing syndrome and may also be secreted by small cell lung cancers. (ii) β-HCG is a well-established marker used for the detection and monitoring of male and female germ cell tumors and trophoblastic tumors. (iii) Gastrin, glucagon, insulin, somatostatin, and the vasoactive intestinal peptide (VIP) are used in the diagnosis and follow-up of gastrinomas, glucagonomas, insulinomas, somatostatinomas, and VIPomas, respectively.

Endocrine Syndromes Secondary to Functioning Neuroendocrine Tumors

Carcinoid tumors of the gastrointestinal (GI) tract are rare slow-growing neoplasms that can secrete various biologically active humoral factors, including 5-HT and 5-HTP, that are responsible for an array of symptoms such as flushing, diarrhea, wheezing, and heart disease. In the vast majority of cases, liver metastasis is a prerequisite for the development of this "carcinoid syndrome." The carcinoid syndrome most frequently occurs secondary to small bowel carcinoid tumors originating from the midgut. This is because carcinoids derived from the hindgut or the foregut are most commonly nonfunctioning. Unlike carcinoid tumors, liver metastasis is not required for the development of the endocrine syndromes associated with insulinomas, glucagonomas, or gastrinomas.

Insulinomas are the most common functioning pancreatic islet cell neoplasms. Hypersecretion of insulin by insulinomas can cause hyperinsulinemic hypoglycemia. The majority of these tumors have benign histology and only 5% to 10% are malignant. Patients with either benign or malignant insulinomas share the same clinical presentation of fasting hypoglycemia. Metastatic lesions may also secrete insulin, and therefore, full enucleation of both the primary and metastatic tumors is required to achieve complete symptom resolution. On the opposite end of the glucose control spectrum, glucagonomas are very rare tumors usually found on the pancreatic tail that secretes the hyperglycemic hormone glucagon which, in addition, has powerful catabolic effects. The majority of glucagonomas are malignant. Although the most common clinical presentation of glucagonomas is unexplained weight loss, the classic endocrine syndrome associated with these tumors is the necrolytic migratory erythema (NME), a characteristic cutaneous skin eruption. About 70% of patients with glucagonoma will initially present with NME, and almost all patients will eventually develop this rash. Approximately 75% of glucagonoma patients will develop diabetes mellitus.

Gastrinomas are the second most common functioning pancreatic islet cell tumors and are the major cause of the Zollinger–Ellison syndrome which is characterized by gastric acid hypersecretion and peptic ulcer disease secondary to excessive gastrin production and release. All gastrinomas should be considered as potentially malignant regardless of histology. Approximately one third of gastrinomas have already metastasized at the time of diagnosis. VIPomas are the third most common functional neuroendocrine tumor of the pancreas and are characterized by hypersecretion of VIP resulting in the watery diarrhea, hypokalemia, and hypochlorhydria or achlorhydria (WDHA) syndrome, also known as Verner–Morrison syndrome or pancreatic cholera syndrome. The majority of VIPomas are malignant and will have metastatic lesions by the time of diagnosis. Somatostatinomas are very rare tumors that secrete excessive amounts of somatostatin and most commonly originate from the pancreas and the duodenum. Approximately 75% of somatostatinomas will have metastasized by the time of the diagnosis. The most common

symptoms at presentation are abdominal pain and weight loss. In addition, approximately 10% of tumors found in the pancreas cause the classic somatostatinoma endocrine syndrome that is characterized by cholelithiasis, diabetes mellitus, and diarrhea with steatorrhea. Patients with duodenal somatostatinomas do not develop the somatostatinoma syndrome and may not always have abnormal circulating somatostatin levels.

Endocrine Paraneoplastic Syndromes

Endocrine paraneoplastic phenomena are produced by the ectopic secretion of hormones by cancers. Neuroblastoma and small cell lung cancer (SCLC) are the tumors most commonly associated with paraneoplastic syndromes in children and in adults, respectively. The syndrome of inappropriate antidiuresis (SIADH) is the most frequent cause of euvolemic hyponatremia and can occur in a variety of cancers. The majority of cases are caused from the ectopic production and secretion of arginine vasopressin (AVP). Paraneoplastic SIADH is most commonly developed in patients with lung cancer, especially in SCLC. It also frequently occurs in patients with intracranial malignancies. Paraneoplastic ectopic ACTH syndrome is most commonly seen in patients with aggressive tumors, such as SCLC, resulting in rapid onset, severe Cushing syndrome. In such cases, cortisol secretion is not suppressed by high-dose dexamethasone. More indolent tumors such as bronchial carcinoid tumors may also produce paraneoplastic ACTH syndrome that is less severe and more slowly progressive. Of note, close to one third of cases will show suppression of cortisol secretion by high-dose dexamethasone as these tumors demonstrate a more pituitary-like behavior.

Hypercalcemia in patients with malignancies may be associated with metastatic bone disease (especially in osteolytic disease), with extensive bone destruction secondary to multiple myeloma bone disease and with ectopic production of parathyroid hormone-related protein (PTHrP) that mimics the effects of parathyroid hormone (PTH) on bone resorption. Ectopic PTHrP secretion accounts for up to 80% of malignancy-associated hypercalcemia. The most common cancers associated with ectopic PTHrP secretion are squamous cell cancers (most commonly of the lungs, head, and neck); renal, bladder, breast, and ovarian cancers; as well as non-Hodgkin lymphomas, chronic myelogenous leukemia (CML), and other leukemias. Non-islet cell tumor hypoglycemia is a paraneoplastic phenomenon resulting from the secretion by a non-pancreatic cancer of a hypoglycemic substance such as the insulin-like growth factor I (IGF-I) and IGF-II.

Hormonal Influences on Tumor Growth

A variety of hormones may influence tumor growth and perhaps even tumor initiation directly by increasing tumor cell proliferation and inhibiting apoptosis as well as indirectly by supporting tumor neoangiogenesis. Sex steroids have thus been associated with tumor progression in hormone-sensitive cancers such as breast and prostate cancers and may also influence other nonclassical hormone-sensitive neoplasias including endometrial, ovarian, and colon cancers. There is also growing evidence for a role of the insulin–IGF axis in the progression of a variety of neoplasias including pancreatic, colon, kidney, and endometrial cancers.

Hormone-Sensitive Tumors

Prostate and breast cancers are the two classic hormone-sensitive cancers. Worldwide, prostate cancer is the second most common cancer among men. Endocrine treatment of prostate cancer is based on the fact that prostate cancer cells are androgen dependent and undergo apoptosis following androgen deprivation. Accordingly, androgen deprivation therapy is the cornerstone of advanced prostate cancer management. Androgen deprivation regimens include gonadotropin-releasing hormone receptor (GnRH) agonists or antagonists with or without concomitant antiandrogens. The vast majority of patients (80–90%) will initially respond to this hormonal manipulation. However, cancer cells will typically adapt and become resistant to androgen blockade after 18–24 months of treatment. The common practice currently in this setting is to begin palliative chemotherapy which can achieve short-term responses. However, the role of further hormonal manipulations after the failure of initial androgen ablation is an area of intense investigation. Such strategies may include the administration of less commonly used antiandrogens such as nilutamide, adrenal suppressants such as ketoconazole, estrogens, corticosteroids, or newer compounds such as abiraterone, an inhibitor of androgen biosynthesis.

Breast cancer is the most common cancer and the leading cause of cancer death in women worldwide. Estrogen and progesterone receptor positivity of breast malignancies is both prognostic of favorable outcome and a predictor of response to endocrine therapy. The most commonly used hormonal modalities include the selective estrogen receptor modulator (SERM) tamoxifen or aromatase inhibitors such as anastrozole, letrozole, and exemestane.

Late Effects of Cancer Therapy on the Endocrine System

Both chemotherapy and radiation therapy can produce endocrine abnormalities on cancer survivors. These late effects can occur even decades after completion of the cytotoxic therapies. Survivors of childhood malignancies in particular may experience disruption of growth and pubertal onset resulting in reduced growth velocity and impaired somatic development. Cranial radiation therapy affecting the hypothalamo–pituitary axis can also cause precocious puberty. Radiation-induced hypopituitarism may also result in gonadotropin, ACTH, and thyroid-stimulating hormone (TSH) deficiencies. Furthermore, cytotoxic chemotherapy and radiation treatment regimens can directly damage the gonads. Various strategies are currently being explored to prevent such gonadal sequelae by using regimens such as doxorubicin, bleomycin, vinblastine, and dacarbazine (ABVD) in dosages that have less impact on the gonadal axis. Furthermore, a number of strategies for the cryopreservation of sperm, oocytes, and embryos are being used to preserve the fertility of cancer survivors.

Patients who receive neck irradiation treatment are at increased risk of developing hypothyroidism, thyroid adenomas, Graves disease, thyroiditis, and thyroid fibrosis. These patients also have significantly higher incidence of thyroid cancer (mainly papillary thyroid cancer and, less frequently, follicular cancer) starting 5–10 years after the radiation treatment and remain at increased risk for several decades. Neck irradiation also increases the incidence of hyperparathyroidism after a long latency period of 25–47 years.

Conclusion

Endocrine oncology is a broad medical discipline covering a heterogeneous group of tumors ranging from rare but frequently dramatic endocrine malignancies and hereditary syndromes to some of the most common malignancies afflicting humankind. Ongoing investigations to elucidate the endocrine pathways and feedback mechanisms related to tumor pathophysiology will improve the detection, prognosis, and treatment of endocrine tumors and hormone-sensitive malignancies.

References

Hay ID, Wass JAH (2008) Clinical endocrine oncology, 2nd edn. Wiley-Blackwell, Malden
Stephen PE (2000) Endocrine oncology. Humana Press, Totowa
Sturgeon C (2009) Endocrine neoplasia. Springer, New York

Endocrine Therapy

Synonyms

Antihormone therapy; Hormone therapy

Definition

Therapy used in ▶ breast cancer, or ▶ ovarian cancer, based on the blockade of steroid hormone receptor activity. The manipulation of hormones in order to treat a disease or condition.

Cross-References

▶ Breast Cancer
▶ Fulvestrant
▶ Ovarian Cancer
▶ Progestin

Endocrine Therapy in Breast Cancer

Claudia Fumarola
Department of Experimental Medicine, Unit of Experimental Oncology, University of Parma, Parma, Italy

Synonyms

Breast cancer antiestrogen therapy; Breast cancer hormonal therapy; Breast cancer hormone therapy

Definition

Endocrine therapy in breast cancer is one of the earliest targeted therapy developed for cancer and is used to treat estrogen receptor alpha (ERα)-positive breast tumors that depend on estrogens for their growth. It works by blocking estrogen signaling through two main modalities: (1) targeting the ERα protein and (2) depriving cancer cells of estrogen supply either by ovarian suppression in premenopausal women or by inhibition of extraovarian estrogen production in postmenopausal women. Endocrine therapy is currently used both in early-stage as well as in advanced/metastatic breast cancer.

Characteristics

ERα-positive (ERα+) breast cancer accounts for approximately 70% of all breast cancers and is typically characterized by positivity for progesterone receptor (PgR) expression, less frequent amplification/overexpression of human epidermal growth factor receptor 2 (HER2), lower proliferation rate and grade, and lower risk of recurrence as compared with other breast cancer subtypes. Based on gene expression profiling, ERα + breast cancers can be further divided into two distinct molecular subtypes, luminal A and luminal B, with the second showing relatively lower expression of ERα and ERα-related genes like PgR, higher proliferation, and poorer prognosis.

Estrogen Receptors

Estrogen signaling is mediated by two distinct receptors, ERα and ERβ, that are members of the nuclear hormone receptor superfamily and distinctly regulate transcription driving growth, proliferation, and differentiation, among many cellular processes.

ERα is well characterized as a crucial driver of breast carcinogenesis. It is activated by binding of estrogens, among which 17β-estradiol (E2) is the most potent ligand. The ERα presents an N-terminal domain with a region called AF1 (hormone-independent transcriptional activation function 1), a central DNA-binding domain, and a C-terminal domain with a region called AF2 (hormone-dependent activation function 2). Estrogen-mediated activation of ERα occurs through both genomic and non-genomic mechanisms. Genomic mechanisms include the "classical" and the "nonclassical" pathways. In the "classical" pathway, estrogen-activated ERα proteins form dimers that directly bind to specific sequences called estrogen-responsive elements (EREs), located in the promoter regions of target genes. Transcription of these genes is enhanced by recruitment of coactivator proteins such as amplified in breast cancer 1 (AIB1), nuclear receptor coactivator 1 (NCOA-1/SRC1), and p300/CBP-associated factor (PCAF). On the other hand, the binding of corepressor proteins such as nuclear receptor corepressor (N-CoR) and silencing mediator of retinoid and thyroid receptor (SMRT) to the ERα complex results in the downregulation of its transcriptional activity. Therefore, the recruitment of coactivators or corepressors affects the magnitude of ERα transcriptional activation. In the "nonclassical" pathway, ERα regulates gene transcription without directly binding to DNA via protein-protein interactions with other DNA-binding transcription factors such as AP-1, SP-1, and NF-kB. This mechanism is involved in the regulation of a large number of estrogen-responsive genes, including insulin-like growth factor I receptor (IGF-IR), c-Myc, cyclin D1, and c-fos. In addition to genomic pathways, estrogens exert rapid non-genomic actions (membrane-initiated steroid signaling) mediated by a subpopulation of membrane-bound ERs that can directly interact with membrane kinase receptors such as IGF-IR, epidermal growth factor receptor (EGFR), and HER2 leading to activation of various signaling cascades including the Ras/Raf/mitogen-activated protein kinase (MAPK) pathway and the phosphatidylinositol 3-kinase PI3K/AKT/mTOR pathway. It is worth noting that ERα can be directly activated in the absence of estrogens through phosphorylation at multiple sites by a variety of kinases, including MAPK, AKT, p90 ribosomal S6 kinase (Rsk), protein kinase A (PKA), and c-Src. Of these sites, serine 118 and serine 167 located in the AF1 region and phosphorylated by MAPK and AKT, respectively, play a major role in this pathway.

The role of ERβ in breast cancer is less clear and its prognostic value is still controversial. In vitro and in vivo findings suggest that ERβ is a potential tumor suppressor that opposes ERα, promoting cell differentiation and inhibiting ER-α-mediated proliferation. It is estimated that approximately 50% of human primary breast cancers are positive for ERβ, but its expression is lost during breast cancer progression, most likely due to promoter hypermethylation. Therefore, positive expression of ERβ appears to correlate with a favorable prognosis.

Therapeutic Strategies Based On ERα Targeting

Selective Estrogen Receptor Modulators (SERMs)
SERMs are a class of compounds that act by directly competing with E2 for binding to the ER and exert estrogen antagonist action in some target tissues while acting as estrogen agonists in

others. SERMs bind to both isoforms of ER receptor with different effects. Indeed they function as pure antagonists when acting through ERβ but can function as partial agonists when acting through ERα. The most common and successfully used member of this group of drugs is the nonsteroidal triphenylethylene tamoxifen (Nolvadex). First described in 1966 and approved for use in women with advanced breast cancer in the 1970s, tamoxifen is currently indicated in the advanced disease and in neoadjuvant and adjuvant setting and is still the antihormonal treatment of choice for premenopausal patients. In addition, it is used as endocrine therapy for ductal carcinoma in situ (DCIS) and remains the appropriate treatment for breast cancer prevention in premenopausal women at elevated risk. Tamoxifen requires metabolic activation by cytochrome P450 (CYP) enzymes to generate active metabolites (4-hydroxytamoxifen and endoxifen), which have 30- to 100-fold higher affinity for ERα than the parental drug. The antiestrogenic activity of these molecules is due to their ability to induce a conformational change of ER that leads to recruitment of corepressors, rather than coactivators, to the promoter regions of estrogen-responsive genes, thus blocking AF2-mediated transcription. In addition to the antagonist properties in the breast, tamoxifen exhibits strong estrogen agonist activities in the uterine endometrium, bone, and cardiovascular system. The underlying mechanism involves the transactivation of responsive genes through the AF1 function of ERα that is not inhibited by tamoxifen. These effects account for increased risk of thromboembolic complications or endometrial cancer especially in postmenopausal women. Moreover, they are associated with increased bone mineral density of the axial skeleton; although in premenopausal women, there may be a decrease in bone density.

It is worth noting that premenopausal women treated with tamoxifen do not experience increases in endometrial cancer risk and blood clots, so the risk:benefit ratio is strongly in favor of tamoxifen treatment.

Nonsteroidal tamoxifen-like triphenylethylenes such as toremifene, droloxifene, and idoxifene, as well as the benzothiophenes raloxifene and arzoxifene, are SERMs that inhibit breast cancer but do not appear to be more efficacious than tamoxifen. Moreover, patients who relapse on tamoxifen do not appear to respond to these SERMs.

Among these SERMs, raloxifene is used for the prevention and treatment of osteoporosis in postmenopausal women. This drug is also approved by the Food and Drug Administration (FDA) for reducing the risk of invasive breast cancer in postmenopausal women.

Selective Estrogen Receptor Downregulators (SERDs)

SERDs are a class of compounds that selectively downregulate ERα cellular levels. They are also known as "pure antiestrogens" since they exert estrogen antagonist activities without showing agonist effects. Among them, fulvestrant (ICI 182780, Faslodex) is the only treatment entered in the clinic and is currently used in postmenopausal women with ERα+, locally advanced or metastatic breast cancer who have progressed on prior endocrine therapy. Fulvestrant acts by competitively binding to ERα with a much stronger affinity as compared with tamoxifen. Due to its steroidal structure containing a long bulky side chain, fulvestrant induces a distinct conformational change of ERα that results in rapid receptor immobilization, polyubiquitination, and degradation in the nuclear matrix. Although preclinical studies have demonstrated the efficacy of fulvestrant at inhibiting breast cancer cell growth either in vitro or in animal models, clinical data are less encouraging. The reason for this is the extremely poor fulvestrant bioavailability that renders difficult to achieve the intratumoral levels required for effective ERα turnover. Indeed, results from clinical trials indicate that higher doses of fulvestrant compared with the currently approved dose are associated with improved response rate, as a result of increased serum steady-state levels of the drug and consequent increased turnover of ERα. Importantly, a high-dose regimen remains equally well tolerated. Nevertheless fulvestrant use continues to be limited to second-line therapy in postmenopausal setting.

Therapeutic Strategies Based On Estrogen Deprivation

Ovarian Ablation/Suppression

Suppression of ovarian function is one of the therapeutic strategies to treat ERα + breast tumors in premenopausal women where estrogens are mainly produced by the ovaries under the control of the hypophysis-hypothalamus axis. It can be achieved by permanent intervention via surgical removal or direct radiation therapy of the ovaries, or by reversible approaches based on administration of luteinizing hormone-releasing hormone (LHRH) analogs, such as goserelin, triptorelin, and leuprolide, that downregulate gonadotropin and hence estrogen to postmenopausal levels. Permanent and temporary approaches have comparable therapeutic effects; however, the second are currently preferred by most patients as well as physicians, although they present the drawback of higher economic cost. All forms of ovarian suppression cause a rapid onset of menopause symptoms (hot flashes, night sweats, mood swings, vaginal dryness), which can be severe. Ovarian ablation/suppression has been proved to be effective as adjuvant therapy for women under 50 years and to significantly reduce the risk of recurrence and breast cancer mortality. However, its use as monotherapy is currently not recommended. Its combination with CMF (cyclophosphamide, methotrexate, and 5-fluorouracil) chemotherapy does not provide any additional benefit because of the suppressive effects of this chemotherapy regimen on ovarian function. By contrast, ovarian suppression may improve outcomes when combined with anthracycline- and taxane-based treatments that tend to be less toxic to the ovaries than CMF; however, no clinical data proving the efficacy of these combinations are so far available. Clinical trials testing the benefit of ovarian suppression with tamoxifen compared to tamoxifen alone with and without chemotherapy are currently ongoing. In conclusion, the current standard adjuvant endocrine therapy recommended by International Consensus Statements for premenopausal women with ERα + breast cancer is tamoxifen over 5 years (with or without chemotherapy),

and ovarian suppression should be considered only for those patients that cannot tolerate or refuse the standard therapy.

Aromatase Inhibitors

The majority of ERα + breast cancer patients are postmenopausal women. In these patients, estrogens are no longer produced in the ovaries. Instead they are synthesized in a number of extragonadal tissues including adipose and breast tissues, where they act as paracrine or even intracrine factors. Thus, circulating levels of estrogens in postmenopausal women are low, whereas their concentration in breast cancer tissue can reach levels as high as those of premenopausal women, indicating that estrogens produced locally are those implicated in breast cancer development. Extragonadal production of estrogens is mediated by aromatase, a cytochrome P450 that catalyzes the conversion of androgens to estrogens. In particular, androstenedione and testosterone are converted to estrone and estradiol, respectively. Upregulated expression of aromatase in breast tissue plays a central role in the progression of postmenopausal ERα + breast cancer; therefore, selective inhibition of aromatasic function has become an effective therapeutic strategy in this setting. So far, three generations of aromatase inhibitors (AIs) have been developed. The first-generation nonsteroidal inhibitor aminoglutethimide (AG) was removed from the clinic because of its inhibitory effects on aldosterone, progesterone, and corticosterone biosynthesis. The second-generation steroid analog formestane (4-hydroxyandrostenedione) was more selective and well tolerated, but its clinical use was limited due to the requirement of intramuscular administration. The third-generation inhibitors, developed in the early 1990s, include anastrozole (Arimidex), letrozole (Femara), and exemestane (Aromasin). Anastrozole and letrozole are nonsteroidal triazole derivatives that interact with the heme prosthetic group of aromatase, acting as competitive inhibitors with respect to the androgen substrates. Exemestane is a steroidal inhibitor catalytically converted by aromatase into active intermediates that irreversibly bind to the enzyme, causing its inactivation.

By eradicating estrogens, AIs suppress both genomic and non-genomic action of ERα, and unlike tamoxifen they do not display agonist activity. All three agents are orally administered and show near-complete specificity and improved potency with respect to suppression of aromatase activity and circulating estrogen levels as compared with previous generations of AIs. In addition, they have been shown to offer a considerable clinical benefit with better tolerability as compared with tamoxifen standard treatment, being associated with a lower incidence of endometrial cancer, thromboembolic events, and contralateral breast cancer occurrence. AIs have challenged tamoxifen as first-line therapy in advanced/metastatic breast cancer as well as in neoadjuvant setting. As adjuvant treatment, current guidelines suggest different options for AIs: upfront therapy for 5 years as an alternative to tamoxifen, sequential therapy following 2–3 years use of tamoxifen, or extended therapy following 5 years of tamoxifen. AIs are not prescribed for premenopausal women, unless used in combination with ovarian suppression.

Resistance to Endocrine Therapy

Despite the documented benefits of ER-targeted therapy in breast cancer, not all patients respond to endocrine manipulation (de novo resistance) and a large number of patients initially responders will eventually develop disease progression or recurrence while on therapy (acquired resistance). A variety of molecular mechanisms have been involved in resistance to endocrine therapy. One of these mechanisms may be the loss of ERα either by clonal selection of ERα-negative cells or by transcriptional suppression of ERα gene expression. Actually, increased signaling from overexpressed growth factor receptors such as EGFR and HER2 has been demonstrated to contribute to downregulation of ERα protein expression. However, experimental and clinical evidence indicates that development of endocrine resistance is more frequently associated with maintenance or enhancement of ERα signaling. Increased ligand-independent phosphorylation and activation of ERα may occur as a consequence of EGFR, HER2, or IGF-IR

overexpression. Non-genomic crosstalk between ERα and growth factor receptors ultimately leads to activation of intracellular signaling cascades involved in cell proliferation/survival, such as the MAPK and the PI3K/AKT/mTOR pathways. On the other hand, both MAPK and AKT can directly augment the genomic activities of ERα, even in the absence of estrogens, further contributing to endocrine resistance. In tumor cells that have developed resistance to aromatase inhibition, crosstalk between ERα and growth factor signaling may contribute to adaptation to a sustained low-estrogen environment enhancing cell sensitivity to very low levels of residual estrogens. Another mechanism of de novo and acquired resistance to endocrine therapy involves changes in the levels of ERα co-regulator proteins. In particular, these changes may affect sensitivity to tamoxifen due to its mixed agonist/antagonist properties. For example, both overexpression of the coactivator AIB1 and reduced expression of the corepressor N-CoR may enhance the agonist effects of tamoxifen on ERα, contributing to resistance. The metabolism of tamoxifen mediated by CYP enzymes has been implicated in intrinsic resistance to therapy. In particular, CYP2D6 genetic variation has been shown to affect the plasma concentration of tamoxifen active metabolites and consequently the clinical outcomes of patients treated with tamoxifen.

A number of therapeutic approaches are in various stages of clinical development to either delay or overcome endocrine resistance. Great efforts have been directed toward assessing the efficacy of strategies based on simultaneous targeting of estrogen and growth factor signaling pathways through combinations of endocrine therapy with drugs that inhibit the HER family receptors, such as the monoclonal anti-HER2 antibody trastuzumab and small-molecule tyrosine-kinase inhibitors (TKIs) against EGFR or HER2 (lapatinib, gefitinib). In addition, combinations of endocrine therapy with multiple novel targeted agents directed against downstream signaling pathways are currently under clinical investigation. These agents include farnesyltransferase inhibitors, Raf and MEK inhibitors, as well as PI3K, AKT, and mTOR inhibitors.

Cross-References

References

Griggs JJ, Somerfield MR, Anderson H, Henry NL, Hudis CA, Khatcheressian JL, Partridge AH, Prestrud AA, Davidson NE (2011) American Society of Clinical Oncology Endorsement of the Cancer Care Ontario practice guideline on adjuvant ovarian ablation in the treatment of premenopausal women with early-stage invasive breast cancer. J Clin Oncol 29(29):3939–3942

Johnston SR, Martin LA, Leary A, Head J, Dowsett M (2007) Clinical strategies for rationale combinations of aromatase inhibitors with novel therapies for breast cancer. J Steroid Biochem Mol Biol 106(1–5):180–186

McDonnell DP, Wardell SE (2010) The molecular mechanisms underlying the pharmacological actions of ER modulators: implications for new drug discovery in breast cancer. Curr Opin Pharmacol 10(6):620–628

Puhalla S, Bhattacharya S, Davidson NE (2012) Hormonal therapy in breast cancer: a model disease for the personalization care. Mol Oncol 6:222–236

Ring A, Dowsett M (2004) Mechanisms of tamoxifen resistance. Endocr Relat Cancer 11(4):643–658

Endocrine-Related Cancers

Hao Zhang
The University of Texas MD Anderson Cancer Center, Houston, TX, USA

Synonyms

Cancers of hormone-responsive organs or tissues; Endocrine-responsive cancers; Hormone-induced cancers; Hormone-related cancers; Malignant neoplasias originating in endocrine tissues or their target organs; Malignant tumors of the endocrine glands; Reproductive system cancers

Definition

The term endocrine-related cancers refers mainly to malignant tumors in either the endocrine glands or the endocrine target tissues. The endocrine system comprises the glands in the body that make and secrete hormones. All endocrine glands, such as the ovaries and testes, secrete hormones directly into the bloodstream, where they travel to a target organ or cell, such as the breast and prostate, to trigger a specific reaction. These glands are primarily involved in controlling a wide range of activities and functions in the body, such as metabolism, growth, and reproduction.

As is true for other types of malignant neoplasias, endocrine-related cancers are characterized by potentially out-of-control growth that either expands locally by invasion or spreads systemically by ▶ metastasis. In many cases, hormone production is abnormal when cancers are found in the endocrine glands, and hormone response can become disrupted when cancers occur in target tissues.

Other malignancies are also associated with the endocrine system. Carcinoids (▶ carcinoid tumors), a group of slow-growing but often malignant tumors, originate in the hormone-producing cells of the diffuse neuroendocrine. A large variety of other cancers that are not primary endocrine cancers also occur in endocrine tissues, including meningioma, ▶ astrocytoma, and lymphoma. ▶ Metastases from other cancers can be found in such endocrine organs as the adrenal glands, pituitary, and thyroid gland. Malignancies such as ▶ lung cancer can also produce hormones and cause patients to display clinically ectopic hormone syndromes.

Characteristics

The subject of endocrine-related cancers encompasses a vast area of clinical medicine and cancer biology. This entry mainly describes endocrine gland cancers. ▶ Breast cancers and prostate cancers, hormone target tissue cancers with a very high prevalence, will be described elsewhere.

Pathogenesis
The special natures of the origin and pathogenesis of endocrine tumors derive from a number of special properties of endocrine cells. As a result of environment risk and the stepwise accumulation of genetic defects, normal cells are transformed into malignant cells by a failure of normal growth regulation. The great majority of endocrine cells are stable, having a low growth rate and a long life span in contrast to the renewing tissues that are maintained by stem cells and have a short intermitotic life. In endocrine tissues, both function and growth are closely linked and are stimulated by one trophic hormone. In contrast to the renewing tissues, which undergo frequent cell divisions during which mutations commonly occur, endocrine tissues divide slowly and mutations are unlikely, but those that do occur are retained for a long time. In most types of endocrine cells, the mutation rate is low, but mutations accumulate; the chance of mutations is increased by any stimulus leading to mitotic growth in the endocrine tissues.

There are a number of possible explanations for the pathogenesis of endocrine cancers.

One possibility is the activation of ▶ oncogenes or inactivation of tumor suppressors. Defects in a variety of oncogenes and ▶ tumor suppressor genes such as *ras*, p53, TGFβ, and IGF-I have been described in thyroid carcinomas.

Another possibility is that abnormal physiologic regulation disrupts the balance between stimulation and inhibition, and the resultant hyperfunction in the gland results in adenoma.

Abnormal stimulation may also induce gene dysregulation, including oncogene activation or tumor suppressor deactivation or damage, which has been described in the carcinogenesis of a number of human endocrine tumors.

Prevalence
Endocrine gland tumors are not common. There is a relative paucity of epidemiological investigations of endocrine cancers mainly because of the rarity of individual cases and the occult nature of the diseases (including asymptomatic cases and their responses to therapy). In addition the reported incidence is also affected by variable levels of recognition or registration of the tumors in different clinical settings, including endocrinology, oncology, surgery, neurology, and even outpatient settings. In the United States, approximately 25,520 new cases of endocrine system cancer were diagnosed in 2004, and endocrine cancers have been estimated to account for approximately 4% of all new cancer cases each year. The most common malignancies of the endocrine system are gonadal tumors and thyroid tumors. In the United States, ovarian cancer accounts for 44% of all endocrine system cancers, with an annual incidence of 8 per 100,000 women. Testicular cancer has an annual incidence of 2 per 100,000 men. In the United States, thyroid cancer has an annual incidence of approximately 6 per 100,000 persons (accounting for 36% of all endocrine cancers), whereas the annual incidence in the United Kingdom is around 0.6–1.5 per 100,000. The other endocrine tumors are less common. ▶ Adrenocortical cancer and other ▶ carcinoid tumors are all extremely rare, having incidences as low as 0.5 per 100,000 per year in the United Kingdom. Parathyroid tumors are relatively more common, with an incidence of 28 per 100,000 per year in the United Kingdom, but the malignancy is still rare.

Clinical Presentation and Pathology
Most endocrine tumors are not malignant, although certain types when they do occur are

likely to be cancerous. Endocrine tumors affect all age groups.

The clinical manifestations vary. Some patients are asymptomatic, with tumors incidentally discovered at autopsy or during sophisticated imaging techniques; others have multiple hormone-related symptoms or severe symptoms at late malignancy. Severity of symptoms depends on the site and the malignancy. Most endocrine tumors of childhood are benign or are low-grade malignancies. A small percentage of gonadal and germ cell tumors, thyroid cancers, and adrenocortical tumors are high grade.

Clinical manifestations of endocrine gland tumors are often associated with abnormalities of hormone secretion. For example, patients with insulin-producing islet cell tumors of the pancreas may have severe effects of low blood sugar due to the excessive insulin in the blood secreted by tumors. In addition to the hormone-producing tumors or functional tumors, there is another category of nonfunctional tumors that do not secrete hormones. Both types of tumors have the potential to be malignant.

Endocrine system tumors involve all anatomic areas from the head to the pelvic area and include tumors in single and multiple sites: pituitary cancer, thyroid cancer, parathyroid cancer, adrenal gland cancer, and endocrine ▶ pancreatic cancers, as well as multiple endocrine neoplasia (MEN), ectopic endocrine tumors, neuroendocrine tumors, and metastases in the endocrine organs from other primary cancers.

Pituitary Tumors

Pituitary tumors are classified by the type of hormone they secrete. They are rarely malignant but, because of the production of a diverse spectrum of excess hormones varying with the cell type of origin, can cause a variety of health problems including ▶ Cushing syndrome, acromegaly, disturbance of milk secretion (prolactinoma), and local symptoms such as visual complication, headache, and injury to cranial nerves. In certain cases, tumors may cause pituitary hormone insufficiency due to pituitary tissue destruction.

Thyroid Tumors

Although only 5% of the tumors found on the thyroid are malignant, their biological behaviors may vary substantially, ranging from well-differentiated tumors to aggressive anaplastic cancers. A classification of thyroid tumors proposed by the World Health Organization (WHO) takes into account the histological features and clinical behavior of the tumor and classifies malignant thyroid tumors into four types (▶ thyroid carcinogenesis).

Parathyroid Tumors

Around 5% of parathyroid tumors are malignant. Overproduction of parathyroid hormone, a condition known as hyperparathyroidism, is a common condition associated with both benign and malignant tumors. Untreated hyperparathyroidism can result in osteoporosis (causing bones to become brittle and fracture easily), kidney stones, peptic ulcers, and nervous system problems.

Endocrine Pancreatic Tumors

Pancreatic islet cell tumors are rare, with malignant tumors even more rare. The most common types of tumors are gastrinomas, which are associated with Zollinger–Ellison syndrome. Pancreatic islet cell tumors include functioning and nonfunctioning varieties. Islet cells are multipotential in respect to peptide production, so functioning tumors can be classified into entopic hormone tumors (e.g., insulinoma, glucagonoma, and somatostatinoma) and ectopic hormone tumors (gastrinoma, VIPoma). Benign tumors are often treatable with surgery combined with medicine. For treatment of malignant tumors, in addition to surgery and radiation, cytotoxic ▶ chemotherapy has also been used.

Adrenal Tumors

There are two adrenal glands, one above each kidney in the back of the upper abdomen. The adrenal gland consists of two layers: an outside layer (the adrenal cortex) and an inner layer (the adrenal medulla).

Cancer of the adrenal cortex is also called ▶ adrenocortical cancer. The cells in the adrenal cortex make hormones that help the body work properly. When cells in the adrenal cortex become cancerous, they may make too much of one or more hormones, which can cause symptoms such as high blood pressure, weakening of the bones, and diabetes. If male or female hormones are affected, the person may undergo changes such as a deepening of the voice, growth of facial hair, swelling of the genitals, or swelling of the breasts.

Tumors that start in the adrenal medulla are called pheochromocytomas. About 10% of pheochromocytomas are malignant.

Most adrenal tumors are disorders of hormonal excess, but rarely malignant disease can cause adrenal insufficiency.

Ovarian Tumors

▶ Ovarian cancer can develop in the egg cells inside the ovary (▶ ovarian stromal and germ cell tumors), but most occur in the cells lining the outside of the ovary (epithelial ovarian cancer), and most of these tumors are benign.

Testicular Tumors

▶ Testicular cancer can occur in one or both of the testes. Most testicular tumors are malignant and account for approximately 1% of all cancers in men. Over 90% develop in the germ cells and are generally malignant. Only 4% involve the endocrine cells of the testes, and these tumors are rarely malignant.

Multiple Endocrine Gland Tumors

Some disorders result in the simultaneous occurrence of tumors on several endocrine glands. Many of these are inherited disorders, including multiple endocrine neoplasia syndromes, von Hippel–Lindau syndrome, and von Recklinghausen disease (neurofibromatosis types).

Carcinoids

▶ Carcinoid tumors are a group of slow-growing but often malignant tumors that originate in hormone-producing cells of the diffuse neuroendocrine system and have been found in almost every organ of the human body. Approximately 10% of carcinoids secrete excess amounts of hormones, most notably the peptide serotonin

(5-HT). Presenting symptoms include flushing, diarrhea, wheezing, abdominal cramping, and peripheral edema. The only curative therapy for carcinoid tumor is surgery.

Secondary Endocrine Malignancies

In addition to primary endocrine cancers, a large variety of other cancers – including meningioma, ▶ astrocytoma, and lymphoma – occur in endocrine tissues. Metastases from other cancers can be found in such endocrine organs as the adrenal glands, pituitary, and thyroid gland.

Ectopic Hormone Syndromes

Nonendocrine tumors can also produce hormones, causing patients to display clinically ectopic hormone syndromes. For example, lung cancers produce parathyroid hormone-related protein and ACTH. Ectopic hormone syndromes are important because of several clinical reasons. Such syndromes may appear as the first signal of a malignancy. Ectopic hormone production may serve as a tumor marker even before the tumor mass is apparent and may serve as follow-up evidence for a response to therapy.

Treatment

Many endocrine cancers can be treated and some can be cured. However, given the complex nature of endocrine tumors, therapy for such cancers is unlike treatment for many other types of cancer. Targeting of these cancers as well as normalizing hormone levels requires the cooperation of medical endocrinologists and endocrine surgeons to apply both surgical and biochemical therapies. Surgery is generally the treatment of choice for most endocrine cancers. Chemotherapy is frequently used and is most effective in germ cell tumors of the ovary and testis. Although some tumors are resistant to radiation, radiation therapy may nonetheless decrease the chance of recurrence and halt the spread of cancer cells. ▶ Hormone replacement therapy is frequently used.

Prognosis

Many benign and malignant endocrine tumors are treatable with a combination of surgery and medication, and the survival rate for many patients with endocrine cancers is good. The best outcome is for patients with ovarian or testicular germ cell tumors, with most achieving disease-free long-term survival. Prognosis for patients with pituitary and thyroid tumors and adrenal adenoma and MEN is generally good. The survival rate for patients with adrenal carcinoma is poor, with an average survival of approximately 1.5 years.

In the United States, 25,520 new cases of endocrine system cancer were diagnosed in 2004, with an estimated 2,440 deaths, for a ratio of deaths to incidence of 0.096 for endocrine system cancers.

Cancers of Endocrine Target Tissues

Hormones have wide effects on a variety of cellular processes, including cell growth and differentiation, metabolic activity, and the metabolism of substances. In fact, the endocrine system directly or indirectly influences all mammalian organs and tissues. Unbalance of endocrine hormones is an endogenous factor that may sensitize the cells to the carcinogenic insult, thereby promoting cancer development. It has been well established that sex hormone-induced cell proliferation plays a critical role in corresponding hormone-initiated carcinogenesis, as is reflected in the fact that the many types of human cancers, including breast cancer, prostate cancer, and ▶ endometrial cancer, occur in sex hormone target tissues.

More than 25% of human tumors originate in the endocrine organs or their target tissues. However, with aspect to prevalence, the major entity in the spectrum of endocrine-related cancers is endocrine target organs such as the breast and prostate. In the Western world, ▶ breast cancer is the most common tumor in women, and prostate cancer is the most prevalent malignancy in men and the second most common cause of male cancer deaths. They have attracted great attention in endocrinology because of not only their extremely high incidence but also their endocrinological nature. Specifically, breast cancer and prostate cancer, for instance, start as estrogen- and androgen-dependent noninvasive disease, respectively, and are sensitive to endocrine agents and treatable at an early stage with hormone therapy

and surgery. However, if left undetected or untreated, these cancers eventually develop into a more aggressive, hormone-independent, highly invasive disease, and the patients die of their diseases after tumor cells metastasize to distant sites in the bodies.

Cross-References

- ▶ Adrenocortical Cancer
- ▶ Astrocytoma
- ▶ Breast Cancer
- ▶ Carcinoid Tumors
- ▶ Chemotherapy
- ▶ Cushing Syndrome
- ▶ Endometrial Cancer
- ▶ Hormone Replacement Therapy
- ▶ Lung Cancer
- ▶ Metastasis
- ▶ Multiple Endocrine Neoplasia Type 1
- ▶ Neurofibromatosis 1
- ▶ Oncogene
- ▶ Ovarian Cancer
- ▶ Ovarian Stromal and Germ Cell Tumors
- ▶ Pancreatic Cancer
- ▶ Pancreatic Cancer Basic and Clinical Parameters
- ▶ Prostate Cancer
- ▶ *RAS* Genes
- ▶ Testicular Cancer
- ▶ Thyroid Carcinogenesis
- ▶ Tumor Suppressor Genes
- ▶ Von Hippel-Lindau Disease

References

Beuschlein F, Reincke M (2006) Adrenocortical tumorigenesis. Ann N Y Acad Sci 1088:319–334

Caplin ME, Buscombe JR, Hilson AJ et al (1998) Carcinoid tumor. Lancet 352:799–805

Epstein R (1996) Hormones, growth factors and tumor growth. In: Sheaves R, Jenkins P, Wass J (eds) Clinical endocrine oncology. Raven, New York, pp 39–43

Green S, Furr B (1999) Prospects for the treatment of endocrine-responsive tumors. Endocr Relat Cancer 6:349–371

Monson JP (2000) The epidemiology of endocrine tumors. Endocr Relat Cancer 7:29–36

See Also

(2012) Acromegaly. In: Schwab M (ed) Encyclopedia of Cancer, 3rd edn. Springer Berlin Heidelberg, p 18. doi:10.1007/978-3-642-16483-5_39

(2012) Adrenocortical Tumors. In: Schwab M (ed) Encyclopedia of Cancer, 3rd edn. Springer Berlin Heidelberg, pp 89-90. doi:10.1007/978-3-642-16483-5_121

(2012) Epithelial Ovarian Cancer. In: Schwab M (ed) Encyclopedia of Cancer, 3rd edn. Springer Berlin Heidelberg, p 1292. doi:10.1007/978-3-642-16483-5_6952

(2012) Germ Cell Tumors. In: Schwab M (ed) Encyclopedia of Cancer, 3rd edn. Springer Berlin Heidelberg, p 1541. doi:10.1007/978-3-642-16483-5_6905

(2012) Lymphoma. In: Schwab M (ed) Encyclopedia of Cancer, 3rd edn. Springer Berlin Heidelberg, p 2124. doi:10.1007/978-3-642-16483-5_3463

(2012) P53. In: Schwab M (ed) Encyclopedia of Cancer, 3rd edn. Springer Berlin Heidelberg, p 2747. doi:10.1007/978-3-642-16483-5_4331

(2012) Tumor Suppressor. In: Schwab M (ed) Encyclopedia of Cancer, 3rd edn. Springer Berlin Heidelberg, p 3803. doi:10.1007/978-3-642-16483-5_6056

(2012) Zollinger-Ellison Syndrome. In: Schwab M (ed) Encyclopedia of Cancer, 3rd edn. Springer Berlin Heidelberg, p 3980. doi:10.1007/978-3-642-16483-5_6303

Endocrine-Responsive Cancers

▶ Endocrine-Related Cancers

Endocurietherapy

▶ Brachytherapy

Endocytosis

Pier Paolo Di Fiore[1] and Simona Polo[2]
[1]IFOM, the FIRC Institute of Molecular Oncology, Milan, Italy
[2]University of Milan, Medical School, Milan, Italy

Synonyms

Clathrin-mediated endocytosis; Fluid phase endocytosis; Internalization; Pinocytosis; Receptor-mediated endocytosis

Definition

A process in eukaryotic cells consisting of a progressive invagination of a small region of the plasma membrane which is subsequently pinched off to form a cytoplasmic vesicle.

Characteristics

Eukaryotic cells use endocytosis to internalize plasma membrane, surface receptors and their bound ligands, nutrients, bacterial toxins, immunoglobulins viruses, and various extracellular soluble molecules. The molecular machinery of endocytosis is also largely overlapping that of synaptic vesicle recycling. Once an intracellular vesicle is formed, following the endocytic process, its content is trafficked through the ▶ endosomal compartment and normally destined to either degradation in the lysosomal compartment or to recycling to the cell surface. The definition of endocytosis should be limited to the process leading to the formation of vesicles of 100–200 nm, which is also known with the general name of pinocytosis.

On the contrary, phagocytosis is the intake of large particles, such as bacteria or parts of broken cells. It is used by many protozoans to ingest food particles and by blood cells (macrophages) to take in and destroy pathogens and dead-cell debris. After the binding of the target particle to the cell surface, the plasma membrane expands along the surface of the particle and eventually engulfs it. Vesicles formed by this process are much larger than those formed by endocytosis (1–2 μm).

Four different types of endocytic processes are described below:

1. Fluid phase endocytosis or macropinocytosis is the simpler and nonspecific form of endocytosis in which any fluid extracellular material is taken up at a rate that is simply proportional to its concentration in the extracellular fluid. Fluid phase endocytosis does not require a specific interaction of extracellular material with surface-bound receptor structures. Macropinocytosis shares similarities with

phagocytosis as both require activation of GTPases that, in turn, stimulates actin polymerization and depolymerization events necessary for membrane protruding. After the collapse of the ruffles onto the plasma membrane, macropinosomes are formed with a mechanism that is independent of the fission-inducing protein dynamin.

2. Clathrin-mediated endocytosis occurs through the formation of clathrin-coated pits at the plasma membrane, followed by the generation of clathrin-coated vesicles of 100–150 nm in diameter. The process requires, in addition to clathrin, the presence of the adaptor protein complex AP2, or of other adaptors. The general function of the adaptors is that of establishing a physical link between plasma membrane receptors and clathrin. The formation of clathrin-coated pits at the plasma membrane shares structural and molecular similarities with the budding of clathrin-coated vesicles from the Golgi apparatus, which also requires clathrin and a different adaptor complex, AP1. Clathrin-mediated endocytosis requires interaction of an extracellular ligand with a surface receptor and takes the name of receptor-mediated endocytosis (or internalization). Two types of receptor-mediated endocytosis are known: (a) constitutive and (b) ligand-induced.

   (a) In constitutive endocytosis, membrane receptors are continuously internalized and after sorting in the endosomal compartment, they are recycled back to the cell surface. When a ligand binds to the receptor, the ligand is also internalized and can undergo different metabolic destinies. Two paradigmatic examples are provided by the constitutive endocytosis of the low-density lipoprotein (LDL) receptor and of the transferrin receptor. In the former case, LDL complexed to cholesterol is internalized with its receptor. In the endosomes, the LDL receptor dissociates from the LDL–cholesterol complex and it is redirected to the cell surface for more cycles of internalization. The LDL–cholesterol complex is routed to the lysosomes, where LDL is degraded and free cholesterol is made available to the cell. The cycle of the transferrin receptor is more complex. Transferrin, bound to iron, is also internalized together with its receptor. In the endosomal compartment, the acidic pH causes the dissociation of iron. Iron-free transferrin (apotransferrin) remains, however, bound to the receptor, and it is recycled to the plasma membrane. The process of constitutive internalization is thus used by the cell mostly for the uptake of nutrients.

   (b) In the ligand-regulated process, internalization is triggered by the interaction of a ligand with its surface receptor. Both ligand and receptors are normally routed to the lysosomal compartment with ensuing degradation. However, a fraction of the internalized receptor can be redelivered to the plasma membrane, in a recycling process not dissimilar to that of constitutive endocytosis. The process of induced internalization therefore serves, in the majority of cases, as a downregulation mechanism to extinguish signals originating on the plasma membrane from signaling receptors, for instance ► Receptor Tyrosine Kinases (Tyrosine kinase receptors). It has, however, become increasingly clear that endocytosis is also required to propagate signals originated from surface receptors, thus not merely constituting an attenuation mechanism (see below).

3. Caveolae-mediated endocytosis occurs through flask-shaped, nonclathrin-coated surface structures called caveolae. The vesicles formed in this process are of 50–85 nm in diameter and caveolin-1 is an essential component, as caveolin-1-null mice lack caveolar structures. By this process, cells are able to internalize certain glycosylphosphatidylinositol-linked proteins, albumin, bacterial toxins as well as membrane-associated receptors like MHC class I, TGFβR, or E-cadherin. Caveolae remain for long periods at the plasma membrane, but their internalization can be

stimulated by various agents. These include the SV40 virus which uses caveolae for entry into cells and stimulates caveolar budding, as well as sterols and glycosphingolipids. Although the molecular circuitries underlying caveolar internalization are not completely understood, critical requirements for dynamin, Src kinases, protein kinase C (PKC), and actin recruitment have been demonstrated. It is important to point out that caveolae represent one type of cholesterol-rich microdomain on the plasma membrane and that in the absence of caveolin, lipid-raft-dependent trafficking occurs (see below).

4. Clathrin- and caveolin-independent endocytosis probably encompasses more than one pathway. Given the poor molecular definition, the mechanisms that govern caveolin- and clathrin-independent endocytosis remain in large part unknown, as illustrated by the fact that these pathways are described only in negative terms. These internalization routes have been shown to mediate the internalization of interleukin-2 receptor-β (IL2Rβ), some GPI-anchored proteins, and receptor tyrosine kinases.

It is likely that these different pathways have evolved so that pinocytosis can be coordinated with more complex aspects of cell physiology, such as signal transduction, development, and modulation of the cell responses to, and interaction with, its environment.

Endocytosis and Signaling

One aspect of endocytosis that has attracted much attention is the relationship between internalization of surface receptors and their ability to transduce signals within the cell. Many surface receptors (including, but not limited to, receptor tyrosine kinases) act as molecular transducers, by binding to extracellular ligands and delivering their signal to the cell, with various mechanisms. By and large these receptors are removed from the cell surface through endocytosis and destined to lysosomal degradation. Thus, for many years, endocytosis has been considered the major form of receptor attenuation leading to signal

extinction. It has, however, become clear that signaling receptors continue to signal while en route through the endosomal compartment, and that in many cases they gain access, in this compartment, to interactors and/or substrates, not available at the plasma membrane. In addition, the recycling of a fraction of the receptor to the cell surface might have a profound effect on signaling. It is becoming apparent that recycled signaling receptors are not simply redelivered to the plasma membrane, but might, in many cases, be delivered to specific regions of the membrane where local signaling is necessary. Therefore, endocytosis provides spatial and temporal dimension to signaling, ensuring prolonged signaling in the endosomes and local redistribution of signaling molecules. In this framework, endocytosis should not be considered merely as an attenuator of ▶ signal transduction but rather as an integrator of signaling and attenuation. This complex interplay is at the basis of numerous cellular functions, including motility, proliferation, and determination of cellular fate/differentiation.

Molecular Mechanisms

The structural and regulatory mechanisms of the clathrin-mediated endocytosis are being elucidated. Four major structural components are required in the formation of an endocytic vesicle (see Fig. 1): clathrin, AP2, receptor tails, and dynamin. Polymerization of clathrin into a hexagonal/pentagonal array forms a cagelike lattice around the internalizing pit and provides the organizing framework of the pit. The AP2 complex drives the polymerization of clathrin and serves as a recruiter of receptors to the forming pit, due to its ability to simultaneously interact with the receptor intracytoplasmic tails and with clathrin. Receptor tails contain endocytic codes, i.e., amino acid sequences capable by themselves of sustaining internalization. Endocytic codes are thought to be cryptic in receptor tyrosine kinases and to be unmasked by conformational changes that follow receptor activation and autophosphorylation. In other types of receptors, such as the transferrin receptor, endocytic codes are probably continuously exposed, thus determining constitutive internalization. In many cases, endocytic codes

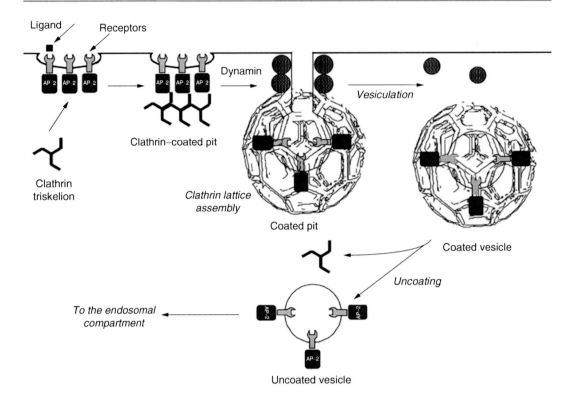

Endocytosis, Fig. 1 Schematic of clathrin-mediated endocytosis. From *left* to *right*, the various temporal phases of endocytosis are depicted: recruitment of receptors into pits, clathrin assembly, vesiculation, uncoating, and trafficking to the endosomal compartment

contain a critical tyrosine residue, thus being known as "tyrosine-based" signals. There is ample evidence that tyrosine-based signals bind directly to the clathrin adaptor protein complex AP2, thus allowing receptor recruitment into the pit. Once a clathrin-coated pit is formed, dynamin is responsible for its release into the cytosol as a vesicle (vesiculation). Vesiculation is caused by conformational changes of dynamin, which require its intrinsic GTPase activity, and which lead to pinch off of the vesicle from the plasma membrane. Once a vesicle is formed, it appears as a coated vesicle, still being surrounded by a clathrin lattice. Subsequent shedding of clathrin and AP2 leads to the formation of an uncoated vesicle. Uncoated vesicles fuse with other vesicles in the endosomal compartment, where decisions are made whether to recycle the content of the vesicles (or part of it) to the plasma membrane or to further process it in the lysosome. Several

other proteins (AP180/CALM, epsins, Eps15 and Eps15R, amphiphysins, synaptojanins, etc.), globally referred to as accessory or regulatory proteins, participate to the various phases of the endocytic process, frequently entering in contact with a forming pit through AP2, in a precise hierarchical fashion. Finally posttranslational modifications, such as phosphorylation, dephosphorylation, and ▶ ubiquitination, are known to play a role in the various cycles of assembly and disassembly of structural and regulatory endocytic proteins.

Clinical Relevance

Pathogenesis of a number of membrane receptor-linked diseases can be directly traced to primary defects in endocytosis. In Familial Hypercholesterolemia, genetic defects in the LDL receptor result in the absence of cholesterol internalization into the cell. Both receptor absent and receptor-defective mutants occur. Among the latter, internalization-defective mutants can be caused by nonsense or frameshift mutations or by single amino acid substitutions. In insulin-resistant diabetes mellitus, mutations of the insulin receptor gene occur that fall into various categories, among which are those involving accelerated receptor degradation and impaired transport of receptors to the cell surface. Another genetic disease that might be linked to the processes of endocytosis and of vesicle transport is Huntington disease. The disease is associated with increased length of a glutamine stretch, located in the NH2-terminal region of the protein huntingtin. Abnormal protein interactions have been proposed as a pathogenetic mechanism. Huntingtin localizes to vesicles of the secretory and endocytic pathway and interacts with proteins involved in vesicle trafficking. Data suggest that huntingtin is a regulator of these pathways, through association with clathrin-coated vesicles.

A remarkable overlap of molecular events between endocytosis and vesicle recycling in synapses is emerging with increasing relevance for those conditions, of neurological interest, in which alterations affecting components of the synaptic vesicles were shown. Among those are the Lambert–Eaton myasthenic syndrome, in which synaptotagmin might be a target of autoimmunity; the Stiff-Man syndrome, in which amphiphysin and the glutamic acid decarboxylase (GAD) (both of which are associated with the cytoplasmic surface of synaptic vesicles) are targets of autoimmunity; and possibly even Alzheimer disease, in which the progressive cognitive loss is associated with synaptic loss in the cortex and decrease in the levels of synaptobrevin and synaptophysin.

Alterations of the processes of endocytosis and intracellular vesicular sorting might play a role in ▶ cancer. For instance, several receptor tyrosine kinases, including Met, EGFR, and ErbB-2, display alteration of their endocytosis in human cancers. In addition, translocations are described in leukemia, which involve accessory endocytic proteins. The *Eps15* and *EEN* genes (*extra eleven-nineteen genes*) are translocated with the *MLL* gene (myeloid/lymphoid, or mixed lineage, leukemia), resulting in the production of fusion proteins. The *CALM* gene is rearranged with the *AF-10 gene* (ALL1 fused gene from chromosome 10), which in turn is a partner in *MLL*-involving translocations. Tumor suppressor and/or cancer predisposition genes are also linked to endocytic/sorting pathways, including *ATM*, the cancer predisposition gene mutated in ataxia-teleangiectasia, and the two genes *TSC1* and *TSC2*, encoding hamartin and tuberin, respectively, identified in tuberous sclerosis.

Another emerging field in cancer research concerns plasma membrane receptors that signal to the cell in a negative fashion, by transducing antiproliferative signals. In such cases, endocytosis could represent a mechanism to attenuate these signals, leading to increased proliferation. This is exemplified by studies of E-cadherin, a transmembrane adhesion molecule that, by establishing homophilic interactions at sites of cell-to-cell contacts, downregulates the β-catenin pathway and participates in the tight control of epithelial tissue homeostasis. There is strong evidence that loss of the tumor-suppressor function of E-cadherin represents a common event in human cancers and causes cell invasiveness in vitro and tumor progression in vivo.

Finally, data demonstrated that endocytic proteins are involved in cell-fate determination. This is particularly relevant considering the emerging concept of cancer stem cell compartments. Signaling by the Notch receptor affects cell-fate specification, proliferation, apoptosis, and migration, and it is frequently deregulated in human malignancies. Both the Notch receptor and its ligands undergo ubiquitin-regulated internalization and degradation, and functional subversion of the endocytic/trafficking events is predicted to have profound impact in the regulation of this receptor. Another endocytic protein, involved in tumors, is Numb, which is also linked to the Notch pathway. Numb is a negative regulator of Notch, and loss of Numb expression is a frequent event in human breast tumors. This, in turn, results in increased Notch signaling, an event that contributes to cellular transformation.

Cross-References

- ► Cancer
- ► Caveolins
- ► Endosomal Compartments
- ► Huntingtin Interacting Protein 1
- ► Signal Transduction
- ► Ubiquitination

References

Conner SD, Schmid SL (2003) Regulated portals of entry into the cell. Nature 422:37–44

Le Roy C, Wrana J (2005) Clathrin- and non-clathrin-mediated endocytic regulation of cell signalling. Nat Rev Mol Cell Biol 2:112–126

Polo S, Di Fiore PP (2006) Endocytosis conducts the cell signaling orchestra. Cell 124:897–900

Polo S, Pece S, Di Fiore PP (2004) Endocytosis and cancer. Curr Opin Cell Biol 2:156–161

Sorkin A (2004) Cargo recognition during clathrin-mediated endocytosis: a team effort. Curr Opin Cell Biol 4:392–399

See Also

(2012) AP2. In: Schwab M (ed) Encyclopedia of cancer, 3rd edn. Springer Berlin Heidelberg, p 230. doi:10.1007/978-3-642-16483-5_342

(2012) Clathrin. In: Schwab M (ed) Encyclopedia of cancer, 3rd edn. Springer Berlin Heidelberg, p 880. doi:10.1007/978-3-642-16483-5_1207

(2012) Dynamin. In: Schwab M (ed) Encyclopedia of cancer, 3rd edn. Springer Berlin Heidelberg, p 1177. doi:10.1007/978-3-642-16483-5_1758

(2012) Extra Eleven-Nineteen gene. In: Schwab M (ed) Encyclopedia of cancer, 3rd edn. Springer Berlin Heidelberg, p 1361. doi:10.1007/978-3-642-16483-5_2065

(2012) Mixed lineage leukemia protein. In: Schwab M (ed) Encyclopedia of cancer, 3rd edn. Springer Berlin Heidelberg, p 2348. doi:10.1007/978-3-642-16483-5_3784

(2012) Recycling. In: Schwab M (ed) Encyclopedia of cancer, 3rd edn. Springer Berlin Heidelberg, pp 3208-3209. doi:10.1007/978-3-642-16483-5_4999

(2012) Synaptic vesicle recycling. In: Schwab M (ed) Encyclopedia of cancer, 3rd edn. Springer Berlin Heidelberg, p 3592. doi:10.1007/978-3-642-16483-5_5619

(2012) Tyrosine kinase receptors. In: Schwab M (ed) Encyclopedia of cancer, 3rd edn. Springer Berlin Heidelberg, p 3824. doi:10.1007/978-3-642-16483-5_6081

Endolysosomal Pathway

- ► Endosomal Compartments

Endometrial Cancer

Alexey Davydov[1] and V. Craig Jordan[2]
[1]Fox Chase Cancer Center, Philadelphia, PA, USA
[2]Breast Medical Oncology, MD Anderson Cancer Center, Houston, TX, USA

Definition

Endometrial carcinoma is the most common malignancy of the female genital tract, and it is estimated to account for approximately 40,880 new cases and more than 7,310 deaths in 2005 in the USA and a similar figure in Europe. It is more frequent than ► ovarian cancer but can be treated successfully more often by surgery and radiation therapy. Endometrial cancer refers to one of the several types of malignancy that arise from the endometrium or lining of the uterus. It may sometimes be referred to as uterus cancer.

Characteristics

Risk Factors

Several risk factors for the development of endometrial carcinoma are related to an extended exposure to unopposed estrogen action, such as in nulliparity, late menopause, obesity, diabetes mellitus, estrogen replacement therapy, and tamoxifen treatment (► Estrogenic hormones and cancer; Hormones and Cancer; ► obesity and cancer risk; ► progestin and cancer; ► tamoxifen). Tamoxifen has proven to be highly effective in the treatment of all stages of ► estrogen receptor (ER) positive breast cancer; however, it has a partial estrogen agonist effect in the human uterus. In the uterus of postmenopausal women, the estrogen-agonistic activity of tamoxifen results in an increased risk for the development of endometrial hyperplasia and endometrial cancer. The National Surgical Adjuvant Breast and Bowel Project-P1 study showed that tamoxifen increases the risk of endometrial cancer in postmenopausal women by four- to fivefold. It is important to stress, however, that the stage and grade of endometrial cancers observed in postmenopausal women were the same as in the general population. There is also evidence that the increased risk of endometrial cancer continues for years after tamoxifen therapy is discontinued.

Classification

Endometrial carcinomas are classified into two types on the basis of biological and histopathological variables:

- Type I tumors, are usually well differentiated and endometrioid in histology and are associated with a history of unopposed estrogen exposure or other hyperestrogenic risk factors such as obesity.
- Type II tumors, which often poorly differentiated, nonendometrioid, and are not associated with hyperestrogenic factors. These tumors are more likely to be metastatic and can recur even after clinical intervention.

Pathogenesis Mechanisms

The pathogenic mechanisms of endometrial cancer are poorly understood. However, as in other malignancies, accumulation of genetic abnormalities and ▶ epigenetic alterations is thought to cause the transformation of normal endometrium to cancerous tissue. These changes disrupt cellular signaling networks that govern processes such as cell proliferation, ▶ apoptosis, and ▶ angiogenesis. So far, no specific gene or genes have been linked to the majority of cases of endometrial cancer. However, molecular analyses have implicated several well-characterized ▶ oncogenes and ▶ tumor suppressor genes in endometrial carcinogenesis. Current data indicate that type I tumors are more commonly associated with abnormalities in the DNA-mismatch repair genes (▶ Mismatch Repair in Genetic Instability), ▶ *KRAS*, PTEN (phosphatase and tensin homologue), and β-catenin (Beta-Catenin), whereas type II tumors seem to be linked to abnormalities in ▶ TP53 and ERBB2 (also known as ▶ HER-2/neu). These can be mutation, deletion and ▶ amplification/overexpression of genes, and/or ▶ epigenetic deregulation.

Approximately 10% of EC cases are associated with hereditary nonpolyposis colorectal cancer (HNPCC), a dominantly inherited syndrome with germ-line abnormalities in one of five DNA-mismatch repair genes with resultant microsatellite instability. Females with HNPCC have a ten-fold increased lifetime risk of EC compared with that of the general population and the lifetime risk of EC (42%) is higher than that for colorectal carcinoma.

Treatment

Treatment for uterus cancer depends on the stage of the disease and the overall health of the patient. Removal of the tumor (surgical resection) is the primary treatment. Radiation therapy, hormone therapy, and/or chemotherapy may be used as adjuvant treatment (i.e., in addition to surgery) in patients with metastatic or recurrent disease.

Surgery

Treatment for uterine cancer usually involves removal of the uterus, including the cervix (called total hysterectomy), and removal of the fallopian tubes and ovaries (called bilateral salpingo-oophorectomy). Surgery may be performed through an incision in the abdomen or through the vagina (called transvaginal hysterectomy).

Surgical treatment is required to determine the degree of myometrial invasion. The following surgical staging has been adopted by the International Federation of Gynecology and Obstetrics (FIGO) and by the American Joint Committee on Cancer (AJCC):

- Stage I endometrial cancer is carcinoma confined to the corpus uteri:
  ## Stage IA: tumor limited to endometrium
  ## Stage IB: invasion to less than one half of the myometrium
  ## Stage IC: invasion to greater than one half of the myometrium
- Stage II endometrial cancer involves the corpus and the cervix but has not extended outside the uterus:
  ## Stage IIA: endocervical glandular involvement only
  ## Stage IIB: cervical stromal invasion

E

- Stage III endometrial cancer extends outside of the uterus but is confined to the true pelvis:
  - ## Stage IIIA: tumor invades serosa and/or adnexa and/or positive peritoneal cytology
  - ## Stage IIIB: vaginal metastases
  - ## Stage IIIC: metastases to pelvic and/or paraaortic lymph nodes
- Stage IV endometrial cancer involves the bladder or bowel mucosa or has metastasized to distant sites:
  - ## Stage IVA: tumor invasion of bladder and/or bowel mucosa
  - ## Stage IVB: distant metastases, including intraabdominal and/or inguinal lymph nodes

Radiation Therapy

Radiation uses high-energy x-rays to destroy cancer cells and shrink tumors. This treatment may be used prior to surgery (called ▶ neoadjuvant therapy) or after surgery to destroy remaining cancer cells. Radiation also may be used in patients who are unable to undergo surgery.

Hormonal Therapy

Endometrioid endometrial cancer has long been associated with states of estrogen excess. As a result, a variety of antiestrogens have been used for systemic treatment.

Progestogens

Following the theory of estrogen excess as a carcinogenic promoter, progestogens have been used in the treatment of endometrial cancer for their antiproliferative effects on the endometrium.

Aromatase Inhibitors

Another approach to reducing the estrogen stimulation of the tumor is to use aromatase inhibitors. Aromatase inhibitors (▶ Aromatase and its Inhibitors) are known to reduce levels of circulating estrogen by reducing estrogen production.

Cross-References

- ▶ Amplification
- ▶ Angiogenesis
- ▶ Apoptosis
- ▶ Aromatase and its Inhibitors
- ▶ Endocrine-Related Cancers
- ▶ Epigenetic
- ▶ Estrogen Receptor
- ▶ Estrogenic Hormones
- ▶ HER-2/neu
- ▶ *KRAS*
- ▶ Mismatch Repair in Genetic Instability
- ▶ Neoadjuvant Therapy
- ▶ Obesity and Cancer Risk
- ▶ Oncogene
- ▶ Ovarian Cancer
- ▶ Progestin
- ▶ Tamoxifen
- ▶ TP53
- ▶ Tumor Suppressor Genes

References

Amant F, Moerman P, Neven P et al (2005) Endometrial cancer. Lancet 366:491–505

Fisher B, Costantino JP, Wickerham DL et al (2005) Tamoxifen for the prevention of breast cancer: current status of the National Surgical Adjuvant Breast and Bowel Project P-1 study. J Natl Cancer Inst 97(22):1652–1662

Kloos I, Delalage S, Pautier P et al (2002) Tamoxifen-related uterine carcinosarcomas occur under/after prolonged treatment: report of five cases and review of the literature. Int J Gynecol Cancer 12:496–500

Shang Y (2006) Molecular mechanisms of estrogen and SERMs in endometrial carcinogenesis. Nat Rev Cancer 6:360–368

See Also

(2012) Beta-catenin. In: Schwab M (ed) Encyclopedia of cancer, 3rd edn. Springer, Berlin/Heidelberg, p 385. doi:10.1007/978-3-642-16483-5_889

(2012) Deletion. In: Schwab M (ed) Encyclopedia of cancer, 3rd edn. Springer, Berlin/Heidelberg, p 1080. doi:10.1007/978-3-642-16483-5_1553

(2012) Endometrium. In: Schwab M (ed) Encyclopedia of cancer, 3rd edn. Springer, Berlin/Heidelberg, p 1239. doi:10.1007/978-3-642-16483-5_1886

(2012) Estrogens. In: Schwab M (ed) Encyclopedia of cancer, 3rd edn. Springer, Berlin/Heidelberg, p 1333. doi:10.1007/978-3-642-16483-5_2019

(2012) Neoadjuvant. In: Schwab M (ed) Encyclopedia of cancer, 3rd edn. Springer, Berlin/Heidelberg, p 2472. doi:10.1007/978-3-642-16483-5_4003

(2012) Progestogens. In: Schwab M (ed) Encyclopedia of cancer, 3rd edn. Springer, Berlin/Heidelberg, p 2994. doi:10.1007/978-3-642-16483-5_4757

(2012) Proliferation. In: Schwab M (ed) Encyclopedia of cancer, 3rd edn. Springer, Berlin/Heidelberg, p 3004. doi:10.1007/978-3-642-16483-5_4766

(2012) Uterus cancer. In: Schwab M (ed) Encyclopedia of cancer, 3rd edn. Springer, Berlin/Heidelberg, p 3866. doi:10.1007/978-3-642-16483-5_6452

Endometrioma

Definition

A benign ovarian cyst containing endometrium-like cells. It is commonly referred to as a chocolate cyst. It has the potential to become malignant so it is generally removed after diagnosis.

See Also

(2012) Endometrium. In: Schwab M (ed) Encyclopedia of cancer, 3rd edn. Springer, Berlin/Heidelberg, p 1239. doi:10.1007/978-3-642-16483-5_1886

Endoplasmic Reticulum Stress

Alex van Vliet and Patrizia Agostinis
Department of Cellular and Molecular Medicine, Cell Death Research and Therapy Lab, KU Leuven Campus Gasthuisberg, Leuven, Belgium

Synonyms

Endoplasmic reticulum stress response

Definition

The endoplasmic reticulum (ER) is an organelle with several essential functions in eukaryotic cells. The ER is both a major intracellular calcium store and the place where proteins are synthesized, folded, modified, and delivered to their final cell surface or extracellular destination. Moreover, in mammalian cells, the ER is the site of sterols and lipids synthesis. Disturbance in any of these functions, which results in the disruption of the proper folding and secretory capacity of the ER and increased load of unfolded proteins in its lumen, defines a condition known as "ER stress." ER stress activates a complex and multifaceted intracellular signal transduction pathway that is essentially designed to re-establish ER homeostasis. Inability to restore ER functions induces cell death, which is usually in the form of **apoptosis**. ER stress contributes to the etiology of several human pathologies, including diabetes, neurodegeneration, and cancer.

Characteristics

Proteins that are expressed at the cell surface or secreted extracellularly, as well as resident proteins of the organelles along the secretory pathway, are first co-translationally translocated into the lumen of the ER as unfolded polypeptide chains. The unique highly oxidizing ER environment, which is equivalent to the extracellular space, is required to sustain a variety of posttranslational and co-translational modifications to which proteins are subjected after entering the ER. These modifications include the formation of intra- and intermolecular disulphide bonds and N-linked glycosylation. In the Ca^{2+}-rich ER lumen several resident chaperones and Ca^{2+} binding proteins, including the glucose-regulated proteins (GRPs) and the lectins, calreticulin and calnexin critically assist and monitor proper folding, maturation and stabilization of the nascent protein. These ATP-requiring processes are part of the stringent ER quality-control mechanisms that allow only proteins adopting a correctly folded or native conformation, or a proper oligomeric assembly in case of multisubunit proteins, to be exported to the Golgi complex and move on through the secretory pathway. Terminal failure in protein folding or in oligomer assembly results in ER-associated protein degradation (ERAD), in which nonnative conformers are retrotranslocated to the cytosol and degraded by the 26S proteasome. ER stress is set off by various intracellular and extracellular perturbations, which alter the protein folding capacity of the ER. These include glucose deprivation, which interferes with N-linked glycosylation and affects ATP levels, redox imbalance and depletion of the ER Ca^{2+} store. Perturbation of ER homeostasis activates an evolutionarily conserved stress response, collectively called the Unfolded Protein Response or UPR. The UPR is primarily a pro-survival response initiated to restore normal ER homeostasis.

Molecular Mechanisms

The UPR consists of three major mechanisms: (1) translational attenuation to limit the biosynthetic load of the ER; (2) transcriptional activation of cytoprotective genes encoding ER resident chaperones to increase the ER folding capacity; and (3) increased clearance of unfolded proteins through the upregulation of ERAD. When these mechanisms fail to restore normal ER homeostasis, ER stress promotes the cell death program, which is likely activated to protect the host from the accumulation of dysfunctional cells.

At the molecular level, the UPR is characterized by the activation of three ER transmembrane proteins: the pancreatic ER kinase (PKR)-like ER kinase (PERK), the inositol-requiring enzyme 1 (IRE1), and the activating transcription factor 6 (ATF6). These transmembrane proteins contain a lumenal domain, which functions as a sensor of the ER folding capacity, and a cytosolic effector domain that provides a signaling bridge connecting the ER to other cellular compartments. In unstressed cells, the luminal domain of the transmembrane receptors is bound to a 78-kDa ER resident chaperone referred to as the glucose-regulated protein (GRP) 78 or immunoglobulin binding protein, BiP. BiP belongs to the heat-shock protein (HSP) 70 class of ▶ molecular chaperones and can form complexes with heterologous proteins that are processed through the ER. Similar to other HSP70 members, BiP binds both ADP and ATP, which serve to regulate, through ADP ribosylation, its binding, and release from nascent polypeptide chains. BiP cycles between an oligomeric and monomeric state, the latter of which is thought to bind preferentially to unfolded proteins. Association of the ER transmembrane receptors

with BiP lock them into an inactive, monomeric conformation that prevents their oligomerization or trafficking to other compartments. On accumulation of unfolded proteins, which promotes a dramatic increase in the luminal pool of monomeric BiP, BiP is competitively titrated away from the luminal domains of the three receptors. BiP dissociation triggers the activation of these proximal ER stress sensors and initiates the first molecular event of the complex transcriptional and translational program defining the UPR.

Signal Transduction by the Three Arms of the UPR

Adaptation to ER stress involves the sequential activation of three major arms of the UPR: the PERK- eIF2α-ATF4, ATF-6, and IRE1 pathways.

- PERK is a type I transmembrane Ser/Thr protein kinase. Dissociation of BiP from PERK drives PERK homo-oligomerization and self-activation by *trans*-autophosphorylation. PERK mediates the specific phosphorylation of nuclear factor (erythroid-derived 2)-like 2 (Nrf2) and the eukaryotic initiation factor-2 α (eIF2α) at Ser51. Phosphorylation of eIF2α interrupts the translation of most mRNAs, thereby reducing the load of newly synthesized proteins in the ER. However, while global translation is transiently repressed, the PERK-eIF2α pathway results in the preferential translation of a subset of mRNAs containing regulatory sequences in their 5' untranslated regions that allow them to bypass the eIF2α translational block. One of these mRNAs encodes ATF4, a member of the bZIP family of transcription factors. ATF4 promotes cell survival through the induction of several genes involved in restoring ER homeostasis. Under conditions of sustained ER stress, ATF4 prompts the expression of the transcription factor C/EBP homologous protein (CHOP) and of one of its target genes, the growth arrest, and DNA damage-inducible gene 34 (GADD34). GADD34 is a protein phosphatase 1 (PP1) regulatory subunit that promotes eIF2α dephosphorylation, which ends the translational block and restores protein synthesis in the ER. Phosphorylation of Nrf2 promotes its dissolution from the Nrf2/Keap1 complex and leads to Nrf2 nuclear import and causes activation of its gene expression program. This leads to the upregulation of proteins mainly involved in cellular redox homeostasis, like glutathione-S-transferase (GST).

- ATF6 is a type II transmembrane protein that contains a transcription factor in its cytoplasmic domain. Under ER stress conditions, the dissociation of BiP frees ATF6 to translocate to the Golgi apparatus, where it is cleaved by Golgi-resident proteases. The limited proteolysis of ATF6 releases the transcription factor domain into the cytosol and allows its migration into the nucleus, where it binds DNA and activates gene expression. ATF6-responsive genes are activated rapidly after the induction of ER stress and include the ER molecular chaperones BiP and GRP94, as well as protein disulphide isomerase (PDI). ATF-6 also leads to the transcriptional upregulation of the X-box binding protein (XBP1) mRNA, which is converted into a stable transcription factor by the endonuclease activity of IRE1.

- IRE1 contains a Ser/Thr kinase domain and an endonuclease domain facing the cytosol. Like PERK, IRE1 is a type I transmembrane receptor that is activated by oligomerization-induced *trans*-autophosphorylation in the ER membrane upon BiP dissociation. IRE1 autophosphorylation stimulates its endonuclease activity and does not result in the propagation of a phosphorylation cascade. Active IRE1 is able to degrade specific mRNAs based on localization and the amino acid sequences that they encode, termed regulated IRE1-dependent decay of mRNA (RIDD). An important result of this mRNA cleavage is the removal of a 26-nucleotide intron from the XBP1 mRNA, generating the XBP1^s frameshift splice variant which encodes a stable and active transcription factor. XBP1^s translocates to the nucleus and transactivates several cytoprotective genes involved in ER quality-control. XBP1^s can also induce a negative feedback loop which, by relieving the PERK-mediated translational block, returns the ER

function to normal once the UPR has been successful. The IRE1 pathway is thought to be the final branch of the UPR to be activated, subsequent to the rapid induction of PERK and ATF6 signaling which have a major cytoprotective role. The IRE1 signal plays a dual role in the UPR and if the ER stress persists, XBP1^s initiates pro-apoptotic protein synthesis, thereby facilitating the induction of apoptosis.

Cell death after ER stress. The adaptive responses initiated by the UPR restrain cell death to the extent that ER homeostasis and protein folding capacity can be re-established. If, in spite of this, protein aggregation persists, the UPR shifts into a proapoptotic response, whose molecular effectors are still poorly defined.

- *CHOP and IRE1.* Sustained CHOP induction, resulting from the late phase of ATF-6 activation and the PERK-eIF2α-ATF4 axis, tips the balance towards apoptosis under conditions of persistent ER stress. Several CHOP target genes contribute to the proapoptotic effect of CHOP induction. One mechanism entails CHOP-mediated suppression of *bcl-2* transcription, which increases the sensitivity of the cell to apoptotic cell death. Also, restoration of protein synthesis by the CHOP-mediated induction of GADD34 can promote the expression of proapoptotic effectors when the ER folding capacity cannot be restored. Additionally, CHOP, in tandem with increased levels of ATF4, also causes the block on protein translation to be lifted, causing increased protein folding which overloads the already stressed ER leading to ATP depletion and oxidative stress, contributing to cell death. The IRE1 pathway leads to activation of the MAPK family member c-Jun N-terminal kinase (JNK) through the adaptor molecule TNF receptor-associated factor 2 (TRAF2) and apoptosis signal-regulating kinase-1 (ASK1). JNK can favor apoptosis induction following ER stress by increasing the proapoptotic activity of some BCL2 proteins, such as BIM. Furthermore, IRE1 can contribute to cell death through RIDD function by selectively degrading mRNAs, for example, those encoding growth-promoting proteins.

- Ca^{2+} *signaling and BCL2 proteins.* The physiological ER Ca^{2+} content is maintained by the balance between the active Ca^{2+} uptake, controlled by SERCA (sarco-endoplasmic reticulum Ca^{2+}-ATPase) pumps, and its release, through opening of IP_3R (inositol 1,4,5-triphosphate receptor) or RyR (ryanodine receptor) Ca^{2+}-release channels. Various stimuli that cause depletion of ER Ca^{2+} stores and result in cytosolic Ca^{2+} overload also stimulate the UPR, since the vast majority of ER resident chaperones involved in ER quality-control are Ca^{2+} binding proteins and can induce cell death. Excessive Ca^{2+} uptake by mitochondria induces changes in the permeability of the inner mitochondrial membrane, which can result in the breakdown of the mitochondria membranes triggering apoptosis. Increased Ca^{2+} leak from the ER activates Ca^{2+}-dependent enzymes in the cytosol, including the Ca^{2+}/calmodulin-dependent phosphatase calcineurin and calpains, which can regulate apoptosis induction by modulating the activity of BCL-2 proteins. In addition to having a crucial role as regulators of mitochondria function during apoptosis, proapoptotic BAX, BAK and antiapoptotic BCL2 proteins regulate ER Ca^{2+} homeostasis and the UPR machinery. For instance, interaction of BAX and BAK with IRE1 during ER stress is required for IRE1-mediated JNK activation and apoptosis induction.

- *Caspases.* The execution phase of apoptosis following ER stress involves the activation of caspases. Both murine caspase 12 and human caspase 4 associate with the ER and are proteolytically processed after ER stress. The general relevance of this event for the initiation of apoptosis following ER stress is however still dubious as caspase 12, which is found only in rodents, and caspase 4 belong to the subclass of inflammatory caspases with primary role in inflammation and innate immunity rather than in ER stress (Fig. 1).

Clinical Relevance

Activation of the UPR is observed in several conformational diseases, including Alzheimer and Creutzfeldt–Jakob diseases, and involves the accumulation of protein aggregates in the ER. In general, it is becoming increasingly clear that a range of pathological conditions are associated with ER stress, including diabetes, ischemia, lysosomal storage disease, cardiovascular diseases, atherosclerosis, neurodegeneration, viral infection, as well as cancer.

With respect to cancer, tumors often exhibit activated molecular effectors of the UPR. Indeed, conditions found in the tumor microenvironment – which, due to insufficient vascularization, often include glucose starvation, hypoxia and acidosis – can promote ER stress in tumors. Because of their heightened proliferation, increased metabolic and biosynthetic demand, and the need to plastically adapt to the hostile microenvironment, tumors also rely heavily on an efficient secretory machinery to cope with the increased load of proteins destined for secretion, which include various protumorigenic or immunosuppressive chemokines and cytokines, or changes in their surface proteome. However, although the UPR is generally considered a protective response, prolonged ER stress can activate cell death and therefore its role in the disease process can be dualistic. The activated UPR could either limit tumor growth (by the induction of cell death in conditions of chronic stress), or instead facilitate it (by increasing the survival of cancer cells). Additionally, therapeutic induction of ER stress incited by ▶ reactive oxygen species (ROS) can elicit immunogenic cell death (ICD), a form of cell death associated with the trafficking through the secretory pathway and extracellular emission of damage associated molecular patterns (DAMPs), which engage the host immune system and stimulate anti-tumor immunity. Further research will be necessary to determine the extent to which ER stress and UPR induction in tumors modulate chemotherapy responses and therapy resistance, which is currently a major obstacle for the treatment of human cancer.

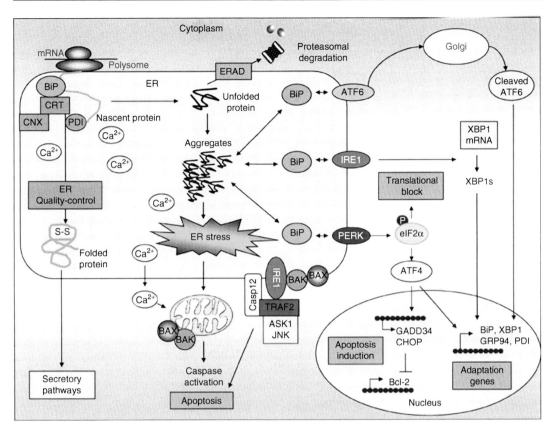

Endoplasmic Reticulum Stress, Fig. 1 **ER stress signaling pathways** After import in the ER lumen a nascent unfolded protein undergoes different posttranslational modifications monitored by molecular chaperones and folding enzymes (*CRT*; calreticulin, *CNX*; calnexin, BiP, PDI), prior to its export from the ER through the secretory pathways. Incorrectly folded proteins are retrotraslocated to the cytosol and degraded by the proteasome through the ERAD. Perturbations in the ER environment resulting in the accumulation of unfolded proteins in the ER lumen, promotes the dissociation of BiP from three proximal ER-stress sensors. This triggers PERK and IRE1 oligomerization/activation and ATF6 release from the ER membranes. The PERK-eiF2α-ATF4 signal is responsible for the initial translational block, whereas the ATF-6 and IRE1 axis, in addition to the phosphorylation of Nrf2 by PERK mediate the rapid transcriptional activation of cytoprotective genes, including several ER resident chaperones. If ER stress persist, apoptosis is activated through several mechanisms involving Ca^{2+} signaling and the transcriptional upregulation of CHOP, which represses *bcl-2* expression. BAX and BAK proteins regulate the IRE1-TRAF2-ASK1 pathway which leads to JNK activation and processing of murine caspase-12

Glossary

Ca^{2+}-release channels Membrane receptors localized in the sarcoplasmic/endoplasmic reticulum that once activated mediates the release of Ca^{2+} from the ER to the cytosol.

Calnexin A Ca^{2+} binding integral protein of the ER that interacts and assists the folding of proteins that carry monoglucosylated N-linked glycans.

Conformational diseases A group of heterologous disorders, which include Alzheimer's, Parkinson's and Creutzfeldt-Jakob diseases, which arise from the dysfunctional aggregation of proteins in non-native conformations.

Damage associated molecular patterns Subset of endogenous molecules that have physiological roles within living cells and which acquire immunomodulatory functions when exposed to the extracellular milieu during the cell death process. Once surface exposed or released, DAMPs are sensed by the innate immune system and act as activators of antigen-presenting cells (APCs) to stimulate adaptive immunity.

ER quality-control The molecular apparatus that monitors the maturation and transport of proteins in the ER and targets nonnative conformers for degradation.

ER-associated protein degradation The mechanism whereby unfolded or misfolded proteins are re-exported from the ER lumen into the cytosol to undergo proteasomal degradation.

Eukaryotic initiation factor-2 α A subunit of eIF2 that recruits Met-tRNAi to the mRNA-40S ribosome complex in its active GTP-binding form. Stress induced eIF2α phosphorylation by a family of protein kinases including PERK, PKR, GCN2, and HRI, prevents the assembly of the eIF2-initiation complex, thereby inhibiting global translation.

Glutathione-S-Transferase Conjugates reduced glutathione to toxic electrophile that have arisen through oxidative stress, for example.

Immunogenic cell death A cell death modality that, in contrast to the physiological caspase mediated-tolerogenic apoptosis, stimulates immunogenicity of the cancer cell along with a protective anticancer immune response in vivo.

Lectins Proteins that bind carbohydrates.

Molecular chaperones A family of cellular proteins that mediate the correct assembly and disassembly of nascent polypeptides, by the reversible interaction with nascent polypeptide chains.

Native conformation (protein) The most energetically favorable state adopted by a protein, which usually corresponds to the conformation with the lowest free energy.

Nuclear factor (erythroid-derived 2)-like 2 Transcription factor that can induce the expression of anti-oxidant proteins.

Proteasome A 26S multiprotein complex that catalyses the degradation of polyubiquitylated proteins.

Protein disulphide isomerase (PDI) An enzyme belonging to the thiol-disulphide oxidoreductases family of proteins that catalyzes the oxidation, isomerization, and reduction of disulphide bonds.

Reactive Oxygen Species A group of chemically reactive molecules and free radical molecules deriving from the incomplete reduction of molecular oxygen. Examples of which are superoxide and peroxide.

Regulated Ire1-dependent decay Activated IRE1 is able to utilize its endoribonuclease domain in the degradation of mRNA which harbor a specific XBP-1-like consensus site.

Secretory pathway A complex network of eukaryotic cell organelles that is central to and mediates the folding, maturation, and trafficking of secreted and transmembrane proteins.

SERCA A pump situated in the membrane of the sarcoplasmic/endoplasmic reticulum that couples ATP hydrolysis to the import of Ca^{2+} from the cytosol to the ER lumen.

Type I transmembrane protein A transmembrane protein that exposes its carboxyl terminus into the cytosol.

Type II transmembrane protein A transmembrane protein that exposes its amino terminus into the cytosol.

References

Berridge MJ (2002) The endoplasmic reticulum: a multifunctional signaling organelle. Cell Calcium 32:235–249

Cullinan SB, Zhang D, Hannink M, Arvisais E, Kaufman RJ, Diehl JA (2003) Nrf2 is a direct PERK substrate and effector of PERK-dependent cell survival. Mol Cell Biol 23:7198–7209

Dejeans N, Manié S, Hetz C, Bard F, Hupp T, Agostinis P, Samali A, Chevet E (2014) Addicted to secrete – novel concepts and targets in cancer therapy. Trends Mol Med 20:242–250

Ellgaard L, Helenius A (2003) Quality control in the endoplasmic reticulum. Nat Rev Mol Cell Biol 4:181–191

Han J, Back SH, Hur J, Lin YH, Gildersleeve R, Shan J, Yuan CL, Krokowski D, Wang S, Hatzoglou M, Kilberg MS, Sartor MA, Kaufman RJ (2013) ER-stress-induced transcriptional regulation increases protein synthesis leading to cell death. Nat Cell Biol 15:481–490

Hollien J, Lin JH, Li H, Stevens N, Walter P, Weissman JS (2009) Regulated Ire1-dependent decay of messenger RNAs in mammalian cells. J Cell Biol 186:323–331

Ma Y, Hendershot LM (2004) The role of the unfolded protein response in tumour development: friend of foe? Nat Rev Cancer 4:966–977

Maurel M, Chevet E, Tavernier J, Gerlo S (2014) Getting RIDD of RNA: IRE1 in cell fate regulation. Trends Biochem Sci 39:245–254

Schroder M, Kaufman RJ (2005) ER stress and the unfolded protein response. Mut Res 569:29–63

Walter P, Ron D (2011) The unfolded protein response: from stress pathway to homeostatic regulation. Science 334:1081–1086

Wang M, Kaufman RJ (2014) The impact of the endoplasmic reticulum protein-folding environment on cancer development. Nat Rev Cancer 14:581–597

Endoplasmic Reticulum Stress Response

▶ Endoplasmic Reticulum Stress

Endosomal Compartments

Vincent Dammai
Dammai-Morgan Scientific Consultants LLC, Mount Pleasant, SC, USA

Synonyms

Endolysosomal pathway; Endosomal protein sorting; Receptor-mediated endocytosis

Definition

In eukaryotic cells, receptor-mediated ▶ endocytosis releases cargo-loaded vesicles at the plasma membrane. The vesicle content pass through a series of discontinuous closed membrane systems called endosomal compartments. In strict sense, endosomal compartments are temporal sorting stations where the fate of endocytosed cargo is determined.

Characteristics

Eukaryotic cells internalize a variety of extracellular material (cargo) either nonspecifically, by phagocytosis and pinocytosis, or selectively through receptor-mediated endocytosis. In phagocytosis, specialized cells (e.g., ▶ macrophages and neutrophils) engulf large particles such as cell debris, bacteria, and viruses. Pinocytosis, exhibited by all cell types, involve uptake of extracellular fluid continually as tiny droplets (micropinocytosis) or sometimes as large droplets (macropinocytosis). By contrast, receptor-mediated endocytosis involves selective uptake of cargo by cognate cell surface receptors. Pinocytosis and receptor-mediated endocytosis are closely related, while phagocytosis is mechanistically distinct and therefore is not considered as endocytosis. In phagocytosis, plasma membrane evaginates to wrap around the particle. Additionally, exocytosis adds membrane material to phagocytic vacuoles and compensates for loss of plasma membrane. In phagocytosis and pinocytosis, hydrolytic enzymes packed in Golgi-derived vesicles are directly delivered to the vesicle/vacuole, and subsequent fusion with lysosomes results in degradation of contents. On the other hand, in receptor-mediated endocytosis, not all cargo is degraded. To illustrate with familiar examples, transferrin (Tf, carry iron) and low-density lipoprotein (LDL, carry cholesterol) are endocytosed along with their respective receptors, TfR and LDL-R. After internalization, iron is released inside the cell, and both Tf and TfR are recycled to the plasma membrane for additional iron uptake. However, LDL is degraded to release cholesterol but LDL-R is recycled to the plasma membrane for further LDL uptake. Ubiquitinated growth factor receptors, on the other hand, are degraded to abolish receptor signaling. To achieve this, the incoming cargo is delivered to endosomal compartments where they are sorted either away from (to be recycled) or toward lysosomes (to be degraded). Besides their roles in differentiated cells, endosomal compartments are important during development, for example, in interpretation of morphogen gradients. In light of the details provided later, the illustration in Fig. 1 is self-explanatory and offers a basic view of the endosomal compartments.

The endosomal compartments are (i) early endosome (EE), (ii) recycling endosome (RE),

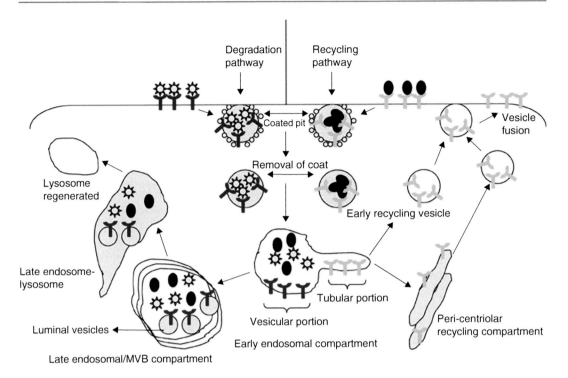

Endosomal Compartments, Fig. 1 Protein cargo entering cells are shunted through either the degradation pathway (diverted to lysosomes for degradation) or recycling pathway (back to the cell surface for re-use). Proteins destined to be degraded are separated away from the recycled proteins within endosomal compartments. Dynamic changes occurring in the endosomal compartments (described in the text) facilitate smooth flow of the incoming cargo and ensure their correct sorting to appropriate destination/fate

(iii) late endosome (LE)/multivesicular bodies (MVBs), and (iv) lysosomes. Endosomal compartments are distinguished by (i) morphology in electron and fluorescence microscopy, (ii) differential density in Ficoll gradients, (iii) pH differences, and (iv) molecular signature provided by marker proteins. Morphologically, EEs are recognizable as tubulovesicular. LEs are tubular, multilamellar, and multivesicular (with internal vesicles). MVB contains internal vesicles (up to 200) because of protein sorting into the lumen by invagination of outer endosome membrane. Electron tomography studies suggest that the internal vesicles are free-floating and discontinuous with outer endosomal membrane. Distinct endosomal subpopulations can be enriched by density centrifugation and has allowed study of membrane fusion events between endosomal compartments. Also, molecular probes sensitive to pH changes can identify individual endosomal compartment as the luminal pH decreases along the degradative but increase along the recycling pathway (cytosolic pH is $\sim$7.2). EEs are mildly acidic (pH $\sim$ 6.0), recycling compartment is toward neutral (pH 6.4–6.5), and LE/MVB is at pH $\sim$ 5.0–5.5 and lysosomes at pH < 5.0. Finally, proteins recruited to specific endosomal compartments, such as Rab GTPases, serve as markers for identification of the compartment in fixed or live-cell imaging experiments. The maturation theory suggests that EE progress toward lysosome by sequential addition of some material and removal of other material. The vesicular transfer theory proposes permanent existence of discrete compartments through which passage of cargo occurs by vesicle-mediated transfer.

Our current knowledge of endosomal compartments is mostly from studies of clathrin-dependent endocytosis (100–200 nm). Other mechanisms, classified as clathrin independent, exist and involve smaller vesicles (40–80 nm). These are mediated by ▶ caveolins (caveolae) and lipid rafts. Upon cargo binding, receptor adaptor proteins mediate polymerization of clathrin or caveolin on the cytosolic side causing plasma membrane invaginations (pits). The receptors and their cargo concentrate in these pits. Receptors and the initially assembled proteins provide binding sites for several proteins, such as Eps15, epsin, amphiphysin, intersectin, endophilin, SNX9, dynamin, and Rab5. Mostly, dynamin activity constricts and snips such invaginations to release receptor/cargo-loaded primary endocytic vesicles. The released vesicles are coated with clathrin or caveolin. Studies of cholera toxin-B (CTB) and simian virus 40 (▶ SV40) endocytosis revealed clathrin- and caveolin-independent pathways. Interestingly, dynamin and Arf6 are not required in this pathway.

After vesicle release, chaperone-mediated removal of the bulky clathrin or caveolin coat enables vesicles to deliver cargo to EEs. The EE compartment (also called sorting endosome) is the first compartment to receive all incoming cargo and is characterized by marker proteins such as Rab5 and EEA1 and the absence of coat. GTP-bound Rab5 interacts with at least 20 different proteins and along with EEA1, Rabex-5, and SNAREs mediates homotypic fusion and fusion of primary vesicles with early endosomal compartment. It was thought that all endocytic cargos fuse with a homogenous early endosomal compartment, where individual cargo is separated via recognition of specific determinants (protein domains, an ubiquitin moiety, or other modifications) and shunted accordingly. However, EEs have been subdivided into dynamic and static EEs (DEE and SEE). This so-called preendosomal sorting mechanism suggests that the type of protein coat and adaptor proteins assembled around the primary vesicle (at the plasma membrane) predetermines whether the cargo is delivered to DEE or SEE. The cargo to be degraded is delivered to DEE and the cargo to be recycled effectively concentrates in the more abundant SEE.

Within the early endosomal compartment, most receptors disengage from their ligands and the receptors to be recycled move into tubular portions, while proteins to be degraded are retained in the vesicular portion. Although specific sorting mechanisms remain unknown for many receptors, ▶ ubiquitinated growth factor receptors are retained in the vesicular portion by sequential assembly of ubiquitin-binding Hrs and Escrt-1, Escrt-II, and Escrt-III complexes and later sorted into intraluminal vesicles (ILV) for degradation by lysosomes. Additional proteins such as STAM, AMSH, and UBPY may positively and negatively regulate the ILV sorting steps. From the tubular portions, Rab4 and Rab11 generate recycling vesicles that fuse with plasma membrane, returning receptors (mostly nonubiquitinated) directly back to cell surface. Such recycling vesicles are called early (fast) REs. A tubular compartment (Rab11-positive but Rab4-negative) exists, which is known as pericentriolar recycling endosomal compartment. It performs similar functions, i.e., returning endocytosed proteins to cell surface. However, its significance is apparent where recycling is important for membrane domain maintenance and membrane protein mobilization. For instance, in migrating fibroblasts, recycling TfRs are routed through the pericentriolar area and reach the plasma membrane of the leading lamella. In neurons, with distinct axonal and somatodendritic plasma membrane domains, polarized sorting of TfR is mediated in pericentriolar recycling compartment. In polarized epithelia, apical–basal polarity is reinforced by endocytosis-mediated recovery of missorted proteins or transcytosis and localized to correct membrane domain by traversing through the pericentriolar recycling compartment.

Beginning from EE, a process of endosome maturation ensues. This involves increasing acidification of the compartments and changes that permit fusion with next endosomal compartment, thereby progressively moving nonrecycled cargo toward lysosomes. LEs acquire Rab7 and lose Rab5, a process called Rab5–Rab7 conversion. Rab7 defines the functional identity of LE and makes the compartment competent for fusion with lysosome (degradation competent). Live-cell imaging shows that Rab5 is either replaced

E

by Rab7 in early endosomal microdomains or Rab7 initially assembles on Rab5-positive EE and bud-off, taking cargo with it to become late endosomal compartment. Thus, the early compartment gradually converts/matures to LEs. An important feature is that LE membranes are highly tubular or multilamellar and contain an unusual lipid, called lysobisphosphatidic acid (LBPA). LEs become MPR positive, allowing vesicles from the *trans*-Golgi network to deliver degradative enzymes to the compartment. Resident lysosomal proteins are also delivered by this mechanism through LEs. Some protein and lipid cargo are degraded in LEs as the pH-sensitive degradative enzymes are not fully activated. MPR is eventually carried away by vesicles back to *trans*-Golgi (to be used again later) following which fusion with lysosome occur. LEs are identified by presence of both Lamp-1 and MPR.

Sorting of ubiquitinated cargo into ILVs results in the characteristic appearance of MVBs. A clear distinction between MVB and LE is difficult and thus mentioned as LE/MVB. In specialized cell types or under specific conditions, a few typical characteristics of MVB assume prominence. In MVB that have a nondegradative function, the LBPA-containing ILVs undergo retrofusion with outer endosomal membrane, releasing their contents into cytosol (such as vascular somatitis virus and anthrax lethal factor). Also, MVBs serve as precursors for antigen-processing compartments (nonendosomal compartment), exosome release, T-cell secretory granules, and melanosomes. Interestingly, ubiquitinated proteins are sorted to MVBs that are distinct from MPR- or LBPA-positive MVB. Since MPR is recycled and LBPA-containing internal vesicles can undergo retrofusion, this mechanism may represent separation of distinct classes of receptors and explain why not all proteins and lipids are degraded in LE/MVB. Finally, the cargo in LE/MVB is delivered to lysosomes. Lysosomes are the endpoint of endocytosis and contain all enzymes necessary for degradation of proteins and lipids. Syntaxin 7 and Rab7 appear to be critical for LE/MVB to deliver lipid and protein material to lysosomes. The cargo

is transferred by both complete fusion and transient membrane interactions (kiss and run). In case of complete fusion, a hybrid LE–lysosome is observed. The resultant products are utilized by cells and in mammals; lysosomes reform after endosome–lysosome fusion and degradation of cargo.

Clinical Aspects

Defects in protein trafficking influence many aspects of tumor growth and ▶ metastasis. Tumorigenic mutations in receptor domains (EGFR, FGFR, PDGFR), adaptor proteins (AP-2), or E3 ubiquitin ligases (oncogenic forms of Cbl) that prevent receptor endocytosis may not represent direct defects in the endosomal compartment, but nevertheless influence normal trafficking of activated membrane receptors through the endolysosomal pathway. By contrast, activation of ▶ oncogenes such as V-Src and K-Ras promote increased growth factor uptake by stimulating macropinocytosis, contributing to tumor growth and metastasis. Hip1, an endocytosis regulator, is directly implicated in tumor formation but its precise role remains unknown. Defects in the endosomal compartments are known to promote tumor progression by multiple mechanisms. Genetic defects leading to fusion of ALL/HRX, AF-10, and PDGFR with protein components of the endocytic pathway (Eps15, CALM, and Hip1, respectively) create abnormal oncogenic protein fusion forms and are detected in certain forms of leukemia. Increased recycling of metastasis-promoting integrins (integrin β1, integrin α6β4, and integrin αvβ3) was directly attributed to Rab11 overexpression. As Rab11 expression is regulated by HIF, a transcription factor found elevated in many human cancers; this phenomenon may have significant implications. Not surprisingly, Rab11 and one of its family member, Rab11c, is linked to skin carcinogenesis, Barrett dysplasia, and aggressiveness of breast and ovarian cancers. Additionally, mutants of Rab5 prevent receptor endocytosis, while mutations in ubiquitin sorting machinery (Hrs, Tsg101, GGA, Escrt complexes) inhibit sorting of ubiquitinated receptors and prolong ▶ receptor tyrosine kinase activity of EGFR, PVR, and Torso by preventing degradation. Cadherins are Ca^{2+}-dependent homophilic cell–cell adhesion molecules (forming ▶ adheren junctions, AJ) and are critical determinants of tissue architecture. ▶ Epithelial to mesenchymal transition (EMT), hallmark of malignant tumors, is promoted by AJ disruption, prominently seen in breast, colon, and gastric carcinomas. Cadherin internalization-recycling cycle has emerged as a major route for maintenance of AJ adhesion strength. MDM2, a E3 ubiquitin ligase, is found overexpressed in brain, breast, lung, and prostate cancers; interferes with this cycle by promoting sorting of E-cadherin toward lysosome for degradation; and thus inhibits cadherin recycling and compromises AJ integrity. Finally, Rab GTPases control both endocytic and exocytic pathways. Abnormal activities of Rabs or their regulatory proteins and effectors lead to several human diseases, including many types of cancers.

Cross-References

▶ Endocytosis

References

Lakadamyali M, Rust MJ, Zhuang X (2006) Ligands for clathrin-mediated endocytosis are differentially sorted into distinct populations of early endosomes. Cell 124:997–1009

Rink J, Ghigo E, Kalaidzidis Y et al (2005) Rab conversion as a mechanism of progression from early to late endosomes. Cell 122:735–749

Russell MR, Nickerson DP, Odorizzi G (2006) Molecular mechanisms of late endosome morphology, identity and sorting. Curr Opin Cell Biol 18:422–428

Stein MP, Dong J, Wandinger-Ness A (2003) Rab proteins and endocytic trafficking: potential targets for therapeutic intervention. Adv Drug Deliv Rev 55:1421–1437

Endosomal Protein Sorting

▶ Endosomal Compartments

Endostatin

Sundaram Ramakrishnan
Department of Pharmacology, University of Minnesota, Minneapolis, MN, USA

Definition

Endostatin is a proteolytic fragment derived from the carboxyterminal, non-collagenous domain 1 (NC1) of collagen type XVIII. Endostatin inhibits ▶ angiogenesis by blocking endothelial cell proliferation and ▶ migration.

Characteristics

Discovery

Endostatin was originally identified in the culture medium of a murine endothelioma cell line by Michael S. O'Reilly et al., from the laboratory of Dr. Judah Folkman. The same group of investigators has also discovered another angiogenesis inhibitor, angiostatin. The search for angiogenesis inhibitors began with the premise that the primary tumor secretes an inhibitor which suppresses the growth of metastatic cells at secondary sites. Biochemical characterization of endostatin showed a molecular mass of 20 KDa containing 184 amino acid residues. Endostatin selectively inhibits endothelial cells and does not affect the growth of non-endothelial cells, including tumor cells. Recombinant mouse endostatin was initially produced in *E. coli* in insoluble form. Treatment of mice with recombinant endostatin inhibited >99% of tumor growth. Soluble forms of both mouse and human endostatins are currently produced in yeast, insect, and mammalian cells.

Structure and Function

The precursor for endostatin is collagen type XVIII, a special kind of collagen found at the basement membrane of endothelium and epithelium. Collagen XVIII along with collagen XV

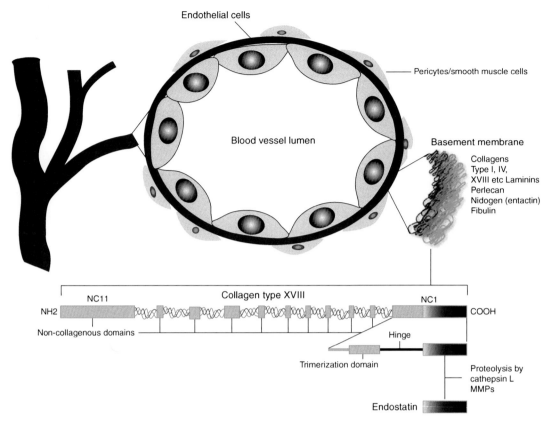

Endostatin, Fig. 1 Generation of endostatin from the vascular basement membrane

belongs to a subfamily called multiplexin. Collagens normally have a contiguous, long stretch of triple-helical region (collagenous domain) which ends at the carboxyterminus with a globular, non-collagenous domain (NC1). Collagenous domains are characterized by repeated sequence of glycine–hydroxylysine–hydroxyproline amino acid residues. The multiplexin subfamily of collagens has multiple interruptions in the collagenous domain. Collagen type XVIII, for example, has 11 non-collagenous (NC1–NC11) domains flanking ten collagenous domains (Fig. 1). Endostatin is generated from the NC1 domain, which contains a trimerization domain at the amino terminus followed by a protease-sensitive hinge region which culminates at the carboxyterminus as endostatin. Cathepsin L and, to some extent, matrix metalloproteinases cleave at the hinge region of NC1 domain to release endostatin.

X-ray crystallographic studies have shown that endostatin is a compact globular molecule.

Schematic diagram shows the complex architecture of vascular basement membrane/extracellular matrix. In addition to providing the scaffold for vascular assembly, the matrix contains a number of sequestered growth factors which can be locally released to modulate angiogenic response. The domain organization of collagen type XVIII is shown. Endostatin is generated by the proteolysis of the carboxyterminal NC1 domain.

A number of positively charged arginine residues are located at the major alpha-helix aligning toward the surface of endostatin. Substitution of arg residues affects heparin binding but does not alter the biological activity. A single point mutation at proline 125 to alanine has been found to increase the biological activity of human endostatin. Comparison of amino acid sequences between various organisms reveals substitutions of proline to a different amino acid residue at this site. Endostatin sequence from chicken and *Xenopus* shows proline to alanine substitution. Japanese puffer fish and drosophila (fruit fly) have glycine at this position. *C. elegans* has an aspartic acid at this location. However, no polymorphism has been found in human collagen XVIII/endostatin at this position. Some studies suggest even shorter fragments of endostatin are biologically active and inhibit angiogenesis.

Mechanism of Action

Endostatin binds to α5β1 integrin and induces clustering within the lipid rafts. In addition, glypican 1, a heparan sulfate glycosaminoglycan, serves as a low-affinity binding site for endostatin. Two phenylalanine residues (F31 and F34) are found to be critical for glypican binding. Mutation of these two residues affects endostatin binding to glypican 1. The integrin-binding site of endostatin has not yet been identified. Clustering of integrin-bound endostatin activates ▶ *Src* kinase and inhibits RhoA activity leading to the disruption of focal adhesion. Associated changes in the actin stress fibers ultimately affect endothelial cell migration. Endostatin is also known to affect a second signaling pathway, the ▶ Wnt signaling pathway, leading to the proteosomal degradation of β-catenin. Reduced levels of cytoplasmic β-catenin block T-cell factor (TCF)-mediated transcription of ▶ cyclin D1 and c-myc. Repression of cyclin D1, one of the critical mediators of cell cycle progression, arrests endothelial cells at the G1 phase of the cell cycle. Other reports have identified cell surface-bound tropomyosin, ▶ vascular endothelial growth factor (VEGF) receptor 2, and matrix metalloproteinase-2 (MMP-2) as potential targets for endostatin.

Genetic Variations

Mutations in collagen XVIII/endostatin gene have been identified in humans. Absence of collagen XVIII/endostatin is associated with Knobloch syndrome (an autosomal recessive disease that is characterized by ocular defects leading to retinal detachment, macular degeneration, and blindness). In mice, collagen XVIII/endostatin gene knockout did not affect developmental angiogenesis but however showed progressive loss of vision. Other mutation, such as D104N substitution in endostatin, has been observed in Knobloch syndrome patients (heterozygous, with one allele truncated and the other with a mutation leading to D104N substitution) as well as in many cancer patients. D104N endostatin does not affect the biological activity of endostatin but showed reduced affinity to laminin. Cancer patients having D104N mutation in endostatin did not show any correlation with the disease progression and survival. These studies suggest that loss of collagen XVIII/endostatin does not paralyze the entire vascular system but leads to some selective changes associated with the eye. In contrast, increased expression of collagen XVIII/endostatin has been postulated to reduce the risk of cancer development. Normal serum levels of endostatin in human are between 25 and 30 ng/ml. Higher levels (about 70%) of endostatin have been observed in Down syndrome patients who have an extra copy of chromosome 21 (trisomy). Collagen XVIII gene is located in chromosome 21. In comparison to age matched, control healthy population, a decreased risk for cancer development has been noticed in the Down syndrome patients. Folkman and Kalluri have hypothesized a physiological tumor suppressive role for endostatin based on these observations.

Preclinical Studies

Endostatin inhibits tumor growth in several model systems. More than fifteen different types of human tumor cell lines (e.g., ▶ ovarian cancer, ▶ breast cancer, glioblastoma, and renal cancer) transplanted into immunodeficient mice were inhibited, to varying degrees by endostatin treatment. Endostatin treatment was successful when initiated at an early stage of tumor growth. In a few studies, endostatin treatment induced the regression of established tumors. Furthermore, twice daily injections of endostatin were found to be more effective than daily bolus injections. Serum half-life (pharmacokinetics) and bioavailability are responsible for the schedule-dependent differences seen in the efficacy of endostatin treatment. This notion was further supported by the observation that tumor growth was better inhibited when endostatin was delivered by mini-osmotic pumps and in slow-release formulations. Endostatin is well tolerated in mice and does not show any toxicity.

Clinical Trials

In 1999, human trials were initiated at three institutions, Dana-Farber Cancer Institute, Boston; University of Wisconsin, Madison; and MD Anderson Cancer Center at Houston. Recombinant endostatin expressed in yeast was used in these phase I trials. A total of 61 patients with advanced disease of different tumor types were enrolled into these studies. Recombinant endostatin was administered either as a 20 min or 1 h intravenous infusion. Dosages ranged from 15 to 600 mg/m^2/day. There was no dose-limiting toxicity observed in all the three trials. Serum half-life of endostatin was found to be between 10 and 11 h. In two of the 25 patients treated at the MD Anderson Cancer Center, there was evidence of minor antitumor effect. Out of the total of 15 patients enrolled in Dana-Farber Cancer Institute, there was a minor response in a patient with a pancreatic neuroendocrine tumor, while two of the other patients showed stabilization of the disease. In the University of Wisconsin trial, there was no objective clinical response in any of the 25 patients treated with endostatin. Dynamic CT scans of the patients treated with endostatin showed some evidence of changes in the microvessel density. Other imaging methods to study functional changes in tumor vasculature indicated changes in blood flow and metabolism following endostatin treatment. A definitive antiangiogenic effect in treated patients could not be documented due to the lack of suitable biomarkers to validate changes in tumor angiogenesis. Another phase I study was undertaken at the Vrije Universiteit Medical Center in the Netherlands. Thirty-two patients received recombinant endostatin as a continuous infusion for 4 weeks.

Treatment was continued after 1 week of rest by twice daily subcutaneous injections. This trial also noted no adverse side effects in the treated patients, and again there was no objective clinical response. However, two patients had a long-lasting stable disease.

Since one of the patients with pancreatic neuroendocrine tumor showed partial response in phase I trial, a phase II trial was initiated at multiple centers. Forty patients with advanced neuroendocrine tumors were treated with recombinant endostatin by daily subcutaneous injections. Even though a steady-state level of potentially effective serum concentration of endostatin was achieved in these patients, no partial or clinical response was observed. While it is disappointing to note that the antitumor effects seen in experimental animals could not be replicated in these early clinical trials, a lot has been learned on the pharmacological properties of endostatin. Future studies will focus on combining endostatin treatment with other modalities to improve the antitumor effects. Indeed, the inhibition of tumor growth by radiotherapy and ► chemotherapy is potentiated by endostatin treatment. Furthermore, ► gene therapy approaches using viral vectors have shown promising effects in experimental animals. In fact, adeno-associated virus (AAV)-mediated expression of endostatin is in the early phases of clinical development. The surgical removal of tumors followed by chemotherapy in combination with endostatin treatment is a promising strategy for cancer treatment.

References

Benezra R, Rafii S (2004) Endostatin's endpoints – deciphering the endostatin antiangiogenic pathway. Cancer Cell 13:205–206

Folkman J (2006) Antiangiogenesis in cancer therapy – endostatin and its mechanism of action. Exp Cell Res 312:594–607

Kalluri R (2003) Basement membranes: structure, assembly and role in tumour angiogenesis. Nat Rev Cancer 3:422–433

Marneros AG, Olsen BR (2005) Physiological role of collagen XVIII/endostatin. FASEB J 19:716–728

Endothelial Cell-Specific Molecule-1

▶ Endocan

Endothelial Transglutaminase

▶ Transglutaminase-2

Endothelial-Derived Gene-1

Mai N. Brooks
Surgical Oncology, School of Medicine,
University of California, Los Angeles, CA, USA

Synonyms

Magicin; Med28

Definition

Endothelial-derived gene EG-1 was discovered in 2002. Although first cloned from a human endothelial cell cDNA library (▶ Angiogenesis), EG-1's transcript has been shown to be present in other cell types as well, particularly in epithelial cells. The calculated mass of EG-1 is 19,520.2 Da based on amino acid sequence alone, whereas the native complete protein is ~22 kDa. The homology between the human EG-1 peptides to its mouse counterpart is 94.9%, 95.5% to its rat counterpart, and 31% to the *Drosophila* one.

Characteristics

EG-1 is strongly associated with cellular proliferation. Overexpression of EG-1 achieved by transfection results in increased proliferation of multiple human cell lines in culture. When these cells are injected subcutaneously into immunodeficient mice, EG-1 overexpressed cells develop into larger xenograft tumors (Sato et al. 2004).

Mechanisms

Overexpression of EG-1 results in activation of the mitogen-activated protein kinase (MAPK) pathway (▶ MAP kinase), which has been shown to be crucial in promoting cellular proliferation. This manifests as increased levels of phosphorylated p44/42 MAP kinase, phosphorylated JNK (Jun-terminal kinase) (▶ JNK Subfamily), and phosphorylated p38 kinase.

EG-1 overexpression also results in c-Src activation. c-Src is a member of the Src family of cytoplasmic tyrosine kinases that regulate cell growth, differentiation, cell shape, migration, and survival (▶ Src). c-Src has been reported to be overexpressed and to play a role in human carcinomas of the breast, colon, and others. Src family tyrosine kinases are often activated by ▶ receptor tyrosine kinases, such as EGF-R (▶ epidermal growth factor receptor) (EGR ligands) or PDGF-R (▶ platelet-derived growth factor receptor) (PDGF). Via its proline-rich region, EG-1 binds to the Src family of protein tyrosine kinases c-Src and Yes and possibly FYN and Hck (hemopoietic cell kinase). As a result of EG-1 binding, c-Src then becomes catalytically active. However, EG-1 is not a direct substrate of c-Src, nor does it increase c-Src expression (Lu et al. 2005).

Two proteins with identical sequences to EG-1 have been identified: Magicin for merlin and Grb2-interacting cytoskeletal protein in 2004 (Wiederhold et al. 2004). Magicin is described to associate with the actin cytoskeleton and is proposed to have a role in receptor-mediator signaling at the cell surface. Magicin binds directly to Grb2 (growth factor receptor bound 2 protein). Med28 is a member of the Mediator, a multiprotein transcriptional coactivator that is expressed ubiquitously in eukaryotes for induction of RNA polymerase II transcription by DNA-binding transcription factors. As Med28, this protein is one subunit of the "adaptor" that bridges RNA polymerase II with its DNA-binding regulatory proteins and transduces both positive and negative signals.

In summary, EG-1 is an important protein that has multiple interactions with crucial cellular pathways involved in cellular proliferation, cytoskeletal function, and transcriptional regulation.

EG-1 in Human Cancer

Immunohistochemistry of human samples demonstrates that EG-1 is present in the nucleus as well as in the cytoplasm and possibly in the cell membrane. There are significantly higher levels of EG-1 peptides in cancer specimens of the breast, colon (Colon cancer), and prostate (Prostate carcinoma), in comparison with their benign counterparts (Zhang et al. 2004) (▶ Cancer). EG-1 has been detected at elevated levels in sera from breast cancer patients (▶ Serum biomarkers). In human urine, EG-1 appears primarily at a larger molecular weight, suggesting that these peptides may co-aggregate or associate with other moieties in urine (Biomarkers). Thus, EG-1 may be secreted or it may be shed with cell death/turnover.

Translational Aspects

Studies have shown that endogenous EG-1 can be targeted to inhibit breast tumor growth (Lu et al. 2007). This inhibition, whether delivered via siRNA lentivirus (▶ siRNA) or polyclonal antibody, results in decreased cellular proliferation in culture and smaller xenograft tumors in mice (▶ Mouse models). The effects are shown in both ER (estrogen receptor)-positive human breast cancer MCF-7 cells (▶ Estrogen receptor) and in ER-negative MDA-MB-231 cells. As breast cancer is the most common malignancy diagnosed in women and as one-third of these patients will die of their disease, a novel target for breast cancer therapeutic development such as EG-1 would be very useful (▶ Breast cancer). Because EG-1 is unique, its use may not be redundant to other gene products/potential targets involved in other molecular pathways (▶ Molecular therapy; ▶ small molecule drugs). Further preclinical studies are warranted to explore the usefulness of targeting EG-1 for future cancer therapy.

Cross-References

- ▶ Angiogenesis
- ▶ Breast Cancer
- ▶ Cancer
- ▶ Chemotherapy
- ▶ Epidermal Growth Factor Receptor
- ▶ Epithelial Tumorigenesis
- ▶ Estrogen Receptor
- ▶ JNK Subfamily
- ▶ MAP Kinase
- ▶ Molecular Therapy
- ▶ Mouse Models
- ▶ Neoadjuvant Therapy
- ▶ Platelet-Derived Growth Factor
- ▶ Prostate Cancer Clinical Oncology
- ▶ Receptor Tyrosine Kinases
- ▶ Serum Biomarkers
- ▶ SiRNA
- ▶ Small Molecule Drugs
- ▶ Src

References

Lu M, Zhang L, Maul RS et al (2005) The novel gene EG-1 stimulates cellular proliferation. Cancer Res 65:6159–6166

Lu M, Zhang L, Sartippour MR et al (2006) EG-1 interacts with c-Src and activates its signaling pathway. Int J Oncol 29:1013–1018

Lu M, Sartippour MR, Zhang L et al (2007) Targeted inhibition of EG-1 blocks tumor growth. Cancer Biol Ther 6:936–941

Sato S, Tomomori-Sato C, Parmely TJ et al (2004) A set of consensus mammalian mediator subunits identified by multidimensional protein identification technology. Mol Cell 14:685–691

Wiederhold T, Lee MF, James M et al (2004) Magicin, a novel cytoskeletal protein associates with the NF2 tumor suppressor merlin and Grb2. Oncogene 23:8815–8825

Zhang L, Maul RS, Rao J et al (2004) Expression pattern of the novel gene EG-1 in cancer. Clin Cancer Res 10:3504–3508

See Also

(2012) Biomarkers. In: Schwab M (ed) Encyclopedia of Cancer, 3rd edn. Springer Berlin Heidelberg, pp 408–409. doi:10.1007/978-3-642-16483-5_6601

Endothelial-Leukocyte Adhesion Molecule 1 (ELAM1)

▶ E-Selectin-Mediated Adhesion and Extravasation in Cancer

Endothelin-2

▶ Endothelins

Endothelins

Matthew J. Grimshaw
Breast Cancer Biology Group, King's College
London School of Medicine, Guy's Hospital,
London, UK

Synonyms

EDN; Endothelin-2; ET; ET-2; vasoactive intestinal contractor; VIC

Definition

Endothelins (ETs) are a family of three similar, small peptides that are among the strongest vasoconstrictors known and play a key part in vascular homeostasis. Endothelins have numerous roles in tumors including modulating ▶ angiogenesis and blood flow, inducing mitogenesis and ▶ invasion of tumor cells, immune activation, and protecting cells from ▶ apoptosis.

Characteristics

Endothelins (ETs) are a family of small, structurally related, vasoactive peptides that have a variety of physiological roles in many tissues, notably vascular homeostasis. The "ET axis" consists of three peptides, two receptors, and two activating enzymes (Table 1). Some examples of the roles of ETs in both normal physiology and pathological conditions are shown below:

1. *Blood vessels*: Maintain basal level of vasoconstriction ("contraction of blood vessels" which controls blood pressure). Involved in the development of hypertension and atherosclerosis.
2. *Heart*: Affect the force and rate of contraction of the heart. Mediate hypertrophy and remodeling in congestive heart failure.
3. *Lungs*: Regulate the tone of airways and blood vessels. Involved in pulmonary hypertension.
4. *Kidney*: Controls water and sodium excretion and acid–base balance. Participate in renal failure.
5. *Brain*: Modulates cardiorespiratory centers and hormone release.
6. *Cancer*: Numerous tumors – including carcinomas of the breast, lung, prostate, and ovary – produce one or more of the ETs and their receptors, and there are many potential roles of the "ET axis" in cancer.

Mitogenesis: Endothelins have a mitogenic ("growth promoting") effect on both tumor and stromal (i.e., the noncancer cell component of a tumor including blood vessels, immune cells, and fibroblasts) (stroma) cells and enhance tumor growth. *Tumor angiogenesis*: Angiogenesis is the growth of new blood vessels and is critical for the growth of a solid tumor. ETs stimulate angiogenesis within solid tumors by acting directly on endothelial cells to modulate proliferation, migration, invasion, and morphogenesis. ETs also modulate angiogenesis indirectly through induction of the angiogenic cytokine ▶ vascular endothelial growth factor (VEGF) expression. ETs also affect blood flow through the established tumor vasculature due to their vasoactive nature. However, the effect of ETs on blood flow appears to be tissue and tumor specific. *Tumor invasion and metastasis*: Invasion is the process by which cancer cells spread beyond the border of the tumor enabling the tumor to grow and spread. ETs stimulate invasion of

Endothelins, Table 1 Genes and peptide sequences of ET axis members. Amino acids that differ in ET-2 and/or ET-3 from ET-1 are marked in bold/red

Gene	Mapping position	Peptide or protein	ET isoform peptide sequence
EDN1	6p24	ET-1	CSCSSLMDKECVYFCHLDIIW
EDN2	1p34	ET-2	CSCSS**W**L**DKECVVF**CHLDIIW
EDN3	20q13	ET-3	C**TC**F**TYK**DKECV**YY**CHLDIIW
EDNRA	4q31	ET-RA	–
EDNRB	13q22	ET-RA	–
ECE1	1p36	ECE-1	–
ECE2	3q28-29	ECE-2	–

several types of tumor cells including ▶ P-Glycoprotein family cells, Ewing's sarcoma and neuroblastoma cells, and ▶ breast cancer cells. Stimulation of, for instance, breast tumor cell lines with ETs leads to an invasive phenotype via several autocrine and paracrine mechanisms including induction of ▶ matrix metalloproteinases (MMPs) and ▶ chemokine receptors and the activation of macrophages. *Protection from apoptosis*: ETs can protect several cell types – including tumor cells, macrophages, and endothelial cells – from apoptosis ("programmed cell death") induced by cellular stresses including ▶ hypoxia, serum starvation, and chemotherapeutic agents. *Immune modulation*: Trafficking, differentiation, and activation of tumor-infiltrating immune cells are all modulated by ETs.

The Endothelin Axis. The ET axis consists of three 21 amino acid (aa) peptides (ET-1, ET-2, and ET-3) (Fig. 1), two G-protein-coupled receptors (ET-RA and ET-RB), and two membrane-bound endothelin-converting enzymes (ECE-1 and ECE-2). ET-1 was initially found in the conditioned medium of cultured endothelial cells, and its activity as a potent vasoconstrictive peptide was described. ET-2 and ET-3 were rapidly described following ET-1's discovery, and further roles in a variety of tissues have been described.

The three ET isoforms – which are highly conserved in human, rat, and mouse – derive from three separately regulated genes yet have a similar structure. Human ET-1 derives from a 212 aa precursor, preproendothelin-1. The removal of the signal sequence generates the 195 aa proendothelin-1, which is further processed to release the intermediate 38 aa "big

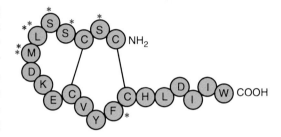

Endothelins, Fig. 1 *Structure of ET-1.* ET-1 is a 21 amino acid peptide with a hydrophobic C-terminus and two disulfide bonds at the N-terminus. ET-2 and ET-3 are structurally similar to ET-1, differing by two and six amino acids, respectively. The amino acids which differ in the ET-2 sequence are indicated by "*," while those which differ in ET-3 marked by "*"

ET-1." ECEs hydrolyze big ET-1 to yield the active 21 aa ET-1. The gene for each ET has a distinct pattern of tissue expression: ET-1 is expressed by endothelial cells of many organs, ET-2 is in the ovary and intestine, and ET-3 is found in the brain. There is a relatively low basal level of synthesis of ETs, but these genes are readily inducible by inflammatory stimuli.

Endothelin Receptors. Two receptors for ETs have been characterized: ET-RA (also known as EDNRA or ET_AR) and ET-RB (EDNRB, ET_BR). Both receptors are expressed in a wide variety of tissue types. ETs bind these receptors with varying affinity: ET-RA binds ET-1 $\geq$ ET-2 $>$ ET-3, but ET-RB shows no selective affinity for any ET subtype. Binding of the ligands to these G-protein-coupled receptors (GPCRs) modulates several overlapping signaling pathways resulting in the activation of phospholipase C and MAPK pathways, an increase in intracellular calcium and the induction of immediate early genes.

Induction of Endothelin Expression by the Tumor Microenvironment. One region of tumor, compared to another, may differ in the levels of hypoxia, cytokine concentration, immune infiltrate, vascularization, necrosis, etc. The "▶ tumor microenvironment" – particularly hypoxia and soluble factors such as cytokines – modulates expression of numerous "pro-tumor" genes, including those of the ET axis. Transcriptional regulation of numerous hypoxia-responsive genes is via the hypoxia-induced transcription factor, HIF-1, which initiates transcription of genes whose promoter contains a hypoxia response element (HRE). Hypoxia induces ET axis transcription in several cell types including endothelial and tumor cells. There is a functioning HRE in the antisense strand of the promoter of ET-1, and induction of ET expression by hypoxia is via HIF-1.

Endothelin Receptor Antagonists. The role of ETs in vasoconstriction has led to the development of several antagonists of the ET receptors that are currently under investigation for the treatment of hypertension, heart failure, and renal disease. Several are now in phase II and III trials for the treatment of various neoplasms, particularly prostate cancer. ET receptor antagonists hold the attractive possibility that they will "hit" several different cell types and mechanisms of cancer progression.

Of the antagonists available, it is the modified peptide-based antagonists BQ123 (ET-RA antagonist) and BQ788 (ET-RB antagonist) that have been used extensively both in vitro and in vivo. Small molecule antagonists such as atrasentan, a highly selective ET-RA antagonist, have been used clinically. These antagonists can be administered orally, are well tolerated, and have few toxic side effects. ET receptor antagonists inhibit proliferation of Kaposi sarcoma cells, Ewing's sarcoma and neuroblastoma cells, melanoma cells, and ovarian carcinoma cells.

As well as the commercially produced antagonists, it is of interest that ET activity may be modified by dietary factors. An extract of red wine polyphenols causes inhibition of ET-1 synthesis in endothelial cells; this is associated with modifications in phosphotyrosine staining, indicating that the active components of red wine cause specific modifications of tyrosine kinase signaling. Green tea polyphenol epigallocatechin-3-gallate inhibits the ET axis and downstream signaling pathways in ovarian carcinoma.

Endothelin Expression in Cancer

Numerous types of tumors produce one or more of the ETs and their receptors. However, the expression and actions of ETs in cancer are incompletely described and are tumor-type specific. ET axis expression is increased in many types of tumor, yet in several types of tumor, expression of the ET axis – particularly the receptors – is *decreased* in neoplastic tissue. For instance, in carcinomas of the breast, both ET-RA and -RB are increased, yet in prostate cancer, ET-RB is decreased, while in lung cancer, ET-RA is downregulated. The function of the ET-RB receptor in tumors is particularly enigmatic; in some cases, such as breast cancer, ligand binding to ET-RB initiates several pro-tumor actions, such as promoting invasion, yet in prostate cancer, the loss of ET-RB expression is postulated to increase ET-1 peptide in the tumor due to the loss of ET-RB's ET clearance function.

Ovarian Carcinoma. ETs have several roles in ovarian tumors including promoting growth and invasion, and ETs stimulate both tumor and stromal cells. Ovarian carcinoma cells secrete ET-1, which acts as an autocrine growth factor via ET-RA and also has a paracrine growth effect on the fibroblastic cells via both receptors. ETs acting through ET-RA promote invasion of the tumor cells by upregulating secretion and activation of MMPs.

ET-RA is found in both tumor cells and intratumoral vessels, whereas ET-RB is expressed mainly in endothelial cells, and ETs induce angiogenesis via hypoxia-inducible factor (HIF)-1α and VEGF. Atrasentan decreases growth of ovarian xenografts in mice, and this is associated with decreased angiogenesis and MMP expression and increased apoptotic tumor cells.

Prostate Cancer. The role of ETs has been studied in both the normal and transformed prostate. ETs are produced in the normal prostate gland by epithelial cells and are found in high

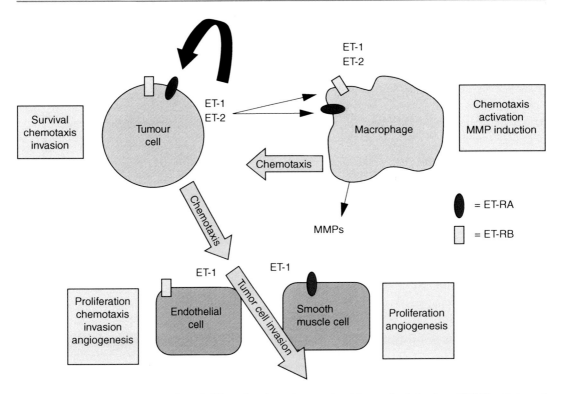

Endothelins, Fig. 2 *Putative roles of ETs in breast tumors.* TAMs express inflammatory cytokines that may induce ET expression by both tumor cells and macrophages, which release ETs and express both receptors. Microenvironmental factors, including hypoxia and the ETs themselves, further stimulate the ET axis. Stimulation of tumor cells and macrophages with ETs leads to chemotaxis of these cells, induction of MMP activity and cytokine expression, and invasion of the tumor cells. ET-2 also protects tumor cells from apoptosis and activates macrophages. Endothelial cells and VSMCs express ET-RB and ET-RA, respectively. Stimulation of endothelial cells and VSMCs with ETs may stimulate angiogenesis

concentrations in seminal fluid (up to 5 μg/l). ET-RA and ET-RB are found in normal prostate tissue, but in the malignant prostate, there is a loss of ET-RB and increased levels of ET-1. Roles for ETs in prostate cancer include growth promotion, apoptosis inhibition, and bone formation. Atrasentan delays time to progression in prostate cancer in phase III clinical trials. ETs modulate nociception ("sense of pain"), and ET-RA antagonism attenuates prostate carcinoma-induced pain.

Breast Cancer. There are numerous potential consequences of ET expression in breast tumors that may lead to a more aggressive tumor cell phenotype (Fig. 2). These include the induction of invasion and angiogenesis via stimulation of tumor cells and ▶ tumor-associated macrophages (TAMs). There is increased expression of the ET axis in invasive ductal carcinoma (IDC) of the breast compared to the normal breast; lymph node metastases (▶ metastasis) have a higher degree of ET staining still. Cells expressing ETs and their receptors in IDC include the tumor cells, the CD68[+] macrophage infiltrate, and the endothelial cells.

ETs have a role in recruiting TAMs – macrophages express both ET receptors and chemotax toward ETs via ET-RB and a MAPK-mediated signaling pathway. Exposure of macrophages to ETs leads to an "activated" phenotype and cytokine secretion. Macrophages not only react to ETs but also produce ETs themselves, and the TAMs contribute to the ETs in the breast tumor microenvironment.

ETs induce expression of chemokine receptors including CCR7 and potentiate the response of breast tumor cells to chemokines including CXCL12 and CCL21, which modulate the organ specificity of breast cancer metastasis. A further potential function of ETs in breast cancer is the modulation of angiogenesis. Boyden chamber (a mixed ET-RA/B antagonist) inhibits tumor vascularization and bone metastasis in a murine model of breast carcinoma cell metastasis. Expression of ET-RA predicts unfavorable response to neoadjuvant chemotherapy in locally advanced breast cancer.

▶ *Melanoma.* The ET axis may be a promising therapeutic target for the treatment of melanomas. Activation of ET-RB promotes melanocyte precursor cell proliferation while inhibiting differentiation, two hallmarks of malignant transformation. In melanoma cell lines, ETs prevent apoptosis, and ET-RB antagonists cause an increase in cell death. ETs are also involved in angiogenesis in mouse models of melanoma. In vivo, BQ788 slows growth of human melanoma tumors in nude mice. A phase II study of bosentan as monotherapy in patients with stage IV metastatic melanoma showed disease stabilization in 6 of 32 patients.

▶ *Lung Cancer.* ET-1 has been proposed as a prognostic marker in non-small cell lung carcinoma (NSCLC). There is higher expression of ET-1, ET-RA, and ECE-1 in lung tumors compared to the normal tissue, while ET-RB is decreased. Interestingly, ET-1 is increased in the breath condensate of NSCLC patients, and this could potentially be used as a noninvasive test for early detection of NSCLC.

▶ *Bladder Cancer.* The ET axis, particularly ET-RB, is overexpressed in bladder cancer. Patients with ET-RB expression tend to have organ-confined tumors and no vascular invasion, and as such, ET-RB is associated with *favorable* disease-free survival. When metastatic bladder carcinoma cells were injected into mice treated with atrasentan, there was a dramatic reduction of metastases to the lungs.

▶ *Nasopharyngeal Carcinoma.* Elevated plasma big ET-1 is associated with distant failure in patients with advanced-stage nasopharyngeal carcinoma.

▶ *Cervical Cancer.* In human papillomavirus-positive cervical cancer cells, ET-RA mediates an ET-induced mitogenic effect. Atrasentan inhibits growth and angiogenesis in cervical cancer xenografts.

Cross-References

▶ Angiogenesis
▶ Apoptosis
▶ Bladder Cancer
▶ Breast Cancer
▶ Cervical Cancers
▶ Chemokines
▶ Hypoxia
▶ Invasion
▶ Lung Cancer
▶ Matrix Metalloproteinases
▶ Metastasis
▶ Nasopharyngeal Carcinoma
▶ P-Glycoprotein
▶ Tumor-Associated Macrophages
▶ Tumor Microenvironment
▶ Vascular Endothelial Growth Factor

References

Grimshaw MJ (2005) Endothelins in breast tumour cell invasion. Cancer Lett 222(2):129–138

Kedzierski RM, Yanagisawa M (2001) Endothelin system: the double-edged sword in health and disease. Annu Rev Pharmacol Toxicol 41:851–876

Nelson J, Bagnato A, Battistini B et al (2003) The endothelin axis: emerging role in cancer. Nat Rev Cancer 3(2):110–116

See Also

(2012) Atrasentan. In: Schwab M (ed) Encyclopedia of Cancer, 3rd edn. Springer Berlin Heidelberg, p 303. doi:10.1007/978-3-642-16483-5_444

(2012) Boyden Chambers. In: Schwab M (ed) Encyclopedia of Cancer, 3rd edn. Springer Berlin Heidelberg, p 465. doi:10.1007/978-3-642-16483-5_696

(2012) Endothelin Converting Enzyme. In: Schwab M (ed) Encyclopedia of Cancer, 3rd edn. Springer Berlin Heidelberg, p 1254. doi:10.1007/978-3-642-16483-5_1903

(2012) ET-RA. In: Schwab M (ed) Encyclopedia of Cancer, 3rd edn. Springer Berlin Heidelberg, p 1339. doi:10.1007/978-3-642-16483-5_2024

(2012) ET-RB. In: Schwab M (ed) Encyclopedia of Cancer, 3rd edn. Springer Berlin Heidelberg, p 1339. doi:10.1007/978-3-642-16483-5_2025

(2012) G-protein Couple Receptor. In: Schwab M (ed) Encyclopedia of Cancer, 3rd edn. Springer Berlin Heidelberg, p 1587. doi:10.1007/978-3-642-16483-5_2294

(2012) Invasive Ductal Carcinoma. In: Schwab M (ed) Encyclopedia of Cancer, 3rd edn. Springer Berlin Heidelberg, p 1906. doi:10.1007/978-3-642-16483-5_3134

(2012) Lymph Node Metastases. In: Schwab M (ed) Encyclopedia of Cancer, 3rd edn. Springer Berlin Heidelberg, p 2116. doi:10.1007/978-3-642-16483-5_3444

(2012) Prostate Cancer. In: Schwab M (ed) Encyclopedia of Cancer, 3rd edn. Springer Berlin Heidelberg, p 3009–3010. doi:10.1007/978-3-642-16483-5_6576

(2012) Stroma. In: Schwab M (ed) Encyclopedia of Cancer, 3rd edn. Springer Berlin Heidelberg, p 3541. doi:10.1007/978-3-642-16483-5_5532

(2012) Stromal Cells. In: Schwab M (ed) Encyclopedia of Cancer, 3rd edn. Springer Berlin Heidelberg, p 3544. doi:10.1007/978-3-642-16483-5_5535

Endotoxin-Induced Factor in Serum

▶ Tumor Necrosis Factor

Engineered Antibody

▶ Diabody

Enlarged Breast Male

▶ Gynecomastia

ENPP2

▶ Autotaxin

Enteropeptidase (TMPRSS15)

▶ Serine Proteases (Type II) Spanning the Plasma Membrane

Enzymes

▶ Antioxidant Enzymes

Enzymic Mouth to Mouth Feeding

▶ Substrate Channeling

Ep

▶ Erythropoietin

EpCAM

M. Asif Chaudry
University Department of Surgery, Royal Free and University College London Medical School, London, UK

Synonyms

17-1A; EGP-2; EGP34; EpCAM; Epithelial cell adhesion molecule; ESA; GA733-2; HEA125; KSA; MK-1; TROP-1

Definition

A membrane protein found on all simple epithelia to varying degrees is a type I membrane protein. It is a pan-epithelial differentiation antigen expressed on the basolateral surface of all carcinomas, varying in density and glycosylation.

As a homotypic cell adhesion molecule, it is intimately integrated within the cadherin-catenin and ▶ Wnt signaling pathways. It modulates the expression of proto-oncogenes such as ▶ Myc oncogene. Its status as a pan-carcinoma antigen has rendered it an attractive target for cancer ▶ immunotherapy.

Characteristics

Structure and Function

Structure

This 37 kDa protein is formed from 314 amino acids (aa) of which only 26 aa face the cytoplasm.

The extracellular component contains three domains: the first is novel and is the site to which most of the antibodies developed are targeted (323 ∼ A3, 17-1A and others). The second is similar to EGF-binding proteins 1 and 6 and thyroglobulin. The third has a novel structure that also has similarities with EGF. The intracellular portion of the antigen has a tyrosine phosphorylation site, the significance of which is uncertain.

Tissue Morphogenesis

EpCAM is essential for stable adhesion formation and tissue morphogenesis similar to adhesion molecules: ▶ carcino-embryonic antigen (CEA) and ICAM-1. The mechanism by which cytoskeletal and intracellular elements mediate this function is being characterized. EpCAM inhibits intercellular adhesion mediated by E-cadherins, in turn interacts with α-, β-, and γ-catenins forming the cadherin-catenin complex.

Catenins link cadherins with the actin cytoskeleton and form complexes with other proteins. Cadherins are crucial for the establishment and maintenance of epithelial cell polarity, morphogenesis of epithelial tissues, and regulation of cell proliferation and apoptosis. Their association with β-catenin is particularly interesting as this is a component in the ▶ Wnt signaling pathway that regulates the expression of proto-oncogenes such as c-Myc: fundamentally associated with tumor development. Wnt glycoproteins are signaling molecules that regulate cell-to-cell interaction during embryogenesis. Wnt proteins bind to receptors of the Frizzled family. Through several cytoplasmic relay components, the signal is transduced to β-catenin, which is stabilized, accumulates in the cytoplasm, and enters the nucleus, where it binds a lymphoid enhancer factor/T-cell factor transcription factor. Together, β-catenin and lymphoid enhancer factor/T-cell factor activate expression of many target genes, such as Myc oncogene, VEGF, and *cyclooxygenase-2*, all associated with neoplasia.

EpCAM directly impacts the cell cycle by upregulating c-Myc and cyclin A/E. Human epithelial cells expressing EpCAM reduce growth factors' dependency and increase metabolism and colony formation. Inhibition of EpCAM expression with antisense nucleic acid reduces proliferation and metabolism in human carcinoma cells. The intracellular domain is essential for these effects.

EpCAM adhesive properties promote calcium-independent homotypic cell sorting. Cells transfected to express EpCAM are sorted from cells of the same line that do not normally express EpCAM. It also inhibits invasive growth in cell colonies. Both activities are inhibited by anti-EpCAM antibodies.

The function of EpCAM thyroglobulin domain is being actively investigated. These domains commonly inhibit cathepsins: cysteine proteases frequently produced by tumor cells and known to be involved in metastasis.

Pattern of Tissue Expression

Normal Tissue

EpCAM is present on all normal epithelia excluding stratified squamous epithelia. Within the gastrointestinal (GI) tract, colonic expression is greatest and gastric lowest. Glandular GI epithelium displays a marked expression gradient from crypts to the apex of villae.

Abnormal Tissue

Carcinomas and actively proliferating tissues show increased and differential expression of EpCAM.

Expression correlates with differentiation in gastric lesions. Immunochemical and mRNA studies show that well-differentiated tumors are more expressive than those less differentiated. Normal background mucosa shows weak expression, but interestingly areas of ▶ Barrett esophagus or metaplasia are highly expressive.

Ninety percent of colorectal carcinoma cells express EpCAM but in a differential form. Modifications include variable glycosylation

E

analogous to tumor-specific antigens such as colonic tumor antigen MUC1. EpCAM exists in the cell membrane of colon carcinoma cells as a high affinity noncovalent *cis*-dimer. Dimers on opposing membranes can associate via a head-to-head interaction to form tetramers with moderate affinity consistent with reversible intercellular associations. It is not known how exactly antibody binding correlates with variable glycosylation or oligomerization in a functional or structural sense; further investigation is required.

Tissue microarray assessment of EpCAM expression in 3,900 tissues of tumor of stratified stages and grades of 134 different histological subtypes sourced from head and neck, lung, gastrointestinal, breast, urogenital, and mesenchymal tumors showed 75% tumor categories expressed EpCAM. At least weak EpCAM expression in >10% of tumors was observed in 87 of 131 different tumor categories. Colon cancer (81%), ▶ gastric cancer, pancreas cancer (78%), and ▶ lung cancer revealed a high proportion of strongly positive tumors suggesting EpCAM is an attractive target for pan-carcinoma ▶ immunotherapy.

Paradox of Expression with Advance in Carcinomas

A simple linear pattern of an increase in EpCAM expression with the progression of all tumors is not seen. The exact nature of EpCAM temporal expression vis-a-vis the grade of different tumor types remains to be stratified.

The functionally paradoxical upregulation of EpCAM with disease progression in colorectal, breast, prostate, and upper GI carcinomas remains unexplained. This is intriguing in metastatic carcinoma in which degradation of intercellular adhesions is a primary feature. Perhaps, EpCAM is upregulated in response to other intra- and extracellular processes that promote destruction of tissue adhesion and morphology, maintaining its constitutional stabilizing function. Conversely in some tumors, e.g., colorectal cancer, a loss of EpCAM expression is associated with increased local recurrence risk and a diffusely infiltrative morphology but not distant recurrence. In prostate cancer, reduced EpCAM expression correlates

with a higher Gleason score, but expression is higher on hormone-refractory tumor tissue than at earlier stages. In cholangiocarcinomas, squamous cell carcinoma of the head and neck and esophagus increased expression correlates with reduced survival.

The relative loss of EpCAM expression in patients with gastric cancer is associated with a significant reduction in survival indicating that loss of EpCAM expression identifies aggressive tumors especially in patients with stages I and II disease. Data from a Dutch study compared p53, ▶ CD44, E-cadherin, EpCAM, and c-erB2/neu in tumors of 300 patients, investigating the extent of lymph node clearance. Patients without loss of EpCAM expression of tumor cells (19%) had a significantly better 10-year survival compared to patients with any loss: 42% versus 22%. The prognostic value was stronger in stages I and II and independent of the TNM stage. Similarly, in breast cancer, a relative reduction in EpCAM expressing disseminated tumor cells in the bone marrow of patients is associated with a relatively poorer prognosis, whereas an increase in EpCAM-positive tumor cells in lymph nodes and peripheral blood is associated with reduced survival.

In prostate cancer patients staged as M0 (no metastasis), BR (biochemical PSA relapse), and M1 (established metastasis), the presence of bone marrow EpCAM expressing tumor cells significantly increased with progression from M0 to BR and M1 stages from 9 to 16–33%. These cells had double the chromosomal aberrations compared to cytokeratin-positive tumor cells. There was only a small overlap between EpCAM$^+$ and CK$^+$ DTC populations of 9.5%. EpCAM marked a cohort of DTC in ca prostate patients that unlike CK$^+$ DTC expanded during biochemical relapse and had a phenotype different from that of CK$^+$ tumor cells. This differential expression of EpCAM as compared to cytokeratin indicates a specific functional role for EpCAM in the development of metastatic precursors over and above the simple role of an adhesion molecule once thought. The possible interaction of EpCAM with immune escape remains to be elucidated.

Once neoplastic transformation has taken place, a reduced EpCAM expression is an indicator of a more aggressive tumor phenotype with increased ▶ invasion, ▶ metastasis, and mortality. This seemingly contradicts a study that suggested EpCAM silencing leads to reduced invasive potential of tumor cells. Breast cancer cell lines were grown in Matrigel invasion/migration chambers. Cells in which EpCAM expression was silenced with ▶ SiRNAs showed a reduction of 35–80% in proliferation, 92% in cell migration, and 96% in cell invasion without increase in cell death or apoptosis. There was however an increase in E-cadherin, α-catenin, and β-catenin. This may be due to silencing of the inhibition that EpCAM exerts on E-cadherin. Alternatively, EpCAM gene silencing may lead to decreased cytoplasmic β-catenin through an increase in its association with the E-cadherin adhesion complex. Hence, reducing EpCAM may decrease β-catenin availability for the Wnt pathway and activation of its target genes downstream.

A process whereby EpCAM expression increases up to a point after which destabilizing factors predominate and its expression is no longer stimulated is plausible. Once this point is reached, which may be variable according to the particular tumor type, tumors are less stable and display greater invasive and metastatic potential. A comprehensive study of the temporal expression of EpCAM during tumor progression is currently lacking. This would be useful as a predictor of the efficacy of immunotherapy in individual patients according to their tumor stage.

A tenfold reduction in expression of EpCAM is seen in ▶ circulating tumor cells compared to primary tumors from whence they emerged and established metastases. Absence of homotypic adhesions stimulating EpCAM expression in the vascular microenvironment may be the causal link. EpCAM-targeted immunotherapy may be more effective in established tumors or metastases as opposed to fluid borne disease. This does not preclude ascites due to peritoneal metastases for which treatment with trifunctional antibodies is effective. However, the efficacy of destroying blood-borne circulating tumor cells in a hope of eradicating minimal residual disease may be limited.

EpCAM-Targeted Immunotherapy

Immunotherapy manipulates competent host immune system inducing tumor growth inhibition, regression, or cytolysis. Approaches include the use of monoclonal antibodies and their derivatives, hybrid bispecific (trifunctional) antibodies, tumor cell vaccines, anti-idiotypic antibodies, and dendritic cell vaccines. Pure immunomodulatory cytokines have been used to enhance the effect of MAbs.

Mechanism of Tumor Inhibition

The mechanisms by which anti–EpCAM antibodies exert tumor inhibition in vivo remain controversial. Cytotoxic mechanisms include antibody-dependent cellular cytotoxicity (ADCC) mediated by natural killer cells and T lymphocytes, complement-mediated cytolysis (CMC), and opsonization promoting phagocytosis mediated by PMNs.

The question of whether anti- EpCAM antibodies directly inhibit tumor cell proliferation remains unanswered. It could be postulated that EpCAM antibodies directly interfere with the activation of the Wnt pathway causing downregulation of c-Myc: this remains untested. The majority of anti-EpCAM antibodies produced are specific for epitopes within the first of two EGF-like domains in the extracellular segment of EpCAM (Fig. 1); none have been shown to mimic the dimerization/tetramerization that EpCAM undergoes on ligation or to interfere with downstream gene activation or cell proliferation in vivo. EpCAM antibodies do obliterate EpCAM-mediated homotypic cell-sorting activity in vitro; this effect may be a competitive event preventing dimerization alone. Although it is unlikely that a similar competitive effect takes place in established tumors, an investigation to see any effect on the establishment of metastases would be interesting.

A comparison between any differences in the cytotoxicity of antibodies according to the functional EpCAM domain targeted is awaited. It is possible that the majority of these antibodies work to opsonize cells alone – inducing the cytolytic mechanisms mentioned above – particularly as no physiological ligands for the extracellular domain

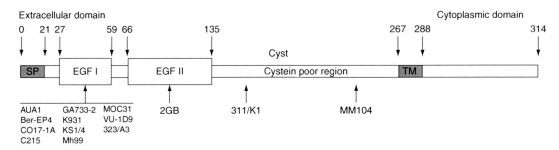

Extracellular domain

Cytoplasmic domain

EpCAM, Fig. 1 Protein domain structure

of EpCAM other than EpCAM itself have been identified.

EpCAM forms a complex with the tight junction protein Claudin-7 within its intramembranous segment; the physiological significance of this is not yet known although an effect on apoptosis resistance in tumors is intriguing.

A flurry of interest followed the assertion that LAIR-1, a member of the inhibitory group of the immunoglobulin-like receptors, was a novel receptor for EpCAM. Speculation that neoplastic cells escape immunological surveillance and clearance by interacting with LAIR-1 via EpCAM gaining selective advantage for their growth, spread, and dissemination was nullified when the original paper by Meyaard et al. was retracted because the observed binding of the LAIR-1 to EpCAM transfected cells was an artifact, attributed to the contamination of the LAIR-1 fusion protein preparation with an antihuman EpCAM monoclonal antibody.

Clinical Trials (Table 1)

Monospecific Murine Antibody

The murine Ig2a antihuman 17-1A monoclonal antibody edrecolomab was the first immunotherapeutic agent licensed for use in large-scale human antitumor immunotherapy trials. Initial trials in patients with advanced colorectal cancer showed little improvement in morbidity or mortality. Augmentation with interferm and GM-CSF increased ADCC with associated tumor lymphocyte infiltration and complement deposition. Patients with greater ADCC survived longer.

In 1994, 189 patients with Dukes C CRC were randomly assigned to adjuvant therapy with edrecolomab or resection alone. Survival at 3 years was 72% for the edrecolomab cohort and 62% for surgery alone. Further follow-up at 7 years showed significantly reduced mortality (32%), disease recurrence (23%), and metastases leading to further phases II and III trials.

In 2002, Punt published results of a trial of 2,761 patients randomized to MAb 17-1A monotherapy, 5-FU and folinic acid or 5-FU + edrecolomab. No additional benefit was seen by adding immunotherapy to the standard chemotherapy regimen at 26 months. Immunotherapy alone was associated with significantly shorter disease-free survival. Edrecolomab was removed from circulation.

The discrepancy between preclinical and clinical findings has led to much debate. What are the reasons for this discrepancy?

EpCAM expression density varies at different stages of tumor growth suggesting patient antigen positivity should be assessed prior to clinical use. EpCAM density is a proven predictor of survival in breast cancer patients.

As a murine antibody, edrecolomab induces a neutralizing humoral response in humans resulting in a short serum half-life. Foreign MAbs are rapidly cleared as immune complexes depositing in the liver, greatly reducing bioavailability. Reduced compatibility with human effector cells may also be significant.

EpCAM-targeted immunotherapy to date has targeted advanced disease: its value weighed against classic adjuvant treatments. The effect of such immunotherapy on earlier, less established disease or cancer models, is unknown.

EpCAM, Table 1 Trials to assess efficacy of **EpCAM**-targeted immunotherapy for intra-abdominal carcinomas

Author	Patients	Treatment	Results	Conclusions
Weiner et al. (1986)	27 metastatic adenocarcinoma of the colon or pancreas	Passive MAb17-1A preceded by 4-day γIFN	No objective clinical markers. Serum tumor markers reduced in 36%. 11 developed Ab3 response	MAb 17-1A safe for clinical use. Evidence of anti-idiotypic response
Herlyn et al. (1994)	Nine CRC	Active anti-idiotypic CO17-1A aluminum hydroxide precipitated	Three patients developed Ab3 response to Ab2 determinants	Marginal success
Herlyn et al. (1994)	54 CRC	Active polyclonal goat and monoclonal rat anti-idiotypic CO17-1A	Majority developed Ab3 response; 30% developed delayed type hypersensitivity	Anti-idiotypic CO17-1A effective in stimulating long-term immunity in cohort
Fagerberg et al. (1995)	Six CRC	Active anti-idiotypic CO17-1A	Six patients developed T-cell immunity; 5 mounted Ab3 response	Small study evidence of anti-idiotypic response
Ragnhammar et al. (1995)	86 Adv CRC	Passive murine MAb17-1A (76) or chimeric MAb17-1A (10)	All patients developed anti-idiotypic Abs increased by GM-CSF; c-MAb less response and more allergic side effects than MAb	Patients with Ab2 response – median survival 9/12
Riethmuller et al. (1998)	189 Dukes C	Passive observation or MAb17-1A adjuvant	7-year evaluation, mortality decreased by 32% and recurrence by 23%	Therapeutic effect maintained after 7 years, mortality/recurrence reduced
Shetye et al. (1998)	20 Adv CRC	Passive single infusion MAb17-1A + GM-CSF	Increased tumoral PMN, monocytes, and T lymphocytes	Increased TILs representing ADCC and CTLs
Hjelm et al. (1999)	20 Adv CRC	Passive MAb17-1A + IL-2 + GM-CSF	One patient partial remission, 2 patients stable disease for 7 and 4 months	No augmentation of effect of MAb 17-1A
Punt et al. (2002)	2761st III CRC	Passive multicenter; (1) 17-1A MAb/5FU/LV or (2) 5 FU/LV or (3) 17-1A MAb	3-year surv DFS (1) 74.7% 63.8% (2) 76.1% 65.5% (3) 70.1% 53.0%	Addition of edrecolomab to standard therapy does not improve the disease outcome. Panorex withdrawn
TRION Pharma, Fresenius (2003)	23 symptomatic ascites Ca ovary	Passive trifunctional multicenter open label intraperitoneal Removab	Well-tolerated 22 of 23 patients ascites free at day 37	Effective treatment of malignant ascites phase III for all-cause malignant ascites underway
Heiss (2005)	Eight peritoneal carcinomatosis	Passive trifunctional, 4–6 applications intraperitoneal	Seven of eight patients no further paracentesis needed. Eradication of tumor cells in ascites	

ADCC antibody-dependent cell cytotoxicity; *CRC* colorectal cancer; *CTL* cytotoxic T cells; *DFS* disease-free survival; *GM-CSF* granulocyte-macrophage colony-stimulating factor; *IFN* interferon; *MAB* monoclonal antibody; *PMN* polymorphonuclear cells; *TIL* tumor infiltrating lymphocytes

E

What Are the Solutions?

Humanized Antibody

A human IgG antibody, MT201 (adecatumumab) combines binding affinity similar to edrecolomab with considerably enhanced ADCC potency with human gastric carcinoma cell lines. Addition of human serum containing IgG or human peripheral blood monocytes halves MT201 ADCC but abolishes that of edrecolomab: indicating the importance of human anti-mouse antibodies (HAMA) and compatibility of syngeneic effector cells.

MT201 reduces tumor growth in xeno-transplanted HT-29 CRC cells in nude mice but only to a level similar to edrecolomab. It is hoped that human effector cells with greater type specificity of Fcγ receptors will facilitate amplified tumor inhibition clinically. Three clinical trials are currently underway: two phase II studies with metastatic breast cancer and early-stage prostate cancer patients, respectively, and a phase I study testing the safety of a combination with Taxotere.

Bispecific Antibodies

Structure and Rationale for Development (Fig. 2)

Normal IgG molecules compose of Fc and FAb segments. The monospecific FAb segment binds to specific epitopes on antigens, whereas the Fc portion recruits cells expressing Fc receptors (e.g., FcγR) such as macrophages. These are described as being bifunctional and monospecific. In trifunctional antibodies, the two halves of the FAb segment have different specificity: they are bispecific and trifunctional.

Both edrecolomab and MT201 are bifunctional antibodies: IgG1 and IgG2a, respectively, with active components being the anti-light chains and Fc portions. Zeidler (1999) successfully constructed a bispecific/trifunctional targeting both and CD3 (BiUII or Removab). The rationale being that ADCC is complemented by the presence of CD3 + T lymphocytes in addition to macrophage/monocytes, NK, and dendritic cells, known to express FcγR binding to the Fc portion

of the antibody. This antibody consists of a murine IgG2a associated with an anti-light chain and rat IgG2b associated with anti-CD3.

In Vitro Cytotoxicity

In vitro experiments of ADCC with cell lines, effector cells, and BiUII demonstrated increased production of interleukins IL-1β, IL-2, IL-6, IL-12, and DC-CK1. Simultaneous stimulation of accessory cells and T lymphocytes leads to antigen presentation to T lymphocytes inducing immunomodulation and cytotoxicity. BiUII induces production of IL-2 in the presence of + cells activating accessory and T-cells without the requirement of exogenous IL-2. An immunologically self-supporting tri-cell complex is formed which is efficient for immune cell activation.

Cytolysis occurs within 1–3 days. The mode of cell death is characteristically necrotic and not apoptotic. Lymphocytes with pore-forming perforin proteins surround the tumor cells causing cytolysis.

Prolonged Antitumor Immunity In Vivo

Another bispecific antibody: BiLu induces long-lasting antitumor immunity consisting of both humoral and cell-mediated responses when administered intraperitoneally in a murine syngeneic model. It targets *murine* CD3 and human EpCAM. The Fc portion is identical to.

The human CD3 counterpart of BiLu: Catumaxomab/Removab has shown promising results.

Clinical Trials (Table 2)

A Phase I/II study for the treatment of ovarian cancer patients with symptomatic ascites has now been completed (23 patients) showing that Removab was safe and effectively reduced ascitic flow and tumor cell content. A substantial Phase II/III trial assessing efficacy in patients with all causes of malignant ascites including primary gastrointestinal tumors commenced in September 2004 (250 patients), and a Phase IIa study of platinum refractory ovarian cancer patients is also underway. Finally, a Phase I/II study of patients with peritoneal carcinomatosis due to GI

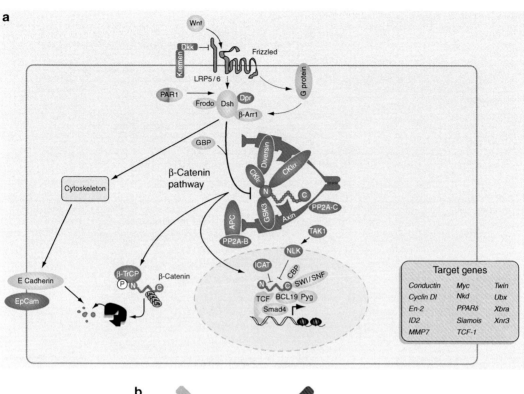

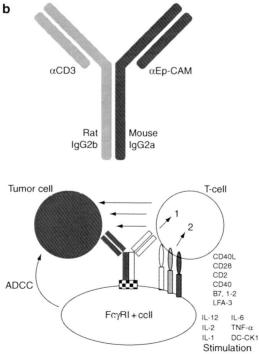

EpCAM, Fig. 2 Structure rationale for the development of bispecific antibodies

EpCAM, Table 2 EpCAM targeting trials underway

Therapeutic (alternative name)	Class	Ongoing or completed trials	Company
Catumaxomab (Removab®)	Trispecific antibody; mouse IgG2a/ rat IgG2b hybrid	Phase II/III in ovarian cancer	Trion Pharma/ Fresenius Biotech
		Phase II in gastric cancer	
Proxinium® Vivendium® (VB4-845)	Immunotoxin; single-chain antibody pseudomonas exotoxin fusion	Phase II/III in head and neck cancer	Viventia
		Phase I/II in bladder cancer	
IGN-101 (edrecolomab)	Vaccine for induction of anti-idiotypic antibody response	Phase II in various adenocarcinomas	Aphton
		Phase II/III in non-small cell lung cancer	
Adecatumumab (MT201)	Fully human IgG1 MAb	Phase II in metastatic breast and early-stage prostate cancer	Micromet, Inc./ Serono
		Phase I in metastatic breast, plus Taxotere	
EMD 273066 (huKS-IL2)	Fusion of humanized MAb KS1/4 with human IL-2	Phase I in hormone-refractory prostate cancer	Lexigen, Inc./ Merck KGaA

EpCAM epithelial cell adhesion activating molecule; *IP* intraperitoneal; *IV* intravenous; *MTD* maximum tolerated dose; *SC* subcutaneous

tumors but without symptomatic ascites is underway. Preliminary results are promising indicating the utility of Removab in the treatment of minimal fluid borne disease and micrometastases.

Cross-References

▶ Barrett Esophagus
▶ Bispecific Antibodies
▶ Carcinoembryonic Antigen
▶ CD44
▶ Circulating Tumor Cells
▶ E-Cadherin
▶ Gastric Cancer
▶ Immunotherapy
▶ Invasion
▶ Lung Cancer
▶ Metastasis
▶ MYC Oncogene
▶ Pancreatic Cancer
▶ Prostate Cancer
▶ Prostate-Specific Antigen
▶ SiRNA
▶ Wnt Signaling

References

Baeuerle PA, Gires O (2007) EpCAM (CD326) finding its role in cancer. Br J Cancer 96(3):417–423

Chaudry MA, Sales K, Ruf P, Lindhofer H, Winslet MC (2007) EpCAM an immunotherapeutic target for gastrointestinal malignancy: current experience and future challenges. Br J Cancer 96:1013–1019

Trzpis M et al (2007) Epithelial cell adhesion molecule: more than a carcinoma marker and adhesion molecule. Am J Pathol 171:386–395

See Also

(2012) Antibody. In: Schwab M (ed) Encyclopedia of Cancer, 3rd edn. Springer Berlin Heidelberg, p 208. doi:10.1007/978-3-642-16483-5_312

(2012) Antisense nucleic acid. In: Schwab M (ed) Encyclopedia of cancer, 3rd edn. Springer Berlin Heidelberg, pp 220–221. doi:10.1007/978-3-642-16483-5_336

(2012) Beta-catenin. In: Schwab M (ed) Encyclopedia of cancer, 3rd edn. Springer Berlin Heidelberg, p 385. doi:10.1007/978-3-642-16483-5_889

(2012) Cadherins. In: Schwab M (ed) Encyclopedia of cancer, 3rd edn. Springer Berlin Heidelberg, pp 581-582. doi:10.1007/978-3-642-16483-5_770

(2012) Chromosomal aberrations. In: Schwab M (ed) Encyclopedia of cancer, 3rd edn. Springer Berlin Heidelberg, p 838. doi:10.1007/978-3-642-16483-5_1138

(2012) Cyclooxygenase-2. In: Schwab M (ed) Encyclopedia of cancer, 3rd edn. Springer Berlin Heidelberg, p 1035. doi:10.1007/978-3-642-16483-5_1435

(2012) EGF. In: Schwab M (ed) Encyclopedia of cancer, 3rd edn. Springer Berlin Heidelberg, p 1211. doi:10.1007/978-3-642-16483-5_1824

(2012) Humanized antibodies. In: Schwab M (ed) Encyclopedia of cancer, 3rd edn. Springer Berlin Heidelberg, p 1760. doi:10.1007/978-3-642-16483-5_2863

(2012) ICAMs. In: Schwab M (ed) Encyclopedia of cancer, 3rd edn. Springer Berlin Heidelberg, p 1803. doi:10.1007/978-3-642-16483-5_2938

(2012) Interferon. In: Schwab M (ed) Encyclopedia of cancer, 3rd edn. Springer Berlin Heidelberg, p 1888. doi:10.1007/978-3-642-16483-5_3090

(2012) Interleukin. In: Schwab M (ed) Encyclopedia of cancer, 3rd edn. Springer Berlin Heidelberg, p 1892. doi:10.1007/978-3-642-16483-5_3094

(2012) Matrigel. In: Schwab M (ed) Encyclopedia of cancer, 3rd edn. Springer Berlin Heidelberg, p 2182. doi:10.1007/978-3-642-16483-5_3551

(2012) MUC1. In: Schwab M (ed) Encyclopedia of cancer, 3rd edn. Springer Berlin Heidelberg, pp 2384–2386. doi:10.1007/978-3-642-16483-5_3870

(2012) P53. In: Schwab M (ed) Encyclopedia of cancer, 3rd edn. Springer Berlin Heidelberg, p 2747. doi:10.1007/978-3-642-16483-5_4331

(2012) Phagocytosis. In: Schwab M (ed) Encyclopedia of cancer, 3rd edn. Springer Berlin Heidelberg, p 2840. doi:10.1007/978-3-642-16483-5_4493

(2012) PSA. In: Schwab M (ed) Encyclopedia of cancer, 3rd edn. Springer Berlin Heidelberg, pp 3111–3112. doi:10.1007/978-3-642-16483-5_6738

(2012) Thyroglobulin. In: Schwab M (ed) Encyclopedia of cancer, 3rd edn. Springer Berlin Heidelberg, p 3687. doi:10.1007/978-3-642-16483-5_5805

(2012) Trifunctional antibody. In: Schwab M (ed) Encyclopedia of cancer, 3rd edn. Springer Berlin Heidelberg, p 3783. doi:10.1007/978-3-642-16483-5_5976

Eph Receptors

Diego Arango
CIBBIM - Nanomedicina Oncologia Molecular, Vall d'Hebron Hospital Research Institute, Barcelona, Spain

Definition

Eph receptors are tyrosine kinases (RTK) bound to the extracellular membrane that function as "switches" that upon activation by ephrin (EFN) ligands initiate signaling cascades that regulate numerous developmental processes, particularly in the vasculature and nervous system.

Characteristics

The name of the Eph receptors is derived from the name of the cell line used to characterize them initially (erythropoietin-producing hepatocellular carcinoma cell line). These receptors and their ephrin (EFN) ligands constitute the largest family of ▶ receptor tyrosine kinases (RTK) known to date. Eph receptors are integral membrane proteins with a conserved N-terminal domain responsible for ligand binding, followed by a cysteine-rich region and two fibronectin type III repeats which are essential for dimerization and interactions with other proteins. The intracellular region of these receptors contains a juxtamembrane domain, a conserved kinase domain, a sterile alpha motif (SAM), and a PDZ-binding motif. There are at least 15 Eph receptors in the human genome. Based on sequence homology and the preferred type of ephrin ligands that they bind, the Eph receptors can be divided into two subclasses, EphA and EphB. The ligands of the Eph receptors (ephrins) are also bound to the extracellular membrane. Ephrins of the A subclass are bound to the membrane through a GPI anchor (*glycosylphosphatidylinositol*), whereas members of the B subclass have a transmembrane domain. At least five ephrins of the A subclass and three of the B subclass have been described.

The signaling cascade is initiated by the binding of a membrane-bound ephrin and an Eph receptor in neighboring cells. This leads to the autophosphorylation of the Eph receptor on several tyrosine residues and the activation of the tyrosine kinase activity. In addition to the "forward" signaling elicited by the Eph receptors, the ephrin ligands are also able to transduce a signal upon interaction with their cognate Eph receptor. This is often referred to as "reverse" signaling. The signaling cascade initiated upon ligand binding regulates many developmental processes and has an important role in tumor initiation and progression in different tissues.

Normal Eph Signaling

The role of Eph signaling in the nervous and vascular systems during normal development has been characterized in detail. During normal embryogenesis, Eph signaling has an important role in the development of the nervous system. Eph signaling regulates the migration of neural crest cells, the formation of the corticospinal tract, the boundary formation between hindbrain segments (rhombomeres), the establishment of neural topographic maps, and the formation and functional properties of neuronal synapses. Eph signaling also plays an important role in the angiogenic process by restricting arterial and venous endothelial mixing. In addition, interactions between Eph receptors and ephrin ligands mediate cytoskeleton organization, cell migration, and substrate attachment. The Eph family also has an important physiologic role in the normal intestinal epithelium. EphB2 and EphB3, together with their ligand ephrin-B1, regulate proliferation and cell positioning within the intestinal crypts.

Eph Signaling in Cancer

Deregulation of the levels of expression and normal Eph signaling are commonly observed in tumors of various origins. This is not surprising since aberrant Eph signaling can interfere with processes that are crucial during malignant transformation such as cell attachment, ▶ migration, proliferation, cytoskeleton organization, and ▶ angiogenesis. Eph/ephrin activation has been implicated in several ▶ signal transduction pathways contributing to the tumorigenic process. For instance, Akt/PI3K (phosphatidylinositol 3-kinase) has been shown to be implicated in the increase in proliferation and migration of endothelial cells after activation of EphB4 with its ligand ephrin-B2. In addition, Eph signaling can regulate cell ▶ motility by modulating the activity of other signal transduction proteins such as ▶ focal adhesion kinase (FAK) and ▶ Rho. The

important role of Eph signaling in the modulation of the cellular attachment to the extracellular matrix seems to be regulated through the modulation of integrin-mediated adhesion.

Most of the studies in the literature report increased levels of expression of Eph receptors and/or ephrin ligands in most of the tumor types studied, compared with the respective normal tissue. For instance, EphA2 is overexpressed in melanomas; EphA1 shows elevated levels in breast, liver, and lung tumors; and EphB4 has been reported to be overexpressed and significantly contribute to tumor progression in prostate, bladder, breast, and head and neck tumors. Although the general view emerging is that Eph receptors and ephrins may function as ▶ oncogenes in the sense that elevated levels or kinase activity promotes tumor formation and/or progression, Eph receptors may be important ▶ tumor suppressor genes in some tissues. This is the case of EphB2 and EphB4 in the intestine. The expression of these two EphB receptors is reduced in colorectal tumors compared to the normal intestinal cells and the premalignant lesions. The levels of expression of EphB2 and EphB4 negatively correlate with tumor progression, and mechanisms of inactivation of these receptors include somatic mutations and hypermethylation of ▶ CpG islands situated within the promoter regions regulating their expression. Moreover, animal studies clearly support the tumor suppressor role of EphB2 in colorectal tumors, and germ line mutations in this gene seem to predispose to prostate and possibly to colorectal cancer.

References

Alazzouzi H, Davalos V, Kokko A et al (2005) Mechanisms of inactivation of the receptor tyrosine kinase EPHB2 in colorectal tumors. Cancer Res 65:10170–10173

Committee EN (1997) Unified nomenclature for Eph family receptors and their ligands, the ephrins, Eph Nomenclature Committee. Cell 90:403–404

Davalos V, Dopeso H, Castano J et al (2006) EPHB4 and survival of colorectal cancer patients. Cancer Res 66:8943–8948

Holder N, Klein R (1996) Eph receptors and ephrins: effectors of morphogenesis. Development 126. 2033–2044

Surawska H, Ma PC, Salgia R (2004) The role of ephrins and Eph receptors in cancer. Cytokine Growth Factor Rev 15:419–433

Ephelide

▶ Carney Complex

Epidemiology of Cancer

Elizabeth B. Claus
Department of Epidemiology and Public Health, Yale University School of Medicine, New Haven, CT, USA

Synonyms

Cancer epidemiology

Definition

Is the study of the incidence, distribution, and, ultimately, the prevention and control of cancer within the general population.

Characteristics

The discipline of cancer epidemiology is a relatively young one, with much of the methodology developed over the past 50 years. Prior to the advent of formal methods of collection and analysis of cancer incidence and risk factor data, associations were generally the result of reports or observations of astute clinicians or scientists. The literature is full of fascinating stories of early attempts at epidemiologic cancer studies such as that of the nineteenth-century physician,

Alfred Haviland, who created elaborate maps of cancer deaths in England and Wales using national mortality statistics. One of the first and probably most well-known reports of a relationship between a risk factor and the occurrence of cancer occurred in 1950, with the publication of several case/control studies detailing the association between cigarette smoking and the development of lung cancer.

Study Design

There are a number of basic study designs in cancer epidemiology. Descriptive cancer epidemiology examines how cancer incidence and mortality rates vary according to demographic characteristics of the study population such as geographic location, race, and sex. One may further analyze such information by categories of age as well as by birth cohort and time period. Ecological studies generally examine aggregate measures of risk and cancer outcome such as median income and cancer incidence across counties of a given state in an effort to identify an association between the two. Alternatively, many epidemiologic studies are based on the use of individuals as the study unit rather than larger groups, or populations of study subjects. Within this category of analysis, there are essentially three study designs (with several variations) including the cross-sectional, case/control, and cohort study design. Cross-sectional study designs allow for the consideration of a reference population at a given point in time. In a case/control study, the frequency of a particular risk factor among individuals with a cancer of interest (cases) is compared with that among individuals without cancer (controls). Case/control studies have been instrumental in the identification of numerous important cancer risk factors including the association between family history and breast cancer as well as between tobacco and lung cancer. In a prospective or cohort study, researchers assemble a cohort of healthy individuals who provide information on risk factors of interest at a baseline point in time. The study subjects are then followed prospectively until they develop cancer or the study is completed. An advantage of this type of study design is that risk factor data is collected before any cancer is diagnosed, thus reducing the amount of recall bias associated with disease status in the reporting of risk factor data such as family history information. There are many well-known examples of cancer cohort studies including the Atomic Bomb Casualty Commission established in 1947 to study the effects of exposure to radiation with outcomes such as leukemia, the British physicians cohort from the 1950s that examined the association between smoking and lung cancer, and the Nurse's Health Studies I and II, started in 1976 and 1991, respectively, which have examined a wide range or hormonal and dietary risk factors (among others) in the development of breast and other cancers.

Risk Factors

Categories of cancer risk factors are many and include infectious agents, diet and lifestyle factors, endogenous and exogenous hormonal components, and genetic factors, to name a few. It is currently popular to divide cancer risk factors into two broad categories defined as genetic and environmental. This has come about because of the many laboratory-based advances in the identification of genetically transmitted or regulated diseases such as cancer, leading to the emergence of a new field of investigation, that is, the genetic epidemiology of cancer. For cancers, a small subset of cases exist that are attributable to rare inherited cancer susceptibility genes. The majority of cases appear to be due to sporadic mutations that may be a result of genetic or environmental events or the result of an interaction between genetic and environmental factors. Much of traditional epidemiologic methodology has been adopted for use in genetic epidemiology. In addition, new methods specific to genetic epidemiology have been developed including the twin, adoption, and pedigree study designs as well as segregation and linkage analyses.

The many advances in genetic testing, as well as in some instances, prevention, or treatment options associated with particular cancer diagnoses, has led to a surge of interest in the availability of personalized risk estimates in the clinical setting and hence the development of cancer risk assessment models used to generate these estimates. Cancer risks may be presented in both relative and absolute terms and may define risk for a discrete period of time or over a lifetime. A wide variety of statistical methods exist to estimate cancer risk, with the most fully developed existing in the area of breast and ovarian cancer. Although the concept of risk assessment is not new to the fields of medicine or genetics, the use of detailed genetic information on a large population-based scale is, with all the associated difficulties of presentation and interpretability.

The field of cancer epidemiology is an exciting scientific discipline, which is able to adapt well to new information and technology. New developments in the field include the integration of biomarkers into exposure data, the inclusion of both molecular genetics and environmental risk factor data into study designs in an effort to explore the complex interaction between genotype and the environment, the creation of international databases via the Internet, and the merging of large databases that combine risk factor information with cancer incidence and mortality data. All of these advances should continue to assist scientists and health-care professionals in the identification of individuals at increased risk of developing cancer so that screening and prevention regimes as well as treatment plans may be developed.

Cross-References

▶ Alcoholic Beverages Cancer Epidemiology
▶ Obesity and Cancer Risk

References

Claus EB (2000) Risk models in genetic epidemiology. Stat Methods Med Res 9(6):589–601
Samet JM, Munez A (1998) Epidemiologic reviews: cohort studies. Am J Epidemiol 20(1):1–136
Szklo M, Nieto FJ (2000) Epidemiology. Beyond the basics. Aspen, Gaithersburg

Epidermal Growth Factor Inhibitors

Deepa S. Subramaniam[1], Marion Hartley[2], Michael Pishvaian[3], Ruth He[3] and John L. Marshall[3]
[1]Georgetown University Hospital, Washington, DC, USA
[2]Ruesch Center for the Cure of Gastrointestinal Cancers, Lombardi Comprehensive Cancer Center, Georgetown University, Washington, DC, USA
[3]Lombardi Comprehensive Cancer Center, Georgetown University, Washington, DC, USA

Definition

The EGFR receptor (EGFR) is a member of the HER family of receptor tyrosine kinases that includes EGFR itself (ErbB1/HER1), ErbB2 (*HER-2/Neu*), ErbB3 (HER3), and ErbB4 (HER4). These proteins are classic membrane-bound tyrosine kinase receptors whose activation is typically ligand dependent. The principal ligands for EGFR are EGF and TGF-α. Other ligands include amphiregulin, heparin-binding EGF, the poxvirus mitogens, epiregulin, and β-cellulin.

Characteristics

EGFR Activation and Downstream Signaling

Receptor activation results in homo- or heterodimerization and autophosphorylation of c-terminal tyrosine residues. Receptor activation enables the docking of cytoplasmic proteins that bind to specific phosphotyrosine residues and initiate several cell signaling pathways. These pathways include the Ras-Raf-MAPK pathway, the PI3K-AKT pathway, the protein kinase C pathway, the

Modified version of Pishvaian M, Marshall JL, He R, Wang D (2012) Epidermal growth factor inhibitors. In: Schwab M (ed) Encyclopedia of cancer, 3rd edn. Springer, Berlin Heidelberg, pp 1271–1275. doi:10.1007/978-3-642-16483-5_1931

STAT pathway, and the src kinase pathway, all of which play important roles in tumor cell proliferation, invasion, migration, and inhibition of apoptosis. Reports have also demonstrated that EGFR can be found in the nucleus where it can act as a transcription factor. EGFR activation does not initiate linear downstream pathway signaling, but rather can activate multiple pathways that cross-connect intracellularly. The pattern of activation is often cell/tissue specific and likely contributes to the rich variety of biological responses to EGFR activation. Ligand activation of EGFR also results in receptor downregulation, mediated through endocytosis and ultimately Cbl-mediated EGFR ubiquitination and degradation (Marshall 2006). Markedly, EGFR can also be activated in a ligand-independent manner. This aberrant EGFR activation can result from receptor overexpression, gene amplification, activating mutations, or loss of regulatory mechanisms.

Molecular Mechanisms of Targeted Therapies

Anti-EGFR therapies include monoclonal antibodies (mAbs) that recognize EGFR and small molecule inhibitors of EGFR tyrosine kinase activity (TKIs). Cetuximab is the first anti-EGFR mAb to be developed and eventually US Food and Drug Administration (FDA) approved. This agent prevents receptor dimerization through steric inhibition of the extracellular domain of EGFR. Cetuximab also promotes receptor internalization and degradation without receptor activation, resulting in receptor downregulation and reduced cell surface expression levels of EGFR. Cetuximab also blocks the transport of EGFR into the cell nucleus, thus inhibiting any direct effects on DNA transcription and/or repair. Finally, cetuximab has the potential to kill its target cells by mediating antibody-dependent cell-mediated cytotoxicity (ADCC) and complement fixation.

The TKIs are competitive inhibitors of adenosine triphosphate (ATP). They block the enzymatic activity of the intracellular domain of EGFR. Because of their mechanism of action, these TKIs can block EGFR mutants that lack an extracellular domain and block ligand-independent receptor activation.

EGFR Mutations

A number of genetic mutations of EGFR have been found in cancer cells. These mutations generally result in constitutive activation of receptor signaling and, in this way, become independent oncogenic drivers. Many of these mutations predict responsiveness to targeted therapies, especially in non-small cell lung cancer.

Many somatic gene mutations – called activating or sensitizing mutations – are associated with an increased response to TKIs and are clustered within exons 18–21 of the receptor kinase domain, near the ATP-binding region. Two of the most prevalent sensitizing mutations include an Exon 19 deletion and a substitution mutation in Exon 21 (L858R). These, and other, sensitizing mutations result in structural changes that confer exquisite sensitivity to TKIs, wherein the presence of EGFR mutations in a patient's tumor results in increased responsiveness and significantly improved progression-free survival compared with a patient whose tumor does not harbor such EGFR mutations. Conversely, certain mutations, such as an Exon 20 insertion, can confer primary resistance to targeting by first-generation TKIs, and other mutations, such as the T790M point mutation in Exon 20, can confer secondary (acquired) resistance after initial TKI therapy. Furthermore, TKI resistance may also result from activation of previously redundant growth signaling pathways, such as ras, raf, and PI3K; loss of the tumor suppressor PTEN pathway; or activation of type I growth factor receptor pathways, such as the IGF-R (insulin growth factor receptor) pathway.

EGFRvIII is a tumor-specific oncogene expressed in a third of all primary or de novo glioblastomas (GBM), but not in normal tissues. The mutation arises from an in-frame deletion, resulting in loss of the majority of the extracellular ligand-binding domain, leading to a constitutively active ligand-independent tyrosine kinase. EGFRvIII-positive tumor cells may induce growth in EGFRvIII-negative cells via paracrine signaling. Expression of EGFRvIII in primary glioblastoma is linked to a poorer long-term patient survival (Schuster et al. 2015).

Only 10–20% of all patients with GBM respond to EGFR inhibitors, despite the fact that most GBMs overexpress EGFR. In a study involving 59 patients, 30 (approximately 50%) expressed EGFRvIII, but only 13 of those 30 patients were seen to respond to TKIs (43% of EGFRvIII-expressing GBM). If patients had tumors that expressed EGFRvIII *together* with PTEN (14 out of 59 patients), 11 of these 14 (78%) responded to TKIs. PTEN is a tumor suppressor whose expression is frequently lost in GBM and acts as an inhibitor of the PI3K/AKT pathway. Loss of PTEN confers resistance to EGFR inhibitors, presumably because of the persistent activation of the PI3K pathway *downstream* of EGFR. Nevertheless, only about 25% of the patients (14/59) actually exhibited this coexpression of EGFRvIII and PTEN. In fact, mutations are often cancer-type specific. For example, while approximately 30–50% of GBMs and head and neck cancers exhibit expression of EGFRvIII, only about 1.5% of non-small cell lung cancers (NSCLC) and 0% of metastatic colorectal cancers express EGFRvIII. By contrast, 10–25% of patients with NSCLC, very few head and neck cancers, and no GBMs contain EGFR tyrosine kinase (EGFR-TK) gene mutations.

Rindopepimut consists of an EGFRvIII-peptide vaccine conjugated to keyhole limpet hemocyanin (KLH) and generates a specific immune response against EGFRvIII-expressing glioblastoma. A randomized phase II trial of rindopepimut or control (KLH alone) with bevacizumab in recurrent EGFRvIII-expressing glioblastoma GBM demonstrated an overall survival advantage in the vaccine-treated population, with a hazard ratio of 0.47 ($p = 0.0208$) and 30% of patients in the treatment arm being alive at 18 months, compared with 15% in the control arm (Reardon et al. 2015). Additionally, a large randomized trial of the current standard of care (temozolomide) with or without rindopepimut is ongoing in patients with EGFRvIII-expressing GBM [NCT01480479].

Surprisingly, there is a lack of reported assessments of EGFR-TK mutations in colorectal cancers.

EGFR Inhibitors in GI Tumors

EGFR is implicated in the pathogenesis of several cancers including CRC, pancreatic cancers, NSCLC, head and neck cancers, breast cancers, and brain cancers, and overexpression of EGFR has been associated with metastasis, chemotherapy resistance, and poor outcome. The receptor has emerged as a rational target for anticancer treatment in these listed tumors. MAbs to EGFR and tyrosine kinase inhibitors to EGFR kinase have been developed with the aim of inhibiting EGFR signaling.

Cetuximab and panitumumab are two mAbs to EGFR that are approved by the FDA in the treatment of CRC. Since the original publication of this book chapter, a great deal has been learned about the incorporation of EGFR receptor-targeting mAbs in the treatment of mCRC. Both cetuximab and panitumumab are now approved as first-line agents, as well as in the refractory setting, and are guideline approved for use in the second-line setting. The most important breakthrough has been the recognition of Ras mutations conferring resistance to these anti-EGFR antibodies. Originally, the only presumptive biomarker for selection of patients for treatment with EGFR antibodies was the presence of EGFR receptors on the patient's colon cancer cells. This proved not to be useful in discerning benefit. Important clinical trials using the EGFR antibodies alone, as well as in combination with chemotherapy, have consistently demonstrated that patients whose tumors have one of several *RAS* mutations fail to respond to this treatment. In fact several studies have demonstrated potential harm when the antibodies are given to patients whose tumors harbor these mutations. This research field moved more rapidly than clinical uptake. It is important to ensure that adequate tumor has been tested for *RAS* mutations prior to considering one of these agents in any line of therapy.

As a consequence of this improved enrichment, the benefit seen when using anti-EGFR mAbs in the properly selected *RAS* wild-type patients has significantly improved. This has resulted in improved outcomes in both progression-free survival and overall survival and response rate in patients appropriately selected. However, much remains to be learned about the appropriate utilization of these compounds in the sequential therapy of mCRC. Increasingly, B-raf is becoming recognized as a potential pathway of resistance, but is less firmly established as a marker.

Cetuximab and panitumumab have been compared head to head in a randomized clinical trial involving patients with chemotherapy-refractory mCRC, and both mAbs demonstrated very similar activity, clinical benefit, and toxicity (Price et al. 2014). Cetuximab has been tested in a randomized phase III clinical trial in the stage III adjuvant setting and failed to demonstrate benefit (Taieb et al. 2014). In addition, cetuximab was tested in the perioperative setting for patients with liver metastasis and was likewise found to be of no additional benefit (Primrose et al. 2014). It is clear that these compounds have a significant role in the traditional metastatic setting, but their role in the neoadjuvant or adjuvant setting has not been fully established and is currently negative.

A survival advantage of adding cetuximab to radiation therapy was demonstrated in a randomized phase III study comparing concurrent cetuximab/radiation therapy with radiation alone in locoregionally advanced head and neck squamous cell carcinoma. Since then, the role of cetuximab in concurrent chemoradiation has been tested in esophageal/gastric cancer, rectal cancer, and pancreas cancer in phase II clinical trials. Patients with localized esophageal/gastric cancer received radiation and concurrent chemotherapy of cetuximab, carboplatin, and paclitaxel. Approximately 67% of the patients had a complete clinical response, and 43% were found to have a complete pathologic response at surgery. A confirmatory phase III study is being planned. Neoadjuvant chemotherapy and chemoradiation in rectal cancer treatment decreases local relapse rate of rectal cancer with higher sphincter preservation rate compared with postoperative chemoradiation. Cetuximab has been evaluated in combination with a capecitabine/oxaliplatin (CapOx)-based regimen in concurrent neoadjuvant chemoradiotherapy of patients with rectal cancer in phase II studies. The combination therapy was found to be feasible and resulted in pathologic complete response (CR) in some patients. In the EXPERT-C trial, which compared "neoadjuvant oxaliplatin, capecitabine, and pre-operative radiotherapy with or without cetuximab, followed by total mesorectal excision in patients with high-risk rectal cancer," the presence of cetuximab was found to significantly increase patient response rate (RR) and overall survival (OS), but not CR, as long as patients had KRAS/BRAF wild-type rectal cancer (Dewdney et al. 2012).

Management of borderline resectable pancreas cancer is challenging, and neoadjuvant chemo-therapy or chemoradiation is controversial in those patients. The role of cetuximab in combina-tion with gemcitabine in concurrent neoadjuvant chemoradiation was examined in a phase II clin-ical trial. Two of ten patients exhibited partial response. Six patients went on to margin (−) resection, including one patient each with border-line resectable and unresectable disease prior to therapy. The role of cetuximab with neoadjuvant chemoradiation in downstaging pancreas cancer for resection needs to be confirmed by a phase III randomized study.

More than 80% of patients treated with cetuximab experience an acneiform rash. EGFR is expressed in the epidermis, sebaceous glands, and hair follicle epithelium. EGFR plays a role in the normal differentiation and development of skin follicles and keratinocytes. Results following the study of mice with EGFR gene knockout, or dominant negative mutational status, predict that EGFR inhibition in humans is feasible but may be associated with cutaneous toxicity. Indeed skin rash is the most common toxicity following EGFR inhibitor therapy (including anti-EGFR mAbs and small molecule inhibitors of EGFR tyrosine kinase activity (TKIs)). The ability of the presence and intensity of a skin rash to predict the response to EGFR inhibitors was investigated in a subgroup analysis from the BOND study. An increase in RR, median time to progression (mTTP), and median OS was all found to correlate with a higher grade of skin toxicity to cetuximab.

The same phenomena were observed with panitumumab treatment. To take a step further, the question was posed as to whether to dose escalate anti-EGFR antibody therapy until a skin rash is detectable. In a phase II study, patients who had no, or mild, skin reactions in response to a standard cetuximab dose were subjected to dose escalation (up to 500 mg/m^2), after which an improved tumor response rate was generally observed.

TKIs did not mirror the clinical activities and toxicity profiles of anti-EGFR antibodies (cetuximab and panitumumab) in CRC; when administered as single agents, TKIs showed min-imal activity in mCRC. When patients with mCRC were treated with a TKI plus fluoropyrimidine-, oxaliplatin-, and irinotecan-based chemotherapy regimens, clinical response rates were observed to range from 24% to 74% in phase II studies. However, TKIs were found to increase grade 3 and 4 toxicities, and some of the trials had to close prematurely due to intolerable adverse effects. To confirm the clinical benefit of TKI therapy in mCRC, a phase III study was carried out that administered bevacizumab-based induction chemotherapy to patients and random-ized those who were subsequently free of disease progression or the need for surgery to mainte-nance therapy with bevacizumab, with or without erlotinib. The target accrual was 640 patients. On final analysis, median overall survival from the beginning of maintenance therapy was signifi-cantly greater following bevacizumab plus erlotinib compared with bevacizumab therapy alone. Median progression-free survival (PFS) was also improved (Tournigand et al. 2015).

Erlotinib in combination with gemcitabine demonstrated benefit in patients with pancreatic cancer in a phase III study. To the best of our knowledge, results from this trial were the first to demonstrate a clinical benefit following the use of a TKI in combination with chemotherapy.

Cetuximab was found to have encouraging activity when combined with gemcitabine and concurrent radiation in localized pancreatic can-cer. In addition, cetuximab showed clinical effi-cacy in combination with gemcitabine/oxaliplatin (GEMOXCET) in the treatment of previously untreated patients with metastatic pancreatic cancer. Further evaluation in a phase III trial is warranted.

Treatment choice for hepatocellular carcinoma (HCC) is limited due to low response rate and transient response following most cytotoxic chemotherapy. Although cetuximab showed very modest activity in HCC, erlotinib demonstrated clinical efficacy. Thirty-eight patients with HCC were treated with erlotinib at a dose of 150 mg daily, and 32% of these patients were found to have PFS at 6 months. Disease control was seen in 59% of the patients. Median overall survival time was 13 months. The role of TKIs in HCC treatment needs to be confirmed in a phase III study.

EGFR Inhibitors in Lung Cancer

The development of EGFR inhibitors as a therapeutic option in NSCLC has been moving at a rapid pace. Gefitinib (Iressa, AstraZeneca), a first-generation reversible small molecule inhibitor of EGFR, initially received accelerated approval from the FDA after phase II clinical studies of this agent showed a promising clinical response in NSCLC patients (Lynch et al. 2004). However, subsequent large phase III trials in the United States failed to confirm these findings, leading the FDA to restrict gefitinib use to only those patients who had already been on this medication with good response. However, in Europe and China, multiple randomized clinical trials have demonstrated the superiority of gefitinib over standard platinum-doublet chemotherapy in EGFR-mutant NSCLC (I-PASS trial), allowing the reapproval of gefitinib in the United States in 2015. Erlotinib (Tarceva, Genentech) is also a first-generation reversible TKI, which, likewise, competes with ATP for the ATP-binding site of the TK domain of EGFR. Erlotinib's initial approval came after the BR.21 phase III trial demonstrated an overall survival benefit of 2 months compared with placebo in second- and third-line NSCLC (Shepherd Frances et al. 2005). Subsequently, erlotinib has demonstrated a statistically significant improvement in progression-free survival compared with platinum-doublet chemotherapy in the frontline treatment of advanced EGFR-mutant NSCLC (EURTAC and OPTIMA). Both

gefitinib and erlotinib are oral medications and are usually well tolerated. The most common adverse effects of these TKIs are acneiform rash and diarrhea; less commonly reported are transaminitis and stomatitis, with rare cases of interstitial lung disease. Initially, it appeared that certain demographic features, such as female sex, Asian ethnicity, never-smoker status, and adenocarcinoma histology correlated with higher response rates to TKIs. However, it is now known that these demographic factors only serve as surrogates for a higher prevalence of sensitizing EGFR mutations, which are identified even in the absence of such demographics. As these mutations are exceptionally rare in squamous NSCLC, the current recommendations of the National Comprehensive Cancer Network (NCCN) are to perform EGFR mutation testing routinely in adenocarcinoma tumors, tumors with mixed adenosquamous histology, or very poorly differentiated NSCLC tumors without clearly defined histology.

Second-generation TKIs have been developed that are irreversible inhibitors that bind to the TK domain of EGFR and ErbB family members. Such inhibitors include afatinib and dacomitinib.

Afatinib (Gilotrif, Boehringer Ingelheim) is FDA approved for patients with tumors harboring Exon 19 deletions or Exon 21 L858R substitution mutations. There are no head-to-head comparisons of these second-generation TKIs with first-generation TKIs in EGFR-mutant NSCLC. LUX-Lung 3 and LUX-Lung 6 are two phase III randomized trials that compare afatinib with cisplatin-doublet chemotherapy (a pemetrexed combination in LUX-Lung 3 and a gemcitabine combination in LUX-Lung 6) in frontline treatment of EGFR-mutant NSCLC. These trials demonstrated a clear PFS advantage following afatinib therapy, leading to this agent's FDA approval. In early phase clinical trials, dacomitinib also showed promise as an active agent in EGFR-mutant NSCLC. This agent is being explored in Her2-mutated NSCLC; one study involving 26 subjects demonstrated a 12% response rate (3 of 26 subjects). However, dacomitinib does not appear to have much activity in Her2-amplified NSCLC. A risk of diarrhea, skin rash, and stomatitis limit the use of dacomitinib and afatinib.

Third-generation EGFR TKIs include the irreversible and potent, pyrimidine-based EGFR inhibitors such as rocelitinib (Clovis Oncology), AZD9291 (Astra Zeneca), and WZ4002 and HM61763 (Hanmi Pharmaceuticals/co-licensed with Boehringer Ingelheim). These drugs appear to target both activating and resistance mutations, such as T790M, which accounts for over 50–60% of cases of acquired resistance to the first-generation TKIs (e.g., erlotinib and gefitinib). The third-generation drugs do not target wild-type EGFR and thus avoid the toxicities-associated wild-type receptor targeting, including skin rash and diarrhea. Rocelitinib demonstrates an objective response rate of 60% and a disease control rate of 90% in patients with an acquired T790M resistance mutation. In patients with T790M-negative tumors, a response rate of 37% is observed (Sequist et al. 2015). Key side effects include hyperglycemia (due to inhibition of IGF-1R/IR), diarrhea, nausea, fatigue, QT prolongation, and cataract formation. A phase I dose expansion cohort study of AZD9291 in 60 subjects demonstrated an objective response rate of 83% at a dose of 160 mg and 63% at a dose of 80 mg (Ramalingam et al. 2015). Approximately 79% of patients remained disease progression-free at one year; the disease control rate was 93% at the 80-mg dose, which is now being explored in the frontline phase III FLAURA trial, comparing AZD9291 with erlotinib in EGFR-mutant NSCLC. It appears that a newly identified EGFR mutation, namely, C797S in Exon 20 of EGFR, may mediate resistance to AZD9291, at least in select cases.

A major drawback of all three generations of EGFR inhibitors is the lack of central nervous system (CNS) penetration: low cerebrospinal fluid (CSF) and plasma concentrations are achieved at standard systemic doses. Therefore, patients with good control of systemic disease may fail therapy if they have brain metastases or leptomeningeal carcinomatosis. A couple of newer EGFR inhibitors with promising CNS penetration in preclinical models are entering clinical trials, including tesavatinib (Kadmon Pharma) and AZD3759 (Astra Zeneca).

In addition to TKIs, cetuximab has been tested in NSCLC with cytotoxic chemotherapy. The role of anti-EGFR mAbs in lung cancer is not very clear as these studies are still ongoing.

NSCLC patients respond differently to EGFR inhibitors, especially to TKIs, as suggested by several clinical trials. K-ras is a downstream molecule in the EGFR-signaling pathway. K-ras mutations are consistently shown to be related to TKI resistance (Eberhard et al. 2005). Other processes such as Her2 gene amplification and Akt phosphorylation have been reported to correlate with clinical response to TKIs (Stasi and Cappuzzo 2014). As with other targeted therapy in oncology, the molecular target of EGFR inhibition needs to be further defined in NSCLC. If this is accomplished, we can individualize anti-EGFR therapy to achieve the best possible outcomes.

References

Dewdney A, Cunningham D, Tabernero J, Capdevila J, Glimelius B, Cervantes A, Tait D, Brown G, Wotherspoon A, Gonzalez de Castro D, Chua YJ, Wong R, Barbachano Y, Oates J, Chau I (2012) Multicenter randomized phase II clinical trial comparing neoadjuvant oxaliplatin, capecitabine, and preoperative radiotherapy with or without cetuximab followed by total mesorectal excision in patients with high-risk rectal cancer (EXPERT-C). J Clin Oncol 30(14): 1620–1627. doi:10.1200/JCO.2011.39.6036. Epub 2 Apr 2012

Eberhard DA, Johnson BE, Amler LC et al (2005) Mutations in the epidermal growth factor receptor and in KRAS are predictive and prognostic indicators in patients with non-small-cell lung cancer treated with chemotherapy alone and in combination with erlotinib. J Clin Oncol 23:5900

Lynch TJ, Bell DW, Sordella R et al (2004) Activating mutations in the epithelial growth factor receptor underlying responsiveness of non-small-cell lung cancer to gefitinib. N Engl J Med 350:2129

Marshall J (2006) Clinical implications of the mechanism of epidermal growth factor receptor inhibitors. Cancer 107(6):1207–1218

Price TJ et al (2014) Panitumumab versus cetuximab in patients with chemotherapy-refractory wild-type KRAS exon 2 metastatic colorectal cancer (ASPECCT): a randomised, multicentre, open-label, non-inferiority phase 3 study. Lancet Oncol 15(6):569–579

Primrose J et al (2014) Systemic chemotherapy with or without cetuximab in patients with resectable colorectal liver metastasis: the New EPOC randomised controlled trial. Lancet Oncol 15(6):601–611

Ramalingam SS et al (2015) AZD9291, a mutant-selective EGFR inhibitor, as first-line treatment for EGFR mutation-positive advanced non-small cell lung cancer (NSCLC): results from a phase 1 expansion cohort. J Clin Oncol 33 (suppl): abstr 8000

Reardon DA, Schuster J, Tran DD, Fink KL, Nabors LB, Li G, Bota DA, Lukas RV, Desjardins A, Ashby LS, Duic JP, Mrugala MM, Werner A, Hawthorne T, He Y, Green JA, Yellin MJ, Turner CD, Davis TA, Sampson JH, The ReACT Study Group (2015) ReACT: Overall survival from a randomized phase II study of rindopepimut (CDX-110) plus bevacizumab in relapsed glioblastoma. J Clin Oncol 33(15):2009

Schuster J, Lai RK, Recht LD, Reardon DA, Paleologos NA, Groves MD, Mrugala MM, Jensen R, Baehring JM, Sloan A, Archer GE, Bigner DD, Cruickshank S, Green JA, Keler T, Davis TA, Heimberger AB, Sampson JH (2015) A phase II, multicenter trial of rindopepimut (CDX-110) in newly diagnosed glioblastoma: the ACT III study. Neuro Oncol 17(6):854–861

Sequist LV et al (2015) Rociletinib in EGFR-mutated non–small-cell lung cancer. N Engl J Med 372:1700–1709. doi:10.1056/NEJMoa1413654

Shepherd Frances A et al (2005) Erlotinib in previously treated non–small-cell lung cancer. N Engl J Med 353:123–132

Stasi I, Cappuzzo F (2014) Second generation tyrosine kinase inhibitors for the treatment of metastatic non-small-cell lung cancer. Transl Resp Med 2:2. doi:10.1186/2213-0802-2-2

Taieb J et al (2014) Oxaliplatin, fluorouracil, and leucovorin with or without cetuximab in patients with resected stage III colon cancer (PETACC-8): an open-label, randomised phase 3 trial. Lancet Oncol 15(8):862–873

Tournigand C et al (2015) Bevacizumab with or without erlotinib as maintenance therapy in patients with metastatic colorectal cancer (GERCOR DREAM; OPTIMOX3): a randomised, open-label, phase 3 trial. Lancet Oncol. doi:10.1016/S1470-2045(15)00216-8. Published online 13 Oct 2015

Epidermal Growth Factor Receptor

Christina L. Addison
Cancer Therapeutics Program, Ottawa Hospital Research Institute, Ottawa, ON, Canada

Synonyms

EGFR; ERBB; ErbB-1; HER1; PLG61

Definition

The epidermal growth factor receptor (EGFR) is the prototype cell surface receptor for the epidermal growth factor (EGF) family of soluble protein ligands. The receptor itself is part of a family of homologous transmembrane tyrosine kinase receptors that includes EGFR (ErbB-1), HER2/c-neu (ErbB-2), Her 3 (ErbB-3), and Her 4 (ErbB-4). EGFR is a 1186aa, 134 kD protein and was the first receptor described to have intrinsic tyrosine kinase activity directed toward not only to itself (autophosphorylation) but also to various downstream target proteins. Following binding of its ligands, EGFR can elicit intracellular signals that modulate a wide variety of cellular functions including cell proliferation, survival, migration, and differentiation. The EGFR gene maps to chromosome 7p11-13, and it is often amplified or rearranged in glioblastoma multiforme, non-small cell lung cancer, and prostate cancer.

Characteristics

EGFR Structure

EGFR is transcribed from a single 26 exon gene on chromosome 7p11-13, and upon translation, it generates a 1186 amino acid mature transmembrane glycoprotein (Fig. 1). The amino terminal 622 amino acids form the extracellular domain which contains two cysteine-rich domains (CR1 and CR2), which comprise the ligand binding domain of the protein. Following this, the transmembrane domain separates the extracellular domain from the intracellular 542 amino acid carboxy-terminus of the protein. The intracellular domain contains three separate motifs, the juxtamembrane domain, the kinase domain, and a carboxy-terminal tail. The juxtamembrane domain primarily plays a role in feedback attenuation of receptor signaling. The kinase domain becomes activated following ligand binding and receptor dimerization, and it then phosphorylates numerous substrates on tyrosine residues. The kinase domain of the EGFR family of receptors is highly conserved with the exception of ErbB3

Epidermal Growth Factor Receptor, Fig. 1 EGFR structure. EGFR is a transmembrane receptor tyrosine kinase whose domain structure is illustrated herein. In the extracellular region, the CR1 and CR2 domains form the ligand binding domain. Intracellularly, the tyrosine kinase domain and a number of tyrosine residues which can become phosphorylated following receptor activation play important roles in EGFR signal transduction. EGFR ligands are grouped into three classes depending on their abilities to bind various EGFR family members in addition to binding EGFR

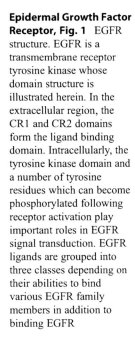
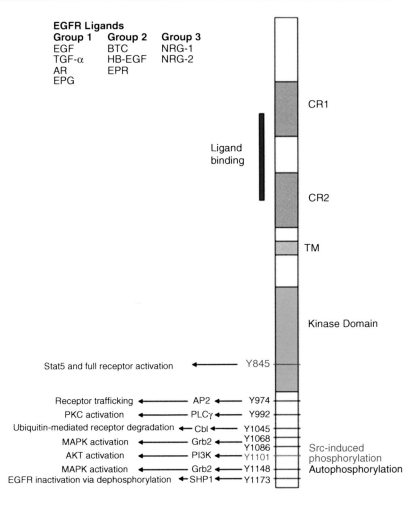

which has amino acid substitutions at critical sites which results in it lacking tyrosine kinase activity. Within the carboxy-terminal tail of EGFR, there are a number of tyrosine residues that are autophosphorylated, which following their phosphorylation can serve as binding sites to recruit additional signaling proteins leading to activation of a number of downstream signal transduction pathways (Fig. 1). The carboxy tail also has additional tyrosine residues that are known to be phosphorylated by other signaling molecules, for example, Src kinase, which also plays a role in modulation of receptor signaling.

EGFR Ligands

EGFR family ligands can be primarily divided into three groups based on their affinity for binding the various EGFR family of receptors. As EGFR receptors can homodimerize or heterodimerize, binding of ligands and subsequent signaling that occurs following this binding will be influenced by the expression of the combination of different EGFR receptors in cells. Group 1 EGFR ligands represent those that bind specifically to the prototypical family member EGFR. These include EGF, tumor growth factor-α (TGF-α), epigen (EPG), and amphiregulin (AR). The second group of ligands includes those that bind EGFR but additionally bind ErbB4. Group 2 ligands include betacellulin (BTC), heparin-binding epidermal growth factor (HB-EGF), and epiregulin (EPR). Group 3 comprises the neuregulin (NRG) ligands, which is further divided into two subgroups. Subgroup 1 is comprised of NRG-1 and NRG-2 which bind to both EGFR and ErbB4, while subgroup

2 contains only those neuregulins that bind ErbB4, namely, NRG-3 and NRG-4. It should also be noted that no known ligands bind to ErbB2.

EGFR Signal Transduction

Upon ligand binding, EGFR homodimerizes or heterodimerizes and undergoes subsequent internalization and activation of the kinase domain of the receptor. This results in phosphorylation of the receptor at a number of tyrosine residues in its cytoplasmic tail, most notably at Y974, Y992, Y1045, Y1068, Y1086, Y1148, and Y1173. These phosphorylation events create docking sites for a number of other proteins that contain Src homology 2 (SH2) domains or phosphotyrosine binding (PTB) domains (Fig. 1). One of the most important downstream signaling events following EGFR activation is the subsequent activation of the Ras/Raf/MEK/MAPK pathway via association of phosphorylated Y1068, Y1086, or Y1148 residues with the adaptor protein growth factor receptor-bound protein-2 (Grb2). The SH3 domains of Grb2 are constitutively associated with SOS (son of sevenless), an exchange factor of Ras GTPase. Besides interaction with SOS, Grb2 SH3 domains are capable of association with several additional proteins, including dynamin and Cbl, which both play a role in the regulation of EGFR endocytosis. Binding of the Grb2 and SOS complex to the EGFR places SOS in proximity to Ras, thus leading to GTP-loading of Ras and subsequent activation of Ras effectors, including Raf kinases and PI3K (phosphatidylinositol 3-kinase). Raf initiates a cascade of phosphorylation events including the phosphorylation and activation of the MEKs (MAPK/ERK kinases) and ERKs (extracellular signal-regulated kinases).

Other important interactions include phosphorylated Y992 of EGFR with phospholipase Cγ (PLCγ), which upon binding becomes activated and in turn activates protein kinase C (PKC), a serine-threonine kinase. This enzyme, which has two SH2 domains, catalyzes the hydrolysis of PIP2, generating the second messengers DAG (1,2-diacylglycerol) and IP3 (inositol trisphosphate). IP3 diffuses through the cytosol and releases stored Ca^{2+} (Calcium) ions from the ER (endoplasmic reticulum). DAG is the physiological activator of PKC (protein kinase C), which in turn leads to phosphorylation of various substrate proteins that are involved in an array of cellular events including enhanced cellular migration.

Phosphorylated EGFR has also been shown to recruit and activate the p85 subunit of PI3-K. Recruitment of PI3K may be mediated by the docking protein GAB1 (Grb2-associated binder-1). Once PI3K is recruited and becomes activated, it phosphorylates membrane bound PIP2 (phosphatidylinositol (4,5)-bisphosphate) to generate PIP3 (phosphatidylinositol-3,4,5-trisphosphate). The binding of the released PIP3 to the PH domain of Akt anchors Akt to the plasma membrane and allows its phosphorylation and activation by PDK1 (phosphoinositide-dependent kinase-1). Akt then phosphorylates several substrates thereby regulating cell survival.

A number of other autophosphorylation sites play a role in regulating the duration of activation of EGFR. For example, the autophosphorylation at Y1045 allows association with the Cbl protein, an E3 ubiquitin ligase which subsequently modifies EGFR and targets it for ubiquitin-mediated proteasomal degradation. Another mechanism by which EGFR signaling may be attenuated is via interaction of phosphorylated Y1173 with the protein tyrosine phosphatase SHP1, which dephosphorylates the autophosphorylated tyrosine residues on EGFR and hence halts its continual interaction with other signaling molecules.

EGFR also activates several other proteins including FAK (focal adhesion kinase), paxillin, caveolin, E-cadherin, and CTNN-beta (catenin-beta). FAK activation links EGFR with regulation of cell motility, and caveolin, cadherin, and CTNN-beta are involved in cytoskeletal regulation.

EGFR Transactivation

In addition to activation of EGFR via ligand binding and subsequent autophosphorylation, EGFR may be activated by other signaling pathways in a process known as transactivation. For example, Janus tyrosine kinase 2 (JAK2) can phosphorylate specific residues on the cytoplasmic tail of EGFR following its own activation by growth hormone or prolactin interactions with their receptors. The intracellular kinase Src has also been shown to directly phosphorylate EGFR on specific residues such as Y845 and Y1101 following Src activation by a number of pathways including other growth factor receptors and integrins. A number of G - protein-coupled receptors have also been shown to transactivate EGFR following binding their own agonists, such as endothelin-1, bombesin, thrombin, or lysophosphatidic acid.

EGFR in Cancer

Given that EGFR signaling results in modulation of cellular proliferation and migration, it is not surprising that EGFR has been shown to play a significant role in cancer. There are a number of mechanisms that result in the dysregulation of EGFR signaling in cancer including: (1) increased production of ligands, (2) increased expression of EGFR protein, (3) acquisition of EGFR mutations which give rise to constitutively active EGFR variants, (4) impairment of pathways which control downregulation of EGFR signals, and (5) increased cross talk with other growth factor receptor systems.

Increased EGFR Ligand Production

In many tumor types, increased expression of EGF and TGF-α has been noted, and these ligands act in a paracrine and autocrine manner to further drive tumor cell proliferation in an EGFR-dependent manner. In fact, stimulation of fibroblasts overexpressing EGFR with EGF or TGF-α resulted in their enhanced proliferation and transformation. Transgenic mice engineered to overexpress TGF-α under the mammary-specific MMTV promoter developed mammary hyperplasia and adenocarcinoma of mammary tissues. It has also been shown that co-expression of TGF-α with EGFR in many human tumors correlates with higher proliferative indices and an overall worse patient survival. The EGFR ligand amphiregulin has also been shown to be upregulated in tumor tissues in patients, and its overexpression in tumor cells modulates proliferation, anchorage independent growth, and response to chemotherapy drugs.

Increased EGFR Protein Levels

Increased EGFR expression in fibroblasts led to increased cellular transformation and migration independent of ligand binding, likely as a result of increased spontaneous receptor dimerization and subsequent activation. Subsequently, in vivo transgenic mouse models overexpressing EGFR under control of mammary-specific promoters resulted in the development of mammary hyperplasia that led to dysplasia and tubular adenocarcinomas in lactating animals. It should be noted, however, that forced overexpression of EGFR in other tissues including urothelium, glial cells, or esophageal keratinocytes resulted in increased cellular proliferation but not carcinoma. The family member ErbB-2, however, has been shown to have more potent transforming ability compared to EGFR, as its forced overexpression in a tissue-specific manner resulted in carcinoma of the breast and the skin.

In human cancer, both EGFR and ErbB-2 have been implicated in the development and progression in a variety of different tumor types. The overexpression of this family of receptors has been shown in the majority of solid tumors and is reviewed elsewhere (Normanno et al. 2006; Arteaga 2002; Mendelsohn and Baselga 2000; Spaulding and Spaulding 2002). On average, 50–70% of lung, colon, and breast carcinomas have been shown to overexpress EGFR. In most cases, receptor overexpression is a result of gene amplification; however, this varies according to tumor type. For example, EGFR amplification or overexpression occurs in ~40–60% of glioblastoma multiforme tumors and ~20–35% of non-small cell lung carcinoma (NSCLC); however, in other tumor types, this incidence of amplification may not be as high.

EGFR Mutations

Mutations in EGFR in tumor cells are quite frequent. EGFR mutations can be divided into three main groups, namely, (1) extracellular mutations, (2) intracellular mutations, and (3) mutations within the kinase domain. Extracellular mutations predominantly involve deletions of specific exons encoding all or part of the extracellular domain of EGFR; however, some characterized mutations have also included duplications of some regions. Most of the deletion mutations give rise to truncated versions of the receptor which are constitutively active and usually escape normal receptor downregulatory mechanisms. Similar to the extracellular mutations, the characterized intracellular mutations also predominantly arise from deletions or duplications. These usually result in truncation of the c-terminus of the receptor. Finally, a number of mutations that alter the tyrosine kinase domain have been identified. Although exons 18–24 are known to encode the tyrosine kinase domain, the identified kinase domain mutations are all restricted to exons 18–21. The most frequent mutations here include: (1) small deletions of 6–7 codons in exon 19 that affect amino acids 746–753 of the EGFR protein, (2) point mutations such as the missense mutation L858R in exon 21 or G719A/C in exon 18, and (3) small duplications or insertions in exon 20. Generally, these mutations are centered around the ATP binding site of the kinase domain and are associated with increased tyrosine kinase activity of the mutant receptors.

Impairment of EGFR Downregulation

In normal cells, EGFR downregulation and attenuation of its signaling occur as a result of receptor internalization and subsequent degradation of ligand-activated receptors. It is known that this downregulation is dependent on the cytoplasmic tail of EGFR, and mutant EGFR lacking portions of the cytoplasmic tail is able to transform fibroblasts in a ligand-dependent manner. The E3 ubiquitin ligase c-Cbl plays a significant role in this process, and both EGFR variants that lack ability to bind c-Cbl or mutant c-Cbl variants that cannot ubiquitinate EGFR both result in increased cellular proliferation. Co-expression of ErbB-2 also results in decreased downregulation of EGFR, as it promotes formation of EGFR/ErbB-2 heterodimers which are not readily downregulated by the same mechanism that controls degradation of EGFR homodimers. Differential ligand stimulation can also modulate EGFR degradation, as it has been shown that EGF-induced EGFR activation results in efficient EGFR degradation, while TGF-α-induced EGFR activation predominantly results in recycling of EGFR back to the cell surface where it can then be restimulated.

EGFR Cross Talk with Other Receptors

In addition to the modulation of EGFR activity following interaction with other EGFR family receptors, EGFR signaling is also affected by the activity of other receptor tyrosine kinases, cell adhesion molecules, cytokine receptors, ion transport channel proteins, and G protein-coupled receptors (GPCR). The most well-characterized EGFR interaction with other receptors is that with its family member ErbB2. This heterodimer is the most potent inducer of cellular transformation and mitogenic signaling compared to other EGFR family heterodimers. Activation of another important growth factor receptor in cancer, c-met, has also been shown to modulate the activity of EGFR and production of EGFR ligands in tumor cells. EGFR activity has also been shown to be modulated by cell adhesion molecules such as the integrins following their binding to extracellular matrix (ECM) proteins. The exact mechanism of this is unclear, however, likely involves their association with EGFR in cell surface molecule clusters along with other adaptor and signaling proteins. Common downstream signaling targets between EGFR and integrins may also facilitate this interaction, such as PI3K or Src. EGFR transactivation has also been shown to be modulated by GPCR agonists in an EGFR ligand-independent manner. This EGFR transactivation results in enhanced cell proliferation and migration. GPCRs may also contribute to EGFR ligand-dependent EGFR activation, as GPCR stimulation by agonists may lead to enhanced membrane bound matrix metalloproteinase (MMP) activity which then cleaves membrane bound EGFR ligand precursors that subsequently bind and activate EGFR.

Therapies Targeting EGFR

Given the established role of EGFR in a number of human cancers, it has become a main focus of targeted cancer therapy in the past 10–15 years. Anti-EGFR agents include mAbs (monoclonal antibodies) targeting the EGFR extracellular receptor domain and small molecule TKIs (tyrosine kinase inhibitors) targeting the EGFR intracellular kinase domain. Both mAbs and TKIs have demonstrated encouraging results as monotherapies and in combination with chemotherapy and radiotherapy. More detailed discussion on anti-EGFR therapies and their results can be found elsewhere (Ganti and Potti 2005; Baselga 2001; Khalil et al. 2003; El-Rayes and LoRusso 2004; Johnston et al. 2006).

A number of mAbs to EGFR have been developed and tested in clinical trials, but the most extensively tested one to date is cetuximab (aka Erbitux). Cetuximab has a higher affinity for binding EGFR than EGFR ligands, hence, acts as a competitive inhibitor to block EGFR activation, in addition to facilitating EGFR internalization and degradation following mAb binding. Preclinical studies with cetuximab suggested that it could inhibit tumor cell proliferation and induce apoptosis following a G1 cell cycle arrest. Cetuximab further inhibited tumor growth in xenograft models where it additionally showed an ability to inhibit tumor vascularization. These preclinical studies supported its use in clinical trials, where early Phase I studies demonstrated excellent patient tolerability with relatively minor side effects. Following Phase III clinical analyses, cetuximab demonstrated antitumor efficacy as a single agent or in combination with other anticancer regimens, which has led to its approval for use clinically in certain advanced cancers. Other mAbs to EGFR are presently being tested and are in various stages of clinical studies.

Similar to mAbs, a number of small molecule TKIs have been developed to target EGFR. The primary mechanism of action is via the ability of the small molecule to act as an ATP mimetic for EGFR, thereby resulting in selective inhibition of its autophosphorylation and subsequent signaling activity. The two EGFR small molecule inhibitors most extensively studied are gefitinib (aka Iressa or ZD1839) and erlotinib (aka Tarceva or OSI774). Both have been shown to effectively inhibit tumor cell proliferation in vitro and tumor growth in vivo in preclinical studies. Following testing in Phase I studies where clinical activity

was most favorable, the majority of advanced clinical testing was performed in patients with NSCLC. Both gefitinib and erlotinib were shown to increase the survival of advanced NSCLC patients. However, it appears that a small subset of patients are sensitive to these agents while the vast majority do not derive significant antitumor benefit from these drugs. The majority of sensitive patients appear to have either EGFR amplification or activating point mutations that confer sensitivity. However, acquisition of other point mutations in EGFR have also been associated with gained resistance to these agents. To circumvent these issues, a new class of "second-generation" EGFR inhibitors has been generated and is currently under clinical evaluation. These "irreversible" EGFR receptor inhibitors have the added benefit of binding to the ATP binding site of the kinase domain in a covalent and irreversible manner following reaction of a nucleophilic cysteine residue. Additionally, as the first-generation inhibitors such as gefitinib and erlotinib were selective for inhibition of EGFR, most of the second-generation inhibitors target more than one EGFR family member. CI-1033 (aka canertinib or PD183805) is one such irreversible inhibitor which actually targets all three of the kinase domain-containing members of the EGFR family. CI-1033 is more effective in inhibiting tumor cell proliferation in vitro than the first-generation EGFR inhibitors, and it has demonstrated antitumor efficacy in EGFR- and ErbB2-dependent tumor xenograft models in preclinical studies. While still in clinical testing, early studies have shown that the drug is reasonably tolerated by patients; however, a maximum tolerated dose was established as a result of some grade 3 toxicities. A number of Phase II and III studies are currently underway with this agent in a variety of tumor types; however, we will have awaited the outcome of the mature clinical data to determine whether acquired tumor resistance to these inhibitors also develops.

Summary

EGFR is a critically important mediator of tumor growth and remains an important therapeutic target. However, regulation of its activity and signal transduction by other cellular receptors or adhesion molecules require further analysis in order to determine what role this receptor cross talk may play in mediating the response of tumor to anti-EGFR therapeutic strategies. It will also be important to understand how different EGFR ligands affect not only EGFR signaling and receptor trafficking but also how they may modulate tumor response to EGFR targeting agents. Finally, understanding the tumor characteristics that are important for response to EGFR-targeting agents will facilitate identification of cancer patients most likely to benefit from these agents in future treatment strategies.

Cross-References

▶ HER-2/neu
▶ Herceptin

References

Arteaga CL (2002) Overview of epidermal growth factor receptor biology and its role as a therapeutic target in human neoplasia. Semin Oncol 29:3–9

Baselga J (2001) The EGFR as a target for anticancer therapy–focus on cetuximab. Eur J Cancer 37(Suppl 4):S16–S22

El-Rayes BF, LoRusso PM (2004) Targeting the epidermal growth factor receptor. Br J Cancer 91:418–424

Ganti AK, Potti A (2005) Epidermal growth factor inhibition in solid tumours. Expert Opin Biol Ther 5:1165–1174

Johnston JB, Navaratnam S, Pitz MW, Maniate JM, Wiechec E, Baust H, Gingerich J, Skliris GP, Murphy LC, Los M (2006) Targeting the EGFR pathway for cancer therapy. Curr Med Chem 13:3483–3492

Khalil MY, Grandis JR, Shin DM (2003) Targeting epidermal growth factor receptor: novel therapeutics in the management of cancer. Expert Rev Anticancer Ther 3:367–380

Mendelsohn J, Baselga J (2000) The EGF receptor family as targets for cancer therapy. Oncogene 19:6550–6565

Normanno N, De Luca A, Bianco C, Strizzi L, Mancino M, Maiello MR, Carotenuto A, De Feo G, Caponigro F, Salomon DS (2006) Epidermal growth factor receptor (EGFR) signaling in cancer. Gene 366:2–16

Spaulding DC, Spaulding BO (2002) Epidermal growth factor receptor expression and measurement in solid tumors. Semin Oncol 29:45–54

Epidermal Growth Factor Receptor (EGFR) Inhibitors

▶ Receptor Tyrosine Kinase Inhibitors

Epidermal Growth Factor-like Ligands

Aleksandra Glogowska and Thomas Klonisch
Department of Human Anatomy and Cell Science, College of Medicine, Faculty of Health Sciences, University of Manitoba, Winnipeg, MB, Canada

Synonyms

EGF family; EGF-like ligands; Growth Factors

Definition

Epidermal growth factor (EGF-)-like family members bind to and activate EGF receptor tyrosine kinases ErbB-1, −2, −3, and −4, also named HER1-4. This triggers the activation of intracellular signal transduction pathways, resulting in cellular proliferation and differentiation. All members of the EGF-like family are produced as membrane-anchored precursors and are processed and released through the action of specific membrane-bound proteolytic enzymes of the sheddase family.

Epidermal Growth Factor- (EGF-) like family consists of at least twelve members: Epidermal growth factor (EGF), transforming growth factor alpha (TGFα), heparin binding-EGF like growth factor (HB-EGF), amphiregulin(AR), betacellulin (BTC), epiregulin (EPR), epigen, cripto, and neuregulins 1–4 (NRG1-4). These EGF-like members are defined by three characteristics: (a) they display high affinity binding to membrane-bound epidermal growth factor tyrosine kinases receptors ErbB1-4, (b) they are eliciting a mitogenic response in EGF-sensitive cells, and (c) they possess at least one primary EGF structural motif of 50–60 amino acid (aa) residues with the general sequence $X_nCX_7CX_{2-3}GXCX_{10-13-}CX CX_3YXGXRCX_4LX_n$ embedded in the transmembrane precursor molecule. Usually present as multiple structural units in the extracellular domain of EGF-like ligands, this conserved cysteine-rich EGF-like motif is crucial for the binding to ErbB receptors. The EGF-like ligands are proteolytically cleaved and liberated from the extracellular region of the transmembrane precursor. Ligands, either membrane-bound or soluble, can bind to homo- and heterodimer combinations of the four currently known membrane-anchored tyrosin kinase receptors ErbB1 (EGFR, HER-1), ErbB2 (Neu antigen, HER-2), ErbB 3 (HER-3), and ErbB4 (HER-4). Of all possible combinations, ErbB2 homodimers appear not to be stable. EGF, AR, and TGFα bind preferentially to EGFR, while BTC, HB-EGF, and EPR bind to both EGFR and ErbB4. NRG1 and NRG2 preferentially bind to ERbB3 and ERbB4, whereas NRG3 and NRG4 bind to ErbB4. Growth factor binding induces homo- or heterodimerization of the receptors and stimulates intracellular protein-tyrosine kinase activity resulting in auto- and cross-phosphorylation of key tyrosine residues in the C-terminal domain in the ErbB homo- or heterodimer. This autophosphorylation sites are docking stations for several proteins that associate with the phosphorylated tyrosine residues through SH2 domains (Src homology 2). This initiates downstream ▶ signal transduction cascades, among them being the ▶ MAPK, AKT, and ▶ JNK pathway, which leads to DNA synthesis, cell proliferation, alterations in apoptotic pathways, and cell migration and adhesion (Jorissen et al. 2003; Yarden 2001).

Epidermal Growth Factor, EGF, was originally isolated from the mouse submaxillary gland as a stimulatory of eyelid opening and tooth eruption in newborn rodents. Other sources of (pro) EGF are the kidney*, epidermis* (*mainly membrane-bound proEGF), pancreas, small intestine, and brain. Human EGF precursor (preproEGF) is a large protein consisting of 1207 aa containing nine EGF motifs in the extracellular proEGF region. The EGF motif closest to the transmembrane domain corresponds to soluble EGF (Cohen 1962). The enzyme responsible for cleavage of the extracellular proEGF domain is ADAM 10 (A Disintegrin And Metalloprotease domain 10) (Dong and Wiley 2000). EGF null mice do not show a significant phenotype. Transgenic overexpression of mouse EGF targeted to villous enterocytes of the jejunum and ileum causes increased villous height and crypt depth in these intestinal parts. Transgenic mice overexpressing human proEGF-driven by the β -actin promoter have revealed that (pro)EGF is the active ligand for the EGFR in germ cells and proper EGF expression is important for the completion of spermatogenesis (Harris et al. 2003; Wong et al. 2001).

Transforming Growth Factor alpha, TGFα, derives from a 160 aa precursor (preproTGFα). This protein was originally characterized by its capacity to induce oncogenic transformation in rat kidney fibroblasts. TGFα is targeted preferentially to the basolateral compartment of polarized epithelial cells where it is cleaved by TNF alpha converting enzyme (TACE)/ADAM17. TGFα mRNA and protein have been detected in various adult tissues, including pituitary gland, brain, skin keratinocytes, and macrophages (Burdick et al. 2000). Transgenic animals with overexpression of TGFα display hypertrophy of skin and hyperkerotosis with alopecia. These animals also have stunted hair growth and psoriasis-like lesions. TGF α null mice show a very mild phenotype which includes wavy fur and whiskers (Harris et al. 2003; Mann et al. 1993).

Heparin binding-EGF like growth factor, HB-EGF, derives from a 204 aa molecule. Human HB-EGF was identified in conditioned media of human monocytes/macrophages and was purified from phorbol ester-treated human U-937 macrophages as a 22-kDa factor with strong affinity for heparin (Higashiyama et al. 1991, 1992). HB-EGF is expressed in a wide variety of hematopoietic cells, endothelial cells, vascular smooth muscle, and epithelial cells. Cleavage of proHB-EGF may involve ADAM9, 10, 12, and 17 (Suzuki et al. 1997). Overexpression of human HB-EGF decreases the weight and growth rate in these transgenic mice, and this may involve alternation of insulin-like growth factor binding protein (IGFBP) pathways. Moreover, histological analysis showed overexpression of HB-EGF exclusively in kidney, liver, lung, and stomach. Employing several mutant mice lacking HB-EGF expression revealed a critical role for HB-EGF in cardiac valve formation and normal heart function. HB-EGF null mice develop heart failure as a result of grossly enlarged ventricular chambers (Asakura et al. 2002; Harris et al. 2003).

Amphiregulin, AR, is synthesized as a 252 aa transmembrane precursor. AR was originally isolated from the phorbol ester-treated human breast adenocarcinoma cell line MCF7. ProAR is expressed in many human tissue including ovary, placenta, pancreas, testes, lung, cardiac muscle, breast, kidney, spleen, and colon. In polarized epithelial cells, cleavage of basolateral proAG is facilitated by TACE/ADAM17 to release a 78–84 aa growth factor. Posttranslational modification results in several different cell surface and soluble isoforms (Brown et al. 1998). Keratin-14 promoter-driven overexpression of AR in the basal epidermal cell layer in transgenic mice induces severe, early-onset skin pathology, resembling psoriasis. Histological examination of the skin in these AR overexpressing mice revealed hyperkeratosis, focal parakeratosis and acanthosis, mixed leukocyte infiltration, including CD3-positive T cells and neutrophils in the epidermis and dermis. These mice also display tortuous dermal vasculature. AR null mice similar to EGF null mice do not display any gross or histological abnormalities (Inui et al. 1997).

Betacellulin (BTC) has been isolated from conditioned media from a pancreatic β cell tumor cell line. BTC is also expressed in a variety of mesenchymal and epithelial cell lines and in many tissue including pancreas, liver, kidney, and small intestine. Membrane-bound proBTC is processed by ADAM10 to release soluble BTC (Dunbar and Goddard 2000; Sasada et al. 1993). Transgenic chicken actin promoter-driven BTC overexpression in mice causes bony deformations of the skull, pulmonary hemorrhage syndrome, and complex eye pathology. Transgenic animals showed decrease in the weight of pancreas and increase weight of the eye, lung, and spleen (Schneider et al. 2008, 2009).

Epiregulin, EPR, was initially purified from conditioned medium of the mouse fibroblast-derived tumor cell line NIH-3 T3 clone T7. EPR is expressed in adult pancreas, liver, kidney, and small intestine. Human epiregulin encodes a 46 aa growth factor which exhibits 24–50% identity with the sequences of other EGF-like ligands. Epiregulin exhibits a bifunctional regulatory property in that it inhibits the growth of several epithelial cell lines, like NIH-3 T3, and stimulates the growth of fibroblasts, primary rat hepatocytes, human smooth muscle cells, and various other cell types, including COS-7 (monkey) and Swiss 3 T3 (mouse) (Shirakata et al. 2000). Epiregulin can bind to and activate EGFR and ErbB4. EPR can also transactivate ErBB2 and ErbB3. Although EPR displays lower affinity towards EGFR, EPR-mediated receptor activation is more sustained when compared with EGF or BTC. The lower affinity of EPR to EGFR results in enhanced dissociation of EPR from the ligand-receptor complex during receptor endocytosis and renewed EGFR activation by recycled EPR. EPR-deficient mice develop chronic dermatitis (Komurasaki et al. 1997).

Neu differentiation factors (NDF, heregulins, Neuregulins, Glia Growth Factors-GGFs, Acetylocholine Receptor Inducing Activity-ARIA) were isolated from mesenchymal cells based on their ability to elevate phosphorylation of ErbB proteins. At least four different genes of neuregulins are known. NRGs are widely expressed in neurons of the central and peripheral nervous system. They are known to play important functions during neuronal development and migration (Britsch 2007). Furthermore, NRG-1 is also involved in heart development. Mice defective in genes encoding either NRG-1 or the receptors ErbB-2 or ErbB-4 display identical failure of trabecular formation in the embryonic heart, consistent with the notion that trabeculation requires NRG-1-mediated activation of ErbB-2/ErbB-4 heterodimers (Kuramochi et al. 2004).

Epigen encodes a protein of 152 aa that contains EGF-like features. Epigen has 24–37%

E

identity with EGF, TGF α, and Epiregulin. In epithelial cells, Epigen stimulates the phosphorylation of ErbB1 and causes activation of MAP kinase signaling pathways. Epigen also activates genes under the control of the SRE (Serum Response Element). Epigen is a mitogen for HaCaT cells and this activity can be significantly reduced by a blocking antibody to the receptor ErbB-1 (Strachan et al. 2001).

Cripto-1 (Cr-1) is a glycoprotein member of 188 aa. Cr-l has been implicated in development. Expression of Cr-l was reported in trophoblast and embryoblast of 4-day-old mouse blastocysts, and the myocardium of developing heart tubes in 8.5-day-old mice embryos. Transgenic murine embryos lacking Cr-l are devoid of cardiac specific gene expression (α- and β- myosin heavy chain, myosin light chains 2A and 2B) (Ozcelik et al. 2006). Antisense RNA silencing studies showed that inhibition of Cr-1 expression inhibits the growth of human colon carcinoma cell lines expressing this factor (Normanno et al. 2004).

EGF-like Ligands and Cancer

Members of the EGF-like ligand receptor system are often amplified in various cancers. EGFR overexpression in brain tumors (oligodendroglioma, glioblastoma) and in carcinoma of the stomach, thyroid, lung, and breast (for the latter two also Neu/ErbB2 amplification) makes these tumors targets for the actions of EGF-like ligands and anti-cancer drugs directed at inhibiting ligand-mediated ErbB1/2 activation. EGF-like ligands affect tumor cell growth, differentiation, and metastasis. Amplification and ligand-induced activation of EGFR correlates with increased tumor cell migration, matrix degradation in vitro, and enhanced tissue invasiveness in vivo. EGFR activation as a result of the prior activation of G protein coupled receptors (GPCRs) involves the proteolytic cleavage of membrane-anchored EGF-like ligands by GPCR-mediated activation of members of the metalloproteinase super family. The GPCR-mediated EGFR activation in COS-7, HEK-293, and breast cancer cells involves cleavage of proHB-EGF, while in colon epithelial cells and head and neck squamous cell carcinomas proTGFα has been implicated (Olayioye et al. 2000; Yarden 2001).

Brain Tumor

Malignant human gliomas are the most common form of primary tumors of the central nervous system and display an invasive growth pattern. Among the genetic alternations found in these tumors, p53 inactivation and PDGF/PDGFR activation represent early events, whereas the loss of chromosome 10, gene amplification, and rearrangement of genomic EGFR are late events in glioma carcinogenesis. Coamplification of TGFα and EGFR in human glioma cell lines and primary glioma tissues coincides with sustained glioma proliferation and suggests an autocrine growth loop promoted by this EGF-like ligand-receptor pair, especially in high grade glioma. Antisense-mediated downregulation of TGFα expression was shown to inhibit glioma growth (Harris et al. 2003; Nair 2005; Perry et al. 2002; Tang et al. 1997).

Carcinoma of the Lung

EGF-like growth factors play an important role in the pathogenesis and progression of nonsmall cell lung cancers (NSCLC) comprising large cell cancer, squamous cancer and adenocarcinoma, and small cell lung cancer (SCLC). Both the presence of cytoplasmic EGFR, but not its membrane-anchored form, and/or higher EGFR expression, either in its cytoplasmic or membrane-bound form, together with the amplification of TGFα are all significantly associated with poor patient survival rates (Nair 2005; Olayioye et al. 2000; Rusch et al. 1997; Wu et al. 2007).

Head and Neck Squamous Cell Carcinoma (HNSCC)

GPCR-EGFR transactivation via GPCRs for gastrin releasing peptide and lysophosphatidic acid (LPA) or carbachol results in a matrix-metalloproteinase-dependent enhanced invasiveness and growth of HNSCC cells in vitro. This coincides with the release of TGFα and AR, but not EGF or HB-EGF, into the supernatant of HNSCC. Enzymatic processing of proamphiregulin and proTGFα upon treatment with lysophosphatidic acid (LPA) or carbachol involves the activation of TNF-converting enzyme (TACE/ADAM-17) and downstream activation of EGFR-induced MAPK signaling pathways in HNSSC (Grandis and Tweardy 1993; Gschwind et al. 2003).

Breast Cancer

TGFα stimulates growth and differentiation of mammary epithelial cells and is implicated in the pathogenesis of human breast cancer (Zheng 2009). High expression of AR has been detected in several human breast cancer cell lines and primary human breast carcinomas. AR acts as an autocrine/juxtacrine growth factor in human mammary epithelial cells transformed with an activated c-Ha-*ras* proto-oncogene or overexpressing Neu/ErbB2 (Normanno et al. 1994). EGF and AR may modulate invasion of metastatic breast cancer cells by increasing the expression of matrix-metalloproteinases. EGFR activation by GPCRs in breast cancer cells appears to be mainly facilitated by HB-EGF. Furthermore, estradiol treatment of breast cancer cells causes EGFR activation by the MMP-2- and MMP-9-mediated release of HB-EGF (Olayioye et al. 2000).

Colorectal Cancer

Most human colon cancer cell lines express TGFα, AR, and cripto (Wu et al. 2009). Cripto and AR appear to be suitable markers for human colorectal cancer tissues since transcripts for both EGF-like ligands are detected in 60–70% of primary or metastatic human colorectal cancers but are present in only 2–7% of normal human colonic mucosa. Also, immunoreactive AR was reported in primary and metastatic colorectal tumors but not in normal colon (Normanno et al. 2004; Ohchi et al. 2012). Agonists to the GPCR prostaglandin E_2 receptor and the M3 muscarinic receptor lead to a metalloproteinase-/

TACE-dependent processing of TGFα resulting in EGF-R transactivation and proliferation of colon cancer (Caco-2, LoVo, and HT-29) cell lines (Fiske et al. 2009).

Cross-References

▶ Adhesion
▶ Amphiregulin
▶ Cripto-1
▶ JNK Subfamily
▶ Pleiotrophin
▶ SH2/SH3 Domains
▶ Signal Transduction
▶ Transforming Growth Factor-Beta

References

Asakura M, Kitakaze M, Takashima S, Liao Y, Ishikura F, Yoshinaka T, Ohmoto H, Node K, Yoshino K, Ishiguro H et al (2002) Cardiac hypertrophy is inhibited by antagonism of ADAM12 processing of HB-EGF: metalloproteinase inhibitors as a new therapy. Nat Med 8:35–40

Britsch S (2007) The neuregulin-I/ErbB signaling system in development and disease. Adv Anat Embryol Cell Biol 190:1–65

Brown CL, Meise KS, Plowman GD, Coffey RJ, Dempsey PJ (1998) Cell surface ectodomain cleavage of human amphiregulin precursor is sensitive to a metalloprotease inhibitor. Release of a predominant *N*-glycosylated 43-kDa soluble form. J Biol Chem 273:17258–17268

Burdick JS, Chung E, Tanner G, Sun M, Paciga JE, Cheng JQ, Washington K, Goldenring JR, Coffey RJ (2000) Treatment of Menetrier's disease with a monoclonal antibody against the epidermal growth factor receptor. N Engl J Med 343:1697–1701

Cohen S (1962) Isolation of a mouse submaxillary gland protein accelerating incisor eruption and eyelid opening in the new-born animal. J Biol Chem 237:1555–1562

Dong J, Wiley HS (2000) Trafficking and proteolytic release of epidermal growth factor receptor ligands are modulated by their membrane-anchoring domains. J Biol Chem 275:557–564

Dunbar AJ, Goddard C (2000) Structure-function and biological role of betacellulin. Int J Biochem Cell Biol 32:805–815

Fiske WH, Threadgill D, Coffey RJ (2009) ERBBs in the gastrointestinal tract: recent progress and new perspectives. Exp Cell Res 315:583–601

E

Grandis JR, Tweardy DJ (1993) TGF-alpha and EGFR in head and neck cancer. J Cell Biochem Suppl 17F:188–191

Gschwind A, Hart S, Fischer OM, Ullrich A (2003) TACE cleavage of proamphiregulin regulates GPCR-induced proliferation and motility of cancer cells. EMBO J 22:2411–2421

Harris RC, Chung E, Coffey RJ (2003) EGF receptor ligands. Exp Cell Res 284:2–13

Higashiyama S, Abraham JA, Miller J, Fiddes JC, Klagsbrun M (1991) A heparin-binding growth factor secreted by macrophage-like cells that is related to EGF. Science 251:936–939

Higashiyama S, Lau K, Besner GE, Abraham JA, Klagsbrun M (1992) Structure of heparin-binding EGF-like growth factor. Multiple forms, primary structure, and glycosylation of the mature protein. J Biol Chem 267:6205–6212

Inui S, Higashiyama S, Hashimoto K, Higashiyama M, Yoshikawa K, Taniguchi N (1997) Possible role of coexpression of CD9 with membrane-anchored heparin-binding EGF-like growth factor and amphiregulin in cultured human keratinocyte growth. J Cell Physiol 171:291–298

Jorissen RN, Walker F, Pouliot N, Garrett TP, Ward CW, Burgess AW (2003) Epidermal growth factor receptor: mechanisms of activation and signalling. Exp Cell Res 284:31–53

Komurasaki T, Toyoda H, Uchida D, Morimoto S (1997) Epiregulin binds to epidermal growth factor receptor and ErbB-4 and induces tyrosine phosphorylation of epidermal growth factor receptor, ErbB-2, ErbB-3 and ErbB-4. Oncogene 15:2841–2848

Kuramochi Y, Cote GM, Guo X, Lebrasseur NK, Cui L, Liao R, Sawyer DB (2004) Cardiac endothelial cells regulate reactive oxygen species-induced cardiomyocyte apoptosis through neuregulin-1beta/erbB4 signaling. J Biol Chem 279:51141–51147

Mann GB, Fowler KJ, Gabriel A, Nice EC, Williams RL, Dunn AR (1993) Mice with a null mutation of the TGF alpha gene have abnormal skin architecture, wavy hair, and curly whiskers and often develop corneal inflammation. Cell 73:249–261

Nair P (2005) Epidermal growth factor receptor family and its role in cancer progression. Curr Sci 88:890–898

Normanno N, Ciardiello F, Brandt R, Salomon DS (1994) Epidermal growth factor-related peptides in the pathogenesis of human breast cancer. Breast Cancer Res Treat 29:11–27

Normanno N, De Luca A, Maiello MR, Bianco C, Mancino M, Strizzi L, Arra C, Ciardiello F, Agrawal S, Salomon DS (2004) CRIPTO-1: a novel target for therapeutic intervention in human carcinoma. Int J Oncol 25:1013–1020

Ohchi T, Akagi Y, Kinugasa T, Kakuma T, Kawahara A, Sasatomi T, Gotanda Y, Yamaguchi K, Tanaka N, Ishibashi Y et al (2012) Amphiregulin is a prognostic factor in colorectal cancer. Anticancer Res 32:2315–2321

Olayioye MA, Neve RM, Lane HA, Hynes NE (2000) The ErbB signaling network: receptor heterodimerization in development and cancer. EMBO J 19:3159–3167

Ozcelik C, Bit-Avragim N, Panek A, Gaio U, Geier C, Lange PE, Dietz R, Posch MG, Perrot A, Stiller B (2006) Mutations in the EGF-CFC gene cryptic are an infrequent cause of congenital heart disease. Pediatr Cardiol 27:695–698

Perry SW, Dewhurst S, Bellizzi MJ, Gelbard HA (2002) Tumor necrosis factor-alpha in normal and diseased brain: conflicting effects via intraneuronal receptor crosstalk? J Neurovirol 8:611–624

Rusch V, Klimstra D, Venkatraman E, Pisters PW, Langenfeld J, Dmitrovsky E (1997) Overexpression of the epidermal growth factor receptor and its ligand transforming growth factor alpha is frequent in resectable non-small cell lung cancer but does not predict tumor progression. Clin Cancer Res 3:515–522

Sasada R, Ono Y, Taniyama Y, Shing Y, Folkman J, Igarashi K (1993) Cloning and expression of cDNA encoding human betacellulin, a new member of the EGF family. Biochem Biophys Res Commun 190:1173–1179

Schneider MR, Antsiferova M, Feldmeyer L, Dahlhoff M, Bugnon P, Hasse S, Paus R, Wolf E, Werner S (2008) Betacellulin regulates hair follicle development and hair cycle induction and enhances angiogenesis in wounded skin. J Invest Dermatol 128:1256–1265

Schneider MR, Mayer-Roenne B, Dahlhoff M, Proell V, Weber K, Wolf E, Erben RG (2009) High cortical bone mass phenotype in betacellulin transgenic mice is EGFR dependent. J Bone Miner Res 24:455–467

Shirakata Y, Komurasaki T, Toyoda H, Hanakawa Y, Yamasaki K, Tokumaru S, Sayama K, Hashimoto K (2000) Epiregulin, a novel member of the epidermal growth factor family, is an autocrine growth factor in normal human keratinocytes. J Biol Chem 275:5748–5753

Strachan L, Murison JG, Prestidge RL, Sleeman MA, Watson JD, Kumble KD (2001) Cloning and biological activity of epigen, a novel member of the epidermal growth factor superfamily. J Biol Chem 276:18265–18271

Suzuki M, Raab G, Moses MA, Fernandez CA, Klagsbrun M (1997) Matrix metalloproteinase-3 releases active heparin-binding EGF-like growth factor by cleavage at a specific juxtamembrane site. J Biol Chem 272:31730–31737

Tang P, Steck PA, Yung WK (1997) The autocrine loop of TGF-alpha/EGFR and brain tumors. J Neurooncol 35:303–314

Wong WC, Dong M, Mak KL, Chan SY (2001) Prospects of EGF transgenic mice researches. Sheng Wu Hua Xue Yu Sheng Wu Wu Li Xue Bao (Shanghai) 33:473–476

Wu W, O'Reilly MS, Langley RR, Tsan RZ, Baker CH, Bekele N, Tang XM, Onn A, Fidler IJ, Herbst RS (2007) Expression of epidermal growth factor (EGF)/transforming growth factor-alpha by human lung cancer cells determines their response to EGF receptor tyrosine kinase inhibition in the lungs of mice. Mol Cancer Ther 6:2652–2663

Wu WK, Tse TT, Sung JJ, Li ZJ, Yu L, Cho CH (2009) Expression of ErbB receptors and their cognate ligands in gastric and colon cancer cell lines. Anticancer Res 29:229–234

Yarden Y (2001) The EGFR family and its ligands in human cancer. signalling mechanisms and therapeutic opportunities. Eur J Cancer 37(Suppl 4):S3–S8

Zheng W (2009) Genetic polymorphisms in the transforming growth factor-beta signaling pathways and breast cancer risk and survival. Methods Mol Biol 472:265–277

See Also

(2012) AKT. In: Schwab M (ed) Encyclopedia of Cancer, 3rd edn. Springer Berlin Heidelberg, p 115. doi:10.1007/978-3-642-16483-5_163

(2012) Apoptosis Pathways. In: Schwab M (ed) Encyclopedia of Cancer, 3rd edn. Springer Berlin Heidelberg, p 244. doi:10.1007/978-3-642-16483-5_365

(2012) Betacellulin. In: Schwab M (ed) Encyclopedia of Cancer, 3rd edn. Springer Berlin Heidelberg, p 385. doi:10.1007/978-3-642-16483-5_591

(2012) Cell Migration. In: Schwab M (ed) Encyclopedia of Cancer, 3rd edn. Springer Berlin Heidelberg, p 738. doi:10.1007/978-3-642-16483-5_1006

(2012) DNA. In: Schwab M (ed) Encyclopedia of Cancer, 3rd edn. Springer Berlin Heidelberg, p 1129. doi:10.1007/978-3-642-16483-5_1663

(2012) Domain. In: Schwab M (ed) Encyclopedia of Cancer, 3rd edn. Springer Berlin Heidelberg, p 1150. doi:10.1007/978-3-642-16483-5_1702

(2012) Epigen. In: Schwab M (ed) Encyclopedia of Cancer, 3rd edn. Springer Berlin Heidelberg, p 1283. doi:10.1007/978-3-642-16483-5_1939

(2012) Epiregulin. In: Schwab M (ed) Encyclopedia of Cancer, 3rd edn. Springer Berlin Heidelberg, p 1291. doi:10.1007/978-3-642-16483-5_1954

(2012) G-protein Couple Receptor. In: Schwab M (ed) Encyclopedia of Cancer, 3rd edn. Springer Berlin Heidelberg, p 1587. doi:10.1007/978-3-642-16483-5_2294

(2012) MAPK. In: Schwab M (ed) Encyclopedia of Cancer, 3rd edn. Springer Berlin Heidelberg, p 2167. doi:10.1007/978-3-642-16483-5_3532

(2012) Phosphorylation. In: Schwab M (ed) Encyclopedia of Cancer, 3rd edn. Springer Berlin Heidelberg, p 2870. doi:10.1007/978-3-642-16483-5_4544

(2012) Tyrosine. In: Schwab M (ed) Encyclopedia of Cancer, 3rd edn. Springer Berlin Heidelberg, p 3822. doi:10.1007/978-3-642-16483-5_6078

Epidermoid Carcinoma

▶ Squamous Cell Carcinoma

Epigallocatechin

Ann M. Bode and Zigang Dong
The Hormel Institute, University of Minnesota, Austin, MN, USA

Synonyms

Catechin; Green tea polyphenol

Definition

Epigallocatechin is one of the several biologically active ingredients that make up the bulk of the potent antioxidant polyphenols known as catechins, which are found in green tea ▶ Green Tea Cancer Prevention.

Characteristics

Next to water, tea is the second most widely consumed beverage in the world. Green tea, like oolong and black teas, is derived from the *Camellia sinensis* plant, but is processed immediately from fresh leaves and is protected from oxidation. The biologically active ingredients in green tea are a family of polyphenols (catechins) and flavonols, which are very strong antioxidants. The catechins comprise about 90% of the bulk of green tea and include epicatechin (EC), epigallocatechin (EGC), epicatechin gallate (ECG), and epigallocatechin gallate (EGCG) (Fig. 1). The catechins are characterized by the presence of a di- or trihydroxyl group substitution on the "B"-ring and the meta-5.7-dihydroxyl substitution at the "A"-ring (Fig. 1). EGCG appears to account for 50–80% of the total catechins found in green tea and is considered to be the most biologically active. A cup of green tea (2.5 g of dried green tea leaves brewed in 200 mL of water) usually contains about 90 mg of EGCG. In addition, green tea contains a similar or slightly smaller amount (65 mg) of EGC, about 20 mg each of ECG and

Epigallocatechin,
Fig. 1 Chemical structures of the major green tea catechins

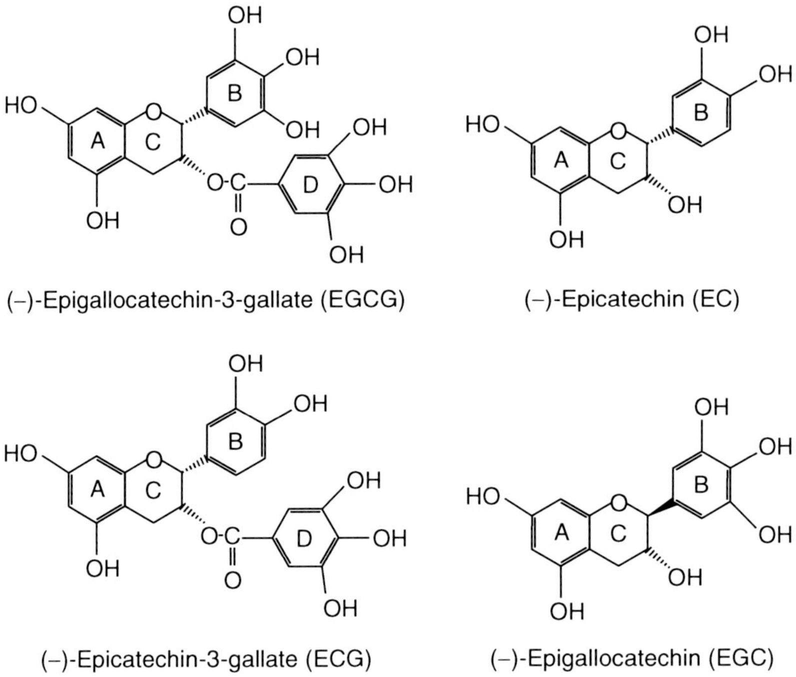

(–)-Epigallocatechin-3-gallate (EGCG)

(–)-Epicatechin (EC)

(–)-Epicatechin-3-gallate (ECG)

(–)-Epigallocatechin (EGC)

EC, and about 50 mg of caffeine. However, physiologically achievable tissue levels of EGCG appear to be in the low micromolar range (i.e., 1–7 μM).

Epidemiologic and laboratory studies suggest that consumption of green tea might be associated with a decreased risk of developing skin, lung, bladder, esophageal, stomach, liver, duodenum and small intestine, pancreatic, colorectal, prostate, or breast ▶ cancer. Many chronic diseases, including cancer, are associated with ▶ oxidative DNA damage produced by free radicals. Much of the effectiveness of green tea has been attributed to its potent antioxidant activity, which is suggested to be greater than that of vitamin C or E or equivalent servings of most vegetables and fruits. Research data also suggest that some of the effects of green tea might be due to its ability to generate ▶ reactive oxygen species as a prooxidant molecule. However, direct unequivocal evidence for either mechanism of action in vivo is lacking. On the other hand, an accumulating number of research studies suggest that EGCG may specifically target and modulate distinct cancer genes or proteins, with little or no

effect on normal molecules, in order to exert its anticancer effects.

Cellular and Molecular Targets of EGCG

Cells respond to their environment through a process known as signal transduction in which information from a stimulus outside a cell is transmitted from the cell membrane into the cell and along an intracellular chain of signaling molecules to perpetuate a response. The development of cancer is a complex, several-step process (i.e., initiation, promotion, progression) that affects innumerable genes and proteins, including signaling molecules that are critical in the regulation of many cellular functions but especially proliferation or growth. The initiation step is an irreversible period involving a process in which a normal cell becomes changed so that it has the capacity to form a tumor (i.e., preneoplastic). The change results from DNA damage that can be induced by a number of "initiators," including radiation, ultraviolet irradiation, chemical carcinogens, and retroviruses. Under normal conditions, the DNA damage is either repaired or, if the damage is too extreme, the cell is eliminated by a tightly controlled process of cell death called "▶ apoptosis." If the damaged DNA is allowed to duplicate, the cell may be predisposed to cancer. The promotion step is a potentially reversible process by which actively proliferating preneoplastic cells accumulate and begin to develop characteristics, including growth factor independence, lack of contact inhibition, and resistance to apoptosis, all of which enhance the cells' ability to proliferate by escaping normal control mechanisms. This step generally occurs over a period of many years, and environmental tumor promoters or host factors may play roles in cancer promotion and latency. The final step of cell transformation is progression in which preneoplastic cells acquire increased metastatic potential and the ability to spread to other tissue sites of the afflicted organism.

Research findings have shown that the dysfunction or deregulation of various cellular signaling molecules is a major factor in cancer development and prevention. The prevailing idea today is that cancer may be prevented or treated by targeting and modulating the activity of specific cancer genes or signaling proteins. Each step of cancer development could be a potential target for anticancer agents, but especially the promotion step because of its length and reversible nature. In addition, an increased interest in discovering and developing natural, nontoxic compounds as chemopreventive agents now exists. The molecular mechanisms explaining how normal cells undergo transformation induced by tumor promoters are rapidly being clarified, and the mechanisms by which natural compounds such as the green tea polyphenol EGCG can act as chemopreventive agents are also being elucidated.

General Anticancer Effects of Green Tea Polyphenols

Green tea polyphenols have been reported to suppress cancer cell proliferation, enhance apoptosis, decrease ▶ angiogenesis, and suppress oncoprotein activation. In particular, the mitogen-activated protein or ▶ MAP kinase signaling pathways are activated differentially by various tumor promoters. The MAP kinases generally transmit signals initiated by tumor promoters such as 12-O-tetradecanoylphorbol-13-acetate (TPA), epidermal growth factor (EGF), and platelet-derived growth factor (PDGF) and are also strongly stimulated by stresses such as ultraviolet (UV) irradiation and arsenic. The activation of these signaling cascades can result in a multitude of cellular responses including apoptosis, proliferation, inflammation, differentiation, and development. MAP kinases activate a variety of target proteins that are important in tumor development, including activator protein-1 (▶ AP-1) and nuclear factor-kappa B (NF-κB), which in turn may promote transcription of a variety of cancer-related genes such as *cyclooxygenase-2* (*cox-2*).

A substantial body of evidence suggests that EGCG or other green tea polyphenols inhibit the phosphorylation and activation of the MAP kinases and various components of another critical cancer-associated pathway, the phosphatidylinositol-3 kinase (▶ PI3K)/▶ Akt pathway. Green tea polyphenols have also been reported to suppress tumor promoter- or growth factor-induced cell transformation and AP-1, NF-κB, or COX-2 activation. EGCG has also been shown to inhibit the phosphorylation of the upstream epidermal growth factor family of proteins. Furthermore, consumption of green tea polyphenols has been associated with suppression of numerous markers of angiogenesis and ▶ metastasis, including expression of ▶ matrix metalloproteinases 2 and 9 and ▶ vascular endothelial growth factor. Other studies indicate that EGCG also inhibits ▶ telomerase activity to induce cell senescence and suppress DNA methyltransferase resulting in the reactivation of the ▶ tumor suppressor gene *p16* [INK4a].

Direct Molecular Targets of EGCG

Identifying EGCG receptors or high-affinity proteins that bind to EGCG is a key step in understanding the molecular and biochemical mechanism of this polyphenol's anticancer effects. The structure of EGCG (Fig. 1) facilitates its ability to bind with varying affinity to a number of proteins. Proteins that have thus far been reported to directly bind with EGCG include ▶ fibronectin, ▶ laminin and the 67-kDa laminin receptor, ▶ insulin growth factor-1 receptor (IGF-1R), Pin1, ZAP-70 kinase, Fyn, ▶ BCL-2 and Bcl-x_L, vimentin, the glucose-regulated protein 78 (GRP78) chaperone protein, apoptosis-associated Fas, and fatty acid synthase. However, not all of these proteins (i.e., Fas, 400 μM; fatty acid synthase, 52 μM) show a high binding affinity, and the consequences or details of the binding with EGCG are not fully understood.

EGCG was first reported to inhibit cancer cell ▶ adhesion to fibronectin by directly binding with this protein. However, the inhibition of cancer cell adhesion was not caused by EGCG binding to the cell-binding domain because EGCG interacted with the adjoining domain. EGCG was also reported to bind to another extracellular matrix protein, laminin, and to inhibit ▶ melanoma cell adhesion. EGCG was later found to inhibit cell growth by binding with the invasion- and metastasis-associated 67-kDa laminin receptor (67LR), which is expressed on various cancer cell types. Importantly, the K_d value for binding was 39.9 nM suggesting possible in vivo relevance. The intermediate filament protein, vimentin, which has an important functional involvement in cell division and proliferation, was also identified as an EGCG-binding protein. Vimentin displayed an even higher affinity ($K_i = 3.3$ nM) for binding with EGCG, and the association also appeared to have a regulatory role in controlling cell proliferation. Another work has suggested that EGCG (20 μg/mL) blocks anchorage-independent growth and human breast and cervical cancer cell phenotype expression through inhibition of IGF-1R downstream signaling. Thus, EGCG appears to interact with surface proteins to suppress cancer cell proliferation.

EGCG may also interact with key proteins to modulate apoptosis. The pro-survival Bcl-2 proteins are overexpressed in many cancer types and thus contribute to resistance of cancer cells to apoptosis. EGCG was reported to directly bind to the BH3 pocket of Bcl-x_L ($K_i = 490$ nM) or Bcl-2 ($K_i = 335$ nM), resulting in the suppression of the antiapoptotic activity of these proteins. EGCG was also shown to directly interact with GRP78, which is associated with the multidrug resistance phenotype of many types of cancer cells. EGCG suppressed GRP78's function and caused an increased etoposide-induced apoptosis in cancer cells. These results strongly suggest that EGCG can enhance apoptosis of cancer cells. Data suggest that EGCG has multiple targets and identification of novel EGCG-binding proteins could facilitate the design of new strategies to prevent cancer and hopefully help translate the effectiveness of EGCG observed in cell and animal models to humans.

Clinical Relevance

Tea polyphenols have attracted a great deal of interest because of their perceived ability to act as highly effective chemopreventive agents. EGCG has been reported to cause growth inhibition, G1-phase arrest, and apoptosis in a variety of human cancer cells. Notably, EGCG's effects appear to target only cancer cells with little or no effect on normal cells. This apparent specificity suggests that EGCG can be used in combination with traditional chemotherapeutic agents to enhance cancer cell death without harming normal cells. For example, treatment of lung cancer cells with EGCG plus ▶ celecoxib, a COX-2 inhibitor, has been shown to synergistically induce apoptosis. EGCG has also been shown to increase the toxicity of the chemotherapeutic drug, ▶ cisplatin, by severalfold in ovarian cancer cells and showed IC$_{50}$ values in the μM range even for ovarian cancers that are known to be resistant to cisplatin. Furthermore, the combination of EGCG with radiotherapy has been suggested to improve the efficacy of ionizing radiation in treating glioblastoma cells.

Although animal and cell culture data suggest a potent anticancer effect for EGCG, reports of anticancer activity of tea polyphenols in humans are less dramatic. Phase I and II clinical trials have been performed to test the anticancer effects of oral administration of green tea but results are still inconclusive. Laboratory data clearly indicate that EGCG and other green tea polyphenols are very unlikely to have only a single target or receptor to account for all its observed activities and effects. Furthermore, based on limited bioavailability, experimental concentrations of EGCG greater than 20 μM may not be relevant to the in vivo situation. Understanding the molecular mechanisms of tea in antitumor promotion may reveal additional high-affinity molecular targets for the development of more effective agents with fewer side effects for the chemoprevention of cancer. A continuing emphasis on obtaining rigorous research data and critical analysis of those data regarding tea polyphenols and other food factors is vital to determine the molecular basis and long-term effectiveness and safety of these compounds as chemopreventive agents. Large-scale and comprehensive studies using combined approaches of biochemical, molecular, animal, and clinical studies are needed to address the bioavailability, toxicity, molecular target, signal transduction pathways, and side effects of tea polyphenols for translation to humans.

Cross-References

► Adhesion
► Angiogenesis
► Apoptosis
► AP-1
► Bcl2
► Cancer
► Celecoxib
► Cisplatin
► DNA Oxidation Damage
► Fibronectin
► Green Tea Cancer Prevention
► Insulin-Like Growth Factors
► MAP Kinase
► Matrix Metalloproteinases
► Metastasis
► PI3K Signaling
► Reactive Oxygen Species
► Telomerase
► Tumor Suppressor Genes
► Vascular Endothelial Growth Factor

References

Bode AM, Dong Z (2003) Signal transduction pathways: targets for green and black tea polyphenols. J Biochem Mol Biol 36:66–77

Bode AM, Dong Z (2006) Molecular and cellular targets. Mol Carcinog 45:422–430

Na HK, Surh YJ (2006) Intracellular signaling network as a prime chemopreventive target of (−)-epigallocatechin gallate. Mol Nutr Food Res 50:152–159

Yang CS, Lambert JD, Hou Z et al (2006a) Molecular targets for the cancer preventive activity of tea polyphenols. Mol Carcinog 45:431–435

Yang CS, Sang S, Lambert JD et al (2006b) Possible mechanisms of the cancer-preventive activities of green tea. Mol Nutr Food Res 50:170–175

See Also

(2012) AKT. In: Schwab M (ed) Encyclopedia of cancer, 3rd edn. Springer, Berlin/Heidelberg, p 115. doi:10.1007/978-3-642-16483-5_163

(2012) Laminin. In: Schwab M (ed) Encyclopedia of cancer, 3rd edn. Springer, Berlin/Heidelberg, pp 1971–1972. doi:10.1007/978-3-642-16483-5_3268

(2012) Senescence. In: Schwab M (ed) Encyclopedia of cancer, 3rd edn. Springer, Berlin/Heidelberg, p 3370. doi:10.1007/978-3-642-16483-5_5236

Epigenetic

Definition

Heritable changes in genome function that occur without a change in DNA sequence. DNA ► methylation at ► CpG-islands and posttranslational histone modifications are the molecular basis for epigenetic information. Epigenetic changes of DNA influence the phenotype without altering the genotype. They consist of changes in the properties of a cell that can be transmitted through cell cycle divisions but do not represent a change in genetic information. The main epigenetic changes that target gene regulatory regions and modulate gene expression are methylation and acetylation.

Cross-References

► CpG Islands
► Epigenetic Gene Silencing
► Epigenetic Therapy
► Methylation

Epigenetic Biomarker

Tina Bianco-Miotto
Robinson Research Institute and School of
Agriculture, Food and Wine, The University of
Adelaide, Adelaide, SA, Australia

Definition

Epigenetic biomarker refers to the measurement of epigenetic modifications in tissues or peripheral fluids like urine, blood, plasma, serum, and stool samples, as markers of disease detection, progression, and therapy response. In particular, most research into epigenetic biomarkers has focused on assessing epigenetic modifications as markers of diagnosis, prognosis, and therapy response in cancers. However, epigenetic biomarkers have also been used as markers of disease (e.g., Alzheimer, Parkinson, diabetes, obesity), and other uses have included markers of response to environmental exposures and toxicology.

Characteristics

Epigenetic Modifications

▶ Epigenetic changes include DNA methylation, ▶ histone modifications, and noncoding RNA which includes ▶ microRNAs. Epigenetic changes result in changes in gene expression without alterations to the DNA sequence. DNA methylation is the addition of a methyl group to the 5′ carbon of a cytosine, which occurs predominantly at cytosine and guanine (CpG) dinucleotides and is catalyzed by DNA methyltransferases. Histone modifications are posttranslational modifications to the amino acids (lysine, arginine, serine, threonine, and proline) within histone tails and include acetylation, phosphorylation, ▶ ubiquitination, and methylation. Histone acetylation is associated with active gene transcription and is a reversible modification regulated by the opposing actions of histone acetyltransferases and histone deacetylases. Histone methylation is associated with both active and inactive gene expression

and is regulated by histone methyltransferases and histone demethylases. Small noncoding RNAs, such as microRNAs, consist of approximately 21 nucleotides, exist naturally in the genome, are involved in the regulation of cellular functions, and can bind to partial or complete complementary mRNA targets to induce gene silencing.

Epigenetic Biomarkers in Cancer

In cancer, biomarkers are used for early detection (diagnosis), prognosis, and therapy response. Since epigenetic modifications such as DNA methylation and microRNAs are easy to measure in tumor tissue and peripheral fluids, they are extensively investigated as biomarkers. Since the 1990s, assessing DNA methylation as a biomarker for cancer diagnosis, prognosis, and therapy response has been extensively pursued. This area has continued to grow to include the assessment of histone modifications as markers of patient outcome and microRNAs and other noncoding RNAs in tumor tissues and other peripheral samples. This has been greatly aided with the newest sequencing technologies and developments.

Prostate Cancer

Prostate cancer is one of the most commonly diagnosed cancers in men of developed Western countries, and globally, it is the 2nd most commonly diagnosed and 6th leading cause of cancer death in men. However, the majority of men diagnosed with prostate cancer have insignificant disease that will not cause their mortality. Current tests do not differentiate between these men and those who have clinically significant disease and life-threatening prostate cancer, for whom early detection and treatment are necessary and whose disease is potentially curable. Several studies have shown that epigenetic modifications play a major role in prostate carcinogenesis, are capable of differentiating between benign and malignant disease, and are useful markers of response to epigenetic therapy. The only biomarker currently used for prostate cancer is serum ▶ prostate-specific antigen (PSA) levels. Studies have shown that a proportion of men without prostate cancer have high levels of serum PSA, and 22% of men with prostate cancer have low serum PSA. Serum PSA levels cannot distinguish between patients with insignificant disease and those with clinically significant cancer at diagnosis. Therefore, there is still a great need to identify biomarkers that differentiate insignificant from clinically significant prostate cancers.

GSTP1 DNA Methylation in Prostate Cancer

DNA methylation is the most frequently studied epigenetic modification in cancer. DNA methylation-based biomarkers for cancer are very appealing due to the high stability of DNA, the ease of analysis with the current techniques available, and the ability to assess the biomarker in body fluids such as blood, urine, and saliva. The most studied epigenetic biomarker for prostate cancer is DNA methylation of the glutathione-S-transferase P1 (GSTP1) gene, which encodes an enzyme required for detoxification and protection of DNA from oxidants and electrophilic metabolites. DNA methylation of GSTP1 has several characteristics which make it an ideal biomarker for prostate cancer: it can be easily measured in body fluids, it has a higher specificity for prostate cancer compared to serum PSA, it can differentiate prostate cancer from other prostatic diseases, and different levels are associated with different stages of prostate cancer and recurrence of the disease following treatment. Although GSTP1 has a much higher specificity than serum PSA, it is still not 100% prostate cancer specific as DNA methylation of GSTP1 has been detected in other cancers. However, several studies have improved the overall specificity and sensitivity by combining GSTP1 with a panel of genes. This may be a better option as a biomarker for cancer and other diseases.

Conclusion

Although there has been extensive research and discovery in the field of epigenetic modifications and their use as biomarkers, further studies are required. An exciting advancement in the development of epigenetic biomarkers is the improvements in technology, which now allow profiling of epigenetic alterations at a much higher sensitivity and genomic scale previously not possible. With the advancement of these new technologies, such as next-generation sequencing, and with the development of platforms for global epigenome analyses, the critical epigenetic alterations involved in tumorigenesis and other diseases and developmental processes will be identified. As we advance our ability to define and elucidate how epigenetic modifications are involved in the etiology of diseases, this will enhance our ability to use the assessment of epigenetic changes as biomarkers of diagnosis, prognosis, and therapy response and to bring epigenetic biomarkers into clinical use.

References

Baylin SB, Jones PA (2011) A decade of exploring the cancer epigenome – biological and translational implications. Nat Rev Cancer 11:726–734

Enokida H, Shiina H, Urakami S, Igawa M, Ogishima T, Li LC, Kawahara M, Nakagawa M, Kane CJ, Carroll PR, Dahiya R (2005) Multigene methylation analysis for detection and staging of prostate cancer. Clin Cancer Res 11:6582–6588

Jemal A, Bray F, Center MM, Ferlay J, Ward E, Forman D (2011) Global cancer statistics. CA Cancer J Clin 61:69–90

Jerónimo C, Henrique R, Hoque MO, Mambo E, Ribeiro FR, Varzim G, Oliveira J, Teixeira MR, Lopes C, Sidransky D (2004) A quantitative promoter methylation profile of prostate cancer. Clin Cancer Res 10:8472–8478

Lee WH, Morton RA, Epstein JI, Brooks JD, Campbell PA, Bova GS, Hsieh WS, Isaacs WB, Nelson WG (1994) Cytidine methylation of regulatory sequences near the pi-class glutathione S-transferase gene accompanies human prostatic carcinogenesis. Proc Natl Acad Sci U S A 91:11733–11737

Nakayama M, Bennett CJ, Hicks JL, Epstein JI, Platz EA, Nelson WG, De Marzo AM (2003) Hypermethylation of the human glutathione S-transferase-pi gene (GSTP1) CpG island is present in a subset of proliferative inflammatory atrophy lesions but not in normal or hyperplastic epithelium of the prostate: a detailed study using laser-capture microdissection. Am J Pathol 163:923–933

Neal DE, Donovan JL (2000) Prostate cancer: to screen or not to screen? Lancet Oncol 1:17–24

Seligson DB, Horvath S, Shi T, Yu H, Tze S, Grunstein M, Kurdistani SK (2005) Global histone modification patterns predict risk of prostate cancer recurrence. Nature 435:1262–1266

Wu T, Giovannucci E, Welge J, Mallick P, Tang WY, Ho SM (2011) Measurement of GSTP1 promoter methylation in body fluids may complement PSA screening: a meta-analysis. Br J Cancer 105:65–73

Epigenetic Gene Silencing

Kenneth P. Nephew
School of Medicine, Indiana University,
Bloomington, IN, USA

Definition

Epigenetic: A heritable change in gene expression that is not accompanied by changes in DNA sequence. The two most studied ▶ epigenetic phenomena are DNA methylation and modifications of histone tails.

Characteristics

The term epigenetics was originally coined to describe the development of phenotype from genotype. The field of epigenetics now encompasses DNA methylation, covalent modifications of histones, nucleosome–DNA interactions, and small inhibitory RNA molecules. While it is well documented that genetic changes, such as mutations and deletions, play a functional role in silencing tumor suppressor genes in cancer cells, the functional importance of covalent epigenetic modifications, such as DNA methylation and histone posttranslational modifications, is becoming increasingly recognized as an important early event during carcinogenesis and tumor development.

Epigenetics and Cancer

Epigenetic alterations are widely observed in cancer, and it has been shown that epigenetic mechanisms can be important to all phases of the cancer process, including tumor initiation, tumor progression, and maintenance of the malignant state of cancer cells (Fig. 1). Tumors exhibit two characteristic changes in DNA methylation patterns. One is genome-wide hypomethylation, primarily in repeat elements and pericentromeric regions, which is responsible for genomic instability. The other is promoter hypermethylation of normally protected ▶ CpG islands near the transcription start site of genes, which is responsible for transcriptional inactivation. Another characteristic of tumors includes hypoacetylation of histones, and HATs have been reported as up- or downregulated in a number of tumors. In this entry, both DNA methylation and histone modifications are discussed in the context of cancer.

Epigenetic Gene Silencing, Fig. 1 Schematic representation of cancer progression. Characteristic epigenetic features of cancer are listed. During tumor progression, accumulation of both genetic and epigenetic abnormalities contributes to carcinogenesis

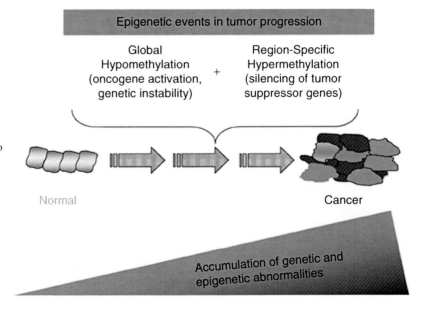

Epigenetic events in tumor progression

Global Hypomethylation (oncogene activation, genetic instability) + Region-Specific Hypermethylation (silencing of tumor suppressor genes)

Normal   Cancer

Accumulation of genetic and epigenetic abnormalities

DNA Methylation

The transfer of a methyl group to the carbon 5 position of a cytosine residue within the context of a CpG dinucleotide is the only known epigenetic modification of DNA itself. The 5-methylcytosine is the best-studied epigenetic modification and often referred to as the "fifth base" present in DNA. Although 5-methylcytosine comprises approximately 1% of the human genome, it is underrepresented in the bulk of the genome due to spontaneous deamination to thymine, which is not recognized by DNA mismatch repair systems. The CpG dinucleotides that are present are almost always methylated in normal cells, as 5-methylcytosine is widely believed to act to "silence" expression and/or retrotransposition of parasitic repeat sequences, such as Alu repeats and long-interspersed elements (LINEs) found throughout the genome. Aberrations in DNA methylation patterns are firmly associated with cancer, as evidenced by vast alterations in methylation patterns that occur during tumorigenesis, including global loss of methylation at CpG dinucleotides and region-specific hypermethylation of distinct regions of 5-methylcytosine, located within specific CG-rich sequences known as CpG islands, often found within the promoter and associated with active genes. About 60% of human genes are associated with unique CpG islands, and it has been estimated that the human genome contains about 29,000 CpG islands. These normally unmethylated CpG islands may become methylated in cancer cells, and the event is associated with loss of expression of flanking genes. It has also been estimated that aberrant methylation can accumulate in as many as 10% of the CpG islands during tumor development. Thus, abnormal de novo methylation of CpG islands in human cancer cells represents one of the most prevalent molecular markers yet identified, and the list of methylated genes identified in various tumor types continues to grow. DNA is methylated by DNMTs, a family of proteins. DNMT1 is responsible for the maintenance of DNA methylation after each round of replication. De novo methylation of DNA is the responsibility of DNMT3a and DNMT3b.

Histone Modifications and Chromatin Remodeling

While DNA methylation is currently the best-studied epigenetic modification, other well-known chromatin alterations play critical roles in normal and aberrant physiological processes, including cancer. Histones are the major components of chromatin, the complex of DNA and protein found within the nucleus of eukaryotic cells. Early structural studies of chromatin revealed that 146 base pairs of DNA are "wrapped" around a core nucleosome, the fundamental unit of chromatin. A nucleosome consists of an octamer of histones, with 50 base pairs of "linker" DNA between repeating octamers. The N-terminal "tail" regions of histones extend from the nucleosome core octamer and are subject to various posttranslational modifications. Histone tails can be acetylated and deacetylated by HATs and HDACs, respectively. These enzymes are gene activators and repressors, and histone acetylation/deacetylation is important in transcriptional regulation. Additional histone modifications include methylation, by HMTs, phosphorylation, sumoylation, and ubiquitination. Overall, these and other modifications of histone tails play a critical role in chromatin compaction (e.g., tightly vs. loosely compacted chromatin), which can determine the access of various factors involved in transcription and gene expression and whether a gene is switched on (activated) or off (repressed). Furthermore, establishing transcriptional silencing of a gene involves a close interplay between histone modifications and DNA methylation, and loss and gain of DNA methylation, histone acetylation, and methylation are observed in cancer cells. The sum total of these covalent alterations in the epigenome has been referred to as the "histone code," which can be "written" by the various modifying enzymes and "read" by various binding proteins that act to further modify chromatin and/or alter gene expression. According to the histone code hypothesis, the combination of DNA and histone modifications allows genes to go from the activate or inactive state interchangeably. The relatively new discipline of ▶ epigenomics promises to reveal novel insights into the histone code and a better understanding of normal development and human disease, including cancer.

Clinical Relevance

Unlike genetic changes, the epigenetic changes in cancer are potentially reversible. The ability to reactivate epigenetically silenced tumor suppressor genes and key control pathways and reverse the cancer cell phenotype is a promising strategy. Epigenetic drugs include both DNMT inhibitors ("demethylating" agents) and HDAC inhibitors (agents that cause "hyperacetylation" of histones). These agents can be used singly or in combination with currently available cancer chemotherapies. Epigenetic therapies themselves are now approved for various hematological malignancies and currently being studied in clinical trials for solid tumors. Furthermore, the combination of DNMT inhibitors with HDAC inhibitors has shown synergistic re-expression of epigenetically silenced genes and inhibition of tumor growth. Moreover, many investigators have shown that DNMT inhibitors with HDAC inhibitors can act to resensitize drug-resistant cancer cells to standard chemotherapeutic and hormonal agents. The ability to measure biochemical responses to epigenetic drugs, such as demethylation of previously hypermethylated genes, and correlate this with clinical responses has also been shown, and epigenetic therapies for chemoprevention in individuals with aberrant epigenetic alterations but have not yet developed cancer are an exciting possibility.

Distinct CpG island methylation profiles for various cancers continue to emerge; consequently, aberrant DNA methylation and histone modification patterns are now being investigated as potential biomarkers and pathway-specific therapeutic targets. Thus, epigenetic profiling of tumors could provide new therapeutic targets and epigenetic biomarkers for prognosis, such as predicting therapy outcome in patients with cancer or, ideally, early cancer detection. Clearly, attractive and promising clinical possibilities exist for epigenetic-based therapies.

In summary, perhaps all human cancers are at least partially associated with epigenetic dysregulation of gene expression, forming a rational basis for future treatment strategies designed to alter this fundamental processes in cancer. Comprehensive elucidation of epigenetic modifications, in both normal and diseased tissues, will allow for an extensive understanding of gene regulatory networks that control both normal and cancer phenotypes. The realization of the Human Epigenome Project (HEP), proposed as an exhaustive annotation of all histone and deoxycytosine modifications throughout the human genome, should have a major impact on cancer.

Cross-References

▶ CpG Islands
▶ Epigenetic
▶ Epigenomics

References

American Association for Cancer Research Human Epigenome Task Force and European Union, Network of Excellence, Scientific Advisory Board (2008) Moving AHEAD with an international human epigenome project. Nature 454(7205):711–715. doi:10.1038/454711a

Egger G, Liang G, Aparicio A, Jones PA (2004) Epigenetics in human disease and prospects for epigenetic therapy. Nature 429:457–463

Feinberg AP, Tycko B (2004) The history of cancer epigenetics. Nat Rev Cancer 4:143–153

Herman JG, Baylin SB (2003) Gene silencing in cancer in association with promoter hypermethylation. N Engl J Med 349:2042–2054

Nephew KP, Huang TH (2003) Epigenetic gene silencing in cancer initiation and progression. Cancer Lett 190:125–133

Toyota M, Issa JP (2005) Epigenetic changes in solid and hematopoietic tumors. Semin Oncol 32:521–530

See Also

(2012) DNA methylation. In: Schwab M (ed) Encyclopedia of cancer, 3rd edn. Springer, Berlin/Heidelberg, p 1140. doi:10.1007/978-3-642-16483-5_1682

(2012) DNMTs. In: Schwab M (ed) Encyclopedia of cancer, 3rd edn. Springer, Berlin/Heidelberg, p 1147. doi:10.1007/978-3-642-16483-5_1698

(2012) Epigenome. In: Schwab M (ed) Encyclopedia of cancer, 3rd edn. Springer, Berlin/Heidelberg, p 1290. doi:10.1007/978-3-642-16483-5_1949

(2012) HAT. In: Schwab M (ed) Encyclopedia of cancer, 3rd edn. Springer, Berlin/Heidelberg, pp 1632–1633. doi:10.1007/978-3-642-16483-5_2573

(2012) HDACs. In: Schwab M (ed) Encyclopedia of cancer, 3rd edn. Springer, Berlin/Heidelberg, p 1635. doi:10.1007/978-3-642-16483-5_2592

Epigenetic Modifications

Definition

Reversible, heritable changes in gene regulation that occur without a change in DNA sequence.

Cross-References

► Epigenetic

Epigenetic Therapy

Debby Hellebrekers and Manon van Engeland
Department of Pathology, GROW-School for
Oncology and Developmental Biology,
Maastricht University Hospital, Maastricht, The
Netherlands

Definition

Epigenetic therapy refers to therapy using inhibitors of DNA ► methylation and histone deacetylation, which reverse these epigenetic modifications. These compounds inhibit tumor growth in vitro and in vivo, which is thought to be due to reactivation of epigenetically silenced ► tumor suppressor genes by inhibition of promoter DNA methylation and histone deacetylation of these genes.

Characteristics

DNA Methylation and Histone Deacetylation
► Epigenetic modifications regulate heritable changes in gene expression without changing the primary DNA sequence. The best-studied epigenetic mechanisms are DNA methylation and posttranslational histone modifications. DNA methylation is the covalent addition of a methyl group to the DNA, predominantly to the base

cytosine 5' to guanine, also called a CpG dinucleotide. CpG dinucleotides are clustered in small stretches of DNA called ► CpG islands, often located in or near the promoter region of approximately half of all human genes. Methylation of CpG dinucleotides, which occurs nonrandomly, is an important ► epigenetic gene silencing mechanism. Most methylation in the human genome occurs in the noncoding DNA, preventing the transcription of repeat elements, inserted viral sequences, and transposons. In contrast, CpG islands are largely unmethylated in both expressing and nonexpressing tissues under normal conditions. Exceptions to this unmethylated state of CpG islands involve the silenced gene alleles for imprinted genes and genes located on the inactive X chromosome of females. DNA methylation is catalyzed by DNA methyltransferases (DNMTs). DNA methylation can induce gene silencing through several mechanisms. By sterically hindering the binding of activating transcription factors to gene promoters, DNA methylation can directly repress gene transcription. Another mechanism is through recruitment of several methyl-binding domain proteins (MBDs) that recognize methylated DNA, including MeCP2, MBD1–4, and Kaiso. These proteins themselves can repress gene transcription or bind proteins which cause gene silencing. The DNA helix is wrapped around a core of histone proteins. The basic amino-terminal tails of histones are subject to various posttranslational modifications, including acetylation, methylation, phosphorylation, ubiquitination, sumoylation, ADP-ribosylation, glycosylation, biotinylation, and carbonylation. The best-characterized histone modification is histone acetylation, which is controlled by histone acetyltransferases (HATs) and histone deacetylases (HDACs). Histone acetylation generally correlates to active gene transcription, whereas histone deacetylation is associated with transcriptional repression, by blocking accessibility of transcription factors to their binding sites.

DNA methylation and histone deacetylation are interconnected in gene silencing. Methyl-binding domain proteins are components of HDAC complexes or recruit these complexes to methylated DNA, resulting in transcriptional silencing. Furthermore, a much more direct connection between DNA methylation and histone deacetylation exists by direct interactions between DNMTs and HDACs. DNA methylation and histone deacetylation are pivotal in X chromosome inactivation, ▶ imprinting, and establishment of tissue-specific gene expression. However, aberrant epigenetic regulation of gene expression also plays a major role in the development of human cancer.

Epigenetic Abnormalities in Cancer

Aberrant epigenetic silencing of tumor suppressor genes by promoter DNA hypermethylation and histone deacetylation plays an important role in the pathogenesis of cancer. According to Knudson's two-hit model, complete loss of function of a tumor suppressor gene requires loss of function of both gene copies. Epigenetic silencing of the wild-type allele of a tumor suppressor gene by aberrant promoter hypermethylation and histone deacetylation can be considered as the second hit in this model, resulting in complete loss of function of the gene. It has become apparent that many genes, located across all chromosome locations, are epigenetically silenced in cancer cells. Examples are genes involved in cell cycle regulation and apoptosis (*p14ARF*, *p15INK4b*, *p16INK4a*, *APC*, *RASSF1A*, and *HIC1*), DNA repair genes (*hMLH1*, *GSTP1*, *MGMT*, and *BRCA1*), and genes related to ▶ metastasis and invasion (*CDH1*, *TIMP-3*, *DAPK*, *p73*, *maspin*, *TSP1*, and *VHL*).

DNA Methyltransferase and Histone Deacetylase Inhibitors

In contrast to genetic alterations, which are almost impossible to reverse, ▶ chromatin remodeling in cancers is potentially reversible. This resulted in the development of pharmacologic inhibitors of DNA methylation and histone deacetylation. By inducing DNA demethylation and histone acetylation, DNMT and HDAC inhibitors can reverse epigenetic silencing of tumor suppressor genes, resulting in reactivation of these genes in tumor cells and restoring of crucial cellular pathways. The most extensively studied DNMT inhibitors

are 5-azacytidine and 5-aza- 2′-deoxycytidine [5-aza-2′deoxycytidine and Cancer], which were initially developed as chemotherapeutic agents. These nucleoside analogs are incorporated into DNA in place of the natural base cytosine during DNA replication and are therefore only active during S phase. Once incorporated into the DNA, a complex is formed with active sites of DNMTs, thereby covalently trapping these enzymes. This results in the depletion of active enzymes and the demethylation of DNA after several cell divisions. 5-Aza- 2′-deoxycytidine is the most commonly used DNMT inhibitor in assays with cultured cells. This compound reactivates dormant tumor suppressor genes by demethylation of their hypermethylated promoter, thereby restoring their normal function. This seems to be a widespread effect of 5-aza-2′-deoxycytidine, because all cancer cell lines studied so far are sensitive to the DNA demethylating effects of this agent. Reactivation of silenced tumor suppressor genes might be the mechanism by which this compound suppresses growth and induces differentiation of human tumor cell lines. Examples of other DNMT inhibitors are the cytidine analogs 5,6-dihydro-5-azacytidine, 5-fluoro-2′-deoxycytidine, and zebularine, the small molecule RG108, which blocks the DNMT active site, and MG98, an antisense oligonucleotide that specifically inhibits DNMT1 mRNA.

By inhibiting histone deacetylation, HDAC inhibitors cause accumulation of acetylated histones, leading to increased transcription of previously silenced tumor suppressor genes in malignant cells. Both naturally existing and synthetic HDAC inhibitors have been characterized. The effects of HDAC inhibitors on gene expression in transformed cells are selective; only about 2–10% of all known genes are affected by these agents. One gene most consistently induced by HDAC inhibition is *CDKN1A*, which encodes the cell cycle inhibitor ▶ p21. HDAC inhibitors can also relieve inappropriate transcriptional repression mediated by chimeric oncoproteins, such as PML-RARα, thereby inducing differentiation in cells harboring these translocations. HDAC inhibitors have many antitumor effects including induction of cell cycle arrest, differentiation, and/or apoptosis in virtually all cultured transformed cell types and in cells from different tumors.

Epigenetic Therapy in Cancer

It is clear from in vitro and preclinical studies that the clinical application of reversing epigenetic aberrations in tumor cells, called epigenetic therapy, is an exciting strategy for cancer treatment. Many agents have been discovered that inhibit DNA ▶ methylation or histone deacetylation, and the value of these compounds will be established by ongoing ▶ clinical trials.

5-Azacytidine (Vidaza) and 5-aza-2'-deoxycytidine (Decitabine) represent the two most prominent inhibitors of DNA methyltransferases that are being used in clinical practice, and these drugs have been approved by the FDA for the treatment of myelodysplastic syndrome (MDS). The use of DNMT inhibitors in the treatment of MDS results from the knowledge that epigenetic gene silencing of, in particular, *p15INK4b* is present in poor-risk MDS subtypes and often predicts transformation to acute myeloid leukemia (AML). Multiple HDAC inhibitors are currently being tested in patients through intravenous or oral administration. Suberoylanilide hydroxamic acid (SAHA) is one of the HDAC inhibitors most advanced in development. Encouraging results were obtained in Phase I, II, and III clinical trials for patients with both hematologic and solid tumors. Other examples of HDAC inhibitors undergoing clinical testing in a range of solid and hematological malignancies are ▶ valproic acid, PXD101, NVP-LAQ824, LBH589, depsipeptide, MS-275, and CI-994. Complete targeting of epigenetic gene regulation might require a combination of chromatin modifying agents. The synergy between demethylating drugs and HDAC inhibitors in reactivation of epigenetically silenced tumor suppressor genes in vitro makes combined treatment with DNMT and HDAC inhibitors a promising epigenetic therapy. Reduction of individual doses should minimize toxic effects and optimize the therapeutic response of such combination. Encouraging anticancer activity of epigenetic therapy has been shown particularly in the treatment of hematologic disorders, but their effectiveness in solid tumors largely remains to be determined.

In addition to the inhibitory effects of DNMT and HDAC inhibitors on tumor cells, by reactivation of epigenetically silenced tumor suppressor genes, these compounds have also been described to inhibit tumor endothelial cell growth and tumor angiogenesis. ▶ Angiogenesis is required for tumor growth to a size of approximately 2 mm^3 but is also instrumental for tumor cells to metastasize to other locations in the body. Angiogenesis is considered to be a promising target of anticancer treatment. By influencing the gene expression profile of tumor cells, DNMT and HDAC inhibitors target genes, which are regulating angiogenesis. Among the epigenetically silenced tumor suppressor genes in tumor cells are genes with angiogenesis inhibiting properties. By reexpression of these genes in tumor cells, DNMT and HDAC inhibitors might indirectly – via the tumor cells – exhibit angiostatic effects in vivo. An example of these genes is *p16INK4a*, which can regulate angiogenesis by modulating ▶ vascular endothelial growth factor (VEGF) expression by the tumor. Another epigenetically silenced tumor suppressor gene that inhibits angiogenesis by downregulation of VEGF is *p73*. The tumor suppressor ▶ maspin, which is often silenced in tumors by epigenetic promoter modifications, is also an effective inhibitor of angiogenesis. Methylation-associated inactivation of the angiogenesis inhibiting factors tissue inhibitor of metalloproteinase-2- and -3 (*TIMP-2/3*) is frequent in many human tumors. Thrombospondin-1 (*TSP-1*) has been described to be repressed by epigenetic promoter modifications in several adult cancers and can be reactivated by 5-aza- 2'-deoxycytidine. The secreted protease ADAMTS-8 (METH-2) has anti-angiogenic properties, which can specifically suppress endothelial cell proliferation, and significant downregulation of *ADAMTS-8* by promoter hypermethylation, which has been described in different tumor types. Besides the indirect effects of DNMT and HDAC inhibitors on tumor angiogenesis, these compounds also directly inhibit endothelial cell growth and angiogenesis in vitro

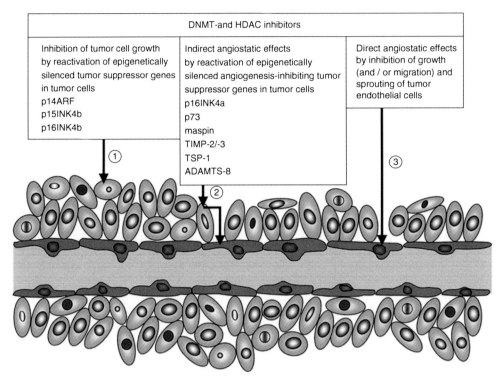

Epigenetic Therapy, Fig. 1 Antitumor effects of DNA methyltransferase and histone deacetylase inhibitors in vivo. (1) DNMT and HDAC inhibitors decrease tumor cell growth by reactivation of epigenetically silenced tumor suppressor genes in tumor cells. (2) Release of transcriptional repression of angiogenesis inhibiting tumor suppressor genes in tumor cells might result in indirect angiostatic effects of DNMT and HDAC inhibitors. (3) DNMT and HDAC inhibitors directly decrease endothelial cell growth and angiogenesis, thereby exhibiting direct angiostatic effects

and in vivo. Potent anti-angiogenic activity has been described for several HDAC inhibitors, such as trichostatin A (TSA), SAHA, depsipeptide, valproic acid, butyrate, apicidin, LBH589, and NVP-LAQ824, as well as for the DNMT inhibitors 5-aza- 2′-deoxycytidine and zebularine. These drugs suppress spontaneous or VEGF-induced angiogenesis in different in vitro, ex vivo, and in vivo angiogenesis assays.

Clearly, the dual targeting of epigenetic therapy in cancer treatment, inhibiting both tumor cells as well as tumor angiogenesis, makes them suitable combinatorial anticancer therapeutics (Fig. 1). By targeting multiple genes and pathways in tumor cells, as well as endothelial cell biology and angiogenesis, DNMT and HDAC inhibitors decrease the development of resistance that is associated with many of the current chemotherapeutic drugs and anti-angiogenic drugs.

Despite the promising data from clinical trials, there are several pitfalls regarding the clinical application of epigenetic therapy. An important side effect that should be taken into account in the use of these drugs is induction of global hypomethylation, which might induce tumorigenesis by aberrant activation of repetitive DNA sequences, transposons and ▶ oncogenes, induction of chromosomal instability, and mutagenesis. Furthermore, the existence of many different DNMTs and HDACs makes the development of selective inhibitors that target individual enzymes imperative.

Cross-References

▶ Angiogenesis
▶ Chromatin Remodeling

- ▸ Clinical Trial
- ▸ CpG Islands
- ▸ Epigenetic
- ▸ Epigenetic Gene Silencing
- ▸ Imprinting
- ▸ Knudson Hypothesis
- ▸ Maspin
- ▸ Metastasis
- ▸ Methylation
- ▸ Oncogene
- ▸ p21
- ▸ Tumor Suppressor Genes
- ▸ Valproic Acid
- ▸ Vascular Endothelial Growth Factor

References

Egger G, Liang G, Aparicio APA et al (2004) Epigenetics in human disease and prospects for epigenetic therapy. Nature 429:457–463

Folkman J (2006) Antiangiogenesis in cancer therapy – endostatin and its mechanisms of action. Exp Cell Res 312:594–607

Hellebrekers DM, Griffioen AW, van Engeland M (2007) Dual targeting of epigenetic therapy in cancer. Biochim Biophys Acta 1775:76–91

Herman JG, Baylin SB (2003) Gene silencing in cancer in association with promoter hypermethylation. N Engl J Med 349:2042–2054

Minucci S, Pelicci PG (2006) Histone deacetylase inhibitors and the promise of epigenetic (and more) treatments for cancer. Nat Rev Cancer 6:38–51

See Also

(2012) Anti-Angiogenic Drugs. In: Schwab M (ed) Encyclopedia of Cancer, 3rd edn. Springer Berlin Heidelberg, pp 207–208. doi:10.1007/978-3-642-16483-5_302

(2012) Chimeric Oncoproteins. In: Schwab M (ed) Encyclopedia of Cancer, 3rd edn. Springer Berlin Heidelberg, p 811. doi:10.1007/978-3-642-16483-5_1095

(2012) Chromatin. In: Schwab M (ed) Encyclopedia of Cancer, 3rd edn. Springer Berlin Heidelberg, p 825. doi:10.1007/978-3-642-16483-5_1125

(2012) DNA Methyltransferases. In: Schwab M (ed) Encyclopedia of Cancer, 3rd edn. Springer Berlin Heidelberg, p 1140. doi:10.1007/978-3-642-16483-5_6997

(2012) FDA. In: Schwab M (ed) Encyclopedia of Cancer, 3rd edn. Springer Berlin Heidelberg, p 1386. doi:10.1007/978-3-642-16483-5_2136

(2012) Histone Deacetylation. In: Schwab M (ed) Encyclopedia of Cancer, 3rd edn. Springer Berlin Heidelberg, p 1702. doi:10.1007/978-3-642-16483-5_2753

(2012) Histone Proteins. In: Schwab M (ed) Encyclopedia of Cancer, 3rd edn. Springer Berlin Heidelberg, p 1706. doi:10.1007/978-3-642-16483-5_2761

(2012) Nucleoside Analogs. In: Schwab M (ed) Encyclopedia of Cancer, 3rd edn. Springer Berlin Heidelberg, p 2577. doi:10.1007/978-3-642-16483-5_4164

(2012) PML-RARa. In: Schwab M (ed) Encyclopedia of Cancer, 3rd edn. Springer Berlin Heidelberg, p 2931. doi:10.1007/978-3-642-16483-5_4645

(2012) Tumor Endothelial Cell. In: Schwab M (ed) Encyclopedia of Cancer, 3rd edn. Springer Berlin Heidelberg, p 3795. doi:10.1007/978-3-642-16483-5_6026

E

Epigenomics

William Chi-Shing Cho
Department of Clinical Oncology, Queen Elizabeth Hospital, Kowloon, Hong Kong

Definition

Epigenomics is the study based on the comprehensive analyses of the entire epigenome using high-throughput technologies, including physical modifications, associations and conformations of genomic DNA sequences. The complete set of the epigenetic landscape in a cell is known as epigenome, which is tissue specific, developmentally regulated, and highly dynamic. This variability offers a potential explanation for individual differences in phenotype. Aberrant epigenetic marks are associated with a range of complex pathologies, including ▸ cancer. The field of epigenomics involves chromatin, the three-dimensional complex of DNA, protein, and ▸ noncoding RNAs that determines the accessibility of DNA by the transcriptional machinery.

Characteristics

The Epigenome and Cancer

The term ▸ epigenetics was first used by Prof. Conrad H. Waddington in 1942 as part of his model of how cell fates are established during development. It usually refers to reversible biochemical alterations of DNA and related proteins.

Although it represents memories of molecular decisions that can be perpetuated through cell divisions, it does not change the DNA sequence. As epigenetic factors (such as DNA ▶ methylation, ▶ histone modification, histone variant, ▶ chromatin remodeling, and noncoding RNA) can influence chromatin structure and dynamics, they play a significant role in determining cell fate and are important epigenetic mechanisms that regulate gene expressions. Posttranslational modification of chromatin facilitates conformational transition between transcriptionally permissive and suppressive chromatin structures. Dysregulation of chromatin modification leads to pathologic alteration in gene transcription, hence to human disease.

DNA methylation is binary at the cellular level, i.e., a CpG site in a given allele in a single cell is either methylated or unmethylated. Methylation of cytosine in the context of CpG dinucleotides is an epigenetic phenomenon in eukaryotes that plays a vital role in genome function and transcription regulation, including host defense of endogenous parasitic sequences, embryonic development, transcription, X chromosome inactivation, and ▶ genomic imprinting. Aberrant change in DNA methylation is a main feature of several human diseases, such as cancer and neurological disorders.

Epigenetic modifications in the human genome can regulate and impact the expression of coding genes and noncoding genes, such as ▶ microRNA (miRNA), at the transcriptional level; they thus play a critical role in cancer initiation and development. Some of the epigenetic changes have been well-characterized, in which a range of modifications (including methylation, acetylation, phosphorylation, ▶ ubiquitination, deamination, citrullination, sumoylation, ADP-ribosylation and isomerization) of histones' N-terminal tails are involved. The enzymatic activities of chromatin modifiers contribute to the epigenetic regulation of gene expression through the remodeling of chromatin. In addition, there are at least 16 distinct classes of histone modifications and more than a hundred different modification sites in the major core histones.

Since anomalous DNA methylation and histone modification are often seen in cancer, epigenetic change is considered to be a characteristic of tumorigenesis. It is believed that changes in DNA methylation are molecular signatures of the tumor itself. As DNA methylation heterogeneity may differentiate the cancer group from the unaffected individuals even before other histopathological changes appear, DNA methylation ▶ biomarkers circulating in blood can be used for early screening and the follow-up of high-risk subjects.

miRNAs can transcriptionally cleave or translationally repress the expressions of major enzymes participating in epigenetic processes, such as DNA methylation and histone modifications. With the regulatory effect on gene expression, miRNAs involve in many vital biological and cellular processes, including proliferation, differentiation, ▶ apoptosis, metabolism and viral infection. They are thus promising therapeutic targets for cancer interventions which may influence the epigenetic states associated with cancer.

Technologies

Bisulfide treatment converting unmethylated cytosine to uracil followed by cloning and Sanger sequencing is the gold standard for DNA methylation analysis of individual genes, which is a quantitative, allelic, contiguous, and base-pair resolution of CpG methylation approach. Development of high-throughput technology and the systematic assessment of accumulated data allow the identification of previously unknown biological processes and disease states in terms of whole-genome profiles of epigenetic signatures at a high resolution.

Focusing on the discovery of specific genetic risk factors, genome-wide association studies have identified many disease-associated loci. It helps us to understand the genetic basis of complex traits and motivates us to explore the epigenetic contribution to interindividual variation in complex phenotypes. The application of genome-wide methylation arrays has proved very informative to investigate both clinical and biological questions in human epigenomics. There are a number of informative functional genomics and epigenomics assays available in cancer research, such as DNA methylome for profiling genome-wide DNA methylation status, HELP tagging using massively parallel sequencing technology for genome-wide methylation profiling, DNase-seq for identifying genome-wide accessible chromatin regions that are hypersensitive to DNase I cleavage, ChIP-seq for investigating genome-wide transcription factor binding or histone modifications, as well as GRO-seq for studying genome-wide nascent RNA and transcriptional rate.

Utilization of high-throughput sequencing also facilitates many applications to understand the transcriptional and epigenetic gene regulations. Rapid advances in the profiling of sequencing-based DNA methylation have enabled comprehensive comparison of the complete DNA methylome between cancer cells and normal tissues. The sequencing-based methods permit the measurement of DNA methylation in interspersed repeat sequences, which is unattainable using microarray platform. The advancement in next-generation sequencing is proceeding in a breathtaking pace. Notable increases in sequence reads per lane and read length, along with the development of paired-end sequencing, have created optimism for the future of sequencing-based methylome mapping.

Epigenomic Researches and Therapies

Aiming to catalog the epigenomic patterns across different cell and tissue types using next-generation sequencing-based methodologies, the National Institute of Health Roadmap Epigenomics was launched in 2008. A new epigenomics database has been established at the National Center for Biotechnology Information to serve as a comprehensive public resource for epigenetic and epigenomic data sets (www.ncbi.nlm.nih.gov/epigenomics). This database collects data from several large-scale projects, including the National Institute of Health Roadmap Epigenomics project, the ENCODE and modENCODE projects, as well as from smaller single laboratory studies. It enables the epigenomic data to be readily available and easily accessed by the scientific community.

Growing data demonstrate that environmental and lifestyle factors can affect epigenetic patterns, such as pollution, alcohol consumption, depression, smoking, diet, obesity, and physical activity. In addition, aging is likely to drift with a loss of DNA methylation and an increased hypermethylation of CpG islands. For example, small but statistically significant age-related demethylation of CpG motifs in the tumor necrosis factor promoter is detected in human peripheral blood cells and primary monocyte-derived macrophages from healthy subjects using pyrosequencing. Unlike genetic modifications, epigenetic alterations are potentially reversible. Exciting new developments in medicinal chemistry deliver an extensive collection of potential therapeutic compounds against chromatin regulators, which may target the catalytic activity of diverse chromatin modifiers/remodelers and chromatin-reading/chromatin-binding modules. There are a number of DNA demethylation agents emerging as therapeutics in the clinic.

The epigenetic treatments targeting against the catalytic activity of chromatin-modifying enzymes have proved to be successful, and they have become a new anticancer therapy. As DNA demethylation therapeutics is emerging as potential clinical agents, there is a pressing need to identify DNA demethylation resistant genes and clarify their role in cancer treatment. Rapid advancements in high-throughput epigenomic technologies have led us to a new era in epigenetic biology which enhances our understanding of cancer cells. Along with the promising anticancer effects shown in some inhibitors of enzymes controlling epigenetic changes, the development of epigenetic-based therapies possesses great potential in the clinical applications against cancer.

Cross-References

► Apoptosis
► Cancer Epigenetics
► Carcinogenesis
► Chromatin Remodeling
► Epigenetic
► Epigenetic Gene Silencing
► Genomic Imprinting

► Histone Modification
► Methylation
► MicroRNA
► Noncoding RNA
► Ubiquitination

References

Baylin SB, Jones PA (2011) A decade of exploring the cancer epigenome – biological and translational implications. Nat Rev Cancer 11:726–734

Cancer Genome Atlas Research Network (2013) Genomic and epigenomic landscapes of adult de novo acute myeloid leukemia. N Engl J Med 368:2059-2074

Cho WC (2010) MicroRNAs in cancer – from research to therapy. Biochim Biophys Acta 1805:209–217

Jiang W, Liu N, Chen XZ, et al (2015) Genome-wide identification of a methylation gene panel as a prognostic biomarker in nasopharyngeal carcinoma. Mol Cancer Ther 14:2864–2873

Mullard A (2015) The Roadmap Epigenomics Project opens new drug development avenues. Nat Rev Drug Discov 14:223–225

Roadmap Epigenomics Consortium, Kundaje A, Meuleman W, et al (2015) Integrative analysis of 111 reference human epigenomes. Nature 518:317–330

Taudt A, Colomé-Tatché M, Johannes F (2016) Genetic sources of population epigenomic variation. Nat Rev Genet 17:319–332

See Also

(2012) Acetylation. In: Schwab M (ed) Encyclopedia of cancer, 3rd edn. Springer, Berlin/Heidelberg, p 17. doi:10.1007/978-3-642-16483-5_24

(2012) Biomarkers. In: Schwab M (ed) Encyclopedia of cancer, 3rd edn. Springer, Berlin/Heidelberg, pp 408–409. doi:10.1007/978-3-642-16483-5_6601

(2012) Hypermethylation. In: Schwab M (ed) Encyclopedia of cancer, 3rd edn. Springer, Berlin/Heidelberg, p 1784. doi:10.1007/978-3-642-16483-5_2910

(2012) Phosphorylation. In: Schwab M (ed) Encyclopedia of cancer, 3rd edn. Springer, Berlin/Heidelberg, p 2870. doi:10.1007/978-3-642-16483-5_4544

(2012) Sumoylation. In: Schwab M (ed) Encyclopedia of cancer, 3rd edn. Springer, Berlin/Heidelberg, p 3562. doi:10.1007/978-3-642-16483-5_5572

(2012) Transcription factor. In: Schwab M (ed) Encyclopedia of cancer, 3rd edn. Springer, Berlin/Heidelberg, p 3752. doi:10.1007/978-3-642-16483-5_5901

Epithelial Cadherin

► E-Cadherin

Epithelial Carcinogenesis

► Epithelial Tumorigenesis

Epithelial Cell Adhesion Molecule

► EpCAM

Epithelial Tumorigenesis

Vassilis Gorgoulis and Eleni A. Georgakopoulou
Department of Histology and Embryology,
Faculty of Medicine, National and Kapodistrian
University of Athens, Athens, Greece

List of Abbreviations

AKT	Nonspecific name abbreviation for a group of serine/threonine kinases involved in many cellular processes
APC	Adenomatous polyposis coli
EGFR	Epidermal growth factor receptor
GSTP	Glutathione S-transferase P enzymes
HPV	Human papillomavirus
HRAS	Harvey sarcoma viral oncogene
KRAS	Kirsten rat sarcoma viral oncogene homologue
LOH	Loss of heterozygosity
MDR1	Multidrug resistance 1 gene
MGMT	O^6-Methylguanine DNA methyltransferase
MLF1	Myeloid leukemia factor 1
MYC	Cellular myelocytomatosis viral oncogene
PTEN	Phosphatase and tensin homologues
RARRES	Retinoic acid receptor responder protein
RASSF2	Ras association domain family member 2
TGF-β	Transforming growth factor-beta
WNT	Wingless-type MMTV (mouse mammary tumor virus) integration site family

Synonyms

Carcinoma pathogenesis; Epithelial carcinogenesis

Definition

Epithelial tumorigenesis relates to the process of developing and/or progressing of a tumor, originating from epithelial cells. Malignant tumors of this origin are known as carcinoma, whereas malignant tumors derived from cells of the connective tissue are known as sarcoma (Fig. 1).

Characteristics

Introduction

Epithelial cells are found in most organs of the human body, and 90% of human cancers originate from epithelial cells. Epithelial cancers (carcinomas) are the most common cancers among elderly, in contrast to mesenchymal cancers and ► hematological malignancies that are more common in younger patients and children. Characteristic epithelial cancers are ► skin cancer and cancers that originate from cells of the epithelial lining of the aerodigestive and genitourinary system.

Epithelial cancers in a simple classification may be of three types:

- Squamous cell carcinomas (e.g., skin carcinoma, mouth carcinoma)
- Adenocarcinomas (e.g., colon adenocarcinoma, lung adenocarcinoma)
- Transitional cell carcinomas (e.g., bladder cancer, renal carcinoma)

Predisposing Factors

Age is the common risk factor for all epithelial cancers as their incidence rises with age. Genetic predisposition in epithelial cancer may be associated with syndromes where inherited mutations lead to cancer development (Table 1). Usually, in the cases of inherited cancer syndromes, the genes affected are those controlling DNA damage repair mechanisms. An example is hereditary nonpolyposis colorectal cancer (HNPCC) in which inherited germline mutations affect the DNA mismatch repair system (*MMR*). Various agents and conditions have been associated with predisposition to epithelial cancers. Most studied are tobacco and ► alcohol consumption for aerodigestive tract cancers, HPV infection for cervical cancer, ultraviolet radiation for skin cancer, ► H. pylori infection for ► gastric cancer, and ► obesity and inflammatory bowel disease for colorectal cancer.

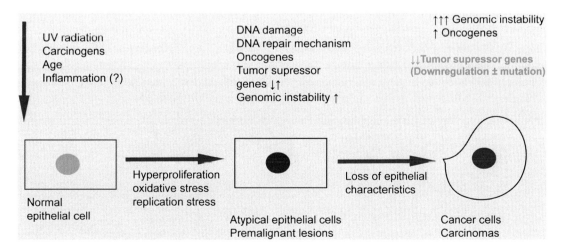

Epithelial Tumorigenesis, Fig. 1 Illustration of the main identified events that characterize epithelial tumorigenesis

Epithelial Tumorigenesis, Table 1 Some hereditary cancer syndromes, associated genetic mutations, and carcinomas

Cancer syndrome	Genes	Carcinomas
Hereditary breast and ovarian cancer syndrome	BRCA1, BRCA2	Breast, ovarian, prostate
Cowden syndrome	PTEN	Breast, thyroid, uterus, kidney
Hereditary nonpolyposis colon cancer	MLH1, MSH2, MSH6	Colorectal ovarian, hepatobiliary
Familial adenomatous polyposis	▸ APC	Colorectal
Xeroderma pigmentosum	XP A-D/ ERCC 4-6	Skin basal and squamous cell carcinoma (and melanoma)

Epithelial Tumorigenesis, Table 2 Common genetic changes in some epithelial cancers

Cancer	Genetic alterations in
Lung	▸ p53, MYC, EGFR
Colon	▸ APC, ▸ *KRAS*, ▸ TP53
Oral	MYC, ▸ TP53
Gastric	MLF1, MGMT, p16, RASSF2
Bladder	▸ *HRAS*, ▸ *KRAS*, RB
Breast	BRCA1, BRCA2
Cervical	EGFR, AKT, PTEN
Prostate	GSTP, RARRES, MDR1

The Multistage Pathogenesis Model of Epithelial Tumorigenesis

In almost all cancers of epithelial origin, a "pattern" of multistage pathogenesis has been identified. This model comprises an initiation phase, a promotion phase, a premalignant phase, and finally malignant progression. It requires the interaction of normal cells with carcinogenic agents which lead to genetic alterations. Alterations in the affected cells are the cause of their clonal expansion and accumulation of further alterations. (Common genetic alterations in epithelial cancers are summarized in Table 2.) A critical point of genetic alterations may lead to autonomy in growth and invasive phenotype. Such cancer models have been described in several carcinomas including oral cancer, colon cancer, and skin cancer. They have been reproduced under laboratory conditions by the repeated application of chemical carcinogens in mouse models.

Example of Multistage Carcinogenesis in Colorectal Cancer

Initiation → promotion → premalignancy → cancer

Normal cell → hyperproliferation → early and late adenoma → carcinoma

Changes in APC/KRAS → LOH in 18q → LOH 17p

Multistage carcinogenesis requires the accumulation of mutations that either enhance the expression of oncogenes or block the expression of tumor suppressor genes and is in accordance with the "mutation theory" for epithelial tumorigenesis. Commonly altered genes in epithelial cancers are the tumor suppressor genes encoding the proteins ▸ p53 and RB (chromosome location 17p and 13q, respectively). P53, often referred to as the "guardian of the genome," is one of the most studied tumor suppressor genes. P53 is inactivated in many cancers, and its role is crucial in cell cycle control. The RB gene is a tumor suppressor gene whose protein is a crucial cell cycle regulator. The Rb protein interacts with ▸ cyclin D and ▸ cyclin-dependent kinases 6 and 4 to allow the cell cycle to progress. Rb is also an E2F transcription factor regulator. The Rb protein is a target of the HPV viral protein E7, which binds to Rb and inhibits E2F control, thus allowing for malignant progression.

Multiple "hits" in different genes are required for the development of malignant cell clones. According to the "mutation theory," mutations are the events responsible for cancer development. However, the role of the epithelial microenvironment in the pathogenesis of epithelial cancer has been highlighted. Nevertheless, mutations especially in the genes that control DNA repair and cell cycle are key events for genomic instability, one of the hallmarks of malignant and premalignant lesions.

Apart from mutations, another characteristic of the premalignant phase and cancer in comparison to normal epithelium is the presence of ▸ DNA damage as well as ▶ DNA damage response activation. In premalignant lesions it is possible that oncogenes may cause DNA replication stress (unprogrammed and defective DNA replication) resulting in DNA damage (e.g., in the form of DNA double-strand breaks). This concept is supported by the fact that DNA double-strand breaks and DNA repair mechanisms are present in cancers and precancerous lesions. P53 activation in premalignant lesions could be a result of DNA damage checkpoint or ▶ *arf* action, and this activation suppresses tumorigenesis. P53 mutations relate to progression of premalignant lesions to cancer.

The Role of Stem Cells in Epithelial Tumorigenesis

The initial cellular origin of epithelial malignant clones is a matter of debate. Adult stem cells have been identified in carcinomas and are named ▶ cancer stem cells. Their role is uncertain. It has been proposed that cancer stem cells are the only cells that can develop clonal populations when they accumulate a critical amount of genetic alterations, in contrast to the concept that any cell may acquire clonal properties. Furthermore, cancer stem cells have been blamed for the resistance of many epithelial cancers to chemotherapy and radiotherapy.

Epigenetic Changes and Epithelial Tumorigenesis

Epigenetic changes are the changes that affect the expression of genes without changing the DNA sequence. Such changes are hypermethylation and hypomethylation of promoter regions and histone deacetylation. Hypermethylation of tumor suppressor genes is a characteristic of many epithelial cancers. DNA hypermethylation may be the result of exogenous carcinogens and may be conserved during mitosis. Also, histone deacetylation results in different histone types that have various functions, e.g., as DNA damage marker (▶ H2AX variant) or gene promoter regions (H2AZ variant). Epigenetic changes may also be identified in DNA areas containing genes that control cell death, cell cycle regulation, or DNA repair. Accumulation of epigenetic changes

contributes to genomic instability and sequential carcinogenesis.

The Interaction of Cellular Microenvironment in Epithelial Tumorigenesis

The epithelial microenvironment is composed of non-epithelial cells that form the supportive stroma of the epithelium (fibroblasts, lymphocytes, endothelial cells in vessels), the extracellular matrix, and produce a variety of growth factors and cytokines. The interaction between stroma and epithelial cells is important during embryogenesis for the organization and differentiation of the various epithelia in different systems of the human body. These interactions have also been proven important for the development and progression of epithelial cancer.

Stroma fibroblasts are involved in epithelial tumor development by remodeling the extracellular matrix and by providing the epithelial tumor cells with the required growth factors signals to enhance their autonomy in proliferating signals.

Epithelial Mesenchymal Transition (EMT)

An important role in the epithelial integrity is played by the junctional structures (gap junctions, desmosomes, hemidesmosomes) that adhere epithelial cells to each other and to the stromal structures, as well as the apicobasal polarity of epithelial cells and the stable direction of their division in parallel with the epithelial sheet. These features are characteristic of the epithelial phenotype. Changes in these cellular characteristics and development of a mesenchymal phenotype are a characteristic of advanced epithelial cancers. This phenomenon described as ▶ epithelial mesenchymal transition enables epithelial malignant cells to invade the connective tissue and also to metastasize in distant organs through the lymphatic or hematogenous route. The most studied epithelial mesenchymal transition molecules are ▶ TGF-β, Twist, ▶ Snail/Slug, ZEB, and ▶ Wnt. In the same sense, epithelial cells also accumulate motility and plasticity as they transform to a mesenchymal phenotype which enables malignant epithelial cells to migrate to surrounding stroma. A characteristic of EMT is the downregulation of the cellular adhesion

E

molecule E-cadherin. E-cadherin is normally located in the cellular membrane of epithelial cells and functions as a blockage for the development and infiltration of malignant cells.

Chronic Inflammation and Epithelial Cancers

Several epithelial cancers develop on the basis of previous chronic inflammation. Such carcinomas are gastric cancers following chronic gastritis, esophageal cancer following chronic esophageal inflammation, and colorectal cancer on the basis of inflammatory bowel disease. The chronic inflammation causes oxidative stress and damage to DNA that result in malignant transformation. The inflammatory mediators produced by the inflammatory cellular populations support the cancer cells with growth signals and enable tumor progression.

Summary

Epithelial tumorigenesis is a complex multifactorial process. Cellular alterations that result in malignant changes are a combination of mutations, DNA damage epigenetic changes, and interactions of malignant cells with their microenvironment.

Cross-References

► Alcohol Consumption
► Cyclin D
► Cyclin-Dependent Kinases
► DNA Damage
► DNA Damage Response
► Epidermal Growth Factor Receptor
► Epithelial-to-Mesenchymal Transition
► Gastric Cancer
► Helicobacter Pylori in the Pathogenesis of Gastric Cancer
► *HRAS*
► *KRAS*
► Lynch Syndrome
► Mismatch Repair in Genetic Instability
► MYC Oncogene
► Skin Cancer
► Tobacco Carcinogenesis
► Tobacco-Related Cancers

► TP53
► Transforming Growth Factor-Beta
► Xeroderma Pigmentosum

References

Fearon ER (2011) Molecular genetics of colorectal cancer. Annu Rev Pathol 6:479–507
Frank SA (2004) Genetic predisposition to cancer – insights from population genetics. Nat Rev Genet 5:764–772
Halazonetis TD, Gorgoulis VG, Bartek J (2008) An oncogene-induced DNA damage model for cancer development. Science 7:1352–1355
McCaffrey LM, Macara IG (2011) Epithelial organization, cell polarity and tumorigenesis. Trends Cell Biol 21:727–735
Shan W, Yang G, Liu J (2009) The inflammatory network: bridging senescent stroma and epithelial tumorigenesis. Front Biosci 14:4044–4057

See Also

(2012) AKT. In: Schwab M (ed) Encyclopedia of cancer, 3rd edn. Springer, Berlin/Heidelberg, p 115. doi:10.1007/978-3-642-16483-5_163
(2012) APC. In: Schwab M (ed) Encyclopedia of cancer, 3rd edn. Springer, Berlin/Heidelberg, p 234. doi:10.1007/978-3-642-16483-5_347
(2012) ARF. In: Schwab M (ed) Encyclopedia of cancer, 3rd edn. Springer, Berlin/Heidelberg, p 265. doi:10.1007/978-3-642-16483-5_383
(2012) Cancer stem cells. In: Schwab M (ed) Encyclopedia of cancer, 3rd edn. Springer, Berlin/Heidelberg, p 626. doi:10.1007/978-3-642-16483-5_815
(2012) Carcinoma. In: Schwab M (ed) Encyclopedia of cancer, 3rd edn. Springer, Berlin/Heidelberg, p 657. doi:10.1007/978-3-642-16483-5_848
(2012) Colorectal cancer. In: Schwab M (ed) Encyclopedia of cancer, 3rd edn. Springer, Berlin/Heidelberg, p 916. doi:10.1007/978-3-642-16483-5_1265
(2012) EGFR. In: Schwab M (ed) Encyclopedia of cancer, 3rd edn. Springer, Berlin/Heidelberg, p 1211. doi:10.1007/978-3-642-16483-5_1828
(2012) Epigenetic changes. In: Schwab M (ed) Encyclopedia of cancer, 3rd edn. Springer, Berlin/Heidelberg, p 1283. doi:10.1007/978-3-642-16483-5_1942
(2012) Epithelial cell. In: Schwab M (ed) Encyclopedia of cancer, 3rd edn. Springer, Berlin/Heidelberg, pp 1291–1292. doi:10.1007/978-3-642-16483-5_1958
(2012) Genomic instability. In: Schwab M (ed) Encyclopedia of cancer, 3rd edn. Springer, Berlin/Heidelberg, p 1539. doi:10.1007/978-3-642-16483-5_2391
(2012) Helicobacter pylori. In: Schwab M (ed) Encyclopedia of cancer, 3rd edn. Springer, Berlin/Heidelberg, p 1639. doi:10.1007/978-3-642-16483-5_2606

(2012) Loss of heterozygosity. In: Schwab M (ed) Encyclopedia of cancer, 3rd edn. Springer, Berlin/Heidelberg, pp 2075–2076. doi:10.1007/978-3-642-16483-5_3415

(2012) MSH2-6. In: Schwab M (ed) Encyclopedia of cancer, 3rd edn. Springer, Berlin/Heidelberg, p 2383. doi:10.1007/978-3-642-16483-5_3860

(2012) MYC family. In: Schwab M (ed) Encyclopedia of cancer, 3rd edn. Springer, Berlin/Heidelberg, p 2426. doi:10.1007/978-3-642-16483-5_3922

(2012) Obesity. In: Schwab M (ed) Encyclopedia of cancer, 3rd edn. Springer, Berlin/Heidelberg, p 2595. doi:10.1007/978-3-642-16483-5_4185

(2012) P53. In: Schwab M (ed) Encyclopedia of cancer, 3rd edn. Springer, Berlin/Heidelberg, p 2747. doi:10.1007/978-3-642-16483-5_4331

(2012) Sarcoma. In: Schwab M (ed) Encyclopedia of cancer, 3rd edn. Springer, Berlin/Heidelberg, p 3335. doi:10.1007/978-3-642-16483-5_5161

(2012) Slug. In: Schwab M (ed) Encyclopedia of cancer, 3rd edn. Springer, Berlin/Heidelberg, p 3439. doi:10.1007/978-3-642-16483-5_5354

(2012) Tobacco. In: Schwab M (ed) Encyclopedia of cancer, 3rd edn. Springer, Berlin/Heidelberg, pp 3716–3717. doi:10.1007/978-3-642-16483-5_5844

(2012) Wnt. In: Schwab M (ed) Encyclopedia of cancer, 3rd edn. Springer, Berlin/Heidelberg, p 3953. doi:10.1007/978-3-642-16483-5_6255

Knowlton CA, Mackay MK, Huth BJ, Roedel C (2013) Hereditary nonpolyposis colorectal cancer (HNPCC). In: Brady LW, Yaeger TE (eds) Encyclopedia of radiation oncology. Springer, Berlin/Heidelberg, p 312. doi:10.1007/978-3-540-85516-3_481

Epithelial-to-Mesenchymal Transition

Ayyappan K. Rajasekaran and Sigrid A. Langhans
Nemours Center for Childhood Cancer Research, Alfred I duPont Hospital for Children, Wilmington, DE, USA

List of Abbreviations

ECM	Extracellular matrix
EMT	Epithelial-mesenchymal transition
MET	Mesenchymal-epithelial transition
NF-κB	Nuclear factor κB
SMA	Smooth muscle actin
TGF-β_1	Transforming growth factor-β_1

Definition

Phenotypic alterations in which epithelial cells adopt characteristics of mesenchymal cells. Epithelial-mesenchymal transition (**EMT**) is a physiological process during normal development and a pathological process during cancer progression and fibrosis.

Characteristics

Epithelial cells line external surfaces and internal cavities of the body. A distinguishing characteristic of epithelial cells is the presence of junctional complexes, such as ▶ tight junctions, ▶ adherens junctions, and desmosomes, and segregation of plasma membrane into apical and basolateral domains. These features promote adhesion, restrict motility, facilitate intercellular communication, and permit individual cells to function as a cohesive unit. The phenotype of epithelial cells cultured in vitro or in tissues is often described as well differentiated (Fig. 1, left panel).

Mesenchymal cells are spindle shaped with fibroblast-like morphology, lack adhesiveness, and are highly motile. They do not have junctional complexes and specialization of the plasma membrane into apical and basolateral domains. When cells of epithelial origin show a mesenchymal phenotype under in vitro culture conditions or in tissues, they are often described as poorly differentiated (Fig. 1, right panel).

The plasticity of epithelial cells enables them to convert between the epithelial and mesenchymal phenotypes. These phenotypic transformations are highly regulated by specific signaling events and molecules. Conversion of the epithelial cell to a mesenchymal phenotype is known as epithelial-mesenchymal transition (EMT) and vice versa as mesenchymal-epithelial transition (**MET**). EMT and MET occur during normal development as well as in cancer progression.

EMT provides a mechanism for epithelial cells to overcome the physical constraints imposed upon them by intercellular junctions and to adopt a motile phenotype. The process was originally identified during specific stages of

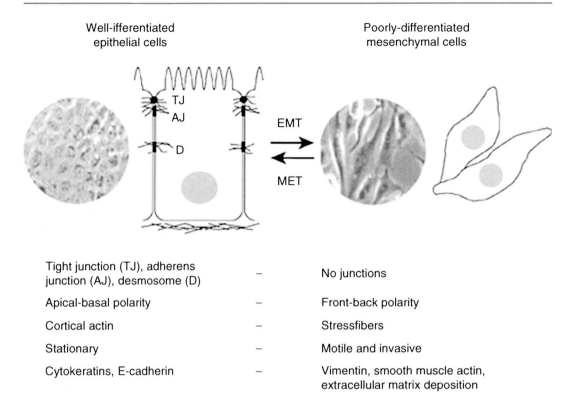

Well-ifferentiated epithelial cells		Poorly-differentiated mesenchymal cells
Tight junction (TJ), adherens junction (AJ), desmosome (D)	–	No junctions
Apical-basal polarity	–	Front-back polarity
Cortical actin	–	Stressfibers
Stationary	–	Motile and invasive
Cytokeratins, E-cadherin	–	Vimentin, smooth muscle actin, extracellular matrix deposition

Epithelial-to-Mesenchymal Transition, Fig. 1 Trans-itions between well-differentiated epithelial cells and poorly differentiated mesenchymal cells during EMT. *Left panel*, phase contrast microscopy and schematic dia-gram of well-differentiated polarized epithelial cells with characteristic apical-basolateral polarity and junctional complexes. *Right panel*, cells that have undergone EMT display mesenchymal morphology with no junctions and are highly motile and invasive. The process of reversion of mesenchymal cells to an epithelial phenotype is called mesenchymal-epithelial transition (*MET*)

embryogenesis in which epithelial cells migrate and colonize various embryonic territories to form different organs. EMT is critical for the formation of ectoderm, mesoderm, and endoderm during gastrulation as well as for the differentiation of neural crest cells into neurons and glia of the peripheral nervous system. During embryogene-sis EMT is spatially and temporally regulated in a subtle manner that is essential for normal organ development. Activation of the EMT program depends on the convergence of multiple cues that are both intrinsic to the cell and received from the **microenvironment**.

EMT during cancer progression occurs in an aggressive and uncontrolled fashion and might facilitate the invasive and metastatic potentials of cancer cells. The phenotypic conversion of epi-thelial cells to mesenchymal cells involves a series of events that includes dissolution of ▶ tight junc-tions, ▶ adherens junctions, and desmosomes, the suppression of molecules involved in restricting invasiveness and motility, and induction of factors that promote invasiveness and motility and gain of **stem cell** attributes. EMT can also occur as a partial transition when the phenotypic conversion is not complete. A characteristic feature of cells undergoing EMT in culture is the change in the organization of the **actin cytoskeleton**. In most cases stress fibers are induced with a concomitant loss of the cortical actin ring. Although it is established that such morphological changes accompany EMT, the chronology of these events is still not deciphered. It is also not known whether all these changes are essential for induction of EMT and the metastatic potential of cancer cells.

Mechanisms

EMT is in part achieved by downregulation of epithelial-specific molecules and induction of proteins expressed in mesenchymal cells. One of the epithelial cell molecules extensively studied that change during EMT is the cell-cell adhesion molecule ► E-cadherin. During epithelial morphogenesis, ► E-cadherin regulates the establishment of ► adherens junctions, which form a continuous adhesive belt below the apical surface. The extracellular domain of ► E-cadherin mediates calcium-dependent homotypic interactions with ► E-cadherin molecules on adjacent cells, and the intracellular domain binds cytosolic catenins and links the ► E-cadherin complex to the actin cytoskeleton. A stable ► E-cadherin complex at the plasma membrane is essential for the cell-cell adhesion function of this protein.

Several studies have shown that expression of ► E-cadherin is reduced during EMT, associated with the loss of junctional complexes and the induction of a mesenchymal phenotype of carcinoma cells. It is believed that the decrease in adhesive force following reduced expression of ► E-cadherin facilitates invasion and dispersion of carcinoma cells from the primary tumor mass. Methods to abolish ► E-cadherin function promote epithelial cell invasion into a variety of substrates, as determined by a number of in vitro and in vivo experimental systems. Loss or reduced expression of ► E-cadherin is also accompanied by expression of mesenchymal markers such as vimentin, smooth muscle actin (**SMA**), γ-actin, β-filamin, and talin and extracellular matrix (**ECM**) components such as fibronectin and collagen precursors. Upregulation of these proteins facilitates cytoskeletal remodeling and promotes cell motility (Table 1).

The diverse molecular mechanisms mediated by growth factors and extracellular matrix proteins contribute to EMT. Growth factors such as epidermal growth factor (EGF), hepatocyte growth factor (HGF), or insulin growth factor II (IGF-II) promote signaling cascades through their cognate receptor tyrosine kinases, which in turn signal through various downstream effector molecules such as Ras, Src, phosphatidylinositol-3-

Epithelial-to-Mesenchymal Transition, Table 1 Markers of EMT

	Increased abundance/activity
Epithelial-mesenchymal markers	N-Cadherin
	Vimentin
	Fibronectin
	Smooth muscle actin
	γ-Actin
	β-Filamin
	Talin
	Collagen precursors
	MMPs
	Stress fibers
Signal transduction molecules/ pathways	Epidermal growth factor (EGF)
	Fibroblast growth factor (FGF)
	Hepatocyte growth factor (HGF)
	Insulin growth factor (IGF)
	Platelet-derived growth factor (PDGF)
	Transforming growth factor (TGF-β)
	Ras
	Src
	PI3K
	Wnt
	Notch
	Hedgehog
	GSK-3β
	MAPK
	TNFα
	NFκB
	Smurf-1
	miR-138
	miR-200
Transcriptional regulators	Snail
	Slug
	ZEB-1
	ZEB-2
	TCF/LEF
	Smads
	Twist1/2
	E12/E47
	DNA methylation
	Histone acetylation
	Decreased abundance/activity
Epithelial-mesenchymal markers	E-cadherin
	Cytokeratin
	Claudin
	Occludin
	Desmoplakin
	Desmoglein

kinase (PI3K), and MAPK leading to EMT. In addition, signaling pathways essential for stem cell function during development such as the Wnt, ► Notch, and ► Hedgehog signaling pathways are activated during EMT. The role of ▶ Wnt signaling has been well established during normal development as well as in EMT. Binding of the soluble ligand Wnt to its receptor frizzled inhibits β-catenin degradation and facilitates its nuclear translocation together with the TCF/LEF transcription factors to activate transcription of target genes such as cyclin D1 and Myc. The transcriptional activity of β-catenin is increased in a wide variety of cancers as well as in growth factor-induced EMT of cultured cells.

A key molecule involved in the induction of EMT and extensively studied is the ▶ transforming growth factor-$β_1$ **(TGF-$β_1$)**. The TGF-β growth factor superfamily comprises TGF-βs, bone morphogenetic proteins (BMPs), activins, and other related proteins. TGF-β induces EMT in epithelia either through transcriptional- or transcription-independent mechanisms. Cooperation between TGF-β and Ras/Raf/MEK/MAPK signaling is involved in the induction and maintenance of EMT. TGF-β has been shown to stimulate ERK1/2 activity in cell culture models of EMT that is required for the disassembly of junctional complexes and the induction of motility. TGF-β also activates phosphatidylinositol-3-kinase (PI3K) in a RhoA-dependent manner, which has been implicated in the disassembly of ▶ tight junctions. In keratinocytes and several epithelial cell types, TGF-β treatment activates the ▶ Notch pathway by inducing the ▶ Notch ligand jagged1 and the ▶ Notch target genes TLE3, HEY1, HEY2, and HES1 at the onset of EMT. TGF-β, in cooperation with oncogenic Ras, induces EMT by the activation of the transcription factor ▶ nuclear factor κ B **(NF-κB)**. Constitutive activation of NF-κB induces EMT and metastasis, whereas inhibition of NF-κB by inhibitory IκBa suppresses EMT and metastasis in a breast tumor model. Although in different cell types TGF-β induces various signaling pathways, these signals subsequently target ▶ E-cadherin expression and the disassembly of epithelial junctional complexes to induce EMT. For example, the TGF-β type I receptor is localized to ▶ tight junctions through the ▶ tight junction protein occludin allowing for efficient TGF-β-dependent dissolution of ▶ tight junctions during EMT. The epithelial polarity protein Par-6 interacts with the TGF-β type I receptor and TGF-β binding initiates Par-6 phosphorylation and activation of the E3-ubiquitin ligase, Smurf-1. Activated Smurf-1 promotes degradation of local RhoA resulting in ▶ tight junction dissociation, inhibition of cell adhesion, and transition to a mesenchymal phenotype.

The signaling cascades described above induce two major types of transcriptional regulators that mediate EMT, zinc finger (Snail, Slug, ZEB-1, ZEB-2) and basic helix-loop-helix (Twist, E12/E47) proteins. The transcription suppressors SNAI1 (Snail) and SNAI2 (Slug) play a central role in the induction of EMT. These zinc-finger proteins recognize E-box elements in the cognate target promoters, and SNAI1 represses the transcription of the ▶ E-cadherin gene during EMT as well as embryonic development. Factors that regulate SNAI1 by phosphorylation, subcellular localization, and transcription have been well described in development and EMT. While phosphorylation of SNAI1 in the two GSK3β phosphorylation consensus motifs targets it for export from the nucleus (motif 2) and ubiquitinylation and degradation (motif 1), phosphorylation of SNAI1 at Ser^{246} by p21-activated kinase (PAK1) results in its accumulation in the nucleus and induction of EMT. LIV-1, an estrogen-regulated member of the LZT subfamily of zinc transporters, is activated by STAT3, which is essential for nuclear localization of SNAI1 and suppression of ▶ E-cadherin expression during gastrulation in zebrafish embryos. Further, SNAI1 expression is transcriptionally suppressed by metastasis-associated gene 3 (MTA3), a subunit of the Mi-2/NuRD transcriptional corepressor, thereby establishing a mechanistic link between estrogen receptor status and invasive growth of breast cancers. While there is great deal of knowledge about SNAI1 regulation, much less is known about SNAI2. It has been shown that SNAI2 suppresses ▶ E-cadherin expression when ectopically expressed in well-differentiated epithelial cells.

HGF and FGF induce SNAI2 to suppress desmoplakin and desmoglein, thereby destabilizing desmosomes. The SMAD-interacting repressors SIP-1/ZEB2 and δEF1/ZEB1 that can be induced by TGF-β bind to the ► E-cadherin promoter to suppress its transcription. The basic helix-loop-helix transcription factors involved in the induction of EMT are E12/E47 (E2A gene product) and Twist, both of which have been shown to repress ► E-cadherin expression and induce EMT. The mechanisms by which these factors suppress ► E-cadherin expression are not well established but recent work emphasized the importance of microRNAs and **epigenetic** factors.

Clinical Relevance

Although EMT represents a fundamentally important process for tumor dissemination and is widely believed to be an essential event involved in cancer metastasis, there are several lines of evidence to suggest that many invasive and metastatic carcinomas have not undergone a complete transition to a mesenchymal phenotype. Many advanced carcinomas of prostate, breast, squamous cell carcinomas derived from a variety of origins, including the esophagus, oral epithelium, lung, cervix, and salivary neoplasms, possess molecular and morphological characteristics of well-differentiated epithelial cells, with the presence of epithelial junctions and apical-basolateral plasma membrane asymmetry. High ► E-cadherin expression was also observed in a wide variety of carcinomas and ► E-cadherin levels did not correlate with invasiveness and metastasis. These results are consistent with the idea that complete EMT might not be necessary for cancer cell metastasis or that cancer cells redifferentiate to an epithelial phenotype following metastasis.

While EMT is well established in cultured cells, there is little evidence for EMT in vivo and if EMT occurs it is not known at what stage of tumor progression. There are several possibilities by which cancer cells could spread without undergoing complete EMT: (1) incomplete EMT by which epithelial cells partially convert to a mesenchymal phenotype acquiring invasive and metastatic potential, (2) cohort migration in which well-differentiated epithelial cells migrate as a cluster and cause metastasis, and (3) reversion of poorly differentiated cells to a well-differentiated phenotype by mesenchymal-epithelial transition (MET) at the site of metastasis. These diverse mechanisms might be regulated by the tumor microenvironment and/or signaling pathways distinct from the molecular machinery of EMT. Thus, there are several mechanisms by which cancer cells could metastasize and EMT may represent one of the global changes associated with malignant transformation of epithelial cells.

Recognizing EMT as a fundamentally important process for tumor dissemination together with the increasing knowledge about the molecular pathways leading to EMT may offer new targets for therapeutic intervention. Indeed, inhibitors of the TGF-β, ERK1/2, and PI3K/Akt pathways have shown encouraging results in the suppression of tumor progression. Further understanding of the molecular requirements of EMT will allow for more effective approaches for future therapeutic intervention.

Glossary

Actin cytoskeleton The actin cytoskeleton is a dynamic structure of actin bundles and networks in the cytoplasm that provides a framework to maintain cell shape, protects the cell, and enables cell locomotion. It also plays an important role in intracellular transport.

Epigenetics Mechanisms that impose a cellular phenotype without a change in its nucleotide sequence and largely achieved by covalent modification of DNA and histone proteins through methylation and acetylation.

Epithelial cell Epithelial cells line and protect both the outside and the inside cavities and lumen of the body. They regulate selective permeability and transcellular transport between the compartments they separate and are involved in secretion absorption and sensation detection.

Microenvironment Biophysical and biochemical factors in the immediate vicinity of a cell that directly or indirectly affect the behavior of a cell. The microenvironment is composed of extracellular matrix homotypic and heterotypic cells, soluble factors including cytokines, hormones and other bioactive agents, and mechanical forces.

Stem cell Undifferentiated cell capable of dividing and renewing itself for long time periods and with the potential to develop into different cell types.

Cross-References

▶ Adherens Junctions
▶ E-Cadherin
▶ Notch/Jagged Signaling
▶ Nuclear Factor-κB
▶ Tight Junction
▶ Wnt Signaling

References

Chapman HA (2011) Epithelial-mesenchymal interactions in pulmonary fibrosis. Annu Rev Physiol 73:413–435
De Craene B, Berx G (2013) Regulatory networks defining EMT during cancer initiation and progression. Nat Rev Cancer 13:97–110
Kalluri R, Weinberg RA (2009) The basics of epithelial-mesenchymal transition. J Clin Invest 119:1420–1428
Lamouille S, Xu J, Derynck R (2014) Molecular mechanisms of epithelial-mesenchymal transition. Nat Rev Mol Cell Biol 15:178–196
Lindsey S, Langhans SA (2014) Crosstalk of oncogenic signaling pathways during epithelial-mesenchymal transition. Front Oncol 4:358
Moreno-Bueno G, Portillo F, Cano A (2008) Transcriptional regulation of cell polarity in EMT and cancer. Oncogene 27:6958–6969
Tam WL, Weinberg RA (2013) The epigenetics of epithelial-mesenchymal plasticity in cancer. Nat Med 19:1438–1449
Thiery JP, Acloque H, Huang RY, Nieto MA (2009) Epithelial-mesenchymal transitions in development and disease. Cell 139:871–890

Epitheliasin (TMPRSS 2)

▶ Serine Proteases (Type II) Spanning the Plasma Membrane

Epithelium

Isabelle Gross
INSERM U1113, Université de Strasbourg, Strasbourg, France

Definition

The epithelium is one of the basic tissue types. It lines the cavities or covers surfaces of structures throughout the body and also forms many glands. Functions of epithelial cells include protection, secretion, selective absorption, transcellular transport, and detection of sensation.

Epithelia are classified by the morphology of their cells (columnar, squamous, cuboidal, or transitional) and the number of layers they are composed of (simple, stratified). They have the following structural and functional characteristics:

- Epithelia form continuous sheets of cells densely packed together through the presence of numerous specialized intercellular junctions. Tight junctions are specific to epithelia and contribute to their barrier function. Desmosomal and adherens junctions ensure mechanic stability through connections with cytokeratin intermediate filaments specific to epithelia and actin microfilaments of the cytoskeleton.
- Epithelia rest on a basement membrane called the basal lamina that acts as a scaffold and connects them to the underlying connective tissue that contains the blood vessels required for nutrient delivery and waste product disposal.
- Epithelial cells exhibit polarity, meaning that they have an asymmetric organization of the cell surface, intracellular organelles, and the cytoskeleton. The apical region is defined as the area lying above the tight junctions and contains the apical membrane that faces the lumen or the outer surface. The basolateral region is the side that is below the tight junctions and contains the basolateral membrane that is in contact with the basal lamina.

Cross-References

▸ Adherens Junctions

See Also

(2012) Epithelial cell. In: Schwab M (ed) Encyclopedia of cancer, 3rd edn. Springer, Berlin/Heidelberg, pp 1291–1292. doi:10.1007/978-3-642-16483-5_1958

Epo

▸ Erythropoietin

Epoetin

▸ Erythropoietin

Epothilone B Analogue

Robert P. Whitehead
Nevada Cancer Institute, Las Vegas, NV, USA

Synonyms

Azaepothilone B; BMS-247550; Ixabepilone; NSC-710428

Definition

The epothilone B analogue, ixabepilone, is the term used to denote one specific agent of a new class of anticancer drugs, the epothilones. The epothilones A and B are derived from fermentation of the myxobacteria *Sorangium cellulosum*. They have been found to have potent cytotoxic activity, which, like that of the taxanes, has been linked to the stabilization of cellular microtubules resulting in the blocking of cell division at the ▸ G2/M Transition portion of the cell cycle. The epothilone B analogue ixabepilone is a semisynthetic analogue of the natural product epothilone B, made by replacing the lactone oxygen of epothilone B with a lactam (azoepothilone B) designed to overcome the metabolic instability of the natural product.

Chemical name: [1S-[1R*,3R*(E),7R*, 10S*,11R*,16S*]]-7,11-Dihydroxy-8,8,10,12,16-pentamethyl-3-[1-methyl-2-(2-methyl-4-thiazolyl) ethenyl]-17-oxa-4-azabicyclo[14.1.0] heptadecane-5,9-dione

Molecular formula: $C_{27}H_{42}N_2O_5S$ M.W. (506.7 g/mole)

Characteristics

Introduction

The epothilones A and B are a new class of anticancer agents isolated in the mid-1990s. They are in the macrolide class of drugs but have a mechanism of action similar to the taxanes. ▸ Paclitaxel, the first taxane to be widely used, was found in 1971 in a screening assay for antitumor agents. It is a complex diterpene compound. Its cytotoxic activity derives from its binding to cellular microtubules stabilizing them and preventing the dynamic growth and shrinkage that occurs during normal cellular processes. In dividing cells, this leads to mitotic arrest and cell death by causing a block at the transition between G_2 and M phases of the cell cycle. Paclitaxel has been found to have important clinical activity against breast, ovarian, lung, and head and neck cancers. However, some tumors such as colorectal cancer have innate resistance to this agent. Other tumors develop resistance to paclitaxel through the multidrug resistance mechanism in which ▸ P-glycoprotein removes cytotoxic agents from the tumor cell or by genetic mutations leading to altered tubulin protein. Side effects such as ▸ neutropenia and ▸ peripheral neuropathy can limit its activity and usefulness in the clinic. Its low solubility requires that paclitaxel be administered in a Cremophor vehicle which itself can induce hypersensitivity reactions. Modifications to paclitaxel to improve its solubility or reduce side effects are difficult because of its complex ring structure. Therefore, when the epothilones A and B were found to have a mechanism of action similar to the taxanes, it was hoped that they might lead to more effective anticancer agents. It was found that the epothilones did show important cytotoxic activity against tumor cells when tested in in vitro cell assays, but when tested in in vivo models of cancer, only modest antitumor activity was present. This was found to be due to poor metabolic stability and other unfavorable characteristics. To overcome these problems, multiple semisynthetic analogues of the epothilones were made and tested. One of these analogues, BMS-247550, epothilone B analogue, or ixabepilone, was found to be the most effective epothilone in a variety of laboratory assays and was also active in paclitaxel-resistant tumor models. It has now been tested in the clinic and has been found to have activity in a variety of clinical cancers.

Preclinical Testing

Ixabepilone has shown potent cytotoxic activity when tested against a broad panel of tumor cell lines in vitro, including human P-glycoprotein family; colon carcinoma; breast, prostate, and lung cancers; ▶ squamous cell carcinoma; human leukemia; and mouse lung carcinoma. The concentration of drug at which 50% of the tested cells were killed (IC_{50}) ranged from 1.4 to 34.5 nM. Included in this panel were cell lines resistant to paclitaxel by either multidrug resistance (MDR) due to P-glycoprotein overexpression or due to mutations of β-tubulin, the two most common mechanisms of resistance to paclitaxel. Ixabepilone is more potent than paclitaxel in causing tubulin polymerization and has similar activity to that of the parental compounds, epothilones A and B. Ixabepilone maintained its activity in whole animal systems. Using paclitaxel-sensitive human ovarian or colon tumor cell lines in the nude mouse models, ixabepilone produces comparable log cell kill and tumor growth delay as paclitaxel, while in nude mouse or rat tumor models using paclitaxel-resistant human ovarian, breast, colon, or ▶ pancreatic cancer cell lines, ixabepilone produced much greater tumor log cell kill and tumor growth delay than did paclitaxel. In contrast to paclitaxel which is usually ineffective when given orally, ixabepilone was also active in paclitaxel-resistant human ovarian cancer and paclitaxel-sensitive human colon cancer nude mouse models when given by the oral route.

Phase I Studies in Cancer Patients

In animal studies, the main toxicities of ixabepilone were related to the gastrointestinal tract, peripheral neuropathy, and bone marrow toxicity. Similar to the taxanes, when tested in cancer patients, ixabepilone required the same solvent for i.v. use, a Cremophor-based formulation (ethanol plus polyoxyethylated castor oil) which can lead to hypersensitivity reactions. After this type of reaction occurred in a patient, subsequent patients were prophylactically treated with histamine-1 (H1) and histamine-2 (H2) blockers. Schedules tested included an every 21-day cycle, weekly administration, a daily times 5 every 21-day cycle, and a daily times 3 every 21-day cycle. In each of these schedules, the drug was administered by the intravenous route. All of the phase I trials showed antitumor responses. These occurred in patients with breast, non-small cell lung, and ovarian cancers and melanoma. Some of these patients had previous treatment with paclitaxel or ▶ docetaxel.

The 21-day cycle trials consisted of a 60-min i.v. infusion on day 1 repeated every 21 days. The dose-limiting toxicities for this schedule were neutropenia and sensory neuropathy. One study recommended a phase II dose of 50 mg/m^2 and another trial of the same schedule suggested 40 mg/m^2 as the phase 2 dose. Most subsequent trials have used the lower dose. Other toxicities commonly seen with this schedule included fatigue, arthralgias, myalgias, and vomiting.

With treatments using weekly infusions or daily times five or daily times three infusions, maximum tolerated doses were lower. Neutropenia, sensory neuropathy, fatigue, and hypersensitivity reactions were seen with the weekly schedules. With the daily times five or three schedules, dose-limiting toxicity was neutropenia with fewer hypersensitivity reactions and less severe neurotoxicity.

In patients given a 1-h infusion of ixabepilone, there was found to be bundling of microtubules in the peripheral blood mononuclear cells and this correlated with the plasma area under the curve, the concentration of drug measured in the blood multiplied by the time it is present. Similar microtubule bundle formation was seen in breast tumor cells obtained from a chest wall mass in a patient who showed a partial response after receiving drug on the 1-h infusion schedule. This patient was taxane refractory and the tumor expressed multidrug resistance protein. Cell death occurred in these tumor cells 23 h after the peak formation of microtubule bundles.

Phase II Studies in Cancer Patients
There is data from some phase II trials of ixabepilone in cancer patients. With ixabepilone given by a 1- or 3-h infusion on an every 21-day schedule, responses have been seen in a modest number of gastric or ► breast cancer patients previously treated with a taxane or ► non-small cell lung cancer patients who had previously received a platinum-based regimen. In breast cancer patients who were previously treated but had not received a taxane, a higher response rate of 34% was seen. A trial in colorectal cancer patients who had previously received an ► irinotecan-based regimen did not show any responses.

Multicenter phase II trials in the Southwest Oncology Group of ixabepilone in previously untreated patients with advanced pancreatic cancer or chemotherapy-naive patients with hormone-refractory prostate cancer have shown encouraging results that suggest that further testing is warranted.

Summary and Future Outlook
The epothilone B analogue ixabepilone is a cancer therapeutic agent with a mechanism of action similar to the taxanes, stabilization of cellular microtubules leading to mitotic arrest and cell death. However, it demonstrates antitumor effects against both taxane-sensitive and taxane-resistant tumors and is clinically active against a broad spectrum of tumor types. Further testing as a single agent or in combinations in previously untreated patients is needed. Phase III trials in which it is compared to standard regimens will further define its role in cancer treatment. Ixabepilone is felt to be an important new anticancer agent that may surpass the taxanes in usefulness.

Cross-References

► Breast Cancer
► Docetaxel
► Gastric Cancer
► G_2/M Transition
► Irinotecan
► Log-Kill Hypothesis
► Lung Cancer
► Neutropenia
► Non-Small-Cell Lung Cancer
► Paclitaxel
► Pancreatic Cancer
► Pancreatic Cancer Basic and Clinical Parameters
► Peripheral Neuropathy
► P-Glycoprotein
► Prostate Cancer
► Prostate Cancer Clinical Oncology
► Squamous Cell Carcinoma

References

Bollag DM, McQueney PA, Zhu J et al (1995) Epothilones, a new class of microtubule-stabilizing agents with a taxol-like mechanism of action. Cancer Res 55:325–2333

Goodin S, Kane MP, Rubin EH (2004) Epothilones: mechanism of action and biologic activity. J Clin Oncol 22:2015–2025

Lee FYF, Borzilleri R, Fairchild CR et al (2001) BMS-247550: a novel epothilone analog with a mode of action similar to paclitaxel but possessing superior antitumor efficacy. Clin Cancer Res 7:1429–1437

McDaid HM, Mani S, Shen HJ et al (2002) Validation of the pharmacodynamics of BMS-247550, an analogue of epothilone B during a phase I clinical study. Clin Cancer Res 8:2035–2043

Whitehead RP, McCoy S, Rivkin SE et al (2006) A phase II trial of epothilone B analogue BMS-247550 (NSC #710428) ixabepilone, in patients with advanced pancreas cancer: a Southwest Oncology Group study. Invest New Drugs 24:515–520

E

Epothilones

Definition

A group of microtubule-targeting agents, like taxanes, and vinca alkaloids. Epothilones were originally identified as metabolites produced by the common soil myxobacterium *Sorangium cellulosum*. They were found initially to have a narrow antifungal spectrum, but they also were found too toxic for use as an antifungal. Subsequently, their anticancer properties were detected. They are important and powerful options in the management of breast cancer and prostate cancer. The epothilones are a new class of cytotoxic molecules, including epothilone A, epothilone B, and epothilone D, identified as potential chemotherapy drugs. Early studies in cancer cell lines and in human cancer patients indicate superior efficacy to the taxanes. Their mechanism of action is similar to that of the taxanes, but their chemical structure is simpler and they are more soluble in water. Although taxane-based therapy has been used successfully, its effectiveness is often compromised by the emergence of ▶ drug resistance. Efforts to overcome drug resistance have led to the discovery of several novel anti-microtubule agents, including the epothilones. Epothilones exhibit broad antitumor activity similar to that of the taxanes, but they are less sensitive to known resistance mechanisms. The ongoing development of microtubule-targeting agents provides new strategies for overcoming taxane resistance and may improve clinical efficacy and patient outcomes.

Cross-References

▶ Drug Resistance

See Also

(2012) Microtubule. In: Schwab M (ed) Encyclopedia of cancer, 3rd edn. Springer, Berlin/Heidelberg, p 2308. doi:10.1007/978-3-642-16483-5_3734

(2012) Taxanes. In: Schwab M (ed) Encyclopedia of cancer, 3rd edn. Springer, Berlin/Heidelberg, pp 3614–3615. doi:10.1007/978-3-642-16483-5_6648

(2012) Vinca alkaloids. In: Schwab M (ed) Encyclopedia of cancer, 3rd edn. Springer, Berlin/Heidelberg, p 3908. doi:10.1007/978-3-642-16483-5_6187

Epstein-Barr Virus

Evelyne Manet
CIRI-International Center for Infectiology Research, INSERM U1111, Université Lyon 1, ENS de Lyon, Lyon, France

Synonyms

EBV; HHV4; Human herpesvirus 4

Definition

Epstein-Barr virus (EBV) was the first virus isolated from a human tumor, Burkitt lymphoma (BL). EBV is a lymphotropic γ-herpesvirus widely spread in the human population: 90–95% of adults have antibodies against the virus. In the majority of cases, the primary infection occurs within the first 3 years of life and is asymptomatic. When EBV infection occurs later in life, usually during adolescence, it results in the symptomatic illness known as infectious mononucleosis (IM). Infected individuals carry the virus all their life, in a very low number of lymphoid B cells (probably resting B cells) in their peripheral blood and lymphatic organs. Intermittent viral shedding occurs into the saliva, due to viral replication in the oropharyngeal lymphoid or epithelial tissues: saliva is the main transmission route of the virus. Since its first discovery in 1964 in Burkitt lymphoma tumor, EBV has been found to be associated with several other human malignancies including the undifferentiated ▶ nasopharyngeal carcinoma (NPC), ▶ Hodgkin disease, rare T-cell and natural killer (NK)-cell lymphomas, gastric carcinomas, and B- and T-cell lymphomas in immunocompromised individuals.

A characteristic unique to EBV is its capacity to induce the indefinite proliferation or immortalization of quiescent B lymphocytes, upon their infection in vitro.

Characteristics

In vitro Immortalization of Primary B Cells by EBV: The Growth Transcription Program

In vitro, infection of B cells by EBV is not productive but results in the outgrowth of latently infected lymphoblastoid cell lines (LCL) (Fig. 1). Such cell lines can also be obtained by culture of peripheral blood lymphocytes (PBL) from naturally infected individuals. The phenotype of LCLs (i.e., morphology and cell surface markers) is very similar to that of antigen-activated B cells, and a limited set of viral gene products is expressed: six nuclear proteins, the Epstein-Barr nuclear antigens (EBNA-1, EBNA-2, EBNA-3A (or EBNA-3), EBNA-3B (or EBNA-4), EBNA-3C (or EBNA-6), and EBNA-LP (or EBNA-5)), three integral membrane proteins (LMP-1, LMP-2A (or TP-1), and LMP-2B (or TP-2), two small non-polyadenylated nuclear RNAs (EBER-1 and EBER-2). This expression profile is referred to as the growth transcription program or latency III (Table 1) and over 25 miRNAs. In such immortalized cell lines, the viral DNA (a 172 kpb double-stranded DNA molecule) is maintained in the nucleus as multiple extrachromosomal copies of the viral episome. Among the nine proteins expressed in latency III, EBNA-1, EBNA-2, EBNA-3A, EBNA-3C, and EBNA-LP and

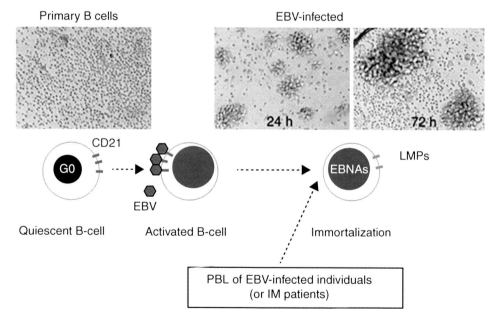

Epstein-Barr Virus, Fig. 1 Immortalization of B cells by EBV. EBV infection of resting primary human B cells in vitro causes cell cycle entry of the infected cells, establishment of latent viral infection, and conversion of the cell culture into permanently growing lymphoblastoid cell lines (*LCLs*), which progress to a fully immortalized state. Such LCLs can also be obtained by culture of PBL from individuals infected by EBV (mostly IM patients). The first step of infection is the interaction of the EBV gp350 glycoprotein with its receptor, the CD21. This induces a decondensation of the chromatin, a prerequisite for expression of a subset of the viral genes that cooperate to induce B-cell proliferation and the expression of several cellular activation markers and adhesion molecules. These latency genes code for three nuclear proteins, the Epstein-Barr virus nuclear antigens (EBNA1, −2, −3A, −3B, −3C, −LP), three integral membrane proteins (LMP1, LMP2A, LMP2B), plus two non-polyadenylated small nuclear RNAs (EBERs). The EBV genome is maintained in these proliferating cells as multiple extrachromosomal episomes and is replicated by cellular proteins, concomitantly with the cellular genome, via a DNA replication origin called *oriP*

Epstein-Barr Virus, Table 1 EBV gene latency transcription programs

Transcription program	Type of latency	EBV genes expressed	Occurrence
Growth	III	EBNA1, 2, 3a, 3b, 3c, LP, LMP1, LMP2a and LMP2b	PTLD, immunoblastic lymphoma (HIV)
Default	II	EBNA1, LMP1 and LMP2a	HD, NPC, gastric carcinoma
EBNA 1 only	I	EBNA1	BL, PEL
Latency	0	None	Memory B cells in peripheral blood

PTLD posttransplant lymphoma disease, *HD* Hodgkin disease, *NPC* nasopharyngeal carcinoma, *BL* Burkitt lymphoma, *PEL* primary effusion lymphoma

LMP-1 are essential for efficient transformation/immortalization of B lymphocytes in vitro. These proteins are thought to cooperate for the initiation and maintenance of B-cell proliferation in vitro.

EBNA-1 is a sequence-specific DNA-binding protein which binds to the EBV origin of replication (*OriP*). This interaction is required for viral DNA replication and for equilibrated distribution of the EBV episomes to the daughter cells during cell division. EBNA-1 may have other roles in EBV-induced oncogenesis as EBNA-1 transgenic mice display an increased incidence of B-cell lymphoma.

EBNA-2 is a transcriptional activator which regulates the expression of all EBV genes expressed in latency III as well as certain cellular genes including CD21, CD23, and cfgr. EBNA-2 does not bind DNA directly but is recruited to EBNA2-responsive elements by the cellular sequence-specific DNA-binding factor RBP-Jκ (also called CBF-1). As RBP-Jκ is part of the Notch signaling pathway, EBNA-2 may mimic part of Notch signal transduction (▶ NOTCH/JAGGED signaling in neoplasia). Although the exact function of EBNA-LP is still unknown, this nuclear factor has been found to cooperate with EBNA-2 for the transcriptional activation of the LMP-1 gene. Furthermore, co-expression of EBNA-2 and EBNA-LP in primary resting B cells previously activated through binding of the EBV glycoprotein gp350 to the EBV receptor CD21 induces cyclin D2 expression and drives resting B lymphocytes into the G1 phase of the cell cycle.

EBNA-3A, EBNA-3B, and EBNA-3C are related proteins but only EBNA-3A and EBNA-3C are essential for B-cell immortalization by EBV in vitro. However, these proteins have at least one common function: repression of EBNA2-activated transcription by directly contacting RBP-Jκ and inhibiting its binding to DNA. Furthermore, EBNA-3C is able to cooperate with activated (Ha)-Ras (RAS), to induce the proliferation of primary rat fibroblasts.

LMP-1 (▶ Epstein–Barr virus latent membrane protein 1) is an integral membrane protein with a very short 24 aa N-terminal cytoplasmic domain, six membrane spanning hydrophobic segments, and a 200 aa cytoplasmic C-terminal domain. LMP-1 transforms rodent fibroblast cell lines and Rat-1 cells expressing LMP-1 are tumorigenic in nude mice. LMP-1 acts as a constitutively activated member of the tumor necrosis factor receptor (TNRF) superfamily. It activates NF-κB transcription factor activity through a pathway that involves the recruitment of TNF-RI receptor-associated factors (TRAFs). It also induces ▶ AP-1 transcription factor activity via triggering of the c-jun N-terminal kinase (JNK). The STAT-1 transcription factor has also been shown to be a target of LMP1 which induces STAT-1 phosphorylation and thus its subsequent transfer to the nucleus.

LMP-2A is an integral membrane protein with a 119 aa hydrophilic N-terminal cytoplasmic tail which contains immunoreceptor tyrosine-based activation motifs (ITAMs) followed by 12 transmembrane domains and a 27 aa hydrophilic C-terminus. LMP-2B differs from LMP-2A by the lack of the N-terminal cytoplasmic domain. Although dispensable for in vitro immortalization of B lymphocytes, LMP-2A has an important role in the biology of EBV in vivo (see below) in mimicking the presence of a B-cell receptor (BCR) and providing important survival signals for B cells.

EBV Biological Cycle In Vivo

Although understanding of EBV infection biology in vivo is still rudimentary, it is believed to mimic the normal B-cell response to environmental antigen. EBV transits the epithelium and infects naive B cells in the underlying tissue. Expression of the latency genes (transcription growth program: Table 1) causes the cell to become activated, proliferate, and migrate to the follicle. After this initial clonal expansion, some EBV-infected cells undergo germinal center reactions. EBV transcription is then limited to EBNA-1, LMP-1, and LMP-2 expression (the default transcription program: Table 1). These germinal center B cells will then differentiate into memory B cells in which EBV expression is turned off (latency program: Table 1). These cells constitute the long-term reservoir of EBV. The lytic viral cycle can be reactivated in these cells by signals that cause B cells to differentiate into antibody-secreting plasma cells (through antigen stimulation) and migrate to the mucosal epithelium allowing the release of viral particles in the saliva, the main route for transmission of the virus between individuals.

Although EBV is considered to be a B-lymphotropic virus, it can also infect epithelial cells in vitro and is found in several EBV-associated carcinomas in vivo. A role of the epithelial cells of the oropharynx in the amplification of virus production in vivo has long been suspected but not yet demonstrated.

Viral Productive Cycle

In vitro, the lytic productive cycle can be induced in EBV-infected cell lines (either LCL or cell lines established from Burkitt lymphoma) by treatments of the cells with various agents such as the phorbol ester TPA, butyric acid (BA) or cross linking of surface immunoglobulin, etc. A key mediator of the entry into the productive cycle is the viral-encoded transcription factor BZLF1 (also called EB1, Zta, and Zebra) which activates both transcription of all the EBV early genes and DNA

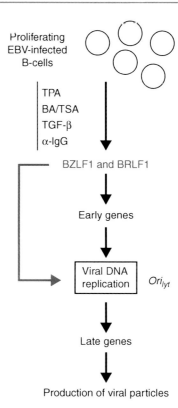

Epstein-Barr Virus, Fig. 2 The EBV replicative cycle. The EBV lytic replication program can be induced in vitro by treatment of proliferating EBV-infected B cells with various chemicals, by cross-linking of the surface immunoglobulins, or by expression of the EBV transcriptional activator EB1 (also called Zta or ZEBRA), product of the BZLF1 gene. In cooperation with another viral transcriptional factor (also called R or Rta), product of the BRLF1 gene, EB1 activates the expression of all the EBV early genes. EB1 also directly stimulates viral DNA replication which is dependent on viral proteins (a DNA-polymerase, a polymerase processivity factor, a single-stranded DNA-binding protein, a primase, and a helicase/primase associated protein). This viral DNA replication is initiated at replication origins, called *ori*$_{lyt}$, which are different from the one used during latency. This lytic replication program leads to amplification of the viral genome, synthesis of structural viral proteins, and the assembly of infectious virus particles

replication from the replication origins (Ori_{lyt}) active during the lytic cycle (Fig. 2).

Clinical Relevance

EBV is associated with several human malignancies in both immunocompromised and immunocompetent individuals.

In immunodepressed individuals – post-transplant patients or AIDS (acquired immunodeficiency syndrome) patients – EBV is probably directly involved in the appearance of immunoblastic B lymphomas (in AIDS patients) or posttransplantation lymphoproliferative diseases (PTLDs) (in patients undergoing organ transplantation) due to the loss of normal cytotoxic T-cell surveillance. These lymphomas are monoclonals or polyclonals, and the cells usually express the full set of EBV genes found in LCLs proliferating in culture (latency III, Table 1).

In immunocompetent individuals EBV is associated with several cancers. Endemic Burkitt lymphoma (BL) is found in certain parts of Africa and South America where malaria – which appears to act as a cofactor – is also endemic and affects mainly children from 7 to 9 years old. In these regions, BL is associated with EBV in more than 90% of cases. In lower-incidence regions, the association is only found in 20–30% of cases. BL is a monoclonal tumor characterized by a translocation of the c-myc gene (▶ Myc oncogene) to one of the immunoglobulin loci which results in an altered regulation of c-myc. The expression of EBV in the tumor cells is limited to EBNA-1, the EBER RNAs, plus several miRNAs. This profile of expression is defined as latency I (Table 1). Burkitt lymphoma cell lines can be readily established from tumor biopsies. On the contrary to LCLs, these cells are tumorigenic in nude mice. However, after several passages of these cells in culture, the expression profile of the EBV genes has been shown to derive towards a latency III profile.

In ▶ Hodgkin disease, EBV is present in the Reed-Sternberg cells (▶ Hodgkin and Reed/Sternberg Cell), in about 40% of cases, mostly of the mixed cellularity type. Expression of EBV in Hodgkin disease is characteristic of latency II (Table 1).

EBV is also found associated with rare but specific types of nasal T-cell lymphomas more common in Southeast Asian populations and also natural killer (NK)-cell (▶ Natural Killer Cell Activation) lymphomas. These types of lymphomas seem to arise either after acute primary infection or in some cases of chronic active EBV infection. The EBV expression profiles in these tumors are characteristic of latency I/II (Table 1).

EBV is also associated with a variety of carcinoma particularly the undifferentiated nasopharyngeal carcinoma (▶ nasopharyngeal carcinoma) (NPC). NPC is associated with EBV in almost 100% of cases worldwide and is particularly common in areas of China and Southeast Asia. Genetic disposition as well as environmental cofactors such as dietary components is thought to be important in the etiology of NPC. EBV gene expression in NPC epithelial cells consists of EBNA-1, the EBER RNAs, and LMP-1, LMP-2A/-2B (in 65% of cases), plus the miRNAs. This profile of expression is similar to latency II (Table 1). Several factors suggest that a reactivation of EBV (i.e., entry into the lytic cycle) precedes or accompanies the development of NPC. EBV is also found in about 10% of gastric adenocarcinomas (▶ gastric cancer) with a pattern of expression of EBV genes similar to that observed in NPC.

The exact role of EBV in the development of these different tumors is not yet understood and both environmental and genetic cofactors also contribute. However, the fact that the EBV genome is present in the great majority of the cells in EBV-associated malignancies and the demonstration that the virus is present in the tumor cells at a very early stage argue for a causative role for EBV in these cancers.

Cross-References

- ▶ AP-1
- ▶ Burkitt Lymphoma
- ▶ Epstein–Barr Virus Latent Membrane Protein 1
- ▶ Gastric Cancer
- ▶ Hodgkin and Reed/Sternberg Cell
- ▶ Hodgkin Disease
- ▶ MYC Oncogene
- ▶ Nasopharyngeal Carcinoma
- ▶ Natural Killer Cell Activation
- ▶ Notch/Jagged Signaling
- ▶ *RAS* Genes

References

Münz C (ed) (2015a) Epstein Barr virus volume 1. One herpes virus: many diseases. In: Current topics in microbiology and immunology, vol 390. Springer International Publishing

Münz C (ed) (2015b) Epstein Barr virus volume 2. One herpes virus: many diseases. In: Current topics in microbiology and immunology, vol 391. Springer International Publishing

Rickinson AB (2014) Co-infections, inflammation and oncogenesis: future directions for EBV research. Semin Cancer Biol 26:99–115

Thorley-Lawson DA, Hawkins JB, Tracy SI, Shapiro ME (2013) The pathogenesis of Epstein-Barr virus persistent infection. Curr Opin Virol 3(3):227–232

See Also

(2012) Burkitt Lymphoma Cell Lines. In: Schwab M (ed) Encyclopedia of Cancer, 3rd edn. Springer Berlin Heidelberg, p 575. doi:10.1007/978-3-642-16483-5_754

(2012) Lymphoblastoid Cell Lines. In: Schwab M (ed) Encyclopedia of Cancer, 3rd edn. Springer Berlin Heidelberg, pp 2122-2123. doi:10.1007/978-3-642-16483-5_3454

Epstein–Barr Virus Latent Membrane Protein 1

Martina Vockerodt
Department of Pediatrics I, Children's Hospital, Georg-August University of Gottingen, Gottingen, Germany

Definition

The latent membrane protein 1 (LMP1) of the ► Epstein-Barr virus (EBV) is an ► oncogene that is expressed during latent EBV infection. LMP1 is sufficient for the transformation of rodent fibroblast cells and is essential for efficient B cell transformation. LMP1 mimics a constitutively active tumor necrosis factor (TNF) receptor and interacts with and deregulates the ► signal transduction network of the host cell leading to altered cell survival, differentiation, and phenotypic changes.

Characteristics

EBV belongs to the family of gamma herpes viruses; it infects humans and establishes a latent infection for the lifetime of the individual. Infection in childhood is usually asymptomatic, but in young adults can lead to infectious mononucleosis. Due to its association with ► B cell tumors as Burkitt lymphoma (BL), ► Hodgkin disease/lymphoma (HL), posttransplant lymphoproliferative disorder, and epithelial tumors as nasopharyngeal carcinoma (NPC) and gastric carcinoma, EBV was the first human tumor virus to be discovered and is now classified as a group 1 carcinogen by the WHO. In vivo, EBV does usually not replicate in B lymphocytes, but instead establishes a latent infection with defined expression of the virus latent genes. In vitro, EBV infection leads to continuously proliferating lymphoblastoid cell lines (LCLs) and a different expression pattern of virus latent genes. The virus genes influence both viral and cellular transcription in the host cell. Among them LMP1 is essential for B cell transformation in vitro and behaves as a classical oncogene in rodent fibroblast transformation assays. When expressed in the B cell compartment of transgenic mice, LMP1 induces the development of B cell lymphoma, whereas its expression in the murine epidermis results in hyperplasia. It is interesting to note that expression of LMP1 is variable among EBV-associated tumors, being always present in EBV-associated Hodgkin lymphoma, almost always absent in virus-positive Burkitt lymphoma, and variably present in NPC.

Structurally, LMP1 is a 66 kDa integral membrane protein consisting of 386 amino acids. It can be divided into three domains: a short cytoplasmic amino-terminus of 24 amino acids, a transmembrane domain consisting of six transmembrane helices which oligomerize to form membrane patches, and a carboxy-terminal cytoplasmic domain of 200 amino acids. The cytoplasmic domain contains two carboxy-terminal activating regions (CTARs), also known as transformation effector sites (TES). The CTARs are critical for EBV's transforming activity; while CTAR1/TES1

E

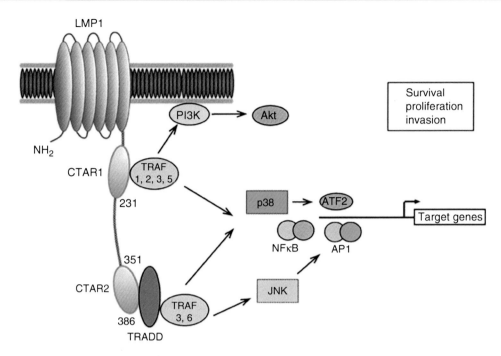

Epstein–Barr Virus Latent Membrane Protein 1, Fig. 1 Structure of LMP1 and its functional domains. The signaling domains recruit TRAFs and TRADD which results in the activation of host cell signal transduction pathways and the regulation of host cell target genes

is essential for initial B cell transformation, CTAR2/TES2 mediates growth factor-like signals which are required for long-term outgrowth of EBV-infected cells. The region between CTAR1 and CTAR2, sometimes referred to as CTAR3, is not required for B cell ▶ immortalization.

Mechanisms

Cell transformation requires activation of the host cell signaling machinery. LMP1 mediates this activation by recruiting molecules of the TNF receptor and ▶ toll-like receptor family (Fig. 1). CTAR1 interacts with the TNF receptor-associated factors (TRAF) 1, 2, 3, and 5 through the PxQxT-binding motif of LMP1. Recruitment of TRAF molecules to CTAR2 may be mediated indirectly through binding of the TNF receptor-associated death domain protein (TRADD), although much of CTAR2-mediated ▶ nuclear factor kappa B (NF-κB) activity has been shown to be TRADD independent (Fig. 1).

These adaptor proteins subsequently recruit multiprotein complexes containing the NF-κB-inducing kinase (NIK) and the I-κB kinases (IKKs). This results in the activation of the canonical I-κB-dependent NF-κB pathway (involving p50-p65 heterodimers) and the noncanonical pathway, leading to the processing of p100 to generate p52-p65 heterodimers. While activation through the CTAR1 domain of LMP1 is mediated by the noncanonical NF-κB pathway, CTAR2 appears to activate the canonical pathway by utilizing TRAF6 and TAK1. CTAR2 is also important in activating the c-Jun N-terminal kinase (▶ JNK subfamily) leading to activation of the transcription factor ▶ AP-1. The phosphatidy-linositol 3-kinase (▶ PI3K signaling) pathway is triggered through CTAR1 leading in epithelial cells to actin polymerization and cell ▶ motility.

Other signaling pathways activated by LMP1 are the p38 ▶ MAP kinase pathway and the ▶ signal transducers and activators of transcription (STAT) pathway.

The NF-κB pathway plays a key role in the activation of many genes and is essential for the transformation of B cells by EBV. NF-κB mediates induction of antiapoptotic genes (▶ *bcl-2*, *bfl-1*, *A20*, *c-IAPs*, *c-FLIP*) and downregulates proapoptotic genes as *Bax* (*Bcl-2*-associated protein x). LMP1 expression in B cells also results in the upregulation of activation markers and ▶ adhesion molecules.

LMP1 also modulates the communication between EBV-infected cells and its cellular environment by upregulating the expression of a large number of cytokines (▶ interleukin-6, −10, TNF-α) and ▶ chemokines (RANTES, IP-10, interleukin-8). It is also involved in migration and invasion processes as it activates proangiogenic factors as ▶ vascular endothelial growth factor, ▶ matrix metalloproteinases, and modulators of the cytoskeleton.

Although LMP1 shows no homology with any cellular protein, it functionally mimics an activated CD40 receptor, which is a costimulatory receptor required for B cell proliferation. Although LMP1 and CD40 both recruit TRAF molecules and regulate an overlapping pattern of signaling pathways and target genes, they have divergent roles in B cell development. For example, in CD40-deficient LMP1 transgenic mice, LMP1, like CD40, can induce extrafollicular B cell differentiation, but in contrast to CD40, LMP1 leads to a defective germinal center reaction, characterized by splenomegaly and lymphadenopathy.

Cross-References

▶ Hodgkin Lymphoma, Clinical Oncology

References

Young LS, Rickinson AB (2004) Epstein-Barr virus: 40 years on. Nat Rev Cancer 4:757–768

ER

▶ Estrogen Receptor

ER Stress

▶ Endoplasmic Reticulum Stress

ERBB

▶ Epidermal Growth Factor Receptor

ErbB-1

▶ Epidermal Growth Factor Receptor

ERBB2

▶ HER-2/neu

Erlotinib

Bassel El-Rayes[1,2], Shirish Gadgeel[3], Shadan Ali[3], Philip A. Philip[3] and Fazlul H. Sarkar[3]
[1]Department of Hematology and Medical Oncology, Emory University School of Medicine, Atlanta, GA, USA
[2]Winship Cancer Institute of Emory University, Atlanta, GA, USA
[3]Karmanos Cancer Institute, Wayne State University, Detroit, MI, USA

Synonyms

Tarceva

Definition

Erlotinib is a potent and selective ▶ epidermal growth factor receptor (EGFR) tyrosine kinase inhibitor. Erlotinib is commercially available in tablets of 25, 100, and 150 mg formulations. Erlotinib is currently approved for use in previously treated non-small cell ▶ lung cancer and in frontline management of ▶ pancreatic cancer in combination with gemcitabine chemotherapy.

Characteristics

Rationale for Targeting the EGFR. The EGFR is frequently dysregulated in epithelial cancers. Overexpression of EGFR can result in malignant transformation of cells. Patients whose tumors have overexpressed or dysregulated EGFR and/or ligand expression may have a worse prognosis. Activation of the EGFR initiates dimerization of the receptors leading to activation of the tyrosine kinase domain. The kinase in turn phosphorylates (Phosphorylation) and activates proteins in the signal transduction cascade promoting cell proliferation, angiogenesis, invasion, and survival. In preclinical models, erlotinib selectively inhibited EGFR tyrosine kinase activity in human cancer cell lines and resulted in inhibition of various tumor growths and induction of apoptosis. Erlotinib potentiated the activity of cytotoxic agents and radiation in cancer cell line as well as in animal models.

Phase I Trials. The maximal tolerated dose of erlotinib was 150 mg once daily in phase I trial. The most commonly observed toxicities were ▶ acneiform rash and diarrhea.

Experience in Non-Small Cell Lung Cancer (NSCLC). Three randomized trials evaluated the efficacy of erlotinib in advanced NSCLC. In the first trial, patients with previously treated advanced NSCLC were randomized to erlotinib or best supportive care. The results revealed a significant improvement in median survival (6.7 vs. 4.7 months, $p = 0.001$) and survival at 1 year (31.2% vs. 21.5%) in favor of erlotinib. The median time to progression and response rates in the erlotinib and best supportive care arms were 9.7 vs. 8.0 weeks ($p < 0.001$) and 8.9 vs. 1.0%, respectively. The incidence of grade 3 and 4 rash and diarrhea in the erlotinib arm was 9% and 6%, respectively. A very important finding, in a multivariate analysis, was that nonsmokers and those with adenocarcinoma histology benefited most from erlotinib. Based on this trial, erlotinib was approved in patients with previously treated NSCLC.

Two trials compared conventional chemotherapy to the same chemotherapy and erlotinib in patients with newly diagnosed advanced NSCLC. The chemotherapy regimens evaluated were ▶ gemcitabine/carboplatin (TALENT) and ▶ paclitaxel/▶ cisplatin (TRIBUTE). No significant difference was observed between patients receiving chemotherapy or erlotinib and chemotherapy with respect to objective response rate, survival, or time to progression. Therefore, erlotinib is not used in combination with chemotherapy to treat patients with NSCLC.

Experience in Pancreatic Cancer. A randomized trial compared gemcitabine to gemcitabine and erlotinib in patients with advanced pancreatic cancer. The trial resulted in a significant improvement in median survival (6.37 vs. 5.91 months, $p = 0.01$) and 1-year survival (24 vs. 17%) in favor of erlotinib. Significant predictors for favorable outcome were performance status 0–1, locally advanced disease, and normal albumin. Treatment with erlotinib resulted in an improvement of progression-free survival (3.75 vs. 3.55 months, $p = 0.009$). The incidence of grade 3 and 4 toxicities was higher in the erlotinib arm with respect to rash (6 vs. 1%) and diarrhea (7 vs. 2%). The Food and Drug Administration (FDA) approved erlotinib in combination with gemcitabine for previously untreated advanced pancreatic cancer. A similar approval was granted in Europe for this particular indication.

Trials in Other Tumor Types. Erlotinib has been evaluated in bile duct, gastric, ▶ esophageal, ▶ hepatocellular, ▶ colorectal (CRC), and head and neck cancers. In bile duct cancer, a multi-institutional phase II trial evaluated single-agent erlotinib in 42 patients with advanced disease. The trial met its primary end point with a progression-free survival at 6 months of 17%. Three patients had a partial response. The Southwest Oncology Group (SWOG) performed a phase II trial of erlotinib in advanced gastric and gastroesophageal (GEJ) tumors. The response rate was 9%. All responses were observed in the GEJ patients. Interestingly, no responses were seen in the patients with gastric cancer. Philip et al. reported the results of 38 patients with advanced hepatocellular cancer treated with erlotinib. Forty-seven percent of the patients had received prior chemotherapy. The study met its primary end point with a 6-month progression-free survival of 32%. Three patients had partial response. Erlotinib has been evaluated in CRC as a single agent and in combination with FOLFIRI, capecitabine, capecitabine/oxaliplatin, and FOLFOX. As a single agent, erlotinib was evaluated in patients with previously treated CRC. Thirty-nine percent of patients had stable disease and no responses were observed. The combination of FOLFIRI and erlotinib resulted in excessive toxicity, and the trial was discontinued. Erlotinib and capecitabine were well tolerated, and in the phase I trial, two of the nine patients with CRC had a partial response. The phase II trial is still ongoing. Two trials evaluated erlotinib and capecitabine/oxaliplatin in patients with previously treated CRC. The partial response rates and stable disease were 20–22% and 61–64%, respectively. There is still uncertainty whether erlotinib plays any role in CRC. Currently, research in CRC is focused on the use of monoclonal antibodies that target the EGFR. Erlotinib in combination with FOLFOX is currently in a phase II trial. Modest activity was observed with erlotinib in head and neck cancer patients with a response rate of 4% and disease stabilization rate of 38%. In conclusion, erlotinib has demonstrated promising activity in cholangiocarcinoma, hepatocellular, and GEJ tumors. The role of erlotinib in these diseases requires further randomized trials.

Future Directions. Erlotinib has demonstrated a significant but modest activity in a number of carcinomas. The future challenge is how to improve the activity of erlotinib. Two approaches are being evaluated in clinical trials. The first is to select patients with a higher likelihood of benefit from erlotinib. For example, patients with NSCLC and either activating mutations of the EGFR or no smoking history have demonstrated an increased benefit from erlotinib therapy. Additional trials are evaluating the efficacy of erlotinib in previously untreated patients whose tumors have EGFR mutations. On the other hand, a trial is evaluating the combination of erlotinib with chemotherapy in previously untreated patients with NSCLC and no smoking history. The results

Erlotinib, Table 1 Ongoing trials of erlotinib in combination with targeted agents

Agent	Target	Phase (disease)
Isoflavone	Akt/NF-kB	II (Pancreatic)
Sorafenib	VEGFR, PDGFR, Raf	I
Dasatinib	Src	I
Cetuximab	EGFR	I
Celecoxib	Cyclooxygenase 2	II (NSCLC)
RAD001	mTOR	Phase I

of these trials will determine whether erlotinib should be used in the frontline management of NSCLC in selected patient populations.

The second approach is based on combining erlotinib with other targeted agents because of the redundancy of the signaling pathways and existence of independently activated survival pathways. Since the inhibition of the EGFR results in inhibition of angiogenesis, the combination of erlotinib and ► bevacizumab is being evaluated in a number of malignancies including hepatocellular, cholangiocarcinoma, NSCLC, and pancreatic cancer. The combinations of erlotinib with other agents targeting the signaling cascade are currently at different stages of development (Table 1).

Cross-References

► Achneiform Rash
► Bevacizumab
► Cisplatin
► Colorectal Cancer
► Epidermal Growth Factor Receptor
► Esophageal Cancer
► Gemcitabine
► Hepatocellular Carcinoma
► Lung Cancer
► Paclitaxel
► Pancreatic Cancer

References

El-Rayes BF, LoRusso PM (2004) Targeting the epidermal growth factor receptor. Br J Cancer 91:418–424
Giaccone G (2005) Targeting HER1/EGFR in cancer therapy: experience with erlotinib. Future Oncol 1:449–460

Mendelsohn J, Baselga J (2006) Epidermal growth factor receptor targeting in cancer. Semin Oncol 33:369–385

Philip PA, Mahoney MR, Allmer C et al (2005) Phase II study of Erlotinib (OSI-774) in patients with advanced hepatocellular cancer. J Clin Oncol 23:6657–6663

Philip PA, Mahoney MR, Allmer C et al (2006) Phase II study of erlotinib in patients with advanced biliary cancer. J Clin Oncol 24:3069–3074

See Also

(2012) Bile Duct. In: Schwab M (ed) Encyclopedia of Cancer, 3rd edn. Springer Berlin Heidelberg, p 399. doi:10.1007/978-3-642-16483-5_616

(2012) Capecitabine. In: Schwab M (ed) Encyclopedia of Cancer, 3rd edn. Springer Berlin Heidelberg, p 640. doi:10.1007/978-3-642-16483-5_828

(2012) FOLFIRI. In: Schwab M (ed) Encyclopedia of Cancer, 3rd edn. Springer Berlin Heidelberg, p 1440. doi:10.1007/978-3-642-16483-5_2232

(2012) FOLFOX. In: Schwab M (ed) Encyclopedia of Cancer, 3rd edn. Springer Berlin Heidelberg, p 1441. doi:10.1007/978-3-642-16483-5_2233

(2012) Phosphorylation. In: Schwab M (ed) Encyclopedia of Cancer, 3rd edn. Springer Berlin Heidelberg, p 2870. doi:10.1007/978-3-642-16483-5_4544

Erlotinib (Tarceva)

▶ Receptor Tyrosine Kinase Inhibitors

ERM Proteins

Ling Ren[1] and Chand Khanna[2]
[1]Pediatric Oncology Branch, National Cancer Institute, Center for Cancer Research, Bethesda, MD, USA
[2]Comparative Oncology Program, Center for Cancer Research, National Cancer Institute, Bethesda, MD, USA

Synonyms

Cytovillin; Villin 2

Definition

The ERM family consists of three closely related proteins, ezrin, radixin, and moesin. ERM proteins are cell membrane and cytoskeleton linker proteins.

Characteristics

History, Structure, and Sequence

Ezrin, the prototype ERM protein, is a 585-amino acid polypeptide, first identified as a constituent of microvilli and shown to be present in actin-containing surface structure on a wide variety of cells. ERM proteins share homology in sequence structure and function. They are composed of three domains: an N-terminal globular domain, an extended α-helical domain, and a charged C-terminal domain. The N-terminal domain of ERM proteins is highly conserved and is also found in ▶ merlin, band 4.1 proteins, and members of the band 4.1 superfamily. This domain is called FERM (fourpointone protein, ezrin, radixin, moesin) domain. The crystal structure of moesin revealed that the FERM domain is composed of three structural modules that, together, form a compact clover-shaped structure. The C-terminal domain can extend across the FERM domain surface, potentially masking recognition sites of other proteins. Ezrin and radixin also contain a polyproline region between the helical and C-terminal domains. The cDNA sequence of radixin encodes a protein of 583 amino acids with 77% identity to ezrin. Moesin, isolated as a heparin-binding protein, consists of 577 amino acids with 74% identity to ezrin.

Regulation

ERM proteins are conformationally regulated. ERM proteins exist in proposed dormant forms in which the C-terminal tail binds to and mask the N-terminal FERM domain (Fig. 1). The activation of ERM protein is mediated by both C-terminal threonine phosphorylation (T567 in ezrin, T564 in radixin, T558 in moesin) and exposure to PIP2. It is likely that phosphorylation at other residues in ERM proteins is needed to maintain an open

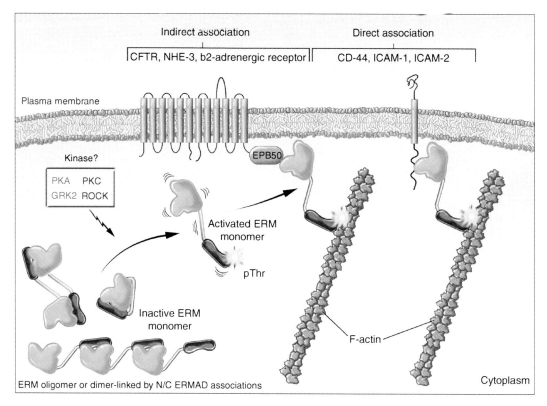

ERM Proteins, Fig. 1 Activation of ERM proteins. The dormant ERM proteins exist as monomers, dimers and oligomers with a closed conformation. The activation of ERM proteins is mediated by both exposure to PIP2 and phosphorylation of the C-terminal threonine. The C-terminal of activated ERM proteins bind to F-actin filaments. The N-terminal domains of activated ERM proteins are associated directly with the adhesion molecules such as CD44 and ICAM-1, -2 and -3 or indirectly with other transmembrane proteins such as NHE3 through EPB50.

activated conformation for ezrin and to direct ezrin-specific effects in cells. It is also unclear what functions are ascribed to the so-called inactive closed conformation of ERM proteins. Several protein kinases have been found to phosphorylate the C-terminal threonine residue of the ERM proteins. Examples include PKCα (▶ protein kinase C family), PKCθ (protein kinase C family), rho kinases/ROCK, G protein-coupled receptor kinase 2 (GRK2), and myotonic dystrophy kinase-related Cdc42-binding kinase (MRCK).

Dephosphorylation of C-terminal Thr of moesin has been suggested to be a crucial step for lymphocyte adhesion and transendothelial migration. The disassembly of microvilli on lymphocyte cell surfaces caused by dephosphorylation of moesin facilitates the cell–cell (lymphocyte–endothelium) contact. Protein phosphatase 2C is involved in the dephosphorylation of moesin through the activation of Rac1 small ▶ GTPase.

Function, Distribution, Localization

ERM proteins either directly associate with the cytoplasmic domains of adhesive type I membrane proteins, such as ▶ CD44, CD43, ICAM-1, ICAM-2, and ICAM-3, or indirectly associate with membrane proteins via PDZ-containing adaptors EBP50 and E3KA-P. Regulated attachment of membrane proteins to F-actin is essential for many fundamental cellular processes, including the determination of cell shape, polarity, and surface structure, cell ▶ adhesion, ▶ motility, cytokinesis, phagocytosis, and integration of membrane transport with signaling pathways. There is functional redundancy between ERM proteins. This is best exemplified by the phenotype of the ezrin knockout mouse. This mouse is viable at birth, suggesting the ability of radixin and moesin to fill the role of ezrin during development. Interestingly, the fatal phenotype of this mouse is characterized by intestinal villous malformations seen at day 13 postpartum. The normal intestinal epithelial cells nearly exclusively express ezrin. Although ezrin, radixin, and moesin are coexpressed in most cultured cells, they exhibit a tissue-specific expression patterns. Ezrin is highly concentrated in the intestine, stomach, lung, and kidney although moesin is prominent in the lung and spleen, and radixin in the liver and intestine. Ezrin is expressed in epithelial and mesothelial cells, while moesin is expressed in endothelial cells. As indicted, the brush border of intestinal epithelial cells expresses only ezrin, and hepatocytes express only radixin.

The Expression and Functions of ERM Proteins in Cancer

Ezrin has been shown to be expressed in most human cancers and linked to progression in several cancers, including carcinomas of endometrium, breast, colon, ovary, in uveal and cutaneous ► melanoma, ► brain tumors, and soft tissue sarcomas. In a cDNA array (► microarray (cDNA) technology) analysis of highly and poorly metastatic ► rhabdomyosarcoma and ► osteosarcoma, ezrin was indicated as a key metastatic regulator. In several murine and human cancer models, suppression of ezrin protein and disruption of ezrin function significantly reduced the metastatic phenotype despite the expression of other ERM proteins. This suggests that the redundancy provided by the other ERM proteins for ezrin does not extend to ► metastasis and that ezrin contributes a unique and necessary function to cells undergoing metastasis. Comparing lung adenocarcinoma with normal lung tissue, the expressions of ezrin, radixin, and moesin were decreased on mRNA level as well as the protein level. Interestingly, the high expression of ezrin was observed in the invading tumor cells in lung adenocarcinoma.

Cross-References

► Adhesion
► Brain Tumors
► CD44
► GTPase
► Merlin
► Metastasis
► Microarray (cDNA) Technology
► Motility
► Osteosarcoma
► Protein Kinase C Family
► Rhabdomyosarcoma

References

Bretscher A, Chambers D, Nguyen R et al (2000) ERM-Merlin and EBP50 protein families in plasma membrane organization and function. Annu Rev Cell Dev Biol 16:113–143

Bretscher A, Edwards K, Fehon RG (2002) ERM proteins and merlin: integrators at the cell cortex. Nat Rev Mol Cell Biol 3:586–599

McClatchey AI (2003) Merlin and ERM proteins: unappreciated roles in cancer development? Nat Rev Cancer 3:877–883

Tsukita S, Yonemura S (1999) Cortical actin organization: lessons from ERM (ezrin/radixin/moesin) proteins. J Biol Chem 274:34507–34510

Vaheri A, Carpen O, Heiska L et al (1997) The ezrin protein family: membrane-cytoskeleton interactions and disease associations. Curr Opin Cell Biol 9:659–666

ERp60

► Calreticulin

Erythrocytosis

► Polycythemia

Erythroid Colony-Stimulating Activity

► Erythropoietin

Erythroid Differentiation Factor

▶ Activin

Erythroleukemia

Definition

A form of acute myeloid leukemia where the myeloproliferation is of abnormal, immature red blood cells.

Cross-References

▶ Erythropoietin

Erythropoiesis-Stimulating Factor

▶ Erythropoietin

Erythropoietin

Jiuwei Cui[1] and Yaacov Ben-David[2]
[1]Jilin University, Changchun, Jilin, China
[2]Division of Molecular and Cellular Biology, Sunnybrook Health Sciences Centre, Toronto, ON, Canada

Synonyms

ECSA; Ep; Epo; Epoetin; Erythroid colony-stimulating activity; Erythropoiesis-stimulating factor; ESF

Definition

Erythropoietin (Epo) (from Greek erythro for red and poietin to make) is a small glycoprotein hormone that is essential for the production of red blood cells. Epo promotes the survival, proliferation, and differentiation of erythroid progenitor cells (BFU-E, CFU-E) to mature erythrocytes and initiates hemoglobin synthesis.

Characteristics

The Epo gene contains at least five exons and resides on chromosome 7q21-q22 in humans and chromosome 5 in mice. DNA sequences from monkey and mouse display 90% and 80% homology to human Epo, respectively. Epo is produced primarily in the kidney and to a lesser extent in the liver. It is an acidic glycoprotein hormone with a molecular weight of 34–37 kD and circulates in the blood plasma at a very low concentration (about 5 pmol/l). It is composed of a single chain polypeptide and is resistant to denaturation by heat, alkali, or reducing agents. Epo is synthesized as a 193-amino acid precursor that is cleaved to yield an active protein of 165 amino acids. It is N-glycosylated at asparagine residues 24, 36, and 83 and O-glycosylated at serine 126. Epo is also sialylated and contains two disulfide bonds at positions 7/161 and 29/33. The alpha form of the hormone consists of 31% carbohydrates, while the beta form consists of 24%. These two forms of Epo have similar biological and antigenic properties. The carbohydrate moiety of Epo plays an important role in the mediation of its full biological effect and the pharmacokinetic behavior of the protein in vivo; non-glycosylated Epo has a very short biological half-life. Epo is fully synthesized in its active form prior to secretion into circulation. Epo, already known as the stimulating hormone for erythropoiesis, has displayed different and interesting pleiotropic actions. It not only affects erythroid cells but also myeloid cells, lymphocytes, and megakaryocytes. This hormone can enhance phagocytic function of polymorphonuclear cells and reduce the activation of macrophages, thus modulating the inflammatory process. Epo also exerts diverse biological effects in many nonhematopoietic tissues and is involved in the wound-healing cascade, functions as a proangiogenic cytokine during physiological

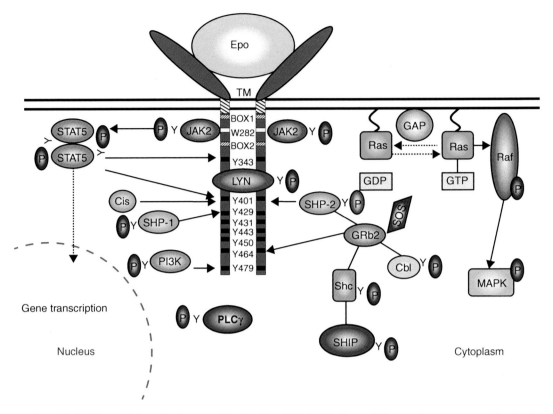

Erythropoietin, Fig. 1 Schematic diagram of the EpoR depicting the positions of tyrosine (Y) residues (*black bars*; Y) in the cytoplasmic domain and attachment sites of signal transduction proteins such as STAT5, SHP-1, and SHP-2. Binding of Epo to its receptor results in the autophosphorylation and activation of JAK2, which in turn phosphorylates eight tyrosine residues in the cytoplasmic domain of the EpoR

angiogenesis in the embryo and uterus, and exerts tissue-protective effects as part of the innate response to stressors.

Cellular and Molecular Regulation

The synthesis of Epo in the kidney is under the control of an oxygen-sensing mechanism. Transcriptional response of the Epo gene to hypoxia is mediated partly by promoter sequences but mainly by a 24 bp hypoxia-response element located at the 3′ flanking region of the Epo gene bound to the hypoxia-inducible factor-1(HIF-1). Epo production is also modulated by several other factors such as hypoglycemia, increased intracellular calcium, insulin release, estrogen, androgenic steroids, and various cytokines.

The biological activity of Epo is mediated by its specific receptors present at 300–3,000 copies per cell that undergo phosphorylation in response to Epo. The Epo receptor (EpoR) belongs to the class I cytokine receptor superfamily. The mouse EpoR consists of 507 amino acids with an extracellular domain, a single hydrophobic transmembrane domain, and a cytoplasmic domain. The human EpoR is a 66 kD protein comprised of 508 amino acids. It consists of eight exons spanning some 6 kb on human chromosome 19p13.3.

The interaction of Epo with its receptor results in the formation of a homodimer and its subsequent internalization (Fig. 1). Dimerization of the receptor results in the autophosphorylation of Janus kinase 2 (JAK2), a protein kinase that is tightly associated with the EpoR. Once activated, JAK2 phosphorylates eight tyrosine residues located in the cytoplasmic domain of the EpoR. Phosphorylation of the EpoR leads to the recruitment and phosphorylation of a number of signal transduction proteins. One such protein is

STAT5, a transcription factor that plays an important role in the regulation of in vivo erythropoiesis. Once phosphorylated by binding to tyrosine 343 and 401 of the EpoR, STAT5 translocates to the nucleus to activate the expression of several downstream target genes. Other signaling cascades triggered by Epo binding to its receptor include phosphatidylinositol 3-kinase (PI3K) that binds to tyrosine 479 and is involved in erythroblast survival and Grb2 that binds to tyrosine 464 and is involved in the activation of the Ras pathway. Ras pathway may be required for the synergistic expansion of erythroid progenitors and precursor cells in response to Epo and stem cell factor (SCF). EpoR-mediated activation of phospholipase A2 and C also leads to the release of membrane phospholipids, the synthesis of diacylglycerol, and the increase in intracellular calcium levels and pH. Since phosphorylation of the EpoR by Epo is diminished after 30 min of stimulation, a number of tyrosine phosphatases have been identified that are involved in attenuating the signal. The tyrosine phosphatase SHP-2 binds to tyrosine 401 of the Epo receptor and stimulates erythroid proliferation, while SHP-1 binds to tyrosine 429 and inhibits proliferation.

Expression of the EpoR is not restricted to hematopoietic cells and exhibits a multi-tissue distribution that includes vascular endothelial cells, muscle cells, and neurons; therefore, Epo is believed to play a physiological role in angiogenesis and cardiac and brain development. Abnormal regulation of Epo-EpoR signaling in hematopoietic cells has been associated with proliferative disorders of the bone marrow, such as polycythemia vera, a disorder characterized by erythrocytosis, as a consequence of an active mutation in the EpoR. Additionally, prolonged activation of STAT5 has been observed in cells transfected with mutant (tyrosine 429) EpoR, suggesting that STAT5 DNA binding activity may play a role in the pathogenesis of erythrocytosis. A point mutation at position 129 of the mouse EpoR gene results in constitutive activation of the receptor without stimulation with Epo. Mice infected with a retrovirus expressing this aberrant receptor develop ▶ erythroleukemia and splenomegaly. Taken together, these provide evidences that the precise control of Epo-EpoR signaling is critical for the normal proliferation and differentiation of erythroid progenitor cells.

Clinical Relevance

The synthesis of Epo is subject to a complex circuit that links the bone marrow and kidney in a feedback loop. Its reference interval in the blood plasma ranges between 3.3 and 16.6 mIU/ml. Patients suffering from most anemias display higher than normal concentrations of serum Epo, whereas those suffering from anemia associated with chronic renal disease have values either low or within the normal range. Epo levels are disproportionately low in anemic patients with chronic disorders as well, such as rheumatoid arthritis, AIDS, and cancer, in which inhibition of Epo production and erythroid progenitor proliferation by inflammatory cytokines, such as IL-1 and TNF, are thought to play major causative roles. Abnormally high concentrations may also be induced by renal neoplasms, benign tumors, polycystic kidney disease, renal cysts, and hydronephrosis. The pathophysiological excess of Epo leads to erythrocytosis that is accompanied by an increase in blood viscosity and may cause heart failure and pulmonary hypertension.

Chronic kidney disease causes the destruction of Epo-producing cells resulting in hyporegenerative normochromic normocytic anemias. Epo is therefore clinically used for the treatment of patients with severe kidney insufficiency. In uremic patients, treatment with recombinant human Epo (rhEpo) effectively reactivates the bone marrow to produce erythrocytes and also improves platelet adhesion and aggregation. Hypertension is an important complication in the treatment of renal anemia with rhEpo. rhEpo is also used to treat nonrenal forms of anemia caused by chronic infections, inflammation, radiation therapy, and chemotherapy. Beyond ameliorating anemia, rhEpo has been shown to restore radiosensitivity and increase cytotoxicity of chemotherapy in the treatment of cancer-related anemia. However, clinical trials have shown increase in the relative risk of thromboembolic complications and lower survival, which raises concerns about the potential adverse effects of rhEpo in cancer patients. Additional studies show that Epo and EpoR expression also occurs in tumor cells, suggesting the potential for the generation of an autocrine or paracrine growth-stimulator Epo-EpoR loop in cancer cells. Further studies will be required to investigate the effects, if any, of rhEpo therapy on disease progression and survival. For its role in stimulating the production of erythrocytes, an important application of Epo is the presurgical activation of erythropoiesis allowing for the collection of autologous donor blood. rhEpo has emerged as a novel anti-inflammatory and cytoprotective agent, as evidenced by its physiological response to various forms of tissue injury. Accordingly, the therapeutic potential of Epo has been shown in acute renal failure, diabetic neuropathy, myocardial infarction, and cerebral ischemia. The characterization of Epo variants, such as asialo-Epo and carbamylated Epo, that retain nonhematopoietic, tissue-protective properties of Epo without stimulating erythropoiesis has uncovered new areas of research into the mechanisms of Epo-mediated signaling in nonhematopoietic tissues as well as novel clinical applications for rhEpo and its derivatives in disorders other than anemia.

References

Hardee ME, Acrasoy MO, Blackwell KL et al (2006) Erythropoietin biology in cancer. Clin Cancer Res 12:332–339

Heuser M, Ganser A (2006) Recombinant human erythropoietin in the treatment of nonrenal anemia. Ann Hematol 85:69–78

Jelkman W (1992) Erythropoietin, structure, control of production and function. Physiol Rev 72:449–489

Lipton S (2004) Erythropoietin for neurological protection and diabetic neuropathy. N Eng J Med 350:2516–2517

Wojchowski DM, Gregory RC, Miller CP et al (1999) Signal transduction in the erythropoietin receptor system. Exp Cell Res 253:143–156

ESA

► EpCAM

ESE (EH Domain and SH3 Domain Regulator of Endocytosis)

► Intersectin

E-Selectin-Mediated Adhesion and Extravasation in Cancer

Liang Zhong, Bryan Simoneau, Pierre-Luc Tremblay, Stéphanie Gout, Martin J. Simard and Jacques Huot
Le Centre de recherche du CHU de Québec-Université Laval: axe Oncologie, Le Centre de recherche sur le cancer de l'Université Laval, Québec, QC, Canada

Keywords

Angiogenesis; death receptor-3; extravasation; endothelium; MAP kinases; metastasis; sialyl Lewis determinants

Synonyms

CD62 antigen-like family member E (CD62E); Endothelial-leukocyte adhesion molecule 1 (ELAM1); Leukocyte-endothelial cell adhesion molecule 2 (LECAM2)

Definition

► Adhesion of circulating cancer cells to the endothelium is a prerequisite for ► metastasis. It requires specific interactions between adhesion molecules such as E- and P- selectins that are present on endothelial cells and their counter-receptors on cancer cells. The specificity of this interaction constitutes the basis of the organ selectivity of metastatic colonization. Subsequently to adhesion, cancer cells form metastasis either by growing locally in the capillaries or by invading the surrounding tissues following extravasation.

Characteristics

Structure

E-selectin (64 kDa) is a transmembrane receptor of the selectin family that also contains L- and P-selectins. Two glycosylated forms of E-selectin are detected at 100 and 115 kDa. The extracellular part of selectins is constituted of three domains: an N-terminal C-type lectin domain, which is calcium-dependent and mediates ligand interaction; an epidermal growth factor (EGF) domain, which also regulates ligand interaction; and consensus complement regulatory protein (CRP) repeats of ~60 amino acids each, which serve as spacers to hold the other two domains away from cell surface and mediate the rolling of adhering cells (see below). The number of CRP repeats distinguishes the extracellular domain of different selectins. E-selectin has six CRP repeats. Selectins are anchored in the membrane through a single helicoidal transmembrane domain followed by a short cytoplasmic tail (Fig. 1). The cytoplasmic tail of E-selectin can trigger signaling in the endothelial cell and is connected to the actin cytoskeleton via actin-binding proteins, which are important mediators of extravasation.

Expression

E-selectin is expressed exclusively by endothelial cells. Its constitutive expression has been detected in the skin and parts of bone marrow

microvasculature. However, in most vessels, the de novo synthesis of E-selectin is induced by proinflammatory molecules such as tumor necrosis factor α (TNFα), interleukin 1β (IL-1β), endothelial monocyte activating polypeptide II (EMAPII), and bacterial lipopolysaccharide (LPS). Following stimulation by TNFα, E-selectin relies on PI3K-Akt-NFκB and JNK-c-Jun pathways for its transcription. In physiological conditions such as inflammation, the expression of E-selectin is transient and often reaches its peak 2–6 h after stimuli. E-selectin is gradually internalized by endocytosis by clathrin-coated pits and degraded in the lysosomes. In the endothelium areas of chronic inflammation, E-selectin may remain upregulated. Several cancer cells have the ability to induce E-selectin. For instance, Lewis lung carcinoma cells induce E-selectin in liver sinusoidal endothelium. Moreover, highly metastatic human colorectal and mouse lung carcinoma cells, upon their entry into the hepatic microcirculation, induce TNFα production by resident Kupffer cells, triggering E-selectin expression. Clinically, patients with various cancers including breast, colorectal, lung, bladder, head and neck, and melanoma have elevated galectin-3 in their serum. In turn, galectin-3 induces secretion of proinflammatory cytokines by blood vascular endothelium, which triggers the expression of E-selectin. A TNFα-inducible microRNA, miR-31, directly targets the mRNA of E-selectin

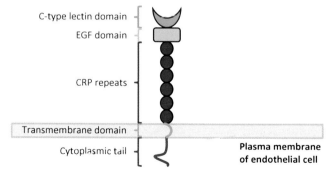

E-selectin-mediated adhesion and extravasation in cancer

E-Selectin-Mediated Adhesion and Extravasation in Cancer, Fig. 1 Structure of E-selectin. The extracellular part of E-selectin is divided into three domains: an N-terminal C-type lectin domain, an epidermal growth factor (EGF) domain, and consensus complement regulatory protein (CRP) repeats. E-selectin is anchored in the membrane through a single helicoidal transmembrane domain, followed by a short cytoplasmic tail

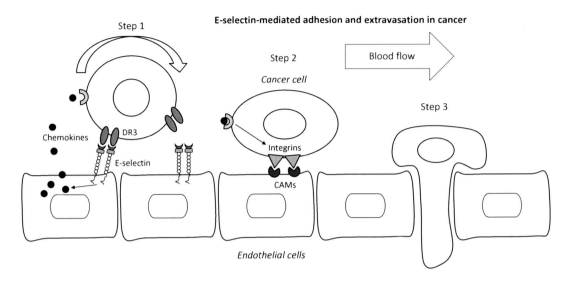

E-Selectin-Mediated Adhesion and Extravasation in Cancer, Fig. 2 E-selectin-mediated adhesion and transendothelial migration of cancer cells. Extravasation of cancer cells is a multistep process. The first step involves the transient adhesion of cancer cells to the endothelium. It requires endothelial E- and P-selectins, and their counter-receptors (such as DR3 and CD44 for E-selectin) on cancer cells. This step is associated with the rolling of cancer cells on the endothelium. The second step involves a firmer adhesion of cancer cells to the endothelium, which is mediated by cell adhesion molecules (CAMs) on the endothelium and integrins on cancer cells. The third step is the extravasation of cancer cells through endothelial cell-cell junctions

and downregulates its expression, suggesting its involvement in carcinoma dissemination.

Function

E-selectin recognizes the sialyl Lewis-a/x tetrasaccharide borne by glycoproteins and glycolipids on the surface of leukocytes and tumor cells. Its glycoprotein ligands include: E-selectin ligand-1 (ESL-1), P-selectin glycoprotein ligand-1 (PSGL-1), β2 integrin, L-selectin, CD43/44, lysosomal-associated membrane protein-1/2 (LAMP-1/2), mucin-16 (MUC16), Mac-2, podocalyxin (PODXL), and death receptor-3 (DR3). Malignant transformation is often associated with abnormal glycosylation such as increased sialyl Lewis-a/x structures. On carcinoma cells, sialyl Lewis-a/x are mostly carried by mucins, making them major E-selectin ligands on carcinoma cells. The physiological role of E-selectin is to mediate the adhesion of leukocytes to the endothelium. In pathological conditions, cancer cells "hijack" the inflammatory system to interact with E-selectin. On the surface of endothelial cell, E-selectin molecules cluster in clathrin-coated pits and lipid rafts. This

distribution pattern of E-selectin enhances its ability to mediate rolling in flow conditions.

The sequence of events is as follows: the C-type lectin domain of E-selectin binds its ligand on cancer cells. This primary adhesion is unstable under shear stress, which allows the rolling of cancer cells along the endothelium. In response to the E-selectin-mediated attachment, chemokines are produced and released by endothelial cells, which activate integrins on cancer cells. Integrins are capable of firmly binding to cell adhesion molecules (ICAM-1/2 and VCAM-1) on endothelial cells, allowing the extravasation of cancer cells into tissues (Fig. 2). Breast, bladder, gastric, and pancreatic carcinoma, leukemia, and lymphoma form metastasis in an E-selectin-dependent manner in organs as various as liver, bone marrow, skin, and lung.

The interaction between E-selectin and its ligand triggers signals in both endothelial cells bearing E-selectin and cancer cells bearing the ligand. When E-selectin binds to DR3 on colon carcinoma cells, on one hand, this interaction activates not only the prosurvival ERK MAP kinase and PI3K pathways but also the

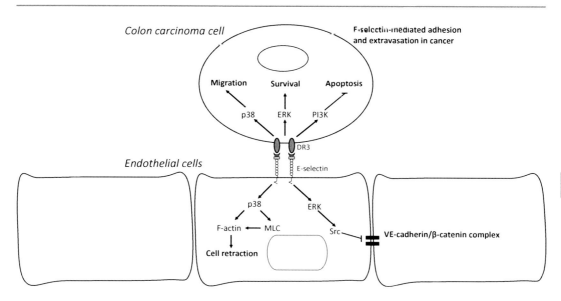

E-Selectin-Mediated Adhesion and Extravasation in Cancer, Fig. 3 E-Selectin-Mediated Bidirectional Signaling in colon carcinoma cells and endothelial cells. The adhesion of colon carcinoma cells to endothelial cells involves the binding of E-selectin on endothelial cells to counter-receptors such as death receptor-3 (DR3) on colon carcinoma cells. The interaction between E-selectin and DR3 induces activation of PI3K, p38 and ERK MAP kinases in cancer cells, which increase their motile and survival potentials. Reciprocally, the interaction triggers the activation of p38 and ERK MAP kinases in endothelial cells, which results in myosin-light chain (MLC)-mediated cell retraction, and dissociation of the VE-cadherin-β-catenin complex, and thereby destruction of adherens junctions leading to increased endothelial permeability and extravasation of cancer cells

promigratory p38 MAP kinase pathway in colon carcinoma cells; on the other hand, in endothelial cells, the interaction activates p38 and ERK MAP kinase pathways to increase the permeability of the endothelium (Fig. 3). Similar mechanism has also been observed with ESL-1, where E-selectin binds to ESL-1 on the circulating prostate cancer cell, and activates the pro-metastatic RAS-ERK-cFos signal cascade in the cancer cell. Moreover, CD44 on melanoma cells can bind to E-selectin on endothelial cells and activate PKCα-p38-SP-1 pathway to up-regulate ICAM-1 on endothelial cells. This bidirectional signal transduction also characterized the tethering of leukocytes: E-selectin binding to PSGL-1 on neutrophils activates β2 integrin through the Syk-Src pathway. At the same time, E-selectin transduces signals into endothelial cells through p38 and p42/p44 MAP kinase pathways. Overall, E-selectin-mediated adhesion of cancer cells increases their metastatic potential by inducing bidirectional signaling that enhances their intrinsic motile and survival

abilities, as well as the permeability of the endothelium.

E-selectin is a double-edged sword in cancer therapy, as it also allows lymphocyte infiltration into tumor. Some cancer cells are able to reduce E-selectin to evade immune detection: squamous cell carcinomas can recruit nitric oxide (NO)–producing myeloid-derived suppressor cells, and this local production of NO inhibits vascular E-selectin expression, preventing T cells from entering in squamous cell carcinomas. In this case, lower E-selectin level is correlated with lower survival. On the other hand, many types of cancer cells benefit from E-selectin in a variety of ways. The recruitment of leukemia cells by E-selectin sequesters leukemia cells in a quiescent state, rendering them immune to chemotherapy. Given that leukemia cells can stimulate endothelial cells by themselves, they promote their own survival through E-selectin. In addition, proliferating hemangioma endothelial cells from infantile hemangioma constitutively express E-selectin, which enhances hemangioma stem cell adhesion and vasculogenesis.

A new role in stem cell proliferation has been identified for E-selectin on bone marrow vascular endothelial cells: it recruits hematopoietic stem cells that express appropriate ligands, and this attachment wakes hematopoietic stem cells up by inducing their proliferation, self-renewal, chemosensitivity, and radiosensitivity.

Clinical Relevance of E-Selectin in Cancer

The finding that cancer cells are recruited by E-selectin-expressing endothelial cells is of significant clinical importance and opens several therapeutic avenues.

Targeting E-selectin and Its Ligands
Various strategies targeting E-selectin and its ligands are promising to suppress E-selectin-mediated cancer cell adhesion. For example, antibodies against E-selectin can impair lung metastasis of colon carcinoma in mice. ESTA, an aptamer targeting E-selectin, is able to reduce metastasis of breast cancer in mice. ESTA is safe as an antagonist as it can be applied at high doses without causing overt side effects. It is of particular interest for the prevention of metastasis of ER (−)/CD44(+) breast cancers. Similarly, SDA, a DNA aptamer antagonizing E- and P-selectins, also exhibits antiadhesive effect for colorectal cancer and leukemia in vitro. In mice, encouraging results have been obtained with colon cancer metastasis by using cimetidine to inhibit E-selectin expression. In clinical trials, cimetidine treatment dramatically improves the 10-year cumulative survival of colorectal cancer patients. Moreover, atrial natriuretic peptide (ANP) is capable of reducing E-selectin expression and preventing recurrence in patients with non-small cell lung cancer. E-selectin is also a target for inhibition of angiogenesis. Knocking down vascular E-selectin in mice inhibited the recruitment of endothelial progenitor cells to the tumor, thus reducing angiogenesis and tumor growth in human melanoma xenograft murine model. Another approach for reducing E-selectin-mediated adhesion is to target ligands of E-selectin. Antibodies against sialyl Lewis-a/x determinants inhibited the formation of metastasis by human pancreatic and gastric cancers in nude mice. The antisense-cDNA for fucosyltransferase genes (FUT III/VI), enzymes producing sialyl Lewis saccharides, suppressed metastatic colonization by colon cancer cells in mice. Along the same lines, celecoxib, an inhibitor of cyclooxygenase-2, impaired the expression of sialyl Lewis-a on colon cancer cells and reduced metastasis. Still in mice, re-introduction of the glycosyltransferase B4GALNT2, which synthesizes "normal" saccharides instead of sialyl Lewis-a/x, prevented dissemination of gastric carcinoma.

Soluble E-selectin as a Diagnostic Marker
A soluble form of E-selectin (sE-selectin) is generated by enzymatic cleavage or when activated endothelial cells shed their damaged parts. The concentration of sE-selectin is directly correlated with its cell surface expression. sE-selectin limits E-selectin-mediated rolling by competing for binding sites on the leukocyte, thus downregulates the inflammatory response. sE-selectin can be used as a marker of activation of endothelium and is therefore useful for diagnosis of acute inflammation and metastasis. Specifically, for breast cancer patients, high sE-selectin level is associated with liver metastasis. In patients with non-small cell lung cancer, high sE-selectin level correlates with poor prognosis when cancer cells express sialyl Lewis-a/x. Increased sE-selectin also characterizes patients with chronic lymphatic leukemia. For patients suffering from oral cavity cancer, higher level of E-selectin correlates with higher risk of cancer transformation and relapse. Hence, the determination of blood sE-selectin on tumor biopsies is of prognostic value.

E-selectin-Mediated Capture
Highly metastatic circulating cancer cells express mucins with increased sialyl Lewis-a/x (such as MUC1), and these mucins consistently expose their core epitope. Nanostructured surface coated with E-selectin offers a method to selectively

capture viable cancer cells from blood samples. This E-selectin-mediated capture allows fast analysis and elimination of circulating cancer cells. For instance, microtube surface with E-selectin-functionalized liposomal doxorubicin specifically captured breast adenocarcinoma MCF7 cells from the perfusion, and induced significant cancer cell death.

E-selectin as a Receptor for Targeted Delivery

E-selectin can serve as a receptor for the delivery of anti-inflammatory drugs, anticancer drugs, and imaging markers in endothelial cells. For this purpose, antibodies against E-selectin, or artificial ligands of E-selectin are conjugated to the surface of polymeric particles. These immunoparticles are used to encapsulate the agent, so they can selectively bind to E-selectin-expressing endothelial cells and get internalized, together with the agent inside them. This technique allows the specific delivery of drugs to the proinflammatory microenvironment harboring tumor cells. Naturally, immunoparticles targeting E-selectin can also directly compete with cancer cells to bind to E-selectin. Based on these principles, intravenous injections of two E-selectin-targeting drug-carrying immunoparticles, P-(Esbp)-DOX and P-(Esbp)-KLAK, inhibited primary tumor growth and metastasis of lung carcinoma in mice. Moreover, the "drug free" immunoparticle P-(Esbp) also exhibited antimetastatic effects by competing with circulating lung carcinoma cells. By targeting E-selectin, we can also carry out targeted gene therapy, if viral vectors are encapsulated. E-selectin thioaptamer-conjugated multistage vector (ESTA-MSV) can carry therapeutic anti-STAT3 siRNA to bone marrow vascular endothelium of mice, and infect breast cancer cells there. In vitro, anti-E-selectin lipoplexes can deliver anti-VE-cadherin siRNAs to inflamed primary vascular endothelial cells originating from different vascular beds, which are generally difficult to transfect.

Overall, E-selectin-mediated endothelial adhesion plays a key role in metastasis, which opens new avenues for therapeutic interventions aiming at inhibiting this fatal complication of cancer.

Glossary

Angiogenesis Angiogenesis is a multistep process that refers to the formation of new blood vessels from pre-existing ones. In cancer, angiogenesis is required to feed the tumors. It is initiated by an hypoxic signal generated by cancer cells and that leads to the expression of angiogenic agents such as vascular endothelial growth factor (VEGF) that will switch on angiogenesis and activate endothelial cells.

Extravasation Extravasation refers to the passage of cancer cells from blood or lymph vessels to the surrounding tissues.

MAP kinases The MAPK cascades are highly conserved signaling networks that transduce the signals elicited by stress and physiological stimuli. The ERK pathway is the best known of these cascades. It is activated, for example, through the binding of agonists to tyrosine kinase (TK) receptors, which results into auto-phosphorylation on tyrosyl residues, hence creating docking sites for adapter proteins and enzymes, followed by the activation in cascade of the GTPase Ras, the MAP kinase kinase kinase (MAPKKK) Raf, the MAP kinase kinases (MAPKK) MEK-1/2, and finally the MAP kinase (MAPK) ERK. In turn, ERK phosphorylates a number of cytoplasmic and nuclear proteins to regulate cell functioning. The functions of the signaling molecules along the MAPK pathways are regulated by scaffolding proteins such as CNK in the ERK pathway. The scaffolding proteins facilitate or restrict the enzyme/substrate interactions by modulating the mutual availability of the signaling components.

Metastasis Metastasis consists in the formation of secondary tumor sites distant from the primary site. The capacity to metastasize is a characteristic of all malignant tumors.

Cross-References

▶ Adherens Junctions
▶ Adhesion
▶ Cell Adhesion Molecules

► MAP Kinase
► Metastasis
► microRNA

References

Bird NC, Mangnall D, Majeed AW (2006) Biology of colorectal liver metastases. A review. J Surg Oncol 94:68–80

Gout S, Morin C, Houle F et al (2006) Death receptor-3, a new E-selectin counter-receptor that confers migration and survival advantages to colon cancer cells by triggering p38 and ERK MAPK activation. Cancer Res 66:9117–9124

Gout S, Tremblay PL, Huot J (2008) Selectins and selectin ligands in extravasation of cancer cells and organ selectivity of metastasis. Clin Exp Metastasis 25:335–344

Jubeli E, Moine L, Vergnaud-Gauduchon J, Barratt G (2012) E-selectin as a target for drug delivery and molecular imaging. J Control Release 158: 194–206

Kannagi R, Izawa M, Koike T et al (2004) Carbohydrate-mediated cell adhesion in cancer metastasis and angiogenesis. Cancer Sci 95:377–384

Khatib AM, Auguste P, Fallavollita L et al (2005) Characterization of the host proinflammatory response to tumor cells during the initial stages of liver metastasis. Am J Pathol 167:749–759

Läubli H, Borsig L (2010) Selectins promote tumor metastasis. Semin Cancer Biol 20(3):169–177

Suárez Y, Wang C, Manes TD, Pober JS (2009) Cutting edge: TNF-induced microRNAs regulate TNF-induced expression of E-selectin and intercellular adhesion molecule-1 on human endothelial cells: feedback control of inflammation. J Immunol 184(1):21–25

Tremblay PL, Auger FA, Huot J (2006) Regulation of transendothelial migration of colon cancer cells by E-selectin-mediated activation of p38 and ERK MAP kinases. Oncogene 25:6563–6573

See Also

(2012) Cyclooxygenase-2. In: Schwab M (ed) Encyclopedia of cancer, 3rd edn. Springer, Berlin/Heidelberg, p 1035. doi:10.1007/978-3-642-16483-5_1435

(2012) Death receptor-3. In: Schwab M (ed) Encyclopedia of cancer, 3rd edn. Springer, Berlin/Heidelberg, p 1065. doi:10.1007/978-3-642-16483-5_1537

(2012) Endothelium. In: Schwab M (ed) Encyclopedia of cancer, 3rd edn. Springer, Berlin/Heidelberg, p 1258. doi:10.1007/978-3-642-16483-5_1905

(2012) Extravasation. In: Schwab M (ed) Encyclopedia of cancer, 3rd edn. Springer, Berlin/Heidelberg, p 1370. doi:10.1007/978-3-642-16483-5_2080

(2012) Galectin-3. In: Schwab M (ed) Encyclopedia of cancer, 3rd edn. Springer, Berlin/Heidelberg, p 1490. doi:10.1007/978-3-642-16483-5_2306

(2012) Integrin. In: Schwab M (ed) Encyclopedia of cancer, 3rd edn. Springer, Berlin/Heidelberg, p 1884. doi:10.1007/978-3-642-16483-5_3084

(2012) Sialyl Lewis-a/x determinants. In: Schwab M (ed) Encyclopedia of cancer, 3rd edn. Springer, Berlin/Heidelberg, pp 3402–3403. doi:10.1007/978-3-642-16483-5_5293

ESF

► Erythropoietin

ESM-1

► Endocan

ESO1

► NY-ESO-1

Esophageal Adenocarcinoma

Landon Inge
Norton Thoracic Institute, St. Joseph's Hospital and Medical Center, Phoenix, AZ, USA

Synonyms

Barrett adenocarcinoma; Oesophageal adenocarcinoma

Definition

Esophageal adenocarcinoma (EAC) is an epithelial malignancy of the lining of the esophagus. EAC is distinct from squamous cell carcinoma of the esophagus, arising from the glandular, metaplastic columnar epithelium characteristic of Barrett esophagus (BE).

Characteristics

Esophageal cancer is the sixth most frequent malignancy worldwide. Squamous cell carcinoma accounts for the majority of esophageal cancers; however, epidemiological data reveal increased incidence of EAC within Western countries. In particular, EAC now accounts for the majority of esophageal cancers in the USA and is one of the few malignancies whose incidence is continuing to increase in the USA (Simard et al. 2012). Five-year relative survival rates (2000–2007) are 47.8% for localized cancer, 8.9% for cancer with regional metastasis, and 2.9% for cancer with distant metastasis. Overall survival for EAC is poor, a consequence of the frequent presence of metastasis at diagnosis (71% of diagnoses).

Epidemiology, Risk Factors, and Clinical Treatment

Within the USA, analyses of data from the National Cancer Institute Surveillance, Epidemiology, and End Results program show that Caucasian males and females carry the highest risk of EAC relative to other ethnic groups. These findings are paralleled by high incidence of EAC in Western Europe and Australia, and increases in EAC incidence have been reported in Japan and Singapore.

Epidemiological analyses have defined several risk factors for EAC. Obesity and symptomatic gastroesophageal reflux disease (GERD) are currently the strongest risk factors for EAC. Case-control and cohort studies show that obesity, defined as a body mass index >30 kg/m^2, increases the relative risk of EAC by 2.4–2.8. Similarly, symptomatic GERD increases the relative risk of EAC fourfold. Additional risk factors (poor diet and smoking) have also been identified. Barrett esophagus (BE), a metaplastic response to GERD, carries a low (~0.5% per year) but significant risk of EAC and is thought to affect 1.5–1.6% of the population in the Western world. Despite the low risk of EAC within overall BE patient population, subsets of BE patients undergo pathological progression, developing neoplastic morphology (dysplasia or intraepithelial neoplasia) within their BE lesion. In addition, histological diagnosis of dysplasia frequently occurs concomitantly with a diagnosis of invasive EAC. As such, current treatment practices are designed to identify and remove dysplasia in patients with BE, despite the understanding that the bulk of these patients will likely never progress to dysplasia and EAC. Current treatment guidelines for BE patients in the USA advocate endoscopic biopsy, combined with pathological review at 1–2-year intervals in order to diagnose dysplasia or noninvasive EAC early, allowing for endoscopic removal of the dysplastic BE/EAC lesion via radio-frequency ablation or resection. For patients diagnosed with invasive/metastatic EAC, treatment is limited to surgery and chemotherapy combine with radiation therapy. Staging determines the treatment approach with patients with early stage (I–III) treated with surgery or preoperative chemoradiation followed by surgery and chemoradiation without surgery for advanced (stage IV, metastatic) disease. Preferred chemotherapies for EAC are platinum-containing DNA cross-linkers (cisplatin, oxaliplatin, carboplatin) combined either with an antimetabolite (5-fluorouracil, capecitabine), a topoisomerase inhibitor (irinotecan, epirubicin), or a mitotic inhibitor (paclitaxel). For EAC patients whose tumors display HER2-neu (ERBB2) overexpression, the addition of the HER2-neu inhibitor, trastuzumab, to standard chemotherapy has been shown to be beneficial.

Mechanisms of Esophageal Adenocarcinoma Pathogenesis

Chronic inflammation has a central role in EAC carcinogenesis. Retrospective analyses of human samples harboring the pathological progression from BE to EAC, as well as murine models of EAC pathogenesis, reveal increases in several inflammatory molecules (IL-1β, IL-8, IL-6, COX-2) and activation of the pro-inflammatory signaling (NF-κB, STAT3). Supplementation with nonsteroidal anti-inflammatory agents (NSAIDs) reduces progression to EAC in surgical murine models of BE to EAC pathogenesis, while retrospective epidemiological analysis found that BE patients taking NSAIDs had a lower risk of EAC. Finally, establishment of a chronic inflammatory microenvironment within the murine esophagus via ectopic, constitutive expression of the pro-inflammatory cytokine, IL-1β, induces EAC disease pathogenesis comparable to the human disease process (esophagitis-BE-EAC) (Quante et al. 2012). Development of chronic inflammation is a consequence of GERD. The repeated exposure to the gastric refluxate, comprised of gastric acid, digestive enzymes, bile salts, and ingested food and their metabolites, results in acute and chronic inflammation and consequently oxidative stress and DNA/tissue damage. Introduction of bile salts or reduced pH in cell culture replicates the inflammatory state within BE and EAC cell lines. Obesity is thought to also contribute to the chronic inflammatory state by either increasing the episodes of GERD or via increased systematic chronic inflammation associated with an obese state. Other contributors, such as diet and smoking have also been demonstrated; however, exactly how they contribute to EAC pathogenesis is still unclear. Some evidence suggests that increased intake of nitrosamines, found at high levels in processed meats as well as in tobacco, is a possible mechanism. Concurrently, there is limited evidence of genetic susceptibility for EAC, based upon studies of families with hereditary history of BE and EAC, as well as larger population-based studies of BE and EAC patients for genetic variants.

Mutational Landscape of Esophageal Adenocarcinoma

Several high-resolution genetic analyses of human EAC tumors have been completed (Nones et al. 2014; Weaver et al. 2014; Dulak et al. 2013). Of particular note, these studies show a high mutation frequency (8.0–9.9 mutations/Mb of DNA) and prevalence of T:A>G:C transversions. A single study found evidence of chromothripsis (Nones et al. 2014), a massive genomic rearrangement during a single catastrophic event, present in 32% of EAC tumors. Combined, these characteristics are suggested to be reflective of the deleterious microenvironment (GERD, chronic inflammation) EAC arises in. Frequent mutational inactivation of several tumor suppressor genes (*CDKN2A*, *ARID1A*, and *SMAD4*) are also reported. *TP53* is the most frequently mutated gene across all three datasets. *TP53* mutations were also found to designate dysplastic BE, and identification of *TP53* had value as a diagnostic marker of high-grade dysplasia (Weaver et al. 2014). In addition to mutational inactivation, methylation of *CDKN2A* is frequently present in EAC. Gain-of-function mutations in known oncogenes have not been found at a high incidence; however, other studies have reported amplification in the *ERBB2*, *KRAS*, *GATA4*, *ERBB1*, and *CCNE1* genes in EAC.

Cross-References

► CDKN2A
► HER-2/Neu
► *KRAS*
► Obesity and Cancer Risk
► TP53

References

Dulak AM et al (2013) Exome and whole-genome sequencing of esophageal adenocarcinoma identifies recurrent driver events and mutational complexity. Nat Genet 45:478

Nones K et al (2014) Genomic catastrophes frequently arise in esophageal adenocarcinoma and drive tumorigenesis. Nat Commun 5:5224

Quante M et al (2012) Bile acid and inflammation activate gastric cardia stem cells in a mouse model of Barrett-like metaplasia. Cancer Cell 21:36

Simard EP, Ward EM, Siegel R, Jemal A (2012) Cancers with increasing incidence trends in the United States: 1999 through 2008. CA: Cancer J Clin

Weaver JM et al (2014) Ordering of mutations in preinvasive disease stages of esophageal carcinogenesis. Nat Genet 46:837

See Also

Barrett esophagus (2012) In: Schwab M (ed) Encyclopedia of cancer, 3rd edn. Springer, Berlin/Heidelberg, p 342. doi:10.1007/978-3-642-16483-5_527

Esophageal Cancer

Ruggero Montesano
International Agency for Research on Cancer, Lyon, France

Definition

Esophageal cancer comprises two main types of malignant epithelial neoplasms: squamous cell carcinoma, originating from the lining squamous epithelium of the esophagus, and adenocarcinoma (also called Barrett adenocarcinoma), originating from metaplastic columnar epithelium in the lower part of the esophagus.

Characteristics

Esophageal cancer is the sixth most frequent cancer worldwide. In 2001, the estimated number of deaths due to esophageal cancer amounted to about 338,000 out of a total of 6.2 million cancer deaths. Of those, more than 80% occurred in developing countries, the majority being squamous cell carcinomas. The occurrence of this cancer varies greatly in different parts of the world, with areas of high mortality rate per year in regions of South Africa, Northeast of Iran and China (30 or more per 100,000 in males and 10 per 100,000 in females). In Europe or USA, the age standardized annual mortality of squamous cell carcinoma is no more than 5 in males and 1 in females per 100,000. There are, however, areas in Europe, namely in Normandy and Brittany in France and North east of Italy, where the

mortality rates, at least in males, are as high as those observed in China.

Adenocarcinoma of the esophagus is less frequent and occurs mainly in industrial countries. However, epidemiological data show increasing numbers of adenocarcinoma cases and this type of cancer accounts for more than 50% of all esophageal cancer in the USA.

The 5-year survival rate for patients with squamous cell carcinoma and adenocarcinoma of the esophagus is similarly poor (~10%), with no difference between industrial and developing countries. This is mainly due to their late detection and the poor therapy efficacy. No reliable prognostic markers are available.

Epidemiological studies have clearly shown that tobacco smoke (▶ Tobacco Carcinogenesis) and alcohol, together with a low intake of fresh fruits, vegetables, and meat, is causally associated with squamous cell carcinoma. It is estimated that in industrial countries, approximately 90% of this cancer is attributable to tobacco and alcohol consumption. Other risk factors are chewing of betel, consumption of pickled vegetables, and hot mate drink in Southeast Asia, China, and South America, respectively.

Adenocarcinoma of the esophagus arises from Barrett esophagus, a condition in which the normal squamous epithelium is replaced by metaplastic columnar epithelium. This condition is frequently present in patients with chronic gastroesophageal reflux and these patients have a more than 100-fold higher risk than the general population to develop adenocarcinoma.

Squamous cell carcinoma and adenocarcinoma of the esophagus show multiple genetic alterations (point mutations, allelic loss, and gene amplification) of several oncogenes and tumor suppressor genes. The most interesting observation in both cancer types is the high prevalence of mutations (up to 80%) of the tumor suppressor gene *p53*. In addition a distinct pattern of p53 mutations, namely a high prevalence of G>A transitions at CpG sites in adenocarcinoma and a higher prevalence of G>T transversions and mutations at A:T base pairs in squamous cell carcinoma. There is good evidence that the

This entry was first published in the 2nd edition of the Encyclopedia of Cancer in 2009.

Esophageal Cancer,
Fig. 1 Incidence of
esophageal cancer in males
from selected world regions
(From Parkin et al. (1999))

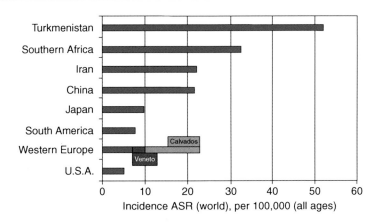

Incidence ASR (world), per 100,000 (all ages)

mutations in squamous cancer types are attributable to carcinogens present in tobacco smoke. In both types of cancers, p53 mutations occur very early and are followed by the accumulation of other genetic alterations during the process of esophageal carcinogenesis. It is evident that these genetic alterations are relevant not only in the understanding of the multifocal monoclonal origin of this cancer but also to the elucidation of its multifactorial etiology (Fig. 1).

Cross-References

► Tobacco Carcinogenesis

References

Devesa SS, Blot WJ, Fraumeni JF Jr (1998) Changing patterns in the incidence of esophageal and gastric carcinoma in the United States. Cancer 83:2049–2053
Montesano R, Hainaut P (1998) Molecular precursor lesions in oesophageal cancer. Cancer Surv 32:53–68
Montesano R, Hollstein M, Hainaut P (1996) Genetic alterations in esophageal cancer and their relevance to etiology and pathogenesis: a review. Int J Cancer 69:225–235
Munoz N, Day N (1997) Esophageal cancer. In: Schottenfeld D, Fraumeni JF (eds) Cancer epidemiology and precvention, 2nd edn. Oxford University Press, Oxford, pp 681–706
Parkin DM, Bray FI, Devesa SS (2001) Cancer burden in the year 2000. The global picture. Eur J Cancer 37(Suppl 8):S4–S66
Pisani P, Parkin DM, Bray F et al (1999) Estimates of the worldwide mortality from 25 cancers in 1990. Int J Cancer 83:18–29

ESR1

► Estrogen Receptor

ESR2

► Estrogen Receptor

Estradiol

Jose Russo and Irma H. Russo
Breast Cancer Research Laboratory, Fox Chase
Cancer Center, Philadelphia, PA, USA

Definition

E2 (estradiol) or estradiol-17β is biologically the most active estrogen, and circulating estrogens are mainly originated from ovarian steroidogenesis in premenopausal women and peripheral aromatization of ovarian and adrenal androgens in postmenopausal women.

Characteristics

Mechanism of Action

It is generally accepted that the biological activities of estrogens are mediated by nuclear ► estrogen receptors (ER) which, upon activation by cognate ligands, form homodimers with another ER–ligand complex and activate the transcription of specific genes containing the estrogen response elements (ERE). According to this classical model, the biological responses to estrogens are mediated by either of the two estrogen receptors, ERα and ERβ. The presence of ERα in target tissues or cells is essential to their responsiveness to estrogen action. The expression levels of ERα in a particular tissue have been used as an index of the degree of estrogen responsiveness. For example, human ► breast carcinomas are initially positive for ERα, and their growth can be stimulated by estrogens and inhibited by antiestrogens. ERβ can be activated by estrogen stimulation and blocked with antiestrogens. Upon activation, ERβ can form homodimers as well as heterodimers with ERα. The existence of two ER subtypes and their ability to form DNA-binding heterodimers suggest three potential pathways of estrogen signaling: via the ERα or ERβ subtype in tissues exclusively expressing each subtype or via the formation of heterodimers in tissues expressing both ERα and ERβ. In addition, estrogens and antiestrogens can induce differential activation of ERα and ERβ to control transcription of genes that are under the control of an ► AP-1 element.

Sources of Estrogens

Circulating estrogens are mainly originated from ovarian steroidogenesis in premenopausal women and peripheral aromatization of ovarian and adrenal androgens in postmenopausal women. Three main enzyme complexes are involved in the synthesis of biologically active estrogen (i.e., estradiol-17β) (Fig. 1): (i) ► aromatase that converts androstenedione to estrone, (ii) estrone sulfatase that hydrolyzes the estrogen sulfate to estrone, and (iii) estradiol-17β hydroxysteroid dehydrogenase that preferentially reduces estrone to estradiol-17β in tumor tissues.

Role of Estrogens in Human Breast Carcinogenesis

There are three mechanisms that have been considered to be responsible for the carcinogenicity of estrogens:

- Receptor-mediated hormonal activity, which has generally been related to stimulation of cellular proliferation, resulting in more opportunities for accumulation of genetic damages leading to ► carcinogenesis
- ► Cytochrome P450 (CYP)-mediated metabolic activation, which elicits direct genotoxic effects both by increasing mutation rates and the induction of ► aneuploidy
- Reduction of the ► DNA repair capability, which allows accumulation of lesions in the genome

Receptor-Mediated Pathway

The receptor-mediated activity of estrogen is generally related to induction of expression of the genes involved in the control of ► cell cycle progression and growth of human breast epithelium. The biological response to estrogen depends upon the local concentrations of the active hormone and its receptors. The proliferative activity and the percentage of ERα-positive cells are highest in Lob 1 in comparison with the various lobular structures composing the normal breast.

Even though it is now generally believed that alterations in the ER-mediated signal transduction pathways contribute to breast cancer progression toward hormonal independence and more aggressive phenotypes, there is also mounting evidence that a membrane receptor coupled to alternative second messenger signaling mechanisms is operational and may stimulate the cascade of events leading to cell proliferation. This knowledge suggests that ERα-negative cells found in the human breast may respond to estrogens through this or other pathways. The biological responses elicited by estrogens are mediated, at least in part, by the production of autocrine and paracrine growth factors from the epithelium and the stroma in the breast. In addition, evidence has accumulated over the last decade supporting the existence of ER variants, mainly a truncated ER and an exon-deleted ER. It has been suggested that expression of ER variants may contribute to breast cancer progression toward hormone independence.

Estradiol, Fig. 1 Steroidogenic pathways leading to the biosynthesis of estrogens (Reprinted from Russo J, Russo IH (2004) Biological and molecular basis of breast cancer. Springer-Verlag, Heidelberg)

Oxidative Metabolism of Estrogen

There is evidence that the oxidative catabolism of estrogens, which is mediated by various CYP complexes, constitutes a pathway of their metabolic activation to reactive free radicals and intermediate metabolites that can cause oxidative stress. Estradiol-17β and estrone, which are continuously interconverted by estradiol-17β hydroxysteroid dehydrogenase (or 17β-oxidoreductase), are the two major endogenous estrogens (Fig. 2). They are generally metabolized via two major pathways: hydroxylation at C-16α position and at the C-2 or C-4 positions (Fig. 2). The carbon position of the estrogen molecules to be hydroxylated differs among

various tissues, and each reaction is probably catalyzed by various CYP isoforms. For example, in MCF-7 human breast cancer cells, which produce catechol estrogens in culture, CYP 1A1 catalyzes hydroxylation of estradiol-17β at C-2, C-15α, and C-16α, CYP 1A2 predominantly at C-2, and a member of the CYP 1B subfamily is responsible for the C-4 hydroxylation of estradiol-17β. CYP3A4 and CYP3A5 have also been shown to play a role in the 16-α-hydroxylation of estrogens in human.

The hydroxylated estrogens are catechol estrogens that will easily be autoxidized to semiquinones and subsequently quinones, both of which are electrophiles capable of covalently

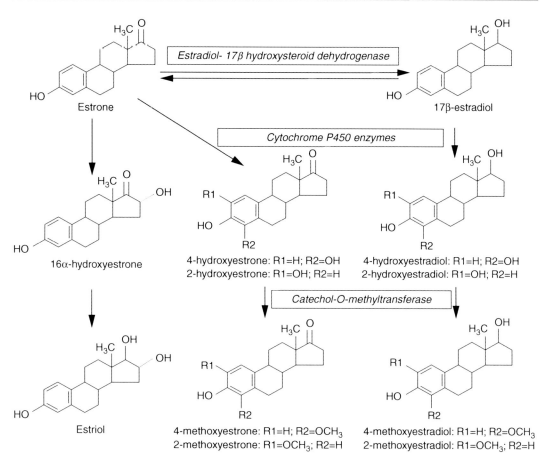

Estradiol, Fig. 2 Biosynthesis and steady-state control of catechol estrogens in human breast tissues (Reprinted from Russo J, Russo IH (2004) Biological and molecular basis of breast cancer. Springer-Verlag, Heidelberg)

binding to nucleophilic groups on DNA via a Michael addition and, thus, serve as the ultimate carcinogenic reactive intermediates in the peroxidatic activation of catechol estrogens. In addition, a redox cycle consisting of the reversible formation of the semiquinones and quinones of catechol estrogens catalyzed by microsomal P450 and CYP-reductase can locally generate superoxide and hydroxyl radicals to produce additional DNA damage. Furthermore, catechol estrogens have been shown to interact synergistically with nitric oxide present in human breast generating a potent oxidant that induces DNA strand breakage.

Steady-state concentrations of catechol estrogens are determined by the CYP-mediated hydroxylations of estrogens and monomethylation of catechols catalyzed by blood-borne catechol O-methyltransferase (Fig. 2). Increased formation of catechol estrogens as a result of elevated hydroxylations of estradiol-17β at C-4 and C-16α positions occurs in human breast cancer patients and in women at a higher risk of developing this disease. There is also evidence that lactoperoxidase, present in milk, saliva, tears, and mammary glands, catalyzes the metabolism of estradiol-17β to its phenoxyl radical intermediates, with subsequent formation of superoxide and hydrogen peroxide that might be involved in estrogen-mediated oxidative stress. A substantial increase in base lesions observed in the DNA of invasive ductal carcinoma of the breast has been postulated to result from the oxidative stress associated with metabolism of estradiol-17β.

The breast is an endocrine organ and can synthesize E2 in situ from precursor androgens via the enzyme ▶ aromatase. Breast tissue contains aromatase and produces amounts of E2 that exert biologic effects on proliferation. The effects of local production exceed those exerted in a classical endocrine fashion by uptake of E2 from plasma.

Estrogens as Inducers of Aneuploidy

Breast cancer is considered the result of sequential changes that accumulate over time. DNA content changes, i.e., loss of heterozygosity (LOH) and ▶ aneuploidy, can be detected at early stages of morphological atypia, supporting the hypothesis that aneuploidy is a critical event driving neoplastic development and progression. Aneuploidy is defined as the gain or loss of chromosomes; it is a dynamic, progressive, and accumulative event that is almost universal in solid tumors. The extensive array of altered gene expression observed in tumors and the numerous altered chromosomes detected by comparative genomic hybridization provide evidence that aneuploidy can disrupt cell homeostatic control. The main question is whether aneuploidy is a consequence of neoplastic development or a cause of neoplastic development. One of the several mechanisms proposed for the development of aneuploidy is the failure to appropriately segregate chromosomes, for example, interference with mitotic spindle dynamics, abnormal centrosome duplication, altered chromosome condensation and cohesion, defective centromeres, and loss of mitotic checkpoints. Functional consequences of centrosome defects may play a role during neoplastic transformation and tumor progression, increasing the incidence of multipolar mitoses that lead to chromosomal segregation abnormalities and aneuploidy. In considering estrogen as a carcinogenic agent, there is evidence that it affects microtubules. The importance of these findings is magnified with publications that demonstrate women on ▶ hormone replacement treatment that include progesterone have increased ▶ mammographic breast density

and increased breast cancer risk than women taking only estrogen.

In the center stage of the research endeavor on aneuploidy are the ▶ centrosomes that are organelles that nucleate microtubule growth and organize the mitotic spindle for segregating chromosomes into daughter cells, establishing cell shape and cell polarity. Centrosomes also coordinate numerous intracellular activities, in part by providing a site enriched for regulatory molecules, including those that control cell cycle progression, centrosome and spindle function, and ▶ cell cycle checkpoints. Although the underlying mechanisms for the formation of abnormal centrosomes are not clear, several possibilities have been proposed and implicated in the development of cancer such as alterations of checkpoint controls initiating multiple rounds of centrosome replication within a single cell cycle and failure of cytokinesis, cell fusion, and cell cycle arrest in S-phase uncoupling DNA replication from centrosome duplication.

References

Chakravarti D, Mailander P, Franzen J et al (1998) Detection of dibenzo[a, l]pyrene-induced H-ras codon 61 mutant genes in preneoplastic SENCAR mouse skin using a new PCR-RFLP method. Oncogene 16:3203–3210

Hu Y-F, Russo IH, Russo J (2001) Estrogen and human breast cancer. In: Matzler M (ed) Endocrine disruptors. Springer, Heidelberg, pp 1–26

Jefcoate CR, Liehr JG, Santen RJ et al (2000) Tissue-specific synthesis and oxidative metabolism of estrogens. In: Cavalieri E, Rogan E (eds) JNCI monograph 27: estrogens as endogenous carcinogens in the breast and prostate. Oxford Press, Oxford, pp 95–112

Liehr JG (2000) Is estradiol a genotoxic mutagenic carcinogen? Endocr Rev 21:40–54

Russo J, Russo IH (2000) Biological and molecular basis of breast cancer. Springer, Heidelberg

Estren-Dameshek Syndrome (= Variant Form)

▶ Fanconi Anemia

Estrogen Receptor

Gwendal Lazennec
INSERM, Montpellier, France

Synonyms

ER; ESR1; ESR2; Estrogen receptor alpha; Estrogen receptor beta

Definition

Estrogen receptors are represented by two main members (estrogen receptor alpha, ERα, and estrogen receptor beta, ERβ), which bind estrogens. These receptors are mainly nuclear, even though membrane receptors are also suspected to exist. Nuclear estrogen receptors belong to a large family of nuclear receptors, which are ligand-activated transcription factors able to modulate the expression of different genes.

Characteristics

History

The mediators of estrogens, namely, estrogen receptors, remained elusive until the synthesis of radiolabeled ▶ estradiol by Jensen and Jacobsen in 1960, which allowed the identification of such receptors. But it took more than 20 years, with the development at a high rate of molecular biology techniques, to have the first cDNA encoding estrogen receptor (α) cloned by the group of Chambon in Strasbourg. In 1996, when most of nuclear receptors had already been cloned, a great surprise arose with the fortuitous isolation of a second estrogen receptor (β) cloned from a ▶ Prostate Cancer Clinical Oncology library.

Molecular Mechanisms of Action

The genes coding for estrogen receptors ERα and ERβ are located on two distinct chromosomes (ERα on chromosome 6q25.1 and ERβ on chromosomes 14q22–14q24). ERα and ERβ proteins have a respective size of 595 and 530 amino acids (Fig. 1). As indicated by their membership to nuclear receptors, estrogen receptors are mainly present in the nucleus of cells, even in the absence of estrogens. In addition, a small pool of ERs localize to the plasma membrane and signal mainly through coupling to G-proteins, but we will not discuss this issue here. Upon binding to estrogens, nuclear estrogen receptors dimerize, bind coactivators, and interact with DNA or DNA-bound proteins to modulate the transcription of estrogen target genes (Fig. 2). Estrogen receptors share the common structure of nuclear receptors composed of 6 domains (A, B, C, D, E, and F) (Fig. 1). The C domain (DBD) is responsible for DNA binding. Two distinct synergistic transcriptional activation functions (AFs) have been identified: the ligand-independent AF-1 located in the N-terminal A/B region and the ligand-dependent AF-2 encompassing region

Estrogen Receptor, Fig. 1 Schematic representation of ERα and ERβ proteins

Estrogen Receptor,
Fig. 2 Mechanism of action of estrogen receptors. *CoA* coactivator, *E2* ▶ estradiol, *ERE* estrogen responsive element

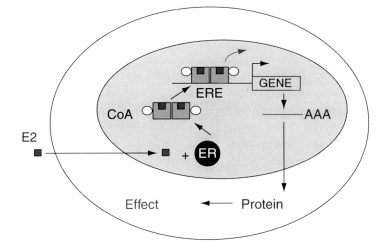

E (the LBD). Both ER AF-1 and AF-2 were found to act in promoter context and cell-specific fashions. Coactivators are mainly interacting with the E domain, but an increasing number of coactivators are also able to interact with the AB domain and the DBD. Estrogen receptors modulate the transcription either by binding to classical estrogen receptor element (ERE) or by interacting with ▶ AP-1, SP-1, or NF-kB-bound transcription factors. Many EREs consist of two inverted, palindromic half sites and are frequently present in the upstream regulation region of estrogen target genes. On the majority of ERE promoters, ERβ is less potent activator than ERα. Interestingly, ERα and ERβ form preferentially heterodimers rather than homodimers when they are both expressed in the same cell, which enables ERβ to decrease ERα transcriptional potential. Both estrogen-bound receptors mediate gene transcription similarly through the ERE pathway but have opposite effects when signaling through the AP-1 pathway. Indeed, ERα activates AP-1 gene transcription, whereas ERβ inhibits this pathway. Moreover, tamoxifen-liganded ERα is inactive on AP-1 elements, but ERβ is able to activate the transcription response.

What Have We Learned from ER Knockout Mice?

Disruption of the ERα gene (αERKO animals) is not lethal, but rather the animals develop normally and exhibit a life span comparable to their wild-type littermates. αERKO mice exhibit several abnormalities and deficiencies, most notable of which are the phenotypic syndromes that result in infertility in both sexes. The mammary phenotype of αERKO female mice demonstrates that embryonic mammary gland development is independent of ERα, but ERα is required for ductal elongation during puberty and complete mammary gland development in the mature mouse. The female reproductive tract of αERKO undergoes normal pre- and neonatal development but is insensitive to estrogens during adulthood. Ovaries undergo normal pre- and neonatal development, but are anovulatory during adulthood, exhibit multiple hemorrhagic cysts, and have no corpora lutea. There is also a 30–40% incidence of ▶ Ovarian Cancer tumors by 18 months of age. Male reproductive tract of αERKO displays dilation of rete testis, atrophy of the seminiferous epithelium, and decreased sperm counts, leading to infertility.

Mice lacking ERβ (βERKO) display a much less pronounced phenotype than αERKO. They develop normally and are indistinguishable grossly and histologically from their littermates. Sexually mature βERKO females are fertile and exhibit normal sexual behavior, but they produced substantially fewer litters as well as significantly less number of pups per litter when compared with their wild-type littermates. This reduction in fertility is the result of reduced ovarian efficiency. The mutant females have normal breast development and lactate normally. However, lactating glands display alveoli which are larger, and there is less secretory epithelium in ERβ than in wild-type mice. Ovaries undergo normal pre- and neonatal development, but do not exhibit normal frequency of spontaneous ovulations during adulthood, and exhibit a severely attenuated response to superovulation treatment with reduced number of oocytes and multiple trapped preovulatory follicles. In addition, male βERKO mice develop prostate hyperplasia.

ER disruption has also distinct effects on sexual behavior of the animals. αERKO male mice, although they rarely ejaculate and are infertile, show almost normal levels of mounts and just reduced levels of intromissions. In contrast, all three components of sexual behaviors are present and robust in βERKO males. Aggressive behavior is greatly reduced in αERKO male mice. In particular, male-typical offensive attacks are almost completely abolished, whereas lunge and bite attacks are still present. On the other hand, the aggressive behavior of βERKO is not reduced but rather elevated depending on age and social experiences.

Involvement of Estrogen Receptors in Physiology and Cancer

Estrogens play a critical role in many physiologic processes, including reproduction, cardiovascular health, bone integrity, immune system, cognition, and behavior. But estrogens are also involved in many pathologies including osteoporosis, endometriosis, or neurodegenerative diseases, but also cancer (breast, ► Ovarian Cancer, ► Endometrial Cancer, ► Prostate Cancer Clinical Oncology, ► Colorectal Cancer Premalignant Lesions).

ERα and ERβ display disparity in their tissue distribution. Studies on rat have shown that ERα mRNA is predominant in the uterus, mammary gland, testis, pituitary, liver, kidney, heart, and skeletal muscle, whereas ERβ transcripts are significantly expressed in the ovary and prostate. In humans, ERα and ERβ are both present in the brain, breast, cardiovascular system, and bone. In addition, ERα is much more abundant than ERβ in the uterus and liver, whereas the opposite is true for ERβ in the ovary, prostate, and gastrointestinal tract. This differential distribution

suggests that the two receptors could play distinct roles.

The most characterized cancer location involving estrogen receptors is definitely the breast. The cumulative exposure of the breast epithelium to estrogens is one of the first risk factors for ► breast cancer incidence. Estrogens are not only believed to be involved in the development and growth of breast tumors but also the late course of the disease, in their ► metastasis. The potential carcinogenic properties of estrogens are also suspected for endometrial and ► ovarian cancers. Two current hypothesis could account for estrogen carcinogenic action. First, estrogen metabolism could lead to the production of genotoxic products which directly alter DNA integrity. Second, by stimulating cell proliferation, estrogens could increase the number of cell divisions and thus the risk of replication errors, leading possibly to selective advantages for the mutated cells, which would exhibit defects in terms of apoptosis, proliferation, and DNA repair. It is interesting to note that this second hypothesis can be further divided into two subhypothesis. Indeed, ERα is generally expressed at low levels in normal breast epithelium cells (approximately 10–20% depending on the phase of menstrual cycle). Moreover, in the normal breast, it seems that there is a discordance between the cells which express ERα and the one that proliferate. For this reason, some people believe that estrogens act directly on epithelial cells to stimulate the proliferation, and others suggest that estrogens act on stromal cells, which in turn secrete soluble growth factors stimulating epithelial cell growth.

In contrast to the situation observed in the normal breast, breast cancer cells express frequently ERα, and the same cells are actively proliferating. Breast cancers are first divided into two subtypes based on whether or not the tumor cells express ERα. ERα-positive cancers represent about two thirds of all breast cancers. The reason for analyzing ERα status is based on the facts that estrogens constitute one of the major mitotic signals for ERα-positive breast cancers. Moreover, antiestrogen drugs such as tamoxifen are commonly used as first-line therapy against ERα-positive breast cancers. Unfortunately though, the use of tamoxifen is rarely associated with long-term remissions in metastatic diseases.

In contrast to the situation described for ERα, it has been reported by several groups that ERβ expression was lower in cancerous tissues compared with normal tissues. This is true for breast, ▶ Ovarian Cancer, ▶ Prostate Cancer Clinical Oncology, lung, and colon cancers, suggesting that this could be a general trend. Moreover, the exogenous delivery of ERβ to breast and prostate cancer leads to an inhibition of proliferation of tumor cells both in vitro and in vivo. In addition, ERβ is able to reduce the ▶ invasion potential of tumor cells. Moreover, ERβ knockout animals display prostate hyperplasia suggesting the possible proliferation gatekeeper role of ERβ. Unlike ERβ, ERα levels are frequently correlated to tumor grade. Patients whose tumors express ERα have a longer interval to recurrence and an improved survival. Concerning ERβ, the prognostic value of this receptor is more controversial, even though several studies have shown that ERβ was associated with a longer disease-free survival.

At the present time, it is not completely understood why tumor cells in some instances lose the expression of ERα or ERβ. Several studies suggest that promoter hypermethylation of both genes could silence the activity of ERα and ERβ promoters. On the other hand, the hyperexpression of ERα in the vast majority of breast cancers is not at all clarified. In addition, the role of the numerous splice variants identified for ERα and ERβ transcripts is not well established. The alternative splicing of ERα mRNA occurs frequently in breast tumors, but the expression of these transcripts is also usually conserved not only in primary tumors but also in distant metastases. In the same line, several single nucleotide polymorphisms (SNPs) in ERα gene have been associated either with an increased or a decreased risk of breast cancer, whereas the situation is much less clear for SNPs found in ERβ gene.

Cross-References

▶ AP-1
▶ Breast Cancer

▶ Colorectal Cancer Premalignant Lesions
▶ Endometrial Cancer
▶ Estradiol
▶ Invasion
▶ Metastasis
▶ Ovarian Cancer
▶ Prostate Cancer Clinical Oncology

References

Deroo BJ, Korach KS (2006) Estrogen receptors and human disease. J Clin Invest 116:561–570
Jensen EV (2005) The contribution of "alternative approaches" to understanding steroid hormone action. Mol Endocrinol 19:1439–1442
Levin ER (2005) Integration of the extranuclear and nuclear actions of estrogen. Mol Endocrinol 19:1951–1959

See Also

(2012) Coactivators. In: Schwab M (ed) Encyclopedia of Cancer, 3rd edn. Springer Berlin Heidelberg, p 886. doi:10.1007/978-3-642-16483-5_1238
(2012) Estrogens. In: Schwab M (ed) Encyclopedia of Cancer, 3rd edn. Springer Berlin Heidelberg, p 1333. doi:10.1007/978-3-642-16483-5_2019
(2012) Nuclear Receptor. In: Schwab M (ed) Encyclopedia of Cancer, 3rd edn. Springer Berlin Heidelberg, p 2571. doi:10.1007/978-3-642-16483-5_4157
(2012) Transcription Factor. In: Schwab M (ed) Encyclopedia of Cancer, 3rd edn. Springer Berlin Heidelberg, p 3752. doi:10.1007/978-3-642-16483-5_5901

Estrogen Receptor Alpha

▶ Estrogen Receptor

Estrogen Receptor Beta

▶ Estrogen Receptor

Estrogen Receptor, Progesterone Receptor, and HER-2 Negative

▶ Triple-Negative Breast Cancer

Estrogen Signaling

Rosamaria Lappano and Marcello Maggiolini
Department of Pharmacy and Health and
Nutritional Sciences, University of Calabria,
Rende, Italy

Definition

Estrogen signaling refers to the transduction pathways which mediate the multifaceted biological actions of estrogens. Estrogen signaling triggers both rapid and genomic responses mainly through: (1) the nuclear estrogen receptor (ER)α and ERβ which act as transcription factors, (2) the membrane-localized ERs, and (3) the G protein-coupled estrogen receptor (GPER), formerly named GPR30 (Fig. 1).

Characteristics

Estrogens, the main female sex steroids, regulate many cellular processes involved in the growth, differentiation, and function of the reproductive systems and exert various biological effects in the cardiovascular, musculoskeletal, immune, and nervous systems in both men and women. In addition to their role in physiology, estrogens may influence the development of hormone-dependent tumors, like breast, endometrial, ovarian, thyroid, and prostate cancer. Endogenous

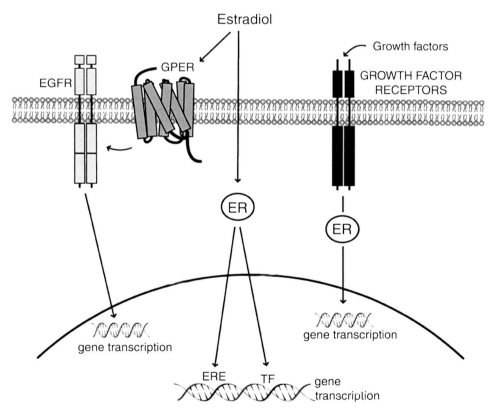

Estrogen Signaling, Fig. 1 Schematic representation of the estrogen signaling. The binding of 17β-estradiol (E2) to the classical ER modulates gene transcription by binding to specific DNA sequences (estrogen response elements, ERE) and other transcription factors complexes (TR) like AP-1 and Sp-1. ER can also be activated in a ligand-independent manner by growth factors. In addition, E2 activates the GPR30/GPER-mediated signaling, which involves the epidermal growth factor receptor (EGFR) and diverse transduction pathways leading to gene transcription

estrogens consist of estrone (E1), estriol (E3), and 17β-estradiol (E2) which is commonly recognized as the most potent female sex hormone. The biological responses to estrogens are mainly mediated by the ERs that initiate a complex array of cellular events upon ligand binding. Generally, these responses are divided into two categories: (i) activation of gene transcription referred to as genomic responses, which requires at least some hours, and (ii) rapid events that occur within seconds or minutes once cells are exposed to estrogens. Although these quick effects are referred to as nongenomic responses, they can lead to gene transcription confounding the aforesaid distinction.

ER-Mediated Signaling

Both the genomic and rapid signaling events initiated by estrogens were ascribed solely to the classical ERs. Estrogens bind to and activate the two thoroughly characterized members of the nuclear receptor superfamily, ERα and ERβ, which are products of two separate genes. Like all steroid receptors, the ERs consist of structurally and functionally distinct domains. The N-terminal domain (A/B region) contains the activating function 1 (AF-1) and several phosphorylation and sumoylation sites. The DNA binding domain (DBD) (C region), which is highly conserved among all nuclear receptors, is responsible for the binding to specific DNA sequences and is involved in protein dimerization. The D region is a hinge domain which contains the nuclear localization signal and different sites involved in posttranslational modifications. The C-terminal domain (E/F region) corresponds to the activating function 2 (AF-2), which includes the dimerization and the ligand binding domain (LBD). The E/F region is also responsible for the binding to chaperones (heat shock proteins 70 and 90). ERα and ERβ, which share a high degree of sequence homology except in their N-terminal domains, have similar affinity for E2 and bind to the same DNA response elements. However, triggering distinct gene transcriptions and signaling pathways, ERα and ERβ mediate specific responses and exert opposite effects on important cellular processes, including proliferation, migration, and apoptosis. Upon ligand activation, ERα and ERβ regulate the transcription through the binding to estrogen-responsive elements (ERE) located within the regulatory regions of target genes. The consensus ERE consists of a 5-bp palindrome with a 3-bp spacer (GGTCAnnnTGACC), although many natural EREs deviate substantially from the consensus sequence. For instance, the well-studied estrogen-responsive genes pS2, cathepsin-D, and progesterone receptor do not contain the perfect consensus ERE sequence. The cell-specific transcriptional response to estrogens depends on multiple factors, like the characteristics of gene promoters and coregulatory proteins. Coactivators which generally do not bind to the DNA can enhance the transcriptional activity once recruited to the promoter of target genes, whereas corepressors negatively regulate gene expression. In addition, posttranslational modifications, namely, acetylation, glycosylation, myristoylation, nitrosylation, palmitoylation, phosphorylation, sumoylation, and ubiquitination, have been reported to regulate ER functions.

Estrogens interacting with other transcription factors modulate the expression of genes whose promoters do not harbor EREs. For instance, the stimulating protein 1 (Sp-1) binds to the estrogen-responsive DNA regulatory regions, then ER enhances the binding of Sp1 to the DNA contributing to coactivator recruitment. Moreover, ERα and ERβ interact with the fos/jun transcription factor complex on the activator protein 1 (AP-1) sites to regulate gene expression.

Ligand-independent ER activation occurs by epidermal growth factor (EGF), insulin, insulin-like growth factor 1 (IGF-1), and transforming growth factor (TGF)-β. In addition, heregulin, interleukin 2, and dopamine can also modulate ER activity. Noteworthy, the growth factor-activated pathways contribute to the hormone-independent tumor progression toward a more aggressive phenotype.

As mentioned above, estrogens induce biological effects in a rapid manner that cannot be accounted for genomic actions. Intriguingly, these effects which are insensitive to inhibitors of transcription and translation occur via membrane-localized ERs. Estrogens activate several membrane initiated transduction cascades, like cAMP/PKA and PKC pathways, endothelial nitric oxide synthase (eNOS) promoting NO release, phospholipase (PL)C-dependent inositol trisphosphate (IP3) production, calcium influx, MAPK and p38 signaling, and phosphatidylinositol-3-kinase (PI3K)/AKT pathway. These estrogen-activated signals which occur in seconds or minutes are typically associated with plasma membrane receptors. In this vein, the palmitoylation of steroid receptors plays a key role as this biological process is required for their membrane localization and the rapid transduction of signals into cell biology. It should be also mentioned the ability of ligand-activated ERα in interacting directly with the IGF-1 receptor, which in turn triggers the MAPK-dependent pathway. As it concerns the interaction of membrane ERα with ErbB2 (HER-2/neu), it was involved in the resistance to tamoxifen-induced apoptosis in breast cancer cells. In addition, E2-activated ERα has been reported to transactivate the EGF receptor by a mechanism which involves the activation of G proteins, Src kinase, and matrix metalloproteinases that in turn activate the MAPK and AKT transduction pathways.

GPER-Mediated Signaling

In the last years, a member of the seven-transmembrane G protein-coupled receptor family, named GPER/GPR30, has been implicated in mediating both rapid and transcriptional responses to estrogens. GPER exhibits many of the expected characteristics of a membrane estrogen receptor, as it binds to estrogens generating biochemical signals mediated by plasma membrane-associated enzymes. For instance, GPER stimulates adenylyl cyclase through G proteins leading to increased cAMP levels. In addition, GPER signaling occurs via the EGFR transactivation which involves a Gβγ-dependent pathway and Src family tyrosine kinases. In particular, ligand-stimulated GPER activates metalloproteinases and induces the extracellular release of heparin-bound epidermal growth factor (HB-EGF), which binding to EGFR triggers

downstream transduction pathways, like ERK and PI3K signaling cascades as well as intracellular calcium mobilization. Likewise, GPER regulates the transcription of several genes which play relevant roles in diverse biological processes such as cell growth, migration, and differentiation. As GPER expression and function are regulated by EGF and insulin/IGF system, it may be considered as a further player involved in the cross-talk among estrogen signaling and these growth factors, toward important biological responses in both normal and cancer cells.

Estrogen Signaling in Cancer

Estrogens influence many physiological processes in mammals, including reproduction, cardiovascular health, bone integrity, cognition, and behavior. Given this widespread role for estrogens in human physiology, it is not surprising that these steroids are also implicated in the development and progression of numerous diseases, including osteoporosis, neurodegenerative and cardiovascular diseases, insulin resistance, endometriosis, and a variety of tumors. In particular, estrogens are critical mediators of breast cancer initiation, progression, and metastasis. At least two current hypotheses may explain this relationship. In the first, estrogens binding to ERα stimulate the proliferation of mammary cells, increasing the target cell number within the tissue. The increase in cell division and DNA synthesis elevates the risk for replication errors, which may result in the acquisition of detrimental mutations that disrupt normal cellular processes such as apoptosis, DNA repair, or cellular proliferation. In the second hypothesis, the metabolism of estrogens leads to the production of genotoxic products that could directly damage DNA resulting in mutations and other adverse effects. In addition to breast cancer, other malignancies including ovarian, endometrial, colon, pancreatic, prostate, and thyroid tumors have been associated with estrogens and estrogen receptor-mediated action.

On the basis of the main role elicited by estrogens in cancer and the existence of multiple estrogen-activated transduction pathways, the inhibition of estrogen signaling still represents one main strategy for targeting hormone-dependent cancer, like breast, ovarian, and endometrial tumors. Indeed, endocrine therapy by using antiestrogens such as the selective estrogen receptor modulator (SERM) tamoxifen and aromatase inhibitors, which ablate peripheral estrogen synthesis, has been shown to significantly improve disease-free survival. Despite the anticancer effects, initial or acquired resistance to endocrine therapies frequently occurs still representing a main concern. Cumulatively, a better understanding on the mechanisms by which estrogens may drive cancer progression is needed in order to identify novel biological targets and therapeutic strategies in hormone-sensitive tumors. In this regard, the effects of estrogens mediated by GPER should be considered in the development of novel SERMs, and the use of GPER antagonists could be included among the current pharmacological approaches in these types of cancer.

References

Ascenzi P, Bocedi A, Marino M (2006) Structure-function relationship of estrogen receptor alpha and beta: impact on human health. Mol Aspects Med 27:299–402

De Marco P, Cirillo F, Vivacqua A, Malaguarnera R, Belfiore A, Maggiolini M (2015) Novel Aspects Concerning the Functional Cross-Talk between the Insulin/IGF-I System and Estrogen Signaling in Cancer Cells. Front Endocrinol (Lausanne) 6:30

Deroo BJ, Korach KS (2006) Estrogen receptors and human disease. J Clin Invest 116:561–570

Hammes SR, Levin ER (2007) Extranuclear steroid receptors: nature and actions. Endocr Rev 28:726–741

Katzenellenbogen BS, Choi I, Delage-Mourroux R, Ediger TR, Martini PG, Montano M, Sun J, Weis K, Katzenellenbogen JA (2000) Molecular mechanisms of estrogen action: selective ligands and receptor pharmacology. J Steroid Biochem Mol Biol 74:279–285

Lappano R, De Marco P, De Francesco EM, Chimento A, Pezzi V, Maggiolini M (2013) Cross-talk between GPER and growth factor signaling. J Steroid Biochem Mol Biol 137:50–6

Lewis JS, Jordan VC (2005) Selective estrogen receptor modulators (SERMs): mechanisms of anticarcinogenesis and drug resistance. Mutat Res 591:247–263

Maggiolini M, Picard D (2010) The unfolding stories of GPR30, a new membrane-bound estrogen receptor. J Endocrinol 204:105–114

McDonnell DP, Wardell SE (2010) The molecular mechanisms underlying the pharmacological actions of ER

modulators: implications for new drug discovery in breast cancer. Curr Opin Pharmacol 10:620–628

Prossnitz ER, Barton M (2011) The G-protein-coupled estrogen receptor GPER in health and disease. Nat Rev Endocrinol 7:715–726

Thomas C, Gustafsson JÅ (2011) The different roles of ER subtypes in cancer biology and therapy. Nat Rev Cancer 11:597–608

Estrogen Synthase

▶ Aromatase and Its Inhibitors

Estrogenic Hormones

Elwood V. Jensen
National Institute of Health, Bethesda, MD, USA

Definition

Estrogens are steroid sex hormones produced chiefly in the ovary and responsible for development and function of the female reproductive tissues such as the uterus and mammary gland. Smaller amounts are also produced by the testis in the male and the adrenal gland in both sexes and contribute to maintenance of bone density and function of cardiovascular and neurological tissues. Estrogenic hormones play a role in both the genesis and treatment of several types of cancer. In general, estrogen-related tumors are those involving tissues of the female reproductive tract, although estrogens produce liver cancers in the hamster and are probably a factor in their occurrence in humans.

Characteristics

Etiology

Carcinogenesis is known to be a multistep process (▶ multistep development), involving both initiation (alteration of DNA) and promotion (proliferation of the altered cells). It is generally agreed that the principal effect of estrogenic hormones is on the promotion stage, especially in tissues where growth and function are normally regulated by estrogen. It has been controversial whether estrogens, especially in physiological amounts, also cause genetic changes in a manner similar to the action of chemical carcinogens such as dimethylbenzanthracene or nitrosourea.

Breast Cancer

The human malignancy most studied in relation to estrogen is ► breast cancer, both because of its high incidence and because its involvement with estrogens is especially striking. Much evidence indicates that estrogenic hormones play an important role in the appearance of mammary cancer, both in experimental animals and in the human. It has long been known that early menarche and/or late menopause increases the risk of ► breast cancer and that artificial menopause induced by ovariectomy or radiation reduces the risk, suggesting a cumulative effect of the number of ovulatory cycles on the incidence of the disease. It is also known that full-term pregnancy before the age of 20 years confers a significant protective effect, whereas nulliparous women have an increased susceptibility to breast cancer, but the basis of this phenomenon is not clear. The effect of hormone replacement therapy has been the subject of much investigation and some controversy, but from studies, it appears that for breast cancer, the risk from unopposed estrogen is small, and the addition of progestin to the regimen makes little difference.

The putative involvement of estrogens in the genesis of breast cancer has afforded an approach to its prevention that is currently under investigation. The antiestrogen tamoxifen was shown to prevent both the induction of mammary tumors by dimethylbenzanthracene in the rat and the appearance of cancer in the contralateral breast after mastectomy in the human. Following this, a large clinical trial has demonstrated that this agent, and related antiestrogens, can lower the incidence of breast cancer in women who are at high risk for developing this disease.

Uterine, Cervical, and Ovarian Cancer

Exposure to estrogen, unopposed by progestin, is a major factor in the occurrence of cancer of the uterine endometrium. Unopposed estrogen replacement therapy for more than 5 years results in an elevated risk of endometrial cancer, which persists for several years after the medication has been discontinued. The addition of progestin to the estrogen replacement regimen substantially reduces this risk. In comparison with cancers of the breast and uterus, involvement of estrogens in the etiology of cervical neoplasia is less clear. Most studies have been of the effect of oral contraceptives and generally have shown some correlation between the prolonged use of these agents and the incidence of cervical cancer. As in the case of breast cancer, late menopause, which gives rise to a longer period of ovulatory activity, results in an increased risk of ovarian cancer. Similarly, pregnancy and the use of oral contraceptives, which decrease the number of ovulatory cycles, are protective.

Vaginal Adenocarcinoma

During the period 1945–1955, large doses of estrogenic hormones were often administered to pregnant women with a history of miscarriage in the belief that this would protect against spontaneous abortion. Because orally active steroidal hormones were not available at that time, a synthetic estrogen called ► diethylstilbestrol (DES) was used. In the early 1970s, a previously rare cancer, clear cell ► adenocarcinoma of the vagina, began to appear in the daughters born to the DES-treated mothers, leading to the impression that estrogens in general, and DES in particular, are carcinogens and should not be used for human medication. However, the amounts administered (0.5–1.0 g) were 5,000–10,000 times the hormonally active dose, and the cancers produced were not in those persons receiving the estrogen but in their offspring, indicating that this is an in utero phenomenon and the action of the hormone is better described as teratogenic than carcinogenic. Longer follow-up has demonstrated a slightly increased incidence of breast cancer among the DES-treated mothers, but not what might be expected from such high doses of a true carcinogen. Genital abnormalities were produced in the male offspring, in keeping with a teratogenic phenomenon.

Hepatoma

In certain animal species such as the hamster, liver cancer can be induced by the simple administration of estrogens. In the Western world, primary liver cancer in humans is a relatively rare phenomenon, except for individuals with cirrhosis. However, it is a major cause of death in Asia and South Africa. The introduction of oral contraceptives has led to an increased incidence of liver tumors after long-term use of preparations containing substantial amounts of estrogen. It has been reported that hepatomas can arise from the use of prolonged ▶ diethylstilbestrol (DES) therapy for prostatic cancer.

Therapy

When cancer occurs in tissues where growth and function depends on estrogenic hormones, in some instances, the malignant cells retain their hormone dependency, while others lose the need for continued stimulation. It is not clearly established what is the exact basis for hormone dependency and whether escape from this regulation takes place on neoplastic transformation or during subsequent tumor progression. For cancers that retain hormone dependency, depriving them of estrogen provides an effective palliative treatment, less traumatic than cytotoxic chemotherapy, which is the only recourse for the majority of non-hormone-dependent metastatic cancers.

Breast Cancer

Hormone-dependent mammary tumors can be deprived of supporting estrogen either by removing the organs in which the hormone is produced, administration of substances that inhibit estrogen biosynthesis, or by giving a so-called antiestrogen that prevents the hormone from exerting its growth-stimulating effect in the cancer cells. More than a century ago, before it was known what estrogens are or that they are produced in the ovary, it was found that removal of the ovaries

from young women with advanced breast cancer caused remission of the disease in some patients. But the majority of breast cancers occur in postmenopausal women, where the ovaries are no longer functional, and it was long suspected that in the older patient the adrenal glands are the source of supporting estrogen. When cortisone became available, first for the treatment of inflammatory diseases, it became possible to remove the adrenal glands or the pituitary gland which controls them and maintain the patient on glucocorticoid replacement therapy. Subsequent clinical experience showed that about one-third of all the patients have mammary tumors that undergo remission when deprived of supporting hormone by any of these procedures, and endocrine ablation (the surgical removal of organs that produce hormones) became first-line therapy for advanced breast cancer, especially after methods were developed to predict which patients will or will not respond to endocrine manipulation.

When it was demonstrated that estrogens, like steroid hormones in general, exert their physiological actions in combination with specific receptor proteins, it was established that patients whose tumors contain low or negligible amounts of ▶ estrogen receptor (ER) rarely respond to any kind of endocrine therapy, whereas most, but not all, patients with ER-rich cancers benefit from such treatment. Determination of estrogen receptor on excised breast cancer specimens, either by immunological or hormone-binding procedures, is now a standard clinical practice.

As an alternative to endocrine ablation, hormone deprivation can be effected by inhibiting the enzymes involved in estrogen biosynthesis. This approach has the advantage that it eliminates not only estrogen arising from the ovary or adrenal gland but also that which, in some cases, appears to be produced by the tumor itself. The first successful agent for this purpose was aminoglutethimide, which inhibits the key enzyme, aromatase, but its clinical utility has been limited by undesirable side effects. Several improved compounds have been developed including fadrozole, letrozole, vorozole, and arimidex, which show promise of increased activity with reduced toxicity.

With the advent of ▶ tamoxifen, the first antiestrogen to be tolerated on prolonged administration, this reversible treatment has largely replaced the irreversible endocrine ablation as first-line therapy for endoplasmic reticulum (ER)-rich breast cancers. Although there are side effects from prolonged treatment as well as a slightly increased risk of endometrial cancer, the benefits greatly outweigh the drawbacks. Tamoxifen and related nonsteroidal compounds such as toremifene, raloxifene, and droloxifene show curious pharmacology in that depending on species, tissue, and dose, they can act either as stimulators or inhibitors. A limitation of tamoxifen therapy is the development in many patients of an "acquired tamoxifen resistance" in which the medication no longer inhibits but actually stimulates the growth of the cancer. Steroidal antiestrogens such as Faslodex (ICI 182,780) and RU 58668 have been developed, which show only inhibitory (antagonist) but not stimulatory (agonist) action.

Uterine and Cervical Cancers

Because growth and development of the uterus are stimulated by estrogen, attempts have been made to treat ▶ endometrial cancer with ▶ tamoxifen in a manner analogous to mammary cancer, but the response rate is low and variable. The most widely used hormonal therapy for this malignancy is treatment with ▶ progestin. Cervical cancer is especially sensitive to radiation and does not metastasize aggressively, and so surgery and radiotherapy are the usual therapeutic procedures, and endocrine therapy has found little application.

Cross-References

- ▶ Adenocarcinoma
- ▶ Breast Cancer
- ▶ Diethylstilbestrol
- ▶ Endometrial Cancer
- ▶ Estrogen Receptor
- ▶ Metastatic Colonization
- ▶ Multistep Development
- ▶ Progestin
- ▶ Tamoxifen

References

Holland JF, Frei E III, Bast RC Jr et al (2000) Cancer medicine, 5th edn. BC Decker, Hamilton

Jensen EV (1999) Oncology. In: Oettel M, Schillinger E (eds) Handbook of experimental pharmacology vol 135/II, estrogens and antiestrogens II pharmacology and clinical application. Springer, Berlin/Heidelberg/New York, pp 195–203

Li JJ, Li SA, Gustafsson J-Å et al (1996) Hormonal carcinogenesis II. Springer, Berlin/Heidelberg/New York

Lindsay R, Dempster DW, Jordan VC (1997) Estrogens and antiestrogens: basic and clinical aspects. Lippincott-Raven, Philadelphia/New York

Parl FF (2000) Estrogens, estrogen receptor and breast cancer. Ios Press, Amsterdam

See Also

(2012) Antiestrogens. In: Schwab M (ed) Encyclopedia of cancer, 3rd edn. Springer Berlin Heidelberg, p 209. doi:10.1007/978-3-642-16483-5_318

(2012) Aromatase. In: Schwab M (ed) Encyclopedia of cancer, 3rd edn. Springer Berlin Heidelberg, p 276. doi:10.1007/978-3-642-16483-5_394

(2012) Biosynthesis. In: Schwab M (ed) Encyclopedia of cancer, 3rd edn. Springer Berlin Heidelberg, p 415. doi:10.1007/978-3-642-16483-5_647

(2012) Cytotoxic chemotherapy. In: Schwab M (ed) Encyclopedia of cancer, 3rd edn. Springer Berlin Heidelberg, p 1058. doi:10.1007/978-3-642-16483-5_1499

(2012) Endocrine ablation. In: Schwab M (ed) Encyclopedia of cancer, 3rd edn. Springer Berlin Heidelberg, p 1223. doi:10.1007/978-3-642-16483-5_1871

(2012) Endometrium. In: Schwab M (ed) Encyclopedia of cancer, 3rd edn. Springer Berlin Heidelberg, p 1239. doi:10.1007/978-3-642-16483-5_1886

(2012) Endoplasmic reticulum. In: Schwab M (ed) Encyclopedia of cancer, 3rd edn. Springer Berlin Heidelberg, p 1240. doi:10.1007/978-3-642-16483-5_1887

(2012) Glucocorticoids. In: Schwab M (ed) Encyclopedia of cancer, 3rd edn. Springer Berlin Heidelberg, p 1558. doi:10.1007/978-3-642-16483-5_2429

(2012) Hormone replacement therapy. In: Schwab M (ed) Encyclopedia of cancer, 3rd edn. Springer Berlin Heidelberg, pp 1733-1734. doi:10.1007/978-3-642-16483-5_2815

(2012) Mastectomy. In: Schwab M (ed) Encyclopedia of cancer, 3rd edn. Springer Berlin Heidelberg, p 2177. doi:10.1007/978-3-642-16483-5_3548

(2012) Teratogenic. In: Schwab M (ed) Encyclopedia of cancer, 3rd edn. Springer Berlin Heidelberg, p 3651. doi:10.1007/978-3-642-16483-5_5731

Estrogen-Replacement Therapy

▶ Hormone Replacement Therapy

E

ET

▶ Endothelins

ET-2

▶ Endothelins

ET-743

▶ Trabectedin

Eta-1

▶ Osteopontin

Ether à-go-go Potassium Channels

Javier Camacho
Department of Pharmacology, Centro de Investigación y de Estudios Avanzados del I.P.N., Mexico City, D.F., Mexico

Synonyms

Eag; Eag1; KCNH1; Kv10.1

Definition

Ether à-go-go potassium channels are transmembrane proteins opening in response to changes in membrane potential and allowing the movement of potassium ions.

Characteristics

Ion transport is crucial for maintaining proper cellular function; this important task is performed by several ▶ membrane transporters including ▶ ion channels. Among other functions, ion channels play key roles in neurotransmission, muscle contraction, metabolism, sensory transduction, ▶ apoptosis, and cell cycle progression. Because of their pivotal role in cellular function, altered expression and/or activity of ion channels leads to several diseases including cardiac arrhythmias, diabetes, and epilepsy, turning these proteins into major drug targets. Participation of ion channels in cell proliferation and cell death makes them not only fundamental elements for the understanding of ▶ cancer but also potential clinical tools both for diagnosis and therapy of cancer.

Ether à-go-go (Eag) comprises a family of voltage-gated potassium channels opening in response to changes in membrane potential. Some members of the Eag family, namely, *human Eag1* (h-Eag1) and *human-Eag-related gene* (h-Erg), are linked to major diseases. h-Erg channels have an essential role in the cardiac action potential; h-Erg mutations produce the long Q-T syndrome type 2 leading to cardiac arrhythmias and eventually – in many cases – death of the patient. A huge number of very different drugs inhibit h-Erg channels producing cardiac arrhythmias as a nondesirable side effect; thus, h-Erg has become an intensively studied ion channel when designing new drugs. In addition, both Eag1 and Erg have been found to be overexpressed in a variety of tumors. h-Erg channels have been found in leukemic cells and biopsies from ▶ colorectal cancer and ▶ endometrial cancer; hence, h-Erg expression has been suggested as a molecular marker for human neoplastic hematopoietic cells and a potential prognostic factor for colorectal cancer. h-Eag1 is the other member of the EAG family involved in cancer. In addition to being overexpressed in many human tumors, h-Eag1 possesses oncogenic properties and has a more restricted distribution in normal healthy tissues in comparison with h-Erg channels, and specific inhibition of h-Eag1 gene expression reduces tumor cell proliferation. In the following, only oncogenic Eag1 channels will be discussed.

Oncogenic Potential of h-Eag1 Channels

Eag1 channels display oncogenic properties. Some characteristics of tumor cells are that they are able to grow in very low serum concentration, lose contact inhibition, and induce tumor formation when injected into immune-deficient mice. Cell lines that normally do not display these characteristics acquire properties of tumor cells when forced to express h-Eag1 channels. The oncogenic potential given to the cells by Eag1 channels is specific because the expression of another type of voltage-gated potassium channel in the same cell type does not induce tumor properties like those induced by Eag1. In addition, cell lines forced to express Eag1 have a higher metabolic activity and DNA synthesis than cells lacking Eag1 or expressing a different potassium channel. Findings of these oncogenic properties of Eag1 raised immediately research on the distribution of Eag1 in normal human tissues and tumor samples and its potential use as a cancer biomarker.

Eag1 as a Diagnostic Marker

Eag1 mRNA is distributed in a very restricted manner in healthy tissues. It is mainly expressed in the brain, although few amounts are found in the testis, adrenal gland, and placenta, and is also transiently expressed in myoblasts. Eag1 is expressed before myoblast fusion, and channel activity has been proposed to modulate the shape of the action potential in some cerebellar cells. Mutations in the Eag1 gene have been associated to epilepsy. In sharp contrast, Eag1 mRNA, protein expression (detected by specific antibodies), and protein activity (studied with the patch-clamp technique) have been found in several tumor cell lines and in a wide variety of biopsies from human tumors including lung, mammary gland, prostate, colon, and uterine-cervix.

Eag1 mRNA expression has been also suggested as a potential early indicator of tumor formation. Cervical cancer studies revealed Eag1 mRNA expression in all of the samples from carcinomas but also in some control samples from patients with normal cervix (diagnosed by pap smears studies);

these patients presented either human papillomavirus (HPV) infection or alteration in nearby regions (e.g., an ovarian cystadenoma or endometrial hyperplasia) suggesting Eag1 as a potential sign of early cellular alterations. In accordance to this idea, several cancer etiological factors induce Eag gene expression (see Fig. 1). Human papillomavirus oncogenes and hormones including estrogen and progesterone upregulate Eag mRNA and protein expression suggesting a novel mechanism by which HPV and hormones promote proliferation. In addition, chemical carcinogens induce Eag expression in a mouse model of colon cancer. Besides, Eag1 mRNA is present not only in mammary gland tumors but also in some free-tumor tissues from the vicinity of the tumors, while Eag1 is not found in commercially available RNA from normal mammary epithelium. Eag is also expressed in human diverticulitis which has the potential to turn into colonic cancer.

Diagnostic methods based on Eag1 expression are promising because in some cases (e.g., cervical cancer), cells can be obtained from the patients and tested for Eag1 protein expression with specific antibodies. On the other hand, it has been shown that fluorescent-labeled antibodies against Eag1 show the presence of the protein in lymph nodes that had not been clinically evident; this approach represents a noninvasive optical technique to detect Eag1. The restricted distribution of Eag1 in normal tissues, the abundant and ubiquitous expression in tumors, and the potential regulation by cancer-associated factors convert Eag1 expression in an attractive cancer biomarker.

Eag1 as a Therapeutic Target and Prognostic Marker

Several approaches have demonstrated that Eag1 expression in tumor cells has an important role in cell proliferation. Eag1 channels are inhibited by several nonspecific potassium channel blockers; however, at the moment there are no drugs inhibiting specifically Eag1 channel activity. Imipramine (a common antidepressant drug) and astemizole (a second-generation antihistamine) both inhibit several ion channels including Eag1 and decrease tumor cell proliferation of different cell lines expressing Eag1. These data strongly

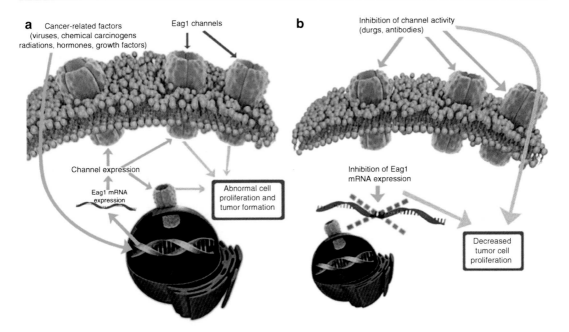

Ether à-go-go Potassium Channels, Fig. 1 Clinical relevance of Eag1 channels. (**a**) Eag1 channels are overexpressed in tumor cells favoring cell proliferation; some cancer-related factors are potential inducers of Eag1 channels which are expressed at the plasma membrane but also probably in the nucleus. (**b**) Inhibition of either channel activity or RNA expression decreases tumor cell proliferation. Therefore, Eag1 channels represent powerful tools both for cancer diagnosis and therapy

suggest that Eag1 channel activity is necessary in proliferation of human tumor cells. Actually, non-conducting Eag1 channels still produce tumor formation.

Molecular biology strategies like RNA silencing have been proved to be very useful for inhibiting specific gene expression. This approach has been used to study the role of Eag1 in proliferation of tumor cells. Specific inhibition of Eag1 gene expression by small interfering RNA (siRNA) caused a marked decrease in proliferation of several tumor cell lines. In some cases, specific Eag1 RNA silencing inhibited tumor cell proliferation in more than 80%. Very specific monoclonal anti-Eag antibodies have been developed. Some of them do not only detect the protein but also inhibit channel activity and display antitumor properties both in vitro and in vivo.

Targeting Eag1 for cancer therapy (see Fig. 1) offers at least two important advantages:

1. Because of Eag1 restricted distribution in normal tissues, targeted cells would be mainly cancer cells and the side effects in normal cells should be almost insignificant. In healthy tissues Eag1 is mainly expressed in the brain which is already protected by the blood-brain barrier.

2. The well-known drug resistance showed by some cancer cells via membrane transporters would be bypassed when using specific blockers of channel activity targeting the plasma membrane Eag1 channels. Eag is also a promising tool in cancer prognosis. Eag gene amplification was found in some samples from human colorectal adenocarcinoma and was significantly associated with adverse outcome. Eag amplification emerged as an independent prognostic marker for colon cancer. Eag expression has also been found in some types of leukemia. In the case of acute myeloid leukemia, Eag expression has been proposed as an independent predictive factor for reduced disease-free and overall survival expression strongly correlated with increasing age, higher relapse rates, and shorter survival. Besides,

high Eag1 expression is associated to poor survival in ovarian cancer patients. In summary, in addition to serving as a diagnostic marker and therapeutic target, Eag might serve also as a cancer prognostic marker.

Mechanisms of Oncogenicity

Molecular mechanisms explaining the oncogenic potential of Eag1 channels remain unknown; however, there are several current hypotheses which can be divided into two groups, general and particular. General hypothesis include the following:

1. Eag1 channels establish a negative membrane potential required for cell cycle progression from the G1 to the S phase. Potassium channel inhibition arrests the cells in the G1 phase of the cell cycle.
2. The negative membrane potential increases the electromotive force for calcium entry which in turn might trigger several transduction pathways.
3. Potassium movement regulates cell volume and volume changes are associated to cell proliferation, for instance, by altering nutrient concentration.

Particular current hypotheses are based on more structural features. Like other voltage-gated potassium channels, Eag1 protein is composed of four subunits (see Fig. 1) having six transmembrane-spanning segments numbered S1–S6; the S4 segment forms the voltage sensor due to its positively charged amino acid residues, and the loop between the segments S5–S6 forms the pore of the channel. Nevertheless, Eag1 has special sequences not shared as a whole with other potassium channels that provide potential clues on its oncogenic mechanisms and include the following:

Eag1 has a nuclear targeting signal presumably to direct the channel to the nucleus. It has been demonstrated for a calcium channel that a segment of the channel encodes a transcription factor regulating the expression of several genes. Something similar should be expected to happen with Eag1 if the channel or at least part of the protein

was expressed in the nucleus (see Fig. 1). Epsin is a protein participating in the endocytosis of growth factor receptors. Epsin binds to Eag1; probably this binding changes the free-epsin levels deregulating endocytosis of growth factor receptors and allowing growing signals to proceed. Calmodulin binding sites (integrating calcium signals), a nucleotide binding domain and a Pern-Arnt-Sim domain (PAS domain involved in responses to hypoxia), as well as putative phosphorylation sites by protein kinase C and mitogen-activated protein kinase (MAP kinase) described for Eag1 might also act in concert to regulate channel activity or interaction with other proteins and favor cell proliferation. Non-conducting channels produced by mutations that abolish ion permeation have been produced. These non-conducting channels fail to completely eliminate xenograft tumor formation by transfected cells. Thus, Eag oncogenic properties seem to not rely exclusively on the role of Eag as an ion channel. In addition, it is observed that Eag-expressing cells increase hypoxia-inducible factor 1 activity, vascular endothelial growth factor secretion, and tumor vascularization. Because of different regions of the protein, Eag channels might have several oncogenic mechanisms. Eag1 knockout mice have been produced. It will be very important to know how tumor development occurs in these mice.

Outline

Expression of Eag1 potassium channels confers oncogenic properties to mammalian cells. Because of their very restricted distribution in normal human tissues but more general distribution in tumor samples, Eag1 mRNA and/or protein expression offers potential tools for the diagnosis of a wide variety of neoplasms. Moreover, the regulation of Eag1 by cancer etiological factors like human papillomavirus, hormones, and carcinogens suggests Eag1 as a potential indicator of early cellular transformation. Specific inhibition of Eag1 produces a drastic decrease in tumor cell proliferation, making Eag1 a promising target for cancer therapy. Important advances have been achieved on the potential molecular mechanisms of Eag1 oncogenicity. In conclusion, Eag1 represents a hopeful tool for cancer diagnosis, prognosis, and therapy.

Cross-References

- ▶ Apoptosis
- ▶ Cancer
- ▶ Colorectal Cancer
- ▶ Endometrial Cancer
- ▶ Ion Channels
- ▶ Membrane Transporters

References

Agarwal JR, Griesinger F, Stuehmer W et al (2010) The potassium channel Ether à go-go is a novel prognostic factor with functional relevance in acute myeloid leukemia. Mol Cancer 9:18

Asher V, Khan R, Warren A, Shaw R, Schalkwyk GV, Bali A, Sowter HM (2010) The Eag potassium channel as a new prognostic marker in ovarian cancer. Diagn Pathol 5:78

Diaz L, Ceja-Ochoa I, Restrepo-Angulo I et al (2009) Estrogens and human papillomavirus oncogenes regulate human Ether à go-go-1 potassium channel expression. Cancer Res 69:3300–3306

Downie BR, Sánchez A, Knötgen H et al (2009) Eag1 expression interferes with hypoxia homeostasis and induces angiogenesis in tumors. J Biol Chem 238:6234–6240

Farias LMB, Ocaña DB, Díaz L et al (2004) Ether à go-go potassium channels as human cervical cancer markers. Cancer Res 64:6996–7001

García-Quiroz J, García-Becerra R, Santos-Martínez N, Barrera D, Ordaz-Rosado D, Avila E, Halhali A, Villanueva O, Ibarra-Sánchez MJ, Esparza-López J, Gamboa-Domínguez A, Camacho J, Larrea F, Díaz L (2014) In vivo dual targeting of the oncogenic Ether-à-go-go-1 potassium channel by calcitriol and astemizole results in enhanced antineoplastic effects in breast tumors. BMC Cancer 14:745

Gómez-Varela D, Zwick-Wallasch E, Knötgen H et al (2007) Monoclonal antibody blockade of the human Eag1 potassium channel function exerts antitumor activity. Cancer Res 67:7343–7349

Hemmerlein B, Weseloh RM, Mello de Queiroz F et al (2006) Overexpression of Eag1 potassium channels in clinical tumours. Mol Cancer 5:41

Mortensen LS, Schmidt H, Farsi Z, Barrantes-Freer A, Rubio ME, Ufartes R, Eilers J, Sakaba T, Stühmer W, Pardo LA (2015) KV 10.1 opposes activity-dependent increase in Ca2+ influx into the presynaptic terminal of the parallel fibre-Purkinje cell synapse. J Physiol 593 (1):181–196

Ousingsawat J, Spitzner M, Puntheeranurak S et al (2007) Expression of voltage-gated potassium channels in human and mouse colonic carcinoma. Clin Cancer Res 13:824–831

Pardo LA, Del Camino D, Sánchez A et al (1999) Oncogenic potential of EAG K channels. EMBO J 18:5540–5547

Ramírez A, Hinojosa LM, Gonzales J d J, Montante-Montes D, Martínez-Benítez B, Aguilar-Guadarrama R, Gamboa-Domínguez A, Morales F, Carrillo-García A, Lizano M, García-Becerra R, Díaz L, Vázquez-Sánchez AY, Camacho J (2013) KCNH1 potassium channels are expressed in cervical cytologies from pregnant patients and are regulated by progesterone. Reproduction 146: 615–623

Rodríguez-Rasgado JA, Acuña-Macías I, Camacho J (2012) Eag1 channels as potential cancer biomarkers. Sensors 12:5986–5995

Schönherr R (2005) Clinical relevance of ion channels for diagnosis and therapy of cancer. J Membr Biol 205:175–184

Simons C, Rash LD, Crawford J, Ma L, Cristofori-Armstrong B, Miller D, Ru K, Baillie GJ, Alanay Y, Jacquinet A, Debray FG, Verloes A, Shen J, Yesil G, Guler S, Yuksel A, Cleary JG, Grimmond SM, McGaughran J, King GF, Gabbett MT, Taft RJ (2015) Mutations in the voltage-gated potassium channel gene KCNH1 cause Temple-Baraitser syndrome and epilepsy. Nat Genet 47(1):73–77

Ufartes R, Schneider T, Mortensen LS, de Juan Romero C, Hentrich K, Knoetgen H, Beilinson V, Moebius W, Tarabykin V, Alves F, Pardo LA, Rawlins JN, Stuehmer W (2013) Behavioural and functional characterization of Kv10.1 (Eag1) knockout mice. Hum Mol Genet 22(11):2247–2262

Etidronate

Definition

Etidronic acid (INN) or 1-hydroxyethane 1,1-diphosphonic acid (HEDP) is a ▶ bisphosphonate used in detergents, water treatment, cosmetics, and pharmaceutical treatment. Etidronic acid is a chelating agent and may be added to bind or, to some extent, counter the effects of substances, such as calcium, iron, or other metal ions, which may be discharged as a component of gray wastewater and could conceivably contaminate groundwater supplies. As a phosphonate it has corrosion inhibiting properties on unalloyed steel.

Clinically, etidronate disodium acts primarily on bone. It can inhibit the formation, growth, and dissolution of hydroxyapatite crystals and their amorphous precursors by chemisorption to calcium phosphate surfaces. Inhibition of crystal resorption occurs at lower doses than are required to inhibit crystal growth. Both effects increase as the dose increases.

Etidronate is used in radionuclide therapy, for instance, of ▶ prostate cancer (▶ Prostate Cancer Radionuclide Therapy). For patients with advanced prostate cancer efficient therapy of painful bony lesions is the primary goal of interdisciplinary treatment strategies. Preservation of quality of life appears to be the main aim rather than prolongation of life. Apart from oral pain relief and local irradiation systemic treatment with radionuclides offers low-risk radiotherapeutic strategies for the palliation of painful, multifocal osteoplastic bone metastases. Depending on the radiopharmaceutical substance chosen response and reduction of pain are described in 65–80%. The duration of pain relief lasts 6–12 weeks. During this time the morphine-based medication can be reduced and in some cases withdrawn, which positively affects quality of life. After improvement of myelosuppression treatment with radionuclides can be repeated. Patients have to be hospitalized for 2 days because of protection from radiation procedures.

Cross-References

- ▶ Bisphosphonates
- ▶ Prostate Cancer
- ▶ Prostate Cancer Radionuclide Therapy

Etiology of Prostate Adenocarcinoma

- ▶ Prostate Cancer Genetic Toxicology

Etoposide

Synonyms

VP-16

Definition

With common trade names Etopophos (etoposide phosphate) and VePesid, etoposide is a cytotoxic chemotherapeutic. Most commonly it is used to treat ▶ Ewing sarcoma, testicular cancer, ▶ lung cancer, lymphoma, nonlymphocytic leukemia, and glioblastoma multiforme. Etoposide inhibits DNA synthesis, through inhibition of DNA topoisomerase II.

Cross-References

- ▶ Ewing Sarcoma
- ▶ Lung Cancer

See Also

(2012) DNA Topoisomerases II. In: Schwab M (ed) Encyclopedia of Cancer, 3rd edn. Springer Berlin Heidelberg, p 1141. doi:10.1007/978-3-642-16483-5_1691.

Ets Transcription Factors

Jürgen Dittmer
Klinik für Gynäkologie, Universität
Halle-Wittenberg, Halle (Saale), Germany

Definition

Ets transcription factors are defined by a unique DNA-binding domain, the ETS domain, which specifically interacts with an ~10 bp long DNA sequence containing a 5′-GGAA/T-3′ core motif (Fig. 1) Ets stands for E26 transformation specific or E26, as the Ets sequence (v-ets) was first

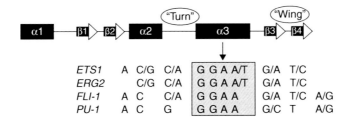

ETS1	A	C/G	C/A	G G A A/T	G/A	T/C	
ERG2		C/G	C/A	G G A A/T	G/A	T/C	
FLI-1	A	C	C/A	G G A A	G/A	T/C	A/G
PU-1	A	C	G	G G A A	G/C	T	A/G

Ets Transcription Factors, Fig. 1 The ETS domain. This winged helix-turn-helix domain binds DNA by a loop-helix-loop scaffold, composed of the helix(α2)-turn-helix(α3) motif and the loops between α2 and α3 (turn) and between the β strands β3 and β4 (wing). All direct contacts with specific bases of the DNA are made by residues in the α3 recognition helix while residues of the two loops contact the phosphate backbone. The resulting neutralization of the phosphate charges is likely to induce DNA bending, as observed in Ets protein-DNA complexes. In contrast to the helices, the loops are not strictly conserved among members of the Ets family. They may, therefore, be responsible for the preference of an individual Ets protein for the sequences flanking the conserved GGAA/T binding motif.

identified in the genome of the avian retrovirus E26. c-Ets1, closely related to v-ets, was the first cellular Ets protein that was discovered. More than 30 different Ets proteins have been identified, found throughout the metazoan world including mammals, sea urchins, worms, and insects. Currently, 27 human Ets proteins are known. The Ets family is subdivided into subfamilies based on the similarity in the ETS domain (Fig. 2).

Characteristics

In contrast to many other transcription factors, Ets proteins bind to DNA as monomers. Most eukaryotic cells express a variety of Ets proteins at the same time. To achieve functional specificity, Ets proteins display differences in preference for certain nucleotides flanking the core motif in the Ets-responsive DNA element and, more important, for certain cooperating partners. A strong interaction with a cooperating partner may even force Ets proteins to bind to an unfavorable DNA-binding site, such as GGAG (Pax5/Ets1 partnership). In many cases, interactions with other proteins depend upon particular protein domains, e.g., for the cooperation with SRF, the so-called B domain is required, which is found in the proteins of the TCF subfamily and the Fli-1 protein. The Pointed domain, named after the Drosophila Ets protein Pointed and shared by many Ets proteins of different subfamilies, shows similarities to the sterile alpha motif (SAM) domain and is an interface for homotypic and heterotypic protein-protein interactions. In Ets1 and Ets2 proteins, the Pointed domain is the docking site for ERK1/2 allowing these kinases to phosphorylate Ets1 and Ets2 at an N-terminal threonine. In contrast, the Pointed domain of the TEL protein mediates homo-oligomerization. Most Ets proteins are transcriptional activators; others (ERF, NET, Tel, Drosophila YAN, Caenorhabditis lin-1) act as repressors. Some, such as Elk-1 and NET, can undergo activator-repressor switching (Fig. 2).

Ets proteins play an important role in transcriptional regulation. Many eukaryotic genes contain Ets DNA-binding sites and are responsive to Ets proteins. Ets-responsive genes are found among critical genes that regulate fundamental cellular processes such as proliferation, differentiation, ▶ invasion, and ▶ adhesion.

Ets Factors and Development
Some Ets factors, including Ets2, Esx, Ese-2, Fli-1, Pu.1, GABPα, and Tel, are essential for embryonic development. Disruption of the Ets2, Ese-2, Fli-1, Pu.1, GABPα, or Tel gene in mice results in early death of the embryo. Lack of Ets2 or Ese-2 leads to defects in trophoblast development and to the absence of extraembryonic ectoderm markers. Ese-2 is also involved in mammary alveolar morphogenesis. Tel null mutant embryos fail to develop a vascular network in the yolk

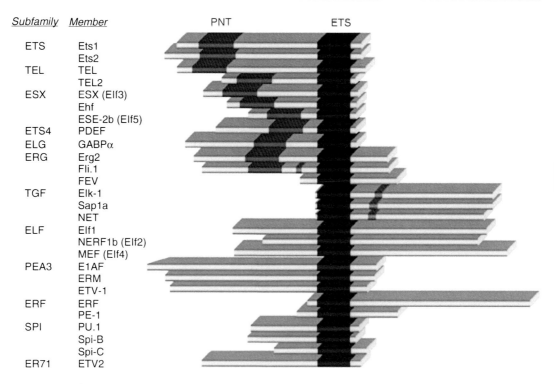

Ets Transcription Factors, Fig. 2 Members of the Ets transcription factor family in humans. The DNA-binding domain (ETS), the Pointed (PNT) domain, and the SRF-interacting B domain are marked. Note that most Ets proteins have several different names. The Esx and Elf proteins are grouped into two separate subfamilies. Splicing variants of the different Ets proteins are not listed. *Tel* translocation, *Ets* leukemia, *Esx* epithelial-restricted with serine box, *Ehf* Ets homologous factor, *ESE* epithelium-specific Ets, *PDEF* prostate-derived Ets factor, *ELG* ets-like gene, *GABP* GA-binding protein, *Erg* ets-related gene, *Fli* Friend leukemia integration, *FEV* fifth Ewing variant, *TCF* ternary complex factor, *Elk* ets-like gene, *Sap* SRF accessory protein, *NET* new ets transcription factor, *Elf* E74-like factor, *NERF* new ets-related factor, *MEF* myeloid Elf-1 like factor, *PEA3* polyoma enhancer A3, *E1AF* adenovirus E1A factor, *ERM* Ets-related molecule, *ETV* Ets translocation variant, *ER81* Ets-related clone 81, *ERF* Ets2 repressor factor, *PE* PU-Ets-related, *Spi* SFFV provirus integration site, *PU* recognizes purine-rich sequences

sac. Pu.1 is necessary for B- and T-cell development, erythropoiesis, terminal myeloid cell differentiation, and maintenance of hematopoietic stem cells. Fli-1 null embryos die of aberrant hematopoiesis and hemorrhaging. Deficiency of GABPα which is expressed in embryonal stem cells leads to embryonic death prior to implantation. GABPα is also required for the function of neuromuscular junctions. Mice lacking Esx die early after birth. Their intestinal epithelial cells fail to differentiate and polarize as result of reduced levels of the TGFβII receptor (▶ transforming growth factor β). Ets1 deficiency leads to defects in B- and T-cell development. In ER81 null mice, two types of mechanoreceptors, muscle spindles and the Pacinian corpuscles, are either absent or degenerated. MEF is involved in osteogenic differentiation.

Regulation of Ets Protein Activities

The activities of Ets proteins are controlled transcriptionally and posttranslationally. The expression of many Ets genes are restricted to certain cell types and/or can be induced by specific extracellular stimuli, e.g., the transcription from the Ets1 gene can be activated by a variety of factors including phorbol ester, ▶ AP-1, ▶ TP53, ▶ retinoic acid, ERK1/2 (▶ MAP kinase), and HIF-1 (▶ hypoxia-inducible factor-1). Many Ets proteins undergo posttranslational modifications, which have an impact on their activities. The most common posttranslational modification of Ets proteins is phosphorylation by MAP kinases, such as ERK1/2. Phosphorylation by MAPK leads to activation of activating Ets proteins, such as Ets1, Ets2, ER81, ERM, Sap1, Elk1, PEA3, or GABPα, and loss of activity of repressing Ets proteins, such as Tel or ERF. When phosphorylated by ERK1/2, NET even switches from a repressor to an activator phenotype. It seems that MAPK-dependent phosphorylation (phosphorylation of proteins) shifts the balance between Ets-dependent activation and repression toward activation. In the case of Ets1, MAPK-dependent phosphorylation enhances the transcriptional activity by recruitment of the coactivator CBP/p300 (▶ P300/CBP coactivators). Some Ets proteins are also targets of PKA, PKC (▶ protein kinase C family), CaMKII (▶ Calcium-Binding Proteins), CKII, and cyclin A-dependent cdk2 (▶ cyclin-dependent kinase). CKII increases the activity of Pu.1 and Spi-B, and PKCα activates Ets1. In contrast, PKA inhibits the DNA-binding activities of ER81 and ERM, whereas CaMKII and cdk2 act as inhibitory on Ets1 and GABPα, respectively. CaMKII phosphorylates Ets1 on serines of a serine-rich region flanking an autoinhibitory module that regulates Ets1 DNA-binding activity. The inhibitory effect of CaMKII on the Ets1 protein increases with each serine that is phosphorylated within the serine-rich region allowing fine-tuning of Ets1-dependent transcription.

A few Ets proteins, Ets1, Elk-1, and Tel, have been shown to undergo sumoylation. This post-translational modification inhibits transcriptional activity of Ets1 and Elk-1 and abrogates the repressing activity of Tel. When sumoylated, the activating Elk-1 protein even transforms to a repressor. Acetylation is another means nature uses to modify the activities of Ets proteins, such as Ets1 and ER81. ER81 becomes acetylated and phosphorylated in response to Her2/neu-stimulated signaling. Acetylation takes place on two lysines within the transactivation domain of ER81 increasing its DNA-binding affinity and protein stability. Elf-1 is an example of an Ets protein that becomes glycosylated. ▶ Glycosylation affects the subcellular localization and DNA-binding activity of Elf-1.

Ets Proteins and Cancer

The Ets proteins Ets1, Ets2, Fli-1, and Erg are able to transform murine cells. These and other Ets proteins are also involved in human carcinogenesis and/or tumor progression. This is in line with the fact that many of these Ets proteins are targets of the ▶ Ras/Raf/MEK/ERK signaling pathway which is often deregulated in human tumors. The Ras-responsive Ets1 protein is found in different types of solid tumors, including carcinomas and sarcomas. Its overexpression often correlates with increased invasion, higher tumor microvessel density, higher grading, and unfavorable prognosis. Ets1 has been linked to the regulation of key proteases, such as ▶ matrix metalloproteases, involved in the degradation of the ▶ extracellular matrix. In tumors, Ets1 is expressed by tumor cells as well as by stromal cells. By its ability to convert endothelial cells to an angiogenic phenotype, Ets1 is involved in tumor-dependent ▶ angiogenesis. A number of other Ets proteins, such as Erg, PEA3, and E1AF, are capable of upregulating proteases and supposed to be involved in tumor progression. PEA3 has particularly been linked to mammary gland development and oncogenesis. Fli-1 and Ets1 has been shown to regulate ▶ tenascin C, an extracellular matrix protein, associated with tumor progression. In some tumors, Ets genes are subject to mutations and recombinations. The inhibitory Ets protein Tel2 has been shown to induce myeloproliferative diseases in mice by cooperating with the ▶ Myc oncogene and stimulating proliferation. Elf-1 has been implicated in tumor-associated angiogenesis. A target of Elf-1 is Tie2 (▶ receptor tyrosine kinases), a receptor tyrosine kinase involved in the activation of endothelial cells. Chromosomal translocations leading to fusion proteins containing Ets proteins are observed in Ewing tumors and certain types of leukemias. EWS-Ets fusion proteins (▶ EWS-FLI (ets) fusion transcripts), most often containing Fli-1 or Erg, rarely ETV-1, EIAF, or FEV, are critically involved in the development of Ewing tumors. The fusion protein presumably acts as a transcription factor that binds through the Ets domain to Ets-responsive genes. In addition, EWS-Ets proteins have been suggested to interfere with RNA splicing. Ets fusion proteins, as found in leukemias, harbor either Tel or Erg2. Erg2 is fused to TLS, a protein structurally related to EWS. Hence, TLS-Erg2 chimeric proteins are supposed to have similar functions as EWS-Ets proteins. Tel is frequently fused to tyrosine kinases, such as PDGFRβ (▶ platelet-derived growth factor), Abl (▶ BCR-ABL1), or Jak2 (▶ signal transducer and activators of transcription in oncogenesis). The Pointed domain of Tel mediates homodimerization resulting in constitutively active kinases. In another fusion protein, Tel is linked to the DNA-binding factor AML-1 (▶ runx1) which together with CBFβ forms the transcription factor CBF. CBF activity is often inhibited in leukemic cells. As a result of Tel-dependent dimerization, CBF function is also blocked when AML-1 is fused to Tel.

References

Dittmer J (2003) The biology of the Ets1 proto-oncogene. Mol Cancer 2:29

Foos G, Hauser CA (2004) The role of Ets transcription factors in mediating cellular transformation. In: Handbook of experimental pharmacology, vol 166, Transcription factors. Springer, Berlin/Heidelberg/New York, pp 259–275

Seth A, Watson DK (2005) ETS transcription factors and their emerging roles in human cancer. Eur J Cancer 41:2462–2478

Tootle TL, Rebay I (2005) Post-translational modifications influence transcription factor activity: a view from the ETS superfamily. Bioessays 27:285–298

ETS Variant Gene 6

▶ E T V6

Everolimus

Daniel C. Cho
Beth Israel Deaconess Medical Center, Boston, MA, USA

Synonyms

Afinitor® (marketed by NOVARTIS); RAD001; 42-O-(2-hydroxyethyl) rapamycin

Definition

Everolimus is an orally administered inhibitor of ▶ mammalian target of rapamycin (mTOR). This agent has significant activity in renal cell carcinoma (RCC) and is approved by the Food and Drug Administration (FDA) for treatment of patients with RCC who have failed treatment with ▶ sorafenib or sunitinib.

Characteristics

Everolimus (RAD001, Afinitor®) is an ester of the immunosuppressant ▶ rapamycin which binds with high affinity to FK506-binding protein 12 (FKBP12), forming a complex that inhibits the kinase activity of mTOR. mTOR is a highly conserved serine/threonine kinase which is activated downstream of Akt (protein kinase B; ▶ Akt signal transduction pathway) and regulates cell growth and metabolism in response to environmental factors (Fig. 1). Once activated, mTOR executes its biologic functions as a critical component of two distinct complexes, TORC1 and TORC2, which have differential sensitivities to the rapalogues (i.e., everolimus, ▶ temsirolimus). TORC1, which includes mTOR, LST8 (GβL), and raptor (regulatory-associated protein of mTOR), is inhibited by the rapalogues, whereas TORC2, which includes mTOR, LST8, Sin1, and rictor, is insensitive to the rapalogues. Thus,

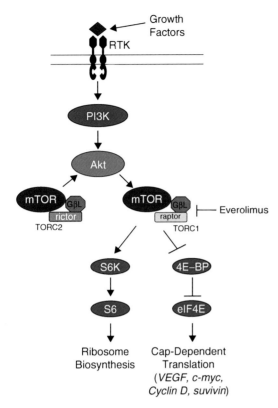

Everolimus, Fig. 1 mTOR signaling pathway and inhibition of TORC1 activity by everolimus

everolimus primarily inhibits the activity of TORC1, which typically acts through its downstream effectors, the eukaryotic translation initiation factor 4E binding protein (4E-BP) and the 40S ribosomal protein p70 S6 kinase (S6K), to stimulate protein synthesis and entrance into G1 phase of the cell cycle.

Mechanism of Action in Renal Cell Carcinoma
While the exact mechanism of action of everolimus in RCC remains unknown, inhibitors of TORC1 likely function by attenuating the translation of critical mRNA. Everolimus inhibits the phosphorylation of 4E-BP by TORC1, allowing 4E-BP1 to remain associated with eukaryotic translation initiation factor 4E (eIF4E). Thus sequestered, eIF4E is hindered from interacting with the 5′ untranslated region (UTR) of capped mRNA and initiating translation. It is now known that certain mRNA,

characterized by lengthy 5′UTR containing stem loop structures, are more dependent upon the availability of eIF4E and are therefore preferentially sensitive to the suppressive effects of TORC1 inhibition. This group of mRNA includes several gene products critical for malignant transformation and tumor progression such as ► vascular endothelial growth factor (VEGF), ► Myc oncogene, ► cyclin D, ► survivin, and ornithine decarboxylase (ODC). It is likely that suppression of translation of one or more of these critical mRNA is central to the clinical efficacy of TORC1 inhibitors. While it has been advocated that the efficacy of TORC1 inhibitors in RCC is primarily achieved through the inhibition of translation of hypoxia-inducible factors (HIF)-1α and HIF-2α (► hypoxia; ► hypoxia-inducible factor-1), evidence that the translation of HIF-2α, believed by many to be the more relevant HIF in RCC, is completely dependent upon the activity of TORC2 has cast some doubt upon this hypothesis.

Clinical Activity in Renal Cell Carcinoma
After demonstrating promising activity in a phase II trial in patients with predominantly clear cell RCC (► Renal Cancer Clinical Oncology), everolimus was assessed in a double-blind, randomized, *placebo*-controlled phase III trial in patients with advanced RCC who had failed treatment with either ► sorafenib or sunitinib. Patients were randomized in a 2:1 fashion to receive either everolimus 10 mg once daily ($n = 272$) or *placebo* ($n = 138$) in conjunction with best ► supportive care. Patients randomized to everolimus experienced a significantly longer progression-free survival (4.0 months, 95% CI 3.7–5.5) compared with those randomized to *placebo* (1.9 months, 95% CI 1.8–1.9) with a hazard ratio of 0.30 (95% CI 0.22–0.40; $p < 0.0001$). Toxicities, which were more common in the everolimus group, included stomatitis, rash, diarrhea, hyperglycemia, hypercholesterolemia, hyperlipidemia, and noninfectious pneumonitis. Based on these findings, everolimus was approved by the FDA on March 30, 2009, for the treatment of patients with advanced RCC who have failed prior therapy with sorafenib or sunitinib.

Conclusion

Having demonstrated activity in a randomized, *placebo*-controlled, phase III trial, everolimus is considered a standard therapeutic option for patients who have failed either sorafenib or sunitinib and is approved by the FDA for this indication. Based on its tolerability, everolimus is now being explored in combination with other active agents in RCC as well as in the adjuvant (▶ adjuvant therapy) setting.

Cross-References

▶ Adjuvant Therapy
▶ Akt Signal Transduction Pathway
▶ Cyclin D
▶ Hypoxia
▶ Hypoxia-Inducible Factor 1
▶ Mammalian Target of Rapamycin
▶ MYC Oncogene
▶ Rapamycin
▶ Renal Cancer Clinical Oncology
▶ Sorafenib
▶ Supportive Care
▶ Survivin
▶ Temsirolimus
▶ Vascular Endothelial Growth Factor

References

Amato RJ, Jac J, Giessinger S, Saxena S, Willis JP (2009) A phase 2 study with a daily regimen of the oral mTOR inhibitor RAD001 (Everolimus) in patients with metastatic clear cell RCC. Cancer 115:2438–2445

Graff JR, Konicek BW, Carter JH, Marcusson EG (2008) Targeting the eukaryotic translation initiation factor 4E for cancer therapy. Cancer Res 68:631–634

Meric-Bernstam F, Gonzalez-Angulo AM (2009) Targeting the mTOR signaling network for cancer therapy. J Clin Oncol 27:2278–2287

Motzer RJ, Escudier B, Oudard S, Hutson TE, Porta C, Bracarda S, Grunwald V, Thompson JA, Figlin RA, Hollaender N, Urbanowitz G, Berg WJ, Kay A, Lebwohl D, Ravaud A (2008) Efficacy of everolimus in advanced renal cell carcinoma: a double-blind, randomized, placebo-controlled phase III trial. Lancet 372:449–456

Toschi A, Lee E, Gadir N, Ohh M, Foster DA (2008) Differential dependence of hypoxia-inducible factors 1 alpha and 2 alpha on TORC1 and TORC2. J Biol Chem 283:34495–34499

See Also

(2012) Adjuvant. In: Schwab M (ed) Encyclopedia of cancer, 3rd edn. Springer Berlin Heidelberg, p 75. doi:10.1007/978-3-642-16483-5_107

(2012) Cell cycle. In: Schwab M (ed) Encyclopedia of Cancer, 3rd edn. Springer Berlin Heidelberg, p 737. doi:10.1007/978-3-642-16483-5_994

(2012) Ornithine decarboxylase. In: Schwab M (ed) Encyclopedia of cancer, 3rd edn. Springer Berlin Heidelberg, p 2656. doi:10.1007/978-3-642-16483-5_4259

(2012) Renal cancer. In: Schwab M (ed) Encyclopedia of cancer, 3rd edn. Springer Berlin Heidelberg, pp 3225–3226. doi:10.1007/978-3-642-16483-5_6575

(2012) Sunitinib. In: Schwab M (ed) Encyclopedia of cancer, 3rd edn. Springer Berlin Heidelberg, p 3562. doi:10.1007/978-3-642-16483-5_5575

(2012) Translation. In: Schwab M (ed) Encyclopedia of cancer, 3rd edn. Springer Berlin Heidelberg, p 3770. doi:10.1007/978-3-642-16483-5_5936

Ewing Sarcoma

Heinrich Kovar
Children's Cancer Research Institute, Vienna, Austria

Synonyms

Ewing sarcoma family tumors; Ewing tumor; Neuroepithelioma; Peripheral primitive neuroectodermal tumor

Definition

Aggressive small round cell tumor affecting bone and soft tissue in children and young adults. Ewing's sarcoma (ES) and peripheral primitive neuroectodermal tumor (pPNET), also called neuroepithelioma, are currently defined as biologically closely related tumors along a gradient of limited neuroglial differentiation. ▶ Askin tumor is the historical designation of Ewing's sarcoma of the chest wall. Today, all these neoplasms are summarized as Ewing tumors (ET) or Ewing sarcoma family tumors (ESFT). Although first described in 1866 and 1890 by Lücke and Hildebrand, respectively, the disease carries the name of the American pathologist James Ewing who, in 1921, was the first to recognize the tumor as a separate entity, which he defined as diffuse endothelioma of the bone.

On the genetic level, Ewing sarcoma family tumors are defined by the consistent presence of a reciprocal ► chromosomal translocation between the long arms of chromosome 22 and either chromosome 11 (85%) or 21 (10%). In rare cases, alternative rearrangements of chromosome 22 with either chromosome 7, 17, or 2 have been reported. These aberrations result in a gene fusion that serves as a diagnostic criterion allowing to discriminate Ewing sarcoma family tumors from osteomyelitis and childhood malignancies with a similar small round cell phenotype, including ► neuroblastoma, ► rhabdomyosarcoma, non-Hodgkin lymphoma, and small cell osteosarcoma. Immunohistochemically, Ewing sarcoma family tumor cells are defined by the abundant presence of the cell surface marker CD99. The exact histogenesis of the disease is not known. An origin from a very primitive, ectodermally derived, migrating cell from the neuroepithelium has been suggested based on ultrastructural and histological signs of limited neural differentiation. Experimental evidence indicates that at least part of the neural resemblance is a functional consequence of the characteristic gene fusion in Ewing sarcoma family tumors. Based on these molecular biological findings, a mesenchymal origin is currently discussed for the disease.

Characteristics

Ewing sarcoma family tumors comprise about 10–15% of malignant bone tumors with a yearly incidence of 0.6 per million in the Caucasian population. The disease is rarely observed among black Africans, African Americans, and Chinese people. It typically occurs in adolescence, in the second decade of life (average age at diagnosis is 13.5 years) with a slight male prevalence. Nevertheless, infants less than 5 years of age as well as adults up to 60 years of age are occasionally diagnosed with Ewing

sarcoma family tumors. The tumor usually presents as a painful swelling, rapidly increasing in size. The duration of symptoms prior to the definitive diagnosis can be weeks to months, or rarely even years, with a median of 3–9 months. In patients with metastatic disease, nonspecific symptoms such as malaise and fever may resemble symptoms of septicemia. The most frequent tumor localizations are the pelvis, the long bones of the extremities, the ribs, the scapula, and the vertebrae. Less frequently, Ewing sarcoma family tumors arise from extraosseous locations. Unlike ► osteosarcomas, Ewing sarcomas tend to arise from the diaphyseal rather than the metaphyseal portion of the bones and are frequently accompanied by tumor-related osteolysis, detachment of the periosteum from the bone, and spiculae of calcification in soft tissue tumor masses.

About 20–25% of Ewing sarcoma patients present at diagnosis with gross, clinically detectable metastases in the lung and/or in bone and/or bone marrow. Metastases to lymph nodes or other sites like liver or central nervous system (CNS) are rare. In contrast to patients with localized disease, cure of this group of patients is very difficult to achieve.

Classic Ewing sarcoma is composed of a monotonous population of small round cells with high nuclear to cytoplasmic ratios arrayed in sheets. The cells have scant, faintly eosinophilic to amphophilic cytoplasm, indistinct cytoplasmic borders, and round nuclei with evenly distributed, finely granular chromatin and inconspicuous nucleoli. Mitotic activity is usually low. By means of immunohistochemistry, the tumor cells occasionally stain positive for neuroglial markers, such as neuron specific enolase, S100 protein, chromogranin A and B, or the gene product PGP9.5. This is in addition to CD99, which is highly expressed in Ewing sarcoma family tumors with consistency. The inclusion of glycogen can frequently be observed in the tumor cells. Infrequently, Ewing sarcoma is focally immunoreactive for cytokeratins.

Cytogenetics and Gene Alterations

The most consistent marker of Ewing sarcoma family tumors is the rearrangement of the Ewing sarcoma gene EWS on chromosome 22 (band q12) with a gene encoding for an ETS transcription factor. These proteins are characterized by a unique structure of their DNA-binding domain determining target gene specificity. In the majority (85%) of Ewing sarcoma family tumors EWS is rearranged with Fli-1, which is located on chromosome 11 (band q24). The second most frequent translocation partner of EWS in this disease is the Ets family member ERG on chromosome 21 (band q22) (10%). In rare cases of Ewing sarcoma family tumors, complex or interstitial chromosomal rearrangements fuse EWS to related Ets transcription factor genes located on chromosomes 7 (band p22), 17 (band q12), and 2 (band q36). These gene rearrangements are currently monitored for diagnostic purposes, either on the chromosomal level using ► fluorescence in situ hybridization (FISH) or on the RNA level using reverse transcriptase polymerase chain reaction (PCR) (RT-PCR). The latter method also allows for high sensitivity detection of minimally disseminated disease in blood, bone marrow, and peripheral blood progenitor cell (PBPC) collections. The prognostic impact of RT-PCR detectable tumor cells in these samples is currently under prospective evaluation in several clinical studies.

As a result of the gene fusion, a potent novel transcription factor with altered structural and functional features is expressed in the tumor cells. EWS-Fli-1 and EWS-ERG fusion proteins have been shown to render mouse fibroblast cell lines and bone marrow-derived mesenchymal progenitor cells tumorigenic in animal models. Using antagonistic agents, generated by means of gene technology, involvement of these aberrant gene products in Ewing sarcoma tumor cell proliferation has experimentally been demonstrated. It is commonly assumed that the EWS-Ets chimeric transcription factors mediate their transforming properties by inappropriately activating or repressing other genes. In vitro gene transfer experiments into a number of different cell types indicated that the spectrum of genes responsive to EWS-Ets fusion proteins is context dependent. EWS-Fli-1 is toxic to most primary human cell types with only very few exceptions.

Among tissues tolerant of the expression of the Ewing sarcoma oncogene are bone marrow-derived mesenchymal progenitor cells. Here, EWS-Fli-1 interferes with the differentiation potential of these cells. With the advent of new technologies to modulate EWS-Ets expression in Ewing sarcoma cells, such as ► RNA interference (RNAi), we are just starting to understand the mechanisms underlying malignant transformation by this oncogene. Large scale gene expression profiling studies have identified a specific signature of EWS-Fli1 within the Ewing sarcoma transcriptome. Chromatin and epigenetic studies indicate that EWS-Fli-1 deregulates gene expression by large scale re-programing of distal gene regulatory elements and aberrant regulation of proximal gene promoters. Current functional studies attempt to separate malfunctions essential for tumorigenesis from collateral damage.

No reliable genetic indicators of prognosis have been identified for Ewing sarcoma family tumors, so far. Cytogenetically, trisomy 8 and 12 accompany the characteristic rearrangement of chromosome 22 in about 44% and 12% of tumors, respectively. Additional structural changes affect chromosomes 1 and 16 in about 20% of tumors, most frequently leading to a gain of 1q and a loss of 16q and the formation of a derivative chromosome. A possible prognostic impact of these cytogenetic alterations has been discussed controversially. Genetic aberrations associated with unfavorable disease in many human malignancies, such as mutations of the tumor suppressor gene p53 and of the Ras oncogene, are infrequent in Ewing sarcoma family tumors. Besides the EWS-Ets gene rearrangement, the only molecularly defined recurrent genomic alterations each occurring in 20–30% of primary Ewing sarcoma family tumors are the homozygous and the loss of STAG2 on Xq25 ► INK4A gene located on chromosome 9 (band p21). Preliminary retrospective data suggest an adverse prognostic impact of these aberration and of rare p53 mutations. However, in the absence of any prospectively confirmed molecular prognostic marker, the extent of disease monitored by clinical imaging techniques at the time of diagnosis (computed tomography of the chest to document or exclude intrathoracic metastases and 99m-Technetium whole body radionuclide bone scans to search for skeletal metastases), microscopically detectable bone marrow micrometastases, and the histopathologically determined tumor response to initial chemotherapy still serve as the only accepted criteria for treatment stratification.

Etiology

The etiology of Ewing sarcoma family tumors is not known. Neither is there evidence for genetic predisposition nor for a role of environmental exposure. A genome-wide association study identified small nucleotide polymorphisms in three regions on chromosomes 1p36, 10q21 and 15q15 displaying linkage disequilibrium in Ewing sarcoma patients potentially responsible for differences in disease incidence between Caucasians, Asians and Africans. Due to its tight association with the disease, the EWS-Ets gene rearrangement is considered the primary event during Ewing sarcoma pathogenesis. No specific recombinogenic activity has been identified as responsible for this aberration and although involvement of a viral agent in generating the chromosomal translocation has been suggested, it has not been confirmed.

Therapy

In the pre-chemotherapy era, less than 10% of Ewing sarcoma patients survived the disease despite the well-known radiosensitivity of the tumor and despite its radical resection, calling for systemic treatment to eradicate disseminated tumor cells. Today, patients with Ewing sarcoma family tumors are treated by multimodal therapeutic regimens including radiotherapy and chemotherapy (combinations of vincristine, actinomycin D, cyclophosphamide, doxorubicin, ifosfamide, and etoposide) as well as surgical resection whenever possible. By using this treatment strategy together with optimized schedules and dose intensities, the result for patients with localized disease was improved to an overall survival rate of 60–70%. The treatment of Ewing sarcoma patients worldwide is organized in cooperative trials, aiming to further improve treatment outcome. In contrast, the management of primary metastatic disease and early relapse remains a clinical challenge that is currently assessed by myeloablative approaches, combining high-dose chemotherapy and total-body irradiation with stem cell reinfusion. The efficacy of this therapeutic approach for high-risk Ewing sarcoma patients remains to be established. To avoid the toxic side effects of chemotherapy, future biologically tailored therapy may target the EWS-Ets fusion protein or genes downstream of this tumor specific aberration.

References

Bernstein M, Kovar H, Paulussen M et al (2006) Ewing's sarcoma family of tumors: current management. Oncologist 11:503–519

Kovar H (2005) Context matters: the hen or egg problem in Ewing's sarcoma. Semin Cancer Biol 15:189–9656

Kovar H (2014) Blocking the road, stopping the engine or killing the driver? Advances in targeting EWS/FLI-1 fusion in Ewing sarcoma as novel therapy. Expert Opin Ther Targets. 18:1315–28.

Lawlor ER and Sorensen PH (2015) Twenty Years on: What Do We Really Know about Ewing Sarcoma and What Is the Path Forward? Crit Rev Oncog 20:155–171.

Gaspar N, Hawkins DS, Dirksen U et al. (2015) Ewing Sarcoma: Current Management and Future Approaches Through Collaboration. J Clin Oncol 33:3036–46.

Sand LG, Szuhai K, Hogendoorn PC (2015) Sequencing Overview of Ewing Sarcoma: A Journey across Genomic, Epigenomic and Transcriptomic Landscapes. Int J Mol Sci 16:16176–215.

Ewing Sarcoma Family Tumors

▶ Ewing Sarcoma

Ewing Tumor

▶ Ewing Sarcoma

EWS-FLI (ets) Fusion Transcripts

Enrique de Alava[1] and Santiago Ramón y Cajal[2]
[1]Institute of Biomedicine of Sevilla (IBiS), Virgen del Rocio University Hospital /CSIC/University de Sevilla, Seville, Spain
[2]Department of Pathology, Vall d'Hebron University Hospital, Barcelona, Spain

Definition

The *EWS-FLI1* fusion transcript is the result of a balanced reciprocal chromosomal *translocation* between chromosomes 11 and 22, which fuses the *EWS* gene in chromosome 22 to the *FLI1* gene in chromosome 11. This fusion transcript is detected in approximately 85% of cases of the ► Ewing sarcoma family of tumors and is considered a tumor-specific molecular rearrangement, therefore useful for diagnosis, prognosis, and presumably specific therapeutics. Approximately 10% of Ewing tumors have fusions involving *EWS* and *ERG* genes. An additional 5% bear fusions between EWS and other less frequent genes.

Characteristics

Structure
Chromosomal translocations result in the genesis of chimeric genes, encoding hybrid transcripts and novel fusion proteins. Many fusion proteins contain juxtaposed functional domains usually found in separate proteins. The EWS-FLI1 fusion protein contains the aminoterminal domain of EWS and the carboxyterminal region of FLI1. *EWS* gene is an RNA-binding protein, which is believed to mediate mRNA transcription probably through its interaction with RNA polymerase II complex. Several forms (at least 12 types) of EWS-FLI1 exist because of variations in the location of the EWS and FLI1 genomic breakpoints. They contain different combinations of exons from EWS and FLI1, the most frequent being the fusion of EWS exons 1–7 to FLI1 exon 6–9 (type 1) and fusion of EWS exons 1–7 to FLI1 exon 5–9 (type 2).

Properties of the *EWS-FLI1* Fusion Transcript and Protein
EWS is widely expressed in most tissues, and because of the genomic structure of the fusion, the EWS promoter drives the expression of EWS-FLI1. The aminoterminal domain of EWS, included in the fusion, has strong transactivating properties. The *FLI1* gene encodes a member of the ETS family of transcription factors and its expression is highly restricted to hematopoietic, endothelial, and mesodermal cells as well as to neural crest cells. The ETS DNA-binding domain of FLI1 is included in the fusion. The resulting EWS-FLI1 protein is therefore an aberrant transcription factor. The chimeric product EWS-FLI1 can transform some cell lines in culture and can inhibit or activate diverse cellular pathways. For example, EWS-FLI1 protein can suppress transcription of transforming growth factor beta type 2 receptor gene leading to TGF-beta resistance and can activate MFNG, a member of fringe family, related to somatic development, and other genes. The action of EWS-FLI1 as a transcription factor is probably related to the cell context in which this fusion is detected. This cellular context is probably influenced by the cell type, stage of differentiation, and microenvironment and could include expression of several growth factors, for example, IGF1 and its receptor. On the other hand, it is likely that *FLI1* gene is developmentally regulated, being expressed at certain times and places, and this can confer some site and tissue specificity to the transcriptional activity generated by the EWS-FLI1 fusion protein (Fig. 1).

EWS and Other Specific Human Translocations
EWS can fuse after chromosomal translocation to other genes, including several members of the ETS transcription factor family. In an analogous way to FLI1, EWS can be detected fused to *ERG* gene through a t(21;22), to *ETV* gene through a t

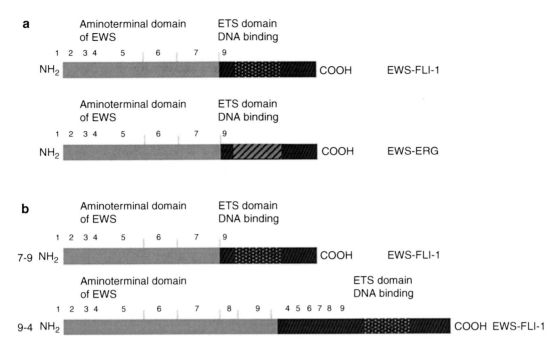

EWS-FLI (ets) Fusion Transcripts, Fig. 1 The figure illustrates two sources of variability of the fusion transcripts in Ewing sarcoma, which include (**a**) the EWS gene partner, and (**b**) the specific number of exons from each gene involved in the fusion

(7;22), to *E1A-F* gene through a t(17;22), or to *FEV* gene through a t(2;22) ► Chromosomal Translocation. All of these *EWS-ETS* gene fusion transcripts can confer a common tumorigenic phenotype of small round cells and can be found in the Ewing family of tumors. But, interestingly, EWS can fuse to other genes and be detected in other tumor types. For example, in desmoplastic small round cell tumor, EWS is fused to the tumor suppressor gene *WT1* through a t(11;22); in clear cell sarcoma of soft tissue, is fused to the ATF1 through a t(12;22); in myxoid and round cell liposarcoma, can also be found fused to the CHOP gene; and to CHN gene, located in chromosome 9, in extraskeletal myxoid chondrosarcoma.

Genesis of the Translocation

The mechanism by which the translocation is generated is another field worth to be studied. In other words why do the breakpoints always occur in the same introns? Is that a random event or are there certain areas of these particular genes particularly prone to recombine? There is some

evidence that several chromosomal translocations may not be random events, but may be specifically promoted by the presence of certain DNA sequence motifs at or around certain target genes. A number of recombinogenic sequences have been described, like topoisomerase-binding sequences in leukemias or lymphomas and translin sequences in alveolar rhabdomyosarcoma and in myxoid liposarcoma. Furthermore it has been suggested that mobile elements or endogenous retroviruses may take part in gene rearrangements. In Ewing tumors, in which illegitimate recombination has been reported to occur, recombinogenic sequences have not been described in a large study of genomic breakpoints.

The EWS-FLI1 Fusion and the Pathogenesis of Ewing Sarcoma

A problem to study sarcoma pathogenesis, in general, and Ewing tumor genesis, in particular, is the absence of preneoplastic lesions, as well as the lack of animal transgenic models. Therefore, to study the mechanism by which EWS-FLI1

induces ▶ Ewing sarcoma, there are two complementary approaches:

1. Induced expression of the fusion in different
   cellular models. For example, induced expression in NIH3T3 fibroblastic cells accelerates
   tumor growth in immunodeficient mouse
   models, while expression in alveolar rhabdomyosarcoma or neuroblastoma cell lines
   induces a shift in differentiation, which
   becomes closer to that of Ewing tumor. The
   same experiment performed in other cell lines
   induces, however, cell death, showing that the
   cellular context matters in the pathogenesis of
   Ewing tumor.
2. By small interference RNA studies blocking
   EWS-FLI1 mRNA, cell cycle arrest and a
   decrease in tumor growth in animal models
   are observed.

Both types of experiments confirm that
EWS-FLI1 functions as an aberrant transcription
factor, although the target genes are still relatively
unknown. The availability of expression
microarrays in the last years, coupled to the experiments described earlier, has been useful to suggest possible target genes. EWS-FLI1 participates
in Ewing tumor pathogenesis controlling cell proliferation and survival by promoting expression of
IGF1, MYC, CCND-1, PDGFC, DAX1-NR0B1,
and NKX2-2 and by repressing genes such as *p21*
^{WAF1}, *p57* ^{kip}, *TGFbRII*, and *IGFBP3*. The genes
related to the induction of the undifferentiated
phenotype of Ewing tumor are largely unknown.

In addition, EWS-FLI1 exerts its function with
the help of other proteins such as RNA helicase A,
which binds to the promoters targeted also by
EWS-FLI1, enhancing its function.

Clinical Relevance

EWS-FLI1 (ets) in Diagnosis

The detection of the chimeric product *EWS-FLI1*,
in the adequate morphologic and immunophenotypic context, is diagnostic of a Ewing family tumor. The fused product can be studied by
fluorescence in situ hybridization (FISH) or by
RT-PCR in very small samples of tissue or cytological specimens. Frozen tissue is the preferred
source of RNA, which can also be extracted from
formalin-fixed paraffin-embedded tissue with variable success. *EWS-FLI1* detection is particularly
useful in cases in which a Ewing family tumor
arises in unusual locations (kidney, skin, lung,
ovary, pancreas), or in patients over 40 years, or
shows atypical features (epithelial differentiation,
no CD99/MIC2 expression).

EWS-FLI1 (ets) Fusion Type and Prognosis

Ewing tumors display a moderate level of molecular heterogeneity. The variability in the chimeric
transcript structure may help to define clinically
distinct risk groups of Ewing tumors. In fact, two
independent groups have found that the type
1 *EWS-FLI1* fusion transcript is associated with
less aggressive clinical behavior than the patients
carrying tumors with other *EWS-FLI1* fusion
types (Fig. 2), regardless of stage at diagnosis,
tumor location, or tumor volume. A study has
shown that this particular gene fusion (type 1)
encodes for a chimeric protein that functions as a
weaker transcription factor than chimeric proteins
encoded by other fusion types. In fact tumors with
EWS-FLI1 type 1 fusions have a lower proliferative rate than their counterparts with other fusion
types. So far, no clinical differences have been
seen between patients with tumors bearing *EWS-
FLI1* fusions and those having *EWS-ERG* transcripts. These results are still pending of confirmation; a European prospective study is being
conducted to assess for clinical differences
among tumors having the most prevalent fusion
types.

EWS-FLI1 (ets) and Detection of Minimal Residual Disease

EWS-ETS fusion transcripts can be detected by
RT-PCR in peripheral blood and bone marrow.
That demonstration may contribute to patient
management and staging, although clinical relevance is still unclear. A French study has shown in
2003 that presence of circulating Ewing tumor
cells and/or Ewing tumor cells in the bone marrow
at the time of diagnosis is a significant predictor of
tumor relapse.

EWS-FLI (ets) Fusion Transcripts, Fig. 2 Early studies showed that patients with Type 1 transcripts were associated to a better overall survival than all others

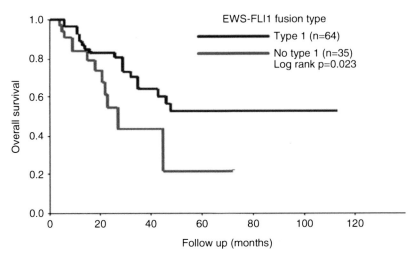

EWS-FLI1 (ets) and Therapeutics

Current efforts are focused on directly inhibiting chimeric proteins (or their downstream targets) and on immunotherapy directed at tumor cell-specific epitopes derived from chimeric products. A potentially useful approach is the delivery of siRNA in the appropriate vectors. The results of an animal model of metastatic Ewing tumor in which EWS-FLI1 siRNA has been injected intravenously are encouraging and could potentially be used in the future treatment of patients with this neoplasm.

Cross-References

▶ Chromosomal Translocations
▶ Ewing Sarcoma

References

Alava E, de Gerald WL (2000) Molecular biology of the Ewing's sarcoma/primitive neuroectodermal tumor family. J Clin Oncol 18:204–213

Hu-Lieskovan S, Heidel JD, Bartlett DW et al (2005) Sequence-specific knockdown of EWS-FLI1 by targeted, nonviral delivery of small interfering RNA inhibits tumor growth in a murine model of metastatic Ewing's sarcoma. Cancer Res 65:8984–8992

Kovar H (2005) Context matters: the hen or egg problem in Ewing's sarcoma. Semin Cancer Biol 15:189–196

Riggi N, Suva ML, Stamenkovic I (2006) Ewing's Sarcoma-like tumors originate from EWS-FLI-1-expressing mesenchymal progenitor cells. Cancer Res 66:9786

Schleiermacher G, Peter M, Oberlin O et al (2003) Increased risk of systemic relapses associated with bone marrow micrometastasis and circulating tumor cells in localized Ewing tumor. J Clin Oncol 21:85–91

See Also

(2012) CD99. In: Schwab M (ed) Encyclopedia of cancer, 3rd edn. Springer, Berlin/Heidelberg, p 704. doi:10.1007/978-3-642-16483-5_941

(2012) Chromosomal translocation t(9;22). In: Schwab M (ed) Encyclopedia of cancer, 3rd edn. Springer, Berlin/Heidelberg, pp 845–846. doi:10.1007/978-3-642-16483-5_1142

(2012) Formalin-fixed Paraffin-embedded Tissue. In: Schwab M (ed) Encyclopedia of cancer, 3rd edn. Springer, Berlin/Heidelberg, p 1446. doi:10.1007/978-3-642-16483-5_2249

(2012) Illegitimate recombination. In: Schwab M (ed) Encyclopedia of cancer, 3rd edn. Springer, Berlin/Heidelberg, p 1808. doi:10.1007/978-3-642-16483-5_2960

(2012) RT-PCR. In: Schwab M (ed) Encyclopedia of cancer, 3rd edn. Springer, Berlin/Heidelberg, p 3322. doi:10.1007/978-3-642-16483-5_5129

(2012) Transcript. In: Schwab M (ed) Encyclopedia of cancer, 3rd edn. Springer, Berlin/Heidelberg, p 3752. doi:10.1007/978-3-642-16483-5_5898

(2012) Transcription factor. In: Schwab M (ed) Encyclopedia of cancer, 3rd edn. Springer, Berlin/Heidelberg, p 3752. doi:10.1007/978-3-642-16483-5_5901

(2012) Translocation. In: Schwab M (ed) Encyclopedia of cancer, 3rd edn. Springer, Berlin/Heidelberg, p 3773. doi:10.1007/978-3-642-16483-5_5942

Exfoliation of Cells

Alexandre Loktionov
DiagNodus Ltd, Babraham Research Campus,
Cambridge, UK

Synonyms

Shedding of cells; Sloughing of cells

Definition

Cell exfoliation (Latin – *exfoliare* – to strip of leaves) is a process of spontaneous or induced complete detachment of single epithelial cells or groups of cells from an epithelial layer.

Characteristics

Exfoliation of cells is one of the main mechanisms of cell loss participating in the homeostatic control of cell population size. Cell exfoliation is a characteristic feature of epithelial tissues forming epithelial layers covering both external body surface (skin epidermis and skin appendages) and surfaces of internal cavities and passages (gastrointestinal tract, respiratory system, and genitourinary system) as well as major exocrine glands and glandular ducts (mammary gland, exocrine pancreas, biliary system, etc.). Cell exfoliation process in normal physiological conditions is closely associated with terminal differentiation and orderly loss of dying cells compensated by permanent cell population renewal. This relationship is exemplified by the structure of stratified squamous (epidermal-type) epithelia where layers of maturing cells are being moved toward the surface due to continuous arrival of younger counterparts produced by proliferation in the basal layers of the epithelium. Eventual death, keratinization (in the epidermis), and inevitable exfoliation of cell remnants is the normal destiny of these terminally differentiated cells. Less is known about cell exfoliation in the epithelia of internal organs; however, studies of well-structured columnar colonic epithelium can be chosen as the most reliable source of information. Obvious links between cell differentiation and eventual shedding exist in this dynamic cell population with extremely high rates of cell proliferation and loss. The progeny of stem cells located at the base of the colonic crypt constantly migrate in the direction of the lumen. During this migration colonocytes gradually lose their proliferative capacity and undergo differentiation, thus terminally differentiated cells reaching surface (luminal) epithelium should be promptly eliminated to be replaced by future generations of their counterparts. Two ways of cell loss coexist in the colonic epithelium under normal conditions (Fig. 1).

1. Exfoliation of single colonocytes or colonocyte groups into the lumen of the gut. Strictly speaking, the shed cells enter the protective mucocellular layer covering colonic mucosa rather than the gut contents. The characterized protective mucus consists of two layers (dense inner and looser outer) formed by Mucin 2 molecules.
2. Apoptosis in situ followed by the engulfment of apoptotic cells by adjacent colonocytes or subepithelial macrophages.

Physiological changes or development of pathological processes can significantly shift proportions of cells eliminated by each of these mechanisms. There are two principal variants of postexfoliation cell fate (Fig. 1). Some colonocytes undergo immediate apoptosis (detachment-induced apoptosis or ▶ anoikis) following exfoliation, whereas other exfoliated cells and especially groups of cells maintain their structural integrity and possibly viability for minutes or even hours following exfoliation. In the human colon, exfoliated cells enter the mucocellular layer overlaying colonic mucosa and can be transferred distally, protected by the mucus, without being incorporated

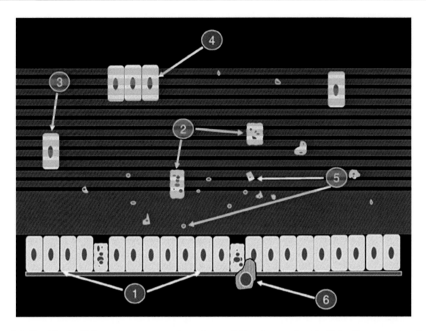

Exfoliation of Cells, Fig. 1 Schematic representation of cell exfoliation in normal physiological conditions. Colon model, surface epithelium, and crypts are not shown. *1* Normal epithelium (two apoptotic cells shown). *2* Anoikis immediately following exfoliation. *3* Exfoliated normal epithelial cell preserving its structure. *4* Group of exfoliated normal epithelial cells preserving their structure. *5* Cellular and nuclear debris. *6* Subepithelial macrophage. Semitransparent horizontal strips above the epithelium correspond to protective colorectal mucus layer (dense inner and looser outer layers)

into the feces. The movement of the cell-containing mucocellular layer is mostly driven by its close contact with moving fecal flow as well as by constant peristaltic movements of the gut. There is no doubt that some proportion of exfoliated colonocytes can reach the gut contents, but these cells should be rapidly destroyed in the anaerobic bacteria-dominated fecal milieu rich in bile acids and other cytolytic agents. Observations of the presence of well-preserved epithelial cells in human fecal samples mostly concern colonocytes excreted together with the mucocellular layer fragments or exfoliated squamous cells of the anal epithelium sometimes misidentified as colonocytes.

Similar exfoliation models with tissue-specific corrections can be applied to the epithelial populations of other internal organs. Preservation of cellular structure after exfoliation is described for cells of bronchial epithelium in sputum samples, bladder, urethral and prostate epithelium in urine samples, gastric epithelium in gastric lavage liquid, cervical, vaginal, and endometrial epithelium in cervical smears, mammary gland epithelium in nipple aspirates, and pancreatic duct epithelium obtained endoscopically. Exfoliated cell migration or passive transport occurs in all internal passages or ducts, being facilitated by contents flow, peristaltic movements, or cell-driven mucus transport (*cilia* of the bronchial epithelium). Shed cells can be commonly found in human body excretions (cells of bladder, urethra, and occasionally prostate epithelium in urine, colonocytes and squamous anal epithelium in feces, bronchial, nasopharyngeal, and oral cavity epithelium in sputum, etc.). Exfoliated epithelial cells should be distinguished from nonepithelial free cells (leukocytes, macrophages, lymphocytes, etc.) that are also often present in the excreted materials but are unrelated to the process of cell exfoliation.

Mechanisms

Molecular mechanisms underlying cell exfoliation in normal conditions are poorly investigated and remain largely unknown. Extrapolation of the

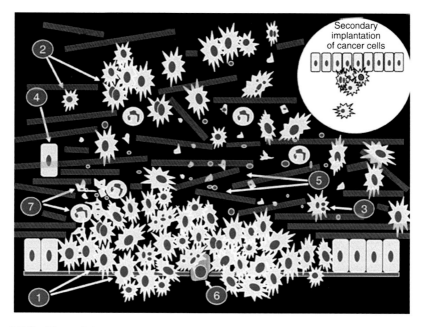

Exfoliation of Cells, Fig. 2 Schematic representation of cell exfoliation from tumor surface. Colon model, surface epithelium, and crypts are not shown. Inset demonstrates secondary implantation of malignant cells distally from the primary tumor. *1* Tumor (malignant cells). *2* Exfoliated tumor cells and cell groups preserving their structure. *3* Occasional anoikis of a tumor cell immediately following exfoliation. *4* Exfoliated normal epithelial cell preserving its structure. *5* Cellular and nuclear debris. *6* Subepithelial macrophage. *7* Neutrophilic leukocytes and necrotic tumor fragments (focal necrosis and inflammatory reaction). Disorderly oriented semitransparent strips of different length indicate disease-related deterioration of the protective mucus layer

information obtained in vitro might be misleading; therefore, only tentative suggestions can be made. At the cellular level, pathways leading to cell exfoliation should involve considerable structural and functional changes of the cytoskeleton, cell membrane, and membrane-bound subcellular structures responsible for the preservation of contacts between neighboring cells (▶ adherens junctions, ▶ tight junctions, ▶ gap junctions) as well as between cells and basal membranes (focal adhesions). A number of proteins associated with these structures (adherens junction–associated cadherins and catenins, tight junction–associated claudins and occludin, gap junction–associated connexins and focal adhesion–associated integrins and focal adhesion kinase) are likely to participate in exfoliation induction. In particular, the integrin system appears to be intimately involved in the homeostatic regulation of cell differentiation, providing communication between epithelial cells and the underlying stromal extracellular matrix. The loss of contact with the extracellular matrix can trigger anoikis through integrin-mediated signaling, but it is evident that many exfoliated cells are able to evade this fate. Mechanisms of this phenomenon remain obscure and need profound investigation as well as other aspects of cell exfoliation in vivo.

Cell Exfoliation in Neoplasia

Cancer development is usually associated with a dramatic increase in cell exfoliation from tumor surface (Fig. 2). The increased exfoliation may reflect a compensatory response to the impairment of the apoptotic mechanism of cell elimination in situ commonly observed in malignant tumors. Anoikis induction appears to be the ultimate homeostatic goal of the enhanced exfoliation. Although this mechanism of cell elimination may be partially functional in early tumors, resistance to anoikis is now regarded as a hallmark of metastatic cancer cells, which tend to survive exfoliation. Loss of cell adhesion (often associated with E-cadherin function impairment) and disruption of normal interactions with underlying stroma associated with malignant progression also strongly contribute to the enhancement of exfoliation from tumors. Disease-related deterioration of the protective mucus layer can also facilitate cell exfoliation. Cancer cells are better adapted to survival in the conditions of oxygen deficiency; therefore their prolonged postexfoliation persistence on the surfaces of internal cavities, passages, and glandular ducts can be expected. These cells shed by tumors are often immature and retain proliferative capacity. Therefore their metastatic potential and ability for secondary implantation can be manifested. Reports describing the occurrence of secondary cancers or peritoneal carcinomatosis following surgical interventions on tumors of several sites (colorectum, ovary, gallbladder) attribute these secondary metastatic cancers to the direct reimplantation of malignant cells exfoliated from the primary tumors. There is a strong probability that cell exfoliation without immediate anoikis in physiological conditions presents a normal "prototype" of metastatic behavior.

Clinical Aspects

The possibility of using exfoliated epithelial cells for various clinical and research purposes is highly attractive because analysis of this material often allows avoiding highly invasive sampling (biopsy) of normally inaccessible tissues. Exfoliated cells can be collected either noninvasively (from sputum, urine, feces, etc.) or by simple and minimally invasive procedures (cervical smears, buccal swabs, direct cell collection from the surface of rectal mucosa during proctoscopy, etc.). Although exfoliated cells became widely employed for some specific medical, forensic, and research purposes (e.g., buccal mucosal cells are routinely used for DNA isolation for genotyping), clinical oncology remains the main area where their use is already common and looks even more promising with the introduction of modern molecular methodologies.

Cytological diagnostic approaches based on the examination of exfoliated cells were developed first. Exfoliative cytology analysis of smears prepared from cervical epithelium (examination of PAP smears) has become a standard screening procedure for cervical cancer. Diagnostic cytology is used for endometrial carcinoma, bronchogenic lung cancer, tumors of the bladder and prostate, and gastric cancer.

The introduction of molecular assays targeting cancer biomarkers created another major direction, aiming to employ exfoliated cells as the material of choice for the molecular diagnosis and screening of oncological conditions. Approaches to colorectal cancer screening occupy a leading position among problems addressed in this area. Significant efforts were concentrated on the analysis of exfoliated colonocytes (or DNA derived from these cells) present in human stool samples. Cancer-specific molecular changes are often detected by such analysis, especially when multimarker panels of PCR-based assays are applied. The sensitivity of these methods has considerably improved, but endoscopic confirmation of the diagnosis is still required. The main reason for remaining problems is likely to be the relatively low presence of exfoliated colonocytes in the fecal samples. Human DNA found in this material can also derive from exfoliated anal epithelium or free cells (leukocytes, lymphocytes, macrophages, etc.), which certainly do not contain tumor markers. Moreover, strong fecal presence normally decreases PCR efficiency. Direct collection of colorectal mucus containing exfoliated cells from the surface of rectal mucosa or the anal area immediately following defecation appear to provide much more abundant and contamination-free material that can be used for a range of analytical procedures. Even simple detection of unusually high DNA yield in samples of exfoliated material directly collected at standardized conditions can be interpreted as a warning signal indicating a high probability of colorectal tumor presence.

Multiple other applications of exfoliated cells for cancer diagnosis by molecular biomarker detection are being developed, but these studies remain research projects rather than proven clinical approaches. The present absence of a single biomarker allowing 100% positive identification of cancer presence is the most important obstacle interfering with the development of molecular diagnostic procedures based upon the use of exfoliated cells.

Cross-References

▸ Adherens Junctions
▸ Anoikis
▸ Gap Junctions
▸ Integrin Signaling
▸ Tight Junction

References

Bogenrieder T, Herlyn M (2003) Axis of evil: molecular mechanisms of cancer metastasis. Oncogene 22:6524–6536
Hanahan D, Weinberg RA (2000) The hallmarks of cancer. Cell 100:57–70
Johansson MEV, Sjövall H, Hansson GC (2013) The gastrointestinal mucus system in health and disease. Nat Rev Gastroenterol Hepatol 10:352–361
Loktionov A (2007) Cell exfoliation in the human colon: myth, reality and implications for colorectal cancer screening. Int J Cancer 120:2281–2289
Osborn NK, Ahlquist DA (2005) Stool screening for colorectal cancer: molecular approaches. Gastroenterology 128:192–206

See Also

(2012) Biopsy. In: Schwab M (ed) Encyclopedia of cancer, 3rd edn. Springer, Berlin/Heidelberg, p 415. doi:10.1007/978-3-642-16483-5_644
(2012) Cilia. In: Schwab M (ed) Encyclopedia of cancer, 3rd edn. Springer, Berlin/Heidelberg, p 857. doi:10.1007/978-3-642-16483-5_1168
(2012) Focal adhesion. In: Schwab M (ed) Encyclopedia of cancer, 3rd edn. Springer, Berlin/Heidelberg, pp 1436–1437. doi:10.1007/978-3-642-16483-5_2227
(2012) Mucocellular layer. In: Schwab M (ed) Encyclopedia of cancer, 3rd edn. Springer, Berlin/Heidelberg, p 2389. doi:10.1007/978-3-642-16483-5_3877
(2012) Multi-marker panels. In: Schwab M (ed) Encyclopedia of cancer, 3rd edn. Springer, Berlin/Heidelberg, p 2394. doi:10.1007/978-3-642-16483-5_3884

Exjade®

▸ Deferasirox

Exobiotics

▸ Xenobiotics

Exosomal MicroRNA

▸ Exosomal miRNA

Exosomal miRNA

Shuai Jiang[1] and Wei Yan[2]
[1]Department of Biology and Biological Engineering, California Institute of Technology, Pasadena, CA, USA
[2]Department of Cancer Biology, Beckman Research Institute of City of Hope, Duarte, CA, USA

Synonyms

Exosomal microRNA; Exosomal shuttle microRNA

Definition

▸ Exosomes, shuttling from donor cells to recipient cells, are cell-derived extracellular vesicles (EV) with 30–100 nm in diameter and dish- or classic cup-shaped morphology. Exosomes transport bioactive cargo carrying selective proteins, lipids, DNA, messenger RNAs (mRNAs), and small and large ▸ noncoding RNAs such as ▸ microRNAs (miRNAs). miRNA embedded in exosomes is termed as exosomal miRNA. Exosomal miRNAs regulate diverse biological processes in recipient cells.

Characteristics

Biogenesis of Exosomes and Exosomal miRNA Packaging

Exosomes play a fundamental role in cell-cell communication and have been found in various

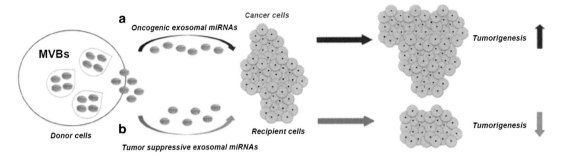

Exosomal miRNA, Fig. 1 Exosomal miRNAs participate in tumorigenesis. Exosomes are secreted by donor cells through the fusion of multivesicular bodies (MVBs) with cell membrane. miRNAs are carried in exosomes and functionally delivered to recipient cancer cells. Oncogenic exosomal miRNAs (**a**) and tumor-suppressive exosomal miRNAs (**b**) promote and inhibit tumorigenesis, respectively

cell types, especially in tumor cells. Cluster of differentiation (CD) 63, CD81, heat shock protein (HSP) 70, Alix (ALG-2-interacting protein X), and TSG101 (tumor susceptibility gene 101) serve as exosomal marker proteins. Exosomes are derived from the multivesicular bodies (MVBs), which are also known as late endosomes. The endosome origin is the hallmark of exosomes distinct from other larger kinds of extracellular vesicles. They are formed by endosomal membrane inward budding in MVB, and this process is controlled by endosomal sorting complex required for transport (ESCRT), ceramide, or tetraspanin complex (Miller and Grunewald 2015; Zhang et al. 2015). When MVBs fuse with plasma membrane, biologically active exosomes are secreted into extracellular environment (Fig. 1). However, the exact mechanism how exosomal contents, especially miRNAs, are loading into MVBs is still under investigation. In fact, miRNA sorted into exosome is a selective process. Profiling studies show that some miRNAs including miR-150, miR-142, miR-320, and miR-451 are preferably enriched in exosomes. Additionally, specific exosomal miRNAs express differently under different conditions. For instance, exosomal Let-7 miRNAs are much more expressed in gastric cell line AZ-P7a compared with other cancer cells, while exosomal miR-21 is abundant in the sera of glioblastoma patients compared with healthy donors.

As for today, there are several models to clarify the mechanisms of the package of exosomal miRNA. Firstly, the specific short sequence located on miRNAs can guide their loading into exosomes. For example, in B lymphocytes, 3′ ends of uridylated endogenous miRNAs are secreted into B lymphocyte-derived exosomes, while 3′ ends of adenylated endogenous miRNAs remain in B lymphocytes. Secondly, the package of miRNAs into exosomes requires the assistance of some intracellular functional proteins. For instance, the overexpression of neural sphingomyelinase 2 (nSMase2), the first reported protein functioning in the miRNA selection into exosomes, can increase the exosomal miRNA number. In addition, argonaute 2 (AGO2) prefers to bind uracil or adenine on the 5′ end of miRNAs. AGO2 deletion can decrease the expression of exosomal miRNAs including miR-150 and miR-451. Moreover, some modified RNA-binding protein contributes to the recognition of specific sequence on miRNAs. Sumoylated protein heterogeneous nuclear ribonucleoprotein A2B1 (hnRNPA2B1) binds to the specific 4 nt motifs (GGAG) in the 3′ region of miRNA to mediate miRNA transporting to MVBs and then packaging into exosomes. KRAS is also engaged in the process of miRNA package into exosomes. ▶ *KRAS* mutant colorectal cancer (CRC) cells have more miR-100 sorted into exosomes, while wild type cells have more exosomal miR-10. Thus, specific motif located on miRNAs and intracellular functional molecules are both critical to the package of exosomal miRNAs. Apart from those two kinds of elements, the affinity between MVB membrane and cellular miRNAs as well as cell-activation-dependent miRNA-targeted transcript level changes in donor cells also contribute to sorting of miRNAs into exosomes (Squadrito et al. 2014).

Exosomal miRNA is secreted into extracellular environment by employing exosome as vesicle. To date, the investigation of the underlying secretion mechanism is still going on. The secretion of exosomes in parent cells is dependent on important regulatory molecules such as small ▸ GTPase Rab family including Rab27, Rab28, Rab31, and Rab11 and Rab effector molecules (SYTL4 and SLAC2B). The tumor suppressor protein p53 modulates transcription of the downstream genes such as *tsap6* and *chmp4c* that regulate the endosomal compartment and lead to elevated exosome secretion. Additionally, nSMase2 and calcium ionophore also regulate exosome secretion.

Bio-functions of Exosomal miRNA in Cancer

Exosomes utilize surface receptor such as MHC interaction or plasma membrane fusion to transfer contents in recipient cells, where exosomal miRNAs can bind to the $3'$ untranslated region ($3'$ UTR) of target mRNAs. By repressing the expression of direct targets in recipient cells at posttranscription level, exosomal miRNAs carry out their functions in acceptor cells, most of which are closely linked to human cancers. Exosomal miRNAs can function as oncogenic exosomal miRNAs in tumor invasion, metastases, tumor angiogenesis, and immune suppression in various cancers. In breast cancer, exosomal miRNAs can promote breast cancer cell metastasis by direct targeting downstream genes in recipient cells. For example, exosomal miR-105, characteristically overexpressed and produced by metastatic breast cancer cell line MDA-MB-231, can directly repress downstream target tight junction-related gene (*ZO-1*) in endothelial cells, damaging the natural barrier integrity against breast cancer metastasis. IL4-activated macrophages-derived exosomal miR-223 can promote invasion of human breast cancer cell line SKBR3. In addition, exosomal miRNAs can also modulate tumor microenvironment. Stromal cell, an important component in tumor microenvironment, can

produce exosomal miRNAs to assist breast cancer cells to be resistant to cancer therapy through activating STAT1 signaling pathway and NOTCH3 on breast cancer cells (Miller and Grunewald 2015). Also, exosomal miRNAs can promote angiogenesis. Blood vessel formation is essential for tumor cell growth. Melanoma cell-derived miR-9 induces the migration of endothelial cells and promotes cancer angiogenesis by regulating JAK-STAT pathway. Exosomal miR-135b derived from hypoxic multiple myeloma cells can block its target factor-inhibiting hypoxia-inducible factor 1 (HIF-FIH-1) when delivered to human umbilical vein endothelial cells (HUVECs). Hypoxic exosomal miRNAs enhance angiogenesis via the HIF-FIH signaling pathway under the condition of hypoxia. Besides, human endothelial cell line HMEC-1 s-derived exosomal miR-214 stimulates migration program and angiogenesis in recipient HMEC-1 cells. Exosomal miRNAs might also assist cancer cells to escape from immune cell detection. For instance, TW03-derived exosomes can block T-cell proliferation and T helper cell differentiation, leading to inhibit the function of T cells in human nasopharyngeal carcinoma (NPC) (Ye et al. 2014). Furthermore, five exosomal miRNAs have been identified to be abundant in the patient sera, and they modulate the MARK1 signaling pathway to affect cell proliferation. By contrast, some exosomal miRNAs are underexpressed in multiple cancers. These are known as tumor-suppressive exosomal miRNAs. Tumor-suppressive exosomal miRNAs have characteristics similar with tumor-suppressive genes that can inhibit cancer cell proliferation, metastasis, and induce apoptosis. For instance, it is well known that Let-7 miRNAs usually play a tumor-suppressive role by repressing oncogenes such as RAS and HMGA2, a metastatic gastric cancer cell line AZ-P7a could keep their oncogenic capacity through releasing tumor-suppressive Let-7 miRNAs via exosomes into the extracellular environment. In addition, tumor-suppressive exosomal miR-143 that derived from noncancerous cells can suppress the growth of prostate cancer cells by inhibiting its target genes including *KRAS* and *ERK5* both in vitro and in vivo. Thus, exosomal miRNAs participate in tumorigenesis. Oncogenic exosomal miRNAs and tumor-suppressive exosomal miRNAs can promote and inhibit process of tumorigenesis, respectively (Fig. 1).

Exosomal miRNAs in Cancer Diagnostic and Clinical Use

Exosomal miRNAs play fundamental characters during cancer progression. So far, more and more evidence shows exosomal miRNAs could be used as diagnostic biomarkers for various cancers. For colorectal cancer, exosomal levels of seven miRNAs including miR-21 and miR-223 are significantly hyperactivated in primary cancer patients compared with healthy controls, with significantly downregulation after surgical resection of colorectal tumors. Thus, exosomal miRNA signatures could be utilized to mirror pathological process of colorectal cancer. For lung cancer, the expression of some exosomal miRNAs in patients' sera is much higher than non-lung cancer tissues, indicating circulating exosomal miRNAs might be useful as biomarker for lung adenocarcinoma as well. For breast cancer, exosomal miRNAs can also be used as a diagnostic biomarker. Certain serum miRNAs are also highly correlated with breast cancer tissues. For instance, oncogenic miR-21 and miR-155 are significantly abundant in breast cancer specimens, whereas miR-126 is dramatically under-expressed. In ovarian cancer, tumor-derived exosomal miRNA signatures exhibit dramatically different profiles compared with that from benign tissues. So, exosomal miRNAs can be applied as diagnostic markers of ovarian cancer and references for tumor stages. Additionally, expression of exosomal miR-21 is correlated with tumor progression stages in esophageal squamous carcinoma, indicating that it may be a useful target for cancer therapy. Exosomal miR-1290 and miR-375 can be used as prognostic marker in castration-resistant prostate cancer as well.

Exosomes have already been employed by human virus to transfer miRNAs to noninfected cells, thereby assisting virus spread, which tells us exosomes may be applied as therapeutic vesicles and function as a good delivery system for tumor-suppressive exosomal miRNAs in cancer therapy. For example, delivering tumor-suppressive exosomal miR-143 leads to the shrink of development of prostate cancer cells in nude mice (Kosaka et al. 2013). Tumor-suppressive miRNA Let-7a can inhibit breast cancer cell growth when introduced into EGFR-expressing cells. Moreover, exosomes can be able to cross the blood–brain barrier, which could enhance the efficiency of delivery of medications to cancer cells, whereas there are still some questions need to be addressed when employing tumor-suppressive exosomal miRNAs in cancer therapy. For instance, how to facilitate more specific tumor-suppressive exosomal miRNA loading and packaging into exosomes? How to enhance the uptake efficiency of exosome by recipient cells? Secondly, apart from tumor-suppressive exosomal miRNAs, oncogenic exosomal miRNAs can also be applied in tumor therapy. Due to their functions in tumor angiogenesis and tumor invasion, oncogenic exosomal miRNAs might be utilized as cancer vaccines, whereas their cancer-promoting effects should be monitored during cancer research trials. Lastly, the small size of exosomal miRNAs also benefits them as an ideal target for drug designing. Collectively, exosomal miRNAs could potentially be served as clinical tools for cancer diagnostic and clinical use for various cancers. However, any side effects when using bioengineered or naturally occurring exosomal miRNAs need to be considered. In a word, using exosomal miRNAs might be an exciting but challenging application in cancer therapy in the near future.

Cross-References

▶ Exosome
▶ GTPase
▶ *KRAS*
▶ MicroRNA
▶ Noncoding RNA

References

Kosaka N et al (2013) Exosomal tumor-suppressive microRNAs as novel cancer therapy "exocure" is another choice for cancer treatment. Adv Drug Deliv Rev 65(3):376–382

Miller IV, Grunewald TG (2015) Tumor-derived exosomes: tiny envelopes for big stories. Biol Cell 107:1–19

Squadrito ML et al (2014) Endogenous RNAs modulate microRNA sorting to exosomes and transfer to accept cells. Cell Rep 8(5):1432–1446

Ye SB et al (2014) Tumor-derived exosomes promote tumor progression and T-cell dysfunction through the regulation of enriched exosomal microRNAs in human nasopharyngeal carcinoma. Oncotarget 5(14):5439–5452

Zhang J et al (2015) Exosome and exosomal microRNA: trafficking, sorting, and function. Genomics Proteomics Bioinformatics 13(1):17–24

See Also

(2012) P53. In: Schwab M (ed) Encyclopedia of cancer, 3rd edn. Springer, Berlin/Heidelberg, p 2747. doi:10.1007/978-3-642-16483-5_4331

Exosomal Shuttle MicroRNA

▶ Exosomal miRNA

Exosome

Peter Kurre[1] and Ben Doron[2]
[1]Department of Pediatrics, Oregon Health and Science University, Portland, OR, USA
[2]Oregon Health and Science University, Portland, OR, USA

Synonyms

Extracellular vesicles

Definition

Exosomes are membrane-enclosed vesicles that are derived from the endocytic compartment and released at the plasma membrane into the extracellular space. The plasma membrane is not the source of the lipid bilayer of exosomes; rather, exosomes originate from luminal pinocytosis of early endosomes. These vesicles range from 30 nm to 100 nm in diameter and traffic cargo in an autocrine, paracrine, and endocrine fashion. Exosomes contain a subset of biologically active macromolecules present in the cell: protein, lipids, and multiple RNA species. The composition of exosomal cargo is unique to the producing cells, giving healthy and diseased cells a specific exosomal signature. The trafficking of these molecules into neighboring cells alters their behavior, a process involved in neuronal signaling, fetal development, tissue homeostasis and repair, adaptive immunity, and cancer progression.

Characteristics

Mechanism

In contrast to the shedding of microvesicles which bud at the plasma membrane, exosomes are contained as vesicles within endosomal compartments, termed multivesicular bodies (MVB). Their unique biogenesis is also reflected in their lipid composition resembling that of the early endosomes. Exosomes originate as inward buddings of the endosomal membrane to create intraluminal vesicles (ILVs). ILVs accrue during the transition from early to late endosomes, also known as multivesicular bodies. Movement toward the plasma membrane is controlled by the cytoskeleton and small GTPases. Endosomes move along microtubule tracks via molecular motors dictated by Rab GTPases and the phosphoinositide profile on the outer lipid leaflet of the vesicles. Secretion occurs when endosomes fuse with the plasma membrane and release their exosomes. This allows the cell to manage exosomal output in a temporally and spatially controlled fashion by using multiple cytoskeletal and membrane proteins to mediate fusion and secretion. Subject to an active process, cells change both the output of exosomes and the cargo within them under various stress-inducing conditions.

Cargo

Protein

Exosomes contain internal maturation proteins and the proteins bound for recipient cells within the lumenal compartment. To date, close to 4,600 different proteins have been shown to associate with exosomes. The cellular origin of exosomes accounts for the fact that many exosomal proteins are involved in endosomal pathways. The most common proteins include the tetraspanins CD9, CD63, and CD81, which act as protein scaffolds; flotillin, which aids in vesicle formation; Alix, an adaptor protein required for endosomal trafficking; and TSG101, a regulator of vesicular trafficking. These proteins are frequently used as markers for the classification as exosomes. Exosomes also contain cytoplasmic proteins that act in recipient cells. These include metabolic enzymes, signal transducers, and transcription factors.

One mechanism by which proteins are sorted into exosomes relies upon ESCRT (endosomal sorting complexes required for transport), a complex of proteins that coordinates both budding and the sorting of proteins into ILVs. This complex was initially described as being required for the sorting of ubiquitinated proteins destined for the lysosomal degradation, but it has been shown to sort proteins into exosomes. The posttranslational modifications that ESCRT recognizes for sorting are still unclear, but seem to be primarily orchestrated by a combination of the ubiquitin profile and association of other "guide" proteins. This work on sorting was performed on viral proteins and may vary for endogenous protein exportation.

Lipid

Exosomes are enriched in cholesterol, ceramides, sphingomyelin, and saturated species of phosphatidylcholine and phosphatidylethanolamine. The lipid composition of exosomes contributes to both ILV formation and trafficking within the cell of origin. In studies where components of the ESCRT complex are knocked out, ILVs are still generated in a mechanism that seems to be aided by the increased incorporation of ceramides, which increase membrane curvature.

Sphingosine-1-phosphate was also shown to contribute to ILV formation, as a reduction in flotillin$^+$ exosomes was observed in sphingosine kinase 2 knockout cell lines. Importantly, in contrast to microvesicles whose bilayer reflects that of the plasma membrane, exosomes contain additional lipid moieties of endosomal origin.

RNA

Deep sequencing of exosomes has revealed the existence of a vast array of RNA species in exosomes. Importantly, the RNA content of exosomes is not a proportional reflection of the transcriptome of the cell, but exhibits enrichment and exclusion of specific transcripts. The mechanism by which selection is accomplished remains under investigation, but may utilize miRNA sequence motifs. This "EXOmotif," in tandem with a sumoylated heterogenous ribonucleoprotein A2B1 (hnRNPA2B1), selectively sorts miRNA into exosomes. Similarly, cis elements on mRNA transcripts have also been identified to be correlated with exosomal sorting. One study demonstrated interplay between miRNA and mRNA in the sorting efficiency of transcripts, and another study showed the presence of miRNA/RISC complex within exosomes. Currently, more research needs to be performed to elucidate these sorting mechanisms, but it is clear that RNA sorting is dependent upon both cis- and trans-elements.

Trafficking

Exosomes are relatively resistant to strong shearing forces and enzymatic degradation, making them suitable candidates for the delivery of fragile or cell impermeable molecules. Alternatively, exosomes in the extracellular space can traffic back to the cell of origin, neighboring cells, or into the bodily fluids for transmission to other organs and tissues. Several mechanisms have been described for cellular entry of exosomes. Recipient cells take up exosomes in a fashion similar to cell-cell adhesion, through the use of ICAMs and integrins. Surface receptors on the recipient cells recognize exosome proteins, glycoproteins, and lipid moieties, leading to cell-type specific trafficking. This is of particular importance in signaling within the immune system. For example, CD169$^+$ macrophages recognize B-cell-derived exosomes particularly by the decoration of 2–3 linked sialic acid and glycoproteins containing specific mannose residues. This is important for communication between cells within the spleen and lymph nodes. Lectins on the surface of recipient cells contribute to preferential exosome uptake, as do clathrins, dynamin, and caveolae. Expression of these surface receptors modulates the specificity and affinity for exosome reception. Further, exosomes can either fuse with the recipient cell's plasma membrane, releasing their cargo into the cytosol, or become engulfed and enter the endocytic pathway. Release of cargo into the cellular cytosol affects recipient cell behavior. Of interest within the exosome field is the trafficking of functional RNA molecules between cells. Messenger RNA transcripts generated in a donor cell have been shown to be translated to functional protein in recipient cells. Furthermore, miRNA generated in a producer cell can exhibit suppression on recipient cell gene expression. This type of cell-cell communication contributes to tissue development and homeostasis and is especially important in the cross talk between the stromal and parenchymal components within organs.

Exosomes in Cancer

The development of cancer is a dynamic process, and exosomes contribute to a variety of events that enable cancer to progress. Almost all tumor types appear to exhibit an increase in exosomal output upon transformation. The cell-autonomous accumulation of mutations and the subsequent signaling dysregulation allow tumors to overcome cellular checkpoints and growth restrictions, whereas intercellular signaling shapes their microenvironment by manipulating neighboring cells and tissue. Exosomal-mediated cell signaling appears to contribute to a tumor-proliferative microenvironment through interactions with neighboring stroma and via immune evasion.

Microenvironment

Tumors are capable of transforming their surroundings into a state that supports malignancy. This includes modulation of the extracellular matrix (ECM) and induction of angiogenesis. ECM is altered in part by the manipulation of fibroblasts near the tumor. These cells are referred to as cancer-associated fibroblasts (CAF) and remodel the architecture and composition of surrounding ECM to promote metastasis and vascularization. Exosome-mediated transport of the signaling molecules TGF-β1 and FGF-2 contributes to the altered fibroblast phenotype. Multiple studies have also shown that exosomal output of stromal cells is altered in a tumor environment, which reinforces tumor proliferation by providing growth factors and signals back to the tumor cells. Angiogenesis is the process by which vascularization is introduced to a tumor. This process is required for solid tumors, as their size and growth rate require the delivery of oxygen to counteract their hypoxic environment. Hypoxia and nutrient depletion provoke the release of exosomes containing angiogenic miRNA and stimulatory signaling molecules that induce the neovascular formation of blood vessels in the tumor.

Immune System

While the tumor microenvironment contains cells from both the innate and adaptive arms of the immune system, most tumors are able to suppress the local immune response. The communication between immune cells, mediated by cellular contact or the release of chemokines and cytokines, establishes the balance between pro- or anti-inflammatory responses. Exosomes influence this equilibrium through several mechanisms. Exosome-mediated immune suppression was shown in experiments demonstrating increased tumor proliferation and decreased immune response when transplanted tumors were accompanied by injections of exosomes derived from the same tumor. Candidate mechanisms in support of this observation include delivery of proapoptotic ligands to reduce tumor infiltrating lymphocytes in the microenvironment, delivery of anti-inflammatory cytokines, and induction of regulatory T cells that increase the immune tolerance of the tumor.

Biomarkers

The unique content of tumor-derived exosomes allows for the use of these vesicles as biomarkers for the detection cancer relapse. As they equilibrate with the bloodstream, circulating exosomes provide a minimally invasive source of substrate for cancer detection from solid tumors and hematologic malignancies. Sampling of body fluids (e.g., blood, urine, saliva, CSF) provides minimally invasive sources of exosomes, allowing more frequent screening and potentially earlier detection. The exosomal miRNA profile has provided a promising platform for disease surveillance, as this RNA population shows increased enrichment in cancer-specific transcripts. Accordingly, the utility as biomarkers is not limited to cancer, but includes infectious disease and degenerative neurological conditions.

Cross-References

▶ Angiogenesis
▶ Ceramide
▶ Chemokines
▶ Endosomal Compartments
▶ GTPase
▶ Hypoxia
▶ Innate Immunity
▶ MicroRNA

References

Lötvall J, Hill AF et al (2014) Minimal experimental requirements for definition of extracellular vesicles and their functions: a position statement from the International Society for Extracellular Vesicles. J Extracell Vesicles 3. doi:10.3402/jev.v3.26913

Mittelbrunn M, Vicente Manzanares M, Sánchez-Madrid F (2015) Organizing polarized delivery of exosomes at synapses. Traffic. doi:10.1111/tra.12258

Roma-Rodriguez C, Fernandes AR, Baptisa PV (2014) Exosome in tumour microenvironment: overview of the crosstalk between normal and cancer cells. Biomed Res Int. doi:10.1155/2014/179486. Epub 2014 May 21

Villarroya-Beltri C, Baixauli F et al (2014) Sorting it out: regulation of exosome loading. Semin Cancer Biol (28) doi:10.1016/j.semcancer.2014.04.009

See Also

(2012) Adaptive immunity. In: Schwab M (ed) Encyclopedia of cancer, 3rd edn. Springer, Berlin/Heidelberg, pp 42–43. doi:10.1007/978-3-642-16483-5_74

(2012) Biomarkers. In: Schwab M (ed) Encyclopedia of cancer, 3rd edn. Springer, Berlin/Heidelberg, pp 408–409. doi:10.1007/978-3-642-16483-5_6601

(2012) Checkpoint. In: Schwab M (ed) Encyclopedia of cancer, 3rd edn. Springer, Berlin/Heidelberg, pp 754–755. doi:10.1007/978-3-642-16483-5_1049

(2012) Cholesterol. In: Schwab M (ed) Encyclopedia of cancer, 3rd edn. Springer, Berlin/Heidelberg, p 821. doi:10.1007/978-3-642-16483-5_1116

(2012) Cytokine. In: Schwab M (ed) Encyclopedia of cancer, 3rd edn. Springer, Berlin/Heidelberg, p 1051. doi:10.1007/978-3-642-16483-5_1473

(2012) Extracellular matrix. In: Schwab M (ed) Encyclopedia of cancer, 3rd edn. Springer, Berlin/Heidelberg, p 1362. doi:10.1007/978-3-642-16483-5_2067

(2012) ICAMs. In: Schwab M (ed) Encyclopedia of cancer, 3rd edn. Springer, Berlin/Heidelberg, p 1803. doi:10.1007/978-3-642-16483-5_2938

(2012) Integrin. In: Schwab M (ed) Encyclopedia of cancer, 3rd edn. Springer, Berlin/Heidelberg, p 1884. doi:10.1007/978-3-642-16483-5_3084

(2012) Lymphocytes. In: Schwab M (ed) Encyclopedia of cancer, 3rd edn. Springer, Berlin/Heidelberg, p 2123. doi:10.1007/978-3-642-16483-5_3455

(2012) Messenger RNA. In: Schwab M (ed) Encyclopedia of cancer, 3rd edn. Springer, Berlin/Heidelberg, p 2250. doi:10.1007/978-3-642-16483-5_6616

(2012) Mutation. In: Schwab M (ed) Encyclopedia of cancer, 3rd edn. Springer, Berlin/Heidelberg, p 2412. doi:10.1007/978-3-642-16483-5_3911

(2012) Phosphoinositides. In: Schwab M (ed) Encyclopedia of cancer, 3rd edn. Springer, Berlin/Heidelberg, p 2867. doi:10.1007/978-3-642-16483-5_4535

(2012) Plasma membrane. In: Schwab M (ed) Encyclopedia of cancer, 3rd edn. Springer, Berlin/Heidelberg, p 2900. doi:10.1007/978-3-642-16483-5_4599

(2012) Rab. In: Schwab M (ed) Encyclopedia of cancer, 3rd edn. Springer, Berlin/Heidelberg, p 3133. doi:10.1007/978-3-642-16483-5_4890

(2012) RISC. In: Schwab M (ed) Encyclopedia of cancer, 3rd edn. Springer, Berlin/Heidelberg, p 3309. doi:10.1007/978-3-642-16483-5_5110

(2012) Signal-transducer proteins. In: Schwab M (ed) Encyclopedia of cancer, 3rd edn. Springer, Berlin/Heidelberg, p 3411. doi:10.1007/978-3-642-16483-5_5299

(2012) Stromal cells. In: Schwab M (ed) Encyclopedia of cancer, 3rd edn. Springer, Berlin/Heidelberg, p 3544. doi:10.1007/978-3-642-16483-5_5535

(2012) Transcription factor. In: Schwab M (ed) Encyclopedia of cancer, 3rd edn. Springer, Berlin/Heidelberg, p 3752. doi:10.1007/978-3-642-16483-5_5901

(2012) Ubiquitin. In: Schwab M (ed) Encyclopedia of cancer, 3rd edn. Springer, Berlin/Heidelberg, p 3825. doi:10.1007/978-3-642-16483-5_6083

Experimental Carcinogenesis

▶ Toxicological Carcinogenesis

Extracellular Matrix Remodeling

Malgorzata Matusiewicz
Department of Medical Biochemistry, Wroclaw
Medical University, Wroclaw, Poland

Definition

Extracellular matrix (ECM) remodeling is a series of quantitative and qualitative changes in ECM during neoplastic transformation facilitating tumor growth and ▶ metastasis.

Characteristics

ECM is produced and assembled by the cells it is surrounding. The main components of ECM include glycosaminoglycans (with predominant hyaluronic acid) and proteoaminoglycans (e.g., perlecan, aggrecan), noncollagenous glycoproteins (such as ▶ fibronectin, laminins, tenascin), collagens, and many other biologically important molecules involved in cell–cell and cell–matrix interactions as well as in matrix remodeling. ECM provides not only the mechanical support for the attachment and organization of cellular structures but is also actively involved in the exchange of information with cells and therefore in the regulation of many important processes such as cell proliferation, migration, differentiation, and survival. Cell–cell and cell–matrix interactions are mediated via adhesive proteins such as cadherins and integrins – adhesive membrane receptors localized on cell surface.

Matrix anchors cells; however, it is also physically confining them. This restriction becomes a problem as organs are growing. To some extent, it can be endured due to the inherent plasticity of a matrix, but after a certain point, structural changes are necessary for a proper functioning of the organ and organism. The cells of a given organ must therefore possess an ability to change/remodel its surroundings. Hence, during physiological processes, such as embryonic development, tissue morphogenesis, ▶ angiogenesis, and cartilage and bone remodeling, a degradation of old elements of matrix and a synthesis of new ones are taking place. The process is governed by signals from regulatory proteins of ECM that are ensuring the divisions and differentiation of cells according to the needs of the organism. The mechanisms that are designated for physiological processes are adapted for the needs of tumor cells. Tumors can be viewed as functional tissues with cells surrounded by the microenvironment of ECM. Tumor cells must remodel the matrix to establish the communication between tumor cells and ECM and break barriers of the controlling mechanisms of the host cells. Local host stroma plays an important role in the transition from normal to malignant tissue. It has been established that stroma and tumor cells can interchange growth factors, cytokines, angiogenic factors, and proteases to activate surrounding ECM and facilitate the expansion of tumor cells. Stromal cells by the release of specific molecules can change cell phenotype and induce neoplastic transformation in the neighboring cells. It has been demonstrated that tumor-associated fibroblasts have altered properties in comparison to fibroblasts from normal epithelial cells. Tumor cells are changing the composition of ECM either by forcing the production of ECM components in an altered form or by stimulation or inhibition of the expression of some other compounds. Hyaluronic acid, which is promoting cell ▶ migration via its surface receptor, is very often overexpressed in malignant tissues. Laminin, which is essential for the integrity of the tissue, is produced in an altered form and in lower quantities. A desmoplastic reaction, which accompanies many solid tumors, is characterized by altered expression of many proteins (such as α-smooth muscle actin, smooth muscle myosin, and desmin) in desmoplastic fibroblasts as well as altered production of some ECM components (such as collagen types III and IV, tenascin, ▶ matrix metalloproteinases (MMPs), tissue inhibitors of metalloproteinases (TIMPs), other proteases involved in ECM degradation, and growth factors).

One of the ways tumors are facilitating their migration is by suppressing cell–cell adhesion. A key molecule in maintaining cell–cell adhesion is ▶ E-cadherin. In human epithelial cancers, a downregulation of E-cadherins and upregulation of another form called N-cadherin has been observed. Downregulation of E-cadherin promotes the invasive and metastatic phenotype of transformed epithelial cells.

Other adhesive receptors – integrins – are engaged in specific binding of the cells and components of ECM. They are heterodimeric proteins composed of α- and β-subunits. There are several types of both α- and β-subunits and therefore a certain number of combinations. Overexpression of many of them has been observed in a number of cancers. Integrins are engaged in signal transmission from the ECM into the cells and regulation of gene expression, for example, of enzymes participating in ECM degradation. They also mediate cell migration. Some integrin receptors are recognizing an RGD sequence (arginine–glycine–asparagine) – a conserved element – which is present in many ECM components. Peptides containing this sequence are implicated in inhibition of metastasis. The manner in which it is conducted is still unclear. It is speculated that they can selectively inhibit either the adhesion of tumor cells to structural proteins, the production of proteases, or the migration of tumor cells.

Degradation of ECM is a key event in ECM remodeling. It is conducted by a number of hydrolytic enzymes such as metalloproteinases, cysteine proteases (cathepsins B and L), aspartic proteases (▶ cathepsin D), serine proteases (elastase), sulfatases, and glycosidases. These enzymes function both under physiological and pathological conditions, and most of them is produced in a form of inactive zymogens activated by proteolytic processes. Some of the proteases (e.g., cathepsins) are considered lysosomal enzymes, but in the case of cancer cells, a change in cellular distribution with significantly elevated expression in cytosolic fraction has been observed.

Cathepsins are working at acidic pH and are involved in intracellular proteolysis, while serine proteases and MMPs act at neutral pH and are mostly responsible for extracellular proteolysis. The proteases can either directly degrade ECM components or indirectly by activation of other proteases, which in turn will also degrade ECM. It seems that the enzymes act in a determined order resulting in a cascade of proteolytic processes. Cathepsin D is produced in inactive form as procathepsin D. The zymogen undergoes autocatalytic activation in an acidic environment. Compared to normal tissue, the extracellular pH in tumors is usually more acidic. The second cathepsin, cathepsin B, also produced in a form of zymogen, can be activated either by cathepsin D or other proteases (elastase, cathepsin G, uPA, tPA). Active cathepsin B can in turn activate prourokinase-type plasminogen activator (pro-uPA). uPA activates plasminogen into plasmin. Both cathepsin B and plasmin are subsequently ready to cleave zymogens of MMPs, producing their active forms. The sequence of the degradation of ECM components seems to be determined as well. Glycoproteins that surround collagen molecules and protect them from proteolysis are degraded first by the action of cathepsins and plasmins. This permits the degradation of collagens by MMPs. As a result, ECM becomes destabilized, and the barriers preventing migration of neoplastic cells are removed. Especially difficult to penetrate by cancer cells is a basement membrane with collagen type IV as a main component. This obstacle can be, however, removed with the help of leukocytes. Leukocyte proteases act on basement membrane, degrade it, and thus facilitate cancer cell migration. Additionally, the contact of leukocytes with neoplastic cells influences the synthesis of other proteases, which further promote the degradation of ECM.

Among other enzymes implicated in ECM remodeling are also ▶ heparanase and sulfatases, which together with MMPs participate in the alternations of heparin sulfate proteoglycans (HSPGs). HSPGs are interacting with many effector molecules such as FGF, IL-8, and ▶ VEGF acting as coreceptors and therefore involved in the regulation of biological activities of cells. HSPGs are overexpressed in many cancers. Additionally, the changes occurring in proteoglycan structure upon the action of the three mentioned groups of enzymes result in altered affinity for growth factors and growth factor receptors dramatically affecting transmission of signals. The accelerated hydrolysis observed in some conditions, including cancers, leads to such changes in cell surface proteoglycans that hinder their ability to mediate cell adhesion. Moreover, the shedded fragments were demonstrated to promote tumor growth and metastasis.

The elevated activity of enzymes is an interplay between enzymes and their inhibitors – it can result either from enhanced expression of enzymes or from the reduction of the available inhibitors. Cathepsin B, for example, not only directly activates metalloproteinases but also further enhances their activity by cleaving and thus inactivating their inhibitors TIPM-1 and TIMP-2. Cathepsin B itself can be inhibited by cystatins and ▶ stefins, and it has been demonstrated that in many pathological conditions, including cancer, the concentration of cystatins has been reduced. An interesting example of interplay between proteases is ▶ plasminogen activator system. Besides direct degradation of ECM components, it seems to be implicated in tumor cell mobility and ▶ invasion. It is composed of proactivators, plasminogens, their cell surface receptors, inhibitors, and antiplasmins. uPA activates plasmin, and pro-uPA – its precursor – is activated not only by cathepsin B and elastase but also by plasmin. uPA and plasmin are inhibited by serpins. The principal role in degradation of ECM is played by MMPs. They are a group of 28 Ca^{2+}- and Zn^{2+}-dependent proteases. Based on their structure and substrate specificity, MMPs were originally classified as collagenases, gelatinases, stromelysins, and matrilysins. Taking into account common functional domains, they are currently divided into eight groups. Under physiological conditions, MMPs are produced in low quantities in zymogen forms. Their expression is induced during ECM remodeling processes by cytokines (e.g., IL-4, IL-10), growth factors (e.g., TGF-α, TGF-β,

FGF), and cell–cell or cell–matrix interactions. The activation of transcription of MMPs genes can involve either of the three mitogen-activated protein kinases pathways: extracellular signal-regulated kinase (ERK), stress-activated protein kinase/Jun N-terminal kinases (SAPK/JNKs), and p38. Zymogens are activated either by autoproteolysis or by another MMP or by a serine protease. MMPs are specifically inhibited by TIMPs and small molecules containing TIMP-like domains such as NC1 domain of collagen type IV. Besides inhibiting MMPs, TIMPs themselves express various antioncogenic properties, and TIMP-3 is suggested to participate in tumor cell death. A list of the main proteases participating in ECM degradation and their inhibitors is presented in Table 1.

Degradation of ECM is necessary for the migration of neoplastic cells, but it also serves other purposes important in tumor cell expansion. It results in the unmasking of cryptic sites, in the production of functional fragments, and in the release of signaling factors. Proteolysis of ECM components reveals new binding sites for the interaction with cell surface receptors and in this way increases tumor metastatic potential. The cleavage of ECM and cell surface molecules produces active fragments influencing tumor growth and spread. It has been documented that the degradation of ECM components by MMPs leads to the production of proangiogenic molecules. MMPs are also cleaving E-cadherin producing a fragment that is inhibiting E-cadherin and thus induces tumor cell invasion. On the other hand, the degradation of collagen XVIII by elastase releases a C-terminal fragment, which is a potent inhibitor of angiogenesis and tumor growth. Signaling factors, such as TGF-β, PDGF, or b-FGF, are in many cases stored in an inactive form, bound to ECM components. Hence, elevated activity of matrix proteases results in the release of increased number of growth factors, which can after binding to their receptors activate various signaling pathways. These processes, initiated by tumor cells, are taking place in host tissue and result in altered regulation of intracellular signaling facilitating tumor growth and metastasis.

Extracellular Matrix Remodeling, Table 1 Proteases participating in the degradation of ECM components

Protease family	Protease	Protease function	Protease inhibitors
Aspartyl protease	Cathepsin D	Degradation of ECM components	
		Conversion of cysteine procathepsins into cathepsins	
Cysteine proteases	Cathepsins B, L, H, K	Degradation of ECM components	Cystatins, stefins, kininogen
		Conversion of pro-MMPs into MMPs	
Serine proteases	Plasmin	Degradation of ECM components	α_2-Antiplasmin
		Activation of uPA	α_2-Macroglobulin
	Urokinase-type plasminogen activator (uPA)	Conversion of inactive elastase into elastase	PAI-1, PAI-2, PAI-3
		Conversion of plasminogen into plasmin	
Neutrophil serine proteases	Elastase	Degradation of ECM	α_2-Antiplasmin
	Cathepsin G	Components	α_2-Macroglobulin secretory leukoprotease inhibitor
Matrix metalloproteinases		Degradation of ECM components	TIMP-1, TIMP-2, TIMP-3, TIMP-4
	Collagenases [MMP-1, MMP-8, MMP-13]	Activation of other pro-MMPs into MMPs	α_2-Macroglobulin
		Degradation of collagens (I, II, III, VII, X) and gelatins	
	Stromelysins [MMP-3, MMP-10]	Degradation of proteoglycans, laminin, gelatins, collagens (III, IV, V, IX), fibronectin, entactin, and collagenase-1	
	Gelatinases [MMP-2, MMP-9]	Degradation of gelatins, collagens (I, IV, V, VII, X), fibronectin, elastin, and procollagenase-3	
	Membrane type [MMP-14, MMP-15, MMP-16, MMP-17, MMP-24, MMP-25]	Degradation of collagens (I, II, III), gelatins, aggregan, fibronectin, laminin, MMP-2, MMP-13, tenascin, nidogen	
	Others [MMP-7, MMP-11, MMP-12, MMP-19, MMP-20, MMP-23]	Degradation of proteoglycans, laminin, gelatins, fibronectin, collagen IV, elastin, entactin, tenascin, α_1-antiproteinase	

From Skrzydlewska et al. (2005)

References

Holmbeck K, Szabowa L (2006) Aspects of extracellular matrix remodeling in development and disease. Birth Defects Res C Embryo Today 78:11–23

Pupa SM, Menard S, Forti S et al (2002) New insights into the role of extracellular matrix during tumor onset and progression. J Cell Physiol 192:259–267

Sanderson RD, Yang Y, Kelly T et al (2005) Enzymatic remodeling of heparin sulfate proteoglycans within the tumor microenvironment: growth regulation and the prospect of new cancer therapies. J Cell Biochem 96:897–905

Skrzydlewska E, Sulkowska M, Koda M et al (2005) Proteolytic-antiproteolytic balance and its regulation in carcinogenesis. World J Gastroenterol 11:1251–1266

Zigrino P, Löffek S, Mauch C (2005) Tumor-stroma interactions: their role in the control of tumor cell invasion. Biochimie 87:321–328

Extracellular Nucleic Acids

▶ Circulating Nucleic Acids

Extracellular Signal-Regulated Kinases 1 and 2

Lars-Inge Larsson and Susanne Holck
Department of Pathology, Copenhagen University
Hospital, Hvidovre, Denmark

Synonyms

Extracellular signal-regulated kinases 1 and 2;
Mitogen-activated protein kinases p42 and p44

Definition

The extracellular signal-regulated kinases (ERKs)
1 and 2 (also referred to as mitogen-activated
protein kinases 1 and 2; MAPK1/2) constitute
major regulators of cell proliferation and survival
and also regulate motility, differentiation, and
senescence. They phosphorylate a multitude of
cytoplasmic and nuclear substrates, including
transcription factors. Mutations or overexpression
of upstream ERK activators are implicated in
oncogene-induced signaling in a wide variety of
tumors. Inhibitors of ERK activation are used in
cancer therapy.

Characteristics

The ERK Activation Cascade
ERKs 1 and 2 are activated through dual phosphorylation by a kinase cascade involving the
upstream signaling entities RAS (rat sarcoma
proto-oncogene), RAF (rat fibrosarcoma proto-
oncogene), and MEK1/2 (MAPK/ERKs 1 and 2)
(Fig. 1). There are three forms of RAS, KRAS
(Kirsten RAS), HRAS (Harvey RAS), and NRAS
(neuroblastoma RAS), and three forms of RAF
(ARAF, BRAF, and CRAF, also referred to as
RAF-1), which can all activate the ERK signaling
cascade.

Multiple factors, including growth factors,
integrin engagement, and activation of G protein-
coupled receptors (GPCRs), activate ERK1/2.
Binding of growth factors to cell surface recep
tors, like the human epidermal growth factor
receptor (HER1), results in receptor homo- or
heterodimerization and triggers its receptor tyrosine kinase (RTK) activity. RTK-dependent
autophosphorylation creates binding sites for
adaptor and docking proteins like growth factor
receptor binding protein 2 (GRB2), which recruits
the guanine nucleotide exchange factor SOS (son
of sevenless) (Fig. 1). SOS exchanges GDP for
GTP in RAS, thus converting it to its active form.
Subsequently, RAS is deactivated through
GTPase activating proteins (GAPs), which convert it to the inactive GDP-bound form.
GTP-bound RAS activates RAF through a complex procedure involving the induction of RAF
dimers and phosphorylation. RAF phosphorylates
and activates MEKs 1 and 2, and these activate
ERKs 1 and 2 through dual threonine/tyrosine
phosphorylation (ERK 1, T202/Y204, and ERK
2, T185/Y187). GPCRs may activate the cascade
either through transactivation of growth factor
receptors or through diacylglycerol-induced activation of protein kinase C (PKC). Pharmacologically, phorbol esters activate PKC and the ERK
cascade through mimicking diacylglycerol.
Exactly how PKC activates the ERK cascade
remains to be resolved.

While it initially was believed that the
RAF-MEK-ERK cascade depended upon stochas-
tic interactions, studies have uncovered the exis-
tence of multiple forms of scaffolding proteins.
Such scaffolds, like IQGAP1 (IQ[isoleucine-
glutamine domain]-guanosine triphosphatase activating protein 1), KSR1/2 (kinase suppressor of
RAS 1/2), paxillin, β-arrestin1/2, and MP1 (MEK
partner 1), bring two or all three of the different
components of the ERK cascade into proximity.
Although its name suggests that IQGAP1 is a
GAP, it is devoid of such activity. Interestingly,
IQGAP1 concentrations appear to be critical to
ERK activation and interference with the ERK
docking site in IQGAP1 inhibits cancer growth
in experimental systems. β-arrestins serve as scaffolds in the context of GPCR stimulation, whereas
paxillin acts as a scaffold for ERK activation at
cell-matrix interaction sites. Paxillin binds RAF,
MEK, and ERK and is a substrate for SRC

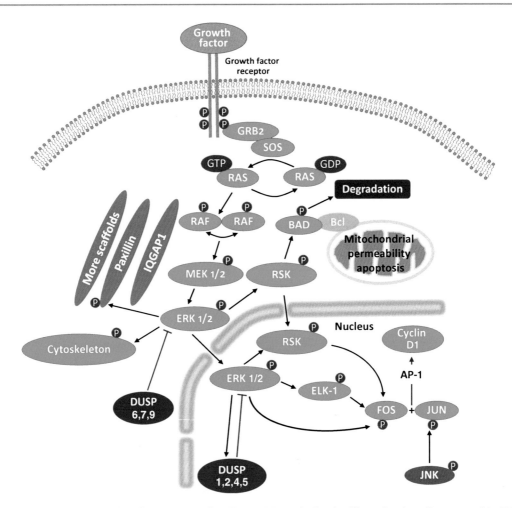

Extracellular Signal-Regulated Kinases 1 and 2, Fig. 1 Schematic drawing illustrating the salient parts of the ERK activation cascade and some of its substrates. Please refer to the text for details

(sarcoma proto-oncogene). SRC-mediated tyrosine phosphorylation increases the affinity of paxillin for ERK. The affinity of paxillin for focal adhesion kinase (FAK) is increased by ERK-mediated phosphorylation. FAK is, in turn, activated by SRC-mediated tyrosine phosphorylation and regulates cell spreading and motility. Thus, scaffolding proteins may both regulate the activation kinetics and compartmentalize ERK activation to specific subcellular sites.

It is debated whether the existence of ERKs 1 and 2 reflects subtle differences in biological roles or represents redundancy. Knockout of ERK2, but not of ERK1, is embryonically lethal. It has been argued that this reflects a higher expression level of ERK2 in most cell types,

making it difficult for ERK1 to compensate. However, other studies have implicated distinct roles for ERKs 1 and 2 so a final consensus has yet to be reached. Through alternative splicing, two distinct molecular forms of MEK (MEK1b) and ERK (ERK1c) arise. These forms do not participate in the canonical ERK activation cascade. However, MEK1b activates ERK1c and both play a role for Golgi fragmentation during mitosis.

ERK Substrates

ERKs are serine/threonine kinases preferring the substrate sequence Ser/Thr-Pro. They possess a multitude of cytoplasmic and nuclear substrates, and immunohistochemistry for active, dually phosphorylated ERK1/2 (henceforth referred to as pERKs) reveals staining of both cytoplasmic and nuclear compartments in most cells. Important pERK substrates include the family of ribosomal S6 kinases (RSK1-4) and the mitogen and stress-activated kinases (MSK1-2), which modulate transcription, apoptosis, and motility. RSKs are phosphorylated by pERKs both in the nucleus and in the cytoplasm (Fig. 1). They appear to differ in biological activities, and opposing actions on tumor cell motility and invasiveness have been reported. The RSKs are expressed in most tissues studied but differ in relative expression levels. They activate nuclear transcription events and also phosphorylate several cytoplasmic proteins, including factors controlling mitochondrial permeability and apoptosis like BAD. Additional substrates for pERKs include ELK-1 (E26-like kinase 1) – a member of the TCF (ternary complex factor) family of ETS transcription factors, which, in complex with the serum response factor, increases transcription of the FOS gene. The FOS protein forms a substrate for pERKs, which phosphorylate it at Ser374 and stabilizes it against degradation. RSKs also phosphorylate FOS (at Ser362) and stabilize it. Together, FOS and JUN family members form the AP-1 transcription complex, which stimulates transcription of the cyclin D1 gene and cell cycle progression (Fig. 1). In addition, pERKs phosphorylate numerous other proteins, including c-MYC, the actin-associated proteins palladin and paxillin, apoptosis-regulating proteins, and several others. Interactions with actin cytoskeletal regulation also involve pERK-mediated phosphorylation of myosin light-chain kinase (MLCK).

Termination of pERK Activity

Efficient systems for terminating the activity of pERKs exist in normal cells but may be at fault in some types of cancer cells. MAPK phosphatases (MKPs, also referred to as dual-specificity phosphatases; DUSPs) remove the activating phosphorylations on both threonines and tyrosines in pERKs. Of ERK-targeting MKPs, class I enzymes (MKP-1/DUSP1, DUSP2/PAC-1, MKP-2/DUSP 4, and DUSP5) are nuclear, and class II enzymes (MKP-3/DUSP6, MKP-X/DUSP7, and MKP-4/ DUSP9) are cytoplasmic. MKP-1 expression is

stimulated by pERKs, which, hence, participate in a negative feedback loop. Importantly, the expression pattern of several MKPs is altered in cancer, and this has been linked to cancer progression and response to chemotherapy. Another important aspect of the MKPs is their interaction and cross talk with stress-activated protein kinases (SAPKs). SAPKs (JUN N-terminal kinases; JNKs and p38 isoforms) are induced by stressful cellular events and by cytokine signaling but may also be stimulated by mitogens. SAPKs are activated through a three-stage kinase cascade similar to ERKs, but involving other enzymes. SAPKs induce a plethora of cellular effects, including phosphorylation of transcription factors like JUN and ATF-2. They regulate cytokine production and immune functions and may either stimulate cell cycle progression and survival or, more commonly, induce apoptosis in a cell-specific context. SAPKs are substrates for several MKPs, including MKP-1. Importantly, activated (SAPK-phosphorylated) ATF-2 increases transcription of the MKP-1 gene, and SAPKs may thus control dephosphorylation of pERKs. A number of additional cross talks between the ERK and other pathways, including the PI3K-AKT-MTOR pathway, exist. Additional phosphatases, including protein phosphatase 2A, also participate in dephosphorylating pERKs. Finally, pERKs exert feedback inhibition of their own activation cascade by phosphorylating and decreasing the activity of CRAF and additional activators like SOS1.

Nuclear Translocation and Localization of ERK

Immunohistochemistry has demonstrated that, following growth factor stimulation, a rapid nuclear translocation of pERKs occurs. These results are consonant with subcellular fractionation studies, which, however, often are bedeviled by diffusion and dephosphorylation of pERKs. The same applies to immunohistochemical studies if the material is not fixed rapidly enough. Thus, localization of pERKs in clinical material requires freshly fixed biopsies, and studies of surgically removed material are often not possible due to unavoidable delays in fixation. However, in freshly fixed material, pERKs localize both to tumor cells and to stromal cells, and in both cell types, both nuclear and cytoplasmic localizations are usually observed. Further support for this localization comes from studies of RSKs, phosphorylated on the pERK site, which also localizes to both cytoplasm and nucleus. A number of additional immunohistochemical control procedures are necessary for establishing credibility of the results. Regrettably, such controls are often not reported, which limits interpretations of results on the immunolocalization of pERKs and other phosphoproteins.

Oncogene-Driven ERK Activation: A Therapeutic Target

Upstream kinases that activate ERK1/2 are often constitutively activated (through mutations, translocations, amplifications, and other mechanisms) in cancer. This includes RTKs, their dimerization partners (like HER2), non-receptor tyrosine kinases (like ABL), and RAS and RAF. Therefore, ERK1/2 activation has been linked to cancer initiation and progression, and inhibitors of the upstream kinases are used in cancer treatment. Such inhibitors include monoclonal antibodies and kinase inhibitors targeting HER1 and/or HER2 as well as imatinib, inhibiting the BCR-ABL oncogene, KIT, and the PDGF (platelet-derived growth factor) receptor. Paradoxically, while RAF inhibitors show clinical effects on melanomas with activating BRAF mutations, they are much less effective in colorectal carcinomas harboring the same mutation. Overall, BRAF mutations occur in 5–10% of all cancers and are more common in some tumors (like melanomas and papillary thyroid carcinomas) than in others. The reason as to why BRAF, and not ARAF or CRAF, is constitutively activated in human cancers may reflect that it only takes a single point mutation to activate it. The by far most common BRAF mutation results in a protein in which Val600 is substituted with Glu (V600E). Point mutations of the *RAS* genes generate constitutively activated GTP-bound forms that are unresponsive to GAPs. Although constitutively active RAS mutations occur in about 20–30% of cancers, RAS inhibitors have so far shown limited usefulness in clinical trials. However, it is believed that activating mutations of

RAS make the use of inhibitors or antibodies targeting upstream kinases fruitless. Thus, CRC patients, eligible for HER1-directed therapies, are routinely screened for activating mutations of KRAS (and also NRAS and BRAF). Presently, the presence of activating KRAS mutations contraindicates HER1-directed therapies.

Much effort and money has gone into developing inhibitors that could target cancer cells like "magic bullets" – a term coined by Paul Ehrlich. Unfortunately, cancer cells frequently develop resistance to these inhibitors. Mechanisms include development of drug-resistant mutations, amplifications of key signaling kinases, and activation of alternative signaling pathways. Possibly, activation of the latter may buy cancer cells time to develop drug-resistant mutations or gene amplifications. Treatment of tumors with RAF inhibitors may lead to a paradoxical activation of ERK1/2. This may result from the induction of RAF heterodimers or through activation of upstream kinases including receptor tyrosine kinases and RAS. Combined treatment with MEK1/2 inhibitors and RAF inhibitors has been shown to produce a better effect in cultured cancer cells. Currently, efforts are devoted toward developing inhibitors, which simultaneously target multiple kinases.

Radiation Therapy and ERK Activation

Not only kinase inhibitors but also radiation therapy may induce resistance in tumor cells by activating ERK1/2. Thus, ERK activation has been associated with radioresistance, which, in experimental systems, is alleviated by MEK1/2 inhibitors. One mechanism by which ionizing radiation induces ERK activation is through free radicals, which inhibit protein tyrosine phosphatases (PTPs). The PTPs constitute a large family of enzymes, most of which work by removing activating tyrosine phosphorylations on RTKs and thus terminate their activation. Radiation-induced activation of, e.g., HER1 has been documented, and this may reflect free radical-induced deactivation of the active sites in HER1-inactivating PTPs. Additionally, data suggesting that also chemotherapeutics like adriamycin/doxorubicin and 5-fluorouracil may activate ERK signaling are accumulating.

Future Directions

It stands to reason that the complex pattern of ERK activation through multiple kinases as well as drugs has created an interest in inhibitors that work downstream in the cascade. Inhibitors of MEK1/2 are in clinical trial, but results from monotherapies have, so far, not been universally encouraging. However, pERKs not only play roles in cell proliferation, motility, invasiveness, and survival but also in induction of differentiation and senescence. Although inhibition of ERK activation may be fruitful in scenarios, where other cancer therapies induce its inappropriate activation, it may be useful to ponder whether such inhibition is desirable in all cases. Additional pathways such as the PI3K-AKT-MTOR and STAT-activating pathways, which drive cell proliferation and survival and interact with the ERK cascade, may constitute targets for simultaneous inhibition, e.g., through inhibitors targeting multiple RTKs.

Cross-References

- ► AP-1
- ► BRaf-Signaling
- ► Epidermal Growth Factor Receptor
- ► HER-2/neu
- ► JNK Subfamily
- ► MAP Kinase
- ► Raf Kinase
- ► RAS Transformation Targets

References

Deschênes-Simard X, Kottakis F, Meloche S, Ferbeyre G (2014) ERKs in cancer: friends or foes? Cancer Res 74:412–419

Keyse SM (2008) Dual-specificity MAP kinase phosphatases (MKPs) and cancer. Cancer Metastasis Rev 27:253–261

Lito P, Rosen N, Solit DB (2013) Tumor adaptation and resistance to RAF inhibitors. Nat Med 19:1401–1409

Mendoza MC, Er EE, Blenis J (2011) The Ras-ERK and PI3K-mTOR pathways: cross-talk and compensation. Trends Biochem Sci 36:320–328

Roskoski R Jr (2012) ERK1/2 MAP kinases: structure, function, and regulation. Pharmacol Res 66:105–143

See Also

(2012) Phorbol ester. In: Schwab M (ed) Encyclopedia of cancer, 3rd edn. Springer, Berlin, p 2865. doi:10.1007/978-3-642-16483-5_4522

Extracellular Vesicles

- ► Exosome

Extrahepatic Bile Duct Carcinoma

- ► Klatskin Tumors

Extrahepatic Cholangiocarcinoma

- ► Klatskin Tumors

Extrahepatic Cholangiocellular Carcinoma

- ► Klatskin Tumors

Extrapulmonary Small Cell Cancer

Rabia K. Shahid[1] and Shahid Ahmed[2]
[1]Department of Medicine, University of Saskatchewan, Saskatoon, SK, Canada
[2]Department of Oncology, University of Saskatchewan, Saskatoon, SK, Canada

Synonyms

Carcinoma with amine precursor uptake decarboxylation cell differentiation; Kulchitsky cell carcinoma; Microcytoma; Oat cell carcinoma; Reserve cell carcinoma; Small cell neuroendocrine carcinoma

Definition

Extrapulmonary small cell carcinoma (EPSCC) is a high-grade epithelial cancer of neuroendocrine origin composed of small, round to fusiform cells with minimal cytoplasm that arises at various anatomical sites in the absence of a primary lung neoplasm.

Small cell carcinoma (SCC) is a distinct clinicopathological entity first described in the lung. It represents approximately 20% of all bronchogenic carcinoma. Extrapulmonary small cell carcinoma (EPSCC) indistinguishable from small cell ► lung cancer was first reported in 1930. Since its first description, EPSCC has been reported in virtually all anatomical sites. The primary sites most frequently involved are gynecologic organs, especially the cervix; genitourinary organs, especially the urinary bladder; the gastrointestinal tract, especially the esophagus; and the head and neck region.

EPSCC often represents a diagnostic and therapeutic challenge. Limited data is available about its clinical behavior and outcome. The available literature is predominantly based on reviews of published cases or analysis of institutional data. The clinicopathological features and general management of EPSCC will be reviewed here, followed by a brief description of SCC specific to the more common sites.

Characteristics

Epidemiology

SCC arising from extrapulmonary sites represents 2–4% of all SCC. Approximately 1,000 cases per year have been reported in the United States, which represents an overall incidence of between 0.1% and 0.4% of all cancer. Patients with EPSCC are generally middle-aged or older similar to SCC of the lung; however, women with SCC of cervix tend to be younger. Both genders are affected and predominance of either gender varies according to the primary site of involvement. For example, SCC of the esophagus, urinary bladder, and head and neck region are more common in men,

whereas female preponderance has been noted in patients with SCC of gallbladder. Although cigarette smoking appears to be associated with EPSCC especially of the head and neck region, it has not been clearly identified as a risk factor for EPSCC, and the role of smoking in the development of malignancy remains speculative.

Pathology

The histological criteria are the same as those for the pulmonary neoplasm. SCC is composed of sheets and nests of round to fusiform cells with minimal amounts of cytoplasm and granular nuclear chromatin. Nucleoli are absent or inconspicuous. The typical organoid architectural patterns of low-grade neuroendocrine neoplasms such as ► carcinoid tumor are generally absent. Mitotic rates are high and necrosis of individual malignant cell is common. It may contain non-SCC elements, varying in type depending on the location.

The pathogenesis of SCC is largely unknown and remained speculative. It exhibits several neuroendocrine features characterized by the presence of enzymes such as of DOPA decarboxylase, ► calcitonin, neuron-specific enolase, chromogranin A, and CD56 (neural cell ► adhesion molecule). SCC is thought to originate from totipotent stem cells present in all tissues. Others have suggested that it may arise from more differentiated tumors during the clonal evolution of a carcinoma as a late-stage phenomenon.

Clinical Features

The clinical presentation is determined by the site of involvement and extent of the disease. Systemic symptoms, such as anorexia and weight loss, are common especially in patients with advanced disease. Focal symptoms are mostly site specific and are usually indistinguishable from those of other neoplasms arising from that anatomical site. Though uncommon, similar to SCC of the lung paraneoplastic syndromes such as ectopic ACTH production or inappropriate antidiuretic hormone secretion may be the dominant presenting feature. Eaton–Lambert syndrome (a disease seen in patients with ► lung cancer and characterized by weakness and fatigue of hip and thigh muscles and an aching back and caused by antibodies directed against the neuromuscular junctions), thyroxine intoxication, and hyperglucagonemia have also been reported but are rare.

Diagnosis and Staging

The diagnosis of SCC is primarily rested on morphological assessment. However, ▶ immunocytochemistry plays an important role and electron microscopy can be of value in difficult cases. The malignant cells are immunoreactive for keratin and epithelial membrane antigen in virtually all cases. Thyroid transcription factor-1 (TTF-1) immunostaining has been proposed by several investigators to differentiate small cell lung carcinoma from EPSCC. However, TTF-1 expression is not specific for SCC of pulmonary origin and should not be used to distinguish primary from metastatic SCC in extrapulmonary sites.

Although no specific staging system for EPSCC has been established, most authors have adopted "two-stage system" originally introduced by the Veterans' Administration Lung Study Group. This staging system consists of two categories: limited disease (LD) defined as tumor contained within a localized anatomic region, with or without locoregional lymphadenopathy, and extensive disease (ED), defined as tumor outside the locoregional boundaries. Information provided by the "tumor–node–metastases (TNM)" staging system may be valuable in certain anatomical sites such as SCC of the large bowel.

The diagnosis of EPSCC requires a normal computed tomographic (CT) study of the chest. A primary lung tumor should be excluded. Some investigators have suggested routine bronchoscopy, but this is not widely adopted. Abdominal and pelvic CT scan is a useful test to determine primary site and to assess the extent of the disease. Although there is a lack of data regarding the role of ▶ positron emission tomography (PET) scan in the management of EPSCC, it can be a useful tool for the detection of primary tumor in SCC of unknown primary site. In the absence of neurologic symptoms, CT scan of the head is not routinely performed. Bone marrow biopsy is indicated if there is cytopenia without other evidence of disseminated disease. Other studies such as endoscopic examination are aimed to assess the affected sites and vary accordingly.

Management

Limited data are available about the optimal management of EPSCC. As in pulmonary SCC, the survival of untreated patients is poor. Treatment goals for extensive-stage and limited-stage diseases are different and they should be treated differently. Treatment for LD is potentially curative, whereas that of ED is palliative.

SCC is sensitive to both radiation therapy and ▶ chemotherapy. The unfavorable prognosis and the chemosensitivity of its pulmonary counterpart have persuaded many clinicians to use combined-modality therapy including surgery, radiation, and chemotherapy. The response rate varies from 48% to 100%. However, the optimal integration of these modalities and precise sequence remained to be defined. Whereas chemotherapy can induce major regression of localized disease and concurrently treat occult metastases, surgery and/or radiation therapy represents the best option for ▶ locoregional therapy at majority of the anatomical sites. In carefully selected patients with LD and small tumor volume, surgery can be curative. Although the role of ▶ adjuvant therapy remains to be defined, platinum-based adjuvant chemotherapy may be beneficial given the chemoresponsiveness of the disease and the high rate of systemic recurrence. The possible synergism between chemotherapy and radiotherapy supports combined ▶ chemoradiotherapy, and for many patients with LD at various anatomical sites, the combination of chemotherapy and radiation therapy can be an effective treatment. Cranial irradiation is not routinely used in the management of these patients who achieved a complete response.

Patients with ED of any site are best managed with systemic chemotherapy. Responses to therapy occur in 60–90% of patients; however, most responses are partial and of short duration. The use of surgery or radiotherapy in these patients is restricted for palliation of local symptoms.

Prognosis

SCC follows an aggressive course with early propensity for metastases. EPSCC of various anatomical sites behaves differently and outcome varies according to the primary site of the disease involvement. In general the prognosis of EPSCC is comparable to SCC of the lung, and the extent of disease is an important factor predicting survival. Poor performance status and abnormal white blood cell count are the other important variables that correlate with survival.

In reported series patients with EPSCC are not uniformly treated or comparably staged. The median overall survival of all patients with EPSCC is 9–15 months, and 5-year survival is 10–15%. Patients with LD have median overall survival of 25–34 months and 5-year survival of 31%, whereas patients with ED have a median overall survival of 2–12 months and 5-year survival of 2%. Despite the generally aggressive behavior of SCC, long-term remission or cure can be achieved in selected patients with a tailored therapy.

Genitourinary Tract

Urinary Bladder

SCC has been reported in the urinary bladder, prostate, and kidney. Although the urinary bladder is the most common site of EPSCC in the genitourinary tract, it accounts for less than 1% of all ▶ bladder cancers. Patients are usually between the ages of 40 and 60 years, and it is three times more common in men than women. SCC may coexist with ▶ transitional cell carcinoma and other types of bladder tumors. Majority of patients present with locally advanced or disseminated disease. Surgery is generally recommended for patients with localized disease often followed by adjuvant chemotherapy. Combination of chemotherapy and radiation has been given concurrently in an effort to preserve the bladder in many cases. The overall median survival in most reported series is about 2 years.

Prostate

The incidence of prostate SCC is less than 1% of the total of prostate cancer. The median age of the patients is approximately 65 years, which is similar to that of patients with adenocarcinoma of prostate. Prostate SCC may present at initial diagnosis or appear later in the evolution of an adenocarcinoma. Approximately 30% of patients present initially with prostatic adenocarcinoma, 20% present with combined adenocarcinoma and SCC, and 50% of patients presented with SCC. ▶ Prostate-specific membrane antigen (PSMA) is not elevated in majority of the patients with prostate SCC, and they respond poorly to antiandrogen therapy. Most of the patients have advanced disease at diagnosis, and median survival is about 15 months. Patients presenting initially with an adenocarcinoma have a median survival of 25 months compared with a median survival of 5 months for patients presenting with SCC.

Gynecological Sites

Cervix

SCC most commonly involve the cervix but may also develop in the endometrium, ovary, vagina, and vulva. SCC represents 0.4–1.4% of all ▶ cervical cancer. The cervix should always be considered as the site of origin in a woman with a SCC of unknown primary site. Women with cervical SCC tend to be younger with median age of diagnosis being about 40–50 years. The prognosis varies with the stage of the disease. Survival is poor with hysterectomy alone, and most patients are treated with a multimodality approach, using chemotherapy regimens that are typically used for small cell lung cancer. Patients with cervical SCC treated with combination of chemotherapy and radiotherapy had 3-year survival of 60%.

Gastrointestinal Tract

Esophagus

Primary SCC involving the esophagus appears to be the most frequently reported digestive tract site of EPSCC. The stomach, pancreas, ampulla of Vater, gallbladder, small intestine, and colon and rectum are the other sites in gastrointestinal tract where SCC has been reported. SCC of the esophagus is rare and incidence has been estimated to range from 0.8% to 2.4% of all ► esophageal cancers. Most cases occur between the ages of 50 and 70 years, and EPSCC is twice as common in men compared with women. Combination chemoradiotherapy is effective against esophageal SCC and may improve survival. Adjuvant systemic chemotherapy is recommended following surgery for localized disease although long-term survival has been reported in a few cases. The reported overall median survival is approximately 5 months, with a median survival for patients with LD being about 8 months and for patients with ED being about 3 months.

Colon and Rectum

SCC arising in the colon and rectum is rare making up approximately 0.2% of all colorectal neoplasms. The epidemiology is somewhat similar to that of adenocarcinoma with a slight male predominance. The majority of the cases are diagnosed between the ages of 50 and 70. Within the large bowel, the most frequent site is the rectum, followed by the cecum and sigmoid. Tumors with mixed histology are often present. The overall prognosis is poor with a median survival of 6 months. Surgery is the primary treatment for localized disease. Adjuvant radiation for local control and systemic chemotherapy to treat ► micrometastases are recommended; however, in the absence of clinical trials, the individual contribution of each component to the survival cannot be determined.

Head and Neck Region

Larynx

Although the larynx is one of the most common extrapulmonary sites, laryngeal SCC accounts for only 0.5% of all primary ► laryngeal carcinoma. Most patients are between the ages of 60 and 80 years, and there is a male predominance.

Smoking, tobacco chewing, and excess ► alcohol consumption have been associated with SCC of the larynx. The supraglottic region is the most commonly reported site. The majority of patients of localized SCC of the head and neck have been treated with local modality of treatment. Although optimal management for these patients is undefined, several investigators have reported that the use of concurrent chemoradiotherapy regimens for limited-stage disease offers potential for long-term survival. Median survival of patients with primary SCC of the larynx, hypopharynx, and trachea is between 7 months and 11 months.

References

Galanis E, Frytak S, Lioyd RV (1997) Extrapulmonary small cell carcinoma. Cancer 79:1729–1736

Haider K, Shahid RK, Finch D et al (2006) Extrapulmonary small cell cancer: a Canadian province's experience. Cancer 107:2262–2269

Remick SC, Ruckdeschel JC (1992) Extrapulmonary and pulmonary small-cell carcinoma: tumor biology, therapy, and outcome. Med Pediatr Oncol 20:89–99

Remick SC, Hafez GR, Carbone PP (1987) Extrapulmonary small cell carcinoma. A review of the literature with emphasis on therapy and outcome. Medicine 66:457–471

Shahid RK, Haider K, Sami A, et al (2008) Extra-pulmonary small cell cancer: diagnosis, treatment, and prognosis. In: Hayat MA (ed) Methods of cancer diagnosis, therapy and prognosis. Springer Netherlands, pp 207–16

Vrouvas J, Ash DV (1995) Extrapulmonary small cell cancer. Clin Oncol 7:377–381

Extreme Hypoxia

► Anoxia